MANUAL OF
Industrial Microbiology and Biotechnology
THIRD EDITION

MANUAL OF
Industrial Microbiology
and Biotechnology

THIRD EDITION

EDITORS IN CHIEF

Richard H. Baltz
CognoGen Biotechnology Consulting, Indianapolis, Indiana

Julian E. Davies
University of British Columbia, Vancouver, British Columbia, Canada

Arnold L. Demain
Charles A. Dana Research Institute (R.I.S.E.), Drew University, Madison, New Jersey

EDITORS

Alan T. Bull
School of Biosciences, University of Kent, Canterbury, Kent, United Kingdom

Beth Junker
Bioprocess Research and Development, Merck Research Laboratories, Rahway, New Jersey

Leonard Katz
SynBERC - Synthetic Biology Engineering Research Center, University of California-Berkeley, Emeryville, California

Lee R. Lynd
Thayer School of Engineering, Dartmouth College, Hanover, New Hampshire

Prakash Masurekar
Department of Plant Biology & Plant Pathology, School of Environmental and Biological Sciences, Rutgers University, New Brunswick, New Jersey

Christopher D. Reeves
Amyris Biotechnologies, Emeryville, California

Huimin Zhao
Departments of Chemical and Biomolecular Engineering, Chemistry, and Bioengineering, Institute for Genomic Biology, Center for Biophysics and Computational Biology, University of Illinois at Urbana-Champaign, Urbana, Illinois

ASM
PRESS

Washington, DC

Library of Congress Cataloging-in-Publication Data

Manual of industrial microbiology and biotechnology / editors in chief, Richard H. Baltz, Julian E.
Davies, and Arnold L. Demain ; editors, Alan T. Bull ... [et al.]. -- 3rd ed.
       p. ; cm.
  Includes bibliographical references and index.
  ISBN 978-1-55581-512-7 (alk. paper)
  1. Industrial microbiology--Handbooks, manuals, etc. 2. Industrial microbiology--Handbooks,
manuals, etc. 3. Biotechnology--Handbooks, manuals, etc. I. Baltz, Richard H. II. Davies, Julian E.
III. Demain, A. L. (Arnold L.), 1927- IV. American Society for Microbiology.
  [DNLM: 1. Biotechnology. 2. Industrial Microbiology. QW 75 M294 2010]
  QR53.M33 2010
  660.6'2--dc22

                                    2010001077

ISBN 978-1-55581-512-7

10 9 8 7 6 5 4 3 2 1

Address editorial correspondence to: ASM Press, 1752 N St., N.W., Washington,
DC 20036-2904, U.S.A.

Send orders to: ASM Press, P.O. Box 605, Herndon, VA 20172, U.S.A.
Phone: 800-546-2416; 703-661-1593
Fax: 703-661-1501
Email: Books@asmusa.org
Online: estore.asm.org

Contents

v

Editors in Chief

Richard H. Baltz
CognoGen Biotechnology Consulting, Indianapolis,
IN 46220

Julian E. Davies
University of British Columbia, Vancouver,
British Columbia, V6R 1Z3 Canada

Arnold L. Demain
Charles A. Dana Research Institute (R.I.S.E.),
Drew University, Madison, NJ 07940

Editors

Alan T. Bull
School of Biosciences, University of Kent, Canterbury,
Kent CT2 7NJ, United Kingdom

Beth Junker
Bioprocess Research and Development,
Merck Research Laboratories, Rahway, NJ 07065

Leonard Katz
SynBERC - Synthetic Biology Engineering Research Center,
University of California-Berkeley, Emeryville, CA 94608

Lee R. Lynd
Thayer School of Engineering, Dartmouth College,
Hanover, NH 03755

Prakash Masurekar
Department of Plant Biology & Plant Pathology,
School of Environmental and Biological Sciences,
Rutgers University, New Brunswick, NJ 08901-8520

Christopher D. Reeves
Amyris Biotechnologies, Emeryville, CA 94608

Huimin Zhao
Departments of Chemical and Biomolecular Engineering,
Chemistry, and Bioengineering, Institute for
Genomic Biology, Center for Biophysics and Computational
Biology, University of Illinois at Urbana-Champaign,
Urbana, IL 61801

Contributors

CHARLES A. ABBAS
James R. Randall Research Center,
Archer Daniels Midland Co., Decatur, IL 62521

CARL B. ABULENCIA
Verenium Corporation, San Diego, CA 92121

SPIROS N. AGATHOS
Unit of Bioengineering (GEBI), Université Catholique
de Louvain, Place Croix du Sud 2/19, B-1348,
Louvain-la-Neuve, Belgium

DYLAN C. ALEXANDER
Cubist Pharmaceuticals, Inc., Lexington, MA 02421

ZHIQIANG AN
Epitomics, Inc., 863 Mitten Road, Burlingame, CA 94010

ADEYMA Y. ARROYO
Process Research & Development, Process Development
Engineering Department, Genentech Inc., South San
Francisco, CA 94080

YASUHISA ASANO
Biotechnology Research Center and Department of
Biotechnology, Toyama Prefectural University,
5180 Kurokawa, Imizu, Toyama 939-0398, Japan

RICHARD H. BALTZ
CognoGen Biotechnology Consulting, Indianapolis,
IN 46220

WU LI BAO
James R. Randall Research Center,
Archer Daniels Midland Co., Decatur, IL 62521

ANDREAS BECHTHOLD
Albert-Ludwigs-Universität, Institut für Pharmazeutische
Wissenschaften, Pharmazeutische Biologie und Biotechnologie,
Stefan-Meier-Strasse 19, 79104 Freiburg, Germany

KYLE E. BEERY
James R. Randall Research Center,
Archer Daniels Midland Co., Decatur, IL 62521

RAMUNAS BIGELIS
Biosynthetic Chemistry and Infectious Diseases,
Pfizer, Pearl River, NY 10965

GERALD BILLS
Fundación MEDINA, Avenida Conocimiento s/n, Parque
Tecnológico Ciencias de la Salud, Armilla, Granada 18100, Spain

WESLEY P. BLACK
Department of Biological Sciences, Virginia Tech,
Blacksburg, VA 24061

MARITE BRADSHAW
Botulinum Toxins Laboratory, Dept. of Bacteriology,
University of Wisconsin, 1550 Linden Dr., 6340 MSB,
Madison, WI 53706

KATIE M. BRANSCUM
Natural Products Discovery Group, Department of Chemistry
and Biochemistry, 620 Parrington Oval, Room 208,
University of Oklahoma, Norman, OK 73019

MATTHEW CHASE
AMRI, 21 Corporate Circle, Albany, NY 12203

C. PERRY CHOU
Department of Chemical Engineering, University of
Waterloo, Waterloo, Ontario, Canada N2L 3G1

ROBERT H. CICHEWICZ
Natural Products Discovery Group, Department of Chemistry
and Biochemistry, 620 Parrington Oval, Room 208,
University of Oklahoma, Norman, OK 73019

PATRICK C. CIRINO
Department of Chemical Engineering, The Pennsylvania
State University, University Park, PA 16802

DOUGLAS S. CLARK
Department of Chemical Engineering, University of
California, Berkeley, 201 Gilman Hall, Berkeley, CA 94720

LOUIS A. CLARK
Codexis Inc., 200 Penobscot Drive, Redwood City, CA 94063

PAMELA CORRINGTON
James R. Randall Research Center,
Archer Daniels Midland Co., Decatur, IL 62521

DON A. COWAN
Institute for Microbial Biotechnology and Metagenomics,
Department of Biotechnology, University of the Western
Cape, Bellville 7535, Cape Town, South Africa

JAMES M. CREGG
Keck Graduate Institute of Applied Life Sciences,
Claremont, CA 91711

CONSUELO CRUZ
James R. Randall Research Center,
Archer Daniels Midland Co., Decatur, IL 62521

SAMUN K. DAHOD
Valent BioSciences Corp., Libertyville, IL 60048

STEFANO DONADIO
KtedoGen, Via Fantoli 16/15, 20138 Milano, Italy, and
NAICONS, Via Fantoli 16/15, 20138 Milano, Italy

WOUTER A. DUETZ
Enzyscreen, Biopartner Center Leiden,
Wassenaarseweg 72, 2333 AL Leiden, The Netherlands

ALEXANDRU DUMITRACHE
Dept. of Chemistry and Biology, Ryerson University,
350 Victoria St., Toronto, Ontario, Canada, M5B 2K3

SLAVA S. EPSTEIN
Department of Biology, Northeastern University,
Boston, MA 02115

DAWN T. ERIKSEN
Department of Chemical and Biomolecular Engineering
and Institute for Genomic Biology, University of Illinois
at Urbana-Champaign, 600 South Mathews Ave.,
Urbana, IL 61801

GIOVANNA E. FELIS
Department of Science, Technology and Market of Vine and
Wine, University of Verona, 37134 Verona, Italy

MANUEL FERRER
Institute of Catalysis, CSIC, Marie Curie 2, Campus
Cantoblanco, 28049 Madrid, Spain

JENS C. FRISVAD
Center for Microbial Biotechnology, Department of Systems
Biology, Building 221, Søltofts Plads, Technical University of
Denmark, DK-2800 Kgs. Lyngby, Denmark

EKATERINA GAVRISH
Department of Biology, Northeastern University,
Boston, MA 02115

PETER N. GOLYSHIN
School of Biological Sciences, Bangor University,
Deiniol Road, LL57 2UW Gwynedd, United Kingdom

OLGA V. GOLYSHINA
School of Biological Sciences, Bangor University,
Deiniol Road, LL57 2UW Gwynedd, United Kingdom

MICHAEL GOODFELLOW
School of Biology, University of Newcastle,
Newcastle upon Tyne, NE1 7RU, United Kingdom

KEVIN A. GRAY
Verenium Corporation, San Diego, CA 92121

RANDOLPH GREASHAM
Independent Consultant, 460 Bayberry Lane,
Mountainside, NJ 07092

MELINDA J. GRIFFITHS
Centre for Bioprocess Engineering Research (CeBER),
Department of Chemical Engineering, University of Cape
Town, Private Bag X3, Rondebosch 7701, South Africa

MARIA-EUGENIA GUAZZARONI
Institute of Catalysis, CSIC, Marie Curie 2, Campus
Cantoblanco, 28049 Madrid, Spain

SUSAN T. L. HARRISON
Centre for Bioprocess Engineering Research (CeBER),
Department of Chemical Engineering, University of Cape
Town, Private Bag X3, Rondebosch 7701, South Africa

JON C. HENRIKSON
Natural Products Discovery Group, Department of Chemistry
and Biochemistry, 620 Parrington Oval, Room 208,
University of Oklahoma, Norman, OK 73019

WILLIAM G. HERRICK
Dept. of Chemical Engineering, University of Massachusetts-
Amherst, 686 N. Pleasant St., Amherst, MA 01003

LUTZ HILTERHAUS
Institute of Technical Biocatalysis, University of Technology
Hamburg-Harburg, D-21071 Hamburg, Germany

DAVID A. HOGSETT
Mascoma Corporation, Lebanon, NH 03655

FLORIAN HOLLFELDER
Department of Biochemistry, University of Cambridge,
Cambridge CB2 1GA, United Kingdom

DAVID HOPKINS
BioProcess Engineering, Genzyme Corporation,
Framingham, MA 01701

WEI-SHOU HU
Department of Chemical Engineering and Materials Science,
University of Minnesota, Minneapolis, MN 55455

ELTON P. HUDSON
Department of Chemical Engineering, University of
California, Berkeley, 201 Gilman Hall, Berkeley, CA 94720

MASATO IKEDA
Department of Bioscience and Biotechnology, Faculty of
Agriculture, Shinshu University, Nagano 399-4598, Japan

THOMAS W. JEFFRIES
USDA, Forest Service, Forest Products Laboratory and
Department of Bacteriology, University of Wisconsin-
Madison, Madison, WI 53726

SUSAN E. JENSEN
Dept. of Biological Sciences, University of Alberta, Edmonton, AB Canada T6G 2E9

ERIC A. JOHNSON
Botulinum Toxins Laboratory, Dept. of Bacteriology, University of Wisconsin, 1550 Linden Dr., 6303 MSB, Madison, WI 53706

BRYAN JULIEN
Allylix, Inc., 1500 Bull Lea Rd., Lexington, KY 40511

PAVAN K. R. KAMBAM
Dept. of Chemical Engineering, University of Massachusetts-Amherst, 686 N. Pleasant St., Amherst, MA 01003

ANNE KANTARDJIEFF
Alexion Pharmaceuticals, Cheshire, CT 06410

JAMES T. KEALEY
Amyris Biotechnologies, 5885 Hollis St., Suite 100, Emeryville, CA 94608

JAY D. KEASLING
Joint BioEnergy Institute (JBEI), Physical Biosciences Division, Berkeley National Laboratory; Dept. of Chemical Engineering, Dept. of BioEngineering, Berkeley Center for Synthetic Biology (SynBERC), University of California, Berkeley, CA 94720

MARTIN KELLER
Biosciences Division, Oak Ridge National Laboratory, Oak Ridge, TN 37831-6026

MAX KENNEDY
Biolighthouse Ltd., Wellington, New Zealand

WAN-SEOP KIM
Bristol-Myers Squibb Company, 6000 Thompson Road, East Syracuse, NY 13057

BRONWYN M. KIRBY
Institute for Microbial Biotechnology and Metagenomics, Department of Biotechnology, University of the Western Cape, Bellville 7535, Cape Town, South Africa

JOEL A. KREPS
Verenium Corporation, 4955 Directors Place, San Diego, CA 92121

NICHOLAS LANGLEY
Centre for Bioprocess Engineering Research (CeBER), Department of Chemical Engineering, University of Cape Town, Private Bag X3, Rondebosch 7701, South Africa

KIM LEWIS
Department of Biology, Northeastern University, Boston, MA 02115

ANDREAS LIESE
Institute of Technical Biocatalysis, University of Technology Hamburg-Harburg, D-21071 Hamburg, Germany

MICHAEL J. LISZKA
Department of Chemical Engineering, University of California, Berkeley, 201 Gilman Hall, Berkeley, CA 94720

LUCAS LOVELESS
James R. Randall Research Center, Archer Daniels Midland Co., Decatur, IL 62521

LEE R. LYND
Thayer School of Engineering and Dept. of Biological Sciences, Dartmouth College, Hanover, NH 03755

R. SCOTT MCIVOR
Department of Genetics, Cell Biology and Development, University of Minnesota, Minneapolis, MN 55455

TRACY L. MEIRING
Institute for Microbial Biotechnology and Metagenomics, Department of Biotechnology, University of the Western Cape, Bellville 7535, Cape Town, South Africa

VERA MEYER
Institute of Biology, Department Molecular Microbiology and Biotechnology, Leiden University, 2333 BE Leiden, The Netherlands

JONATHAN R. MIELENZ
BioEnergy Science Center, Biosciences Division, Oak Ridge National Laboratory, Oak Ridge, TN 37831

UFFE H. MORTENSEN
Department of Systems Biology, Technical University of Denmark, DK-2800 Kgs. Lyngby, Denmark

NIKHIL U. NAIR
Department of Chemical and Biomolecular Engineering, University of Illinois at Urbana-Champaign, 600 South Mathews Ave., Urbana, IL 61801

KIEN T. NGUYEN
Cubist Pharmaceuticals, Inc., Lexington, MA 02421

DOMINICA NICHOLS
Department of Biology, Northeastern University, Boston, MA 02115

JENS NIELSEN
Department of Chemical and Biological Engineering, Chalmers University of Technology, SE-41296 Gothenburg, Sweden

AHARON OREN
Department of Plant and Environmental Sciences, Institute of Life Sciences, and the Moshe Shilo Minerva Center for Marine Biogeochemistry, The Hebrew University of Jerusalem, 91904 Jerusalem, Israel

NANCY L. PAIVA
Dept. of Chemistry, Computer and Physical Sciences, Southeastern Oklahoma State University, Durant, OK 74701

NICOLAI S. PANIKOV
Thayer School of Engineering, Dartmouth College, 8000 Cummings Hall, Hanover, NH 03755

CHRISTOPHER J. PETZOLD
Joint BioEnergy Institute (JBEI), Physical Biosciences Division, Lawrence Berkeley National Laboratory, Berkeley, CA 94720

JACK PRIOR
BioProcess Engineering, Genzyme Corporation,
Framingham, MA 01701

KATHARINA PROBST
Albert-Ludwigs-Universität, Institut für Pharmazeutische
Wissenschaften, Pharmazeutische Biologie und Biotechnologie,
Stefan-Meier-Strasse 19, 79104 Freiburg, Germany

PETER J. PUNT
TNO Quality of Life, 3700 AJ Zeist, The Netherlands

MICHAEL E. PYNE
Department of Chemical Engineering, University of
Waterloo, Waterloo, Ontario, Canada N2L 3G1

ARTHUR F. J. RAM
Institute of Biology, Department Molecular Microbiology
and Biotechnology, Leiden University, 2333 BE Leiden,
The Netherlands

MANFRED T. REETZ
Department of Synthetic Organic Chemistry, Max-Planck-Institut
für Kohlenforschung, 45476 Mülheim an der Ruhr, Germany

URSULA RINAS
Helmholtz Center for Infection Research, Inhoffenstr. 7,
D-38124 Braunschweig, Germany

EDUARDO RODRIGUEZ
Instituto de Biología Molecular y Celular de Rosario (IBR-
CONICET) and Departamento de Microbiología, Facultad de
Ciencias Bioquímicas y Farmacéuticas, Universidad Nacional
de Rosario, Suipacha 531, S2002LRK, Rosario, Argentina

DANIEL J. SAYUT
Dept. of Chemical Engineering, University of Massachusetts-
Amherst, 686 N. Pleasant St., Amherst, MA 01003

GARGI SETH
Genentech, Inc., South San Francisco, CA 94080

JOSEPH SHILOACH
Biotechnology Core Laboratory, National Institute of
Diabetic, Digestive and Kidney Diseases, National Institutes
of Health, Bethesda, MD 20892

VERENA SIEWERS
Department of Chemical and Biological Engineering, Chalmers
University of Technology, SE-41296 Gothenburg, Sweden

MARGHERITA SOSIO
KtedoGen, Via Fantoli 16/15, 20138 Milano, Italy, and
NAICONS, Via Fantoli 16/15, 20138 Milano, Italy

MARTIN SPARKS
James R. Randall Research Center,
Archer Daniels Midland Co., Decatur, IL 62521

MELISSA ST. AMAND
Dept. of Chemical Engineering, University of Delaware,
Newark, DE 19711

ULRICH STRYCH
Dept. of Biology and Biochemistry, University of Houston,
4800 Calhoun Rd., Houston, TX 77204-5001

KARAN S. SUKHIJA
Department of Chemical Engineering, University of
Waterloo, Waterloo, Ontario, Canada N2L 3G1

JIBIN SUN
Chinese Academy of Sciences, Tianjin Institute of
Industrial Biotechnology, 300308 Tianjin, China, and
Hamburg University of Technology, Institute of Bioprocess
and Biosystems Engineering, Denickestr. 15, 21073
Hamburg, Germany

LIANHONG SUN
Dept. of Chemical Engineering, 221 Goessmann Lab,
University of Massachusetts-Amherst,
686 N. Pleasant St., Amherst, MA 01003

SEIKI TAKENO
Department of Bioscience and Biotechnology, Faculty of
Agriculture, Shinshu University, Nagano 399-4598, Japan

WENG LIN TANG
Department of Chemical and Biomolecular Engineering,
University of Illinois at Urbana-Champaign, 600 South
Mathews Ave., Urbana, IL 61801

SANDRA TORRIANI
Department of Science, Technology and Market of Vine and
Wine, University of Verona, 37134 Verona, Italy

KELLI TREI
James R. Randall Research Center,
Archer Daniels Midland Co., Decatur, IL 62521

ROBERT P. VAN HILLE
Centre for Bioprocess Engineering Research (CeBER),
Department of Chemical Engineering, University of Cape
Town, Private Bag X3, Rondebosch 7701, South Africa

JOHAN E. T. VAN HYLCKAMA VLIEG
Danone Research, Gut and Microbiology Platform, R.D. 128,
91767 Palaiseau Cedex, France

BERT VAN LOO
Department of Biochemistry, University of Cambridge,
Cambridge CB2 1GA, United Kingdom

CARYN VENGADAJELLUM
Centre for Bioprocess Engineering Research (CeBER),
Department of Chemical Engineering, University of Cape
Town, Private Bag X3, Rondebosch 7701, South Africa

XIAORU WANG
Natural Products Discovery Group, Department of Chemistry
and Biochemistry, 620 Parrington Oval, Room 208,
University of Oklahoma, Norman, OK 73019

STEVEN M. WELLS
Verenium Corporation, San Diego, CA 92121

RICHARD C. WILLSON
Dept. of Chemical and Biomolecular Engineering, University
of Houston, 4800 Calhoun Rd., Houston, TX 77204-4004

GIDEON M. WOLFAARDT
Dept. of Chemistry and Biology, Ryerson University, 350
Victoria St., Toronto, Ontario, Canada, M5B 2K3

FENG XU
Novozymes, 1445 Drew Avenue, Davis, CA 95618

XIAOMING YANG
Process and Product Development, Amgen Inc.,
Thousand Oaks, CA 91320

ZHAOMIN YANG
Department of Biological Sciences, Virginia Tech,
Blacksburg, VA 24061

AN-PING ZENG
Hamburg University of Technology, Institute of Bioprocess
and Biosystems Engineering, Denickestr. 15, 21073
Hamburg, Germany

JINYOU ZHANG
Immunomedics, Inc., 300 American Road,
Morris Plains, NJ 07950

NINGYAN ZHANG
Merck Research Laboratories, Merck & Co. Inc.,
Sumneytown Pike, P.O. Box 4, West Point, PA 19486

HUIMIN ZHAO
Department of Chemical and Biomolecular Engineering
and Institute for Genomic Biology, and Departments of
Chemistry, Biochemistry, and Bioengineering, University of
Illinois at Urbana-Champaign, 600 South Mathews Ave.,
Urbana, IL 61801

Preface

Industrial microorganisms have contributed enormously to the pharmaceutical, food, beverage, and chemical industries in the past, and are rapidly advancing in the production of renewable energy; they have been central in the development of an ever-expanding field of biotechnology.

It is hard to imagine what life would be like without the fruits of industrial microbes. Some aspects of industrial microbiology have not changed much over the years. The production of fine wines has been refined and perfected, and there is probably not much yet to be learned by studying the molecular biology of the specific yeast cultivars. Chemical mutagens still work just as they did when the first volume of *Manual of Industrial Microbiology and Biotechnology* (MIMB) was published in 1986, and they still provide powerful empirical approaches for strain improvement. However, most of the methodologies that can be applied to industrial microbes and biotechnology have advanced dramatically, even since the second volume of MIMB was published in 1998. It is because of these advances that it is now important to provide the scientific community, including academic and industrial scientists and science administrators, an updated and totally new 3rd edition of MIMB.

Many of the chapters in the first two editions continue to be relevant resources for important methods, and these have not been replicated in this edition. Rather, the 3rd edition has focused on advancements that have brought us to the current state of the art and science in 2010. As such, it contains a totally new section VI devoted to "Microbial Fuels (Biofuels) and Fine Chemicals," and a much heavier emphasis on mammalian cell culture methods. Genomic and other "omic" approaches are widely applied in the biological sciences, and these methodologies are incorporated in many chapters of the current edition. Methods to enhance the capacity of microbes for use in bioremediation are becoming ever more critical to the maintenance of a clean environment.

Microbes continue to yield novel secondary metabolites and novel enzymes, and the current status of this field is covered in section I. Industrial strain improvement remains a very productive approach to improve the economics of microbial processes, and section III covers methods applicable to corynebacteria, *Clostridium* species, myxobacteria, *Escherichia coli*, *Saccharomyces cerevisiae*, other yeasts and filamentous fungi, and mammalian cells. Even wine yeasts can be improved! Combinatorial biosynthetic and focused genetic engineering methodologies directed at the production of novel secondary metabolites have advanced substantially in recent years, and section IV covers approaches to engineer glycosylation, poyketide assembly, and nonribosomal peptide assembly in *Streptomyces*, myxobacteria, and *E. coli* among others. Microbial enzymes continue to be important industrial catalysts which are amenable to directed evolution and refinement, and these subjects are covered in section V. Finally, fermentation and cell culture methods and biological engineering and scale-up methodologies continue to be key elements in industrial microbiology and biotechnology, and are updated in sections II and VII.

We sincerely hope that the current edition will be of use to researchers, administrators, and students throughout the world.

RICHARD H. BALTZ
ARNOLD L. DEMAIN
JULIAN E. DAVIES

ISOLATION AND SCREENING FOR SECONDARY METABOLITES

I

THE OPENING SECTION OF THE 2ND EDITION OF THE MANUAL OF *Industrial Microbiology and Biotechnology* (MIMB2) described the initial stages of establishing an industrial microbiological process: culture isolation and preservation, screening, small-scale fermentation, and process optimization. Much of what was recommended in those discussions remains valid and relevant for developing modern industrial processes; what the contributors to the present Third Edition (MIMB3) provide is new vision and options with which to approach such objectives. The current rapid pace of scientific discovery and technological innovation is such that completely novel options have become available to the industrial microbiologist in the quest for new products and processes since the last edition was prepared. Arguably the most dramatic event changing the face of search and discovery has been the accelerating move from a culture-based to a genomics-based strategy. This is set, not least, against the background of a failure of combinatorial chemistry to deliver exploitable drug leads. The challenge, therefore, rests on the evolutionary power of natural biological processes to fulfill the requirement for new drugs and other products of this increasingly biotechnology-dependent age.

The contributions in this section focus broadly on two themes, namely, some central issues pertaining globally to industrial microbiology (isolation and characterization of organisms), and the routes to biodiscovery. Although each chapter addresses specific questions and provides appropriate intellectual and practical support for their resolution, a number of recurring themes can be found in the advice offered by authors having different professional backgrounds and interests. Examples include the increasing dependence on integrated skills in order to develop successful screening protocols or uncover and exploit silent biosynthetic pathways (SBPs), and the strong desirability of using efficient dereplication procedures at the chemical, organism, and genomic levels so as to avoid unnecessary and cost-ineffective redundancy during the early stages of process or product development.

A recurrent source of embarrassment for microbiologists is the fact that only a minute proportion of the estimated numbers of, in particular, bacteria have been isolated and maintained in the laboratory. Slava Epstein and his colleagues in chapter 1 tackle this dilemma head-on and argue that because such a major part of microbial diversity remains "missing" it hampers progress in basic and applied microbiology: cultures enable the comprehensive study of whole organisms and the elucidation of complete genomes. These authors have developed techniques with which to target microorganisms in their natural habitats such that they are grown in situ and subsequently isolated and domesticated under laboratory conditions. All-importantly, the approach recognizes the impact of naturally occurring growth conditions and microbial interactions in enabling cultivation in ways that attempts at in vitro cultivation rarely achieve. In the following chapter, Michael Goodfellow presents methods for selectively isolating members of one particular group of bacteria—the actinobacteria. Although these organisms are viewed traditionally as typical members of soil and fresh-water ecosystems, their distribution in the biosphere now is known to be among the widest of all bacteria, and they remain a prime choice in the search for novel industrially important natural products and biocatalysts. Goodfellow shows that intelligent manipulation of standard media components and incubation conditions, often guided by ecological and taxonomic database information and allied to the probing of poorly studied and neglected habitats, continues to yield isolates of new species and genera having biotechnological value. In this context, the isolation of members of deep lineage actinobacteria remains as a timely challenge.

Two essential tasks need to be undertaken following the isolation of an organism(s) that possesses the required properties: culture curation (described in MIMB2) and taxonomic characterization. All too often, the latter exercise is ignored or treated superficially, but as Giovanna Felis and her coauthors opine in chapter 3, such characterization is essential for organism identification, rigorous scientific communication, securing intellectual property rights, and establishing taxonomic road maps to genes and products. The importance of a polyphasic taxonomic approach is reiterated in this chapter, and "how-to" details are provided with regard to prokaryotes (similar strategies being entirely apt to eukaryotic microorganisms). Felis et al. also consider the likely impact of whole-genome sequencing on taxonomy, which, given the speed and affordable cost of genome sequencing, might be seen as the "ultimate source" for taxonomic characterization. However, they rightly caution that phenotypic analysis remains valid in order to complete the organism's characterization, to provide a more precise idea of how it interacts with its environment, and to give clues to its industrially exploitable biological activities.

Turning to the question of biodiscovery, the sine qua non is to design search methods that avoid, or at least minimize, rediscovery of known entities. To this end it is desirable to isolate new organisms, explore new ecological niches, design new screens, and apply new methods for detecting natural products. These elements of the discovery process are clearly revealed in chapter 4 by Don Cowan and colleagues, who focus on obtaining biocatalysts

from extreme environments, the so-called extremozymes. An important part of this contribution is devoted to the sampling of extreme environments and extracting DNA from such extreme samples. There follow detailed protocols for producing metagenomic libraries, host-vector systems for use with extremophiles and the expression of extremozymes, and activity screens. In chapter 5, Stefano Donadio and Margherita Sosio describe cell-based assays in screening for anti-infective compounds and argue the case for these screens vis-à-vis cell-free systems. They guide the reader along the critical path from target identification and validation to assay development and assay implementation, and introduce a range of options that include reporter, antisense, and reversion assays. Important recommendations are made with regard to screening algorithms and the evaluation of target "hits."

The final chapters in section I examine two rapidly evolving components of the search and discovery toolbox, namely, detecting the metabolic products of expressed genes (metabolomics) and probing genomes for the potential to synthesize new products (genome mining of SBPs). Metabolomics is discussed by Jens Frisvad, who at the outset reminds us that the metabolome is a snapshot of an organism's metabolic capacity under a specific set of growth conditions, and that its definition is crucially dependent on the analytical techniques used for metabolite detection. Thus, maximizing expression and detecting compounds again emphasize the necessity of an integrated skills platform. Frisvad demonstrates the steps in metabolome analysis with reference to microfungi, but the protocols are pertinent to all microorganisms. Finally, strategies for accessing microbial secondary metabolites from SBPs are presented in detail in chapter 7, by Robert Cichewicz and his coworkers. While acknowledging the value of culture condition manipulation as a means of triggering SBPs, these authors show that a new suite of genomic-driven, mechanism-based, and hybrid approaches is providing an unprecedented opportunity for identifying SBPs. Such approaches include genome scanning, heterologous and in situ expression of SBP genes, and chemical epigenetic and genoisotopic methods. Yet again, the integration of microbiological and chemical techniques is a key element in the successful application of these techniques. Especially helpful features of this chapter include an analysis of the major strengths and weaknesses of the various methodological approaches, and case studies that guide the reader on such points.

New Approaches to Microbial Isolation

SLAVA S. EPSTEIN, KIM LEWIS, DOMINICA NICHOLS, AND
EKATERINA GAVRISH

1

1.1. INTRODUCTION

It has been known for over a century that most microorganisms do not grow under standard laboratory conditions (2, 7, 20, 32, 34–36, 39). At first, this appeared to be a cell number phenomenon: cells in the sample grossly outnumbered the number of colonies they produced on artificial media. The introduction of the rRNA approach to the study of microbial diversity (25) revealed that it was also a species diversity phenomenon: the number of microbial species in nature proved much larger than the richness of established culture collections (10, 12, 14, 16, 18, 22, 29, 33). In 1985, a term was coined to describe these observations: the Great Plate Count Anomaly (30).

Metagenomics approaches promised to provide access to the uncultivated species while bypassing the problem of uncultivability (11, 17). Many of these promises have been fulfilled, resulting in remarkable findings (4). The application of these and high-throughput sequencing approaches also showed that the microbial diversity in nature is so staggering that even the largest metagenomic efforts still sample only a small fraction of the metagenomes of microbial communities (18). Metagenomics libraries are dominated by gene sequences with unknown functions, and this limits the knowledge gained from the community genome. In practical application, metagenomic approaches have not yet produced a robust pipeline of bioactive compounds from uncultivated species. Microbial culture thus remains a basic necessity for the study of microbes in nature, and a prerequisite for drug discovery from natural sources. However, over a hundred years after its discovery, the anomaly remains unresolved, and more than half of microbial divisions are still without a single cultivable representative (27, 28). The fact that the major part of microbial diversity remains "missing" hampers progress in both basic and applied microbiology, and is widely recognized as one of the most important challenges facing microbiology today (19, 37).

This state of affairs set the stage for a recent resurgence of cultivation efforts (3, 5, 6, 8, 9, 13, 15, 21, 24, 26, 31, 38; D. Nichols, N. Cahoon, E. M. Trakhtenberg, L. Pham, A. Mehta, A. Belanger, T. Kanigan, K. Lewis, and S. S. Epstein, submitted for publication). Informing these efforts was the principal lesson from the past: traditional manipulation of the standard media components likely brings only incremental success. Significant departures from conventional techniques were clearly in order, and indeed, the new technologies substantially diverged from the cultivation tradition by adopting single-cell and high-throughput strategies (8, 27, 39; Nichols et al., submitted), better mimicking the natural milieu (3, 6, 13, 31) and increasing the length of incubation and lowering the concentration of nutrients (9). Our research group contributed to the effort by developing three cultivation methodologies (15, 21; Nichols et al., submitted). Their gist is to replace the in vitro cultivation by growing microorganisms in situ, and thus provide target microorganisms with their natural growth conditions, followed by domestication of the isolates for growth under laboratory conditions. This chapter describes the principle of these methods, details their application and the initial results of this application, and summarizes the lessons learned.

1.2. METHOD 1: DIFFUSION CHAMBER

1.2.1. Principle of the Method

Designing this method to cultivate environmental microorganisms, we considered that natural environments by definition provide the necessary growth conditions to the microorganisms inhabiting these environments. No matter how transient such conditions may be for any given species, they do occur, and the species grows. It follows that by placing a sample of environmental cells into a diffusion chamber, and incubating this chamber in the natural environment of these cells, we could ensure that at least some of them would be supplied by the required growth factors via diffusion. If so, such cells should grow inside the diffusion chamber just as their kin cells would outside the diffusion chamber. The key element of this strategy is incubation of the diffusion chamber in situ, which allows diffusion to occur while preventing cells from migrating in and out of the chamber. Reducing this idea to practice, we tested a number of different membranes and different ways to construct the chamber, and designed a simple device (21) described below.

1.2.2. Diffusion Chamber Design

We construct the diffusion chambers (Fig. 1A) using stainless steel washers (70 mm outside diameter, 33 mm interior diameter, 3 mm thickness; Bruce Watkins Supply, Inc., Wilmington, NC) and 0.03-μm-pore-size membranes

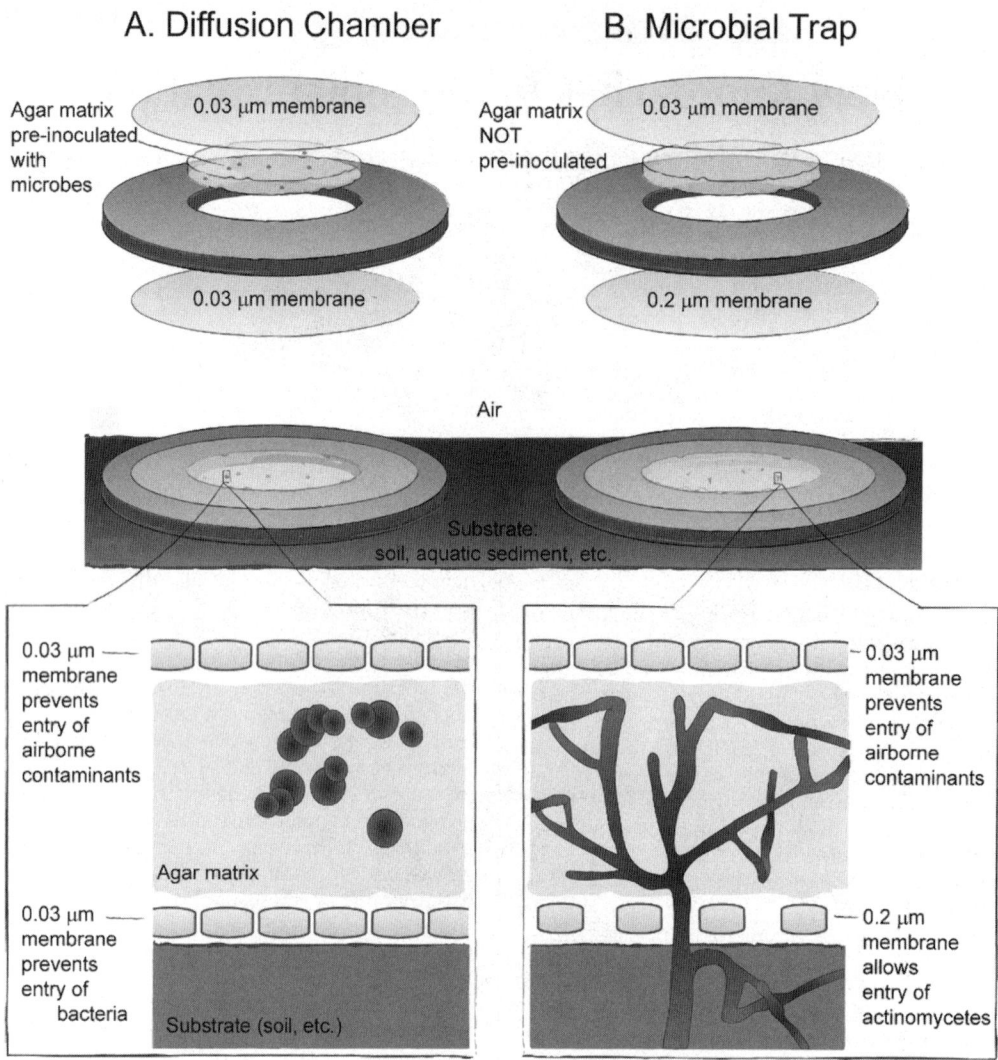

FIGURE 1 Design and application of diffusion chamber (A) and microbial trap (B). Explanations in the text.

(Osmonics, Inc., Westborough, MA). After a membrane is glued to one side of the washer using Silicone Glue II (General Electric, Waterford, NY), the half-assembled chamber is kept in a laminar flow hood until bonding is complete. Meanwhile, we prepare the inoculum by mixing environmental cells with warm (and still liquid) agar (0.7 to 1.5% final concentration). Note that the agar does not have to be supplemented with nutrients, as they are supposed to diffuse into the chamber during in situ incubation. However, when targeting specific microorganisms, their selection may be enhanced by relevant supplements, for example, with cellulose when species capable of cellulose degradation are of interest.

The diffusion chamber accommodates approximately 3 ml of the cell/agar mix; in a typical experiment this inoculum contains between 10^4 and 10^5 cells. Our experience with soil and aquatic environments shows that such inocula usually produce a number of colonies that is optimal for observations and subculturing. Once the 3-ml inoculum is administered into the chamber by pipetting, it fills the space formed by the bottom membrane and the opening in the washer. We then seal the chamber by

gluing the second membrane on the upper surface of the washer and wait for the glue to bond in a laminar flow hood. It is critical to provide extra moisture to the chamber during this drying period and we accomplish this by placing the drying chambers on moist paper. This second membrane sandwiches the microbe sample/agar mixture within the chamber between two membranes, and prepares the chamber for incubation.

The chamber can be incubated in two different ways. The first is to simply return the chamber to the place of the cells' origin. This is convenient for environments that are both close to the home laboratory and experience little recreational use. Often this approach is impractical because one or both of these conditions are not met. In these cases, the chamber can be incubated in a simulated natural environment, such as a block of freshly collected aquatic sediment or soil kept in an aquarium at ambient temperature. In our experience, this mimics the natural environment well for 2 to 4 weeks after the sample collection. As the sample ages, it exhibits signs of a shift to a new community represented by few abundant species. This results in an impoverished collection of species grown inside the diffusion chamber, dominated by strains of uncertain environmental relevance. We note that this observation is strictly qualitative, and we have not examined the phenomenon in detail.

The length of incubation largely depends on the nature of the target environment, and could be best determined empirically. Applying the method to boreal aquatic environments and soils, we typically incubate the chambers for 3 to 4 weeks if the ambient temperature is between 10 and 20°C, and for 2 to 3 weeks at higher temperatures. When the method is applied to samples from human microbiome or sewage treatment facilities, incubations under 1 week appear sufficient to obtain microbial growth adequate for observation and subculturing.

After incubation, the diffusion chambers are retrieved and opened by cutting off the upper membrane with a sterile razor blade run along the interior diameter of the washer. Direct examination of the grown material, e.g., under a dissecting microscope, is not an efficient way to quantify growth, because most of the colonies formed during incubation are of microscopic size (Fig. 2A). In addition, the translucent nature of the bottom membrane interferes with such observation. Instead, we subsample the agar material by taking randomly distributed, measured cores using Pasteur pipettes as coring devices. As a rule, cores 1 to 30 μl in volume, depending on the density of growth, are sufficient for further examination and colony counting. Placed between a glass slide and coverslip, the flattened cores are examined

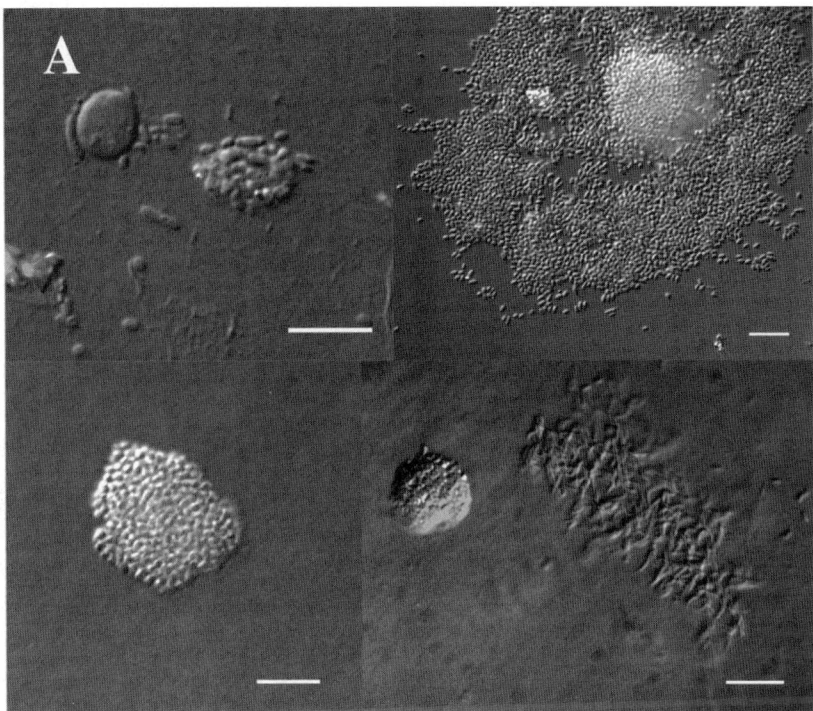

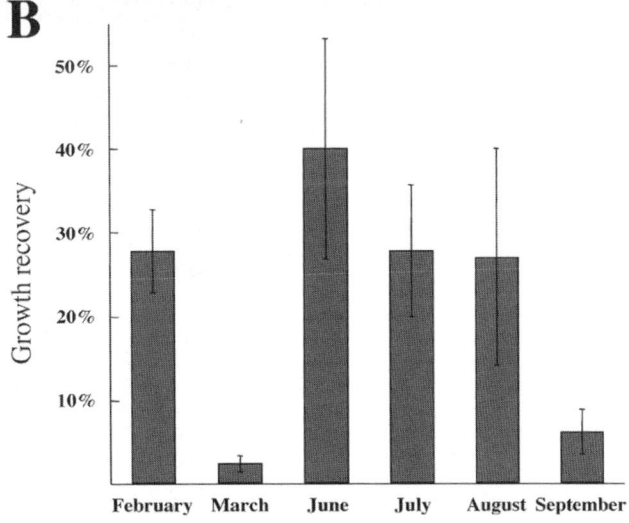

FIGURE 2   (A) Microcolonies of marine microorganisms grown inside a diffusion chamber incubated in simulated natural environment. Bars, 5 μm. (B) Growth recovery of environmental cells in diffusion chambers plotted as percentage of inoculated cells forming colonies.

in their entirety at 400 to 1,000× magnification under a compound microscope equipped for differential interference contrast (DIC), the latter being crucial for colony visualization. In a typical application, we record between 1 to 5 and several hundred microcolonies per replicate. The counts are then used to estimate the total number of colonies grown in the chamber, and these estimates are compared with the total number of cells inoculated to calculate percentage of recovery.

Subsampling grown colonies for subculturing represents one of the more time- and effort-consuming aspects of the overall protocol. If the colonies are rare in the diffusion chamber agar, they may be difficult to locate. However, increasing their number by inoculating more cells leads to a new difficulty: frequent failure to sample cells from one—and only one—colony. The next round of growth thus results in a mixed culture, necessitating additional purification efforts. One recommended way to pick up colonies from the diffusion chamber-grown material is to place a sample of this material under a compound microscope, verify the target colony under high magnification (e.g., using a dry 40× objective with a long working distance), and then sample this colony using tungsten wires (75-μm shaft, <1 μm at tip; FHC, Inc., Bowdoinham, ME). This material is then syringe mixed in a small volume of sterile water (seawater in marine application) and inoculated into a new diffusion chamber. If this produces a mixed culture, a series of chamber-to-chamber transfers may be necessary to achieve purity. Purity is confirmed by a combination of light microscopy at high magnification, scanning electron microscopy, and analyses of PCR-amplified 16S rRNA gene fragments.

1.2.3. Diffusion Chamber Applications

We have applied the method across a variety of different environments to quantify microbial growth inside the chambers, and assessed what part of grown species could be domesticated (subcultured under conventional laboratory conditions).

1.2.3.1. Marine Sediments Application

In this application, we grew microorganisms inhabiting a sandy tidal flat in Massachusetts Bay, close to one of our home laboratories at the Marine Science Station of Northeastern University, Nahant, MA (21). Microorganisms from the uppermost layer of these sediments were detached from sediment grains by vortexing 1-g sediment samples in 5 ml of autoclaved seawater for 5 min. The heaviest sediment particles in the supernatant were allowed to settle, and subsamples of the supernatant were mixed with 2.5 ml of warm (40°C) agar supplemented with 0.01% technical grade casein (Sigma-Aldrich, St. Louis, MO). The cells/agar mix was inoculated into the diffusion chambers as described above. Some chambers received agar but not cells; these served as a contamination control. Inoculated growth chambers were incubated in an aquarium containing freshly collected intertidal sediments and recirculating natural seawater. After 1 to 3 weeks of incubation, the chambers were opened and microbial colonies were examined, counted, and subcultured as described above. This experiment was repeated 13 times over a period of 7 months; its main results are given in Fig. 2B. The number of cells forming colonies in diffusion chambers averaged approximately 22%, exceeding hundredfold what could be achieved using standard petri dishes. This strongly indicated that the diffusion chamber-based approach to microbial cultivation

likely provides access to some of the previously uncultivated species. Indeed, in this and follow-up research (5, 24) we isolated numerous species that did not grow in petri dishes inoculated with environmental samples, but could be grown and maintained in the diffusion chambers.

Cultivation of novel species inside diffusion chambers is a welcome development, but their properties and abilities could be fully studied and utilized only if they could be adapted for growth in vitro. We therefore examined whether repetitive cultivation in a series of generations of diffusion chambers facilitated domestication of the grown strains, and discovered a positive correlation between the number of cultivation rounds in situ and the probability of obtaining a variant capable of growth in vitro (24). Of the 23 strains tested, approximately a quarter acquired the ability to grow on standard media after just one round, and this proportion steadily grew to 70% after four cultivation rounds (24). The nature of the domestication process remains unclear, but the empirical observations suggest that rounds of in situ cultivation adapt a significant number of otherwise "uncultivable" strains to growth under conventional conditions of standard petri dishes. This important observation was confirmed in follow-up studies conducted in different environments (see below).

1.2.3.2. Freshwater Sediment Application

We applied the diffusion chamber-based method to freshwater sediments to cultivate microbial species from the Turtle Pond, a small freshwater pond in Boston, MA (5). Pond sediment material was diluted in agar supplemented with Casamino Acids (0.01%), yeast extract (0.01%), and hot water extract from the sediment (0.1%), in various combinations, and inoculated into diffusion chambers. These were incubated on top of a freshly collected block of sediment kept under a layer of aerated extant water in an aquarium. After 4 weeks of incubation, the chambers were opened, the material was homogenized by passaging the material through a syringe equipped with a gauge 25 needle and diluted with sterile, filtered pond water, and part of the material was incubated in the next generation of diffusion chambers. The remaining material was used to prepare standard pour plates supplemented with diluted (10% of manufacturer recommended) Luria-Bertani broth (LB; BD, Franklin Lakes, NJ). Reincubation in new diffusion chambers was performed once again, to a total of three rounds of in situ cultivation, each time with parallel subculturing in standard petri dishes. In this way, we could assess if new species would appear by repetitive incubations of the chambers, and what species from thus grown diversity could be grown in vitro. Indeed, 70% of the 438 strains isolated came from the diffusion chamber-reared material, and several strains unique to this approach only appeared after two or more rounds of cultivation in situ; notably, these represented rarely cultivated phyla Verrucomicrobia and Acidobacteria. Overlap between species composition grown in standard petri dishes inoculated with pond material directly versus petri dishes inoculated with cells first grown in diffusion chambers was minimal (7%). This confirmed a similar observation made in a different environment (24) that one or more rounds of growth inside diffusion chambers leads to domestication of novel microbial species that are difficult to grow otherwise.

1.2.3.3. Application to Contaminated Environments

We also used this method in contaminated soils and groundwaters in several areas around the Field Research Center,

Oak Ridge National Laboratory, TN. These data are being prepared for publication, but the principal results mirror those from the above-mentioned studies of marine and freshwater environments: the diffusion chamber-based method allows for growth and domestication of novel species from contaminated soils and groundwater. It also brings into culture organisms previously known only from the rRNA gene and metagenomics surveys of the sampled areas (data not shown). This indicates that some microbial species, including the likely bioremediation players, do not grow under standard conditions unless first preincubated in the diffusion chambers deployed in situ.

1.3. METHOD 2: ICHIP

1.3.1. Principles of the Method

In a typical application, the diffusion chamber described above is inoculated with a mix of environmental cells, and so the first growth event leads to a mixed culture. This necessitates additional purification and isolation efforts, which are often substantial and limit the throughput. To resolve this bottleneck, we developed a variant of the diffusion chamber method that transforms it into a high-throughput platform for massively parallel microbial isolation. While based on the same principle (placing microorganisms into diffusion chambers for incubation in vivo), it differs significantly from the original chamber design described above. The key element of this new method is the Isolation Chip, or ichip for short (Fig. 3; Nichols et al., submitted), which is a combination of hundreds of miniature diffusion chambers loaded with an average of one cell per chamber. The ichip thus represents a tool helping to achieve microbial growth and isolation in a single step.

1.3.2. Ichip Design

The ichip consists of several elements: the central plate, the main function of which is to house growing microorganisms; semipermeable membranes on each side of the plate, which separate the plate from the environment; and two side panels, playing mainly a structural role (Fig. 3). The central plate and side panels have multiple matching through holes. When the central plate is dipped into a suspension of cells in molten agar (Fig. 3A), its through holes capture miniature volumes of this suspension that

solidify in the form of small agar plugs. These plugs are sealed from both each other and the outside environment by membranes pressed against the central plate by the side panels, fastened with screws (Fig. 3C). The seal transforms the assembly into an array of small diffusion chambers, each filled with agar and cells. The number of cells inside each through hole is the function of the initial dilution rate in molten agar, and thus can be controlled. When an average agar plug contains a single cell, the majority of through holes will contain individual cells (Fig. 3B). The assembled ichip is incubated in nature (or simulated natural environment) similarly to the diffusion chambers of the original design (Fig. 1A). The growth of individual cells automatically leads to pure cultures inside the through holes, minimizing isolation efforts.

The ichips currently in use in our laboratories were manufactured by HI-TECH Manufacturing (Schiller Park, IL). The plastic elements are machined from blocks of hydrophobic plastic polyoxymethylene, known as Delrin®. All the through holes are uniform in their diameter (1 mm), and are arranged in two arrays with 192 through holes per array. The size of the array is such that it can be completely covered by standard 25- or 47-mm-diameter membranes. It is also possible to use commercially available sheet membranes that could be cut to the appropriate size in advance.

Prior to assembly, we sterilize the plastic components in ethanol, dry them in a laminar flow hood, and rinse the parts in particle-free DNA grade water (Fisher Scientific, Hampton, NH). After incubation, ichips are washed vigorously in particle-free DNA grade water, and disassembled. The central plate is then examined under a compound or high-power dissecting microscope for colony count. The colonies are removed from the ichip's central plate by touching agar plugs with unwound and sterile #1 gauge paper clips for further analyses and/or subculturing (e.g., in microtiter plates).

In a series of preliminary tests, we checked whether the pressure applied by tightening the screws was sufficient to prevent cells from migrating in and out of the agar plugs. Ichips were loaded with sterile 1% agar (BD, Franklin Lakes, NJ), assembled, submerged into *Escherichia coli* K-12 culture growing in LB, and incubated for 24 h. After incubation, the content of ichips was examined for growth under a compound microscope. In parallel, ichips loaded with *E. coli* K-12 cells mixed with 1% warm LB

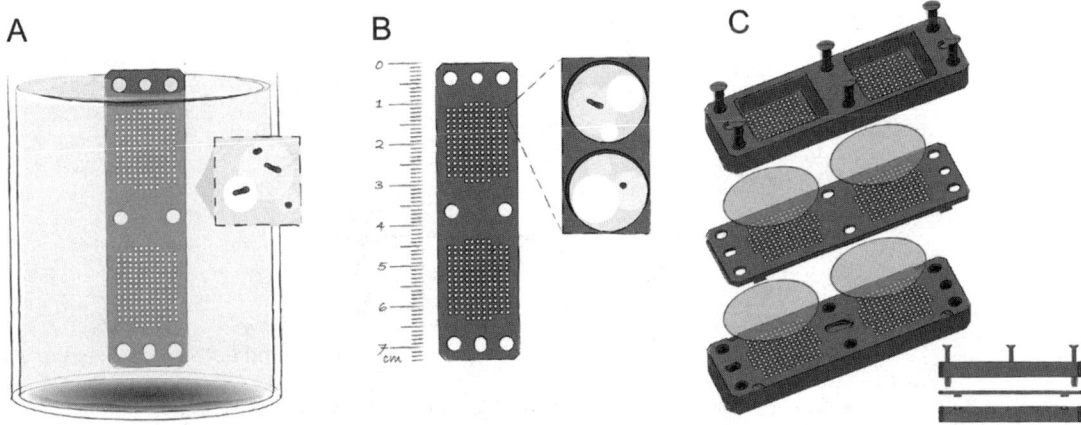

FIGURE 3 Ichip design. Explanations in the text.

agar were incubated for 24 h while bathed in sterile LB. The outside medium was then examined for growth. In neither case were microbial cells observed crossing the barrier established by the membranes (Nichols et al., submitted).

1.3.3. Ichip Applications

In a proof-of-concept study, we used the ichip to grow marine water column and soil microorganisms, recorded the colony count, and compared the rRNA gene diversity of ichip-grown microorganisms and their colony count with microorganisms from parallel incubations in standard petri dishes (Nichols et al., submitted). Seawater and waterlogged soil samples were obtained from the respective environments in the vicinity of the Marine Science Center of Northeastern University, Nahant, MA. Microbial cells were dislodged from soil particles by two 10-s-long sonication pulses (Sonics Vibra-Cell VC130; 3-mm stepped microtip; Sonics & Materials, Inc., Newtown, CT) and collected after larger particles settled for 60 s. Seawater samples were used without sonication. All cells were enumerated with 4′,6-diamidino-2-phenylindole (DAPI) (Sigma-Aldrich, St. Louis, MO) and diluted in warm diluted (0.1% wt/vol) LB agar to a concentration of 10^3 cells/ml, with or without supplementation with 4% sea salts (Sigma-Aldrich, St. Louis, MO). In the present ichip configuration (Fig. 3), each through hole captures approximately 1.25 µl of agar, or 500 µl of cell/agar mix per assembly. Given the dilution, each loaded through hole contained 500/384 = 1.25 cells (on average). To approximate the species diversity inoculated into ichips, for conventional cultivation we established miniature petri dishes with 500 µl of the cell/agar mixes in 24-well culture plates (Corning Costar, Corning, NY). The ichips were returned to the respective environments of cells within,

incubated for 2 weeks, and disassembled. Approximately 50 agar plugs from each ichip were removed for a detailed microscopic examination and colony counting, whereas the rest served as a source of DNA for 16S rRNA gene sequence-based identification of grown species. Agar material from the conventional petri dishes was processed in an identical manner.

This study led to three important observations. First, 40 to 50% of cells incubated in ichips formed colonies, which exceeded the petri dish recovery fivefold. Second, the ichip- and petri dish-grown collections of microorganisms were dramatically different and shared only one species (*Vibrio* sp.). This lack of overlap between the two species lists was not due to undersampling, because petri dishes inoculated with seawater and soil samples did share several species, as did ichips inoculated with microorganisms from these two sources. Third, the novelty of isolates grown in ichips significantly exceeded that of petri dish-reared species (Fig. 4). Our general conclusion is that the ichip-based method enables growth of a significant fraction of microbial diversity, and produces unique microbial collections of high phylogenetic novelty; it does so in a time- and effort-efficient manner, compatible with contemporary tools for high-throughput analyses, such as microtiter plates.

We note that the flexibility of the ichip design makes it useful for a variety of applications beyond those described here. For example, a single incubation of the device currently in use (Fig. 3) affords tens of microbial cultures, and cultivation efforts can be easily scaled up via minimal changes in configuration of the ichip. Matching the position of through holes with wells of microtiter plates paired with commercially available replicators that have pins spaced to match the wells will allow massively parallel growth and isolation of hundreds of cultures per incubation, with minimal labor involved. Further modifications of the ichip may make

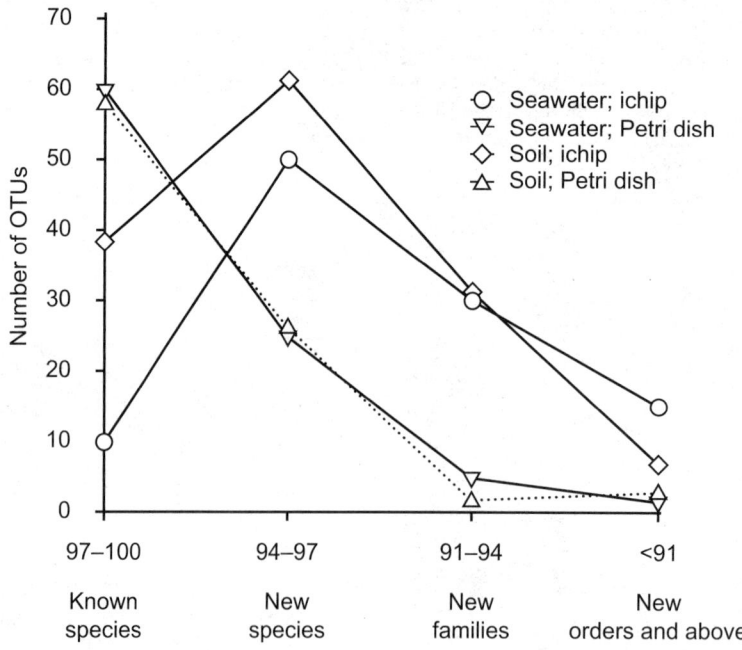

FIGURE 4 Novelty of seawater and soil microbial strains grown in ichip and petri dish. Equation of sequence novelty, in percentage of diversion from the known species, and taxonomic rank of novelty (genus, family level, etc.) is approximate. OTU, operational taxonomic unit.

it suitable to grow previously uncultivated microorganisms, not only from natural environments, but also from the human microbiome. For example, miniaturization of the ichip design opens a possibility of its incubation in the oral cavity. In this application, microorganisms from a dental plaque can be loaded into a mini-ichip (Fig. 5A and B), which is then incubated on a retainer inside the volunteer mouth (Fig. 5C). This is important because >50% of oral microflora has not been cultivated yet, and the uncultivated fraction contains species implicated in oral diseases (1, 23). Equally simple modifications can make the ichip useful to grow microorganisms from other parts of the human microbiome.

1.4. METHOD 3: MICROBIAL TRAP

1.4.1. Principles of the Method

Several types of microorganisms that are particularly important for drug discovery, microscopic fungi and actinomycetes, grow by forming filaments capable of penetrating soft substrates. Considering that actinomycetes can pass through filters with pores as small as 0.2 μm, we hypothesized that the reverse use of the diffusion chamber (Fig. 1A) could selectively enrich for these microorganisms by capitalizing on their unique abilities (trap method, Fig. 1B [15]). Visually, the trap is indistinguishable from the diffusion chamber, but

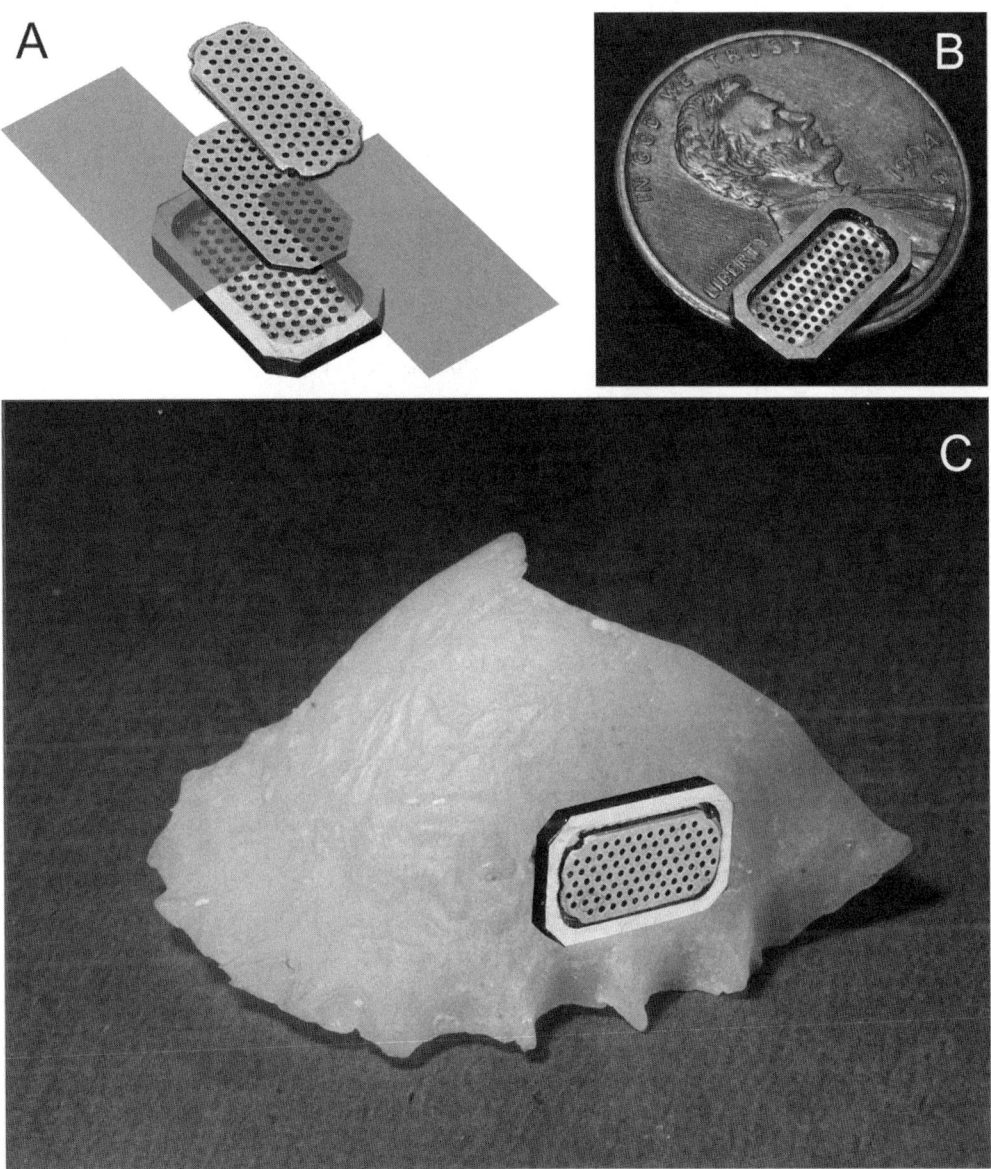

FIGURE 5 Prospective ichip to grow oral microorganisms. (A) The assembly of the device is similar to the ichip currently in use (Fig. 3): the central plate is loaded with target (dental plaque) microorganisms; two additional panels with matching through holes press the 0.03-μm-pore-size membranes against the central plate, creating multiple diffusion minichambers. (B) Relative size of the prospective ichip for oral microflora cultivation. (C) Assembled ichip is fastened in the opening of a mold of the volunteer's upper palate, to be incubated in the mouth in touch with the molar used to sample the plaque.

the main principle of its application is very different. The trap is built from similar components except the membranes have larger pores, and it is *not* inoculated with environmental cells (contrast Fig. 1A and B). Instead, the trap is incubated in the environment with sterile agar inside, with the expectation that diffusion through membranes will establish conditions inside the trap that closely mimic the natural conditions. The pore size above 0.2 µm should allow filamentous organisms active at the time of the experiment to penetrate the membranes, grow into the agar inside the trap, and establish colonies there. Though selected motile microorganisms could in principle cross the membranes and grow inside the trap as well, we hypothesized that the method should enrich for target (filamentous) organisms. These could be further purified by subculturing in diffusion chamber or petri dishes. The test study described below supported these expectations.

1.4.2. Trap Design

Traps are built using 0.2- to 0.6-µm-pore-size polycarbonate membranes, glued to the bottom of washers as in the diffusion chamber method. The trap is filled with 3 ml of sterile 1% agar or 1.2% gellan gum, with or without vitamin supplement. After the medium solidifies, the top polycarbonate membrane, with a 0.03-µm pore size, is glued to the washer, sealing the trap. After the glue dries, the traps are placed on top of the moist soil or marine sediment, ensuring that the bottom filter (with larger pores) is in good contact with the substrate. The small pore size of the upper membranes prevents contamination from the air. In aquatic application, the two membranes can be the same, with 0.2- to 0.6-µm pore sizes, allowing penetration by filamentous organisms from both sides.

1.4.3. Trap Application

We tested the trap performance (15) by incubating these devices on top of garden soil kept in large petri dishes, sealed with Parafilm to prevent evaporation. After 14 to 21 days of incubation at room temperature in the dark, the traps were opened and the slabs of solid agar or gellan

gum were removed, inverted, and placed into a sterile petri dish. The solid disks were examined for growth under a stereomicroscope at 20 to 100× magnification, and visible microcolonies were sampled with sterile needles. In several experiments, the agar slabs were incubated undisturbed for 5 to 7 days prior to colony sampling to allow actinomycetes to form aerial mycelia to facilitate colony picking. All sampled material was streaked on plates with agar or gellan gum medium supplemented with 0.1% Casamino Acids and 0.1% nutrient broth (BD Difco, Detroit, MI), or Actinomycete Isolation Agar (BD Difco, Detroit, MI). Because the experiment aimed to isolate actinomycetes, nystatin (50 µg/ml) and cycloheximide (100 µg/ml) were added to the media in petri dishes to prevent growth of fungi. Subcultivation was repeated to obtain pure cultures. For comparative purposes, we also prepared parallel conventional petri dishes by inoculating them with the air-dried soil material directly, followed by 2-week-long incubation in the laboratory. Similarly to subculturing trap-derived material, conventional petri dishes were established with agar or gellan gum and were supplemented with the same nutrient additives. Upon incubation and purification, 90 colonies from traps and 90 colonies from conventional petri dishes were identified by sequencing their 16S rRNA genes.

Visual inspection of traps revealed an abundance of actinobacterial colonies (Fig. 6). Traps with 0.4- to 0.6-µm-pore-size bottom membranes contained a significant number of fungal filaments, but cutting the pore size to 0.2 µm effectively excluded fungi. The majority of organisms grown in the traps proved to be actinomycetes, some of which represented rare and unusual species from the genera *Dactilosporangium*, *Catellatospora*, *Catenulispora*, *Lentzea*, and *Streptacidiphilus*. In general, gellan gum-based traps appeared to have captured more diverse assemblage of actinomycetes; conventional petri dishes produced more nonfilamentous actinobacteria. We concluded that the trap method allows for a selective capture of filamentous actinomycetes enriched for new and rare species.

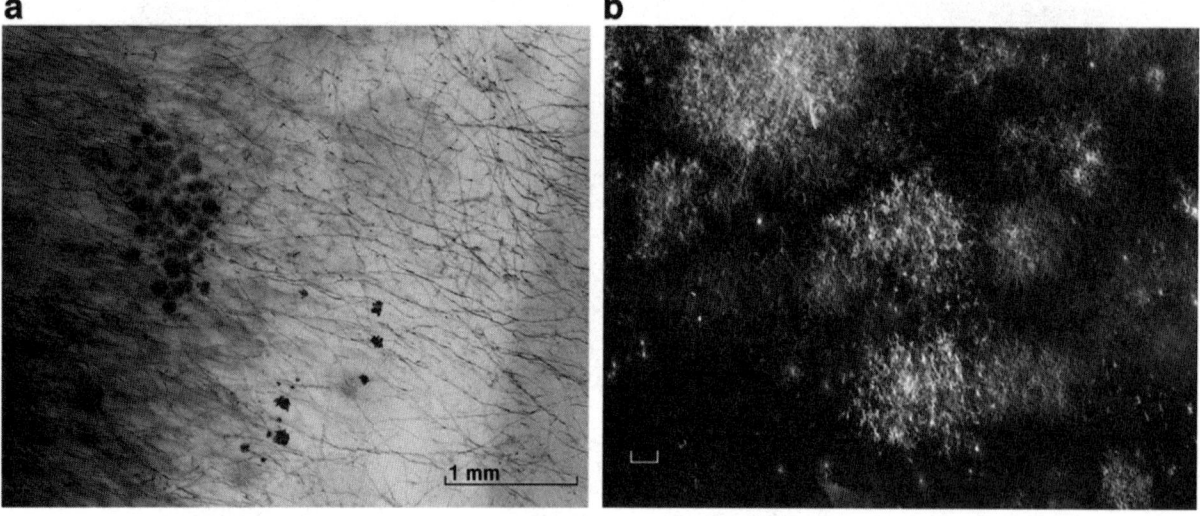

FIGURE 6 Bacterial colonies and fungal hyphae (a) and actinomycetes-like colonies (b; scale, 0.1 mm) grown in a trap after 2 weeks of incubation in garden soil.

1.5. CONCLUSIONS

There is a growing appreciation of microbial cultivation because microbial culture remains central not only to the study and utilization of microbial properties and capabilities, but also to the placement of molecular surveys and metagenomics data into a larger context. With the majority of microbial species remaining uncultivated, there is an urgent need to develop novel methodologies to grow previously uncultivated species from the environment and human and animal microbiomes. One way to gain access to "missing" species is to grow environmental cells in their natural habitat, capitalizing on a naturally occurring suite of growth factors. This can be accomplished by placing target cells in diffusion chambers incubated back in the cells' habitat. We developed several variants of such diffusion chambers, each with its own set of advantages. The diffusion chamber method is a simple and inexpensive tool to grow mixes of novel environmental microorganisms. The ichip design allows for massively parallel, high-throughput isolation of novel microorganisms that may be difficult or impossible to grow by use of conventional techniques. Miniaturization of the ichip makes it suitable to grow presently uncultivated microorganisms from human microbiome. A reverse use of the diffusion chamber leads to a different type of cultivation (trap method), which selectively enriches for new and rare filamentous actinomycetes. This makes the trap method particularly useful for the discovery of novel secondary metabolites.

REFERENCES

1. Aas, J. A., B. J. Paster, L. N. Stokes, I. Olsen, and F. E. Dewhirst. 2005. Defining the normal bacterial flora of the oral cavity. *J. Clin. Microbiol.* **43:**5721–5732.
2. Amann, J. 1911. Die direkte Zählung der Wasserbakterien mittels des Ultramikroskops. *Zentralbl. Bakteriol.* **29:**381–384.
3. Aoi, Y., T. Kinoshita, T. Hata, H. Ohta, H. Obokata, and S. Tsuneda. 2009. Hollow-fiber membrane chamber as a device for in situ environmental cultivation. *Appl. Environ. Microbiol.* **75:**3826–3833.
4. Beja, O., L. Aravind, E. V. Koonin, M. T. Suzuki, A. Hadd, L. P. Nguyen, S. B. Jovanovich, C. M. Gates, R. A. Feldman, J. L. Spudich, E. N. Spudich, and E. F. DeLong. 2000. Bacterial rhodopsin: evidence for a new type of phototrophy in the sea. *Science* **289:**1902–1906.
5. Bollmann, A., K. Lewis, and S. S. Epstein. 2007. Incubation of environmental samples in a diffusion chamber increases the diversity of recovered isolates. *Appl. Environ. Microbiol.* **73:**6386–6390.
6. Bruns, A., H. Cypionka, and J. Overmann. 2002. Cyclic AMP and acyl homoserine lactones increase the cultivation efficiency of heterotrophic bacteria from the central Baltic Sea. *Appl. Environ. Microbiol.* **68:**3978–3987.
7. Cholodny, N. 1929. Zur Methodik der quantitativen Erforschung des bakteriellen Planktons. *Zentralbl. Bakteriol. Parasitenkd. Infektionskr. Hyg. A* **77:**179–193.
8. Connon, S. A., and S. J. Giovannoni. 2002. High-throughput methods for culturing microorganisms in very-low-nutrient media yield diverse new marine isolates. *Appl. Environ. Microbiol.* **68:**3878–3885.
9. Davis, K. E., S. J. Joseph, and P. H. Janssen. 2005. Effects of growth medium, inoculum size, and incubation time on culturability and isolation of soil bacteria. *Appl. Environ. Microbiol.* **71:**826–834.
10. DeLong, E. F. 1992. Archaea in coastal marine environments. *Proc. Natl. Acad. Sci. USA* **89:**5685–5689.
11. DeLong, E. F. 2005. Microbial community genomics in the ocean. *Nat. Rev. Microbiol.* **3:**459–469.
12. Dojka, M. A., J. K. Harris, and N. R. Pace. 2000. Expanding the known diversity and environmental distribution of an uncultured phylogenetic division of bacteria. *Appl. Environ. Microbiol.* **66:**1617–1621.
13. Ferrari, B. C., S. J. Binnerup, and M. Gillings. 2005. Microcolony cultivation on a soil substrate membrane system selects for previously uncultured soil bacteria. *Appl. Environ. Microbiol.* **71:**8714–8720.
14. Fuhrman, J. A., K. McCallum, and A. A. Davis. 1992. Novel major archaebacterial group from marine plankton. *Nature* **356:**148–149.
15. Gavrish, E., A. Bollmann, S. Epstein, and K. Lewis. 2008. A trap for in situ cultivation of filamentous actinobacteria. *J. Microbiol. Methods* **72:**257–262.
16. Giovannoni, S. J., T. B. Britschgi, C. L. Moyer, and K. G. Field. 1990. Genetic diversity in Sargasso Sea bacterioplankton. *Nature* **345:**60–63.
17. Handelsman, J. 2004. Metagenomics: application of genomics to uncultured microorganisms. *Microbiol. Mol. Biol. Rev.* **68:**669–685.
18. Huber, J. A., D. B. Mark Welch, H. G. Morrison, S. M. Huse, P. R. Neal, D. A. Butterfield, and M. L. Sogin. 2007. Microbial population structures in the deep marine biosphere. *Science* **318:**97–100.
19. Hurst, C. J. 2005. Divining the future of microbiology. *ASM News* **71:**262–263.
20. Jannasch, H. W., and G. E. Jones. 1959. Bacterial populations in seawater as determined by different methods of enumeration. *Limnol. Oceanogr.* **4:**128–139.
21. Kaeberlein, T., K. Lewis, and S. S. Epstein. 2002. Isolating "uncultivable" microorganisms in pure culture in a simulated natural environment. *Science* **296:**1127–1129.
22. Liesack, W., and E. Stackebrandt. 1992. Occurrence of novel groups of the domain Bacteria as revealed by analysis of genetic material isolated from an Australian terrestrial environment. *J. Bacteriol.* **174:**5072–5078.
23. Marsh, P. D. 2003. Are dental diseases examples of ecological catastrophes? *Microbiology* **149:**279–294.
24. Nichols, D., K. Lewis, J. Orjala, S. Mo, R. Ortenberg, P. O'Connor, C. Zhao, P. Vouros, T. Kaeberlein, and S. S. Epstein. 2008. Short peptide induces an "uncultivable" microorganism to grow in vitro. *Appl. Environ. Microbiol.* **74:**4889–4897.
25. Olsen, G. J., D. J. Lane, S. J. Giovannoni, N. R. Pace, and D. A. Stahl. 1986. Microbial ecology and evolution: a ribosomal RNA approach. *Annu. Rev. Microbiol.* **40:**337–365.
26. Rappe, M. S., S. A. Connon, K. L. Vergin, and S. J. Giovannoni. 2002. Cultivation of the ubiquitous SAR11 marine bacterioplankton clade. *Nature* **418:**630–633.
27. Rappe, M. S., and S. J. Giovannoni. 2003. The uncultured microbial majority. *Annu. Rev. Microbiol.* **57:**369–394.
28. Schloss, P. D., and J. Handelsman. 2004. Status of the microbial census. *Microbiol. Mol. Biol. Rev.* **68:**686–691.
29. Sogin, M. L., H. G. Morrison, J. A. Huber, D. Mark Welch, S. M. Huse, P. R. Neal, J. M. Arrieta, and G. J. Herndl. 2006. Microbial diversity in the deep sea and the underexplored "rare biosphere." *Proc. Natl. Acad. Sci. USA* **103:**12115–12120.
30. Staley, J. T., and A. Konopka. 1985. Measurement of in situ activities of nonphotosynthetic microorganisms in aquatic and terrestrial habitats. *Annu. Rev. Microbiol.* **39:**321–346.
31. Stevenson, B. S., S. A. Eichorst, J. T. Wertz, T. M. Schmidt, and J. A. Breznak. 2004. New strategies for cultivation and detection of previously uncultured microbes. *Appl. Environ. Microbiol.* **70:**4748–4755.
32. Waksman, S. A., and M. Hotchkiss. 1937. Viability of bacteria in sea water. *J. Bacteriol.* **33:**389–400.

33. **Ward, D. M., R. Weller, and M. M. Bateson.** 1990. 16S rRNA sequences reveal numerous uncultured microorganisms in a natural community. *Nature* **345:**63–65.

34. **Wilson, G. S.** 1922. The proportion of viable bacteria in young cultures with especial reference to the technique employed in counting. *J. Bacteriol.* **7:**405–446.

35. **Winslow, C.-E. A., and G. E. Willcomb.** 1905. Tests of a method for the direct microscopic enumeration of bacteria. *J. Infect. Dis.* Suppl. **1:**273–283.

36. **Winterberg, H.** 1898. Zur Methodik der Bakterienzahlung. *Zeitschr. Hyg.* **29:**75–93.

37. **Young, P.** 1997. Major microbial diversity initiative recommended. *ASM News* **63:**417–421.

38. **Zengler, K., G. Toledo, M. Rappe, J. Elkins, E. J. Mathur, J. M. Short, and M. Keller.** 2002. Cultivating the uncultured. *Proc. Natl. Acad. Sci. USA* **99:**15681–15686.

39. **ZoBell, C. E.** 1946. *Marine Microbiology: a Monograph on Hydrobacteriology.* Chronica Botanica Co., Waltham, MA.

Selective Isolation of Actinobacteria

MICHAEL GOODFELLOW

2

2.1. INTRODUCTION

The choice of biological material for exploitable biotechnology is a daunting one given the sheer scale of microbial diversity. Molecular ecological studies, for instance, show that only a tiny fraction of prokaryotic diversity in natural habitats has been cultivated (18, 20, 70, 138), an unseen majority that encompasses enormous genetic diversity for biotechnology (15, 16, 182). Nevertheless, it is encouraging that innovative selective isolation and characterization studies are providing a way of making previously uncultivated taxa available for ecological and biotechnological studies (34, 153). 16S rRNA gene sequence data provide a particularly effective way of detecting prokaryotic taxa and resolving deep-rooted phylogenetic lineages (49, 201).

Among prokaryotes, members of the suborders *Micromonosporineae*, *Pseudonocardineae*, *Streptomycineae*, and *Streptosporangineae* of the order *Actinomycetales* of the class *Actinobacteria* (Fig. 1) are currently the most prolific sources of novel secondary metabolites, notably antibiotics (11, 105, 162). *Streptomyces* strains have a unique capacity to produce bioactive compounds and remain the richest source of novel antibiotics (33, 72, 73, 162). The significance of streptomycetes in this respect is underpinned by results from whole-genome-sequencing studies which show that the whole genomes of model *Streptomyces* strains contain more then 20 naturally produced biosynthetic gene clusters (10, 81). In contrast, whole-genome sequence data show that most other prokaryotes lack the capacity to synthesize antibiotics.

There is a continued interest in screening members of the order *Actinomycetales* because they remain the richest source of novel bioactive compounds. However, it has become increasingly difficult to find commercially useful secondary metabolites from well-known actinomycetes, because this practice leads to the wasteful rediscovery of known bioactive metabolites, especially antibiotics. There is, therefore, an urgent need to isolate and characterize representatives of the many poorly studied actinobacterial taxa, and to cultivate the astonishingly rich diversity of novel actinobacteria diversity present in natural habitats (2, 36, 51, 154, 202). The principles and practices currently used to selectively isolate and recognize previously uncultivated taxa is the subject of this chapter.

2.2. SELECTIVE ISOLATION: THE GROUND RULES

Most actinobacteria are saprophytes in aquatic and terrestrial habitats, notably in self-heating plant material, soils, and freshwater and marine habitats (13, 17, 20, 46, 47, 84). It is not possible to recommend a single procedure for the selective isolation of the multiplicity of the different kinds of actinobacteria present in environmental samples given their diverse growth and incubation requirements. Consequently, innumerable approaches have been recommended for the isolation of specific actinobacterial taxa (102, 124, 183). The most selective strategies have tended to focus on the isolation of members of a particular taxonomic group or related taxonomic groups based on the biological properties of the organisms that are being sought (28, 42, 46, 179, 195). Such approaches require some prior knowledge of the growth and incubation requirements of the target organisms. Most selective isolation procedures involve the extraction of propagules from selected environmental samples, pretreatment(s) of samples, choice of selective media, incubation, and colony selection.

2.2.1. Selection of Environmental Samples

Actinobacteria mainly occur as saprophytes in diverse natural habitats, including soil, the initial focus of selective isolation studies. The numbers and kinds of actinobacteria found in soil and other substrates are greatly influenced by primary ecological factors, such as aeration, pH, temperature, salinity, and moisture and organic matter content (46, 185, 190, 193). Indeed, the success in isolating large numbers of specific actinobacterial taxa can be highly dependent on the choice of environmental samples, as illustrated by the selective isolation of acidophilic and alkaliphilic streptomycetes (4, 9, 39, 185). Similarly, self-heating materials like bagasse, composts, grains, and fodders are useful sources of thermophilic, filamentous, spore-forming actinobacteria (103, 195).

Soil remains a fruitful source of novel actinobacteria (168, 199), although it is becoming increasingly clear that un- and underexplored habitats, notably marine ecosystems, are a rich source of novel taxa that have the capacity to produce new bioactive compounds, including the discovery of first-in-the-class drug candidates (18, 33, 74). Indeed, an encouraging stream of anticancer and anti-infective compounds have been derived from marine sources, as

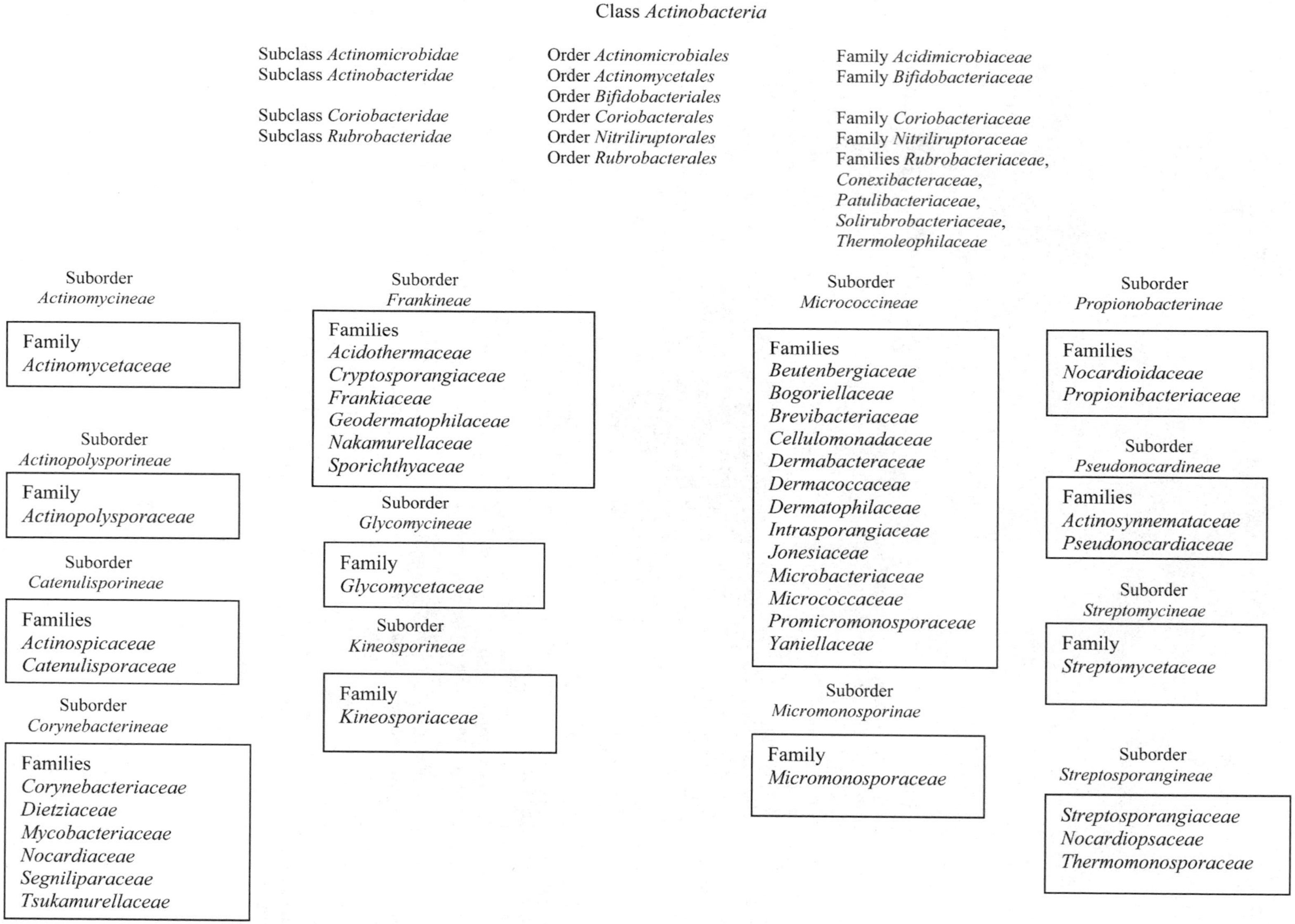

Class *Actinobacteria*

Subclass *Actinomicrobidae*   Order *Actinomicrobiales*   Family *Acidimicrobiaceae*
Subclass *Actinobacteridae*   Order *Actinomycetales*   Family *Bifidobacteriaceae*

Subclass *Coriobacteridae*   Order *Bifidobacteriales*   Family *Coriobacteriaceae*
Subclass *Rubrobacteridae*   Order *Coriobacterales*   Family *Nitriliruptoraceae*
                                           Order *Nitriliruptorales*   Families *Rubrobacteriaceae,*
                                           Order *Rubrobacterales*   *Conexibacteraceae,*
                                                                            Patulibacteriaceae,
                                                                            Solirubrobacteriaceae,
                                                                            Thermoleophilaceae

Suborder
Actinomycineae

Family
Actinomycetaceae

Suborder
Actinopolysporineae

Family
Actinopolysporaceae

Suborder
Catenulisporineae

Families
Actinospicaceae
Catenulisporaceae

Suborder
Corynebacterineae

Families
Corynebacteriaceae
Dietziaceae
Mycobacteriaceae
Nocardiaceae
Segniliparaceae
Tsukamurellaceae

Suborder
Frankineae

Families
Acidothermaceae
Cryptosporangiaceae
Frankiaceae
Geodermatophilaceae
Nakamurellaceae
Sporichthyaceae

Suborder
Glycomycineae

Family
Glycomycetaceae

Suborder
Kineosporineae

Family
Kineosporiaceae

Suborder
Micrococcineae

Families
Beutenbergiaceae
Bogoriellaceae
Brevibacteriaceae
Cellulomonadaceae
Dermabacteraceae
Dermacoccaceae
Dermatophilaceae
Intrasporangiaceae
Jonesiaceae
Microbacteriaceae
Micrococcaceae
Promicromonosporaceae
Yaniellaceae

Suborder
Micromonosporineae

Family
Micromonosporaceae

Suborder
Propionobacterineae

Families
Nocardioidaceae
Propionibacteriaceae

Suborder
Pseudonocardineae

Families
Actinosynnemataceae
Pseudonocardiaceae

Suborder
Streptomycineae

Family
Streptomycetaceae

Suborder
Streptosporangineae

Streptosporangiaceae
Nocardiopsaceae
Thermomonosporaceae

FIGURE 1 Hierarchic classification of the phylum *Actinobacteria* based on 16S rRNA gene sequences and taxon-specific 16S rDNA signature nucleotides (adapted from reference 201).

exemplified by the discovery of abyssomicin C, a potent polycyclic polyketide active against methicillin-resistant *Staphylococcus aureus* and synthesized by "*Verrucosispora maris*" (12, 151), and salinosporamide A, an anti-cancer agent produced by *Salinispora tropica* (32, 112). Novel endophytic actinobacteria are also a promising source of anticancer and antitumor agents (23, 80). Niche habitats, such as beehives (147), copper-polluted sediments (1), hyperarid soils (137), and root nodules of legumes (173, 174), are useful sources of novel actinomycetes.

2.2.2. Representative Sampling of Actinobacteria
Physicochemical interactions of bacterial propagules with particulate substrates affect their recovery from environmental samples. Traditional methods used to separate bacteria from organic matter, sediment, and soil particles, such as shaking in water or weak buffers (e.g., one-quarter strength Ringer's solution), are not effective (75). It is particularly important to thoroughly break up soil aggregates because many microorganisms, notably those showing mycelial growth, may be bound within them (54, 123, 148). Procedures used to promote the dissociation of microorganisms from particulate material include the use of chelating agents (108), buffered diluents (122), elutriation (76), and ultrasonication (148). All of these procedures address the problem of quantitative and representative sampling to varying degrees.

The dispersion and differential centrifugation (DDC) technique, a multistage procedure introduced by Hopkins et al. (75), combines several physicochemical treatments and has been found to be effective in representative sampling of bacteria, including actinomycetes, from freshwater and marine sediments, as well as from diverse soils (5, 76, 109, 119, 155). The procedure involves the use of a mild detergent (sodium cholate), Tris buffer, attenuated physical disruption (mild ultrasonication), and ionic shock (distilled water).

The DDC procedure has been shown to be much more effective in extracting propagules of streptomycetes from soil and marine sediment samples than traditional, but still widely used, reciprocal shaking techniques (5, 113, 155). Indeed some kinds of streptomycetes were only isolated by use of the DDC procedure. These observations suggest that associations between taxonomically different kinds of streptomycetes and particulate components in soil and sediments may be factors that limit representative sampling of streptomycetes, and that the DDC procedure may be effective in breaking down such associations.

2.2.3. Pretreatment of Samples
Various pretreatments can be used to select for different fractions of actinobacterial communities present in environmental samples. In general, pretreatment regimes select for the target actinobacteria by inhibiting or eliminating unwanted microorganisms. Several physical pretreatments have been used for the isolation of actinobacteria. Actinobacterial spores are more resistant to desiccation than most bacteria; hence, simply air-drying soil/sediment samples at room temperature will eliminate most unwanted gram-negative bacteria that might otherwise overrun isolation plates (102, 118, 193). A pretreatment based on alternate drying and wetting of soil has been used to enrich for sporangia (spore vesicles)-forming actinobacteria (110). Rare spore-forming actinobacteria (e.g., *Actinomadura*, *Microtetraspora*, *Pseudonocardia*, and *Streptosporangium*) have been isolated from irradiated soil samples and soil suspensions (14, 169). Cesium chloride density gradient centrifugation has been used to isolate *Micromonospora* species from soil (89). Similarly, *Nocardia* species have been selectively isolated from soil using sucrose gradient centrifugation (200).

Resistance of actinobacterial propagules to desiccation is usually accompanied by some measure of resistance to heat. The basis of this resistance is not clear, but it is apparent that many actinobacterial spores (e.g., *Micromonospora* and *Microtetraspora*), spore vesicles (*Dactylosporangium* and *Streptosporangium*), and hyphal fragments (e.g., *Rhodococcus*) are more resistant to heat than gram-negative cells. Actinobacteria are more sensitive to wet than to dry heat; hence, much lower temperatures are used to isolate these organisms from suspensions of environmental samples.

Nocardia asteroides strains have been isolated from soil following pretreatment of 10^{-1} suspensions at 55°C for 6 min (141); modifications of this procedure were used to isolate novel nocardiae from soil, water, and deteriorating natural rubber rings (111). The same mild pretreatment regime was used to isolate *Rhodococcus coprophilus* strains from aquatic and terrestrial habitats (152). More stringent heat pretreatment regimes select for actinobacteria that produce more heat-resistant spores, such as those formed by *Micromonospora*, *Microbispora*, and *Microtetraspora* strains (130–132). A number of additional heat pretreatment regimes are shown in Table 1. Heat pretreatment protocols usually lead to a decrease in the ratio of bacteria to actinomycetes on isolation plates, although actinobacterial counts may also be reduced (193).

A simple and effective technique for the selective isolation of filamentous actinobacteria on soil dilution plates was introduced by Hirsch and Christensen (71). This procedure involves the spread of soil or soil suspensions over the surface of membrane filters placed on the surface of agar plates. The plates are incubated for several days to allow actinobacterial hyphae to grow through pores on the filters and grow on the agar surfaces below. In contrast,

TABLE 1 Selective heat pretreatments for the isolation of actinobacteria

Pretreatment	Targets	Reference(s)
Air-dried soil heated at 120°C for 1 h	*Microbispora* and *Streptosporangium* spp.	128, 129
Air-dried soil heated at 100°C for 15 min	*Actinomadura* spp.	7
Water or soil suspensions heated at 45°C or 50°C for 10 min	*Streptomyces* spp.	28
Water or soil suspensions heated at 60°C for 30 min	*Micromonospora* spp.	28
Air-dried soil heated at 120°C for 1 h	*Dactylosporangium* and *Streptosporangium* spp.	59
Air-dried soil heated at 28°C for 1 wk	*Herbidospora cretea*	93
Soil suspension heated at 110°C for 1 h	*Microtetraspora glauca*	61

nonfilamentous bacteria are confined to the surface of the membrane. The filters are removed after this initial growth phase and plates incubated further to allow the growth of actinobacteria (see also chapter 1, this volume).

Another ingenious pretreatment method involves the addition of a cocktail of polyvalent phage to soil suspensions to reduce the growth of unwanted bacteria, notably aerobic, endospore-forming bacteria and neutrophilic streptomycetes, thereby raising the probability of isolating members of rare or novel actinobacterial genera (97, 99, 196). This approach has been especially useful for the isolation of rare filamentous actinobacteria, such as *Amycolatopsis*, *Catellatospora*, *Couchiplanes*, *Kibdelosporangium*, *Thermomonospora*, and *Thermobifida* strains, from diverse environmental samples (96, 98). An impressive phage library is now available for such isolation experiments (96).

Chemical pretreatments of mixed inocula are used to isolate specific actinobacterial taxa, notably members of the genera classified in the family *Streptosporangiaceae* (57, 59, 65, 68). The selective chemical procedures introduced by Hayakawa and his colleagues are based on the differential ability of actinobacterial spores to withstand treatment with chemical germicides such as benzethonium chloride, chlorhexidine gluconate, and phenol (55). Treatment with these agents for 30 min at 30°C kills vegetative cells of aerobic, endospore-forming bacilli and pseudomonads.

The simultaneous use of more than one chemical germicide can further enhance selectivity, as exemplified by the use of chlorheximide gluconate and phenol for the isolation of *Microbispora* strains (65). Sodium dodecyl sulfate and yeast extract broth inhibit the growth of unwanted bacteria and enhance the germination of actinobacterial spores (125). Mycobacteria are routinely isolated from clinical and environmental samples by treating them with dilute acids or alkali with or without quaternary ammonium compounds, followed by neutralization, and decontaminated material is inoculated to agar or inspissated egg medium supplemented with malachite green (178). Sodium hypochloride and osmium tetroxide are routinely used to decontaminate root surfaces/root nodules before isolating endophytic actinobacteria, including *Frankia* (22, 175). Other effective chemical pretreatment procedures are shown in Table 2.

Low numbers of actinobacteria in environmental samples can be concentrated, then plated out onto nutrient or selective media to facilitate their isolation. Actinobacterial propagules in freshwater and seawater can be concentrated by filtration through membrane filters, which are then placed over agar surfaces to promote growth of constituent strains (21, 136). Similarly, actinobacterial spores can be separated from bacteria and fungal propagules by centrifuging soil suspensions for 20 min at 3,000 rpm before

seeding isolation media with supernatants containing the actinobacterial spores (150). Differential centrifugation has also been used to separate *Frankia* bacteroids from root nodule homogenates (8, 149) and to selectively isolate motile actinobacteria from soil and plant litter (64).

Nutrient amendment and baiting techniques are used to increase the populations of specific fractions of actinobacterial communities in environmental samples to facilitate their isolation on nutrient and selective media. Soil amendments with substrates such as chitin and keratin were first used by Jensen (82, 83) to boost the numbers of streptomycetes. This approach has fallen into neglect to some extent, but does provide an effective way of enhancing specific components of actinobacterial communities in soil (125, 192).

Actinobacteria with a motile spore phase in their life cycle have been isolated from environmental samples baited with natural substrates, such as hair and pollen grains, ever since the pioneering studies on actinoplanetes by Couch (24–27). The method is based on the ability of zoospores to migrate and colonize baits floated on the surfaces of soil suspended in distilled water in petri dishes. Following incubation for 1 to 2 weeks, the baits are removed and used to inoculate growth media to isolate the organisms. This procedure is still widely used for the isolation of actinobacteria, such as members of the genera *Actinokineospora*, *Actinoplanes*, *Actinosynnema*, *Catenuloplanes*, *Cryptosporangium*, *Dactylosporangium*, *Geodermatophilus*, *Kineospora*, *Pilimelia*, *Sporichthya*, and *Spirillospora* (35, 64, 67, 134, 171).

Motile actinobacteria can also be isolated by taking advantage of their ability to show chemotactic responses (55). Palleroni (143) found that actinoplanete zoospores present in soil suspensions migrated into glass capillaries containing potassium chloride within about 40 min and could then be isolated on nutrient agar plates seeded with contents of the capillaries. An improved chemotactic method based on the use of γ-collidine as chemoattractant was used to isolate *Actinoplanes* and *Dactylosporangium* species from various soils (68). Similarly, vanillin has been used as a chemoattractant for the isolation of members of the genera *Catenuloplanes* (56) and *Virgosporangium* (165).

An interesting departure from the dilution plate procedure involves the isolation of spore-forming actinobacteria from self-heating plant material. The latter is placed either in a wind tunnel (48, 103) or in a simple sedimentation chamber (104) and the resultant spore cloud passed through filters with progressively smaller pores onto plates of agar media held in an Andersen sampler (3). This method has been used to selectively isolate spore-forming taxa from composts and overheated cereal grains, including carboxydotrophic streptomycetes (31), saccharopolysporae (41), and thermomonosporae (116).

TABLE 2 Chemical pretreatments for the selective isolation of actinobacteria

Chemical	Substrate	Targets	Reference(s)
Benzethonium chloride (0.03%, wt/vol)	Soil	*Streptosporangium* spp.	59
Benzethonium chloride (0.01%, wt/vol)	Soil	*Dactylosporangium* spp.	59
Benzethonium chloride	Soil	M. *glauca*	61
Chloramine-T	Soil	*Herbidospora*, *Microbispora*, *Microtetraspora*, *Nonomuraea*, and *Streptosporangium*	57
Chlorine	Water	*Streptomyces* spp.	21
γ-Collidine	Soil	*Actinoplanes* and *Dactylosporangium*	68
Phenol (1.5%, wt/vol)	Soil	*Micromonospora* spp.	58, 65

2.2.4. Choice of Selective Media

Innumerable media have been recommended for the isolation of actinobacteria, in general, and for selected families, genera, and species, in particular. Most of the "general" or "nonselective" media were formulated empirically, notably without reference to either the nutritional or tolerance preferences of the target organisms. Some recommended isolation media are nutritionally poor because many actinobacteria have a tendency to survive and grow, albeit poorly, on nutrients scavenged from tap water or even purified agar. However, most selective media have high carbon-to-nitrogen ratios because they contain complex carbon and nitrogen sources (e.g., casein, humic acid, starch, and xylan) that favor the growth of actinobacteria over the generality of bacteria that grow better on media with low carbon-to-nitrogen ratios and are usually unable to metabolize high-molecular-weight, resistant organic polymers.

Nonselective media, such as colloidal chitin-mineral salts and starch-casein-nitrate agars (78, 100, 107), are now known to select for a relatively narrow range of *Streptomyces* species (177, 188) and do not support the growth of actinomycetes with different growth requirements (29, 188). These media are still widely used even though their value is mainly restricted to detecting the presence of streptomycetes in extreme environments (137, 144). However, starch-casein-nitrate agar adjusted to pH 4.5 or 10.5 has been used to good effect in the isolation of acidophilic and alkaliphilic streptomycetes, respectively (4, 9, 39, 91, 185), and when prepared with seawater can be used to recover streptomycetes from marine habitats (40, 74). The presence of seawater can be crucial for the selective isolation of marine actinomycetes (15, 84), notably for indigenous marine taxa such as *Salinispora* (85, 112, 120).

Antimicrobial agents, notably antibiotics, provide an effective means of increasing the selectivity of isolation media. It is standard practice to control or eliminate fungal contaminants by using antifungal antibiotics, such as actidione, nystatin, and pimaricin (48, 145, 184). Similarly, penicillin G and polymyxin B have been used to select filamentous actinobacteria from competing soil-inhabiting bacteria (128, 184), as have nalidixic acid and trimethoprim (66). Antibacterial antibiotics are widely used for the selective isolation for specific actinomycete taxa, as exemplified in Table 3.

2.2.5. Incubation

Gaseous regimes, temperature, and incubation times also contribute to selectivity. Incubation at 25 to 30°C favors mesophilic actinobacteria, whereas their thermophilic counterparts require higher temperatures. Members of commonly isolated genera, such as *Micromonospora*, *Nocardia*, and *Streptomyces*, may be selected from isolation plates after 14 days. In contrast, thermophilic actinomycetes, such as *Pseudonocardia thermophila* and *Saccharopolyspora rectivirgula*, only require incubation for 2 to 3 days at 45 to 50°C; plates incubated at these temperatures should be incubated within polythene bags or moist chambers to prevent desiccation. However, incubation times of up to 5 weeks may be required to isolate some genera, such as those classified in the family *Streptosporangiaceae*. Nonomura and Ohara, for instance, incubated plates at 30°C for a month to selectively isolate *Microbispora* and *Microtetraspora* strains (133). Prolonged incubation was also one of the factors that contributed to the initial isolation of the slow-growing, mutualistic symbiont *Frankia* (106).

Relatively little attention has been paid to the selective isolation of anaerobic, barophilic, halophilic, or psychrophilic actinobacteria from a biotechnological perspective. However, large numbers of diverse, facultatively autotrophic actinobacteria have been isolated by use of gaseous C_1 compounds, including carbon monoxide (31, 135). Some of these isolates were formally described as new *Streptomyces* species (92).

2.2.6. Selection of Colonies

Selection of colonies remains the most time-consuming and subjective stage in the isolation procedure. Colonies on isolation plates can be selected randomly or with some degree of choice. Random selection of colonies is especially laborious if it is performed manually in the absence of automated systems backed up by image analysis. When selective isolation media are used, the target organism(s) can often be provisionally identified either on the basis of colony morphology or by the examination of colonies for distinctive features, such as the presence of spore vesicles and the nature of spore chains, using a long working distance objective. However, the ability to accurately identify spore-forming actinomycetes directly on primary isolation plates requires considerable taxonomic prowess and experience. Chromogenic or fluorogenic substrates, which indicate the synthesis of taxon-specific enzymes, coupled with the use of highly specific isolation media might help in the rapid recognition of particular taxa.

It is not usually possible to distinguish between species of the same actinobacterial genus on selective isolation plates. In such instances, the selection of large numbers

TABLE 3 Antibiotics used in the selective isolation of actinobacterial taxa

Antibiotic(s)	Target genera	Reference
Bruneomycin and streptomycin	*Actinomadura*	146
Gentamicin	*Actinomadura* and *Streptosporangium*	157
Josamycin, kanamycin, and nalidixic acid	*A. viridis*	60
Kanamycin, nalidixic acid, and nofloxacin	*Microtetraspora*	61
Fradiomycin, kanamycin, nalidixic acid, and trimethoprim	*Actinokineospora*	142
Leucomycin and nalidixic acid	*Streptosporangium*	59
Nalidixic acid and penicillin	*Saccharothrix*	157
Nalidixic acid and tunicamycin	*Micromonospora*	125
Neomycin sulfate	*Amycolatopsis*	168
Novobiocin and streptomycin	*Glycomyces*	101

of colonies is laborious and can lead to serious duplication of strains and hence effort, especially in low-throughput screening systems. However, this problem can be overcome by grouping isolates based on pigmentation characters and molecular fingerprinting patterns (4, 168). Surface spread plates, dried thoroughly before or after inoculation, provide a satisfactory way of recognizing and enumerating filamentous actinobacteria, including streptomycetes (176).

2.3. RATIONAL DESIGN OF SELECTIVE ISOLATION MEDIA

Selective isolation media can be formulated in an objective way by using information held in taxonomic databases (42, 188). One of the early successes of this approach was the development of a medium selective for members of the genus *Nocardia* (111, 139–141). The discovery that diagnostic sensitivity agar supplemented with tetracyclines was selective for *Nocardia* species was based on information held in an antibiotic sensitivity database (43). A logical extension of this work was the visual scanning of phenotypic databases to highlight antibiotics that might form the basis of selective isolation media (Table 4). The number of verticillate streptomycetes (formerly members of the genus *Streptoverticillium*), for example, was increased by reducing the number of competing neutrophilic streptomycetes by supplementing isolation media with neomycin and oxytetracycline (52); verticillate streptomycetes were known to be resistant and other neutrophilic streptomycetes were sensitive to these antibiotics (187, 191). The addition of lysozyme to this medium further increased its selectivity (53).

Phenotypic databases generated in numerical taxonomic studies are ideal resources for the formulation of novel selective isolation media because they are rich in data on the antibiotic, nutritional, and physiological properties of members of numerically defined clusters (42, 188). Such databases are available for several industrially significant actinobacterial taxa, including the genera *Actinomadura* (6, 172), *Gordonia* (45), *Nocardia* (140), *Rhodococcus* (45), *Streptomyces* (88, 114, 186, 187, 189), and *Thermomonospora* (117), and for members of the family *Pseudonocardiaceae* (180). Several computer-generated selective media have been designed for the isolation of specific fractions of the actinobacterial community from natural habitats (Table 4).

The most extensive studies on computer-based selective media formulations have been performed by Stan Williams and his colleagues (177, 188, 194). They used combinations of amino acids and carbohydrates, with or without antibacterial antibiotics, to favor the isolation of uncommon streptomycetes known to be a promising source of new antibiotics or to discourage the growth of *Streptomyces albidoflavus*, which is ubiquitous in terrestrial habitats. The selective agents were chosen after examining a neutrophilic streptomycete database (189) by use of the DIACHAR program (158), which selects for the most diagnostic characters of individual numerically defined clusters. The highest scores obtained by using this procedure are given by properties that are either all positive or all negative for strains in one cluster when compared with all of the associated numerically defined taxa (177).

The most widely used selective isolation formulation arising from the pioneering studies of Williams and his colleagues is raffinose-histidine agar (177). This medium was designed to reduce the growth of *S. albidoflavus* and related strains, as the latter grow poorly on raffinose and histidine as sole carbon and nitrogen sources, in favor of the isolation of rare and novel streptomycetes. Raffinose-histidine agar has been shown to be effective in the selective isolation of

TABLE 4 Selective media based on information mined from phenotypic databases generated in numerical taxonomic studies

Selective agent(s)	Basal medium	Target organisms	Reference(s)
A. Visual scanning of databases			
Kanamycin	Half-strength tryptic soy agar supplemented with casein hydrolysate	*T. chromogena*	116
Neomycin and oxytetracycline	Semidefined agar medium	Verticillate streptomycetes	52
Rifampicin	Glucose-yeast extract agar[a]	*Actinomadura* spp.	7
		Saccharomonospora and *Thermomonospora* spp.	
Tetracyclines	Diagnostic sensitivity test agar	*Nocardia* spp.	139
B. Selective agents highlighted using the DIACHAR program			
Aminobutyric acid and rhamnose	Aminobutyric acid-rhamnose agar	*S. albidoflavus* and *Streptomyces violaceusniger*	194
Dextran, L-histidine, and penicillin G	Mineral salts agar	*S. violaceoruber* clade	30
D-Melezitose	SM2 agar	*Amycolatopsis* spp.	168
Nalidixic acid and novobiocin	Gauze's medium 2 agar	*Amycolatopsis* spp.	
Neomycin and D-sorbitol	SM1 agar	*Amycolatopsis* spp.	
Raffinose and histidine	Raffinose-histidine agar	"Rare" *Streptomyces* spp.	5, 177, 191
Rifampicin		*Streptomyces atroolivaceus* and *S. diastaticus*	191
Oleandomycin and sodium chloride	Bennett's agar	*S. atroolivaceus*	194

[a]Isolation plate incubated at 30 and 50°C.

Streptomyces cyaneus, *Streptomyces platensis*, and *Streptomyces rochei* and related taxa from a sand dune soil; *Streptomyces chromofuscus*, *S. cyaneus*, and *S. rochei* and similar organisms from a rose bed soil; *S. chromofuscus*, *S. cyaneus*, and *Streptomyces diastaticus* and related species from a grassland soil (177); and for the isolation of two novel species, *S. aureus* and *Streptomyces sanglieri*, from a hay meadow soil (5, 115). These studies show that the analysis of the growth requirements of different streptomycete taxa using the DIACHAR program provides an objective strategy for designing novel selective isolation media.

The approaches to the selective isolation of streptomycetes outlined above have been refined by Duangmal et al. (30), who found that a number of antibiotics, sole-carbon, and sole-nitrogen compounds favored the growth of members of the *Streptomyces violaceoruber* 16S rRNA gene clade; the putative selective agents had been highlighted following the application of the DIACHAR program (158) to a neutrophilic streptomycetes database modified from the one devised by Williams et al. (186). Various combinations of the putative selective agents were added to a basal mineral salts medium and the resultant agar plates were inoculated with spore suspensions prepared from members of the *S. violaceoruber* gene clade and representative *Streptomyces* species. All of the *S. violaceoruber* clade strains formed characteristic red-pigmented colonies when the basal medium was supplemented with dextran, L-histidine, sodium chloride, penicillin G, and rifampicin; the sodium chloride and rifampicin were added to the selective agents chosen using output from the DIACHAR program, because they are considered to enhance the growth of soil streptomycetes at appropriate concentrations (177, 188). None of the remaining strains were able to grow on this medium, apart from *Streptomyces rutgersensis* subspecies *castelasensis* NRRL B-1567^T.

A similar approach to the one described above was used to formulate media designed to select isolate members of the genus *Amycolatopsis* from soil samples (168). The absence of an effective selective isolation procedure had limited the biotechnological exploitation of members of this taxon, a potentially rich source of new secondary metabolites, notably antibiotics (77, 166, 197). Three of 20 putatively selective agar media, namely SM1 (Stevenson's basal medium supplemented with D-sorbitol and neomycin sulfate), SM2 (Stevenson's basal medium supplemented with D-melezitose and neomycin sulfate), and SM3 (Gauze's medium no. 2 supplemented with nalidixic acid and novobiocin), supported the growth of all of the representatives of the genus *Amycolatopsis*. These strains formed abundant crusty colonies covered with white aerial hyphae on the SM1 and SM2 agar plates, and leathery colonies covered by white to yellow aerial hyphae on the corresponding SM3 plates. Colonies presumptively assigned to the genus *Amycolatopsis* based on these properties were seen on all three putative selective media after incubation with soil suspensions and incubation at 28°C for 3 weeks. Representatives of strains taken from the selective isolation plates were shown to be related to *Amycolatopsis mediterranea* and *Amycolatopsis orientalis* using 16S rRNA gene sequence data, although many others formed distinct phyletic lines in the *Amycolatopsis* gene tree. Members of two of these novel phyletic branches were subsequently described as new species, that is, as *Amycolatopsis australiensis* (167) and *Amycolatopsis regifaucium* (166). Additional putatively novel *Amycolatopsis* species have been isolated from Atacama Desert soil by using the selective isolation and characterization strategy introduced by Tan et al. (168).

It is good practice to evaluate, and if necessary, to modify computer-generated media formulations by altering the combinations of selective agents to increase the recovery of target actinobacteria from environmental samples. Vickers and her colleagues identified streptomycetes growing on a range of objectively defined isolation media with use of a computer-assisted identification procedure (189) and found that it was possible to achieve increased yields of some species, although not with all of the soils. Some species were found in greater or lesser numbers than could be anticipated from information held in the streptomycete database. Such results are not altogether surprising because the surfaces of isolation plates can be sites of intensive competition between the different kinds of bacteria found in mixed inocula. The outcome of such struggles for survival will be influenced not only by the selectivity of isolation media, but also by the mix of species in the inoculum that are able to grow on it. This unpredictability can be turned to good effect when selective pressures allow the growth of novel actinobacteria that are present in environmental samples and of interest to biotechnologists (42).

Huck et al. (79) used stepwise discriminant analysis to highlight the most useful selective substrates that could be used to formulate media able to enhance the recovery of antibiotic-producing actinobacteria from soil. Initially a range of soil actinobacteria and eubacteria were examined for their ability to produce antibacterial antibiotics and for their growth responses to a range of nutritional and physiological tests. Characters that were selective for actinobacteria relative to eubacteria included growth on proline and humic acid as sole sources of both carbon and nitrogen, growth on nitrate as a nitrogen source, and growth at pH 7.7 to 8.0. Similarly, antibiotic- and non-antibiotic-producing actinobacteria were distinguished by the ability of the former to grow on proline and humic acid as sole carbon and nitrogen sources and to grow on asparagine as a nitrogen source, and by their growth in the presence of vitamins. Several simple isolation media that contained selective substrates identified by stepwise discriminant analysis were found to increase the proportion of actinobacteria isolated from soil samples. In addition, a considerable increase was found in the percentage of isolates able to produce antibiotics.

2.4. RECOGNITION OF TARGET AND NOVEL TAXA

Until recently, relatively little effort was made to evaluate the effectiveness of procedures recommended for the isolation of specific fractions of actinobacterial communities present in natural habitats. The failure to establish whether or not target organisms had been isolated not only limited the value of such procedures, but also meant that members of potentially novel taxa went undetected. Consequently, the selection of strains for screening programs owed more to custom and practice than to scientific rigor. This problem was partly historical because the need to characterize and identify representative strains growing on isolation plates had rarely been seen as a relevant task for those involved in search and discovery programs, although it also reflected the dearth of simple and accurate procedures for the classification and identification of commercially significant actinomycetes. This situation is now dramatically different because powerful taxonomic tools are available for these purposes (see chapter 4, this volume).

The assignment of unknown actinobacteria to either formally described or novel taxa is essentially a two-stage process. Reliable criteria are needed to allocate isolates to genera and families prior to the selection of diagnostic tests for identification of new or established species. Assignment of isolates at or above the generic level can readily be achieved by comparing their 16S rRNA gene sequences with corresponding sequences retrieved from GenBank (for details, see chapter 4, this volume) and by evaluating the resultant data in light of complementary chemotaxonomic and morphological information. Similarly, assignment to established or new species can be realized by using an appropriate combination of phenotypic tests. The effectiveness of such a polyphasic approach is apparent from the steady increase in the number of actinobacterial genera, as witnessed by descriptions of the novel genera *Glacibacter* (90), *Gordonibacter* (198), *Iami* (95), *Haloglycomyces* (50), *Marinoactinospora* (170), *Ferrimicrobium* (87), *Ferrithrix* (87), and *Paraeggerthella* (198).

It is not usually possible to identify colonies growing on isolation plates unless a particular taxon is sought, as in the case of *Amycolatopsis* and *Nocardia* strains growing on SM1 and diagnostic sensitivity agar supplemented with tetracyclines, respectively (139–141, 168), or painstaking examination of individual colonies for key morphological markers is undertaken (74). However, it is often possible to group colonies with similar, distinctive growth forms and pigmentation following random transfer of individual colonies to appropriate media. Streptomycetes, for instance, can be grouped based on aerial spore mass, colony reverse

and diffusible pigment colors on oatmeal agar, and the formation of melanin pigments on peptone-yeast extract-iron agar; isolates taken to represent such taxa key out either to formally described or to novel *Streptomyces* species based on computer-assisted identification (5, 40, 115, 194), Curie-point mass spectrometry (5), and polyphasic taxonomic procedures (115). Excellent congruence has been found between the assignment of alkaliphilic streptomycetes to color and molecular fingerprint groups (4). The identity of representative isolates can also be achieved by using other molecular systematic procedures, notably by the use of taxon-specific 16S rRNA oligonucleotide primers (94, 121, 168).

2.5. INTEGRATED SELECTIVE ISOLATION PROCEDURES

It is apparent from the examples shown in Table 5 that the isolation of specific actinobacterial taxa can be achieved by combining suitable pretreatment methods with appropriate selective media. The selective isolation of some genera in the family *Streptosporangiaceae*, for instance, can be achieved by dry-heat treatment of air-dried soil samples and dilution plate culture with several selective media (62, 63, 126–130, 181). To this end, soil samples are initially dried at room temperature for a week to promote sporulation of the indigenous sporoactinomycetes. The samples

TABLE 5 Integrated procedures for the selective isolation of actinobacterial taxa

Substrate	Pretreatment	Medium[a]	Target	Reference(s)
Beach and dune sand	10^{-1} suspensions heated at 55°C for 6 min	Starch-casein + cycloheximide	Alkaliphilic streptomycetes	4
Composite arid soils	10^{-1} suspensions of air-dried soil heated at 50°C for 6 min	Dextran-histidine-sodium chloride mineral salts agar + nystatin, penicillin G, and rifampicin	*S. violaceoruber* clade	30
Composite arid soils	Air-dried soil	SM1-SM3 agars + cycloheximide, nystatin, and antibacterial antibiotics	*Amycolatopsis* spp.	168
Marine sediments	Wet sediment heated at 55°C for 6 min	M2-M5 agars + cycloheximide and rifampicin	*Salinispora* spp.	112, 120
Mushroom compost	Dilution in air using a sedimentation chamber and Andersen sampler	Half-strength TSA + casein hydrolysate and kanamycin (50°C)	*T. chromogena*	116
Soil	$CaCO_3$ enrichment, rehydration, and centrifugation	HV agar + fradiomycin, kanamycin, nalidixic acid, and trimethoprim	*Actinokineospora* spp.	142
Soil	Chloramine-T	HV agar	*Herbidospora, Microbispora, Microtetraspora, Nonomuraea,* and *Streptosporangium* spp.	57
Soil	Suspension of air-dried soil treated with phenol (1.5%, wt/vol) at 30°C for 10 min	HV agar + nystatin	*S. violaceusniger* phenotypic cluster	69
Soil	Suspension of air-dried soil treated with phenol (1.5%, wt/vol) at 30°C for 10 min	HV agar + chlortetracycline, cycloheximide, and nalidixic acid	*Nocardia* spp.	200

[a]HV agar, humic acid-vitamin agar; TSA, trypticase soy agar.

are then sieved, gently ground in a mortar, and heated in a hot-air oven at 100 to 120°C. The numbers of bacteria and streptomycetes in soil samples are greatly reduced by this process, leading to the enhanced recovery of *Microbispora*, *Microtetraspora*, *Nonomuraea*, and *Streptosporangium* strains on media such as arginine-vitamin, chitin-V, and humic acid-vitamin (HV) agars. These media are usually supplemented with antifungal antibiotics and sometimes with penicillin and polymyxin to further increase their selectivity. Inoculated plates are incubated at 30°C for 4 to 6 weeks and colonies examined for the presence of spore vesicles by using a light microscope fitted with a long working distance objective. Thermophilic species of the genera *Microbispora* and *Streptosporangium* can be recovered by incubating the isolation plates at 50°C for 2 to 3 weeks.

A combination of heat and chemical pretreatments have proved to be highly selective for the isolation of several genera, notably *Microbispora* (59, 65, 66), and for species such as *Actinomadura viridis* (60). The improved method for the selective isolation of *Microbispora* spp. involved making a suspension of heat-pretreated, air-dried soil in phenol (1.5%) and chlorhexidine gluconate (0.03%) prior to dilution and plating onto HV agar supplemented with nalidixic acid. The dry-heat pretreatment regime led to a sharp reduction in the nonfilamentous bacterial community and a significant fall in the number of *Streptomyces* spp. without a reduction in the *Microbispora* component. Indeed, over 90% of colonies growing on isolation plates seeded with treated suspensions of various field soils proved to be *Microbispora* strains. Similarly, heat pretreatments (110 or 120°C) together with the use of benzethonium chloride (0.05 or 0.01%) proved to be highly selective for the isolation of *Dactylosporangium*, *Microtetraspora*, and *Streptosporangium* strains on HV agar supplemented with antibacterial antibiotics (59, 66). Innovative developments from such approaches have proved to be highly selective for the isolation of some members of the family *Streptosporangiaceae* that form motile spores, notably for *Planobispora* and *Planomonospora* (163, 164).

The genus-specific isolation procedures outlined above, while effective, have provided relatively little information on taxonomic diversity of the target organisms growing on isolation plates. This omission has been addressed in more recent studies as representative isolates taken from isolation plates have been characterized by using a combination of genotypic and phenotypic procedures (30, 168). Tan et al. (168) assigned 175 isolates shown to belong to the genus *Amycolatopsis* to 18 multimembered and 20 single-membered color groups based on their ability to form pigments on Bennett's agar supplemented with mannitol and soybean flour. Representatives of the color groups formed distinct phyletic lines in the *Amycolatopsis* gene tree, showing that the genus *Amycolatopsis* is grossly underspeciated and, hence, forms an attractive target for search and discovery programs.

2.6. CONCLUDING REMARKS

The isolation and characterization of specific actinobacterial taxa are procedures of vital importance in actinobacterial biology, the more so in light of culture-independent surveys that show that the taxonomic diversity of actinobacteria is very much greater than previously recognized (36, 159, 160, 202). It is especially important to design effective procedures for the selective isolation of commercially significant actinobacterial taxa to obtain high-quality biological material for biotechnological purposes (18, 20).

Improved selective isolation and characterization procedures show that actinobacterial taxa once considered rare are both common and widely distributed in natural habitats, as exemplified by studies on acidophilic actinomycetes (39, 91, 185), alkaliphilic streptomycetes (4, 9), and motile actinomycetes (56, 67, 163, 164). Reliable procedures are available for the selective isolation of members of specific taxa, such as the genera *Actinomadura* (7, 172), *Actinoplanes* (62, 67), *Amycolatopsis* (168), *Gordonia* (45), *Nocardia* (111, 139, 141), and *Rhodococcus* (152) and for individual species like *A. viridis* (60), *Catenuloplanes japonicus* (56), *S. violaceoruber* (30), and *Thermomonospora chromogena* (116). The availability of rapid and reliable taxonomic tools, such as the use of genus-specific probes as selective amplification primers, offers a practical way of detecting target actinobacteria taken from isolation plates (121, 168).

The application of reliable selective isolation and characterization procedures to surveys of the cultivable actinobacteria in poorly studied and neglected habitats is yielding a steady flow of novel taxa (19, 74, 137). In this context the marine environment is proving to be an especially rich source of novel taxa, members of which have the capacity to produce new anticancer and antimicrobial products, such as salinosporamide A from *S. tropica* (32) and abyssomicin C from "*V. maris*" (12, 151). Integrated selective isolation strategies can be expected to yield additional novel, commercially significant taxa, especially if neglected but effective physicochemical procedures like the dispersion and differential centrifugation technique (75) are used. Other approaches that require care and patience, such as the use of long incubation times (37) and dilution-to-extinction culturing (161), can also be expected to yield dividends.

It can be argued that conventional manipulation of standard media components is only likely to bring incremental success (see chapter 1, this volume). This may well be so; but, nevertheless, more certainly does not mean less as witnessed by the addition of the genera *Salinispora* and *Verrucosispora* to the family *Micromonosporaceae* (112, 113). Indeed, there is a compelling need to develop integrated procedures for the selective isolation of a host of well-known taxa of potential biotechnological value, not least the genera *Actinopolyspora*, *Herbidospora*, *Lechevalieria*, *Pseudonocardia*, and *Saccharopolyspora*. The detection of members of such taxa in natural habitats by the use of culture-independent methods can be used to guide the development and application of selective isolation strategies along the lines used by Maldonado et al. (113).

There is, however, an even greater need to focus on the selective isolation of understudied actinobacterial taxa, such as members of the families *Conexibacteriaceae*, *Coriobacteriaceae*, and *Rubrobacteriaceae*, which form deep branches in the 16S rRNA actinobacterial gene tree (201). It seems likely that the successful isolation of such taxa will require a profound understanding of the physicochemical-biological features of their habitats to unravel their ecophysiology and thereby help inform isolation strategies. The impressive ecophysiological studies performed by Seviour and his colleagues on actinobacteria in activated sludge systems is a model of what needs to be done in this respect (156). Other innovative approaches, such as the use of in situ procedures, are also relevant in this respect (see chapter 1, this volume).

It can be anticipated that the pace of technological change, notably in biosystematics, genomics, and metagenomics, will have a profound impact on approaches to selective isolation, characterization, and selection of strains for future screening programs. These aspects are beyond the scope of the present article, although they are touched upon in some recent broadly based reviews on aspects of search and discovery of novel natural products (18, 20).

REFERENCES

1. Albarracin, V.H., B. Wink, E. Kothe, M.J. Amoroso, and C.M. Abate. 2008. Copper bioaccumulation by the actinobacterium *Amycolatopsis* sp. ABO. *J. Basic Microbiol.* 48:323–330.

2. Allgaier, M., and H.-P. Grossart. 2006. Diversity and seasonal dynamics of *Actinobacteria* populations in four lakes in Northeastern Germany. *Appl. Environ. Microbiol.* 72:3489–3497.

3. Andersen, A. A. 1958. New sampler for the collection, sizing and enumeration of viable airborne particles. *J. Bacteriol.* 76:471–484.

4. Antony-Babu, S., and M. Goodfellow. 2008. Biosystematics of alkaliphilic streptomycetes isolated from seven locations across a beach and dune sand system. *Antonie van Leeuwenhoek* 94:581–591.

5. Atalan, E., G. P. Manfio, A. C. Ward, R. M. Kroppenstedt, and M. Goodfellow. 2000. Biosystematic studies on novel streptomycetes from soil. *Antonie van Leeuwenhoek* 77:337–353.

6. Athalye, M., M. Goodfellow, J. Lacey, and R. P. White. 1985. Numerical classification of *Actinomadura* and *Nocardiopsis*. *Int. J. Syst. Bacteriol.* 35:86–98.

7. Athalye, M., J. Lacey, and M. Goodfellow. 1981. Selective isolation and enumeration of actinomycetes using rifampicin. *J. Appl. Bacteriol.* 51:289–299.

8. Baker, D., J. G. Torrey, and G. H. Kidd. 1979. Isolation and sucrose density fractionation and cultivation *in vitro* of actinomycetes from nitrogen-fixing root nodules. *Nature* (London) 281:76–78.

9. Basilio, A., I. Gonzales, M. F. Vicente, J. Gorrochategui, A. Cabello, and A. Gonzalez. 2003. Patterns of antimicrobial activities from soil actinomycetes under different conditions of pH and salinity. *J. Appl. Microbiol.* 95:814–823.

10. Bentley, S. D., K. F. Chater, A-M. Cerdeno-Tarraga, G. L. Challis, N. R. Thompson, K. D. Jones, D. E. Harris, M. A. Quail, H. Kieser, D. Harper, A. Bateman, S. Brown, G. Chandra, C. W. Chen, M. Collins, A. Cronin, A. Fraser, A. Goble, J. Hidalgo, T. Hornsby, S. Howarth, C. H. Huang, T. Kieser, L. Larke, L. Murphy, K. Oliver, S. O'Neil, E. Rabbinowitsch, M. A. Rajandream, K. Rutherford, S. Rutter, K. Seeger, D. Saunders, S. Sharp, R. Squares, S. Squares, K. Taylor, T. Warren, A. Wietzorrek, J. Woodward, B. G. Barrell, J. Parkhill, and D. A. Hopwood. 2002. Complete genome sequence of the model actinomycete *Streptomyces coelicolor* A3(2). *Nature* 417:141–147.

11. Bérdy, J. 2005. Bioactive microbial metabolites. *J. Antibiot.* 58:1–26.

12. Bister, B., D. Bischoff, M. Ströbele, J. Riedlinger, A. Reichke, F. Walter, A. T. Bull, H. Zähner, H-P. Fielder, and R. D. Süssmuth. 2004. Abyssomicin C – a polyketide antibiotic from a marine *Verrucosispora* strain as an inhibitor of the *p*-aminobenzoic acid/tetrahydrofolate biosynthetic pathway. *Angew. Chem. Int. Ed.* 43:2574–2576.

13. Bredholt, H., E. Fjaervik, G. Johnsen, and S. B. Zotchev. 2008. Actinomycetes from sediments in the Trondheim Fjord, Norway: diversity and biological activity. *Mar. Drugs* 6:12–24.

14. Bulina, T. I., I. V. Alferova, and L. P. Terekhova. 1997. A new method for the isolation of actinomycetes with the use of microwave irradiation of soil samples. *Mikrobiologiya* 66:278–282.

15. Bull, A. T. 2004. Biotechnology, the art of exploiting biotechnology, p. 3–13. *In* A. T. Bull (ed.), *Microbial Diversity and Bioprospecting.* ASM Press, Washington, DC.

16. Bull, A. T. 2004. Microbial diversity: the resource, p. 13–15. *In* A. T. Bull (ed.), *Microbial Diversity and Bioprospecting.* ASM Press, Washington, DC.

17. Bull, A. T., M. Goodfellow, and J. H. Slater. 1992. Biodiversity as a source of innovation in biotechnology. *Annu. Rev. Microbiol.* 46:219–252.

18. Bull, A. T., and J. E. M. Stach. 2007. Marine actinobacteria: new opportunities for natural product search and discovery. *Trends Microbiol.* 15:491–499.

19. Bull, A. T., J. E. M. Stach, A. C. Ward, and M. Goodfellow. 2005. Marine actinobacteria: perspectives, challenges, future directions. *Antonie van Leeuwenhoek* 87:65–79.

20. Bull, A. T., A. C. Ward, and M. Goodfellow. 2000. Search and discovery strategies for biotechnology: the paradigm shift. *Microbiol. Mol. Biol. Rev.* 64:573–606.

21. Burman, N. P., C. W. Oliver, and J. K. Stevens. 1969. Membrane filtration techniques for the isolation from water of coli-aerogenes, *Escherichia coli*, faecal streptococci, *Clostridium perfringens*, actinomycetes and microfungi, p. 127–134. *In* D. A. Shapton and G. W. Gould (ed.), *Isolation Methods for Microbiologists.* Academic Press, London, United Kingdom.

22. Caruso, M., A. L. Colombo, L. Fedele, A. Pavesi, S. Quaroni, M. Saracchi, and G. Ventrella. 2000. Isolation of endophytic fungi and actinomycetes taxane producers. *Ann. Microbiol.* 50:3–13.

23. Castillo, U. F., G. A. Strobel, E. J. Ford, W. H. Hess, H. Porter, J. B. Jensen, H. Albert, R. Robinson, M. A. M. Condron, D. B. Teplow, D. Stevens, and D. Yaver. 2002. Munumbicins, wide spectrum antibiotics produced by *Streptomyces* NRRL 30562, endophytic on *Kennedia nigrisca*. *Microbiology* 148:2675–2685.

24. Couch, J. N. 1949. A new group of organisms related to *Actinomyces*. *J. Elisha Mitchell Sci. Soc.* 65:315–318.

25. Couch, J. N. 1954. The genus *Actinoplanes* and its relatives. *Trans. N. Y. Acad. Sci.* 16:315–318.

26. Couch, J. N. 1955. A new genus of the family *Actinomycetales* with a revision of the genus *Actinoplanes*. *J. Elisha Mitchell Sci. Soc.* 71:148–155.

27. Couch, J. N. 1963. Some new genera and species of the *Actinoplanaceae*. *J. Elisha Mitchell Sci. Soc.* 79:53–70.

28. Cross, T. 1982. Actinomycetes: a continuing source of new metabolites. *Dev. Ind. Microbiol.* 33:1–18.

29. Cross, T., T. J. Rowbotham, E. N. Mishustin, E. Z. Tepper, F. Antoine-Portaels, K. P. Schaal, and H. Bickenbach. 1976. The ecology of nocardioform actinomycetes, p. 337–371. *In* M. Goodfellow, G. H. Brownell, and J. A. Serrano (ed.), *The Biology of Nocardiae.* Academic Press, London, United Kingdom.

30. Duangmal, K., A. C. Ward, and M. Goodfellow. 2005. Selective isolation of members of the *Streptomyces violaceoruber* clade from soil. *FEMS Microbiol. Lett.* 245:321–327.

31. Falconer, C. S. 1988. The isolation, physiology and taxonomy of carboxydotrophic actinomycetes. Ph.D. thesis. University of Newcastle, Newcastle upon Tyne, United Kingdom.

32. Feling, R. H., G. O. Buchanan, T. J. Mincer, C. A. Kaufman, P. R. Jensen, and W. Fenical. 2003. Salinosporamide A: a highly cytotoxic proteasome inhibitor from a novel microbial source, a marine bacterium of the new genus *Salinospora*. *Angew. Chem. Int. Ed.* 42:355–357.

33. Fiedler, H. P., C. Bruntner, A. T. Bull, A. C. Ward, M. Goodfellow, and G. Mihm. 2005. Marine actinomycetes as a source of new secondary metabolites. *Antonie van Leeuwenhoek* 87:37–42.

34. Fry, J. S. 2004. Culture dependent microbiology, p. 80–97. *In* A. T. Bull (ed.), *Microbial Diversity and Bioprospecting.* ASM Press, Washington, DC.

35. Garrity, G. M., B. K. Heimbuch, and M. Gaghardi. 1996. Isolation of zoosporogenous actinomycetes from desert soils. *J. Ind. Microbiol.* **17**:260–267.

36. Glockner, F. O., E. Zaichikov, N. Belkova, L. Denissova, J. Pernthaler, A. Pernthaler, and R. Amann. 2000. Comparative 16S rRNA analysis of the lake bacterioplankton reveals global distributed phylogenetic clusters including an abundant group of actinobacteria. *Appl. Environ. Microbiol.* **66**:5053–5065.

37. Gontang, E. A., W. Fenical, and P. R. Jensen. 2007. Phylogenetic diversity of Gram-positive bacteria cultured from marine sediments. *Appl. Environ. Microbiol.* **73**:3272–3282.

38. Goodfellow, M., A. R. Beckham, and M. D. Barton. 1982. Numerical classification of *Rhodococcus equi* and related actinomycetes. *J. Appl. Bacteriol.* **53**:199–207.

39. Goodfellow, M., and D. Dawson. 1978. Qualitative and quantitative studies of bacteria colonizing *Picea sitchensis* litter. *Soil Biol. Biochem.* **10**:303–307.

40. Goodfellow, M., and J. A. Haynes. 1984. Actinomycetes in marine sediments, p. 453–472. *In* L. Ortiz-Ortiz, L. F. Bojalil, and Y. Yakoleff (ed.), *Biological, Biochemical and Biomedical Aspects of Actinomycetes.* Academic Press, Orlando, FL.

41. Goodfellow, M., J. Lacey, M. Athalye, T. M. Embley, and T. Bowen. 1989. *Saccharopolyspora gregorii* and *Saccharopolyspora hordei*: two new actinomycete species from fodder. *J. Gen. Microbiol.* **135**:2125–2139.

42. Goodfellow, M., and A. G. O'Donnell. 1989. Search and discovery of industrially-significant actinomycetes, p. 343–383. *In* S. Baumberg, I. S. Hunter, and P. M. Rhodes (ed.), *Microbial Products: New Approaches.* Cambridge University Press, Cambridge, United Kingdom.

43. Goodfellow, M., and V. A. Orchard. 1974. Antibiotic sensitivity of some nocardioform bacteria and its value as a criterion for taxonomy. *J. Gen. Microbiol.* **33**:375–387.

44. Goodfellow, M., L. J. Stanton, K. E. Simpson, and D. E. Minnikin. 1990. Numerical classification of *Actinoplanes* and related genera. *J. Gen. Microbiol.* **136**:19–36.

45. Goodfellow, M., F. M. Stainsby, R. Davenport, J. Chun, and T. P. Curtis. 1998. Activated sludge foaming: the true extent of actinomycete diversity. *Water Sci. Technol.* **37**:511–519.

46. Goodfellow, M., and S. T. Williams. 1983. Ecology of actinomycetes. *Annu. Rev. Microbiol.* **37**:189–216.

47. Goodfellow, M., and E. Williams. 1986. New strategies for the selective isolation of industrially important bacteria. *Biotechnol. Genet. Eng. Rev.* **4**:213–262.

48. Gregory, P. H., and M. E. Lacey. 1963. Mycological examination of dust from mouldy hay associated with farmers lung disease. *J. Gen. Microbiol.* **30**:75–88.

49. Gribaldo, S., and H. Phillipe. 2002. Ancient phylogenetic relationships. *Theor. Popul. Biol.* **61**:391–408.

50. Guan, T.-W., S.-K. Tang, J.-Y. Wu, X.-Y. Zhi, L.-H. Xu, L.-L. Zhang, and W.-J. Li. 2009. *Haloglycomyces albusi* gen. nov., sp. nov., a halophilic, filamentous actinomycete of the family *Glycomycetaceae*. *Int. J. Syst. Evol. Microbiol.* **59**:1297–1301.

51. Hahn, M. W. 2009. Description of seven candidate species affiliated with the phylum *Actinobacteria*, representing planktonic freshwater bacteria. *Int. J. Syst. Evol. Microbiol.* **59**:112–117.

52. Hanka, L. J., P. W. Rueckert, and T. Cross. 1985. A method for isolating strains of the genus *Streptoverticillium* from soil. *FEMS Microbiol. Lett.* **30**:365–368.

53. Hanka, L. J., and R. D. Schaadt. 1988. Methods for isolation of streptoverticillia from soil. *J. Antibiot.* **41**:576–578.

54. Hattori, T., and R. Hattori. 1976. The physical environment in soil microbiology: an attempt to extend principles of microbiology to soil microorganisms. *CRC Crit. Rev. Microbiol.* **4**:423–461.

55. Hayakawa, M. 2003. Selective isolation of rare actinomycete genera using pretreatment techniques, p. 55–81. *In* I. Kurtböke (ed.), *Selective Isolation of Rare Actinomycetes.* Queensland Complete Printing Services, Nambour, Queensland, Australia.

56. Hayakawa, M., M. Ariizumi, T. Yamazaki, and H. Nonomura. 1995. Chemotaxis in the zoosporic actinomycete *Catenuloplanes japonicus*. *Actinomycetologica* **9**:152–163.

57. Hayakawa, M., H. Iino, S. Takeuchi, and T. Yamazaki. 1997. Application of a method incorporating treatment with chloramine-T for the selective isolation of *Streptosporangiaceae* from soil. *J. Ferment. Bioeng.* **84**:599–602.

58. Hayakawa, M., K. Ishizawa, and H. Nonomura. 1988. Distribution of rare actinomycetes in Japanese soils. *J. Ferment. Technol.* **66**:367–373.

59. Hayakawa, M., J. Kajiura, and H. Nonomura. 1991. New methods for the highly selective isolation of *Streptosporangium* and *Dactylosporangium* from soil. *J. Ferment. Bioeng.* **12**:327–333.

60. Hayakawa, M., Y. Momose, T. Kajiura, T. Yamazaki, T. Tamura, K. Hatano, and H. Nonomura. 1995. A selective isolation method for *Actinomadura viridis* in soil. *J. Ferment. Bioeng.* **79**:287–289.

61. Hayakawa, M., Y. Momose, T. Yamazaki, and H. Nonomura. 1996. A method for the selective isolation of *Microtetraspora glauca* and related four spored actinomycetes from soil. *J. Appl. Bacteriol.* **80**:375–386.

62. Hayakawa, M., and H. Nonomura. 1987. Humic acid-vitamin agar, a new medium for the selective isolation of soil actinomycetes. *J. Ferment. Technol.* **65**:501–509.

63. Hayakawa, M., and H. Nonomura. 1987. Efficacy of artificial humic acid as a selective nutrient in HV agar and for the isolation of soil actinomycetes. *J. Ferment. Technol.* **65**:609–616.

64. Hayakawa, M., M. Otoguro, T. Takeuchi, T. Yamazaki, and Y. Iimura. 2000. Application of a method incorporating differential centrifugation for selective isolation of motile actinomycetes in soil and plant litter. *Antonie van Leeuwenhoek* **78**:171–185.

65. Hayakawa, M., T. Sadakata, T. Kajiura, and H. Nonomura. 1991. New methods for the highly selective isolation of *Micromonospora* and *Microbispora* from soil. *J. Ferment. Bioeng.* **72**:320–326.

66. Hayakawa, M., T. Takeuchi, and T. Yamazaki. 1996. Combined use of trimethoprim and nalidixic acid for the selective isolation of actinomycetes from soil. *Actinomycetologia* **10**:80–90.

67. Hayakawa, M., T. Tamura, H. Iino, and H. Nonomura. 1991. Pollen-baiting and drying method for the highly selective isolation of *Actinoplanes* spp. from soil. *J. Ferment. Bioeng.* **72**:433–438.

68. Hayakawa, M., T. Tamura, and H. Nonomura. 1991. Selective isolation of *Actinoplanes* and *Dactylosporangium* from soil using γ-collidine as the chemoattractant. *J. Ferment. Bioeng.* **72**:426–432.

69. Hayakawa, M., Y. Yoshida, and Y. Iimura. 2004. Selective isolation of bioactive soil actinomycetes belonging to the *Streptomyces violaceusniger* phenotypic cluster. *J. Appl. Microbiol.* **96**:973–981.

70. Head, I. M., J. R. Saunders, and R. W. Pickup. 1988. Microbial evolution, diversity and ecology: a decade of ribosomal RNA analysis of uncultivated microorganisms. *Microb. Ecol.* **35**:1–21.

71. Hirsch, C. F., and D. L. Christensen. 1983. Novel method for the selective isolation of actinomycetes. *Appl. Environ. Microbiol.* **46**:925–929.

72. Hohmann, C., K. Schneider, C. Bruntner, R. Brown, A. L. Jones, M. Goodfellow, M. Kramer, J. F. Imhoff, G. Nicholson, H-P. Fiedler, and R. D. Süssmuth. 2009. Albidopyrone, a new α-pyrone-containing metabolite

from marine-derived *Streptomyces* sp. NTK 227. *J. Antibiot.* **62**:75–79.

73. **Hohmann,C., K. Schneider, C. Bruntner, E. Irran, G. Nicholson, A. T. Bull, A. L. Jones, R. Brown, J. E. M. Stach, M. Goodfellow, W. Beil, M. Kramer, J. F. Imhoff, R. D. Süssmuth, and H. P. Fiedler.** 2009. Caboxamycin, a new antibiotic of the benzoxazole family produced by the deep-sea strain *Streptomyces* sp. NTK 937. *Antonie van Leeuwenhoek* **62**:99–104.

74. **Hong, K., A.-H. Gao, Q.-Y. Xie, H. Gao, L. Zhuang, H.-P. Lin, H.-P. Yu, X.-S. Yao, M. Goodfellow, and J.-S. Ruan.** 2009. Actinomycetes for marine drug discovery isolated from mangrove soils and plants in China. *Mar. Drugs* **7**:24–44.

75. **Hopkins, D. W., S. J. MacNaughton, and A. G. O'Donnell.** 1991. A dispersion and differential centrifugation technique for representatively sampling microorganisms from soil. *Soil Biol. Biochem.* **23**:217–225.

76. **Hopkins, D. W., A. G. O'Donnell, and S. J. MacNaughton.** 1991. Evaluation of a dispersion and elutriation technique for sampling microorganisms from soil. *Soil Biol. Biochem.* **23**:227–232.

77. **Hopmann, C., M. Kurz, J. Brönstrup, J. Wink, and D. Le Beller.** 2002. Isolation and structure elucidation of vancoresmycin – a new antibiotic from *Amycolatopsis* sp. ST101170. *Tetrahedron Lett.* **43**:435–438.

78. **Hsu, S. C., and J. L. Lockwood.** 1975. Powdered chitin as a selective medium for enumeration of actinomycetes in water and soil. *Appl. Microbiol.* **29**:422–426.

79. **Huck, T. A., N. Porter, and M. E. Bushell.** 1991. Positive selection of antibiotic producing actinomycetes. *J. Gen. Microbiol.* **137**:2310–2329.

80. **Igarashi, Y., S. Miura, T. Fujita, and T. Furumai.** 2006. A cytotoxic compound from the endophyte *Streptomyces hygroscopicus*. *J. Antibiot.* **59**:193–195.

81. **Ikeda, H., J. Ishikawa, A. Hanamoto, M. Shinose, H. Kikuchi, T. Shiba, Y. Sakaki, M. Hattori, and S. Ōmura.** 2003. Complete genome sequence and comparative analysis of the industrial microorganism *Streptomyces avermitilis*. *Nat. Biotechnol.* **21**:526–531.

82. **Jensen, H. L.** 1930. Actinomycetes in Danish soils. *Soil Sci.* **30**:59–77.

83. **Jensen, H. L.** 1932. Contributions to our knowledge of the actinomycetes. II. The definition and subdivision of the genus *Actinomyces*, with a preliminary account of Australian soil actinomycetes. *Proc. Linnean Soc. N. S. W.* **59**:19–61.

84. **Jensen, P. R., R. Dwight, and W. Fenical.** 1991. Distribution of actinomycetes in near-shore tropical marine sediments. *Appl. Environ. Microbiol.* **57**:1102–1108.

85. **Jensen, P. R., E. Gontang, C. Mafnas, T. J. Mincer, and W. Fenical.** 2005. Culturable marine actinomycete diversity from tropical Pacific Ocean sediments. *Environ. Microbiol.* **7**:1039–1048.

86. **Jensen, P. R., and F. M. Lauro.** 2008. An assessment of actinobacterial diversity in the marine environment. *Antonie van Leeuwenhoek* **94**:51–62.

87. **Johnson, D. B., P. Bacelar-Nicolau, N. Okibe, A. Thomas, and K. B. Hallberg.** 2009. *Ferrimicrobium acidiphilum* gen. nov., sp. nov. and *Ferrithrix thermotolerans* gen. nov., sp. nov. heterotrophic, iron-oxidizing, extremely acidophilic actinobacteria. *Int. J. Syst. Evol. Microbiol.* **59**:1082–1989.

88. **Kämpfer, P., R. M. Kroppenstedt, and D. Wolfgang.** 1991. A numerical classification of the genera *Streptomyces* and *Streptoverticillium* using miniaturized physiological tests. *J. Gen. Microbiol.* **137**:1831–1891.

89. **Karawowski, J.** 1986. The selective isolation of *Micromonospora* from soil by cesium chloride density gradient ultracentrifugation. *J. Ind. Microbiol.* **1**:181–186.

90. **Katayama, T., T. Kato, M. Tanaka, T. A. Douglas, A. Brouchkova, M. Fukuda, F. Tomito, and K. Asano.** 2009. *Glacibacter superstetes* gen. nov., sp. nov., a novel member of the family *Microbacteriaceae* isolated from a permafrost ice wedge. *Int. J. Syst. Evol. Microbiol.* **59**:482–486.

91. **Khan, M. R., and S. T. Williams.** 1975. Studies on the ecology of actinomycetes in soil. VIII. Distribution and characteristics of acidophilic actinomycetes. *Soil Biol. Biochem.* **7**:345–348.

92. **Kim, S. B., C. Falconer, E. Williams, and M. Goodfellow.** 1998. *Streptomyces thermocarboxydovorans* sp. nov. and *Streptomyces thermocarboxydus* sp. nov., two moderately thermophilic carboxydotrophic species from soil. *Int. J. Syst. Bacteriol.* **48**:59–68.

93. **Kudo, T., T. Itoh, S. Miyadoh, T. Shamura, and A. Seino.** 1993. *Herbidospora* gen. nov., a new genus of the family *Streptosporangiaceae* Goodfellow et al.1990. *Int. J. Syst. Bacteriol.* **43**:319–328.

94. **Kumar, Y., P. Aiemsun-ang, A. C. Ward, and M. Goodfellow.** 2007. Diversity and geographical distribution of members of the *Streptomyces violaceusniger* 16S rRNA gene clade detected by clade specific primers. *FEMS Microbiol. Ecol.* **62**:54–63.

95. **Kurahashi, M., Y. Tukunaga, Y. Sakiyama, S. Harayama, and A. Yokota.** 2009. *Iamia majanohamensis* gen. nov., sp. nov., an actinobacterium isolated from sea cucumber *Halothuria edulis*, and proposal of family *Iamiaceae* fam. nov. *Int. J. Syst. Evol. Microbiol.* **59**:869–873.

96. **Kurtböke, D.I.** 2003. Use of bacteriophage for the selective isolation of rare actinomycetes, p. 10–54. *In* I. Kurtböke (ed.), *Selective Isolation of Rare Actinomycetes*. Queensland Complete Printing Services, Nambour, Queensland, Australia.

97. **Kurtböke, D. I., G. F. Chen, and S. T. Williams.** 1992. Use of polyvalent phage for reduction of streptomycetes on soil dilution plates. *J. Appl. Bacteriol.* **72**:103–111.

98. **Kurtböke, D. I., N. E. Murphy, and K. Swasithamparam.** 1993. Use of bacteriophage for the selective isolation of thermophilic actinomycetes from composted *Eucalyptus* bark. *Can. J. Microbiol.* **39**:46–51.

99. **Kurtböke, D. I., and S. T. Williams.** 1991. Use of actinophages for selective isolation purposes: current problems in actinomycetes, p. 31–34. *In* R. Locci, H. Lechevalier, and S. T. Williams (ed.), *Actinomycetes*. International Centre for Theoretical and Applied Ecology, Gorizia, Italy.

100. **Küster, E., and S. T. Williams.** 1964. Selective media for isolation of streptomycetes. *Nature (London)* **202**: 928–929.

101. **Labeda, D. P., and R. M. Kroppenstedt.** 2005. *Stackebrandtia nassauensis* gen. nov., sp. nov. and emended description of the family *Glycomycetaceae*. *Int. J. Syst. Evol. Microbiol.* **55**:1687–1691.

102. **Labeda, D. P., and M. C. Shearer.** 1991. Isolation of actinomycetes for biotechnological applications, p. 1–19. *In* D. P. Labeda (ed.), *Isolation of Biotechnological Organisms from Nature*. McGraw-Hill Publishing Co., New York, NY.

103. **Lacey, J.** 1971. The microbiology of moist barley storage in unsealed silos. *Ann. Appl. Biol.* **69**:187–212.

104. **Lacey, J., and J. Dutkiewicz.** 1976. Isolation of actinomycetes and fungi from mouldy hay using a sedimentation chamber. *J. Appl. Bacteriol.* **41**:315–319.

105. **Lazzarini, A., L. Cavaletti, G. Toppo, and F. Marinelli.** 2000. Rare genera of actinomycetes as potential producers of new antibiotics. *Antonie van Leeuwenhoek* **78**: 399–405.

106. **Lechevalier, M. P.** 1981. Ecological associations involving actinomycetes, p. 159–166. *In* K. P. Schaal and G. Pulverer (ed.), *Actinomycetes*. Gustav Fischer Verlag, Stuttgart, Germany.

107. Lingappa, Y., and J. L. Lockwood. 1961. A chitin medium for isolation, growth and maintenance of actinomycetes. *Nature* **189:**158–159.

108. MacDonald, R. M. 1986. Sampling soil microfloras: dispersion of soil by ion exchange and extraction of specific microorganisms by elutriation. *Soil Biol. Biochem.* **18:**399–406.

109. MacNaughton, S. J., and A. G. O'Donnell. 1994. Tuberculostearic acid as a means of estimating the recovery (using dispersion and differential centrifugation) of actinomycetes from soil. *J. Microbiol. Methods* **20:** 69–77.

110. Makkar, N. S., and T. Cross. 1982. Actinoplanetes in soil and on plant litter from freshwater habitats. *J. Appl. Bacteriol.* **52:**209–218.

111. Maldonado, L., J. V. Hookey, A. C. Ward, and M. Goodfellow. 2000. The *Nocardia salmonicida* clade, including descriptions of *Nocardia cummidelens* sp. nov., *Nocardia fluminea* sp. nov. and *Nocardia soli* sp. nov. *Antonie van Leeuwenhoek* **78:**367–377.

112. Maldonado, L. A., W. Fenical, P. R. Jensen, C. A. Kaufman, T. J. Mincer, A. C. Ward, A. T. Bull, and M. Goodfellow. 2005. *Salinispora avenicola* gen. nov., sp. nov. and *Salinispora tropica* sp. nov., obligate marine actinomycetes belonging to the family *Micromonosporaceae*. *Int. J. Syst. Evol. Microbiol.* **55:**1759–1766.

113. Maldonado, L. A., J. E. Stach, W. Pathom-aree, A. C. Ward, A. T. Bull, and M. Goodfellow. 2005. Diversity of cultivable actinobacteria in geographically widespread marine sediments. *Antonie van Leeuwenhoek* **87:**11–18.

114. Manfio, G. P., J. Zakrzewska-Czerwinska, E. Atalan, and M. Goodfellow. 1995. Towards minimal standards for the description of *Streptomyces* species. *Biotechnologia* **7–8:**242–253.

115. Manfio, G. P., J. Zakrzewska-Czerwinska, M. Mordarski, C. Rodriguez, M. D. Collins, and M. Goodfellow. 2003. Classification of novel soil streptomycetes as *Streptomyces aureus* sp. nov., *Streptomyces laceyi* sp. nov., and *Streptomyces sanglieri* sp. nov. *Antonie van Leeuwenhoek* **83:**245–255.

116. McCarthy, A. J., and T. Cross. 1981. A note on a selective isolation medium for the thermophilic actinomycete *Thermomonospora chromogena*. *J. Appl. Bacteriol.* **51:**299–302.

117. McCarthy, A. J., and T. Cross. 1984. A taxonomic study of *Thermomonospora* and other monosporic actinomycetes. *J. Gen. Microbiol.* **130:**5–25.

118. Meiklejohn, J. 1957. Number of bacteria and actinomycetes in a Kenya soil. *J. Soil Sci.* **8:**240–247.

119. Mexson, J., M. Goodfellow, and A. T. Bull. 2000. Preliminary studies of the diversity of micromonosporae in Indonesian soils and lake sediments. *Indones. J. Biotech.* Special Issue, **June:**384–388.

120. Mincer, T. J., P. R. Jensen, C. A. Kaufmann, and W. F. Fenical. 2002. Widespread and persistent populations of a major new marine actinomycete taxon in ocean sediments. *Appl. Environ. Microbiol.* **68:**5005–5011.

121. Monciardini, P., M. Sosio, L. Cavaletti, C. Chiochini, and S. Donadio. 2002. New PCR primers for the selective amplification of 16S rDNA from different groups of actinomycetes. *FEMS Microbiol. Ecol.* **42:**419–429.

122. Niepold, F., R. Conrad, and H. G. Schlegel. 1979. Evaluation of the efficiency of extract for quantitative estimation of hydrogen bacteria in soil. *Antonie van Leeuwenhoek* **45:**485–497.

123. Nishiyama, M., K. Senoo, H. Wada, and S. Matsumato. 1992. Identification of soil micro-habitats for growth, death and survival of a bacterium, γ-1, 2, 3, 4, 5, 6, -hexachlorocyclohexane assimilating *Sphingomonas paucimobilis*, by fractionation of soil. *FEMS Microbiol. Ecol.* **101:**145–150.

124. Nolan, R. D., and T. Cross. 1988. Isolation and screening of actinomycetes, p. 1–32. *In* M. Goodfellow, S. T. Williams, and M. Mordarski (ed.), *Actinomycetes in Biotechnology*. Academic Press, San Diego, CA.

125. Nonomura, H., and M. Hayakawa. 1988. New methods for the selective isolation of soil actinomycetes, p. 288–293. *In* Y. Okami, T. Beppu, and K. Ogawara (ed.), *Biology of Actinomycetes '88*. Japan Scientific Societies Press, Tokyo, Japan.

126. Nonomura, H., and Y. Ohara. 1957. Distribution of actinomycetes in soil. II. *Microbispora*, a new genus of *Streptosporangiaceae*. *J. Ferment. Technol.* **35:**307–311.

127. Nonomura, H., and Y. Ohara. 1960. Distribution of actinomycetes in soil. V. The isolation and classification of the genus *Streptosporangium*. *J. Ferment. Technol.* **38:**405–409.

128. Nonomura, H., and Y. Ohara. 1969. Distribution of actinomycetes in soil. VI. A culture method effective for preferential isolation and enumeration of *Microbispora* and *Streptosporangium* strains in soil (Part 1). *J. Ferment. Technol.* **47:**463–469.

129. Nonomura, H., and Y. Ohara. 1969. Distribution of actinomycetes in soil. VII. A culture method effective for preferential isolation and enumeration of *Microbispora* and *Streptosporangium* strains in soil (Part 2). *J. Ferment. Technol.* **47:**701–709.

130. Nonomura, H., and Y. Ohara. 1971. Distribution of actinomycetes in soil. IX. New species of the genera *Microbispora* and *Microtetraspora* and their isolation methods. *J. Ferment. Technol.* **49:**887–894.

131. Nonomura, H., and Y. Ohara. 1971. Distribution of actinomycetes in soil. XI. Some new species of the genus *Actinomadura* Lechevalier *et al*. *J. Ferment. Technol.* **49:** 904–912.

132. Nonomura, H., and Y. Ohara. 1971. Distribution of actinomycetes in soil. VIII. Green spore group of *Microtetraspora*, its preferential isolation and taxonomic characteristics. *J. Ferment. Technol.* **49:**1–7.

133. Nonomura, H., and Y. Ohara. 1971. Distribution of actinomycetes in soil. IX. New species of the genera *Microbispora* and *Microtetraspora* and their isolation method. *J. Ferment. Technol.* **48:**887–894.

134. Nonomura, H., and S. Takagi. 1977. Distribution of actinomycetes in soil of Japan. *J. Ferment. Technol.* **55:** 423–428.

135. O'Donnell, A. G., C. Falconer, M. Goodfellow, A. C. Ward, and E. Williams. 1993. Biosystematics and diversity amongst novel carboxydotrophic actinomycetes. *Antonie van Leeuwenhoek* **64:**325–340.

136. Okami, Y., and T. Okazaki. 1972. Studies on marine microorganisms. I. *J. Antibiot.* **25:**456–460.

137. Okoro, C. K., R. Brown, A. L. Jones, B. A. Andrews, J. A. Asenjo, M. Goodfellow, and A. T. Bull. 2009. Diversity of culturable actinomycetes in hyper-arid soils of the Atacama Desert, Chile. *Antonie van Leeuwenhoek* **95:**121–133.

138. Olsen, G. J., D. J. Lane, S. J. Giovannoni, and N. P. Pace. 1986. Microbial ecology and evolution: a ribosomal RNA approach. *Annu. Dev. Microbiol.* **40:** 337–365.

139. Orchard, V. A., and M. Goodfellow. 1974. The selective isolation of *Nocardia* from soil using antibiotics. *J. Gen. Microbiol.* **85:**160–162.

140. Orchard, V. A., and M. Goodfellow. 1980. Numerical classification of some named strains of *Nocardia asteroides* and related isolates from soil. *J. Gen. Microbiol.* **118:** 295–312.

141. Orchard, V. A., M. Goodfellow, and S. T. Williams. 1977. Selective isolation and occurrence of nocardiae in soil. *Soil Biol. Biochem.* **9:**233–238.

142. Otoguro, M., M. Hayakawa, T. Yamazaki, T. Tamura, K. Hatano, and Y. Iimura. 2001. Numerical phenetic and phylogenetic analyses of *Actinokineospora* isolates with a description of *Actinokineospora auranticolor* sp. nov. and *Actinokineospora enzanensis* sp. nov. *Actinomycetologia* **15**:30–39.

143. Palleroni, N. J. 1980. A chemotactic method for the isolation of *Actinoplanaceae*. *Arch. Microbiol.* **128**:53–55.

144. Pathom-aree, W., J. E. M. Stach, A. C. Ward, K. Horikoshi, A. T. Bull, and M. Goodfellow. 2006. Diversity of actinomycetes isolated from Challenger Deep sediment (10,898m) from the Mariana Trench. *Extremophiles* **10**:181–189.

145. Porter, J. N., J. J. Wilhelm, and H. D. Tresner. 1960. Method for preferential isolation of actinomycetes from soils. *Appl. Microbiol.* **8**:174–178.

146. Preobrazhenskaya, T. P., N. V. Lavrova, R. S. Ukholina, and N. P. Nechaeva. 1975. Isolation of new species of *Actinomadura* on selective media with bruneomycin and streptomycin. *Antibiotiki* **20**:404–409.

147. Promnuan, Y., T. Kudo, and P. Chatawannakul. 2009. Actinomycetes isolated from beehives in Thailand. *World J. Microbiol. Biotechnol.* DOI 10.1007/s11274-009-0051-1.

148. Ramsey, A. J. 1984. Extraction of bacteria from soil: efficiency of shaking or ultrasonication as indicated by direct counts and autoradiography. *Soil Biol. Biochem.* **16**:475–481.

149. Reibach, P. H., P. Mask, and J. G. Streeter. 1981. A rapid one-step method for the isolation of bacteroids from root nodules of soybean plants, utilizing self-heating Percoll gradients. *Can. J. Microbiol.* **27**:491–495.

150. Rehacêk, Z. 1959. Isolation of actinomycetes and determination of the number of their spores in soil. *Microbiology* **28**:220–225.

151. Riedlinger, J., A. Reike, H. Zähner, B. Krismer, A. T. Bull, L. A. Maldonado, A. C. Ward, M. Goodfellow, B. Bister, D. Bischoff, R. D. Süssmuth, and H.-P. Fiedler. 2004. Abyssomicins, inhibitors of the *para*-aminobenzoic acid pathway produced by the marine *Verrucosispora* strain AB-18-032. *J. Antibiot.* **57**:271–279.

152. Rowbotham, T. J., and T. Cross. 1977. Ecology of *Rhodococcus coprophilus* and associated actinomycetes in freshwater and agricultural habitats. *J. Gen. Microbiol.* **100**:231–240.

153. Sait, M., P. Hugenholtz, and P. H. Janssen. 2002. Cultivation of globally distributed soil bacteria from phylogenetic lineages previously only detected in cultivation-independent surveys. *Environ. Microbiol.* **4**:654–666.

154. Seker, R., A. Pernthaler, J. Pernthaler, F. Warnecke, T. Posch, and R. Amann. 2003. An improved protocol for quantification of freshwater *Actinobacteria* by fluorescence *in situ* hybridization. *Appl. Environ. Microbiol.* **69**:2928–2935.

155. Sembiring, L., A. C. Ward, and M. Goodfellow. 2000. Selective isolation and characterisation of members of the *Streptomyces violaceusniger* clade associated with the roots of *Paraserianthes falcataria*. *Antonie van Leeuwenhoek* **78**:353–366.

156. Seviour, R. J., C. Kragelund, Y. Kong, K. Eales, J. L. Nielsen, and P. H. Nielsen. 2008. Ecophysiology of the *Actinobacteria* in activated sludge systems. *Antonie van Leeuwenhoek* **94**:21–33.

157. Shearer, M. C. 1987. Methods for the isolation of non-streptomycete actinomycetes. *Dev. Ind. Microbiol.* **28**: 91–97.

158. Sneath, P. H. A. 1980. BASIC program for the most diagnostic properties of groups from an identification matrix of percentage positive characters. *Comput. Geosci.* **6**:21–26.

159. Stach, J. E. M., L. A. Maldonado, D. G. Masson, A. C. Ward, M. Goodfellow, and A. T. Bull. 2003. Statistical approaches for estimating actinobacterial diversity in marine sediments. *Appl. Environ. Microbiol.* **69**:6189–6200.

160. Stach, J. E. M., L. A. Maldonado, A. C. Ward, M. Goodfellow, and A. T. Bull. 2003. New primers for the class *Actinobacteria*: application to marine and terrestrial environments. *Environ. Microbiol.* **5**:828–841.

161. Stingl, U., J.-C. Cho, W. Foo, K. L. Vergin, B. Lanoil, and S. J. Giovannoni. 2008. Dilution-to-extinction culturing of psychrotolerant planktonic bacteria from permanently ice-covered lakes in the McMurdo Dry Valleys, Antarctica. *Microb. Ecol.* **55**:395–405.

162. Strohl, W. 2004. Antimicrobials, p. 288–313. *In* A. T. Bull (ed.), *Microbial Diversity and Bioprospecting*. ASM Press, Washington, DC.

163. Suzuki, S.-I., T. Okuda, and S. Komatsubara. 2001. Selective isolation and distribution of the genus *Planomonospora* in soils. *Can. J. Microbiol.* **47**:253–263.

164. Suzuki, S.-I., T. Okuda, and S. Komatsubara. 2001. Selective isolation and study of the global distribution of the genus *Planobispora* in soils. *Can. J. Microbiol.* **47**:979–986.

165. Tamura, T., M. Hayakawa, and K. Hatano. 2001. A new genus of the order *Actinomycetales*, *Virgosporangium* gen. nov., with descriptions of *Virgosporangium ochraceum* sp. nov. and *Virgosporangium auranticolor* sp. nov. *Int. J. Syst. Evol. Bacteriol.* **51**:1809–1816.

166. Tan, G. Y. A., S. Robinson, E. Lacey, R. Brown, W. Kim, and M. Goodfellow. 2007. *Amycolatopsis regifaucium* sp. nov., a novel actinomycete that produces kigamicins. *Int. J. Syst. Evol. Microbiol.* **57**:2562–2567.

167. Tan, G. Y. A., S. Robinson, E. Lacey, and M. Goodfellow. 2006. *Amycolatopsis australiensis* sp. nov., a novel actinomycete isolated from arid soils. *Int. J. Syst. Evol. Microbiol.* **56**:2297–2301.

168. Tan, G. Y. A., A. C. Ward, and M. Goodfellow. 2006. Exploration of *Amycolatopsis* diversity in soil using genus specific primers and novel selective media. *Syst. Appl. Microbiol.* **29**:557–569.

169. Terekhova, L. 2003. Isolation of actinomycetes with the use of microwaves and electric pulses. *In* I. Kurtböke (ed.), *Selective Isolation of Rare Actinomycetes*. Queensland Complete Printing Services, Nambour, Queensland, Australia.

170. Tian, X.-P., S.-K. Tang, J.-D. Dong, Y.-Q. Zhang, L.-H. Xu, S. Zhang, and W.-J. Li. 2009. *Marinactinospora thermotolerans* gen. nov., sp. nov., a marine actinomycete isolated from a sediment in the northern South China Sea. *Int. J. Syst. Evol. Microbiol.* **59**:948–952.

171. Tribe, H. T., and S. M. Abu El-Souod. 1979. Colonization of hair in soil-water cultures, with special reference to the genera *Pilimelia* and *Spirillospora* (*Actinomycetales*). *Nova Hédwigia* **31**:789–805.

172. Trujillo, M. E., and M. Goodfellow. 2003. Numerical phenetic classification of clinically significant aerobic sporoactinomycetes and related organisms. *Antonie van Leeuwenhoek* **84**:39–68.

173. Trujillo, M. W., R. M. Kroppenstedt, C. Fernandez-Molinero, P. Schumann, and E. Martinez-Molina. 2007. *Micromonospora lupini* sp. nov. and *Micromonospora saelicerisensis* sp. nov., isolated from root nodules of *Lupinus angustifolius*. *Int. J. Syst. Evol. Microbiol.* **57**:2799–2804.

174. Trujillo, M. E., R. M. Kroppenstedt, P. Schumann, L. Carro, and E. Martinez-Molina. 2006. *Micromonospora coriariae* sp. nov., isolated from root nodules of *Coriaria myrtifolia*. *Int. J. Syst. Evol. Microbiol.* **56**:2381–2385.

175. Verma, V. C., S. K. Gond, A. Kumar, A. Mishra, R. N. Kharwar, and A. C. Gange. 2009. Endophytic actinomycetes from *Azadirachia indica* A. Juss.: isolation, diversity, and antimicrobial activity. *Microb. Ecol.* **57**:749–756.

176. **Vickers, J. C., and S. T. Williams.** 1987. An assessment of plate inoculation procedures for the enumeration and isolation of soil streptomycetes. *Microbios Lett.* **35:**113–117.

177. **Vickers, J. C., S. T. Williams, and G. W. Ross.** 1984. A taxonomic approach to selective isolation of streptomycetes from soil, p. 553–561. *In* L. Ortiz-Ortiz, L. F. Bojalil, and Y. Yakoleff (ed.), *Biological, Biochemical and Biomedical Aspects of Actinomycetes.* Academic Press, Orlando, FL.

178. **Wayne, L. G., and G. P. Kubica.** 1986. Genus *Mycobacterium* Lehmann and Neumann 1896, 363[AL], p. 1436–1457. *In* P. H. H. Sneath, N. S. Mair, M. E. Sharpe, and J. G. Holt (ed.), *Bergey's Manual of Systematic Bacteriology*, vol. 2. Williams & Wilkins, Baltimore, MD.

179. **Wellington, E. M. H., and T. Cross.** 1983. Taxonomy of antibiotic producing actinomycetes and new approaches to their selective isolation, p. 7–36. *In* M. E. Bushell (ed.), *Progress in Industrial Microbiology.* Elsevier, Amsterdam, The Netherlands.

180. **Whitehead, D.** 1989. Classification, selective isolation of members of the family *Pseudonocardiaceae.* Ph.D. thesis. Newcastle University, Newcastle upon Tyne, UK.

181. **Whitham, T. S., M. Athalye, D. E. Minnikin, and M. Goodfellow.** 1993. Numerical and chemical classification of *Streptosporangium* and related actinomycetes. *Antonie van Leeuwenhoek* **64:**387–429.

182. **Whitman, W. B., D. C. Coleman, and W. J. Wiebe.** 1998. Prokaryotes: the unseen majority. *Proc. Natl. Acad. Sci. USA* **95:**6578–6583.

183. **Williams, S. T., and T. Cross.** 1971. Isolation, purification, cultivation and preservation of actinomycetes. *Methods Microbiol.* **4:**295–334.

184. **Williams, S. T., and F. L. Davies.** 1965. Use of antibiotics for selective isolation and enumeration of actinomycetes in soil. *J. Gen. Microbiol.* **38:**251–261.

185. **Williams, S. T., F. L. Davies, C. I. Mayfield, and M. R. Khan.** 1971. Studies on the ecology of actinomycetes in soil. II. The pH requirements of streptomycetes from two acid soils. *Soil Biol. Biochem.* **3:**187–195.

186. **Williams, S. T., M. Goodfellow, and G. Alderson.** 1989. Genus *Streptomyces* Waksman and Henrici 1943, 339[AL], p. 2452–2492. *In* S. T. Williams, M. E. Sharpe, and J. G. Holt (ed.), *Bergey's Manual of Systematic Bacteriology*, vol. 4. Williams and Wilkins, Baltimore, MD.

187. **Williams, S. T., M. Goodfellow, G. Alderson, E. M. H. Wellington, P. H. A. Sneath, and M. J. Sackin.** 1983. Numerical classification of *Streptomyces* and related genera. *J. Gen. Microbiol.* **129:**1743–1813.

188. **Williams, S. T., M. Goodfellow, and J. C. Vickers.** 1984. New microbes from old habitats? p. 481–528. *In* D. P. Kelley and N. G. Carr (ed.), *The Microbe 1984*, vol. 2. Cambridge University Press, Cambridge, United Kingdom.

189. **Williams, S. T., M. Goodfellow, E. M. H. Wellington, J. C. Vickers, G. Alderson, P. H. A. Sneath, M. J. Sackin, and A. M. Mortimer.** 1983. A probability matrix for the identification of streptomycetes. *J. Gen. Microbiol.* **129:**1815–1830.

190. **Williams, S. T., S. Lanning, and E. M. H. Wellington.** 1984. Ecology of actinomycetes, p. 481–528. *In* M. Goodfellow, M. Mordarski, and S. T. Williams (ed.), *The Biology of the Actinomycetes.* Academic Press, London, United Kingdom.

191. **Williams, S. T., R. Locci, J. Vickers, G. M. Schofield, P. H. A. Sneath, and A. M. Mortimer.** 1985. Probabilistic identification of *Streptoverticillium* species. *J. Gen. Microbiol.* **131:**1681–1689.

192. **Williams, S. T., and C. I. Mayfield.** 1971. Studies on the ecology of actinomycetes in soil. III. The behaviour of neutrophlic streptomycetes in soil. *Soil Biol. Biochem.* **3:**197–208.

193. **Williams, S. T., M. Shameemullah, E. T. Watson, and C. I. Mayfield.** 1972. Studies on the ecology of actinomycetes in soil. VI. The influence of moisture tension on growth and survival. *Soil Biol. Biochem.* **4:**215–225.

194. **Williams, S. T., and J. C. Vickers.** 1988. Detection of actinomycetes in the natural environment—problems and perspectives, p. 265–270. *In* Y. Okami, T. Beppu, and K. Ogawara (ed.), *Biology of Actinomycetes '88.* Japan Scientific Societies Press, Tokyo, Japan.

195. **Williams, S. T., and E. M. H. Wellington.** 1982. Principles and problems of selective isolation of microbes, p. 9–26. *In* J. D. Bu'Lock, L. J. Nisbet, and D. J. Winstanley (ed.), *Bioactive Products: Search and Discovery.* Academic Press, London, United Kingdom.

196. **Williams, S. T., and E. M. H. Wellington.** 1982. Actinomycetes, p. 969–987. *In* A. L. Page, R. H. Miller, and D. R. Keeney (ed.), *Methods of Soil Analysis.* Part 2. *Chemical and Microbiological Properties*, 2nd ed. American Society of Agronomy/Soil Science Society of America, Madison, WI.

197. **Wink, J. M., R. M. Kroppenstedt, N. Ganguli, S. R. Nadkari, P. Schumann, G. Leibert, and E. Stackebrandt.** 2003. Three new antibiotic producing species of the genus *Amycolatopsis*: *Amycolatopsis balhimycina* sp. nov., *Amycolatopsis tolypomycina* sp. nov., *Amycolatopsis vancoresmycina* sp. nov., and description of *Amycolatopsis keratophila* subsp. *keratophila* subsp. nov. and *Amycolatopsis keratophila* subsp. *nogabecina* subsp. nov. *Syst. Appl. Microbiol.* **26:**38–46.

198. **Würdemann, D., B. J. Tindall, R. Pukall, H. Lunsdorf, C. Strompl, T. Namuth, H. Nahrstedt, M. Wos-Oxley, S. Ott, S. Schreiber, K. N. Timmis, and A. P. A. Oxley.** 2009. *Gordonibacter pamelaceae* gen. nov., sp. nov., a new member of the *Coriobacteriaceae* isolated from a patient with Crohn's Disease, and reclassification of *Eggerthella hongkongensis* Lau *et al.* 2006 as *Paraeggerthella hongkongensis* gen. nov., comb. nov. *Int. J. Syst. Evol. Microbiol.* **59:**1405–1415.

199. **Xu, C., L. Wang, Q. Cui, Y. Huang, Z. Liu, G. Zhang, and M. Goodfellow.** 2006. Four novel neutrotolerant acidophilic *Streptomyces* species isolated from acidic soils in China: *Streptomyces guanduensis* sp. nov., *Streptomyces paucisporeus* sp. nov., *Streptomyces rubidus* sp. nov. and *Streptomyces yanglinensis* sp. nov. *Int. J. Syst. Evol. Microbiol.* **56:**1109–1115.

200. **Yamamura, H., M. Hayakawa, and I. Iimura.** 2003. Application of sucrose-gradient centrifugation for selective isolation of *Nocardia* spp. from soil. *J. Appl. Microbiol.* **95:**677–685.

201. **Zhi, X.-Y., W.-J. Li, and E. Stackebrandt.** 2009. An update of the structure and 16S rRNA gene sequence-based definition of higher ranks of the class *Actinobacteria*, with the proposal of two new suborders and four new families and emended descriptions of the existing higher taxa. *Int. J. Syst. Evol. Microbiol.* **59:**589–608.

202. **Zwart, G., B. C. Crump, M. P. Kamst-van Agterveld, F. Hagen, and S.-K. Han.** 2002. Typical freshwater bacteria: an analysis of available 16S rRNA gene sequences from plankton of lakes and rivers. *Aquat. Microb. Ecol.* **28:**141–155.

Taxonomic Characterization of Prokaryotic Microorganisms

GIOVANNA E. FELIS, SANDRA TORRIANI,
JOHAN E. T. VAN HYLCKAMA VLIEG, AND AHARON OREN

3

3.1. INTRODUCTION

Prokaryote microorganisms constitute a major part of the genetic diversity of our planet and they play a pivotal role in the biochemical processes of the biosphere. In addition, they find application in a broad array of industrial and biotechnological applications (18, 108, 156). In the search for new industrial strains and catalysts, prospecting of natural diversity has greatly intensified over the last decades. Improved cultivation techniques and newly emerging (meta)genomic tools have provided access to sources of microbial diversity that were previously uncovered.

The interest of biotechnologists in what microorganisms can do is not always accompanied by a serious investigation on who they are. Identification of the isolates is of primary importance, not only for the exchange of scientific knowledge, but also in an applied context, because names are crucial for regulations in biotechnology and biosafety. A survey of species names associated with published genome sequences has, in several cases, revealed incorrect, out-of-date, or incomplete strain names (44). This suggests that reductionism brought about by molecular biology (159) and the interests of single research groups could significantly affect the coherence of names, with negative consequences on accessibility of microbiological data.

Taxonomy provides the framework for correct identification of prokaryotic strains. The aim of this chapter is to present microbiologists in biotechnology and molecular biology with the basic conceptual framework and the currently accepted procedures for characterization and consistent naming of newly isolated prokaryotic strains.

3.2. THE CONCEPTS OF PROKARYOTE TAXONOMY

3.2.1. What Is Taxonomy

Taxonomy is the systematic study of biological diversity, aimed at arranging organisms in an ordered scheme. The scheme is hierarchical, and attempts to reflect the evolutionary relationships among organisms. Classification, identification, and nomenclature are three separate, but interrelated subdisciplines of taxonomy. Classification is the process of clustering organisms into taxonomic groups (taxa) on the basis of similarities or relationships. Nomenclature is the assignment of names to taxonomic groups according to internationally accepted rules (for prokaryotes, the International Code of Nomenclature of Bacteria) (84). These rules are very clear to avoid misunderstandings. Finally, identification is the process of determining whether a new isolate belongs to one of the established and named taxa (80, 136). An isolate that cannot be assigned to a known taxon should be described and classified as a new taxon (species, genus, etc.). Therefore, classification precedes identification, since it would be impossible to assign new isolates to a taxon without prior arrangements of individuals into groups (113, 126).

Prokaryote taxonomy is a highly dynamic science. There is no official classification of prokaryotes, and classification schemes change in accordance with the development of new techniques and approaches. The sections below present an overview of the current concepts for characterization and identification of isolates and their practical application.

3.2.2. The Species Concept for Prokaryotes

The basic unit of classification and nomenclature of all living organisms, prokaryotes included, is the species. Species are grouped in genera, genera in families, families in orders, orders in classes, and classes in phyla or divisions. Species are categories based on the observations of recurrent patterns within the diversity of living organisms. Those patterns are different for plants, animals, and microorganisms, and arise in different taxonomic schemes, because, while all taxonomies are based on the category species, no universal species concept exists, and prokaryote systematics still lacks a firm theoretical basis (120).

The species concept for prokaryotes is a long-debated issue; its in-depth discussion is outside the scope of this chapter, but it has been discussed and reviewed in many publications (e.g., 1, 2, 15, 25, 26, 52, 68, 82, 121, 135, 153). The most accepted concept, considered to be pragmatic and useful for species definition, is the so-called phylophenetic species concept: a monophyletic and genomically coherent cluster of individual organisms (strains) that show a high degree of overall similarity in many independent characteristics, and is diagnosable by one or more discriminative phenotypic properties (121). It is therefore recommended that a distinct genospecies (i.e., a species discernible only by nucleic acid comparisons) that cannot be differentiated from another genospecies on the basis of any known phenotypic property should not be named as a new species until some phenotypic differentiating property is found.

The development of a natural species concept is more difficult for prokaryotes than for higher organisms, where the biological species concept proposed by Ernst Mayr (99) is widely accepted. Periodic technological improvements in molecular biology and the availability of new types of data (e.g., genomic data) progressively allow a deeper understanding of prokaryote population biology, and have affected the definition of a natural species concept for prokaryotes. They allow the delineation of bacterial diversity at higher resolution, but these types of data and analysis have not been completely elaborated and incorporated in the taxonomic framework to devise a novel classification scheme that could be applied to all microorganisms (120).

Based on the experience of the past 25 years a pragmatic definition of the prokaryote species has emerged, based on the recommendations published in 1987 by a committee of experts (155) and updated a few years ago (133). The species concept is based on a "polyphasic" approach, which includes description of diagnostic phenotypic features combined with genomic properties. Many of the phenotypic and chemotaxonomic characters used individually as diagnostic properties are insufficient to delineate the species, but together they provide sufficient descriptive information to allow the definition of a species. The most widely used genomic property included in descriptions of new species, often (and not always rightfully) used to delineate species, is the sequence of the small-subunit rRNA gene (16S rRNA for prokaryotes), and many classification schemes are based on this.

3.2.3. The Definition of Species and of Higher Taxonomic Ranks

The gold-standard technique for species delineation is total DNA relatedness: a species is conventionally defined as a group of strains sharing at least 70% total genome DNA-DNA similarity and having less than 5°C ΔT_m (the difference in the melting temperature between the homologous and the heterologous hybrids formed under standard conditions, indicating the thermal stability of DNA duplexes) (132, 134). These values should not be taken as absolute limits for circumscribing the species; the concept allows more relaxed DNA-DNA similarity frontiers, and an internal genomic heterogeneity is permitted (121). The delineation value of 70%, as introduced around 1987, is artificial, but has proven satisfactory in most cases. DNA relatedness values between 30 and 70% indicate a moderate degree of relationship, often parallel to the extent of the genus. At the level of genera or higher taxa, the resolving power of DNA-DNA similarity assays is limited. There are cases of species and even genera that share more than 70% DNA-DNA similarity. For example, the enteric bacterial genera *Escherichia* and *Shigella* share more than 85% similarity. However, for pragmatic reasons, the separation into two genera is maintained. Despite the great advances made in recent years in genomic analysis and the development of other modern approaches to characterization of prokaryote isolates, DNA-DNA hybridization combined with a thorough phenotypic characterization is still the best criterion for species delineation (133).

Compared with the species concept in the plant and the animal world, the bacterial species concept is exceptionally broad. When applying the "70% DNA-DNA similarity" criterion to delineate species of higher eukaryotes, the number of recognized species would decrease dramatically, because, for example, humans would belong to the same species as chimpanzees (98.4% DNA-DNA relatedness), gorillas (97.7%), orangutans (96.5%), and even lemurs (78%).

The technique to determine DNA-DNA relatedness is laborious and can be used only to define relatedness between pairs of strains. Therefore the first step in characterization is usually determination of the almost complete 16S rRNA gene sequence (at least 1,300 nucleotides, sequenced on both strands after cloning, to guarantee sequence reliability and no influence of sequence heterogeneity among different operons). Empirical correlations have been observed between DNA-DNA hybridization values and 16S rRNA gene sequence similarity; as a general indication, if two strains share less than 97–98% 16S rRNA gene sequence identity, they share less than 70% DNA-DNA relatedness, and they thus should belong to different species (2, 63, 121). There is good congruence between DNA-DNA relatedness and genome sequence-derived parameters, such as the average nucleotide identity of common genes and the percentage of conserved DNA (57). Many species are further divided into subspecies, which are based on the observation of minor but consistent phenotypic variations within the species or on genetically determined clusters of strains within the species (136), although phenotypic differences are usually considered more reliable (149). Other infraspecific clusters can be recognized for practical reasons, such as pathovar, phagovar, etc., but those have no standing in official nomenclature. Given that there is no proper concept of the prokaryotic species, the guidelines for the delineation of higher taxa (genera, families, orders, classes, etc.) are even less clear. The genus may be defined as "a collection of species with many characters in common," but how many and what kind of characters should be in common for species to be classified in a single genus is largely a matter of personal judgment (131).

There is a general consensus that the classification of species into higher taxonomic levels should reflect phylogenetic relationships. Sequence analysis of small-subunit rRNA (16S rRNA in prokaryotes) has provided much insight in the phylogenetic relationships among microorganisms (87, 88, 157, 158). A DNA-DNA similarity level of less than 70% (the operative delineation of a species), in general, corresponds to less than 97% 16S rRNA sequence identity. In general, species classified in a single genus share at least 93 to 95% identity in their 16S rRNA gene sequence. However, at the species level, the 16S rRNA-based methods lack the necessary resolving power, and in these instances DNA-DNA reassociation experiments are still required.

3.2.4. Type Strains and Culture Collections of Prokaryotes

The terms "strain" and "isolate" refer to the descendants of a single isolation in pure culture. They are usually made up of a succession of cultures derived from an initial single colony. Identification of unknown isolates should use type strains of species with validly published names for comparison.

For each species (or subspecies) a type strain has been designated that is the name bearer of that species and is the reference specimen for the name. The type strain, which is permanently linked to the species name, represents the connecting point between the existing biodiversity and the conventional, artificial scheme of classification. Similarly to the type strain for a species, type genera are indicated for families and for orders, and type orders are indicated for classes.

When describing a novel species, a type strain has to be indicated and included in the subsequent taxonomic studies. One of the conditions of valid publication of a name of a new species is that the strain must be deposited by the authors in at least two public culture collections located in different countries (e.g., the American Type Culture Collection [ATCC], the Deutsche Sammlung von Mikroorganismen und Zellkulturen [DSMZ], the Japan Collection of Microorganisms [JCM], etc.) so that subcultures will be available to any interested scientist for further study. A comprehensive list of such culture collections is found in www.bacterio.cict.fr/collections.html. Culture collections generally preserve bacterial strains either in dry, lyophilized form, to be revived by wetting and suspension in suitable growth medium, or frozen in liquid nitrogen. These culture collections ensure the availability of type strains for scientific studies without restrictions (144).

3.2.5. How Many Species Have Been Described?

The number of species of prokaryotes with names validly published under the rules of the Bacteriological Code is surprisingly small compared with the numbers of plant and animal species that have been described, which are in the order of hundreds of thousands or even millions. The number of validly published names of different prokaryote species on November 3, 2009 was approximately 8,226, classified in 1,732 genera, 253 families, 116 orders, and 70 classes (see http://www.bacterio.cict.fr/number.html). This number is updated monthly after the publication of each new issue of the *International Journal of Systematic and Evolutionary Microbiology*, the only journal in which names of new prokaryote taxa can be validly published (see section 3.3.2 below).

Experimental data based mainly on sequencing of 16S rRNA genes retrieved directly from the natural environment without prior cultivation suggest existence of at least two orders of magnitude more species. Today it is generally accepted that less than 1% of the prokaryote species that inhabit our planet have been named (37, 108).

Description of a new species should ideally be based on a comparative study of a large number of independent isolates to define the degree of variation of certain properties within the boundaries of the species. In practice, however, over 90% of the new species descriptions published in recent years have been based on single isolates. Unfortunately, such species descriptions can lead to improper phenotypic circumscription of taxa, making the identification of new isolates as members of a taxon problematic. From time to time formal proposals have been made to the International Committee for Systematics of Prokaryotes/International Committee for Systematic Bacteriology (ICSP; see section 3.3.1) to define a minimum number of isolates necessary for description of a new species (21, 43). Such proposals have never been formally approved or fully implemented. If they had, the names of many prokaryotes described as new species in recent years could not have been validly published because it had not been feasible to obtain more isolates.

3.2.6. Bacteria and Archaea—the Two Domains of Prokaryotes

The Linnaean hierarchical scheme has been applied to the bacterial world from the beginning of classification attempts, but only in the past 30 years has the availability of sequencing techniques for nucleic acids led to the establishment of an evolutionary framework for prokaryotes (159). A comprehensive picture of the phylogenetic diversity of prokaryotic microorganisms has recently been presented (161).

Phylogenetic analysis based on rRNAs has been the first successfully applied method to investigate evolutionary relationships among prokaryotic organisms (157). This has led to the description of two domains or kingdoms for the prokaryotes (160). The Archaea (formerly known as Archaebacteria) and the Bacteria (formerly called Eubacteria), despite both being prokaryotes, differ in many characteristics such as the structure of the cell wall, the nature of the membrane phospholipids, the properties of the protein-synthesizing system, sensitivity to different antibiotics, RNA metabolism, etc. It is intriguing to observe that several components of archaeal informational processes (e.g., primases, helicases, and polymerases) are more similar to eukaryotic than to their bacterial homologues, if such homologues exist (60). On the other hand, Archaea and Bacteria resemble each other in size and organization of chromosomes, polycistronic transcription, and presence of Shine-Dalgarno sequences for gene transcription. "To sum up, Archaea look like organisms that use eukaryotic-like proteins in a bacterial like context" (60).

The recognition of Archaea as a third domain of life has been questioned (62), with the suggestion that both the Archaea and the gram-positive bacteria should be described as monoderm prokaryotes (surrounded by a single membrane), while all true gram-negative bacteria, containing both an inner cytoplasmic membrane and an outer membrane, should be classified as diderm prokaryotes. However, genomic data have confirmed the coherence of the three domains as distinct groups (60).

Archaea and Bacteria are widespread in different environments. Archaea are not uniquely extremophiles as originally thought, but occur in the open sea, in soil, and in other "conventional" environments as well (60).

For matters concerning nomenclature, Archaea and Bacteria are both prokaryotes and their nomenclature follows the same rules fixed in the International Code of Nomenclature of Prokaryotes. The conceptual framework outlined below for the taxonomic characterization of prokaryotes applies to both groups. However, the types of phenotypic tests performed during the taxonomic characterization of isolates may differ, because, for example, determination of the chemical structure of peptidoglycan is relevant for Bacteria only. Likewise, due to the different types of lipids present in the cytoplasmic membranes, different tests should be applied in the characterization of the lipids of Bacteria and of Archaea.

3.3. PROKARYOTE NOMENCLATURE

Nomenclature is the assignment of names to the taxonomic groups defined during classification. The rules of nomenclature of prokaryotes are laid down in the International Code of Nomenclature of Bacteria (the Bacteriological Code) (see section 3.3.2).

An entirely new start was made in prokaryote nomenclature in 1980 with the publication of the "Approved Lists of Bacterial Names" (127). Rule 24a of the Bacteriological Code (see below) states: "Priority of publication dates from 1 January 1980. On that date all names published prior to 1 January 1980 and included in the Approved Lists of Bacterial Names of the ICSB are treated for all nomenclatural purposes as though they had been validly published for the first time on that date." With the publication of the Approved Lists, which encompassed approximately 2,500 species names (127, 128), all earlier published names lost their validity. Central registration and indexing of new names validly published after January 1, 1980, according to the rules of the International Code of Nomenclature of Prokaryotes (84; see below) ensured that the number of validly named species, genera, families, and orders of prokaryotes is known at any time (130).

The nomenclature of prokaryotes is subject to changes, and there are many examples of species that have been renamed, moved to other existing genera, or reclassified in newly established genera in accordance with new insights. For example, when based on an increased understanding of the phylogenetic relationships within the aerobic gram-positive endospore-forming bacteria it became desirable to split the genus *Bacillus* into a number of genera, the species formerly known as *Bacillus polymyxa* and *Bacillus stearothermophilus* were renamed *Paenibacillus polymyxa* and *Geobacillus stearothermophilus*, respectively. From the moment these new names were validly published, they obtained status in the nomenclature of prokaryotes.

Practical guidelines for the naming of new taxa of prokaryotes have been given by Trüper (147, 148).

3.3.1.   The International Committee for Systematics of Prokaryotes

The body that oversees the nomenclature of prokaryotes is the International Committee for Systematics of Prokaryotes (ICSP) (before 2000, the International Committee of Systematic Bacteriology). This committee is a constituent part of the International Union of Microbiological Societies (IUMS). The committee meets once every 3 years to discuss nomenclatural problems that have arisen in different groups of prokaryotes and to propose changes and amendments to the rules of the Bacteriological Code.

The ICSP has established several subcommittees. One important subcommittee is the Judicial Commission, which deals with problematic cases in bacterial nomenclature and renders judicial decisions in instances of controversy about the validity of a name, identity of type strains, and cases of emerging problems with the interpretation of the rules of the Bacteriological Code. It may propose amendments to the Code and consider exceptions that may be needed to certain rules. Specific subcommittees exist that deal with the systematics and nomenclature of specific groups: "Genera *Agrobacterium* and *Rhizobium*," "*Bacillus* and related organisms," "*Bifidobacterium*, *Lactobacillus* and related organisms," "*Campylobacter* and related bacteria," "*Clostridia* and *Clostridium*-like organisms," "*Flavobacterium*- and *Cytophaga*-like bacteria," "Gram-negative anaerobic rods," "Family *Halobacteriaceae*," "Family *Halomonadaceae*," "Genus *Leptospira*," "Genus *Listeria*," "Methanogens," "Suborder *Micrococcineae*," "Families *Micromonosporaceae*, *Streptosporangiaceae* and *Thermomonosporaceae*," "Class *Mollicutes*," "Genus *Mycobacterium*," "*Nocardia* and related organisms," "Family *Pasteurellaceae*," "Photosynthetic prokaryotes," "*Pseudomonas*, *Xanthomonas* and related organisms," "Suborder *Pseudonocardineae*," "Staphylococci and streptococci," "Family *Streptomycetaceae*," and "Family *Vibrionaceae*."

The minutes of the meetings of the ICSP, the Judicial Commission, and the taxonomic subcommittees are published in the *International Journal of Systematic and Evolutionary Microbiology*. Information on the ICSP, its current officers, and its subcommittees can be found at http://www.the-icsp.org.

3.3.2.   The International Code of Nomenclature of Prokaryotes

The first Code of Nomenclature for prokaryotes was published in 1948 as "The International Code of Nomenclature of Bacteria and Viruses." The current version of the International Code of Bacterial Nomenclature (the Bacteriological Code) was approved at the Ninth International Congress of Microbiology, Moscow, 1966, revised in 1990 (84), and amended at the meetings of the ICSP (see section 3.3.1). The Bacteriological Code presents the formal framework according to which prokaryotes are named and according to which existing names can be changed or rejected. It covers the rules for the naming of species (and subspecies), genera, families, orders, and classes of prokaryotes. No provisions are made by the Code for the naming of the higher taxa: phylum and kingdom.

The Code contains general considerations, principles, rules, and recommendations referring to ranking and naming of taxa, nomenclatural types and their designation, priority and publication of names, citation of authors and names, changes in names of taxa as a result of transference, union, or change in rank, rules about illegitimate names and epithets, replacement, rejection, and conservation of names and epithets. When taxonomic re-evaluations are performed and changes in names occur, the existence of rules and the availability of lists of names allow the association of new name characteristics linked with older ones, avoiding loss of information. There are also appendices that clarify different aspects of nomenclature. An online version of the Code is available at http://www.ncbi.nlm.nih.gov/books/bv.fcgi?rid=icnb.TOC&depth=2.

The Code was renamed in 2000 as the International Code of Nomenclature of Prokaryotes, and the current version will soon be published in book form as well as electronically.

It must be stressed that prokaryote nomenclature is thus governed by internationally approved rules and regulations, but that there is no "official classification of prokaryotes." General Consideration 4 of the Code states: "Rules of nomenclature do not govern the delimitation of taxa nor determine their relations. The rules are primarily for assessing the correctness of the names applied to defined taxa; they also prescribe the procedures for creating and proposing new names." The Code deals with all prokaryotes: the nomenclatures of the Archaea and the Bacteria are governed by the same rules.

3.3.3.   Effective and Valid Publication of Names of Prokaryotes

Since the publication of the Approved Lists (127, 128), the only framework in which new names of taxa can be published to obtain standing in the nomenclature of prokaryotes ("valid publication" according to Rules 27 to 32 of the Code) is the *International Journal of Systematic Bacteriology* (before 2000), now the *International Journal of Systematic and Evolutionary Microbiology*.

One way to achieve valid publication of the new name of species, genera, etc. is by publication of the description of the new taxa in that journal, while meeting the different criteria as determined by the Code. For new species a nomenclatural type has to be designated, which is the name bearer of that species and is the reference specimen for the name. As already mentioned (see section 3.2.3), when a new species is described, the type strain must be made available to the scientific community through deposit in at least two publicly accessible culture collections. For an example of the desired format of a description of a new species, see reference 72.

Descriptions of new taxa and publication of new names are also possible in other journals. This is known as "effective publication" of the names (Rule 25 of the Code). Names thus published can only obtain standing in the nomenclature when they are included in the "Validation Lists" (lists of new names and new combinations previously effectively, but not validly, published) that appear periodically in the *International Journal of Systematic and Evolutionary Microbiology*. Authors of papers in which names have been effectively published and who wish to validate the names must submit copies of the original publication, together with proof that all criteria for valid publication have been met, such as deposition of the type strain in two culture collections, availability of the type strain without restrictions to the scientific community, etc. The request for validation of the name is then handled by one of the associate editors of the journal together with the list editor. The name becomes validly published when it is featured in a Validation List (145).

The *International Journal of Systematic and Evolutionary Microbiology* also publishes articles that address taxonomy, phylogeny, and evolution of prokaryotes (as well as of fungi and other groups of eukaryotic protists). In addition, it contains the minutes of the meetings of the ICSP, its Judicial Commission, and its taxonomic subcommittees.

3.4. CHARACTERIZATION OF PROKARYOTES—CURRENT PRACTICE AND PROCEDURES

Characterization of isolates of prokaryotes and description of new species is based on a "polyphasic approach," because it uses a combination of a variety of different genotypic and phenotypic properties (54, 151). The more properties are included in the descriptions, the more robust and stable the resulting classification will be. Different properties have different resolving power; some are species specific, while others are valuable to discriminate genera, families, and orders. The properties tested generally include morphological characters (cell shape and size, the Gram reaction, cell inclusions, presence and nature of the surface layers, including extracellular capsules), information on motility (presence of flagella, their number and the way they are inserted into the cell, gliding movement), the mode of nutrition (assimilatory metabolism) and energy generation (dissimilatory metabolism), the cells' relationship to molecular oxygen, temperature, pH, tolerance toward and requirement for salt, and many others. Miniaturized tests such as the BIOLOG® and the API system are often very useful. Additional tests of value toward the identification of the isolate may be its sensitivity toward different antibiotics, as well as immunological properties. Genotypic information includes 16S rRNA gene sequence information, which is a very valuable tool for rapidly placing any isolate within the current classification scheme, at least down to the family and genus level. For final identification of the species, DNA-DNA hybridization tests are the ultimate tool to decide whether two isolates should be classified in the same species. More specific tests such as serotyping or phage typing may be necessary for certain groups of microorganisms to obtain a reliable identification.

For several groups of prokaryotes, the taxonomic subcommittees of the ICSP have established recommended minimal standards for the description of new isolates. Groups of prokaryotes for which such minimal standards have been published are the methanogenic Archaea (11), the class *Mollicutes* (16), root- and stem-nodulating bacteria (59), anoxygenic phototrophic bacteria (67), the order

Halobacteriales (109), the families *Halomonadaceae* (3), *Flavobacteriaceae* (7), *Pasteurellaceae* (21), and *Campylobacteraceae* (150), and the genera *Moraxella* and *Acinetobacter* (12), *Brucella* (30), *Helicobacter* (33), *Staphylococcus* (50), and *Mycobacterium* (85).

A prerequisite for characterization and identification of prokaryotes is the isolation in pure culture. This means that appropriate growth media have to be set up and balanced in terms of the physical and chemical parameters such as macro- and microelements to be supplied, other growth factors (e.g., vitamins), pH, osmolarity, and temperature chosen for isolation and propagation. An in-depth discussion on the procedures for isolation of microorganisms was given by Overmann (110).

When dealing with multiple isolates from the same niche, a dereplication step is convenient to roughly determine the diversity of the sample and focus on representative members of the collection, to save time and effort. On the other hand, the more strains included in a study, the better the circumscription. Considering that a massive isolation procedure for the selection of strains potentially interesting for an industrial application could result in the isolation of hundreds of isolates, the first thing to do is to assess how many different strains are present, before their identity can be ascertained (Fig. 1). Many techniques could be used successfully for the discrimination of strains below species level: genetic fingerprinting methods (e.g., pulsed-field gel electrophoresis, restriction fragment length polymorphism, randomly amplified polymorphic DNA, and ribotyping) or phenotype-based fingerprints (such as sodium dodecyl sulfate-polyacrylamide gel electrophoresis of proteins). Some of those techniques are extremely sensitive, but they are also time- and labor-consuming; therefore, the adoption of automated techniques is really helpful. Clearly, techniques useful for dereplication also have taxonomic value, because of the discrimination they can provide; therefore short protocols are reported in the sections below. Typically, techniques chosen for dereplication provide the scientist with a complex pattern of bands (a fingerprint) characteristic of each isolate. Several software packages are available, either commercial or freely available (see below), for primary data analysis and clustering isolates according to similarity.

Description of a species should ideally be based on a comparative study of a large number of isolates to define also the degree of variation of certain properties within the defined boundaries of the species. In practice, however, most new species descriptions published in recent years have been based on the study of single isolates.

3.4.1. Genotypic Characterization

Genotype-based methods used in the characterization of prokaryotes include the sequencing of single genes (16S rRNA, 23S rRNA, the intergenic transcribed spacer [ITS] between these two genes, different housekeeping genes) and molecular fingerprinting techniques, as well as methods that provide information on the entire genome: the determination of the mol% guanine + cytosine, DNA-DNA hybridization to compare genomes, and in some cases even complete genome sequencing. Many prokaryotes contain plasmids, and these plasmids often contain genes of interest for biotechnological purposes. Therefore, characterization and analysis of plasmid DNA may provide important information.

Different DNA isolation procedures exist, aimed at obtaining as pure and undegraded DNA as possible. The

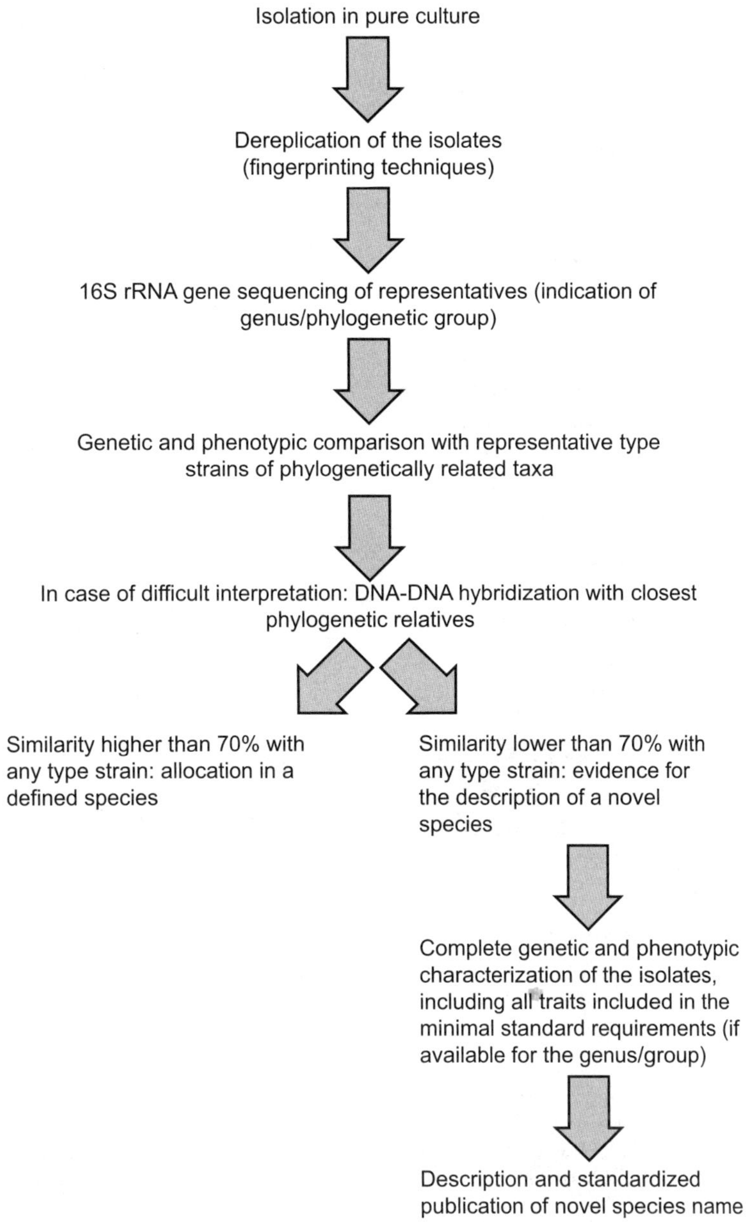

Isolation in pure culture

Dereplication of the isolates
(fingerprinting techniques)

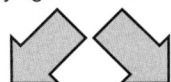

16S rRNA gene sequencing of representatives (indication of
genus/phylogenetic group)

Genetic and phenotypic comparison with representative type
strains of phylogenetically related taxa

In case of difficult interpretation: DNA-DNA hybridization with closest
phylogenetic relatives

Similarity higher than 70% with
any type strain: allocation in a
defined species

Similarity lower than 70% with
any type strain: evidence for
the description of a novel
species

Complete genetic and phenotypic
characterization of the isolates,
including all traits included in the
minimal standard requirements (if
available for the genus/group)

Description and standardized
publication of novel species name

FIGURE 1 Flow sheet showing the different stages in the taxonomic characterization of new isolates of prokaryotes.

protocol proposed by Marmur (96) is still widely used in different modifications: it involves lysis of the cells by enzymatic and/or mechanical means, inactivation of nucleases, separation of the nucleic acids from proteins, and use of organic solvents such as chloroform and isoamyl alcohol to separate the DNA from other cellular components. Precipitation with ethanol or isopropyl alcohol concludes the isolation procedure. Commercial kits are available for general-purpose DNA extractions that enable efficient isolation of DNA for PCR and other applications.

3.4.1.1. rRNA-Based Methods
For sequencing the 16S rRNA gene (83), universal primers have been designed that amplify the gene in all

known prokaryotes, Archaea as well as Bacteria (4) and are reported in Table 1. In cases where the approximate phylogenetic affiliation of the organism is known, more specific primers aimed at the taxonomic group can be used. Cloning of amplicons before sequencing should be performed, because different ribosomal operons could be characterized by sequence heterogeneity, and therefore sequencing of mixed populations could generate chimeras and mistakes. There are many cases in which one organism harbors multiple copies of the 16S rRNA gene that may differ in sequence by approximately 5% (105).

After almost complete sequences (>1,400 nucleotides) have been obtained, comparison with database information

TABLE 1 Sequences of universal primers available for amplification of ribosomal genes from Bacteria and Archaea[a]

Gene	Primer sequence	Annealing temp (°C)	Reference(s)
Bacteria			
E8F	AGAGTTTGATCCTGGCTCAG	55	98, 116, 117
E9F	GAGTTTGATCCTGGCTCAG	53	42, 65, 100
Eb246F*	AGCTAGTTGGTGGGGT	45	31
E334F	CCAGACTCCTACGGGAGGCAGC	65	122
E341F	CCTACGGGIGGCIGCA	51	154
U341F*	CCTACGGGRSGCAGCAG	54	64
U515F*	GTGCCAGCMGCCGCGGTAA	60	115, 116
U519F*	CAGCMGCCGCGGTAATWC	54	137
Ab779F*	GCRAASSGGATTAGATACCC	56	17
E786F	GATTAGATACCCTGGTAG	47	29
Ab789F*	TAGATACCCSSGTAGTCC	51	6
U1053F*	GCATGGCYGYCGTCAG		31
E533R	TIACCGIIICTICTGGCAC	56	154
E926R	CCGICIATTIITTTIAGTTT	50	154
E939R	CTTGTGCGGGCCCCCGTCAATTC	71	122
E1115R	AGGGTTGCGCTCGTTG	47	116
E1541R	AAGGAGGTGATCCANCCRCA	57	137
Archaea			
A571F*	GCYTAAAGSRICCGTAGC	52	4
A751F*	CCGACGGTGAGRGRYGAA	54	4
Ab909R*	TTTCAGYCTTGCGRCCGTAC	49	17
E926R	CCGICIATTIITTTIAGTTT	50	154
UA1204R*	TTMGGGGCATRCIKACCT	53	4
UA1406*	ACGGGCGGTGWGTRCAA	52	4

[a]Primers are named according to *Escherichia coli* numbering position or to the position with respect to the 1,300-bp archaeal alignment (the latter indicated with an asterisk) (4). F and R indicate forward and reverse primer, respectively.

allows a tentative identification of the strain. Tools available for sequence comparison include the classical Blast in GenBank, the RDP-II database (91), the EzTaxon tool (23), and the most recent updates of ARB (89) and SILVA (114). This first comparison in general allows identification of a genus to which the strain may belong. Then, a refined phylogenetic analysis can be performed that includes all related species. The data set should also include sequences from a few unrelated species to obtain a more complete picture and to better evaluate the branching pattern (134). Sequences have to be aligned before the phylogenetic tree can be reconstructed and statistically evaluated (5). Careful data selection and optimal alignment of the sequences, including removal of ambiguous positions, are prerequisites for reliable phylogenetic conclusions. Many programs are available for alignment (142, 143), and different approaches can be applied for phylogenetic analysis, such as distance, maximum parsimony, and maximum likelihood methods of phylogeny (88); finally, verification of the statistical support of branching patterns can be obtained by bootstrap analysis (46). Available software for phylogenetic analysis include PHYLIP (47), MEGA (81, 141), and ARB software (89). Different programs allow tree inference with different methods and models (e.g., references 45, 46, 70, 73, and 123).

If the isolate under study shares less than 97 to 98% sequence identity with any known species, it might well represent a novel species. However, 16S rRNA sequence identity is not sufficient to guarantee species identity (49). Where 16S rRNA gene sequence values are above 97%, other methods such as DNA-DNA hybridization or sequence comparison of genes with a greater resolution should be used to test the affiliation of the isolate with a known species. If the sequence under study shows very low similarity with related taxa, it might represent a novel genus. Sequence variation between members of a genus may range from small (for instance, above 97% for *Micrococcus*), to moderate (92% for *Streptococcus*), to significant (approximately 79% for *Spirochaeta*) (134).

The 16S rRNA gene is most widely used as a phylogenetic marker for the prokaryotes. In some cases, the 23S rRNA gene of the large ribosome subunit is included in phylogenetic analyses as well. In many taxonomic studies the ITS region separating the 16S and the 23S rRNA genes is used, because this region is much more variable than the rRNA molecules themselves, and therefore comparison of ITS sequences can detect differences at a much higher resolution than 16S rRNA sequence comparisons.

3.4.1.2. Other Genes as Taxonomic Markers

Multilocus sequence typing (MLST) or multilocus sequence analysis (MLSA) are powerful tools to obtain taxonomic information on the species level and the interspecies variability (52, 92). The method was first developed to characterize bacterial pathogens. Sequences are determined of an internal portion (usually 450 to 700 bp) of a small number (most often seven) of housekeeping genes. Examples of genes used for this purpose are GMP synthase, methionine aminopeptidase, pyruvate formate lyase, inorganic pyrophosphatase, pyruvate kinase, RNA polymerase beta subunit, superoxide dismutase, elongation factor Tu, *recA*, and ATPase subunit genes (9, 38). Comparisons of protein sequence-based phylogenies have by and large confirmed the general structure of the 16S rRNA gene phylogenetic tree. There is a growing database available of gene sequences relevant to multilocus sequence analysis (see www.eMLSA.net). For each housekeeping gene, different sequences are assigned as distinct alleles. A dendrogram generated from the matrix of pairwise differences between the allelic profiles will display the relatedness of isolates (34, 39).

3.4.1.3. Molecular Fingerprinting Techniques

When large numbers of strains are to be compared, individual sequencing of the 16S rRNA genes of all isolates is often not feasible. Molecular fingerprinting techniques can then be used. Such methods are also useful for typing at the infraspecific level, i.e., to subdivide species into more homogeneous groups that could represent subspecies. These techniques, which include DNA- and protein-based methods, can be used also for dereplication of isolates. The choice of the most suitable technique depends on the taxon analyzed; therefore, it is difficult to give general information on the rationale to be applied for selection. Most fingerprinting techniques use PCR-based methods (151, 152). The methods differ according to the primers used, e.g., oligonucleotides of approximately 20 bases in arbitrarily primed PCR, or oligonucleotides of approximately 10 bases in randomly amplified polymorphic DNA assays (RAPD-PCR). Primers complementary to repetitive elements dispersed throughout the genomes or to tRNA gene regions could result in the production of patterns of amplification useful for discrimination (BOX- and ERIC-PCR) (151).

The discriminatory power of given primer sequences depends on the taxon investigated, e.g., arbitrary primers with a given GC content will give different results on the basis of the genome GC content of the microorganism under study, because the probability of finding a certain sequence in the genome is higher if the GC content of the genome and the primer are similar.

PCR-based DNA-typing methods can be combined with restriction enzyme analysis in the so-called amplified-rDNA restriction analysis method (ARDRA) (151): 16S or 23S rRNA genes can be amplified and then digested with restriction enzymes and compared. The results are often species specific. AFLP is also a combination of PCR and restriction enzyme methodologies: its basic principle is a restriction analysis followed by PCR-mediated amplification to select particular DNA fragments from the pool of restriction fragments. PCR-restriction fragment length polymorphism–single-strand conformational polymorphism (PCR-RFLP-SSCP; PRS) was also used successfully to analyze large numbers of strains, and it shows good congruence with chemometric fingerprinting (13, 14).

The ribotyping method consists of a restriction enzyme digestion of genomic DNA, which is afterward transferred to a membrane and hybridized with a labeled probe targeting rRNA. This DNA-typing technique has also been reported for dereplication (118).

Pulsed-field gel electrophoresis is a powerful and discriminatory technique, based on enzymatic digestion of chromosomal DNA with a rare cutter enzyme and migration of the bands in a variably oriented electric field. This allows the separation of large fragments of DNA and the generation of highly discriminative strain-specific fingerprints. However, the technique is labor- and time-consuming and requires a sophisticated instrument.

A phenotype-targeting fingerprinting technique is standardized sodium dodecyl sulfate-polyacrylamide gel electrophoresis of whole-cell proteins, which can be used effectively for dereplication because it has discriminative information at or below the species level (69, 112, 151).

3.4.1.4. DNA-DNA Hybridization and G+C Content Analyses

When collected data are not sufficient to clearly determine identification, DNA-DNA hybridization should be performed, as a hybridization value of 70% was proposed as the boundary for the delineation of prokaryote species (61, 155). If a new isolate shows a high degree of similarity to more than one species, DNA-DNA hybridization experiments should be performed with all relevant type strains to ensure that there is enough dissimilarity to warrant classification of the strain(s) as a new taxon. A number of techniques are available; most of these have been validated and show comparable results (32, 41, 57, 58, 66, 120).

Most laboratories do not routinely perform DNA-DNA hybridization assays. The protocol includes labeling of the DNA (in general, with radioisotopes), shearing, and denaturing the sheared DNA, after which the labeled denatured DNA is mixed with excess of unlabeled DNA from the second organism. The mixture is then cooled and allowed to reanneal under carefully controlled conditions. Duplex DNA is separated from any unhybridized DNA remaining, and the amount of bound DNA is quantified. A control experiment with homologous DNA is included and its results of the heterologous DNA binding are normalized with respect to the homologous control. Nonradioactive methods have also been introduced. The procedure is time-consuming and allows for pairwise comparisons only, making comparisons of large numbers of strains cumbersome.

Another analysis performed historically is the determination of genome G+C base content. The G+C base ratio varies from approximately 20 to 80% in the prokaryote world. While an identical G+C content of two species does not prove the existence of any relationship, a difference of 5 mol% or more shows that the isolates cannot be closely related. Techniques for the experimental determination of the G+C percentage include thermal denaturation profiles, centrifugation methods that assess the buoyant density of the DNA, and high-performance liquid chromatography methods that determine the amount of each nucleotide after hydrolysis of the DNA (20, 55, 94, 95, 97, 101, 102, 111, 139, 140). Whatever method is used, different types of reference DNA of known G+C content should be included in the tests for calibration, and information about the method used should be provided.

3.4.2. Phenotypic Characterization

Phenotypic properties important in prokaryote taxonomy comprise morphological, physiological, and biochemical features (10, 107, 129, 131, 151). Phenotypic physiological and biochemical tests should be carried out under standardized conditions. Type strains of the most closely related taxa should be tested at the same time for comparison, and known positive and negative controls should always be included.

3.4.2.1. Morphological Properties

Morphological characteristics commonly used in the taxonomic characterization of prokaryotes include:

- Cell shape and size, with use of phase-contrast microscopy, transmission electron microscopy, and other microscopy techniques.
- Gram-staining behavior, providing information on the structure of the cell wall of the Bacteria. For halophilic prokaryotes, a special modification of the fixation protocol used is necessary (35).
- Other staining methods are used to identify other cellular components, such as acid-fast staining, used to detect the presence of mycolic acids in the cell wall (106).
- Presence of extracellular capsular material. Extracellular polymers can often be visualized by suspending cells in a solution of India ink and observing wet mounts under the microscope.
- Colony morphology and size when grown on solid surfaces.
- Pigmentation.
- Formation, location, and shape of endospores.
- Presence of intracellular structures such as gas vesicles, intracellular membranes, granules of storage materials such as poly(β-hydroxyalkanoates), elemental sulfur, and others.
- Motility (swimming, gliding), and presence and location of flagella.

3.4.2.2. Physiological Properties

Physiological traits commonly tested for prokaryotes are given below. Not all tests are equally relevant for all groups of organisms. For example, tests for the consumption of different organic compounds are irrelevant for chemoautotrophic sulfur bacteria; for such specialized groups special tests should be added such as the range of sulfur compounds used as electron donor. Fermentation products should be determined for obligate or facultative anaerobes that grow in the absence of an external electron acceptor, etc. Among the most relevant and generally recorded characteristics are:

- The temperature range and optimum for growth.
- The pH range and pH optimum for growth.
- The influence of salt concentration on growth rate.
- The relation to oxygen: obligate aerobic, facultative anaerobic, obligate anaerobic.
- The use of electron acceptors other than oxygen for respiration, e.g., anaerobic respiration with nitrate as electron acceptor (denitrification) or reduction of sulfate and other oxidized sulfur compounds.
- The ability to grow heterotrophically, phototrophically, and chemoautotrophically.
- For chemoautotrophs, the range of inorganic electron donors and energy sources used.
- The range of organic compounds used as carbon and energy source by aerobic heterotrophs. Miniaturized standardized tests such as those provided by the BIOLOG® system are often helpful. The latter consists of microtiter plates with wells that contain potential growth substrates and a redox indicator. Utilization of the carbon source causes a color change of the indicator, enabling the simultaneous determination of the utilization of a large number of substrates within a short time. The procedure can also be easily automatized.
- For fermentative prokaryotes: the nature of the fermentation products made and their ratio.
- Different enzymatic tests (oxidase, catalase, dissimilatory nitrate reduction, and others) (65). In addition, commercially available test kits (e.g., the API system for identification of enteric bacteria, which consists of miniaturized test tubes with different media and reagents) are helpful to routinely test isolates under standardized conditions.
- Excretion of extracellular enzymes such as amylases, proteases, and lipases.
- Susceptibility to different antibiotics and other antibacterial compounds.

3.4.2.3. Chemotaxonomic Properties

Chemical properties of prokaryotic cells useful in polyphasic taxonomy studies include the composition of the cell wall, membrane lipids, respiratory quinones, polyamines, and others. Not all chemical tests should be performed on all prokaryotes. Determination of the detailed structure of the amino acid bridges in peptidoglycan is important only for certain groups of gram-positive Bacteria that possess a peptidoglycan cell wall and for which variations in peptidoglycan structure have been documented. For Archaea, which lack peptidoglycan, different chemical assays can yield information on the structure of the wall (glycoproteins, pseudomurein, etc.) (75). Analysis of teichoic acids is important only for gram-positive Bacteria, and mycolic acids are present only in mycobacteria and relatives. Likewise, fatty acid profiles are not relevant for the Archaea, which lack ester-type lipids that yield fatty acids on hydrolysis. Here, special methods are needed to obtain information on the presence of a lipid bilayer with 20- and/or 25-carbon isoprenoid branched chains (71) or, alternatively, a lipid monolayer membrane in which two glycerol moieties are linked by 40-carbon isoprenoid (biphytanyl) chains, providing a highly stable membrane with covalent bonds spanning over its whole width. The list below, therefore, provides general information only, and the nature of the tests to be performed depends to a large extent on the group to which the organisms to be studied belong.

Among the properties often tested are:

- The type of peptidoglycan, and especially the nature of the peptides linking the polysaccharide chains. The discriminatory power of the structure of peptidoglycan is apparently restricted to gram-positive bacteria, whereas no variation has been reported among Proteobacteria and Bacteroidetes. The simplest analysis is the determination of the characteristic diamino acid (lysine or diaminopimelic acid) in the cross-linking peptide, and more complex tests for a complete structure elucidation are available as well (74, 124, 125).
- Teichoic acids, which are polyols consisting predominantly of glycerol, ribitol, and mannitol, and are covalently linked to the peptidoglycan through phosphodiester bonds, can be useful for characterization of gram-positive Bacteria, and can be extracted, purified, and analyzed by gas-liquid chromatography (151).

- Polar lipids, both for Bacteria and Archaea, can be extracted with organic solvents, separated by one- or two-dimensional thin-layer chromatography, and detected by different spray reagents, including reagents that specifically stain phospholipids, glycolipids, amino lipids, etc. (74, 146).

- Fatty acid profiles (for Bacteria only) are made by acid methanolysis of the lipids, followed by gas chromatographic separation of the fatty acid methyl esters and detection and identification by mass spectrometry. For such studies, growth conditions must be strictly standardized, because factors such as temperature and age of the culture may have a strong influence on the fatty acid pattern (74).

- The structure of the outer membrane of gram-negative Bacteria shows considerable variation in the nature of the sugar present in lipid A and in the fatty acids in the lipopolysaccharide (104).

- The presence and length of the carbon chain of the mycolic acids that occur in certain members of the high-GC gram-positive Bacteria (19, 56).

- The presence and types of respiratory quinones. Respiratory lipoquinones are widely distributed in both anaerobic and aerobic organisms within the Bacteria and Archaea. These can be divided into two structural classes: naphthoquinones and benzoquinones (which include ubiquinones, rhodoquinones, and plastoquinones). A third class includes the benzothiophene derivatives, such as *Sulfolobus* quinone and *Caldariella* quinone, which are probably restricted to members of the order *Sulfolobales*. The side-chain lengths recorded to date range from 6 to 15 isoprenoid units, and different variations in the structure of the quinones and the degree of saturation of the side chain have been reported for specific groups of prokaryotes. Different techniques of thin-layer chromatography, high-performance liquid chromatography, mass spectrometry, and nuclear magnetic resonance should be combined to fully characterize the respiratory quinones present (27, 28, 103, 138).

- Many prokaryotes are colored by carotenoids, flexirubins, (bacterio)chlorophyll, pyocyanin, and other pigments. Extraction of pigments in suitable solvents and spectrophotometric characterization add much important information that can be used in polyphasic taxonomic characterization (67).

- The types of polyamines present in the cells is another chemotaxonomically relevant property.

3.5. CONCLUDING REMARKS: WHOLE-GENOME SEQUENCING IN PROKARYOTE TAXONOMY

The availability of the methodology to determine complete genome sequences of prokaryotes has initiated discussions on changing the rules and concepts used in delineating species in the near future. Several papers have already been published on the impact of genomics in taxonomy (e.g., references 53 and 76–79). A full genome sequence provides a view of the genetic information in a microbe at the highest possible resolution. It can therefore be considered the "ultimate source" of molecular information for taxonomic and phylogenetic studies. In addition, whole genomic sequences contain the information required for all of the molecular tools currently applied for genotyping; therefore, these data can easily be correlated

to existing data sets in this field. Classical Sanger sequencing is a time-consuming and costly process and does not allow the throughput needed for most taxonomic studies. Yet, it has proven a valuable source of information for phylogenetic studies at the species, genera, and higher levels for microbes that are of key importance to industrial microbiology (24, 93).

The emergence of a new generation of sequencing technologies is rapidly changing this notion. It addresses the two major limitations of Sanger sequencing because it provides a dramatic increase in throughput at a much lower cost (90). The costs of sequencing per base pair has already decreased by a factor of 1,000 or more, and, as a result, it is anticipated that very soon a full genome sequence will be obtained via commercial providers for a few hundred to a thousand euros or dollars. These costs may become even lower if low-coverage sequence data are sufficient, as may be the case for taxonomic studies. These costs are comparable or even lower than the costs of many of the molecular and biochemical typing techniques described above.

Therefore, it can be anticipated that publication of the complete genome sequence of the type strain will soon become obligatory for the description of a new species. However, the impact that this could have on the general taxonomic scheme is less trivial. First, genome sequencing provides a framework of the potentialities of the organism, but the phenotypic analysis necessarily completes the characterization of the organism and can give a more precise idea on how the organism itself interacts with the environment. Moreover, a careful analysis will be needed to determine new cutoffs and methodologies for comparison of taxa. Approaches based on whole-genome sequences could successfully integrate previous "traditional" knowledge, only if bioinformatics pipelines become available to rapidly extract the information relevant for taxonomic and phylogenetic purposes (119), and only if methods for comparison are evaluated and carefully chosen, because more than 20 methods to compare genomes are currently available (8). Therefore, there still is considerable uncertainty regarding how complete genome sequence data could best be evaluated in a taxonomic framework.

APPENDIX: IMPORTANT RESOURCES OF INFORMATION RELEVANT TO PROKARYOTE TAXONOMY

Information on taxonomy of prokaryotes can be found in many sources: handbooks, review articles, public databases, and other sources of information available online. A selection of those sources follows here.

- The International Committee for Systematics of Prokaryotes (ICSP) and its taxonomic subcommittees (www.the-icsp.org).

- The International Code of Nomenclature of Bacteria (the Bacterial Code), 1990 version (84). Available at: http://www.ncbi.nlm.nih.gov/books/bv.fcgi?rid=icnb.TOC&depth=2. An updated version entitled International Code of Nomenclature of Prokaryotes is in preparation.

- The *International Journal of Systematic and Evolutionary Microbiology* (http://ijs.sgmjournals.org/), the official journal of the ICSP, and the only journal where names of new taxa of prokaryotes can be validly published.

- *Bergey's Manual of Systematic Bacteriology*, the major handbook with descriptions of prokaryote species and higher taxa, that provides a systematic framework for their classification (http://www.taxonomicoutline.org). (Note that this classification has no official status.) *Bergey's Manual of Systematic Bacteriology* and its predecessor, *Bergey's Manual of Determinative Bacteriology*, have served microbiologists since 1923. Eight editions of *Bergey's Manual of Determinative Bacteriology* were published between 1923 and 1974. The first edition of *Bergey's Manual of Systematic Bacteriology* was published in four volumes between 1984 and 1990. The first volumes of the five-volume second edition appeared in 2001 and 2005 (51), and the third volume is due to be published in 2009.

- *The Prokaryotes*, a comprehensive seven-volume handbook on the biology of Bacteria and Archaea, now in its third edition (36).

- The list of prokaryotic names with standing in nomenclature, compiled by Dr. Jean Euzéby of the University of Toulouse, France (www.bacterio.cict.fr) (40). This website contains a wealth of information on prokaryote nomenclature, and it is updated monthly with the publication of the latest issue of the *International Journal of Systematic and Evolutionary Microbiology*.

- For information on culture collections from which type strains of prokaryotes and other isolates can be obtained, see the web pages of the World Federation for Culture Collections (http://www.wfcc.info/datacenter.html). A list of culture collections that maintain cultures of prokaryotes can also be found at www.bacterio.cict.fr/collections.html.

- Some public culture collections maintain websites that, in addition to the strain catalogs and technical details about depositing and ordering strains, provide information about the history and the nomenclature of the strains, recipes for media in which the isolates can be grown, and more. The sites of the Deutsche Sammlung von Mikroorganismen und Zellkulturen (DSMZ, Braunschweig, Germany; www.dsmz.de) and the American Type Culture Collection (ATCC, Manassas, VA; www.atcc.org) are especially recommended.

- Sequence information on 16S rRNA genes can be retrieved from the GenBank (http://www.ncbi.nlm.nih.gov) and results can be easily formatted to visualize the phylogenetic placement of the newly obtained sequence with respect to the closest relatives. The European database EMBL (http://www.ebi.ac.uk) and the Japanese DDBJ (http://www.ddbj.nig.ac.jp) share sequences with GenBank on a daily basis.

- The Ribosomal Database Project (http://rdp.cme.msu.edu/) is a database devoted to storage and analysis of ribosomal sequences. Other databases related to the ribosomal operon sequences and numbers are the 5S rRNA database (http://biobases.ibch.poznan.pl/5SData/) and the RRNDB, on the rRNA operon numbers in different prokaryotes (http://ribosome.mmg.msu.edu/rrndb/index.php) (86).

- A very promising resource is the SILVA rRNA database project, a "comprehensive on-line resource for quality checked and aligned ribosomal RNA sequence data" (http://www.arb-silva.de/) (161).

- For comparison of 16S rRNA gene sequences with the type strains of prokaryotes with validly published names, the EzTaxon tool has been developed (http://147.47.212.35:8080) (23).

- Multilocus sequence typing data can be found at www.eMLSA.net (9) and at http://pubmlst.org.

- For complete prokaryote genome sequences online, http://www.genomesonline.org and http://www.jcvi.org provide a wealth of information.

REFERENCES

1. **Achtman, M., and M. Wagner.** 2008. Microbial diversity and the genetic nature of microbial species. *Nat. Rev. Microbiol.* **6:**431–440.
2. **American Academy of Microbiology.** 2006. *Reconciling Microbial Systematics and Genomics.* http://academy.asm.org/images/stories/documents/reconcilingmicrobialsystematicsandgenomicsfull.pdf. Accessed December 2, 2009.
3. **Arahal, D. R., R. H. Vreeland, C. D. Litchfield, M. R. Mormile, B. J. Tindall, A. Oren, V. Bejar, E. Quesada, and A. Ventosa.** 2007. Recommended minimal standards for describing new taxa of the family *Halomonadaceae*. *Int. J. Syst. Evol. Microbiol.* **57:**2436–2446.
4. **Baker, G. C., J. J. Smith, and D. A. Cowan.** 2003. Review and re-analysis of domain-specific 16S primers. *J. Microbiol. Methods* **55:**541–555.
5. **Baldauf, S. L.** 2003. Phylogeny for the faint of heart: a tutorial. *Trends Genet.* **19:**345–351.
6. **Barns, S. M., R. E. Fundyga, M. W. Jeffries, and N. R. Pace.** 1994. Remarkable archaeal diversity detected in Yellowstone National Park hot spring environment. *Proc. Natl. Acad. Sci. USA* **91:**1609–1613.
7. **Bernardet, J.-F., Y. Nakagawa, and B. Holmes.** 2002. Proposed minimal standards for describing new taxa of the family *Flavobacteriaceae* and emended description of the family. *Int. J. Syst. Evol. Microbiol.* **52:**1049–1070.
8. **Binnewies, T. T., Y. Motro, P. F. Hallin, O. Lund, D. Dunn, T. La, D. J. Hampson, M. Bellgard, T. M. Wassenaar, and D. W. Ussery.** 2006. Ten years of bacterial genome sequencing: comparative-genomics-based discoveries. *Funct. Integr. Genom.* **6:**165–185.
9. **Bishop, C. J., D. M. Aanensen, G. E. Jordan, M. Kilian, W. P. Hanage, and B. G. Spratt.** 2009. Assigning strains to bacterial species via the internet. *BMC Biol.* **7:**3.
10. **Bochner, B. R.** 2009. Global phenotypic characterization of bacteria. *FEMS Microbiol. Rev.* **33:**191–205.
11. **Boone, D. R., and W. B. Whitman.** 1988. Proposal of minimal standards for describing new taxa of methanogenic bacteria. *Int. J. Syst. Bacteriol.* **38:**212–219.
12. **Bøvre, K., and S. D. Hendriksen.** 1976. Minimal standards for description of new taxa within the genera *Moraxella* and *Acinetobacter*: proposal by the subcommittee on *Moraxella* and *Acinetobacter*. *Int. J. Syst. Bacteriol.* **26:**92–96.
13. **Brandão, P. F. B., J. P. Clapp, and A. T. Bull.** 2002. Discrimination and taxonomy of geographically diverse strains of nitrile-metabolizing actinomycetes using chemometric and molecular sequencing techniques. *Environ. Microbiol.* **4:**262–276.
14. **Brandão, P. F. B., M. Torimura, R. Kurane, and A. T. Bull.** 2002. Dereplication for biotechnology screening: PyMS analysis and PRC-RFLP-SSCP (PRS) profiling of 16S rRNA genes of marine and terrestrial actinomycetes. *Appl. Microbiol. Biotechnol.* **58:**77–83.
15. **Brenner, D. J., J. T. Staley, and N. R. Krieg.** 2001. Classification of procaryotic organisms and the concept of bacterial speciation, p. 27–31. *In* G. M. Garrity, D. R. Boone and R. W. Castenholz (ed.), *Bergey's Manual of Systematic Bacteriology*, 2nd ed., vol. 1, *The Archaea and the Deeply Branching and Phototrophic Bacteria*. Springer-Verlag, New York, NY.
16. **Brown, D. R., R. F. Whitcomb, and J. M. Bradbury.** 2007. Revised minimal description of new species of the class *Mollicutes* (division *Tenericutes*). *Int. J. Syst. Evol. Microbiol.* **57:**2703–2719.

17. **Brunk, C. F., and N. Eis.** 1998. Quantitative measure of small subunit rRNA gene sequences of the Kingdom *Korarchaeota. Appl. Environ. Microbiol.* **64:**5064–5066.

18. **Bull, A. T., A. C. Ward, and M. Goodfellow.** 2000. Search and discovery strategies for biotechnology: the paradigm shift. *Microbiol. Mol. Biol. Rev.* **64:**573–606.

19. **Butler, W. R., and L. S. Guthertz.** 2001. Mycolic acid analysis by high-performance liquid chromatography for identification of *Mycobacterium* species. *Clin. Microbiol. Rev.* **14:**704–726.

20. **Cashion, P., M. A. Holder-Franklin, J. McCully, and M. Franklin.** 1977. A rapid method for the base ratio determination of bacterial DNA. *Anal. Biochem.* **81:**461–466.

21. **Christensen, H., M. Bisgaard, W. Fredricksen, R. Mutters, P. Kuhnert, and J. E. Olsen.** 2001. Is characterization of a single isolate sufficient for valid publication of a new genus or species? Proposal to modify Recommendation 30b of the *Bacteriological Code* (1990 Revision). *Int. J. Syst. Evol. Microbiol.* **51:**2221–2225.

22. **Christensen, H., P. Kuhnert, H.-J. Busse, W. C. Frederiksen, and M. Bisgaard.** 2007. Proposed minimal standards for the description of genera, species and subspecies of the *Pasteurellaceae. Int. J. Syst. Evol. Microbiol.* **57:**166–178.

23. **Chun, J., J. H. Lee, Y. Jung, M. Kim, S. Kim, B. K. Kim, and Y. W. Lim.** 2007. EzTaxon: a web-based tool for the identification of prokaryotes based on 16S ribosomal RNA gene sequences. *Int. J. Syst. Evol. Microbiol.* **57:**2259–2261.

24. **Claesson, M. J., D. van Sinderen, and P. W. O'Toole.** 2008. *Lactobacillus* phylogenomics—towards a reclassification of the genus. *Int. J Syst. Evol. Microbiol.* **58:**2945–2954.

25. **Cohan, F. M.** 2001. Bacterial species and speciation. *Syst. Biol.* **50:**513–524.

26. **Cohan, F. M.** 2002. What are bacterial species? *Annu. Rev. Microbiol.* **56:**457–487.

27. **Collins, M. D.** 1985. Analysis of isoprenoid quinones. *Methods Microbiol.* **18:**329–366.

28. **Collins, M. D.** 1994. Isoprenoid quinones, p. 265–310. *In* M. Goodfellow and A. G. O'Donnell (ed.), *Chemical Methods in Prokaryotic Systematics.* Wiley, Chichester, United Kingdom.

29. **Coloqhoun, J. A.** 1997. Discovery of deep-sea Actinomycetes. Ph.D. Dissertation. Research School of Biosciences, University of Kent, Canterbury, United Kingdom.

30. **Corbel, M. J., and W. J. Brinley-Morgan.** 1975. Proposal for minimal standards for description of new species and biotypes of the genus *Brucella. Int. J. Syst. Bacteriol.* **25:**83–89.

31. **DasSarma, S., and E. F. Fleischmann.** 1995. *Archaea: A Laboratory Manual—Halophiles*, p. 269–272. Cold Spring Harbor Laboratory Press, Cold Spring Harbor, NY.

32. **de Ley, J., H. Cattoir, and A. Reynaerts.** 1970. The quantitative measurement of DNA hybridization from renaturation rates. *Eur. J. Biochem.* **12:**133–142.

33. **Dewhirst, F. E., J. G. Fox, and S. L. W. On.** 2000. Recommended minimal standards for describing new species of the genus *Helicobacter. Int. J. Syst. Evol. Microbiol.* **50:**2231–2237.

34. **Diancourt, L., V. Passet, C. Chervaux, P. Garault, T. Smokvina, and B. Brisse.** 2007. Multilocus sequence typing of *Lactobacillus casei* reveals a clonal population structure with low levels of homologous recombination. *Appl. Environ. Microbiol.* **73:**6601–6611.

35. **Dussault, H. P.** 1955. An improved technique for staining red halophilic bacteria. *J. Bacteriol.* **70:**484–485.

36. **Dworkin, M., S. Falkow, E. Rosenberg, K.-H. Schleifer, and E. Stackebrandt (ed.).** 2006. *The Prokaryotes*, 3rd ed. Springer-Verlag, Berlin, Germany.

37. **Dykhuizen, D. E.** 1998. Santa Rosalia revisited: why are there so many species of bacteria? *Antonie van Leeuwenhoek* **73:**25–33.

38. **Eisen, J. A.** 1995. The RecA protein as a model molecule for molecular systematic studies of bacteria: comparison of trees of RecA and 16S rRNA from the same species. *J. Mol. Evol.* **41:**1105–1123.

39. **Enright, M. C., and B. G. Spratt.** 1999. Multilocus sequence typing. *Trends Microbiol.* **7:**482–487.

40. **Euzéby, J.** 1997. List of bacterial names with standing in nomenclature: a folder available on the internet. *Int. J. Syst. Bacteriol.* **47:**590–592. http://www.bacterio.cict.fr

41. **Ezaki, T., Y. Hashimoto, and E. Yabuuchi.** 1989. Fluorimetric deoxyribonucleic acid-deoxyribonucleic acid hybridization in microdilution wells as an alternative to membrane filter hybridization in which radioisotopes are used to determine genetic relatedness among bacterial strains. *Int. J. Syst. Bacteriol.* **39:**224–229.

42. **Farelly, V., F. A. Rainey, and E. Stackebrandt.** 1995. Effect of genome size and rrn gene copy number on PCR amplification of 16S rRNA genes from a mixture of bacterial species. *Appl. Environ. Microbiol.* **61:**2798–2801.

43. **Felis, G. E., and F. Dellaglio.** 2007. On species descriptions based on a single strain: proposal to introduce the status species proponenda (sp. pr.). *Int. J. Syst. Evol. Microbiol.* **57:**2185–2187.

44. **Felis, G. E., D. Molenaar, F. Dellaglio, and J. E. van Hylckama Vlieg.** 2007. Dichotomy in post-genomic microbiology. *Nat. Biotechnol.* **25:**848–849.

45. **Felsenstein, J.** 1981. Evolutionary trees from DNA sequences: a maximum likelihood approach. *J. Mol. Evol.* **17:**368–376.

46. **Felsenstein, J.** 1985. Confidence limits on phylogenies: an approach using the bootstrap. *Evolution* **39:**783–791.

47. **Felsenstein, J.** 2008. PHYLIP (phylogeny inference package), version 3.68. Department of Genome Sciences, University of Washington, Seattle, WA.

48. **Fitch, W. M., and E. Margoliash.** 1967. Construction of phylogenetic trees: a method based on mutation distances as estimated from cytochrome *c* sequences is of general applicability. *Science* **155:**279–284.

49. **Fox, G. E., J. D. Wisotzkey, and P. Jurtshuk, Jr.** 1992. How close is close: 16S rRNA sequence identity may not be sufficient to guarantee species identity. *Int. J. Syst. Bacteriol.* **42:**166–170.

50. **Freney, J., W. E. Kloos, V. Hajek, J. A. Webster, M. Bes, Y. Brun, and C. Vernozy-Rozand.** 1999. Recommended minimal standards for description of new staphylococcal species. *Int. J. Syst. Bacteriol.* **49:**489–502.

51. **Garrity, G. M., et al. (ed.).** *Bergey's Manual of Systematic Bacteriology*, 2nd ed. Springer-Verlag, New York, NY. Vol. 1, 2001; vol. 2, 2005; vol. 3, 2009; vol. 4 to 5, pending.

52. **Gevers, D., F. M. Cohan, J. G. Lawrence, B. G. Spratt, T. Coenye, E. J. Feil, E. Stackebrandt, Y. Van de Peer, P. Vandamme, F. L. Thompson, and J. Swings.** 2005. Opinion: re-evaluating prokaryotic species. *Nat. Rev. Microbiol.* **3:**733–739.

53. **Gevers, D., P. Dawyndt, P. Vandamme, A. Willems, M. Vancanneyt, J. Swings, and P. de Vos.** 2006. Stepping stones towards a new prokaryotic taxonomy. *Philos. Trans R. Soc. Lond. B Biol. Sci.* **361:**1911–1916.

54. **Gillis, M., P. Vandamme, P. de Vos, J. Swings, and K. Kersters.** 2001. Polyphasic taxonomy, p. 43–48. *In* G. M. Garrity, D. R. Boone, and R. W. Castenholz (ed.), *Bergey's Manual of Systematic Bacteriology*, 2nd ed., vol. 1, *The Archaea and the Deeply Branching and Phototrophic Bacteria.* Springer-Verlag, New York, NY.

55. **Gonzalez, J. M., and C. Saiz-Jimenez.** 2002. A fluorimetric method for the estimation of G + C mol% content in microorganisms by thermal denaturation temperature. *Environ. Microbiol.* **4:**770–773.

56. **Goodfellow, M., M. D. Collins, and D. E. Minnikin.** 1976. Thin-layer chromatographic analysis of mycolic acid and other long-chain components in whole-organism methanolysates of coryneform and related taxa. *J. Gen. Microbiol.* **96:**351–358.

57. **Goris, J., K. T. Konstantinidis, J. A. Klappenbach, T. Coenye, P. Vandamme, and J. M. Tiedje.** 2007. DNA-DNA hybridization values and their relationship to whole-genome sequence similarities. *Int. J. Syst. Evol. Microbiol.* **57:**81–91.

58. **Goris, J., K. Suzuki, P. de Vos, T. Bakase, and K. Kersters.** 1998. Evaluation of a microplate DNA-DNA hybridization method compared with the initial renaturation method. *Can. J. Microbiol.* **44:**1148–1153.

59. **Graham, P. H., M. J. Sadowsky, H. H. Keyser, Y. M. Barnet, R. S. Bradley, J. E. Cooper, D. J. de Ley, B. D. W. Jarvis, E. B. Roslycky, B. W. Strijdom, and J. P. W. Young.** 1991. Proposed minimal standards for the description of new genera and species of root- and stem-nodulating bacteria. *Int. J. Syst. Bacteriol.* **41:**582–587.

60. **Gribaldo, S., and C. Brochier-Armanet.** 2006. The origin and evolution of Archaea: a state of the art. *Philos. Trans. R. Soc. Lond. B Biol. Sci.* **361:**1007–1022.

61. **Grimont, P. A. D.** 1981. Use of DNA reassociation in bacterial classification. *Can. J. Microbiol.* **34:**541–546.

62. **Gupta, R. S.** 1998. Life's third domain (Archaea): an established fact or an endangered paradigm? *Theor. Popul. Biol.* **54:**91–104.

63. **Hagström, Å., J. Pinhassi, and U. L. Zweifel.** 2000. Biogeographical diversity among marine bacterioplankton. *Aquat. Microb. Ecol.* **21:**231–244.

64. **Hansen, M. C., T. Tolker-Neilson, M. Givskov, and S. Molin.** 1998. Biased 16S rDNA PCR amplification caused by interference from DNA flanking template region. *FEMS Microbiol. Ecol.* **26:**141–149.

65. **Holding, A. J., and J. G. Collee.** 1971. Routine biochemical tests. *Methods Microbiol.* **6:**1–32.

66. **Huß, V. A. R., H. Festl, and K-H. Schleifer.** 1983. Studies on the spectrophotometric determination of DNA hybridization from renaturation rates. *Syst. Appl. Microbiol.* **4:**184–192.

67. **Imhoff, J. F., and P. Caumette.** 2004. Recommended standards for the description of new species of anoxygenic phototrophic bacteria. *Int. J. Syst. Evol. Microbiol.* **54:**1415–1421.

68. **Istock, C. A., J. A. Bell, N. Ferguson, and N. L. Istock.** 1996. Bacterial species and evolution: theoretical and practical perspectives. *J. Indust. Microbiol.* **17:**137–150.

69. **Jackman, P. J. H.** 1988. Microbial systematics based on electrophoretic whole-cell protein patterns. *Methods Microbiol.* **19:**209–225.

70. **Jukes, T. H., and C. R. Cantor.** 1969. Evolution of protein molecules, p. 21–132. *In* H. N. Munro (ed.), *Mammalian Protein Metabolism.* Academic Press, New York, NY.

71. **Kamekura, M.** 1993. Lipids of extreme halophiles, p. 135–161. *In* R. H. Vreeland and L. I. Hochstein (ed.), *The Biology of Halophilic Bacteria.* CRC Press, Boca Raton, FL.

72. **Kämpfer, P., S. Buczolits, A. Albrecht, H.-J. Busse, and E. Stackebrandt.** 2003. Towards a standardized format for the description of a novel species (of an established genus): *Ochrobactrum gallinifaecis* sp. nov. *Int. J. Syst. Evol. Microbiol.* **53:**893–896.

73. **Kimura, M.** 1980. A simple method for estimating evolutionary rates of base substitutions through comparative studies of nucleotide sequences. *J. Mol. Evol.* **16:**111–120.

74. **Komagata, K., and K. Suzuki.** 1987. Lipid and cell-wall analysis in bacterial systematics. *Methods Microbiol.* **19:**161–207.

75. **König, H.** 1995. Isolation and analysis of cell walls from methanogenic *Archaea*, p. 315–328. *In* K. R. Sowers and H. T. Schreier (ed.), *Archaea: A Laboratory Manual*, vol. 2,

Methanogens. Cold Spring Harbor Laboratory Press, Cold Spring Harbor, NY.

76. **Konstantinidis, K. T., A. Ramette, and J. M. Tiedje.** 2006. The bacterial species definition in the genomic era. *Philos. Trans. R. Soc. Lond. B* **361:**1929–1940.

77. **Konstantinidis, K. T., and J. M. Tiedje.** 2005. Towards a genome-based taxonomy for prokaryotes. *J. Bacteriol.* **187:**6258–6264.

78. **Konstantinidis, K. T., and J. M. Tiedje.** 2005. Genomic insights that advance the species definition for prokaryotes. *Proc. Natl. Acad. Sci. USA* **102:**2567–2572.

79. **Konstantinidis, K. T., and J. M. Tiedje.** 2007. Prokaryotic taxonomy and phylogeny in the genomic era: advancements and challenges ahead. *Curr. Opin. Microbiol.* **10:**504–509.

80. **Krieg, N. R.** 2001. Identification of Procaryotes, p. 33–38. *In* G. M. Garrity, D. R. Boone, and R. W. Castenholz (ed.), *Bergey's Manual of Systematic Bacteriology*, 2nd ed., vol. 1, *The Archaea and the Deeply Branching and Phototrophic Bacteria.* Springer-Verlag, New York, NY.

81. **Kumar, S., J. Dudley, M. Nei, and K. Tamura.** 2008. MEGA: a biologist-centric software for evolutionary analysis of DNA and protein sequences. *Brief. Bioinf.* **9:**299–306.

82. **Lan, R., and P. R. Reeves.** 2001. When does a clone deserve a name? A perspective on bacterial species based on population genetics. *Trends Microbiol.* **9:**419–424.

83. **Lane, D. J.** 1991. 16S/23S rRNA sequencing, p. 115–175. *In* E. Stackebrandt and M. Goodfellow (ed.), *Nucleic Acid Techniques in Bacterial Systematics.* Wiley, Chichester, United Kingdom.

84. **Lapage, S. P., P. H. A. Sneath, E. F. Lessel, Jr., V. B. D. Skerman, H. P. R. Seelinger, and W. A. Clark (ed).** 1992. *International Code of Nomenclature of Bacteria (1990 Revision).* American Society for Microbiology, Washington, DC.

85. **Levy-Frebault, V. V., and F. Portales.** 1992. Proposed minimal standards for the genus *Mycobacterium* and for description of new slowly growing *Mycobacterium* species. *Int. J. Syst. Bacteriol.* **42:**315–323.

86. **Ludwig, L., K.-H. Schleifer, and E. Stackebrandt.** 2006. Databases, p. 24–28. *In* M. Dworkin, S. Falkow, E. Rosenberg, K.-H. Schleifer, and E. Stackebrandt (ed.), *The Prokaryotes*, 3rd ed., vol. 1, *Symbiotic Associations, Biotechnology, Applied Microbiology.* Springer-Verlag, Berlin, Germany.

87. **Ludwig, W., and H.-P. Klenk.** 2001. Overview: a phylogenetic backbone and taxonomic framework for procaryotic systematics, p. 49–65. *In* G. M. Garrity, D. R. Boone, and R. W. Castenholz (ed.), (editor-in-chief), *Bergey's Manual of Systematic Bacteriology*, 2nd ed., vol. 1, *The Archaea and the Deeply Branching and Phototrophic Bacteria.* Springer-Verlag, New York, NY.

88. **Ludwig, W., and K.-H. Schleifer.** 1994. Bacterial phylogeny based on 16S and 23S rRNA sequence analysis. *FEMS Microbiol. Rev.* **15:**155–173.

89. **Ludwig, W., O. Strunk, R. Westram, L. Richter, H. Meier, Yadhukumar, A. Buchner, T. Lai, S. Steppi, G. Jobb, W. Förster, I. Brettske, S. Gerber, A. W. Ginhart, O. Gross, S. Grumann, S. Hermann, R. Jost, A. König, T. Liss, R. Lüssmann, R. May, B. Nonhoff, B. Reichel, R. Strehlow, A. Stamatakis, N. Stuckmann, A. Vilbig, M. Lenke, T. Ludwig, A. Bode, and K.-H. Schleifer.** 2004. ARB: a software environment for sequence data. *Nucleic Acids Res.* **32:**1363–1371.

90. **MacLean, D., J. D. Jones, and D. J. Studholme.** 2009. Application of 'next-generation' sequencing technologies to microbial genetics. *Nat. Rev. Microbiol.* **7:**287–296.

91. **Maidak, B. L., J. R. Cole, T. G. Lilburn, C. T. Parker, Jr., P. R. Saxman, R. J. Farris, G. M. Garrity, G. J. Olsen, T. M. Schmidt, and J. M. Tiedje.** 2001. The RDP-II (Ribosomal Database Project). *Nucleic Acids Res.* **29:**173–174.

92. **Maiden, M. C. J., J. A. Bygraves, E. Feil, G. Morelli, J. E. Russell, R. Urwin, Q. Zhang, J. Zhou, K. Zurth, D. A. Caugant, I. M. Feavers, M. Achtman, and B. G. Spratt.** 1998. Multilocus sequence typing: a portable approach to the identification of clones within populations of pathogenic organisms. *Proc. Natl. Acad. Sci. USA* **95:**3140–3145.

93. **Makarova, K. S., and E. V. Koonin.** 2006. Evolutionary genomics of lactic acid bacteria. *J Bacteriol.* **189:** 1199–1208.

94. **Mandel, M., and J. Marmur.** 1968. Use of ultraviolet absorbance-temperature profile for determining the guanine plus cytosine content of DNA. *Methods Enzymol.* **12B:**195–206.

95. **Mandel, M., C. L. Schildkraut, and J. Marmur.** 1968. Use of CaCl gradient analysis for determining the guanine plus cytosine content of DNA. *Methods Enzymol.* **12B:**184–195.

96. **Marmur, J.** 1961. A procedure for the isolation of deoxyribonucleic acid from microorganisms. *J. Mol. Biol.* **3:**208–218.

97. **Marmur, J., and P. Doty.** 1962. Determination of the base composition of deoxyribonucleic acid from its thermal denaturation temperature. *J. Mol. Biol.* **5:**109–118.

98. **Martinez-Murcia, A. J., S. G. Acinas, and F. Rodriguez-Valera.** 1995. Evaluation of prokaryotic diversity by restrictase digestion of 16S rDNA directly amplified from hypersaline environments. *FEMS Microbiol. Ecol.* **17:**247–256.

99. **Mayr, E.** 1942. *Systematics and the Origin of Species—From the Viewpoint of a Zoologist.* Harvard University Press, Cambridge, MA.

100. **McInnery, J. O., M. Wilkinson, J. W. Patching, T. M. Embley, and R. Powell.** 1995. Recovery and phylogenetic analysis of novel archaeal rRNA sequences from deep-sea deposit feeder. *Appl. Environ. Microbiol.* **61:**1646–1648.

101. **Mesbah, M., U. Premachandran, and W. B. Whitman.** 1989. Precise measurement of the G+C content of deoxyribonucleic acid by high-performance liquid chromatography. *Int. J. Syst. Bacteriol.* **39:**159–167.

102. **Mesbah, M., and W. B. Whitman.** 1989. Measurement of deoxyguanosine/thymidine ratios in complex mixtures by high-performance liquid chromatography for determination of the mole percentage guanine + cytosine of DNA. *J. Chromatogr.* **479:**297–306.

103. **Minnikin, D. E., A. G. O'Donnell, M. Goodfellow, G. Alderson, M. Athalye, K. Schaal, and J. H. Parlett.** 1984. An integrated procedure for the extraction of bacterial isoprenoid quinones and polar lipids. *J. Microbiol. Methods* **8:**303–310.

104. **Morrison, D.C., and J. L. Ryan (ed.).** 1992. *Bacterial Endotoxin Lipopolysaccharides.* Vol. I. *Molecular Biochemistry and Cellular Biology.* CRC Press, Boca Raton, FL.

105. **Mylvaganam, S., and P. P. Dennis.** 1992. Sequence heterogeneity between the two genes encoding 16S rRNA from the halophilic archaeobacterium *Haloarcula marismortui. Genetics* **130:**399–410.

106. **Norris, J. R., and H. Swain.** 1971. Staining bacteria. *Methods Microbiol.* **6A:**105–134.

107. **O'Brien, M., and E. Colwell.** 1987. Characterization tests for numerical taxonomic studies. *Methods Microbiol.* **19:**69–104.

108. **Oren, A.** 2004. Prokaryote diversity and taxonomy: present status and future challenges. *Philos. Trans. R. Soc. B* **359:**623–638.

109. **Oren, A., A. Ventosa, and W. D. Grant.** 1997. Proposed minimal standards for description of new taxa in the order *Halobacteriales. Int. J. Syst. Bacteriol.* **47:**233–238.

110. **Overmann, J.** 2006. Principles of enrichment, isolation, cultivation and preservation of prokaryotes, p. 80–136.

In M. Dworkin, S. Falkow, E. Rosenberg, K.-H. Schleifer, and E. Stackebrandt (ed.), *The Prokaryotes*, 3rd ed., vol. 1, *Symbiotic Associations, Biotechnology, Applied Microbiology.* Springer-Verlag, Berlin, Germany.

111. **Owen, R. J., and D. Pitcher.** 1985. Current methods for estimating DNA base composition and levels of DNA-DNA hybridization, p. 67–93. *In* M. Goodfellow and D. E. Minnikin (ed.) *Chemical Methods in Bacterial Systematics.* Academic Press, London, United Kingdom.

112. **Pot, B., P. Vandamme, and K. Kersters.** 1994. Analysis of electrophoretic whole-organism protein fingerprints, p. 493–521. *In* M. Goodfellow and A. G. O'Donnell (ed.), *Chemical Methods in Prokaryotic Systematics.* Wiley, Chichester, United Kingdom.

113. **Priest, F., and B. Austin.** 1993. *Modern Bacterial Taxonomy.* Chapman & Hall, London, United Kingdom.

114. **Pruesse, E., C. Quast, K. Knittel, B. M. Fuchs, W. Ludwig, J. Peplies, and F. O. Glockner.** 2007. SILVA: a comprehensive online resource for quality checked and aligned ribosomal RNA sequence data compatible with ARB. *Nucleic Acids Res.* **35:**7188–7196.

115. **Reysenbach, A. L., L. J. Giver, G. S. Wickham, and N. R. Pace.** 1992. Differential amplification of rRNA genes by polymerase chain reaction. *Appl. Environ. Microbiol.* **58:**3417–3418.

116. **Reysenbach, A.-L., and N. R. Pace.** 1995. Reliable amplification of hyperthermophilic archaeal 16S rRNA genes by PCR, p. 101–107. *In* F. T. Robb and A. R. Place (ed.), *Archaea: A Laboratory Manual—Thermophiles.* Cold Spring Harbor Laboratory Press, Cold Spring Harbor, NY.

117. **Reysenbach, A.-L., G. S. Wickham, and N. R. Pace.** 1994. Phylogenetic analysis of the hyperthermophilic pink filament community in Octopus Spring, Yellowstone National Park. *Appl. Environ. Microbiol.* **60:**2113–2119.

118. **Ritacco, F. V., B. Haltli, J. E. Janso, M. Greenstein, and V. S. Bernan.** 2003. Dereplication of *Streptomyces* soil isolates and detection of specific biosynthetic genes using an automated ribotyping instrument. *J. Ind. Microbiol. Biotechnol.* **30:**472–479.

119. **Rokas, A., and P. Abbot.** 2009. Harnessing genomics for evolutionary insights. *Trends Ecol. Evol.* **24:**192–200.

120. **Rosselló-Mora, R.** 2005. Updating prokaryotic taxonomy. *J. Bacteriol.* **187:**6255–6257.

121. **Rosselló-Mora, R., and R. Amann.** 2001. The species concept for prokaryotes. *FEMS Microbiol. Rev.* **25:**39–67.

122. **Rudi, K., O. M. Skulberg, F. Larsen, and K. S. Jacoksen.** 1997. Strain classification of oxyphotobacteria in clone cultures on the basis of 16S rRNA sequences from variable regions V6, V7 and V8. *Appl. Environ. Microbiol.* **63:**2593–2599.

123. **Saitou, N., and M. Nei.** 1987. The neighbor-joining method: new method for reconstructing phylogenetic trees. *Mol. Biol. Evol.* **4:**406–425.

124. **Schleifer, K.-H.** 1985. Analysis of the chemical composition and primary structure of murein. *Methods Microbiol.* **18:**123–156.

125. **Schleifer, K.-H., and O. Kandler.** 1972. Peptidoglycan types of bacterial cell walls and their taxonomic implications. *Bacteriol. Rev.* **36:**407–477.

126. **Schleifer, K.-H., and W. Ludwig.** 1994. Molecular taxonomy: classification and identification, p. 1–15. *In* F. G. Priest, A. Ramos-Cormenzana, and B.J. Tindall (ed.), *Bacterial Diversity and Systematics.* Plenum Press, New York, NY.

127. **Skerman, V. B. D., V. McGowan, and P. H. A. Sneath.** 1980. Approved lists of bacterial names. *Int. J. Syst. Bacteriol.* **30:**225–420.

128. **Skerman, V. B. D., V. McGowan, and P. H. A. Sneath (ed.).** 1989. *Approved Lists of Bacterial Names*

(Amended Edition). American Society for Microbiology, Washington, DC.

129. **Smibert, R. M., and R. N. Krieg.** 1994. Phenotypic characterization, p. 607–654. *In* P. Gerhardt, R. G. E. Murray, W.A. Wood, and N. R. Kried (ed.), *Manual of Methods for General Microbiology*. American Society for Microbiology, Washington, DC.

130. **Sneath, P. H. A.** 2001. Bacterial nomenclature, p. 83–88. *In* G. M. Garrity, D. R. Boone, and R. W. Castenholz (ed.), *Bergey's Manual of Systematic Bacteriology*, 2nd ed., vol. 1, *The Archaea and the Deeply Branching and Phototrophic Bacteria*. Springer-Verlag, New York, NY.

131. **Stackebrandt, E.** 2006. Defining taxonomic ranks, p. 29–57. *In* M. Dworkin, S. Falkow, E. Rosenberg, K.-H. Schleifer, and E. Stackebrandt (ed.), *The Prokaryotes*, 3rd ed., vol. 1, *Symbiotic Associations, Biotechnology, Applied Microbiology*. Springer-Verlag, Berlin, Germany.

132. **Stackebrandt, E., and J. Ebers.** 2006. Taxonomic parameters revisited: tarnished gold standards. *Microbiol. Today* **33:**152–155.

133. **Stackebrandt, E., W. Frederiksen, G. M. Garrity, P. A. D. Grimont, P. Kämpfer, M. C. J. Maiden, X. Nesme, R. Rosselló-Mora, J. Swings, H. G. Trüper, L. Vauterin, A. C. Ward, and W. B. Whitman.** 2002. Report of the ad hoc committee for the re-evaluation of the species definition in bacteriology. *Int. J. Syst. Evol. Microbiol.* **52:**1045–1047.

134. **Stackebrandt, E., and B. M. Goebel.** 1994. Taxonomic note: a place for DNA:DNA reassociation and 16S rRNA sequence analysis in the present species definition in bacteriology. *Int. J. Syst. Bacteriol.* **44:**846–849.

135. **Staley, J. T.** 2006. The bacterial species dilemma and the genomic-phylogenetic species concept. *Philos. Trans. R. Soc. Lond. B* **361:**1899–1909.

136. **Staley, J. T., and N. R. Krieg.** 1989. Classification of procaryotic organisms: an overview, p. 1601–1603. *In* J. T. Staley, M. P. Bryant, N. Pfennig, and J. G. Holt (ed.), *Bergey's Manual of Systematic Bacteriology*, vol. 3. Williams & Wilkins, Baltimore, MD.

137. **Suzuki, M. T., and S. J. Giovannoni.** 1996. Bias caused by template annealing in the amplification of mixtures of 16S rRNA genes by PCR. *Appl. Environ. Microbiol.* **62:**625–630.

138. **Tamaoka, J.** 1986. Analysis of bacterial menaquinone mixtures by reverse-phase high-performance liquid chromatography. *Methods Enzymol.* **123:**251–256.

139. **Tamaoka, J.** 1994. Determination of DNA base composition, p. 463–470. *In* M. Goodfellow and A. G. O'Donnell (ed.), *Chemical Methods in Prokaryotic Systematics*. Wiley, Chichester, United Kingdom.

140. **Tamaoka, J., and K. Komagata.** 1984. Determination of DNA base composition by reversed-phase high-performance liquid chromatography. *FEMS Microbiol. Lett.* **25:**125–128.

141. **Tamura, K., J. Dudley, M. Nei, and S. Kumar.** 2007. MEGA4: molecular evolutionary genetics analysis (MEGA) software version 4.0. *Mol. Biol. Evol.* **24:**1596–1599.

142. **Thompson, J. D., D. G. Higgins, and T. J. Gibson.** 1994. Clustal W: improving the sensitivity of progressive multiple sequence alignments through sequence weighing position-specific gap penalties and weight matrix choice. *Nucleic Acids Res.* **22:**4673–4680.

143. **Thompson, J. D., T. J. Gobson, F. Plewniak, F. Jeanmougin, and D. G. Higgins.** 1997. The CLUSTAL_ X windows interface: flexible strategies for multiple sequence alignment aided by quality analysis tools. *Nucleic Acids Res.* **25:**4876–4882.

144. **Tindall, B. J., P. Kämpfer, J. Euzéby, and A. Oren.** 2006. Valid publication of names of prokaryotes according to the rules of nomenclature: past history and current practice. *Int. J. Syst. Evol. Microbiol.* **56:**2715–2720.

145. **Tindall, B. J., and G. M. Garrity.** 2008. Proposals to clarify how type strains are deposited and made available to the scientific community for the purpose of systematic research. *Int. J. Syst. Evol. Microbiol.* **58:**1987–1990.

146. **Tornabene, T. G.** 1985. Lipid analysis and the relationship to chemotaxonomy. *Methods Microbiol.* **18:**209–234.

147. **Trüper, H. G.** 1999. How to name a prokaryote? Etymological considerations, proposals and practical advice in prokaryote nomenclature. *FEMS Microbiol. Rev.* **23:**231–249.

148. **Trüper, H. G.** 2001. Etymology in nomenclature of prokaryotes, p. 89–99. *In* G. M. Garrity, D. R. Boone and R. W. Castenholz (ed.), *Bergey's Manual of Systematic Bacteriology*, 2nd ed., vol. 1, *The Archaea and the Deeply Branching and Phototrophic Bacteria*. Springer-Verlag, New York, NY.

149. **Trüper, H. G., and K.-H. Schleifer.** 2006. Prokaryote characterization and identification, p. 58–79. *In* M. Dworkin, S. Falkow, E. Rosenberg, K.-H. Schleifer, and E. Stackebrandt (ed.), *The Prokaryotes*, 3rd ed., vol. 1, *Symbiotic Associations, Biotechnology, Applied Microbiology*. Springer-Verlag, Berlin, Germany.

150. **Ursing, J. B., H. Lior, and R. J. Owen.** 1994. Proposal of minimal standards for describing new species of the family *Campylobacteraceae*. *Int. J. Syst. Bacteriol.* **44:**842–845.

151. **Vandamme, P., B. Pot, M. Gillis, P. de Vos, K. Kersters, and J. Swings.** 1996. Polyphasic taxonomy, a consensus approach to bacterial systematics. *Microbiol. Rev.* **60:**407–438.

152. **Vaneechoutte, M.** 1996. DNA-fingerprinting techniques for microorganisms. *Mol. Biotechnol.* **6:**114–143.

153. **Ward, D. M.** 1998. A natural species concept for prokaryotes. *Curr. Opin. Microbiol.* **1:**271–277.

154. **Watanabe, K., Y. Kodama, and S. Harayama.** 2001. Design and evaluation of PCR primers to amplify 16S ribosomal DNA fragments used for community fingerprinting. *J. Microbiol. Meth.* **44:**253–262.

155. **Wayne, L. G., D. J. Brenner, R. R. Colwell, P. A. D. Grimont, O. Kandler, M. I. Krichevsky, L. H. Moore, W. E. C. Moore, R. G. E. Murray, E. Stackebrandt, M. P. Starr, and H. G. Trüper.** 1987. Report of the ad hoc committee on reconciliation of approaches to bacterial systematics. *Int. J. Syst. Bacteriol.* **37:**463–464.

156. **Whitman, W. B., D. C. Coleman, and W. J. Wiebe.** 1998. Prokaryotes: the unseen majority. *Proc. Natl. Acad. Sci. USA* **95:**6578–6583.

157. **Woese, C. R.** 1987. Bacterial evolution. *Microbiol. Rev.* **51:**221–271.

158. **Woese, C. R.** 1992. Prokaryote systematics: the evolution of a science, p. 3–18. *In* A. Balows, H. G. Trüper, M. Dworkin, W. Harder, and K.-H. Schleifer (ed.), *The Prokaryotes. A Handbook on the Biology of Bacteria: Ecophysiology, Isolation, Identification, Applications*, 2nd ed., vol. 1. Springer-Verlag, Berlin, Germany.

159. **Woese, C. R.** 2006. How we do, don't and should look at bacteria and bacteriology, p. 3–23. *In* M. Dworkin, S. Falkow, E. Rosenberg, K.-H. Schleifer, and E. Stackebrandt (ed.), *The Prokaryotes*, 3rd ed., vol. 1, *Symbiotic Associations, Biotechnology, Applied Microbiology*. Springer-Verlag, Berlin, Germany.

160. **Woese, C. R., O. Kandler, and M. L. Wheelis.** 1990. Towards a natural system of organisms: proposal of the domains Archaea, Bacteria, and Eucarya. *Proc. Natl. Acad. Sci. USA* **87:**4576–4579.

161. **Yarza, P., M. Richter, J. Peplies, J. Euzéby, R. Amann, K.-H. Schleifer, W. Ludwig, F. O. Glöckner, and R. Rosselló-Mora.** 2008. The All-Species Living Tree Project: a 16S rRNA-based phylogenetic tree of all sequenced type strains. *Syst. Appl. Microbiol.* **31:**241–250.

Enzymes from Extreme Environments

DON A. COWAN, BRONWYN M. KIRBY, TRACY L. MEIRING,
MANUEL FERRER, MARIA-EUGENIA GUAZZARONI,
OLGA V. GOLYSHINA, AND PETER N. GOLYSHIN

4

4.1. INTRODUCTION

Environmental samples are colonized by complex microbial communities whose composition and functions may differ significantly. Some organisms are overrepresented in a sample, whereas others can be found at concentrations below 10 cells per g of soil (106). However, at present, we simply do not know the extent of the functional diversity that microbes encompass: a classical theoretical analysis estimates that the prokaryotic population on Earth is approximately 10^{30} bacteria, a few orders of magnitude higher than the number of stars in the known universe (estimated 10^{22} to 10^{24}), with most microbes being members of complex communities (106). Any individual survey to study such diversity is limited because of our relatively poor ability to grow most microorganisms under standard laboratory conditions (53). This problem is exacerbated in the case of organisms living in habitats with extreme conditions (e.g., low or high temperatures, low and high pH, high pressure, and high salinity), the so-called extremophiles that operate under conditions generally considered to be hostile to life. These environments often contain very few microbes. Even though the genetic diversity of such environments may in some cases be low, it may nevertheless be particularly interesting, warranting the special effort required to access it. To do so, along with traditional cultivation methods, a wide range of approaches collectively described as "metagenomics" have been developed to study communities through the analysis of their genetic material without culturing individual organisms (40, 41). Metagenomics is analogous to genomics with the difference that it does not deal with the single genome from a clone or microbe cultured or characterized in the laboratory, but rather with that from the entire microbial community present in an environmental sample, the so-called community genome. Metagenomics represents a strategic concept that includes investigations at three major interconnected levels (sample processing, DNA sequencing, and functional analysis), with an ultimate goal of obtaining a comprehensive view of the functioning of the microbial world.

One of the major problems associated with metagenomic analysis is that the total DNA extracted from environmental samples in many cases may not contain an even representation of the population genome, meaning that rare organisms would contribute less to the overall DNA diversity, with the library being dominated by the most abundant organisms (32, 33). It is thus of great priority to adapt sample and DNA extraction methods and cloning strategies for normalization of the sample, because the relative abundance of representatives of a certain group of microbes is not necessarily linked to the importance of that group in the community functioning: common organisms may not necessarily play a critical role in a community despite their numbers, and organisms that comprise only 0.1% of the total population (e.g., nitrogen fixers) can be of pivotal importance (106). To date, much of the research has been focused on the sequence-centered analysis of community DNA. This analysis has a great throughput in data production and can access much greater genomic information of microbial communities, but sequence-based functional reconstruction and predictions of whole-community metabolic processes is reminiscent of crystal ball gazing: most of the genes have no clear function; and there is no support through functional verification, albeit this method can suggest some candidates/targets for expression and further characterization. Since new technologies allow dramatic expansion of sequencing data production, the current development of new analytical approaches and techniques will help to make more sense of these data and provide a holistic insight into the global ecosystem functioning (26). Nevertheless, in contrast to this sequence-centered metagenomics, another function-based, or activity-based approach currently shows better options to link specific microbes to specific ecological functions (12, 58, 105).

Below we provide a view of technical issues to illustrate the potential of getting appropriate sample material to create representative gene libraries as the first step for analysis of community genomes. We also provide an extensive description of tools and protocols for DNA extraction and cloning, as well as of host/vector expression systems developed for specific groups of microorganisms, with a special focus on extremophiles.

4.2. SAMPLING PROTOCOLS

Little of the technology relevant to environmental sampling of extremophilic habitats is unique to these organisms, and where specialized methods are required, they are normally the obvious consequences of the properties of the specific group of organisms and their favored habitats. Nevertheless, general protocols used in "normal" environmental sampling should be equally valid when sampling extreme habitats. In particular, sampling protocols that take account of potential problems such as microheterogeneous distribution should be used. The use of such protocols is critical where sampling is for the purpose of comparative microbial ecology or where representative metagenomic coverage is required. Although there is no standardized sampling protocol, and most field biologists will adapt procedures to the requirements of their sampling program, the physical characteristics of their sampling site, and the technology available to them, the following protocol can be used as a guideline for sampling where issues of microheterogeneity are potentially important.

For open soils and the margins of aquatic systems:

- Lay out a 50-m transect line, marked at the center point (to increase coverage, a second 50-m line may be laid out at right angles, intersecting at the midpoint).
- Establish 1-m² quadrants at the center and the ends of the transects (to increase coverage, quadrants may also be sited at the 10-m points).
- Sample an appropriate volume (using sterile procedure as appropriate) at four points (typically the corners) of each quadrant.
- Label and identify each sample separately.
- Subsequently, recover and mix approximately 50% of each of the four samples from each quadrant into a fifth sample.

The purpose of this approach is to pre-prepare samples for subsequent analysis of homogeneous (or otherwise) microbial distribution. This would typically be done by denaturing gradient gel electrophoresis or automated ribosomal interspacer sequence analysis (ARISA). The results of this analysis can be used as a guide for subsequent separation (or mixing) of the samples.

There is no standard technology for recovery and retention of samples. However, sampling devices and sample containers should be compatible with the physical and chemical properties of the sample, and compatible with the requirements of postsampling transport and storage. Sterile 50-ml plastic Greiner® tubes and wide-mouth Nalgene® containers (for volumes of >100 ml) are ideal for liquid samples, while WhirlPak® bags conserve both weight and space for collection of solid/dry samples.

4.2.1. Rules for Sampling Extreme Environments

In general, the more detailed and sophisticated the microenvironmental data obtained, the better! Parameters such as temperature, pH, percentage of relative humidity, salinity, dO_2, redox state, moisture content, light, local geology, weather (cloud, precipitation, wind speed), and so on, may be critical or irrelevant. Date and time of sampling together with accurate Global Positioning System coordinates should be recorded. Further, as a general rule, good sterile technique should be used in any sampling protocol, the objective being to minimize contamination of both the sample and the sampling site. Avoiding cross-contamination between multiple sites may be important. Sampling tools (metal or wooden spatulas, trowels, etc.) can be presterilized in tinfoil and opened on-site, or (for larger tools where one-off use is not feasible) can be cleaned and sterilized between samples by using sterilizing wipes. The following are issues for consideration in the sampling of various specific extreme environments.

4.2.1.1. Thermophilic Samples

In hydrothermal systems, the solid phase (sinter surfaces, sand and gravel deposits, sediments, etc.) typically have much higher cell loads than the aqueous phase (99). Given that even a gloved hand cannot be immersed for more than fractions of a second in thermal waters of above approximately 60°C, a sampling device is essential for recovering sediment material. The simplest effective device is a pair of tongs, preferably long-handled, with which a wide-mouthed plastic bottle can be grasped and manipulated. More sophisticated devices can be engineered—a 500-ml aluminium or titanium collection vessel with a threaded detachable handle is practical for both transport and sampling.

For recovery of liquid samples, handheld suction devices ("turkey basters") are adequate for recovering small volumes, but they are readily blocked by particulates. Microbial cell yields from bulk liquids may be no more than 10^4 to 10^5 ml^{-1}, so filtration may be preferable for recovery of larger volumes of biomass. A handheld suction device (such as the Nalgene Mityvac®), used in conjunction with a plastic Buchner flask and a Millipore® filtration unit (for 7-cm diameter 0.2-μm filters), is a slow but effective method for scaling up the recovery of biomass from bulk liquids. Filters will block when fully coated with cellular material: after filtration of a few hundred milliliters (for very high biomass fluids such as algal blooms) or quite a few liters for low biomass (e.g., hydrothermal) fluids. Once blinded, filters can be transferred with sterile tweezers to sealable plastic bags, tubes of growth media, or nucleic acid-stabilizing solutions.

To the authors' knowledge, there is no strong evidence that the survival of thermophilic species in environmental samples is improved by maintaining samples at the in vivo temperature during transport from site to laboratory. If it is deemed essential to maintain sample temperatures, volumes of a few liters can be transported in Thermos® flasks with good temperature retention over 24 to 36 h.

In thermal environments of above approximately 75°C, most microbial species are facultative or obligate anaerobes. The maintenance of cell viability for this group of organisms thus critically depends on the retention of anaerobic status. The minimum necessary action is to ensure that sample containers are "brim-full"—i.e., that air is effectively excluded. Note also that some plastics are highly permeable to oxygen! Ideally, where anaerobic status is important, addition of a sterile sodium sulfide solution (e.g., sodium sulfide nonahydrate ACS reagent; Sigma-Aldrich, St. Louis, MO) after collection, together with a drop of the redox indicator resazurin (e.g., 0.1% solution prepared in distilled water), will result in the maintenance of anaerobic conditions, and allow the researcher to visually confirm that status.

4.2.1.2. Psychrophilic Samples

True psychrophiles are highly sensitive to elevated temperatures and may be inactivated by temperatures as low as 30°C (87). All cold-active microorganisms probably maintain some level of metabolic activity at temperatures well below 0°C. For these reasons, temperature control is particularly important in the handling of samples from psychrophilic environments. Ideally, field samples collected in psychrophilic environments (mountain, polar, etc.) should be stored at or just below 0°C for as short a time as possible before transfer to colder storage—domestic freezer temperature (−12 to −18°C) at least. Even at these temperatures, it is likely that many psychrophiles maintain some metabolic activity, and ideally, 70°C storage, in particular for longer-term maintenance of samples, is preferable. However, note that freeze-thaw cycling is extremely damaging to both viable cells and polynucleotides, and large samples should be subfractionated before long-term storage.

4.2.1.3. Halophilic Samples

High-salt environments frequently carry very high biomass loads. This usually facilitates the recovery of biomass, in particular in the filtration of water samples (see above).

The cytoplasm of extremely halophilic organisms is generally hyperosmotic, and cells will lyse very rapidly if exposed to a low-ionic-strength environment. Consequently, buffers or media used to stabilize filters should be adjusted (or pre-prepared) to the approximate salt concentration of the sampling site.

4.2.1.4. Acidophilic Samples

Field researchers report that samples recovered from acidophilic sites, in particular high-temperature (thermoacidophilic) sites, show rapid loss of viable biomass and degradation of high-molecular-weight DNA. This is believed to be the result of acidification of cellular cytoplasm under conditions where the organisms are metabolically inactive, and the resulting accelerated acid hydrolysis of ester, amide, and phosphodiester bonds. Some field researchers prefer to neutralize acidophilic samples at the time of recovery; typically by addition of sterile Na_2CO_3, NaOH, or $CaCO_3$, and the judicious use of pH paper. Others, concerned by the possible impact of large shifts in pH and ionic strength on cell viability, do nothing. To the authors' knowledge, this problem has been neither rigorously investigated nor satisfactorily resolved.

4.2.1.5. "Normal" Superficial Sea- and Freshwater Samples

Immediately after collection, samples (approximately 2 to 20 liters) are filtered onto a 500-kDa NMWL (nominal molecular weight limit or membrane cutoff in kilodaltons) ultrafiltration disc (Biomax polyethersulfone; Millipore). After this, the filters are cut into strips (1 cm × 2 mm) and used directly for DNA extraction (see the protocols described below). Alternatively, in the case of large sample amounts (e.g., 100 to 200 liters) a tangential flow filtration device can be used, such as Pellicon TFF 0.1 μm (Millipore), to separate solid particles and picoeukaryotic organisms (2 to 4 μm). Part of the overconcentrated product (retentate solution; e.g., 10 ml) can be used for the enrichment of cells with a desired supplement (e.g., minimal medium supplemented with petroleum, to enrich for microbes able to metabolize it) or this retentate solution can be filtered onto a 500-kDa pore size with the Amicon® system (Millipore), and the filter cut into strips and used directly for DNA extraction.

4.2.1.6. Soil Samples

Soil or sediment samples (1 to 5 g) are resuspended in 5 to 40 ml (depending on the sample properties) of Tris-EDTA (TE) buffer, pH 8.0, and mixed by inverting the tube 10 to 15 times. The samples are mixed with moderate shaking, to release cells from the solid matrix, and centrifuged at a low speed (approximately 200 to 400 × g for 1 to 5 min) to eliminate bulky soil particles. The supernatant is transferred to a fresh sterile tube and is centrifuged at 6,000 × g for 15 to 30 min at 4°C. The supernatant is discarded and the pellet is retained for DNA extraction. In some cases, for example, dealing with the samples containing particulate materials (soils, sediments), a Nycodenz® extraction technique is highly recommended. During the physical separation of the bacterial fraction with use of a Nycodenz® cushion, a whitish band of microbial biomass is obtained at the interface between the Nycodenz® and the aqueous layer by using a high-speed density gradient.

This method has been used successfully to isolate DNA from freshwater, compost, rhizosphere-associated soils, as well as from pristine- and contaminated sediments. The procedure is outlined below.

- Prepare sample suspension: to 15 g of sample, add disruption buffer (35 ml total volume: 0.2 M NaCl, 50 mM Tris-HCl, pH 8.0) and mix (preferably overnight with orbital shaking).
- Centrifuge at a low speed (approximately 200 to 400 × g for 1 to 5 min) to eliminate large soil particles and then use supernatant for biomass separation via Nycodenz®.
- Transfer 25 ml of the soil homogenate to an ultracentrifuge tube and carefully pipette 9 to 11 ml of Nycodenz® (0.8 to 1.3 g ml^{-1}) to form a layer below the homogenate.
- Centrifuge at 10,000 × g for 20 to 40 min at 4°C.
- A faint, whitish band containing bacterial cells is resolved at the interface between the Nycodenz® and the aqueous layer. Transfer this band into a sterile tube. Note that some soils, in particular clay soils, contain small particles that do not separate from the microbial biomass, forming a layer mixed on the Nycodenz® surface.
- Add approximately 35 ml of phosphate-buffered saline, and pellet the cells by centrifugation at 10,000 × g for 20 min. Discard the supernatant and resuspend the cell pellet in 0.5 to 2.0 ml of TE buffer, pH 8.0.

4.3. DNA EXTRACTION

4.3.1. Wet-Lab Techniques for DNA Extraction

Two major extraction protocols may be used: (i) phenol extraction, where the gentle hydrophobicity of phenol makes it a good solvent for DNA extraction, and (ii) the use of commercial kits (in this case, it is highly recommend to separate cells from the matrix to allow efficient DNA recovery yields). The protocols are as follows.

Isolation of High-Quality DNA by the Phenol/ Chloroform Method Followed by DNA Cleaning

The following procedure is recommended when a large quantity of humic acids is present in the soil samples and for Nycodenz®-separated biomass to minimize the volume of solvents required for large-scale soils.

1. To the solution obtained as above (cell pellet in TE buffer), add 25 μl of lysozyme solution (10% wt/vol, prepared before use), and incubate for 2 h at 37°C (1,500 rpm).
2. Add 6 μl of RNase solution (1% wt/vol) free of DNase, and incubate for approximately 30 min at 37°C.
3. Add 8 μl of solution of proteinase K (1% wt/vol) and 60 μl of sodium dodecyl sulfate solution (10% wt/vol) and incubate 30 min at 50°C, until it becomes clear and viscous.
4. Add 100 μl of 5 M NaCl and mix gently by inverting the tube four to six times.
5. Add 80 μl of a cetyltrimethylammonium bromide solution prewarmed at 65°C (10% wt/vol in 0.7 M NaCl), mix gently by inverting the tube four to six times, and incubate 10 min at 65°C.
6. The final volume should be approximately 748 μl.
7. Add 750 μl of CHCl$_3$/isoamyl alcohol (24:1), mix gently by inverting the tube four to six times, then centrifuge for 3 min at 12,000 × g and quickly transfer the upper aqueous phase into a new 2-ml Eppendorf tube.

8. Add 700 μl of phenol/chloroform/isoamyl alcohol (25:24:1); mix gently by inverting the tube four to six times. Centrifuge for 3 min at 12,000 × g, and quickly transfer the upper aqueous phase to a new 2-ml Eppendorf tube.

9. Add 650 μl of chloroform/isoamyl alcohol (24:1) and mix gently by inverting the tube four to six times. Centrifuge for 3 min at 14,000 × g, and quickly transfer the upper aqueous phase to a new 1.5-ml Eppendorf tube.

10. Finally, add 0.6 volumes of isopropyl alcohol and mix gently by inverting the tube four to six times. Place the tubes on ice for 10 min, followed by centrifugation for 30 min in a benchtop centrifuge at maximal speed (>12,000 × g). The supernatant can be removed by gentle suction with a vacuum pump.

11. Add 500 μl of ethanol (70% vol/vol) to the DNA pellet, centrifuge in a microcentrifuge at the maximal speed (>12,000 relative centrifugal force), and using a gentle suction with a vacuum pump remove the supernatant.

12. Quickly transfer the tube into the Speed-Vac and dry for approximately 5 min without heating.

13. Resuspend the DNA pellet in 500 μl of TE buffer, pH 8.0 (typically, at room temperature, overnight).

14. If the quality of DNA from the previous step is good enough (of satisfactory size, as can be judged with agarose gel electrophoresis) and does not contain inhibitors (as can be judged by the enzymatic endonuclease trial digestion of a small DNA aliquot and its subsequent gel electrophoresis), then proceed directly to the digestion and cloning step. If not, an ultrapure DNA protocol should be applied. DNA Clean & Concentrator® (Zymo Research Corp., Orange, CA) has been shown to yield a high-purity DNA. By using this product one could purify DNA by using a single-buffer system that allows efficient and selective DNA adsorption onto a matrix. Here, it is important to use at least 200 μg of DNA since the recovery of DNA is, in most samples, lower than 40%.

Important note: in case high-molecular-weight DNA is needed, avoid pipetting samples, because this may shear the DNA.

Isolation of High-Quality DNA by Commercial Kits

Commercial kits such as UltraClean MegaPrep (MoBio Laboratories, Inc., Carlsbad, CA) and GNOME® DNA Extraction Kit (BIO101) may be used for isolation of metagenomic DNA from eukaryotic or prokaryotic cells and tissues in less than 2 h with no organic extractions. Preliminary separation of cellular biomass from soil homogenate via Nycodenz® gradient is recommended to achieve maximal DNA recovery per gram of soil. DNA purification kits from other manufacturers presumably work equally well, but they have not been tested in our laboratory. The GNOME® DNA kit uses RNase Mix to eliminate RNA immediately after lysis, and Protease Mix to rapidly degrade cellular proteins. This is followed by a proprietary "salting out" technique that precludes the need for phenol, chloroform, or other organic extractions. Preparation of metagenomic DNA using this kit is described below.

1. Immediately after collection, samples are either stored in 95% ethanol at 4°C or are shock-frozen in liquid nitrogen, followed by storage at −80°C. Alternatively, 10 g of soil is directly homogenized with 40 ml of a 0.2 M NaCl, 50 mM Tris-HCl, pH 8.0, buffer, mixed (overnight, 4°C, orbital shaking), and then centrifuged at low speed (approximately 200 to 500 × g for 1 to 5 min) to eliminate large soil particles. Cells (plus small particles) are separated by centrifugation at high speed (9,000 × g for 15 min). Freshwater samples (>600 ml) are subjected to cell separation either by centrifugation (9,000 × g for 15 min) or filtration through a 0.20-μm filter (the filter is removed and used directly). DNA is isolated from these samples after steps 2 to 10. To each 0.1 g of cells (from pellets obtained by centrifugation, or cells separated by filtration), add 1.85 ml of Cell Suspension Solution (use a 15-ml clear-plastic tube for efficient mixing). Mix until the solution appears homogeneous.

2. Add 50 μl of RNase Mix and mix thoroughly. Add 100 ml of Cell Lysis/Denaturing Solution and mix well.

3. Incubate at 55°C for 15 min.

4. Add 25 μl of Protease Mix and mix thoroughly.

5. Incubate at 55°C for 30 to 120 min (the longer time will result in minimal protein carryover and will also allow for substantial reduction in residual protease activity).

6. Add 500 μl of "Salt-Out" Mixture and mix gently, yet thoroughly. Divide sample into 1.5-ml tubes. Refrigerate at 4°C for 10 min.

7. Spin for 10 min at maximum speed in a microcentrifuge (at least 10,000 × g). Carefully collect the supernatant, avoiding the pellet. If a precipitate remains in the supernatant, spin again until it is clear. Pool the supernatants in a 15-ml (or larger) clear-plastic tube.

8. To this supernatant, add 2 ml of TE buffer and mix. Then add 8 ml of 100% ethanol. If spooling the DNA, add the ethanol slowly and spool the DNA at the interphase with a clean glass rod. If centrifuging the DNA, add the ethanol and gently mix the solution by inverting the tube.

9. Spin for 15 min at 1,000 to 1,500 × g. Eliminate the excess ethanol by blotting or air drying the DNA.

10. Dissolve the genomic DNA in TE (10 mM Tris, pH 7.5, 1 mM EDTA).

4.3.2. DNA Extraction from Extreme Samples

Many of the methods that are used routinely to extract genomic DNA from mesophiles can be used to extract DNA from extremophiles. However, from the authors' experience, extracting good-quality, high-molecular-weight DNA from halophilic organisms can be problematic, due to the high salt concentration of the culture medium necessary for the growth of obligate halophiles. In our laboratory, the method of Yu et al. (109) has been adapted successfully to extract genomic DNA from high-salt soil samples or halophilic bacterial cultures. The method involves direct cell lysis of soil or cell pellets embedded in agarose plugs, with the agarose plugs being cast in 96-well microtiter plates. Yu and coworkers used pulsed-field gel electrophoresis followed by dialysis to remove soil contaminants and cellular debris. However, our laboratory has replaced the pulsed-field gel electrophoresis with electrophoresis in low-melt agarose. Relatively pure, high-molecular-weight DNA can be obtained by this method, which can be used directly for downstream applications, including cloning into fosmids.

1. One gram of soil/200 μl of cell pellet is mixed with 1 ml of suspension buffer (10 mM Tris-HCl, 1 M NaCl, 50 mM EDTA) and 1 ml of 1% low-melting-point agarose. Keep at 37°C to prevent the agarose from solidifying.

2. Several (at least five per sample) 100-μl aliquots of the slurry are pipetted into a mold (we use a 96-well microtiter plate) and placed at 4°C to solidify. Soil that has a lot of plant litter or compost is not well suited to this

method, because the agarose plugs are not firm and tend to break apart during the extraction process.

3. For each sample, place three plugs in a 2-ml Eppendorf tube and add 1.5 ml of lysis buffer (50 mM Tris-HCl, 100 mM EDTA, 100 mM NaCl, 0.2% [wt/vol] sodium deoxycholate, 0.5% [wt/vol] 20 cetyl ether, 0.5% [wt/vol] lauroylsarcosine sodium salt, and 5 mg/ml lysozyme). Place tubes at 37°C for 24 h with gentle agitation (50 rpm).

4. Remove the lysis buffer and add 1.5 ml of digestion buffer (100 mM Tris-HCl, 500 mM EDTA, 100 mM sodium phosphate, 1.5 M NaCl, 1.5 mg/ml proteinase K, and 0.5% [wt/vol] sodium dodecyl sulfate). Place tubes at 50°C for 24 h with gentle agitation.

5. Collect the plugs and wash in 50 volumes of wash buffer (10 mM Tris-HCl, pH 8, 100 mM EDTA) and then several times with ice-cold 0.5× TBE buffer.

6. Secure the washed plugs with molten agarose to the teeth of an electrophoresis comb and allow to solidify. Pour molten 0.7% low-melt agarose carefully around the plugs and allow the gel to solidify. The gel is electrophoresed for 18 h at 35 V. The gel is then stained, viewed under UV (360 nm). DNA of the desired size is excised from the gel, destained, and stored at −20°C.

7. Remove agarose by agarase digestion with standard methods and purify the DNA by ethanol precipitation.

4.4. CONSTRUCTION OF METAGENOMIC LIBRARIES

Two distinct strategies are taken in metagenomics, according to the primary goal. First, large-insert libraries (cosmid, fosmid, or bacterial artificial chromosomes) are constructed for archiving and sequence homology screening purposes, to capture the largest amount of the available genetic resources available in the sample and archive it for further studies/interrogation. Second, small-insert expression libraries, especially those made in lambda phage vectors, are constructed for activity screening. The small size of the cloned fragments means that most genes present in the appropriate orientation will be under the influence of the extremely strong vector expression signals, and thus have a good chance of being expressed and detected by activity screens (32, 35). Although these two strategies may differ in some technical aspect, both are increasingly used together because of their complementarity. Activity mining often reveals novel enzymes, but the nature of the organism from which they originate can rarely be determined, nor can their genetic context, which may harbor equally or even more interesting similar or related enzymes. Primary enzyme discovery in an expression library, followed by identification of the same gene in a large-insert library and genome walking on the identified fragment, constitutes a powerful means of maximizing the discovery process and identifying the interesting new organisms that are producing such enzymes. Below we describe the methods used to construct small- and large-insert libraries.

4.4.1. Libraries in pCC1FOS Vectors

The CopyControl™ Fosmid Library Production kit (Epicentre, Madison, WI) uses a strategy based on the cloning of randomly physically sheared and end-repaired DNA with an average insert size of 40 kbp. Physical shearing of the DNA into approximately 35- to 40-kb-long fragments results in the generation of more random DNA fragments compared with those from partial endonuclease digestion of the genomic/metagenomic DNA. Genomic DNA is frequently sheared sufficiently, as a result of pipetting

during the purification process, so additional shearing is not necessary. Test the extent of shearing of the DNA by first running a small amount of it (approximately 100 ng). Run the sample on a 20-cm-long 1% agarose gel at 30 to 35 V overnight at 4°C and stain with ethidium bromide. If 10% or more of the genomic DNA migrates with the Fosmid control DNA provided with the kit (36 kb size), then one can proceed to the end-repair protocol. If the genomic DNA has a higher molecular mass than the 36-kb DNA fragment supplied with the kit, then the DNA needs to be sheared. The DNA (2.5 μg) is sheared by passing through a 200-μl small-bore pipette tip 50 to 100 times. The degree of shearing must be controlled through gel electrophoresis: if the (meta)genomic DNA migrates faster than the 36-kb fragment, then it has to be reisolated; the kit uses phage-assisted packaging and size selection of the cloned fragment, and in case the fragments are too short, it could lead to the hybrid insertions. For the end-repair protocol, take into account the following suggestions.

1. Thaw and thoroughly mix all of the following reagents before dispensing, and place on ice. Combine the following on ice: 8 μl of 10× End-Repair Buffer, 8 μl of 2.5 mM dNTP Mix, 8 μl of 10 mM ATP, 32 μl of sheared insert DNA (approximately 4.3 μg), 20 μl of sterile water, 4 μl of End-Repair Enzyme Mix, 80 μl of total reaction volume. (The end-repair reaction can be scaled up or down depending on the quantity of DNA available.)

2. Incubate at room temperature for 45 min.

3. Add gel-loading buffer and incubate at 70°C for 10 min to inactivate the End-Repair Enzyme Mix.

4. Select the size of the end-repaired DNA by low-melting-point (LMP) agarose gel electrophoresis. Run the sample on a 20-cm-long 1% agarose gel at 30 to 35 V overnight at 4°C. Do not stain the DNA with ethidium bromide, and do not expose it to UV. Use ethidium bromide-stained DNA marker lanes as a ruler to cut out the agarose region corresponding to 25- to 60-kb DNA and trim excess agarose, then proceed with the agarose gel-digesting assay using the "GELase (Epicentre) Agarose Gel-Digesting protocol" described in steps 5 to 11 below.

5. Thoroughly melt the gel slice by incubating at 70°C for 3 min for each 200 mg of gel.

6. Transfer the molten agarose immediately to 45°C and equilibrate 2 min for each 200 mg of gel.

7. Digest the agarose with 1 U of gelase for 30 min at 45°C.

8. Centrifuge the tubes in a microcentrifuge at maximum speed (15,000 × g) for 15 min at 4°C to pellet any insoluble oligosaccharides. Carefully remove the upper 90% to 95% of the supernatant, which contains the DNA, to a sterile 1.5-ml tube. Be careful to avoid the gelatinous pellet.

9. Precipitate the DNA by adding 1/10 volume of 3 M sodium acetate, pH 7.0, and 2.5 volumes of ethanol, and mix gently.

10. Wash the pellet with 70% ethanol. Gently resuspend the DNA pellet in TE buffer (approximately 200 μl).

11. Concentrate the DNA in a Microcon-100 (Millipore) concentrator membrane (100 kDa cutoff) at 4°C to a final volume of 20 to 30 μl. The DNA concentration can be approximately 100 ng/μl. This concentrated DNA will serve as the insert to ligate with the pCC1FOS vector.

The next step is the ligation of DNA fragment into the pCC1FOS fosmid vector. A single ligation reaction will produce 10^3 to 10^6 clones depending on the quality of the insert DNA. Based on this information, calculate the number of ligation reactions that you will need to perform. The ligation reaction can be scaled up as needed. A 10:1 molar ratio of pCC1FOS vector to insert DNA is optimal. If we use 0.5 μg of a 50-kb DNA insert, we need approximately 0.1 μg of vector. Combine the following reagents in the order listed and mix thoroughly after each addition: 1 μl of 10× Fast-Link Ligation Buffer, 1 μl of pCC1FOS (0.5 μg/μl), 1 μl of 10 mM ATP, up to 6.8 μl of concentrated insert DNA (75 ng/μl), 0.2 μl of MilliQ water, 1 μl of Fast-Link DNA ligase, and 10 μl of total reaction volume.

Incubate at room temperature for 2 h and then transfer the reaction to 70°C for 10 min to inactivate the Fast-Link DNA ligase; proceed with in vitro packaging as follows:

1. Thaw on ice one tube of the MaxPlax Lambda Packaging Extract for every ligation reaction performed in the step above.
2. When thawed, immediately transfer 25 μl (one-half) of each package extract to a second 1.5-ml microfuge tube and place on ice.
3. Add 10 μl of the ligation reaction to each 25 μl of the thawed extracts being held on ice.
4. Mix by pipetting the solutions several times. Avoid the introduction of air bubbles. Briefly centrifuge the tubes to get all liquid to the bottom.
5. Incubate the packaging reactions at 30°C for 90 min. After the 90-min packaging reaction is complete, add the remaining 25 μl of MaxPlax Lambda Packaging Extract to each tube.
6. Incubate the reactions for an additional 90 min at 30°C.
7. At the end of the second 90-min incubation, add Phage Dilution buffer (PD buffer: 10 mM Tris-HCl, pH 8.3, 100 mM NaCl, 10 mM MgCl₂) to 1 ml of final volume in each tube and mix gently. Add 25 μl of chloroform to each. Mix gently and store at 4°C (up to a month). A viscous precipitate may form after addition of the chloroform. This precipitate will not interfere with library production. Determine the titer of the phage particles (packaged fosmid clones) and then plate the fosmid library.

The day of the packaging reactions, inoculate 50 ml of LB broth + 10 mM MgSO₄ with 5 ml of the EPI300-T1^R overnight culture. Shake at 37°C to an optical density at 600 nm (OD$_{600nm}$) of 0.8 to 1.0. Store the cells at 4°C until needed for titer determination. The cells may be stored for up to 72 h at 4°C if necessary.

Before plating the library, we recommend that the titer of packaged fosmid clones be determined. This will aid in determining the number of plates and dilutions to make to obtain a library that meets the needs of the user (first, to determine the clone densities before picking the colonies).

1. Make serial dilutions of the 1 ml of packaged phage particles into PD buffer in sterile microcentrifuge tubes. For example, use dilutions 1:10^1, 1:10^2, 1:10^4, and 1:10^5.
2. Add 10 μl of each of these dilutions, individually, to 100 μl of the prepared EPI300-T1^R host cells. Incubate each for 20 min at 37°C.
3. Spread infected EPI300-T1^R cells on an LB plate plus 12.5 μg/ml chloramphenicol and incubate at 37°C overnight to select for the fosmid clones.

4. Count colonies and calculate the titer of the packaged clones as follows: if there were 200 colonies on the plate streaked with the 1:10^4 dilution, then the titer in CFU/ml of this reaction would be: (# of colonies) (dilution factor) (1,000 μl/ml) / (volume of phage plated [μl]).

That is, (200 CFU) (10^4) (1,000 μl/ml)/(10 μl) = 2×10^8 CFU/ml. Note that the surface area of a standard 90-mm petri dish corresponds to one-eighth of a standard 22.5 × 22.5-cm square plate that is typically used.

Based on the titer of the phage particles determined previously, dilute the phage particles with PD buffer to obtain the desired number of clones and clone density on the plate. Mix the diluted phage particles with EPI300-T1^R cells prepared in the ratio of 100 μl of cells (prepared as above) for every 10 μl of diluted phage particles. Spread the infected bacteria on an LB plate plus 12.5 μg/ml chloramphenicol and incubate at 37°C overnight to select for the fosmid clones. Subsequently these clones are plated with the help of a colony-picker robot, in 384-well plates (LB, 12.5 μg/ml chloramphenicol and 15% of glycerol). Plates are incubated overnight without shaking at 37°C. The colony-picker robot is again used to produce copies of the 384-well plates.

4.4.2. Libraries in the Cosmid pLAFR3

Of particular interest is the mining and further reconstitution of natural-product biosynthetic pathways where large multienzyme assemblies should be functionally expressed and where the choice of a suitable heterologous host is critical (108). In this case, the generation of broader host range vectors mobilized in different gram-negative species—such as pLAFR3 vector, which is able to replicate in *Pseudomonas* strains—has been proposed (97). Strains of *Pseudomonas* are known to be ubiquitous and metabolically versatile, with a great ability to metabolize toxic organic chemicals, such as aliphatic and aromatic hydrocarbons (28). This ability, along with their genetic plasticity, makes these bacteria very attractive for cloning of DNA libraries rich in biodegradation pathways. To this end, below we provide the protocol to generate metagenomic libraries in the pLAFR3 vector, which allows the cloning of ca. 23 kbp of insert DNA and their transfer into different gram-negative expression hosts. To prepare the pLAFR3 shoulders:

1. Inoculate 200 ml of LB, tetracycline 10 μg/ml with a single colony of *Escherichia coli* S17-3 (bearing pLAFR3 cosmid), and grow overnight with orbital shaking (250 rpm) at 30°C. Pellet cells for 10 min at 7,000 × g and isolate pLAFR3 cosmid according to the Large Construct Kit (Qiagen), treating the sample with ATP-dependent exonuclease to eliminate the chromosomal DNA of *E. coli*.
2. Take two aliquots of approximately 3 μg of pLAFR3 and cut one with HindIII (shoulder 1) and the other with EcoRI (shoulder 2) at 37°C overnight. Run small aliquots in a 0.75% agarose gel to check that the digestion worked properly, then incubate samples at 65°C for 20 min to inactivate restriction enzymes.
3. Add 3 μl of Shrimp Alkaline Phosphatase (SAP; e.g., from Biotec ASA) to dephosphorylate DNA and incubate for 1 h at 37°C. To prevent DNA shearing, avoid pipetting; just stir the tube content with a pipette tip to mix. Then incubate samples at 65°C for 20 min to inactivate SAP.
4. To remove the endonuclease buffers, mix the pLAFR3 shoulders from the above two reactions 1:1, add 400 μl of water, transfer into a Microcon-100 (Millipore) concentration device, and wash the sample via centrifugation. Concentrate to a final volume of approximately 30 to 40 μl.

5. To a volume of 37 µl of Microcon-concentrated DNA add 5 µl of 10× New England Biolabs Buffer 3 (NEB3), 5 µl of bovine serum albumin (BSA) 10×, 2 µl of MilliQ water, and 1 µl of BamHI enzyme, and digest overnight at 37°C.

6. Run small aliquots in a 0.75% agarose electrophoresis gel to ensure that the size of the fragments (22 kbp) has not been altered by digestion with BamHI.

7. Use the GeneClean Kit (BIO101) to inactivate BamHI and to concentrate the pLAFR3 shoulders.

8. Ligate the shoulders with insert DNA overnight at 4°C (see the preparation procedure of the insert DNA partially digested with Sau3AI below).

9. Perform the packaging with Gigapack XL (Stratagene, La Jolla, CA) as described by the supplier and select on LB, tetracycline 10 µg/ml, X-Gal 40 µg/ml.

To obtain DNA fragments of 25 to 50 kbp partially digested with Sau3AI it is recommended to do some trial reactions using different amounts of enzyme. Set up a series of reactions starting, for example, with 2 U of enzyme per 1 µg of DNA, 1 U/µg, 0.5 U/µg, 0.25 U/µg, and 0.125 U/µg. Incubate for 20 min at 37°C. Stop reactions by adding 65 mM EDTA (pH 8) from 0.5 M stock solution, and by heating the samples at 65°C for 15 min. Then run a 20-cm-long 1% agarose gel at 30 to 35 V overnight at 4°C and stain with ethidium bromide. Monitor the partial digestion conditions that optimally should result in a majority of the DNA migrating in the desired size range (25 to 50 kbp) in the agarose gel. To make the preparative (larger-scale) digestion, scale up the enzyme quantity, amount of DNA, and reaction volume, proportionally. Choose two different restriction conditions, as in the following example:

Reaction 1: 0.14 U/µl - 20 µl of concentrated insert DNA (12 µg), 5 µl of buffer NEB1 10×, 5 µl of BSA 10×, 2 µl of MilliQ water, 18 µl of Sau3AI 0.4 U/µl. Total reaction volume: 50 µl.

Reaction 2: 0.21 U/µl - 20 µl of concentrated insert DNA (12 µg), 7 µl of buffer NEB1 10×, 7 µl of BSA 10×, 2 µl of MilliQ water, 36 µl of Sau3AI, 0.4 U/µl. Total reaction volume: 70 µl.

Incubate 20 min at 37°C. Stop reactions by adding 1.5 µl of 0.5 M EDTA (pH 8) to each 10-µl reaction volume and heat the samples to 65°C for 15 min. Then mix both reactions and load samples on a 20-cm-long preparative (seal together the comb teeth with an adhesive band) 1% LMP agarose gel, run it at 30 to 35 V overnight at 4°C and cut and stain the slots with the DNA marker with ethidium bromide. Do not stain the part of the gel containing your DNA for cloning. Under UV light cut out the part of the gel blocks with the DNA markers in the range of ca. 20 kbp and above to use them as a marker to cut the gel part containing the DNA to be cloned. Cut out the desired gel region (25 to 40 kbp gel) and trim the excess of agarose. Then proceed to the agarose gel digestion following the GELase (Epicentre) protocol and concentrate the insert DNA on a Microcon-100 as described above.

Further, ligate overnight at 14°C partially digested DNA and pLAFR3 shoulders in a ratio 1:2 or 1:1. The ligation volume must be as low as possible, preferably up to 5 µl. If you take 100 ng of both shoulders together, then add 50 or 100 ng of the insert. Depending on the amount of DNA available, you may consider doing two separate ligation reactions (50 and 100 ng of DNA, respectively) to determine which reaction generates the best results. It is highly recommended to run small aliquots (e.g., 1 µl) of all your samples after any manipulation, and after ligation in an agarose gel. Then, package with Gigapack XL (Stratagene, La Jolla, CA) as described in the Packaging Protocol:

1. Remove the appropriate number of packaging extracts from a −80°C freezer and place the extracts on dry ice.

2. Quickly thaw the packaging extract by holding the tube between your fingers until the contents of the tube just begin to thaw.

3. Add the experimental DNA immediately (1 to 4 µl containing 0.1 to 1.0 µg of ligated DNA) to the packaging extract.

4. Stir the tube with a pipette tip to mix well. Gentle pipetting is allowable provided that air bubbles are avoided.

5. Spin the tube quickly (for 3 to 5 s), if desired, to ensure that all contents are at the bottom of the tube.

6. Incubate the tube at room temperature (22°C) for 2 h.

7. Add 500 µl of SM buffer (50 mM Tris-HCl, pH 7.5, 0.1 M NaCl, 8.5 mM $MgSO_4$, and 0.01% [wt/vol] gelatin) to the tube. The gelatin in SM buffer stabilizes lambda phage particles during storage.

8. Add 20 µl of chloroform and mix the contents of the tube gently.

9. Spin the tube briefly to sediment the debris.

10. The supernatant containing the phage is ready for titering. The supernatant may be stored at 4°C for up to 1 month.

The next step is the titration of the cosmid-packaging reaction to check the library size:

1. Streak the bacterial glycerol stock (*E. coli* DH5α or XL1Blue) onto the LB agar plates. Incubate the plates overnight at 37°C. Do not add antibiotic to the medium in the following step. The antibiotic will bind to the bacterial cell wall and will inhibit the ability of the phage to infect the cell.

2. Inoculate 50 ml of LB, supplemented with 10 mM $MgSO_4$ and 0.2% (wt/vol) maltose, with a single colony.

3. Grow at 37°C, shaking for 4 to 6 h (do not grow past an OD_{600} of 1.0). Alternatively, grow overnight at 30°C, shaking at 200 rpm.

4. Pellet the bacteria at 500 × g for 10 min.

5. Gently resuspend the cells in half the original volume with sterile 10 mM $MgSO_4$.

6. Dilute the cells to an OD_{600} of 0.5 with sterile 10 mM $MgSO_4$. The bacteria should be used immediately after dilution.

7. Prepare a 1:10 and a 1:50 dilution of the cosmid-packaging reaction in SM buffer.

8. Mix 25 µl of each dilution with 25 µl of the appropriate bacterial cells at an OD_{600} of 0.5 in a microcentrifuge tube and incubate the tube at room temperature for 30 min.

9. Add 200 µl of LB broth to each sample and incubate for 1 h at 37°C, shaking the tube gently once every 15 min. This incubation will allow time for expression of the antibiotic resistance.

10. Spin the microcentrifuge tube for 1 min and resuspend the pellet in 50 µl of fresh LB broth.

11. Using a sterile spreader, plate the cells on LB agar plus 10 µg/ml tetracycline and incubate at 37°C overnight to select for the fosmid clones. Incubate the plates overnight at 37°C.

12. Count colonies and calculate the titer of the packaged phage particles as described above.

Based on the titer of the phage particles, dilute the phage particles with SM buffer to obtain the desired number of clones and clone density on the plate. Mix the diluted phage particles with *E. coli* DH5α or XL1Blue cells prepared in the ratio of 100 μl of cells for every 10 μl of diluted phage particles. Spread the infected bacteria on LB agar, tetracycline 10 μg/ml, X-Gal 40 μg/ml plates and incubate at 37°C overnight to select for the plasmid clones. These clones are subsequently picked with a colony-picker robot and transferred into 384-well plates containing liquid LB, tetracycline 10 μg/ml, and 15% of glycerol. Plates are incubated overnight without shaking at 37°C. The colony-picker robot or the 384-pin replicators can be used to produce multiple copies of the primary library arrayed in 384-well plates.

4.4.3. Libraries in Lambda Zap® Express

Small insert expression libraries, especially those made in lambda phage vectors, are specially constructed for activity screens; however, in contrast with cosmid or fosmid vectors, the Zap Express pBK vector (Stratagene, La Jolla, CA) allows cloning of up to 15 kbp (usually, approximately 6.5- to 8.5-kbp-long DNA inserts). To obtain DNA fragments of approximately 8.5 to 9.5 kbp partially digested with Sau3AI it is recommended to conduct some trial reactions using different amounts of enzyme. Set up a series of reactions starting, for example, from 0.1 to 0.04 U of enzyme per 1 μg of DNA:

Reaction 1: 0.04 U (0.004 U/μl) - 0.683 μl of DNA (585.5 ng/μl), 1.0 μl of NEB buffer ×10, 1.0 μl of BSA ×10, 7.23 μl of H$_2$O, 0.1 μl of Sau3AI 0.4 U/μl.

Reaction 2: 0.06 U (0.006 U/μl) - 0.683 μl of DNA (585.5 ng/μl), 1.0 μl of NEB buffer ×10, 1.0 μl of BSA ×10, 7.17 μl of H$_2$O, 0.15 μl of Sau3AI 0.4 U/μl.

Reaction 3: 0.1 U (0.01 U/μl) - 0.683 μl of DNA (585.5 ng/μl), 1.0 μl of NEB buffer ×10, 1.0 μl of BSA ×10, 7.07 μl of H$_2$O, 0.25 μl of Sau3AI 0.4 U/μl.

Reaction 4: 0.5 U (0.05 U/μl) - 0.683 μl of DNA (585.5 ng/μl), 1.0 μl of NEB buffer ×10, 1.0 μl of BSA ×10, 6.07 μl of H$_2$O, 1.25 μl of Sau3AI 0.4 U/μl.

Reaction 5: 1 U (0.1 U/μl) - 0.683 μl of DNA (585.5 ng/μl), 1.0 μl of NEB buffer ×10, 1.0 μl of BSA ×10, 4.82 μl of H$_2$O, 2.5 μl of Sau3AI 0.4 U/μl.

Incubate for 20 min at 37°C. Stop reactions by adding 65 mM EDTA, pH 8, and by heating the samples at 65°C for 15 min. Then run a 20-cm-long 1% agarose gel at 30 to 35 V overnight at 4°C and stain with ethidium bromide. Use the partial digestion conditions that result in a majority of the DNA migrating in the desired size range (5 to 15 kbp). For the partial digestion of the DNA, you should scale up the Sau3AI enzyme amount for at least 2 μg of DNA. The two best restriction conditions to be selected and scaled up are as follows:

Reaction 1: 20 μl of DNA (11.7 μg), 5 μl of buffer NEB ×10, 5 μl of BSA ×10, 2 μl of H$_2$O MilliQ, 18 μl of Sau3A 0.4 U/μl (7.2 U; 0.144 U/μl of reaction; 0.61 U/μg of DNA)

Reaction 2: 20 μl of DNA (11.7 μg), 7 μl of buffer NEB ×10, 7 μl of BSA ×10, 0 μl of H$_2$O MilliQ, 36 μl of Sau3A 0.4 U/μl (14.4 U; 0.206 U/μl of reaction; 1.23 U/μl DNA).

Incubate for 20 min at 37°C. Stop reactions by adding 65 mM EDTA, pH 8 (1.5 μl of 0.5 M, pH 8.0 EDTA for each 10-μl reaction volume) and heat the samples to 65°C for 15 min. Then mix both reactions and load samples on a 20-cm-long preparative 1% LMP agarose gel, run it at 30 to 35 V overnight at 4°C, and cut out and stain marker lanes with ethidium bromide. On the latter mark the desired gel region (10- to 15-kbp gel region) and cut the corresponding region from the nonstained part of the gel, and trim the excess of the agarose. Do not expose the gel slice to UV light to minimize the formation of pyrimidine dimers. Then proceed with the agarose gel digestion following the GELase (Epicentre) protocol and concentrate DNA as above.

The next step is the ligation of pBK-CMV vector predigested with BamH1 (Stratagene, La Jolla, CA) with the insert DNA prepared above. Ligate overnight at 4°C Sau3AI-digested DNA and pBK-CMV, using the following ligation conditions: 1 μl of Zap Express Vector, 0.6 μl of T4 ligase buffer (×10), 4 μl of concentrated insert, 0.6 μl of T4 DNA ligase. The final volume should not exceed 5.0 to 5.5 μl.

Further, the ligation reaction is packaged with Gigapack XL (Stratagene, La Jolla, CA), as described above, and titrated as follows:

1. Streak the bacterial glycerol stock (*E. coli* XL1 MRF′) onto the LB agar plates and incubate overnight at 37°C. Do not add antibiotic to the medium in the next step. The antibiotic will bind to the bacterial cell wall and inhibit the ability of the phage to infect the cell.
2. Inoculate 50 ml of LB, supplemented with 10 mM MgSO$_4$ and 0.2% (wt/vol) maltose with a single colony.
3. Grow at 37°C with shaking for 4 to 6 h (do not grow past an OD$_{600}$ of 1.0). Alternatively, grow overnight at 30°C, shaking at 200 rpm.
4. Pellet the bacteria at 500 × g for 10 min.
5. Gently resuspend the cells in half the original volume with sterile 10 mM MgSO$_4$.
6. Dilute the cells to an OD$_{600}$ of 0.5 with sterile 10 mM MgSO$_4$. The bacteria should be used immediately after dilution.
7. Prepare dilutions from 1:1 to 1:10^5 of the packaging reaction in SM buffer.
8. Mix 1 μl of each dilution with 200 μl of the appropriate bacterial cells at an OD$_{600}$ of 0.5 in a microcentrifuge tube and incubate the tube at 37°C for 15 min, shaking the tube gently.
9. Add 500 μl of NZY soft agar to each sample plate on NZY agar plates. Incubate the plates overnight at 37°C.
10. Count phage particles and calculate the titer of the packaged phage particles as described above.

After the titration, used to calculate the library size, the library is further amplified. Amplification can be performed both in liquid medium or agar plates. For amplification in liquid culture use the following protocol:

1. Mix 2 ml of a fresh, overnight bacterial culture (OD$_{600}$, 0.95) with approximately 10^6 PFU of bacteriophage in a sterile culture tube.
2. Incubate for 15 min at 37°C to allow the bacteriophage particles to adsorb.
3. Add 8 ml of prewarmed LB medium (or NZY) and incubate for 6 to 12 h at 37°C with vigorous shaking until lysis occurs: the cell suspension becomes less dense.
4. After lysis has occurred, add 2 drops of chloroform and continue incubation for 15 min at 37°C.
5. Centrifuge at 4,000 × g for 10 min at 4°C.
6. Recover the supernatant, add 1 drop of chloroform, and store at 4°C. The titer of the stock should be approximately 10^{10} PFU/ml, and this usually remains unchanged as long as the stock is stored at 4°C.

For the amplification in solid agar, *E. coli* XL1 MRF′cells are prepared as described above in MgSO$_4$ 10 mM and with an OD$_{600}$ of 0.5. Then proceed as follows:

1. Two aliquots are prepared, each of them containing approximately 5×10^4 PFU and 600 μl of *E. coli* cells. Do not exceed 300 μl of phage suspension per 600 μl of cells.
2. Incubate for 15 min at 37°C with gentle shaking, after which 3 ml of NZY broth is added and further spread over NZY agar plate (22.5 × 20 cm) prewarmed at 37°C.
3. Incubate the plates at 37°C for about 8 to 10 h, after which 8 to 10 ml of SM buffer is added. Shake the plates at 50 rpm for 10 h at 4°C.
4. Decant into a Falcon (50 ml) tube. Add additional 2 ml of SM buffer to the agar, shake the plate briefly, and decant to the same Falcon tube.
5. Add 5% (vol/vol) chloroform and incubate 15 min at 4°C.
6. Centrifuge at 500 × *g* for 10 min at 4°C.
7. The supernatant is collected and stored: one small aliquot at 4°C for lab use and the remaining supernatant is stored at −70°C after the addition of 7% dimethyl sulfoxide (DMSO). The library is then ready to use.

4.5. HOST-VECTOR SYSTEMS FOR "COMMON" ENZYMES AND "EXTREMOZYME" EXPRESSION

The production of enzymes of industrial interest is usually accomplished through the overexpression of the corresponding genes in well-characterized bacterial or eukaryotic hosts such as *E. coli*, *Bacillus subtilis*, and *Saccharomyces cerevisiae*. Numerous sophisticated genetic tools for heterologous gene cloning and expression in these hosts are commercially available. However, the success of this approach for the production of extremozymes has been limited.

Expression in an enzymatically active form is commonly achieved only for simple monomeric or homo-oligomeric soluble extremozymes that do not require *trans*-acting factors that need to be provided by the host. Host-specific factors include transcription factors, tRNAs for rare codons, cofactors, posttranslational modification systems, proteolytic processing, chaperones, and components for multisubunit assembly. In addition, factors such as temperature may affect the folding of enzymes into active forms. For thermophilic bacteria and archaea it is currently estimated that a large proportion (more than 40%) of the proteins encoded within the genome cannot be expressed or would be expressed in enzymatically inactive forms in common mesophilic hosts (54).

For a significant number of extremozymes large-scale production in heterologous nonextremophile hosts will therefore probably not be achievable. Similarly, the use of these hosts for activity-based screening of genomic and metagenomic libraries from extremophilic organisms limits the success rate for assessing enzymatic diversity and isolating novel extremozymes. The development of shuttle vectors with wider host ranges, such as the *E. coli*-*Actinomycetes* shuttle vector (18) and a bacterial artificial chromosome shuttle vector system capable of replicating in *E. coli*, *Streptomyces lividans*, and *Pseudomonas putida* (69), should increase the probability of detecting genes encoding novel extremozymes. Expression in extremophilic hosts is however likely to prove the ultimate solution to these problems. The next section will therefore focus on the recent development and use of genetic tools for extremophilic organisms.

4.5.1. *Thermus thermophilus* Expression Systems

The thermophilic gram-negative bacterium *T. thermophilus* has become the model of choice for the development of genetic tools for the expression of thermozymes (reviewed in reference 14). The genome sequences of the two commonly used strains, *T. thermophilus* HB8 and HB27, are available (42, 70). Well-adapted laboratory strains grow aerobically at optimum temperatures ranging from 62 to 75°C with high growth rates and good cell yields on complex medium (TB medium, 8 g/liter trypticase, 4 g/liter yeast extract, and 3 g/liter NaCl, pH 7.5) without the addition of specific amino acids or vitamins (24). *Thermus* spp. are readily accessible to genetic manipulation due to the constitutive expression of a natural competence apparatus, with high transformation efficiencies typically achieved (91). *Thermus* vector systems have been developed for the expression of homologous or heterologous genes, genomic or metagenomic libraries, and the tailoring of the biocatalytic properties of thermozymes.

Numerous cloning and expression vectors have been described for *T. thermophilus* (23, 61, 63, 71, 84). However, in most cases, these systems achieved only moderate protein overexpression levels. Moreno and coworkers (74), however, report the construction of a shuttle expression vector, pMKE2, facilitating overexpression in *T. thermophilus* to levels greater than that obtained in *E. coli* expression systems. Plasmid pMKE2 (Biotools, Madrid, Spain; http://www.biotools.eu) contains several restriction sites for cloning, a selection gene encoding a thermostable resistance to kanamycin under the control of a bifunctional *E. coli-Thermus* promoter, and replicative origins for *E. coli* and *Thermus* (75). Inserted genes are placed under the control of an inducible modified promoter from the respiratory nitrate reductase operon (*Pnar*) of HB8. Plasmids are transformed into HB27::*nar*, a derivative of HB27 that contains the gene cluster for nitrate respiration. This system allows for inducible expression based on the presence of nitrate and absence of oxygen. Transformed HB27::*nar* cells are grown aerobically in TB medium supplemented with kanamycin to an OD$_{550}$ of 0.2 to 0.3 at 60 or 70°C under mild stirring. Transcription from the *narp* promoter is then initiated through the simultaneous addition of KNO$_3$ (40 mM) and induction of anoxia through incubation without shaking. Thermophilic enzymes expressed in this system include β-galactosidase (75), Mn-dependent catalase (46), DNA polymerase (76), and Omp85 (78).

A *T. thermophilus*–*E. coli* fosmid system for functional screening of genomic and metagenomic libraries from thermophiles has recently been described (3). The fosmid integration vector, pCT3FK, allows for the construction of large-insert libraries in *E. coli* and the transfer of these recombinant libraries to *T. thermophilus* for functional screening. pCT3FK (Fig. 1) is a derivative of pCC1FOS (Epicentre), with the addition of a gene encoding thermostable resistance to kanamycin and two fragments that flank the *pyrE* locus derived from the *T. thermophilus* HB27 genome. Fosmid libraries are generated by shearing the genomic DNA of interest to an average size of 40 kb, endrepairing and ligating to Eco721 linearized dephosphorylated pCT3FK. Fosmids are then prepared by use of lambda phage packaging extracts and propagated in *E. coli* EPI300 (Epicentre). Fosmid DNA is isolated using commercial DNA isolation kits and used to transform *T. thermophilus* HB27. The incorporation of the sequences flanking the

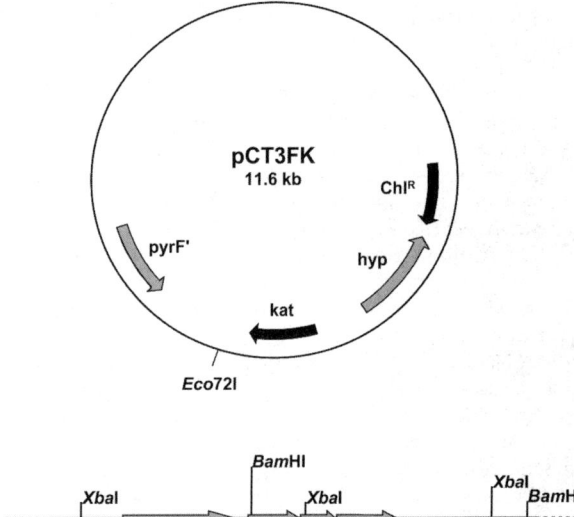

FIGURE 1 Vector map of the shuttle fosmid pCT3FK (Chl, chloramphenicol resistance; kat, thermostable kanamycin resistance) and the region of *T. thermophilus* HB27 genome containing the *pyr* genes. G3PDH, glycerol-3-phosphate dehydrogenase; hyp, hypothetical protein (3).

pyrE locus on either side of the Eco721 site in pCT3FK then guides the homologous integration of the insert into the *pyrE* locus of HB27, creating site-specific *Thermus* libraries. In a xylanase activity screen test of a *Spirochaeta thermophila* genomic library created in pCT3FK, 50% of the fosmids that conferred xylanase activity in *T. thermophilus* did not give rise to xylanase-positive *E. coli* clones (3).

Recently, new technologies for the generation and identification of thermostable mutants in *T. thermophilus*, named Massive Mutagenesis and THR, have been developed by Biométhodes (Evry, France; http://www.biomethodes.com). THR is based on an *E. coli-Thermus* spp. shuttle plasmid, pNCK, developed by Chautard and coworkers (15). This plasmid contains both *Thermus* and *E. coli* origins of replication, an ampicillin resistance gene, and a gene encoding thermostable resistance to kanamycin (*kat*^tv). Two unique restriction enzyme sites (NcoI and NotI) inserted between the *kat*^tv promoter and open reading frame facilitate the cloning and expression of the gene of interest as an N-terminal fusion construct with *kat*^tv. The recombinant pNCK plasmid is introduced into *E. coli* and mutant libraries (representing complete diversity introduced in a combinatorial manner at each amino acid position) generated using Massive Mutagenesis technology (88). Libraries are then transformed into *T. thermophilus* by electroporation and transformants grown at high temperature on TB plates supplemented with kanamycin. Because improper folding of the N-terminal portion of a protein has a negative impact on the folding of the C terminus (in this case, the reporter Kat^tv), the stability of the protein of interest will influence the antibiotic resistance of transformed cells. In the presence of kanamycin and high temperature, only *Thermus* transformants encoding thermostable proteins will have increased antibiotic resistance, thus allowing direct overnight plate selection (15). Biométhodes' Massive Mutagenesis and THR technology should allow for the rapid tailoring of the thermal stability and activity of enzymes.

4.5.2. *Geobacillus* spp. Expression System

The thermophilic bacilli of the genus *Geobacillus* have emerged as having considerable potential in biotechnological applications. Strains can be cultivated under aerobic or anaerobic conditions on a wide range of growth substrates, their metabolism has been researched extensively, and they are amenable to genetic manipulation. Full genomes for *Geobacillus kaustophilus* and *Geobacillus thermodenitrificans* are available, and the genome of *Geobacillus stearothermophilus* is currently being sequenced (*Bacillus* [*Geobacillus*] *stearothermophilus* Genome Sequencing Project: http://www.genome.ou.edu/bstearo.html).

An improved versatile shuttle vector, pUCG18, has been developed for heterologous gene expression in thermophilic *Geobacillus* spp. (101). This vector can replicate in *E. coli* and *Geobacillus* spp., encodes ampicillin and thermostable kanamycin resistance, has multiple cloning sites, and permits LacZ blue-white screening for the selection of recombinants in *E. coli*. The vector has been used for the expression of *Zymomonas palmae* pyruvate decarboxylase in *Geobacillus thermoglucosidasius*.

4.5.3. *Sulfolobus* spp. Expression Systems

Sulfolobus spp. belong to the class *Crenarchaeota* and are thermoacidophiles with an optimal growth temperature above 70°C (up to 82°C) and an optimum pH of 3. *Sulfolobus* strains are grown in the laboratory with relative ease. Strains can be grown to high cell density heterotrophically on organic substrates such as tryptone and a variety of sugars. During liquid cultivation in complex media, doubling times of 3 to 6 h are achieved during exponential growth. Strains can be grown on plates prepared with gellan gum (Gelzan or Gelrite; Sigma), a gelling agent that, in the presence of divalent cations, does not melt at 80°C. *Sulfolobus* can be transformed by electroporation (90). Complete genome sequences for the commonly used laboratory strains *Sulfolobus solfataricus* P2 and *Sulfolobus acidocaldarius* are available (16, 93).

A variety of virus- and plasmid-derived *Sulfolobus-E. coli* shuttle vectors have been developed with variable success and reproducibility (reviewed in reference 9). Of these the pA to pN shuttle vector series appears the most straightforward and promising in terms of vector size (approximately 9 kbp), unique cloning sites, lack of repetitive sequence, and stability in *E. coli*. In addition, the performance of this vector has been confirmed independently (10). The pA to pN vector series were constructed by the insertion of an *E. coli* replicon and the marker genes *pyrEF* into the *Sulfolobus* plasmid pRN1 at a number of different sites. These vectors therefore differ only in the orientation and location of these inserted genes. Transformation is achieved by electroporation of stable *S. acidocaldarius* or *S. solfataricus pyrE* mutants and selection on uracil-deficient media. Cloning of an *S. solfataricus* expression cassette encoding *lacS* and expression of this enzyme in an active form indicated that the system is suitable for gene expression.

4.5.4. *Pyrococcus abyssi* Expression System

Pyrococcus spp. are hyperthermophilic anaerobic euryarchaea that grow optimally at 100°C. Complete genome sequences for *P. abyssi*, *P. furiosis*, and *P. horikoshii* are available (17). *Pyroccocus* spp. can be grown to high cell densities and are relatively easy to handle. Transformation is achieved by use of a polyethylene glycol-mediated spheroplast method (66). An *E. coli-P. abyssi* shuttle vector, pYS2, was derived from the endogenous pGT5 plasmid of the *P. abyssi* strain

GE5 and includes the *pyrE* gene from *S. acidocaldarius* as a selectable auxotrophic marker. Transformation of a *pyrE* mutant *P. abysii* strain GE9 with pYS2 restored growth on uracil to wild-type levels. Under selective conditions pYS2 was shown to be stable and maintained at high copy numbers in both *E. coli* and *P. abysii* (66, 86).

4.5.5. *Thermococcus kodakaraensis* Expression System

Thermococcus kodakaraensis is a hyperthermophilic euryarchaeon that is naturally competent for DNA uptake. An *E. coli-T. kodakaraensis* shuttle vector pLC70 was recently constructed by Santangelo and coworkers (89). This vector can replicate stably and express genes in both *E. coli* and *T. kodakaraensis* (at 85°C). A selectable marker for Δ*trpE* complementation in *T. kodakaraensis* KW128 (Δ*pyrF* Δ*trpE::pyrF*) allows for the selection of transformants on plates in the absence of tryptophan. An additional mevinolin resistance marker allows for transformant selection in *T. kodakaraensis* without the Δ*trpE::pyrF* mutation. Recombinant expression of a subunit (RpoL) of the *T. kodakaraensis* RNA polymerase using this system revealed incorporation into functionally active RNA polymerase holoenzymes (89).

4.5.6. Haloarchaeal Expression Systems

Haloarchaea are adapted to high-salt environments (typically requiring molar salt concentrations for growth) and accumulate equivalent cytoplasmic concentrations of salt. For an excellent technical resource for haloarchaea, detailing methods used in growth media preparation, cultivation, transformation, and DNA extraction procedures, see *The Halohandbook* (29). Haloarchaea were the first archaea to be successfully transformed, allowing for the development of a number of genetic tools. Genetic manipulation systems include vectors, selection systems, or reporter genes used primarily for targeted gene knockout, exchange, or mutation, and the analysis of *cis*-acting factors involved in transcription and translation in archaea (e.g., see references 7, 44, and 49). Several *E. coli-Halobacterium* shuttle vectors have been constructed based on, for example, the endogenous plasmids pHV2 (62, 79, 85) and pHK2 (48, 49). Shuttle plasmids pMDS24, and derivatives pBAP5009 and pJAM202, all include the P2 promoter of the rRNA operon of *Halobacterium cutirubrum* for recombinant expression and ampicillin, mevinolin, or novobiocin resistance genes. These plasmids have been used successfully for the overexpression of a number of enzymes in *Halobacterium volcanii* DS70 (22, 55–57, 107).

4.5.7. Acidophiles and Alkalophiles

Native plasmids have been isolated from acidophiles (4, 94) and alkalophiles (34). To our knowledge no suitable expression or shuttle vectors have been generated, this is probably due to the large size and cryptic nature of most plasmids isolated. Increased interest in the biotechnological applications of acid- and alkaline-resistant proteins should promote the development of genetic tools for these hosts.

4.6. ENZYMES FROM EXTREMOPHILES

4.6.1. Thermophiles

Thermophiles are defined as organisms with an upper growth limit between 50°C and 70°C, that are able to grow, albeit slowly, at mesophilic temperatures (25 to 40°C). Hyperthermophiles, which include representatives from the domains *Bacteria* and *Archaea*, are defined as microorganisms that grow fastest between 80°C and 110°C (98). Many of the enzymes produced by thermophilic organisms are thermostable (20). These enzymes are of much interest to the biotechnology industry, not only for their stability at elevated temperatures, but also because of their resistance to other denaturants. It has been well documented that some mechanisms that impart resistance to denaturation by heat also confer resistance to other denaturants such as detergents and organic solvents (19). One advantage of working with thermophilic enzymes is that, in general, they can be purified by simple heat precipitation, even when they have been expressed in a mesophilic host (27).

When performing an assay with a thermostable enzyme, one should be aware of several factors that may affect the outcome of the experiment. First, although most enzymes are relatively thermostable at the optimal growth temperature for the producing strain, some enzymes may be more or less thermostable at the optimal growth temperature, and the best temperature at which to perform the assay should be determined empirically. For some thermostable enzymes optimal thermostability may be achieved at temperatures from 10 to 20°C above or below the optimal growth temperature (19). The thermostability of an enzyme should be determined before selecting at which temperature to perform an assay (38, 72). Second, some reactants or products may become unstable at elevated temperatures and, therefore, it is not always feasible to carry out the reaction at an enzyme's thermal optimum. The loss of catalytic efficiency caused by running a reaction at a lower temperature should be balanced out with the increased stability of all the reactants. However, some substrates, including polymeric substrates such as starch, are in fact more susceptible to catalysis at high temperatures (19, 45). Finally, because many of these enzymes need to be assayed at elevated temperatures, researchers should consider the effect that temperature has on the pH of a buffer. Table 1 lists the pK_a of some commonly used biological buffers at 20, 25, and 37°C. Table 2 shows the change in pH of Trizma buffer versus temperature.

4.6.2. Psychrophiles

While we live in a world dominated by "cold" environments, where the ambient temperature is permanently below 5°C, the amount of published work on cold-adapted proteins is limited. For several decades the most commonly cited definition of a psychrophile has been any organism that has a growth optimum below 15°C, with an upper limit for growth at ~20°C (77). However, this broad classification is highly ambiguous and could be extended to include many microorganisms, plants, and invertebrates living in colder habitats. Recently, researchers have started to define true psychrophiles as any organism that permanently "*thrives*" at temperatures close to the freezing point of water (31). Many psychrophilic microorganisms are completely adapted to living in extremely cold environments. This is evident because, when some psychrophiles are cultured in rich medium at 4°C, they have a doubling time similar to that of *E. coli* when cultured at 37°C; however, these same psychrophilic organisms will be unable to grow at elevated temperatures (31).

Two striking adaptive features of cold-adapted proteins are their weak stability and high activity at low temperatures. Because low temperatures inhibit the rate of catalytic reactions, one of the main challenges faced by psychrophilic organisms is the need to maintain the rate of enzyme-catalyzed reactions that are essential for their survival.

TABLE 1 Useful pH range of common biological buffers (25°C, 0.1 M)[a]

Buffers[b]	Useful pH range	pK$_a$ at 20°C	pK$_a$ at 25°C	pK$_a$ at 37°C
Bis-Tris	5.8–7.2	n/a	6.50	6.36
CABS	10.0–11.4	n/a	10.70	n/a
CAPS	9.7–11.1	10.56	10.40	10.02
HEPES	6.8–8.2	7.55	7.48	7.31
MES	5.5–6.7	6.16	6.10	5.97
MOPS	6.5–7.9	7.28	7.20	7.02
MOPSO	6.2–7.6	n/a	6.90	6.75
PIPES	6.1–7.5	6.80	6.76	6.66
TES	6.8–8.2	7.50	7.40	7.16

[a]See reference 21 for details.
[b]CABS, 4-[cyclohexylamino]-1-butanesulfonic acid; CAPS, N-cyclohexyl-3-aminopropanesulfonic acid; MES, morpholineethanesulfonic acid; MOPS, morpholinepropanesulfonic acid; MOPSO, 3-(N-morpholino)-2-hydroxypropanesulfonic acid; PIPES, piperazine-N,N′-bis(2-ethanesulfonic acid); TES, N-tris(hyrdoxymethyl)methyl-2-aminoethanesulfonic acid.

Therefore, the activity of many cold-active, heat-labile proteins is 10-fold greater at low temperatures compared with the activity of the homologous mesophilic proteins at mesophilic temperatures. The increased activity of cold-adapted proteins is achieved by destabilizing either the active site of the enzyme or the entire protein, rendering the catalytic site more flexible (31). Two general considerations when assaying psychrophilic enzymes is that, while these heat-labile proteins may have thermal optima that are significantly lower than those of the corresponding mesophilic proteins, these thermal optima may be significantly higher than the optimal growth temperature, and may even exceed the upper growth limits, for the producing strain (86). Second, some cold-adapted proteins are inactivated at a

TABLE 2 Trizma buffer: pH versus temperature[a]

pH at 5°C	pH at 25°C	pH at 37°C
7.76	7.20	6.91
7.89	7.30	7.02
7.97	7.40	7.12
8.07	7.50	7.22
8.18	7.60	7.30
8.26	7.70	7.40
8.37	7.80	7.52
8.48	7.90	7.62
8.58	8.00	7.71
8.68	8.10	7.80
8.78	8.20	7.91
8.88	8.30	8.01
8.98	8.40	8.10
9.09	8.50	8.22
9.18	8.60	8.31
9.28	8.70	8.42
9.36	8.80	8.51
9.47	8.90	8.62
9.56	9.00	8.70

[a]See reference 21 for details.

temperature below which the protein itself denatures (30). The thermodynamics of psychrophilic proteins is discussed in detail by Feller (30).

4.6.3. Halophiles

While many prokaryotes can survive over a wide range of salinities, true halophilic organisms require NaCl for growth. Halotolerant organisms can grow over a salt range of 0 to 1.0 M, moderate halophiles are capable of growth from 0.4 to 3.5 M salt, while an extreme halophile has a growth range from 2.0 to 5.2 M and typically displays optimal growth at >3.0 M (39). Because halophiles need to contend with high intra- and extracellular ionic concentrations, the enzymes they produce must retain functionality under conditions of low water activity.

The purification of halophilic proteins can be problematic because of their requirement for salt to maintain structural integrity. In general, halophilic proteins have acidic surfaces and the associated negative electrostatic forces may potentially provide a stabilizing force. These proteins may unfold at low salt concentrations due to the repulsive force exerted by the acidic residues. Many halophilic proteins also have a low thermal stability (95). Ion-exchange chromatography with an increasing salt gradient to elute the protein is a technique that is routinely used to purify mesophilic proteins. However, because a high salt concentration must be maintained to retain halophilic proteins in their native, active conformation, ion-exchange chromatography is not a suitable method to purify these proteins. Possible methods that can be used to purify halophilic proteins include gel filtration and hydroxylapatite chromatography using high-salt buffers, hydrophobic chromatography with a decreasing gradient of ammonium sulfate on carboxymethylcellulose or Sepharose 4B columns, and affinity chromatography (73). Electrophoresis of halophilic proteins may also be problematic because of the high electrical conductivity of the suspending solution required to maintain the protein in its native configuration (82). Structural studies of halophilic proteins may also prove to be problematic because of the difficulty in crystallizing proteins at high ion concentrations (95).

4.6.4. pH Extremes: Alkaliphiles and Acidophiles

Horikoshi (51) defined an alkaliphile as "an organism whose optimum rate of growth is observed at least two pH units above neutrality." Alkali-tolerant organisms have an optimum growth rate around neutrality but can grow or survive at pH ≥9. The current consensus is that a true alkaliphile grows optimally at pH ≥9, but is unable to grow or grows very slowly at pH 6.5 (51). Obligate alkaliphiles require alkaline conditions and sodium ions for growth, sporulation, and germination; in fact, many alkaliphiles display sodium ion-dependent nutrient uptake.

Acidophiles have a growth optimum between pH 0 and 5.5 and cannot grow at neutral pH. The cytoplasmic membrane of an obligate acidophile requires a high concentration of hydrogen ions to maintain membrane stability, and at neutral pH the membrane begins to dissolve and the cell lyses (68). The internal pH of acidophiles is typically between pH 5 and pH 7 (81). Serour and Antranikian (92) isolated heat- and acid-stable glucoamylases from three thermoacidophilic *Archaea* species. The enzymes displayed optimal activity at 90°C and pH 2.0. These authors encountered a problem in that, when they ran the concentrated supernatant on native polyacrylamide gels, they were unable to detect the proteins. They surmise that the acidophilic proteins failed to migrate into the gel under nondenaturing conditions because of the aggregation of the proteins at low pH (92). When performing an assay on acidophilic proteins the pH optima of the enzyme may not necessarily mimic that of the natural environment and one should consider the intracellular pH of the host strain, the extracellular pH, and the stability of a substrate at an acidic pH when selecting the assay conditions. The pH profile of an acidophilic protein may also shift by the addition of NaCl or CaCl$_2$. The intracellular glucoamylase produced by the extreme thermoacidophilic archaeon *Thermoplasma acidophilum* displayed maximal amylolytic activity at pH 5. With the addition of NaCl or CaCl$_2$ the pH optimum of the enzyme was pH 5.5 (27).

4.6.5. Archaeal Enzymes

Archaea are seen by many scientists as an ideal source of thermophilic and/or acidophilic enzymes. However, the large-scale production of archaeal proteins has several disadvantages. The biomass yield from archaea may be low (less than 10 g/liter wet weight versus 5 to 30 g for many common eubacteria) (19). While a possible solution to this obstacle would be to clone the gene of interest into a suitable host, some archaeal proteins can not be expressed in eukaryotic or bacterial expression hosts. At the protein level there are a number of explanations as to why an expression host may be unable to express archaeal proteins, including the absence of appropriate cellular chaperones, some strains that possess the correct translational machinery lack the protease Lon, or the formation of inclusion bodies. Possible ways to prevent or overcome the formation of inclusion bodies include moderate heat shock, decreased temperature during expression, and use of growth media that are enriched with amino acids (25, 45, 52).

4.7. ENZYMES FROM "COMMON" AND EXTREME ENVIRONMENTS

4.7.1. Lignocellulose-Degrading Extremozymes

It has become increasingly apparent that the production of fuel from traditional, nonrenewable sources is unable to match the global population's demand for fuel. Therefore, there is an incentive to discover alternative, readily available sources. The production of biofuels from lignocellulosic biomass is an attractive option, because the processing costs are low, large quantities of the raw materials are available, and production is environmentally benign (almost carbon neutral) (67). Most of the methods currently used to produce biofuels from lignocelluloses require enzymes that function at high temperature *and* under acidic conditions (30). The elevated temperatures are needed to solubilize the starch and maintain a low viscosity (27). Therefore, extremophiles are an obvious source of novel enzymes that may revolutionize the biofuels industry.

Much research has focused on thermoacidophilic species. When screening for enzymes from these organisms it is important to note that some small sugars (i.e., maltose) are not suitable substrates because the combination of low pH and high temperature may result in the occurrence of Maillard's reaction. (Maillard's reaction is a chemical reaction between the carbonyl group of reducing sugars and an amino group [92].)

4.7.1.1. Cellulose-Degrading Enzymes

Cellulases (EC 3.2.1) are enzymes that catalyze the hydrolysis of cellulose by cleaving the β-1,4-D-glycosidic linkage between glucose monomers. Examples of enzymes that degrade cellulose include β-1,4-endoglucanase, carboxymethyl cellulase, and endo-1,4-D-glucanase (25). See Table 3 for examples of lignocellulosic-degrading enzymes that have been isolated from extremophiles.

4.7.1.2. Amylases

Glucoamylases are exohydrolases that attack the substrate from the nonreducing end, producing glucose monomers with a β-anomeric configuration. Amylases have been characterized from various extremophiles. Heat- and acid-stable glucoamylases have been isolated from the thermoacidophilic archaeon species *T. acidophilum*, *Picrophilus torridus*, and *Picrophilus oshimae*. All three species grow optimally at 60°C and at a pH of ≤2, and the enzymes they produce have optimal activity at 90°C and at a pH of 2.0 (92). α-Amylases have also been isolated from psychrophilic strains, including the Antarctic bacterium *Pseudoalteromonas haloplanktis* (30). Horikoshi (50) was the first to describe an alkaline-stable α-amylase from an alkaliphilic *Bacillus* species. This enzyme was most active between pH 10 and 10.5, and hydrolyzed 70% of starch yielding glucose, maltose, and maltotriose (50). γ-Amylases (EC 3.2.1.3) cleave the α-1,4-glycosidic linkage at the nonreducing end of amylase and amylopectin, as well as cleaving α-1,6-glycosidic linkages. γ-Amylases are most effective in acidic environments, and many have optimal activity at pH 3.

4.7.1.3. Hemicellulose-Degrading Enzymes

Hemicellulose is one of the most abundant polysaccharides in nature, and unlike cellulose, which contains only glucose monomers, hemicellulose contains an array of different of sugar monomers, including galactose, glucose, mannose, and rhamnose, with xylose often being present in large quantities.

4.7.1.4. Ethanol Production from Xylose

Theoretically, thermophilic anaerobes can produce large quantities of ethanol from xylose, with yields of 0.51 g of ethanol per g of xylose being reported (96). Sommer et al. (96) investigated the potential of thermophilic anaerobic bacterial strains and anaerobic enrichment cultures (all with optimal growth occurring at 70 to 80°C) to produce ethanol from D-xylose and hemicellulose hydrolysate from wheat straw. The authors used pretreated wet, oxidized hemicelluloses (13) and detected the end products of fermentation

TABLE 3 Lignocellulosic-degrading enzymes produced by extremophiles: producing strains and assays used

Enzyme	EC no.	Assay substrate	Assay and reference	Producing organisms	Reference
Endocellulase	EC 3.2.1.4	Carboxymethyl cellulose	Carboxymethyl cellulase activity (100)	*Chaetomium thermophilum* (thermophilic fungi)	64
Endo-β-1,4-xylanase	EC 3.2.1.8	Xylan	0.4% Remazo Brilliant Blue-Xylan (98)	*Thermomyces lanuginosus* (thermophilic fungi)	98
		Xylan	Release of reducing sugar from OSX[a] (74)	*Thermomonospora fusca* (thermophilic actinomycete)	103
Glucoamylase	EC 3.2.1.3	Amylopectin (potato)	3,5-Dinitrosalicyclic DNS assay (11)	*Thermoplasma acidophilum* (thermoacidophilic archaeon)	27
Cellobiose dehydrogenase	EC 1.1.99.18	Cellobiose or lactose	Reduction of benzoquinone or dichlorophenol indophenols (43)		43
Endo-β-1,4-endoglucanase	EC 3.2.1.4	Carboxymethyl cellulose	Modified Somogi-Nelson method (47)	*Pyrococcus horikoshii* (hyperthermophilic archaeon)	59
Laccase	EC 1.10.3.2	ABTS	Oxidation of ABTS (80)	*Melanocarpus albomyces* (thermophilic fungi)	60
Peroxidase	EC 1.11.1.x	2,4-Dichlorophenol	(103)	*Thermomonospora fusca* (thermophilic actinomycete)	103
1,4-β-D-Glucan 4-glucanohydrolase	EC 3.2.1.4	Carboxymethyl cellulose	Release of reducing sugar from OSX[a] (74)	*Thermomonospora fusca* (thermophilic actinomycete)	103
β-Xylosidase	EC 3.2.1.37	*p*-Nitrophenyl-β-D-xylopyranoside	Release of *p*-nitrophenol (5)	*Thermomonospora fusca* (thermophilic actinomycete)	103
α-L-Arabinofuranosidase	EC 3.2.1.55	*p*-Nitrophenyl-α-L-arabinofuranoside	Hydrolysis of *p*-nitrophenyl-α-L-arabinofuranoside (37)	*Bacillus stearothermophilus* (thermophilic eubacteria)	37

[a]OSX, oat-spelt xylan.

including ethanol, lactate, and acetate by high-pressure liquid chromatography (96). Conclusions drawn from this study confirmed previously published results, in that while some strains may be able to produce ethanol from pure xylose in well-defined growth media, they are unable to ferment lignocellulosic hydrolysate effectively. This is possibly due to the presence of inhibitors such as weak organic acids (including acetate) (104) or phenolic monomers generated from the degradation of lignin or sugars present in the hydrolysate (83).

4.7.1.5. Xylanases

Xylanases (EC 3.2.1.8) are produced by a range of fungi, bacteria, and yeasts. Xylanases break down β-1,4-xylan into xylose and can be used to digest hemicelluloses. Xylanases can be produced from a number of different substrates (for a comprehensive list, see reference 102). Many fungal species produce xylanases when cultured in the presence of

1% (wt/vol) xylan, *Bacillus* species on 0.5 to 5.0% wheat bran, while some actinomycete species (including thermophilic strains) have detectable activity when grown in media containing 1% oat-spelt xylan.

4.7.1.6. Lignin-Degrading Enzymes

Several bacterial and fungal species produce ligninases, with the enzymes produced by brown- and white-rot fungi being studied the most extensively. Enzymes with lignolytic activity include the redox enzymes manganese peroxidase and the copper-containing phenol oxidase, laccase, as well as lignin peroxidase and cellobiose dehydrogenase (CDH) (60).

4.7.1.7. CDH

CDHs oxidize soluble cellodextrins, lactose, and mannodextrins into their corresponding lactones via a ping-pong mechanism. The most extensively studied CDHs were isolated from the white-rot fungi *Trametes versicolor* and

Phanerochaete chrysosporium. The assay most commonly used to detect CDHs is based on the reduction of benzoquinone or dichlorophenol indophenols (DCPIPs) (43). However, Henriksson et al. (43) identified two possible sources of error with this assay that should be considered. First, some sugar oxidases, such as glucose oxidase, might show activity with the DCPIP assay, thereby generating a false-positive reaction (43). Second, many white-rot fungi produce laccases that may reoxidize all the electron acceptors available, thereby masking the CDH activity (42). Baminger et al. (6) reported that the inclusion of 4 mM NaF, prior to performing a DCPIP assay, inactivated all laccases (6).

4.7.1.8. Laccases

Thermomonospora fusca BD25, a thermophilic actinomycete, produces several extracellular lignocellulose-degrading enzymes, including endoglucanase, endoxylanase, and peroxidase (90). Studies have found that the production of these enzymes depends on the growth phase. The synthesis of endoglucanases and endoxylanases occurs during the growth phase (0 to 96 h), while the production of peroxidases is nongrowth phase associated and is likely to be produced as a secondary metabolite. (Tuncer et al. [103] reported that peroxidase production peaked within 36 h, after which it rapidly decreased [1, 65]). Tuncer and coworkers found that the production of all three enzymes was highest on growth media containing oat-spelt xylan or ball-milled wheat straw, and that the addition of straw (0.6% or 0.8% wt/vol) induced the production of all three enzymes. The authors also found that the ratio of carbon/nitrogen affected enzyme production and that a C/N ratio of 4:1 to 5.3:1 resulted in maximal enzyme production. This article also highlighted the need to grow all test strains on a range of different carbon sources, because the amount of extracellular enzymes produced varies depending on the carbon source, and it is essential to find the best medium for enzyme production (90).

4.8. ACTIVITY SCREENS

The activity-based mining of enzymes performing a particular chemical conversion seems to be a better option, in contrast to the "(meta)genome gazing" approach, to look for new proteins that do not share any homology with those that are available in the public databases (8). Numerous assays enable the detection of enzymatic activities in colonies cultured on agar or crude cell lysates by the production of a fluorophore or chromophore (see examples in reference 2). Assays on agar-plated colonies typically enable the screening of $>10^4$ variants in a matter of days, but are often limited in sensitivity. The range of assays that are applicable for crude cell lysates is obviously much wider, but their throughput is rather restricted to only 10^3 to 10^4 clones (36). These low- to medium-throughput screens have proved effective for the isolation of enzymes from natural or preenriched metagenomic libraries, some of which has been successfully applied at industrial scale (for extensive review, see reference 2). A number of examples are described below.

4.8.1. Esterase Screens

Plates are incubated for 12 h at 37°C and then are covered with a second layer containing the substrate (20 ml of buffer HEPES 50 mM, pH 7.5, 0.4% [wt/vol] agarose, 320 μl of Fast Blue RR [Sigma] solution in dimethyl sulfoxide [80 mg/ml], and 320 μl

of α-naphthyl acetate solution in acetone 20 mg/ml). Positive clones appear due to the formation of a brown precipitate.

4.8.2. Cellulase-Like Screens

Plates are incubated for 12 h at 37°C in appropriate solid medium containing 0.5 to 1.0% (wt/vol) of the substrate (e.g., crystalline cellulose). The plates are subsequently stained for 20 min with 0.5% Congo red. Transparent halos in Congo red are formed around the positive colonies/clones.

4.8.3. P450 Oxidoreductase Screens

Plates are incubated for 12 h at 37°C and then are covered with a second layer containing the substrate (100 ml of Tris-HCl 50 mM, pH 7.5, 0.4% agarose, 50 μl of *p*-nitrophenoxy-carboxylic acids, e.g., 12-*p*NC$_{10}$ in DMSO [15 mM]). Positive clones appear visible due to the formation of a yellow color.

4.8.4. Laccase-Like Screens

Laccase production by library clones is screened by plating hybrid phage-infected cells or clones on appropriate soft agar containing 50 μM syringaldazine. Positive clones are identified by a purple halo, produced by the oxidation of syringaldazine, on agar plate.

4.8.5. Screens for Sugar-Fermenting Enzymes Able to Produce Alcohol

The presence of ethanol produced by an active clone is determined by adding a solution of 0.4% agarose made up in a sodium phosphate buffer (0.1 M, pH 8), containing 50 μl of 0.05 M 2,6-dichlorophenolindophenol and 100 μl of 0.15 M NAD solution. The solution must be well shaken, poured over the agar plate, and allowed to solidify. The plates are incubated at room temperature for 30 min, after which 5 to 20 ml of 0.005 M 5-methyl-phenazinium methyl sulfate is spread over each plate by flooding or spraying and allowed further incubation at 30°C for 30 min. A yellow color indicates the reaction zone on a blue background.

4.8.6. Alcohol Oxidoreductase Screens

The screening is performed on indicator plates that contain 1,2-ethanediol, 2,3-butanediol, or a mixture of 1,2-propanediol and glycerol (as test substrates, although other alcohols may be used) and a mixture of *p*-ararosaniline and bisulfite for the detection of carbonyl compounds formed by the *E. coli* clones. In brief, indicator plates are prepared by adding 8 ml of *p*-ararosaniline (2.5 mg/ml of 95% ethanol; nonautoclaved) and 100 mg of sodium bisulfite (nonsterile dry powder) to 400-ml batches of precooled (45°C) Luria agar lacking added carbohydrate. Most of the dye is immediately converted to the *leuko* form by reaction with the bisulfite to produce a rose-colored medium. Plates with colonies are then stored at room temperature, away from fumes that contain aldehydes (cigarette smoke, many plastic containers, etc.) and light, both of which promote increased background color. Upon production of carbonyls from the test substrates, an intensely red Schiff base is formed. Thus, colonies capable of carbonyl formation appear red on the indicator medium and are surrounded by a red zone, whereas colonies failing to produce carbonyl compounds remain uncolored.

4.8.7. Catechol Dioxygenase Screen

Agar plates, containing the medium that depends on the strain growth requirement (e.g., M9 supplemented with biphenyl crystals placed on the plate lid), are incubated for 12 h at 37°C and then are covered with a second layer of 100 ml of Tris-HCl 50 mM, pH 7.5 buffer containing the substrate (5 μl of catechol; 0.5 mM final concentration) and 0.4% (wt/vol) agarose, and the reaction plates are incubated at 25°C. Positive clones are normally identified after incubation for 1 h for 16 h, by the development of intense yellow color.

4.8.8. Polyol Oxidase-Like Screens

Screening of enzymes able to oxidize methyl-β-D-galactopyranoside (or other polyol such as 2-butanol, to cite some) is performed in a 0.4% (wt/vol) agarose solution prepared in sodium phosphate solution (50 mM, pH 7.0) that contains substrate (300 mM), catalase (700 U; Sigma), and $CuSO_4$ (0.5 mM). The screening is performed at room temperature and detection of a dark-blue color indicates a positive clone.

4.9. CONCLUSIONS

This chapter presented a number of methods that have been successfully applied in the screening for enzymes from extremophilic microorganisms and their communities. Nevertheless, the mining of enzymes from extremophiles and their communities will never be a trivial business and will require a number of parallel strategies, depending on the nature of the sample to be analyzed. There is certainly a need for a standardized set of methods for performing metagenomics projects, from physical-chemical description of sampling sites and sampling procedures down to the data interpretation and integration.

REFERENCES

1. Adhi, T. D., R. A. Korus, and D. L. Crawford. 1989. Production of major extracellular enzymes during lignocellulose degradation by two streptomycetes in agitated submerged culture. *Appl. Environ. Microbiol.* **55**:1165–1168.
2. Andexer, J., J. K. Guterl, M. Pohl, and T. Eggert. 2006. A high-throughput screening assay for hydroxynitrile lyase activity. *Chem. Commun. (Camb.)* **40**:4201–4203.
3. Angelov, A., M. Mientus, S. Liebl, and W. Liebl. 2009. A two-host fosmid system for functional screening of (meta)genomic libraries from extreme thermophiles. *Syst. Appl. Microbiol.* **32**:177–185.
4. Aparicio, T., P. Lorenzo, and J. Perera. 2000. pT3.2I, the smallest plasmid of *Thiobacillus* T3.2. *Plasmid* **44**:1–11.
5. Bachmann, S. L., and A. J. McCarthy. 1989. Purification and characterisation of a thermostable β-xylosidase from *Thermomonospora fusca*. *J. Gen. Microbiol.* **135**:293–299.
6. Baminger, U., B. Nidetzky, K. D. Kulbe, and D. Haltrich. 1999. A simple assay for measuring cellobiose dehydrogenase in the presence of laccase. *J. Microbiol. Methods* **35**:253–259.
7. Bauer, M., L. Marschaus, M. Reuff, V. Besche, S. Sartorius-Neef, and F. Pfeifer. 2008. Overlapping activator sequences determined for two oppositely oriented promoters in halophilic *Archaea. Nucleic Acids Res.* **36**:598–606.
8. Beloqui, A., P. D. de María, P. N. Golyshin, and M. Ferrer. 2008. Recent trends in industrial microbiology. *Curr. Opin. Microbiol.* **11**:204–248.
9. Berkner, S., D. Grogan, S. V. Albers, and G. Lipps. 2007. Small multicopy, non-integrative shuttle vectors based on the plasmid pRN1 for *Sulfolobus acidocaldarius* and *Sulfolobus solfataricus*, model organisms of the (cren-) archaea. *Nucleic Acids Res.* **35**:e88.
10. Berkner, S., and G. Lipps. 2008. Genetic tools for *Sulfolobus* spp.: vectors and first applications. *Arch. Microbiol.* **190**:217–230.
11. Bernfeld, P. 1955. Amylases alpha and beta. *Methods Enzymol.* **1**:149–158.
12. Biddle, J. F., S. Fitz-Gibbon, S. C. Schuster, J. E. Brenchley, and C.H. House. 2008. Metagenomic signatures of the Peru Margin subseafloor biosphere show a genetically distinct environment. *Proc. Natl. Acad. Sci. USA* **105**:10583–10588.
13. Bjerre, A. B., A. Bjerring Olesen, T. Fernqvist, A. Ploger, and A. Skammelsen Schmidt. 1996. Pretreatment of wheat straw using combined wet oxidation and alkaline hydrolysis resulting in convertible cellulose and hemicellulose. *Biotechnol. Bioeng.* **49**:568–577.
14. Cava, F., A. Hidalgo, and J. Berenguer. 2009. *Thermus thermophilus* as biological model. *Extremophiles* **13**:213–231.
15. Chautard, H., E. Blas-Galindo, T. Menguy, L. Grand'Moursel, F. Cava, J. Berenguer, and M. Delcourt. 2007. An activity-independent selection system of thermostable protein variants. *Nat. Methods* **4**:919–921.
16. Chen, L., K. Brugger, M. Skovgaard, P. Redder, Q. She, E. Torarinsson, B. Greve, M. Awayez, A. Zibat, H. P. Klenk, and R. A. Garrett. 2005. The genome of *Sulfolobus acidocaldarius*, a model organism of the *Crenarchaeota*. *J. Bacteriol.* **187**:4992–4999.
17. Cohen, G. N., V. Barbe, D. Flament, M. Galperin, R. Heilig, O. Lecompte, O. Poch, D. Prieur, J. Querellou, R. Ripp, J. C. Thierry, J. Van der Oost, J. Weissenbach, Y. Zivanovic, and P. Forterre. 2003. An integrated analysis of the genome of the hyperthermophilic archaeon *Pyrococcus abyssi*. *Mol. Microbiol.* **47**:1495–1512.
18. Courtois, S., C. M. Cappellano, M. Ball, F. X. Francou, P. Normand, G. Helynck, A. Martinez, S. J. Kolvek, J. Hopke, M. S. Osburne, P. R. August, R. Nalin, M. Guerineau, P. Jeannin, P. Simonet, and J. L. Pernodet. 2003. Recombinant environmental libraries provide access to microbial diversity for drug discovery from natural products. *Appl. Environ. Microbiol.* **69**:49–55.
19. Cowan, D. 1992. Enzymes from thermophilic archaebacteria: current and future applications in biotechnology, p. 149–169. In M. J. Danson, D. W. Hough, and G. G. Lunt (ed.), *The Archaebacteria: Biochemistry and Biotechnology*. Portland Press, London, United Kingdom.
20. Daniel, R. M., M. J. Danson, D. W. Hough, C. K. Lee, M. E. Peterson, and D. A. Cowan. 2008. Enzyme stability and activity at high temperatures, p. 1–34. In K. S. Siddiqui and T. Thomas (ed.), *Protein Adaptation in Extremophiles*. Nova Science Publishers Inc., New York, NY.
21. Dawson, R. M. C., D. C. Elliot, W. H. Elliot, and K. M. Jones. 1986. *Data for Biochemical Research*, 3rd ed. Oxford Science Publications, Oxford, United Kingdom.
22. De Castro, R. E., D. M. Ruiz, M. I. Gimenez, M. X. Silveyra, R. A. Paggi, and J. A. Maupin-Furlow. 2008. Gene cloning and heterologous synthesis of a haloalkaliphilic extracellular protease of *Natrialba magadii* (Nep). *Extremophiles* **12**:677–687.
23. de Grado, M., P. Castán, and J. Berenguer. 1999. A high-transformation efficiency cloning vector for *Thermus thermophilus*. *Plasmid* **42**:241–245.
24. de Grado, M., I. Lasa, and J. Berenguer. 1998. Characterization of a plasmid replicative origin from an extreme thermophile. *FEMS Microbiol. Lett.* **165**:51–57.
25. de Groot, N. S., and S. Ventura. 2006. Effect of temperature on protein quality in bacterial inclusion bodies. *FEBS Lett.* **580**:6471–6476.
26. DeLong, E.F. 2009. The microbial ocean from genomes to biomes. *Nature* **459**:200–206.
27. Dock, C., M. Hess, and G. Antranikian. 2008. A thermoactive glucoamylase with biotechnological relevance from the thermoacidophilic Euryarchaeon *Thermoplasma acidophilum*. *Appl. Microbiol. Biotechnol.* **78**:105–114.

28. **Dos Santos, V. A., S. Heim, E. R. Moore, M. Stratz, and K. N. Timmis.** 2004. Insights into the genomic basis of niche specificity of *Pseudomonas putida* KT2440. *Environ. Microbiol.* **6:**1264–1286.

29. **Dyall-Smith, M.** 2008. *The Halohandbook: Protocols for Haloarchaeal Genetics.* http://www.haloarchaea.com. [Online.] Accessed 30 April 2009.

30. **Feller, G.** 2008. Enzyme function at low temperatures in psychrophiles, p. 35–70. *In* K. S. Siddiqui and T. Thomas (ed.), *Protein Adaptation in Extremophiles.* Nova Science Publishers Inc., New York, NY.

31. **Feller, G., and C. Gerday.** 2003. Psychrophilic enzymes: hot topics in cold adaptation. *Nat. Rev. Microbiol.* **1:** 200–208.

32. **Ferrer, M., A. Beloqui, K. N. Timmis, and P. N. Golyshin.** 2008. Metagenomics for mining new genetic resources of microbial communities. *J. Mol. Microbiol. Biotechnol.* **16:**109–123.

33. **Ferrer, M., A. Beloqui, J. M. Vieites, M. E. Guazzaroni, I. Berger, and A. Aharoni.** 2009. Interplay of metagenomics and *in vitro* compartmentalization. *Microbial Biotechnol.* **2:**31–39.

34. **Fish, S. A., A. W. Duckworth, and W. D. Grant.** 1999. Novel plasmids from alkaliphilic halomonads. *Plasmid* **41:**268–273.

35. **Gabor, E. M., E. J. de Vries, and D. B. Janssen.** 2004. Construction, characterization, and use of small-insert gene banks of DNA isolated from soil and enrichment cultures for the recovery of novel amidases. *Environ. Microbiol.* **6:**948–958.

36. **Geddie, M. L., L. A. Rowe, O. B. Alexander, and I. Matsumura.** 2004. High throughput microplate screens for directed protein evolution. *Methods Enzymol.* **388:** 134–145.

37. **Gilead, S., and Y. Shoham.** 1995. Purification and characterization of a-L-arabinofuranosidase from *Bacillus stearothermophilus* T-6. *Appl. Environ. Microbiol.* **61:**170–174.

38. **Giver, L., A. Gershenson, P. O. Freskgard, and F. H. Arnold.** 1998. Directed evolution of a thermostable esterase. *Proc. Natl. Acad. Sci. USA* **95:**12809–12813.

39. **Grant, W. D., R. T. Gemmell, and T. J. McGenity.** 1998. Halophiles, p. 93–132. *In* K. Horikoshi and W. D. Grant (ed.), *Extremophiles—Microbial Life in Extreme Environments.* Wiley-Liss Inc., New York, NY.

40. **Handelsman, J.** 2004. Metagenomics: application of genomics to uncultured microorganisms. *Microbiol. Mol. Biol. Rev.* **68:**669–685.

41. **Handelsman, J.** 2008. Metagenomics is not enough. *DNA Cell Biol.* **27:**219–221.

42. **Henne, A., H. Bruggemann, C. Raasch, A. Wiezer, T. Hartsch, H. Liesegang, A. Johann, T. Lienard, O. Gohl, R. Martinez-Arias, C. Jacobi, V. Starkuviene, S. Schlenczeck, S. Dencker, R. Huber, H. P. Klenk, W. Kramer, R. Merkl, G. Gottschalk, and H. J. Fritz.** 2004. The genome sequence of the extreme thermophile *Thermus thermophilus. Nat. Biotechnol.* **22:**547–553.

43. **Henriksson, G., G. Johansson, and G. Pettersson.** 2000. A critical review of cellobiose dehydrogenases. *J. Biotechnol.* **78:**93–113.

44. **Hering, O., M. Brenneis, J. Beer, B. Suess, and J. Soppa.** 2009. A novel mechanism for translation initiation operates in haloarchaea. *Mol. Microbiol.* **71:**1451–1463.

45. **Hess, M.** 2008. Thermoacidophilic proteins for biofuel production. *Trends Microbiol.* **16:**414–419.

46. **Hidalgo, A., L. Betancor, R. Moreno, O. Zafra, F. Cava, R. Fernandez-Lafuente, J. M. Guisan, and J. Berenguer.** 2004. *Thermus thermophilus* as a cell factory for the production of a thermophilic Mn-dependent catalase which fails to be synthesized in an active form in *Escherichia coli. Appl. Environ. Microbiol.* **70:**3839–3844.

47. **Hiromi, K., Y. Takahashi, and S. Ono.** 1963. Kinetics of hydrolytic reaction catalyzed by crystalline bacterial α-amylase. The influence of temperature. *Bull. Chem. Soc. Jpn.* **36:**563–569.

48. **Holmes, M. L., S. D. Nuttall, and M. L. Dyall-Smith.** 1991. Construction and use of halobacterial shuttle vectors and further studies on *Haloferax* DNA gyrase. *J. Bacteriol.* **173:**3807–3813.

49. **Holmes, M., F. Pfeifer, and M. Dyall-Smith.** 1994. Improved shuttle vectors for *Haloferax volcanii* including a dual-resistance plasmid. *Gene* **146:**117–121.

50. **Horikoshi, K.** 1971. Production of alkaline enzymes by alkalophilic microorganisms. II. Alkaline amylase production by *Bacillus* No. A-40-2. *Agric. Biol. Chem.* **35:** 1783–1791.

51. **Horikoshi, K.** 1998. Alkaliphiles, p. 155–180. *In* K. Horikoshi and W. D. Grant (ed.), *Extremophiles—Microbial Life in Extreme Environments.* Wiley-Liss Inc., New York, NY.

52. **Hunke, S., and J. M. Betton.** 2003. Temperature effect on inclusion body formation and stress response in the periplasm of *Escherichia coli. Mol. Microbiol.* **50:**1579–1589.

53. **Ingham, C. J., A. Sprenkels, J. Bomer, D. Molenaar, A. van den Berg, J. E. van Hylckama Vlieg, and W. M. de Vos.** 2007. The micro-Petri dish, a million-well growth chip for the culture and high-throughput screening of microorganisms. *Proc. Natl. Acad. Sci. USA* **104:** 18217–18222.

54. **Jenney, F. E., Jr., and M. W. Adams.** 2008. The impact of extremophiles on structural genomics (and vice versa). *Extremophiles* **12:**39–50.

55. **Jolley, K. A., E. Rapaport, D. W. Hough, M. J. Danson, W. G. Woods, and M. L. Dyall-Smith.** 1996. Dihydrolipoamide dehydrogenase from the halophilic archaeon *Haloferax volcanii:* homologous overexpression of the cloned gene. *J. Bacteriol.* **178:**3044–3048.

56. **Jolley, K. A., R. J. Russell, D. W. Hough, and M. J. Danson.** 1997. Site-directed mutagenesis and halophilicity of dihydrolipoamide dehydrogenase from the halophilic archaeon, *Haloferax volcanii. Eur. J. Biochem.* **248:** 362–368.

57. **Kaczowka, S. J., C. J. Reuter, L. A. Talarico, and J. A. Maupin-Furlow.** 2005. Recombinant production of *Zymomonas mobilis* pyruvate decarboxylase in the haloarchaeon *Haloferax volcanii. Archaea* **1:**327–334.

58. **Kalyuzhnaya, M. G., A. Lapidus, N. Ivanova, A.C. Copeland, A. C. McHardy, E. Szeto, A. Salamov, I. V. Grigoriev, D. Suciu, S. R. Levine, V. M. Markowitz, I. Rigoutsos, S. G. Tringe, D. C. Bruce, P. M. Richardson, M. E Lidstrom, and L. Chistoserdova.** 2008. High-resolution metagenomics targets specific functional types in complex microbial communities. *Nat. Biotechnol.* **26:**1029–1034.

59. **Kashima, Y., K. Mori, H. Fukada, and K. Ishikawa.** 2005. Analysis of the function of a hyperthermophilic endoglucanase from *Pyrococcus horikoshii* that hydrolyzes crystalline cellulose. *Extremophiles* **9:**37–43.

60. **Kiiskinen, L.-L., L. Viikari, and K. Kruuk.** 2002. Purification and characterisation of a novel laccase from the ascomycete *Melanocarpus albomyces. Appl. Microbiol. Biotechnol.* **59:**198–204.

61. **Kobayashi, H., A. Kuwae, H. Maseda, A. Nakamura, and T. Hoshino.** 2005. Isolation of a low-molecular-weight, multicopy plasmid, pNHK101, from *Thermus* sp. TK10 and its use as an expression vector for *T. thermophilus* HB27. *Plasmid* **54:**70–79.

62. **Lam, W. L., and W. F. Doolittle.** 1992. Mevinolin-resistant mutations identify a promoter and the gene for a eukaryote-like 3-hydroxy-3-methylglutaryl-coenzyme A reductase in the archaebacterium *Haloferax volcanii. J. Biol. Chem.* **267:**5829–5834.

63. Lasa, I., M. de Grado, M. A. de Pedro, and J. Berenguer. 1992. Development of *Thermus-Escherichia* shuttle vectors and their use for expression of the *Clostridium thermocellum celA* gene in *Thermus thermophilus*. *J. Bacteriol.* **174:** 6424–6431.

64. Li, D.-C., M. Lu, Y.-L. Li, and J. Lu. 2003. Purification and characterization of an endocellulase from the thermophilic fungus *Chaetomium thermophilum* CT2. *Enzyme Microb. Technol.* **33:**932–937.

65. Lodha, S. J., A. R. Korus, and D. L. Crawford. 1991. Synthesis and properties of lignin peroxidases from *Streptomycetes viridosporus* T7A. *Appl. Biochem. Biotechnol.* **28:**411–420.

66. Lucas, S., L. Toffin, Y. Zivanovic, D. Charlier, H. Moussard, P. Forterre, D. Prieur, and G. Erauso. 2002. Construction of a shuttle vector for, and spheroplast transformation of, the hyperthermophilic archaeon *Pyrococcus abyssi*. *Appl. Environ. Microbiol.* **68:**5528–5536.

67. Lynd, L.R., W. H. van Zyl, J. E. McBride, and M. Laser. 2005. Consolidated bioprocessing of cellulosic biomass: an update. *Curr. Opin. Biotechnol.* **16:**577–583.

68. Madigan, M. T., J. M. Martinko, and J. Parker. 2000. *Brock Biology of Microorganisms*, 9th ed. Prentice Hall International Inc., London, United Kingdom.

69. Martinez, A., S. J. Kolvek, C. L. Yip, J. Hopke, K. A. Brown, I. A. MacNeil, and M. S. Osburne. 2004. Genetically modified bacterial strains and novel bacterial artificial chromosome shuttle vectors for constructing environmental libraries and detecting heterologous natural products in multiple expression hosts. *Appl. Environ. Microbiol.* **70:**2452–2463.

70. Masui, R., K. Kurokawa, N. Nakagawa, F. Tokunaga, Y. Koyama, T. Shibata, T. Oshima, S. Yokoyama, T. Yasunaga, and S. Kuramitsu. 2005. *Thermus thermophilus* HB8, complete genome. http://www.ncbi.nlm.nih.gov/nuccore/AP008226. (Online.) Accessed 30 April 2009.

71. Mather, M. W., and J. A. Fee. 1992. Development of plasmid cloning vectors for *Thermus thermophilus* HB8: expression of a heterologous, plasmid-borne kanamycin nucleotidyltransferase gene. *Appl. Environ. Microbiol.* **58:**421–425.

72. Matsuura, T., K. Miyai, S. Trakulnaleamsai, T. Yomo, Y. Shima, S.Miki, K. Yamamoto, and I. Urabe. 1999. Evolutionary molecular engineering by random elongation mutagenesis. *Nat. Biotechnol.* **17:**58–61.

73. Mevarech, M., F. Frolow, and L. M. Gloss. 2000. Halophilic enzymes: proteins with a grain of salt. *Biophys. Chem.* **86:**155–164.

74. Miller, G. 1959. Use of dinitrosalisylic acid reagent for determination of reducing sugar. *Anal. Chem.* **31:**426–428.

75. Moreno, R., A. Haro, A. Castellanos, and J. Berenguer. 2005. High-level overproduction of His-tagged Tth DNA polymerase in *Thermus thermophilus*. *Appl. Environ. Microbiol.* **71:**591–593.

76. Moreno, R., O. Zafra, F. Cava, and J. Berenguer. 2003. Development of a gene expression vector for *Thermus thermophilus* based on the promoter of the respiratory nitrate reductase. *Plasmid* **49:**2–8.

77. Morita, P. 1975. Psychrophilic bacteria. *Bacteriol. Rev.* **39:**144–167.

78. Nesper, J., A. Brosig, P. Ringler, G. J. Patel, S. A. Muller, J. H. Kleinschmidt, W. Boos, K. Diederichs, and W. Welte. 2008. Omp85(Tt) from *Thermus thermophilus* HB27: an ancestral type of the Omp85 protein family. *J. Bacteriol.* **190:**4568–4575.

79. Nieuwlandt, D. T., and C. J. Daniels. 1990. An expression vector for the archaebacterium *Haloferax volcanii*. *J. Bacteriol.* **172:**7104–7110.

80. Niku-Paavola, M.-L., E. Karhunen, P. Salola, and V. Raunio. 1988. Ligninolytic enzymes of the white-rot fungus *Phlebia radiata*. *Biochem. J.* **254:**877–884.

81. Norris, P. R., and D. B. Johnson. 1998. Acidophilic microorganisms, p. 133–154. *In* K. Horikoshi and W. D. Grant (ed.), *Extremophiles—Microbial Life in Extreme Environments*. Wiley-Liss Inc., New York, NY.

82. Oren, A. 2002. *Halophilic Microorganisms and Their Environments*. Springer, Berlin, Germany.

83. Parekh, S. R., S. Yu, and M. Wayman. 1989. Adaptation of *Candida shehatae* and *Pichia stipitis* to wood hydrolysates for increased ethanol production. *Appl. Microbiol. Biotechnol.* **25:**300–304.

84. Park, H. S., K. J. Kayser, J. H. Kwak, and J. J. Kilbane II. 2004. Heterologous gene expression in *Thermus thermophilus*: beta-galactosidase, dibenzothiophene monooxygenase, PNB carboxy esterase, 2-aminobiphenyl-2,3-diol dioxygenase, and chloramphenicol acetyl transferase. *J. Ind. Microbiol. Biotechnol.* **31:**189–197.

85. Pfeifer, F., S. Offner, K. Krüger, P. Ghahraman, and C. Englert. 1994. Transformation of halophilic archaea and investigation of gas-vesicle synthesis. *Syst. Appl. Microbiol.* **16:**569–577.

86. Prieur, D., G. Erauso, C. Geslin, S. Lucas, M. Gaillard, A. Bidault, A. C. Mattenet, K. Rouault, D. Flament, P. Forterre, and M. Le Romancer. 2004. Genetic elements of *Thermococcales*. *Biochem. Soc. Trans.* **32:**184–187.

87. Russell, N. J., and T. Hamamoto. 1998. Psychrophiles, p. 25–46. *In* K. Horikoshi and W. D. Grant (ed.), *Extremophiles—Microbial Life in Extreme Environments*. Wiley-Liss Inc., New York, NY.

88. Saboulard, D., V. Dugas, M. Jaber, J. Broutin, E. Souteyrand, J. Sylvestre, and M. Delcourt. 2005. High-throughput site-directed mutagenesis using oligonucleotides synthesized on DNA chips. *Biotechniques* **39:**363–368.

89. Santangelo, T. J., L. Cubonova, and J. N. Reeve. 2008. Shuttle vector expression in *Thermococcus kodakaraensis*: contributions of cis elements to protein synthesis in a hyperthermophilic archaeon. *Appl. Environ. Microbiol.* **74:**3099–3104.

90. Schleper, C., K. Kubo, and W. Zillig. 1992. The particle SSV1 from the extremely thermophilic archaeon *Sulfolobus* is a virus: demonstration of infectivity and of transfection with viral DNA. *Proc. Natl. Acad. Sci. USA* **89:**7645–7649.

91. Schwarzenlander, C., and B. Averhoff. 2006. Characterization of DNA transport in the thermophilic bacterium *Thermus thermophilus* HB27. *FEBS J.* **273:**4210–4218.

92. Serour, E., and G. Antranikian. 2002. Novel thermoactive glucoamylases from the thermoacidophilic Archaea *Thermoplasma acidophilum*, *Picrophilus torridus* and *Picrophilus oshimae*. *Antonie van Leeuwenhoek* **81:**73–83.

93. She, Q., R. K. Singh, F. Confalonieri, Y. Zivanovic, G. Allard, M. J. Awayez, C. C. Chan-Weiher, I. G. Clausen, B. A. Curtis, A. De Moors, G. Erauso, C. Fletcher, P. M. Gordon, I. Heikamp-de Jong, A. C. Jeffries, C. J. Kozera, N. Medina, X. Peng, H. P. Thi-Ngoc, P. Redder, M. E. Schenk, C. Theriault, N. Tolstrup, R. L. Charlebois, W. F. Doolittle, M. Duguet, T. Gaasterland, R. A. Garrett, M. A. Ragan, C. W. Sensen, and J. Van der Oost. 2001. The complete genome of the crenarchaeon *Sulfolobus solfataricus* P2. *Proc. Natl. Acad. Sci. USA* **98:**7835–7840.

94. Singh, S. K., and P. C. Banerjee. 2007. Nucleotide sequence analysis of cryptic plasmid pAM5 from *Acidiphilium multivorum*. *Plasmid* **58:**101–114.

95. Sivakumar, N., N. Lia, J. W. Tang, B. K. C. Patel, and K. Swaminathan. 2006. Crystal structure of AmyA lacks acidic surface and provide insights into protein stability at poly-extreme condition. *FEBS Lett.* **580:**2646–2652.

96. Sommer, P., T. Georgieva, and B. K. Ahring. 2004. Potential for using thermophilic anaerobic bacteria for bioethanol production from hemicellulose. *Biochem. Soc. Trans.* **32:**283–289.

97. **Staskawicz, B., D. Dahlbeck, N. Keen, and C. Napoli.** 1987. Molecular characterization of cloned avirulence genes from race 0 and race 1 of *Pseudomonas syringae* pv. *glycinea. J. Bacteriol.* **169:**5789–5794.

98. **Stephens, D. E., K. Rumbold, K. Permaul, B. A. Prior, and S. Singh.** 2007. Directed evolution of the thermostable xylanase from *Thermomyces lanuginosus. J. Biotechnol.* **127:**348–354.

99. **Stetter, K. O.** 1998. Hyperthermophiles: isolation, classification and properties, p. 1–24. *In* K. Horikoshi and W. D. Grant (ed.), *Extremophiles—Microbial Life in Extreme Environments.* Wiley-Liss Inc., New York, NY.

100. **Stewart, B. J., and J. M. Leatherwood.** 1976. Derepressed synthesis of cellulase by *Cellulomonas. J. Bacteriol.* **128:**609–615.

101. **Taylor, M. P., C. D. Esteban, and D. J. Leak.** 2008. Development of a versatile shuttle vector for gene expression in *Geobacillus* spp. *Plasmid* **60:**45–52.

102. **Techapun, C., N. Poosaran, M. Watanabe, and K. Sasaki.** 2003. Thermostable and alkaline-tolerant microbial cellulase-free xylanases produced from agricultural wastes and the properties required for use in pulp bleaching bioprocesses: a review. *Process Biochem.* **38:**1327–1340.

103. **Tuncer, M, A. S. Ball, A. Rob, and M. T. Wilson.** 1999. Optimization of extracellular lignocellulolytic enzyme production by a thermophilic actinomycete *Thermomonospora fusca* BD25. *Enzyme Microb. Technol.* **25:**38–47.

104. **Van Zyl, C., B. A. Prior, and J. C. du Preez.** 1991. Acetic acid inhibition of d-xylose fermentation by *Pichia stipitis. Enzyme Microb. Technol.* **13:**82–86.

105. **Venter, J. C., K. Remington, J. F. Heidelberg, A. L. Halpern, D. Rusch, J. A. Eisen, D. Wu, I. Paulsen, K. E. Nelson, W. Nelson, D. E. Fouts, S. Levy, A. H. Knap, M. W. Lomas, K. Nealson, O. White, J. Peterson, J. Hoffman, R. Parsons, H. Baden-Tillson, C. Pfannkoch, Y. H. Rogers, and H. O. Smith.** 2004. Environmental genome shotgun sequencing of the Sargasso Sea. *Science* **304:**66–74.

106. **Vieites, J. M., M. E. Guazzaroni, A. Beloqui, P. N. Golyshin, and M. Ferrer.** 2008. Metagenomics approaches in systems microbiology. *FEMS Microbiol. Rev.* **33:**236–255.

107. **Wendoloski, D., C. Ferrer, and M. L. Dyall-Smith.** 2001. A new simvastatin (mevinolin)-resistance marker from *Haloarcula hispanica* and a new *Haloferax volcanii* strain cured of plasmid pHV2. *Microbiology* **147:**959–964.

108. **Wenzel, S. C., and R. Muller.** 2005. Recent developments towards the heterologous expression of complex bacterial natural product biosynthetic pathways. *Curr. Opin. Biotechnol.* **16:**594–606.

109. **Yu, W.-H., S.-C. Su, and C.-Y. Lee.** 2008. A novel retrieval system for nearly complete microbial genomic fragments from soil samples. *J. Microbiol. Methods* **72:**197–205.

Cell-Based Screening Methods for Anti-Infective Compounds

STEFANO DONADIO AND MARGHERITA SOSIO

5

5.1. INTRODUCTION

This chapter describes selected cell-based assays, and the selection of the corresponding target(s), amenable for the discovery of novel antibacterial and antifungal agents from microbial products, screened as complex mixtures resulting from partial processing of fermentation broths. We also provide a framework to view target choice and assay design in the context of the microbial diversity available for screening, highlighting the importance of designing an overall strategy for an effective drug discovery program. A related chapter present in the previous edition of this manual (18) described some aspects of microbial product screening, which will not be covered here.

5.1.1. Microbial Products in Anti-Infective Screening

Aging, immunosuppression, and invasive surgical procedures are determining an increase in the number of people at risk of contracting a bacterial infection, while currently available antibiotics are becoming less effective because antibiotic resistance is spreading among microbial pathogens. Despite the increasing medical need, the pharmaceutical industry as a whole has manifested a diminished interest in the anti-infective field, with a limited number of antibiotics in clinical development and just three new chemical classes of antibacterial and antifungal agents approved for life-threatening infections during the past four decades (3).

Despite this grim scenario, the sequenced genomes of microbial pathogens provide a multitude of targets (28), while effective assays can be designed and implemented to discover new leads by high-throughput screening (HTS). Along with these new opportunities, there is also the increasing recognition that microbial products represent one of the best sources for screening programs aimed at identifying novel antibacterial and antifungal compounds (29). However, because screening of microbial products for antibiotics has been extremely intensive and successful, with several thousands of bioactive metabolites described, effective screening strategies must be devised to increase the probability of discovering novel compounds.

In the anti-infective field, a drug discovery program based on microbial products has the ultimate goal of discovering a new, patentable chemical entity possessing desired properties, such as antimicrobial spectrum, molecular weight, solubility, and preferred route of administration.

Obviously, it would be quite difficult and prohibitively expensive to design and implement assays that can measure all the properties sought in a drug candidate at the screening stage. Therefore, HTS programs aim at identifying *hits*, compounds, or mixtures thereof possessing one or more desired properties. Hits undergo further evaluation and characterization to possibly become *leads* (Fig. 1).

The effectiveness of an HTS program lies in devising a series of assays (the *screening algorithm*) that can rapidly identify samples possessing desired properties. When screening microbial products, it must also be realized that some classes of antibiotics have been extensively investigated, either because many variants are naturally produced by different microbial strains or because they have been intensively explored by chemical derivatization. There is thus a small probability of discovering improved variants of highly explored classes through HTS. As a consequence, screening programs based on microbial products should aim at discovering either a novel class or an improved variant of a poorly explored class.

5.1.2. Strategies for Screening Programs

Most HTS programs are target based, i.e., they involve evaluating an appropriate number of samples for their ability to modulate the activity of one or more targets through use of a series of assays. Samples passing the threshold criteria defined in the screening algorithm become *hits*. In microbial product screening, hits emerging from HTS must be evaluated in two parallel paths: for their biological properties, using further tests that complement those employed in HTS; and for their potential novelty, comparing the newly identified bioactive compounds with previously discovered molecules possessing similar chemical and biological properties, a process designated *novelty evaluation* (Fig. 1). The target and assay type determine which classes of known microbial metabolites can potentially be detected. Furthermore, since most classes of microbial metabolites are restricted to particular taxa, there is also a strong influence on this frequency from the microbial strains used as a source of chemical diversity. Therefore, the selectivity and sensitivity of the assays and the microbial diversity used in screening determine the number of known chemical classes and their expected frequency (see section 5.1.4). In any case, appropriate tools must be available to rapidly recognize the expected known microbial products. It is our experience that any HTS program will usually detect additional classes of known

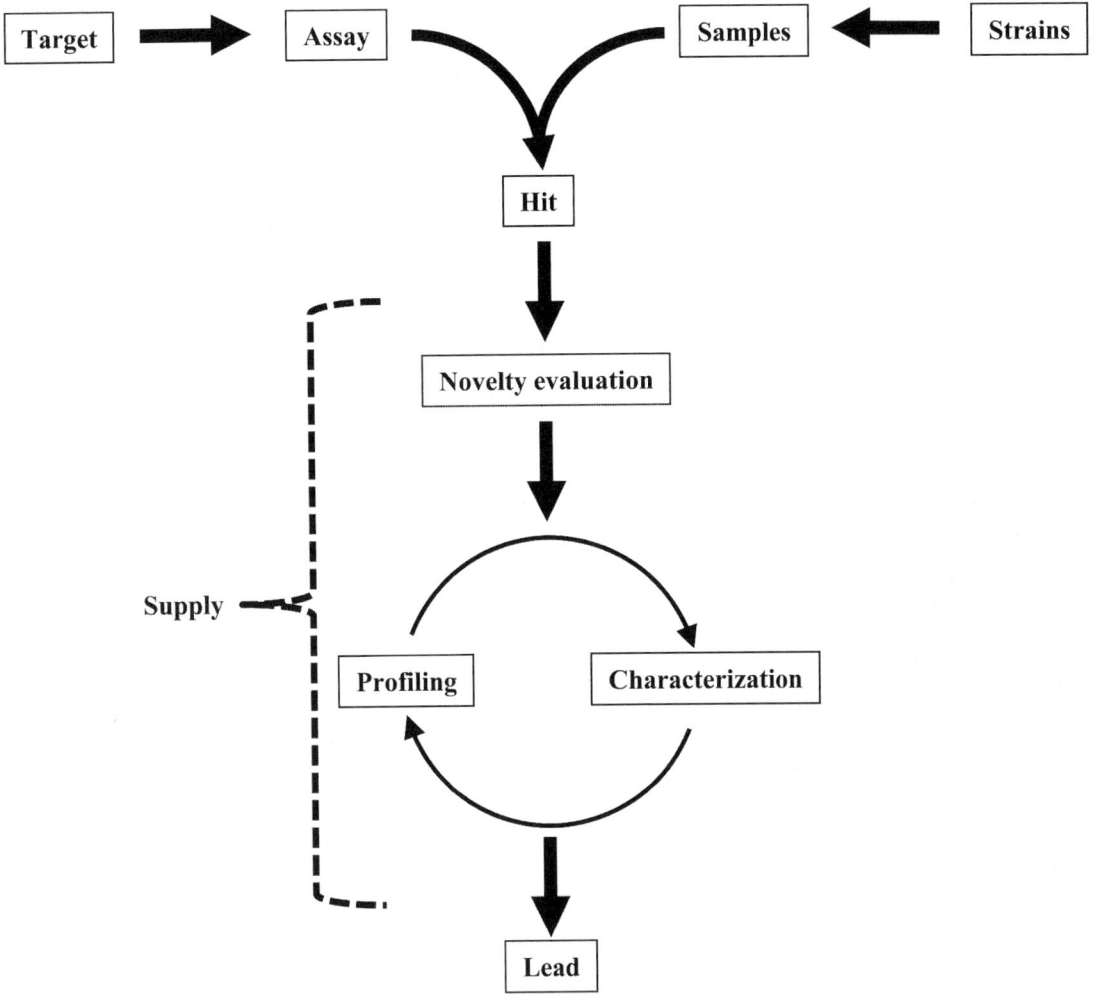

FIGURE 1 Scheme of a screening process.

metabolites, either because they represent *false positives* in the assay, or because their target was not yet known at the time of assay design.

5.1.3. Cell-Based and Cell-Free Assays

Cell-free assays directly measure the effect of a sample on the biological activity of one or more relevant targets. In general, these types of assays are more specific and sensitive than cell-based assays. Because most cell-free assays were unconceivable before the advent of recombinant DNA, they rarely were used during the golden age of antibiotic discovery and are thus more likely to discover new classes of inhibitors. However, a serious limitation of cell-free assays is that they also detect compounds that are unable to enter a microbial cell. While these types of inhibitors might be expected to offer little selective advantage to the producer strain, they are in fact relatively frequent among microbial metabolites. Furthermore, some of these inhibitors may be prohibitively hard to modify to allow them to enter a microbial cell while retaining activity and selectivity on the target. Cell-based assays, instead, do not suffer from this limitation, and progress in microbial genetics can ameliorate limitations in their sensitivity and specificity. The possible approaches and types of cell-based assays are discussed in section 5.2.

5.1.4. Novelty of the Program

A key factor for success in any HTS program based on microbial metabolites is to introduce elements of novelty. Although many details of previous screening campaigns are not known, a reasonable assumption is that easy-to-isolate strains (e.g., streptomycetes and other easily retrieved actinomycetes) were screened for antibacterial and antifungal agents by simple growth inhibition tests, prioritizing the many positives by potency, selective effect on one macromolecular synthesis, and/or lack of toxicity. Consequently, most discovered compounds are produced by relatively abundant species within the sampled genera, act on canonical targets (e.g., cell wall synthesis, transcription, and translation), and are rather potent. Thus, the novelty of a program can result from one or more of the following elements: (i) a relatively unexplored source of microbial diversity, which is expected to produce novel compounds at a higher frequency than an explored microbial group; (ii) an assay directed at canonical targets that is more sensitive than growth inhibition of commonly used test strains, allowing detection of compounds of lower potency or produced at lower levels; and (iii) an assay directed at "novel" targets, which is expected to detect compounds with a lower probability of having been pursued in the past. These elements of novelty must be appropriately combined and evaluated to increase the probability of identifying novel compounds at a reasonable cost.

5.2. TARGETS AND ASSAYS

5.2.1. Target Identification and Validation

In general, it is accepted that an antibacterial or antifungal target must respond to three criteria: (i) it must be a cellular component essential for cell viability or required for infection and disease; (ii) it must be present and well conserved in the desired range of pathogens, where it must play a similar essential function; and (iii) it must be specific to bacteria or fungi, or at least significantly different from similar components present in humans. The first criterion is generally dichotomous (i.e., a target is either essential or not essential), while the second and third criteria, by definition, are quantitative and require a definition of similarity thresholds of the same target among different pathogens or between a microbial and a human target. Indeed, because of the existence of common protein domains and folds, virtually any target will have some relatedness to one or more human proteins. A recent example highlights that a chemical library biased for inhibitors of human targets can afford selective inhibitors of a bacterial enzyme (26), through the use of an effective screening algorithm. In any case, when using targets with related human counterparts, it is essential to have comparable assays in place to evaluate selectivity.

Current clinically used antibiotics inhibit approximately 25 to 30 distinct cellular components, while the number of essential genes in a microbial cell is about 1 order of magnitude higher, thus offering ample choice of possible targets. However, while most antibacterial targets show limited relatedness to human proteins, fungal pathogens have a limited number of essential components sufficiently different from higher eukaryotic counterparts to represent promising drug targets. Therefore, in the antifungal field target choice is limited and appropriate assays for measuring selectivity must be always considered.

As mentioned above, target validation requires experimental demonstration of its essentiality in a given set of microbial cells. This can be achieved by the inability to perform a gene knockout, by conditional antisense-RNA expression, or by other experimental demonstrations (Fig. 2). Advances in genomic technologies have allowed the systematic investigation of essential genes in many strains, including the prokaryotes

Bacillus subtilis (23), *Escherichia coli* (14), *Haemophilus influenzae* (31), *Helicobacter pylori* (21), *Salmonella enterica* (22), and *Staphylococcus aureus* (20), and the eukaryotes *Aspergillus fumigatus* (17), *Saccharomyces cerevisiae* (15), and *Candida albicans* (33). The ample literature on the topic has resulted in the availability of catalogs of essential genes for many microbial strains. Comparative genomics (Fig. 2) can be used for predicting the presence of chosen genes in desired microbial pathogens and for establishing the level of relatedness to human proteins. A key factor in choosing "esoteric" targets is to ensure their essentiality in different pathogens. Examples of genes essential in one species but not in another are known (39). We refer the reader to a previously reported discussion on the opportunity and limitations of novel targets (9).

New anti-infective agents must act on drug-resistant pathogens. Drug resistance mechanisms are often class specific and, as mentioned, current drugs affect few targets. Furthermore, most targets present multiple sites for potential inhibition. Therefore, it is extremely unlikely that a new antibiotic class will exhibit cross-resistance, either because it acts on a target not affected by currently available drugs or because it acts at a different site. This may not be the case, however, when drug efflux is the prevalent mechanism operating in multiresistant pathogens.

5.2.2. Assay Development and Validation

An identified target must be transformed into an assay concept (how inhibitors will be detected), which is strictly connected to the assay format (Fig. 2). Although this is the logical flow, a favorite assay format, which has proven effective in HTS, can often be adapted to different targets. In HTS, assays are designed to test a large number of samples (on the order of 100,000) and must be performed robotically in small volumes, to reduce labor and reagent cost. Furthermore, they must be sufficiently sensitive to allow detecting bioactive compounds present also at low concentrations, and specific for the target to avoid detecting compounds acting on other targets. Finally, the assay must be robust, i.e., reproducible and relatively unaffected by the type of samples used in screening (see below). Assays can be set up and performed in multiple ways, but the implementation of a target into a screening program requires common steps as described below and shown in the flow chart in Fig. 2.

Cell-based assays detect, by definition, compounds able to exert their action on an intact microbial cell, but they present the challenge of being made specific, i.e., responsive to inhibition of the desired target(s) only, since signal is

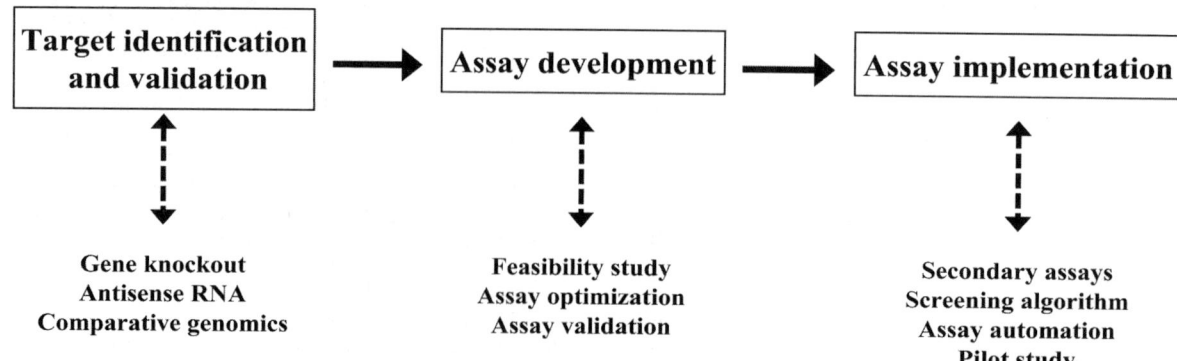

FIGURE 2 From target to screens. Main conceptual passages required (top) and the experimental steps required at each stage (bottom).

measured in the presence of all cellular targets and operating networks. A further challenge is to make cell-based assays more sensitive than simple growth inhibition assays. In the past decade, general formats have become available for cell-based assays that are amenable to multiple targets: reporter (11, 37) and antisense assays (6, 20) represent the two most notable examples. Other assay formats are also possible, and selected examples will be provided.

During assay selection, it is important to be aware of the likely positivity rate, which depends on the assay type and sensitivity, on the number of affected targets, and on the samples used. For example, an assay in which multiple targets can be inhibited (e.g., a B. subtilis reporter assay responding to any inhibitor of translation; 37) may have a positivity rate 2 orders of magnitude higher than an assay based on the inhibition of a single target within a gram-negative cell (see, e.g., reference 13). Therefore, the former assay may be used with a sample library obtained from Streptomyces strains (which at high frequency produce compounds able to inhibit B. subtilis growth) only in the presence of an effective screening algorithm to decrease the hit rate.

Assay development can be conceptually simplified as consisting of three phases: a feasibility study, assay optimization, and assay validation (Fig. 2). However, it is more an iterative than a linear process, because results obtained during a later phase may require reevaluating earlier parameters. The feasibility study includes a determination of whether the test is specific for the desired target. To do this, it is necessary to perform the test in the presence of positive (compounds acting on the target) and negative controls (compounds inhibiting the growth of the test strain but acting on different targets) to determine the threshold of sensitivity and the presence of possible false positives. For many novel targets, appropriate small-molecule inhibitors may not be available as positive controls, and one solution is to generate an inhibitor within the test strain itself by producing a peptide or nucleotide inhibitor after addition of an inducer (13). The assay should also be executable in miniaturized format and be sufficiently reproducible and simple enough for automated handling, generating a readout amenable to automated analysis. The ideal test is performed in a single well, requires only addition of reagents, and does not require separating the reaction product(s) before detection (a so-called mix-and-measure assay). Any general interference from the samples (i.e., residues from the fermentation broths or the solvent used to dissolve them) must be evaluated at this stage. Finally, the availability and stability of the assay reagents must be also taken into consideration.

The objective of the assay optimization phase is to ensure that the assay is more sensitive than growth inhibition of the same test strain. This usually involves varying the assay parameters (such as inoculum, incubation times, and temperatures) to establish the best signal-to-noise ratio, with use of positive and negative controls. Samples representative of the extract bank are also used, and a few samples are spiked with the positive control(s) to ensure lack of interference. Further negative controls are tested at this stage. Depending on the results obtained, secondary assays are devised at this stage, and the screening algorithm is implemented. In this respect, it is also important to consider that a primary assay, performed on the entire library, must be robust and inexpensive, while the cost of secondary assays, which are performed on a subset of samples only, can be higher.

After establishing the protocol, the assay is validated by executing it on a relevant number of samples (from a few hundred to a few thousand), which must be representative of the entire collection, both in terms of producing strains and extraction methods. This step is actually a first "road test" of the assay, evaluating parameters such as performance, variability, and positivity rate, which should be within expectations. In some cases, the assay is validated directly during a pilot study. However, we prefer to perform assay validation manually, before spending resources in assay automation. In some cases, an assay may actually turn out to have an unexpectedly high positivity rate and require going back to the drawing board.

Once the newly developed assay has been validated, it is adapted to the available HTS instrumentation, automated, adjusted for expected throughput and personnel working hours, and implemented into a pilot study. At this stage, the screening algorithm is also definitively established, including the numerical threshold for accepting or discarding a sample. Because in microbial product screening the identity of the bioactive compound(s) present in a hit is, in most cases, unknown, an effective screening algorithm is almost as critical as the primary assay for pursuing only the most promising hits. Therefore, secondary assays are directed at recognizing known antibiotic classes, at highlighting undesirable biological properties (e.g., selectivity for the microbial target), or, more rarely, at independently confirming the activity on the selected target. Secondary assays usually involve evaluating the potency (the lowest dilution still giving a signal above threshold) of the positives on the primary assay and on selected secondary assays, to establish selectivity.

It is important to keep in mind that the screening algorithm must be designed with consideration of the complexity of the samples to be used (e.g., a single microbial extract may contain two or more classes of bioactive metabolites), and that the primary and secondary assays must have comparable windows. For example, a screening algorithm consisting of an antisense assay to detect inhibitors of a single fungal target (primary test) and of cytotoxicity against a mammalian cell line (secondary test) assumes the existence of only one bioactive substance in the samples and may be appropriate when screening prefractionated microbial extracts.

5.2.3. General Types of Cell-Based Assays

As mentioned above, reporter and antisense assays can in principle be applied to multiple targets, but they require validation to ensure that the assay is truly responsive to inhibitors of the chosen target(s), with acceptable rates of false positives and false negatives. The choice of assay format is linked to the anticipated screening strategy. Antisense assays can in principle be applied to any target, but a separate assay is required for each target. Instead, a single reporter assay can detect inhibitors of multiple targets, thus increasing the probability of discovering hits. However, a suitable reporter assay may not be available in a desired cell for the chosen target(s). Furthermore, technical implementation and troubleshooting strongly depend on the assay format, so it may be preferable to apply a favorite assay format to many different targets rather than devising different formats for each target.

Important decisions to be made at the onset are whether to identify inhibitors of a single target or of multiple targets (a cellular pathway); and the microbial strain in which to perform the assay, including the possible use of hyperpermeable strains or strains resistant to particular classes of compounds. These decisions are part of the general strategy and are intimately connected with the samples to be screened, the size of the screening effort, and, often, the idiosyncrasies of the screening team.

One important issue relates to the expected positivity rate. In a given microbial strain, an assay directed at a cellular pathway consisting of multiple targets is expected to have a positivity rate higher than an equivalent assay involving a single target. Similarly, independently of the individual cellular response, the same assay is expected to have different sensitivity according to the intrinsic permeability of the microbial cells employed. As an example, permeability generally decreases along the following list of bacterial strains: *B. subtilis*, *Enterococcus faecium*, and *E. coli*. Permeability can be modulated by employing hypersensitive strains (e.g., *E. coli* strains devoid of the major porin TolC; 27) or strains constitutively expressing promiscuous efflux pumps (e.g., *Pseudomonas aeruginosa* MexAB-OprM, *E. coli* AcrAB, and *S. aureus* NorA; 40). Employing microbial strains with defined resistance mechanisms can also affect the positivity rate. Thus, no screening assay is ideal, and each choice results in opportunities and caveats. Our general advice is to use hypersensitive strains when employing assays based on just one or a few well-defined molecular targets, while relying on strains of wild-type permeability when using assays based on multiple targets such as "pathway assays." The use of strains expressing defined resistance markers is always an advantage, provided that there is no fitness cost associated with the resistance mechanism. In any case, the resistance markers employed should be relevant to the microbial diversity employed in screening. For example, the use of a vancomycin-resistant strain is unlikely to affect the positivity rate when screening samples derived from fungi or mycobacteria.

Finally, while it may be preferable to utilize a microbial strain representative of the pathogens of interest, difficulties in its genetic manipulation or safety issues while handling pathogens in HTS may tip the scale in favor of using a related microbial strain.

5.2.3.1. Reporter Assays

Reporter assays are conceptually based on the observation that microbial cells respond to perturbations in many cellular pathways by specific transcriptional responses (16). These perturbations can be brought about by sub-MIC concentrations of an antibiotic inhibiting a particular pathway. Hence, bacterial cells can be engineered so that the expression of a convenient reporter gene is under the control of a promoter specifically induced upon perturbation of a cellular pathway, as diagrammed in Fig. 3. Many such assays have been described, which respond to inhibition of one or more targets: DNA replication (e.g., reference 36), translation (e.g., reference 37), cell wall (e.g., reference 25), fatty acid formation (e.g., reference 11), or secretion (1). In general, the promoter(s) responding to the stimulus is determined empirically, and systematic evaluations of antibiotics within a single microbial species can identify potential promoters to drive expression of a desired reporter gene (19).

These assays can be effectively adapted to microwell format and usually provide advantages such as increased sensitivity and relatively short incubation times, since transcriptional responses are sensitive (sub-MIC) and rapid. Two important considerations must be made: one is that transcriptional responses can be species specific, and different microbial strains are likely to switch on different genes in response to the same physiological insult; the other is that the molecular mechanism through which a microbial cell senses an antibiotic-caused insult is often not known, and thus a reporter assay must be validated to understand the extent of possible false positives and false negatives (e.g., reference 7).

For example, a reporter assay for detecting antimicrobial compounds affecting DNA replication takes advantage of the ability of DNA-damaging compounds to induce SOS response genes. Thus, an *E. coli* strain bearing the highly SOS-inducible promoter of the cell division inhibitor *sulA* fused to β-galactosidase gene can be used (36). The lower limit of detection for nalidixic acid, used as positive control, is 1.25 μg/ml, or 0.25× MIC. This amount is sufficient to cause an about fourfold increase in β-galactosidase activity relative to untreated cultures, while maximum signal-to-noise ratio is obtained at 10 μg/ml. The test, originally validated using six different antibiotics acting on other macromolecular synthesis (36), responds only to SOS-inducing agents. The assay can be implemented in a Δ*tolC* background to increase sensitivity and is readily adapted to 96-well microtiter plates. Midexponential cultures of *E. coli sulA::lacZ* Δ*tolC* grown in LB medium at 30°C are added to the wells of a 96-well microplate, containing 10 μl of 80 different samples, leaving in each plate 8 wells with 10 μl of 2.5 μg/ml nalidixic acid and of 10% dimethyl sulfoxide as positive and negative controls, respectively. After 2 h at 30°C, the A_{600} is measured in a thermostated microplate reader. Next, 25 μl per well of B-PER II bacterial protein extraction reagent (Pierce, Rockford, IL) is added and the plate is agitated for 15 s before addition of 50 μl of ZOB buffer (36). The A_{420} is read immediately after ZOB buffer addition and after 30 min at room temperature. β-Galactosidase activity can be expressed as $\Delta A_{420}/A_{600}$. (If growth of individual wells is sufficiently reproducible, the A_{600} measure can be skipped.) Results are normalized to those of positive controls using the formula

$$I_i = 100 \times (A_i - A_c)/(A_p - A_c)$$

where I_i represents the relative β-galactosidase activity of sample i, with a ΔA_{420} of A_i, while A_c and A_p represent the average ΔA_{420} values of negative and positive controls, respectively.

5.2.3.2. Antisense Assays

Antisense assays are based on the principle that sensitivity of a microbial cell to an antibiotic is inversely related to the concentration of the target affected by that antibiotic. The level of a cellular target can be progressively diminished by increasing the expression levels of an antisense-RNA, which sequesters the target's mRNA, leading to decreased translation. When no mRNA is available for translation, the cell is no longer viable. Therefore, the trick with antisense assays is to find a level of antisense-RNA expression high enough to result in hypersensitivity to a specific inhibitor of the chosen target, but not sufficient to affect significantly the duplication time, which may render the cell hypersensitive to many different antibiotics. This is achieved by placing the antisense-RNA under the control of a promoter whose activity can be modulated by increasing levels of an exogenously added inducer.

The antisense approach has been used for the systematic identification of essential genes in microbial strains (20). In *S. aureus*, an inducible antisense-RNA system has been used to identify genes essential or critical for growth, leading also to a collection of "antisense strains," each suitable as a hypersensitive, single-target-based whole-cell assay (12). For example, expression from a xylose-inducible promoter of the RNA antisense to *fabF* conferred hypersensitivity to the FabF inhibitor cerulenin. In a similar experiment, antisense-RNA expression led to reduction of FabI levels with a corresponding hypersensitivity of the antisense strain to the FabI inhibitor triclosan.

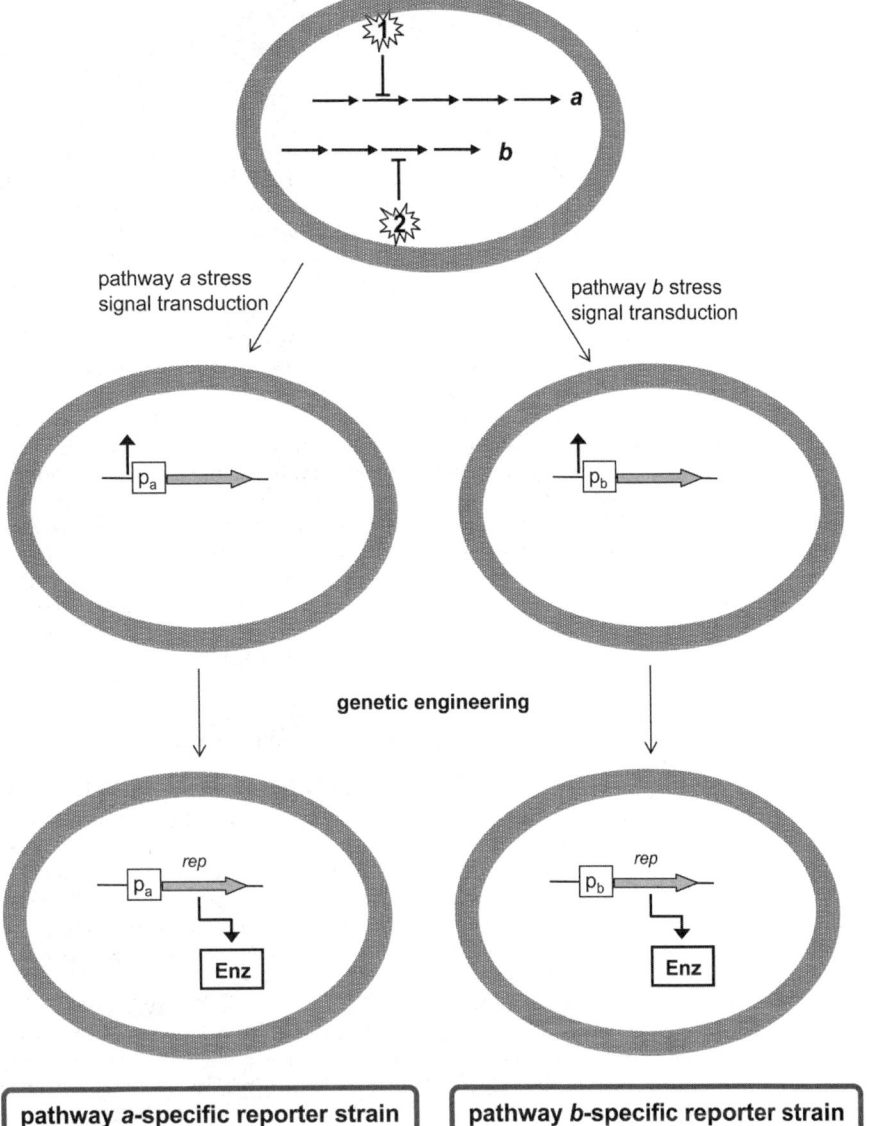

FIGURE 3 Concept of reporter assays. In a bacterial cell, the antibiotics 1 and 2 act on the different pathways a and b, respectively (top panel), leading to pathway perturbation due to specific stress-response signals and upregulation of genes under the same stimulon, as exemplified by the promoters p_a and p_b, which responds to perturbations in pathways a and b, respectively (middle panel). Two reporter assays can be designed by engineering cells so that expression of the reporter gene rep is under the control of the inducible promoter p_a or p_b, resulting in increased activity of the reporter enzyme Enz (bottom panel).

In contrast to reporter assays, antisense assays are intrinsically *differential* assays in which the response of the antisense test must be compared with that of an isogenic control. The standard approach to measure differential activity is to execute the assay in microtiter format, testing first all samples of the library for growth inhibition of the antisense strain. Then, the positive samples are retrieved, serially diluted, and simultaneously assayed on the antisense and control strains. This approach involves standard liquid-handling procedures, but identifies as positives all samples inhibiting the growth of the antisense strain, and only a small fraction of these will be preferentially active

on the antisense strain. If the positivity rate is relatively high (for example, microbial fermentation extracts from actinomycetes may contain detectable antistaphylococcal compounds at a frequency of approximately 10%), this approach would be extremely labor-intensive. An alternative approach is to perform the assay by agar diffusion, comparing the inhibition haloes observed on two agar plates, one seeded with antisense strain (the AS plate) and one with the control strain (42). The positive sample will show a larger halo on the AS plate compared with the control plate. An advantage of this approach is that it identifies directly the samples preferentially inhibiting the antisense strain, but it requires preparing agar plates and measuring inhibition haloes, procedures once the staple of microbial product screening, but somehow alien to highly automated HTS operations.

As an example we report the discovery of FabF/FabH inhibitors. Samples (20 μl) are applied to wells (prepared with well casters placed into the agar) on both the control and AS plates, which are then incubated at 37°C for 18 h. A clear, circular zone is formed in the presence of samples inhibiting cell growth, whose size depends on the concentration, potency, and diffusion rate of the inhibitor. Different concentrations of cerulenin and thiolactomycin (1, 0.5, 0.25, and 0.125 μg/ml), selective FabH/FabF inhibitors, are used as positive controls, while any other antibiotic can be used as negative control. At those concentrations, cerulenin and thiolactomycin show a larger inhibition halo only on the AS plate, while the negative controls show no differences in halo size between the two plates. Samples showing a difference in zone size between the AS plate and the control plate of ≥15 mm are retested as serial dilutions on the two-plate system, thus defining the ratio between the concentrations giving identical inhibition haloes on the AS and the control strains. This approach led to the identification of a novel class of antibiotics (38).

5.2.3.3. Other Assays

A conceptual variation of the antisense assay modulates the level of an essential target by placing the corresponding gene under the control of a promoter whose activity can be controlled by adjusting the concentration of a transcription inducer. When target expression is decreased, the cell is expected to become hypersensitive to inhibitors of that target, while increased expression levels should lead to reduced sensitivity. This principle was validated in *E. coli* with the following target-inhibitor pairs: fosfomycin and MurA, triclosan and FabI, and kirromycin and EF-Tu (8). The test is performed in 96-well microtiter plates, reading A_{600} after overnight incubation at 37°C. Positives (samples showing A_{600} values of <0.1 during the primary test) are then evaluated using the engineered strain grown under different concentrations of inducer. The potency of a target-specific hit is expected to decrease upon upregulation of the target. In this case, assay execution is conceptually similar to an antisense assay, as described above.

Another widely used assay type consists of the so-called *reversion assays*, in which the activity of a growth-inhibiting compound is reverted upon the extracellular addition of an excess of the inhibitor's target (*target reversion assays*) or of an essential nutrient whose synthesis is impaired by the inhibitor (*nutrient reversion assays*). Target reversion assays have been widely used to identify cell wall-active antibiotics by observing lack of activity in the presence of an excess of a bacterial cell wall preparation (24) or portions thereof (30). Those screens actually identified compounds binding to an intermediate in peptidoglycan formation rather than to enzyme inhibitors; thus, they employed nonprotein targets that, for example, are not amenable to antisense assays.

In a variation of this concept, a screen was implemented to detect inhibitors of the translation factor EF-Tu (35). Kirromycin, a known EF-Tu inhibitor, is used to validate the assay (positive control). The test is performed by using two agar plates, each inoculated with the test microorganism (i.e., 10^5 CFU/ml *S. aureus*). An appropriate number of 6-mm paper discs is then laid on the agar plates. On each disc, a 20-μl drop of 5 mg/ml *E. coli* EF-Tu is loaded on the reversion plate, while just 20 μl of H_2O is placed on the control plate. Samples are then spotted on the paper discs and plates incubated at 37°C overnight. Scoring of inhibition halos proceeds as described above. This screen led to the identification of two novel antibiotics representative of a novel class of EF-Tu inhibitors (35). It should be noted, however, that this assay is possible insofar as the chosen target is relatively stable under the incubation conditions.

Nutrient reversion assays can be applied when (i) the essential nutrient is not available in an animal host, and (ii) the chosen microbial cell can effectively uptake the exogenously added nutrient. An example of this type of assay is the screening for inhibitor of the biosynthetic pathway transforming chorismate into p-aminobenzoic acid (32). The assay is again a differential assay in which a *B. subtilis* strain is grown on two separate agar plates, one containing minimal medium and the other the same minimal medium supplemented with 5 mM L-tryptophan, 5 mM L-phenylalanine, 5 mM L-tyrosine, and 5 mM p-aminobenzoic acid. As above, 6-mm paper discs are laid on the agar plates and spotted with 10 μl of fermentation broth extracts. After overnight incubation at 37°C, inhibition haloes are scored, and the samples able to inhibit *B. subtilis* growth only in the unsupplemented medium are further evaluated. This screening led to the identification of a new class of antibiotics from a marine actinomycete (32).

5.3. SCREENING

5.3.1. Choice of Strains and Samples

As mentioned above, the choice of cell-based assays must also consider the chemical diversity available for screening. If a representative collection of samples is already available, target selection and assay format should be chosen on the basis of previous experience related to the frequency of known microbial metabolites and the performance of different assay formats.

When a sample library is not available, focus should be on the retrieval of strains offering a higher probability of producing new compounds. Historically, secondary metabolites with anti-infective activity have been identified from a variety of strains, with fungi and actinomycetes representing the most intensively screened groups. Strain novelty and diversity are key elements for identifying diverse molecules, but the ability to produce bioactive metabolites is not uniformly present in the microbial world. Furthermore, certain classes of metabolites are observed at higher frequencies than others, while ease of strain isolation and cultivation can greatly vary. Thus, a newly assembled collection should consist of strains that produce anti-infective compounds at an appreciable frequency, are underrepresented as producers of the frequent metabolites acting on the chosen target(s), and can be cultivated with existing methodologies and equipment.

The corresponding collection of samples can be generated by culturing the strains in appropriate media under suitable conditions, followed by sample processing, if required. While sample processing is essential when using cell-free assays (to remove interfering substances such as enzymes, salts, and nutrilites), it is not necessary with cell-based assays. Nonetheless, sample processing can be used to reduce the chemical complexity of the extracts, for example, by performing a prefractionation. By reducing the probability that a single sample contains more than one bioactive compound, sample prefractionation facilitates the design of effective screening algorithms and the step of hit deconvolution (see below), but adds a substantial cost to sample preparation. Additional considerations on strain selection and sample processing can be found in a recent review (10).

5.3.2. Devising the Screening Algorithm

The examples provided above often illustrate that a screening program rarely consists of a single assay, but it requires a logical sequence of tests with well-defined threshold criteria. With differential assays, one secondary assay is dictated by the assay concept. In other cases, the selection of a secondary assay derives from experience or from results observed during assay development. The screening algorithm may include additional tests aimed at recognizing known classes of metabolites. As a rule of thumb, the higher the number of targets covered by the assay, the larger the probability of detecting known compounds and thus the need for additional tests in the screening algorithm.

Two parameters must be taken into consideration when designing and implementing the screening algorithm: the threshold level for accepting and rejecting a sample, and the concentration ratio of a sample active in experimental versus control tests.

With differential assays that rely on growth inhibition, threshold criteria are easy to set (i.e., a clear inhibition halo or 90% inhibition when measuring turbidimetry). In these types of assays, where the anticipated result is a different minimal concentration of sample required to inhibit growth in the experimental versus the control system, a concentration ratio of ≥ 8 is usually selected. However, the potency of each sample is not known, and a 1:2 dilution may be sufficient to dilute the bioactive compound(s) to result in a concentration insufficient to elicit a signal above the set threshold. Consequently, the screening team must decide whether such samples should be considered hits, bearing in mind that true specificity can be established only when a more concentrated batch of the sample has become available (which may require a new fermentation of the producer strain), or abandon them, with the risk of losing possible hits. Although there is no straightforward answer to this dilemma, a useful rule of thumb is to consider the actual positivity rate of the screen: in the presence of a few hits, it may be worthwhile to include samples with a twofold ratio.

In other cases, the assay readout may not just be growth of the test strain, and the threshold criteria for positivity relies on a combination of theoretical expectations and experimental validation. For example, reporter assays ultimately measure the activity of a reporter enzyme, and different metabolites may induce its expression to different levels, sometimes higher than the positive control. In some cases, a control reporter assay may be employed to recognize unwanted events, and desired metabolites may elicit partial responses in the control assay too. With the primary and control reporter assay threshold, criteria can be empirically set at levels that allow detection of desired compounds at reasonable concentrations, but not of unwanted compounds.

Before performing an HTS campaign, a pilot study is run on a significant number of samples (a few thousand) to ensure that no unexpected results are encountered (Fig. 2). This is particularly relevant for microbial product screening, because different strains, growth media, and extraction procedures may result in heterogeneous samples and lower signal-to-noise ratios. The pilot study performs all the steps of a true HTS program: execution of primary assays; identification of positives; cherry-picking of an independent sample of the positives; repetition of the primary assay and execution of the secondary assay(s); data analysis; and, finally, hit declaration. Often, after the pilot study, the threshold criteria for the primary and secondary assays are adjusted, using appropriate statistical analysis (43). Furthermore, comparison with other screening programs may help to identify possible false positives and suggest strategies to recognize them. Often, the hits identified during the pilot study are processed (section 5.3.3) to identify the active molecules. This last step may strengthen the rationale for a full-scale HTS campaign or suggest additional secondary assays to recognize unwanted compounds.

5.3.3. Evaluating Hits

The ultimate goal of microbial product screening is to establish the chemical identity of the hits and their biological properties. Since this is a lengthy and expensive process, further filters are introduced to dedicate resources only to the most promising hits. These filters involve *dereplication* and *novelty evaluation* (Fig. 1).

Different hits discovered within a single screening program may actually contain the same bioactive principle. For example, they may represent different samples derived from the same strain (through the use of different production media or extraction procedures), or highly related strains may have been independently added to the library, or the same compound may be produced by unrelated strains. The process of choosing a single representative for each supposedly identical hit is designated *dereplication* in HTS jargon. Depending on the information available and the likely relatedness of the hits, dereplication may be performed querying the strain/sample database (in the case of hits derived from the same or similar strains), or it may require actual processing of the hits (see below). After dereplication, the sample showing the highest potency or specificity ratio is selected for further analysis. If the hit then shows interesting properties, it is advisable to reanalyze the strains/samples discarded during dereplication because they may produce lower amounts of different congeners; this will help to understand the structure-activity relationships of newly identified compound(s).

Novelty evaluation is a continuous process that involves comparing the properties of the bioactive compound(s) in each hit with those of natural products described in the literature, until its novelty is firmly established. The process at the beginning is based on negative evidence: lack of resemblance to known molecules is taken as an indication of possible novelty, and this comparison is repeated until convincing proof about the compound identity is established. Routinely, novelty evaluation involves a high-performance liquid chromatography separation of the sample, establishing which fractions contain the bioactivity observed with the primary assay. Simultaneously, data are acquired on molecular mass and UV spectrum of the active peaks. This information, together with additional chemical (chemical class, presence of particular functional groups, molecular formula, etc.) and biological (antimicrobial spectrum, target, producer strain) data, is used to query natural product databases. Detailed procedures for novelty evaluation have been described recently (10).

Depending on sample availability, complexity (weight per culture volume after extraction), and potency (i.e., highest dilution giving a measurable signal), novelty evaluation may be performed directly on the sample bank. On the other hand, when dealing with low-potency or high-complexity samples, it may be preferable to confirm the activity by reprocessing the producing strain before performing novelty evaluation. Since different batches are likely to yield different concentrations of the active compound and of interfering substances, each sample preparation must be validated with respect to the biological selection criteria through the screening algorithm, and consistency of the results must be thoroughly checked.

Eventually, the activity observed during screening must be reconfirmed by growing the producer strain and preparing a new sample under conditions identical to those that yielded the original sample. Along with these filters goes biological profiling and the supply of increasing amounts of samples for analysis (Fig. 1).

In antibacterial and antifungal screening programs, hits are characterized for their activity on different microbial strains. When limited information on a hit is available, a microbial panel may include just a few isogenic antibiotic-resistant and -sensitive strains to assist in novelty evaluation. As the interest in the hit increases, panels include relevant pathogens and cytotoxicity tests. These activities are common to any anti-infective discovery program, independently of the origin of the inhibitor, and will not be described here. However, for low-purity samples care must be taken when evaluating spectrum and cytotoxicity, as well as actual MIC data, since samples may contain more than one bioactive substance and the molecule of interest may be present at low purity.

HTS requires investments in automation, data acquisition, and data management. Relevant to microbial product is also the availability of databases linking the producer strains to the screened samples and to the identified compounds, as well as queryable databases of known microbial products for novelty evaluation. When pharmaceutical companies were actively involved in natural product screening, each had built its customized database, containing relevant chemical and biological properties on microbial products described in the literature. These databases are particularly valuable for old compounds discovered when analytical tools were less sophisticated than they are today.

5.4. CONCLUSIONS

In this chapter, we have provided an overview of the possible cell-based assays for discovering novel antibacterial and antifungal agents. Because of the almost unlimited number of possibilities, we have focused our attention on assay formats that can be applied to different targets, because technical challenges in assay development are similar for tests of the same type. When performing several HTS programs using different cell-based assays, the use of one or a few formats facilitates assay development, execution, and hit dereplication.

We have tried to provide an overview focusing on key elements of the process and combining strategic considerations with selected detailed procedures. There are significant challenges in the screening for novel anti-infective agents from microbial sources, but also opportunities offered by newly discovered strains, by a better understanding of the large genomic potential of the known strains to produce metabolites, and by the possibility of designing effective assays to identify novel compounds. The recent discoveries of novel chemical classes (4, 5, 32, 34, 38) bode well for the future.

A successful screening program depends on the integration of different skills and experiences, and on the humbleness of not pretending to reinvent the wheel. Before embarking on a screening program, we advise that valuable insights can be obtained from the detailed retrospective analyses by scientists who have witnessed several decades of microbial product screening (2, 24, 41).

REFERENCES

1. **Alksne, L. E., P.Burgio, W. Hu, B. Feld, M.P. Singh, M. Tuckman, P.J. Petersen, P. Labthavikul, M. McGlynn, L. Barbieri, L. McDonald, P. Bradford, R. G. Dushin, D. Rothstein, and S. J. Projan.** 2000. Identification and analysis of bacterial protein secretion inhibitors utilizing a SecA-LacZ reporter fusion system. *Antimicrob. Agents Chemother.* **43:**1418–1427.
2. **Baltz, R. H.** 2005. Antibiotic discovery from actinomycetes: will a renaissance follow the decline and fall? *SIM News* **55:**186–196.
3. **Boucher, H. W., G. H. Talbot, J. S. Bradley, J. E. Edwards, D. Gilbert, L. B. Rice, M. Scheld, B. Spellberg, and J. Bartlett.** 2009. Bad bugs, no drugs: no ESKAPE! An update from the Infectious Diseases Society of America. *Clin. Infect. Dis.* **48:**1–12.
4. **Brandi, L., A. Fabbretti, A. La Teana, M. Abbondi, D. Losi, S. Donadio, and C. O. Gualerzi.** 2006. Specific, efficient, and selective inhibition of prokaryotic translation initiation by a novel peptide antibiotic. *Proc. Natl. Acad. Sci. USA* **103:**39–44.
5. **Brötz-Oesterhelt, H., D. Beyer, H. P. Kroll, R. Endermann, C. Ladel, W. Schroeder, B. Hinzen, S. Raddatz, H. Paulsen, K. Henninger, J. E. Bandow, H. G. Sahl, and H. Labischinski.** 2005. Dysregulation of bacterial proteolytic machinery by a new class of antibiotics. *Nat. Med.* **11:**1082–1087.
6. **DeBacker, M. D., B. Nelissen, M. Logghe, J. Viaene, I. Loonen, S. Vandoninck, R. de Hoogt, S. Dewaele, F. A. Simons, P. Verhasselt, G. Vanhoof, R. Contreras, and W. H. Luyten.** 2001. An antisense-based functional genomics approach for identification of genes critical for growth of *Candida albicans*. *Nat. Biotechnol.* **19:**235–241.
7. **De Pascale, G., C. Grigoriadou, D. Losi, I. Ciciliato, M. Sosio, and S. Donadio.** 2007. Validation for high-throughput screening of a VanRS-based reporter gene assay for bacterial cell wall inhibitors. *J. Appl. Microbiol.* **103:**133–140.
8. **DeVito, J. A., J. A. Mills, V. G. Liu, A. Agarwal, C. F. Sizemore, Z. Yao, D. M. Stoughton, M. G. Cappiello, M. D. Barbosa, L. A. Foster, and D. L. Pompliano.** 2002. An array of target-specific screening strains for antibacterials discovery. *Nat. Biotechnol.* **20:**478–483.
9. **Donadio, S., L. Brandi, S. Serina, M. Sosio, and S. Stinchi.** 2005. Discovering novel antibacterial agents by high throughput screening. *Front. Drug Design Discov.* **1:**3–16.
10. **Donadio, S., P. Monciardini, and M. Sosio.** 2009. Approaches to discovering novel antibacterial and antifungal agents. *Methods Enzymol.* **458:**3–28.
11. **Fischer, H. P., N. A. Brunner, B. Wieland, J. Paquette, L. Macko, L. Ziegelbauer, and C. Freiberg.** 2004. Identification of antibiotic stress-inducible promoters: a systematic approach to novel pathway-specific reporter assays for antibacterial drug discovery. *Genome Res.* **14:**90–98.
12. **Forsyth, R. A., R. J. Haselbeck, K. L. Ohlsen, R. T. Yamamoto, H. Xu, J. D. Trawick, D. Wall, L. Wang, V. Brown-Driver, J. M. Froelich, G. C. Kedar, P. King, M. McCarthy, C. Malone, B. Misiner, D. Robbins, Z. Tan, Z.-Y. Zhu, G. Carr, D. A. Mosca, C. Zamudio, J. G. Foulkes, and J. W. Zyskind.** 2002. A genome-wide strategy for the identification of essential genes in *Staphylococcus aureus*. *Mol. Microbiol.* **43:**1387–1400.

13. Fossum, S., G. De Pascale, C. Weigel, W. Messer, S. Donadio, and K. Skarstad. 2008. A robust screen for novel antibiotics: specific knockout of the initiator of bacterial DNA replication. *FEMS Microbiol. Lett.* **281:**210–214.

14. Gerdes, S. Y., M. D. Scholle, J. W. Campbell, G. Balázsi, E. Ravasz, M. D. Daugherty, A. L. Somera, N. C. Kyrpides, I. Anderson, M. S. Gelfand, A. Bhattacharya, V. Kapatral, M. D'Souza, M. V. Baev, Y. Grechkin, F. Mseeh, M. Y. Fonstein, R. Overbeek, A. L. Barabási, Z. N. Oltvai, and A. L. Osterman. 2003. Experimental determination and system level analysis of essential genes in *Escherichia coli* MG1655. *J. Bacteriol.* **185:**5672–5684.

15. Giaever, G., A. M. Chu, L. Ni, C. Connelly, L. Riles, and S. Véronneau. 2002. Functional profiling of the *Saccharomyces cerevisiae* genome. *Nature* **418:**387–391.

16. Goh, E. B., G. Yim, W. Tsui, J. McClure, M. G. Surette, and J. Davies. 2002. Transcriptional modulation of bacterial gene expression by subinhibitory concentrations of antibiotics. *Proc. Natl. Acad. Sci. USA* **99:**17025–17030.

17. Hu, W., S. Sillaots, S. Lemieux, J. Davison, S. Kauffman, A. Breton, A. Linteau, C. Xin, J. Bowman, J. Becker, B. Jiang, and T. Roemer. 2007. Essential gene identification and drug target prioritization in *Aspergillus fumigatus.* *PLoS Pathog.* **3:**e24.

18. Huang, L., S. Stevens-Miles, and R. B. Lingham. 1999. Screening for bioactivities, p. 21–28. *In* A. L. Demain and J. E. Davies (ed.), *Manual of Industrial Microbiology and Biotechnology*, 2nd ed. ASM Press, Washington, DC.

19. Hutter, B., C. Fischer, A. Jacobi, C. Schaab, and H. Loferer. 2004. Panel of *Bacillus subtilis* reporter strains indicative of various modes of action. *Antimicrob. Agents Chemother.* **48:**2588–2594.

20. Ji, Y., B. Zhang, S. F. Van Horn, P. Warren, G. Woodnutt, M. K. Burnham, and M. Rosenberg. 2001. Identification of critical staphylococcal genes using conditional phenotypes generated by antisense RNA. *Science* **293:**2266–2269.

21. Kavermann, H., B. P. Burns, K. Angermüller, S. Odenbreit, W. Fischer, K. Melchers, and R. Haas. 2003. Identification and characterization of *Helicobacter pylori* genes essential for gastric colonization. *J. Exp. Med.* **197:**813–822.

22. Knuth, K., H. Niesalla, C. J. Hueck, and T. M. Fuchs. 2004. Large-scale identification of essential Salmonella genes by trapping lethal insertions. *Mol. Microbiol.* **51:**1729–1744.

23. Kobayashi, K., S. D. Ehrlich, A. Albertini, G. Amati, K. K. Andersen, M. Arnaud, K. Asai, S. Ashikaga, S. Aymerich, P. Bessieres, F. Boland, S. C. Brignell, S. Bron, K. Bunai, J. Chapuis, L. C. Christiansen, A. Danchin, M. Débarbouille, E. Dervyn, E. Deuerling, K. Devine, S. K. Devine, O. Dreesen, J. Errington, S. Fillinger, S. J. Foster, Y. Fujita, A. Galizzi, R. Gardan, C. Eschevins, T. Fukushima, K. Haga, C. R. Harwood, M. Hecker, D. Hosoya, M. F. Hullo, H. Kakeshita, D. Karamata, Y. Kasahara, F. Kawamura, K. Koga, P. Koski, R. Kuwana, D. Imamura, M. Ishimaru, S. Ishikawa, I. Ishio, D. Le Coq, A. Masson, C. Mauël, R. Meima, R. P. Mellado, A. Moir, S. Moriya, E. Nagakawa, H. Nanamiya, S. Nakai, P. Nygaard, M. Ogura, T. Ohanan, M. O'Reilly, M. O'Rourke, Z. Pragai, H. M. Pooley, G. Rapoport, J. P. Rawlins, L. A. Rivas, C. Rivolta, A. Sadaie, Y. Sadaie, M. Sarvas, T. Sato, H. H. Saxild, E. Scanlan, W. Schumann, J. F. Seegers, J. Sekiguchi, A. Sekowska, S. J. Séror, M. Simon, P. Stragier, R. Studer, H. Takamatsu, T. Tanaka, M. Takeuchi, H. B. Thomaides, V. Vagner, J. M. van Dijl, K. Watabe, A. Wipat, H. Yamamoto, M. Yamamoto, Y. Yamamoto, K. Yamane, K. Yata, K. Yoshida, H. Yoshikawa, U. Zuber, and N. Ogasawara. 2003. Essential *Bacillus subtilis* genes. *Proc. Natl. Acad. Sci. USA* **100:**4678–4683.

24. Lancini, G. 2006. Forty years of antibiotic discovery at Lepetit: a personal journey. *SIM News* **56:**192–212.

25. Mascher, T., S. L. Zimmer, T. A. Smith, and J. D. Helmann. 2004. Antibiotic-inducible promoter regulated by the cell envelope stress-sensing two-component system LiaRS of *Bacillus subtilis*. *Antimicrob. Agents Chemother.* **48:**2888–2896.

26. Miller, J. R., S. Dunham, I. Mochalkin, C. Banotai, M. Bowman, S. Buist, B. Dunkle, D. Hanna, H. J. Harwood, M. D. Huband, A. Karnovsky, M. Kuhn, C. Limberakis, J. Y. Liu, S. Mehrens, W. T. Mueller, L. Narasimhan, A. Ogden, J. Ohren, J. V. Prasad, J. A. Shelly, L. Skerlos, M. Sulavik, V. H. Thomas, S. Vanderroest, L. Wang, Z. Wang, A. Whitton, T. Zhu, and C. K. Stover. 2009 A class of selective antibacterials derived from a protein kinase inhibitor pharmacophore. *Proc. Natl. Acad. Sci. USA* **106:**1737–1742.

27. Nikaido, H. 2003. Molecular basis of bacterial outer membrane permeability revisited. *Microbiol. Mol. Biol. Rev.* **67:**593–656.

28. Payne, D. J., M. N. Gwynn, D. J. Holmes, and D. L. Pompliano. 2007. Drugs for bad bugs: confronting the challenges of antibacterial discovery. *Nat. Rev. Drug Discov.* **6:**29–40.

29. Peláez, F. 2006. The historical delivery of antibiotics from microbial natural products—can history repeat? *Biochem. Pharmacol.* **71:**981–990.

30. Rake, J. B., R. Gerber, R. J. Mehta, D. J. Newman, Y. K. Oh, C. Phelen, M. C. Shearer, R. D. Sitrin, and L. J. Nisbet. 1986. Glycopeptide antibiotics: a mechanism-based screen employing a bacterial cell wall receptor mimetic. *J. Antibiot.* **39:**58–67.

31. Reich, K. A., L. Chovan, and P. Hessler. 1999. Genome scanning in *Haemophilus influenzae* for identification of essential genes. *J. Bacteriol.* **181:**4961–4968.

32. Riedlinger, J., A. Reicke, H. Zähner, B. Krismer, A. T. Bull, L. A. Maldonado, A. C. Ward, M. Goodfellow, B. Bister, D. Bischoff, R. D. Süssmuth, and H. P. Fiedler. 2004. Abyssomicins, inhibitors of the para-aminobenzoic acid pathway produced by the marine *Verrucosispora* strain AB-18-032. *J. Antibiot.* **57:**271–279.

33. Roemer, T., B. Jiang, J. Davison, T. Ketela, K. Veillette, A. Breton, F. Tandia, A. Linteau, S. Sillaots, C. Marta, N. Martel, S. Veronneau, S. Lemieux, S. Kauffman, J. Becker, R. Storms, C. Boone, and H. Bussey. 2003. Large-scale essential gene identification in *Candida albicans* and applications to antifungal drug discovery. *Mol. Microbiol.* **50:**167–181.

34. Scott, J. J., D. C. Oh, M. C. Yuceer, K. D. Klepzig, J. Clardy, and C. R. Currie. 2008. Bacterial protection of beetle-fungus mutualism. *Science* **322:**63.

35. Selva, E., N. Montanini, S. Stella, A. Soffientini, L. Gastaldo, and M. Denaro. 1997. Targeted screening for elongation factor Tu binding antibiotics. *J. Antibiot.* **50:**22–26.

36. Shapiro, E., and F. Baneyx. 2002. Stress-based identification and classification of antibacterial agents: second-generation *Escherichia coli* reporter strains and optimization of detection. *Antimicrob. Agents Chemother.* **46:**2490–2497.

37. Urban, A., S. Eckermann, B. Fast, S. Metzger, M. Gehling, K. Ziegelbauer, H. Rübsamen-Waigmann, and C. Freiberg. 2007. Novel whole-cell antibiotic biosensors for compound discovery. *Appl. Environ. Microbiol.* **73:**6436–6443.

38. Wang, J., S. M. Soisson, K. Young, W. Shoop, S. Kodali, A. Galgoci, R. Painter, G. Parthasarathy, Y. S. Tang, R. Cummings, S. Ha, K. Dorso, M. Motyl, H. Jayasuriya, J. Ondeyka, K. Herath, C. Zhang, L. Hernandez, J. Allocco, A. Basilio, J. R. Tormo, O. Genilloud, F. Vicente, F. Pelaez, L. Colwell, S. H. Lee, B. Michael, T. Felcetto, C. Gill, L. L. Silver, J. D. Hermes, K. Bartizal, J. Barrett, D. Schmatz, J. W. Becker, D. Cully, and S. B. Singh.

2006. Platensimycin is a selective FabF inhibitor with potent antibiotic properties. *Nature* **441:**358–361.

39. **Washburn, R. S., A. Marra, A. P. Bryant, M. Rosenberg, and D. R. Gentry.** 2001. Rho is not essential for viability or virulence in *Staphylococcus aureus. Antimicrob. Agents Chemother.* **45:**1099–1103.

40. **Webber, M. A., and L. J. Piddock.** 2003. The importance of efflux pumps in bacterial antibiotic resistance. *J. Antimicrob. Chemother.* **51:**9–11.

41. **Weinstein, M. P.** 2004. *Micromonospora* antibiotic discovery at Schering-Plough (1961–1973): a personal reminiscence. *SIM News* **54:**56–66.

42. **Young, K., H. Jayasuriya, J. G. Ondeyka, K. Herath, C. Zhang, S. Kodali, A. Galgoci, R. Painter, V. Brown-Driver, R. Yamamoto, L. L. Silver, Y. Zheng, J. I. Ventura, J. Sigmund, S. Ha, A. Basilio, F. Vicente, J. R. Tormo, F. Pelaez, P. Youngman, D. Cully, J. F. Barrett, D. Schmatz, S. B. Singh, and J. Wang.** 2006. Discovery of FabH/FabF inhibitors from natural products. *Antimicrob. Agents Chemother.* **50:**519–526.

43. **Zhang, J. H., T. D. Chung, and K. R. Oldenburg.** 1999. A simple statistical parameter for use in evaluation and validation of high throughput screening assays. *J. Biomol. Screen.* **4:**67–73.

Metabolomics for the
Discovery of Novel Compounds

JENS C. FRISVAD

6

6.1. BACKGROUND

"Extrolites" is an ecological term for exported, potentially functional molecules produced by living organisms. These can be biopolymers and small molecules, also called secondary metabolites. Secondary metabolites are mostly found in bacteria, fungi, algae, and plants, and many of these have been shown to be bioactive, i.e., they can be effective antibiotic, anticancer, antiviral, cholesterol-lowering, antidiabetic, immunosuppressant, anti-Parkinson, antimalarial, antistrongyloidiasis compounds, animal growth promoters, etc. (32). Even though new interesting lead compounds are constantly found, it is rare that a secondary metabolite makes it all the way to a commercial drug. For this reason screening a large number of organism extracts for new compounds is necessary to have a chance of finding new efficient drugs and drug leads. In the plant kingdom, this may be guided by human experience with use of plant extracts, but in microorganisms the search for new drugs is of a more fortuitist nature, like, for example, the discovery of penicillin by Fleming. Since a large number of secondary metabolites have already been discovered, dereplication of known compounds is a very important part of any screening campaign for new bioactive compounds in nature. In many screenings for bioactive compounds, a very large number of microorganisms have been examined to increase the probability of discovering new compounds. These microorganisms have often been obtained from animals, plants (endophytes), macrofungi, algae, or the natural environments of these organisms, including soil, freshwater, seawater (marine microorganisms), and extreme habitats (8, 45).

Metabolomics is, by definition, based on a full description of all metabolites taken up by, contained in, and secreted by an organism, often both in qualitative and quantitative terms. However, because of the many factors affecting the quantitative amounts of different metabolites at any one moment, i.e., a quantitative snapshot of the metabolic status of an organism, quantitative metabolomics is a very complex discipline, heavily depending on the analytical techniques used for detection of all the metabolites. Exometabolomics is a more restricted discipline, only concerned with detection of all secreted or cell wall-deposited secondary metabolites. Restricted this way, exometabolomics is a more approachable technique, often performed using secondary metabolite profiling (13, 16, 39), and is often, in the screening phase, a matter of qualitative determination of secondary metabolites produced by any one microorganism under different nutritional and environmental conditions. However, even the less ambitious task of secondary metabolite profiling or metabolic footprinting (19, 39, 55, 56) depends on the growth media and environmental conditions, and on extraction, separation, detection, and identification techniques; no single method can cover all secondary metabolites produced by a microorganism. In addition, many microorganisms require a chemical signal from another organism to produce secondary metabolites that are usually not expressed because of histone silencing.

Combinations of secondary metabolites are occasionally much more effective than a metabolite alone, and therefore metabolite profiling can be of even greater importance than originally thought (62).

This chapter outlines the procedures for optimal discovery of new compounds, based on experience with filamentous fungi. However, the microscale technique recommended here is also useful for other culturable microorganisms, especially bacteria.

6.2. TAXONOMY AND IDENTIFICATION OF THE PRODUCING MICROORGANISMS

An often neglected aspect of screening for new compounds is the chemoconsistency of isolates of a particular species (31). It has been shown that filamentous fungi produce species-specific profiles of secondary metabolites (13, 21, 36), so to find new bioactive compounds from microorganisms, it is better to screen a large number of species than a large number of isolates. This will require correct identification of the microorganisms (see chapter 3, this volume), a goal that is achievable in well-known genera of microorganisms that have a high chemical diversity, such as *Streptomyces* (4, 5, 7, 9, 53), *Alternaria* (2), *Daldinia* (52), *Fusarium* (54), and *Penicillium* and *Aspergillus* (13, 18), but that is more difficult, for example, in endophytic microorganisms, which are notoriously difficult to identify. In bacterial genera such as *Salinispora* it has recently been recognized that these bacteria also produce species-specific profiles of secondary metabolites (28).

Sequencing of ribosomal genes is often used for identification in bacteriology and mycology, but experience within the fungi has shown that sequencing one or two housekeeping genes gives a much more accurate identification at the species level than ribosomal genes alone. In a study by Skouboe et al. (47) of common secondary metabolite-producing *Penicillium* species, it was shown that only a few species could be identified accurately by ITS sequencing, and it was later shown that β-tubulin sequencing was much more effective in separating the individual species (42). Correct species identification will also help in the dereplication of already known compounds, because the profile of secondary metabolites produced is species specific (13, 18), and at least some of the secondary metabolites are known from literature data. On the other hand, many producers of bioactive compounds have been associated with a producer that was misidentified, so it is best to rely only on detailed and well-documented chemotaxonomic studies rather than on individual reports of secondary metabolite production by a known or new species. It is always advisable to identify microorganisms by use of a polyphasic approach, but for screening purposes sequencing of (parts of) rDNA in conjunction with sequencing of at least two housekeeping genes is very useful.

In mycology, a polyphasic approach has been used among others on the *Aspergillus* section *Nigri* (34, 40), *Aspergillus* section *Clavati* (58), *Aspergillus* section *Candidi* (59), *Aspergillus* section *Usti* (26), and *Aspergillus* section *Fumigati* (17, 41). The methods used included sequencing of the ITS, β-tubulin, and calmodulin genes or spacers, micro- and macromorphology, physiology, especially growth at different temperatures, and water activities and secondary metabolite profiling.

6.3. CONDITIONS FOR PRODUCTION OF SECONDARY METABOLITES

Many screening procedures have been suggested, but if a large number of microorganisms have to be examined, a microscale technique is very valuable. Some industrial firms have used liquid culture (shake or still culture) for screening, because the results from this screening can more easily be the basis of scale-up than results from agar cultures. However, liquid cultures require elaborated extraction and rinse-up procedures. In contrast, a few agar plugs cut out of an agar colony from a microorganism can easily be extracted without any rinse-up procedures and the filtered extract used for any separation method available. This so-called agar plug method was originally proposed for thin-layer chromatography (TLC) where the agar plug was placed directly on a TLC plate (10, 11), and later modified to be used in conjunction with high-performance liquid chromatography (HPLC) or mass spectrometry (MS) methods (48, 50). An additional advantage of using colonies of microorganisms grown on agar is that the growth and development of the colony can be monitored, and several differentiation phases can be analyzed. Furthermore, contaminations can easily be detected. By analyzing the colony itself, cell-associated secondary metabolites can be detected, and by analyzing the agar surrounding the colony, secreted secondary metabolites can be detected.

Bacteria and filamentous fungi grow and differentiate well on different kinds of agar media, but in general the media should contain trace metals, potassium, magnesium, sulfate, phosphate, a nitrogen source, and a carbon source. Often yeast extract, corn steep liquor, or enzyme-digested proteins (peptone, tryptone) are good nitrogen sources, and sucrose or glucose are good carbon sources. However, expression of some secondary metabolites requires other simple carbohydrates or complex ones such as starch or cellulose. Plant extracts often contain these carbohydrates, and in addition, some plant secondary metabolites may stimulate secondary metabolite production in the microorganisms. Examples of these are malt extract, potato extract, and carrot extract. Particular metabolites may require special media. For example, siderophores are best expressed on a medium deficient of iron, and β-lactams such as penicillin are produced best if there is not too much carbohydrate in the medium. No medium can support production of all secondary metabolites produced by microorganisms, so several media should be used to detect as close to the full potential of secondary metabolites as possible. A new way of having silent gene clusters expressed is to use histone inhibitors, which have been shown to induce secondary metabolites that were otherwise only produced after triggering by plant secondary metabolites (12, 46, 60). A combination of genome mining (23) and chemical induction of silent biosynthetic pathways by suberoylanilide hydroxamic acid and 5-azacitidine in potato dextrose broth did show that several secondary metabolites were upregulated in *Aspergillus niger* ATCC 1015 and CBS 513.88, both full-genome-sequenced strains (12; see also chapter 7, this volume). However, many of these secondary metabolites indicated by expressed sequence tags were actually produced by a direct approach using yeast extract sucrose (YES) agar (34).

Incubation temperature, atmosphere, redox potential, pH, and water activity optimal for growth are not always the optimal conditions for secondary metabolite production, but microorganisms rarely produce secondary metabolites optimally far away from their growth optima. Some initial experimentation may be necessary to establish the optimal conditions for production of as many different secondary metabolites as possible. Incubation time is also important. For many microorganisms, 5 to 10 days of incubation will be optimal.

Miniaturized methods have been used for screening of enzymes in mutants of fungi (1), and the same kind of miniaturized methods and microwell methods can also be used for screening for secondary metabolites. Such methods are mostly used in biotechnological firms, but with the increased sensitivity in separation and detection methods, this is also an approach that can be used in exometabolomics.

6.4. EXTRACTION OF CULTURES

A very efficient extraction mixture for most secondary metabolites is ethylacetate/dichlormethane/methanol (3:2:1) with some added formic acid (48). Ultrasonication helps in extracting a large proportion of the secondary metabolites present in the agar plugs. This mixture is especially well suited for apolar and medium polar compounds. Usually the extraction mixture is then evaporated and the remaining dry extract redissolved in methanol. By use of the 3:2:1 mixture many "primary metabolites" are not included in the final extract being analyzed. For very polar compounds 75% methanol or 84% acetonitrile are also effective extraction liquids, and the resulting extract can be analyzed directly without any rinse-up. That rinse-up is not necessary is one of the advantages in the agar plug methods. However, many other extraction solvents appear to be effective, including dichlormethane, ethylacetate, and acetone. All these extraction solvents have been used extensively to produce crude plant or microorganism extracts.

6.5. SEPARATION OF SECONDARY METABOLITES

Volatile secondary metabolites are best separated by use of gas chromatography (29). However, most bioactive secondary metabolites are not volatile, so other separation methods are used. Even though micellar electrokinetic capillary chromatography appears to be a very effective separation technique (37), HPLC or ultra performance liquid chromatography (UPLC) are the methods of choice for general separation and simultaneous detection of secondary metabolites. UPLC has only been used recently (22, 57), but HPLC methods are still those used by most researchers. Many types of columns are effective, but reversed-phase columns have usually been used and acetonitrile/water gradients are nearly always used for elution (19, 20, 35). An alkylphenone retention index can be calculated and used in conjunction with retention time to partially identify the different secondary metabolites eluted (19). The more authentic standards of known secondary metabolites that can be analyzed together with the extracts of the microorganisms, the better.

Direct-inlet MS can be used without a need to separate the different secondary metabolites (22, 39, 49, 50, 51). By using electrospray conditions and weak ionization, many compounds will appear as their M + 1 mass peak (for positive ionization) or as their M − 1 peak (for negative ionization) and few fragments will be seen. If a high-accuracy time-of-flight mass spectrometer is used, it is often possible to calculate the molecular formula for each compound. Such an analysis takes 1 to 2 min, compared with an HPLC separation, which may take 15 to 30 min. Smedsgaard et al. (50) incubated isolates from 54 species of *Penicillium* on Czapek yeast autolysate agar and YES agar for 1 week at 25°C, and extracted three small agar plugs from these cultures by use of a two-stage extraction procedure using first ethylacetate/dichlormethane/methanol (3:2:1) with some added formic acid and then 2-propanol. After evaporation and redissolving in methanol, these extracts could be injected directly into the mass spectrometer. By using this method nearly all 54 species could be separated and several M + 1 ions could be directly related to different secondary metabolites produced by the isolates analyzed, as accurate mass detection was used.

6.6. DETECTION AND IDENTIFICATION OF SECONDARY METABOLITES

In addition to retention time, other informative spectrometric data will increase the chance of correctly identifying the known secondary metabolites. The two major ways of identifying the different peaks eluted from the HPLC are a hyphenated diode array detector (UV-visible spectra) or a hyphenated mass spectrometric detector. Many secondary metabolites have conjugated double bonds, and thereby they have more or less specific UV spectra. These UV spectra are often alike in biosynthetically related secondary metabolites, and new compounds with the same chromophore can be found, or known compounds can be deselected (3, 25, 31, 38). This can be accomplished by comparing UV spectra with use of image analysis techniques (25, 30). The same kind of image analysis techniques can be used for high-resolution mass spectra to deselect known compounds or to discover probable new compounds (24). Databases on secondary metabolites are not very detailed concerning UV spectra, often only providing the absorption maxima, but data for fungal metabolites can be found in Frisvad and Thrane (20) and Nielsen and Smedsgaard (35) and in the

original literature. Mass spectra are easier to use for identification, because most databases, such as ANTIBASE, have accurate monoisotopic masses listed for all compounds. The combination of taxonomy of the producer, retention time, UV spectrum, and high-resolution mass spectrum of the secondary metabolite under question, and available databases, will thus be sufficient to identify many known compounds, so valuable time is not wasted rinsing up and elucidating the structure of a known compound.

Nuclear magnetic resonance (NMR) techniques can also be used for screening secondary metabolites from fungi. Isaka et al. (27) used a combination of TLC and ^{1}H NMR for insect-associated fungi. Characteristic proton signals were used to indicate certain steroids and these were later separated by use of TLC, chemical color reactions, and repeated NMR confirmation of identity. This technique was used for certain hopane steroids that did not have a chromophore or had very poor mass sensitivity. However the TLC-NMR method is not very sensitive, and a new method based on microcoil probes (33, 43) and two-dimensional NMR spectra is very promising. The latter method was used for *Tolypocladium cylindrosporum* to find new terpene indol alkaloids (44). New chemoinformatic methods can then be used to identify the metabolites (61).

6.7. CHARACTERIZATION OF NEW SECONDARY METABOLITES

Data from the screening of microorganisms are also useful to elucidate the structures of new compounds. The UV and mass spectra, in addition to taxonomic and biosynthetic reasoning, can in some cases help in proposing a chemical structure for a new secondary metabolite, but in most cases the new compound has to be purified, and characterized further by NMR, X-ray crystallography, infrared spectra, and circular dichroism. Closely related fungi often produce the same secondary metabolites; hence, a correct identification can be of value. For example, most, but not all, species in the *Aspergillus ochraceus* group (*Aspergillus* section *Circumdati*) produce five types of polyketides: penicillic acid, ochratoxins, xanthomegnins, aspyrones, and melleins (15), so if a new species in this group is discovered, it will probably produce some or all of those metabolites. New metabolites from a new species in that group would then probably be a terpene or a nonribosomal peptide, even though a fifth kind of polyketide could also be the bioactive compound. Many compounds of mixed biosynthetic origin have also been found, and the structure of those may be more difficult to predict based on taxonomic and biosynthetic reasoning. After deselection of known compounds, using, for example, UV and mass spectra, the effort can be concentrated on what appear to be new compounds. These are often purified based on the analytical results and separated by use of preparative reversed-phase HPLC, alternatively preparative TLC, or high-speed countercurrent chromatography (6). Dalsgaard et al. (6) compared reversed-phase HPLC and high-speed countercurrent chromatography to purify and elucidate the structure of two new cyclic peptides from *Penicillium*.

6.8. CONCLUSION

The discovery of new bioactive compounds requires a combination of methods and a systems biology approach, using both molecular and phenotypic data. It is recommended to use a combination of taxonomy, bioinformatics, optimal growth conditions, high-resolution separation and identification techniques, and biosynthetic reasoning in the efforts to discover new lead drugs. Such a polyphasic approach needs to be made in such a way that it can be miniaturized and used in high-throughput screening; the use of microscale techniques is recommended. One of the most important parts of finding new compounds is to be able to quickly deselect already known compounds. The use of fast methods such a direct-inlet MS combined with databases is one possibility, while UPLC-UV-MS may also prove to be the chosen chemical screening technique for new bioactive natural products in the future.

REFERENCES

1. **Alberto, F., D. Navarro, R. P. de Vries, M. Asther, and E. Record.** 2009. Technical advances in fungal biotechnology: development of a miniaturized culture method and an automated high-throughput screening. *Lett. Appl. Microbiol.* **49:**278–282.
2. **Andersen, B., A. Dongo, and P. M. Pryor.** 2008. Secondary metabolite profiling of *Alternaria dauci*, *A. porri*, *A. solani* and *A. tomatophila*. *Mycol. Res.* **112:**241–250.
3. **Andersen, B., M. E. Hansen, and J. Smedsgaard.** 2005. Automated and unbiased classification of chemical profiles from fungi using high performance liquid chromatography. *J. Microbiol. Methods* **61:**295–304.
4. **Challis, G. L.** 2008. Genome mining for novel natural product discovery. *J. Med. Chem.* **51:**2618–2628.
5. **Challis, G. L. and D. A. Hopwood.** 2003. Synergy and contingency as driving forces of the evolution of secondary metabolite production by *Streptomyces* species. *Proc. Natl. Acad. Sci. USA* **100:**14555–14561.
6. **Dalsgaard, P. W., K. F. Nielsen, and T. O. Larsen.** 2005. UV-guided isolation of fungal metabolites by HSCCC. *J. Liq. Chromatogr. Relat. Technol.* **28:**2029–2039.
7. **Denery, J. R., M. J. Cooney, and Q. X. Li.** 2006. Metabolic profiling to reflect gene expression in *Streptomyces tenjimariensis*. *Ind. Biotechnol.* **2:**51–54.
8. **de Vita-Marques, A. M., S. P. Lira, R. G. S. Berrlinck, M. H. R. Seleghim, S. R. P. Sponchiado, S. M. Tauk-Tornisielo, M. Barata, C. Pessoa, M. O. de Moraes, B. C. Cavalcanti, G. G. F. Naschimento, A. O. de Souza, F. C. S. Galetti, C. L. Silva, M. Silva, E. F. Pimenta, O. Thiermann, M. R. Z. Passarini, and L. D. Sette.** 2008. A multi-screening approach for marine-derived fungal metabolites and the isolation of cyclodepsipeptides from *Beauveria feline*. *Quim. Nova* **31:**1099–1103.
9. **Fiedler, H.-P., C. Bruntner, A. Bull, A. Ward, M. Goodfellow, O. Potterat, C. Puder, and G. Mihm.** 2005. Marine actinomycetes as a source of novel secondary metabolites. *Antonie van Leeuwenhoek* **87:**37–42.
10. **Filtenborg, O., and J. C. Frisvad.** 1980. A simple screening method for toxigenic fungi in pure cultures. *Lebensm.-Wiss. Technol.* **13:**128–130.
11. **Filtenborg, O., J. C. Frisvad, and J. A. Svendsen.** 1983. Simple screening method for moulds producing intracellular mycotoxins in pure cultures. *Appl. Environ. Microbiol.* **45:**581–585.
12. **Fisch, K. M., A. F. Gillaspy, M. Gipson, J. C. Henrikson, A. R. Hoover, L. Jackson, F. Z. Najar, H. Wägele, and R. H. Cichewicz.** 2009. Chemical induction of silent biosynthetic pathway transcription in *Aspergillus niger*. *J. Ind. Microbiol. Biotechnol.* **36:**1199–1213.
13. **Frisvad, J. C., B. Andersen, and U. Thrane.** 2008. The use of secondary metabolite profiling in fungal taxonomy. *Mycol. Res.* **112:**231–240.
14. **Frisvad, J. C., and O. Filtenborg.** 1983. Classification of terverticillate Penicillia based on profiles of mycotoxins and other secondary metabolites. *Appl. Environ. Microbiol.* **46:**1301–1310.
15. **Frisvad, J. C., J. M. Franks, J. A. M. P. Houbraken, A. F. A. Kuijpers, and R. A. Samson.** 2004. New ochratoxin producing species of *Aspergillus* section *Circumdati*. *Stud. Mycol.* **50:**23–43.
16. **Frisvad, J. C., T. O. Larsen, R. de Vries, M. Meijer, J. Houbraken, F. J. Cabañes, K. Ehrlich, and R. A. Samson.** 2007. Secondary metabolite profiling, growth profiles and other tools for species recognition and important *Aspergillus* mycotoxins. *Stud. Mycol.* **59:**31–37.
17. **Frisvad, J.-C., C. Rank, K. F. Nielsen, and T. O. Larsen.** 2009. Metabolomics of *Aspergillus fumigatus*. *Med. Mycol.* **47:**S53–S71.
18. **Frisvad, J. C., J. Smedsgaard, T. O. Larsen, and R. A. Samson.** 2004. Mycotoxins, drugs and other extrolites produced by species in *Penicillium* subgenus *Penicillium*. *Stud. Mycol.* **49:**201–241.
19. **Frisvad, J. C., and U. Thrane.** 1987. Standardized high-performance liquid chromatography of 182 mycotoxins and other fungal metabolites based on alkylphenone indices and UV-VIS spectra (diode-array detection). *J. Chromatogr.* **404:**195–214.
20. **Frisvad, J. C., and U. Thrane.** 1993. Liquid column chromatography of mycotoxins, p. 253–372. *In* V. Betina (ed.), *Chromatography of Mycotoxins: Techniques and Applications.* Journal of Chromatography Library, 54. Elsevier, Amsterdam, The Netherlands.
21. **Frisvad, J. C., U. Thrane, and O. Filtenborg.** 1998. Role and use of secondary metabolites in fungal taxonomy, p. 289–319. *In* J. C. Frisvad, P. D. Bridge, and D. K. Arora (ed.), *Chemical Fungal Taxonomy.* Marcel Dekker, New York, NY.
22. **Geng, R., X. Meng, G. Bai, and G. Luo.** 2008. Profiling of acarviostatin family secondary metabolites secreted by *Streptomyces coelicoflavus* ZG0656 using ultra-performance liquid chromatography with electrospray ionization mass spectrometry. *Anal. Chem.* **80:**7554–7561.
23. **Gross, H.** 2009. Genomic mining, a concept for the discovery of new bioactive natural products. *Curr. Opin. Drug Discov. Dev.* **12:**207–219.
24. **Hansen, M. E., and J. Smedsgaard.** 2006. A new matching algorithm for high resolution mass spectra. *J. Am. Soc. Mass Spectrom.* **15:**1173–1180.
25. **Hansen, M. E., J. Smedsgaard, and T. O. Larsen.** 2005. X-hitting: an algorithm for novelty detection and dereplication by UV-spectra in complex mixtures of natural products. *Anal. Chem.* **77:**6805–6817.
26. **Houbraken, J., M. Due, J. Varga, M. Meijer, J. C. Frisvad, and R. A. Samson.** 2007. Polyphasic taxonomy of *Aspergillus* section *Usti*. *Stud. Mycol.* **59:**107–128.
27. **Isaka, M., N. L. Hywel-Jones, M. Sappan, S. Mongkolsamrit, and S. Saidaengkham.** 2009. Hopane triterpenes as chemotaxonomic markers for the insect pathogens *Hypocrella s. lat.* and *Aschersonia*. *Mycol. Res.* **113:**491–497.
28. **Jensen, P. R., P. G. Williams, D. C. Oh, L. Zeigler, and W. Fenical.** 2007. Species-specific secondary metabolite production in marine actinomycetes of the genus *Salinispora*. *Appl. Environ. Microbiol.* **73:**1146–1152.
29. **Larsen, T. O., and J. C. Frisvad.** 1995. Comparison of different methods for collection of volatile chemical markers from fungi. *J. Microbiol. Methods* **24:**135–144.
30. **Larsen, T. O., B. O. Petersen, D. Sørensen, J. Ø. Duus, J. C. Frisvad, and M. E. Edberg.** 2005. Discovery of new natural products by application of X-hitting, a novel algorithm for automated comparison of full UV-spectra, combined with structural determination by NMR spectroscopy. *J. Nat. Prod.* **68:**871–874.
31. **Larsen, T. O., J. Smedsgaard, K. F. Nielsen, M. E. Hansen, and J. C. Frisvad.** 2005. Phenotypic taxonomy and metabolite profiling in microbial drug discovery. *Nat. Prod. Rep.* **22:**672–695.

32. **Meinwald, J.** 2009. The chemistry of biotic interactions in perspective: small molecules take center stage. *J. Org. Chem.* **74:**1813–1825.

33. **Molinski, T. F.** 2009. Nanomole-scale natural products discovery. *Curr. Opin. Drug Discov. Dev.* **12:**197–206.

34. **Nielsen, K. F., J. M. Mogensen, M. Johansen, T. O. Larsen, and J. C. Frisvad.** 2009. Review of secondary metabolites and mycotoxins from the *Aspergillus niger* group. *Anal. Bioanal. Chem.* doi: 10.1007/s00216-009-3081-5.

35. **Nielsen, K. F., and J. Smedsgaard.** 2003. Fungal metabolite screening: database of 474 mycotoxins and fungal metabolites for dereplication by standard liquid chromatography-UV-mass spectrometry methodology. *J. Chromatogr. A* **1002:**111–136.

36. **Nielsen, K. F., J. Smedsgaard, T. O. Larsen, F. Lund, U. Thrane, and J. C. Frisvad.** 2004. Chemical identification of fungi—metabolite profiling and metabolomics, p. 19–35. *In* D. K. Arora (ed.), *Fungal Biotechnology in Agricultural, Food and Environmental Applications.* Marcel Dekker, New York, NY.

37. **Nielsen, M. S., P. V. Nielsen, and J. C. Frisvad.** 1996. Micellar electrokinetic capillary chromatography of fungal metabolites: resolution optimized by experimental design. *J. Chromatogr. A* **721:**337–344.

38. **Nielsen, N.-P. V., J. Smedsgaard, and J. C. Frisvad.** 1999. Full second-order chromatographic/spectrometric data matrices for automated sample identification and component analysis by non data reducing image analysis. *Anal. Chem.* **71:**727–735.

39. **Pope, G. A., D. A. MacKenzie, M. Defernez, and I. N. Roberts.** 2009. Metabolic footprinting for the study of microbial diversity. *Cold Spring Harbor Protoc.* doi: 10.1101/pdb.prot5222.

40. **Samson, R. A., S. Hong, S. W. Peterson, J. C. Frisvad, and J. Varga.** 2007. Diagnostic tools to identify black aspergilli. *Stud. Mycol.* **59:**129–145.

41. **Samson, R. A., S. Hong, S. W. Peterson, J. C. Frisvad, and J. Varga.** 2007. Polyphasic taxonomy of *Aspergillus* section *Fumigati* and its teleomorph *Neosartorya. Stud. Mycol.* **59:**147–207.

42. **Samson, R. A., K. A. Seifert, A. F. A. Kuijpers, J. A. M. P. Houbraken, and J. C. Frisvad.** 2004. Phylogenetic analysis of *Penicillium* subgenus *Penicillium* using partial β tubulin sequences. *Stud. Mycol.* **49:**175–200.

43. **Schroeder, F. C., and M. Gronquist.** 2006. Extending the scope of NMR spectroscopy with microcoil probes. *Angew. Chem. Int. Ed.* **45:**7122–7133.

44. **Schroeder, F. C., D. M. Gibson, A. C. L. Churchill, P. Sojikul, E. J. Wursthorn, S. B. Krasnoff, and J. Clardy.** 2007. Differential analysis of 2D NMR spectra: new natural products from a pilot-scale fungal extract library. *Angew. Chem. Int. Ed.* **46:**901–904.

45. **Schulz, B., S. Draeger, T. E. de la Cruz, J. Rheinheimer, K. Siems, S. Loesgen, J. Bitzer, O. Schloerke, A. Zeeck, I. Kock, H. Hussain, J. Q. Dai, and K. Krohn.** 2008. Screening strategies for obtaining novel, biologically active, fungal secondary metabolites from marine habitats. *Bot. Mar.* **51:**219–234.

46. **Schwab, E. K., J. W. Bok, M. Tribus, J. Galehr, S. Grassle, and N. P. Keller.** 2007. Histone deacetylase activity regulates chemical diversity in *Aspergillus. Eukaryot. Cell* **6:**1656–1664.

47. **Skouboe, P., J. C. Frisvad, D. Lauritsen, M. Boysen, J. W. Taylor, and L. Rossen.** 1999. Nucleotide sequences from the ITS region of *Penicillium* species. *Mycol. Res.* **103:**873–881.

48. **Smedsgaard, J.** 1997. Micro-scale extraction procedure for standardized screening of fungal metabolite production in cultures. *J. Chromatogr. A* **760:**264–270.

49. **Smedsgaard, J., and J. C. Frisvad.** 1996. Using direct electrospray mass spectrometry in taxonomy and secondary metabolite profiling of crude fungal extracts. *J. Microbiol. Methods* **25:**5–17.

50. **Smedsgaard, J., M. E. Hansen, and J. C. Frisvad.** 2004. Classification of terverticillate Penicillia by electrospray mass spectrometric profiling. *Stud. Mycol.* **49:**243–251.

51. **Smedsgaard, J., and J. Nielsen.** 2005. Metabolite profiling of fungi and yeast: from phenotype to metabolome by MS and informatics. *J. Exp. Bot.* **56:**273–286.

52. **Stadler, M., H. Wollweber, A. Mühlbauer, T. Henkel, Y. Asahawa, T. Hashimoto, Y. M. Ju, J. D. Rogers, H. G. Welzstein, and H. V. Tichy.** 2001. Secondary metabolite profiles, genetic fingerprints and taxonomy of *Daldinia* and allies. *Mycotaxon* **77:**379–429.

53. **Thomas, R.** 2001. A biosynthetic classification of fungal and *Streptomyces* fused-ring aromatic polyketides. *Chem-BioChem* **2:**612–627.

54. **Thrane, U.** 1993. *Fusarium* species and their specific profiles of secondary metabolites, p. 199–225. *In* J. Chełkowski (ed.), *Fusarium: Mycotoxins, Taxonomy and Pathogenicity.* Elsevier, Amsterdam, The Netherlands.

55. **Thrane, U., B. Andersen, J. C. Frisvad, and J. Smedsgaard.** 2007. The exo-metabolome of filamentous fungi, p. 235–252. *In* M. Jewitt and J. Nielsen (ed.), *Metabolomics. A Powerful Tool in Systems Biology.* Topics in Current Chemistry, 276. Springer, Berlin, Germany.

56. **Tormo, J. R., and J. B. Garcia.** 2005. Automated analysis of HPLC profiles of microbial extracts: a new tool for drug discovery and production, p. 33–56. *In* L. Zhang and A. L. Demain (ed.), *Natural Products. Drug Discovery and Therapeutic Medicine.* Humana Press, New York, NY.

57. **Urbain, A., A. Marston, E. Marsden-Edwards, and K. Hostettmann.** 2008. Ultra-performance liquid chromatography/time-of-flight mass spectrometry as a chemotaxonomic tool for the analysis of Gentianaceae species. *Phytochem. Anal.* **20:**134–138.

58. **Varga, J., M. Due, J. C. Frisvad, and R. A. Samson.** 2007. Polyphasic taxonomy of *Aspergillus* section *Candidi* based on molecular, morphological and physiological data. *Stud. Mycol.* **59:**75–88.

59. **Varga, J., J. C. Frisvad, and R. A. Samson.** 2007. Polyphasic taxonomy of *Aspergillus* section *Fumigati* and its teleomorph *Neosartorya. Stud. Mycol.* **59:**147–207.

60. **Williams, R. B., J. Henrikson, A. R. Hoover, A. E. Lee, and R. H. Cichewicz.** 2008. Epigenetic remodeling of fungal secondary metabolism. *Org. Biomol. Chem.* **6:**1895–1897.

61. **Xia, J., T. C. Bjorndal, P. Tang, and D. S. Wishart.** 2008. MetaboMiner—semi-automated identification of metabolites from 2D NMR spectra of complex biofluids. *BMC Bioinformatics* **9:**507.

62. **Zhang, L. X., K. Z. Yan, Y. Zhang, R. Huang, J. Bian, C. S. Zheng, H. X. Sun, Z. H. Chen, N. Sun, R. An, F. G. Min, W. B. Zhao, Y. Zhuo, J. L. You, Y. J. Song, Z. Y. Yu, Z. H. Liu, K. Q. Yang, H. Gao, H. Q. Dai, Z. L. Zhang, J. Wang, C. Z. Fu, G. Pei, J. T. Liu, S. Zhang, M. Goodfellow, Y. Y. Jiang, J. Kuai, G. C. Zhou, and X. P. Chen.** 2007. High-throughput synergy screening identifies microbial metabolites as combination agents for the treatment of fungal infections. *Proc. Natl. Acad. Sci. USA* **104:**4606–4611.

Strategies for Accessing Microbial Secondary Metabolites from Silent Biosynthetic Pathways

ROBERT H. CICHEWICZ, JON C. HENRIKSON, XIAORU WANG, AND KATIE M. BRANSCUM

7

7.1. INTRODUCTION

The theory and practice of microbial natural products research has been revolutionized by genomic technologies. The proliferation of fully sequenced microbial genomes during the past half-decade, many of which possess remarkable numbers of genes and gene clusters encoding for the production of unknown secondary metabolites, demonstrates that only a fraction of Nature's chemical bounty has been explored. This discovery has sparked significant interest in probing silent biosynthetic pathways (SBPs) for new and bioactive natural products and led to the rapid development of several innovative technologies for investigating these compounds. In this review, we examine the growing array of methods used to access microbial SBPs and discuss the challenges of effectively exploring this chemical resource.

7.2. SCOPE OF THE CHAPTER

Given the substantial scientific interest in this field, several review articles have been published within the past 5 years about microbial SBPs and we refer the reader to these excellent treatises for further discussion and alternative perspectives on the topic (5, 13, 18, 22, 23, 39, 45, 78, 80, 81, 85, 93, 95, 99). Unique to this chapter is a concentration on the methodological approaches used to access the products of SBPs, and we present a number of important case studies that are intended to help the reader evaluate the advantages and disadvantages of each technique. While we have attempted to provide a broad coverage of contemporary methods for accessing SBPs, we have found it necessary to exclude a number of studies to (i) avoid unnecessary redundancy describing investigations utilizing the same experimental approach and (ii) maintain the chapter's focus on techniques that have been successfully applied toward the exploration of SBPs.

7.3. BACKGROUND

The concept of SBPs was born out of the realization that many microorganisms express only a fraction of their secondary-metabolite-encoding pathways under typical laboratory culture conditions (Table 1). In each of these cases, the numbers of secondary metabolites observed from laboratory-grown cultures fall short of their genetically encoded biosynthetic potential. Several terms have arisen in the natural products literature as descriptors of this phenomenon. The words "silent," "cryptic," and "orphan" (45, 78) have all been used in reference to the refractory nature of these genes and the general inaccessibility of their biosynthetic products. Our group prefers the use of the word "silent" since this terminology offers a rather liberal definition for broadly encapsulating the concept of muted gene pathways and their corresponding enzymes and secondary metabolites. Our use of the term "silent" throughout this chapter is intended to include the broad range of suppressive transcriptional, translational, and biosynthetic mechanisms responsible for inhibiting secondary-metabolite formation. It is interesting to consider that some secondary-metabolite-encoding pathways that appear functionally silent might actually be transcribed at rather low levels, resulting in the generation of relatively minor concentrations of their respective biosynthetic products. Consequently, this low-level production could preclude the straightforward detection of a gene's or gene cluster's associated natural products (56). In other cases, SBPs have demonstrated complete transcriptional suppression under laboratory culture (38, 56). Irrespective of the cause, SBPs occupy the vast gulf that exists between our knowledge of what microorganisms are seemingly *genetically* able to produce and the natural products that researches are capable of *chemically* detecting.

During the preparation of this chapter, we have found it extremely useful to consider each method in its chronological context (Fig. 1). An important benefit of this approach is that it provides a unique window through which the evolution of SBP studies can be viewed, and suggests the future trajectory for this field. Consequently, we have presented the methods for probing SBPs in their approximate chronological order of development to highlight significant research trends. In general, the reader will note a pronounced shift away from processes based strictly on culture manipulation toward more genomic-driven technologies. An exciting new dimension in SBP studies is the emergence of novel mechanism-based and other hybrid techniques that meld the technological simplicity of culture manipulation strategies with recent breakthroughs in our understanding of microbial molecular biology. This has yielded a new generation of approaches that are anticipated to greatly enhance SBP-mining practices.

TABLE 1 Numbers of PKS, NRPS, and HPN gene clusters in selected bacteria and fungi

Organism	Genome size (Mb)	Number of gene clusters			Reference(s)
		PKS	NRPS	HPN	
Aspergillus fumigatus Af293	29.4	14	13	1	65
Aspergillus nidulans FGSC A4	30.0	27	13	1	42
Aspergillus niger ATCC 1015	37.2	31	15	9	38
Aspergillus niger CBS 513.88	33.7	34	17	7	72
Aspergillus oryzae RIB40	37.0	30	18	0	59
Frankia sp. Ccl3	5.4	6	3	1	89
Magnaporthe grisea 70–15	37.9	23	6	8	34
Mycobacterium tuberculosis H5N1	4.4	11	2	0	89
Neurospora crassa N150	38.6	7	2	1	41
Nocardia farcinica IFM 10152	6.0	6	7	1	89
Penicillium chrysogenum Wisconsin54-1255	32.2	20	10	2	92
Salinispora tropica CNB-440	5.2	6	3	4	89
Streptomyces avermitilis	9.0	12	4	0	50, 68
Streptomyces coelicolor A3(2)	8.7	7	3	0	8
Ustilago maydis 521	20.5	3	4	9	16, 52

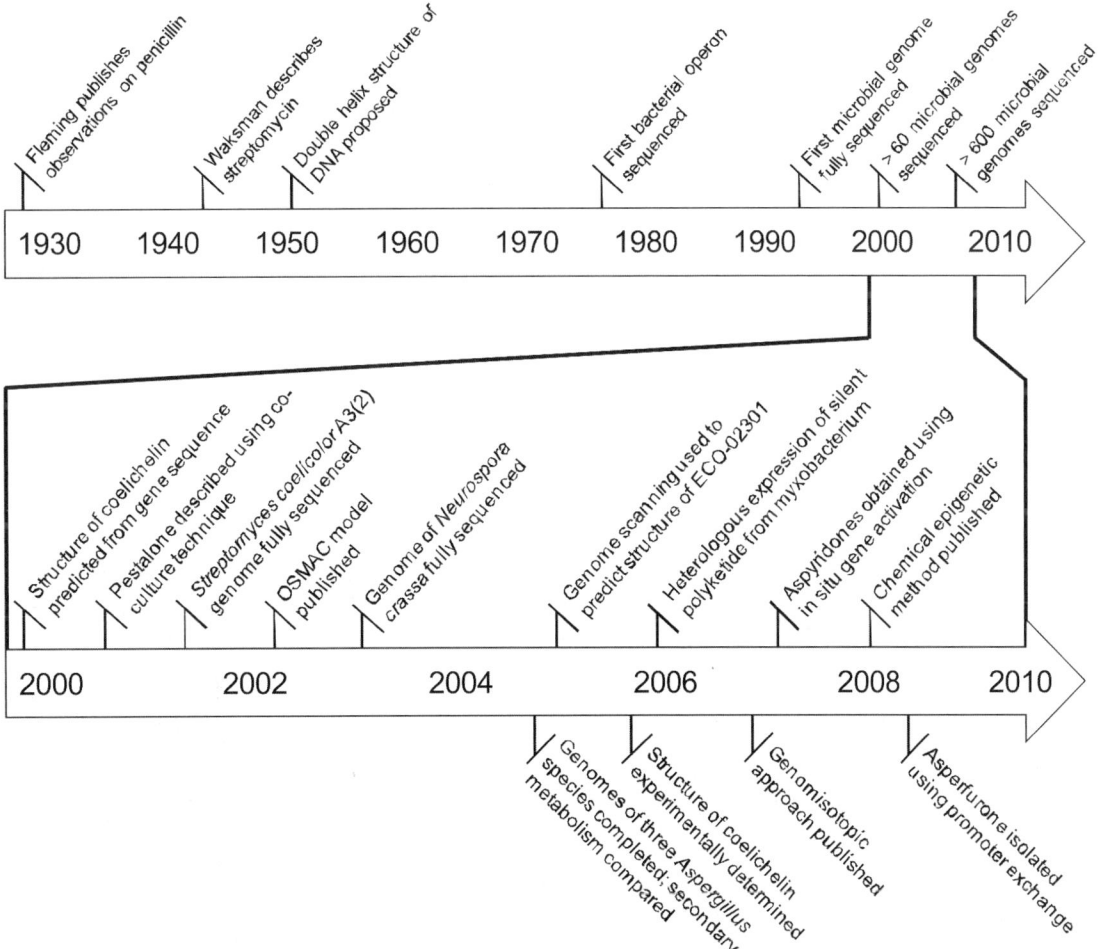

FIGURE 1 Timeline illustrating significant events related to the field of SBP studies. Note the substantial surge in the number of experimental methods and techniques for investigating SBPs that were developed during the past several years.

7.4. MANIPULATION OF ABIOTIC AND BIOTIC CULTURE CONDITIONS

7.4.1. Malleability of Microbial Chemical Phenotypes

It has long been recognized that efforts to manipulate microbial culture conditions can have a profound influence on the production of microbial secondary metabolites. This understanding has served as the backdrop of numerous studies describing the variable impact that alterations of abiotic or biotic factors have on the biosynthesis of natural products. A key advantage of this approach is that efforts to alter a microorganism's phenotypic (chemical) traits can be made independent of genomic knowledge. That is to say, if one assumes that a microbe harbors SBPs, then altering its growth conditions is a viable way to trigger production of its hidden natural product potential. The significant disadvantage of this technique is the lack of rationale for which growth factors should be manipulated to elicit production of new natural products. Moreover, conditions facilitating SBP expression for one microbe are not necessarily conducive for other species (i.e., microorganisms respond nonuniformly to stimulants). Consequently, unique culturing methods must be developed independently for each set of SBP-derived natural products in each microbial strain.

The ability to activate SBPs by altering microbial culture methods was well known long before researchers were knowledgeable of the functional roles that DNA, RNA, and proteins played in their production. To demonstrate this point, it is informative to note some of the early accounts describing altered chemical profiles of microorganisms following the modification of their respective culturing conditions. At the turn of the twentieth century, there existed significant debate as to the roles that various nutrients and chemical "stimulants" contributed to the phenotypic traits of microbes (86, 87). While most of these studies focused heavily on features such as cell mass accumulation and spore formation, it was also observed that culture manipulation resulted in significant qualitative and quantitative changes in an organism's chemical composition (76). For example, the addition of 0.01% ZnSO₄ to *Aspergillus niger* was shown to result in a 41% increase in the production of citric acid (Fig. 2), along with the appearance and greater yields of many other unidentified extractable small organic substances. Similarly, the effects of varying media/culture conditions were observed to appreciably alter the production of pigments and antibiotics including fumigacin (*Aspergillus fumigatus*) (94), clavacin (*Aspergillus clavatus*) (94), rubrofusarin and aurofusarin (*Fusarium culmorum*) (4), phoenicine (*Penicillium phoeniceum*) (31), and penicillins (*Penicillium* spp.) (21, 32, 51, 75, 84, 88) (Fig. 2). While this list of examples is by no means exhaustive, it does provide important evidentiary support for understanding the conceptual basis on which modern culture manipulation strategies have been designed.

7.4.2. Current Approaches to SBPs Using Culture Manipulation

In 2002, the idea of using culture manipulation techniques to probe SBPs for new natural products was formally reviewed and presented as the OSMAC theory (One Strain – MAny Compounds) (12). Although this approach had been widely used for decades, the compilation of data presented by the authors offered a compelling summary of disparate abiotic and biotic manipulation techniques for eliciting new natural products from bacteria and fungi (Table 2). For example, the fungus *Sphaeropsidales* sp. F-24′707 was noted to be metabolically responsive to a broad range of different media, cultivation vessels, solid versus liquid fermentation

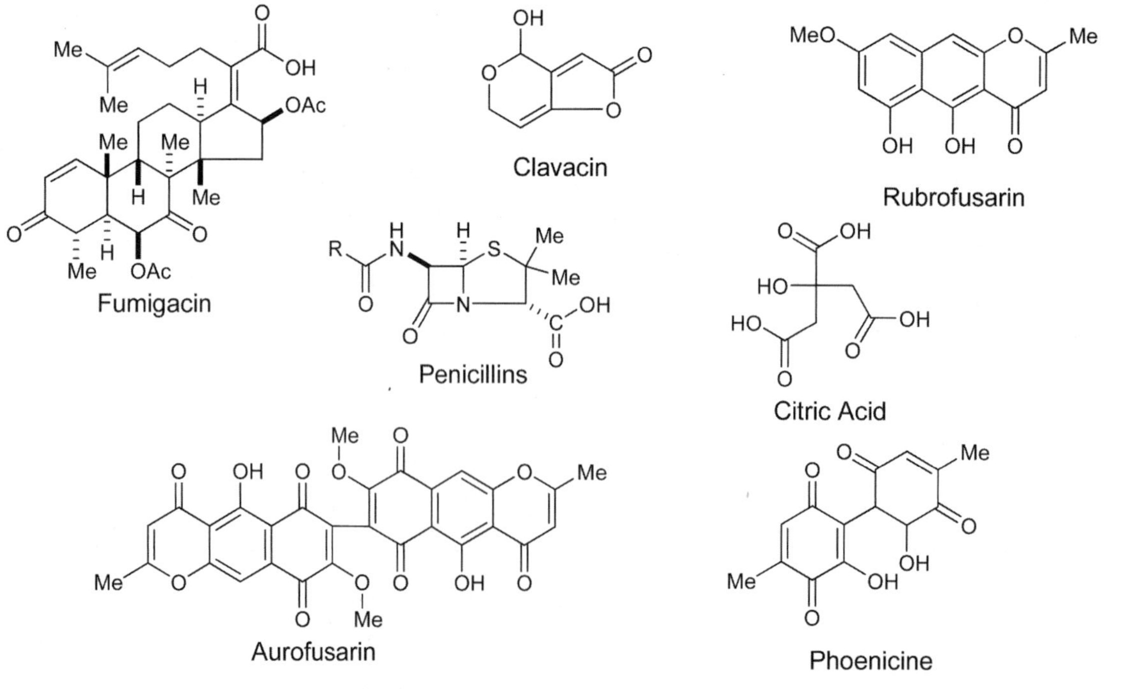

FIGURE 2 Representative microbial compounds identified by use of early culture manipulation techniques.

TABLE 2 Examples of some reported culture manipulation techniques

Microbe	Elicitation	Compounds	Reference
Aspergillus ochraceus	Variation of media composition, vessel, and oxygenation	Total of 15 compounds produced including seven new pentaketides	40
Chaetomium chiversii	Switch from solid to liquid medium	Chaetochromin A (liquid), radicicol (solid)	71
Gymnascella dankaliensis	Carbon source of media varied	Dankasterones A and B; gymnasterones A, B, C, and D	2
Halomonas sp. GWS-BW-H8hM	Addition of anthranilic acid	Aminophenoxazinones	11
Paraphaeosphaeria quadriseptata	Changing from tap H_2O to distilled H_2O	Cytosporones F-I, quadriseptin A, 5′-hydroxymonocillin III	71
Phomopsis asparagi	Addition of jasplakinolide (F-actin inhibitor)	Chaetoglobosins 510, 540, and 542	29
Sphaeropsidales sp. F-24′707	Variation of culture conditions	Cladospirones B-I	14
Sphaeropsidales sp. F-24′707	Addition of tricyclazole (inhibitor of 1,8-dihydroxynaphthalene pathway)	Sphaerolone, dihydrosphaerolone	15
Spicaria elegans	10 media types with and without shaking	Spicochalasin A, aspochalasins M–Q	58
Streptomyces spp.	Addition of dimethyl sulfoxide and other organic solvents to cultures	Chloramphenicol, tetracenomycin C, thiostrepton	26
Streptomyces spp.	Addition of scandium and other rare earth elements	Increased production of actinorhodin and other metabolites	53
Streptomyces sp. Gö 40/14	Variation of media composition, oxygenation, and pH	Several metabolites from different structure classes	12, 79
Streptomyces cellulosae S1013	Changing culture vessels	Hexacyclinic acid	48
Streptomyces fradiae	Addition of dimethyl sulfoxide	Increase the production of tylosin	20
Streptomyces venezuelae ISP5230	Shifting temperature from 37°C to 42°C	Jadomycin B	36

techniques, and enzyme inhibitors resulting in the generation of 19 new and known spirobisnaphthalene, bisnaphthalene, naphthoquinone, and macrolide metabolites from this single strain (Fig. 3). Similarly, *Aspergillus ochraceus* DSM7428 yielded an impressive array of nine structurally diverse natural products of polyketide synthase (PKS) and nonribosomal peptide synthetase (NRPS) origins following culture modifications (shake versus static fermentations and different vessel types) (Fig. 4).

Over the past few years, several new compounds have been described from fungi and bacteria by the use of culture manipulation techniques (Table 2). Ascospiroketals A and B from *Ascochyta salicorniae* (82); chaetoglobosins 510, 540, and 542 from *Phomopsis asparagi* (29); cytosporones F to I from *Paraphaeosphaeria quadriseptata* (71); daldinins A to C from *Daldinia concentrica* (83); and dankasterones A and B from *Gymnascella dankaliensis* (2) represent pertinent examples demonstrating the value that culture manipulation strategies have for enhancing metabolite yields or inducing the expression of new natural products from microorganisms (Fig. 5).

7.4.3. Microbial Coculture Techniques

Based on the premise that bacteria and fungi live in complex communities involving a wide range of competitive, commensalistic, predatory, mutualistic, parasitic, and synergistic interactions, it has been proposed that cohabitating microorganisms might serve as eliciting agents for the production of natural products (43). Consequently, it was envisioned that agents capable of triggering SBP

production might include a variety of undefined signaling molecules (e.g., secondary metabolites, primary metabolites, complex cell wall materials, proteins, etc.), making coculture a highly promising approach for natural product discovery (19). Published accounts describe a wide variety of approaches that have been applied in coculture studies ranging from the simple mixing of multiple live organisms to the addition of purified or semipurified microbial-derived components to cultures of a different microorganism (Table 3).

Several outstanding examples using coculture methods point to the exceptional opportunities that this technique provides. For example, the microbial metabolite pestalone was reported in 2001 from a marine-derived *Pestalotia* sp. (Fig. 6) (30). Pestalone is a unique chlorinated benzophenone possessing significant antimicrobial activity against several bacterial strains. It is noteworthy that pestalone was not detected in appreciable amounts from monocultures of the *Pestalotia* sp. However, addition of live bacteria cells (an unidentified gram-negative unicellular bacterium described as strain CNJ-328) to the growing fungus resulted in accumulation of the metabolite. It is interesting that neither an organic extract of the bacteria nor a cell-free aliquot of CNJ-328 culture broth were capable of eliciting pestalone production. Later, these authors expanded their application of this cocultures approach to include the production of libertellenones A to D from a marine-derived *Libertella* sp. fermented with CNJ-328 (66) and emericellamide from the fungus *Emericella* sp. grown in combination with a marine actinomycete (*Salinispora*

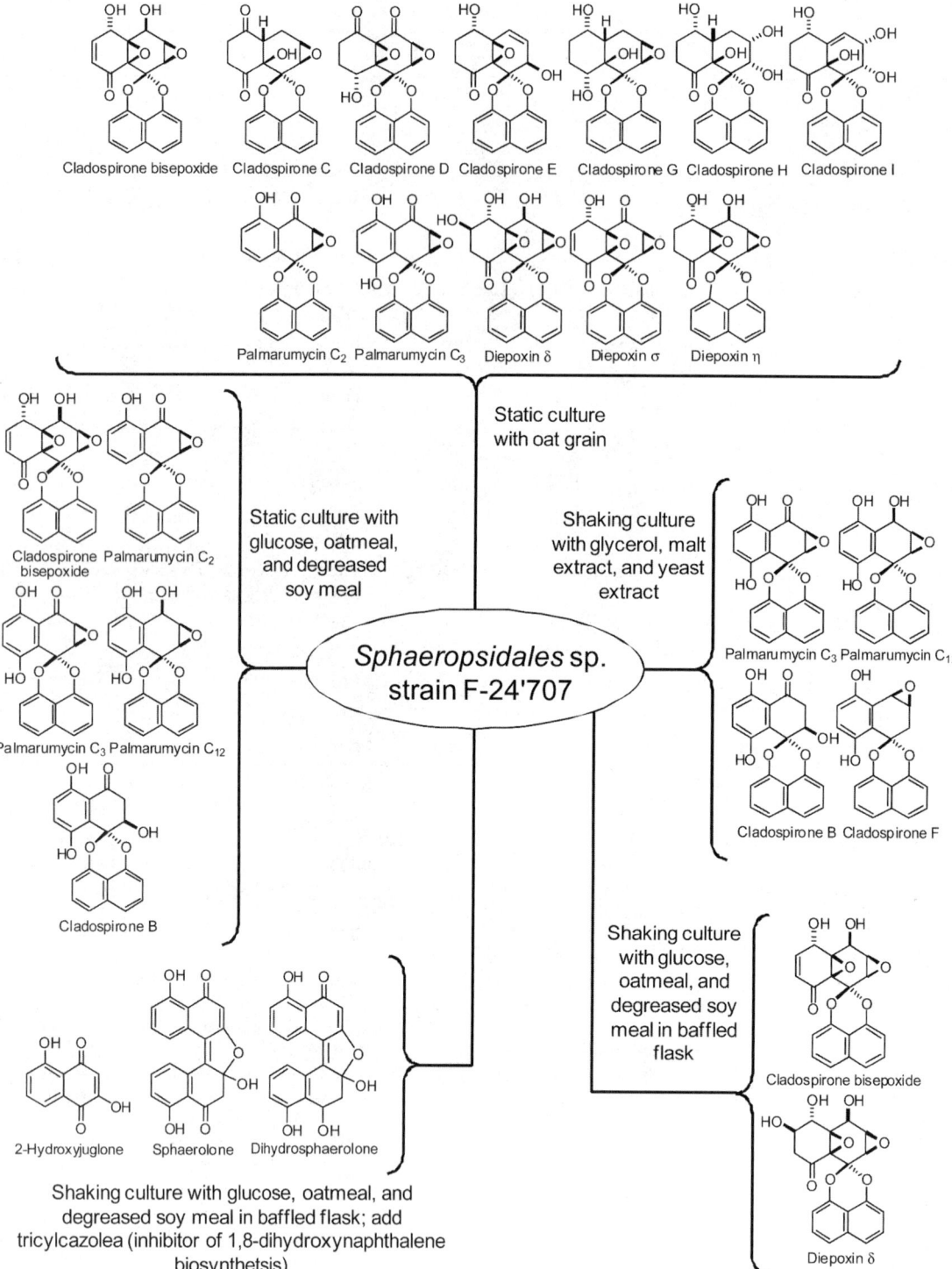

FIGURE 3 Compounds obtained from *Sphaeropsidales* sp. F-24'707 by use of culture manipulation.

arenicola) (Fig. 6) (67). The production of pyocyanin from mixed fermentations of marine-sediment-derived *Pseudomonas aeruginosa* and *Enterobacter* sp. (Fig. 6) was described recently (3). In liquid culture, neither isolate alone was found capable of generating the blue metabolite. By using a series of Boyden chamber experiments, the authors were able to demonstrate that a small-molecule membrane-diffusible factor(s) was generated by *Enterobac-*

ter sp., leading to the induction of pyocyanin production in *P. aeruginosa*.

7.4.4. Final Thoughts and New Directions for Using Culture Manipulation Techniques

An important point to note is that each of the described culture manipulation methods involves changes in the inclusion/deletion of single components (e.g., add or

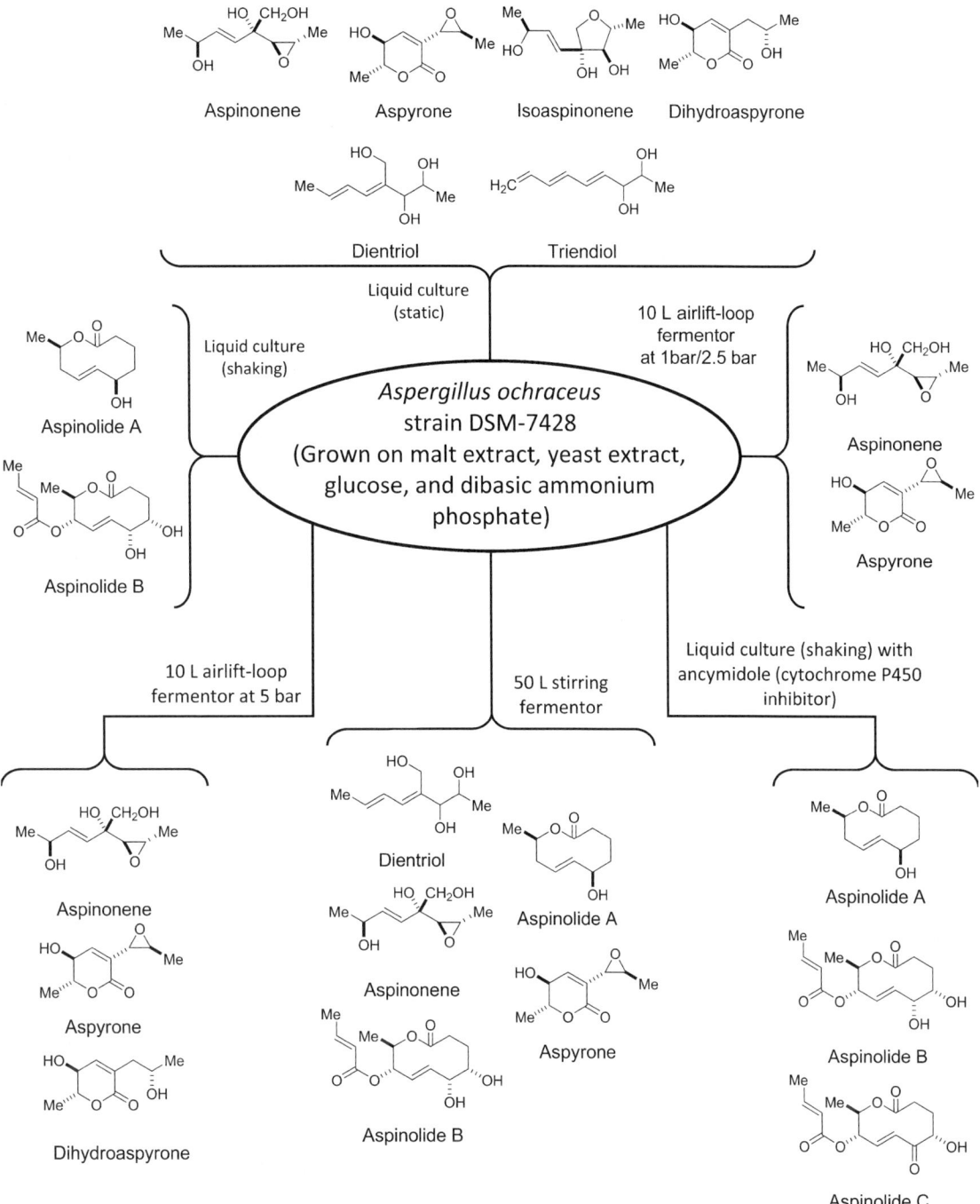

FIGURE 4 Compounds obtained from *A. ochraceus* DSM7428 by use of culture manipulation.

increase levels of NaCl, alter pH, add or increase various metals, etc.) or component mixtures (e.g., carbon source, media type, coculture species, etc.), or alteration in fermentation conditions (e.g., shake versus static cultures, culture vessel type, solid versus liquid media, etc.). It quickly becomes apparent that an unlimited assortment of culture parameters exist that can be altered singly or in combination. This makes the selection of which abiotic and biotic features to modify subject to considerable conjecture. Despite attempts to rationalize this process, there are no empirical rules or guidelines as to which culture conditions should be routinely evaluated. The arbitrary probing of abiotic and biotic factors until a suitable parameter set is discovered for inducing SBP activation can create a tremendous workload for natural products researchers. In light of these difficulties, Bills and colleagues (10) have proposed the use of "nutritional arrays," which allow for the parallel screening of hundreds of media conditions. Although conceptually appealing, this technique is likely to have only limited practical value because of the tremendous workload entailed in generating arrays and analyzing the large numbers of resulting samples.

The most detrimental factor limiting the effective use of culture manipulation techniques is the general lack

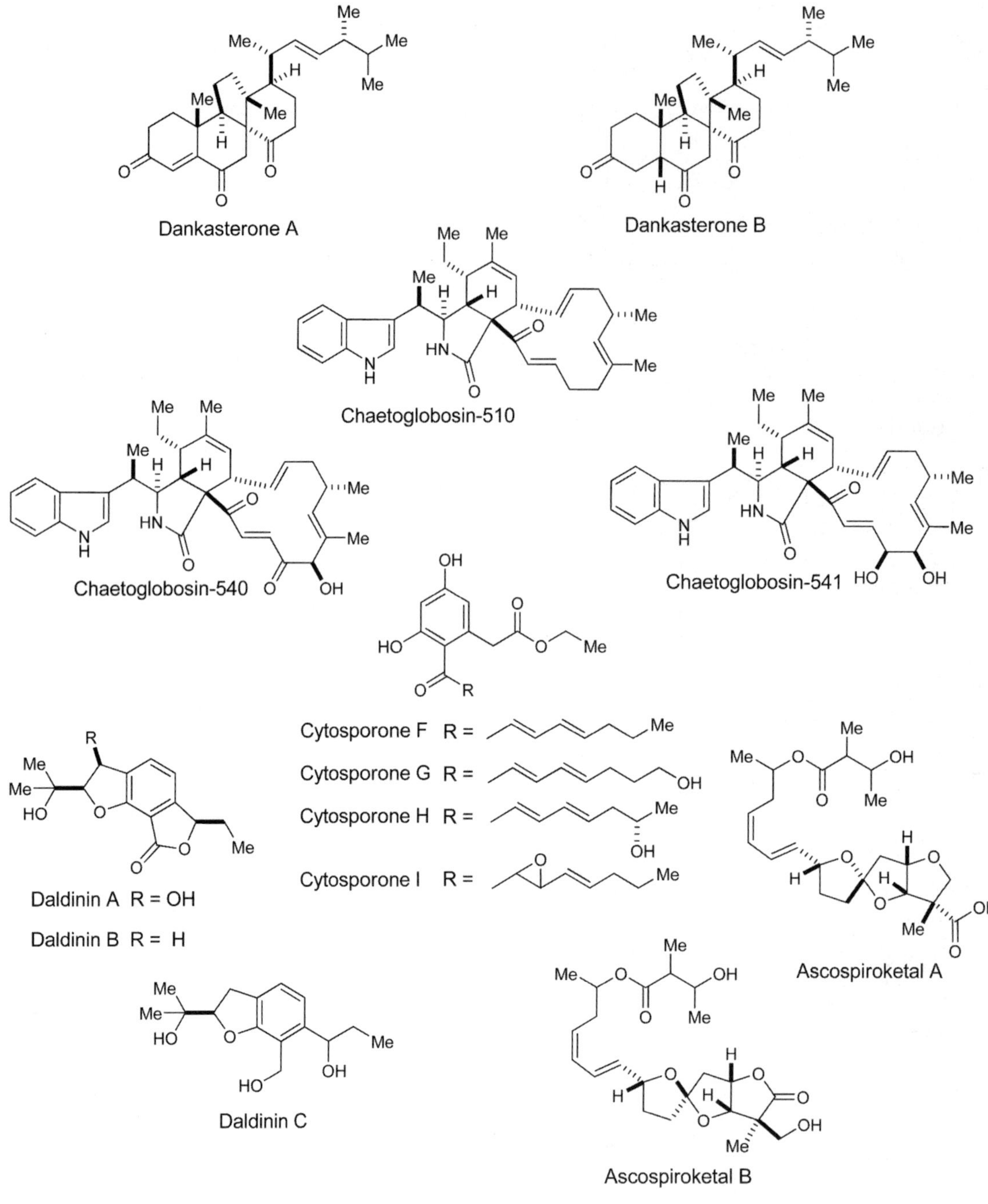

FIGURE 5 Examples of natural products obtained by manipulating microbial culture conditions.

of knowledge concerning which environmental cues microorganisms rely on to induce SBP expression. Coupled to this is the natural product research community's imprecise knowledge concerning the functional roles that SBP-derived secondary metabolites play under natural conditions (24, 33, 37). Conceivably, understanding the native functions of SBP-derived compounds could help researchers judiciously select which conditions should elicit metabolite biosynthesis. An exceptional example of utilizing an informed approach to identifying the product of an SBP is the discovery of coelichelin from *Streptomyces coelicolor* M145 (Fig. 7) (23, 25, 55). Key to this compound's discovery were (i) the realization that the targeted natural product was predicted to possess hydroxamate functionalities (these are frequently found in siderophores) and (ii) the fact that the gene cluster controlling coelichelin's biosynthesis was positioned adjacent to a presumed iron-binding repressor. This permitted Challis and colleagues (55) to accurately predict that use of iron-deficient media would facilitate the production of coelichelin. Although significant insight from genome mining was used to advance this discovery (discussed

TABLE 3 Examples of cocultivation methods for investigating SBPs in microorganisms

Methods	Outcome	Reference
Marine bacteria were cocultured with live and heat-killed terrestrial bacteria	Marine bacteria strains produced uncharacterized antibiotics	64
Marine bacteria were cocultured with live and/or cell-free supernatants of marine and terrestrial bacteria	Marine bacteria strains produced uncharacterized antibiotics	19
Panels of up to 76 live *Streptomyces* species were cocultured together in a pairwise fashion	Many species were able to elicit and produce uncharacterized antibiotics depending on the coculture partner	90
Two live fungi species (*Acremonium* sp. Tbp-5 and *Mycogone rosea*) were cocultured together	Induced production of acremostatins A, B and C	35
A bacterium (*Lactobacillus plantarum* NC8) was cocultured with live and heat-killed bacteria (*Lactococcus lactis* and *Pediococcus pentosaceus*)	Induced production of plantaricin NC8	61
A bacterium (*Streptomyces californicus*) was cocultured with a fungus (*Stachybotrys chartarum*)	Induced production of an uncharacterized cytostatic immunotoxin	73

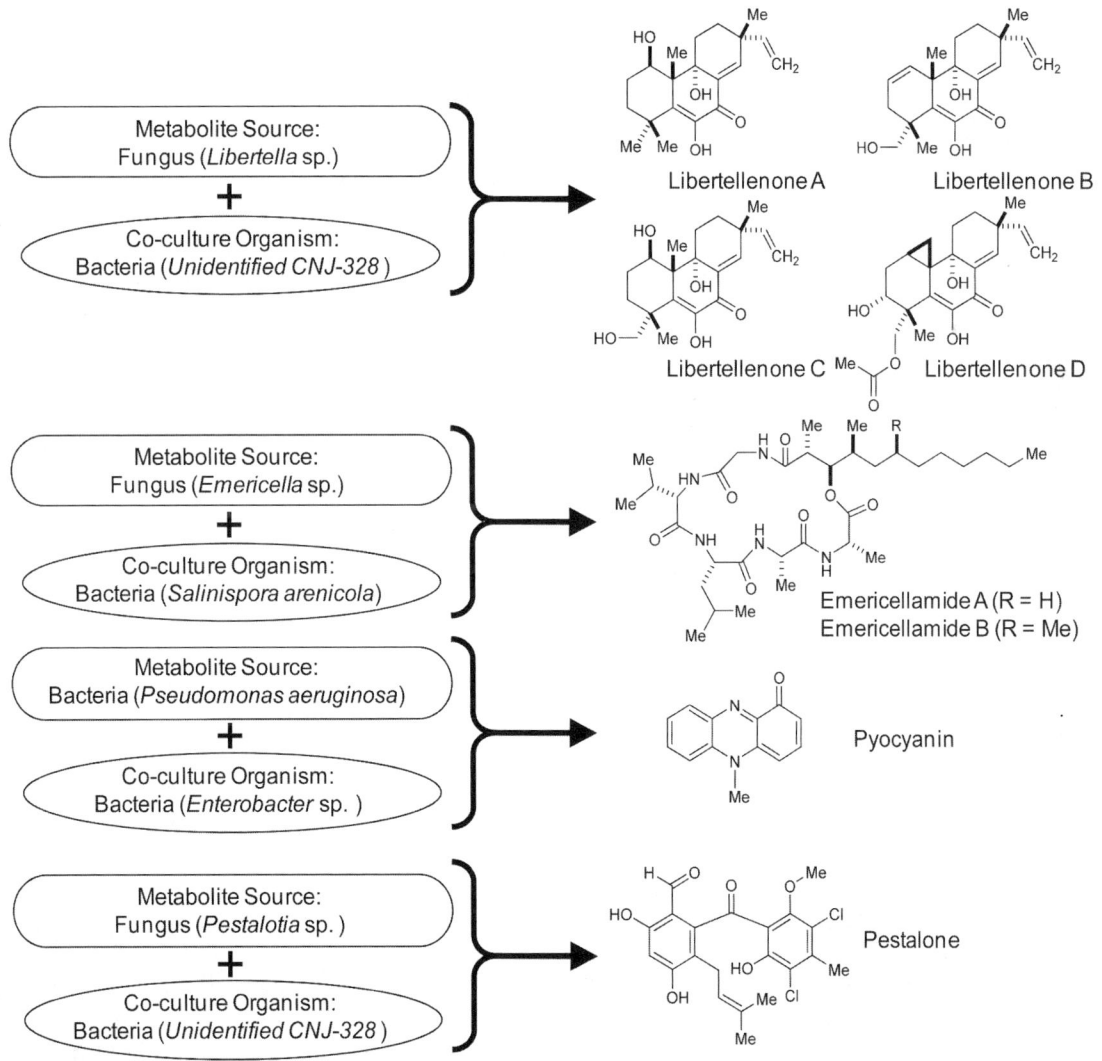

FIGURE 6 Examples demonstrating the use of coculture techniques to induce the production of microbial natural products.

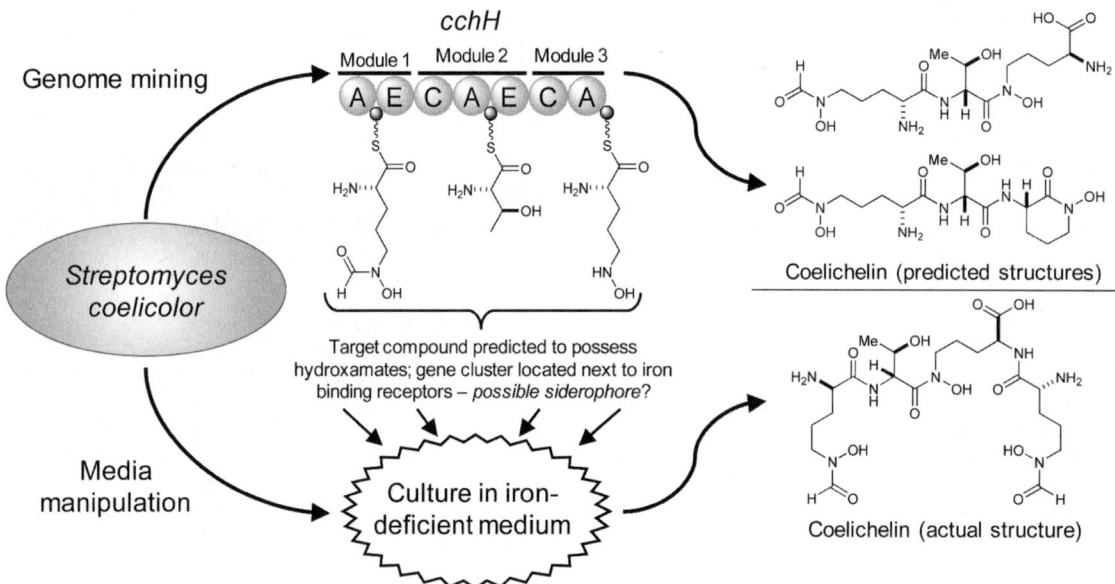

FIGURE 7 Combined approach leading to the discovery of coelichelin from *S. coelicolor*.

below in section 7.5.2), this work serves as a testament to the fact that, in the absence of similar guiding factors, culture modification strategies will continue to involve significant guesswork regarding which culture amendments will provide access to SBPs.

7.5. GENOMICS-DRIVEN APPROACHES TO INVESTIGATING SBPS

7.5.1. Overview of Genomic Technologies

Since the first report of a fully sequenced microbial genome (Fig. 1), growth in the number of bacteria and fungi whose genomes have been determined has been exponential. Given the recent emergence of a new generation of sequencing methods such as Applied Biosystems SOLiD, Helicos HeliScope, Illumina Solexa, Roche 454 GS FLX, and Pacific Biosciences SMRT (60, 62), the time and cost required to sequence microbial species has diminished significantly. These technologies are expected to facilitate the sequencing of microbial genomes at even greater rates, which is anticipated to provide natural products researchers with unprecedented capacity for identifying SBPs (13, 23, 93). Several resources have been created to address the critical need for sharing the massive quantities of data emerging from microbial genome-sequencing projects (Table 4). The centralized archiving of sequencing data will come to play increasingly important roles for the mining of SBPs. It is also likely that the evolving speed and efficiency of these techniques will lead to the further compilation of privately held (industry) generated genomes that will be largely inaccessible to the natural products research community. Hopefully, meaningful data-sharing plans will help to loosen restricted access to these resources and help foster additional discoveries.

7.5.2. Genome Scanning for SBPs

As a testament to the power of genome-scanning methods, Ecopia Biosciences, Inc. (now Thallion Pharmaceuticals, Inc.) developed a commercial enterprise based on the rapid generation and screening of bacterial genomic

information for the presence of secondary-metabolite-encoding gene clusters (98). Two principal features were key to facilitating the success of this method: (i) secondary-metabolite-encoding gene clusters are readily recognizable in microbial genomes because of the inclusion of distinctive conserved motifs, and (ii) gene cluster sequence information can be used in certain circumstances to predict the final biosynthetic protein's substrate specificity and to tailor functions with reasonable accuracy (this is particularly true for bacterial modular PKS and NRPS enzymes). Use of this approach led to the discovery of several new secondary metabolites, some of which are illustrated in Fig. 8. This includes the furanone-containing polyketides E-492, E-837, and E-975 (6); the glycosylated polyketide ECO-0501 (7); and a rather lengthy polyketide, ECO-02301 (63).

Beyond these discoveries, academic laboratories appear to be responsible for spearheading the bulk of genomics-driven discoveries of SBP-derived secondary metabolites. Many excellent examples can be found describing how the identification of silent genes in bacteria and fungi resulted in the discovery of novel microbial metabolites (13, 22, 23, 45, 78). A particularly elegant example of applying the genome-scanning approach has been described for coelichelin from *S. coelicolor* M145 (see above and Fig. 7) (23, 25, 55). It is important to note that, in each of these cases, the biosynthetic products of the investigated genes were ultimately obtained from the organisms by use of traditional culture techniques. What does one need to do to access metabolites if standard culture conditions prove completely incapable of facilitating the production of a target compound? This situation requires a different set of methodological tools, including heterologous expression and in situ gene manipulation, which are discussed in the following examples.

7.5.3. Heterologous Expression of SBPs

Heterologous expression techniques provide the opportunity to selectively transfer an SBP into a new host possessing a nontranscriptionally suppressive regulatory environment to produce the enzymatic machinery for silent natural

TABLE 4 Examples of online resources providing archiving of microbial genomic data[a]

Database	Description and link
Broad Institute Argo Genome Browser	Downloadable tool for visualizing and manually annotating genomic data http://www.broad.mit.edu/annotation/argo/
DOE Joint Genome Institute – Integrated Microbial Genomes	Comparative analysis and annotation of publicly available genomes including bacteria and eukaryotic microorganisms http://img.jgi.doe.gov/cgi-bin/pub/main.cgi
EMBL-EBI Ensembl Genome Browser	Website that maintains annotated eukaryotic genomes http://www.ebi.ac.uk/ensembl
Genome Channel Organism Navigation	Displays annotated microbial genomes sequenced by the DOE Joint Genome Institute http://compbio.ornl.gov/channel/
Gold Genomes Online Database	Collection of publicly available genomic data including many microorganisms http://genomesonline.org/index2.htm
JCVI Comprehensive Microbial Resource	Free website used to display publicly available prokaryotic genomes http://cmr.jcvi.org/tigr-scripts/CMR/CmrHomePage.cgi
Kyoto Encyclopedia of Genes and Genomes	Annotated and integrated system for comparisons of selected microbial genomic data http://www.genome.jp/kegg/
Microbial Genome Database	Collection of published microbial genomic data for comparative analyses http://mbgd.genome.ad.jp/
Microbial Genomics Resources	List of web-based resources for microbial genomics http://microbialgenomics.energy.gov/databases.shtml
Microbes Online	Website for browsing and comparing prokaryotic genomes http://www.microbesonline.org
NCBI GenBank	National Institutes of Health genetic sequence database, annotated collection of publicly available DNA sequences http://www.ncbi.nlm.nih.gov/Genbank/index.html
Sanger Institute Bacterial Genomes	Bacterial genomes and list of current genome projects http://www.sanger.ac.uk/Projects/Microbes/

[a]Note: specialized databases limited to one or a few related genera have been excluded from this list.

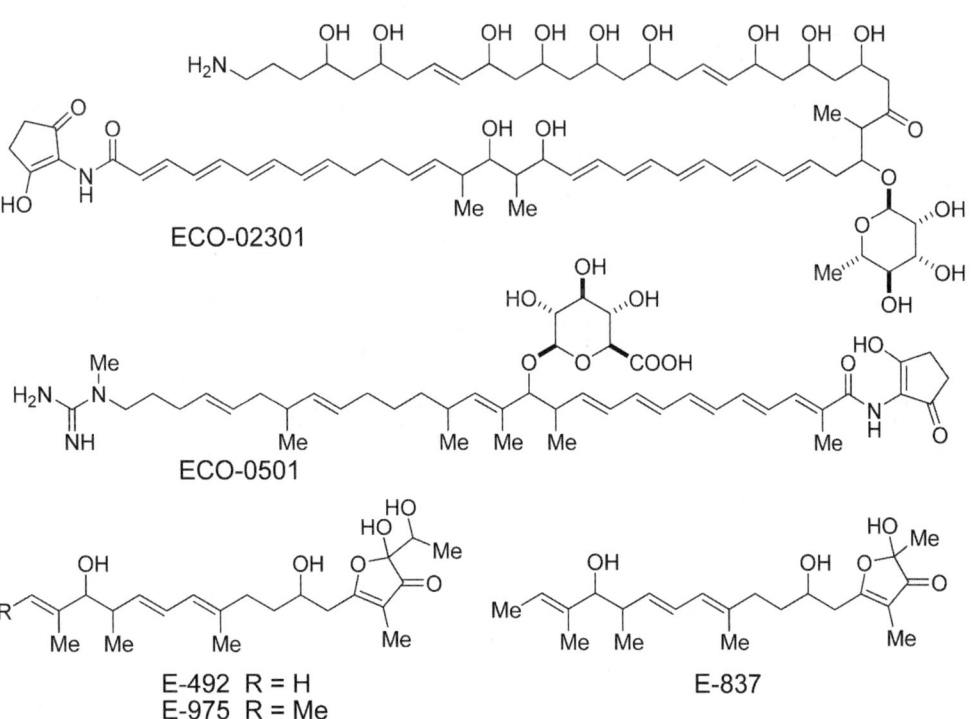

FIGURE 8 Examples of microbial metabolites whose structures were predicted based on genome scanning. The structure of each metabolite was later confirmed by isolation and characterization using NMR and mass spectrometric analyses.

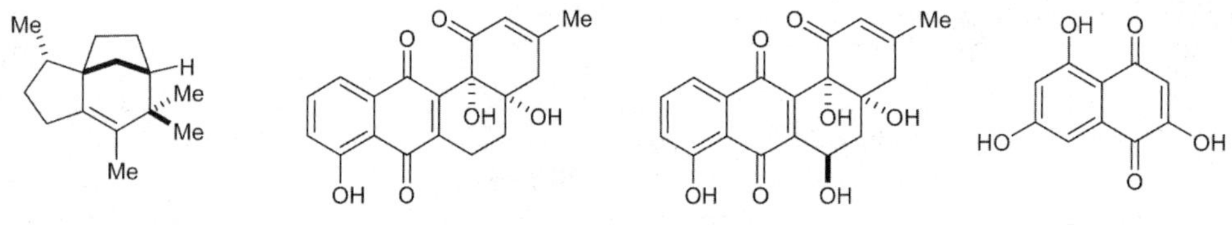

FIGURE 9 Examples of SBP-derived natural products that were identified by use of heterologous expression techniques.

product production. While this technology has been widely used for investigating numerous non-secondary-metabolite-encoding genes, its application to the study of natural products and SBPs has evolved dramatically in recent years. In 2006, genome scanning by Müller and colleagues led to the discovery of a type III PKS in the myxobacterium *Sorangium cellulosum* for which no corresponding secondary metabolite could be identified (44). The lack of production of the unknown metabolite was confirmed through gene activation by inserting a pOPB20 plasmid into the PKS coding sequence, resulting in no detectable changes to the microbial secondary metabolome. Cloning and subsequent transfection of the type III PKS into *Escherichia coli* and *Pseudomonas* sp. demonstrated that only the pseudomonad was able to function as an appropriate host for expression. The transfected *Pseudomonas* sp. culture turned visibly red and high-performance liquid chromatography profiling confirmed the presence of a new metabolite. The structure of the compound was determined to be the naphthoquinone flaviolin, which is readily generated from the oxidation of the actual biosynthetic precursor product, 1,3,6,8-tetrahydroxynaphthalene (Fig. 9).

An important advantage of heterologously expressing SBPs is the opportunity it affords to manipulate biosynthetic genes and enzymes, as well as study secondary-metabolite production in a cell-free environment. For example, the Canes group noted a terpene synthase of unknown function in *S. coelicolor* A3(2) and transfected the cloned gene into *E. coli*. Using gene manipulation, the researchers were able to produce a modified His$_6$-tagged translational product allowing the transgenic protein to be readily purified by Ni^{2+} affinity chromatography (57). In vitro incubation of the recombinant protein with farnesyl diphosphate in the presence of Mg^{2+} yielded the new sesquiterpene (+)-epi-isozizaene (Fig. 9). The importance of this approach is that utilization of the recombinant protein product allowed the group to carefully analyze the reaction mechanisms and kinetics of the new terpene synthase, which is generally not possible in a cellular host.

Palmu et al. (70) have taken the idea of SBP manipulation one step further, leading to the production of new angucy-cline analogs. This group observed that both *Streptomyces* sp. PGA64 and *Streptomyces* sp. H021 possessed PKS gene clusters, exhibiting significant homology to each other. Heterologous expression of varying portions of the gene clusters in *Streptomyces lividans* TK24 resulted in the generation of natural products bearing an angucyclic carbon skeleton (Fig. 9). These and other examples demonstrate the promise of using heterologous expression systems to manipulate SBPs. The ability to generate unique pools of natural and "unnatural" secondary metabolites presents an exceptional opportunity to identify new drugs and engineer novel chemical scaffolds.

7.5.4. Controlling SBP Expression by Means of In Situ Gene Manipulation

Despite its promising features, the routine use of heterologous expression systems for activating SBPs faces a number of significant difficulties. These challenges include handling large, multicomponent gene clusters, as well as identifying suitable expression hosts (81, 93). In light of these problems, methods to manipulate the in situ expression of SBPs hold great potential for rationally activating bacterial and fungal biosynthetic pathways. Although the concept of using promoter regulation to control natural product biosynthesis has been in use for years (1, 27, 69, 91), it was not until recently that this technique was applied to the activation of SBPs. The Hertweck group has presented evidence of a new and simple method for inducing SBP expression in a fungus (9). Having identified a silent hybrid PKS-NRPS (HPN) gene cluster in *Aspergillus nidulans*, the researchers cloned an associated putative activator gene (*apdR*) into a pAL4 vector and placed it under the control of an inducible alcohol dehydrogenase promoter (*alcAp*). Ectopic insertion of the *alcA-apdR* complex resulted in the inducible expression of *apdR*, which was confirmed by Northern blot analysis. Moreover, expression of *apdR* led to the transcriptional activation of the target HPN gene cluster and subsequent production of two new pyridone metabolites: aspyridones A and B (Fig. 10).

More recently, Chiang and colleagues noted the presence of two additional silent PKS gene clusters adjacent to one another in the *A. nidulans* genome (28). A homolog of the citrinin biosynthetic transcriptional activator was noted

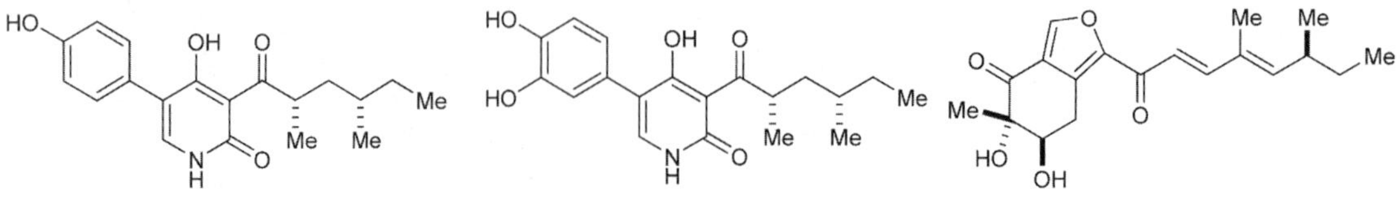

FIGURE 10 Compounds obtained from SBPs following activation using promoter/activator manipulation.

adjacent to the PKS cluster and it was rationalized that this gene was responsible for controlling the transcription of one or both of the silent PKS gene clusters. Replacement of the putative promoter with an inducible *alc*Ap resulted in the accumulation of two new (one major and one minor) metabolites under inducing conditions. Purification of the major metabolite yielded the new polyketide asperfuranone, which is structurally reminiscent of the azaphilones.

7.5.5. Final Thoughts and New Directions for Using Genome-Based Techniques

The power afforded by genome-scanning and molecular mining methods for investigating SBPs offers unprecedented opportunities to the natural products community. By use of genomic data, it is now feasible to identify transcriptionally suppressed secondary-metabolite-encoding gene clusters and to manipulate their expression in situ or transfect them into novel hosts. The ability to directly link gene clusters to their resultant biosynthetic products is expected to provide important new insights into the evolution and function of SBP-derived natural products. While methods for working with bacterial SBPs have advanced significantly (5, 13, 23, 45), corresponding advances in the exploration of fungal SBPs have developed more slowly. This is due in part to the greater complexity of fungal natural-product-encoding pathways (e.g., incorporation of introns, strain-specific activation of polyketide acyl carrier protein domains, instability of large gene clusters in heterologous hosts, etc.) compared with their bacterial counterparts (81).

Despite the advantageous features and potential of genome-based methods, these techniques suffer from a variety of significant shortcomings that limit their utility. These problems include the lack of "standard" methods that would be widely applicable to a majority of microorganisms and diverse types of SBPs. Consequently, significant effort must be devoted to generating appropriate heterologous hosts or engineering suitable microbial strains for in situ manipulation. These problems appreciably reduce the speed and efficiency with which researchers are able to investigate SBPs. It is remarkable that the swiftness with which a microbial genome can be fully sequenced (<1 day), reasonably well annotated (days to weeks), and SBPs identified (<1 min using predetermined BLAST search parameters) far outpaces the speed of cloning and/or expressing a targeted biosynthetic gene cluster (typically months in academic laboratories). Until these limitations can be overcome, access to SBPs by use of genome-based methods will continue to be severely limited. New and widely applicable tools (e.g., universal heterologous hosts, novel vectors for routinely handling large gene clusters, etc.) are needed for probing the overwhelmingly large number of SBPs that are predicted to exist.

7.6. MECHANISM-BASED AND HYBRID STRATEGIES FOR ACCESSING SBPs

7.6.1. Advances in the Speed of Detection of SBP-Derived Natural Products

Up to this point, we have focused on two distinct sets of methods for accessing microbial SBPs: culture manipulation techniques and genomics-guided approaches. While both sets of methods have important strengths, they also present a number of limitations that restrict their usefulness. These weaknesses are particularly critical when considering each method's applications in high-throughput discovery settings where the speed

of preparing new bioactive substances from SBPs is crucial. Two new strategies offer great potential for rationally procuring SBP-derived natural products at an accelerated pace. These techniques, chemical epigenetics and genomisotopic methods, represent the hybridization of genomic- and molecular-based strategies with culture-based applications. It is anticipated that these methods will find widespread use for the discovery of SBPs and investigation of SBP-derived natural products.

7.6.2. Chemical Epigenetic Approach

As part of our research group's search for salient molecular mechanisms involved in controlling SBP expression, we were inspired by the work of Keller and colleagues, which provided tantalizing evidence for an epigenetic component to the transcriptional regulation of secondary-metabolite-encoding genes and gene clusters (17, 54, 74). To test this assumption, we had assembled a library of small-molecule epigenetic modifying agents that inhibit histone deacetylase and DNA methyltransferase/histone methyltransferase functions. Screening this library against a panel of filamentous fungi revealed that in 11 of 12 cases the addition of epigenetic modifiers to the culture medium resulted in the production of new and/or enhanced levels of secondary metabolites (96). Closer examination of two of the fungi yielded several new epigenetically induced metabolites that were not biosynthesized under typical culture conditions. These included the production of new oxylipins and perylenequinones from *Cladosporium cladosporioides* using 5-azacytidine (DNA methyltransferase/histone methyltransferase inhibitor) and suberoylanilide hydroxamic acid (histone deacetylase inhibitor), respectively (Fig. 11). Similarly, a *Diatrype* sp. yielded two new glycosylated polyketides, lunalides A and B, following treatment with 5-azacytidine (Fig. 11).

To further investigate the extent to which epigenetic modifying agents were able to alter the transcription of secondary-metabolite-encoding pathways, we treated *A. niger* with suberoylanilide hydroxamic acid and 5-azacytidine, and examined changes in the transcription of its 31 PKSs, 15 NRPSs, and 9 HPN gene clusters (38). Whereas less than 30% of biosynthetic genes were transcribed under typical laboratory culture conditions utilizing an extensive range of media types, cultures treated with histone deacetylase and DNA methyltransferase/histone inhibitors exhibited elevated transcription levels for the majority of these pathways (38). We have also begun the task of characterizing the new SBP-derived metabolites that are produced by *A. niger* following epigenetic modifier treatment. This has led to the identification of the new *N*-phenylpyridine metabolite nygerone A following addition of suberoylanilide hydroxamic acid to the culture medium (Fig. 11) (47). The chemical epigenetic approach holds substantial promise for activating a multitude of fungal SBPs in a mechanistically rational manner and providing a means for rapidly screening SBP-derived compounds for biological activities.

7.6.3. Genomisotopic Method

Beginning with an NRPS whose biosynthetic product was unknown, the Gerwick group tested the use of a hybrid genomisotopic method to identify the structure of the secondary metabolite produced by the gene cluster (46). The genomisotopic technique utilizes a combined approach consisting of structure prediction (based on a gene sequence) and isotopic labeling (based on feeding of isotopically labeled precursors expected to be incorporated into the metabolite) for detection of the target natural product (Fig. 12). Using this

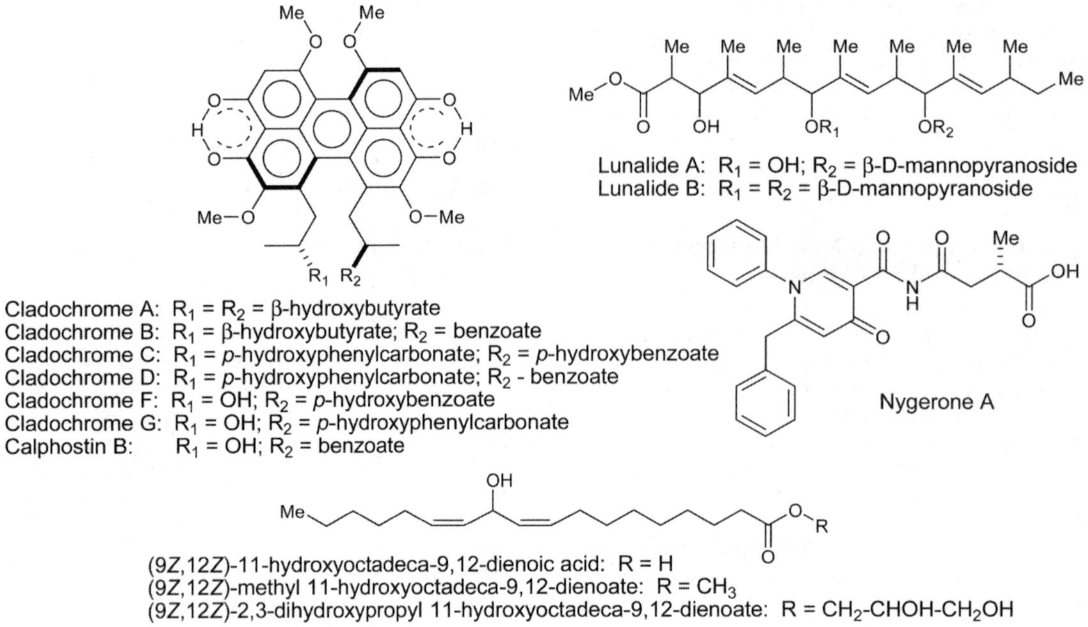

Lunalide A: R1 = OH; R2 = β-D-mannopyranoside
Lunalide B: R1 = R2 = β-D-mannopyranoside
Cladochrome A: R1 = R2 = β-hydroxybutyrate
Cladochrome B: R1 = β-hydroxybutyrate; R2 = benzoate
Cladochrome C: R1 = p-hydroxyphenylcarbonate; R2 = p-hydroxybenzoate
Cladochrome D: R1 = p-hydroxyphenylcarbonate; R2 - benzoate
Cladochrome F: R1 = OH; R2 = p-hydroxybenzoate
Cladochrome G: R1 = OH; R2 = p-hydroxyphenylcarbonate
Calphostin B: R1 = OH; R2 = benzoate
Nygerone A
(9Z,12Z)-11-hydroxyoctadeca-9,12-dienoic acid: R = H
(9Z,12Z)-methyl 11-hydroxyoctadeca-9,12-dienoate: R = CH3
(9Z,12Z)-2,3-dihydroxypropyl 11-hydroxyoctadeca-9,12-dienoate: R = CH2-CHOH-CH2OH

FIGURE 11 Natural products obtained by using the chemical epigenetic induction of SBPs.

approach, *Pseudomonas fluorescens* Pf-5 was noted to contain a large NRPS cluster encoding for a decapeptide product. Four of the NRPS's modules exhibited highly conserved domains that were presumed responsible for the addition of leucine residues to metabolite. Using ^{1}H-^{15}N heteronuclear multiple bond correlation nuclear magnetic resonance (HMBC NMR), the researchers were able to track the incorporation of ^{15}N-labeled leucine residues into a group of related secondary metabolites, which led to the isolation of orfamide A and its related congeners (Fig. 12). This technique is well suited for the exploration of SBPs, and it should provide an important new tool for microbial secondary-metabolite investigation.

Orfamide gene cluster
ofaA
ofaB
ofaC
Genome-scanning reveals NRPS gene cluster of undetermined function
Gene sequence used to predict partial product structure
Partial predicted structure of NRPS product
Pseudomonas fluorescens
Orfamide A (actual structure)
Identify conditions conducive for gene expression and biosynthesis of target compound
15N-labeled leucine fed to P. fluorescens
Isolation of isotopically-labeled product guided by 1H-15N HMBC NMR

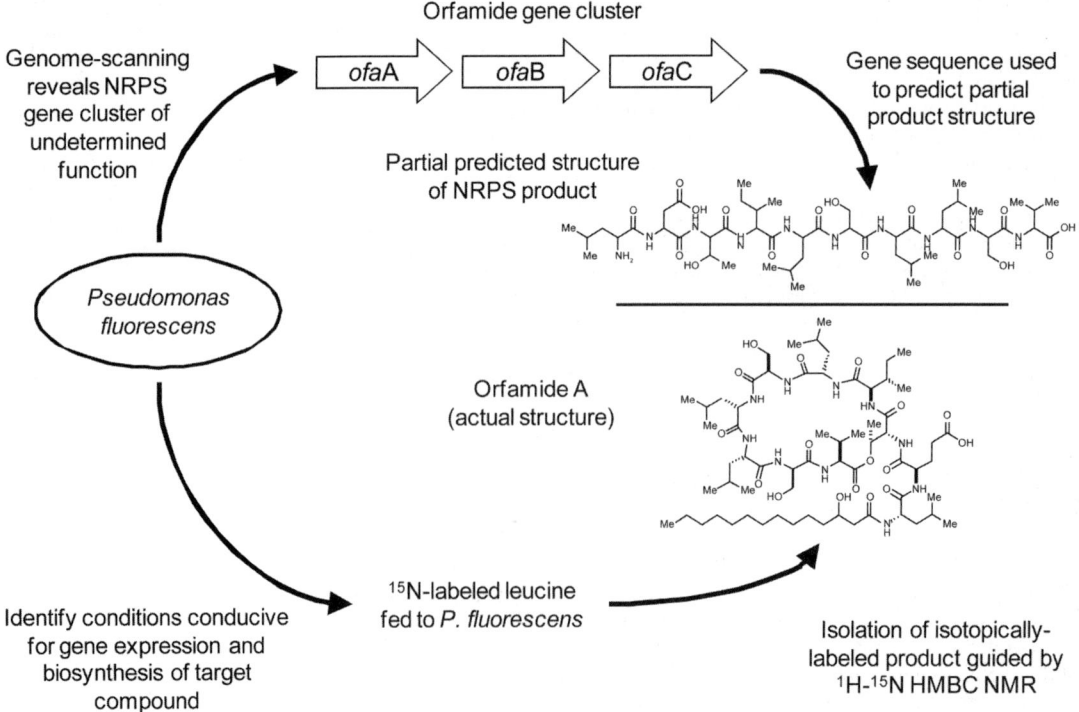

FIGURE 12 Overview of the genomisotopic method and its application to the study of the orfamide gene cluster in *P. fluorescens* Pf-5.

TABLE 5 Summary of the major strengths and weaknesses of methods for investigating microbial SBPs

Method	Technique	Advantages	Disadvantages
Culture-dependent approaches	Culture manipulation	Ease of use Low cost Technical simplicity Genomic information not required	Time- and labor-intensive No molecular targets defined a priori No direct link established between metabolites of interest and biosynthetic genes Same metabolites repeatedly observed; substantial dereplication efforts required
	Coculture	Ease of use Low cost Ability to utilize complex intercellular interactions to induce secondary-metabolite production Genomic information not required	Time- and labor-intensive Generation of metabolites by target organism not predictable Purported ecological relevance is uncertain Metabolite source difficult to identify in mixtures of live organisms
Genomics-driven approaches	Genome scanning and mining	In silico method enables rapid gene identification Full biosynthetic potential of microorganisms revealed (i.e., all secondary-metabolite-encoding pathways can be detected)	Additional efforts needed to produce and identify metabolite of interest Prediction of metabolite structure presents several challenges, especially for fungi Biological activity of target metabolite is unknown Predicted physiochemical properties of target compound may match other endogenous metabolites
	Heterologous expression	Offers potential for exceptional control over gene/gene cluster transcription Genes can be readily subjected to manipulation to create new "unnatural" secondary metabolites Biosynthetic pathways can be transferred to suitable hosts for ease of fermentation	Significant molecular biology experience required; may limit widespread use Thorough understanding of gene cluster function, tailoring enzymes, and essential metabolic precursors required Difficulties in handling and manipulating of large gene clusters (particularly fungi) Identification of suitable transgenic host can be challenging
	Promoter/ activator manipulation	Handling of large gene clusters is eliminated All host-specific biosynthetic factors are intact and presumably functional Successful integration of relatively small "activator" gene into host microbe is less challenging compared with heterologous expression approach	Heterologous expression of promoter/activator may be problematic in poorly studied species Not applicable to high-throughput investigations Generalized applicability of this method has not been established Transcription-level manipulation may not directly correlate with changes in translation and metabolite biosynthesis
Mechanism-based and hybrid approaches	Chemical epigenetics	Rational approach targeting regulatory control of secondary-metabolite-encoding biosynthetic pathways Ease of implementation in high-throughput settings for bioactive compound screening Relatively inexpensive and rapid process for screening microbes without need for a priori genome mining	Transcription-level manipulation may not directly correlate with changes in translation and metabolite biosynthesis Limited to fungal systems; not all fungi are responsive Rapid biotransformation and degradation of epigenetic modifiers can limit the effectiveness of the epigenetic agents in certain cases
	Genomisotopic	Does not require transfer of target gene/gene cluster to heterologous host Potential for rapid metabolite identification when judiciously applied to appropriate biosynthetic targets	Extensive genomic-level knowledge of target gene required Testing of various media/culture conditions required to induce compound biosynthesis Accurate predictions of target compound's structure and biosynthetic precursors required Suitable conditions for precursor incorporation must be identified and isotopically labeled precursors must be unique substrate for target metabolite

7.6.4. Final Thoughts and New Directions for Using Mechanism-Based and Hybrid Approaches

Mechanism-based and hybrid techniques represent a powerful new generation of investigational tools that will help address the growing needs of natural products researchers to rapidly access SBPs and their secondary metabolites. Key to the significant utility of these methods is their cross-disciplinary blending of chemical and biological techniques to efficiently probe SBPs. However, if further methodological advances employing other translational approaches are to occur, new information concerning the molecular underpinnings of SBP transcription, translation, and biosynthesis are needed. Despite the fact that there have been many significant advances toward understanding the cellular events responsible for controlling secondary-metabolite production in microorganisms (49, 77, 97), there is still a woeful lack of information concerning universally applicable processes that are exploitable for accessing SBPs. Further molecular-based studies will be needed to help fill this gap in our current knowledge and provide new approaches to SBP discovery.

7.7. CONCLUSIONS AND PERSPECTIVES ON METHODS FOR ACCESSING SBPS

The investigation of SBPs has advanced significantly in the past 5 years, and it is expected that further advances in genome-based technologies will continue to fuel this trend. Currently available techniques for investigating SBPs fall into three discrete categories: (i) culture-dependent, (ii) genomics-driven, and (iii) mechanism-based and hybrid approaches. Each of these method sets present unique advantages and disadvantages for studying SBPs (Table 5). With so many new and exceedingly beneficial techniques becoming available, we anticipate that traditional culture-based approaches will have significantly diminished value as part of the repertoire of contemporary SBP research methods. Replacing these approaches will be a variety of new tools that offer increased speed and efficiency for evaluating SBPs in a high-throughput fashion. Consequently, we expect that microbial natural products will undergo a renaissance of renewed interest for utilization as a resource to discover new bioactive molecules.

One of the foreseeable outcomes of current trends in the field of SBP studies is the necessity for increased cooperation between chemists and biologists. Driving this need will be an exponentially increasing pool of genomic data containing enormous numbers of candidate genes for testing. Sifting through these massive quantities of data and identifying key parameters for choosing which genes to target will become increasingly important for efficiently mining this resource. In addition, new discoveries related to the molecular mechanisms contributing to SBP suppression are certain to require the need for novel biologically based solutions. Likewise, increasingly sensitive and sophisticated mass spectrometry and NMR tools will aid in the detection and structure confirmation of SBP-derived natural products. By focusing the complementary talents of multidisciplinary teams on SBPs, researchers are certain to uncover fascinating new insights into the structures, functions, and commercialization potential of the microbial world's hidden natural products.

REFERENCES

1. **Aceti, D. J., and W. C. Champness.** 1998. Transcriptional regulation of *Streptomyces coelicolor* pathway-specific antibiotic regulators by the *absA* and *absB* loci. *J. Bacteriol.* **180:**3100–3106.

2. **Amagata, T., M. Tanaka, T. Yamada, M. Doi, K. Minoura, H. Ohishi, T. Yamori, and A. Numata.** 2007. Variation in cytostatic constituents of a sponge-derived *Gymnascella dankaliensis* by manipulating the carbon source. *J. Nat. Prod.* **70:**1731–1740.

3. **Angell, S., B. J. Bench, H. Williams, and C. M. H. Watanabe.** 2006. Pyocyanin isolated from a marine microbial population: synergistic production between two distinct bacterial species and mode of action. *Chem. Biol.* **13:**1349–1359.

4. **Ashley, J. N., B. C. Hobbs, and H. Raistrick.** 1937. Studies in the biochemistry of micro-organisms: the crystalline colouring matters of *Fusarium culmorum* (W. G. Smith) Sacc. and related forms. *Biochem. J.* **31:**385–397.

5. **Baltz, R. H.** 2008. Renaissance in antibacterial discovery from actinomycetes. *Curr. Opin. Pharmacol.* **8:**557–563.

6. **Banskota, A. H., J. B. McAlpine, D. Sorensen, M. Aouidate, M. Piraee, A.-M. Alarco, S. Omura, K. Shiomi, C. M. Farnet, and E. Zazopoulos.** 2006. Isolation and identification of three new 5-alkenyl-3,3(2H)-furanones from two *Streptomyces* species using a genomic screening approach. *J. Antibiot.* **59:**168–176.

7. **Banskota, A. H., J. B. McAlpine, D. Sorensen, A. Ibrahim, M. Aouidate, M. Piraee, A.-M. Alarco, C. M. Farnet, and E. Zazopoulos.** 2006. Genomic analyses lead to novel secondary metabolites. *J. Antibiot.* **59:**533–542.

8. **Bentley, S. D., K. F. Chater, A. M. Cerdeno-Tarraga, G. L. Challis, N. R. Thomson, K. D. James, D. E. Harris, M. A. Quail, H. Kieser, D. Harper, A. Bateman, S. Brown, G. Chandra, C. W. Chen, M. Collins, A. Cronin, A. Fraser, A. Goble, J. Hidalgo, T. Hornsby, S. Howarth, C. H. Huang, T. Kieser, L. Larke, L. Murphy, K. Oliver, S. O'Neil, E. Rabbinowitsch, M. A. Rajandream, K. Rutherford, S. Rutter, K. Seeger, D. Saunders, S. Sharp, R. Squares, S. Squares, K. Taylor, T. Warren, A. Wietzorrek, J. Woodward, B. G. Barrell, J. Parkhill, and D. A. Hopwood.** 2002. Complete genome sequence of the model actinomycete *Streptomyces coelicolor* A3(2). *Nature* **417:**141–147.

9. **Bergmann, S., J. Schumann, K. Scherlach, C. Lange, A. A. Brakhage, and C. Hertweck.** 2007. Genomics-driven discovery of PKS-NRPS hybrid metabolites from *Aspergillus nidulans*. *Nat. Chem. Biol.* **3:**213–217.

10. **Bills, G. F., G. Platas, A. Fillola, M. R. Jiménez, J. Collado, F. Vicente, J. Martín, A. González, J. Bur-Zimmermann, J. R. Tormo, and F. Peláez.** 2008. Enhancement of antibiotic and secondary metabolite detection from filamentous fungi by growth on nutritional arrays. *J. Appl. Microbiol.* **104:**1644–1658.

11. **Bitzer, J., T. Grosse, L. Z. Wang, S. Lang, W. Beil, and A. Zeeck.** 2006. New aminophenoxazinones from a marine *Halomonas* sp.: fermentation, structure elucidation, and biological activity. *J. Antibiot.* **59:**86–92.

12. **Bode, H. B., B. Bethe, R. Höfs, and A. Zeeck.** 2002. Big effects from small changes: possible ways to explore nature's chemical diversity. *ChemBioChem* **3:**619–627.

13. **Bode, H. B., and R. Müller.** 2005. The impact of bacterial genomics on natural product research. *Angew. Chem. Int. Ed.* **44:**6828–6846.

14. **Bode, H. B., M. Walker, and A. Zeeck.** 2000. Secondary metabolites by chemical screening. 42. Cladospirones B to I from *Sphaeropsidales* sp F-24'707 by variation of culture conditions. *Eur. J. Org. Chem.* **2000:**3185–3193.

15. **Bode, H. B., and A. Zeeck.** 2000. Sphaerolone and dihydrosphaerolone, two bisnaphthyl-pigments from the fungus *Sphaeropsidales* sp. F-24'707. *Phytochemistry* **54:**597–601.

16. **Boelker, M., C. W. Basse, and J. Schirawski.** 2008. *Ustilago maydis* secondary metabolism: from genomics to biochemistry. *Fungal Genet. Biol.* **45:**S88–S93.

17. **Bok, J. W., and N. P. Keller.** 2004. LaeA, a regulator of secondary metabolism in *Aspergillus* spp. *Eukaryot. Cell* **3:**527–535.

18. Brakhage, A. A., J. Schuemann, S. Bergmann, K. Scherlach, V. Schroeckh, and C. Hertweck. 2008. Activation of fungal silent gene clusters: a new avenue to drug discovery. *Prog. Drug Res.* **66:**3–12.

19. Burgess, J. G., E. M. Jordan, M. Bregu, A. Mearns-Spragg, and K. G. Boyd. 1999. Microbial antagonism: a neglected avenue of natural products research. *J. Biotechnol.* **70:**27–32.

20. Butler, A. R., and E. Cundliffe. 2001. Influence of dimethylsulfoxide on tylosin production in *Streptomyces fradiae*. *J. Ind. Microbiol. Biotechnol.* **27:**46–51.

21. Calam, C. T., and D. J. D. Hockenhull. 1949. The production of penicillin in surface culture, using chemically defined media. *J. Gen. Microbiol.* **3:**19–31.

22. Challis, G. L. 2008. Genome mining for novel natural product discovery. *J. Med. Chem.* **51:**2618–2628.

23. Challis, G. L. 2008. Mining microbial genomes for new natural products and biosynthetic pathways. *Microbiology* **154:**1555–1569.

24. Challis, G. L., and D. A. Hopwood. 2003. Synergy and contingency as driving forces for the evolution of multiple secondary metabolite production by *Streptomyces* species. *Proc. Natl. Acad. Sci. USA* **100:**14555–14561.

25. Challis, G. L., and J. Ravel. 2000. Coelichelin, a new peptide siderophore encoded by the *Streptomyces coelicolor* genome: structure prediction from the sequence of its non-ribosomal peptide synthetase. *FEMS Microbiol. Lett.* **187:**111–114.

26. Chen, G. H., G. Y. S. Wang, X. Li, B. Waters, and J. Davies. 2000. Enhanced production of microbial metabolites in the presence of dimethyl sulfoxide. *J. Antibiot.* **53:**1145–1153.

27. Chen, L., Y. Lu, J. Chen, W. Zhang, D. Shu, Z. Qin, S. Yang, and W. Jiang. 2008. Characterization of a negative regulator AveI for avermectin biosynthesis in *Streptomyces avermitilis* NRRL8165. *Appl. Microbiol. Biotechnol.* **80:**277–286.

28. Chiang, Y.-M., E. Szewczyk, A. D. Davidson, N. Keller, B. R. Oakley, and C. C. C. Wang. 2009. A gene cluster containing two fungal polyketide synthases encodes the biosynthetic pathway for a polyketide, asperfuranone, in *Aspergillus nidulans*. *J. Am. Chem. Soc.* **131:**2965–2970.

29. Christian, O. E., J. Compton, K. R. Christian, S. L. Mooberry, F. A. Valeriote, and P. Crews. 2005. Using jasplakinolide to turn on pathways that enable the isolation of new chaetoglobosins from *Phomopsis asparagi*. *J. Nat. Prod.* **68:**1592–1597.

30. Cueto, M., P. R. Jensen, C. Kauffman, W. Fenical, E. Lobkovsky, and J. Clardy. 2001. Pestalone, a new antibiotic produced by a marine fungus in response to bacterial challenge. *J. Nat. Prod.* **64:**1444–1446.

31. Curtin, T., G. Fitzgerald, and J. Reilly. 1940. Production of phoenicine on synthetic media: *Penicillium phoeniceum* Van Beyma. 2. *Penicillium rubrum* Grasberger-Stoll. *Biochem. J.* **34:**1605–1610.

32. Davey, V. F., and M. J. Johnson. 1953. Penicillin production in corn steep media with continuous carbohydrate addition. *Appl. Microbiol.* **1:**208–211.

33. Davies, J. 2006. Are antibiotics naturally antibiotics? *J. Ind. Microbiol. Biotechnol.* **33:**496–499.

34. Dean, R. A., N. J. Talbot, D. J. Ebbole, M. L. Farman, T. K. Mitchell, M. J. Orbach, M. Thon, R. Kulkarni, J. R. Xu, H. Q. Pan, N. D. Read, Y. H. Lee, I. Carbone, D. Brown, Y. Y. Oh, N. Donofrio, J. S. Jeong, D. M. Soanes, S. Djonovic, E. Kolomiets, C. Rehmeyer, W. X. Li, M. Harding, S. Kim, M. H. Lebrun, H. Bohnert, S. Coughlan, J. Butler, S. Calvo, L. J. Ma, R. Nicol, S. Purcell, C. Nusbaum, J. E. Galagan, and B. W. Birren. 2005. The genome sequence of the rice blast fungus *Magnaporthe grisea*. *Nature* **434:**980–986.

35. Degenkolb, T., S. Heinze, B. Schlegel, G. Strobel, Gr. Auml, and U. Fe. 2002. Formation of new lipoaminopeptides, acremostatins A, B, and C, by co-cultivation of *Acremonium* sp. Tbp-5 and *Mycogone rosea* DSM 12973. *Biosci. Biotechnol. Biochem.* **66:**883–886.

36. Doull, J. L., S. W. Ayer, A. K. Singh, and P. Thibault. 1993. Production of a novel polyketide antibiotic, jadomycin B, by *Streptomyces venezuelae* following heat-shock. *J. Antibiot.* **46:**869–871.

37. Firn, R. D., and C. G. Jones. 2000. The evolution of secondary metabolism—a unifying model. *Mol. Microbiol.* **37:**989–994.

38. Fisch, K. M., A. F. Gillaspy, M. Gipson, J. C. Henrikson, A. R. Hoover, L. Jackson, F. Z. Najar, H. Wägele, and R. H. Cichewicz. 2009. Chemical induction of silent pathway transcription in *Aspergillus niger*. *J. Ind. Microbiol. Biotechnol.* **36:**1199–1213.

39. Fox, E. M., and B. J. Howlett. 2008. Secondary metabolism: regulation and role in fungal biology. *Curr. Opin. Microbiol.* **11:**481–487.

40. Fuchser, J., and A. Zeeck. 1997. Secondary metabolites by chemical screening. 34. Aspinolides and aspinonene/aspyrone co-metabolites, new pentaketides produced by *Aspergillus ochraceus*. *Liebigs Ann. Rec.* **1997:**87–95.

41. Galagan, J. E., S. E. Calvo, K. A. Borkovich, E. U. Selker, N. D. Read, D. Jaffe, W. FitzHugh, L. J. Ma, S. Smirnov, S. Purcell, B. Rehman, T. Elkins, R. Engels, S. G. Wang, C. B. Nielsen, J. Butler, M. Endrizzi, D. Y. Qui, P. Ianakiev, D. B. Pedersen, M. A. Nelson, M. Werner-Washburne, C. P. Selitrennikoff, J. A. Kinsey, E. L. Braun, A. Zelter, U. Schulte, G. O. Kothe, G. Jedd, W. Mewes, C. Staben, E. Marcotte, D. Greenberg, A. Roy, K. Foley, J. Naylor, N. Stabge-Thomann, R. Barrett, S. Gnerre, M. Kamal, M. Kamvysselis, E. Mauceli, C. Bielke, S. Rudd, D. Frishman, S. Krystofova, C. Rasmussen, R. L. Metzenberg, D. D. Perkins, S. Kroken, C. Cogoni, G. Macino, D. Catcheside, W. X. Li, R. J. Pratt, S. A. Osmani, C. P. C. DeSouza, L. Glass, M. J. Orbach, J. A. Berglund, R. Voelker, O. Yarden, M. Plamann, S. Seller, J. Dunlap, A. Radford, R. Aramayo, D. O. Natvig, L. A. Alex, G. Mannhaupt, D. J. Ebbole, M. Freitag, I. Paulsen, M. S. Sachs, E. S. Lander, C. Nusbaum, and B. Birren. 2003. The genome sequence of the filamentous fungus *Neurospora crassa*. *Nature* **422:**859–868.

42. Galagan, J. E., S. E. Calvo, C. Cuomo, L. J. Ma, J. R. Wortman, S. Batzoglou, S. I. Lee, M. Basturkmen, C. C. Spevak, J. Clutterbuck, V. Kapitonov, J. Jurka, C. Scazzocchio, M. Farman, J. Butler, S. Purcell, S. Harris, G. H. Braus, O. Draht, S. Busch, C. D'Enfert, C. Bouchier, G. H. Goldman, D. Bell-Pedersen, S. Griffiths-Jones, J. H. Doonan, J. Yu, K. Vienken, A. Pain, M. Freitag, E. U. Selker, D. B. Archer, M. A. Penalva, B. R. Oakley, M. Momany, T. Tanaka, T. Kumagai, K. Asai, M. Machida, W. C. Nierman, D. W. Denning, M. Caddick, M. Hynes, M. Paoletti, R. Fischer, B. Miller, P. Dyer, M. S. Sachs, S. A. Osmani, and B. W. Birren. 2005. Sequencing of *Aspergillus nidulans* and comparative analysis with A. *fumigatus* and A. *oryzae*. *Nature* **438:**1105–1115.

43. Gloer, J. B. 2007. Applications of fungal ecology in the search for new bioactive natural products, p. 257–283. *In* C. P. Kubicek and I. S. Druzhinina (ed.), *The Mycota*. Springer, Berlin, Germany.

44. Gross, F., N. Luniak, O. Perlova, N. Gaitatzis, H. Jenke-Kodama, K. Gerth, D. Gottschalk, E. Dittmann, and R. Müller. 2006. Bacterial type III polyketide synthases: phylogenetic analysis and potential for the production of novel secondary metabolites by heterologous expression in pseudomonads. *Arch. Microbiol.* **185:**28–38.

45. Gross, H. 2007. Strategies to unravel the function of orphan biosynthesis pathways: recent examples and future prospects. *Appl. Microbiol. Biotechnol.* **75:**267–277.

46. Gross, H., V. O. Stockwell, M. D. Henkels, B. Nowak-Thompson, J. E. Loper, and W. H. Gerwick. 2007. The genomisotopic approach: a systematic method to isolate products of orphan biosynthetic gene clusters. *Chem. Biol.* **14:**53–63.

47. Henrikson, J. C., A. R. Hoover, P. M. Joyner, and R. H. Cichewicz. 2009. A chemical epigenetics approach for

engineering the *in situ* biosynthesis of a cryptic natural product from *Aspergillus niger*. *Org. Biomol. Chem.* **7:**435–438.

48. **Hofs, R., M. Walker, and A. Zeeck.** 2000. Hexacyclinic acid, a polyketide from *Streptomyces* with a novel carbon skeleton. *Angew. Chem. Int. Ed.* **39:**3258–3261.

49. **Horinouchi, S.** 2007. Mining and polishing of the treasure trove in the bacterial genus *Streptomyces*. *Biosci. Biotechnol. Biochem.* **71:**283–299.

50. **Ikeda, H., J. Ishikawa, A. Hanamoto, M. Shinose, H. Kikuchi, T. Shiba, Y. Sakaki, M. Hattori, and S. Omura.** 2003. Complete genome sequence and comparative analysis of the industrial microorganism *Streptomyces avermitilis*. *Nat. Biotechnol.* **21:**526–531.

51. **Jarvis, F. G., and M. J. Johnson.** 1947. The role of the constituents of synthetic media for penicillin production. *J. Am. Chem. Soc.* **69:**3010–3017.

52. **Kamper, J., R. Kahmann, M. Bolker, L. J. Ma, T. Brefort, B. J. Saville, F. Banuett, J. W. Kronstad, S. E. Gold, O. Muller, M. H. Perlin, H. A. B. Wosten, R. de Vries, J. Ruiz-Herrera, C. G. Reynaga-Pena, K. Snetselaar, M. McCann, J. Perez-Martin, M. Feldbrugge, C. W. Basse, G. Steinberg, J. I. Ibeas, W. Holloman, P. Guzman, M. Farman, J. E. Stajich, R. Sentandreu, J. M. Gonzalez-Prieto, J. C. Kennell, L. Molina, J. Schirawski, A. Mendoza-Mendoza, D. Greilinger, K. Munch, N. Rossel, M. Scherer, M. Vranes, O. Ladendorf, V. Vincon, U. Fuchs, B. Sandrock, S. Meng, E. C. H. Ho, M. J. Cahill, K. J. Boyce, J. Klose, S. J. Klosterman, H. J. Deelstra, L. Ortiz-Castellanos, W. X. Li, P. Sanchez-Alonso, P. H. Schreier, I. Hauser-Hahn, M. Vaupel, E. Koopmann, G. Friedrich, H. Voss, T. Schluter, J. Margolis, D. Platt, C. Swimmer, A. Gnirke, F. Chen, V. Vysotskaia, G. Mannhaupt, U. Guldener, M. Munsterkotter, D. Haase, M. Oesterheld, H. W. Mewes, E. W. Mauceli, D. DeCaprio, C. M. Wade, J. Butler, S. Young, D. B. Jaffe, S. Calvo, C. Nusbaum, J. Galagan, and B. W. Birren.** 2006. Insights from the genome of the biotrophic fungal plant pathogen *Ustilago maydis*. *Nature* **444:**97–101.

53. **Kawai, K., G. Wang, S. Okamoto, and K. Ochi.** 2007. The rare earth, scandium, causes antibiotic overproduction in *Streptomyces* spp. *FEMS Microbiol. Lett.* **274:**311–315.

54. **Keller, N. P., G. Turner, and J. W. Bennett.** 2005. Fungal secondary metabolism—from biochemistry to genomics. *Nat. Rev. Microbiol.* **3:**937–947.

55. **Lautru, S., R. J. Deeth, L. M. Bailey, and G. L. Challis.** 2005. Discovery of a new peptide natural product by *Streptomyces coelicolor* genome mining. *Nat. Chem. Biol.* **1:**265–269.

56. **Lee, B.-N., S. Kroken, D. Y. T. Chou, B. Robbertse, O. C. Yoder, and B. G. Turgeon.** 2005. Functional analysis of all nonribosomal peptide synthetases in *Cochliobolus heterostrophus* reveals a factor, NPS6, involved in virulence and resistance to oxidative stress. *Eukaryot. Cell* **4:**545–555.

57. **Lin, X., R. Hopson, and D. E. Cane.** 2006. Genome mining in *Streptomyces coelicolor*: molecular cloning and characterization of a new sesquiterpene synthase. *J. Am. Chem. Soc.* **128:**6022–6023.

58. **Lin, Z., T. Zhu, H. Wei, G. Zhang, H. Wang, and Q. Gu.** 2009. Spicochalasin A and new aspochalasins from the marine-derived fungus *Spicaria elegans*. *Eur. J. Org. Chem.* **2009:**3045–3051.

59. **Machida, M., K. Asai, M. Sano, T. Tanaka, T. Kumagai, G. Terai, K. I. Kusumoto, T. Arima, O. Akita, Y. Kashiwagi, K. Abe, K. Gomi, H. Horiuchi, K. Kitamoto, T. Kobayashi, M. Takeuchi, D. W. Denning, J. E. Galagan, W. C. Nierman, J. J. Yu, D. B. Archer, J. W. Bennett, D. Bhatnagar, T. E. Cleveland, N. D. Fedorova, O. Gotoh, H. Horikawa, A. Hosoyama, M. Ichinomiya, R. Igarashi, K. Iwashita, P. R. Juvvadi, M. Kato, Y. Kato, T. Kin, A. Kokubun, H. Maeda, N. Maeyama, J. Maruyama, H. Nagasaki, T. Nakajima, K. Oda, K. Okada, I. Paulsen, K. Sakamoto, T. Sawano, M. Takahashi, K. Takase, Y. Terabayashi, J. R. Wortman, O. Yamada, Y. Yamagata, H.** Anazawa, Y. Hata, Y. Koide, T. Komori, Y. Koyama, T. Minetoki, S. Suharnan, A. Tanaka, K. Isono, S. Kuhara, N. Ogasawara, and H. Kikuchi. 2005. Genome sequencing and analysis of *Aspergillus oryzae*. *Nature* **438:**1157–1161.

60. **MacLean, D., J. D. Jones, and D. J. Studholme.** 2009. Application of "next-generation" sequencing technologies to microbial genetics. *Nat. Rev. Microbiol.* **7:**287–296.

61. **Maldonado, A., J. L. Ruiz-Barba, and R. Jimenez-Diaz.** 2003. Purification and genetic characterization of plantaricin NC8, a novel coculture-inducible two-peptide bacteriocin from *Lactobacillus plantarum* NC8. *Appl. Environ. Microbiol.* **69:**383–389.

62. **Mardis, E. R.** 2008. The impact of next-generation sequencing technology on genetics. *Trends Genet.* **24:**133–141.

63. **McAlpine, J. B., B. O. Bachmann, M. Piraee, S. Tremblay, A.-M. Alarco, E. Zazopoulos, and C. M. Farnet.** 2005. Microbial genomics as a guide to drug discovery and structural elucidation: ECO-02301, a novel antifungal agent, as an example. *J. Nat. Prod.* **68:**493–496.

64. **Mearns-Spragg, A., M. Bregu, K. G. Boyd, and J. G. Burgess.** 1998. Cross-species induction and enhancement of antimicrobial activity produced by epibiotic bacteria from marine algae and invertebrates, after exposure to terrestrial bacteria. *Lett. Appl. Microbiol.* **27:**142–146.

65. **Nierman, W. C., A. Pain, M. J. Anderson, J. R. Wortman, H. S. Kim, J. Arroyo, M. Berriman, K. Abe, D. B. Archer, C. Bermejo, J. Bennett, P. Bowyer, D. Chen, M. Collins, R. Coulsen, R. Davies, P. S. Dyer, M. Farman, N. Fedorova, N. Fedorova, T. V. Feldblyum, R. Fischer, N. Fosker, A. Fraser, J. L. Garcia, M. J. Garcia, A. Goble, G. H. Goldman, K. Gomi, S. Griffith-Jones, R. Gwilliam, B. Haas, H. Haas, D. Harris, H. Horiuchi, J. Huang, S. Humphray, J. Jimenez, N. Keller, H. Khouri, K. Kitamoto, T. Kobayashi, S. Konzack, R. Kulkarni, T. Kumagai, A. Lafton, J.-P. Latge, W. Li, A. Lord, C. Lu, W. H. Majoros, G. S. May, B. L. Miller, Y. Mohamoud, M. Molina, M. Monod, I. Mouyna, S. Mulligan, L. Murphy, S. O'Neil, I. Paulsen, M. A. Penalva, M. Pertea, C. Price, B. L. Pritchard, M. A. Quail, E. Rabbinowitsch, N. Rawlins, M.-A. Rajandream, U. Reichard, H. Renauld, G. D. Robson, S. R. de Cordoba, J. M. Rodriguez-Pena, C. M. Ronning, S. Rutter, S. L. Salzberg, M. Sanchez, J. C. Sanchez-Ferrero, D. Saunders, K. Seeger, R. Squares, S. Squares, M. Takeuchi, F. Tekaia, G. Turner, C. R. V. de Aldana, J. Weidman, O. White, J. Woodward, J.-H. Yu, C. Fraser, J. E. Galagan, K. Asai, M. Machida, N. Hall, B. Barrell, and D. W. Denning.** 2005. Genomic sequence of the pathogenic and allergenic filamentous fungus *Aspergillus fumigatus*. *Nature* **438:**1151–1156.

66. **Oh, D.-C., P. R. Jensen, C. A. Kauffman, and W. Fenical.** 2005. Libertellenones A-D: induction of cytotoxic diterpenoid biosynthesis by marine microbial competition. *Bioorg. Med. Chem.* **13:**5267–5273.

67. **Oh, D. C., C. A. Kauffman, P. R. Jensen, and W. Fenical.** 2007. Induced production of emericellamides A and B from the marine-derived fungus *Emericella* sp. in competing coculture. *J. Nat. Prod.* **70:**515–520.

68. **Omura, S., H. Ikeda, J. Ishikawa, A. Hanamoto, C. Takahashi, M. Shinose, Y. Takahashi, H. Horikawa, H. Nakazawa, T. Osonoe, H. Kikuchi, T. Shiba, Y. Sakaki, and M. Hattori.** 2001. Genome sequence of an industrial microorganism *Streptomyces avermitilis*: deducing the ability of producing secondary metabolites. *Proc. Natl. Acad. Sci. USA* **98:**12215–12220.

69. **Onaka, H., N. Ando, T. Nihira, Y. Yamada, T. Beppu, and S. Horinouchi.** 1995. Cloning and characterization of the A-factor receptor gene from *Streptomyces griseus*. *J. Bacteriol.* **177:**6083–6092.

70. **Palmu, K., K. Ishida, P. Mäntsälä, C. Hertweck, and M. Metsä-Ketelä.** 2007. Artificial reconstruction of two

cryptic angucycline antibiotic biosynthetic pathways. *ChemBioChem* **8:**1577–1584.

71. **Paranagama, P. A., E. M. K. Wijeratne, and A. A. L. Gunatilaka.** 2007. Uncovering biosynthetic potential of plant-associated fungi: effect of culture conditions on metabolite production by *Paraphaeosphaeria quadriseptata* and *Chaetomium chiversii. J. Nat. Prod.* **70:**1939–1945.

72. **Pel, H. J., J. H. de Winde, D. B. Archer, P. S. Dyer, G. Hofmann, P. J. Schaap, G. Turner, R. P. de Vries, R. Albang, K. Albermann, M. R. Andersen, J. D. Bendtsen, J. A. E. Benen, M. van den Berg, S. Breestraat, M. X. Caddick, R. Contreras, M. Cornell, P. M. Coutinho, E. G. J. Danchin, A. J. M. Debets, P. Dekker, P. W. M. van Dijck, A. van Dijk, L. Dijkhuizen, A. J. M. Driessen, C. d'Enfert, S. Geysens, C. Goosen, G. S. P. Groot, P. W. J. de Groot, T. Guillemette, B. Henrissat, M. Herweijer, J. P. T. W. van den Hombergh, C. A. M. J. J. van den Hondel, R. T. J. M. van der Heijden, R. M. van der Kaaij, F. M. Klis, H. J. Kools, C. P. Kubicek, P. A. van Kuyk, J. Lauber, X. Lu, M. J. E. C. van der Maarel, R. Meulenberg, H. Menke, M. A. Mortimer, J. Nielsen, S. G. Oliver, M. Olsthoorn, K. Pal, N. N. M. E. van Peij, A. F. J. Ram, U. Rinas, J. A. Roubos, C. M. J. Sagt, M. Schmoll, J. Sun, D. Ussery, J. Varga, W. Vervecken, P. J. J. van de Vondervoort, H. Wedler, H. A. B. Wosten, A.-P. Zeng, A. J. J. van Ooyen, J. Visser, and H. Stam.** 2007. Genome sequencing and analysis of the versatile cell factory *Aspergillus niger* CBS 513.88. *Nat. Biotechnol.* **25:**221–231.

73. **Penttinen, P., J. Pelkonen, K. Huttunen, and M.-R. Hirvonen.** 2006. Co-cultivation of *Streptomyces californicus* and *Stachybotrys chartarum* stimulates the production of cytostatic compound(s) with immunotoxic properties. *Toxicol. Appl. Pharmacol.* **217:**342–351.

74. **Perrin, R. M., N. D. Fedorova, J. W. Bok, R. A. Cramer, J. R. Wortman, H. S. Kim, W. C. Nierman, and N. P. Keller.** 2007. Transcriptional regulation of chemical diversity in *Aspergillus fumigatus* by LaeA. *PLoS Pathog.* **3:**508–517.

75. **Pirt, S. J., and R. C. Righelato.** 1967. Effect of growth rate on the synthesis of penicillin by *Penicillium chrysogenum* in batch and chemostat cultures. *Appl. Microbiol.* **15:**1284–1290.

76. **Porges, N.** 1932. Chemical composition of *Aspergillus niger* as modified by zinc sulphate. *Bot. Gaz.* **94:**197.

77. **Rokem, J. S., A. E. Lantz, and J. Nielsen.** 2007. Systems biology of antibiotic production by microorganisms. *Nat. Prod. Rep.* **24:**1262–1287.

78. **Scherlach, K., and C. Hertweck.** 2009. Triggering cryptic natural product biosynthesis in microorganisms. *Org. Biomol. Chem.* **7:**1753–1760.

79. **Schiewe, H. J., and A. Zeeck.** 1999. Cineromycins, gamma-butyrolactones and ansamycins by analysis of the secondary metabolite pattern created by a single strain of *Streptomyces. J. Antibiot.* **52:**635–642.

80. **Schneider, P., M. Misiek, and D. Hoffmeister.** 2008. *In vivo* and *in vitro* production options for fungal secondary metabolites. *Mol. Pharm.* **5:**234–242.

81. **Schumann, J., and C. Hertweck.** 2006. Advances in cloning, functional analysis and heterologous expression of fungal polyketide synthase genes. *J. Biotechnol.* **124:**690–703.

82. **Seibert, S. F., A. Krick, E. Eguereva, S. Kehraus, and G. M. Konig.** 2007. Ascospiroketals A and B, unprecedented cycloethers from the marine-derived fungus *Ascochyta salicorniae. Org. Lett.* **9:**239–242.

83. **Shao, H.-J., X.-D. Qin, Z.-J. Dong, H.-B. Zhang, and J.-K. Liu.** 2008. Induced daldinin A, B, C with a new skeleton from cultures of the ascomycete *Daldinia concentrica. J. Antibiot.* **61:**115–119.

84. **Shwartzman, G.** 1944. Enhanced production of penicillin in fluid medium containing cellophane. *Science* **100:**390–392.

85. **Singh, S. B., and F. Pelaez.** 2008. Biodiversity, chemical diversity and drug discovery, p. 141–174. *In* F. Peterson and R. Amstutz (ed.), *Natural Compounds as Drugs*, vol. I. Birkhäuser, Basel, Switzerland.

86. **Steinberg, R. A.** 1919. A study of some factors in the chemical stimulation of the growth of *Aspergillus niger. Am. J. Bot.* **6:**330–356.

87. **Steinberg, R. A.** 1919. A study of some factors in the chemical stimulation of the growth of *Aspergillus niger. Am. J. Bot.* **6:**357–372.

88. **Stone, R. W., and M. A. Farrell.** 1946. Synthetic media for penicillin production. *Science* **104:**445–446.

89. **Udwary, D. W., L. Zeigler, R. N. Asolkar, V. Singan, A. Lapidus, W. Fenical, P. R. Jensen, and B. S. Moore.** 2007. Genome sequencing reveals complex secondary metabolome in the marine actinomycete Salinispora tropica. *Proc. Natl. Acad. Sci. USA* **104:**10376–10381.

90. **Ueda, K., S. Kawai, H. Ogawa, A. Kiyama, T. Kubota, H. Kawanobe, and T. Beppu.** 2000. Wide distribution of interspecific stimulatory events on antibiotic production and sporulation among *Streptomyces* species. *J. Antibiot.* **53:**979–982.

91. **Uguru, G. C., K. E. Stephens, J. A. Stead, J. E. Towle, S. Baumberg, and K. J. McDowall.** 2005. Transcriptional activation of the pathway-specific regulator of the actinorhodin biosynthetic genes in *Streptomyces coelicolor. Mol. Microbiol.* **58:**131–150.

92. **van den Berg, M. A., R. Albang, K. Albermann, J. H. Badger, J.-M. Daran, A. J. M Driessen, C. Garcia-Estrada, N. D. Fedorova, D. M. Harris, W. H. M. Heijne, V. Joardar, J. A. K. W Kiel, A. Kovalchuk, J. F. Martin, W. C. Nierman, J. G. Nijland, J. T. Pronk, J. A. Roubos, I. J. van der Klei, N. N. M. E. van Peij, M. Veenhuis, H. von Dohren, C. Wagner, J. Wortman, and R. A. L. Bovenberg.** 2008. Genome sequencing and analysis of the filamentous fungus *Penicillium chrysogenum. Nat. Biotechnol.* **26:**1161–1168.

93. **Van Lanen, S. G., and B. Shen.** 2006. Microbial genomics for the improvement of natural product discovery. *Curr. Opin. Microbiol.* **9:**252–260.

94. **Waksman, S. A., E. S. Horning, and E. L. Spencer.** 1943. Two antagonistic fungi, *Aspergillus fumigatus* and *Aspergillus clavatus*, and their antibiotic substances. *J. Bacteriol.* **45:**233–248.

95. **Wilkinson, B., and J. Micklefield.** 2007. Mining and engineering natural-product biosynthetic pathways. *Nat. Chem. Biol.* **3:**379–386.

96. **Williams, R. B., J. C. Henrikson, A. R. Hoover, A. E. Lee, and R. H. Cichewicz.** 2008. Epigenetic remodeling of the fungal secondary metabolome. *Org. Biomol. Chem.* **6:**1895–1897.

97. **Yu, J.-H., and N. Keller.** 2005. Regulation of secondary metabolism in filamentous fungi. *Annu. Rev. Phytopathol.* **43:**437–458.

98. **Zazopoulos, E., K. Huang, A. Staffa, W. Liu, B. O. Bachmann, K. Nonaka, J. Ahlert, J. S. Thorson, B. Shen, and C. M. Farnet.** 2003. A genomics-guided approach for discovering and expressing cryptic metabolic pathways. *Nat. Biotechnol.* **21:**187–190.

99. **Zerikly, M., and G. L. Challis.** 2009. Strategies for the discovery of new natural products by genome mining. *ChemBioChem* **10:**625–633.

FERMENTATION AND CELL CULTURE

THIS SECTION COVERS RECENT ADVANCES IN CULTIVATING MICROBIAL, plant, mammalian, and insect cells and their use for production of therapeutic compounds. The eight chapters include miniaturization of methods used to grow microorganisms, technologies used for cultivating them in solid phase, and culturing plant, insect, and mammalian cells. The use of plant cells to produce secondary metabolites is described, and that of bacterial, yeast and fungal, insect, and mammalian cell cultures to produce recombinant proteins is also illustrated.

Chapter 8 by Wouter Duetz and coauthors provides an excellent review of the history of small-scale cultures and the basic principles of culture miniaturization. Due to the very small size of the culture vessels, the problems encountered are quite different from those faced in larger-scale cultures. For example, surface tension and gas bubble size significantly affect performance of a miniaturized culture. Various types of microtiter plates used for this purpose, and the development of appropriate well closure systems and of the miniaturized sensors for oxygen and pH measurements, are described. An exhaustive discussion of use of these systems for screening for bioactive molecules can also be found in this chapter.

Chapter 9 by Ramunas Bigelis covers aerobic and anaerobic solid-phase fermentations, with their advantages and limitations. The author details variables that affect microbial growth and thereby product formation, such as temperature, concentrations of oxygen and carbon dioxide, moisture content, water activity, and physical properties of the solid substrate. Examples are given on the use of this technology to produce enzymes, primary metabolites like organic acids and ethanol, and secondary metabolites such as antibiotics, gibberellic acid, flavor compounds, and ergot alkaloids. Bigelis also discusses the use of food fermentations to enhance the flavor of food, change its physical characteristics, or increase its shelf life. Solid-phase fermentations used to prepare biocontrol agents are illustrated as well. A new development, mixed-phase (solid and liquid) fermentation, is described. The chapter concludes with an operating protocol for solid-phase fermentations that will be very useful as a starting point for process development.

The use of bacterial cultures for the production of proteins and other biological products is described in chapter 10 by Joseph Shiloach and Ursula Rinas. Bacteria have been used for a long time to produce small molecules, but in recent years they have also been used to make recombinant proteins. This chapter provides detailed information on all aspects of developing and operating a bacterial process to accomplish this, focusing on the processes that use *Escherichia coli*. The preparation of medium and inoculum, bioreactors and sterilization, instruments, necessary infrastructure, and procedures for initial product recovery are described in detail. The chapter also discusses products from bacteria other than *E. coli*.

Ninyan Zhang and Zhiqiang An give in chapter 11 a comprehensive report on heterologous protein expression in yeast and filamentous fungi. Just like *E. coli* among the bacteria, *Saccharomyces cerevisiae* has been the member of the fungi most commonly used for the production of heterologous proteins. One of the advantages that *S. cerevisiae* has over bacterial systems is its ability to glycosylate, as many antibodies require glycosylation for activity. The chapter notes the recent advances in fungal genetic technology and their effect on the enhancement of production. The authors describe the criteria for choosing an expression system and provide information on the vectors and promoters used and on the transcriptional and translational regulation of heterologous gene expression. Another yeast which has been widely used is *Pichia pastoris*, a methylotropic yeast that is capable of growing to very high cell density, e.g., a dry cell weight of 100 g/liter. Fermentation technology for this yeast is also well established. Many mammalian proteins need posttranslational modification, and both *S. cerevisiae* and *P. pastoris* are capable of doing this. Protein folding and glycosylation receive special attention in this chapter. The chapter concludes with a discussion of the use of *Aspergillus* species for heterologous protein production, including a description of the host strains, vectors, and promoters used as well as the regulation of posttranslational modification.

The two chapters by Jinyou Zhang on production of heterologous proteins in mammalian cells cover the entire process from selection of cells to downstream processing and regulatory considerations. Chapter 12 deals with the upstream part of the process, and chapter 13 with the downstream part. In the first chapter, Zhang surveys the biopharmaceutical industry to introduce the reader to the subject, and follows with a discussion of the suitability of mammalian cell cultures for the production of heterologous proteins. Coverage of the construction and selection of high-producing cell lines includes details on expression systems, host cell lines, screening for desired cell lines, and cell line engineering. Medium development for mammalian cell culture is also explored, including general features of cell culture media, animal component-free media, development of cell culture media, and development of feeds. Later, a discussion of process development comprises reactor systems, monitoring and control, optimization of culture conditions, scale-up and scale-down, and the relationship between process improvement and product quality.

The second chapter by Zhang dovetails seamlessly with the previous one, as it begins with the detailed description of downstream processing. This section comprehensively examines recovery separation, primary and additional purification, polishing chromatography, virus removal, and final formulation. Next, formulation development and fill/finish issues such as preformulation, formulation stability, drug delivery, and fill and finish are considered in detail. Another important aspect of the production of biopharmaceuticals is quality control. For this purpose, Zhang discusses analytical method development and qualification. Specific items covered include lot release testing, glycosylation analysis, and assay development and validation. This chapter concludes with consideration of Good Manufacturing Procedures requirements.

Plant cell cultures have been used for several years. For chapter 14, Nancy Paiva has updated her contribution to the second edition of this book to describe various medium components and their function in supporting the growth of plant cells. The section on plant growth regulators deserves special mention. The need for extra care during the preparation of the medium is highlighted. It is emphasized that operating parameters such as the absolute necessity of a sterile transfer facility, and control of temperature, aeration, and exposure to light, are critical for good plant cell growth and maximum product formation. Special attention is given to the sterilization procedures to be used. Appropriately, the section on process development concludes with a description of issues pertaining to the use of bioreactor and scale-up. Most of the considerations with respect to the scale-up of microbial processes also apply to plant cell cultures, with the exception of high shear sensitivity. As a result, impellers capable of low shear or airlift fermentors have to be used. The chapter concludes with descriptions of procedures used for regeneration of plants from cell cultures and procedures for DNA transformation.

Although the development of insect cell cultures for biomanufacturing began later than the other systems mentioned, it is now mastered to an extent similar to mammalian cell culture. The application of baculoviruses as vectors for the production of recombinant proteins or for manufacturing of viral biopesticides is safe and is amenable to scale-up. Chapter 15 by Spiros N. Agathos describes the characteristics of various cell lines and the development of the cell lines commonly used for stable expression. Procedures for small-scale (250 ml) and large-scale (several hundred liters) cultures along with potential pitfalls are highlighted. A mathematical analysis of different strategies used for the operation—i.e., batch, fed-batch, and continuous cultures—is presented. Infection of insect cell lines with baculoviruses is thoroughly covered, including vector generation and amplification and infection of cell lines. For this last, important parameters are cell line selection, time and multiplicity of infection, cell density at infection, and time of harvest. The chapter concludes with a discussion of product recovery.

Miniaturization of Fermentations

WOUTER A. DUETZ, MATTHEW CHASE, AND GERALD BILLS

8

8.1. HISTORICAL CONTEXT AND APPLICATIONS OF SMALL CULTURES

Well into the 20th century, a large gap in cultivation scale continued to exist between agar colonies on the one hand and Erlenmeyer flask cultures on the other. Most laboratories had little incentive to reduce culture volumes below 50 ml, simply because the number of "house strains" and their mutants was rather limited. During the 1980s and 1990s, a number of new developments spurred the demand for and utilization of high-quality and parallel small-scale growth systems. First, several large-scale fermentative bioprocesses (such as those for antibiotic production) became fully mature, and it became increasingly difficult to find significantly improved mutants. In practice, this meant screening ever larger mutant libraries, and accepting that the improvement in the productivity of the best mutants would be less than 5 or 10% compared to the parent strain, and required an evaluation process with very low variability. This in turn created a demand for high-quality cultivation systems. The same trends held for medium optimization projects, where increasingly large numbers of medium parameters were being tested. Second, analytical techniques including chromatographic techniques such as liquid chromatography-mass spectrometry (LC-MS) and gas chromatography (GC), but also receptor binding assays and cell-based inhibition assays, were being improved to an extent that they could be performed with picogram or nanogram amounts of cell products, which in most cases is compatible with submilliliter culture volumes. The more sensitive bioassays, e.g., mode-of-action-based assays to detect anticancer or antibiotic activities of microbial metabolites, also enabled screening for new drugs and antibiotics, which are often secondary metabolites from actinomycetes or fungi, to be performed with much smaller amounts of extracts. Third, screening projects with extremely expensive medium components, such as ^{13}C-labeled substrates as used in flux-analysis studies, were financially feasible only when fermented in small culture volumes. Other application areas included the screening for new biocatalysts and the improvement of specific enzymes by screening large mutant libraries.

Initial ad hoc efforts on miniaturization were done in-house by industry in the 1980s, but systematic knowledge in this area started to accumulate only after several universities, including those in Aachen, Zurich, Munich, and London, started dedicated research projects in the second half of the 1990s. Much research was focused on 96- and 24-well microtiter plates (MTPs), which were a logical choice because they were standardized and being widely applied in other research areas. Furthermore, a broad range of compatible equipment—pipetting stations, microplate readers, etc.—was available, which saved money and time. The relatively large size of the part of this chapter dedicated to MTPs (section 8.3) reflects their current importance in industry and academia alike. Sections 8.4 and 8.5 are dedicated to test tubes and other types of miniature cultivation systems, respectively.

8.2. BASICS OF CULTURE MINIATURIZATION

The knowledge that the average microbial cell is still more than 1,000- to 10,000-fold smaller than a culture vessel of 10-mm diameter leads microbiologists easily into believing that the effects of miniaturization on an individual cell are bound to be minimal. This intuition may be largely true for growth vessels with diameters above 10 mm, but not for smaller vessels, mainly as a result of surface tension effects and/or different behavior of gas bubbles.

8.2.1. Surface Tension Effects

Surface tension interferes with the flow and movement of the culture medium due to bubbling, stirring, or *g*-forces arising from orbital shaking. The importance of culture size becomes evident when comparing the movement of fluids inside vessels with various diameters during orbital shaking (Fig. 1). From these photographic studies it became evident that the effect of surface tension rapidly increases at vessel diameters below 8 mm: even at a *g*-force of 2.5, no significant movement of the culture fluid can be established in culture vessels of 3 or 4 mm. A moderating, although also complicating, factor (it may give rise to false positives during mutant screenings) is the reducing effect that culture broth components (e.g., cell membrane components, proteins, and extracellular glycolipids) exert on the surface tension. This lower surface tension does significantly improve the situation with vessel sizes around 6 mm (compare Fig. 1 and 3), but not at 3 or 4 mm (17). The absence of fluid movement as a result of surface tension effects generally results in low oxygen transfer rates (OTRs) and other problems (see also section 8.3).

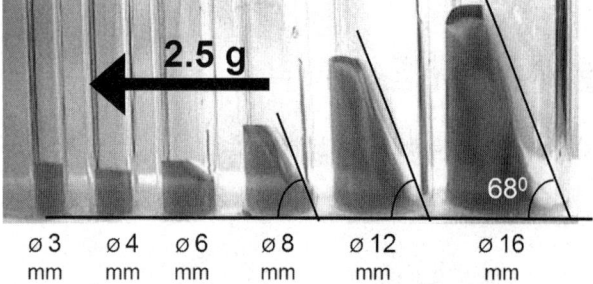

FIGURE 1 High-speed photograph showing the ability of a g-force of 2.5 (generated by orbital shaking at 300 rpm and a shaking diameter of 50 mm) to induce a rotating movement of an aqueous solution of bromocresol blue (0.1% [wt/wt], surface tension of 72 dynes/cm) in round polyacrylate vessels with various diameters. The theoretical angle of the liquid/air surface with the horizontal plane of 68° is only reached at vessel diameters of 12 mm and higher. The angle of more than 68° at a vessel diameter of 16 mm can be explained by the situation that the shaking diameter for the bulk of the liquid is in effect approximately 60 instead of 50 mm. The practical absence of any movement of the liquid at 3 and 4 mm is due to the high surface tension.

8.2.2. Gas Bubble Size Effects

The behavior of gas bubbles becomes increasingly unpredictable at smaller bubble sizes. In the first place, small bubbles tend to ascend more slowly and are more easily deviated from an exactly vertical movement than is the case with large bubbles. The large variety of bubble patterns in duplicate miniature growth vessels may lead to large differences in hydrodynamics and OTRs and indirectly to poor reproducibility of process performance.

8.2.3. Methods for Determining OTRs

When assessing the applicability of a novel miniature cultivation system—especially if one wants to mimic the physicochemical situation in larger cultures—it is advisable to determine OTRs from the gas phase to the liquid phase. Various methods are available to determine OTRs in cultures smaller than 1 ml. The traditional and best-validated method is the cobalt-catalyzed oxidation of sulfite (22). In this method, the headspace of the vessel is flushed slowly (and at a constant rate) with an oxygen-containing gas (e.g., air). The medium in the vessel contains sulfite and cobalt as a catalyst. All oxygen diffusing into the medium is consumed in the oxidation of sulfite to sulfate so that the concentration of oxygen in the liquid phase remains close to zero. The OTR can be calculated from the difference in oxygen concentration in the gas inlet and gas outlet, or from the pH drop (using a pH indicator) associated with the oxidation of sulfite (22). A slight disadvantage of this method is the high salt concentration applied (usually 0.5 M), which reduces the maximal solubility of oxygen to a level well below that in low-salt culture media, which generally contains 0.1 to 0.2 M salts. Therefore, the maximal OTR is systematically underestimated, and a 20 to 30% correction factor is necessary when translating the results to microbial cultures (J. Buchs, personal communication). Another drawback of this method is the high surface tension as compared to microbial cultures (see earlier), leading to an underestimate of potentially achievable OTRs in small-scale cultures with well diameters of less than 8 mm.

Enzymatic methods (using oxygen-consuming enzymes such as oxygenases and oxidases) can also be used to determine OTRs in small vessels. Glucose oxidase in combination with diammonium 2,2′-azino-di-(3-ethylbenzthiazoline sulfonate) (ABTS) (16) and catechol 2,3-dioxygenase (C23O) are applicable, as they both give rise to colored products that are easy to measure spectrophotometrically. The C23O method (35) is preferred because the colored compound (a semialdehyde) is formed in a single-step reaction in a 1:1 proportion to the amount of oxygen consumed. The glucose oxidase method (16) shows a less clear stoichiometry and therefore requires calibration with another method (such as the cobalt-catalyzed oxidation of sulfite), but has the advantage that all reagents are available commercially.

A third possibility is to derive OTRs from the oxygen-limited growth phase of growth curves of aerobic strains such as *Pseudomonas putida* (15) or *Bacillus subtilis* (13). These latter methods provide a more realistic comparison with "real-life" conditions (also in terms of surface tension) but are more laborious and time-consuming, because it is necessary to sample the cultures (in order to be able to measure the optical density) at regular time intervals for 10 to 20 h.

8.3. CULTIVATION IN MTPs

Orbital shaking is currently the major method applied to obtain satisfactory OTRs in MTPs. It is popular because the force driving the increase of the gas-liquid interface, as determined by the centrifugal force, is applied identically to all wells in all MTPs mounted on a shaking platform, and therefore eliminates a potential source of well-to-well variation. In contrast, aeration by air sparging and magnetic stirring almost inevitably gives rise to well-to-well variations, especially at lower culture volumes.

During orbital shaking of microbial cultures in MTPs, two main factors determine the quality and reproducibility of the culture: (i) the well-closure system and (ii) the combination of orbital shaking conditions and the culture volume, which in turn determine the OTR and the degree of mixing.

8.3.1. Well-Closure Systems

The well-closure system for the individual wells should fulfill three functions: (i) prevent (cross)-contamination, (ii) permit the exchange of headspace air, and (iii) limit evaporation. Equivalent physical conditions in all wells is a further requirement: the wells in the corners should have exactly the same characteristics as the wells in the middle of the MTP, and uniformity is especially important in quantitative work, e.g., the comparison of certain (enzyme) activities in mutant libraries.

8.3.1.1. Sealing Tapes

Gas-permeable membrane filters (sealing tapes) are available from many suppliers, and are well suited when used in combination with humidified growth cabinets since the water vapor permeability is high for all sealing tapes (47) as long as the culture fluid does not come in direct contact with the sealing tape during shaking. The sealing tape will remain dry when deep-well MTPs are used in combination with relatively small culture volumes and mild shaking conditions and when no condensation appears. When sealing tapes are used in combination with high shaking speeds and/or shallow-well MTPs are used, however, culture fluid may settle on the foil and cause a number of adverse effects on the quality of the culture. First, glue components may partially dissolve, and if a droplet falls back into the culture, it may influence the growth and/or subsequent analytical procedures. Second, the proportion of the membrane area covered by the droplets/aerosol will differ from well to well, contributing to well-to-well variation, both with respect to evaporation losses and to gas concentrations (oxygen, CO_2) in the headspace. Third, depending on the quality of the sealing tape, the glue may not hold when wetted, risking cross-contamination and airborne contamination. Apart from aerosols, sealing tapes can also become wet (and so hamper gas exchange) as a result of water condensation. Condensation readily occurs when the membrane is colder than the culture. This, in turn, may happen when the shaker table is (slightly) warmer than the air in the shaker cabinet, which often occurs when the shaker motor warms the shaker table. It can also result from heat generation by fast-growing cells. For further practical information on various types of sealing tapes and their relative permeabilities for water and oxygen, we refer to reference 47. In conclusion, sealing tapes are often well suited for qualitative work under relatively mild shaking conditions, but not under harsh conditions.

8.3.1.2. Sandwich Covers

A more generally applicable, though more complex, closure system consists of a silicone layer with small holes above each well in combination with suitable layers of bacterial filters, pressed onto the MTP at a sufficient force to prevent cross-contamination (15; Fig. 2; see also Fig. 1). Because the only connection with the ambient air is via the small hole in the silicone layer above the center of each well, the degree of exchange of headspace air (via passive diffusion) is fully determined by the hole's dimensions (diameter and length). Depending on the culture volume and the type of microbial strain, one can choose suitable hole dimensions. For fast-growing strains, the rule of thumb borrowed from the field of stirred-tank bio-

reactors (1 to 2 culture volumes of air supply per minute) has proven to give satisfactory results for small cultures as well. For culture volumes of 0.5 ml per well, the required headspace exchange rate of 1 ml air per minute can be achieved with holes that are 1.5 mm in diameter and 6 mm long (15). Higher or lower exchanges of headspace can be achieved by using larger or smaller holes, respectively. The exact rates of exchange of headspace air corresponding to varying hole size can be experimentally determined by measuring the water loss from a well after several days of incubation in a shaker cabinet with a defined temperature and humidity.

Such a rate of exchange of headspace air of 1 to 2 culture volumes per minute will keep the oxygen concentration in the headspace above 18% (vol/vol) under all conditions, and will limit culture volume loss through evaporation to a few percent per day, even at 50% ambient humidity in the shaking cabinet. The evaporation rate can be further lowered by increasing the humidity in the shaking cabinet, though one has to realize that long-term humidities over 75% may result in mold growth inside the shaking cabinet, which in turn can harm the health of laboratory workers who open the shaking cabinets and are exposed to spores, as well as increasing contamination levels.

The holes in the silicone layer of such covers may be obstructed when heavy splashing occurs inside the well, or when the culture volume is too high relative to the total well volume and the shaking conditions. It is therefore recommended to perform a range of tests first, such as the use of different volumes, different shaking conditions, etc., and inspect the sandwich cover for blockage of the holes. As with the sealing tapes discussed above, condensation on the silicone rubber may occur when the shaker table is warmer than the air inside the shaking cabinet or when the culture produces heat. Insulating the MTP from the shaker table, e.g., by using a layer of air or rubber sponge, may alleviate this problem.

A general advantage of sandwich covers is that shaking rates and thus OTRs can be higher: direct contact

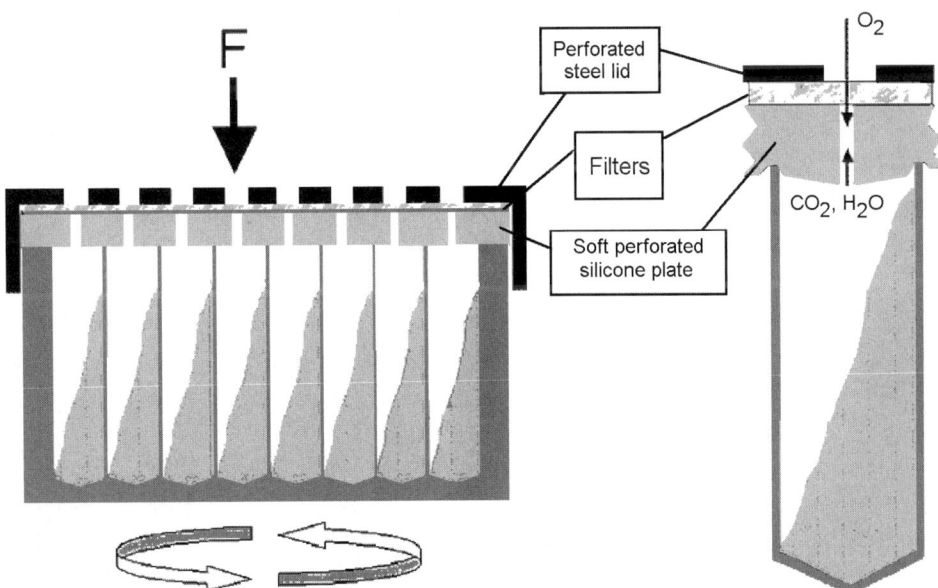

FIGURE 2 Schematic view of the sandwich cover (section 8.3.1.2) used for (i) the prevention of (cross)-contamination of wells of a 96-square-deep-well MTP during vigorous orbital shaking and (ii) the limitation of evaporation.

between the culture fluid and the inert silicone layer does not have any adverse consequences. Cross-contamination will still not occur as long as the sandwich cover is not taken off during growth. However, if intermediate sampling (e.g., for optical density measurements) is required, one should preferably choose a suitable combination of culture volume, type of MTP, and shaking conditions so that the culture fluid does not touch the silicone layer; otherwise cross-contamination may occur when the sandwich cover is placed back onto the MTP after sampling.

8.3.2. OTRs

A sufficient degree of exchange of headspace air, as can be achieved using sealing tapes or sandwich covers as described above, is a prerequisite for a good oxygen supply, but does not automatically ensure that the oxygen in the headspace reaches the cells in suspension. To achieve this, a sufficient level of gas-liquid transfer needs to be established. Table 1 lists OTRs in various types of MTPs, at various culture volumes and shaking conditions. The current consensus in literature (especially on the basis of the work at universities in Aachen and Zurich) is that sufficient levels of OTRs (20 to 50 mmol liter^{-1} h^{-1}) can be most easily reached when using a relatively high shaking amplitude of 25 or 50 mm in combination with a shaking frequency of 225 to 300 rpm (17). For 24-well plates (both those with square wells and round wells), good results can be obtained with either a 50-mm amplitude (even at 225 rpm) or a 25-mm amplitude (at 300 rpm). For all types of 96-well MTPs, a shaking amplitude of 50 mm gives far better results than 25 mm (up to threefold higher OTRs; see Table 1). This initially counterintuitive conclusion, i.e., the smaller a vessel is, the higher the shaker amplitude should be in order to achieve a satisfactory OTR, becomes logical when one realizes that with large vessels (e.g., Erlenmeyer flasks), the largest part of the diameter at which the bulk of the liquid is swirling actually stems from the vessel itself, and not from the shaker. The contribution of the vessel diameter to the effective throw of the bulk of the liquid is insignificant in 96-well MTPs. A second explanation for the apparent need for a high g-force (as can be most readily achieved using a high shaking amplitude) is that this leads to an increase of the angle of the fluid with the horizontal plane (see also Fig. 1), which leads to a larger percentage increase of the gas-liquid interface area with 96-well MTPs than with, for example, 24-well MTPs.

High OTRs (above 20 mmol liter^{-1} h^{-1}) can also be achieved in 96-well MTPs at shaking diameters of 6 or 3 mm, but only at shaking frequencies exceeding 800 and 1,000 rpm, respectively (23). Such high frequencies, however, are not achievable with most standard shaking equipment, and are therefore less practical for standard applications.

8.3.3. Turbulence and Mixing

In the process of choosing a suitable combination of a specific type of MTP, culture volume, and shaking conditions, it is important to realize that a satisfactory OTR does not guarantee a good degree of mixing. If thorough mixing is crucial, it is often advisable to employ MTPs with square wells; these generally give rise to a more turbulent hydrodynamic pattern (the walls act as a sort of baffle), and in addition allow a higher "filling height" for a given required OTR. Roughly, it can be concluded (see Table 1) that for an identical OTR, MTPs with square wells allow three- to fourfold higher culture volumes per well than MTPs with

TABLE 1 OTRs in various types of MTPs at various volumes during orbital shaking at 300 rpm and a shaking diameter of 50 and 25 mm

Type of MTP	Cells present?	Vol (liquid height before start of shaking)	OTR at 300 rpm (mmol of O$_2$ liter^{-1} h^{-1}) 50 mm	25 mm	Reference
96 shallow round wells, 6-mm diameter	No	0.2 ml (6.2 mm)	20[a]	9[a]	23
	Yes	0.2 ml (6.2 mm)	26[b]	ND[c]	17
	Yes	0.15 ml (4.7 mm)	32[b]	ND[c]	17
96 square deep wells, 8 × 8 mm	Yes	0.5 ml (8.5 mm)	38[b]	13[b]	15
	Yes	0.75 ml (13 mm)	24[b]	7[b]	15
	Yes	1 ml (17 mm)	18[b]	4[b]	15
48 shallow round wells, 11-mm diameter	No	0.4 ml (4 mm)	38[a]	22[a]	28
	No	0.6 ml (6 mm)	27[a]	16[a]	28
24 shallow round wells, 16-mm diameter	No	0.75 ml (4 mm)	40	25	Estimated on basis of reference 28
	No	1 ml (5 mm)	30	20	Estimated on basis of reference 28
24 square deep wells, 17 × 17 mm	No	2.5 ml (8.7 mm)	51[d]	39[d]	16

[a]Determined by the cobalt-catalyzed oxidation of sulfite but corrected for the lower oxygen solubility (original values multiplied by 1.3 to compensate for low oxygen solubility; see text).
[b]Derived from the oxygen-limited growth phase of growth curves of *Pseudomonas putida*.
[c]ND, not determined.
[d]Determined enzymatically using glucose oxidase.

round wells. The higher turbulence in MTPs with square wells can be either a disadvantage where shear forces may affect growth negatively, or an advantage in a situation where the mass transfer between different phases is limiting (e.g., when working with two-phase systems or solid medium components). Figure 3 illustrates the higher degree of turbulence that can be achieved in MTPs with square wells in comparison to round wells.

8.3.4. Toxicants Leaching from the Plastic of MTPs
A trivial problem that occurs quite frequently is the negative effects on cell growth exerted by toxic compounds leaching from polypropylene (as used for both 24- and 96-square-deep-well MTPs). The compounds responsible are believed to be UV absorbers and antioxidants (e.g., phenols and phosphites), whiteners (coumarins and benzoxazoles), and heavy metals (such as cobalt), which are often used as additives in

polypropylene (4). In order to alleviate these toxic effects, it is advisable to soak newly acquired polypropylene MTPs in boiling NaOH (1 M) for 4 h and HCl (1 M, 70°C) for 4 h (15). Because this procedure is quite laborious, it is advisable to reuse such treated MTPs for as long as possible. Direct extraction of fermentation broth in MTPs with organic solvents, e.g., acetone or methyl ethyl ketone, may also result in extraction of plastic components along with metabolites in the culture extract. Such compounds may be detected by analytical techniques, e.g., LC-MS, and in some cases may interfere with analysis of microbial metabolites in the extracts. If planning direct extraction in the plates, we recommend extracting control MTPs with solvents to determine their background.

Polystyrene MTPs are generally nontoxic when used off-the-shelf, and are therefore often preferred, especially in the case of high-throughput projects. However, the

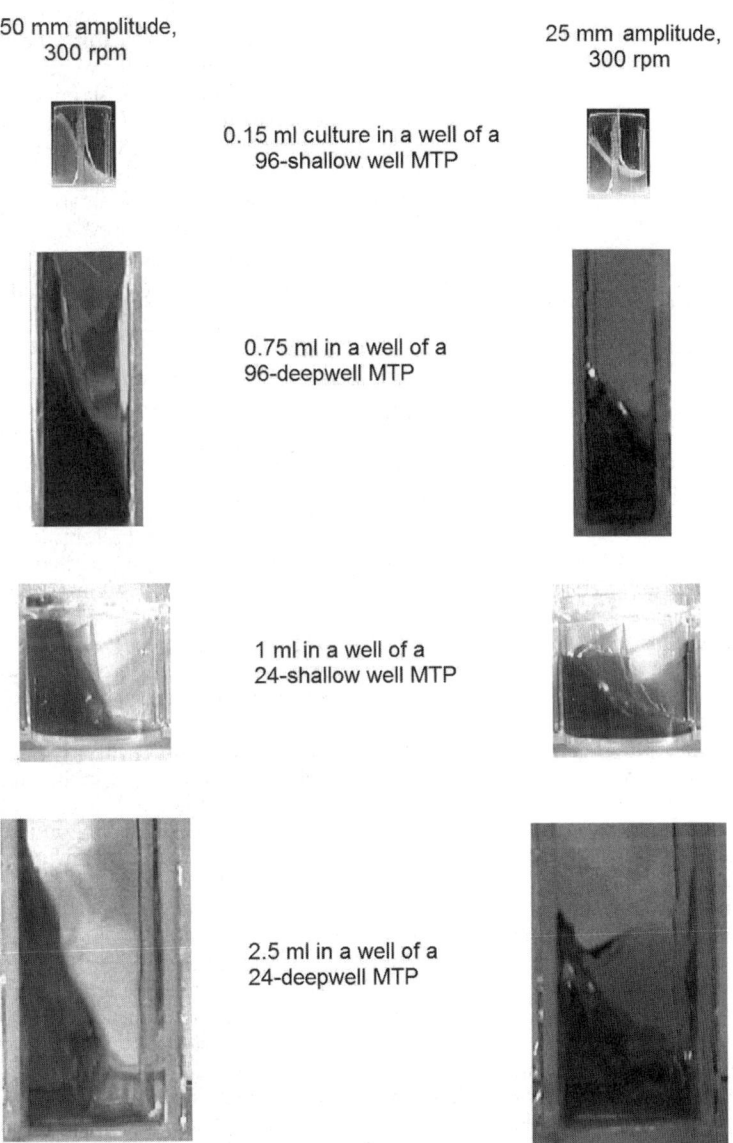

FIGURE 3 Photographs illustrating the hydrodynamic flow pattern inside wells of various MTPs during orbital shaking at 300 rpm, at a shaking amplitude of either 50 mm (left) or 25 mm (right).

choice in polystyrene MTPs is limited; currently no polystyrene square-deep-well plates are commercially available, although MTPs with 96 square shallow wells are available from various suppliers.

8.3.5. Applicability of Various MTPs for Different Purposes

In the process of choosing a suitable type of MTP for a specific application, a range of factors should be considered. As discussed in section 8.3.2 (and as can be seen in Table 1), for all types of 96-, 48-, and 24-well MTPs, OTRs exceeding 30 mmol liter^{-1} h^{-1} can be achieved, so this is generally not an important selection criterion. In most cases, the more relevant criterion is the minimum culture volume necessary for subsequent assays and analysis. However, even if only a few microliters are required for analysis, e.g., for GC or high-performance liquid chromatography (HPLC) analysis, it does not automatically suggest that the standard 96-shallow-well plates are the correct choice. It is often advisable to start performing tests with a range of different types of MTPs, and to specifically compare the variation between independent duplicates with regard to the relevant analyte concentration or other parameter to be considered for the different cultures (see protocol 1). As a rule, much lower standard deviations can be achieved with the relatively high culture volumes that are feasible in 24-well MTPs. The underlying reasons for the impact of culture volume on standard deviation are poorly understood, but pipetting errors (percentagewise larger with smaller volumes) and a larger biological variation in small cultures, especially when cells do not grow fully suspended, have a substantial impact at smaller volumes. Also, the smaller amounts of biomass used for inoculation (which preferably should be exactly the same when comparing various cultures) will inevitably show a larger variation. The final choice for a certain type of MTP is often a trade-off between the throughput on the one hand and the standard deviation between independent duplicates on the other.

Protocol 1. Preparation of a Screening: Optimization of the Reproducibility

1. Decide what standard deviation between independent duplicates is acceptable. This will among other things depend on the expected percentagewise improvement of the best mutants or culture condition.
2. Start with optimizing the reproducibility of the analytical procedure. If possible, apply HPLC-UV (in combination with relatively large injection volumes), infrared-based methods, or colorimetric/spectroscopic methods, since they generally give rise to the lowest standard deviations (1 to 3% once optimized). Methods dependent on internal standards (e.g., GC and LC-MS) generally result in relatively poor reproducibilities.
3. Test the standard deviations between duplicate cultures inoculated with the same (overnight) suspended culture (e.g., a shake flask culture) using a range of different MTPs and culture volumes, each at three- to eightfold. If applicable, also vary the cultivation/incubation times. On the basis of the resulting standard deviations, choose an MTP-culture volume combination that will allow you to detect the sort of mutants you are looking for (as defined in step 1).
4. Test the standard deviation between independent duplicates (inoculated with colonies from the same agar plate) using the MTP-culture volume combination selected in step 3. In order to achieve a standard deviation close to the standard deviation achieved in step 3, it is often necessary to synchronize the cultures as described in protocol 2. It may be important to optimize the culture lengths of the primary and secondary cultures as well as the amounts of inoculation.
5. Now that the general framework of a suitable screening protocol has been established, it may take another few man-months to further reduce the individual sources of error that contribute to the overall standard deviation. Focus in this stage generally lies on pipetting methods (pipetting/robotic stations versus manual pipetting; manual pipettes versus electronic pipettes) and the minimization of biological variation by optimizing the inoculation procedures.

8.3.5.1. MTPs with 384 or 1,536 Wells

As a result of surface tension effects (see section 8.2.1), water-based medium in 384- and 1,536-well MTPs (well dimension in the horizontal plane 3 to 4 mm and 1 to 2 mm, respectively) cannot be significantly moved by the centrifugal forces that can be achieved with standard orbital shakers. The resulting absence of active mixing often causes cells to sink to the bottom, and OTRs are bound to stay relatively low. Therefore, the application of 384- and 1,536-well MTPs is mainly limited to projects with anaerobic strains or when the generation of small amounts of DNA is the sole objective.

8.3.5.2. MTPs with 96 Shallow Round Wells

Standard polystyrene 96-shallow-well MTPs (with round wells of 350-μl volume) are widely applied for the cultivation of large libraries (3,000 strains or more). Using appropriate sandwich covers (see section 8.3.1.2) with spacers, multiple MTPs can be stacked, and up to 80 MTPs (7,680 cultures) can be mounted on a single standard shaker. In practice, the application of polystyrene 96-shallow-well MTPs is mainly limited to strains that grow fully suspended, such as *Escherichia coli* or *Bacillus* species. The typical culture volume applied is 150 μl, which allows shaking at 300 rpm and a 50-mm amplitude at OTRs of 30 mmol liter^{-1} h^{-1} (see Table 1). It is advisable to use a growth medium with a relatively low surface tension containing peptides and/or proteins, such as nutrient broth or Luria-Bertani (LB) medium. Such a medium will allow a thorough mixing up to the bottom of the wells (17).

Culture volumes of 100 μl or less can be used when high OTRs are desired (40 mmol liter^{-1} h^{-1} or more). With such small culture volumes, limiting evaporation becomes increasingly important, which can be achieved by the application of sandwich covers with smaller holes for the exchange of headspace air and/or the humidification of the air inside the shaker cabinet. With culture volumes of 100 μl or less, it becomes increasingly difficult to aspirate the supernatant fluid after centrifugation without disturbing the cell pellet. A practical solution to this issue is to add sterile buffer or water (e.g., 100 or 200 μl) after growth but prior to centrifugation. Of course, the resulting dilution will require correction of the subsequently measured analyte concentrations or other bioactivities.

The application of polystyrene 96-shallow-well MTPs for strains that grow in pellets or aggregates, such as actinomycetes, fungi (e.g., *Trichoderma*), etc., is generally not advisable for quantitative comparisons, due to poor well-to-well reproducibility.

8.3.5.3. MTPs with 96 Square Deep Wells

In comparison to the standard shallow 96-well MTPs discussed above, MTPs with 96 square deep wells have the advantage that culture volumes can be fourfold higher at the same OTR (up to 0.75 to 1 ml). Higher culture volumes are also associated with better reproducibility and an easier separation of pellet and supernatant fluid after centrifugation. In addition, the square shape of the wells gives rise to a more turbulent hydrodynamic shaking pattern, which may be a distinct advantage when working with actinomycetes (pellets are better dispersed) or when a sufficient mass transfer is required between two phases (solid particles) or with a solvent as a second phase (30). To achieve good mixing (up to the bottom) and high OTRs, it is strongly advisable to use an orbital shaker with a shaking amplitude of at least 50 mm, in combination with a frequency of 300 rpm (15; see also Table 1).

8.3.5.4. MTPs with 48 Round Shallow Wells

The advantage of polystyrene MTPs with 48 shallow wells is that at their well diameter of 11 mm, surface tension effects are almost absent (see section 8.2.1, Fig. 1), even when using mineral medium with a high surface tension, in contrast to 96-shallow-well MTPs. Compared to 24-shallow-well MTPs, an advantage is the higher ratio between the height of the wells (approximately 16 mm) and the diameter of the wells (11 mm), which allows a higher g-force to be applied (and so a higher shaking frequency) without the culture fluid reaching the tops of the wells. At culture volumes of 300 to 600 μl, OTRs of 30 mmol liter^{-1} h^{-1} and higher (up to 280 mmol liter^{-1} h^{-1} at 1,450 rpm, a shaking diameter of 3 mm, and a working volume of 300 μl) are readily attained (28). The poor compatibility of the 48-well format with standard MTP side equipment, such as microplate readers and multipipettes, is a distinct disadvantage, most particularly for smaller research groups that do not have the means to obtain custom-made tools and equipment.

8.3.5.5. MTPs with 24 Round Shallow Wells

Polystyrene MTPs with 24 round shallow wells with optimal culture volumes of 0.75 or 1 ml per well are very versatile when it comes to their application as growth vessels. For bacterial cultures, sufficient degrees of mixing and OTRs (30 mmol liter^{-1} h^{-1}) are readily achievable, even at the shaker speed of approximately 225 rpm in combination with a 50-mm amplitude, or at higher shaker speeds of preferably 300 rpm with shakers that only have a maximal shaker amplitude of 25 mm. In addition, MTPs with 24 shallow wells are also well suited for eukaryotic cell cultures that are sensitive to shear forces, because in contrast to 96-well MTPs, even at shaking frequencies of 150 to 200 rpm the mixing is still reasonable, and because surface tension effects play no role. Polystyrene MTPs with 24 shallow wells with various surfaces in terms of hydrophobicity (high binding, low binding) are commercially available from a range of suppliers.

8.3.5.6. MTPs with 24 Square Deep Wells

Polypropylene MTPs with 24 square deep wells (11-ml total volume per well) are the preferred choice when large culture volumes (up to 3 ml per well) and/or high OTRs (up to 60 mmol liter^{-1} h^{-1}) are required (16). As with the 24-shallow-well MTPs, good results in terms of mixing and OTR can be achieved with shaker amplitudes of either 25 or 50 mm (see Table 1 and Fig. 3). Also, shaking frequencies can be lowered to 150 or 200 rpm without the risk of a poor degree of mixing, which might result in the movement of cells to the bottom. Important application areas include medium optimization projects and the screening of mutant banks where only slightly better activities/titers are expected (10% improvement or less). Standard deviations for independent duplicates of less than 3% have been shown to be achievable, even with pellet-forming actinomycetes. The low standard deviations make MTPs with 24 deep wells well suited for secondary screenings: the initial screening (of mutant banks of a few thousand strains) is performed with 96-well MTPs, and the secondary screening (on a selected number of mutants) is performed in MTPs with 24 square deep wells. This sequence often allows the rejection of false positives from the primary screening and achievement of a better quantitative insight into the real mutants.

Twenty-four-square-well MTPs are also often used in the discovery of new secondary metabolites. The relative ease with which high OTRs are achieved (13) ensures high biomass and sufficient secondary-metabolite production by anabolic pathways requiring oxygenases. Furthermore, culture volumes of 2 to 4 ml facilitate separation of cells, pellets, or mycelia from the supernatant fluid by centrifugation and provide adequate amounts of extracted compounds for multiple bioassays, such as those for antibiotic or anticancer activities.

8.3.5.7. MTPs with Six Shallow Wells

Polystyrene MTPs with six shallow round wells most prominently differ from all MTPs described above with regard to the ratio between their diameter and height, i.e., 35 mm and 18 mm, respectively. This high ratio sets limits to the shaking frequency and thus to the g-force that can be applied. On the positive side, shaking frequencies of 60 rpm already result in reasonable mixing, at a very low shear force. The liquid swirls gently, which makes such MTPs applicable for sensitive eukaryotic cells, such as plant cells. Accumulation of cells in the middle of the wells may occur, which may be counteracted by small baffles inside the wells.

8.3.6. Procedures for Inoculation and "Synchronization" of Cultures

When screening libraries of mutants, the inoculation procedure is crucial for reproducible results. In many cases, the mutant library will be initially available as single colonies on an agar plate. A robotic colony picker may be used to inoculate liquid medium dispensed in the wells of a 96-well MTP. Alternatively for small libraries, one may use toothpicks for this initial inoculation. After an appropriate well-closure system has been applied (see section 8.3.1.2), this primary MTP may be incubated in an orbital shaker for 1 to 2 days. In the case of a qualitative screening, e.g., to screen for the presence or absence of a certain gene, product, or enzyme, the resulting cell suspension can be used immediately for analysis or bioassay. For quantitative screenings, e.g., when searching for high-activity mutants, it is often advisable to use this primary MTP to inoculate a second MTP, by transferring 1 to 5 μl with a multichannel pipette. The rest of the primary MTP can be stored at -80°C, after addition of glycerol, for later use if desired (see also protocol 2). As can be seen in Fig. 4, the cultures in this secondary MTP will be more or less "synchronized," i.e., the cells grow in parallel and reach the stationary phase at the same time. The latter is especially relevant if the product is unstable or is prominently formed in the stationary phase. Nonsynchronized cultures (such as from direct inoculation starting from colonies) often give rise

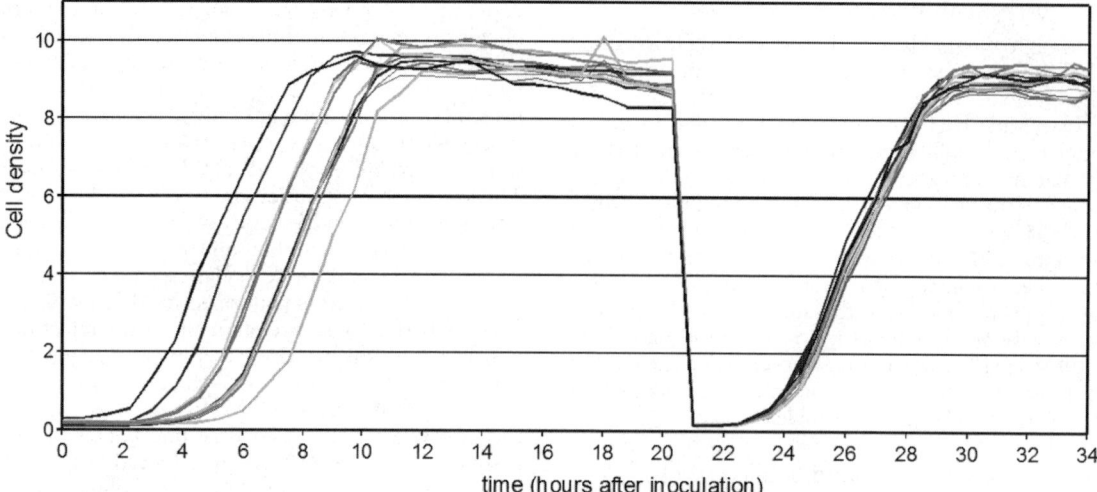

FIGURE 4 Synchronization of 12 isogenic cultures (0.75 ml OF LB medium) of *E. coli* TOP10 in wells of a 24-shallow-well MTP during orbital shaking (250 rpm, 50-mm amplitude, 30°C). The initial cultures ($t = 0$) were inoculated manually with toothpicks from colonies on an LB agar plate. At $t = 20$ h, 5 μl of each of the cultures (all in stationary phase) was used as an inoculum for 12 fresh cultures: the 12 cultures now grow in a "synchronized" pattern (reach the stationary phase at the same time), which will—in most cases—result in a smaller standard deviation of the parameter under study (e.g., an enzyme activity) and in relative ease of identifying high-activity mutants.

to large numbers of false positives, and possibly false negatives. In practice, synchronizing cultures becomes increasingly difficult at smaller culture volumes, most notably as a result of a larger variation in the size of the inoculum (see also section 8.3.5).

When starting with a library in 96-well MTP format (cells frozen in the presence of 15% [vol/vol] glycerol), one can use a 96-pin replicator with fixed pins to sample the library after melting the master plate. Alternatively, one can press a sterile spring-loaded replicator onto the frozen cultures (no melting required, so there is no viability loss of the remaining frozen cultures; see reference 15) to get a small film of cell suspension on the tips of the pins. In either case, one can subsequently transfer the sampled cells either directly into a liquid culture or onto an agar surface in a rectangular petri dish. The choice depends on the viability of the host strain in liquid medium after the cells have been frozen at −80°C. Many strains grow much better in liquid medium after having been precultured on an agar-based medium.

Protocol 2. Screening of a Mutant/Construct Library in *E. coli* or *Bacillus*

1. Fill all 96 wells of a sterile standard polystyrene shallow-well MTP (round wells with a total volume of approximately 0.35 ml) with 0.15 ml of an appropriate sterile rich medium (e.g., LB medium).
2. Inoculate each well with a single colony from an agar plate. Use sterile toothpicks or, alternatively, a colony-picking robotic system.
3. Cover the inoculated MTP with a sterile sealing tape or a sandwich cover (see section 8.3.1).
4. Incubate the MTP plus cover in an orbital shaker at 30 to 37°C for 16 to 24 h. Shaker conditions: 300 rpm, 50-mm shaking amplitude in case a sandwich cover is applied (see section 8.3.1.2, Fig. 2), or 200 rpm, 50-mm amplitude in case a sealing tape is applied (alternatively, 300 rpm, 25-mm amplitude).

5. Prepare a second MTP (as in step 1) in case synchronized cultures are desired (see section 8.3.6, Fig. 4).
6. Inoculate this second MTP by transfer of 5 μl from each well of the first MTP. For this purpose, use either a 12-channel multipipette, a 96-channel pipetting machine, or a pipetting robot.
7. Repeat steps 3 and 4.
8. Optional: use the first MTP to prepare a frozen master plate for possible rescreening at a later time. Add 150 μl of a 30% (vol/vol) glycerol in water solution to each well. Use a 12-channel multipipette (move up and down) for thorough mixing. Use wide-orifice tips in case cultures are viscous or are not fully homogeneous. Put on a polystyrene lid. Freeze at −80°C and store in a cryobox.
9. Harvest the cells and/or supernatant by centrifugation of the second MTP. Optionally: add 150 μl sterile water or buffer prior to centrifugation; this will make it easier to take off the supernatant after centrifugation (see section 8.3.7).
10. Perform the assay on either the supernatant (in case of an extracellular product or enzyme) or lysed cells (in case of a cytoplasmatic enzyme product).

8.3.7. Centrifugation and Filtration of Cultures in MTPs

Filters in MTP format, available from a range of suppliers in combination with a vacuum unit, can often be used effectively for the separation of cells and supernatant fluid, as long as the cell density is not too high, i.e., does not cause clogging of the membrane filter. Their use is not recommended for relatively dense cultures (more than 1 to 2 g dry wt/liter), which may cause clogging of filters, or if the cost of such filters is inhibitory. In such cases, centrifugation is the method of choice. Tabletop centrifuge rotors are available that can accommodate deep-well MTPs or multiple shallow-well MTPs. Depending on the type of strain, and the strength of the MTP, a minimum speed of 2,000 to 4,000 rpm for 10 to 20 min is required, preferably under refrigeration. Removal of the supernatant using a robotic system or a multichannel pipette can be problematic, especially with small culture volumes. As already mentioned in section 8.3.5.2, addition of extra buffer prior to centrifugation may alleviate this problem (see also the 96-well plunger device described in section 8.3.10).

8.3.8. Glucose Feeding Systems

Many large-scale bioprocesses are fed-batch systems: when the glucose initially present (maximally 100 to 200 mM for toxicity reasons) is fully consumed, the continuous feeding of a highly concentrated glucose solution is started. Mutant or medium screenings aimed at an increased productivity of such fed-batch systems are ideally also done in a similar fed-batch mode. This may be a realistic option for 24 or 96 cultures using peristaltic pumps or syringe-based feeding systems, but such external dosing systems are not practically feasible for thousands of cultures. For large numbers of cultures in MTPs, two internal glucose delivery systems are applicable.

First, small silicone elastomer disks containing glucose crystals may be added to each well (26). The glucose will slowly diffuse from the disks into the medium. A second option is the addition of a combination of glucosidases and starch or cyclodextrin to each well. By varying the amount of glucosidase in the medium, the glucose supply rate can be adjusted to a level that mimics the large-scale bioprocess. This second strategy was patented by Green and Rheinwald from the Massachusetts Institute of Technology in 1975 (20), and recently further elaborated upon by Panula-Perälä et al. (36). In the latter paper, the starch is added into an agar gel on the bottom of each well, while glucoamylase is added to the growth medium itself. Distinct advantages of the silicone elastomer disks are that compounds other than glucose can also be used and that it can also be applied for strains that produce proteases (which would destroy the glucosidase used in the enzymatic method). A practical and logistic advantage of the enzymatic method is that the necessary components can be added as liquids.

The substantially higher cell densities that can be reached with these systems also make it more challenging to keep the pH within certain limits in the absence of an active pH control system. A relatively strong buffer (e.g., 0.15 to 0.2 M phosphate buffer) is recommended. Also, ammonia is preferably not used as a nitrogen source since with the consumption of each molecule of NH_4^+ one proton is released, and the medium may thus acidify rapidly. Use of an acid carbon and energy source (e.g., succinic acid or acetic acid) leads to a pH rise when consumed and may thus counteract a pH drop due to the consumption of NH_4^+. If a sugar is used as a carbon and energy source, applying high OTR shaking conditions will keep the fermentative formation of acids to a minimum for many microbial strains.

8.3.9. Measurement of pH, O_2, and Biomass during Growth

It is obvious that the small size of the cultures in MTPs sets limits to the methods used for monitoring culture parameters. The application of electrodes or other probes is not very practical, not least because of the required wiring. For these reasons, noninvasive methods such as optical sensors are generally preferred.

Frequently used for pH and oxygen determinations are fluorescence-based sensors made of thin, hydrophilic sensing films consisting of an analyte-sensitive indicator and a reference fluorophore that are deposited on the bottom of the MTP. With the help of special readers, the change in pH and/or dissolved oxygen tension can be followed in time for all wells (3).

For the quantification of biomass inside the wells of MTPs, optical methods based on light scattering (reflection and refraction by the cells) can be applied. Essentially three approaches can be distinguished. In the first approach, a regular MTP reader is used to follow the optical density in time, either in the presence or absence of low-orbit shaking. During the measurement, shaking is interrupted. A limitation of this approach is that OTRs are generally low and that the unhindered transmission of light (the principle on which optical density measurements are based) becomes almost zero at optical densities higher than 3, resulting in a low accuracy at high cell densities. In the second approach (38), the bottoms of the individual wells of an MTP are illuminated using an optical fiber bundle (under an angle of 20 to 30°). The intensity of the light reflected by the cells growing inside the wells is measured continuously using a photodetector at the other end of the fiber bundle. As long as the shaking diameter is not too large (typically 3 mm), the shaking does not need to be stopped during the measurement; thus, mixing and O_2 supply are not interrupted. As one is free to choose different wavelengths for the excitation light and the emission light, this approach is also suitable for fluorescence measurements. The third approach consists of the mounting of a flatbed scanning device between the shaker and the MTPs (18). At regular time intervals, e.g., every hour, the shaker is stopped and the bottoms of the MTPs are scanned, immediately followed by an automatic restart of the orbital shaker. Subsequently, image analysis software automatically generates growth curves. A disadvantage of this third approach is the low oxygen supply during the scanning phase (20 to 40 s), which can have unpredictable effects and can cause stress reactions for certain strains. An advantage, however, is that the user is free to choose a large shaking diameter (e.g., 50 mm) if desired. The availability of high-resolution graphics of each well enables the user to study morphological aspects (formation of pellets, aggregates, wall growth, etc.). Both the second and the third approach (in contrast to the first approach) allow the quantification of relatively high biomass concentrations up to 20 to 50 g/liter cell dry weight.

8.3.10. Specific Aspects of Growth and Screening of Fungi in MTP Format

Parallel microfermentations of bacteria and yeasts in MTPs have gained widespread acceptance, but few cases have been reported that suggest that filamentous fungi could be effectively grown and that they would produce their enzymes or secondary metabolites in MTPs. MTP fermentation systems and pin tool inoculations were designed and developed for use with heterogeneous strain collections of aerobic bacteria and actinomycetes and clonal libraries in *E. coli* and yeasts, but protocols were not available for filamentous fungi. Several features of filamentous growth complicate manipulation of fungi in MTP systems. Strains that lack spores and that grow only as multicellular filaments cannot be transferred as homogeneous microliter volumes. Second, in heterogeneous collections of fungi, rates and patterns of hyphal growth vary enormously among species. Some species may grow very rapidly and produce enough biomass to overflow and escape from a microwell. When one is trying to move inoculum or pipette extracted mycelia, mycelial masses clog and interfere with pipette tips. Furthermore, aeration of microwell cultures by orbital shaking may be ineffective because mycelia can quickly form solid masses. Therefore, new protocols were needed to adapt MTPs to the distinctive biology of filamentous fungi, especially those fungi that only grow as hyphae in culture.

The first obstacle to use of fungi in MTPs is difficulty in achieving consistent inoculation. MTPs can be used to grow fungi that sporulate abundantly in culture because spore suspensions can be reliably sampled from frozen inoculum using a 96-well pin tool or with tips of a multichannel pipette. This approach may be adequate when screening mutants or libraries of transformants in a heavily sporulating host, e.g., *Aspergillus niger*. However, in the context of secondary-metabolite discovery, using MTPs to only grow sporulating fungi limits screening experiments to only a small fraction of the culturable fungi diversity. Within a few well-studied orders of fungi, e.g., the Eurotiales (e.g., *Penicillium* and *Aspergillus*) and Hypocreales (e.g., *Fusarium* and *Trichoderma*), sporulation in vitro is common; however, the vast majority of culturable fungi only produce vegetative hyphae in culture. The best-known uses of MTPs for fungi are the phenotype microarray systems that measure patterns of carbon and nitrogen utilization and responses to cellular inhibitors which are predispensed in 100-μl cultures in 96-well MTPs (9). Fresh conidial suspensions have been used in MTP assays of fungi in phenotype microarrays to inoculate plates for profiling substrate utilization or screening for enzymatic activity (11, 29, 40). In some cases, fungal growth characteristics were comparable among agar, shake flask, and 100-μl cultures in MTPs, such as with *Hypocrea jecorina* (14).

The second obstacle in adaptation of fungi to MTPs is manipulation of filamentous growth in microwells. In an effort to obtain fungal growth compatible with microplate technologies, the filamentous fungus *Chrysosporium lucknowense* has been mutated for nonfilamentous growth and reduced medium viscosity so that it could serve as a eukaryotic host in high-throughput DNA library screening (43). However, agitated aerated growth in 96-well plates is likely unobtainable with most fungi because of the restricted space and tendency of fungi to adhere to the container walls, although 24-well plates permit some degree of agitation of fungal mycelium. For example, transformants of *Aspergillus nidulans* were screened for modified secondary-metabolite profiles by growing them in a 24-well MTP system (1). However, the space limitation in 96-well MTPs dictated that screening of filamentous fungi should adopt a solid-state fermentation strategy (6). Mycotoxin and secondary-metabolite production is routinely evaluated from agar growth in petri plates (19, 39). The successful application of agar plug microextraction techniques to fungi suggested that secondary-metabolite production in small-scale static fermentations in microwells would be feasible.

Recently the spring-loaded replicator (section 8.3.6) and sandwich covers (section 8.3.1.2) have been adapted for inoculation and growth of filamentous fungi in 96-well and 24-well formats. A technique was developed for efficiently generating inoculum from fungal hyphae and transferring inocula among microwells as hyphal suspensions (6). A simple screening system was established using heterogeneous collections of 80 fungi grown in a 96-well format. Fungal metabolites could be efficiently extracted from microwell plates, and metabolite production was adequate for detection of biological activity. Growth in 96-well plates can yield 500 to 1,000 μl of mycelial extract

and supernatant, enough to carry out several assays, repeat them if necessary, and evaluate the extract via HPLC–LC-MS. Active strains discovered by high-throughput screening of MTP fermentations usually can be scaled up in shake flasks or static flask cultures in order to isolate and characterize active components (6, 21, 34). The 96-well or 24-well MTP format permits creative and high-throughput experimentation with fungi that normally is resource limited and cumbersome with conventional shake flask systems (1, 6–8).

Protocol 3. Inoculation and Growth of Filamentous Fungi in MTPs

1. *Equipment.* The equipment and materials for growth of fungi for secondary-metabolite production in static MTPs have been described in detail (6). An environmental chamber equipped with shaking platforms (5-cm displacement) and inclinable (approximately 75° angle) steel tube racks (25-mm diameter, 16 tubes per rack) is used to incubate and agitate seed inoculum tubes. Production cultures are grown in 96-well (2.5 ml per well) or 24-well (10.0 ml per well) deep-well polypropylene plates and incubated in environmental cabinets. Culture manipulations, master plate preparation, and plate replications should be carried out in biosafety cabinets.

2. *Preparation of inoculum: replication and growth of 80 fungi in microplate fermentations.* Select sets of 90 to 100 strains of freshly isolated fungi or fungi from a culture collection and grow for 2 to 3 weeks in 60-mm petri dishes. Grow fungi in YM agar (malt extract, 10 g; yeast extract, 2 g; bacteriological agar, 20 g, H$_2$O, 1,000 ml). Avoid heavily sporulating species with dry airborne spores (e.g., *Aspergillus* spp.) or species with extremely fast radial growth (e.g., *Trichoderma* spp.) to minimize the possibility of cross-well contamination. When using screening applications that employ fungi with dry airborne spores (*Apergillus, Penicillium*) or aggressive growth (*Botrytis, Trichoderma*), consider using inoculation patterns that skip alternative rows or wells. Prepare seed medium in 25-by-150-mm glass tubes. Add two 22-mm^2 cover glasses to each tube. Agitation of the tubes on an orbital shaker causes the cover glasses to continually shear hyphae and mycelial pellets, resulting in a homogeneous hyphal suspension (Fig. 5). Dispense 8 ml of melted and homogeneously stirred Sabouraud's maltose yeast extract medium with 0.35% agar (maltose, 40 g; yeast extract, 10 g; neopeptone, 10 g; agar, 3.5 g; distilled H$_2$O, 1 liter) into each tube. The additional agar in the medium aids adherence of mycelial fragments to the replicator tips. Cap the tubes and autoclave. To inoculate the tubes, remove two 5-mm-diameter agar disks from each culture with a Transfer Tube (Spectrum Laboratories, Rancho Dominguez, CA). Push the Transfer Tube's plunger against the bottom of the tube to mash and extrude the agar disks against the glass, thus increasing the number of mycelial growing points. Agitate the tubes (200 rpm, 5-cm displacement) for 3 to 6 days. Most strains exhibit moderate to dense growth. However, tubes that appear to the unaided eye to have little growth, upon microscopic inspection often reveal a dense suspension of fine hyphal fragments.

3. *Preparation of master plate for replication.* From among the set of 100 tubes, select 80 strains with adequate growth to prepare master plates for replication (Fig. 6). A sterilized master plate holds the hyphal inoculum for replication. Use sterilized Transfer Tubes to transfer 1 ml

FIGURE 5 Preparation of fungal inoculum for master plate replication. Two microscope cover glasses are agitated with growing mycelium to produce a homogeneous mycelial suspension.

of hyphal suspensions to wells. After loading, the master plate is closed until replicated in fermentation plates. Freeze additional aliquots of each inoculum tube in 20% glycerol in 1.8-ml cryovials at −80°C.

4. *Preparation of fermentation media in multiwell plates.* Prepare fermentation media in advance in aliquots of 100 ml in Erlenmeyer flasks. Use 80 ml of the aliquot to fill the 80 central wells of the deep-well growth plate at 1 ml per well. Leave the first and last columns of the plates empty so those positions can be used for evaluation of

assay positive and negative controls. Media formulations are usually limited to soluble components; media with insoluble components or precipitates are mixed with a magnetic stirrer during dispensing.

Prepare seed-based solid media by filling an MTP with 200-μl wells with seeds (wheat, rice, millet, cracked corn) and leveling the seed depth to the tops of the wells. Align the open wells of an inverted growth plate on top of the seed-containing MTP, and flip the two plates rapidly so that a measured seed volume falls into each well. Add a liquid basal medium (700 μl) to the seeds; close the plates with the sandwich lids and autoclave the assembled plates.

5. *Replication of master plates across the nutritional array.* Sterilize the cryoreplicator pin tool (section 8.3.6) by immersing the pins in 70% ethanol for 1 min and then evaporating the ethanol on a hot plate. Lower the cooled cryoreplicator pins several millimeters into the hyphal suspensions of the master plate, and then raise them. Switch the master plate for a medium-filled growth plate, and lower the inoculum-covered pins to the bottom of the liquid medium. Switch the growth plate again for the master plate, and repeat the inoculation three to five times. Repeated inoculation from the master plate ensures that sufficient fungal cells are transferred to initiate growth. Repeat the process again for each different medium plate. Sterilization of the cryoreplicator during multiple transfers from the same master plate is unnecessary because the cryoreplicator press prevents accidental lateral movements, and the inoculum source for each well remains constant. Change and sterilize cryoreplicators between consecutive inoculations of master plates.

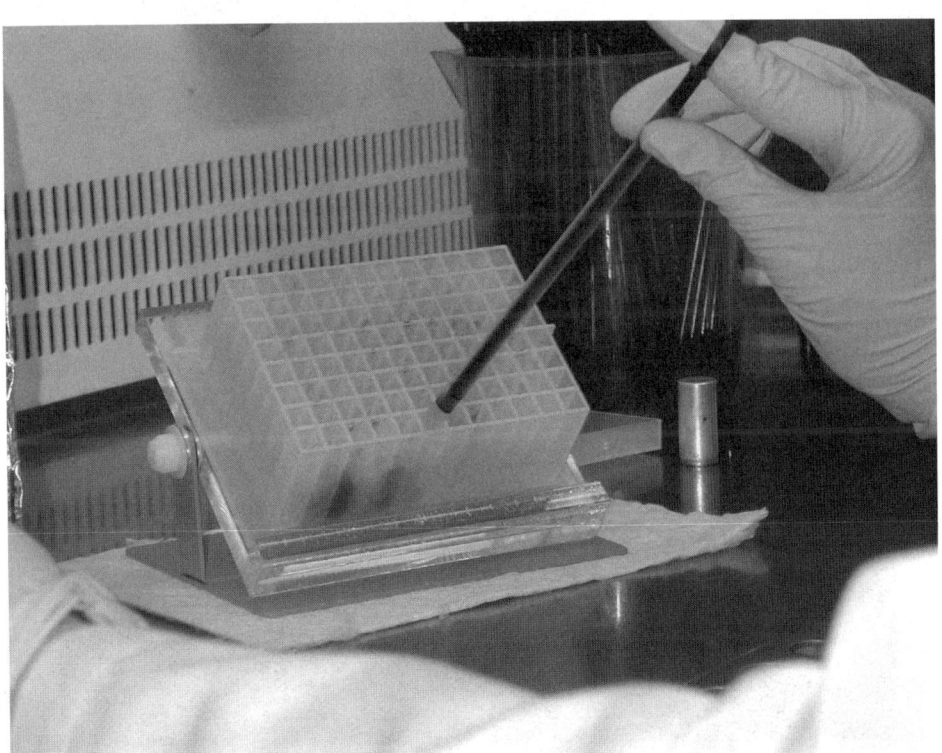

FIGURE 6 Loading the master plate with fungal mycelium suspensions. Afterwards, the cryoreplicator tool is used to replicate each well across multiple growth conditions in the nutritional array.

6. *Incubation.* The clamps used to press the sandwich cover onto growth plates and attach them to shaker platforms are awkward for static incubation and therefore are not used. Instead, incubate plates statically at about an 85° angle (Fig. 7). Inclined incubation increases the surface area of the liquid media and improves aeration. Plates are incubated for 2- to 3-week growth cycles at 22°C. During early experiments, the sandwich covers were held in place on growth plates with heavy rubber bands. A second-generation closure used modified carpenter's wood clamps to clamp lids securely to growth plates (Fig. 7, upper shelf). Finally, stainless steel plate holders that simultaneously clamped plates shut with two lateral wing nuts and supported them at an 85° angle were custom built (Fig. 7, lower shelf).

7. *Contamination and viability check.* After inoculating the medium plates, inoculate Omnitray plates filled with YM agar from the same master plate and incubate at 22°C. Inoculate a second Omnitray plate filled with Luria broth agar and incubate at 28°C to check for bacterial contamination. Replication of master plates onto agar verifies that each well contains viable inoculum and that strains are not contaminated.

8. *Extract preparation.* After growth, open plates and inspect for contamination. Replace the sandwich cover's 96-hole silicone mat with a solid silicone mat, and remove the breathable cotton layer. Pump 1 ml of a water-miscible solvent (acetone or isopropanol) into each fungal culture. Gently dislodge mycelia adhering to the well walls, and crush and mix mycelial masses with the solvent with the cryoreplicator tool. Reassemble sandwich covers with the solid silicone mats, clamp them on a shaker board, and agitate for 1 h. After about 30 min, reverse the shaker board to change the direction of agitation. Finally, centrifuge plates for 5 min to settle

mycelia to the bottom of the wells. To facilitate removal of extracts from growth plates, use a custom-built aluminum 96-well hollow plunger (Fig. 8) to push mycelium to the bottom of wells while the extract and supernatant fill the plunger's hollow columns. The hollow center of the plunger columns and open windows at the sides prevent spilling and overflow of the extracts and largely eliminate mycelial interference and facilitate pipetting with a liquid-handling robot. Transfer a total of 900 μl of the acetone- or isopropanol-medium supernatant from each well to wells of a 2-ml recipient plate. To minimize metabolite precipitation during solvent evaporation, add 100 μl of dimethyl sulfoxide (DMSO) to the 900 μl of the culture extract. The culture-solvent-DMSO mixture is reduced to 50% of its original volume under a stream of N_2 from a 96-tip manifold or by vacuum evaporation. After solvent removal, approximately 500 μl of aqueous extract with 20% DMSO remains per well, and extracts can be stored at −20°C until assayed. Extract plates can be thawed and briefly shaken on a plate mixer prior to assay.

8.3.11. Specific Aspects of Growth and Screening of Actinomycetes in MTP Format

Actinomycetes provide an array of opportunities to the industrial microbiologist. From an industrial point of view, actinomycetes are biosynthetic factories capable of producing a wide range of complex secondary metabolites. They are also efficient producers of enzymes useful for biocatalytic reactions, including many useful in mimicry of human metabolic reactions. In contrast to many nonactinomycete strains, which exhibit more homogeneous growth, the ability to screen hundreds of actinomycete strains for these capabilities can be complicated given the different growth

FIGURE 7 Static incubation of fungal growth plates for the nutritional array. The plates on the upper shelf are closed with carpenter's clamps and supported at an 85° angle by wood tongue depressors. The plates on the lower shelf are supported by a custom-made stainless steel support.

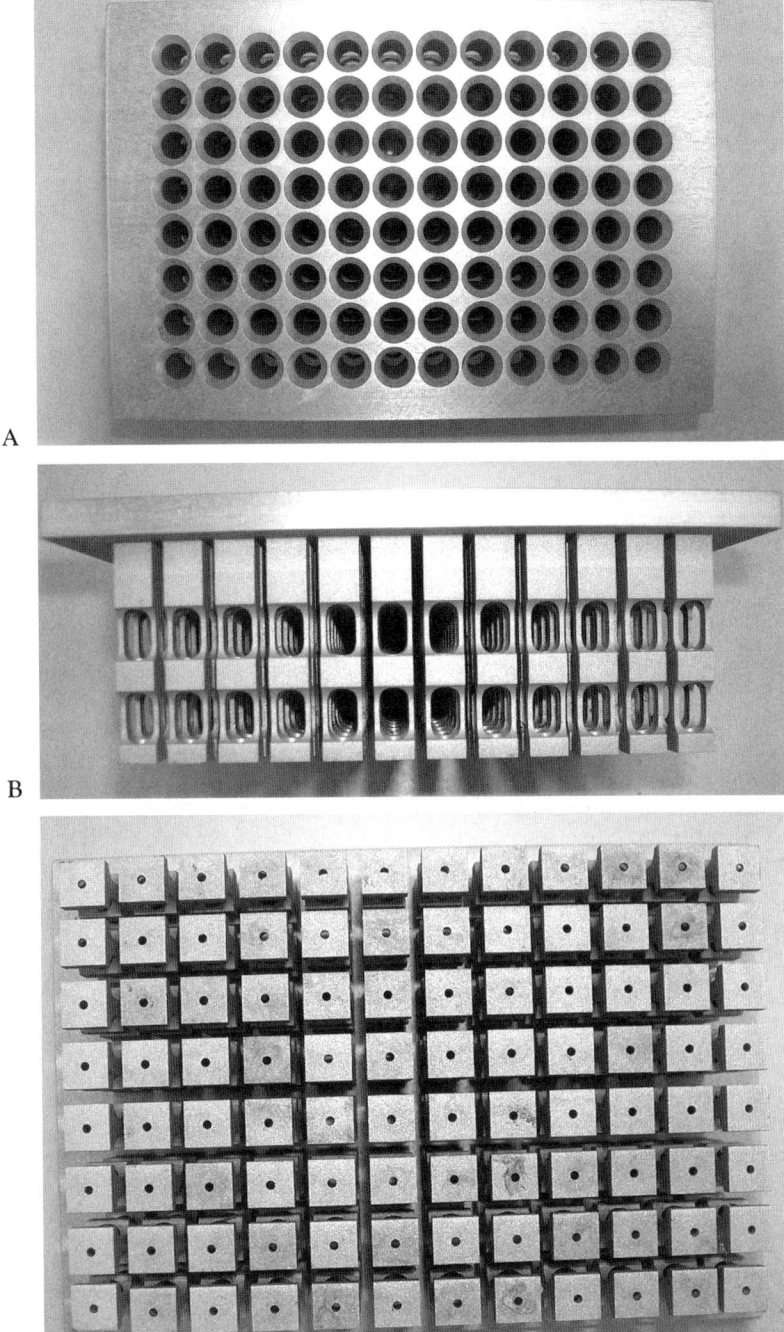

FIGURE 8 Custom-built 96-well aluminum plunger tool used to compress and separate fungal mycelium to bottom of each well while permitting solvent and culture supernatant mixture to rise to top of plate for pipetting. (A) Top view. (B) Side view. (C) Bottom view.

morphologies, lengths of culture periods, and varying medium requirements for growth and induction of pathways. However, given the need for new natural products in pharmaceutical discovery and the utility of biocatalytic derivatization of complex structures in, as examples, medicinal chemistry structure-activity relationship studies, metabolite syntheses, organic syntheses, and biotechnology initiatives, screening of actinomycete cultures is a tool industrial microbiologists must continue to utilize and improve upon. The

ability to quickly and effectively screen banks of mutant populations for improvements in these various activities is another effective use of the miniaturized format.

Growth of most actinomycete strains in liquid culture generally results in dispersed mycelial growth, clumped mycelial growth, or pelleted growth of some type. The reasons that determine submerged liquid growth morphology of one culture as compared to another may include the genus of the strain, inoculum quality, quantity of growth points such as spores or mycelia fragments in the inoculum, medium composition, medium viscosity, and physical culture conditions. Physical culture conditions include factors such as the ratio of liquid volume to vessel volume, and vessel configurations such as baffling and agitation rate. As discussed previously in this chapter, physical culture conditions in miniaturized growth systems can typically be dictated by the revolutions-per-minute set point of an orbital shaker, the use of round or square well plates, the geometries of the wells of the plates, the volume ratio of medium in a particular well geometry, and the viscosity of the medium. Maximizing these conditions for a diverse population of strains can be difficult, but several steps can be taken to optimize multiple parameters around a set of distinct strains for particular purposes.

To begin any screening effort with actinomycetes, there must be a good working knowledge of the strains to be screened. Actinomycete strains will differ in growth rate, optimal growth temperature, medium requirements, and morphology, so decisions about strain selection for the purposes of populating the wells of a particular MTP may be driven by attempting to synchronize culture populations and growth requirements. In addition, the culture behavior as it relates to the needs of postfermentation processing will influence this design. For example, it might make sense to separate those strains typically growing as fragmented or dispersed mycelial cultures, such as many members of the nocardioforms, from those typically observed as pelleted, clumped, or thick and viscous cultures, such as many members of the streptomycetes. One may then be able to more effectively select postgrowth activities such as liquid-handling steps, filtration steps, extraction steps, or centrifugation steps between the more fragmented, filamentous strains and the pelleted or clumped strains. It is difficult to process 24 or 96 wells of actinomycete cultures simultaneously if 1 or 2 or 10 cultures do not effectively filter, clog filters of a solid-phase extraction plate, or are difficult to pipette, etc. If these difficult-to-process strains are absolutely required in the screen, then either more work must be done to understand the issues involved or it may be beneficial to segregate these problem strains to MTPs with those which behave similarly, or perhaps even screen them separately in flasks.

As mentioned above, distribution of actinomycete populations in MTPs for miniaturized growth and screening may have more to do with the practical aspects of physical manipulations and morphological characteristics than with taxonomy. With regard to morphology, it is known that in specific cases greater antibiotic productivity can result from cultures growing in a nonpelleted or dispersed mycelial morphology, so efforts may be made to reduce the pelleting of particular cultures (12, 42). Similarly, reducing the extent of formation of pellets may be required in biocatalyst screening since in such cases mixing, oxygen transfer, substrate availability, and effective active-site populations may be enhanced. In addition, and in consideration of scale-up of successful reactions or fermentations to larger bioreactors or fermentors, fragmented culture morphology may be desired for the same reasons just presented. Medium composition variations and addition of components to medium formulations known to increase the viscosity of liquids as well as that of several polymeric compounds have also been used with some success to limit culture pelleting (24, 32, 46). An excellent review of morphological forms of actinomycetes in submerged cultures lists strains that normally grow as pellets, as fragmented mycelium, or both, under shaken-flask conditions (45). This review is a great

source of information on the factors that may be influenced to control pellet formation.

Methods for preparation, preservation, and use of culture inoculum vary from laboratory to laboratory, and in most cases these methods are determined experimentally for each strain. Whether or not a particular strain will sporulate in a laboratory setting using standard solidified medium formulations is a factor in the procedures followed for preservation of the strain. Harvested spore suspensions containing a cryopreservative such as glycerol can be frozen in cryogenic vials and stored under liquid nitrogen or in liquid nitrogen vapors for extended periods or at −80°C. Others have reported cryopreservation of spore suspensions prepared directly from cultures grown on solidified medium in MTP wells (30). For strains that do not sporulate or are difficult to sporulate in a typical lab setting, one may choose to cryopreserve harvested mycelia either from homogenized scrapings from growth on appropriate agar plates or from mycelia obtained from submerged cultures harvested via centrifugation or filtration. Several techniques and procedures for preparing, preserving, and working with various types of strains are available (31).

The steps used in the preservation of inoculum are important for many reasons, including reproducibility. It has been reported for several filamentous microorganisms that the quality of a seed inoculum and the quantity of points of growth such as spores or mycelia fragments can affect the morphology of the culture in culture vessels (27, 45). These observed differences in morphology resulting from quality and quantity of inoculum are similar in miniaturized growth systems. One may observe increased incidence of pelleting with several actinomycete strains in 96-deep-well plates versus 24-deep-well plates in similar medium formulations and under similar inoculum conditions. Differences in mixing and turbulence, as discussed earlier in this chapter, between the different vessels may in fact affect the interaction of particles such as spores or mycelial fragments and thus pellet formation. Variations in either inoculum or inoculation conditions may therefore lead to wide-ranging culture morphology changes. These morphological changes may in turn affect culture viscosity and overall culture behavior. These effects may impact scalability, as well as reproducibility, with a strain from experiment to experiment or even between independent replicates.

Analysis of various culture preservation techniques and inoculum development techniques, with special attention to reproducibility of resulting cultures in terms of morphology (in MTPs and flasks), natural product yields and enzyme activity, and growth rates and nutrient utilization, is important. Techniques resulting in optimum performance in miniaturized culturing devices are generally successful in seed culturing flasks and therefore can be used for both. Cryopreserved inoculum can then be used to seed multiple identical "master plates" to be used to inoculate screening plates in future initiatives. These master plates are typically produced in sterile, labeled polystyrene 96-shallow-well plates with sterile polystyrene lids and stored in racks at −80°C or colder. These master plates can also be used to inoculate agar plates or MTPs containing solidified media. One may observe irreproducible results using 96-pin inoculation devices with diverse actinomycete strains, and thus attention should be paid to developing alternative techniques such as utilizing up to several hundred microliters of thawed cryopreserved inoculum per well from master plates, depending on the strain and on the MTP geometry. Growing strains on solidified medium in MTP wells can be used for various biocatalyst screens when the substrate is known to be cytotoxic. One can overlay growth in each well with an appropriate buffer, dose with substrate, and shake in a shaking incubator at appropriate temperatures and speeds. One of the benefits of miniaturized growth systems is that one can perform these steps with multiple replicates to test various buffers and different pH values fairly easily and reproducibly. An alternative to this is to harvest mycelia from growing plate cultures by filtration or centrifugation and resuspend cell material as "resting cells" in an appropriate biocatalysis buffer system followed by dosing and incubation. The latter "resting cells" strategy may be more pragmatic from a scale-up perspective.

Many filamentous strains tend to adhere to the walls of MTPs at the "tide line" created by the medium during shaking. This wall growth will then typically continue to grow, and one may even observe sporulation with some strains. This presents a situation that can be very different from those conditions encountered in many shaking flask configurations as well as in fermentors and thus presents a potential scale-up issue, not to mention a potential source of irreproducibility. In addition, if the strains in these wells are to be screened for potential bioconversion capability and thus dosed with an organic compound, neat or suspended in a carrier solvent, the dosed material may get held up or absorbed by the wall growth and never completely make contact with the submerged culture. There is also the possibility of the absence of submerged growth in the medium because of adherence and wall growth. This phenomenon may happen to a greater extent in one miniaturized growth system configuration versus another, for example, 96-square-deep-well systems versus 24-square-deep-well formats. Medium formulation may also play a role in wall growth, as complex media formulations with insoluble components such as starch or Pharmamedia tend to facilitate more wall growth.

Other beneficial uses of the high-density and parallel format of miniaturized culture techniques include screening of mutant populations and optimization of medium components. Classical methods for strain improvement generally involve selection from large numbers of mutants, isolates exhibiting improved strain characteristics (44). Often, random mutagenesis is used to improve production strains with regard to yields of secondary natural product. Thousands of colonies of surviving mutants may be picked and patched into wells of 96-deep-well plates containing 0.5 ml of growth medium and grown up (with sandwich covers; see section 8.3.1.2) for cryopreservation and crossing into screen plates. Mutant cultures exhibiting improved productivity may then be screened in shaking flasks and ultimately in 20-liter fermentor studies over a 9-month period. Concurrently, medium optimization studies can be done in 96-deep-well plates utilizing the wild-type production strain. This wild-type strain can then be used to assay effectiveness of varying medium formulations and conditions such as carbon-to-nitrogen ratio as well as the selection and concentrations of various nitrogen and carbon sources. Use of the 96-well formats, once conditions for growth and natural product formation were identified, allows for testing of very large numbers of medium formulations, multiple controls, and culturing conditions, such as additions of sequestering agents like adsorbent resins, addition of various oils and fatty acid methyl esters, as well as potential inducers and pathway intermediates.

As can be surmised from the preceding, it is quite challenging to make generalizations concerning the miniaturization of hundreds of cultures of actinomycete strains. However, what is hopefully very clear is that the success of miniaturized growth of actinomycete strains with respect to screening and reproducibility depends on several factors that must be determined using a strong, basic understanding of the microbiology and physiology of the strains of interest, combined with an experimental strategy. Important factors include the scheme employed for populating strains into the miniaturized format, the medium used for synchronous growth among the populations of strains in any one device, appropriate inoculation protocols, optimum strain preservation protocols, and choice of MTP format.

Protocol 4.   Screening for an Improved Actinomycete Production Strain

1. Fill all 96 wells of sterile 96-deep-well MTPs (square wells with a total volume of approximately 2.0 ml) with 0.5 ml of an appropriate sterile rich liquid growth medium (e.g., yeast extract-malt extract medium [YEME]).

2. Inoculate each well with a single colony from an agar plate spread with an appropriately diluted mutagenized population. Due to the nature of many actinomycete strains growing on agar, picking of colonies may involve taking "core samples" from the agar plate.

3. Cover the inoculated MTPs with sterile sandwich covers (see section 8.3.1).

4. Incubate the MTP with cover in an orbital shaker at an appropriate temperature, typically 25 to 28°C, for 48 h. Shaker conditions: 300 rpm, 50-mm shaking amplitude (other conditions may be appropriate based on experimental determination with specific production strains; however, these conditions approximate those in "standard" lab shaking flask conditions with actinomycete strains).

5. Prepare a second MTP (as in step 1) with an appropriately optimized and tested production medium for the production strain.

6. Inoculate this second MTP by transfer of 25 μl from each well of the first MTP. For this purpose, use an 8- or 12-channel pipette with wide-orifice tips. Use of robotics may be possible at this step depending on laboratory setup with respect to aseptic practices; culture morphology and medium viscosity are additional factors potentially affecting use of robotics.

7. Optional (recommended): use the first MTP to prepare a frozen master plate for possible rescreening at a later time. Add 100 μl each of a 50% (vol/vol) glycerol solution and of the primary plate culture to each well of a sterile polystyrene 96-well plate. Use an 8- or 12-channel pipette (move up and down) for thorough mixing. Put on a sterile, labeled polystyrene lid and secure lid. Freeze at −80°C and store in a cryobox.

8. Incubate production culture MTPs in an orbital shaker at an appropriate temperature, typically 25 to 28°C, for an appropriate length of time for the particular production strain. As in step 4, shaker conditions: 300 rpm, 50-mm shaking amplitude.

9. Assay populations for titer, activity, or other targeted criteria:
   a. Cell-associated natural products: perform extractions of cells/fermentation solids as necessary from centrifuged or filtered mass to extract compound of interest. Use extracts directly for LC-MS, HPLC, GC-MS, or cellular-based assays of titer or activity.
   b. Extracellular products: collect supernatant by centrifugation or filtration. Transfer supernatant to 96-well

collection/storage plates. Use supernatant directly for LC-MS, HPLC, GC-MS, or cellular-based assays of titer or activity.

c. Alternatives:

i. Extract entire well using liquid-liquid extraction techniques with an appropriate water-immiscible solvent such as ethyl acetate or dichloromethane.

ii. Apply entire well contents to an appropriate solid support such as Celite-filled 96-well filter plate devices followed by liquid-liquid extraction techniques with an appropriate water-immiscible solvent such as ethyl acetate or dichloromethane.

iii. Apply supernatant samples to appropriate 96-well solid-phase extraction devices followed by appropriate extraction and, if needed, concentration steps.

8.4. CULTIVATION IN TEST TUBES

While they are not an inherently parallelized system, one should not rule out the utility of test tubes for the culturing of small to large numbers of microbial strains, especially if one wishes to handle all cultures individually (which is difficult when using MTPs). Test tubes are used for various purposes by the microbiologist, but in the context of this chapter, focus is placed on use of test tubes for high-throughput culturing of microbes in small volumes. Ohta et al. screened 1,237 actinomycete strains for the ability to enzymatically hydroxylate cyclosporin A, thus mimicking human metabolism of this complex structure (33). In this case, all strains were inoculated from frozen vegetative mycelial stocks into 2 ml of medium in test tubes, and the tubes were grown on a rotary shaker at 210 rpm. In the end, 89 of the strains grown and screened in this manner were found to produce the hydroxylated metabolites of interest and several were successfully scaled up to 30-liter jar fermentors. Tormo et al. reported screening 250 diverse actinomycete strains in test tube cultures utilizing 10 different fermentation media for the purposes of identifying HPLC profiles of metabolites produced. This was intended to identify those media typically supportive of wider chemical diversity of the products (41). Each strain in this work was inoculated into 10 ml of each production medium in 25-by-150-mm test tubes and the cultures were grown in tubes at a 35° angle. The results of this screening of strains and various production media were ultimately supportive of the "one strain-many compounds" approach to microbial natural product screening (10) and also demonstrated the utility of test tubes in high-throughput microbial screening.

8.5. CONTROLLED MINI-BIOREACTOR SYSTEMS

The use of mini-bioreactor systems such as MTPs has aided scientists in a diverse range of microbiological applications. This chapter has primarily focused on MTPs of various well geometries and volumes, in the absence of medium, pH, and/or O_2 control. However, there exists a range of alternative mini-bioreactor systems each having potential for various applications. From the simple to the exquisite, the literature contains several examples of these other mini-bioreactor and fermentation systems. Betts and Baganz present an extensive review of miniaturized bioreactors and explore advantages and disadvantages of several of the current and emerging technologies (5). As explained in the review, there may be an advantage to selecting each

individual type of mini- or micro-bioreactor based on the application. Two specific examples of mini-reactor systems are described below.

Puskeiler et al. from the Technical University of Munich developed a reaction block with 48 mini-reactors with working volumes of 8 to 12 ml each, and free-floating and self-centering magnetic impellers (37). The shape of the impeller in combination with baffles caused the air to be sucked into the medium, leading to maximal OTRs of around 300 mmol liter^{-1} h^{-1}. A robotic liquid-handling system (using syringes that can take samples, but also inject medium, acid or base) allows the system to function as a set of 48 fully controlled fed-batch bioreactors. In an automated, pH-controlled, fed-batch cultivation, an E. coli strain could be grown to a density of more than 20 g dry wt/liter.

The miniature bioreactor system described by Isett et al. (25) consists of a modified polystyrene 24-well MTP (with higher wells). pH and oxygen are continuously monitored using fluorescence-based sensors as described in section 8.3.9. The pH can be automatically adjusted to a set level by either CO_2 or ammonia via the gas phase. Orbital shaking at 500 to 800 rpm enables OTRs in the range of 10 to 15 mmol liter^{-1} h^{-1}. Optionally, sparging of air can be used to supply extra O_2 to the cultures (leading to OTRs of approximately 25 mmol liter^{-1} h^{-1}) (25). The system was evaluated with Saccharomyces cerevisiae, E. coli, and Pichia pastoris strains.

Although it was not miniaturized to the extent as the two examples above, we also want to mention here the development of a shaken-flask parallel system (2). Use of this system allows for growth of microorganisms in 14 parallel shaking flasks: 6 standard laboratory flasks for sampling as well as 8 specialized flasks for online measurements of OTRs, carbon dioxide transfer rates, and the respiratory quotient. This system allows for screening of various conditions, such as medium composition, while gaining information on the metabolism of the organism(s).

8.6. SUMMARY

Miniaturized cultivation systems are attractive not only in reducing demands for incubation space and medium but also in making the parallel handling of large numbers of strains more practicable. Provided the user is willing to invest a sufficient amount of time and effort on the optimization for their specific purpose, the presently available systems allow the cultivation at a 0.1- to 10-ml scale at a reproducibility and quality approaching those of larger-scale fermentation equipment. The presently available systems are almost without exception based on the MTP standard, which contributes—in combination with the use of other MTP-based equipment—to a continuous improvement of the time efficiency of screenings and medium optimization projects.

REFERENCES

1. **An, Z., G. Harris, D. Zink, R. Giacobbe, P. Lu, R. Sangari, G. Bills, V. Svetnik, B. Gunter, A. Liaw, P. Masurekar, J. Liesch, S. Gould, and W. Strohl.** 2005. Expression of cosmid-size DNA of slow-growing fungi in *Aspergillus nidulans* for secondary metabolite screening, p. 167–187. *In* Z. An (ed.), *Handbook of Industrial Mycology.* Marcel Dekker, New York, NY.

2. **Anderlei, T., W. Zang, M. Papaspyrou, and J. Buchs.** 2004. Online respiration activity measurement (OTR, CTR, RQ) in shake flasks. *Biochem. Eng. J.* **17:**187–194.

3. **Arain, S., G. T. John, C. Krause, J. Gerlach, O. S. Wolfbeis, and I. Klimant.** 2006. Characterization of microtiter plates with integrated optical sensors for oxygen and pH, and their applications to enzyme activity screening, respirometry, and toxicological assays. *Sensors Actuators B* **113:**639–648.

4. **Becker, R. F., L. P. J. Burton, and S. E. Amos.** 1996. Additives, p. 177–210. *In* E. P. Moore, Jr. (ed.), *Polypropylene Handbook.* Hanser-Gardner, Cincinnati, OH.

5. **Betts, J. I., and F. Baganz.** 2006. Miniature bioreactors: current practices and future opportunities. *Microb. Cell Fact.* **5:**21.

6. **Bills, G., G. Platas, A. Fillola, M. R. Jiménez, J. Collado, F. Vicente, J. Martín, A. González, J. Bur-Zimmermann, J. R. Tormo, and F. Peláez.** 2008. Enhancement of antibiotic and secondary metabolite detection from filamentous fungi by growth on nutritional arrays. *J. Appl. Microbiol.* **104:**1644–1658.

7. **Bills, G. F., J. Martín, J. Collado, G. Platas, D. Overy, J. R. Tormo, F. Vicente, G. Verkleij, and P. Crous.** 2009. Measuring the distribution and diversity of antibiosis and secondary metabolites in the filamentous fungi. *Soc. Ind. Microbiol. News,* **59:**133–146.

8. **Bills, G. F., G. Platas, D. P. Overy, J. Collado, A. Fillola, M. R. Jiménez, J. Martín, A. González del Val, F. Vicente, J. R. Tormo, F. Peláez, K. Calati, G. Harris, C. Parish, D. Xu, and T. Roemer.** 2009. Discovery of the parnafungins, antifungal metabolites that inhibit mRNA polyadenylation, from the *Fusarium larvarum* complex and other Hypocrealean fungi. *Mycologia* **101:**449–472.

9. **Bochner, B. R.** 2009. Global phenotypic characterization of bacteria. *FEMS Microbiol. Rev.* **33:**191–205.

10. **Bode, H. B., B. Bethe, R. Höfs, and A. Zeeck.** 2002. Big effects from small changes: possible ways to explore nature's chemical diversity. *Chembiochem* **3:**619–627.

11. **De la Cruz, T. E. E., B. E. Schulz, C. P. Kubicek, and I. S. Druzhinina.** 2006. Carbon source utilization by the marine *Dendryphiella* species *D. arenaria* and *D. salina. FEMS Microbiol. Ecol.* **58:**343–353.

12. **Dobson, L. F., and D. G. O'Shea.** 2008. Antagonistic effect of divalent cations Ca^{2+} and Mg^{2+} on the morphological development of *Streptomyces hygroscopicus* var. *geldanus. Appl. Microbiol. Biotechnol.* **81:**119–126.

13. **Doig, S. D., A. Diep, and F. Baganz.** 2005. Characterisation of a novel miniaturised bubble column bioreactor for high throughput cell cultivation. *Biochem. Eng. J.* **23:**97–105.

14. **Druzhinina, I. S., M. Schmoll, B. Seiboth, and C. P. Kubicek.** 2006. Global carbon utilization profiles of wild-type, mutant, and transformant strains of *Hypocrea jecorina. Appl. Environ. Microbiol.* **72:**2126–2133.

15. **Duetz, W. A., L. Rüedi, R. Hermann, K. O'Connor, J. Büchs, and B. Witholt.** 2000. Methods for intense aeration, growth, storage, and replication of bacterial strains in microtiter plates. *Appl. Environ. Microbiol.* **66:**2641–2646.

16. **Duetz, W. A., and B. Witholt.** 2004. Oxygen transfer by orbital shaking of square vessels and deepwell microtiter plates of various dimensions. *Biochem. Eng. J.* **17:**181–185.

17. **Duetz, W. A.** 2007. Microtiter plates as mini-bioreactors: miniaturization of fermentation methods. *Trends Microbiol.* **15:**470–475.

18. **Duetz, W. A.** October, 2008. An apparatus and a method for investigation of microtiter plates subjected to orbital shaking. Netherlands Patent 86185NL00.

19. **Frisvad, J. C., and R. A. Samson.** 2004. Polyphasic taxonomy of *Penicillium* subgenus *Penicillium,* a guide to identification of food and air-borne terverticillate penicillia and their mycotoxins. *Stud. Mycol.* **49:**1–173.

20. **Green, H., and J. G. Rheinwald.** December, 1975. Method of controllably releasing glucose into a cell culture medium. U.S. patent 3,926,723.

21. **Herath, K. B., G. H. Harris, H. Jayasuriya, D. L. Zink, S. K. Smith, F. Vicente, G. F. Bills, J. Collado, A. González del Val, B. Jiang, J. N. Kahn, S. Galuska, R. A. Giacobbe, G. K. Abruzzo, E. J. Hickey, P. A. Liberator, T. Roemer, and S. B. Singh.** 2009. Isolation, structure and biological activity of phomafungin, a cyclic lipodepsipeptide from a widespread tropical *Phoma* sp. *Bioorg. Med. Chem.* **17:**1361–1369.

22. **Hermann, R., N. Walther, U. Maier, and J. Büchs.** 2001. Optical method for the determination of the oxygen-transfer capacity of small bioreactors based on sulfite oxidation. *Biotechnol. Bioeng.* **74:**355–363.

23. **Hermann, R., M. Lehmann, and J. Büchs.** 2003. Characterization of gas-liquid mass transfer phenomena in microtiter plates. *Biotechnol. Bioeng.* **81:**178–186.

24. **Hobbs, G., C. M. Frazer, D. C. J. Gardner, J. A. Cullum, and S. G. Oliver.** 1989. Dispersed growth of *Streptomyces* in liquid culture. *Appl. Microbiol. Biotechnol.* **31:**272–277.

25. **Isett, K., H. George, W. Herber, and A. Amanullah.** 2007. Twenty-four-well plate miniature bioreactor high-throughput system: assessment for microbial cultivations. *Biotechnol. Bioeng.* **98:**1017–1028.

26. **Jeude, M., B. Dittrich, H. Niederschulte, T. Anderlei, C. Knocke, D. Klee, and J. Buchs.** 2006. Fed-batch mode in shake flasks by slow-release technique. *Biotechnol. Bioeng.* **95:**433–445.

27. **Junker, B. H., M. Heese, B. Burgess, P. Masurekar, N. Connors, and A. Seely.** 2004. Early phase process scale-up challenges for fungal and filamentous bacterial cultures. *Appl. Biochem. Biotechnol.* **119:**241–277.

28. **Kensy, F., H. F. Zimmermann, I. Knabben, T. Anderlei, H. Trauthwein, U. Dingerdissen, and J. Büchs.** 2005. Oxygen transfer phenomena in 48-well microtiter plates: determination by optical monitoring of sulfite oxidation and verification by real-time measurement during microbial growth. *Biotechnol. Bioeng.* **89:**698–708.

29. **Kubicek, C. P., J. Bissett, I. Druzhinina, C. Kullnig-Gradinger, and G. Szakacs.** 2003. Genetic and metabolic diversity of *Trichoderma*: a case study on South-East Asian isolates. *Fungal Genet. Biol.* **38:**310–319.

30. **Minas, W., J. E. Bailey, and W. Duetz.** 2000. Streptomycetes in micro-cultures: growth, production of secondary metabolites, and storage and retrieval in the 96-well format. *Antonie van Leeuwenhoek* **78:**297–305.

31. **Monaghan, R. L., M. M. Gagliardi, and S. L. Streicher.** 1999. Culture preservation and inoculum development, p. 29–48. *In* A. L. Demain and J. E. Davies (ed.), *Manual of Industrial Microbiology and Biotechnology,* 2nd ed. ASM Press, Washington, DC.

32. **O'Cleirigh, C., J. T. Casey, P. K. Walsh, and D. G. O'Shea.** 2005. Morphological engineering of *Streptomyces hygroscopicus* var. *geldanus*: regulation of pellet morphology through manipulation of broth viscosity. *Appl. Microbiol. Biotechnol.* **68:**305–310.

33. **Ohta, K., H. Agematu, T. Yamada, K. Kaneko, and T. Tsuchida.** 2005. Production of human metabolites of cyclosporin A, AM1, AM4N and AM9, by microbial conversion. *J. Biosci. Bioeng.* **99:**390–395.

34. **Ondeyka, J., G. Harris, D. Zink, A. Basilio, F. Vicente, G. Bills, G. Platas, J. Collado, A. González, M. de la Cruz, J. Martín, J. N. Kahn, S. Galuska, R. Giacobbe, G. Abruzzo, E. Hickey, P. Liberator, B. Jiang, D. Xu, T. Roemer, and S. B. Singh.** 2009. Isolation, structure elucidation and biological activity of virgineone from *Lachnum virgineum* using the genome-wide *Candida albicans* fitness test. *J. Nat. Prod.* **72:**136–141.

35. **Ortiz-Ochoa, K., S. D. Doig, J. M. Ward, and F. Baganz.** 2005. A novel method for the measurement of oxygen mass transfer rates in small-scale vessels. *Biochem. Eng. J.* **25:**63–68.

36. **Panula-Perälä, J., J. Šiurkus, A. Vasala, R. Wilmanowski, M. G. Casteleijn, and P. Neubauer.** 2008. Enzyme controlled glucose auto-delivery for high cell density cultivations in microplates and shake flasks. *Microb. Cell Fact.* **7:**31–42.

37. **Puskeiler, R., A. Kusterer, G. T. John, and D. Weuster-Botz.** 2005. Miniature bioreactors for automated high-throughput bioprocess design (HTBD): reproducibility of parallel fed-batch cultivations with *Escherichia coli*. *Biotechnol. Appl. Biochem.* **42:**227–235.

38. **Samorski, M., G. Muller-Newen, and J. Buchs.** 2005. Quasi-continuous combined scattered light and fluorescence measurements: a novel measurement technique for shaken microtiter plates. *Biotechnol. Bioeng.* **92:** 61–68.

39. **Smedsgaard, J.** 1997. Micro-scale extraction procedure for standardized screening of fungal metabolite production in cultures. *J. Chromatogr. A* **760:**264–270.

40. **Talbot, N. J., P. Vincent, and H. G. Wildman.** 1996. The influence of genotype and environment on the physiological and metabolic diversity of *Fusarium compactum*. *Fungal Genet. Biol.* **20:**254–267.

41. **Tormo, J. R., J. B. Garcia, M. DeAntonio, J. Feliz, A. Mira, M. T. Diez, P. Hernandez, and F. Pelaez.** 2003. A method for the selection of production media for actinomycete strains based on their metabolite HPLC profiles. *J. Ind. Microbiol. Biotechnol.* **30:**582–588.

42. **Van Wezel, G. P., P. Krabben, B. A. Traag, B. J. F. Keijser, R. Kerste, E. Vijgenboom, J. J. Heijnen, and B. Kraal.** 2006. Unlocking *Streptomyces* spp. for use as sustainable industrial production platforms by morphological engineering. *Appl. Environ. Microbiol.* **72:**5283–5288.

43. **Verdoes, J. C., P. J. Punt, R. Burlingame, J. Bartels, R. van Dijk, E. Slump, M. Meens, R. Joosten, and M. Emalfarb.** 2007. A dedicated vector for efficient library construction and high throughput screening in the hyphal fungus *Chrysosporium lucknowense*. *Indust. Biotechnol.* **3:**48–57.

44. **Vinci, V. A., and G. Byng.** 1999. Strain improvement by nonrecombinant methods, p. 103–113. *In* A. L. Demain and J. E. Davies (ed.), *Manual of Industrial Microbiology and Biotechnology*, 2nd ed. ASM Press, Washington, DC.

45. **Whitaker, A.** 1992. Actinomycetes in submerged culture. *Appl. Biochem. Biotechnol.* **32:**23–35.

46. **Yang, W., E. A. Hartwieg, A. Fang, and A. L. Demain.** 2003. Effects of carboxymethylcellulose and carboxypolymethylene on morphology of *Aspergillus fumigatus* NRRL 2346 and fumagillin production. *Curr. Microbiol.* **46:**24–27.

47. **Zimmermann, H. F., G. T. John, H. Trauthwein, U. Dingerdissen, and K. Huthmacher.** 2003. Rapid evaluation of oxygen and water permeation through microplate sealing tapes. *Biotechnol. Prog.* **19:**1061–1063.

Solid-Phase Fermentation: Aerobic and Anaerobic

RAMUNAS BIGELIS

9

The science of solid-phase fermentation (SPF) draws on the same ancient origins that are the foundation of the field of biotechnology and of the modern methods used to grow microorganisms. This approach to culturing microorganisms has been used to produce chemicals, commodity enzymes, pharmaceuticals, high-value biochemicals, and diverse foods and beverages. It also serves as a valuable research tool for the discovery of new natural products. An earlier article by Sato and Sudo (131) covers small-scale solid-state fermentation, as do other reviews on SPF (4, 12, 34, 49, 65, 72, 80, 93, 96, 98, 102, 104, 106, 109, 110, 120, 125, 150).

9.1. GENERAL CONSIDERATIONS

9.1.1. History

The origins of fermentation date to prehistory and gave rise to the first, perhaps accidental, methods used to preserve food and prevent them from spoiling and causing illness (16). SPF is as old as leavened bread making in ancient Egypt (2600 BC) and beer making in Babylonia (6000 BC), centuries-old soy fermentations in south Asian cultures, and a variety of traditional food fermentations that spread worldwide alongside primitive agriculture.

The koji process, a form of SPF and a landmark development, originated in China and was brought to Japan in the seventh century as a fermentation with *Aspergillus oryzae* on steamed rice. Used as a "starter," it still is the first stage in a second fermentation that produces miso, tempeh, soy sauce, sake, and other traditional Oriental foods. It is fitting that the word "koji" is an abbreviation of the Japanese word *kabi-tachi*, which means "bloom of the mold." Its development and spread were seminal events in the history of SPF.

Significant commercial exploitation of SPF beyond food fermentations, however, occurred late in the 19th century (85, 91). In 1894, Jokichi Takamine patented a process for producing amylase by culturing *A. oryzae* on solid substrates via what is still called the Takamine process (143). These pioneering efforts evolved into industries that employed SPF for production of a variety of fungal and bacterial enzymes, citric acid, gluconic acid, and kojic acid. And, in part, they inspired the development of modern biotechnology (15, 17, 47, 52, 85, 91). Today, the impact of industrial SPF in Europe and North America is small; it

is more common in regions that lack modern fermentation equipment. However, interest in this method of culturing microorganisms has been renewed (40, 135).

9.1.2. Characteristics of SPF

As defined inclusively, SPF under either aerobic or anaerobic conditions is a process of culturing microorganisms on solid materials in the presence of little or no free water. The solid materials on which the organisms proliferate, and which they may penetrate, typically contain water (and often soluble nutrients) or have the capacity to absorb it. These substrates are commonly cereal grains, legumes, and various lignocellulosics. It can be said that SPF encompasses both solid-substrate and solid-state fermentation since solid-substrate fermentation employs a natural material as a carbon/energy source in the presence of little or no free water, but the broader designation, solid-state fermentation, includes the possible application of an inert, degradable, or poorly degradable solid support in a suitable medium. The inclusion of inert supports (26, 104) can play an important role in SPF and shape its definitions and classification. Krishna (80) discusses a number of such definitions of solid-state and solid-substrate fermentation as put forth by Moo-Young et al. (98), Pandey and Ramachandran (111), Aidoo et al. (4), Raimbault and Alazard (121), Raimbault (120), Lonsane et al. (86), Soccol (137), Hesseltine (72), Viniegra-Gonzalez (150), and others, and relates this terminology to the presence, absence, or irrelevance of added inert solid supports. Additionally, SPF can be subcategorized as to the nature of the microorganisms applied to the fermentation: natural-indigenous culture, mixed culture-cofermentation, and pure monoculture fermentation. Pure, defined cultures are generally used in most industrial SPFs since this approach is associated with greater quality control and reproducibility, though some fermentations with agricultural residues and food applications employ mixed cultures with benefit. All common definitions of SPF differentiate its various forms from submerged fermentation, termed liquid fermentation or liquid-phase fermentation (LPF), which is carried out in dilute solutions or slurries, in contrast to SPF, where little or no free liquid is present (26, 80, 86, 121, 137, 151). LPF is of major significance in industrial microbiology (23, 47) and is used to produce numerous and varied commercial products. Another approach, termed mixed-phase fermentation (MPF) by Bigelis et al. (26), bridges SPF and LPF, and employs absorbent inert solid supports in abundant liquid medium to facilitate the discovery of new natural products.

The physicochemical nature of the substrate and the growth environment differentiate it from LPF. The solid substrate integral to this microbial process is *generally* heterogeneous; inherently low in moisture content; susceptible to heat buildup and moisture loss due to evaporation during microbial growth; and often difficult to agitate, mix, and, at times, aerate. Many SPFs are stationary and nonagitated, though rotary drum or fluidized-bed fermentors can be configured to allow substrate mixing. Moreover, the solid substrate in SPF predominantly supports surface growth rather than the homogeneously dispersed growth that typically occurs in an agitated liquid environment, and in the case of molds, it is quite often associated with aerial mycelia rather than pellet morphologies (74, 83, 104, 144, 155). Similar patterns have been observed with actinomycetes (37, 60). Generally, the nonuniform nature of the solid substrate in SPF is correlated with a nonuniform growth environment, resulting in conditions that can influence productivity and yield. Owing to these general characteristics, process control and optimization of SPF present considerably more challenges than do those of LPF.

The final processing steps of SPF may generate valuable by-products such as animal feed, compost, or fertilizer and significantly enhance the economics of the process.

9.1.3. Advantages and Disadvantages of SPF

9.1.3.1. Advantages

SPF may offer advantages, especially cost savings, over submerged fermentation (10, 42, 74, 75, 120, 125, 151). In a number of cases, SPF has been shown to be a simpler technology. It is (i) more productive; (ii) less susceptible to contamination by undesired organisms due to the reduced water activity (a_w); (iii) independent of agitation and mixing; and (iv) cost-effective since (a) substrate sterility (and pasteurization or autoclaving) may be nonessential, (b) culture media may be simple, consisting of agro-industrial residues, and (c) aeration needs may be reduced (39, 40, 42, 61, 86, 131, 148). Furthermore, SPF can use fermentors of reduced size and volume with concomitant reduced water needs and effluent production (4, 86), compact fermentation facilities, and undemanding culture conditions resembling a natural environment (128), requiring reduced processing and recovery with reagents and solvents (148). Processing of the fermented residues is usually very simple, only requiring quick drying of the low-moisture product to produce animal feed, fertilizer, or other dry commercial products (131). Finally, convenient production of conidiophores by the koji method provides readily available starter cultures. Such spores are reportedly more stable, robust after cryostorage, and resistant to dehydration than spores obtained by LPF (75, 83).

SPF is especially advantageous in fungal applications, and is more common with this group of microorganisms. The low moisture content of many solid media used for SPF makes it especially suitable for culturing molds and other filamentous fungi rather than bacteria, whose growth may be restricted by low a_w (131). SPF, which is commonly a stationary process, often avoids the shear stress and mechanical forces characteristic of LPF which disrupt mycelia. As a result, SPF may promote conditions for maximum production. Moreover, fungal SPF has a record of good production of secondary metabolites (11–13, 26, 45, 80, 125).

9.1.3.2. Disadvantages

The disadvantages of SPF may outweigh the advantages and force alternative fermentation strategies. The lack of scalability of this fermentation technology frequently limits its application. Another major drawback is the low thermal conductivity of the solid substrate, which makes temperature control difficult. Temperature increases, resulting from metabolic heat, can create local hot spots despite forced aeration and the application of thermal regulatory devices (132). Other disadvantages are linked to agitation and process control: difficulties in mixing of the substrate bed may lead to nonhomogeneity and uneven distribution of water, nutrients, biomass, heat, or pH control agents (131). As a consequence of poor mixing, product diffusion may also be limited and interfere with the desired physiological state and pathway regulation of the production organism. Various other disadvantages may involve (i) difficulty in measuring and directly monitoring microbial cell growth and fermentation parameters, (ii) inability to control critical fermentation parameters, (iii) uneconomical requirements for pretreatment of solid materials and substrates, (iv) impurity of the final product elevating product recovery costs, (v) significant moisture loss during long-lasting fermentations, (vi) the need for large inocula, and (vi) limitations due to the critical role of water and a_w (42, 131).

9.2. MICROBIAL GROWTH AND PRODUCTS

9.2.1. Microbial Growth

The growth characteristics of microorganisms during SPF are strongly dependent on the physicochemical nature of the solid substrate and its interaction with water and nutrients, and are associated with growth habits and morphologies that are distinct from those observed during LPF (7, 113). Mathematical models and empirical growth equations that describe fungal growth patterns in SPF have been developed for various molds that characterize growth kinetics in response to key environmental variables (43, 95, 123, 129, 149).

Complex microbial growth phenomena are associated with solid substrates that are typically heterogeneous and porous with void spaces. Fungal surface growth promotes formation of aerial mycelia in this ideal environment for differentiation, leading to sporangiophores bearing conidiospores or other sporulation structures (83, 144), somewhat comparable to sporulation processes in actinomycetes (37, 60). Hyphal penetration of interior spaces, however, can lead to different morphologies that may depend on changes in oxygen availability in the void spaces and nutrient availability in the substrate bed (75, 97, 129). Sudo et al. (140) have reported differences in amylase production associated with submerged and aerial mycelia in rice koji inoculated with *Aspergillus kawachii*. Akao et al. (5) have compared SPF cultures of *A. oryzae* with LPF cultures by subtractive cloning of cDNA, and shown differential gene expression, as well as higher enzyme levels and hyphal differentiation during SPF. Yamane et al. (156) have developed a model SPF system ("cellulose agar cube") to simulate and study the mycelial morphology associated with glucoamylase production by *A. oryzae* during koji culture. A good match between the predictions of the model and the experimental results was observed.

Generally, upon germination of the spores, fungal hyphae form a mycelial mat that proliferates on the solid surface and then invades gaseous spaces and liquid-filled pores.

Moisture control determines the proportion of each of these two environments. Metabolic activities occur primarily on or near the surface and within void spaces, though such activity is also associated with aerial mycelia, which can exchange metabolites with penetrative mycelia. Secreted hydrolytic enzymes infuse into the substrate bed and promote digestion of its components, yielding nutrients. Gradients of nutrients, O_2, CO_2, pH, and secreted primary or secondary metabolites may develop during the course of the SPF, together with an uneven distribution of heat. The behavior of microorganisms during SPF as a microscale process has been discussed by Holker and Lenz (75), and as a bioreactor system by Mitchell et al. (96, 97) and Viccini et al. (149). Both microscale and macroscale processes are illustrated in Fig. 1 (149).

It should be noted that sporulation, a hallmark of SPF, is a key safety concern. Without proper safeguards fungal and bacterial spores can be a major health hazard (82), as can improperly handled grain-based materials that form dust (70).

9.2.1.1. Temperature

Temperature is a key factor in SPF (14, 96, 109, 142). The thermotolerance of the organism must be clearly established, since a buildup or uneven distribution of temperature may interfere with growth. In addition, hot spots can cause denaturation of products, especially thermolabile substances (75, 130). The systems for heat removal, recognizing the generation of considerable metabolic heat and the presence of temperature gradients in the substrate bed, must take into account bed dimensions, heat conductivity, and changes that may occur during the fermentation. Additionally, the effects of temperature on both product

formation and growth must be considered together since each stage of the process may have different requirements for each.

9.2.1.2. Moisture Content and a_w

Like temperature, the moisture content of the solid substrate strongly influences microbial growth (39, 62, 63, 96, 109). In fact, the two variables are closely related in that water evaporation increases as more metabolic heat is evolved. Water can be replaced with periodic additions or humidification. Excessive moisture loss reduces substrate swelling and microbial growth. On the other hand, excessive moisture reduces the porosity of the substrate, lowering gaseous diffusion and exchange.

When considering parameters critical to SPF, a_w is a more relevant representation of moisture limitations than is water content, since a_w represents the amount of water that is available to microorganisms. a_w is the ratio of the vapor pressure of water in a material (p) to the vapor pressure of pure water (p_o) at the same temperature and, being a ratio, has no units. Thus, it ranges from 0.0 (completely dry) to 1.0 (pure water). Relative humidity and a_w are related: multiplication of a_w by 100 gives the equilibrium relative humidity in percent. Many microorganisms thrive at an a_w of 0.99 and most need an a_w higher than 0.91 to grow (39, 131). Bacteria are typically less tolerant of low a_w than are yeasts and molds.

Since water is chemically bound by components of the substrate bed, making it unavailable to microbes, a sensor that measures a_w or equilibrium relative humidity (14, 96) provides critical data on the nature of the substrate, expected growth characteristics, a_w stress tolerance of the production organism (61, 148), and susceptibility of the

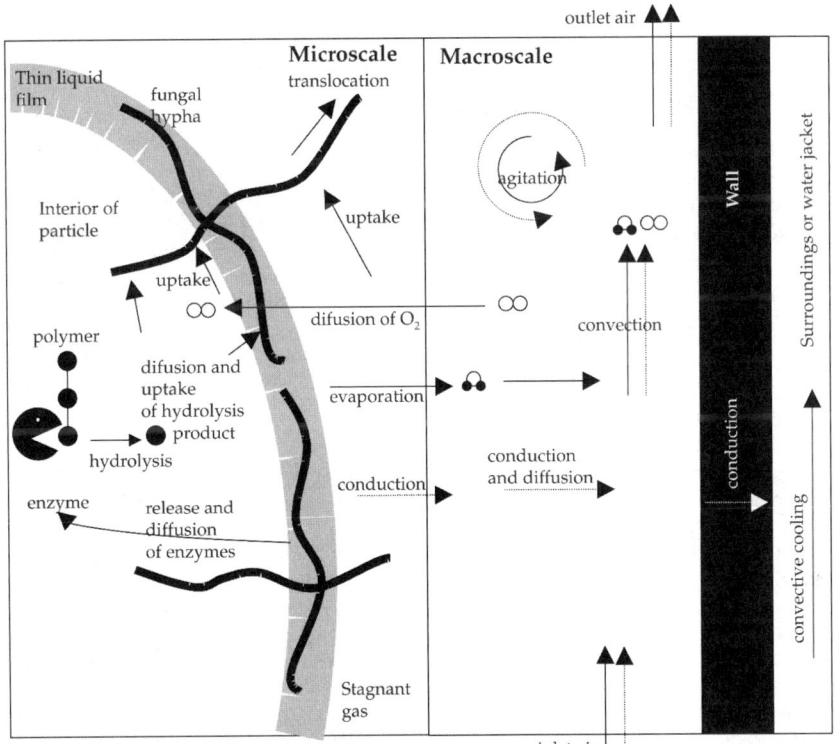

FIGURE 1 The macroscale and microscale processes that occur within an SPF bioreactor. (Adapted from reference 149.)

SPF to bacterial contamination. In fact, a full characterization of the production organism's tolerance to low a_w may be necessary (67). Astoreca et al. (8) have illustrated this point by showing how the interaction of a_w, temperature, and lag phase influenced three ochratoxigenic isolates of *Aspergillus niger* grown in various media. Lag times increased considerably at low a_ws, especially at low temperatures, for all studied isolates.

9.2.1.3.   Oxygen and Carbon Dioxide

The monitoring of oxygen uptake and carbon dioxide evolution, coupled with aeration, is important for characterization and optimization of SPF (105). Such essential knowledge allows the estimation of biomass and growth rate, and a correlation with fermentation performance.

Efficient SPF aeration systems enable accurate maintenance of appropriate oxygen levels, desorption of carbon dioxide, regulation of moisture level, distribution of volatile metabolites, heat dissipation, and temperature control (63, 66, 80, 96, 97, 109). Mixing may improve oxygen transfer to pockets of moisture and water films around substrate particles, increasing gas exchange between the substrate bed and aerial mycelia. In addition, secreted digestive enzymes, especially amylases, may increase the microporosity of substrate particles and even cause some disintegration, influencing the availability of gaseous oxygen to mold mycelia in stationary and agitated SPF. However, excessive porosity and water in void spaces risk reduction of oxygen transfer.

Oostra et al. (105) explored intraparticle oxygen transfer with *Rhizopus oligosporus* grown on the surface of a solid medium. After direct measurement of oxygen levels in the fungal mat with microelectrodes, they found no depletion in the upper aerial layer, but steep oxygen gradients in the wet bottom layer. Oxygen consumption occurred mainly in the wet portion of the fungal mat. Since the contribution of aerial mycelia to overall oxygen consumption was minimal, it was concluded that optimal oxygen transfer in SPF depends on the available interfacial gas-liquid surface area and on the thickness of the wet fungal layer. Moreover, it was apparent that during SPF, moisture levels in the substrate matrix influence both of these parameters and interact with them to influence oxygen transfer.

Some undesirable characteristics of the solid substrate bed may restrict the diffusion of oxygen critical to an aerobic fermentation and limit performance. As would a medium of low porosity, an excessively deep bed may deter mycelial penetration of even high-moisture grains, kernels, beans, or other enriched substrates since oxygen transfer could be limited or prevented. In such cases, deviation from the theoretical respiratory quotient of 1.0 for an aerobic organism to lower values may indicate insufficient oxygen transfer. Such effects have been observed with molds in aerobic SPF. Illustrating these deleterious effects (95, 131), researchers have reported that (i) deeply submerged mold mycelia grow poorly in high-moisture rice koji, (ii) low oxygen levels increase lag time (but not growth rate) of *A. oryzae* grown on steamed rice, (iii) poor oxygen diffusion limits growth of *R. oligosporus* on cassava starch/κ-carrageenan, (iv) low-moisture wheat bran affects oxygen transfer even during high aeration of an *A. niger* culture, (v) reduced oxygen concentrations reduce α-amylase production by *A. oryzae*, and (vi) a thickening of *A. niger* growth around starch-rich particles reduces porosity together with oxygen transfer to the bed.

In aerobic SPF, substrate mixing or agitation can significantly change productivity by oxygenating the solid substrate, as in the case of citric acid or ochratoxin production. In contrast, increased agitation (drum rotation) and aeration reduce ethanol production by *Saccharomyces cerevisiae* (131).

9.2.1.4.   Physical Properties of the Solid Substrate and Particle Size

The SPF substrate is typically a natural raw material that serves as the carbon and energy source and is rich in polysaccharides, such as starch, cellulose, pectin, or lignocellulose. Such substrates are generally cheap agro-industrial by-products, often heterogeneous waste material (40). Solid substrates that have a large surface area (in the range of 10^3 to 10^6 m^2/cm^3) are chosen most often in order to enhance exposure to microbial action (111).

The particle size of the SPF substrate is an important variable that influences growth patterns of the production microorganism, similar to critical parameters such a heat and mass transfer (93, 96, 149). All of these parameters involve the exchange of oxygen and carbon dioxide and the movement of microbial products with respect to the solid surface (Fig. 1). They may change during the course of the fermentation in conjunction with particle degradation and changes in the substrate bed. The surface-to-volume ratio of the particle, as well as its packing density, influences the invasiveness of the microorganism and the exposure of the substrate to biodegradation (79, 109). The geometry of the particle also affects void spaces and, consequently, gas transfer. In general, small particles, owing to their larger specific surface area, are advantageous for prolific mold growth, heat exchange, and gas transfer—but only up to a point. Very small particles may reduce aeration and respiration by limiting the interparticle space. Very small particles may also lead to substrate agglomeration, and their use or appearance risks compression and contraction of the substrate bed, leading to channeling of airflow and thus to reduced respiration and poor growth. In some instances, large particles may be favored, with some drawbacks (110). Most SPF processes use a mixed range of particle sizes for optimal production and cost-effectiveness. For example, the commonly used SPF substrate wheat bran, available in two sizes, fine (<500 to 600 μm) and coarse, can be used in different ratios (111).

Particle size and/or packing density has been shown to influence SPF with *Fusarium oxysporum* on rice, SPF with *Penicillium chrysogenum* on sugarcane bagasse, SPF with *A. niger* on barley, SPF with *Pleurotus ostreatus* strains on sugarcane bagasse, and fermentations for a variety of enzymes (90, 107, 111, 131).

9.2.2.   Production of Enzymes

Enzymes that have been produced using SPF technology include the major classes of hydrolytic enzymes such as α-amylases, glucoamylases, β-fructo-furanosidases, cellulases, hemicellulases, pectinases, galactosidases, proteases, and lipases (19–22, 40, 49, 65, 66, 80, 109, 130). More recently, xylanases (40, 153), phytases (81, 114), and ligninases (41) have been added to this list, with more additions expected in the future (111). These enzymes have a history of use in the food, feed, fuel, textile, detergent, and pharmaceutical industries (23, 66, 151). SPF competes with LPF approaches for enzyme production; however, the latter fermentation method prevails in large-scale commercial enzyme manufacture. While many types of enzymes have been produced by SPF and employed in industrial applications (36, 111), currently only phytases (81), proteases (65), and galactosidases (109, 134) are known to be manufactured on a commercial scale by such means (80). Reported advantages for enzyme production associated with SPF over LPF are (i) higher enzyme yields with selected strains (3, 11, 57, 77, 109, 149, 151), (ii) enhanced enzyme stability (2, 56), (iii) reduced levels of proteolytic degradation of the enzyme product (151), (iv) reduced catabolite repression (100, 119, 120, 131), (v) suitability of crude product with minimal or no downstream processing (103, 109), (vi) efficient growth and biomass production (126), and (vii) good biosynthetic efficiency at high levels of oxygen demand with limited mixing (in tray reactors) (126, 151). Additional advantages point to (viii) simpler fermentation equipment (109), (ix) reduced costs for dewatering and product concentration (109), and (x) lower expenditures for raw materials (40).

9.2.3. Production of Metabolites

SPF has a history of use in industrial mycology to manufacture food-processing aids and ingredients, fuels, chemicals, and pharmaceuticals.

9.2.3.1. Organic Acids

The worldwide industrial output of citric acid surpasses that of other primary metabolites derived by fermentation (24, 25, 111). Citric acid is used predominantly in the food/beverage and detergent industries. Today, virtually all production is by LPF with A. *niger* in large, computer-controlled, aerated fermentors of stirred or airlift design, although up to 20% of world output was obtained by surface fermentation and SPF late in the last century. Continuous processes with *Candida* yeast strains are performed only in small volumes (6).

Variations of the traditional koji process with *Aspergillus* species have been applied to citric acid production in Japan and Southeast Asia on a relatively small scale (85). Owing to economic advantages, the raw materials are residual pulps from starch manufacture or food processing, or similar by-products mixed with a solid support (40). The starchy material, bearing considerable cellulosic substrate, is soaked to a final water content of 65 to 70% and excess water is removed, steamed, cooled to yield a sterilized (or pasteurized) paste, loaded into trays or shallow pans, and spray inoculated with conidia of selected strains. Frequently, A. *niger* strains that are less responsive to metal ions are preferred. Separate or simultaneous enzyme pretreatment with commercial amylases may be used in the saccharification of the starch-based substrate (22, 25). The pH is maintained at about 5.5 (usually with $CaCO_3$) and temperature at 28 to 30°C for 7 to 10 days. Nutrients are often supplied to enrich the fermentation medium. Aeration is optimized in accordance with strain requirements, culture conditions, and the ratio of surface area to volume. Citric acid yields are generally lower than those obtained by LPF owing to difficulties in controlling process parameters and the presence of trace metals, which interfere with citric acid production. Yields of 60% of the theoretical are common; however, some reported yields reach the neighborhood of 80 to 85% of the weight of the initial carbohydrate with up to 90% recovery efficiency (85). A variety of additives and stimulants such as alcohols, metabolic inhibitors, trace elements, and ferrocyanide have been employed to improve citric acid yield (24, 25, 85). Recent attention has focused on SPF for citric acid with

A. *niger* using inexpensive raw materials and agroresidues, diverse food-processing wastes, and inert polymeric support systems (42, 111, 117). SPF for citric acid in flasks and bioreactors has been compared (147), with a focus on production and respiration rate (117). These optimization studies reported that strain selection was an important factor in achieving high citric acid yields, and that strains with high requirements for nitrogen and phosphorus were disfavored in SPF, while strains with low requirements were favored.

The production of citric acid, lactic acid, fumaric acid, gallic acid, and other organic acids by SPF has been reviewed elsewhere (24, 25, 40, 85, 91, 110, 111).

9.2.3.2. Ethanol

Justifications for bioethanol production by SPF have been proposed, and advantages claimed are diminished requirements for extraction and purification of carbon sources, reduced water demand and fermentor capacity, and lowered capital expenses for distillation and energy. However, most reported SPF processes are small-scale and many are still in the experimental stage. All remain linked to petroleum prices, labor costs, and other economic factors. As an alternative to LPF, SPF with yeast has been investigated using plant crops favoring substrates such as corn, beets, sweet sorghum, cassava, and other tubers, in addition to starchy substances such as cereal flours, cereal starches, and potato starch (131, 138). Also, solid wastes and biomass, such as fruit pomace, cassava bagasse, sugarcane bagasse, and other food-processing and agricultural by-products (40); wood residues, such as forestry, paper-mill, and wood-product manufacturing by-products; or even municipal solid waste, each offer unique economic advantages as substrates for SPF. Cellulase technology, perhaps with enzyme produced by SPF (35), will be an important component of any future biomass-to-ethanol process.

9.2.3.3. Natural Products

Natural products have extended the human life span by improving health and nutrition, and have advanced agriculture and animal health, with a major impact on the well-being and economics of society (46, 47, 52, 101). For these reasons, microbial secondary metabolites continue to attract the attention of applied microbiologists long after the end of one era, the Golden Age of Antibiotics (15, 48), and now well into another, the *new* era of natural products. Current natural products research blends more disciplines with modern biotechnology and also adds high-throughput screening for novel therapeutic targets and bioapplications (51, 60). An important additional discipline in natural products research involves SPF and innovative approaches for growing microorganisms.

9.2.3.3.1. Microbial Secondary Metabolites and SPF

The wealth of biodiversity in the microbial world is matched by its associated chemodiversity, and this remarkable variety is a driving force in natural products research. The value of microbial secondary metabolites for pharmaceutical discovery (46, 48, 49, 52, 101, 116) has inspired bioprospectors to investigate new species, unexplored environments, new culturing methods, and novel approaches such as metagenomics.

SPF is a valuable tool in the quest for new bioactive compounds (Table 1) (110) and offers technologies for scaled-up production (10–12, 29, 38, 76, 110, 125, 145). The mycelial state associated with SPF is well suited

TABLE 1 Production of bioactive natural products by SPF[a]

Compound	Source	Substrate	Function
Aflatoxin	A. oryzae, Aspergillus parasiticus	Wheat, oats, rice, maize, peanuts	Mycotoxin
Ochratoxin	Aspergillus ochraceus, Aspergillus carbonarius, Penicillium viridicatum	Wheat, rice, corn, corn kernels	Mycotoxin
Bacterial endotoxins	Bacillus thuringiensis	Coconut waste	Insecticide
Gibberellic acid	G. fujikuroi, Fusarium moniliforme	Wheat bran, corncob, cassava flour, sugarcane bagasse[b]	Plant growth hormone
Zearalenone	F. moniliforme	Corn	Growth promoter
Ergot alkaloids	Claviceps purpurea, C. fusiformis	Sugarcane bagasse[b]	Disease treatment
Penicillin	P. chrysogenum	Sugarcane bagasse[b]	Antibiotic
Cephalosporin	Cephalosporium acremonium	Barley	Antibiotic
Cephamycin C	Streptomyces clavuligerus	Wheat rawa with cottonseed cake and sunflower cake	Antibiotic
Tetracycline, chlortetracycline	Streptomyces viridifaciens	Sweet potato residue	Antibiotic
Oxytetracycline	Streptomyces rimosus	Corncob	Antibiotic
Iturin	Bacillus subtilis	Okara, wheat bran	Antibiotic
Actinorhodin, methylenomycin	Streptomyces coelicolor	Agar medium[b]	Antibiotic
Surfactin	B. subtilis	Soybean residue, okara	Antibiotic
Monorden	Humicola fuscoatra	Agar medium[b]	Antibiotic
Cyclosporine	Tolypocladium inflatum	Wheat bran	Immunosuppressive drug
Ustiloxins	Ustilaginoidea virens	Rice panicles	Antimitotic cyclic peptide
Antifungal volatiles	B. subtilis	Impregnated loam-based compost	Antifungal compounds
Destruxins A and B	M. anisopliae	Rice, rice bran, rice husk	Cyclodepsipeptides
Clavulanic acid	S. clavuligerus	Wheat rawa with soy flour and sunflower cake	β-Lactamase inhibitor, antibacterial
Mycophenolic acid	Penicillium brevicompactum	Wheat bran	

[a]Adapted from reference 110.
[b]As inert solid support.

for growth and production of such metabolites. SPF can elevate the yield of desired metabolites or facilitate the discovery of new ones undetectable after LPF (12, 26). The range of SPF applications is broad. Antibiotic production with fungi under such conditions has frequently been associated with higher yields compared to LPF methods (11, 12, 38, 125). SPF even overcomes current limitations in microbial flavor production, as well as widening the spectrum of metabolites useful for food technology (59). SPF has been successfully applied to the production of gibberellic acid with *Gibberella fujikuroi*, a source of this potent plant growth stimulant widely used in agriculture (145). A superior yield of 250 mg gibberellin/kg dry solid medium was obtained after SPF. Parallel observations with *Claviceps fusiformis* have reported enhanced yields (three to nine times higher) of ergot alkaloids, and production of a broader range of compounds of this class (146). SPF has been proposed as an effective means of cephalosporin (76) and cephamycin C (30) production with actinomycetes. Figure 2 compares lovastatin production by *Aspergillus terreus* under conditions of LPF and SPF with a polyurethane foam support. Superior productivity under SPF conditions was correlated with more intense transcription of lovastatin biosynthetic genes (13).

9.2.3.3.2. Microbial Secondary Metabolites and New Approaches—MPF

Innovative fermentation methods and medium formulations complement bioprospecting for natural products and offer new approaches for screening programs (29, 78). Variations of SPF and LPF can enhance the value of known and less-studied microorganisms as sources of new natural products. MPF, one such alternative approach, bridges LPF and SPF by employing inert solid supports in *abundant* liquid, and facilitates the discovery of new natural products from actinomycetes (27) and fungi (26, 28). Moreover, these approaches take advantage of increased surface area, promoting the diffusion of nutrients, metabolites, and gases—and thereby microbial growth. Solid support systems used in such new ways exploit the tendency of some microorganisms to adhere to solid surfaces, possibly simulating natural habitats. These environmental stimuli can affect differentiation and developmental programs that are linked to secondary metabolism in fungi and actinomycetes (32, 37, 50, 60). Some growth and metabolic responses may be linked to the microarchitecture of the support or substrate, initial steps involved in plant infection or pathogenesis, or responses to native environments that resemble soils or decaying plant matter.

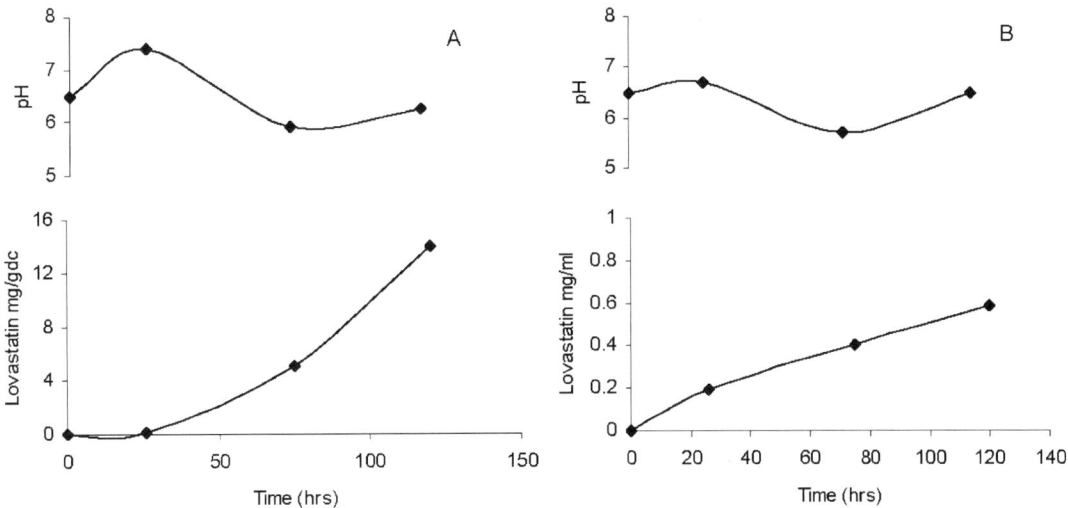

FIGURE 2 Time course of lovastatin production by *A. terreus* in SPF (A) and LPF (B). Note the different scales in lovastatin concentration. pH evolution is shown in the upper part. (Adapted from reference 13.)

Three versions of MPF serve as examples of this fermentation approach: (i) stationary supports on agar surfaces, (ii) shaken supports in abundant liquid medium, and (iii) milled cereal grains or processed solid substrates dispersed on polyester cellulose carrying abundant liquid with or without dissolved nutrient supplements. Bigelis et al. (26) have used MPF with various inert polymeric substances (polyurethane foam, polypropylene, polypropylene cellulose, polyester cellulose) to facilitate the discovery of antimicrobial compounds from fungi (26) and to enhance the production of secondary metabolites (27, 28). In their reports, parallel studies with LPF revealed low or undetectable levels of these secondary metabolites. The MPF method generated low levels of unwanted compounds and medium components (after removal of biomass attached to the polyester cellulose support) in final extracts, and simplified harvesting, extraction, and purification.

A second approach with discs, cylinders, pulps, or other configurations has utilized solid supports in agitated liquid and revealed similar advantages (26, 28). Concentrated extracts of different fungi or actinomycetes grown by MPF in agitated flasks showed greater secondary-metabolite yield or diversity after liquid chromatography-mass spectrometry analysis than did conventional LPF extracts, thus identifying more candidate cultures for further screening. In addition, most fungi and many actinomycetes that were examined were immobilized and exhibited growth morphologies that differed from those observed after conventional LPF (26, 28). These distinctive growth patterns were possibly linked to group behavior, a sociomicrobiology trait (115) of mycelial aggregates firmly attached to each other and to the solid support, and to modified differentiation patterns (37, 50, 60). The ecological role of secondary metabolites in such aggregates may be related to cell-cell microbial communication, a proposed function that has been reviewed by Davies (44), Demain (45), Demain and Fang (50), and Yim et al. (157) and also discussed by Parsek and Greenberg (115) from the perspective of biofilm formation. Cell density-dependent regulatory networks in microorganisms control processes related to cell-cell interactions and may play a role in growth and metabolic pathways that influence secondary metabolism during SPF or MPF (73).

A third MPF approach has been applied to natural product screening programs with fungi in stationary flasks. It utilized coarsely milled cereal grains dispersed on cut strips of polyester cellulose saturated with water or media (28). A Schnitzer Grano 200™ electric grain mill performs the milling function very well. Liquid may be in excess or even be replenished during the fermentation. MPF with milled/dispersed cereal grains was superior to conventional SPF performed with whole cereal grains (28). The absorbent, porous solid support greatly increased accessible surface area (when combined with dispersed milled substrate), water availability and retention, nutrient and metabolite diffusion, as well as exposure to the gaseous phase in interstitial spaces. Fungal differentiation and exposure of substrate to mycelial infiltration were pronounced, and differed from growth patterns observed with compacted, unmilled, and nondispersed moistened grain. Typically, growth and secondary-metabolite yields were enhanced, residual unused solid substrate was significantly reduced, and carryover of unwanted substrate constituents into the final extract was diminished, benefiting high-performance liquid chromatography analysis and purification. Table 2 compiles formulations for conventional SPF fungal media and MPF versions that incorporate substrates dispersed on absorbent polyester cellulose (or other polymeric support). Other media useful for SPF and MPF have been compiled by Atlas (9).

9.2.4. Food Fermentations

Fermented foods are products of SPF that have been influenced by microbial action to produce desired biochemical changes (16, 18, 33). Such favorable modifications with edible fungi and bacteria may make the food material more nutritious and health-beneficial, increase digestibility and processibility, enhance flavor and other sensory aspects, modify physical characteristics, and even improve safety. Food fermentation is a cost-effective means of preservation that can prolong product life and reduce the dependence on refrigeration, food additives, specialized packaging or processing, or other energy-intensive food technological methods. SPFs with foods vary based on geography and dietary customs, but most often are classified according to substrate such

TABLE 2 Medium formulations for small-scale fermentation with solid-substrate system in 2.8-liter Fernbach flasks[a]

Medium	Solid substrate (g/flask)	Liquid supplement[b] (ml/flask)	Support system[c] (g/flask: size of p-c supports)	Water[d] (ml/flask)
SPF corn (cracked)	81	90, 0.1% yeast extract	None	None
SPF rice (white)	102	90, 0.1% yeast extract	None	None
SPF alfalfa (dried)	27	65, 0.1% yeast extract	None	None
SPF milled Fiber One™ food	61	90, 0.1% yeast extract	None	None
MPF coarsely milled corn	81	90, 0.1% yeast extract	15 g (8 × 60 mm) p-c	75
MPF coarsely milled white rice	102	90, 0.1% yeast extract	15 g (8 × 60 mm) p-c	75
MPF homogenized alfalfa	27	90, 0.1% yeast extract	15 g (8 × 60 mm) p-c	75
MPF milled Fiber One™	61	90, 0.1% yeast extract	15 g (8 × 60 mm) p-c	75
Vermiculite	None	70, Czapek-Dox medium	21 g vermiculite	None

[a]For 250-ml flasks divide additions by 7. SPF inoculation: 20 ml of seed broth per flask distributed on substrate surface. MPF inoculation: 20 ml of seed broth per flask dispersed on substrate and support by manual shaking of flask culture; periodic manual shaking may accelerate growth.
[b]Liquid medium supplement added aseptically after sterilization.
[c]p-c, cut polyester-cellulose, KenAG milk-filter (KenAG Animal Care Group, Ashland, OH), dispersal system for milled grain.
[d]To promote adhesion of solids to p-c support, water is added to the support first before addition of processed solid substrate. Solids are dispersed on p-c by manual shaking of the flask, and then the medium is sterilized.

as cereal, dairy, fish, fruit and vegetable, meat, starch crop, and legume products. Mixtures of fermentation substrates and cofermentations with more than one microorganism are common. Fermented beverage production may involve SPF in whole or in part to produce beers (and lagers, ales, and malts), wines, spirits, liqueurs, and nonalcoholic beverages such as coffee, cocoa, and tea (16, 33). Fermented foods continue to provide consumers throughout the world with a significant portion of their nutritional needs, as they have over centuries. Mushroom biotechnology as a large-scale microbial technology has been reviewed elsewhere (154), along with composting processes for substrate production (71, 88).

9.2.5. Biocontrol Agents

SPF is used for inoculum production in processes yielding alternative biocontrol agents against fungal diseases that constrain crop production. The mass production of fungal spores is key to the manufacture of such agents, and SPF offers advantages for their production. De Vrije et al. (55) have discussed the relevance of SPF to the production of commercial products containing *F. oxysporum* (nonpathogenic), *Gliocladium catenulatum*, *Phlebiopsis gigantea*, and the *Trichoderma* spp. *T. harzianum*, *T. polysporum*, and *T. viride*, as well as that for the promising biocontrol agent *Coniothyrium minitans*. Other fungal biocontrol agents used in crop protection, such as *Beauveria bassiana* and *Metarhizium anisopliae*, can be readily produced as either aerial conidia or mycelia and blastospores. SPF allows production of aerial conidia, which are similar to those produced naturally on the surface of insect cadavers and superior to those produced by LPF.

9.3. PROCESS CONTROL

Key SPF process variables are related to the microorganism and its preparation, the choice of substrate and its pretreatment, and environmental factors which determine how the microorganism and the substrate interact, often in a large-scale bioreactor (80, 97). Successful process control depends on model-based strategies descriptive of mass and thermal transfer processes (43, 93, 131).

9.3.1. Preparation of the Organism and Inoculation

Inoculation protocols take into account the organism and its cellular state (spores or vegetative cells), medium choice,

density of the inoculum and its mode of application, and the nature and state of the substrate (temperature, pH, a_w, sterility, mixability). Compared to vegetative mycelia, spores are often preferred since they offer convenience; viability and storage advantages; and ease of quantitation, manipulation, and transfer, often by spray inoculation (83). They also provide multiple growth centers. Despite their hardiness, spores may display lag times as they grow out of dormancy, and require larger inocula and growth conditions quite different from those for vegetative growth.

Unlike LPF, where sterility is often required, pure culture techniques may be unnecessary in SPF, especially if the process organism is able to outcompete the contaminating organisms. While cleanliness is essential, aseptic operation may not be necessary and that may lower the cost (110).

9.3.2. Preparation of Materials

SPF substrates and starting materials may be pretreated before inoculation, when necessary, to facilitate mycelial penetration, and even conditioned to promote mixing, moistening, aeration, or diffusion of gases and nutrients. Dehulled rice, a common substrate for the koji process, for example, may be polished to remove up to 40 to 70% of its weight, and then steamed, raising its water content by 10 to 15% by weight (131). Steam pretreatment is used to sterilize or pasteurize the substrate. Steam may also enhance susceptibility of the substrate to degradation by carbohydrases (22). In addition to steaming or cooking, substrates such as cereal grains may be crushed, cracked, milled, powdered, or blended to meet the requirements of the SPF—or even roasted to enrich for flavor compounds in food fermentations.

Inexpensive raw agro-industrial wastes may be chopped or ground to decrease size and increase substrate exposure; cleansed and/or sanitized; cooked or steamed; partially degraded via additional combinations of physical, chemical, and/or enzymatic pretreatments; and enriched with nutrient supplements to promote microbial growth (120). The inclusion of an inert support in the substrate bed may aid in the retention of favorable physical and spatial characteristics during microbial action on its constituents, and improve the management of heat and mass transfer during the course of the SPF (104). The use of inert supports may also have processing advantages during product recovery (92, 104, 152).

9.3.3. Temperature, pH, and Moisture Control

The temperature optima for product formation and stability during SPF may differ from culture conditions most suitable for common fungi and bacteria, which typically grow within the range of 20 to 55°C. Some SPF substrates, especially starchy or lignocellulosic solid substrates, may not accommodate these optima owing to poor thermal conductivity and low water content. For example, heat buildup as a result of microbial metabolism can reach high levels, even as high as 80°C in a composting environment (71, 88), with thermal gradients extending vertically and horizontally. Thus, monitoring and control of temperature and humidity are essential to avoid drying of the medium, diminished a_w, and an associated reduction in nutrient availability (64, 69). Solutions to heat buildup focus on forced aeration, stirring, or evaporative cooling, although the last approach risks moisture loss and drying of solid materials. The temperature is generally measured in the solid layer, and is also monitored at entry and exit ports of a bioreactor. Thermosensors are usually spaced and inserted radially from the center of the fermentor. The sensors commonly employed are thermocouples, metallic probes, or thermistors. Thermistors offer the advantage of higher sensitivity and short response times, but may show lower long-term stability than metallic probes. Metallic probes, while simple and precise, may be more expensive and slower to respond. Thermocouples provide good fidelity, but drawbacks are associated with the measurement of relative temperature rather than absolute temperature via this technique. Various temperature-sensing techniques have been discussed by Bellon-Maurel et al. (14).

Control of pH is another essential aspect of SPF requiring management of excessive acidity (during accumulation of organic acids) or undesirable alkalinity (during protein hydrolysis). Owing to substrate nonhomogeneity, the lack of free water, and the limitations in pH electrode design, pH adjustment during SPF is generally inefficient. Thus, pH control is often accomplished by incorporating medium components (urea, ammonium salts, calcium salts) that introduce a buffering capacity without diminishing yield or productivity (87). The pH can also be managed by incorporating acids or bases in the cooling water added to the fermentation.

Water influences growth and metabolism during SPF, the activity of intracellular and extracellular enzymes, and transport processes such as the entry of nutrients, exit of metabolic products, and gas exchange. As a consequence of its central role, moisture levels must be carefully maintained. Fluctuations in water content during SPF occur as a result of evaporation caused by metabolic heat, and also via metabolic water production. Excessive moisture risks the development of impediments to gas transfer, particle agglomeration, and a reduced deterrence of bacterial contamination (via elevation of a_w). Inadequate water content (1, 34, 124) leads to suboptimal microbial growth and restricts metabolic activities, besides influencing the physical properties of the substrate bed. Moisture levels in a large-scale SPF bioreactor are often controlled by aeration with water-saturated air, while small-scale laboratory SPF may employ a culture chamber with saturated saline solution or spraying of sterilized cooled water. A model has been proposed (119) for diffusion of water in the substrate that takes into account the significant water retention of the fungal biomass layer (e.g., 2 kg/kg/dry substrate) and the shrinkage of the substrate matrix caused by the removal of water (99).

During SPF, water content is commonly assayed by measuring dry weight; however, this approach does not discriminate between water content and a_w, which reflects water unavailable for microbial action. The most common online technique to measure a_w employs the capacitive sensor, a device with a slow response time often unable to respond to rapid changes in biological processes. Alternative approaches employ a system with a faster response time based on optical absorbance using a mirror that captures dew condensate. Bellon-Maurel et al. (14) have reviewed such techniques that measure a_w, together with online methods for controlling humidity and dew temperature, studying mass transfer limitations, and estimating the effect of humidity on productivity.

9.3.4. Monitoring of Microbial Growth

The mass of microbial cells is central to the control, characterization, and full evaluation of any fermentation process. The direct estimation of microbial biomass during SPF is prevented owing to its intimate association with the substrate, and with the solid support, if present, and the difficulty in separating these entangled fibrous or granular structures. Consequently, indirect measurements that assay cellular constituents are sometimes employed (53). However, the correlation of biomass with cellular components or activities can be inaccurate (53); i.e., the levels of biomolecules or activities can depend on the stage of fungal development or medium composition.

The quantitative determination of DNA, protein, total sugar, cell components like chitin or glucosamine, ergosterol, or ATP has provided an index of biomass levels, as have methods for other cell components (14, 53, 54, 89, 96, 131). Similarly, biological activities may be directly proportional to biomass, permitting its estimation by measurement, for example, of hydrolytic enzymes, nutrient consumption, reactivity to specific antibodies, or respiratory rates (14). The last-named method is commonly employed to analyze fermentor off-gases via respirometry and is used to measure microbial growth (106, 150).

The determination of CER (CO_2 evolution rate in g CO_2/g solid/h) via infrared spectroscopy permits estimation of cumulative cell concentration, which is accurate at high growth rates typical at the beginning of a fermentation. However, stationary-phase cells continue to produce CO_2 and may cause misleading biomass estimates by this method (120). Measurement of OUR (O_2 uptake rate in g O_2/g solid/h) also permits a convenient estimate of the biomass of aerobic microorganisms (14). Versatile online systems for monitoring of CO_2 and O_2 and control of gases during aerobic or anaerobic SPF have been developed. They have proved reliable for obtaining information on the physiological state and respiration rate of cultures in real-time processes, as well as providing estimates of specific growth rates for aerobic cultures (58, 133). New techniques have been proposed for biomass measurement that rely on image analysis, near-infrared spectroscopy, and Fourier transform infrared spectroscopy, but superior and more reliable online methods are still being sought (14, 68, 118).

General Procedure for SPF

1. *Culture.* Choose appropriate SPF microorganism, selecting a natural, mixed, or pure culture. Develop an archival culture storage plan. Define the seed preparation protocol and medium for generating dependable inocula. Evaluate the special needs of the organism, as well as its sensitivity to storage, handling, and mechanical forces of the SPF, aiming for culture stability and reproducibility.

2. *Safety.* Perform a safety evaluation and develop biohazard control for the SPF process and facility, complying with regulatory requirements and good laboratory practices.
3. *Pilot studies.* Perform preliminary studies with small-scale SPF to define the best conditions for growth, product formation, and eventual scale-up. Understand the growth kinetics in response to environmental variables, aiming for models that describe growth and production. Complete an economic assessment considering the cost of raw materials. Verify the stability of the product.
4. *Substrate, nutrients, solid support, and scale.* Choose a suitable solid substrate and nutrient formulation, and solid support if beneficial, based on medium studies. Confirm scalability. Understand the physicochemical characteristics of the medium, e.g., moisture content and retention, thermal conductivity, porosity, mixability, heat and mass transfer, and possible changes that may occur during the SPF, together with the micro and macro characteristics of the solid substrate.
5. *Pretreatment of solid substrate.* Review and evaluate pretreatment protocols for the solid substrate and its preparation, e.g., physical, chemical, or enzymatic processing; steaming or cooking; pasteurization; autoclaving; mixing with a support; and enrichment.
6. *Inoculation.* Initiate the SPF with a spore or mycelial preparation dispersed by spraying or mixing using a quantified inoculum of appropriate density to accelerate the fermentation.
7. *Fermentation and monitoring of parameters.* Perform the SPF under stationary or agitated growth conditions, with an appropriate bioreactor design that permits monitoring and process control of environmental factors on the desired scale. Manage moisture, a_w, temperature, aeration, O_2, CO_2, pH, mixing efficiency, and growth—and correlate with yield and productivity. Reduce the negative effect of by-products on yield, and enhance the value of potential coproducts.
8. *Processing and recovery of product.* Define product specifications and purity requirements, and execute the recovery plan with processing steps that maximize efficiency and yield, recovering/recycling solvents and processing materials whenever possible.

Example: Small-scale pilot fermentation with cereal grain

Place 100 g of rice in a 2.8-liter Fernbach flask with an air-permeable cover. Add 90 ml of 0.1% yeast extract before (for cooking) or after autoclaving. Cool and then inoculate with 10 to 20 ml of liquid seed by dispersal on sterile substrate surface. Incubate stationary for 1 to 3 weeks at 22 to 28°C, or as suitable for product formation. Process after lyophilization or drying, or extract as is, to recover product. Perform larger-scale SPF and compare results.

9.4. BIOREACTORS/FERMENTORS

Bioreactors, or fermentors, have been used in small- and large-scale SPF to produce diverse metabolites and enzyme products, as well as applied to processes involving fermented food products, feeds, edible mushrooms, biofertilizers, biopesticides, and bioremediation (93, 96, 109, 118). While laboratory-scale SPF employs flasks, jars, roller bottles, columns, and petri dishes, common bioreactor/fermentor systems used for industrial scale-up are tray fermentors, packed-bed fermentors, rotary drum fermentors,

and fluidized-bed fermentors. Most of these systems are aerobic and often require aeration, either intermittent or continuous. A few SPF processes are anaerobic and require no agitation or aeration. Some SPF processes do not require bioreactors: spreading the substrate on the ground will suffice, as with some forms of silage fermentation and composting (71).

9.4.1. Tray Bioreactor

A tray bioreactor contains an unmixed substrate bed without forced aeration. Multiple trays may be stacked in a chamber with a controlled environment, where air is circulated around the array and temperature and relative humidity are regulated to maximize efficiency and production. An open top and perforated bottom promote maximum surface area for gas exchange. Sterile water can be sprayed to maintain moisture levels. Manual mixing can be used to promote microbial growth and cooling. This traditional SPF method employing flat stationary trays (often on mobile trolleys) requires the allotment of considerable work space and labor for large-scale production and is often designed for continuous operation (31, 80, 93, 131). Automated handling systems have been described and some have been applied to the production of rice koji for sake, shochu, and soy sauce (31, 131). The PlaFractor™ automated technology developed by Biocon (Bangalore, India) (141) is a contained modular system that incorporates sealed trays, mixing arms with blades, and superior temperature control, all with space and energy savings.

The tray design is very simple and has limitations. The uncovered tray is usually filled with substrate to a depth of about 5 to 15 cm. The bed height cannot be increased because this results in overheating; thus the only option for scale-up is an increase in the number of trays or the tray area. Heat and mass transfer limitations, as well as temperature gradients, restrict the application of this method and add to a list of disadvantages for tray bioreactors, which also includes contamination risk and inefficient substrate utilization (122). The tray system is suitable mainly for circumstances with low-volume needs, and regions where technology is unavailable and labor costs are low (152).

The tray bioreactor is of historical importance since it is based on the koji process (91), and it was introduced to the Western world in 1891 by J. Takamine to manufacture commercial enzymes with *A. oryzae* grown on wheat bran ("Taka-koji") (143). This type of reactor system was scaled up to manufacture citric acid in the 20th century (85).

9.4.2. Packed-Bed Bioreactor

The packed-bed bioreactor usually consists of a static bed in a cylindrical column with forced aeration through the substrate, which rests on a perforated base plate (93, 96, 131). Alternate versions of this system adapt a forced aeration device inserted into the center of the bed for improved performance (97). Packed-bed bioreactors are hampered by the formation of axial gradients, owing to end-to-end aeration (7). Temperature gradients, which reduce bed porosity, have a more negative impact on performance than does uneven distribution of oxygen (64). Water-saturated air can alleviate some of the effects of temperature and successfully remove metabolic heat, but this approach also increases evaporation and drying, thereby restricting microbial growth. A jacket for water circulation may serve as an alternate means for temperature control. A specially designed bioreactor termed the Zymotis packed-bed reactor (127), configured to remove heat by radial conduction, employs closely spaced internal heat-transfer plates and has achieved superior enzyme yields and productivities as a result of improved removal of metabolic heat (93–95). Bioreactors with intermittently stirred beds resemble packed-bed systems, but contain an agitator, and are favored for large-scale operations that tolerate mixing (42).

Packed-bed bioreactors have limitations besides temperature gradients. Pressure drop (0.1 to 0.5 cm water per cm bioreactor) is a characteristic of packed-bed bioreactors that limits performance by reducing airflow, as well as promoting channeling of airflow and cracking of the substrate bed along the wall of the reactor (93). Additional drawbacks involve nonuniform growth, bed caking, scale-up difficulties, problematic continuous operation, and difficulties in product recovery (42).

9.4.3. Rotary Drum Bioreactor

A rotary drum fermentor employs a complex operational system that rotates a drum horizontally (or nearly so), continuously or intermittently (96). Air is blown across the top of the substrate through the headspace, unlike packed-bed systems. The rotating drum design may incorporate dampers to vary airflow or internal lifters, paddles, or baffles to enhance mixing. A major disadvantage to this type of bioreactor is the low fill volume; the substrate bed is mixed poorly if filled beyond 30% capacity. Other issues are related to poor mixing (slumping flow), heat and mass transfer, shear in the substrate bed, and axial temperature gradients (139). While the rotary drum bioreactor was used to produce α-amylase early in the 19th century and penicillin in the mid-20th century, currently this bioreactor system is rarely used owing to difficult operation and complicated reactor construction (152).

9.4.4. Fluidized-Bed Bioreactor

Fluidized-bed bioreactors employ both continuous mixing and forced aeration (93, 97, 112, 152). Gas is vigorously directed upwards through a perforated base plate to fluidize substrate particles, working best with fine particles. A bottom mixer may be included to prevent substrate agglomeration by large, cohesive particles, which may be difficult to fluidize. A high flow rate of gas ensures good heat and mass transfer between substrate particles and the gas phase. The gas-solid fluidized system has attracted research in recent years focusing on yeast biomass and ethanol production (84, 136). Despite the advantages of good heat and mass transfer, more research is necessary using these types of bioreactors to overcome the disruption of microbial integrity and function caused by continuous-agitation systems with high shear forces. Moreover, productivity gains achieved with this type of fluidized bioreactor must offset high energy costs associated with continuous aeration and high air velocity essential for fluidization (93, 136).

9.5. PROSPECTS FOR THE FUTURE

While the industrial application of SPF has waned in the last 50 years, research focusing on this technology has been renewed in the new century, as productivity and cost advantages are confirmed. With the advent of innovative "green" applications relevant to agricultural/industrial products, wastes, and hazardous by-products, together with progress in bioreactor design and mathematical modeling of biological phenomena, these efforts can be expected to accelerate (49, 108–110, 135, 152). New "bioproducts" and "bioapplications" related to SPF, e.g., biopesticides, bioherbicides,

biocontrol agents, biofertilizers, and bioremediation, are already offering unique biological solutions, and new opportunities for this traditional yet modernized fermentation technology. SPF continues as a versatile microbial technology: the "bloom of the mold" retains its allure.

REFERENCES

1. **Acuna-Arguelles, M., M. Gutierrez-Rojas, G. Viniegra-Gonzalez, and E. Favela-Torres.** 1994. Effect of water activity on exo-pectinase production by *Aspergillus niger* CH4 on solid-state fermentation. *Biotechnol. Lett.* **16:**23–28.
2. **Acuna-Arguelles, M. E., M. Gutierrez-Rojas, G. Viniegra-Gonzalez, and E. Favela-Torres.** 1995. Production and properties of three pectinolytic activities produced by *Aspergillus niger* in submerged and solid-state fermentation. *Appl. Microbiol. Biotechnol.* **43:**808–814.
3. **Aguilar, C. N., J. C. Contreras-Esquivel, R. Rodriguez, L. A. Prado, and O. Loera.** 2004. Differences in fungal enzyme productivity in submerged and solid state cultures. *Food Sci. Biotechnol.* **13:**109–113.
4. **Aidoo, K. E., R. Hendry, and B. J. B. Wood.** 1982. Solid substrate fermentation. *Adv. Appl. Microbiol.* **28:**201–237.
5. **Akao, T., K. Gomi, K. Goto, N. Okazaki, and O. Akita.** 2002. Subtractive cloning of cDNA from *Aspergillus oryzae* differentially regulated between solid-state culture and liquid (submerged) culture. *Curr. Genet.* **41:**275–281.
6. **Anastassiadis, S., and H.-J. Rehm.** 2006. Citric acid production from glucose by yeast *Candida oleophila* ATCC 20177 under batch, continuous and repeated batch cultivation. *Electronic J. Biotechnol.* **9:**26–39.
7. **Ashley, V. M., D. A. Mitchell, and T. Howes.** 1999. Evaluating strategies for overcoming overheating problems during solid-state fermentation in packed bed bioreactors. *Biochem. Eng. J.* **3:**141–150.
8. **Astoreca, A., C. Magnoli, M. L. Ramirez, M. Combina, and A. Dalcero.** 2007. Water activity and temperature effects on growth of *Aspergillus niger, A. awamori* and *A. carbonarius* isolated from different substrates in Argentina. *Int. J. Food Microbiol.* **119:**314–318.
9. **Atlas, R. M.** 1993. *Handbook of Microbiological Media.* CRC Press, Boca Raton, FL.
10. **Balakrishnan, K., and A. Pandey.** 1996. Production of biologically active secondary metabolites in solid state fermentation. *J. Sci. Ind. Res.* **55:**365–372.
11. **Barrios-González, J., T. E. Castillo, and A. Mejia.** 1993. Development of high penicillin producing strains for solid state fermentation. *Biotechnol. Adv.* **11:**525–537.
12. **Barrios-González, J., and A. Mejia.** 1996. Production of secondary metabolites by solid-state fermentation. *Biotechnol. Annu. Rev.* **2:**85–121.
13. **Barrios-González, J., J. G. Baños, A. A. Covarrubias, and A. Garay-Arroyo.** 2008. Lovastatin biosynthetic genes of *Aspergillus terreus* are expressed differentially in solid-state and in liquid submerged fermentation. *Appl. Microbiol. Biotechnol.* **79:**179–186.
14. **Bellon-Maurel, V., O. Orliac, and P. Christen.** 2003. Sensors and measurements in solid state fermentation: a review. *Process Biochem.* **38:**881–896.
15. **Bennett, J. W.** 1998. Mycotechnology: the role of fungi in biotechnology. *J. Biotechnol.* **66:**101–107.
16. **Beuchat, L. R.** 1987. Traditional fermented food products, p. 269–306. *In* L. R. Beuchat (ed.), *Food and Beverage Mycology,* 2nd ed. Avi Publishing Co., Westport, CT.
17. **Bigelis, R.** 1985. Primary metabolism and industrial fermentations, p. 357–401. *In* J. W. Bennett and L. L. Lasure (ed.), *Gene Manipulations in Fungi.* Academic Press, New York, NY.

18. **Bigelis, R.** 1991. Fungal metabolites in food processing, p. 415–443. *In* D. K. Arora, K. G. Mukerji, and E. H. Marth (ed.), *Handbook of Applied Mycology,* vol. 3. *Foods and Feeds.* Marcel Dekker, New York, NY.

19. **Bigelis, R.** 1991. Fungal enzymes in food processing, p. 445–498. *In* D. K. Arora, K. G. Mukerji, and E. H. Marth (ed.). *Handbook of Applied Mycology,* vol. 3. *Foods and Feeds.* Marcel Dekker, New York, NY.

20. **Bigelis, R.** 1992. Flavor metabolites and enzymes from filamentous fungi. *Food Technol.* **46:**151, 154–156, 158, 161.

21. **Bigelis, R.** 1992. Food enzymes, p. 361–415. *In* D. B. Finkelstein and C. Ball (ed.), *Biotechnology of Filamentous Fungi. Technology and Products.* Butterworth-Heinemann, Stoneham, MA.

22. **Bigelis, R.** 1993. Carbohydrases, p. 121–158. *In* T. W. Nagodawithana and G. Reed (ed.), *Enzymes in Food Processing,* 3rd ed. Academic Press, New York, NY.

23. **Bigelis, R.** 1999. Fungal fermentation: industrial, p. 1–10. *In Encyclopedia of Life Sciences.* Nature Publishing Group, London, United Kingdom. www.els.net. (Online.)

24. **Bigelis, R., and D. K. Arora.** 1992. Organic acids of fungi, p. 357–376. *In* D. K. Arora, R. P. Elander, and K. G. Mukerji (ed.), *Handbook of Applied Mycology,* vol. 4. *Fungal Biotechnology.* Marcel Dekker, New York, NY.

25. **Bigelis, R., and S.-P. Tsai.** 1995. Microorganisms for organic acid production, p. 239–280. *In* Y. H. Hui and G. G. Khachatourians (ed.), *Food Biotechnology.* Microorganisms, Food Science and Technology Series. VCH Publishers, Inc., New York, NY.

26. **Bigelis, R., H. He, H. Y. Yang, L. P. Chang, and M. Greenstein.** 2006. Production of fungal antibiotics using polymeric solid supports in solid-state and liquid fermentation. *J. Ind. Microbiol. Biotechnol.* **33:**815–826.

27. **Bigelis, R., H. He, H. Y. Yang, S. W. Luckman, L. P. Chang, and D. M. Roll.** 2007. New approaches to natural product discovery: mixed-phase fermentation for secondary metabolites from actinomycetes and fungi. Abstract P68. *Ann. Mtg. Soc. Ind. Microbiol.,* July 30, 2007, Denver, CO.

28. **Bigelis, R., H. Y. Yang, and H. He.** 2009. Production of fungal natural products by mixed phase fermentation with milled cereal grains dispersed on polyester cellulose. Abstr. P68. *Ann. Mtg. Soc. Ind. Microbiol.,* July 27, 2009, Toronto, Ontario.

29. **Bills, G. F., G. Platas, A. Fillola, M. R. Jimenez, J. Collado, F. Vicente, J. Martin, A. Gonzalez, J. Bur-Zimmermann, J. R. Tormo, and F. Pelaez.** 2008. Enhancement of antibiotic secondary metabolite detection from filamentous fungi by growth on nutritional arrays. *J. Appl. Microbiol.* **104:**1644–1658.

30. **Bussari, B., P. S. Saudagar, N. S. Shaligram, S. A. Survase, and R. S. Singhal.** 2008. Production of cephamycin C by *Streptomyces clavuligerus* NT4 using solid-state fermentation. *J. Ind. Microbiol. Biotechnol.* **35:**49–58.

31. **Byndoor, M. G., N. G. Karanth, and G. V. Rao.** 1997. Efficient and versatile design of a tray type solid state fermentation bioreactor, p. 113–119. *In* S. Roussos, B. K. Lonsane, M. Raimbault, and G. Viniegra-Gonzalez (ed.), *Advances in Solid State Fermentation.* Kluwer Academic Publishers, Dordrecht, The Netherlands.

32. **Calvo, A. M., R. A. Wilson, J. W. Bok, and N. P. Keller.** 2002. Relationship between secondary metabolism and fungal development. *Microbiol. Mol. Biol. Rev.* **66:**447–459.

33. **Campbell-Platt, G.** 1987. *Fermented Foods of the World. A Dictionary and Guide.* Butterworths, Cambridge, UK.

34. **Cannel, E., and M. Moo-Young.** 1980. Solid-state fermentation systems. *Process Biochem.* **15:**2–7.

35. **Cen, P., and L. Xia.** 1999. Production of cellulase by solid-state fermentation. *Adv. Biochem. Eng. Biotechnol.* **65:**69–72.

36. **Chisti, Y.** 1999. Solid substrate fermentations, enzyme production, food enrichment, p. 2446–2462. *In* M. C. Flickinger and S. W. Drew (ed), *Encyclopedia of Bioprocess Technology: Fermentation, Biocatalysis, and Bioseparation,* vol. 5. John Wiley & Sons, Inc., New York, NY.

37. **Claessen, D., W. de Jong, L. Dijkhuizen, and H. A. B. Wosten.** 2006. Regulation of *Streptomyces* development: reach for the sky! *Trends Microbiol.* **14:**313–319.

38. **Compos, C., F. J. Fernandez, E. C. Sierra, F. Fierro, A. Gray, and J. Barrios-Gonzalez.** 2008. Improvement of penicillin yield in solid state and submerged fermentation of *Penicillium chrysogenum* by amplification of the penicillin biosynthetic gene cluster. *World J. Microbiol. Biotechnol.* **24:**3017–3022.

39. **Corry, J. E. L.** 1987. Relationships of water activity to fungal growth, p. 51–99. *In* L. R. Beuchat (ed.), *Food and Beverage Mycology,* 2nd ed. Avi Publishing Co., Westport, CT.

40. **Couto, S. R.** 2008. Exploitation of biological wastes for the production of value-added products under solid-state fermentation conditions. *Biotechnol. J.* **3:**859–870.

41. **Couto, S. R., and M. A. Sanromán.** 2005. Application of solid-state fermentation to ligninolytic enzyme production—review. *Biochem. Eng. J.* **22:**211–219.

42. **Couto, S. R., and M. A. Sanromán.** 2006. Application of solid-state fermentation to food industry—a review. *J. Food Eng.* **76:**291–302.

43. **Dalsenter, F. D. H., G. Viccini, M. C. Barga, D. A. Mitchell, and N. Krieger.** 2005. A mathematical model describing the effect of temperature variations on the kinetics of microbial growth in solid-state culture. *Process Biochem.* **40:**801–807.

44. **Davies, J.** 2006. Are antibiotics naturally antibiotics? *J. Ind. Microbiol. Biotechnol.* **33:**496–499.

45. **Demain, A. L.** 1998. Induction of microbial secondary metabolism. *Int. Microbiol.* **1:**259–264.

46. **Demain, A. L.** 1999. Pharmaceutically active secondary metabolites of microorganisms. *Appl. Microbiol. Biotechnol.* **52:**455–463.

47. **Demain, A. L.** 2000. Microbial biotechnology. *Trends Biotechnol.* **13:**26–31.

48. **Demain, A. L.** 2006. From natural products discovery to commercialization: a success story. *J. Ind. Microbiol. Biotechnol.* **33:**486–495.

49. **Demain, A. L.** 2007. The business of biotechnology. *Ind. Biotechnol.* **3:**269–283.

50. **Demain, A. L., and A. Fang.** 2000. The natural functions of secondary metabolites. *Adv. Biochem. Eng. Biotechnol.* **69:**1–39.

51. **Demain, A. L., and L. Zhang.** 2005. Natural products and drug discovery, p. 3–32. *In* L. Zhang and A. Demain (ed.), *Natural Products: Drug Discovery and Therapeutic Medicines.* Humana Press, Totowa, NJ.

52. **Demain, A. L., and J. L. Adrio.** 2008. Contributions of microorganisms to industrial biology. *Mol. Biotechnol.* **38:**41–55.

53. **Desgranges, C., C. Vergoignan, M. Georges, and A. Durand.** 1991. Biomass estimation in solid-state fermentation. I. Manual biochemical methods. *Appl. Microbiol. Biotechnol.* **35:**200–205.

54. **Desgranges, C., M. Georges, C. Vergoignan, and A. Durand.** 1991. Biomass estimation in solid state fermentation. II. On-line measurements. *Appl. Microbiol. Biotechnol.* **35:**206–209.

55. **de Vrije, T., N. Antoine, R. M. Buitelaar, S. Bruckner, M. Dissevelt, A. Durand, M. Gerlagh, E. E. Jones, P. Lüth, J. Oostra, W. J. Ravensberg, R. Renaud, A. Rinzema, F. J. Weber, and J. M. Whipps.** 2001. The fungal biocontrol agent *Coniothyrium minitans:* production by solid-state fermentation, application and marketing. *Appl. Microbiol. Biotechnol.* **56:**58–68.

56. Diaz, J. C. M., J. A. Rodriguez, S. Roussos, J. Cordova, A. Abousalham, F. Carriere, and J. Baratti. 2006. Lipase from the thermotolerant fungus *Rhizopus homothallicus* is more thermostable when produced using solid state fermentation than liquid fermentation procedures. *Enzyme Microb. Technol.* **39:**1042–1050.

57. Díaz-Godínez, G., J. Soriano-Santos, C. Augur, and G. Viniegra-González. 2001. Exopectinases produced by *Aspergillus niger* in solid state and submerged fermentation: a comparative study. *J. Ind. Microbiol. Biotechnol.* **26:**271–275.

58. Dominguez, M., A. Mejia, and J. Barrios-Gonzalez. 2000. Respiration studies of penicillin solid-state fermentation. *J. Biosci. Bioeng.* **89:**409–413.

59. Feron, G., P. Bonnarme, and A. Durand. 1996. Prospects for the microbial production of food flavours. *Trends Food Sci. Technol.* **7:**285–293.

60. Flärdh, K., and M. J. Buttner. 2009. *Streptomyces* morphogenetics: dissecting differentiation in a filamentous bacterium. *Nat. Rev. Microbiol.* **7:**36–49.

61. Gervais, P., P. Molin, W. Grajek, and M. Bensoussan. 1988. Influence of the water activity of a solid substrate on the growth rate and sporogenesis of filamentous fungi. *Biotechnol. Bioeng.* **31:**457–463.

62. Gervais, P., P. A. Marrchal, and P. Molin. 1996. Water relations of solid state fermentation. *J. Sci. Ind. Res.* **55:**343–357.

63. Gervais, P., and P. Molin. 2003. The role of water in solid-state fermentation. *Biochem. Eng. J.* **13:**85–101.

64. Ghildyal, N. P., M. K. Gowthaman, K. S. Rao, and N. G. Karanth. 1994. Interaction of transport resistances with biochemical reaction in packed-bed solid-state fermentors: effect of temperature gradients. *Enzyme Microb. Technol.* **16:**253–257.

65. Gowthaman, M. K., C. Krishna, and M. Moo-Young. 2001. Fungal solid state fermentation—an overview, p. 305–352. *In* G. G. Khachatourians and D. K. Arora (ed.), *Agriculture and Food Production. Applied Mycology and Biotechnology*, vol. 1. Elsevier, Amsterdam, The Netherlands.

66. Graminha, E. B. N., A. Z. L. Gonçalves, R. D. P. B. Pirota, M. A. A. Balsalobre, R. Da Silva, and E. Gomes. 2008. Enzyme production by solid-state fermentation: application to animal nutrition. *Anim. Feed Sci. Technol.* **144:**1–22.

67. Grayek, W., and P. Gervais. 1987. Influence of water activity on the enzyme biosynthesis and enzyme activities produced by *Trichoderma viride* TS in solid-state fermentation. *Enzyme Microb. Technol.* **9:**658–662.

68. Greene, R. V., S. N. Freer, and S. H. Gordon. 1988. Determination of solid-state fungal growth by Fourier transform infrared-photoacoustic spectroscopy. *FEMS Microbiol. Lett.* **52:**73–77.

69. Gutierrez-Rojas, M., S. Amar Aboul Hosn, R. Auria, S. Revah, and E. Favela Tomes. 1996. Heat transfer in citric acid production by solid state fermentation. *Process Biochem.* **31:**363–369.

70. Halstensen, A. S., K. C. Nordby, I. M. Wouters, and W. Eduard. 2007. Determinants of microbial exposure in grain farming. *Ann. Occup. Hyg.* **51:**581–592.

71. Haug, R. T. 1993. *The Practical Handbook of Compost Engineering.* CRC Press, Boca Raton, FL.

72. Hesseltine, C. W. 1977. Substrate fermentation. *Process Biochem.* **12:**24–27.

73. Hogan, D. A. 2006. Talking to themselves, autoregulation and quorum sensing in fungi. *Eukaryotic Cell* **5:**613–619.

74. Holker, U., M. Hofer, and J. Lenz. 2004. Biotechnological advantages of laboratory-scale solid-state fermentation with fungi. *Appl. Microbiol. Biotechnol.* **64:**175–186.

75. Holker, U., and J. Lenz. 2005. Solid-state fermentation—are there any biotechnological advantages? *Curr. Opin. Microbiol.* **8:**301–306.

76. Jermini, M. F. G., and A. L. Demain. 1989. Solid state fermentation for cephalosporin production by *Streptomyces clavuligerus* and *Cephalosporium acremonium*. *Experientia* **4:**1061–1065.

77. Kashyap, D. R., S. K. Soni, and R. Tewari. 2003. Enhanced production of pectinase by *Bacillus* sp. DT7 using solid state fermentation. *Bioresour. Technol.* **88:**251–254.

78. Koehn, F. E., and G. T. Carter. 2005. The evolving role of natural products in drug discovery. *Nat. Rev. Drug Discov.* **4:**206–220.

79. Krishna, C. 1999. Production of bacterial cellulases by solid state bioprocessing of banana wastes. *Bioresour. Technol.* **69:**231–239.

80. Krishna, C. 2005. Solid-state fermentation systems—an overview. *Crit. Rev. Biotechnol.* **25:**1–30.

81. Krishna, C., A. Pandey, and A. Mohandas. 2004. Microbial phytases, p. 569–576. *In* A. Pandey (ed.), *Concise Encyclopedia of Bioresource Technology.* Haworth Press, Binghamton, NY.

82. Lacey, J., and B. Crook. 1988. Fungal and actinomycete spores as pollutants of the workplace and occupational allergens. *Ann. Occup. Hyg.* **32:**515–533.

83. Larroche, D. 1996. Microbial growth and sporulation behaviour in solid state fermentation. *J. Sci. Ind. Res.* **55:**408–423.

84. Liu, C.-Z., F. Wang, and F. Ou-Yang. 2009. Ethanol fermentation in a magnetically fluidized bed reactor with immobilized *Saccharomyces cerevisiae* in magnetic particles. *Bioresour. Technol.* **100:**878–882.

85. Lockwood, L. B. 1979. Production of organic acids by fermentation, p. 355–387. *In* H. J. Peppler and D. Perlman (ed.), *Microbial Technology*, vol. 1. Academic Press, New York, NY.

86. Lonsane, B. K., N. P. Ghildyal, S. Budiatman, and S. V. Ramakrishna. 1985. Engineering aspects of solid-state fermentation. *Enzyme Microb. Technol.* **7:**258–265.

87. Lonsane, B. K., O. Saucedo-Castaneda, M. Raimbault, S. Roussos, O. Viniegra-Gonzalez, N. P. Ghildyal, M. Ramakrishna, and M. M. Krishnaiah. 1992. Scale-up strategies for solid state fermentation systems. *Process Biochem.* **27:**259–273.

88. Mason, I. G., and M. W. Milke. 2005. Physical modelling of the composting environment: a review. Part 1: reactor systems. *Waste Manag.* **25:**489–500.

89. Matcham, S. E., B. R. Jordan, and D. A. Wood. 1985. Estimation of fungal biomass in a solid substrate by three independent methods. *Appl. Microbiol. Biotechnol.* **21:**108–112.

90. Membrillo, I., C. Sánchez, M. Meneses, E. Favela, and O. Loera. 2008. Effect of substrate particle size and additional nitrogen source on production of lignocellulolytic enzymes by *Pleurotus ostreatus* strains. *Bioresour. Technol.* **99:**7842–7847.

91. Miall, L. M. 1978. Organic acids, p. 47–119. *In* A. H. Rose (ed.), *Primary Products of Metabolism, Economic Microbiology*, vol. 2. Academic Press, New York, NY.

92. Mitchell, D. A., M. Berovic, and N. Krieger. 2000. Biochemical engineering aspects of solid state bioprocessing. *Adv. Biochem. Eng. Biotechnol.* **68:**61–138.

93. Mitchell, D. A., N. Krieger, D. M. Stuart, and A. Pandey. 2000. New developments in solid state fermentation. II. Rational approaches to the design, operation and scale-up of bioreactors. *Process Biochem.* **35:**1211–1225.

94. Mitchell, D. A., O. F. von Meien, L. F. Luiz, Jr., and N. Krieger. 2002. Evaluation of productivity of Zymotis solid-state bioreactor. *Food Technol. Biotechnol.* **40:**135–144.

95. Mitchell, D. A., O. F. von Meien, N. Krieger, and F. D. H. Dalsenter. 2004. A review of recent developments in modeling of microbial growth kinetics and intraparticle phenomena in solid-state fermentation. *Biochem. Eng. J.* **17:**15–26.

96. **Mitchell, D. A., N. Krieger, and M. Berovič (ed.).** 2006. *Solid-State Fermentation Bioreactors: Fundamentals of Design and Operation.* Springer, Berlin, Germany.

97. **Mitchell, D. A., M. Berovič, M. Nopharatana, and N. Krieger.** 2006. The bioreactor step of SSF: a complex interaction of phenomena, p. 13–32. *In* D. A. Mitchell, N. Krieger, and M. Berovič (ed.), *Solid-State Fermentation Bioreactors: Fundamentals of Design and Operation.* Springer, Berlin, Germany.

98. **Moo-Young, M., A. R. Moriera, and R. P. Tengerdy.** 1983. Principles of solid state fermentation, p. 117–144. *In* J. E. Smith, D. R. Berry, and B. Kristiansen (ed.), *The Filamentous Fungi,* vol. 4. *Fungal Biotechnology.* Edward Arnold Publishers, London, United Kingdom.

99. **Nagel, F. J. I., J. Tramper, M. S. N. Bakker, and A. Rinzema.** 2001. Model for on-line moisture-content control during solid-state fermentation. *Biotechnol. Bioeng.* **72:**231–243.

100. **Nandakumar, M. P., M. S. Thakur, K. S. M. S Raghavarao, and N. P. Ghildyal.** 1999. Studies on catabolite repression in solid state fermentation for biosynthesis of fungal amylases. *Lett. Appl. Microbiol.* **29:**380–384.

101. **Newman, D. J., and G. M. Cragg.** 2007. Natural products as sources of new drugs over the last 25 years. *J. Nat. Prod.* **70:**461–477.

102. **Nigam, P., and D. Singh.** 1994. Solid-state (substrate) fermentation systems and their applications in biotechnology. *J. Basic Microbiol.* **6:**405–423.

103. **Nigam, P., and D. Singh.** 1996. Processing of agricultural wastes in solid state fermentation for cellulolytic enzymes production. *J. Sci. Ind. Res.* **55:**457–463.

104. **Ooijkaas, L. P., F. J. Weber, R. M. Buitelaar, J. Tramper, and A. Rinzema.** 2000. Defined media and inert supports: their potential as solid-state fermentation production systems. *Trends Biotechnol.* **18:**356–360.

105. **Oostra, J., E. P. le Comte, J. C. van den Heuvel, J. Tramper, and A. Rinzema.** 2001. Intra-particle oxygen diffusion limitation in solid-state fermentation. *Biotechnol. Bioeng.* **74:**13–24.

106. **Oriol, E., B. Schettino, and G. Viniegra-Gonzales.** 1988. Solid-state culture of *Aspergillus niger* on support. *J. Ferment. Technol.* **66:**57–62.

107. **Pandey, A.** 1991. Effect of particle size of substrate on enzyme production in solid-state fermentation. *Bioresour. Technol.* **37:**169–172.

108. **Pandey, A.** 2003. Solid-state fermentation. *Biochem. Eng. J.* **13:**81–84.

109. **Pandey, A., P. Selvakumar, C. R. Soccol, and P. Nigam.** 1999. Solid state fermentation for the production of industrial enzymes. *Curr. Sci.* **77:**149–162.

110. **Pandey, A., C. R. Soccol, and D. Mitchell.** 2000. New developments in solid state fermentation: I—bioprocesses and products. *Process Biochem.* **35:**1153–1169.

111. **Pandey, A., and S. Ramachandran.** 2006. Process development for food applications, p. 87–110. *In* K. Shetty, G. Paliyath, A. Pometto, and R. E. Levin (ed.), *Food Biotechnology,* 2nd ed. CRC Press, Boca Raton, FL.

112. **Pandey, A., C. R. Soccol, and C. Larroche.** 2008. *Current Developments in Solid-State Fermentation.* Springer, Berlin, Germany.

113. **Papagianni, P.** 2004. Fungal morphology and metabolite production in submerged mycelial processes. *Biotechnol. Adv.* **22:**189–259.

114. **Papagianni, M., S. E. Nokes, and K. Filer.** 1999. Production of phytase by *Aspergillus niger* in submerged and solid state fermentation. *Process Biochem.* **35:**397–402.

115. **Parsek, M. P., and E. P. Greenberg.** 2005. Sociomicrobiology: the connections between quorum sensing and biofilms. *Trends Microbiol.* **13:**27–33.

116. **Pearce, C.** 1997. Biologically active fungal metabolites. *Adv. Appl. Microbiol.* **44:**1–80.

117. **Pintado, J., A. Torrado, M. P. Gonzalez, and M. A. Murado.** 1998. Optimization of nutrient concentration for citric acid production by solid-state culture of *Aspergillus niger* on polyurethane foams. *Enzyme Microb. Technol.* **23:**149–156.

118. **Raghavarao, K. S. M. S., T. V. Ranganathan, and N. G. Karanth.** 2003. Some engineering aspects of solid-state fermentation. *Biochem. Eng. J.* **13:**127–135.

119. **Rahardjo, Y. S. P., J. Tramper, and A. Rinzema.** 2006. Modeling conversion and transport phenomena in solid-state fermentation: a review and perspectives. *Biotechnol. Adv.* **24:**161–179.

120. **Raimbault, M.** 1998. General and microbiological aspects of solid substrate fermentation. *Electronic J. Biotechnol.* **1:**1–15.

121. **Raimbault, M., and D. Alazard.** 1980. Culture method to study fungal growth in solid fermentation. *Eur. J. Appl. Microbiol. Biotechnol.* **9:**199–202.

122. **Rajagopalan, S., and J. M. Modak.** 1995. Modeling of heat and mass transfer for solid state fermentation process in tray bioreactor. *Bioprocess Biosyst. Eng.* **13:**161–169.

123. **Rajagopalan, R., and J. M. Modak.** 1995. Evaluation of relative growth limitation due to depletion of glucose and oxygen during fungal growth on a spherical solid particle. *Chem. Eng. Sci.* **50:**803–811.

124. **Ramesh, M. V., and B. K. Lonsane.** 1990. Critical importance of moisture content of the medium in alpha-amylase production by *Bacillus licheniformis* M27 in a solid-state fermentation system. *Appl. Microbiol. Biotechnol.* **3:**201–205.

125. **Robinson, T., D. Singh, and P. Nigam.** 2001. Solid-state fermentation: a promising microbial technology for secondary metabolite production. *Appl. Microbiol. Biotechnol.* **55:**284–289.

126. **Romero-Gomez, R. J., C. Augur, and G. Viniegra-Gonzalez.** 2000. Invertase production by *Aspergillus niger* in submerged and solid-state fermentation. *Biotechnol. Lett.* **22:**1255–1258.

127. **Roussos, S., M. Raimbault, J.-P. Prebois, and B. K. Lonsane.** 1993. Zymotis, a large scale solid state fermenter. *Appl. Biochem. Biotechnol.* **42:**37–52.

128. **Roussos, S., E. Bresson, G. Saucedo-Castaneda, P. Martinez, J. Guymberteau, and J.-M. Olivier.** 1997. Production of mycelial cells of *Pleurotus opuntiae* on natural support in solid state fermentation, p. 483–498. *In* S. Roussos, B. K. Lonsane, M. Raimbault, and G. Viniegra-Gonzalez (ed.), *Advances in Solid State Fermentation.* Kluwer Academic Publishers, Dordrecht, The Netherlands.

129. **Sangsurasak, P., M. Nopharatana, and D. A. Mitchell.** 1996. Mathematical modeling of the growth of filamentous fungi in solid state fermentation. *J. Sci. Ind. Res.* **55:**333–342.

130. **Santos, M. M., A. S. Rosa, S. Dal'boit, D. A. Mitchell, and N. Krieger.** 2004. Thermal denaturation: is solid state fermentation really a good technology for the production of enzymes? *Bioresour. Technol.* **93:**261–268.

131. **Sato, K., and S. Sudo.** 1999. Small-scale solid-state fermentations, p. 61–89. *In* A. L. Demain and J. E. Davies (ed.), *Manual of Industrial Microbiology and Biotechnology,* 2nd ed. ASM Press, Washington, DC.

132. **Saucedo-Castaneda, G., M. Gutierrez-Rojas, G. Bacquet, M. Raimbault, and G. Viniegra-Gonzalez.** 1990. Heat transfer simulation in solid substrate fermentation. *Biotechnol. Bioeng.* **5:**802–808.

133. **Saucedo-Castaneda, G., M. R. Trejo-Hernandez, J. M. Lonsane, J. M. Navarro, S. Roussos, D. Dufour, and M. Raimbault.** 1994. On-line automated monitoring and control systems for CO_2 and O_2 in aerobic and anaerobic solid-state fermentations. *Process Biochem.* **29:**13–24.

134. **Shankar, S. K., and V. H. Mulimani.** 2007. α-Galactosidase production by *Aspergillus oryzae* in solid-state fermentation. *Bioresour. Technol.* **98:**958–961.

135. **Singhania, R. R., C. R. Soccol, and A. Pandey.** 2009. Recent advances in solid-state fermentation. *Biochem. Eng. J.*, **44:**13–18.

136. **Smith, P. G.** 2007. *Gas-Solid Fluidized Bed Fermentation.* Wiley Interscience, New York, NY.

137. **Soccol, C. R.** 1996. Biotechnology products from cassava root by solid state fermentation. *J. Sci. Ind. Res.* **55:**358–364.

138. **Sree, N. K., M. Sridhar, K. Suresh, and L. V. Rao.** 1999. High alcohol production by solid substrate fermentation from starchy substrates using thermotolerant *Saccharomyces cerevisiae. Bioprocess Biosyst. Eng.* **20:**561–563.

139. **Stuart, D. M., D. A. Mitchell, M. R. Johns, and J. D. Litster.** 1999. Solid-state fermentation in rotating drum bioreactors: operating variables affect performance through their effects on transport phenomena. *Biotechnol. Bioeng.* **63:**383–391.

140. **Sudo, S., S. Kobayashi, A. Kaneko, K. Sato, and T. Oba.** 1995. Growth of submerged mycelia of *Aspergillus kawachii* in solid-state culture. *J. Ferment. Bioeng.* **79:**252–256.

141. **Suryanarayan, S.** 2003. Current industrial practice in solid state fermentations for secondary metabolite production: the Biocon India experience. *Biochem. Eng. J.* **13:**189–195.

142. **Szewczyk, K. W., and L. Myszka.** 1994. The effect of temperature on the growth of *A. niger* in solid state fermentation. *Bioprocess Eng.* **10:**123–126.

143. **Takamine, J.** September, 1894. Preparing and making taka-koji. U.S. patent 525,820.

144. **Talbot, N. J.** 2003. Aerial morphogenesis: enter the chaplins. *Curr. Biol.* **13:**R696–R698.

145. **Tomasini, A., C. Fajardo, and J. Barrios-Gonzalez.** 1997. Gibberellic acid production using different solid-state fermentation systems. *World J. Microbiol. Biotechnol.* **13:**203–206.

146. **Trejo-Hernandez, M. R., M. Raimbault, S. Roussos, and B. K. Lonsane.** 1992. Potential of solid state fermentation for production ergot alkaloids. *Lett. Appl. Microbiol.* **14:**156–159.

147. **Vandenberghe, L. P. S., C. R. Soccol, F. C. Prado, and A. Pandey.** 2004. Comparison of citric acid production by solid-state fermentation in flask, column, tray, and drum bioreactors. *Appl. Biochem. Biotechnol.* **118:**293–303.

148. **Varzakas, T. H., S. Roussos, and I. S. Arvanitoyannis.** 2008. Glucoamylases production of *Aspergillus niger* in solid state fermentation using a continuous counter-current reactor. *Int. J. Food Sci. Technol.* **43:**1159–1168.

149. **Viccini, G., D. A. Mitchell, S. D. Boit, J. C. Gern, A. S. da Rosa, R. M. Costa, F. D. H. Dalsenter, O. F. von Meien, and N. Krieger.** 2001. Analysis of growth kinetic profiles in solid-state fermentation. *Food Technol. Biotechnol.* **39:**271–294.

150. **Viniegra-Gonzalez, G.** 1996. Solid state fermentation: definition, characteristics, limitations and monitoring, p. 5–22. *In* S. Roussos, B. K. Lonsane, M. Raimbault and G. Viniegraz-Gonzalez (ed.), *Advances in Solid State Fermentation.* Kluwer Academic Publishers, Dordrecht, The Netherlands.

151. **Viniegra-González, G., E. Favela-Torres, C. N. Aguilar, S. de Jesus Rómero-Gomez, G. Díaz-Godínez, and C. Augur.** 2003. Advantages of fungal enzyme production in solid state over liquid fermentation systems. *Biochem. Eng. J.* **13:**157–167.

152. **Wang, L., and S.-T. Yang.** 2007. Bioprocessing for value-added products from renewable resources. New technologies and applications, p. 465–489. *In* S.-T. Yang (ed.), *Solid State Fermentation and Its Applications.* Elsevier, Amsterdam, The Netherlands.

153. **Wiacek-Zychlinska, A., J. Czajak, and R. Sawicka-Zukowska.** 1994. Xylanase production by fungal strains in solid-state fermentations. *Bioresour. Technol.* **49:**13–16.

154. **Wood, D. A.** 1989. Mushroom biotechnology. *Int. Ind. Biotechnol.* **9:**5–8.

155. **Wosten, H. A. B., M. A. van Wetter, L. G. Lugones, H. C. van der Mei, H. J. Busscher, and J. G. H. Wessels.** 1999. How a fungus escapes the water to grow into the air. *Curr. Biol.* **9:**85–88.

156. **Yamane, Y., M. Yoshii, S. Mikami, H. Fukuda, and Y. Kizaki.** 2000. A solid-state culture system using a cellulose carrier containing defined medium as a useful tool for investigating characteristics of koji culture. *J. Biosci. Bioeng.* **89:**33–39.

157. **Yim, G., H. H. Wang, and J. Davies.** 2007. Antibiotics as natural signaling molecules. *Philos. Trans. R. Soc. Lond. B Biol. Sci.* **362:**1195–1200.

Bacterial Cultivation for Production of Proteins and Other Biological Products

JOSEPH SHILOACH AND URSULA RINAS

10

10.1. INTRODUCTION

Large numbers of biological products are currently being produced on an industrial scale from microorganisms such as filamentous fungi, yeast, and bacteria. These products can be divided into several groups: primary metabolites such as acetic acid, ethanol, and amino acids; secondary metabolites such as antibiotics; and recombinant products, especially proteins that are produced for pharmaceutical purposes and technical applications. In this chapter, we concentrate on the production of biological products from bacteria. Since the basic steps of the production processes of the various products mentioned above are similar, we describe one of these processes in detail to give the reader the necessary information. Based on this information, the reader will be able to design production processes for different types of products from various types of microorganisms. The process we describe in detail is the production of recombinant proteins from *Escherichia coli* (34, 36). This process includes the following steps: (i) preparation of the bacterial strain (not described in this chapter); (ii) determination of the growth and production parameters such as growth strategy, production strategy, medium composition, pH, and optimal concentration of dissolved oxygen (DO) and temperature (not described in this chapter); (iii) preparation of the growth vessels; (iv) preparation of the starter culture (inoculum); (v) bacterial growth and product formation; (vi) process termination and preparation for the protein recovery step; and (vii) protein recovery and purification (not described in this chapter).

10.2. RECOMBINANT PROTEIN PRODUCTION FROM *E. COLI*

10.2.1. General Process Description

Following the determination of the batch production size, the proper growth vessels are prepared: this includes the bioreactors used for growth of the starter culture (in some cases, the volume required is large and more than one transfer is needed) and for production. The preparation involves (i) making the culture medium for growth and production, (ii) electrode calibration and installation, (iii) assembly of hoses to transfer the culture from one growth vessel to another (in cases where there is no permanent pipe connection between the various growth vessels), and

(iv) sterilization of the bioreactor. In addition, there is the need to prepare and/or calibrate various follow-up instruments such as pH meter, spectrophotometer, conductivity meter, glucose analyzer, and microscope. The recombinant bacteria are removed from storage (working cell bank) and transferred to the starter culture shake flask or bioreactor containing the proper medium and are grown to the required density (which was determined when the process was developed). In some cases, cells are first transferred to an agar plate and a single colony is used to inoculate the starter culture; in other cases, freshly transformed cells are needed for inoculation of the starter culture. When the starter culture reaches its designated density, it is transferred to the production bioreactor and then the production process commences. In most cases, the production process involves two phases, i.e., a growth phase and a production phase. In the growth phase, the bacteria are grown to a high density by implementing a fed-batch growth procedure (see section 10.2.6.3). The second phase is associated with the production of the recombinant protein by inducing its biosynthesis. In some cases, the growth conditions after the induction are different from those before the induction. After the bacteria have synthesized the desired recombinant protein to the expected level, the culture is cooled down and the cells (if the product is accumulated in the cells) or the supernatant fluid (if the product is secreted into the outside medium) are collected and processed. The process flow is summarized in Fig. 1.

To give the reader a better feeling for the general process described above, we provide here details on the production process of recombinant *Pseudomonas aeruginosa* exotoxin A in *E. coli* (5), which has been adapted for routine production. For a batch production size of 50 liters, a bioreactor with 50 liters of medium is prepared and sterilized. One liter of starter culture is prepared by inoculating a 2.8-liter Fernbach flask containing 1 liter of medium with a frozen culture stock. After 12 h of growth at 37°C, the starter culture is transferred to the bioreactor and the bacteria are grown at 37°C at a 30% DO concentration and a pH of 6.8. When the culture's optical density reaches the value of 40 (measured at 600 nm), the inducer isopropyl-β-D-thiogalactopyranoside (IPTG) is added to a final concentration of 100 mmol liter^{-1} and growth is continued for 30 to 60 min; the culture is then cooled down and the cells are collected and processed.

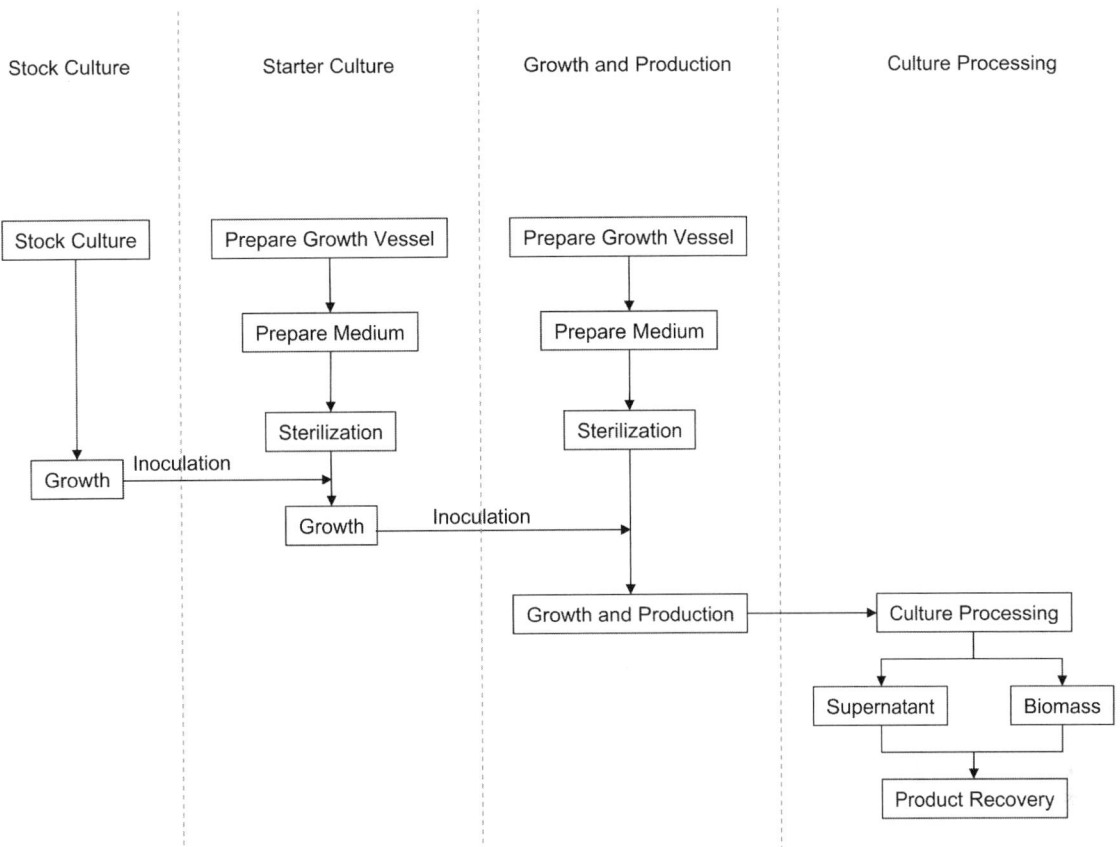

FIGURE 1 General layout of the bacterial cultivation process for production of proteins and other biological products.

10.2.2. Instrumentation and Infrastructure Required

To conduct a process of recombinant protein production with *E. coli*, the investigator needs to have access to the following instrumentation.

1. Cold storage equipment: a −80°C freezer to store the bacterial master and working cell banks; a −20°C freezer to store collected samples; a 4°C cold room; and 4°C refrigerators to store samples, medium, agar plates, etc.
2. Sterilization equipment: a steam-operated autoclave that can hold a benchtop stirred-tank bioreactor in a volume up to 10 liters.
3. Propagation equipment: incubators in the range of 20 to 45°C to grow the bacteria on agar plates; incubator shakers that can accommodate different sizes of shake flasks, from 50-ml Erlenmeyer flasks to 2.8-liter Fernbach flasks, which are needed for the initial growth of the culture from the plates; and stirred-tank bioreactors in various sizes to be used for both starter culture growth and protein production. Depending on their size, these bioreactors can be divided into two types: one group includes bioreactors up to 10 liters. These are called benchtop reactors and, in most cases, are sterilized in the autoclave. The other group includes bioreactors with higher volumes that are sterilized-in-place. In most cases, the bioreactors used for bacterial growth are stirred-tank reactors equipped with air inlet, air outlet, impellers, baffles, air sparger, and numerous inlets and outlets for removal of samples and medium and for adding various solutions to the growing culture. These include nutrients and growth factors to support growth and production, acid or base for pH control, and antifoam for foam control. The bioreactor is also equipped with openings for installation of various probes, especially for pH and DO. A general scheme of the stirred-tank bioreactor can be seen in Fig. 2. To add various solutions to the bioreactor, it should also be equipped with variable-speed pumps, either sterilizable or outfitted with sterilizable tubing. The bioreactor is supported by instrumentation to measure and control agitation, airflow, temperature, pressure, pH, DO, and foam. In some cases, it can also include instruments to analyze the concentrations of CO_2 and O_2 in the off-gas. The measurements of all these variables are collected by a digital control unit that can also be used for process control based on the analysis of one or more process variables. The bioreactor is connected to a source of water, air, oxygen, and steam and also to a drain. Since during the process foam can be generated by the growing culture, the bioreactor should be equipped with a foam probe that detects the foam level and triggers the addition of antifoam solution (see section 10.2.4). Another instrument is a level probe that can detect the liquid level in the bioreactor. General information on bioreactor principles and operation can be found in several comprehensive books (3, 15, 32).
4. Analytical instrumentation: to control and follow bacterial growth and protein production, access is needed to the following analytical instruments: (i) optical microscope

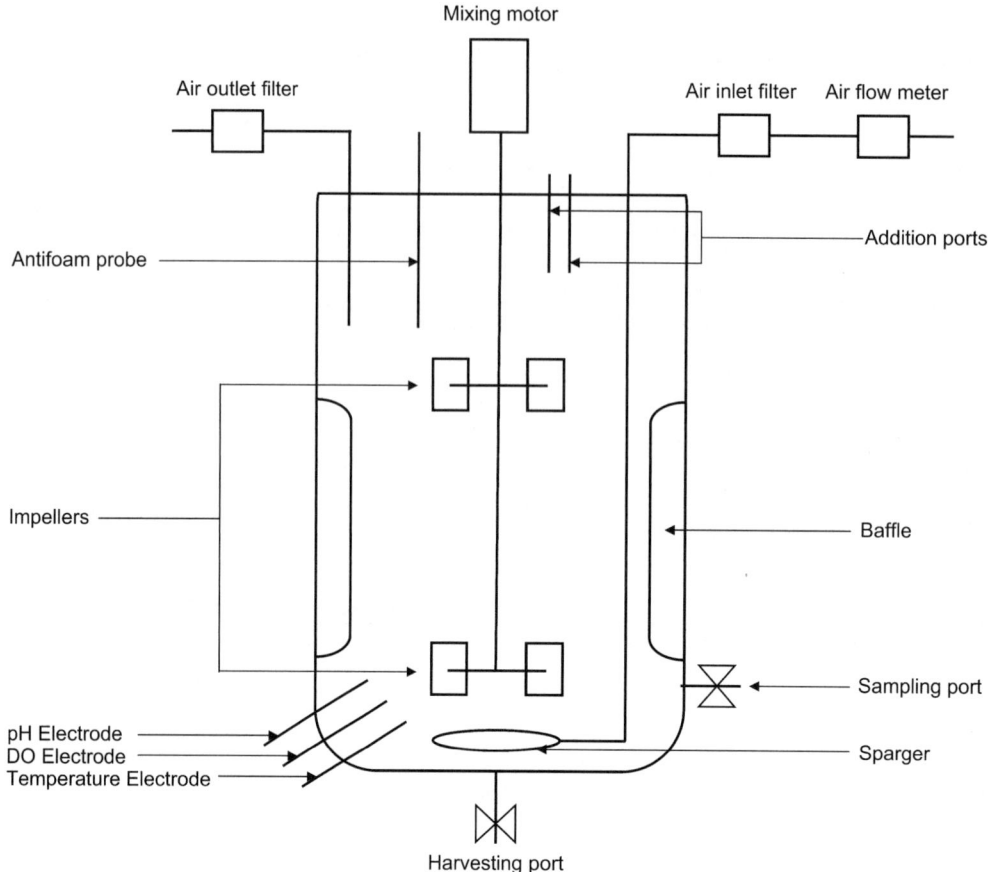

FIGURE 2 General scheme of a stirred-tank bioreactor.

to check the condition of the bacterial culture; (ii) spectrophotometer for the measurement of bacterial density; (iii) pH meter; (iv) benchtop centrifuge to separate the bacterial mass from the supernatant fluid; and (v) additional instrumentation required for product measurements, such as gel electrophoresis, enzyme-linked immunosorbent assay apparatus, and in some cases, equipment to measure the composition of the off-gas (e.g., mass spectrometer).

5. Processing equipment: equipment is needed to separate the bacterial biomass from the medium. The separation can be done with a continuous centrifuge or a tangential flow device (4).

10.2.3. Seed Culture: Preparation and Storage

Following the research and development stage, a bacterial producer strain or strains are selected. These strains are usually stored frozen at −80°C or stored as freeze-dried samples (lyophilized). The currently accepted procedure is to grow the selected strains in their specified medium and to collect the cells at the mid-logarithmic phase. Once collected, the cells are prepared for storage. The cells (e.g., *E. coli*) are suspended in an equal volume of a special freezing medium with the following composition: 6.3 g K_2HPO_4, 1.8 g KH_2PO_4, 0.45 g sodium citrate, 0.09 g $MgSO_4$, 0.9 g $(NH_4)_2SO_4$, 44 g glycerol, and 450 ml water. The suspended cells are divided into 0.5-ml aliquots in cryogenic tubes and stored in the −80°C freezer. Preparation of freeze-dried aliquots of other bacteria requires special equipment and is not described in this chapter. Detailed descriptions

of preservation methods for different bacteria can be found in *ATCC Preservation Methods* (30) and in *Maintenance of Microorganisms* (11). The aliquots prepared from the grown culture are kept as the master cell bank. An aliquot from the master cell bank is than used to prepare a working cell bank in the same way. When there is a need to start a production process, a sample from the working cell bank is taken out and used for inoculum preparation that later will be used for the production process.

10.2.4. General Information on Medium Composition, Preparation, and Sterilization

Chemotrophic bacteria need various chemical compounds designated as substrates for cell maintenance, cell growth, and production. Most bacteria used for the production of low- and high-molecular-weight organic compounds (e.g., acids, DNA, proteins) are chemoorganotrophic; thus, in addition to inorganic substrates, they also need organic substrates for cell maintenance, growth, and production. The substrates required by these bacteria can be grouped into two categories: (i) substrates that serve as an energy source and as a building unit to generate more cells or product, and (ii) substrates that only serve as building units for generating more cells and product, but their transformation by the bacterial cells does not generate energy.

Carbon substrates such as glucose, glycerol, and other compounds containing carbon atoms can be used by these bacteria for the generation of biomass and product and for the generation of energy through substrate-level phosphorylation, and during complete oxidation using the respiratory pathway. Other substrates, such as salts containing nitrogen, phosphor, and sulfur atoms, only serve as building units for generating more cells and/or product. For example, the elemental composition of *E. coli* grown on a defined medium is $CH_{1.85}O_{0.574}N_{0.22}$ plus 12% ash (10). This elemental composition does not vary significantly with the growth rate. In addition to these elements, cells need phosphorus for the formation of RNA and DNA, and sulfur for the formation of the amino acids methionine and cysteine. Other trace elements, such as metal ions, are required by various enzymes for their proper function. Some metal ions are required in high concentrations, such as iron needed for the heme-containing enzymes of the respiratory chain. Other trace metals are required in small amounts, such as copper, as there are few copper-containing enzymes. In some cases, bacteria have specific requirements for compounds that they cannot synthesize; these compounds must be added to the medium in order to allow cell growth. An example is the protein producer *E. coli* K-12 strain TG1, which is a thiamine auxotroph and therefore requires supplementation with thiamine when grown on a chemically defined medium (10, 12).

The composition of a defined medium that can support both small-scale (test tube or shake flask) and large-scale batch cultivations of *E. coli* is described in Table 1.

Preparation of the Medium

1. Dissolve KH_2PO_4, $(NH_4)_2HPO_4$, and citric acid in ~800 ml of deionized water in a beaker.
2. Add trace element solutions.
3. Adjust the pH of this solution to 6.8 using 5 mol liter^{-1} NaOH and fill it up to 900 ml using deionized water. Transfer to a 1-liter bottle.
4. Add $MgSO_4$ into a 50-ml bottle and fill up to 20 ml.
5. Add glucose into a 100-ml bottle and fill up to 80 ml.
6. Sterilize these three bottles in an autoclave for 30 min at 120°C.
7. Mix all three components under sterile conditions. You can store the sterile medium for approximately 4 weeks at room temperature.
8. Add the volume of medium you need to sterile flasks.
9. If necessary, add filter-sterilized thiamine (4.5 mg liter^{-1}) and antibiotics.

When this medium is used in larger-scale bioreactors, KH_2PO_4, $(NH_4)_2HPO_4$, and citric acid are added directly to the bioreactor and heat-sterilized. The other solutions are prepared in concentrated form in separate containers. It is important to note that some compounds should be heat-sterilized separately. For example, magnesium and phosphate should not be heat-sterilized together as they will form a precipitate that will not dissolve again. Also, glucose should be heat-sterilized separately, as it forms brown Maillard products when heat-sterilized with other compounds, e.g., amino acids. Another group of medium components are heat-labile and therefore need to be filter-sterilized. Most antibiotics and compounds such as thiamine and IPTG, a common inducer used to initiate recombinant protein production, are not heat-sterilized and need to be filter-sterilized through a 0.22-μm-pore-size filter.

For growth of *E. coli* to high cell densities, it is important to supply the required substrates in such a way that their concentrations are below growth-inhibitory values in the bioreactor. For example, cells will not grow when the glucose concentration exceeds 50 g liter^{-1}, the phosphate concentration 10 g liter^{-1}, and the ammonium concentration 4 g liter^{-1}. On the other hand, magnesium phosphate has a very low solubility, and therefore for high-cell-density cultivations it is recommended to add the majority of

TABLE 1 Defined medium using glucose as carbon substrate[a]

A. Components of medium

Component	Concn (g liter^{-1})	In vol of H_2O
Glucose·H_2O	12.00	80 ml
$MgSO_4$·$7H_2O$	1.20	20 ml
KH_2PO_4	13.30	
$(NH_4)_2HPO_4$	4.00	900 ml
Citric acid·H_2O	1.70	

B. Trace elements

Trace element	Concn (mg liter^{-1})	Concn in stock solution (mg ml^{-1})	Add vol (ml)
Fe(III)citrate	100.80	12.00	8.40
$CoCl_2$·$6H_2O$	2.50	2.50	1.00
$MnCl_2$·$4H_2O$	15.00	15.00	1.00
$CuCl_2$·$2H_2O$	1.50	1.50	1.00
H_3BO_3	3.00	3.00	1.00
Na_2MoO_4·$2H_2O$	2.10	2.50	0.84
$Zn(CH_3COOH)_2$·$2H_2O$	33.80	13.00	2.60
Titriplex III (EDTA)	14.10	8.40	1.68

[a]This medium can be used to grow *E. coli* in test tubes, shake flasks, and batch bioreactor cultures.

TABLE 2 Defined medium to grow *E. coli* to high cell density using fed-batch culture technique or to grow *E. coli* in continuous culture

Medium components	Batch culture	Feeding solution, fed-batch culture	Feeding solution, continuous culture
Glucose·H$_2$O	27.5 g liter^{-1}	875.0 g liter^{-1}	11.0 g liter^{-1}
KH$_2$PO$_4$	13.3 g liter^{-1}		2.7 g liter^{-1}
(NH$_4$)$_2$HPO$_4$	4.0 g liter^{-1}		0.8 g liter^{-1}
(NH$_4$)$_2$SO$_4$			8.0 g liter^{-1}
MgSO$_4$·7H$_2$O	1.2 g liter^{-1}	20.0 g liter^{-1}	1.0 g liter^{-1}
Citric acid·H$_2$O	1.7 g liter^{-1}		0.35 g liter^{-1}
Fe(III) citrate	100.8 mg liter^{-1}	40.0 mg liter^{-1}	12.0 mg liter^{-1}
CoCl$_2$·6H$_2$O	2.5 mg liter^{-1}	4.0 mg liter^{-1}	0.5 mg liter^{-1}
MnCl$_2$·4H$_2$O	15.0 mg liter^{-1}	23.5 mg liter^{-1}	3.0 mg liter^{-1}
CuCl$_2$·2H$_2$O	1.5 mg liter^{-1}	2.3 mg liter^{-1}	0.3 mg liter^{-1}
H$_3$BO$_3$	3.0 mg liter^{-1}	4.7 mg liter^{-1}	0.6 mg liter^{-1}
Na$_2$MoO$_4$·2H$_2$O	2.1 mg liter^{-1}	4.0 mg liter^{-1}	0.5 mg liter^{-1}
Zn(CH$_3$COOH)$_2$·2H$_2$O	33.8 mg liter^{-1}	16.0 mg liter^{-1}	1.6 mg liter^{-1}
Titriplex III (EDTA)	14.1 mg liter^{-1}	813.0 mg liter^{-1}	1.7 mg liter^{-1}

needed phosphate at the beginning of the cultivation and the needed magnesium continuously during the cultivation (12, 27). Nitrogen addition to high-cell-density cultures is done in a similar way, since adding all the required nitrogen at the beginning will inhibit growth. The nitrogen is added continuously to the growing culture as ammonium hydroxide in response to the change of the pH. An example of defined medium used for growing nonrecombinant *E. coli* to high cell densities of 128 g liter^{-1} dry cell mass using glucose (12) and 165 g liter^{-1} dry cell mass using glycerol as carbon source (23), and for production of recombinant proteins (1, 7, 27), is given in Table 2. This medium has also been successfully applied to produce 27.5 g liter^{-1} amorpha-4,11-diene, a precursor of the antimalarial drug artemisinin, with a genetically engineered strain of *E. coli* in high-cell-density culture (37). The feeding solution of this medium can be adapted for usage in carbon-limited continuous culture experiments by decreasing the phosphor and increasing the nitrogen content (10).

10.2.5. Starter Culture Preparation

The starter culture (inoculum) preparation for the recombinant protein production process can be divided into the following steps. (i) Choose the proper growth vessel to accommodate production batch size. The starter culture volume is usually between 0.25 and 1% of the initial production volume. For starter culture volumes of up to 4 to 5 liters, shake flasks are sufficient, but for higher volumes, it is better to grow the starter culture in a bioreactor. (ii) Prepare the medium and the growth vessels; for details refer to section 10.2.6.1. (iii) Inoculation of the starter culture is usually done in two phases. In the first phase, an aliquot of the working cell bank is removed from the −80°C freezer and transferred to a small shake flask, containing usually between 50 and 100 ml of medium. In some cases, the first culture will be inoculated from a single colony of freshly transformed cells or from a single colony of a fresh agar plate generated from the working cell bank. In the second phase, this culture, when it reaches the desired density, is used to inoculate the starter culture vessel or vessels.

10.2.6. Growth and Production

10.2.6.1. Bioreactor Preparation

The bioreactor preparation process can be divided into the following steps. (i) Making sure that the bioreactor is clean, that it is equipped with an inlet and an outlet air filter in good shape, and that all the valves controlling the addition ports, the sampling ports, and the harvest port are working satisfactorily. (ii) Installing the *on-line* probes and calibrating them. In most cases, the only probes required are those for pH and DO. (iii) Medium preparation: the medium is usually composed of heat-stable and heat-sensitive reagents (see section 10.2.4). The heat-stable reagents can be sterilized directly in the bioreactor and the heat-sensitive reagents are filter-sterilized as concentrated solutions in a separate container that can be connected aseptically to the bioreactor. Sometimes, it is not possible to heat-sterilize certain reagents together, and there is a need to separately prepare concentrated solutions of those heat-stable ingredients and to heat-sterilize them in the autoclave in a separate container that can be aseptically connected to the bioreactor following the sterilization.

As indicated above, there are two types of bioreactors: those that are sterilized in the autoclave (not more than 10 liters in working volume) and those that are sterilized-in-place (above 10 liters). To sterilize the bioreactor in the autoclave, it is important to ensure that the air outlet is open and that all the inlet or outlet ports (which are submersed in the growth medium) are either plugged or connected to a port that is not submersed. The air inlet filter is usually sterilized separately and is hooked to the sparger port after sterilization. A 10-liter bioreactor should be sterilized for an hour. Following the sterilization, the bioreactor is placed next to its controlling instruments, the air source is connected to the inlet of the air filter, and the outlet of the air filter is connected to the sparger. The bioreactor is allowed to cool to the growth temperature and is ready for inoculation. When dealing with sterilized-in-place bioreactors, the sterilization process has several steps that are coordinated and monitored, in most cases, by a programmed controller. The following is a description of the process. (i) Agitation is turned on, and the bioreactor is heated up by steam that flows into the bioreactor jacket. (ii) When the temperature reaches 100°C, the steam is allowed to go directly into the medium through the air inlet filter, sterilizing the air filter at this time and raising the medium temperature to 121.5°C. (iii) At this point, it is advisable to sterilize all the auxiliary ports. Each port is equipped with its own steam inlet and condensate outlet. The bioreactor is kept at this temperature for at least 20 min and is allowed to cool down to the growth temperature by circulating cold water through the bioreactor jacket. (iv) It is essential to confirm that when the medium temperature reaches below 100°C, air can flow into the bioreactor to compensate for pressure loss due to condensation. (v) When the vessel reaches the growth temperature, it is ready for inoculation.

Specific details of the recombinant *P. aeruginosa* exotoxin A production process adapted to routine production are based on procedures described before (5).

Preparation of a Bioreactor Containing a 50-Liter Working Volume

1. DO and pH electrodes are installed; the bioreactor is filled with 45 liters of distilled water containing 250 g of yeast extract (Difco), 500 g of tryptone (Difco), 250 g of NaCl, 250 g of K_2HPO_4, and 5 ml of antifoam P-2000 (Fluka).
2. The bioreactor is then heat-sterilized. Three solutions are heat-sterilized separately: (i) 4 liters of 50% glucose solution in a 5-liter transferring bottle, (ii) a 500-ml solution of 123.24 g (1 mol) $MgSO_4 \cdot 7H_2O$ in a bottle, and (iii) 50 ml of trace element solution (27.0 g liter^{-1} $FeCl_3 \cdot 6H_2O$, 2.0 g liter^{-1} $ZnCl_2 \cdot 4H_2O$, 2.0 g liter^{-1} $CoCl_2 \cdot 6H_2O$, 2.0 g liter^{-1} $Na_2MoO_4 \cdot 2H_2O$, 1.0 g liter^{-1} $CaCl_2 \cdot 2H_2O$, 1.0 g liter^{-1} $CuCl_2$, 0.5 g liter^{-1} H_3BO_3, 100 ml liter^{-1} concentrated HCl). In addition, a 500-ml solution of 5 g ampicillin is filter-sterilized.
3. Following the sterilization of the bioreactor, the above solutions (glucose, $MgSO_4$, trace element solution, and ampicillin) are added into the bioreactor.

4. In addition, a solution of 50% ammonium hydroxide (approximately 2 liters) for pH control is prepared in an aspirator bottle equipped with silicone tubing suitable for peristaltic pumping and connected to the bioreactor. At this point, the bioreactor is ready for the inoculation.

10.2.6.2. Inoculation

The volume and density of the starter culture (inoculum) depend on the process development parameters and on the volume of the production bioreactor. In general, the starter culture should be in the middle of the logarithmic growth phase, and the volume can be somewhere between 0.25 and 1% of the initial production volume. After verifying that the starter culture is not contaminated (using a light microscope) and that it is in its proper growth phase, it is transferred aseptically to the production vessel. To ensure that the starter culture is not contaminated, it is advisable to streak an agar plate for later visual colony inspection. If the starter culture grew in another bioreactor, it is transferred directly from that bioreactor to the production bioreactor via a sterilized hose using a pump or by pressurizing the starter culture bioreactor. If the starter culture is grown in shake flasks, it is transferred first to a transfer container and from this container into the production bioreactor by either pressure or pump.

The details associated with recombinant *P. aeruginosa* exotoxin A production are as follows. (i) One liter inoculum is grown for 12 h at 37°C in the following medium: 5 g liter^{-1} yeast extract, 10 g liter^{-1} peptone, and 5 g liter^{-1} NaCl. (ii) After overnight growth, the pH and the optical density (OD) are analyzed, and if they are in the accepted limits, the culture is transferred to the bioreactor to start the growth and production process.

10.2.6.3. Growth Strategies

There are three major strategies to grow bacteria—batch, fed-batch, and continuous culture—shown schematically in Fig. 3. When cells are grown in a batch procedure, all nutrients are added at the beginning of the cultivation, and the cell growth and production process ends when the essential nutrients are depleted. The limiting essential nutrient, in

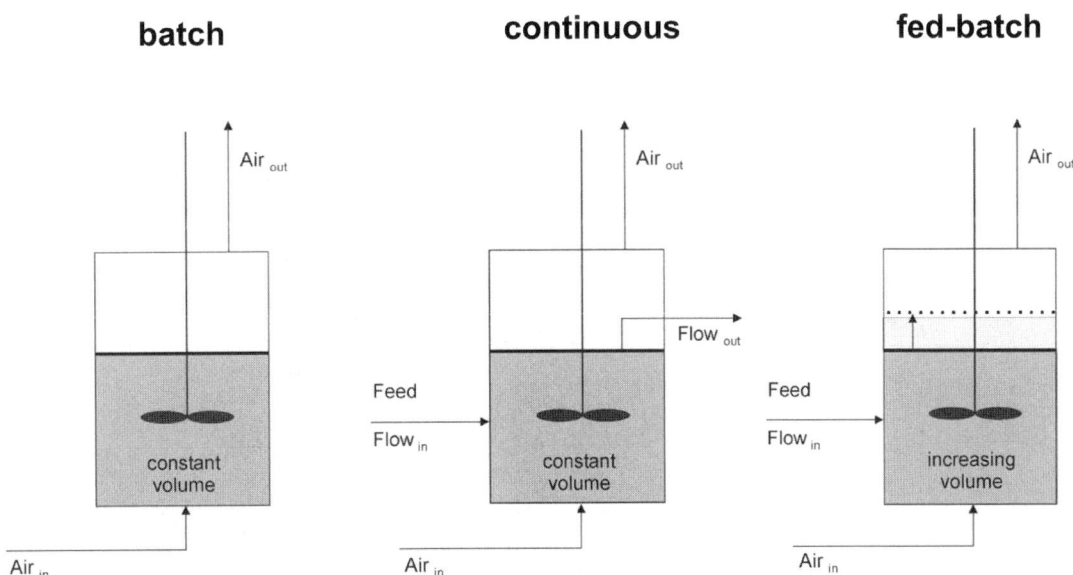

FIGURE 3 General scheme of cultivation strategies.

most cases, is the carbon source, and in the *E. coli* growth process it is usually glucose. Since most *E. coli* strains are sensitive to high glucose concentrations, the final cell density and the final product concentration are relatively low. *E. coli* strain B is exceptional in its capability to tolerate glucose concentrations as high as 40 g liter^{-1} without excessive acetate formation and therefore can grow to a relatively high density and consequently produce a higher level of product while growing in a batch mode strategy (21). The batch culture technique is simple to implement and can be handled in laboratories that cannot accommodate sophisticated growth strategies. For recombinant protein production in batch culture, it is recommended to use either a complex medium, e.g., Luria broth (LB) or terrific broth, supplemented with either glucose or glycerol as an additional carbon substrate or a defined medium as described in section 10.2.4. With the defined medium described in section 10.2.4, cell densities of approximately 10 g liter^{-1} dry cell mass (corresponding to an optical density of approximately 20 at 600 nm) can be obtained in batch culture for *E. coli* K-12 strain TG1 at 20 g liter^{-1} glucose (12). The maximum protein concentration that can be reached is affected by the properties of the protein and the expression vector used for production.

In order to obtain higher productivities, fed-batch culture strategies are being used. In these growth strategies, one of the essential growth components, usually the carbon source, is added continuously to the growing culture. As stated above, in batch cultivations the final cell concentration is limited by the initial glucose concentration. However, high glucose concentrations usually cause acetate formation, which will decrease the biomass yield on glucose or even completely inhibit bacterial growth (28, 29). Fed-batch cultivation eliminates acetate formation by adding the glucose continuously into the bioreactor but keeping its concentration below a detectable level. When using this growth strategy, it is important that all other nutrients are in excess, so that the growth is controlled only by the available carbon source. To allow growth at a defined but restricted growth rate under carbon-limiting conditions, the glucose (or another carbon substrate such as glycerol) is added to the bioreactor as follows (12):

$$m_s(t) = F(t)S_F(t) = \left(\frac{\mu(t)}{Y_{X/S}} + m\right)V(t)X(t) \qquad (1)$$

where m_s is the mass flow of substrate (g h^{-1}), F is the volumetric feeding rate (liters h^{-1}), S_F is the concentration of the substrate in the feeding solution (g liter^{-1}), μ is the specific growth rate (h^{-1}), $Y_{X/S}$ is the biomass/substrate yield coefficient (g g^{-1}), m is the specific maintenance coefficient (g g^{-1} h^{-1}), X is the biomass concentration (g liter^{-1}), and V is the cultivation volume (liters). In a fed-batch system, the following growth equation applies:

$$\frac{d(XV)}{dt} = \mu XV \qquad (2)$$

Assuming μ (growth rate) does not change with time, one obtains on integration of equation 2, when starting the feeding at time t_F:

$$X(t) V(t) = X_{t_F} V_{t_F} e^{\mu(t - t_F)} \qquad (3)$$

Thus, by introducing equation 3 into equation 1, the substrate mass feeding rate for a constant specific growth rate (μ_{set}) follows as

$$m_s(t) = \left(\frac{\mu_{set}}{Y_{X/S}} + m\right) V_{t_F} X_{t_F} e^{\mu_{set}(t - t_F)} \qquad (4)$$

To allow *E. coli* to grow to high cell densities, a growth rate must be chosen (μ_{set}) that does not lead to the formation of acetic acid. It is generally the case when the specific growth rate μ is below 0.15 h^{-1}. This feeding strategy allows exponential growth at a constant specific growth rate when the yield coefficient $Y_{X/S}$ and the maintenance coefficient m do not change with time. It implies that the same amount of biomass is generated per amount of carbon consumed and that the cells always use the same amount of carbon per biomass and time unit for maintaining vital cell functions. This exponential feeding strategy, called "predetermined feeding protocol," does not depend on the measurement of any growth variables. There is no need to continuously measure the bacterial concentration, the oxygen consumption, or the carbon dioxide production. It is only required to know the time for starting or changing the feeding, the culture volume and the biomass concentration at these time points, and the yield and maintenance coefficients, and to choose the desired growth rate (μ_{set}). As a rough assumption, a yield coefficient $Y_{X/S}$ of 0.5 and 0.45 for glucose and glycerol, respectively, and a maintenance coefficient of m = 0.025 g g^{-1} h^{-1} for both substrates can be considered. If a desired specific growth rate (μ_{set}) of 0.12 h^{-1} is chosen, formation of acetic acid should not be observed. For temperature-induced production of recombinant proteins, the desired specific growth temperature should be reduced to 0.08 h^{-1} after raising the temperature to 42°C (1, 25, 27).

Fed-batch cultivations are normally preceded by batch culture growth, and the fed-batch phase of the cultivation is started when the glucose of the batch phase is consumed. This can be followed by monitoring the DO concentration, which will sharply rise after all glucose has been consumed. It is possible to start the feeding according to equation 4 after this rise in the DO concentration. Another option is to wait until the acetate that accumulated in the batch phase has been consumed by the cells. This can also be followed by monitoring the DO concentration. After the sharp rise in the DO concentration as a result of glucose depletion, the bacteria will start to consume the accumulated acetate; this will be indicated by a slow decline of the DO concentration. When the DO concentration rises again, all acetate has been consumed by the cells and feeding can be started.

The fed-batch procedure described above is a simple feed-forward strategy allowing exponential growth at a constant specific growth rate as long as the carbon source yield and maintenance coefficients do not change with time, as assumed in equations 1 to 4. In cases when programmable pumps are not available, it is possible to manually adjust the carbon source feeding rate by stepwise increases in such a way that it follows a pseudoexponential increase as predetermined by that same equation.

Another, simpler fed-batch strategy involves linear feeding. In this case, the amount of glucose or any other carbon source added per unit time does not vary with the culture time. A drawback of this linear feeding strategy is that it leads to declining growth rates with the increase in biomass. In some cases, mixed fed-batch strategies are applied. First, an exponential feeding strategy is implemented to allow growth at a constant specific growth rate. The culture is grown in this way until the supply of DO reaches its limitation. At this point, the feeding is switched to linear feeding. Fed-batch cultivation strategies can also be based on the metabolic activities of the growing culture, such as oxygen consumption or pH changes. For example, the change in the DO concentration activates a pump that delivers the carbon source in such a way that a certain DO concentration is maintained (17).

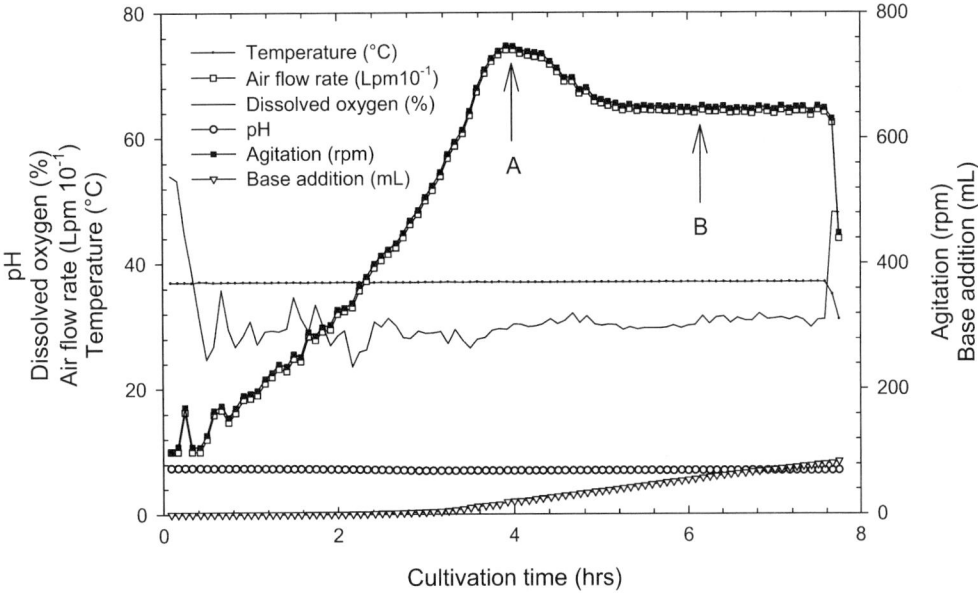

FIGURE 4 On-line data on batch cultivation process for production of recombinant exotoxin A from *E. coli* in 5-liter bioreactor. Arrow A indicates the point of introducing oxygen-enriched air to the culture; arrow B indicates the time when IPTG was added to the culture.

More-detailed discussions on the pros and cons of different fed-batch strategies can be found elsewhere (13, 22).

Fed-batch processes are the most common strategies in industrial settings. However, in some cases, continuous culture techniques are used. In continuous culture the carbon source is added to the growing culture at a certain rate, but unlike the fed-batch technique, where the culture volume increases with time, in the continuous culture the total volume of the culture is kept constant and the excess culture volume, containing cells and product, is collected at the same rate. Theoretically, the highest productivities are reachable in continuous cultures, provided the bacteria exhibit sufficient genetic and physiological stability (16). In most cases, high-producer strains designed by genetic engineering or traditional mutagenesis will lose their production capabilities in long-term continuous cultures.

10.2.6.4. Following Growth and Production

The growth and production process starts after the bioreactor has been inoculated with the starter culture. The follow-up of the process is done by monitoring both on-line and off-line data. The common on-line data are DO concentration, pH, agitation (revolutions per minute), airflow (liters per minute), temperature, bioreactor pressure, and the accumulative volume of acid or base added to keep the pH at its predetermined value. In some cases, other on-line data are monitored, such as the CO_2 and O_2 concentrations in the outlet air, and the turbidity. The DO concentration and the pH are important variables that have direct effects on the growth and production process and therefore must be continuously monitored and controlled. In most protein production processes by recombinant *E. coli*, the DO concentration is kept around 20 to 30% air saturation by varying the agitation, airflow, and pressure independently, sequentially, simultaneously, or based on a specific control strategy. The pH is controlled usually at around 7 by the addition of acid or base depending on the medium. In the case of an *E. coli* recombinant protein production process, the carbon source is usually glucose and the pH is controlled by the addition

of ammonium hydroxide, and seldom by the addition of sodium hydroxide. As was mentioned in section 10.2.1, this process has usually two phases: in the first phase the cells are grown to the desired density, and in the second phase the recombinant protein production is induced. An example of pH and DO control together with the measurement of other on-line variables during an *E. coli* protein production run is shown in Fig. 4. It is important to note that a specific control algorithm is implemented to keep the DO at 30% saturation by increasing the agitation and the airflow with time. In addition, base is added to keep the pH at a value of 7.

The off-line data are measured using a sample removed from the bioreactor through a special sampling port. These data include the bacterial concentration, concentrations of various substrates such as glucose or metabolites such as acetic acid, and the product concentration. Bacterial concentration is usually evaluated by measuring the turbidity of the culture using a spectrophotometer. This method provides quick information on the bacterial concentration when the medium is clear. If the medium is not clear, it is not possible to assess the bacterial concentration by turbidity measurement, and in such cases, the packed cell volume can be an alternative. Other methods, such as dry cell mass measurements and cell counting, are time-consuming and do not provide data in real time. Glucose concentration can be measured by high-pressure liquid chromatography or by a glucose analyzer based on the enzyme glucose oxidase. In most cases, the amount of the product cannot be determined during the production process itself due to the time required for analysis and therefore is done later on stored samples. An example of measuring off-line bacterial concentration by OD at 600 nm and glucose concentration by glucose analyzer (YSI Inc., Yellow Springs, OH) is shown in Fig. 5.

Additional details of the process for recombinant *P. aeruginosa* exotoxin A production are listed here. The following on-line variables are monitored during the process: DO concentration, pH, airflow, agitation, pressure, amount of base added, and temperature (Fig. 4). Throughout the process, the DO concentration is kept at 30% air saturation and

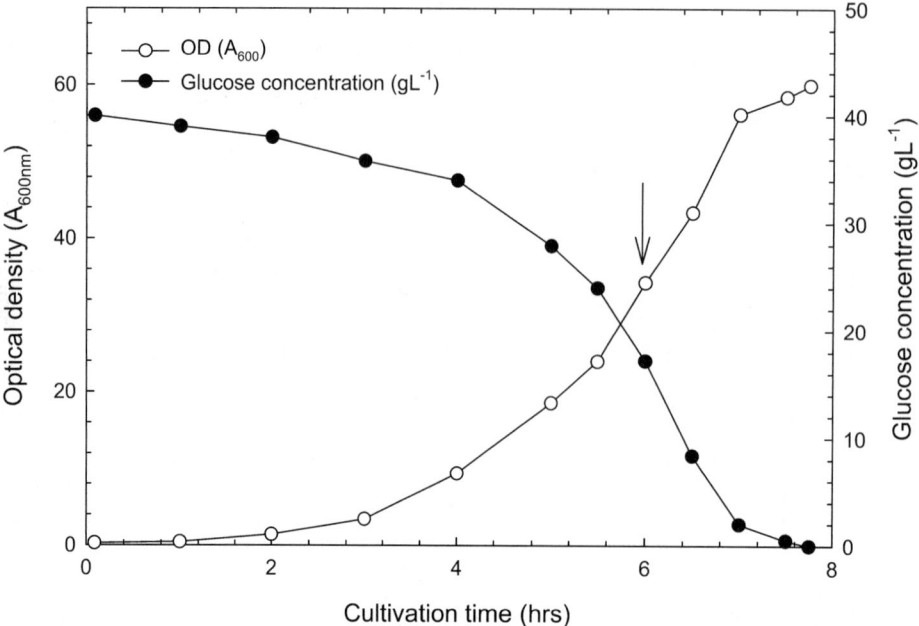

FIGURE 5 Off-line data on batch cultivation process for production of recombinant exotoxin A from *E. coli* in 5-liter bioreactor. The arrow indicates the time when IPTG was added to the culture.

is controlled by simultaneously increasing the agitation and the airflow. The pH is kept at 6.8 by adding 50% ammonium hydroxide solution automatically, and the amount of the base added is monitored continuously. The following variables are measured off-line: bacterial concentration is determined by the OD value at 600 nm and the glucose concentration is measured using the glucose analyzer made by YSI (Fig. 5). *E. coli* BL21, the strain used for the production, can tolerate glucose concentrations as high as 40 g liter^{-1} (21); therefore, the cultivation is carried out as a batch process.

10.2.6.5. Bioprocess Calculations

Growth of most bacteria, including *E. coli*, occurs by cell division. Thus, during unlimited growth (as found in the batch phase when all nutrients are in excess), growth can be described as follows:

$$\frac{dX}{dt} = \mu X \tag{5}$$

or in the integrated form:

$$X = X_0 e^{\mu(t - t_0)} \tag{6}$$

where X is the biomass (g) at time t (h), X_0 the biomass at t_0 (usually the beginning of the cultivation; $t_0 = 0$ h), μ the specific growth rate (h^{-1}), and e the Euler number (2.718281828. . .).

The specific growth rate can be determined from measurements of the bacterial biomass (e.g., analysis of the optical density) at different time points as follows:

$$\mu = \frac{\ln X - \ln X_0}{t - t_0} \tag{7}$$

During carbon-limited fed-batch cultivation when no acids are produced, the growth of bacterial cells can also be determined on-line by following the ammonia consumption (26). Thus, when ammonia is used for pH control, it does not only serve as a base for pH control but also as a nitrogen source. In contrast to the biomass yield coefficient for carbon substrates such as glucose, $Y_{X/C_6H_{12}O_6}$, which is not

constant, the biomass yield coefficient for nitrogen, such as ammonia, Y_{X/NH_3}, is constant and is not affected by the metabolic status of the cells.

When Y_{X/NH_3} (g g^{-1}) and the concentration of ammonia in the feeding/base solution are constant, their absolute values are not required and the actual specific growth rate, μ, (h^{-1}), can be calculated from the time-dependent change of the natural logarithm of the dimensionless signal of the ammonia balance, M_{NH_3}(g g^{-1}), according to equation 8.

$$\mu = \frac{d\ln(M_{NH_3})}{dt} \tag{8}$$

The actual biomass in the bioreactor, X (g), can be calculated according to equation 9:

$$X = M_{NH_3} C_{NH_3} Y_{X/NH_3} \tag{9}$$

where M_{NH_3} (g) is the mass of the ammonia solution added into the bioreactor (g), C_{NH_3} is the concentration of ammonia in this solution (g g^{-1}), and Y_{X/NH_3} is the average biomass yield coefficient with respect to ammonia (7 g g^{-1}).

The volumetric oxygen and carbon dioxide transfer rates (OTR and CTR, respectively) (g liter^{-1} h^{-1}) can be calculated from the mass balance of the gas phase as follows (10):

$$OTR = \frac{M_{O_2} F_G^{in}}{V(t) V_M} \left| x_{O_2}^{in}(t) - x_{O_2}^{out}(t) \frac{1 - x_{O_2}^{in}(t) - x_{CO_2}^{in}(t)}{1 - x_{O_2}^{out}(t) - x_{CO_2}^{out}(t)} \right| \tag{10}$$

and

$$CTR = \frac{M_{CO_2} F_G^{in}}{V(t) V_M} \left| x_{CO_2}^{out}(t) \frac{1 - x_{O_2}^{in}(t) - x_{CO_2}^{in}(t)}{1 - x_{O_2}^{out}(t) - x_{CO_2}^{out}(t)} - x_{CO_2}^{in}(t) \right| \tag{11}$$

where M_{O_2} and M_{CO_2} are the molecular mass of oxygen and carbon dioxide (g mol^{-1}), respectively; F_G^{in} is the volumetric inlet airflow (liters h^{-1}) at standard conditions; $V(t)$ is the working volume of the bioreactor (liters); V_M is the mol volume of the ideal gas (liters mol^{-1}) at standard conditions; $x_{O_2}^{in}$ and $x_{CO_2}^{in}$ are the molar fractions of oxygen and carbon dioxide (mol mol^{-1}), respectively, in the inlet air; and $x_{O_2}^{out}(t)$ and $x_{CO_2}^{out}(t)$ are the molar fractions of oxygen and carbon dioxide (mol mol^{-1}), respectively, in the outlet air of the bioreactor. For calculation of specific rates, the convective flow of oxygen and carbon dioxide can be neglected and the transfer rates OTR and CTR can be considered to be identical to the oxygen uptake and carbon dioxide evolution rates. Specific rates can then be calculated by dividing volumetric rates by cell concentration.

10.2.6.6. Initial Product Recovery

Depending on the process and the product, the recombinant protein can accumulate inside the cells or can be secreted into the growth medium. When the product is secreted into the medium, the bacterial biomass is separated from the medium by either centrifugation or filtration and the protein is recovered from the supernatant. When the product accumulates in the cells, it is possible to recover the protein either from the biomass after its separation from the medium or, especially when the cell concentration is high, directly from the complete broth containing both the cells and the growth medium without further separation. Details on continuous-flow centrifuges and tangential filtration systems can be found in several books dealing with downstream processing and protein purification (4, 33).

For example, the following steps are required for the downstream processing of *P. aeruginosa* exotoxin A after termination of its production in the bioreactor. The protein is secreted into the periplasmic space, thus accumulating inside the cells. The initial downstream processing involves the separation of the cells from the medium, followed by lysis of the cells by osmotic shock, which releases the accumulated protein into the supernatant. The recovery and clarification of the supernatant and final purification of the protein are described in detail elsewhere (8).

10.3. OTHER PRODUCTS FROM *E. COLI*

Recombinant protein production from *E. coli* started in the late 1970s and early 1980s with insulin as the first recombinant protein product (9). In addition to recombinant proteins for biopharmaceutical use, *E. coli* is also used to produce other recombinant proteins such as enzymes for technical applications, polysaccharides and other biopolymers such as plasmid DNA, amino acids, and primary metabolites such as organic acids. The production principles and the overall procedures are similar to the recombinant protein production procedure as described in section 10.2. The differences are in the medium composition, the growth strategy, and production variables such as pH, DO concentration, temperature, cell density, and length of cultivation. Below are some examples of processes for production of biological products other than recombinant proteins. One is the production of the amino acid L-alanine by genetically engineered *E. coli*, the second is the conversion of ferulic acid to vanillin by recombinant *E. coli*, and the third is the production of polysialic acid by a selected *E. coli* K1 strain. Two more examples are associated with two new products: one is plasmid DNA and the other is a precursor of artemisinin, a promising antimalarial drug, which is being produced in a two-phase partitioning bioreactor, a novel approach for the biosynthesis of organic compounds.

- *Production of the amino acid L-alanine by metabolically engineered* E. coli *(39)*. The modified *E. coli* strain is grown anaerobically on a defined medium of the following composition (14), in mmol liter^{-1}: 19.92 (NH$_4$)$_2$HPO$_4$, 7.56 NH$_4$H$_2$PO$_4$, 2.0 KCl, 1.5 MgSO$_4$·7H$_2$O, and 1.0 betaine·KCl. The following concentrations are in μmol liter^{-1}: 8.8 FeCl$_3$·6H$_2$O, 1.26 CoCl$_2$·6H$_2$O, 0.88 CuCl$_2$·H$_2$O, 2.2 ZnCl$_2$, 1.24 Na$_2$MoO$_4$·2H$_2$O, 1.21 H$_3$BO$_3$, and 2.5 MnCl$_2$·4H$_2$O. The trace metal stock solution (1,000 times concentrated) is prepared in 120 mmol liter^{-1} HCl. The concentration of the carbon source glucose is 120 g liter^{-1}. The growth is carried out at 37°C and the pH is maintained at 7 by automatic addition of 5 mol liter^{-1} NH$_4$OH. The product accumulates in the medium. The major differences between this process and the recombinant *P. aeruginosa* exotoxin A production process described above are the utilization of defined medium, the anaerobic growth, and the high concentration of glucose.

- *Production of vanillin.* Vanillin, an organic compound with the formula C$_8$H$_8$O$_3$, is a flavoring agent used in foods and beverages. The compound is produced by genetically engineered *E. coli* by conversion of ferulic acid (2). The process has two phases: in the first phase, the engineered *E. coli* strain is grown in a complex medium to generate the required biomass, and in the second step, the cells are collected and resuspended in a buffer containing ferulic acid. The suggested process is as follows. (i) The cells are grown in LB medium (10 g liter^{-1} tryptone, 5 g liter^{-1} yeast extract, and 5 g liter^{-1} NaCl), containing 25 mg liter^{-1} tetracycline, at 37°C in an aerated bioreactor. (ii) At the end of this growth phase, the cells are collected, washed, and resuspended at a concentration of 4 g liter^{-1} in M9 saline/phosphate buffer (4.2 mmol liter^{-1} Na$_2$HPO$_4$, 2.2 mmol liter^{-1} KH$_2$PO$_4$, 0.9 mmol liter^{-1} NaCl, and 1.9 mmol liter^{-1} NH$_4$Cl) containing 0.5 mg liter^{-1} yeast extract and 5 mmol liter^{-1} ferulic acid. (iii) The cells, now in a resting phase, convert the ferulic acid to vanillin. This way, 2.5 g liter^{-1} vanillin can be produced (2).

- *Production of long-chain polysialic acid.* Polysialic acid is a polymer that is being investigated as a potential additive to different biomedical applications such as tissue engineering. In the following example, the production process of this polymer from *E. coli* is described (24). In this process, the selected *E. coli* K1 strain is grown in a defined medium containing 1.2 g liter^{-1} NaCl, 1.1 g liter^{-1} K$_2$SO$_4$, 13 mg liter^{-1} CaCl$_2$, 0.15 g liter^{-1} MgSO$_4$·7H$_2$O, 1 mg liter^{-1} FeSO$_4$·7H$_2$O, 1 mg liter^{-1} CuSO$_4$·5H$_2$O, 6.67 g liter^{-1} K$_2$HPO$_4$, and 0.25 g liter^{-1} KH$_2$PO$_4$. Additionally, the medium includes 13.3 g liter^{-1} glucose and 10 g liter^{-1} (NH$_4$)$_2$SO$_4$. The cells are grown aerobically (the DO is measured throughout the cultivation) at pH 7.5 and 37°C. The overall process lasts 25 h, the bacteria grow exponentially for 10 h, and the polysialic acid accumulates during the following 15 h. Following production, the cells are removed by continuous centrifugation and the polysialic acid is recovered from the supernatant fluid.

- *Production of plasmid DNA.* Plasmid DNA is produced by *E. coli* in a process similar to recombinant protein production. The plasmid DNA is amplified when the culture temperature is increased to 42°C. The cells are initially grown at 30°C to high cell density. After the completion of this growth phase, plasmid DNA production is commenced by increasing the temperature to 42°C for several more hours (20, 38).

- *Production of amorpha-4,11-diene.* This compound is a precursor of the new antimalarial drug artemisinin (6). This natural product is made in *E. coli* containing a heterologous nine-gene pathway (18). The production procedure has two phases. In the first phase, the culture is grown in the medium containing 12 g liter^{-1} tryptone, 24 g liter^{-1} yeast extract, 14.9 g liter^{-1} phosphate, and 10 g liter^{-1} glycerol. The medium also contains three different antibiotics (ampicillin at 100 mg liter^{-1}, tetracycline at 5 mg liter^{-1}, and chloramphenicol at 25 mg liter^{-1}). During this phase, the DO level is kept around 30% air saturation and the pH at 7 by the addition of 10% NaOH. In the second phase, production of amorpha-4,11-diene is induced by adding 0.5 mmol liter^{-1} IPTG. At 30 min following induction, 20% (vol/vol) dodecane is added. This represents a cultivation process called "two-phase partitioning bioreactor," where the product accumulates in the organic phase and is recovered continuously by condensation (18). Growing the cells in the complex medium described above in a batch procedure to low cell density resulted in 0.5 g liter^{-1} amorpha-4,11-diene (18). Further improvement of the strain, utilization of a defined growth medium, and the application of a carbon-limited fed-batch procedure increased the product concentration to 27 g liter^{-1} (37).

10.4. PRODUCTS FROM BACTERIA OTHER THAN *E. COLI*

The list of biological products that can be produced from different bacteria is long. It includes enzymes such as amylases, proteases, and pectinases; metabolites such as amino acids, ethanol, and acetone; organic acids such as acetic, citric, lactic, and glutamic acids; and nucleotides, vitamins, antibiotics, insecticides, polysaccharides, vaccines, and bacterial biomass that is needed for processes such as biotransformation. The list of the producing bacteria is also long and includes genera such as *Bacillus* for enzymes and insecticide production, *Streptomyces* for antibiotics, *Corynebacterium* and *Brevibacterium* for amino acids, *Clostridium* for acetone and butanol, *Lactobacillus* for organic acids, and *Acetobacter* for polysaccharides. The production principles are similar to those described for recombinant protein production using genetically engineered *E. coli*. The differences are in the medium compositions and the growth and production variables. Following are three examples of production of biological products using different types of bacteria. The first two examples describe classical production processes using genetically unmodified bacteria: production of the antibiotic streptomycin by *Streptomyces griseus* and production of glutamic acid by *Corynebacterium glutamicum*. The third example is the description of succinic acid production by a novel, genetically engineered strain of *Mannheimia succiniciproducens*.

- *Production of the antibiotic streptomycin using* S. griseus. The streptomycete can be grown in a defined or complex medium. A typical complex medium composition is 1% glucose, 1% soybean meal, and 0.5% sodium chloride. The aerobic cultivation takes place at a temperature between 25 and 30°C and at a pH in the range of 7 to 8. The production process takes about 80 h and has two phases. The first phase is biomass production, and the second phase is streptomycin production. The process is terminated when there is no further increase in streptomycin production. Increasing the glucose concentration and adding ammonium sulfate prolong the production phase and increase the final concentration of streptomycin (31).

- *Glutamic acid production from* C. glutamicum. The aerobic production process can last 70 h, the pH is controlled at 7, and the growth temperature is in the range of 30 to 35°C. Typical medium composition is glucose 4.75%, calcium carbonate 1.25%, urea 0.07%, KH$_2$PO$_4$ 0.05%, MgSO$_4$ 0.01%, and ferric sulfate 8 ppm. It was found that maintaining a low biotin level increased glutamic acid production. Moreover, controlling the pH by feeding urea increased biomass and glutamic acid production (35).

- *Succinic acid production from* M. succiniciproducens. A genetically engineered strain of M. *succiniciproducens*, deficient in several catabolic genes leading to unwanted by-product formation, was employed for succinic acid production (19). The cultivation was carried out in a batch procedure using a semisynthetic medium containing (per liter) glucose 18 g, yeast extract 5 g, NaCl 1 g, K$_2$HPO$_4$ 8.708 g, CaCl$_2$·2H$_2$O 0.02 g, and MgCl$_2$·6H$_2$O 0.2 g. The cultivation is carried out at anaerobic conditions, leading to final concentrations of 15.5 g liter^{-1} succinic acid, corresponding to a final product yield of 0.86 grams of succinic acid per gram of glucose. In this case, the agitation speed was identified as an important variable affecting the final product yield.

10.5. SUMMARY

Bacterial cultivation for production of proteins and other biological products is a broad topic. It is not possible to cover all the variations of these processes in this chapter since they depend on the microorganism on the one hand and on the products on the other hand. However, the general principles of the process, as outlined in Fig. 1, are similar for all the different processes. They involve the following phases: (i) preparation of the growth vessels both for the starter culture and the production culture, (ii) preparation of the growth and the production medium, (iii) sterilization of the medium and the growth vessels, (iv) preparation of the auxiliary equipment, and (v) performing the production process itself. A large amount of work has to be done on development and optimization of the production process before commercial production can be initiated. This includes optimization of the medium composition, growth conditions (such as temperature, pH, and oxygen saturation level), growth strategies (batch, fed-batch, or continuous), and the length of the process. Production of recombinant proteins from *E. coli* was chosen as an example and described in detail, but as was shown in section 10.2, the optimized parameters and methodologies are different for different processes. For example, in some cases the process is carried out anaerobically, and in other cases the process has two distinct phases where after the growth phase the bacteria are given the opportunity to produce the desired product from a specific precursor. Thus, when dealing with a specific production process, it is clearly advisable to consult the vast scientific literature related to a specific product.

Funding was provided by the intramural program at the NIDDK, NIH and HZI. We thank D. Livant for proofreading the manuscript.

REFERENCES

1. Babu, K. R., S. Swaminathan, S. Marten, N. Khanna, and U. Rinas. 2000. Production of interferon-alpha in high-cell density cultures of recombinant *Escherichia coli*

and its single step purification from refolded inclusion body proteins. *Appl. Microbiol. Biotechnol.* **53:**655–660.

2. **Barghini, P., D. Di Giola, F. Fava, and M. Ruzzi.** 2007. Vanillin production using metabolically engineered *Escherichia coli* under non-growing conditions. *Microb. Cell Fact.* **6:**13.

3. **El-Mansi, E. M. T., and C. F. A. Bryce (ed.).** 2007. *Fermentation Microbiology and Biotechnology,* 2nd ed. CRC, Boca Raton, FL.

4. **Endo, I., T. Nagamune, S. Katoh, and T. Yonemoto.** 2001. *Bioseparation Engineering.* John Wiley & Sons, Inc., New York, NY.

5. **Fass, R., M. van de Walle, A. Shiloach, A. Joslyn, J. Kaufman, and J. Shiloach.** 1991. Use of high density cultures of *Escherichia coli* for high level production of recombinant *Pseudomonas aeruginosa* exotoxin A. *Appl. Microbiol. Biotechnol.* **36:**65–69.

6. **Hale, V., J. D. Keasling, N. Renninger, and T. T. Diagana.** 2007. Microbially derived artemisinin: a biotechnology solution to the global problem of access to affordable antimalarial drugs. *Am. J. Trop. Med. Hyg.* **77:**198–202.

7. **Hoffmann, F., J. van den Heuvel, N. Zidek, and U. Rinas.** 2004. Minimizing inclusion body formation during recombinant protein production in *Escherichia coli* at bench and pilot plant scale. *Enzyme Microb. Technol.* **34:**235–241.

8. **Johansson, H. J., C. Jägersten, and J. Shiloach.** 1996. Large scale recovery and purification of periplasmic recombinant protein from *E. coli* using expanded bed adsorption chromatography followed by new ion exchange media. *J. Biotechnol.* **48:**914.

9. **Johnson, I. S.** 1983. Human insulin from recombinant DNA technology. *Science* **219:**632–637.

10. **Kayser, A., J. Weber, V. Hecht, and U. Rinas.** 2005. Metabolic flux analysis of *Escherichia coli* in glucose-limited continuous culture: I. Growth rate dependent metabolic efficiency at steady state. *Microbiology* **151:**693–706.

11. **Kirsop, B. E., and J. J. S. Snell.** 1984. *Maintenance of Microorganisms.* Academic Press, London, United Kingdom.

12. **Korz, D. J., U. Rinas, K. Hellmuth, E. A. Sanders, and W. D. Deckwer.** 1995. Simple fed-batch technique for high cell density cultivation of *Escherichia coli. J. Biotechnol.* **39:**59–65.

13. **Lee, J., S. Y. Lee, S. Park, and A. P. J. Middelberg.** 1999. Control of fed-batch fermentations. *Biotechnol. Adv.* **17:**29–48.

14. **Martinez, A., T. B. Grabar, K. T. Shanmugam, L. P. Yomano, S. W. York, and L. O. Ingram.** 2007. Low salt medium for lactate and ethanol production by recombinant *Escherichia coli* B. *Biotechnol. Lett.* **29:**397–404.

15. **McNeil, B., and L. M. Harvey (ed.).** 2008. *Practical Fermentation Technology.* John Wiley & Sons Ltd., Chichester, United Kingdom.

16. **Mori, H., T. Yamane, T. Kobayashi, and S. Shimizu.** 1983. Comparison of cell productivities among fed-batch, repeated fed-batch and continuous cultures at high cell concentration. *J. Ferment. Technol.* **61:**391–401.

17. **Mori, H., T. Yano, T. Kobayashi, and S. Shimizu.** 1979. High density cultivation of biomass in fed-batch system with DO-stat. *J. Chem. Eng.* **12:**313–319.

18. **Newman, J. D., J. Marshall, M. Chang, F. Nowroozi, E. Paradise, D. Pitera, K. L. Newman, and J. D. Keasling.** 2006. High-level production of amorpha-4,11-diene in a two-phase partitioning bioreactor of metabolically engineered *Escherichia coli. Biotechnol. Bioeng.* **95:**684–691.

19. **Oh, I. J., D. H. Kim, E. K. Oh, S. Y. Lee, and J. Lee.** 2009. Optimization and scale-up of succinic acid production by *Mannheimia succiniciproducens* LPK7. *J. Microbiol. Biotechnol.* **19:**167–171.

20. **Phue, J. N., S. J. Lee, L. Trinh, and J. Shiloach.** 2008. Modified *Escherichia coli* B (BL21), a superior producer of plasmid DNA compared with *Escherichia coli* K (DH5α). *Biotechnol. Bioeng.* **101:**831–836.

21. **Phue, J. N., S. B. Noronha, R. Hattacharyya, A. J. Wolfe, and J. Shiloach.** 2005. Glucose metabolism at high density growth of *E. coli* B and *E. coli* K: differences in metabolic pathways are responsible for efficient glucose utilization in *E. coli* B as determined by microarrays and Northern blot analyses. *Biotechnol. Bioeng.* **90:**805–820.

22. **Posten, C., and U. Rinas.** 2000. Control strategies for high-cell density cultivation of *Escherichia coli,* p. 374–390. *In* K. Schügerl and K. H. Bellgardt (ed.), *Bioreaction Engineering. Modelling and Control.* Springer-Verlag, Berlin, Germany.

23. **Rinas, U., K. Hellmuth, R. Kang, A. Seeger, and H. Schlieker.** 1995. Entry of *Escherichia coli* into stationary phase is indicated by endogenous and exogenous accumulation of nucleobases. *Appl. Environ. Microbiol.* **61:**4147–4151.

24. **Rode, B., C. Endres, C. Ran, F. Stahl, S. Beutel, C. Kasper, S. Galuska, R. Geyer, M. Muhlenhoff, R. Gerardy-Schahn, and T. Scheper.** 2008. Large-scale production and homogenous purification of long chain polysialic acids from *E. coli* K1. *J. Biotechnol.* **135:**202–209.

25. **Schmidt, M., K. R. Babu, N. Khanna, S. Marten, and U. Rinas.** 1999. Temperature-induced production of recombinant human insulin in high-cell density cultures of recombinant *Escherichia coli. J. Biotechnol.* **68:**71–83.

26. **Schmidt, M., E. Viaplana, F. Hoffmann, S. Marten, A. Villaverde, and U. Rinas.** 1999. Secretion-dependent proteolysis of heterologous protein by recombinant *Escherichia coli* is connected to an increased activity of the energy-generating dissimilatory pathway. *Biotechnol. Bioeng.* **66:**61–67.

27. **Seeger, A., B. Schneppe, J. E. G. McCarthy, W.-D. Deckwer, and U. Rinas.** 1995. Comparison of temperature- and isopropyl-β-D-thiogalacto-pyranoside-induced synthesis of basic fibroblast growth factor in high-cell-density cultures of recombinant *Escherichia coli. Enzyme Microb. Technol.* **17:**947–953.

28. **Shiloach, J., and R. Fass.** 2005. Growing *E. coli* to high cell density—a historical perspective on method development. *Biotechnol. Adv.* **23:**345–357.

29. **Shiloach, J., and U. Rinas.** 2009. Glucose and acetate metabolism in *E. coli*—system level analysis and biotechnological applications in protein production processes, p. 377–400. *In* S. Y. Lee (ed.), *Systems Biology and Biotechnology of Escherichia coli.* Springer, Berlin, Germany.

30. **Simione, F. P., and E. M. Brown.** 1991. *ATCC Preservation Methods.* American Type Culture Collection, Manassas, VA.

31. **Singh, A., E. Bruzelius, and H. Heading.** 1976. Streptomycin, a fermentation study. *Eur. J. Appl. Microbiol. Biotechnol.* **3:**97–101.

32. **Stanbury, P. F., A. Whitaker, and S. J. Hall.** 1995. *Principles of Fermentation Technology,* 2nd ed. Butterworth-Heinemann, Burlington, MA.

33. **Subramanian, G. (ed.).** 1998. *Bioseparation and Bioprocessing: a Handbook.* Wiley-VCH Verlag GmbH and Co. KGaA, Weinheim, Germany.

34. **Swartz, J. R.** 2001. Advances in *Escherichia coli* production of therapeutic proteins. *Curr. Opin. Biotechnol.* **12:**195–201.

35. **Tanaka, K., T. Iwasaki, and S. Kinoshita.** 1960. Studies on L-glutamic acid fermentation. Part V. Biotin and L-glutamic acid accumulation by bacteria. *J. Agric. Chem. Soc. Jpn.* **34:**583.

36. **Terpe, K.** 2006. Overview of bacterial expression systems for heterologous protein production: from molecular and

biochemical fundamentals to commercial systems. *Appl. Microbiol. Biotechnol.* **72:**211–222.

37. **Tsuruta, H., C. J. Paddon, D. Eng, J. R. Lenihan, T. Horning, L. C. Anthony, R. Regentin, J. D. Keasling, N. S. Renninger, and J. D. Newman.** 2009. High-level production of amorpha-4,11-diene, a precursor of the antimalarial agent artemisinin, in *Escherichia coli. PLoS ONE* **4:**e4489.

38. **Williams, J. A., A. E. Carnes, and C. P. Hodgson.** 2009. Plasmid DNA vaccine vector design: impact on efficacy, safety and upstream production. *Biotechnol. Adv.* **27:** 353–370.

39. **Zhang, X., K. Jantama, J. C. Moore, K. T. Shanmugam, and L. O. Ingram.** 2007. Production of L-alanine by metabolically engineered *Escherichia coli. Appl. Microbiol. Biotechnol.* **77:**355–366.

Heterologous Protein Expression in Yeasts and Filamentous Fungi

NINGYAN ZHANG AND ZHIQIANG AN

11

11.1. INTRODUCTION

Development of recombinant DNA technologies in the late 1970s laid the foundation for the advancement of various heterologous protein expression systems, including those based on microbial platforms. In the early 1980s, *Escherichia coli* and *Saccharomyces cerevisiae* expression systems were developed and used for heterologous protein expression (47). Many comprehensive reviews of these two expression systems are available (12, 29, 41, 48, 60, 72, 74, 110, 134). Development of other microbial expression systems, such as the methylotrophic yeast *Pichia pastoris* (72) and filamentous fungi (74), followed quickly, and the broad utility of these systems is becoming increasingly apparent.

Heterologous protein expression technologies are widely used in both academic research and industrial applications. In basic research, heterologous gene expression serves as an important tool to prepare proteins and enzymes that are either at limited levels or difficult to purify in their native context. Heterologously expressed proteins and enzymes are used across the entire spectrum of biology. Many enzymes and proteins are of commercial importance, and therefore improved processes for large-scale production of industrial enzymes and proteins are of great interest. Since different enzymes and proteins possess different molecular and biochemical characteristics, different expression systems with specific features are required to meet the growing need for heterologous protein expression at the level of both basic and industrial research.

With the steady advancement in gene expression technology, such as new vectors for gene delivery, improvement in throughput of screening for strain selection, and genetically engineered hosts, production of recombinant proteins has become more efficient with broader applications. In addition, introduction of fusion tags such as hexa-histidine (6× His) and inducible promoters has allowed rapid detection and purification of recombinant proteins and increased expression titers. Since enzymes and proteins have diverse physical and biochemical properties, expression yields vary widely from microgram to low-milligram (37, 57) to multi-gram levels (62). Choosing a suitable expression system for a protein of interest is one of the most critical steps for the development of a high-expression process. Some advantages and drawbacks for commonly used heterologous protein expression systems are listed in Table 1 (138). *E. coli* expression systems provide quick and easy molecular manipulation and, in general, require low-cost growth media. However, the protein expression in *E. coli* cannot be used if foreign proteins require transcriptional and/or posttranslational processing and modifications. By contrast, fungal protein expression systems including yeast and filamentous fungi can provide high expression levels and posttranscriptional and posttranslational processes and modifications of eukaryotic proteins. Similar to *E. coli* protein expression systems, fungal expression systems need low-cost culture media, the technologies needed for scale-up of fungal fermentation processes are well developed, and the organisms are Generally Regarded As Safe (GRAS) (24, 127). In contrast, insect and mammalian expression systems require special culture conditions and relatively high-cost culture media and often have lower expression titers.

When choosing a heterologous protein expression system, the following general criteria should be considered: (i) requirement of translational modification and processing of the protein; (ii) authenticity of the expressed proteins; (iii) protein expression level and scale-up requirement; (iv) localization, e.g., intracellular, membrane-bound, or secreted proteins; (v) cofactor requirement; and (vi) cost, safety, and other regulatory-related issues. Each protein expression system offers a unique set of properties, and for successful production of a heterologous protein, it is often required to tailor a process specifically based on the properties of the protein to be produced and the expression system used.

Many filamentous fungi and yeast species have been used for heterologous protein expression (32, 34, 53, 54, 89, 94, 101, 105, 110, 112, 132, 133). There are numerous reviews addressing foreign protein expressions in each of the fungal protein expression systems (14, 20, 21, 44, 63, 101, 107, 110). Most reviews are focused on one specific system. Furthermore, some address specific aspects of the expression system, such as protein secretion in *S. cerevisiae* (121) and glycosylation of proteins in *P. pastoris* (11). Significant progress has been made in recent years in the production of therapeutic proteins and antibodies using humanized yeast (40). This chapter covers the most commonly used fungal expression systems (*S. cerevisiae*, *P. pastoris*, and *Aspergillus* species), with a focus on vector systems, promoter and leader sequences, posttranslational modifications, and fermentation scalability.

TABLE 1 Pros and cons for some commonly used heterologous protein expression systems

Expression hosts	Pros	Cons
E. coli	Easy genetic manipulation, rapid growth, simple media requirements and low cost to grow, no human virus carried	Inclusion bodies, no secretion into media, lack of posttranslational process
S. cerevisiae	Low-cost culture and easy to grow, genetic regulation well known, secretion protein, no risk for human virus infection	Overglycosylation at N-linked sites, cell difficult to break, expression level limited
P. pastoris	High-level expression, low cost and easy to grow, genetic regulation well known, secretion protein, no risk for human virus infection	Royalties can be expensive, potential for overglycosylation, cell difficult to break
Filamentous fungi	High level of expression, good secretion system, some level of posttranslational modification, culture medium is inexpensive	No good commercial vectors are available, longer culture time, spores are concerns for health problems, not well documented for production of therapeutic proteins
Insect cells	Easy to grow up to 400 liters, easy to manipulate genetically for baculovirus, can grow in the same-style tanks as mammalian cells	Relatively low expression, serum-free medium is expensive, royalties for the commercially available systems can be costly, few contractors have current Good Manufacturing Practice experience
Mammalian cell lines	Authentic protein expression and proper posttranslational processing and assembly for therapeutic proteins and other mammalian proteins of interest, extensively used, reasonable expression level, endotoxin is not a concern	Medium is expensive, special tank, require CO_2, potential risk of human virus in culture, longer expression time

11.2. *S. CEREVISIAE* FOR HETEROLOGOUS PROTEIN EXPRESSION

S. cerevisiae, a unicellular yeast, is one of the most studied eukaryotic microorganisms. Its genetics, physiology, biochemistry, metabolism, and fermentation are well researched and a wealth of information is readily available. Thousands of genes from *S. cerevisiae* have been characterized (43) and its genome sequence has been determined (33). *S. cerevisiae* has been extensively used as a host for heterologous gene expression of eukaryotic proteins for research and industrial applications (44, 110).

S. cerevisiae is a GRAS organism and has been used for centuries in the brewing and baking industries. Similar to *E. coli*, *S. cerevisiae* requires simple culture media for growth and is easily scaled up for large-volume fermentations. On the other hand, unlike *E. coli*, *S. cerevisiae* is capable of accomplishing many posttranslational modifications, such as proteolytic processing, disulfide bond formation, glycosylation, and other posttranscriptional and translational processing unique to eukaryotic organisms. Many vector systems containing a variety of promoters and auxotrophic or dominant selectable markers have been developed, allowing constitutive or regulated gene expression. Several methods for transformation of foreign DNA into *S. cerevisiae* with high efficiency have also been developed (31, 85). As a result, *S. cerevisiae* is one of the most widely used microbial hosts for heterologous gene expression. However, there are limitations to using *S. cerevisiae* for heterologous gene expression; for example, the specific rate of production can vary from protein to protein, ranging from as low as 0.03 g/kg/h to as high as 4.5 g/kg/h (44). The use of episomal vector systems for heterologous protein expression in *S. cerevisiae* often results in plasmid instability during fermentation, which may lead to a lower growth rate and reduced overall protein yield. Furthermore, since many proteins of therapeutic interest are secreted glycoproteins, the tendency of *S. cerevisiae* to hyperglycosylate proteins not only reduces the efficiency of protein secretion but also may lead to undesired changes in the immunogenic properties or biological activities of the expressed proteins (23, 39, 40, 44, 84).

11.2.1. Vectors

S. cerevisiae expression vectors usually are shuttle plasmids that contain sequences for propagation and selection in both *S. cerevisiae* and *E. coli*. Two types of vectors have been described based on their mode of replication: episomal and integrating vectors (Table 2) (16, 18, 27, 45, 73, 86, 108, 109, 111, 138). Episomal vectors can be characterized according to their copy numbers, mode of replication, and stability. There are three major types of episomal vectors: YRp, YCp, and YEp. YRp-type vectors contain an autonomous replication sequence from the *S. cerevisiae* genome and have an average copy number of 1 to 10 per cell, even though higher copy numbers (up to 100 copies per cell) have been reported (44). These vectors are unstable, and plasmid loss rates can be as high as 10% per generation (44). This instability greatly limits their use in large-scale fermentation processes. YCp-type vectors are derived from the incorporation of *S. cerevisiae* centromeres into YRp plasmids and have improved plasmid stability. However, YCp-type vectors have lower copy numbers, typically at 1 to 2 copies per cell. The most commonly used episomal vectors are of the YEp type derived from the naturally occurring plasmid called 2μ circle in *S. cerevisiae*. These vectors are present at an average of 40 copies per cell and exhibit higher stability than the YRp and YCp vectors (5, 16). Consequently, YEp vectors are the best-developed vectors for heterologous gene expression in *S. cerevisiae*. These vectors are especially useful for controlled expression of toxic proteins. More than 80 compact expression vectors have been developed based

TABLE 2 Vector systems used for heterologous protein expression in *S. cerevisiae*

Vector	Copy number per cell	Reference
Episomal		
YEp: 2μ-based	25–200	27
YCp: centromere	1–2	18
YRp: replicating	1–20	86
Regulated copy number	3–100	16
Integrating		
YIp	>1	45
rDNA-integrating	100–200	73
Tyδ	<20	111
Transplacement	1	109

on the pRS series of centromeric and 2μ plasmids (26). The GATEWAY vectors (Invitrogen Life Technologies, Carlsbad, CA) incorporated some of these features into their systems in order to provide fast cloning and transfer of genes to different vectors for heterologous expression in most strains of *S. cerevisiae*. Other 2μ-based expression vectors available from commercial sources include YEpFLAG-1 from Sigma (St. Louis, MO), YES vectors from Invitrogen, and pESC vectors from Stratagene (La Jolla, CA).

Integrative vectors contain selectable *S. cerevisiae* genes and lack sequences for autonomous replication. These vectors are highly stable but usually present at low copy numbers. They include the YIp-type vectors, and their integration can be directed by homologous recombination between sequences carried on the plasmid and its homologous counterpart in the *S. cerevisiae* genome. Some integration vector systems use repetitive elements such as delta sequences, Ty elements, or tRNA genes for heterologous gene integration (119). A system for multiple-site integration of a heterologous gene into the ribosomal DNA (rDNA) 9.1-kb segment has also been developed (73). Stability of the genome-integrated expression cassette can be maintained under selection. However, a decrease of copy numbers of the integrated expression cassette was observed under nonselective culture conditions (44).

11.2.2. Promoters

An array of *S. cerevisiae* promoters has been used for heterologous gene expression. There are constitutive promoters derived from genes such as *CYC1* (cytochrome C oxidase), *TEF2* (translation elongation factor), and *GAPDH* (glyceraldehyde-3-phosphate dehydrogenase), as well as regulated promoters such as *GAL1* (galactokinase/galactose epimerase 1), *ADH2* (alcohol dehydrogenase 2), *CUP1* (metallothionein), and *PHO5* (acid phosphatase) (6, 10, 15, 46, 51, 55, 56, 58, 65, 66, 80, 97, 98, 124). It is often advantageous for improving expression titer to couple the protein production phase with cell growth phases if there is no toxicity issue for the protein of interest. Some inducible promoter systems that work well at the shake flask scale, such as temperature-regulated promoters, do not always function well in large-scale fermentation due to operational restraints (44). Hybrid promoters such as *GAP/GAL* and *GAL10/CYC1* have also been reported (6). Several reviews on promoters and their applications in *S. cerevisiae* are available (108, 128, 138).

11.2.3. Transcriptional and Translational Regulation

In order to fully utilize the transcriptional, translational, and posttranslational regulatory machinery and to get efficient heterologous gene expression in *S. cerevisiae*, it is very important to match the expressed protein with the proper expression vector system. There have been many studies and extensive reviews on these aspects (22, 44, 82, 123). Generally, the use of strong promoters and robust transcriptional terminators of *S. cerevisiae* is essential for maximal expression since most prokaryotic or higher eukaryotic transcription terminators are not active in *S. cerevisiae*. More importantly, mRNA levels are controlled by the rate of transcription initiation and the transcript turnover rate (mRNA stability). In addition to upstream regulatory elements, downstream activation sequences are also required for maximum transcription initiation in *S. cerevisiae* (123). Expression levels of some foreign genes with high AT content (>60%) in *S. cerevisiae* can be low or absent due to incomplete transcript elongation (108). Transcript-destabilizing sequence elements have been reported in the 5′ nontranslated regions, coding regions, and 3′ nontranslated regions of different mRNAs (22). Therefore, if the titer is low due to the instability of mRNA, it may be necessary to use a stronger promoter, to delete the transcript-destabilizing sequence elements, and to introduce alternative codons. It has been shown that secondary structure of mRNA is one of the most important factors affecting the rate of initiation and translational elongation (123). For secreted proteins, posttranslational processes such as proteolytic cleavage, N-terminal modification, and glycosylation are critical to both proper protein folding and high protein yields. Consequently, it is worthwhile to evaluate the effect of the posttranslational steps when strong promoters and different strains fail to produce desired product yields.

S. cerevisiae has been used for expression of many eukaryotic proteins in both intracellular and secreted forms. For production of secreted proteins, secretion signals can come either from native protein signal sequences or from *S. cerevisiae* secretion protein signal sequences. Due to the limited studies available in the literature on the effects of different signal peptides on the yield of heterologous protein expression in *S. cerevisiae* (120), it is a good starting point to include the signal sequence of the foreign protein in the initial test of expression cassettes. Alternatively, *S. cerevisiae* signal sequences from invertase (SUC2, 19 amino acids), acid phosphatase (PHO5, 17 amino acids), and the most widely used α-factor pheromone (MFα1, 20 amino acids) can be incorporated into the recombinant construct(s).

An important step among posttranslational modification processes is glycosylation. Heterologous glycoproteins expressed in *S. cerevisiae* are glycosylated at both N-linked and O-linked sites (64). Little is known about O-linked glycosylation, but more information is available on the process of N glycosylation. In *S. cerevisiae*, a core sugar moiety consisting of two *N*-acetylglucosamines (GlcNAc), nine mannoses, and three glucoses is added to the N-amide of asparagine at the Asn-X-Ser/Thr sequence in the endoplasmic reticulum (ER) in the early process of N glycosylation (67). Three glucose residues and one mannose residue are subsequently removed in the early Golgi apparatus. It is in the Golgi apparatus where further modification takes place that results in major differences between *S. cerevisiae* and higher eukaryotes in oligosaccharide structures. In higher eukaryotes, additional mannose residues are removed and several other sugars such as galactose, sialic acid, and fucose are added (39, 40). *S. cerevisiae* maintains the core mannose scaffold and is often extended with a large number of additional mannose residues. This often results in a hyperglycosylated outer chain containing more than 50 mannose residues with many branch chains. Many examples of hyperglycosylation of heterologous proteins in *S. cerevisiae* have been reported (23, 68, 84). Since specific glycosylation patterns are important for the immune response in mammalian systems, wild-type *S. cerevisiae* is not a good system for the production of therapeutic proteins such as monoclonal antibodies that have complex sugar structures.

11.2.4. Industrial Protein Production in *S. cerevisiae*

Many heterologous proteins have been produced in *S. cerevisiae* and large-scale fermentation processes have been developed (43, 108). Nevertheless, most heterologous protein production in *S. cerevisiae* remains in shake flask cell cultures. The first therapeutic recombinant protein expressed in *S. cerevisiae* was human alpha interferon (47). Since then, many other therapeutic proteins have been expressed in *S. cerevisiae*. These include the hepatitis B surface antigen (126), human insulin (122), the human papillomavirus (HPV) vaccine Gardasil™ (93, 135), and many others. In the case of Gardasil, the four vaccine components were expressed in *S. cerevisiae*, which were transformed by pGAL110-based yeast expression vectors, each containing the gene of interest coding for HPV L1 types 6, 11, 16, or 18 (75, 104). An excellent review by Vasavada (128) details a comprehensive list of foreign proteins expressed in *S. cerevisiae* at small and large industrial scale. Various strategies have been reported to improve the expression levels of different heterologous proteins in *S. cerevisiae*, such as screening of various host strains with different expression vectors, altering host genetics by molecular engineering, and optimization of fermentation conditions (83, 115, 128). Since there are no defined conditions that can be used for all protein production, it is a general rule of thumb to examine different parameters to develop an optimized process for heterologous protein expression in *S. cerevisiae*.

11.3. *P. PASTORIS* FOR HETEROLOGOUS PROTEIN EXPRESSION

Pichia expression systems have been widely used for expression of heterologous proteins from a diverse array of organisms, including bacteria, fungi, protists, plants, invertebrates, nonhuman vertebrates, and humans (72). Similar to *S. cerevisiae*, protein expression systems in *Pichia* have the advantages of well-established, cost-effective fermentation technology and comprehensive understanding of the host molecular genetics. *P. pastoris* is a methylotrophic yeast and has strong preference for respiratory growth (72). Cell cultures can reach extremely high density, with dry weight mass reaching as high as 100 g/liter (44). Most of the *Pichia*-based protein expression systems use a tightly controlled and highly inducible promoter derived from the alcohol oxidase 1 gene (*AOX1*). Under methanol-induced conditions, a single protein such as alcohol oxidase can reach as high as 30% of the total cell proteins (44, 72). Similar methods used for the molecular genetic manipulation for *S. cerevisiae* can be applied to *P. pastoris*. Extensive studies have been published on the use of *Pichia* systems in heterologous protein expression. A recent review article by Cregg

et al. (14, 20) listed 241 individual studies of heterologous proteins expressed in *P. pastoris*, more than 80 of which were human proteins. Expression levels for heterologous proteins differ greatly, ranging from microgram-per-liter to gram-per-liter levels. For example, a hydroxynitrile lyase from a tropical rubber plant (*Hevea brasiliensis*) was expressed at levels as high as 22 g/liter (42). There are numerous reports on heterologous protein expression in *P. pastoris* (9, 91, 94, 107, 113, 114, 130, 138).

11.3.1. Expression Vectors and Host Strains

A fully developed *P. pastoris* expression system is commercially available (Invitrogen). The *Pichia* vectors can be grouped into two categories based on their promoters used: (i) inducible promoters derived from the *AOX1* gene or the formaldehyde dehydrogenase gene (*FLD1*) (117), and (ii) a strong constitutive promoter derived from the *P. pastoris* glyceraldehyde-3-phosphate dehydrogenase gene (*pGAP*). The methanol-inducible *AOX1* promoter provides tighter control of gene expression. The *FLD1* promoter can be induced by either methanol as the sole carbon source or certain methylated amines such as methylamine as the sole nitrogen source. Alternatively, the constitutive *GAP* promoter does not require cultures to be shifted from one carbon source to another for the induction of heterologous gene expression. However, if the foreign proteins to be expressed are toxic to the host cells, constitutive expression of the enzyme or proteins may cause inhibition of cell growth or result in loss of the expression vector from the cell.

There are no stable episomal expression vectors available in *P. pastoris*, and the expression genes need to be integrated into the genome to obtain stable expression strains. The simplest way of integration is single crossover homologous recombination by digesting the vector at a unique site with a restriction enzyme within either the marker gene (e.g., *HIS4*) or the promoter sequences (e.g., *AOX1*) (117). In the case of *HIS4* as the selectable marker, transformants of auxotrophic mutant strain (*HIS4*) can be selected for His+ cell growth. Alternatively, expression strains can be generated with *AOX1* flanking sequences at both the 5′ and 3′ terminals by cutting the vector in the *AOX1* sequence region. The flanking *AOX1* sequences at both ends can stimulate gene replacement events at the *AOX1* site in the *Pichia* genome, and this can result in a gene replacement integration. Heterologous proteins expressed in *P. pastoris* can be retained inside cells (intracellular expression) or secreted into medium depending on signal sequences in the expression vector systems. Commercial vectors for secreted protein expression often carry a leader sequence from MFα1 (mating factor) of *S. cerevisiae* as a secretion signal (117). Other leader sequences, such as the alkaline phosphatase (PHO) signal sequence from *P. pastoris* and some native protein leader sequences, have also been used for secretion of foreign proteins in *Pichia* (79). To facilitate protein detection and purification, most vectors have 6× His or *c-myc* tag sequences fused at the C-terminal or N-terminal end.

A variety of selection markers are available for *Pichia* protein expression. Some can be used for selecting multicopy gene integration, such as the drug resistance markers zeocin and blasticidin, while others provide auxotrophic selection such as his4− mutant strains. Commercially available *Pichia* host strains for protein expression include X-33, GS115, SMD1168, and KM71. Additional host strains for

P. pastoris heterologous protein expression are reviewed by Cregg et al. (20).

11.3.2. Protein Folding and Glycosylation in *P. pastoris*

P. pastoris, similar to *S. cerevisiae* and other fungi species, can perform protein posttranslational modifications, such as cleavage of signal sequences for secreted proteins, N- and O-linked glycosylation, and formation of disulfide bonds (71, 72). It has been well documented that overexpression of heterologous proteins in *P. pastoris* can have a deleterious effect on its secretion pathway. Protein folding in the ER and Golgi is often the rate-limiting step for heterologous protein expression. Overloading the secretion pathway can lead to accumulation of partial or misfolded heterologous proteins in cells. Existence of high levels of unfolded or misfolded proteins in cells may induce a stress reaction termed the "unfolded protein response" (UPR), which triggers the expression of folding chaperones and enzymes in addition to other physiological responses. Both heat shock protein 70-type chaperone (BiP) and a protein disulfide isomerase (Pdi1) were reported to be induced when cells face UPR in *Pichia* (44, 72). BiP and Pdi1 showed synergistic effects on improving secretory capacity (136).

Glycosylation is one of the most important posttranslational modifications for heterologous protein expression, since it not only affects proper folding and secretion of the protein but also has a direct impact on the biological activities of proteins. Both N- and O-linked glycosylation have been reported using *Pichia* systems for heterologous protein expression (20). The consensus sequence for N-linked glycosylation in *Pichia* is the same as that in *S. cerevisiae* (Asn-X-Thr/Ser), in which Ser or Thr provides a hydroxyl group for glycosylation. Similar to *S. cerevisiae* and other fungal organisms, *Pichia* transfers a lipid-linked oligosaccharide unit, $Glc_3Man_9GlcNAc_2$, to an amino acid (Asn) in the ER. This oligosaccharide core is further modified in the Golgi. Unlike hyperglycosylation (>50 Man) occurring in *S. cerevisiae*, the most commonly found N glycosylation structure in *P. pastoris* is $Man_{14}GlcNAc_2$. There has been no α-1,3-mannosyltransferase activity detected in *P. pastoris*; only α-1,6 and α-1,2 linkages were found in glycosylated proteins in *Pichia* (11). The O-linked oligosaccharides in *Pichia* are also reported, but a consensus sequence for O-linked glycosylation is not available. The O-linked oligosaccharides in *P. pastoris* are generally short (<5 residues) and contain only α-1,2-linked mannose units. Phosphorylation of oligosaccharide units has been found in *P. pastoris* expression systems, but the exact sites of phosphate linkage on saccharides are unknown (72). In order to produce antibodies for therapeutic use, extensive glycoengineering efforts have been carried out to make humanlike glycoforms in *Pichia* (39, 40). Fully humanized sialylated antibody was produced successfully in a humanized *Pichia* strain (39). It is noteworthy that unlike antibodies produced in mammalian cells, antibodies produced in humanized *Pichia* do not contain fucose in the core carbohydrate structures and the afucosylated antibodies showed increased binding affinity to Fcγ receptor IIIA, which in turn results in higher antibody-dependent cellular cytotoxicity (70).

11.3.3. Heterologous Protein Production in *Pichia*

Expression levels of heterologous proteins in *P. pastoris* vary greatly depending on copy numbers of the foreign gene integrated and the stability of both mRNA and protein.

Since expression levels are correlated with gene copy number, it is important to achieve high copy numbers of gene integration. To obtain multiple copies of the target gene integrated in *Pichia*, one can screen large numbers of transformants in a 96-well plate format, applying high concentrations of the selection agents (e.g., zeocin and blasticidin), or construct multiple copies of the gene in tandem repeats in vitro.

In addition to the high-copy-number transformants, optimization of the expression process is vital to achieve high-level expression of heterologous proteins in *Pichia*. Although some proteins can be expressed well in flask cultures, it is common that expression levels are greatly enhanced in a controlled environment. Cultures in fermentor vessels offer several advantages over flask cultures, such as control of pH and dissolved oxygen levels, and continuous feeding of carbon and other components during cell growth and protein expression (49, 72). Since *P. pastoris* cells have a preference for respiratory growth, cultures can be grown to extremely high cell densities (up to 500 absorbance units at 600 nm) (20, 21). This is particularly important for secreted proteins since the concentration of the expressed proteins is often proportional to the concentration of cells in culture. Common defined media for *P. pastoris* protein expression are composed of glucose and glycerol as the carbon source and Casamino Acids and/or ammonium salts as the nitrogen source in addition to biotin and trace elements. The *Pichia* fermentation for heterologous protein expression can be separated into two phases. The first phase is that of growth using glycerol as a carbon source, and at this stage the *AOX1* promoter is fully repressed. After depletion of glycerol, a transitional phase may be added to feed more glycerol at a growth-limiting rate. The second phase of feeding methanol starts induction of the *AOX1* promoter and heterologous gene expression. Detailed protocols for fermentation of *P. pastoris* can be found in various articles (30, 50, 72, 90, 96).

Similar to other heterologous protein expression systems, proteolytic hydrolysis often causes instability of foreign proteins in *Pichia*. In addition to use of proteinase-deficient strains, several fermentation strategies can be applied to control proteolysis of foreign proteins. Some common approaches that have been used successfully to reduce hydrolysis of heterologous proteins in *Pichia* include (i) adding amino acid-rich supplements such as peptone or Casamino Acids; (ii) culturing at more acidic pH values (*Pichia* cultures grow well between pH 3 and 7); (iii) using a protease (*PEP4*, vacuolar protease gene)-deficient strain; and (iv) controlling culture temperature (44, 72). Another limitation of heterologous protein expression in *Pichia* is inefficient secretion of some mammalian proteins. In these cases, alternative protein secretion sequences can be tested first, followed by the consideration of more extensive molecular engineering approaches to improve protein chaperone and folding machinery.

11.4. HETEROLOGOUS PROTEIN EXPRESSION IN FILAMENTOUS FUNGI

Filamentous fungi, most notably *Aspergillus* species, can produce high levels of proteins and are capable of secreting proteins into the media. Many industrial enzymes, such as amylase, invertase, pectinase, and cellulases, that are used widely in the food industry for production of high-fructose corn syrup, fruit juice, and wine are produced in *Aspergillus* species. The most well-known species for producing commercial enzymes include *A. oryzae* and *A. niger*; both have been given GRAS status by the U.S. FDA (116) for production of enzymes and proteins (35). The GRAS status makes them more attractive for use in the production of enzymes and proteins for the food industry and for pharmaceutical products such as therapeutic proteins and antibodies (29, 131). However, in comparison with *S. cerevisiae* and *P. pastoris*, filamentous fungi are less studied for heterologous protein production with respect to their transcription and translation control, mRNA stability, glycosylation, and regulation of protein secretion. With growing demand for the production of therapeutic proteins including antibodies, there is renewed interest in developing cost-effective fungal expression systems including yeast and *Aspergillus* species for heterologous protein production. In recent years, many advances have been made in the use of filamentous fungi for protein production (74, 87, 129).

11.4.1. Host Strains and Selection Markers for Transformation

Unlike *Pichia* and *Saccharomyces*, there are no commercially available protein expression systems for *Aspergillus*. However, there are several Web-accessible organizations such as the ATCC (American Type Culture Collection), the FGSC (Fungal Genetics Stock Center), the NRRL (Agricultural Research Service Culture Collection), and the BCCP (Belgium Culture Collection) from which the necessary tools for heterologous protein expression in filamentous fungi can be obtained.

For heterologous protein expression in *Aspergillus* species, the first step is to select a suitable host strain. The expression host must be either auxotrophic or antibiotic susceptible so that it can be used for efficient transformation and selection of transformants. The most genetically accessible strain is *A. nidulans*, but *A. niger*, *A. awamori*, and *A. oryzae* are more extensively used because of their capacity for high-level expression of commercial enzymes. *A. niger* can efficiently secrete many enzymes including food-related enzymes. *A. oryzae* has become more amenable to molecular analysis, and current research is focused on the identification of strong promoters to improve expression of industrially valuable enzymes. Due to the limited number of studies on *Aspergillus* strains and patent restrictions on the use of some strains and markers, more studies are warranted to fully utilize the potential of *Aspergillus* host strains for heterologous gene expression.

Transformation of host strains can result in chromosomal integration of plasmids, either through a random mechanism or through site-directed integration. Markers commonly used for transformation of auxotrophic strains are listed in Table 3 (7, 13, 59, 81, 100–102, 125, 137, 138). The *pyrG* gene (uracil/uridine auxotrophs) from *A. niger* is a common marker found in many plasmid vectors. The *A. nidulans argB* gene is also used extensively for efficient transformation of suitable auxotrophic host strains. For example, Buxton et al. (13) described the complementation of an auxotrophic *A. niger* strain with the *A. nidulans argB* gene. More recently, there was a report on the complete disruption of the *argB* gene on chromosome I in *A. niger* (69). The *A. niger argB* deletion strain can be complemented by the *argB* genes from both *A. niger* and *A. nidulans* and can serve as a useful host strain for heterologous protein expression in *A. niger*.

Although plasmids can be readily integrated into the chromosome (stable and multicopy integration) during transformation of *Aspergillus* species, it is desirable to have a replicative plasmid for heterologous protein expression in filamentous fungi. Sequences have been isolated that allow one to perform replicative transformation in *A. nidulans* (2). Initial work with linear plasmids containing human telomeric elements was encouraging, but later these were shown to be unstable (3). To date, a synthetic *Aspergillus* replicative plasmid of wide utility is not yet available (74).

11.4.2. Expression Vectors and Promoters

Promoters commonly used for heterologous protein expression in *Aspergillus* species are listed in Table 4 (17, 25, 35, 38, 61, 99, 103). Constitutive promoters include the *trpC* and the *gpdA* promoters from *A. nidulans*. Other frequently used promoters are the starch-inducible promoters, α-amylase and the glucoamylase promoters from *A. oryzae* and *A. niger*. The glucoamylase promoter responds to maltose as a carbon source and is repressed by glucose. More recently the *TEF1-α* promoter of *A. oryzae* has been studied for use as a strong constitutive promoter (61).

TABLE 3 Selectable markers used for heterologous gene expression in *Aspergillus* species

Gene markers	Enzyme	Cellular function	Gene host	Reference
Auxotrophic				
argB	Ornithine carbamoyltransferase	Arginine biosynthesis	*A. nidulans*	13
pyr4	Orotidine 5′-phosphate decarboxylase	Uracil biosynthesis	*Neurospora crassa*	7
pyrG	Orotidine 5′-phosphate decarboxylase	Uracil biosynthesis	*A. niger*	81
trpC	Phosphoribosylanthranilate isomerase	Tryptophan biosynthesis	*A. nidulans*	137
amdS	Acetamidase	Acetamide or acrylamide assimilation, N$_2$ source	*A. nidulans*	59
niaD	Nitrate reductase	Nitrate assimilation, N$_2$ source	*A. niger*	125
Dominant				
hph (HmB^r)	Hygromycin B phosphotransferase		*E. coli*	100
phleo^r	*E. coli* transposon Tn5 phleo^r DNA			102

TABLE 4 Promoters used in heterologous gene expression in *Aspergillus* species

Gene	Enzyme	Species	Reference(s)
Strong inducible			
alcA	Alcohol dehydrogenase 1 (ADH1) (ethanol inducible)	*A. nidulans*	25
α-Amylase	Starch inducible	*A. oryzae*	17
glaA	Glucoamylase (starch and maltose inducible)	*A. nidulans*	36
Constitutive			
gpdA	Glyceraldehyde-3-phosphate dehydrogenase	*A. nidulans*	99, 103
trpC	Phosphoribosylanthranilate isomerase	*A. nidulans*	38
TEF1-α	Translation elongation factor 1-α	*A. oryzae*	61

Another suitable promoter for the expression of heterologous proteins in filamentous fungi is the ethanol-regulated promoter *alcA* for alcohol dehydrogenase from *A. nidulans*. Genes for ethanol utilization in *A. nidulans* are efficiently transcribed. The *alcA* promoter is tightly regulated through *alcR*, a *trans*-activating gene, and is subject to repression by the major carbon catabolite repressor CreA (25). In a recent report, expression of antisense *creA* in *A. nidulans* increased levels of expression for glucose-repressible enzymes in the presence of glucose (8). The derepression effect by the *creA* antisense DNA did not alter cell morphology and growth parameters, as observed in a *creA* null mutant. The *alcR-alcA* regulatory system from *A. nidulans* has been transformed into *A. niger*, which does not possess an inducible *alc* system (88). The *alcR-alcA*-transformed strain of *A. niger* showed similar levels of heterologous expression of the reporter enzyme β-glucuronidase, as in *A. nidulans*.

Ishida et al. (52) reported that the tyrosinase gene promoter (*melO*) was four times stronger for the heterologous expression of β-glucuronidase from *E. coli* in a submerged culture as compared with other gene promoters, such as *amyB*, *glaA*, and *agdA*, previously used for heterologous gene expression in *A. oryzae*. The *melO* promoter was used to produce the *glaB*-type glucoamylase in a submerged culture of *A. oryzae* and achieved maximum yield of 3.3 g/liter and 99% purity. This work establishes the *melO* promoter for high-level expression and high-purity production of heterologous proteins in *A. oryzae*.

11.4.3. Posttranslational Regulations

Large numbers of mammalian enzymes and proteins have been expressed through genetic engineering; however, secretion of heterologous proteins is often much less efficient than the secretion of native fungal proteins. During the last 10 years, the production levels of heterologous proteins in *Aspergillus* have been improved by fusing the protein of interest to the 3′ end of a highly expressed gene encoding a protein that is efficiently secreted from the host. The gene fusion or carrier protein strategy has been used with the glucoamylase gene from *A. niger* or *A. awamori*. Increases of 5- to 1,000-fold in protein production levels have been observed (87). Gouka et al. (36) extensively reviewed the production levels of different proteins fused to the glucoamylase carrier gene. They also performed detailed experiments on the production levels of various proteins with carrier fusions to the glucoamylase protein in isogenic, single-copy strains of *A. awamori*. The glucoamylase gene

can be divided into three distinct domains: the catalytic N-terminal domain, the C-terminal starch-binding domain, and a flexible linker region that separates the two domains. The C-terminal starch-binding domain can be replaced by a heterologous protein. The glucoamylase carrier proteins are believed to improve translocation in the ER, to enhance proper protein folding, and possibly to afford protection against proteolysis. In the secretory pathway, the carrier protein can be cleaved by a KEX2-like protease at a specific proteolytic cleavage site separating the glucoamylase carrier and the protein of interest. The amino acid sequence (NVISKR) of the cleavage site is derived from the propeptide of glucoamylase. The carrier approach appears to be the first choice in attempts to produce secreted proteins in *Aspergillus* species. There are many reports on the use of carriers for heterologous protein secretion. A recent study reported a successful secretion of laccase (*lac1*) (an industrially useful enzyme for the oxidation of *p*-diphenols) from *Pycnoporus cinnabarinus* that used either the native laccase signal sequence or the *A. niger* glucoamylase pre-pro-sequence as a carrier for the heterologous enzyme expression in *A. niger* (106). The fusion of the glucoamylase pre-pro-sequence increased expression levels 80-fold in comparison to the levels observed with the laccase signal peptide sequence. Another advantage of using the glucoamylase pre-pro-sequence as a carrier for laccase in *A. niger* is the production of the 70-kDa protein, which is similar to the molecular mass of the native protein (106). However, it was reported that laccase expressed in *P. pastoris* had an estimated 110-kDa molecular mass due to hyperglycosylation in the yeast (76, 77, 106). Human interleukin-6 (IL-6) secretion was also improved using the native glucoamylase as a carrier when expressed in *A. niger* (74). It has been reported (4) that the UPR, i.e., the deliberate upregulation of "foldases" and chaperones, can also improve the secretion of heterologous proteins.

Efficient heterologous protein production in fungi cannot always be achieved by alleviating the limitations of transcription. Posttranslational regulations such as glycosylation, protein folding, and secretion often pose challenges for successful expression of heterologous proteins in *Aspergillus* species. Another major limiting factor in protein production in filamentous fungi is proteolytic degradation by high levels of intracellular and extracellular proteases. *A. niger* has four predominant extracellular proteases, two aspartyl proteases (PEPA and PEPB) and two serine carboxypeptidases (PEPF and PEPG). In *A. nidulans*, neutral and alkaline proteases often are the dominant ones affecting the yields of heterologous proteins. Efforts to eliminate specific protease activities through mutagenesis or gene disruption have improved protein production levels. Van den Hombergh et al. (127) reviewed fungal proteases and described the development of a set of tester strains of *A. niger* that may be used to identify the appropriate genetic background for a particular protein.

Posttranslational modifications such as N glycosylation are particularly important for many pharmaceutically relevant proteins because of their impact on biological activities (19). High mannose-type glycosylation is observed in filamentous fungi, but it is much less extensive than that seen in proteins produced in *Saccharomyces* and *Pichia* species. Maras et al. (78) have nicely reviewed the fungal glycosylation pathway and approaches and compared the carbohydrate structures with proteins produced in mammalian cells.

Successful expression of *Drosophila melanogaster* xanthine dehydrogenase in *A. nidulans* is a recent example of proper posttranslational modification of a heterologous enzyme in filamentous fungi (1). In comparison to the baculovirus expression system, which produced only about 10% of active xanthine dehydrogenase enzyme due to incomplete incorporation of the cofactor molybdenum and two iron-sulfur clusters, *A. nidulans* produced 40% of active enzyme with relatively more complete cofactor incorporation.

11.4.4. Scale-Up of Fermentation and Protein Production in *Aspergillus*

In addition to the molecular genetic approaches, a combination of host strain selection, strain development, and production process development including media optimization have resulted in commercially attractive protein yields in *Aspergillus* protein expression systems (101). For example, IL-6 gene carrier fusion in *A. nidulans* in combination with host strain development and growth optimization resulted in a 200-fold increase in production of active IL-6 (101). A combination of gene dosage and growth condition optimization for the improvement of protein expression in *A. oryzae* resulted in a high production titer of phospholipase A_1 (PLA1) at the industrial scale (118). Transformants with 15 to 20 copies of the *PLA1* gene were integrated in the chromosome of *A. oryzae* and produced 60- to 7,500-fold higher activity than the native expression level. In addition, improvement of the mycelial morphology from a pellet form to a filamentous form improved production of PLA1 in scaled-up fermentations and resulted in production of PLA1 at 1.8 g/liter consistently. This is consistent with the observation that improvements in oxygen transfer and broth rheology obtained with the pellet form of growth often resulted in increased protein production.

It has been recognized that fungal morphology plays an important role in determining process productivity in mycelial fermentations. This has led to investigations into alternative ways to conduct fermentations to optimize production yields. Papagianni et al. investigated the effect of cell immobilization on protein production and secretion (92). They reported on cells of *A. niger* that were immobilized by attachment to metal surfaces or by spore entrapment, and subsequently grown on Celite beads. They noted that cell attachment on metal surfaces decreased secretion of proteases and increased glucoamylase production. It was observed that growth on Celite beads in shake flasks reduced the secreted proteases by half and almost doubled the specific activity of glucoamylase. In bioreactor cultures, conditions can be better controlled than in flask culture and effects of culture conditions on protein yields are much greater.

11.5. FUTURE PROSPECTS

The demand for production of a variety of proteins and enzymes in heterologous hosts for research and industrial applications is growing continuously. It is of great interest to develop and optimize diverse host systems for heterologous protein expression to meet the growing demand. Fungal expression systems offer many advantages over other expression systems and have been widely used for production of heterologous proteins from many organisms. Discovery of new promoters and development of new vectors will provide more choices for different protein expression needs. Greater understanding of transcriptional control, mRNA stability, glycosylation, protein folding, and other posttranslational processes will allow us to overcome the difficulties associated with heterologous protein expression in fungi, such as hyperglycosylation in yeasts and poor secretion associated with some mammalian proteins. It has been documented that the use of protease-deficient strains can increase the yield of recombinant protein production (28, 95) and that the use of specific mutants may eliminate hyperglycosylation problems with some proteins expressed in *Saccharomyces* (115). The exponential increase in fungal genomics and proteomics in recent years will no doubt fuel the advancement of filamentous fungi as heterologous protein expression hosts (74). Collectively, fungal protein expression systems have been and will continue to serve as a choice for heterologous protein production in small or large quantities.

We want to express our gratitude to Prakash Masurekar and Arnold Demain for spending time editing and reviewing this article.

REFERENCES

1. **Adams, B., D. J. Lowe, A. T. Smith, C. Scazzocchio, S. Demais, and R. C. Bray.** 2002. Expression of *Drosophila melanogaster* xanthine dehydrogenase in *Aspergillus nidulans* and some properties of the recombinant enzyme. *Biochem. J.* **362:**223–229.
2. **Aleksenko, A., and A. J. Clutterbuck.** 1997. Autonomous plasmid replication in *Aspergillus nidulans*: AMA1 and MATE elements. *Fungal Genet. Biol.* **21:**373–387.
3. **Aleksenko, A., and L. Ivanova.** 1998. In vivo linearization and autonomous replication of plasmids containing human telomeric DNA in *Aspergillus nidulans. Mol. Gen. Genet.* **260:**159–164.
4. **Archer, D. B., D. J. Jeenes, and D. A. Mackenzie.** 1994. Strategies for improving heterologous protein production from filamentous fungi. *Antonie van Leeuwenhoek* **65:**245–250.
5. **Armstrong, K. A., T. Som, F. C. Volkert, A. Rose, and J. R. Broach.** 1989. Propagation and expression of genes in yeast using 2-micron circle vectors. *Biotechnology* **13:**165–192.
6. **Bae, J. Y., J. Laplaza, and T. W. Jeffries.** 2008. Effects of gene orientation and use of multiple promoters on the expression of *XYL1* and *XYL2* in *Saccharomyces cerevisiae. Appl. Biochem. Biotechnol.* **145:**69–78.
7. **Ballance, D. J., F. P. Buxton, and G. Turner.** 1983. Transformation of *Aspergillus nidulans* by the orotidine-5′-phosphate decarboxylase gene of *Neurospora crassa. Biochem. Biophys. Res. Commun.* **112:**284–289.
8. **Bautista, L. F., A. Aleksenko, M. Hentzer, A. Santerre-Henriksen, and J. Nielsen.** 2000. Antisense silencing of

the *creA* gene in *Aspergillus nidulans*. *Appl. Environ. Microbiol.* **66:**4579–4581.

9. **Berrin, J. G., W. R. McLauchlan, P. Needs, G. Williamson, A. Puigserver, P. A. Kroon, and N. Juge.** 2002. Functional expression of human liver cytosolic β-glucosidase in *Pichia pastoris*. Insights into its role in the metabolism of dietary glucosides. *Eur. J. Biochem.* **269:**249–258.

10. **Brake, A. J., J. P. Merryweather, D. G. Coit, U. A. Heberlein, F. R. Masiarz, G. T. Mullenbach, M. S. Urdea, P. Valenzuela, and P. J. Barr.** 1984. β-Factor-directed synthesis and secretion of mature foreign proteins in *Saccharomyces cerevisiae*. *Proc. Natl. Acad. Sci. USA* **81:**4642–4646.

11. **Bretthauer, R. K., and F. J. Castellino.** 1999. Glycosylation of *Pichia pastoris*-derived proteins. *Biotechnol. Appl. Biochem.* **30:**193–200.

12. **Buckholz, R. G., and M. A. Gleeson.** 1991. Yeast systems for the commercial production of heterologous proteins. *Biotechnology* (New York) **9:**1067–1072.

13. **Buxton, F. P., D. I. Gwynne, and R. W. Davies.** 1985. Transformation of *Aspergillus niger* using the *argB* gene of *Aspergillus nidulans*. *Gene* **37:**207–214.

14. **Cereghino, J. L., and J. M. Cregg.** 2000. Heterologous protein expression in the methylotrophic yeast *Pichia pastoris*. *FEMS Microbiol. Rev.* **24:**45–66.

15. **Chen, Y., D. Pioli, and P. W. Piper.** 1994. Overexpression of the gene for polyubiquitin in yeast confers increased secretion of a human leucocyte protease inhibitor. *Biotechnology* (New York) **12:**819–823.

16. **Chlebowicz-Sledziewska, E., and A. Z. Sledziewski.** 1985. Construction of multicopy yeast plasmids with regulated centromere function. *Gene* **39:**25–31.

17. **Christensen, T., H. Woeldike, E. Boel, S. B. Mortensen, K. Hjortshoej, L. Thim, and M. T. Hansen.** 1988. High-level expression of recombinant genes in *Aspergillus oryzae*. *Bio/Technology* **6:**1419–1422.

18. **Clarke, L., and J. Carbon.** 1980. Isolation of a yeast centromere and construction of functional small circular chromosomes. *Nature* **287:**504–509.

19. **Conesa, A., G. Weelink, C. A. van den Hondel, and P. J. Punt.** 2001. C-terminal propeptide of the *Caldariomyces fumago* chloroperoxidase: an intramolecular chaperone? *FEBS Lett.* **503:**117–120.

20. **Cregg, J. M., J. L. Cereghino, J. Shi, and D. R. Higgins.** 2000. Recombinant protein expression in *Pichia pastoris*. *Mol. Biotechnol.* **16:**23–52.

21. **Cregg, J. M., T. S. Vedvick, and W. C. Raschke.** 1993. Recent advances in the expression of foreign genes in *Pichia pastoris*. *Biotechnology* (New York) **11:**905–910.

22. **Day, D. A., and M. F. Tuite.** 1998. Post-transcriptional gene regulatory mechanisms in eukaryotes: an overview. *J. Endocrinol.* **157:**361–371.

23. **De Nobel, J. G., and J. A. Barnett.** 1991. Passage of molecules through yeast cell walls: a brief essay-review. *Yeast* **7:**313–323.

24. **Devchand, M., and D. I. Gwynne.** 1991. Expression of heterologous proteins in *Aspergillus*. *J. Biotechnol.* **17:**3–9.

25. **Felenbok, B., M. Flipphi, and I. Nikolaev.** 2001. Ethanol catabolism in *Aspergillus nidulans*: a model system for studying gene regulation. *Prog. Nucleic Acid Res. Mol. Biol.* **69:**149–204.

26. **Funk, M., R. Niedenthal, D. Mumberg, K. Brinkmann, V. Ronicke, and T. Henkel.** 2002. Vector systems for heterologous expression of proteins in *Saccharomyces cerevisiae*. *Methods Enzymol.* **350:**248–257.

27. **Futcher, A. B., and B. S. Cox.** 1984. Copy number and the stability of 2-micron circle-based artificial plasmids of *Saccharomyces cerevisiae*. *J. Bacteriol.* **157:**283–290.

28. **Gardell, S. J., T. R. Hare, J. H. Han, H. Z. Markus, B. J. Keech, C. E. Carty, R. W. Ellis, and L. D. Schultz.** 1990. Purification and characterization of human plasminogen activator inhibitor type I expressed in *Saccharomyces cerevisiae*. *Arch. Biochem. Biophys.* **278:**467–474.

29. **Gasser, B., and D. Mattanovich.** 2007. Antibody production with yeasts and filamentous fungi: on the road to large scale? *Biotechnol. Lett.* **29:**201–212.

30. **Gellissen, G.** 2000. Heterologous protein production in methylotrophic yeasts. *Appl. Microbiol. Biotechnol.* **54:**741–750.

31. **Gietz, R. D., and R. A. Woods.** 2002. Transformation of yeast by lithium acetate/single-stranded carrier DNA/polyethylene glycol method. *Methods Enzymol.* **350:**87–96.

32. **Giga-Hama, Y., and H. Kumagai.** 1999. Expression system for foreign genes using the fission yeast *Schizosaccharomyces pombe*. *Biotechnol. Appl. Biochem.* **30**(Pt. 3)**:**235–244.

33. **Goffeau, A., B. G. Barrell, H. Bussey, R. W. Davis, B. Dujon, H. Feldmann, F. Galibert, J. D. Hoheisel, C. Jacq, M. Johnston, E. J. Louis, H. W. Mewes, Y. Murakami, P. Philippsen, H. Tettelin, and S. G. Oliver.** 1996. Life with 6000 genes. *Science* **274:**546, 563–567.

34. **Gouka, R. J., P. J. Punt, J. G. Hessing, and C. A. van den Hondel.** 1996. Analysis of heterologous protein production in defined recombinant *Aspergillus awamori* strains. *Appl. Environ. Microbiol.* **62:**1951–1957.

35. **Gouka, R. J., P. J. Punt, and C. A. van den Hondel.** 1997. Efficient production of secreted proteins by *Aspergillus*: progress, limitations and prospects. *Appl. Microbiol. Biotechnol.* **47:**1–11.

36. **Gouka, R. J., P. J. Punt, and C. A. van den Hondel.** 1997. Glucoamylase gene fusions alleviate limitations for protein production in *Aspergillus awamori* at the transcriptional and (post) translational levels. *Appl. Environ. Microbiol.* **63:**488–497.

37. **Hackel, B. J., D. G. Huang, J. C. Buboz, X. X. Wang, and E. V. Shusta.** 2006. Production of soluble and active transferrin receptor-targeting single-chain antibody using *Saccharomyces cerevisiae*. *Pharm. Res.* **23:**790–797.

38. **Hamer, J. E., and W. E. Timberlake.** 1987. Functional organization of the *Aspergillus nidulans trpC* promoter. *Mol. Cell. Biol.* **7:**2352–2359.

39. **Hamilton, S. R., R. C. Davidson, N. Sethuraman, J. H. Nett, Y. W. Jiang, S. Rios, P. Bobrowicz, T. A. Stadheim, H. J. Li, B. K. Choi, D. Hopkins, H. Wischnewski, J. Roser, T. Mitchell, R. R. Strawbridge, J. Hoopes, S. Wildt, and T. U. Gerngross.** 2006. Humanization of yeast to produce complex terminally sialylated glycoproteins. *Science* **313:**1441–1443.

40. **Hamilton, S. R., and T. U. Gerngross.** 2007. Glycosylation engineering in yeast: the advent of fully humanized yeast. *Curr. Opin. Biotechnol.* **18:**387–392.

41. **Harris, T. J., and J. S. Emtage.** 1986. Expression of heterologous genes in *E. coli*. *Microbiol. Sci.* **3:**28–31.

42. **Hasslacher, M., M. Schall, M. Hayn, R. Bona, K. Rumbold, J. Luckl, H. Griengl, S. D. Kohlwein, and H. Schwab.** 1997. High-level intracellular expression of hydroxynitrile lyase from the tropical rubber tree *Hevea brasiliensis* in microbial hosts. *Protein Expr. Purif.* **11:**61–71.

43. **Heinisch, J. J.** 1993. Expression of heterologous phosphofructokinase genes in yeast. *FEBS Lett.* **328:**35–40.

44. **Hensing, M. C., R. J. Rouwenhorst, J. J. Heijnen, J. P. van Dijken, and J. T. Pronk.** 1995. Physiological and technological aspects of large-scale heterologous-protein production with yeasts. *Antonie van Leeuwenhoek* **67:**261–279.

45. **Hinnen, A., J. B. Hicks, and G. R. Fink.** 1978. Transformation of yeast. *Proc. Natl. Acad. Sci. USA* **75:**1929–1933.

46. **Hitzeman, R. A., C. Y. Chen, F. E. Hagie, E. J. Patzer, C. C. Liu, D. A. Estell, J. V. Miller, A. Yaffe, D. G. Kleid, A. D. Levinson, and H. Oppermann.** 1983. Expression of hepatitis B virus surface antigen in yeast. *Nucleic Acids Res.* **11:**2745–2763.

47. Hitzeman, R. A., F. E. Hagie, H. L. Levine, D. V. Goeddel, G. Ammerer, and B. D. Hall. 1981. Expression of a human gene for interferon in yeast. *Nature* **293:**717–722.

48. Hockney, R. C. 1994. Recent developments in heterologous protein production in *Escherichia coli. Trends Biotechnol.* **12:**456–463.

49. Homann, M. J., W.-C. Suen, N. Zhang, and A. Zaks. 2006. Comparative analysis of chemical and biocatalytic syntheses of drug intermediates, p. 647–661. *In* R. Patel (ed.), *Biocatalysis in the Pharmaceutical and Biotechnology Industries.* Marcel Dekker, New York, NY.

50. Hong, F., N. Q. Meinander, and L. J. Jonsson. 2002. Fermentation strategies for improved heterologous expression of laccase in *Pichia pastoris. Biotechnol. Bioeng.* **79:**438–449.

51. Hottiger, T., J. Kuhla, G. Pohlig, P. Furst, A. Spielmann, M. Garn, S. Haemmerli, and J. Heim. 1995. 2-Micron vectors containing the *Saccharomyces cerevisiae* metallothionein gene as a selectable marker: excellent stability in complex media, and high-level expression of a recombinant protein from a CUP1-promoter-controlled expression cassette in cis. *Yeast* **11:**1–14.

52. Ishida, H., K. Matsumura, Y. Hata, A. Kawato, K. Suginami, S. Abe, S. Imayasu, and E. Ishishima. 2001. Establishment of a hyper-protein production system in submerged *Aspergillus oryzae* culture under tyrosinase-encoding gene (*melO*) promoter control. *Appl. Microbiol. Biotechnol.* **57:**131–137.

53. Janowicz, Z. A., K. Melber, A. Merckelbach, E. Jacobs, N. Harford, M. Comberbach, and C. P. Hollenberg. 1991. Simultaneous expression of the S and L surface antigens of hepatitis B, and formation of mixed particles in the methylotrophic yeast, *Hansenula polymorpha. Yeast* **7:**431–443.

54. Jeenes, D. J., B. Marczinke, D. A. MacKenzie, and D. B. Archer. 1993. A truncated glucoamylase gene fusion for heterologous protein secretion from *Aspergillus niger. FEMS Microbiol. Lett.* **107:**267–271.

55. Johnston, S. A., J. M. Salmeron, Jr., and S. S. Dincher. 1987. Interaction of positive and negative regulatory proteins in the galactose regulon of yeast. *Cell* **50:**143–146.

56. Karin, M., R. Najarian, A. Haslinger, P. Valenzuela, J. Welch, and S. Fogel. 1984. Primary structure and transcription of an amplified genetic locus: the CUP1 locus of yeast. *Proc. Natl. Acad. Sci. USA* **81:**337–341.

57. Kariya, Y., K. Ishida, Y. Tsubota, Y. Nakashima, T. Hirosaki, T. Ogawa, and K. Miyazaki. 2002. Efficient expression system of human recombinant laminin-5. *J. Biochem.* **132:**607–612.

58. Kaslow, D. C., and J. Shiloach. 1994. Production, purification and immunogenicity of a malaria transmission-blocking vaccine candidate: TBV25H expressed in yeast and purified using nickel-NTA agarose. *Biotechnology* (New York) **12:**494–499.

59. Kelly, J. M., and M. J. Hynes. 1985. Transformation of *Aspergillus niger* by the amdS gene of *Aspergillus nidulans. EMBO J.* **4:**475–479.

60. Kingsman, S. M., A. J. Kingsman, M. J. Dobson, J. Mellor, and N. A. Roberts. 1985. Heterologous gene expression in *Saccharomyces cerevisiae. Biotechnol. Genet. Eng. Rev.* **3:**377–416.

61. Kitamoto, N., J. Matsui, Y. Kawai, A. Kato, S. Yoshino, K. Ohmiya, and N. Tsukagoshi. 1998. Utilization of the TEF1-alpha gene (*TEF1*) promoter for expression of polygalacturonase genes, *pgaA* and *pgaB*, in *Aspergillus oryzae. Appl. Microbiol. Biotechnol.* **50:**85–92.

62. Kobayashi, K., S. Kuwae, T. Ohya, T. Ohda, M. Ohyama, H. Ohi, K. Tomomitsu, and T. Ohmura. 2000. High-level expression of recombinant human serum albumin from the methylotrophic yeast *Pichia pastoris* with minimal protease production and activation. *J. Biosci. Bioeng.* **89:**55–61.

63. Kruszewska, J. S. 1999. Heterologous expression of genes in filamentous fungi. *Acta Biochim. Pol.* **46:**181–195.

64. Kukuruzinska, M. A., M. L. Bergh, and B. J. Jackson. 1987. Protein glycosylation in yeast. *Annu. Rev. Biochem.* **56:**915–944.

65. Lai, M. T., D. Y. T. Liu, and T. H. Hseu. 2007. Cell growth restoration and high level protein expression by the promoter of hexose transporter, HXT7, from *Saccharomyces cerevisiae. Biotechnol. Lett.* **29:**1287–1292.

66. Lau, W. W., K. R. Schneider, and E. K. O'Shea. 1998. A genetic study of signaling processes for repression of PHO5 transcription in *Saccharomyces cerevisiae. Genetics* **150:**1349–1359.

67. Lehle, L., and E. Bause. 1984. Primary structural requirements for N-glycosylation and O-glycosylation of yeast mannoproteins. *Biochim. Biophys. Acta* **799:**246–251.

68. Lemontt, J. F., C. M. Wei, and W. R. Dackowski. 1985. Expression of active human uterine tissue plasminogen activator in yeast. *DNA* **4:**419–428.

69. Lenouvel, F., P. J. van de Vondervoort, and J. Visser. 2002. Disruption of the *Aspergillus niger* argB gene: a tool for transformation. *Curr. Genet.* **41:**425–432.

70. Li, H., N. Sethuraman, T. A. Stadheim, D. Zha, B. Prinz, N. Ballew, P. Bobrowicz, B. K. Choi, W. J. Cook, M. Cukan, N. R. Houston-Cummings, R. Davidson, B. Gong, S. R. Hamilton, J. P. Hoopes, Y. Jiang, N. Kim, R. Mansfield, J. H. Nett, S. Rios, R. Strawbridge, S. Wildt, and T. U. Gerngross. 2006. Optimization of humanized IgGs in glycoengineered *Pichia pastoris. Nat. Biotechnol.* **24:**210–215.

71. Li, M., F. Hubalek, P. Newton-Vinson, and D. E. Edmondson. 2002. High-level expression of human liver monoamine oxidase A in *Pichia pastoris*: comparison with the enzyme expressed in *Saccharomyces cerevisiae. Protein Expr. Purif.* **24:**152–162.

72. Li, P., A. Anumanthan, X. G. Gao, K. Ilangovan, V. V. Suzara, N. Duzgunes, and V. Renugopalakrishnan. 2007. Expression of recombinant proteins in *Pichia pastoris. Appl. Biochem. Biotechnol.* **142:**105–124.

73. Lopes, T. S., J. Klootwijk, A. E. Veenstra, P. C. van der Aar, H. van Heerikhuizen, H. A. Raue, and R. J. Planta. 1989. High-copy-number integration into the ribosomal DNA of *Saccharomyces cerevisiae*: a new vector for high-level expression. *Gene* **79:**199–206.

74. Lubertozzi, D., and J. D. Keasling. 2009. Developing *Aspergillus* as a host for heterologous expression. *Biotechnol. Adv.* **27:**53–75.

75. Mach, H., D. B. Volkin, R. D. Troutman, B. Wang, Z. Luo, K. U. Jansen, and L. Shi. 2006. Disassembly and reassembly of yeast-derived recombinant human papillomavirus viruslike particles (HPV VLPs). *J. Pharm. Sci.* **95:**2195–2206.

76. Madzak, C., M. C. Mimmi, E. Caminade, A. Brault, S. Baumberger, P. Briozzo, C. Mougin, and C. Jolivalt. 2006. Shifting the optimal pH of activity for a laccase from the fungus *Trametes versicolor* by structure-based mutagenesis. *Protein Eng. Des. Sel.* **19:**77–84.

77. Madzak, C., L. Otterbein, M. Chamkha, S. Moukha, M. Asther, C. Gaillardin, and J. M. Beckerich. 2005. Heterologous production of a laccase from the basidiomycete *Pycnoporus cinnabarinus* in the dimorphic yeast *Yarrowia lipolytica. FEMS Yeast Res.* **5:**635–646.

78. Maras, M., I. van Die, R. Contreras, and C. A. van den Hondel. 1999. Filamentous fungi as production organisms for glycoproteins of bio-medical interest. *Glycoconj. J.* **16:**99–107.

79. Martinez-Ruiz, A., A. Martinez del Pozo, J. Lacadena, J. M. Mancheno, M. Onaderra, C. Lopez-Otin, and J. G. Gavilanes. 1998. Secretion of recombinant pro- and mature fungal alpha-sarcin ribotoxin by the methylotrophic yeast *Pichia pastoris*: the Lys-Arg motif is required for maturation. *Protein Expr. Purif.* **12:**315–322.

80. **Matsuyama, A., A. Shirai, and M. Yoshida.** 2008. A series of promoters for constitutive expression of heterologous genes in fission yeast. *Yeast* **25:**371–376.

81. **Mattern, I. E., S. Unkles, J. R. Kinghorn, P. H. Pouwels, and C. A. van den Hondel.** 1987. Transformation of *Aspergillus oryzae* using the *A. niger pyr*G gene. *Mol. Gen. Genet.* **210:**460–461.

82. **McCarthy, J. E.** 1998. Posttranscriptional control of gene expression in yeast. *Microbiol. Mol. Biol. Rev.* **62:**1492–1553.

83. **Mendoza-Vega, O., J. Sabatie, and S. W. Brown.** 1994. Industrial production of heterologous proteins by fed-batch cultures of the yeast *Saccharomyces cerevisiae. FEMS Microbiol. Rev.* **15:**369–410.

84. **Moir, D. T., and D. R. Dumais.** 1987. Glycosylation and secretion of human alpha-1-antitrypsin by yeast. *Gene* **56:**209–217.

85. **Morawski, B., Z. Lin, P. Cirino, H. Joo, G. Bandara, and F. H. Arnold.** 2000. Functional expression of horseradish peroxidase in *Saccharomyces cerevisiae* and *Pichia pastoris. Protein Eng.* **13:**377–384.

86. **Murray, A. W., and J. W. Szostak.** 1983. Pedigree analysis of plasmid segregation in yeast. *Cell* **34:**961–970.

87. **Nevalainen, K. M., V. S. Te'o, and P. L. Bergquist.** 2005. Heterologous protein expression in filamentous fungi. *Trends Biotechnol.* **23:**468–474.

88. **Nikolaev, I., M. Mathieu, P. van de Vondervoort, J. Visser, and B. Felenbok.** 2002. Heterologous expression of the *Aspergillus nidulans alcR-alcA* system in *Aspergillus niger. Fungal Genet. Biol.* **37:**89–97.

89. **Nyyssonen, E., M. Penttila, A. Harkki, A. Saloheimo, J. K. Knowles, and S. Keranen.** 1993. Efficient production of antibody fragments by the filamentous fungus *Trichoderma reesei. Biotechnology* (New York) **11:**591–595.

90. **O'Callaghan, J., M. M. O'Brien, K. McClean, and A. D. Dobson.** 2002. Optimisation of the expression of a *Trametes versicolor* laccase gene in *Pichia pastoris. J. Ind. Microbiol. Biotechnol.* **29:**55–59.

91. **Outchkourov, N. S., W. J. Stiekema, and M. A. Jongsma.** 2002. Optimization of the expression of equistatin in *Pichia pastoris. Protein Expr. Purif.* **24:**18–24.

92. **Papagianni, M., N. Joshi, and M. Moo-Young.** 2002. Comparative studies on extracellular protease secretion and glucoamylase production by free and immobilized *Aspergillus niger* cultures. *J. Ind. Microbiol. Biotechnol.* **29:**259–263.

93. **Park, M. A., H. J. Kim, and H. J. Kim.** 2008. Optimum conditions for production and purification of human papillomavirus type 16 L1 protein from *Saccharomyces cerevisiae. Protein Expr. Purif.* **59:**175–181.

94. **Paus, E. J., J. Willey, R. J. Ridge, C. R. Legg, M. A. Finkelman, T. J. Novitsky, and P. A. Ketchum.** 2002. Production of recombinant endotoxin neutralizing protein in *Pichia pastoris* and methods for its purification. *Protein Expr. Purif.* **26:**202–210.

95. **Pohlig, G., W. Zimmermann, and J. Heim.** 1991. Influence of yeast proteases on hirudin expression in *Saccharomyces cerevisiae. Biomed. Biochim. Acta* **50:**711–716.

96. **Porres, J. M., M. J. Benito, and X. G. Lei.** 2002. Functional expression of keratinase (*ker*A) gene from *Bacillus licheniformis* in *Pichia pastoris. Biotechnol. Lett.* **24:**631–636.

97. **Price, V. L., W. E. Taylor, W. Clevenger, M. Worthington, and E. T. Young.** 1990. Expression of heterologous proteins in *Saccharomyces cerevisiae* using the ADH2 promoter. *Methods Enzymol.* **185:**308–318.

98. **Proudfoot, A. E., G. Turcatti, T. N. Wells, M. A. Payton, and D. J. Smith.** 1994. Purification, cDNA cloning and heterologous expression of human phosphomannose isomerase. *Eur. J. Biochem.* **219:**415–423.

99. **Punt, P. J., M. A. Dingemanse, A. Kuyvenhoven, R. D. M. Soede, P. H. Pouwels, and C. A. van den Hondel.** 1990. Functional elements in the promoter region of the *Aspergillus nidulans gpd*A gene encoding glyceraldehyde-3-phosphate dehydrogenase. *Gene* **93:**101–109.

100. **Punt, P. J., R. P. Oliver, M. A. Dingemanse, P. H. Pouwels, and C. A. van den Hondel.** 1987. Transformation of *Aspergillus* based on the hygromycin B resistance marker from *Escherichia coli. Gene* **56:**117–124.

101. **Punt, P. J., N. van Biezen, A. Conesa, A. Albers, J. Mangnus, and C. van den Hondel.** 2002. Filamentous fungi as cell factories for heterologous protein production. *Trends Biotechnol.* **20:**200–206.

102. **Punt, P. J., and C. A. van den Hondel.** 1992. Transformation of filamentous fungi based on hygromycin B and phleomycin resistance markers. *Methods Enzymol.* **216:**447–457.

103. **Punt, P. J., N. D. Zegers, M. Busscher, P. H. Pouwels, and C. A. van den Hondel.** 1991. Intracellular and extracellular production of proteins in *Aspergillus* under the control of expression signals of the highly expressed *Aspergillus nidulans gpd*A gene. *J. Biotechnol.* **17:**19–33.

104. **Ramqvist, T., K. Andreasson, and T. Dalianis.** 2007. Vaccination, immune and gene therapy based on virus-like particles against viral infections and cancer. *Expert Opin. Biol. Ther.* **7:**997–1007.

105. **Raymond, C. K., T. Bukowski, S. D. Holderman, A. F. Ching, E. Vanaja, and M. R. Stamm.** 1998. Development of the methylotrophic yeast *Pichia methanolica* for the expression of the 65 kilodalton isoform of human glutamate decarboxylase. *Yeast* **14:**11–23.

106. **Record, E., P. J. Punt, M. Chamkha, M. Labat, C. A. van den Hondel, and M. Asther.** 2002. Expression of the *Pycnoporus cinnabarinus* laccase gene in *Aspergillus niger* and characterization of the recombinant enzyme. *Eur. J. Biochem.* **269:**602–609.

107. **Reddy, S. T., and N. M. Dahms.** 2002. High-level expression and characterization of a secreted recombinant cation-dependent mannose 6-phosphate receptor in *Pichia pastoris. Protein Expr. Purif.* **26:**290–300.

108. **Romanos, M. A., C. A. Scorer, and J. J. Clare.** 1992. Foreign gene expression in yeast: a review. *Yeast* **8:**423–488.

109. **Rothstein, R. J.** 1983. One-step gene disruption in yeast. *Methods Enzymol.* **101:**202–211.

110. **Russell, C., J. Mawson, and P. L. Yu.** 1991. Production of recombinant products in yeasts: a review. *Aust. J. Biotechnol.* **5:**48–55.

111. **Sakai, A., F. Ozawa, T. Higashizaki, Y. Shimizu, and F. Hishinuma.** 1991. Enhanced secretion of human nerve growth factor from *Saccharomyces cerevisiae* using an advanced delta-integration system. *Biotechnology* (New York) **9:**1382–1385.

112. **Sakai, Y., T. Rogi, R. Takeuchi, N. Kato, and Y. Tani.** 1995. Expression of *Saccharomyces* adenylate kinase gene in *Candida boidinii* under the regulation of its alcohol oxidase promoter. *Appl. Microbiol. Biotechnol.* **42:**860–864.

113. **Sarramegna, V., F. Talmont, M. Seree de Roch, A. Milon, and P. Demange.** 2002. Green fluorescent protein as a reporter of human μ-opioid receptor overexpression and localization in the methylotrophic yeast *Pichia pastoris. J. Biotechnol.* **99:**23–39.

114. **Schilling, B. M., J. C. Goodrick, and N. C. Wan.** 2001. Scale-up of a high cell density continuous culture with *Pichia pastoris* X-33 for the constitutive expression of rh-chitinase. *Biotechnol. Prog.* **17:**629–633.

115. **Schultz, L. D., H. Z. Markus, K. J. Hofmann, D. L. Montgomery, C. T. Dunwiddie, P. J. Kniskern, R. B. Freedman, R. W. Ellis, and M. F. Tuite.** 1994. Using molecular genetics to improve the production of recombinant proteins by the yeast *Saccharomyces cerevisiae. Ann. N. Y. Acad. Sci.* **721:**148–157.

116. Schuster, E., N. Dunn-Coleman, J. C. Frisvad, and P. W. M. van Dijck. 2002. On the safety of *Aspergillus niger*: a review. *Appl. Microbiol. Biotechnol.* **59:**426–435.

117. Shen, S., G. Sulter, T. W. Jeffries, and J. M. Cregg. 1998. A strong nitrogen source-regulated promoter for controlled expression of foreign genes in the yeast *Pichia pastoris*. *Gene* **216:**93–102.

118. Shiba, Y., C. Ono, F. Fukui, I. Watanabe, N. Serizawa, K. Gomi, and H. Yoshikawa. 2001. High-level secretory production of phospholipase A1 by *Saccharomyces cerevisiae* and *Aspergillus oryzae*. *Biosci. Biotechnol. Biochem.* **65:**94–101.

119. Shuster, J. R., D. Moyer, and H. Lee. 1992. A DNA vector system for the integration and amplification of heterologous DNA into dispersed chromosomal locations in the yeast, *Saccharomyces cerevisiae*, genome. *Abstr. Pap. Am. Chem. Soc.* **203:**1198.

120. Sleep, D., G. P. Belfield, and A. R. Goodey. 1990. The secretion of human serum albumin from the yeast *Saccharomyces cerevisiae* using five different leader sequences. *Biotechnology* (New York) **8:**42–46.

121. Smith, J. D., and A. S. Robinson. 2002. Overexpression of an archaeal protein in yeast: secretion bottleneck at the ER. *Biotechnol. Bioeng.* **79:**713–723.

122. Stepien, P. P., R. Brousseau, R. Wu, S. Narang, and D. Y. Thomas. 1983. Synthesis of a human insulin gene. VI. Expression of the synthetic proinsulin gene in yeast. *Gene* **24:**289–297.

123. Tucker, M., and R. Parker. 2000. Mechanisms and control of mRNA decapping in *Saccharomyces cerevisiae*. *Annu. Rev. Biochem.* **69:**571–595.

124. Tuite, M. F., M. J. Dobson, N. A. Roberts, R. M. King, D. C. Burke, S. M. Kingsman, and A. J. Kingsman. 1982. Regulated high efficiency expression of human interferon-alpha in *Saccharomyces cerevisiae*. *EMBO J.* **1:**603–608.

125. Unkles, S. E., E. I. Campbell, P. J. Punt, K. L. Hawker, R. Contreras, A. R. Hawkins, C. A. Van den Hondel, and J. R. Kinghorn. 1992. The *Aspergillus niger niaD* gene encoding nitrate reductase: upstream nucleotide and amino acid sequence comparisons. *Gene* **111:**149–155.

126. Valenzuela, P., A. Medina, and W. J. Rutter. 1982. Synthesis and assembly of hepatitis-B virus surface-antigen particles in yeast. *Nature* **298:**347–350.

127. van den Hombergh, J. P., P. J. van de Vondervoort, L. Fraissinet-Tachet, and J. Visser. 1997. *Aspergillus* as a host for heterologous protein production: the problem of proteases. *Trends Biotechnol.* **15:**256–263.

128. Vasavada, A. 1995. Improving productivity of heterologous proteins in recombinant *Saccharomyces cerevisiae* fermentations. *Adv. Appl. Microbiol.* **41:**25–54.

129. Wang, L., D. Ridgway, T. Gu, and M. Moo-Young. 2005. Bioprocessing strategies to improve heterologous protein production in filamentous fungal fermentations. *Biotechnol. Adv.* **23:**115–129.

130. Wang, S. H., T. S. Yang, S. M. Lin, M. S. Tsai, S. C. Wu, and S. J. Mao. 2002. Expression, characterization, and purification of recombinant porcine lactoferrin in *Pichia pastoris*. *Protein Expr. Purif.* **25:**41–49.

131. Ward, M., C. Lin, D. C. Victoria, B. P. Fox, J. A. Fox, D. L. Wong, H. J. Meerman, J. P. Pucci, R. B. Fong, M. H. Heng, N. Tsurushita, C. Gieswein, M. Park, and H. Wang. 2004. Characterization of humanized antibodies secreted by *Aspergillus niger*. *Appl. Environ. Microbiol.* **70:**2567–2576.

132. Ward, P. P., J. Y. Lo, M. Duke, G. S. May, D. R. Headon, and O. M. Conneely. 1992. Production of biologically active recombinant human lactoferrin in *Aspergillus oryzae*. *Biotechnology* (New York) **10:**784–789.

133. Ward, P. P., G. S. May, D. R. Headon, and O. M. Conneely. 1992. An inducible expression system for the production of human lactoferrin in *Aspergillus nidulans*. *Gene* **122:**219–223.

134. Wilcox, G., and G. M. Studnicka. 1988. Expression of foreign proteins in microorganisms. *Biotechnol. Appl. Biochem.* **10:**500–509.

135. Woo, M. K., J. M. An, J. D. Kim, S. N. Park, and H. J. Kim. 2008. Expression and purification of human papillomavirus 18 L1 virus-like particle from *Saccharomyces cerevisiae*. *Arch. Pharmacal Res.* **31:**205–209.

136. Xu, P., D. Raden, F. J. Doyle, and A. S. Robinson. 2005. Analysis of unfolded protein response during single-chain antibody expression in *Saccharomyces cerevisiae* reveals different roles for BiP and PDI in folding. *Metab. Eng.* **7:**269–279.

137. Yelton, M. M., J. E. Hamer, and W. E. Timberlake. 1984. Transformation of *Aspergillus nidulans* by using a *trpC* plasmid. *Proc. Natl. Acad. Sci. USA* **81:**1470–1474.

138. Zhang, N., D. L. Daubaras, and W.-C. Suen. 2005. Heterologous protein expression in yeasts and filamentous fungi, p. 667–687. *In* Z. An (ed.), *Handbook of Industrial Mycology*. Marcel Dekker, New York, NY.

Mammalian Cell Culture for Biopharmaceutical Production

JINYOU ZHANG

12

12.1. OVERVIEW OF THE BIOPHARMACEUTICAL INDUSTRY

Combining various definitions (178–180, 235), biopharmaceuticals are medical drugs produced using biotechnology, that is, therapeutic products created through the genetic manipulation of living cells or organisms. They are pharmaceuticals with complex structures that are used for therapeutic, prophylactic, and even in vivo diagnostic purposes, including proteins, peptides, nucleic acids, and other large molecules, and are produced by means other than chemical synthesis or direct extraction from natural biological sources. Their production employs modern biotechnological tools, such as recombinant DNA and hybridoma technology, and is carried out in engineered biological systems. Included in biopharmaceutical categories are recombinant forms of natural proteins (insulin, interferon, interleukins, cytokines, erythropoietin, etc.), biologics derived from natural sources (most vaccines, enzymes, polypeptides, coagulation factors, etc.), and nucleic acid-based products (DNA, RNA, gene therapy, plasmids, etc.). Monoclonal antibodies (MAbs) are the single largest group of biopharmaceuticals.

In recent years, while the pharmaceutical industry as a whole has suffered from only single-digit growth, thinner pipelines, and pressure from an ever aggressive generics industry, biopharmaceuticals have represented the fastest-growing sector of the pharmaceutical industry, with more than 200 marketed products and hundreds more in development (203). During the period from 2003 to 2006, regulators in Europe and the United States approved 32 biopharmaceuticals for human use (44), and the growth rate of this market was expected to be at 12% annually for the next decade (108). The biopharmaceuticals market in 2007 reached ~$106 billion out of the $650 billion global pharmaceuticals market (172). This market segment has expanded rapidly owing to the significantly higher clinical success rate of biopharmaceuticals compared with new small-molecule chemical entities, their greater potential for curing diseases rather than just treating symptoms, and their greater efficacy and reduced side effects (219).

While biopharmaceutical segments such as replacement hormones and traditional vaccines are mature, segments such as MAbs and modern vaccines are maturing. In particular, the MAb therapeutics segment is thriving and boasts one of the most active and promising pipelines in the industry. MAbs are by far the predominant modality and leading segment of biopharmaceuticals, being increasingly used in therapy because of their high specificity, low toxicity, long half-life, predictable pharmacokinetics, and high dose demand. With global sales of $21 billion in 2006 (59, 140), $27 billion in 2007, and $33 billion in 2008 based on one report (141), MAbs represent the fastest-growing sector within the biopharmaceutical industry (161, 187). They are expected to achieve a compound annual growth of 14% between 2006 and 2012, easily outstripping the 0.6% growth rate of the more traditional small-molecules market (18, 28). Currently, MAbs play an important role in the treatment of cancer and autoimmune and inflammatory diseases. In 2004, 6 of the 12 new biopharmaceuticals approved in the United States and European Union were MAb-based products (236), and by 2009 a total of 26 have been approved by the FDA for the market (Table 1; www.fda.gov), with many new MAbs expected to be approved within the coming years. Newer MAbs, which include chimeric and humanized MAbs, had the greatest success with FDA approval. For example, 26% of the chimeric MAbs in development reach the market, while the approval rate for murine products, by comparison, is 4.5% (30).

12.2. MAMMALIAN CELLS AS DESIRED EXPRESSION SYSTEMS FOR PROTEIN BIOPHARMACEUTICALS

The majority of biopharmaceuticals are of protein nature and are expressed in recombinant hosts. For expressing certain biopharmaceutical proteins, microbial systems are attractive because of their low cost, high productivity, and rapid implementation. In addition, there is no adventitious virus concern to regulatory authorities. Indeed, relatively simple recombinant proteins, such as insulin (35, 226) and bovine growth hormone (226), have been successfully produced in *Escherichia coli* or *Saccharomyces cerevisiae*. Antibody fragments are readily produced in microbial systems (27). For example, *E. coli* is used to produce antibody fragments, followed by PEGylation to decrease rate of clearance in vivo (104). A nonglycosylated whole antibody can be produced in *E. coli* that is secreted into the periplasm (215). Filamentous fungi such as *Aspergillus* are also used to produce antibodies (239). The considerations in selecting an expression platform (85), the process development for microbial biopharmaceuticals (46), and the manufacturing of recombinant therapeutic

TABLE 1 FDA-approved monoclonal antibody products

Product	Manufacturer	Date	Indication/*target*[a]	Host cell
Orthoclone OKT3	J & J	1986	Transplant rejection/*T-cell CD3 receptor*	Hybridoma
ReoPro	Centocor/Lilly	1994	Cardiovascular disease/*glycoprotein IIb/IIIa*	Sp2/0
Rituxan	Genentech	1997	Non-Hodgkin's lymphoma and diffuse large B-cell/*CD20*	CHO
Zenapax	Roche	1997	Organ rejection/*IL-2 receptor a*	NS0
Simulect	Novartis	1998	Organ rejection/*IL-2 receptor a*	Sp2/0
Remicade	Centocor	1998	Rheumatoid arthritis and Crohn's disease/*TNF-α signaling*	NS0
Herceptin	Genentech	1998	Breast cancer/*ErbB2*	CHO
Synagis	MedImmune	1998	Respiratory syncytial virus/*F protein epitope*	NS0
Mylotarg	Wyeth	2000	Acute myelogenous leukemia/*CD33*	NS0
Campath-1H	Genzyme	2001	Lymphocytic leukemia/*CD52*	CHO
Zevalin	Spectrum Pharmaceuticals	2002	Non-Hodgkin's lymphoma/*CD20*	CHO
Humira	Abbott	2002	Rheumatoid arthritis and Crohn's disease/*TNF-α signaling*	CHO
Xolair	Genentech	2003	Asthma/*immunoglobulin E*	CHO
Bexxar	GlaxoSmithKline	2003	Non-Hodgkin's lymphoma/*CD20*	Hybridoma
Raptiva	Genentech	2003	Plaque psoriasis/*CD11a*	CHO
Erbitux	Imclone/BMS	2004	Cancer (colorectal)/*epidermal GF receptor*	Sp2/0
Avastin	Genentech	2004	Cancer (colon or rectum)/*endothelial GF*	CHO
Tysabri	Biogen Idec	2004	Multiple sclerosis and Crohn's disease/*T-cell VLA4 receptor*	NS0
Vectibix	Amgen	2006	Cancer (colorectal)/*epidermal GF receptor*	CHO
Lucentis[b]	Genentech	2006	Macular degeneration/*vascular endothelial GF*	E. coli
Soliris	Alexion	2007	PNH (hemolysis)/*complement protein C5*	NS0
Cimzia[c]	UCB-Celltech	2008	Rheumatoid arthritis and Crohn's disease/*TNF-α*	E. coli
Simponi	Centocor	2009	Rheumatoid and psoriatic arthritis, ankylosing spondylitis/*TNF-α*	CHO
Ilaris	Norvatis	2009	Cryopyrin-associated periodic syndrome/*IL-1*	Sp2/0
Stelara	Centocor	2009	Psoriasis/*IL-12 and IL-23*	Sp2/0
Arzerra	GSK/Genmab	2009	Chronic lymphocytic leukemia/*CD20*	CHO

[a]IL, interleukin; TNF-α, tumor necrosis factor alpha; GF, growth factor.
[b]Human antibody Fab fragment.
[c]PEGylated anti-TNF-α antibody Fab fragment.

proteins in microbial systems (91) have been well established, and are covered in other chapters of this book.

However, there are things microbes simply cannot do. Many biopharmaceutical molecules are too large and complex to be made by simple prokaryotic bacteria, or even the lower eukaryotic forms such as fungi and yeasts. Complex biomolecules, such as functional MAbs or highly glycosylated proteins, require the posttranscriptional metabolic machinery only available in mammalian cells (35). Glycosylation of MAbs is important for their biological function and pharmacokinetics (17, 110). Proper posttranslational modifications convey higher or "humanlike" quality and efficacy to the protein when compared with proteins produced by microbes. Aglycosylated forms of glycoproteins tend to be misfolded, biologically inactive, or rapidly cleared from circulation (45). Eukaryotic fungi and yeast can glycosylate proteins, but the resultant glycans are not the structures normally found on human proteins (83). While this inability of microbes is being tackled by a genetic engineering approach that humanizes the glycosylation machinery of the yeast *Pichia pastoris* for desired glycosylation of recombinant protein products (86, 95, 245), the best yeast

production systems like *P. pastoris* are proprietary and require license to use (115). As for the eukaryotic insect cells, they possess only limited glycosylation capabilities and are mostly used as vehicles for production of genetically engineered viral vaccines using baculovirus expression systems (193).

As the need for producing properly folded protein molecules including MAbs with appropriate glycosylation patterns grew, the industry shifted in the 1980s toward the use of mammalian cells. Mammalian cells are the preferred hosts for the production of most complex protein therapeutics, since their functionally and pharmacokinetically relevant posttranslational modifications are highly human compatible. Stable cell lines, the prerequisite for manufacturing biopharmaceutical proteins, can be created by introducing exogenous DNA containing the product gene into host cells by transfection of a nonviral vector system. The economic concern about low productivities from mammalian cells has been eased by the 1- to 2-g/liter yield commonly achieved and 10-g/liter yield recently claimed on the production scale (157). Therefore, during the past 2 decades, despite the well-known lower production yield, higher manufacturing cost, and other challenges compared with microbial systems, cultured mammalian cells have become a widely used platform for producing recombinant proteins (250), and cell culture capacity has grown from less than 300,000 liters to around 3 million liters today, driven by the demands for MAbs (219).

About one-half (212, 237) to two-thirds (152) of all the approved biopharmaceuticals are glycosylated proteins produced by mammalian cell culture, such as MAbs, blood factors, anticoagulants, thrombolytics, erythropoietin, specific interferons, hormones, and various therapeutic enzymes. In 2007, there were 15 biopharmaceuticals approved in the United States and European Union (238), including 6 erythropoietin-based products, 3 antibodies, 2 hormones, 1 recombinant vaccine, 1 growth factor, and a fusion product. Eleven of these 15 are produced in mammalian cell culture systems, 2 in *E. coli*, 1 in *S. cerevisiae*, and 1 in a baculovirus/insect cell-based system, confirming the prominence of mammalian-based expression systems in the biopharmaceutical sector. The rapid increase of MAbs as product candidates in recent years has been enabled by improvements made in the recombinant expression technologies that use mammalian cells. While there are 2 nonglycosylated MAb Fab fragment products from *E. coli* fermentations, Lucentis (for macular degeneration) and Cimzia (for Crohn's disease), which recently were approved by the FDA (18, 180), the other 24 MAb products currently on the market are produced in mammalian cells (Table 1) (186, 237, 263), of which all but 2 are produced by recombinant technology. In addition, in the last 2 decades, recombinant therapeutic protein yields from mammalian cells in batch and fed-batch processes have increased dramatically due to the utilization of highly productive cell lines, improved media, and better bioprocess conditions that enhance cell viabilities in high-density suspension cultures (94), thus easing the major concern of using mammalian production systems. It is obvious that mammalian cell culture will continue to be the cornerstone of large-scale production for biopharmaceuticals, especially for MAbs.

This review will summarize the recent development of biopharmaceutical production using mammalian cell culture. Since it is not practical to attempt to cover all categories of biopharmaceutical products, this chapter will focus on MAbs, which is the most dominant and most rapidly developing sector in the biopharmaceutical industry, and its large-scale production the most established. The concepts and methods described below should nevertheless be applicable to the production of other biopharmaceutical molecules derived from mammalian cells. Specific development and manufacturing information for vaccines can be found in the published literature (204, 227).

Mammalian cells are fragile because they are approximately 10- to 50-fold larger than microbial cells, without the tough cell walls of microbes, and they are sensitive to impurities. Simply getting these cells to thrive and reproduce in culture is a challenge in itself. The fundamental and unique aspects of mammalian cell culture characteristics, growth kinetics, cell metabolism, bioreactor development, process monitoring and control, and process issues in large-scale mammalian cell cultures were well reviewed in the last edition of this manual (174). The history of the discovery, development, and production of therapeutic antibodies is covered in a recent review (114). In the present review, the recently published literature from peer-reviewed journals, but not conference presentations or Web seminars, etc., is summarized, including each of the major mammalian culture process development steps for MAbs such as cell line, medium, and cell culture. Other important aspects

such as purification, formulation, fill/finish, analytics, and current Good Manufacturing Practices (cGMPs) will be covered in another chapter. Only the systems currently commonly used in the industry are described, but references are given for unique or emerging technologies. Where multiple sources exist, the more recent ones are usually cited.

12.3. CONSTRUCTION AND SELECTION OF HIGH-PRODUCING CELL LINES

Mammalian cells have become the dominant system for producing recombinant protein products for clinical application because of their capacity to properly fold and assemble proteins and add humanlike posttranslational modifications. In fact, all cell lines used for biopharmaceutical protein production so far have originated from mammals (44). However, mammalian cell line development is often very time-consuming. In addition, the mammalian cell culture process is hampered by low yields and unstable expression by the cells (130). Productivity and stability of expression are the prerequisites for developing commercially viable processes. So the ultimate goal of cell line development is to obtain clonal cell lines that secrete the protein of interest with high specific productivity (Qp), and at consistently high levels over an extended number of cell generations, allowing for scale-up and cost-efficient manufacturing. Expression vector and cell line engineering are the keys for achieving this goal.

Advances in cell line generation include new expression vectors and transfection technology to introduce the genes into cells, novel parental cell lines that have been selected or designed to grow to maximum density, and robust screening technologies that in combination can enable rapid generation of production cell lines. Fine-tuning of vector construction, identification and use of selectable markers, optimization of transfection in production media, gene targeting, and high-throughput screening have led to high Qp ranging from approximately 20 to 60 pg/cell/day (20, 192). The challenge is to create cell lines that not only have high Qp but also have the growth characteristics that will lead to high volumetric productivity in the manufacturing process. This can be achieved by selecting cell lines with the combination of high Qp and ability to achieve a high integral of viable cell concentration (IVC) and consequent high volumetric production rate in a screen designed to mimic the final production process (24). The resulting cell line also needs to be selected for stability, growth in suspension culture using an animal-free medium, and the ability to make the desired posttranslational modifications such as glycosylation. In addition, the cell line must be able to grow to high density at a reasonable growth rate.

Despite the progress, protein expression levels in mammalian cells are relatively low and often unstable during the time of development (4, 250), resulting in high development and production costs for therapeutic proteins. Although ~30-g/liter antibody titers have been achieved at the laboratory scale with the introduction of certain novel technologies (173, 207), the highest reported yields at industrial scale in mammalian cells are about 5 g/liter (35, 212, 250), which are still about threefold lower than those claimed in bacteria and yeast (8, 23, 86, 247). In addition, cell line construction and selection, which ideally should create a high enough producer that could be used for manufacturing throughout, in reality often yields to the demand for a shorter timeline to go to clinical trial to prove the concept and to obtain the safety profiles. Therefore, there is often a need to create an improved second-generation cell line at a later stage of product development. Also, as with any other steps, the potential benefit of utilizing proprietary expression and engineering technologies need to be carefully weighed against the costs, in particular that for intellectual property rights, which may lead to downstream royalty stacking.

12.3.1. Expression Systems

Many component technologies are included in an expression system, such as vector construction, host cell, transfection method, selection and amplification approach, and screening tool (181). The most important factors in choosing an optimal expression system are productivity, process economics, product quality and safety, production stability, lead times, scalability, and regulatory acceptance. For glycoproteins, expression titers are mainly determined by the promoter construct, its copy number and site of integration into the chromosome, and the type of proteins in question (43). Despite the fact that a variety of new expression systems have been developed for recombinant protein production, the basic aspects and components of the expression systems used in large-scale biopharmaceutical manufacturing have changed little over recent decades (181). The situation is exemplified by a handful of companies controlling a great majority of the world's large-scale (10,000- to ~20,000-liter) biopharmaceutical manufacturing capacity and using only established systems, mostly Chinese hamster ovary (CHO) and murine myeloma (NS0, Sp2/0) expression systems. This is particularly true for manufacture of MAbs, which requires high capacity because of administration in repeated high dosages. These high-Qp cell lines and their high-density growth significantly favor process economics in cost of goods and use of plant capacity, and their well-established posttranslational modification patterns help to ensure product qualities, including glycosylation and protein folding, which largely determine the molecules' biological activities. In addition, these two expression systems are very familiar to regulatory authorities (Table 1).

12.3.1.1. Vectors

Biopharmaceutical production processes predominantly use systems in which the gene of interest is randomly integrated into a host chromosome, and the protein product is almost always secreted into an extracelluar medium for ease of downstream processing. The cell line's ability to deliver high volumetric productivities [= Qp (μg/cell/day) × IVC (cell · day/ml)] lies in the efficient transcription of the antibody gene by a well-designed expression vector (promoter construct, intron, copy number, and site of integration into chromosome, etc.), as well as the efficient translation of the antibody mRNAs with sufficient capacity in assembling, modifying, and secreting the complete antibody molecules. To create a production cell line for an antibody, expression vectors containing the heavy- and light-chain genes under control of strong mammalian promoters are introduced into the host cell line. Usually, a selectable marker is included, so that cells carrying the genes can be easily selected in the presence of an inhibitor that would kill the cells not expressing the selectable marker activity. Other genetic elements are also included, such as transcription termination sequences [poly(A)] and translation control sequences. Two expression vector systems dominate antibody production in mammalian cell culture, one based on dihydrofolate reductase (DHFR) genes (198) and the other on glutamine synthetase (GS) genes (16). Additional expression vectors and their selectable markers are available (24). To achieve

high levels of MAb gene expression, DHFR and GS vectors usually have strong promoters to drive expression of the immunoglobulin (Ig) genes, while the GS and DHFR genes themselves are under control of relatively weak promoters. The strong promoters are typically of viral origin (e.g., human cytomegalovirus) or they are derived from genes that are highly expressed in a mammalian cell (49). The details of vector elements were reviewed (145) and the use of their new combinations were studied (124).

In a DHFR expression system, the folate analogue methotrexate (MTX) is used to inhibit the function of DHFR, an essential metabolic enzyme for purine and pyrimidine, which produces a cofactor for thymidylate synthetase. Transfection with an expression vector containing the DHFR gene prevents MTX from poisoning transfected cells. Selection occurs in the absence of glycine, hypoxanthine, and thymidine. Frequently, an antibiotic resistance gene is used in DHFR expression vectors to act as the selectable marker, turning the primary function of the DHFR gene to facilitating vector amplification. A weak promoter, such as one from simian virus 40, is usually used to control the DHFR gene to reduce promoter interference due to read-through from the upstream promoter, which inhibits expression from the downstream promoter. Despite the usually long amplification time (up to five cycles, with up to 12 weeks per cycle), the DHFR system is effective and has been used in conjunction with other aspects of cell line development to achieve MAb expression levels of multiple grams per liter (114).

GS synthesizes glutamine from glutamate and ammonium. Since glutamine is an essential amino acid, transfection of cells lacking endogenous GS, such as NS0, with the GS vector confers the ability to grow in glutamine-free media (12, 23). Selection occurs in the absence of glutamine. Even for cells that do express functional levels of GS activities, such as CHO, including in the medium the GS inhibitor methionine sulfoximine (MSX), an analogue of glutamate, enables use of the GS expression vectors. GS expression vectors also contain the GS gene downstream of a weak promoter from simian virus 40, minimizing the promoter interference, thus increasing expression of the Ig genes. GS selection can be used to select high-expressing cell lines without amplification, which reduces the time compared with the DHFR selection approach. The GS system has enabled the rapid identification and selection of production cell lines expressing up to 20 to ~50 pg/cell/day and productivity of multiple grams per liter. According to Lonza (Basel, Switzerland) (www.lonza.com), more than 85 global pharmaceutical/biotechnology companies are currently using this technology and several products using the GS system are on the market, including Synagis (179) and Zenapax (12).

12.3.1.2. Stable Expression

Efficient delivery of a vector to host cells is variable depending on the techniques used, which should be based on the requirement for transient, episomal, or stable transfection of the cells. Compared with other physical and chemical methods, electroporation has been traditionally used for the generation of stable transfectants, because it requires smaller quantities of DNA (241), can be used on a variety of cell lines, and has a much lower cost but higher transfection efficiency than chemical reagents (166).

Transcription can be increased by several approaches, such as gene amplification and insertion into a transcriptionally active region. Gene amplification is usually achieved by linking genes of interest to an amplifiable gene such as GS or DHFR. Transfected cells are then gradually exposed to increasing levels of specific enzyme inhibitors at concentrations substantially higher than those used for selection. Only the cells that overproduce these essential enzymes can survive. The overproduction of the enzyme (e.g., GS or DHFR) is usually caused by the increased levels of its particular mRNA, resulting from either an increase in the integrated vector copy number in host chromosomes (amplification) or from more efficient transcription (88). During this process, other tightly linked sequences on the vector, including the genes of interest, are often coamplified, leading to high copy numbers of transgene, which may increase the Qp of the cells, especially with the DHFR expression system (24). However, it can also have a detrimental effect on other cellular properties, resulting in poor growth performance of the cell population and/or altered cellular metabolism. In addition, amplification and the varied copy number can change the inherent stability of expression; thus the continued presence of the selective agent may be required. Some of the selective agents are toxic and thus require safety measures in operation and demonstration of their removal during purification. Also, the Qp will vary among clones, so the identification of high producers will require the screening of hundreds or thousands of cell lines.

Some systems do not rely on gene amplification to achieve high productivities, including GS and certain DHFR variants (24). Instead, they rely on chromatin remodeling technology. In these cases, insertion of plasmid DNA containing antibody genes into an actively transcribed region of the genome (euchromatin), instead of a condensed and thus transcriptionally inactive region of the genome (heterochromatin), is desired since the former favors high and stable expression of the recombinant genes, whereas the latter results in the repression of gene expression. The transcriptionally active regions are easily available to the enzymes that transcribe all the genes integrated at these sites into RNA, thereby increasing the rate of transcription. Recent improvements include incorporation into transgenes of the constructs of DNA elements that can either act as an enhancer or prevent expression silencing by integration into heterochromatin, improving stability and possibly productivity while reducing the number of clones to be screened (78, 130). For example, expression vectors can be constructed that contain a specific targeting sequence that will direct the vector to integrate by homologous recombination into a particular active site. Such a sequence was identified in the Ig locus of NS0 (101), and most high-producing NS0 cell lines use vectors containing this sequence. Logically, the sequence flanking the transcriptionally active locus can also be incorporated into the vector, which will cause the random integration site to become transcriptionally active and available to the enzymes that transcribe the genes, creating a favorable environment for expression that is independent of its integration site into the genome. Methods of target integration of transfected plasmid DNA to transcriptionally active sites of the genome have been developed to generate high-producing cell lines (7), but the consistency of such methods and the long-term expression stabilities of the cell lines generated have yet to be established (94). Vectors incorporating matrix attachment regions (MARs) (38, 67, 120) from Selexis (Marlborough, MA) and ubiquitous chromatin opening element (UCOE) (19) from Millipore (Billerica, MA) are two major types of elements currently used that function to create a region of transcriptionally active chromatin. The MAR elements are inserted into the vector to surround the IgG genes, thus imposing an open chromatin configuration, allowing RNA polymerase and other transcription factors to access promoters and enhancers found in the expression vector. This increases the number of independently transformed cells that express the desired protein and makes possible expression levels in the initial transfectants of as high as 50 to 70 pg/cell/day (114). Success was reported (www.selexis.com) (67, 207) in using MAR elements (currently known as Genetic Elements™) to boost productivity in various recombinant proteins, including MAbs in CHO, HEK-293, BHK, and other cell lines. UCOE consists of regions that are rich in the sequence CpG, which increases the accessibility of the surrounding chromatin and thus the expression levels of linked genes (19). UCOE also exhibits non-tissue-specific preventive activity against transcriptional silencing (14) and can be incorporated into expression vectors to increase transgene expression (250). Stimulating and anti-repressor elements (STAR) (129) from Crucell (Leiden, The Netherlands), flanking sequences of the CHO elongation factor-1α gene (49), and expression augmenting sequence elements (EASE) (160) from Amgen (Thousand Oaks, CA) have also been shown to increase and maintain transgene expression. These are being applied to immunoglobulin (Ig) genes, enabling higher transcription levels in a higher percentage of transfectants. In addition, use of the artificial chromosomal expression system (ACE) from Chromos Molecular Systems (Burnaby, Canada), which functions as stable, nonintegrating vectors with large capacity and allows multiple copies of the same transgene, has resulted in fourfold higher expression levels over the DHFR-CHO system (60). Transgene expression can also be increased by enhancing promoter strength using engineered zinc-finger protein transcription factors that bind a DNA sequence within a promoter (188).

Loss of recombinant gene copy number and the appearance of nonproducing populations of cells appear to be the predominant causes of instability of production, but there are several other factors that may affect expression levels and stability of production, as molecular loci exist at which instability may most likely be engendered (14). Some of the expression instabilities are caused by gene silencing at the level of chromatin—the so-called epigenetic gene silencing, which can be interfered with to enhance and stabilize transgene expression by employing epigenetic gene regulation tools, such as targeting histones, including specific DNA elements, and targeting sites of high expression (130).

12.3.1.3. Transient Expression

Before a stable manufacturing cell line becomes available, rapid and cost-effective expression technologies are frequently employed to quickly deliver milligrams to grams of protein materials for development needs. The methods include the use of uncloned transfectant pools and the use of transient gene expression (251). HEK-293 and CHO cells are the two major host lines for producing recombinant proteins using the transient gene expression or extrachromosomal expression technology (11). The current scale for this method goes up to 100-liter (55). The particular transfection reagents and methods, as well as cell culture parameters, differ from those for production from stably engineered cell lines.

12.3.2. Host Cell Lines

Current biomanufacturing processes require cell lines to be capable of achieving unclumped and robust growth in suspension culture, stably and productively integrating heterologous DNA, producing high antibody concentrations in a given system, and performing desired posttranslational modifications with uniform product characteristics (40). In addition, the availability of a suitable expression system and the speed at which a high-yielding clone can be obtained may also influence the cell line choice. For regulatory approval, production cell lines must be well characterized and genetically stable; thus, using a cell type familiar to regulators will limit the intensity of their scrutiny. Some cell line-specific differences can significantly affect the performance of a production system; for example, glycosylation of a given protein varies with the type of mammalian cells used (111), and even two subclones from the same parental line can differ greatly in metabolic requirements (121).

Over two-thirds, or 140, of the recombinant therapeutic proteins currently on the market are produced in suspension culture-adapted CHO fibroblast cells (237). In addition, over 60% of all new target proteins currently in clinical trials are produced using CHO cells (152). CHO was chosen as the host for tissue plasminogen activator, the first recombinant therapeutic protein from mammalian cells to gain regulatory approval. Other manufacturers continued to use CHO cells because these were considered an acceptable host system for products intended for intravenous administration in humans. With time, other mammalian cell lines such as murine myeloma lymphoblastoid-like cells (NS0 and Sp2/0-Ag14), baby hamster kidney fibroblast cells (BHK-21), and human embryonic kidney epithelial cells (HEK-293) became acceptable (250). BHK-21 was once used for monoclonal antibody production (106), and HEK-293 was initially the producer for Humira. But BHK and HEK cells are now used to a lesser extent, mainly for virus and vaccine production (42, 176). CHO and NS0 remain the tested and most commonly used workhorses, with known safety and productivity profiles and capabilities (40, 44). These parental cell lines are also the ones commonly used for the antibodies currently in clinical trials. Among the 24 full MAb products currently approved by the FDA (including 8 blockbuster drugs) (Table 1), over 40% are produced in CHO, 25% in NS0, and 25% in Sp2/0 cell lines, despite the well-known differences in glycosylation between CHO and NS0 (10, 213). Murine hybridomas and some new cell lines such as the human PER.C6 are also being used (113, 114). It should be noted that the MAb products produced in hybridoma cell lines generally have lower dose requirements and are also older than those produced in highly engineered systems such as CHO, NS0, and Sp2/0. Since glycosylation of IgGs is "species specific," it is necessary to select the proper cell line to express recombinant IgGs for therapy in humans (182, 183).

12.3.2.1. CHO

CHO, established from a Chinese hamster's ovary biopsy over 40 years ago, is the most widely used and the best-understood mammalian host cell in the industry. Because of the widespread adoption of this host cell line, the growth characteristics, metabolism, behavior in bioreactors, virulence factors, and likely host cell-related impurities that might be in a process or product are well understood. Moreover, because there is a strong regulatory history of CHO cells, more and more products in development are now made using CHO cells.

Most biopharmaceuticals on the market are produced in CHO-K1 or CHO-derived cell lines such as DHFR⁻, DG44, and DUK-B11 cells (9, 40, 92), because they enable amplification of specific genes through selection with an amplifiable marker such as DHFR or GS. The most commonly used CHO strains with DHFR-based vectors are DUK-B11 and

DG44, both of which lack DHFR, while the GS system uses CHO-K1 or its suspension-adapted derivative, CHO-K1SV (189). Although CHO-K1 expresses functional GS, inclusion of the GS inhibitor MSX in the media allows use of GS expression vectors. MSX is cytotoxic at 3 μM, at which endogenous GS is inhibited. At 50 μM, only those cell lines that have stably incorporated the expression vector into a transcriptionally active locus will form transfectant colonies, producing enough GS enzyme both to titrate out MSX and to synthesize glutamine from glutamate and ammonium for cellular needs. CHO cells offer the advantage of producing and secreting recombinant proteins with efficient humanlike posttranslational modifications, as exemplified by N-linked carbohydrates that are compatible with prolonged in vivo activities in humans (112). The glycosylation patterns of native and CHO-derived recombinant proteins are similar (24). For example, glycosylation of the IgG1 form of Campath expressed in CHO cells is consistent with normal human IgG (213).

Wild-type CHO strains have adherent cell morphology and require serum supplementation for growth. It would take up to 9 months to adapt the recombinant CHO cell lines from adherent to suspension culture and further to simple serum-free, animal-free media. The current industry practice is to preadapt the parental CHO cell line to suspension growth in the desired media, thus cutting short the overall cell line development timeline to about 6 months (189, 217).

12.3.2.2. NS0 and Sp2/0

The murine lymphoid cell lines NS0 and Sp2/0-Ag14 were both derived over 20 years ago from a plasmacytoma cell line originated from a BALB/c mouse (12, 214). The starting cell lines underwent numerous rounds of cloning and, in the case of Sp2/0, fusion with spleen cells from another BALB/c mouse, to generate these two parental cell lines. Since the parental cell type is a differentiated B cell that is inherently capable of high levels of Ig production, while these cell lines lack the ability to synthesize and secrete Ig protein due to evolution, these two cell lines are favored for manufacturing antibodies at high yields. In addition, these two myeloma cell lines tend to grow well in suspension when adapted to serum-free media. Sp2/0 has an extensive record as a null parent for hybridomas and transfectomas. Half of the antibody products on the market use these two myeloma cells (Table 1). However, it should be noted that instability and relatively slow growth have been reported for murine myeloma cell lines (13).

NS0 and Sp2/0 cells contain very low levels of endogenous GS compared with CHO cells and are obligate auxotrophs for glutamine, which means the glutamine independence can be conferred upon transfection with a functional GS gene. Therefore, a GS expression system is often used for NS0 cells. Unlike NS0 cells, mutants from Sp2/0 cells that no longer require glutamine occur spontaneously with relatively high frequency, raising some concerns about using the GS expression vector in this cell line (24). It should be pointed out that unlike CHO and human cells, NS0 is known to add the nonhuman antigenic sugar residue Galα1-3Gal (170) onto the N-glycans of proteins it makes; however, there is no evidence yet that its presence on recombinant IgG is immunogenic in humans. For example, IgG1 and IgG4 forms of Campath expressed in NS0 cells include two potentially immunogenic glycoforms which contain either one or two nonreducing terminal α-linked galactose residues (213).

12.3.2.3. PER.C6

In recent years, the human PER.C6 cell line attracted attention (113) because of its human glycosylation and other posttranslational modification machinery. The cell line was specifically designed for biopharmaceutical production. Rather than by viral transformation, PER.C6 was derived from a single healthy human embryonic retinal cell by transfection with the adenovirus E1 region followed by selection for transfectants with an immortal phenotype; thus the cells can replicate indefinitely, allowing them to be cultured in single-cell suspension under serum-free conditions. It was shown (122) that the IgG1 antibody produced in PER.C6 had few mannose or hybrid structures in its glycosylation pattern, which is biantennary with core fucose, very much like the IgG1 found in human serum. These cells do not synthesize potentially immunogenic MAb glycan structures. This expression platform requires no amplification of inserted genes or selection agent to deliver stable clones with high levels of protein expression within a few months. A low copy number is sufficient to retain stable and efficient protein expression. The cell line has been adapted to grow to very high density, and productivity of >2 g/liter in fed-batch culture can be routinely achieved. Lately, with the use of this cell line, a very high production titer of 27 g/liter was reported using a unique combination of fed-batch and perfusion approaches (173). It appears that PER.C6 and other human cell lines may become reliable, safe, scalable, and economical alternatives to the widely used CHO and myeloma cell lines.

12.3.2.4. Hybridoma

MAbs were first produced from hybridomas consisting of a murine B cell producing a specific antibody fused to an immortal murine lymphoid cell line (125), which are specifically designed to secrete MAbs. After being injected into the abdomens of mice, the hybridoma cells grow and MAb-rich ascites fluid accumulates, which can be withdrawn by needles at intervals. The collected ascites fluid is very complex and highly contaminated, but can often reach the 1-g/liter level. One commercial MAb is produced in this way. In vitro culture is later developed, and improved cell density, longevity, and antibody titer are achieved by fed-batch system. Still, these cell lines are generally difficult to engineer for high levels of protein expression and usually grow to only moderate densities in bioreactors. In addition, hybridomas are not suitable for obtaining high-affinity antibodies (159, 224). Therefore, hybridomas are mostly used for creation and production of murine MAbs for research and analytical use, or as limited-dose therapeutics. It is also possible to make human antibodies using murine hybridoma technology (47), in which the spleen cells are taken from a transgenic mouse that has the murine Ig locus replaced by the human genes, or by directly employing primary human B cells for generating hybridoma cell lines (136).

12.3.3. Screening for Desired Cell Lines

After transfection with vectors bearing the antibody light- and heavy-chain genes, cells are screened for highly productive cell lines through growth recovery, serum-free and suspension adaptation, amplification, single-cell cloning, and final clone selection. These steps can vary in order and are as needed. To improve outgrowth and transfection efficiency, transfection is predominantly performed in serum-containing media, but using host or parental cell lines preadapted to grow in suspension and serum-free media reduces adaptation times and increases the likelihood of reaching high cell densities and high product yields in serum-free suspension culture. Recent progress in using serum-free, protein-free transfection conditions (133)—thus the fact that all steps listed above can now be executed under serum-free conditions—has significantly speeded up cell line development while complying with regulatory requirements (201). Usually, the transfected cells in 96-well plates are first screened for fast growers, then their productivities are analyzed to identify the candidates with the best production potential and growth characteristics, followed by amplification with increasing concentrations of inhibitors to the selective markers. The key parameters, such as secretion rates, intracellular amounts of the product, and gene copy numbers, are followed (127). The surviving transfectants are single-cell cloned by methods such as limited dilution, and the promising clones are adapted to grow in the enriched or large-scale production medium, with a similar feed scheme, in shake flasks. The leading candidates are cryopreserved as the pre-master cell bank (pre-MCB) and are evaluated for passaging stability that should cover the longest possible manufacturing duration. In addition to phenotypic stability, the amplified cell lines need to be evaluated for genetic stability (14) in the absence of selective pressure, which generally is not applied at the production stage. Once the stable and highly productive cell line is identified, the MCB is generated under cGMP conditions (71, 72) before initiation of process development for commercial production. Mammalian MCB cells must be characterized to be free of adventitious agents, be demonstrated suitable for large-scale culture, offer high-yield production, be genetically stable, and provide integrity of the protein product, which is assessed by a battery of protein and carbohydrate analytical methods (142, 197). Then the MCB needs to be subcultured to provide seed stock that produces working cell banks (WCB) for all future production runs.

Transfectants with high Qp are rarely generated, so stringent selection is needed for an increased frequency of high producers. Levels of about 15 to 20 pg/cell/day are considered appropriate for initial transfectants for antibodies (114), with greater productivity arising from optimized cell culture conditions or amplification using selective pressure. Since a transfectant with high Qp does not necessarily perform well in the production process, a sufficient number of cell lines need to be generated to allow for the attrition when screening for other desired characteristics. It is suggested (23) that a few thousand transfectants should be screened after enrichment by a stringent selection system in order to be confident of getting multiple cell lines with the desired productivity characteristics. A large number of high-producing transfectants is also needed to screen against the additional growth criteria that contribute to high productivity in a manufacturing process.

Selecting the final clone can be complicated and difficult because of the harsh environment in which single cells are generated through limited dilution or facilitated cloning. The survivor cells after selection are often subjected to serial dilution in secondary containers like 96-well plates, and the resulting cells that show as single colonies are subsequently expanded in larger containers like 12- or 24-well plates. The process is lengthy primarily due to the fact that many cell lines have a transfection efficiency of <1% (54), and most cells need to communicate with surrounding cells to enable growth. Supplementing with serum or preconditioned medium can improve the probability of a single cell multiplying successfully (54), but during serum-free scale-up some cell clones will be lost, as they cannot survive without serum. The conventional manual limited-dilution cloning from a transfection pool is labor-intensive and time-consuming, as it involves initial analysis and consequent monitoring of single-cell wells. Typical timelines from single-cell seeding to clone productivity evaluation can be about 6 to 12 weeks, and the effort required is proportional to the number of clones selected. To make things more challenging, regulatory guidelines recommend at least two rounds of dilution cloning.

The recovery of cell lines with high Qp has been improved through the development of high-throughput screening tools (31). Combining image processing and highly mechanized robotics to identify and pick mammalian colonies of interest, many automated systems have been developed that can aid cell line development by assessing cell growth from a single-cell clone and isolating higher expressers for further expansion. This allows a much greater number of clones to be analyzed without the need for multiple operators. Some systems, like fluorescence-activated cell sorting (FACS), can identify cells secreting high levels of antibody and sort them away from the low producers. The system employs a fluorescence-tagged antibody against the product expressed either on the cell surface or in a microgel bead (31, 102), or an artificial affinity matrix on the surface of viable cells (25), followed by isolation using FACS. The feasibility of using FACS in conjunction with a nonfluorescent reporter protein molecule that can be detected by a fluorescent antibody has also been demonstrated for identifying the high producers from the sorted single-cell clones (51). Other advanced systems include ClonePix (Genetix, Boston, MA) and Cell Xpress (SAFC Biosciences, Lenexa, KS), both of which employ fluorescence technology to detect high-expressing cells from among thousands. To shorten the time for clone selection, miniaturized bioreactors or shaking flasks are also being used that can simulate standard production bioreactor conditions (246). This technology usually employs high-speed automation to streamline liquid handling and analytical capability for early evaluation of growth and productivity profiles of candidate clones. Despite their high expenses and lack of a track record, at the present time these high-throughput screening methods can reduce the time taken to create a robust cell line from 12 to 18 months to 6 to 12 months and provide high-throughput analysis while maintaining and selecting clones (54).

The fundamental mechanisms that dictate clonal productivity and stability are still unknown. The most likely molecular loci at which instability may be engendered were highlighted for CHO and NS0 (14). DNA microarrays and two-dimensional electrophoresis on 11 GS-NS0 cell lines showed that major functional class changes between low and high producers are in protein synthesis and cell death/growth (211). Another report suggested there is a positive correlation between the Qp and the ratio of heavy-to light-chain mRNA expression of an antibody produced by a GS-NS0 cell line (15). In GS-CHO, it was found that the optimal MAb production represents a compromise between heavy-chain abundance and the requirement for excess light chain to render MAb folding and assembly more efficient (200). Obviously, in addition to high Qp, growth characteristics such as the IVC, harvest titer and viability, clone stability, and product quality are all important selection criteria for production cell lines. A single-cell secretion assay coupled with cell sorting was used to screen subclones with altered production kinetics or improved stability (25).

As an alternative to cell line engineering for improved performances (see below), the parental cell line or its recombinant derivatives are often selected from many different phenotypes obtained during initial screening that show characteristics applicable to bioreactor scale-up

scenarios and growth under high-density situations during the production phase (94). For example, the cell sensitivity to osmolality and pCO_2 may need to be considered as well because the cells may have to counteract high osmotic pressure and elevated dissolved carbon dioxide (dCO_2) levels at large scale. A CHO-DG44 variant with 500-mOsm/kg osmolality resistance was stable for at least 75 generations while exhibiting higher Qp (163). The success of these screening schemes depends on the use of a model that is proven predictive of the manufacturing process. In many cases, shake-flask fed-batch culture serves well as the tool of choice.

12.3.4. Cell Line Engineering

The manufacturing cost can be very high because of unpredictable stability and productivity issues in MAb and non-antibody therapeutic protein production by mammalian cell lines. To lower both the cost and risk, most development programs now include cell engineering strategies toward the ultimate production of stable and high-yield cGMP cell banks. These strategies include vector optimization, chromosomal DNA and mRNA elements engineering, integration method improvement, genetic or pathway enhancement, and others (118). Since the recombinant cell lines currently used were mostly constructed using expression vectors developed to give high mRNA levels, transcription is probably not limiting antibody secretion (24). It is likely that with modern cell lines, events downstream of transcription are the limiting factor for productivity; thus, only a few of them can be addressed by vector engineering. For a panel of GS-NS0 cell lines at Qp from 1.2 to 23 pg/cell/day, there was no correlation observed between Qp and mRNA levels (221).

To enhance Qp, targeted approaches to affect posttranslational modifications of antibodies (229) and antibody secretion (26) have been demonstrated. Since antibody production is likely limited at folding and assembly reactions, an approach is to modify the translational or secretory pathways to increase Qp (57). Based on the fact that foldases and chaperones exist as large multiprotein complexes, it was proposed that global expansion of all components of the secretory pathway is required for generic improvement of antibody secretion rather than overexpression of selected proteins. It has also been proposed to overexpress proteins known or suspected to have a functional role in modulating signaling pathways, such as BLIMP-1 (57). Alternatively, zinc-finger protein transcription factor libraries, which can modulate the expression of any genes, can be randomized, so that a specific phenotype can potentially be created with enhanced recombinant protein production (171).

The combination of high Qp (pg/cell/day) and high IVC (cell · day/ml) leads to high antibody titers (g/liter) in cell culture. Maximum IVC can be achieved by screening for a cell line capable of growing to a high density, then maintaining the cell viability for as long as possible, which requires the cell death rate to be minimized. For essentially all industrially used cell lines, such as CHO, myelomas, hybridomas, and HEK-293, apoptosis, or programmed death, is the major cause of cell death, which is the obstacle for these cells to survive high cell densities. Apoptosis is induced by various chronic insults and is mediated by several mechanisms (2, 131). Various strategies were implemented to inhibit apoptosis (6, 23), including the prevention of nutrient limitation by fed-batch operation, which can delay the onset and extent of apoptosis, and engineering of resistance into the cell line, such as the regulatory Bcl-2 family of proteins, which affect the activation of apoptosis pathways in cells (144). However, while the overexpression of anti-apoptotic Bcl-2 family members can be employed to protect industrially important cell lines from stresses typically experienced during cell culture operation, productivities are often compromised (65, 107, 216), and the behaviors of the engineered cell lines are rarely reported. It can thus be concluded that inhibiting apoptosis, whether through the Bcl-2 method or simply by limiting cell culture nutrients, will enhance cellular viability but will not necessarily increase productivity (44), because overexpression of exogenous genes may have unintended consequences with regard to other cellular functions, resulting in cells with undesirable phenotypes. Still, the integration of the antiapoptotic genes *E1B-19K* and *Aven* has recently been reported to improve performance of CHO cell culture (65). The coexpression of cell proliferation genes and antiapoptotic genes into CHO cells has also been reported to improve cell culture performance (105). Alternatively, parental cell lines or recombinant derivatives with elevated resistance to apoptosis may be selected, based on survival under cultivation conditions with increased cell stress. This approach may be more likely to succeed than the host engineering strategy, given the complexity of apoptotic pathways (94).

In addition to limiting apoptotic death, recent studies have also examined regulating the life cycle for metabolic control of cell growth. The cyclin-dependent kinase family of proteins is a good target for control of cell cycle transitions from growth to cell death. The binding of cyclin-dependent kinases with cyclins leads to formation of complexes with cyclin-dependent inhibitors such as p21 and p27 (79, 80); overexpression of p21 and p27 causes cell cycle arrest at G_1 phase, and an increase in recombinant productivity as a consequence. It was shown that overexpression of p21 increased recombinant productivity in CHO cells by fourfold because of cell cycle lengthening, which increased mitochondrial mass and activity and improved ribosome biogenesis (20). The application of recently available "-omics" tools has started to provide a better understanding of the cell biology that defines the desired cell line phenotype. When GS-NS0 cell lines with various antibody production rates were compared in their proteomes, it was demonstrated that changes in productivity were associated with the changes in the abundance of several proteins (221).

Product quality is also the subject of cell line engineering. One such example is to alter the cell's ability to perform posttranslational modifications, particularly glycosylation, which is believed to play a role in effector functions of the antibody, such as antibody-dependent cellular cytotoxicity (ADCC). Reports show that oligosaccharide engineering may optimize ADCC, and thus antibody potency, because the degree of galactosylation, fucosylation, and sialylation and the proportion of bisecting GlcNAc residues have all been implicated in modulating effector functions (110). In a CHO-DG44 host cell with deleted α-1,6-fucosyltransferase activity, the ADCC of the resulting antibody was 100-fold higher than the fucosylated form (256). Also, overexpression of *N*-acetylglucosaminyltransferase III in CHO cells increased the proportion of bisecting GlcNAc residues, increasing ADCC substantially compared with the parent molecule (229). It is also possible to screen for NS0 clones that exhibit a low incidence of Galα1-3Gal because of their genetic makeup. Targeted knockout of mammalian host cell genes is now possible using RNA interference (66) or engineered zinc fingers (230). This technology was used to modify the glycosylation machinery in CHO cells targeting fucosyltransferase, resulting in MAbs with reduced fucosylation and thus increased ADCC (158).

12.4. MEDIUM DEVELOPMENT FOR MAMMALIAN CELL CULTURE

Obtaining a good-producing clone through cell line construction and selection makes it possible to achieve high productivity and the desired product quality. For the cell line to achieve its full potential, it is important to create a suitable culture environment that provides the optimal chemical and physical conditions for cells to grow and produce. A culture medium determines the chemical environment and thus can have a dramatic effect on cell growth and product yield, and even on product quality. Consequently, when development reaches the late clinical phases, as in the case of the cell line, change in culture medium should be avoided. This requires the selection of the right components and careful formulation and optimization of the medium before product development reaches the final stage. Medium development is an important part of process optimization, which has led to the improvement in biopharmaceutical manufacturing for several decades (93); therefore this topic is reviewed here as a separate section. It should be pointed out that detailed information on medium development and medium compositions is usually proprietary and not available in the public domain.

12.4.1. General Features of Cell Culture Media

Whereas most microbes easily flourish in a simple and defined mixture of carbon, nitrogen, phosphorus sources, and mineral salts, mammalian cells in addition frequently need vitamins, amino acids, nucleotides, lipids, precursors, protectants, reducing agents, and even growth factors. Based on this author's own experiences, there are several key aspects of cell culture media that are different from the traditional microbial fermentation media. A cell culture medium is usually highly cell line and clone specific in the absence of serum, and individual optimization can normally improve production levels two- to fivefold (32). The presence of animal-derived components such as serum creates a more natural environment for better cell growth, but results in serious safety concerns about the biopharmaceuticals produced (148); yet the common non-animal replacements such as recombinant proteins are usually very expensive and carry their own impurities (196). Mammalian cells are unique in their ability to "adapt" to new medium formulations, voluntarily or involuntarily selecting variants. Another unique feature is that the medium performance is culture history dependent, and a conclusion about a new medium can only be drawn after multiple passages. Additional difficulties include the large variations in medium evaluation methods, such as growth assessment by cell counting, which can vary by about 10 to 20%; thus, quantum change in the measured parameters is needed to draw conclusions about an impact. In addition, there can easily be over 50 components in a typical cell culture medium, making it practically very difficult to fully optimize the composition. Finally, optimized batch medium is rarely optimal when a fed-batch culture is employed, and optimized growth medium is frequently nonoptimal for production.

12.4.2. Animal Component-Free Media

As more MAbs are being employed as therapeutic agents, their production methods are coming under increasing regulatory scrutiny. Among the concerns is the presence of animal-derived components in the media employed to grow mammalian cells, because animal-derived substances can potentially introduce contaminants (adventitious agents) into the process and thus potentially into the final product. There are clearly risks associated with using animal-sourced materials, as these materials could be a source for prion protein transmission (81) and viral contamination (149); therefore, the regulations have become more stringent in order to ensure that the biopharmaceutical production process is free of potential contamination by these adventitious agents. For example, regulatory authorities specifically require that non-animal alternatives be used where available and demand that "manufacturers shall always justify the use of materials of animal origin" (62).

For better composition definition, reduced contamination by adventitious and infectious agents, and lowered cost, serum-free media are now commonly used in industry. However, to preserve the multifaceted beneficial functions of serum, many of the serum-free media still contain some bovine-derived large-molecule components, particularly bovine proteins such as albumin, insulin, transferrin, and lipoproteins, some of which act as carriers of certain essential nutrients such as lipids while others facilitate cell metabolism (29, 52). Because any animal-derived proteins could carry a theoretical risk of introducing prions (191) or other adventitious agents (97), it is highly desirable to develop animal protein-free media for cell culture processes (148). Although very often recombinant proteins can be employed to replace their animal-sourced counterparts, using organic or inorganic small molecules where possible may dramatically lower the medium cost and minimize even the indirect risks, such as those potentially from fermentation raw materials used for production of these recombinant protein medium components. An entire medium formulation can now be made animal free, because all components, including small molecules such as amino acids that are usually animal derived, are available from fermentation, synthetic, or plant sources (109, 263). Not only animal protein-free media but also protein-free media for CHO and NS0 have also become commercially available, and some can be found in the literature (202, 260, 263). In addition, a host of chemically defined media were reported to support high-density cell culture and to facilitate small-scale development activities (54, 222, 263, 266). Chemically defined media are inherently less expensive and make downstream processing more straightforward as there are fewer, and known, contaminants to monitor and remove. Process optimization is also easier in a defined environment. It should be pointed out that the definition for chemically defined medium varies in the literature, with some included recombinant proteins such as insulin-like growth factor-1, transferrin, and insulin, etc., as signals for cell growth (114). In this author's view, a chemically defined medium should only contain well-characterized components, with known identity and structure and consistent purity/impurity profiles; it can contain animal-derived components such as cholesterol, galactose, and amino acids, but should not include any protein components that are generally less defined. Those that contain minimal quantities of small-mass proteins should be termed as low-protein media (29).

Significant progress has been made in substituting animal components in media with non-animal materials for mammalian cell growth and recombinant protein production (90, 109, 119). Specifically, (i) bovine serum albumin could be replaced by recombinant human albumin, although costly and not straightforward (29), plus an organic carrier for lipids (cyclodextrin) where needed (109, 263); (ii) bovine lipoprotein by synthetic cholesterol and plant-derived fatty acids (90); (iii) bovine insulin by recombinant human insulin or zinc ions (249, 263); and (iv) bovine transferrin by small-molecule tropolone (150) or other iron-chelating agents (196, 264) for iron delivery. Animal protein-free media are currently the most prevalent platforms being developed by many companies, and can contain complex ingredients like hydrolysates (157). All the popular industrial cell lines mentioned above (CHO, NS0, Sp2/0, and PER.C6) can now be cultured efficiently in these serum-free, animal component-free media.

12.4.3. Development of Cell Culture Media

Cell culture media are usually developed by selecting a commercially available, well-established, and chemically defined basal formulation and supplementing it with additives specific to the cell line or recombinant construct. In addition to the difference in host cell lines, most clonal derivatives present their own metabolic phenotypes, so cell culture optimization is often regarded as cell line or clone dependent. Therefore, the basis of medium development is a better quantitative understanding of the nutritional requirements of the cells, often helped by analysis of nutrient depletion followed by supplementation of the relevant nutrient in the medium or feed. The development process is usually iterative until the desired growth and production levels are achieved at large scale, together with satisfactory product purification and product quality. The commonly used strategies for medium formulation and optimization include (i) literature survey to identify possible growth requirements based on related cell lines or constructs, (ii) component swapping and titration to improve performance in a simple and straightforward way, and (iii) medium blending to select the best composition in an evolutionary manner (69). Other approaches are also used, such as (i) material balance to estimate minimum nutritional needs based on cell compositions and available intracellular stoichiometry (253, 255), (ii) spent medium analysis to identify and fortify limiting nutrients (82), (iii) statistically designed experimentation to gain rich data including component interactions without exhaustive experiments (228), and (iv) microarray analysis to identify ligands to the expressed receptors that can be included as medium components (1). The latter few methods have been employed less frequently in cell culture medium development because of the incomplete understanding of metabolic pathways and the huge numbers of medium components, in addition to their high complexity. However, high-throughput systems are being developed and their implementation has begun to speed up medium development (50, 84). This is occurring in conjunction with the promising advancement in high-throughput analytical measurements, such as measuring antibody yields in 96-well format.

An example of medium development involves the delivery of iron to cells. Iron is essential for cell respiration and metabolism, and without iron cells stop growing and eventually die. Iron is traditionally delivered by transferrin, but the bovine form carries the risk of viral contamination, while the recombinant form is often costly. As for the non-animal small-molecule iron delivery vehicles, some have been patented, such as tropolone (64). Selenium is an essential trace element for cell growth and development in vivo and in vitro and can protect cells from oxidative damage (259). Accordingly, selenium (as selenite salts) has been extensively used as a supplement in mammalian cell culture media (98, 123). A recent finding revealed that selenite also serves as an effective carrier in animal protein-free media to deliver the essential iron to mammalian cells for high-level cell growth and production (264). Another example that deserves mentioning is the delivery of water-insoluble cholesterol and fatty acids to cell lines such as NS0 and PER. C6, which require them to grow. For animal protein-free and defined media, since the natural carriers like bovine serum albumin cannot be used, carrier molecules like cyclodextrin were found effective in delivering these molecules (234). Such media require careful preparation in order not to affect other aspects of the cell culture (169). Interestingly, it has been reported recently that NS0 cells, which were thought to be cholesterol auxotrophs, can be successfully grown in the absence of cholesterol (33). Overexpression of 17β-hydroxysteroid dehydrogenase type 7 in NS0 cells also led to cholesterol-independent growth (210).

Non-animal-sourced hydrolysates have been commonly added to defined basal media to increase cell density, culture viability, and productivity (34, 155). Plant or yeast hydrolysates are protein digests composed of amino acids, short peptides, carbohydrates, vitamins, nucleosides, and minerals, providing a variety of nutritional supplements to media. The inclusion of hydrolysates has resulted in significant improvements in MAb yields in various industrially important cell lines. The exact mechanism for this hydrolysate-induced increase is yet unknown. Initial investigations reported that certain hydrolysates provided nutritional benefits, which resulted in increased cell numbers and final titers (168, 199). But further studies suggested that the benefits from hydrolysates were not just nutritional, since some peptides may exhibit antiapoptotic effects (34, 76, 225), affect the metabolism of the cells (75, 168), affect the cell cycle (77), and/or improve specific antibody productivity (34, 77, 225). However, the exact mechanisms of how peptides act as antiapoptotic agents or improve specific productivity are not fully understood. Different sources of hydrolysates may be preferred for different expression systems in order to maintain production stability, with yeast-derived hydrolysates less suitable to DHFR and plant-derived hydrolysates less suitable to GS, based on considerations of the relative amounts of nucleosides or glutamine-containing peptides in these hydrolysates (unpublished results). Due to their complex composition and possible lot-to-lot variations, hydrolysates can be a significant source of medium variation, so prescreening is necessary to ensure that the lots used meet the performance criteria (138). There are efforts toward developing a chemically defined replacement for hydrolysates (48, 223) and toward the use of more defined size-excluded fractions of hydrolysates (41, 63).

Irrespective of the characteristics of the medium developed, for suspension culture it is common to include a synthetic polymer such as Pluronic F-68 to act as a protectant against the damage to cells by sparging gases (139). To prevent foaming caused by sparging, antifoam may be used, for which the non-animal-sourced FoamAway™ Irradiated AOF is available from Invitrogen (Carlsbad, CA; www.invitrogen.com). In addition, some supplementation to a standard medium is often required at the time of use to provide components that are too labile to include in the basic formulations, or optional to standard use, or vector-determined selection agents or inducing agents, or to support cell line- or application-specific requirements (243).

12.4.4. Development of Feeds

Common goals for feed supplements include replenishing depleted nutrients, adding a particular substrate to drive an alternative metabolic pathway, introducing materials to specifically influence apoptosis, and altering culture metabolism from growth into production mode (244). This last category may often be achieved by changing pH or overall tonicity, increasing specific ion complements (e.g., acetate), specifically inhibiting cell division or DNA replication (190, 242), and introducing toxic or cytostatic agents (e.g., butyrate). Apoptosis can be controlled by directly adding antiapoptotic agents (6) or supplementing cultures at the appropriate time with identified nutritional components, antioxidants, or growth factors.

High amino acid and carbohydrate concentrations can be used in basal serum-free media for convenient batch culture, with NaCl omitted to maintain desired osmolality. But in fed-batch culture, beginning with a much leaner base medium can offer even greater efficiency. High initial levels of glucose, while contributing to an early boost in culture expansion, can induce many cultures to shift to "Crabtree effect," which inhibits oxidative metabolism and results in production of lactate, leading to premature cellular stress or apoptosis. High concentrations of glutamine can cause ammonium levels over 10 mM, impeding not only cell growth but also aspects of product formation such as protein glycosylation (39). Therefore, an approach is to use low glucose and/or glutamine in basal media and leave those substrates as part of the controlled nutrient feeding (265), allowing cultures to detoxify themselves from inhibitory metabolites and reducing accumulation of amino acid by-products, thus increasing both culture and product quality (3). Despite the fact that the accumulation of inhibitory metabolites such as lactate and ammonium can be minimized by maintaining low glucose and glutamine concentrations (254), frequent or continuous feeding is practically less desirable in large-scale manufacturing due to its operational complexity. Stepwise bolus addition of feed solution is commonly used in industry because of its simplicity and scalability.

For feed solutions used in large-scale fed-batch culture, the most common approach is to develop concentrated basal medium, with salts lowered to avoid high osmolality (103). In addition to glucose and glutamine, a feed formulation may include (i) nutrient concentrates (amino acids, vitamins, minerals) to increase both biomass and product, (ii) lipid preparations for specific needs of the cell line or clone, (iii) product enhancers such as butyrate to increase specific productivity, (iv) acid/base/buffers to restore pH to growth optimum or adjust to production optimum, (v) antioxidants or growth factors to promote culture longevity and thus production, and (vi) silicone-based surfactants to reduce shear stress and foaming (244). Certain key components such as phosphate may need to be added at extra strength to reach high density (56). Non-animal protein hydrolysates are commonly added as part of a feeding supplement and serve as a nutritive replenishment of free amino acids, carbohydrates, vitamins, minerals, nucleotides, and others. The bioactive small peptides in the hydrolysates can potentially activate specific pathways within the cells, act as survival signals, and suppress apoptotic death (76). Asparagine, arginine, and cysteine are among those amino acids that often have disproportionately higher use rates. Others, such as alanine, proline, and isoleucine, tend to vary greatly in use between cell lines or even individual clones. Some vitamins and metals provide significant benefits, and some can be tolerated at many times the levels found in commercial basal media. At mid culture, the cell density is much higher than the seeding density; thus high molar levels of feed components can be brought in to promote special pathways or cellular responses without detrimental effect (35). More specific approaches include influencing (even controlling) particular cellular metabolic pathways or activities, such as feeding with nucleotide sugars or their precursors to enhance product glycosylation (10). It is important to also optimize the timing and rates of feed administration.

Like media, feed solution development is an iterative process, based on the analyses of nutrient consumption, by-product accumulation, and the balance between growth and production. Usually the first one or two rounds will show the most improvement, but eventually no increase in cell mass or product will be obtained even when every measurably depleted component is replenished, because nutrient depletion is not the only limiting factor in a cell culture system (244). A fully optimized feed often exists as two or more separate solutions for storage stability reasons. Lipids and very hydrophobic ingredients usually require their own separate concentrate solutions (243). During the concentrated feed solution preparation, pH and/or temperature may need to be adjusted to completely dissolve some low-solubility components. Both the physical stability of a concentrated feed solution and the chemical stability of its components are practical issues. Varying solution pH can drastically increase the solubility of the amino acid constituents. Blending concentrates of previously solubilized individual components, rather than sequential addition of powder to one solution, can lead a higher total concentration of each.

12.5. PROCESS DEVELOPMENT FOR MAMMALIAN CELL CULTURE

In the last 2 decades, volumetric yields from recombinant mammalian cell lines have increased dramatically (94). In addition to the generation of high-specific-productivity cell lines and formulation of media and feeds to support high-density cell cultivation, as described above, there have also been significant advances in understanding cell culture processes and in sustaining cell viability in high-density cultures for optimal production (5, 94, 157). There are two crucial issues facing modern process development: to minimize the time taken to provide material for clinical studies and to develop a process that can deliver sufficient drug product to meet market demands at an acceptable price per dose (24). For the first goal, platform technologies are often employed, which in upstream processes means the same host cell line, generic media and feeds, and standard growth or operation conditions. For the second goal, individual optimization of the cell culture medium and process is needed in order to fully realize the potential of the final clone selected, aiming to deliver maximum productivity at manufacturing scale, along with desired product quality and safety in a consistent and economical way.

A typical cell culture process starts with (i) thawing a working cell bank vial; (ii) expanding the culture through increased size or number of vessels at about 1:5 to 1:20 split ratio to reach sufficient cell viability, density, and volume; (iii) seeding the culture to production bioreactors of up to 20,000 liters; (iv) operating fed-batch systems to continue cell growth and accumulate product in culture medium; and (v) harvesting to deliver clarified broth to downstream purification. There are some fundamental considerations in developing mammalian cell culture processes, including (i) cell generation number (affecting product yield and quality); (ii) production bioreactor seeding age and density (affecting productivity during the given cycle time); (iii) medium performance (making a direct difference in product titer or even quality); (iv) feed or perfusion rate (determining how long cell growth and production will be sustained); and (v) parameters such as dissolved oxygen (DO), dCO_2,

pH, osmolality, lactate and ammonium level, etc. A step-by-step process development approach is illustrated in the literature (74), which led to titer progression by 30-fold to 4.3 g/liter for an antibody.

12.5.1. Reactor Systems for Biopharmaceutical Production

Anchorage-dependent cells generally require large surface areas to generate sufficient biomass for the required productivity, which can be achieved by using multiple tissue culture flasks, multitray bioreactors, roller bottles, microcarrier culture, and hollow-fiber bioreactors. Cell growth in suspension culture is not dependent on two-dimensional surface areas; it is thus far more straightforward to cultivate than adherent cells, because no enzymatic treatment is necessary, meaning less manual handling and contamination risk. For these reasons suspension culture dominates the biopharmaceutical industry and will be the focus of this review.

There are several fundamental differences between suspension mammalian cells and microbial cells that can affect bioreactor configuration and operation (116). The time needed for cells to double, which reflects growth rate, can be counted in minutes for microbes but upwards of a day for animal cells. Mammalian cell culture bioreactors therefore require added contamination safeguards because of the long culture cycle, such as the use of magnetic-coupled bearing housings and threadless welded ports. In contrast to microbial cells, mammalian cells often overreact to small changes in temperature and pH. They can only tolerate a very narrow temperature range, typically around 37°C; therefore mammalian cell bioreactors require a regulation capability of ±0.1°C, unlike the ±1°C for microbial fermentors. While specific microbes thrive in conditions with a typical range of pH 2 to 10, a pH drop of just 0.4 can lower antibody titers from >500 mg/liter to 50 mg/liter (205). Mammalian cells require a narrow pH range of 6.8 to 7.4, so indirect addition of acid for pH control by CO_2 sparging must be used to avoid a harsh environment, and nitrogen can be sparged to neutralize the effect of CO_2 and thus raise pH. Due to the fact that the DO will be influenced by the amount of sparged CO_2 and N_2 (if used), an interactive four-gas controller may be ideally used to balance each gas component. Due to the low growth rate and the limited growth mass reached by mammalian cell culture, its oxygen uptake is much lower than in microbial fermentation, so the airflow-control capabilities of mammalian bioreactors must be more refined (delivering up to 0.5 vvm) than those for microbial fermentors (delivering up to 2 vvm). Unlike microbial cells, which can allow an agitation rate of 800 rpm or higher by radial-type impellers (such as Rushton turbines) at large scale, mammalian cells are extremely shear sensitive, in that just stirring the liquid medium can break up these fragile cells. Thus, usually no more than 150 rpm of agitation can be used even with specialized low-shear impellers (such as marine or pitched-blade), and forced-air sparging is the major means of introducing air into the mix. Finally, direct sparging of gas and high protein content in culture medium are the two main culprits in foaming, but the addition of antifoam via a peristaltic pump controlled by a foam sensor, as used in microbial culture, is usually not adopted in mammalian cell culture. Instead, manual addition of antifoam based on need is generally the practice.

To meet the rapidly growing demands of large production capacity, two types of suspension culture systems remain suitable for large-scale manufacturing: fed-batch and continuous-perfusion culture (21, 40, 100). During perfusion culture (or repeated-batch), fresh medium is added continuously, which is in balance with the volumetric removal rate of the product-containing, cell-free spent medium. The culture time in a perfusion system is thus long and the harvest volume high, yet the product titer is relatively low per liter of medium consumed. The operation is also complex, involving the use of internal or external cell-retention devices (233). The continuous changing of medium conditions and long culture times frequently lead to inconsistent processes, higher contamination risks, and variable glycosylation and other posttranslational modifications in the product over time in culture, making the process difficult to scale up and generally less robust (137, 151). The key advantage of such a system is that smaller reactors can be used for production of large quantities of product. As an example, a 500-liter industrial perfusion system was claimed to operate for about 15 to 35 days, generating multi-kilogram quantities of a MAb, at a throughput about 10 times higher than that achieved in a batch or fed-batch system (53). Recently a successful scale-up from 200-liter to 1,000-liter perfusion cell culture in disposable stirred-tank bioreactors has been demonstrated (167, 252). However, since per-cell yields or specific productivities have dramatically improved of late, which leads to high product titers in bioreactors (250), the above advantage of perfusion systems has been reduced. Another use of perfusion is for those sensitive protein products whose accumulation can inhibit or be toxic to cell growth (201) or that are themselves particularly labile. The cell viability was often found to be high, while residual host cell protein and DNA were often low in harvests of perfusion culture.

Different from batch and perfusion systems, fed-batch cultures are supplied with a concentrated nutrient solution at particular intervals to replenish the depleting nutrients, with no spent medium removed, leading to increased broth volume, high harvest titer, and easier scale-up, and ultimately to greatly increased IVC and volumetric productivity (244). The advantages of a fed-batch system include (i) reduced direct cost (e.g., reduced medium consumption, waste generation, and personnel requirement); (ii) ease of process validation and characterization; (iii) reduced footprint and turnaround time; (iv) greater lot consistency and definition; (v) ease of downstream clarification, harvest concentration, and storage; and (vi) overall reduced time to product approval. Despite the continued debate over fed-batch versus perfusion for large-scale production (143), fed-batch processes are currently by far the most commonly used in the biopharmaceutical industry because of their robust, reproducible, scalable, and reliable manufacturing outcomes (244), and they are now being operated at scales as high as 20,000-liter working volumes. While the capital investments in a manufacturing facility using fed-batch culture are higher than those for a perfusion-based facility, the overall costs of goods for the two are similar (114). The development of improved feeds in particular has made a significant contribution to increased antibody yields. The industry is now converging on using fed-batch suspension culture in stirred-tank bioreactors with controlled feeding.

The stainless bioreactor system is robust in design and is proven in small- and large-scale production. However, it has high capital costs and high maintenance needs and requires sterilization and cleaning validation. As an alternative, the use of disposable-bag bioreactor systems has become an integral part of the bioprocess in the last decade, with 20- to 1,000-liter working volume systems commercially available (61, 162, 232). The systems are particularly useful for seed culture development (153) and for rapidly meeting early development needs, because they are flexible and quick to implement, eliminating the need for cleaning-in-place and sterilizing-in-place, along with significant economic benefits (220). This is particularly advantageous for GMP manufacturing and can reduce the expensive investment in a stainless steel bioreactor facility. A wide range of small to semilarge disposable bioreactors are currently available (218), including Wave (175), Xcellerex (68, 167, 252), and SUB (single-use bioreactor) (61). Critical components for the practical use of disposable bioreactors are now available, including disposable mixing technologies, aseptic connections, and sensing technology. In-process monitoring and control conferred by traditional stirred-tank bioreactors have been incorporated into disposable systems (185).

12.5.2. Monitoring and Control of Cell Culture Processes

While disposable bioreactors are gaining popularity for smaller-scale production, and while up to 5,000-liter airlift reactors have also been used (22), large-scale cell culture has been and will continue to be typically carried out in stirred stainless steel tanks. Electrochemical sensors for pH and DO remain the most commonly used in bioprocess monitoring, and optical sensors for DO and dCO_2 are now commercially available (96). Online monitoring of intrinsic chemical and biological parameters other than the traditional pH, DO, and temperature is limited due to the lack of reliable online electrochemical sensors. Cell density, major nutrients, metabolites, and antibody products are mostly measured offline through daily sampling, using instruments such as Cedex, Guava, and Vi-Cell (for cell counting); BioProfile, YSI, and high-performance liquid chromatography (for nutrients and metabolites); blood gas analyzers or radiometers (for CO_2, O_2, and pH); and enzyme-linked immunosorbent assay, high-performance liquid chromatography, and Biacore (for products). Systems from YSI Inc. (Yellow Springs, OH; www.ysi.com) and Nova Biomedical (Waltham, MA; www.novabiomedical.com) allow real-time monitoring of about 10 nutrients and metabolites. Recently, several novel online monitoring biosensors have become available, mostly for development scale, including a biomass probe (36) and a glucose probe (117). There have also been great advances in online sampling for at-line analyses of nutrients, metabolites, and amino acids, etc. (132). For example, both YSI and BioProfile analyzers can now take samples directly from bioreactors. The next phase will likely involve the implementation of online probes for measuring viable cell density for commercial processes, which is important because multiple decisions such as seed transfer and feed additions are dependent on the viable cell density of the culture. In addition, the advances in instrumentation and data analysis algorithms have made nondestructive near-infrared spectroscopy a useful tool for measuring multiple components simultaneously from a single probe, including glucose, glutamine, lactate, ammonium, and even protein product concentration in cell culture broth (258). These in-process, online, at-line, and offline measurements are applied to improve process control and performances, such as optimizing nutrient feeding strategy, identifying critical process parameters, and initiating appropriate control actions to main process and product consistency.

Bioreactor control methods compare either measured or derived values with mathematical simulations to determine the timing and interval of feed introduction to mammalian cell culture (177). In the standard "closed-loop" strategy, the orchestrated nutrient supplementation is based on precise mathematical formulas, while the "open-loop" methods use real-time process measurements to provide feedback to controlling devices and guide their actions (146). When a simple formula cannot describe a process due to lack of accurate data, the nonlinearity of many functions, or delays in biological response to control inputs, recent statistical and nonformulaic techniques such as "fuzzy logic" and "neural networks" are becoming prevalent (194). A proportional-integral-derivative type of controller is frequently used to integrate the control algorithms into electronic and mechanical hardware for reactor control, which brings values to their set points and maintains them without overshooting by controlling various actuators (e.g., peristaltic pumps for glucose feeding). Control features different from microbial fermentors are reflected in the reactor design for mammalian cells (231), in order to ensure minimum stress to cells while maintaining adequate mixing and mass transfer of not only oxygen but also CO_2. DO level is regulated by the controlled sparging of air, and oxygen gas if needed. The pH control is achieved by CO_2 and N_2 sparging, or alkaline agents such as $NaHCO_3$, Na_2CO_3, or NaOH. The latter should be added in a well-mixed zone of the reactor to avoid high local departure from the set pH point (240).

12.5.3. Optimization of Culture Conditions

Enhancements in recombinant MAb production on a per-cell basis are often the result of combined improvements of cell line, culture medium, and culture conditions. The almost 10-fold improvement in titer every decade since 1980 (250) has been partially due to improvements in expression technology and clone selection, as noted above, and partially due to the optimization of cell culture processes (52), which include media and feeds and culture conditions. As a result, much higher cell concentrations can now be achieved in bioreactors (easily >1 to 2×10^7 viable cells/ml), and the cells can be maintained in a viable state for a much longer period, leading to dramatically enhanced productivity or titer (= Qp × IVC). If culture duration cannot be extended due to limitations of the cell line or plant throughput considerations, efforts should then be focused on increasing Qp and maximum cell density. Strategies for developing media and feed solutions were described in the previous section, as well as in published literature (247). As for the optimization of a fed-batch process, it involves a delicate interplay among such variables as feed solution constituents and concentrations, the timing and duration of feed introduction, and control limits for reactor operating parameters (134, 195, 248). The overall aim of the fed-batch approach is to increase maximum cell concentrations in the bioreactor and to prolong the production phase without decreasing Qp. By this approach, the productivities of NS0 (52) and Sp2/0 processes (195) were improved by about 1 order of magnitude. It should be noted that a preferred cell growth condition may not be the optimal condition for productivity and vice versa; therefore, optimization of culture conditions needs to balance cell growth with antibody production.

In addition to maintaining nutritional supply to cells through feeding, a key factor to consider is the toxic effect of accumulated catabolites such as ammonium and lactate. It has been demonstrated that these metabolites can be inhibitory to mammalian cell growth and production (39). It is possible to use stoichiometric feeding of key energy substrates (glutamine and glucose) to reduce these metabolites (195, 254). Accumulation of ammonium can be substantially reduced by eliminating the main source, glutamine, in the medium, which can be the case when cell lines have endogenous GS or are constructed with GS-containing vectors; thus the growth does not require glutamine in the medium. An ~45% increase in the specific production rate of both lactate

and ammonium was observed when osmolality was elevated from 316 to 450 mOsm/kg in a CHO culture (267).

For most mammalian cell culture, pH is normally controlled within the range of 6.8 to 7.4. High pH is preferred for initial cell growth, whereas low pH facilitates antibody production (170, 195). However, high pH is usually associated with increased anaerobic cell metabolism, which converts more glucose to lactate and drives down the medium pH. The control mechanism for pH in bioreactors then causes more base addition to the medium, resulting in increased osmolality, which can delay cell growth and accelerate cell death. So the suggested control mechanism is that once fast cell growth is achieved, the pH set point is shifted to a lower level to minimize base addition and to allow the best antibody production (135). For most cell culture processes, lactate is consumed in the later stage, so pH can rise by itself.

In addition to pH control agents, addition of feeds can lead osmolality to reach growth-inhibitory levels. However, due to the fact that elevated osmolality also enhances Qp (261), the final titer may not be much affected (267). CO_2 concentration in large-scale bioreactors can also reach inhibitory levels (>120 mm Hg) in cell culture, decreasing the specific growth rate and specific production rate (164, 267), which is why sufficient air sparging needs to be maintained in large-scale bioreactor operation—not only to supply enough DO but also to strip out the accumulated CO_2 adequately. The challenge is that a high sparging rate often leads to a tendency to foaming, so control of a bioreactor run is an act of balancing various interacting factors. By using either antifoam with sparge stone or an open pipe sparging in a 1,000-liter CHO culture, the dCO_2 was reduced from ~180 to ~70 mm Hg, resulting in twofold higher titer (164). DO is usually considered a less significant factor in mammalian cell growth, so it can be controlled in a wide range of 20 to 100% air saturation (135). But DO can change product quality: it was reported that a reduction in DO caused decreased glycosylation of a MAb's N-glycan chains (128).

The maintenance of cells at a high density without growth can be achieved by different means, including temperature reduction or addition of chemicals that interfere with cell physiology, resulting in enhancement and stabilization of viability. For example, manipulating the cell cycle in CHO cell culture through a temperature shift can extend culture longevity. It was found that lowering temperature from 37 to 30°C at 48 h postinoculation can retain cells in the G_1 phase longer and thus delay the onset of apoptosis (156). It was also found (126) that growth arrest by lowering temperature following 37°C growth of CHO-K1 cells leads to increased recombinant protein productivity, and a number of key regulatory proteins and pathways were identified as modulating the cell response to mild hypothermia.

12.5.4. Process Scale-Up and Scale-Down

The scale-up of cell culture processes remains poorly understood, which often leads to altered yields and product quality attributes. While the exact criteria are still unknown, a successful process scale-up depends on determining, measuring, and monitoring critical scale-independent process parameters, then designing and operating equipment appropriately to deliver those same parameters at large scale (87). The least directly scalable process parameters in cell culture production are fluid mixing and oxygen transfer. The shear sensitivity of mammalian cells is no longer considered a serious issue in scale-up, because with no gas-liquid interface present, the local energy dissipation rate the mammalian cells would experience is orders of magnitude lower than that demonstrated to be catastrophically damaging to them

(154). Successful examples of scale-up of antibody cell culture processes can be found in the literature (206, 257). A systematic approach was proposed (58) to address the incompatibility between lab procedures and pilot scale and manufacturing, such as batch versus continuous operations, and theory versus reality. Other factors include raw material differences and dissimilar bioreactor hydrodynamics at different scales. Even scheduling and work-shift difference needs to be taken into consideration (89).

Using a good laboratory model is an easy and sometimes the only practical way to troubleshoot and improve a large-scale production process. Much of the feed and process optimization today is accomplished in scaled-down physical models built by imitating conditions existing in full-scale production reactors (50). Often the best small-scale conditions are sacrificed in favor of approximating those that can be actually achieved in a large-scale system. Newer technologies such as SimCell miniaturized bioreactors (www.bioprocessors.com) promise to accelerate optimization with more accurate scaled-down conditions. Scale-down models are also needed for process validation studies, in which a lab-scale version is validated for its ability to accurately represent a development process, then used in setting performance ranges and specifications for ultimate implementation in a manufacturing plant (184). Worst-case conditions, which pose the greatest chance of process or product failure (37), are tested to establish the operating ranges.

12.5.5. Process Improvement versus Product Quality

Frequently, change in productivity is not the only response cells exhibit to process modifications. It is possible that product characteristics alter as well, e.g., increased aggregation, different glycosylation, or faster degradation, potentially resulting in clinical implications for efficacy and safety. Glycosylation variation, which can affect in vivo IgG functions and product stability, is one of the most sensitive quality attributes that is dependent on cell line and cell culture conditions (110). Because the biosynthesis of glycans does not follow a genetic "code" like protein biosynthesis does, many variations are possible, thus leading to glycan heterogeneity. The biosynthesis of glycoproteins in mammalian cells and their characterization are reviewed in a recent paper (147). Due to the sensitivity of MAb glycosylation profiles to host cell enzymatic systems and bioreactor conditions (182, 209), selection of host cells and culture conditions must consider the requirement for stably producing each specific glycosylation pattern. Differences among the mammalian expression systems are well documented, particularly for CHO and NS0 cells (10). The effect of cell culture media and conditions on antibody glycosylation has been extensively studied (110, 112). Factors such as medium serum, glucose, ammonium, nutrient availability, particular sugar component, DO, dCO_2, and broth osmolality were reported to cause glycosylation changes in different cell lines (73, 208, 209, 262). The reason for this is that these factors likely affect monosaccharide transferase activities and/or sugar transport to the Golgi apparatus, the major glycosylation site in mammalian cells. Culture pH difference changed the galactosylation of glycans for a murine antibody (165). Different medium components can reduce Galα1-3Gal content on NS0 glycoproteins (10, 99). Some modifications can occur even on the antibody backbone under certain cell culture conditions. For example, the IgG heavy-chain C-terminal lysine is normally cut off through posttranscriptional modification by carboxypeptidase B. The activity of this type of metal enzyme can be affected by the trace element concentrations in the medium, resulting in charge variants (135). It is thus important to monitor product quality attributes during process optimization in order to ensure comparability across various clinical stages and commercial manufacturing. A recent regulation has provided comparability guidelines for biotechnological and biological products (70).

12.6. SUMMARY

Biopharmaceuticals, particularly MAbs, have been growing at a much faster rate than the traditional pharmaceutical industry in recent years, and the trend is continuing. Mammalian cells are the dominant production system due to their capacity for adding humanlike posttranslational modifications to complex protein therapeutics. High productivities and desired product qualities have been achieved by the development of high-yielding and stable cell lines, animal-free and productive media, high-density fed-batch cell culture conditions, and successful scale-up to large manufacturing bioreactors. Novel technologies such as enhanced expression systems, automated screening methods, cell line engineering, improved process monitoring, and disposable apparatuses are leading to more productive and efficient production of biopharmaceuticals by mammalian cell culture.

REFERENCES

1. **Allison, D. W., K. A. Aboytes, D. K. Fong, S. L. Leugers, T. K. Johnson, and H. N. Loke.** 2005. Development and optimization of cell culture media: genomic and proteomic approaches. *BioProcess Int.* **3:**38–45.
2. **Al-Rubeai, M., and R. P. Singh.** 1998. Apoptosis in cell culture. *Curr. Opin. Biotechnol.* **9:**152–156.
3. **Altamirano, C., C. Paredes, A. Illanes, J. J. Cairo, and F. Godia.** 2004. Strategies for fed-batch cultivation of t-PA producing CHO cells: substitution of glucose and glutamine and rational design of culture medium. *J. Biotechnol.* **110:**171–179.
4. **Anderson, D. C., and L. Krummen.** 2002. Recombinant protein expression for therapeutic applications. *Curr. Opin. Biotechnol.* **13:**117–123.
5. **Anderson, D. C., and D. E. Reilly.** 2004. Production technologies for MAbs and their fragments. *Curr. Opin. Biotechnol.* **15:**456–462.
6. **Arden, N., and M. J. Betehbaugh.** 2004. Life and death in mammalian cell culture: strategies for apoptosis inhibition. *Trends Biotechnol.* **22:**174–180.
7. **Baer, A., and J. Bode.** 2001. Coping with kinetics and thermodynamic barriers: RMCE, an effective strategy for the targeted integration of transgenes. *Curr. Opin. Biotechnol.* **12:**473–480.
8. **Baez, J., D. Olson, and J. W. Polarek.** 2005. Recombinant microbial systems for the production of human collagen and gelatin. *Appl. Microbiol. Biotechnol.* **69:**245–252.
9. **Bailey, C. G., A. S. Tait, and N. A. Sunstrom.** 2002. High-throughput clonal selection of recombinant CHO cells using a dominant selectable and amplifiable metallothionein-GFP fusion protein. *Biotechnol. Bioeng.* **80:**670–676.
10. **Baker, K. N., M. H. Rendall, A. E. Hills, M. Hoare, R. B. Freedman, and D. C. James.** 2001. Metabolic control of recombinant protein N-glycan processing in NS0 and CHO cells. *Biotechnol. Bioeng.* **73:**188–202.
11. **Baldi, L., N. Muller, S. Picasso, R. Jacquet, P. Girard, H. P. Thanh, E. Derow, and F. M. Wurm.** 2005. Transient gene expression in suspension HEK-293 cells: application to large-scale protein production. *Biotechnol. Prog.* **21:**148–153.

12. **Barnes, L. M., C. M. Bentley, and A. J. Dickson.** 2000. Advances in animal cell recombinant protein production: GS-NS0 expression system. *Cytotechnology* **32:**109–123.

13. **Barnes, L. M., C. M. Bentley, and A. J. Dickson.** 2001. Characterization of the stability of recombinant protein production in GS-NS0 expression system. *Biotechnol. Bioeng.* **73:**261–270.

14. **Barnes, L. M., C. M. Bentley, and A. J. Dickson.** 2003. Stability of protein production from recombinant mammalian cells. *Biotechnol. Bioeng.* **81:**631–639.

15. **Barnes, L. M., C. M. Bentley, N. Moy, and A. J. Dickson.** 2007. Molecular analysis of successful cell line selection in transfected GS-NS0 myeloma cells. *Biotechnol. Bioeng.* **96:**337–348.

16. **Bebbington, C. R., G. L. Renner, S. Thomson, D. King, D. Abrams, and G. T. Yarranton.** 1992. High level expression of a recombinant antibody from myeloma cells using a glutamine synthetase gene as an amplifiable selectable marker. *Bio/Technology* **10:**169–175.

17. **Beck, A., E. Wagner-Rousset, M. C. Bussat, M. Lokteff, C. Klinguer-Hamour, J. F. Haeuw, L. Goetsch, T. Wurch, A. van Dorsselaer, and N. Corvaia.** 2008. Trends in glycosylation, glycoanalysis and glycoengineering of therapeutic antibodies and Fc-fusion proteins. *Curr. Pharm. Biotechnol.* **9:**482–501.

18. **Beck, A., T. Wurch, and N. Corvaia.** 2008. Therapeutic antibodies and derivatives: from the bench to the clinic. *Curr. Pharm. Biotechnol.* **9:**421–422.

19. **Benton, T., T. Chen, M. McEntee, B. Fox, D. King, R. Crombie, and T. C. Thomas.** 2002. The use of UCOE vectors in combination with a preadapted serum free, suspension cell line allows for rapid production of large quantities of protein. *Cytotechnology* **38:**43–46.

20. **Bi, J. X., J. Shuttleworth, and M. Al-Rubeai.** 2004. Uncoupling of cell growth and proliferation results in enhancement of productivity in p21^{CIP1}-arrested CHO cells. *Biotechnol. Bioeng.* **85:**741–749.

21. **Birch, J. R.** 2000. Cell products—antibodies, p. 411–424. *In* R. E. Spier (ed.), *Encyclopedia of Cell Technology.* John Wiley & Sons, Inc., Hoboken, NJ.

22. **Birch, J. R., J. Bonnerjea, and S. Flatman.** 1995. The production of monoclonal antibodies, p. 231–265. *In* J. R. Birch and E. S. Lennox (ed.), *Monoclonal Antibodies: Principles and Applications.* John Wiley & Sons, Inc., Hoboken, NJ.

23. **Birch, J. R., D. O. Mainwaring, and A. J. Racher.** 2005. Use of the glutamine synthetase (GS) expression system for the rapid development of highly productive mammalian cell processes, p. 809–832. *In* J. Knablein (ed.), *Modern Biopharmaceuticals.* Wiley-VCH Verlag GmbH and Co. KGaA, Weinheim, Germany.

24. **Birch, J. R., and A. J. Racher.** 2006. Antibody production. *Adv. Drug Delivery Rev.* **58:**671–685.

25. **Bohm, E., R. Voglauer, W. Steinfellner, R. Kunert, N. Borth, and H. Katinger.** 2004. Screening for improved cell performance: selection of subclones with altered production kinetics of improved stability by cell sorting. *Biotechnol. Bioeng.* **88:**699–706.

26. **Borth, N., D. Mattanovich, R. Kunert, and H. Katinger.** 2005. Effect of increased expression of protein disulfide isomerase and heavy chain binding protein on antibody secretion in a recombinant CHO cell line. *Biotechnol. Prog.* **21:**106–111.

27. **Bowering, L. C.** 2004. Microbial systems for the manufacture of therapeutic antibody fragments. *BioProcess Int.* **2:**40–47.

28. **BPI Reports.** 2007. MAb market predicted to continue dominating. *BioProcess Int.* **5:**8.

29. **Broedel, S., Jr., and S. M. Papciak.** 2003. The case for serum-free media. *BioProcess Int.* **1:**56–58.

30. **Brower, A.** 2005. Number of monoclonal antibodies on market nearly doubles by 2008. *Biotechnol. Healthcare* **June:**64.

31. **Browne, S. M., and M. Al-Rubeai.** 2007. Selection methods for high-producing mammalian cell lines. *Trends Biotechnol.* **25:**425–432.

32. **Burgener, A., and M. Butler.** 2006. Medium development, p. 41–80. *In* S. S. Ozturk and W.-S. Hu (ed.), *Cell Culture Technology for Pharmaceutical and Cell-Based Therapies* (Biotechnology and Bioprocessing Series). CRC Press, Boca Raton, FL.

33. **Burky, J. E., M. C. Wesson, A. Young, S. Farnsworth, B. Dionne, and Y. Zhu.** 2007. Protein-free fed-batch culture of non-GS NS0 cell lines for production of recombinant antibodies. *Biotechnol. Bioeng.* **96:**281–293.

34. **Burteau, C. C., F. R. Verhoeye, J. F. Mols, J. S. Ballez, S. N. Agathos, and Y. J. Schneider.** 2003. Fortification of a protein-free cell culture medium with plant peptones improves cultivation and productivity of an interferon-g-producing CHO cell line. *In Vitro Cell. Dev. Biol. Anim.* **39:**291–296.

35. **Butler, M.** 2005. Animal cell cultures: recent achievements and perspectives in the production of biopharmaceuticals. *Appl. Microbiol. Biotechnol.* **68:**283–291.

36. **Carvell, J. P.** 2008. Viable biomass sensor for bioreactors. *BIO Forum Europe* **9:**54–55.

37. **Center for Drug Evaluation and Research.** 1993. *Guideline on General Principles of Process Validation.* U.S. Food and Drug Administration, Rockville, MD.

38. **Chattopadhyay, S., and L. Pavithra.** 2007. MARs and MARBPs: key modulators of gene regulation and disease manifestation. *Subcell. Biochem.* **41:**213–230.

39. **Chen, P., and S. W. Harcum.** 2007. Identification of genes sensitive to ammonium in CHO cell cultures using a differential display. *Appl. Biochem. Biotechnol.* **141:**349–360.

40. **Chu, L., and D. Robinson.** 2001. Industrial choices for protein production by large-scale cell culture. *Curr. Opin. Biotechnol.* **12:**180–187.

41. **Chun, B. H., J. H. Kim, H. J. Lee, and N. Chung.** 2007. Usability of size-excluded fractions of soy protein hydrolysates for growth and viability of Chinese hamster ovary cells in protein-free suspension culture. *Bioresour. Technol.* **98:**1000–1005.

42. **Clark, R. G., D. Mortensen, D. Reifsynder, M. Mohler, T. Etcheverry, and V. Mukku.** 1993. Recombinant human insulin-like growth factor binding protein-3 (rhIGFBP-3): effects on the glycemic and growth promoting activities of rhIGF-1 in the rat. *Growth Regul.* **3:**50–52.

43. **Coco-Martin, J. M.** 2004. Mammalian expression of therapeutic proteins—a review of advancing technology. *BioProcess Int.* **2:**32–40.

44. **Coco-Martin, J. M., and M. M. Harmsen.** 2008. A review of therapeutic protein expression by mammalian cells. *BioProcess Int.* **6:**S28–S33.

45. **Coloma, M. J., A. Clift, L. Wims, and S. L. Morrison.** 2000. The role of carbohydrate in the assembly and function of polymeric IgG. *Mol. Immunol.* **37:**1081–1090.

46. **Cui, L.** 2004. Microbial process development for biopharmaceuticals. *BioProcess Int.* **2:**32–39.

47. **Davies, C. G., M. L. Gallo, and J. R. F. Corvalan.** 1999. Transgenic mice as a source of fully human antibodies for the treatment of cancer. *Cancer Metastasis Rev.* **18:**421–425.

48. **Deeds, Z. W., S. Updike, B. J. Cutak, and M. V. Caple.** 2007. Working towards a chemically-defined replacement for hydrolysates, p. 649–651. *In* R. Smith (ed.), *Cell Technology for Cell Products. Proceedings of the 19th ESACT Meeting, Harrogate, UK.* Springer, Dordrecht, The Netherlands.

49. **Deer, J. R., and D. S. Allison.** 2004. High-level expression proteins in mammalian cells using transcription regulatory sequences from the Chinese hamster EF1-α gene. *Biotechnol. Prog.* **20:**880–889.

50. **De Jesus, M. J., and F. M. Wurm.** 2009. Medium and process optimization for high yield, high density suspension cultures: from low throughput spinner flasks to high throughput milliliter reactors. *BioProcess Int.* **7:**S12–S17.

51. **DeMaria, C. T., V. Cairnes, C. Schwarz, J. Zhang, M. Guerin, and E. Zuena.** 2007. Accelerated clone selection for recombinant CHO cells using a FACS-based high-throughput screen. *Biotechnol. Prog.* **23:**465–472.

52. **Dempsey, J., S. Ruddock, M. Osborne, A. Ridley, S. Sturt, and R. Field.** 2003. Improved fermentation processes for NS0 cell lines expressing human antibodies and glutamine synthetase. *Biotechnol. Prog.* **19:**175–178.

53. **Deo, Y. M., M. D. Mahadevan, and R. Fuchs.** 1996. Practical considerations in operation and scale-up of spin-filter based bioreactors for monoclonal antibody production. *Biotechnol. Prog.* **12:**57–64.

54. **DePalma, A.** 2008. Strengthening mammalian cell culture. *Genet. Eng. Biotechnol. News* **28**(18).

55. **Derouazi, M.** 2004. Serum-free large-scale transient transfection of CHO cells. *Biotechnol. Bioeng.* **87:**537–545.

56. **DeZengotita, V. M., P. Girard, F. Van Tilborgh, K. Iglesias, N. Muller, M. Bertschinger, and F. M. Wurm.** 2000. Phosphate feeding improves high-cell-concentration for monoclonal antibody production. *Biotechnol. Bioeng.* **69:**566–576.

57. **Dinnis, D. M., and D. C. James.** 2005. Engineering mammalian cell factories for improved recombinant monoclonal antibody production: lessons from nature? *Biotechnol. Bioeng.* **91:**180–189.

58. **Douglas, A. S.** 2005. Efficient product development—a systems approach. *BioProcess Int.* **3:**26–27.

59. **Dubel, S.** 2007. Recombinant therapeutic antibodies. *Appl. Microbiol. Biotechnol.* **74:**723–729.

60. **Duncan, A., and G. Hadlaczky.** 2007. Chromosomal engineering. *Curr. Opin. Biotechnol.* **18:**420–424.

61. **Eibl, R., and D. Eibl.** 2009. Disposable bioreactors in cell culture-based upstream processing. *BioProcess Int.* **7:**S18–S23.

62. **European Medicines Evaluation Agency.** 2002. Note for guidance on minimizing the risk of transmitting animal spongiform encephalopathies via human and veterinary medicinal products. EMEA/410/01, Rev. 2. EMEA, London, United Kingdom.

63. **Farges-Haddani, B., B. Tessier, S. Chenu, I. Chevalot, C. Harscoat, I. Marc, J. L. Goergen, and A. Marc.** 2006. Peptide fractions of rapeseed hydrolysates as an alternative to animal proteins in CHO cell culture media. *Process Biochem.* **41:**2297–2304.

64. **Field, R. P.** July, 2003. Animal cell culture. U.S. patent 6,593,140 (Assignee: Lonza Group AG).

65. **Figueroa, B., Jr., E. Ailor, D. Osborne, J. M. Hardwick, M. Reff, and M. J. Betenbaugh.** 2007. Enhanced cell culture performance using inducible anti-apoptotic genes *E1B-19K* and *Aven* in the production of a monoclonal antibody with Chinese hamster ovary cells. *Biotechnol. Bioeng.* **97:**877–892.

66. **Fire, A., S. Xu, M. K. Montgomery, S. A. Kostas, S. E. Driver, and C. C. Mello.** 1998. Potent and specific genetic interference by double-stranded RNA in *Caenorhabditis elegans*. *Nature* **391:**806–811.

67. **Fisch, I.** 2007. The role of matrix-attachment regions in increasing recombinant protein expression. *BioProcess Int.* **5:**66–72.

68. **Fisher, M.** 2006. A stirred-tank bioreactor delivered in eight weeks and one hour. *BioProcess Int.* **4:**S28–S30.

69. **Fletcher, T.** 2005. Designing culture media for recombinant protein production: a rational approach. *BioProcess Int.* **3:**30–36.

70. **Food and Drug Administration.** 2005. Guidance for industry: Q5E comparability of biotechnological/biological products subject to changes in their manufacturing process (ICH). Food and Drug Administration, Rockville, MD. (Online.) http://www.fda.gov/downloads/Drugs/GuidanceCompliance-RegulatoryInformation/Guidances/ucm073476.pdf.

71. **Food and Drug Administration.** 2009. Current good manufacturing practice in manufacturing, processing, packing, or holding of drugs; general (21 CFR, Part 210) Food and Drug Administration, Rockville, MD. (Online.) http://www.accessdata.fda.gov/scripts/cdrh/cfdocs/cfCFR/CFRSearch.cfm?CFRPart=210&showFR=1.

72. **Food and Drug Administration.** 2009. Current good manufacturing practices for finished pharmaceuticals (21 CFR, Part 211). Food and Drug Administration, Rockville, MD. (Online.) http://www.accessdata.fda.gov/scripts/cdrh/cfdocs/cfcfr/CFRSearch.cfm?CFRPart=211&showFR=1.

73. **Forno, G., M. Bollati Fogolin, M. Oggero, R. Kratje, M. Etcheverrigaray, H. S. Conradt, and M. Nimtz.** 2004. N- and O-linked carbohydrates and glycosylation site occupancy in recombinant human granulocyte-macrophage colony-stimulating factor secreted by a Chinese hamster ovary cell line. *Eur. J. Biochem.* **271:**907–919.

74. **Fox, S.** 2004. Maximizing outsourced biopharma production. *Contract Pharma.* **June:**72–78..

75. **Franek, F.** 2004. Gluten of spelt wheat (*Triticum aestivum* subspecies *spelta*) as a source of peptides promoting viability and product yield of mouse hybridoma cell cultures. *J. Agric. Food Chem.* **52:**4097–4100.

76. **Franek, F.** 2004. Peptide modulates growth and productivity of mammalian cell cultures and suppress apoptosis. *BioProcess Int.* **2:**48–52.

77. **Franek, F., T. Eckschlager, H. Katinger.** 2003. Enhancement of monoclonal antibody production by lysine-containing peptides. *Biotechnol. Prog.* **19:**169–174.

78. **Fukuda, Y., and S. Nishikawa.** 2003. Matrix attachment regions enhance transcription of a downstream transgene and the accessibility of its promoter region to micrococcal nuclease. *Plant Mol. Biol.* **51:**665–675.

79. **Fussenegger, M., and J. E. Bailey.** 1998. Molecular regulation of cell-cycle progression and apoptosis in mammalian cells: implications for biotechnology. *Biotechnol. Prog.* **14:**807–833.

80. **Fux, C., S. Moser, S. Schlatter, M. Rimann, J. E. Bailey, and M. Fussenegger.** 2001. Streptogramin- and tetracycline-responsive dual regulated expression of p27^{Kip1} sense and antisense enables positive and negative growth control of Chinese hamster ovary cell. *Nucleic Acids Res.* **29:**E19.

81. **Galbraith, D. N.** 2002. Transmissible spongiform encephalopathies and tissue cell culture. *Cytotechnology* **39:**117–124.

82. **Gambhir, A., C. Zhang, A. Europa, and W.-S. Hu.** 1999. Analysis of the use of fortified medium in continuous culture of mammalian cells. *Cytotechnology* **31:**243–254.

83. **Gasser, B., and D. Mattanovich.** 2007. Antibody production with yeasts and filamentous fungi: on the road to large scale? *Biotechnol. Lett.* **29:**201–212.

84. **Ge, X., M. Hanson, H. Shen, Y. Kostov, K. A. Brorson, D. D. Fray, A. Moreira, and G. Rao.** 2006. Validation of an optical sensor-based high-throughput bioreactor system for mammalian cell culture. *J. Biotechnol.* **122:**293–306.

85. **Gellissen, G., A. W. M. Strasser, and M. Suckow.** 2005. Key and criteria to the selection of an expression platform, p. 1–6. *In* G. Gellissen (ed.), *Production of Recombinant Proteins: Novel Microbial and Eucaryotic Expression Systems.* Wiley-VCH Verlag GmbH and Co. KGaA, Weinheim, Germany.

86. **Gerngross, T.** 2004. Advances in the production of human therapeutic protein in yeasts and filamentous fungi. *Nat. Biotechnol.* **22:**1409–1414.

87. **Gerson, D. F., and B. Mukherjee.** 2005. Manufacturing process development for high-volume, low-cost vaccines. *BioProcess Int.* **3:**42–50.

88. **Gillies, S. D., H. Dorai, J. Wasolowski, G. Majeau, D. Young, J. Boyd, J. Gardner, and K. James.** 1989. Expression of human anti-tetanus toxoid antibody in transfected murine myeloma cells. *Bio/Technology* **7:**799–804.

89. **Goochee, C. F.** 2002. The role of a process development group in biopharmaceutical process startup. *Cytotechnology* **38:**63–76.

90. **Gorfien, S., R. M. Fike, J. L. Dzimian, J. P. Godwin, P. J. Price, D. A. Epstein, D. Gruber, and C. McClure.** 2000. Growth of NS0 cells in protein-free, chemically-defined medium. *Biotechnol. Prog.* **16:**682–687.

91. **Graumann, K., and A. Premstaller.** 2006. Manufacturing of recombinant therapeutic proteins in microbial systems. *Biotechnol. J.* **1:**164–168.

92. **Grillari, J., K. Fortschegger, R. M. Grabherr, O. Hohenwarter, R. Kunert, and H. Katinger.** 2001. Analysis of alterations in gene expression after amplification of recombinant genes in CHO cells. *J. Biotechnol.* **87:**59–65.

93. **Grosvenor, S.** 2008. The role of media development in process optimization: an historical perspective. *In* Guide to Protein Production. *BioPharm Int.* **21:**S28–S36.

94. **Hacker, D. L., S. Nallet, and F. M. Wurm.** 2008. Recombinant protein production yields from mammalian cells: past, present, and future. *In* Guide to Protein Production. *BioPharm Int.* **21:**S6–S14.

95. **Hamilton, S. R., and T. U. Gerngross.** 2007. Glycosylation engineering in yeasts: the advent of fully humanized yeast. *Curr. Opin. Biotechnol.* **18:**387–392.

96. **Harms, P., Y. Kostov, and G. Rao.** 2002. Bioprocess monitoring. *Curr. Opin. Biotechnol.* **13:**124–127.

97. **Hellman, K. B., J. P. Honstead, and C. K. Vincent.** 1996. Adventitious agents from animal-derived raw materials and production systems. *Dev. Biol. Stand.* **88:**231–234.

98. **Hewlett, G.** 1991. Strategies for optimizing serum-free media. *Cytotechnology* **5:**3–14.

99. **Hills, A. E., A. Patel, P. Boyd, and D. C. James.** 2001. Metabolic control of recombinant monoclonal antibody N-glycosylation in GS-NS0 cells. *Biotechnol. Bioeng.* **75:**239–251.

100. **Ho, L., S. E. Lee, and A. E. Humphrey.** 2003. Industrial cell culture: principles, processes and products, p. 1046–1072. *In* J. A. Kent (ed.), *Reigel's Handbook of Industrial Chemistry*, 10th ed. Springer, Berlin, Germany.

101. **Hollis, G. F., and G. E. Mark.** June 2004. Homologous recombination antibody expression system for murine cells. U.S. Patent 6,750,041.

102. **Holmes, P., and M. Al-Rubeai.** 1999. Improved cell line development by a high throughput affinity capture surface display technique to select for high secretors. *J. Immunol. Methods* **230:**141–147.

103. **Huang, E. P., C. P. Marquis, and P. P. Gray.** 2004. Process development for a recombinant Chinese hamster ovary (CHO) cell line utilizing a metal-induced and amplified metallothionein expression system. *Biotechnol. Bioeng.* **88:**437–450.

104. **Humphreys, D. P.** 2003. Production of antibodies and antibody fragments in *Escherichia coli* and a comparison of their functions, uses and modifications. *Curr. Opin. Drug Discov. Devel.* **6:**188–196.

105. **Ifandi, V., and M. Al-Rubeai.** 2005. Regulation of cells by the coexpression of c-Myc and Bcl-2. *Biotechnol. Prog.* **21:**671–677.

106. **Inoue, Y., L. B. Lopez, S. Kawamoto, N. Arita, K. Teruya, M. K. Seki, M. Shoji, S. Karmei, Y. Hashizume, H. Shiozawa, H. Tachibana, K. Ohashi, T. Yasumoto, T. Suzuki, K. Imai, M. Nomoto, Y. Takenoyama, and S. Katakura.** 1996. Production of recombinant human monoclonal antibody using ras-amplified BHK-21 cells in a protein-free medium. *Biosci. Biotechnol. Biochem.* **60:**811–817.

107. **Ishaque, A., and M. Al-Rubeai.** 2002. Role of vitamins in determining apoptosis and extent of suppression by bcl-2 during hybridoma cell culture. *Apoptosis* **7:**231–239.

108. **Jakovcic, K.** 2007. Biomanufacturing strategies: market drivers, build-vs-buy decisions and opportunities in contract relationship management. Business Insights Ltd., London, UK. (Online.) www.globalbusinessinsights.com.

109. **Jayme, D. W.** 1999. An animal origin perspective of common constituents of serum-free medium formulations. *Dev. Biol. Stand.* **99:**181–187.

110. **Jefferis, R.** 2005. Glycosylation of recombinant antibody therapeutics. *Biotechnol. Prog.* **21:**11–16.

111. **Jenkins, N.** 2003. Analysis and manipulation of recombinant glycoproteins manufactured in mammalian cell culture, p. 3–20. *In* V. A. Vinci and S. R. Parekh (ed.), *Handbook of Industrial Cell Culture: Mammalian, Microbial, and Plant Cells.* Humana Press, Totowa, NJ.

112. **Jenkins, N., R. B. Parekh, and D. C. James.** 1996. Getting the glycosylation right: implications for the biotechnology industry. *Nat. Biotechnol.* **14:**975–981.

113. **Jones, D., N. Kroos, R. Anema, B. van Montfoort, A. Vooys, S. van der Kraats, E. van der Helm, S. Smits, J. Schouten, K. Brouwer, F. Lagerwerf, P. van Berkel, D. J. Opstelten, T. Logtenberg, and A. Bout.** 2003. High-level expression of recombinant IgG in the human cell line PER.C6. *Biotechnol. Prog.* **19:**163–168.

114. **Jones, S. D., F. J. Castillo, and H. L. Levine.** 2007. Advances in the development of therapeutic monoclonal antibodies. *BioPharm Int.* **20:**96–114.

115. **Julien, C.** 2006. Production of humanlike recombinant proteins in *Pichia pastoris:* from expression vector to fermentation strategy. *BioProcess Int.* **4:**22–33.

116. **Julien, C.** 1998. Scaling-up from spinners, T-flasks and shakers: a versatile bioreactor for mammalian and microbial cells. *Am. Biotechnol. Lab.* **16**(6):12–13.

117. **Jung, B., S. Lee, I. H. Yang, T. Good, and G. L. Cote.** 2002. Automated on-line noninvasive optical glucose monitoring in a cell culture system. *Appl. Spectrosc.* **56:**51–57.

118. **Kayser, K., N. Lin, D. Allison, L. Donahue, and M. Caple.** 2006. Cell line engineering methods for improving productivity. *BioProcess Int.* **4:**S6–S13.

119. **Keen, K. J., and C. Hale.** 1996. The use of serum-free medium for the production of functionally active humanized monoclonal antibody from NS0 mouse myeloma cells engineered using glutamine synthetase as a selectable marker. *Cytotechnology* **18:**207–217.

120. **Kim, J. M., J.-S. Kim, D.-H. Park, H. S. Kang, J. Yoon, K. Baek, and Y. Yoon.** 2004. Improved recombinant gene expression in CHO cells using matrix attachment regions. *J. Biotechnol.* **107:**95–105.

121. **Kim, N. S., S. J. Kim, and G. M. Lee.** 1998. Clonal variability within dihydrofolate reductase-mediated gene amplified Chinese hamster ovary cells: stability in the absence of selective pressure. *Biotechnol. Bioeng.* **60:**679–688.

122. **Kirschweger, G.** 2003. Crucell: biopharmaceuticals—as human as they get. *Mol. Ther.* **7:**5–6.

123. **Kisiday, J. D., B. Kurz, M. A. DiMicco, and A. J. Grodzinsky.** 2005. Evaluation of medium supplemented with insulin-transferrin-selenium for culture of primary bovine calf chondrocyte in three-demensional hydrogel scaffolds. *Tissue Eng.* **11:**141–151.

124. **Kobr, M., P. Chatellard, and M. O. Imhof.** 2008. Expression vector engineering for cell line development—new roles for "old" sequences. *BioProcess. J.* **7:**16–20.

125. **Kohler, G., and C. Milstein.** 1975. Continuous cultures of fused cells secreting antibody of predefined specificity. *Nature* **256:**495–497.

126. **Kumar, N., P. Gammell, P. Meleady, M. Henry, and M. Clynes.** 2008. Differential protein expression following low temperature culture of suspension CHO-K1 cells. *BMC Biotechnol.* **8:**42–54.

127. **Kunert, R., S. Wolbank, M. Chang, R. Voglauer, N. Borth, and H. Katinger.** 2004. Control of key parameters in developing mammalian production clones. *BioProcess Int.* **2:**54–59.

128. **Kunkel, J. P., D. C. H. Jan, J. C. Jamieson, and M. Butler.** 1998. Dissolved oxygen concentration in serum-free continuous culture affects N-linked glycosylation of a monoclonal antibody. *J. Biotechnol.* **62:**55–71.

129. **Kwaks, T. H., P. Barnett, W. Hemrika, T. J. Siersma, R. G. Sewalt, D. P. Satijn, J. F. Brons, R. Van Blokland, P. Kwakman, A. L. Kruckeberg, A. Kelder, and A. P. Otte.** 2003. Identification of anti-repressor elements that confer high and stable protein production in mammalian cells. *Nat. Biotechnol.* **21:**553–558.

130. **Kwaks, T. H., and A. P. Otte.** 2006. Employing epigenetics to augment the expression of therapeutic proteins in mammalian cells. *Trends Biotechnol.* **24:**137–142.

131. **Laken, H. A., and M. W. Leonard.** 2001. Understanding and modulating apoptosis in industrial cell culture. *Curr. Opin. Biotechnol.* **12:**175–179.

132. **Larson, T. M.** 2002. Chemometric evaluation of on-line HPLC in mammalian cell cultures: analysis of amino acids and glucose. *Biotechnol. Bioeng.* **77:**553–563.

133. **Lattenmayer, C., M. Loeschel, K. Schriebl, W. Steinfellner, T. Sterovsky, E. Trummer, K. Vorauer-Uhl, D. Müller, H. Katinger, and R. Kunert.** 2007. Protein-free transfection of CHO host cells with an IgG-fusion protein: selection and characterization of stable high-producers and comparison of conventionally transfected clones. *Biotechnol. Bioeng.* **96:**1118–1126.

134. **Lavric, V., I. Ofiteru, and A. Woinaroschy.** 2005. Sensitivity analysis of the fed-batch animal-cell bioreactor with respect to some control parameters. *Biotechnol. Appl. Biochem.* **41:**29–35.

135. **Li, F., J. Z. Zhou, X. Yang, T. Tressel, and B. Lee.** 2006. Current therapeutic antibody production and process optimization. *BioProcess. J.* **4:**16–25.

136. **Li, J., T. Sai, M. Berger, Q. Chao, D. Davidson, G. Deshmukh, B. Drozdowski, W. Ebel, S. Harley, M. Henry, S. Jacob, B. Kline, E. Lazo, F. Rotella, E. Routhier, K. Rudolph, J. Sage, J. Yao, Y. Zhou, M. Kavuru, T. Bonfield, M. J. Thomassen, P. M. Sass, N. C. Nicolaides, and L. Grasso.** 2006. Human antibodies for immunotherapy development generated via a human B cell hybridoma technology. *Proc. Natl. Acad. Sci. USA* **103:**3557–3562.

137. **Lim, A. C., J. Washbrook, N. J. Titchener-Hooker, and S. S. Farid.** 2006. A computer-aided approach to compare the production economics of fed-batch and perfusion culture under uncertainty. *Biotechnol. Bioeng.* **93:**687–697.

138. **Luo, Y., and G. Chen.** 2007. Combined approach of NMR and chemometrics for screening peptones used in the cell culture medium for the production of a recombinant therapeutic protein. *Biotechnol. Bioeng.* **97:**1654–1659.

139. **Ma, N., J. J. Chalmers, J. G. Aunins, W. Zhou, and L. Xie.** 2004. Quantitative studies of cell-bubble interactions and cell damage at different Pluronic F-68 and cell concentrations. *Biotechnol. Prog.* **20:**1183–1191.

140. **Maggon, K.** 2007. Monoclonal antibody "gold rush." *Curr. Med. Chem.* **14:**1978–1987.

141. **Maggon, K.** 2009. Global monoclonal antibodies market review 2008 (world top ten MoAbs). (Online.) http://knol.google.com/k/krishan-maggon/global-monoclonal-antibodies-market.

142. **Magil, S. G.** 2005. Biopharmaceutical characterization techniques for early phase development of proteins. *In* Guide to Bioanalytical Advances. *BioPharm Int.* **18:**S34–S42.

143. **Mahadevan, M. D.** 2003. Bioreactor process selection for large-scale manufacture of monoclonal antibodies: tradeoffs and recommendations. *BioProcess. J.* **2:**25–31.

144. **Majors, B. S., M. J. Betenbaugh, and G. G. Chiang.** 2007. Links between metabolism and apoptosis in mammalian cells: application for anti-apoptosis engineering. *Metab. Eng.* **9:**473–480.

145. **Makrides, S. C.** 1999. Components for vectors for gene transfer and expression in mammalian cells. *Protein Expr. Purif.* **17:**183–202.

146. **Mandenius, C. F.** 2004. Recent developments in the monitoring, modeling and control of biological production systems. *Bioprocess Biosyst. Eng.* **26:**347–351.

147. **Manzi, A. E.** 2008. Carbohydrates and their analysis, part two: glycoprotein characterization. *BioProcess Int.* **6:**50–67.

148. **Merten, O. W.** 1999. Safety issues of animal products used in serum-free medium. *Dev. Biol. Stand.* **99:**167–180.

149. **Merten, O. W.** 2002. Virus contaminations of cell cultures—a biotechnological view. *Cytotechnology* **39:**91–116.

150. **Metcalfe, H., R. P. Field, and S. J. Froud.** 1994. The use of 2-hydroxy-2,4,6-cycloheptarin-1-one (tropolone) as a replacement for transferrin, p. 88–90. *In* R. E. Spier, J. B. Griffiths, and W. Berthold (ed.), *Animal Cell Technology: Products of Today, Prospects of Tomorrow.* Butterworth-Heinemann, Woburn, MA.

151. **Meuwly, F., U. Weber, T. Ziegler, A. Gervais, R. Mastrangeli, C. Crisci, M. Rossi, A. Bernard, U. von Stockar, and A. Kadouri.** 2006. Conversion of a CHO cell culture from perfusion to fed-batch technology without altering product quality. *J. Biotechnol.* **123:**106–116.

152. **Meyer, H. P., J. Brass, C. Jungo, J. Klein, J. Wenger, and R. Mommers.** 2008. An emerging start for therapeutic and catalytic protein production. *BioProcess Int.* **6:**S10–S21.

153. **Mirasol, F.** 2008. Disposable bioreactor use grows in commercial production: a disposable future. *ICIS Chem. Bus.* **273**(10). (Online.) http://www.icis.com/Articles/2008/03/10/9105767/Disposable-bioreactor-use-grows-in-commercial-production.html.

154. **Mollet, M., N. Ma, Y. Zhao, R. Brodkey, R. Taticek, and J. J. Chalmers.** 2004. Bioprocess equipment: characterization of energy dissipation rate and its potential to damage cells. *Biotechnol. Prog.* **20:**1437–1448.

155. **Mols, J., C. Peeters-Joris, R. Wattiez, S. N. Agathos, and Y. J. Schneider.** 2005. Recombinant interferon-gamma secreted by Chinese hamster ovary-320 cells cultivated in protein-free media is protected against extracellular proteolysis by the expression of natural protease inhibitors and by the addition of plant protein hydrolysates to the culture medium. *In Vitro Cell. Dev. Biol. Anim.* **41:**83–91.

156. **Moore, A., J. Mercer, G. Dutina, C. J. Donahue, K. D. Bauer, J. P. Mather, T. Etcheverry, and T. Ryll.** 1997. Effects of temperature shift on cell cycle, apoptosis, and nucleotide pools in CHO cell batch culture. *Cytotechnology* **23:**47–54.

157. **Moreira, A. R.** 2007. The evolution of protein expression and cell culture. *BioPharm Int.* **20:**56–68.

158. **Mori, K., R. Kuni-Kamochi, N. Yamane-Ohnuki, M. Wakitani, K. Yamano, H. Imai, Y. Kanda, R. Niwa, S. Iida, K. Uchida, K. Shitara, and M. Satoh.** 2004. Engineering Chinese hamster ovary cells to maximize effector function of produced antibodies using FUT8 siRNA. *Biotechnol. Bioeng.* **88:**901–908.

159. **Moroney, S., and A. Pluckthun.** 2005. Modern antibody technology: the impact on drug development, p. 1147–1186. *In* J. Knablein (ed.), *Modern Biopharmaceuticals.* Wiley-VCH Verlag GmbH and Co. KGaA, Weinheim, Germany.

160. **Morris, A. E., C.-C. Lee, and J. N. Thomas.** October, 2001. Expression augmenting sequence elements (ease) for eukaryotic expression system. U.S. patent 6,312,951.

161. **Morrow, K. J.** 2004. Antibody technology highlighted in Europe. *Genet. Eng. Biotechnol. News* **24**(1).

162. **Morrow, K. J.** 2006. Disposable bioreactors gaining favor. *Genet. Eng. Biotechnol. News* **26**(12).

163. **Morrow, K. J., Jr.** 2008. Method for maximizing antibody yields—new technologies could help usher in lower costs and increased availability. *Genet. Eng. Biotechnol. News* **28**(12).

164. **Mostafa, S., and X. Gu.** 2003. Strategies for improved dCO_2 removal in large-scale fed-batch cultures. *Biotechnol. Prog.* **19**:45–51.

165. **Muthing, J., S. E. Kemminer, H. S. Conradt, D. Sagi, M. Nimtz, U. Karst, J. Peter-Katalinic.** 2003. Effects of buffering conditions and culture pH on production rates and glycosylation of clinical Phase I anti-melanoma mouse IgG3 monoclonal antibody R24. *Biotechnol. Bioeng.* **83**:321–334.

166. **Nickoloff, J. A.** 1995. Preface, p. v–vi. *In* J. A. Nickoloff (ed.), *Electroporation Protocols for Microorganisms.* Humana Press, Totowa, NJ.

167. **Niloff, M.** August 2008. Scale-up to 1000-liter perfusion cell culture in a single-use, stirred tank bioreactor. *ISPE Boston Area Chapter Newsletter.* International Society for Pharmaceutical Engineering, Tampa, FL.

168. **Nyberg, G. B., R. R. Balcarcel, B. D. Follstad, G. Stephanopoulos, and D. I. C. Wang.** 1999. Metabolism of peptide amino acids by Chinese hamster ovary cells grown in a complex medium. *Biotechnol. Bioeng.* **62**:324–335.

169. **Okonkowski, J., U. Balasubramanian, C. Seamans, S. Fries, J. Zhang, P. Salmon, D. Robinson, and M. Chartrain.** 2007. Cholesterol delivery to NS0 cells: challenges and solutions in disposable liner low-density polyethylene-based bioreactors. *J. Biosci. Bioeng.* **103**:50–59.

170. **Osman, J., J. Birch, and J. Varley.** 2001. The response of GS-NS0 myeloma cells to pH shifts and pH perturbations. *Biotechnol. Bioeng.* **75**:63–73.

171. **Park, K. S., W. Seol, H.-Y. Yang, S. I. Lee, S. K. Kim, E.-J. Kwon, Y.-H. Kim, B. L. Roh, and J. S. Seong.** 2005. Identification and use of zinc finger transcription factors that increase production of recombinant proteins in yeast and mammalian cells. *Biotechnol. Prog.* **21**:664–670.

172. **Pendlebury, D.** 2008. Disposable systems in biopharmaceutical manufacturing. *Fut. Pharm.* **Quarter 2**:98–100.

173. **Percivia.** 2008. Press release: DSM and Crucell announce record achievement in PER.C6 technology. June 16. Percivia, Cambridge, MA. (Online.) http://investors.crucell.com/C/132631/PR/200806/1227870_5_5.html.

174. **Peshwa, M. V.** 1999. Mammalian cell culture, p. 181–191. *In* A. L. Demain and J. E. Davies (ed.), *Manual of Industrial Microbiology and Biotechnology,* 2nd ed. ASM Press, Washington, DC.

175. **Pierce, L. N., and P. W. Shabram.** 2004. Scalability of a disposable bioreactor from 25 L–500 L run in perfusion mode with a CHO-based cell line: a tech review. *BioProcess. J.* **4**:51–56.

176. **Pitti, R. M., S. A. Marsters, M. Haak-Frendscho, G. C. Osaka, J. Mordenti, S. M. Chamow, and A. Ashkenazi.** 1994. Molecular and biological properties of an interleukin-1 receptor immunoadhesin. *Mol. Immunol.* **31**:1345–1351.

177. **Portner, R., J. O. Schwabe, and B. Frahm.** 2004. Evaluation of selected control strategies for fed-batch cultures of a hybridoma cell line. *Biotechnol. Appl. Biochem.* **40**:47–55.

178. **Rader, R. A.** 2005. Biopharmaceutical terminology: what is a biopharmaceutical? *BioExecutive Internat.* Part 1, March; Part 2, May.

179. **Rader, R. A.** 2005. *Biopharmaceutical Products in the US and European Markets,* 4th ed. BioPlan Associates, Inc., Rockville, MD.

180. **Rader, R. A.** 2006. Biopharmaceutical approval review. *Genet. Eng. Biotechnol. News* **26**(14).

181. **Rader, R. A.** 2008. Expression systems for process and product improvement. *BioProcess Int.* **6**:S4–S9.

182. **Raju, T. S.** 2003. Glycosylation variations with expression systems and their impact on biological activity of therapeutic immunoglobulins. *BioProcess Int.* **1**:44–53.

183. **Raju, T. S., J. B. Briggs, S. M. Borge, and A. J. S. Jones.** 2000. Species-specific variation in glycosylation of IgG: evidence for the species-specific sialylation and branch-specific galactosylation and importance for engineering recombinant glycoprotein therapeutics. *Glycobiology* **10**:477–486.

184. **Ramelmeier, R. A., B. D. Kelley, and C. Van Horn.** 1998. Historical, current, and future trends for validating biological processes, p. 1–11. *In* B. D. Kelley and R. A. Ramelmeier (ed.), *Validation of Biopharmaceutical Manufacturing Processes.* American Chemical Society, Washington, DC.

185. **Rao, G., Y. Kostov, A. Moreira, D. Frey, M. Hanson, M. Jornitz, O.-W. Reif, R. Baumfalk, and J. Qualitz.** 2009. Non-invasive sensors as enablers of "smart" disposables. *BioProcess Int.* **7**:S24–S27.

186. **Reichert, J. M.** 2008. Monoclonal antibodies as innovative therapeutics. *Curr. Pharm. Biotechnol.* **9**:423–430.

187. **Reichert, J. M., C. J. Rosensweig, L. B. Faden, and M. C. Dewitz.** 2005. Monoclonal antibodies successes in the clinic. *Nat. Biotechnol.* **23**:1073–1078.

188. **Reik, A., Y. Zhou, T. N. Collingwood, M. Warfe, V. Bartsevich, Y. Kong, K. A. Henning, B. K. Fallentine, L. Zhang, X. Zhong, Y. Jouvenot, A. C. Jamieson, E. J. Rebar, C. C. Case, A. Korman, X.-Y. Li, A. Black, D. J. King, and P. D. Gregory.** 2007. Enhanced protein production by engineered zinc finger proteins. *Biotechnol. Bioeng.* **97**:1180–1189.

189. **Rendall, M. H., A. Maxwell, D. Tatham, P. Khan, R. D. Gay, R. C. Kallmeier, J. R. Wayte, and A. J. Racher.** 2005. Transfection to manufacturing: reducing timelines for high yielding GS-CHO processes, p. 701–704. *In* F. Godia and M. Fussenegger (ed.), *Animal Cell Technology Meets Genetics.* Springer, Dordrecht, The Netherlands.

190. **Rodriguez, J., M. Spearman, N. Huzel, and M. Butler.** 2005. Enhanced production of monomeric interferon-β by CHO cells through the control of culture conditions. *Biotechnol. Prog.* **21**:22–30.

191. **Rohwer, R. G.** 1996. Analysis of risk to biomedical products developed from animal sources (with special emphasis on the spongiform encephalopathy agents, scrapie and BSE). *Dev. Biol. Stand.* **88**:247–256.

192. **Ryu, J. S.** 2001. Effects of cloned gene dosage on the response of recombinant CHO cells to hyperosmotic pressure in regard to cell growth and antibody production. *Biotechnol. Prog.* **17**:993–999.

193. **Safdar, A., and M. M. Cox.** 2007. Baculovirus-expressed influenza vaccines: a novel technology for safe and expeditious vaccine production for human use. *Expert Opin. Investig. Drugs* **16**:927–934.

194. **Sarkar, D., and J. M. Modak.** 2004. Algorithms with filters for optimal control problems in fed-batch bioreactors. *Bioprocess Biosyst. Eng.* **26**:295–306.

195. **Sauer, P. W., J. E. Burky, M. C. Wesson, H. D. Sternard, and L. Qu.** 2000. A high-yielding, generic fed-batch cell culture process for production of recombinant antibodies. *Biotechnol. Bioeng.* **67**:585–597.

196. **Schenerman, M. A., J. Casas-Finet, M. J. Axley, and C. N. Oliver.** 2003. Characterization of alternatives to animal-derived raw materials. *BioProcess Int.* **1**:42–49.

197. **Schenerman, M. A., B. R. Sunday, S. Kozlowski, K. Webber, H. Gazzano-Santoro, and A. Mire-Sluis.** 2004. CMC

strategy forum report—analysis and structure characterization of monoclonal antibodies. *BioProcess Int.* **2**:42–52.

198. **Schimke, R. T., D. S. Roos, and P. C. Brown.** 1987. Amplification of genes in somatic mammalian cells. *Methods Enzymol.* **151**:85–104.

199. **Schlaeger, E. J.** 1996. The protein hydrolysate, Primatone RL, is a cost-effective multiple growth promoter of mammalian cell culture in serum containing and serum-free media and displays anti-apoptosis properties. *J. Immunol. Methods* **194**:191–199.

200. **Schlatter, S., S. H. Stanfield, D. M. Dinnis, A. J. Racher, J. R. Birch, and D. C. James.** 2005. On the optimal ratio of heavy to light chain genes for efficient recombinant antibody production by CHO cells. *Biotechnol. Prog.* **21**:122–133.

201. **Schonfelder, M., and P. Schlenke.** 2007. Keep the whole process in view: from stable cell lines to robust manufacturing processes. *BioProcess Int.* **5**:70–77.

202. **Schroder, M., K. Matischak, and P. Friedl.** 2004. Serum- and protein-free media formulations for the Chinese hamster ovary cell line DUKXB11. *J. Biotechnol.* **108**:279–292.

203. **Scott, C.** 2008. In the bioprocess zone: the state of the art is science. *In* Bio International Convention Preshow Planner. *BioProcess Int.* **6**:S8–S19.

204. **Scott, C.** 2008. Biotech leads a revolution in vaccine manufacturing. *BioProcess Int.* **6**:S12–S18.

205. **Scott, C., S. A. Montgomery, and L. J. Rosin.** 2007. Genetic engineering leads to microbial, animal cell, and transgenic expression systems. *In* Bio International Convention: Official Pre-Event Planner. *BioProcess Int.* **5**:S27–S34.

206. **Seamans, T. C., S. Fries, A. Beck, T. Wurch, S. Chenu, C. Chan, M. Ushio, J. Bailey, A. Kejariwal, C. Ranucci, N. Villani, S. Ozuna, L. Goetsch, N. Corvaia, P. Salmon, D. Robinson, and M. Chartrain.** 2008. Cell cultivation process transfer and scale-up in support of production of early clinical supplies of an anti-IGF-1R antibody, part 2. *BioProcess Int.* **6**:34–42.

207. **Selexis SA and Artelis SA.** 2008. Press release: Artelis and Selexis report record antibody production levels in CHO cells of over 31 grams per liter. Sept. 20. Artelis SA, Brussels, Belgium. (Online.) http://www.artelis.be/uploads//ARTELISSELEXIS%20SEPT08.pdf.

208. **Senger, R. S., and M. N. Karim.** 2003. Effect of shear stress on intrinsic CHO culture state and glycosylation of recombinant tissue-type plasminogen activator protein. *Biotechnol. Prog.* **19**:1199–1209.

209. **Senger, R. S., and M. N. Karim.** 2007. Optimization of fed-batch parameters and harvest time of CHO cell cultures for a glycosylated product with multiple mechanism of inactivation. *Biotechnol. Bioeng.* **98**:378–390.

210. **Seth, G., R. S. McIvor, and W.-S. Hu.** 2006. 17β-Hydroxysteroid dehydrogenase type 7 (Hsd17b7) reverts cholesterol auxotrophy in NS0 cells. *J. Biotechnol.* **121**:241–252.

211. **Seth, G., R. J. Philp, A. Lau, K. Y. Jiun, M. Yap, and W.-S. Hu.** 2007. Molecular portrait of high productivity in recombinant NS0 cells. *Biotechnol. Bioeng.* **97**:933–951.

212. **Sharfstein, S. T.** 2008. Advances in cell culture process development: tools and techniques for improving cell line development and process optimization. *Biotechnol. Prog.* **24**:727–734.

213. **Sheeley, D. M., B. M. Merrill, and L. C. E. Taylor.** 1997. Characterization of monoclonal antibody glycosylation: comparison of expression systems and identification of terminal α-linked galactose. *Anal. Biochem.* **247**:102–110.

214. **Shulman, M., C. D. Wilde, and G. A. Kohler.** 1978. A better cell line for making hybridoma secreting specific antibodies. *Nature* **276**:269–270.

215. **Simmons, L.C., D. Reilly, L. Klimowski, T. S. Raju, G. Meng, P. Sims, K. Hong, R. L. Shields, L. A. Damico, and P. Rancatore.** 2002. Expression of full-length immunoglobulins in *Escherichia coli*: rapid and efficient production of aglycosylated antibodies. *J. Immunol. Methods* **263**:133–147.

216. **Simpson, N. H., R. P. Singh, A. Perani, C. Goldenzon, and M. Al-Rubeai.** 1998. In hybridoma cultures, deprivation of any single amino acid leads to apoptotic death, which is suppressed by the expression of the *bcl-2* gene. *Biotechnol. Bioeng.* **59**:90–98.

217. **Sinacore, M. S., T. S. Charlebois, S. Harrison, S. Brennan, T. Richards, M. Hamilton, S. Scott, S. Brodeur, P. Oakes, M. Leonard, M. Switzer, A. Anagnostopoulos, B. Foster, A. Harris, M. Jankowski, M. Bond, S. Martin, and S. R. Adamson.** 1996. CHO DUKX cell lineages preadapted to growth in serum-free suspension culture enable rapid development of cell culture processes for the manufacture of recombinant proteins. *Biotechnol. Bioeng.* **52**:518–528.

218. **Sinclair, A.** 2008. Disposable bioreactors: the next generation. *BioPharm Int.* **21**:38–40.

219. **Sinclair, A.** 2009. An industry in transition. *BioProcess Int.* **7**:S1.

220. **Sinclair, A., and M. Monge.** 2002. Quantitative economic evaluation of single use disposables in bioprocessing. *Pharm. Eng.* **22**:20–34.

221. **Smales, C. M., D. M. Dinnis, S. H. Stansfield, D. E. Alete, E. A. Sage, J. R. Birch, A. J. Racher, C. T. Marshall, and D. C. James.** 2004. Comparative proteomic analysis of GS-NS0 murine myeloma cell lines with varying recombinant monoclonal antibody production rate. *Biotechnol. Bioeng.* **88**:474–488.

222. **Spens, E., and L. Haggstrom.** 2005. Defined protein-free NS0 myeloma cell cultures: stimulation of proliferation by conditioned medium factors. *Biotechnol. Prog.* **21**:87–95.

223. **Stein, A.** 2007. Decreasing variability in your cell culture. *Biotechniques* **43**:228–229.

224. **Stockwin, L. H., and S. Holmes.** 2003. Antibodies as therapeutic agents: vive la renaissance. *Expert Opin. Biol. Ther.* **3**:1133–1152.

225. **Sung, Y. H., S. W. Lim, J. Y. Chun, and J. M. Lee.** 2004. Yeast hydrolysate as a low-cost additive to serum-free medium for the production of human thrombopoietin in suspension cultures of Chinese hamster ovary cells. *Appl. Microbiol. Biotechnol.* **63**:527–536.

226. **Swartz, J. R.** 2001. Advances in *Escherichia coli* production of therapeutic proteins. *Curr. Opin. Biotechnol.* **12**:195–201.

227. **Thalen, M.** 2008. The next generation of biologicals and their production systems. *BioProcess Int.* **6**:S20–S23.

228. **Tsao, Y. S., S. L. Gould, and D. K. Robinson.** 2000. Animal cell culture media, p. 31–35. *In* R. E. Spier (ed.), *Encyclopedia of Cell Technology.* John Wiley & Sons, Inc., Hoboken, NJ.

229. **Umana, P., P. Jean-Mairet, R. Moudry, H. Amstutz, and J. E. Bailey.** 1999. Engineered glycoforms of an anti-neuroblastoma IgG1 with optimized antibody-dependent cellular cytotoxic activity. *Nat. Biotechnol.* **17**:176–180.

230. **Urnov, F. D., J. C. Miller, Y.-L. Lee, C. M. Beausejour, J. M. Rock, S. Augustus, A. C. Jamieson, M. H. Porteus, P. D. Gregory, and M. C. Holmes.** 2005. Highly efficient endogenous human gene correction using designed zinc-finger nucleases. *Nature* **435**:646–651.

231. **Varley, J., and J. R. Birch.** 1999. Reactor design for large-scale suspension animal cell culture. *Cytotechnology* **29**:177–205.

232. **Vicki, G.** 2005. Disposable bioreactors become standard fare. *Genet. Eng. Biotechnol. News* **25**(14).

233. **Voisard, D., F. Meuwly, P. A. Ruffieux, G. Baer, and A. Kadouri.** 2003. Potential of cell retention techniques for large-scale high-density perfusion culture of suspended mammalian cells. *Biotechnol. Bioeng.* **82**:751–765.

234. **Walowitz, J. L., R. M. Fike, and D. W. Jayme.** 2003. Efficient lipid delivery to hybridoma culture by use of cyclodextrin in a novel granulated dry-form medium technology. *Biotechnol. Prog.* **19:**64–68.

235. **Walsh, G.** 2003. Pharmaceuticals, biologics and biopharmaceuticals. *In* G. Walsh (ed.), *Biopharmaceuticals: Biochemistry and Biotechnology,* 2nd ed. John Wiley & Sons, Inc., Hoboken, NJ.

236. **Walsh, G.** 2005. Biopharmaceuticals; recent approvals and likely directions. *Trends Biotechnol.* **23:**553–558.

237. **Walsh, G.** 2006. Biopharmaceutical benchmark 2006. *Nat. Biotechnol.* **24:**769–776.

238. **Walsh, G.** 2008. Biopharmaceuticals: approval trends in 2007. *BioPharm Int.* **5:**52–65.

239. **Ward, M., L. Cherry, D. C. Victoria, B. P. Fox, J. A. Fox, D. L. Wong, H. J. Meerman, J. P. Pucci, R. B. Fong, M. H. Heng, N. Tsurushita, C. Gieswein, M. Park, and H. Wang.** 2004. Characterization of humanized antibodies secreted by *Aspergillus niger. Appl. Environ. Microbiol.* **70:**2567–2576.

240. **Wayte, M., R. Boraston, H. Bland, J. Varley, and M. Brown.** 1997. pH: effects on growth and productivity of cell lines producing monoclonal antibodies: control in large-scale fermentors. *Genet. Eng. Biotechnol. News* **17:**125–132.

241. **Weaver, J. C.** 1995. Electroporation theory: concepts and mechanisms, p. 1–26. *In* J. A. Nickoloff (ed.), *Electroporation Protocols for Microorganisms.* Humana Press, Totowa, NJ.

242. **Whitford, W. G.** 2004. Large-scale exogenous protein production in higher animal cells. *In* S. R. Parekh (ed.), *The GMO Handbook: Genetically Modified Animals, Microbes, and Plants in Biotechnology.* Humana Press, Totowa, NJ.

243. **Whitford, W. G.** 2005. Supplementation of animal cell culture media. *BioProcess Int.* **3:**S28–S36.

244. **Whitford, W. G.** 2006. Fed-batch mammalian cell culture in bioproduction. *BioProcess Int.* **4:**30–44.

245. **Wildt, S., and T. U. Gerngross.** 2005. The humanization of *N*-glycosylation pathways in yeast. *Nat. Rev. Microbiol.* **3:**119–128.

246. **Wilson, G.** 2007. Evaluation of a novel micro-bioreactor system for cell culture optimization, p. 611. *In* R. Smith (ed.), *Cell Technology for Cell Products. Proceedings of the 19th ESACT Meeting, Harrogate, UK.* Springer, Dordrecht, The Netherlands.

247. **Winder, R.** 2005. Cell culture changes gear. *Chem. Ind.* October:18–20.

248. **Wong, D. C. F., K. T. K. Wong, L. T. Goh, C. K. Heng, and M. G. S. Yap.** 2005. Impact of dynamic on-line fed-batch strategies on metabolism, productivity and *N*-glycosylation quality in CHO cell cultures. *Biotechnol. Bioeng.* **89:**164–177.

249. **Wong, V. T., K. W. Ho, and M. G. S. Yap.** 2004. Evaluation of insulin-mimetic trace metals as insulin replacements in mammalian cell cultures. *Cytotechnology* **45:**107–115.

250. **Wurm, F. M.** 2004. Production of recombinant protein therapeutics in cultivated mammalian cells. *Nat. Biotechnol.* **22:**1393–1398.

251. **Wurm, F. M., and A. R. Bernard.** 1999. Large scale transient expression in mammalian cells for recombinant protein production. *Curr. Opin. Biotechnol.* **10:**156–159.

252. **Xcellerex.** 2008. Application brief: scale-up to 1000-L perfusion cell culture in a disposable stirred-tank bioreactor. April 7. Xcellerex, Marlborough, MA. (Online.) www.xcellerex.com/pdf/perfusion.pdf.

253. **Xie, L., and D. I. C. Wang.** 1994. Application of improved stoichiometric model in medium design and

254. **Xie, L., and D. I. C. Wang.** 1996. High cell density and high monoclonal antibody production through medium design and rational control in a bioreactor. *Biotechnol. Bioeng.* **51:**725–729.

255. **Xie, L., and D. I. C. Wang.** 1997. Integrated approaches to the design of media and feeding strategies for fed-batch cultures of animal cells. *Trends Biotechnol.* **15:**109–113.

256. **Yamane-Ohnuki, N., S. Kinoshita, M. Inoue-Urakubo, M. Kusunoki, S. Iida, R. Nakano, M. Wakitani, R.Niwa, M. Sakurada, K. Uchida, K. Shitara, and M. Satoh.** 2004. Establishment of *FUT8* knockout Chinese hamster ovary cells: an ideal host cell line for producing completely defucosylated antibodies with enhanced antibody-dependent cellular cytotoxicity. *Biotechnol. Bioeng.* **87:**614–622.

257. **Yang, J., C. Lu, B. Stasny, J. Henley, W. Guinto, C. Gonzalez, M. Fung, B. Collopy, M. Benjamino, J. Gangi, M. Hanson, and E. Ille.** 2007. Fed-batch bioreactor process scale-up from 3-L to 2,500-L scale for monoclonal antibody production from cell culture. *Biotechnol. Bioeng.* **98:**141–154.

258. **Yeung, K. S., M. Hoare, N. F. Thornhill, T. Williams, and J. D. Vaghjiani.** 1999. Near-infrared spectroscopy for bioprocess monitoring and control. *Biotechnol. Bioeng.* **63:**684–693.

259. **Yoon, S. O., M. M. Kim, S. J. Park, D. Kim, J. Chung, and A. S. Chung.** 2002. Selenite suppresses hydrogen peroxide-induced cell apoptosis through inhibition of ASK1/JNK and activation of PI3-K/Akt pathways. *FASEB J.* **16:**111–113.

260. **Zang, M., H. Trautmann, C. Gandor, F. Messi, F. Asselbergs, C. Leist, A. Fiechter, and J. Reiser.** 1995. Production of recombinant protein in Chinese hamster cells using a protein-free cell culture medium. *Bio/Technology* **13:**389–392.

261. **Zeng, A., and J. Bi.** 2006. Cell culture kinetics and modeling, p. 299–348. *In* S. S. Ozturk and W.-S. Hu (ed.), *Cell Culture Technology for Pharmaceutical and Cell-Based Therapies* (Biotechnology and Bioprocessing Series). CRC Press, Boca Raton, FL.

262. **Zhang, F., M. A. Saarinen, L. J. Itle, S. C. Lang, D. W. Murhammer, and R. J. Linhardt.** 2002. The effect of dissolved oxygen (DO) concentration on the glycosylation of recombinant protein produced by the insect cell-baculovirus expression systems. *Biotechnol. Bioeng.* **77:**219–224.

263. **Zhang, J., and D. Robinson.** 2005. Development of animal-free, protein-free and chemically-defined media for NS0 cell culture. *Cytotechnology* **48:**59–74.

264. **Zhang, J., D. Robinson, and P. Salmon.** 2006. A novel function for selenium in biological system: selenite as a highly effective iron carrier for Chinese hamster ovary cell growth and monoclonal antibody production. *Biotechnol. Bioeng.* **95:**1188–1197.

265. **Zhang, L., H. Shen, and Y. Zhang.** 2004. Fed-batch culture of hybridoma cells in serum-free medium using an optimized feeding strategy. *J. Chem. Technol. Biotechnol.* **79:**171–181.

266. **Zhao, D., R. Fike, B. Naumovich, and M. Stramaglia.** 2008. Improving protein production in CHO cells. *In* Guide to Protein Production. *BioPharm Int.* **21:**S22–S27.

267. **Zhu, M. M., A. Goyal, D. L. Rank, S. K. Gupta, T. V. Boom, and S. S. Lee.** 2005. Effects of elevated pCO_2 and osmolality on growth of CHO cells and production of antibody-fusion protein B1: a case study. *Biotechnol. Prog.* **21:**70–77.

fed-batch cultivation of animal cells in bioreactor. *Cytotechnology* **15:**17–29.

Manufacture of Mammalian Cell Biopharmaceuticals

JINYOU ZHANG

13

Following the discussion in the previous chapter on cell line development and cell culture production for biopharmaceuticals, the present chapter reviews other important aspects of the manufacturing of biopharmaceuticals in mammalian systems. These include recovery, purification, formulation, fill and finish, bioanalytics, and current Good Manufacturing Practices (cGMP). As in the previous chapter, the dominant sector in the industry, monoclonal antibody (MAb) production, is used as an example where available. Only the common systems currently used in the biopharmaceutical industry are described, but references are given for unique and emerging technologies. When multiple citations exist, the more recent ones are usually listed.

13.1. DOWNSTREAM PROCESS DEVELOPMENT FOR BIOPHARMACEUTICALS

Historically, the major bottleneck in production of biopharmaceuticals was cell culture capacity. As described earlier, the industry responded to capacity shortage by developing processes with ever higher titers from both new and novel expression systems, developing and optimizing new and existing growth media, and increasing mammalian cell densities by bioreactor technology improvement. All of these advancements have helped to increase the productivity of upstream processes, with the best examples of optimized MAb cell culture achieving yields of approximately 27 to 31 g/liter at development scale (4, 122). Just since the turn of the century, the industry has gone from considering 1 g/liter as impressive to seeing that as the baseline. Now about 60 to 70% of the industrial titers are in the 1- to 3-g/liter range (38), and 15,000- to 20,000-liter bioreactors are used in sets of four or six. This means that severalfold more products are coming from mammalian cell culture, which necessitates severalfold-higher downstream processing capability.

However, the chromatography resin capacity has not been increased dramatically, so larger filter and column sizes, higher product and buffer volumes, and longer processing times are needed and are becoming issues. Therefore, the bottleneck of manufacturing has been pushed downstream (99), mostly with respect to the first product capture step. At the same time, in-process impurities have increased, mainly due to the use of more complex media and higher cell densities with lower cell viabilities, thus introducing elevated levels of host cell proteins (HCPs), lipids, and nucleic acids into the process (133). In addition to the non-product impurities, large amounts of undesirable product-related isomers are now also present. As a result, there are three major challenges in developing an efficient downstream process for MAb manufacturing with highly productive cell culture: (i) process capacity, (ii) removal of process impurities, and (iii) removal of product-related impurities. Unless radically new product recovery methods become available and accepted quickly, product concentrations higher than 2 to 5 g/liter may not even be considered as desirable (74).

Over the past few years, to cope with the much higher cell culture titers, an increasing emphasis has been placed on developing effective and economic downstream purification processes. Downstream recovery has become a significant portion of total manufacturing cost and also has begun to limit overall facility throughput. Rather than investing in greater infrastructure, which can be costly and risky and ultimately is limited by column sizes, the MAb industry is moving more toward using current capacity more efficiently (24). This is being achieved by (i) altering the order of operations, (ii) integrating and combining unit operations to reduce overall processing steps, (iii) introducing improved chromatography matrices such as membrane separation, (iv) using new resins with enhanced dynamic binding capacity (71, 138, 186), and (v) employing disposable technologies as opposed to fixed equipment (32, 115, 177). In addition, a platform approach that begins with a boilerplate purification process that is adjusted and optimized for a given project is gaining popularity. This maximizes the efficiency of the purification process and product purity while minimizing the development time for early-phase therapeutic antibodies (39, 72). New techniques and technologies such as single-use apparatus, cation exchange membrane adsorbers, simulated moving bed chromatography, single-pass tangential flow filtration, high-performance tangential flow filtration, precipitation, and crystallization are being introduced, which may help utility use and overall cost of goods, and overcome bottlenecks in existing facilities (28, 143). The improved efficiency also leads to other cost benefits, including reductions in raw material use, buffer consumption, waste disposal, and downstream processing facility requirements.

13.1.1. Overview of Downstream Processing

In the biopharmaceutical industry, downstream processing is a series of unit operations that starts with a large volume of a complex biological mixture and refines it to a small volume containing the purified product in bulk form (134). The stepwise process begins with cell separation, followed by concentration and purification steps of increasing specificity or resolution based on molecular size, electric charge, solubility, and/or binding affinities. For economic reasons, volume reduction is the first concern. For regulatory compliance and patient safety, inactivation and/or removal of contaminants takes the highest priority. For the sake of product, solubility and structural integrity are of greatest importance.

Three major types of operations are used in biopharmaceutical downstream processing. (i) Fluid filtration is vital to the clarification, concentration, and refinement of process streams, whether using tangential flow filters (cross flow) or depth filters (normal flow). Ultrafiltration (UF) and microfiltration (MF) are most often used in bioprocessing, with UF as a pressure-driven convective process using semipermeable membranes to separate or concentrate soluble species by molecular size or shape (109, 141), and MF as a lower-pressure process to separate micron-level particles (including cells) for solution concentration or clarification (135, 168). One advantage of this technique is the creation of a particle-free harvest stream that requires minimal additional filtration. Normal flow filters are used where clarification and/or bioburden reduction is desired in streams with relatively modest solid contents. They protect or enhance downstream operations or final polishing to achieve sterility. Positively charged depth filters have been used for a variety of applications, including the removal of endotoxin and DNA (35), HCPs (165, 181), and retrovirus as well as parvovirus (165). Membrane filters often find their use by supplementing chromatography steps, by providing clean feed streams, buffer exchange, product concentration, virus removal, and sterile filtration. But membrane fouling due to filter clogging or accumulating layers of cake- or jelly-like sedimentation can cause protein precipitation and aggregation, leading to product loss. Chemical additives and periodic backwashing can keep it from happening. Air must be kept to a minimum in process streams to prevent foaming, and low flow rates and pressures must be used for delicate proteins. (ii) Liquid chromatography now forms the basis of industrial purification processes after 5 decades of innovation (26). Usually two or more different types are employed, and a process chromatographic capture-purify-polish regimen is ubiquitous. The basic chemistries involved for large-scale application are mostly based on liquid adsorption, i.e., molecular affinities, ion exchange, and hydrophobic interaction. The steps typically include column packing, sample load, and product elution. Elution at large scale is mostly achieved by using displacement, while pH or salt gradient is the common choice for analytical application. Ion exchange chromatography (IEC) is the "workhorse" method, and most bioseparations can be assumed to include at least one IEC step (146), because of its high binding capacity, high selectivity, fairly inexpensive resins, and straightforward scale-up, ideal for relatively large volumes. (iii) Centrifugation separation is most often engaged at the front end of a downstream process, where it separates nonsoluble material from cell culture supernatant, preventing filter fouling at the later steps. Disk-stack desludging centrifuges offer continuous operation, making their throughput consistent with the desire to limit the time for harvest, and therefore have been popular for removing cells and cell debris (146, 156). But the operating conditions (flow rate, revolutions

per minute, etc.) for disk-stack centrifuges frequently require some empirical optimization (156). In addition to these three operations, other techniques have been or are being developed for bioseparations, such as protein precipitation with concomitant virus kill (16), perfusion chromatography using monoliths (163), and displacement chromatography (76).

By and large, however, the regulatory acceptance of filtration and chromatography in their various forms supports their dominance in the industry. The latest trend is in fact the combination of chromatographic chemistries with filtration media for what is known as "membrane chromatography," a niche application that is ideally suited to capture large molecules from diluted feed streams (187). In flow-through polishing steps, these targets are critical impurities such as nucleic acids, viruses, endotoxins, and host proteins. A disposable membrane is typically oversized for these highly diluted contaminants and is particularly attractive for virus clearance validation (26). In addition, unlike the traditional column chromatography, membrane chromatography can be used in a "plug-and-play" manner. Despite the concerns about its cost, relatively low binding capacity, and issues related to integrity testing, due to the current focus on disposable processing solutions, this technology is establishing a position in the bioseparations arsenal, particularly in MAb purification (103).

For MAb production, the industry has moved to the development of platform technologies to shorten the timeline (43, 72, 155), although case-by-case optimization still remains a challenge for manufacturers with diverse product types. After clarification from cell culture medium, the vast majority of MAbs on the market and in clinical trials are first purified with protein A affinity chromatography for product capture, followed by anion-exchange (AEX) chromatography in flow-through mode to extract negatively charged contaminants such as HCPs, endotoxins, host DNA, and leached protein A, and then cation-exchange (CEX) chromatography or hydrophobic interaction chromatography (HIC) in bind-and-elution mode to remove positively charged contaminants, including residual HCPs and undesired product-related derivatives such as aggregates or degradation products (13, 71). Size exclusion chromatography can also be used for polishing. The bulk product is filtered through a 20-nm filter for removal of viruses, both large and small (139, 164); concentrated to target concentration in the final formulation buffer using an ultrafiltration/diafiltration (UF/DF) device; and finally, sterile filtered through a terminal 0.2-μm membrane filter. Typical yields from an overall downstream process for MAbs range from 60 to 80%.

Purification process optimization must examine the operation of units in concert with one another, and the ultimate goal is to make as much purified product as possible by the cheapest, quickest, most reproducible, and most robust and efficient route. Process time is critical because, due to the limited stability of biomolecules and the high likelihood of an adventitious agent growing in the spent medium environment, any decrease in time will improve process throughput. In addition, the optimization efforts should also focus on system capacity, resin cost, endotoxin removal method, aggregation at higher antibody concentration, worldwide regulatory compliance, and chemical and equipment compatibility (80).

Process impurities include HCPs, nucleic acids (DNA, RNA), leached protein A, potential viral contaminants, and certain medium components (e.g., insulin, methotrexate).

The accepted levels according to current guidelines from regulatory authorities are <5 ppm for HCPs, <10 ng/dose for ribosomal DNA (rDNA), and <5 ppm for leached protein A (43). The most critical concern in mammalian cell processes is viral clearance (142). The virus removal steps should achieve a log reduction sufficient to ensure theoretical viral counts no higher than 1 per million doses, and the ability of purification steps has to be demonstrated for sufficient removal of a range of virus types through virus clearance validation studies (42, 54).

13.1.2. Recovery Separation—Removal of Cells and Cellular Debris

The harvest process separates the antibody products released in the culture supernatant from the cells. A primary clarification step is usually achieved by centrifugation or a membrane process such as tangential flow (or cross-flow) microfiltration (133, 156). Although depth filtration can be used as a primary harvest method, it is commonly seen to follow the primary harvest step to provide additional clarification (180). The unavoidable bioreactor variability has little effect on sizing of primary harvest operations because they are scaled based on bioreactor volume, not product concentration. Due to improved centrifuge design by leading vendors (Westfalia, Oelde, Germany, and Alfa Laval, Lund, Sweden) and the changeover flexibility (171), continuous centrifugation has become the most common harvest method for large-scale cell culture broth volumes of 2,000 and 20,000 liters (156). The cell turbidity of the cell-free supernatant can vary with the total number of shots (full or partial) and the buffer volume per shot (100); therefore the supernatant needs further clarification through depth filtration. Because depth filters do not have an absolute pore size cutoff rating, the clarification needs to be followed by an in-line 0.2-μm dead-ended microfilter to effectively remove any residual particulates for the subsequent capture chromatographic step. The optimal depth filtration is dependent on filter type, flux, and total membrane area. Large-scale depth filtration can now be readily scaled up with process-scale housings and disposable filter modules. Often, the filter can not only reduce the turbidity, but also remove up to 50% of DNA impurities and 15% of HCPs at neutral pH (100). Another widely used harvest technique is a tangential flow MF-based strategy, particularly at scales at or smaller than 2,000-liter, because centrifugation is considered a significant capital investment. The development of an MF harvest method is outlined step by step (135), and its optimization for cell culture harvest has been well studied (34, 82, 128, 172). Benefiting from the evolution and evaluation carried out in the past 15 years, harvest techniques for mammalian cell culture systems can now be expected to operate with high yields (>98%) and minimal cell disruption (156). The very high titers that can be achieved in cell culture operation, as described above, mean that the current cell culture scales are likely to stay with us for some time, so the improvements in harvest and clarification are likely to be in the form of gains in efficiency and throughput and improved understanding of existing unit operations.

13.1.3. Primary Purification—Initial Product Capture and Concentration

At present, MAbs or Fc fusion proteins dominate the biopharmaceutical industry pipeline, and the use of affinity chromatography resin continues to be the mainstay of antibody purification, which offers a simple and effective way of reducing the number of processing steps by incorporating both a capture and concentration step with a high degree of purification. Protein A is an affinity ligand that binds specifically the Fc region on immunoglobulin G (IgG) molecules, allowing everything else to wash through; then a high-salt or low-pH buffer is applied to desorb the antibody from the column. Matrix fouling can happen if a large number of impurities are present. Wider columns can be used to accommodate larger sample volumes, and regeneration and reuse are critical. It can be repeatedly used to capture and purify the product to up to 98% purity and to remove most process impurities including protease in one step, with step yield easily over 95%. Also, protein A resin can offer greater vendor reliability, improved resin packing permeability, or lower nonspecific binding than cheaper alternatives (80). Regulatory authorities also have familiarity with protein A affinity chromatography. Therefore, despite the growing trend of searching for alternatives due to its very high cost (about 10 times that for the less specific ion exchange resins) (8, 118), stability concern related to cleaning-in-place regimes, ligand leakage, and limited capacity (<50 g/liter) (97), protein A affinity chromatography remains the most efficient purification step for antibodies and is unlikely to be replaced for the foreseeable future in the primary capture step, which is the most significant bottleneck of all (26, 89).

Two major types of resin dominate the industry—agarose-based (GE Healthcare, Amersham) and glass bead-based (Millipore, Billerica, MA) resins. Both are robust enough to handle high flux and resist most cleaning chemicals including alkaline. Dynamic binding capacity is in the range of 20 to 50 g of antibody per liter of resin, and both can be reused up to 200 times. The difference is that agarose-based resin has very low nonspecific binding, but it is less mechanically rigid, so a linear flow rate of ≤300 cm/h and a bed height of ≤20 cm should be employed. On the other hand, glass bead resin can tolerate high flow rate and bed height without significant back-pressure, but it has higher nonspecific binding and is also sensitive to caustic solutions. To minimize proteolytic degradation of antibody product and protein A resin, a low loading temperature (12 to 18°C) should be used. Other operation conditions should be optimized to achieve maximal binding capacity (through resin type and loading residence time), minimal residual HCP and rDNA levels (through composition, pH, and volume of wash buffer), and minimal turbidity, conductivity, and aggregation in eluate (through composition and pH of elution buffer). High-capacity resins are needed to keep column and tank sizes low, to minimize buffer consumption and disposal, and to make the best use of facility space. The dramatically improved cell culture titers have put downstream processing as the rate-limiting step for antibody production, but the "true bottleneck in recovery processes is the first adsorptive column" (105).

Usually after protein A affinity chromatography, a viral inactivation step at low pH follows. This is because viruses are best removed early in the first purification steps so they don't complex with antibodies to get a "free ride," which can ease quality and regulatory concerns. At pH of approximately 3.4 to 3.8 for 60 to 120 min at room temperature, the inactivation is effective on enveloped RNA viruses such as the retrovirus murine leukemia virus (MuLV), while nonenveloped DNA viruses such as murine minute virus (MMV) are not efficiently killed. In some cases, antibody is captured onto a protein A matrix, followed by low-pH hold and elution, which is effective as a combined purification and virus inactivation step (81). After low-pH inactivation, the subsequent pH up-adjustment sometimes results in turbidity, which can be removed by a 0.2-μm membrane filter or a depth filter, a step that may also reduce HCP and rDNA levels, and even some MuLV and MMV viruses at 2 to 4 log reduction units (165).

13.1.4. Additional Purification—Removal of Bulk In-Process Impurities

Post-protein A, among several chromatography methods, AEX is perhaps the most powerful in removing residual amounts of negatively charged impurities such as endotoxin, rDNA, HCPs, and viruses. AEX is generally carried out in flow-through mode under a pH lower than the pI of the MAb but higher than most impurities, so that the negatively charged impurities bind to the resin while the positively charged antibody product flows through. The Q Sepharose (strong) and the DEAE Sepharose (weak) anion exchangers (GE Healthcare) are commonly used. To permit high volumetric flow rates, a large-diameter column with specific bed height is required for proper flow distribution (43, 182). This leads to a large column volume, which is optimized for fast flow but not for binding capacity. To overcome this disadvantage, membrane chromatography has been developed. Current Q membrane technology offers major advantages over packed-bed resin chromatography for antibody purification, including fast operation, low buffer consumption, disposability (thus no requirement for cleaning and storage validation), easy and quick qualification, good scalability, and high antibody recovery (>98%) (75, 92).

13.1.5. Polishing Chromatography—Removal of Specific, Low-Level In-Process Impurities

Typical polishing objectives are to remove high-molecular-weight aggregates, product isomers, leached protein A, and remaining trace amounts of HCPs and rDNA. CEX in bind-and-elution mode is a powerful tool for removing product-related impurities left behind after protein A affinity chromatography and also those that cannot be taken care of by AEX. While it is not considered a significant mechanism for viral clearance and HCP/rDNA removal compared to AEX, an ideal CEX resin can leave the majority of residual HCPs, DNA/RNA, and endotoxin in the flow-through fraction. Deamidated or acidic by-products can be separated in the front peak, while the amidated or basic materials and dimers/aggregates can be isolated in the tailing part. CEX is the major mechanism to remove antibody aggregates. CEX resins should be screened for low HCP binding, high dynamic binding capacity under high conductivity, and high resolution to remove product variants (100). Linear gradient elution provides better purity, process monitoring and control, and reproducibility, while stepwise elution obviously is mechanically simpler with higher product concentration. Compared to the commonly used salt gradient, a pH gradient elution can support similar product purity but higher yield at <50% pool volume, which was shown at large scale with an overall recovery of >95% (100). The process reliability of pH-conductivity hybrid gradient CEX with high-step purity and yield at multiple 2,000-liter runs was also reported (185). The SP Sepharose (strong) and the CM Sepharose (weak) cation exchangers (GE Healthcare) are commonly used.

HIC and ceramic hydroxyapatite chromatography can also serve as efficient methods to remove dimers and aggregates (68). However, HIC has relatively low yield and leads to high salt concentration in the elution pools. So instead of bind-and-elution, HIC is used more in a flow-through mode to remove a large percentage of aggregates, in which slow flow rate is necessary to allow sufficient time for aggregates to bind to the resin. For ceramic hydroxyapatite chromatography, remaining technical issues include lot-to-lot variability, extractables, and the short resin half-life.

13.1.6. Virus Removal Filtration

Viral contamination is a risk to all biopharmaceutical products (including MAbs) derived from cell lines of human or animal origin (42, 129). Viral safety should be considered throughout the life of a biopharmaceutical manufacturing process, from process design to selection of source materials to setting process limits (51, 53, 54). Contamination of a product with endogenous viruses from cell banks or adventitious viruses from production processes can have serious clinical implications, so regulatory guidelines (25, 42, 54) mandate that therapeutic biological product manufacturers implement adequate technologies in their manufacturing processes to remove or inactivate known or adventitious contaminants based on a process-specific virus clearance strategy. At least two virus removal/inactivation steps are needed, one based on filtration and another on inactivation, which is typically incorporated between the earlier steps such as protein A and AEX, as described above. The entire purification process for clinical manufacturing must be validated to achieve a 16- to 20-log reduction of viral load using two (for early clinical phase) to four (for late clinical phase) different and representative viruses.

A statistically independent combination of methods (orthogonal approach) is required for removing enveloped and nonenveloped viruses, with the methods based on different physical principles of removal and inactivation and yet complementary to each other (160). Viral inactivation and filtration are specifically designed and dedicated steps that effectively clear viruses. Solvents, detergents, chemicals, low pH, etc., can be used for virus inactivation. Low-pH inactivation is demonstrated to be effective against enveloped viruses, and its effectiveness is highly dependent on time, temperature, and pH (15). Nanofiltration such as using 20-nm retentive filters has traditionally been accepted as a robust method for clearance of both large and small viruses (139, 164), despite the fact that it is the most expensive downstream step and needs a lot of optimization (74). However, the combined log reduction values of these steps rarely exceed 12 logs due to the limited sensitivity of virus assays, which means that chromatography adsorption must provide additional clearance to achieve a typical 16- to 20-log reduction requirement. This can be mainly achieved by AEX (about 4 to 6 logs) plus protein A (about 2 to 3 logs). The relatively new membrane chromatography technique is efficient in removing small nonenveloped viruses (188), meeting or exceeding the viral clearance performance of AEX (7, 129). Its disposable feature not only reduces capital costs, but also eliminates post-use cleaning, sterilization, validation, and risk of contamination carryover, thereby simplifying adsorptive virus clearance (46, 188).

Viral clearance studies are used to measure the relative safety of purification processes, and it is essential to ensure that the studies are carefully designed to exactly reflect the scale-down of the manufacturing conditions, or the worst-case scenarios if established, thus providing key data rather than becoming just a box to check on the way to licensure of a product (21, 167). A virus spiking trial needs to be performed to validate the efficiency of the purification process steps (78, 164). Duplicate runs are needed (42). Depending on the feasibility and purpose, relevant/specific model viruses or nonspecific model viruses should be selected (54). For cell lines that express retroviral particles, such as CHO, NS0, and Sp2/0, two model viruses are used to assess the ability of downstream processes to clear virus (54). MMV or porcine parvovirus (PPV) can be used as a model of parvoviruses, which are a known contaminant of bioreactors and are relatively small and resistant to physical inactivation. MMV or PPV can be readily removed by a 20-nm filter (>4 logs in this one step). The other is a specific model of retroviruses for cells of murine origin, MuLV. Retrovirus-like particles theoretically pose safety concerns to humans because of their morphological and biochemical resemblance to tumorigenic retroviruses (33, 153). The target of clearance is 6 logs in excess of retroviral load in the unprocessed cell culture harvest broth, which is typically in the range of 10^5 to 10^9 virus-like particles/ml broth. For example, if there are 9 logs of counts in broth, a production process of 2-g/liter cell culture titer needs to achieve an 18-log reduction factor during purification to ensure no more than 10^{-6} viral counts per dose (54) for a typical antibody dosing of ~2 g. Higher titer or lower dosage can reduce the log reduction factor requirement.

13.1.7. Final Formulation—Final Dosage Concentration and Buffer Formulation Including Bioburden Reduction Filtration

After the antibody is purified out from the mixture, the "bulk" drug goes into final processing—formulation. The development of formulation buffer is discussed in the next section. Usually at this stage, the antibody solution from the virus removal filtration step is concentrated by UF, then exchanged into the desired formulation buffer by DF. The two operations are performed on the same UF/DF membrane filter with a molecular mass cutoff of 30 to 50 kDa, which retains the IgG molecules of about 150 kDa. The UF and DF steps can alternate based on an optimized scheme to minimize buffer consumption while avoiding product aggregation, until the desired final MAb concentration is reached, with the buffer thoroughly exchanged into that of the final formulation. While final concentrations are normally about 110 mg/ml (145), a robust, scalable UF/DF formulation process was developed which produced up to 183 mg/ml MAb as clinical study supplies (109). The formulated solution is then subject to the terminal sterile filtration using a 0.2-μm membrane to give rise to the bulk or drug substance.

13.2. FORMULATION DEVELOPMENT AND FILL/FINISH FOR BIOPHARMACEUTICALS

Just as the physiochemical properties of an antibody make all the difference in how it is purified, they also determine what type of formulation will suit it best. One of the most challenging tasks of developing protein biopharmaceuticals such as MAbs is to deal with and overcome their inherent physical and chemical instabilities (147), which have the potential to alter the state of the molecule from a desired (native) to an undesirable form upon storage, compromising patient safety and drug efficacy. Formulation development is to overcome this inherent instability to changes in pH, osmolality, tonicity, pressure, and temperature, and it is a major area of biopharmaceutical development (144). Various factors can affect the decision on a formulation, including the reliability, cost, and availability of analytical methods; the availability of qualified excipients and stabilizers on the

market; methods and devices of delivery; product form, dosage frequency, and concentrations; patient preferences and behavior; biology of the disease; transportation and storage; packaging; manufacturing process; and considerations of legal, sales, and marketing issues (145).

By far the most common delivery method used in the biopharmaceutical industry is parenteral dosing, that is, liquid formulations for subcutaneous (SC) injection or intravenous (IV) drip, because of the assured bioavailability compared to oral route (70). Although an IV liquid formulation is most commonly used for toxicology and phase I studies, potential target indications and market competition often lead different dosage forms to be evaluated. For example, for non-life-threatening diseases that require systemic administration, the convenience of an SC formulation is preferred to an IV formulation. Similarly, for administration convenience and patient compliance, a stable liquid formulation is a better choice than a lyophilized product. Further, a higher product concentration is desirable if less frequent administration provides the same therapeutic effect, and in some cases such as SC administration it is a must because of the limit on the injection volume.

13.2.1. Preformulation

Preformulation is an initial development of formulation, which is an exploratory activity early in development that characterizes the antibody and its compatibility with potential excipients toward safe, stable, beneficial, and marketable product. For example, administration by the SC route requires that the MAb is available at high concentration (over 100 mg/ml) in the vial. Therefore, a concentration study in various buffers needs to be considered to investigate solubility behavior, concentration effect on viscosity, and increased potential for aggregation. These studies have the potential to strongly influence the target product profile as well as the design of clinical trials.

While there are numerous ways for a protein product to lose its stability, three are the most common: aggregation (physical), oxidation, and deamidation (chemical) (120). Physical degradation changes only the higher structures (secondary, tertiary, and quaternary) of a polypeptide, resulting in aggregation but not necessarily creating a brand-new molecule. Such degradation can also occur as adsorption, unfolding, and precipitation. Chemical degradation changes the primary structure of a protein, which is usually preceded by a causal physical process, typically unfolding, that makes available residues that are usually inaccessible for chemical reactions with their environment. Certain amino acids in protein molecules are susceptible to oxidation, and metal ions, high pH, fluorescent light, and hydrogen peroxide can all cause or aggravate oxidation, in addition to atmospheric O_2 itself. Deamidation leads to the hydrolysis of a side-chain amide on a polypeptide's glutamine or asparagine residue to yield a carboxylic acid, which is facilitated by elevated temperature and pH (73).

Several approaches have been identified to prevent or inhibit chemical degradation, including effective excipients, reducing agents, optimal pH, and lyophilization. Sucrose was shown to inhibit the conformation mobility of a serine protease, thus inhibiting oxidation (31). It was observed that addition of an antioxidant such as methionine thiosulfate significantly reduced the oxidation rate of an anti-HER2 antibody (96). The improved stability of a therapeutic protein was also demonstrated by pretreatment with a reducing agent (184). Most antibodies require a slightly acidic to neutral pH for maximum stability because of their slightly basic isoelectric points (pH of approximately 8 to 9). For those protein products not sufficiently stable as aqueous solutions, the lyophilized formulation can be used. The key is to protect the protein from unfolding during the freezing and vacuum-drying stages (19). Thermodynamic stabilizers can be used to inhibit unfolding of a protein in the presence of low concentrations of denaturants (90). Sucrose and trehalose are the most commonly used excipients to serve as cryoprotectant and lyoprotectant. The former excludes water and protects proteins from freezing stress (147), while the latter substitutes for water and protects protein from drying stress (22). In addition, nonionic surfactants such as polysorbate (plant-derived) or Tween (animal-derived) can be employed to protect proteins against freeze-thaw and agitation-induced aggregation, primarily by competing with stress-induced soluble aggregates for interfaces, thus inhibiting subsequent transition to insoluble aggregates (20). Polysorbate 80 is also widely used for stabilizing antibody IgGs in liquid formulation because it prevents small aggregates from progressing (27) and can be used to reduce the detrimental surface absorption at low protein concentrations. In addition, polysorbate 80 decreases the opalescence in a NaCl-containing formulation for MAbs (173). By far the most widely accepted methods of protein stabilization are additives and excipients, cryopreservation, and freeze-drying (12).

There are practical issues that need to be considered during formulation development (23). Chemicals used in a manufacturing process can contribute to product instabilities. For example, chloride ion in a low-pH formulation can corrode stainless steel storage tanks for buffers and/or products, thus compromising protein stability. Increase in ionic strength can result in higher viscosity and opalescence (173). High viscosity of a high-concentration MAb solution is difficult to scale because of poor recovery and instrument constraints. Formulations containing sodium phosphate can experience significant pH drop during lyophilization because of precipitation of Na_2HPO_4 upon ice crystallization. A citrate buffer in an SC dosage form may cause injection-site pain. A near neutral pH and isotonic formulation is preferred for SC dosage forms. New excipients require extensive toxicity studies. Finally, cost of certain excipients may be a factor; for example, trehalose is much more expensive than sucrose and may present supply issues.

13.2.2. Formulation Stability

Formulation development is primarily about stability and the mode of use of antibody. The fragile proteins need to be protected from denaturation and degradation until they can be delivered to their site of action in a patient's body. The antibody development process does not allow enough time to confirm stability requirements for a final formulation, which could take 2 years, before it is ready to apply for marketing. But testing of the structure and innate stability of the antibody using various analytical methods should begin as soon as a final formulation is decided, and the results should be updated frequently (59). Accelerated stability tests subject products to various stresses over a period of about 3 to 6 months, including a range of pH values, heat, light, freezing and thawing conditions, additives, and surface materials and interfaces (112, 161). The signs of degradation are observed and used to predict long-term storage behavior. Although accelerated tests are needed, real-time tests are the ultimate proof (110, 111). During stability studies, in addition to physical and chemical assays

to monitor the physiochemical properties of the protein, bioassays are used to evaluate antibody activity, identity, and critical pathways. If any of these parameters is changing with storage time, the formulation will need to be improved in order to meet the current market demand of 1 to 2 years of storage for most products. In general, formulation decisions are made primarily on product stability (shelf life), followed by convenience of use and cost-effectiveness.

Because of the extreme sensitivity of antibody stability to multiple physical and chemical factors as discussed above, in addition to the testing of purified bulk substance and drug product, testing the stability of process intermediates using various analytical methods should take place wherever there is potentially a storage period during a purification process. Physical and chemical assays are employed to monitor the physiochemical properties of IgG, such as molecular integrity, charge variation, aggregation, concentration, and pH in the solution. Microbiological assays are used to demonstrate the absence of microbial growth, which can otherwise introduce contaminants to the preparation, potentially changing the protein structure and even jeopardizing the safety profile of the product. If any of these parameters change with storage time, the in-process hold step will need to be redefined, either by shortening or eliminating the storage period or by improving the storage environment such as buffer formulation or temperature or container. A degraded or aggregated process intermediate cannot lead to a final drug product that meets the quality requirements.

13.2.3. Drug Delivery

How a biopharmaceutical is formulated may also depend on the intended mode of delivery. The most common delivery method used in the antibody/biopharmaceutical industry is parenteral dosing of liquid formulations (70). High-concentration formulation (>100 mg/ml) enables less than 1 ml per SC injection, for which gentler fill technology than traditional vial-fill is required to avoid sheared, precipitated, or aggregated product during the filling process. Lyophilization or UF/DF can be used to concentrate the formulation (109). For better patient comfort, convenience, and dosage compliance, other notable technologies under development include needle-free injection by means of high-pressure air or liquid that forces tiny droplets through the skin, allowing self-administration of medications without a hypodermic needle, and controlled delivery that makes it possible to control the duration (sustained delivery) and the site of action (targeted delivery) of a biopharmaceutical (12). The sustained delivery is achieved by a depot of drug created in the body to release the drug over a specified time, which can be implanted by emulsification, coacervation, extrusion, and polymerization; targeted delivery has been practiced for some time using the high specificity of MAbs and surgical implantation. Currently, needles and syringes—especially with the new trend toward prefilled syringe packaging—remain the most practical option for protein delivery (145).

13.2.4. Fill and Finish

Fill and finish is the actual process of dividing a biopharmaceutical bulk formulation into dosage forms for sale or clinical study use. The product container, the cap, the label, the packaging, the patient instruction, and the transportation are all part of filling and finishing a product. The operation is often at least partly automated, with aseptic conditions being the priority. Unit processes in manufacturing finished dosage forms include compounding and mixing, filtration, filling, lyophilization, closing and sealing, sorting and inspection, labeling, and final packaging for distribution. Each of these steps faces challenges and must be controlled carefully to ensure quality of the relatively unstable and interactive biopharmaceuticals (1).

The protein nature of MAbs requires more care in handling and preservation than classical small-molecule pharmaceuticals, because proteins have limited stability in their liquid state. To preserve the bulk liquid therapeutics before fill so as to meet the current market demand of 1 to 2 years of storage for most products, the most effective method currently is freezing, which also adds flexibility to the manufacturing process. But the subsequent thawing operation stresses protein molecules, and each additional thaw reportedly increases degradation (174). Parameters such as processing time, temperature, pressure, solute distribution, and pH need to be carefully controlled, and the product must not change physicochemical characteristics nor lose bioactivity during freezing and thawing.

For protein products not sufficiently stable as aqueous solutions post-fill, lyophilized formulations are often used. A lyophilized biopharmaceutical drug maintains its potency over time, thus adding flexibility to the entire manufacturing operation and extending the shelf life of the drug product. About half of the biopharmaceuticals currently sold are shipped and stored in freeze-dried form at room temperature (145). To achieve this formulation, the protein needs to be protected from unfolding/aggregation during the freezing and the subsequent drying stage (2), and thermodynamic stabilizers can be used to inhibit the unfolding of a protein (90). As stated above, sucrose and trehalose are common excipients to serve as cryoprotectant and lyoprotectant, and polysorbate 80 can protect proteins against freeze-thaw and agitation-induced aggregation. Operationally, control considerations in a freezing process include minimizing freezing speed, reducing convection and cryoconcentration, and preventing mechanical stresses in the vessel. One recent trend is toward use of organic solvents such as ethanol for increased sublimation rates and improved product stability (145). For a thawing process, the important control factors include avoiding overheating, maximizing convection in liquid phase, preventing interaction among formulation components, and avoiding internal melting, which creates a negative-pressure cavity that can degrade proteins. The options for reconstitution of lyophilized biopharmaceuticals were reviewed recently (131).

As parenteral drugs, MAbs must be free of all microorganisms; therefore, they must be packaged aseptically in a class 100 clean room with strict limits on viable and particulate contamination, and with production personnel thoroughly gowned up to keep the drug from being exposed to human-borne contaminants (60). There are many considerations for aseptic filling of parenteral biopharmaceuticals, including the product's physical characteristics and its intended market, which determine the most appropriate facility, filling equipment, delivery system, and container-closure system (77, 158). An automated parenteral production line usually consists of a washer, a depyrogenation tunnel, and a filler (with companion check-weigher). The filled vials are stoppered and secured with an aluminum cap. The final product is inspected, labeled, boxed with patient insert information, and shipped. Lyophilized products are filled in liquid form and then freeze-dried before completing the packaging. Alternatives to vials include ampoules,

syringes, and cartridges. Recent trends in fill and finish include the development of barrier systems (123), which enclose the fill operation and thus reduce the chances of opportunistic contamination and the need for a clean room. Barrier isolators, which are gaining rapid acceptance by industry and regulatory authorities, have the major advantage of sterilizing equipment and enable sterile filling instead of the traditional aseptic filling, but are specific to the product and process used, difficult to clean, and complex to design. Blow-fill-seal production is an alternative approach that fills and seals protein drug solutions into plastic containers in a sterile enclosed area inside a machine, minimizing human intervention (88).

13.3. BIOANALYTICAL METHOD DEVELOPMENT AND QUALIFICATION

A broad variety of bioassays and analytical testing are required for a biopharmaceutical before it reaches the market. The relation between the kinds of results needed and the type of test that must be performed, as well as regulatory requirements and how to ensure consistency of results over time and at different locations, are topics of various surveys, reviews, and publications (132, 137, 157). The analytical methods, procedures, and instruments need to be validated to qualify the use (55, 56, 61, 93, 94). No company can develop and perform every kind of testing in-house, so there need to be strategies for deciding which tests to outsource.

13.3.1. Lot Release Testing

The development of complex biological materials such as MAbs as therapeutic agents requires comprehensive assessment with regard to regulatory guidelines. Reflecting the importance and complexity of MAb products, several statutory and guidance documents have been issued by the U.S. Food and Drug Administration (FDA) recommending approaches for characterization and lot release (47, 48, 53, 55). Generic test methods for MAbs cannot be recommended because the "critical quality attributes" that affect product potency and safety are product specific. However, a basic set of lot release test methods should be included for each individual MAb. The view generally accepted is to perform release tests on the quality attributes that are related to product performance (safety or efficacy) or stability (137), including aggregation and size, molecular weight, free sulfhydryl groups, heavy-chain glycosylation, C- and N-terminal heavy-chain heterogeneity, and fragments/IgG4 half molecules.

The most commonly used analytical characterization and lot release techniques for development of biopharmaceutical proteins are spectrophotometric, chromatographic, electrophoretic, and mass spectrometric (113), although the current preference is for assays easier to perform and control than mass spectrometry. For MAb products, the following test methods often need to be developed as the analytical tools for characterization and lot release (119, 137): (i) identity (for product identification) by isoelectric focusing (IEF), capillary gel electrophoresis, specific binding, or peptide mapping (36); (ii) native size (for aggregates under native condition) by size exclusion chromatography, analytical centrifugation, and electrospray ionization mass spectrometry (ESI-MS); (iii) denatured size (for fragments/half molecules) by denaturing sodium dodecyl sulfate-polyacrylamide gel electrophoresis, size exclusion chromatography, and capillary gel electrophoresis; (iv) charge heterogeneity (for C-terminal lysine truncation and other charge variation) by IEF and capillary IEF, IEC, and capillary zone electrophoresis; (v) N-terminal heavy-chain heterogeneity (for structure integrity) by peptide mapping, ESI-MS, enzyme digestion, and high-performance liquid chromatography (HPLC); (vi) heavy-chain glycosylation (for oligosaccharide pattern) by ESI-MS, enzyme digestion, and HPLC; (vii) free sulfhydryl groups (for antibody stability) by fluorescent technique (183); and (viii) potency (for functional capability or biological effect) by cell- or animal-based bioassays, enzymatic assays, and binding assays. In addition to the product itself, process-specific analytical tools need to be established and validated to ensure that impurities introduced during production of antibodies are eliminated to the desired extent, for example, components added to the medium such as methotrexate for the dihydrofolate reductase expression system or methionine sulfoximine for the glutamine synthetase expression system, insulin, transferrin, etc.; protein A leached from the resin; and host cell DNA and proteins (157). Immunological and enzymatic methods are widely used in the detection of biological impurities due to their increased sensitivity and specificity over analytical methods (e.g., HPLC).

During release, stability, and formulation studies, cell-based biological assays are a critical and ubiquitous component of MAb testing, because the product's biological function, or potency, is a quality issue. From candidate selection and product release to product stability assessment and process change proposal, as well as during pharmacokinetics/pharmacodynamics (PK/PD) evaluation and drug response measurement, cell-based bioassays for functional and binding measurements are required. Since binding assay results do not necessarily correlate with functional assay results, the correlation of a binding assay with a cellular/functional end point must be validated before it can be used for lot release (137). Despite the fact that it is only in clinical trials that the clinical potency is determined and the dose range defined, regulatory guidelines state that "a correlation between the expected clinical response and the activity in biological assay should be established" (55). Thus, there needs to be an understanding of the mechanism of action of the product. It is not always possible to completely understand the mechanism, in which case more than one potency assay may be necessary. It is also often the case that in vivo assays are established later in the product development cycle when the "go" status has been reached, so at early stages only the binding assay is available for simple quantification of materials and for pharmacokinetics analyses. The various practical aspects of potency assay development have been reviewed (137). The current industry status and trends on bioassays are summarized in a recent survey (132). It should be pointed out that the complexity of the materials and procedures involved in the biological assays results in significantly increased variability as compared to analytical methods, and bioassays are often difficult to develop for use in lot release.

13.3.2. Glycosylation Analysis

Among the many characteristics of MAbs, glycosylation is particularly critical and highly sensitive to upstream process variations. A detailed description of the glycol structure features is increasingly part of the expected content of a New Drug Application or comparability protocol (61, 114). Glycosylation of MAbs is important not only for their biological function (11, 14, 83, 101) but also for pharmacokinetics (37). Because the biosynthesis of glycans does not follow a genetic "code" like protein biosynthesis does, many variations are possible, thus leading to glycan heterogeneity.

The biosynthesis of glycoproteins in mammalian cells and their characterization are well documented in literature (114). Due to the sensitivity of MAb glycosylation profiles to host cell enzymatic systems and bioreactor conditions (124, 148), selection of host cells and culture conditions must consider the requirement for stably producing each specific glycosylation pattern. Appropriate analytical tools must therefore be developed for in-process monitoring to determine any changes due to culture conditions. A high-throughput analysis of the N-glycans of NS0 cell-secreted antibodies is described (29). Peptide mapping, HPLC, CEX, IEF, capillary electrophoresis oligosaccharide mapping, nuclear magnetic resonance, various types of MS, and specific potency assays can all be used to detect glycol variants. A combination of tandem MS, methylation analysis, and exoglycosidase digestion was employed to characterize the oligosaccharide structures of Campath IgG1 from CHO and NS0 (152). A combination of tryptic digestion, HPLC with fluorescence detection, and MS analysis was used to characterize N-linked oligosaccharides of MAbs (91, 107).

13.3.3. Assay Development and Validation

When developing a new assay method for a new antibody/biopharmaceutical product, optimizing an existing method for an existing product, or changing a release method for a licensed product, many validation elements should be considered (56, 57, 61). The development and optimization process can improve an analytical method, but validation does not. Validation is the final proof that regulations and expectations are met. A list of regulations and guidance governing analytical method development and validation are provided by Kanarek (87). The minimum requirements for the development and validation of a new method are summarized in ICH Q2(R1) (61), and method validation approaches can be found in recently published literature (87, 95). Current GMP regulations also require analytical instruments to be qualified to demonstrate suitability for the intended use (57, 166). Similar to analytical method validation, analytical instrument qualification is to ensure the quality of each analysis before tests are conducted, which includes qualifications on design, installation, operation, and performance (10). A standardized DQ (Design Qualification), IQ (Installation Qualification), OQ (Operation Qualification), and PQ (Performance Qualification) approach was discussed (178).

Many new analytical technologies are now available for development of biopharmaceuticals in general and MAbs in particular (108, 119, 137), resulting in shorter testing times, increased throughput, greater ease of use, and improved sensitivity, selectivity, and precision. But changing biological assays, such as impurity testing by enzyme-linked immunosorbent assay, to automated or more sensitive ones requires extra care, because any bias in results must be reflected in the release specifications (55). In-process and product specifications should reflect production consistency and analytical capabilities, unless otherwise dictated by regulation (55). Current GMP guidelines state that GMP documentation and the detail of validation activities should increase as the production process progresses (58). Testing upstream stages may actually be more critical than many final container release tests because it provides evidence of cell culture quality and the efficiency of impurity removal, despite the fact that the tests are more uncertain and variable. Final container testing attests that active and inactive formulation components remain at predicted levels with little variability. The analytical methods should detect lot-to-lot variations, while whether the measurements are extremely accurate (100% recovery) is not as important.

13.4. GMP MANUFACTURING OF BIOPHARMACEUTICALS FROM MAMMALIAN CELL CULTURE SYSTEMS

Having developed a robust and productive cell culture process, a reproducible and high-yielding purification method, and reliable analytical tests for manufacturing controlled lots to be used for clinical trials and eventually for commercial sale, successful implementation of large-scale manufacturing in compliance with cGMP guidelines becomes the ultimate test for biopharmaceutical companies. Some of the cGMP manufacturing concepts have been introduced in the above sections and in the previous chapter. Additional regulatory compliance requirements for cGMP manufacturing are included in this section.

The goals of cGMP can be summarized as (143): protect the product from contamination; prevent mix-ups; know what you are to do before you do it; document what really happens; strive for consistency and control; have an independent group make the final decisions; and solve problems, learn from mistakes, monitor, and continuously improve. It should be pointed out that the regulatory requirements of the FDA and European Medicines Evaluation Agency (EMEA) can differ, with the FDA focusing on process and the EMEA on results (175). For investigational new drugs, the FDA issued "Guidance for industry: cGMP for phase 1 investigational drugs" (cGMP1) (65), whose EMEA counterpart is "Good automated manufacturing practices 5" (GAMP 5) (79). Differences can be reflected in procedures, facilities, and release requirements, etc. For example, different clean-room settings are expected for fill/finish. It is important to reconcile the guidelines from both agencies, to understand the legal framework, and thus to implement corresponding strategies under the new European Union GMP environment for investigational medicinal products (117).

13.4.1. Characteristics of Desired Manufacturing Processes

Most antibody therapy requires multiple doses at dosages 10 to 100 times greater than those for successful recombinant protein drugs, such as erythropoietin and human growth hormone (126); thus, large amounts of pure product are needed, which requires highly productive and consistent manufacturing processes derived from optimization and validation. On the other hand, most biopharmaceuticals that enter development will fail to make it to the market, so the high cost and resource commitment of a thorough process development effort for a specific drug candidate is difficult to justify until the risk of failure has been decreased by successful results in early-stage clinical trials. Furthermore, speed to market is critical to quickly deliver health benefit to patients and achieve business success. Therefore, for early-stage clinical studies (phase I and phase IIa), a standardized platform process is usually employed despite its relatively low productivity and relatively high variability. Only when the product candidate is proven to be safe and the initial therapeutic efficacy is demonstrated is the process individually optimized and validated to become high-yielding, robust, and economical, suitable to late-phase clinical studies and commercialization. It is important to demonstrate through analytical, nonclinical, and even clinical comparability studies that the manufacturing process changes will not have an adverse effect on the quality, safety, and efficacy of the drug product (62).

It is critical for a biopharmaceutical company to determine what constitutes a good manufacturing process and factor those attributes into the early development platform process, which is much simpler and more effective than trying to change an approved, validated commercial process. Characteristics of a good antibody manufacturing process are twofold: product specific and generic. The former includes maximum yield, high purity, desired log of viral clearance, expected bulk product concentration, adequate product stability, low production cost, sufficient campaign delivery, and adherence to manufacturing schedule. The latter includes safety, compliance, plant capacity, ease of operation, production cost, process characterization/validation, and robustness (179). In general terms, a good manufacturing process should have a safe working environment that minimizes personnel exposure to hazardous chemicals, process waste or by-products, and dangerous working conditions. It should be in compliance with regulatory guidelines on quality (67), which requires the process running as designed, with product specifications guarding against deviations from normal practice. It should maximize plant capacity, and thus productivity, by using fewer steps and/or tanks and/or fewer buffers, but still achieving higher yield and higher success rates. Its operation should use simple and reliable process equipment and solution preparation. The manufacturing process should be cost-effective based on a combined consideration of raw materials, utilities, labor, and facility, with a high batch success rate. A key feature of a good manufacturing process is adequate process characterization and validation, which is required by cGMP regulations (50, 66). Critical process parameters are identified by process characterization, and then the acceptable ranges are validated, before filing of a Biologics License Application. Finally, process robustness is the most important consideration when deciding whether it is a good manufacturing process, which reflects the relative insensitivity to small changes in process parameters or conditions and the simplicity and reliability in operation.

13.4.2. Process Validation

According to the FDA, process validation is "establishing documented evidence which provides a high degree of assurance that a specific process will consistently produce a product meeting its predetermined specifications and quality characteristics" (50). This is required by worldwide regulatory authorities, because validation helps to provide a better understanding of procedures and processes, thus making them more controllable for consistent product quality. It also helps prevent costly problems for companies.

Validation begins in process development, continues through clinical trials, and resides in the final manufacturing process (125). In the development stage, statistically designed experimentation and high-throughput methodologies may be employed to high thresholds for optimization in relatively short timelines and using small amounts of starting material (140). It is necessary to have validated analytical tools available to assess any impact on product yield and quality (95). Changes to a process resulting from these investigations must be exemplified in representative systems, and any impact on downstream processing addressed early. Prior to transferring a process to a cGMP production facility, it is vital to demonstrate scalability using pilot-scale versions of the production equipment. Developing a sound process validation strategy early in clinical studies is critical to the success of a validation program (49) and is consistent with the FDA's Quality by Design initiative (63).

MAb manufacturers prepare written validation plans or protocols that specify procedures, tests, and necessary data and analyses. That includes in-process monitoring of important process variables. Test protocols specify a number of process runs to demonstrate reproducibility and variability. All variables cannot be challenged to their extremes, but those that affect product quality are subjected to stressed conditions to demonstrate process robustness (143). The first step in process validation is process characterization, including process, controls, variables, changes, equipment, and assays; performance qualification comes next to ensure the process performs as designed, with process robustness demonstrated at lab or pilot scale; the final phase of validation requires successful completion of several consecutive manufacturing lots that are properly documented following a validation protocol (125).

Although the use of disposables for biopharmaceutical manufacturing may reduce validation efforts, especially in cleaning validation, it does not eliminate them, because there may still be many validation-related issues, such as raw material identity and consistency, leachables and extractables, interactions between antibodies and bag/tubing materials, stability, installation and operation qualification, process validation, disposal, and even cleaning and sanitization (159). Even for using disposable biocontainers for sterile liquid transportation, the system needs to be evaluated and validated (121).

13.4.3. Raw Materials for GMP Manufacturing

In each of upstream, downstream, and formulation steps, biopharmaceutical products come in contact with a variety of raw materials (RMs), starting with culture media for production; followed by filters, buffers, and resins for purification; and finally excipients such as sugars, salts, amino acids, and surfactants for final drug formulation. To illustrate the many considerations in cGMP manufacturing, the RM aspect is discussed here.

An RM is a component that GMP regulation refers as "any ingredient intended for use in the manufacture of a drug product, including those that may not appear in the drug product" (64). The scope of RMs thus includes chemicals, biological materials, liquids, solids—anything that contacts the product during the manufacturing process, including all inactive ingredients contained in final formulations (150). The biopharmaceutical industry uses compendial RMs, which are those that have monographs in major compendia. The grade for an RM can be USP (United States Pharmacopoeia) or NF (National Formulary). For the worldwide market, compliance with multinational regulatory rules is required; thus, multicompendial grades (MC) are needed, or EP (European Pharmacopoeia) or JP (Japanese Pharmacopoeia) specifically.

RM selection, testing, and control strategies form an important part of biopharmaceutical manufacturing tasks. It is critical to understand the relationship among RMs, process design, process settings, and end-product results (106), because biological product characteristics are defined by their production processes, which are affected by the RM quality. Therefore, the primary objective during RM selection and evaluation is quality. Since manufacturing processes employ a variety of RMs that can vary widely in quality (84), an important task is to develop methodologies and strategies that take this variation into account for manufacturing a biopharmaceutical product in a consistent manner.

In the manufacture under cGMP guidelines, RMs need to be characterized and released to ensure that they meet predefined specifications on their quality. Characterization is an important check of the physical, chemical, and biological attributes of an RM, while release ensures that the RM is suitable for its intended use (136, 151). The guidelines for release (40, 58) govern from the time an RM is procured until it is finally consumed in a biopharmaceutical manufacturing process.

As an example (9), a vendor-made cell culture medium, treated as a single RM, can be characterized and qualified through in-process testing based on established release criteria. The quality attributes include identity, purity, suitability, and traceability. The need for traceability can be exemplified by the requirement in EMEA guidelines (41) that the use of any animal-derived RMs should be justified based on the source animals and their geographical origin, the nature of the animal material, and the production process(es) and the quality assurance system. Properties of a medium that need to be checked include chemical (grade, potency, purity, impurity profiles, etc.), physical (appearance, identity, etc.), physiochemical (pH, osmolality, etc.), and biological (functional capacity in supporting growth and production, etc.). In addition, qualification tests on sterility, endotoxin, and mycoplasma need to be performed to demonstrate lack of bioburden. For liquid media, release criteria from vendors typically are appearance, pH, osmolality, sterility, endotoxin, and in some cases, growth promotion. For powder media, release criteria can also include moisture. Additional quality tests such as measurement of trace elements or heavy metals may also be performed. The test results are supplied on the vendor's Certificate of Analysis.

Upon receipt of the medium from a vendor, most biopharmaceutical manufacturers would repeat some or all of the tests. Assays on pH, osmolality, sterility, and endotoxin are standardized procedures that are well established in the industry, and can also be contracted out to specialized test labs. In addition, according to 21 CFR, Part 211 (64), identity must be confirmed. This can be done by spectroscopy like infrared or nuclear magnetic resonance to see if the RM matches the label claim based on established fingerprints of functional groups. But most of all, the medium needs to be able to support cell growth. A "Pass" or "Acceptable for Use" release criterion by the vendor does not always ensure that the medium can meet the customer's specific process requirements. Growth promotion testing of media is considered unique and needs to be performed by the user (customer) itself, since frequently the criteria for release by vendors do not match the customer's requirements, for the following reasons: (i) the vendor usually does not possess the actual production cell line for real performance evaluation; (ii) the vendor cannot test the medium matrix effect because the customer may supplement additional components to the medium at use; (iii) there may be container effects, such as cholesterol/lipid depletion during medium storage in plastic bags (85, 176); (iv) filtration prior to use in a bioreactor vessel may retain filter-extractable substances; (v) there can be an impact on cell culture performance that can only be detected by a sensitive cell line and when carryover is completely eliminated; and finally, (vi) in-house growth promotion testing at small scale can better take scale-up factors into consideration. Case studies were described to illustrate the development and application of growth promotion testing, along with the influencing factors (9).

13.4.4. Operation and Release

Standard, fixed manufacturing remains a good investment for high-demand drugs. But modular manufacturing facilities let companies easily change and optimize operations and floor space as needs evolve and technology improves. A current trend is to use multiple production bioreactors in a cell culture facility, basing the size and quantity of production bioreactors on the forecasted annual demand of product, the titer that can be achieved, the bioreactor success rate, and overall recovery and purification yields. Batch processing makes chromatography columns sit idle at times. Staggering batches could increase throughput and relieve bottlenecks by approximating continuous processing. A flexible facility can process more production campaigns, handle a wide range of products and process yields, and ultimately increase the facility output dramatically. Also, productivity improvements can be gained by using smaller equipment or the similar-size equipment in a more time-efficient manner (104).

In addition to widely used disposable bioreactors, single-use apparatuses like filters, tubings, bags, mixing systems, and aseptic connectors can now be incorporated in every process step. Single-use technologies such as noninvasive sensors (127), fully disposable cross-flow cassettes or systems for cell harvest, concentration, and DF, and disposable membrane chromatography capsules have proven effective and cost- and time-efficient for certain steps, replacing the traditional packed-bed chromatography in flow-through operations, including polishing for removal of viruses and contaminants (69, 189). Closed, sterile mixing bags are used for column eluate collection, dilution, titration, and viral inactivation. The single-use disposable process has also been integrated into the critical aseptic processing (fill) operations (17). One application increasing in use is transportation of sterile liquids (121). Single-use systems can be scaled up faster than their stainless steel counterparts and offer process flexibility every step of the way because they eliminate cleaning, sterilization, and their validations; reduce use of caustic chemicals and detergents as well as consumption of water for injection; allow efficient use of space; minimize cross-contamination; reduce downtime associated with stainless steel assembly, labor, cleaning-in-place, and sterilizing-in-place; provide the ability and flexibility to increase production capacity; and decrease capital investment (18). The growth in use of disposables will be a function of economics, which is justified in biopharmaceutical manufacturing (98, 102). Phase I and II production operations can use disposable manufacturing to increase flexibility and avoid tying up capital in stainless steel (32). However, completely disposable manufacturing plants for the large-scale commercial production of specific molecules is thought unlikely to become a reality because of the natural technology limitations (169).

Release testing is a major aspect of biopharmaceutical manufacturing, particularly viral safety (42, 51, 53, 54). The cell banks, unprocessed bulk, final bulk, and finished product should all be characterized and tested before being released for use. According to regulatory guidelines (42, 54), the viral safety of an investigational medicinal product should be ensured by three complementary approaches involving (i) selecting and testing cell lines and other RMs of human or animal origin for viral contaminants, (ii) assessing the capacity of downstream processing to clear infectious viruses, and (iii) testing the product at appropriate steps for contaminating viruses. To fulfill these requirements, Master Cell Bank/Working Cell Bank and all biologically originated RMs should be tested for viral safety; the purification process needs to be validated for virus clearance; and the unprocessed bulk materials need to be evaluated for the presence of adventitious viruses and the count of retroviral particles.

13.4.5. Production Cost

Mammalian cell culture on the industrial scale is not to be taken lightly. It involves expensive growth media, complicated purification processes, and expensive cell line characterization and documentation. In addition, the cost and startup times for large-scale protein production facilities

are substantial, typically ranging from $500 million to $1 billion and taking about 3 to 5 years to achieve full regulatory approval (3). Therefore, biopharmaceuticals such as MAbs are among the most expensive drugs, as the annual cost per cancer patient can reach $35,000 (44), and final products from mammalian cell culture may cost $500 to $5,000 per gram (146), of which about half is due to purification and the other half is shared by cell culture and quality control/quality assurance. Despite the attractive returns, potential losses in revenue due to product approval delay have made companies focus more on speed to market rather than on improving process economics, and the competition and pressure will force MAb therapeutics to increasingly depend on economic factors (116). As a result, production costs and capacity use are becoming critical factors for the industry (45, 86).

The cost structure of manufactured lots depends on process time and productivity. To shorten the time, recent technologies include using perfusion cell cultures to amplify dihydrofolate reductase-based expression cell lines by incrementally increasing methotrexate in the perfusion medium for the faster generation of stable and high-productivity clones, and performing perfusion processes in single-use bioreactors to prepare cell banks at high density, which minimizes the inoculation time from vial thaw to seed bioreactor, leading to cost-effective facility utilization (5). In addition, all phases of cell line development have been integrated using a single medium—the production medium—thus reducing the cell culture development time and minimizing intermittent adaptation processes, which saves both time and cost (5). To enhance productivity, a perfusion process with compact packed-bed bioreactors achieved reported CHO cell densities of ~250 × 10^6/ml and as high as 31 g/liter of bed-volume MAb productivity at development scale (4). Another example used perfusion-fed batch technology to enable both high cell density and antibody concentration (~25 g/liter) in low-volume cell culture medium, while integrating recovery operations (122). These high titers minimize equipment sizes and processing volumes. The assembly of individual process units becomes easy and flexible because of the application of single-use systems, which can lead to significant cost reduction in upstream and recovery unit operations. For downstream processing, practically feasible approaches are also emerging to control economics and reduce purification cost, which has been the major cost driver for MAb manufacturing (6, 30). It is claimed that the future of therapeutic MAbs lies in the development of economically feasible downstream processing (44).

Outsourcing of biopharmaceutical product manufacturing has grown significantly in the past few years, mainly because of the increase in the number of biopharmaceuticals being developed and the availability of highly qualified contract manufacturing organizations (CMOs) (149, 162). Such a transformation is gradually taking place in the biopharmaceutical industry, as exemplified by the fact that the total number of related articles published in 2005 is more than in the entire decade of the 1990s (120). Before 1995, companies were required to manufacture phase II clinical materials at the final commercial scale. In 1995 the FDA authorized the use of pilot facilities, allowing companies to make clinical materials at smaller scale or license those smaller GMP facilities later for commercial manufacturing (52). This incurred less financial risk for companies that do not have the capacity, capability, or in-house expertise to manufacture the clinical products in the event of phase II failure, as they did not have to construct expensive large-scale facilities. A sound Chemistry, Manufacturing and Control (CMC) strategic plan is the key to the selection of the right CMO (149), and well-executed technical transfer of manufacturing processes to CMOs leads to the success of custom manufacturing (154). It should be pointed out that a change in production scale and/or site is considered a change in manufacturing process, so analytical, nonclinical, and even clinical studies are needed to demonstrate the comparability in quality, safety, and efficacy of the drug product (62).

Mammalian cell culture requires more testing than microbial fermentation systems. Generally the more advanced the bioactive entity, the greater the problem in paying for discovery, development, scale-up, facilities, regulatory compliance, and overall investment. These economic aspects of the biotechnology industry are described in detail in an excellent article in the last edition of this manual (130). A recent review discussed the key process economic drivers and the impact of scale and titer on trade-offs in antibody purification processes (44). To address these two critical factors, the industry is looking to improve the overall downstream processing yield and reduce the batch duration by using platform processes based on chromatography steps without intermediate buffer exchange steps, to increase capacity by improving resins to allow increased throughput over shorter times, and to lower buffer demands and validation cost using new technologies such as membrane chromatography. Overall assay development and validation for biomolecular analysis can cost around $1.5 million—about $150,000 for three immunological assays, about $1 million for maybe a dozen chemical/biochemical assays, and almost $0.5 million for just three bioassays (143). The cost of one lot of MAb product release assays can be $50,000 to $100,000, and that for systematic stability testing $250,000 to $500,000. As for the preclinical development, the overall cost estimate 10 years ago for developing a protein molecule into the clinical phase was already about $1.7 million to $2.7 million (170).

13.5. SUMMARY

The dramatic advancement in mammalian cell line and cell culture development has shifted the biopharmaceutical manufacturing bottleneck to downstream processing. In response, the industry is moving toward using current purification capacity more efficiently by improved operations and new technologies such as membrane chromatography and disposable apparatuses, and toward using platform purification approaches until late phase for products such as MAbs. However, the first step using protein A affinity resin remains rate limiting due to its low capacity and high cost despite great purification. Formulation is critical to the stability and application of therapeutic proteins, and its development remains mostly the combination of art and science. Fill and finish of these products require more care than classical small-molecule drugs because of their limited stability in liquid and their parenteral administration. Modern bioanalytical tools have been established to characterize and release biopharmaceuticals, and the methods need to be validated. Manufacture of biopharmaceutical is governed by various cGMP regulations, requiring validation of process, qualification/release of all RMs, and thorough testing/release of cell banks and products to ensure safety and potency. Costs are high for production using mammalian cell systems due to expensive facilities, equipment, RMs, testing, validation, etc. Alternative operation approaches are emerging, including incorporation of more single-use devices and outsourcing.

REFERENCES

1. **Akers, M.** 2006. Special challenges in production of biopharmaceutical dosage forms. *BioProcess Int.* **4:**36–43.

2. **Andya, J. D., C. C. Hsu, and S. J. Shire.** 2003. Mechanisms of aggregate formation and carbohydrate excipient stabilization of lyophilized humanized monoclonal antibody formulations. *AAPS PharmSci.* **5:**E10.

3. **Anicetti, V.** 2009. Biopharmaceutical processes: a glance into the 21st century. *BioProcess Int.* **7:**S4–S11.

4. **Artelis SA and Selexis SA.** 2008. Artelis and Selexis announces 31 g/L yields in multiple perfusion harvests. *BioPharm. Bull.* **Sept.** (Online.) http://biopharminternational.findpharma.com/biopharm/issue/issueDetail.

5. **Arunakumari, A.** 2009. Implementing cost reduction strategies for HuMab manufacturing processes. *BioProcess Int.* **7:**S48–S54.

6. **Arunakumari, A., J. M. Wang, and G. Ferreira.** 2007. Alternative to Protein A: improved downstream process design for human monoclonal antibody production. *BioPharm Int.* **20:**S36–S40.

7. **Arunakumari, A., J. M. Wang, and G. Ferreira.** 2007. Improved downstream process design for human monoclonal antibody production. *BioPharm Int.* **20:**S6–S10.

8. **Arunakumari, A., J. M. Wang, and G. Ferreira.** 2009. Advances in non-Protein A purification processes for human monoclonal antibodies. *BioPharm Int.* **22:**S22–S26.

9. **Balasubramanian, U., P. Salmon, D. Robinson, and J. Zhang.** 2006. Characterization and release of raw materials used in upstream processes for production of monoclonal antibodies by mammalian cell culture. *BioProcess. J.* **5:**7–13.

10. **Bansal, S. K., T. Layloff, E. D. Bush, M. Hamilton, E. A. Hankinson, J. S. Landy, S. Lowes, M. M. Nasr, P. A. St. Jean, and V. P. Shah.** 2004. Qualification of analytical instruments for use in the pharmaceutical industry: a scientific approach. *AAPS PharmSci.* **5:**E22.

11. **Beck, A., E. Wagner-Rousset, M. C. Bussat, M. Lokteff, C. Klinguer-Hamour, J. F. Haeuw, L. Goetsch, T. Wurch, A. van Dorsselaer, and N. Corvaia.** 2008. Trends in glycosylation, glycoanalysis and glycoengineering of therapeutic antibodies and Fc-fusion proteins. *Curr. Pharm. Biotechnol.* **9:**482–501.

12. **BioPharm International.** 2004. Guide to formulation, fill, and finish. *BioPharm Int.* **17:**S1–S42.

13. **Birch, J. R., J. Bonnerjea, and S. Flatman.** 1995. The production of monoclonal antibodies, p. 231–265. *In* J. R. Birch and E. S. Lennox (ed.), *Monoclonal Antibodies: Principles and Applications.* John Wiley & Sons, Inc., Hoboken, NJ.

14. **Boyd, P. N., A. C. Lines, and A. K. Patel.** 1995. The effect of the removal of sialic acid, galactose, and total carbohydrate on the functional activity of CAMPATH-1H. *Mol. Immunol.* **32:**1311–1318.

15. **Brorson, K., S. Krejci, K. Lee, E. Hamilton, K. Stein, and Y. Xu.** 2003. Bracketed generic inactivation of rodent retroviruses by low pH treatment for monoclonal antibodies and recombinant proteins. *Biotechnol. Bioeng.* **82:**321–329.

16. **Buchacher, A., and G. Iberer.** 2006. Purification of intravenous immunoglobulin G from human plasma—aspects of yield and virus safety. *Biotechnol. J.* **2:**148–163.

17. **Cappia, J. M., and N. B. T. Holman.** 2004. Integrating single-use disposable processes into critical aseptic processing operations. *BioProcess Int.* **2:**S56–S63.

18. **Cardona, M., and B. Allen.** 2006. Incorporating single-use systems in biopharmaceutical manufacturing. *BioProcess Int.* **4:**S10–S14.

19. **Carpenter, J. F., B. S. Chang, W. Garzon-Rodriguez, and T. W. Randolph.** 2002. Rational design of stable lyophilized protein formulations: theory and practice, p. 109–133. *In* J. F. Carpenter and M. C. Manning (ed.), *Rational Design of Stable Protein Formulations: Theory and Practice.* Springer, Berlin, Germany.

20. **Carpenter, J. F., M. J. Pikal, B. S. Chang, and T. W. Randolph.** 1997. Rational design of stable lyophilized

protein formulations: some practical advice. *Pharm. Res.* **14:** 965–975.

21. **Carter, J., and H. Lutz.** 2003. Validation of virus filtration—ensuring regulatory compliance. *BioProcess Int.* **1:**52–62.

22. **Chang, L., D. Shepherd, J. Sun, D. Ouellette, K. L. Grant, X. Tang, and M. J. Pikal.** 2005. Mechanism of protein stabilization by sugars during freeze-drying and storage: native structure preservation, specific interaction, and/or immobilization in a glassy matrix? *J. Pharm. Sci.* **94:**1427–1444.

23. **Chen, B., G. Zapata, and S. M. Chamow.** 2003. Strategies for rapid development of liquid and lyophilized antibody formulations. *BioProcess Int.* **2:**46–55.

24. **Clutterbuck, A.** 2008. Integrating and streamlining biopharm purification processes. *Innovations Pharm. Technol.* October:42–46.

25. **Committee for Human Medicinal Products/Biotechnology Working Party.** 1995. Note for guidance on virus validation studies: the design, contribution and interpretation of studies validating the inactivation and removal of viruses. CHMP/BWP/268/95. EMEA, London, United Kingdom.

26. **Curling, J., and U. Gottschalk.** 2007. Process chromatography: five decades of innovation. *BioPharm Int.* **20:** 70–93.

27. **Daugherty, A., and R. J. Mrsny.** 2006. Formulation and delivery issues for monoclonal antibody therapeutics. *Adv. Drug Deliv. Rev.* **58:**686–706.

28. **Davies, J.** 2009. A purification platform for the production of MAbs from fermentors with titers of 5 g/L and beyond. *BioPharm Int.* **22:**S28–S31.

29. **Defrancq, L., N. Callewaert, J. Zhu, W. Laroy, and R. Contreras.** 2004. High-throughput analysis of the N-glycans of NS0 cell-secreted antibodies. *BioProcess Int.* **2:**60–68.

30. **DePalma, A.** 2008. Reducing downstream purification cost. *Genet. Eng. Biotechnol. News* **28**(13).

31. **DePaz, R. A., C. C. Barnett, D. A. Dale, J. F. Carpenter, A. L. Gaertner, and T. W. Randolph.** 2000. The excluding effect of sucrose on a protein chemical degradation pathway: methionine oxidation in subtilisin. *Arch. Biochem. Biophys.* **384:**123–132.

32. **Diblasi, K., M. W. Jornitz, U. Gottschalk, and P. M. Priebe.** 2007. Disposable biopharmaceutical processes—myth or reality? *BioPharm Int.* **20:**S19–S24.

33. **Dinowitz, M., Y. S. Lie, M. A. Low, R. Lazar, C. Fautz, B. Potts, J. Sernatinger, and K. Anderson.** 1992. Recent studies on retrovirus-like particles in Chinese hamster ovary cells. *Dev. Biol. Stand.* **76:**201–207.

34. **Dominguez, C. A., E. Rivera, C. Escaboar, and J. Weidner.** 2008. Improving tangential flow filtration yield. *BioPharm Int.* **21:**42–59.

35. **Dorsey, N., J. Eschrich, and G. Cyr.** 1997. The role of charge in the retention of DNA by charged cellulose based depth filters. *BioPharm Int.* **10:**46–49.

36. **Dougherty, J., R. Mhatre, and S. Moore.** 2003. Using peptide maps as identity and purity tests for lot release testing of recombinant therapeutic proteins. *BioPharm Int.* **16:**54–58.

37. **Drickamer, K.** 1991. Clearing up glycoprotein hormones. *Cell* **67:**1029–1932.

38. **Dutton, G.** 2008. Downstream bottlenecks: are they myth or reality? *Genet. Eng. Biotechnol. News* **28**(8).

39. **Eppink, M. H. M., R. Schreurs, A. Gusen, and K. Verhoeven.** 2009. Platform technology for developing purification processes. *BioPharm Int.* **22:**S32–S36.

40. **European Commission.** 2009. EudraLex. The rules governing medicinal products in the European Union, vol. 4. Good manufacturing practice (GMP) guidelines, annex 8. Sampling of starting and packaging materials. European Commission, Brussels, Belgium. (Online.) http://ec.europa.eu/enterprise/pharmaceuticals/eudralex/vol-4/pdfs-en/anx08en.pdf.

41. **European Medicines Evaluation Agency.** 2002. Note for guidance on minimizing the risk of transmitting animal spongiform encephalopathies via human and veterinary medicinal products. EMEA/410/01, Rev. 2. EMEA, London, United Kingdom.

42. **European Medicines Evaluation Agency.** 2008. Guideline on virus safety evaluation of biotechnological investigational medicinal products. EMEA/CHMP/BWP/398498/2005. EMEA, London, United Kingdom.

43. **Fahrner, R. L., H. L. Knudsen, C. D. Basey, W. Galan, D. Feuerheim, M. Vanderlaan, and G. S. Blank.** 2001. Industrial purification of pharmaceutical antibodies: development, operation, and validation of chromatography processes. *Genet. Eng. Biotechnol. Rev.* **18:**301–327.

44. **Farid, S. S.** 2008. Economic drivers and trade-offs in antibody purification processes. *BioPharm. Int.* **21:**S37–S42.

45. **Farid, S. S., J. Washbrook, and N. J. Titchener-Hooker.** 2005. Decision-support tool for assessing bio-manufacturing strategies under uncertainty: stainless steel versus disposable equipment for clinical trial material preparation. *Biotechnol. Prog.* **21:**486–497.

46. **Farshid, M., R. E. Taffs, D. Scott, D. M. Asher, and K. Brorson.** 2005. The clearance of viruses and transmissible spongiform encephalopathy agents from biologicals. *Curr. Opin. Biotechnol.* **16:**561–567.

47. **Federal Register.** 1995. Interim definition and elimination of lot-by-lot release for well-characterized therapeutic recombinant DNA-derived and monoclonal antibody biotechnology products. *Fed. Regist.* **60:**63048.

48. **Federal Register.** 1996. Well-characterized biotechnology products: elimination of establishment license application. *Fed. Regist.* **61:**2733.

49. **Fetterolf, D. M.** 2007. Developing a sound process validation strategy. *BioPharm Int.* **20:**38–46.

50. **Food and Drug Administration.** 1987. Guideline on general principles of process validation. Food and Drug Administration, Rockville, MD. (Online.) http://www.fda.gov/Drugs/GuidanceComplianceRegulatoryInformation/Guidances/ucm124720.htm

51. **Food and Drug Administration.** 1993. Points to consider in the characterization of cell lines used for biological products. Food and Drug Administration, Rockville, MD. (Online.) http://www.fda.gov/downloads/BiologicsBloodVaccines/GuidanceComplianceRegulatoryInformation/OtherRecommendationsforManufacturers/UCM062745.pdf

52. **Food and Drug Administration.** 1995. FDA guidance document concerning use of pilot manufacturing facilities for the development and manufacture of biological products. Food and Drug Administration, Rockville, MD. (Online.) http://www.fda.gov/downloads/BiologicsBloodVaccines/GuidanceComplianceRegulatoryInformation/Guidances/General/UCM168111.pdf

53. **Food and Drug Administration.** 1997. Points to consider in the manufacture and testing of monoclonal antibody products for human use. Food and Drug Administration, Rockville, MD. (Online.) http://www.fda.gov/downloads/BiologicsBloodVaccines/GuidanceComplianceRegulatoryInformation/OtherRecommendationsforManufacturers/UCM153182.pdf

54. **Food and Drug Administration.** 1998. Guidance for industry: Q5A viral safety evaluation of biotechnological products derived from cell lines of human or animal origin (ICH). Food and Drug Administration, Rockville, MD. (Online.) http://www.fda.gov/downloads/RegulatoryInformation/Guidances/UCM129101.pdf

55. **Food and Drug Administration.** 1999. Guidance on specifications: Q6B test procedures and acceptance criteria for biotechnological/biological products (ICH). Food and Drug Administration, Rockville, MD. (Online.)

http://www.fda.gov/downloads/RegulatoryInformation/Guidances/UCM129100.pdf

56. **Food and Drug Administration.** 2000. Guidance for industry: analytical procedures and methods validation. Food and Drug Administration, Rockville, MD. (Online.) http://www.fda.gov/downloads/Drugs/GuidanceComplianceRegulatoryInformation/Guidances/ucm122858.pdf

57. **Food and Drug Administration.** 2001. Guidance for industry: bioanalytical method validation. Food and Drug Administration, Rockville, MD. (Online.) http://www.fda.gov/downloads/Drugs/GuidanceComplianceRegulatoryInformation/Guidances/UCM070107.pdf

58. **Food and Drug Administration.** 2001. Guidance for industry: Q7A good manufacturing practice for active pharmaceutical ingredients (ICH). Food and Drug Administration, Rockville, MD. (Online.) http://www.fda.gov/downloads/RegulatoryInformation/Guidances/UCM129098.pdf

59. **Food and Drug Administration.** 2003. Guidance for industry: Q1A(R2) stability testing of new drug substances and products (ICH). Food and Drug Administration, Rockville, MD. (Online.) http://www.fda.gov/downloads/RegulatoryInformation/Guidances/ucm128204.pdf

60. **Food and Drug Administration.** 2004. Guidance for industry: sterile drug products produced by aseptic processing—current good manufacturing practice. Food and Drug Administration, Rockville, MD. (Online.) http://www.fda.gov/downloads/Drugs/GuidanceComplianceRegulatoryInformation/Guidances/ucm070342.pdf

61. **Food and Drug Administration.** 2005. Guidance for industry: Q2(R1) validation of analytical procedures: test and methodology (ICH). Food and Drug Administration, Rockville, MD. (Online.) http://www.fda.gov/downloads/RegulatoryInformation/Guidances/UCM128048.pdf; http://www.fda.gov/downloads/RegulatoryInformation/Guidances/UCM128049.pdf

62. **Food and Drug Administration.** 2005. Guidance for industry: Q5E comparability of biotechnological/biological products subject to changes in their manufacturing process (ICH). Food and Drug Administration, Rockville, MD. (Online.) http://www.fda.gov/downloads/RegulatoryInformation/Guidances/ucm128076.pdf

63. **Food and Drug Administration.** 2006. Guidance for industry: Q9 quality risk management (ICH). Food and Drug Administration, Rockville, MD. (Online.) http://www.fda.gov/downloads/Drugs/GuidanceComplianceRegulatoryInformation/Guidances/ucm073511.pdf

64. **Food and Drug Administration.** 2008. Current good manufacturing practices for finished pharmaceuticals (21 CFR, Part 211). Food and Drug Administration, Rockville, MD. (Online.) http://www.accessdata.fda.gov/scripts/cdrh/cfdocs/cfcfr/CFRSearch.cfm?CFRPart=211&showFR=1

65. **Food and Drug Administration.** 2008. Guidance for industry: cGMP for phase 1 investigational drugs. Food and Drug Administration, Rockville, MD. (Online.) http://www.fda.gov/downloads/Drugs/GuidanceComplianceRegulatoryInformation/Guidances/UCM070273.pdf

66. **Food and Drug Administration.** 2008. Guidance for industry: process validation: general principles and practices. Food and Drug Administration, Rockville, MD. (Online.) http://www.fda.gov/downloads/Drugs/GuidanceComplianceRegulatoryInformation/Guidances/UCM070336.pdf

67. **Food and Drug Administration.** 2007. Guidance for industry: Q10 pharmaceutical quality system (ICH). Food and Drug Administration, Rockville, MD. (Online.) http://www.fda.gov/downloads/RegulatoryInformation/Guidances/ucm128031.pdf

68. **Frankin, S.** 2003. Removal of aggregate from an IgG4 product using CHT ceramic hydroxyapatite. *BioProcess Int.* **1:**50–51.

69. **Fraud, N.** 2008. Membrane chromatography: an alternative to polishing column chromatography. *BioProcess. J.* **7:**34–48.

70. **Ganderton, D.** 1991. The development of peptide and protein pharmaceuticals, p. 211–277. *In* R. C. Hider and D. Barlow (ed.), *Polypeptide and Protein Drugs: Production, Characterization, and Formulation.* Ellis Horwood Ltd., Chichester, United Kingdom.

71. **Giovannoni, L., M. Ventani, and U. Gottschalk.** 2008. Antibody purification using membrane adsorbers. *BioPharm Int.* **21:**48–52.

72. **Glynn, J., T. Hagerty, T. Pabst, G. Annathur, K. Thomas, P. Johnson, N. Ramasubramanyan, and P. Mensah.** 2009. The development and application of a monoclonal antibody purification platform. *BioPharm Int.* **22:**S16–S20.

73. **Goolcharran, C., J. L. Cleland, R. Keck, A. J. Jones, and R. T. Borchardt.** 2000. Comparison of the rates of deamidation, diketopiperazine formulation and oxidation in recombinant human vascular endothelial growth factor and model peptides. *AAPS PharmSci.* **2:**E5.

74. **Gottschalk, U.** 2007. The renaissance of protein purification. *BioPharm Int.* **20:**S41–S42.

75. **Gottschalk, U., and S. Fishcher-Fruehholz.** 2004. A cutting edge process technology at the threshold. *BioProcess Int.* **2:**56–65.

76. **Guhan, J., Y. F. Li, J. A. Moore, and S. M. Cramer.** 1995. Ion-exchange displacement chromatography of proteins: dendritic polymers as novel displacers. *J. Chromatogr. A* **702:**143–155.

77. **Imensek, M.** 2003. Sterile fill facilities: problems and resolutions. *Biopharm Int.* **16:**44–54.

78. **Immelmann, A., K. Kellings, O. Stamm, and K. Tarrach.** 2005. Validation and quality procedures for virus and prion removal in biopharmaceuticals. *BioProcess Int.* **3:**S26–S31.

79. **International Society for Pharmaceutical Engineering.** 2008. GAMP 4 to GAMP 5 summary. International Society for Pharmaceutical Engineering, Tampa, FL. (Online.) http://www.ispe.org/galleries/publications-files/GAMP_4_to_GAMP_5_Summary.pdf.

80. **Iyer, H., F. Henderson, E. Cunningham, J. Welbb, J. Hanson, C. Bork, and L. Conley.** 2002. Considerations during development of a Protein A-based antibody purification process. *BioPharm Int.* **15:**14–20.

81. **Jacob, L. R., and M. Frech.** 2004. Scale up of antibody purification: from laboratory scale to production, p. 101–132. *In* G. Subramanian (ed.), *Antibodies*, vol. 1. *Production and Purification.* Kluwer Academic/Plenum Publishers, New York, NY.

82. **Jacobsen, M, and C. Sandstrom.** 2006. An approach to optimizing large-scale cell harvest. *BioProcess Int.* **4:**28–35.

83. **Jefferis, R.** 2005. Glycosylation of recombinant antibody therapeutics. *Biotechnol. Prog.* **21:**11–16.

84. **Jorgensen, K., and T. Naes.** 2004. A design and analysis strategy for situations with uncontrolled raw material variation. *J. Chemometrics* **18:**45–52.

85. **Kadarusman, J., R. Bhatia, J. McLaughlin, and W. R. Lin.** 2005. Growing cholesterol-dependent NS0 myeloma cell line in the Wave bioreactor system: overcoming cholesterol-polymer interaction by using pretreated polymer or inert fluorinated ethylene propylene. *Biotechnol. Prog.* **21:**1341–1346.

86. **Kamarck, M. E.** 2006. Building biomanufacturing capacity—the chapter and verse. *Nat. Biotechnol.* **24:**503–505.

87. **Kanarek, A. D.** 2005. Method validation guidelines. *In* Guide to Bioanalytical Advances. *BioPharm Int.* **18:**S28–S33.

88. **Kania, K.** 2002. Blow-fill-seal packaging: a sterile environment. *Pharm. Med. Packaging News* September.

89. **Kelley, B.** 2007. Very large scale monoclonal antibody purification: the case for conventional unit operations. *Biotechnol. Prog.* **23:**995–1008.

90. **Kendrick, B. S., J. F. Carpenter, J. L. Cleland, and T. W. Randolph.** 1998. A transient expansion of the native state

precedes aggregation of recombinant human interferon-gamma. *Proc. Natl. Acad. Sci. USA* **95**:14142–14146.

91. **Kilgore, B. R., A. D. Lucka, R. Patel, B. A. Andrien, Jr., and S. T. Dhume.** 2008. Comparability and monitoring immunogenic N-linked oligosaccharides from recombinant monoclonal antibodies from two different cell lines using HPLC with fluorescence detection and mass spectrometry. *Methods Mol. Biol.* **446**:333–346.

92. **Knudsen, H. L., R. L. Fahrner, Y. Xu, L. A. Norling, and G. S. Blank.** 2001. Membrane ion-exchange chromatography for process-scale antibody purification. *J. Chromatogr. A* **907**:145–154.

93. **Krause, S. O.** 2004. Development and validation of analytical methods for biopharmaceuticals, part I: development and optimization. *BioPharm Int.* **17**:52–61.

94. **Krause, S. O.** 2007. *Validation of Analytical Methods for Biopharmaceuticals—a Guide to Risk Based Validation and Implementation Strategies.* PDA/DHI Publishing, Bethesda, MD.

95. **Krause, S. O.** 2008. Formal method validation. *In* Guide to Analytical Methods. *BioPharm Int.* **21**:S14–S22.

96. **Lam, X. M., J. Y. Yang, and J. L. Cleland.** 1997. Antioxidants for prevention of methionine oxidation in recombinant monoclonal antibody HER2. *J. Pharm. Sci.* **86**:1250–1255.

97. **Langer, E. S.** 2008. Quantifying trends toward alternative to Protein A. *BioProcess Int.* **6**:72.

98. **Langer, E. S., and J. Ranck.** 2005. The ROI case—economic justification for disposables in biopharmaceutical manufacturing. *BioProcess Int.* **3**:S46–S50.

99. **Langer, E. S., and J. Ranck.** 2006. Capacity bottleneck squeezed by downstream processes. *BioProcess Int.* **4**:14–18.

100. **Li, F., J. Z. Zhou, X. Yang, T. Tressel, and B. Lee.** 2006. Current therapeutic antibody production and process optimization. *BioProcess. J.* **4**:16–25.

101. **Lifely, M. R., C. Hale, S. Boyce, M. J. Keen, and J. Phillips.** 1995. Glycosylation and biological activity of CAMPATH-1H expressed in different cell lines and grown under different culture conditions. *Glycobiology* **5**:813–822.

102. **Lim, J. A. C., A. Sinclair, M. Hirai, and U. Gottschalk.** 2007. Disposable membrane chromatography—counting the cost. *BioPharm Int.* **20**:S34–S40.

103. **Lim, J. A. C., A. Sinclair, D. S. Kim, and U. Gottschalk.** 2007. Economic benefits of single-use membrane chromatography for process-scale antibody purification. *J. Chromatogr. A* **907**:145–154.

104. **Liu, X., M. Collins, N. Fraud, J. Campbell, I. Lowenstein, K. M. Lacki, and A. S. Rathore.** 2009. Upcoming technologies to facilitate more efficient biologics manufacturing. *BioPharm Int.* **22**:38–51.

105. **Low, D., R. O'Leary, and N. S. Pujar.** 2007. Future of antibody purification. *J. Chromatogr. B* **848**:48–63.

106. **Lubiniecki, A. S., and P. J. Shadle.** 1997. Raw material considerations. *Dev. Biol. Stand.* **91**:65–72.

107. **Lucka, A. W., B. R. Kilgore, R. Patel, B. A. Andrien, Jr., and S. T. Dhume.** 2008. Mass spectrometry and HPLC with fluorescent detection-based orthogonal approaches to characterize N-linked oligosaccharides of recombinant monoclonal antibodies. *Methods Mol. Biol.* **446**:347–361.

108. **Lucy, P. K., and R. G. Beri.** 2003. Key considerations in process transfer. *BioProcess Int.* **1**:36–43.

109. **Luo, R., R. Waghmare, M. Krishnan, C. Adams, E. Poon, and D. Kahm.** 2006. High-concentration UF/DF of a monoclonal antibody—strategy for optimization and scale-up. *BioProcess Int.* **4**:44–48.

110. **Magari, R. T.** 2002. Estimating degradation in real time and accelerated stability tests with random lot-to-lot variation: a simulation study. *J. Pharm. Sci.* **91**:893–899.

111. **Magari, R. T.** 2003. Assessing shelf life using real-time and accelerated stability tests. *BioPharm Int.* **16**:36–48.

112. **Magari, R. T., K. P. Murphy, and T. Fernandez.** 2002. Accelerated stability model for predicting shelf-life. *J. Clin. Lab. Anal.* **16**:221–226.

113. **Magil, S. G.** 2005. Biopharmaceutical characterization techniques for early phase development of proteins. *In* Guide to Bioanalytical Advances. *BioPharm Int.* **18**:S34–S42.

114. **Manzi, A. E.** 2008. Carbohydrates and their analysis, part two: glycoprotein characterization. *BioProcess Int.* **6**:50–67.

115. **Meyeroltmanns, F., J. Schmitz, and M. N. Kamal.** 2005. Disposable bioprocess components and single-use concepts for optimized process economy in biopharmaceutical production. *BioProcess Int.* **3**:S60–S66.

116. **Mitchell, P.** 2005. Next-generation monoclonals less profitable than trailblazers? *Nat. Biotechnol.* **23**:906.

117. **Moritz, A.** 2005. The new GMP environment for investigational medicinal products in the European Union. *BioProcess Int.* **3**:28–38.

118. **Morrow, K. J., Jr.** 2008. Method for maximizing antibody yields—new technologies could help usher in lower costs and increased availability. *Genet. Eng. Biotechnol. News* **28**(12).

119. **Nguyen, L. T., J. M. Wiencek, and L. E. Kirsch.** 2003. Characterization methods for the physical stability of biopharmaceuticals. *PDA J. Pharm. Sci. Technol.* **57**:429–445.

120. **Niazi, S. K., and T. L. Flynn III.** 2006. A practical model for outsourced biomanufacturing. *BioProcess Int.* **4**:S10–S16.

121. **Norris, D.** 2008. Method for evaluating bio-container transportation applications. *BioProcess Int.* **6**:142–143.

122. **Percivia.** 2008. Press release: DSM and Crucell announce record achievement in PER.C6 technology. June 16. Percivia, Cambridge, MA. (Online.) http://investors.crucell.com/C/132631/PR/200806/1227870_5_5.html.

123. **Polin, J. B.** 2000. Doing aseptic filling in barrier isolators. *Pharm. Med. Packaging News* November.

124. **Raju, T. S.** 2003. Glycosylation variations with expression systems and their impact on biological activity of therapeutic immunoglobulins. *Bioprocess Int.* **1**:44–53.

125. **Ramelmeier, R. A., B. D. Kelley, and C. Van Horn.** 1998. Historical, current, and future trends for validating biological processes, p. 1–11. *In* B. D. Kelley and R. A. Ramelmeier (ed.), *Validation of Biopharmaceutical Manufacturing Processes.* American Chemical Society, Washington, DC.

126. **Ransohoff, T. C.** 2007. Successful strategies for production optimization. *BioPharm. Int.* **20**:165–174.

127. **Rao, G., Y. Kostov, A. Moreira, D. Frey, M. Hanson, M. Jornitz, O.-W. Reif, R. Baumfalk, and J. Qualitz.** 2009. Non-invasive sensors as enablers of "smart" disposables. *BioProcess Int.* **7**:S24–S27.

128. **Rathores, A. S., A. Wang, M. Menon, J. Martin, J. Campbell, and E. Goodrich.** 2004. Optimization, scale-up, and validation issues in filtration of biopharmaceuticals. *BioPharm Int.* **17**:42–50.

129. **Ray, S., and K. Tarrach.** 2008. Virus clearance strategy using a three-tier orthogonal technology platform. *BioPharm Int.* **21**:50–59.

130. **Reisman, H. B.** 1999. Economics, p. 273–288. *In* A. L. Demain and J. E. Davies (ed.), *Manual of Industrial Microbiology and Biotechnology*, 2nd ed. ASM Press, Washington, DC.

131. **Reynolds, G.** 2006. The market need for reconstitution systems. *BioProcess Int.* **4**:18–21.

132. **Robinson, C. J., L. E. Little, and H.-J. Wallny.** 2008. Bioassay survey 2006–2007. *BioProcess Int.* **6**:38–49.

133. **Roush, D. J., and Y. Lu.** 2008. Advances in primary recovery: centrifugation and membrane technology. *Biotechnol. Prog.* **24:**488–495.

134. **Russell, E., A. Wang, and A. S. Rathore.** 2007. Harvest of a therapeutic protein product, p. 1–58. *In* A. A. Shukla, M. R. Etzel, and S. Gadam (ed.), *Process Scale Bioseparations for the Biopharmaceutical Industry*. CRC Press, Boca Raton, FL.

135. **Russotti, G., and K. Goklen.** 2001. Cross-flow membrane filtration of fermentation broth, p. 85–159. *In* W. K. Wang (ed.), *Membrane Separations in Biotechnology*, 2nd ed. Marcel Dekker, New York, NY.

136. **Schenerman, M. A., J. Casas-Finet, M. J. Axley, and C. N. Oliver.** 2003. Characterization of alternatives to animal-derived raw materials. *BioProcess Int.* **1:**42–49.

137. **Schenerman, M. A., B. R. Sunday, S. Kozlowski, K. Webber, H. Gazzano-Santoro, and A. Mire-Sluis.** 2004. CMC startegy forum report—analysis and structure characterization of monoclonal antibodies. *BioProcess Int.* **2:**42–52.

138. **Schimdt, M. T., S. Sze-Khoo, A. R. Cothran, B. Q. Thai, S. Sargis, B. Lebreton, B. Kelley, and G. S. Blank.** 2009. Purification strategies to process 5 g/L titers on monoclonal antibodies. *BioPharm Int.* **22:**S8–S15.

139. **Schmidt, S., J. Mora, S. Dolan, and J. Kauling.** 2005. An integrated concept for robust and efficient virus clearance and contaminant removal in biotech process. *BioProcess Int.* **3:**S26–S31.

140. **Schreyer, H. B., S. E. Miller, and S. Rodgers.** 2007. Application note: high-throughput process development. *Genet. Eng. Biotechnol. News* **27**(17).

141. **Schwartz, L.** 2003. Diafiltration for desalting or buffer exchange. *BioProcess Int.* **1:**43–49.

142. **Scott, C.** 2005. Methodologies for viral safety. *BioProcess Int.* **3:**S8–S14.

143. **Scott, C.** 2006. Process development: turning science into technology. *BioProcess Int.* **4:**S24–S41.

144. **Scott, C.** 2006. Formulation development: making the medicine. *BioProcess Int.* **4:**S42–S56.

145. **Scott, C.** 2008. In the drug delivery zone: patients are the priority. *In* Bio International Convention Preshow Planner. *BioProcess Int.* **6:**S34–S41.

146. **Scott, C., S. A. Montgomery, and L. J. Rosin.** 2007. Advances in separation methods for increasing process streams. *In* Bio International Convention Preshow Planner. *BioPharm Int.* **5:**S43–S58.

147. **Sellers, S. P., and Y.-F. Maa.** 2005. Principles of biopharmaceutical protein formulation. *Methods Mol. Biol.* **38:**243–263.

148. **Senger, R. S., and M. N. Karim.** 2003. Effect of shear stress on intrinsic CHO culture state and glycosylation of recombinant tissue-type plasminogen activator protein. *Biotechnol. Prog.* **19:**1199–1209.

149. **Seymour, P. M., H. L. Levine, and S. D. Jones.** 2006. CMC strategies for outsourcing biopharmaceutical product manufacturing. *BioProcess Int.* **4:**S26–S29.

150. **Shadle, P. J.** 2004. The art of raw materials and supplier qualification. *BioProcess. J.* **3:**43–48.

151. **Shadle, P. J.** 2004. Qualification of raw materials for biopharmaceutical use. *BioPharm Int.* **17:**28–35.

152. **Sheeley, D. M., B. M. Merrill, and L. C. E. Taylor.** 1997. Characterization of monoclonal antibody glycosylation: comparison of expression systems and identification of terminal α-linked galactose. *Anal. Biochem.* **247:**102–110.

153. **Shi, L., Q. Chen, L. A. Norling, A. S. Lau, S. Krejci, and Y. Xu.** 2004. Real-time quantitative PCR as a method to evaluate xenotropic murine leukemia virus removal during pharmaceutical protein purification. *Biotechnol. Bioeng.* **87:**884–896.

154. **Shipston, N.** 2006. Technical transfer of manufacturing processes from client sites to a CMO. *BioProcess Int.* **4:**S40–S42.

155. **Shukla, A. A., B. Hubbard, T. Tressel, S. Guhan, and D. Low.** 2007. Downstream processing of monoclonal antibodies—application of platform technologies. *J.Chromatgr. B* **848:**28–39.

156. **Shukla, A. A., and J. R. Kandula.** 2008. Harvest and recovery of monoclonal antibodies from large-scale mammalian cell culture. *BioPharm Int.* **21:**34–46.

157. **Simmerman, H., and R. Donnelly.** 2005. Defining your product profile and maintaining control over it, part 1. *BioProcess Int.* **3:**32–38.

158. **Smith, K. A.** 2006. Considerations for aseptic filling of parenterals. *BioProcess Int.* **4:**12–17.

159. **Sofer, G.** 2004. Validation issues for disposable manufacturing. *BioProcess Int.* **2:**S18–S20.

160. **Sofer, G., D. C. Lister, and J. A. Boose.** 2003. Inactivation methods grouped by virus: virus inactivation in the 1990s—and into the 21st century. *BioPharm Int.* **16:**S37–S42.

161. **Some, I., P. Bogaerts, R. Hanus, M. Hanocq, and J. Dubois.** 2001. Stability parameter estimation at ambient temperature from studies at elevated temperatures. *J. Pharm. Sci.* **90:**1759–1766.

162. **Steffy, C. P.** 2003. What are the options? *BioProcess Int.* **1:**2S–11S.

163. **Strancar, A., A. Podgornik, M. Barut, and R. Necina.** 2002. Short monolithic columns as stationary phases for biochromatography. *Adv. Biochem. Eng. Biotechnol.* **76:**49–85.

164. **Tarrach, K., A. Meyer, J. E. Dathe, and H. Sun.** 2007. The effect of flux decay on a 20-nm nanofilter for virus retention. *BioPharm Int.* **20:**S15–S18.

165. **Tipton, B., J. A. Boose, W. Larsen, J. Beck, and T. O'Brien.** 2002. Retrovirus and parvovirus clearance from an affinity column product using adsorptive depth filtration. *BioPharm Int.* **15:**43–50.

166. **United States Pharmacopeia.** 2005. Analytical instrument qualification (draft). *Pharmacopeial Forum* **31:**1–5.

167. **Valax, P., E. Charbaut, J. E. Dathe, K. Tarrach, A. Lamproye, and H. Broly.** 2009. Robustness of parvovirus-retentive membranes and implications for virus clearance validation requirements. *BioProcess Int.* **7:**S56–S62.

168. **Van Reis, R., and A. Zydney.** 2001. Membrane separations in biotechnology. *Curr. Opin. Biotechnol.* **12:**208–211.

169. **Vogt, R., and T. Paust.** 2009. Disposable factory or tailor made integration of single-use systems? *BioProcess Int.* **7:**S72–S77.

170. **Wallace, K. K., and A. R. Moreira.** 1998. Changes in biologics regulation: impact on the development and validation of the manufacturing processes for well-characterized products, p. 170–179. *In* B. D. Kelley and R. A. Ramelmeier (ed.), *Validation of Biopharmaceutical Manufacturing Processes*. American Chemical Society, Washington, DC.

171. **Wang, A., R. Lewus, and A. Rathore.** 2006. Comparison of different options for harvest of a therapeutic protein product from high cell density yeast fermentation broth. *Biotechnol. Bioeng.* **94:**91–104.

172. **Wang, J., T. Diehl, M. Watkins-Fischl, D. Perkins, D. Aguiar, and A. Arunakumari.** 2008. Optimizing the primary recovery step in nonaffinity purification schemes of HuMAbs. *Suppl. BioPharm Int.* **21:**S4–S10.

173. **Wang, N., B. Hu, R. Ionescu, H. Mach, J. Sweeney, C. Hamm, M. J. Kirchmeier, and B. K. Meyer.** 2009. Opalescence of an IgG1 monoclonal antibody formulation is mediated by ionic strength and excipients. *BioPharm Int.* **22:**36–47.

174. **Webb, S. D., J. N. Webb, T. G. Hughes, D. F. Sesin, and A. C. Kincaid.** 2002. Freezing bulk-scale biopharmaceuticals

using common techniques—and the magnitude of freeze-concentration. *BioPharm Int.* **15**:22–34.

175. **Weinberg, S., and S. J. Advant.** 2009. Serving two masters: reconciling EMEA/GAMP 5 and FDA/cGMP Phase 1. *BioPharm Int.* **22**:24–25.

176. **Whitford, W. G.** 2005. Lipids in bioprocess fluids. *BioProcess Int.* **3**:46–56.

177. **Wilson, J. S.** 2006. A fully disposable monoclonal antibody manufacturing train. *BioProcess Int.* **4**:S34–S36.

178. **Winter, W.** 2006. Analytical instrument qualification: standardization on the 4Q model. *BioProcess Int.* **4**:46–50.

179. **Wolk, B., P. Bezy, R. Arnold, and G. Blank.** 2003. Characteristics of a good antibody manufacturing process. *BioProcess Int.* **1**:50–58.

180. **Yavorsky, D., R. Blanck, C. Lambalot, and R. Brunkow.** 2003. The clarification of bioreactor cell cultures for biopharmaceuticals. *Pharm. Technol.* **27**:62–74.

181. **Yigzaw, Y., M. Tranh, R. Riper, and A. Shukla.** 2006. Exploitation of the adsorptive properties of depth filters for host cell protein removal during monoclonal antibody purification. *Biotechnol. Prog.* **22**:288–296.

182. **Yuan, Q. S., A. Rosenfeld, T. W. Root, D. J. Klingenberg, and E. N. Lightfoot.** 1999. Flow distribution in chromatographic columns. *J. Chromatogr.* A **831**:149–165.

183. **Zhang, W., and M. J. Czupryn.** 2002. Free sulfhydryl in recombinant monoclonal antibodies. *Biotechnol. Prog.* **18**:509–513.

184. **Zhang, L., M. Moo-Young, and C. P. Chou.** 2008. Stability improvement of a therapeutic protein by reducing agent treatment. *Chinese J. Biotechnol.* **24**:2142–2143.

185. **Zhou, J. X., S. Dermawan, F. Solamo, G. Flynn, R. Stenson, T. Tressel, and S. Guhan.** 2007. pH-conductivity hybrid gradient cation-exchange chromatography for process-scale monoclonal antibody purification. *J. Chromatogr.* A **1175**:69–80.

186. **Zhou, J. X., and T. Tressel.** 2005. Membrane chromatography as a robust purification system for large-scale antibody production. *BioProcess Int.* **3**:S32–S37.

187. **Zhou, J. X., and T. Tressel.** 2006. Basic concepts in Q membrane chromatography for large-scale antibody production. *Biotechnol. Prog.* **22**:341–349.

188. **Zhou, J. X., T. Tressel, U. Gottschalk, F. Soalmo, A. Pastor, and S. Dermawan.** 2006. New Q membrane scale-down model for process-scale antibody purification. *J. Chromatogr.* A **1134**:66–73.

189. **Zhou, J. X., T. Tressel, and S. Guhan.** 2007. Disposable chromatography. *BioPharm Int.* **20**:S25–S33.

Plant Cell Culture

NANCY L. PAIVA

14

Plant cell culture has found wide applications ranging from studies on basic plant biochemistry and molecular biology to mass propagation and genetic engineering of crop species. The media, conditions, culture vessels, and other critical parameters vary widely depending on the intended use of the culture. This chapter covers the basic requirements for common procedures in plant cell culture. For work with a specific species, it is recommended that one survey the literature to see if procedures and media for the chosen species or a close relative have already been developed, since optimum conditions can be species or genus dependent. Several books and reviews containing detailed protocols have been published (2, 13, 18, 21, 28, 36, 38, 41, 46, 48, 50, 52, 53, 55), and several journals are dedicated to reporting the latest breakthroughs in plant tissue culture.

In general, plant tissue culture begins with the initiation of callus cultures, i.e., dedifferentiated masses of rapidly dividing cell cultures, on solid, agar-based nutrient media. The callus is generated by exposing sterile pieces of plant tissue to plant growth regulators. Normally, cell division in an intact plant is restricted to various meristematic regions (for example, root and shoot tip meristems) by the plant regulating its natural hormone distribution. The plant growth regulators are either natural plant hormones or synthetic analogs that bind to receptors in the plant's cells, causing all cells to resume division and growth.

Callus cultures can be maintained indefinitely by transferring clumps of cells to fresh agar media at regular intervals. Alternatively, callus pieces can be placed in shake flasks containing liquid media to eventually form suspension cultures. These grow faster than callus cultures, probably due to better nutrient exchange with the medium. Plant cell suspensions consist of some single cells and many clumps of a few to hundreds of cells; the size of the clumps depends upon a number of factors including the species, the medium, and the length of time since the culture was initiated. Suspension cultures can be used as a uniform, year-round source of plant cells for biochemical and plant physiological studies or as a source of valuable plant secondary metabolites following appropriate treatments.

The callus cultures can also be allowed to differentiate (regenerate) back into plants; this process is often accelerated by the omission of or changes in the types of plant growth regulators. Each piece of callus can give rise to one to hundreds of plantlets, each with the same genome as the parent, giving rise to the industry of micropropagation. Either early or late in the callus stage, foreign DNA can be introduced into the dividing plant cells. If the cells are then allowed to regenerate, a portion of the plantlets will be transgenic and can be selected for in a number of ways. At one time, the main way of introducing foreign DNA and organelles into plant cells involved the production of protoplasts (wall-less plant cells released by the action of cellulases and pectinases). Currently, several methods of DNA delivery are employed, including delivery by the "tamed" plant pathogen *Agrobacterium tumefaciens* or delivery by direct shooting of DNA-coated particles into the cells.

In the last 2 decades, transformation of important crop plant species such as corn, soybean, cotton, alfalfa, and wheat has become increasingly routine. Transgenic soybean, corn, and cotton varieties are now being planted on more than 50 to 90% of the farmland in the United States dedicated to those crops, and global use of transgenic crops is increasing rapidly. Transgenic plants and cell cultures are also used widely in reverse genetics experiments to determine the function of genes discovered in large-scale plant genome sequencing and mutational studies. Given good sterile technique and some patience for the comparatively slow growth rates, much can be accomplished today with plant cell culture.

14.1. CULTURE MEDIA

The composition of a plant cell culture medium resembles that of an all-purpose plant fertilizer, with the addition of a carbon source, since most cultures cannot photosynthesize enough sugars for growth, and plant growth regulators to stimulate cell growth (14, 18, 20, 33, 45, 52, 53, 55). Most plant cell lines can be cultured in completely defined media, although for some applications complex ingredients such as coconut milk or yeast extract were used in some early formulations and are still used today. Compositions of two of the most commonly used media are shown in Table 1. Murashige and Skoog medium (MS) (33) was one of the first developed for the growth of plant cells, using tobacco as the test species; it is still used today for the growth of callus and suspension cultures, as well as the growth of axenic cultures of plants. Gamborg's B5 medium (B5) (20) was an adaptation of MS, optimized for the growth of soybean

TABLE 1 Composition of common plant cell culture media

Component	Concn in:			
	MS		B5	
	mM	mg/liter	mM	mg/liter
Macronutrients				
NH_4NO_3	20.6	1,650		
KNO_3	18.8	1,900	25	2,500
$CaCl_2 \cdot 2H_2O$	3.0	440	1.0	150
$MgSO_4 \cdot 7H_2O$	1.5	370	1.0	250
KH_2PO_4	1.25	170		
$(NH_4)_2SO_4$			1.0	134
$NaH_2PO_4 \cdot H_2O$			1.1	150
Micronutrients				
KI	0.005	0.83	0.0045	0.75
H_3BO_3	0.100	6.3	0.050	3.0
$MnSO_4 \cdot 4H_2O$	0.100	22.3		
$MnSO_4 \cdot H_2O$			0.060	10.0
$ZnSO_4 \cdot 7H_2O$	0.030	8.6	0.007	2.0
$Na_2MoO_4 \cdot 2H_2O$	0.001	0.25	0.001	0.25
$CuSO_4 \cdot 5H_2O$	0.0001	0.025	0.0001	0.025
$CoSO_4 \cdot 6H_2O$	0.0001	0.025	0.0001	0.025
Disodium EDTA	0.100	37.3	0.100	37.3
$FeSO_4 \cdot 7H_2O$	0.100	27.8	0.100	27.8
Vitamins/amino acids				
Inositol	0.49	100	0.49	100
Nicotinic acid	0.0047	0.5	0.0094	1.0
Pyridoxine·HCl	0.0024	0.5	0.0048	1.0
Thiamine·HCl	0.0003	0.1	0.0300	10.0
Glycine	0.0005	2.0		
Carbon source				
Sucrose		30 g/liter		20 g/liter
pH		5.8		5.5

cells, and is used for many species. There is some confusion in the terminology used in the literature, but generally a medium is defined by its formulation of inorganic salts and organic compounds (the macro- and micronutrients), often excluding the major carbon sources (usually sucrose). Some authors include either the carbon source or growth regulators or both in their definitions, so it is important to check formulations in the literature carefully.

14.1.1. Macronutrients

The macronutrients provide the sources of N, P, K, S, and Ca, all required for healthy plant growth. The Na and Cl included as counterions are not essential; MS medium contains no Na source other than a trace amount included with the micronutrients, and studies have indicated that plants can synthesize organic anions to take the place of chloride ions as needed. Na, K, and Cl ions can be tolerated at levels up to 50 to 60 mM and can be used as counterions for additives or for pH adjustment without concern. Nitrogen is mainly provided as nitrate, although some media contain both ammonium and nitrate. Utilization of ammonium causes the culture pH to drop, while utilization of nitrate causes the pH to rise. Some species have a requirement for ammonium or greatly benefit from the inclusion of ammonium or glutamine, but high concentrations of ammonium can be toxic either directly or by causing extremely low pH shifts. Phosphate salts provide both a source of P and the majority of the buffer capacity of the media.

14.1.2. Micronutrients

The micronutrients are sources of various trace elements required for growth. These are similar to those found in microbial defined media, with the addition of iodide. Studies have indicated that iodide is not essential, but greatly stimulates cell growth. Iron is always added in a chelated form, either as an EDTA salt or as agricultural chelates (e.g., Sequestrene 330 Fe [Novartis Crop Protection, Inc., Greensboro, NC]).

14.1.3. Carbon Sources

The most common carbon source is sucrose, generally in the range of 20 to 30 g/liter; the original formulation for MS called for 30 g/liter, while B5 called for 20 g/liter. Glucose is used less often, and fructose only occasionally. In many cultures, much of the sucrose is rapidly taken up by the cells in the first 1 or 2 days of culture and converted to starch granules inside the cells. If different carbon sources are to be tested for growth, one must take into account that the cells previously cultured on sucrose may have substantial starch reserves, allowing growth for several days. Cultures of other species often hydrolyze the sucrose to fructose and glucose and utilize the glucose first and then the fructose. Very high sugar concentrations (40 to 100 g/liter) have been used in specialized secondary-metabolite production media (21, 23, 24, 28, 36) and to adjust the osmotic potential of the media in short-term treatments for regeneration (2, 18).

14.1.4. Vitamins

Although intact, soil-grown plants normally synthesize all of the vitamins required for growth, vitamins are generally added to all plant culture media. Very early work indicated an absolute requirement for thiamine (vitamin B_1) in the growth of tomato cells (56), but this has not been determined for all species. Pyridoxine, nicotinic acid, and *myo*-inositol have each been shown to increase the growth of certain species and should also be included, especially when establishing new callus cultures. In many cases, however, these may not be required for growth, and some species have been reported to be unable to take up exogenous *myo*-inositol. All of the common vitamins listed here (thiamine, pyridoxine, nicotinic acid, and *myo*-inositol) are sufficiently stable during autoclaving at common medium pH values. Other vitamins have been added (1 mg/liter or less) in some instances, especially when cells or protoplasts are cultured at very low densities (21, 55); these include biotin, pantothenic acid, folic acid, choline chloride, *p*-aminobenzoic acid, riboflavin, vitamin B_{12}, and ascorbic acid. Ascorbic acid and cysteine·HCl are often added at very high concentrations (25 to 100 mg/liter) as antioxidants to prevent browning of cultures (21, 52).

14.1.5. pH

Most cell culture media are adjusted to pH 5.5 to 5.8 before autoclaving. Following inoculation, the cells may rapidly shift the pH of the media by the release of stored ions or by the uptake and utilization of nutrients such as ammonium. After a week or more of growth, suspension cell culture media may exceed pH 6.0. Rapid alkalinization of the medium is also common after elicitation (see below), with values reaching as high as pH 7.0.

14.1.6. Plant Growth Regulators

Plant growth regulators include both the naturally occurring plant hormones (phytohormones) and synthetic compounds, most of which are structural analogs of the natural hormones and bind to the same receptors. Commonly used regulators are listed in Table 2, along with their abbreviations. One class of regulators are called auxins, which cause plant cells to grow, the naturally occurring plant auxin being indoleacetic acid (IAA). IAA is unstable in solution and is easily oxidized and conjugated to inactive forms by plant cells. However, IAA and the slightly more stable indolebutyric acid (IBA) are still used when low doses or short pulses of hormone are needed, as in certain regeneration systems or in rooting micropropagated plants or cuttings. Naphthaleneacetic acid (NAA) and 2,4-dichlorophenoxyacetic acid (2,4-D) are the most frequently used auxins. 2,4-D is more potent than NAA, followed by IBA and then IAA. IAA and IBA must be added after autoclaving, and the media must be used immediately thereafter; stock solutions should be frozen or made weekly and refrigerated. 2,4-D and NAA can be added before autoclaving and will not break down in the media during storage; stock solutions can be stored indefinitely at 4°C. Many callus cultures can grow with auxins alone; the original formulation for B5 contained 1 ppm of 2,4-D (i.e., 1 mg/liter, or 4.5 μM).

A second class of regulators, the cytokinins, promote cell division and are sometimes required for culture initiation

TABLE 2 Growth regulators and additives commonly used in plant tissue cultures

Class of additives	Compound name	Abbreviation(s)
Auxins	Indole-3-acetic acid	IAA
	Indole-3-butyric acid	IBA
	1-Naphthaleneacetic acid	NAA
	(2,4-Dichlorophenoxy)acetic acid	2,4-D
	(2,4,5-Dichlorophenoxy)acetic acid	2,4,5-T
	(*para*- or 4-Chlorophenoxy)acetic acid	PCPA or CPA
	Picloram (4-amino-3,5,6-trichloropicolinic acid)	PIC
	2-Naphthoxyacetic acid	NOA
Cytokinins	6-Benzylaminopurine	BA or BAP
	Zeatin	ZEA
	Kinetin (6-furfurylaminopurine)	KIN
	(2-Isopentenyl)adenine	2iP
	Adenine	ADE
Other hormones	Gibberellic acid (gibberellin A3)	GA or GA3
	Abscisic acid	ABA
Other additives	Casein hydrolysate (NZ Amine)	CH or C
	Coconut water	CW
	Ethylenediaminetetraacetic acid	EDTA

or long-term growth. Cytokinins are often added to induce shoot initiation during plant regeneration from callus (18), at which time the auxins may be decreased or omitted from the medium. The first naturally occurring cytokinin to be identified was zeatin, a purine derivative (22). Zeatin is occasionally used in culture media, but the most common cytokinins in use today are kinetin (6-furfurylaminopurine) and 6-benzylaminopurine (also known as N6-benzyladenine); adenine and adenosine can also have cytokinin activity, particularly at high levels. Most cytokinins, except for zeatin and its derivatives, can be sterilized by autoclaving.

Gibberellins (usually GA3, gibberellic acid) and abscisic acid are sometimes used to stimulate embryo or shoot development (2, 18). Other plant growth regulators such as the recently discovered brassinosteroids are still being evaluated in culture systems.

Auxins and cytokinins are used in the range of 1 to 50 μM. Most plant growth regulators can be dissolved in a small volume of reagent-grade ethanol and then diluted in warm water to the final stock concentration; the ethanol will evaporate during autoclaving. Alternatively, NaOH is used to dissolve most auxins and HCl can be used to dissolve cytokinins, but these should be kept to a minimum to avoid altering the pH of the media.

14.1.7. Other Additives

Complex ingredients such as coconut milk, yeast extract, fruit juices, and casein hydrolysates (such as NZ Amine) were used in many early culture experiments and are still used today. They provide plant growth regulators, vitamins, and amino acids that can enhance plant cell growth rates and survival of some cultures. However, they are expensive, sometimes variable in quality, and sometimes detrimental, such as when yeast extract induces stress responses in dilute cell suspensions.

Mannitol, sorbitol, or combinations of these two sugar alcohols are routinely added as an osmoticum (i.e., a supplemental agent increasing osmolarity). A concentration of 100 g/liter (0.6 M) sugar alcohol is used in most media for the isolation of plant protoplasts, to prevent the cells from bursting once the rigid cellulose cell walls are removed (55). The addition of 0.2 to 0.5 M total osmoticum (300 to 900 mOsm/kg of H_2O, added as an equimolar mixture of sorbitol and mannitol) to the culture medium increases the number of transformants produced by bombardment techniques (43).

Organic acids such as citric, fumaric, malic, or succinic acids, or synthetic buffers such as Tris, morpholineethanesulfonic acid (MES), or HEPES, have been used to help maintain culture pH, especially when ammonium salts are used as the nitrogen source (21). Neutralized activated charcoal is occasionally added to young regenerating cultures to remove toxic phenolics released by the stressed plant cells, or to help remove plant growth regulators introduced at an earlier stage. Silver thiosulfate serves as an ethylene biosynthesis inhibitor, preventing this gaseous plant hormone from accumulating to detrimental concentrations.

"Nurse" medium or "conditioned" medium is the liquid medium removed from a suspension of fast-growing cells. It contains uncharacterized growth factors released by growing cells, or it may be depleted of growth-inhibiting substances, such as high levels of ammonium ions. Conditioned medium is usually removed aseptically from a culture and never autoclaved. A nurse culture or "feeder" layer is usually a thin layer of fine, fast-growing cells spread on the agar surface and covered with a moist sterile filter paper.

Conditioned media and nurse cultures are sometimes used in the culture of regenerating protoplasts, single cells, or very dilute cell suspensions (29, 49). The nurse media or cultures are often not from the same species as the species of interest; e.g., a tobacco cell culture may be used as the nurse culture for carrot cells.

14.2. MEDIUM PREPARATION

Only high-purity (reagent-grade) chemicals should be used for plant cell culture, especially when new cultures are being initiated or sensitive cell lines are being used. Water should be double-distilled (glass stills preferred) or deionized, such as 18 MΩ water produced by MilliQ devices (Millipore, Bedford, MA). Glassware must be free of residual detergents. If glassware is cleaned with chromic acid (or any solution containing heavy metal ions), it must be treated with EDTA or another chelator to remove all traces of chromium, since some cultures are very sensitive. In fermentors or bioreactors, all parts in contact with the medium must be made of high-quality stainless steel, glass, or plastic for work with sensitive cell lines. In studies with opium poppy cells (*Papaver somniferum*), the use of bioreactors containing even a small amount of silver solder or brass screws results in the browning and death of the cultures, while periwinkle (*Catharanthus roseus*) and tobacco (*Nicotiana tobacum*) cells grow well (personal observation).

The options for gelling agents for plate cultures include bacteriological-grade agar (such as Bacto Agar [Difco, Detroit, MI]), Noble agar (agar reduced in ionic impurities [Difco]), tissue culture-grade Phytagar (specially prepared agar [Gibco-BRL, Gaithersburg, MD]), agarose (highly purified agar), and Phytagel (Gellam gum agar substitute [Sigma, St. Louis, MO] or Gelrite [Merck/Kelco Division, San Diego, CA]). For routine work, our laboratory prefers Phytagar at 0.6 to 0.8% or agarose at 0.55%, although many cultures can tolerate the impurities in bacteriological agar. Carrageenan (vegetable gelatin) and calcium alginate have been used for cell immobilization, particularly in bioreactor studies.

A number of strategies have been used to handle medium preparation. One traditional method is to prepare four liquid stock solutions, containing (i) the five or six macroelements (20× stock), (ii) the microelements omitting the iron source (200×), (iii) the iron chelate (200×), and (iv) the vitamins (200 to 1,000×). The first three are stored at 4°C, while the vitamins are stored frozen. The carbon source is added as a solid. If hormones are to be included (Table 2), heat-stable hormones are predissolved and added as a 200 to 1,000× stock; otherwise, they are filter sterilized and added after autoclaving. Agar, if required, is added after pH adjustment. The use of multiple concentrated stocks of N and P sources make medium variations easier.

Once an acceptable medium is identified and large quantities are needed, a convenient practice is to prepare a 10× concentrated stock and freeze it in appropriately sized aliquots. An example of this strategy for the preparation of an SH stock (supplemented with three growth regulators and 30 g/liter of sucrose for use with alfalfa cell cultures [2, 14, 45]) is provided in Table 3. A 4-liter volume of 10× concentrate, including three growth regulators but excluding sucrose, is dissolved, distributed into plastic Whirlpak bags (Nasco, Grand Prairie, TX) in 100-ml aliquots, and frozen at −20°C or lower. To prepare the medium, one bag of concentrate is thawed for every liter of liquid or agar needed, and the contents of the bag are transferred with rinsing to a large flask or beaker and dissolved by stirring

TABLE 3 Preparation of SH frozen stock[a]

Medium component	Concn (mM) in final medium	Amt (g/40 liters) per batch of 10× stock
KNO_3	25.0	101.0
$MgSO_4 \cdot 7H_2O$	1.5	14.8
$NH_4H_2PO_4$	2.5	11.6
$CaCl_2 \cdot 2H_2O$	1.5	8.8
myo-Inositol	5.0	40.0
$MnSO_4$	0.059	0.358
H_3BO_3	0.13	0.200
$ZnSO_4 \cdot 7H_2O$	0.0035	0.040
KI	0.0060	0.040
$CuSO_4 \cdot 5H_2O$	0.0008	0.008
$FeSO_4 \cdot 7H_2O$	0.0054	0.600
Na_2EDTA	0.0054	0.800
Thiamine·HCl	0.0150	0.200
Nicotinic acid	0.0410	0.200
Pyridoxine·HCl	0.0024	0.020

[a]For 10× concentrate without sugar or pH adjustment. The method is as follows.
1. Dissolve all of the above solids in 2.5 liters of distilled H_2O.
2. Add 8 ml of Mo/Co microstock: 50 mg of Na_2MoO_4 and 50 mg of $CoCl_2 \cdot 6H_2O$ dissolved together in 100 ml. (Store at 4°C.)
3. Add 20 ml of kinetin stock: 43 mg/200 ml.
4. Add 80 ml of 2,4-D stock: 225 mg/liter.
5. Add 400 ml of PCPA stock: 186.6 mg/liter.
6. Adjust volume to 4 liters. Mix well. Divide into 100-ml lots in Whirlpak bags. Store at −20 to −40°C.
Note: each 100-ml bag is equivalent to 1 liter of SH medium. To use, thaw 1 bag/liter needed, add 30 g of sucrose per liter, adjust volume, and titrate to pH 5.7 with 1 N NaOH. For plates, add 8 g of agar per liter.

for 30 to 60 min at room temperature. Sucrose (30 g/liter) is added, the pH is adjusted to 5.7, and the medium is distributed into flasks (with or without agar) for autoclaving. We find it advantageous to add the sugar at the time the concentrate is thawed for use, because otherwise the stock would contain 30% sucrose, which is difficult to dissolve, distribute, and freeze. The pH of the stock is slightly acidic and should not be adjusted prior to freezing. A slight precipitate may form in the bags during freezing, but this will redissolve with extended stirring.

A few companies produce a variety of premixed plant cell culture media. Usually, the medium is divided into "basal salts" (the macro- and micronutrients), sold as a dried powder, and vitamins, sold separately as a solution. These components may also be sold as a combined dried powder ("basal salts with minimal organics"), with or without sugar included. The medium is usually dispensed into preweighed packets such that each makes 1 to 10 liters of medium; one can also purchase a jar of well-blended powder that makes 50 liters and weigh the powder as needed. Plant growth regulators and agar are rarely included. Due to inconsistencies in formulations and terminology in the literature, manufacturers usually provide very detailed tables of their formulations; one manufacturer supplied 15 variations of MS salts. If one is careful to select the correct preparation and labor is limited, these commercial powders offer a slightly expensive but reliable option.

For many liquid media including MS, a small amount of white precipitate may be visible in shake flasks, especially after extended autoclaving or storage. For routine culture maintenance, this medium will work well, since the

precipitate will gradually be used by the cells. Media should be autoclaved at 121°C for a minimum of 15 min for small volumes and for longer times for larger volumes or thick-walled vessels. The required times may need to be evaluated on-site; 30 min for 500 ml of agar or 1 h for a 2-liter bioreactor vessel is not unusual (6).

14.3. FACILITIES

Successful plant cell culture requires extremely close attention to proper sterile technique and temperature control. The growth of plant cell cultures is relatively slow compared to common microbial cultures, necessitating very long culture times for experiments; a low level of contamination which would be outgrown and perhaps unnoticed in a bacterial culture would have time to overwhelm plant cultures. It may take months to establish a high-producing cell line or to regenerate transgenic plants from the original explants.

14.3.1. Sterile Transfer Facilities

Laminar-flow hoods are strongly recommended. An even flow of HEPA (high-efficiency particulate air) filter-sterilized air comes from the back of the hood, protecting the cultures when the vessels are opened. Unlike primary animal cell cultures, no biohazard containment of the cultures is required. Some laboratories use containment hoods when introducing microbes into plant cell cultures, such as during genetic transformation or inoculation with plant pathogens. Each hood should be equipped with a natural gas outlet for a Bunsen burner and lighting (usually overhead). A UV germicidal lamp is strongly recommended for use in decontaminating the hood work surface after the HEPA filter is replaced, if the hood is switched off for more than 30 min, or if the hood is accidentally contaminated. A selection of stainless steel forceps, spatulas, and scalpels with replaceable blades are required for culture initiation, callus subculture, and transformation and regeneration. For quantitative transfers, an electronic (digital) top-loading balance (400-g to 4-kg capacity) with auto-tare feature is essential.

14.3.2. Temperature

Most cell cultures grow well only in a narrow temperature range. The most common growth temperature is 25°C. Some plant species, especially heat-tolerant plants such as tobacco or *C. roseus*, can continue to grow at temperatures of 30°C and above, while heat-sensitive cultures like opium poppy will cease to grow above 27°C and will die at higher temperatures.

Cyclic temperature changes may occur during the day-night cycles used for some culture procedures. Localized heating of shelves occurs during the day, and they then cool during the night. Although not directly lethal to cultures, this contributes to the formation of condensation inside the lids of petri dishes and other culture vessels, which can dry out the cultures or contribute to contamination. Very careful room design and high airflows can reduce these temperature fluctuations, but these features may add to the cost of culture rooms. Stacking an uninoculated petri dish containing water agar on top of the important cultures will greatly reduce condensation yet still allow sufficient light for regeneration.

14.3.3. Light

Callus and shake flask cultures generally do not require light for growth. Usually they are grown in dark rooms, equipped with lights which are turned on when inspecting or subculturing the cultures. Absolute darkness is not usually required. High light intensities or low levels of continuous illumination can cause cultures to turn green (form small numbers of chloroplasts) or brown (accumulate polymerized phenolics). Enough light to induce either color change can be sufficient to inhibit secondary-metabolite accumulation, and browning may greatly slow the growth of cultures. If callus and shake flasks must be incubated in lighted rooms (such as rooms shared with regenerating cultures or aseptically grown plants), they can be shielded with foil or boxes.

For regeneration, cool white fluorescent lights are recommended. Interspersing incandescent bulbs will greatly increase the quality of the light, providing a spectrum close to that of sunlight. Lights are generally suspended 12 to 14 in. above the culture vessels. Sufficient distance must be maintained between the lights and vessels to allow cooler air to circulate and avoid overheating the plates. Wire shelving, positioned a few inches from the culture room walls, is often preferred, providing good air circulation at a reasonable cost. The shelves can be partially covered with aluminum foil or Plexiglas to provide a more stable, easily cleaned surface. If several layers of closely lit shelves are used, heat from a lower shelf may warm the shelf above; plates can be partially insulated by placing them on foil-covered sheets of styrofoam, or lights should be suspended by chains to allow heat dissipation.

Too much light can actually be detrimental to culture growth and regeneration. One indication of a high light level is the accumulation of anthocyanins (red to purple pigments) in callus or differentiated tissues. Continuous light is sometimes detrimental to whole-plant growth and is rarely used in plant cell cultures. The use of household timers is adequate for generating day-night cycles (16 to 18 h of light, 6 to 8 h of dark). One exception is the culture of photoautotrophic cells, which require high-intensity and high-quality light to enable the cells to photosynthesize sufficient sugars from CO_2 (36).

14.3.4. Aeration

Under most conditions, shake flask cultures of plant cell suspension have a lower volumetric owxygen demand than do microbial cultures. The respiration rates (Q_{O_2}) reported for plant cell cultures range from 0.2 to 3.6 mmol of O_2/h/g (dry weight), compared with 2 to 16 mmol/h/g for *Escherichia coli*, 39 to 570 mmol/h/g for *Bacillus* spp., and 1.9 to 2.8 mmol/h/g for *Saccharomyces* spp. (48). Shaker speeds (100 to 150 rpm with a 1-in. throw) are generally lower than those used for microbial cultures, and the fill ratios in the flasks (volume of medium/volume of flask generally 0.2 to 0.4) are generally higher. Baffled flasks are rarely used to increase aeration but are occasionally used for breaking up clumps of cells, especially when a suspension is first being initiated.

Generally, shake flask cultures can be removed from a shaker for subculturing and observation for 1 h or more without killing the cells. However, this procedure will affect the growth rate and can affect the ability of the culture to produce secondary metabolites. High oxygen demands have been reported immediately following elicitation (see below), since a short-term burst of O_2 utilization occurs as a result of activation of many defense processes. If elicitation is used to induce secondary-metabolite accumulation, one should prevent even brief interruptions in aeration of shake flask cultures, and it may be necessary to increase the aeration of production bioreactors.

A wide variety of closures have been used for shake flask cultures. Traditional cotton plugs in Erlenmeyer flasks can work well, particularly if the cotton plugs "mushroom" over the rim of the flask, keeping it free of microbial contamination. Since the cultures may be on the shakers for weeks, particulates in the culture room air will tend to collect around the rim if unprotected, making it impossible to open the flask for subculture without contamination. A preferred option is straight-walled Delong-style flasks with metal culture tube closures; a small piece of folded cheesecloth or gauze is placed in the top of the closure to act as a sterile filter, and the prongs on the closures hold it securely on the flask. These caps have the advantage of providing the same sterility barrier and good aeration as cotton plugs while covering the rim and upper edge of the flasks. Any dust from the shaker rooms or media storage areas can be easily removed by a brief flaming before the flask is opened.

Petri dishes containing agar media and callus cultures or regenerating plants are incubated agar side down, not inverted like microbial cultures. Plates are always sealed with Parafilm or porous medical tape (such as Urgopore tape [Karlan Research Products Co., Santa Rosa, CA]) or can be placed inside a sterilized box or new plastic self-sealing freezer bag for short-term incubation. Otherwise, contaminants will enter the plate during the long cultures or the agar media will dehydrate.

Closures on culture vessels must not only allow oxygen and CO_2 to pass but, in sensitive species, must also allow ethylene and other volatiles to escape. Ethylene, a gas, is a plant hormone often released by plant cells in response to wounding and infection or during senescence. It can have a number of effects on axenically grown plants and cell cultures (18). In very sensitive species, such as *Arabidopsis* and potato, the addition of ethylene biosynthesis inhibitors such as silver ions (as silver thiosulfate) greatly increases the yield of regenerated transgenic plants. Without these, the wounded tissue produces enough ethylene to inhibit the development or lead to browning and death of nonwounded cells.

Dedicated growth rooms provide the most efficient and flexible use of space. Commercial growth chambers equipped with lights, temperature control, and (optional) humidity control are available; several growth chambers allow many different conditions to be maintained, which may be important if diverse species are to be cultivated. For important cultures, it is recommended that several separate cultures be maintained, preferably in separate culture rooms or incubators. Temperature recorders are strongly recommended to diagnose possible problems in temperature control. Motion sensors and temperature monitors with audible alarms can also be employed to minimize the damage caused by heating and cooling and by shaker equipment breakdowns. Even brief overheating to 35°C or more can greatly reduce regeneration and secondary-metabolite production if it occurs at critical times, while regenerated plantlets are relatively resistant. Long-term exposures (hours) to high temperatures can stop the growth or rapidly kill suspension cultures.

14.4. CULTURE INITIATION

Rapidly growing plant tissues are generally best for rapidly producing callus cultures, although any living plant cell can in theory be induced to divide. While authors often indicate the plant organ from which a callus was derived (i.e., root, leaf, or hypocotyl), after several generations the cultures are fully dedifferentiated and there is little evidence that the original tissue source influences the behavior of the culture, such as the products it will accumulate. In contrast, for short-term cultures such as for transformation and quick regeneration, the explant source can be critical; for wheat and some recalcitrant legumes, immature seeds or cotyledonary nodes will produce easily regenerable callus while leaf or root callus cultures will not.

14.4.1. Sterile Explants

There are two basic sources for obtaining surface-sterilized plant tissues. One is to surface-sterilize seeds, which are dormant with tough, dry seed coats and can therefore tolerate very harsh sterilization conditions. The seeds are then germinated under sterile conditions, and the various tissues are placed on callus-inducing media. The other is to surface-sterilize undamaged tissue from a growing plant, which may be the only option for vegetatively propagated species or tree species that are slow to set seed. The same sterilization strategies can be used in each case. Most commonly, a brief ethanol wash serves to remove and kill spores and bacteria on the plant surface and to thoroughly wet the tissue; it is followed by a hypochlorite soak to kill any remaining microbes.

14.4.1.1. Seeds

The following protocol works well for soybeans and other medium-sized to large seeds. The seeds are placed in a clean plastic petri dish (25 × 100 mm preferred) or beaker and covered with 70% ethanol for 5 min. The dish is stirred or gently swirled occasionally to ensure that the ethanol contacts all the seed surfaces. The seeds are transferred using a sterile (flamed) spoon-shaped spatula to a sterile dish with enough bleach solution (20% Chlorox in distilled water with 1 drop of Tween 20 or 2 drops of 10% sodium dodecyl sulfate solution per 100 ml; equivalent to 1% sodium hypochlorite [final concentration]) to cover the seeds. They are soaked for 15 to 20 min with periodic stirring. Any cracked, floating, or noticeably damaged seeds are removed. The seeds may begin to swell during this time. The seeds are transferred with a sterile spatula to a sterile dish containing sterile distilled water and are rinsed two or three more times by transferring to new dishes of sterile water. They are then transferred to sugar-water agar plates (5 g of sucrose and 8 g of Phytagar/liter; 50 ml/25-mm-deep plate) and sealed with Parafilm. If possible, the seeds are spread out so they do not touch each other, using several plates, so that any unsterilized seeds will not contaminate clean seeds. The seeds are incubated at approximately 25°C.

Germinating the seeds on media containing sugar or other nutrients helps locate contaminants by stimulating their growth on the agar surface. Seeds may also be germinated on damp filter paper or cotton. Tiny seeds, such as those of tobacco, lettuce, and *Arabidopsis* spp., may be more easily sterilized by being treated in sterile 1.5-ml plastic snap-cap tubes, briefly centrifuging them to the bottom of the tubes to remove solutions. The seeds can be transferred to new tubes by being suspended and pipetted. To make tiny seeds easier to distribute, after sterilization they can be suspended in a viscous 0.1% agarose solution and dripped onto the agar surface. If a special treatment is recommended prior to planting in soil, the same treatment should be applied before sterilization. For example, some species require extended cold treatments, exposure to light, or scarification (scratching the seed surface with sandpaper or nicking the seed coat with a razor blade).

14.4.1.2. Leaves

Young leaves, free of insect and microbial damage, should be selected. They are rinsed with distilled water to remove dirt and kept in a moist container to prevent wilting. For leaves that are thick or have a thick waxy cuticle, the above seed sterilization protocol should work well. For very thin leaves or for delicate leaves such as those grown in a very humid environment, less harsh conditions may be required. For example, for tender alfalfa leaves, the ethanol concentration may be reduced to 50% and the exposure time to 1 to 2 min to avoid killing the leaf cells.

14.4.1.3. Other Tissues

Depending upon the species and environment, other plant parts may be available. For species that make tubers or rhizomes, these organs should be first scrubbed free of soil and may be surface-sterilized, peeled, then sterilized briefly again. Young stems and petioles are often more resistant to damage during sterilization than are leaves.

Other sterilizing agents have been used, such as $HgCl_2$ or antibiotics, especially for seeds that have a natural bacterial population under their seed coats. A procedure using chlorine gas is sometimes recommended for dry seeds such as soybeans. The seeds are placed in a dish in a larger, sealable glass container such as a desiccator, together with a beaker containing bleach. The bleach is acidified (3 to 5 ml of concentrated HCl per 100 ml of Chlorox), which releases chlorine gas, and the container is quickly sealed. After 6 h to overnight, the container is opened and the seeds are rinsed with sterile water and transferred to plates for germination. Procedures have also recommended acidifying the bleach solution in the leaf and seed procedures described above to cause less damage to the tissues and to "activate" the chlorine. Any procedures in which bleach and acid are mixed should be carried out in a chemical fume hood, not a sterile transfer hood, due to human toxicity of the chlorine gas.

14.4.2. Callus Culture Initiation

Once surface-sterilized seedlings or explants are available, these are cut into small pieces (0.5- to 1-cm segments or squares) with a sterile scalpel and placed on callus-inducing media. Hypocotyl and root segments are most commonly used. The edges of leaves and any cut surface that was exposed to bleach should be trimmed away and discarded, since these will not yield callus. The tissues will increase in size and should show visible callus formation within 1 month. Some tissues, such as older leaf and stem segments, may show callus development only at the cut edges, while younger tissues, such as roots, hypocotyls, and very young leaves, may completely dedifferentiate. If tissues curl and lose contact with the medium, they should be cut and pushed gently into the agar surface.

Much time can be saved by consulting the literature and finding medium formulations that have been previously developed for the species of interest (2, 18, 21, 52). If this information is not available, one should try media that have been successful with closely related species or try a number of different basal salt and hormone combinations. MS and B5 salts with 1 mg of 2,4-D per liter and SH medium with the two auxins and one cytokinin described above (Table 3) provide a wide range of media for dicots. While callus may form on several media, the texture and growth rate may vary greatly. For example, peanut callus on B5 salts with 1 mg of 2,4-D per liter is very soft and friable, while the callus after several subcultures on SH medium (Table 3) is extremely dense and hard. Monocots often require very high concentrations of auxins (up to 20 mg of 2,4-D per liter) for callus initiation but require reduced levels for extended rapid callus growth (2). If a particular cultivar or variety of a species is not required, several should be tried, since callus formation may vary with the genotype. Several plates of each type should be initiated, due to variations in callus formation response between individual explants and to allow for possible losses to contamination.

The callus should be subcultured every 2 weeks to 1 month by simply moving clumps of rapidly dividing cells to fresh plates with a cooled, sterilized spatula. During the first transfers, it may be beneficial to cut any remaining intact explants to expose fresh tissue to the medium, or to break up large callus pieces. If sufficient material is available, the softest, most friable calli should be selected for subculturing.

14.4.3. Suspension Culture Initiation

Suspension cultures are initiated by inoculating liquid media in shake flasks with callus clumps. Approximately 1 to 2 in^3 of callus (5 to 15 g [fresh weight]) should be suspended in 50 ml of medium. Generally, the same medium used for callus culture is used for suspension cultures. Large clumps can be gently broken or cut into smaller pieces. Softer, friable, rapidly growing callus works best, since it will more quickly disperse into a fine suspension, whereas hard callus clumps may simply enlarge without dispersing. If only a small amount of callus is available, less medium should be used, since high liquid-to-cell ratios may inhibit growth.

The growth rate of suspension cultures will gradually increase as faster-growing cells take over the culture. Initially, growth may be slow, and it may take weeks for the culture to double or triple in biomass. At this time, the culture should be divided into two new medium flasks. This is often accomplished by simply swirling the culture and pouring cells and old medium into the new flasks. If the major cell aggregates are white but the medium is brown and contains much cell debris, it may be beneficial to pour off the old medium and transfer the cells with a spatula. As the growth rate increases, the cultures will have to be divided more frequently. Also, the aggregate size should decrease, and the ratio of old culture to new medium can be reduced.

A portion of the original callus should be reserved and subcultured. If the initial suspension does not grow well, after additional months of subculturing a callus that was too hard to form good suspensions may improve in texture and can be tried again. After a good suspension culture is established, a portion of the cells may be returned to agar to form a callus culture again; this callus will usually grow faster than the original. If the shaker stops and/or the shaker room overheats, suspension cultures can be killed or may lose desirable properties such as the ability to produce secondary metabolites; healthy callus cultures serve as a backup in case such problems arise.

14.4.4. Bioreactors and Scale-Up

The same principles regarding scale-up that apply to microbial or mammalian cell cultures also apply to the scale-up of plant cell cultures (9, 13, 27, 28, 34, 36, 48). Because aeration is usually not limiting for growth in shake flasks, scale-up to larger shake flasks (up to 4 liters) is often easily accomplished for biomass production, keeping the ratio of medium volume to flask volume constant.

Plant cell suspension cultures have been grown successfully in a wide range of bioreactor designs, but there are two important design considerations. First, plant cells tend to be very shear sensitive, and impeller damage can greatly impair the growth of a culture. Strategies to decrease impeller damage include decreasing the angle of the impeller, changing to a marine-style impeller, or using airlift fermentors. By manipulating the hormones or medium components, the culture may be induced to form larger aggregates, which can be less shear sensitive. Second, the cultures can be very heterogeneous. Large cell aggregates tend to settle to the bottom of the reactor vessels, while small aggregates may be carried out of the medium to stick to the reactor walls or float on foam, forming a "meringue" layer above the useful area of the reactor.

To reduce the problems of shear and cell floating or settling, many immobilization matrices have been tested. Cells have been embedded in beads of calcium alginate and carrageenan or entrapped in porous matrices such as foam rubber or polyester fabric (13, 29, 48). For cells that release products into the medium or carry out biotransformation reactions, immobilization allows the cells to be recycled. Unfortunately, new problems arise, such as decreased mass transfer to and from the cells and decreased volumetric productivity of the bioreactor, plus the added cost associated with immobilization.

Plant cell suspension cultures have been grown in industrial-scale fermentors. An extremely fast-growing cell line of tobacco (TBY-2) was grown in a 369-liter, then a 1,500-liter, and finally a specially designed 20,000-liter culture tank (27, 34). For the commercial production of shikonin, cells are first cultured in a 250-liter fermentor in a growth medium and then transferred to a production medium in a 750-liter fermentor (9).

14.5. GROWTH QUANTITATION

Many methods have been used to estimate the growth of plant cell cultures both directly and indirectly. The method chosen will depend on the requirements for maintaining the sterility of the culture and the speed and accuracy of the measurement. One must keep in mind that the doubling time for most plant cell cultures exceeds 12 h and that a typical culture experiment will continue for 1 to 4 weeks.

14.5.1. Fresh Weight

The growth of callus cultures can be measured by simply moving the callus mass to a preweighed sterile container and measuring the weight of the callus at regular intervals. The new agar medium plates can be weighed before and after inoculation at each subculture.

Cells can be collected from suspension cultures by filtration using Miracloth (CalBiochem, San Diego, CA), nylon mesh filters (40- to 200-μm mesh), or sintered glass filters. Filter paper may also be used, but it may clog quickly due to the abundant cell debris or polysaccharides present in most cultures. Due to their extremely high water content, the cells will quickly become too dry, so the length of time the vacuum is applied should be closely monitored, either by timing the filtration or by stopping it as soon as the liquid stops dripping from the filter. The weight of cells collected in this manner is often referred to as "filtered fresh weight."

An alternative measurement is "wet fresh weight" or "drained fresh wet." By placing a pipette loosely against the bottom of a flask, much of the liquid medium can be withdrawn from a suspension culture while the majority of the cells will remain in the flask. The cell mass will retain a high percentage of medium (often about 50%), but this percentage will be fairly constant if the cell aggregate size remains constant. The wet fresh weight is calculated by subtracting the weight of the culture flask. This procedure can be carried out aseptically with cotton-plugged pipettes, preweighed flasks, and a clean top-loading digital balance in a laminar-flow transfer hood. These cells can then be weighed into new culture flasks, giving fairly reproducible inoculum levels without risking damage to the cells by filtering them dry. This measurement has also been adapted to immobilized cells, for which the weight of the medium-soaked immobilization matrix is subtracted from the weight of the inoculated, drained matrix.

14.5.2. Dry Weight

Collected fresh cells can be dried on preweighed filters or dishes to obtain the dry weight. Due to their high water content, the dry weight can be as low as 1 to 2% of the fresh weight. Cells should be dried at low temperatures (55 to 65°C), preferably either under vacuum or in a convection oven, to avoid being caramelized. Filter-collected cells should be washed briefly with distilled water to remove medium solids before being dried. Cells that are drained but not filtered can also be dried for a relative comparison of biomass, but the solids in the medium retained by the cells will contribute greatly to the dry weight. If a more accurate estimate of the dry weight is needed, the percentage of the wet fresh weight contributed by the medium, multiplied by the solids content (largely the dissolved sugars), can be subtracted from the dry weight (29).

Increases in dry weight can be misleading, particularly soon after cells are subcultured. Cell cultures of many species will quickly absorb the sugars from the medium and convert them to starch granules inside the cells. This starch reserve will then be slowly utilized during the growth of the culture. If a growth curve of the culture is determined, starch formation will cause the dry weight to increase dramatically for 1 to 2 days and then to decrease slowly, while the fresh weight is still increasing. (The presence of starch granules can be confirmed by examining cells under a high-power microscope; starch granules will polarize light and will stain blue with 1% iodine solution.) Thus, for initial studies, researchers will often monitor more than one parameter and then determine which is the best indicator of growth for the particular culture. Some cell lines will also concentrate and store substantial amounts of phosphate. Each new cell line must be evaluated for starch and phosphate accumulation, since some cell lines will accumulate neither, both, or only one or of these substances.

14.5.3. Packed Cell Volume

Cell suspensions can be transferred to calibrated conical screw-cap tubes and gently centrifuged at 200 to 500 × g for 10 min. The volume of cells and medium can be read directly from the tube.

14.5.4. Indirect Measurements

In some instances, the entire cell mass cannot be harvested. For example, in long-term bioreactor experiments or immobilized-cell culture systems, only a small sample of the culture or medium may be obtained. If the suspension is homogeneous and a representative sample can be obtained, any of the above methods can be used to estimate the entire contents of a bioreactor. At other times, the biomass can be estimated indirectly by monitoring the consumption of nutrients. A correlation coefficient will have to be established for each cell line, usually from small-scale shake flask experiments. If the culture does not accumulate starch, the growth of the culture can be correlated with the disappearance of sugars from the medium. If sucrose is the carbon source, many cultures will release invertase, which hydrolyzes sucrose to fructose and glucose, and utilize the glucose before the fructose. In such cases, the combination of all three carbohydrates must be monitored by a method such as ion-exchange high-pressure liquid chromatography with refractive index detection. If glucose is used as the sole carbon source, a simple reducing sugar assay may be sufficient. If the culture does not accumulate phosphate reserves (immediately depleting the medium of phosphate), phosphate uptake can be correlated with growth by using either colorimetric or chromatographic detection.

Nitrate is the major nitrogen source in most cell culture media. Very few cultures accumulate and store nitrate, and therefore depletion of nitrate from the medium is often well correlated with the growth of the culture. Several methods of nitrate quantitation have been developed, including colorimetric methods, high-pressure liquid chromatography, and capillary electrophoresis. Since nitrate (NO_3^-) is also the highest-molarity ion (40 mM in MS and 25 mM in B5), it is the major source of conductivity in the medium. It is often possible to correlate a decrease in conductivity with cell growth (14). For some experiments, the exact biomass is not required. For example, if the same inoculation and culture regimen is used each time, a certain conductivity value can be simply correlated with the correct time for harvest or elicitation.

An example of a typical plant suspension cell culture growth curve and the corresponding pattern of sugar hydrolysis and consumption in shake flasks is given in Fig. 1. The poppy cell line used in this experiment does not accumulate starch, and the total sugar consumption closely parallels the increases of dry and fresh weight. Depletion of free carbohydrates in the medium also correlated well with the end of the growth phase.

14.5.5. Viability Assays

At times, the percentage of live cells in a culture must be determined. This is often needed when cells have been subjected to harsh treatments as part of an experiment, such

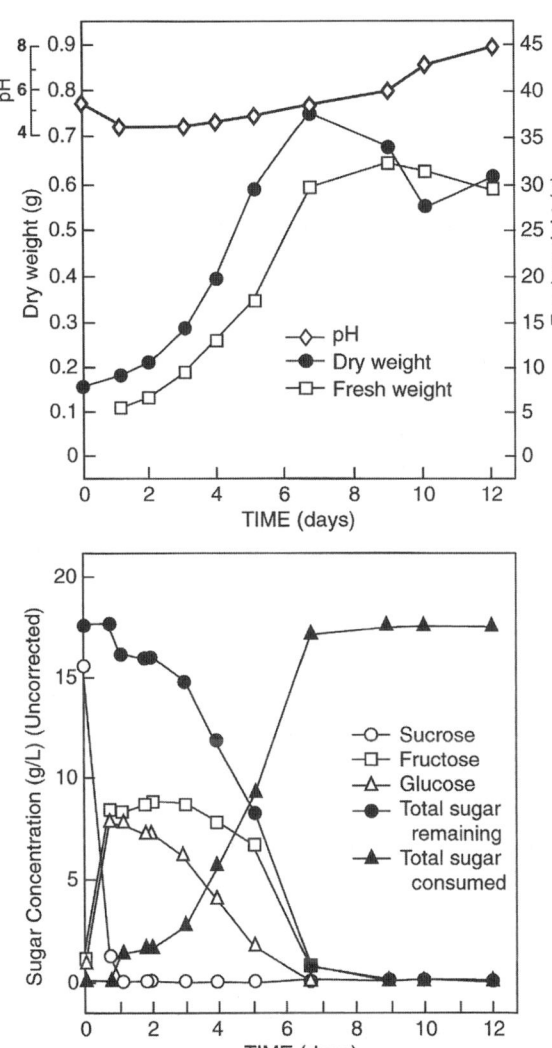

FIGURE 1 Growth of opium poppy cell suspension cultures in 250-ml shake flasks. The inoculum (*Papaver somniferum* L. cv. Marianne, PBI 2009) (29) was 10 g (wet fresh weight) (13% [wt/vol]) per 75 ml of medium per 250-ml flask. Cells were cultured at 25°C and 150 rpm. The medium was 1B5C (B5 salts supplemented with 20 g of sucrose per liter, 1 mg of 2,4-D per ml, and 1 g of casein [NZ-Amine] per liter; pH 5.5). The initial sucrose level was less than 20 g/liter, due to dilution by sucrose-depleted medium in the inoculum.

as heat shock, chemical or pathogen addition, or freezing during the development of cryopreservation conditions. It is also useful when cells have been unintentionally stressed such as during a culture room malfunction. Callus cultures can be dead but not brown or collapse for months.

Several assays have been used, depending on the equipment available and the texture of the tissue. The best assay for viability is to subculture the cells and watch for a visible increase in cell mass. Unfortunately, this may take several days or weeks. Assays in which the stain (such as fluorescein diacetate [FDA] or triphenyltetrazolium chloride [TTC]) is meant to be taken up only by living cells are known to sometimes give false-positive results with cells that no longer divide.

14.5.5.1. Cytoplasmic Streaming

A quick estimate of viability can be made by examining individual cells under a high-power microscope. The cytoplasm and some organelles can be seen circulating inside live cells. This procedure works well for fine cell suspensions but not for large aggregates or regenerating embryos and plantlets. If a researcher is unfamiliar with this phenomenon, both positive and negative controls (a sample of a known well-growing culture and a heat- or freezing-killed sample, respectively) should be used. False-positive results can come from mistaking Brownian motion for cytoplasmic streaming.

14.5.5.2. FDA Staining

The diester FDA will not fluoresce. However, it can pass through the plant cell membrane, whereas esterases will remove the acetate groups. The charged product molecule is unable to leave the intact cell, due to a pH gradient, and fluoresces strongly. Only live cells should emit fluorescence, since dead cells will leak the stain or will not have active esterases. A 0.5% stock solution of FDA in anhydrous acetone can be stored at −20°C indefinitely. When needed, the stock is diluted 50-fold (final concentration, 0.01%) in ice-cold culture medium. One drop of FDA solution is mixed with one drop of culture on a microscope slide, covered with a coverslip, and monitored for 5 to 30 min. The slide is observed under a microscope with a UV/fluorescence attachment (excitation filter, 450 to 490 nm; barrier filter, 520 nm). If relevant, the total number of fluorescing cells can be counted and divided by the total number of cells (counted under visible light) to calculate the percent viability (55).

14.5.5.3. Evans Blue Staining

The blue stain is able to enter dead cells but cannot penetrate intact cell membranes. A solution of Evans blue (5 mg/ml) in culture medium is prepared. The cells are mixed with stain solution and incubated for 5 min, and then live (colorless) versus dead (blue) cells are counted under a microscope with bright-field illumination (57).

14.5.5.4. TTC Reaction

Viable cells will reduce TTC to a pink formazan product, which can diffuse out of the cells and into the reaction medium. Viability can be assessed visually or quantitatively in a spectrophotometer. This procedure can be used for both fine and coarse cells and callus, as well as whole embryos or leaf pieces (4). A 0.6% solution of TTC in 0.5 M phosphate buffer (pH 7.0) is prepared. The tissue (100 mg of cells or plant tissue/3 ml of solution) is suspended and incubated overnight (15 h) at 30°C. The TTC solution is removed, the cells are washed with distilled water, and 7 ml of 95% ethanol is added. The mixture is incubated at 80°C for

5 min. The extract is cooled to room temperature, and the volume is adjusted to 10 ml with 95% ethanol. The absorbance at 530 nm is measured.

14.5.5.5.   Phenosafranin Staining

Dead cells will be stained red with phenosafranin, but the dye cannot penetrate living cells, similar to Evans blue. A 0.1% solution dissolved in culture medium is used, and the cells are observed under bright-field microscopy (14).

14.6.   SECONDARY-METABOLITE PRODUCTION

One application of plant cell cultures is the study of plant secondary-metabolite biosynthesis and the commercial production of such valuable natural products. Plant secondary products have been used for centuries in human medicine, such as codeine and morphine (painkillers) from opium poppy (*P. somniferum*) latex, atropine and scopolamine (neuroregulatory), quinine (antimalarial) from *Chinchona* sp., and diosgenin and digoxin (cardioactive) from the foxglove plant (*Digitalis* sp.). Modern bioactivity-guided screens have identified a range of new important compounds, especially the anticancer compounds vincristine and vinblastine from periwinkle (*C. roseus*), Taxol from the yew tree (*Taxus* sp.), and camptothecin (*Camptotheca acuminata*). Some medicinal plant parts are consumed whole or as a complex extract, such as ginseng root. There are also valuable food ingredients including coloring agents such as anthocyanins from many species and flavor extracts such as vanilla from the vanilla bean (a species of orchid). Fragrant oils such as mint, rose, and jasmine have always commanded high prices in the perfume industry. Some plant extracts are used as pesticides in organic gardening, including rotenone from tropical legumes, pyrethrins from the flowers of the pyrethrum daisy (*Chrysanthemum coccineum* and other species), and azadirachtin from the neem tree (*Azadirachta indica*).

Many studies have investigated the production of these and hundreds of other secondary metabolites in plant tissue culture (9, 13, 14, 21, 23, 28, 36, 38, 46, 48, 50, 51, 53). In some cases, cell cultures provide a source of homogeneous, highly active cells, ideal for studying the biosynthetic pathways by labeled-precursor feeding, enzyme purification, and cloning of the genes encoding pathway enzymes. Other studies have focused on increasing the productivity of cultures, with a goal of profitably producing these metabolites. Many of the above products are currently produced by growing the plants, often in tropical regions, and harvesting and processing the appropriate parts. This means that production is subject to seasonal and environmental variation, disease, competition with other crops and land uses, and political pressures such as environmental concerns or international embargoes. The growth of the source plant may be slow, such as yew trees (for Taxol) or ginseng roots. Occasionally, cell cultures make high levels of one or a few compounds, which would not normally accumulate in the intact plant, decreasing downstream purification costs. For example, elicited *P. somniferum* cultures produce very high levels of sanguinarine and few other trace alkaloids, while the most competitive whole-plant system (*Macleaya cordata*) accumulates more sanguinarine but only as a complex mixture with other undesirable alkaloids, adding several expensive chromatographic steps to the purification.

In some cases, plant tissue cultures might appear to be an economical source of metabolites. Unfortunately, due to their slow growth and often low volumetric productivity, only extremely valuable compounds may be profitably produced; one analysis estimated that a product must be worth over $1,000 per kg, given average productivities and growth rates (50). Cell cultures are also often capable of degrading the desired products, and cell lines can suddenly lose their productivity, making industrial production difficult. Many strategies have been used to improve production, including selecting for higher-producing cell lines, developing production media, and inducing production by elicitation.

14.6.1.   Constitutive Production and Strain Improvement

In certain species, media that allow good growth of the cultures will also allow good accumulation of secondary metabolites. In general, compounds that accumulate in root tissues or throughout all organs of a plant have been most easily produced at high levels in fully dedifferentiated cell cultures. This may in part be due to the lack of light or the hormone regimes; it is often said that cell culture metabolism most closely resembles that of plant roots.

Production can vary greatly among cultures of the same species. When initiating cell cultures for secondary-metabolite production, many separate cultures should be initiated from individual seedlings, and plants or seedlings should be obtained from as many sources (companies, botanical gardens, and research laboratories) as possible. These should be evaluated separately to determine if there is a variation in production between individuals, cultivars, varieties, or other genotype designations.

In some cases, product accumulation decreases either rapidly or gradually as the culture is subcultured, possibly due to natural selection for faster-growing cells, especially in suspension (12, 15). This natural variation can also be used advantageously, in that occasionally a subpopulation of highly producing cells can be selected. Visual selection was successfully applied to the production of shikonin, a red alkaloid pigment with antibacterial activity. For initial rounds of selection, portions of callus which were darker red were picked by hand and subcultured separately; for later rounds of selection, quantitative chemical analysis was used (9). Fluorescence microscopy was employed to select cell lines with high berberine production. The berberine content was increased from approximately 5% in unselected cells to 13.2% in a selected line (44).

14.6.2.   Production Media

Many cell lines will accumulate only low levels of secondary metabolites in standard growth media, but production can be increased severalfold by altering one or more medium components. A wide range of production media have been developed, and a few trends have been established. In general, the growth of the culture may slow or stop completely. Production media often contain lower levels of growth regulators or none at all, or they contain weaker auxins such as NAA or IBA instead of 2,4-D. For anthocyanins (colored phenolic compounds), limiting the growth of the culture by restricting either the nitrate or phosphate supply while simultaneously providing very high concentrations of sugars has greatly improved production. Increasing the sucrose concentration from 20 to 80 g/liter and decreasing the nitrate concentration from 25 to 2.5 mM in B5 increases anthocyanin accumulation 10-fold (24), while phosphate limitation together with a high sucrose concentration is also successful (60). In the production of alkaloids (nitrogen-containing compounds), low phosphate levels are beneficial, but high nitrate levels increase accumulation in some cases and are inhibitory in others.

In many cases, fully dedifferentiated tissue cultures will not accumulate the same valuable product as the parent plant or will accumulate only small amounts, sometimes only when cells are allowed to differentiate back into shoots or roots (58). For example, thiophenes are light-activated insecticidal and antimicrobial compounds that are highly toxic to both plant pests and host plant cells; they are accumulated only in specialized veins running longitudinally inside the roots (28). For unknown reasons, opium alkaloids accumulate only in specialized lacticifers in the aboveground portions of plants, and production in tissue culture has been associated only with redifferentiation into shoots (30). Similar observations have been made for the famous anticancer compounds vincristine and vinblastine from *C. roseus*. These molecules are heterodimers derived from the indole alkaloids catharanthine and vindoline. While high levels of catharanthine (5 to 20 times that in the plant) accumulate, no significant levels of vindoline, vincristine, or vinblastine have been reported in homogeneous cell cultures, despite the fact that this has been one of the most thoroughly investigated systems (5). One of the enzymes in vindoline biosynthesis is light activated, while another may require the differentiation of functional chloroplasts, which are absent under standard cell culture conditions.

14.6.3. Elicitation

Many economically valuable plant secondary metabolites serve as defense compounds for the producing species, protecting them against attacks by insect herbivores and microbial pathogens. Some of these compounds are absent or present only at very low levels in healthy tissues, but they accumulate to very high levels in a narrow zone surrounding the site of microbial infections; these are referred to as phytoalexins (from the Greek for "plant warding-off agents"). In 1981, it was observed that contamination of a plant callus culture by a fungus resulted in the accumulation of a pigmented antifungal alkaloid (47). Later, it was determined that live fungus was not required to cause the response but that either autoclaved fungal culture filtrates or a number of purified compounds could "elicit" phytoalexin production and other defense responses, in both callus and suspension cultures. The process was referred to as "elicitation," because the regulatory mechanisms involved were not known at the time; it has subsequently been shown that in most cases, the increase in the production of secondary metabolites is the result of an increase in the transcription of the genes encoding the corresponding biosynthetic pathway enzymes.

Elicitors are classed as biotic or abiotic. Abiotic elicitors include heavy metals such as copper or mercury salts, and UV light. Biotic elicitors include many components of fungal cell walls (chitin, chitosan, β-glucan, peptidoglycans, and arachidonic acid), compounds excreted by plant pathogens (cellulase, ribonuclease, pectinase, antibiotics, various sugar polymers, and proteins), and compounds released from plant cell walls during degradation by plant pathogens (pectin fragments) (11). The type of elicitor that is most effective in inducing secondary metabolism in a particular plant cell culture varies widely. The elicitor does not have to be derived from a pathogen of the source plant. In general, members of the same plant genus will respond well to the same elicitors but members of the same family may not. For example, cultures of several legume species (alfalfa, soybean, and green bean) are highly elicited by a crude β-glucan derived from *Saccharomyces cerevisiae* (28,

47) but are unresponsive to chitin. Solanaceous species (tobacco, potato, and tomato) are all elicited by cellulases, arachidonic acid, and β-glucans, but while two members of the poppy family, *P. somniferum* and *Eschscholtzia californica*, are both elicited by β-glucans, only the former is responsive to chitin hydrolysates (29, 34).

In the *Eschscholtzia* system, in which elicitation was first studied, the control cultures accumulate a low level of one alkaloid. Following addition of the elicitor, the level of this alkaloid increases and five related alkaloids appear; the total benzophenanthridine alkaloid content of suspension cultures increases more than 20-fold compared to that in unelicited cultures (34). In some cultures, such as soybean and green bean, no pterocarpan phytoalexins are present before elicitor addition but reach maximum levels (16, 19) within 24 to 48 h after elicitation. In an attempt to use elicitation to stimulate the accumulation of opium (morphinan) alkaloids in *P. somniferum* cultures, the authors were surprised to induce instead the accumulation of sanguinarine, a member of a different (benzophenanthridine) class of alkaloids but one which is also inducible by pathogen attack on poppy plants (17).

The optimal time for addition of the elicitor is usually in the middle to late growth phase of the culture, often 3 to 7 days after subculture. Elicitor addition before or after this time results in little or no response from the plant cells. The optimal time may shift; for example, as a cell suspension is repeatedly subcultured and the net growth rate increases, the optimal time will become earlier by as much as 2 or 3 days. Elicitors are often effective when added to standard growth media, although combining a production medium with elicitation may result in even higher production levels (8). Many elicitors cause a browning of the cell cultures and may result in the death of a high percentage of the cells.

The first secondary metabolite commercially produced by plant cell cultures was shikonin, a pigmented antibacterial alkaloid that was normally extracted from the older (3- to 7-year-old) roots of *Lithospermum erythrorhizon*. By using a combination of strain selection and production medium, the investigators found that the cultures could accumulate 12% dry weight of shikonin, versus 1 to 2% in older roots (9). While other products have come close to commercialization, in general, producers have defaulted to using field-grown plants as the source. The most recent promising candidate for tissue culture production is Taxol, due to its high value and environmental concerns regarding harvesting of wild trees. While plantation production and semisynthetic methods are currently the favored production routes, recent manipulations in explant sources and culture conditions have resulted in the accumulation of Taxol in plant cell cultures at levels several times higher than that found in the tree bark (25).

14.7. REGENERATION

Even individual plant cells or protoplasts, if cultured properly, can regenerate into fully developed, fertile plants. This property is referred to as "totipotency" and was first demonstrated in plants in 1965 (18). In current applications, small explants (rather than individual cells) are first induced to form calli, which are then induced to regenerate into a number of plants. There are two modes of regeneration, organogenesis and somatic embryogenesis.

In organogenesis, leaf explants, callus cultures, or suspension cultures differentiate directly to form shoots, which can then be induced to form roots. Generally the procedures involve induction of rapid cell division and/or callus formation on media containing high auxin levels, followed by a medium containing low auxin and high cytokinin levels to induce shoot formation. For example, for tobacco, young leaves are surface-sterilized, cut into 1-cm squares, and placed onto MS agar containing 4.5 μM 2,4-D and 30 g of sucrose per liter for 2 to 4 weeks to induce callus formation. The explants are then transferred to MS agar supplemented with 5 μM 6-benzylaminopurine and placed under lights; shoots appear within a few weeks. Young shoots are cut off with a scalpel and rooted in MS with 30 g of sucrose per liter without hormones. Organogenesis is easily accomplished for most solanaceous species (tobacco, tomato, potato, petunia, etc.) and has been reported for many dicot species but is rarer among monocots.

In somatic embryogenesis, embryos or embryolike structures are produced from somatic cells, as opposed to the maturation of zygotic cells in plant flowers. Somatic embryogenesis was first demonstrated for carrot cultures in 1964 but has since been demonstrated for many species, both dicots and monocots. For carrot (18, 55), callus cultures are simply initiated on MS agar supplemented with 4.5 μM 2,4-D and 30 g of sucrose per liter for 1 month; then a fine, rapidly growing suspension is initiated in the same medium without agar. After 2 to 4 weeks, the cells are washed with the same medium without 2,4-D and incubated in the dark to induce embryo formation. Embryos appear after 8 to 10 days and continue to increase in size for 2 weeks. The embryos are then transferred to MS agar medium (with sucrose and no hormones) for plantlet formation; this last step is referred to as conversion or germination of the embryos. In some systems, many embryos are formed but not all of them will germinate. The addition of abscisic acid may aid in the development of normal embryos, and supplementation of the medium with proline and glutamine can greatly increase embryogenesis.

14.8. MICROPROPAGATION

In both organogenesis and somatic embryogenesis systems, hundreds of plantlets can be generated from a single explant. This is one method by which plants can be multiplied for commercial purposes. Another method is simply establishing sterile shoot cultures of plants, from which axenic cuttings are made and rooted in culture; cuttings can be made at any time of the year and will usually grow more quickly under plant tissue conditions than in the field. Either method is referred to as micropropagation. Many houseplants and ornamentals are propagated in this manner, allowing producers to introduce new varieties in a short time. Crop species such as blueberries, strawberries, garlic, and potatoes are also propagated by micropropagation to produce virus-free plants. The regeneration step can also be used to induce secondary-metabolite accumulation (see the section on production media, above).

14.9. TRANSFORMATION

Genetic transformation of plant cells, and their subsequent regeneration into transgenic plants, was first accomplished in the early 1980s for tobacco and is now routine for many species. Important crop species such as corn, wheat, rice, and soybean, which were deemed impossible to transform or regenerate only a few years ago, are now routinely manipulated. Usually, breakthroughs in transformation protocols of so-called recalcitrant species are first made with a highly regenerable cultivar. This transformed cultivar can then be crossed with elite high-producing cultivars to introduce the transgene into a number of plant lines.

In very early versions of plant transformation protocols, foreign DNA was introduced into plant protoplasts (plant cells treated with cellulase and other digestive enzymes to remove the cellulose cell walls) by using electroporation or polyethylene glycol (39). Since many plants are regenerated from protoplasts only with great difficulty, this method has not been widely applied. Most plant transformation has been carried out by A. tumefaciens-mediated DNA delivery (41). Wild-type A. tumefaciens causes crown gall disease by transferring a segment of plasmid DNA (called the T-DNA, for transfer DNA) containing bacterial genes encoding auxin biosynthesis and other pathways into wounded plants, which in turn cause dedifferentiated callus tumors to form, releasing nutrients to the surrounding A. tumefaciens organisms in the soil. A. tumefaciens strains have been modified (disarmed) such that the auxin biosynthetic genes have been deleted so they can no longer cause disease but the strains still contain the vir genes required for T-DNA transfer. Foreign genes of interest, including a selectable marker, are inserted into binary vectors, such that the foreign genes are between the ends of the T-DNA region. When in contact with cultivated plant cells, usually at the beginning of the callus induction phase of either organogenesis or somatic embryogenesis, the disarmed A. tumefaciens strain transfers the foreign gene construct. Selection pressure is applied (usually an antibiotic or herbicide in the medium), so that the only cells that are able to grow are those that received the foreign T-DNA (including the selectable marker, usually a drug resistance or herbicide resistance gene). The A. tumefaciens strain is then killed by antibiotics (such as penicillin derivatives) that do not act against plant cells. The surviving transformed plant cells are then induced to regenerate into transgenic plants.

A. tumefaciens-mediated transformation has been used for many dicot species but has been less effective with monocots. A more recently developed DNA delivery system is called particle bombardment or the biolistic approach (42, 43). Early versions of biolistic equipment used the explosive force of gunpowder to propel the 1-μm tungsten or gold particles, coated with the desired foreign gene construct. The current version of the equipment (patented by Dupont but marketed through Bio-Rad [Hercules, CA]) uses high-pressure helium gas and calibrated membranes (rupture disks) as a safer and more reproducible driving force. Transgenic plants have been produced by "shooting" leaf explants, embryogenic suspensions, or young embryos followed by applying drug or herbicide selection pressures, as described for the A. tumefaciens system, and then by regeneration. The biolistic approach was the first system to allow routine transformation of many monocot species such as corn, wheat (54), and rice (1), but it is also popular for dicot transformation (42).

Plant transformation and regeneration have resulted in the release and marketing of many genetically engineered plants, worth billions of dollars annually, and their use and the available genotypes are increasing rapidly. Transgenic soybeans have been the most widely adopted, with herbicide-tolerant transgenic soybeans being planted on over 90% of the U.S. and over 50% of the worldwide soybean farmland in 2008, based on U.S. Department of Agriculture public reports. Glyphosate-resistant (or Round-Up Ready, from Monsanto Co. [St. Louis, MO]) soybeans are engineered to be resistant to the herbicide glyphosate (an inhibitor of aromatic amino acid synthesis), which is highly toxic to most plants but relatively harmless to people and animals and which is rapidly destroyed in the environment. Farmers using these varieties achieve much better weed suppression with fewer herbicide applications and can adopt "no till" planting strategies, which further reduce fuel consumption and soil erosion. Over 80% of U.S. cotton and corn planted in 2008 contained either herbicide tolerance (resistant to glyphosate, bialophos, or acetolactate synthase-inhibiting herbicides), insect tolerance, or both types of transgenes. The most common method of conferring insect resistance is the expression of the insecticidal *Bacillus thuringiensis* (Bt) toxin protein; this protein only kills lepidopteran insect pests that eat the transgenic tissues, thus sparing beneficial and nontarget insects. Some products have been developed but are not being marketed, such as ripening-delayed tomatoes and transgenic wheat, due to lack of consumer acceptance or fear of import bans from countries outside the United States.

Many more transgenic plant products are under development, including plants with altered nutritional properties (e.g., high-lysine corn, increased vitamins, or improved protein quality), fatty acid or oil content (for nutritional, industrial, or fuel applications), and secondary-metabolite accumulation. Studies have indicated that consumption of certain plant secondary metabolites such as isoflavonoids, anthocyanins, resveratrol, lignans, and carotenoids may help reduce the incidence of human ailments such as cancer, heart disease, osteoporosis, and sight loss. Genes from the biosynthetic pathways of these so-called nutraceuticals have been cloned and overexpressed in transgenic plants either to increase their levels in native hosts or to introduce them into more widely consumed foods (32, 37). Several researchers joined forces to develop "golden rice," a carotenoid-accumulating transgenic rice containing multiple foreign genes, to provide a cheap source of dietary vitamin A for tropical developing countries with high levels of childhood blindness (caused by vitamin A deficiency) (61). In the case of soybean oil, researchers have introduced constructs expressing plant and microbial enzymes to increase the overall fatty acid unsaturation level or to introduce desirable omega-3 and omega-6 fatty acids (7, 10). Since transformation and regeneration of soybean relies on somatic embryogenesis, and the oil profiles in transgenic somatic embryos in tissue culture closely predict those that will accumulate in developing soybean embryos in pods in the field, companies are able to test many transformation constructs quickly by analyzing embryos from early-in-transformation cultures, without waiting several months for transgenic plants to set viable seed (7).

Plant tissue culture and plant transformation are becoming routine tools in both basic research and applied crop development projects and have yielded many important discoveries. Early plant studies were supported by the floral industry, and in the process of trying to overexpress anthocyanin pigment-related genes in an attempt to produce darker-red petunia flowers, the studies found that some transgenic plants produced only white flowers (35). Later studies with plant RNA viruses demonstrated that transgenic plants overexpressing certain viral RNAs were resistant to infection by the same virus (31, 40). Years later, additional work analyzing these perplexing results led to the widely applied tools of posttranscriptional gene silencing or RNA interference, now used in the plant, animal, and microbial fields. In recent years, RNA interference has become a quick way to knock out a specific gene in both whole plants and tissue cultures, producing the same phenotype as if the gene had been deleted or suppressed. One can then attempt to determine the role of the targeted gene in plants. Now that the genomes or large cDNA pools from several plant species have been sequenced, the demand has increased for quick plant transformation and efficient culture methods for screening transformed cells to identify and characterize new genes.

Plant transformation itself has become a highly developed mutagenesis and gene discovery tool. When foreign DNA (often referred to as T-DNA) is introduced into a plant cell, it is integrated into chromosomes at one or more seemingly random sites. It may insert into intergenic spaces with no additional phenotype, but it may also disrupt coding or regulatory regions of endogenous genes; such T-DNA-tagged populations of model plant species have yielded many useful mutants, and the altered gene is easily recovered and sequenced using the foreign gene sequences as a marker (59). A variation on this is "activation tagging," wherein the introduced DNA bears strong promoters (to activate the expression of nearby endogenous genes after insertion) or easily detected marker genes (to be activated by nearby endogenous genes after insertion) (3, 26).

In closing, plant tissue cuture began in the early 1900s primarily as basic laboratory experiments and a means to readily propagate expensive tropical houseplants. While the media and growth conditions still vary widely depending on which of the thousands of plant species are being cultured and the purpose of the culture (e.g., biochemical analysis, regeneration, transformation), the principles behind their selection are well understood. In the past 25 years, plant tissue culture has become a major tool in plant biochemistry, plant gene discovery and functional analysis, and plant genetic modification, with a huge economic impact on agriculture and human nutrition worldwide.

REFERENCES

1. **Aldemita, R. R., and T. K. Hodges.** 1996. *Agrobacterium tumefaciens*-mediated transformation of *japonica* and *indica* rice varieties. *Planta* **199:**612–617.
2. **Ammirato, P. V., D. A. Evans, W. R. Sharp, and Y. Yamada.** 1984. *Handbook of Plant Cell Culture*, vol. 3. *Crop Species.* MacMillan Publishing Co., New York, NY.
3. **An, G., S. Lee, S. H. Kim, and S. R. Kim.** 2005. Molecular genetics using T-DNA in rice. *Plant Cell Physiol.* **46:**14–22.
4. **Bajaj, Y. P. S., and J. Reinert.** 1977. Cryobiology of plant cell cultures and establishment of gene banks, p. 757–777. *In* J. Reinert and Y. P. S. Bajaj (ed.), *Applied and Fundamental Aspects of Plant Cell, Tissue, and Organ Culture.* Springer-Verlag, Berlin, Germany.
5. **Balsevich, J., and G. Bishop.** 1989. Distribution of catharanthine, vindoline and 3′,4′-anhydrovinblastine in the aerial parts of some *Catharanthus roseus* plants and the significance thereof in relation to alkaloid production in cultured cells, p. 149–153. *In* W. G. W. Kurz (ed.), *Primary and Secondary Metabolism of Plant Cell Cultures II.* Springer-Verlag, New York, NY.
6. **Burger, D. W.** 1988. Guidelines for autoclaving media used in plant tissue culture. *Hort. Sci.* **23:**1066–1068.
7. **Cahoon, E. B., T. J. Carlson, K. G. Ripp, B. J. Schweiger, G. A. Cook, S. E. Hall, and A. J. Kinney.** 1999. Biosynthetic origin of conjugated double bonds: production of fatty acid components of high-value drying oils in transgenic soybean embryos. *Proc. Natl. Acad. Sci. USA* **96:**12935–12940.
8. **Cline, S. D., and C. J. Coscia.** 1988. Stimulation of sanguinarine production by combined fungal elicitation and hormonal deprivation in cell suspension cultures of *Papaver bracteatum*. *Plant Physiol.* **86:**161–165.
9. **Curtin, M. E.** 1983. Harvesting profitable products from plant tissue culture. *Bio/Technology* **1:**649–657.

10. **Damude, H. G., H. Zhang, L. Farrall, K. G. Ripp, J.-F. Tomb, D. Hollerbach, and N. S. Yadav.** 2006. Identification of bifunctional Δ12/ω3 fatty acid desaturases for improving the ratio of ω3 to ω6 fatty acids in microbes and plants. *Proc. Natl. Acad. Sci. USA* **103:**9446–9451.

11. **Darvill, A. G., and P. Albersheim.** 1984. Phytoalexins and their elicitors—a defense against microbial infection in plants. *Annu. Rev. Plant Physiol.* **35:**243–275.

12. **Deus-Neumann, B., and M. H. Zenk.** 1984. Instability of indole alkaloid production in *Catharanthus roseus* cell suspension cultures. *Planta Med.* **50:**427–431.

13. **DiCosmo, F., and M. Misawa.** 1996. *Plant Cell Culture Secondary Metabolism: Toward Industrial Application.* CRC Press, Inc., Boca Raton, FL.

14. **Dixon, R. A., and R. A. Gonzales.** 1994. *Plant Cell Culture. A Practical Approach.* Oxford University Press, Oxford, United Kingdom.

15. **Dougall, D. K., and D. L. Vogelien.** 1987. The stability of accumulated anthocyanin in suspension cultures of the parental line and high and low accumulating subclones of wild carrot. *Plant Cell Tissue Org. Cult.* **8:**113–123.

16. **Ebel, J., A. R. Ayers, and P. Albersheim.** 1976. Host-pathogen interactions XII. Response of suspension-cultured soybean cells to the elicitor isolated from *Phytophthora megasperma* var. *sojae*, a fungal pathogen of soybeans. *Plant Physiol.* **57:**775–779.

17. **Eilert, U., W. G. W. Kurz, and F. Constabel.** 1984. Stimulation of sanguinarine accumulation in *Papaver somniferum* cell cultures by fungal elicitors. *J. Plant Physiol.* **119:**65–76.

18. **Evans, D. A., W. R. Sharp, P. V. Ammirato, and Y. Yamada.** 1983. *Handbook of Plant Cell Culture*, vol. 1. *Techniques for Propagation and Breeding.* Macmillan Publishing Co., New York, NY.

19. **Funk, C., K. Gügler, and P. Brodelius.** 1987. Increased secondary product formation in plant cell suspension cultures after treatment with a yeast carbohydrate preparation (elicitor). *Phytochemistry* **26:**401–405.

20. **Gamborg, O. L., R. A. Miller, and K. Ojima.** 1968. Nutrient requirements of suspension cultures of soybean root cells. *Exp. Cell Res.* **50:**151–158.

21. **George, E. F., D. J. M. Puttock, and H. J. George.** 1987. *Plant Culture Media*, vol. 1. *Formulations and Uses.* Exegetics Ltd., Edington, United Kingdom.

22. **Goodwin, T. W., and E. I. Mercer.** 1983. *Introduction to Plant Biochemistry.* Pergamon Press, Oxford, United Kingdom.

23. **Hay, C. A., L. A. Anderson, M. F. Roberts, and J. D. Phillipson.** 1988. Alkaloid production by plant cell cultures, p. 97–140. *In* A. Mizrahi (ed.), *Biotechnology in Agriculture.* Alan R. Liss, Inc., New York, NY.

24. **Hirasuna, T. J., M. L. Shuler, V. K. Lackney, and R. M. Spanswick.** 1991. Enhanced anthocyanin production in grape cell cultures. *Plant Sci.* **78:**107–120.

25. **Hirasuna, T. J., L. J. Pestchanker, V. Srinivasan, and M. L. Shuler.** 1996. Taxol production in suspension cultures of *Taxus baccata*. *Plant Cell Tissue Org. Cult.* **44:**95–102.

26. **Ichikawa, T., M. Nakazawa, M. Kawashima, S. Muto, K. Gohda, K. Suzuki, A. Ishikawa, H. Kobayashi, T. Yoshizumi, Y. Tsumoto, Y. Tsuhara, H. Iizumi, Y. Goto, and M. Matsui.** 2003. Sequence database of 1172 T-DNA insertion sites in *Arabidopsis* activation tagging lines that showed phenotypes in T1 generation. *Plant J.* **36:**421–429.

27. **Kato, A., S. Kawazoe, M. Iijima, and Y. Shimizu.** 1976. Continuous culture of tobacco cells. *J. Ferment. Technol.* **54:**82–87.

28. **Kurz, W. G. W.** 1989. *Primary and Secondary Metabolism of Plant Cell Cultures II.* Springer-Verlag, New York, NY.

29. **Kurz, W. G. W., N. L. Paiva, and R. T. Tyler.** 1990. Biosynthesis of sanguinarine by elicitation of surface-immonilized cells of *Papaver somniferum* L, p. 682–688. *In* H. J. J. Nikamp, L. H. W. VanDerPlas, and J. Van Aartrijk (ed.), *Progress in Plant Cellular and Molecular Biology.* Kluwer, Boston, MA.

30. **Kutchan, T. M., S. Ayabe, R. J. Krueger, E. M. Coscia, and C. J. Coscia.** 1983. Cytodifferentiation and alkaloid accumulation in cultured cells of *Papaver bracteatum.* *Plant Cell Rep.* **2:**281–284.

31. **Lindbo, J. A., and W. G. Dougherty.** 2005. Plant pathology and RNAi: a brief history. *Annu. Rev. Phytopathol.* **43:**191–204.

32. **Maxwell, C. A., M. A. Restrepo-Hartwig, A. O. Hession, and B. McGonigle.** 2004. Secondary metabolism in model systems: chapter eight. Metabolic engineering of soybean for improved flavor and health benefits. *Recent Adv. Phytochem.* **38:**153–176.

33. **Murashige, T., and F. Skoog.** 1962. A revised medium for rapid growth and bioassays with tobacco tissue cultures. *Physiol. Plant.* **15:**473–497.

34. **Nagata, T., Y. Nemoto, and S. Hesazawa.** 1992. Tobacco BY-2 cell line as the "HeLa" cell in the cell biology of higher plants. *Int. Rev. Cytol.* **132:**1–30.

35. **Napoli, C., C. Lemieux, and R. Jorgensen.** 1990. Introduction of a chalcone synthase gene into *Petunia* results in reversible co-suppression of homologous genes in trans. *Plant Cell* **2:**279–289.

36. **Neumann, K.-H., W. Barz, and E. Reinhard.** 1985. *Primary and Secondary Metabolism of Plant Cell Cultures.* Springer-Verlag, Berlin, Germany.

37. **Paiva, N. L.** 2002. Engineering resveratrol glucoside accumulation into alfalfa: crop protection and nutraceutical applications, p. 118–130. *In* K. Rajasekaran, T. J. Jacks, and J. W. Finley (ed.), *Crop Biotechnology* (American Chemical Society Symposium Series 829). Oxford University Press, New York, NY.

38. **Parr, A. J.** 1988. Secondary products from plant cell culture, p. 1–34. *In* A. Mizrahi (ed.), *Biotechnology in Agriculture.* Alan R. Liss, Inc., New York, NY.

39. **Paszkowski, J., and M. W. Saul.** 1986. Direct gene transfer to plants. *Methods Enzymol.* **118:**668–684.

40. **Register, J. C., III, and R. N. Beachy.** 1988. Resistance to TMV in transgenic plants results from interference with an early event in infection. *Virology* **166:**524–532.

41. **Rogers, S. G., R. B. Horsch, and R. T. Fraley.** 1986. Gene transfer in plants: production of transformed plants using Ti plasmid vectors. *Methods Enzymol.* **118:**627–640.

42. **Russell, D. R., K. M. Wallace, J. H. Bathe, B. J. Martinell, and D. E. McCabe.** 1993. Stable transformation of *Phaseolus vulgaris* via electric-discharge mediated particle acceleration. *Plant Cell Rep.* **12:**165–169.

43. **Sanford, J. C., F. D. Smith, and J. A. Russell.** 1993. Optimizing the biolistic process for different biological applications. *Methods Enzymol.* **217:**483–509.

44. **Sato, F., and Y. Yamada.** 1984. High berberine-producing cultures of *Coptis japonica* cells. *Phytochemistry* **23:**281–285.

45. **Schenk, R. U., and A. C. Hildebrandt.** 1972. Medium and techniques for induction and growth of monocotyledonous and dicotyledonous plant cell cultures. *Can. J. Bot.* **50:**199–204.

46. **Schripsema, J., and R. Verpoorte.** 1995. *Primary and Secondary Metabolism of Plants and Cell Cultures III.* Kluwer Academic Publishers, Dordrecht, The Netherlands.

47. **Schumacher, H.-M., H. Gundlach, F. Fiedler, and M. H. Zenk.** 1987. Elicitation of benzophenanthridine alkaloid synthesis in *Eschscholtzia* cell cultures. *Plant Cell Rep.* **6:**410–413.

48. **Shargool, P. D., and T. T. Ngo (ed.).** 1994. *Biotechnological Applications of Plant Cultures.* CRC Press, Inc., Boca Raton, FL.

49. **Shneyour, Y., A. Zelcer, S. Izhar, and J. S. Beckmann.** 1984. A simple feeder-layer technique for the plating of plant cells and protoplasts at low densities. *Plant Sci. Lett.* **33:**293–302.

50. **Staba, E. J.** 1980. *Plant Tissue Culture as a Source of Biochemicals.* CRC Press, Inc., Boca Raton, FL.

51. **Staba, E. J.** 1985. Milestones in plant tissue culture systems for the production of secondary products. *J. Nat. Prod.* **48:**203–209.

52. **Thorpe, T. A.** 1981. *Plant Tissue Culture. Methods and Applications in Agriculture.* Academic Press, New York, NY.

53. **Trigiano, R. N., and D. J. Gray.** 1996. *Plant Tissue Culture Concepts and Laboratory Exercises.* CRC Press, Inc., Boca Raton, FL.

54. **Weeks, J. T., O. D. Anderson, and A. E. Blechl.** 1993. Rapid production of multiple independent lines of fertile transgenic wheat (*Triticum aestivum*). *Plant Physiol.* **102:**1077–1084.

55. **Wetter, L. R., and F. Constabel (ed.).** 1982. *Plant Tissue Culture Methods,* 2nd ed. National Research Council of Canada, Prairie Regional Laboratories, Saskatoon, Saskatchewan, Canada.

56. **White, P. R.** 1937. Vitamin B1 in the nutrition of excised tomato roots. *Plant Physiol.* **12:**803–811.

57. **Widholm, J. M.** 1972. The use of fluorescein diacetate and phenosafranine for determining viability of cultured plant cells. *Stain Technol.* **47:**189–194.

58. **Wierman, R.** 1981. Secondary plant products and cell and tissue differentiation. p. 85–116. *In* E. E. Conn (ed.), *The Biochemistry of Plants,* vol. 7. Academic Press, New York, NY.

59. **Winkler, R. G., M. R. Frank, D. W. Galbraith, R. Feyereisen, and K. A. Feldmann.** 1998. Systematic reverse genetics of transfer-DNA-tagged lines of *Arabidopsis:* isolation of mutations in the cytochrome P450 gene superfamily. *Plant Physiol.* **118:**743–750.

60. **Yamakawa, T., S. Kato, K. Ishida, T. Kodama, and Y. Minoda.** 1983. Formation and identification of anthocyanins in cultured cells of *Vitis* sp. *Agric. Biol. Chem.* **47:**997–1001.

61. **Ye, X., S. Al-Babili, A. Klöti, J. Zhang, P. Lucca, P. Beyer, and I. Potrykus.** 2000. Engineering provitamin A (β-carotene) biosynthetic pathway into (carotenoid-free) rice endosperm. *Science* **287:**303–305.

Insect Cell Culture

SPIROS N. AGATHOS

15

15.1. INTRODUCTION

Insect cells were first isolated and put into culture in the form of continuous cell lines in the late 1950s and early 1960s for the study of insect metabolism and physiology but also for the in vitro synthesis of baculoviruses (a group of viruses that infect invertebrates) as biological control agents against insect pests. However, insect cell culture received a major boost when the genetic manipulation of baculoviruses in the early 1980s made it possible to use them as vectors for the production of recombinant proteins. Today the in vitro cultivation of insect cells coupled with infection with baculovirus vectors (the insect cell-baculovirus expression vector system, or IC-BEVS) constitutes a major biomanufacturing platform for the commercial production of occluded viruses as biopesticides and of recombinant proteins as human and animal vaccines and therapeutics. This is the result of a combination of advantages of both the cell culture component and the vectors ensuring the transfer of heterologous genes: the culture of insect cells in vitro is now mastered to a similar extent as mammalian cell culture, and the application of baculoviruses as vectors for recombinant protein production or for manufacturing of viral biopesticides is safe and readily amenable to scale-up (1, 36, 60). The IC-BEVS is a highly versatile system because it can express gene products of practically any origin (from bacteria to human tissue), and in contrast to most industrial mammalian cell culture systems, it is based on engineering only the vector and not the host cell line. As a result, the development time needed to progress from gene cloning to protein overproduction is much shorter (weeks instead of months). Compared to other biomanufacturing platforms, the IC-BEVS offers other significant advantages, including typically high product titers, a range of posttranslational modifications, and the possibility to express multimeric proteins or even several distinct proteins using the same vector. The ready adaptation of insect cells to suspension culture and the continuous improvement of cell culture media and additives (3) are contributing to reliable and robust scale-up practices for commercial applications.

15.2. CELL GROWTH

15.2.1. Characteristics of Cell Lines

Several hundreds of insect cell lines have been established from more than 100 different species encompassing seven orders and originate from eggs or adult tissues like ovaries, imaginal disks, midgut, etc. (50). The IC-BEVS makes use mostly of lepidopteran cell lines, in particular from *Bombyx mori* (silkworm), *Mamestra brassicae*, *Spodoptera frugiperda* (fall army worm), and *Trichoplusia ni* (cabbage looper). Among them, Sf-9 and Sf-21, isolated from *S. frugiperda*, and Tn-368 and BTI-TN-5B1-4, isolated from *T. ni*, are the cell lines most commonly used in industrial applications. These cell lines are very susceptible to infection by *Autographa californica* multiple nucleopolyhedrosis virus (AcMNPV) and other baculoviruses that form the basis for the construction of vectors in the IC-BEVS. A detailed, comprehensive, and up-to-date reference on all the techniques and methodologies involving the use of insect cell culture and baculoviruses for research and industrial applications is given by Murhammer (55).

The first line to be intensively used in research and technological applications was Sf-21, isolated from ovarian tissues of *S. frugiperda*. The Sf-9 cell line, derived from Sf-21 (68), remains probably the most widely used of all insect cell lines, thanks to its improved growth and high susceptibility to baculovirus infection, leading to excellent viral particle and protein yields. Tn-368 was obtained from ovarian tissues of *T. ni*, and BTI-TN-5B1-4 is a clone of the embryonic Tn-5 cell line isolated from *T. ni* (31). BTI-TN-5B1-4, patented by Granados (30), has been commercialized under the name High Five™ (Invitrogen, Carlsbad, CA) because of its superior capacity for secreted glycoprotein production compared to Sf cell lines (20, 66) and now rivals Sf-9 as the most popular cell line for heterologous gene expression. Conversely, the Sf-9 cell line is used more frequently for extracellular virus production (e.g., biopesticides) because of its higher yield of assembled viral particles. A genetically modified Sf cell line commercialized under the name Mimic™ Sf-9 (Invitrogen) is used for the stable expression of complex glycosylated heterologous proteins and is the same as the SfSWT-1 cell line developed by the group of Jarvis (32).

The Sf lines are adapted to suspension cultivation and are easily detached from T-flask (or other recipient) surfaces by gentle agitation without trypsinization (57). Tn lines were originally anchorage dependent, but today they are fully adapted to suspension cultivation. Moreover, upon microcarrier cultivation, High Five cells are more easily detached than Sf-9 cells (34). Among Sf lines, Sf-21 cells are more fragile than Sf-9; less tolerant to osmotic, pH, and shear stress than Sf-9; and have a lower growth rate (57). High Five cells are more robust to shear stress and osmotic shocks than Sf-9 (42), although Sf-9 cells are more resistant to thermal shock (28). High Five cells (15 μm) are larger and have higher protein content than Sf-9 cells (13 μm), and they have wider cell size distribution than Sf-9 cells (J. C. Drugmand, Y. J. Schneider, and S. N. Agathos, unpublished data). However, the cell size depends on medium osmolarity, shear stress, and cell state (viable, apoptotic, etc.) (58). In general, Sf and Tn cells are bigger when infected (49). Maximal cell densities in serum-free batch culture can reach up to 9.6×10^6 cells/ml with High Five cells in YPR medium (Drugmand et al., unpublished) and 8.1×10^6 cells/ml with Sf-9 cells (63) in Sf900-II medium. Insect cell lines can grow over long-term passaging, which, however, can give rise to morphological and physiological changes: decrease of productivity, increase of growth rate and cell diameter (22), and lower susceptibility to growth enhancement by conditioned medium (11, 12). Finally, the tendency of baculovirus-infected cultured insect cells toward production of defective interfering particles ("passage effect") increases with higher cell passage number (43, 61). It is recommended to renew the working subcultivation after 30 passages (60).

The industrially and biotechnologically important cell lines of *S. frugiperda* and *T. ni* grow in the typical pattern of other animal cells in vitro. Following a short lag phase, the cells grow exponentially with doubling times between 18 and 30 h. Starting from seeding densities of 2 to 4×10^5 cells/ml, maximal cell densities between 5 and 10×10^6 cells/ml for Sf-9 and High Five cells are typically obtained on day 5 to 7 postinoculation in batch shake flask culture and in suspension bioreactors, depending on the medium formulation and the passage number of the inoculum. These values can easily increase by 100% or more in cases of fed-batch culture, especially when medium replacement or specific nutrient supplementation is practiced (36). The optimal temperature of insect cell cultivation is 27 to 28°C and it does not require sparging of CO_2. The optimal pH for the growth of most insect cells in culture lies between 6.2 and 6.3, somewhat more acidic compared to the growth pH required by cultured mammalian cells.

The biotechnological exploitation of cultured insect cells is based on their infection either with a recombinant baculovirus to produce a recombinant protein or with a baculovirus (mostly wild-type but also recombinant) to produce biopesticides in the form of budded virus. The production of the recombinant protein or of the viral particles occurs late in the infection phase, when the infected cells are lysed due to the pathogenic effect of the baculovirus. Thus the IC-BEVS production platform is a lytic system and the formation of the viral or protein product is transient, as a new batch of insect cells must be grown and infected for a new cycle of production.

15.2.2. Insect Cell Culture Media Development

Insect cell media contain the same basal ingredients as mammalian cell media (carbohydrates, amino acids, and salts) but at concentrations adapted to insect cell metabolism. In contrast to mammalian media, they are supplemented with specific additives such as a lipid mixture to supply lipids that insect cells are unable to synthesize. Other ingredients added include the shear-protective agent Pluronic® F-68 (Sigma-Aldrich, St. Louis, MO) (typically at 0.1 to 0.2% [wt/vol]) and antifoam for agitated and sparged reactor cultivation.

Insect cells have been cultured for decades in noncommercial basal media, such as Grace's, TNM-FH, TC-100, or IPL-41, that have been supplemented with 5 to 20% fetal bovine serum. In the last decade serum-free media have become dominant in insect cell culture in order to overcome the drawbacks of serum (cost, lot-to-lot variability, potential adventitious agents or contaminants, interference with product purification, etc.). Serum-free media are formulated from the above basal media (typically IPL-41 or TNM-FH) upon supplementation with yeastolate (ultrafiltered yeast extract), a lipid mixture emulsified in Pluronic F-68 (51) and other, mostly proprietary complements. Commercial serum- and protein-free media, including BD BaculoGold Max-XP (BD Biosciences, Franklin Lakes, NJ), ESF 921 (Expression Systems, Woodland, CA), EX-CELL® 405 and 420 (Sigma-Aldrich, St. Louis, MO), Express Five® and Sf900III™ (Invitrogen), HyClone SFX-Insect™ (Thermo Scientific, Logan, UT), and Insect-XPRESS™ (Cambrex Bio Science, Walkersville, MD), are used routinely for insect cell cultivation. These media, although able to support high cell densities and recombinant protein titers, are typically expensive, cell line specific, and of proprietary composition. A comprehensive listing of most serum-free media currently available for insect cell culture has been compiled by Agathos (3). A dominant feature of medium design today in response to stringent regulations is the elimination of animal-derived components. Thus, supplements designed to replace serum and other animal-derived ingredients are protein hydrolysates (peptones) of plant or microbial origin. Hydrolysates may have a nutritional role if basal media with lower amino acid content or no amino acids are used, but mostly they are assumed to perform many serum functions (protection from oxidative stress, shear damage, or apoptosis, etc.), as shown by the use of yeastolate, which moreover facilitates the purification of the recombinant proteins produced in the IC-BEVS (3, 36).

In addition to their high cost, the lack of information on the detailed composition of commercial serum-free media can also limit their attractiveness for specific biotechnological applications. Hence, there is a need for formulating serum-free media based on a rational assessment of the metabolic requirements of the lepidopteran cells during both the growth and production phases. The development of low-cost, in-house serum- and protein-free media can be based on empirical, statistical, or advanced genetic algorithm approaches (3).

A typical empirical path toward formulating an in-house, cost-effective medium starts with a basal medium requiring serum supplementation, such as IPL-41 (72), which has a simple composition and is easy to filter. Levels of glucose and of glutamine are fixed first, according to the requirements of the cell line to be cultured. Next, the lipid mixture emulsified in Pluronic F-68 (51). Alternatively, if the level of Pluronic F-68 is to be varied, the lipid mixture (1,000×, liquid; Sigma-Aldrich) can be mixed directly with Pluronic F-68 (10% solution; Sigma-Aldrich or Invitrogen). Serum and amino acids are replaced with hydrolysates (or, alternatively, with different defined peptides and vitamins). All components of the serum-free medium under development are mixed together, followed by filtration through a 0.2-μm filter. Each component is tested by individual titration to determine its optimal level of supplementation. For this, criteria of growth rate, viability, and growth extent (maximal or final cell density) are used for the growth phase, and growth extent, viability, and product virus or protein titer for the infection (production) phase. The last steps after serum substitution must be repeated as different components are added, in order to exclude combinations that have an adverse effect on growth and/or production.

A statistical approach involves factorial experimental design, which enables the rational formulation of media through screening of many ingredients at the same time while keeping the number of experimental runs required low and manageable (54). This approach can result in great time and cost savings. Ikonomou et al. (33) used a 2^{7-4} fractional factorial experiment for the screening of seven different hydrolysates and a subsequent 2^2 full factorial experiment for the optimization of the concentrations of the two selected hydrolysates (yeastolate and Primatone RL). This led to the formulation of the serum-free medium YPR, which performed equally well with commercial serum-free media costing 10- to 20-fold more (33). A variation of this serum-free medium that we have found to perform equally well and to be as cost-effective is given in Table 1. Designated YSD, this versatile serum-free medium is an evolution of YPR to which a low level of dextran has been added and in which Primatone RL, a meat hydrolysate, has been replaced by soy protein hydrolysate to ensure that the medium is free of animal-derived components.

In contrast to factorial experiments, which use only a small number of levels, genetic algorithms may enable the comprehensive optimization of insect cell media and cultivation conditions because they allow the screening of a wide variety of components and each of them at a wide range of concentrations. For instance, in the case of a fed-batch culture of *Helicoverpa zea* insect cells, a feed was developed from 11 different medium components, each used at a range of up to 31 concentrations. The feed was then optimized within four sets of 20 experiments (52).

15.2.3. Small-Scale Culture

Small bench-scale culture involves volumes ranging from a total of a few milliliters in wells of multiple-well plates or T-flasks for adherent culture to shake flasks or spinners up to 250 ml for suspension culture. The equipment required is limited to an incubator and an orbital shaker. All addition and removal of medium should be done in a biological safety cabinet or laminar-flow hood by using aseptic conditions and scrupulously sterilized equipment. The incubator is held at an optimal temperature of $27 \pm 0.5°C$ for Sf-9, Sf-21, and High Five cells, which, moreover, do not need CO_2, since most media are buffered.

Most types of insect cells tend to clump and lose viability when they are transferred abruptly to suspension from monolayer culture or to serum-free from serum-supplemented medium. It is recommended that one adapt the cells to suspension culture before attempting to grow them in serum-free medium. The increased sensitivity of the cells in serum-free culture requires reduced use or total elimination of antibiotics, avoidance of trypsinization for dislodging adherent cells from culture vessels, and consistent cryopreservation (freezing) and recovery (thawing) procedures. Cells propagated in serum-free medium are also more fragile than cells grown in serum-containing medium. Therefore, upon subculturing, insect cells grown in serum-free medium must be separated by low-speed centrifugation. Dislodging of adherent cells from culture vessels should be done with mild tapping or shaking or with a gentle stream of medium.

For routine monolayer culture in polystyrene T-flasks, dishes, or multiwell plates, cells are inoculated at 2 to 5 × 10^5 cells/ml (e.g., 0.8 to 2 × 10^6 cells/25-cm^2 T-flask). The recipient container is incubated at $27 \pm 0.5°C$ for 3 days, whereupon spent supernatant medium is removed from the monolayer and the culture is refed with fresh medium gently introduced to the side of the recipient container. Subculturing to the next passage is done when the monolayer reaches 90 to 100% confluency. Exponential growth can be maintained by splitting cells at a 1:5 dilution. After about 30 passages (2 to 3 months in culture) and after increased doubling times (>28 h) have been reached, insect cells tend to lose their viability and it is time to thaw new cell cryovials to initiate new monolayer cultures.

Cell viability should be monitored at each passage and should exceed 90%. Most insect cell lines do not adhere to monolayer substrata (e.g., T-flasks, dishes) as strongly as do anchorage-dependent mammalian cell lines. In the case of High Five cells, which have already been adapted to suspension in many laboratories, an occasional tendency toward adhering to substratum and clumping can be counteracted by addition of heparin or dextran sulfate (20a) in order to obtain single-cell suspensions. This is not necessary when an already suspension-adapted cell line is obtained from a commercial supplier (e.g., High Five cells from Invitrogen have been adapted to grow individually in suspension). Thus, monolayer cultures of all types of lepidopteran cells can be used to initiate suspension cultures.

A typical protocol of adaptation requires six to eight confluent 25-cm^2 T-flasks to start a 30-ml suspension culture (or six to ten 75-cm^2 T-flasks to start a 100-ml suspension). After cells are dissociated from the bottom of the flasks with a gentle stream of medium or by a sharp shaking motion, the suspensions from the various T-flasks are pooled into a shake flask or spinner and the cells counted. Fresh medium is added to the cell suspension to reach 2 to 5 × 10^5 cells/ml. The shake flasks are incubated on the shaker at 90 to 100 rpm and the spinners are operated at a stirring rate of 40 to 60 rpm in an incubator (or constant-temperature room) at $27 \pm 0.5°C$. Subculturing into fresh medium is done when the viable cell count reaches 1 to 2 × 10^6 cells/ml,

---

TABLE 1 Formulation of YSD, a generic all-purpose serum-free medium for insect cell culture

Component	Content	Supplier(s)
Basal IPL medium[a]		HyClone
Glucose supplement[b]	41 mM	
Glutamine supplement[c]	3.5 mM	
NaHCO₃	4 mM	
Pluronic® F-68	0.1% (vol/vol)	Sigma-Aldrich
Lipid mixture for insect cell culture	0.25% (vol/vol)	Sigma-Aldrich
Dextran 67 kDa	0.1% (wt/vol)	Sigma-Aldrich
Yeastolate	6 g/liter	Sigma-Aldrich
Soy hydrolysate	5 g/liter	DMV International Nutritionals (SE 50 MAF), Kerry Bio-Science (HySoy)

[a]Preparation in Milli-Q® water (Millipore), pH 6.2 to 6.3.
[b]Final glucose concentration, 55 mM.
[c]Final glutamine concentration, 10 mM. For certain applications a higher supplementation with free glutamine may be needed (final concentration, 15 to 17 mM).

and the stirring or shaking speed is increased by increments of 5 rpm as long as viability is maintained above 90%. This last step is repeated until a constant agitation speed of 90 rpm for spinner flasks or 130 rpm for shake flasks is reached and a consistent viability above 90% is maintained. Further subculturing from this point can be done at 0.3 to 1×10^6 cells/ml (i.e., every 1 to 3 days). If cell clumping is observed, the larger cell aggregates can be left to settle and only single cells and small clumps are selected from the upper third of the suspension for subsequent subculture. In a final step, a sufficient quantity of cells fully adapted to suspension culture should be frozen into cell banks to initiate future suspension cultures.

Erlenmeyer shake flask culture is by far the easiest way to grow S. *frugiperda* and *T. ni* insect cells in suspension. The most common configurations are 100-ml (working volume, 30 to 40 ml) or 250-ml (working volume, 80 to 100 ml) flasks. Loosening the cap approximately one quarter turn enables unhampered aeration to the cells, which can grow at their maximum specific growth rate. In a representative sequence of operations, the desired number of 250-ml shake flasks is inoculated with 80 to 100 ml of medium containing 3×10^5 viable cells/ml at a viability of at least 90%. The flasks are placed on an orbital shaker agitated at 125 to 140 rpm (depending on the cell line and the medium) and maintained in an environment of $27 \pm 0.5°C$. When the culture reaches 1 to 3×10^6 viable cells/ml, the shake flask contents are split to subculture the cells to about 3×10^5 viable cells/ml into similar-size shake flasks with fresh medium. The subculturing should be done when the cells are in mid-exponential growth and at maximum viability (i.e., above 90%). Every 3 weeks, cultures are centrifuged (100 to 200 $\times$ *g* for 5 min) and the cell pellet is resuspended into fresh medium to reduce cell debris and toxic by-product accumulation.

Spinner culture is less common than shake flask culture but better simulates the environment of a stirred bioreactor for eventual scale-up. A spinner flask with a vertical impeller is preferred over that with a hanging stir bar because it supplies better aeration. The liquid culture volume in a spinner should not exceed one-half the graduated volume of the vessel, to ensure proper aeration (e.g., a 500-ml spinner should contain no more than 250 ml of culture fluid). A minimum volume should also be respected, so the impeller stays submerged (e.g., 200 ml for a 500-ml spinner). In a spinner culture protocol for S. *frugiperda*, exponential-phase monolayer culture cells can be inoculated at a cell density of 0.3 to 1×10^6 viable cells/ml (four or five confluent 75-cm² T-flasks can supply this biomass for a 100-ml culture). Each spinner is incubated at $27 \pm 0.5°C$ with constant stirring at 75 to 90 rpm. The top of the impeller should be slightly above the medium surface to provide additional aeration, and the sidearm caps of the vessels are kept loose to equilibrate the headspace with the atmosphere. When the cells reach a density of 2 to 3×10^6 viable cells/ml (i.e., once every 1 to 3 days), a sufficient quantity of fresh medium is added to the spinner(s) to bring the cell density back to 10^6 cells/ml. Alternatively, new, sterile spinners containing fresh medium can be reseeded to about 3×10^5 cells/ml from a growing suspension culture. As in all growth protocols, insect cells should be monitored daily for density and viability.

Growing shake flask or spinner cultures of sufficient volume and good physiological characteristics (in mid-exponential phase and with >90% viability) can be used for maintenance culture or for inoculating a higher-scale vessel, e.g., a stirred Celligen™ bioreactor (New Brunswick Scientific, Edison, NJ).

15.2.4. Large-Scale Culture

The capacity of insect cells to grow in suspension and the wide availability of serum-free media and additives, including hydrodynamic shear-protective agents, together with the vast prior experience using mammalian cells in biotechnological applications, have opened the way for the development of reliable and efficient large-scale insect cell culture processes. The successful scale-up from spinners or shake flasks of a few hundred milliliters to a large bioreactor of several hundred liters is done in steps, passing through progressively larger vessels. Scale-up requires a good understanding and control of fundamental process parameters, such as adequate oxygen delivery, good mixing, and appropriate modes of feeding, i.e., parameters whose consequences are crucial for optimal cell growth and efficient baculovirus infection. For instance, the crucial phenomena of insect cell damage in an aerated and stirred bioreactor can be assessed and mitigated by understanding and controlling the complex interplay among the aeration rate through the orifices of a sparger, the agitation rate with a mechanical impeller, and the existence (and extent) of gas-liquid interface for air bubble bursting and its control by additives such as Pluronic F-68 (14).

Generally speaking, the culture conditions allowing the attainment of high cell densities in the growth phase are different from conditions compatible with high production during the infection phase. Although the pH, temperature, dissolved oxygen (DO), and medium conditions to sustain high cell densities are well established, the corresponding parameters favoring an optimal outcome of the infection process and hence high recombinant protein productivity are less well documented. In addition to these, productivity of infected cells depends upon several more factors such as virus construction (62), number of passages of the virus (45), cell density at infection (CDI), multiplicity of infection (MOI) (16), "culture age" (36), cell state (73), and state of medium depletion (37). An important phenomenon underlying the distinct optima between the conditions for growth and the conditions for efficient virus infection and productivity is the "cell density effect," i.e., the consistently observed drop in cell-specific productivity when CDI exceeds a certain value (36). Its consequences and practices aiming at overcoming it are discussed below.

15.2.4.1. Selection of Bioreactors

A number of bioreactor systems have been investigated and proven their merits in terms of robustness and scalability (2, 36, 60). The selection of a production-scale bioreactor system from the point of view of design (stirred-tank, airlift, packed-bed, or Wave) and from that of cultivation mode (batch, fed-batch, or continuous perfusion) should be guided by considerations of product titer, overall productivity, and product quality. Although in the 1990s there was a tendency to experiment with a wide variety of bioreactor designs including prototypes, the last few years have seen a reaffirmation of the selection of mainstream reactors, mostly of the stirred-tank type. For instance, a side-by-side comparison of the bioprocess behavior of Sf-9 cells in an 8-liter stirred-tank reactor and in a 6-liter airlift reactor showed similar growth characteristics and productivity of a recombinant kinase (65). The flexibility and wide availability of stirred-tank reactors make them more attractive than airlifts, despite the latter's recognized potential for adequate oxygen transfer rates with low and homogeneously distributed shear compared to conventional stirred vessels. At several hundred liters, airlift scale-up rules impose a constant height-to-circumference ratio that can result in very tall, customized vessels. In contrast, stirred-tank fermentors widely used in microbial applications can be readily retrofitted with low-shear impellers (e.g., marine propellers or hydrofoils) and adapted spargers for mammalian and insect cell cultivation.

Packed-bed reactors make use of a cell immobilization matrix (nonwoven fibers, foamy polymeric materials, and macroporous microcarriers) compactly loaded into a retention basket within the vessel (29). The main advantage of the fixed bed is its ability to retain the cells and perfuse the medium across the reactor under low shear stress. In this design, high cell densities can be reached in small bioreactors, potentially leading to high productivity if the cell density effect can be managed. However, the density of the immobilized cells in the bed is difficult to estimate, cell distribution inside the matrix can be heterogeneous, and problems of oxygen limitation with gas exchange and metabolic byproduct accumulation are possible. In addition, the packed cells must remain viable and productive for prolonged periods during the culture. Although insect cell immobilization studies have some promise, as seen with microcarriers in a stirred reactor (34), insect cell cultivation in packed-bed reactors (5, 17, 18, 74) seems potentially more attractive for constitutive rather than transient protein production.

The Wave bioreactor was conceived for the suspension culture of shear-sensitive animal cells without the need for protective additives like Pluronic F-68 and is part of the current trend toward single-use biomanufacturing systems (67). It consists of disposable, presterilized polyethylene bags (from 1 to 1,000 liters with working volumes from 100 ml to 500 liters) placed on a rocking platform. The bags are filled with medium and cells and the rest of the chamber is inflated with air (67). The rocking motion of the platform generates waves, which enable good oxygen transfer and mixing. A 20-liter Wave bioreactor was used in a side-by-side comparison with a stirred vessel for the production of a secreted adhesion molecule (intercellular adhesion molecule-1) by baculovirus-infected Sf-9 cells (71). Similar product yields were obtained in both the Wave bioreactor and the stirred-tank reactor, but operational costs were 40% lower for the disposable bioreactor. The Wave reactor is well adapted to the fed-batch mode of operation. On the other hand, despite generating a constantly renewed liquid-gas interface, oxygen may be limiting at high-density infections where oxygen uptake rate (OUR) increases transiently after infection (2). The introduction of a floating filter and a perfusion control system in the Wave bioreactor (56) has opened the way to its use in insect cell perfusion culture. In recent versions of this system it is possible to fit DO and pH probes and to connect sample, input, and exit lines to a variety of analyzers for online monitoring (41).

15.2.4.2. Batch, Fed-Batch, and Continuous Culture

Batch culture remains the most common method for large-scale IC-BEVS processing because of its inherent simplicity and flexibility in bioreactor equipment. In a typical protocol adapted from Elias et al. (27), the batch bioreactor is inoculated at a density of between 3 and 5×10^6 cells/ml by pooling cells from exponentially growing smaller-scale cultures (see section 15.2.3). The volume of the inoculum should represent approximately 10% of the working volume. Cells are grown to 2 to 3×10^6 cells/ml (a CDI that is considered safe for infection, i.e., is not linked to a cell density effect) and are then infected by adding the baculovirus stock. If a synchronous infection is desired in the interest of fast and abundant production, an MOI of 5 to 10 is appropriate. This ensures that a growth cessation due to the baculovirus vector's cytopathic effect on the cells will occur at about 24 h postinfection and the production of the recombinant protein will increase after that time. Based on the definition of MOI as the ratio of PFU per cell at the time of infection, the volume of baculovirus stock, V_b, is given by the equation

$$V_b = \frac{X \times V \times MOI \times 10^3}{P} \qquad (1)$$

where X is the viable cell density (cells/ml), V is the working bioreactor volume (liters), MOI is the target multiplicity of infection (PFU/cell), and P is the baculovirus stock titer. However, as a rule, the volume of virus stock added should not be much higher than 2% of the total culture volume.

A close monitoring of the growth and infection phases by a combination of offline and online techniques such as OUR (see section 15.4) will allow the determination of the best time for harvesting, which, depending on the protein, may be between 36 and 72 h postinfection.

Fed-batch cultivation is today perhaps the most robust bioreactor operation mode, and its success seems to be linked in large measure to its capacity to overcome the cell density effect, which depends on the CDI. This effect represents a consistent reduction of the specific productivity when the cell concentration at infection is above 3 to 4 $\times$ 10^6 Sf-9 cells/ml or above 5 to 6 $\times$ 10^6 High Five cells/ml (26). This is thought to be due to nutrient depletion in the medium during infection, although its understanding remains incomplete. For instance, it has been shown that the drop in productivity of Sf-9 cells occurs prior to nutrient depletion (23). Complete medium replenishment or selected feeding of nutrients counteracts the cell density effect (as reviewed in reference 36), lending support to nutrient depletion as at least one causative factor. Generally, infection is more rapid at high cell density and high MOI, but the temporary burst in metabolic activity may not be sustainable in the absence of fresh nutrients. Cell productivity is also affected by the physiological state of the cells (e.g., apoptotic cells are less productive than exponentially growing cells). Finally, the effect of the time of infection (TOI; also known as culture age), i.e., the duration of the culture from inoculation to infection, depends directly on the specific growth rate of cells, on the inoculation cell density, and on the target CDI. Hence, as stated above in the batch culture protocol, infections that aim to reach high recombinant protein production must be as synchronized as possible and applied when the cells are highly viable in exponential phase, and at that point, medium depletion must be compensated with appropriate nutrient feeding.

Due to the depletion of medium ingredients and the density effect, infection of simple batch cultures is not optimal for high protein production levels. An initial approach to counteract this problem has been the total or partial medium replacement at the time of infection. Medium replacement can be applied before or shortly after the infection and has been found to be efficient for recovering high specific production at high cell density (reviewed in reference 36). Moreover, the use of a cell retention (or centrifugation) system to separate the cells from the exhausted media could serve to concentrate the cells (37). Partial medium replacement is less costly and allows the retention of potential growth-promoting factors secreted by the cells (11, 12, 24, 36a). Complete medium replacement, originally proposed in the early 1990s (47), helps to obtain high production with Sf-9 cells. Today, however, this solution is considered neither economical nor practical for large-scale applications.

Fed-batch operation involves essentially the intermittent addition of nutrients to a batch culture. It enables increased cell density and productivity without medium depletion. It is easier and cheaper to implement than perfusion and avoids exposure of the cells to high shear stress in a cell retention device. However, it requires knowledge of physiological requirements of the cells. Moreover, in fed-batch mode, accumulation of by-products may be considerable. Nevertheless, the low sensitivity of insect cells to by-product accumulation, to osmolarity increases, and to pH variation has made them a better candidate for fed-batch processing than mammalian cells. Fed-batch cultivation of insect cells has been studied intensively in the last several years, especially in the context of increasing both the cell density and the production yield using serum-free media and a suite of feeding strategies. Bédard et al. (6) demonstrated that Sf cells needed a single-pulse addition of yeastolate and amino acid mixture as supplement. Incidentally, the use of a richer supplement consisting of glucose, tyrosine, vitamins, iron, and a trace metal solution led to very high densities of Sf-9 cells (30×10^6 cells/ml) (7). In a report from the same group, (semi)continuous addition was better than single-pulse addition for achieving high cell densities of Sf-9 cells (26). The highest cell density reported with lepidopteran cells in fed-batch culture was 52×10^6 cells/ml using Sf-9 cells, but the cell density effect resulted in unsuccessful infection at such a cell density (26). The highest reported Sf-9 CDIs sustaining high production were 11.5×10^6 cells/ml (7), 12×10^6 cells/ml (15), and 17×10^6 cells/ml (26). There is less published information on the fed-batch cultivation of *T. ni* cells. Infection of High Five cells at a density of 3.8×10^6 cells/ml attained in serum-supplemented fed-batch culture allowed a modest production of β-galactosidase (75). In another report, a serum-free fed-batch strategy was developed based on culture pH as an indicator of a metabolic switch linked to the requirements of High Five cells for glucose and glutamine (70).

A protocol adapted from the developers of high-cell-density culture using fed-batch methodology (27) involves a start-up phase of batch cultivation of Sf-9 cells followed by additions of concentrated nutrient solutions together with close on-line monitoring and control of the integral process. In this sequence, the baculovirus infection is done routinely at cell densities of 10×10^6 cells/ml. The initial batch is inoculated at a density of 3 to 5×10^6 cells/ml with an exponentially growing culture of Sf-9 cells. The nutrient concentrate, which includes glucose, amino acids, yeastolate, lipids, vitamins, and trace metals (6, 7, 26), is added at 10% (vol/vol) to the culture volume at 24-h intervals and the working volume of the culture is maintained constant by removing an equivalent volume of mixed liquor between two consecutive feeds of nutrient concentrate. Finally, the culture is infected when the cell density reaches 10×10^6 cells/ml. As the cell density is progressively increasing, it is important to ensure adequate oxygenation of the culture by using an appropriate airflow rate or, if need be, an O_2-enriched gas mixture. The efficiency of oxygen supply can be checked by monitoring the DO concentration, which should not be allowed to drop below 20% air saturation. Both oxygen supply and nutrient adequacy should be checked throughout both the growth and the infection phases. As suggested in the batch culture protocol above, an MOI of 5 to 10 is also recommended here for synchronous infection. However, if a lower MOI is chosen based on small-scale trials because of better and prolonged product formation, the asynchronous infection will enable significant postinfection growth that will contribute to oxygen and nutrient consumption. Therefore, these substrates should be carefully monitored, and if need be, the feeding strategy should be adapted in a manner ensuring that there is no nutritional limitation during the infection phase, since this may compromise the final production level.

Continuous culture involving the steady-state feed of sterile medium and replacement of culture liquor without insect cell retention has not been used much for production of baculovirus or recombinant protein because the continuous withdrawal of the culture from the continuous stirred-tank reactor (CSTR) causes the dilution of cells and product (25). Moreover, the two distinct phases of the production process (growth and infection) are not easily implemented in a continuous mode in the same reactor due to the transient, lytic nature of BEVS. Thus, continuous operation in a reactor cascade has been proposed, consisting of two (or more) CSTRs in series, where the first is used for cell growth and the next for infection (77). A key limitation in this configuration is the accumulation of defective virus (69).

In contrast to classical CSTR operation, continuous perfusion culture in a reactor equipped with a cell retention system is better adapted to insect cell propagation because, just as in the case of fed-batch operation, it allows increase in the cell density (16). Among the many cell retention systems available, perfusion culture of insect cells makes use of membrane-based modules (internal or external). The lepidopteran cells cultivated in perfusion are reported to attain very high cell densities and production levels. For example, when cultivated in perfusion, Sf-21 cells were found to reach 55×10^6 cells/ml (21), while Sf-9 cells attained 30×10^6 cells/ml (13), and about 500 mg/ml of a recombinant protein was obtained with High Five cells (16).

15.3. BACULOVIRUS INFECTION

15.3.1. Recombinant Baculovirus Vector Generation

Recombinant AcMNPV is generated by cotransfecting into insect cells the linearized AcMNPV genome with a transfer plasmid carrying the gene of interest in an expression cassette controlled by a late viral promoter such as that of the polyhedrin (*polh*) gene or of the *p10* gene. Following recombination of the expression cassette into the AcMNPV genome, the recombinant virus is amplified in an insect cell host and used for productive infections (62). The replacement of either of the above nonessential genes (*polh, p10*) with the foreign gene of choice and the very strong promoter control allow the production of recombinant protein at very high yield (53, 68). Owing to continuous improvements in recent years to the flexibility and efficiency of baculovirus vector-based expression, this system is now adapted for simultaneous expression of several foreign genes using a single recombinant baculovirus (76). The recombinant viruses produced using these transfer vectors have facilitated the study of complex protein assembly, by coexpressing protein subunits in the same host cells. Coinfection with several recombinant viruses is also possible to express different genes (44, 64). The incorporation of the recombinant expression cassette into the AcMNPV genome by site- and orientation-specific Tn7-mediated transposition has improved and simplified the selection procedure for amplified recombinant baculovirus (4). In addition, recombinant protein yield is significantly enhanced by the deletion of the genes coding for the baculoviral enzymes chitinase (*chiA*) and v-cathepsin (V-CATH), which are involved in insect cell host lysis when the cytopathic effect takes hold (40). The above improvements have been combined in a baculovirus expression system named MultiBac, which is used mainly for the expression of heterologous multiprotein complexes (8, 10). Many of the components and reagents needed for the generation of recombinant baculovirus vectors are now commercially available in the form of kits (e.g., from Invitrogen).

15.3.2. Vector Amplification and Viral Titration

Depending on the amount of viral stock required, the baculovirus vector can be amplified in monolayer cultures, shake flasks, or bioreactors, typically using a routine protocol of small-scale culture (see section 15.2.3). Although S. frugiperda and T. ni cells, in addition to several other lepidopteran cell lines, are readily susceptible to infection by wild-type and recombinant AcMNPV baculovirus, the general tendency in practice is to use Sf-9 or Sf-21 cell cultures for routine production of baculovirus stock. It has been observed repeatedly that despite its capacity for overexpression of recombinant proteins (especially secreted ones) T. ni is less prolific than S. frugiperda in the production of extracellular virus. A low MOI of about 0.1 should be used for viral stock production to ensure the purity of the stock and prevent mutant accumulation. Usually a series of amplification steps are carried out until the viral stocks reach 1×10^8 or 1×10^9 PFU/ml. Adding 5 to 10% serum can enhance the stability of baculoviruses stored at 4°C (used as a working viral stock), but cryopreservation of the viral stock at −85°C is also possible without loss of stability for several months (master viral stocks).

An accurate baculovirus titer (in terms of PFU/ml) is essential to determine the strength of the stock for future infections. This effective concentration of infective particles can be obtained upon infection of insect cells in standardized laboratory assays, usually in culture dishes and multiwell plates. The most commonly used methodologies for baculovirus titration are the plaque assay and end-point dilution (57). The plaque assay is carried out on tissue culture-treated dishes and involves the infection of a monolayer of insect cells with a low concentration of virus, followed by an agarose medium overlay. When the infection is complete, a virus plaque (i.e., an isolated region of infected cells) is identified by visual inspection, and an agarose plug is removed at the corresponding plaque location for generating viral stocks. With the end-point dilution method, the culture medium is serially diluted to the point at which any infection event is expected to result from a single infectious viral particle. The direct visual identification of positive wells (i.e., wells containing cells infected by the baculovirus) is difficult for recombinant baculovirus vectors lacking the polyhedrin gene. Because of the exacting and time-consuming nature of both the end-point dilution and plaque assay methods, a number of alternative techniques for determining the viral titer have been proposed. Thermo Scientific has commercialized a kit containing the necessary ingredients for plaque assay, shortening the time of preparation (HyQ BEVS plaKit). Clontech (Mountain View, CA), Expression Systems, and Merck Biosciences (Darmstadt, Germany) offer quick and efficient enzyme-linked immunosorbent assay kits for titrating the virus based on the detection of the gp64 viral capsid protein with a monoclonal antibody. Finally, viral titration can also be accomplished through the detection of green fluorescent protein, whose gene is incorporated in the virus genome and coexpressed together with the recombinant protein of choice (19).

15.3.3. Infection of Insect Cell Cultures

Important considerations involved in the process of infecting an insect cell culture with a baculovirus are the cell line to use, the timing of the infection, the cell density at which the cells should be infected (CDI), the MOI to apply, and the time at which to harvest the product.

15.3.3.1. Cell Line Selection

The choice of insect cell line for use in the IC-BEVS biomanufacturing platform depends primarily on the product to be expressed. Lepidopteran insect cells are able to bring about eukaryotic posttranslational modification of heterologous proteins, such as N and O glycosylation, phosphorylation, fatty acid acylation, sialylation, α-amidation, N-terminal amidation, carboxymethylation, and isoprenylation (9, 32, 38, 39, 59). Moreover, baculovirus-infected insect cells are able to effect folding of heterologous proteins and peptide cleavage, although these recombinant proteins exhibit differences from their authentic counterparts. As a rule, Sf-21 and Sf-9 are the cell lines of choice for overproduction of whole viruses (e.g., as biopesticides) and also for the production of intracellular or periplasmic proteins, whereas High Five has been repeatedly demonstrated to overexpress many glycosylated and secreted proteins. Specifically, High Five cells are supposed to carry out better posttranslational modifications than Sf-9 cells, with a higher percentage of complex glycoforms and even some sialylation (38, 39).

15.3.3.2. TOI, CDI, and MOI

The fraction of the population of insect cells initially infected by the virus is a crucial factor affecting cell productivity. This fraction of the cell population is a function of the MOI, which corresponds to the number of infectious particles (PFU) divided by the number of cells, and follows a Poisson distribution given by equation 2. In this relationship, $p(r)$ is the probability of a cell being infected by r virus particles and MOI is expressed in PFU/cell (57).

$$p(r) = \frac{\text{MOI}^r \cdot \exp(-\text{MOI})}{r!} \qquad (2)$$

The probability $p(r)$ is equivalent to the fraction of cells infected with r baculovirus particles at a given MOI. Since, according to equation 2, the fraction of uninfected cells (i.e., cells infected with $r = 0$ virus particles) at the given MOI would be $p(0) = \exp(-\text{MOI})$, then it follows that the fraction f of cells infected with at least 1 virus particle would be $f = 1 - p(0)$, i.e.:

$$f = 1 - \exp(-\text{MOI}) \qquad (3)$$

Various reports have examined the combined effect of TOI and MOI on productivity. Infection using a high MOI (>1) implies that the infection is synchronized (i.e., all cells are infected at the same time), while with an MOI of <1 all cells are not simultaneously infected and several cycles of infection are required to infect the entire cell population. High MOI is thus predicted to lead to a short production period with high productivity (i.e., product concentration per unit time of process duration). On the other hand, in larger bioreactors infection at a high MOI would also imply a need for large volumes of virus stock (see equation 1). As seen previously (section 15.3.2), a low MOI is used for virus amplification and for applications where viral stocks are low (57). In addition, infections at low MOI and low CDI can be successful and cost-effective in cell lines like Sf-9 that are very prolific in extracellular virus production or when it has been empirically found that the accumulation of the product is not adversely affected by a long infection phase. However, as a rule, asynchronous infections increase the time of harvest, lead to the production of defective viral particles, and expose the recombinant protein to proteases (2). Nonetheless, different opinions exist on this point. Some authors consider that an MOI between 1 and 20 leads to essentially similar levels of productivity (51), while others contend that an MOI close to 1 is better (2). Still others recommend an optimization of both the TOI and MOI in an effort to ensure that all cells are infected by the baculovirus and that there are sufficient concentrations of nutrients left for product (virus or recombinant protein) formation (73). Although previous attempts to predict the effect of MOI (and TOI) on recombinant protein production based on probabilistic mathematical simulations (46) have been borne out experimentally (73), the phenomena are complex and MOI optimization is not straightforward. In conclusion, MOI optimization depends on the virus, the cell line, the media/nutritional conditions, and the physiological state of the insect cells at the time of infection.

15.3.3.3. Time of Harvesting
The time of harvesting is an important consideration in the management of any biomanufacturing process involving IC-BEVS. The recombinant protein must be harvested before potential degradation by proteases that are secreted by the insect cells, which may occur during lysis and disintegration of the cells (35, 48). This is especially critical in the case of a secreted protein product. Therefore, as a general guideline, a protein product should be harvested when the cell viability falls below 80%.

15.4. QUANTITATIVE ASSESSMENT OF GROWTH AND INFECTION PHASE, AND PROCESS MONITORING
Insect cell growth is evaluated in terms of population doubling time, specific growth rate (μ), growth extent (maximal cell density or CDI), length of the stationary/decline phase, and viability. Infection at a given MOI is assessed by the time course of the postinfection cell density and viability and, most importantly, by the evaluation of the viral or recombinant protein titer. CDI, MOI, and time of harvesting must be optimized on an individual basis for different recombinant products, even though some general rules can be deduced (see section 15.3.3.2).

Cell growth and infection in insect cell culture are commonly monitored by counting total and viable cell numbers with a hemacytometer and using trypan blue staining/exclusion. A more accurate but more time-consuming alternative is the 3-(4,5-dimethyl-thiazol-2-yl)-2,5-diphenyltetrazolium bromide cell proliferation assay, which can be applied using a commercial kit (e.g., from the American Type Culture Collection [Manassas, VA]). Other less labor-intensive or invasive techniques are also possible for monitoring cell growth and infection, both offline and online (2). Offline measurements of growth make use of optical density or of a Coulter particle counter. Online monitoring of growth through respiratory activity is based on determinations of cellular OUR or CO_2 production rate using mass spectrometry or individual analyzers (O_2 paramagnetic analyzer or CO_2 infrared analyzer). Specific OUR can also be determined using a DO electrode. With proper calibration it is possible to monitor the infection phase via OUR because of the transient increase in oxygen consumption at the onset of infection. Epifluorescence microscopy of live and dead cells using fluorescence stains (Molecular Probes, Invitrogen) is

another technique for tracking the infection phase. Finally, protein production can be followed by a variety of assays, including fluorometric staining, Western blots, enzyme-linked immunosorbent assay, and enzyme activity.

15.5. PRODUCT RECOVERY
The recovery of the recombinant protein or viral product makes use of various well-established downstream processes that are commonly used in many biomanufacturing platforms (microbial fermentation, mammalian cell culture, etc.) but can vary widely as a function of the nature of the product (intracellular or extracellular protein, molecular size, state of ionization, hydrophobicity, etc.). The separation of the insect cells may be a prerequisite in the case of intracellular, periplasmic, or membrane-associated recombinant proteins. This cell recovery step and all associated operations should be carried out at 4°C. Insect cells can be recovered by centrifugation at low speeds (500 to 1,000 × g for 5 to 10 min) or by tangential flow membrane filtration, e.g., using a Pellicon tangential filtration unit (Millipore Corp., Billerica, MA) or a Filtron membrane unit (Pall Corp., East Hills, NY). An alternative possibility is cross-flow microfiltration using hollow-fiber cartridges (GE Healthcare, Piscataway, NJ). The release of the intracellular or cell-associated protein may be aided by a prior cell lysis step (mechanical milling or using a detergent mix in buffer). In the case of a protein product secreted in the culture supernatant, the cells can be centrifuged at 1,500 × g for 10 min. If further processing is to be done several days after harvesting, the cell paste and supernatant must be stored at −80°C until the next phase of separation and purification. It is recommended that large volumes of supernatant be concentrated by ultrafiltration through an appropriate membrane cutoff before being frozen for further processing.

15.6. SUMMARY
The practice of insect cell culture is becoming more and more widespread as an essential component of the IC-BEVS, perhaps one of the most successful and far-reaching expression systems available today for the production of recombinant proteins and viral particles. In addition to its established applications in the production of ecological biopesticides and of recombinant proteins as diagnostics and veterinary vaccines, this technology platform is attracting increasing attention because of emerging applications in human health, including new-generation vaccines against human papillomavirus and influenza virus. Furthermore, in the era of proteomics, the IC-BEVS is an obvious choice for the ample and rapid expression of hitherto unknown or poorly understood proteins of interest in fundamental biological research or with potential applications in drug discovery, sustainable agriculture practice, etc. Therefore, the understanding of insect cell physiology in vitro and the constant improvement of insect cell culture techniques using some of the tools and concepts outlined above will ensure the industrial maturity of the IC-BEVS through further development of efficient, scalable, and safe production processes.

REFERENCES
1. **Agathos, S. N.** 1991. Production scale insect cell culture. *Biotechnol. Adv.* **9:**51–68.
2. **Agathos, S. N.** 1996. Insect cell bioreactors. *Cytotechnology* **20:**173–189.

3. **Agathos, S. N.** 2007. Development of serum-free media for lepidopteran insect cell lines, p. 155–185. *In* D. W. Murhammer (ed.), *Baculovirus Expression Protocols*, 2nd ed. Humana Press, Totowa, NJ.

4. **Airenne, K. J., E. Peltomaa, V. P. Hytonen, O. H. Laitinen, and S. Ylä-Herttuala.** 2003. Improved generation of recombinant baculovirus genomes in *Escherichia coli. Nucleic Acids Res.* **31:**e101.

5. **Archambault, J., J. Robert, and L. Tom.** 1994. Culture of immobilized insect cells. *Bioprocess Eng.* **11:**189–197.

6. **Bédard, C., A. Kamen, R. Tom, and B. Massie.** 1994. Maximization of recombinant protein yield in the insect cell/baculovirus system by one-time addition of nutrients to high-density batch cultures. *Cytotechnology* **15:**129–138.

7. **Bédard, C., S. Perret, and A. A. Kamen.** 1997. Fed-batch culture of Sf-9 cells supports 3×10^7 cells per ml and improves baculovirus-expressed recombinant protein yields. *Biotechnol. Lett.* **19:**629–632.

8. **Berger, I., D. J. Fitzgerald, and T. J. Richmond.** 2004. Baculovirus expression system for heterologous multiprotein complexes. *Nat. Biotechnol.* **22:**1583–1587.

9. **Betenbaugh, M. J., E. Ailor, E. Whiteley, P. Hinderliter, and T.-A. Hsu.** 1996. Chaperone and foldase coexpression in the baculovirus-insect cell expression system. *Cytotechnology* **20:**149–156.

10. **Bienossek, C., T. J. Richmond, and I. Berger.** 2008. Multi-Bac: multigene baculovirus-based eukaryotic protein complex production. *Curr. Protoc. Protein Sci.* **51:**5.20.1–5.20.26.

11. **Calles, K., I. Svensson, E. Lindskog, and L. Häggström.** 2006. Effects of conditioned medium factors and passage number on Sf9 cell physiology and productivity. *Biotechnol. Prog.* **22:**394–400.

12. **Calles, K., U. Eriksson, and L. Häggström.** 2006. Effect of conditioned medium factors on productivity and cell physiology in *Trichoplusia ni* insect cell cultures. *Biotechnol. Prog.* **22:**653–659.

13. **Cavegn, C., H. D. Blasey, M. A. Payton, B. Allet, J. Li, and A. R. Bernard.** 1992. Expression of recombinant protein in high density insect cell cultures, p. 569–578. *In* R. E. Spier, J. B. Griffiths, and C. MacDonald (ed.), *Animal Cell Technology: Developments, Processes and Products.* Butterworth-Heinemann, Oxford, United Kingdom.

14. **Chalmers, J. J.** 1996. Shear sensitivity of insect cells. *Cytotechnology* **20:**163–171.

15. **Chan, L. C. L., P. F. Greenfield, and S. Reid.** 1998. Optimising fed-batch production of recombinant proteins using the baculovirus expression vector system. *Biotechnol. Bioeng.* **59:**178–188.

16. **Chico, E., and V. Jäger.** 2000. Perfusion culture of baculovirus-infected BTI-Tn-5B1-4 insect cells: a method to restore cell-specific β-trace glycoprotein productivity at high cell density. *Biotechnol. Bioeng.* **70:**574–586.

17. **Chiou, T.-W., Y.-C. Wang, and H.-S. Liu.** 1998. Utilizing the macroporous media for insect cell/baculovirus expression. Part 1: the entrapment kinetics of insect cells within porous packed bed. *Bioprocess Eng.* **18:**45–53.

18. **Chiou, T.-W., Y.-C. Wang, and H.-S. Liu.** 1998. Utilizing the macroporous media for insect cell/baculovirus expression. Part 2: the production of human interleukin-5 in polyurethane foam and cellulose foam packed bed bioreactors. *Bioprocess Eng.* **18:**91–100.

19. **Dalal, N. G., W. E. Bentley, and H. J. Cha.** 2005. Facile monitoring of baculovirus infection for foreign protein expression under very late polyhedron promoter using green fluorescent protein reporter under early-to-late promoter. *Biochem. Eng. J.* **24:**27–30.

20. **Davis, T. R., M. L. Shuler, R. R. Granados, and H. A. Wood.** 1993. Comparison of oligosaccharide processing among various insect cell lines expressing a secreted glycoprotein. *In Vitro Cell. Dev. Biol. Anim.* **29A:**842–846.

20a. **Dee, K. U., M. L. Shuler, and H. A. Wood.** 1997. Inducing single-cell suspension of BRI-TnB1-4 insect cells. 1. The use of sulfated polyanions to prevent cell aggregation and enhance recombinant protein production. *Biotechnol. Bioeng.* **54:**191–205.

21. **Deutschmann, S. M., and V. Jäger.** 1994. Optimization of the growth conditions of Sf21 insect cells for high-density perfusion culture in stirred-tank bioreactors. *Enzyme Microb. Technol.* **16:**506–512.

22. **Donaldson, M. S., and M. L. Shuler.** 1998. Effects of long-term passaging of BTI-Tn5B1-4 insect cells on growth and recombinant protein production. *Biotechnol. Prog.* **14:**543–547.

23. **Doverskog, M., J. Ljunggren, L. Öhman, and L. Häggström.** 1997. Physiology of cultured animal cells. *J. Biotechnol.* **59:**103–115.

24. **Doverskog, M., E. Bertram, J. Ljunggren, L. Öhman, R. Sennerstam, and L. Häggström.** 2000. Cell cycle progression in serum-free cultures of Sf9 insect cells: modulation by conditioned medium factors and implication for proliferation and productivity. *Biotechnol. Prog.* **16:**837–846.

25. **Drews, M., T. Paalme, and R. Vilu.** 1995. The growth and nutrient utilization of the insect cell line *Spodoptera frugiperda* Sf9 in batch and continuous culture. *J. Biotechnol.* **40:**187–198.

26. **Elias, C. B., A. Zeiser, C. Bédard, and A. A. Kamen.** 2000. Enhanced growth of Sf-9 cells to a maximum density of 5.2×10^7 cells per mL and production of β-galactosidase at high cell density by fed batch culture. *Biotechnol. Bioeng.* **68:**381–388.

27. **Elias, C. B., B. Jardin, and A. Kamen.** 2007. Recombinant protein production in large-scale agitated bioreactors using the baculovirus expression vector system, p. 225–245. *In* D. W. Murhammer (ed.), *Baculovirus Expression Protocols*, 2nd ed. Humana Press, Totowa, NJ.

28. **Gerbal, M., P. Fournier, P. Barry, M. Mariller, F. Odier, G. Devauchelle, and M. Duonor-Cerutti.** 2000. Adaptation of an insect cell line of *Spodoptera frugiperda* to growth at 37°C: characterization of an endodiploid clone. *In Vitro Cell. Dev. Biol. Anim.* **36:**117–124.

29. **Goosen, M. F. A.** 1993. Insect cell immobilization, p. 69–104. *In* M. F. A. Goosen, A. J. Daugulis, and P. Faulkner (ed.), *Insect Cell Culture Engineering.* Marcel Dekker, New York, NY.

30. **Granados, R. R.** April, 1994. *Trichoplusia ni* cell line which supports replication of baculoviruses. U.S. patent 5,300,435. Boyce Thompson Institute for Plant Research, Ithaca, NY.

31. **Hink, W. F.** 1970. Established insect cell line from the cabbage looper, *Trichoplusia ni. Nature* **226:**466–467.

32. **Hollister, J. R., E. Grabenhorst, M. Nimtz, H. O. Conradt, and D. L. Jarvis.** 2002. Engineering the protein N-glycosylation pathway in insect cells for production of biantennary, complex N-glycans. *Biochemistry* **41:**15093–15104.

33. **Ikonomou, L., G. Bastin, Y.-J. Schneider, and S. N. Agathos.** 2001. Design of an efficient medium for insect cell culture and recombinant protein production. *In Vitro Cell. Dev. Biol. Anim.* **37:**549–559.

34. **Ikonomou, L., J.-C. Drugmand, G. Bastin, Y.-J. Schneider, and S. N. Agathos.** 2002. Microcarrier culture of lepidopteran cell lines: implications for growth and recombinant protein production. *Biotechnol. Prog.* **18:**1345–1355.

35. **Ikonomou, L., C. Peeters-Joris, Y.-J. Schneider, and S. N. Agathos.** 2002. Supernatant proteolytic activities of High-Five insect cells grown in serum-free culture. *Biotechnol. Lett.* **24:**965–969.

36. Ikonomou, L., Y.-J. Schneider, and S. N. Agathos. 2003. Insect cell culture for industrial production of recombinant proteins. *Appl. Microbiol. Biotechnol.* **62:**1–20.

36a. Ikonomou, L., G. Bastin, Y.-J. Schneider, and S. N. Agathos. 2004. Effect of partial medium replacement on cell growth and protein production for the High Five™ insect cell line. *Cytotechnology* **44:**67–76.

37. Jäger, V. 1996. Perfusion bioreactors for the production of recombinant proteins in insect cells. *Cytotechnology* **20:**191–198.

38. Joosten, C. E., and M. L. Shuler. 2003. Production of a sialylated N-linked glycoprotein in insect cells: role of glycosidase and effect of harvest time on glycosylation. *Biotechnol. Prog.* **19:**193–201.

39. Joshi, L., T. R. Davis, T. S. Mattu, P. M. Rudd, R. A. Dwek, M. L. Shuler, and H. A. Wood. 2000. Influence of baculovirus-host cell interactions on complex glycosylation of a recombinant human protein. *Biotechnol. Prog.* **16:**650–656.

40. Kaba, S. A., A. M. Salcedo, P. O. Wafula, J. M. Vlak, and M. M. van Oers. 2004. Development of a chitinase and v-cathepsin negative bacmid for improved integrity of secreted recombinant proteins. *J. Virol. Methods* **122:**113–118.

41. Kadwell, S. H., and P. I. Hardwicke. 2007. Production of baculovirus-expressed recombinant proteins in Wave bioreactors, p. 247–266. *In* D. W. Murhammer (ed.), *Baculovirus Expression Protocols*, 2nd ed. Humana Press, Totowa, NJ.

42. Kioukia, N., A. W. Nienow, A. N. Emery, and M. Al-Rubeai. 1995. Physiological and environmental factors affecting the growth of insect cells and infection with baculovirus. *J. Biotechnol.* **38:**243–251.

43. Kool, M., J.-W. Voncken, F. L. van Lier, J. Tramper, and J. M. Vlak. 1991. Detection and analysis of *Autographa californica* nuclear polyhedrosis virus mutants with defective interfering properties. *Virology* **183:**739–746.

44. Kost, T. A., J. P. Condreay, and D. L. Jarvis. 2005. Baculovirus as versatile vectors for protein expression in insect and mammalian cells. *Nat. Biotechnol.* **23:**567–575.

45. Krell, P. J. 1996. Passage effect of virus infection in insect cells. *Cytotechnology* **20:**125–137.

46. Licari, P., and J. E. Bailey. 1992. Modelling the population dynamics of baculovirus-infected insect cells: optimizing infection strategies for enhanced recombinant protein yields. *Biotechnol. Bioeng.* **39:**432–441.

47. Lindsay, D. A., and M. J. Betenbaugh. 1992. Quantification of cell culture factors affecting recombinant protein yield in baculovirus-infected insect cells. *Biotechnol. Bioeng.* **39:**614–618.

48. Lindskog, E., I. Svensson, and L. Häggström. 2006. A homologue of cathepsin L identified in conditioned medium from Sf9 insect cells. *Appl. Microbiol. Biotechnol.* **71:**444–449.

49. Lloyd, D. R., P. Holmes, L. P. Jackson, A. N. Emery, and M. Al-Rubeai. 2000. Relationship between cell size, cell cycle and specific recombinant protein productivity. *Cytotechnology* **34:**59–70.

50. Lynn, D. E. 2001. Novel techniques to establish new insect cell lines. *In Vitro Cell. Dev. Biol. Anim.* **37:**319–321.

51. Maiorella, B., D. Inlow, A. Shauger, and D. Harano. 1988. Large-scale insect cell-culture for recombinant protein production. *Bio/Technology* **6:**1406–1410.

52. Marteijn, R. C. L., O. Jurrius, J. Dhont, C. D. de Gooijer, J. Tramper, and D. E. Martens. 2003. Optimization of a feed medium for fed-batch culture of insect cells using a genetic algorithm. *Biotechnol. Bioeng.* **81:**269–278.

53. Merrington, C. L., J. E. Bailey, and R. D. Possee. 1997. Protocol: manipulation of baculovirus vectors. *Mol. Biotechnol.* **8:**283–297.

54. Montgomery, D. C., and G. C. Runger. 2006. *Applied Statistics and Probability for Engineers*, 4th ed. John Wiley & Sons, Inc., New York, NY.

55. Murhammer, D. W. (ed.). 2007. *Baculovirus and Insect Cell Expression Protocols*, 2nd ed. Humana Press, Totowa, NJ.

56. Ohashi, R., V. Singh, and J.-F. P. Hamel. 2001. Perfusion cell culture in disposable bioreactors, p. 403–409. *In* E. Lindner-Olsson, N. Chatzissavidou, and E. Lüllau (ed.), *Animal Cell Technology: from Target to Market.* Kluwer Academic, Dordrecht, The Netherlands.

57. O'Reilly, D. R., L. K. Miller, and V. A. Luckow. 1992. *Baculovirus Expression Vectors: a Laboratory Manual.* W. H. Freeman & Co., New York, NY.

58. Palomares, L. A., J. C. Pedroza, and O. T. Ramirez. 2001. Cell size as a tool to predict the production of recombinant protein by the insect-cell baculovirus expression system. *Biotechnol. Lett.* **23:**359–364.

59. Palomares, L. A., C. E. Joosten, P. R. Hughes, R. R. Granados, and M. L. Shuler. 2003. Novel insect cell line capable of complex N-glycosylation and sialylation of recombinant proteins. *Biotechnol. Prog.* **19:**185–192.

60. Palomares, L. A., S. Estrada, and O. T. Ramirez. 2006. Principles and applications of the insect cell-baculovirus expression vector system, p. 627–692. *In* S. S. Ozturk and W.-S. Hu (ed.), *Cell Culture Technology for Pharmaceutical and Cell-Based Therapies.* Taylor & Francis, New York, NY.

61. Pijlman, G. P., E. van den Born, D. E. Martens, and J. M. Vlak. 2001. *Autographa californica* baculoviruses with large genomic deletions are rapidly generated in infected insect cells. *Virology* **283:**132–138.

62. Possee, R. D. 1997. Baculoviruses as expression vectors. *Curr. Opin. Biotechnol.* **8:**569–572.

63. Rhiel, M., C. M. Mitchell-Logean, and D. W. Murhammer. 1997. Comparison of *Trichoplusia ni* BTI-Tn-5B1-4 (High Five™) and *Spodoptera frugiperda* Sf-9 insect cell line metabolism in suspension cultures. *Biotechnol. Bioeng.* **55:**909–920.

64. Rhodes, D. J. 1996. Economics of baculovirus-insect cell production systems. *Cytotechnology* **20:**291–297.

65. Rice, J. W., N. B. Rankl, T. M. Gurganus, C. M. Marr, J. B. Barna, M. M. Walters, and D. J. Burns. 1993. A comparison of large-scale Sf9 insect cell growth and protein production: stirred vessel vs. airlift. *BioTechniques* **15:**1052.

66. Saarinen, M. A., K. A. Troutner, S. G. Gladden, C. M. Mitchell-Logean, and D. W. Murhammer. 1999. Recombinant protein synthesis in *Trichoplusia ni* BTI-TN-5B1-4 insect cell aggregates. *Biotechnol. Bioeng.* **63:**612–617.

67. Singh, V. 1999. Disposable bioreactor for cell culture using wave-induced agitation. *Cytotechnology* **30:**149–158.

68. Smith, G. E., M. D. Summers, and M. J. Fraser. 1983. Production of human interferon in insect cells infected with a baculovirus expression vector. *Mol. Cell. Biol.* **3:**2156–2165.

69. van Lier, F. L., J. P. van den Hombergh, C. D. de Gooijer, M. M. den Boer, J. M. Vlak, and J. Tramper. 1996. Long-term semi-continuous production of recombinant baculovirus protein in a repeated (fed-)batch two-stage reactor system. *Enzyme Microb. Technol.* **18:**460–466.

70. Wang, M.-Y., and S.-R. Doong. 2000. A pH-based fed-batch process for the production of a chimeric recombinant infectious bursal disease virus (IBDV) structural protein (rVP2H) in insect cells. *Process Biochem.* **35:**877–884.

71. Weber, W., E. Weber, S. Geisse, and K. Memmert. 2001. Catching the Wave: the BEVS and the Biowave, p. 335–337. *In* E. Lindner-Olsson, N. Chatzissavidou,

and E. Lüllau (ed.), *Animal Cell Technology: from Target to Market*. Kluwer Academic, Dordrecht, The Netherlands.

72. **Weiss, S.A., G. C. Smith, S. S. Kalter, and J. L. Vaughn.** 1981. Improved method for the production of insect cell cultures in large volume. *In Vitro* **17:**495–502.

73. **Wong, K. T. K., C. H. Peter, P. F. Greenfield, S. Reid, and L. K. Nielsen.** 1996. Low multiplicity infection of insect cells with a recombinant baculovirus: the cell yield concept. *Biotechnol. Bioeng.* **49:**659–666.

74. **Wu, J.-Y., and M. F. A. Goosen.** 1996. Immobilization of insect cells. *Cytotechnology* **20:**199–208.

75. **Wu, J., Q. Ruan, and H. Y. P. Lam.** 1998. Evaluation of spent medium recycle and nutrient feeding strategies for recombinant protein production in the insect cell-baculovirus process. *J. Biotechnol.* **66:**109–116.

76. **Zavodszky, P., and S. Cseh.** 1996. Production of multidomain complement glycoproteins in insect cells. *Cytotechnology* **20:**279–288.

77. **Zhang, J., N. Kalogerakis, and K. Iatrou.** 1993. A two stage bioreactor system for the production of recombinant proteins using a genetically engineered baculovirus/insect cell system. *Biotechnol. Bioeng.* **42:**357–366.

GENETICS, STRAIN IMPROVEMENT, AND RECOMBINANT PROTEINS

NUMEROUS DIVERSE MICROORGANISMS, INCLUDING CULTURED CELLS derived from multicellular organisms, are used in industry to produce products ranging from therapeutic proteins, enzymes, and biopolymers to antibiotics, pharmaceuticals, and biofuels. Almost without exception, the strains used in such bioprocesses have been genetically altered, either through mutagenesis and screening for higher titers or through strain engineering using modern technology. Since the previous edition of this manual was published in 1999, the genetic toolbox for most organisms has grown significantly. The chapters in this section describe contemporary methods and strategies for engineering and improving some of the most widely used industrial microorganisms. Each is focused on a particular taxonomic grouping that shares similar methods of genetic manipulation and culturing. Although it was not possible to cover the genetic manipulation of every industrial microorganism in this section, much can be found here that should apply more widely. The genetic manipulation of industrial microorganisms in the context of natural products was recently covered in "Complex enzymes in microbial natural product biosynthesis" (*Methods Enzymol.*, volumes 458 and 459, edited by D. A. Hopwood [Academic Press, 2009]). Also, the useful "Guide to Yeast Genetics and Molecular Cell Biology" was recently updated (*Methods Enzymol.*, volume 464, edited by J. Weissman, C. Guthrie, and G. R. Fink [Academic Press, 2009]).

In chapter 16, "Genetic Engineering of Corynebacteria," Masato Ikeda and Seiki Takeno provide a historical context for the extensive work on the gram-positive genus *Corynebacterium*, which is most famous for its use in large-scale production of amino acids. The authors describe how genomics and other "-omics" technologies have been applied to strains derived from classical mutagenesis and screening programs, allowing the rational reconstruction of industrial production strains. They discuss many strategies for metabolic pathway engineering in corynebacteria and the methods used to apply those strategies.

In chapter 17, "Genetics of *Clostridium*," Marite Bradshaw and Eric Johnson discuss the latest tools for genetic manipulation of this gram-positive bacterial genus known for its many strictly anaerobic species. *Clostridium* is diverse and important in both industry and medicine, but has been relatively difficult to engineer. However, as this chapter attests, significant progress has been made during the past decade.

In chapter 18, "Genetic Manipulation of Myxobacteria," Wesley P. Black and colleagues describe new genetic tools available for these fascinating gram-negative bacteria, which have been of interest to academia for the morphogenesis of complex, multicellular fruiting bodies. Myxobacteria are valuable to industry because they produce a wide diversity of novel biologically active compounds. This chapter describes many methods that have been developed in the last decade, which hopefully will facilitate increased use of myxobacteria in biotechnology and drug discovery.

Michael Pyne and coworkers, in chapter 19, "Strain Improvement of *Escherichia coli* To Enhance Recombinant Protein Production," have focused on the latest developments for achieving high-level production of heterologous proteins from *E. coli*. They discuss the construction of host/vector systems and the various strategies for achieving efficient transcription and translation, as well as high cell densities. They describe strategies for overcoming many of the obstacles encountered at the different stages of a bioprocess (e.g., expression vector construction, strain engineering, cell cultivation, and downstream processing), thereby improving recovery or activity of the recombinant protein. Because there is a plethora of protocols for genetic manipulation of *E. coli*, this chapter takes a higher-elevation view of the major technical issues.

Chapter 20, "Genetic Engineering Tools for *Saccharomyces cerevisiae*" by Verena Siewers and coworkers, describes both new and established methods used to engineer this powerful and versatile cell factory. The authors focus on those methods for gene deletion, promoter replacement, heterologous gene expression, etc., that can be applied in high-throughput mode to allow the engineering of complex metabolic pathways, where it is necessary to interrogate the expression of several interacting genes in combinatorial fashion to maximize yield.

In "Protein Expression in Nonconventional Yeasts" (chapter 21), Thomas W. Jeffries and James M. Cregg discuss the advantages and disadvantages of yeasts, particularly species other than *S. cerevisiae*. They focus primarily on heterologous expression of single protein products, although the strategies also apply to metabolic pathway engineering, which requires the balanced expression of multiple proteins under complex regulatory conditions. Because transformation and expression techniques for non-*Saccharomyces* yeasts have been well covered in recent reviews, this chapter emphasizes more general features of expression systems in these yeasts.

Chapter 22, "Genetics, Genetic Manipulation, and Approaches to Strain Improvement of Filamentous Fungi" by Vera Meyer and colleagues, discusses the recently developed genetic tools for strain improvement and optimization of protein production in filamentous fungi. The authors focus mainly on *Aspergillus niger*, although most of the methods described should be applicable to other industrially important filamentous fungi.

Finally, in chapter 23, "Genetic Manipulations of Mammalian Cells for Protein Expression," Anne Kantardjieff and her coauthors provide a thorough discussion of the methods for both stable and transient expression of heterologous genes. Mammalian cells are handled industrially as though they were microorganisms, and they are used to produce biologically active proteins that, in many cases, would be difficult or impossible using any other system. In addition to presenting methods for engineering efficient heterologous gene expression, this chapter also discusses methods for engineering cell physiology to achieve optimum performance at scale.

Genetic Engineering of Corynebacteria

MASATO IKEDA AND SEIKI TAKENO

16

Corynebacteria are gram-positive, high-G+C-content microorganisms that inhabit diverse ecological niches, such as soil, sewage, manure, skin, vegetables, and fruits. The genus *Corynebacterium* formerly included a diverse collection of morphologically similar species, both pathogenic and nonpathogenic. However, modern phylogenetic approaches based on 16S ribosomal DNA (rDNA) sequence analysis have redefined the genus *Corynebacterium* and reclassified some of its former members. In the new phylogenetic classification system, the genus *Corynebacterium* belongs to the class *Actinobacteria*, the order *Actinomycetales*, the suborder *Corynebacterineae*, and the family *Corynebacteriaceae* (23). The suborder *Corynebacterineae* includes the genera *Mycobacterium*, *Nocardia*, and *Rhodococcus*, which are characterized by the presence of mycolic acids, long-chain α-alkyl, β-hydroxy fatty acids, in the cell wall (23).

The genus *Corynebacterium* includes some biotechnologically relevant species, like *Corynebacterium glutamicum*, *Corynebacterium ammoniagenes*, and *Corynebacterium efficiens*. *C. glutamicum* is an especially important industrial microorganism with an annual production of more than 2 million tons of amino acids (33). Due to its importance as an amino acid producer and potential producer of other commodity chemicals and heterologous proteins, *C. glutamicum* is also the best-investigated species of the genus. There is an extensive body of literature on the physiology, genetics, biochemistry, genomics, and metabolic engineering of *C. glutamicum*, including many patents, which emphasizes the commercial importance of this microbe (16, 23).

During the last decade, genomic and other "-omics" data have accumulated for *C. glutamicum*, profoundly affecting strain development methods and providing a global understanding of this microbe. Such work has revealed new regulatory networks and functions that had not previously been identified in this microbe. A novel methodology that merges genomics with classical strain improvement has been developed and used to rationally reconstruct classically derived production strains. This chapter describes the technology and provides strategies for molecular strain improvement using current genetic engineering tools, global analysis techniques, and genome-based engineering approaches, with particular emphasis on the industrially important *C. glutamicum*. The fundamental techniques and tools for genetic engineering have been omitted because they can be found in the 2nd edition of this book and many other publications (16, 23).

16.1. GENOME ANALYSIS

Several corynebacterial genomes have been entirely sequenced: *C. glutamicum* ATCC 13032 (35, 44), *C. glutamicum* R (134), *C. efficiens* YS-314 (82), *C. diphtheriae* NCTC 13129 (19), and *C. jeikeium* K411 (122). The two different genome sequences for *C. glutamicum* ATCC 13032 (BA000036 and BX927147) revealed a genome of 3.3 Mbp with a G+C content of 53.8% and approximately 3,000 predicted genes. Similarity searches have predicted the function of more than 60% of the genes.

The more thermotolerant glutamic acid-producing bacterium *C. efficiens* YS-314, which could potentially reduce cooling costs during fermentation, has a genome of 3.1 Mbp with a G+C content of 63.4%. Comparative genome sequence analysis between *C. glutamicum* and *C. efficiens* has revealed a rather high frequency of certain amino acid substitutions in *C. efficiens* (K to R, S to A or T, and I to V), which might play a significant role in its thermotolerance (82). Detailed features of the corynebacterial genomes and their comparative genomics have been reviewed (16, 23, 134).

16.2. GENOME ENGINEERING

16.2.1. Transposon Mutagenesis

Although the genome sequences of three *C. glutamicum* strains have been completely determined, a large number of the genes are still annotated as hypothetical proteins or proteins with unknown function. The role of each gene may be elucidated by phenotypic analyses following targeted gene disruption, but this approach is laborious, because each disruption requires the construction of a deletion vector and a transformation. On the other hand, transposon (Tn) mutagenesis can generate a mutant library composed of a large number of single-gene disruptants. One can screen for mutants with a desired phenotype and then identify which genes were disrupted.

The Tn mutagenesis experiments described below were performed using special Tn delivery vectors containing insertion sequences (ISs). These vectors cannot replicate in *C. glutamicum*, contain an antibiotic resistance gene between IS elements, and express a transposase gene. Identification of the disrupted gene is done by direct sequencing upstream and downstream of the Tn insertion or by using a standard plasmid rescue technique (121).

The Tn vector pCGL0040 has been used for in vivo mutagenesis of the *C. glutamicum* strain (1, 12). The vector contains a Tn*5531* formed by two IS*1207* sequences isolated from *C. glutamicum* subsp. *lactofermentum* ATCC 21086. The vector is used only for mutagenesis of *C. glutamicum* strains that lack IS*1207* or close homologs in the chromosome, because homologous recombination will occur between the vector and the host chromosome. For this reason, the applicable strains are *C. efficiens* YS-314, *C. glutamicum* ATCC 14752, and *Corynebacterium* sp. strain 2262. Genes encoding the phosphoserine aminotransferase (SerC) and the phosphoserine phosphatase (SerB) were identified by analyzing the serine auxotrophs isolated from a Tn*5531*-mediated mutant library of ATCC 14752 (92). The threonine exporter gene (*thrE*) was also identified using the same Tn vector (108).

Several IS elements, including IS*31831*, IS*14751*, IS*14999*, and IS*13655*, have been isolated from *C. glutamicum* strains, and Tn vectors containing these ISs have been constructed (39, 124, 126, 127). In particular, the Tn vector miniTn*31831* and its derivatives, which are based on IS*31831* isolated from *C. glutamicum* ATCC 31831, have been proposed for high-throughput mutagenesis and used in a reduced genome project described later (115). Vector pCRA732, which is based on IS*14751* isolated from *C. glutamicum* ATCC 14751, was also constructed (39). It has been suggested that the pCRA732-based mutagenesis is applicable to *C. glutamicum* strains R, ATCC 13058, ATCC 13761, ATCC 13826, ATCC 14020, ATCC 14306, ATCC 14752, ATCC 14999, ATCC 15455, and ATCC 15990, but not to type strain ATCC 13032, as in the case of pCGL0040. The transpositional activities of IS*14999*- and IS*13655*-based Tn vectors have been evaluated only in the *C. glutamicum* R strain (124, 126).

As a tool for the mutagenesis of *C. glutamicum* ATCC 13032, Tn vector pAT6100 has been constructed (121). The vector has an IS*6100* sequence found in R-plasmids pCG4 of ATCC 31830 (46) and pTET3 of *C. glutamicum* LP-6 (121). The histidinol-phosphate phosphatase gene involved in histidine biosynthesis has been identified by analyzing the histidine-auxotrophic mutants from an ATCC 13032 Tn mutant library comprising about 10,000 independent clones (77).

16.2.2. Reduced Genome

It has been assumed that a strain with a minimal set of essential genes required for production and cell survival would serve as the preferable host in bioindustrial applications, because minimizing the genome should increase productivity by reducing the metabolic burden caused by maintenance and expression of unnecessary genes (65). Moreover, such a strain is expected to exhibit less regulation and, therefore, be an ideal platform for production (75). It is also thought that a minimal genome should result in decreased by-product outputs. Based on these points of view, the concept of a minimum genome factory, which is defined as an engineered strain with an optimal minimum gene set needed for the intended process, has been proposed and is being tested in several biotechnologically relevant microbes, including *Escherichia coli* (59, 75), *Bacillus subtilis* (2), and *C. glutamicum* (113).

In *C. glutamicum*, comparative genomic analysis between two *C. glutamicum* strains, R and ATCC 13032, revealed that 11 strain-specific islands are scattered in the genomes (116). Such strain-specific islands are thought to be composed of dispensable genes acquired by horizontal gene transfer (59). Thus, these islands were selected as

the first candidates for deletion. Individual excision experiments demonstrated that these 11 regions, comprising 233 open reading frames, are nonessential for cell growth under normal laboratory conditions, indicating that a total genomic region of 250 kb (7.5% of the *C. glutamicum* R genome) can be deleted (116).

The most common excision method is shown in Fig. 1. The Cre/*loxP* excision system is based on Cre recombinase and two 34-bp *loxP* sequences (116, 117). When two *loxP* sites are integrated in the same orientation in the genome, Cre recombinase catalyzes recombination at those sites, resulting in excision of the region between the two sites. The major disadvantage of this method is that one *loxP* site is left on the genome after the deletion, thereby interfering with subsequent rounds of excision. To remedy this, two improved methods have been proposed. One is based on I-*SceI* meganuclease and its cleavage site. In this case, the *loxP* site left on the genome is designed so that it is flanked by two foreign I-*SceI* cleavage sites and short regions of homology. When I-*SceI* meganuclease cleaves the chromosome at such sites, the segment containing the *loxP* site is released and the resultant broken chromosomal ends are subject to RecA-mediated repair by intramolecular recombination, leaving only a single copy of the short homology (114). The other method is based on *loxP* characteristics. Two mutant *loxP* sequences that contain different mutations can be subjected to Cre recombinase-mediated recombination, leaving a *loxP* sequence with both mutations in the chromosome. The double-mutated *loxP* site shows dramatically reduced binding affinity for Cre recombinase and therefore does not interfere with subsequent rounds of excision. The latter method was successfully applied to achieve a 190-kb reduction of the genome (5.7% of the *C. glutamicum* R genome) (113).

Recently, random segment deletion was performed in *C. glutamicum* using IS*31831*-based transposon vectors and the Cre/*loxP* system (125). The principle is that two *loxP* sites can be randomly inserted into the genome by using transposon vectors. The system is independent of prior knowledge about which genes are dispensable. A schematic diagram of the method appears in Fig. 2. Using this method, 331 genes were identified to be nonessential for growth under normal laboratory conditions by analyzing the growth properties of 42 random excision mutants, and a 394-kb region (11.9% of the *C. glutamicum* R genome) was deleted in total (125). To date, no studies have been performed to test the practical utility of *C. glutamicum* strains having a reduced genome.

16.3. POSTGENOMIC TECHNOLOGIES FOR COMPREHENSIVE ANALYSIS

Determining the whole genome sequence of *C. glutamicum* is aimed at gaining sufficient information to manipulate the metabolism or physiology on a global scale, eventually integrating the information for development of more efficient production strains. The sequenced and annotated genomic information alone, however, is not sufficient to achieve this goal. However, the genomic data enable powerful DNA array technologies and proteomics, which are currently undergoing rapid development in this microbe.

16.3.1. DNA Arrays for Transcriptomics

Techniques for the design, construction, and validation of DNA arrays for *C. glutamicum* have been established and include selected gene subsets (28, 69) and arrays that monitor virtually all of the predicted genes (78, 130). Recent articles on the use of DNA arrays for *C. glutamicum* research have already shown how the technologies can be

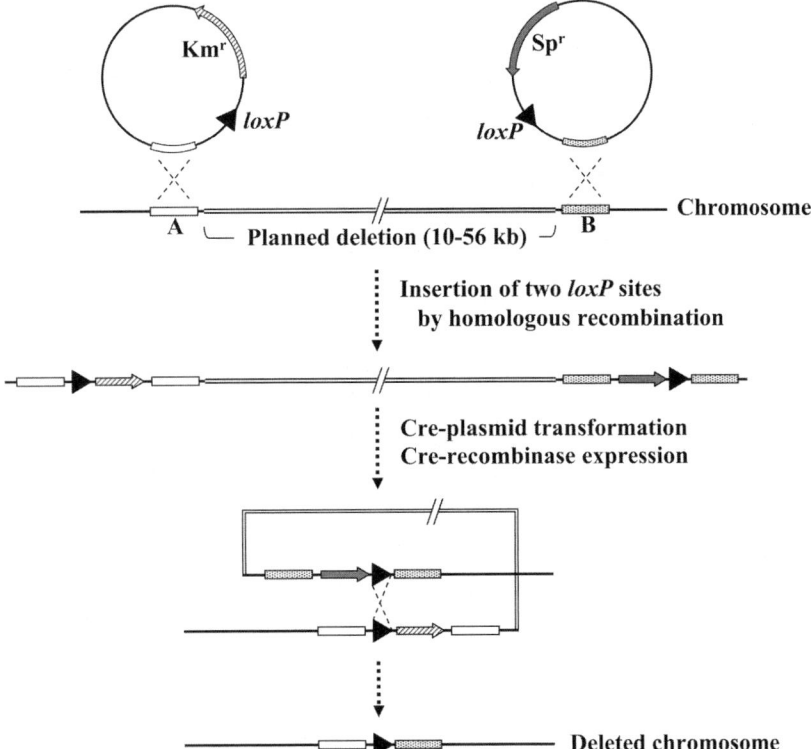

FIGURE 1 Cre/*loxP*-mediated deletion of the C. *glutamicum* genome. Boxes A and B are short segments of the C. *glutamicum* genome. These PCR-amplified segments are integrated into the genome by homologous recombination using two separate vectors. After integration, the Cre-containing plasmid is introduced into the recombinant cell in order to excise the target region. Km[r], kanamycin resistance gene; Sp[r], spectinomycin resistance gene.

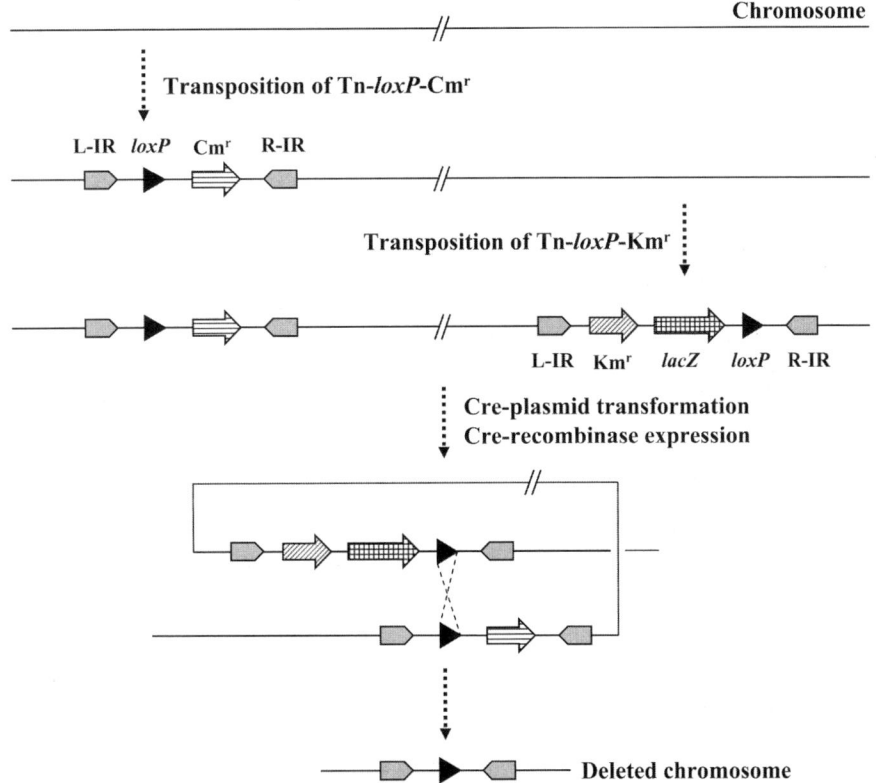

FIGURE 2 Random segment deletion of the C. *glutamicum* genome. Two IS*31831*-based transposon vectors are serially introduced into the C. *glutamicum* cell, randomly integrating two *loxP* sites into the genome, followed by Cre-mediated recombination. Cm[r], chloramphenicol resistance gene; L-IR, left inverted repeat; R-IR, right inverted repeat.

applied to study the metabolic and regulatory properties of this bacterium on a genomic scale (14, 26, 28, 78, 94, 107, 130). Importantly, previously unknown regulatory systems of central metabolism are being discovered, leading to the design of new metabolic engineering strategies. The array technologies can also be used to detect mutations that result in altered transcript levels. One example of using DNA arrays to map a spontaneous mutation in *C. glutamicum* was the characterization of a suppressor mutation in a pyruvate kinase-defective mutant (80).

Besides providing a basic understanding of the biology, these technologies can be used to engineer *C. glutamicum* strains toward improved production. For example, DNA arrays have been used to identify targets to improve valine production (63) or achieve higher productivity of pantothenate (31). Furthermore, the technologies can be used to identify which mutations confer beneficial traits for high-level production in strains from random mutagenesis and screening programs. This has been exemplified by the DNA array-based prediction of target genes for increased production of amino acids such as lysine, arginine, and citrulline (27, 34, 109).

16.3.2. Proteomics

Proteome analysis is also a powerful technology for understanding global regulatory networks by defining sets of coregulated proteins. The technology allows analysis of protein localization, quantification, and posttranslational modifications, which are not accessible by transcriptome analysis. Protocols for proteome analysis of *C. glutamicum* have been established, and high-resolution proteome reference maps have been developed for the cytoplasmic proteins (30, 67, 101), membrane proteins (102), phosphoproteins (9), and secreted proteins (30). Currently, proteomics is recognized as a promising tool for understanding both basic and applied aspects of this bacterium, including the global responses to various environmental conditions such as nitrogen limitation (107), heat shock (3), pH changes (5), and an excess amount of valine (63); the global effects of specific genetic modifications such as an H^+-ATPase-defective mutation (66); and the comparative proteomic studies between an amino acid producer and a wild-type strain (103).

16.3.3. Toward Systems Biology

In addition to transcriptomics and proteomics, technologies for metabolomics (112) and fluxomics (133), combined with metabolic modeling (70), have been developed in *C. glutamicum*. Thus, systems biology is emerging for *C. glutamicum* (131). Very recently, a genome-scale model of the *C. glutamicum* metabolic network has been constructed, based on the annotated genome, available literature, and various "-omics" data (56). The constructed metabolic model consisted of 446 reactions and 411 metabolites, and the predicted metabolic fluxes during lysine production and growth under various conditions were highly consistent with experimental values. Such predictions of the metabolic state for maximum production yield can be used to guide strain engineering. This strategy is exemplified by the rational design of high-lysine-producing strains of *C. glutamicum* (7, 61, 131).

16.4. STRAIN IMPROVEMENT STRATEGIES

16.4.1. Overview

In the last 2 decades, metabolic engineering of *C. glutamicum* has seen rapid development. The targets of metabolic engineering have expanded beyond the core biosynthetic pathways leading to products of interest and now include central metabolism, cofactor-regeneration systems, uptake and export systems, energy metabolism, global regulation, and stress responses. In short, strain development is beginning to achieve optimization of whole cellular systems. In addition, the product spectrum of this microbe has also expanded and metabolic engineering has been applied to the production of chemicals not naturally biosynthesized in *C. glutamicum* (e.g., cadaverine and poly-3-hydroxybutyrate) or amino acids that previously could not be produced cost-effectively from glucose (e.g., serine and methionine). Furthermore, strains are being engineered from the environmental point of view to utilize alternative feedstocks, such as whey, lignocellulose-derived xylose and arabinose, and glycerol, that do not compete with human food or energy sources.

16.4.2. General Strategies with Lysine Production as a Model

Exhaustive studies have been directed to metabolic engineering of *C. glutamicum* for lysine production, resulting in a large body of literature. Thus, lysine production is used here as a model to illustrate the overall strategies of rational strain improvement.

16.4.2.1. Fundamentals of Pathway Engineering

An essential first step for overproduction of lysine is the elimination of bottlenecks in the biosynthetic pathway from aspartic acid to lysine. This can be accomplished by mutations that reduce the expression or activity of homoserine dehydrogenase or desensitize aspartokinase to feedback inhibition by lysine (45, 93, 99, 105). These two modifications, when combined, appear to be synergistic for lysine production (86). *C. glutamicum* mutants with either or both modifications generally show lysine production yields of 10 to 30% from glucose (45, 86, 99). Overexpression of the dihydrodipicolinate synthase gene also can increase lysine production (20).

16.4.2.2. Strategies to Modify Central Metabolism

Once the core pathway is optimized, further gains can be made by increasing the precursor supply, in this case the pathways leading to aspartate. Several strategies have proved beneficial for this objective. Increasing carbon flux from pyruvate to oxaloacetate by overexpression of the pyruvate carboxylase gene or by deletion of the phosphoenolpyruvate carboxykinase gene resulted in significantly increased production of lysine (90, 91, 96). Increasing the availability of pyruvate by decreasing or abolishing pyruvate dehydrogenase activity can also improve lysine production (11, 106). As described later, disruption of malate:quinone oxidoreductase in the tricarboxylic acid cycle also has a beneficial effect, probably due to increased availability of oxaloacetate (74).

Although overexpression of key enzyme activities, as just discussed, can increase lysine production, this can be at the cost of reduced cell growth due to perturbation of the natural homeostatic mechanisms of the cell. This can be partially avoided by expressing additional activities that counterbalance the negative effect on growth. For example, a *C. glutamicum* lysine producer was recently engineered by compensating for the negative consequences of an overexpressed aspartate kinase by simultaneously increasing the pyruvate carboxylase anaplerotic activity that replenishes the tricarboxylic acid cycle intermediates (58).

16.4.2.3. Increased Availability of NADPH

Aside from pathway engineering to direct carbon toward lysine, the NADPH supply is crucial for efficient production of lysine because this cofactor is used either directly or indirectly at four steps in lysine biosynthesis. Augmentation of NADPH supply has been accomplished by engineering the redirection of carbon from glycolysis into the pentose phosphate pathway. Strategies have included disruption of the phosphoglucose isomerase gene (72), overexpression of the fructose 1,6-bisphosphatase gene (7) or the glucose 6-phosphate dehydrogenase gene (6), and introduction of a mutant allele of the 6-phosphogluconate dehydrogenase gene encoding an enzyme less sensitive to feedback inhibition (85). In addition, expression of the membrane-bound transhydrogenase genes from *E. coli* provided an alternative source of NADPH in *C. glutamicum* (42).

16.4.2.4. Engineering of Product Export

With the discovery of active export systems for several kinds of amino acids in *C. glutamicum*, this area has become another focus for further strain improvement. The exporters so far identified for *C. glutamicum* are LysE, which exports the basic amino acids lysine and arginine (128); ThrE, which exports threonine and serine (108); and BrnFE, which exports the branched-chain amino acids and methionine (49, 123). Recently, the NCgl1221 gene product, a mechanosensitive channel homolog, was identified as a possible glutamic acid exporter (79).

Sometimes the export step is critical for efficient amino acid production (17). Identification of the amino acid specificities of exporters, which generally have a limited capacity for amino acid efflux, has allowed the enhancement of efflux by increasing the dosage of genes encoding key exporters. It was revealed that overexpression of the *lysE* gene resulted in a fivefold enhancement of the excretion rate for lysine compared to the rate of the control strain (128). The functions of such exporters also can be transferred to heterologous bacterial species. For example, a mutant allele of the *C. glutamicum lysE* gene has been successfully used for improved lysine production in the methylotroph *Methylophilus methylotrophus* (25). Limited threonine production by *C. glutamicum* was also improved by heterologous expression of an *E. coli* threonine exporter (22).

16.4.2.5. Engineering of Respiratory Energy Efficiency

The supply of ATP is also critical to efficient amino acid production. Improving the efficiency of ATP synthesis is thus another strategy for increasing amino acid production. *C. glutamicum* possesses a branched respiratory chain with two terminal oxidases (13). One branch is composed of the cytochrome bc_1-aa_3 supercomplex, which has a threefold-higher bioenergetic efficiency than the other cytochrome *bd* branch. Disruption of the inefficient cytochrome *bd* branch caused increased lysine production and, perhaps unexpectedly, no marked effect on growth or glucose consumption (43).

16.4.2.6. Other Potential Targets of Metabolic Engineering

16.4.2.6.1. Global Regulation

Amino acid biosynthesis in *C. glutamicum* is subject to both pathway-specific and global regulation (14, 60). Thus, global regulation is also important in strain improvement.

An interesting finding is that global induction of amino acid biosynthesis genes occurs in a classically derived industrial lysine producer (29). The *lysC* gene, encoding the key enzyme aspartokinase, is upregulated severalfold in this strain, even though a repression mechanism for lysine biosynthesis is not known in *C. glutamicum*. Although the genetic elements responsible for these changes remain to be elucidated, introduction of a mutant allele of the *leuC* gene into a defined lysine producer can trigger a stringent-like global response and thereby lead to a significant increase in lysine production (27). Recently, the use of DNA microarray technology and the genome sequence of *C. glutamicum* has identified a variety of global regulators. These include GlxR (53) and SugR (24), controlling carbon metabolism; AmtR (8), controlling nitrogen metabolism; PhoR (100), controlling phosphorus metabolism; McbR (94, 95) and SsuR (57), controlling sulfur metabolism; DtxR (132), controlling iron homeostasis; and FarR (26) and LtbR (15), controlling amino acid metabolism, all of which are potential targets of future metabolic engineering.

16.4.2.6.2. Stress Responses

In the industrial fermentation of *C. glutamicum*, cells must adapt to various suboptimal conditions due to the considerable heterogeneity within large-scale fermentors. The cells are assumed to possess a variety of mechanisms for adaptation to such conditions. For example, cells maintain the pH of their cytoplasm relative to the external environment. Although the mechanisms of pH homeostasis in *C. glutamicum* are not fully understood, it has been shown that a putative transporter of the cation diffusion facilitator family is responsible for alkaline pH homeostasis (118). Only limited studies have so far been conducted on other stress response proteins of *C. glutamicum*. Some examples include SigH (54) and WhcE (55), involved in heat and oxidative stress responses, and BetP (88), EctP (89), ProP (89), LcoP (110), MtrB (76), and MtrA (76), involved in osmotic stress response. These are potential targets for future engineering.

16.4.3. Metabolic Engineering for Serine Production

Serine is among the few amino acids where it has been difficult to engineer strains that give high production yields by fermentation on glucose. Classical mutagenesis and screening for a strain producing serine from glucose had not been successful. Recently, a systematic metabolic engineering strategy resulted in a *C. glutamicum* strain that accumulated considerable amounts of serine (92). In this case it was necessary to intervene at several points in the complicated metabolism of serine. Initial overexpression of serine biosynthesis genes coding for deregulated enzymes did not lead to significant serine accumulation, nor did the further deletion of the serine dehydratase gene, which catalyzes serine degradation to pyruvate. However, by reducing the *glyA*-encoded serine hydroxymethyltransferase (SHMT) activity, considerable serine accumulation was ultimately achieved. Since SHMT is essential for growth of *C. glutamicum* as the unique route leading to C_1 supply, a precise level of reduction of this enzyme activity was the key. This was first achieved by replacing the *glyA* promoter with the *tac* promoter, which caused *glyA* expression to be reduced in the absence of isopropyl-thio-β-D-galactopyranoside. However, this method of *glyA* control was unstable, with mutations in *lacI*q restoring expression. To overcome this inconvenient *glyA* expression control, a better physiological method was developed in which the strain was made auxotrophic for folate and SHMT activity could be controlled by the availability of 5,6,7,8-tetrahydrofolate. This resulted in serine accumulation to 345 mM in a 20-liter controlled fed-batch culture (111).

16.4.4. Metabolic Engineering for Methionine Production

As with serine, it has also been difficult to engineer strains that produce high yields of methionine. Recently, *C. glutamicum* was shown to possess both transsulfuration and direct sulfhydrylation pathways, in contrast to most microorganisms, including *E. coli*, which utilize only one of these two pathways (32, 64). Metabolic engineering to redirect carbon from the lysine pathway into the methionine pathway led to a *C. glutamicum* strain that produced 2.9 g/liter of methionine, together with 23.8 g/liter of lysine (87). In *C. glutamicum*, two regulatory genes relevant to methionine biosynthesis have been identified: *mcbR* (cg3253) and NCgl2640. Inactivation of either in wild-type *C. glutamicum* resulted in increased methionine production (71, 94, 95).

While progress has been made in creating improved methionine producers, the yields remain low compared with those attained for other amino acids. Metabolic pathway analysis has been used to evaluate the theoretical maximum yields of methionine production on the substrates glucose, sulfate, and ammonia in *C. glutamicum* and *E. coli*. The theoretical yield (mol-C methionine per mol-C glucose) of *C. glutamicum* was 0.49, while *E. coli* displayed a higher value of 0.52. This analysis also showed that introduction of the *E. coli* glycine cleavage system into *C. glutamicum* would increase the theoretical maximal methionine yield to 0.57 (62).

16.4.5. Engineering Production of Other Products

In the past decade, metabolic engineering approaches have been used to develop strains that produce chemicals not naturally biosynthesized in *C. glutamicum*. For example, cadaverine (1,5-diaminopentane), an expected raw material of polyamines, and poly-3-hydroxybutyrate, a biodegradable polyester, have received considerable research attention as alternatives to petroleum-derived chemicals. Production of cadaverine and poly-3-hydroxybutyrate by *C. glutamicum* has been achieved through heterologous expression of lysine decarboxylase from *E. coli* (73) and the poly-3-hydroxybutyrate biosynthetic pathway from *Ralstonia eutropha* (40), respectively. In the latter case, poly-3-hydroxybutyrate accumulation could be observed as granules in the cytoplasm of the cell. Not only gene dosage but also codon optimization of the poly-3-hydroxybutyrate biosynthetic genes has been shown to further improve the product content (41).

C. glutamicum also has potential for the production of organic acids such as lactic acid and succinic acid, which can serve as intermediates for the synthesis of a variety of chemicals and polymers. Since *C. glutamicum* is aerobic and does not grow under anaerobic conditions, little work has been done on anaerobic growth and fermentation. However, it was recently shown that this bacterium retains its primary metabolic capacity under growth-arrested anaerobic conditions (38), and production of lactic acid, succinic acid, and acetic acid from glucose was demonstrated (38). A significant increase of succinic acid production was obtained by deleting the *ldhA* gene for L-lactate dehydrogenase and overexpressing the endogenous *pyc* gene for

pyruvate carboxylase. The succinic acid yield of this strain was approximately 1.2 mol per mol of glucose.

More recently, it was shown that *C. glutamicum* can grow anaerobically by nitrate respiration (81, 119). In the presence of nitrate, lysine and arginine production occurred anaerobically, albeit at a very low level, indicating the potential of this bacterium for anaerobic amino acid production (119).

16.4.6. Engineering Feedstock Utilization

Presently, the primary feedstocks for industrial fermentations are sugars from agricultural crops. However, there is an increasing need to engineer the use of alternative raw materials, especially those that are not used for human food or energy supply. Wild-type *C. glutamicum* cannot utilize lactose, galactose, starch, xylose, arabinose, or glycerol for growth, but recently strains have been engineered that can. For example, by overexpressing both *lacYZ* from *Lactobacillus delbrueckii* subsp. *bulgaricus* and *galMKTE* from *Lactococcus lactis* subsp. *cremoris* in a lysine-producing strain of *C. glutamicum*, the strain was able to produce lysine at up to 2 g/liter when fed whey, which contains lactose and galactose (4). Another lysine-producing *C. glutamicum* strain expressing the α-amylase gene from *Streptomyces griseus* utilized soluble starch for lysine production, albeit at a lower efficiency than that obtained for glucose (104). More-efficient lysine production from soluble starch by *C. glutamicum* has been achieved by displaying the α-amylase from *Streptococcus bovis* on the cell surface. As the anchor protein, PgsA from *B. subtilis* was fused to the N terminus of α-amylase. A lysine producer displaying the fusion protein on its cell surface produced 6.04 g/liter of lysine with the conversion yield of 18.89% on starch. The titer and yield were higher than those obtained in glucose medium (120). This was the first study on a cell surface display system of *C. glutamicum*.

The use of lignocellulose as a feedstock is limited in part by poor catabolism of the xylose component. A xylose-utilizing *C. glutamicum* strain was engineered by expressing the *xylA* and *xylB* genes from *E. coli* on a high-copy plasmid. It is interesting that the *E. coli xylB* gene contributed to improved growth performance on xylose despite the existence of a functional *xylB* gene on the *C. glutamicum* wild-type genome. This engineered strain produced lactic acid and succinic acid from xylose anaerobically (48). As another example, heterologous expression of the *E. coli* arabinose-utilizing pathway in *C. glutamicum* gave a strain that was able to grow on arabinose, another component of lignocellulose (47).

Glycerol, the main by-product of biodiesel production, is also a potential carbon source in biotechnological processes. *C. glutamicum* was engineered to grow on glycerol by expressing the *E. coli* glycerol utilization genes, *glpF*, *glpK*, and *glpD*. This engineering allowed production of glutamate and lysine from glycerol with yields of 11% and 19% (grams of product per grams of glycerol), respectively (97).

16.5. GENOME-BASED STRAIN RECONSTRUCTION

The classical approach to strain improvement involves random mutagenesis and screening, but this technology cannot avoid accumulating detrimental or unnecessary mutations. Genome-based strain reconstruction is a technology that identifies the useful mutations in such strains through genome sequencing and then systematically and precisely

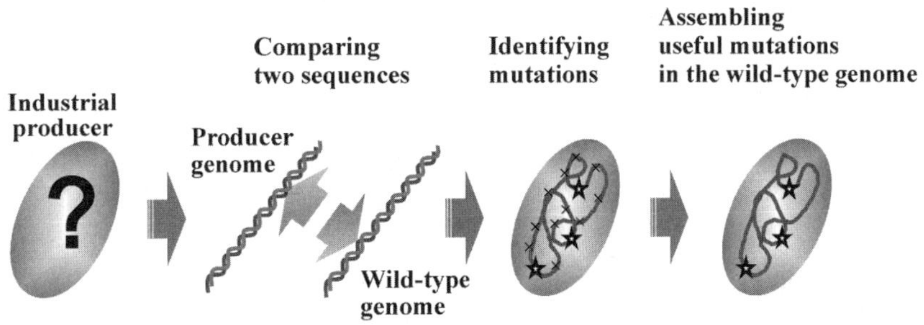

FIGURE 3 Methodology to create a minimally mutated strain. Useful mutations relevant to amino acid production are indicated (stars), together with unnecessary mutations ($\times$).

engineers those mutations into the wild-type genome, creating a strain with only the useful mutations (Fig. 3). Strains derived by this "reverse engineering" method are expected to be more robust, give higher fermentation yields in a shorter time, and resist stressful conditions. The technology is more fully described in the following sections.

16.5.1. Reverse Engineering of a Lysine Producer

Reconstruction of a *C. glutamicum* industrial lysine producer is carried out in the following way. (i) First, the genome sequences of the industrial lysine producers are compared with wild-type sequences to identify the mutational differences. (ii) The mutations are then sequentially introduced by allelic replacement into the wild-type genome (37). Mutations in the relevant terminal pathways are first introduced, followed by those in central metabolism, and finally genes involved in global regulation. (iii) Each strain is evaluated to determine the contribution of each mutation to production. When the mutation is beneficial, the resulting strain is used as the parent to introduce and evaluate further mutations. This iterative cycle makes it possible to generate a minimally mutated strain having only relevant mutations (27, 36, 74, 85, 86). It should be noted that the combination of mutations chosen for introduction and the host strain used at the beginning of the process can have a significant impact on the ultimate outcome. In the case of lysine fermentation, a key mutation, *lysC311*, that confers high-level lysine production on wild-type *C. glutamicum* was used to screen different wild-type strains of *C. glutamicum* to identify the best background to begin the process (84). Although this procedure is simple to describe, it is a challenge to execute in a meaningful time frame.

An industrial lysine producer that had undergone years of mutagenesis and screening was found to have more than 1,000 mutations accumulated in its genome (36). Among the mutations that resulted in increased production, two (*hom59* and *lysC311*) were identified in the terminal pathway to lysine (86), three (*pyc458*, *gnd361*, and *mqo224*) were identified in central metabolism (74, 85, 86), and one (*leuC456*) was identified that caused global induction of the amino acid-biosynthetic genes and thereby contributed to further increased production (27). Subsequent assembly of these six useful mutations in a robust wild-type strain was shown to substantially improve producer performance, resulting in a final titer of near 100 g/liter after only 30 h with 5-liter jar fermentor cultivation at a suboptimal temperature of 40°C (27, 36, 83).

16.5.2. Reengineering of an Arginine and Citrulline Producer

With accumulated knowledge of mutations relevant to production, it becomes possible to combine positive muta-

tions derived from different lines of classical producers in a single wild-type background. Such an advanced approach has recently led to an impressive result in the production of arginine and citrulline (arg+cit) by *C. glutamicum* (34). The final strain was generated according to the strategy described below.

The first step was to identify the most important mutation(s) causing arg+cit overproduction in wild-type *C. glutamicum*. Three independently derived industrial arg+cit producer strains were sequenced and compared to their natural ancestors. This identified a variety of mutations potentially associated with arginine biosynthesis. Five were located within *arg* operons (*argB26*, *argB31*, *argR123*, *argG92*[up], and *argG45*), and their relevance to arg+cit production was therefore examined in a wild-type background. Consequently, *argB26* and Δ*argR* (*argR123*-derived deletion mutation) were found to be the most important mutations. The second step was to screen for the wild-type background giving the best performance, as discussed for lysine production above. Thus, the two mutations, *argB26* and Δ*argR*, were introduced into six different *C. glutamicum* wild-type strains to generate isogenic mutants, which were then screened for their ability to produce arg+cit under suboptimal, 38°C conditions. This revealed that strain ATCC 13032 had the highest potential for arg+cit production at elevated temperatures. By combining those mutations in the best host, a robust producer was obtained, but its production was still only one-third of that of the best classically derived strain.

Thus, a third step was to identify what was limiting in the new strain by conducting transcriptome analyses. This revealed that the *arg* operon in the classically derived strain was much more highly expressed than it was in the new strain. From this analysis, *leuC456*, a mutation in an industrial lysine producer that provoked global induction of the amino acid-biosynthetic genes, including the *arg* operon, was thought to be a good target for engineering. When this mutation was introduced into the wild-type strain along with *argB26* and Δ*argR*, production was increased, but the strain also showed a growth defect. On the other hand, replacing the chromosomal *argB* with the heterologous *E. coli argB*, which is insensitive to arginine inhibition, increased production threefold without retarding growth, revealing that a prime target for engineering was the properties of the *argB* product.

Accordingly, the final step was to engineer the *argB* product, N-acetyl-L-glutamate kinase, so that it would not be feedback inhibited by arginine. To this end, in addition to *argB26*, the other *argB* mutation, *argB31*, was introduced into the new strain, causing more complete deregulation of the enzyme and resulting in dramatically increased production. This reconstructed strain, designated strain RBid, displayed enhanced performance, which is clearly reflected

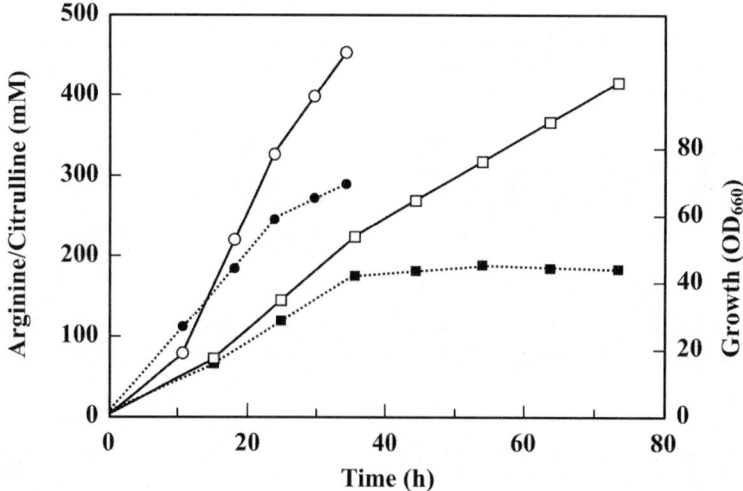

FIGURE 4 Fermentation kinetics of the newly developed strain RBid at 38°C in 5-liter jar fermentor cultivation. For comparison, the profiles of the best classical producer, A-27, which was cultured under its optimal 30°C conditions, are shown as controls. ○, arg+cit of strain RBid; ●, growth of strain RBid; □, arg+cit of strain A-27; ■, growth of strain A-27.

in the differences in the fermentation profiles between strain RBid and the best classical producer (Fig. 4). Figure 5 summarizes this entire strain reconstruction process.

16.6. ENGINEERING FOR HETEROLOGOUS PROTEIN SECRETION

C. glutamicum is nonpathogenic and produces no hazardous toxins, and thus is also a potential host for protein production. Various bacterial proteins have been secreted by *C. glutamicum*, including a staphylococcal nuclease (68), a protease from *Dichelobacter nodosus* (10), subtilisin from *Bacillus* (10), and a fibronectin-binding protein from *Mycobacterium tuberculosis* (98). *C. glutamicum* has at least two advantages as a protein production host. First, no proteolytic activity is detected in its culture supernatants. Second, it naturally secretes only a limited number of extracellular proteins, thus facilitating downstream purification of secreted heterologous proteins. *C. glutamicum* may be a

favorable host for producing food enzymes and pharmaceutical proteins.

A recent notable example is secretion of the *Streptoverticillium mobaraense* transglutaminase, a food enzyme, in an active form and in a significant amount by *C. glutamicum* (50). In this protein expression system, *C. glutamicum* can secrete the pro-transglutaminase efficiently by fusing the protein with a signal peptide derived from a cell surface protein of corynebacteria. When a subtilisin-like protease from *Streptomyces albogriseolus* is cosecreted, the pro-domain is processed by the protease to yield 142 mg/liter of active transglutaminase. It also has been demonstrated that the *C. glutamicum* protein expression system can be used to secrete a pharmaceutically important protein efficiently and in an active form, namely human epidermal growth factor (21). These successes relied on the signal peptide-dependent general secretion (Sec) pathway (18). However, heterologous proteins are often poorly secreted by the Sec pathway. Another food enzyme, *Chryseobacterium proteolyticum* protein-glutaminase, is an example where secretion by the Sec pathway failed. Very recently, this problem has been successfully overcome by using a second signal peptide dependent on the twin-arginine translocation (Tat) pathway (51). It has been shown that overexpression of TatABC, the components of the Tat pathway, significantly improves protein secretion in this organism (52).

Comprehensive screening of secretion signal sequences has been conducted in *C. glutamicum* R by using bioinformatic analysis and a high-throughput secretion assay, which identified a total of 108 candidate signal sequences that could secrete heterologous α-amylase in the organism (129). These secretion signals may be used for secretion of other heterologous proteins.

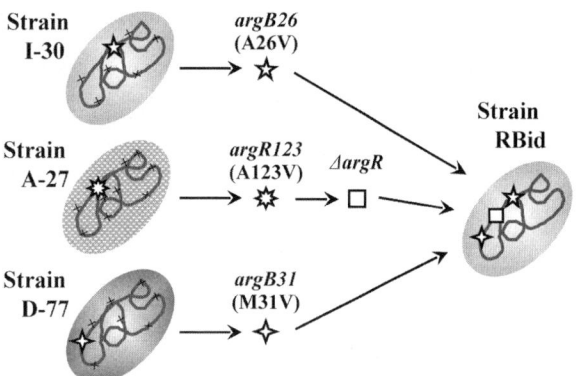

FIGURE 5 Schematic diagram of the creation of new strain RBid. Useful mutations identified in classical producers I-30, A-27, and D-77 are indicated (stars), together with unnecessary mutations (×).

16.7. CONCLUSIONS AND OUTLOOK

This chapter has discussed the technology and strategies for molecular strain improvement of corynebacteria, with special focus on the biotechnologically important *C. glutamicum*. The genetic engineering tools and global analysis techniques for this organism are well developed and have been successfully applied. Understanding of the whole cellular metabolism of corynebacteria continues to improve. In parallel, in silico modeling and simulation approaches are being used to help identify new targets for further engineering and strain improvement. The power of such systems-level approaches will surely increase as modeling is combined with the ever-accumulating data from global analysis techniques such as transcriptomics, proteomics, metabolomics, and fluxomics. The rapid progress in genomic analysis technology and *C. glutamicum* genomics has already allowed the reengineering of classically derived industrial strains. The technology will continue to have an important impact on industrial fermentation.

C. glutamicum has a long history of classical strain improvement, providing many industrially useful strains. Now the beneficial mutations in these strains are being identified and are widely available to the amino acid industry. As a result, the conventional mutagenesis and screening approach is rapidly being replaced by the reengineering of strains using these identified mutations. The two examples of lysine and arg+cit production described herein will be a paradigm for future strain development in the fermentation industry.

REFERENCES

1. **Ankri, S., I. Serebrijski, O. Reyes, and G. Leblon.** 1996. Mutations in the *Corynebacterium glutamicum* proline biosynthetic pathway: a natural bypass of the *proA* step. *J. Bacteriol.* **178:**4412–4419.
2. **Ara, K., K. Ozaki, K. Nakamura, K. Yamane, J. Sekiguchi, and N. Ogasawara.** 2007. *Bacillus* minimum genome factory: effective utilization of microbial genome information. *Biotechnol. Appl. Biochem.* **46:**169–178.
3. **Barreiro, C., E. González-Lavado, S. Brand, A. Tauch, and J. F. Martín.** 2005. Heat shock proteome analysis of wild-type *Corynebacterium glutamicum* ATCC 13032 and a spontaneous mutant lacking GroEL1, a dispensable chaperone. *J. Bacteriol.* **187:**884–889.
4. **Barrett, E., C. Stanton, O. Zelder, G. Fitzgerald, and R. P. Ross.** 2004. Heterologous expression of lactose- and galactose-utilizing pathways from lactic acid bacteria in *Corynebacterium glutamicum* for production of lysine in whey. *Appl. Environ. Microbiol.* **70:**2861–2866.
5. **Barriuso-Iglesias, M., D. Schluesener, C. Barreiro, A. Poetsch, and J. F. Martín.** 2008. Response of the cytoplasmic and membrane proteome of *Corynebacterium glutamicum* ATCC 13032 to pH changes. *BMC Microbiol.* **8:**225.
6. **Becker, J., C. Klopprogge, A. Herold, O. Zelder, C. J. Bolten, and C. Wittmann.** 2007. Metabolic flux engineering of L-lysine production in *Corynebacterium glutamicum*—over expression and modification of G6P dehydrogenase. *J. Biotechnol.* **132:**99–109.
7. **Becker, J., C. Klopprogge, O. Zelder, E. Heinzle, and C. Wittmann.** 2005. Amplified expression of fructose 1,6-bisphosphatase in *Corynebacterium glutamicum* increases in vivo flux through the pentose phosphate pathway and lysine production on different carbon sources. *Appl. Environ. Microbiol.* **71:**8587–8596.
8. **Beckers, G., J. Strösser, U. Hildebrandt, J. Kalinowski, M. Farwick, R. Krämer, and A. Burkovski.** 2005. Regulation of AmtR-controlled gene expression in *Corynebacterium glutamicum*: mechanism and characterization of the AmtR regulon. *Mol. Microbiol.* **58:**580–595.
9. **Bendt, A. K., A. Burkovski, S. Schaffer, M. Bott, M. Farwick, and T. Hermann.** 2003. Towards a phosphoproteome map of *Corynebacterium glutamicum*. *Proteomics* **3:**1637–1646.
10. **Billman-Jacobe, H., L. Wang, A. Kortt, D. Stewart, and A. Radford.** 1995. Expression and secretion of heterologous proteases by *Corynebacterium glutamicum*. *Appl. Environ. Microbiol.* **61:**1610–1613.
11. **Blombach, B., M. E. Schreiner, M. Moch, M. Oldiges, and B. J. Eikmanns.** 2007. Effect of pyruvate dehydrogenase complex deficiency on L-lysine production with *Corynebacterium glutamicum*. *Appl. Microbiol. Biotechnol.* **76:**615–623.
12. **Bonamy, C., J. Labarre, L. Cazaubon, C. Jacob, F. L. Bohec, O. Reyes, and G. Leblon.** 2003. The mobile element IS*1207* of *Brevibacterium lactofermentum* ATCC 21086: isolation and use in the construction of Tn*5531*, a versatile transposon for insertional mutagenesis of *Corynebacterium glutamicum*. *J. Biotechnol.* **104:**301–309.
13. **Bott, M., and A. Niebisch.** 2003. The respiratory chain of *Corynebacterium glutamicum*. *J. Biotechnol.* **104:**129–153.
14. **Brockmann-Gretza, O., and J. Kalinowski.** 2006. Global gene expression during stringent response in *Corynebacterium glutamicum* in presence and absence of the *rel* gene encoding (p)ppGpp synthase. *BMC Genomics* **7:**230.
15. **Brune, I., N. Jochmann, K. Brinkrolf, A. T. Hüser, R. Gerstmeir, B. J. Eikmanns, J. Kalinowski, A. Pühler, and A. Tauch.** 2007. The IclR-type transcriptional repressor LtbR regulates the expression of leucine and tryptophan biosynthesis genes in the amino acid producer *Corynebacterium glutamicum*. *J. Bacteriol.* **189:**2720–2733.
16. **Burkovski, A.** 2008. *Corynebacteria: Genomics and Molecular Biology.* Caister Academic Press, Norfolk, United Kingdom.
17. **Burkovski, A., and R. Krämer.** 2002. Bacterial amino acid transport proteins: occurrence, functions, and significance for biotechnological applications. *Appl. Microbiol. Biotechnol.* **58:**265–274.
18. **Caspers, M., and R. Freudl.** 2008. *Corynebacterium glutamicum* possesses two *secA* homologous genes that are essential for viability. *Arch. Microbiol.* **189:**605–610.
19. **Cerdeño-Tárraga, A. M., A. Efstratiou, L. G. Dover, M. T. Holden, M. Pallen, S. D. Bentley, G. S. Besra, C. Churcher, K. D. James, A. De Zoysa, T. Chillingworth, A. Cronin, L. Dowd, T. Feltwell, N. Hamlin, S. Holroyd, K. Jagels, S. Moule, M. A. Quail, E. Rabbinowitsch, K. M. Rutherford, N. R. Thomson, L. Unwin, S. Whitehead, B. G. Barrell, and J. Parkhill.** 2003. The complete genome sequence and analysis of *Corynebacterium diphtheriae* NCTC13129. *Nucleic Acids Res.* **31:**6516–6523.
20. **Cremer, J., L. Eggeling, and H. Sahm.** 1991. Control of the lysine biosynthesis sequence in *Corynebacterium glutamicum* as analyzed by overexpression of the individual corresponding genes. *Appl. Environ. Microbiol.* **57:**1746–1752.
21. **Date, M., H. Itaya, H. Matsui, and Y. Kikuchi.** 2006. Secretion of human epidermal growth factor by *Corynebacterium glutamicum*. *Lett. Appl. Microbiol.* **42:**66–70.
22. **Diesveld, R., N. Tietze, O. Fürst, A. Reth, B. Bathe, H. Sahm, and L. Eggeling.** 2008. Activity of exporters of *Escherichia coli* in *Corynebacterium glutamicum*, and their use to increase L-threonine production. *J. Mol. Microbiol. Biotechnol.* **16:**198–207.
23. **Eggeling, L., and M. Bott.** 2005. *Handbook of Corynebacterium glutamicum.* CRC Press, Inc., Boca Raton, FL.
24. **Engels, V., S. N. Lindner, and V. F. Wendisch.** 2008. The global repressor SugR controls expression of genes of glycolysis and of the L-lactate dehydrogenase LdhA in *Corynebacterium glutamicum*. *J. Bacteriol.* **190:**8033–8044.
25. **Gunji, Y., and H. Yasueda.** 2006. Enhancement of L-lysine production in methylotroph *Methylophilus methylotrophus* by introducing a mutant LysE exporter. *J. Biotechnol.* **127:**1–13.
26. **Hänssler, E., T. Müller, N. Jessberger, A. Völzke, J. Plassmeier, J. Kalinowski, R. Krämer, and A. Burkovski.** 2007. FarR, a putative regulator of amino acid metabolism

in *Corynebacterium glutamicum*. *Appl. Microbiol. Biotechnol.* **76:**625–632.

27. **Hayashi, M., H. Mizoguchi, J. Ohnishi, S. Mitsuhashi, Y. Yonetani, S. Hashimoto, and M. Ikeda.** 2006. A *leuC* mutation leading to increased L-lysine production and *rel*-independent global expression changes in *Corynebacterium glutamicum*. *Appl. Microbiol. Biotechnol.* **72:**783–789.

28. **Hayashi, M., H. Mizoguchi, N. Shiraishi, M. Obayashi, S. Nakagawa, J. Imai, S. Watanabe, T. Ota, and M. Ikeda.** 2002. Transcriptome analysis of acetate metabolism in *Corynebacterium glutamicum* using a newly developed metabolic array. *Biosci. Biotechnol. Biochem.* **66:**1337–1344.

29. **Hayashi, M., J. Ohnishi, S. Mitsuhashi, Y. Yonetani, S. Hashimoto, and M. Ikeda.** 2006. Transcriptome analysis reveals global expression changes in an industrial L-lysine producer of *Corynebacterium glutamicum*. *Biosci. Biotechnol. Biochem.* **70:**546–550.

30. **Hermann, T., W. Pfefferle, C. Baumann, E. Busker, S. Schaffer, M. Bott, H. Sahm, N. Dusch, J. Kalinowski, A. Pühler, A. K. Bendt, R. Krämer, and A. Burkovski.** 2001. Proteome analysis of *Corynebacterium glutamicum*. *Electrophoresis* **22:**1712–1723.

31. **Hüser, A. T., C. Chassagnole, N. D. Lindley, M. Merkamm, A. Guyonvarch, V. Elisáková, M. Pátek, J. Kalinowski, I. Brune, A. Pühler, and A. Tauch.** 2005. Rational design of a *Corynebacterium glutamicum* pantothenate production strain and its characterization by metabolic flux analysis and genome-wide transcriptional profiling. *Appl. Environ. Microbiol.* **71:**3255–3268.

32. **Hwang, B. J., S. D. Park, Y. Kim, P. Kim, and H. S. Lee.** 2007. Biochemical analysis on the parallel pathways of methionine biosynthesis in *Corynebacterium glutamicum*. *J. Microbiol. Biotechnol.* **17:**1010–1017.

33. **Ikeda, M.** 2003. Amino acid production processes, p. 1–35. *In* R. Faurie and J. Thommel (ed.), *Advances in Biochemical Engineering/Biotechnology*, vol. 79. *Microbial Production of L-Amino Acids*. Springer-Verlag, Berlin, Germany.

34. **Ikeda, M., S. Mitsuhashi, K. Tanaka, and M. Hayashi.** 2009. Reengineering of a *Corynebacterium glutamicum* L-arginine and L-citrulline producer. *Appl. Environ. Microbiol.* **75:**1635–1641.

35. **Ikeda, M., and S. Nakagawa.** 2003. The *Corynebacterium glutamicum* genome: features and impacts on biotechnological process. *Appl. Microbiol. Biotechnol.* **62:**99–109.

36. **Ikeda, M., J. Ohnishi, M. Hayashi, and S. Mitsuhashi.** 2006. A genome-based approach to create a minimally mutated *Corynebacterium glutamicum* strain for efficient L-lysine production. *J. Ind. Microbiol. Biotechnol.* **33:**610–615.

37. **Ikeda, M., J. Ohnishi, and S. Mitsuhashi.** 2005. Genome breeding of an amino acid-producing *Corynebacterium glutamicum* mutant, p. 179–189. *In* J. L. S. Barredo (ed.), *Microbial Processes and Products*. Humana Press, Totowa, NJ.

38. **Inui, M., S. Murakami, S. Okino, H. Kawaguchi, A. A. Vertes, and H. Yukawa.** 2004. Metabolic analysis of *Corynebacterium glutamicum* during lactate and succinate production under oxygen deprivation conditions. *J. Mol. Microbiol. Biotechnol.* **7:**182–196.

39. **Inui, M., Y. Tsuge, N. Suzuki, A. A. Vertès, and H. Yukawa.** 2005. Isolation and characterization of a native composite transposon, Tn*14751*, carrying 17.4 kilobases of *Corynebacterium glutamicum* chromosomal DNA. *Appl. Environ. Microbiol.* **71:**407–416.

40. **Jo, S. J., M. Maeda, T. Ooi, and S. Taguchi.** 2006. Production system for biodegradable polyester polyhydroxybutyrate by *Corynebacterium glutamicum*. *J. Biosci. Bioeng.* **102:**233–236.

41. **Jo, S. J., K. Matsumoto, C. R. Leong, T. Ooi, and S. Taguchi.** 2007. Improvement of poly(3-hydroxybutyr-

ate) [P(3HB)] production in *Corynebacterium glutamicum* by codon optimization, point mutation and gene dosage of P(3HB) biosynthetic genes. *J. Biosci. Bioeng.* **104:**457–463.

42. **Kabus, A., T. Georgi, V. F. Wendisch, and M. Bott.** 2007. Expression of the *Escherichia coli pntAB* genes encoding a membrane-bound transhydrogenase in *Corynebacterium glutamicum* improves L-lysine formation. *Appl. Microbiol. Biotechnol.* **75:**47–53.

43. **Kabus, A., A. Niebisch, and M. Bott.** 2007. Role of cytochrome *bd* oxidase from *Corynebacterium glutamicum* in growth and lysine production. *Appl. Environ. Microbiol.* **73:**861–868.

44. **Kalinowski, J., B. Bathe, D. Bartels, N. Bischoff, M. Bott, A. Burkovski, N. Dusch, L. Eggeling, B. J. Eikmanns, L. Gaigalat, A. Goesmann, M. Hartmann, K. Huthmacher, R. Krämer, B. Linke, A. C. McHardy, F. Meyer, B. Möckel, W. Pfefferle, A. Pühler, D. A. Rey, C. Rückert, O. Rupp, H. Sahm, V. F. Wendisch, I. Wiegräbe, and A. Tauch.** 2003. The complete *Corynebacterium glutamicum* ATCC 13032 genome sequence and its impact on the production of L-aspartate-derived amino acids and vitamins. *J. Biotechnol.* **104:**5–25.

45. **Kase, H., and K. Nakayama.** 1974. Mechanism of L-threonine and L-lysine production by analog-resistant mutants of *Corynebacterium glutamicum*. *Agr. Biol. Chem.* **38:**993–1000.

46. **Katsumata, R., A. Ozaki, T. Oka, and A. Furuya.** 1984. Protoplast transformation of glutamate-producing bacteria with plasmid DNA. *J. Bacteriol.* **159:**306–311.

47. **Kawaguchi, H., M. Sasaki, A. A. Vertès, M. Inui, and H. Yukawa.** 2008. Engineering of an L-arabinose metabolic pathway in *Corynebacterium glutamicum*. *Appl. Microbiol. Biotechnol.* **77:**1053–1062.

48. **Kawaguchi, H., A. A. Vertès, S. Okino, M. Inui, and H. Yukawa.** 2006. Engineering of a xylose metabolic pathway in *Corynebacterium glutamicum*. *Appl. Environ. Microbiol.* **72:**3418–3428.

49. **Kennerknecht, N., H. Sahm, M. R. Yen, M. Patek, M. H. Saier, Jr., and L. Eggeling.** 2002. Export of L-isoleucine from *Corynebacterium glutamicum*: a two-gene-encoded member of a new translocator family. *J. Bacteriol.* **184:**3947–3956.

50. **Kikuchi, Y., M. Date, K. Yokoyama, Y. Umezawa, and H. Matsui.** 2003. Secretion of active-form *Streptoverticillium mobaraense* transglutaminase by *Corynebacterium glutamicum*: processing of the pro-transglutaminase by a cosecreted subtilisin-like protease from *Streptomyces albogriseolus*. *Appl. Environ. Microbiol.* **69:**358–366.

51. **Kikuchi, Y., H. Itaya, M. Date, K. Matsui, and L. F. Wu.** 2008. Production of *Chryseobacterium proteolyticum* protein-glutaminase using the twin-arginine translocation pathway in *Corynebacterium glutamicum*. *Appl. Microbiol. Biotechnol.* **78:**67–74.

52. **Kikuchi, Y., H. Itaya, M. Date, K. Matsui, and L. F. Wu.** 2009. TatABC overexpression improves *Corynebacterium glutamicum* Tat-dependent protein secretion. *Appl. Environ. Microbiol.* **75:**603–607.

53. **Kim, H. J., T. H. Kim, Y. Kim, and H. S. Lee.** 2004. Identification and characterization of *glxR*, a gene involved in regulation of glyoxylate bypass in *Corynebacterium glutamicum*. *J. Bacteriol.* **186:**3453–3460.

54. **Kim, T. H., H. J. Kim, J. S. Park, Y. Kim, P. Kim, and H. S. Lee.** 2005. Functional analysis of *sigH* expression in *Corynebacterium glutamicum*. *Biochem. Biophys. Res. Commun.* **331:**1542–1547.

55. **Kim, T. H., J. S. Park, H. J. Kim, Y. Kim, P. Kim, and H. S. Lee.** 2005. The *whcE* gene of *Corynebacterium glutamicum* is important for survival following heat and oxidative stress. *Biochem. Biophys. Res. Commun.* **337:**757–764.

56. Kjeldsen, K. R., and J. Nielsen. 2009. In silico genome-scale reconstruction and validation of the *Corynebacterium glutamicum* metabolic network. *Biotechnol. Bioeng.* **102:**583–597.

57. Koch, D. J., C. Rückert, A. Albersmeier, A. T. Hüser, A. Tauch, A. Pühler, and J. Kalinowski. 2005. The transcriptional regulator SsuR activates expression of the *Corynebacterium glutamicum* sulphonate utilization genes in the absence of sulphate. *Mol. Microbiol.* **58:**480–494.

58. Koffas, M. A. G., G. Y. Jung, and G. Stephanopoulos. 2003. Engineering metabolism and product formation in *Corynebacterium glutamicum* by coordinated gene overexpression. *Metab. Eng.* **5:**32–41.

59. Kolisnychenko, V., G. Plunkett III, C. D. Herring, T. Fehér, J. Pósfai, F. R. Blattner, and G. Pósfai. 2002. Engineering a reduced *Escherichia coli* genome. *Genome Res.* **12:**640–647.

60. Krömer, J. O., C. J. Bolten, E. Heinzle, H. Schröder, and C. Wittmann. 2008. Physiological response of *Corynebacterium glutamicum* to oxidative stress induced by deletion of the transcriptional repressor McbR. *Microbiology* **154:**3917–3930.

61. Krömer, J. O., O. Sorgenfrei, K. Klopprogge, E. Heinzle, and C. Wittmann. 2004. In-depth profiling of lysine-producing *Corynebacterium glutamicum* by combined analysis of the transcriptome, metabolome, and fluxome. *J. Bacteriol.* **186:**1769–1784.

62. Krömer, J. O., C. Wittmann, H. Schröder, and E. Heinzle. 2006. Metabolic pathway analysis for rational design of L-methionine production by *Eschrichia coli* and *Corynebacterium glutamicum*. *Metab. Eng.* **8:**353–369.

63. Lange, C., D. Rittmann, V. F. Wendisch, M. Bott, and H. Sahm. 2003. Global expression profiling and physiological characterization of *Corynebacterium glutamicum* grown in the presence of L-valine. *Appl. Environ. Microbiol.* **69:**2521–2532.

64. Lee, H. S., and B. J. Hwang. 2003. Methionine biosynthesis and its regulation in *Corynebacterium glutamicum*: parallel pathways of transsulfuration and direct sulfhydrylation. *Appl. Microbiol. Biotechnol.* **62:**459–467.

65. Lee, J. H., B. H. Sung, M. S. Kim, F. R. Blattner, B. H. Yoon, J. H. Kim, and S. C. Kim. 2009. Metabolic engineering of a reduced-genome strain of *Escherichia coli* for L-threonine production. *Microb. Cell Fact.* **8:**2–13.

66. Li, L., M. Wada, and A. Yokota. 2007. A comparative proteomic approach to understand the adaptations of an H⁺-ATPase-defective mutant of *Corynebacterium glutamicum* ATCC14067 to energy deficiencies. *Proteomics* **7:**3348–3357.

67. Li, L., M. Wada, and A. Yokota. 2007. Cytoplasmic proteome reference map for a glutamic acid-producing *Corynebacterium glutamicum* ATCC 14067. *Proteomics* **7:**4317–4322.

68. Liebl, W., A. J. Sinskey, and K. H. Schleifer. 1992. Expression, secretion, and processing of staphylococcal nuclease by *Corynebacterium glutamicum*. *J. Bacteriol.* **174:**1854–1861.

69. Loos, A., C. Glanemann, L. B. Willis, X. M. O'Brien, P. A. Lessard, R. Gerstmeir, S. Guillouet, and A. J. Sinskey. 2001. Development and validation of *Corynebacterium* DNA microarrays. *Appl. Environ. Microbiol.* **67:**2310–2318.

70. Magnus, J. B., D. Hollwedel, M. Oldiges, and R. Takors. 2006. Monitoring and modeling of the reaction dynamics in the valine/leucine synthesis pathway in *Corynebacterium glutamicum*. *Biotechnol. Prog.* **22:**1071–1083.

71. Mampel, J., H. Schröder, S. Haefner, and U. Sauer. 2005. Single-gene knockout of a novel regulatory element confers ethionine resistance and elevates methionine production in *Corynebacterium glutamicum*. *Appl. Microbiol. Biotechnol.* **68:**228–236.

72. Marx, A., S. Hans, B. Mockel, B. Bathe, and A. A. de Graaf. 2003. Metabolic phenotype of phosphoglucose

73. Mimitsuka, T., H. Sawai, M. Hatsu, and K. Yamada. 2007. Metabolic engineering of *Corynebacterium glutamicum* for cadaverine fermentation. *Biosci. Biotechnol. Biochem.* **71:**2130–2135.

74. Mitsuhashi, S., M. Hayashi, J. Ohnishi, and M. Ikeda. 2006. Disruption of malate:quinone oxidoreductase increases L-lysine production by *Corynebacterium glutamicum*. *Biosci. Biotechnol. Biochem.* **70:**2803–2806.

75. Mizoguchi, H., H. Mori, and T. Fujio. 2007. *Escherichia coli* minimum genome factory. *Biotechnol. Appl. Biochem.* **46:**157–167.

76. Möker, N., J. Krämer, G. Unden, R. Krämer, and S. Morbach. 2007. In vitro analysis of the two-component system MtrB-MtrA from *Corynebacterium glutamicum*. *J Bacteriol.* **189:**3645–3649.

77. Mormann, S., A. Lömker, C. Rückert, L. Gaigalat, A. Tauch, A. Pühler, and J. Kalinowski. 2006. Random mutagenesis in *Corynebacterium glutamicum* ATCC 13032 using an IS6100-based transposon vector identified the last unknown gene in the histidine biosynthesis pathway. *BMC Genomics* **7:**205–224.

78. Muffler, A., S. Bettermann, M. Haushalter, A. Hörlein, U. Neveling, M. Schramm, and O. Sorgenfrei. 2002. Genome-wide transcription profiling of *Corynebacterium glutamicum* after heat shock and during growth on acetate and glucose. *J. Biotechnol.* **98:**255–268.

79. Nakamura, J., S. Hirano, H. Ito, and M. Wachi. 2007. Mutations of the *Corynebacterium glutamicum* NCgl1221 gene, encoding a mechanosensitive channel homolog, induce L-glutamic acid production. *Appl. Environ. Microbiol.* **73:**4491–4498.

80. Netzer, R., M. Krause, D. Rittmann, P. G. Peters-Wendisch, L. Eggeling, V. F. Wendisch, and H. Sahm. 2004. Roles of pyruvate kinase and malic enzyme in *Corynebacterium glutamicum* for growth on carbon sources requiring gluconeogenesis. *Arch. Microbiol.* **182:**354–363.

81. Nishimura, T., A. A. Vertès, Y. Shinoda, M. Inui, and H. Yukawa. 2007. Anaerobic growth of *Corynebacterium glutamicum* using nitrate as a terminal electron acceptor. *Appl. Microbiol. Biotechnol.* **75:**889–897.

82. Nishio, Y., Y. Nakamura, Y. Kawarabayasi, Y. Usuda, E. Kimura, S. Sugimoto, K. Matsui, A. Yamagishi, H. Kikuchi, K. Ikeo, and T. Gojobori. 2003. Comparative complete genome sequence analysis of the amino acid replacements responsible for the thermostability of *Corynebacterium efficiens*. *Genome Res.* **13:**1572–1579.

83. Ohnishi, J., M. Hayashi, S. Mitsuhashi, and M. Ikeda. 2003. Efficient 40°C fermentation of L-lysine by a new *Corynebacterium glutamicum* mutant developed by genome breeding. *Appl. Microbiol. Biotechnol.* **62:**69–75.

84. Ohnishi, J., and M. Ikeda. 2006. Comparisons of potentials for L-lysine production among different *Corynebacterium glutamicum* strains. *Biosci. Biotechnol. Biochem.* **70:**1017–1020.

85. Ohnishi, J., R. Katahira, S. Mitsuhashi, S. Kakita, and M. Ikeda. 2005. A novel *gnd* mutation leading to increased L-lysine production in *Corynebacterium glutamicum*. *FEMS Microbiol. Lett.* **242:**265–274.

86. Ohnishi, J., S. Mitsuhashi, M. Hayashi, S. Ando, H. Yokoi, K. Ochiai, and M. Ikeda. 2002. A novel methodology employing *Corynebacterium glutamicum* genome information to generate a new L-lysine-producing mutant. *Appl. Microbiol. Biotechnol.* **58:**217–223.

87. Park, S. D., J. Y. Lee, S. Y. Sim, Y. Kim, and H. S. Lee. 2007. Characteristics of methionine production by an engineered *Corynebacterium glutamicum* strain. *Metab. Eng.* **9:**327–336.

88. **Peter, H., A. Burkovski, and R. Krämer.** 1996. Isolation, characterization, and expression of the *Corynebacterium glutamicum betP* gene, encoding the transport system for the compatible solute glycine betaine. *J. Bacteriol.* **178:**5229–5234.

89. **Peter, H., B. Weil, A. Burkovski, R. Krämer, and S. Morbach.** 1998. *Corynebacterium glutamicum* is equipped with four secondary carriers for compatible solutes: identification, sequencing, and characterization of the proline/ectoine uptake system, ProP, and the ectoine/proline/glycine betaine carrier, EctP. *J. Bacteriol.* **180:**6005–6012.

90. **Petersen, S., C. Mack, A. A. de Graaf, C. Riedel, B. J. Eikmanns, and H. Sahm.** 2001. Metabolic consequences of altered phosphoenolpyruvate carboxykinase activity in *Corynebacterium glutamicum* reveal anaplerotic mechanisms in vivo. *Metab. Eng.* **3:**344–361.

91. **Peters-Wendisch, P. G., B. Schiel, V. F. Wendisch, E. Katsoulidis, B. Möckel, H. Sahm, and B. J. Eikmanns.** 2001. Pyruvate carboxylase is a major bottleneck for glutamate and lysine production by *Corynebacterium glutamicum*. *J. Mol. Microbiol. Biotechnol.* **3:**295–300.

92. **Peters-Wendisch, P., M. Stolz, H. Etterich, N. Kennerknecht, H. Sahm, and L. Eggeling.** 2005. Metabolic engineering of *Corynebacterium glutamicum* for L-serine production. *Appl. Environ. Microbiol.* **71:**7139–7144.

93. **Pfefferle, W., B. Möckel, B. Bathe, and A. Marx.** 2003. Biotechnological manufacture of lysine. *Adv. Biochem. Eng. Biotechnol.* **79:**59–112.

94. **Rey, D. A., S. S. Nentwich, D. J. Koch, C. Rückert, A. Pühler, A. Tauch, and J. Kalinowski.** 2005. The McbR repressor modulated by the effector substance S-adenosylhomocysteine controls directly the transcription of a regulon involved in sulphur metabolism of *Corynebacterium glutamicum* ATCC 13032. *Mol. Microbiol.* **56:**871–887.

95. **Rey, D. A., A. Pühler, and J. Kalinowski.** 2003. The putative transcriptional repressor McbR, member of the TetR-family, is involved in the regulation of the metabolic network directing the synthesis of sulfur containing amino acids in *Corynebacterium glutamicum*. *J. Biotechnol.* **103:**51–65.

96. **Riedel, C., D. Rittmann, P. Dangel, B. Möckel, H. Sahm, and B. J. Eikmanns.** 2001. Characterization, expression, and inactivation of the phosphoenolpyruvate carboxykinase gene from *Corynebacterium glutamicum* and significance of the enzyme for growth and amino acid production. *J. Mol. Microbiol. Biotechnol.* **3:**573–583.

97. **Rittmann, D., S. N. Lindner, and V. F. Wendisch.** 2008. Engineering of a glycerol utilization pathway for amino acid production by *Corynebacterium glutamicum*. *Appl. Environ. Microbiol.* **74:**6216–6222.

98. **Salim, K., V. Haedens, J. Content, G. Leblon, and K. Huygen.** 1997. Heterologous expression of the Mycobacterium tuberculosis gene encoding antigen 85A in *Corynebacterium glutamicum*. *Appl. Environ. Microbiol.* **63:**4392–4400.

99. **Sano, K., and I. Shiio.** 1971. Microbial production of L-lysine. IV. Selection of lysine-producing mutants from *Brevibacterium flavum* by detecting threonine sensitivity or halo-forming method. *J. Gen. Appl. Microbiol.* **17:**97–113.

100. **Schaaf, S., and M. Bott.** 2007. Target genes and DNA-binding sites of the response regulator PhoR from *Corynebacterium glutamicum*. *J. Bacteriol.* **189:**5002–5011.

101. **Schaffer, S., B. Weil, V. D. Nguyen, G. Dongmann, K. Günther, M. Nickolaus, T. Hermann, and M. Bott.** 2001. A high-resolution reference map for cytoplasmic and membrane-associated proteins of *Corynebacterium glutamicum*. *Electrophoresis* **22:**4404–4422.

102. **Schluesener, D., F. Fischer, J. Kruip, M. Rögner, and A. Poetsch.** 2005. Mapping the membrane proteome of *Corynebacterium glutamicum*. *Proteomics* **5:**1317–1330.

103. **Schluesener, D., M. Rögner, and A. Poetsch.** 2007. Evaluation of two proteomics technologies used to screen the membrane proteomes of wild-type *Corynebacterium glutamicum* and an L-lysine-producing strain. *Anal. Bioanal. Chem.* **389:**1055–1064.

104. **Seibold, G., M. Auchter, S. Berens, J. Kalinowski, and B. J. Eikmanns.** 2006. Utilization of soluble starch by a recombinant *Corynebacterium glutamicum* strain: growth and lysine production. *J. Biotechnol.* **124:**381–391.

105. **Shiio, I., and R. Miyajima.** 1969. Concerted inhibition and its reversal by end products of aspartate kinase in *Brevibacterium flavum*. *J. Biochem.* **65:**849–859.

106. **Shiio, I., Y. Toride, and S. Sugimoto.** 1984. Production of lysine by pyruvate dehydrogenase mutants of *Brevibacterium flavum*. *Agric. Biol. Chem.* **48:**3091–3098.

107. **Silberbach, M., M. Schäfer, A. T. Hüser, J. Kalinowski, A. Pühler, R. Krämer, and A. Burkovski.** 2005. Adaptation of *Corynebacterium glutamicum* to ammonium limitation: a global analysis using transcriptome and proteome techniques. *Appl. Environ. Microbiol.* **71:**2391–2402.

108. **Simic, P., H. Sahm, and L. Eggeling.** 2001. L-Threonine export: use of peptides to identify a new translocator from *Corynebacterium glutamicum*. *J. Bacteriol.* **183:**5317–5324.

109. **Sindelar, G., and V. F. Wendisch.** 2007. Improving lysine production by *Corynebacterium glutamicum* through DNA microarray-based identification of novel target genes. *Appl. Microbiol. Biotechnol.* **76:**677–689.

110. **Steger, R., M. Weinand, R. Krämer, and S. Morbach.** 2004. LcoP, an osmoregulated betaine/ectoine uptake system from *Corynebacterium glutamicum*. *FEBS Lett.* **573:**155–160.

111. **Stolz, M., P. Peters-Wendisch, H. Etterich, T. Gerharz, R. Faurie, H. Sahm, H. Fersterra, and L. Eggeling.** 2007. Reduced folate supply as a key to enhanced L-serine production by *Corynebacterium glutamicum*. *Appl. Environ. Microbiol.* **73:**750–755.

112. **Strelkov, S., M. von Elstermann, and D. Schomburg.** 2004. Comprehensive analysis of metabolites in *Corynebacterium glutamicum* by gas chromatography/mass spectrometry. *Biol. Chem.* **385:**853–861.

113. **Suzuki, N., H. Nonaka, Y, Tsuge, M. Inui, and H. Yukawa.** 2005. New multiple-deletion method for the *Corynebacterium glutamicum* genome, using a mutant *lox* sequence. *Appl. Environ. Microbiol.* **71:**8472–8480.

114. **Suzuki, N., H. Nonaka, Y. Tsuge, S. Okayama, M. Inui, and H. Yukawa.** 2005. Multiple large segment deletion method for *Corynebacterium glutamicum*. *Appl. Microbiol. Biotechnol.* **69:**151–161.

115. **Suzuki, N., N. Okai, H. Nonaka, Y. Tsuge, M. Inui, and H. Yukawa.** 2006. High-throughput transposon mutagenesis of *Corynebacterium glutamicum* and construction of a single-gene disruptant mutant library. *Appl. Environ. Microbiol.* **72:**3750–3755.

116. **Suzuki, N., S. Okayama, H. Nonaka, Y. Tsuge, M. Inui, and H. Yukawa.** 2005. Large-scale engineering of the *Corynebacterium glutamicum* genome. *Appl. Environ. Microbiol.* **71:**3369–3372.

117. **Suzuki, N., Y. Tsuge, M. Inui, and H. Yukawa.** 2005. Cre/loxP-mediated deletion system for large genome rearrangements in *Corynebacterium glutamicum*. *Appl. Microbiol. Biotechnol.* **67:**225–233.

118. **Takeno, S., M. Nakamura, R. Fukai, J. Ohnishi, and M. Ikeda.** 2008. The Cgl1281-encoding putative transporter of the cation diffusion facilitator family is responsible for alkali-tolerance in *Corynebacterium glutamicum*. *Appl. Microbiol. Biotechnol.* **190:**531–538.

119. **Takeno, S., J. Ohnishi, T. Komatsu, T. Masaki, K. Sen, and M. Ikeda.** 2007. Anaerobic growth and potential for amino acid production by nitrate respiration in *Corynebacterium glutamicum*. *Appl. Microbiol. Biotechnol.* **75:**1173–1182.

120. **Tateno, T., H. Fukuda, and A. Kondo.** 2007. Production of L-lysine from starch by *Corynebacterium glutamicum* displaying α-amylase on its cell surface. *Appl. Microbiol. Biotechnol.* **74:**1213–1220.

121. **Tauch, A., S. Götker, A. Pühler, J. Kalinowski, and G. Thierbach.** 2002. The 27.8-kb R-plasmid pTET3 from *Corynebacterium glutamicum* encodes the aminoglycoside adenyltransferase gene cassette *aadA9* and the regulated tetracycline efflux system Tet 33 flanked by active copies of the widespread insertion sequence IS*6100*. *Plasmid* **48:**117–129.

122. **Tauch, A., O. Kaiser, T. Hain, A. Goesmann, B. Weisshaar, A. Albersmeier, T. Bekel, N. Bischoff, I. Brune, T. Chakraborty, J. Kalinowski, F. Meyer, O. Rupp, S. Schneiker, P. Viehoever, and A. Pühler.** 2005. Complete genome sequence and analysis of the multiresistant nosocomial pathogen *Corynebacterium jeikeium* K411, a lipid-requiring bacterium of the human skin flora. *J. Bacteriol.* **187:**4671–4682.

123. **Trötschel, C., D. Deutenberg, B. Bathe, A. Burkovski, and R. Krämer.** 2005. Characterization of methionine export in *Corynebacterium glutamicum*. *J. Bacteriol.* **187:**3786–3794.

124. **Tsuge, Y., K. Ninomiya, N. Suzuki, M. Inui, and H. Yukawa.** 2005. A new insertion sequence, IS*14999*, from *Corynebacterium glutamicum*. *Microbiology* **151:**501–508.

125. **Tsuge, Y., N. Suzuki, M. Inui, and H. Yukawa.** 2007. Random segment deletion based on IS*31831* and Cre/*loxP* excision system in *Corynebacterium glutamicum*. *Appl. Microbiol. Biotechnol.* **74:**1333–1341.

126. **Tsuge, Y., N. Suzuki, K. Ninomiya, M. Inui, and H. Yukawa.** 2007. Isolation of a new insertion sequence, IS*13655*, and its application to *Corynebacterium glutami-* cum genome mutagenesis. *Biosci. Biotechnol. Biochem.* **71:** 1683–1690.

127. **Vertès, A. A., Y. Asai, M. Inui, M. Kobayashi, Y. Kurusu, and H. Yukawa.** 1994. Transposon mutagenesis of coryneform bacteria. *Mol. Gen. Genet.* **245:**397–405.

128. **Vrljié, M., H. Sahm, and L. Eggeling.** 1996. A new type of transporter with a new type of cellular function: L-lysine export from *Corynebacterium glutamicum*. *Mol. Microbiol.* **22:**815–826.

129. **Watanabe, K., Y. Tsuchida, N. Okibe, H. Teramoto, N. Suzuki, M. Inui, and H. Yukawa.** 2009. Scanning the *Corynebacterium glutamicum* R genome for high-efficiency secretion signal sequences. *Microbiology* **155:**741–750.

130. **Wendisch, V. F.** 2003. Genome-wide expression analysis in *Corynebacterium glutamicum* using DNA microarrays. *J. Biotechnol.* **104:**273–285.

131. **Wendisch, V. F., M. Bott., J. Kalinowski, M. Oldiges, and W. Wiechert.** 2006. Emerging *Corynebacterium glutamicum* systems biology. *J. Biotechnol.* **124:**74–92.

132. **Wennerhold, J., and M. Bott.** 2006. The DtxR regulon of *Corynebacterium glutamicum*. *J. Bacteriol.* **188:**2907–2918.

133. **Wittmann, C., and E. Heinzle.** 2002. Genealogy profiling through strain improvement by using metabolic network analysis: metabolic flux genealogy of several generations of lysine-producing corynebacteria. *Appl. Environ. Microbiol.* **68:**5843–5859.

134. **Yukawa, H., C. A. Omumasaba, H. Nonaka, P. Kós, N. Okai, N. Suzuki, M. Suda, Y. Tsuge, J. Watanabe, Y. Ikeda, A. A. Vertès, and M. Inui.** 2007. Comparative analysis of the *Corynebacterium glutamicum* group and complete genome sequence of strain R. *Microbiology* **153:**1042–1058.

Genetic Manipulation of *Clostridium*

MARITE BRADSHAW AND ERIC A. JOHNSON

17

17.1. INTRODUCTION

The clostridia have a rich history in medical and industrial microbiology (40, 77, 82, 166, 204, 233). Clostridia were probably first alluded to about 30 centuries ago in the writings of Hippocrates, who reported on wound infections in his *Epidemics III* (13). Pasteur noted in 1861 that the anaerobic butyric acid fermentation was due to a rod-shaped organism that he called *Vibrion butyrique*, which likely corresponds to *Clostridium butyricum*, the type species of the genus *Clostridium* (233). Throughout history, clostridia have also had positive properties and utilities in biotechnology, the chemical industry, and medicine (40, 77, 166, 204). Certain nonpathogenic clostridia have been used as a source of solvents, including butanol, as a precursor for butadiene for synthetic rubber, and acetone, used in the synthesis of munitions (39, 82, 230). Solventogenic and cellulosic clostridia are seeing a renewed interest as energy sources for production of solvents and for partial replacement of oil (39). Clostridia are a rich source of enzymes and vaccines, and are ecologically important in the degradation of various pesticides and xenobiotics (40, 77). A remarkable medical application is the development of clostridial toxins for the treatment of a myriad of neurological disorders (73, 185). The study of clostridia has had an important impact on the development of microbiological techniques, including the development of anaerobic media and methods for cultivation of anaerobes from the rumen, human intestine, and other anaerobic environments (77, 233). Despite the medical and industrial importance of this genus, the genetic manipulation of *Clostridium* is still relatively elemental, although considerable progress has been made since the second edition of this manual was published in 1999 (126).

The genus *Clostridium* is classified in the family *Clostridiaceae* of the phylum *Firmicutes* of *Bacteria* (35, 43, 208). The genus comprises a large and diverse group of at least 12 lineages of anaerobic and some aerotolerant gram-positive bacteria (26, 63). Most of the species are rod shaped and form endospores. Traditionally, the clostridia have been recognized to have common properties (35, 43, 208): (i) a gram-positive cell wall structure, (ii) formation of endospores, (iii) an anaerobic and fermentative metabolism, and (iv) the inability to reduce sulfate to sulfide. The genus *Clostridium* has a wide range of 21 to 54 mol% G+C content of DNA, supporting the heterogeneity of this group of bacteria.

Currently, clostridia are mainly classified on the basis of morphology and motility, spore formation, physiological properties including anaerobiosis and fermentative metabolism, phenotypic properties such as protein toxin formation, and rRNA gene sequence homologies. On the basis of sequencing of genes encoding 16S rRNA, 19 homology groups have been proposed (26). The type species, *Clostridium butyricum*, and most of the clinically important clostridia cluster within homology group I. The DSMZ (Deutsche Sammlung von Mikroorganismen und Zellkulturen, the German Collection of Microorganisms and Cell Cultures) has over 170 clostridial species listed in its collection. From the 16S rRNA gene sequence analyses, several species have been proposed as type species of new genera (209), including several nonsporulating genera. Although endospore formation is known to occur only in the low-mol% G+C branch of *Firmicutes*, it is a very complex developmental process and endosporulation requires more than 150 gene products, with at least 75 acting sequentially (211). Disruption of any of these genes could result in lack of sporulation, which may explain the relatedness of nonsporulating genera to *Clostridium* (149).

The genetic manipulation of clostridia is still in an early stage of development, but significant advances have been made in recent years, and this chapter emphasizes newly developed genetic methods and strategies. Considering the diversity of clostridial species that are important industrially and in medicine, it is not feasible to describe genetic protocols for all the various groups. Most genetic methods are described in detail in the literature, and the reader is referred to the pertinent sources cited in this chapter and in authoritative reviews. Furthermore, certain traditional methods of genetic manipulation, including transduction, protoplast formation and fusion, and chemical mutagenesis, have mainly been transcended by newer techniques, and these are described in earlier reviews (31, 110, 126, 133, 134, 219, 239, 240).

17.2. PHYSIOLOGY OF *CLOSTRIDIUM* RELATED TO GENETIC MANIPULATION

The clostridia have considerable diversity in metabolism and physiology (6, 58). Certain physiological properties of clostridia are integral for genetic manipulation and gene expression (6, 41, 79, 136). Most clostridia are strict

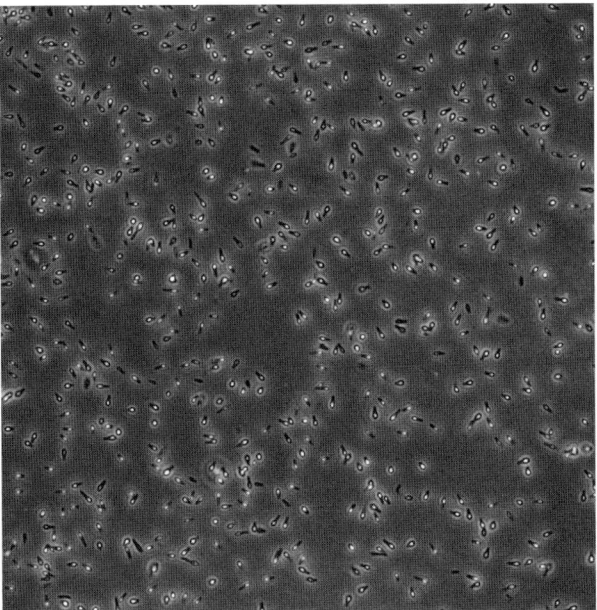

FIGURE 1 Characteristic spindle morphology of *C. botulinum* and presence of endospores. The photograph shows a phase-contrast visual micrograph (750×) of a culture of *C. botulinum* type A.

anaerobes and only grow in the absence of oxygen. Their manipulation requires anaerobic conditions such as the use of glove boxes, specialized culture tubes, and fermentors; reduced media; and other necessary conditions. Systems for manipulation of anaerobic organisms have been described (76). In addition, many clostridia tend to grow as mixed cultures with other anaerobes and facultative organisms, and strict attention is needed to affirm the purity of the culture under study or during culture growth and solvent, enzyme, or toxin production. Clostridia produce endospores, which are resistant to heat, radiation, drying, and chemicals. Spores can be visualized by phase-contrast microscopy (Fig. 1). Safe laboratory practices and special training of personnel are needed for the study of toxigenic species such as *Clostridium botulinum* and *Clostridium tetani*.

Clostridia are classified as proteolytic or saccharolytic depending on their preference for energy-yielding substrates. Many clostridial species are nutritionally demanding and are commonly isolated and grown in rich media (38, 63, 67, 81, 142). The optimal isolation of different species requires specialized media and conditions. The availability of defined media is useful for the isolation of certain classes of mutants such as auxotrophs (193). Chemically defined media have been described for several clostridia (80, 108, 179, 193, 228, 232). Media void of animal constituents have also been used for development of inocula for production of toxins for vaccines and other medicinal uses (48, 49). Rich media are commonly used for spore production (67, 196, 235) although spore production of certain clostridia has been obtained in defined media (179). For production of master cell banks and working cell banks for industrial processes, bacterial stocks should be prepared from mid-log cultures, particularly for processes where nonsporulating cultures are used (185). On the other hand, spore stocks are often valuable for maintaining physiological properties and preventing strain degeneration (89, 128).

17.3. NATIVE PLASMIDS, TRANSPOSONS, AND BACTERIOPHAGES

17.3.1. Native Plasmids

Several studies in the 1970s and 1980s showed that *Clostridium* spp. possess extrachromosomal plasmids (reviewed in reference 100). Most of the plasmids are considered cryptic, but plasmids carrying antibiotic resistance genes, bacteriocins, virulence factors, or genes for solvent production have been demonstrated (110, 175, 176). With the advent of new genetic tools, the function of some cryptic plasmids will be elucidated. Transferability of *Clostridium perfringens* plasmids was demonstrated in the 1970s (16). Plasmid states are also observed in pseudolysogenic phage cycles and as intermediates in transfer of conjugative transposons (19, 21). Native plasmids range in size from several thousand to several hundred thousand base pairs. The most extensively studied are *C. perfringens* plasmids, and this was the first *Clostridium* species that was found to carry conjugative plasmids (recently reviewed in reference 110).

Large plasmids (>50 kb) have also been found in several clostridia. Analysis of the large plasmids has lagged behind studies of gram-negative plasmids mostly due to difficulties encountered in purification and characterization of such large plasmids from clostridia (175). In the late 1980s large native clostridial plasmids encoding virulence genes were identified in neurotoxigenic clostridia. The gene encoding tetanus neurotoxin in *C. tetani* was discovered on an 84-kb plasmid in *C. tetani* (51). The *C. botulinum* serotype G neurotoxin gene cluster was also found on a 114-kb plasmid in *C. botulinum* (45, 243). In *C. botulinum* serotypes C and D the botulinum neurotoxin genes were determined to be located on pseudolysogenic bacteriophages, which may exist as a pseudolysogenic circular extrachromosomal state when not integrated into the chromosome (180).

Plasmid analyses using standard purification approaches did not initially identify these very large plasmids in clostridia. Large plasmids in clostridia were discovered using the pulsed-field gel electrophoresis (PFGE) method of total DNA preparations (27, 90, 122). PFGE analysis of nondigested preparations of total DNA from solvent-producing *Clostridium acetobutylicum* strain 824 identified an ~210-kb megaplasmid, pSOL1, that possesses genes for acetone and butanol production (27, 28). During sequencing it was established that the actual size of the plasmid is 192 kb and its nucleotide sequence was determined (147), and the role of the plasmid in *C. acetobutylicum* physiology and its relation to chromosomal genes have been studied (85, 153, 154). Katayama and coworkers (90) employed PFGE analysis of total DNA from different *C. perfringens* strains cleaved with the intron-encoded endonuclease I-CeuI from *Chlamydomonas eugametos*. This enzyme was shown to cleave only within the rRNA operons, and therefore this method was used to determine the localization of the virulence genes on extrachromosomal plasmids since I-CeuI only cuts chromosomal DNA. Based on these two approaches, the PFGE analysis has been used by several laboratories to identify large plasmids in *C. perfringens* (102, 138, 184) and *C. botulinum* (52, 122, 181). Virulence plasmids ranging in size from 47 to 270 kb have been identified in *C. botulinum* serotypes A, B, Ab, Ba, and Bf and in *C. tetani* (52, 122, 203). The presence of large plasmids in *C. botulinum* and *C. tetani* encoding neurotoxin genes has been confirmed by genomic sequencing (18, 203). Thus, large plasmids appear to be widely distributed in the clostridia, and it is anticipated that more will be discovered that encode metabolic and virulence properties.

The genetics, structure, and function of plasmids have been most extensively studied in *C. perfringens*. The nucleotide sequences of several *C. perfringens* plasmids have been determined, including a 10.2-kb plasmid, pIP404 (53); a 54-kb plasmid, pCP13 (strain 13) (200); the 47-kb pCW3 (strain CW92) (9); the 75.3-kb pCPF5603 (strain F5603) (138); the 70.5-kb pCPF4969 (strain F4969) (138); 12.4- and 12.2-kb plasmids from strain SM101 (143); and the 64.7-kb pCT8533ext (strain NCTC 8533) (139).

Plasmid pIP404 is considered a paradigm plasmid in *C. perfringens* and has been extensively studied (176). The plasmid contains 10 open reading frames (ORFs), and several of the genes are involved in plasmid replication and maintenance, while others are associated with bacteriocin production. The pIP404 replicon has been used in construction of clostridial shuttle vectors (see section 17.6.2). Sequencing and diversity analysis of *C. perfringens* plasmids has revealed extensive similarities among antibiotic resistance plasmids (175), as well as within different toxin-encoding plasmids in *C. perfringens* types A, B, and D (184) and E strains (102, 138, 139, 184). Among conjugative antibiotic resistance plasmids, the 47-kb tetracycline resistance plasmid pCW3 has been studied in detail. The regions responsible for the conjugation and replication functions of pCW3 have been identified (9), and the role of the *tcp* (transfer clostridial plasmid) locus in the transfer of pCW3 has been confirmed (9, 156, 210, 213).

Interestingly, the replication region identified in pCW3 is conserved in all conjugative plasmids from *C. perfringens* (9). It appears that antibiotic resistance and toxin plasmids in *C. perfringens* share a common but poorly understood mechanism for plasmid transfer, and that the *C. perfringens* *tcp* locus is unique to this organism and differs from the transfer regions of conjugative plasmids from other gram-positive bacteria (9). Conjugal transfer of the toxin plasmids to normal flora *C. perfringens* strains possibly has given rise to pathogenic isolates and is implicated in disease dissemination (139, 210). These suggestions are based on the sequence similarities observed between different *C. perfringens* plasmids encoding antibiotic resistance and toxins, as well as the presence of the same *tcp* locus on these plasmids and mobile insertion sequence (IS) elements that carry different toxin genes on the plasmids or on the chromosomes in pathogenic *C. perfringens*. The pCPF5603 enterotoxin plasmid of *C. perfringens* is conjugative (69) and has extensive homologies to the epsilon-toxin-encoding plasmid in *C. perfringens* (139).

The plasmids found in proteolytic *C. botulinum* serotypes are highly related and contain large regions of homology (203). A 47-kb plasmid found in nonproteolytic type B *C. botulinum* strain 17B is unrelated to the proteolytic serotype plasmids (157). The only regions revealing homology between the plasmids from proteolytic and nonproteolytic strains are their BoNT gene clusters (157). In addition, the gene for C2 toxin has recently been identified on a 107-kb plasmid in *C. botulinum* serotype C (181). Although functional studies of the genes located on *C. botulinum* plasmids are not yet described, putative genes involved in conjugal transfer have been identified by genome annotation (see the Pathema website at http://pathema.jcvi.org/cgi-bin/Clostridium/PathemaHomePage.cgi). Our laboratory has recently demonstrated that certain of these plasmids are conjugative in *C. botulinum* (K.M. Marshall, M. Bradshaw, and E. A. Johnson, unpublished data). Identification of conjugative virulence plasmids and availability of sequences from 13 *C. botulinum* genomes including several large virulence plasmids should facilitate future studies of this pathogenic clostridial species.

17.3.2. Clostridial Transposons and IS Elements

Several transposable elements have been demonstrated in clostridia (114). Conjugative transposable elements have been found in *C. perfringens* and *Clostridium difficile* (reviewed in references 4 and 114). Recently, a Tn*5382*-like transposon was demonstrated in a hospital isolate of *Clostridium bolteae*, a common inhabitant of the human intestine (37). Transposable elements carry antibiotic resistance determinants, particularly tetracycline and erythromycin. The Tc^r determinant of *C. perfringens* could be transferred to other *Clostridium* spp., including *C. difficile* (192), but the transconjugants were often unstable. Transposon Tn*1549* has been shown to transfer vancomycin resistance from *Clostridium symbiosum* to *Enterococcus* in a mouse model (99).

IS elements, which possess only genes involved in their transposition, have been found in clostridia (15, 36, 61, 62, 70, 106, 109, 115, 203). These elements have been located not only on the chromosomes but also on plasmids and bacteriophages (115, 180, 181, 203). They mainly belong to the IS3, IS6, IS200, IS256, and IS481 families. In contrast to many other bacterial species such as *Escherichia coli*, IS elements appear to be present in relatively small numbers in clostridia, and many are truncated or contain deletions and are presumably nonfunctional. Active IS elements are likely involved in mobile gene transfer and recombination of chromosomal regions of certain clostridia, such as *C. difficile* (191), and in recombination of highly diverse loci such as the neurotoxin and flagellin gene clusters in *C. botulinum* (22, 36, 203). Interestingly, IS elements also appear to be involved in cellulosome activity and associated microcrystalline cellulose degradation (245). The roles of transposons and IS elements warrant further study in clostridia in relation to gene movement, recombination, and metabolic activities, and for development of genetic tools.

17.3.3. Bacteriophages

Bacteriophages are present in many clostridia (3, 83, 148). The only group of *Clostridium* spp. that appear to lack phages are certain thermophiles (83). Phages are associated with toxin formation in *C. botulinum* types C and D, production of alpha-toxin by *Clostridium novyi* (44), and possibly sporulation of *C. perfringens* (244). Clostridial phages have also caused lytic infections of production strains, resulting in low yields of important industrial products such as acetone-butanol (84) and vaccines. Recently, bacteriophages from *C. botulinum* type C and *C. difficile* have been characterized at a molecular level (59, 71, 127, 180, 181).

Bacteriophages have not been routinely used for genetic analysis and manipulation by transduction in *Clostridium* spp. This is due in part to the unstable pseudolysogenic nature of many clostridial phages (4, 44), where the phage has a nonstable association with the chromosome and appears throughout much of its life cycle as an unstable element in the cytoplasm (78, 83). Phage manipulations have been hindered by practical aspects of working with the anaerobic clostridia (44, 83). However, more recent studies have demonstrated the ability to induce and assay for phage infection (59, 127). The development of transduction methods could have much utility in genetic analysis of *Clostridium* spp., but these methods require development.

17.4. DNA ISOLATION

17.4.1. Genomic DNA

Traditional protocols used for the preparation of clostridial genomic DNA consist of cell lysis with enzymes in the presence of detergents, followed by series of extractions with phenol and chloroform prior to precipitation of the nucleic acids with ethanol or isopropanol. A detailed protocol using this approach has been published in the *Handbook on Clostridia* (31). However, in the past decade an increasing number of publications on clostridia have utilized various commercial DNA extraction kits for purification of genomic DNA. For example, researchers have used the MasterPure Gram-Positive DNA Purification Kit (Epicentre Biotechnologies, Madison, WI) for DNA isolation from *C. perfringens* (139) and *C. botulinum* (171); Qiagen DNeasy Minikit for gram-positive organisms (Qiagen, Valencia, CA) for *C. difficile* (121); InstaGene Matrix kit (Bio-Rad, Hercules, CA) for *C. perfringens* (33) and *C. difficile* (14); Puregene Genomic DNA Purification kit (Gentra Systems, Minneapolis, MN) for *C. acetobutylicum* (189); and High Pure PCR Template Preparation kit (Roche, Mannheim, Germany) for *Clostridium septicum* (145). Our laboratory uses the ChargeSwitch Genomic DNA Purification kit (Invitrogen, Carlsbad, CA) for *C. botulinum*. These kits allow purification of the genomic DNA in concentrations sufficient for different applications, as well as minimizing the time required and the use of caustic organic solvents. Prior treatment of bacterial cells with lysozyme at 37°C for 15 to 60 min has been included as a modification for some kits to achieve complete lysis of bacterial cells and thus increase DNA yield (31, 145; our laboratory, unpublished data).

For some PCR applications, crude cell lysates are prepared by resuspending bacterial cells in distilled water and boiling (29, 86) or heating them in a microwave oven (231). The protocol described below was used to prepare template DNA for the multiplex PCR of the *C. perfringens* toxin genotyping assay (231). Four or five *C. perfringens* colonies from the brain heart infusion plate are suspended in 200 μl of sterilized phosphate-buffered saline. The cell suspensions are microcentrifuged at 14,000 × *g* for 5 min, and the cell pellets are washed twice with phosphate-buffered saline and resuspended in 100 μl of PCR-grade H₂O. Microcentrifuge tubes containing washed *C. perfringens* cells are sealed with Parafilm and subjected to heating in a microwave (700 W) for a total of 20 min (administered as four separate 5-min heat treatments with 1-min cooling intervals) to induce bacterial lysis. The resultant lysates are cleared by microcentrifugation at 14,000 × *g* for 5 min, and 5 μl of each supernatant is used directly as template DNA in a multiplex PCR assay.

An automated method for the extraction of high-quality DNA from clostridial strains has been developed using the Nuclisens EasyMag bio-robot (Biomérieux, Marcy l'étoile, France) for testing the specimens of fat meat samples for contamination with *C. botulinum*, *Clostridium baratii*, and *C. butyricum* (47). According to the protocol, 25 or 50 g of meat samples (foie gras) were diluted 10-fold (wt/vol) in prereduced tryptone-yeast extract-glucose broth containing 200 μg/ml of D-cycloserine under a gas flow of N₂/H₂ (95:5 [vol/vol]) and then incubated anaerobically for 48 h at 37°C. After the enrichment step, the broth was collected from the extracts, a 2-ml aliquot was centrifuged at 9,000 × *g* for 5 min, and the supernatant was discarded. The cell pellet was mixed with 1 ml of the EasyMag lysis buffer (Biomérieux), and DNA extraction was performed with the Nuclisens EasyMag bio-robot following the instructions of the manufacturer. Extracted DNA (5 μl) was used in the PCR reactions.

17.4.2. Plasmid DNA

Protocols for isolation of plasmid DNA from clostridia are mainly based on the standard alkaline lysis method (182),

followed by extraction of nucleic acids from the lysates with phenol-chloroform and precipitation of plasmid DNA with ethanol or isopropanol. The lysis step of bacterial cells with lysozyme is recommended prior to the denaturation step (31, 126). Different commercial plasmid kits are also frequently used for plasmid isolation from clostridia, and preincubation of the cells with lysozyme is also strongly recommended (31).

A method for isolation of large plasmids (~50 kb) from *C. perfringens* has been developed (2). *C. perfringens* cells were cultured overnight on nutrient agar containing 5 μg/ml of tetracycline and then subcultured in 100 ml of Trypticase-peptone-glucose broth. Mid-log-phase cells were washed and suspended in 2 ml of 25% sucrose in 30 mM TES (Tris-hydrochloride buffer [pH 8.0] containing 5 mM EDTA and 50 mM NaCl). Then, 0.4 ml of a lysozyme solution (10 mg/ml in TES) was added and the cell suspension was incubated for 15 min at 37°C. This was followed by addition of 0.8 ml of 250 mM EDTA (pH 8.0), and incubation continued for an additional 5 min. Complete lysis of the cells was achieved by adding 0.5 ml of 10% (wt/vol) sodium dodecyl sulfate and incubating the preparation at 37°C for 5 min. The lysate was then treated with addition of 5 M NaCl solution to the viscous cell lysate with gentle mixing to a final concentration of 1 M. The lysates were stored at 4°C overnight and centrifuged at 17,000 × *g* for 30 min. This procedure removed essentially all chromosomal DNA and left plasmid DNA in the supernatant fluid. The resultant lysates were then subjected to cesium chloride-ethidium bromide density gradient centrifugation for 20 h. Plasmid bands were removed, dialyzed twice against 0.1 × TES, and stored at 4°C.

Routine methods for isolation of purified high-quality DNA from clostridial megaplasmids ranging in size between 100 and 300 kb are not available. The PFGE method of undigested total chromosomal samples has been employed for identification of these large plasmids (see section 17.3.1). The following protocol is routinely used in our laboratory for preparation of PFGE samples from clostridial strains. Bacterial cultures were grown anaerobically in anaerobic tubes in 10 ml of Trypticase-peptone-glucose-yeast extract medium until the optical density at 600 nm reached approximately 0.6. Bacterial cells were fixed with formaldehyde to inhibit DNase activity by injection of 1 ml of formaldehyde solution through the septum into the 10-ml bacterial cultures. The contents were mixed, the tubes were incubated on ice for 30 min, and cells were harvested by centrifugation. Cell pellets were washed twice with 10 ml of 0.85% NaCl solution; resuspended in 1.5 ml of solution containing 20 mM Tris-HCl (pH 8.0), 2 M NaCl, and 200 mM EDTA; and incubated for 10 min in a 50°C water bath. Equal volumes of the cell suspension and 1.5% InCert agarose (BioWhitaker Molecular Applications, Rockland, ME) prepared in distilled water were mixed, and the cell-agarose mixtures were loaded into plug-forming molds. The gel molds were chilled in a refrigerator for 15 min, and the agarose plugs were removed and incubated in 4 volumes of lysis buffer (10 mM Tris-HCl [pH 8.0], 1 M NaCl, 100 mM EDTA, 0.5% Brij 58, 0.2% deoxycholate, 0.5% sarcosyl, 100 μg/ml RNase A, 5 mg/ml lysozyme, and 40 U/ml mutanolysin) at 37°C for 12 h. The lysis buffer was removed, and the plugs were rinsed with 50 mM EDTA solution to remove lysozyme, RNase, and mutanolysin. Two volumes of proteinase K (PK) buffer (10 mM Tris-HCl [pH 8.0], 0.5 M EDTA, 1% sarcosyl, and 0.5 mg/ml proteinase K) were added, and the agarose plugs were incubated for 48 h with gentle agitation in a 50°C water bath. Following proteinase K treatment, the plugs were washed with Tris-EDTA (TE) buffer and incubated for 2 h at 37°C, with shaking in TE buffer containing 1 mM phenylmethylsulfonyl fluoride to inactivate proteinase K. The agarose plugs were washed three times for 2 h each with 10 volumes of TE buffer to remove phenylmethylsulfonyl fluoride, then inserted into the agarose gel; the samples were separated using a clamped homogenous electric field system (CHEF-DRII [Bio-Rad, Hercules, CA]). Several methods for extraction of the plasmid DNA from the agarose gels following PFGE have been tried; however, the recovered DNA was significantly degraded during the extraction process and the yields decreased. More improved methods are necessary for isolation of these megaplasmids.

17.4.3. Bacteriophage DNA

Bacteriophages from *C. botulinum* type C, *C. difficile*, and *C. perfringens* have been investigated at the genomic level (59, 180, 244). In these studies, phage growth was induced by treatment of the bacterial cultures with mitomycin C or UV to induce lysis, and culture lysates were cleared by filtration or centrifugation. The cell-free supernatants were then treated with DNase and RNase and washed with NaCl solution, and phage particles were isolated by precipitation with polyethylene glycol followed by cesium chloride ultracentrifugation. The DNA from the phages can be isolated using DNA isolation kits available from various manufacturers following standard techniques (182). Another strategy for phage DNA isolation from *C. botulinum* type C strain was used by Sakaguchi and coworkers (180). These investigators prepared phage particles from the bacterial culture as described above, then embedded the phages into the agarose plugs, and after treatment with proteinase K and *N*-lauroylsarcosine at 50°C subjected the phage preparations to PFGE. The phage DNA migrating at the 186-kb position was recovered from the gel using a DNA extraction kit (Prep-A-Gene DNA purification kit [Bio-Rad]), and the DNA was sequenced.

17.5. GENE ISOLATION

17.5.1. Direct Cloning

The cloning and analysis of clostridial genes has led to several important discoveries of gene structure and function. The construction of clostridial gene banks began in the 1980s and generally used multicopy plasmids such as pBR322 and the pUC series (239, 240). The chromosomal DNA fragments were generally prepared by partial digestion of total chromosomal DNA with a frequently cutting restriction endonuclease, Sau3AI and/or MboI; both enzymes recognize the same GATC sequence, but Sau3AI cleaves methylated while MboI cleaves unmethylated sites (240). The resulting DNA fragments were separated through the agarose gels, and the fraction containing 4- to 10-kb fragments was eluted from the gels and ligated into *E. coli* cloning vectors. The average size of the inserts was ~4.5 to 5.0 kb. Therefore, 3,000 to 5,000 clones were necessary to cover the entire genome. For generation of larger insert libraries (20 to 40 kb), phage λ or cosmid vectors were employed (240). The screening of genome libraries by immunological methods and hybridization of clones with specific gene probes were used to isolate and further characterize different genes (reviewed in reference 240).

With the development of DNA amplification technologies, PCR methods are now routinely used for gene isolation and cloning. Amplification can be problematic in *Clostridium* due to the high % A+T of the coding regions

in some species. Generally, for amplification of random gene fragments, 20- to 25-mer primers can be designed from the gene regions containing 45 to 55% of the G-C bases. For amplification of the entire genes, which may contain very few G-C bases in their 5′ and 3′ regions, longer PCR primers (30- to 40-mers) are usually designed to ensure proper annealing to the target sequence without a need to use annealing temperatures that are too low in the amplification reactions. The PCR methods also permit addition of appropriate restriction sites during amplification of clostridial genes or their fragments to facilitate cloning into various commercial cloning vectors. The employment of PCR enzymes with high fidelity is strongly recommended to increase the accuracy of the amplified gene products. Generally, PCR reactions for amplification of clostridial genes are performed following the instructions of the enzyme manufacturers. For situations when the initial PCR conditions did not result in a desired product, investigators have optimized their reaction conditions, such as by increasing the purity of the target DNA, altering the stringency of the PCR, using longer primers or different enzymes, or other modifications. In our laboratory, we have observed that increasing the concentration of the dATP/dTTP twofold in the reaction mixes helps to achieve higher yields of the longer PCR products (>3 kb). Interestingly, in 1989 only ~50 genes from 15 different clostridial species were cloned and partially sequenced (239, 240). Currently there are 24,101 sequence entries in GenBank, containing different genes, gene fragments, and entire genomes from 40 clostridial species (see section 17.9).

17.5.2. DNA Transfer Techniques

Several different approaches have been used for introduction of DNA into *Clostridium* spp. Natural competence has not been reported in clostridia (31). The most commonly used approaches are electrotransformation (electroporation) and conjugation of vectors from *E. coli* or *Enterococcus faecalis* to *Clostridium*. Detailed protocols for DNA transfer are described in the second edition of this manual (126) and in *Handbook on Clostridia* (31).

17.5.2.1. Electroporation

Electroporation efficiencies are generally low in *Clostridium* spp. compared to many other organisms. In early studies, gene transfer by electroporation was highly strain dependent. For example, for *C. perfringens*, strain 13 has been used for genetic manipulation in preference to many other strains because it lacks certain restriction/modification systems that prevent the introduction of foreign DNA (176). Similar problems have been encountered in other clostridial species. However, methylation of plasmid DNA prior to electroporation enables plasmid transfer to many *Clostridium* spp. (31, 126). Systematic studies have not been published for increasing transfer efficiencies by electroporation or conjugation, but studies for optimization of DNA uptake may be useful to improve DNA transfer efficiency.

17.5.2.2. Conjugal Transfer between Clostridia

Conjugal transfer between the same and different species of *Clostridium* certainly occurs in the environment and in the gastrointestinal tracts of mammals, but it is not used for routine genetic manipulation. Conjugal plasmids have rarely been identified in clostridia, and so far have only been demonstrated in *C. perfringens* (176) and recently in *C. botulinum* (Johnson laboratory, unpublished data).

Intraspecies conjugal transfer frequencies are generally low, and transfer is strain specific. Therefore, host vector systems have been developed to facilitate gene transfer.

17.6. HOST VECTOR SYSTEMS

17.6.1. Host Strains
Due to a low efficiency of DNA transfer in clostridia, alternative hosts are used for clostridial gene manipulation. *E. coli* has mostly been utilized as a host for generation of gene constructs; thus, shuttle plasmids must contain both clostridial and *E. coli* replication sequences (126). *E. coli* strains are also used as donor strains for conjugative transfer of shuttle plasmids into clostridial strains. Shuttle plasmids designed for conjugal transfer carry the origin of transfer (*oriT*) from the IncP plasmid RK2 (237). Other conjugation functions are provided in *trans* from another IncP plasmid present in the donor strain (237) or from the chromosomal copy in the donor strain (111). In early studies *Bacillus subtilis* was also used as the host strain for plasmid transfer in *C. acetobutylicum* and *C. difficile* (132, 134).

17.6.2. Vector Systems
Most of the shuttle vectors described in the literature were originally developed for studies of *C. perfringens* and *C. acetobutylicum*. These vectors were later modified and adapted for use in other clostridial species. However, systematic comparisons of various plasmid components have not been performed in different clostridial strains, and modifications of these vectors would require considerable cloning and gene manipulation. Construction of various vectors and detailed descriptions of vector components have been reviewed (31, 66, 110, 126, 133–135, 176, 177, 219, 237–239) and therefore will not be described here. In this chapter, we will only describe the replicons (Table 1), selective markers (Table 2), and reporter genes (Table 3) most frequently used for construction of clostridial shuttle vectors and then describe the latest developments in the field.

Recently the Minton laboratory designed and constructed a standardized modular system for generation of *Clostridium*/*E. coli* shuttle plasmids (66). This system enables construction of specific shuttle vectors suitable for a particular host and application by combining standardized plasmid components or modules; it is simple and rapid and the exchange of the modules does not require extensive cloning steps. A schematic presentation of this modular vector system appears in Fig. 2.

In this system, the necessary shuttle plasmid elements are organized into four different modules. The modules include (i) a gram-positive/*Clostridium* replicon, (ii) a selectable marker, (iii) a gram-negative replicon with or without a conjugal transfer function, and (iv) an application-specific module. Each module is flanked by two designated restriction sites of rare cutting type II restriction endonucleases: AscI, FseI, PmeI, and SbfI, which recognize 8-bp palindromes.

TABLE 1 Replicons used for construction of clostridial shuttle vectors

Plasmid	Organism	Comments	Reference(s)
Heterologous replicons			
pBC161	*Bacillus cereus*	Unstable in *C. acetobutylicum*	217
pPIM13	*B. subtilis*	ssDNA[a] replication	167
pAMβ1	*E. faecalis*	Theta replicating, broad host range	212
pWV01	*L. lactis*	Functional in *E. coli* and clostridia	95
pUB110	*Staphylococcus aureus*	ssDNA replication, thermostable in *S. aureus*	206
pT127	*S. aureus*	Unstable in *C. acetobutylicum*	217
Clostridial replicons			
pCAK1	*C. acetobutylicum*	Phagemid ssDNA intermediate	93
pSC86	*C. acetobutylicum*		236
pBP1	*C. botulinum*	Stable in *C. botulinum*	66
pCB101	*C. butyricum*	ssDNA replication	17, 240
pCB102	*C. butyricum*		135
pCB103	*C. butyricum*		135
pCBU2	*C. butyricum*		101
pCD6	*C. difficile*		168
pCPA1	*Clostridium paraputrificum*		135
pHB101	*C. perfringens*		94
pIP404	*C. perfringens*		54, 168
pJU121	*C. perfringens*	Replicates in *C. beijerinckii*	12
pJU122	*C. perfringens*	Replicates in *C. beijerinckii*	12

[a]ss, single-stranded.

TABLE 2 Antibiotic resistance genes for construction of clostridial shuttle vectors

Marker	Gene/organism	Reference(s)
catP	Chloramphenicol acetyltransferase/pIP401 C. perfringens	113
ermB	Erythromycin/pAMβ1 E. faecalis	150
ermBP	Macrolide-lincosamide-streptogramin B/pIP402 C. perfringens	10
ermC	Erythromycin/pPIM13 (B. subtilis)	141, 217
aphA-3	3′-Aminoglycoside phosphotransferase of type III E. faecalis strain BM4127 (resistance to kanamycin)	123
aad(9)	Spectinomycin adenyl transferase/E. faecalis	66
tetA	Tetracycline/C. perfringens	112
tetM	Tetracycline/C. perfringens	112
tetP	Tetracycline/pCW3 C. perfringens	1
tetC	Tetracycline/pT127 S. aureus	217

These restriction sites are not present in any of the plasmid modules. In this system every shuttle vector contains one module of each of the four types, and the elements are always present in the same order. For certain applications a functional module can be replaced with a short spacer.

The elements for the modules have been selected from known plasmid components with a wide range of properties suitable for use in different clostridial hosts (Table 4). The first module contains four different gram-positive replicons flanked by AscI and FseI restriction sites. These include a novel replicon from indigenous plasmid pBP1 identified in *C. botulinum* strain NCTC 2916, from *B. subtilis* plasmid pIM13, from *C. butyricum* plasmid pCB102, and from *C. difficile* plasmid pCD6. The second module contains selectable marker genes flanked by FseI and PmeI restriction sites. Two of the most commonly used markers, *catP* and *ermB*, were selected along with the *tetA* gene from *C. perfringens* plasmid pCW3 and the *aad9* gene encoding resistance to spectinomycin from an *E. faecalis* plasmid. All four genes are also functional in *E. coli*. The third module contains gram-negative replicons and is flanked by PmeI and SbfI restriction sites. Two of the replicons selected are the high-copy-number ColE1 replicon, suitable for most applications, and p15a replicon

for low-copy-number plasmids for cloning and expression of potentially toxic products. The two replicons are also available in combination with a conjugal transfer function from RK2, which can be easily deleted by ApaI digestion. Thus, four different elements are available for this module. The last module is designed for insertion of application-specific elements, such as a promoter or reporter gene. This module is flanked by SbfI and AscI restriction sites, and it contains a multiple cloning site (MCS) and two bidirectional transcription terminators on each side of the MCS (Fig. 2B). Three additional elements have been constructed for this module. Two of them contain the promoter and ribosome-binding site of the thiolase gene (*thl*) from *C. acetobutylicum* ATCC 824 (55) or the ferrodoxin gene (*fdx*) from *Clostridium sporogenes* NCIMB 10696 (158). These are inserted between the NotI and NdeI sites, resulting in modules suitable for gene expression. In the third derivative, the entire *lacZα* ORF is replaced by the *catP* ORF with an NdeI site overlapping the start codon, to serve as a reporter module. Each of the module elements is given a number, and therefore constructed vectors can be described by a standard nomenclature. The investigators have reserved plasmid designations pMTL80000 to pMTL89999, where the new shuttle plas-

TABLE 3 Reporter genes used for construction of clostridial shuttle vectors

Reporter gene, organism	Clostridium sp. studied	Reference(s)
amyP, C. acetobutylicum	C. acetobutylicum	178
catP, C. perfringens	C. perfringens	20, 75, 87, 125
englA, C. acetobutylicum	C. beijerinckii	169
gusA, E. coli	C. acetobutylicum	55
	C. beijerinckii	173
	C. perfringens	224, 227
	C. difficile	119, 120
lacZ, Thermanaerobacterium sulfurigenes	C. acetobutylicum	32, 188, 189, 215, 222
	C. botulinum	32
luxAB, Vibrio fischeri	C. perfringens	163, 164
	C. botulinum	32
luxB, Photinus pyralis	C. acetobutylicum	50
	C. botulinum	186

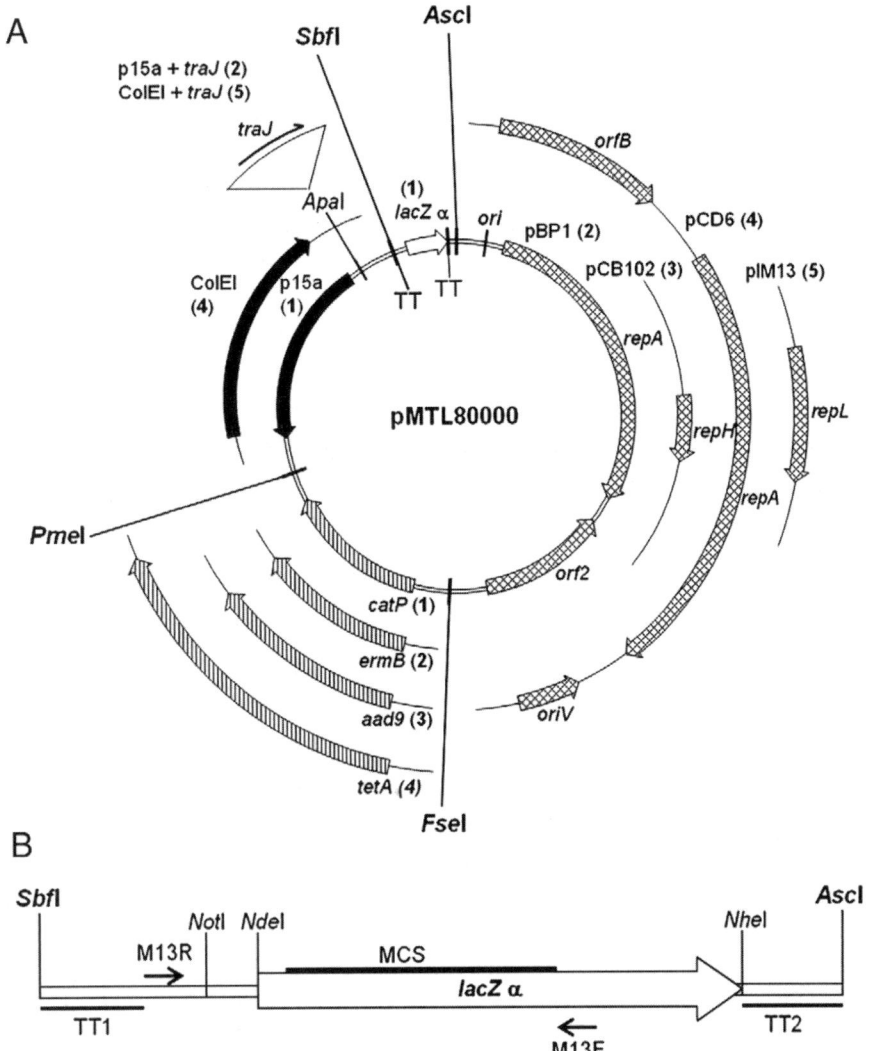

FIGURE 2 (A) Schematic presentation of pMTL80000-series modular plasmids, showing most available modules. (B) The MCS module. TT1, transcriptional terminator from downstream of the CD0164 ORF of *C. difficile* 630; TT2, transcriptional terminator from the *fdx* gene of *Clostridium pasteurianum*. (Reprinted from reference 66 with permission of the publisher.)

mids are named by the combination of different elements in the determined order, as shown in Table 4.

It is possible to construct 400 different vectors by combining the 18 modules generated so far. The five vectors presented in Table 4 contain each of these 18 modules at least once, and therefore the system is very portable. New modules can be constructed using the same specifications described above. Construction of the modules, their properties, and testing of different module combinations in several clostridial species have been described in detail (66), and additional information is available at http://clostron.com/.

17.7. GENE MANIPULATION AND GENE INACTIVATION METHODOLOGIES

17.7.1. Classical Mutagenesis

One of the first demonstrations of deliberate isolation of mutants in clostridia was described by Sebald and Costilow

in 1975 (193). Nutritional auxotrophs were isolated using *N*-methyl-*N'*-nitro-*N*-nitrosoguanidine and a chemically defined medium for selection. Strain development of *Clostridium thermocellum* for increased cellulase formation was also performed in the early 1980s using the chemical mutagen ethyl methanesulfonate (129). Chemical mutagenesis using *N*-methyl-*N'*-nitro-*N*-nitrosoguanidine has also been applied to *C. butyricum*, *C. acetobutylicum*, and other clostridia for industrial process improvement (137). With the development of molecular and directed mutagenesis systems, chemical mutagenesis has largely been transcended in clostridial strain development, and especially in genetic analyses.

17.7.2. Transposon Mutagenesis

Transposons have played a crucial role in defining microbial gene functions. A revised nomenclature for transposable genetic elements has been described, and a registry for designation of Tn numbers is available (http://www.ucl.

TABLE 4 Numbering scheme for pMTL80000-series modular plasmids[a]

Plasmid	Gram[+] replicon	Marker	Gram[−] replicon	Application-specific
pMTL80110	0. Spacer	1. *catP*	1. p15a	0. Spacer
pMTL82254	2. pBP1	2. *ermB*	5. ColE1 + *tra*	4. *catP* reporter
pMTL83353	3. pCB102	3. *aad9*	5. ColE1 + *tra*	3. P*fdx* + MCS
pMTL84422	4. pCD6	4. *tetA*	2. p15a + *tra*	2. P*thl* + MCS
pMTL85141	5. pIM13	1. *catP*	4. ColE1	1. MCS

[a]The core distribution set of five plasmids, which include all 18 modules constructed to date. GenBank accession numbers: pMTL80110, FJ797644; pMTL82254, FJ797646; pMTL83353, FJ797648; pMTL84422, FJ797650; pMTL85141, FJ797651. (Reprinted from reference 66 with permission of the publisher.)

ac.uk/eastman/tn/) (174). It is envisaged that the website will include sequence and ORF data and will be interfaced with ISfinder (174). In clostridia, generation of random mutants has been restricted to the use of the conjugative transposons Tn916 and Tn1545 (reviewed in reference 64). However, their utility has several significant limitations. These elements themselves are extremely large and may affect expression of genes near the integration site, and their transfer efficiencies are low. In some clostridial species Tn916 inserted randomly, while in others "hot spots" were identified. Moreover, a significant proportion of isolated transconjugants contained multiple copies of Tn916. In addition, Tn916 remains active after insertion, which can result in unstable mutants, and deletions of genome regions upon insertion of Tn916 have also been reported (7, 79, 104). Thus a smaller, nonconjugative element would provide much greater utility for generation of random mutants in clostridia. Until recently such elements had not been available for mutagenesis in clostridia.

Random mutagenesis methods based on transposition of a phage Mu and Tn5 have been recently reported for *C. perfringens* (98, 225). This approach appears to be more efficient and random than previously used conjugative transposons and, most importantly, generated mutants with only single-copy transposon insertions (98, 225).

The phage Mu and Tn5 transposon systems are based on the same principle and consist of similar components. The strategy involves an initial assembly of the transposition machinery in vitro, with subsequent introduction of these transposomes into the recipient cells by electroporation, resulting in genomic integration of the delivered transposon DNA within the recipient genome (57, 96). A transposition complex, or transposome, is formed between a transposase and the transposon by covalently linking the enzyme to the inverted repeat ends of the transposon. The transposome is catalytically inactive in the absence of divalent cations, but becomes activated for transposition chemistry in the presence of Mg^{2+} in the recipient cells (http://www.epibio.com).

Both Ez-Tn5- and bacteriophage Mu-based transposition systems are commercially available from Epicentre Biotechnologies and are described as simple, rapid, and straightforward methods for in vivo mutagenesis (http://www.epibio.com). This approach avoids the need to generate species-compatible transposase expression systems and specific suicide vectors, since transposition of the transposome does not require specific host factors. Therefore, this methodology is recommended for generation of mutant libraries in species that have poorly described genetic systems or lack adequate molecular tools but for which electroporation protocols have been established and reasonable electroporation efficiencies can be achieved. The higher the transformation efficiency of the target strain, the more

clones will be produced. Considering the low electroporation efficiencies reported for different clostridial species, it is not surprising that this mutagenesis strategy has been successful for generation of mutants in *C. perfringens* strain 13, where the highest electroporation efficiencies can be achieved. Interestingly, phage Mu-based mutants were generated not only in *C. perfringens* strain 13 and its derivative JIR25, but also in a field isolate after optimization of the electroporation protocol (98). In *C. perfringens* strain 13 derivative JIR25, 239 transformant colonies per 1 μg of Mu-transposome were obtained, as were 134 transformants in field isolate 56 (98), while ~50-fold more transformants (11,200) per 1 μg of EZ-Tn5 DNA were generated in *C. perfringens* strain 13 (225). Thus, a 2- to 3-log reduction in the number of transformants obtained in *C. perfringens* is observed compared to the number of transformants reported for *E. coli* target strains (http://www.epibio.com). Interestingly, 43% of the mutants obtained by the Mu system were inserted into rRNA genes (98), while in only 18% of the Tn5 mutants the element was inserted into the rRNA genes (225). However, for comparison of these two mutagenesis elements, more data are required, since 200 Mu clones and only 11 Tn5 clones were analyzed. Similar preferential insertion of the Mu-transposon into the rRNA cluster has also been described for *Saccharomyces cerevisiae* (151). It has been suggested that phage Mu insertions favor GC-rich sequences; this would explain insertion into rRNA genes. Detailed protocols have been described for construction of the EZ-Tn5 and Hyper-Mu transposons containing erythromycin resistance genes functional in clostridia for mutant selection, as well as for preparation of transposomes, electroporation into *C. perfringens* target strains, and clone analysis (98, 225). These studies are of considerable significance and should encourage research groups to adapt this mutagenesis technique to other clostridial species by optimizing and improving their electroporation efficiencies, as described by Lanckriet and coworkers (98).

17.7.3. Directed Gene Inactivation

17.7.3.1. Directed Gene Inactivation Using Host-Mediated Recombination

Generation of directed mutants using classical methods based on host-mediated recombination has been very inefficient in clostridia due to rare recombination events observed between the mutated and the wild-type gene, the lack of conditional shuttle vectors, and the relatively low efficiencies observed for DNA transfer by electroporation and conjugation. Over the past decades only a limited number of mutants have been isolated from a few *Clostridium* species, such as *C. perfringens*, *C. difficile*, *C. septicum*, *C. acetobutylicum*, *C. beijerinckii*, and *C. tyrobutyricum* (reviewed in reference 64). Most of these mutants were derived by a single crossover integration of a replication-defective (suicide) plasmid by homologous recombination. Such integrants are segregationally unstable because the entire plasmid carrying the modified gene has integrated into the genome at the target site by a Campbell-like mechanism. Ideally, stable integrants resulting from allelic exchange are preferred; however, double-crossover mutants of several genes have been generated primarily in *C. perfringens*, mostly in strain 13, where the highest transformation frequencies have been observed (64), and for a single gene in *C. septicum* (91). In general, these experiments are tedious and require substantial time, effort, and a fortuitous outcome. In a recent review, different approaches used and problems encountered during generation of these single- and double-crossover mutants in clostridia are discussed (64).

17.7.3.2. Directed Gene Inactivation Using a Retargeted Group II Intron

An alternative approach to achieve gene inactivation in clostridia is to use DNA integration mechanisms that are independent of host recombination factors. One such system is based on mobile bacterial group II introns (88), which are found in bacteria and eukaryotic organelles, where they function as catalytic RNAs and retrotransposable elements (97). The mobilization mechanisms and target specificities of the group II intron from the *ltrB* gene of *Lactococcus lactis* have been described (reviewed in reference 97). The L1.LtrB is a site-specific retroelement that uses a mobility mechanism termed "retrohoming," by which the excised intron lariat RNA inserts directly into a DNA target site and is then reverse transcribed by the associated intron-encoded enzyme protein (IEP) (97). The insertion target site is recognized primarily by base pairing between the excised intron lariat RNA and the target-site DNA. Therefore, modification of the intron sequence at a small number of positions recognized by the protein can lead to altered target specificity (140). Based on these observations, Lambowitz and colleagues developed group II introns into a new class of gene-targeting vectors ("targetrons"), which can be reprogrammed to insert into desired DNA targets simply by modifying the intron RNA. Detailed analysis of the nucleotide frequencies at different active target sites resulted in generation of an algorithm that predicts optimal L1.LtrB intron-insertion sites and designs primers for modifying the intron to target any gene of interest (161). The IEP gene in the targetron is removed from the intron and placed downstream from the intron on the same plasmid. Such arrangement of the intron components (as a minitransposon) does not affect the intron mobility but results in stable mutants. By removal of the IEP-encoding plasmid after insertion of the element in the desired position, the intron is not able to splice out. Furthermore, to facilitate positive selection of inactivated genes, an antibiotic resistance marker designated retrotransposition-activated marker (RAM) was engineered into the group II intron (241). RAM is an antibiotic resistance gene interrupted by another intron, a self-splicing group I intron. The group II intron and RAM are positioned and oriented in such a way that only after insertion of the group II intron into the target site in the genome can the group I intron be removed by self-splicing, thus restoring the integrity of the antibiotic resistance gene and allowing positive selection for integration events. The L1.LtrB mobile group II intron has been commercially developed as a genetic tool (TargeTron) for

insertional inactivation of genes. Several vectors for targeting both gram-positive and gram-negative bacterial genes with or without positive selection (containing RAM) are available from Sigma-Aldrich (St. Louis, MO; http://www.sigmaaldrich.com/life-science/functional-genomics-and-rnai/targetron.html).

The targetron technology has been adapted and successfully used for gene inactivation in clostridia by several research laboratories (23–25, 66). Construction of specific vectors and further improvements of this technology for inactivation of genes in different clostridial species have been recently reviewed (64).

The first report of inactivation of the *plc* gene in *C. perfringens* strain 13 using the group II intron technology was described by Chen and coworkers (24). Researchers used the basic TargeTron vector pACD3 (Sigma-Aldrich) and made modifications by including a clostridial replicon in the vector and replacing a gram-negative antibiotic resistance marker with a functional marker in *C. perfringens*. Additionally, the T7 promoter for transcription of the group II intron was replaced with a promoter region from the *C. perfringens* gene *cpb2*, generating the vector pJIR750ai. This vector was further modified for targeting the *plc* gene of *C. perfringens*. The targeting vector was transferred by electroporation into *C. perfringens* strain 13, and *plc* knockout clones were selected using a simple phenotypic screen followed by clone analysis with PCR and sequencing.

Plasmid pJIR750ai has also been used to construct a vector, pSY6, that was used for targeting two genes in *C. acetobutylicum*: *buk* (encoding the butyrate kinase) and *solR* (a putative repressor of solvent formation genes) (197). The targetron vector was constructed by inserting the group II intron from pJIR750ai into the *E. coli-C. acetobutylicum* shuttle vector pIMP1 (131), which is under control of the *C. acetobutylicum* phosphotransbutyrylase gene (*ptb*) promoter. The vector contains *B. subtilis* replication sequences. Transformants were screened by PCR in order to isolate *buk* and *solR* gene knockout clones.

Significant improvements of the group II intron technology for use in clostridia have been developed by the Minton laboratory (64, 65). Their system, designated ClosTron, is designed for generation of mutants in different clostridial species, and it enables positive selection of the mutants (64, 65). Positive selection was not used for mutant screening in the experiments described above for isolation of the *plc* mutant in *C. perfringens* and *buk* and *solR* mutants in *C. acetobutylicum*. The two RAM elements available for use with the L1.LtrB group II introns are based on kanamycin or trimethoprim selection (241); however, these cannot be used in clostridial species. Therefore, a new RAM gene was constructed based on the *ermB* gene from *E. faecalis* plasmid pAMβ1, which has been widely used for construction of *E. coli-Clostridium* shuttle vectors (126). The self-splicing group I intron from the *td* gene of phage T4 was inserted between the *ermB* ORF and its promoter region, and the new RAM was placed under control of a stronger promoter derived from the *C. acetobutylicum* thiolase gene (*thlA*) to ensure that expression of *ermB* from a single-copy gene on the chromosome was sufficient to provide erythromycin resistance in the host cells (65). The group II intron was transcribed from an isopropyl-β-D-thiogalactopyranoside-inducible clostridial *fac* promoter in pMTL007, and under constitutive *fdx* promoter in the second-generation ClosTron vector pMTL007C-E2 (64). The ClosTron vectors contain the *catP* gene, and these constructs are maintained in *E. coli* in the presence of chloramphenicol and in clostridia in the presence of thiamphenicol. To ensure the loss of the vector upon intron insertion, ClosTron vectors contain a pCB102 replicon from *C. butyricum* that functions with various degrees of efficiency in different clostridial species. Therefore, the plasmid can be removed from the mutant strains by streaking out the integrant clones on media without thiamphenicol.

The ClosTron vectors can be transferred to clostridial strains by conjugation from a suitable *E. coli* donor strain (such as CA434) since the vectors carry the *oriT* region and *traJ* gene, as well as by electroporation. A new methylation plasmid, pAN2, compatible with pMTL007, also has been constructed. This plasmid is used to methylate ClosTron vectors prior to electroporation into clostridial strains to circumvent their type II restriction/modification systems. Lastly, to eliminate possible polar effects due to the presence of a powerful *thlA* promoter inside the intron, which can lead to increased expression of downstream and upstream genes, the *ermB* gene in the second-generation vector pMTL007C-E2 is flanked by FLP recombination target sites. This modification permits excision of the *ermB* gene in the presence of the FLP recombinase, which can be expressed from a segregationally unstable plasmid introduced into the mutant strains. Removal of the *thlA-ermB* gene from the mutants would significantly reduce potential polar effects and facilitate generation of multiple mutations by employing the ClosTron vectors with the same positive selection of mutants using erythromycin resistance. Detailed protocols for construction of retargeting intron plasmids for gene inactivation and other useful tools and information can be found at the TargeTron website (http://www.sigmaaldrich.com/life-science/functional-genomics-and-rnai/targetron.html) and at the ClosTron website (http://clostron.com/).

The group II intron mutagenesis system can be used not only for gene inactivation (knock-out) but also for gene transfer (knock-in). Such experiments have been successfully performed in clostridia. For example, Chen and coworkers (23) used the TargeTron vector pJIR750ai for insertion of the simian immunodeficiency virus (SIV) p27 gene into the theta-toxin gene (*pfoA*) on the *C. perfringens* chromosome, which simultaneously inactivated the theta-toxin gene and introduced the SIV p27 gene onto the bacterial chromosome. The mutant strain produced a large amount of SIV p27 protein and did not produce theta-toxin. This reengineered *C. perfringens* strain could provide a safe and efficient oral vehicle for delivery of antigens or therapeutic compounds to the gastrointestinal tract. The *virR* gene and its mutated form, *virRD57N*, were introduced into a unique MluI site in the TargeTron vector pJIR750ai (designed to target the *C. perfringens plc* gene) in such a way that the genes were transcribed in the same direction as the intron (25). The targeting plasmids pJIR3326 (*virR*) and pJIR3243 (*virRD57N*) were introduced into *C. perfringens* VirR mutant strain TS133 and single-copy integrants (in the *plc* gene) were isolated for both *virR* and mutant *virRD57N* genes. The mutant strains were used to further study the VirSR two-component signal transduction system and to elucidate the role of the phosphorylation of the VirR response regulator.

The ability to create desired mutations in genes using the group II intron mutagenesis system opens new horizons for functional genomic studies in clostridia. Considering that it takes as little as 10 to 14 days to generate mutants (64) and that the method is highly efficient and reproducible, new and interesting studies can be expected in the near future.

17.8. MODULATION OF GENE EXPRESSION WITH ANTISENSE RNA TECHNOLOGY

Antisense RNA (asRNA) technology offers an attractive alternative to gene knockout as a means of evaluation of gene functions as well as for modulation of gene expression. Most natural regulatory asRNAs are small molecules that bind to the complementary 5′ untranslated regions of the target RNA, including the translation initiation site and/or coding regions, thus preventing RNA translation and facilitating its degradation (226). Generation of mutants using asRNA has several advantages over gene inactivation. First, asRNA technology is faster than isolation of gene insertion mutants, and the gene inactivation is not absolute. This is especially important for studies of genes with essential cellular functions, as their disruption would result in a lethal phenotype. Also, asRNA is invaluable for studies of genes organized in polycistronic transcription units, because complete inactivation of some genes in these units may prevent synthesis of the gene products located downstream from the mutated gene. Construction of an artificial asRNA is usually performed by insertion of the asRNA gene containing the ribosome binding site and a portion of the coding region of the target gene between a strong promoter and a transcription terminator in a high-copy-number vector (reviewed in reference 220). It has been shown that by including the ribosome binding region and 33 to 100% of the target gene's coding region for construction of asRNA, over 80% downregulation of the target gene can be achieved (220). Detailed experiments for designing efficient asRNA molecules have been performed in *C. acetobutylicum* in E. T. Papoutsakis' laboratory (34, 218, 220, 221). Initially, investigators used a software program, Gene Construction Kit (Textco, Inc., West Lebanon, NH), to generate the asRNA constructs, and then an analysis was performed of the secondary structure of these constructs using MFold (http://mfold.bioinfo.rpi.edu/). Only the asRNA constructs with the most energetically favorable structure as predicted by MFold were used for experiments. Use of this strategy enabled these investigators to generate significantly more efficient asRNA constructs compared to their previous constructs made without MFold analysis. The asRNA technology was used in combination with overexpression of certain genes for detailed studies of butanol, acetone, and ethanol production in *C. acetobutylicum* (reviewed in reference 216). Furthermore, these technologies have been employed for metabolic engineering of *C. acetobutylicum* to create strains of industrial importance for increased production of these organic solvents (34, 152, 201, 202, 216). The asRNA technology has also been used for metabolic engineering of other clostridial species, such as *C. beijerinckii* (107) and *C. saccharoperbutylacetonicum* (144).

asRNA technology has been useful for studies of functional genomics in clostridia. The enzymatic composition of cellulosomes produced by *Clostridium cellulolyticum* was modified by inhibiting synthesis of the major enzymatic protein Cel48F (159). The asRNA technology has also been used to decrease synthesis in *C. perfringens* of small acid-soluble spore proteins, enabling an analysis of resistance to moist heat and UV radiation (170). In toxigenic clostridia, asRNA was employed to demonstrate the role of BotR in the regulation of the expression of botulinum neurotoxin and its nontoxigenic components (124).

Further improvements of asRNA technology have been suggested by the Papoutsakis laboratory, including development of efficient computational methods for designing efficient asRNA constructs, generation of constructs that permit modulation of multiple target genes, and the use of different promoters to control efficient transcription of these asRNAs (220). These researchers have proposed construction of efficient asRNAs for each of the ORFs, thus generating a library of strains that can be used to downregulate individual genes or gene operons in any clostridial strain for which the genomic sequence has been determined. Such libraries would facilitate molecular analysis to elucidate complex regulatory networks in these study organisms.

17.9. GENOME SEQUENCING AND ANALYSIS

Currently GenBank lists 91 genome sequencing projects of strains from 40 different *Clostridium* spp. (Table 5). Of these genome projects, 25 are completed genome sequences, 46 are draft sequences, and the sequencing of an additional 20 clostridial genomes is in progress. The complete sequences of 28 clostridial plasmids ranging in size from 2.4 kb to 270 kb have been determined and are also included in the table. Several strains have been identified that contain more than one plasmid. The most completed genomes are for *C. botulinum* (10); five genomes are at the draft stage, and the sequencing of two additional genomes is in progress. In *C. perfringens*, three genomes are complete and six draft genomes are available, while for *C. difficile* only one genome is completed, nine are in draft, and an additional five are being sequenced. While most of these sequences have been deposited in GenBank, only a few of the genomes have been analyzed in detail and published. These include *C. acetobutylicum* strain ATCC 824 (147); *C. botulinum* strain ATCC 3502 (190); *C. difficile* strain 630 (191); *C. kluyveri* strain DSM 555 (194); *C. novyi* strain NT (11); *C. perfringens* strains 13 (200) and SM101 and ATCC 13124 (143); *C. tetani* strain E88 (18); and *Moorella thermoacetica* (formerly *C. thermoaceticum*) (165).

Based on completed genome sequences, DNA microarrays have been designed using several different platforms and used for the comparison of strains within the same species or related clostridia (see, for example, references 22, 74, 105, 172, and 207) as well as for characterization of transcriptomes (46, 199, 220, 242). The most comprehensive and detailed transcriptional analysis including the life cycle of *C. acetobutylicum* has been performed in the Papoutsakis laboratory (85, 152). This study described major cellular regulatory systems; sigma and sporulation factors, including activity assays for major sporulation factors of the canonical sets of genes from their regulons; assessment of expression intensities; and identification of putative histidine kinases that may phosphorylate Spo0A (85).

Microarrays have also been developed for detection of food-borne bacteria based on their 23S rRNA variable region (68), based on their virulence genes (195), or for biological warfare agent detection (60, 195, 205). Specialized arrays have been designed for genotyping *C. perfringens* toxins (5), as well as for screening of gram-positive bacteria for 90 antibiotic resistance genes (160). Availability of a large number of completed genome sequences from clostridia, including multiple genomes from the same species, will enable detailed studies of the evolution of the genus *Clostridium* and will facilitate functional studies of genes at the genome level.

17.10. HOMOLOGOUS AND HETEROLOGOUS EXPRESSION OF CLOSTRIDIAL PROTEINS

While not as well developed as expression systems in other gram-positive bacteria such as *Bacillus*, *Staphylococcus*, *Streptococcus*, and *Lactococcus*, homologous and heter-

ologous expression systems are available and are being improved in *Clostridium*. The clostridia produce more extracellular protein antigens than any other bacterial genus (223), and several species are prolific producers of toxins, enzymes, and other proteins (42). *Clostridium* has certain advantages for production of proteins, including efficient secretion systems and lack of endotoxins. However, compared to other systems such as *B. subtilis* (187), many species are pathogenic, and clostridia have not been granted "Generally Regarded As Safe" status by the U.S. Food and Drug Administration. Certain clostridia also have high A+T coding bias, which can create considerable difficulties in expression, and relatively little information is known regarding transcription and translation mechanisms.

The high A+T codon content exhibits strongly selected codon usage bias (198), and the disparity in codon usage causes inefficient translation by heterologous hosts; furthermore, expression of clostridial proteins is difficult in such hosts as *E. coli* and *B. subtilis*. This is particularly apparent for proteins of relatively high mass, i.e., <50 kDa. The secretion systems in other gram-positive organisms, including the Sec system, the Tat pathway, and signal recognition particle systems, have not been established in *Clostridium*. While *B. subtilis* can obtain very high cell densities and accumulate recombinant proteins to up to 10 to 30% of total cellular protein, the fermentative metabolism of clostridia limits the energy capacity for protein accumulation. The codon bias can be partially alleviated by supplementation with rare tRNAs or by use of synthetic genes for optimized *Clostridium* codon usage (117, 198, 234).

Attempted production of clostridial enzymes and toxins in heterologous hosts has often resulted in very low yields, insolubility, or truncated or full-length inactive proteins (see, e.g., reference 56). However, hyperexpression of glucanase B from *C. thermocellum* was obtained using an amplication unit, pE194 (162). Expression of beta-toxoid of *C. perfringens* at 200 mg per liter was achieved, but the beta-toxoid was poorly secreted (146). Eukaryotic hosts including baculovirus and *Pichia pastoris* have also been used to achieve high-level expression of clostridial proteins (8).

17.11. TARGETED EXPRESSION OF *CLOSTRIDIUM* PRODUCTS FOR HUMAN DISEASE

Clostridium toxins and spores have been used experimentally for the study of eukaryotic physiology using an intriguing targeting expression approach (reviewed in reference 130). It has been known for more than 60 years that *Clostridium* endospores, when administered systemically to humans, locate and germinate within the anaerobic environment of solid tumors (118, 155). Targeted vector expression systems have been developed that express antitumor factors from *Clostridium* spores (229). In one exemplary case, intravenously injected *C. novyi* spores germinated within the avascular regions of tumors in mice and destroyed surrounding viable tumor cells (30). When *C. novyi* spores were administered together with conventional chemotherapeutic drugs, hemorrhagic necrosis of tumors was observed within 24 h. Spores are nonimmunogenic and can be administered for long periods. In a species comparison, *C. novyi* and *Clostridium sordellii* had the most antitumor activity, probably due to their high mobility within the tumors giving maximum spread. A *C. novyi* strain made nontoxigenic by genetic modification is in clinical trials for tumor ablation.

TABLE 5 Clostridial genome sequences[a]

Species and strain	Chromosome and/or plasmid	Length (nt)	ORFs	rRNAs	GC content (%)	Accession no.	Status	Reference
C. acetobutylicum ATCC 824	Chromosome	3,940,880	4,067	107	30	AE001437	Complete	147
	Plasmid pSOL1	192,000	178	None	30	AE001438	Complete	147
C. asparagiforme DSM 15981	Chromosome	6,224,391	6,881	63	55	ACCJ00000000	Draft assembly	
C. bartlettii DSM 19795	Chromosome	2,971,856	2,871	84	28	ABEZ00000000	Draft assembly	
C. beijerinckii NCIMB 8052	Chromosome	6,000,632	5,243	141	29	CP000721	Complete	
C. bolteae ATCC BAA-613	Chromosome	6,556,988	7,349	65	49	ABCC00000000	Draft assembly	
C. botulinum								
Type A1 ATCC 3502	Chromosome	3,886,916	3,776	114	28	AM412317	Complete	190
	Plasmid pBOT3502	16,344	19	None	26	AM412318	Complete	190
Type A1 ATCC 19397	Chromosome	3,863,450	3,692	98	28	CP000726	Complete	203
Type A1 Hall	Chromosome	3,760,560	3,569	106	28	CP000727	Complete	203
Type A2 Kyoto	Chromosome	4,155,278	4,035	105	28	CP001581	Complete	203
Type A3 Loch Maree	Chromosome	3,992,906	3,776	108	28	CP000962	Complete	203
	Plasmid pCLK	266,785	329	None	25	CP000963	Complete	203
Type B Eklund 17B	Chromosome	3,800,327	3,586	116	27	CP001056	Complete	203
	Plasmid pCLL	47,642	54	None	24	CP001057	Complete	203
Type B1 Okra B	Chromosome	3,958,233	3,780	109	28	CP000939	Complete	203
	Plasmid pCLD	148,780	195	None	25	CP000940	Complete	203
Type Ba4 657	Chromosome	3,977,794	3,940	115	28	CP001083	Complete	203
	Plasmid pCLJ	270,022	310	1	25	CP001081	Complete	203
	Plasmid pCLJ2	9,953	6	None	24	CP001082	Complete	203
Type Bf	Chromosome	4,217,754	4,217	127	28	ABDP00000000	Draft assembly	
Type C Eklund	Chromosome	2,961,186	2,954	126	28	ABDQ00000000	Draft assembly	
Type D 1873	Chromosome	2,379,404	2,330	119	27	ACSJ00000000	Draft assembly	
	Plasmid pCLG1	107,690	128	None	26	CP001659	Complete	
	Plasmid pCLG2	54,152	48	None	25	CP001660	Complete	
Type E1 Beluga	Chromosome	3,999,201	3,830	116	27	ACSC00000000	Draft assembly	
Type E3 Alaska E43	Chromosome	3,659,644	3,381	117	27	CP001078	Complete	
Type F Langeland	Chromosome	3,995,387	3,832	148	28	CP000728	Complete	
	Plasmid pCLI	17,531	24	None	26	CP000729	Complete	
Type G	Chromosome						In progress	
	Plasmid							
Type B(A) NCTC 2916	Chromosome	4,031,357	3,877	129	28	ABDO00000000	Draft assembly	
Type E Iwanii							In progress	
Type C2 203U28	Plasmid pC2C203U28	106,981	122	None	26	AP010934	Complete	

Species / Strain	Replicon	Size (bp)				Accession no.	Status	Ref.
C. butyricum								
Strain 2CR37							In progress	
Strain 5521		4,540,699	4,210	125	28	ABDT00000000	Draft assembly	
Strain BoNT/E4 BL5262		4,758,422	4,418	116	28	ACOM00000000	Draft assembly	
Strain MIYAIRI 588	Plasmid pCBM588	8,060	9	None	27	AB365348	Complete	
C. carboxidivorans								
Strain P7[b]							In progress	
Strain P7[c]	Chromosome	5,592,854	5,517	57	29	ACVI00000000	Draft assembly	
C. celatum DSM 1785							In progress	
C. cellulolyticum strain H10	Chromosome	4,068,724	3,578	88	37	CP001348	Complete	
C. cellulovorans								
Strain 743B[b]	Chromosome						In progress	
Strain 743B[c]	Chromosome	5,122,778	4,552	51	31	ACPD00000000	Draft assembly	
C. difficile								
Strain 630	Chromosome	4,290,252	3,971	185	29	AM180355	Complete	191
	Plasmid pCD630	7,881	11	None	27	AM180356	Complete	191
Strain ATCC 43255	Chromosome	4,204,780	3,959	54	28	ABKJ00000000	Draft assembly	
Strain CD196							In progress	
Strain CIP 107932	Chromosome	4,032,580	3,686	43	28	ABKK00000000	Draft assembly	
Strain NAP07							In progress	
Strain NAP08							In progress	
Strain QCD-23m63	Chromosome	3,396,085	3,611	29	28	ABKL00000000	Draft assembly	
Strain QCD-32g58	Chromosome	4,108,089	4,071	NA	28	AAML00000000	Draft assembly	
Strain QCD-37×79	Chromosome	4,329,888	4,042	40	28	ABHG00000000	Draft assembly	
Strain QCD-63q42	Chromosome	4,440,437	4,243	55	28	ABHD00000000	Draft assembly	
Strain QCD-66c26	Chromosome	4,126,050	3,769	46	28	ABFD00000000	Draft assembly	
Strain QCD-76w55	Chromosome	4,392,595	4,094	38	28	ABHE00000000	Draft assembly	
Strain QCD-97b34	Chromosome	4,059,010	3,748	37	28	ABHF00000000	Draft assembly	
Strain R20291							In progress	
Strain gs							In progress	
Strain CD (R8375)	Plasmid pCD6	6,830	5	None	24	AY350745	Complete	168
C. hathewayi DSM 13479	Chromosome	7,163,884	NA	NA	46	ACIO00000000	Draft assembly	
C. hiranonis DSM 13275	Chromosome	2,423,348	2,304	33	31	ABWP00000000	Draft assembly	
C. hylemonae DSM 15053	Chromosome	3,885,459	4,032	56	48	ABYI00000000	Draft assembly	

(Continued on next page)

TABLE 5 Clostridial genome sequences[a] (Continued)

Species and strain	Chromosome and/or plasmid	Length (nt)	ORFs	RNAs	GC content (%)	Accession no.	Status	Reference
C. innocuum SB23							In progress	
C. kluyveri								
Strain DSM 555	Chromosome	3,964,618	3,919	81	32	CP00673	Complete	194
	Plasmid pCKL555A	59,182	75	None	33	AP009050	Complete	194
Strain NBRC 12016	Chromosome	3,896,121	3,549	81	31	AP009049	Complete	
	Plasmid pCKL1	59,182	55	None			Complete	
C. leptum DSM 753	Chromosome	3,270,109	3,979	56	50	ABCB00000000	Draft assembly	
C. ljungdahlii							In progress	
C. methylpentosum DSM 5476	Chromosome	3,406,326	3,964	57	51	ACEC00000000	Draft assembly	
C. nexile DSM 1787	Chromosome	3,861,016	4,301	62	40	ABWO00000000	Draft assembly	
C. novyi NT	Chromosome	2,547,720	2,427	112	28	CP00382	Complete	11
C. orbiscindens DSM 6740							In progress	
C. papyrosolvens DSM 2782		4,874336	4,423	56	36	ACXX00000000	Draft assembly	
C. pasteurianum BC1							In progress	
C. perfringens								
Strain ATCC 13124	Chromosome	3,256,683	3,017	118	28	CP00246	Complete	143
Strain SM101	Chromosome	2,897,393	2,701	124	28	CP00312	Complete	143
	Plasmid 1	12,397	10	None	26	CP000313	Complete	143
	Plasmid 2	12,206	11	None	25	CP000314	Complete	143
Strain 13	Chromosome	3,031,430	2,786	126	28	BA000016	Complete	200
	Plasmid pCP13	54,310	63	None	25	AP003515	Complete	200
B strain ATCC 3626	Chromosome	3,896,305	3,710	120	28	ABDV00000000	Draft assembly	
C strain JGS1495	Chromosome	3,661,329	3,447	114	28	ABDU00000000	Draft assembly	
CPE strain F4969	Chromosome	3,510,272	3,290	139	28	ABDX00000000	Draft assembly	
	Plasmid	70,480	62	None	26	AB236336	Complete	138
D strain JGS1721	Chromosome	4,045,016	3,904	125	28	ABOO00000000	Draft assembly	
E strain JGS1987	Chromosome	4,127,102	4,034	138	28	ABDW00000000	Draft assembly	
Strain NCTC 8239	Chromosome	3,324,319	3,068	132	28	ABDY00000000	Draft assembly	
Strain CW92	Plasmid pCW3	47,263	51	None	27	DQ366035	Complete	9
Strain F5603	Plasmid pBCNF5603	36,695	36	None	25	AB189671	Complete	139
Strain NCTC 8533	Plasmid pCP8533ext	64,753	63	None	25	AB444205	Complete	139
C. phytofermentans ISDg	Chromosome	4,847,594	4,023	89	35	CP000885	Complete	
C. ramosum DSM 1402	Chromosome	3,234,795	3,169	51	31	ABFX00000000	Draft assembly	

Organism/strain	Replicon	Size				Accession	Status	Ref
C. scindens ATCC 35704	Chromosome	3,619,905	4,078	83	46	ABFY00000000	Draft assembly	
Clostridium spp.								
Strain 7 2 43FAA	Chromosome	3,779,307	3,512	57	27	ACDK00000000	Draft assembly	
Strain A2-232							In progress	
Strain ATCC BAA-442							In progress	
Strain D5							In progress	
Strain L2-50	Chromosome	2,954,116	3,004	55	41	AAYW00000000	Draft assembly	
Strain M62/1	Chromosome	3,781,854	4,328	62	50	ACFX00000000	Draft assembly	
Strain SS2/1	Chromosome	3,141,381	3,243	76	37	ABGC00000000	Draft assembly	
Strain MCF-1	Plasmid pMCF-1	2,450	1	None	39	U59416	Complete	
C. spiroforme DSM 1552	Chromosome	2,207,485	2,465	68	28	ABIK00000000	Draft assembly	
C. sporogenes ATCC 15579	Chromosome	4,102,125	3,873	106	28	ABKW00000000	Draft assembly	
C. sticklandii DSM 519							In progress	
C. symbiosum ATCC 14940							In progress	
C. tetani E88	Chromosome	2,799,251	2,373	72	28	AE15927	Complete	18
	Plasmid pE88	74,082	61	None	24	AF528097	Complete	18
C. thermocellum								
Strain ATCC 27405	Chromosome	3,843,301	3,305	71	38	CP000568	Complete	
Strain DSM 2360	Chromosome	3,454,608	3,146	55	39	ACVX00000000	Draft assembly	
Strain DSM 4150	Chromosome	3,321,980	3,021	42	38	ABVG00000000	Draft assembly	
Moorella thermoacetica (formerly C. thermoaceticum)	Chromosome	2,628,784	2,615	51	56	CP000232	Complete	165
Thermoanaerobacter ethanolicus CCSD1 (thermophilic anaerobe related to Clostridium)	Chromosome	2,204,850	2,367	56	34	ACXY00000000	Draft assembly	
Thermoanaerobacter pseudethanolicus ATCC 33223 (Clostridium thermohydrosulfuricum strain 39E)	Chromosome	2,362,816	2,243	75	34	CP000924	Complete	
Thermoanaerobacter thermosaccharolyticum DSM 571 (formerly Clostridium thermosaccharolyticum)	Chromosome	2,706,457	2,770	55	33	ACVG00000000	Draft assembly	

[a]Source: http://www.ncbi.nlm.nih.gov.ezproxy.library.wisc.edu/sites/entrez (accessed August 25, 2009) and http://pathema.jcvi.org/cgi-bin/Clostridium/shared/Genomes.cgi?crumbs=genomes. NA, not available.
[b]Sequencing performed at DOE Joint Genome Institute.
[c]Sequencing performed at Mississippi State University.

Clostridium spp. have shown enhanced oncolytic activity compared to other targeted organisms such as *Salmonella* and *Bifidobacterium*. The further development of genetic systems for *Clostridium* spp. will enhance their efficacy as oncolytic treatments for human diseases (214).

17.12. MOLECULAR TYPING AND GENETICS OF *CLOSTRIDIUM* POPULATIONS

The identification of *Clostridium* spp. has classically been performed on the basis of source of isolation, morphology, physiological and metabolic characteristics, pathogenicity and disease association, staining reactions, and serologic properties (63, 67, 142). Sequencing of 16S rRNA genes has become a powerful tool to identify and determine the species of the clostridia (26, 63, 81). However, 16S rRNA gene sequencing alone has certain limitations, including the inability to resolve recent speciations, and therefore it is recommended to combine ribosomal sequencing with determination of physiological properties and other typing methods such as multilocus sequence typing (see, e.g., references 72, 92, 103, 116, 145, and 183). Although many typing methods have been investigated, they vary in efficacy depending on their intended use, e.g., strain differentiation or understanding population structure.

Only recently, the population genetics and interactions in the environment of *Clostridium* spp. have begun to be investigated. The availability of genomic sequences (see section 17.9) provides insights into population genetics and evolution of *Clostridium* (see, e.g., reference 22). It is anticipated that considerable knowledge of the evolution of clostridia and novel methods for improvement of industrial processes will be gained in the near future.

17.13. CONCLUSIONS AND PERSPECTIVES

Considerable progress has been achieved in the analysis and genetic manipulation of *Clostridium* since the publication of the second edition of this manual in 1999. In particular, the availability of numerous clostridial genome sequences has initiated studies for evaluation of functional genomics at the genome level. There is an opportunity and a need for in-depth bioinformatic analyses of clostridial genomes to improve understanding of pathogenesis, metabolism, and evolution. A second major breakthrough has been the development of gene disruption systems. There is a need for further improvement of the knockout systems to achieve multiple gene deletions to facilitate studies of gene functions. Ideally, the development of further genetic tools will foster fruitful collaborations within the clostridial community to understand this complex and difficult organism.

Research in our laboratory has been supported by industry, the University of Wisconsin, and grants from the National Institutes of Health. We are grateful to our collaborators in this field.

REFERENCES

1. Abraham, L. J., D. I. Berryman, and J. I. Rood. 1988. Hybridization analysis of the class P tetracycline resistance determinant from the *Clostridium perfringens* R-plasmid, pCW3. *Plasmid* **19:**113–120.
2. Abraham, L. J., and J. I. Rood. 1985. Molecular analysis of transferable tetracycline resistance plasmids from *Clostridium perfringens*. *J. Bacteriol.* **161:**636–640.
3. Ackermann, H. W. 1974. Classification of *Bacillus* and *Clostridium* bacteriophages. *Pathol. Biol.* **22:**909–917.
4. Adams, V., D. Lyras, K. A. Farrow, and J. I. Rood. 2002. The clostridial mobilisable transposons. *Cell. Mol. Life Sci.* **59:**2033–2043.
5. Al-Khaldi, S. F., K. M. Myers, A. Rasooly, and V. Chizhikov. 2004. Genotyping of *Clostridium perfringens* toxins using multiple oligonucleotide microarray hybridization. *Mol. Cell. Probes* **18:**359–367.
6. Andreesen, J. R., H. Bahl, and G. Gottschalk. 1989. Introduction to the physiology and biochemistry of the genus *Clostridium*, p. 27–62. *In* N. P. Minton and D. J. Clarke (ed.), *Clostridia*. Plenum Press, New York, NY.
7. Awad, M. M., and J. I. Rood. 1997. Isolation of α-toxin, θ-toxin and κ-toxin mutants of *Clostridium perfringens* by Tn916 mutagenesis. *Microb. Pathog.* **22:**275–284.
8. Band, P. A., S. Blais, T. A. Neubert, T. J. Cardozo, and K. Ichtchenko. Recombinant derivatives of botulinum neurotoxin A engineered for trafficking studies and neuronal delivery. *Protein Expr. Purif.*, in press.
9. Bannam, T. L., W. L. Teng, D. Bulach, D. Lyras, and J. I. Rood. 2006. Functional identification of conjugation and replication regions of the tetracycline resistance plasmid pCW3 from *Clostridium perfringens*. *J. Bacteriol.* **188:**4942–4951.
10. Berryman, D. I., and J. I. Rood. 1995. The closely related *ermB-ermAM* genes from *Clostridium perfringens*, *Enterococcus faecalis* (pAMβ1), and *Streptococcus agalactiae* (pIP501) are flanked by variants of a directly repeated sequence. *Antimicrob. Agents Chemother.* **39:**1830–1834.
11. Bettegowda, C., X. Huang, J. Lin, I. Cheong, M. Kohli, S. A. Szabo, X. Zhang, L. A. Diaz, Jr., V. E. Velculescu, G. Parmigiani, K. W. Kinzler, B. Vogelstein, and S. Zhou. 2006. The genome and transcriptomes of the anti-tumor agent *Clostridium novyi*-NT. *Nat. Biotechnol.* **24:**1573–1580.
12. Birrer, G. A., W. R. Chesbro, and R. M. Zsigray. 1994. Electro-transformation of *Clostridium beijerinckii* NRRL B-592 with shuttle plasmid pHR106 and recombinant derivatives. *Appl. Microbiol. Biotechnol.* **41:**32–38.
13. Bleck, T. P. 1991. Tetanus: pathophysiology, management, and prophylaxis. *Dis. Mon.* **37:**547–603.
14. Bouvet, P. J., and M. R. Popoff. 2008. Genetic relatedness of *Clostridium difficile* isolates from various origins determined by triple-locus sequence analysis based on toxin regulatory genes *tcdC*, *tcdR*, and *cdtR*. *J. Clin. Microbiol.* **46:**3703–3713.
15. Braun, V., M. Mehlig, M. Moos, M. Rupnik, B. Kalt, D. E. Mahony, and C. von Eichel-Streiber. 2000. A chimeric ribozyme in *Clostridium difficile* combines features of group I introns and insertion elements. *Mol. Microbiol.* **36:**1447–1459.
16. Brefort, G., M. Magot, H. Ionesco, and M. Sebald. 1977. Characterization and transferability of *Clostridium perfringens* plasmids. *Plasmid* **1:**52–66.
17. Brehm, J. K., A. Pennock, H. M. Bullman, M. Young, J. D. Oultram, and N. P. Minton. 1992. Physical characterization of the replication origin of the cryptic plasmid pCB101 isolated from *Clostridium butyricum* NCIB 7423. *Plasmid* **28:**1–13.
18. Brüggemann, H., S. Baumer, W. F. Fricke, A. Wiezer, H. Liesegang, I. Decker, C. Herzberg, R. Martinez-Arias, R. Merkl, A. Henne, and G. Gottschalk. 2003. The genome sequence of *Clostridium tetani*, the causative agent of tetanus disease. *Proc. Natl. Acad. Sci. USA* **100:**1316–1321.
19. Brynestad, S., and P. E. Granum. 1999. Evidence that Tn5565, which includes the enterotoxin gene in *Clostridium perfringens*, can have a circular form which may be a transposition intermediate. *FEMS Microbiol. Lett.* **170:**281–286.

20. **Bullifent, H. L., A. Moir, and R. W. Titball.** 1995. The construction of a reporter system and use for the investigation of *Clostridium perfringens* gene expression. *FEMS Microbiol. Lett.* **131:**99–105.

21. **Burrus, V., G. Pavlovic, B. Decaris, and G. Guedon.** 2002. Conjugative transposons: the tip of the iceberg. *Mol. Microbiol.* **46:**601–610.

22. **Carter, A. T., C. J. Paul, D. R. Mason, S. M. Twine, M. J. Alston, S. M. Logan, J. W. Austin, and M. W. Peck.** 2009. Independent evolution of neurotoxin and flagellar genetic loci in proteolytic *Clostridium botulinum*. *BMC Genomics* **10:**115.

23. **Chen, Y., L. Caruso, B. McClane, D. Fisher, and P. Gupta.** 2007. Disruption of a toxin gene by introduction of a foreign gene into the chromosome of *Clostridium perfringens* using targetron-induced mutagenesis. *Plasmid* **58:**182–189.

24. **Chen, Y., B. A. McClane, D. J. Fisher, J. I. Rood, and P. Gupta.** 2005. Construction of an alpha toxin gene knockout mutant of *Clostridium perfringens* type A by use of a mobile group II intron. *Appl. Environ. Microbiol.* **71:**7542–7547.

25. **Cheung, J. K., M. M. Awad, S. McGowan, and J. I. Rood.** 2009. Functional analysis of the VirSR phosphorelay from *Clostridium perfringens*. *PLoS One* **4:**e5849.

26. **Collins, M. D., P. A. Lawson, A. Willems, J. J. Cordoba, J. Fernandez-Garayzabal, P. Garcia, J. Cai, H. Hippe, and J. A. Farrow.** 1994. The phylogeny of the genus *Clostridium*: proposal of five new genera and eleven new species combinations. *Int. J. Syst. Bacteriol.* **44:**812–826.

27. **Cornillot, E., R. V. Nair, E. T. Papoutsakis, and P. Soucaille.** 1997. The genes for butanol and acetone formation in *Clostridium acetobutylicum* ATCC 824 reside on a large plasmid whose loss leads to degeneration of the strain. *J. Bacteriol.* **179:**5442–5447.

28. **Cornillot, E., and P. Soucaille.** 1996. Solvent-forming genes in clostridia. *Nature* **380:**489.

29. **Dahlsten, E., H. Korkeala, P. Somervuo, and M. Lindstrom.** 2008. PCR assay for differentiating between Group I (proteolytic) and Group II (nonproteolytic) strains of *Clostridium botulinum*. *Int. J. Food Microbiol.* **124:**108–111.

30. **Dang, L. H., C. Bettegowda, D. L. Huso, K. W. Kinzler, and B. Vogelstein.** 2001. Combination bacteriolytic therapy for the treatment of experimental tumors. *Proc. Natl. Acad. Sci. USA* **98:**15155–15160.

31. **Davis, I. J., G. Carter, M. Young, and N. P. Minton.** 2005. Gene cloning in clostridia, p. 37–52. *In* P. Dürre (ed.), *Handbook on Clostridia*. Taylor & Francis Group, Boca Raton, FL.

32. **Davis, T. O., I. Henderson, J. K. Brehm, and N. P. Minton.** 2000. Development of a transformation and gene reporter system for group II, non-proteolytic *Clostridium botulinum* type B strains. *J. Mol. Microbiol. Biotechnol.* **2:**59–69.

33. **Deguchi, A., K. Miyamoto, T. Kuwahara, Y. Miki, I. Kaneko, J. Li, B. A. McClane, and S. Akimoto.** 2009. Genetic characterization of type A enterotoxigenic *Clostridium perfringens* strains. *PLoS One* **4:**e5598.

34. **Desai, R. P., and E. T. Papoutsakis.** 1999. Antisense RNA strategies for metabolic engineering of *Clostridium acetobutylicum*. *Appl. Environ. Microbiol.* **65:**936–945.

35. **De Vos, P., G. Garrity, D. Jones, N. R. Krief, W. Ludwig, F. A. Rainey, K.-H. Schleifer, and W. B. Whitman (ed.).** 2009. *Bergey's Manual of Systematic Bacteriology*, 2nd ed., vol. 3. *The Firmicutes*. Springer-Verlag, New York, NY.

36. **Dineen, S. S., M. Bradshaw, and E. A. Johnson.** 2003. Neurotoxin gene clusters in *Clostridium botulinum* type A strains: sequence comparison and evolutionary implications. *Curr. Microbiol.* **46:**345–352.

37. **Domingo, M. C., A. Huletsky, A. Bernal, R. Giroux, D. K. Boudreau, F. J. Picard, and M. G. Bergeron.** 2005. Characterization of a Tn5382-like transposon containing the *vanB2* gene cluster in a *Clostridium* strain isolated from human faeces. *J. Antimicrob. Chemother.* **55:**466–474.

38. **Dowell, V. R., and G. L. Lombard.** 1977. Media for isolation, characterization and identification of obligately anaerobic bacteria. US Department of Health and Human Services, Public Health Service, Centers for Disease Control, Atlanta, GA.

39. **Dürre, P.** 2008. Fermentative butanol production: bulk chemical and biofuel. *Ann. N. Y. Acad. Sci.* **1125:**353–362.

40. **Dürre, P.** 2001. From Pandora's box to cornucopia, p. 1–13. *In* H. Bahl and P. Dürre (ed.), *Clostridia. Biotechnology and Medical Applications*. Wiley-VCH, Weinheim, Germany.

41. **Dürre, P. (ed.).** 2005. *Handbook on Clostridia*. Taylor & Francis Group, Boca Raton, FL.

42. **Dürre, P.** 2005. Formation of solvents in clostridia, p. 671–693. *In* P. Dürre (ed.), *Handbook on Clostridia*. Taylor & Francis Group, Boca Raton, FL.

43. **Dworkin, M. (ed. in chief).** 2006. *The Prokaryotes*, 3rd ed. Springer, New York, NY.

44. **Eklund, M. W., F. T. Pousky, and W. H. Habig.** 1989. Bacteriophages and plasmids in *Clostridium botulinum* and *Clostridium tetani* and their relationship to production of toxins, p. 25–51. *In* L. L. Simpson (ed.), *Botulinum Neurotoxin and Tetanus Toxin*. Academic Press, San Diego, CA.

45. **Eklund, M. W., F. T. Poysky, L. M. Mseitif, and M. S. Strom.** 1988. Evidence for plasmid-mediated toxin and bacteriocin production in *Clostridium botulinum* type G. *Appl. Environ. Microbiol.* **54:**1405–1408.

46. **Emerson, J. E., R. A. Stabler, B. W. Wren, and N. F. Fairweather.** 2008. Microarray analysis of the transcriptional responses of *Clostridium difficile* to environmental and antibiotic stress. *J. Med. Microbiol.* **57:**757–764.

47. **Fach, P., P. Micheau, C. Mazuet, S. Perelle, and M. Popoff.** 2009. Development of real-time PCR tests for detecting botulinum neurotoxins A, B, E, F producing *Clostridium botulinum*, *Clostridium baratii* and *Clostridium butyricum*. *J. Appl. Microbiol.* **107:**465–473.

48. **Fang, A., D. F. Gerson, and A. L. Demain.** 2006. Menstrum for culture preservation and medium for seed preparation in a tetanus toxin production process containing no animal or dairy products. *Lett. Appl. Microbiol.* **43:**360–363.

49. **Fang, A., D. F. Gerson, and A. L. Demain.** 2009. Production of *Clostridium difficile* toxin in a medium totally free of both animal and dairy proteins or digests. *Proc. Natl. Acad. Sci. USA* **106:**13225–13229.

50. **Feustel, L., S. Nakotte, and P. Dürre.** 2004. Characterization and development of two reporter gene systems for *Clostridium acetobutylicum*. *Appl. Environ. Microbiol.* **70:**798–803.

51. **Finn, C. W., Jr., R. P. Silver, W. H. Habig, M. C. Hardegree, G. Zon, and C. F. Garon.** 1984. The structural gene for tetanus neurotoxin is on a plasmid. *Science* **224:**881–884.

52. **Franciosa, G., A. Maugliani, C. Scalfaro, and P. Aureli.** 2009. Evidence that plasmid-borne botulinum neurotoxin type B genes are widespread among *Clostridium botulinum* serotype B strains. *PLoS ONE* **4:**e4829.

53. **Garnier, T., and S. T. Cole.** 1988. Complete nucleotide sequence and genetic organization of the bacteriocinogenic plasmid, pIP404, from *Clostridium perfringens*. *Plasmid* **19:**134–150.

54. **Garnier, T., and S. T. Cole.** 1988. Identification and molecular genetic analysis of replication functions of the bacteriocinogenic plasmid pIP404 from *Clostridium perfringens*. *Plasmid* **19:**151–160.

55. **Girbal, L., I. Mortier-Barriere, F. Raynaud, C. Rouanet, C. Croux, and P. Soucaille.** 2003. Development of a sensitive gene expression reporter system and an inducible promoter-repressor system for *Clostridium acetobutylicum*. *Appl. Environ. Microbiol.* **69:**4985–4988.

56. **Girbal, L., G. von Abendroth, M. Winkler, P. M. Benton, I. Meynial-Salles, C. Croux, J. W. Peters, T. Happe, and P. Soucaille.** 2005. Homologous and heterologous overexpression in *Clostridium acetobutylicum* and characterization of purified clostridial and algal Fe-only hydrogenases with high specific activities. *Appl. Environ. Microbiol.* **71:**2777–2781.

57. **Goryshin, I. Y., and W. S. Reznikoff.** 1998. Tn5 *in vitro* transposition. *J. Biol. Chem.* **273:**7367–7374.

58. **Gottschalk, G.** 1989. *Bacterial Metabolism*, 2nd ed. Springer-Verlag, New York, NY.

59. **Govind, R., J. A. Fralick, and R. D. Rolfe.** 2006. Genomic organization and molecular characterization of *Clostridium difficile* bacteriophage PhiCD119. *J. Bacteriol.* **188:**2568–2577.

60. **Hashsham, S. A., L. M. Wick, J. M. Rouillard, E. Gulari, and J. M. Tiedje.** 2004. Potential of DNA microarrays for developing parallel detection tools (PDTs) for microorganisms relevant to biodefense and related research needs. *Biosens. Bioelectron.* **20:**668–683.

61. **Hasselmayer, O., V. Braun, C. Nitsche, M. Moos, M. Rupnik, and C. von Eichel-Streiber.** 2004. *Clostridium difficile* IStron CdISt1: discovery of a variant encoding two complete transposase-like proteins. *J. Bacteriol.* **186:**2508–2510.

62. **Hasselmayer, O., C. Nitsche, V. Braun, and C. von Eichel-Streiber.** 2004. The IStron CdISt1 of *Clostridium difficile*: molecular symbiosis of a group I intron and an insertion element. *Anaerobe* **10:**85–92.

63. **Hatheway, C. L., and E. A. Johnson.** 1998. *Clostridium*: the spore-bearing anaerobes, p. 731–782. *In* L. Collier, A. Balows, and M. Sussman (ed.), *Topley and Wilson's Microbiology and Microbial Infections*, 9th ed., vol. 2. *Systematic Bacteriology*. Arnold, London, United Kingdom.

64. **Heap, J. T., S. T. Cartman, O. J. Pennington, C. M. Cooksley, J. C. Scott, B. Blount, D. A. Burns, and N. P. Minton.** 2009. Development of genetic knock-out systems for clostridia, p. 179–198. *In* H. Brüggemann and G. Gottschalk (ed.), *Clostridia. Molecular Biology in the Post-Genomic Era*. Caister Academic Press, Norfolk, United Kingdom.

65. **Heap, J. T., O. J. Pennington, S. T. Cartman, G. P. Carter, and N. P. Minton.** 2007. The ClosTron: a universal gene knock-out system for the genus *Clostridium*. *J. Microbiol. Methods* **70:**452–464.

66. **Heap, J. T., O. J. Pennington, S. T. Cartman, and N. P. Minton.** 2009. A modular system for *Clostridium* shuttle plasmids. *J. Microbiol. Methods* **78:**79–85.

67. **Holdeman, L. V., E. P. Cato, and W. E. C. Moore.** 1977. *Anaerobe Laboratory Manual*, 4th ed. Department of Anaerobic Microbiology, Virginia Polytechnic Institute and State University, Blacksburg, VA.

68. **Hong, B. X., L. F. Jiang, Y. S. Hu, D. Y. Fang, and H. Y. Guo.** 2004. Application of oligonucleotide array technology for the rapid detection of pathogenic bacteria of foodborne infections. *J. Microbiol. Methods* **58:**403–411.

69. **Hughes, M. L., R. Poon, V. Adams, S. Sayeed, J. Saputo, F. A. Uzal, B. A. McClane, and J. I. Rood.** 2007. Epsilon-toxin plasmids of *Clostridium perfringens* type D are conjugative. *J. Bacteriol.* **189:**7531–7538.

70. **Hussein, H. M., A. L. Cookson, and G. T. Attwood.** 2008. A new growth medium for rapid selection and purification of *Clostridium proteoclasticum* transposon mutants. *J. Microbiol. Methods* **73:**203–207.

71. **Hwang, H. J., J. C. Lee, Y. Yamamoto, M. R. Sarker, T. Tsuchiya, and K. Oguma.** 2007. Identification of structural genes for *Clostridium botulinum* type C neurotoxin-converting phage particles. *FEMS Microbiol. Lett.* **270:**82–89.

72. **Jacobson, M. J., G. Lin, T. S. Whittam, and E. A. Johnson.** 2008. Phylogenetic analysis of *Clostridium botulinum* type A by multi-locus sequence typing. *Microbiology* **154:**2408–2415.

73. **Jankovic, J., A. Albanes, M. Souhair Atassi, J. O. Dolly, M. Hallett, and N. H. Mayer.** 2009. *Botulinum Toxin: Therapeutic Clinical Practice & Science.* Saunders Elsevier, Philadelphia, PA.

74. **Janvilisri, T., J. Scaria, A. D. Thompson, A. Nicholson, B. M. Limbago, L. G. Arroyo, J. G. Songer, Y. T. Grohn, and Y. F. Chang.** 2009. Microarray identification of *Clostridium difficile* core components and divergent regions associated with host origin. *J. Bacteriol.* **191:**3881–3891.

75. **Johanesen, P. A., D. Lyras, T. L. Bannam, and J. I. Rood.** 2001. Transcriptional analysis of the *tet*(P) operon from *Clostridium perfringens*. *J. Bacteriol.* **183:**7110–7119.

76. **Johnson, E. A.** 1999. Anaerobic fermentations, p. 139–150. *In* A. L. Demain and J. E. Davies (ed.), *Manual of Industrial Microbiology and Biotechnology*, 2nd ed. ASM Press, Washington, DC.

77. **Johnson, E. A.** 2009. Clostridia, p. 87–93. *In* J. Lederberg (ed.), *Encyclopedia of Microbiology*. Elsevier, Oxford, United Kingdom.

78. **Johnson, E. A.** 1997. Extrachromosomal virulence determinants in the clostridia, p. 35–48. *In* J. I. Rood, B. A. McClane, J. G. Songer, and R. W. Titball (ed.), *The Clostridia: Molecular Genetics and Pathogenesis*. Academic Press, London, United Kingdom.

79. **Johnson, E. A., and M. Bradshaw.** 2001. *Clostridium botulinum* and its neurotoxins: a metabolic and cellular perspective. *Toxicon* **39:**1703–1722.

80. **Johnson, E. A., A. Madia, and A. L. Demain.** 1981. Chemically defined minimal medium for growth of the anaerobic cellulolytic thermophile *Clostridium thermocellum*. *Appl. Environ. Microbiol.* **41:**1060–1062.

81. **Johnson, E. A., P. Summanen, and S. M. Finegold.** 2006. Clostridium. *Manual of Clinical Microbiology*, 9th ed. ASM Press, Washington, DC.

82. **Jones, D. T.** 2001. Applied acetone-butanol fermentation, p. 125–168. *In* H. Bahl and P. Dürre (ed.), *Clostridia. Biotechnology and Medical Applications*. Wiley-VCH, Weinheim, Germany.

83. **Jones, D. T.** 2005. Bacteriophages in *Clostridium*, p. 697–717. *In* P. Dürre (ed.), *Handbook on Clostridia*. Taylor & Francis Group, Boca Raton, FL.

84. **Jones, D. T., M. Shirley, X. Wu, and S. Keis.** 2000. Bacteriophage infections in the industrial acetone butanol (AB) fermentation process. *J. Mol. Microbiol. Biotechnol.* **2:**21–26.

85. **Jones, S. W., C. J. Paredes, B. Tracy, N. Cheng, R. Sillers, R. S. Senger, and E. T. Papoutsakis.** 2008. The transcriptional program underlying the physiology of clostridial sporulation. *Genome Biol.* **9:**R114.

86. **Jost, B. H., H. T. Trinh, and J. G. Songer.** 2006. Clonal relationships among *Clostridium perfringens* of porcine origin as determined by multilocus sequence typing. *Vet. Microbiol.* **116:**158–165.

87. **Kaji, M., O. Matsushita, E. Tamai, S. Miyata, Y. Taniguchi, S. Shimamoto, S. Katayama, S. Morita, and A. Okabe.** 2003. A novel type of DNA curvature present in a *Clostridium perfringens* ferredoxin gene: characterization and role in gene expression. *Microbiology* **149:**3083–3091.

88. **Karberg, M., H. Guo, J. Zhong, R. Coon, J. Perutka, and A. M. Lambowitz.** 2001. Group II introns as controllable gene targeting vectors for genetic manipulation of bacteria. *Nat. Biotechnol.* **19:**1162–1167.

89. **Kashket, E. R., and Z. Y. Cao.** 1995. Clostridial strain degeneration. *FEMS Microbiol. Rev.* **17:**307–315.

90. **Katayama, S., B. Dupuy, G. Daube, B. China, and S. T. Cole.** 1996. Genome mapping of *Clostridium perfringens* strains with I-CeuI shows many virulence genes to be plasmid-borne. *Mol. Gen. Genet.* **251:**720–726.

91. **Kennedy, C. L., E. O. Krejany, L. F. Young, J. R. O'Connor, M. M. Awad, R. L. Boyd, J. J. Emmins, D. Lyras, and J. I. Rood.** 2005. The α-toxin of *Clostridium septicum* is essential for virulence. *Mol. Microbiol.* **57:**1357–1366.

92. **Killgore, G., A. Thompson, S. Johnson, J. Brazier, E. Kuijper, J. Pepin, E. H. Frost, P. Savelkoul, B. Nicholson, R. J. van den Berg, H. Kato, S. P. Sambol, W. Zukowski, C. Woods, B. Limbago, D. N. Gerding, and L. C. McDonald.** 2008. Comparison of seven techniques for typing international epidemic strains of *Clostridium difficile*: restriction endonuclease analysis, pulsed-field gel electrophoresis, PCR-ribotyping, multilocus sequence typing, multilocus variable-number tandem-repeat analysis, amplified fragment length polymorphism, and surface layer protein A gene sequence typing. *J. Clin. Microbiol.* **46:**431–437.

93. **Kim, A. Y., and H. P. Blaschek.** 1993. Construction and characterization of a phage-plasmid hybrid (phagemid), pCAK1, containing the replicative form of viruslike particle CAK1 isolated from *Clostridium acetobutylicum* NCIB 6444. *J. Bacteriol.* **175:**3838–3843.

94. **Kim, A. Y., and H. P. Blaschek.** 1989. Construction of an *Escherichia coli*-*Clostridium perfringens* shuttle vector and plasmid transformation of *Clostridium perfringens*. *Appl. Environ. Microbiol.* **55:**360–365.

95. **Kok, J., J. M. van der Vossen, and G. Venema.** 1984. Construction of plasmid cloning vectors for lactic streptococci which also replicate in *Bacillus subtilis* and *Escherichia coli*. *Appl. Environ. Microbiol.* **48:**726–731.

96. **Lamberg, A., S. Nieminen, M. Qiao, and H. Savilahti.** 2002. Efficient insertion mutagenesis strategy for bacterial genomes involving electroporation of in vitro-assembled DNA transposition complexes of bacteriophage Mu. *Appl. Environ. Microbiol.* **68:**705–712.

97. **Lambowitz, A. M., and S. Zimmerly.** 2004. Mobile group II introns. *Ann. Rev. Genet.* **38:**1–35.

98. **Lanckriet, A., L. Timbermont, L. J. Happonen, M. I. Pajunen, F. Pasmans, F. Haesebrouck, R. Ducatelle, H. Savilahti, and F. Van Immerseel.** 2009. Generation of single-copy transposon insertions in *Clostridium perfringens* by electroporation of phage Mu DNA transposition complexes. *Appl. Environ. Microbiol.* **75:**2638–2642.

99. **Launay, A., S. A. Ballard, P. D. Johnson, M. L. Grayson, and T. Lambert.** 2006. Transfer of vancomycin resistance transposon Tn*1549* from *Clostridium symbiosum* to *Enterococcus* spp. in the gut of gnotobiotic mice. *Antimicrob. Agents Chemother.* **50:**1054–1062.

100. **Lee, C. K., P. Dürre, H. Hippe, and G. Gottschalk.** 1987. Screening for plasmids in the genus *Clostridium*. *Arch. Microbiol.* **148:**107–114.

101. **Lee, S. Y., L. D. Mermelstein, and E. T. Papoutsakis.** 1993. Determination of plasmid copy number and stability in *Clostridium acetobutylicum* ATCC 824. *FEMS Microbiol. Lett.* **108:**319–323.

102. **Li, J., K. Miyamoto, and B. A. McClane.** 2007. Comparison of virulence plasmids among *Clostridium perfringens* type E isolates. *Infect. Immun.* **75:**1811–1819.

103. **Li, K., B. Chen, Y. Zhou, R. Huang, Y. Liang, Q. Wang, Z. Xiao, and J. Xiao.** 2009. Multiplex quantification of 16S rDNA of predominant bacteria group within human fecal samples by polymerase chain reaction-ligase detection reaction (PCR-LDR). *J. Microbiol. Methods* **76:**289–294.

104. **Lin, W. J., and E. A. Johnson.** 1991. Transposon Tn*916* mutagenesis in *Clostridium botulinum*. *Appl. Environ. Microbiol.* **57:**2946–2950.

105. **Lindström, M., K. Hinderink, P. Somervuo, K. Kiviniemi, M. Nevas, Y. Chen, P. Auvinen, A. T. Carter, D. R. Mason, M. W. Peck, and H. Korkeala.** 2009. Comparative genomic hybridization analysis of two predominant Nordic group I (proteolytic) *Clostridium botulinum* type B clusters. *Appl. Environ. Microbiol.* **75:**2643–2651.

106. **Liyanage, H., P. Holcroft, V. J. Evans, S. Keis, S. R. Wilkinson, E. R. Kashket, and M. Young.** 2000. A new insertion sequence, ISCb1, from *Clostridium beijerinckii* NCIMB 8052. *J. Mol. Microbiol. Biotechnol.* **2:**107–113.

107. **Liyanage, H., M. Young, and E. R. Kashket.** 2000. Butanol tolerance of *Clostridium beijerinckii* NCIMB 8052 associated with down-regulation of gldA by antisense RNA. *J. Mol. Microbiol. Biotechnol.* **2:**87–93.

108. **Lundie, L. L., Jr., and H. L. Drake.** 1984. Development of a minimally defined medium for the acetogen *Clostridium thermoaceticum*. *J. Bacteriol.* **159:**700–703.

109. **Lyras, D., V. Adams, S. A. Ballard, W. L. Teng, P. M. Howarth, P. K. Crellin, T. L. Bannam, J. G. Songer, and J. I. Rood.** 2009. tISCpe8, an IS*1595*-family lincomycin resistance element located on a conjugative plasmid in *Clostridium perfringens*. *J Bacteriol.* **191:**6345–6351.

110. **Lyras, D., and J. I. Rood.** 2006. Clostridial genetics, p. 672–687. *In* V. A. Fischetti (ed.), *Gram-Positive Pathogens*, 2nd ed. ASM Press, Washington, DC.

111. **Lyras, D., and J. I. Rood.** 1998. Conjugative transfer of RP4-*oriT* shuttle vectors from *Escherichia coli* to *Clostridium perfringens*. *Plasmid* **39:**160–164.

112. **Lyras, D., and J. I. Rood.** 1996. Genetic organization and distribution of tetracycline resistance determinants in *Clostridium perfringens*. *Antimicrob. Agents Chemother.* **40:**2500–2504.

113. **Lyras, D., and J. I. Rood.** 1997. Transposable genetic elements and antibiotic resistance determinants from *Clostridium perfringens* and *Clostridium difficile*, p. 73–92. *In* J. I. Rood, B. A. McClane, J. G. Songer, and R. W. Titball (ed.), *The Clostridia: Molecular Biology and Pathogenesis*. Academic Press, London, United Kingdom.

114. **Lyras, D., and J. I. Rood.** 2005. Transposable genetic elements of clostridia, p. 631–643. *In* P. Dürre (ed.), *Handbook on Clostridia*. Taylor & Francis Group, Boca Raton, FL.

115. **Maamar, H., P. de Philip, J. P. Belaich, and C. Tardif.** 2003. ISCce1 and ISCce2, two novel insertion sequences in *Clostridium cellulolyticum*. *J. Bacteriol.* **185:**714–725.

116. **Macdonald, T. E., C. H. Helma, L. O. Ticknor, P. J. Jackson, R. T. Okinaka, L. A. Smith, T. J. Smith, and K. K. Hill.** 2008. Differentiation of *Clostridium botulinum* serotype A strains by multiple-locus variable-number tandem-repeat analysis. *Appl. Environ. Microbiol.* **74:**875–882.

117. **Makoff, A. J., M. D. Oxer, M. A. Romanos, N. F. Fairweather, and S. Ballantine.** 1989. Expression of tetanus toxin fragment C in *E. coli*: high level expression by removing rare codons. *Nucleic Acids Res.* **17:**10191–10202.

118. **Malmgren, R. A., and C. C. Flanigan.** 1955. Localization of the vegetative form of *Clostridium tetani* in mouse tumors following intravenous spore administration. *Cancer Res.* **15:**473–478.

119. **Mani, N., B. Dupuy, and A. L. Sonenshein.** 2006. Isolation of RNA polymerase from *Clostridium difficile* and characterization of glutamate dehydrogenase and rRNA gene promoters in vitro and in vivo. *J. Bacteriol.* **188:**96–102.

120. **Mani, N., D. Lyras, L. Barroso, P. Howarth, T. Wilkins, J. I. Rood, A. L. Sonenshein, and B. Dupuy.** 2002. Environmental response and autoregulation of *Clostridium difficile* TxeR, a sigma factor for toxin gene expression. *J. Bacteriol.* **184:**5971–5978.

121. **Marsh, J. W., M. M. O'Leary, K. A. Shutt, A. W. Pasculle, S. Johnson, D. N. Gerding, C. A. Muto, and L. H. Harrison.** 2006. Multilocus variable-number tandem-repeat analysis for investigation of *Clostridium difficile* transmission in hospitals. *J. Clin. Microbiol.* **44:** 2558–2566.

122. **Marshall, K. M., M. Bradshaw, S. Pellett, and E. A. Johnson.** 2007. Plasmid encoded neurotoxin genes in *Clostridium botulinum* serotype A subtypes. *Biochem. Biophys. Res. Commun.* **361:**49–54.

123. **Martin, P., P. Trieu-Cuot, and P. Courvalin.** 1986. Nucleotide sequence of the *tetM* tetracycline resistance determinant of the streptococcal conjugative shuttle transposon Tn*1545*. *Nucleic Acids Res.* **14:**7047–7058.

124. **Marvaud, J. C., M. Gibert, K. Inoue, Y. Fujinaga, K. Oguma, and M. R. Popoff.** 1998. *botR/A* is a positive regulator of botulinum neurotoxin and associated non-toxin protein genes in *Clostridium botulinum* A. *Mol. Microbiol.* **29:**1009–1018.

125. **Matsushita, C., O. Matsushita, M. Koyama, and A. Okabe.** 1994. A *Clostridium perfringens* vector for the selection of promoters. *Plasmid* **31:**317–319.

126. **Mauchline, M. L., T. O. Davis, and N. P. Minton.** 1999. Clostridia, p. 475–490. *In* A. L. Demain and J. E. Davies (ed.), *Manual of Industrial Microbiology and Biotechnology*, 2nd ed. ASM Press, Washington, DC.

127. **Mayer, M. J., A. Narbad, and M. J. Gasson.** 2008. Molecular characterization of a *Clostridium difficile* bacteriophage and its cloned biologically active endolysin. *J. Bacteriol.* **190:**6734–6740.

128. **McCoy, E., and E. B. Fred.** 1941. The stability of a culture for industrial fermentation. *J. Bacteriol.* **41:**90–91.

129. **Mendez, B. S., and R. F. Gomez.** 1982. Isolation of *Clostridium thermocellum* auxotrophs. *Appl. Environ. Microbiol.* **43:**495–496.

130. **Mengesha, A., L. Dubois, K. Paesmans, B. Wouters, P. Lambin, and J. Theys.** 2009. Clostridia in anti-tumour therapy, p. 199–214. *In* H. Brüggemann and G. Gottschalk (ed.), *Clostridia. Molecular Biology in the Post-Genomic Era.* Caister Academic Press, Norfolk, United Kingdom.

131. **Mermelstein, L. D., N. E. Welker, G. N. Bennett, and E. T. Papoutsakis.** 1992. Expression of cloned homologous fermentative genes in *Clostridium acetobutylicum* ATCC 824. *Biotechnology* **10:**190–195.

132. **Minton, N., G. Carter, M. Herbert, T. O'Keeffe, D. Purdy, M. Elmore, A. Ostrowski, O. Pennington, and I. Davis.** 2004. The development of *Clostridium difficile* genetic systems. *Anaerobe* **10:**75–84.

133. **Minton, N. P., J. K. Brehm, T. J. Swinfield, S. M. Whelan, M. L. Mauchline, N. Bodsworth, and J. D. Oultram.** 1993. Clostridial cloning vectors, p. 119–156. *In* D. R. Woods (ed.), *The Clostridia and Biotechnology.* Butterworth-Heinemann, Stoneham, MA.

134. **Minton, N. P., and J. D. Oultram.** 1988. Host: vector systems for gene cloning in *Clostridium. Microbiol. Sci.* **5:**310–315.

135. **Minton, N. P., T. J. Swinfield, J. K. Brehm, S. M. Whelan, and J. D. Oultram.** 1993. Vectors for use in *Clostridium acetobutylicum*, p. 120–140. *In* M. Sebald (ed.), *Genetics and Molecular Biology of Anaerobic Bacteria.* Springer-Verlag, New York, NY.

136. **Mitchell, W. J.** 2001. Biology and physiology, p. 49–104. *In* H. Bahl and P. Dürre (ed.), *Clostridia. Biotechnology and Medical Applications.* Wiley-VCH, Weinheim, Germany.

137. **Mitchell, W. J.** 1998. Physiology of carbohydrate to solvent conversion by clostridia. *Adv. Microb. Physiol.* **39:**31–130.

138. **Miyamoto, K., D. J. Fisher, J. Li, S. Sayeed, S. Akimoto, and B. A. McClane.** 2006. Complete sequencing and diversity analysis of the enterotoxin-encoding plasmids in *Clostridium perfringens* type A non-foodborne human gastrointestinal disease isolates. *J. Bacteriol.* **188:**1585–1598.

139. **Miyamoto, K., J. Li, S. Sayeed, S. Akimoto, and B. A. McClane.** 2008. Sequencing and diversity analyses reveal extensive similarities between some epsilon-toxin-encoding plasmids and the pCPF5603 *Clostridium perfringens* enterotoxin plasmid. *J. Bacteriol.* **190:**7178–7188.

140. **Mohr, G., D. Smith, M. Belfort, and A. M. Lambowitz.** 2000. Rules for DNA target-site recognition by a lactococcal group II intron enable retargeting of the intron to specific DNA sequences. *Genes Dev.* **14:**559–573.

141. **Monod, M., C. Denoya, and D. Dubnau.** 1986. Sequence and properties of pIM13, a macrolide-lincosamide-streptogramin B resistance plasmid from *Bacillus subtilis. J. Bacteriol.* **167:**138–147.

142. **Moore, L. V. H., E. P. Cato, and W. E. C. Moore.** 1987. *Anaerobe Manual Update. Supplement to the VPI Anaerobe Manual*, 4th ed. Department of Anaerobic Microbiology, Virginia Polytechnic Institute and State University, Blacksburg, VA.

143. **Myers, G. S., D. A. Rasko, J. K. Cheung, J. Ravel, R. Seshadri, R. T. DeBoy, Q. Ren, J. Varga, M. M. Awad, L. M. Brinkac, S. C. Daugherty, D. H. Haft, R. J. Dodson, R. Madupu, W. C. Nelson, M. J. Rosovitz, S. A. Sullivan, H. Khouri, G. I. Dimitrov, K. L. Watkins, S. Mulligan, J. Benton, D. Radune, D. J. Fisher, H. S. Atkins, T. Hiscox, B. H. Jost, S. J. Billington, J. G. Songer, B. A. McClane, R. W. Titball, J. I. Rood, S. B. Melville, and I. T. Paulsen.** 2006. Skewed genomic variability in strains of the toxigenic bacterial pathogen, *Clostridium perfringens. Genome Res.* **16:**1031–1040.

144. **Nakayama, S., T. Kosaka, H. Hirakawa, K. Matsuura, S. Yoshino, and K. Furukawa.** 2008. Metabolic engineering for solvent productivity by downregulation of the hydrogenase gene cluster *hupCBA* in *Clostridium saccharoperbutylacetonicum* strain N1-4. *Appl. Microbiol. Biotechnol.* **78:**483–493.

145. **Neumann, A. P., and T. G. Rehberger.** 2009. MLST analysis reveals a highly conserved core genome among poultry isolates of *Clostridium septicum. Anaerobe* **15:**99–106.

146. **Nijland, R., C. Lindner, M. van Hartskamp, L. W. Hamoen, and O. P. Kuipers.** 2007. Heterologous production and secretion of *Clostridium perfringens* β-toxoid in closely related Gram-positive hosts. *J. Biotechnol.* **127:**361–372.

147. **Nolling, J., G. Breton, M. V. Omelchenko, K. S. Makarova, Q. Zeng, R. Gibson, H. M. Lee, J. Dubois, D. Qiu, J. Hitti, Y. I. Wolf, R. L. Tatusov, F. Sabathe, L. Doucette-Stamm, P. Soucaille, M. J. Daly, G. N. Bennett, E. V. Koonin, and D. R. Smith.** 2001. Genome sequence and comparative analysis of the solvent-producing bacterium *Clostridium acetobutylicum. J. Bacteriol.* **183:**4823–4838.

148. **Ogata, S., and M. Hongo.** 1979. Bacteriophages of the genus *Clostridium. Adv. Appl. Microbiol.* **25:**241–273.

149. **Onyenwoke, R. U., J. A. Brill, K. Farahi, and J. Wiegel.** 2004. Sporulation genes in members of the low G+C Gram-type-positive phylogenetic branch (*Firmicutes*). *Arch. Microbiol.* **182:**182–192.

150. **Oultram, J. D., and M. Young.** 1985. Conjugal transfer of plasmid pAMβ1 from *Streptococcus lactis* and *Bacillus subtilis* to *Clostridium acetobutylicum. FEMS Microbiol. Lett.* **27:**129–134.

151. **Paatero, A. O., H. Turakainen, L. J. Happonen, C. Olsson, T. Palomaki, M. I. Pajunen, X. Meng, T. Otonkoski, T. Tuuri, C. Berry, N. Malani, M. J. Frilander, F. D. Bushman, and H. Savilahti.** 2008. Bacteriophage Mu integration in yeast and mammalian genomes. *Nucleic Acids Res.* **36:**e148.

152. **Papoutsakis, E. T.** 2008. Engineering solventogenic clostridia. *Curr. Opin. Biotechnol.* **19:**420–429.

153. **Paredes, C. J., K. V. Alsaker, and E. T. Papoutsakis.** 2005. A comparative genomic view of clostridial sporulation and physiology. *Nat. Rev. Microbiol.* **3:**969–978.

154. **Paredes, C. J., I. Rigoutsos, and E. T. Papoutsakis.** 2004. Transcriptional organization of the *Clostridium acetobutylicum* genome. *Nucleic Acids Res.* **32:**1973–1981.

155. **Parker, R. C., H. C. Plummer, et al.** 1947. Effect of histolyticus infection and toxin on transplantable mouse tumors. *Proc. Soc. Exp. Biol. Med.* **66:**461–467.

156. **Parsons, J. A., T. L. Bannam, R. J. Devenish, and J. I. Rood.** 2007. TcpA, an FtsK/SpoIIIE homolog, is essential for transfer of the conjugative plasmid pCW3 in *Clostridium perfringens. J. Bacteriol.* **189:**7782–7790.

157. **Peck, M. W.** 2009. Biology and genomic analysis of *Clostridium botulinum. Adv. Microb. Physiol.* **55:**183–265, 320.

158. **Pennington, O. J.** 2006. The development of molecular tools for the expression of prodrug converting enzymes in *Clostridium sporogenes.* Ph.D. thesis. The University of Nottingham, Nottingham, United Kingdom.

159. **Perret, S., H. Maamar, J. P. Belaich, and C. Tardif.** 2004. Use of antisense RNA to modify the composition of cellulosomes produced by *Clostridium cellulolyticum. Mol. Microbiol.* **51:**599–607.

160. **Perreten, V., L. Vorlet-Fawer, P. Slickers, R. Ehricht, P. Kuhnert, and J. Frey.** 2005. Microarray-based detection of 90 antibiotic resistance genes of gram-positive bacteria. *J. Clin. Microbiol.* **43:**2291–2302.

161. **Perutka, J., W. Wang, D. Goerlitz, and A. M. Lambowitz.** 2004. Use of computer-designed group II introns to disrupt *Escherichia coli* DExH/D-box protein and DNA helicase genes. *J. Mol. Biol.* **336:**421–439.

162. **Petit, L., M. Gibert, and M. R. Popoff.** 1999. *Clostridium perfringens*: toxinotype and genotype. *Trends Microbiol.* **7:**104–110.

163. **Phillips-Jones, M. K.** 1993. Bioluminescence (lux) expression in the anaerobe *Clostridium perfringens. FEMS Microbiol. Lett.* **106:**265–270.

164. **Phillips-Jones, M. K.** 2000. Use of a lux reporter system for monitoring rapid changes in alpha-toxin gene expression in *Clostridium perfringens* during growth. *FEMS Microbiol. Lett.* **188:**29–33.

165. **Pierce, E., G. Xie, R. D. Barabote, E. Saunders, C. S. Han, J. C. Detter, P. Richardson, T. S. Brettin, A. Das, L. G. Ljungdahl, and S. W. Ragsdale.** 2008. The complete genome sequence of *Moorella thermoacetica* (f. *Clostridium thermoaceticum*). *Environ. Microbiol.* **10:**2550–2573.

166. **Prescott, S. C., and C. G. Dunn.** 1940. *Industrial Microbiology.* McGraw-Hill Company, New York, NY.

167. **Projan, S. J., M. Monod, C. S. Narayanan, and D. Dubnau.** 1987. Replication properties of pIM13, a naturally occurring plasmid found in *Bacillus subtilis*, and of its close relative pE5, a plasmid native to *Staphylococcus aureus. J. Bacteriol.* **169:**5131–5139.

168. **Purdy, D., T. A. O'Keeffe, M. Elmore, M. Herbert, A. McLeod, M. Bokori-Brown, A. Ostrowski, and N. P. Minton.** 2002. Conjugative transfer of clostridial shuttle vectors from *Escherichia coli* to *Clostridium difficile* through circumvention of the restriction barrier. *Mol. Microbiol.* **46:**439–452.

169. **Quixley, K. W., and S. J. Reid.** 2000. Construction of a reporter gene vector for *Clostridium beijerinckii* using a *Clostridium* endoglucanase gene. *J. Mol. Microbiol. Biotechnol.* **2:**53–57.

170. **Raju, D., P. Setlow, and M. R. Sarker.** 2007. Antisense-RNA-mediated decreased synthesis of small, acid-soluble spore proteins leads to decreased resistance of *Clostridium perfringens* spores to moist heat and UV radiation. *Appl. Environ. Microbiol.* **73:**2048–2053.

171. **Raphael, B. H., and J. D. Andreadis.** 2007. Real-time PCR detection of the nontoxic nonhemagglutinin gene as a rapid screening method for bacterial isolates harboring the botulinum neurotoxin (A-G) gene complex. *J. Microbiol. Methods* **71:**343–346.

172. **Raphael, B. H., C. Luquez, L. M. McCroskey, L. A. Joseph, M. J. Jacobson, E. A. Johnson, S. E. Maslanka, and J. D. Andreadis.** 2008. Genetic homogeneity of *Clostridium botulinum* type A1 strains with unique toxin gene clusters. *Appl. Environ. Microbiol.* **74:**4390–4397.

173. **Ravagnani, A., K. C. Jennert, E. Steiner, R. Grunberg, J. R. Jefferies, S. R. Wilkinson, D. I. Young, E. C. Tidswell, D. P. Brown, P. Youngman, J. G. Morris, and M. Young.** 2000. Spo0A directly controls the switch from acid to solvent production in solvent-forming clostridia. *Mol. Microbiol.* **37:**1172–1185.

174. **Roberts, A. P., M. Chandler, P. Courvalin, G. Guedon, P. Mullany, T. Pembroke, J. I. Rood, C. J. Smith, A. O. Summers, M. Tsuda, and D. E. Berg.** 2008. Revised nomenclature for transposable genetic elements. *Plasmid* **60:**167–173.

175. **Rood, J. I.** 2004. Virulence plasmids of spore-forming bacteria, p. 413–422. *In* B. E. Funnell and G. J. Phillips (ed.), *Plasmid Biology.* ASM Press, Washington, DC.

176. **Rood, J. I., and S. T. Cole.** 1991. Molecular genetics and pathogenesis of *Clostridium perfringens. Microbiol. Rev.* **55:**621–648.

177. **Rood, J. I., and M. Lyristis.** 1995. Regulation of extracellular toxin production in *Clostridium perfringens. Trends Microbiol.* **3:**192–196.

178. **Sabathe, F., C. Croux, E. Cornillot, and P. Soucaille.** 2002. amyP, a reporter gene to study strain degeneration in *Clostridium acetobutylicum* ATCC 824. *FEMS Microbiol. Lett.* **210:**93–98.

179. **Sacks, L. E., and P. A. Thompson.** 1978. Clear, defined medium for the sporulation of *Clostridium perfringens. Appl. Environ. Microbiol.* **35:**405–410.

180. **Sakaguchi, Y., T. Hayashi, K. Kurokawa, K. Nakayama, K. Oshima, Y. Fujinaga, M. Ohnishi, E. Ohtsubo, M. Hattori, and K. Oguma.** 2005. The genome sequence of *Clostridium botulinum* type C neurotoxin-converting phage and the molecular mechanisms of unstable lysogeny. *Proc. Natl. Acad. Sci. USA* **102:**17472–17477.

181. **Sakaguchi, Y., T. Hayashi, Y. Yamamoto, K. Nakayama, K. Zhang, S. Ma, H. Arimitsu, and K. Oguma.** 2009. Molecular analysis of an extrachromosomal element containing the C2 toxin gene discovered in *Clostridium botulinum* type C. *J. Bacteriol.* **191:**3282–3291.

182. **Sambrook, J., P. MacCallum, and D. Russell.** 2001. *Molecular Cloning: a Laboratory Manual*, 3rd ed. Cold Spring Harbor Laboratory Press, Cold Spring Harbor, NY.

183. **Sawires, Y. S., and J. G. Songer.** 2005. Multiple-locus variable-number tandem repeat analysis for strain typing of *Clostridium perfringens. Anaerobe* **11:**262–272.

184. **Sayeed, S., J. Li, and B. A. McClane.** 2007. Virulence plasmid diversity in *Clostridium perfringens* type D isolates. *Infect. Immun.* **75:**2391–2398.

185. **Schantz, E. J., and E. A. Johnson.** 1992. Properties and use of botulinum toxin and other microbial neurotoxins in medicine. *Microbiol. Rev.* **56:**80–99.

186. **Schmidt, J. A.** 1998. M.S. thesis. University of Wisconsin-Madison, Madison, WI.

187. **Schumann, W.** 2007. Production of recombinant proteins in *Bacillus subtilis. Adv. Appl. Microbiol.* **62:**137–189.

188. **Scotcher, M. C., and G. N. Bennett.** 2008. Activity of abrB310 promoter in wild type and spo0A-deficient strains of *Clostridium acetobutylicum. J. Ind. Microbiol. Biotechnol.* **35:**743–750.

189. Scotcher, M. C., and G. N. Bennett. 2005. SpoIIE regulates sporulation but does not directly affect solventogenesis in *Clostridium acetobutylicum* ATCC 824. *J. Bacteriol.* **187:**1930–1936.

190. Sebaihia, M., M. W. Peck, N. P. Minton, N. R. Thomson, M. T. Holden, W. J. Mitchell, A. T. Carter, S. D. Bentley, D. R. Mason, L. Crossman, C. J. Paul, A. Ivens, M. H. Wells-Bennik, I. J. Davis, A. M. Cerdeno-Tarraga, C. Churcher, M. A. Quail, T. Chillingworth, T. Feltwell, A. Fraser, I. Goodhead, Z. Hance, K. Jagels, N. Larke, M. Maddison, S. Moule, K. Mungall, H. Norbertczak, E. Rabbinowitsch, M. Sanders, M. Simmonds, B. White, S. Whithead, and J. Parkhill. 2007. Genome sequence of a proteolytic (Group I) *Clostridium botulinum* strain Hall A and comparative analysis of the clostridial genomes. *Genome Res.* **17:**1082–1092.

191. Sebaihia, M., B. W. Wren, P. Mullany, N. F. Fairweather, N. Minton, R. Stabler, N. R. Thomson, A. P. Roberts, A. M. Cerdeno-Tarraga, H. Wang, M. T. Holden, A. Wright, C. Churcher, M. A. Quail, S. Baker, N. Bason, K. Brooks, T. Chillingworth, A. Cronin, P. Davis, L. Dowd, A. Fraser, T. Feltwell, Z. Hance, S. Holroyd, K. Jagels, S. Moule, K. Mungall, C. Price, E. Rabbinowitsch, S. Sharp, M. Simmonds, K. Stevens, L. Unwin, S. Whithead, B. Dupuy, G. Dougan, B. Barrell, and J. Parkhill. 2006. The multidrug-resistant human pathogen *Clostridium difficile* has a highly mobile, mosaic genome. *Nat. Genet.* **38:**779–786.

192. Sebald, M. 1994. Genetic basis for antibiotic resistance in anaerobes. *Clin. Infect. Dis.* **18**(Suppl. 4):S297–S304.

193. Sebald, M., and R. N. Costilow. 1975. Minimal growth requirements for *Clostridium perfringens* and isolation of auxotrophic mutants. *Appl. Microbiol.* **29:**1–6.

194. Seedorf, H., W. F. Fricke, B. Veith, H. Brüggemann, H. Liesegang, A. Strittmatter, M. Miethke, W. Buckel, J. Hinderberger, F. Li, C. Hagemeier, R. K. Thauer, and G. Gottschalk. 2008. The genome of *Clostridium kluyveri*, a strict anaerobe with unique metabolic features. *Proc. Natl. Acad. Sci. USA* **105:**2128–2133.

195. Sergeev, N., M. Distler, S. Courtney, S. F. Al-Khaldi, D. Volokhov, V. Chizhikov, and A. Rasooly. 2004. Multipathogen oligonucleotide microarray for environmental and biodefense applications. *Biosens. Bioelectron.* **20:**684–698.

196. Setlow, M., and E. A. Johnson. 2007. Spores and their significance, p. 35–67. *In* M. P. Doyle and L. R. Beuchat (ed.), *Food Microbiology: Fundamentals and Frontiers*, 3rd ed. ASM Press, Washington, DC.

197. Shao, L., S. Hu, Y. Yang, Y. Gu, J. Chen, W. Jiang, and S. Yang. 2007. Targeted gene disruption by use of a group II intron (targetron) vector in *Clostridium acetobutylicum*. *Cell Res.* **17:**963–965.

198. Sharp, P. M., E. Bailes, R. J. Grocock, J. F. Peden, and R. E. Sockett. 2005. Variation in the strength of selected codon usage bias among bacteria. *Nucleic Acids Res.* **33:**1141–1153.

199. Shi, Z., and H. P. Blaschek. 2008. Transcriptional analysis of *Clostridium beijerinckii* NCIMB 8052 and the hyper-butanol-producing mutant BA101 during the shift from acidogenesis to solventogenesis. *Appl. Environ. Microbiol.* **74:**7709–7714.

200. Shimizu, T., K. Ohtani, H. Hirakawa, K. Ohshima, A. Yamashita, T. Shiba, N. Ogasawara, M. Hattori, S. Kuhara, and H. Hayashi. 2002. Complete genome sequence of *Clostridium perfringens*, an anaerobic flesheater. *Proc. Natl. Acad. Sci. USA* **99:**996–1001.

201. Sillers, R., M. A. Al-Hinai, and E. T. Papoutsakis. 2009. Aldehyde-alcohol dehydrogenase and/or thiolase overexpression coupled with CoA transferase downregulation lead to higher alcohol titers and selectivity in *Clostridium acetobutylicum* fermentations. *Biotechnol. Bioeng.* **102:**38–49.

202. Sillers, R., A. Chow, B. Tracy, and E. T. Papoutsakis. 2008. Metabolic engineering of the non-sporulating, non-solventogenic *Clostridium acetobutylicum* strain M5 to produce butanol without acetone demonstrate the robustness of the acid-formation pathways and the importance of the electron balance. *Metab. Eng.* **10:**321–332.

203. Smith, T. J., K. K. Hill, B. T. Foley, J. C. Detter, A. C. Munk, D. C. Bruce, N. A. Doggett, L. A. Smith, J. D. Marks, G. Xie, and T. S. Brettin. 2007. Analysis of the neurotoxin complex genes in *Clostridium botulinum* A1-A4 and B1 strains: BoNT/A3, /Ba4 and /B1 clusters are located within plasmids. *PLoS One* **2:**e1271.

204. Smyth, H. F., and W. L. Obold. 1930. *Industrial Microbiology: the Utilization of Bacteria, Yeasts, and Molds in Industrial Processes*. The Williams & Wilkins Company, Baltimore, MD.

205. Song, L., S. Ahn, and D. R. Walt. 2006. Fiber-optic microsphere-based arrays for multiplexed biological warfare agent detection. *Anal. Chem.* **78:**1023–1033.

206. Soutschek-Bauer, E., L. Hartl, and W. L. Staudenbauer. 1985. Transformation of *Clostridium thermohydrosulfuricum* DSM 568 with plasmid DNA. *Biotechnol. Lett.* **7:**705–710.

207. Stabler, R. A., D. N. Gerding, J. G. Songer, D. Drudy, J. S. Brazier, H. T. Trinh, A. A. Witney, J. Hinds, and B. W. Wren. 2006. Comparative phylogenomics of *Clostridium difficile* reveals clade specificity and microevolution of hypervirulent strains. *J. Bacteriol.* **188:**7297–7305.

208. Stackebrandt, E., and H. Hippe. 2001. Taxonomy and systematics, p. 19–48. *In* H. Bahl and P. Dürre (ed.), *Clostridia. Biotechnology and Medical Applications*. Wiley-VCH, Weinheim, Germany.

209. Stackebrandt, E., I. Kramer, J. Swiderski, and H. Hippe. 1999. Phylogenetic basis for a taxonomic dissection of the genus *Clostridium*. *FEMS Immunol. Med. Microbiol.* **24:**253–258.

210. Steen, J. A., T. L. Bannam, W. L. Teng, R. J. Devenish, and J. I. Rood. 2009. The putative coupling protein TcpA interacts with other pCW3-encoded proteins to form an essential part of the conjugation complex. *J. Bacteriol.* **191:**2926–2933.

211. Steil, L., M. Serrano, A. O. Henriques, and U. Volker. 2005. Genome-wide analysis of temporally regulated and compartment-specific gene expression in sporulating cells of *Bacillus subtilis*. *Microbiology* **151:**399–420.

212. Swinfield, T. J., J. D. Oultram, D. E. Thompson, J. K. Brehm, and N. P. Minton. 1990. Physical characterisation of the replication region of the *Streptococcus faecalis* plasmid pAM beta 1. *Gene* **87:**79–90.

213. Teng, W. L., T. L. Bannam, J. A. Parsons, and J. I. Rood. 2008. Functional characterization and localization of the TcpH conjugation protein from *Clostridium perfringens*. *J. Bacteriol.* **190:**5075–5086.

214. Theys, J., A. W. Landuyt, S. Nuyts, L. Van Mellaert, P. Lambin, and J. Anne. 2001. *Clostridium* as a tumor-specific delivery system of therapeutic proteins. *Cancer Detect. Prev.* **25:**548–557.

215. Thormann, K., L. Feustel, K. Lorenz, S. Nakotte, and P. Dürre. 2002. Control of butanol formation in *Clostridium acetobutylicum* by transcriptional activation. *J. Bacteriol.* **184:**1966–1973.

216. Tomas, C. A., S. B. Tummala, and E. T. Papoutsakis. 2005. Metabolic engineering of solventogenic clostridia, p. 813–830. *In* P. Dürre (ed.), *Handbook on Clostridia*. Taylor & Francis Group, Boca Raton, FL.

217. Truffaut, N., J. Hubert, and G. Reysset. 1989. Construction of shuttle vectors useful for transforming *Clostridium acetobutylicum*. *FEMS Microbiol. Lett.* **58:**15–20.

218. **Tummala, S. B., S. G. Junne, and E. T. Papoutsakis.** 2003. Antisense RNA downregulation of coenzyme A transferase combined with alcohol-aldehyde dehydrogenase overexpression leads to predominantly alcohologenic *Clostridium acetobutylicum* fermentations. *J. Bacteriol.* **185:**3644–3653.

219. **Tummala, S. B., C. Tomas, L. M. Harris, N. E. Welker, F. B. Rudolph, G. N. Bennett, and E. T. Papoutsakis.** 2001. Genetic tools for solventogenic clostridia, p. 105–123. *In* H. Bahl and P. Dürre (ed.), *Clostridia. Biotechnology and Medical Applications.* Wiley-VCH, Weinheim, Germany.

220. **Tummala, S. B., C. Tomas, and E. Papoutsakis.** 2005. Gene analysis in clostridia, p. 53–70. *In* P. Dürre (ed.), *Handbook on Clostridia.* Taylor & Francis Group, Boca Raton, FL.

221. **Tummala, S. B., N. E. Welker, and E. T. Papoutsakis.** 2003. Design of antisense RNA constructs for downregulation of the acetone formation pathway of *Clostridium acetobutylicum. J. Bacteriol.* **185:**1923–1934.

222. **Tummala, S. B., N. E. Welker, and E. T. Papoutsakis.** 1999. Development and characterization of a gene expression reporter system for *Clostridium acetobutylicum* ATCC 824. *Appl. Environ. Microbiol.* **65:**3793–3799.

223. **Van Heyningen, W. E.** 1950. *Bacterial Toxins.* Blackwell Scientific Publications, Oxford, United Kingdom.

224. **Varga, J., V. L. Stirewalt, and S. B. Melville.** 2004. The CcpA protein is necessary for efficient sporulation and enterotoxin gene (*cpe*) regulation in *Clostridium perfringens. J. Bacteriol.* **186:**5221–5229.

225. **Vidal, J. E., J. Chen, J. Li, and B. A. McClane.** 2009. Use of an EZ-Tn5-based random mutagenesis system to identify a novel toxin regulatory locus in *Clostridium perfringens* strain 13. *PLoS One* **4:**e6232.

226. **Wagner, E. G., and R. W. Simons.** 1994. Antisense RNA control in bacteria, phages, and plasmids. *Annu. Rev. Microbiol.* **48:**713–742.

227. **Walters, D. M., V. L. Stirewalt, and S. B. Melville.** 1999. Cloning, sequence, and transcriptional regulation of the operon encoding a putative N-acetylmannosamine-6-phosphate epimerase (*nanE*) and sialic acid lyase (*nanA*) in *Clostridium perfringens. J. Bacteriol.* **181:**4526–4532.

228. **Watt, B., and R. Brown.** 1975. A defined medium for the growth of *Clostridium tetani* and other anaerobes of clinical interest. *J. Med. Microbiol.* **8:**167–172.

229. **Wei, M. Q., A. Mengesha, D. Good, and J. Anne.** 2008. Bacterial targeted tumour therapy—dawn of a new era. *Cancer Lett.* **259:**16–27.

230. **Weizmann, C.** 1949. *Trial and Error: the Autobiography of Chaim Weizmann.* Harper & Brothers, New York, NY.

231. **Wen, Q., and B. A. McClane.** 2004. Detection of enterotoxigenic *Clostridium perfringens* type A isolates in American retail foods. *Appl. Environ. Microbiol.* **70:**2685–2691.

232. **Whitmer, M. E., and E. A. Johnson.** 1988. Development of improved defined media for *Clostridium botulinum* serotypes A, B, and E. *Appl. Environ. Microbiol.* **54:**753–759.

233. **Willis, A. T.** 1969. *Clostridia of Wound Infection.* Butterworths, London, United Kingdom.

234. **Wu, S. C., J. C. Yeung, Y. Duan, R. Ye, S. J. Szarka, H. R. Habibi, and S. L. Wong.** 2002. Functional production and characterization of a fibrin-specific single-chain antibody fragment from *Bacillus subtilis*: effects of molecular chaperones and a wall-bound protease on antibody fragment production. *Appl. Environ. Microbiol.* **68:**3261–3269.

235. **Yang, W. W., E. N. Crow-Willard, and A. Ponce.** 2009. Production and characterization of pure *Clostridium* spore suspensions. *J. Appl. Microbiol.* **106:**27–33.

236. **Yoshino, S., T. Yoshino, S. Hara, S. Ogata, and S. Hayashida.** 1990. Construction of shuttle vector plasmid between *Clostridium acetobutylicum* and *Escherichia coli. Agric. Biol. Chem.* **54:**437–441.

237. **Young, M.** 1993. Development and exploitation of conjugative gene transfer in clostridia, p. 99–117. *In* D. R. Woods (ed.), *The Clostridia and Biotechnology.* Butterworth-Heinemann, Stoneham, MA.

238. **Young, M., and S. T. Cole.** 1993. *Clostridium*, p. 35–52. *In* A. L. Sonenshein, J. A. Hoch, and R. Losick (ed.), *Bacillus subtilis and Other Gram-Positive Bacteria: Biochemistry, Physiology and Molecular Genetics.* ASM Press, Washington, DC.

239. **Young, M., N. P. Minton, and W. L. Staudenbauer.** 1989. Recent advances in the genetics of the clostridia. *FEMS Microbiol. Rev.* **5:**301–325.

240. **Young, M., W. L. Staudenbauer, and N. P. Minton.** 1989. Genetics of *Clostridium*, p. 63–103. *In* N. P. Minton and D. J. Clarke (ed.), *Biotechnology Handbooks: Clostridia.* Plenum Publishing Corporation, New York, NY.

241. **Zhong, J., M. Karberg, and A. M. Lambowitz.** 2003. Targeted and random bacterial gene disruption using a group II intron (targetron) vector containing a retrotransposition-activated selectable marker. *Nucleic Acids Res.* **31:**1656–1664.

242. **Zhou, H., J. Gong, J. Brisbin, H. Yu, A. J. Sarson, W. Si, S. Sharif, and Y. Han.** 2009. Transcriptional profiling analysis of host response to *Clostridium perfringens* infection in broilers. *Poult. Sci.* **88:**1023–1032.

243. **Zhou, Y., H. Sugiyama, H. Nakano, and E. A. Johnson.** 1995. The genes for the *Clostridium botulinum* type G toxin complex are on a plasmid. *Infect. Immun.* **63:**2087–2091.

244. **Zimmer, M., S. Scherer, and M. J. Loessner.** 2002. Genomic analysis of *Clostridium perfringens* bacteriophage φ3626, which integrates into *guaA* and possibly affects sporulation. *J. Bacteriol.* **184:**4359–4368.

245. **Zverlov, V. V., M. Klupp, J. Krauss, and W. H. Schwarz.** 2008. Mutations in the scaffoldin gene, *cipA*, of *Clostridium thermocellum* with impaired cellulosome formation and cellulose hydrolysis: insertions of a new transposable element, IS*1447*, and implications for cellulase synergism on crystalline cellulose. *J. Bacteriol.* **190:**4321–4327.

Genetic Manipulation of Myxobacteria

WESLEY P. BLACK, BRYAN JULIEN, EDUARDO RODRIGUEZ,
AND ZHAOMIN YANG

18

18.1. INTRODUCTION

Myxobacteria, a coherent group of the *Deltaproteobacteria*, are recognized as important producers of secondary metabolites of industrial and medical importance. These gram-negative bacteria have fascinated scientists for over half a century because of their surface-gliding motility and their starvation-induced multicellular development (40, 84, 95). In more recent years, they have emerged as prominent producers of secondary metabolites, as discussed in several recent reviews (5, 6, 22, 23, 69). Myxobacteria as a group produce more bioactive compounds than any other gram-negative phylum and rank third after actinomycetes and bacilli (69, 70). Currently there are over 100 different known natural products from myxobacteria with about 500 structural variants (5, 6). *Sorangium cellulosum* accounts for nearly 50% of these and *Myxococcus* and *Stigmatella* account for about 25% (22, 23). Importantly, the bioactive compounds from myxobacteria tend to be novel and distinct (5). For example, *S. cellulosum* produces epothilone (33), which is currently marketed as an anticancer drug (Ixempra® by Bristol-Myers Squibb [Princeton, NJ]). Although epothilones act as microtubule stabilizers like paclitaxel (8), they remain effective against paclitaxel-resistant cancer cells (46). Genome sequencing of a few myxobacteria has revealed genes for the biosynthesis of unidentified secondary metabolites (6, 24, 74). Considering the need for new drugs to treat a variety of diseases, further exploration of the myxobacteria is imperative.

Tools for genetic manipulation of myxobacteria are important for identifying and engineering strains to maximize production of secondary metabolites. Fortunately, a considerable number of such tools have been developed over the years, including methods for mutagenesis, transduction, conjugation, and transformation. Several developments have enabled heterologous gene expression in myxobacteria, including the construction and development of regulated promoters and the identification of strong constitutive promoters. Recently, the first autonomously replicating plasmid was identified. While most genetic tools were first developed for the model myxobacterium *Myxococcus xanthus*, many have been applied to *S. cellulosum*, *Stigmatella aurantiaca*, and the lesser-known myxobacteria *Chondromyces crocatus*, *Cystobacter fuscus*, *Angiococcus disciformis*, and *Corallococcus macrosporus*. We describe these tools in this chapter and hope to facilitate the increased use of myxobacteria for applications in biotechnology and drug discovery.

18.2. TRANSFER OF DNA INTO MYXOBACTERIA

18.2.1. Transduction of Myxobacteria

18.2.1.1. Transduction of DNA from *Escherichia coli* to *M. xanthus*

The first reported transfer of DNA from *E. coli* to *M. xanthus* was associated with the coliphage P1 (41). P1 harboring a chloramphenicol resistance gene was shown to inject its DNA into *M. xanthus*, producing unstable drug-resistant transformants. Because P1 is unable to replicate in or lysogenize *M. xanthus*, it became the earliest suicide vector for delivery of Tn5 for transposon mutagenesis (48). Because P1 is a generalized transducing phage for *E. coli*, it has the ability to randomly package DNA. This was exploited to package plasmid DNA for transfer to *M. xanthus*. Because of the large size requirement, approximately 100 kb, to produce stable infecting P1 phage particles, the transfer of larger plasmids is more efficient than smaller ones (62). Using P1 for plasmid DNA delivery before the discovery of electroporation allowed for mutant complementation and strain construction. Two other *E. coli* phages, P2 and P4, bind to the same cell surface receptor as P1. P4 has been shown to inject DNA into *M. xanthus* using a P4::miniTn5Kan^r derivative. Like P1, P4 is unable to replicate or lysogenize in *M. xanthus* (B. Julien, unpublished results). Although P4 does not package DNA randomly, it can package plasmid DNA that contains the 125-bp recognition sequence for DNA packaging (94). With the development of electroporation protocols (see below), P1 is used less frequently to introduce plasmid DNA into *M. xanthus*. However, with the construction of bacterial artificial chromosome libraries harboring large polyketide synthase biosynthetic gene clusters, such as the ~70-kb ambruticin gene cluster, the ability to move them into *M. xanthus* with P1 may prove useful since electroporation is not efficient with such large DNA (Julien, unpublished). For methods using P1 to transfer DNA from *E. coli* to *M. xanthus*, refer to O'Connor and Zusman (62).

18.2.1.2. Transducing Phages of *M. xanthus*

Currently, only four phages have been identified that infect myxobacteria, all of which were isolated using *M. xanthus* as a host. The first myxophage isolated was Mx1 (12, 13). It is a lytic phage and does not transduce chromosomal DNA.

The other three myxophages, Mx4, Mx8, and Mx9 (14, 21, 55), are temperate phages. Mx4 and Mx8 have been shown to transfer genetic markers between different strains of M. xanthus. They have been developed as generalized transducing phages for mapping mutations and strain constructions. Mutants of both have been isolated that carry out general transduction with improved efficiency (14, 21, 55). For those requiring transduction using Mx4 or Mx8, please refer to Campos et al. (14) or Martin et al. (55), respectively, for more details.

18.2.2. Transformation of Myxobacteria by Electroporation

Electroporation has become the most common technique to introduce DNA into myxobacteria, including the species M. xanthus, S. aurantiaca, C. fuscus, A. disciformis, and C. macrosporus (43, 63, 68, 72, 77). In addition to plasmids, it can be used with genomic DNA fragments, which is useful as a substitute for transduction to transfer mutations between M. xanthus strains (82). More recently, genomic DNA electroporation has been used to establish genetic linkages in M. xanthus. In our experience, markers that are about 10 kb apart can be cotransformed easily using genomic DNA transformation. Although electroporation of DNA into S. cellulosum has been reported (45), it has not been duplicated due to lack of details provided in the original publication.

Protocol 1. Electroporation Protocol for M. xanthus

This protocol is based on Kashefi and Hartzell (43), with all steps performed at room temperature unless otherwise noted.

1. Grow an overnight culture at 32°C in Casitone-Yeast Extract (CYE) to an optical density at 600 nm (OD_{600}) of ~0.5 to 1.0.
2. For each electroporation, pellet 1.5 ml of culture at 15,000 × g in a microcentrifuge for 4 min. Aspirate off medium.
3. Wash cells by resuspending in 1 ml of sterile distilled deionized water (ddH_2O).
4. Pellet cells as in step 2. Aspirate ddH_2O (pellet may become loose).
5. Repeat washing with sterile ddH_2O two or three times.
6. Resuspend cells in 40 μl of sterile ddH_2O.
7. Add 1 to 5 μl of salt-free DNA to the cell suspension and mix by vortexing. The amount of DNA to use for each electroporation is as follows.
   a. 100 to 300 ng of replicative plasmid or plasmid with phage attachment site
   b. 0.5 to 1.5 μg of plasmid for integration by homologous recombination
   c. 3 to 5 μg of genomic DNA
8. Transfer cells and DNA to an electroporation cuvette with a 1-mm gap and perform electroporation with the following settings.
   a. Resistance at 400 Ω
   b. Voltage at 0.65 kV
   c. Capacitance at 25 μF
   d. Optimal time constant usually between 8 and 9.5
9. Quickly remove the cells from the cuvette with 1 ml of CYE and transfer to a tube with an additional 1.5 ml of CYE.
10. Incubate at 32°C for 4 to 5 h with shaking at 300 rpm for recovery. A 2- to 3-h incubation may suffice for phage attachment plasmids or replicative plasmids, but for homologous recombination with sequences less than 500 bp an overnight recovery may be necessary.

11. Plate cells on selective medium as follows.
    a. For attachment or replicative plasmids, plate 100 μl directly and 100 μl of a 1/10 dilution.
    b. For homologous recombination and genomic DNA transformations, plate 100 to 150 μl directly. Concentrate the rest and plate on two plates.
12. Incubate plates at 32°C for 5 to 6 days to obtain single colonies.

18.2.3. Conjugation of Myxobacteria

Conjugation is the most frequently used technique to introduce DNA into S. cellulosum, and several different protocols have been described (35, 44, 66, 75). Strain-specific modifications appear necessary to achieve maximal efficiency (44). Protocol 2 is a general conjugation procedure for S. cellulosum. This protocol is applicable to M. xanthus except that the cells do not need to be heat shocked and the medium used for growth of M. xanthus is CYE. Conjugation has also been used to create knockout mutations in C. crocatus (67). For M. xanthus and S. aurantiaca, electroporation is highly favored over conjugation for small plasmids. However, for large plasmids, ~100 kb, it is necessary to perform conjugation with M. xanthus (Julien, unpublished). For conjugation protocols using M. xanthus or S. aurantiaca, please refer to Breton et al. (11), Glomp et al. (23a), and Jaoua et al. (34) for more details.

Protocol 2. General Conjugation Protocol for S. cellulosum

1. Grow the myxobacterial strain in 50 ml of 307 medium to an OD_{600} of no more than 1.0 (~1 × 10^8 cells/ml) but aiming for an OD_{600} of ~0.5.
2. The night before the myxobacterial strain is ready, grow the E. coli donor strain in 20 ml of LB containing 0.5% NaCl in a 125-ml flask at 37°C overnight without shaking.
3. The next day, heat shock 1 × 10^9 S. cellulosum cells at 50°C for 10 min and then pellet cells. When performing conjugation to deliver the mariner transposon, this heat shock step is not necessary. Pellet 0.1 to 10 ml of donor E. coli. Decant supernatants, leaving about 400 μl of 307 medium for myxobacteria, and remove all LB from E. coli. Resuspend S. cellulosum in the remaining medium, add to the E. coli pellet, and resuspend those cells too.
4. Pipette the combined cells onto the center of an S42 plate. Use a dry plate to ensure that the liquid will be absorbed. Incubate at 30°C for 24 h.
5. Scrape off the cells, suspend in 1 ml of 307 medium, and plate various dilutions onto S42 plates containing kanamycin (Kan) at 50 μg/ml (to inhibit E. coli) and the appropriate antibiotic for selection in S. cellulosum. For integration of plasmid DNA by homologous recombination, plate 10 to 100 μl/plate from the 1-ml suspension. If conjugating the mariner transposon, suspend cells in 10 ml of 307 medium and plate 5 to 25 μl/plate.

18.2.4. Selectable Markers in Myxobacteria

A variety of antibiotic selection markers are used in myxobacteria. The most commonly used marker is nptII from Tn5 (2), which confers kanamycin resistance (Kan^r) and has been used with M. xanthus, S. aurantiaca, C. fuscus, C. macrosporus, and A. disciformis (48, 63, 67, 68, 72, 77). Kanamycin is typically used at 50 to 100 μg/ml. For S. cellulosum strains, which are naturally resistant to kanamycin, the best alternatives are hygromycin (Hyg) using the hyg resistance gene or phleomycin (Phleo) and its analogue bleomycin (Bleo) using the bleo resistance gene from Tn5. Hygromycin is used at 60 μg/ml (66) and phleomycin at 20 to 40 μg/ml (37, 38). The *bleo* resistance gene also works well in M. *xanthus* using phleomycin concentrations of 20 to 50 μg/ml (Julien, unpublished). When the *hyg* resistance gene is used with C. *crocatus*, selection requires 100 μg/ml hygromycin (67). Tetracycline resistance (Tet^r), which is conferred by *tetA* from pSC101 or pBR322 (7, 18), has been used for selection in M. *xanthus* and S. *aurantiaca* (1). Due to the long incubations typical for myxobacteria, the more stable tetracycline derivative oxytetracycline is generally used at 15 μg/ml. Another good marker for M. *xanthus* is resistance to apramycin (Apra) conferred by the *aacC4* gene driven by the promoter from the Kan resistance gene *nptII* (4). Apramycin is typically used at 60 μg/ml (Julien, unpublished). Although rarely used, trimethoprim (Tp) (250 μg/ml) can be used for selection in M. *xanthus* (19, 90).

18.3. MUTAGENESIS OF MYXOBACTERIA

Mutagenesis is fundamental to any genetic manipulation. Classical techniques utilizing UV irradiation and chemical mutagens have been used to introduce random point mutations in M. *xanthus*, especially in the early days of basic research (14, 30, 31, 58). Although it is laborious to identify a point mutation in myxobacteria, UV or chemical mutagenesis may prove valuable for the improvement of industrial strains. Interested readers are directed to Hodgkin and Kaiser (31) and Morrison and Zusman (58) for related protocols of chemical and UV mutagenesis. Here we will focus on mutagenesis procedures based on transposition and homologous recombination.

18.3.1. Transposon Mutagenesis of Myxobacteria

Transposon mutagenesis is critical for myxobacterial genetics due to the ease of mutation identification. As mentioned earlier, Tn5 was the first transposon utilized in myxobacteria (48). It proved powerful not only for mutagenesis, but also for mapping point mutations from previous studies. Several derivatives of Tn5 have been useful. Tn5-132 and Tn-Tp allowed in situ replacement of Kan^r with either Tet^r or trimethoprim resistance, respectively (1). It also made it possible to combine multiple Tn5 insertions in the same strain. Tn5*lac*, which contains a promoterless *lacZ*, has been useful in dissecting the temporal regulation of gene expression during M. *xanthus* development (47). Another useful derivative is Tn*phoA* (54), which creates translational fusions to the extracellular phosphatase, PhoA. Since PhoA is only active when exported to the periplasm, Tn*phoA* has proved useful in the identification of secreted proteins from M. *xanthus* (10, 42). A serious drawback of Tn5 is its lack of true randomness in transposition. It was determined that it transposes near sequences matching or resembling the consensus A-GNTYWRANC-T (25). As such, insertions in certain genes may never occur even when insertions in other genes are saturated or overrepresented.

A welcome development was the demonstration that *mariner*-based transposons are functional in M. *xanthus*. Transposons of the *mariner* family, which require only a TA dinucleotide for insertion, are nearly random in transposition (71). Several *mariner* derivatives have been used in myxobacteria. *magellan4*, which was initially constructed for mutagenesis in mycobacteria (71), was first used in M. *xanthus* by Youderian et al. (92) to identify genes required for motility. Another is *miniHimar1*-Tet, which was constructed by replacing the Kan^r from *magellan4* with Tet^r (H. Kaplan, unpublished data). This enables transposon mutagenesis in a strain that is already Kan^r. Julien and Fehd (38) developed

another *mariner* derivative for mutagenesis of S. *cellulosum*. This transposon, which can be conjugated into S. *cellulosum*, contains both kanamycin and bleomycin resistance for selection. The conjugative delivery plasmid also encodes a mutated transposase that increases the frequency of transposition. Another *mariner*-based transposon that can be conjugated into S. *cellulosum* (44) is similar to *magellan4* except that Kan^r was replaced with the hygromycin resistance marker.

Identification of the insertion site of *magellan4* and its derivatives is relatively easy due to the inclusion of the R6Kγ replication origin in these transposons. Genomic DNA of a transposon insertion mutant is isolated and digested with a restriction enzyme that does not cut within the transposon. The DNA is self-ligated and transformed into a λ*pir* E. *coli* strain. Plasmids isolated from these transformants can be sequenced using primers complementary to the left and right insertion elements.

Protocol 3. Genomic DNA Preparation

This small-scale genomic DNA prep (65) is sufficient for genomic DNA transformations of M. *xanthus* and for the isolation of *mariner* insertions (see next section).

1. Pellet 3 to 4 ml of an overnight culture at an OD_{600} of ~1.
2. Resuspend pellet in 500 μl of STE buffer (20 mM Tris-HCl [pH 7.5], 75 mM NaCl, 25 mM EDTA [pH 8.0]).
3. Add lysozyme to a final concentration of ~1 mg/ml (10 μl of a 50-mg/ml stock), mix by inversion, and incubate at 37°C for 30 min.
4. Add 50 μl of 10% sodium dodecyl sulfate (about 0.1 volume) and mix by inversion.
5. Add 15 μl of a 20-mg/ml stock of proteinase K, mix by inversion, and incubate at 55°C for 30 to 60 min.
6. Add 1/3 volume of 5 M NaCl (~190 μl) and mix by inversion.
7. Add 1 volume of chloroform (~760 μl) and mix by inversion until milky. Allow layers to separate for 2 to 3 min and mix once more.
8. Centrifuge for 5 to 10 min at 5,000 × g in a microcentrifuge.
9. Carefully transfer the upper aqueous layer (~700 μl) to a new tube (avoid cell debris at the interface). Add an equal volume of isopropanol (~700 μl) and mix by inversion until DNA precipitates.
10. Centrifuge for 1 min at 15,000 × g in a microcentrifuge and keep the white DNA pellet.
11. Wash DNA pellet with 750 μl of 70% ethanol and repeat step 10.
12. Air-dry the pellet on the bench or at 37°C for ~5 to 15 min.
13. Gently resuspend DNA pellet in ~100 μl of sterile ddH₂O or other suitable buffer (Tris-EDTA or 10 mM Tris-HCl [pH 8.0]). Typical yield should be ~1 to 2 μg/μl.

Protocol 4. Isolation and Identification of *mariner* Insertions

This procedure is based on that reported by Youderian et al. (92).

1. Isolate genomic DNA according to protocol 3.
2. Digest 1 μg of genomic DNA in 20 μl using 10 to 20 units of an appropriate restriction enzyme. (Generally SacII or BssHII is chosen because they do not cleave within *magellan4* and cleave the high-GC myxobacterial DNA more frequently.)

3. Heat-inactivate enzyme following instructions from supplier.
4. Use 10 μl of the digestion to set up a 25-μl ligation reaction. Incubate at 15°C for 4 h to overnight.
5. Use 4 to 5 μl of the ligation mixture to transform λpir E. coli such as DH5α λpir (71). Use electrocompetent cells or cells of high competence if prepared chemically.
6. Plate cells on LB plates with appropriate antibiotic.
7. Prepare plasmid DNA from the transformants.
8. Identify disrupted gene by sequencing using primers complementary to the left and right arms of *mariner*:
   a. Mar1 (left): 5′ CGCCATCTATGTGTCAGACC-GGGG
   b. Mar2 (right): 5′ TGTGTTTTTCTTTGTTAGACCG

18.3.2. Random Mutagenesis through Homologous Recombination

A method based on homologous recombination provides an alternative to transposons for generating random insertions in myxobacteria (17). Cho and Zusman developed this method to identify genes important for development, and it has been used successfully to identify other genes (16, 17, 53). A plasmid library is constructed by cloning random genomic DNA fragments (~500 bp) into the *E. coli* plasmid pZErO-2 (Table 1), which is unable to replicate in *M. xanthus*. The library is electroporated into *M. xanthus* to create insertion mutations through homologous recombination. The mutations are identified by cloning the plasmid with flanking DNA in a similar fashion to that described for *mariner* insertions (see protocol 4).

TABLE 1 Selected plasmids for myxobacteria

Plasmid	Relevant characteristics	Reference
For integration by homologous recombination		
pZErO-2	Cloning vector, Kan^r	33a
pBGS18	Cloning vector, Kan^r	76
pBR322	Cloning vector, Tet^r	7
Phage attachment for *M. xanthus*		
pWB200	*intP/attP* (Mx8), Kan^r	91
pSWU19	*intP/attP* (Mx8), Kan^r	87
pSWU30	*intP/attP* (Mx8), Tet^r	9
pYC274	*intP/attP* (Mx8), Kan^r	29
pKOS249-31	*intP/attP* (Mx9), Bleo^r	37
Allelic exchange		
pBJ113	Kan^r-*galK* cassette	39
pSWU35	Kan^r-*sacB* cassette	88
Replicative		
pZJY41	*M. xanthus* replicative, Kan^r	93
pZJY156	*M. xanthus* replicative, Kan^r	93
pRP-GFP	*S. cellulosum* replicative, Tet^r, P$_{epoA}$-GFP	80
Expression for *M. xanthus*		
pDAH217	Light-inducible: P$_{carQRS}$-*lacZ*, Kan^r	50
pLOJ4	IPTG-inducible: P$_{pilA-lacO}$, Kan^r (requires pLOJ28)	36
pLOJ28	P$_{pilA}$-*lacI*, *intP/attP* (Mx8), Tet^r	36
pCT2	Tet-inducible: P$_{tetA}$, *galK*, Kan^r	57

18.3.3. Targeted Insertion Mutagenesis through Homologous Recombination

Homologous recombination can also be used to inactivate a specific gene by insertion. A short internal fragment of a gene is cloned into an *E. coli* plasmid, and chromosomal integration of the plasmid into myxobacteria by homologous recombination leaves two copies of the target gene truncated at the 3′ and the 5′ ends, respectively (Fig. 1). In practice, PCR is often used to amplify the internal fragment, which is then cloned into a plasmid with Kan^r such as pZErO-2 (Table 1). Essentially any *E. coli* plasmid with a selectable marker for myxobacteria will serve this purpose, since none can replicate in myxobacteria. Insertional mutant alleles can be verified by Southern blot or PCR. One limitation of this method is the difficulty of disrupting small genes due to the length requirement for homologous recombination in myxobacteria. Fragments larger than 500 bp are generally not a problem, but fragments less than 500 bp can be very troublesome, with success requiring greater amounts of DNA for electroporation and longer recovery periods after transformation (60). There have been no reports of success with fragments smaller than 250 bp. One should also be mindful of the polar effects of an insertional mutation on downstream genes when they are in an operon.

18.3.4. Allelic Exchange through Homologous Recombination

An alternative to insertion is to construct an in-frame deletion through allelic exchange, which avoids the drawbacks of polar effects of an insertion. In principle, an in-frame deletion of a gene is first constructed on a plasmid with a cassette that enables both a positive and negative selection in myxobacteria. The first such positive-negative cassette developed for *M. xanthus* is based on Kan^r and the *E. coli galK* gene (39, 81). Initial plasmid integration is selected by Kan^r (Fig. 2). The replacement of the wild-type allele with the in-frame deletion is achieved by a second homologous recombination event that leads to the loss of the plasmid DNA from the chromosome. The second event is selected for with medium containing galactose, because galactose is toxic

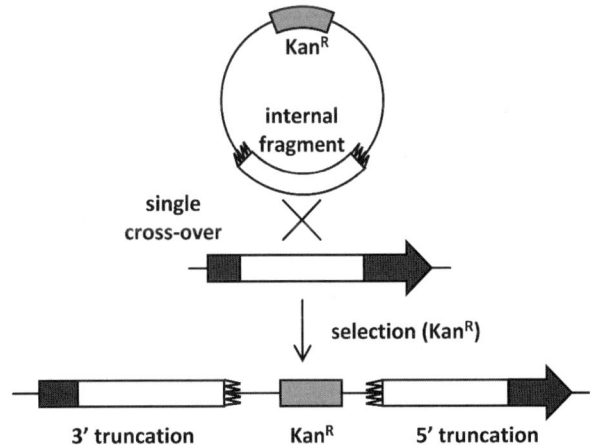

FIGURE 1 Schematic for insertional inactivation. An internal fragment of the gene to be disrupted is cloned into a nonreplicative plasmid for integration by homologous recombination (Table 1). Single homologous crossover between the fragment and the chromosome results in plasmid integration and a partial merodiploid with a 3′ and a 5′ truncation of the gene.

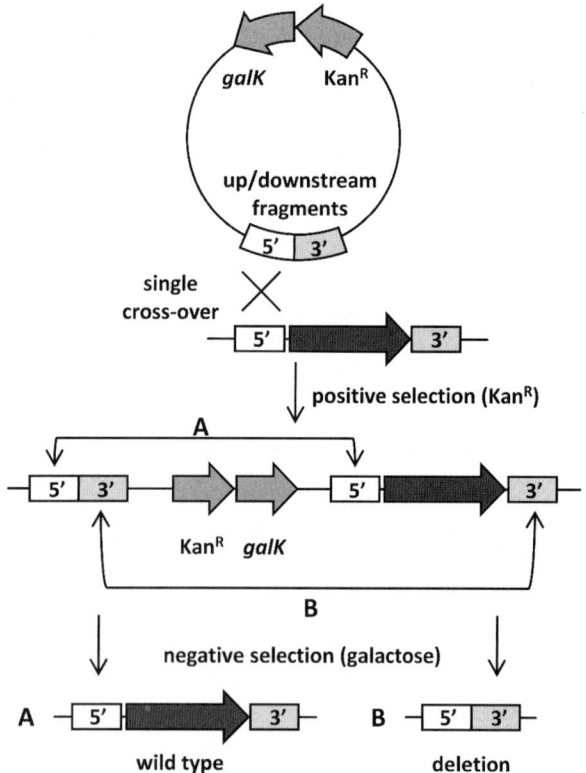

FIGURE 2 Diagram of allelic exchange to construct a deletion. A fragment containing the upstream and the downstream DNA of a gene (see text for details) is cloned into an allelic exchange plasmid (Table 1). The first round of homologous recombination, selectable by Kan^r, results in integration of the plasmid onto the chromosome. This may happen through the 5′ or the 3′ fragment. For simplicity, only integration through the 5′ end is shown here. Kan^r transformants are plated for the selection of galactose resistance (negative selection) and the loss of the plasmid by a second round of recombination. Kanamycin-sensitive and galactose-resistant colonies may contain either the wild-type allele (A) or the deletion allele (B), depending on whether the second recombination occurred through the 5′ or the 3′ end, respectively.

to myxobacterial cells containing the *galK* product (galactokinase). Two outcomes are possible after plasmid loop-out (Fig. 2): restoration of the wild-type allele or its replacement by the deletion allele. PCR or Southern analysis can be used to identify which one occurred. In addition to *galK*, *sacB* from *Bacillus subtilis* has been used for negative selection (88), with sucrose used in place of galactose. The *sacB* system has been used in both M. *xanthus* and S. *aurantiaca*, while the *galK* system has only been used in M. *xanthus* (39, 83, 88). Both systems are likely applicable to S. *cellulosum* except that the Kan^r gene should be replaced with another marker suitable for S. *cellulosum*. Protocol 5 is a detailed protocol for making in-frame deletions in M. *xanthus*. A cautionary note here is that problems have been experienced by the authors and others with both the *galK*- and the *sacB*-based systems (88). The major issue is the emergence of galactose or sucrose resistance, apparently due to spontaneous mutations in the *galK/sacB* genes. It is therefore critical to test sensitivity to kanamycin before proceeding to mutant allele identification.

In-frame deletion alleles of a gene may be constructed in different ways. If a fragment with the full-length gene is available, an internal fragment may be removed by restriction digestion and religation. A second method is to PCR amplify the upstream and the downstream regions of a gene, excluding the sequence to be deleted. For this method, the same restriction enzyme site is engineered into both the 3′ end of the upstream fragment and the 5′ end of the downstream fragment. These two can be digested with the same restriction enzyme, ligated, and cloned into a vector with the appropriate positive-negative selection cassette (Table 1). A third method that we use almost exclusively is to join the upstream and the downstream fragments by overlap PCR, as illustrated in Fig. 3 (61). The two fragments are PCR amplified first. Instead of a restriction site, the PCR primers are designed such that the 3′ primer (R1) for the upstream fragment and the 5′ primer (F2) for the downstream fragment are complementary for about 20 bp in sequence. The two PCR products are joined by a second-round PCR, which gives rise to the final PCR product with the designed deletion. The in-frame deletion alleles constructed by these methods are cloned into either the *galK*- or *sacB*-based selection plasmid (Table 1). It is desirable to ensure that the upstream and downstream fragments are of similar lengths. Otherwise, recombination will preferentially occur at the longer homologous sequence and result in a high percentage of wild-type alleles after the negative selection.

These positive-negative selection systems can also be used for the construction of point mutations through allelic exchange. This is similar to the construction of in-frame deletions described above. That is, the mutation is first constructed on a plasmid with the positive-negative selection cassette. The mutant allele is then used to replace the wild-type allele through two rounds of recombination. For the construction of the point mutation, an upstream and a downstream fragment with the desired point mutation are PCR amplified using complementary primers containing the desired mutation (R1 and F2 in Fig. 4). A second round of PCR is used to join the fragments, and the final PCR product containing the desired mutation is cloned into either the *galK*- or *sacB*-based selection plasmid (Table 1). It is always helpful to engineer a restriction site along with the point mutation to facilitate mutant identification in subsequent steps. When changing codons in myxobacteria, it is desirable to maintain a high-GC codon bias to ensure proper expression. Protocol 5 is a general protocol for performing allelic exchange in M. *xanthus* and should be applicable to other myxobacteria.

Protocol 5. Protocol for Allelic Exchange in *M. xanthus*

1. Transform ~0.5 to 1.5 μg of the plasmid with the mutant allele by electroporation as described in protocol 1. Allow recovery and plate on CYE with kanamycin. Incubate at 32°C for 5 to 7 days for colony formation.

2. Patch 5 to 10 colonies to a fresh CYE plate with kanamycin, and incubate for 1 to 3 days.

3. Transfer cells from the patches into 5 ml of CYE without kanamycin and thoroughly disperse cells. Cells may be incubated at 32°C with shaking at 300 rpm for several hours for better dispersion.

4. Plate cells using a series of 10-fold dilutions (i.e., undiluted to 10^{-5}) onto CYE agar with 1% galactose (assuming a *galK*-based plasmid), and incubate at 32°C.

5. Allow 7 to 10 days for the appearance of galactose-resistant colonies.

6. Patch 20 to 30 galactose-resistant colonies onto CYE plates with galactose and patch these same colonies onto CYE plates with kanamycin.

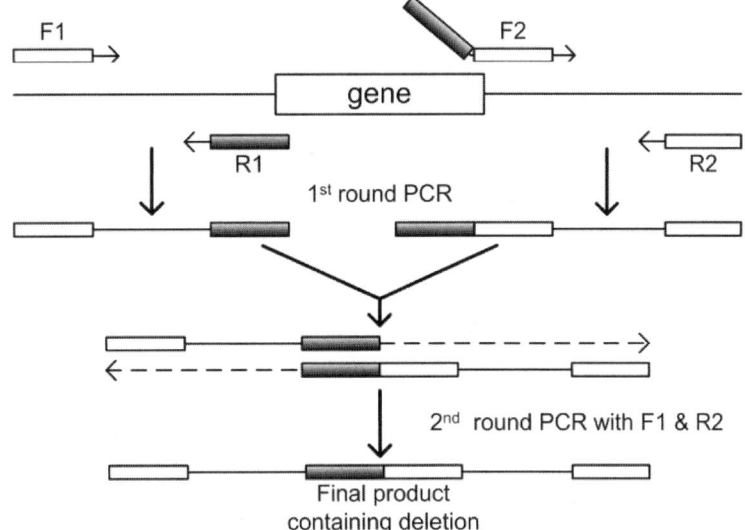

FIGURE 3 PCR-based method for creating a deletion fragment. In the first-round PCR, primer pairs F1-R1 and F2-R2 are used to amplify the upstream and the downstream fragments individually. The 20 bases at the 5′ end of primer F2 are complementary to primer R1. The fragments are gel purified, mixed, and subjected to the second round of PCR using only F1 and R2 to produce the deletion allele. For convenience, restriction sites may be engineered into primers F1 and R2 to facilitate cloning.

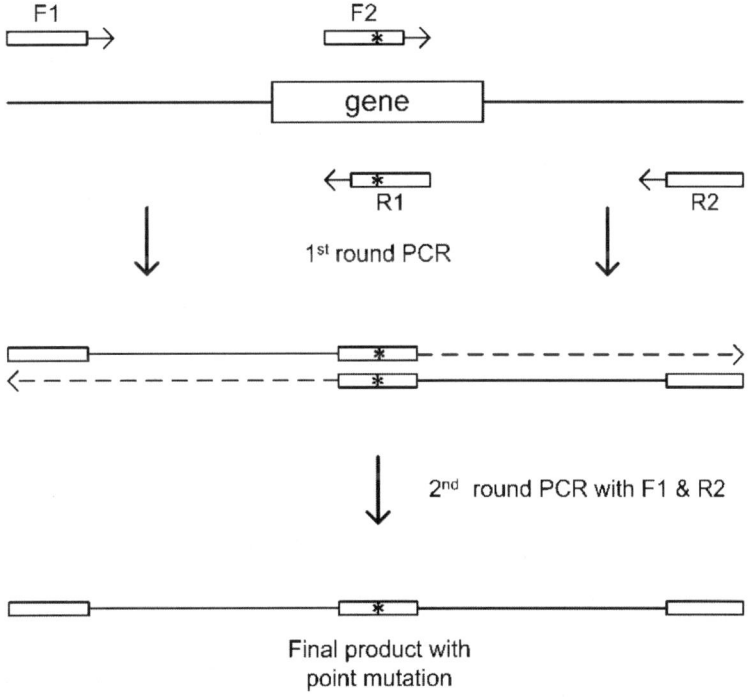

FIGURE 4 PCR-based method for introducing point mutations. In the first-round PCR, primer pairs F1-R1 and F2-R2 are used to amplify the upstream and the downstream fragments individually. Primers R1 and F2 contain the desired point mutation at the center and are fully complementary. It is desirable to engineer a restriction site within R1 and F2 for screening purposes in later steps. The fragments are gel purified, mixed, and subjected to the second round of PCR using only F1 and R2. For convenience, restriction sites may be engineered into primers F1 and R2 to facilitate cloning.

7. Analyze the galactose-resistant and kanamycin-sensitive isolates by PCR or Southern blot to identify mutants.

Note: If the mutant allele is constructed with a *sacB*-based plasmid, use 5% sucrose starting at step 4 instead of 1% galactose.

18.4. GENETIC MAPPING AND LINKAGE ANALYSIS IN MYXOBACTERIA

Genetic mapping and linkage analysis may be performed using generalized transduction or genomic DNA transformation by electroporation in *M. xanthus*. This is especially critical for the identification of point mutations from chemical or UV mutagenesis. Kuner and Kaiser used Mx4-mediated generalized transduction to link point mutations to Tn5 insertions (48). In principle, Tn5 or a *mariner*-based transposon is randomly inserted onto the chromosome. If an insertion occurs near the mutation of interest, the mutation will cotransduce with the insertion at a detectable frequency. The frequency of cotransduction is proportional to the distance between the mutation and the transposon insertion. That is, the shorter the distance between two markers, the more likely they will be cotransduced. Given that Mx4 and Mx8 package about 40 kb of DNA, the distance between two markers can be calculated from their cotransduction efficiency using the following Wu equation (89):

$$\text{frequency of cotransduction} = [1 - (\text{distance between markers}/40 \text{ kb})]^3$$

As mentioned in section 18.2.2, transformation by electroporation using genomic DNA may be a viable alternative to transduction for linkage analysis (82). However, in this case, the integrity of the genomic DNA or lack thereof will influence the cotransformation frequency. It is not yet possible to calculate the distance between two markers from genomic DNA transformations, due to the lack of systematic studies.

18.5. HETEROLOGOUS GENE EXPRESSION IN MYXOBACTERIA

There are two areas of development that have made it possible to express desired genes in myxobacteria. The first is the ability to stably maintain exogenous DNA. The second is the availability of regulated promoters. The following section will focus on these two aspects, with an emphasis on heterologous gene expression.

18.5.1. Integrative Plasmids

Integration onto the chromosome is an effective way to maintain heterologous genes in myxobacteria. This can be achieved in two ways. The first is integration via phage attachment sites. The second is by homologous recombination.

The first phage attachment site used for the construction of integrative plasmids for *M. xanthus* was from the temperate phage Mx8 (51, 55). It was determined that the Mx8 *intP* gene itself contains all the necessary components required for integration, including the *attP* attachment site and the integrase gene (79). One problem with Mx8 is that integration of genes at the Mx8 *attB* site results in reduced gene expression (51, 85). More recently, an integrative plasmid containing the Mx9 phage integrase and attachment site was constructed (37). Mx9, a relative of Mx8, also contains all the necessary components for integration within the *intP* gene. Unlike Mx8, integration of genes at either of the two Mx9 *attB* sites does not result in an obvious

reduction in gene expression (37). Integration of Mx8 *attP* plasmids in *S. aurantiaca* has been reported, although it appears to result in developmental defects (59). See Table 1 for a list of plasmids with Mx8 and Mx9 attachment sites.

In addition to using native phage attachment sites, the φC31 and φBT1 attachment sites from *Streptomyces* phages (3, 26, 52) have been successfully utilized for integration of plasmids into the chromosome of *M. xanthus* (Julien, unpublished). Because integration of these phages requires only the Int protein, *attP* and *attB*, and no host factors, they have been developed as integration systems for a variety of *Streptomyces* (3) and other organisms, including eukaryotes (27, 28, 78). Insertion of the φC31 and φBT1 *attB* site into the chromosome of *M. xanthus* results in highly efficient integration of vectors harboring the corresponding *attP* sites and *int* genes. For φC31, integration can occur at a low frequency in the *M. xanthus* chromosome without *attB*. This nonnative integration occurs at pseudosites, four of which have been identified in *M. xanthus* (Julien, unpublished). These sites can be engineered into any myxobacteria to provide a highly efficient means of integrating DNA into the chromosome for stable maintenance.

In some cases, it may be advantageous to integrate genes of interest at a site other than *attB*. This can be accomplished through homologous recombination by providing myxobacterial sequences on a plasmid. Strategies for both single crossover and allelic exchange have been devised for this purpose. Section 18.5.3 provides some examples.

18.5.2. Replicative Plasmids for Use in Myxobacteria

Self-replicating plasmids are now available for *M. xanthus*. Zhao et al. recently isolated a cryptic plasmid from *Myxococcus fulvus* (93). An ~2-kb fragment from this plasmid was found to contain the minimal region with an origin of replication. This fragment was used to successfully construct a couple of *E. coli*-*M. xanthus* shuttle vectors (Table 1). Although the initial constructs appeared somewhat unstable in *M. xanthus*, stable derivatives were isolated after passages through *M. xanthus*. These plasmids, which are maintained at approximately 10 copies per cell, provide promising leads for the development of heterologous expression systems in myxobacteria.

In addition, a broad-host-range plasmid (Table 1) that allegedly replicates autonomously in *S. cellulosum* strain So ce90 has been reported (80). This plasmid was able to express green fluorescent protein (GFP) in *S. cellulosum* and, as such, could be useful as a heterologous gene expression plasmid. However, this plasmid belongs to the IncP (RK4 derivative) incompatibility group with no additional origin of replication. Other IncP group (RP4 derivatives) plasmids were previously used as suicide vectors in both *M. xanthus* and *S. cellulosum* (11, 34, 35, 73). Since the 900-bp promoter driving GFP expression in the RK4 derivative (80) is from *S. cellulosum*, this plasmid has the potential to integrate by homologous recombination. This claim of autonomous replication may need further verification.

18.5.3. Regulated Gene Expression Systems for Use in Myxobacteria

Several systems have been used for regulated gene expression in *M. xanthus* (Table 1). The light-inducible *carQSR* promoter (P_{carQRS}) was the first described (32, 50, 56). P_{carQRS}, native to *M. xanthus*, was discovered through studies of carotenoid biogenesis in this bacterium. This promoter has relatively low basal activity and can be induced approximately 50- to

100-fold by light (32, 50). For maximal induction, genes to be expressed should be cloned downstream of P_{carQRS} and integrated at the carQRS locus, not at the Mx8 attB site (85). The major limitation of this system is the difficulty of exposing fermentation-scale cultures to light for the induction. The regulation of P_{carQRS} requires at least two proteins, CarQ (sigma factor) and CarR (anti-sigma factor), which may limit its use in species other than M. xanthus.

Another inducible promoter for M. xanthus was constructed using the lac regulatory element from E. coli (36). In this system, a lac operator was placed downstream of the strong and constitutive pilA promoter to make it LacI responsive. lacI is expressed from an additional pilA promoter integrated at the Mx8 attB site using pLOJ28 (Table 1) (36). For heterologous expression, a gene can be cloned downstream of the engineered pilA promoter in pLOJ4 (Table 1) (36). The construct with the foreign gene can be integrated at the pil locus through homologous recombination. This system allowed very high levels of induction, although the basal level of expression is relatively high. In addition, this dual plasmid design requires integrations at both the Mx8 attB site and the M. xanthus pil locus, which may limit its use to M. xanthus.

The tetR-tetA regulatory system from E. coli (86) was used recently to engineer another regulated promoter in M. xanthus. Mignot et al. (57) constructed pCT2 (Table 1), an integrative plasmid carrying a fragment with the tetR repressor and the tetA promoter (P_{tetA}) followed by a multiple cloning site. This fragment is flanked on both sides by sequences from the car locus to allow integration by homologous recombination at the M. xanthus car locus. The basal level of expression appears low, and expression can be induced by anhydrotetracycline at nanomolar concentrations without any toxic effects on the cells (57). Two nice features of this system are particularly noteworthy. The first is that correct integration at the car locus can be easily identified because the disruption of car genes leads to colony color changes on plates. Second, pCT2 carries galK for negative selection for the removal of the plasmid backbone along with Kan^r from the chromosome by selection on galactose. This eliminates the Kan^r marker and leaves behind the expression system intact at the car locus. On the other hand, the requirement for integration at the car locus may limit its use to M. xanthus. However, the E. coli tetR repressor and P_{tetA} are functional in M. xanthus with minimum modification. Since both elements are carried on the same DNA fragment, it is probably relatively easy to adapt this TetR-P_{tetA} system for use in other myxobacteria.

An additional inducible system utilizes the strong T7A1 promoter from the E. coli phage T7. This promoter utilizes the host RNA polymerase and not the T7 RNA polymerase. To regulate the T7A1 promoter, a version was engineered to contain two LacI binding sites to allow regulation by isopropyl-β-D-thiogalactopyranoside (IPTG) in E. coli (20, 49). To use this promoter in myxobacteria, a plasmid has been constructed with this modified T7A1 promoter along with the lacI gene under the control of the strong constitutive promoter from the epothilone biosynthetic gene cluster (P_{epo}). Fusion of this inducible promoter to lacZ reveals very little expression of β-galactosidase in the absence of induction in M. xanthus, while in the presence of IPTG, very high levels of expression are achieved (Julien, unpublished). This plasmid has been used to regulate the expression of epoK in both M. xanthus and S. cellulosum that produce epothilone D. In the presence of IPTG, the induced EpoK protein can effectively convert epothilone D to epothilone B (Julien, unpublished).

APPENDIX: STANDARD GROWTH MEDIA FOR SELECTED MYXOBACTERIA

M. xanthus

CYE medium (15) (per liter)
Casitone .... 10 g
Yeast extract .... 5 g
MgSO$_4$·7H$_2$O .... 1 g
MOPS (morpholinepropanesulfonic acid) ...... 10 mM

A MOPS stock at 1 M can be prepared and adjusted to pH 7.6 with KOH. Standard CYE plates contain 1.5% agar.

S. aurantiaca

Tryptone medium (64) (per liter)
Tryptone .... 10 g
MgSO$_4$·7H$_2$O .... 1 g

Adjust pH to 7.2. Standard tryptone media plates contain 1.5% agar.

S. cellulosum

307 medium (per liter)
Casitone .... 9 g
Fructose .... 3 g
MgSO$_4$·7H$_2$O .... 0.5 g
CaCl$_2$·2H$_2$O .... 0.5 g
HEPES .... 12 g

Adjust pH to 7.6.

S42 plates (per liter)
Tryptone .... 0.5 g
MgSO$_4$·7H$_2$O .... 1.5 g
CaCl$_2$·2H$_2$O .... 1 g
HEPES .... 12 g
K$_2$HPO$_4$ .... 0.06 g
Fe-EDTA .... 0.008 g
Glucose .... 3.5 g
(NH$_4$)$_2$SO$_4$ .... 0.5 g
3.5% spent liquid medium
Sodium hydrosulfite .... 0.1 g
Agar .... 15 g

Adjust pH to 7.4.

REFERENCES

1. **Avery, L., and D. Kaiser.** 1983. In situ transposon replacement and isolation of a spontaneous tandem genetic duplication. Mol. Gen. Genet. 191:99–109.
2. **Beck, E., G. Ludwig, E. A. Auerswald, B. Reiss, and H. Schaller.** 1982. Nucleotide sequence and exact localization of the neomycin phosphotransferase gene from transposon Tn5. Gene 19:327–336.
3. **Bierman, M., R. Logan, K. O'Brien, E. T. Seno, R. N. Rao, and B. E. Schoner.** 1992. Plasmid cloning vectors for the conjugal transfer of DNA from Escherichia coli to Streptomyces spp. Gene 116:43–49.
4. **Blondelet-Rouault, M. H., J. Weiser, A. Lebrihi, P. Branny, and J. L. Pernodet.** 1997. Antibiotic resistance gene cassettes derived from the omega interposon for use in E. coli and Streptomyces. Gene 190:315–317.
5. **Bode, H. B., and R. Muller.** 2006. Analysis of myxobacterial secondary metabolism goes molecular. J. Ind. Microbiol. Biotechnol. 33:577–588.
6. **Bode, H. B., and R. Muller.** 2008. Secondary metabolism in myxobacteria, p. 259–282. In D. E. Whitworth (ed.), Myxobacteria: Multicellularity and Differentiation. ASM Press, Washington, DC.

7. Bolivar, F., R. L. Rodriguez, P. J. Greene, M. C. Betlach, H. L. Heyneker, and H. W. Boyer. 1977. Construction and characterization of new cloning vehicles. II. A multipurpose cloning system. *Gene* **2:**95–113.

8. Bollag, D. M., P. A. McQueney, J. Zhu, O. Hensens, L. Koupal, J. Liesch, M. Goetz, E. Lazarides, and C. M. Woods. 1995. Epothilones, a new class of microtubule-stabilizing agents with a taxol-like mechanism of action. *Cancer Res.* **55:**2325–2333.

9. Boysen, A., E. Ellehauge, B. Julien, and L. Sogaard-Andersen. 2002. The DevT protein stimulates synthesis of FruA, a signal transduction protein required for fruiting body morphogenesis in *Myxococcus xanthus*. *J. Bacteriol.* **184:**1540–1546.

10. Breton, A. M., I. Buon, and J. F. Guespin-Michel. 1990. Use of Tn *phoA* to tag exported proteins in *Myxococcus xanthus*. *FEMS Microbiol. Lett.* **67:**179–186.

11. Breton, A. M., S. Jaoua, and J. Guespin-Michel. 1985. Transfer of plasmid RP4 to *Myxococcus xanthus* and evidence for its integration into the chromosome. *J. Bacteriol.* **161:**523–528.

12. Brown, N. L., D. W. Morris, and J. H. Parish. 1976. DNA of *Myxococcus* bacteriophage MX-1: macromolecular properties and restriction fragments. *Arch. Microbiol.* **108:**221–226.

13. Burchard, R. P., and M. Dworkin. 1966. A bacteriophage for *Myxococcus xanthus*: isolation, characterization and relation of infectivity to host morphogenesis. *J. Bacteriol.* **91:**1305–1313.

14. Campos, J. M., J. Geisselsoder, and D. R. Zusman. 1978. Isolation of bacteriophage MX4, a generalized transducing phage for *Myxococcus xanthus*. *J. Mol. Biol.* **119:**167–178.

15. Campos, J. M., and D. R. Zusman. 1975. Regulation of development in *Myxococcus xanthus*: effect of 3′:5′-cyclic AMP, ADP, and nutrition. *Proc. Natl. Acad. Sci. USA* **72:**518–522.

16. Cho, K., and D. R. Zusman. 1999. AsgD, a new two-component regulator required for A-signalling and nutrient sensing during early development of *Myxococcus xanthus*. *Mol. Microbiol.* **34:**268–281.

17. Cho, K., and D. R. Zusman. 1999. Sporulation timing in *Myxococcus xanthus* is controlled by the *espAB* locus. *Mol. Microbiol.* **34:**714–725.

18. Cohen, S. N., A. C. Chang, H. W. Boyer, and R. B. Helling. 1973. Construction of biologically functional bacterial plasmids *in vitro*. *Proc. Natl. Acad. Sci. USA* **70:**3240–3244.

19. Crawford, E. W., Jr., and L. J. Shimkets. 2000. The stringent response in *Myxococcus xanthus* is regulated by SocE and the CsgA C-signaling protein. *Genes Dev.* **14:**483–492.

20. Deuschle, U., W. Kammerer, R. Gentz, and H. Bujard. 1986. Promoters of *Escherichia coli*: a hierarchy of in vivo strength indicates alternate structures. *EMBO J.* **5:**2987–2994.

21. Geisselsoder, J., J. M. Campos, and D. R. Zusman. 1978. Physical characterization of bacteriophage MX4, a generalized transducing phage for *Myxococcus xanthus*. *J. Mol. Biol.* **119:**179–189.

22. Gerth, K., O. Perlova, and R. Muller. 2008. *Sorangium cellulosum*, p. 329–348. *In* D. E. Whitworth (ed.), *Myxobacteria: Multicellularity and Differentiation*. ASM Press, Washington, DC.

23. Gerth, K., S. Pradella, O. Perlova, S. Beyer, and R. Muller. 2003. Myxobacteria: proficient producers of novel natural products with various biological activities—past and future biotechnological aspects with the focus on the genus *Sorangium*. *J. Biotechnol.* **106:**233–253.

23a. Glomp, I., P. Saulnier, J. Guespin-Michel, and H. U. Schairer. 1988. Transfer of IncP plasmids into *Stigmatella aurantiaca* leading to insertional mutants affected in spore development. *Mol. Gen. Genet.* **214:**213–217.

24. Goldman, B. S., W. C. Nierman, D. Kaiser, S. C. Slater, A. S. Durkin, J. A. Eisen, C. M. Ronning, W. B. Barbazuk, M. Blanchard, C. Field, C. Halling, G. Hinkle, O. Iartchuk, H. S. Kim, C. Mackenzie, R. Madupu, N. Miller, A. Shvartsbeyn, S. A. Sullivan, M. Vaudin, R. Wiegand, and H. B. Kaplan. 2006. Evolution of sensory complexity recorded in a myxobacterial genome. *Proc. Natl. Acad. Sci. USA* **103:**15200–15205.

25. Goryshin, I. Y., J. A. Miller, Y. V. Kil, V. A. Lanzov, and W. S. Reznikoff. 1998. Tn5/IS50 target recognition. *Proc. Natl. Acad. Sci. USA* **95:**10716–10721.

26. Gregory, M. A., R. Till, and M. C. Smith. 2003. Integration site for *Streptomyces* phage φBT1 and development of site-specific integrating vectors. *J. Bacteriol.* **185:**5320–5323.

27. Groth, A. C., M. Fish, R. Nusse, and M. P. Calos. 2004. Construction of transgenic *Drosophila* by using the site-specific integrase from phage φC31. *Genetics* **166:**1775–1782.

28. Groth, A. C., E. C. Olivares, B. Thyagarajan, and M. P. Calos. 2000. A phage integrase directs efficient site-specific integration in human cells. *Proc. Natl. Acad. Sci. USA* **97:**5995–6000.

29. Guo, D., M. G. Bowden, R. Pershad, and H. B. Kaplan. 1996. The *Myxococcus xanthus rfbABC* operon encodes an ATP-binding cassette transporter homolog required for O-antigen biosynthesis and multicellular development. *J. Bacteriol.* **178:**1631–1639.

30. Hagen, D. C., A. P. Bretscher, and D. Kaiser. 1978. Synergism between morphogenetic mutants of *Myxococcus xanthus*. *Dev. Biol.* **64:**284–296.

31. Hodgkin, J., and D. Kaiser. 1977. Cell-to-cell stimulation of movement in nonmotile mutants of *Myxococcus*. *Proc. Natl. Acad. Sci. USA* **74:**2938–2942.

32. Hodgson, D. A. 1993. Light-induced carotenogenesis in *Myxococcus xanthus*: genetic analysis of the *carR* region. *Mol. Microbiol.* **7:**471–488.

33. Hofle, G., and H. Reichenbach. 2005. Epothilone, a myxobacterial metabolite with promising antitumor activity, p. 413–450. *In* G. M. L. Cragg, D. Kingston, and D. J. Newman (ed.), *Anticancer Agents from Natural Products*. Taylor & Francis/CRC Press, Boca Raton, FL.

33a. Invitrogen. 1998. pZErO-2, Zero Background Cloning Kit. Invitrogen, Carlsbad, CA. (Online.) http://tools.invitrogen.com/content/sfs/manuals/pZero2_plus_man.pdf.

34. Jaoua, S., J. F. Guespin-Michel, and A. M. Breton. 1987. Mode of insertion of the broad-host-range plasmid RP4 and its derivatives into the chromosome of *Myxococcus xanthus*. *Plasmid* **18:**111–119.

35. Jaoua, S., S. Neff, and T. Schupp. 1992. Transfer of mobilizable plasmids to *Sorangium cellulosum* and evidence for their integration into the chromosome. *Plasmid* **28:**157–165.

36. Jelsbak, L., and D. Kaiser. 2005. Regulating pilin expression reveals a threshold for S motility in *Myxococcus xanthus*. *J. Bacteriol.* **187:**2105–2112.

37. Julien, B. 2003. Characterization of the integrase gene and attachment site for the *Myxococcus xanthus* bacteriophage Mx9. *J. Bacteriol.* **185:**6325–6330.

38. Julien, B., and R. Fehd. 2003. Development of a *mariner*-based transposon for use in *Sorangium cellulosum*. *Appl. Environ. Microbiol.* **69:**6299–6301.

39. Julien, B., A. D. Kaiser, and A. Garza. 2000. Spatial control of cell differentiation in *Myxococcus xanthus*. *Proc. Natl. Acad. Sci. USA* **97:**9098–9103.

40. **Kaiser, D.** 2003. Coupling cell movement to multicellular development in myxobacteria. *Nat. Rev. Microbiol.* **1**:45–54.

41. **Kaiser, D., and M. Dworkin.** 1975. Gene transfer to myxobacterium by *Escherichia coli* phage P1. *Science* **187**:653–654.

42. **Kalos, M., and J. Zissler.** 1990. Transposon tagging of genes for cell-cell interactions in *Myxococcus xanthus*. *Proc. Natl. Acad. Sci. USA* **87**:8316–8320.

43. **Kashefi, K., and P. L. Hartzell.** 1995. Genetic suppression and phenotypic masking of a *Myxococcus xanthus frzF⁻* defect. *Mol. Microbiol.* **15**:483–494.

44. **Kopp, M., H. Irschik, F. Gross, O. Perlova, A. Sandmann, K. Gerth, and R. Muller.** 2004. Critical variations of conjugational DNA transfer into secondary metabolite multiproducing *Sorangium cellulosum* strains So ce12 and So ce56: development of a *mariner*-based transposon mutagenesis system. *J. Biotechnol.* **107**:29–40.

45. **Kopp, M., H. Irschik, S. Pradella, and R. Muller.** 2005. Production of the tubulin destabilizer disorazol in *Sorangium cellulosum*: biosynthetic machinery and regulatory genes. *Chembiochem* **6**:1277–1286.

46. **Kowalski, R. J., P. Giannakakou, and E. Hamel.** 1997. Activities of the microtubule-stabilizing agents epothilones A and B with purified tubulin and in cells resistant to paclitaxel (Taxol®). *J. Biol. Chem.* **272**:2534–2541.

47. **Kroos, L., and D. Kaiser.** 1984. Construction of Tn5 *lac*, a transposon that fuses *lacZ* expression to exogenous promoters, and its introduction into *Myxococcus xanthus*. *Proc. Natl. Acad. Sci. USA* **81**:5816–5820.

48. **Kuner, J. M., and D. Kaiser.** 1981. Introduction of transposon Tn5 into *Myxococcus* for analysis of developmental and other nonselectable mutants. *Proc. Natl. Acad. Sci. USA* **78**:425–429.

49. **Lanzer, M., and H. Bujard.** 1988. Promoters largely determine the efficiency of repressor action. *Proc. Natl. Acad. Sci. USA* **85**:8973–8977.

50. **Letouvet-Pawlak, B., C. Monnier, S. Barray, D. A. Hodgson, and J. F. Guespin-Michel.** 1990. Comparison of beta-galactosidase production by two inducible promoters in *Myxococcus xanthus*. *Res. Microbiol.* **141**:425–435.

51. **Li, S. F., and L. J. Shimkets.** 1988. Site-specific integration and expression of a developmental promoter in *Myxococcus xanthus*. *J. Bacteriol.* **170**:5552–5556.

52. **Lomovskaya, N. D., K. F. Chater, and N. M. Mkrtumian.** 1980. Genetics and molecular biology of *Streptomyces* bacteriophages. *Microbiol. Rev.* **44**:206–229.

53. **Lu, A., K. Cho, W. P. Black, X. Y. Duan, R. Lux, Z. Yang, H. B. Kaplan, D. R. Zusman, and W. Shi.** 2005. Exopolysaccharide biosynthesis genes required for social motility in *Myxococcus xanthus*. *Mol. Microbiol.* **55**:206–220.

54. **Manoil, C., and J. Beckwith.** 1985. Tn*phoA*: a transposon probe for protein export signals. *Proc. Natl. Acad. Sci. USA* **82**:8129–8133.

55. **Martin, S., E. Sodergren, T. Masuda, and D. Kaiser.** 1978. Systematic isolation of transducing phages for *Myxococcus xanthus*. *Virology* **88**:44–53.

56. **McGowan, S. J., H. C. Gorham, and D. A. Hodgson.** 1993. Light-induced carotenogenesis in *Myxococcus xanthus*: DNA sequence analysis of the *carR* region. *Mol. Microbiol.* **10**:713–735.

57. **Mignot, T., J. P. Merlie, Jr., and D. R. Zusman.** 2007. Two localization motifs mediate polar residence of FrzS during cell movement and reversals of *Myxococcus xanthus*. *Mol. Microbiol.* **65**:363–372.

58. **Morrison, C. E., and D. R. Zusman.** 1979. *Myxococcus xanthus* mutants with temperature-sensitive, stage-specific defects: evidence for independent pathways in development. *J. Bacteriol.* **140**:1036–1042.

59. **Muller, S., H. Shen, D. Hofmann, H. U. Schairer, and J. R. Kirby.** 2006. Integration into the phage attachment site, *attB*, impairs multicellular differentiation in *Stigmatella aurantiaca*. *J. Bacteriol.* **188**:1701–1709.

60. **Murphy, K., and A. Garza.** 2008. Genetic tools for studying *Myxococcus xanthus* biology, p. 491–501. In D. E. Whitworth (ed.), *Myxobacteria: Multicellularity and Differentiation*. ASM Press, Washington, DC.

61. **Newton, C. R., and A. Graham.** 1997. *PCR*, 2nd ed. BIOS Scientific Publishers; Springer-Verlag New York, Inc., New York, NY.

62. **O'Connor, K. A., and D. R. Zusman.** 1983. Coliphage P1-mediated transduction of cloned DNA from *Escherichia coli* to *Myxococcus xanthus*: use for complementation and recombinational analyses. *J. Bacteriol.* **155**:317–329.

63. **Perlova, O., K. Gerth, S. Kuhlmann, Y. Zhang, and R. Muller.** 2009. Novel expression hosts for complex secondary metabolite megasynthetases: production of myxochromide in the thermopilic isolate *Corallococcus macrosporus* GT-2. *Microb. Cell Fact.* **8**:1.

64. **Plaga, W., I. Stamm, and H. U. Schairer.** 1998. Intercellular signaling in *Stigmatella aurantiaca*: purification and characterization of stigmolone, a myxobacterial pheromone. *Proc. Natl. Acad. Sci. USA* **95**:11263–11267.

65. **Pospiech, A., and B. Neumann.** 1995. A versatile quick-prep of genomic DNA from gram-positive bacteria. *Trends Genet.* **11**:217–218.

66. **Pradella, S., A. Hans, C. Sproer, H. Reichenbach, K. Gerth, and S. Beyer.** 2002. Characterisation, genome size and genetic manipulation of the myxobacterium *Sorangium cellulosum* So ce56. *Arch. Microbiol.* **178**:484–492.

67. **Rachid, S., D. Krug, B. Kunze, I. Kochems, M. Scharfe, T. M. Zabriskie, H. Blocker, and R. Muller.** 2006. Molecular and biochemical studies of chondramide formation—highly cytotoxic natural products from *Chondromyces crocatus* Cm c5. *Chem. Biol.* **13**:667–681.

68. **Rachid, S., F. Sasse, S. Beyer, and R. Muller.** 2006. Identification of StiR, the first regulator of secondary metabolite formation in the myxobacterium *Cystobacter fuscus* Cb f17.1. *J. Biotechnol.* **121**:429–441.

69. **Reichenbach, H.** 2001. Myxobacteria, producers of novel bioactive substances. *J. Ind. Microbiol. Biotechnol.* **27**:149–156.

70. **Reichenbach, H., and G. Hofle.** 1993. Production of bioactive secondary metabolites, p. 347–397. *In* M. Dworkin and D. Kaiser (ed.), *Myxobacteria II*. American Society for Microbiology, Washington, DC.

71. **Rubin, E. J., B. J. Akerley, V. N. Novik, D. J. Lampe, R. N. Husson, and J. J. Mekalanos.** 1999. In vivo transposition of *mariner*-based elements in enteric bacteria and mycobacteria. *Proc. Natl. Acad. Sci. USA* **96**:1645–1650.

72. **Sandmann, A., F. Sasse, and R. Muller.** 2004. Identification and analysis of the core biosynthetic machinery of tubulysin, a potent cytotoxin with potential anticancer activity. *Chem. Biol.* **11**:1071–1079.

73. **Saulnier, P., J. Hanquier, S. Jaoua, H. Reichenbach, and J. F. Guespin-Michel.** 1988. Utilization of IncP-1 plasmids as vectors for transposon mutagenesis in myxobacteria. *J. Gen. Microbiol.* **134**:2889–2895.

74. **Schneiker, S., O. Perlova, O. Kaiser, K. Gerth, A. Alici, M. O. Altmeyer, D. Bartels, T. Bekel, S. Beyer, E. Bode, H. B. Bode, C. J. Bolten, J. V. Choudhuri, S. Doss, Y. A. Elnakady, B. Frank, L. Gaigalat, A. Goesmann, C. Groeger, F. Gross, L. Jelsbak, J. Kalinowski, C. Kegler, T. Knauber, S. Konietzny, M. Kopp, L. Krause, D. Krug, B. Linke, T. Mahmud, R. Martinez-Arias, A. C. McHardy, M. Merai, F. Meyer, S. Mormann, J. Munoz-Dorado, J. Perez, S. Pradella, S. Rachid, G. Raddatz, F. Rosenau, C. Ruckert, F. Sasse, M. Scharfe, S. C. Schuster, G. Suen, A. Treuner-Lange, G. J. Velicer,**

F. J. Vorholter, K. J. Weissman, R. D. Welch, S. C. Wenzel, D. E. Whitworth, S. Wilhelm, C. Wittmann, H. Blocker, A. Puhler, and R. Muller. 2007. Complete genome sequence of the myxobacterium *Sorangium cellulosum*. *Nat. Biotechnol.* **25:**1281–1289.

75. **Schupp, T., C. Toupet, B. Cluzel, S. Neff, S. Hill, J. J. Beck, and J. M. Ligon.** 1995. A *Sorangium cellulosum* (myxobacterium) gene cluster for the biosynthesis of the macrolide antibiotic soraphen A: cloning, characterization, and homology to polyketide synthase genes from actinomycetes. *J. Bacteriol.* **177:**3673–3679.

76. **Spratt, B. G., P. J. Hedge, S. te Heesen, A. Edelman, and J. K. Broome-Smith.** 1986. Kanamycin-resistant vectors that are analogues of plasmids pUC8, pUC9, pEMBL8 and pEMBL9. *Gene* **41:**337–342.

77. **Stamm, I., A. Leclerque, and W. Plaga.** 1999. Purification of cold-shock-like proteins from *Stigmatella aurantiaca*—molecular cloning and characterization of the *cspA* gene. *Arch. Microbiol.* **172:**175–181.

78. **Thyagarajan, B., E. C. Olivares, R. P. Hollis, D. S. Ginsburg, and M. P. Calos.** 2001. Site-specific genomic integration in mammalian cells mediated by phage φC31 integrase. *Mol. Cell. Biol.* **21:**3926–3934.

79. **Tojo, N., K. Sanmiya, H. Sugawara, S. Inouye, and T. Komano.** 1996. Integration of bacteriophage Mx8 into the *Myxococcus xanthus* chromosome causes a structural alteration at the C-terminal region of the IntP protein. *J. Bacteriol.* **178:**4004–4011.

80. **Tu, Y., G. P. Chen, and Y. L. Wang.** 2007. Autonomously replicating plasmid transforms *Sorangium cellulosum* So ce90 and induces expression of green fluorescent protein. *J. Biosci. Bioeng.* **104:**385–390.

81. **Ueki, T., S. Inouye, and M. Inouye.** 1996. Positive-negative KG cassettes for construction of multi-gene deletions using a single drug marker. *Gene* **183:**153–157.

82. **Vlamakis, H. C., J. R. Kirby, and D. R. Zusman.** 2004. The Che4 pathway of *Myxococcus xanthus* regulates type IV pilus-mediated motility. *Mol. Microbiol.* **52:**1799–1811.

83. **Weinig, S., T. Mahmud, and R. Muller.** 2003. Markerless mutations in the myxothiazol biosynthetic gene cluster: a delicate megasynthetase with a superfluous nonribosomal peptide synthetase domain. *Chem. Biol.* **10:**953–960.

84. **Whitworth, D. E. (ed.).** 2008. *Myxobacteria: Multicellularity and Differentiation*. ASM Press, Washington, DC.

85. **Whitworth, D. E., S. J. Bryan, A. E. Berry, S. J. McGowan, and D. A. Hodgson.** 2004. Genetic dissection of the light-inducible *carQRS* promoter region of *Myxococcus xanthus*. *J. Bacteriol.* **186:**7836–7846.

86. **Wray, L. V., Jr., R. A. Jorgensen, and W. S. Reznikoff.** 1981. Identification of the tetracycline resistance promoter and repressor in transposon Tn*10*. *J. Bacteriol.* **147:**297–304.

87. **Wu, S. S., and D. Kaiser.** 1995. Genetic and functional evidence that Type IV pili are required for social gliding motility in *Myxococcus xanthus*. *Mol. Microbiol.* **18:** 547–558.

88. **Wu, S. S., and D. Kaiser.** 1996. Markerless deletions of *pil* genes in *Myxococcus xanthus* generated by counterselection with the *Bacillus subtilis sacB* gene. *J. Bacteriol.* **178:**5817–5821.

89. **Wu, T. T.** 1966. A model for three-point analysis of random general transduction. *Genetics* **54:**405–410.

90. **Xu, D., C. Yang, and H. B. Kaplan.** 1998. *Myxococcus xanthus sasN* encodes a regulator that prevents developmental gene expression during growth. *J. Bacteriol.* **180:**6215–6223.

91. **Xu, Q., W. P. Black, S. M. Ward, and Z. Yang.** 2005. Nitrate-dependent activation of the Dif signaling pathway of *Myxococcus xanthus* mediated by a NarX-DifA interspecies chimera. *J. Bacteriol.* **187:**6410–6418.

92. **Youderian, P., N. Burke, D. J. White, and P. L. Hartzell.** 2003. Identification of genes required for adventurous gliding motility in *Myxococcus xanthus* with the transposable element *mariner*. *Mol. Microbiol.* **49:**555–570.

93. **Zhao, J. Y., L. Zhong, M. J. Shen, Z. J. Xia, Q. X. Cheng, X. Sun, G. P. Zhao, Y. Z. Li, and Z. J. Qin.** 2008. Discovery of the autonomously replicating plasmid pMF1 from *Myxococcus fulvus* and development of a gene cloning system in *Myxococcus xanthus*. *Appl. Environ. Microbiol.* **74:**1980–1987.

94. **Ziermann, R., and R. Calendar.** 1990. Characterization of the *cos* sites of bacteriophages P2 and P4. *Gene* **96:**9–15.

95. **Zusman, D. R., A. E. Scott, Z. Yang, and J. R. Kirby.** 2007. Chemosensory pathways, motility and development in *Myxococcus xanthus*. *Nat. Rev. Microbiol.* **5:**862–872.

Strain Improvement of *Escherichia coli* To Enhance Recombinant Protein Production

MICHAEL E. PYNE, KARAN S. SUKHIJA, AND C. PERRY CHOU

19

Among many host systems developed for recombinant protein production, the gram-negative bacterium *Escherichia coli* still retains high popularity due to several technical advantages. First, its fast growth rate and superior environmental adaptability allow *E. coli* to be cultured easily, inexpensively, and if necessary, to a high cell density. Second, its genome-encoded proteins and metabolic pathways are well characterized. Third, numerous technologies and protocols for engineering *E. coli* strains are well developed. Fourth, heterologous proteins of interest can be expressed to high levels (up to 50% of total cellular protein). However, *E. coli* also has technical limitations in recombinant protein production. For example, many human therapeutic proteins require specific glycosylation for full bioactivity. These proteins cannot be produced in *E. coli*, which does not perform such glycosylation, although in some cases it has been possible to engineer a glycosylation pathway in *E. coli* (28). Also, many eukaryotic proteins have multiple domains or disulfide bonds and cannot be functionally expressed in *E. coli* primarily because the organelles and mechanisms in eukaryotic cells for assembling such highly complex protein structures are missing in *E. coli*.

The wild-type *E. coli* cell is not optimally designed for industrial applications. Hence, developing a bioprocess often starts with the construction of the host/vector system so that the *E. coli* strain can be genetically transformed and phenotypically suitable for biomanufacturing. Since the gene of interest is heterologously expressed along with cell growth during the cultivation stage for recombinant protein production, high-level gene expression and high-cell-density cultivation have to be performed simultaneously in order to optimize the culture performance. After cultivation for recombinant protein purification, either the cell pellet (for recombinant proteins retained intracellularly) or the extracellular medium (for recombinant proteins released extracellularly) is harvested and a series of downstream bioprocessing steps are carried out. Experience over the past few decades has shown that most biomanufacturing issues arising at various bioprocessing stages can be addressed through genetic manipulation of the host/vector system. It is the most effective and economic approach to enhance recombinant protein production in terms of mediating functional expression, increasing protein yield, and facilitating downstream purification. However, factors associated with the genetic mechanisms (i.e., replication and gene expression steps) and bioprocessing conditions are often tangled in an unpredictable manner, resulting in complications in strategy development.

Various technical issues that potentially limit recombinant protein production in *E. coli* and the corresponding strategies to address them are summarized in Table 1. Many of these strategies involve alleviating (or even eliminating) the obstacle or limitation associated with a particular bioprocessing stage (i.e., gene expression, cell cultivation, or downstream processing) or with the recombinant protein itself. Typically, the biotechnological basis for strain improvement to enhance recombinant protein production relies on the permanent implementation of desirable traits into the production strain to stimulate both cell growth and functional expression of the target gene during the cultivation (Fig. 1). This is done by genetic restructuring of the production strain, involving both expression vector construction and genetic modifications to the host genome using either recombinant DNA technology, classical mutagenesis and screening technology, or both. Since genetic manipulation of *E. coli* host/vector systems is extensively described in the literature, this chapter reviews the major technical issues associated with *E. coli* strain engineering to enhance recombinant protein production and directs the reader to protocols appropriate for specific applications.

19.1. CONSTRUCTION OF EXPRESSION VECTORS

19.1.1. Genetic Elements of Expression Vectors

Heterologous gene expression in *E. coli* typically involves two major intracellular molecular mechanisms, namely, (i) replication of the expression vector and (ii) expression of the target gene, which includes transcription, translation, and posttranslational processing steps (Fig. 2). To improve a strain for recombinant protein production, effective replication should be ensured such that a high dosage of the target gene is maintained in all the cells of the population. Various genetic elements found in a typical expression vector used for recombinant protein production in *E. coli* are shown schematically in Fig. 3. The functions of each element and a list of common features are summarized in Table 2. Basically, these genetic elements are in place to increase the gene dosage and the efficiency of transcription, translation,

TABLE 1 Factors limiting recombinant protein production in *E. coli* and corresponding strategies for overcoming these limitations

Stage	Limiting events	Strategies for strain improvement	Reference(s)
Gene expression			
Replication	Plasmid instability	Genetically stabilize the plasmid, apply proper culture conditions	7, 39
	Gene dosage	Use a high-copy-number plasmid	35, 99
Transcription	Transcriptional efficiency	Use a strong inducible promoter	11, 95
	Promoter leakage	Use a tightly regulated promoter or other genetic configurations	37, 91
Posttranscriptional stage	mRNA secondary structure near the 5′ terminus	Optimize gene sequence	53, 108
	mRNA stability	Use an RNase-deficient mutant	48, 64
Translation	Translational efficiency	Optimize the Shine-Dalgarno sequence for optimal binding of ribosome, optimize coding sequence	26, 53, 108
	Ribosome availability	Enhance the production of functional ribosome	104, 109
	tRNA availability (codon bias)	Coexpress the rare tRNA gene, optimize the codon	53, 108
Posttranslational stage	Protein translocation	Use a proper signal peptide	18
	Protein folding	Coexpress chaperones, use a low culture temperature, use protein fusion technology, secrete the protein into the extracellular medium	43, 51, 56, 82
	Protein disulfide bond formation	Secrete the protein into the periplasm or extracellular medium, coexpress factors involved in disulfide bond formation, use a mutant that can form disulfide bonds in the cytoplasm	24, 51
	Proteolysis	Use a protease-deficient mutant, use protein fusion technology, perform extracellular protein secretion	4, 32, 97
	Glycosylation	Use a genetically engineered strain	57, 84
Cultivation	Physiological deterioration	Eliminate various deterioration factors	8, 30
	Heat shock responses	Manipulate heat shock responses	15, 110
	Nutrient uptake	Avoid nutrient overfeeding	61
	Toxic metabolite accumulation	Reduce the accumulation	17, 23
	Growth inhibition	Eliminate inhibitory factors	54, 76
Downstream processing	Extracellular protein secretion	Use a proper secretion strategy	32
	Protein recovery	Use protein fusion technology	97
	Protein solubility	Use protein fusion technology, modify the protein sequence	66, 97
Protein feature	Toxicity	Use a tightly regulated promoter to prevent expression leakage	37, 91
	High complexity and difficult expression	Use a special strain for functional expression	71, 105
	Proteolysis	Modify protein sequence	75

and posttranslational processing so that the target gene can be functionally expressed at a high rate. The copy number of the expression vector is intrinsically determined by the replication origin. A strong inducible promoter is required to increase the transcriptional efficiency and boost the mRNA level. Because of the coupling of transcription and translation, recombinant protein production is often improved simply by using a strong promoter. It may be necessary that the promoter be tightly regulated to avoid leaky expression

under noninduced conditions, which is particularly important for proteins with a high toxicity. The molecular details of the transcriptional regulation mechanism and whether it involves an activator or repressor can be very important for achieving maximum production. The terminator used is also important to ensure proper termination of transcription of the target gene, since transcription running past the target gene can interfere with other functions of the expression vector. The ribosome binding site and sequences near

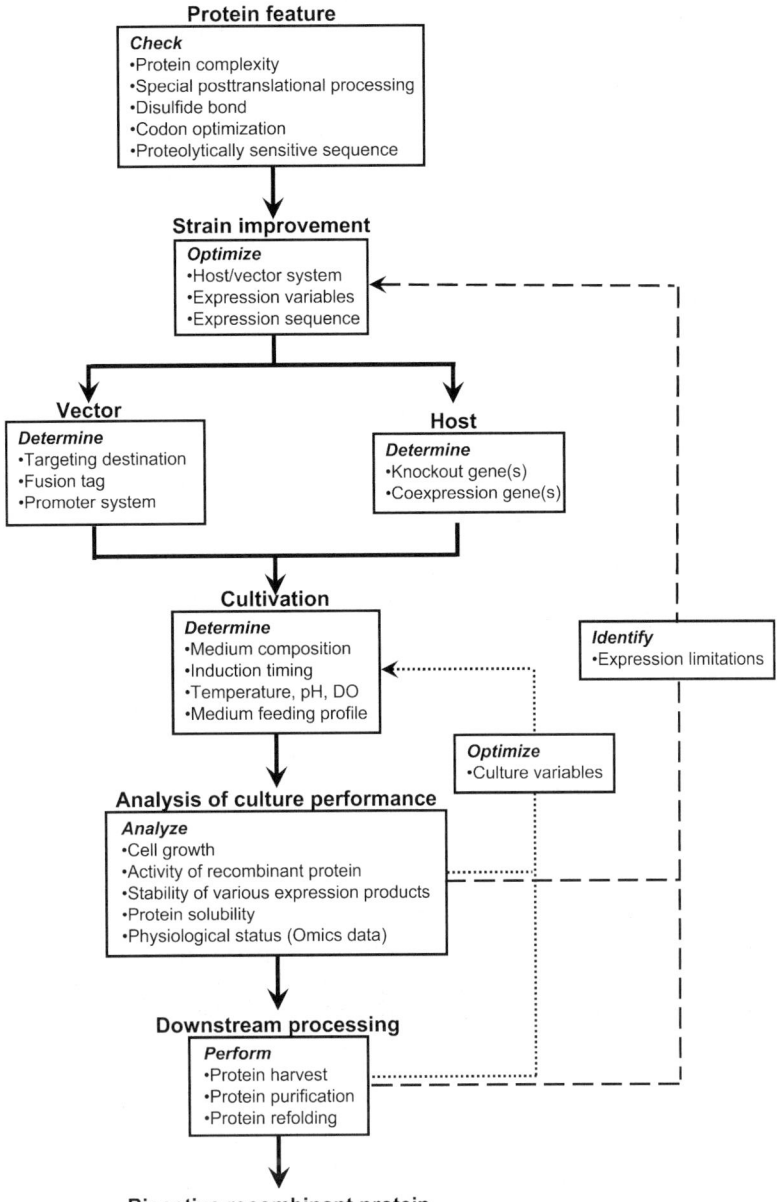

FIGURE 1 Flowchart for *E. coli* strain improvement to enhance recombinant protein production. The strategies depend on identification of the specific factors limiting the overall recombinant protein yield and include optimization of the host/vector system, expression variables, and expression sequence. Refer to Table 1 for a list of technical limitations and how to deal with them. DO, dissolved oxygen.

the start codon affect the efficiency of ribosome binding to the target mRNA and the rate of translation initiation. The antibiotic resistance marker enables selection for cells harboring the expression vector. The fusion tag, situated either at the 5′ or 3′ end of the target gene (so that the expressed tag is fused with the N or C terminus of the target protein), offers several expression advantages, particularly for soluble expression and protein purification. Most expression vectors have multiple cloning sites to enable flexible cloning of various target genes with transcriptional or translational fusions for different expression scenarios (Fig. 3). Many expression vectors with a selection of these genetic elements are commercially available and significantly facilitate the

construction work. While the construction of the expression vector is considered a key step for recombinant protein production, developing an effective bioprocess also relies on a proper design of the expression host, cultivation, and downstream processing that are compatible with the expression vector.

19.1.2. Limiting Steps in Gene Expression

Recombinant protein production involves a series of molecular mechanisms that must be optimized, regulated, and coordinated. A malfunction at any of the steps (replication, transcription, translation, posttranslational processing, etc.) or the presence of various undesired events (such

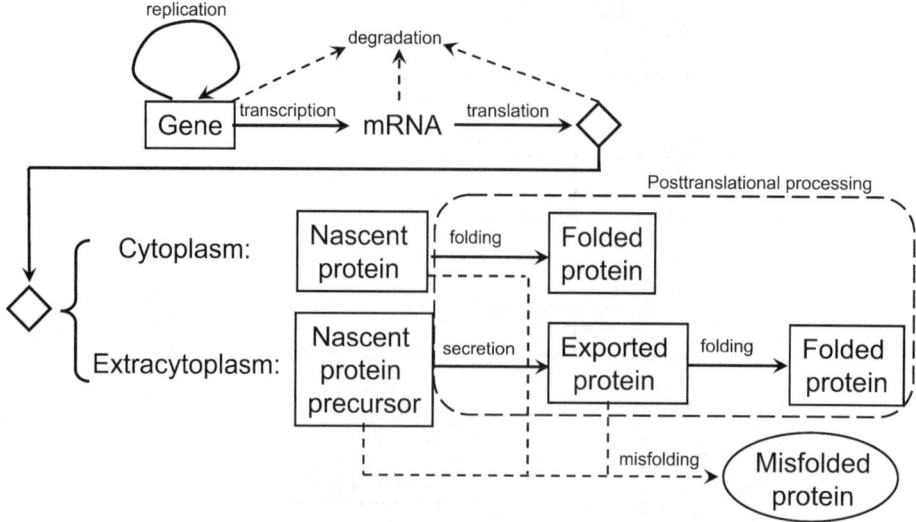

FIGURE 2   Molecular events associated with recombinant protein production in *E. coli*. Recombinant protein production involves a series of complex molecular mechanisms, such as replication of the expression vector, transcription and translation of the gene of interest, and various posttranslational processing steps (including protein secretion, folding, and disulfide bond formation). Production can be limited by low efficiency at any one of these steps or by an abnormal event that diverts protein synthesis into a nonproductive pathway (e.g., protein misfolding or degradation of DNA, mRNA, or protein).

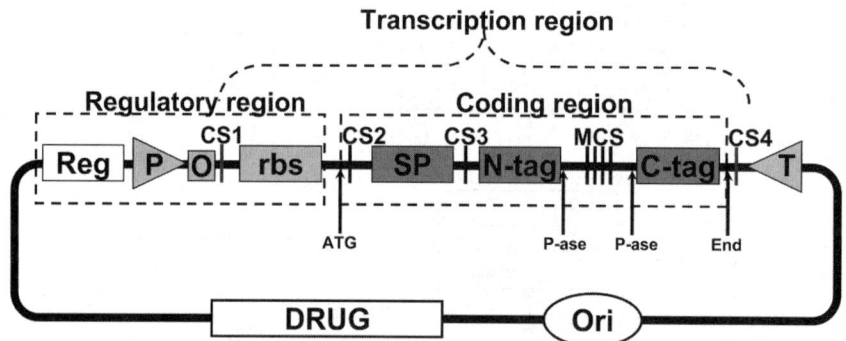

Cloning sites	Expression product	Comments
CS2(or CS1)/CS4	ORF	1. Another rbs is needed when CS1 is used.
CS2(or CS1)/MCS	ORF::C-tag	2. ORF might contain a SP for secretion.
CS3/CS4	SP::ORF	1. SP might not be present for cytoplasmic expression.
CS3/MCS	SP::ORF::C-tag	2. P-ase (protease cleavage site) might be
MCS/CS4	SP::N-tag::ORF	available to release the target protein moiety.
MCS/MCS	SP::N-tag::ORF::C-tag	

FIGURE 3   A typical *E. coli* expression vector (i.e., plasmid) for recombinant protein production. Several expression and cloning features are shown, including Reg (gene encoding the regulator, either transcriptional activator or repressor), P (promoter), O (operator), rbs (ribosome binding site), SP (signal peptide), N-tag (N-terminal fusion tag), C-tag (C-terminal fusion tag), T (terminator), CS (cloning site), MCS (multiple cloning sites), Ori (replication origin), DRUG (drug resistance gene), ATG (initiation codon encoding methionine), P-ase (protease cleavage site), and End (stop codon). Note that, depending on the cloning site(s) for insertion of a target gene (i.e., open reading frame [ORF]), a transcriptional or translational fusion vector can be constructed to express a gene product (either ORF or ORF-fusion) containing various feature domains and targeting in the cytoplasm or extracytoplasmic compartment.

TABLE 2 Genetic elements found in a typical expression vector used for recombinant protein production

Element	Common features	Function(s)
Origin	pUC, pMB1, ColE1, p15A, pSC101	Regulate gene dosage
Promoter	T7, *araB*, *lacUV5*, *tac*, *trc*, *tetA*, λ$_L$, λ$_R$	Enhance transcriptional efficiency
Terminator	*rrnB* T1, T2	Ensure transcriptional termination
Regulator	Transcriptional activators or repressors	Regulate transcriptional efficiency
Ribosome binding site	Shine-Dalgarno sequence, translational enhancer	Improve translational initiation
Drug marker	Ampicillin, chloramphenicol, tetracycline, kanamycin	Select the transformed cell
Signal sequence	Tat, TorA, OmpA, PhoA, PelB, SpA	Translocate the protein precursor across the cytoplasmic membrane
Tag	Streptavidin-binding peptide (Strep), poly-histidine, FLAG peptide, DsbA, DsbC, thioredoxin (TRX), glutathione *S*-transferase (GST), maltose-binding protein (MBP), N-utilization substance protein A (NusA), calmodulin binding peptide (CBP), small ubiquitin-like modifier (SUMO)	Facilitate protein purification, improve protein solubility

as degradation or misfolding of the target protein) can jeopardize the production process (Fig. 2). A high yield of bioactive protein implies not only a high rate for all the expression steps (i.e., fluxes), but also coordination of various intracellular events to balance these expression fluxes. Typically, for a particular host/vector system under a specific growth condition, overall performance can be limited by a single expression step, which can result in the accumulation of an inactive intermediate prior to the limiting step or a reduced level of the intermediate after that step. Theoretically, strategies based on enhancing the limiting step can lead to an overall improvement in recombinant protein production. For example, transcription is recognized as a common step limiting the abundance of mRNA for subsequent translation. A stronger promoter can boost the mRNA level through increased transcriptional efficiency, and the recombinant protein yield is often increased as a result. In addition to the high expression rate, all the expression intermediates (i.e., mRNA and protein precursors) and the final gene product should remain stable in order to achieve a high protein yield during the production. There are cases where one or more of these intermediates are subject to intracellular degradation and recombinant protein production is reduced as a result. Strategies to stabilize these labile biomolecules can be developed for strain improvement.

While the limiting step is determined by the intracellular genetic makeup of the host/vector system, it is prone to change depending on the transient physiological state. This implies that the strategies for genetic construction of the production strain and for its biological cultivation should be simultaneously configured to optimize the culture performance for recombinant protein production. Technically, it is important that the limiting step be properly identified since improving the efficiency of that step should enhance the overall expression performance. However, improving a single expression step can simply shift the limitation to a different step, giving little to no improvement in the overall recombinant protein yield. For example, though the transcriptional efficiency can be enhanced by using a strong inducible promoter, the boosted mRNA level does not necessarily lead to the increased production of the target protein due to the limitation from another expression step, such as translation or protein folding. Even more troublesome,

the genetic strategy used to resolve one limiting step might actually cause the development of another limitation. For example, the presence of high-copy-number plasmids to increase the gene dosage or the induction of strong promoters to enhance the transcriptional efficiency could negatively affect the physiological condition of the producing cells, resulting in growth inhibition and a reduced recombinant protein yield. Such complications often limit the applicability of these genetic strategies.

19.1.3. Localization of Recombinant Protein

Another key issue for recombinant protein production in *E. coli* is the location to which the final protein product should be targeted, including the cytoplasm, periplasm, cell surface, or extracellular medium. This is determined by various factors associated with the recombinant protein, such as its susceptibility to intracellular proteolysis, its tendency to misfold, the requirement for disulfide bond formation, and the process used for downstream protein purification. While the cytoplasm is the intracellular compartment where all the nascent proteins are first synthesized and where most of them reside, it might not be a suitable destination for a recombinant protein. Therefore, the target gene product might require the secretion to either the periplasm or extracellular medium for functional expression (18, 32). The periplasm provides an oxidative environment that allows disulfide bond formation, as well as a separate compartment from the cytoplasm for facilitating protein purification. The expression of a recombinant protein on the *E. coli* cell surface is performed only for special applications other than mass production (62). Secretion to the extracellular medium is an alternative strategy for functional expression of recombinant proteins that are otherwise difficult to express intracellularly due to technical limitations such as protein misfolding or disulfide bond formation. With the host/vector system acting as a whole-cell biocatalyst, extracellular secretion enables the continuous production of recombinant protein and facilitates downstream purification. However, recombinant protein yields are often low when secretion is involved.

There are two major structural barriers to protein secretion, the cytoplasmic membrane and the outer membrane. Protein translocation across the cytoplasmic membrane is usually driven by the Sec-dependent type II secretion pathway under the direction of a cleavable signal peptide at the N terminus of the secreted polypeptide. Technically, a DNA sequence encoding a signal peptide can be attached at the 5′ end of the open reading frame encoding the target protein to form a translational fusion. The expressed polypeptide precursor is expected to be translocated across the cytoplasmic membrane under the direction of the signal peptide. Several signal peptides (see Table 2) have been adopted successfully for exporting recombinant proteins into the periplasm or onto the outer membrane of *E. coli*. Very few *E. coli* proteins pass the outer membrane and are secreted extracellularly. Two other secretion pathways (i.e., type I and type III) are known to be responsible for the extracellular secretion of certain *E. coli* proteins without involving a periplasmic intermediate (10, 98). Although biochemical and genetic strategies (36, 90, 102, 106) based on permeabilizing the outer membrane have been developed for the extracellular release of periplasmic proteins (which are typically exported from the cytoplasm via the type II secretion system), they often impair cell physiology and reduce productivity. The type I and type III secretion systems have been successfully used to secrete target proteins extracellularly, though the host/vector system often requires a specific design (9, 22, 29, 101).

19.1.4. Protein Fusion Technology

Protein fusion technology serves as a powerful tool with multiple applications for recombinant protein production (97). Basically, the target gene product is expressed as a fusion protein, where the fusion tag does not affect the bioactivity of the target protein. The typical fusion tag is one that can be captured by affinity chromatography, greatly facilitating downstream purification. In addition, a specific cleavage site can be engineered at the junction between the two fusion partners so that the target protein can be released as the native protein. While such tags theoretically can be fused with the target protein at either the N or C terminus, the N terminus is more often used because many N-terminal tags contain local DNA sequences optimized for translational initiation. With an appropriate design, the production of fusion protein can be enhanced. Furthermore, several fusion tags have been demonstrated to promote protein folding because they can rapidly attain their native conformation. As a result, the solubility and stability of a fusion protein are often superior to the heterologous protein alone (107). In some cases, particularly with a small tag, the target protein retains its bioactivity and can be used without removing the tag.

19.1.5. Protein Misfolding

Another common technical issue for recombinant protein production in *E. coli* is protein misfolding, which results in loss of bioactivity. The problem can be viewed as the expression flux imbalance between polypeptide formation and polypeptide processing at any posttranslational stage (including folding). Such imbalance diverts protein synthesis into a nonproductive pathway and results in the accumulation of misfolded protein species, which typically aggregate into insoluble inclusion bodies that are inactive. Inclusion bodies can be formed in either the cytoplasm or periplasm. Factors that can cause this expression flux imbalance and protein misfolding include heat shock, high protein concentration, certain amino acid sequences within the expressed protein, and physiological stress on the cells. Some of the strategies that can be used to alleviate this problem include lowering the cultivation temperature or

coexpressing particular chaperones. However, in some expression scenarios, the formation of inclusion bodies is considered advantageous, particularly for toxic proteins, since inclusion bodies are often easy to harvest and are less toxic to the production cell. The use of inclusion bodies as an expression strategy hinges primarily on the ease of recovering bioactivity through in vitro protein refolding.

19.2. DEVELOPMENT OF THE EXPRESSION HOST

19.2.1. Genetic Modification

Enhancing recombinant protein production relies on the construction of a compatible host/vector system for effective expression of the target gene. Wild-type *E. coli* cells have several mechanisms to protect them from environmental stress, and these can be activated during high-level recombinant protein production. These natural safeguards can potentially limit performance in industrial applications. For example, several nucleases and proteases can be induced to digest foreign DNA and protein, respectively, in response to the physiological stress caused by recombinant protein production. Hence, the host strain requires genetic modification to improve its suitability for biomanufacturing through the implementation of desirable traits on either the host chromosome or expression vector. Thus, obstacles to recombinant protein production due to the natural safeguard mechanisms mentioned above can be eliminated or at least mitigated through appropriate genetic engineering.

19.2.2. Identifying the Targets that Require Genetic Modification

Essentially, genetic modifications involve changing the expression level of certain key genes. At one extreme native *E. coli* genes can be knocked out, and at the other extreme heterologous genes encoding a useful activity can be expressed. Expression vectors provide a flexible way to coexpress multiple genes and can be used for such genetic modification (100). If necessary, the expression level can be fine-tuned with tunable promoters (25) or gene-silencing mechanisms (48, 49). Identifying the key genes that can directly or indirectly affect recombinant protein production is one of the most critical and difficult aspects in developing the genetic engineering strategy. While individual genes of potential value are constantly being identified in many research programs and large-scale random overexpression projects (65), recent progress in transcriptomic and proteomic analyses offers a systematic approach for rapid identification of hundreds of such candidates (1, 31, 38, 83, 111). However, proper selection of the key genes among the candidates that might facilitate optimal engineering of the production strain remains challenging.

General methods for engineering microbial strains to derive desired phenotypes have been extensively reviewed (78, 88). Whole-organism mutagenesis followed by phenotypic screening and selection has been the classical strategy for obtaining desirable traits for biomanufacturing, such as increased product titer (yield or productivity) and greater physiological robustness under the overproduction conditions. Random mutations can be introduced by exposing the production strain to UV light or a chemical mutagen such as *N*-methyl-*N*′-nitro-*N*-nitrosoguanidine, or by random genome-wide transposon insertions (2). Multiple rounds of mutagenesis and screening are often required to optimize the strain. This approach can potentially identify multiple mutation targets leading to the desired phenotype, and the targets can be related to the production pathway in obscure ways. Given the increasing accessibility of genomic sequencing services, such mutations can be identified and novel mechanisms relevant to recombinant protein production can be discovered. A major drawback with multiple rounds of mutagenesis and screening is that the process can be time-consuming and resource intensive. Furthermore, the result of identifying the manipulation targets might not be reproducible. While mutagenesis is easy to perform, establishing an effective screening protocol can be challenging. Recent development in high-throughput screening technologies with advanced analytical tools such as flow cytometry, chromatography, and mass spectrometry significantly improves the screening efficiency, particularly in the area of metabolic engineering applications. However, these approaches are often product specific.

19.2.3. Physiological Improvement

The metabolic burden arising from recombinant protein production can induce physiological stress and in turn reduce recombinant protein yield. One strategy to alleviate the impact of this burden is to mimic or boost the natural stress response, thus maintaining healthy cell physiology during recombinant protein overproduction (20). Several types of genes can be useful for this strategy, one example being those encoding chaperones, which are often induced during the heat shock response and assist with folding of damaged or misfolded proteins in the cytoplasm (5) or periplasm (72). These chaperones can enhance the solubility, structural stability, translocation/secretion efficacy, or even disulfide bond formation of the expressed gene products (41, 55, 87). Various proteases (cytoplasmic, periplasmic, or membrane-bound) are also induced during stress responses that degrade damaged or abnormal proteins. Such proteases can also degrade the desired recombinant protein (42) and therefore represent major targets for gene knockout. While there are reports of using protease-deficient mutants to improve recombinant protein production (69), knocking out a large number of protease genes for this purpose will result in a sick strain because *E. coli* requires these proteases for regulated protein turnover and physiological adaptation. Some have reported improvement in both cell physiology and recombinant protein yield via protease coexpression, which is presumably mediated by selective proteolysis of misfolded proteins that may be toxic (47, 77). Various stress-sensing pathways have been identified (27, 45, 63), and genes involved in this process can be logical targets for genetic modification. Stationary-phase genes encode proteins that may lead to a reduction in cellular and metabolic activity, which can negatively affect recombinant protein production, and as such, these genes are targets for strain improvement (19, 46). Other metabolic engineering strategies that have been used to improve a strain's adaptation to harsh production conditions include reducing the secretion of toxic metabolites (34, 103), boosting the nutrient assimilation rate (67), or increasing the supply of limiting cofactors (86).

19.2.4. Genotypes Suitable for Recombinant Protein Production

E. coli strains engineered for the application of recombinant DNA technology have been derived from both K-12 and B strains. The K-12-derived strains (e.g., DH5α and JM109) often carry several endogenous mutations (e.g., *endA*, *recA*, and *lacZ*) specifically designed for molecular cloning pur-

poses, though they can also be used as a host for recombinant protein production. On the other hand, *E. coli* B strains, e.g., BL21 F⁻ *ompT dcm lon gal hsd*S$_B$(r$_B$⁻, m$_B$⁻) (95) and its derivatives, are probably the most common production strains primarily owing to the two mutations in protease genes, i.e., *lon*, encoding intracellular ATP-dependent protease, and *ompT*, encoding outer membrane protease, which can potentially alleviate the degradation of heterologous protein products. In addition, BL21 is less sensitive to glucose than K-12 strains (80) and can tolerate higher levels of metabolic stress associated with recombinant protein production (44), making it suitable for high-cell-density cultivation. These technical advantages may be related to its more integrated outer membrane structure, since it is observed that BL21 cannot be used as a suitable expression host for cell-surface display or extracellular secretion of recombinant protein (N. Narayanan and C. P. Chou, unpublished data).

Several genetic traits have been introduced into BL21 to improve its utility as a protein overproducer. The most popular is the genomic integration of a DE3 lysogen that contains the T7 RNA polymerase gene, expression of which is regulated by the *lacUV5* promoter. The resulting strain, BL21(DE3), can be used with expression vectors containing the T7 promoter system for overexpression of recombinant protein (96). Other genetic features have also been included in BL21 to enhance its tolerance against physiological impacts associated with recombinant protein production (particularly for toxic proteins). For example, the basal level arising from leaky expression during noninducing conditions is minimized by replacing the isopropyl-β-D-thiogalactopyranoside (IPTG)-inducible *lacUV5* promoter in DE3 with the *araB* promoter for tighter transcriptional regulation (14) (e.g., BL21-AI™, available from Invitrogen [Carlsbad, CA]). Alternatively, a plasmid (pLysS or pLysE with a p15A replication origin) or a mini-F plasmid containing the gene encoding T7 lysozyme, which is a natural inhibitor of T7 RNA polymerase (94), can be included to minimize leaky expression [e.g., BL21(DE3) (pLysS) and BL21(DE3) (pLysE) from Novagen (Madison, WI) or NEB Express and T7 Express from NEB (Ipswich, MA)].

Another interesting approach is to challenge the expression host with a plasmid that overexpresses a toxic protein and then screen for mutants that gain the ability to produce the protein. C41(DE3) and C43(DE3) (available from Lucigen Corp. [Middleton, WI]) contain mutations that facilitate overexpression of certain membrane proteins and were derived from BL21(DE3) by this strategy (71). The mutations resulting in the improved expression performance of these strains were recently identified to be located in the *lacUV5* promoter, and the discovery led to the construction of a BL21(DE3) derivative, i.e., Lemo21(DE3), with a precise T7 RNA polymerase activity for tunable protein expression (105). Precise control of expression has also been achieved with a *lacY* mutation (e.g., Tuner™ and Origami™ B from Novagen). It appears that precise control of the expression level of the heterologous gene of interest is critical for achieving maximum protein production.

Since T7 RNA polymerase synthesizes mRNA rapidly, resulting in the uncoupling of transcription and translation, transcripts are subject to degradation by endogenous RNases. Mutation of the *rne* gene, encoding RNase E, the major enzyme for RNA degradation (64), stabilizes such highly expressed mRNA transcripts, and BL21 strains with this mutation are available (e.g., Star™ from Invitrogen).

The *endA* gene encodes a nonspecific endonuclease, endonuclease I, which can degrade plasmid DNA prepared by the typical mini-prep protocols. In addition, *recA* encodes an ATPase involved in homologous DNA recombination in vivo (6). K-12 strains often carry mutations in these two genes to maintain the structural stability and product quality of plasmid DNA for molecular cloning. These two mutations have been introduced into BL21 strains for large-scale production of plasmid DNA (79). Though the functions associated with the *recA* and *endA* gene products might seem to be required for healthy cell physiology, protein overexpression using such double-mutant BL21 strains (e.g., Acella™ from EdgeBio [Gaithersburg, MD]) does not seem to encounter any problems associated with recombination or DNA degradation. Furthermore, such strains allow both protein expression and molecular cloning to be conducted in a single host.

Overexpression of heterologous proteins can potentially result in transient shortage of amino acids and the corresponding charged tRNAs for translation, which in turn can lead to the problems discussed above due to uncoupling of transcription and translation. This physiological regulation, known as the stringent response (12, 40), can be relaxed by the *relA* mutation such that RNA is still synthesized in the absence of protein synthesis. On the other hand, protein overexpression can be limited by depletion of certain rare tRNAs, particularly for expression of genes having a codon bias significantly different from that of *E. coli* (92). This limitation can be alleviated by coexpressing genes for those rare tRNAs, such as *argU*, *proL*, *ileY*, and *leuW*. Several BL21-CodonPlus® (Stratagene [La Jolla, CA]) and Rosetta™ (Novagen) strains expressing these tRNA genes are available, and their use for recombinant protein production appears to be justified (50). Alternatively, the heterologous gene can be modified or completely synthesized to have the optimal codon usage of *E. coli* (53, 74, 108).

For certain eukaryotic or therapeutic proteins, heterologous overexpression is limited by posttranslational processing, such as disulfide bond formation, which can be crucial for a protein's bioactivity. Though *E. coli* contains the oxidative milieu of the periplasm, where disulfide bonds can be formed in vivo, targeting of proteins to this compartment can be limited by the efficiency of translocation or periplasmic folding. Hence, there is a desire for strains that can produce proteins containing disulfide bonds in the cytoplasm. However, the *E. coli* cytoplasm contains two thioredoxins and three glutaredoxins, all of which are maintained in a reduced state by thioredoxin reductase (TrxB) and glutathione. Consequently, disulfide bond formation hardly occurs in this reduced compartment. Introducing mutations in the genes for both thioredoxin reductase (*trxB*) and glutathione reductase (*gor*) (or glutathione synthetase [*gshB*]) can transform the reduced cytoplasm into an oxidized one (93). This double mutation in the BL21 background is found in the Origami™ B strains (Novagen), which offer an alternative for the overexpression of proteins containing disulfide bonds in the oxidative cytoplasm. However, cell growth appears to be impaired by this double mutation.

Another host feature that can be useful for downstream processing is genomic inclusion of the λ R gene, encoding the λ lysozyme (i.e., endolysin). The resulting strain (XJb Autolysis™ from Zymo Research Corp. [Orange, CA]) is efficiently lysed by arabinose induction.

Note that most of the above mutations are implemented with an antibiotic resistance marker, which limits the choice of markers for the expression vector. In practice, it will be more convenient to construct mutants without drug markers using chromosomal engineering strategies developed recently (21).

19.2.5. Chromosomal Engineering of *E. coli*

Chromosomal engineering refers to the techniques for making site-specific insertions, replacements, or deletions in the chromosome. With an extensive ability to knock out an existing gene or express a foreign gene, the technique becomes a powerful tool for strain improvement. The most common one, known as recombineering, is based on homologous recombination (Fig. 4). In general, exogenous DNA containing an allele of interest is first delivered into the recipient cell directly or via a shuttle vector (either plasmid or virus based) or conjugation. Because *E. coli* is artificially transformable, conjugation is seldom used for delivering the exogenous DNA. The efficiency for homologous recombination in vivo can be enhanced (e.g., by expressing exonuclease and recombinase) so that the key allele in the delivered DNA can be site-specifically exchanged with the corresponding genomic allele or can be inserted into the chromosome. A cotransduced drug marker is often used for selection of the mutant derivative, whose genotype can be verified by colony PCR or genomic sequencing. Several methods based on transposon mutagenesis and homologous recombination (21, 33, 73, 112–114) have been developed for genomic modification to construct *E. coli* strains. Very large fragments carried on bacterial artificial chromosomes can be inserted into the *E. coli* genome (58, 85). This is particularly useful for metabolic engineering of *E. coli*, since novel strains can be constructed that express entire pathways for nonnative metabolites. A comprehensive *E. coli* mutant library was made using the chromosomal engineering protocol involving the λ Red-mediated recombination system (21). Each gene in the chromosome was replaced with a kanamycin resistance cassette flanked by homologous regions to the target allele. This library of single-gene knockout strains is called the Keio collection (3) (www.shigen.nig.ac.jp/ecoli/strain/top/top.jsp). Each mutant gene in the Keio collection can be transferred via P1-phage transduction (70) from its original BW25113 background to other genetic backgrounds using kanamycin for selection. Furthermore, the kanamycin resistance cassette between two flanking FLP recombination target (FRT) sites can be removed via FLP recombination using a helper plasmid (16). This will allow the construction of drug-resistance-free strains with multiple gene knockouts. The establishment of the Keio collection and its availability to researchers worldwide have marked an important biotechnological milestone by significantly facilitating *E. coli* mutant construction. The recently commercialized recombineering protocol from Gene Bridges (Heidelberg, Germany; www.genebridges.com), also based on λ Red-mediated recombination, allows versatile chromosomal engineering, including gene disruption, deletion, insertion, point mutation, modification, and even promoter fine-tuning, and can serve as a versatile manipulation tool for strain improvement and even optimization.

Protocol 1. Gene Knockouts in *E. coli* Using λ Red-Mediated Recombination

This protocol is based on Datsenko and Wanner (21) using the *E. coli* K-12 strain. There are other similar recombineering methods that use plasmids containing different promoters and markers, but these protocols are also based on the recombinases from either the Rac prophage or bacteriophage λ or both (58). Genetic modification using recombineering is not limited to chromosomal genes, but can

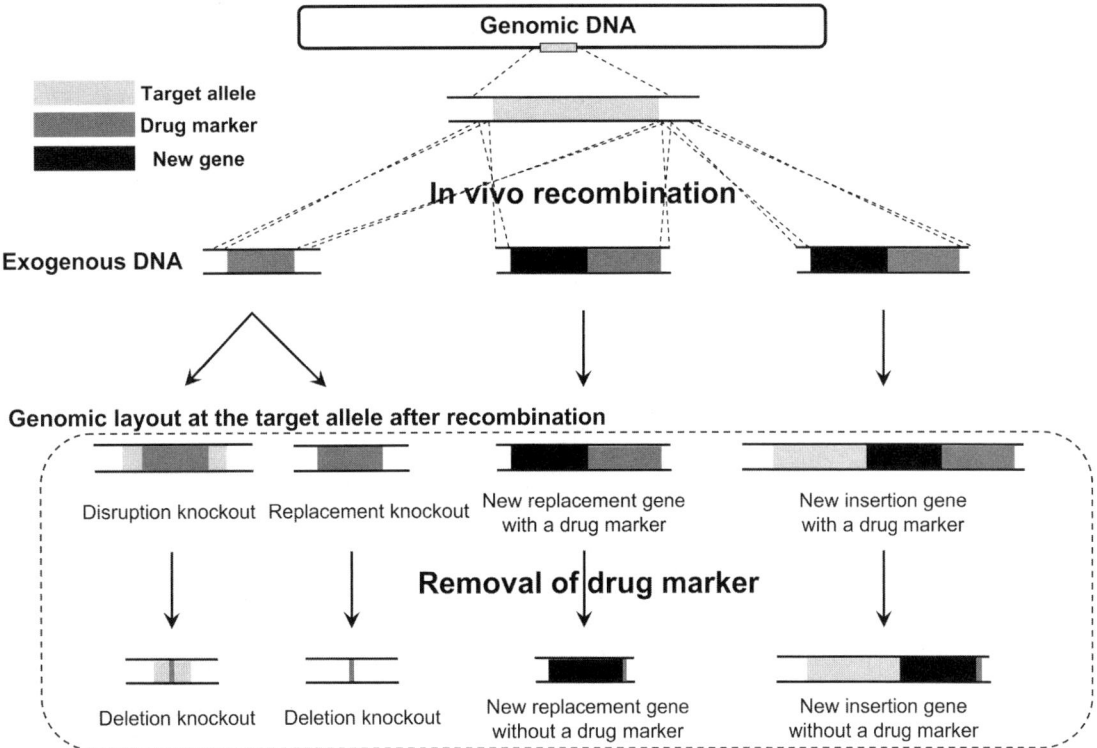

FIGURE 4 Chromosomal engineering of *E. coli* based on homologous recombination for either site-specific gene knockout or gene insertion. The target allele/site is first selected and the exogenous segment is prepared in vitro (e.g., by PCR). Because *E. coli* is artificially transformable, the exogenous DNA can be delivered into the recipient cell through electroporation. The efficiency of in vivo recombination can be enhanced by expressing key enzyme(s) associated with the recombination. A drug resistance marker is often introduced as the major replacing cassette or cotransduced with a new gene for selection of transformed cells, and can be subsequently deleted in vivo (e.g., by FLP recombination).

also be applied to extrachromosomal DNA molecules, such as plasmids and bacterial artificial chromosomes.

1. Obtain the set of recombineering vectors denoted pKD46, pKD3 (or pKD4), and pCP20.
2. Transform pKD46 into the *E. coli* strain containing the gene you wish to modify. The plasmid has a low copy number and a temperature-sensitive origin of replication, repA101. Thus, the cells should be incubated at 30°C during the transformation.
3. Design the primers for PCR amplification of the insertion cassette on a template plasmid (pKD3 or pKD4). The insertion cassette is flanked by approximately 35- to 50-bp homology extensions. These extensions are homologous to the sequences surrounding the chromosomal region to be replaced. The sense and antisense primers for the PCR should be designed such that the last 20 bp of each primer anneal to the insertion cassette to be amplified. The PCR product is gel purified, DpnI digested, and then repurified to ensure that all template plasmid is removed. The insertion cassette will contain the chloramphenicol resistance gene (for pKD3) or kanamycin resistance gene (for pKD4) between two FRT sites, which are used to remove the antibiotic resistance genes after recombination.
4. Make an electrocompetent cell suspension by growing a colony obtained from step 2 in LB medium containing 10% arabinose at 30°C to induce the expression of the three main λ Red genes on pKD46, *exo*, *bet*, and *gam*.

5. Electroporate approximately 0.5 to 1.0 μg of the insertion cassette prepared in step 3 into the electrocompetent cells by selection of the transformants on agar plates containing 34 μg/ml chloramphenicol or 50 μg/ml kanamycin at 37°C.
6. Verify the absence of pKD46 in the transformants by streaking and growing a few colonies on agar plates containing 100 μg/ml ampicillin as well as agar plates containing 34 μg/ml chloramphenicol (or 50 μg/ml kanamycin). The cells that have lost pKD46 will only grow on the chloramphenicol (or kanamycin) plates, and not on the ampicillin plates. Insertion of the antibiotic resistance genes can be further verified using colony PCR.
7. Remove the antibiotic resistance gene via FLP-FRT recombination. Transform the recombineered cells with pCP20, which contains a temperature-sensitive origin of replication, repA101. Plate the cells on LB agar plates containing 100 μg/ml ampicillin at 30°C.
8. Prepare overnight cultures of a few of the colonies obtained from step 7 in LB medium without antibiotic at 37°C to induce the expression the flippase (*flp*) gene for excising the antibiotic resistance gene region between the FRT sites.
9. Verify the loss of all antibiotic resistance genes by streaking the overnight cultures on LB agar plates containing appropriate antibiotics and incubating overnight at 37°C. The loss of the antibiotic resistance genes can be further verified using colony PCR.

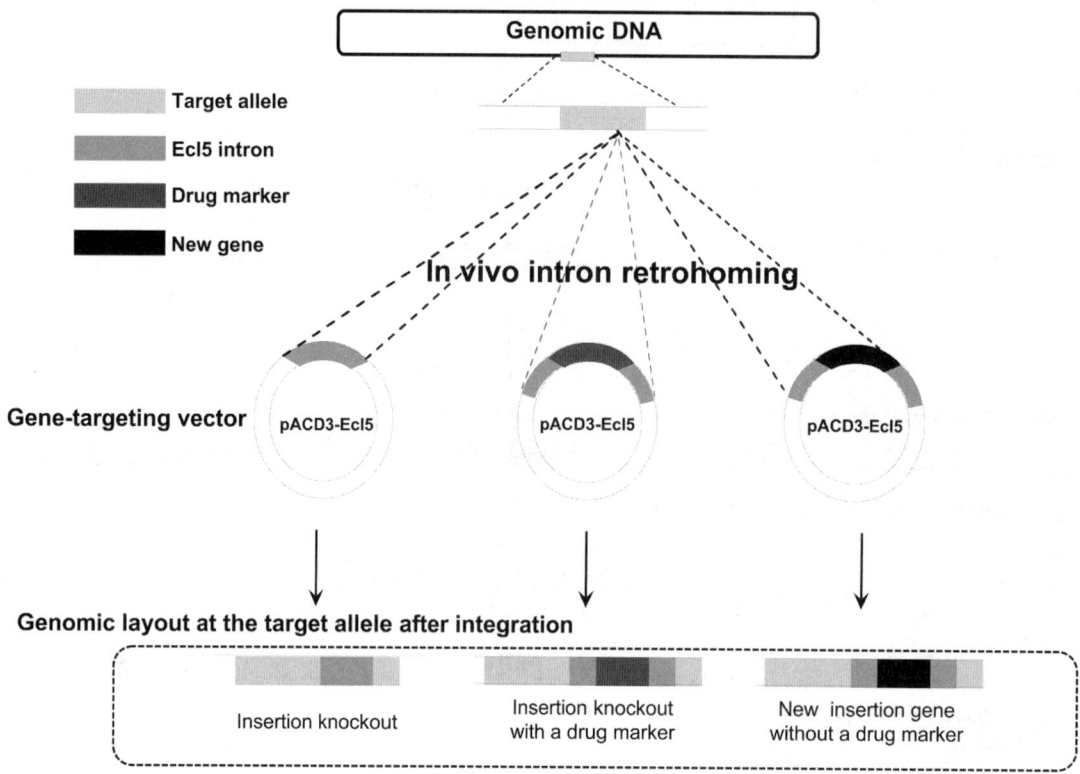

FIGURE 5 Chromosomal engineering of *E. coli* based on the intron gene-targeting system for either site-specific gene knockout or gene insertion. The target allele/site is first selected and the gene sequence is entered into the EcI5 computer algorithm to obtain putative insertion sites and corresponding mutagenesis primers. The intron is then retargeted, ligated into a targetron vector, and expressed within the appropriate host strain. The pACD3-EcI5 vectors contain a convenient MluI restriction site for inserting cargo genes such as a drug marker for knockout selection or a foreign gene for chromosomal expression in *E. coli*.

A chromosomal gene knockout strategy not involving homologous recombination is also available (Fig. 5). The technique utilizes a bacterial group II intron, such as L1.LtrB from *Lactococcus lactis*, and an intron-encoded protein (IEP), for site-specific integration via an in vivo retrohoming process (68). The mechanism begins with the IEP locating certain fixed nucleotide positions within the target DNA sequence, thereby facilitating specific base-pairing interactions between the intron RNA and target DNA. Due to both endonuclease and reverse transcriptase activity of the IEP, the intron RNA is able to hybridize with one strand of the target DNA and serve as a template for cDNA synthesis. Once the integration and reverse transcription events are complete, the nicked insertion site is sealed by host repair mechanisms to generate a stable and permanent 0.9-kb L1.LtrB intron insertion, which contains a stop codon in all six reading frames. Since specific base pairing between the intron RNA and target DNA occurs at only a few key regions (denoted IBS, EBS1d, and EBS2), it is possible to mutate and retarget the L1.LtrB intron for insertion into virtually any chromosomal site. Recently, a new bacterial group II intron isolated from *E. coli* O157:H7, called EcI5, has been shown to possess higher site-specific integration frequencies than the L1.LtrB intron (115), eliminating the need for antibiotic selection markers.

Protocol 2. Gene Knockouts in *E. coli* Using the EcI5 Intron

This protocol, based on the recent work by Zhuang et al. (115), describes gene knockout starting from EcI5 intron retargeting through induction of intron transcription and mutant screening.

1. Obtain the set of EcI5 gene-targeting vectors denoted pACD3-EcI5A, pACD3-EcI5C, pACD3-EcI5G, and pACD3-EcI5T. The vectors differ in just one nucleotide position at the EBS3 site. The appropriate vector must be chosen such that the EBS3 nucleotide is complementary to the +1 position (IBS3) within the predicted EcI5 intron insertion site.

2. Select a gene to be knocked out and input the coding sequence into the computer algorithm developed by Zhuang et al. (115). The algorithm ranks putative intron insertion sites and provides primer sequences corresponding to each insertion site.

3. Select at least two insertion sites and obtain the IBS1/IBS2, EBS1, and EBS2 primers for each site, in addition to the universal EcI5 primer.

4. The EcI5 intron is retargeted by a two-step PCR procedure in which the IBS1/IBS2 and EBS2 primers are used to generate a 336-bp PCR product, while the EBS1 and universal EcI5 primer are used to generate a 230-bp PCR product. These small PCR products are then gel purified and used in a second PCR to generate the fully mutated and retargeted 546-bp product. The internal EBS2 and EBS1 primers contain 20 bp of overlap and are fused together in the second PCR using the flanking IBS1/IBS2 and universal primers. The final product is also gel purified.

5. The retargeted PCR product is flanked by XbaI and AvaII sites, which are present in the IBS1/IBS2 and

universal EcI5 primers, respectively. The retargeted PCR product is swapped for the corresponding region in a previously targeted EcI5 targetron vector. Prepare restriction digests using AvaII and XbaI to digest both the intron PCR product and pACD3-EcI5A/G/C/T vector.

6. Gel purify the digested EcI5 targetron vector.

7. Prepare a ligation reaction containing 15 to 30 ng digested PCR product and 50 ng digested and purified targetron vector. This corresponds to a molar insert-to-vector ratio of approximately 3:1 to 6:1.

8. Inactivate the ligase using heat or phenol extraction.

9. Transform 1 μl of the ligation into an *E. coli* cloning strain using the heat shock or electroporation method. Plate the cells onto LB agar plates containing 25 μg/ml chloramphenicol. All pACD3-EcI5 targetron vectors contain a chloramphenicol antibiotic marker for plasmid selection and propagation.

10. Screen approximately five colonies by restriction analysis or colony PCR to verify the presence of the 546-bp insert.

11. Prepare overnight cultures of two or three colonies containing the 546-bp insert at 37°C for approximately 16 h.

12. Verify the IBS1, IBS2, EBS1, and EBS2 mutations by DNA sequencing using the universal T7 promoter sequencing primer.

13. Transform the retargeted plasmid into the desired knockout strain using the appropriate strain-specific transformation protocol. Since the intron is expressed under the control of the T7 promoter system, the knockout strain should be a DE3 strain of *E. coli*. If the knockout is to be conducted in a non-DE3 strain, the cells must be supplied with a source of T7 RNA polymerase. Alternatively, gene knockouts can be conducted in DE3 strains and then transduced to non-DE3 strains using P1-phage transduction.

14. Revive electroporated cells in 1 ml warm SOC medium (a nutrient-rich broth used to increase transformation efficiency) for 1 to 2 h at 37°C.

15. Dilute the transformation reaction into 4 ml warm LB medium containing 25 μg/ml chloramphenicol and grow for 16 h at 37°C.

16. Inoculate 5 ml of fresh LB medium containing 25 μg/ml chloramphenicol with 50 μl of the overnight culture and grow to an optical density at 600 nm of 0.2 to 0.3.

17. To induce intron transcription, inoculate 5 ml of fresh LB medium containing 100 μM IPTG and grow for 3 h at 37°C.

18. To remove excess IPTG, pellet the culture by centrifugation and resuspend in 1 ml fresh LB medium without antibiotic or IPTG.

19. Dilute a portion of the culture 1,000-fold in LB medium and spread 50 to 100 μl onto LB agar without antibiotic.

20. Perform colony PCR on colonies using primers flanking the intron insertion site or one flanking primer and one intron-specific primer. The intron insertion is ~0.9 kb.

There are also efforts to reduce the *E. coli* genome, thereby removing nonessential regions without affecting the growth behavior (52, 81). With a cleaner and more efficient background, such reduced-genome *E. coli* strains (e.g., Clean Genome® *E. coli* from Scarab Genomics, LLC [Madison, WI; www.scarabgenomics.com]) have been demonstrated to be effective in the production of recombinant proteins, DNAs, and metabolites (13, 60, 89).

19.3. CHALLENGES FOR STRAIN IMPROVEMENT

Proteins having a complex structure or multiple domains or requiring complex folding, multiple disulfide bonds, or other special posttranslational processing steps are difficult to express functionally in *E. coli*, and these technical difficulties still present a great challenge for strain improvement. Once the target protein is determined to be suitable for expression in *E. coli*, strain improvement is aimed at the host/vector system, target protein sequence, and various expression variables (Fig. 1). The general guidelines for strain improvement are (i) to ensure the genetic stability of the host/vector system, (ii) to maximize the synthesis fluxes for all the gene expression steps (i.e., transcription, translation, and posttranslational processing steps), (iii) to ensure the flux balance of these protein synthesis steps, (iv) to stabilize all the expression intermediates and final products, and (v) to minimize the physiological impact associated with high-level gene expression and high-cell-density cultivation. Successful strain improvement often relies on identification of the key steps limiting the overall culture performance. However, these limiting steps tend to be specific to the product of interest, difficult to identify, and subject to change. While certain types of limitations are now relatively easy to overcome, achieving a precise balance of the protein synthesis fluxes through optimization of the host/vector system remains challenging. In other words, we still do not have sufficient knowledge to undertake complete genome restructuring to achieve optimum strain performance in a given situation. Modern approaches based on systems biology and "-omics" studies continue to improve our knowledge, however. The genetic modifications currently adopted in most practical applications are primarily limited to gene knockouts and extrachromosomal overexpression, which cannot be fine-tuned. Modern chromosomal engineering techniques will potentially allow the inclusion of multiple exogenous genes as well as more precise regulation of the expression level of the modified genes in the chromosome. As a result, innovative strains can be more flexibly, effectively, and optimally tailored to enhance recombinant protein production. Despite the power of *E. coli* as an expression system, much work remains to be done to expand the types of protein products that can be functionally expressed and to improve its performance relative to the numerous alternative expression systems.

We greatly appreciate critical comments from George N. Bennett, Chris Reeves, and Richard Baltz.

REFERENCES

1. **Aldor, I. S., D. C. Krawitz, W. Forrest, C. Chen, J. C. Nishihara, J. C. Joly, and K. M. Champion.** 2005. Proteomic profiling of recombinant *Escherichia coli* in high-cell-density fermentations for improved production of an antibody fragment biopharmaceutical. *Appl. Environ. Microbiol.* **71:**1717–1728.

2. **Alexeyev, M. F., and I. N. Shokolenko.** 1995. Mini-Tn*10* transposon derivatives for insertion mutagenesis and gene delivery into the chromosome of gram-negative bacteria. *Gene* **160:**59–62.

3. **Baba, T., T. Ara, M. Hasegawa, Y. Takai, Y. Okumura, M. Baba, K. A. Datsenko, M. Tomita, B. L. Wanner, and H. Mori.** 2006. Construction of *Escherichia coli* K-12 in-frame, single-gene knockout mutants: the Keio collection. *Mol. Syst. Biol.* **2:**2006.0008.

4. **Baneyx, F., and G. Georgiou.** 1991. Construction and characterization of *Escherichia coli* strains deficient in multiple secreted proteases: protease III degrades high-molecular-weight substrates in vivo. *J. Bacteriol.* **173:**2696–2703.

5. **Baneyx, F., and M. Mujacic.** 2004. Recombinant protein folding and misfolding in *Escherichia coli*. *Nat. Biotechnol.* **22:**1399–1408.

6. **Bell, C. E.** 2005. Structure and mechanism of *Escherichia coli* RecA ATPase. *Mol. Microbiol.* **58:**358–366.

7. **Benito, A., M. Vidal, and A. Villaverde.** 1993. Enhanced production of P_L-controlled recombinant proteins and plasmid stability in *Escherichia coli* RecA$^+$ strains. *J. Biotechnol.* **29:**299–306.

8. **Bentley, W. E., N. Mirjalili, D. C. Andersen, R. H. Davis, and D. S. Kompala.** 2009. Plasmid-encoded protein: the principal factor in the "metabolic burden" associated with recombinant bacteria. *Biotechnol. Bioeng.* **102:**1284–1297.

9. **Bergmann, S., D. Wild, O. Diekmann, R. Frank, D. Bracht, G. S. Chhatwal, and S. Hammerschmidt.** 2003. Identification of a novel plasmin(ogen)-binding motif in surface displayed α-enolase of *Streptococcus pneumoniae*. *Mol. Microbiol.* **49:**411–423.

10. **Binet, R., S. Letoffe, J. M. Ghigo, P. Delepelaire, and C. Wandersman.** 1997. Protein secretion by Gram-negative bacterial ABC exporters: a review. *Gene* **192:**7–11.

11. **Boer, H. A., L. J. Comstock, and M. Vasser.** 1983. The *tac* promoter: a functional hybrid derived from the *trp* and *lac* promoters. *Proc. Natl. Acad. Sci. USA* **80:**21–25.

12. **Cashel, M., and K. E. Rudd.** 1987. The stringent response, p. 1410–1438. *In* F. C. Neidhardt, J. L. Ingraham, K. B. Low, B. Magasanik, M. Schaechter, and H. E. Umbarger (ed.), *Escherichia coli and Salmonella typhimurium: Cellular and Molecular Biology*, vol. 2. American Society for Microbiology, Washington, DC.

13. **Chakiath, C. S., and D. Esposito.** 2007. Improved recombinational stability of lentiviral expression vectors using reduced-genome *Escherichia coli*. *BioTechniques* **43:**466, 468, 470.

14. **Chao, Y. P., C. J. Chiang, and W. B. Hung.** 2002. Stringent regulation and high-level expression of heterologous genes in *Escherichia coli* using T7 system controllable by the araBAD promoter. *Biotechnol. Prog.* **18:**394–400.

15. **Chen, J. Q., T. B. Acton, S. K. Basu, G. T. Montelione, and M. Inouye.** 2002. Enhancement of the solubility of proteins overexpressed in *Escherichia coli* by heat shock. *J. Mol. Microbiol. Biotechnol.* **4:**519–524.

16. **Cherepanov, P. P., and W. Wackernagel.** 1995. Gene disruption in *Escherichia coli*: Tc^R and Km^R cassettes with the option of Flp-catalyzed excision of the antibiotic-resistance determinant. *Gene* **158:**9–14.

17. **Cho, S. H., D. Shin, G. E. Ji, S. Heu, and S. Ryu.** 2005. High-level recombinant protein production by overexpression of Mlc in *Escherichia coli*. *J. Biotechnol.* **119:**197–203.

18. **Choi, J. H., and S. Y. Lee.** 2004. Secretory and extracellular production of recombinant proteins using *Escherichia coli*. *Appl. Microbiol. Biotechnol.* **64:**625–635.

19. **Chou, C.-H., G. N. Bennett, and K.-Y. San.** 1996. Genetic manipulation of stationary-phase genes to enhance recombinant protein production in *Escherichia coli*. *Biotechnol. Bioeng.* **50:**636–642.

20. **Chou, C. P.** 2007. Engineering cell physiology to enhance recombinant protein production in *Escherichia coli*. *Appl. Microbiol. Biotechnol.* **76:**521–532.

21. **Datsenko, K. A., and B. L. Wanner.** 2000. One-step inactivation of chromosomal genes in *Escherichia coli* K-12 using PCR products. *Proc. Natl. Acad. Sci. USA* **97:**6640–6645.

22. **Davis, S. J., and R. D. Vierstra.** 1998. Soluble, highly fluorescent variants of green fluorescent protein (GFP) for use in higher plants. *Plant Mol. Biol.* **36:**521–528.

23. **De Anda, R., A. R. Lara, V. Hernandez, V. Hernandez-Montalvo, G. Gosset, F. Bolivar, and O. T. Ramirez.** 2006. Replacement of the glucose phosphotransferase transport system by galactose permease reduces acetate accumulation and improves process performance of *Escherichia coli* for recombinant protein production without impairment of growth rate. *Metab. Eng.* **8:**281–290.

24. **de Marco, A.** 2009. Strategies for successful recombinant expression of disulfide bond-dependent proteins in *Escherichia coli*. *Microb. Cell Fact.* **8:**26.

25. **De Mey, M., J. Maertens, G. J. Lequeux, W. K. Soetaert, and E. J. Vandamme.** 2007. Construction and model-based analysis of a promoter library for *E. coli*: an indispensable tool for metabolic engineering. *BMC Biotechnol.* **7:**34.

26. **de Smit, M. H., and J. van Duin.** 1990. Secondary structure of the ribosome binding site determines translational efficiency: a quantitative analysis. *Proc. Natl. Acad. Sci. USA* **87:**7668–7672.

27. **Diaz-Acosta, A., M. L. Sandoval, L. Delgado-Olivares, and J. Membrillo-Hernandez.** 2006. Effect of anaerobic and stationary phase growth conditions on the heat shock and oxidative stress responses in *Escherichia coli* K-12. *Arch. Microbiol.* **185:**429–438.

28. **Feldman, M. F., M. Wacker, M. Hernandez, P. G. Hitchen, C. L. Marolda, M. Kowarik, H. R. Morris, A. Dell, M. A. Valvano, and M. Aebi.** 2005. Engineering N-linked protein glycosylation with diverse O antigen lipopolysaccharide structures in *Escherichia coli*. *Proc. Natl. Acad. Sci. USA* **102:**3016–3021.

29. **Fernandez, L. A., I. Sola, L. Enjuanes, and V. de Lorenzo.** 2000. Specific secretion of active single-chain Fv antibodies into the supernatants of *Escherichia coli* cultures by use of the hemolysin system. *Appl. Environ. Microbiol.* **66:**5024–5029.

30. **Flores, S., R. de Anda-Herrera, G. Gosset, and F. G. Bolivar.** 2004. Growth-rate recovery of *Escherichia coli* cultures carrying a multicopy plasmid, by engineering of the pentose-phosphate pathway. *Biotechnol. Bioeng.* **87:**485–494.

31. **Franchini, A. G., and T. Egli.** 2006. Global gene expression in *Escherichia coli* K-12 during short-term and long-term adaptation to glucose-limited continuous culture conditions. *Microbiology* **152**(Pt. 7):2111–2127.

32. **Georgiou, G., and L. Segatori.** 2005. Preparative expression of secreted proteins in bacteria: status report and future prospects. *Curr. Opin. Biotechnol.* **16:**538–545.

33. **Goryshin, I. Y., J. Jendrisak, L. M. Hoffman, R. Meis, and W. S. Reznikoff.** 2000. Insertional transposon mutagenesis by electroporation of released Tn5 transposition complexes. *Nat. Biotechnol.* **18:**97–100.

34. **Gosset, G.** 2005. Improvement of *Escherichia coli* production strains by modification of the phosphoenolpyruvate: sugar phosphotransferase system. *Microb. Cell Fact.* **4:**14.

35. **Grabherr, R., and K. Bayer.** 2002. Impact of targeted vector design on ColE1 plasmid replication. *Trends Biotechnol.* **20:**257–260.

36. **Gumpert, J., and C. Hoischen.** 1998. Use of cell wall-less bacteria (L-forms) for efficient expression and secretion of heterologous gene products. *Curr. Opin. Biotechnol.* **9:**506–509.

37. **Guzman, L.-M., D. Belin, M. J. Carson, and J. Beckwith.** 1995. Tight regulation, modulation, and high-level expression by vectors containing the arabinose P_{BAD} promoter. *J. Bacteriol.* **177:**4121–4130.

38. **Haddadin, F. T., and S. W. Harcum.** 2005. Transcriptome profiles for high-cell-density recombinant and wild-type *Escherichia coli*. *Biotechnol. Bioeng.* **90:**127–153.

39. **Hagg, P., J. W. de Pohl, F. Abdulkarim, and L. A. Isaksson.** 2004. A host/plasmid system that is not dependent on antibiotics and antibiotic resistance genes for stable plasmid maintenance in *Escherichia coli*. *J. Biotechnol.* **111:**17–30.

40. **Harcum, S. W., and W. E. Bentley.** 1993. Response dynamics of 26-, 34-, 39-, 54-, and 80-kDa proteases in induced cultures of recombinant *Escherichia coli*. *Biotechnol. Bioeng.* **42:**675–685.

41. **Heo, M. A., S. H. Kim, S. Y. Kim, Y. J. Kim, J. H. Chung, M. K. Oh, and S. G. Lee.** 2006. Functional expression of single-chain variable fragment antibody against

c-Met in the cytoplasm of *Escherichia coli*. *Protein Expr. Purif.* **47:**203–209.

42. **Hoffmann, F., and U. Rinas.** 2004. Roles of heat-shock chaperones in the production of recombinant proteins in *Escherichia coli*. *Adv. Biochem. Eng. Biotechnol.* **89:**143–161.

43. **Hoffmann, F., J. van den Heuvel, N. Zidek, and U. Rinas.** 2004. Minimizing inclusion body formation during recombinant protein production in *Escherichia coli* at bench and pilot plant scale. *Enzyme Microb. Technol.* **34:**235–241.

44. **Hoffmann, F., J. Weber, and U. Rinas.** 2002. Metabolic adaptation of *Escherichia coli* during temperature-induced recombinant protein production: 1. Readjustment of metabolic enzyme synthesis. *Biotechnol. Bioeng.* **80:**313–319.

45. **Ihssen, J., and T. Egli.** 2004. Specific growth rate and not cell density controls the general stress response in *Escherichia coli*. *Microbiology* **150**(Pt. 6)**:**1637–1648.

46. **Jeong, K. J., J. H. Choi, W. M. Yoo, K. C. Keum, N. C. Yoo, S. Y. Lee, and M. H. Sung.** 2004. Constitutive production of human leptin by fed-batch culture of recombinant *rpoS⁻ Escherichia coli*. *Protein Expr. Purif.* **36:**150–156.

47. **Kadokura, H., H. Kawasaki, K. Yoda, M. Yamasaki, and K. Kitamoto.** 2001. Efficient export of alkaline phosphatase overexpressed from a multicopy plasmid requires *degP*, a gene encoding a periplasmic protease of *Escherichia coli*. *J. Gen. Appl. Microbiol.* **47:**133–141.

48. **Kemmer, C., and P. Neubauer.** 2006. Antisense RNA based down-regulation of RNaseE in *E. coli*. *Microb. Cell Fact.* **5:**38.

49. **Kim, J. Y. H., and H. J. Cha.** 2003. Down-regulation of acetate pathway through antisense strategy in *Escherichia coli*: improved foreign protein production. *Biotechnol. Bioeng.* **83:**841–853.

50. **Kleber-Janke, T., and W.-M. Becker.** 2000. Use of modified BL21(DE3) *Escherichia coli* cells for high-level expression of recombinant peanut allergens affected by poor codon usage. *Protein Expr. Purif.* **19:**419–424.

51. **Kolaj, O., S. Spada, S. Robin, and J. G. Wall.** 2009. Use of folding modulators to improve heterologous protein production in *Escherichia coli*. *Microb. Cell Fact.* **8:**9.

52. **Kolisnychenko, V., G. Plunkett, C. D. Herring, T. Feher, J. Posfai, F. R. Blattner, and G. Posfai.** 2002. Engineering a reduced *Escherichia coli* genome. *Genome Res.* **12:**640–647.

53. **Kudla, G., A. W. Murray, D. Tollervey, and J. B. Plotkin.** 2009. Coding-sequence determinants of gene expression in *Escherichia coli*. *Science* **324:**255–258.

54. **Kurland, C. G., and H. J. Dong.** 1996. Bacterial growth inhibition by overproduction of protein. *Mol. Microbiol.* **21:**1–4.

55. **Kurokawa, Y., H. Yanagi, and T. Yura.** 2000. Overexpression of protein disulfide isomerase DsbC stabilizes multiple-disulfide-bonded recombinant protein produced and transported to the periplasm in *Escherichia coli*. *Appl. Environ. Microbiol.* **66:**3960–3965.

56. **Kyratsous, C. A., S. J. Silverstein, C. R. DeLong, and C. A. Panagiotidis.** 2009. Chaperone-fusion expression plasmid vectors for improved solubility of recombinant proteins in *Escherichia coli*. *Gene* **440:**9–15.

57. **Langdon, R. H., J. Cuccui, and B. W. Wren.** 2009. N-linked glycosylation in bacteria: an unexpected application. *Future Microbiol.* **4:**401–412.

58. **Lee, E.-C., D. Yu, J. Martinez de Velasco, L. Tessarollo, D. A. Swing, D. L. Court, N. A. Jenkins, and N. G. Copeland.** 2001. A highly efficient *Escherichia coli*-based chromosome engineering system adapted for recombinogenic targeting and subcloning of BAC DNA. *Genomics* **73:**56–65.

59. Reference deleted.

60. **Lee, J. H., B. H. Sung, M. S. Kim, F. R. Blattner, B. H. Yoon, J. H. Kim, and S. C. Kim.** 2009. Metabolic engineering of a reduced-genome strain of *Escherichia coli* for L-threonine production. *Microb. Cell Fact.* **8:**2.

61. **Lee, S. Y.** 1996. High cell density culture of *Escherichia coli*. *Trends Biotechnol.* **14:**98–105.

62. **Lee, S. Y., J. H. Choi, and Z. Xu.** 2003. Microbial cell-surface display. *Trends Biotechnol.* **21:**45–52.

63. **LeThanh, H., P. Neubauer, and F. Hoffmann.** 2005. The small heat-shock proteins IbpA and IbpB reduce the stress load of recombinant *Escherichia coli* and delay degradation of inclusion bodies. *Microb. Cell Fact.* **4:**6.

64. **Lopez, P. J., I. Marchand, S. A. Joyce, and M. Dreyfus.** 1999. The C-terminal half of RNase E, which organizes the *Escherichia coli* degradosome, participates in mRNA degradation but not rRNA processing in vivo. *Mol. Microbiol.* **33:**188–199.

65. **Lynch, M. D., T. Warnecke, and R. T. Gill.** 2007. SCALEs: multiscale analysis of library enrichment. *Nat. Methods* **4:**87–93.

66. **Malissard, M., and E. G. Berger.** 2001. Improving the solubility of the catalytic domain of human β-1,4-galactosyltransferase 1 through rationally designed amino-acid replacements. *Eur. J. Biochem.* **268:**4352–4358.

67. **March, J. C., M. A. Eiteman, and E. Altman.** 2002. Expression of an anaplerotic enzyme, pyruvate carboxylase, improves recombinant protein production in *Escherichia coli*. *Appl. Environ. Microbiol.* **68:**5620–5624.

68. **Matsuura, M., R. Saldanha, H. W. Ma, H. Wank, J. Yang, G. Mohr, S. Cavanagh, G. M. Dunny, M. Belfort, and A. M. Lambowitz.** 1997. A bacterial group II intron encoding reverse transcriptase, maturase, and DNA endonuclease activities: biochemical demonstration of maturase activity and insertion of new genetic information within the intron. *Genes Dev.* **11:**2910–2924.

69. **Meerman, H. J., and G. Georgoiu.** 1994. Construction and characterization of a set of *E. coli* strains deficient in all known loci affecting the proteolytic stability of secreted recombinant proteins. *Bio/Technology* **12:**1107–1110.

70. **Miller, J. H.** 1992. *A Short Course in Bacterial Genetics*. Cold Spring Harbor Laboratory Press, Cold Spring Harbor, NY.

71. **Miroux, B., and J. E. Walker.** 1996. Over-production of proteins in *Escherichia coli*: mutant hosts that allow synthesis of some membrane proteins and globular proteins at high levels. *J. Mol. Biol.* **260:**289–298.

72. **Mogensen, J. E., and D. E. Otzen.** 2005. Interactions between folding factors and bacterial outer membrane proteins. *Mol. Microbiol.* **57:**326–346.

73. **Mori, H., K. Isono, T. Horiuchi, and T. Miki.** 2000. Functional genomics of *Escherichia coli* in Japan. *Res. Microbiol.* **151:**121–128.

74. **Nakamura, Y., T. Gojobori, and T. Ikemura.** 2000. Codon usage tabulated from international DNA sequence databases: status for the year 2000. *Nucleic Acids Res.* **28:**292.

75. **Narayanan, N., and C. P. Chou.** 2009. Alleviation of proteolytic sensitivity to enhance recombinant lipase production in *Escherichia coli*. *Appl. Environ. Microbiol.* **75:**5424–5427.

76. **Neubauer, P., H. Y. Lin, and B. Mathiszik.** 2003. Metabolic load of recombinant protein production: inhibition of cellular capacities for glucose uptake and respiration after induction of a heterologous gene in *Escherichia coli*. *Biotechnol. Bioeng.* **83:**53–64.

77. **Pan, K.-L., H.-C. Hsiao, C.-L. Weng, M.-S. Wu, and C. P. Chou.** 2003. Roles of DegP in prevention of protein misfolding in the periplasm upon overexpression of penicillin acylase in *Escherichia coli*. *J. Bacteriol.* **185:**3020–3030.

78. **Patnaik, R.** 2008. Engineering complex phenotypes in industrial strains. *Biotechnol. Prog.* **24:**38–47.

79. **Phue, J.-N., S. J. Lee, L. Trinh, and J. Shiloach.** 2008. Modified *Escherichia coli* B (BL21), a superior producer of plasmid DNA compared with *Escherichia coli* K (DH5). *Biotechnol. Bioeng.* **101:**831–836.

80. Phue, J. N., S. B. Noronha, R. Hattacharyya, A. J. Wolfe, and J. Shiloach. 2005. Glucose metabolism at high density growth of *E. coli* B and *E. coli* K: differences in metabolic pathways are responsible for efficient glucose utilization in *E. coli* B as determined by microarrays and Northern blot analyses. *Biotechnol. Bioeng.* **90:**805–820.

81. Posfai, G., G. Plunkett, T. Feher, D. Frisch, G. M. Keil, K. Umenhoffer, V. Kolisnychenko, B. Stahl, S. S. Sharma, M. de Arruda, V. Burland, S. W. Harcum, and F. R. Blattner. 2006. Emergent properties of reduced-genome *Escherichia coli*. *Science* **312:**1044–1046.

82. Rabhi-Essafi, I., A. Sadok, N. Khalaf, and D. M. Fathallah. 2007. A strategy for high-level expression of soluble and functional human interferon α as a GST-fusion protein in *E. coli. Protein Eng. Des. Sel.* **20:**201–209.

83. Raman, B., M. P. Nandakumar, V. Muthuvijayan, and M. R. Marten. 2005. Proteome analysis to assess physiological changes in *Escherichia coli* grown under glucose-limited fed-batch conditions. *Biotechnol. Bioeng.* **92:**384–392.

84. Rich, J. R., and S. G. Withers. 2009. Emerging methods for the production of homogeneous human glycoproteins. *Nat. Chem. Biol.* **5:**206–215.

85. Rong, R., M. M. Slupska, J.-H. Chiang, and J. H. Miller. 2004. Engineering large fragment insertions into the chromosome of *Escherichia coli*. *Gene* **336:**73–80.

86. San, K. Y., G. N. Bennett, S. J. Berrios-Rivera, R. V. Vadali, Y. T. Yang, E. Horton, F. B. Rudolph, B. Sariyar, and K. Blackwood. 2002. Metabolic engineering through cofactor manipulation and its effects on metabolic flux redistribution in *Escherichia coli*. *Metab. Eng.* **4:**182–192.

87. Sandee, D., S. Tungpradabkul, Y. Kurokawa, K. Fukui, and M. Takagi. 2005. Combination of Dsb coexpression and an addition of sorbitol markedly enhanced soluble expression of single-chain Fv in *Escherichia coli*. *Biotechnol. Bioeng.* **91:**418–424.

88. Santos, C. N., and G. Stephanopoulos. 2008. Combinatorial engineering of microbes for optimizing cellular phenotype. *Curr. Opin. Chem. Biol.* **12:**168–176.

89. Sharma, S. S., F. R. Blattner, and S. W. Harcum. 2007. Recombinant protein production in an *Escherichia coli* reduced genome strain. *Metab. Eng.* **9:**133–141.

90. Shokri, A., A. M. Sande'n, and G. Larsson. 2003. Cell and process design for targeting of recombinant protein into the culture medium of *Escherichia coli*. *Appl. Microbiol. Biotechnol.* **60:**654–664.

91. Skerra, A. 1994. Use of the tetracycline promoter for the tightly regulated production of a murine antibody fragment in *Escherichia coli*. *Gene* **151:**131–135.

92. Sorensen, H. P., and K. K. Mortensen. 2005. Advanced genetic strategies for recombinant protein expression in *Escherichia coli*. *J. Biotechnol.* **115:**113–128.

93. Stewart, E. J., F. Aslund, and J. Beckwith. 1998. Disulfide bond formation in the *Escherichia coli* cytoplasm: an in vivo role reversal for the thioredoxins. *EMBO J.* **17:**5543–5550.

94. Studier, F. W. 1991. Use of bacteriophage-T7 lysozyme to improve an inducible T7 expression system. *J. Mol. Biol.* **219:**37–44.

95. Studier, F. W., and B. A. Moffatt. 1986. Use of bacteriophage T7 RNA polymerase to direct selective high-level expression of cloned genes. *J. Mol. Biol.* **189:**113–130.

96. Studier, F. W., A. H. Rosenberg, J. J. Dunn, and J. W. Dubendorff. 1990. Use of T7 RNA polymerase to direct the expression of cloned genes. *Methods Enzymol.* **185:**60–89.

97. Terpe, K. 2003. Overview of tag protein fusions: from molecular and biochemical fundamentals to commercial systems. *Appl. Microbiol. Biotechnol.* **60:**523–533.

98. Thomas, N. A., and B. B. Finlay. 2003. Establishing order for type III secretion substrates—a hierarchical process. *Trends Microbiol.* **11:**398–403.

99. Togna, A. P., M. L. Shuler, and D. B. Wilson. 1993. Effects of plasmid copy number and runaway plasmid replication on overproduction and excretion of β-lactamase from *Escherichia coli*. *Biotechnol. Prog.* **9:**31–39.

100. Tolia, N. H., and L. Joshua-Tor. 2006. Strategies for protein coexpression in *Escherichia coli*. *Nat. Methods* **3:**55–64.

101. Tzschaschel, B. D., C. A. Guzman, K. N. Timmis, and V. de Lorenzo. 1996. An *Escherichia coli* hemolysin transport system-based vector for the export of polypeptides: export of Shiga-like toxin IIeB subunit by *Salmonella typhimurium aroA*. *Nat. Biotechnol.* **14:**765–769.

102. van der Wal, F. J., C. M. ten Hagen-Jongman, B. Oudega, and J. Luirink. 1995. Optimization of bacteriocin-release-protein-induced protein release by *Escherichia coli*: extracellular production of the periplasmic molecular chaperone FaeE. *Appl. Microbiol. Biotechnol.* **44:**459–465.

103. Vemuri, G. N., M. A. Eiteman, and E. Altman. 2006. Increased recombinant protein production in *Escherichia coli* strains with overexpressed water-forming NADH oxidase and a deleted ArcA regulatory protein. *Biotechnol. Bioeng.* **94:**538–542.

104. Vind, J., M. A. Sorensen, M. D. Rasmussen, and S. Pedersen. 1993. Synthesis of proteins in *Escherichia coli* is limited by the concentration of free ribosomes: expression from reporter genes does not always reflect functional mRNA levels. *J. Mol. Biol.* **231:**678–688.

105. Wagner, S., M. M. Klepsch, S. Schlegel, A. Appel, R. Draheim, M. Tarry, M. Hagbom, K. J. van Wijk, D. J. Slotboom, J. O. Persson, and J.-W. de Gier. 2008. Tuning *Escherichia coli* for membrane protein overexpression. *Proc. Natl. Acad. Sci. USA* **105:**14371–14376.

106. Wan, E. W., and F. Baneyx. 1998. TolAIII co-overexpression facilitates the recovery of periplasmic recombinant proteins into the growth medium of *Escherichia coli*. *Protein Expr. Purif.* **14:**13–22.

107. Waugh, D. S. 2005. Making the most of affinity tags. *Trends Biotechnol.* **23:**316–320.

108. Welch, M., A. Villalobos, C. Gustafsson, and J. Minshull. 2009. You're one in a googol: optimizing genes for protein expression. *J. R. Soc. Interface* **6**(Suppl. 4): S467–S476.

109. Wood, T. K., and S. W. Peretti. 1991. Construction of a specialized-ribosome vector for cloned-gene expression in *E. coli*. *Biotechnol. Bioeng.* **38:**891–906.

110. Wu, M. S., K. L. Pan, and C. P. Chou. 2007. Effect of heat-shock proteins for relieving physiological stress and enhancing the production of penicillin acylase in *Escherichia coli*. *Biotechnol. Bioeng.* **96:**956–966.

111. Yoon, S. H., M. J. Han, S. Y. Lee, K. J. Jeong, and J. S. Yoo. 2003. Combined transcriptome and proteome analysis of *Escherichia coli* during high cell density culture. *Biotechnol. Bioeng.* **81:**753–767.

112. Yu, B. J., K. H. Kang, J. H. Lee, B. H. Sung, M. S. Kim, and S. C. Kim. 2008. Rapid and efficient construction of markerless deletions in the *Escherichia coli* genome. *Nucleic Acids Res.* **36:**e84.

113. Yu, D., H. M. Ellis, E.-C. Lee, N. A. Jenkins, N. G. Copeland, and D. L. Court. 2000. An efficient recombination system for chromosome engineering in *Escherichia coli*. *Proc. Natl. Acad. Sci. USA* **97:**5978–5983.

114. Zhang, Y. M., F. Buchholz, J. P. Muyrers, and A. F. Stewart. 1998. A new logic for DNA engineering using recombination in *Escherichia coli*. *Nat. Genet.* **20:**123–128.

115. Zhuang, F. L., M. Karberg, J. Perutka, and A. M. Lambowitz. 2009. EcI5, a group IIB intron with high retrohoming frequency: DNA target site recognition and use in gene targeting. *RNA* **15:**432–449.

Genetic Engineering Tools for
Saccharomyces cerevisiae

VERENA SIEWERS, UFFE H. MORTENSEN, AND JENS NIELSEN

20

The yeast *Saccharomyces cerevisiae* is a widely used cell factory. Since ancient times it has been used for the production of bread, beer, and wine, and with the development of industrial processes for bioethanol production it has also become the cell factory of choice due to its high ethanol tolerance and high fermentative capacity. After the first demonstration of genetic engineering in *Escherichia coli*, techniques for genetic engineering of *S. cerevisiae* were rapidly developed, allowing its use for the production of biopharmaceuticals, e.g., human insulin. Since then many other pharmaceutical proteins have been produced using *S. cerevisiae*, including hepatitis virus and human papillomavirus vaccines. Furthermore, this yeast is currently being exploited as a production host for a number of fuels and chemicals, such as butanols, organic acids, novel antibiotics, and nutraceuticals such as resveratrol (89).

Among the advantages of using *S. cerevisiae* as a cell factory are: (i) it is the best-characterized eukaryotic microorganism in terms of genomics, biochemistry, cell biology, and physiology; (ii) it has a very efficient metabolism that can ensure very high glycolytic fluxes; (iii) many high-throughput experimental techniques were first developed for analysis of *S. cerevisiae*, and very large databases are available for this kind of data; (iv) detailed metabolic models have been developed for this organism; and (v) very efficient genetic engineering tools are available for this organism. This last is particularly important for metabolic engineering of *S. cerevisiae* (87, 95), as it is often necessary to evaluate many different strategies. An efficient metabolic engineering strategy may require many genetic modifications, such as gene deletion, promoter replacement, and/or heterologous gene expression. *S. cerevisiae* is probably matched only by *E. coli* in terms of the genetic engineering tools available.

20.1. YEAST TRANSFORMATION

Genetic engineering requires efficient transformation techniques, and many protocols for transformation of *S. cerevisiae* have been developed. These include methods involving spheroplast generation (48), electroporation (25), and biolistics with DNA-coated microprojectiles (6). The most widely applied method, however, is the lithium acetate/polyethylene glycol method first described by Ito et al. (1983) (50). This method has undergone many improvement steps over the years, one of them being the implementation of single-stranded DNA as carrier. It has also been adapted to different needs (e.g., high-efficiency versus fast and easy protocol) (36, 37).

Fast and Easy Protocol

1. Scrape a 50-μl volume of yeast from a fresh YPAD (1% [wt/vol] yeast extract, 2% [wt/vol] peptone, 2% [wt/vol] glucose, 80 mg/liter adenine hemisulfate, 18 g/liter agar) plate and suspend in 1 ml of sterile water.
2. Pellet cells in a microcentrifuge for 30 s and discard the supernatant.
3. Add 240 μl of PEG 3350 (50% wt/vol), 36 μl of lithium acetate (1 M), 50 μl of single-stranded carrier DNA (2 mg/ml in 10 mM Tris-HCl, 1 mM disodium EDTA, pH 8; denatured by 5 min of boiling), and 34 μl of sterile water containing any plasmid DNA (up to 1 μg). Mix thoroughly (Vortex mixer).
4. Incubate at 42°C for 20 to 180 min (dependent on strain).
5. Pellet cells for 30 s and remove supernatant.
6. Resuspend pellet in 400 to 1,000 μl of sterile water.
7. Spread 100 to 200 μl on appropriate dropout selection plates and let the liquid be absorbed by the plates.
8. Incubate for 3 to 4 days at 30°C.

20.2. PLASMID VECTORS

For expression of heterologous genes or overexpression of homologous genes in *S. cerevisiae*, plasmid vectors are usually employed. Most are shuttle vectors that allow cloning and amplification in *E. coli*. Yeast plasmid vectors can be divided into different types. Yeast integrative plasmids (YIps) do not possess any yeast replication origin and can therefore only be maintained after chromosomal integration (see below). Yeast episomal plasmids (YEps) contain a replication origin derived from the endogenous yeast 2μm plasmid allowing autonomous replication. YEps usually exist at a level of 10 to 80 copies per cell. However, numbers vary between cells of the same population, and copy number as well as mitotic stability of the single plasmid may depend on the marker gene applied (116). Yeast replicating plasmids (YRps), which carry an autonomous replicating sequence instead of the 2μm origin, also provide a high copy number, but are less stable than YEps (128) and therefore not used very frequently. Insertion of centromere sequences into YRps led to the construction of yeast centromeric plasmids

(YCps). YCps have a higher mitotic stability than YRps and YEps, but exist usually only in one to two copies per cell. Additional cloning of telomere sequences into YCps allows the formation of yeast artificial chromosomes after linearization. Yeast articifial chromosomes are used for the construction of DNA libraries rather than for gene overexpression. They allow stable maintenance of inserts larger than 1 Mb, whereas the maximal insert capacity of plasmid vectors is around 20 kb.

20.2.1. Marker Genes

Selection of transformants and stable maintenance of plasmid-containing strains require the use of selectable marker genes. Table 1 lists a number of yeast selectable markers. However, many of them have only been used in a few studies and therefore may not be generally useful.

Probably the most widely used type of selectable markers are ones that confer prototrophy to auxotrophic strains containing mutations in amino acid and/or nucleotide biosynthetic pathways, e.g., *LEU2* and *URA3*. For plasmid maintenance, these systems are dependent on the use of chemically defined media, which are often not practical in industrial-scale fermentations and may lead to suboptimal growth and lower cell densities. Moreover, cross-feeding may result in mixed cultures, in which cells containing the marker gene secrete the required nutrient(s) or a pathway intermediate, thus allowing markerless cells to grow also (99).

An alternative is an autoselection system, where the marker gene is essential for cell viability independent of the medium composition. This allows the use of complex media for optimal growth conditions and avoids the possibility of cross-feeding. Examples are genes involved in cell division (e.g., *CDC4* or *CDC28*) or the glycolytic pathway (e.g., *PGK1*). Because such genes are essential, special procedures for establishing the desired strain are required. One method is plasmid shuffling, where the host strain contains a maintenance plasmid carrying a marker gene that can be counterselected after transformation with the expression vector (35).

Both prototrophic and autoselection markers imply the availability of strains that have the respective genes deleted. These are usually easy to obtain for haploid laboratory strains, but difficult to achieve for industrial strains, which are often diploid. Here, dominant resistance markers are required, which can be divided into three groups: (i) yeast wild-type genes that when overexpressed cause resistance to certain compounds, e.g., *ERG11* conferring resistance to azole antifungals (27); (ii) mutant alleles of yeast genes, such as *AUR1-C*, providing aureobasidin A resistance (47); and (iii) heterologous drug resistance genes, such as the widely used *kan* gene derived from *E. coli* (124). Drawbacks for some of these resistance markers are low transformation efficiency, the occurrence of background colonies, and spontaneous genomic mutant alleles (1). The high cost of some selective compounds limits their industrial applications, and the use of antibiotics is generally restricted for public health reasons, in particular for the production of feed, food, or pharmaceutical products.

Choice of the marker gene also may influence copy number and stability of the vector. For example, *URA3*-containing plasmids have a higher mitotic stability than plasmids carrying *LEU2* as a marker (92, 116). In order to achieve high copy numbers, a low expression level of the marker gene and/or a low activity of the resulting gene products may be advantageous. This may be obtained by using truncated promoters (29, 90) or heterologous promoters/genes (64, 123, 125).

The use of genes derived from other organisms has the additional advantage that they are not present in the *S. cerevisiae* genome, thus avoiding undesirable recombination events. Several heterologous and native marker genes are available as MX cassettes, in which gene expression is controlled by the *Ashbya gossypii TEF1* promoter and terminator (40, 41, 46, 51, 122–124).

20.2.2. Counterselection

The ability to select for the absence of a marker gene can be very useful, allowing an integrated marker to be removed for subsequent reuse (see below) or to eliminate the maintenance plasmid in an autoselection system. Several of the prototrophic markers offer this possibility, the most widely used being the *URA3* marker, which can be selected against using 5-fluoroorotic acid (5-FOA) (14). Further examples are the *LYS2(LYS5)*/α-aminoadipate, the *MET15*/methylmercury, the *TRP1*/5-fluoroanthranilic acid, the *GAP1*/D-histidine, and the *FYC1(FCA1)*/5-fluorocytosine systems (21, 24, 46, 51, 102, 115). Alternatively, an additional marker, which may itself not be selectable, can be employed. This could be a drug sensitivity marker such as *CAN1*, conferring canavanine sensitivity; *CYH2*, conferring cycloheximide sensitivity; or a marker gene that leads to growth inhibition when overexpressed under control of a strong conditional promoter, such as *GIN11* or *PKA3* (2, 93).

20.2.3. Promoters

Promoter choice and copy number are the most important factors for achieving the desired gene transcription levels. The choice of promoter can affect expression levels by up to 4 orders of magnitude. In general, one distinguishes constitutive and regulatable promoters. However, one should bear in mind that even promoters described as constitutive may have different expression levels depending on the environmental conditions, e.g., the applied carbon source. Furthermore, what are generally referred to as regulatable (inducible/repressible) promoters can be leaky; i.e., basal expression may be observed even under repressive conditions. Table 2 provides a list of *S. cerevisiae* promoters that have been used for heterologous gene expression.

Among strong constitutive promoters, the most commonly used for high-level gene expression are P_{TEF1}, P_{ADH1}, and promoters derived from glycolytic pathway genes. The best-studied regulatable promoters are the different P_{GAL} promoters, which are repressed by glucose and induced by galactose and reach high expression levels under induced conditions. Several approaches have been used to improve the P_{GAL} system, since galactose is a relatively expensive carbon source. *gal1*Δ strains are not able to consume galactose and can be grown in media with both glucose for cell growth and galactose (which remains at a constant level) for gene induction (86). Additional deletion of *HXK2* relieves glucose repression of the *GAL* promoters, allowing gene expression even at high glucose consumption rates (68). However, deletion of *HXK2* attenuates the growth rate (and the glycolytic flux), and one may therefore alternatively delete the main transcriptional repressor of the *GAL* genes, Mig1p (56).

Promoters that are induced in late growth phase or stationary phase without requiring supply of an inducer can be useful for industrial fermentation applications. Examples are the glucose-repressed P_{HXT7} and P_{ADH2}, which become induced after glucose is depleted, or P_{SPI1}, which is induced under different stress conditions including stationary phase.

TABLE 1 Marker genes

Marker gene	Gene product	Reference
Prototrophic markers		
ADE2	Phosphoribosylaminoimidazole carboxylase involved in purine biosynthesis	17
ADE8	Phosphoribosyl-glycinamide transformylase involved in purine biosynthesis	106
ECM31	Ketopantoate hydroxymethyltransferase involved in pantothenic acid biosynthesis	109
HIS3	Imidazoleglycerol-phosphate dehydratase involved in histidine biosynthesis	110
LEU2	β-Isopropylmalate dehydrogenase involved in leucine biosynthesis	110
LEU2d	LEU2 with a partially defective promoter conferring a low expression level	29
LYS2	α-Aminoadipate reductase involved in lysine biosynthesis	106
LYS5	Phosphopantetheinyl transferase involved in lysine biosynthesis	51
MET2-CA	L-Homoserine-O-acetyltransferase involved in methionine biosynthesis	93
MET15	O-Acetyl homoserine-O-acetyl serine sulfhydrylase involved in sulfur amino acid biosynthesis	24
TRP1	Phosphoribosylanthranilate isomerase involved in tryptophan biosynthesis	110
TRP1d	TRP1 with a partially defective promoter conferring a low expression level	90
URA3	Orotidine-5′-phosphate decarboxylase involved in pyrimidine biosynthesis	110
KlURA3	K. lactis orotidine-5′-phosphate decarboxylase	64
CaURA3	C. albicans orotidine-5′-phosphate decarboxylase	41
Autoselection systems		
CDC4	F-box protein	55
CDC9	DNA ligase	117
CDC28	Catalytic subunit of the main cell cycle cyclin-dependent kinase	35
FBA1	Fructose-1,6-bisphosphate aldolase	23
MOB1	Component of the mitotic exit network	35
PGI1	Phosphoglucose isomerase	55
PGK1	3-Phosphoglycerate kinase	8
POT	Schizosaccharomyces pombe triose phosphate isomerase	114
PSA1 (SRB1)	GDP-mannose pyrophosphorylase	101
TPI	Aspergilllus nidulans triose phosphate isomerase	55
URA3 fur1	Orotidine-5′-phosphate decarboxylase and uracil phosphoribosyltransferase	85
Dominant and semidominant resistance markers		
CUP1	Metallothionein conferring resistance to copper and cadmium	49
ERG11	Lanosterol 14α-demethylase conferring resistance to azole antifungals	27
GRE2	Methylglyoxal reductase conferring resistance to methylglyoxal	82
SSU1	Plasma membrane sulfite pump conferring sulfite resistance	97
SFA1	Formaldehyde dehydrogenase conferring resistance to formaldehyde	119
YAP1	Transcription factor conferring resistance to cerulenin and cycloheximide	2
ARO4-OFP	Mutated DAHP synthase conferring resistance to fluorophenylalanine	32
AUR1-C	Mutated inositol-phosphoceramide synthase conferring resistance to aureobasidin A	47
cyh2	Mutated ribosomal protein conferring resistance to cycloheximide	26
FZF1-4	Mutated transcription factor conferring sulfite resistance	97
LEU4-1	Mutated α-isopropylmalate synthase conferring resistance to trifluoroleucine	13
pdr3-9	Mutated transcriptional activator conferring multidrug resistance	61
SMR1B	Mutated acetolactate synthase conferring resistance to sulfometuron methyl	126
aroA	5-Enolpyruvylshikimate-3-phosphate synthase conferring resistance to glyphosate	59
ble	Phleomycin binding protein conferring resistance to phleomycin	42
bsd	Blasticidin S deaminase conferring resistance to blasticidin S	33
cat	Acetyltransferase conferring resistance to chloramphenicol	44
dehH1	Fluoroacetate dehalogenase conferring resistance to fluoroacetate	119

(Continued on next page)

TABLE 1 *(Continued)*

Marker gene	Gene product	Reference
dhfr	Dihydrofolate reductase conferring resistance to methotrexate	78
dsdA	D-Serine deaminase conferring resistance to D-serine	122
hph	Phosphotransferase conferring resistance to hygromycin B	40
kan	Phosphotransferase conferring resistance to G418	124
mdr3	Phosphoglycoprotein conferring resistance to FK520	100
nat1	Acetyltransferase conferring resistance to nourseothricin	40
pat	Acetyltransferase conferring resistance to bialaphos	40
Other markers		
GAP1	General amino acid permease allowing growth on L-citrulline as sole nitrogen source	102
FCY1	Cytosine deaminase allowing growth on cytosine as sole nitrogen source	46
FCA1	C. albicans cytosine deaminase allowing growth on cytosine as sole nitrogen source	46

As mentioned before, it may be beneficial to minimize *S. cerevisiae*-derived sequences in an expression vector to avoid recombination. Apart from the *A. gossypii TEF1* promoter, which drives marker gene expression in the MX cassettes (see above), several other heterologous promoters, such as the constitutive GAP promoter from *Pichia pastoris* (121), the *TEF1* promoter from *Arxula adeninivorans* (15), and the regulatable *Candida tropicalis* ICL promoter (54), have successfully been applied in *S. cerevisiae*. The Tet system is widely used, and different versions of it allow either repression or induction of a *tet* promoter by doxycycline (12).

New promoter properties, e.g., altered regulation or enhanced expression, can be achieved by combining regulatory elements derived from different promoters. An example is the introduction of P_{GAL1} or P_{GAL10} upstream activation sequences into the *ENO2* promoter (118). Since highest gene expression levels may not always lead to highest protein production, promoter libraries can be used to fine-tune gene expression. These may be generated by error-prone PCR from existing yeast promoters (4) or by using primers that contain certain conserved regulatory elements, but are otherwise degenerate (53).

20.2.4. Terminators

Proper transcription termination is a requirement for mRNA stability and thus also for high expression levels. The origin of the termination signal is usually of less importance and even the presence of the native terminator, when expressing a heterologous gene, may be sufficient. However, an influence of the termination signal origin on protein expression levels has also been observed (69). The most common *S. cerevisiae* terminators used for heterologous gene expression are derived from the *PGK1*, *ADH1*, and *CYC1* genes.

20.2.5. Protein Secretion Signals

Even though the production of extracellular proteins using *S. cerevisiae* has some drawbacks (often lower efficiency than in other host systems, and problems with hyperglycosylation), it has found many industrial applications. Protein secretion requires the presence of an N-terminal signal sequence. If the protein is also secreted in its natural host, the genuine secretion signal may be sufficient for extracellular targeting in yeast. In case a yeast signal sequence is to be used, the leader of the mating pheromone α-factor (encoded by *MFα1*) is usually chosen (18). It consists of a pre-peptide, a pro-peptide, and a Kex2p processing site. An alternative is the signal sequence of the *S. cerevisiae*

invertase Suc2p (91). Further secretion signals that can be applied are, for example, the leader of the *Kluyveromyces lactis* killer toxin (10) or the glucoamylase signal peptide of *Saccharomyces diastaticus* (65).

20.2.6. Available Expression Vectors

Table 3 lists several plasmid vectors used for gene expression in *S. cerevisiae* that are available from different suppliers and also specifies their main features.

20.3. CHROMOSOMAL INTEGRATION

Once the full genome sequence for *S. cerevisiae* was obtained (39), it became possible to introduce directed genetic changes by gene targeting (104). Using this technique, new strains can be tailor-made to produce new products and to display a preferred metabolic flux or growth pattern, ensuring maximum yield and desired product quality. Most commonly, this has been achieved by inserting new genes and/or by deleting or disrupting genes that influence a given process. However, gene targeting also allows metabolism to be engineered in more sophisticated ways. For example, it is possible to alter transcriptional activity of a selected gene by exchanging its native promoter for one with different strength or induction properties. Even point mutations can be introduced to change specific aspects of gene function. Hence, it is possible to alter the substrate specificity of a given protein or to specifically eliminate a single activity in a multifunctional protein. Furthermore, by using gene targeting, genes can be extended by sequences that encode new functions such as epitope tags, allowing for immunological detection of protein levels; visualization tags like green fluorescent protein (GFP) to address the localization of the protein in the cell; sequences to direct a protein to a specific compartment in the cell; and purification tags to facilitate protein isolation.

Gene targeting is particularly efficient in *S. cerevisiae* because DNA double-strand breaks are preferentially repaired by homologous recombination as opposed to nonhomologous end joining (30). When a linear molecule of DNA is transformed into a yeast cell, the cell perceives the fragment as broken DNA that needs repair. If the fragment contains sequences that are identical to a target site in the genome, it will efficiently integrate at that target site via the homologous recombination repair process (94). Henceforth we refer to DNA fragments made for genomic integration at a specific locus as gene-targeting substrates because they represent substrates for DNA repair. The different types of

TABLE 2 Promoters for heterologous gene expression

Promoter	Gene product	Comment	Reference(s)
Constitutive promoters			
P_{ACT1}	Actin	Strong	79
P_{ADH1}	Alcohol dehydrogenase	Strong on glucose; long version is repressed by nonfermentable carbon sources	79, 81, 105
P_{ENO2}	Enolase II	Strong on glucose; expression level is carbon-source dependent	118
P_{KEX2}	Serine protease	Very weak	84
P_{PGK1}	3-Phosphoglycerate kinase	Strong on glucose; expression level is carbon-source dependent	79, 84, 120
P_{PMA1}	Plasma membrane H$^+$-ATPase	Upregulated on glucose	58, 108
P_{TDH3}	Glyceraldehyde-3-phosphate dehydrogenase (alias GPD)	Strong	9, 81
P_{TEF1}	Translational elongation factor EF-1 alpha	Strong	9, 81
P_{TPI1}	Triose phosphate isomerase	Strong; expression level is carbon-source dependent	55, 98
Regulatable promoters			
P_{ADH2}	Alcohol dehydrogenase II	Repressed by glucose	66
P_{CUP1}	Metallothionein	Induced by copper(II) ions	60, 75
P_{CYC1}	Cytochrome *c*	Regulated by carbon source and oxygen availability	74, 81, 84
P_{DAN1}	Cell wall mannoprotein	Induced under anaerobic conditions	88
P_{FUS1}	Membrane protein	Pheromone-induced	31
P_{GAL1}/P_{GAL10}	Galactokinase/UDP-glucose-4-epimerase	Bidirectional promoter; induced by galactose, repressed by glucose	75, 80
P_{GAL7}	Galactose-1-phosphate uridyl transferase	Induced by galactose, repressed by glucose	75, 111
P_{HXT7}	High-affinity hexose transporter	Low expression at high glucose concentration, high expression at low glucose concentration	62
P_{MET3}	ATP sulfurylase	Repressed by methionine	7
P_{MET25} (P_{MET17})	O-Acetylhomoserine sulfhydrylase	Repressed by methionine	112
P_{PHO5}	Acid phosphatase	Induced at low phosphate concentration, repressed at high phosphate concentration	67
P_{SOR1}	Sorbitol dehydrogenase	Induced by sorbitol and xylose	76
P_{SPI1}	Glycosylphosphatidylinositol-anchored cell wall protein	Induced under stress and at stationary phase	19
P_{SUC2}	Invertase	Repressed at high glucose concentration	20

specific genetic manipulations listed above are obtained by varying the design of the gene-targeting substrate, as described in the following sections.

20.3.1. Integrative Plasmids

Integrative plasmids are most often used to insert a (novel) gene into the genome at a defined location. This is typically achieved by using *E. coli* vectors with an added yeast selectable marker gene (see above examples). The gene-targeting experiment is initiated by inserting Your Favorite Gene (*YFG1*) into such a vector using *E. coli*-based cloning. This vector is linearized by a restriction enzyme that cuts within the yeast marker gene. The resulting gene-targeting

substrate is transformed into yeast, where it integrates at the corresponding marker locus in the yeast genome by what is commonly referred to as an ends-in or loop-in recombination event (113) (Fig. 1). Notably, during integration the selectable marker is duplicated and will flank the inserted *YFG1* as a direct repeat. If insertion of *YFG1* significantly influences the fitness of the host strain, it is important to keep in mind that such insertions are inherently unstable, as direct repeat recombination (also known as pop-out or loop-out recombination) eliminates the inserted gene at a low rate (57) and the resulting recombinants may take over the population. To avoid this, genes can be inserted using gene-targeting substrates that integrate via separate

TABLE 3 Available plasmid expression vectors

Name	Description	Marker genes	Promoters	Reference(s)	Available from[a]
pRS series	Set of YCp, YIp, and YEp shuttle vectors containing different selectable markers	HIS3, TRP1, LEU2, URA3		22, 110	ATCC
p4XX(prom)	Set of YCp and YEp shuttle vectors based on the pRS series containing different promoters and the CYC1 terminator	HIS3, TRP1, LEU2, URA3	P_{CYC1}, P_{TEF} (=P_{TEF1}), P_{ADH1}, P_{GPD} (=P_{TDH3}), P_{MET25}, P_{GAL1}, P_{GALL}, P_{GALS}	80, 81	ATCC
Y(C/E/I)plac series	Set of YCp, YIp, and YEp shuttle vectors based on pUC19 containing different selectable markers	TRP1, LEU2, URA3		38	ATCC
pFL series	Set of YCp, YIp, and YEp shuttle vectors based on pUC19 containing different selectable markers	TRP1, LEU2, URA3		16	ATCC
pIS series	Set of YIp shuttle vectors for single-copy chromosomal integration	ade8, fcy1, his3, leu2, lys2, met15; URA3		106	Euroscarf
pCM series	Set of YCp and YEp shuttle vectors containing tetracycline regulatable promoters	TRP1, URA3	tetO	34	ATCC, Euroscarf
pGREG series	Set of YCp shuttle vectors based on the pRS series; with and without tagging	HIS3, TRP1, LEU2, URA3; kanMX	P_{GAL1}	52	Euroscarf
pVV series	Set of YEp and YCp shuttle vectors for gateway cloning; with and without tagging	TRP1, URA3	tetO, P_{PGK1}	120	Euroscarf
pYES vectors	YEp vectors offering different tagging and cloning strategies	URA3, TRP1	P_{GAL1}		Invitrogen
pYC vectors	YCp vectors	URA3, bsdR	P_{GAL1}		Invitrogen
pESC vectors	YEp vectors for coexpression of two genes from one vector	HIS3, TRP1, LEU2, URA3	P_{GAL1}/P_{GAL10}		Stratagene
pKS vectors	YEp vectors with or without SUC2 export signal	kanMX	P_{ADH2}		Dualsystems
p(prom)-MF	YEp vectors containing MFα leader sequence	URA3	P_{TEF1}, P_{GAL}		Dualsystems
p417-CYC, p427-TEF	YCp and YEp vector	kanMX	P_{CYC1}, P_{TEF1}		Dualsystems

[a] ATCC, American Type Culture Collection; Euroscarf, European *Saccharomyces cerevisiae* Archive for Functional Analysis; Invitrogen, Invitrogen, Carlsbad, CA; Stratagene, Stratagene, La Jolla, CA; Dualsystems, Dualsystems Biotech AG, Schlieren, Switzerland.

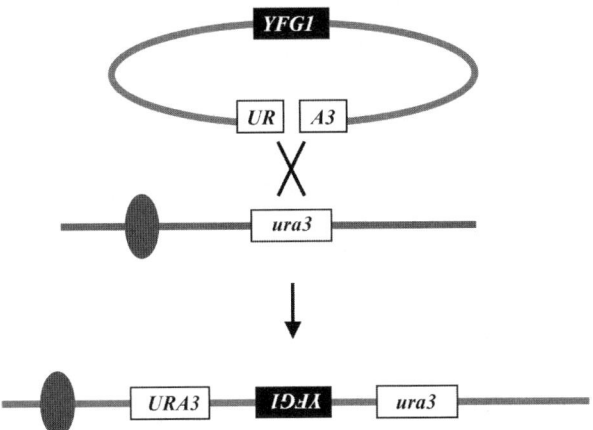

FIGURE 1 Genomic integration using an integrative plasmid. *YFG1* and a selective marker, in this case *URA3*, are cloned into an *E. coli* plasmid. The plasmid is linearized at a site within *URA3* prior to transformation into a *ura3* strain. After integration, the *URA3* gene is duplicated, typically resulting in one functional and one nonfunctional copy. Chromosomal DNA is indicated by the presence of a gray oval representing the centromere.

recombination events at both ends, thus not generating repeats (see below).

Multicopy insertions of the same gene may be achieved if the gene-targeting substrate is designed to target repeated elements—δ-elements or ribosomal DNA (rDNA) repeats—in the genome. In this way, stable strains expressing 20 to 200 gene copies have been reported (71, 96). Selection for multicopy integrations can be performed by choosing a weakly expressed resistance marker such as *neo* (G418/kanamycin resistance). Addition of high concentrations of G418 selects for multiple integration events, as G418 resistance and *neo* copy number are roughly proportional (125).

20.3.2. Gene Disruption/Deletion

Elimination of gene functions is most commonly achieved by disrupting the *YFG1* open reading frame (ORF) or by deleting the entire (or part of) *YFG1* in a one-step recombination procedure. A typical gene-targeting substrate for ORF disruption is simply made by inserting a selectable marker into the *YFG1* ORF, generating a gene-targeting substrate with the selectable marker flanked by *YFG1* ORF up- and downstream sequences, respectively. For *YFG1* ORF deletion, a gene-targeting substrate composed of a sequence upstream of the *YFG1* ORF, a selectable marker, and a sequence downstream of the *YFG1* ORF is constructed. Such gene-targeting substrates integrate at endogenous *YFG1* by double homologous recombination to disrupt/delete *YFG1* in a process that does not generate repeats (Fig. 2).

The starting point for *YFG1* gene disruption is typically a plasmid containing *YFG1*. The gene-targeting substrate is then simply made by inserting a selectable marker into the *YFG1* ORF by conventional cloning. Finally, the gene-targeting substrate is liberated from the vector backbone by restriction enzymes to provide a linear gene-targeting substrate for transformation (Fig. 2A).

Gene-targeting substrates for gene deletion are composed of a selectable marker flanked by two sequences that are identical to sequences up- and downstream of the region to be deleted, respectively. Recombination between

such substrates and the genome results in integration of the marker and loss of the genomic sequence situated between the up- and downstream sequences (Fig. 2B). Deletion substrates can be constructed by conventional cloning or by PCR. However, since construction of gene-targeting substrates by PCR is generally more convenient, the following sections will focus only on procedures using PCR. In the simplest form, a gene-targeting substrate for gene deletion is generated by amplifying the selectable marker using PCR primers with extensions at their 5′ ends homologous to the up- and downstream targeting sequences, respectively (11, 72). However, the efficiency of gene targeting dramatically decreases as the length of the homologous targeting sequences is shortened (73), and therefore it is advisable to use targeting sequences of at least 200 bp (maximum efficiency is achieved when the regions of homology are >500 bp). Since such long targeting sequences cannot be incorporated into a PCR primer, it is necessary to build the gene-targeting substrates in two steps. In a first round of PCR reactions, the two targeting fragments (up- and downstream of *YFG1*) and the marker fragment are amplified in separate PCR reactions. The gene-targeting substrate is completed in a second PCR reaction that fuses the targeting fragments to the ends of the fragment containing the selectable marker, as illustrated in Fig. 2C (5). Fusion of the primary PCR fragments is made possible by using primers in the first round with extensions at their 5′ ends that overlap the pieces to which they are being fused. Alternatively, and using simpler PCR, the substrate can be made as a bipartite gene-targeting substrate (Fig. 2D) (28). In this case, one part of the substrate is made by fusing the upstream targeting sequence to a fragment containing the 2/3 upstream region of the selectable marker by PCR. The other part of the substrate is made by fusing the 2/3 downstream region of the selectable marker to the downstream targeting sequence. After transformation with the bipartite gene-targeting substrate, integration events can be selected because the selectable marker is reconstructed by homologous recombination (Fig. 3).

20.3.3. Iterative Gene Targeting

In many cases, a strain needs to be modified several times by successive gene-targeting steps before it obtains the desired genotype. With the gene-targeting principle described above, the selectable marker is stably integrated at the target site. Since in many strains the number of useful markers is limited, it is often necessary to recycle a favorable marker during a series of gene-targeting experiments. This requires that the selectable marker be eliminated from the targeted site between each round of gene targeting. Marker elimination can be achieved either by employing the Cre/Lox system (107), by spontaneous mutagenesis, or, if the selectable marker in the gene-targeting substrate is flanked by direct repeats, by recombination. In the Cre/Lox system, the selectable marker, typically *kanMX*, is flanked by the short (34-bp) *loxP* sequences (43). After gene targeting, the marker can be excised from the genome by expressing the Cre recombinase, e.g., from a plasmid. Recycling of a marker by spontaneous mutation or recombination requires that the selectable marker also can be counterselected. The most widely used marker for this purpose is *URA3*, but other markers can also be used (see above). In this case, it is useful to employ a functional marker from a related organism, e.g., *URA3* from *K. lactis*, to avoid recombination with the endogenous marker locus during transformation. Marker loss by mutation is rare. For *URA3* at its native location,

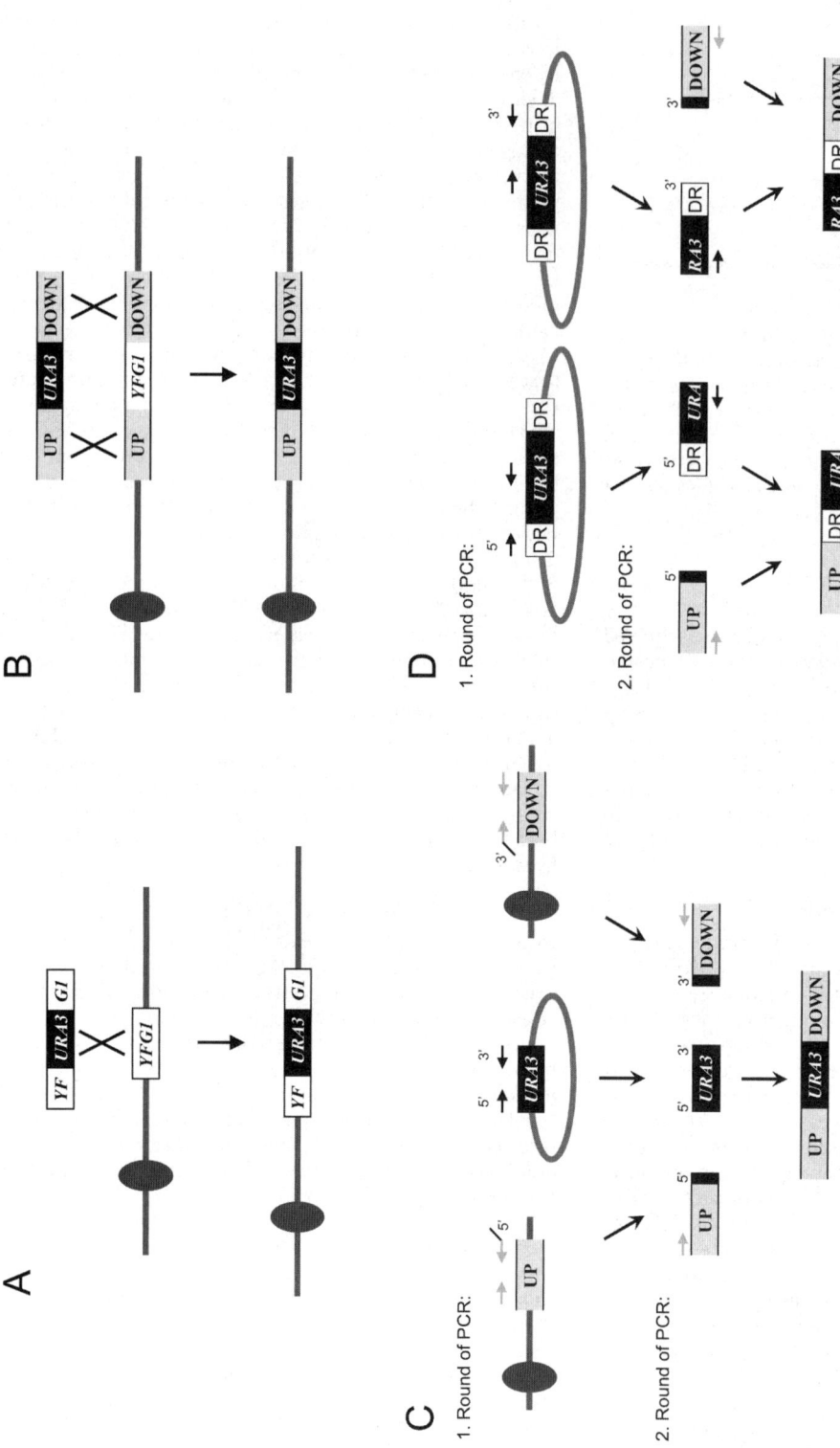

FIGURE 2 Substrates for disruption and deletion of *YFG1*. (A) Disruption of *YFG1* is achieved by using a linear substrate composed of a selectable marker (*URA3*) flanked by sequences up- and downstream of the integration site in *YFG1*. (B) Deletion of *YFG1* is achieved by using a linear substrate composed of a selectable marker (*URA3*) flanked by sequences up- and downstream of *YFG1*. Original sequence between the *YFG1* up- and downstream sequences will be deleted. (C) Construction of a gene deletion substrate by PCR. In a first round of PCR, fragments containing the up- and downstream sequences of *YFG1* and the marker gene (*URA3*) are generated in three independent PCR reactions. Note that one of the primers in the primer pair used to generate the upstream fragment is extended by a sequence identical to the extreme 5′-upstream section of *URA3* as indicated by a black line added to the gray primer. The sequence of the extension is complementary to the 5′-primer used to amplify *URA3*. Similarly, one of the primers used to amplify the downstream fragment is extended by a sequence that is complementary to the 3′-primer used to amplify *URA3*. The two primer extensions produce sequence overlaps that allow all three fragments to be fused and amplified in a second round of PCR. (D) Construction of a bipartite gene deletion substrate. The up- and downstream sections are generated as shown in panel C. In a first round of PCR, two fragments consisting of the 2/3 upstream and 2/3 downstream parts of *URA3* are generated in two individual PCR reactions. The primer extensions added to primers used to generate the up- and downstream fragments to be individually fused to the 5′-2/3 and 3′-2/3 *URA3* fragments in separate PCR reactions.

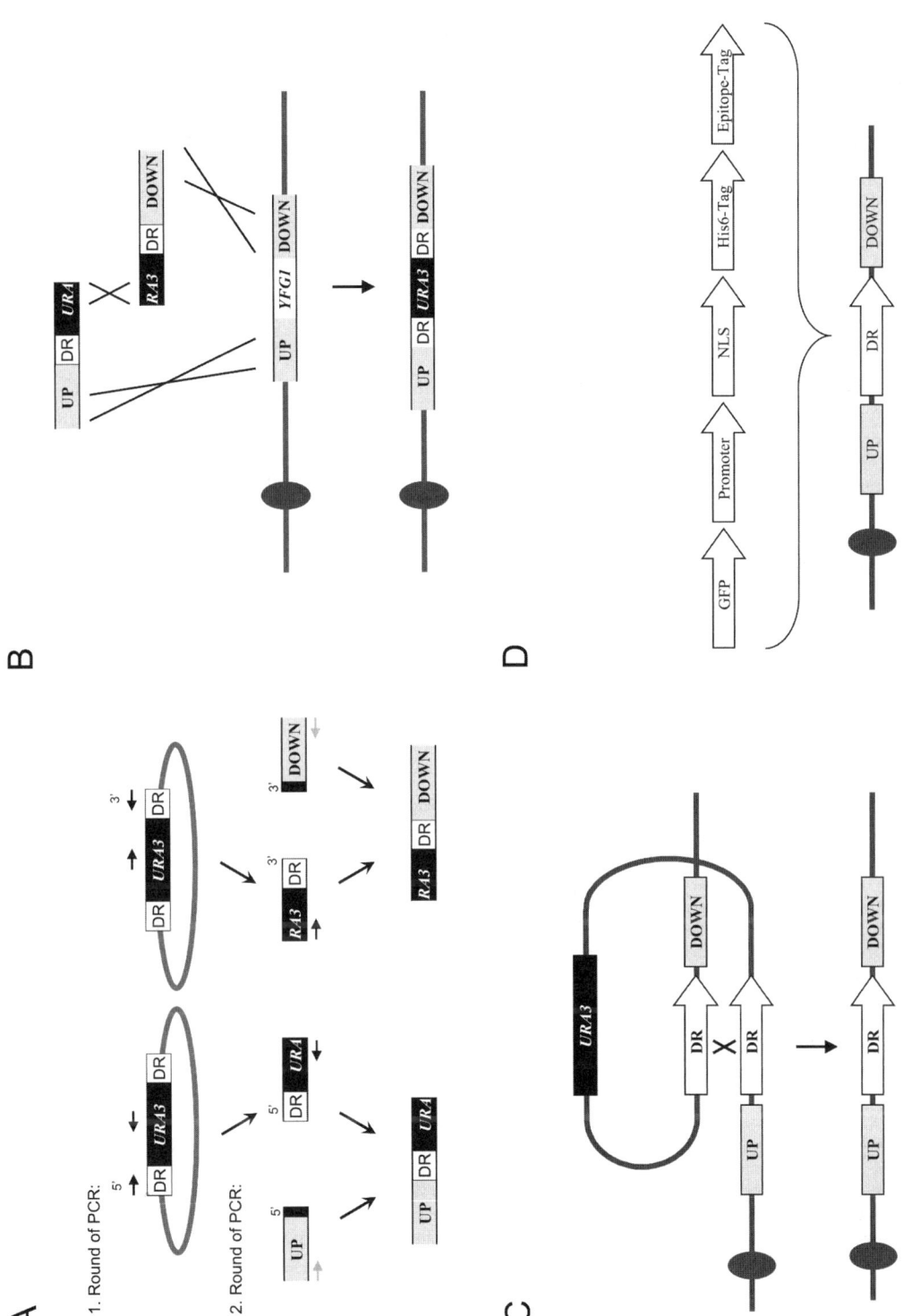

FIGURE 3 Substrates for iterative gene deletion. Direct repeats flanking the selectable marker allow the marker to be eliminated from the genome by recombination. If the selectable marker is *URA3*, recombinants can be selected using 5-FOA. (A) Construction of a bipartite gene deletion substrate with the marker flanked by a direct repeat (labeled DR). The template for the marker fragments is a plasmid that contains direct repeats flanking the marker. The remaining steps are similar to the legend for Fig. 2D. (B) Genomic integration of a bipartite gene-targeting substrate. Note that homologous recombination completes the selectable marker after cotransforming the two substrate fragments into the cell. (C) Direct repeat recombination eliminates the selectable marker from the genome. Note that one repeat remains in the genome. (D) By varying the sequence content of the DR repeats flanking the marker in the template plasmid, different useful sequences may be introduced between defined up and down sequences, as indicated; see text for details.

one 5-FOA-resistant mutant is generated per 5.4×10^8 cell divisions (63), and in some cases survivors may arise from mutations in another gene. For example, *ura5* mutants are also viable on 5-FOA medium (14). To eliminate the marker by recombination, it is necessary to build a gene-targeting substrate in which the selectable/counterselectable marker is flanked by direct repeats (3) (Fig. 3A, B, and C). Marker loss by direct repeat recombination is more frequent (one event per 10^4 to 10^5 cell divisions for repeats longer than 1 kb) (57) than by mutation. To reduce the size of the gene-targeting substrate, a repeat size of 143 bp is sufficient for efficient pop-out recombination (103). A continuous gene-targeting substrate based on a selectable marker flanked by a direct repeat is not easily assembled by PCR since PCR recombination between the repeats often eliminates the central part of the fragment (77). Accordingly, it is advisable to construct the substrate as a bipartite gene-targeting substrate, in which each repeat is contained within separate fragments (Fig. 3A). Selection is possible in such a bipartite gene-targeting substrate because the marker gene is reconstituted by homologous recombination after transformation. It is important to restreak the primary transformants on selective medium before counterselection, because the original transformation colony may contain untransformed cells that are kept alive by cross-feeding. These cells grow on counterselective medium and will provide undesired background.

20.3.4. Gene-Targeting Cassettes for Tagging Proteins

As described above, iterative gene targeting can be performed by flanking the selectable/counterselectable marker, e.g., *URA3*, in the gene-targeting substrate by direct repeats. After pop-out recombination, a single sequence unit of the repeat is left as a scar in the genome. This sequence scar can be highly useful as it allows a simple way to extend proteins with epitope, fluorescence, or purification tags etc. (Fig. 3D). For example, to extend a protein by GFP, the first step is to build two universal vectors that can be used for tagging proteins with GFP at any location in the sequence. In one vector, the DNA sequence encoding GFP is inserted upstream of *URA3*, and in the other, the GFP sequence is inserted downstream of *URA3* (70) (Fig. 4A). If the goal is to C-terminally tag a protein encoded by *YFG1*, a bipartite gene-targeting substrate for this purpose can easily be constructed. First, the two targeting sequences to direct the GFP sequence to the proper position are made by PCR using genomic DNA as template (Fig. 4B). The upstream targeting sequence is a fragment containing the sequence corresponding to the downstream end of *YFG1*, ending just before the stop codon. The downstream targeting sequence is a fragment that contains a stop codon followed by the 3′ nontranslated region of *YFG1*. Both fragments are produced by primers containing extensions that allow them to be fused to the proper *GFP-URA3* fragments generated from the universal GFP-tagging vector set (see the legend to Fig. 4B). One part of the bipartite gene-targeting substrate is made by fusing the *YFG1* upstream targeting fragment to a PCR fragment containing the *GFP* sequence followed by the 2/3 upstream section of *URA3* (the latter fragment has been amplified from one member of the GFP-tagging vector set). The other part of the bipartite substrate is made by fusing a PCR fragment containing the 2/3 downstream section of *URA3* followed by the *GFP* sequence (amplified from the other member of the GFP-tagging vector set) to the *YFG1* downstream fragment.

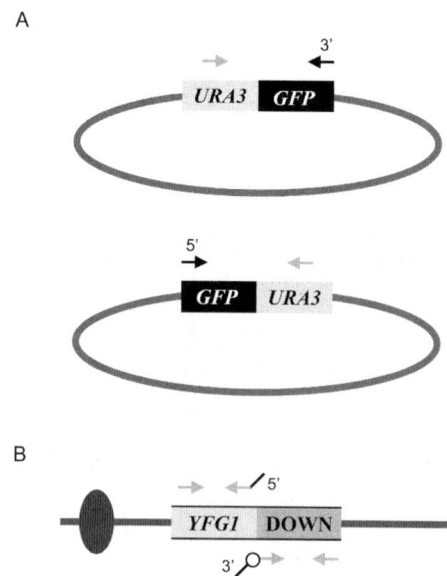

FIGURE 4 Incorporation of GFP DNA sequence into the genome. (A) Universal vector set for incorporating the GFP DNA sequence at any location in the genome using a bipartite gene-targeting substrate. Positions of the primers are indicated. (B) Strategy to extend the protein encoded by *YFG1* C-terminally with GFP. The position of the primer pairs used to amplify the two targeting sequences is indicated. Each fragment will terminate with a sequence tag matching either the 5′- or the 3′-end of GFP as indicated. Note that a sequence complementary to a stop codon, shown as an open circle, must be incorporated in the forward primer used to amplify the downstream targeting fragment. This codon is positioned right after the GFP segment of this primer as indicated.

Hence, in this fragment the stop codon follows right after the GFP sequence. After transformation to integrate the construct, followed by pop-out recombination, the *GFP* sequence will be fused in frame to the *YFG1* sequence just after its last codon.

20.3.5. Genomic Site-Directed Mutagenesis

A simple variation of the gene-targeting methods described above allows the introduction of point mutations or similar small sequence modifications into the genome. The starting point is a sequence fragment that includes the mutation. In two separate PCR reactions, this mutant fragment is fused to both ends of the selectable marker (Fig. 5). These reactions are similar to what is shown in Fig. 2D, except that the up and down sequences in this case are the same mutant gene fragment. As a guideline, the mutant fragment should be around 500 bp or larger. Note that a nonlethal mutation can be introduced in an essential gene if the mutant fragment contains the entire ORF so that it is not disrupted as the result of the gene-targeting experiment (28). When the substrate integrates at the correct genomic locus, the mutation will be integrated as a direct repeat flanking *URA3*. Subsequent pop-out recombination leaves only the point mutation in the genome (Fig. 5). It is important to note that the substrate may occasionally integrate such that only one of the repeats contains the desired mutation (28). In this case, pop-out recombination yields both wild-type and mutant recombinants and a subsequent screen is necessary to identify the desired strain.

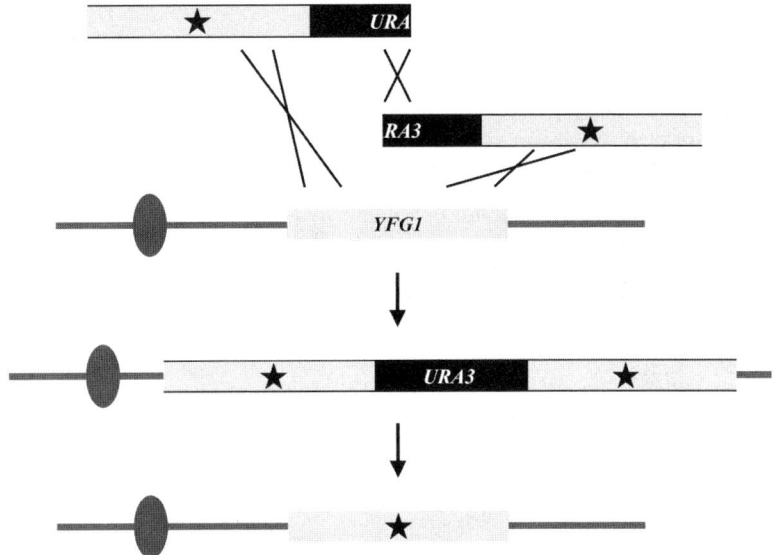

FIGURE 5 Substrates for introduction of point mutations. Identical direct repeats of *YFG1* sequences containing the relevant mutation flank a selectable marker (*URA3*). The substrate is shown as a bipartite gene-targeting substrate. Direct repeat recombination between mutant repeats leaves only the mutation in the genome.

To reduce the amount of screening, it is advisable to position the mutation in a central part of the targeting sequence, as it optimizes the chance of retaining the mutation in the genome.

20.4. FACTORS INFLUENCING TRANSLATION EFFICIENCY

The engineering tools discussed so far mainly aim at regulating gene expression levels through control of copy number and transcription. Although influencing translational processes is usually less often employed, there are important factors to consider.

Translation efficiency is to a certain extent influenced by the nucleotide sequence surrounding the AUG start codon. The consensus sequence for highly expressed yeast genes was determined as (A/U)A(A/C)A(A/C)A$\underline{AUG}$UC(U/C) (45). An adenosine at position −3 relative to the AUG is especially important for high-level translation (127).

Another important factor can be the codon composition of the gene to be expressed. Often, heterologous genes—especially those that contain a high percentage of codons rarely used in yeast—are only poorly expressed in *S. cerevisiae*. Rarely used codons are reflected by a low abundance of the respective tRNA molecules. Therefore, overexpression of selected tRNA genes can improve the expression of difficult genes such as those with a high GC content (83). Generally, the use of codon-optimized synthetic genes is highly recommended. Since the codon composition at the gene's 5′ end seems to be most important, it may be sufficient only to optimize the first 14 to 32 codons to achieve good expression (83). However, due to its ever-decreasing cost, the use of completely synthetic genes has become routine. Many companies offer the design and synthesis of genes based on a given protein sequence.

The algorithms developed for codon optimization follow two different strategies. In the first and most common strategy, the synthetic genes are optimized towards the most frequently used codons, which will result in a high codon adaptation index. The second strategy optimizes towards the overall yeast codon usage. Usually, other factors are included in the optimization process as well, such as the GC content and avoidance of motifs leading to mRNA instability, premature translation termination, or RNA secondary structure. However, current knowledge about the relationship between the gene sequence and its expression level is still relatively limited, leaving room for improvement during the coming years.

REFERENCES

1. **Akada, R.** 2002. Genetically modified industrial yeast ready for application. *J. Biosci. Bioeng.* **94**:536–544.
2. **Akada, R., Y. Shimizu, Y. Matsushita, M. Kawahata, H. Hoshida, and Y. Nishizawa.** 2002. Use of a *YAP1* overexpression cassette conferring specific resistance to cerulenin and cycloheximide as an efficient selectable marker in the yeast *Saccharomyces cerevisiae*. *Yeast* **19**:17–28.
3. **Alani, E., L. Cao, and N. Kleckner.** 1987. A method for gene disruption that allows repeated use of *URA3* selection in the construction of multiply disrupted yeast strains. *Genetics* **116**:541–545.
4. **Alper, H., C. Fischer, E. Nevoigt, and G. Stephanopoulos.** 2005. Tuning genetic control through promoter engineering. *Proc. Natl. Acad. Sci. USA* **102**:12678–12683.
5. **Amberg, D. C., D. Botstein, and E. M. Beasley.** 1995. Precise gene disruption in *Saccharomyces cerevisiae* by double fusion polymerase chain reaction. *Yeast* **11**:1275–1280.
6. **Armaleo, D., G.-N. Ye, T. M. Klein, K. B. Shark, J. C. Sanford, and S. A. Johnston.** 1990. Biolistic nuclear transformation of *Saccharomyces cerevisiae* and other fungi. *Curr. Genet.* **17**:97–103.
7. **Asadollahi, M. A., J. Maury, K. Møller, K. F. Nielsen, M. Schalk, A. Clark, and J. Nielsen.** 2007. Production of plant sesquiterpenes in *Saccharomyces cerevisiae*: effect of *ERG9* repression on sesquiterpene biosynthesis. *Biotechnol. Bioeng.* **99**:666–677.

8. **Ayub, M. A. Z., S. Astolfi-Filho, F. Mavituna, and S. G. Oliver.** 1992. Studies on plasmid stability, cell metabolism and superoxide dismutase production by Pgk⁻ strains of *Saccharomyces cerevisiae. Appl. Microbiol. Biotechnol.* **37:**615–620.

9. **Bae, J. Y., J. Laplaza, and T. W. Jeffries.** 2008. Effects of gene orientation and use of multiple promoters on the expression of *XYL1* and *XYL2* in *Saccharomyces cerevisiae. Appl. Biochem. Biotechnol.* **145:**69–78.

10. **Baldari, C., J. A. H. Murray, P. Ghiara, G. Cesareni, and C. L. Galeotti.** 1987. A novel leader peptide which allows efficient secretion of a fragment of human interleukin 1β in *Saccharomyces cerevisiae. EMBO J.* **6:**229–234.

11. **Baudin, A., O. Ozier-Kalogeropoulos, A. Denouel, F. Lacroute, and C. Cullin.** 1993. A simple and efficient method for direct gene deletion in *Saccharomyces cerevisiae. Nucleic Acids Res.* **21:**3329–3330.

12. **Bellí, G., E. Garí, L. Piedrafita, M. Aldea, and E. Herrero.** 1998. An activator/repressor dual system allows tight tetracycline-regulated gene expression in budding yeast. *Nucleic Acids Res.* **26:**942–947.

13. **Bendoni, B., D. Cavalieri, E. Casalone, M. Polsinelli, and C. Barberio.** 1999. Trifluoroleucine resistance as a dominant molecular marker in transformation of strains of *Saccharomyces cerevisiae* isolated from wine. *FEMS Microbiol. Lett.* **180:**229–233.

14. **Boeke, J. D., F. LaCroute, and G. R. Fink.** 1984. A positive selection for mutants lacking orotidine-5′-phosphate decarboxylase activity in yeast: 5-fluoro-orotic acid resistance. *Mol. Gen. Genet.* **197:**345–346.

15. **Böer, E., G. Steinborn, A. Matros, H.-P. Mock, G. Gellissen, and G. Kunze.** 2007. Production of interleukin-6 in *Arxula adeninivorans, Hansenula polymorpha* and *Saccharomyces cerevisiae* by applying the wide-range yeast vector (CoMed™) system to simultaneous comparative assessment. *FEMS Yeast Res.* **7:**1181–1187.

16. **Bonneaud, N., O. Ozier-Kalogeropoulos, G. Li, M. Labouesse, L. Minvielle-Sebastia, and F. Lacroute.** 1991. A family of low and high copy replicative, integrative and single-stranded *S. cerevisiae/E. coli* shuttle vectors. *Yeast* **7:**609–615.

17. **Brachmann, C. B., A. Davies, G. J. Cost, E. Caputo, J. Li, P. Hieter, and J. D. Boeke.** 1998. Designer deletion strains derived from *Saccharomyces cerevisiae* S288C: a useful set of strains and plasmids for PCR-mediated gene disruption and other applications. *Yeast* **14:**115–132.

18. **Brake, A. J., J. P. Merryweather, D. G. Coit, U. A. Heberlein, F. R. Masiarz, G. T. Mullenbach, M. S. Urdea, P. Valenzuela, and P. J. Barr.** 1984. α-Factor-directed synthesis and secretion of mature foreign proteins in *Saccharomyces cerevisiae. Proc. Natl. Acad. Sci. USA* **81:**4642–4646.

19. **Cardona, F., P. Carrasco, J. E. Pérez-Ortín, M. L. del Olmo, and A. Aranda.** 2007. A novel approach for the improvement of stress resistance in wine yeasts. *Int. J. Food Microbiol.* **114:**83–91.

20. **Cha, H. J., H. J. Chae, S. S. Choi, and Y. J. Yoo.** 2000. Production and secretion patterns of cloned glucoamylase in plasmid-harboring and chromosome-integrated recombinant yeasts employing an *SUC2* promoter. *Appl. Biochem. Biotechnol.* **87:**81–92.

21. **Chattoo, B. B., F. Sherman, D. A. Azubalis, T. A. Fjellstedt, D. Mehnert, and M. Ogur.** 1979. Selection of *lys2* mutants of the yeast *Saccharomyces cerevisiae* by the utilization of α-aminoadipate. *Genetics* **93:**51–65.

22. **Christianson, T. W., R. S. Sikorski, M. Dante, J. H. Shero, and P. Hieter.** 1992. Multifunctional yeast high-copy-shuttle vectors. *Gene* **100:**119–122.

23. **Compagno, C., A. Tura, B. M. Ranzi, L. Alberghina, and E. Martegani.** 1993. Copy number modulation in an autoselection system for stable plasmid maintenance in *Saccharomyces cerevisiae. Biotechnol. Prog.* **9:**594–599.

24. **Cost, G. J., and J. D. Boeke.** 1996. A useful colony colour phenotype associated with the yeast selectable/counterselectable marker *MET15. Yeast* **12:**939–941.

25. **Delorme, E.** 1989. Transformation of *Saccharomyces cerevisiae* by electroporation. *Appl. Environ. Microbiol.* **55:**2242–2246.

26. **del Pozo, L., D. Abarca, M. G. Claros, and A. Jiménez.** 1991. Cycloheximide resistance as a yeast cloning marker. *Curr. Genet.* **19:**353–358.

27. **Doignon, F., M. Aigle, and P. Ribereau-Gayon.** 1993. Resistance to imidazoles and triazoles in *Saccharomyces cerevisiae* as a new dominant marker. *Plasmid* **30:**224–233.

28. **Erdeniz, N., U. H. Mortensen, and R. Rothstein.** 1997. Cloning-free PCR-based allele replacement methods. *Genome Res.* **7:**1174–1183.

29. **Erhart, E., and C. P. Hollenberg.** 1983. The presence of a defective *LEU2* gene on 2μ DNA recombinant plasmids of *Saccharomyces cerevisiae* is responsible for curing and high copy number. *J. Bacteriol.* **156:**625–635.

30. **Featherstone, C., and S. P. Jackson.** 1999. DNA double-strand break repair. *Curr. Biol.* **9:**R759–R761.

31. **Fink, G., J. Trueheart, and E. A. Elion.** November, 1991. DNA fragment containing a pheromone-inducible yeast promoter useful for transforming yeast cells to produce foreign proteins, which may be toxic to yeast cells. U.S. patent 5063154.

32. **Fukuda, H., and Y. Kizaki.** 1999. A new transformation system of *Saccharomyces cerevisiae* with blasticidin S deaminase gene. *Biotechnol. Lett.* **21:**969–971.

33. **Fukuda, K., M. Watanabe, K. Asano, K. Ouchi, and S. Takasawa.** 1992. Molecular breeding of a sake yeast with a mutated *ARO4* gene which causes both resistance to o-fluoro-DL-phenylalanine and increased production of β-phenethyl alcohol. *J. Ferment. Bioeng.* **73:**366–369.

34. **Garí, E., L. Piedrafita, M. Aldea, and E. Herrero.** 1997. A set of vectors with a tetracycline-regulatable promoter system for modulated gene expression in *Saccharomyces cerevisiae. Yeast* **13:**837–848.

35. **Geymonat, M., A. Spanos, and S. G. Sedgwick.** 2007. A *Saccharomyces cerevisiae* autoselection system for optimised recombinant protein expression. *Gene* **399:**120–128.

36. **Gietz, R. D., and R. H. Schiestl.** 2007. High-efficiency yeast transformation using the LiAc/SS carrier DNA/PEG method. *Nat. Protoc.* **2:**31–34.

37. **Gietz, R. D., and R. H. Schiestl.** 2007. Quick and easy yeast transformation using the LiAc/SS carrier DNA/PEG method. *Nat. Protoc.* **2:**35–37.

38. **Gietz, R. D., and A. Sugino.** 1988. New yeast-*Escherichia coli* shuttle vectors constructed with in vitro mutagenized yeast genes lacking six-base pair restriction sites. *Gene* **74:**527–534.

39. **Goffeau, A., B. G. Barrell, H. Bussey, R. W. Davis, B. Dujon, H. Feldmann, F. Galibert, J. D. Hoheisel, C. Jacq, M. Johnston, E. J. Louis, H. W. Mewes, Y. Murakami, P. Philippsen, H. Tettelin, and S. G. Oliver.** 1996. Life with 6000 genes. *Science* **274:**546, 563–567.

40. **Goldstein, A. L., and J. H. McCusker.** 1999. Three new dominant drug resistance cassettes for gene disruption in *Saccharomyces cerevisiae. Yeast* **15:**1541–1553.

41. **Goldstein, A. L., X. Pan, and J. H. McCusker.** 1999. Heterologous URA3MX cassettes for gene replacement in *Saccharomyces cerevisiae. Yeast* **15:**507–511.

42. **Gueldener, U., J. Heinisch, G. J. Koehler, D. Voss, and J. H. Hegemann.** 2002. A second set of loxP marker cassettes for Cre-mediated multiple gene knockouts in budding yeast. *Nucleic Acids Res.* **30:**e23.

43. **Güldener, U., S. Heck, T. Fielder, J. Beinhauer, and J. H. Hegemann.** 1996. A new efficient gene disruption cassette for repeated use in budding yeast. *Nucleic Acids Res.* **24:**2519–2524.

44. Hadfield, C., A. M. Cashmore, and P. A. Meacock. 1986. An efficient chloramphenicol-resistance marker for *Saccharomyces cerevisiae* and *Escherichia coli*. *Gene* **45**:149–158.

45. Hamilton, R., C. K. Watanabe, and H. A. de Boer. 1987. Compilation and comparison of the sequence context around the AUG startcodons in *Saccharomyces cerevisiae* mRNAs. *Nucleic Acids Res.* **15**:3581–3593.

46. Hartzog, P. E., B. P. Nicholson, and J. H. McCusker. 2005. Cytosine deaminase MX cassettes as positive/negative selectable markers in *Saccharomyces cerevisiae*. *Yeast* **22**:789–798.

47. Hashida-Okado, T., A. Ogawa, I. Kato, and K. Takesako. 1998. Transformation system for prototrophic industrial yeasts using the *AUR1* gene as a dominant selection marker. *FEBS Lett.* **425**:117–122.

48. Hinnen, A., J. B. Hicks, and G. R. Fink. 1978. Transformation of yeast. *Proc. Natl. Acad. Sci. USA* **75**:1929–1933.

49. Hottiger, T., J. Kuhla, G. Pohlig, P. Fürst, A. Spielmann, M. Garn, S. Haemmerli, and J. Heim. 1995. 2-μm vectors containing the *Saccharomyces cerevisiae* metallothionein gene as a selectable marker: excellent stability in complex media, and high-level expression of a recombinant protein from a *CUP1*-promoter-controlled expression cassette in *cis*. *Yeast* **11**:1–14.

50. Ito, H., Y. Fukuda, K. Murata, and A. Kimura. 1983. Transformation of intact yeast cells treated with alkali cations. *J. Bacteriol.* **153**:163–168.

51. Ito-Harashima, S., and J. H. McCusker. 2004. Positive and negative selection LYS5MX gene replacement cassettes for use in *Saccharomyces cerevisiae*. *Yeast* **21**:53–61.

52. Jansen, G., C. Wu, B. Schade, D. Y. Thomas, and M. Whiteway. 2005. Drag&Drop cloning in yeast. *Gene* **344**:43–51.

53. Jeppsson, M., B. Johansson, P. R. Jensen, B. Hahn-Hägerdal, and M. F. Gorwa-Grauslund. 2003. The level of glucose-6-phosphate dehydrogenase activity strongly influences xylose fermentation and inhibitor sensitivity in recombinant *Saccharomyces cerevisiae* strains. *Yeast* **20**:1263–1272.

54. Kanai, T., H. Atomi, K. Umemura, H. Ueno, Y. Teranishi, M. Ueda, and A. Tanaka. 1996. A novel heterologous gene expression system in *Saccharomyces cerevisiae* using the isocitrate lyase gene promoter from *Candida tropicalis*. *Appl. Microbiol. Biotechnol.* **44**:759–765.

55. Kawasaki, G. H., and L. Bell. February, 1999. Production of proteins in yeast host cells by transforming cells with DNA which complements a deficiency present in the cells and DNA encoding the proteins. U.S. patent 5871957.

56. Klein, C. J. L., L. Olsson, and J. Nielsen. 1998. Glucose control in *Saccharomyces cerevisiae*: the role of *MIG1* in metabolic functions. *Microbiology* **144**:13–24.

57. Klein, H. L. 1995. Genetic control of intrachromosomal recombination. *Bioessays* **17**:147–159.

58. Kolariková, K., P. Galuszka, I. Sedlárová, M. Šebela, and I. Frébort. 2009. Functional expression of amine oxidase from *Aspergillus niger* (AO-I) in *Saccharomyces cerevisiae*. *Mol. Biol. Rep.* **36**:13–20.

59. Kunze, G., R. Bode, H. Rintala, and J. Hofemeister. 1989. Heterologous gene expression of the glyphosate resistance marker and its application in yeast transformation. *Curr. Genet.* **15**:91–98.

60. Labbé, S., and D. J. Thiele. 1999. Copper ion inducible and repressible promoter systems in yeast. *Methods Enzymol.* **306**:145–153.

61. Lacková, D., and J. Šubík. 1999. Use of mutated *PDR3* gene as a dominant selectable marker in transformation of prototrophic yeast strains. *Folia Microbiol.* **44**:171–176.

62. Lai, M.-T., D. Y.-T. Liu, and T.-H. Hseu. 2007. Cell growth restoration and high level protein expression by the promoter of hexose transporter, *HXT7*, from *Saccharomyces cerevisiae*. *Biotechnol. Lett.* **29**:1287–1292.

63. Lang, G. I., and A. W. Murray. 2008. Estimating the per-base-pair mutation rate in the yeast *Saccharomyces cerevisiae*. *Genetics* **178**:67–82.

64. Längle-Rouault, F., and E. Jacobs. 1995. A method for performing precise alterations in the yeast genome using a recyclable selectable marker. *Nucleic Acids Res.* **23**:3079–3081.

65. Lee, J.-W., D.-O. Kang, B.-Y. Kim, W.-K. Oh, T.-I. Mheen, Y.-R. Pyun, and J.-S. Ahn. 2000. Mutagenesis of the glucoamylase signal peptide of *Saccharomyces diastaticus* and functional analysis in *Saccharomyces cerevisiae*. *FEMS Microbiol. Lett.* **193**:7–11.

66. Lee, K. M., and N. A. DaSilva. 2005. Evaluation of the *Saccharomyces cerevisiae ADH2* promoter for protein synthesis. *Yeast* **22**:431–440.

67. Lee, S. Y., Y. C. Park, H. S. Cho, K. S. Ra, H. S. Baik, S.-Y. Paik, J. W. Yun, H. S. Park, and J. W. Choi. 2003. Expression of an artificial polypeptide with a repeated tripeptide glutamyl-tryptophanyl-lysine in *Saccharomyces cerevisiae*. *Lett. Appl. Microbiol.* **36**:121–128.

68. Lee, T.-H., M.-D. Kim, S.-Y. Shin, H.-K. Lim, and J.-H. Seo. 2006. Disruption of hexokinase II (*HXK2*) partly relieves glucose repression to enhance production of human kringle fragment in gratuitous recombinant *Saccharomyces cerevisiae*. *J. Biotechnol.* **126**:562–567.

69. Lind, K., and J. Norbeck. 2009. A QPCR-based reporter system to study post-transcriptional regulation via the 3′ untranslated region of mRNA in *Saccharomyces cerevisiae*. *Yeast* **26**:407–413.

70. Lisby, M., R. Rothstein, and U. H. Mortensen. 2001. Rad52 forms DNA repair and recombination centers during S phase. *Proc. Natl. Acad. Sci. USA* **98**:8276–8282.

71. Lopes, T. S., J. Klootwijk, A. E. Veenstra, P. C. van der Aar, H. van Heerikhuizen, H. A. Raúe, and R. J. Planta. 1989. High-copy-number integration into the ribosomal DNA of *Saccharomyces cerevisiae*: a new vector for high-level expression. *Gene* **79**:199–206.

72. Lorenz, M. C., R. S. Muir, E. Lim, J. McElver, S. C. Weber, and J. Heitman. 1995. Gene disruption with PCR products in *Saccharomyces cerevisiae*. *Gene* **158**:113–117.

73. Manivasakam, P., S. C. Weber, J. McElver, and R. H. Schiestl. 1995. Micro-homology mediated PCR targeting in *Saccharomyces cerevisiae*. *Nucleic Acids Res.* **23**:2799–2800.

74. Martens, C., B. Krett, and P. J. Laybourn. 2001. RNA polymerase II and TBP occupy the repressed *CYC1* promoter. *Mol. Microbiol.* **40**:1009–1019.

75. Maya, D., M. J. Quintero, M. D. Muñoz-Centeno, and S. Chávez. 2008. Systems for applied gene control in *Saccharomyces cerevisiae*. *Biotechnol. Lett.* **30**:979–987.

76. McGonigal, T., P. Bodelle, C. Schopp, and A. V. Sarthy. 1998. Construction of a sorbitol-based vector for expression of heterologous proteins in *Saccharomyces cerevisiae*. *Appl. Environ. Microbiol.* **64**:793–794.

77. Meyerhans, A., J. P. Vartanian, and S. Wain-Hobson. 1990. DNA recombination during PCR. *Nucleic Acids Res.* **18**:1687–1691.

78. Miyajima, A., I. Miyajima, K.-I. Arai, and N. Arai. 1984. Expression of plasmid R388-encoded type II dihydrofolate reductase as a dominant selective marker in *Saccharomyces cerevisiae*. *Mol Cell. Biol.* **4**:407–414.

79. Monfort, A., S. Finger, P. Sanz, and J. A. Prieto. 1999. Evaluation of different promoters for the efficient production of heterologous proteins in baker's yeast. *Biotechnol. Lett.* **21**:225–229.

80. Mumberg, D., R. Müller, and M. Funk. 1994. Regulatable promoters of *Saccharomyces cerevisiae*: comparison of transcriptional activity and their use for heterologous expression. *Nucleic Acids Res.* **22**:5767–5768.

81. Mumberg, D., R. Müller, and M. Funk. 1995. Yeast vectors for the controlled expression of heterologous proteins in different genetic backgrounds. *Gene* **156**:119–122.

82. **Murata, K., Y. Fukuda, M. Shimosaka, K. Watanabe, T. Saikusa, and A. Kimura.** 1985. Phenotypic character of the methylglyoxal resistance gene in *Saccharomyces cerevisiae*: expression in *Escherichia coli* and application to breeding wild-type yeast strains. *Appl. Environ. Microbiol.* **50:**1200–1207.

83. **Mutka, S. C., S. M. Bondi, J. R. Carney, N. A. Da Silva, and J. T. Kealey.** 2006. Metabolic pathway engineering for complex polyketide biosynthesis in *Saccharomyces cerevisiae*. *FEMS Yeast Res.* **6:**40–47.

84. **Nacken, V., T. Achstetter, and E. Degryse.** 1996. Probing the limits of expression levels by varying promoter strength and plasmid copy number in *Saccharomyces cerevisiae*. *Gene* **175:**253–260.

85. **Napp, S. J., and N. A. Da Silva.** 1993. Enhancement of cloned gene product synthesis via autoselection in recombinant *Saccharomyces cerevisiae*. *Biotechnol. Bioeng.* **41:**801–810.

86. **Napp, S. J., and N. A. Da Silva.** 1994. Enhanced productivity through gratuitous induction in recombinant yeast fermentations. *Biotechnol. Prog.* **10:**125–128.

87. **Nevoigt, E.** 2008. Progress in metabolic engineering of *Saccharomyces cerevisiae*. *Microbiol. Mol. Biol. Rev.* **72:**379–412.

88. **Nevoigt, E., C. Fischer, O. Mucha, F. Matthäus, U. Stahl, and G. Stephanopoulos.** 2006. Engineering promoter regulation. *Biotechnol. Bioeng.* **96:**550–558.

89. **Nielsen, J., and M. C. Jewett.** 2008. Impact of systems biology on metabolic engineering of *Saccharomyces cerevisiae*. *FEMS Yeast Res.* **8:**122–131.

90. **Nieto, A., J. A. Prieto, and P. Sanz.** 1999. Stable high-copy-number integration of *Aspergillus oryzae* α-amylase cDNA in an industrial baker's yeast strain. *Biotechnol. Prog.* **15:**459–466.

91. **Nishizawa, M., F. Ozawa, and F. Hishinuma.** 1989. Construction of yeast secretion vectors designed for production of mature proteins using the signal sequence of yeast invertase. *Appl. Microbiol. Biotechnol.* **32:**317–322.

92. **Novak Frazer, L., and R. T. O'Keefe.** 2007. A new series of yeast shuttle vectors for the recovery and identification of multiple plasmids from *Saccharomyces cerevisiae*. *Yeast* **24:**777–789.

93. **Olesen, K., P. F. Johannesen, L. Hoffmann, S. B. Sørensen, C. Gjermansen, and J. Hansen.** 2000. The pYC plasmids, a series of cassette-based yeast plasmid vectors providing means of counter-selection. *Yeast* **16:**1035–1043.

94. **Orr-Weaver, T. L., J. W. Szostak, and R. J. Rothstein.** 1981. Yeast transformation: a model system for the study of recombination. *Proc. Natl. Acad. Sci. USA* **78:**6354–6358.

95. **Ostergaard, S., L. Olsson, and J. Nielsen.** 2000. Metabolic engineering of *Saccharomyces cerevisiae*. *Microbiol. Mol. Biol. Rev.* **64:**34–50.

96. **Parekh, R. N., M. R. Shaw, and K. D. Wittrup.** 1996. An integrating vector for tunable, high copy, stable integration into the dispersed Ty delta sites of *Saccharomyces cerevisiae*. *Biotechnol. Prog.* **12:**16–21.

97. **Park, H., N. I. Lopez, and A. T. Bakalinsky.** 1999. Use of sulfite resistance in *Saccharomyces cerevisiae* as a dominant selectable marker. *Curr. Genet.* **36:**339–344.

98. **Pirkov, I., E. Albers, J. Norbeck, and C. Larsson.** 2008. Ethylene production by metabolic engineering of the yeast *Saccharomyces cerevisiae*. *Metab. Eng.* **10:**276–280.

99. **Pronk, J. T.** 2002. Auxotrophic yeast strains in fundamental and applied research. *Appl. Environ. Microbiol.* **68:**2095–2100.

100. **Raymond, M., S. Ruetz, D. Y. Thomas, and P. Gros.** 1994. Functional expression of P-glycoprotein in *Saccharomyces cerevisiae* confers cellular resistance to the immunosuppressive and antifungal agent FK520. *Mol. Cell. Biol.* **14:**277–286.

101. **Rech, S. B., L. I. Stateva, and S. G. Oliver.** 1992. Complemetation of the *Saccharomyces cerevisiae srb1-1* mutation: an autoselection system for stable plasmid maintenance. *Curr. Genet.* **21:**339–344.

102. **Regenberg, B., and J. Hansen.** 2000. GAP1, a novel selection and counter-selection marker for multiple gene disruptions in *Saccharomyces cerevisiae*. *Yeast* **16:**1111–1119.

103. **Reid, R. J., M. Lisby, and R. Rothstein.** 2002. Cloning-free genome alterations in *Saccharomyces cerevisiae* using adaptamer-mediated PCR. *Methods Enzymol.* **350:**258–277.

104. **Rothstein, R. J.** 1983. One-step gene disruption in yeast. *Methods Enzymol.* **101:**202–211.

105. **Ruohonen, L., M. K. Aalto, and S. Keränen.** 1995. Modifications to the *ADH1* promoter of *Saccharomyces cerevisiae* for efficient production of heterologous proteins. *J. Biotechnol.* **39:**193–203.

106. **Sadowski, I., T.-C. Su, and J. Parent.** 2007. Disintegrator vectors for single-copy yeast chromosomal integration. *Yeast* **24:**447–455.

107. **Sauer, B.** 1987. Functional expression of the cre-lox site-specific recombination system in the yeast *Saccharomyces cerevisiae*. *Mol. Cell. Biol.* **7:**2087–2096.

108. **Sauer, N., and J. Stolz.** 1994. SUC1 and SUC2: two sucrose transporters from *Arabidopsis thaliana*; expression and characterization in baker's yeast and identification of the histidine-tagged protein. *Plant J.* **6:**67–77.

109. **Shimoi, H., M. Okuda, and K. Ito.** 2000. Molecular cloning and application of a gene complementing pantothenic acid auxotrophy of sake yeast Kyokai no. 7. *J. Biosci. Bioeng.* **90:**643–647.

110. **Sikorski, R. S., and P. Hieter.** 1989. A system of shuttle vectors and yeast host strains designed for efficient manipulation of DNA in *Saccharomyces cerevisiae*. *Genetics* **122:**19–27.

111. **Slibinskas, R., D. Samuel, A. Gedvilaite, J. Staniulis, and K. Sasnauskas.** 2004. Synthesis of the measles virus nucleoprotein in yeast *Pichia pastoris* and *Saccharomyces cerevisiae*. *J. Biotechnol.* **107:**115–124.

112. **Solow, S. P., J. Sengbusch, and M. W. Laird.** 2005. Heterologous protein production from the inducible *MET25* promoter in *Saccharomyces cerevisiae*. *Biotechnol. Prog.* **21:**617–620.

113. **Symington, L. S.** 2002. Role of RAD52 epistasis group genes in homologous recombination and double-strand break repair. *Microbiol. Mol. Biol. Rev.* **66:**630–670.

114. **Thim, L., M. T. Hansen, K. Norris, I. Hoegh, E. Boel, J. Forstrom, G. Ammerer, and N. P. Fiil.** 1986. Secretion and processing of insulin precursors in yeast. *Proc. Natl. Acad. Sci. USA* **83:**6766–6770.

115. **Toyn, J. H., P. L. Gunyuzlu, W. H. White, L. A. Thompson, and G. F. Hollis.** 2000. A counterselection for the tryptophan pathway in yeast: 5-fluoroanthranilic acid resistance. *Yeast* **16:**553–560.

116. **Ugolini, S., V. Tosato, and C. V. Bruschi.** 2002. Selective fitness of four episomal shuttle-vectors carrying *HIS3, LEU2, TRP1,* and *URA3* selectable markers in *Saccharomyces cerevisiae*. *Plasmid* **47:**94–107.

117. **Unternährer, S., D. Pridmore, and A. Hinnen.** 1991. A new system for amplifying 2 μm plasmid copy number in *Saccharomyces cerevisiae*. *Mol. Microbiol.* **5:**1539–1548.

118. **Van Arsdell, S., R. S. Daves, and R. R. Yocum.** July, 1996. Hybrid yeast promoter containing *ENO2* promoter and second upstream activator provides high level expression of heterologous genes in yeast, especially glucoamylase for production of low-calorie beer. U.S. patent 5541084.

119. **van den Berg, M. A., and H. Y. Steensma.** 1997. Expression cassettes for formaldehyde and fluoroacetate resistance, two dominant markers in *Saccharomyces cerevisiae*. *Yeast* **13:**551–559.

120. Van Mullem, V., M. Wery, X. De Bolle, and J. Vandenhaute. 2003. Construction of a set of *Saccharomyces cerevisiae* vectors designed for recombinational cloning. *Yeast* **20:**739–746.

121. Vellanki, R. N., N. Komaravelli, R. Tatineni, and L. N. Mangamoori. 2007. Expression of hepatitis B surface antigen in *Saccharomyces cerevisiae* utilizing glyceraldehyde-3-phosphate dehydrogenase promoter of *Pichia pastoris*. *Biotechnol. Lett.* **29:**313–318.

122. Vorachek-Warren, M. K., and J. H. McCusker. 2004. DsdA (D-serine deaminase): a new heterologous MX cassette for gene disruption and selection in *Saccharomyces cerevisiae*. *Yeast* **21:**163–171.

123. Wach, A., A. Brachat, C. Alberti-Segui, C. Rebischung, and P. Philippsen. 1997. Heterologous *HIS3* marker and GFP reporter modules for PCR-targeting in *Saccharomyces cerevisiae*. *Yeast* **13:**1065–1075.

124. Wach, A., A. Brachat, R. Pöhlmann, and P. Philippsen. 1994. New heterologous modules for classical or PCR-based gene disruptions in *Saccharomyces cerevisiae*. *Yeast* **10:**1793–1808.

125. Wang, X., Z. Wang, and N. A. Da Silva. 1996. G418 selection and stability of cloned genes integrated at chromosomal δ sequences of *Saccharomyces cerevisiae*. *Biotechnol. Bioeng.* **49:**45–51.

126. Xie, Q., and A. Jiménez. 1996. Molecular cloning of a novel allele of *SMR1* which determines sulfometuron methyl resistance in *Saccharomyces cerevisiae*. *FEMS Microbiol. Lett.* **137:**165–168.

127. Yun, D.-F., T. M. Laz, J. M. Clements, and F. Sherman. 1996. mRNA sequences influencing translation and the selection of AUG initiator codons in the yeast *Saccharomyces cerevisiae*. *Mol. Microbiol.* **19:** 1225–1239.

128. Zhang, Z., M. Moo-Young, and Y. Chisti. 1996. Plasmid stability in recombinant *Saccharomyces cerevisiae*. *Biotechnol. Adv.* **14:**401–435.

Protein Expression in Nonconventional Yeasts

THOMAS W. JEFFRIES AND JAMES M. CREGG

21

21.1. INTRODUCTION

Heterologous expression of biopharmaceuticals and enzymes is common, but increasingly, metabolic engineering of biosynthetic or biodegradative pathways requires the balanced expression of multiple proteins under complex regulatory conditions. The transformation and manipulation of well-studied hosts such as *Escherichia coli* or *Saccharomyces cerevisiae* draw on many genetic tools and strains and a plethora of research into regulatory responses and protein functions. These data on familiar workhorses, while essential, represent only a small fraction of the biochemistry and physiology present in the microbial realm. Moreover, with the advent of whole-genome sequencing, it is becoming possible to obtain particular advantages by using different microbial systems for specialized applications.

21.1.1. Yeasts as Expression Hosts

As unicellular organisms, yeasts are relatively easy to isolate, scale up, and manipulate genetically. They can achieve high growth rates. Their thick cell walls make them more resistant to osmotic shock. Because they are larger than bacteria, they are easier to harvest by filtration or centrifugation. They will grow at an acidic pH, which discourages the growth of most contaminants. They are not susceptible to most viral infections, and they lack lipopolysaccharide endotoxins. They generally have minimal nutritional requirements and can be cultivated at very high cell density. Cell and protein yields are often high. Specifically with respect to heterologous proteins for therapeutic use, yeasts have an advantage over bacteria in their capacity to produce N-glycosylated proteins with disulfide linkages (42).

Some of the physiological features of yeasts cause difficulties. Their thick cell walls can make release of intracellular proteins more difficult, and their natural glycosylation systems differ from those of mammals. However, even these problems have been addressed with yeasts that secrete heterologous proteins and by the genetic development of strains having humanized glycosylation patterns.

Within the dikaryotic Ascomycetes to which the budding yeasts belong, at least 20 species of yeasts have been examined for their industrial properties. Several have been used for heterologous protein expression or metabolically engineered for novel properties. *S. cerevisiae* is the best studied of the ascomycetous budding yeasts. Genetically it is the best understood eukaryote, which makes it very easy to manipulate. It is not as useful for heterologous protein production because it lacks respiration complex 1 and hence exhibits relatively low ATP and cell yields. It has relatively limited capacity to metabolize carbon sources other than glucose and sucrose. It tends to hyperglycosylate proteins, and it does not possess a strong secretion pathway, so proteins larger than about 30 kDa occur as intracellular or periplasmic products. Researchers and industrial processes have drawn on the features of nonconventional yeasts to overcome these problems.

21.1.2. Nonconventional Yeasts

Nonconventional yeasts can be roughly defined as any yeast other than *S. cerevisiae*. A number are used in industrial settings. If one says "yeast" without further clarification, *S. cerevisiae* is assumed. The realm of "yeasts," however, is far more diverse. Closely related to *S. cerevisiae* are *Kluyveromyces marxianus* and *Kluyveromyces lactis*. Their expression vectors, selectable markers, and transformation systems are similar to those of *S. cerevisiae*. Nonconventional yeasts such as *Pichia pastoris*, *Hansenula polymorpha*, *Arxula adeninivorans*, and *Yarrowia lipolytica* cannot readily employ 2μm plasmid-based vectors and generally use selectable markers and autonomous replication sequences derived from autologous sources. These yeasts have found wide application in heterologous expression because they either produce very high levels of heterologous proteins, secrete proteins, or exhibit more versatile substrate utilization. The popularity of methylotrophic yeasts and particularly *P. pastoris* and *H. polymorpha* for heterologous protein production stems from the establishment of their Generally Recognized As Safe status, their capacity for cultivation at very high cell densities, and the existence of effective systems for transformation and expression.

Nonconventional yeasts are notable for their use of diverse carbon sources, physiological properties, and unusual biochemical pathways. Utilization of lactose and galactose is particularly important in the fermentation of whey (64), and xylose and cellobiose fermentations are important in the bioconversion of lignocellulosics (86, 141a). However, utilization of alkanes (117, 121) and particularly methanol (35) has proven significant because protein products can accumulate to high levels in these yeasts. Thermotolerance is also a valued trait that can suppress contamination or increase growth rates. *A. adeninivorans* will grow at temperatures up to 48°C (174), and *K. marxianus*, which will ferment cellobiose, will grow at temperatures up to 45°C, making it potentially useful for simultaneous saccharification and fermentation. *Debaryomyces hansenii* shows unusual osmotolerance (129). Production of extracellular enzymes such as lipase (14, 135, 153), amylase (23, 40, 100, 154), and endoglucanases (123, 124) is facilitated in yeasts with effective secretion systems. Finally, nonconventional yeasts are particularly useful in the production of glycerol (27), lactic acid (131), citric acid (73, 99), or carotenoids such as astaxanthin (111, 113, 164, 185). The capacities of nonconventional yeasts to produce useful products other than ethanol or to produce ethanol from substrates other than glucose, sucrose, and fructose have led to efforts that focus on their metabolic engineering for improved product formation.

Several researchers have previously reviewed the general transformation and expression techniques for non-*Saccharomyces* yeasts (15, 20, 37, 60, 61, 108, 127, 171), so our emphasis here will be on more generalized features of expression systems in nonconventional yeast systems.

21.2. ELEMENTS OF TRANSFORMATION VECTORS

All yeast transformation and expression systems require ways to introduce engineered DNA into a cell, enable it to survive, assist its propagation, and promote its transcription. The most essential requirements are a selectable marker and a functional promoter. With these two elements alone, it is possible to introduce modified genes and effect transformations of the DNA. The efficiencies of such simple systems, however, are very low. Linearized DNA is susceptible to degradation by endo- and exonucleases, and the DNA cannot be transcribed until it has become integrated into the genome, which can occur only after the DNA enters the nucleus. To increase the probability of survival, replication, and expression, autonomous replication sequences (ARSs) are often incorporated into the transformation vector. In the case of *S. cerevisiae* and related genera, these are derived from the origin of replication (Ori) sequences of native 2μm plasmids that are found in *Saccharomyces* and *Kluyveromyces* species. While the 2μm origin of replication will function in many yeasts, it is often inefficient or poorly stable in yeasts outside of the *Saccharomyces-Kluyveromyces* clade. Occasionally an ARS identified through functional studies in *S. cerevisiae* can be used in another yeast (137); however, the efficiency of transformation is generally low. It is often necessary to isolate an ARS from the species of interest.

21.3. SELECTABLE MARKERS

A selectable marker enables the isolation of cells that have taken up the exogenous DNA. They can be based on complementation of genetic lesions that create nutritional deficiencies (auxotrophies) or on genes that enable acquired resistance to otherwise toxic medium constituents. The latter are termed dominant selectable markers. These are preferred since they can be employed with commercial yeast strains that are diploid or lacking auxotrophic deficiencies. Drug resistance markers are often used, but they should not be left in the final construct for commercial strains. This is especially true of yeasts that are used in producing food-grade products.

21.3.1. Auxotrophic Markers

By far the most common auxotrophic selectable marker employed is *URA3* (13, 39, 52, 91, 120, 139, 184). Its popularity is attributable to the relative ease of obtaining *ura3* mutations by selection for resistance to the orotic acid homolog 5'-fluoroorotic acid (5'FOA). Also, if *URA3* carried on an episomal vector is used to complement a *ura3* mutation, it is very easy to cure the host by cultivation on medium containing 5'FOA (13). Cells with a complete pathway for uracil biosynthesis take up the 5'FOA antimetabolite and incorporate it into their RNA. The 5'-fluoro group causes mispairing, and cells prototrophic for uracil biosynthesis are killed. 5'FOA selection plates also contain uracil or uridine at a concentration of 50 to 100 μg/ml. If a cell is deficient in either orotidine-5'-phosphate decarboxylase (*URA3*, *pyrG*) or orotate phosphoribosyltransferase (*URA5*) (39, 120), it will not take up the 5'FOA, and the mutant survives by using the uracil or uridine provided in the medium. The *ura3* mutant must be homozygotic for the trait in order to survive in the presence of 5'FOA, so it is sometimes necessary to sporulate the culture prior to selection. Also, residual orotidine-5'-phosphate decarboxylase in the cell cytoplasm can be sufficient to incorporate 5'FOA, so it is useful to cultivate sporulated or mutated cells on medium containing uracil for a few generations prior to selection on 5'FOA plates.

Other popular auxotrophic markers include the multifunctional *HIS4* gene (109, 119), the *LEU2* gene for 3-isopropylmalate dehydrogenase (16, 41, 71a, 112, 142, 175), the *TRP1* gene for phosphoribosylanthranilate isomerase (28, 154, 155), and the *ADE1* gene for *N*-succinyl-5-aminoimidazole-4-carboxamide ribotide (SAICAR) synthetase (24, 71a, 109). Recipient hosts can be obtained either through random mutagenesis and screening, which is not recommended for industrial strains since it can result in other unwanted changes, or by targeted deletion using transposable elements, *URA3*, or some other selectable marker. Auxotrophs in general and *Leu2* and *Ade1* auxotrophs in particular often grow poorly—even when their deficiencies are compensated with nutritional supplements—and are not highly suitable for industrial applications. Backcrossing the mutated cells against a parental strain and subsequent screening and selection for the desired auxotrophic marker are recommended.

21.3.2. Drug Resistance Markers

Drug resistance genes are widely used as dominant selectable markers to transform yeasts, other fungi, and bacteria. Their extreme utility means that this practice will continue, but possible proliferation of drug resistance into human, plant, and animal pathogens requires good practices in biosafety containment. Where possible—and certainly when modified organisms are to be released into nature—dominant resistance markers that do not use or that compromise clinically important drugs should be employed. Alternatively, once the genetic modification is completed, the marker should be excised (see below).

To be effective for transformation, the target yeast must be susceptible to the antibiotic, and the protein must confer resistance even when expressed at a low level. Both of these requirements can present problems when developing transformation systems with nonconventional yeasts. Non-*Saccharomyces* yeasts often exhibit high levels of drug resistance. This means that high concentrations need to be used in the selection plate in order to kill the cells, and the specialized antibiotics used for yeasts are often expensive.

Second, many of the drug resistance markers are derived from bacterial proteins, and expression of functional proteins occurs at only a low level in many yeasts. A significant fraction of yeasts belonging to the genus *Candida*, along with *Debaryomyces hansenii*, *Clavispora lusitaniae*, *Lodderomyces elongisporis*, *Candida rugosa*, and others, use CUG (CTG in DNA) to code for serine rather than leucine (51, 158). While the CUG codon is infrequently used by these species, bacteria use it regularly. The substitution of serine for a leucine can disrupt protein structure, so the CUG yeasts do not express the bacterial drug resistance markers properly. To get around this problem, correct expression of bacterial drug resistance markers in CUG yeasts requires changing CUGs to some other codon for leucine (101, 102, 161) or synthesizing the gene with optimum codon usage (see below).

When employing a drug resistance marker for transformation, it is necessary to add a short recovery incubation period (ca. 4 h) in complete medium without the antibiotic to allow expression of the drug resistance protein before exposing the cells to the selective agent.

The two most widely used resistance markers are genes that code for aminoglycoside-3-phosphotransferase, which confers resistance to Geneticin (G418) or kanamycin. The *kan*^r marker is widely available (53, 67, 145, 171). To inhibit yeast growth, Geneticin is typically used at a concentration of about 50 μg/ml in the final medium. It is soluble in water at 50 mg/ml and stable at room temperature. A 1-mg/ml solution in 0.1 M potassium phosphate buffer (pH 8.0) should be sterilized by filtration, stored at 2 to 8°C, and used within 30 days. Kanamycin is likewise used at about 50 μg/ml. It is much less stable than Geneticin and even less stable than penicillin at 25 to 37°C (50).

The gene for hygromycin phosphotransferase, *hph* (66, 187), confers resistance to hygromycin B. It is widely used as a selectable marker for yeasts and filamentous fungi (3, 7, 21, 68, 137, 152, 165, 186). At a concentration of 0.38 mM, hygromycin completely halts growth of eukaryotic cells even in rich medium by blocking chain elongation in ribosomes (65). Typical concentrations used to inhibit yeasts and fungi are about 0.2 to 1.0 mg/ml. Hygromycin B is stable in solution for about 2 years at 4°C and about 1 month at 37°C (77).

Some yeasts and fungi are notably resistant to gentamicin and hygromycin, so other drugs must be used. In recent years, Zeocin has proven to be very useful as an inhibitory antibiotic for yeasts and other organisms (1, 4, 102). Zeocin effectively inhibits growth at a concentration of 25 to 100 μg/ml. It is stable in a dry powder form for up to a year at −20°C, but in solution, it is light sensitive. Plates containing Zeocin can be stored for several weeks at 4°C in the dark. The mechanism of action of Zeocin is not known, but it is structurally similar to bleomycin and phleomycin, which complex with Fe²⁺ to cleave DNA (18). The resistance factor *ble* codes for a protein that binds bleomycin, thereby eliminating its DNA cleavage activity (46). *ble* genes have been described from a number of bacterial sources (85). In addition, the bacterial gene for chloramphenicol acetyltransferase (*cat*) can impart resistance to phleomycin when it is expressed properly in yeasts (31). It also has been widely used to transform various yeasts (85).

The gene for nourseothricin acetyltransferase (*nat1*) from *Streptomyces noursei* confers resistance to the aminoglycoside antibiotic nourseothricin (63, 144, 167), which is effective at concentrations of 25 to 100 μg/ml and stable for up to 2 years at 20°C. In solution it shows some light sensitivity.

Most eukaryotic cells are sensitive to inhibition by the glutarimide antibiotic cycloheximide, but resistance to it is widespread among yeasts (156). Cycloheximide binds to the 80S ribosomal subunit, thereby inhibiting translation. The resistance mechanism is attributable to a specific amino acid difference in ribosomal protein L41. Yeasts that are susceptible to cycloheximide can be transformed with a mutated form of L41 to confer resistance (92).

21.3.3. Dominant Non-Drug Resistance Markers

With the spread of antibiotic resistance among clinical isolates, concern is rising over the use of drug resistance markers to transform medically important human, animal, or plant pathogens. One useful selectable marker that does not use drug resistance is the *IMH3* gene for inosine monophosphate (IMP) dehydrogenase, which confers resistance to mycophenolic acid (MPA). IMP dehydrogenase catalyzes an NAD-dependent rate-limiting step in the biosynthesis of guanine nucleotides. Growth of *Candida albicans* is inhibited by 1 μg/ml of MPA, which is a specific inhibitor of the *IMH1* form of IMP dehydrogenase. However, transformants overexpressing a different form of IMP dehydrogenase, the *C. albicans IMH3* gene, will resist up to 40 μg/ml of MPA (11, 88, 95). The bioavailable form of MPA, mycophenolate mofetil, has a number of clinical applications, mainly as an immune suppressant for transplantation, but it is not widely used as an antimicrobial. The *CaIMH3* gene has been used successfully as a dominant selectable marker in *Candida parapsilosis* and *C. albicans* (11, 44, 55).

Sulfite resistance has long been studied in yeasts since it is important in wine making. Two genes for sulfite resistance from *S. cerevisiae*, the transcription activator *FZF1-4* (*RSU1-4*) and to a lesser extent its regulated target, *SSU1*, have been used as dominant selectable markers in sulfite-susceptible strains (125). The latter is a plasma membrane sulfite pump, which enables cells to survive in the presence of sulfite.

21.4. EXCISION ELEMENTS

Most nonconventional yeasts have only a few selectable markers for genetic transformation. It is therefore very useful to be able to recover them for reuse. *URA3* is almost unique in exhibiting a bidirectional selection. It is possible to identify *URA3* transformants by selecting for growth on uracil⁻ (Ura⁻) medium, and it is possible to select for *URA3* by plating on 5′FOA. Normally the reverse selection is employed when the *URA3* gene is carried on an autonomous plasmid. However, it is also possible to design *URA3* and other selectable markers so that they will excise readily. Direct flanking repeat sequences on either side of the selectable marker will occasionally (ca. 10⁻⁵) recombine to excise the intervening sequence (169). For example, short (40-bp) flanking direct repeat sequences from λ phage promoted the spontaneous excision of *URA3* after it was used to disrupt *LEU2* (112). The excised *URA3* mutants were recovered on 5′FOA plates. In the case of selectable markers that do not have such sensitive means for identifying spontaneous excision reversions, two recombinase proteins can greatly enhance the frequency.

The Cre recombinase from bacteriophage P1 has been shown to excise intervening DNA that is flanked by direct repeats of its cognate *loxP* sequence. The efficiency of this excision approaches 70%, making identification of the resulting marker-deficient strains relatively easy (67). In a similar manner, the Flp recombinase, encoded by the 2μm plasmid of *S. cerevisiae*, specifically recognizes 13-bp repeat elements found in the 2μm circle. When these are present as flanking direct repeats, the intervening sequence is excised (6). When these or the *loxP* sequences are in opposite orientations, the intervening sequence is inverted.

21.5. PROMOTERS

In the past, promoter selection for heterologous expression has been based on experience with a relatively limited set of genes. In recent years, however, as we have gained powerful tools for large-scale quantification of promoter expression profiles, selection has become tuned to specific applications. When faced with developing promoters for a new yeast system, little is likely to be known about promoter properties since the strength and expression profiles can vary widely from one organism to another. Genome-wide expression arrays provide the most comprehensive way to identify useful promoters. The induction patterns for *S. cerevisiae* genes can be obtained from the *Saccharomyces* genome database (http://www.yeastgenome.org/).

ADH1 has been employed for strong heterologous expression in *S. cerevisiae*, and the *ADH2* promoter has been used for glucose-repressible expression. However, few other yeasts exhibit such high levels of fermentation. The galactokinase (*GAL1*) or the bifunctional UDP-glucose-4-epimerase (Galactowaldenase), aldose-1-epimerase (Mutarotase) (*GAL10*) promoter can be used for regulated (glucose-repressed/galactose-induced) expression. Both of these are widely distributed among yeasts. The *TDH3* (GAP) promoter is often used for strong constitutive expression (25, 34, 71, 74, 81, 104, 146). Triose phosphate dehydrogenase has three isoforms in *S. cerevisiae* and between one and three isoforms in other yeasts. This is also true in *Pichia stipitis* (87). Alternatively, the translation elongation factor (*TEF1*) promoter (2, 15, 161) or the histone H4 (*AHSB4*) promoter has been used for strong constitutive expression (173).

In *P. pastoris* and *H. polymorpha*, promoters that are responsive to the presence of methanol are used. Even though these yeasts are frequently employed for heterologous protein production, only a limited set of promoters is in common use (146). The *AOX1* promoter for alcohol oxidase is most often employed. It is very strong, tightly controlled through glucose repression, and highly induced upon a shift to methanol as a carbon source (38). The promoter for formaldehyde dehydrogenase, *FLD1*, is sometimes used as an alternative to *AOX1* (147). Occasionally more precise expression levels are required for optimal protein production. For this purpose, a library of *P. pastoris* promoters has been characterized (69). The use of a constitutive promoter such as *PMA1* for plasma membrane H$^+$-ATPase can avoid the fire hazard inherent in the use of methanol as a carbon source at large-scale use (35).

21.6. REPORTER GENES

Once a promoter is isolated, it is necessary to determine its relative strength. This is most conveniently done through the use of reporter genes. β-D-Galactosidase (*lacZ*) is commonly used due to the availability of a rapid screen (22, 136). Some yeasts, however, such as *P. stipitis*, *D. hansenii*, *Pichia guilliermondii*, *K. marxianus*, and *K. lactis* possess β-D-galactosidase that can interfere with heterologous expression assays. Also, CUG yeasts will not translate *lacZ* properly. For example, *E. coli lacZ* contains 54 CUG codons.

Even more popular than *lacZ* is green fluorescent protein (GFP). While the quantitative analysis of expression requires a measurement of fluorescence, this marker is particularly useful when expression occurs only in a fraction of the cell population, during a certain phase of the cell cycle, or in a region of the cell. The native GFP does not express well in yeast, and especially in CUG yeasts; however, a codon-optimized form of GFP, in which all of the CUGs have been replaced with TTG to specify leucine, expresses well in both *C. albicans* and *S. cerevisiae* (33).

21.7. INTEGRATED EXPRESSION

Integrated expression (expression of genes after integration into the chromosome) is critical for stable industrial production strains. Even with integrated expression, mutations that enable more rapid growth of the host to the detriment of product formation can proliferate. The use of episomal vectors can result in strain instability and inconsistencies from batch to batch (61).

Integration can be random or targeted to a particular site. Targeted integration is generally preferred since this yields better-defined transformants with potentially higher productivity. Regulatory certification is easier to obtain if the integration site is defined, and by knocking out genes that can form by-products, cell yields are increased.

Integration of exogenous DNA requires double-strand break (DSB) repair, which can occur either through homologous recombination (HR) or through nonhomologous end joining (NHEJ). The former is mediated by Rad52-related proteins (97) and the latter by Ku70/Ku80, Lig4, and a number of other gene products (5, 46, 143). HR requires relatively long regions of homology to repair a DSB, whereas NHEJ requires little or no homology to join two broken strands (Fig. 1).

HR predominates in *S. cerevisiae*, but in *P. stipitis*, *K. lactis* (97), fungi such as *Aspergillus* (160) and *Neurospora* (122) spp., and higher organisms (78), the frequency of NHEJ can be 1 to 4 orders of magnitude higher than targeted integration by HR. Mutating the *KU80* or *KU70* genes in *K. lactis* (97), *P. stipitis* (94, 114), and other fungi (122) greatly increases the proportion of targeted disruption by HR; however, the overall number of recovered transformants drops by more than an order of magnitude. Even though there are fewer total transformants, recovery of site-specific integrants is much easier because they represent 80 to 97% of the recovered mutants (Fig. 2A).

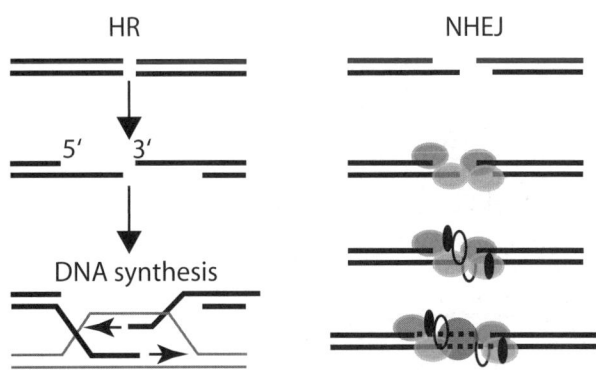

FIGURE 1 Homologous recombination and nonhomologous end-joining mechanisms for recombination.

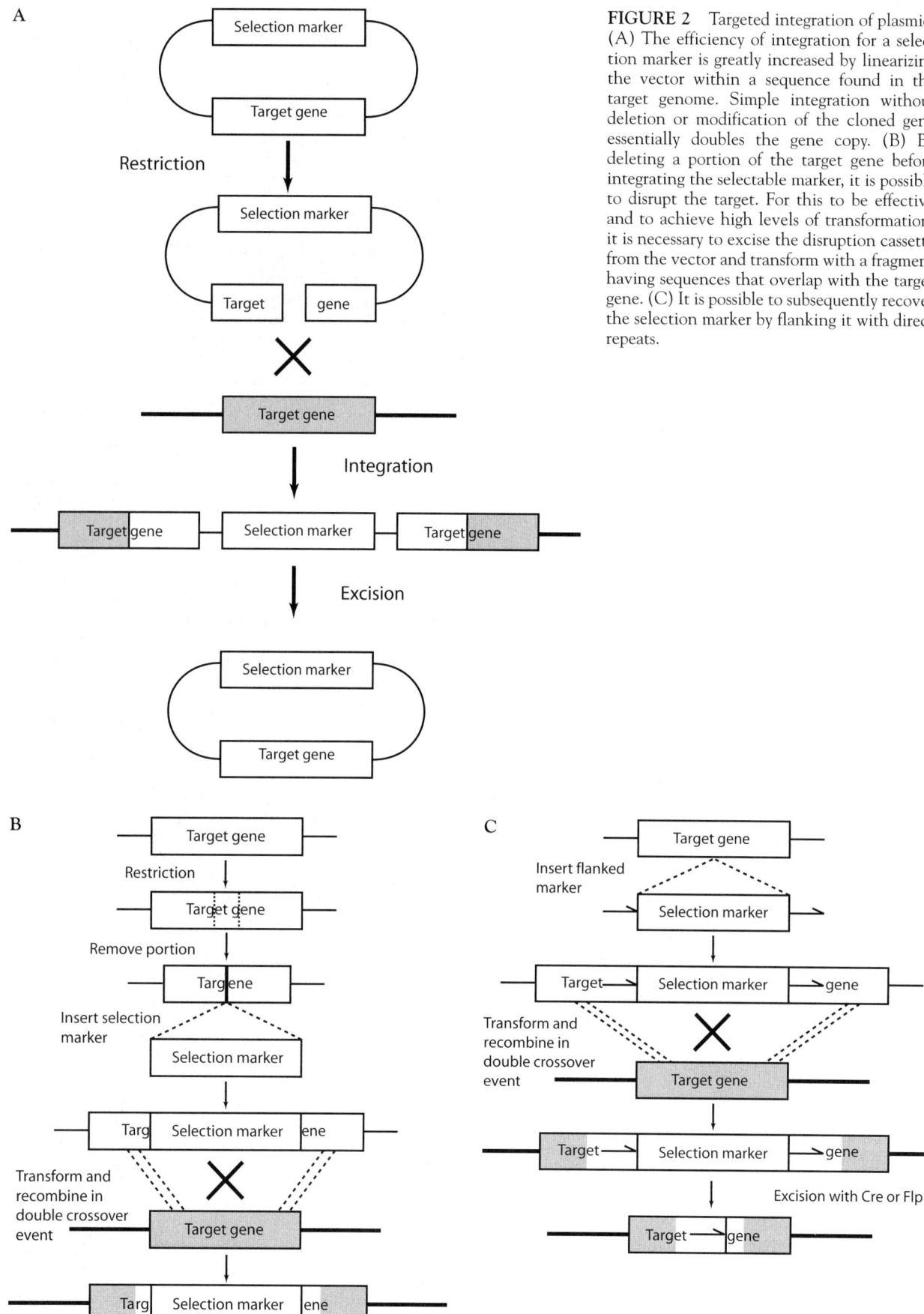

FIGURE 2 Targeted integration of plasmid. (A) The efficiency of integration for a selection marker is greatly increased by linearizing the vector within a sequence found in the target genome. Simple integration without deletion or modification of the cloned gene essentially doubles the gene copy. (B) By deleting a portion of the target gene before integrating the selectable marker, it is possible to disrupt the target. For this to be effective and to achieve high levels of transformation, it is necessary to excise the disruption cassette from the vector and transform with a fragment having sequences that overlap with the target gene. (C) It is possible to subsequently recover the selection marker by flanking it with direct repeats.

Aside from targeted integration, restricting the efficiency of the promoter of the selectable marker will also increase the recovery of transformants with multiple integration events. In *Arxula adeninivorans*, it was possible to obtain transformants with multiple integrational events for amylase production by reducing the length of the *LEU2* promoter from its initial 565 bp to only 56 bp (154). Transformants recovered from the complete promoter contained only a single integration event, but transformants obtained with the abbreviated promoter contained up to eight integrated copies when both were targeted to the 25S RNA gene (Fig. 2B). Presumably, strains with more integrated copies are favored for growth because they achieve a functional level of *LEU2* expression. Madzak et al. have reported similar results with the use of a defective *URA3* marker (115).

Restriction endonuclease-mediated integration has been reported to increase the frequency of site-specific integration (43, 166). While it does improve the overall frequency of integration events, linearization alone is effective as well (Fig. 2C). In fact, Maassen et al. (114) have suggested that the increase in integration events that they observed when transforming *P. stipitis* with plasmids along with the restricting enzymes was attributable to the prevention of recircularization rather than to any specific effect on the target genomic DNA.

The frequency of integration events can also be increased by incubating the host cells with sublethal doses of bleomycin just prior to transformation with linearized DNA. Bleomycin introduces DSBs in DNA, so exposure to sublethal doses of this antibiotic increases opportunities for recombination.

21.7.1. Multicopy Integration

More than 400 copies of the long terminal repeat elements of Ty1 and Ty2 are found in the *S. cerevisiae* genome. These have been used as targets for multiple integration events in order to increase the stable transmission of heterologous genes (106). Transformation frequencies are higher and stabilities are greater when two flanking copies of the δ-region are used (105). This technique has been fairly widely used in *S. cerevisiae* (29, 47, 93, 106), and use of the same sequence has been attempted in *K. lactis* (172).

Ylt1 is a repetitive element that exists in about 35 copies with an additional 30 copies of long terminal repeat elements in the *Yarrowia lipolytica* genome. Relatively high-copy-number integration events could be obtained when a defective *URA3* allele was used as the selectable marker. Most of these were direct tandem integration events (90, 126).

The ribosomal DNA (rDNA) locus has been used as a site for multiple integration events in *Y. lipolytica*, *K. lactis*, *S. cerevisiae*, *Candida utilis*, *Schizosaccharomyces pombe*, and *Phaffia rhodozyma* (19, 21, 33, 37, 39, 44).

21.8. EXPRESSION FROM AUTONOMOUS PLASMIDS

An ARS can greatly increase the efficiency of transformation by enabling replication of plasmid DNA in its target host. An autologous ARS can be obtained from the prospective host by cloning a library of short restriction fragments from the host into a vector bearing a selectable marker. The library is then transformed into the target host. The transformed yeasts are pooled, and a plasmid preparation is made from the yeast cells. Any rescued plasmids are transformed back into *E. coli*,

and a number of clones are selected for characterization. After two or three cycles of transformation and rescue, it is possible to obtain an ARS that allows stable replication in both hosts (1, 17, 30, 96, 119, 157, 168, 184). Alternatively, it is possible to synthesize a library of random AT sequences and screen for ARS activities in the same manner (54). Centromeric sequences are needed for appropriate partitioning (168, 182, 183).

21.9. TRANSFORMATION OF *P. PASTORIS* AND *H. POLYMORPHA*

The logic for expression of recombinant genes in either *P. pastoris* or *H. polymorpha* is much the same. Both yeasts are strong methanol-inducible yeast species, and methanol-regulated promoters are most often utilized for expression. Thus, one clones the recombinant gene under control of a methanol-regulatable promoter such as that from the alcohol oxidase gene (known as the *AOX1* gene in *P. pastoris* and the *MOX* gene in *H. polymorpha*). Then one proceeds to culture the expression strains in glucose for growth and strain maintenance. Under these conditions, expression of the recombinant gene is "off," minimizing selection against a strain that produces large amounts of foreign protein when induced. When production of the recombinant protein is desired, one switches to a methanol medium. A difference between the yeasts is that the *P. pastoris AOX1* promoter is strictly dependent on methanol and requires the alcohol for full expression levels, whereas the *H. polymorpha MOX* promoter can be derepressed to full expression levels by feeding of glycerol as a carbon source or by feeding glucose at a carbon limited rate. If ethanol is a problem for production with *P. pastoris*, then a variety of other promoters are available for this yeast, such as the *GAP* promoter, which is constitutive, or the *ADH* promoter, which can be derepressed to a high level of expression in response to growth rate-limiting amounts of glucose. Another difference between these yeasts is growth temperature. *P. pastoris* does not grow well above 30°C, whereas *H. polymorpha*'s optimal growth temperature is about 37°C. As a result, *H. polymorpha* grows significantly faster than *P. pastoris* and most other yeasts, and processes for *H. polymorpha* take a significantly shorter time. An advantage of *P. pastoris* is that the strains and vectors needed for recombinant gene expression are commercially available through Invitrogen (Carlsbad, CA) and other sources, whereas the *H. polymorpha* system is not sold commercially and is more difficult to obtain. In addition, the *P. pastoris* system has been much more extensively utilized, and therefore there is much more published literature available on expression in this yeast.

Techniques for culturing *P. pastoris* or *H. polymorpha* at the bench level are very similar to those used for *S. cerevisiae*. The most common rich medium for cultivation of either yeast species is YPD (1% yeast extract, 2% peptone, 2% dextrose), and defined medium is YNB (0.67% yeast nitrogen base with ammonium sulfate and without amino acids, 2% dextrose plus any amino acids or nucleotides required for growth at ~50 μg/ml each). Growth of *P. pastoris* or *H. polymorpha* on methanol requires that the dextrose be replaced with methanol to 0.5%. Incubations of *P. pastoris* are typically done at 30°C, whereas those for *H. polymorpha* are generally done at 37°C. In liquid YPD, *P. pastoris* has a generation time of ca. 90 min, and *H. polymorpha* at 37°C in the same medium has a generation time of approximately 1 h. With methanol as sole carbon source and a defined culture medium, the generation time of *P. pastoris* is around 5 h and that for *H. polymorpha* is about 3 h per generation at 37°C.

The first consideration in the selection of a *P. pastoris* or *H. polymorpha* expression vector is whether you intend the yeast to secrete a protein product or produce it intracellularly. A general rule of thumb is to produce a recombinant protein in the same way that it is expressed in its native host: if a protein is produced intracellularly by its native host, one should also produce it intracellularly in the yeast host; if the protein is secreted from its native host, secrete it from the yeast system. Although there have been exceptions to this general rule, it is generally best to follow it since the intracellular and secretory environments are very different from each other and synthesis of a protein in the wrong compartment may result in improper folding or degradation. A number of vectors have been constructed for each yeast species; a list and detailed discussion of these vectors for *P. pastoris* can be found in Lin Cereghino et al. (109) and at www.Biogrammatics.com, and *H. polymorpha* is discussed in reference 61.

These days, genes to be expressed are usually obtained by PCR amplification from genomic DNA or cDNA, or by gene synthesis; either method facilitates the incorporation of convenient restriction sites, signal sequences, purification tags, etc. When the goal is efficient secretion of the gene product, the cloning can be a little trickier. Use of the native secretion signal sequence associated with the heterologous gene is the most straightforward, but those signal sequences may not be efficiently used by the yeast host. Alternatively, a yeast signal sequence, such as that for the *S. cerevisiae* alpha mating factor (αMF), can replace the heterologous signal sequence in the expression vector. The αMF secretion signal is commonly used because it has been proven to give efficient secretion of many types of recombinant proteins. Although the recombinant protein may be successfully secreted using the αMF signal, proper N-terminal processing of the heterologous protein may not occur, and modifications of the αMF signal, or the use of an alternative secretion signal sequence, may ultimately be necessary to obtain a properly processed protein.

21.9.1. Transformation by Electroporation

For transformation of either *P. pastoris* or *H. polymorpha*, electroporation is the most commonly used procedure, and therefore a modified version of that described by Becker and Guarente (10) will be related here. For the other procedures, readers are referred to either of the volumes of *Methods in Molecular Biology: Pichia Protocols* (36, 72); for *H. polymorpha*, please refer to *Hansenula polymorpha: Biology and Applications* by G. Gellissen (59).

21.9.2. DNA Preparation

For all transformation methods, linearized plasmid DNA is most commonly transformed into either yeast for integration into the yeast genome. The DNA sequences at the ends of the linear plasmid DNA stimulate integration by a single crossover recombination event into the locus shared by vector and host genome. Therefore, linearization of an expression plasmid is performed within a yeast sequence such as the promoter sequence. The final vector, prepared in *E. coli*, is cut with a restriction enzyme that linearizes the vector, and then the DNA is purified and concentrated to approximately 100 ng/μl prior to transformation. At this point the vector is ready for transformation into yeast.

Preparation of Electrocompetent Cells for Transformation of *P. pastoris*

The procedure for growth and preparation of electrocompetent *P. pastoris* cells is taken from Lin-Cereghino et al. (110) and is described below. All solutions should be autoclaved except for the dithiothreitol (DTT) and HEPES solutions, which should be filter sterilized.

1. Electroporation instrument: BTX Electro Cell Manipulator 600 (BTX, San Diego, CA). Parameters for electroporation vary considerably depending on the apparatus used. Check the instructions provided by the manufacturer for specifics for each instrument.
2. Inoculate 10 ml of YPD medium with a single fresh *P. pastoris* colony of the strain to be transformed from an agar plate and grow overnight with shaking at 30°C.
3. Use the overnight culture to inoculate a 500-ml YPD culture in a 2.8-liter Fernbach culture flask to a starting optical density at 600 nm (OD_{600}) of 0.01 and grow to an OD_{600} of 1.0 (ca. 12 h).
4. Harvest the cells by centrifugation at 2,000 × g at 4°C, discard the supernatant, and resuspend the cells in 100 ml of fresh YPD medium plus 1 M HEPES buffer (pH 8.0) in a sterile 250-ml centrifuge tube.
5. Add 2.5 ml of 1 M DTT and gently mix.
6. Incubate at 30°C for 15 min with slow rotating.
7. Add 150 ml of cold water to the culture and wash by centrifugation at 4°C with an additional 250 ml of cold water. At this stage and from here on, keep the cells ice cold and do not vortex the cells to resuspend them (slow pipetting is best).
8. Wash cells a final time in 20 ml of cold 1 M sorbitol, then resuspend in 0.5 ml of cold 1 M sorbitol (final volume including cells will be 1.0 to 1.5 ml).
9. Use these cells directly without freezing to achieve the most transformants.
10. To freeze competent cells, distribute in 40-μl aliquots to sterile 1.5-ml minicentrifuge tubes, and place the tubes in a −70°C freezer.

Electroporation Procedure

1. Add up to 1 μg of linearized plasmid DNA sample in no more than 5 μl of water to a tube containing 40 μl of frozen or fresh competent cells and then transfer the entire mixture to a 2-mm gap electroporation cuvette held on ice.
2. Pulse cells according to the parameters suggested for yeast by the manufacturer of the specific electroporation instrument.
3. Immediately add 0.5 ml of cold 1 M sorbitol and 0.5 ml of cold YPD, then transfer the entire cuvette contents to a sterile 1.5- to 2.0-ml minicentrifuge tube.
4. Incubate for 3.5 to 4 h at 30°C with slow shaking (100 rpm).
5. Spread aliquots onto selective agar plates and incubate for 2 to 4 days.
6. To avoid mixed colonies, pick and restreak transformants on selective medium at least once before proceeding with further analysis.

Preparation of Electrocompetent Cells and Electrotransformation of *H. polymorpha*

The procedure for growth and preparation of electrocompetent *H. polymorpha* cells is taken from Gellissen (59) and is described below. All solutions should be autoclaved except for the DTT solution, which should be filter sterilized.

1. Inoculate 200 ml of YPD medium with an appropriate volume of fresh preculture and grow until the cells reach an OD_{600} of 0.8 to 1.2.
2. Harvest the cells by centrifugation and resuspend them in 0.2 volume of prewarmed (37°C) 50 mM potassium phosphate buffer (pH 7.5), then add DTT (1 M) to a final concentration of 25 mM.
3. Incubate the cells at 37°C in a water bath (without shaking) for 15 min.
4. Harvest the cells by centrifugation and wash twice: the first in 1 volume and the second in 0.5 volume of STM buffer (270 mM sucrose, 10 mM Tris-HCl, pH 7.5, 1 mM $MgCl_2$) while being kept at 0°C.
5. Resuspend the final cell pellet in 0.005 volume of STM buffer and dispense the cells in 60-μl aliquots. Freeze the aliquots at −70°C.
6. To transform, thaw aliquots of competent cells in ice, and add plasmid DNA (200 ng to 1 μg). Transfer the cells to a prechilled 2-mm cuvette.
7. After careful drying with a paper towel, place the cuvette in the electroporator and pulse (e.g., 2.0 kV, 25 μF, 200 ohm, for a Bio-Rad Gene Pulser II device). Note: Conditions for electroporation vary considerably depending on the instrument. Be sure to follow the instrument's instructions.
8. Immediately add 1 ml of YPD medium, transfer the cells to a microtube, and incubate at 37°C for 1 h of recovery.
9. Harvest cells by centrifugation, wash them once in 1 ml of selective medium, and suspend them for plating on selective plates. Incubate plates at 37°C for 2 to 3 days.

Protocol for Small-Scale Expression in *P. pastoris*

This protocol describes how to perform a shake flask study of two P. pastoris strains, one secreting a protein, human serum albumin (HSA), and a second expressing an intracellular enzyme, bacterial β-lactamase (β-lac) (see Fig. 3). The HSA gene is expressed from the methanol-inducible *AOX1* promoter, whereas the β-lac gene is expressed constitutively from the *GAP* promoter. (These strains can be obtained from the authors by contacting J.M.C.) For the strains

expressing HSA, samples of culture medium are collected and subjected to sodium dodecyl sulfate-polyacrylamide gel electrophoresis (SDS-PAGE). For strains expressing β-lac, a simple qualitative enzyme assay is employed.

Secretion of HSA

1. Pick a colony of an HSA-secreting strain and inoculate into 50 ml of YPD medium in a sterile Erlenmeyer flask. Incubate with shaking overnight.
2. The next day, shift the cultures by centrifugation. To shift, determine the OD_{600} of each culture and centrifuge 200 OD_{600} units (OD_{600} unit = volume in milliliters × OD_{600} reading) of each culture.
3. After centrifugation, decant the supernatant and suspend each cell pellet in 10 ml of BMMY medium (1% yeast extract, 2% peptone, 100 mM potassium phosphate [pH 6.0], 1.34% YNB with ammonium sulfate and without amino acids, 0.5% methanol). Transfer each culture to a small (50-ml) sterile shake flask and place in a 30°C incubator with vigorous shaking. Before starting the induction, harvest 0.5 ml of each culture for a t = 0 sample.
4. Centrifuge the t = 0 samples in a microcentrifuge for 5 min at maximum speed. Transfer the supernatants to a clean tube. Discard the cell pellets but save the supernatants by storing them frozen at −20°C.
5. Add methanol to each culture to a final concentration of 0.5%, every 12 h.
6. Take additional aliquots of HSA-secreting culture (or negative control) at 12, 24, 48, and 64 h and prepare them for storage at −20°C as described above.
7. Thaw and analyze proteins in each HSA sample by SDS-PAGE to evaluate HSA synthesis and secretion over time. For each sample, mix 25 μl of culture medium supernatant with 4 μl of the 6× SDS-PAGE loading. Load 25 to 30 μl on an SDS-polyacrylamide gel. Follow standard procedures for Coomassie staining of an SDS-polyacrylamide gel.

Expression of β-lac

1. Pick a colony of each of the two β-lac-producing *P. pastoris* strains and inoculate into 50 ml of YPD medium. In one strain, the gene for *bla* is fused to the inducible

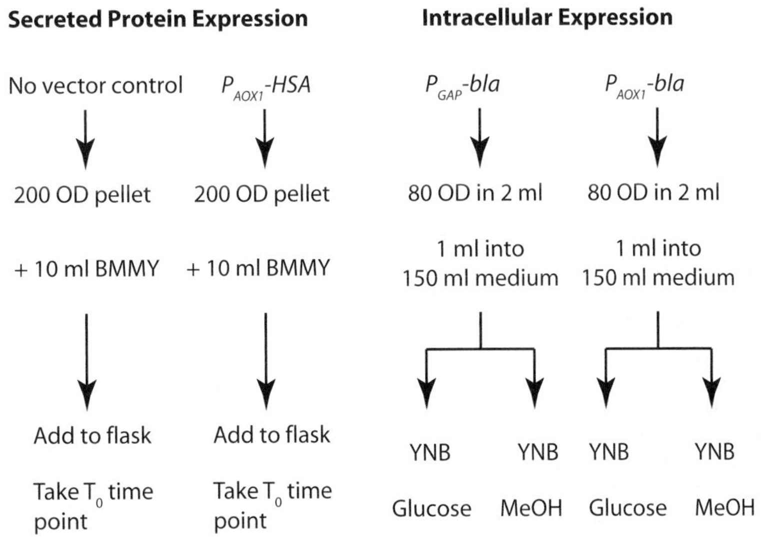

Secreted Protein Expression

No vector control → 200 OD pellet → + 10 ml BMMY → Add to flask → Take T_0 time point

P_{AOX1}-HSA → 200 OD pellet → + 10 ml BMMY → Add to flask → Take T_0 time point

Intracellular Expression

P_{GAP}-bla → 80 OD in 2 ml → 1 ml into 150 ml medium → YNB Glucose / YNB MeOH

P_{AOX1}-bla → 80 OD in 2 ml → 1 ml into 150 ml medium → YNB Glucose / YNB MeOH

FIGURE 3 Protocol for assays of protein expression in *P. pastoris*.

alcohol oxidase (*AOX1*) promoter. In the other, *bla* expression is regulated by the constitutive glyceraldehyde-3-phosphate dehydrogenase (*GAP*) promoter.

2. The next day, remove a 1-ml aliquot of each culture to a clean microcentrifuge tube. Determine the absorbance of each culture at OD_{600} and centrifuge 80 OD_{600} units of each. Then resuspend the 80 OD_{600} units in 2 ml of sterile water.

3. For each culture, add 1 ml of cells to each of two flasks containing 150 ml of medium, one with YNB glucose medium and one with YNB methanol medium.

4. Determine the OD_{600} of each culture and harvest 20 OD_{600} units of each by centrifugation. Decant the supernatant medium and save the pellets at $-20°C$. These are t = 0 samples.

5. Place the remainder of the two cultures at 30°C with shaking. Add additional methanol to each methanol medium culture (i.e., not the glucose medium cultures), to a final concentration of 0.5%, at 12 h.

6. At 6, 12, and 24 h, determine the OD_{600} of each culture and harvest 20 OD_{600} units from each. Centrifuge each sample, decant the supernatants, and store cell pellets in labeled tubes at $-20°C$.

7. Transfer all β-lac 20-OD_{600} cell pellet samples from storage at $-20°C$ (plus at least one nonexpressing negative control sample) to an ice bucket to thaw.

8. Suspend cell pellets in 100 μl of ice-cold breaking buffer (25 mM Tris-HCl, pH 7.5). Add one scoop (~100 μl) of glass beads to each sample.

9. Vortex as vigorously as possible. Multiple samples (up to 12) can be broken simultaneously using a multisample vortex head (Disruptor Genie, Scientific Industries, Inc.) in the cold box by vortexing constantly at 4°C for 10 min.

10. Centrifuge samples for 10 min at full speed in a microcentrifuge at 4°C.

11. Transfer each supernatant to a clean minicentrifuge tube. (Be careful to avoid transferring pellet and cell debris; i.e., transfer only clear cell extract.) Save supernatants (cell-free extracts) in labeled tubes specifying the strain and time point.

12. Determine protein concentration in each sample using the Pierce assay or other method.

13. Carry out assays for β-lac activity in extracts as follows. For each strain and time point, transfer 5 μg of protein to a microcentrifuge tube containing 198 μl of ice-cold breaking buffer. Include a negative control with breaking buffer substituted for extract. Start the assay by adding 2 μl of PADAC substrate solution. Mix and incubate at 30°C for 5 min. β-Lac activity is indicated by a change of color from purple to yellow.

21.10. SYNTHETIC DNA

As synthetic DNA continues to drop in price, it is now routinely used to speed up construction of expression cassettes for metabolic engineering. Gene synthesis is useful to eliminate unwanted restriction and alter specific regions of the protein. More importantly, through codon optimization, it can also increase translation rates. Codon optimization was based initially on the observation that organisms use a nonrandom subset of the 61 possible sense codons available in the genetic code. This can result in a bias in their codon usage. Codon usage varies from one species to another. It is consistent within the genes of any given genome, and the codon usage in open reading frames correlates with the

most abundant aminoacyl-tRNAs present in the cell (150). It is not clear, however, that altering codon usage alone will result in an optimal translation rate. In addition to codon preference based on tRNA usage, it is also necessary to take into account secondary structures and contextual optimization. For example, it is possible to increase protein synthesis by optimizing the initial codons in an open reading frame, thereby avoiding the formation of secondary structures which may interfere with the ribosomal complex (76, 80, 98). Codon optimization can adjust the overall G+C usage to something more appropriate for the producing organisms, and contextual codon optimization can avoid the sequential use of certain acyl-tRNAs that interfere with one another during translation (70). Optimization of codons based on frequency of acyl-tRNA usage alone does not give consistently better results, so care should be taken in considering the various factors in designing synthetic genes.

While limited sequence optimization—such as the elimination of restriction sites or specific codons—might be done manually, the optimization of open reading frames for various codon usage tables can be aided through the use of software (53) or online tools (180). Reducing ambiguity in the overlaps for optimized gene assembly requires elaborate algorithms and significant computational power (103).

Codon optimization can be highly effective in the expression of bacterial (19, 178) or human proteins in yeasts. It can eliminate cryptic polyadenylation signals in native mRNA, such as AT-rich stretches (76), thereby increasing the efficiency of protein synthesis (162). Codon optimization has been used to increase production of several proteins in *P. pastoris* (75, 76, 118, 170, 179).

Synthetic DNA can be used to increase expression of open reading frames, but it is somewhat more difficult to apply to the construction of eukaryotic promoters since these often have high AT content, along with inverted and direct repeats that can interfere with DNA synthesis. The cost of synthetic DNA is dropping rapidly, and genes up to 1,500 bp in length can be obtained from commercial operations with turnaround times of a few weeks or less.

21.11. ENHANCING OVEREXPRESSION OR SECRETION

Many factors besides transcript formation determine the efficiency of protein production. One of the most critical is oxygen supply. During heterologous protein production in large-scale bioreactors, regions of the reactor vessel are relatively poorly mixed. Moreover, even if the whole vessel is mixed, packets of fluid are rapidly depleted in oxygen, generating cycles of aeration and oxygen deprivation. This is particularly true in high-density fermentations for the production of enzymes or other heterologous proteins. The oscillation between aeration and oxygen limitation can disrupt cellular metabolism. One interesting approach to overcome this problem is the heterologous expression of a bacterial hemoglobin (Vhb) from the gram-positive bacterium *Vitreoscilla* (159). Trapping oxygen when cells are exposed to aerobic conditions and releasing it when under less aeration can mitigate the effect of poor mixing. Expression of Vhb increased recombinant β-galactosidase production in *P. pastoris* (181), improved cell growth and secretion of extracellular enzymes in *Y. lipolytica* (12), and enhanced ethanol production by *S. cerevisiae* (26), including the production of ethanol from xylose (141).

Excessive oxygen can be toxic to yeasts due to the production of reactive oxygen species such as superperoxide anions ($O_2^{\cdot-}$). Superperoxide dismutase, which catalyzes the conversion of $O_2^{\cdot-}$ into oxygen and hydrogen peroxide, is the first line of defense (130). In *K. lactis*, overexpression of *KlSOD1* effectively reduced oxygen stress during heterologous protein production.

21.12. CHAPERONE, FOLDASE, CELL WALL, AND ER FACTORS

Overexpression of heterologous proteins in a recombinant host can saturate or overload the endoplasmic reticulum (ER) secretory pathway, thereby resulting in misfolding (57, 148, 151). There are two basic modes of eukaryotic protein translocation. In cotranslational translocation, secretion occurs contemporaneously with peptide synthesis. The growing polypeptide chain moves from the ribosomal tunnel into a membrane channel formed by the Sec61 membrane complex, and the whole process is driven by GTP hydrolysis during translation. In posttranslational translocation, the Sec62 membrane complex forms a pore between the cytosol and the ER membrane. It binds the signal sequence of the preformed protein, which is then drawn through the membrane (132).

Kar2p (the yeast equivalent to the mammalian BiP) is a member of the Hsp70 family of heat shock proteins that serves as a ratchet to draw the preformed protein through the membrane in posttranslational secretion. Further processing such as folding, disulfide cross-linking, and glycosylation then occurs in the lumen of the ER (9, 32, 62).

Several proteins in the ER are involved in this process. Protein disulfide isomerase (PDI) appears to play a role in folding. Overexpression of proteins involved in the ER secretory pathway has been reported to improve secretion in *S. cerevisiae*. Overexpression of PDI in *K. lactis* strongly stimulated secretion of highly disulfide-bonded serum albumin (8) and human lysozyme (82). Likewise, overexpression of PDI in *P. pastoris* was able to increase secretion of a hookworm vaccine protein that contained 20 disulfide linkages (79). Overexpression of the chaperones Kar2, Ssa1, or PDI can improve protein secretion by two- to sevenfold in *P. pastoris* (188).

A systematic study of yeast genes to enhance protein secretion in *S. cerevisiae* showed the cell wall protein Ccw12 and the ER resident thiol oxidase, Ero1, as lending the most general enhancement (176). Gasser et al. (58) used a transcriptomics-based approach to examine a *P. pastoris* strain that would overexpress human trypsinogen, in comparison with one that would not. More than 500 genes were significantly upregulated in the secreting strain. Of these, 13 had potential function in secretion pathways. Based on previous studies of secretion enhancement in *S. cerevisiae*, these 13 genes, along with homologs to genes previously characterized as having roles in secretion, were cloned and overexpressed in *P. pastoris*. Ero1, Kar2, and Sso2 (involved in fusion of secretory vesicles at the plasma membrane), along with Bfr2 and Bmh2 (involved in protein transport), the chaperones Ssa4 and Sse1, the vacuolar ATPase subunit Cup5, and Kin2 (a protein kinase connected to exocytosis), all proved effective at increasing production of antibody fragments.

21.13. HUMANIZED YEAST EXPRESSION SYSTEMS

21.13.1. Market

Monoclonal antibodies (MAbs) represent the fastest growing sector of biotherapeutic proteins (107). The global therapeutic MAb market grew to $5.4 billion in 2002, with chimeric MAbs accounting for 43% of sales (133). From 2004 to 2008, growth actually exceeded expectations (128). The global sales of MAbs have been reported as $33 billion in 2008 as compared to $27 billion in 2007 (116). This explosive growth stems from the tremendous ability of MAbs to bind a plethora of antigens ranging from toxins, viruses, and bacteria to cancer cells. The main applications, however, lie in their potential to boost novel treatments of cancer and inflammatory and infectious diseases. In 2004, the global clinical antibody pipeline had 132 products in development (133). As of 2009, more than 500 antibody-based therapeutics were in development, with 200 more in clinical trials (84, 128, 134). These factors suggest that large-scale antibody production in yeasts and fungi might be a major emerging opportunity (see reference 56).

Clinically proven MAbs have been produced in tissue culture (49), mainly Chinese hamster ovary cells, with its incumbent costs (107). This technology results in MAbs similar but not identical to their native counterparts. Large-scale tissue culture (up to 200,000 liters) has an estimated productivity of only 1 to 2 mg liter^{-1} h^{-1}, with an estimated annual productivity of 8 to 16 kg/year (177). In 2006, the estimated cost for producing a MAb by mammalian tissue culture was between $300 and $3,000 (US) per g, with the major cost centers coming from low titers and downstream purification (49). While such MAbs are valuable as biopharmaceutics, the doses required are typically much larger than for other biologics (89), typically >100 mg, due to their low potency (84). As a result, MAbs are among the most expensive pharmaceuticals available, with annual costs for cancer therapies reaching up to $35,000 US/year per patient (48). Clearly this is an area in need of improved production technology, especially in light of efforts to control medical costs while expanding coverage.

21.13.2. Yeast Production Systems

MAbs must be capable of two binding events if they are to be therapeutically effective. In the first event, the variable domain, which determines the antibody specificity, binds to a specific target protein on the cell surface. In the second event, the immune system's effector cells bind to constant region (Fc) of the antibody and destroy the cell to which the antibody is attached through a process known as antibody-dependent cell cytotoxicity. This cytotoxic process depends on a specific Fc Asn$_{297}$ N glycosylation and is affected by the glycan composition of the constant region (108, 140). Glycosylation patterns vary with different mammalian cell lines, and the pattern elucidating the most potent antibody-dependent cell cytotoxic response for a given cell type is complex, which leaves open the possibilities of further development and optimization.

MAbs are obtained by (i) immunizing mice with a specific antigen, (ii) harvesting the mouse spleen lymphocytes, (iii) hybridizing the lymphocytes with myeloma cells, (iv) screening the resulting hybridoma cells for antibodies to the desired antigen, and (v) carrying the selected potent clones into tissue culture. Antibody production then depends on the efficiency of the scale-up and downstream processing (84). To produce antibodies in yeast cells, it is further necessary to clone the specific variable sequence into a host for which the glycosylation machinery has been altered to match that of humans.

P. pastoris is preferred for the production of N-glycan-containing biopharmaceuticals since it exhibits a low frequency of serine O mannosylation. Moreover, mannoglycosylation is much less extensive in *P. pastoris*, and unlike *S. cerevisiae*, *P. pastoris* lacks the transferase necessary to create terminal α-1,3-linked mannosyl residues that can cause strong antigenic responses in humans (42). The glycosylation pathways for yeasts and mammalian systems have recently been reviewed in excellent detail and do not need to be repeated here (83, 138). In brief, deletion of *OCH1*, which codes for the mannosyltransferase of the *cis*-Golgi apparatus, eliminates the native hypermannosylation machinery. Combining the correct Golgi leader sequence from *P. pastoris* with the appropriate catalytic domains of human Golgi α-(1-2)-mannosidases IA, IB, and IC enabled construction of humanized complex-type *N*-glycans (GlcNAc2Man3GlcNAc2) (138). A complete protocol for production of complex N-glycosylated proteins in *P. pastoris* has recently been published (83).

Posttranslational modification such as N glycosylation is initiated in the ER lumen. Proteins are exported to the Golgi only if they have folded in the correct conformation (57). For heterologous expression, genes for MAb heavy- and light-chain domains are fused to a Kar2 signal peptide to ensure processing and secretion through the ER along with an appropriate promoter (107). When transformed into wild-type *P. pastoris* cells, the antibodies yield high-mannose-type proteins.

REFERENCES

1. **Adamikova, L., P. Griac, L. Tomaska, and J. Nosek.** 1998. Development of a transformation system for the multinuclear yeast *Dipodascus* (*Endomyces*) *magnusii*. *Yeast* **14:**805–812.

2. **Ahn, J., J. Hong, H. Lee, M. Park, E. Lee, C. Kim, E. Choi, and J. Jung.** 2007. Translation elongation factor 1-alpha gene from *Pichia pastoris*: molecular cloning, sequence, and use of its promoter. *Appl. Microbiol. Biotechnol.* **74:**601–608.

3. **Alcaino, J., S. Barahona, M. Carmona, C. Lozano, A. Marcoleta, M. Niklitschek, D. Sepúlveda, M. Baeza, and V. Cifuentes.** 2008. Cloning of the cytochrome p450 reductase (*crtR*) gene and its involvement in the astaxanthin biosynthesis of *Xanthophyllomyces dendrorhous*. *BMC Microbiol.* **8:**169.

4. **Alderton, A. J., I. Burr, F. A. Muhlschlegel, and M. F. Tuite.** 2006. Zeocin resistance as a dominant selective marker for transformation and targeted gene deletions in *Candida glabrata*. *Mycoses* **49:**445–451.

5. **Andaluz, E., A. Ciudad, J. R. Coque, R. Calderone, and G. Larriba.** 1999. Cell cycle regulation of a DNA ligase-encoding gene (*CaLIG4*) from *Candida albicans*. *Yeast* **15:**1199–1210.

6. **Andrews, B. J., G. A. Proteau, L. G. Beatty, and P. D. Sadowski.** 1985. The FLP recombinase of the 2-mu circle DNA of yeast—interaction with its target sequences. *Cell* **40:**795–803.

7. **Bailey, A. M., G. L. Mena, and L. Herrera-Estrella.** 1991. Genetic transformation of the plant pathogens *Phytophthora capsici* and *Phytophthora parasitica*. *Nucleic Acids Res.* **19:**4273–4278.

8. **Bao, W. G., and H. Fukuhara.** 2001. Secretion of human proteins from yeast: stimulation by duplication of polyubiquitin and protein disulfide isomerase genes in *Kluyveromyces lactis*. *Gene* **272:**103–110.

9. **Bao, W. G., K. K. Huo, Y. Y. Li, and H. Fukuhara.** 2000. Protein disulphide isomerase genes of *Kluyveromyces lactis*. *Yeast* **16:**329–341.

10. **Becker, D. M., and L. Guarente.** 1991. High efficiency transformation of yeast by electroporation. *Methods Enzymol.* **194:**182–187.

11. **Beckerman, J., H. Chibana, J. Turner, and P. T. Magee.** 2001. Single-copy IMH3 allele is sufficient to confer resistance to mycophenolic acid in *Candida albicans* and to mediate transformation of clinical *Candida* species. *Infect. Immun.* **69:**108–114.

12. **Bhave, S. L., and B. B. Chattoo.** 2003. Expression of *Vitreoscilla* hemoglobin improves growth and levels of extracellular enzyme in *Yarrowia lipolytica*. *Biotechnol. Bioeng.* **84:**658–666.

13. **Boeke, J. D., F. Lacroute, and G. R. Fink.** 1984. A positive selection for mutants lacking orotidine-5′-phosphate decarboxylase activity in yeast—5-fluoro-orotic acid resistance. *Mol. Gen. Genet.* **197:**345–346.

14. **Boer, E., H. P. Mock, R. Bode, G. Gellissen, and G. Kunze.** 2005. An extracellular lipase from the dimorphic yeast *Arxula adeninivorans*: molecular cloning of the ALIP1 gene and characterization of the purified recombinant enzyme. *Yeast* **22:**523–535.

15. **Boer, E., G. Steinborn, G. Kunze, and G. Gellissen.** 2007. Yeast expression platforms. *Appl. Microbiol. Biotechnol.* **77:**513–523.

16. **Boer, E., G. Steinborn, A. Matros, H. P. Mock, G. Gellissen, and G. Kunze.** 2007. Production of interleukin-6 in *Arxula adeninivorans*, *Hansenula polymorpha* and *Saccharomyces cerevisiae* by applying the wide-range yeast vector (CoMed (TM)) system to simultaneous comparative assessment. *FEMS Yeast Res.* **7:**1181–1187.

17. **Boretsky, Y., A. Voronovsky, O. Liuta-Tehlivets, M. Hasslacher, S. D. Kohlwein, and G. M. Shavlovsky.** 1999. Identification of an ARS element and development of a high efficiency transformation system for *Pichia guilliermondii*. *Curr. Genet.* **36:**215–221.

18. **Bostock, J. M., K. Miller, A. J. O'Neill, and I. Chopra.** 2003. Zeocin resistance suppresses mutation in hypermutable *Escherichia coli*. *Microbiology* **149:**815–816.

19. **Brat, D., E. Boles, and B. Wiedemann.** 2009. Functional expression of a bacterial xylose isomerase in *Saccharomyces cerevisiae*. *Appl. Environ. Microbiol.* **75:**2304–2311.

20. **Buckholz, R. G., and M. A. G. Gleeson.** 1991. Yeast systems for the commercial production of heterologous proteins. *BioTechnology* **9:**1067–1072.

21. **Burland, T. G., J. Bailey, D. Pallotta, and W. F. Dove.** 1993. Stable, selectable, integrative DNA transformation in *Physarum*. *Gene* **132:**207–212.

22. **Casadaban, M. J., A. Martinezarias, S. K. Shapira, and J. Chou.** 1983. Beta-galactosidase gene fusions for analyzing gene expression in *Escherichia coli* and yeast. *Methods Enzymol.* **100:**293–308.

23. **Chang, C. C., D. D. Y. Ryu, C. S. Park, and J. Y. Kim.** 1997. Enhancement of rice alpha-amylase production in recombinant *Yarrowia lipolytica*. *J. Ferment. Bioeng.* **84:**421–427.

24. **Chang, C. C., D. D. Y. Ryu, C. S. Park, J. Y. Kim, and D. M. Ogrydziak.** 1998. Recombinant bioprocess optimization for heterologous protein production using two-stage, cyclic fed-batch culture. *Appl. Microbiol. Biotechnol.* **49:**531–537.

25. **Chen, G. H., L. J. Yin, I. H. Chiang, and S. T. Jiang.** 2007. Expression and purification of goat lactoferrin from *Pichia pastoris* expression system. *J. Food Sci.* **72:**M67–M71.

26. **Chen, W., D. E. Hughes, and J. E. Bailey.** 1994. Intracellular expression of *Vitreoscilla* hemoglobin alters the aerobic metabolism of *Saccharomyces cerevisiae*. *Biotechnol. Progr.* **10:**308–313.

27. **Chen, X. Z., H. Y. Fang, Z. M. Rao, W. Shen, B. Zhuge, Z. X. Wang, and J. Zhuge.** 2008. An efficient genetic transformation method for glycerol producer *Candida glycerinogenes*. *Microbiol. Res.* **163:**531–537.

28. Cheon, S. A., E. J. Han, H. A. Kang, D. M. Ogrydziak, and J. Y. Kim. 2003. Isolation and characterization of the TRPI gene from the yeast *Yarrowia lipolytica* and multiple gene disruption using a TRP blaster. *Yeast* 20:677–685.

29. Choi, E. Y., J. N. Park, H. O. Kim, D. J. Shin, Y. H. Chun, S. Y. Im, S. B. Chun, and S. Bai. 2002. Construction of an industrial polyploid strain of *Saccharomyces cerevisiae* containing *Saprolegnia ferax* beta-amylase gene and secreting beta-amylase. *Biotechnol. Lett.* 24:1785–1790.

30. Chung, H. Y., M. H. Hong, Y. H. Chun, S. Bai, S. Y. Im, H. B. Lee, J. C. Park, D. H. Kim, and S. B. Chun. 1998. Nucleotide analysis of *Phaffia rhodozyma* DNA fragment that functions as ARS in *Saccharomyces cerevisiae*. *J. Microbiol. Biotechnol.* 8:650–655.

31. Cohen, J. D., T. R. Eccleshall, R. B. Needleman, H. Federoff, B. A. Buchferer, and J. Marmur. 1980. Functional expression in yeast of the *Escherichia coli* plasmid gene coding for chloramphenicol acetyl transferase. *Proc. Natl. Acad. Sci. USA* 77:1078–1082.

32. Copeland, C. S., K. P. Zimmer, K. R. Wagner, G. A. Healey, I. Mellman, and A. Helenius. 1988. Folding, trimerization and transport are sequential events in the biogenesis of influenza virus hemagglutinin. *Cell* 53:197–209.

33. Cormack, B. P., G. Bertram, M. Egerton, N. A. R. Gow, S. Falkow, and A. J. P. Brown. 1997. Yeast-enhanced green fluorescent protein (yEGFP): a reporter of gene expression in *Candida albicans*. *Microbiology* 143:303–311.

34. Cos, O., R. Ramón, J. L. Montesinos, and F. Valero. 2006. Operational strategies, monitoring and control of heterologous protein production in the methylotrophic yeast *Pichia pastoris* under different promoters: a review. *Microb. Cell Fact.* 5:17.

35. Cox, H., D. Mead, P. Sudbery, R. M. Eland, I. Mannazzu, and L. Evans. 2000. Constitutive expression of recombinant proteins in the methylotrophic yeast *Hansenula polymorpha* using the *PMA1* promoter. *Yeast* 16:1191–1203.

36. Cregg, J. M. 2007. Pichia Protocols, 2nd ed. *Methods Mol. Biol.* 389. Humana, Totowa, N.J.

37. Cregg, J. M., J. L. Cereghino, J. Y. Shi, and D. R. Higgins. 2000. Recombinant protein expression in *Pichia pastoris*. *Mol. Biotechnol.* 16:23–52.

38. Cregg, J. M., T. S. Vedvick, and W. C. Raschke. 1993. Recent advances in the expression of foreign genes in *Pichia pastoris*. *BioTechnology* 11:905–910.

39. Dave, M. N., and B. B. Chattoo. 1997. A counter selectable marker for genetic transformation of the yeast *Schwanniomyces alluvius*. *Appl. Microbiol. Biotechnol.* 48:204–207.

40. de Alteriis, E., G. Silvestro, M. Poletto, V. Romano, and P. Parascandola. 2006. Heterologous glucoamylase production with immobilised *Kluyveromyces lactis* cells in a fluidised bed reactor operating as a two-(liquid-solid) or a three-(gas-liquid-solid) phases system. *Process Biochem.* 41:2352–2356.

41. De la Rosa, J. M., J. A. Perez, F. Gutierrez, J. M. Gonzalez, T. Ruiz, and L. Rodriguez. 2001. Cloning and sequence analysis of the LEU2 homologue gene from *Pichia anomala*. *Yeast* 18:1441–1448.

42. Demain, A. L., and P. Vaishnav. 2009. Production of recombinant proteins by microbes and higher organisms. *Biotechnol. Adv.* 27:297–306.

43. Dmytruk, K. V., A. Y. Voronovsky, and A. A. Sibirny. 2006. Insertion mutagenesis of the yeast *Candida famata* (*Debaryomyces hansenii*) by random integration of linear DNA fragments. *Curr. Genet.* 50:183–191.

44. Du, W. J., M. Coaker, J. D. Sobel, and R. A. Akins. 2004. Shuttle vectors for *Candida albicans*: control of plasmid copy number and elevated expression of cloned genes. *Curr. Genet.* 45:390–398.

45. Dudásová, Z., A. Dudás, and M. Chovanec. 2004. Nonhomologous end-joining factors of *Saccharomyces cerevisiae*. *FEMS Microbiol. Rev.* 28:581–601.

46. Dumas, P., M. Bergdoll, C. Cagnon, and J. M. Masson. 1994. Crystal structure and site-directed mutagenesis of a bleomycin resistance protein and their significance for drug sequestering. *EMBO J.* 13:2483–2492.

47. Ekino, K., H. Hayashi, M. Moriyama, M. Matsuda, M. Goto, S. Yoshino, and K. Furukawa. 2002. Engineering of polyploid *Saccharomyces cerevisiae* for secretion of large amounts of fungal glucoamylase. *Appl. Environ. Microbiol.* 68:5693–5697.

48. Farid, S. S. 2009. Economic drivers and trade-offs in antibody purification processes. *BioPharm Internat.* 2009:37–42.

49. Farid, S. S. 2006. Established bioprocesses for producing antibodies as a basis for future planning. *Adv. Biochem. Eng. Biotechnol.* 101:1–42.

50. Feigin, R. D., K. S. Moss, and P. G. Shackelford. 1973. Antibiotic stability in solutions used for intravenous nutrition and fluid therapy. *Pediatrics* 51:1016–1026.

51. Fitzpatrick, D. A., M. E. Logue, and G. Butler. 2008. Evidence of recent interkingdom horizontal gene transfer between bacteria and *Candida parapsilosis*. *BMC Evol. Biol.* 8:181.

52. Francois, F., F. Chapeland-Leclerc, J. Villard, and T. Noel. 2004. Development of an integrative transformation system for the opportunistic pathogenic yeast *Candida lusitaniae* using URA3 as a selection marker. *Yeast* 21:95–106.

53. Fuglsang, A. 2003. Codon optimizer: a freeware tool for codon optimization. *Protein Expr. Purif.* 31:247–249.

54. Fukuhara, H. 2006. Random AT library: autonomously replicating sequence (ARS) activity of chemically synthesized random sequences for transformation of nonconventional yeast species. *FEMS Yeast Res.* 6:1281–1287.

55. Gacser, A., S. Salomon, and W. Schafer. 2005. Direct transformation of a clinical isolate of *Candida parapsilosis* using a dominant selection marker. *FEMS Microbiol. Lett.* 245:117–121.

56. Gasser, B., and D. Mattanovich. 2007. Antibody production with yeasts and filamentous fungi: on the road to large scale? *Biotechnol. Lett.* 29:201–212.

57. Gasser, B., M. Saloheimo, U. Rinas, M. Dragosits, E. Rodríguez-Carmona, K. Baumann, M. Giuliani, E. Parrilli, P. Branduardi, C. Lang, D. Porro, P. Ferrer, M. L. Tutino, D. Mattanovich, and A. Villaverde. 2008. Protein folding and conformational stress in microbial cells producing recombinant proteins: a host comparative overview. *Microb. Cell Fact.* 7:11.

58. Gasser, B., M. Sauer, M. Maurer, G. Stadlmayr, and D. Mattanovich. 2007. Transcriptomics-based identification of novel factors enhancing heterologous protein secretion in yeasts. *Appl. Environ. Microbiol.* 73:6499–6507.

59. Gellissen, G. 2002. Hansenula polymorpha: *Biology and Applications*. Wiley-VCH Verlag GmbH and Co., Weinheim, Germany.

60. Gellissen, G. 2000. Heterologous protein production in methylotrophic yeasts. *Appl. Microbiol. Biotechnol.* 54:741–750.

61. Gellissen, G., G. Kunze, C. Gaillardin, J. M. Cregg, E. Berardi, M. Veenhuis, and I. van der Klei. 2005. New yeast expression platforms based on methylotrophic *Hansenula polymorpha* and *Pichia pastoris* and on dimorphic *Arxula adeninivorans* and *Yarrowia lipolytica*—a comparison. *FEMS Yeast Res.* 5:1079–1096.

62. Gething, M. J., and J. Sambrook. 1992. Protein folding in the cell. *Nature* 355:33–45.

63. Goldstein, A. L., and J. H. McCusker. 1999. Three new dominant drug resistance cassettes for gene disruption in *Saccharomyces cerevisiae*. *Yeast* 15:1541–1553.

64. Golubev, W. I., and N. W. Golubev. 2004. Selection and study of potent lactose-fermenting yeasts. *Appl. Biochem. Microbiol.* **40:**280–284.

65. González, A., A. Jimenez, D. Vazquez, J. E. Davies, and D. Schindler. 1978. Studies on the mode of action of hygromycin B, an inhibitor of translocation in eukaryotes. *Biochim. Biophys. Acta* **521:**459–469.

66. Gritz, L., and J. Davies. 1983. Plasmid encoded hygromycin B resistance—the sequence of hygromycin B phosphotransferase gene and its expression in *Escherichia coli* and *Saccharomyces cerevisiae*. *Gene* **25:**179–188.

67. Güldener, U., S. Heck, T. Fiedler, J. Beinhauer, and J. H. Hegemann. 1996. A new efficient gene disruption cassette for repeated use in budding yeast. *Nucleic Acids Res.* **24:**2519–2524.

68. Hara, A., M. Arie, T. Kanai, T. Matsui, H. Matsuda, K. Furuhashi, M. Ueda, and A. Tanaka. 2001. Novel and convenient methods for *Candida tropicalis* gene disruption using a mutated hygromycin B resistance gene. *Arch. Microbiol.* **176:**364–369.

69. Hartner, F. S., C. Ruth, D. Langenegger, S. N. Johnson, P. Hyka, G. P. Lin-Cereghino, J. Lin-Cereghino, K. Kovar, J. M. Cregg, and A. Glieder. 2008. Promoter library designed for fine-tuned gene expression in *Pichia pastoris*. *Nucleic Acids Res.* **36:**e76.

70. Hatfield, G. W., L. Z. Larsen, C. D. Wassman, S. P. Hung, I. B. Chen, and R. H. Lathrop. 2006. Translationally engineering synthetic genes for high level protein production in heterologous host systems. *Mol. Cell. Proteomics* **5:**779.

71. Heo, J. H., W. K. Hong, E. Y. Cho, M. W. Kim, J. Y. Kim, C. H. Kim, S. K. Rhee, and H. A. Kang. 2003. Properties of the *Hansenula polymorpha*-derived constitutive GAP promoter, assessed using an HSA reporter gene. *FEMS Yeast Res.* **4:**175–184.

71a. Hiep, T. T., V. N. Noskov, and Y. I. Pavlov. 1993. Transformation in the methylotrophic yeast *Pichia methanolica* utilizing homologous *ADE1* and heterologous *Saccharomyces cerevisiae ADE2* and *LEU2* genes as genetic markers. *Yeast* **9:**1189–1197.

72. Higgins, D. R., and J. M. Cregg. 1998. Pichia *Protocols. Methods Mol. Biol.* **103.** Humana Press, Totowa, NJ.

73. Holz, M., A. Forster, S. Mauersberger, and G. Barth. 2009. Aconitase overexpression changes the product ratio of citric acid production by *Yarrowia lipolytica*. *Appl. Microbiol. Biotechnol.* **81:**1087–1096.

74. Hong, I. P., S. Anderson, and S. G. Choi. 2006. Evaluation of a new episomal vector based on the GAP promoter for structural genomics in *Pichia pastoris*. *J. Microbiol. Biotechnol.* **16:**1362–1368.

75. Hu, S. Y., L. W. Li, J. J. Qiao, Y. J. Guo, L. S. Cheng, and J. Liu. 2006. Codon optimization, expression, and characterization of an internalizing anti-ErbB2 single-chain antibody in *Pichia pastoris*. *Protein Expr. Purif.* **47:**249–257.

76. Huang, H., P. Yang, H. Luo, H. Tang, N. Shao, T. Yuan, Y. Wang, Y. Bai, and B. Yao. 2008. High-level expression of a truncated 1,3-1,4-beta-D-glucanase from *Fibrobacter succinogenes* in *Pichia pastoris* by optimization of codons and fermentation. *Appl. Microbiol. Biotechnol.* **78:**95–103.

77. Hygromycin.com. 2004. Hygromycin Service. Hygromycin.com, a division of A.G. Scientific, Inc. http://www.hygromycin.com/service.htm. [Online.]

78. Iizumi, S., A. Kurosawa, S. So, Y. Ishii, Y. Chikaraishi, A. Ishii, H. Koyama, and N. Adachi. 2008. Impact of non-homologous end-joining deficiency on random and targeted DNA integration: implications for gene targeting. *Nucleic Acids Res.* **36:**6333–6342.

79. Inan, M., D. Aryasomayajula, J. Sinha, and M. M. Meagher. 2006. Enhancement of protein secretion in *Pichia pastoris* by overexpression of protein disulfide isomerase. *Biotechnol. Bioeng.* **93:**771–778.

80. Ingolia, N. T., S. Ghaemmaghami, J. R. Newman, and J. S. Weissman. 2009. Genome-wide analysis in vivo of translation with nucleotide resolution using ribosome profiling. *Science* **324:**218–223.

81. Iwakiri, R., Y. Noda, H. Adachi, and K. Yoda. 2006. Isolation and characterization of promoters suitable for a multidrug-resistant marker CuYAP1 in the yeast *Candida utilis*. *Yeast* **23:**23–34.

82. Iwata, T., R. Tanaka, M. Suetsugu, M. Ishibashi, H. Tokunaga, M. Kikuchi, and M. Tokunaga. 2004. Efficient secretion of human lysozyme from the yeast, *Kluyveromyces lactis*. *Biotechnol. Lett.* **26:**1803–1808.

83. Jacobs, P. P., S. Geysens, W. Vervecken, R. Contreras, and N. Callewaert. 2009. Engineering complex-type N-glycosylation in *Pichia pastoris* using GlycoSwitch technology. *Nat. Protocols* **4:**58–70.

84. Jain, E., and A. Kumar. 2008. Upstream processes in antibody production: evaluation of critical parameters. *Biotechnol. Adv.* **26:**46–72.

85. Janatova, I., P. Costaglioli, J. Wesche, J. M. Masson, and E. Meilhoc. 2003. Development of a reporter system for the yeast *Schwanniomyces occidentalis*: influence of DNA composition and codon usage. *Yeast* **20:**687–701.

86. Jeffries, T. W. 2006. Engineering yeasts for xylose metabolism. *Curr. Opin. Biotechnol.* **17:**320–326.

87. Jeffries, T. W., and J. H. Van Vleet. 2009. *Pichia stipitis* genomics, transcriptomics, and gene clusters. *FEMS Yeast Res.* **9:**793–807.

88. Jenks, M. H., and D. Reines. 2005. Dissection of the molecular basis of mycophenolate resistance in *Saccharomyces cerevisiae*. *Yeast* **22:**1181–1190.

89. Jones, S. D., F. J. Castillo, and H. L. Levine. 2007. Advances in the development of therapeutic monoclonal antibodies. *BioPharm Internat.* **20:**96–107.

90. Juretzek, T., M. T. Le Dall, S. Mauersberger, C. Gaillardin, G. Barth, and J. M. Nicaud. 2001. Vectors for gene expression and amplification in the yeast *Yarrowia lipolytica*. *Yeast* **18:**97–113.

91. Kato, M., H. Iefuji, K. Miyake, and Y. Iimura. 1997. Transformation system for a wastewater treatment yeast, *Hansenula fabianii* J640: isolation of the orotidine-5′-phosphate decarboxylase gene (*URA3*) and uracil auxotrophic mutants. *Appl. Microbiol. Biotechnol.* **48:**621–625.

92. Kim, I. G., S. K. Nam, J. H. Sohn, S. K. Rhee, G. H. An, S. H. Lee, and E. S. Choi. 1998. Cloning of the ribosomal protein L41 gene of *Phaffia rhodozyma* and its use as a drug resistance marker for transformation. *Appl. Environ. Microbiol.* **64:**1947–1949.

93. Kim, M. D., Y. J. Yoo, S. K. Rhee, and J. H. Seo. 2001. Enhanced transformation efficiency of an anticoagulant hirudin gene into *Saccharomyces cerevisiae* by a double delta-sequence. *J. Microbiol. Biotechnol.* **11:**61–64.

94. Klinner, U., and B. Schafer. 2004. Genetic aspects of targeted insertion mutagenesis in yeasts. *FEMS Microbiol. Rev.* **28:**201–223.

95. Köhler, G. A., T. C. White, and N. Agabian. 1997. Overexpression of a cloned IMP dehydrogenase gene of *Candida albicans* confers resistance to the specific inhibitor mycophenolic acid. *J. Bacteriol.* **179:**2331–2338.

96. Kong, D. C., and M. L. DePamphilis. 2002. Site-specific ORC binding, pre-replication complex assembly and DNA synthesis at *Schizosaccharomyces pombe* replication origins. *EMBO J.* **21:**5567–5576.

97. Kooistra, R., P. J. J. Hooykaas, and H. Y. Steensma. 2004. Efficient gene targeting in *Kluyveromyces lactis*. *Yeast* **21:**781–792.

98. Kudla, G., A. W. Murray, D. Tollervey, and J. B. Plotkin. 2009. Coding-sequence determinants of gene expression in *Escherichia coli*. *Science* **324:**255–258.

99. Kumari, K.S., I. S. Babu, and G. H. Rao. 2008. Process optimization for citric acid production from raw glycerol using response surface methodology. *Indian J. Biotechnol.* **7:**496–501.

100. Laitila, A., A. Wilhelmson, E. Kotaviita, J. Olkku, S. Home, and R. Juvonen. 2006. Yeasts in an industrial malting ecosystem. *J. Indust. Microbiol. Biotechnol.* **33:**953–966.

101. Laplaza, J. M., and T. W. Jeffries. March, 2009. Yeast transformation system. U.S. patent 7,501,275.

102. Laplaza, J. M., B. R. Torres, Y. S. Jin, and T. W. Jeffries. 2006. Sh ble and Cre adapted for functional genomics and metabolic engineering of *Pichia stipitis*. *Enzyme Microb. Technol.* **38:**741–747.

103. Lathrop, R. H., L. Z. Larsen, C. D. Wassman, R. Colman, S. P. Hung, H. Wunsch, J. D. Kittle, and G. W. Hatfield. 2006. Computationally optimized DNA assembly. *Mol. Cell. Proteomics* **5:**776.

104. Lee, C. C., T. G. Williams, D. W. S. Wong, and G. H. Robertson. 2005. An episomal expression vector for screening mutant gene libraries in *Pichia pastoris*. *Plasmid* **54:**80–85.

105. Lee, F. W. F., and N. A. DaSilva. 1997. Improved efficiency and stability of multiple cloned gene insertions at the delta sequences of *Saccharomyces cerevisiae*. *Appl. Microbiol. Biotechnol.* **48:**339–345.

106. Lee, F. W. F., and N. A. DaSilva. 1996. Ty1-mediated integration of expression cassettes: host strain effects, stability, and product synthesis. *Biotechnol. Progr.* **12:** 548–554.

107. Li, H. J., N. Sethuraman, T. A. Stadheim, D. X. Zha, B. Prinz, N. Ballew, P. Bobrowicz, B. K. Choi, W. J. Cook, M. Cukan, N. R. Houston-Cummings, R. Davidson, B. Gong, S. R. Hamilton, J. P. Hoopes, Y. W. Jiang, N. Kim, R. Mansfield, J. H. Nett, S. Rios, R. Strawbridge, S. Wildt, and T. U. Gerngross. 2006. Optimization of humanized IgGs in glycoengineered *Pichia pastoris*. *Nat. Biotechnol.* **24:**210–215.

108. Li, P. Z., A. Anumanthan, X. G. Gao, K. Ilangovan, V. V. Suzara, N. Duzgunes, and V. Renugopalakrishnan. 2007. Expression of recombinant proteins in *Pichia pastoris*. *Appl. Biochem. Biotechnol.* **142:**105–124.

109. Lin Cereghino, G. P., J. Lin Cereghino, A. J. Sunga, M. A. Johnson, M. Lim, M. A. G. Gleeson, and J. M. Cregg. 2001. New selectable marker/auxotrophic host strain combinations for molecular genetic manipulation of *Pichia pastoris*. *Gene* **263:**159–169.

110. Lin-Cereghino, J., W. W. Wong, S. Xiong, W. Giang, L. T. Luong, J. Vu, S. D. Johnson, and G. P. Lin-Cereghino. 2005. Condensed protocol for competent cell preparation and transformation of the methylotrophic yeast *Pichia pastoris*. *BioTechniques* **38:**44–48.

111. Lodato, P., J. Alcaino, S. Barahona, M. Niklitschek, M. Carmona, A. Wozniak, M. Baeza, A. Jimenez, and V. Cifuentes. 2007. Expression of the carotenoid biosynthesis genes in *Xanthophyllomyces dendrorhous*. *Biol. Res.* **40:**73–84.

112. Lu, P., B. P. Davis, J. Hendrick, and T. W. Jeffries. 1998. Cloning and disruption of the beta-isopropylmalate dehydrogenase gene (*LEU2*) of *Pichia stipitis* with *URA3* and recovery of the double auxotroph. *Appl. Microbiol. Biotechnol.* **49:**141–146.

113. Lukacs, G., B. Linka, and I. Nyilasi. 2006. *Phaffia rhodozyma* and *Xanthophyllomyces dendrorhous*: astaxanthin-producing yeasts of biotechnological importance—a minireview. *Acta Aliment.* **35:**99–107.

114. Maassen, N., S. Freese, B. Schruff, V. Passoth, and U. Klinner. 2008. Nonhomologous end joining and homologous recombination DNA repair pathways in integration mutagenesis in the xylose-fermenting yeast *Pichia stipitis*. *FEMS Yeast Res.* **8:**735–743.

115. Madzak, C., C. Gaillardin, and J. M. Beckerich. 2004. Heterologous protein expression and secretion in the non-conventional yeast *Yarrowia lipolytica*: a review. *J. Biotechnol.* **109:**63–81.

116. Maggon, K. 2008. Global Monoclonal Antibodies Market Review 2008 (World Top Ten MAbs). Knol [Online.] http://knol.google.com/k/krishan-maggon/global-monoclonal-antibodies-market/3fy5eowy8suq3/11#

117. Mauersberger, S., H. J. Wang, C. Gaillardin, G. Barth, and J. M. Nicaud. 2001. Insertional mutagenesis in the n-alkane-assimilating yeast *Yarrowia lipolytica*: generation of tagged mutations in genes involved in hydrophobic substrate utilization. *J. Bacteriol.* **183:**5102–5109.

118. Micheelsen, P. O., P. R. Ostergaard, L. Lange, and M. Skjot. 2008. High-level expression of the native barley alpha-amylase/subtilisin inhibitor in *Pichia pastoris*. *J. Biotechnol.* **133:**424–432.

119. Minhas, A., D. Biswas, and A. K. Mondal. 2009. Development of host and vector for high-efficiency transformation and gene disruption in *Debaryomyces hansenii*. *FEMS Yeast Res.* **9:**95–102.

120. Nett, J. H., and T. U. Gemgross. 2003. Cloning and disruption of the *PpURA5* gene and construction of a set of integration vectors for the stable genetic modification of *Pichia pastoris*. *Yeast* **20:**1279–1290.

121. Neuveglise, C., J. M. Nicaud, P. Ross-Macdonald, and C. Gaillardin. 1998. A shuttle mutagenesis system for tagging genes in the yeast *Yarrowia lipolytica*. *Gene* **213:**37–46.

122. Ninomiya, Y., K. Suzuki, C. Ishii, and H. Inoue. 2004. Highly efficient gene replacements in *Neurospora* strains deficient for nonhomologous end-joining. *Proc. Natl. Acad. Sci. USA* **101:**12248–12253.

123. Park, C. S., C. C. Chang, and D. D. Ryu. 2000. Expression and high-level secretion of *Trichoderma reesei* endoglucanase I in *Yarrowia lipolytica*. *Appl. Biochem. Biotechnol.* **87:**1–15.

124. Park, C. S., Y. S. Sohn, C. Crispino, C. C. Chang, and D. D. Y. Ryu. 1998. Isolation of oversecreting mutant strains of the yeast *Yarrowia lipolytica*. *J. Ferment. Bioeng.* **85:**180–184.

125. Park, H., N. I. Lopez, and A. T. Bakalinsky. 1999. Use of sulfite resistance in *Saccharomyces cerevisiae* as a dominant selectable marker. *Curr. Genet.* **36:**339–344.

126. Pignede, G., H. J. Wang, F. Fudalej, M. Seman, C. Gaillardin, and J. M. Nicaud. 2000. Autocloning and amplification of LIP2 in *Yarrowia lipolytica*. *Appl. Environ. Microbiol.* **66:**3283–3289.

127. Porro, D., M. Sauer, P. Branduardi, and D. Mattanovich. 2005. Recombinant protein production in yeasts. *Mol. Biotechnol.* **31:**245–259.

128. Potgieter, T. I., M. Cukan, J. E. Drummond, N. R. Houston-Cummings, Y. W. Jiang, F. Li, H. Lynaugh, M. Mallem, T. W. McKelvey, T. Mitchell, A. Nylen, A. Rittenhour, T. A. Stadheim, D. X. Zha, and M. d'Anjou. 2009. Production of monoclonal antibodies by glycoengineered *Pichia pastoris*. *J. Biotechnol.* **139:**318–325.

129. Prista, C., M. C. Loureiro-Dias, V. Montiel, R. Garcia, and J. Ramos. 2005. Mechanisms underlying the halotolerant way of *Debaryomyces hansenii*. *FEMS Yeast Res.* **5:**693–701.

130. Raimondi, S., E. Zanni, C. Talora, M. Rossi, C. Palleschi, and D. Uccelletti. 2008. SOD1, a new *Kluyveromyces lactis* helper gene for heterologous protein secretion. *Appl. Environ. Microbiol.* **74:**7130–7137.

131. Rajgarhia, V., V. Hatzimanikatis, S. Olson, T. Carlson, J. N. Starr, J. J. Kolstad, and A. Eyal. November, 2002. Production of lactate using Crabtree negative organisms in varying culture conditions. U.S. patent 6,485,947 B1.

132. Rapoport, T. A. 2008. Protein transport across the endoplasmic reticulum membrane. *FEBS J.* **275:**4471–4478.

133. **Reichert, J., and A. Pavlou.** 2004. Monoclonal antibodies market. *Nat. Rev. Drug Discovery* **3**:383–384.

134. **Reichert, J. M.** 2008. Monoclonal antibodies as innovative therapeutics. *Curr. Pharm. Biotechnol.* **9**:423–430.

135. **Resina, D., A. Serrano, F. Valero, and P. Ferrer.** 2004. Expression of a *Rhizopus oryzae* lipase in *Pichia pastoris* under control of the nitrogen source-regulated formaldehyde dehydrogenase promoter. *J. Biotechnol.* **109**:103–113.

136. **Rezaee, A.** 2003. A rapid and sensitive assay of beta-galactosidase in yeast cells. *Ann. Microbiol.* **53**:343–347.

137. **Ricaurte, M. L., and N. S. Govind.** 1999. Construction of plasmid vectors and transformation of the marine yeast *Debaryomyces hansenii*. *Marine Biotechnol.* **1**:15–19.

138. **Rich, J. R., and S. G. Withers.** 2009. Emerging methods for the production of homogeneous human glycoproteins. *Nat. Chem. Biol.* **5**:206–215.

139. **Rose, K., M. Liebergesell, and A. Steinbuchel.** 2000. Molecular analysis of the *Aureobasidium pullulans ura3* gene encoding orotidine-5′-phosphate decarboxylase and isolation of mutants defective in this gene. *Appl. Microbiol. Biotechnol.* **53**:296–300.

140. **Rothman, R. J., B. Perussia, D. Herlyn, and L. Warren.** 1989. Antibody dependent cytotoxicity mediated by natural killer cells is enhanced by castanospermine induced alterations of IGG glycosylation. *Mol. Immunol.* **26**:1113–1123.

141. **Ruohonen, L., A. Aristidou, A. D. Frey, M. Penttila, and P. T. Kallio.** 2006. Expression of *Vitreoscilla* hemoglobin improves the metabolism of xylose in recombinant yeast *Saccharomyces cerevisiae* under low oxygen conditions. *Enzyme Microb. Technol.* **39**:6–14.

141a. **Ryabova, E. B., O. M. Chmil, and A. A. Sibirny.** 2003. Xylose and cellobiose fermentation to ethanol by the thermotolerant methylotrophic yeast *Hansenula polymorpha*. *FEMS Yeast Res.* **4**:157–164.

142. **Sakai, Y., and Y. Tani.** 1992. Directed mutagenesis in an asporogenous methylotrophic yeast—cloning, sequencing and one step gene disruption of the 3-isopropylmalate dehydrogenase gene (*LEU2*) of *Candida boidinii* to derive doubly auxotrophic marker strains. *J. Bacteriol.* **174**:5988–5993.

143. **Santoyo, G., and J. N. Strathern.** 2008. Non-homologous end joining is important for repair of Cr(VI)-induced DNA damage in *Saccharomyces cerevisiae*. *Microbiol. Res.* **163**:113–119.

144. **Sato, M., S. Dhut, and T. Toda.** 2005. New drug-resistant cassettes for gene disruption and epitope tagging in *Schizosaccharomyces pombe*. *Yeast* **22**:583–591.

145. **Scorer, C. A., J. J. Clare, W. R. McCombie, M. A. Romanos, and K. Sreekrishna.** 1994. Rapid selection using G418 of high copy number transformants of *Pichia pastoris* for high level foreign gene expression. *BioTechnology* **12**:181–184.

146. **Sears, I. B., J. O'Connor, O. W. Rossanese, and B. S. Glick.** 1998. A versatile set of vectors for constitutive and regulated gene expression in *Pichia pastoris*. *Yeast* **14**:783–790.

147. **Shen, S. G., G. Sulter, T. W. Jeffries, and J. M. Cregg.** 1998. A strong nitrogen source-regulated promoter for controlled expression of foreign genes in the yeast *Pichia pastoris*. *Gene* **216**:93–102.

148. **Shusta, E. V., R. T. Raines, A. Pluckthun, and K. D. Wittrup.** 1998. Increasing the secretory capacity of *Saccharomyces cerevisiae* for production of single-chain antibody fragments. *Nat. Biotechnol.* **16**:773–777.

149. Reference deleted.

150. **Sinclair, G., and F. Y. M. Choy.** 2002. Synonymous codon usage bias and the expression of human glucocerebrosidase in the methylotrophic yeast, *Pichia pastoris*. *Protein Expr. Purif.* **26**:96–105.

151. **Smith, J. D., B. C. Tang, and A. S. Robinson.** 2004. Protein disulfide isomerase, but not binding protein, overexpression enhances secretion of a non-disulfide-bonded protein in yeast. *Biotechnol. Bioeng.* **85**:340–350.

152. **Smulian, A. G., R. S. Gibbons, J. A. Demland, D. T. Spaulding, and G. S. Deepe.** 2007. Expression of hygromycin phosphotransferase alters virulence of *Histoplasma capsulatum*. *Eukaryot. Cell* **6**:2066–2071.

153. **Song, H. T., Z. B. Jiang, and L. X. Ma.** 2006. Expression and purification of two lipases from *Yarrowia lipolytica* AS 2.1216. *Protein Expr. Purif.* **47**:393–397.

154. **Steinborn, G., G. Gellissen, and G. Kunze.** 2007. A novel vector element providing multicopy vector integration in *Arxula adeninivorans*. *FEMS Yeast Res.* **7**:1197–1205.

155. **Steinborn, G., T. Wartmann, G. Gellissen, and G. Kunze.** 2007. Construction of an *Arxula adeninivorans* host-vector system based on trp1 complementation. *J. Biotechnol.* **127**:392–401.

156. **Stevens, D. R., A. Atteia, L. G. Franzen, and S. Purton.** 2001. Cycloheximide resistance conferred by novel mutations in ribosomal protein L41 of *Chlamydomonas reinhardtii*. *Mol. Gen. Genet.* **264**:790–795.

157. **Struhl, K., D. T. Stinchcomb, S. Scherer, and R. W. Davis.** 1979. High frequency transformation of yeast—autonomous replication of hybrid DNA molecules. *Proc. Natl. Acad. Sci. USA* **76**:1035–1039.

158. **Sugita, T., and T. Nakase.** 1999. Non-universal usage of the leucine CUG codon and the molecular phylogeny of the genus *Candida*. *Syst. Appl. Microbiol.* **22**:79–86.

159. **Suthar, D. H., and B. B. Chattoo.** 2006. Expression of *Vitreoscilla* hemoglobin enhances growth and levels of alpha-amylase in *Schwanniomyces occidentalis*. *Appl. Microbiol. Biotechnol.* **72**:94–102.

160. **Tadashi, T., T. Masuda, and Y. Koyama.** 2006. Enhanced gene targeting frequency in ku70 and ku80 disruption mutants of *Aspergillus sojae* and *Aspergillus oryzae*. *Mol. Genet. Genomics* **275**:460–470.

161. **Tang, S. J., K. H. Sun, G. H. Sun, T. Y. Chang, W. L. Wu, and G. C. Lee.** 2003. A transformation system for the nonuniversal CUG(Ser) codon usage species *Candida rugosa*. *J. Microbiol. Methods* **52**:231–238.

162. **Tokuoka, M., M. Tanaka, K. Ono, S. Takagi, T. Shintani, and K. Gomi.** 2008. Codon optimization increases steady-state mRNA levels in *Aspergillus oryzae* heterologous gene expression. *Appl. Environ. Microbiol.* **74**:6538–6546.

163. Reference deleted.

164. **Ukibe, K., T. Katsuragi, Y. Tani, and H. Takagi.** 2008. Efficient screening for astaxanthin-overproducing mutants of the yeast *Xanthophyllomyces dendrorhous* by flow cytometry. *FEMS Microbiol. Lett.* **286**:241–248.

165. **Van Bogaert, I. N. A., S. L. De Maeseneire, D. Develter, W. Soetaert, and E. J. Vandamme.** 2008. Cloning and characterisation of the glyceraldehyde 3-phosphate dehydrogenase gene of *Candida bombicola* and use of its promoter. *J. Ind. Microbiol. Biotechnol.* **35**:1085–1092.

166. **van Dijk, R., K. N. Faber, A. T. Hammond, B. S. Glick, M. Veenhuis, and J. Kiel.** 2001. Tagging *Hansenula polymorpha* genes by random integration of linear DNA fragments (RALF). *Mol. Genet. Genomics* **266**:646–656.

167. **Van Driessche, B., L. Tafforeau, P. Hentges, A. M. Carr, and J. Vandenhaute.** 2005. Additional vectors for PCR-based gene tagging in *Saccharomyces cerevisiae* and *Schizosaccharomyces pombe* using nourseothricin resistance. *Yeast* **22**:1061–1068.

168. **Vernis, L., L. Poljak, M. Chasles, K. Uchida, S. Casaregola, E. Kas, M. Matsuoka, C. Gaillardin, and P. Fournier.** 2001. Only centromeres can supply the

partition system required for ARS function in the yeast *Yarrowia lipolytica*. *J. Mol. Biol.* **305:**203–217.

169. **Wach, A., A. Brachat, R. Pohlmann, and P. Philippsen.** 1994. New heterologous modules for classical or PCR based gene disruptions in *Saccharomyces cerevisiae*. *Yeast* **10:**1793–1808.

170. **Wang, H., Q. L. Wang, F. F. Zhang, Y. H. Huang, Y. L. Ji, and Y. Hou.** 2008. Protein expression and purification of human Zbtb7A in *Pichia pastoris* via gene codon optimization and synthesis. *Protein Expr. Purif.* **60:**97–102.

171. **Wang, T. T., Y. J. Choi, and B. H. Lee.** 2001. Transformation systems of non-*Saccharomyces* yeasts. *Crit. Rev. Biotechnol.* **21:**177–218.

172. **Wang, Y. C. M., L. L. Chuang, F. W. F. Lee, and N. A. Da Silva.** 2003. Sequential cloned gene integration in the yeast *Kluyveromyces lactis*. *Appl. Microbiol. Biotechnol.* **62:**523–527.

173. **Wartmann, T., C. Bellebna, E. Boer, O. Bartelsen, G. Gellissen, and G. Kunze.** 2003. The constitutive AHSB4 promoter—a novel component of the *Arxula adeninivorans*-based expression platform. *Appl. Microbiol. Biotechnol.* **62:**528–535.

174. **Wartmann, T., and G. Kunze.** 2000. Genetic transformation and biotechnological application of the yeast *Arxula adeninivorans*. *Appl. Microbiol. Biotechnol.* **54:**619–624.

175. **Wartmann, T., R. Stoltenburg, E. Boer, H. Sieber, O. Bartelsen, G. Gellissen, and G. Kunze.** 2003. The *ALEU2* gene—a new component for an *Arxula adeninivorans*-based expression platform. *FEMS Yeast Res.* **3:**223–232.

176. **Wentz, A. E., and E. V. Shusta.** 2007. Novel high-throughput screen reveals yeast genes that increase secretion of heterologous proteins. *Appl. Environ. Microbiol.* **73:**1189–1198.

177. **Werner, R. G.** 2004. Economic aspects of commercial manufacture of biopharmaceuticals. *J. Biotechnol.* **113:**171–182.

178. **Wiedemann, B., and E. Boles.** 2008. Codon-optimized bacterial genes improve L-arabinose fermentation in recombinant *Saccharomyces cerevisiae*. *Appl. Environ. Microbiol.* **74:**2043–2050.

179. **Wu, A. B., H. D. Chen, Z. Z. Tang, B. W. Ye, W. J. Liu, H. Y. Jia, and D. B. Zhang.** 2007. Synthesis of *Drosophila melanogaster* acetylcholinesterase gene using yeast preferred codons and its expression in *Pichia pastoris*. Presented at the 9th International Meeting on Cholinesterases, Suzhou, Peoples Republic of China, 6–10 May 2007.

180. **Wu, G., N. Bashir-Bello, and S. J. Freeland.** 2006. The Synthetic Gene Designer: a flexible web platform to explore sequence manipulation for heterologous expression. *Protein Expr. Purif.* **47:**441–445.

181. **Wu, J. M., T. A. Hsu, and C. K. Lee.** 2003. Expression of the gene coding for bacterial hemoglobin improves beta-galactosidase production in a recombinant *Pichia pastoris*. *Biotechnol. Lett.* **25:**1457–1462.

182. **Yamane, T., T. Ogawa, and M. Matsuoka.** 2008. Derivation of consensus sequence for protein binding site in *Yarrowia lipolytica* centromere. *J. Biosci. Bioeng.* **105:**671–674.

183. **Yamane, T., H. Sakai, K. Nagahama, T. Ogawa, and M. Matsuoka.** 2008. Dissection of centromeric DNA from yeast *Yarrowia lipolytica* and identification of protein-binding site required for plasmid transmission. *J. Biosci. Bioeng.* **105:**571–578.

184. **Yang, V. W., J. A. Marks, B. P. Davis, and T. W. Jeffries.** 1994. High-efficiency transformation of *Pichia stipitis* based on its URA3 gene and a homologous autonomous replication sequence, Ars2. *Appl. Environ. Microbiol.* **60:**4245–4254.

185. **Ye, R. W., H. Yao, K. Stead, T. Wang, L. Tao, Q. Cheng, P. L. Sharpe, W. Suh, E. Nagel, D. Arcilla, D. Dragotta, and E. S. Miller.** 2007. Construction of the astaxanthin biosynthetic pathway in a methanotrophic bacterium, *Methylomonas* sp. strain 16a. *J. Ind. Microbiol. Biotechnol.* **34:**289–299.

186. **Yehuda, H., S. Droby, M. Wisniewski, and M. Goldway.** 2001. A transformation system for the biocontrol yeast, *Candida oleophila*, based on hygromycin B resistance. *Curr. Genet.* **40:**282–287.

187. **Zalacain, M., A. González, M. C. Guerrero, R. J. Mattaliano, F. Malpartida, and A. Jimenez.** 1986. Nucleotide sequence of the hygromycin B phosphotransferase gene from *Streptomyces hygroscopicus*. *Nucleic Acids Res.* **14:**1565–1581.

188. **Zhang, W., H. L. Zhao, C. Xue, X. H. Xiong, X. Q. Yao, X. Y. Li, H. P. Chen, and Z. M. Liu.** 2006. Enhanced secretion of heterologous proteins in *Pichia pastoris* following overexpression of *Saccharomyces cerevisiae* chaperone proteins. *Biotechnol. Progr.* **22:**1090–1095.

Genetics, Genetic Manipulation, and Approaches to Strain Improvement of Filamentous Fungi

VERA MEYER, ARTHUR F. J. RAM, AND PETER J. PUNT

22

22.1. INTRODUCTION

Only a limited number of filamentous fungi are used in industry for the production of enzymes or metabolites. They include *Aspergillus niger*, *Aspergillus oryzae*, *Trichoderma reesei*, and *Penicillium chrysogenum*. An industrial filamentous fungus must meet several demands such as strong and versatile metabolic capacities, good growth characteristics on low-cost substrates, and the ability to be cultivated in large-scale fermentors. In addition, the strain should not be pathogenic or produce any toxic by-products during the fermentation process. Finally, to allow further improvements and optimization of the production process, a successful industrial filamentous fungus should be amenable to genetic manipulation and genomic approaches. The filamentous fungi exploited these days as cell factories all meet these criteria (Table 1). Within the past few years, genome sequences of several industrial fungi have been published (34, 36, 47, 56), clearly documenting the genetic diversity remaining to be understood and exploited for the production of new proteins or metabolites. In addition to the genome sequence data, related "omics" approaches also have the potential to generate leads for further strain improvement by genetic engineering. With the recent breakthroughs in DNA and RNA sequencing methods, the number of fungal genome sequences continues to grow, as does the need for efficient fungal expression platforms to produce new and interesting compounds. In this review, we will focus on some recently developed genetic tools and strain improvement approaches for further optimizing protein production in filamentous fungi in the "omics" era. The technologies and approaches described are mainly focused on A. niger, but in principle could be applied to any of the other industrially relevant filamentous fungi.

22.2. GENETICS OF FUNGAL CELL FACTORIES IN THE "OMICS" ERA

Within the past 5 years, genome sequences for the most important industrial filamentous fungi have become available (Table 1), facilitating four main areas of current research: (i) genome mining and identification of new compounds and enzymes; (ii) in silico reconstitution of metabolic pathways; (iii) systems biology approaches to comprehensively understand the complex regulation of growth, physiology, and metabolism in fungi; and (iv) further development of industrial strains as cell factories for the production of enzymes or compounds.

22.2.1. Genome Mining and Identifications of New Proteins and Metabolites

Without exception, the analysis of the genomes of industrial filamentous fungi has uncovered previously unknown enzymes, especially for carbohydrate metabolism (16, 34–36, 47). In order to predict the functions of these new enzymes, BLAST analyses and searches for carbohydrate-active domains in the CAZy database are combined with expression profiling of the respective genes. A great challenge is to determine substrate specificities and catalytic properties of these enzymes (7). Genome sequences have also revealed many new proteases and lipases, which are interesting candidate products for other industrial applications (27, 34, 47).

Filamentous fungi, including the industrial strains, are well known for their capacity to produce an astonishing wealth of secondary metabolites including both commercially exploited compounds and unwanted mycotoxins (1, 24, 26, 51). Compounds produced by industrial filamentous fungi include penicillin, by *P. chrysogenum*, and lovastatin by *Aspergillus terreus* (6, 56). The availability of genome sequence data and various genomics tools, including DNA microarrays, has allowed the comparison of low- and high-producing strains and identification of the factors involved in metabolite synthesis at the molecular level (6, 56).

The analysis of genomic data from A. niger and A. oryzae has revealed the presence of several secondary metabolite gene clusters. The numbers of polyketide synthase (PKS) (34), nonribosomal peptide synthase (NRPS) (17), and hybrid PKS-NRPS (7) genes in A. niger illustrate its potential for secondary metabolite production (47). Among these secondary metabolites, mycotoxins like fumonisins can potentially be produced (18). However, the structure of most fungal secondary metabolites has not been elucidated, and the growth conditions required for their synthesis are not known. Therefore, much current research is focused on identifying the products of these new secondary metabolite clusters. A straightforward approach to this end is (i) deletion of, e.g., a PKS-encoding gene, (ii) cultivation of the mutant strain and the respective wild-type strain under various growth conditions, and (iii) comparison of their metabolic profiles to identify missing or altered metabolite(s) in the mutant strain (13).

318

TABLE 1 Overview of selected industrial filamentous fungi, their products, and the current "omics" status (spring 2009)

Organism	Main products	Genomic sequence available (reference)	Transcriptomics platforms available	NHEJ-deficient strains available
A. niger	Citric acid Amylases Lipases Proteases Phytases Xylanases Lactoferrin Chymosin Glucose oxidase	Yes[a] (47; www.broad.mit.edu)	Yes (Affymetrix)	Yes (37)
A. oryzae	Kojic acid Amylases Glucose oxidase Phytases	Yes (34)	Yes (Affymetrix)	Yes (53)
A. terreus	Itaconic acid Lovastatin	Yes (www.broad.mit.edu)	Yes (cDNA arrays)	No
C. lucknowense	Hemicellulases	Yes (www.dyadic.com)	No	Yes (59)
P. chrysogenum	Penicillin	Yes (56)	Yes (Affymetrix)	Yes (52)
T. reesei	Cellulases Cellobiohydrolases Xylanases Hydrophobins	Yes (36)	Yes (cDNA arrays, TRAC[b])	Yes (20)

[a]The genome sequences of two A. niger strains have been published, strain CBS 513.88 (47) and ATCC 1015 (www.broad.mit.edu).
[b]TRAC, transcript analysis with aid of affinity capture.

22.2.2. Reconstitution of Metabolic Pathways and Systems Biology Approaches

Using the complete genome sequences of A. niger and A. oryzae, Nielsen and coworkers have reconstructed the metabolic networks of both organisms, leading to genome-scale metabolic models for both fungi (3, 47, 64). These metabolic networks, in combination with newly established tools for filamentous fungi such as transcriptomics, proteomics, and metabolomics (see reference 5 for details), will provide new leads for process development and strain improvements. In addition, in top-down systems biology approaches, large data sets from transcriptomic, proteomic, or metabolomic studies are used in an unbiased way to perform statistical data analysis (preferably multivariate data analysis) and to identify relevant genes, proteins, or metabolites. Two recent publications (10, 46) provide good examples of how the application of systems biology tools helps to understand biological processes in filamentous fungi and generates leads for process and strain improvement programs.

22.3. GENETIC MANIPULATION

One key step in the improvement of fungal production strains is their optimization at the molecular level. Two complementary approaches can be followed: nonrecombinant and recombinant methods. The power of nonrecombinant approaches to achieve genetic alterations, e.g., by mutagenesis or protoplast fusion, was reviewed in the last edition of this manual (62). During the past decade, however, considerable progress has been made in establishing very efficient recombinant methods to introduce and control gene expression in filamentous fungi. In this chapter, we focus on

the most recent key concepts and technologies for genetic engineering of filamentous fungi. To obtain deeper insights into the theoretical basis of these technologies, the reader is directed to a number of recent reviews (39, 40, 45, 50).

One of the first steps in a project aimed at efficient production in a fungal host is the choice of the host strain itself. The suitability of the specific host strain depends on the amenability of the strain for genetic transformation and the availability of large-scale fermentation processes. With respect to efficient production of specific proteins, important considerations include the availability of suitable expression cassettes, appropriate secretion signals, and protease-deficient strains (to ensure low proteolytic activity against the protein of interest). For most of the strains listed in Table 1, good transformation characteristics using a variety of selection markers have been published.

22.3.1. Selection Marker

For fungal transformation, various dominant and auxotrophic selection markers have been developed (Table 2). Very versatile dominant selection markers include hygromycin B and phleomycin resistance based on prokaryotic resistance genes (49). However, for use in industrial hosts, selection based on auxotrophic and nutritional markers is highly preferred, as these markers can also be used in so-called self-cloning approaches. In the European Union (EU), the definition of self-cloning is described in Directive 90/219/EEC: self-cloning is the reintroduction of DNA "into cells of the same species or into cells of phylogenetically closely related species which can exchange genetic material by natural physiological processes where the resulting micro-organism is unlikely to cause disease to humans, animals or plants".

TABLE 2 Selection markers, respective vectors, and the impact of the selection strategy on integration events

Selection marker	Origin	Type	Multicopy selection[a]	Locus-specific integration[b]	Available vector	Reference
Hygromycin B resistance	*Escherichia coli* (*hph*)	Dominant	+	−	pAN7-1	48
Phleomycin resistance	*Streptoalloteichus hindustanus* (*ble*)	Dominant	+	−	pAN8-1	49
Acetamidase	*Aspergillus nidulans* (*amdS*)	Dominant	++	−	p3SR2, pMTL-25synAmdS	15
Orotidine-5′-phosphate decarboxylase	*Aspergillus niger/oryzae* (*pyrG*)	Auxotrophic	−	+	pAB4-1, pAO4.1, pABpyrGNot	17, 58
Orotidine phosphoribosyl transferase	*A. niger* (*pyrE*)	Auxotrophic	−	+	pBluepyrE	61
Acetamidase/orotidine-5′-phosphate decarboxylase	*A. nidulans* (*amdS*) *A. niger* (*pyrG*)	Dominant and auxotrophic	++	+	pAMDSPYRG-M1	23
Hygromycin B resistance/orotidine-5′-phosphate decarboxylase	*E. coli* (*hph*) *A. niger* (*pyrG*)	Dominant and auxotrophic	+	−	pAN7-1pyrG	Punt, unpublished

[a]In *A. niger*, the use of the selection marker results in medium (+; 1–10) or high (++; 10–200) copy numbers.
[b]In *A. niger*, the use of selection marker results in locus-specific integration (+) or heterologous integration (−).

Self-cloning may also "include the use of recombinant vectors with an extended history of safe use in the particular micro-organisms" (15a).

Along these lines, various vectors for a dominant nutritional marker, *amdS*, were developed (15) (Table 2). In relation to protein overproduction, the *amdS* marker in particular allows isolation of transformants carrying multiple vector copies (29). Other versatile auxotrophic selection markers that fulfill the EU requirement are based on complementation of *pyrG*⁻ or *pyrE*⁻ mutant strains (58). The availability of combined *pyrG/amdS* selection markers may be useful in cases where the use of the single dominant selection marker is limited by a high level of background growth and, consequently, low transformation frequencies (P. Punt, unpublished observations). In that case, the *pyrG* marker can be used for initial selection, while the presence of the *amdS* marker allows selection of those transformants carrying multiple vector copies. The advantage of a *pyrG* or *pyrE* complementation approach is that *pyrG*⁻ or *pyrE*⁻ strains can easily be obtained by direct selection on 5′-fluoroorotic acid (FOA) medium without any mutagenic treatment. Another advantage of *pyrG* and *pyrE* markers is that they can be used repeatedly (e.g., after integration of a *pyrG*- or *pyrE*-containing plasmid into the genome, the transformants can be cured from *pyrG* or *pyrE* by cultivating them on FOA plates).

The procedure to obtain *pyrG*⁻ or *pyrE*⁻ strains is based on the observation that mutants resistant to the antimetabolite FOA lack one of the two enzymes required for the conversion of orotic acid into uridine 5′-phosphate. *pyrE* encodes an orotate phosphoribosyltransferase, required for the first step in this conversion (orotic acid into orotine 5′-phosphate), and *pyrG* encodes an orotidine 5′-monophosphate decarboxylase, required for the conversion of orotine 5′-phosphate into uridine 5′-phosphate. Mutations in *pyrE* or *pyrG* prevent the conversion of FOA into a toxic compound, making the strains auxotrophic for uridine or uracil, but resistant to

FOA (8). The following protocol for the generation of *pyrG*⁻ and *pyrE*⁻ mutant strains of *A. niger* can, in our experience, also be applied to other filamentous fungi.

Protocol for Generation of *pyrG*⁻ or *pyrE*⁻ Strains

1. Inoculate 2×10^5 to 2×10^7 cells or spores on an FOA plate containing minimal medium (MM) supplemented with 0.75 mg/ml FOA and 10 mM uridine. Proline at 10 mM is usually used as nitrogen source. Note: Do not autoclave FOA, uridine, or proline; use filter-sterilized solutions. MM contains 1% glucose, 2 mM $MgSO_4$, trace element solution (63), 7 mM KCl, 8 mM KH_2PO_4, and 70 mM $NaNO_3$ and has been adjusted to pH 5.5.
2. Incubate for 1 to 2 weeks at an appropriate cultivation temperature and transfer FOA-resistant mutants onto fresh FOA plates for purification.
3. Test for uridine/uracil auxotrophy. Mutants should not be able to grow on MM lacking uridine or uracil.
4. To discriminate between *pyrE* or *pyrG* auxotrophic mutants, transform the mutant strains with a *pyrG*-containing plasmid (e.g., pAB4-1 [58]) and a *pyrE*-containing plasmid, pBluepyrE (61).

The *amdS* gene is a popular and frequently used selectable marker for fungal transformations, because it is a dominant, nonantibiotic marker. Dominant selectable markers are preferable to auxotrophic markers because they can be used directly without the need to generate mutant strains. The *amdS* gene encodes an acetamidase, enabling the strains to grow on acetamide as the sole nitrogen source. Most interestingly, the *amdS* gene can even be used as a dominant marker in fungi that harbor an endogenous *amdS* gene (54). In these cases, any background growth can be suppressed by the addition of cesium chloride to the selection medium. In addition, multicopy transformants can easily be screened using acrylamide instead of acetamide as nitrogen source. Finally, the *amdS* gene can be used, much like *pyr*G and *pyr*E, as a bidirectional marker by counterselecting for the loss of *amdS* with media containing the antimetabolite fluoroacetamide (FAA). To further simplify cloning procedures using the *amdS* selection marker, we have also designed a synthetic *amdS* selection marker gene in which most of the 6-bp cutter restriction sites have been removed (pMTL25synAmdS; Punt, unpublished).

Protocol for *amdS* Selection Media

1. To select for *amdS* integration in an *amdS⁻* recipient strain, supplement MM (without any N source) with 10 mM acetamide.
2. To select for *amdS* integration in a wild-type recipient strain, supplement MM with 15 mM cesium chloride and 10 mM acetamide.
3. To select for multicopy *amdS* integration, supplement MM with 15 mM cesium chloride and 10 mM acrylamide. Note: Do not autoclave acetamide, acrylamide, or cesium chloride; use filter-sterilized solutions.

Protocol for Curing of the *amdS* Selection Marker from *amdS⁺* Strains

1. Inoculate 2×10^5 to 2×10^7 cells or spores on an FAA plate (MM supplemented with 0.2% FAA; 10 mM urea is usually added as additional nitrogen source). Note: Do not autoclave FAA or urea; use filter-sterilized solutions.
2. Incubate for 1 to 2 weeks at an appropriate cultivation temperature and transfer FAA-resistant mutants onto fresh FAA plates for purification.
3. Test for growth on acetamide medium. Mutants should not be able to grow on MM containing 10 mM acetamide as sole nitrogen source.

22.3.2. Transformation

The establishment of a suitable and effective transformation method for the fungus of interest is of crucial importance for any genetic engineering approach. As summarized elsewhere, four transformation methods have been shown to be valuable tools for filamentous fungi—protoplast-mediated transformation, *Agrobacterium*-mediated transformation, electroporation, and biolistic transformation—each technique having its own advantages and disadvantages (see reference 39 and references therein). Protoplast-mediated and *Agrobacterium*-mediated transformation, however, seem to be the most commonly used techniques for filamentous fungi. In our hands, both of these methods have been successful in transforming different (industrially used) filamentous fungi belonging to the Ascomycetes, Basidiomycetes, or Zygomycetes. As a detailed protocol for *Agrobacterium*-mediated transformation has recently been published (41), we here describe a protocol for the transformation of *A. niger* protoplasts which can be used for many other filamentous fungi. In general, the yield of protoplasts depends on both the quality and quantity of the lytic enzyme batch used and the age of the culture used as starting material.

Protocol for Protoplast-Mediated Transformation

1. Inoculate 2.5×10^8 spores in 250 ml of complete medium (CM) and incubate for 16 to 18 h at 30°C and 0 to 50 rpm (CM is MM supplemented with 0.1% Casamino Acids and 0.5% yeast extract).
2. Prepare protoplastation solution as follows: dissolve 400 mg of Lysing Enzymes (Sigma) or 400 mg of

Vinoflow FCE (Novozymes) in 10 ml of SMC (1.33 M sorbitol, 50 mM CaCl₂, 20 mM morpholineethanesulfonic acid [MES] buffer; pH 5.8). Set pH to 5.6 with 1 M NaOH. Filter-sterilize the solution and transfer it to a sterile 50-ml tube.
3. Harvest the mycelium by filtration through sterile Miracloth (CalBiochem) or a sterile coffee filter and wash 1 time with SMC. Add less than 0.5 to 1 g of mycelium (wet weight) to the protoplastation solution and incubate for 1.5 to 2.5 h at 37°C with gentle shaking at 60 to 75 rpm, adjusting the tube in a shaker horizontally. Check protoplast formation under the microscope after 1 h of incubation and then every 30 min.
4. Before harvesting protoplasts, add 10 ml of STC (1.33 M sorbitol, 50 mM CaCl₂, 10 mM Tris HCl; pH 7.5) to the protoplastation sample and mix carefully by repeatedly pipetting up and down (this additionally releases protoplasts from the mycelium).
5. Collect protoplasts through a sterile Miracloth filter (or silica wool). Mycelial debris is not able to pass through the Miracloth (or silica wool).
6. Centrifuge at $2,000 \times g$ for 10 min at 10°C and decant the supernatant.
7. Gently resuspend the pellet in 1 ml of STC and centrifuge for 5 min at $3,000 \times g$. Discard supernatant and repeat the STC wash step twice.
8. For each transformation sample, add to a sterile 50-ml tube and gently mix: 200 μl of protoplasts (2×10^6 to 2×10^7), 20 μl of DNA solution (8 to 10 μg), and 50 μl of freshly made PEG buffer (per 10 ml: 2.5 g of PEG-6000 in 50 mM CaCl₂, 10 mM Tris HCl; pH 7.5)
9. Add 1 ml of PEG buffer and mix gently. After exactly 5 min of incubation (do not extend this incubation period because PEG is toxic for protoplasts), add 2 ml of STC and mix gently by tipping.
10. Add top agar (0.95 M sucrose in MM, 0.6% Oxoid agar; 47 to 50°C) to a total volume of 30 ml and gently mix by inverting the tube several times. Pour 7 ml onto each of four selective transformation plates (0.95 M sucrose in MM, 1.2% Oxoid agar; use 15-cm petri dishes). An alternative method which does not make use of top agar is to directly plate the protoplasts onto transformation plates. In our experience, this procedure can even improve the transformation efficiency when selecting for *pyr*G/*pyr*E or *amdS*.
11. Incubate selective transformation plates at an appropriate temperature for 3 to 6 days until colonies become visible. Purify primary transformants at least twice by plating diluted spore suspensions on selective medium in order to get homokaryotic transformants.

22.3.3. High-Throughput Screening of Transformants

Two approaches can be used to analyze transformants at the gene level and to verify that the DNA has inserted correctly: diagnostic PCR and Southern analysis of genomic DNA. For both approaches, we have developed high-throughput, rapid, and small-scale methods to isolate genomic DNA from transformants (V. Meyer, unpublished). We routinely use both methods to isolate high-molecular-weight genomic DNA from different *Aspergillus* species or other filamentous fungi.

Protocol for Isolation of Genomic DNA for Diagnostic PCR

1. As starting material, either spores (simply taken off a colony with 5 μl of 0.05% Triton X-100), substrate

mycelium (scraped from a colony with a sterile toothpick), or young mycelium (grown overnight in 2-ml Eppendorf tubes containing 700 μl of CM) can be used. If spores or substrate mycelium are used, transfer them into 2-ml Eppendorf tubes containing 600 μl of H_2O. If young mycelium is used, harvest the mycelium by centrifugation (10,000 × g), wash it twice with 1 ml of H_2O, and resuspend it in 600 μl of H_2O.

2. Disrupt cells by repeated freezing in liquid nitrogen and thawing in a 75°C water bath (three times). It is not necessary to include in this step any DNA extraction buffer or to treat the samples with proteinases or RNases.

3. Extract the genomic DNA once with an equal volume of phenol and once with an equal volume of chloroform/isoamylalcohol (24:1).

4. Precipitate the DNA from the aqueous phase with 2 volumes of 96% ethanol and 1/25 volume of 3 M sodium acetate (pH 6). Incubate the mixture for 1 h at −80°C. (Note: This incubation is crucial for efficient precipitation.)

5. Centrifuge at 10,000 × g for 15 min, discard the supernatant, and wash the DNA pellet with 100 μl of 70% ethanol. (Note: The DNA pellet is not easy to see.)

6. Resuspend the DNA in 20 μl of H_2O and incubate the solution at 65°C for 15 min in order to completely dissolve the DNA. One microliter of the DNA solution can directly be used in a PCR reaction.

Protocol for Isolation of Genomic DNA for Southern Analyses

1. Grow strains overnight in 20 ml of rich CM at 30 to 37°C.

2. Harvest mycelium by filtration through sterile Miracloth or a sterile coffee filter and transfer it without washing into a 2-ml Eppendorf tube (fill the tube up to the 0.5-ml line with mycelium).

3. Freeze samples at −80°C or in liquid nitrogen and dry the samples overnight in a freeze-dryer.

4. Grind the mycelium using small-scale pellet pestles (Fisher Scientific) for 10 to 20 s and resuspend the pulverized cells in 800 μl of extraction buffer (0.5% sodium dodecyl sulfate, 10 mM Tris HCl, 0.1 M EDTA; pH 8.0).

5. Add 1 μl of RNase (10 μg/μl) and incubate the suspension for 30 min at 37°C. Shake the mixture occasionally.

6. Extract the genomic DNA twice with an equal volume of phenol-chloroform-isoamylalcohol (25:24:1) and once with an equal volume of chloroform.

7. Precipitate the DNA from the aqueous phase with 1 volume of isopropanol and 1/10 volume of 3 M sodium acetate (pH 6).

8. Centrifuge at 10,000 × g for 15 min, discard the supernatant, and wash the DNA pellet with 100 μl of 70% ethanol.

9. Resuspend the DNA in 30 to 50 μl of H_2O and incubate the solution at 65°C for 15 min to completely dissolve the DNA. About 2 to 10 μl of the DNA solution (about 5 μg) is usually used for Southern analyses.

To analyze transformants for production of specific proteins or metabolites, we have also developed various growth and detection assays that can be performed in a microtiter plate (MTP) format. The assay can be greatly affected by the cultivation conditions, which in turn depend on the growth characteristics of the specific host strain being studied. Recently, it was found that a newly developed fungal production host, *Chrysosporium lucknowense* C1, can be cultivated in MTPs (61). Consequently, high-throughput screening can be used to analyze large collections of strains of this species in a short time frame (61). Many protocols have been published for MTP-based enzyme activity assays. In our research, we have in particular found the various hydrolase substrates sold by MEGAZYME very useful for development of MTP assays.

Protocol for MTP-Based Hydrolase Activity Assay

1. Use for cultivation Corning no. 3799 MTPs (or other round-bottom MTPs) and add 200 μl of the preferred growth and induction medium per well. Inoculate each well with spores or mycelial fragments using toothpick transfer.

2. Incubate the MTPs in an MTP shaking device until sufficient growth is obtained. In our experience, the Multitron equipment (INFORS) allows efficient submerged growth for filamentous fungi such as *Trichoderma* and *Chrysosporium*. However, for some other species such as *A. niger*, even incubation in the Multitron gyrotory MTP incubator (INFORS) results in an undesirable surface growth. (Note: If longer incubation times or higher growth temperatures are required, the outer wells of the culture plates should be filled with culture medium only, to limit evaporation of the center wells.)

3. After cultivation, centrifuge MTPs for 10 min at 3,500 × g. For surface-grown cultures, this step may not result in sufficient separation of mycelium and supernatant, thus requiring an additional transfer step to a second sample plate.

4. Perform the hydrolase activity assay as follows. Fill the assay-MTP (any MTP type) with 500 μl of 0.2% azo-dyed cross-linked (AZCL) substrate in any appropriate buffer. Using a multichannel pipette, add 10 μl of culture medium (or any other required dilution) to the substrate, followed by appropriate incubation. Depending on the hydrolase expression level, various incubation times and temperatures may be chosen.

5. After incubation, centrifuge the assay MTP for 10 min at 3,500 × g.

6. Transfer 100 μl of the supernatant to the measure-MTP (Corning no. 3596). Measure absorption at 590 nm using a microplate reader (Tecan Infinite 200 or a similar apparatus).

22.4. STRAIN IMPROVEMENT BY GENETIC ENGINEERING

Genetic engineering approaches can be used with both natural and recombinant production strains to increase production levels, to produce novel compounds, or to direct synthesis toward a desired metabolite. Some of the concepts and tools that have proven particularly powerful for optimizing industrial fungal strains will be discussed in the following sections.

22.4.1. Gene Overexpression and Protein Secretion

Different approaches are available for overexpression of a gene of interest. For example, targeted integration of a gene into a genomic locus known to be transcriptionally active is a strategy that provides good protein synthesis in a highly controlled fashion. An alternative method to maximize protein production is to integrate multiple copies of the gene at multiple genomic sites. Here in particular, the use of the *amdS* gene as a selection marker is a favorable option. In general, efficient gene transcription, translation, and (preferably) secretion of the protein can be achieved by the use of appropriate expression cassettes. Several expression vectors have been used to successfully express proteins in different fungal host strains, as summarized in Table 3.

The basic expression vector (type I) consists of an efficient fungal promoter and a suitable fungal terminator sequence. In our research on *Aspergillus*, we have primarily used two promoters: P_{gpdA} (glyceraldehyde-3-phosphate dehydrogenase A promoter) and P_{glaA} (glucoamylase A promoter). In particular, the *gpdA* promoter has been shown to be very useful for a wide variety of fungal species, allowing constitutive expression of the gene of interest under various culture conditions. Type II expression vectors carry, in addition to the basic expression cassette, a sequence encoding a secretion signal from *A. niger*, e.g., from the *glaA* gene. Using this vector type, the gene of interest will be fused downstream of the *glaA* secretion signal sequence, resulting in an efficient secretion of the gene product. A third type of vector, also designed to lead to secretion of the gene product of interest, contains, besides the basic expression cassette and the *glaA* secretion signal sequence, a gene encoding an efficiently secreted *A. niger* protein (*glaA*). The design of this vector is based on results obtained with the so-called gene fusion approach (for a review of this approach, see reference 19).

Although filamentous fungi are very efficient cell factories for the expression of proteins of fungal and nonfungal origin, the expression and secretion levels of heterologous proteins are usually two to three orders of magnitude lower than homologous proteins. Besides aspects of protein secretion, another reason for this discrepancy could be that the heterologous gene uses codons that are rarely used by the fungal host strain. Furthermore, A-T-rich sequences within heterologous mRNA molecules might be recognized as cryptic polyadenylation signals (19). These signals probably cause premature polyadenylation of some heterologous transcripts and rapid degradation by the nonstop mRNA decay system (55). One strategy to overcome this is to synthesize the gene with the codon usage of the fungal host strain. Such "codon optimization" has been shown to improve heterologous gene expression considerably

in different industrial *Aspergillus* strains (12, 28, 55). Most recently, a method applying genetic algorithms to optimize both single codon fitness and codon pair fitness was published (50a), which clearly showed that increased protein expression levels can be achieved, e.g., in *A. niger* by optimized protein coding sequences. During recent years, chemical DNA synthesis has become relatively inexpensive and the length of DNA that can be synthesized has extended from <1 kb to >30 kb (67). Hence, codon optimization and other targeted gene modifications can nowadays easily be designed in silico and manufactured rapidly by use of synthetic DNA technology. Various commercial suppliers offer these services.

22.4.2. Gene Targeting

Whereas overexpression of genes as described in the section above has been successful in many industrial strain development programs, high-level production of a heterologous protein can be hampered by other issues, the most prominent being production of endogenous proteases by fungi. A conceptually straightforward approach to reduce protease production is the targeted deletion of relevant protease genes (9). However, gene targeting has historically been a difficult venture in filamentous fungi. Due to very low frequencies of homologous recombination, any DNA that is introduced predominantly integrates randomly in the genome as a result of a nonhomologous integration event. Therefore, it has not been trivial to modify or delete an endogenous gene, or to target a gene of interest to a particular genomic locus. Fortunately, this limitation has recently been overcome. In 2004, Ninomiya and colleagues reported that *Neurospora crassa* mutant strains, lacking components of the nonhomologous end joining (NHEJ) pathway, became extremely efficient recipient strains for gene targeting (44). Since this breakthrough, many other NHEJ mutant strains, such as Ku70, Ku80, and Lig4 mutants, were generated in other filamentous fungi, and it was demonstrated that they all gave rise to very high homologous recombination frequencies (reference 39 and references therein). For example, deletion of the Ku70 homologue in *A. niger* reached more than 80% homologous integration efficiency, compared to 7% in the wild-type background, when 500-bp homologous flanks were used (37). Thus, targeted genetic alterations are more straightforward and considerably less time-consuming in NHEJ-deficient

TABLE 3 Vectors for improved protein production and secretion

Vector	Promoter	Secretion sequence	Terminator	Cloning sites (5′ → 3′)	Accession no. or reference
Type I					
pAN52-1NotI	P*gpdA*	–	T_{trpC} (long)	NcoI, BamHI	Z32524
pAN52-2	P*gpdA*	–	T_{trpC} (short)	HindIII	Z32688
pAN52-3	P*gpdA*	–	T_{trpC} (short)	NcoI, HindIII	Z32689
pAN52-7Not	P*glaA*	–	T_{trpC} (long)	NcoI, BamHI	60
Type II: pAN52-4NotI	P*gpdA*	S*glaA* (24 aa[a])	T_{trpC} (short)	BssHII, BamHI, HindIII	Z32750
Type III					
pAN56-1	P*gpdA*	S*glaA* (24 aa) + *glaA*	T_{trpC} (short)	NcoI, EheI, BglII	Z32700
pAN56-2	P*glaA*	S*glaA* (24 aa) + *glaA*	T_{trpC} (short)	EheI, BglII	Z32690

[a]aa, amino acids.

strains. One important observation, however, is that NHEJ mutant strains might also show some undesired phenotypes such as reduced fitness and diminished stress resistance (37, 43, 52). A very elegant solution to bypass this drawback has recently been reported for *A. nidulans* (43) and *Chrysosporium* (59). This rationale involves silencing the NHEJ machinery transiently, a concept that can easily be adapted to any other filamentous fungus. Basically, a gene encoding one component of the NHEJ machinery, e.g., *ku70*, is disrupted by inserting a counterselectable marker, e.g., *pyrG*. Most importantly, the *pyrG* marker is flanked on each side by a short *ku70* sequence, forming a direct repeat. After the desired genetic manipulation has been carried out, the *ku70* gene becomes restored by growing the strain on FOA medium, whereby recombination between the direct repeats causes excision of the *pyrG* marker and re-establishment of an intact *ku70* gene copy (43).

22.4.3. Gene Silencing

The knockout of genes encoding enzymes in pathways leading to side products can redirect the flux into the product-forming pathway. Such a gene knockout can be achieved by deleting or disrupting a gene of interest. However, to circumvent the need for gene targeting, RNA-based tools ("gene knockdown" tools) may be attractive alternatives. Several gene knockdown techniques, such as antisense

RNA, hammerhead ribozymes, and RNA interference, have been developed for filamentous fungi. Common to these methods is that they silence gene expression posttranscriptionally and, depending on how it is done, can silence gene expression partially to almost completely (references 2 and 39 and references therein). Basically, gene knockdown tools are superior to gene knockout approaches when (i) gene targeting approaches fail, (ii) multiple copies or close homologues of a gene of interest are present in the genome, (iii) isogenes might compensate for the knockout of the deleted gene, (iv) gene knockouts are lethal, and/or (v) partial reduction of gene expression is required to obtain the preferred phenotype. For more information on the principles of gene knockdown technologies and their applicability for filamentous fungi, we recommend recent reviews such as those in references 39 and 42. Most recently, the robustness and stability of the RNA interference silencing tool were demonstrated in *T. reesei* and *A. niger*, supporting the usefulness of gene silencing tools in industrial applications (11).

22.5. ALTERNATIVE STRAIN IMPROVEMENT APPROACHES

22.5.1. Screening for Protease-Deficient Strains

Besides the various genetic engineering approaches described above, classical strain improvement approaches

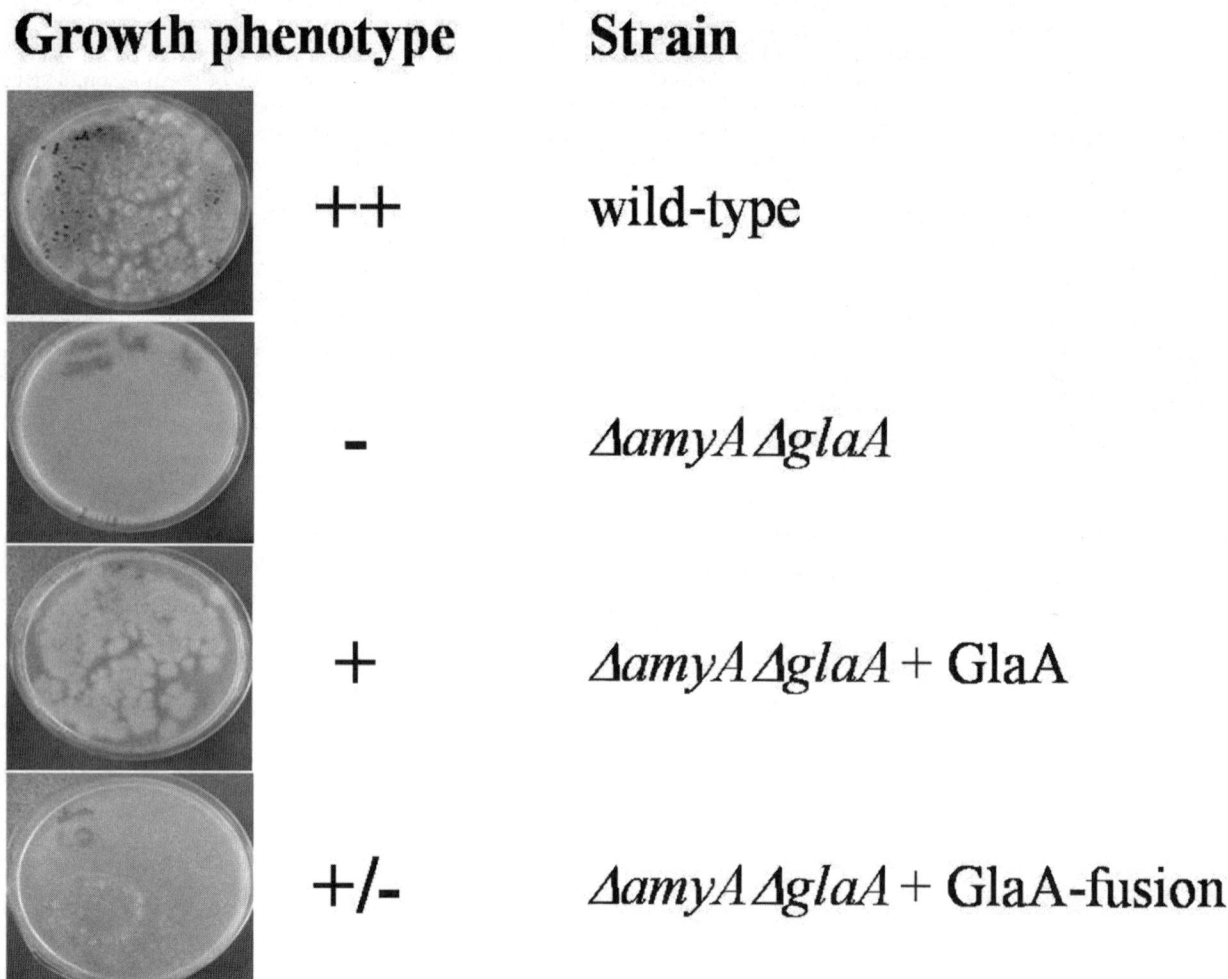

FIGURE 1 Growth phenotypes of different *A. niger* strains obtained on starch plates.

remain important for industrial strain development. Recently, we have developed a very successful selection approach using classical mutagenesis and screening, based on a so-called suicide substrate (9). In this approach, protease-deficient mutants can be selected efficiently from a background of wild-type host strains, resulting in identification of promising protease-deficient mutant strains at a 10,000-fold higher frequency compared to a typical screening approach.

22.5.2. Screening for Hypersecretion Strains

To improve laccase-producing A.*niger* strains, we have also developed a screening approach that combines classical mutagenesis and a glucoamylase-fusion approach (see section 22.4.1). The glucoamylase carrier approach was combined with the use of a specifically designed starch-nonutilizing host strain (65) generated by deleting the two genes encoding the major amylolytic proteins, i.e., acid amylase and glucoamylase (*aamA* and *glaA*, respectively). The resulting host strain (Δ*aamA*, Δ*glaA*) is practically unable to use starch as carbon source. Reintroduction of the native glucoamylase gene results in a strain that regains the ability to grow on starch, whereas introduction of a fusion gene encoding a poorly secreted glucoamylase-laccase (or any other fusion gene) into this strain results in transformants which grow only poorly on medium containing starch as sole carbon source (Fig. 1). One of these poorly growing glucoamylase-laccase transformants was mutagenized with UV light and mutants were selected for better growth on starch medium. This provided strains with increased laccase production levels (65).

22.6. DESIGNING A STRAIN IMPROVEMENT PROGRAM—AN EXAMPLE

One of our research interests is in the production of a separate starch-binding domain (SBD). SBDs are part of the group of carbohydrate-binding modules (CBMs) that are noncatalytic, polysaccharide-recognizing domains present in polysaccharide-degrading enzymes such as cellulases and chitinases. They bind to the insoluble polysaccharide and disrupt its structural organization. CBMs can be applied in various ways, e.g., to lower the tensile index of paper (32) or, if fused to another enzyme, to change the affinity and/or activity of the enzyme (31). In order to overproduce an SBD from the glucoamylase of A. *niger*, we have developed a screening approach to identify SBD-hyperproducing strains based on the starch-nonutilizing A. *niger* host strain described above. In this approach, we used a glucoamylase fusion to directly select for starch growth and thereby to identify the most productive transformants. In this case, no mutagenesis step was applied, and the best-performing transformants were selected directly. A related approach was recently also used for the identification of a novel GH13 amylase-like protein (57). In the following paragraphs, we detail the steps of our approach, providing a "how to" guide for strain improvement programs for filamentous fungi.

Based on publicly available sequence information, the DNA fragment encoding the A. *niger* glucoamylase SBD (CBM20 domain) was identified. The fragment encodes the C-terminally localized SBD preceded by a small threonine-rich linker region (108 amino acids encompassing C^{509}TTPTA.... VTDTWR^{616}; see also reference 30). The DNA-encoding SBD domain was amplified by PCR and subsequently cloned into the expression vector pAN56-1 (Table 3). The resulting vector carries the *gpdA* promoter, which drives expression of

the glucoamylase-SBD fusion product with the glucoamylase sequence separated from the SBD sequence by a dibasic endoprotease cleavage site (NVISKR). This site is recognized by a secretion pathway-specific intracellular protease of A. *niger*, resulting in the release of the SBD as soon as the mature glucoamylase-SBD fusion becomes exported from the cells. To explore the possibility of increasing the SBD yield and possibly adding new cross-linking functionalities, expression of a dimeric SBD module was investigated too. For this purpose, a DNA fragment encoding a dimeric SBD in the form of SBD$_1$-flex-SBD$_2$ was synthesized ("flex" represents the amino acid sequence GGGGSGGGGSGGGGS) and cloned into pAN56-1 (Table 3).

Both SBD expression vectors were introduced into a starch-nonutilizing, protease-deficient A. *niger* strain (Δ*aamA*, Δ*glaA*, *prtT13*) using cotransformation with the *amdS/pyrG* selection vector pAMDSPYRG-M1 (Table 2). For our specific purposes, this strain has an additional advantage over the ones already mentioned above, in that deletion of *glaA* and *aamA* also removed the only two genes encoding proteins with endogenous CBM20 domains. Primary *amdS*$^+$ (co)transformants obtained were subjected to a round of selection by transferring them to starch plates as well as to acetamide-containing plates. (Increased) growth on the former designated the transformants as true cotransformants and allowed selection of the best-growing strains. Cultivation of selected transformants in shake flasks and analysis for the presence of glucoamylase and SBD in the culture supernatants (using relevant antisera) allowed selection of SBD-hyperproducing strains. Finally, cultivation of the best-performing strains in 5-liter bioreactors (applying a fed-batch cultivation protocol) resulted in the production of a few grams of free SBD (P. Punt, unpublished data; Fig. 2).

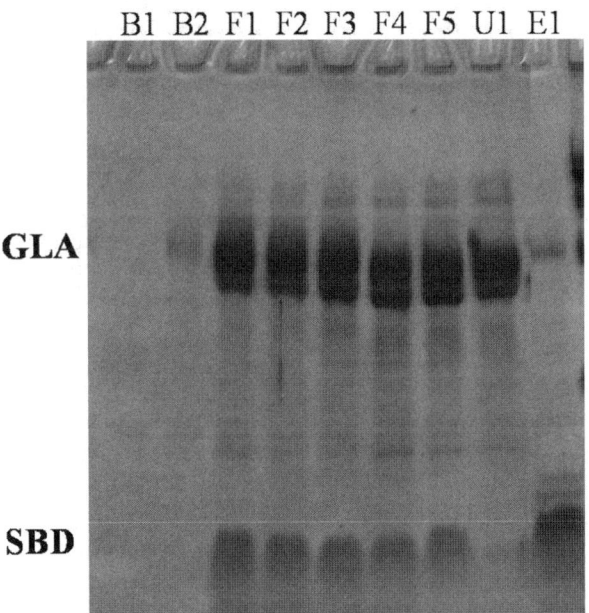

FIGURE 2 Production of SBD in A. *niger* during fed-batch cultivation. Depicted is a protein gel loaded with protein samples from culture supernatants obtained from different time points during cultivation. Indicated are the protein bands of the glucoamylase (GLA) and the starch-binding domain (SBD). B1–B2, medium samples from batch phase; F1–F5, medium samples from feed phase; U1, starch-unbound protein fraction; E1, starch-eluted protein fraction.

22.7. THE ULTIMATE TEST OF CANDIDATE STRAINS: CULTIVATIONS IN LAB-SCALE BIOREACTORS

Promising candidate strains—selected either from MTP-based screening approaches or from shake-flask cultivations—must ultimately perform under controlled cultivation conditions, in a bioreactor. As a standard reactor, we recommend the use of the BioFlo3000 fermentor (New Brunswick Scientific), which allows cultivation in a total working volume of 5 liters. Depending on the final product, batch, fed-batch, or chemostat cultivations should be performed. Depending on the strain (A. niger, C. lucknowense, P. chrysogenum, T. reesei, etc.) and expression cassette used, different conditions should be tested (medium composition, pH, temperature, inducing agents, etc.). It is thus impossible to give here a detailed protocol that would apply to any of the strains or products. However, we summarize in the following protocol some rules of thumb that we found critical for successful cultivation of filamentous fungi in a bioreactor. For more information, e.g., on the respective medium compositions, cultivation time, physiology, and achievable biomass and product yields, refer to references 4, 25, and 38.

Protocol: Guidelines for Bioreactor Cultivations

1. Equip the fermentor vessel with cooling tubing that surrounds the headspace of the vessel to limit wall growth in the headspace.
2. Use two Rushton impellers, one in the bottom of the vessel and one just below the level of the intended culture volume (5-liter maximum). In between these two impellers, it is possible to mount a marine downflow impeller.
3. Mount an air/O_2 sparger near the bottom of the vessel and equip off-gas outlets with a cooling device and overpressure valves.
4. If spores are used for inoculation, inoculate the medium with 10^9 spores/liter. Use in the beginning a low initial agitation speed (such as 250 rpm) and headspace aeration in order to keep spores from being blown out from the medium. After spore germination has been completed, increase impeller speed up to 750 to 800 rpm and switch headspace aeration to sparger aeration.
5. If precultures are used for inoculation, make them by inoculating 100 ml of CM with, e.g., 1×10^6 spores/ml. After 1 to 2 days of growth, use two flasks, each containing 100 ml of preculture, to inoculate the fermentor. Start directly with an impeller speed of 750 to 800 rpm and sparger aeration.
6. Always add antifoam to the medium (if spores are used for inoculation, add the antifoam after they have germinated). Use either 0.01% sterile polypropyleneglycol or 0.01% Struktol J673.
7. Maintain the dissolved oxygen at 20% by controlling the agitation speed. If dissolved oxygen drops below 20%, substitute air sparging for O_2 sparging. In general, the use of an airflow of 1 to 1.2 liters/min is sufficient.
8. Maintain the pH of the medium either by using 8 M KOH and 2 M H_3PO_4 or by using 1 M HCl and 2 M NaOH as titrants.

22.8. FINAL CONSIDERATIONS AND OUTLOOK

The high protein-producing capacity of filamentous fungi such as A. niger, A. oryzae, and T. reesei has prompted their development as cell factories for the production of heterologous proteins (33, 50). Multiple studies, however, have shown that upon expression of these heterologous proteins, the unfolded protein response pathway becomes induced in filamentous fungi (see for example references 14, 66), suggesting that heterologous proteins become misfolded and subsequently degraded during their way through the secretory pathway. Genome-wide transcriptomic approaches have thus been used to identify genes that are induced or repressed during the expression of heterologous proteins and have provided target genes for genetic manipulation and strain improvement approaches (21). Jacobs et al. have taken it one step further by integrating transcriptomic and proteomic data sets for further strain improvement (22). Leads from such studies can rapidly be tested by the use of NHEJ-deficient recipient strains, which have been established for most of the industrial filamentous fungal strains.

Several reports have already emphasized the importance of highly controlled and reproducible growth conditions to perform reliable transcriptomic, proteomic, and metabolomic analyses (6, 9, 22, 25). Genome-wide studies should therefore be designed and performed under defined conditions which can only be ensured by cultivations in bioreactors. Here, medium composition, pH, temperature, and impeller speed can be controlled and physiological data can be obtained—e.g., from off-gas analyses and titrant addition—and product formation can be determined online as well.

Finally, continuing advances in high-throughput DNA sequencing (such as 454 Life Science pyrosequencing and Illumina sequencing technologies) are revolutionizing fungal biotechnology and leading to the identification of many more interesting enzymes (e.g., glycosyl hydrolases, proteases, lipases) and potentially interesting secondary metabolite gene clusters. The filamentous fungi listed in Table 1 will continue to be used as chassis for screening approaches as well as for the production of newly identified proteins. It is likely that industrial fungi with well-known and well-developed fermentation characteristics also will be used as production hosts for secondary metabolites discovered from fungi less suited to industrial exploitation.

REFERENCES

1. **Abe, K., K. Gomi, F. Hasegawa, and M. Machida.** 2006. Impact of *Aspergillus oryzae* genomics on industrial production of metabolites. *Mycopathologia* **162:**143–153.
2. **Akashi, H., S. Matsumoto, and K. Taira.** 2005. Gene discovery by ribozyme and siRNA libraries. *Nat. Rev. Mol. Cell. Biol.* **6:**413–422.
3. **Andersen, M. R., M. L. Nielsen, and J. Nielsen.** 2008. Metabolic model integration of the bibliome, genome, metabolome and reactome of *Aspergillus niger*. *Mol. Syst. Biol.* **4:**178.
4. **Andersen, M. R., W. Vongsangnak, G. Panagiotou, M. P. Salazar, L. Lehmann, and J. Nielsen.** 2008. A tri-species *Aspergillus* microarray: comparative transcriptomics of three *Aspergillus* species. *Proc. Natl. Acad. Sci. USA* **105:**4387–4392.
5. **Andersen, M. R., and J. Nielsen.** 2009. Current status of systems biology in Aspergilli. *Fungal Genet. Biol.* **46**(Suppl. 1)**:**S180–190.
6. **Askenazi, M., E. M. Driggers, D. A. Holtzman, T. C. Norman, S. Iverson, D. P. Zimmer, M. E. Boers, P. R. Blomquist, E. J. Martinez, A. W. Monreal, T. P. Feibelman, M. E. Mayorga, M. E. Maxon, K. Sykes, J. V. Tobin, E. Cordero, S. R. Salama, J. Trueheart, J. C. Royer, and K. T. Madden.** 2003. Integrating transcriptional and metabolite profiles to direct the engineering of lovastatin-producing fungal strains. *Nat. Biotechnol.* **21:**150–156.

7. Bauer, S., P. Vasu, S. Persson, A. J. Mort, and C. R. Somerville. 2006. Development and application of a suite of polysaccharide-degrading enzymes for analyzing plant cell walls. *Proc. Natl. Acad. Sci. USA* **103:** 11417–11422.

8. Boeke, J. D., J. Trueheart, G. Natsoulis, and G. R. Fink. 1987. 5-Fluoroorotic acid as a selective agent in yeast molecular genetics. *Methods Enzymol.* **154:** 164–175.

9. Braaksma, M., and P. J. Punt. 2008. *Aspergillus* as a cell factory for protein production: controlling protease activity in fungal production, p. 441–456. *In* G. H. Goldman and S. A. Osmani (ed.), *The Aspergilli: Genomics, Medial Aspects, Biotechnology and Research Methods*. CRC Press, Boca Raton, FL.

10. Braaksma, M., R. A. van den Berg, M. J. van der Werf, and P. J. Punt. 2009. A top-down systems biology approach for the identification of targets for fungal strain and process development, p. 25–35. *In* K. A. Borkovich and D. J. Ebbole (ed.), *Cellular and Molecular Biology of Filamentous Fungi*. ASM Press, Washington, DC.

11. Brody, H., and S. Maiyuran. 2009. RNAi-mediated gene silencing of highly expressed genes in the industrial fungi *Trichoderma reesei* and *Aspergillus niger*. *Ind. Biotechnol.* **5:**53–60.

12. Cardoza, R. E., S. Gutierrez, N. Ortega, A. Colina, J. Casqueiro, and J. F. Martin. 2003. Expression of a synthetic copy of the bovine chymosin gene in *Aspergillus awamori* from constitutive and pH-regulated promoters and secretion using two different pre-pro sequences. *Biotechnol. Bioeng.* **83:**249–259.

13. Chiang, Y. M., E. Szewczyk, A. D. Davidson, N. Keller, B. R. Oakley, and C. C. Wang. 2009. A gene cluster containing two fungal polyketide synthases encodes the biosynthetic pathway for a polyketide, asperfuranone, in *Aspergillus nidulans*. *J. Am. Chem. Soc.* **131:**2965–2970.

14. Collen, A., M. Saloheimo, M. Bailey, M. Penttila, and T. M. Pakula. 2005. Protein production and induction of the unfolded protein response in *Trichoderma reesei* strain Rut-C30 and its transformant expressing endoglucanase I with a hydrophobic tag. *Biotechnol. Bioeng.* **89:**335–344.

15. Corrick, C. M., A. P. Twomey, and M. J. Hynes. 1987. The nucleotide sequence of the *amdS* gene of *Aspergillus nidulans* and the molecular characterization of 5′ mutations. *Gene* **53:**63–71.

15a. Council of the European Communities. 1990. Directive 90/219/EEC: Council Directive 90/219/EEC of 23 April 1990 on the contained use of genetically modified micro-organisms. Council of the European Communities. (Online.) http://europa.eu/legislation_summaries/agriculture/ food/l21157_en.htm.

16. Coutinho, P. M., M. R. Andersen, K. Kolenova, P. A. vanKuyk, I. Benoit, B. S. Gruben, B. Trejo-Aguilar, H. Visser, P. van Solingen, T. Pakula, B. Seiboth, E. Battaglia, G. Aguilar-Osorio, J. F. de Jong, R.A. Ohm, M. Aguilar, B. Henrissat, J. Nielsen, H. Stålbrand, and R. P. de Vries. 2009. Post-genomic insights into the plant polysaccharide degradation potential of *Aspergillus nidulans* and comparison to *Aspergillus niger* and *Aspergillus oryzae*. *Fungal Genet. Biol.* **46:**S161–S169.

17. de Ruiter-Jacobs, Y. M., M. Broekhuijsen, S. E. Unkles, E. I. Campbell, J. R. Kinghorn, R. Contreras, P. H. Pouwels, and C. A. van den Hondel. 1989. A gene transfer system based on the homologous *pyrG* gene and efficient expression of bacterial genes in *Aspergillus oryzae*. *Curr. Genet.* **16:**159–163.

18. Frisvad, J. C., J. Smedsgaard, R. A. Samson, T. O. Larsen, and U. Thrane. 2007. Fumonisin B2 production by *Aspergillus niger*. *J. Agric. Food Chem.* **55:**9727–9732.

19. Gouka, R. J., P. J. Punt, and C. A. van den Hondel. 1997. Efficient production of secreted proteins by *Aspergillus*: progress, limitations and prospects. *Appl. Microbiol. Biotechnol.* **47:**1–11.

20. Guangtao, Z., L. Hartl, A. Schuster, S. Polak, M. Schmoll, T. Wang, V. Seidl, and B. Seiboth. 2009. Gene targeting in a nonhomologous end joining deficient *Hypocrea jecorina*. *J. Biotechnol.* **139:**146–151.

21. Guillemette, T., N. N. van Peij, T. Goosen, K. Lanthaler, G. D. Robson, C. A. van den Hondel, H. Stam, and D. B. Archer. 2007. Genomic analysis of the secretion stress response in the enzyme-producing cell factory *Aspergillus niger*. *BMC Genomics* **8:**158.

22. Jacobs, D. I., M. M. Olsthoorn, I. Maillet, M. Akeroyd, S. Breestraat, S. Donkers, R. A. van der Hoeven, C. A. van den Hondel, R. Kooistra, T. Lapointe, H. Menke, R. Meulenberg, M. Misset, W. H. Muller, N. N. van Peij, A. Ram, S. Rodriguez, M. S. Roelofs, J. A. Roubos, M. W. van Tilborg, A. J. Verkleij, H. J. Pel, H. Stam, and C. M. Sagt. 2009. Effective lead selection for improved protein production in *Aspergillus niger* based on integrated genomics. *Fungal Genet. Biol.* **46**(Suppl. 1)**:**S141–152.

23. Joosten, V., R. J. Gouka, C. A. van den Hondel, C. T. Verrips, and B. C. Lokman. 2005. Expression and production of llama variable heavy-chain antibody fragments (V(HH)s) by *Aspergillus awamori*. *Appl. Microbiol. Biotechnol.* **66:**384–392.

24. Jorgensen, T. R. 2007. Identification and toxigenic potential of the industrially important fungi, *Aspergillus oryzae* and *Aspergillus sojae*. *J. Food Prot.* **70:** 2916–2934.

25. Jorgensen, T. R., T. Goosen, C. A. Hondel, A. F. Ram, and J. J. Iversen. 2009. Transcriptomic comparison of *Aspergillus niger* growing on two different sugars reveals coordinated regulation of the secretory pathway. *BMC Genomics* **10:**44.

26. Keller, N. P., G. Turner, and J. W. Bennett. 2005. Fungal secondary metabolism—from biochemistry to genomics. *Nat. Rev. Microbiol.* **3:**937–947.

27. Kobayashi, T., K. Abe, K. Asai, K. Gomi, P. R. Juvvadi, M. Kato, K. Kitamoto, M. Takeuchi, and M. Machida. 2007. Genomics of *Aspergillus oryzae*. *Biosci. Biotechnol. Biochem.* **71:**646–670.

28. Koda, A., T. Bogaki, T. Minetoki, and M. Hirotsune. 2005. High expression of a synthetic gene encoding potato alpha-glucan phosphorylase in *Aspergillus niger*. *J. Biosci. Bioeng.* **100:**531–537.

29. Kolar, M., P. J. Punt, C. A. van den Hondel, and H. Schwab. 1988. Transformation of *Penicillium chrysogenum* using dominant selection markers and expression of an *Escherichia coli lacZ* fusion gene. *Gene* **62:**127–134.

30. Le Gal-Coeffet, M. F., A. J. Jacks, K. Sorimachi, M. P. Williamson, G. Williamson, and D. B. Archer. 1995. Expression in *Aspergillus niger* of the starch-binding domain of glucoamylase. Comparison with the proteolytically produced starch-binding domain. *Eur. J. Biochem.* **233:**561–567.

31. Levasseur, A., S. Pages, H. P. Fierobe, D. Navarro, P. Punt, J. P. Belaich, M. Asther, and E. Record. 2004. Design and production in *Aspergillus niger* of a chimeric protein associating a fungal feruloyl esterase and a clostridial dockerin domain. *Appl. Environ. Microbiol.* **70:** 6984–6991.

32. Levy, I., and O. Shoseyov. 2002. Cellulose-binding domains: biotechnological applications. *Biotechnol. Adv.* **20:**191–213.

33. Lubertozzi, D., and J. D. Keasling. 2009. Developing *Aspergillus* as a host for heterologous expression. *Biotechnol. Adv.* **27:**53–75.

34. Machida, M., K. Asai, M. Sano, T. Tanaka, T. Kumagai, G. Terai, K. Kusumoto, T. Arima, O. Akita, Y. Kashiwagi, K. Abe, K. Gomi, H. Horiuchi, K. Kitamoto, T. Kobayashi, M. Takeuchi, D. W. Denning, J. E. Galagan, W. C. Nierman, J. Yu, D. B. Archer, J. W. Bennett, D. Bhatnagar, T. E. Cleveland, N. D. Fedorova, O. Gotoh, H. Horikawa, A. Hosoyama, M. Ichinomiya, R. Igarashi, K. Iwashita, P. R. Juvvadi, M. Kato, Y. Kato, T. Kin, A. Kokubun, H. Maeda, N. Maeyama, J. Maruyama, H. Nagasaki, T. Nakajima, K. Oda, K. Okada, I. Paulsen, K. Sakamoto, T. Sawano, M. Takahashi, K. Takase, Y. Terabayashi, J. R. Wortman, O. Yamada, Y. Yamagata, H. Anazawa, Y. Hata, Y. Koide, T. Komori, Y. Koyama, T. Minetoki, S. Suharnan, A. Tanaka, K. Isono, S. Kuhara, N. Ogasawara, and H. Kikuchi. 2005. Genome sequencing and analysis of Aspergillus oryzae. Nature 438:1157–1161.

35. Martens-Uzunova, E. S., and P. J. Schaap. 2009. Assessment of the pectin degrading enzyme network of Aspergillus niger by functional genomics. Fungal Genet. Biol. 46:S170–S179.

36. Martinez, D., R. M. Berka, B. Henrissat, M. Saloheimo, M. Arvas, S. E. Baker, J. Chapman, O. Chertkov, P. M. Coutinho, D. Cullen, E. G. Danchin, I. V. Grigoriev, P. Harris, M. Jackson, C. P. Kubicek, C. S. Han, I. Ho, L. F. Larrondo, A. L. de Leon, J. K. Magnuson, S. Merino, M. Misra, B. Nelson, N. Putnam, B. Robbertse, A. A. Salamov, M. Schmoll, A. Terry, N. Thayer, A. Westerholm-Parvinen, C. L. Schoch, J. Yao, R. Barabote, M. A. Nelson, C. Detter, D. Bruce, C. R. Kuske, G. Xie, P. Richardson, D. S. Rokhsar, S. M. Lucas, E. M. Rubin, N. Dunn-Coleman, M. Ward, and T. S. Brettin. 2008. Genome sequencing and analysis of the biomass-degrading fungus Trichoderma reesei (syn. Hypocrea jecorina). Nat. Biotechnol. 26:553–560.

37. Meyer, V., M. Arentshorst, A. El-Ghezal, A. C. Drews, R. Kooistra, C. A. van den Hondel, and A. F. Ram. 2007. Highly efficient gene targeting in the Aspergillus niger kusA mutant. J. Biotechnol. 128:770–775.

38. Meyer, V., R. A. Damveld, M. Arentshorst, U. Stahl, C. A. van den Hondel, and A. F. Ram. 2007. Survival in the presence of antifungals: genome-wide expression profiling of Aspergillus niger in response to sublethal concentrations of caspofungin and fenpropimorph. J. Biol. Chem. 282:32935–32948.

39. Meyer, V. 2008. Genetic engineering of filamentous fungi—progress, obstacles and future trends. Biotechnol. Adv. 26:177–185.

40. Michielse, C. B., P. J. Hooykaas, C. A. van den Hondel, and A. F. Ram. 2005. Agrobacterium-mediated transformation as a tool for functional genomics in fungi. Curr. Genet. 48:1–17.

41. Michielse, C. B., P. J. Hooykaas, C. A. van den Hondel, and A. F. Ram. 2008. Agrobacterium-mediated transformation of the filamentous fungus Aspergillus awamori. Nat. Protoc. 3:1671–1678.

42. Nakayashiki, H., and Q. B. Nguyen. 2008. RNA interference: roles in fungal biology. Curr. Opin. Microbiol. 11:494–502.

43. Nielsen, J. B., M. L. Nielsen, and U. H. Mortensen. 2008. Transient disruption of non-homologous end-joining facilitates targeted genome manipulations in the filamentous fungus Aspergillus nidulans. Fungal Genet. Biol. 45:165–170.

44. Ninomiya, Y., K. Suzuki, C. Ishii, and H. Inoue. 2004. Highly efficient gene replacements in Neurospora strains deficient for nonhomologous end-joining. Proc. Natl. Acad. Sci. USA 101:12248–12253.

45. Olmedo-Monfil, V., C. Cortes-Penagos, and A. Herrera-Estrella. 2004. Three decades of fungal transformation: key concepts and applications. Methods Mol. Biol. 267:297–313.

46. Panagiotou, G., M. R. Andersen, T. Grotkjaer, T. B. Regueira, J. Nielsen, and L. Olsson. 2009. Studies of the production of fungal polyketides in Aspergillus nidulans by using systems biology tools. Appl. Environ. Microbiol. 75:2212–2220.

47. Pel, H. J., J. H. de Winde, D. B. Archer, P. S. Dyer, G. Hofmann, P. J. Schaap, G. Turner, R. P. de Vries, R. Albang, K. Albermann, M. R. Andersen, J. D. Bendtsen, J. A. Benen, M. van den Berg, S. Breestraat, M. X. Caddick, R. Contreras, M. Cornell, P. M. Coutinho, E. G. Danchin, A. J. Debets, P. Dekker, P. W. van Dijck, A. van Dijk, L. Dijkhuizen, A. J. Driessen, C. d'Enfert, S. Geysens, C. Goosen, G. S. Groot, P. W. de Groot, T. Guillemette, B. Henrissat, M. Herweijer, J. P. van den Hombergh, C. A. van den Hondel, R. T. van der Heijden, R. M. van der Kaaij, F. M. Klis, H. J. Kools, C. P. Kubicek, P. A. van Kuyk, J. Lauber, X. Lu, M. J. van der Maarel, R. Meulenberg, H. Menke, M. A. Mortimer, J. Nielsen, S. G. Oliver, M. Olsthoorn, K. Pal, N. N. van Peij, A. F. Ram, U. Rinas, J. A. Roubos, C. M. Sagt, M. Schmoll, J. Sun, D. Ussery, J. Varga, W. Vervecken, P. J. van de Vondervoort, H. Wedler, H. A. Wosten, A. P. Zeng, A. J. van Ooyen, J. Visser, and H. Stam. 2007. Genome sequencing and analysis of the versatile cell factory Aspergillus niger CBS 513.88. Nat. Biotechnol. 25:221–231.

48. Punt, P. J., R. P. Oliver, M. A. Dingemanse, P. H. Pouwels, and C. A. van den Hondel. 1987. Transformation of Aspergillus based on the hygromycin B resistance marker from Escherichia coli. Gene 56:117–124.

49. Punt, P. J., and C. A. van den Hondel. 1992. Transformation of filamentous fungi based on hygromycin B and phleomycin resistance markers. Methods Enzymol. 216:447–457.

50. Punt, P. J., N. van Biezen, A. Conesa, A. Albers, J. Mangnus, and C. van den Hondel. 2002. Filamentous fungi as cell factories for heterologous protein production. Trends Biotechnol. 20:200–206.

50a. Roubos, J. A., and N. N. M. van Peij. 2008. A method for achieving improved polypeptide expression. World Patent Application WO2008000632.

51. Schuster, E., N. Dunn-Coleman, J. C. Frisvad, and P. W. Van Dijck. 2002. On the safety of Aspergillus niger—a review. Appl. Microbiol. Biotechnol. 59:426–435.

52. Snoek, I. S., Z. A. van der Krogt, H. Touw, R. Kerkman, J. T. Pronk, R. A. Bovenberg, M. A. van den Berg, and J. M. Daran. 2009. Construction of an hdfA Penicillium chrysogenum strain impaired in non-homologous end-joining and analysis of its potential for functional analysis studies. Fungal Genet. Biol. 46:418–426.

53. Takahashi, T., T. Masuda, and Y. Koyama. 2006. Enhanced gene targeting frequency in ku70 and ku80 disruption mutants of Aspergillus sojae and Aspergillus oryzae. Mol. Genet. Genomics 275:460–470.

54. Tilburn, J., C. Scazzocchio, G. G. Taylor, J. H. Zabicky-Zissman, R. A. Lockington, and R. W. Davies. 1983. Transformation by integration in Aspergillus nidulans. Gene 26:205–221.

55. Tokuoka, M., M. Tanaka, K. Ono, S. Takagi, T. Shintani, and K. Gomi. 2008. Codon optimization increases steady-state mRNA levels in Aspergillus oryzae heterologous gene expression. Appl. Environ. Microbiol. 74:6538–6546.

56. van den Berg, M. A., R. Albang, K. Albermann, J. H. Badger, J. M. Daran, A. J. Driessen, C. Garcia-Estrada, N. D. Fedorova, D. M. Harris, W. H. Heijne, V. Joardar, J. A. Kiel, A. Kovalchuk, J. F. Martin, W. C. Nierman, J. G. Nijland, J. T. Pronk, J. A. Roubos, I. J. van der Klei, N. N. van Peij, M. Veenhuis, H. von Dohren, C. Wagner, J. Wortman, and R. A. Bovenberg. 2008. Genome sequencing and analysis of the filamentous fungus *Penicillium chrysogenum*. *Nat. Biotechnol.* **26:**1161–1168.

57. van der Kaaij, R. M., S. Janecek, M. J. van der Maarel, and L. Dijkhuizen. 2007. Phylogenetic and biochemical characterization of a novel cluster of intracellular fungal alpha-amylase enzymes. *Microbiology* **153:**4003–4015.

58. van Hartingsveldt, W., I. E. Mattern, C. M. van Zeijl, P. H. Pouwels, and C. A. van den Hondel. 1987. Development of a homologous transformation system for *Aspergillus niger* based on the *pyrG* gene. *Mol. Gen. Genet.* **206:**71–75.

59. van Zeijl, C., M. van Muijlwijk, M. Heerikhuisen, J. C. Verdoes, and P. J. Punt. 2009. Gene targeting in *Chrysosporium lucknowense* using a retrievable mutation in the KU70 gene. *Fungal Genet. Rep.* **56S:**92.

60. Verdoes, J. C., P. J. Punt, A. H. Stouthamer, and C. A. van den Hondel. 1994. The effect of multiple copies of the upstream region on expression of the *Aspergillus niger* glucoamylase-encoding gene. *Gene* **145:**179–187.

61. Verdoes, J. C., P. J. Punt, R. Burlingame, J. Bartels, R. van Dijk, E. Slump, M. Meens, R. Joosten, and M. Emalfarb. 2007. A dedicated vector for efficient library construction and high throughput screening in the hyphal fungus *Chrysosporium lucknowense*. *Ind. Biotechnol.* **3:**48–57.

62. Vinci, V. A., and G. Byng. 1999. Strain improvement by nonrecombinant methods, p. 103–113. *In* A. E. Demain and J. E. Davies (ed.), *Manual of Industrial Microbiology and Biotechnology*, 2nd ed. ASM Press, Washington, DC.

63. Vishniac, W., and M. Santer. 1957. The thiobacilli. *Bacteriol. Rev.* **21:**195–213.

64. Vongsangnak, W., P. Olsen, K. Hansen, S. Krogsgaard, and J. Nielsen. 2008. Improved annotation through genome-scale metabolic modeling of *Aspergillus oryzae*. *BMC Genomics* **9:**245.

65. Weenink, X. O., P. J. Punt, C. A. van den Hondel, and A. F. Ram. 2006. A new method for screening and isolation of hypersecretion mutants in *Aspergillus niger*. *Appl. Microbiol. Biotechnol.* **69:**711–717.

66. Wiebe, M. G., A. Karandikar, G. D. Robson, A. P. Trinci, J. L. Candia, S. Trappe, G. Wallis, U. Rinas, P. M. Derkx, S. M. Madrid, H. Sisniega, I. Faus, R. Montijn, C. A. van den Hondel, and P. J. Punt. 2001. Production of tissue plasminogen activator (t-PA) in *Aspergillus niger*. *Biotechnol. Bioeng.* **76:**164–174.

67. Xiong, A. S., R. H. Peng, J. Zhuang, F. Gao, Y. Li, Z. M. Cheng, and Q. H. Yao. 2008. Chemical gene synthesis: strategies, softwares, error corrections, and applications. *FEMS Microbiol. Rev.* **32:**522–540.

Genetic Manipulation of Mammalian Cells for Protein Expression

ANNE KANTARDJIEFF, WEI-SHOU HU, GARGI SETH, AND R. SCOTT McIVOR

23

23.1. INTRODUCTION

Genetic manipulation of mammalian cells is used in a wide spectrum of applications. Foremost, genetically engineered mammalian cells are used in the production of a variety of therapeutic proteins such as factor VIII, protein C, and a number of antibodies. In laboratories, proteins which require extensive posttransitional modification and are difficult to produce in an active form in other systems are also expressed in genetically manipulated mammalian cells. While the majority of protein expression is for secreted proteins, a number of receptors and membrane proteins are also expressed in mammalian systems. In many applications, the genetic manipulation is intended to impart some desirable properties to the cell, rather than producing the protein per se. This is often referred to as cell engineering or metabolic engineering. These metabolically engineered cells are used for research purposes, for in vitro or in vivo studies, or in the bioprocess production of other proteins. For example, the overexpression of heat shock protein 70 (Hsp70) is often used in baby hamster kidney (BHK) cells to enhance the productivity of factor VIII (46). A variety of cells have also been engineered to overexpress antiapoptotic factor(s), thus delaying the onset of apoptotic cell death (4). Depending on the objective, genetic manipulation may also take a different path. For overproduction of a product protein, the protocol of genetic manipulation may include a gene amplification step to maximize cellular content of the product's transcript. In contrast, for cell engineering purposes, an appropriate expression level, which may not be very high, is necessary to confer the desired physiological effect.

For many applications, it is necessary to generate a permanent population of cells for production purposes or research use. In other cases, the genetic element introduced remains inside the cell for only a short duration, during which the protein produced is harvested for study or the cellular properties imparted are characterized. This process is referred to as transient expression. Both stable and transient expression are widely used in research and at the production scale. Transient expression is used in production scale of hundreds of liters or more, although the protein produced is mostly used for chemical and physical characterization or research use, rather than as therapeutic proteins. When stable expression is used, the transfected cells may be used as a population wherein each cell may have integrated the gene of interest in different locations of the genome. As a result, not all cells in the population will have the same properties or productivity. For many applications, especially when generating a commercial protein product, clones originating from a single cell are isolated, characterized, and used in the production. In this chapter, all of the above-mentioned aspects of genetic manipulation will be discussed.

23.2. BASIC STEPS IN GENETIC MANIPULATIONS FOR HETEROLOGOUS PROTEIN PRODUCTION

Expressing genes in mammalian cells has become increasingly important to understand their functional significance. Heterologous expression in mammalian cells is also the primary method of production of proteins for therapeutic use. Only a few cell lines account for the majority of secreted proteins produced in mammalian cells, namely Chinese hamster ovary (CHO) cells, mouse myeloma cells (NS0, Sp2/0), and BHK cells. However, other mammalian expression systems are also available, including human embryonic kidney (HEK-293) cells, African green monkey CV1 kidney (COS) cells, and Madin-Darby canine kidney (MDCK) cells. In this section, the basic steps of genetic manipulation for heterologous protein production are outlined.

23.2.1. Vectors

For heterologous expression, the foreign gene is inserted into a plasmid that includes genetic elements that facilitate genetic manipulation, vector production, and expression in the host cell (Fig. 1). These elements include a bacterial origin of replication and a bacterial selectable marker. The vector introduced into the host cell will need to be translocated into the nucleus for transcription, and in the case of stable transfections, the inserted sequences will be integrated into the chromosome. Only a small fraction of foreign DNA entering the cell reaches the nucleus, whether in plasmid form or as virus genomes; an even smaller fraction is integrated into the chromosomes. In most transfection processes, a very large portion (typically 30 to 90%) of cells will not express the foreign gene, as judged by the expression of a reporter or selectable marker, either because those cells fail to take up the plasmid or because the plasmids taken up fail to reach the nucleus. In expressing a foreign protein, the selection and design of a vector will critically affect the outcome.

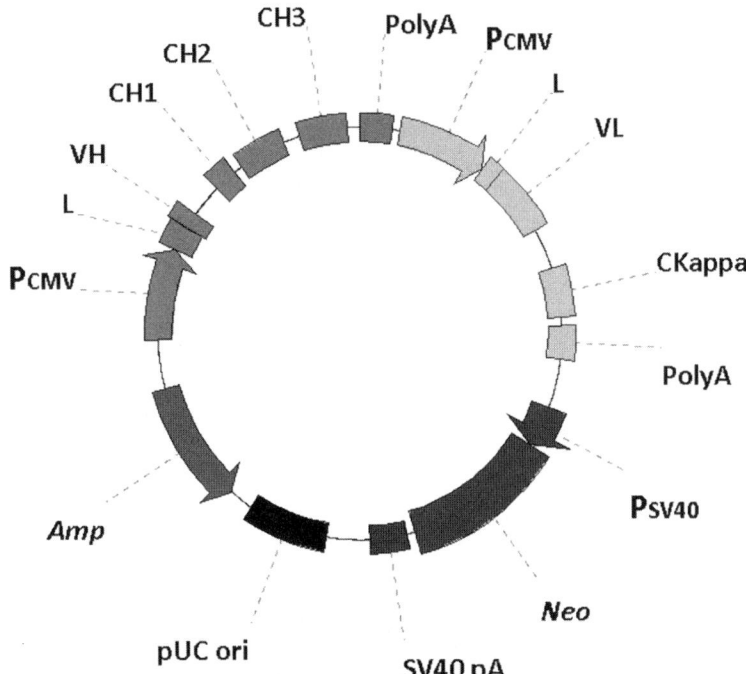

FIGURE 1 Typical vector for expression of antibody product in mammalian cells. This representative plasmid contains both the heavy-chain (VH, CH1, CH2, CH3) and light-chain (VL, CKappa) genes of the IgG gene. Each is driven by its own promoter (P_{CMV}) and has a polyadenylation signal (PolyA) downstream. A leader sequence (L) which serves as a signal for secretion is present for each IgG gene. A mammalian selectable marker, *Neo* in this case, is included with its own promoter (P_{SV40}) and polyadenylation signal (SV40 pA). Sequence elements for replication and selection in *E. coli* are also included, namely, an origin of replication (pUC ori) and selectable marker (*Amp*).

23.2.1.1. Basic Components

Genetic elements of interest (promoter, coding sequence, polyadenylation signal, selectable marker, or reporter gene) are collected and combined in the form of a plasmid, followed by insertion of the gene of interest. There are multiple criteria to consider when choosing a vector, including but not limited to maximum insert size required, commercial availability of expression systems, and availability of a suitable multiple cloning site for insertion of the gene of interest. Multiple genes can also be expressed from the same plasmid, provided that each gene has its own accessory sequences (i.e., promoter, etc.). One common example of such an approach is to include the coding sequences of both the heavy and light chains of the immunoglobulin G (IgG) gene on the same plasmid (Fig. 1). Selectable markers are discussed in section 23.2.3.1, while reporter genes are discussed in section 23.3.3.

Promoters regulate transcription of the gene of interest and any selectable markers, so the appropriate choice of promoters is critical. Promoters are short sequences, typically <1 kb, that bind to endogenous transcription factors. Some promoters are known to be active only in certain cell types, and this should be considered when selecting a promoter. Also, many promoters may function well in different species, yet their expression strength and response to regulatory elements may not be identical in cells of different species. Another consideration is the desired level of expression. Genetic manipulation for heterologous protein production often requires high-level expression, so for this purpose, strong constitutive promoters are typically used. Table 1 shows commonly used constitutive promoters.

The presence of a polyadenylation signal determines the 3′ terminus of mRNA and directs the addition of a poly(A) tail to the nascent mRNA transcript, often enhancing transcript stability and increasing expression. A polyadenylation signal consists of a conserved sequence (AAUAA) 20 to 30 nucleotides upstream of the 3′ end and a GU- or U-rich region downstream. While different sources of polyadenylation signals are used, the most common include those derived from bovine growth hormone, mouse β-globin, the simian virus 40 (SV40) early transcription unit, and the herpes simplex virus thymidine kinase genes (59).

23.2.1.2. Optional Components

In addition to the basic components described above, optional sequence elements can be incorporated into expression vectors with the aim of facilitating transcriptional activity of the foreign DNA inside the cell. These elements can help modulate the influence of neighboring elements at the site of transgene integration. They can also help prevent the decline of transgene expression with time, which could be occurring as a result of structural effects such as gene elimination or epigenetic silencing.

23.2.1.2.1. Enhancer Elements

Enhancer elements work in concert with promoters to recruit transcription factors and stabilize the transcription initiation complex. Many enhancers confer cell- and tissue-specific expression and have been used extensively to boost expression in the engineering of cultured cells for high-level recombinant protein expression. Several classes

TABLE 1 Commonly used constitutive promoters for expression constructs

Promoter	Source
Moderate constitutive expression promoters	
SV40	Simian virus 40
hUbC	Human ubiquitin C gene
mPGK	Mouse phosphoglycerate kinase gene
High-level constitutive expression promoters	
MoMLV-LTR	Moloney murine leukemia virus LTR
hCMV-IE/mCMV-IE	Human and mouse cytomegalovirus immediate-early promoter and enhancer genes
RSV-LTR	Rous sarcoma virus LTR
Ad2MLP-TLP	Adenovirus major late promoter and tripartite leader
hEF-1α	Human translation elongation factor 1α subunit gene
β-Actin	Chicken β-actin gene

of enhancer elements have been identified and applied in expression engineering. The first such class is STAR elements, or stabilizing and antirepressor elements. These elements were first identified based on their ability to block chromatin-associated repression (55). They are typically on the order of 500 to 2,000 bp and are thought to reduce the extent of histone deacetylation and the spread of methylation in the vicinity of the inserted transgene. Their utility has been demonstrated in CHO-K1 cells and a human osteosarcoma cell line (55).

The second class of elements is known as S/MAR elements (scaffold/matrix associated regions). It has been proposed that these AT-rich sequences serve as boundaries between active and inactive chromatin elements, interacting with the nuclear matrix. This interaction creates loops of coordinated gene expression, maintaining the DNA in a lightly wound conformation and insulating sequences from repression (33, 34). However, the exact molecular mechanisms involved are still under investigation. The chicken lysozyme MAR and human β-globin MAR have been applied to successfully increase transgene expression in CHO cells. Identified elements are ~3 kb in size and have been reported to be independent of the cell line or promoter used (53, 96).

A third class of elements was identified from housekeeping genes and is known as UCOE (ubiquitous chromatin opening elements). Housekeeping genes are highly transcriptionally active, with high histone acetylation, and elements within their promoter regions have been successfully used to increase expression in a number of cultured cell lines (3, 8, 91). While these are typically large elements (~16 kb), the use of shorter fragments (~2.5 kb) has been reported.

23.2.1.2.2. Introns

Studies have shown that inclusion of an intronic sequence, although not necessary for expression, can increase transgene expression by as much as 10- to 20-fold (16). This beneficial effect has been attributed to enhanced RNA polyadenylation and/or nuclear transport as a consequence of RNA splicing (45). Note that the intron must be flanked by appropriate 5′ and 3′ sequence signals to ensure effective splicing. However, caution should be exercised when including these elements, as their presence in the 3′ end of transcripts has been reported to lead to aberrant splicing (44, 92). The inclusion of cytomegalovirus intron has been shown to

increase recombinant factor VIII expression in CHO cells (18) and COS cells (21).

23.2.2. Introduction of DNA

There are many different methods for the transfer of expression vectors into cells. The choice of method depends on the host cell to be used, the number of cells to be transfected, and the efficiency of transfection. The different transfer strategies include chemical (DNA-calcium phosphate coprecipitation, cationic polymers, liposomes, molecular conjugates), physical (electroporation, microinjection), and biological (viral vectors; see section 23.3.1). While maximizing transfection efficiency is an objective when introducing foreign DNA into the cell, this usually comes at the cost of increased toxicity. Consequently, a trade-off should be struck which maximizes transfection efficiency while minimizing toxicity. Typically, for the transfection of adherent cells, reagents containing calcium phosphate, cationic polymers, or liposomes are used. For the transfection of suspension cells, electroporation or liposome-mediated transfection are commonly used. For detailed protocols of the methods described below, the reader is referred to references 39, 42, 54, and 75.

In calcium phosphate transfection, a precipitate containing calcium phosphate and DNA is formed by diluting DNA in a solution buffered within a narrow pH range and containing a low concentration of phosphate ions. Calcium ions (as $CaCl_2$) are added and the DNA is coprecipitated with the inorganic salts. These precipitates are then added to the cells (69). Parameters that can be utilized to optimize this process include the amount of DNA in the precipitate, the period of time the precipitate is left on the cells, and the possible inclusion of glycerol or dimethyl sulfoxide treatment, which has been shown to increase transfection efficiency in some cases. One of the main advantages of this method is low cost. Disadvantages include the need to use large amounts of DNA for efficient precipitate formation, and the somewhat variable nature of the procedure. This method is used for both transient and stable transfections.

Polycation-mediated DNA uptake involves the formation of complexes between positively charged polymer and negatively charged DNA. DEAE-dextran is the most commonly used polymer, but protamine and polyethyleneimine are also employed. The mechanism by which cells take up DNA in this method involves interaction of the condensed DNA with the negatively charged cell surface, followed by endocytosis (69). Parameters to optimize for this method include the number of cells transfected, the DNA concentration, and the DEAE-dextran concentration. The major advantages of this technique are its relative simplicity and speed and its limited expense. Disadvantages include cell growth inhibition. This method is most often used for transient transfections or transfections with episomal vectors.

Liposome-mediated transfection involves the formation of lipid-DNA complexes by using cationic lipids. The cationic lipids combine with the negatively charged DNA through electrostatic interactions. The condensed aggregates then interact with the negatively charged and hydrophobic plasma membrane and enter the cell (63). Parameters used to optimize this method include the DNA and lipid concentrations and the incubation time of these complexes with the cells. Advantages of this approach include high transfection efficiencies. Disadvantages include higher reagent costs, as compared to other transfection methods.

Electroporation is the most commonly used physical method of DNA delivery. This method uses a high-voltage electric pulse to introduce DNA into the cell. Cells are placed in an electroporation buffer and transferred to an electroporation cuvette, after which DNA is added. An electrical pulse of defined magnitude and length is applied, most often using a commercial device, and cells are subsequently transferred to normal growth medium (75). Parameters to optimize include the amplitude and length of the electrical pulse. This method is less affected by the amount of DNA used in the transfection. Advantages include high transfection efficiencies and the ability to deliver large DNA molecules, as well as its applicability for a broad range of cell types. Disadvantages include the need for specialized equipment, and difficulty in processing many samples, although industrial-scale continuous-flow electroporation is feasible. This method is used for both transient and stable transfections and has been shown to be very efficient for suspension cells.

23.2.3. Isolation of Transfectants

After introduction of the vector, the resulting host cell population is invariably heterogeneous, including cells not expressing any transgene and others exhibiting a wide range of expression. Typically, selection pressure is applied to enrich for cells that express the selectable marker and thus carry the transfected vector. By cultivating the surviving cells in the presence of a selection agent for a sufficient period of time (usually 2 weeks), the surviving population will contain only cells in which the selectable marker gene has integrated into a chromosome (Fig. 2). Such a cell population remains heterogeneous. For applications in which the primary objective is to characterize the properties of cells expressing the target gene, or to produce proteins for research use, obtaining a heterogeneous population of cells is sufficient. To establish a reproducible process of protein production, it is critical to establish a steady supply of cells that originate from a single cell clone. In such clonal isolates, variability in productivity and even in some key characteristics of the protein product, such as glycosylation, is unavoidable. It is thus critical to isolate a reasonably large but manageable number of clones and to establish a method of screening these clones to identify those candidates with the best combination of important features for protein product expression. Several commonly used selection and screening methods are described below.

23.2.3.1. Selection Methods

Approximately 1 in 10^4 cells in a transfected population will stably integrate foreign DNA carried in plasmids, with varying efficiencies observed in different cell types. It is therefore necessary to include a selectable marker in order to select cells that have integrated the foreign DNA.

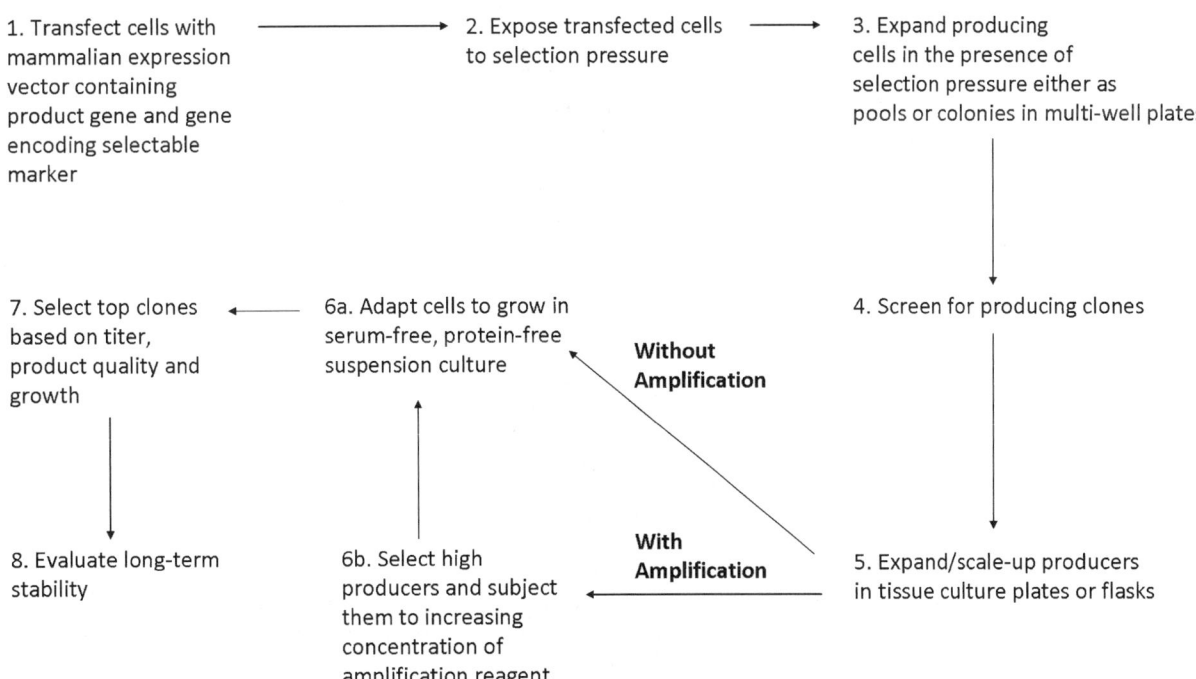

FIGURE 2 Overview of selection of stable and amplified clones.

Selectable markers include recessive markers, in which the recipient cells must have a special phenotype and introduction of the selectable marker provides a selection advantage over the host background, and dominant markers, in which no special phenotype is necessary in the recipient cell population. Commonly used recessive markers include wild-type dihydrofolate reductase (DHFR) (discussed in greater detail in section 23.2.4.1, where the recipient cells lack DHFR activity, and thymidine kinase, where the recipient cells lack thymidine kinase activity. In this latter system, the herpesvirus thymidine kinase gene is introduced into the cells, and positive transfectants are selected in HAT (hypoxanthine aminopterin thymidine) medium. Several dominant selectable markers are available for selection in mammalian cells, and some commonly used markers are shown in Table 2.

23.2.4. Amplification

Certain selectable markers can also be used for gene amplification by applying systematic increases in selective pressure. Gene amplification is commonly used to increase the number of copies of the transgene and its transcript levels, thus generating high levels of recombinant protein expression. In the presence of selective pressure, the selectable marker is coamplified alongside the product gene, and copy numbers between tens and several thousands have been reported (7, 14, 51).

23.2.4.1. DHFR System

One of the most commonly used gene amplification systems is based on the DHFR gene, whose chemical antagonist, methotrexate, can be used for gene amplification. DHFR is an enzyme which catalyzes the conversion of folate to tetrahydrofolate, a compound required for the biosynthesis of glycine, thymidine monophosphate, and purine nucleotides. CHO cells deficient in DHFR were isolated after ethyl methanesulfonate- and UV irradiation-induced mutagenesis (86, 87). These DHFR-deficient cells require the addition of thymidine, glycine, and hypoxanthine to the medium. These cells do not grow in the absence of added nucleosides unless they acquire a functional DHFR gene.

The primary advantage of the DHFR selection system is the ability to select for cells which contain an amplified copy number of DHFR genes. Methotrexate, a folate analogue, binds and inhibits DHFR, thereby leading to cell death. When cells are selected for growth in methotrexate, the surviving population contains increased levels of DHFR, resulting from an amplification of the DHFR gene. This is also accompanied by amplification of 10 to 10,000 kb of DNA surrounding the site of integration (24, 83). Therefore, by introducing a gene of interest (e.g., protein product gene) alongside the DHFR gene, coamplification of the product gene can be achieved. Upon further increases in methotrexate concentration, cells with higher degrees of DHFR gene amplification are obtained. Highly methotrexate-resistant cells have been obtained that contain several thousand copies of the DHFR gene (93). Thus, DHFR-deficient CHO cells have proven a useful host for gene transfer experiments devoted towards high-level expression of heterologous genes, due to their ability to amplify

TABLE 2 Commonly used selectable markers

Selectable marker	Example of selection conditions	Mechanism of action	Comments	References
Neomycin phosphotransferase (neo, G418, APH)	100 to 800 μg/ml G418	G418 inhibits protein synthesis by interfering with ribosomal functions. Gene expression detoxifies G418.	Resistant colonies usually take 1 to 2 weeks to emerge.	82
Hygromycin B phosphotransferase (HPH)	10 to 400 μg/ml hygromycin B	Hygromycin inhibits protein synthesis via translocation disruption and promotion of mistranslation. Gene expression detoxifies hygromycin through phosphorylation.	Selection typically takes 10 days to 3 weeks.	38, 71
Puromycin N-acetyltransferase (PAC, puro)	0.5 to 10 μg/ml puromycin	Puromycin inhibits protein synthesis. Gene expression detoxifies puromycin through acetylation.	Cell death occurs very rapidly and colonies can be obtained in as little as 7 days.	26
Bleomycin (phleo, bleo, zeocin)	0.1 to 50 μg/ml phleomycin	The gene encodes a protein which stoichiometrically binds the drug and inactivates it.	Selection typically takes 10 days to 3 weeks.	67, 85
Blasticidin-S deaminase (bsr)	3 to 50 μg/ml blasticidin	Blasticidin inhibits peptide bond formation by the ribosome, preventing protein synthesis.	Cell death occurs very rapidly and colonies can be obtained in as little as 7 days.	49

the gene copy number of DHFR. It should be noted that methotrexate resistance can also emerge from other cellular modifications, including altered methotrexate transport properties (79) or the development of a methotrexate-resistant DHFR enzyme (29, 64).

23.2.4.2. Glutamine Synthetase System

The glutamine synthetase selection system is based on the biosynthetic pathway of glutamine from the substrates glutamate and ammonia. Most mammalian cell lines require glutamine supplementation in their culture media to grow, and their endogenous glutamine synthetase activity is low. The glutamine synthetase expression vector contains a glutamine synthetase gene along with the gene of interest, allowing for selection by growth in glutamine-free cell culture medium. The glutamine synthetase gene is usually driven by a weaker promoter, typically the SV40 promoter (10). With a high concentration of the glutamine synthetase inhibitor methionine sulfoximine, it is possible to select for transfectants with gene amplification.

Several variations on the systems described above have been reported, aimed at increasing achievable expression levels. The use of DHFR in conjunction with an impaired neomycin resistance gene has been reported, resulting in the isolation of high-level-expressing clones with only a few integrated gene copies (6). In another application, expression of the protein product gene was coupled to that of the selectable marker using bicistronic expression vectors with an internal ribosomal entry site (IRES) (2, 9). It should be noted that in amplified systems, it is common to see a reduction in protein titers upon removal or reduction of the selective pressure.

23.2.4.3. Cell Clones versus Cell Pools

After transfection and selection, a mixture of transfected cells is present in the population. These cells are heterogeneous. Each originally transfected cell has a different chromosome integration site and copy number; each is also likely to exhibit different growth rates, kinetics, and productivity (78). A subpopulation which has a high productivity may be masked by others, thus giving an overall medium productivity level. Over time, one subpopulation may outgrow the others. This mixture of transfected cells is referred to as a cell pool. For the development of industrial recombinant protein-producing cell lines, it is required to isolate clonal lines to ensure a robust production process. However, such clonal lines are not always necessary for research applications or in cell engineering. Cell cloning is a long and tedious process, since each clone must start from a single cell. It takes more than 10 population doublings for a small colony to be visible to the naked eye. In certain applications, it may therefore be desirable to forgo the time-consuming step of generating clones and instead characterize the behavior of a pool of transfectants.

23.2.4.4. Cell Clone Screening

Screening is a crucial component in isolating the desired clones, but the process can be both time-consuming and labor-intensive. For industrial producers, typical protocols involve screening hundreds of clones. Consequently, instrumentation has been developed to facilitate high-throughput screening of cell clones. Below, we discuss some of the available cell line screening methods, a number of which are amenable to high-throughput screening. For a recent review on the subject, the reader is referred to reference 15.

23.2.4.4.1. Screening for Protein Expression Level or Productivity

It is a common practice to combine limiting-dilution cloning with productivity measurements, typically using enzyme-linked immunosorbent assay (ELISA), during cell screening. In this procedure, cells are diluted in microtiter plates to ensure that the probability of the presence of more than one growing cell per well is sufficiently low. Therefore, cells growing out in any well after a sufficient number of cell doublings can be considered to have originated from a single cell. Supernatants from positive wells can be assayed for product titer, usually using ELISA, and a number of producing clones, usually those with higher product concentrations, will be expanded for further characterization. There is a wide distribution in expression levels of the transgene among different clones. A wide range of possible causes may contribute to different productivity, such as the different sites of chromosomal integration of the transgene, and the difference in copy number of the transgene. This broad range of expression levels and productivities is the main driver for screening multiple clones.

In an effort to increase throughput, there has been a growing use of automation at various stages in the process, including the use of liquid handling systems for cell seeding and plating, and automating product quantification assays such as ELISA. One platform in particular has been developed with the aim of automating the entire cell line development platform: the Cello robotic system from the Automation Partnership (http://www.automationpartnership.com). This system incorporates clone screening, cell expansion, and product quantification. Contained within a laminar flow cabinet, it automates cell seeding following transfection, microscopic monitoring of microtiter plates, expansion of individual wells, and supernatant collection for product quantification. The system, though rather expensive, can currently process 600 to 800 plates.

23.2.4.4.2. Screening Based on Fluorescence

The occurrence of a high producer is a rare event in a heterogeneous population of producing cells. The use of flow cytometry can improve the odds of identifying these rare events, due to its ability to screen millions of cells, a scale that is far beyond the capability of the limited-dilution method. We describe here a number of methods that incorporate flow cytometry for sorting high-producing cells.

One approach to identify potential high producers has been to correlate productivity with the presence of product molecules on the cell surface. Upon labeling with fluorescent antibody, cells are sorted based on fluorescence intensity. Since the protein of interest is usually a secreted product, fluorescent staining of product molecules on the cell surface captures only those in secretory vesicles near the cell surface in the process of being excreted. A correlation between cell surface labeling and expression level has been reported in many cases, but does not always hold true (22, 62, 65, 77). Reporter proteins, such as green fluorescent protein (GFP), have been applied for screening potential high producers. In one approach, GFP was coexpressed with the product gene and cells were sorted based on fluorescence intensity (60, 66). The use of two fluorescent reporters for antibody production has also been reported, whereby the heavy-chain and light-chain constructs each express a different fluorescent reporter. Cell populations can then be selected for maximal double fluorescence (80). The use of fluorescent methotrexate has also been applied for cell screening. Cells are sorted based on fluorescence intensity as a result of methotrexate bound to the DHFR enzyme (95).

23.2.4.4.3. Screening Based on Protein Secretion

Another type of screening method relies on confining the secreted product to the vicinity of the cell, thus allowing for detection and isolation of cells secreting a high level of the desired protein product. The final product concentration in a growing culture is the combined result of many factors, including cell growth rate, the ability to reach and sustain a high cell concentration, and the specific productivity of the product. These methods aim to capture cells with the highest final product level, as opposed to the methods described in the above section, which capture the ones possibly having a faster secretion rate or specific productivity, but not necessarily robust growth characteristics.

Gel microdrops have been used to encapsulate single cells in biotinylated agarose droplets, consequently capturing proteins secreted by the cell. To allow protein-specific capture, antibodies are bound to the biotin-derivatized gel matrix and captured protein is then quantified using a fluorescent detection antibody (89, 90). Due to their small size (20 to 50 μm in diameter), gel microdrops can be sorted using fluorescence-activated cell sorting. This technique has been used to select high-level producers in hybridoma cells (37) and CHO cells (41). In matrix-based secretion assays, a product-specific antibody is immobilized on the cell surface, allowing capture of the secreted protein. Quantification of the secreted protein can be accomplished through immunostaining with a detection antibody, which can also be used for cell sorting. In order to prevent the secreted product from binding to neighboring cells, cells are typically cultured in a high-viscosity medium, thereby minimizing diffusion of the secreted product. This approach was originally described in hybridomas and activated T lymphocytes (61), and has since been applied to CHO (11) and NS0 (43) cells. A number of automated systems have recently been developed that make use of these screening approaches to isolate high-level-producing clones. The first is LEAP (laser-enabled analysis and processing). This system combines the use of product capture and individual cell protein secretion measurement with laser-mediated cell removal. Briefly, secreted protein is captured in situ using a capture matrix, and product secretion from individual cells is determined using a fluorescent detection antibody. The system, which incorporates automated image analysis, then secreting identifies highly secreting cells. Nonsecreting or poorly secreting cells surrounding the desired colonies are killed using laser irradiation, allowing the remaining colonies to proliferate.

A number of automated colony pickers have been commercialized that allow for automatic selection of high-level-producing clones. In these systems, cells are immobilized in semisolid medium at sufficiently low densities to allow colony formation. The secreted proteins are retained in the vicinity of the cell and can be visualized using appropriate detection methods. When the method is applied to screen for antibody production using secondary antibodies, colonies with large "halos" of captured secreted antibody are excised and expanded. Commercialized systems include the ClonePix™ from Genetix (http://www.genetix.com) and the CellCelector™ from Aviso (http://www.aviso-gmbh.de).

The cold capture antibody secretion assay is another method which allows retention of the secreted product in the vicinity of the cell. In this method, producing cells are isolated and maintained at low temperature, which is thought to retard plasma membrane turnover and the release of protein being secreted to the plasma membrane. Subsequently, staining with a fluorescent antibody allows cells to be sorted based on fluorescence intensity (13).

23.2.4.4.4. Screening Based on DNA or mRNA Levels

Attempts have been made to correlate productivity and titer with DNA and mRNA levels of product gene or selectable/amplifiable marker (e.g., DHFR). Good correlations have been observed between specific productivity and antibody product mRNA levels in CHO cells (56, 58) and hybridomas (28), with heavy-chain IgG being a better predictor of productivity than light chain. In one study of CHO cells, heavy-chain IgG mRNA levels were measured in a panel of clones both at the 96-well plate stage and further in the expansion process (58). While a good correlation was observed between mRNA levels and specific productivity in the expanded clones (cultured in 125-ml shake flasks or 50-ml tubes), mRNA levels at the 96-well plate stage could only be used for prediction of the top 25% producing clones upon expansion.

23.2.5. Characterization of Transfectants

From the screening and selection stage, a small number of clones are isolated for further study. Those clones can possibly exhibit a wide range of behavior even though they have all been screened for a common property, typically the productivity level. Some may exhibit diminished growth, or produce excess amounts of undesired metabolites, or secrete product molecules that have an altered glycoform. Thus, at the minimum, the expression of the transgene and growth kinetics of these clones should be characterized.

23.2.5.1. Expression Characterization

Following screening and selection of stable transfectants, several characterization steps may be included to verify integration and expression of the protein product. These include measuring mRNA levels of the expressed gene using quantitative real-time PCR, confirming presence of the target protein using Western blots, measuring protein levels using ELISA, and analyzing samples for product quality attributes.

23.2.5.2. Cell Characterization

The growth behavior of the isolated clones should be characterized. Different clones may even have different nutritional requirements in terms of their need for animal serum or special growth factors. Clones used for industrial production of recombinant protein need to be stable in terms of their growth, protein productivity, and quality over a time span that takes into account the duration of a manufacturing campaign, from a frozen vial to the manufacturing-scale bioreactor. The level of protein production can be measured using assays like ELISA and high-pressure liquid chromatography, and needs to be demonstrated to be stable. Even for research use, the stability of the transfectant should be examined for any cell age effects, although not necessarily over a duration as stringent as in industrial production.

23.2.6. Transient Expression

In transient expression, plasmids or virus vectors harboring the protein product gene are introduced into the cell and maintained and/or replicated as an extrachromosomal unit. Transfection efficiency, or the percentage of cells taking up and expressing the DNA of interest, is the most important factor controlling the overall expression level. High transfection efficiencies ($\geq$70%) are desirable in order to attain a manageable protein concentration. The duration of protein production using transient expression systems is usually limited by loss of vector DNA from the cell population with time, and the toxic effect of high copy numbers of foreign DNA.

Protein concentrations attainable in transient systems range from $\geq$1 to 100 mg/liter (corresponding to specific productivities of approximately 0.1 to 1 pg/cell/day) in approximately 5 to 10 days (5). While not suitable for long-term production or commercial manufacturing, these systems are extremely useful for rapid production of research material in small quantities, for drug candidate identification, and for in vivo evaluation, assay development, structural studies, and toxicology studies. Large-scale transient transfection cultures up to 100 liters have been reported, with yields of 75 mg (32).

Both viral and nonviral expression vectors are used for transient expression. Episomal systems, which allow extrachromosomal replication of the expression vector in mammalian cells, are commonly employed. These include the Epstein-Barr virus and SV40, which require expression of the Epstein-Barr virus nuclear antigen 1 (EBNA1) and SV40 large-T antigen, respectively, for replication. Transient expression typically requires a large amount of plasmid DNA, on the order of 1 to 1.25 µg/ml of culture (5). Large-scale production of plasmid DNA is therefore a significant contributor to the overall cost of transient transfection.

As reviewed in section 23.2.2, several methods are available for DNA delivery into the cell. One consideration when selecting a delivery method for transient transfection is cost, especially for large-scale transient transfections. While lipid-mediated transfection is very effective, it is not frequently used at large scales due to its comparatively higher cost. Polyethyleneimine-mediated transfection is more frequently employed at larger scales (5).

The host cells frequently used for transient expression include HEK-293, COS, and BHK cells (5). As CHO cells are a major workhorse for industrial protein production in mammalian cells, there have been developments in their use as a platform for transient expression. However, transient production in CHO cells has been reported to be comparatively lower than in HEK-293 cells (5).

23.3. GENETIC MANIPULATIONS FOR CELLULAR ENGINEERING APPLICATIONS

Within the context of cell engineering applications, genetic manipulations are employed to modify cellular functions. In such applications, the objective may not be to get maximal expression of the gene of interest, but rather to achieve the desired phenotype. Consequently, transfectants will be selected on the basis of their phenotype, rather than maximal recombinant protein production. The steps involved in genetic manipulations for cell engineering remain largely the same as described previously. However, additional options have now become available, including wider choices in vector expression platforms, promoters for transgene expression, and the possibility of using reporter constructs. These are discussed below.

23.3.1. Additional Vector Expression Systems

Plasmid DNA constructs, described in section 23.2.1, can be used as vehicles for gene delivery in cell engineering applications. However, a number of other expression systems are also commonly employed in such applications. These include the use of transposon-mediated gene delivery, such as the Sleeping Beauty (47) and piggyBac (27) systems, reviewed in reference 48. The use of viral expression systems is also common in cell engineering applications and is described below.

While viral vectors are rarely used for stable production of recombinant proteins, they are commonly used in cell engineering and other research applications. A number of viral vectors are available for gene expression, and the choice of system depends mainly on infectivity of the host cells, whether host cells are resting or dividing, and whether stable integration of the gene of interest is desired. Table 3 summarizes some of the viral systems available for gene expression. Retroviral vectors represent one of the most commonly used viral expression platforms, with wide commercial availability. We will therefore focus the discussion solely on this system.

Retroviral vectors are based on the *Retroviridae*, enveloped RNA viruses that contain two identical molecules of linear, single-stranded RNA. These are packaged into virion particles, which consist of an inner core containing the viral genome, an outer shell formed by capsid proteins, and an envelope surrounding the entire particle. This envelope consists of a lipid bilayer and includes virus-specific glycoproteins that are responsible for entry of the virion particle into the host cell through interaction with cell surface receptors (72). While many retrovirus species have been used to generate expression vectors, the most commonly used vectors are based on Moloney murine leukemia virus, whose biology has been extensively studied.

The retroviral genome contains several characteristic genes, including *gag*, *pol*, and *env*. The *gag* gene encodes the Gag polyprotein, which is cleaved into several structural subunits essential for virion assembly (Fig. 3). The *pol* gene encodes the enzymes reverse transcriptase and integrase, responsible for converting the viral RNA genome to DNA and mediating the incorporation of the viral DNA into the host chromosome, respectively. Finally, the *env* gene encodes viral surface glycoprotein, which interacts with host cell surface receptors to mediate infection. The retroviral genome is flanked by sequences known as long terminal repeats (LTRs), which are replicated during reverse transcription of the RNA genome. Each LTR is further divided into three functionally distinct regions: U5 (unique 5'), R (repeat), and U3 (unique 3'). The U3 region contains the viral promoter, including a TATA box and other transcriptional regulatory elements (72).

The life cycle of a retrovirus begins with entry of the virion particles into the host cell, either through direct fusion or through endocytosis-mediated fusion. Both infection processes require the interaction of the viral surface glycoproteins with cell surface receptor proteins. Once inside the cell, the virion sheds its envelope and the viral RNA genome is reverse transcribed to DNA using reverse transcriptase contained within the viral particle. This DNA is then transported into the nucleus and integrated into the host chromosome. This is followed by transcription and translation of the viral genome and assembly of new virions, which bud from the cell surface (45). One virion produces one integrated copy of the viral genome. Retroviral vectors typically contain a modified retrovirus genome to make them replication defective. This is accomplished by removing the viral coding sequences from the vector, i.e., the *gag*, *pol*, and *env* genes. In order for the vector system

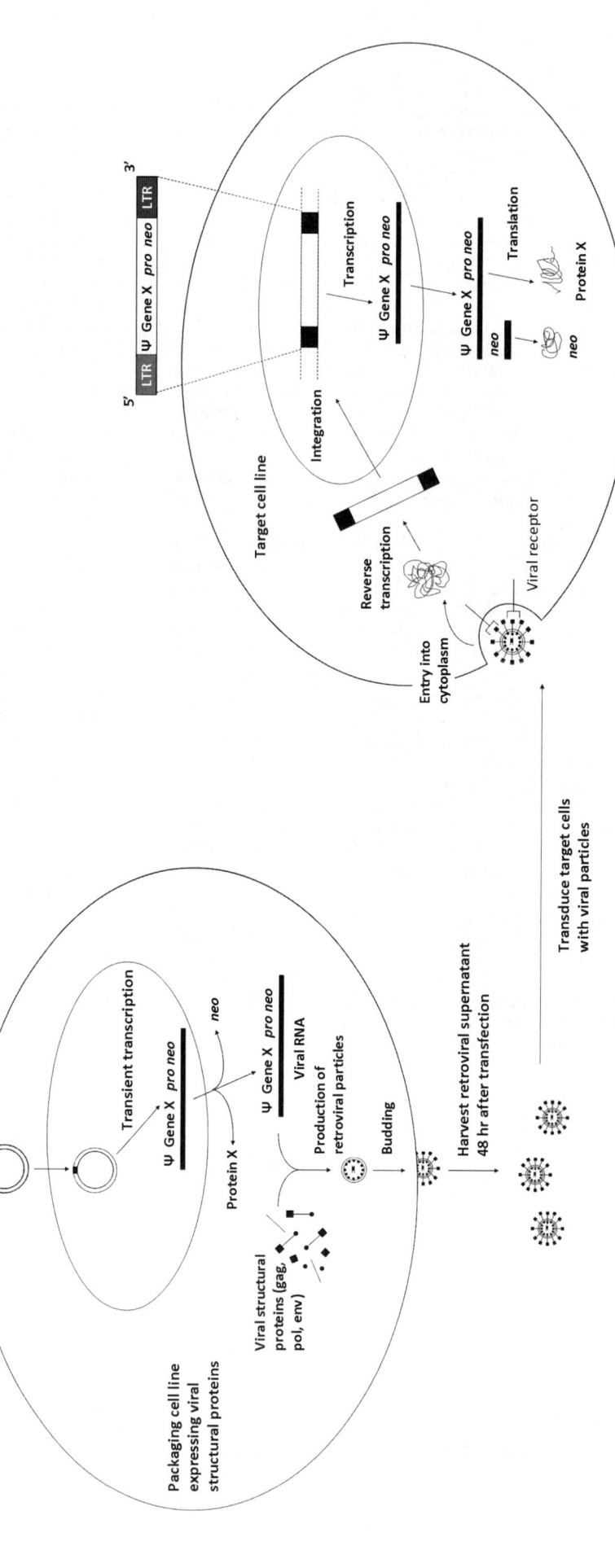

TABLE 3 Common viral expression systems

Virus	Genome[a]	Integration into host chromosome?	Insert size	Comments	Reference
Herpes simplex virus	Enveloped, dsDNA	No	≤30 kb	Broad mammalian host range	17
Epstein-Barr virus	Enveloped, dsDNA	No	~180 kb	This plasmid can be maintained episomally, given the viral origin of replication (*oriP*) and the viral gene encoding the *trans*-acting factor EBNA-1.	52
Simian virus 40 (SV40)	Nonenveloped, dsDNA	Yes	≤5 kb	Can integrate into the host genome and be maintained episomally in the presence of SV40 ori and large-T antigen	84
Adeno-associated virus	Nonenveloped, ssDNA	>90% episomal, <10% integrated	≤5 kb	Can integrate into the host genome and be maintained episomally	94
Adenovirus	Nonenveloped, dsDNA	No	~15 kb	Broad mammalian host range. This plasmid is maintained episomally.	12
Vaccinia virus	Enveloped, dsDNA	No	≤25 kb	Broad mammalian host range. Used for *trans* gene expression and vaccination.	19
Baculovirus	Enveloped, dsDNA	Yes	≤40 kb	Broad host range. Use of mammalian promoters allows *trans* gene expression in mammalian cells.	23
Retrovirus	Enveloped, ssRNA	Yes	≤9 kb	Host range: ecotropic virus replicates solely in cells derived from the host species, while amphotropic virus replicates in a wide range of mammalian cells.	72
Lentivirus	Enveloped, ssRNA	Yes	≤9 kb	Transduces both resting and dividing cells	31

[a]dsDNA, double-stranded DNA; ssDNA, single-stranded DNA; ssRNA, single-stranded RNA.

FIGURE 3 Overview of production of replication-defective retroviral particles using packaging cells and subsequent transduction of target cells. A replication-incompetent vector, which includes the 5′ and 3′ LTR regions, the packaging sequence (ψ), a gene of interest (Gene X), and a mammalian selectable marker (*neo*) with its own promoter (*pro*), is introduced into a packaging cell line which supplies the required viral accessory proteins, namely Gag, Pol, and Env. This bacterial plasmid also contains the required genetic elements for replication (ori) and selection (*Amp*) in E. coli. Upon entry into the packaging cell line (via any desired transfection methodology), the plasmid is transported into the nucleus, where it is transiently transcribed. The viral transcript initiates at the 5′ LTR and terminates at the 3′ LTR, generating a full-length viral transcript. It contains the packaging sequence (ψ), which allows packaging into viral particles. Fully assembled viral particles subsequently bud out of the packaging cell line. These viral particles are harvested from the supernatant 48 h after transfection. The virus-containing supernatant is subsequently incubated with the target cells, and entry of the viral particles occurs through interaction with host-cell receptors. Once the viral core is released into the cytoplasm, reverse transcription of the RNA viral genome begins using reverse transcriptase contained within the viral particle. The DNA copy of the viral genome then integrates into the host genome, and the gene of interest and selectable marker can be expressed. One infectious virion produces one integrated copy of the viral genome. Figure adapted from reference 20.

to be functional, these sequences must be supplied in *trans* (57). This is known as packaging and is typically carried out in 3T3 or HEK-293 cells. The necessary *trans*-retroviral elements are either constitutively expressed by the packaging cell line or cotransfected into the packaging cells with the retroviral vector. Along with the gene of interest and LTR sequences, the retroviral vector must also contain a packaging sequence (ψ) that is required for incorporation of the transcript into the viral particle. Active viral particles can then be produced in these cells and secreted into the growth medium, where they are harvested and used for subsequent transduction of a target cell line (81). These retroviral particles can infect the target cell and integrate the transgene into the host genome, yet are unable to replicate further and infect neighboring cells. The term "transduction" is typically used to refer to viral gene transfer using replication-defective particles, while "infection" refers to the use of replication-competent virus. The choice of viral receptor will determine which cell types the assembled viral particles can infect, and recombinant retroviral vectors can be pseudotyped using envelope proteins from different viruses. A commonly used envelope protein is vesicular stomatitis virus G protein (VSV-G), as it allows for efficient transduction in a broad range of species (72).

Transduction of target cells with retroviral particles is a relatively straightforward procedure. Typically, 12 to 18 h prior to transduction, cells are subcultured in fresh growth medium. For infection, the infectious retroviral particles are added along with a polycation, such as Polybrene, which promotes vector-cell interaction. After 4 to 24 h of incubation, the vector-containing medium is removed and replaced (or the cells are replated) with fresh growth medium. Integration and expression can be expected within 24 to 48 h and generally reach a plateau after 72 h (81). An additional factor to consider when attempting to maximize gene transfer efficiency is the multiplicity of infection (MOI). MOI represents the ratio of infectious particles to infection targets (i.e., cells). Optimal MOI values vary from cell to cell, although an MOI of 10 to 20 infectious vector particles per cell is common (25).

Retroviral vectors have several advantages, including their effectiveness with cell types which are difficult to transfect, as well as providing a single integrant for the gene of interest, a direct consequence of the mechanism of retroviral integration. Furthermore, it should be noted that the use of certain viral expression systems can require additional biosafety considerations, especially when pseudotyping viral particles for broad infectious potential (e.g., the use of VSV-G).

23.3.2. Constitutive and Inducible Promoters

Promoters can be broadly classified as constitutive or inducible. While inducible promoters are widely used for the production of heterologous proteins in bacteria, recombinant protein production in mammalian cells most often involves the use of a strong constitutive promoter, as discussed in section 23.2.1.1. However, cellular engineering applications can potentially benefit from the use of an inducible promoter system so that the desired cell phenotype is only expressed under induced conditions. For a thorough review of inducible expression systems, the reader is referred to reference 30.

Another class of promoters includes tissue-specific and developmental stage-specific promoters, whose expression is specific in some tissues or development stages. A large number of genes exhibit a wide range of dynamics responsive to various events, such as growth rate, for which no single "inducer" has yet been identified. These promoters allow the expression of the transgene to be conditional upon the event or tissue location. For a list of known event-triggered promoters, the reader is referred to reference 68.

We present here a summary of commonly used small-molecule-inducible gene expression systems, of which many are commercially available. Inducible expression systems should ideally have the following features: minimal baseline expression of the transgene under noninduced conditions; adjustable expression of the transgene, subject to dose response of the inducer molecule; the ability to achieve high expression levels that rival those of commonly used constitutive promoters, under induced conditions; rapid transition from noninduced to induced state, and vice versa; and straightforward design and development of inducible cell lines.

The Tet-inducible systems, which are referred to as Tet-OFF (tTA) and Tet-ON (rtTA), are perhaps the most widely used inducible promoter systems and have been extensively studied (40). These systems were originally burdened by high basal expression "leakiness" as well as a low efficiency of regulation (1). However, second-generation systems which address these issues have now been commercialized by a number of vendors (73). In the Tet-OFF system, the *tetO* operator sequence linked to a minimal promoter is used to regulate transgene expression (35). In the absence of tetracycline, a fusion protein consisting of the tetracycline repressor (TetR) and the VP16 transactivation domain (TA) from herpes simplex virus is bound to the operator sequence, allowing expression of the gene of interest. Addition of tetracycline displaces the complex from the operator sequence, thus suppressing transgene expression. A tetracycline derivative, doxycycline, is also commonly used in place of tetracycline. In the Tet-ON system, the fusion protein is based on a reverse Tet repressor (rTetR). In the absence of doxycycline, the fusion protein cannot bind the operator sequence, thus preventing expression of the transgene. Upon addition of doxycycline, the fusion protein can now bind to the operator, thereby allowing expression of the gene of interest (36).

Another inducible system based on the addition of a chemical inducer molecule is the rapamycin-dependent gene expression system. This system relies on the expression of two critical transcription factor binding domains, and these separate polypeptides are cross-linked using a chemical dimerizing drug as an inducer for transcription. Rapamycin binds two peptides, FKBP12 (FK506 binding protein of 21 kDa) and FRAP (FKBP rapamycin-associated protein). These have been modified by fusing the FKBP12 domain to the ZFHD1 (zinc finger homeo domain) DNA binding domain and fusing the rapamycin-binding domain of FRAP to a transcriptional activation domain from the p65 subunit of NF-κB (76). This system has been reported to provide tight expression control, with robust peak expression levels (74). This system is one of many dimerizer systems that are reviewed in reference 74.

A number of steroid receptor-derived inducible gene expression systems are also commonly used. One such class is based on insect steroid receptors. Ecdysteroid hormones, such as ecdysone, regulate a number of developmental processes in insects. Ecdysone binds and activates its receptor, EcR, leading to the formation of a heterodimer with the ultraspiracle protein (USP). This results in activation of gene expression. This system has been adapted and commercialized under the name Rheoswitch® (New England Biolabs). In this system, the original DNA binding site of *Drosophila melanogaster* EcR was replaced with a GAL4 DNA binding domain, while the activation protein was replaced with human fusion protein. Furthermore, induction is now achieved through the addition of a nonsteroidal synthetic ligand (70). This system has been reported to have high sensitivity to ligand induction, low background activity, and high inducibility of target gene expression (88).

23.3.3. Reporter Genes

Reporter systems are frequently used to track the quantity, spatial localization, and dynamics of the promoter by which their expression is controlled. Their applications are thus many, such as quantifying the efficiency of transfection, visualizing the dynamics of a regulatory element, and sorting a subpopulation based on gene expression. A number of reporter genes of both prokaryotic and eukaryotic origin are available. Ideally, a reporter system should exhibit phenotypic characteristics that are absent in the experimental system and for which there is a sensitive detection method

with a broad linear dynamic range that is quantitative, reproducible, and cost-effective.

Depending on the application, different reporter constructs are required. When measuring transfection efficiency or sorting transfected cells, a reporter gene driven by a strong constitutive promoter is typically employed. When characterizing regulatory sequences, expression of the reporter gene is put under the control of the regulatory sequences under study. Finally, for the study of cellular trafficking, protein fusion constructs are used, where the reporter gene coding sequence is linked to the protein of interest, usually through the addition of a short linker sequence. Table 4 presents a summary of commonly used reporter genes along with their advantages and disadvantages. GFP and other fluorescent proteins are most commonly used to assess the frequency of gene transfer in transfected cell populations. Firefly luciferase and secreted alkaline phosphatase are more commonly used to assess overall gene expression. The advantage of secreted alkaline phosphatase is that it does not require cell lysis for sample preparation, while luciferase offers a remarkably broad range of quantification (as much as 10^5-fold).

TABLE 4 Commonly used reporter genes (adapted from reference 50)

Reporter gene	Detection assay	Advantages	Disadvantages	Detection limit (no. of molecules)	Cost/assay (US$)
Green fluorescent protein (GFP)	Fluorescence	Fluorescence is species independent; useful for high-throughput assays performed in 96-well plates	Signal intensity may be too weak for some applications (although can use brighter GFP, e.g., EGFP).	NA	No additional reagents required
Secreted alkaline phosphatase	Colorimetric, bioluminescence, or chemiluminescence	Nonisotopic; secreted reporter protein; various assay formats available; useful for high-throughput assays performed in 96-well plates	May not be suitable for cells expressing low levels of placental-type alkaline phosphatase; may not be suitable if experimental conditions are known to impact cellular secretion	$10^4 - 10^5$	0.01 – 0.04
Firefly luciferase	Bioluminescence	Nonisotopic; good sensitivity and broad linear range; minimal endogenous activity in mammalian cells	Short half-life of protein; conventional assay lacks reproducibility	$1 \times 10^5 - 2.5 \times 10^5$	0.10 – 0.25
Chloramphenicol acetyltransferase	Chromatography, differential extraction, fluorescence, or immunoassay	Minimal endogenous activity in mammalian cells; various assay formats available	Assays are time-consuming and laborious; most require expensive radioactive substrate; relatively low sensitivity and narrow linear range; short half-life of CAT mRNA	$5 \times 10^7 - 10 \times 10^7$	2.00 – 7.00
β-Galactosidase	Colorimetric, fluorescence, or chemiluminescence	Nonisotopic; various assay formats available	Many cell types have high endogenous β-galactosidase activity.	$10^4 - 10^5$	0.01 – 0.02

23.4. CONCLUDING REMARKS

This chapter describes basic methodology used in expressing heterologous proteins in mammalian cells and in engineering their characteristics. Although expression of proteins in mammalian cells is more technically involved than commonly used microbial systems, especially *E. coli*, the wide range of vectors and commercial kits that are readily available has made their genetic manipulation easily accessible to researchers in need of carrying out the task. The development of new technologies for genetic manipulation of mammalian cells and their screening will continue to drive discovery and improve robustness of our protein production processes.

REFERENCES

1. **Agha-Mohammadi, S., M. O'Malley, A. Etemad, Z. Wang, X. Xiao, and M. T. Lotze.** 2004. Second-generation tetracycline-regulatable promoter: repositioned tet operator elements optimize transactivator synergy while shorter minimal promoter offers tight basal leakiness. *J. Gene Med.* **6:**817–828.
2. **Aldrich, T. L., A. Viaje, and A. E. Morris.** 2003. EASE vectors for rapid stable expression of recombinant antibodies. *Biotechnol. Prog.* **19:**1433–1438.
3. **Antoniou, M., L. Harland, T. Mustoe, S. Williams, J. Holdstock, E. Yague, T. Mulcahy, M. Griffiths, S. Edwards, P. A. Ioannou, A. Mountain, and R. Crombie.** 2003. Transgenes encompassing dual-promoter CpG islands from the human TBP and HNRPA2B1 loci are resistant to heterochromatin-mediated silencing. *Genomics* **82:**269–279.
4. **Arden, N., and M. J. Betenbaugh.** 2004. Life and death in mammalian cell culture: strategies for apoptosis inhibition. *Trends Biotechnol.* **22:**174–180.
5. **Baldi, L., D. L. Hacker, M. Adam, and F. M. Wurm.** 2007. Recombinant protein production by large-scale transient gene expression in mammalian cells: state of the art and future perspectives. *Biotechnol. Lett.* **29:**677–684.
6. **Barnett, R. S., K. L. Limoli, T. B. Huynh, E. A. Ople, and M. E. Reff.** 1995. Antibody production in Chinese hamster ovary cells using an impaired selectable marker. *ACS Symp. Ser.* **604:**27–40.
7. **Bebbington, C. R., G. Renner, S. Thomson, D. King, D. Abrams, and G. T. Yarranton.** 1992. High-level expression of a recombinant antibody from myeloma cells using a glutamine synthetase gene as an amplifiable selectable marker. *Bio/Technology* **10:**169–175.
8. **Benton, T., T. Chen, M. McEntee, B. Fox, D. King, R. Crombie, T. C. Thomas, and C. Bebbington.** 2002. The use of UCOE vectors in combination with a preadapted serum free, suspension cell line allows for rapid production of large quantities of protein. *Cytotechnology* **38:**43–46.
9. **Bianchi, A. A., and J. T. McGrew.** 2003. High-level expression of full-length antibodies using trans-complementing expression vectors. *Biotechnol. Bioeng.* **84:**439–444.
10. **Birch, J. R., and A. J. Racher.** 2006. Antibody production. *Adv. Drug Deliv. Rev.* **58:**671–685.
11. **Borth, N., M. Zeyda, and H. Katinger.** 2001. Efficient selection of high-producing subclones during gene amplification of recombinant Chinese hamster ovary cells by flow cytometry and cell sorting. *Biotechnol. Bioeng.* **71:**266–273.
12. **Bourbeau, D., Y. Zeng, and B. Massie.** 2003. Virus-based vectors for gene expression in mammalian cells: adenovirus. *New Compr. Biochem.* **38:**109–124.
13. **Brezinsky, S. C., G. G. Chiang, A. Szilvasi, S. Mohan, R. I. Shapiro, A. MacLean, W. Sisk, and G. Thill.** 2003. A simple method for enriching populations of transfected CHO cells for cells of higher specific productivity. *J. Immunol. Methods* **277:**141–155.
14. **Brown, M. E., G. Renner, R. P. Field, and T. Hassell.** 1992. Process development for the production of recombinant antibodies using the glutamine synthetase (GS) system. *Cytotechnology* **9:**231–236.
15. **Browne, S. M., and M. Al-Rubeai.** 2007. Selection methods for high-producing mammalian cell lines. *Trends Biotechnol.* **25:**425–432.
16. **Buchman, A. R., and P. Berg.** 1988. Comparison of intron-dependent and intron-independent gene expression. *Mol. Cell. Biol.* **8:**4395–4405.
17. **Burton, E. A., Q. Bai, W. F. Goins, D. J. Fink, and J. C. Glorioso.** 2003. Virus-based vectors for gene expression in mammalian cells: herpes simplex virus. *New Compr. Biochem.* **38:**27–54.
18. **Campos-da-Paz, M., C. S. Costa, L. S. Quilici, I. de Carmo Simoes, C. M. Kyaw, A. Q. Maranhao, and M. M. Brigido.** 2008. Production of recombinant human factor VIII in different cell lines and the effect of human XBP1 co-expression. *Mol. Biotechnol.* **39:**155–158.
19. **Carroll, M. W., and G. R. Kovacs.** 2003. Virus-based vectors for gene expression in mammalian cells: vaccinia virus. *New Compr. Biochem.* **38:**125–136.
20. **Cepko, C., and W. Pear.** 1996. Transduction of genes using retrovirus vectors. *Curr. Protocols Mol. Biol.* **36:**9.9.1–9.9.16.
21. **Chapman, B. S., R. M. Thayer, K. A. Vincent, and N. L. Haigwood.** 1991. Effect of intron A from human cytomegalovirus (Towne) immediate-early gene on heterologous expression in mammalian cells. *Nucleic Acids Res.* **19:**3979–3986.
22. **Cherlet, M., S. J. Kromenaker, and F. Srienc.** 1995. Surface IgG content of murine hybridomas: direct evidence for variation of antibody secretion rates during the cell cycle. *Biotechnol. Bioeng.* **47:**535–540.
23. **Condreay, J. P., and T. A. Kost.** 2003. Virus-based vectors for gene expression in mammalian cells: baculovirus. *New Compr. Biochem.* **38:**137–149.
24. **Coquelle, A., E. Pipiras, F. Toledo, G. Buttin, and M. Debatisse.** 1997. Expression of fragile sites triggers intrachromosomal mammalian gene amplification and sets boundaries to early amplicons. *Cell* **89:**215–225.
25. **Cornetta, K.** 2008. Murine leukemia virus-based retroviral vectors, p. 7–17. *In* B. Dropulic and B. Carter (ed.), *Concepts in Genetic Medicine.* John Wiley & Sons, Inc., Hoboken, NJ.
26. **de la Luna, S., I. Soria, D. Pulido, J. Ortin, and A. Jimenez.** 1988. Efficient transformation of mammalian cells with constructs containing a puromycin-resistance marker. *Gene* **62:**121–126.
27. **Ding, S., X. Wu, G. Li, M. Han, Y. Zhuang, and T. Xu.** 2005. Efficient transposition of the piggyBac (PB) transposon in mammalian cells and mice. *Cell* **122:**473–483.
28. **Dorai, H., B. Csirke, B. Scallon, and S. Ganguly.** 2006. Correlation of heavy and light chain mRNA copy numbers to antibody productivity in mouse myeloma production cell lines. *Hybridoma* **25:**1–9.
29. **Flintoff, W. F., and K. Essani.** 1980. Methotrexate-resistant Chinese hamster ovary cells contain a dihydrofolate reductase with an altered affinity for methotrexate. *Biochemistry* **19:**4321–4327.
30. **Fussenegger, M.** 2001. The impact of mammalian gene regulation concepts on functional genomic research, metabolic engineering, and advanced gene therapies. *Biotechnol. Prog.* **17:**1–51.
31. **Gasmi, M., and F. Wong-Staal.** 2003. Virus-based vectors for gene expression in mammalian cells: lentiviruses. *New Compr. Biochem.* **38:**251–264.
32. **Girard, P., M. Derouazi, G. Baumgartner, M. Bourgeois, M. Jordan, B. Jacko, and F. M. Wurm.** 2002. 100-liter transient transfection. *Cytotechnology* **38:**15–21.

33. **Girod, P.-A., M. Zahn-Zabal, and N. Mermod.** 2005. Use of the chicken lysozyme 5′ matrix attachment region to generate high producer CHO cell lines. *Biotechnol. Bioeng.* **91:**1–11.

34. **Girod, P. A., M. M. Zahn-Zabal, and N. Mermod.** 2005. MAR elements as tools to increase protein production by CHO cells, p. 411–415. *In Animal Cell Technology Meets Genomics: Proceedings of the 18th ESACT Meeting, Granada, Spain, May 11–14, 2003*, vol. 2. Springer, Dordrecht, The Netherlands.

35. **Gossen, M., and H. Bujard.** 1992. Tight control of gene expression in mammalian cells by tetracycline-responsive promoters. *Proc. Natl. Acad. Sci. USA* **89:**5547–5551.

36. **Gossen, M., S. Freundlieb, G. Bender, G. Mueller, W. Hillen, and H. Bujard.** 1995. Transcriptional activation by tetracyclines in mammalian cells. *Science* **268:**1766–1769.

37. **Gray, F., J. S. Kenney, and J. F. Dunne.** 1995. Secretion capture and report web: use of affinity derivatized agarose microdroplets for the selection of hybridoma cells. *J. Immunol. Methods* **182:**155–163.

38. **Gritz, L., and J. Davies.** 1983. Plasmid-encoded hygromycin B resistance: the sequence of hygromycin B phosphotransferase gene and its expression in *Escherichia coli* and *Saccharomyces cerevisiae*. *Gene* **25:**179–188.

39. **Gulick, T.** 2003. Transfection using DEAE-dextran. *Curr. Protocols Mol. Biol.* **40:**9.2.1–9.2.10.

40. **Guo, Z. S., Q. Li, D. L. Bartlett, J. Y. Yang, and B. Fang.** 2008. Gene transfer: the challenge of regulated gene expression. *Trends Mol. Med.* **14:**410–418.

41. **Hammill, L., J. Welles, and G. R. Carson.** 2000. The gel microdrop secretion assay: identification of a low productivity subpopulation arising during the production of human antibody in CHO cells. *Cytotechnology* **34:**27–37.

42. **Hawley-Nelson, P., V. Ciccarone, and M. L. Moore.** 2008. Transfection of cultured eukaryotic cells using cationic lipid reagents. *Curr. Protocols Mol. Biol.* **91:**9.4.1–9.4.17.

43. **Holmes, P., and M. Al-Rubeai.** 1999. Improved cell line development by a high throughput affinity capture surface display technique to select for high secretors. *J. Immunol. Methods* **230:**141–147.

44. **Huang, M. T., and C. M. Gorman.** 1990. The simian virus 40 small-t intron, present in many common expression vectors, leads to aberrant splicing. *Mol. Cell. Biol.* **10:**1805–1810.

45. **Huang, M. T., and C. M. Gorman.** 1990. Intervening sequences increase efficiency of RNA 3′ processing and accumulation of cytoplasmic RNA. *Nucleic Acids Res.* **18:**937–947.

46. **Ishaque, A., J. Thrift, J. E. Murphy, and K. Konstantinov.** 2007. Over-expression of Hsp70 in BHK-21 cells engineered to produce recombinant factor VIII promotes resistance to apoptosis and enhances secretion. *Biotechnol. Bioeng.* **97:**144–155.

47. **Ivics, Z., P. B. Hackett, R. H. Plasterk, and Z. Izsvak.** 1997. Molecular reconstruction of Sleeping Beauty, a Tc1-like transposon from fish, and its transposition in human cells. *Cell* **91:**501–510.

48. **Ivics, Z., and Z. Izsvak.** 2006. Transposons for gene therapy! *Curr. Gene Ther.* **6:**593–607.

49. **Izumi, M., H. Miyazawa, T. Kamakura, I. Yamaguchi, T. Endo, and F. Hanaoka.** 1991. Blasticidin S-resistance gene (bsr): a novel selectable marker for mammalian cells. *Exp. Cell Res.* **197:**229–233.

50. **Kain, S., and S. Ganguly.** 1996. Uses of fusion genes in mammalian transfection: overview of genetic reporter systems. *Curr. Protocols Mol. Biol.* **64:**9.0.1–9.0.5.

51. **Kaufman, R. J., L. C. Wasley, A. J. Spiliotes, S. D. Gossels, S. A. Latt, G. R. Larsen, and R. M. Kay.** 1985. Coamplification and coexpression of human tissue-type plasminogen activator and murine dihydrofolate reductase sequences in Chinese hamster ovary cells. *Mol. Cell. Biol.* **5:**1750–1759.

52. **Kennedy, G., and B. Sugden.** 2003. Virus-based vectors for gene expression in mammalian cells: Epstein-Barr virus. *New Compr. Biochem.* **38:**55–70.

53. **Kim, J.-M., J.-S. Kim, D.-H. Park, H. S. Kang, J. Yoon, K. Baek, and Y. Yoon.** 2004. Improved recombinant gene expression in CHO cells using matrix attachment regions. *J. Biotechnol.* **107:**95–105.

54. **Kingston, R. E., C. A. Chen, and J. K. Rose.** 2003. Calcium phosphate transfection. *Curr. Protocols Mol. Biol.* **63:**9.1.1–9.1.11.

55. **Kwaks, T. H., P. Barnett, W. Hemrika, T. Siersma, R. G. Sewalt, D. P. Satijn, J. F. Brons, R. van Blokland, P. Kwakman, A. L. Kruckeberg, A. Kelder, and A.P. Otte.** 2003. Identification of anti-repressor elements that confer high and stable protein production in mammalian cells. *Nat. Biotechnol.* **21:**553–558.

56. **Lattenmayer, C., E. Trummer, K. Schriebl, K. Vorauer-Uhl, D. Mueller, H. Katinger, and R. Kunert.** 2007. Characterisation of recombinant CHO cell lines by investigation of protein productivities and genetic parameters. *J Biotechnol.* **128:**716–725.

57. **Lech, P., and N. V. Somia.** 2008. Retrovirus vectors. *Contrib. Nephrol.* **159:**30–46.

58. **Lee, C. J., G. Seth, J. Tsukuda, and R. W. Hamilton.** 2009. A clone screening method using mRNA levels to determine specific productivity and product quality for monoclonal antibodies. *Biotechnol. Bioeng.* **102:**1107–1118.

59. **Makrides, S. C.** 2003. Vectors for gene expression in mammalian cells. *New Compr. Biochem.* **38:**9–26.

60. **Mancia, F., S. D. Patel, M. W. Rajala, P. E. Scherer, A. Nemes, I. Schieren, W. A. Hendrickson, and L. Shapiro.** 2004. Optimization of protein production in mammalian cells with a coexpressed fluorescent marker. *Structure* **12:**1355–1360.

61. **Manz, R., M. Assenmacher, E. Pflueger, S. Miltenyi, and A. Radbruch.** 1995. Analysis and sorting of live cells according to secreted molecules, relocated to a cell-surface affinity matrix. *Proc. Natl. Acad. Sci. USA* **92:**1921–1925.

62. **Marder, P., R. S. Maciak, R. L. Fouts, R. S. Baker, and J. J. Starling.** 1990. Selective cloning of hybridoma cells for enhanced immunoglobulin production using flow cytometric cell sorting and automated laser nephelometry. *Cytometry* **11:**498–505.

63. **Masson, C., V. Escriou, M. Bessodes, and D. Scherman.** 2003. Lipid reagents for DNA transfer into mammalian cells. *New Compr. Biochem.* **38:**279–289.

64. **McIvor, R. S.** 1996. Drug-resistant dihydrofolate reductases: generation, expression and therapeutic application. *Bone Marrow Transplant.* **18**(Suppl. 3):S50–S54.

65. **McKinney, K. L., R. Dilwith, and G. Belfort.** 1991. Manipulation of heterogeneous hybridoma cultures for overproduction of monoclonal antibodies. *Biotechnol. Prog.* **7:**445–454.

66. **Meng, Y. G., J. Liang, W. L. Wong, and V. Chisholm.** 2000. Green fluorescent protein as a second selectable marker for selection of high producing clones from transfected CHO cells. *Gene* **242:**201–207.

67. **Mulsant, P., A. Gatignol, M. Dalens, and G. Tiraby.** 1988. Phleomycin resistance as a dominant selectable marker in CHO cells. *Somat. Cell Mol. Genet.* **14:**243–252.

68. **Nettelbeck, D. M., V. Jerome, and R. Muller.** 2000. Gene therapy: designer promoters for tumour targeting. *Trends Genet.* **16:**174–181.

69. **Norton, P. A., and C. J. Pachuk.** 2003. Methods for DNA introduction into mammalian cells. *New Compr. Biochem.* **38:**265–277.

70. Palli, S. R., M. Z. Kapitskaya, M. B. Kumar, and D. E. Cress. 2003. Improved ecdysone receptor-based inducible gene regulation system. *Eur. J. Biochem.* **270:**1308–1315.

71. Palmer, T. D., R. A. Hock, W. R. Osborne, and A. D. Miller. 1987. Efficient retrovirus-mediated transfer and expression of a human adenosine deaminase gene in diploid skin fibroblasts from an adenosine deaminase-deficient human. *Proc. Natl. Acad. Sci. USA* **84:**1055–1059.

72. Parolin, C., and G. Palu. 2003. Virus-based vectors for gene expression in mammalian cells: retrovirus. *New Compr. Biochem.* **38:**231–250.

73. Pluta, K., M. J. Luce, L. Bao, S. Agha-Mohammadi, and J. Reiser. 2005. Tight control of transgene expression by lentivirus vectors containing second-generation tetracycline-responsive promoters. *J. Gene Med.* **7:**803–817.

74. Pollock, R., and T. Clackson. 2002. Dimerizer-regulated gene expression. *Curr. Opin. Biotechnol.* **13:**459–467.

75. Potter, H. 2003. Transfection by electroporation. *Curr. Protocols Mol. Biol.* **62:**9.3.1–9.3.6.

76. Rivera, V. M., T. Clackson, S. Natesan, R. Pollock, J. F. Amara, T. Keenan, S. R. Magari, T. Phillips, N. L. Courage, F. Cerasoli, Jr., D. A. Holt, and M. Gilman. 1996. A humanized system for pharmacologic control of gene expression. *Nat. Med.* **2:**1028–1032.

77. Sen, S., W. S. Hu, and F. Srienc. 1990. Flow cytometric study of hybridoma cell culture: correlation between cell surface fluorescence and IgG production rate. *Enzyme Microb. Technol.* **12:**571–576.

78. Seth, G., S. Charaniya, K. F. Wlaschin, and W. S. Hu. 2007. In pursuit of a super producer—alternative paths to high producing recombinant mammalian cells. *Curr. Opin. Biotechnol.* **18:**557–564.

79. Sirotnak, F. M., D. M. Moccio, L. E. Kelleher, and L. J. Goutas. 1981. Relative frequency and kinetic properties of transport-defective phenotypes among methotrexate-resistant L1210 clonal cell lines derived in vivo. *Cancer Res.* **41:**4447–4452.

80. Sleiman, R. J., P. P. Gray, M. N. McCall, J. Codamo, and N. A. Sunstrom. 2008. Accelerated cell line development using two-color fluorescence activated cell sorting to select highly expressing antibody-producing clones. *Biotechnol. Bioeng.* **99:**578–587.

81. Somia, N. 2004. Gene delivery to cells in culture using retroviruses. *Methods Mol. Biol.* **246:**491–498.

82. Southern, P. J., and P. Berg. 1982. Transformation of mammalian cells to antibiotic resistance with a bacterial gene under control of the SV40 early region promoter. *J. Mol. Appl. Genet.* **1:**327–341.

83. Stark, G. R., and G. M. Wahl. 1984. Gene amplification. *Annu. Rev. Biochem.* **53:**447–491.

84. Strayer, D. S., P. Cordelier, J. Landre, A. Matskevitch, H. J. McKee, C. N. Nichols, M. K. White, and M. S. Strayer. 2003. Virus-based vectors for gene expression in mammalian cells: SV40. *New Compr. Biochem.* **38:**71–91.

85. Sugiyama, M., C. J. Thompson, T. Kumagai, K. Suzuki, R. Deblaere, R. Villarroel, and J. Davies. 1994. Characterisation by molecular cloning of two genes from *Streptomyces verticillus* encoding resistance to bleomycin. *Gene* **151:**11–16.

86. Urlaub, G., and L. A. Chasin. 1980. Isolation of Chinese hamster cell mutants deficient in dihydrofolate reductase activity. *Proc. Natl. Acad. Sci. USA* **77:**4216–4220.

87. Urlaub, G., E. Kas, A. M. Carothers, and L. A. Chasin. 1983. Deletion of the diploid dihydrofolate reductase locus from cultured mammalian cells. *Cell* **33:**405–412.

88. Vilaboa, N., and R. Voellmy. 2006. Regulatable gene expression systems for gene therapy. *Curr. Gene Ther.* **6:**421–438.

89. Weaver, J. C. 1986. Gel microdroplets for microbial measurement and screening: basic principles. *Biotechnol. Bioeng. Symp.* **17:**185–195.

90. Weaver, J. C., P. McGrath, and S. Adams. 1997. Gel microdrop technology for rapid isolation of rare and high producer cells. *Nat. Med.* **3:**583–585.

91. Williams, S., T. Mustoe, T. Mulcahy, M. Griffiths, D. Simpson, M. Antoniou, A. Irvine, A. Mountain, and R. Crombie. 2005. CpG-island fragments from the HNRPA2B1/CBX3 genomic locus reduce silencing and enhance transgene expression from the hCMV promoter/enhancer in mammalian cells. *BMC Biotechnol.* **5:**17.

92. Wise, R. J., S. H. Orkin, and T. Collins. 1989. Aberrant expression of platelet-derived growth factor A-chain cDNAs due to cryptic splicing of RNA transcripts in COS-1 cells. *Nucleic Acids Res.* **17:**6591–6601.

93. Wurm, F. M., K. A. Gwinn, and R. E. Kingston. 1986. Inducible overproduction of the mouse c-myc protein in mammalian cells. *Proc. Natl. Acad. Sci. USA* **83:**5414–5418.

94. Xiao, X. 2003. Virus-based vectors for gene expression in mammalian cells: adeno-associated virus. *New Compr. Biochem.* **38:**93–108.

95. Yoshikawa, T., F. Nakanishi, Y. Ogura, D. Oi, T. Omasa, Y. Katakura, M. Kishimoto, and K.-I. Suga. 2001. Flow cytometry: an improved method for the selection of highly productive gene-amplified CHO cells using flow cytometry. *Biotechnol. Bioeng.* **74:**435–442.

96. Zahn-Zabal, M., M. Kobr, P. A. Girod, M. Imhof, P. Chatellard, M. de Jesus, F. Wurm, and N. Mermod. 2001. Development of stable cell lines for production or regulated expression using matrix attachment regions. *J. Biotechnol.* **87:**29–42.

GENETIC ENGINEERING OF SECONDARY METABOLITE SYNTHESIS

IV

"Secondary metabolism" is the term used to describe the biosynthesis of compounds that are not essential to the growth or maintenance of the host organism. Secondary metabolites include thousands of compounds from plants, fungi, and bacteria, including large, complex molecules that have applications in the pharmaceutical, agricultural, chemical, and fuel industries. Examples include phenolics (antiaging agents), terpenoids (antimalarial agents), polyketides (antibiotics, antifungals, anthelminthics, hypercholesterolemics), glycosides (antibiotics), and nonribosomal peptides (antibiotics and anticancer agents). Over the past 2 decades, the importance of secondary metabolites has driven significant efforts in the understanding of the genetics of their biosynthesis. Such efforts have produced the concomitant effects of the development of methodology for the genetic manipulation of many of the heretofore genetically intractable microorganisms, as well as the ability to place the genes responsible for the metabolic pathways into familiar organisms. These developments have enabled two important applications: novel, directed approaches to titer improvements, and the generation of novel derivatives of some of these secondary metabolites through manipulation of the corresponding genes. Both are represented in this section.

In chapter 24, "Glycosylation of Secondary Metabolites To Produce Novel Compounds," Andreas Bechtold and Katharina Probst describe the biosynthesis of various glycosides of aromatic polyketides in various *Streptomyces* hosts, and how the understanding of the genetic bases of biosynthesis has enabled the production of novel, hybrid molecules. Methods to manipulate the described organisms are included in the chapter. In chapter 25, "Metabolic Engineering of *Escherichia coli* for the Production of a Precursor to Artemisinin, an Antimalarial Drug," Christopher J. Petzold and Jay D. Keasling describe the assembly of genes in *E. coli* from bacteria and yeast to produce artemisinic acid, the sesquiterpene precursor to artemisinin. The chapter outlines how an "omics" (proteomics/transcriptomics/metabolomics) approach is being used to overcome bottlenecks in the biosynthesis of the product. In chapter 26, "Heterologous Production of Polyketides in *Streptomyces coelicolor* and *Escherichia coli*," James T. Kealey describes the steps taken to move the genes that determine the polyketide synthase that catalyzes the production of the backbone of erythromycin from the producing organism, *Saccharopolyspora erythraea*, to the genetically tractable hosts *Streptomyces coelicolor* and *E. coli*, and the concomitant strategies undertaken to enable the latter to produce the required precursor, methylmalonyl-CoA.

Richard H. Baltz, Kien T. Nguyen, and Dylan C. Alexander, in their chapter 27, "Genetic Engineering of Acidic Lipopeptide Antibiotics," describe the genetic basis for the production of the lipopetide daptomycin,

345

a recently marketed antibiotic effective against methicillin-resistant *Staphylococcus aureus* infections, and the methodology developed to enable manipulation of the corresponding peptide synthetase to result in the production of novel derivatives in the producing host, *Streptomyces roseosporus*. The chapter includes descriptions of the biosynthesis of other lipopeptides in *Streptomyces* hosts. In "Genetic Engineering To Regulate Production of Secondary Metabolites in *Streptomyces clavuligerus*," chapter 28, Susan E. Jensen describes methodology developed to genetically manipulate *Streptomyces clavuligerus*, which enabled the determination of how the biochemical pathway for the production of cephamycin C, the chemical backbone of numerous marketed cephamycins, is regulated. In addition, the methodologies developed allowed the understanding of the biochemical pathway for this production, as well as the

genes for regulation, in the same host, of clavulanic acid, the natural product used as a codrug to inhibit β-lactamases found in various gram-positive bacteria that cause respiratory infections.

Finally, in chapter 29, "Genetic Engineering of Myxobacterial Natural Product Biosynthetic Genes," Bryan Julien and Eduardo Rodriguez describe the methods developed to manipulate the genomes of the myoxobacterium *Sorangium cellulosum* and the various derivatives of the anticancer polyketide epothilone that were produced. This chapter, together with chapter 18, "Genetic Manipulation of Myxobacteria" (Black et al.) in Section III, presents a comprehensive set of methods and examples for the manipulation of these soil and marine microorganisms, which are rapidly becoming a new and important source of novel secondary metabolites.

Glycosylation of Secondary Metabolites To Produce Novel Compounds

ANDREAS BECHTHOLD AND KATHARINA PROBST

24

24.1. INTRODUCTION

Carbohydrates form the most abundant group of natural products. The central role of sugars and deoxysugars as energy and biosynthetic resources (for glycolysis, the pentose phosphate cycle, and the shikimate pathway), as energy storage devices (for photosynthesis), and as key structural elements in the formation of biological backbones (2-deoxyribose for DNA and N-acetylglucosamine for murein) has been well known for decades (74). But only recently it was recognized that carbohydrates also have crucial functions for signaling and information transfer in biological systems (59). In addition, carbohydrates are important and essential moieties of many pharmaceutical agents (92).

Although sugars and saccharides are involved in many processes within cells, the details of how saccharide structures influence biological functions remain one of the least explored avenues of chemical biology. This lack of knowledge compared to that of proteins and nucleic acids is due to the inaccessibility of numerous saccharide and oligosaccharide structures. Sugars can be connected by a wide range of regio- and stereochemically different linkages. This variability in structure makes classical synthetic chemistry difficult because multiple protection and deprotection steps are required (12). A highly attractive alternative is the use of biological systems or enzymatic tools. In this review, we summarize our own work on cloning and characterizing genes involved in sugar biosynthesis and attachment. We will give an overview of the methods which are available to create novel glycosylated compounds.

24.2. STRUCTURES AND MODES OF ACTION OF SELECTED NATURAL PRODUCTS

24.2.1. Avilamycin A

Avilamycin A (Fig. 1) and avilamycin C are produced by *Streptomyces viridochromogenes* Tü57. Along with curamycins, everninimicins (Fig. 1), and flambamycins, they belong to the orthosomycin class of antibiotics (96). Structural features of the avilamycins are a terminal dichloroisoeverninic acid moiety and a heptasaccharide side chain consisting of D-olivose, 2-deoxy-D-evalose, 4-O-methyl-D-fucose, 2,6-di-O-methyl-D-mannose, L-

lyxose, and eurekanate. Avilamycin A shows excellent activity against a broad range of gram-positive pathogenic bacteria, including glycopeptide-resistant enterococci, methicillin-resistant staphylococci, and penicillin-resistant streptococci, and is therefore an interesting candidate for the generation of new antibiotics (26, 41, 67). In 2003 Treede et al. reported that the methylation of G2535 and U2479 in domain V of the 23S rRNA confers resistance to avilamycin by preventing the antibiotic from binding to the ribosome (Fig. 2) (84). Both rRNA methyltransferases responsible for methylation were crystallized, and their structures were resolved (65, 66). The investigation of the crystal structure of the 50S ribosomal subunit of *Deinococcus radiodurans* in a complex with gavibamycin L1, a desmethyl derivative of avilamycin A, revealed the unique inhibitory mechanism of orthosomycins. The binding sites of gavibamycin L1 are C2480 and C2537 of the 23S rRNA, located in helices 89 and 91, respectively. The binding of gavibamycin L1 probably prevents the binding of aminoacyl-tRNA, initiation factor 2, and the ribosome recycling factor RRF (38). Interestingly, gavibamycin L1 binds to the ribosome in a U form (10). Both ends of the molecule, the methyleurekanate and the orsellinic acid moiety, are involved in the interaction with the rRNA. Although not yet confirmed by X-ray crystallography, it appears that evernimicin interacts with the ribosome in a very similar way. Unfortunately, evernimicin was dropped from clinical trials some years ago due to toxicity problems. Nevertheless, as the binding site of the orthosomycins is distinct from those of other ribosome-targeting antibiotics, it may be a target for new antibiotics.

24.2.2. Saccharomicin A

The heptadecaglycoside antibiotics saccharomicins A and B (Fig. 3) are produced by the rare actinomycete *Saccharothrix espanaensis* and represent an interesting class of antibiotics (45, 82). Both compounds show potent antibacterial activity in vitro and in vivo against multiple-drug-resistant strains of *Staphylococcus aureus* and vancomycin-resistant enterococci. The saccharomicins consist of an oligosaccharide portion and the intriguing aglycone N-(m,p-dihydroxycinnamoyl) taurine, in which caffeic acid is linked to the amino sulfonic acid taurine via an amide bond. All 17 monosaccharide residues of the saccharide chain are 6-deoxyhexoses: 5 are

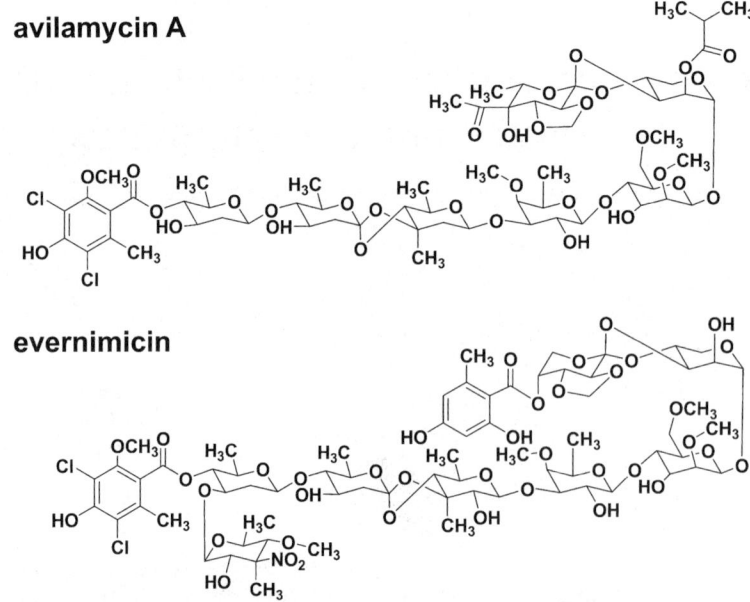

FIGURE 1   Structures of avilamycin A and evernimicin.

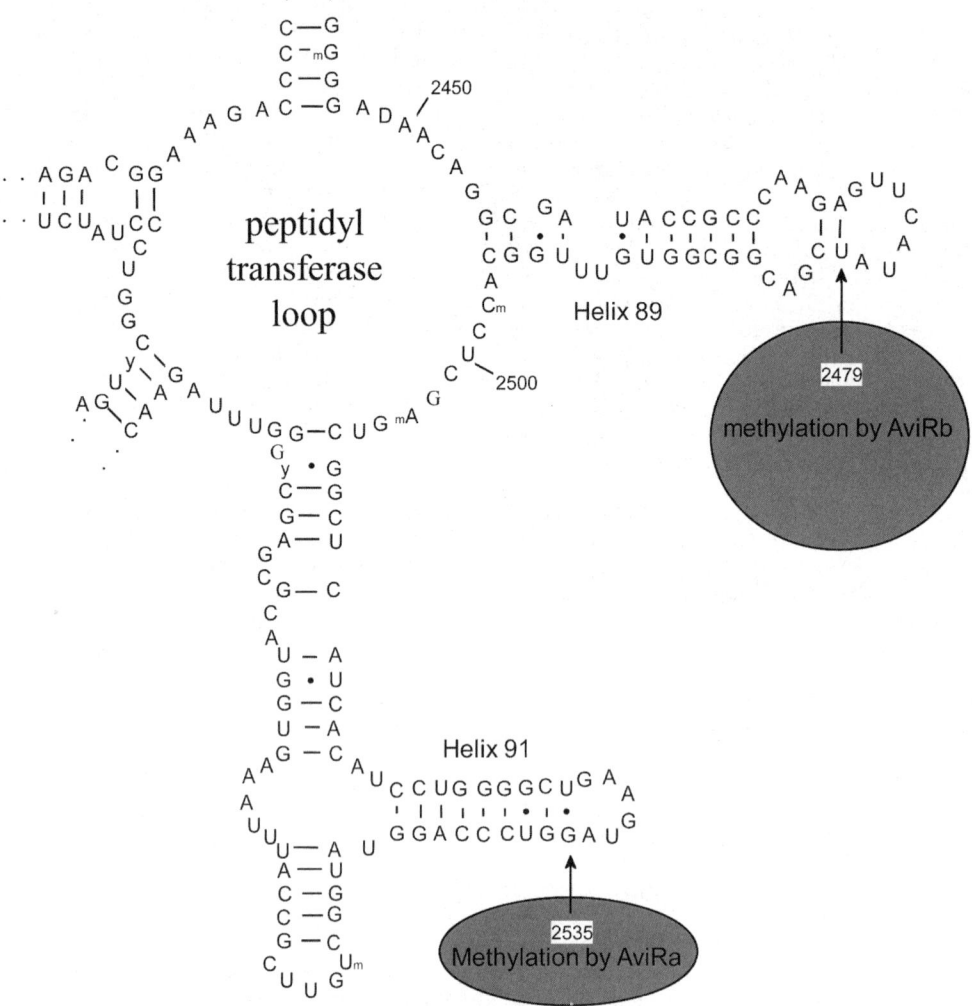

FIGURE 2   Methylation sites of the avilamycin resistance proteins AviRa and AviRb in the peptidyltransferase center.

saccharomicin A

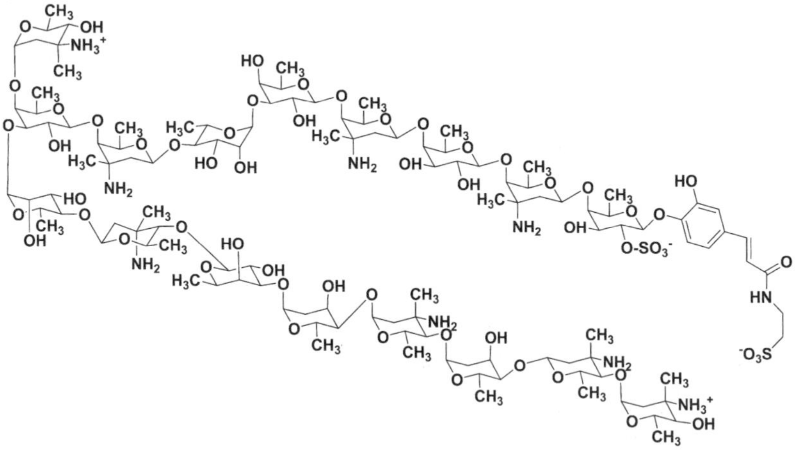

FIGURE 3 Structure of saccharomicin A.

fucoses, 4 are saccharosamines, 4 are 4-*epi*-vancosamines, and 4 are a combination of rhamnose(s) and digitoxoses. The exact mode of action of saccharomicin A is not known. It has been shown that the in vitro activities of the saccharomicins against gram-positive bacteria are unaffected by the presence of Ca^{2+} or Mg^{2+}. In contrast, the activities against gram-negative bacteria are substantially reduced by these divalent cations. Like other antibiotics, saccharomicins self-promote their uptake into bacterial cells by interacting with the binding sites of the divalent cations on the lipopolysaccharide. Thus, the primary cellular target appears to be the lipopolysaccharides of the bacterial membrane.

24.2.3. Landomycins, Urdamycin A, Saquayamycin Z, and Simocyclinone

Landomycins (produced by *Streptomyces cyanogenus* S136 and *Streptomyces globisporus*) (31), urdamycins (produced by *Streptomyces fradiae* Tü2717) (17), saquayamycins (produced by *Micromonospora* sp. strain 6368) (83), and simocyclinones (produced by *Streptomyces antibioticus* Tü6040) (80) are angucycline-type anticancer agents (Fig. 4). All compounds contain an angucyclic polyketide core and one or more sugar side chains consisting of D-olivose and/or L-rhodinose moieties. Saquayamycin Z contains additional 2-deoxy-L-fucose and L-aculose moieties, and simocyclinone D8 contains a tetraene side chain and a halogenated aminocoumarin. The angucyclic polyketides differ in functional groups, oxidation state, and sugars/sugar chains, which are either O-glycosidically or C-glycosidically linked to the polyketide core. Landomycin A and saquayamycin Z in particular show powerful anticancer activities, especially against prostate cancer cell lines (3, 14). Like urdamycins and other members of the angucyclines, both compounds also show cytotoxic activity. In contrast to the other angucyclines, simocyclinone D8 is a potent inhibitor of gyrase supercoiling, having a 50% inhibitory concentration lower than that of novobiocin (23, 69). However, it exhibits no competitive inhibition of the DNA-independent ATPase reaction of GyrB, which is characteristic of other aminocoumarins.

Simocyclinone D8 also inhibits DNA relaxation by gyrase. It does not stimulate cleavage complex formation, as do the quinolones, the other major class of gyrase inhibitors. Instead, it prevents Ca^{2+}- and quinolone-induced cleavage complex formation. Results from binding studies suggest that simocyclinone D8 interacts with the N-terminal domain of GyrA, which is a novel mechanism for a gyrase inhibitor.

24.2.4. Aranciamycin and Polyketomycin

Aranciamycins (Fig. 5), which belong to the anthracycline group of natural products, have been isolated from *Streptomyces echinatus*. Aranciamycin A, the main compound, shows weak antitumor activity and appears to be a specific inhibitor of *Clostridium histolyticum* collagenase (11). Aranciamycin consists of the aglycone aranciamycinone and the sugar 2-O-methyl-L-rhamnose. It differs from the very well-known anthracyclines (e.g., daunomycin, adriamycin, and carminomycin) with respect to the substitution of ring A and the sugar moiety, which does not contain an amino group. The tetracyclic quinone glycoside polyketomycin (Fig. 5) was isolated from *Streptomyces diastatochromogenes* Tü6028 (71). It inhibits the growth of gram-positive bacteria and shows activity against several tumor cell lines. Polyketomycin consists of two polyketide residues, the decaketide polyketomycinone and the tetraketide 3,6-dimethylsalicylic acid. A sugar chain consisting of β-D-amicetose and α-L-axenose connects both moieties. Structurally related to poyketomycin are the antitumor agents dutomycin from *Streptomyces* sp. strain 1725, the DNA methyltransferase inhibitor DMI-2 (a tautomer of dutomycin) produced by *Streptomyces* sp. strain 560, and the cervimycins isolated from *Streptomyces tendae* HKI-179. They share a *p*-quinone in ring D (15).

24.2.5. Lipomycin

α-Lipomycin (Fig. 6) is an acyclic polyene antibiotic isolated from *Streptomyces aureofaciens* Tü117 (47) and was named from the finding that antibiotic activity is antagonized by several lipids, including lecithin and some sterols. Its structure consists of a carbonyl group linking

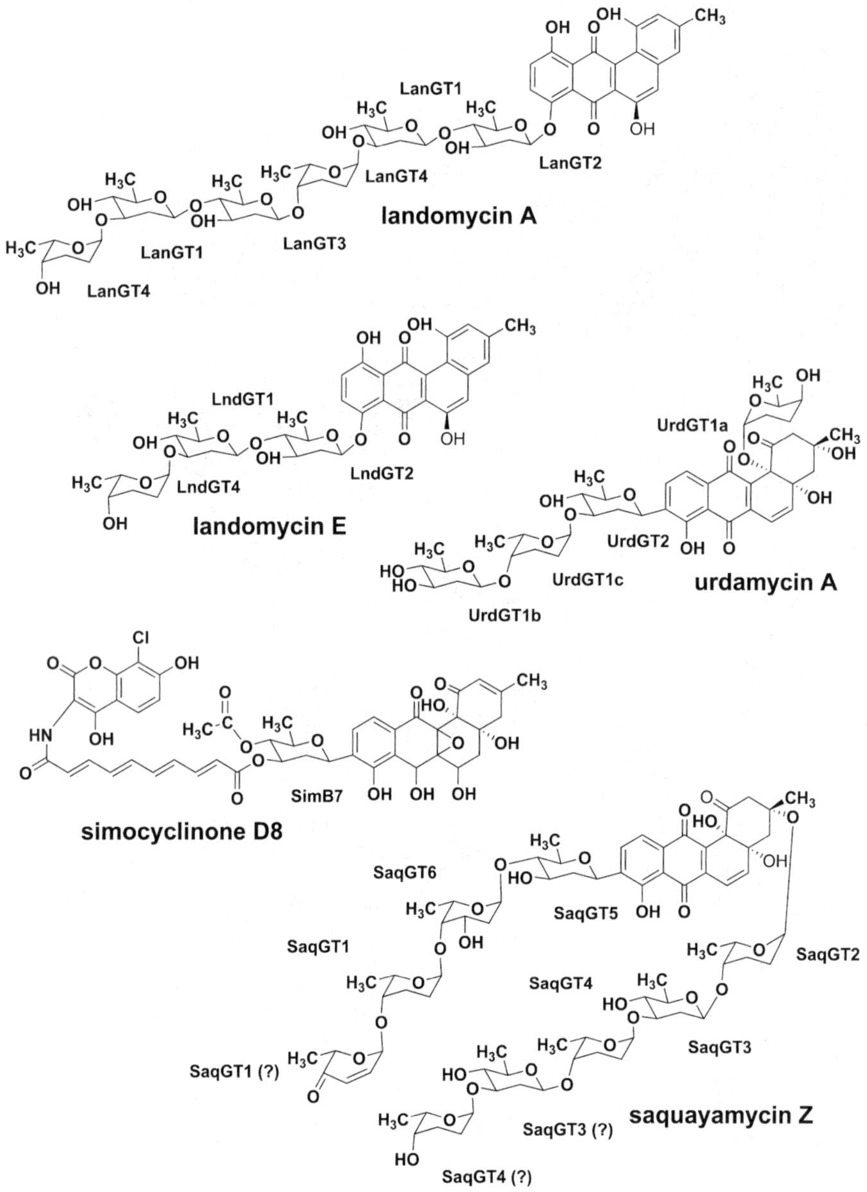

FIGURE 4 Structure of selected angucyclines and functions of GTs involved in their biosynthesis.

a pentaene chain with a derivative of N-methyltetramic acid composing the aglycone. An L-digitoxose molecule is attached to the polyketide part of the aglycone. α-Lipomycin is active against gram-positive bacteria but has no effect upon the growth of fungi, yeast, and gram-negative bacteria. The tetramic acid (2,4-pyrrolidinedione) ring system has been known since the early 20th century, when the first simple derivatives were prepared. Natural products containing the tetramic acid moiety continue to arouse interest due to the range of biological activities they display, including antibiotic and antiviral activities, mycotoxicity, cytotoxicity, and phospholipase A2 inhibition. Other members of this class are responsible for the pigmentation of certain molds and sponges. Examples of natural compounds containing a tetramic

acid moiety are aurantoside B, militarinone B, erythroskyrin, fuligurubin, and oleficin (78).

24.2.6. Phenalinolactone

Phenalinolactone A and its derivatives B through D are terpene glycosides produced by *Streptomyces* sp. strain Tü6071 (Fig. 7). Phenalinolactones are active against some gram-positive bacteria. The tricyclic terpenoid backbone of phenalinolactone is connected to a γ-butyrolactone moiety, an O-glycosidically bound L-amicetose, and a pyrrole-2-carboxylic ester moiety. Additionally, the oxygen function at C-3 contains an acetyl group (19). The structure of the phenalinolactones resembles those of other terpenoids and especially that of brasillicardin from *Nocardia terpenica* (16).

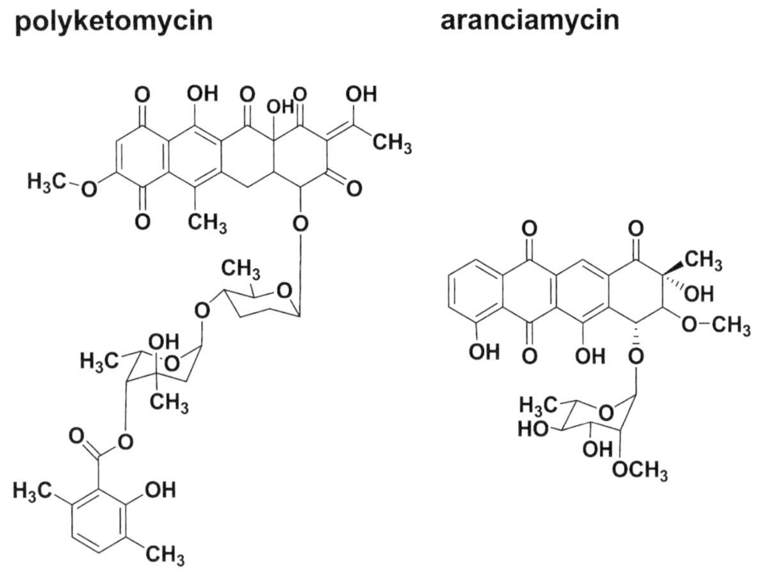

polyketomycin   **aranciamycin**

FIGURE 5   Structures of aranciamycin and polyketomycin.

24.3. CLONING OF BIOSYNTHETIC GENE CLUSTERS OF COMPOUNDS CONTAINING DEOXYSUGARS

24.3.1. Generation and Screening of Cosmid Libraries

One of the most frequently used bifunctional cosmid vectors in *Streptomyces* genetics is pOJ436 (6). This cosmid can replicate in *Escherichia coli* and *Streptomyces* and is therefore very useful for the preparation and screening of cosmid libraries and for heterologous expression. A reliable procedure is described below, although it may require some specific optimization for certain strains.

1. For the preparation of DNA from actinomycetes, the mycelia must be embedded in agarose.
2. After cell disruption, the genomic DNA is partially digested with Sau3AI and ligated into cosmid vector pOJ436, which has been digested with HpaI and BamHI. The DNA is then in vitro packaged with the Gigapack III Gold packaging extract kit according to the instructions of the manufacturer (Stratagene, San Diego, CA).
3. Robotically produced high-density colony arrays (Hybond N+; Amersham Pharmacia, Freiburg, Germany) can be utilized for the screening procedures. In each case, ca. 2,000 cosmid clones should be used.

α-lipomycin

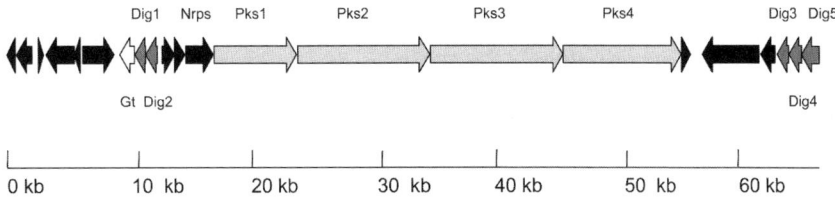

FIGURE 6   Structure of α-lipomycin and organization of the α-lipomycin biosynthetic gene cluster. The GT gene is shown in white, polyketide genes are shown in light gray, sugar biosynthetic genes are shown in dark gray, and all other genes are shown in black. Nrps, nonribosomal peptide synthase gene.

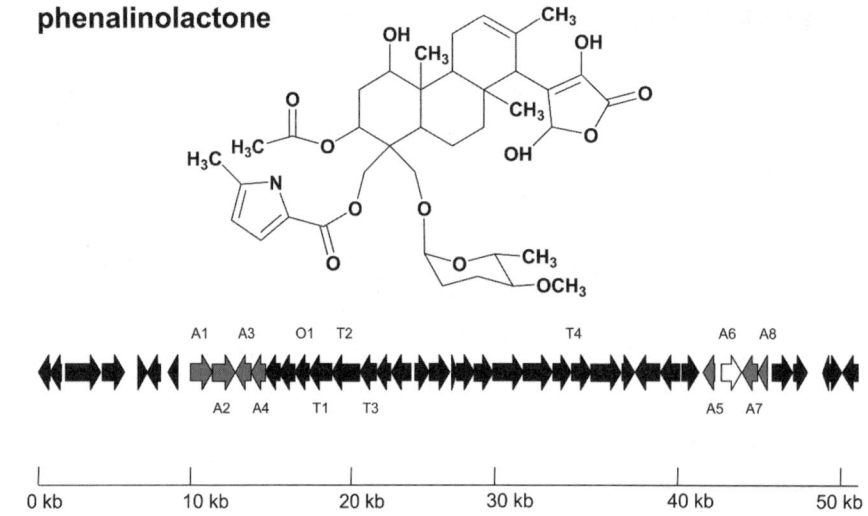

FIGURE 7 Structure of phenalinolactone and organization of the phenalinolactone biosynthetic gene cluster. The GT gene is shown in white, sugar biosynthetic genes are shown in gray, and all other genes are shown in black.

24.3.2. Gene Probes To Screen Cosmid Libraries

The best DNA probes are PCR fragments amplified from each individual strain. The PCR fragments can be obtained with primers directed toward conserved gene sequences for acyltransferase domains of polyketide synthase I (PKSI), polyketide synthase II gene (KS$_\alpha$), nucleoside diphosphate (NDP)-glucose-4,6-dehydratase genes, and NDP-glucose-2,3-dehydratase genes.

The following primers may be used (abbreviations are according to standard International Union of Pure and Applied Chemistry nomenclature for degenerate nucleotides).

For cloning cosmids containing PKSI genes (e.g., the α-lipomycin gene cluster [7]):
5′-CTSGGGSGACCCSATCGAG-3′
5′-CTSGGGSGACCCSATCGAG-3′

For cloning cosmids containing PKSII genes (e.g., the polyketomycin cluster [15]):
5′-TSGCSTGCTTCGAYGCSATC-3′
5′-TGGAANCCGCCGAABCCGCT-3′

For cloning cosmids containing NDP-glucose-4,6-dehydratase genes (e.g., the phenalinolactone gene cluster [19]):
5′-CSGGSGSSGCSGGSTTCATSGG-3′
5′-GGGWRCTGGYRSGGSCCGTAGTTG-3′

For cloning cosmids containing 2,3-dehydratase genes (e.g., the saccharomicin gene cluster [4]):
5′-GAGGCSACSWSSAACTACAC-3′
5′-SWRGAASCGSCCSCCCTCCTC-3′

The following PCR protocol may be used to prepare the desired gene probe (PCR mixture components and concentrations: chromosomal DNA, 0.2 µg/100 µl; primers, 1 µg/100 µl; deoxynucleoside triphosphates, 200 µM [each]; buffer, 25 mM Tris-acetate [pH 7.7], 100 mM potassium acetate, 1mM dithiothreitol; polymerase, 10 U; and dimethyl sulfoxide [DMSO], 1%).

1. Denaturation of DNA at 94°C for 3 min
2. Addition of reaction components and annealing at 60 to 72°C for 1 min
3. Elongation at 72°C for 2 min
4. Thirty cycles of denaturation (94°C for 45 s), annealing (60 to 72°C for 1 min), and elongation (72°C for 2 min)
5. Terminal elongation at 72°C for 7 min

Labeling of DNA probes and colony hybridization are performed by following standard procedures.

24.3.3. Cloning of GT Genes from Actinomycetes

For the cloning of unidentified glycosyltransferase (GT) genes, a specific PCR-based cloning strategy has been developed by Luzhetskyy et al. (53). Hybrid primers (P1 [5″-AGGTCTGCGGCGCCSCNCAYGCNMG-3″] and P2 [5″-GGTGGCGGCGCCNCCRTGRTG-3″]) for the amplification of GT genes involved in natural-product biosynthesis were designed by using sequences of genes (*desVII*, *eryBV*, *eryCIII*, *megDI*, *megCIII*, and *megBV*) encoding GTs catalyzing the deoxysugar attachment to a polyketide moiety. Preselected cosmid DNA hybridizing to one of the probes listed above may be used as a template in the PCR mixture (components and concentrations: chromosomal DNA, 0.5 µg/100 µl; primers, 1 µg/100 µl; deoxynucleoside triphosphates, 200 µM [each]; buffer and polymerase; and DMSO, 1%).

1. Denaturation of DNA at 94°C for 5 min
2. Addition of reaction components and annealing at 45°C for 45 s
3. Elongation at 72°C for 1 min
4. Thirty-five cycles of denaturation (94°C for 1 min), annealing (45°C for 45 s), and elongation (72°C for 1 min)
5. Terminal elongation at 72°C for 5 min

Usually, a 900-bp fragment corresponding to an internal segment of a GT gene is amplified by PCR. The cloning of PCR fragments is performed by following standard procedures. Full-length sequences of the GT genes may be obtained using cosmids as a template.

24.4. GENE CLUSTERS ENCODING NATURAL PRODUCTS CONTAINING DEOXYSUGARS

Most sugar components of natural products from actinomycetes are deoxygenated sugars. These deoxysugars usually derive from glucose 1-phosphate, which is converted by an NDP-D-glucose synthase to NDP-D-glucose. Dehydratases,

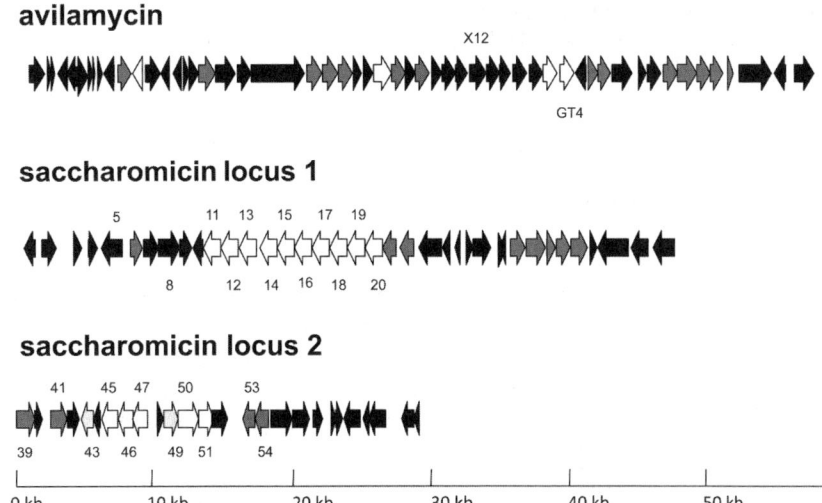

avilamycin
X12
GT4
saccharomicin locus 1
5
11 13 15 17 19
8 12 14 16 18 20
saccharomicin locus 2
41 45 47 50 53
39 43 46 49 51 54
0 kb
10 kb
20 kb
30 kb
40 kb
50 kb

FIGURE 8 Organization of the avilamycin A and saccharomicin gene clusters. Two loci which may be involved in saccharomicin biosynthesis have been identified. GT genes are shown in white, sugar biosynthetic genes are shown in gray, and all other genes are shown in black. The locations of genes for AviX12 (X12) and AviGT4 (GT4), as well as some saccharomicin genes, are shown.

ketoreductases, epimerases, methyltransferases, and other enzymes are used to produce the nucleotide-activated sugar structures that are the substrates for GTs. In many cases, genes encoding these enzymes are located in the biosynthetic gene clusters, but they are usually not clustered into one transcription unit, nor are they necessarily present in an uninterrupted linear array. Here, we give an overview of the gene clusters cloned in our lab and discuss the functions of recently discovered genes and enzymes involved in sugar biosynthesis or attachment.

24.4.1. The Avilamycin Biosynthetic Gene Cluster

The avilamycin biosynthetic gene cluster was cloned in 2001 by using the 4,6-dehydratase gene as a probe (89). The cluster is located within a 60-kb DNA segment of the chromosome. About 50 genes are involved in the biosynthesis of avilamycin A and its regulation (Fig. 8). The functions of several genes were elucidated by gene inactivation experiments (37, 85, 90) (Fig. 9). By deleting methyltransferase genes, novel desmethylated compounds that showed increased hydrophilicity could be generated (37). Biosynthetic enzymes involved in sugar biosynthesis or modification that have recently been investigated are AviGT4, a GT, and AviX12, a very unusual epimerase. AviGT4 catalyzes the attachment of a thus far undefined precursor of the eurekanate moiety to the L-lyxose moiety as part of a saccharide chain (37). The crystal structure of AviGT4 has recently been resolved (58). It contains two

AviG3
AviGT4
AviB1/AviB2/AviO2
AviG4
AviM
AviG2
AviG6
AviX12
AviG5
AviG1
AviH

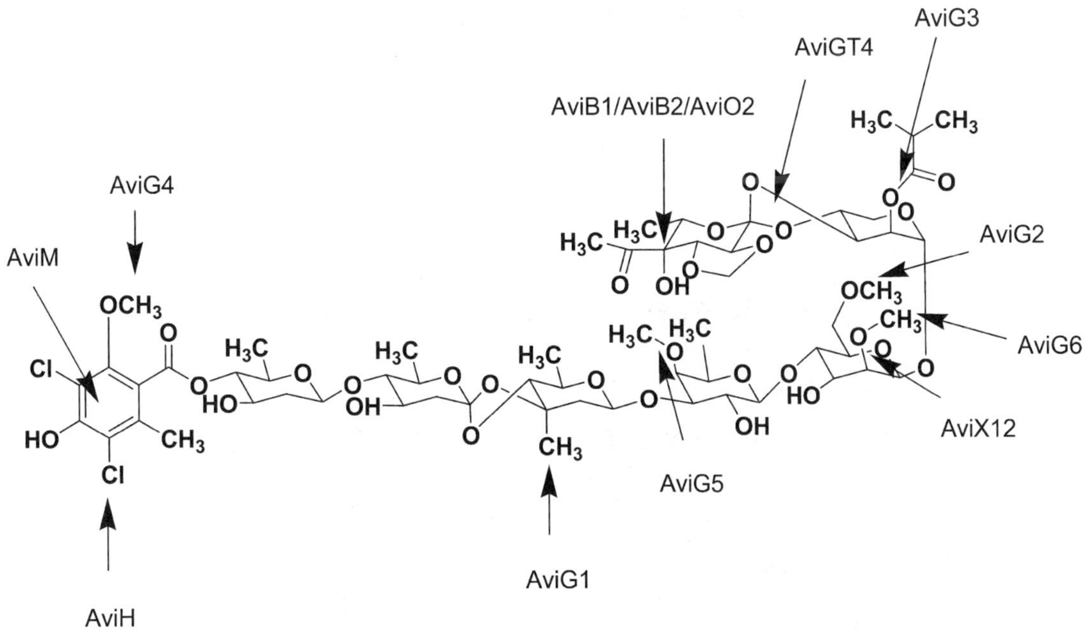

FIGURE 9 Function of genes/enzymes involved in avilamycin A biosynthesis.

"Rossmann-like" (beta/alpha/beta) domains characteristic of the GT-B fold, one of the two folds found in GTs. It was suggested that AviGT4 is an ancestral enzyme, from which inverting and retaining GTs have evolved.

Inactivation of *aviX12* led to the formation of a novel avilamycin derivative named gavibamycin N1, which harbors glucose as a component of the heptasaccharide chain in place of a mannose moiety in avilamycin A. Antibacterial activity tests with a spectrum of gram-positive organisms showed that the new derivative possesses decreased biological activity compared to avilamycin A. Thus, AviX12 seems to be involved in converting avilamycin to its bioactive conformation by catalyzing an unusual epimerization reaction. Sequence comparisons classified AviX12 into the radical *S*-adenosyl-L-methionine protein family. It is the first member of this family known to catalyze an epimerization step (10).

24.4.2. The Saccharomicin Biosynthetic Gene Cluster

In order to clone the saccharomicin biosynthetic genes, the 2,3-dehydratase gene probe and the 4,6-dehydratase gene probe were used. Some cosmids hybridized to the 2,3-dehydratase gene probe (cosmid group 1) and some to the 4,6-dehydratase gene probe (cosmid group 2). No cosmid hybridizing to both probes was detected. Cosmids from both groups were sequenced. Analyses of the final contiguous DNA sequences revealed the presence of 38 open reading frames (ORFs) located on cosmids of group 1 and 26 ORFs located on cosmids of group 2 (Fig. 8). Cosmids of group 1 span a 48.7-kbp genome region; cosmids of group 2 comprise another 33.7 kbp. Twelve potential genes encoding the biosynthesis of the five D-glucose 1-phosphate-derived deoxyhexose moieties were found. Four of these genes (*sam39*, *sam41*, *sam53*, and *sam54*) were found in cosmids of group 2. Downstream of *sam10* (in the group 1 DNA region) are 10 ORFs named *sam11* to *sam20*, all transcribed in the same direction, which encode proteins similar to GTs that are involved

in secondary metabolism in three different actinomycete strains. Cosmids of group 2 carry another seven putative GT genes (*sam43*, *sam45*, *sam46*, *sam47*, *sam49*, *sam50*, and *sam51*). Additional genes in the cluster are regulatory genes, genes with unknown functions, and *sam5* and *sam8*. Sam8 is a tyrosine ammonia-lyase, and Sam5 is a 4-coumarate 3-hydroxylase, as revealed by expression experiments. Both are responsible for the formation of the caffeic acid moiety as part of N-(m,p-dihydroxycinnamoyl)taurine (4, 5).

24.4.3. The Landomycin A, Landomycin E, Urdamycin A, Saquayamycin Z, and Simocyclinone Biosynthetic Gene Clusters

The gene clusters of all five angucyclines have been cloned and sequenced (gene clusters of landomycin A, landomycin E, urdamycin A, and saquayamycin Z are shown in Fig. 10) (21, 22, 27, 86, 91). The amino acid sequences deduced from the PKSII gene sequences in all five biosynthetic gene clusters are similar to one another, with proportions of identical amino acids between 70 and 90%. Furthermore, most of the deduced amino acid sequences corresponding to cyclase genes, oxygenase genes, ketoreductase genes, GT genes, and sugar biosynthetic genes found in the clusters are very similar. Additional genes responsible for the formation and attachment of a PKSI unit and the halogenated coumarin are located in the simocyclinone gene cluster. During the last 10 years, our research has focused on the functions of GTs (Fig. 4). Genes for C-GTs (UrdGT2, SaqGT5, and SimB7) have been detected in the gene clusters for urdamycin A, saquayamycin Z, and simocyclinone (86). These enzymes share a high number of identical amino acids with the O-glycosyltransferases LanGT2 and LndGT2. The first characterized C-GT was UrdGT2. This enzyme is remarkably flexible. It tolerates D- and L-rhodinose and D-mycarose as donor substrates (35) and is also able to catalyze the O

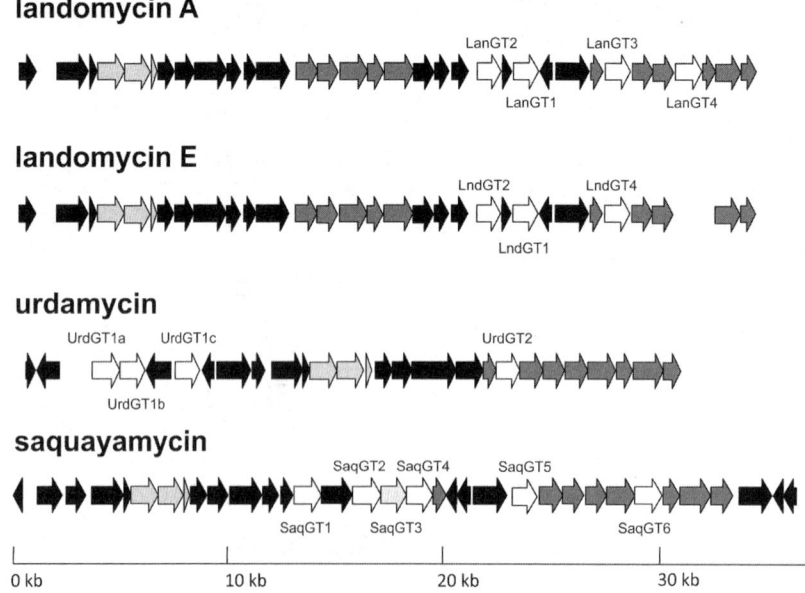

FIGURE 10 Organization of the landomycin A, landomycin E, urdamycin A, and saquayamycin Z biosynthetic gene clusters. GT genes are shown in white, sugar biosynthetic genes are shown in gray, and all other genes are shown in black.

glycosylation of 1,2-dihydroxyanthraquinone (18). The crystal structure of UrdGT2 was resolved. Like almost all other GTs involved in natural-product biosynthesis, it belongs to the GT-B superfamily of GTs (64). Interestingly, the expression of *urdGT2* in the *saqGT5* mutant restored wild-type production of saquayamycin Z and galtamycin B. The expression of *simB7* led only to the production of monoglycosylated galtamycinone but did not restore wild-type production. The results of this experiment may indicate that SimB7 is not used to work in an enzyme complex and that therefore the biosynthesis is blocked (21). Further aspects of our angucycline GT work are outlined below. The biosynthesis processes for the sugar components, mostly D-olivose and L-rhodinose, are nearly identical in all pathways. NDP-4-keto-2,6-di-deoxy-D-glucose is a key intermediate. After 4-ketoreduction, NDP-D-olivose is produced, and after dehydration, 5-epimerization, and 4-ketoreduction, NDP-L-rhodinose is produced. It has been shown that the NDP-sugar concentration and the enzyme expression level influence the substrate specificities of the GTs. When the pathway to D-olivose was blocked by deleting the 4-ketoreductase gene in the urdamycin producer, a novel urdamycin derivative carrying either a C-glycosidically attached D-rhodinose or L-rhodinose instead of a D-olivose was produced (32) (see above).

24.4.4. The Polyketomycin and Aranciamycin Biosynthetic Gene Clusters

The aranciamycin and polyketomycin biosynthetic gene clusters were isolated due to the capacity to hybridize to the PKSII gene probe and to the NDP-glucose-4,6-dehydratase gene probe, respectively. The polyketomycin cluster consists of 41 ORFs, located on 52 kbp of continuous DNA. The aranciamycin cluster is located on one cosmid and contains 24 ORFs in a 35.9-kb region. The arrangement of the minimal PKSII genes of the polyketomycin and aranciamycin gene clusters resembles that of other PKSII-type gene clusters (e.g., actinorhodin and granaticin gene clusters) (15, 54). Only AraGT, the GT involved in aranciamycin biosynthesis, has been investigated in detail. The expression of the aranciamycin gene cluster in a different actinomycete strain resulted in novel aranciamycin derivatives (Fig. 11). Due to the low sugar substrate specificity of AraGT, different sugars were attached to the aranciamycin aglycone (54, 55). Interestingly, the sugar biosynthetic genes are not clustered with the genes responsible for aglycone production.

24.4.5. The Lipomycin Biosynthetic Gene Cluster

The entire lipomycin biosynthetic gene cluster (Fig. 6) was cloned using the PKSI and deoxysugar 1 gene probes. Twenty-two genes are involved in lipomycin biosynthesis. In the center of this cluster is a polyketide synthase locus (*lipPks1* to *lipPks4*) that encodes an eight-module system comprising four multifunctional proteins. In addition, one ORF (*lipNrps*) encodes a product showing homology to non-ribosomal peptide synthetases, indicating that α-lipomycin belongs to hybrid peptide-polyketide natural products. Furthermore, the lipomycin cluster includes genes responsible for the formation and attachment of D-digitoxose (*lipDig1* to *lipDig5* and *lipGt*), as well as ORFs that resemble genes for putative regulatory and export functions. The function of LipGtf as a GT was proven by gene deletion experiments. It is one of the rare GTs glycosylating a linear polyketide moiety (7).

24.4.6. The Phenalinolactone Biosynthetic Gene Cluster

The phenalinolactone cluster (Fig. 7) was isolated using an NDP-glucose-4,6-dehydratase gene. It is located in a 42-kb region consisting of 35 ORFs and, thus far, is the largest identified cluster responsible for the biosynthesis of isoprenoids. The cluster contains a geranylgeranyl diphosphate synthase gene (*plaT4*) responsible for the generation of the polyprenyl chain, a terpene cyclase gene (*plaT2*), and a prenyltransferase gene (*plaT3*). A gene (*plaT1*) encoding an enzyme that is similar to a eukaryotic squalene epoxidase is also located in the cluster. In contrast to many other terpene cyclization processes in bacteria, the phenalinolactone cyclization process is putatively initiated by an epoxidation reaction of the linear terpene chain, as has been described for myxobacterial steroid formation. Five oxygenase genes (*plaO1* to *plaO5*) were also identified in the cluster. One (*plaO1*) codes for an α-ketoglutarate-dependent dioxygenase catalyzing the formation of the γ-butyrolactone moiety. The phenalinolactone gene cluster represents the complete natural pathway for the rare deoxyhexose amicetose. Eight genes can reliably be predicted to encode the biosynthesis and transfer of this sugar moiety. The multistep conversion of glucose 1-phosphate to NDP-L-amicetose includes NDP-D-glucose synthesis (PlaA4), 4,6-dehydratation (PlaA3), 2,3-dehydratation (PlaA2), 3-ketoreduction (PlaA5), and 3-deoxygenation (PlaA1) to obtain NDP-D-cinerulose. The final steps are 5-epimerization (PlaA8) and stereospecific 4-ketoreduction (PlaA7) toward NDP-L-amicetose. Finally, the attachment of NDP-L-amicetose to O-21 is probably catalyzed by PlaA6, which resembles many natural-product GTs. The enzyme for the further modification of the sugar ligand through 4-O-methylation is PlaM1 (19).

24.5. GENERATION OF NOVEL GLYCOSYLATED NATURAL PRODUCTS

The growing importance of natural-product sugar moieties has motivated scientists to develop methods for natural-product glycosylation. These methods range from whole-cell feeding experiments to enzymatic approaches, each having its advantages and disadvantages. Here, the different approaches are introduced with a focus on practicability and limitations.

24.5.1. Generation of Novel Compounds Using Living Cells (Biotransformation)

Biotransformation is the process whereby a substance is changed (transformed) from one chemical to another by a chemical reaction performed by a microorganism, plant, or animal. Several examples in which a variety of mostly aromatic compounds were glycosylated have been described in the literature (Table 1). In most cases, one organism was able to provide one sugar as a donor substrate and accepted several aglycones as acceptor substrates.

The biotransformation procedure can be performed as follows.

1. Use a seed culture to inoculate 100 ml of medium (preculture) (28°C).
2. After 3 to 4 days, use 5 ml of the preculture to inoculate 100 ml of medium (main culture).
3. After 3 days, add 5 mg of the desired aglycone (dissolved in 0.2 ml of DMSO) to the main culture.

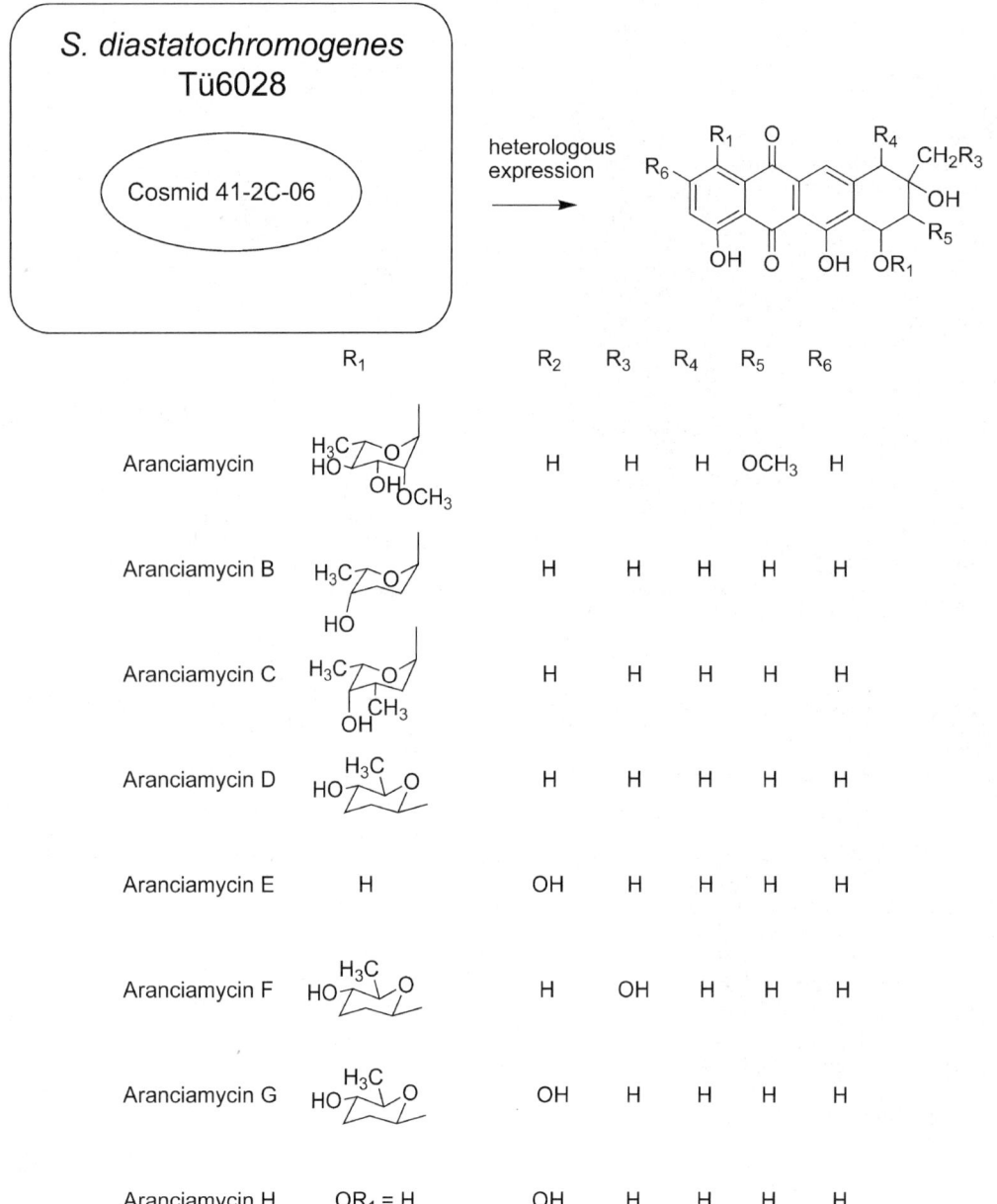

FIGURE 11 Heterologous expression of the aranciamycin gene cluster resulted in novel aranciamycin derivatives.

4. After 7 days, perform ethyl acetate extraction and high-pressure liquid chromatography analysis.

[Control experiments are performed (i) without the strain and (ii) without the aglycone.]

Advantages of this approach are that the yield of glycosylated compounds is often high and biotransformation is easy and quick to perform.

Disadvantages of this approach are that the aglycone substrate flexibility of the GTs seems to be restricted especially to aromatic compounds, and donor substrate variability is restricted to one sugar in most cases.

24.5.2. Generation of Novel Compounds Using GTs and Nucleotide-Activated Sugars (Glycorandomization)

The in vitro approach of glycorandomization is based on two steps. In the first step, natural or chemically derived

sugars are converted by an engineered galactose kinase and a nucleotidyltransferase into nucleotide-activated sugars (NDP-sugars). Then these NDP-sugars are used as substrates by GTs to glycosylate the aglycone. This approach has been used to produce libraries of various natural-product scaffolds, including glycopeptide antibiotics (Table 2) (9, 25).

The glycorandomization procedure can be performed as follows.

1. Provide sugar 1-phosphate "libraries" using chemically synthesized sugars and a genetically modified kinase (e.g., a Y371H GalK mutant).
2. Generate NDP-sugars using sugar 1-phosphates, dTTP, and the α-D-glycopyranosyl phosphate thymidyltransferase from *Salmonella enterica* LT2.
3. Generate glycosylated compounds using a GT, an acceptor molecule, and the dTDP-sugars.

TABLE 1 Glycosylation of compounds by biotransformation

Organism	GT(s)	Attached sugar(s)	Glycosylated compound(s)	Reference(s)
Bacteria				
Bacillus cereus		β-D-Glucose	Flavonoids	43, 76
Streptomyces sp. strain GöM1		6-Desoxy-α-L-talose	Benzoic acid derivates	8
Streptomyces sp. strain LS136		α-L-Rhamnose	Lignane	20
Streptomyces sp. strain Q53834		β-D-Glucose	Brefeldin A	81
Streptomyces aureofaciens B96		β-D-Glucose	Diverse anthraquinones and anthracycline, tetracycline	39, 40, 60
Streptomyces griseoviridis Tü3634		α-L-Rhamnose	Aromatic carbonic acids, emodin	28, 29
Fungi				
Absidia coerulea		β-D-Glucose	Anthraquinones	102
Beauveria bassiana ATCC 7159		β-D-Methylglucose	Anthraquinones, flavonoids	97
Coriolus versicolor		β-D-Xylose, β-D-glucose	Lignin	44
Cunninghamella elegans var. *elegans*		β-D-Glucose	Xanthohumol	42
Plant				
Arabidopsis thaliana	Various GTs	β-D-Glucose	Benzoic acid, caffeic acid, flavonoids, coumarin	49
	UGT78D1	α-L-Rhamnose	Quercetin	50

An advantage of this approach is that it allows for the preparation of a variety of glycosylated compounds.

Disadvantages of this approach are that the yield of glycosylated compounds is low and scale-up experiments are difficult to perform.

24.5.3. Generation of Novel Compounds Using Two GTs and a Nucleotide Diphosphate (Aglycone Exchange)

One of the newest approaches is the use of glycosylated natural products as the source of activated sugars. A GT

TABLE 2 Glycosylation of compounds

Glycosylation method	Lead compound(s)	Reference(s)
Glycorandomization	Novobiocin	2, 95
	Vancomycin	24
	Cardiac glycosides	48
	Colchicine	1
	Oleandomycin	93
	Flavonoids	93
	Coumarins	93
Aglycone exchange method	Amphotericin B	101
	Avermectin	99
	Calicheamicin	98
	Vancomycin	98
	Vicenistatin	63
	Erythromycin	100
Expression of deoxy-sugar gene cassettes	Tetraenomycin	51, 72, 73, 77
	Staurosporins and rebeccamycins	79

catalyzes the NDP-dependent deglycosylation and coupled formation of an NDP-sugar. The NDP-sugar is then used as a donor substrate by a second GT to glycosylate an aglycone. In 2005, Minami and coworkers reported that VinC, a GT that attaches NDP-vicenisamine to vicenilactam, is also able to catalyze the dTDP-dependent deglycosylation of vicenistatin to yield dTDP-vicenisamine (63). Subsequently, Zhang and coworkers expanded this approach by including a second GT, which is able to use the generated NDP-sugar (98) (Table 2).

The aglycone exchange procedure can be performed as follows.

Perform incubations with a glycosylated natural product, the corresponding GT, dTDP, a nonglycosylated natural product, and a second GT.

Advantages of this approach are that there is no need for the preparation of NDP-sugars and that once GTs are available the experiment is easy to perform; a disadvantage of this approach is that scale-up experiments are difficult to perform.

24.5.4. Generation of Novel Compounds by Genetic Engineering of Producer Strains (Pathway Engineering)

Pathway engineering in vivo is a fourth approach to generate novel glycosylated compounds. As the transfer of sugars is often a late biosynthetic step during the biosynthesis of natural products, genetic engineering of producer strains has become a powerful tool. Different strategies have been employed to create compound libraries with nonnatural glycosylation patterns.

24.5.4.1. Modification of the Aglycone Structure

Novel compounds have been generated by the deletion of genes of the aglycone structure and/or the expression of genes involved in modifying the aglycone structure. Many examples of this approach, especially processes resulting in novel macrolides and anthracyclines, can be found in the literature (61). This approach proceeds in two steps. First, the genes involved in the biosynthesis of the aglycone are deleted from a strain. Thereafter, either the expression of other gene clusters (54, 55) or the feeding of other aglycones (mutasynthesis) (30, 88) in this strain finally results in novel glycosylated compounds. When a cosmid containing the aranciamycin biosynthetic gene cluster was expressed in *S. fradiae* A0, a novel product named aranciamycin B was generated. This compound contains an L-rhodinose moiety at the C-7-OH position of aranciamycinone instead of the original L-rhamnose (54, 55). Examples in which mutasynthesis resulted in novel glycosylated compounds are reported by Heide et al. (30) and Weist et al. (88).

24.5.4.2. Manipulation of Deoxysugar Biosynthetic Pathways

By either deletion or expression of sugar biosynthetic genes, a strain's capability to synthesize sugars can be altered, which may also lead to novel derivatives. When a gene encoding an NDP-4-keto-2,6-dideoxyglucose-ketoreductase in *S. fradiae* Tü2717 was deleted, the strain produced NDP-rhodinose in both D- and L-forms. Due to the flexibility of the GT UrdGT2, responsible for attaching D-olivose at C-9 of the aglycone in the wild-type strain, the mutant accumulated two novel urdamycin derivatives with either D-rhodinose or L-rhodinose at position C-9 (35) (Fig. 12). Novel tylosin derivatives were produced by expressing genes of the narbomycin pathway in the tylosin producer. Expression resulted in the additional formation of NDP-D-desosamine, which was used as a substrate instead of NDP-D-mycaminose (13). By combining the deletion of a gene (*dnmV*, encoding a ketoreductase) and the expression of a gene (*eryBIV*, encoding a ketoreductase with different stereospecificity from that of DnmV), Madduri and coworkers generated a strain producing epirubicin instead of doxorubicin. The two compounds differ in the configuration at C-4 of the sugar moiety (57).

The introduction of whole deoxysugar biosynthetic pathways into one strain, followed by the generation of novel compounds, has been very successfully performed by Méndez et al. (62). By this approach, novel tetracenomycin and indolocarbazole compounds have been generated (Table 2).

24.5.4.3. Use of GT Genes as Tools

The most important tools for drug design in glycol biosynthesis are GTs. By deleting GT genes, nonglycosylated compounds or compounds with reduced numbers of sugars can be produced (56). The successful expression of a GT gene in a wild-type host has been described a few times. One example describes the expression of *lanGT4* in *S. fradiae* Tü2717, which resulted in the conversion of the trisaccharide side chain of urdamycin A into a tetrasaccharide chain (36). Due to the competition between the wild-type GT (LanGT2) and the heterologously expressed GT (UrdGT2), *S. cyanogenus* S136 harboring *urdGT2* did not produce novel derivatives or produced them in very small amounts. When *S. cyanogenus* S136 Δ*lanGT2* was used as the host, novel C-glycosylated tetrangulols were produced (Fig. 12) (52). Coexpression of two GT genes (*urdGT2* and *lanGT1*) from two different *Streptomyces* species (*S. fradiae* and *S. cyanogenus*) in a mutant of *Streptomyces argillaceus* in which either one or all GT genes were inactivated resulted in the production of 9-C-olivosylpremithramycinone, 9-C-mycarosylpremithramycinone, and their respective 4-O-demethyl analogues (87) (Fig. 12).

24.5.4.4. GT Engineering

Although the examples described above show that some GTs are sufficiently promiscuous for use in "combinatorial biosyntheses," the strict substrate specificities of others remain a limiting factor. Altering the substrate specificities of GTs and the generation of artificial nonspecific GTs are two important tasks (94). Successful examples have been described previously, but further work needs to be done in this important research field. The first successful results in the engineering of natural-product GTs were obtained with UrdGT1b and UrdGT1c, both involved in urdamycin biosynthesis. The enzymes share 91% identical amino acids, as the corresponding genes most probably derived from gene duplication. Surprisingly, the enzymes show different specificities for both nucleotide sugar donor and acceptor substrates. UrdGT1c accepts strictly dNDP-L-rhodinose and transfers it via an α(1-3)-glycosidic bond to the C-3 hydroxyl of D-olivose, while UrdGT1b transfers D-olivose to the C-4 hydroxyl group of L-rhodinose, forming a β(1-4)-glycosidic bond. A region of 31 amino acids within UrdGT1b and UrdGT1c was identified to control the substrate specificities of both enzymes. By mutating this region, the selectivity for the donor and the acceptor substrates was changed (33, 34). The group of G. J. Williams worked with the macrolide GT OleD, a GT involved in oleandomycin biosynthesis. Sequence variants of OleD showed improved efficiency and accepted additional nucleotide sugar donors compared to OleD (93).

Both the UrdGT1 and OleD studies revealed N-terminal domain residues that control sugar binding. This finding was not expected, as in GTs of the GT-B family, almost all sugar binding residues are in the C-terminal domain (68, 75).

Recently, the swapping of the N- and C-terminal domains of GTs was described. Park et al. fused the N-terminal domain of KanF, involved in kanamycin biosynthesis, and the C-terminal domain of GtfE, involved in vancomycin biosynthesis, producing novel kanamycin derivatives (70).

Another aspect of GT flexibility was recently addressed. Oligosaccharide antibiotics like landomycin A, saquayamycin Z, and saccharomicin are made of disaccharide or trisaccharide units which can be found twice in one molecule. As the numbers of GT genes detected in the biosynthetic gene clusters for these compounds are lower than the numbers of sugar moieties, some GTs are acting iteratively during the biosynthesis of the side chains. This has been shown for LanGT1 (attaching the second and fifth sugars) and LanGT4 (attaching the third and sixth sugars), involved in landomycin A biosynthesis. While landomycin A is produced by *S. cyanogenus* S136, landomycin E is produced by *S. globisporus*. In *S. globisporus*, LndGT1 catalyzes the attachment of the second sugar during landomycin E biosynthesis. This enzyme does not work iteratively; thus, it was not able to attach a fifth sugar when expressed in a *lanGT1* mutant of *S. cyanogenus* S136. By generating several hybrids from LanGT1 and LndGT1 and by introducing mutations into both enzymes, amino acids that specifically contribute to iterative action were identified. All mutations have been made in the N-terminal domain (46).

An advantage of this approach is that once a genetic system for a strain is established, production of a desired product is simple. Disadvantages are that the genetic engineering of strains still requires much time, and the productivity of genetically engineered strains is usually lower than the productivity of the corresponding wild type.

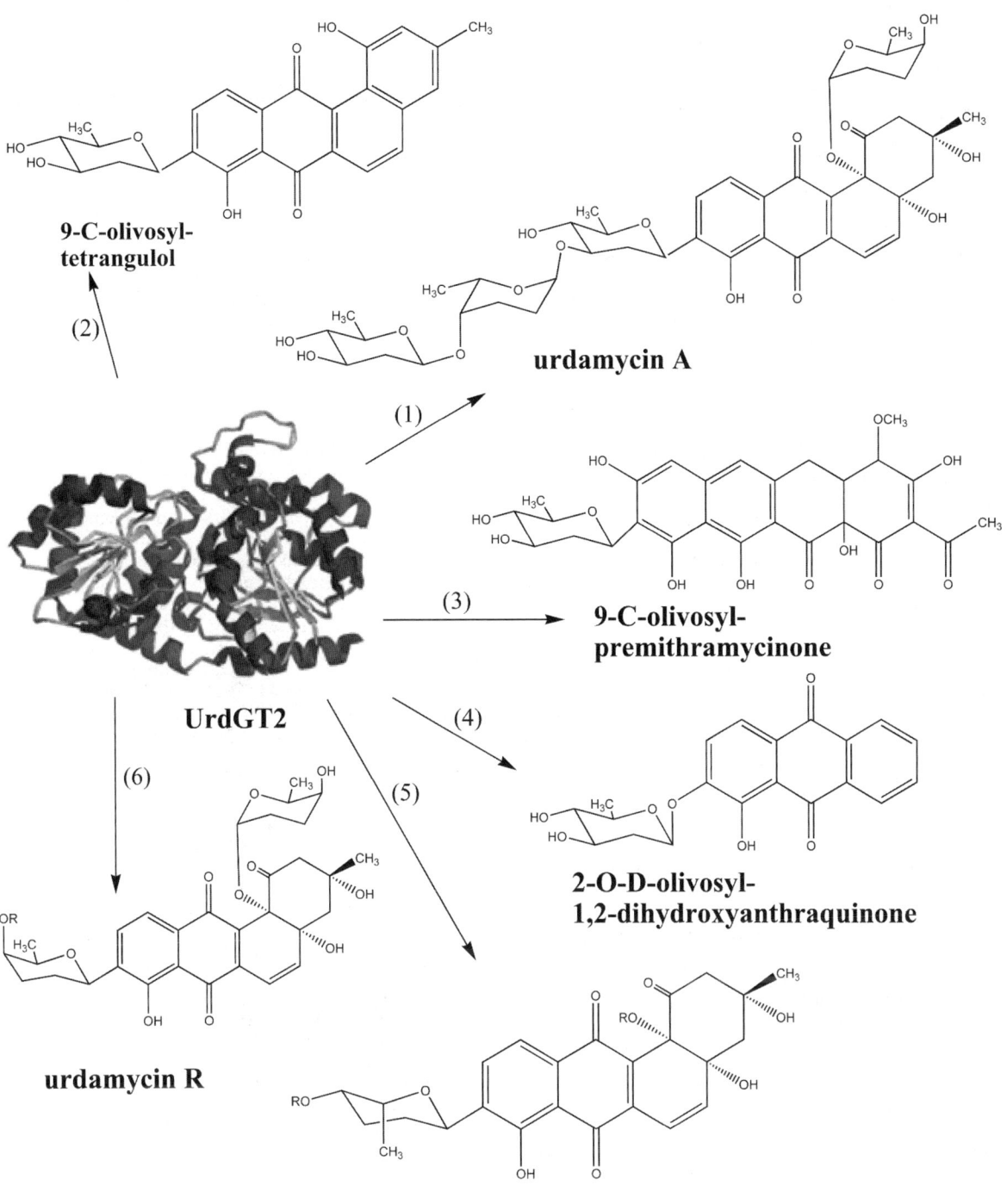

FIGURE 12 Generation of natural products: (1) by UrdGT2 in the wild-type strain; (2 and 3) by combinatorial biosynthesis (*urdGT2* was expressed in *S. cyanogenus* S136 Δ*lanGT2*, a mutant lacking the GT LanGT2 [2], and in *S. argillaceus* Δ*mtmGIV*, a mutant lacking the GT MtmGIV [3]); (4) by mutasynthesis (1,2-dihydroxyanthraquinone was fed to *S. fradiae* XKS containing a deletion in the polyketide synthase genes); and (5 and 6) by manipulating the deoxysugar biosynthetic pathway (compounds were produced by *S. fradiae* Δ*urdR*, a mutant lacking the dTDP-4-keto-2,6-dideoxy-D-glucose 4-ketoreductase UrdR).

24.6. CONCLUSION

In the light of applied pharmaceutical sciences, whole-pathway engineering and single-enzyme engineering have supported the development of novel glycosylated natural products. Different technologies are available for glycosylation. Improving these technologies toward higher product yields will be the most important issue for scientists in the next decade.

REFERENCES

1. **Ahmed, A., N. R. Peters, M. K. Fitzgerald, J. A. Watson, F. M. Hoffmann, and J. S. Thorson.** 2006. Colchicine glycorandomization influences cytotoxicity and mechanism of action. *J. Am. Chem. Soc.* **128:**14224–14225.
2. **Albermann, C., A. Soriano, J. Jiang, H. Vollmer, J. B. Biggins, W.A. Barton, J. Lesniak, D. B. Nikolov, and J. S. Thorson.** 2003. Substrate specificity of NovM: implications for novobiocin biosynthesis and glycorandomization. *Org. Lett.* **5:**933–936.
3. **Antal, N., H. P. Fiedler, E. Stackebrandt, W. Beil, K. Ströch, and A. Zeeck.** 2005. Retymicin, galtamycin B, saquayamycin Z and ribofuranosyllumichrome, novel secondary metabolites from *Micromonospora* sp. Tü 6368. I. Taxonomy, fermentation, isolation and biological activities. *J. Antibiot.* (Tokyo) **58:**95–102.
4. **Berner, M.** 2006. Ph.D. thesis. University of Freiburg, Freiburg, Germany.
5. **Berner, M., D. Krug, C. Bihlmaier, A. Vente, R. Müller, and A. Bechthold.** 2006. Genes and enzymes involved in caffeic acid biosynthesis in the actinomycete *Saccharothrix espanaensis*. *J. Bacteriol.* **188:**2666–2673.
6. **Biermann, M., R. Logan, K. O'Brien, E. T. Seno, R.N. Rao, and B. E. Schoner.** 1992. Plasmid cloning vectors for the conjugal transfer of DNA from *Escherichia coli* in *Streptomyces* spp. *Gene* **116:**43–49.
7. **Bihlmaier, C., E. Welle, C. Hofmann, K. Welzel, A. Vente, E. Breitling, M. Müller, S. Glaser, and A. Bechthold.** 2006. Biosynthetic gene cluster for the polyenoyltetramic acid α-lipomycin. *Antimicrob. Agents Chemother.* **50:**2113–2121.
8. **Bitzer, J., and A. Zeeck.** 2006. 6-Deoxy-alpha-L-talopyranosides from *Streptomyces* sp. *Eur. J. Org. Chem.* **16:**3661–3666.
9. **Blanchard, S., and J. S.Thorson.** 2006. Enzymatic tools for engineering natural product glycosylation. *Curr. Opin. Chem. Biol.* **10:**263–271.
10. **Boll, R., C. Hofmann, B. Heitmann, G. Hauser, S. J. Glaser, T. Koslowski, T. Friedrich, and A. Bechthold.** 2006. The active conformation of avilamycin A is conferred by AviX12, a radical SAM enzyme. *J. Biol. Chem.* **281:**14756–14763.
11. **Bols, M., L. Binderup, J. Hansen, and P. Rasmussen.** 1992. Synthesis and collagenase inhibition of new glycosides of aranciamycinone: the aglycon of the naturally occurring antibiotic aranciamycin. *Carbohydr. Res.* **235:**141–149.
12. **Bongat, A. F., and A. V. Demchenko.** 2007. Recent trends in the synthesis of O-glycosides of 2-amino-2-deoxysugars. *Carbohydr. Res.* **26:**374–406.
13. **Butler, A. R., N. Bate, D. E. Kiehl, H. A. Kirst, and E. Cundliffe.** 2002. Genetic engineering of aminodeoxyhexose biosynthesis in *Streptomyces fradiae*. *Nat. Biotechnol.* **20:**713–716.
14. **Crow, T., B. Rosenbaum, R. Smith, Y. Guo, K. Ramos, and G. Sulikowski.** 1999. Landomycin A inhibits DNA synthesis and G1/S cell cycle progression. *Bioorg. Med. Chem. Lett.* **9:**1663–1666.
15. **Daum, M., I. Peintner, A. Linnenbrink, A. Frerich, M. Weber, T. Paululat, and A. Bechthold.** 2009. Or-ganisation of the biosynthetic gene cluster and tailoring enzymes in the biosynthesis of the tetracyclic quinone glycoside antibiotic polyketomycin. *Chembiochem* **10:**1073–1083.
16. **Daum, M., S. Herrmann, B. Wilkinson, and A. Bechthold.** 2009. Genes and enzymes involved in bacterial isoprenoid biosynthesis. *Curr. Opin. Chem. Biol.* **13:**180–188.
17. **Drautz, H., H. Zähner, J. Rohr, and A. Zeeck.** 1986. Metabolic products of microorganisms. Urdamycins, new angucycline antibiotics from *Streptomyces fradiae*. I. Isolation, characterization and biological properties. *J. Antibiot.* (Tokyo) **39:**1657–1669.
18. **Dürr, C., D. Hoffmeister, S. E. Wohlert, K. Ichinose, M. Weber, U. von Mulert, J. S. Thorson, and A. Bechthold.** 2004. The glycosyltransferase UrdGT2 catalyzes both C- and O-glycosidic sugar transfer. *Angew. Chem.* **43:**2962–2965.
19. **Dürr, C., H. J. Schnell, A. Luzhetskyy, R. Murillo, M. Weber, K. Welzel, A. Vente, and A. Bechthold.** 2006. Biosynthesis of the terpene phenalinolactone in *Streptomyces* sp. Tü6071: analysis of the gene cluster and generation of derivatives. *Chem. Biol.* **13:**365–377.
20. **Eklund, P., T. Holmstrom, L. Al Ubaydy, R. Sjoholm, and J. Hakal.** 2006. Rhamnosylation of lignans by a *Streptomyces* strain. *Tetrahedron Lett.* **47:**1645–1648.
21. **Erb, A., A. Luzhetskyy, U. Hardter, and A. Bechthold.** 2009. Cloning and sequencing of the biosynthetic gene cluster for saquayamycin Z and galtamycin B and the elucidation of the assembly of their saccharide chains. *Chembiochem* **10:**1392–1401.
22. **Faust, B., D. Hoffmeister, G. Weitnauer, L. Westrich, S. Haag, P. Schneider, H. Decker, E. Künzel, J. Rohr, and A. Bechthold.** 2000. Two new tailoring enzymes, a glycosyltransferase and an oxygenase, involved in biosynthesis of the angucycline antibiotic urdamycin A in *Streptomyces fradiae* Tü2717. *Microbiology* **146:**147–154.
23. **Flatman, R. H., A. J. Howells, L. Heide, H. P. Fiedler, and A. Maxwell.** 2005. Simocyclinone D8, an inhibitor of DNA gyrase with a novel mode of action. *Antimicrob. Agents Chemother.* **49:**1093–1100.
24. **Fu, X., C. Albermann, J. Jiang, J. Liao, C. Zhang, and J. S. Thorson.** 2003. Antibiotic optimization via in vitro glycorandomization. *Nat. Biotechnol.* **21:**1467–1469.
25. **Fu, X., C. Albermann, C. Zhang, and J. S. Thorson.** 2005. Diversifying vancomycin via chemoenzymatic strategies. *Org. Lett.* **17:**1513–1515.
26. **Fuchs, P. C., A. L. Barry, and S. D. Brown.** 1999. In vitro activities of SCH27899 alone and in combination with 17 other antimicrobial agents. *Antimicrob. Agents Chemother.* **43:**2996–2997.
27. **Gromyko, O., Y. Rebets, B. Ostash, A. Luzhetskyy, M. Fukuhara, A. Bechthold, T. Nakamura, and V. Fedorenko.** 2004. Generation of *Streptomyces globisporus* SMY622 strain with increased landomycin E production and its initial characterization. *J. Antibiot.* **57:**383–390.
28. **Grond, S., H. J. Langer, P. Henne, I. Sattler, R. Thiericke, S. Grabley, H. Zahner, and A. Zeeck.** 2000. Secondary metabolites by chemical screening. 39. Acyl alpha-L-rhamnopyranosides, a novel family of secondary metabolites from *Streptomyces* sp.: isolation and biosynthesis. *Eur. J. Org. Chem.* **6:**929–937.
29. **Grond, S., I. Papastavrou, and A. Zeeck.** 2000. Studies of precursor-directed biosynthesis with streptomyces. 3. Structural diversity of 1-O-acyl alpha-L-rhamnopyranosides by precursor-directed biosynthesis. *Eur. J. Org. Chem.* **10:**1875–1881.
30. **Heide, L., B. Gust, C. Anderle, and S. M. Li.** 2008. Combinatorial biosynthesis, metabolic engineering and mutasynthesis for the generation of new aminocoumarin antibiotics. *Curr. Top. Med. Chem.* **8:**667–679.

31. Henkel, T., J. Rohr, J. M. Beale, and L. Schwenen. 1990. Landomycins. New angucycline antibiotics from *Streptomyces* sp. I. Structural studies on landomycins A-D. *J. Antibiot.* **43:**492–502.

32. Hoffmeister, D., K. Ichinose, S. Domann, B. Faust, A. Trefzer, G. Dräger, A. Kirschning, C. Fischer, E. Künzel, D. W. Bearden, J. Rohr, and A. Bechthold. 2000. The NDP-sugar co-substrate concentration and the enzyme expression level influence the substrate specificity of glycosyltransferases: cloning and characterization of deoxysugar biosynthetic genes of the urdamycin biosynthetic gene cluster. *Chem. Biol.* **7:**821–831.

33. Hoffmeister, D., K. Ichinose, and A. Bechthold. 2001. Two sequence elements of glycosyltransferases involved in urdamycin biosynthesis are responsible for substrate specificity and enzymatic activity. *Chem. Biol.* **8:**557–567.

34. Hoffmeister, D., B. Wilkinson, G. Foster, P. J. Sidebottom, K. Ichinose, and A. Bechthold. 2002. Engineered urdamycin glycosyltransferases are broadened and altered in substrate specificity. *Chem. Biol.* **9:**287–295.

35. Hoffmeister, D., G. Dräger, K. Ichinose, J. Rohr, and A. Bechthold. 2003. The C-glcosyltransferase UrdGT2 is unselective towards D- and L-configurated nucleotide-bound rhodinose. *J. Am. Chem. Soc.* **125:**4678–4679.

36. Hoffmeister, D., M. Weber, G. Dräger, K. Ichinose, C. Dürr, and A. Bechthold. 2004. Rational saccharide extension by using the natural product glycosyltransferase LanGT4. *Chembiochem* **5:**369–371.

37. Hofmann, C., R. Boll, B. Heitmann, G. Hauser, C. Dürr, A. Frerich, G. Weitnauer, S. J. Glaser, and A. Bechthold. 2005. Identification of genes encoding enzymes responsible for the biosynthesis of L-lyxose and the attachment of methyleurkanate during avilamycin biosynthesis. *Chem. Biol.* **12:**1137–1143.

38. Hofmann, C. 2005. Ph.D. thesis. University of Freiburg, Freiburg, Germany.

39. Hovorkov, N., J. Cudlin, J. Mateju, M. Blumauer, and Z. Vanek. 1974. Microbial glucosidation of alizarin and anthraflavin. *Collect. Czech. Chem. Commun.* **39:**662–667.

40. Hovorkov, N., J. Cudlin, J. Mateju, M. Blumauer, and Z. Vanek. 1974. Microbial glucosidation of monohydroxyanthraquinones. *Collect. Czech. Chem. Commun.* **39:**3568–3572.

41. Jones, R. N., and M. S. Barrett. 1995. Antimicrobial activity of SCH27899, oligosaccharide member of the everninomicin class with a wide gram-positive spectrum. *J. Clin. Microb. Infect.* **1:**35–43.

42. Kim, H. J., and I. S. Lee. 2006. Microbial metabolism of the prenylated chalcone xanthohumol. *J. Nat. Prod.* **69:**1522–1524.

43. Ko, J. H., B. G. Kim, and J. H. Ahn. 2006. Glycosylation of flavonoids with a glycosyltransferase from *Bacillus cereus*. *FEMS Microbiol. Lett.* **258:**263–268.

44. Kondo, R., H. Yamagami, and K. Sakai. 1993. Xylosylation of phenolic hydroxyl groups of the monomeric lignin model compounds 4-methylguaiacol and vanillyl alcohol by *Coriolus versicolor*. *Appl. Environ. Microbiol.* **59:**438–441.

45. Kong, F. M., N. Zhao, M. M. Siegel, K. Janota, J. S. Ashcroft, F. E. Koehn, D. B. Borders, and G. T. Carter. 1998. Saccharomicins, novel heptadecaglycoside antibiotics effective against multidrug-resistant bacteria. *J. Am. Chem. Soc.* **120:**13301–13311.

46. Krauth, C., M. Fedoryshyn, C. Schleberger, A. Luzhetskyy, and A. Bechthold. 2009. Engineering a novel function into a glycosyltransferase. *Chem. Biol.* **16:**28–35.

47. Kunze, B., K. Schabacher, H. Zahner, and A. Zeeck. 1972. Metabolic products of microorganisms. 3. Lipomy-cins. I. Isolation, characterization and first studies of the structure and the mechanism of action. *Arch. Mikrobiol.* **86:**147–174. (In German.)

48. Langenhan, J. M., N. R. Peters, I. A. Guzei, F. M. Hoffmann, and J. S. Thorson. 2005. Enhancing the anticancer properties of cardiac glycosides by neoglycorandomization. *Proc. Natl. Acad. Sci. USA* **102:**12305–12310.

49. Lim, E. K., C. J. Doucet, Y. Li, L. Elias, D. Worrall, S. P. Spencer, J. Ross, and D. J. Bowles. 2002. The activity of *Arabidopsis* glycosyltransferases toward salicylic acid, 4-hydroxybenzoic acid, and other benzoates. *J. Biol. Chem.* **277:**586–592.

50. Lim, E. K., D. A. Ashford, and D. J. Bowles. 2006. The synthesis of small-molecule rhamnosides through the rational design of a whole-cell biocatalysis system. *Chembiochem* **7:**1181–1185.

51. Lombó, F., M. Gibson, L. Greenwell, A. F. Braña, J. Rohr, J. A. Salas, and C. Méndez. 2004. Engineering biosynthetic pathways for deoxysugars: branched-chain sugar pathways and derivatives from the antitumor tetracenomycin. *Chem. Biol.* **11:**1709–1718.

52. Luzhetskyy, A., T. Taguchi, M. Fedoryshyn, C. Dürr, S. Wohlert, V. Novikov, and A. Bechthold. 2005. LanGT2 catalyzes the first glycosylation step during landomycin A biosynthesis. *Chembiochem* **6:**1406–1410.

53. Luzhetskyy, A., H. Weiss, A. Charge, E. Welle, A. Linnebrink, A. Vente, and A. Bechthold. 2007. A strategy for cloning glycosyltransferase genes involved in natural product biosynthesis. *Appl. Microbiol. Biotechnol.* **75:**1367–1375.

54. Luzhetskyy, A., A. Mayer, J. Hoffmann, S. Pelzer, M. Holzenkämper, B. Schmitt, S. E. Wohlert, A. Vente, and A. Bechthold. 2007. Cloning and heterologous expression of the aranciamycin biosynthetic gene cluster revealed a new flexible glycosyltransferase. *Chembiochem* **8:**599–602.

55. Luzhetskyy, A., J. Hoffmann, S. Pelzer, S. E. Wohlert, A. Vente, and A. Bechthold. 2008. Aranciamycin analogs generated by combinatorial biosynthesis show improved antitumor activity. *Appl. Microbiol. Biotechnol.* **80:**15–19.

56. Luzhetskyy, A., and A. Bechthold. 2008. Features and applications of bacterial glycosyltransferases: current state and prospects. *Appl. Microbiol. Biotechnol.* **80:**945–952.

57. Madduri, K., J. Kennedy, G. Rivola, A. Inventi-Solari, S. Filippini, G. Zanuso, A. L. Colombo, K. M. Gewain, J. L. Occi, D. J. MacNeil, and C. R. Hutchinson. 1998. Production of the antitumor drug epirubicin (4′-epidoxorubicin) and its precursor by a genetically engineered strain of *Streptomyces peucetius*. *Nat. Biotechnol.* **16:**69–74.

58. Martinez-Fleites, C., M. Proctor, S. Roberts, D. N. Bolam, H. J. Gilbert, and G. J. Davies. 2006. Insights into the synthesis of lipopolysaccharide and antibiotics through the structures of two retaining glycosyltransferases from family GT4. *Chem. Biol.* **13:**1143–1152.

59. Martin-Rendon, E., and D. J. Blake. 2003. Protein glycosylation in disease: new insights into the congenital muscular dystrophies. *Trends Pharmacol. Sci.* **24:**178–183.

60. Mateju, J., J. Cudlin, N. Hovorkov, M. Blumauer, and Z. Vanek. 1974. Microbial glucosidation of dihydroxyan-thraquinones—general properties of glucosidation system. *Folia Microbiol.* **19:**307–316.

61. Mendez, C., G. Weitnauer, A. Bechthold, and A. Salas. 2000. Structure alteration of polyketides by recombinant DNA technology in producer organisms—prospects for the generation of novel pharmaceutical drugs. *Curr. Pharm. Biotechnol.* **1:**355–395.

62. Méndez, C., A. Luzhetskyy, A. Bechthold, and J. Salas. 2008. Deoxysugars in bioactive natural products: development of novel derivatives by altering the sugar pattern. *Curr. Med. Chem.* **8:**710–724.

63. Minami, A., K. Kakinuma, and T. Eguchi. 2005. Aglycon switch approach toward unnatural glycosides from natural glycoside with glycosyltransferase VinC. *Tetrahedron Lett.* **46:**6187–6190.

64. Mittler, M., A. Bechthold, and G. Schulz. 2007. Structure and action of the C-C bond-forming glycosyltransferase UrdGT2 involved in the biosynthesis of the antibiotic urdamycin. *J. Mol. Biol.* **372:**67–76.

65. Mosbacher, T. G., A. Bechthold, and G. E. Schulz. 2003. Crystal structure of the avilamycin resistance-conferring methyltransferase AviRa from *Streptomyces viridochromogenes. J. Mol. Biol.* **329:**147–157.

66. Mosbacher, T., A. Bechthold, and G. E. Schulz. 2005. Structure and function of the antibiotic resistance-mediating methyltransferase AviRb from *Streptomyces viridochromogenes. J. Mol. Biol.* **345:**535–545.

67. Nakashio, S., H. Iwasawa, F.Y. Dun, K. Kanemitsu, and J. Shimada. 1995. Everninomicin, a new oligosaccharide antibiotic: its antimicrobial activity, post-antibiotic effect and synergistic bactericidal activity. *Drugs Exp. Clin. Res.* **21:**7–16.

68. Offen, W., C. Martinez-Fleites, M. Yang, E. Kiat-Lim, B. G. Davis, C. A. Tarling, C. M. Ford, D. J. Bowles, and G. J. Davies. 2006. Structure of a flavonoid glucosyltransferase reveals the basis for plant natural product modification. *EMBO J.* **25:**1396–1405.

69. Oppegard, L. M., B. L. Hamann, K. R. Streck, K. C. Ellis, H.-P. Fiedler, A. B. Khodursky, and H. Hiasa. 2009. In vivo and in vitro patterns of the activity of simocyclinone D8, an angucyclinone antibiotic from *Streptomyces antibioticus. Antimicrob. Agents Chemother.* **53:**2110–2119.

70. Park, S. H., H. Y. Park, J. K. Sohng, H. C. Lee, K. Liou, Y. J. Yoon, and B. G. Kim. 2009. Expanding substrate specificity of GT-B fold glycosyltransferase via domain swapping and high-throughput screening. *Biotechnol. Bioeng.* **102:**988–994.

71. Paululat, T., A. Zeeck, J. M. Gutterer, and H. P. Fiedler. 1999. Biosynthesis of polyketomycin produced by *Streptomyces diastatochromogenes* Tü 6028. *J. Antibiot.* (Tokyo) **52:**96–101.

72. Pérez, M., F. Lombó, L. Zhu, M. Gibson, A. F. Braña, J. Rohr, J. A. Salas, and C. Méndez. 2005. Combining sugar biosynthesis genes for the generation of L- and D-amicetose and formation of two novel antitumor tetracenomycins. *Chem. Commun.* (Cambridge) **28:**1604–1606.

73. Pérez, M., F. Lombó, I. Baig, A. F. Braña, J. Rohr, J. A. Salas, and C. Méndez. 2006. Combinatorial biosynthesis of antitumor deoxysugar pathways in *Streptomyces griseus:* reconstitution of "unnatural natural gene clusters" for the biosynthesis of four 2,6-D-dideoxyhexoses. *Appl. Environ. Microbiol.* **72:**6644–6652.

74. Pilobello, K. T., and L. K. Mahal. 2007. Deciphering the glycocode: the complexity and analytical challenge of glycomics. *Curr. Opin. Chem. Biol.* **11:**300–305.

75. Ramos, A., C. Olano, A. F. Braña, C. Méndez, and J. A. Salas. 2009. Modulation of deoxysugar transfer by the elloramycin glycosyltransferase ElmGT through site-directed mutagenesis. *J. Bacteriol.* **191:**2871–2875.

76. Rao, K. V., and N. T. Weisner. 1981. Microbial transformation of quercetin by *Bacillus cereus. Appl. Environ. Microbiol.* **42:**450–452.

77. Rodríguez, L., I. Aguirrezabalaga, N. Allende, A. F. Braña, C. Méndez, and J. A. Salas. 2002. Engineering deoxysugar biosynthetic pathways from antibiotic-producing microorganisms. A tool to produce novel glycosylated bioactive compounds. *Chem. Biol.* **9:**721–729.

78. Royles, B. J. L. 1995. Naturally occurring tetramic acids: structure, isolation, and synthesis. *Chem. Rev.* **95:**1981–2001.

79. Salas, A. P., L. Zhu, C. Sánchez, A. F. Braña, J. Rohr, C. Méndez, and J. A. Salas. 2005. Deciphering the late steps in the biosynthesis of the anti-tumour indolocarbazole staurosporine: sugar donor substrate flexibility of the StaG glycosyltransferase. *Mol. Microbiol.* **58:**17–27.

80. Schimana, J., H. P. Fiedler, I. Groth, R. Süssmuth, W. Beil, M. Walker, and A. Zeeck. 2000. Simocyclinones, novel cytostatic angucyclinone antibiotics produced by *Streptomyces antibioticus* Tü 6040. I. Taxonomy, fermentation, isolation and biological activities. *J. Antibiot.* (Tokyo) **53:**779–787.

81. Shibazaki, M., H. Yamaguchi, T. Sugawara, K. Suzuki, and T. Yamamoto. 2003. Microbial glycosylation and acetylation of brefeldin A. *J. Biosci. Bioeng.* **96:**344–348.

82. Singh, M. P., P. J. Petersen, W. J. Weiss, F. Kong, and M. Greenstein. 2000. Saccharomicins, novel heptadecaglycoside antibiotics produced by *Saccharothrix espanaensis:* antibacterial and mechanistic activities. *Antimicrob. Agents Chemother.* **44:**2154–2159.

83. Ströch, K., A. Zeeck, N. Antal, and H. P. Fiedler. 2005. Retymicin, galtamycin B, saquayamycin Z and ribofuranosyllumichrome, novel secondary metabolites from *Micromonospora* sp. Tü 6368. II. Structure elucidation. *J. Antibiot.* (Tokyo) **58:**103–110.

84. Treede, I., L. Jacobsen, F. Kirpekar, B. Vester, G. Weitnauer, A. Bechthold, and S. Douthwaite. 2003. The avilamycin resistance determinants AviRa and AviRb methylate 23S rRNA at the guanosine 2535 base and the uridine 2479 ribose. *Mol. Microbiol.* **49:**309–318.

85. Treede, I., G. Hauser, A. Mühlenweg, C. Hofmann, M. Schmidt, G. Weitnauer, and A. Bechthold. 2005. Genes involved in formation and attachment of a two-carbon chain as a component of eurekanate, a branched-chain sugar moiety of avilamycin A. *Appl. Environ. Microbiol.* **71:**400–406.

86. Trefzer, A., S. Pelzer, J. Schimana, S. Stockert, C. Bihlmaier, H. P. Fiedler, K. Welzel, A. Vente, and A. Bechthold. 2002. The biosynthetic gene cluster of simocyclinone, a natural multihybrid antibiotic. *Antimicrob. Agents Chemother.* **46:**1174–1182.

87. Trefzer, A., G. Blanco, L. Remsing, E. Künzel, U. Rix, F. Lipata, A. F. Braña, C. Méndez, J. Rohr, A. Bechthold, and J. A. Salas. 2002. Rationally designed glycosylated premithramycins: hybrid aromatic polyketides using genes from three different biosynthetic pathways. *J. Am. Chem. Soc.* **124:**6056–6062.

88. Weist, S., C. Kittel, D. Bischoff, B. Bister, V. Pfeifer, G. J. Nicholson, W. Wohlleben, and R. D. Süssmuth. 2004. Mutasynthesis of glycopeptide antibiotics: variations of vancomycin's AB-ring amino acid 3,5-dihydroxyphenylglycine. *J. Am. Chem. Soc.* **126:**5942–5943.

89. Weitnauer, G., A. Mühlenweg, A. Trefzer, D. Hoffmeister, R. Süssmuth, G. Jung, K. Welzel, A. Vente, U. Girreser, and A. Bechthold. 2001. Biosynthesis of the orthosomycin antibiotic avilamycin A: deductions from the molecular analysis of the *avi* biosynthetic gene cluster of *Streptomyces viridochromogenes* Tü57 and production of new antibiotics. *Chem. Biol.* **8:**569–581.

90. Weitnauer, G., G. Hauser, C. Hofmann, U. Linder, R. Boll, K. Pelz, S. J. Glaser, and A. Bechthold. 2004. Novel avilamycin derivatives with improved polarity generated by targeted gene disruption. *Chem. Biol.* **11:**1403–1411.

91. Westrich, L., S. Domann, B. Faust, D. Bedford, D. A. Hopwood, and A. Bechthold. 1999. Cloning and characterization of the landomycin biosynthetic gene cluster of *Streptomyces cyanogenus* S136. *FEMS Microbiol. Lett.* **170:**381–387.

92. Weymouth-Wilson, A. C. 1997. The role of carbohydrates in biologically active natural products. *Nat. Prod. Rep.* **14:**99–110.

93. **Williams, G. J., C. Zhang, and J. S. Thorson.** 2007. Expanding the promiscuity of a natural-product glycosyltransferase by directed evolution. *Nat. Chem. Biol.* **3:** 657–662.

94. **Williams, G. J., R. W. Gantt, and J. S. Thorson.** 2008. The impact of enzyme engineering upon natural product glycodiversification. *Curr. Opin. Chem. Biol.* **12:** 556–564.

95. **Williams, G. J., R. D. Goff, C. Zhang, and J. S. Thorson.** 2008. Optimizing glycosyltransferase specificity via "hot spot" saturation mutagenesis presents a catalyst for novobiocin glycorandomization. *Chem. Biol.* **15:**393–401.

96. **Wright, D. E.** 1979. The orthosomycins, a new family of antibiotics. *Tetrahedron* **35:**1207–1237.

97. **Zhan, J. X., and A. A. L. Gunatilaka.** 2006. Selective 4'-O-methylglycosylation of the pentahydroxy-flavonoid quercetin by *Beauveria bassiana* ATCC 7159. *Biocatal. Biotransformation* **24:**396–399.

98. **Zhang, C., B. R. Griffith, Q. Fu, C. Albermann, X. Fu, I. K. Lee, L. Li, and J. S. Thorson.** 2006. Exploiting the reversibility of natural product glycosyltransferase-catalyzed reactions. *Science* **313:**1291–1294.

99. **Zhang, C., C. Albermann, X. Fu, and J. S. Thorson.** 2006. The in vitro characterization of the iterative avermectin glycosyltransferase AveBI reveals reaction reversibility and sugar nucleotide flexibility. *J. Am. Chem. Soc.* **128:**16420–16421.

100. **Zhang, C., Q. Fu, C. Albermann, L. Li, and J. S. Thorson.** 2007. The in vitro characterization of the erythronolide mycarosyltransferase EryBV and its utility in macrolide diversification. *Chembiochem* **8:**385–390.

101. **Zhang, C., R. Moretti, J. Jiang, and J. S. Thorson.** 2008. The in vitro characterization of polyene glycosyltransferases AmphDI and NysDI. *Chembiochem* **9:**2506–2514.

102. **Zhang, W., M. Ye, J. X. Zhan, Y. J. Chen, and D. Guo.** 2004. Microbial glycosylation of four free anthraquinones by *Absidia coerulea*. *Biotechnol. Lett.* **26:**127–131.

Metabolic Engineering of *Escherichia coli* for the Production of a Precursor to Artemisinin, an Antimalarial Drug

CHRISTOPHER J. PETZOLD AND JAY D. KEASLING

25

25.1. INTRODUCTION

25.1.1. Need

Malaria is a severe human disease caused by protozoan parasites of the genus *Plasmodium* that claims over two million lives annually and threatens hundreds of millions more. Recently, economical malarial treatments currently in use, such as quinine and chloroquine, have become less effective as multidrug-resistant malarial strains have begun to spread. This trend poses little risk to residents of more prosperous nations, who can afford the top-of-the-line therapies, yet the spread of multidrug-resistant malarial strains significantly increases mortality among malarial patients living in developing countries (64). To combat resistance, more advanced drug combination regimens, such as artemisinin-based combination therapies (ACTs), have been developed. ACTs are highly effective, yet the demand for over 100 million ACT drug courses per year (32) greatly exceeds the supply (26). The major component in ACTs is artemisinin, a terpenoid produced naturally by *Artemisia annua*, a plant that has been used in Chinese medicine for over 2,000 years to treat many illnesses, including malaria (7). The main active ingredient in *A. annua*, artemisinin, was not identified until 1971 (45). Since then, the utility of artemisinin to combat many diseases has sparked interest in identifying its chemical structure (34) and determining the biological reactivities of artemisinin and its derivatives (45, 46). Efforts in the 1980s to develop a more stable source resulted in several reports describing the total synthesis of artemisinin (78, 91). Unfortunately, chemical synthesis of artemisinin is complex and inefficient due to side product formation and multiple purification steps. Furthermore, expensive reactants and catalysts are needed in large quantities and are not environmentally friendly. Despite the high demand for artemisinin, total global production is low and highly variable, as demonstrated in 2004, when a seed shortage caused severe artemisinin shortfalls across the world (58). Overall, these issues contribute to the high costs of ACTs, which limit their utility to residents of developing countries. For the people most in need of the drugs, those living in developing countries in temperate areas of the world, ACTs remain prohibitively costly, prompting a need for the application of modern industrial biotechnology methods to reduce costs and stabilize supplies.

25.1.2. Biosynthetic Production of a Precursor to Artemisinin

Biotechnology offers multiple means for improving artemisinin production. Engineering of the native host to overproduce precursors to artemisinin is promising, but genetic tools for *A. annua* manipulation are not well developed, and currently the results have been limited (1, 18, 54, 87, 95). Synthetic biology and metabolic engineering of microbes offer an attractive alternative to natural sources as a means for inexpensive pharmaceutical production. A metabolically engineered microbe is an inherently scalable and robust platform for production that is advantageous compared to natural product extraction and chemical synthesis (41). For example, unlike extraction from plants, production by metabolically engineered microbes is not dependent on weather conditions or the political climate, thus providing a stable supply. Biosynthetic artemisinin can be produced without competing synthesis pathways, or pathways for other isoprenoids common to plants, reducing recovery expenses. Metabolic engineering approaches often utilize native enzymatic biosynthesis pathways to produce the target molecule, offering selectivity for the required stereochemistry and efficient conversion of intermediates to the final product (48). The isoprenoid precursors used in artemisinin biosynthesis are common to both prokaryotes and eukaryotes; consequently, model microbes, such as *Escherichia coli* and *Saccharomyces cerevisiae*, can be used to convert simple, cheap sugar to artemisinin precursors. The challenge associated with this process lies in engineering high carbon flux from sugar to the artemisinin precursors to reduce production costs. Using *E. coli* for artemisinin production is particularly advantageous due to the bacterium's limited isoprenoid metabolism, which minimizes losses to native processes and the production of structurally similar products. However, identification of the enzymes dedicated to artemisinin production and incorporation into and optimization in *E. coli* are necessary.

In the engineering of microbes to produce secondary metabolites, very different challenges from those associated with high-level production of pharmaceutical proteins arise. To optimize system characteristics for metabolite production, the expression of biosynthetic enzymes must be balanced with sufficient metabolite flux and microbial growth (39). Overexpression of biosynthetic enzymes reduces the available carbon for metabolite conversion, whereas too little expression of biosynthetic enzymes reduces the rate of production and final titers. Beyond the need to balance enzyme expression, artemisinin production in *E. coli* is complicated by uncertainty in the native biosynthesis pathway. Indeed, the final steps in artemisinin biosynthesis may be nonenzymatic (18). Given that the conversion of artemisinic acid to artemisinin and its derivatives has been established (32, 74), metabolic engineering of *E. coli* focuses on two fronts, increasing the titer of the isoprenoid precursors via the incorporation of the mevalonate pathway into *E. coli*, and the subsequent conversion of mevalonate to artemisinic acid, a precursor of artemisinin.

25.2. INCORPORATION OF AMORPHA-4, 11-DIENE BIOSYNTHETIC PATHWAY INTO *E. COLI*

One universal microbe is not likely to be appropriate for all metabolic engineering projects. Consequently, careful selection of the host is necessary. Well-characterized hosts such as *E. coli* and *S. cerevisiae* are often the first choices due to their amenable culturing conditions and high growth rates, the wide variety of genetic tools available for these organisms, and their significant history as biosynthetic platforms.

25.2.1. Isoprenoid Biosynthesis in *E. coli*

Prenyl pyrophosphate biosynthesis of isopentenyl diphosphate (IPP) and its isomer dimethylallyl diphosphate (DMAPP) occurs by two known pathways (Fig. 1), the mevalonate pathway in eukaryotes and the deoxyxylulose-5-phosphate (DXP) pathway typically found in prokaryotes (25, 35), although exceptions exist in both cases. There are a variety of detailed reviews of isoprenoid biosynthesis focusing on the mevalonate and DXP pathways (2, 20, 43, 47, 51, 89). In *E. coli*, the native DXP pathway produces isoprenoids for prenylation of tRNAs (17) and conversion into polyisoprenoids such as ubiquinones and menaquinones. However, the flux to these metabolites is significantly lower than what is needed to make biosynthetic artemisinin cost-effective; consequently, modification and improvement of the IPP/DMAPP biosynthesis pathway is necessary. One way increased flux can be achieved is by overexpression of the rate-limiting enzymes in the native (DXP) pathway. An advantage to this method is that utilization of native *E. coli* proteins alleviates the complications of expressing heterologous genes. Indeed, increased isoprenoid production was achieved by overproducing DXP (1-deoxy-D-xylulose-5-phosphate) synthase (encoded by the gene *dxs*) and the DXP reductoisomerase (encoded by the gene *ispC* [*dxr*]) in the DXP pathway and other native proteins (31, 39, 40, 42, 92). Similarly, β-carotene production in *E. coli* was enhanced by integration of additional copies of *dxs*, *ispC* (*dxr*), *idi*, and *ispA* into the chromosome (94). Efforts to optimize the DXP pathway by systematic and combinatorial gene knockouts (4) and promoter engineering (3) also yielded higher isoprenoid titers. Despite significant improvement in flux through the DXP pathway, the highest levels achieved were not feasible for inexpensive artemisinin production.

25.2.2. Addition of an Exogenous Mevalonate Pathway

The limitations associated with the DXP pathway led to the work of Martin et al. (56), whereby the incorporation of the eukaryotic mevalonate pathway into *E. coli* was used to divert more carbon to isoprenoid production. This strategy

has a couple of notable advantages. First, since the mevalonate pathway is exogenous in *E. coli*, regulation machinery that can limit production is absent. Furthermore, the intermediates in the exogenous pathway are less likely to be diverted to other cellular functions, enabling a net higher flux to be directed to artemisinin precursors. Incorporation of an exogenous pathway into *E. coli* has its challenges as well. Diversion of carbon from native processes can have wide-ranging effects on *E. coli* metabolism. Proper expression and folding of nonnative proteins are not certain because the metabolic burden of producing inactive proteins precludes optimal metabolite flux. Thus, careful monitoring of the production levels and enzymatic activity is required. Buildup of toxic intermediates is often immediately obvious, resulting in growth inhibition; however, more subtle impacts on native metabolism may occur as well.

The mevalonate pathway (Fig. 2) was added to *E. coli* on two plasmids. The genes on the first plasmid, pMevT, include *atoB*, encoding an enzyme used to convert acetyl coenzyme A (acetyl-CoA) to mevalonate by means of the native acetyl-CoA thiolase, and two genes from *S. cerevisiae*, encoding 3-hydroxy-3-methylglutaryl (HMG)-CoA synthase (HMGS) and a truncated form of HMG-CoA reductase (HMGR). The second plasmid, pMBIS, consists of *S. cerevisiae* genes encoding enzymes that sequentially convert mevalonate to phosphomevalonate (mevalonate kinase [*ERG12*]), to diphosphomevalonate (phosphomevalonate kinase [*ERG8*]), and via decarboxylation, to IPP (mevalonate diphosphate decarboxylase [*MVD1*]) As will be described below, increasing the expression of two native *E. coli* genes, those for IPP isomerase (*idi*) and the farnesyl diphosphate (FPP) synthase (*ispA*), was also necessary to produce high levels of FPP. To establish that the exogenous mevalonate pathway was active, a strain in which the DXP pathway was inactivated was transformed with the pMBIS plasmid. Strain DYM1 (50) was selected because the gene encoding the second enzyme in the endogenous DXP pathway, *ispC* (*dxr*), was deleted, resulting in a growth defect without an alternate means of isoprenoid production. Because IPP and DMAPP are essential for growth, only a functioning mevalonate pathway in the DYM1 strain with exogenous mevalonate supplementation in the medium will result in active cultures. Both the DYM1 and DYM1(pMBIS) strains grew on plates supplemented with 2-C-methyl-erythritol, the metabolic product of IspC; however, only the DYM1(pMBIS) strain grew on plates supplemented with 1 mM mevalonate. For liquid-phase cultures, when supplemental mevalonate levels exceeded 10 mM, the growth defect returned (Fig. 3), indicating a toxic effect of the mevalonate or one of the downstream intermediates. This observation will be addressed in greater detail in section 25.3.2.

The addition of the entire mevalonate pathway (pMevT and pMBIS) directed a significant amount of metabolic flux to FPP. Because native *E. coli* metabolism utilizes FPP in the biosynthesis of ubiquinone and menaquinone, which are needed in small quantities only, it was necessary to convert the excess FPP to a metabolite that is not naturally consumed by the host. The first committed step toward artemisinin is the conversion of FPP to amorpha-4,11-diene via amorpha-4,11-diene synthase (13). Terpene synthases can be difficult to express in *E. coli* (55); consequently, the amorpha-4,11-diene synthase gene (*ADS*) (59) was codon optimized to aid expression and introduced into *E. coli* on a third plasmid (pADS). With all three plasmids (pMevT, pMBIS, and pADS) incorporated into *E. coli*, complete conversion of glucose to amorpha-4,11-diene was achieved.

DXP Pathway

Mevalonate Pathway

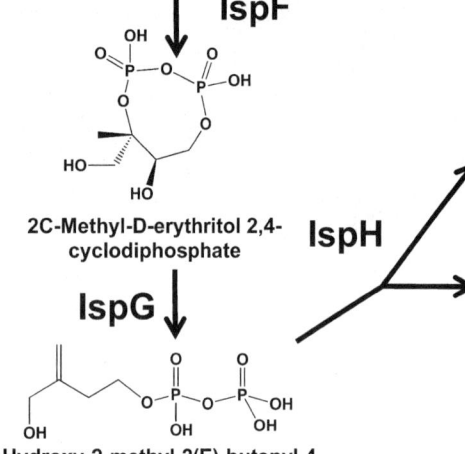

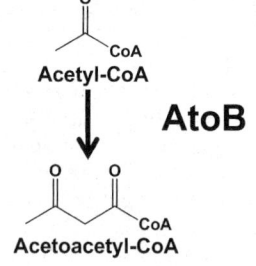

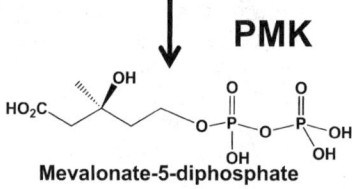

Pyruvate   glyceraldehyde 3-phosphate

Dxs

1-deoxy-D-xylose-5-phosphate

IspC

2-C-methyl-D-erythritol-4-phosphate

IspD

4-Diphosphocytidyl-2C-methyl-D-erythritol

IspE

4-Diphosphocytidyl-2C-methyl-D-erythritol 2-phosphate

IspF

2C-Methyl-D-erythritol 2,4-cyclodiphosphate

IspG

1-Hydroxy-2-methyl-2(E)-butenyl-4-diphosphate

IspH

Acetyl-CoA

AtoB

Acetoacetyl-CoA

HMGS

HMG-CoA

HMGR

Mevalonate

MK

Mevalonate-5-phosphate

PMK

Mevalonate-5-diphosphate

MPD

IPP

Idi

DMAPP

IspA

FPP

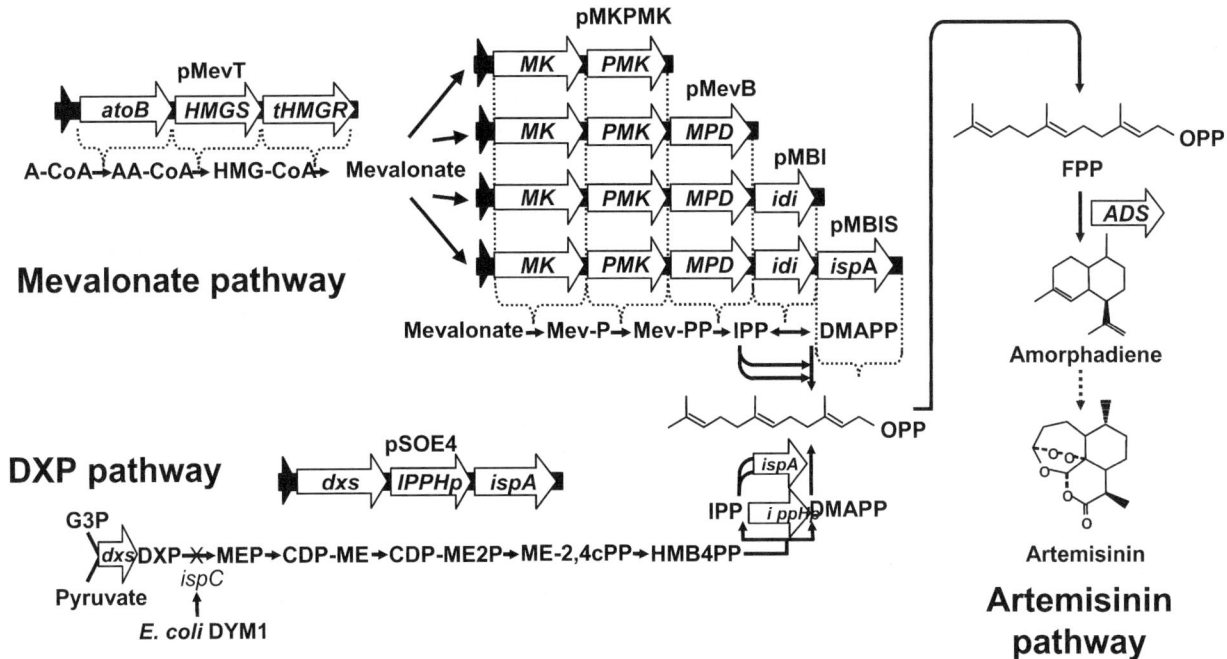

FIGURE 2 Production of amorpha-4,11-diene via the DXP or mevalonate isoprenoid pathway and depiction of the synthetic operons used by Martin et al. (56). The amorpha-4,11-diene pathway was separated into three plasmids, with one plasmid carrying the steps from acetyl-CoA to mevalonate, the second plasmid expressing the steps from mevalonate to FPP, and the third plasmid encoding amorpha-4,11-diene synthase (ADS). Several intermediate plasmids converting mevalonate to mevalonate-5-diphosphate (pMKPMK), to IPP (pMevB), and to DMAPP (pMBI) were constructed for comparison to pMBIS. tHMGR, truncated form of HMGR gene; A-CoA, acetyl-CoA; AA-CoA, acetoacetyl-CoA; MK, mevalonate kinase gene; PMK, phosphomevalonate kinase gene; MPD, mevalonate pyrophosphate decarboxylase gene; Mev-P, mevalonate phosphate; Mev-PP, mevalonate diphosphate; OPP, pyrophosphate; IPPHp, IPP, isopentenyl pyrophosphate; G3P, glyceraldehyde-3-phosphate; MEP, 2-C-methyl-D-erythritol 4-phosphate; CDP-ME, 4-diphosphocytidyl-2-C-methyl-D-erythritol; CDP-ME2P, 4-diphosphocytidyl-2-C-methyl-D-erythritol 2-phosphate; ME-2,4cPP, 2-C-methyl-D-erythritol 2,4-cyclodiphosphate; HMB4PP, 1-hydroxy-2-methyl-2-(E)-butenyl 4-diphosphate.

25.3. AMORPHA-4,11-DIENE PATHWAY OPTIMIZATION

25.3.1. Metabolite Analysis

Metabolic engineering for artemisinin precursor production required accurate, rapid, and sensitive metabolite analysis. Optimal metabolite analysis depends on the detection of all engineered pathway intermediates, the carbon source, and the common *E. coli* metabolites. However, not all intermediates are amenable to analysis, and highly abundant native metabolites preclude facile detection of low-concentration intermediates. To overcome these difficulties, a variety of extraction, separation, and detection techniques were used to identify and quantify native metabolites and pathway intermediates. Chromatography, both gas and liquid, and capillary electrophoresis were the main techniques used to separate the metabolites of interest from the rest of the metabolites present. In general, these methods were coupled to mass spectrometry (MS) for identification.

At the beginning of the work, the amount of mevalonate and/or amorpha-4,11-diene produced was used to measure progress toward the goal of economical artemisinin production. These two intermediates are easily extracted from the medium and analyzed via gas chromatography (GC)-MS. GC-MS analysis is a rapid and robust method

FIGURE 1 Biosynthetic routes to polyprenyl pyrophosphate isoprenoid biosynthetic pathways in *E. coli* and *S. cerevisiae*. Dxs, DXP synthase; IspC, DXP reductoisomerase; IspD, 4-diphosphocytidyl-2-C-methyl-D-erythritol synthase; IspE, 4-diphosphocytidyl-2-C-methyl-D-erythritol kinase; IspF, 2-C-methyl-D-erythritol 2,4-cyclodiphosphate synthase; IspG, 1-hydroxy-2-methyl-2-(E)-butenyl 4-diphosphate synthase; IspH, 1-hydroxy-2-methyl-2-(E)-butenyl 4-diphosphate reductase; AtoB, acetoacetyl-CoA thiolase; HMGS, 3-hydroxymethylglutaryl-CoA synthase; HMGR, 3-hydroxymethylglutaryl-CoA reductase 1; MK, mevalonate kinase; PMK, phosphomevalonate kinase; MPD, mevalonate pyrophosphate decarboxylase; Idi, isopentenyl pyrophosphate isomerase. Adapted from *Applied Microbiology and Biotechnology* (90) with permission of the publisher.

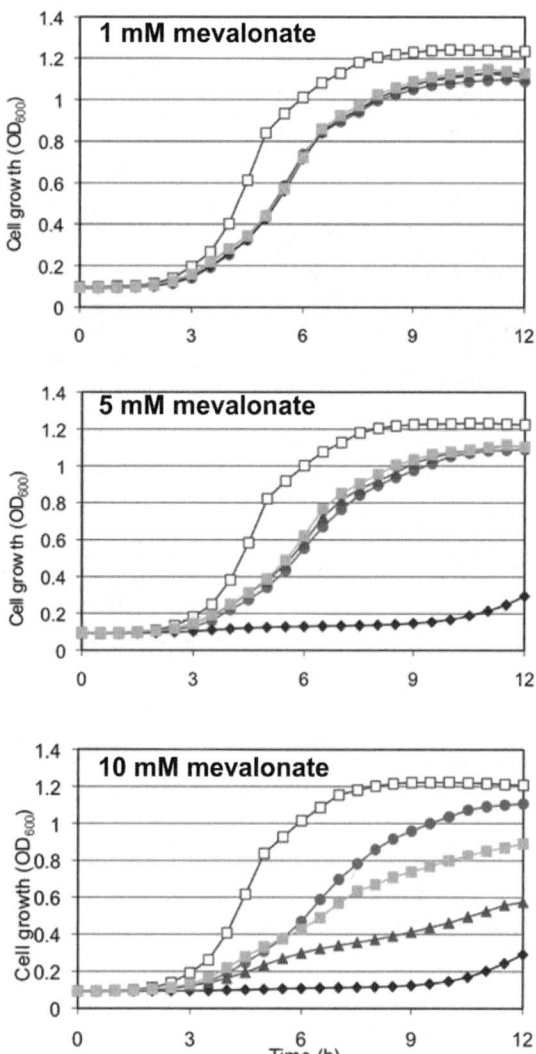

FIGURE 3 Growth curves for *E. coli* showing the inhibition effects caused by increasing concentrations of DL-mevalonate in Luria-Bertani medium. Strains expressing plasmids pBBR1MCS-3 (open squares), pMKPMK (circles), pMevB (diamonds), pMBI (triangles), and pMBIS (solid squares). OD$_{600}$, optical density at 600 nm. Reprinted from *Nature Biotechnology* (56) with permission of the publisher.

by which confident analyte identification is possible via comparison to the National Institute of Standards and Technology database. Additionally, the location of these intermediates in the biosynthesis pathway made monitoring of these molecules useful metrics; however, analytical measurements are only as good as the experimental technologies employed. Initial experiments (56) measuring pathway performance suggested that amorpha-4,11-diene was lost from the cultures due to its volatility, resulting in underestimation of production levels. To address this issue, a two-phase partitioning bioreactor was developed to trap the amorpha-4,11-diene produced (62). The organic partitioning layer effectively trapped amorpha-4,11-diene that had been lost to evaporation under normal conditions and also simplified extraction for GC-MS analysis. Additionally, any toxic effects associated with

amorpha-4,11-diene were minimized by sequestering it from the growing cells. Remarkably, cultures grown with a partitioning layer produced 0.4 g/liter, notably more amorpha-4,11-diene (8.5-fold) than the amount reported for the same culture grown without such a layer (62). Use of the two-phase partitioning bioreactor proved to be a robust method for characterizing amorpha-4,11-diene-producing strains.

25.3.2. Optimization of Metabolic Flux from Acetyl-CoA to Mevalonate

Improving carbon flux to amorpha-4,11-diene requires optimization of the pathway as a whole, but since the mevalonate pathway was separated into two plasmids (pMevT and pMBIS) (Table 1), optimization efforts were divided into mevalonate and amorpha-4,11-diene improvements. Though there were no bottlenecks at the mevalonate and amorpha-4,11-diene stages, the possibility that bottlenecks were present in association with other metabolites or enzymes could not be excluded. Detailed examination showed metabolite imbalances at both the top part of the pathway (acetyl-CoA to mevalonate) and the bottom part (mevalonate to FPP), resulting in significant growth defects (56, 69). Ultimately, optimal production of amorpha-4,11-diene requires careful control of metabolic flux upstream from the desired product. Because supplementation with exogenous mevalonate resulted in higher amorpha-4,11-diene titers than the utilization of the MevT plasmid (the upper mevalonate pathway) alone to produce mevalonate, efforts to improve carbon flux to mevalonate were initiated. The MevT enzymes were expressed from two separate plasmids (pBAD33MevT and pBAD24MevT) with the arabinose-inducible promoter (Table 1), which was stronger than the original *lac* promoter (69). However, as demonstrated in Fig. 4, a growth defect was noted for both strains. This toxicity may be due to increased protein production from the cells with pBAD-33MevT, pathway intermediate-induced stress, or a combination of both. Importantly, identification of the true nature of the toxicity required differentiation between the metabolic burden of protein production and the potential toxicity of intermediary metabolites. To this end, the HMGS was inactivated by mutation of the cysteine-159 residue to alanine (72). Induction of the inactive MevT pathway [carried on pMevT(C159A)] yielded a host that would experience the metabolic burden of protein expression without the related metabolic flux. Cells expressing the plasmid containing an inactivated HMGS gene grew only slightly slower than empty-plasmid controls, whereas cells expressing the active pathway showed severe growth inhibition. Thus, production of a potentially toxic metabolite appeared to be the cause of the stress. Individual expression of each protein from the MevT plasmid showed a growth defect for the HMGS-expressing strain. This result supported the hypothesis that HMG-CoA was the source of the growth defect since acetyl-CoA and acetoacetyl-CoA are native *E. coli* metabolites and high levels of mevalonate supplementation in the growth medium did not cause growth defects.

To verify this conclusion, high-performance liquid chromatography methods were developed to identify the intermediates produced by the MevT enzymes and several similar native metabolites. Direct analysis of acetyl-CoA, acetoacetyl-CoA, HMG-CoA, malonyl-CoA, and free CoA permitted direct determination of which intermediates were accumulating and whether levels of native metabolites

TABLE 1 Plasmids used for metabolic engineering of *E. coli* for production of precursors to artemisinin

Strain or plasmid	Genotype or description[a]	Source or reference
Strains		
DH10B	F⁻ *mcrA* Δ(*mrr-hsdRMS-mcrBC*) φ80*lacZ*ΔM15 Δ*lacX74 recA1 endA1* *araD139* Δ(*ara-leu*)*7697 galU galK* λ⁻ *rpsL* (Str^r) *nupG*	Invitrogen
DP5	DH10B *ispA*::P_LAC-(*ERG12-ERG8-MPD1-idi-ispA*)::*ispA* Δ*ispC*	67
Plasmids		
pBAD33	Cloning vector containing Chl^r marker, pACYC184 origin, *araC*, and P_BAD promoter	29
pBAD24	Cloning vector containing Amp^r marker, modified pBR322 origin with truncation of *rop* gene, *araC*, and P_BAD promoter	29
pJA4	pACYC184 (NEB) derivative containing engineered MCS; Chl^r	5
pMevT	pBAD33 derivative containing *atoB*, HMGS, and tHMG1 genes under control of P_LAC; Chl^r	56
pMBIS	pBBR1MCS-3 containing MK (*ERG12*), PMK (*ERG8*), *MVD1*, *idi*, and *ispA* genes under control of P_LAC; Tet^r	56
pADS	pTrc99A containing synthetic *ADS*; Amp^r	56
pBAD24MevT	pBAD24 containing *atoB*, HMGS, and tHMG1 genes under control of P_BAD promoter; Amp^r	56
pBAD33MevT	pBAD33 containing *atoB*, HMGS, and tHMG1 genes under control of P_BAD promoter; Chl^r	56
pAtoB	pBAD33 containing *atoB* under control of P_BAD promoter; Chl^r	69
pHMGS	pBAD33 containing HMGS gene under control of P_BAD promoter; Chl^r	69
pHMGR	pBAD33 containing tHMG1 gene under control of P_BAD promoter; Chl^r	69
pMevT(C159A)	pBAD33MevT derivative containing HMGS gene with C159A mutation	69
pAM45	pJA4 derivative containing *lacUV5* promoters in 5′ direction from codon-optimized MevT and MBIS operons; Chl^r	5
pAM92	pAM45 derivative containing *lacI*^q-P_trc-*ADS*; Chl^r	5
pAM94	pADS derivative containing codon-optimized MK gene	5
pKLN59	Green fluorescent protein-expressing biosensor plasmid; Amp^r Kan^r *gfp*	63

[a]tHMG1, truncated form of HMGR; NEB, New England Biolabs; MCS, multiple-cloning site; MK, mevalonate kinase; PMK, phosphomevalonate kinase.

such as malonyl-CoA and free CoA were perturbed. To rapidly quench metabolism, the cells and growth medium were centrifuged through silicone oil into a trichloroacetic acid solution by the method described by Shimazu et al. (80). Analysis of the energy charge was used to verify that metabolism was quenched rapidly. These results showed that HMG-CoA accumulated in the strain harboring pBAD33MevT and in the strain expressing HMGS alone. Free-CoA levels were reduced when HMGS was expressed alone, while malonyl-CoA accumulated in both strains. Acyl-CoA analyses prompted efforts directed at achieving balanced enzymatic activity. Strains containing the pBAD-33MevT plasmid and an additional plasmid harboring an extra copy of either *atoB* or the HMGS or HMGR gene were cultured, and the changes in growth and acyl-CoA levels were monitored. The strain harboring pBAD33MevT and the plasmid with an additional copy of the HMGR gene showed restored growth and lower HMG-CoA levels. Characterization of the inhibitory effects of HMG-CoA will be discussed in section 25.4.1. Alternate methods for balancing the enzymatic activity, including varying plasmid

attributes, performing mutagenesis, and tuning intergenic regions to refine the conversion of acetyl-CoA to mevalonate, are discussed below.

25.3.3. Development of a Mevalonate Biosensor for High-Throughput Screening

Mutagenic and combinatorial methods are powerful tools for optimizing metabolically engineered systems. However, their utility is limited by the quality of screens available for the library of interest (49, 71, 86). Chromatographic methods for metabolite analysis, while sensitive and selective, are not readily compatible with high-throughput screening of thousands or tens of thousands of samples. To aid optimization of the top part of the mevalonate pathway (pMevT), an *E. coli* mevalonate auxotroph was engineered for use as a biosensor (67). The sensor reports the mevalonate concentration in the medium via a change in growth rate. The auxotroph was engineered to constitutively produce green fluorescent protein (from pKLN59) (Table 1) so that the change in growth rate could be conveniently monitored by fluorescence. The reporter strain was constructed by deletion of the native *E. coli ispC* (*dxr*)

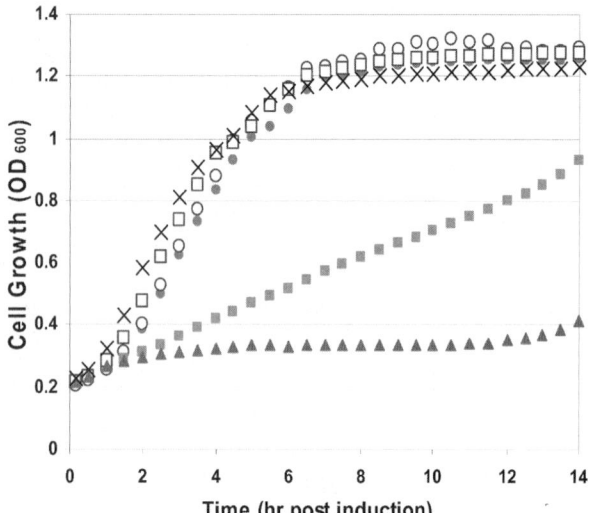

FIGURE 4 Growth of *E. coli* with increasing expression of the MevT operon. *E. coli* DP10 cells harbored the following plasmids (in order of increasing expression of the MevT genes): pMevT (black circles), pBAD33MevT (gray squares), and pBAD24MevT (black triangles). Cells expressing pLac33 (open circles), pBAD33 (open squares), and pBAD24 (multiplication signs) were the respective empty-plasmid controls. Reprinted from *Metabolic Engineering* (69) with permission of the publisher.

gene, the second gene in the DXP isoprenoid biosynthesis pathway. Because both IPP and DMAPP are required for *E. coli* growth, isoprenoid biosynthesis was restored by insertion of the genes from the MBIS plasmid into the *E. coli* chromosome via the recombination method established by Martinez-Morales et al. (57) (yielding strain DP5) (Table 1). Following induction of MBIS with isopropyl-β-D-thiogalactopyranoside (IPTG), growth in the presence of mevalonate was restored. High-throughput screening of mevalonate production by mutant libraries in 96-well plates with the mevalonate auxotroph as a biosensor was implemented (Fig. 5).

This method was applied to measure mevalonate levels associated with a library of posttranscriptional control elements for the modulation of protein levels produced from MevT plasmids. Posttranscriptional engineering offers a second layer of control, beyond transcriptional control, to balance pathway flux (76). To this end, Pfleger et al. integrated a library of tunable intergenic regions between the coding regions of the MevT transcripts to modulate translation initiation, RNA stability, and translation termination (68). Then they used the mevalonate biosensor to screen a library of MevT plasmids containing over 600 tunable intergenic regions for production of increased mevalonate titers. The mevalonate producers were grown for 24 h at 37°C, at which point the cells were pelleted by centrifugation. Subsequently, the supernatant was added to a second 96-well plate containing new media and the mevalonate auxotroph. Following incubation, the wells containing the most mevalonate showed the strongest fluorescent signals.

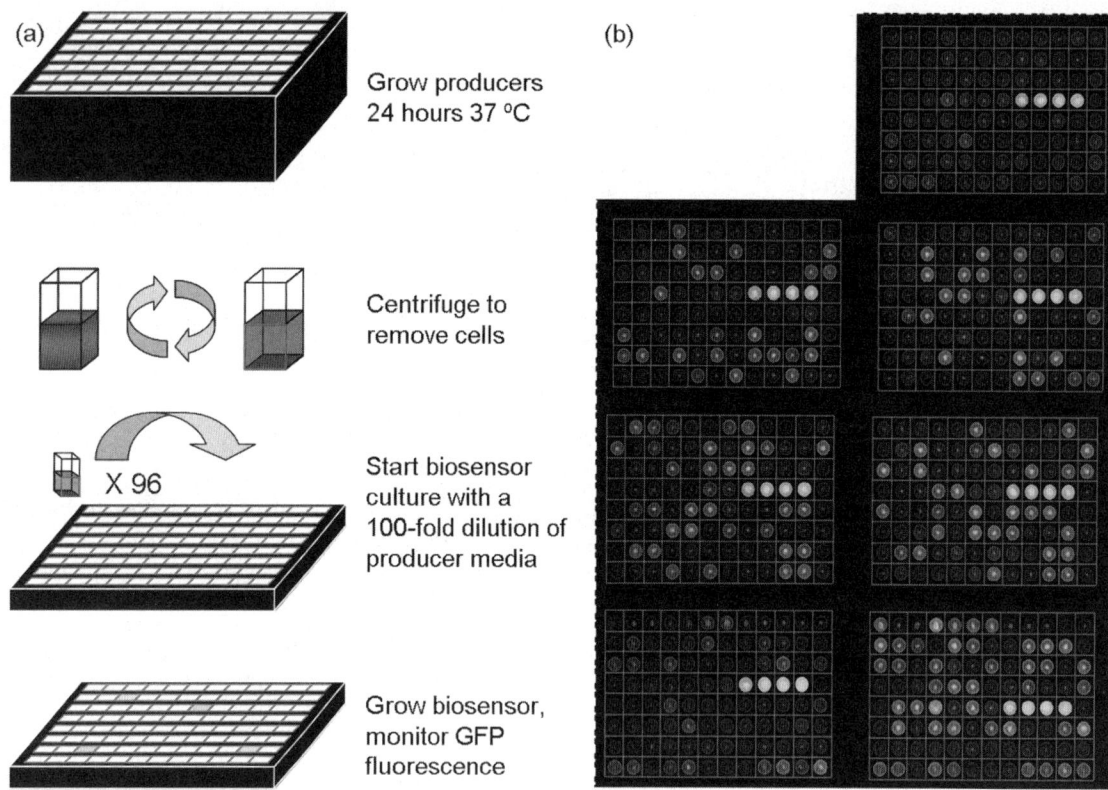

FIGURE 5 Mevalonate biosensor screening strategy. (a) Screening methodology. Mevalonate producers were grown in a 96-well format in C medium with inducers for 24 h. Producer cells were removed by centrifugation, and the spent medium was passed to new cultures inoculated with the biosensor. Wells containing the most mevalonate were the most fluorescent (they showed the greatest intensity of white on the plates). GFP, green fluorescent protein. (b) Mevalonate library. Seven 96-well plates are shown 15 h post-biosensor inoculation. The four white wells on each plate are mevalonate controls. Reprinted from *Metabolic Engineering* (67) with permission of the publisher.

By using this technique, Pfleger et al. isolated four strains that produce seven times more mevalonate than the strain containing the original MevT plasmid (68). The four isolated strains displayed improved growth (improved total production) and flux through the pathway (improved specific production). Notably, acyl-CoA analysis revealed increased acetyl-CoA levels relative to that from the original MevT-containing strain.

25.3.4. Optimization of Metabolic Flux from Mevalonate to Amorpha-4,11-Diene

Attempts to increase flux from mevalonate to FPP revealed growth inhibition when a plasmid (pMevB) containing only the first three genes (*ERG12*, *ERG8*, and *MVD1*) was expressed, as depicted in Fig. 3 (56). This inhibition increased with the concentration of exogenous mevalonate, indicating that the toxicity resulted from metabolite flux and not protein overproduction. Partial growth was restored when the IPP isomerase (*idi*) and FPP synthase (*ispA*) genes were coexpressed with the first three genes. These results suggest that IPP is a toxic intermediate; however, metabolite analysis using [^{3}H]IPP and [^{3}H]DMAPP indicated that FPP accumulation is also toxic. The growth defect persisted for a strain containing both the top (pMevT) and bottom (pMBIS) parts of the mevalonate pathway, confirming that the top part of the pathway produced over 5 mM mevalonate. Complete rescue from the growth defect occurred when the amorpha-4,11-diene synthase gene was expressed on a separate plasmid to consume FPP (Fig. 6). Both of these growth defects were observed as a result of native-metabolite imbalances. Hence, pulling on the pathway via the

amorpha-4,11-diene synthase is just as important as pushing flux through the pathway. Continuing work focuses on analyzing each intermediate in the lower part of the pathway.

25.3.5. Genetic Optimization of the Mevalonate Pathway from Acetyl-CoA to Amorpha-4,11-Diene

As shown by the metabolite analysis work described above, the amount of amorpha-4,11-diene produced by the original three-plasmid (pMevT-pMBIS-pADS) system, while significant, was not optimal. Steps to increase flux to amorpha-4,11-diene involved overexpression of pathway enzymes but did not address the metabolic burden of plasmid maintenance (27, 39), the maintenance of multiple antibiotic resistance genes, and total pathway protein overproduction. Tight control of enzyme concentration is especially important for efficient amorpha-4,11-diene production. Optimal pathway gene expression is dependent on a wide variety of factors (8, 37) including, but not limited to, gene product toxicity, solubility, codon usage, mRNA secondary structure (21, 22), mRNA stability, and translational efficiency. Additionally, plasmid-specific characteristics such as the type of origin of replication used, the ribosomal binding site strength (10), plasmid design, and promoter strength (3, 30, 38) generally can have a profound effect on the amount of protein produced in the cell at any given time.

Thus, in addition to the metabolite studies described above, effort was directed toward reducing the number of plasmids and balancing production of the pathway proteins from the new plasmids. The original plasmids were modified by the incorporation of all the genes required for the conversion of acetyl-CoA to amorpha-4,11-diene via the mevalonate pathway into one- and two-plasmid systems. To aid rapid development, the new plasmids were designed for facile substitution of promoters and operon constructs (5). The initial step in increasing amorpha-4,11-diene production was codon optimization of the MevT plasmid, followed by replacement of the wild-type *lac* promoter with the *lacUV5* promoter. The strain harboring the optimized MevT operon yielded 1.5 times more amorpha-4,11-diene than the original strain (5). This finding led Anthony et al. (5) to design a larger plasmid (pAM45) consisting of all the genes of the (codon-optimized) pMevT and pMBIS plasmids under the control of the stronger *lacUV5* promoter. The strain carrying pAM45, which harbored pADS (Table 1) as well, produced threefold more amorpha-4,11-diene than the original three-plasmid strain (5). Integration of ADS into the pAM45 plasmid created pAM92. By using these variants, two limitations in the pathway were identified. Gene titration experiments involving the genes from the lower part of the pathway (the pMBIS genes) revealed that expressing additional copies of the mevalonate kinase gene improved amorpha-4,11-diene titers by 2.3-fold (5). None of the other genes significantly impacted the amorpha-4,11-diene titers when provided on another plasmid, indicating that they do not represent bottlenecks in the pathway. The second limitation was identified when pAM92 was created. Although cells harboring pAM92 produced more amorpha-4,11-diene than cells harboring the original three-plasmid system, they produced less than cells harboring both pAM45 and pADS and exhibited decreased viability. When ADS was incorporated into pAM45, with the same promoter retained, the most likely cause of lower titers was the decrease in plasmid copy number (pADS was maintained at 20 to 30 copies/cell, while pAM45 was present at 10 to 15 copies/cell). Amorpha-4,11-diene production was increased when the strain containing pAM92 was transformed with the original pADS plasmid, confirming that the bottleneck occurred due to amorphadiene synthase limitation. Ultimately, the highest amorpha-4,11-diene level (~0.3 g/liter/optical density unit at 600 nm in shake

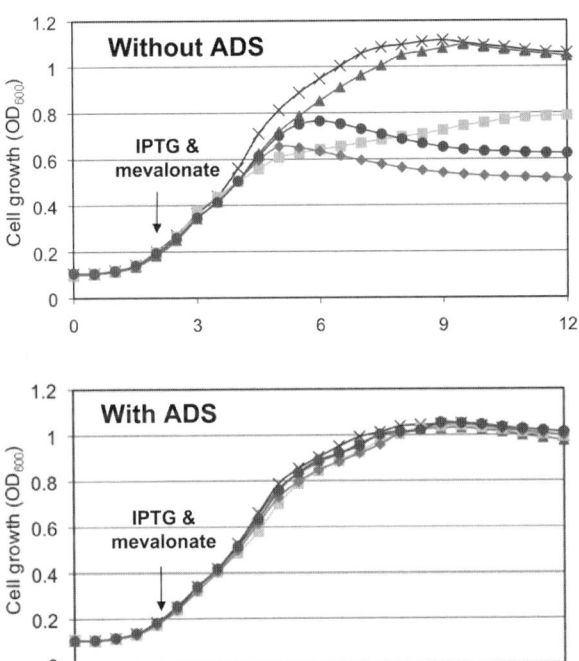

FIGURE 6 Effect of amorphadiene synthase (ADS) expression on the growth of *E. coli* harboring pMBIS. Cells carried pMBIS and the empty expression vector pTrc99A (without the *ADS* gene) (top) or pADS expressing the amorphadiene synthase (bottom). Luria-Bertani medium was supplemented with 0, 5, 10, 20, or 40 mM DL-mevalonate. Reprinted from *Nature Biotechnology* (56) with permission of the publisher. OD$_{600}$, optical density at 600 nm.

flask cultures) was produced by a strain containing pAM45 and pAM94, created via the incorporation of an additional copy of the mevalonate kinase gene into pADS.

25.4. "-OMICS" ANALYSES FOR INCREASED AMORPHA-4,11-DIENE PRODUCTION

The work described above focused on balancing amorpha-4,11-diene-producing pathways and optimizing their flux to meet stringent economic restrictions. As discussed above, biological factors such as gene selection, plasmid design, enzyme expression levels, and isoprenoid intermediate toxicity need to be considered when production of the highest titers of amorpha-4,11-diene and artemisinic acid is desired. Additionally, native host metabolism, especially for enzymes with a particular intermediate as a precursor metabolite, may have an impact on carbon flux through the engineered pathway (81, 83). However, effects that limit final production levels are not al-

ways obvious; perturbation of native regulation and disruption of complex interactions may manifest themselves in unique ways. Despite a vast amount of knowledge about *E. coli* metabolism, such interactions can be highly complex and difficult to elucidate via testable hypotheses. A complete systems-level understanding of the engineered host provides insight into the nonobvious perturbations to the system under amorpha-4,11-diene- and artemisinic acid-producing conditions.

Cell-wide studies of RNA (transcriptomics), proteins (proteomics), and metabolites (metabolomics) reveal the state of the cell under defined conditions, and comparison to the state of the cell under a second set of conditions points to subtle changes often overlooked by hypothesis-driven experiments. By combining the data from transcriptomics, proteomics, and metabolomics experiments, one can develop a greater understanding of the impact of metabolic engineering on the host metabolism and design ways to increase end product titers (Fig. 7). Ultimately, engineering

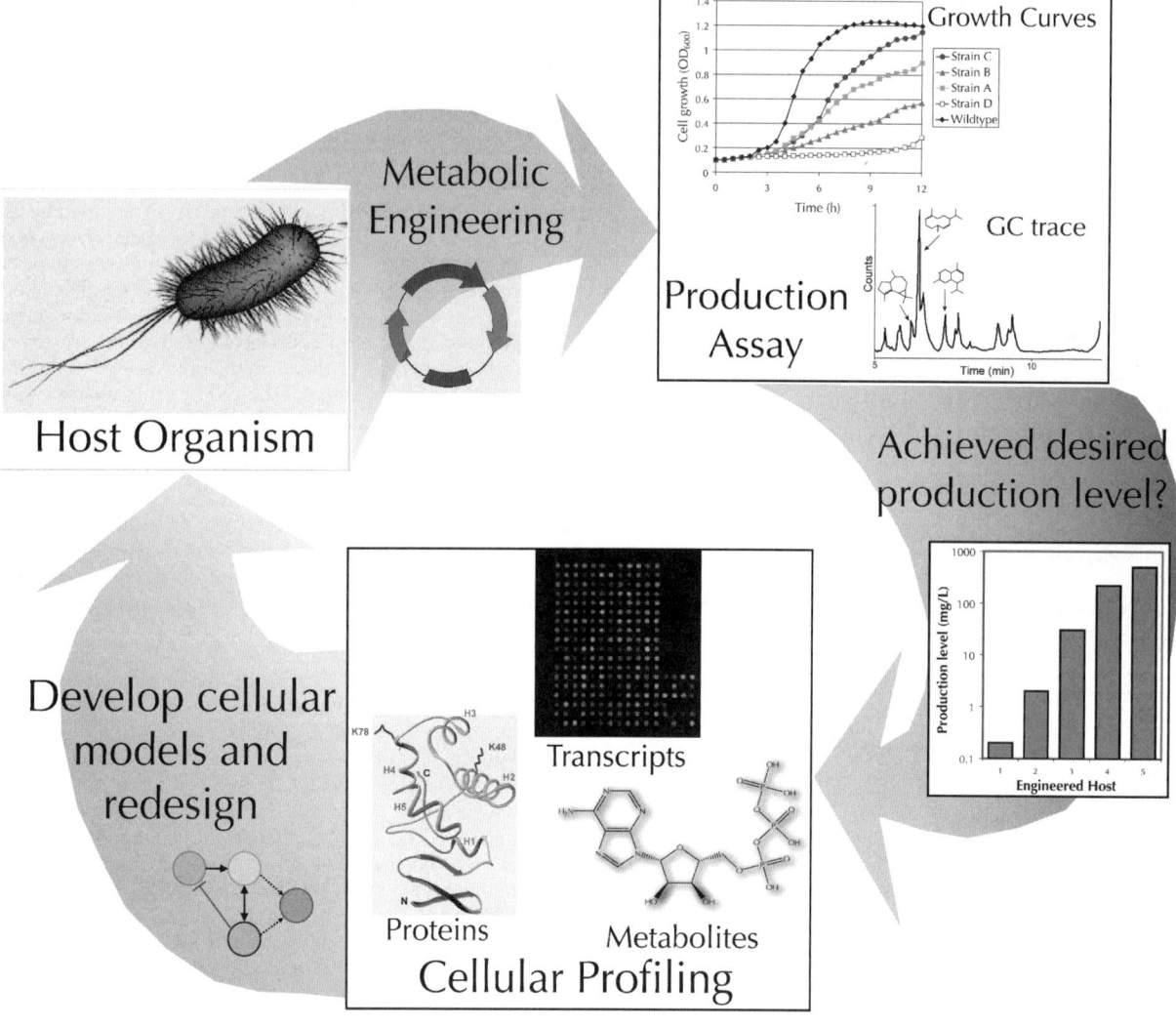

FIGURE 7 Flow chart of systems biology applied to achieving a production target. A host organism is selected and metabolically engineered to produce the molecule(s) of interest. The results of metabolic engineering are evaluated, for example, by monitoring phenotypes demonstrated by growth curves and product titers determined by GC. Cellular profiling is performed using measurements at all levels of the system, including mRNA, proteins, and metabolites, and flux analysis to identify bottlenecks in the pathway. Once bottlenecks are identified, another round of engineering is carried out to overcome the limitations. The cycle is completed once the desired production level has been achieved. OD$_{600}$, optical density at 600 nm. Reprinted from *Current Opinion in Biotechnology* (60) with permission of the publisher.

of microorganisms for efficient production of high levels of high-value compounds requires a comprehensive understanding of both the metabolite biosynthesis pathway and the native metabolism of the host. Combining targeted analyses with systems-level experiments offers the greatest chance to gain such an understanding. "Omics" analyses were used to examine the impacts of the increased metabolic burden of plasmid maintenance or protein expression, the accumulation of pathway intermediates, and the significant consumption of resources (e.g., carbon, ATP, and cofactors) during the production of amorpha-4,11-diene in an attempt to circumvent the corresponding constraints.

25.4.1. Transcriptomics Analysis of a Mevalonate-Producing Strain

Work by Pitera et al. (69), described in section 25.3.2, revealed that the accumulation of HMG-CoA is toxic to *E. coli* and that balancing the levels of HMGS and HMGR expression is necessary for robust growth and high mevalonate titers. Complementary to that work, transcriptomics analysis was used to elucidate which essential aspect of *E. coli* metabolism is impacted by HMG-CoA accumulation (44). To determine the stresses that come from flux through the pathway as opposed to the stress due to protein overproduction alone, the strain harboring a catalytically inactivated HMGS gene [on pMevT(C159A); discussed in section 25.3.2 above] was used as a control. The activated- and inactivated-pathway strains were analyzed separately over a time course and also subjected to cross comparison at the same time points (e.g., the activated-pathway strain versus the inactivated-pathway strain at time point 1). At time zero, just prior to induction, the transcriptomes for the active- and inactive-pathway strains were quite similar; however, several distinct differences were apparent postinduction. Several transcripts encoding fatty acid biosynthesis (FAB) proteins were upregulated over the time course (Fig. 8), as well as in the active-pathway strain compared to the inactive-pathway strain, in a pattern similar to that observed during FAB inhibition in *Mycobacterium tuberculosis* (12). Among the products of upregulated FAB genes (*fabB*, *fabD*, and *fabH*), only FabD interacts with malonyl-CoA, which accumulates along with HMG-CoA. The transcripts for biotin biosynthetic enzymes, which are associated with the initial steps of FAB, were also upregulated. Follow-up analysis of the impact of mevalonate production on FAB via fatty acid methyl ester analysis revealed a shift in fatty acid composition from saturated to unsaturated fatty acids in the active-pathway strain, consistent with observations of FAB inhibition and CoA depletion by Jackowski and Rock (36). Overall, the total amount of fatty acids as a percentage of dry cell weight decreased in the active-pathway strain. Medium supplementation with saturated fatty acids was capable of reversing the growth defect observed for the pMevT active-pathway strain.

25.4.2. Transcriptomics and Proteomics Analyses of an Amorpha-4,11-Diene-Producing Strain

Like genomics before it, proteomics began with relatively small sample sets and targeted analyses, yielding, as technology progressed, analyses of complex mixtures that represent the entire proteome for an organism, organelle, or other biological source (e.g., biofluid or tissue). Proteomics analysis represents the effort to establish the identities of all proteins present in an organism under experimental conditions. Significant strides toward complete proteome analysis have been made in the last 15 years. Liquid chromatography coupled to MS provides a highly sensitive analytical method to analyze proteins from complex cell lysates. To enhance the ability to detect low-abundance proteins, such studies require extensive sample preparation, separation, and/or purification prior to mass spectrometric analysis. The strengths of MS, such as high-throughput capabilities, facile automation, and rapid protein identification, are well suited for the complex mixture analysis that large-scale proteomics requires. For metabolic engineering, quantitative proteomics is useful to compare the proteomes of several strains under a variety of conditions. Such comparisons produce insight into the bacterial responses to metabolite stresses or the expression of exogenous proteins, complementing transcriptomics and metabolite analyses. Transcriptomics analysis of the pMevT strain that experienced HMG-CoA toxicity (see above) revealed clearly different profiles from those of strains not showing toxicity. The differences manifested themselves most clearly via growth defects. A more subtle experiment was the comparison of an amorphadiene-producing strain (harboring pMevT, pMBIS, and pADS) with its inactive analog, which consists of the same plasmids but with a point mutation (C159A) in the HMGS gene such that the gene is catalytically inactivated. Thus, following induction, protein is produced in the active and inactive strains; however, the inactive analog does not feel the stress of metabolic flux through the pathway. The three-plasmid-containing strain produces significant amounts of amorpha-4,11-diene without a growth defect. Yet results from gene titration experiments have shown that the system is not yet optimized. Consequently, cell-wide quantitative proteomics experiments were used to verify the observations. Liquid chromatography-tandem MS analysis of the three-plasmid system confirmed that the levels of production of HMGR and mevalonate kinase over a 24-h time course were significantly lower than those of the other pathway enzymes. By using transcriptomics and proteomics, several native *E. coli* processes (e.g., C1 biosynthesis and amino acid biosynthesis) were observed to change drastically in the active strain compared to those in the inactive strain. Follow-up experiments to characterize the effects of various medium supplements on mevalonate and amorpha-4,11-diene production are currently under way.

25.5. BIOSYNTHETIC OXIDATION OF AMORPHA-4,11-DIENE

25.5.1. Identification of the Native Amorpha-4,11-Diene Oxidase Gene

Production of high levels of amorpha-4,11-diene was an important step toward biosynthetically derived artemisinin. Yet three selective oxidation steps and one reduction step are required to convert amorpha-4,11-diene to dihydroartemisinic acid, the final enzymatic product prior to artemisinin. Chemical synthesis of artemisinin from amorpha-4,11-diene is feasible, yet chemical conversion of artemisinic acid can be achieved in only two steps with yields exceeding 30% (74). Consequently, oxidation of amorpha-4,11-diene to artemisinic acid via tailoring of enzymes is economically desirable. Ro et al. undertook an effort to identify the native enzyme or enzymes that convert amorpha-4,11-diene to artemisinic acid (70). In 2005, Bertea and coworkers (11) proposed that the enzymes responsible for the oxidation were cytochrome P450 monooxygenases. These enzymes catalyze a wide variety of reactions including hydroxylation, epoxidation, dehydrogenation, and other oxidations, yet such reactions are often highly specific for a particular substrate. The high degree of sequence variability found in P450 enzymes complicates

	1 hr	3 hr
accB	2.9 (4.3)	3.3 (4.3)
cfa	1.4 (2.2)	3.1 (5.4)
bioD	1.6 (2.2)	3.1 (5.3)
accC	2.4 (3.1)	3.0 (4.2)
fabD	1.1 (0.9)	2.6 (4.8)
fabB	2.3 (3.4)	2.6 (4.5)
bioA	1.3 (1.4)	2.4 (3.5)
bioB	1.9 (2.7)	2.2 (4.1)
fabH	1.7 (2.0)	2.1 (3.9)

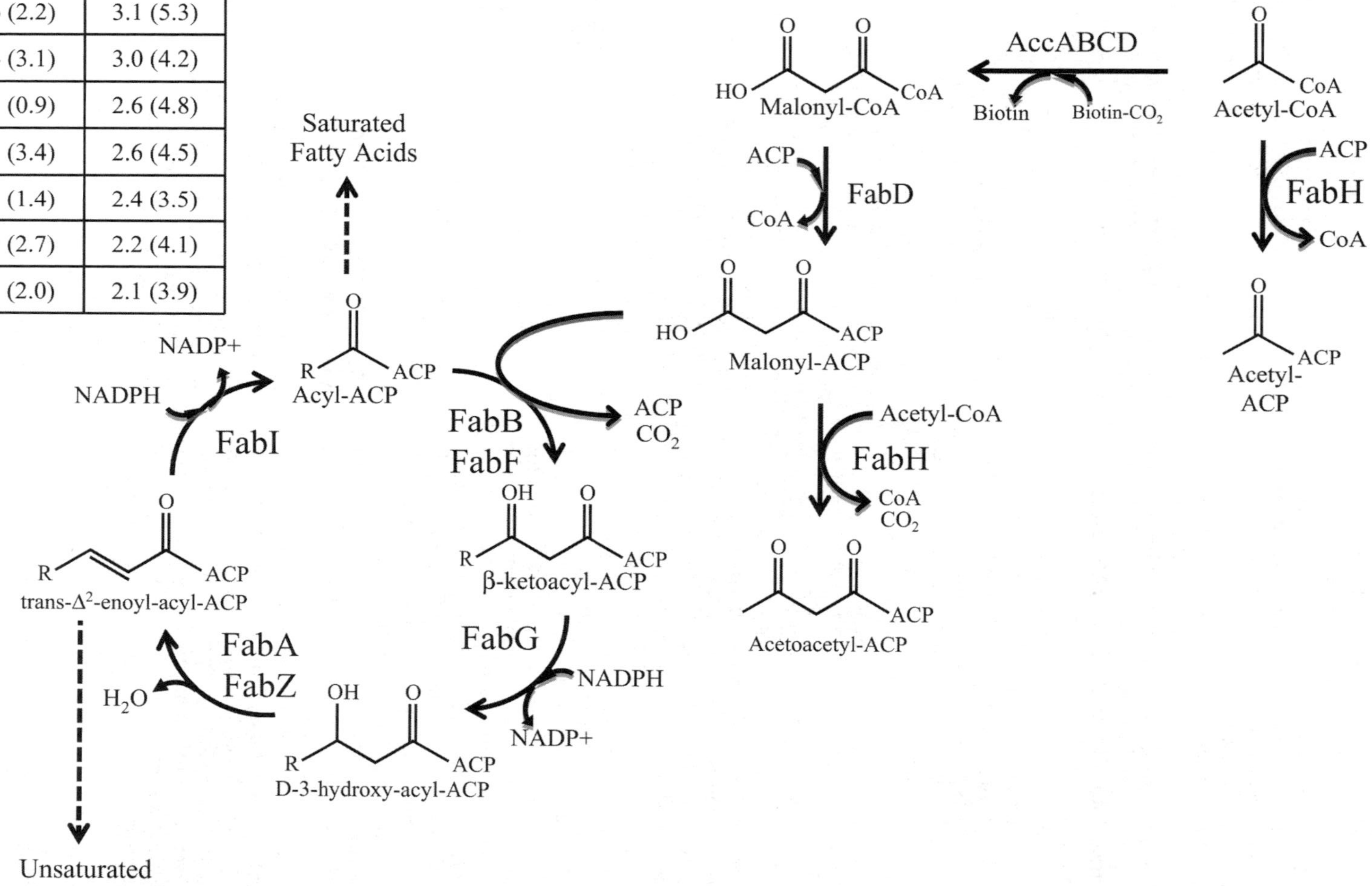

FIGURE 8 Transcript profiles of the initial steps of type II FAB in *E. coli*. Malonyl-CoA is synthesized from acetyl-CoA by the action of acetyl-CoA carboxylase, a heterotetramer composed of subunits encoded by *accABCD*. The malonate moiety is transferred from CoA to the acyl carrier protein (ACP) by the action of malonyl-CoA:ACP transacylase (FabD). Also shown (inset) are the expression values and Z scores (in parentheses) for FAB genes that exhibited biologically significant upregulation in the mevalonate-producing strain (*E. coli* DP10 containing pBAD33MevT and pBAD18) relative to the inactive-pathway control strain [*E. coli* DP10 containing pMevT(C159A) and pBAD18] in the microarray analysis. Values for the control strain were set at 1.0. Adapted from *Applied and Environmental Microbiology* (44).

the identification of specific P450 enzymes by homology. To narrow the search, Ro and coworkers undertook comparative genomic analyses of *A. annua*, chicory, lettuce, and sunflower, all members of the Asteraceae family, because they produce sesquiterpene lactones such as artemisinin. Ro et al. (70) obtained P450 expressed sequence tag libraries to target CYP71 and CYP82 P450 families in lettuce and sunflower. By using CYP-specific primers, several fragments were isolated from an *A. annua* trichome-enriched complementary DNA library. The product of the full-length open reading frame (CYP71AV1) from one of the fragments was found to oxidize amorpha-4,11-diene to artemisinic acid (up to 100 mg/liter) when expressed with its cytochrome P450 reductase (CPR) redox partner in *S. cerevisiae* (70). In vitro experiments showed that the isolated amorphadiene oxidase (CYP71AV1) catalyzed all three oxidations of amorpha-4,11-diene to artemisinic alcohol, artemisinic aldehyde, and artemisinic acid. Concurrently, Teoh et al. used expressed sequence tag libraries generated from *A. annua* trichomes to identify CYP71AV1 for amorpha-4,11-diene oxidation (84). Subsequent work by Zhang and coworkers (96) identified an artemisinic aldehyde Δ11(13) reductase which catalyzes the reduction of artemisinic aldehyde to dihydroartemisinic aldehyde prior to the formation of dihydroartemisinic acid.

25.5.2. Expression of the Native Amorpha-4,11-Diene Oxidase Gene in *E. coli*

Despite successful production of artemisinic acid in *S. cerevisiae*, bio-oxidation of amorpha-4,11-diene in *E. coli* was desirable (16). Yet biotechnological use of plant P450s in bacteria has met with limited success and many difficulties. Functional expression of membrane-bound plant P450s in *E. coli* is hampered by problems associated with protein folding, membrane insertion, posttranslational modification, and cofactor utilization. Indeed, when native CYP71AV1-CPR was expressed in *E. coli*, no oxidation of amorpha-4,11-diene was detectable (15). Consequently, another plant cytochrome P450, 8-cadinene hydroxylase (CAH), was used to optimize functional expression in *E. coli* for application to amorphadiene oxidase. Initial production titers of 8-hydroxycadinene were low (0.4 mg/liter), but following addition of the lower mevalonate pathway and codon optimization for *E. coli*, production was increased to 25 mg/liter. Notably, when CAH was expressed with the CPR redox partner for the amorphadiene oxidase, production decreased 1.8-fold, indicating that interactions between the oxidase and reductase play a significant role in activity.

Increasing the solubility and stability of plant P450s in bacteria is a particular challenge. Previous studies have shown that modification of the membrane anchor can have a dramatic effect on enzymatic activity (9, 75, 77, 82). Consequently, several other anchors that had been previously characterized with respect to bacterial expression (9, 19, 73, 77, 79, 82) were used to replace the N-terminal portion of CAH. The production of 8-hydroxycadinene improved with all of the modifications, up to 105 ± 7 mg/liter with the anchor optimized for the expression of 17α-hydroxylase (9). Chang et al. applied the codon optimization and N-terminal transmembrane engineering methods to amorphadiene oxidase, yielding constructs capable of producing small amounts (5.6 mg/liter) of artemisinic alcohol (15). The oxidized amorpha-4,11-diene metabolite yield was further increased by switching from the pETDUET-1 vector to a P450-specific vector, pCWori (9). Changing to the pCWori vector also resulted in measurable amounts of artemisinic acid; however, artemisinic alcohol and aldehyde remained the dominant products. By removing the dodecane layer, which sequestered the substrate and less-oxidized products, the artemisinic acid level was increased to 105 ± 10 mg/liter.

25.5.3. Engineering of *B. megaterium* P450 BM-3 To Oxidize Amorpha-4,11-Diene

Much of the effort associated with increasing amorphadiene oxidase activity in *E. coli* involved techniques to stabilize the native enzyme. An alternative method to achieving specific oxidation of amorpha-4,11-diene is engineering a bacterial P450 through mutagenesis to produce the desired product. Yoshikuni et al. recently mutagenized (+)-δ-cadinene synthase by using error-prone PCR and site-directed saturation mutagenesis to achieve higher specificity and activity (93). This method is attractive because the issues associated with expressing plant P450 enzymes can be avoided through judicious choice of the starting P450 enzyme. The bacterial P450 enzyme from *Bacillus megaterium*, P450 BM-3, is ideal because it benefits from good solubility and a high catalytic rate and is self-contained; thus, coexpression of a redox partner is not required (66). Additionally, the structure and activity of BM-3 have been extensively studied (33, 61, 65, 88), and the product specificity has been engineered previously for the oxidation of a wide variety of substrates (6, 14, 24, 28, 33, 52, 53). To increase the binding pocket for amorpha-4,11-diene, the wild-type BM-3 was mutated at residue Phe87 by Dietrich and coworkers (23). When the engineered BM-3 was incorporated into a strain that produced amorpha-4,11-diene, they observed a product with an electron impact mass spectrum and a retention time identical to those of an artemisinic-11S,12-epoxide (23). While the artemisinic epoxide is not an intermediate in the native artemisinin biosynthesis pathway, a high-yielding chemical synthetic route exists to convert the epoxide into dihydroartemisinic acid, the final enzymatic step prior to the formation of artemisinin (18, 84). Subsequent engineering of BM-3 production yielded more than 250 mg/liter artemisinic epoxide.

25.6. SUMMARY

The recent increase in multidrug-resistant strains of *Plasmodium falciparum* has pushed ACTs to the front of the worldwide fight against malaria. Development of artemisinin sources that are stable, sufficient to meet the growing need, and affordable represents the most significant challenge for combating malaria. Metabolic engineering of *E. coli* to meet these needs has progressed significantly in the last 8 years. The genes to convert acetyl-CoA to artemisinic acid have been cloned and expressed in *E. coli*. Ultimately, improvements in the titers of the end product (artemisinic acid) as well as the intermediates of the pathway have come from the manipulation of expression levels of the pathway genes or changes in copy numbers of the plasmids on which they are carried. Titers of amorpha-4,11-diene, the first committed step toward artemisinin, have been improved from levels in micrograms per liter in original strains to 0.3 g/liter in an optimized strain (5) and over 0.4 g/liter in cultures in a partitioning bioreactor (62). Furthering this work, Amyris Biotechnologies recently reported that they achieved commercially relevant titers of amorpha-4,11-diene (27.4 g/liter) in fed-batch fermentations (85). Careful selection of the promoter strength, plasmid, and gene copy number led to improved enzyme expression levels, reduced bottlenecks, balanced metabolite flux, and ultimately, increased titers of amorpha-4,11-diene. "Omics" analyses proved to be useful means for identifying bottlenecks in

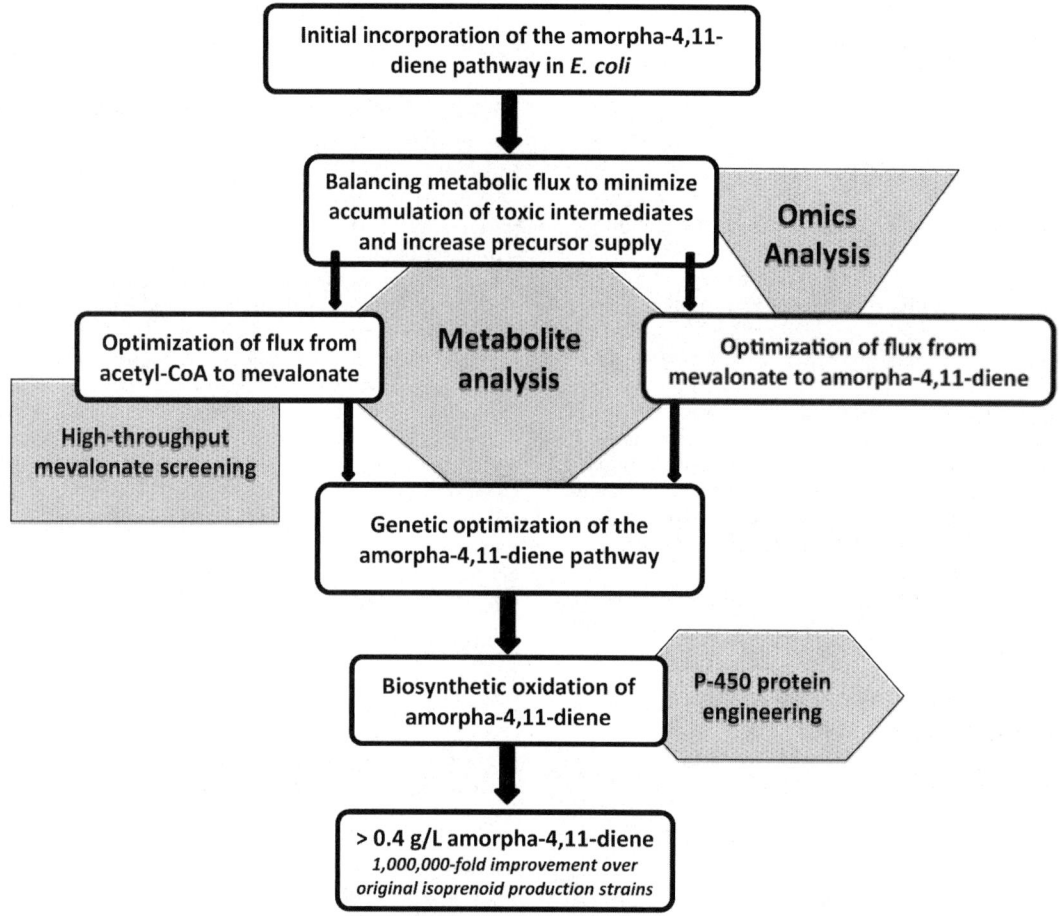

FIGURE 9 Flow chart of metabolic engineering efforts for the high-level production of precursors to artemisinin.

the flux of intermediates through the pathway, enabling a series of subsequent directed genetic manipulations that proved effective in improving the titers. This multifaceted approach is outlined in Fig. 9.

Identification of the native amorpha-4,11-diene oxidase revealed that amorpha-4,11-diene was oxidized in three steps to artemisinic acid. The gene for this P450 enzyme was successfully cloned into *S. cerevisiae*, but unfortunately, expression of amorphadiene oxidase in yeast was not predictive of success in *E. coli*. Improvements in the expression of plant P450 enzymes in *E. coli* were made through codon optimization and N-terminal anchor modification to ultimately produce over 0.1 g/liter of artemisinic acid (15). An alternate means of amorpha-4,11-diene oxidation by engineering a substrate-promiscuous P450 was developed to complement the native-enzyme efforts. Protein engineering of *B. megaterium* P450 BM-3 yielded selective oxidation of amorpha-4,11-diene at a high rate to produce artemisinic epoxide at titers of 250 mg/liter. Taken as a whole, past efforts have led to significant progress in engineering *E. coli* for the production of high levels of artemisinin precursors through a variety of traditional and modern methods.

REFERENCES

1. Abdin, M. Z., M. Israr, R. U. Rehman, and S. K. Jain. 2003. Artemisinin, a novel antimalarial drug: biochemical and molecular approaches for enhanced production. *Planta Med.* **69:**289–299.

2. Ajikumar, P. K., K. Tyo, S. Carlsen, O. Mucha, T. H. Phon, and G. Stephanopoulos. 2008. Terpenoids: opportunities for biosynthesis of natural product drugs using engineered microorganisms. *Mol. Pharm.* **5:**167–190.

3. Alper, H., C. Fischer, E. Nevoigt, and G. Stephanopoulos. 2005. Tuning genetic control through promoter engineering. *Proc. Natl. Acad. Sci. USA* **102:**12678–12683.

4. Alper, H., K. Miyaoku, and G. Stephanopoulos. 2005. Construction of lycopene-overproducing *E. coli* strains by combining systematic and combinatorial gene knockout targets. *Nat. Biotechnol.* **23:**612–616.

5. Anthony, J. R., L. C. Anthony, F. Nowroozi, G. Kwon, J. D. Newman, and J. D. Keasling. 2009. Optimization of the mevalonate-based isoprenoid biosynthetic pathway in *Escherichia coli* for production of the anti-malarial drug precursor amorpha-4,11-diene. *Metab. Eng.* **11:**13–19.

6. Appel, D., S. Lutz-Wahl, P. Fischer, U. Schwaneberg, and R. D. Schmid. 2001. A P450 BM-3 mutant hydroxylates alkanes, cycloalkanes, arenes and heteroarenes. *J. Biotechnol.* **88:**167–171.

7. Balint, G. A. 2001. Artemisinin and its derivatives: an important new class of antimalarial agents. *Pharmacol. Ther.* **90:**261–265.

8. Baneyx, F. 1999. Recombinant protein expression in *Escherichia coli. Curr. Opin. Biotechnol.* **10:**411–421.

9. Barnes, H. J., M. P. Arlotto, and M. R. Waterman. 1991. Expression and enzymatic activity of recombinant cytochrome P450 17 alpha-hydroxylase in *Escherichia coli. Proc. Natl. Acad. Sci. USA* **88:**5597–5601.

10. Barrick, D., K. Villanueba, J. Childs, R. Kalil, T. D. Schneider, C. E. Lawrence, L. Gold, and G. D. Stormo. 1994. Quantitative analysis of ribosome binding sites in *E. coli. Nucleic Acids Res.* **22:**1287–1295.

11. Bertea, C. M., J. R. Freije, H. van der Woude, F. W. Verstappen, L. Perk, V. Marquez, J. De Kraker, M. A. Posthumus, B. J. Jansen, A. de Groot, M. C. Franssen, and H. J. Bouwmeester. 2005. Identification of intermediates and enzymes involved in the early steps of artemisinin biosynthesis in *Artemisia annua. Planta Med.* **71:**40–47.

12. Betts, J. C., A. McLaren, M. G. Lennon, F. M. Kelly, P. T. Lukey, S. J. Blakemore, and K. Duncan. 2003. Signature gene expression profiles discriminate between isoniazid-, thiolactomycin-, and triclosan-treated *Mycobacterium tuberculosis. Antimicrob. Agents Chemother.* **47:**2903–2913.

13. Bouwmeester, H. J., T. E. Wallaart, M. H. Janssen, B. van Loo, B. J. Jansen, M. A. Posthumus, C. O. Schmidt, J. W. De Kraker, W. A. König, and M. C. Franssen. 1999. Amorpha-4,11-diene synthase catalyses the first probable step in artemisinin biosynthesis. *Phytochemistry* **52:**843–854.

14. Carmichael, A. B., and L. L. Wong. 2001. Protein engineering of *Bacillus megaterium* CYP102. The oxidation of polycyclic aromatic hydrocarbons. *Eur. J. Biochem.* **268:**3117–3125.

15. Chang, M. C. Y., R. A. Eachus, W. Trieu, D. Ro, and J. D. Keasling. 2007. Engineering *Escherichia coli* for production of functionalized terpenoids using plant P450s. *Nat. Chem. Biol.* **3:**274–277.

16. Chang, M. C. Y., and J. D. Keasling. 2006. Production of isoprenoid pharmaceuticals by engineered microbes. *Nat. Chem. Biol.* **2:**674–681.

17. Connolly, D. M., and M. E. Winkler. 1989. Genetic and physiological relationships among the *miaA* gene, 2-methylthio-N^6-(Δ^2-isopentenyl)-adenosine tRNA modification, and spontaneous mutagenesis in *Escherichia coli* K-12. *J. Bacteriol.* **171:**3233–3246.

18. Covello, P. S. 2008. Making artemisinin. *Phytochemistry* **69:**2881–2885.

19. Craft, D. L., K. M. Madduri, M. Eshoo, and C. R. Wilson. 2003. Identification and characterization of the CYP52 family of *Candida tropicalis* ATCC 20336, important for the conversion of fatty acids and alkanes to α,ω-dicarboxylic acids. *Appl. Environ. Microbiol.* **69:**5983–5991.

20. Das, A., S. Yoon, S. Lee, J. Kim, D. Oh, and S. Kim. 2007. An update on microbial carotenoid production: application of recent metabolic engineering tools. *Appl. Microbiol. Biotechnol.* **77:**505–512.

21. de Smit, M. H., and J. van Duin. 1994. Control of translation by mRNA secondary structure in *Escherichia coli*. A quantitative analysis of literature data. *J. Mol. Biol.* **244:**144–150.

22. de Smit, M. H., and J. van Duin. 1990. Secondary structure of the ribosome binding site determines translational efficiency: a quantitative analysis. *Proc. Natl. Acad. Sci. USA* **87:**7668–7672.

23. Dietrich, J., Y. Yoshikuni, K. Fisher, F. Woolard, D. Ockey, D. McPhee, N. Renninger, M. Chang, D. Baker, and J. D. Keasling. 2009. A novel semi-biosynthetic route for artemisinin production using engineered substrate-promiscuous P450(BM3). *ACS Chem. Biol.* **4:**261–267.

24. Dietrich, M., T. A. Do, R. D. Schmid, J. Pleiss, and V. B. Urlacher. 2009. Altering the regioselectivity of the subterminal fatty acid hydroxylase P450 BM-3 towards gamma- and delta-positions. *J. Biotechnol.* **139:**115–117.

25. Eisenreich, W., F. Rohdich, and A. Bacher. 2001. Deoxyxylulose phosphate pathway to terpenoids. *Trends Plant Sci.* **6:**78–84.

26. Enserink, M. 2005. Infectious diseases. Source of new hope against malaria is in short supply. *Science* **307:**33.

27. Flores, S., R. de Anda-Herrera, G. Gosset, and F. G. Bolívar. 2004. Growth-rate recovery of *Escherichia coli* cultures carrying a multicopy plasmid, by engineering of the pentose-phosphate pathway. *Biotechnol. Bioeng.* **87:**485–494.

28. Glieder, A., E. T. Farinas, and F. H. Arnold. 2002. Laboratory evolution of a soluble, self-sufficient, highly active alkane hydroxylase. *Nat. Biotechnol.* **20:**1135–1139.

29. Guzman, L. M., D. Belin, M. J. Carson, and J. Beckwith. 1995. Tight regulation, modulation, and high-level expression by vectors containing the arabinose P$_{BAD}$ promoter. *J. Bacteriol.* **177:**4121–4130.

30. Hammer, K., I. Mijakovic, and P. R. Jensen. 2006. Synthetic promoter libraries—tuning of gene expression. *Trends Biotechnol.* **24:**53–55.

31. Harker, M., and P. M. Bramley. 1999. Expression of prokaryotic 1-deoxy-D-xylulose-5-phosphatases in *Escherichia coli* increases carotenoid and ubiquinone biosynthesis. *FEBS Lett.* **448:**115–119.

32. Haynes, R. K. 2006. From artemisinin to new artemisinin antimalarials: biosynthesis, extraction, old and new derivatives, stereochemistry and medicinal chemistry requirements. *Curr. Top. Med. Chem.* **6:**509–537.

33. Hilker, B. L., H. Fukushige, C. Hou, and D. Hildebrand. 2008. Comparison of *Bacillus* monooxygenase genes for unique fatty acid production. *Prog. Lipid Res.* **47:**1–14.

34. Hofheinz, W., H. Bürgin, E. Gocke, C. Jaquet, R. Masciadri, G. Schmid, H. Stohler, and H. Urwyler. 1994. Ro 42-1611 (arteflene), a new effective antimalarial: chemical structure and biological activity. *Trop. Med. Parasitol.* **45:**261–265.

35. Hunter, W. N. 2007. The non-mevalonate pathway of isoprenoid precursor biosynthesis. *J. Biol. Chem.* **282:**21573–21577.

36. Jackowski, S., and C. O. Rock. 1986. Consequences of reduced intracellular coenzyme A content in *Escherichia coli. J. Bacteriol.* **166:**866–871.

37. Jana, S., and J. K. Deb. 2005. Strategies for efficient production of heterologous proteins in *Escherichia coli. Appl. Microbiol. Biotechnol.* **67:**289–298.

38. Jensen, P. R., and K. Hammer. 1998. Artificial promoters for metabolic optimization. *Biotechnol. Bioeng.* **58:**191–195.

39. Jones, K. L., S. W. Kim, and J. D. Keasling. 2000. Low-copy plasmids can perform as well as or better than high-copy plasmids for metabolic engineering of bacteria. *Metab. Eng.* **2:**328–338.

40. Kang, M. J., Y. M. Lee, S. H. Yoon, J. H. Kim, S. W. Ock, K. H. Jung, Y. C. Shin, J. D. Keasling, and S. W. Kim. 2005. Identification of genes affecting lycopene accumulation in *Escherichia coli* using a shot-gun method. *Biotechnol. Bioeng.* **91:**636–642.

41. Keasling, J. D. 2008. Synthetic biology for synthetic chemistry. *ACS Chem. Biol.* **3:**64–76.

42. Kim, S. W., and J. D. Keasling. 2001. Metabolic engineering of the nonmevalonate isopentenyl diphosphate synthesis pathway in *Escherichia coli* enhances lycopene production. *Biotechnol. Bioeng.* **72:**408–415.

43. Kirby, J., and J. D. Keasling. 2008. Metabolic engineering of microorganisms for isoprenoid production. *Nat. Prod. Rep.* **25:**656–661.

44. Kizer, L., D. J. Pitera, B. F. Pfleger, and J. D. Keasling. 2008. Application of functional genomics to pathway optimization for increased isoprenoid production. *Appl. Environ. Microbiol.* **74:**3229–3241.

45. Klayman, D. L. 1985. Qinghaosu (artemisinin): an antimalarial drug from China. *Science* **228:**1049–1055.

46. Klayman, D. L., A. J. Lin, N. Acton, J. P. Scovill, J. M. Hoch, W. K. Milhous, A. D. Theoharides, and A. S. Dobek. 1984. Isolation of artemisinin (qinghaosu) from *Artemisia annua* growing in the United States. *J. Nat. Prod.* **47:**715–717.

47. Klein-Marcuschamer, D., P. K. Ajikumar, and G. Stephanopoulos. 2007. Engineering microbial cell factories for biosynthesis of isoprenoid molecules: beyond lycopene. *Trends Biotechnol.* **25**:417–424.

48. Koeller, K. M., and C. H. Wong. 2001. Enzymes for chemical synthesis. *Nature* **409**:232–240.

49. Kuchner, O., and F. H. Arnold. 1997. Directed evolution of enzyme catalysts. *Trends Biotechnol.* **15**:523–530.

50. Kuzuyama, T., S. Takahashi, and H. Seto. 1999. Construction and characterization of *Escherichia coli* disruptants defective in the yaeM gene. *Biosci. Biotechnol. Biochem.* **63**:776–778.

51. Kuzuyama, T. 2002. Mevalonate and nonmevalonate pathways for the biosynthesis of isoprene units. *Biosci. Biotechnol. Biochem.* **66**:1619–1627.

52. Landwehr, M., L. Hochrein, C. R. Otey, A. Kasrayan, J. Bäckvall, and F. H. Arnold. 2006. Enantioselective alpha-hydroxylation of 2-arylacetic acid derivatives and buspirone catalyzed by engineered cytochrome P450 BM-3. *J. Am. Chem. Soc.* **128**:6058–6059.

53. Li, Q. S., U. Schwaneberg, P. Fischer, and R. D. Schmid. 2000. Directed evolution of the fatty-acid hydroxylase P450 BM-3 into an indole-hydroxylating catalyst. *Chemistry* **6**:1531–1536.

54. Liu, C., Y. Zhao, and Y. Wang. 2006. Artemisinin: current state and perspectives for biotechnological production of an antimalarial drug. *Appl. Microbiol. Biotechnol.* **72**:11–20.

55. Martin, V. J., Y. Yoshikuni, and J. D. Keasling. 2001. The in vivo synthesis of plant sesquiterpenes by *Escherichia coli*. *Biotechnol. Bioeng.* **75**:497–503.

56. Martin, V. J. J., D. J. Pitera, S. T. Withers, J. D. Newman, and J. D. Keasling. 2003. Engineering a mevalonate pathway in *Escherichia coli* for production of terpenoids. *Nat. Biotechnol.* **21**:796–802.

57. Martinez-Morales, F., A. C. Borges, A. Martinez, K. T. Shanmugam, and L. O. Ingram. 1999. Chromosomal integration of heterologous DNA in *Escherichia coli* with precise removal of markers and replicons used during construction. *J. Bacteriol.* **181**:7143–7148.

58. McNeil, D. November 14, 2004. Plant shortage leaves campaigns against malaria at risk. *N. Y. Times* vol. CLIV.

59. Mercke, P., M. Bengtsson, H. J. Bouwmeester, M. A. Posthumus, and P. E. Brodelius. 2000. Molecular cloning, expression, and characterization of amorpha-4,11-diene synthase, a key enzyme of artemisinin biosynthesis in *Artemisia annua* L. *Arch. Biochem. Biophys.* **381**:173–180.

60. Mukhopadhyay, A., A. M. Redding, B. J. Rutherford, and J. D. Keasling. 2008. Importance of systems biology in engineering microbes for biofuel production. *Curr. Opin. Biotechnol.* **19**:228–234.

61. Munro, A. W., D. G. Leys, K. J. McLean, K. R. Marshall, T. W. B. Ost, S. Daff, C. S. Miles, S. K. Chapman, D. A. Lysek, C. C. Moser, C. C. Page, and P. L. Dutton. 2002. P450 BM3: the very model of a modern flavocytochrome. *Trends Biochem. Sci.* **27**:250–257.

62. Newman, J. D., J. Marshall, M. Chang, F. Nowroozi, E. Paradise, D. Pitera, K. L. Newman, and J. D. Keasling. 2006. High-level production of amorpha-4,11-diene in a two-phase partitioning bioreactor of metabolically engineered *Escherichia coli*. *Biotechnol. Bioeng.* **95**:684–691.

63. Newman, K. L., R. P. P. Almeida, A. H. Purcell, and S. E. Lindow. 2004. Cell-cell signaling controls *Xylella fastidiosa* interactions with both insects and plants. *Proc. Natl. Acad. Sci. USA* **101**:1737–1742.

64. Newton, P., and N. White. 1999. Malaria: new developments in treatment and prevention. *Annu. Rev. Med.* **50**:179–192.

65. Noble, M. A., C. S. Miles, S. K. Chapman, D. A. Lysek, A. C. MacKay, G. A. Reid, R. P. Hanzlik, and A. W. Munro. 1999. Roles of key active-site residues in flavocytochrome P450 BM3. *Biochem. J.* **339**(Pt. 2):371–379.

66. Peters, M. W., P. Meinhold, A. Glieder, and F. H. Arnold. 2003. Regio- and enantioselective alkane hydroxylation with engineered cytochromes P450 BM-3. *J. Am. Chem. Soc.* **125**:13442–13450.

67. Pfleger, B. F., D. J. Pitera, J. D. Newman, V. J. J. Martin, and J. D. Keasling. 2007. Microbial sensors for small molecules: development of a mevalonate biosensor. *Metab. Eng.* **9**:30–38.

68. Pfleger, B. F., D. J. Pitera, C. D. Smolke, and J. D. Keasling. 2006. Combinatorial engineering of intergenic regions in operons tunes expression of multiple genes. *Nat. Biotechnol.* **24**:1027–1032.

69. Pitera, D. J., C. J. Paddon, J. D. Newman, and J. D. Keasling. 2007. Balancing a heterologous mevalonate pathway for improved isoprenoid production in *Escherichia coli*. *Metab. Eng.* **9**:193–207.

70. Ro, D., E. M. Paradise, M. Ouellet, K. J. Fisher, K. L. Newman, J. M. Ndungu, K. A. Ho, R. A. Eachus, T. S. Ham, J. Kirby, M. C. Y. Chang, S. T. Withers, Y. Shiba, R. Sarpong, and J. D. Keasling. 2006. Production of the antimalarial drug precursor artemisinic acid in engineered yeast. *Nature* **440**:940–943.

71. Rohlin, L., M. K. Oh, and J. C. Liao. 2001. Microbial pathway engineering for industrial processes: evolution, combinatorial biosynthesis and rational design. *Curr. Opin. Microbiol.* **4**:330–335.

72. Rokosz, L. L., D. A. Boulton, E. A. Butkiewicz, G. Sanyal, M. A. Cueto, P. A. Lachance, and J. D. Hermes. 1994. Human cytoplasmic 3-hydroxy-3-methylglutaryl coenzyme A synthase: expression, purification, and characterization of recombinant wild-type and Cys129 mutant enzymes. *Arch. Biochem. Biophys.* **312**:1–13.

73. Roosild, T. P., J. Greenwald, M. Vega, S. Castronovo, R. Riek, and S. Choe. 2005. NMR structure of Mistic, a membrane-integrating protein for membrane protein expression. *Science* **307**:1317–1321.

74. Roth, R. J., and N. Acton. 1989. A simple conversion of artemisinic acid into artemisinin. *J. Nat. Prod.* **52**:1183–1185.

75. Sandhu, P., Z. Guo, T. Baba, M. V. Martin, R. H. Tukey, and F. P. Guengerich. 1994. Expression of modified human cytochrome P450 1A2 in *Escherichia coli*: stabilization, purification, spectral characterization, and catalytic activities of the enzyme. *Arch. Biochem. Biophys.* **309**:168–177.

76. Santos, C. N. S., and G. Stephanopoulos. 2008. Combinatorial engineering of microbes for optimizing cellular phenotype. *Curr. Opin. Chem. Biol.* **12**:168–176.

77. Schafmeister, C. E., L. J. Miercke, and R. M. Stroud. 1993. Structure at 2.5 A of a designed peptide that maintains solubility of membrane proteins. *Science* **262**:734–738.

78. Schmid, G., and W. Hofheinz. 1983. Total synthesis of qinghaosu. *J. Am. Chem. Soc.* **105**:624–625.

79. Schoch, G. A., R. Attias, M. Belghazi, P. M. Dansette, and D. Werck-Reichhart. 2003. Engineering of a water-soluble plant cytochrome P450, CYP73A1, and NMR-based orientation of natural and alternate substrates in the active site. *Plant Physiol.* **133**:1198–1208.

80. Shimazu, M., L. Vetcher, J. L. Galazzo, P. Licari, and D. V. Santi. 2004. A sensitive and robust method for quantification of intracellular short-chain coenzyme A esters. *Anal. Biochem.* **328**:51–59.

81. Stephanopoulos, G., and J. J. Vallino. 1991. Network rigidity and metabolic engineering in metabolite overproduction. *Science* **252**:1675–1681.

82. Sueyoshi, T., L. J. Park, R. Moore, R. O. Juvonen, and M. Negishi. 1995. Molecular engineering of microsomal P450 2a-4 to a stable, water-soluble enzyme. *Arch. Biochem. Biophys.* **322**:265–271.

83. **Suthers, P. F., A. P. Burgard, M. S. Dasika, F. Nowroozi, S. Van Dien, J. D. Keasling, and C. D. Maranas.** 2007. Metabolic flux elucidation for large-scale models using 13C labeled isotopes. *Metab. Eng.* **9:**387–405.

84. **Teoh, K. H., D. R. Polichuk, D. W. Reed, G. Nowak, and P. S. Covello.** 2006. *Artemisia annua* L. (Asteraceae) trichome-specific cDNAs reveal CYP71AV1, a cytochrome P450 with a key role in the biosynthesis of the antimalarial sesquiterpene lactone artemisinin. *FEBS Lett.* **580:**1411–1416.

85. **Tsuruta, H., C. J. Paddon, D. Eng, J. R. Lenihan, T. Horning, L. C. Anthony, R. Regentin, J. D. Keasling, N. S. Renninger, and J. D. Newman.** 2009. High-level production of amorpha-4,11-diene, a precursor of the antimalarial agent artemisinin, in *Escherichia coli*. *PLoS ONE* **4:**e4489.

86. **Turner, N. J.** 2003. Directed evolution of enzymes for applied biocatalysis. *Trends Biotechnol.* **21:**474–478.

87. **Van Geldre, E., A. Vergauwe, and E. Van den Eeckhout.** 1997. State of the art of the production of the antimalarial compound artemisinin in plants. *Plant Mol. Biol.* **33:**199–209.

88. **Warman, A. J., O. Roitel, R. Neeli, H. M. Girvan, H. E. Seward, S. A. Murray, K. J. McLean, M. G. Joyce, H. Toogood, R. A. Holt, D. Leys, N. S. Scrutton, and A. W. Munro.** 2005. Flavocytochrome P450 BM3: an update on structure and mechanism of a biotechnologically important enzyme. *Biochem. Soc. Trans.* **33:**747–753.

89. **Withers, S. T., S. S. Gottlieb, B. Lieu, J. D. Newman, and J. D. Keasling.** 2007. Identification of isopentenol biosynthetic genes from *Bacillus subtilis* by a screening method based on isoprenoid precursor toxicity. *Appl. Environ. Microbiol.* **73:**6277–6283.

90. **Withers, S. T., and J. D. Keasling.** 2007. Biosynthesis and engineering of isoprenoid small molecules. *Appl. Microbiol. Biotechnol.* **73:**980–990.

91. **Xing-Xiang, X., Z. Jie, H. Da-Zhong, and Z. Wei-Shan.** 1986. Total synthesis of arteannuin and deoxyarteannuin. *Tetrahedron* **42:**819–828.

92. **Yoon, S., H. Park, J. Kim, S. Lee, M. Choi, J. Kim, D. Oh, J. D. Keasling, and S. Kim.** 2007. Increased beta-carotene production in recombinant *Escherichia coli* harboring an engineered isoprenoid precursor pathway with mevalonate addition. *Biotechnol. Prog.* **23:**599–605.

93. **Yoshikuni, Y., T. E. Ferrin, and J. D. Keasling.** 2006. Designed divergent evolution of enzyme function. *Nature* **440:**1078–1082.

94. **Yuan, L. Z., P. E. Rouvière, R. A. Larossa, and W. Suh.** 2006. Chromosomal promoter replacement of the isoprenoid pathway for enhancing carotenoid production in *E. coli*. *Metab. Eng.* **8:**79–90.

95. **Zhang, L., F. Jing, F. Li, M. Li, Y. Wang, G. Wang, X. Sun, and K. Tang.** 2009. Development of transgenic *Artemisia annua* (Chinese wormwood) plants with an enhanced content of artemisinin, an effective anti-malarial drug, by hairpin-RNA-mediated gene silencing. *Biotechnol. Appl. Biochem.* **52:**199–207.

96. **Zhang, Y., K. H. Teoh, D. W. Reed, L. Maes, A. Goossens, D. J. H. Olson, A. R. S. Ross, and P. S. Covello.** 2008. The molecular cloning of artemisinic aldehyde Δ11(13) reductase and its role in glandular trichome-dependent biosynthesis of artemisinin in *Artemisia annua*. *J. Biol. Chem.* **283:**21501–21508.

Heterologous Production of Polyketides in *Streptomyces coelicolor* and *Escherichia coli*

JAMES T. KEALEY

26

Polyketides are structurally diverse secondary metabolites produced in bacteria, fungi, and plants. As a class of natural products, polyketides have contributed to a wealth of valuable pharmaceuticals, including antibiotics (e.g., erythromycin), cholesterol-lowering agents (e.g., lovastatin), immunosuppressants (e.g., rapamycin), and anticancer compounds (e.g., daunorubicin and epothilone). Because advanced genetic and fermentation systems are often not available for native polyketide producers, there has been interest in developing generic and genetically tractable host systems for therapeutically important polyketides.

This chapter will review the development of heterologous polyketide production systems in *Streptomyces coelicolor* and *Escherichia coli*, providing specific examples of polyketides produced in these hosts to illustrate the utility of the approach. The first section covers *S. coelicolor*, an actinomycete that produces a number of polyketides, most notably the blue pigment actinorhodin (3). Heterologous polyketide production in *S. coelicolor* demonstrated the feasibility of the approach, albeit in a host that naturally produces polyketides. The success with *S. coelicolor* provided motivation to develop heterologous production systems in *E. coli* and *Saccharomyces cerevisiae*, user-friendly organisms for which advanced molecular biology tools and fermentation systems are available. In the second section, the development of an *E. coli* production system is described. Demonstration of polyketide production in *E. coli* showed that all activities necessary for polyketide production can be recapitulated in a host that does not naturally make polyketides. The availability of *E. coli* as a heterologous host has accelerated the discovery of new polyketide analogs, as well as facilitated studies of polyketide synthase (PKS) enzymology.

26.1. PRODUCTION OF POLYKETIDES IN *S. COELICOLOR*

26.1.1. Strains, Vectors, and General Methods

Except where noted, the PKS-related genetic engineering described in this chapter generally involves standard molecular biology techniques (50). Intermediate cloning steps are performed with *E. coli* (e.g., XL1-Blue or DH5α), and target DNA fragments are transferred into *Steptomyces*/*E. coli* shuttle vectors (see Table 1) by standard ligation-based methods. Efficient cloning is facilitated by engineering unique and strategic restriction sites into cloning vectors

and target DNA fragments. Large DNA fragments (≥10 bp) can be separated on low-percentage, low-melting-point agarose gels with subsequent manipulations performed "in gel" as outlined in protocol 1.

PKS genes are expressed from the *S. coelicolor actI* promoter, which is activated by *actII*-open reading frame 4 (ORF4), the actinorhodin (*act*) pathway-specific activator gene and positive regulator of transcription. The *actI* promoter-*actII*-ORF4 expression system is upregulated in stationary phase of mycelial growth and therefore couples heterologous expression to the onset of secondary metabolite production in *S. coelicolor*.

Standard *Streptomyces* transformation methods (27) are employed to introduce expression plasmids into *S. coelicolor* CH999, a strain that contains a chromosomal deletion of the actinorhodin gene cluster encoding the activities required for the synthesis of the aromatic polyketide actinorhodin (35). CH999 was constructed by integration via homologous recombination of the *ermE* gene fragment (conferring lincomycin resistance) into the *act* locus. Because *S. coelicolor* has a methyl-specific restriction system, plasmids must first be passed through a methylation-deficient *E. coli* host, such as ET12567. Alternatively, plasmids can be expressed in *Streptomyces lividans* K4-114, a strain with an inactive *act* gene cluster that can be transformed with methylated DNA at high levels of efficiency (63). CH999 and K4-114 are considered "clean hosts" because they contain the necessary ancillary pathways for polyketide production but do not naturally produce the targeted polyketide whose biosynthesis genes are introduced into these hosts.

Protocol 1.   In-Gel Ligation of Large DNA Fragments

1. Prepare low-melting-point agarose gel (0.8 to 1%) in TAE buffer (40 mM Tris, 20 mM acetic acid, 1 mM EDTA, pH 8). To minimize breakage upon handling, place the gel into the refrigerator/cold room and chill for at least 30 min before use. Following electrophoresis, stain the gel with ethidium bromide and visualize *briefly* under long-wavelength UV light to enable the excision of relevant DNA bands. Care should be taken to excise the region of the band where DNA is most concentrated, thereby avoiding excessive dilution of the DNA.

2. Melt agar gel pieces containing target DNA at 65°C and add directly to ligation mixtures. Ligation reaction mixtures (40 μl) contain 15 μl water, 4 μl of 10 × T4 DNA

ligase buffer, molten gel fragments containing insert and vector DNA (20 μl total combined volume), and 1 μl T4 DNA ligase. In practice, a series of ligation reactions are set up to evaluate a range of insert-to-vector ratios (0.5 to 5).

3. After incubating the ligation reaction mixtures for 1 h to overnight at room temperature (~25°C), stop the reactions (and melt the agar) by heating the mixtures at 65°C for 5 min and add 40 μl of TCM buffer (10 mM Tris, pH 7.5, 10 mM calcium chloride, 10 mM magnesium chloride). Use an aliquot (10 to 25 μl) of the resulting mixture to transform 100 μl of chemically competent *E. coli* cells (e.g., XL1-Blue or DH5α). Typically, the entire transformation mixture is plated onto Luria-Bertani (LB) agar plates containing appropriate antibiotics.

26.1.2. 6-MSA

6-Methylsalicylic acid synthase (6-MSAS) from the fungus *Penicillium patulum* catalyzes the synthesis of a simple fungal polyketide, 6-methylsalicylic acid (6-MSA; compound 1) (Fig. 1 and 2), from the acyl coenzyme A (acyl-CoA) precursors acetyl-CoA and malonyl-CoA. The biosynthesis of 6-MSA has been studied extensively and proceeds via three iterative condensation reactions and NADPH-dependent ketoreduction. The single β-ketoacyl acyl carrier protein (ACP) synthase active site of the 6-MSAS monomer is used three times during the synthesis of one molecule of 6-MSA.

The 6-MSAS gene from *P. patulum* has been cloned and sequenced (1). For heterologous expression in *S. coelicolor*, the 6-MSAS ORF was reconstructed from a 7-kb *P. patulum* genomic DNA fragment (2). Exon 2 (5,235 bp) was joined to a synthetic version of exon 1 (87 bp), thereby removing the single 69-bp intron. To facilitate heterologous expression, the synthetic fragment contained a ribosome binding site complementary to the 3' region of *S. coelicolor* 16S rRNA and four of the first seven codons were changed to synonymous codons that better matched codon bias in *S. coelicolor*. Unique PacI and NdeI restriction sites (the lat-

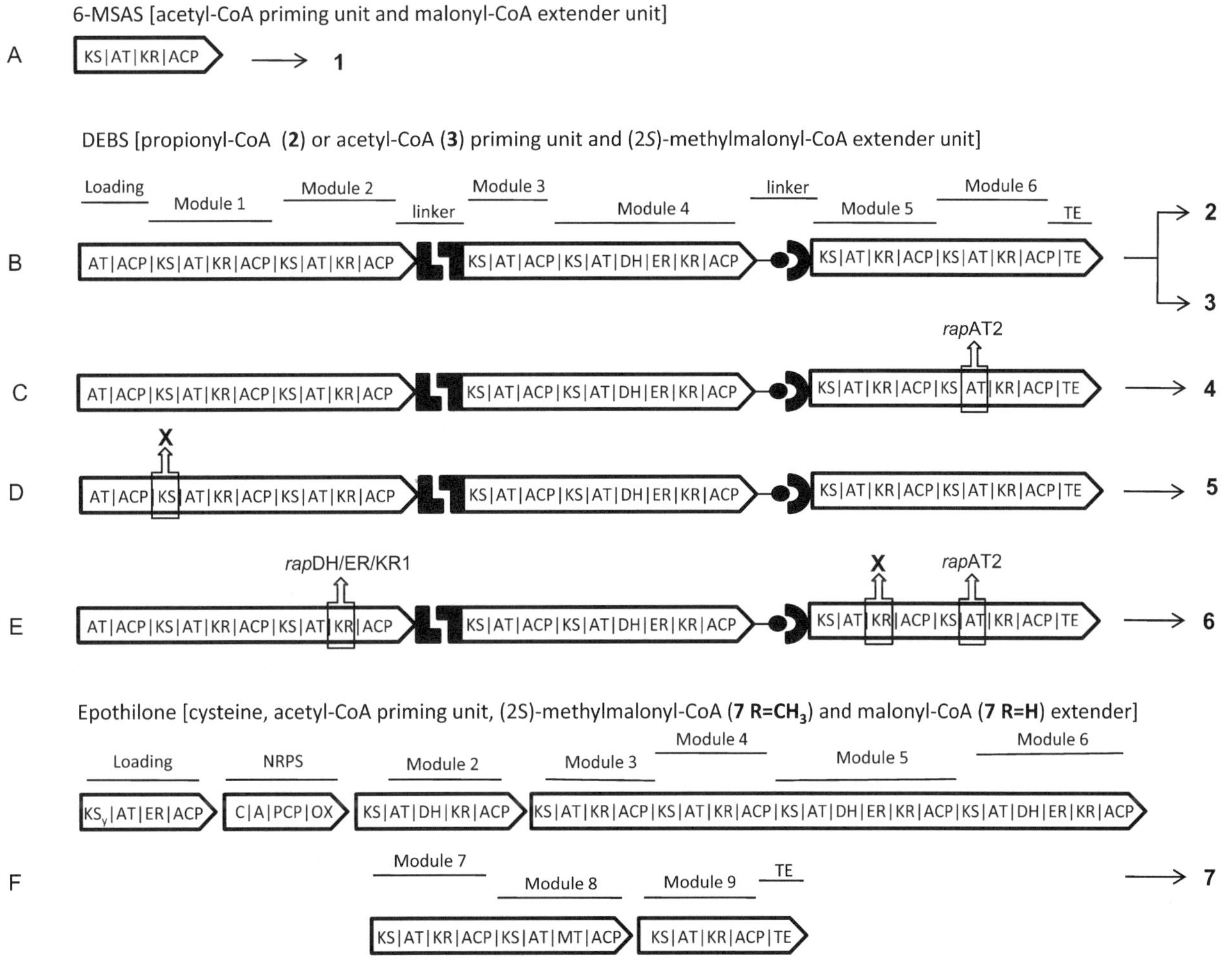

FIGURE 1 Module and domain organization of PKSs expressed in *S. coelicolor* and/or *E. coli*: 6-MSAS (A); DEBS (B); DEBS with AT6 replaced with AT2 (malonyl-CoA specific) from RAPS (C); DEBS with KS1 inactivated by a point mutation (D); DEBS with KR2 replaced with dehydratase (DH)/enoylreductase (ER)/KR1 from RAPS, KR5 inactivated by insertion of a stuffer fragment, and AT6 replaced with AT2 (malonyl-CoA specific) from RAPS (E); and epothilone PKS (F). KSy, inactive ketosynthase; TE, thioesterase; NRPS, nonribosomal peptide synthase; MT, methyltransferase; C, condensation; A, adenylation; PCP, peptidyl carrier protein; Ox, oxidase; linker, interpeptide linker that facilitates interaction between the DEBS proteins.

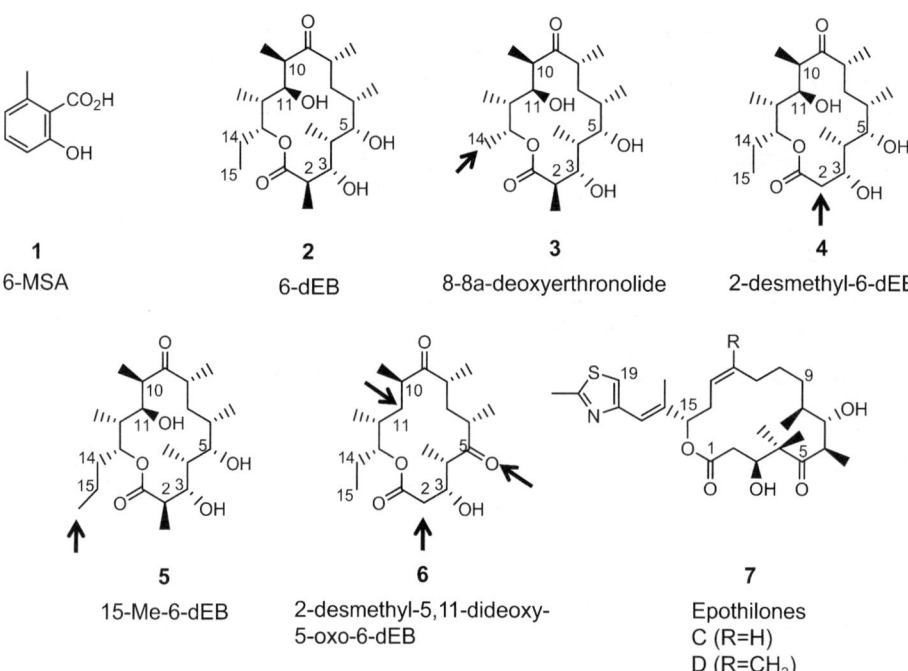

1
6-MSA

2
6-dEB

3
8-8a-deoxyerthronolide

4
2-desmethyl-6-dEB

5
15-Me-6-dEB

6
2-desmethyl-5,11-dideoxy-
5-oxo-6-dEB

7
Epothilones
C (R=H)
D (R=CH₃)

FIGURE 2 Polyketides heterologously produced in *S. coelicolor* and/or *E. coli*. For analogs, arrows indicate the site(s) of functional group modifications with respect to the 6dEB core structure (see the text for details).

ter including the ATG start codon) were introduced at the 5′ end of the gene, and a unique XbaI site was introduced at the 3′ end. The PacI-XbaI cassette containing the 6-MSAS ORF and the *S. coelicolor* ribosome binding site was introduced into PacI- and XbaI-restricted pKOA18′, a modified version of pRM5 (35) (Table 1), placing the 6-MSAS gene under the control of the *actI* promoter and *actII*-ORF4 transcriptional activator, resulting in plasmid pDB106.

The 6-MSAS expression plasmid pDB106 was introduced into protoplasts of *S. coelicolor* CH999, transformants were grown on R5 agar medium (27) for 6 days, and the agar and mycelia were extracted with ethyl acetate–1% acetic acid. Analysis of extracts by thin-layer chromatography revealed the presence of a UV-active product from the CH999(pDB106) fermentation that was not present in CH999 harboring a vector plasmid lacking the 6-MSAS gene. Further characterization by nuclear magnetic resonance confirmed that the UV-active compound was 6-MSA, which was produced at 60 mg/liter, a titer 10-fold lower than that produced by the native host, *P. patulum* (52). Production of 6-MSA in *S. coelicolor* demonstrated that all of the requirements for successful heterologous production of a fungal polyketide were present or facilitated in the bacterial heterologous host, including adequate precursor supply, proper monomer folding and tetramer assembly, and successful posttranslational modification of the ACP domain of the PKS with the 4′-phosphopantetheine moiety.

26.1.3. 6dEB

6-Deoxyerythronolide B synthase (DEBS) from *Saccharopolyspora erythraea* catalyzes the synthesis of the erythromycin precursor polyketide 6-deoxyerythronolide B (6dEB; compound 2) (Fig. 1B and 2). The three DEBS proteins (DEBS1, DEBS2, and DEBS3) are encoded by three large genes, *eryAI*, *eryAII*, and *eryAIII*. The catalytic sites of the DEBS megasynthase are organized into six modules, each

of which catalyzes a two-carbon condensation reaction and subsequent β-carbonyl processing. As the prototypical example of a modular PKS, DEBS has been extensively studied (for reviews, see references 21, 24, 25, 48, and 53). The *ery* genes from *S. erythraea* span 32 kb of contiguous DNA sequence, which posed a formidable challenge for cloning and introduction into *S. coelicolor*. Kao and colleagues used a strategy of recombinational cloning into *E. coli* to assemble the *ery* genes and place them under the control of the *S. coelicolor actI* promoter (20). A donor plasmid contained a temperature-sensitive replicon and carried the 3′ end of *eryAI* and intact *eryAII-eryAIII* genes. The recipient plasmid, a derivative of pRM5 (35), contained the *S. coelicolor actII*-ORF4 transcriptional regulator, the *actI* promoter upstream of the 5′ end of *eryAI*, a tetracycline resistance marker, and the 3′ end of *eryAIII*. Enrichment and isolation of the recombinant plasmid arising from double crossover was accomplished by plating *E. coli* transformants onto agar medium with appropriate antibiotic supplementation at the restrictive temperature. The final plasmid, pCK7, was recovered from *E. coli* and introduced into *S. coelicolor* CH999 by a standard protoplast transformation procedure (27).

CH999(pCK7) transformants were grown on R2YE medium (27), and two macrolides were isolated from the strain. As confirmed by nuclear magnetic resonance spectroscopy, the most abundant macrolide was 6dEB (compound 2; 40 mg/liter) (Fig. 1B and 2), which is the aglycone product of the DEBS PKS primed with a propionate starter unit. The aglycone arising from the incorporation of an acetate starter unit, 8,8a-deoxyoleandolide (compound 3) (Fig. 1B and 2), was also observed as a minor product (at 10 mg/liter). The latter fact demonstrates that the DEBS loading module displays relaxed specificity for the starter unit (see also section 26.2.4), as has been demonstrated in vitro (33, 45, 46). The final ratio of aglycone products (compound 2 to compound 3) produced in the heterologous host is expected to reflect both the intracellular

TABLE 1 *Streptomyces* shuttle vectors and strains

Plasmid or strain	Replicons	Marker(s)[a]	Lineage	Reference(s)	Description
Plasmids					
pRM1	SCP2*, ColE1	*bla, tsr*	pIJ903	36	*Streptomyces/E. coli* shuttle vector
pRM5	SCP2*, ColE1	*bla, tsr*	pRM1	36	*Streptomyces/E. coli* shuttle vector engineered to express PKS genes
pCK7	SCP2*, ColE1	*bla, tsr*	pRM5	21	DEBS expression vector
pDB106	SCP2*, ColE1	*bla, tsr*	pRM5	2	6-MSAS expression vector
JRJ2	SCP2*, ColE1	*bla, tsr*	pCK7	20	DEBS KS1 null vector for precursor-directed biosynthesis
pKOS021-30	SCP2*, ColE1	*bla, tsr*	pRM1	63	Component of 3-plasmid system for expression of *eryAI*
pKOS025-143	SCP2*, ColE1	*bla, hyg*	pRM1	63	Component of 3-plasmid system for expression of *eryAII*
pKOS010-153	Integrating bacteriophage φC31 integrase gene for insertion at φC31 *att* site, pUC18	*apr*	pSET152	4, 63	Component of 3-plasmid system for expression of *eryAIII*
pBoost	pBR322, SCP2*containing 45-bp deletion	*apr*	pHU152/pBR322	18	Forms cointegrates with SCP2* to boost copy number
Strains					
CH999		*ermE*	*S. coelicolor* strain lacking actinorhodin gene cluster	36	Heterologous production strain
K4-114		*ermE*	*S. lividans* heterologous production strain lacking actinorhodin gene cluster	63	Exhibits high-level efficiency of transformation with methylated DNA

[a]*bla*, ampicillin and carbenicillin resistance gene; *apr*, apramycin resistance gene; *tsr*, thiostrepton resistance gene; *hyg*, hygromycin resistance gene; *ermE*, lincomycin resistance gene.

concentrations of propionyl-CoA and acetyl-CoA and the relative specificity of the DEBS loading domain for the corresponding acyl-CoA. Accordingly, the ratio may be changed by providing additional propionate in the growth medium. In the native erythromycin producer, *S. erythraea*, the ratio of propionate-derived erythromycin to acetate-derived erythromycin (15-norerythromycin) is extremely high (26), indicating that host-dependent factors can influence product profiles. For example, Hu et al. found that an editing type II thioesterase from *S. erythraea* functions to remove misprimed acetate units from the DEBS PKS, thereby accounting for the high compound 2/compound 3 ratio observed in the native host (18). Kim et al. have also identified a type II esterase associated with the pikromycin PKS in *Streptomyces venezuelae* (28).

The pioneering work of Khosla and colleagues demonstrated functional expression of a large modular PKS in an actinomycete heterologous clean host (20). When expressed in the heterologous host, the DEBS proteins folded correctly and were properly posttranslationally modified with the phosphopantetheine moiety from CoA. Hence, the *Streptomyces* heterologous host contained an endogenous phosphopantetheinyl (p-pant) transferase that converted the DEBS apo-ACP domains into their holoforms. The heterologous host also contained adequate levels of required acyl-CoA substrates, propionyl-CoA and (2S)-methylmalonyl-CoA.

26.1.4. 6dEB Analogs

There is a linear correspondence between the order of polyketide chain assembly and the order of active sites in DEBS (and other modular PKSs). Consequently, analogs of modular polyketides can be produced by genetic reprogramming of the PKS genes—the "chemistry by genetics" approach (21, 24). Pioneering work in this area involved genetic engineering of the *ery* genes in the native host, *S. erythraea* (6, 9). The cloning strategy developed by Khosla and colleagues for manipulation and heterologous expression of the DEBS genes in *S. coelicolor* simplified the construction of mutated DEBS genes and facilitated the production of 6dEB analogs, many of which were intractable by traditional (chemical) synthetic routes. The construction of mutant genes was carried out in *E. coli*, after which the shuttle vector(s) containing the mutated PKS gene(s) was transferred into the heterologous production host CH999 or an analogous heterologous host, *S. lividans* K4-114 (63). This methodology was initially employed to produce targeted analogs from the manipulation of a single DEBS module. For example, Liu et al. replaced the methylmalonyl-CoA-specific DEBS acyltransferase domain 6 (AT6) with AT2 (specific for malonyl-CoA) from the rapamycin PKS (RAPS) to produce 2-nor-6dEB (34) (compound 4) (Fig. 1C and 2). Analogs with changes at the starter-unit position were produced by a procedure known as chemobiosynthesis. This method involves supplying an exogenous starter unit in the form of a chemically synthesized diketide thioester to a polyketide-biosynthetic host in which the DEBS ketosynthase domain 1 (KS1) is inactivated, usually by mutation of the active-site cysteine to alanine. The synthetic diketide effectively mimics the ACP domain 1 (ACP1)-bound diketide and is incorporated at KS2, where it is subsequently elongated into the full-length polyketide analog. The Cys-to-Ala mutation was introduced into KS1 on plasmid pCK7, and CH999 was transformed with the resulting plasmid (19). When CH999 was cultured in R2YE medium containing (2S,3R)-2-methyl-3-hydroxyhexanoate-N-acetylcysteamine thioester (propyl-SNAc), 15-methyl-6dEB (compound 5) (Fig. 1D and 2) was produced. Optimized diketide feeding protocols and further process modifications led to significant improvements in titers (to >1 g/liter) of 6dEB starter-unit analogs (8, 32, 58).

The technology for 6dEB analog production was further advanced by the development of a system for combining single alterations in DEBS domains to produce analogs with multiple substitutions in the 6dEB core structure (36). To facilitate domain manipulation, strategic restriction sites were introduced between domain boundaries, and the fragments containing the engineered restriction sites were subcloned into *E. coli*. The engineered restriction sites did not adversely affect 6dEB production. Domain cassettes from the RAPS gene were generated by PCR, with appropriate restriction sites incorporated at the domain boundaries. The DNA sequences comprising alternative PKS modules were introduced into the DEBS subclones via appropriate restriction sites. In the final cloning step, unique restriction sites were used to transfer the hybrid PKS DNA sequences into the expression plasmid pCK7. CH999 was transformed with the resulting plasmids for heterologous polyketide production and subsequent analysis. Over 50 6dEB analogs containing changes involving methyl substitution and/or β-carbon processing were produced by individual or combined DEBS-RAPS domain swaps. Compound 6 (Fig. 1E and 2), obtained from a triple substitution in DEBS module 2 (ketoreductase domain 2 [KR2] is replaced by dehydratase, enoylreductase, and KR1 from RAPS), module 5 (KR is removed by replacement of the coding sequence with a "stuffer" DNA sequence), and module 6 (methylmalonyl-specific AT is replaced with malonyl-specific AT), is an example of the type of analog accessible by this approach.

The combinatorial potential of the DEBS system was expanded by the development of a multiplasmid approach for DEBS expression in *S. coelicolor* CH999 or *S. lividans* K4-114 (61). The DEBS1 and DEBS2 genes were carried on separate pRM1-based plasmids with thiostrepton and hygromycin resistance markers, respectively, and the DEBS3 gene was carried on the integrating bacteriophage φC31-based plasmid pSET152 with an apramycin resistance marker (4). In all cases, DEBS genes were expressed from the *S. coelicolor* actI promoter and actII-ORF4 transcriptional activator. The two pRM1-based plasmids, both with SCP2* origins of replication, were stably maintained when the culture medium contained appropriate antibiotic

supplementation, and the titer (50 mg/liter) of 6dEB produced by the strain expressing wild-type DEBS from three different plasmids was similar to that produced by the strain expressing DEBS from a single plasmid. The multiplasmid approach allows for the combination of single mutations by simple transformation, thereby eliminating the need to construct a separate plasmid for each desired analog. The combinatorial potential of this system arising from the combination of major elements is 262,144 polyketides (2 different ATs × 4 different β-keto-modifying activities per module = 8 combinations per module; 8 × 8 per protein; 64 × 64 × 64 = 262,144 per DEBS molecule). To generate such a library by the multiplasmid approach requires the construction of only 192 (64 × 3) different plasmids carrying single mutations. In contrast, the same number of genetic constructions using the single-vector combinatorial approach described above would yield only 192 different polyketides (36).

The utility of the multiplasmid approach was demonstrated by a proof-of-concept experiment: *S. lividans* K4-114 was transformed with eight variants of the DEBS3 gene (*eryAIII*) on the integrating vector. The resulting strains were cotransformed with four DEBS1 variants (*eryAI*) and two DEBS2 variants (*eryAII*) on the replicating plasmids. By starting with 14 different expression constructs, 64 transformants were generated. Among the 64 strains tested, 43 different polyketides were produced, 28 of which had been prepared previously using the single-plasmid approach (36) and 15 of which represented novel structures.

To convert the 6dEB analogs into corresponding erythromycin analogs, the aglycone product is purified from CH999 and fed to a bioconversion strain of the natural host, *S. erythraea* (5, 60), which contains a mutation in a DEBS gene that prevents 6dEB production, but harbors the downstream "tailoring" activities (glycosidation, oxidation, and methylation) necessary for completion of the biosynthesis pathway. The efficiency of bioconversion depends on the tolerance of the tailoring enzymes for the engineered change in the aglycone. Consequently, not all aglycone analogs are efficiently bioconverted by *S. erythraea*.

26.1.5. Epothilone

The epothilones are mixed polyketides/nonribosomal peptides that have microtubule-stabilizing activities similar to those of the taxanes but are active against Taxol-resistant tumors. The natural producer of epothilones, the myxobacterium *Sorangium cellulosum*, is not an ideal host for large-scale production due to its slow growth and genetic intractability. The large epothilone gene cluster (52 kb) comprises seven ORFs with products ranging in size from 47 to 765 kDa (yielding compound 7) (Fig. 1F and 2). Tang et al. expressed the epothilone gene cluster in *S. coelicolor* CH999 from two operons, each carried on a separate plasmid and each under the control of the *actI* promoter/*actII*-ORF4 transcriptional activator (54). The first operon, consisting of the EpoA to EpoD genes, was cloned into a pCK7 plasmid derivative, resulting in a 55-kb expression plasmid. The second operon, consisting of EpoE and EpoF genes and accessory genes, was introduced into a pSET152-based vector, giving a 30-kb expression plasmid. CH999 was transformed sequentially with the plasmids, and transformants were grown in R2YE medium. Following fermentation, ethyl acetate extracts were purified by high-performance liquid chromatography (HPLC) and epothilones A and B were detected by liquid chromatography-mass spectrometry (LC-MS); final titers were estimated to be about 50 μg/liter. While achieving heterologous production of epothilones in *S. coelicolor* was a major accomplishment, significant strain improvement would be required to develop a strain suitable for large-scale production of epothilone.

26.1.6. Boosting Polyketide Production in S. coelicolor

Many polyketides, including epothilone (54) and certain analogs of 6dEB (36), are produced at low levels in the S. coelicolor heterologous production system. Polyketide titers can potentially be increased by increasing the copy number of the SCP2*-based expression plasmid, normally present at 2 to 5 copies per cell. Hu et al. discovered a plasmid (SCP2@) that could recombine with the low-copy-number SCP2* (JRJ2) to form a cointegrate that has a 30-times-higher copy number (100 to 125 copies) and can be stably maintained (10, 17). Moreover, increased plasmid copy number correlates positively with product yield, resulting in ~25-fold-higher production of the polyketide analog 15-methyl-6dEB (compound 5) (Fig. 1D and 2). SCP2@ is identical to SCP2*, except for a 45-bp deletion upstream of repI. A plasmid called pBoost, containing SCP2@, a pBR322 replicon, and an apramycin resistance cassette, was constructed to provide a copy number boost to large pRM5 (SCP2*)-based expression plasmids upon cotransformation (17).

26.2. PRODUCTION OF POLYKETIDES IN E. COLI

26.2.1. 6-MSA

The polyketide 6-MSA was chosen as a paradigm for heterologous polyketide production in E. coli and S. cerevisiae, workhorse prokaryotic and eukaryotic hosts extensively employed in the biotechnology industry. The 6-MSAS ORF was excised as an NdeI-XbaI fragment from pDB106 (Table 1) and introduced into vectors for expression in E. coli or S. cerevisiae (22). The 6-MSAS ORF and sfp, encoding a heterologous p-pant transferase from Bacillus subtilis required for conversion of apo-6-MSAS to holo-6-MSAS, were coexpressed in E. coli from compatible plasmids with different antibiotic resistance markers. In S. cerevisiae, the 6-MSAS ORF and sfp were expressed from 2µm plasmids with different auxotrophic markers (22). Following induction of gene expression in the E. coli and S. cerevisiae production hosts, 6-MSA levels were determined by HPLC analysis of cell-free media by using UV detection (wavelength, 306 nm). The final titer of 6-MSA produced in E. coli was approximately 75 mg/liter, similar to that produced in S. coelicolor (2). The yeast host produced 1.7 g 6-MSA/liter in yeast extract-peptone-dextrose medium in unoptimized shake flask fermentations (22).

Since E. coli and S. cerevisiae do not naturally produce polyketides, the work on 6-MSA production in these hosts defined the minimum requirements for heterologous polyketide production.

1. *Soluble protein expression.* Production of 6-MSA in E. coli required that the cultivation temperature be reduced from 37°C (E. coli temperature optimum) to 30°C during heterologous protein expression, presumably to facilitate proper protein folding.
2. *Phosphopantetheinylation.* Coexpression of a heterologous p-pant transferase was required for phosphopantetheinylation of the apo-ACP domain of the PKS. The p-pant transferase from the B. subtilis surfactin biosynthetic pathway, Sfp, displays broad substrate specificity toward ACP domains (29) and therefore was chosen as a candidate heterologous p-pant transferase for activation of apo-6-MSAS. Since the first demonstration of its utility

for polyketide production in E. coli and S. cerevisiae (22), Sfp has been used extensively for heterologous activation of PKSs expressed in E. coli and/or S. cerevisiae. To construct sfp expression cassettes, the sfp gene (0.5 kb) is amplified from B. subtilis DNA by PCR using amplification primers that introduce convenient restriction sites. Alternatively, sfp can be synthesized from overlapping DNA oligonucleotides (see section 26.3). The sfp gene is placed under the control of an appropriate promoter (e.g., the T5, T7, or lac promoter) for expression in E. coli, and the expression cassette is introduced into the host strain on a plasmid vector or is integrated into the host genome. Other promiscuous p-pant transferases such as Svp (51) can be employed as well.

3. *Adequate levels of acyl-CoA substrates.* The substrates for 6-MSAS—acetyl-CoA, malonyl-CoA, and NADPH—are used for fatty acid biosynthesis and are therefore present in E. coli and S. cerevisiae. For 6-MSA production in E. coli, glycerol was included in the culture medium to increase intracellular malonyl-CoA levels (56), underscoring the importance of adequate provision of acyl-CoA substrates.
4. *Nontoxic products and/or intermediates.* 6-MSA is not toxic to E. coli or yeast. However, for heterologous production of other polyketides or natural products, introduced activities or pathways may cause the accumulation of intermediates or products that are toxic to host cells. In such cases, balancing pathway flux by overexpression of downstream enzymes can potentially relieve bottlenecks and decrease toxicity. For example, heterologous expression of the yeast mevalonate pathway in E. coli causes the toxic buildup of 3-hydroxy-3-methylglutaryl-CoA, which inhibits fatty acid biosynthesis (and results in an increase in the malonyl-CoA pool). Balancing pathway flux by increasing the expression of the downstream enzyme 3-hydroxy-3-methylglutaryl-CoA reductase relieves toxicity (47). Heterologous expression of the Propionibacterium shermanii methylmalonyl-CoA mutase/epimerase pathway in E. coli also results in elevated malonyl-CoA levels, suggesting that the introduced pathway likely impacts fatty acid biosynthesis (7) (see section 26.2.2). If the product itself is toxic, introduction of an appropriate resistance gene may be required. For example, in E. coli strains engineered for erythromycin production, the ermE resistance gene, encoding a 23S rRNA N-methyltransferase that confers resistance to erythromycin, was coexpressed (43) (see section 26.2.4).

26.2.2. 6dEB

Successful production of the simple polyketide 6-MSA in E. coli and S. cerevisiae demonstrated the feasibility of producing polyketides in non-polyketide-producing, user-friendly, genetically tractable heterologous hosts. Pfeifer and colleagues extended the work with E. coli by engineering a strain that produces the complex polyketide 6dEB (44). Like 6-MSA production in E. coli, successful heterologous production of 6dEB resulted when the minimal requirements, as outlined in section 26.2.1, were satisfied.

1. *Soluble protein expression.* Production of 6dEB in E. coli required that the postinduction cultivation temperature be reduced from 37 to ~22°C to facilitate proper folding of the megaenzymes.
2. *Phosphopantetheinylation.* The sfp gene, under the control of the T7 promoter, was integrated into the prpC locus in the E. coli chromosome, inactivating the 2-methylcitrate pathway and propionyl-CoA metabolism (13, 15). Concurrently, a T7 promoter was placed upstream of

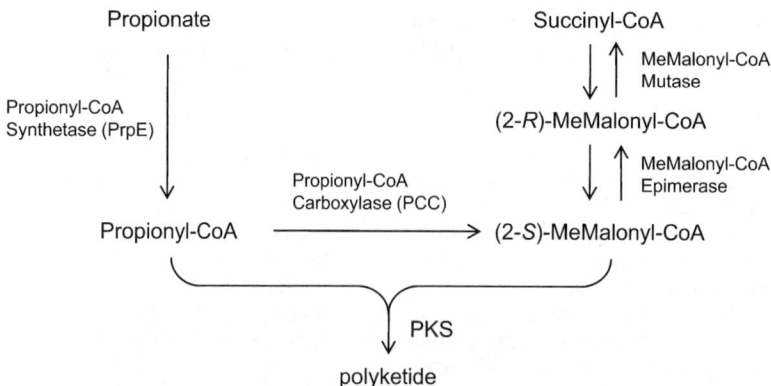

FIGURE 3 Heterologous biosynthetic pathways for (2S)-methylmalonyl-CoA introduced into *E. coli*.

prpE, a gene encoding propionyl-CoA synthetase, which catalyzes the conversion of propionate to propionyl-CoA (14). The genetic manipulation at the *prp* locus by Pfeifer et al. (44) illustrates a common strategy in metabolic engineering: knocking in genes with required functions while knocking out endogenous competing genes or pathways whose products may direct carbon flux away from engineered product pathways.

3. *Adequate levels of acyl-CoA substrates.* The starter unit for 6dEB biosynthesis, propionyl-CoA, was generated from propionate (included in the medium at the induction of gene expression) via priopionyl-CoA synthase. The (2S)-methylmalonyl-CoA extender unit was generated from

propionyl-CoA by the action of the *S. coelicolor* propionyl-CoA carboxylase (PCC), comprising a biotin carrier subunit and a transcarboxylase subunit (49) (Fig. 3).

4. *Nontoxic products and/or intermediates.* At the levels produced, there is no apparent toxicity of 6dEB to the *E. coli* host.

In the initial *E. coli* 6dEB production system developed by Pfeifer et al. (44), the DEBS1 and PCC genes were carried on a pET-based plasmid (pBP144) with kanamycin resistance, and the DEBS2 and DEBS3 genes were present on a second pET-based plasmid (pBP130) with ampicillin resistance (Table 2). An *E. coli* strain (BAP1) containing a single copy of *sfp* integrated at the *prpC* locus was cotransformed

TABLE 2 *E. coli* vectors and strains

Plasmid or strain	Replicon	Marker(s)[a]	Lineage	Reference	Description or genotype
Plasmids					
pBP130	pBR322	*bla*	pET22b	45	Expresses DEBS2 and DEBS3 genes under control of T7 promoter
pBP144	pBR322	Kan	pET26b	45	Expresses DEBS1 and PCC genes under control of T7 promoter
pKOS207-129	RSF1010	Str, Spt	RSF1010	41	Expresses DEBS1 gene under control of T7 promoter
pKOS452-53e	RSF	Kan	pRSF-1b	44	Expresses mycarose and P450 hydroxylase genes under control of T7 promoter
pKOS452-54	CDF	Ble	pCDF-1b	44	Expresses desosamine and P450 hydroxylase genes and *ermE* under control of T7 promoter
pGro7	P15A	Cm	pACYC	43	pBAD-*groES-groEL*
pKOS486-2-1	pBR322	*bla*	pET22b	43	pBAD-*epoABC*
pKOS455-166	RSF	Kan	pRSF-1b	43	pBAD-*epoD₁-epoD₂*
pKOS431-188	CDF	Str, Spt	pCDF-1b	43	pBAD-*epoEF*
Strains					
BAP1			BL21(DE3)	45	*prpC*::T7p-*sfp* T7p-*prpE*
K-207-3			BAP1 *panD*::*panD*(S25A)	41	*ygfG*::T7p-*pcc*
K214-37			BAP1 *panD*::*panD*(S25A)	41	Expresses mutase and epimerase A to C genes under control of *ygfG*::T7p
YW9			BAP1	59	Expresses PCC, DEBS1, DEBS2, and DEBS3

[a]*bla*, ampicillin and carbenicillin resistance gene; Kan, kanamycin resistance; Str, streptomycin resistance; Spt, spectinomycin resistance; Ble, bleomycin resistance; Cm, chloramphenicol resistance.

with the plasmids, as described above. All introduced genes (and endogenous *prpE*) were under the control of the *lac*-inducible T7 promoter. Transformants were cultivated in LB medium with appropriate antibiotic supplementation. Upon induction of gene expression with IPTG (isopropyl-β-D-thiogalactopyranoside) at mid-log phase of growth (optical density at 600 nm of ~0.4 to 0.6), propionate was added to the culture medium. Although the BAP1(pBP144, pBP130) strain produced 6dEB, the system suffered from plasmid instability. Murli et al. developed a second-generation stable *E. coli* strain that produced 6dEB at higher yields (40). In the second-generation strain, K-207-3, the *pcc* genes were integrated at *ygfG*, a gene encoding methylmalonyl-CoA decarboxylase. The DEBS1 gene was introduced into an RSF1010-based plasmid (pKOS207-129) such that it was compatible with the pET-based pBP130. In K-207-3(pKOS207-129, pBP130), the plasmids were stably maintained without antibiotic supplementation and 6dEB was produced at high titers (~1 g/liter) in high-density *E. coli* fermentations (31). Recently, Wang and Pfeifer developed a plasmid-free *E. coli* strain for 6dEB production; however, this strain, YW9, produced low levels of 6dEB (≤0.52 mg/liter) (57).

As an alternative to the PCC pathway for the production of (2S)-methylmalonyl-CoA, the methylmalonyl-CoA mutatase/epimerase pathway has been introduced into *E. coli* (Fig. 3) (7, 23, 40). Since mutase/epimerase produces methylmalonyl-CoA from succinyl-CoA, the extender unit can be produced independently of the starter unit, a potential advantage for the production of analogs of 6dEB with variations at the starter-unit position (see section 26.2.4 below). Moreover, the mutase/epimerase pathway generates the extender unit from an intermediate (succinyl-CoA) derived from central metabolism, potentially providing a more economical route to 6dEB by reducing or eliminating the propionate requirement.

Protocol 2.   Quantification of 6dEB (and Analogs) Produced in *E. coli*

1. Plate *E. coli* transformants onto LB agar with appropriate antibiotics.
2. Pick at least two transformants for each strain analyzed. Inoculate single colonies into LB medium containing appropriate antibiotics and grow cultures overnight at 37°C in an incubator/shaker. Use an aliquot (1/50 final dilution) of each overnight culture to inoculate production flasks containing LB medium with appropriate antibiotic supplementation. Grow cultures at 37°C to mid-log phase (optical density at 600 nm, ~0.6) and then cool the cultures to 22°C. Induce the expression of regulated genes and add medium supplements (for example, add IPTG to a 0.5 mM final concentration and sodium propionate to 5 mM); continue shaking the cultures at 22°C.
3. After appropriate cultivation time, transfer 50 ml of fermentation culture into a 50-ml conical tube and collect cells by centrifuging at 3,500 rpm (2,000 × g) for 15 min.
4. Transfer supernatant into a clean, 50-ml conical (Falcon) tube.
5. In a 15-ml conical tube, mix 5 ml of culture supernatant and 5 ml of ethyl acetate.
6. Shake vigorously for at least 15 min.
7. Centrifuge at 3,500 rpm (2,000 × g) for 15 min.
8. Remove the upper (organic) layer and transfer it into a 16- by 100-mm glass tube.
9. Dry in a vacuum concentrator (approximately 1 h). Note that depending on the yield of polyketide, steps 1 through 7 can be performed in a 1.5-ml Eppendorf tube by mixing 0.5 ml of cell-free culture with 0.5 ml of ethyl acetate.

10. Resuspend the residue in 500 µl of methanol (MeOH) and prepare samples of different concentrations by dilution in 50% MeOH. Analysis of several dilutions of the extract will ensure that the amount of 6dEB analyzed lies within the linear range for the assay. Analyze 6dEB by LC-MS and/or LC–evaporative light-scattering detection (ELSD). Prepare a serially diluted calibration standard of 6dEB. Dilute the standard to 2, 5, 10, 15, and 20 mg/liter in 50% MeOH. Perform analysis and quantitation by LC-MS-ELSD. A typical LC-MS system for analysis of 6dEB is an HPLC system with a C₁₈ column (e.g., 5-µm particle size; 4.1 by 150 mm) and a linear gradient from 35 to 100% methyl cyanide (0.1% acetic acid) at 1 ml/min over 10 min. The eluate is split 1:1 between ELSD and MS detectors. Under these conditions, 6dEB elutes at ~7.4 min. 6dEB titers from sample cultures are determined by comparing the integrated area of the ELSD response to a standard reference curve generated from authentic 6dEB.

26.2.3.   Balancing Heterologous Expression

The *E. coli* production systems described above make use of T7 promoters for expression of genes involved in polyketide production. The exclusive use of the same promoters can result in unbalanced gene expression and reduction of polyketide titers. An extreme example of this phenomenon was described by Menzella et al. for an *E. coli* 6dEB production system that employed synthetic DEBS genes with codons tailored for *E. coli* tRNA frequency bias (39). Expression of the large plasmid-borne PKS genes from T7 promoters at high transcription and translation efficiencies effectively attenuated expression of the chromosomally localized PCC genes, each under the control of a T7 promoter. The resulting strain failed to produce 6dEB, due to inadequate acyl-CoA precursor levels. When the synthetic DEBS genes were expressed from alternative promoters (e.g., pBAD and p*lac*) to better balance PKS and PCC expression, PCC expression increased and polyketide production was restored.

26.2.4.   6dEB Analogs and Erythromycin

The *E. coli* production system also was employed to produce analogs of 6dEB with changes at the starter-unit position. These analogs served as precursors for novel ketolides, a derivative class of macrolides with activities against macrolide-resistant pathogenic bacteria (62). Three approaches were used to produce 6dEB analogs in *E. coli*. The first involved chemobiosynthesis, as described in section 26.1.4, to produce compound 5 (Fig. 2) and related analogs. Synthetic diketides are "fed" to K-207-3 expressing DEBS2, DEBS3, and DEBS1 with nonfunctional KS1 (alternatively, module 2 of DEBS1 can be used in place of DEBS1 with a KS1 knockout). A similar construct has been successfully employed in *S. coelicolor* (58). A potential drawback to the chemobiosynthetic approach is the significant scale-up cost associated with chemical synthesis of the diketide raw material. Murli and coworkers developed an alternative chemobiosynthetic approach that involved feeding more economical monoketide acyl-thioester analogs to K-207-3 strains expressing DEBS2, DEBS3, and DEBS1 lacking the loading didomain (41). A third route to 6dEB starter-unit analogs such as compound 5 takes advantage of the relaxed substrate specificity of the DEBS loading domain toward alternative acyl-CoA starter units. Biosynthesis of polyketide analogs by incorporation of novel acyl-CoAs is cost-effective because the method does not require synthetic precursors, but the spectrum of analogs that can be produced by this strategy is somewhat limited. Major challenges inherent to this methodology include (i) the ability to engineer a pathway to supply the desired acyl-CoA starter unit and (ii) efficient utilization of the alternative acyl-CoA by the DEBS loading domain. As an example of this approach, compound 5 was produced in *E. coli* through the incorporation of butyryl-CoA instead of propionyl-CoA as the primer unit (23). *E. coli* was engineered to produce high levels of butyryl-CoA from butyrate in a strain that generated the (2S)-methylmalonyl-CoA extender unit from the mutase/epimerase pathway (7). By producing (2S)-methylmalonyl-CoA from succinyl-CoA (and not propionyl-CoA via PCC), propionyl-CoA could be effectively removed from the system. Because the DEBS loading domain prefers propionyl-CoA over butyryl-CoA by at least a factor of 10 (23, 30, 33), eliminating competition between propionyl-CoA and butyryl-CoA for the DEBS loading domain was crucial for efficient utilization of butyryl-CoA by DEBS and successful heterologous production of compound 5.

As discussed in section 26.1.4, 6dEB analogs produced in *E. coli* or *S. coelicolor* are fed to a bioconversion strain of *S. erythraea* to produce the corresponding bioactive erythromycin analogs. As a step toward developing a single-host system for the production of erythromycin analogs, Gramajo and colleagues introduced into *E. coli* all of the downstream activities required to convert 6dEB into erythromycin (43, 44). The complete erythromycin production system consisted of 23 heterologous genes and five different plasmids in the K-207-3 strain background. Two plasmids (pBP130 and pKOS207-129) carried the DEBS genes; two other plasmids (pKOS452-53e and pKOS452-54) contained genes required for sugar biosynthesis and glycosyl transfer, P450-dependent hydroxylations, and erythromycin resistance (*ermE*). The fifth plasmid (pGro7) contained the *groEL* and *groES* chaperone genes, the products of which form a complex to promote proper folding of one or more of the heterologously expressed proteins (Table 2). Although the final erythromycin titers in this system were low (~0.4 mg/liter of erythromycin C), the demonstration of heterologous erythromycin production is a crucial first step toward the development of more efficient *E. coli* strains to accelerate the discovery of new antibiotics.

26.2.5. Epothilone

The successful heterologous production of epothilones C and D in *E. coli* serves an illustrative example of the application of synthetic biology to metabolic engineering (see also section 26.3 for further elaboration) (42). The entire epothilone gene cluster (55 kb) was redesigned and synthesized to allow for facile genetic manipulation and optimized gene expression in *E. coli*. Because it was not possible to express the largest epothilone protein, EpoD (765 kDa), this protein was divided into two parts (*epoD₁* and *epoD₂*) containing cognate interaction domains. Such domains facilitate protein-protein interaction between the two halves of EpoD, thereby reconstituting the intact protein. The strategy is employed by natural PKS proteins (such as DEBS) to ensure that the correct molecular assembly line is formed (11, 12, 55). A similar strategy has been employed for production of the ansamycin polyketide precursor in *E. coli* (59). The complete *E. coli* epothilone expression system comprised four plasmids, including a plasmid that carried molecular chaperone genes *groES-groEL*, *dnaK*, *dnaJ*, *grpE*, and *tig* (encoding a trigger factor), in strain K431-37-2A (a derivative of K-207-3). Although the titers of epothi-

lones produced in *E. coli* were low (<1 μg/liter), the work demonstrates proof of concept and provides a basis for titer improvement, as well as a platform for further investigation of epothilone PKS biochemistry and enzymology.

26.3. APPLICATION OF SYNTHETIC BIOLOGY TO POLYKETIDE BIOSYNTHESIS IN *E. COLI*

Recently, custom gene synthesis (yielding >200 bp) has become more affordable, and the lead time for delivery of synthetic fragments has improved significantly. It is now practical to consider a totally synthetic route to the engineering of *E. coli* cell factories for the production of polyketides and other natural products. Indeed, as described above, DEBS and epothilone synthase are two examples of megaenzymes (encoded by DNA sequences of >30 kb) whose genes were resynthesized to facilitate genetic manipulation and expression. Advantages of the synthetic approach include (i) tailoring of the gene's codons to the host's tRNA and G+C biases, (ii) removal and/or addition of restriction sites to facilitate cloning and modular design, (iii) addition of epitope or fusion tags for protein purification and/or protein detection, (iv) the ability to clone any gene starting with only the protein coding sequence, and (v) avoidance of potential recombination between highly homologous sequences often found in PKS genes (for an example, see reference 16).

Custom gene synthesis, together with user-friendly *E. coli* expression systems, has been a powerful tool for studies of PKS specificity, modularity, and flexibility. In the work of Menzella et al., a library of PKS coding sequences were synthesized, unique common restriction sites were engineered in strategic locations between domains to facilitate cloning and modularity, and linker regions were added to mediate module-module assembly (38). Individual modules carried on compatible plasmids with unique antibiotic restriction markers were expressed in *E. coli* K-207-3 in pairwise combinations, and the functionality of bimodular pairs was assessed by measuring the formation of the expected triketide lactone. Those module pairs that were active in the first round of screening were combined with a third module in the next round of screening, and functional trimodular combinations were assessed by assaying for tetraketide formation (37). These experiments laid the foundation for the development of a generic strategy to produce complex polyketides in *E. coli* by modular assembly of novel PKSs.

26.4. SUMMARY AND PROSPECTS

The development of heterologous polyketide production systems in *S. coelicolor* and *E. coli* is a significant step toward establishing generic microbial cell factories for the production of therapeutically important polyketides. The versatility of the platform is underscored by the fact that a number of different polyketides have been successfully produced in these systems—from the simple polyketide 6-MSA to the complex hybrid polyketide/nonribosomal peptide epothilone. Progressing from a generic polyketide production system to one that is tailored to high-level production of an individual polyketide may require additional development, including process improvement, further strain engineering, and/or random mutagenesis. The iterative cycle of random mutagenesis and screening, followed by rapid whole-genome sequencing of improved clones to identify positive causative mutations, is a potentially powerful approach for the rapid engineering of high-producing strains from generic predecessors. The random-mutagenesis/reverse-engineering approach is particularly attractive for hosts such as *E. coli*, a microorganism that is particularly amenable to high-throughput cultivation and rapid genome sequencing.

The availability of heterologous *E. coli* production systems, together with recent advances in synthetic biology, has accelerated studies of PKS modularity and specificity. In recent work, Menzella and coworkers have begun to build a library of synthetic PKS modules and to elucidate a basic set of rules or a code linking polyketide assembly with PKS module-module activity and connectivity (37, 38). With further refinement of the connectivity rules, it is anticipated that virtually any polyketide can be produced in *E. coli* by coexpression of a suitable set of synthetic PKS modules.

REFERENCES

1. Beck, J., S. Ripka, A. Siegner, E. Schiltz, and E. Schweizer. 1990. The multifunctional 6-methylsalicylic acid synthase gene of *Penicillium patulum*. Its gene structure relative to that of other polyketide synthases. *Eur. J. Biochem.* **192:**487–498.

2. Bedford, D. J., E. Schweizer, D. A. Hopwood, and C. Khosla. 1995. Expression of a functional fungal polyketide synthase in the bacterium *Streptomyces coelicolor* A3(2). *J. Bacteriol.* **177:**4544–4548.

3. Bentley, S. D., K. F. Chater, A. M. Cerdeno-Tarraga, G. L. Challis, N. R. Thomson, K. D. James, D. E. Harris, M. A. Quail, H. Kieser, D. Harper, A. Bateman, S. Brown, G. Chandra, C. W. Chen, M. Collins, A. Cronin, A. Fraser, A. Goble, J. Hidalgo, T. Hornsby, S. Howarth, C. H. Huang, T. Kieser, L. Larke, L. Murphy, K. Oliver, S. O'Neil, E. Rabbinowitsch, M. A. Rajandream, K. Rutherford, S. Rutter, K. Seeger, D. Saunders, S. Sharp, R. Squares, S. Squares, K. Taylor, T. Warren, A. Wietzorrek, J. Woodward, B. G. Barrell, J. Parkhill, and D. A. Hopwood. 2002. Complete genome sequence of the model actinomycete *Streptomyces coelicolor* A3(2). *Nature* **417:**141–147.

4. Bierman, M., R. Logan, K. O'Brien, E. T. Seno, R. N. Rao, and B. E. Schoner. 1992. Plasmid cloning vectors for the conjugal transfer of DNA from *Escherichia coli* to *Streptomyces* spp. *Gene* **116:**43–49.

5. Carreras, C., S. Frykman, S. Ou, L. Cadapan, S. Zavala, E. Woo, T. Leaf, J. Carney, M. Burlingame, S. Patel, G. Ashley, and P. Licari. 2002. *Saccharopolyspora erythraea*-catalyzed bioconversion of 6-deoxyerythronolide B analogs for production of novel erythromycins. *J. Biotechnol.* **92:**217–228.

6. Cortes, J., S. F. Haydock, G. A. Roberts, D. J. Bevitt, and P. F. Leadlay. 1990. An unusually large multifunctional polypeptide in the erythromycin-producing polyketide synthase of *Saccharopolyspora erythraea*. *Nature* **348:**176–178.

7. Dayem, L. C., J. R. Carney, D. V. Santi, B. A. Pfeifer, C. Khosla, and J. T. Kealey. 2002. Metabolic engineering of a methylmalonyl-CoA mutase-epimerase pathway for complex polyketide biosynthesis in *Escherichia coli*. *Biochemistry* **41:**5193–5201.

8. Desai, R. P., T. Leaf, E. Woo, and P. Licari. 2002. Enhanced production of heterologous macrolide aglycones by fed-batch cultivation of *Streptomyces coelicolor*. *J. Ind. Microbiol. Biotechnol.* **28:**297–301.

9. Donadio, S., M. J. Staver, J. B. McAlpine, S. J. Swanson, and L. Katz. 1991. Modular organization of genes required for complex polyketide biosynthesis. *Science* **252:**675–679.

10. Fong, R., J. A. Vroom, Z. Hu, C. R. Hutchinson, J. Huang, S. N. Cohen, and C. M. Kao. 2007. Characterization of

11. Gokhale, R. S., and C. Khosla. 2000. Role of linkers in communication between protein modules. *Curr. Opin. Chem. Biol.* **4:**22–27.

12. Gokhale, R. S., S. Y. Tsuji, D. E. Cane, and C. Khosla. 1999. Dissecting and exploiting intermodular communication in polyketide synthases. *Science* **284:**482–485.

13. Horswill, A. R., and J. C. Escalante-Semerena. 1997. Propionate catabolism in *Salmonella typhimurium* LT2: two divergently transcribed units comprise the *prp* locus at 8.5 centisomes, *prpR* encodes a member of the sigma-54 family of activators, and the *prpBCDE* genes constitute an operon. *J. Bacteriol.* **179:**928–940.

14. Horswill, A. R., and J. C. Escalante-Semerena. 1999. The prpE gene of *Salmonella typhimurium* LT2 encodes propionyl-CoA synthetase. *Microbiology* **145**(Pt. 6): 1381–1388.

15. Horswill, A. R., and J. C. Escalante-Semerena. 1999. *Salmonella typhimurium* LT2 catabolizes propionate via the 2-methylcitric acid cycle. *J. Bacteriol.* **181:**5615–5623.

16. Hu, Z., R. P. Desai, Y. Volchegursky, T. Leaf, E. Woo, P. Licari, D. V. Santi, C. R. Hutchinson, and R. McDaniel. 2003. Approaches to stabilization of inter-domain recombination in polyketide synthase gene expression plasmids. *J. Ind. Microbiol. Biotechnol.* **30:**161–167.

17. Hu, Z., D. A. Hopwood, and C. R. Hutchinson. 2003. Enhanced heterologous polyketide production in *Streptomyces* by exploiting plasmid co-integration. *J. Ind. Microbiol. Biotechnol.* **30:**516–522.

18. Hu, Z., B. A. Pfeifer, E. Chao, S. Murli, J. Kealey, J. R. Carney, G. Ashley, C. Khosla, and C. R. Hutchinson. 2003. A specific role of the *Saccharopolyspora erythraea* thioesterase II gene in the function of modular polyketide synthases. *Microbiology* **149:**2213–2225.

19. Jacobsen, J. R., C. R. Hutchinson, D. E. Cane, and C. Khosla. 1997. Precursor-directed biosynthesis of erythromycin analogs by an engineered polyketide synthase. *Science* **277:**367–369.

20. Kao, C. M., L. Katz, and C. Khosla. 1994. Engineered biosynthesis of a complete macrolactone in a heterologous host. *Science* **265:**509–512.

21. Katz, L. 1997. Manipulation of modular polyketide synthases. *Chem. Rev.* **97:**2557–2576.

22. Kealey, J. T., L. Liu, D. V. Santi, M. C. Betlach, and P. J. Barr. 1998. Production of a polyketide natural product in nonpolyketide-producing prokaryotic and eukaryotic hosts. *Proc. Natl. Acad. Sci. USA* **95:**505–509.

23. Kennedy, J., S. Murli, and J. T. Kealey. 2003. 6-Deoxyerythronolide B analogue production in *Escherichia coli* through metabolic pathway engineering. *Biochemistry* **42:**14342–14348.

24. Khosla, C. 1997. Harnessing the biosynthetic potential of modular polyketide synthases. *Chem. Rev.* **97:**2577–2590.

25. Khosla, C., Y. Tang, A. Y. Chen, N. A. Schnarr, and D. E. Cane. 2007. Structure and mechanism of the 6-deoxyerythronolide B synthase. *Annu. Rev. Biochem.* **76:** 195–221.

26. Kibwage, I. O., G. Janssen, R. Busson, J. Hoogmartens, H. Vanderhaeghe, and L. Verbist. 1987. Identification of novel erythromycin derivatives in mother liquor concentrates of *Streptomyces erythraeus*. *J. Antibiot.* (Tokyo) **40:**1–6.

27. Kieser, T., M. J. Bibb, M. J. Buttner, K. F. Chater, and D. A. Hopwood. 2000. *Practical* Streptomyces *Genetics*. John Innes Foundation, Colney, United Kingdom.

28. Kim, B. S., T. A. Cropp, B. J. Beck, D. H. Sherman, and K. A. Reynolds. 2002. Biochemical evidence for an

editing role of thioesterase II in the biosynthesis of the polyketide pikromycin. *J. Biol. Chem.* **277:**48028–48034.

29. **Lambalot, R. H., A. M. Gehring, R. S. Flugel, P. Zuber, M. LaCelle, M. A. Marahiel, R. Reid, C. Khosla, and C. T. Walsh.** 1996. A new enzyme superfamily—the phosphopantetheinyl transferases. *Chem. Biol.* **3:**923–936.

30. **Lau, J., D. E. Cane, and C. Khosla.** 2000. Substrate specificity of the loading didomain of the erythromycin polyketide synthase. *Biochemistry* **39:**10514–10520.

31. **Lau, J., C. Tran, P. Licari, and J. Galazzo.** 2004. Development of a high cell-density fed-batch bioprocess for the heterologous production of 6-deoxyerythronolide B in *Escherichia coli. J. Biotechnol.* **110:**95–103.

32. **Leaf, T., L. Cadapan, C. Carreras, R. Regentin, S. Ou, E. Woo, G. Ashley, and P. Licari.** 2000. Precursor-directed biosynthesis of 6-deoxyerythronolide B analogs in *Streptomyces coelicolor*: understanding precursor effects. *Biotechnol. Prog.* **16:**553–556.

33. **Liou, G. F., J. Lau, D. E. Cane, and C. Khosla.** 2003. Quantitative analysis of loading and extender acyltransferases of modular polyketide synthases. *Biochemistry* **42:**200–207.

34. **Liu, L., A. Thamchaipenet, H. Fu, M. Betlach, and G. Ashley.** 1997. Biosynthesis of 2-nor-6-deoxyerythronolide B by rationally designed domain substitution. *J. Am. Chem. Soc.* **119:**10553–10554.

35. **McDaniel, R., S. Ebert-Khosla, D. A. Hopwood, and C. Khosla.** 1993. Engineered biosynthesis of novel polyketides. *Science* **262:**1546–1550.

36. **McDaniel, R., A. Thamchaipenet, C. Gustafsson, H. Fu, M. Betlach, and G. Ashley.** 1999. Multiple genetic modifications of the erythromycin polyketide synthase to produce a library of novel "unnatural" natural products. *Proc. Natl. Acad. Sci. USA* **96:**1846–1851.

37. **Menzella, H. G., J. R. Carney, and D. V. Santi.** 2007. Rational design and assembly of synthetic trimodular polyketide synthases. *Chem. Biol.* **14:**143–151.

38. **Menzella, H. G., R. Reid, J. R. Carney, S. S. Chandran, S. J. Reisinger, K. G. Patel, D. A. Hopwood, and D. V. Santi.** 2005. Combinatorial polyketide biosynthesis by de novo design and rearrangement of modular polyketide synthase genes. *Nat. Biotechnol.* **23:**1171–1176.

39. **Menzella, H. G., S. J. Reisinger, M. Welch, J. T. Kealey, J. Kennedy, R. Reid, C. Q. Tran, and D. V. Santi.** 2006. Redesign, synthesis and functional expression of the 6-deoxyerythronolide B polyketide synthase gene cluster. *J. Ind. Microbiol. Biotechnol.* **33:**22–28.

40. **Murli, S., J. Kennedy, L. C. Dayem, J. R. Carney, and J. T. Kealey.** 2003. Metabolic engineering of *Escherichia coli* for improved 6-deoxyerythronolide B production. *J. Ind. Microbiol. Biotechnol.* **30:**500–509.

41. **Murli, S., K. S. MacMillan, Z. Hu, G. W. Ashley, S. D. Dong, J. T. Kealey, C. D. Reeves, and J. Kennedy.** 2005. Chemobiosynthesis of novel 6-deoxyerythronolide B analogues by mutation of the loading module of 6-deoxyerythronolide B synthase 1. *Appl. Environ. Microbiol.* **71:**4503–4509.

42. **Mutka, S. C., J. R. Carney, Y. Liu, and J. Kennedy.** 2006. Heterologous production of epothilone C and D in *Escherichia coli. Biochemistry* **45:**1321–1330.

43. **Peiru, S., H. G. Menzella, E. Rodriguez, J. Carney, and H. Gramajo.** 2005. Production of the potent antibacterial polyketide erythromycin C in *Escherichia coli. Appl. Environ. Microbiol.* **71:**2539–2547.

44. **Pfeifer, B. A., S. J. Admiraal, H. Gramajo, D. E. Cane, and C. Khosla.** 2001. Biosynthesis of complex polyketides in a metabolically engineered strain of *E. coli. Science* **291:**1790–1792.

45. **Pieper, R., S. Ebert-Khosla, D. Cane, and C. Khosla.** 1996. Erythromycin biosynthesis: kinetic studies on a fully active modular polyketide synthase using natural and unnatural substrates. *Biochemistry* **35:**2054–2060.

46. **Pieper, R., G. Luo, D. E. Cane, and C. Khosla.** 1995. Cell-free synthesis of polyketides by recombinant erythromycin polyketide synthases. *Nature* **378:**263–266.

47. **Pitera, D. J., C. J. Paddon, J. D. Newman, and J. D. Keasling.** 2007. Balancing a heterologous mevalonate pathway for improved isoprenoid production in *Escherichia coli. Metab. Eng.* **9:**193–207.

48. **Rawlings, B. J.** 2001. Type I polyketide biosynthesis in bacteria. Part A. Erythromycin biosynthesis. *Nat. Prod. Rep.* **18:**190–227.

49. **Rodriguez, E., and H. Gramajo.** 1999. Genetic and biochemical characterization of the alpha and beta components of a propionyl-CoA carboxylase complex of *Streptomyces coelicolor* A3(2). *Microbiology* **145:**3109–3119.

50. **Sambrook, J., and D. W. Russell.** 2001. *Molecular Cloning: a Laboratory Manual*, 3rd ed. Cold Spring Harbor Laboratory Press, Cold Spring Harbor, NY.

51. **Sanchez, C., L. Du, D. J. Edwards, M. D. Toney, and B. Shen.** 2001. Cloning and characterization of a phosphopantetheinyl transferase from *Streptomyces verticillus* ATCC15003, the producer of the hybrid peptide-polyketide antitumor drug bleomycin. *Chem. Biol.* **8:**725–738.

52. **Spencer, J. B., and P. M. Jordan.** 1992. Purification and properties of 6-methylsalicylic acid synthase from *Penicillium patulum. Biochem. J.* **288**(Pt. 3):839–846.

53. **Staunton, J., and K. J. Weissman.** 2001. Polyketide biosynthesis: a millennium review. *Nat. Prod. Rep.* **18:**380–416.

54. **Tang, L., S. Shah, L. Chung, J. Carney, L. Katz, C. Khosla, and B. Julien.** 2000. Cloning and heterologous expression of the epothilone gene cluster. *Science* **287:**640–642.

55. **Tsuji, S. Y., N. Wu, and C. Khosla.** 2001. Intermodular communication in polyketide synthases: comparing the role of protein-protein interactions to those in other multidomain proteins. *Biochemistry* **40:**2317–2325.

56. **Vallari, D. S., S. Jackowski, and C. O. Rock.** 1987. Regulation of pantothenate kinase by coenzyme A and its thioesters. *J. Biol. Chem.* **262:**2468–2471.

57. **Wang, Y., and B. A. Pfeifer.** 2008. 6-Deoxyerythronolide B production through chromosomal localization of the deoxyerythronolide B synthase genes in *E. coli. Metab. Eng.* **10:**33–38.

58. **Ward, S. L., R. P. Desai, Z. Hu, H. Gramajo, and L. Katz.** 2007. Precursor-directed biosynthesis of 6-deoxyerythronolide B analogues is improved by removal of the initial catalytic sites of the polyketide synthase. *J. Ind. Microbiol. Biotechnol.* **34:**9–15.

59. **Watanabe, K., M. A. Rude, C. T. Walsh, and C. Khosla.** 2003. Engineered biosynthesis of an ansamycin polyketide precursor in *Escherichia coli. Proc. Natl. Acad. Sci. USA* **100:**9774–9778.

60. **Weber, J. M., C. K. Wierman, and C. R. Hutchinson.** 1985. Genetic analysis of erythromycin production in *Streptomyces erythreus. J. Bacteriol.* **164:**425–433.

61. **Xue, Q., G. Ashley, C. R. Hutchinson, and D. V. Santi.** 1999. A multiplasmid approach to preparing large libraries of polyketides. *Proc. Natl. Acad. Sci. USA* **96:**11740–11745.

62. **Zhanel, G. G., T. Hisanaga, K. Nichol, A. Wierzbowski, and D. J. Hoban.** 2003. Ketolides: an emerging treatment for macrolide-resistant respiratory infections, focusing on *S. pneumoniae. Expert Opin. Emerg. Drugs* **8:**297–321.

63. **Ziermann, R., and M. C. Betlach.** 1999. Recombinant polyketide synthesis in *Streptomyces*: engineering of improved host strains. *Biotechniques* **26:**106–111.

Genetic Engineering of Acidic Lipopeptide Antibiotics

RICHARD H. BALTZ, KIEN T. NGUYEN, AND DYLAN C. ALEXANDER

27

27.1. INTRODUCTION

Lipopeptide antibiotics are produced by actinomycetes, bacilli, and fungi. The clinically useful lipopeptide antibacterial agents include the basic or cationic polymyxins produced by *Bacillus* sp. (36) and the acidic, Ca^{2+}-dependent daptomycin produced by *Streptomyces roseosporus* (2, 9, 23), which acts as a cationic peptide when bound to Ca^{2+} (8, 59). The antifungal echinocandins are produced by several fungal species (64).

The biosynthesis of lipopeptides has been studied most extensively in the actinomycetes, where the gene clusters for daptomycin, A54145, calcium-dependent antibiotic (CDA), and friulimicin have been cloned, sequenced, and analyzed (6, 31, 44, 45, 48). Other related lipopeptide antibiotics produced by actinomycetes have been reviewed (9, 61) and will not be discussed here. Biosynthesis of the peptide portions of these molecules is carried out by nonribosomal peptide synthetases (NRPSs). The mechanisms of assembly of peptides by NRPSs are now fairly well understood (21, 56). NRPSs are often composed of several very large protein subunits that carry out multiple enzymatic functions. Each protein subunit often contains several modules comprising three or more enzymatic functions required for processing specific amino acids. Typical modules have three enzymatic domains: a condensation (C) domain for coupling amino acids, an adenylation (A) domain for binding and activating specific amino acids, and a thiolation (T) domain for tethering activated amino acids and growing peptides to a phosphopantetheine swinging arm. Some modules contain additional domains, such as an epimerase (E) to convert L-amino acids to their corresponding D-isomers or a methyltransferase (M) to N-methylate specific amino acids. Also, the final module often has a thioesterase (TE) domain to cleave and, in the case of cyclic peptides, cyclize the completed peptide chain. Typical extension modules are structured as C-A-T, C-A-T-E, C-A-M-T, or C-A-T-TE. Some starter modules have A-T didomains. The individual enzymatic domains are connected by interdomain linkers, and individual subunits of NRPS multienzymes are held together by interpeptide docking sequences (21, 56). In cyclic acidic lipopeptides such as daptomycin and A54145 (Fig. 1), the biosynthesis is initiated by the coupling of a long-chain lipid to the N-terminal Trp_1 by a special C^{III} domain (Fig. 2; 44), or $^{F}C_L$, indicating that it couples a fatty acid to an L-amino acid.

Because there are no examples of genetic engineering of polymyxins in bacilli or of echinocandins in fungi, this chapter is restricted to the genetic engineering of the daptomycin and A54145 biosynthetic pathways. These two pathways are of particular note because their biosynthetic machinery may provide a means to generate novel lipopeptides to address the one currently understood shortcoming of daptomycin. Daptomycin is very effective at treating skin and skin structure infections (2) and bacteremia and right-sided endocarditis caused by *Staphylococcus aureus* (23), but is not effective at treating *Streptococcus pneumoniae* pulmonary infections (53) because it most likely becomes sequestered in surfactant in the lung (8, 57). Thus, improving the antibiotic activity in the presence of pulmonary surfactant can be addressed by combinatorial biosynthesis.

In this chapter, we use the terminology of "genetic engineering" and "combinatorial biosynthesis." The latter term embraces a broad set of methodologies—including genetic engineering, the use of mutants blocked in specific biosynthetic steps, and the exploitation of natural variations in substrates and feeding unnatural substrates—to expand the numbers of compounds generated beyond what can be achieved by genetic engineering alone (4, 5). These approaches are particularly useful to generate derivatives of complex cyclic peptides that cannot be generated and scaled up economically by medicinal chemistry approaches.

27.2. CLONING OF LIPOPEPTIDE BIOSYNTHETIC GENES

A common method to clone antibiotic biosynthetic gene clusters is to construct a cosmid library and to screen for overlapping cosmids containing the genes of interest by using probes to conserved sequences as entry points. For relatively small antibiotic gene clusters, it is possible to isolate complete biosynthetic gene clusters on individual cosmids. Another approach is to clone antibiotic gene clusters in bacterial artificial chromosome (BAC) vectors, which can accommodate DNA inserts of >100 kb (24, 44; D. C. Alexander, J. Rock, X. He, V. Miao, P. Brian, and R. H. Baltz, in preparation).

27.2.1. Cosmid and BAC Cloning of Daptomycin Genes

S. roseosporus normally produces A21978C factors (Fig. 1). Daptomycin is the *N*-decanoyl derivative produced by

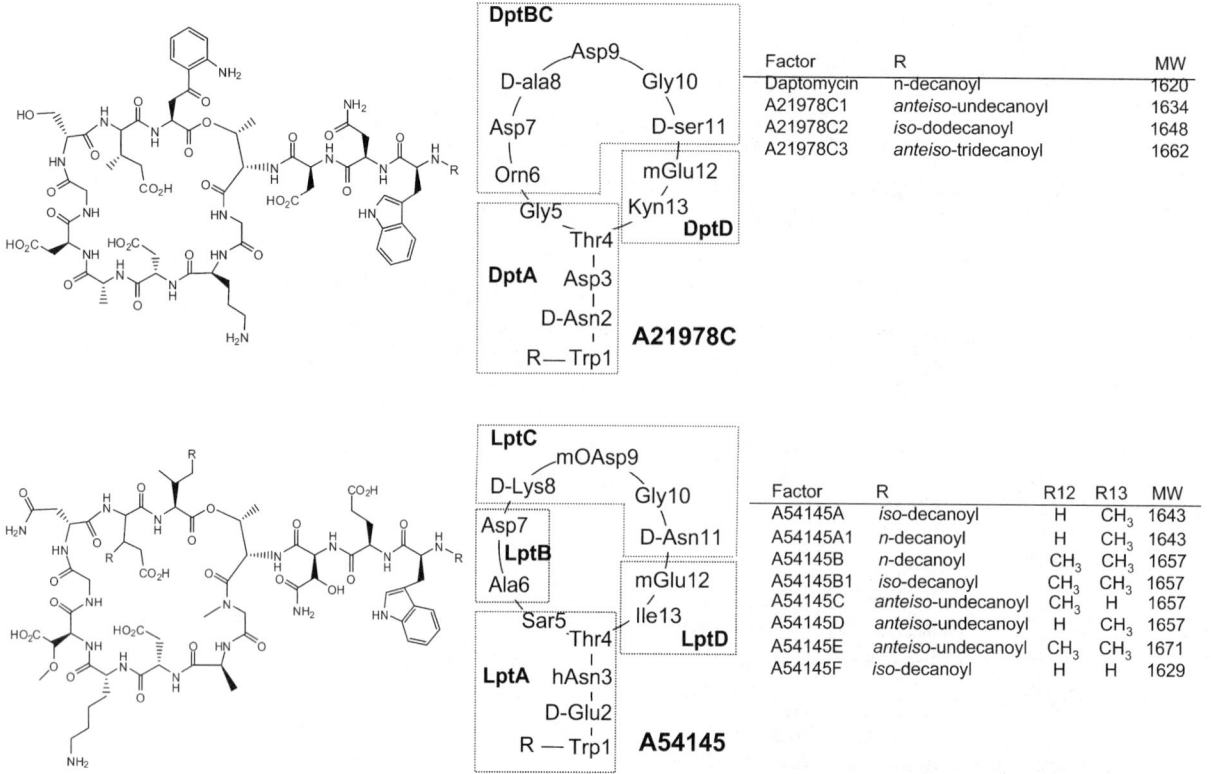

FIGURE 1 Structures of lipopeptide antibiotics and of NRPS subunits. (Top) A21978C factors naturally produced by *S. roseosporus* A21978.6 and A21978.65 (Table 1), and daptomycin, which is produced by feeding decanoic acid during fermentation. (Bottom) A54145 factors naturally produced by *S. fradiae* A54145.

feeding decanoic acid during fermentation (6, 9). The daptomycin biosynthetic gene cluster was localized to one end of the *S. roseosporus* linear chromosome by Tn5099 insertion mutagenesis (42, 43). Tn5099 (30) contains a hygromycin resistance gene (Hm^r) from *Streptomyces hygroscopicus* that can be expressed in streptomycetes and *Escherichia coli*. Thus the DNA segment flanking the Tn5099 insertions in the daptomycin gene cluster was readily cloned in *E. coli* and sequenced to confirm the localization in NRPS domains. The transposon insertions localized the

daptomycin gene cluster to a region centered ca. 450 kb from one end of the linear chromosome and established the direction of transcription of NRPS genes to the chromosome proximal end (43). A cosmid library of *S. roseosporus* DNA prepared in pKC1471 (see Table 1 for a list of key strains and plasmids) was screened using sequences flanking the transposon insertions, and overlapping cosmids were obtained (43). Two overlapping cosmids with large inserts were sequenced, but the inserts did not span the complete daptomycin gene cluster (44).

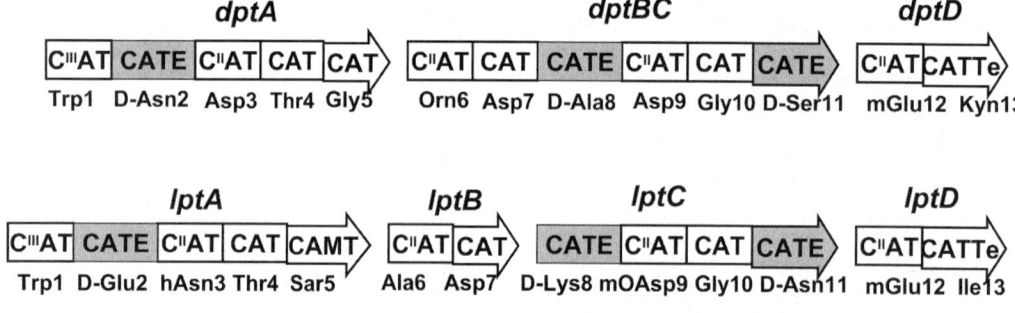

FIGURE 2 Functional organization of NRPS. The condensation domains designated as C (or ^{L}C$_L$), C^{II} (or ^{D}C$_L$), and C^{III} (or ^{F}C$_L$) normally couple L-amino acids, D-amino acids, and long-chain fatty acids to L-amino acids, respectively. The shaded modules show CATE structures followed by C^{II} (^{D}C$_L$) condensation domains. A, adenylation domain; E, epimerase domain; M, methyltransferase domain; T, thiolation domain; TE, thioesterase domain. Figure modified from reference 6.

TABLE 1 Key strains and plasmids used for genetic engineering of lipopeptide biosynthesis

Strains and plasmids	Relevant characteristics[a]	Reference(s)
E. coli		
DH10B	General cloning strain	Invitrogen
XL1-Blue-MR	General cloning strain	Stratagene
S17-1	Conjugation donor	12
ML22	DH10B (pUZ8002) conjugation donor	15
S. ambofaciens		
BES1260	Wild type, Srm$^+$ Np$^+$	ATCC 15154; 39, 54
BES1946	Srm^{++} Np$^-$ from BES1260	22, 54
BES2074	Srm$^-$ (*srmR7*::Tn*10*) Np$^-$ from BES1946	54
S. fradiae		
NRRL 18160	A54145 producer (Lpt$^+$)	13
XH25	Sm^r (*rpsL*) Lpt$^+$, derived from NRRL 18160	In prep[b]
DA613	XH25 Δ*lptI*::*tsr*	In prep
JR397	XH25 Δ*lptI*	In prep
DA728	XH25 Δ*lptD*	In prep
DA895	XH25 Δ*lptD* Δ*lptI*::*tsr*	In prep
DA740	XH25 Δ*lptBCD*	In prep
DA901	XH25 Δ*lptBCD* Δ*lptI*::*tsr*	In prep
DA1187	XH25 Δ*lptEFABCDGHJKLMNPI*	In prep
S. lividans		
TK23-521	Δ*act*::*ermE*	52
S. roseosporus		
A21978.6	Wild type, Dap$^+$	NRRL 11379; 19
UA343 (A21978.65)	Dap^{++}	44
UA117	UA343 *rpsL7* (Sm^r) Dap^{++}	46
UA378	UA117 Δ*dptD*::*ermE*	46
UA474	UA117 Δ*dptA* Δ*dptD*::*ermE*	15
UA431	UA117 Δ*dptEFABCDGHIJ*	50
KN100	UA117 Δ*dptBCD*	50
KN125	UA117 Δ*dptBCDGHIJ*	50
Plasmids		
pKC1471	3.0-kb cosmid containing Spc^r, *rep*^{p15A} MCS	43
pBR322	Tet^r, Amp^r	55
pRHB538	*rpsL rep*^{ts} Am^r	32
pHM11a	Hm^r *oriT ermEp* att/int*^{IS117}	47
pRT802	Kan^r *att/int*$^{\phi BT1}$	27
pRB04	pHM11a::*dptD*	46
pMF26	pHM11a::*cdaPS3*	46
pMF30	pHM11a::*lptDGHJKLMN*	46
pKN38	pHM11a::*dptGHJ*	49
pStreptoBAC V	BAC *oriT att/int*$^{\phi C31}$ Am^r	44
pCV1	pStreptoBAC V::128 kb (*dpt*)	44
pDA300	pStreptoBAC V::*dptMNEFABCDGHIJ*	50
pLT01	pStreptoBAC V::*dptPMNEFA*	15
pLT02	pStreptoBAC V::*dptEF ermEp* dptA*	15
pKN24	pStreptoBAC V::*ermEp* dptEFA*	49
pKN25	pStreptoBAC V::*ermEp* dptBCD*	49
pKN26	pStreptoBAC V::*ermEp* dptBCDGHIJ*	49
pKN54	pRT802:: *ermEp* *	Submitted[c]

(Continued on next page)

TABLE 1 Key strains and plasmids used for genetic engineering of lipopeptide biosynthesis *(continued)*

Strains and plasmids	Relevant characteristics[a]	Reference(s)
pKN55	pKN54::*ermEp** *dptIJ*	Submitted
pCB01	pECBAC1 containing *lptEFABCDGHJKLMNPI*	In prep
pDA2002	pECBAC1 *oriT att/int*$^{\phi C31}$ Am^r::*lptEFABCDGHJKLMNPI*	In prep
pJR153	pECBAC1 *oriT att/int*$^{\phi BT1}$ Am^r::*lptEFABCDGHJKLMNPI*	In prep
pDA2012	pECBAC1 *oriT att/int*$^{\phi C31}$ Am^r::*ermEp** *lptBCDGHJKLMNPI*	In prep
pDA2016	pECBAC1 *oriT att/int*$^{\phi C31}$ Am^r::*ermEp** *lptDGHJKLMNPI*	In prep
pDA2048	pECBAC1 *oriT att/int*$^{\phi C31}$ Am^r::*ermEp** *lptBC*	In prep
pDA2054	pECBAC1 *oriT att/int*$^{\phi C31}$ Am^r::*lptEFABCDGHJKLMNP*	In prep
pDA2106	pECBAC1 *oriT att/int*$^{\phi BT1}$ Hm^r::*ermEp** *lptDGHJKLMNPI*	In prep

[a]Dap$^+$, produces A21978C; Dap^{++}, produces higher levels of A21978C than wild type; Lpt$^+$, produces A54145; Np$^+$, produces netropsin; Np$^-$, defective in netropsin production; Srm$^+$, produces spiramycin; Srm^{++}, stably produces higher levels of spiramycin than wild type.
[b]In prep, Alexander, Rock, He, et al., in preparation.
[c]Submitted, Nguyen et al., submitted.

To complete the sequencing and to facilitate the genetic engineering experiments described below, the complete daptomycin biosynthetic gene cluster was cloned in pStreptoBac V (44), which has the following features: pUC19 (which is removed during the cloning process) for high-copy-number replication in *E. coli*; *sacB* for selection of BACs containing inserts; *oriT* for conjugal transfer from *E. coli* to streptomycetes; φC31 *att/int* for stable insertion into streptomycete chromosomes; and an apramycin resistance (Am^r) gene to select for clones containing BACs in *E. coli* and streptomycetes. To prepare a BAC library, *S. roseosporus* cells were grown overnight in 25 ml of medium F10A (see section 27.7 for media composition), washed, and embedded in 1% Seakem GTG agarose. Agarose plugs were incubated in 2 mg/ml lysozyme for 2 h, then transferred to a solution of 0.2 mg/ml proteinase K and 1% sarcosine (wt/vol) and incubated at 50°C overnight to release high-molecular-weight genomic DNA. Agarose plugs were washed in 0.1 mM EDTA and treated with BamHI. Partially digested DNA was separated by pulsed-field gel electrophoresis, and 200 ng of DNA in the range of 75 to 145 kb was recovered by electroelution and then ligated to BamHI-digested, dephosphorylated pStreptoBAC V. The BamHI treatment of pStreptoBAC V releases the fragment containing pUC19, so clones containing large inserts can stably replicate in *E. coli* from the low-copy F replicon of the BAC. BAC clones were introduced into *E. coli* DH10B and screened for Am^r, and ca. 2,000 clones were archived in microtiter plates. Clones containing daptomycin (*dpt*) biosynthetic genes were identified by PCR amplification of segments containing published NRPS sequences from cosmid clones (43); pCV1 was determined to contain the complete *dpt* gene cluster (Fig. 3A), as confirmed by heterologous expression in *Streptomyces lividans* (44, 52; see section 27.4.1.1).

27.2.2. Cosmid and BAC Cloning of A54145 Genes

A cosmid library was prepared in pStreptoCos X (45). pStreptoCos X is derived from SuperCos (Stratagene) and carries *oriT* and φC31 *attP/int* functions to facilitate conjugal transfer from *E. coli* to streptomycetes and site-specific integration into streptomycete chromosomal φC31 *attB* sites.

Streptomyces fradiae cells were grown in F10A medium and incubated at 30°C with shaking at 200 rpm. Cells were harvested and genomic DNA was released in agarose plugs

as described in section 27.2.1 for *S. roseosporus* (44). DNA was partially digested with BamHI, and fragments of 30 to 60 kb were purified by gel electrophoresis, ligated in the BamHI site of pStreptoCos X, then packaged with Gigapack III Gold packaging extract (Stratagene). Packaged DNA was introduced into *E. coli* XL1-Blue-MR, and a library was screened using a PCR product amplified from an NRPS sequence from the daptomycin gene cluster (43). Cosmid clones flanking the original clone were obtained by walking, and the genes were sequenced and assembled (44).

A pECBAC1 (24) library of *S. fradiae* DNA was constructed in *E. coli* DH10B by Amplicon Express (Pullman, WA), using Sau3AI-digested genomic DNA purified from *S. fradiae* protoplasts embedded in agarose as described in section 27.2.1, and the DNA was inserted into the BamHI-digested pECBAC1. BAC clones believed to contain the complete *lpt* biosynthetic gene cluster were identified by PCR screening with primer sets specific for *orf21* upstream of the cluster and *lptI* gene at the downstream end of the cluster (45). pCB01 was chosen for further work because it had the largest insert (>100 kb; Fig. 3). It was further modified for conjugal transfer and site-specific insertion in streptomycetes by inserting a cassette containing the Am^r gene, *oriT* site, and φC31 *attP/int* functions, generating pDA2002 (Alexander, Rock, He, et al., in preparation) (see section 27.5.1). pDA2002 was used to produce A54145 factors in *Streptomyces ambofaciens* and *S. roseosporus* (see section 27.4.1.2) and as a starting vector to generate a number of additional vectors containing different sets of the A54145 biosynthetic genes (Table 1; Fig. 3). Similarly, pJR153 contains a cassette containing the Hm^r gene, *oriT* site, and φBT1 *attP/int* functions integrated into pCB01 (Fig. 3B).

27.3. ORGANIZATION OF DAPTOMYCIN AND A54145 BIOSYNTHETIC GENE CLUSTERS

In order to design methodology to carry out genetic engineering and combinatorial biosynthesis of novel lipopeptides related to daptomycin and A54145, it was important to understand how the gene clusters are organized and transcribed. Figure 1 shows the structures of daptomycin and A54145. Both are composed of cyclic tridecapeptides with 10 amino acid rings and three exocyclic amino acids coupled to long-chain fatty acids through the N-terminal

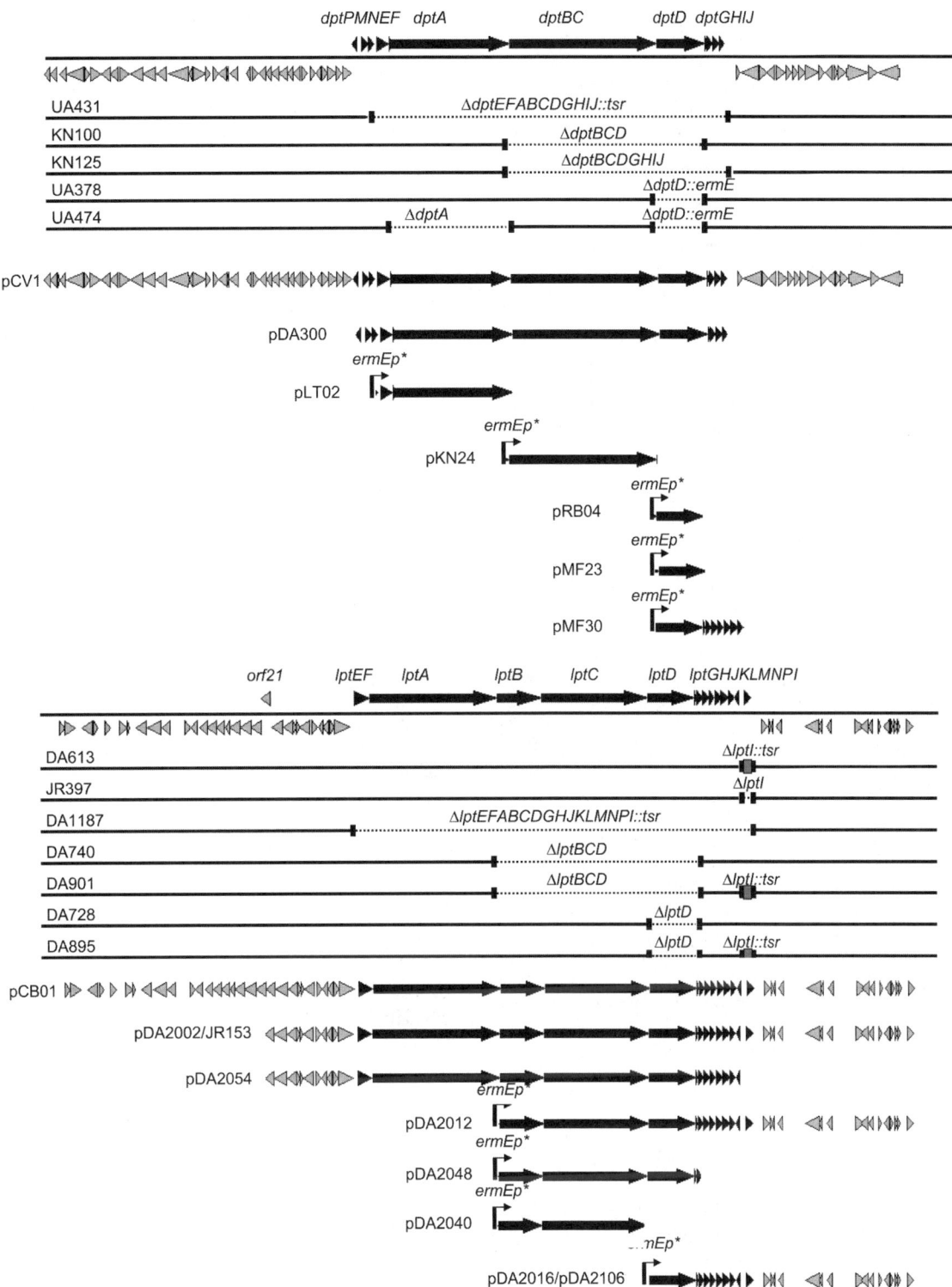

FIGURE 3 Ectopic *trans*-complementation systems for S. *roseosporus* and S. *fradiae*. (Top) The *dpt* genes on BAC pCV1 are shown in black. S. *roseosporus* strains deleted for different *dpt* genes (dotted lines) and plasmids containing different sets of *dpt* genes (pDA300, pLT02, pKN24, and pRB04), *cdaPS3* (pMF23), and *lptD* (pMF30) are also shown (see Table 1 for plasmid details). (Bottom) The *lpt* genes on BAC pCB01 are shown in black, and S. *fradiae* deletion mutants and plasmids for *trans*-complementation are shown below.

Trp_1. Both have D-amino acids at positions 2, 8, and 11 and achiral amino acids, Gly or sarcosine (Sar), at positions 5 and 10. They both have Ca^{2+} binding motifs DXDG at positions 7–10 and 3-methyl-glutamic acid (3mGlu) at position 12, and both cyclize through the hydroxyl group of Thr_4 to form depsipeptides. In spite of their similarities, daptomycin and A54145 have different amino acids at eight positions. In addition to $3mGlu_{12}$, A54145 has the modified amino acids hydroxy-Asn_3 ($hAsn_3$), Sar_5, and methoxy-Asp_9 ($mOAsp_9$).

Figures 2 and 3 show the biosynthetic gene clusters for daptomycin (*dpt* genes) and A54145 (*lpt* genes). The key genes involved in lipopeptide biosynthesis begin with *dptE* and *dptF* and the fused *lptEF* genes, which encode acyl-coenzyme A ligase and acyl carrier protein functions involved in coupling the long-chain fatty acids to the N terminus of Trp_1 to initiate lipopeptide biosynthesis (6, 63). Immediately downstream are the NRPS genes, starting with *dptA* and *lptA*; both encode proteins containing five modules with the enzymatic functions required to incorporate the first five amino acids. Module 1 of each protein has a C^{III} ($^F C_L$) condensation domain required for coupling long-chain fatty acids to L-Trp_1. Both proteins have E domains in module 2 to convert L-amino acids to their corresponding D-isomers. LptA has an additional M domain to N-methylate Gly to Sar_5. The next genes are the related *dptBC* and the *lptB* and *lptC* genes that encode the incorporation of the next six amino acids. These genes have E domains appropriately placed for the D-amino acids at positions 8 and 11. Both DptBC and LptB proteins start with C^{II} ($^D C_L$) condensation domains, which normally follow upstream D-amino acids, even though these follow the achiral Gly and Sar. These $^D C_L$ domains may be vestiges from an ancestral NRPS pathway that encoded D-amino acids at this position in the ring, as observed in the CDA and friulimicin pathways (6). The final NRPS genes are *dptD* and *lptD*, which encode proteins with two modules for the incorporation of the final two amino acids. The modules for the coupling of $3mGlu_{12}$ to the preceding D-amino acids start with $^D C_L$ condensation domains, and each module 13 terminates with a thioesterase (TE) domain responsible for cyclization and release of the completed lipodepsipeptides. All of the other modules in the NRPS genes from both pathways required for the coupling of L-amino acids to L-amino acids have typical CAT domains starting with $^L C_L$ condensation domains.

Downstream of the *dptD* gene are *dptG*, *dptH*, *dptI*, and *dptJ*. DptG is a protein of unknown function required for normal production of daptomycin; DptH is an editing thioesterase; DptI is a methyltransferase involved in two-step conversion of α-ketoglutarate to 3mGlu (38); and DptJ is involved in the formation of Kyn. Deletion of *dptG*, *dptH*, *dptI*, or *dptJ* caused reduced lipopeptide production. Deletion of *dptI* also caused the production of daptomycin analogs containing Glu_{12} and eliminated production of $3mGlu_{12}$-containing lipopeptides (49).

Downstream of *lptD* are *lptG* and *lptH*, with functions similar to *dptG* and *dptH*. Further downstream are the *lptJ*, *lptK*, and *lptL* genes, involved in the modification of amino acids. The *lptJ* and *lptK* genes encode oxygenase and methyltransferase enzymes involved in biosynthesis of $mOAsp_9$, and the *lptL* gene encodes an oxygenase involved in biosynthesis of $hAsn_3$ (D. C. Alexander, J. Rock, J.-Q. Gu, C. Mascio, M. Chu, P. Brian, and R. H. Baltz, in preparation). LptL is related to the AsnO non-heme Fe^{2+}/α-ketoglutarate-dependent L-Asn oxygenase involved in CDA biosynthesis (6, 60). Further downstream are the *lptM*, *lptN*, and *lptP* genes, which have homologs *dptM*, *dptN*, and *dptP* located upstream of the *dptE* gene. These genes are likely to be involved in lipopeptide transport and/or resistance. Downstream of *lptP* is the *lptI* gene involved in methylation of α-ketoglutarate in the biosynthesis of 3mGlu. Deletion of *lptI* eliminated production of $3mGlu_{12}$-containing lipopeptides and resulted in improved yields of Glu_{12}-containing lipopeptides (Alexander, Rock, He, et al., in preparation).

Much of the daptomycin biosynthetic gene cluster is transcribed as a single >50-kb mRNA molecule (15). This is supported by reverse transcription-PCR across the junctions of the *dptEFABCDGH* genes. The transcript is likely to include the *dptI* and *dptJ* genes, which are just downstream of *dptH* and transcribed in the same direction (44; Fig. 3) but were not included in the reverse transcription-PCR analysis (15). Furthermore, the *dptA*, *dptBC*, and *dptD* genes have overlapping stop and start codons, suggesting that they may be translationally coupled to assure the production of the NRPS proteins in stoichiometric amounts. For example, inspection of the region spanning the <u>TGA</u> stop codon in *dptA* and the **GTG** start codon in *dptBC* revealed a nucleotide sequence TG**GTG**AACCGC. The region spanning the <u>TGA</u> stop of *dptBC* and the **ATG** start of *dptD* has the sequence GG**ATG**ACGCAG.

The *lptEFABCDGHJKLMN* genes are transcribed in the same direction, and the *lptA*, *lptB*, *lptC*, and *lptD* genes have overlapping stop and start codons, all using TGA stop codons that overlap with GTG, ATG, and possibly CTG start codons. It is therefore likely that these genes are transcribed as a large multicistronic mRNA, as observed with the *dpt* genes. In this case, however, the *lptP* gene is transcribed in the opposite direction from a different promoter (Fig. 3) and *lptI* is expressed from its own promoter (Alexander, Rock, He, et al., in preparation). The transcriptional organization of the daptomycin and A54145 NRPS genes, with overlapping stop and start codons, poses tactical issues for ectopic expression of specific genes for combinatorial biosynthesis: it is important to excise specific NRPS genes without disrupting the natural transcriptional and translational relationships between the remaining genes.

27.4. EXPRESSION SYSTEMS FOR GENETIC ENGINEERING

There are at least three ways to set up genetic engineering systems for lipopeptides and other complex biosynthetic pathways. The first is to engineer the production of hybrid pathways in a heterologous host by preparing BAC clones containing the complete biosynthetic gene clusters in *E. coli*, engineering the pathway in *E. coli* in any of a number of ways, including λ Red-mediated recombination (18), then transferring the engineered pathway into a heterologous host by conjugation and site-specific integration into the chromosome. The second approach is to delete the complete pathway from a lipopeptide-producing host, engineer the biosynthetic gene cluster on a BAC in *E. coli*, transfer the engineered pathway into the homologous host lacking the genes by conjugation, and insert the genes by site-specific integration into an ectopic chromosomal site. A third approach is to engineer homologous lipopeptide-producing hosts to contain deletions of one or more of the lipopeptide biosynthetic genes, and to express engineered genes from ectopic loci in the chromosome using strong promoters to drive transcription. Each of these approaches is discussed below.

27.4.1. Heterologous Expression of Complete Gene Clusters

27.4.1.1. Daptomycin Production in S. lividans

The pStreptoBac V derivative pCV1 containing the complete daptomycin biosynthetic gene cluster was introduced into S. lividans strains TK23, TK64 (str-6), and TK23-521 (Δact) by protoplast transformation (44, 52). S. lividans strains carrying pCV1 were grown on an A9 agar slant (44; see section 27.7) containing 100 μg/ml apramycin (Am) for 7 to 10 days at 28°C. A mixed mycelial and spore suspension was made in 4 ml of 0.1% Tween 80, and 2 ml was inoculated into 40 ml of seed medium A345 (44) containing 25 μg/ml Am in baffled flasks and shaken at 240 rpm and 5-cm orbit at 30°C for 24 to 28 h. Then, 2.5 ml of culture was added to flasks containing 50 ml of medium B (52). Replicate flasks were sampled for lipopeptide production by high-pressure liquid chromatography for 2 to 6 days.

S. lividans TK64 harbors str-6, an rpsL mutation that causes the K88E substitution in ribosomal protein S12 that enhances actinorhodin (Act) production in S. lividans TK24 (51). The introduction of pCV1 caused the production of A21978C factors in both TK64 and TK23, accompanied by high-level production of Act and CDA factors in both cases. The production of Act interfered with the quantification of A21978C factors. Part of the Act gene cluster was deleted, and the resulting strain, S. lividans TK23-521 (52) containing pCV1, produced about 20 mg/liter of A21978C$_{1-3}$. When medium B was supplemented with 3 g/liter K$_2$HPO$_4$, A21978C$_{1-3}$ levels increased to 55 mg/liter at 72 h and were readily separated from the phosphorylated CDAs. Further development of this system would benefit from the deletion of the CDA gene cluster.

27.4.1.2. A54145 Production in S. ambofaciens and S. roseosporus

S. ambofaciens normally produces the 16-membered macrolide antibiotic spiramycin and the pyrrole-amidine oligopeptide antibiotic netropsin. S. ambofaciens is a potentially useful host for heterologous expression because it is relatively nonrestricting for bacteriophage and plasmid DNA (17, 39), and stable mutants defective in spiramycin production are available (22, 54; Table 1). S. ambofaciens BES2074, which is defective in spiramycin and netropsin biosynthesis (54), and S. roseosporus UA431, which is deleted for the daptomycin biosynthetic gene cluster (50), were used for heterologous expression of A54145.

S. roseosporus and S. ambofaciens strains were grown for 24 to 48 h in Trypticase soy broth (TSB), homogenized, centrifuged at 800 × g for 10 min, resuspended in TSB, centrifuged again at 800 × g for 10 min, then resuspended in TSB and chilled on ice. E. coli donor strains were grown overnight, diluted 1:40 into 10 ml of fresh LB broth plus appropriate antibiotics, and grown for 2 to 3 h. E. coli cultures were pelleted by centrifugation, washed with LB broth, pelleted by centrifugation, resuspended in 2 ml of LB broth, and mixed with mycelial fragments of S. ambofaciens or S. roseosporus in volume ratios of 1:4 and 4:1. Finally, 0.1 ml of the mix was plated onto AS-1 agar (section 27.7) supplemented with 10 mM MgCl$_2$ (12) for S. roseosporus or onto AS-1 agar supplemented with 10 mM MgCl$_2$ and 20 mM TES [N-tris(hydroxymethyl)methyl-2-aminoethanesulfonic acid] (pH 7.6) (1) for S. ambofaciens. Conjugation plates were incubated at 30°C for 18 to 24 h, then overlaid with 2.5 ml of NB soft agar (34) containing nalidixic acid (Nal) (50 μg/ml final bottom agar concentration) to select against E. coli and Am (50 μg/ml final bottom agar concentration) to select for the S. roseosporus and S. ambofaciens strains containing pDA2002. Colonies were patched onto mR2YE agar plates containing Am (50 μg/ml) and Nal (50 μg/ml).

S. roseosporus and S. ambofaciens strains containing chromosomal insertions of pDA2002 containing the A54145 gene cluster were grown for 48 h in TSB as starter cultures, and for 48 h in A355 seed medium. Fermentations were carried out in DSF medium (13) plus 0.79% (wt/vol) L-Ile. Fermentations were harvested by centrifugation of whole broth at 24,000 × g for 30 min at 4°C, and product yields were determined as described (Alexander, Rock, He, et al., in preparation). S. roseosporus DA1155 produced about 100 mg/liter of A54145 lipopeptides and S. ambofaciens DA1164 produced about 390 mg/liter, or 88% of the S. fradiae control (Alexander, Rock, He, et al., in preparation). These results indicated that either of these relatively easily manipulated strains would be a suitable host for the engineering of A54145 biosynthetic genes. It is likely that the yields can be further improved by fermentation medium development, as has been demonstrated with the S. lividans strain that produces daptomycin.

27.4.2. Expression of Individual Genes or Sets of Genes in Lipopeptide Producers

27.4.2.1. Genetic Manipulation of S. roseosporus and S. fradiae

Most streptomycetes produce restriction endonucleases that interfere with bacteriophage plaque formation, and often retard or block the introduction of plasmid DNA by protoplast transformation (3, 17, 39). S. roseosporus appears to exhibit relatively mild restriction of bacteriophage plaque formation. Ten of 11 broad-host-range streptomycete bacteriophages formed plaques on S. roseosporus (42). FP43 did not form plaques, but was able to transduce plasmid DNA containing the FP43 pac sequence (41) into S. roseosporus (42), indicating that FP43 can attach and inject DNA, but that FP43 phage DNA is restricted.

Because S. roseosporus expressed some restriction for double-stranded DNA, protoplast transformation was not pursued. Instead, conjugal transfer of plasmid DNA from E. coli was developed (42, 43). Plasmids were first introduced into E. coli S17-1 by standard transformation. S. roseosporus cells were grown at 29°C for 16 h in TSB from a 2% (vol/vol) inoculum of stationary-phase cells, washed by centrifugation, and resuspended in an equal volume of TSB. E. coli donor strains were grown overnight at 37°C in TY broth, diluted 1:10 into 10 ml of fresh TY broth plus appropriate antibiotics, and grown for 2 to 3 h. Cells were pelleted by centrifugation, washed with an equal volume of TY broth, pelleted again by centrifugation, and resuspended in 2 ml of TY broth. E. coli and S. roseosporus cells were mixed in ratios of 1:9, 5:5, and 9:1, spread on AS-1 agar, incubated at 29°C for 18 h, and overlaid with NB soft agar containing Nal plus Am or Nal plus hygromycin (Hm) to give final bottom agar concentrations of 20 μg/ml Nal, 100 μg/ml Am, or 400 μg/ml Hm to select against E. coli (Nal) and for S. roseosporus transconjugants containing plasmids expressing Am^r or Hm^r. Maximum efficiency (>10^{-5} transconjugants per recipient) was obtained when equal volumes of donor and recipient (~10^7 CFU each) were plated on AS-1 agar. Conjugations were also carried out on B agar with very poor results (42), emphasizing the importance of AS-1 agar for carrying out the intergeneric matings.

Plasmid DNA can also be introduced into *S. fradiae* spore germlings by conjugation from *E. coli* (Alexander, Rock, He, et al., in preparation). *S. fradiae* frozen spores were thawed and centrifuged at 800 × *g* for 10 min, washed with 1 ml of spore germination medium (34), and recentrifuged. Spores were resuspended in 0.5 ml of germination medium, heat shocked at 50°C for 10 min, then added to 10 ml of germination medium and shaken at 30°C for 2 to 4 h. The spore suspension was centrifuged at 800 × *g* for 10 min, washed with 10 ml of LB broth, centrifuged again, resuspended in 10 ml of LB broth, and placed on ice. *E. coli* cells grown in LB broth (section 27.4.1.2) were mixed with *S. fradiae* spore germlings in volumetric ratios of 1:4 and 4:1. The conjugation mix (0.1 ml) was plated onto ISP4 agar (Difco) supplemented with 10 mM MgCl$_2$. Plates were incubated at 30°C for 18 to 24 h, then overlaid with 3 ml of R2 modified soft agar (40) containing trimethoprim (Tmp), to select against *E. coli* donor strains, and thiostrepton (Tsr), neomycin (Nm), Am, or Hm to select for transconjugants. The final antibiotic concentrations in bottom agar were 50 µg/ml (Tmp), 10 µg/ml (Tsr), 10 µg/ml (Am), 10 µg/ml (Nm), or 150 µg/ml (Hm). *S. fradiae* colonies were picked after 4 to 6 days of incubation at 30°C and patched onto mR2YE agar plates (34) containing the appropriate antibiotics.

27.4.2.2. Isolation of Deletion Mutations in *S. roseosporus* and *S. fradiae*

The generation of mutants of *S. roseosporus* and *S. fradiae* deleted for one or more of the daptomycin (*dpt*) and A54145 (*lpt*) biosynthetic genes, respectively, was facilitated by the use of pRHB538 (Table 1), a vector that can be used for dominance selection (32). This method exploits the dominance of streptomycin susceptibility (Sm^s) over resistance (Sm^r). For instance, a strain carrying a chromosomal Sm^r mutation in the *rpsL* gene and a wild-type (Sm^s) allele on a plasmid is phenotypically Sm^s. If the wild-type *rpsL* gene is located on a temperature-sensitive plasmid containing cloned DNA segments flanking a gene to be deleted, then insertion of the plasmid into the chromosome by a single crossover can be facilitated by growing cells at the nonpermissive temperature and selecting for an antibiotic resistance marker on the plasmid. The second crossover, deleting the gene of interest, can be selected by growth on a medium containing Sm at the nonpermissive temperature for plasmid replication. This system has been used to isolate deletion mutations in the daptomycin and A54145 biosynthetic gene clusters. For example (44), a *dptD* deletion was generated in Sm^r *S. roseosporus* UA117 (*rpsL7*) by PCR amplifying 1.1- and 1.5-kb segments of DNA flanking the *dptD* gene and constructing a *dptD-ermE-dptD* cassette for deletion and insertion of the *ermE* (erythromycin resistance [Erm^r]) gene. The cassette was inserted into pRHB538, which has *oriT*, *rep*^{ts}, *rpsL* (Sm^s), and Am^r for selection. The plasmid was introduced into *E. coli* ML22 by electroporation, then transferred to UA117 by conjugation. Transconjugants were selected for Am^r, then grown at 38°C to cure the plasmid and plated on AS-1 agar containing Sm. Sm^r colonies were then screened for Erm^r (gene replacement) and Am^s (loss of plasmid) and strains with the correct phenotype were confirmed by Southern hybridization analysis and loss of daptomycin production. Other mutants containing markerless in-frame deletions of different sets of *dpt* genes (Fig. 3) were isolated directly from colonies containing single crossovers after growth at room temperature (49). In one case, the *dptA* gene was deleted and the strong constitutive promoter *ermEp** was inserted in front of the downstream *dptBC* gene (15).

S. fradiae mutants deleted for different combinations of *lpt* genes were generated in a similar manner, but without the use of positive selection. For example, *lptI* deletion mutants (Δ*lptI::tsr*) were generated using pJC12, a pRHB538 plasmid with ~2-kb DNA segments upstream and downstream of the *lptI* gene flanking the Tsr^r cassette. Recombinants containing pJC12 were grown in CSM broth plus Tsr and Tmp at 30°C. Strains were homogenized, inoculated into fresh CSM plus Tsr, grown at 38°C for 48 h, homogenized again, diluted, and grown for 48 h on mR2YE agar with Tsr. Colonies were screened for loss of pJC12 (Am^s) and presence of *lptI* replacement mutation (Tsr^r) by patching mycelia onto mR2YE agar containing Tsr or Am, and *lptI* deletions were confirmed by PCR. Mutants deleted for *lptD*, *lptBCD*, and *lptEFABCDGHJKLMNPI* (*lptEF-I*) were also generated by this procedure (Alexander, Rock, He, et al., in preparation) (Table 1 and Fig. 3).

27.4.2.3. Ectopic *trans*-Complementation in *S. roseosporus* and *S. fradiae* and Fermentation Analysis of Recombinants

To develop efficient systems for genetic engineering and combinatorial biosynthesis, it was desirable to engineer individual NRPS and other genes separately and to recombine the altered genes or gene cassettes in different combinations, using the *S. roseosporus* and *S. fradiae* deletion mutants (section 27.4.2.2) as expression hosts. Because the daptomycin and A54145 NRPS genes were both likely to be transcribed as single giant polycistronic mRNAs that are translated with overlapping stop and start codons, combinatorial biosynthesis required some additional steps to ensure success. These steps would not be necessary for smaller monocistronic genes each with its own promoter. Due to the possibility of a polar mutation affecting downstream gene function in the operon, markerless in-frame deletions were generated in most cases (50; Alexander, Rock, He, et al., in preparation). It was also necessary to engineer heterologous promoters in front of NRPS genes for complementation experiments because it was not likely that they possessed their own promoters. The individual genes were excised and expressed from different loci in the host chromosomes. To accomplish this, three different site-specific integrating plasmid vectors were used. Two of them utilize the *attP/int* genes from the streptomycete bacteriophages φC31 and φBT1 (12, 27, 35). The other uses the *att/int* functions from IS*117* (34, 47). Another important element in the experimental design was to express the genes from the ectopic loci from the strong constitutive *ermEp** promoter (10, 11). This promoter has been used successfully to drive high-level expression of cloned genes in other streptomycetes (4, 5, 58).

To test the feasibility of expressing different daptomycin and A54145 genes at different chromosomal sites, reconstruction experiments were carried out with the lipopeptide genes being expressed at up to three different loci. *S. roseosporus* recombinants were tested for the production of A21978C factors after fermentation. Fermentations were carried out in 20 ml of broth in 125-ml baffled flasks at 30°C and 200 rpm, 5-cm orbit. Starter cultures were grown in TSB for 48 h, and then 1 ml of culture was transferred to medium A355 (44) and grown for 24 h. One milliliter was transferred to F10A or A346 production medium, and fermentations were carried out for 6 days. Lipopeptide yields were determined by high-pressure liquid chromatography analysis (44). *S. fradiae* strains were grown for 24 h in CSM broth (32) containing the appropriate antibiotics. One milliliter of culture was

TABLE 2 *Trans*-complementation systems in S. *roseosporus* and S. *fradiae* to express daptomycin and A54145 biosynthetic genes from different chromosomal sites

Antibiotic	Genes expressed from locus[a]				Relative yield[b] (%)	Reference
	Nat	φC31	φBT1	IS117		
Daptomycin	dptABCD	–	–	–	100	44
	dptABC	–	–	p*dptD	96	44
	dptA	p*dptBC	–	p*dptD	9	15
	p*dptA	p*dptBC	–	p*dptD	50	15
	dptBC	p*dptA	–	p*dptD	79	15
A54145	lptABCD	–	–	–	100	In prep[c]
	–	lptABCD+	–	–	107	In prep
	–	–	lptABCD+	–	122	In prep
	lptA	p*lptBCD+	–	–	82	In prep
	lptA	p*lptBC	p*lptD+	–	85	In prep

[a]All lipopeptide genes other than those expressed at ectopic loci are expressed from the native locus for daptomycin. For A54145, the complete set of *lpt* genes (−) or all genes downstream of *lptA* (+) were deleted from the native locus and present on complementing plasmids. Genes were expressed on native or *ermEp** (*) promoters.

[b]Control daptomycin and A54145 lipopeptide yields were 340 mg/liter (15) and 480 mg/liter (Alexander, Rock, He, et al., in preparation), in S. *roseosporus* UA117 and S. *fradiae* XH25, respectively.

[c]In prep, Alexander, Rock, He, et al., in preparation.

transferred to a 125-ml baffled flask containing 25 ml of A355 medium (44) and grown at 30°C for 24 h at 200 rpm, 5-cm orbit. Two milliliters of culture was transferred to a 250-ml baffled flask containing 50 ml of DSF production medium (14) plus 0.79% L-Ile. A54145 factor yields were determined by high-pressure liquid chromatography analysis (Alexander, Rock, He, et al., in preparation). Table 2 shows that the strain that expressed the *dptD* gene from the IS117 site under the control of the *ermEp** promoter produced about 96% of the control yield. The strain that expressed the *dptBC* gene from the native locus and the *dptA* and *dptD* genes from the φC31 and IS117 sites, respectively, each transcribed from the *ermEp** promoter, produced about 79% of the control yield. Likewise, when *dptA* was expressed from the native site and *dptBC* and *dptD* from the ectopic sites, all three under the control of *ermEp**, the recombinant produced 50% of control. These results indicate that it is feasible to express the three NRPS genes from three different loci without sacrificing substantial yield, thus providing a system to express independently engineered NRPS genes combinatorially.

In S. *fradiae*, it was shown that the complete A54145 biosynthetic pathway can be deleted from the native locus and expressed from the φC31 or φBT1 *attB* sites without sacrificing lipopeptide yield (Alexander, Rock, He, et al., in preparation). In both cases, the recombinants produced more than 100% of the control yield (Table 2), indicating that both insertion sites are neutral if not positive for lipopeptide production. The suitability of using these sites for combinatorial biosynthesis was confirmed by expressing the *lptA* gene from the native locus and the *lptBC* and *lptD* plus downstream genes from the ectopic loci. The recombinant produced 85% of control lipopeptides (Table 2). The ability to express the complete A54145 biosynthetic pathway efficiently from three different chromosomal sites also suggests the possibility of making pathway duplications and triplications for yield enhancement.

27.5. GENETIC ENGINEERING AND COMBINATORIAL BIOSYNTHESIS OF NOVEL LIPOPEPTIDES

27.5.1. Engineering BAC Clones by λ Red-Mediated Recombination in *E. coli*

Over the past 3 decades, the ability to manipulate antibiotic-producing actinomycetes has become significantly easier and faster, allowing for more sophisticated strain constructions. Some early key milestones were the development of (i) bifunctional plasmids and resistance markers for use in both *E. coli* and actinomycetes; (ii) site-specific integration plasmids; and (iii) intergeneric conjugation from *E. coli* to actinomycetes. All of these properties were represented in a series of plasmids generated at Eli Lilly and Co. (12). The ability to do the majority of the cloning work in *E. coli* allows for significant gains in efficiency because of the much faster growth rate, very robust methods of transformation, and an expanded repertoire of molecular genetic tools (55).

No *E. coli* tool has changed the way actinomycete strain engineering is currently carried out more dramatically than has high-frequency homologous recombination of PCR products using the λ Red or RecET systems (18). Traditional gene replacement experiments in *E. coli* rely on *recA*-dependent homologous recombination, which requires ~1 kb of homologous sequences flanking the region to be exchanged for efficient recombination. This is time-consuming because DNA sequencing is required to confirm the accuracy of cloned PCR products, and multiple steps at nonpermissive growth temperatures or with counterselections are needed to generate mutants with double crossovers. With λ Red-mediated recombination, the number of steps is reduced dramatically, allowing for gene replacements to be completed in *E. coli* within a week. An antibiotic resistance marker is PCR amplified with primers containing 40- to 50-nucleotide extensions that are

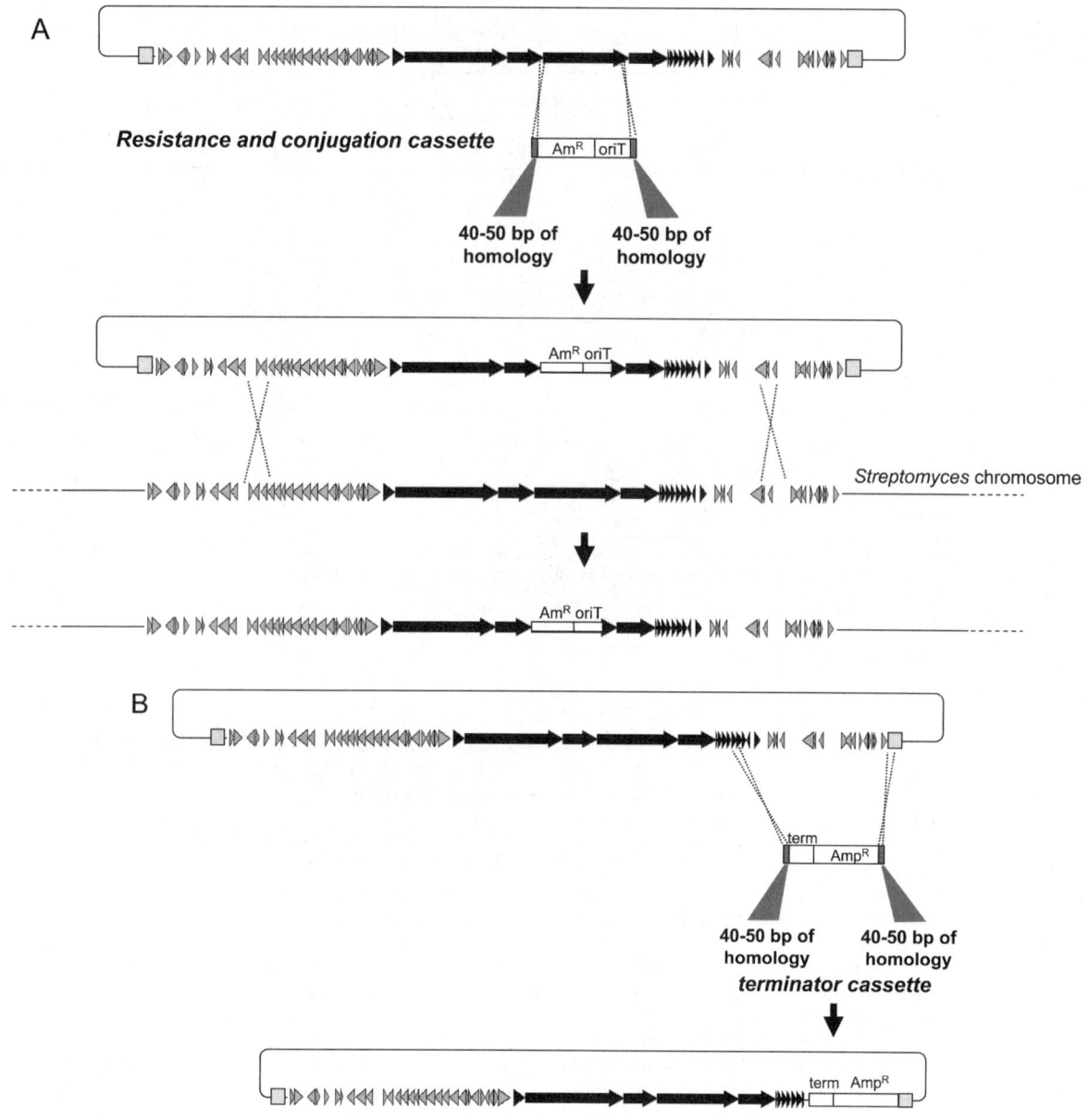

FIGURE 4 λ Red-mediated recombination in *E. coli*. (A) BAC pCB01 containing the complete *lpt* gene cluster is engineered to delete *lptC* by λ Red-mediated insertion of an Am^r-*oriT* cassette. The recombinant BAC can be transferred to *Streptomyces* by conjugation via *oriT*, and recombinants can be selected for Am^r. Other manipulations include (B) the insertion of a terminator cassette, (C) deletion and insertion of a promoter, and (D) insertion of a conjugation/integration cassette. (*continued*)

homologous to the gene to be disrupted. The PCR product is introduced by electroporation into cells induced for λ Red expression, and colonies are selected for antibiotic resistance. Recombination can be carried out in a strain that contains cosmid or BAC vectors containing parts or all of a biosynthetic gene cluster, so it provides a simple and efficient way to disrupt or modify specific genes. For example, λ Red recombination was used to trim BAC pCV1 to generate pDA300, which contains the minimal daptomycin gene cluster (Fig. 3). The first recombination event utilized the Am^r gene flanked by regions of homology for genes downstream of the daptomycin cluster to remove ~26 kb of DNA, thus generating pDA299. The second recombination utilized a spectinomycin resistance (Spc^r) gene flanked by

regions of homology for genes upstream of the daptomycin cluster to remove ~46 kb of DNA, generating pDA300.

Gust and colleagues (29) capitalized on this technology to rapidly generate gene disruptions in *Streptomyces* using cosmids. Their method required the PCR amplification of a bifunctional resistance gene and the origin of transfer (*oriT*) with extensions of homology targeting a specific gene present on a cosmid. This technology has been extended to BAC clones (as illustrated in Fig. 4A, which shows how the resistance cassette was inserted into the *lptC* gene to generate a deletion mutant). The resulting engineered BAC is introduced into *Streptomyces* by conjugation from *E. coli*, but since the BAC lacks *Streptomyces* replication and site-specific insertion functions, it can only confer antibiotic

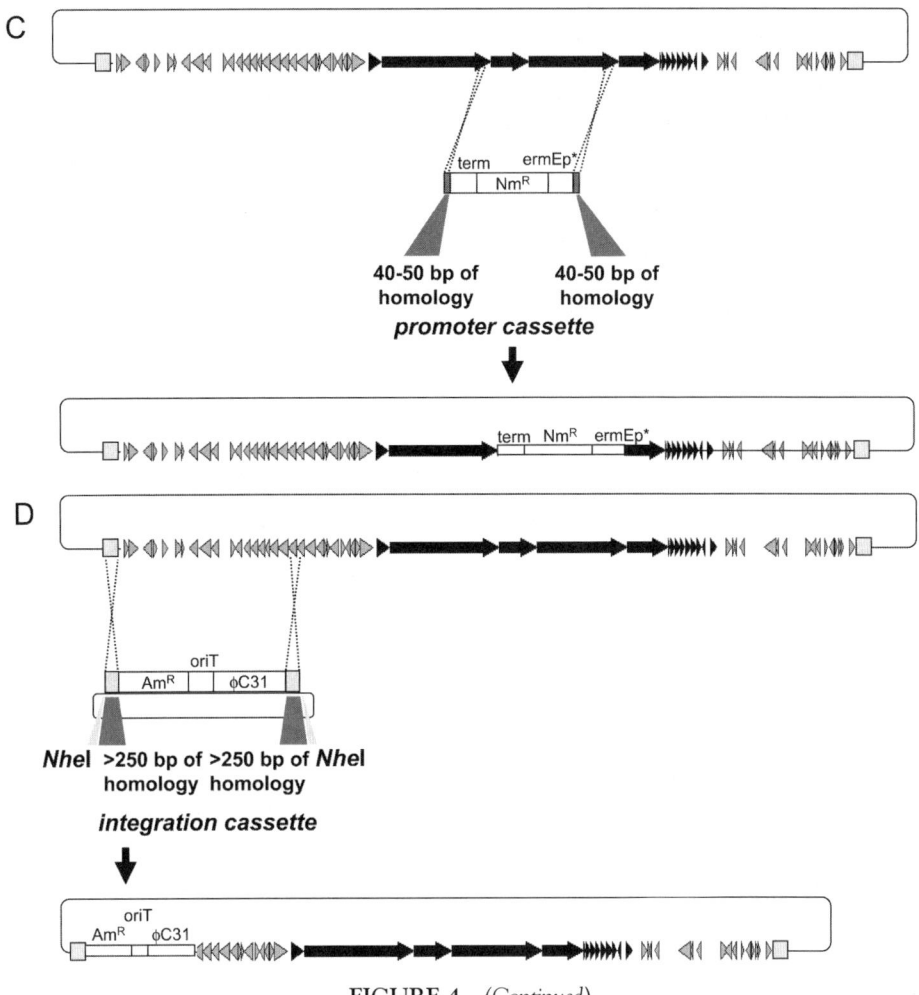

FIGURE 4 (*Continued*)

resistance by recombination between the regions of homology flanking the antibiotic resistance marker in the BAC and the chromosome. A large percentage of the Am^r *Streptomyces* colonies have undergone double crossovers resulting in allelic exchange: these can be identified by loss of the BAC (or cosmid as in reference 29) resistance marker. This system relies on antibiotic resistance as a selection for allelic exchange and cannot be used directly to generate markerless deletions.

The ability to clone entire biosynthetic pathways in *E. coli* in BAC vectors allows for the complex manipulation of genes needed for pathway engineering or module exchanges. For pathway engineering, a number of tools were created to modify BACs containing antibiotic gene clusters, using λ Red-mediated recombination, beyond the simple insertion of a disrupting antibiotic resistance marker. We designed a terminator cassette containing the λ t$_o$ terminator and the Am^r gene, which is flanked by P1 (TGTGTAGGCTGGAGCTGCTTC) and 1233 (AGCG-GATAACAATTTCACACAGGA) primer binding sites (Alexander, Rock, He, et al., in preparation). Primers were designed to contain 40- to 50-nucleotide regions of BAC insert or vector homology plus one of the primer binding sites. PCR-amplified products would recombine and delete genes and place a transcription terminator to stop transcrip-

tion of the biosynthetic gene cluster (Fig. 4B). The terminator cassette has been successfully used to precisely delete small or large regions of DNA for rapid engineering of the A54145 biosynthetic pathway to examine the functions of genes involved in amino acid modifications (Alexander, Rock, Gu, et al., in preparation).

To further extend this concept, a promoter cassette was constructed in which the Nm^r gene or other resistance genes were cloned between the fd phage terminator and the *ermEp** promoter fragments, which are flanked by P1 (TGTGTAGGCTGGAGCTGCTTC) and P3 (CATATGTCCGCCTCCTTTGGTCAC) primer binding sites. The correct positioning of the homologous primer with the start codon in the P3 primer fuses the downstream *lptD* gene with the *ermEp** promoter at the start codon (Fig. 4C). These promoter cassettes have been used successfully to create complementing plasmids for the daptomycin gene cluster (15, 50).

The pCB01 BAC was identified from the *S. fradiae* BAC library by PCR screening and was believed to contain the entire biosynthetic gene cluster. Functional validation of the *lpt* gene cluster on this BAC was not possible because the pECBAC1 vector lacks functions to drive conjugal transfer and chromosomal insertion in *Streptomyces*. To introduce pCB01 into streptomycetes required the addition of an integration cassette containing the Am[r] gene, *oriT* site, and φC31 *attP/int* functions (Alexander, Rock, He, et al., in preparation). The large size of the integration cassette (~4.2 kb) prevented the efficient PCR amplification required to generate a template for λ Red-mediated recombination. As a result, we constructed a recombination cassette by cloning regions of homology needed for recombination on alternate ends of the integration cassette. To generate a BAC containing the entire *lpt* cluster, the integration cassette was flanked with regions of homology to the pECBAC1 backbone (0.26 kb) and the *orf21* region (1.4 kb), both upstream of the *lpt* gene cluster. The entire integration cassette was released from the cloning vector by digestion with NheI followed by gel purification before being used as a template for recombination. Successful λ Red-mediated recombination generated pDA2002 (Fig. 4D), containing the *lpt* cluster that could be transferred to *Streptomyces* by conjugation and site-specific integration. The use of compatible markers on these cassettes allows them to be used in combination to generate BAC vectors modified in very complicated ways.

By using this system, multiple constructs of NRPS domain and module exchanges were rapidly generated. Many of these may not have been feasible using *recA*-dependent recombination. Domains or modules of an NRPS pathway are typically 3 to 5 kb in length and share a high degree of sequence homology, especially in the condensation domains. This sequence homology results in problems for PCR amplification, yielding multiple different products of similar sizes, and *recA*-dependent recombination may also target alternate sites. The application of the λ Red system has simplified and accelerated the genetic manipulation of NRPSs by module and multidomain exchanges (Fig. 5). A specific module(s) present in the BAC can be replaced by

an antibiotic resistance marker flanked by rare restriction sites. The BAC DNA is digested to remove the resistance marker, treated with shrimp alkaline phosphatase, and used directly in a ligation with a replacement DNA fragment containing the new module. The replacing module is cloned directly from the BAC by using the λ Red system in a process called gap repair (37). A PCR product is generated from pBR322 using primers that are homologous with the ends of the module. When electroporated into cells induced for λ Red expression, the pBR322 PCR product will recombine with the BAC to pick up the module fragment and circularize the plasmid, thus allowing for stable replication (Fig. 5). The compatible rare restriction sites are incorporated into primers used for amplification of the pBR322 fragment. Only end sequencing of the cloned 3- to 5-kb module is necessary to check for errors, as the "cloned" module is not exposed to the low level of misincorporation mutagenesis associated with PCR. During the ligation of the module into the BAC, at least one restriction site is destroyed by combining compatible cohesive or blunt ends that are not recleaved upon ligation (e.g., AvrII and NheI), and the resulting ligation mix can be digested with that restriction enzyme to remove BACs resulting from ligation of the original antibiotic resistance gene insert before transformation into *E. coli*. BACs containing the desired module exchanges are identified by PCR screening, restriction mapping, and sequencing of the fusion junctions.

27.5.2. NRPS Subunit *trans*-Complementation

The similarities in size and function of the terminal subunits of the NRPSs of daptomycin, A54145, and CDA made them attractive candidates for subunit exchanges. The *dptD*, *lptD*, and *cdaPS3* genes appear to have evolved from a common ancestral NRPS gene (6). The DptD, LptD, and

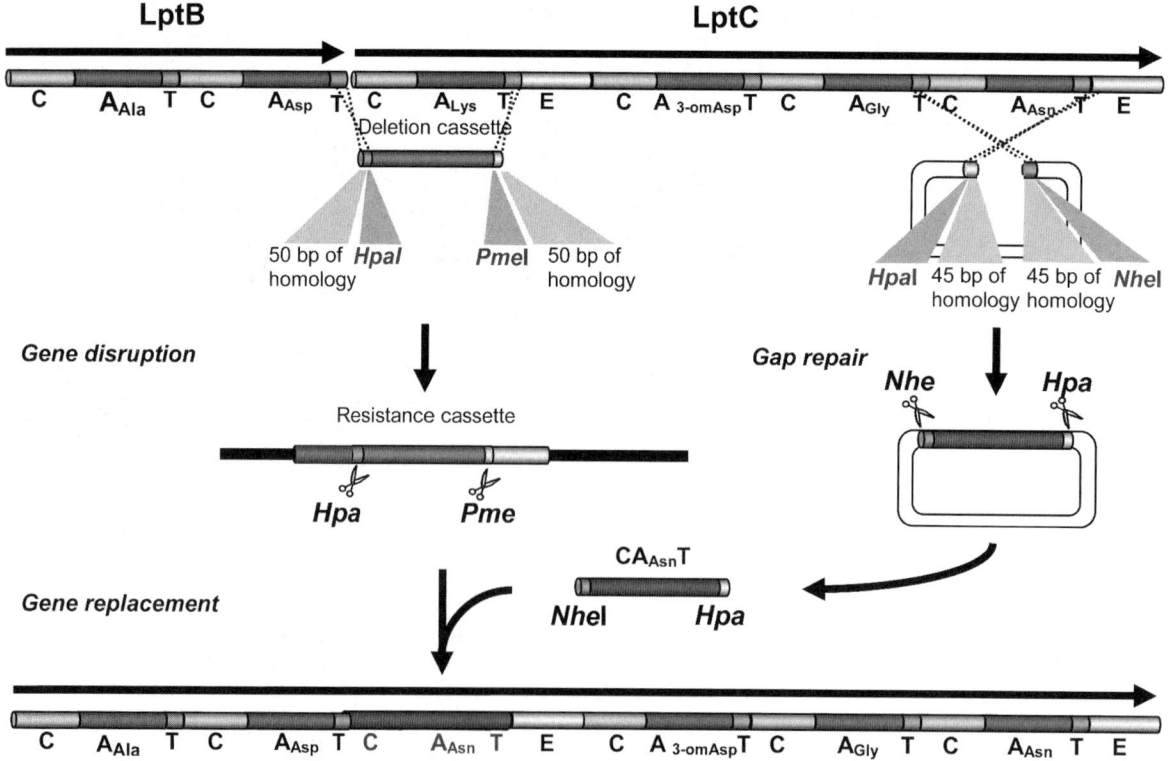

FIGURE 5 λ Red-mediated recombination to generate a C-A-T tridomain exchange and gene fusion in *lptB* and *lptC*.

CdaPS3 proteins share ~50% amino acid identities, and they show some conservation of N-terminal interpeptide docking sequences. The *lptD* and *cdaPS3* genes were cloned into pHM11a under the transcriptional control of *ermEp**, generating pRB04 and pMF26, respectively (Table 1). The plasmids were introduced into *S. roseosporus* UA378 (Δ*dptD::ermE*) by conjugation from *E. coli* and inserted into the IS*117 attB* site. Recombinants containing the *lptD* and *cdaPS3* genes produced A21978C$_{1-3}$ derivatives containing Ile$_{13}$ or Trp$_{13}$ at about 25 and 50% of control levels, respectively (15, 46). Similar experiments have been attempted in an *S. fradiae* strain deleted for the *lptD* genes, but recombinants containing the *dptD* or *cdaPS3* genes inserted in the φC31 site did not produce hybrid lipopeptides (D. C. Alexander, J. Rock, C. Mascio, C. Li, A. Van Praagh, L. Mortin, J.-Q. Gu, M. Chu, J. A. Silverman, P. Brian, and R. H. Baltz, in preparation). These mixed results suggest that subunit exchanges may have limited applicability.

27.5.3. NRPS Domain and Module Exchanges

27.5.3.1. Flexibility of Module Fusions at T-C Linkers

The production of A21978C$_{1-3}$ derivatives containing Ile$_{13}$ or Trp$_{13}$ in *S. roseosporus* can also be achieved by module exchange by fusing the 3mGlu$_{12}$ module from *dptD* with the Ile$_{13}$ or Trp$_{13}$ modules from *lptD* and *cdaPS3*, respectively, as C-A-T::C-A-T-TE fusions (20). In this way, the homologous protein-protein docking relationship between DptBC and DptD is preserved. As a control, the homologous DptD C-A-T::C-A-T-TE fusions were made in the intermodule T-C linker region by inserting a restriction enzyme cleavage site to facilitate subsequent fusions. The flexibility of the system was tested by making three different double-amino-acid substitutions, a four-amino-acid insertion, and a four-amino-acid deletion in the linker region. The reconstructed *dptD* genes were cloned into pHM11a and expressed ectopically from the *ermEp** promoter at the IS*117 attB* site as described above, and all recombinants produced A21978C$_{1-3}$ factors at control yields. Heterologous C-A-T::C-A-T-TE module fusions between *dptD* 3mGlu$_{12}$ and *lptD* Ile$_{13}$ or *cdaPS3* Trp$_{13}$ were constructed, and the recombinant strains produced A21978C$_{1-3}$ derivatives containing Ile$_{13}$ or Trp$_{13}$ in higher yields (~100% of control) than observed with the whole subunit exchanges. The success in this system may be due in part to the maintenance of efficient protein-protein interactions between homologous interpeptide docking sites. Therefore, module fusions that preserve the natural protein-protein docking relationships may provide the most robust approach to generate hybrid NRPSs.

27.5.3.2. Didomain Fusions at A-T Linkers (C-A::T-TE)

Since few C-A-T-TE modules were available to further explore amino acid substitutions at position 13, the ectopic *trans*-complementation system in *S. roseosporus* was used to generate hybrid C-A-T-TE modules retaining the homologous TE for ring closure. The **C-A-T** or **C-A** domains from the LptC **C-A$_{D-Asn11}$-TE** module were exchanged for the C-A-T or C-A domains from the C-A$_{Kyn13}$-T-TE module of DptD. The recombinant expressing the C-A-T::**C-A**::T-TE double fusion produced A21978C$_{1-3}$ derivatives containing Asn$_{13}$ at ~43% of control. The strain expressing the C-A-T::**C-A-T**::TE double fusion produced no lipopeptides. These results point out the importance of maintaining the

homologous interaction between native T and TE domains (25, 62, 65). A C-A-T::**C-A$_{Kyn}$**::T-TE double fusion in *lptD* in *S. fradiae* was used successfully to generate A54145 analogs containing Kyn$_{13}$ from the daptomycin pathway (Alexander, Rock, Mascio, et al., in preparation) (Table 3).

27.5.3.3. Module and Domain Fusions Facilitated by λ Red-Mediated Recombination

Exchanges of C-A-T tridomains and C-A-T-E modules have been carried out at positions 8 and 11 in DptBC at splicing sites in interdomain linkers (20, 28, 50). Recombinants produced the predicted A21978C$_{1-3}$ derivatives (50). The production yields were highest (10 to 50% of control) with the C-A-T exchanges, leaving the homologous E domains intact to interact with downstream ^{D}C$_{L}$ domains. These changes have been coupled with subunit exchanges and deletion of the *dptI* gene to generate larger sets of hybrid compounds (49, 50; Table 3).

In *S. fradiae*, the λ Red-mediated recombination system (18) was used to exchange single or multiple modules, or multidomains in plasmids containing different sets of A54145 biosynthetic genes, using splicing sites located in the interdomain regions (K. Nguyen, X. He, D. Alexander, L. Chen, J.-Q. Gu, C. Mascio, A. Van Praagh, L. Mortin, M. Chu, J. A. Silverman, P. Brian, and R. H. Baltz, submitted) similar to those used above (49, 50). Exchanges at position 8 eliminated the stop codon of *lptB*, generating fused *lptBC* hybrid genes. Exchanges at positions 8 or 11 in the fused *lptBC* gene were carried out in plasmid pDA2048 (Fig. 3) and introduced into strains DA740 (Δ*lptBCD*) and DA901 (Δ*lptBCD* Δ*lptI::tsr*) at the φC31 *attB* site, generating lipopeptides containing 3mGlu$_{12}$ and Glu$_{12}$. Exchanges at positions 2, 2 and 3, or 2–8 were made in plasmid pDA2054 and introduced into strain DA1187 at the φC31 *attB* site to produce novel compounds containing Glu$_{12}$. These amino acid substitutions were coupled with 3mGlu$_{12}$ by introducing plasmid pKN55 containing the *dptIJ* genes into the φBT1 *attB* site in recombinants. Hybrid lipopeptides containing one to three amino acid changes were produced in yields of 3 to 48 mg/liter, whereas the lipopeptides with five or six changes were produced at ~1 mg/liter (Nguyen et al., submitted).

27.5.4. Disruption of Genes Involved in Modification of Amino Acids

As already mentioned, disruption of the *dptI* and *lptI* genes in *S. roseosporus* and *S. fradiae* strains caused the production of lipopeptides containing Glu$_{12}$ rather than 3mGlu$_{12}$. These simple changes doubled the number of potential novel compounds generated by combinatorial biosynthesis. Importantly, as discussed in more detail below, disruption of *lptI* provided a means to generate derivatives of A54145 with low mammalian toxicity and substantially increased antibacterial activity in the presence of surfactant. A54145 has four modified amino acids, and the genes involved in the modifications have been disrupted singly and in combinations (Alexander, Rock, Gu, et al., in preparation), some by the insertion of a terminator cassette via λ Red-mediated recombination of pDA2002. Disruption of the *lptL* asparagine oxygenase gene caused the production of lipopeptides containing Asn$_3$ instead of hAsn$_3$. Disruption of the *lptK* O-methyltransferase gene caused the production of lipopeptides containing hAsp$_9$ rather than mOAsp$_9$, and *lptJ* aspartic acid oxygenase mutants produced Asp$_9$. The properties of mutants containing one or more mutations blocking amino acid modifications are summarized in Table 3 and are discussed in section 27.5.5.

TABLE 3 Lipopeptide antibiotics generated by combinatorial biosynthesis

Compound[a]	Amino acid at position: 2	3	5	6	8	9	11	12	13	S. aureus MIC (µg/ml) −Surfactant	+Surfactant (1%)	Ratio (+/−)
Daptomycin	D-Asn	Asp	Gly	Orn	D-Ala	Asp	D-Ser	3mGlu	Kyn	0.5	64	128
CB-181,220	D-Asn	Asp	Gly	Orn	D-Ala	Asp	D-Ser	3mGlu	Kyn	0.5	64	128
CB-182,098	D-Asn	Asp	Gly	Orn	D-Ala	Asp	D-Ser	3mGlu	*Trp*	1	32	32
CB-182,107	D-Asn	Asp	Gly	Orn	D-Ala	Asp	D-Ser	3mGlu	*Ile*	2	8	4
CB-182,106	D-Asn	Asp	Gly	Orn	D-Ala	Asp	D-Ser	3mGlu	*Val*	4	8	2
A21978C1(Asn₁₃)	D-Asn	Asp	Gly	Orn	D-Ala	Asp	D-Ser	3mGlu	*Asn*	128	ND	ND
CB-182,130	D-Asn	Asp	Gly	Orn	D-Ala	Asp	D-Ser	*Glu*	Kyn	8	16	2
CB-182,166	D-Asn	Asp	Gly	Orn	D-Ala	Asp	D-*Ala*	3mGlu	Kyn	1	16	16
CB-182,290	D-Asn	Asp	Gly	Orn	D-Ala	Asp	D-*Asn*	3mGlu	Kyn	1	16	16
CB-182,123	D-Asn	Asp	Gly	Orn	D-*Ser*	Asp	D-Ser	3mGlu	Kyn	1	32	32
CB-182,257	D-Asn	Asp	Gly	Orn	D-*Asn*	Asp	D-Ser	3mGlu	Kyn	8	ND	ND
CB-182,286	D-Asn	Asp	Gly	Orn	D-Ala	Asp	D-*Asn*	3mGlu	*Ile*	4	ND	ND
CB-182,251	D-Asn	Asp	Gly	Orn	D-Ala	Asp	D-*Asn*	*Glu*	Kyn	32	ND	ND
CB-182,263	D-Asn	Asp	Gly	Orn	D-*Asn*	Asp	D-Ser	*3mGlu*	*Ile*	16	ND	ND
CB-182,269	D-Asn	Asp	Gly	Orn	D-*Asn*	Asp	D-Ser	*Glu*	Kyn	128	ND	ND
CB-182,296	D-Asn	Asp	Gly	Orn	D-*Lys*	Asp	D-*Asn*	3mGlu	Kyn	1	32	32
A54145E	D-Glu	hAsn	Sar	Ala	D-Lys	mOAsp	D-Asn	3mGlu	Ile	1	32	32
A54145D	D-Glu	hAsn	Sar	Ala	D-Lys	mOAsp	D-Asn	*Glu*	Ile	2	4	2
CB-182,548	D-Glu	hAsn	Sar	Ala	D-Lys	mOAsp	D-*Ala*	3mGlu	Ile	1	16	16
CB-182,332	D-Glu	hAsn	Sar	Ala	D-Lys	mOAsp	D-*Ser*	3mGlu	Ile	2	16	8
CB-182,443	D-Glu	hAsn	Sar	Ala	D-Lys	*Asp*	D-Asn	3mGlu	Ile	2	4	2
CB-182,571	D-Glu	hAsn	Sar	Ala	D-*Ala*	mOAsp	D-Asn	3mGlu	Ile	1	32	32
CB-182,549	D-Glu	hAsn	Sar	Ala	D-*Ser*	mOAsp	D-Asn	3mGlu	Ile	1	16	16
CB-182,510	D-Glu	hAsn	Sar	Ala	D-*Asn*	mOAsp	D-Asn	3mGlu	Ile	8	64	8
CB-182,363	D-Glu	*Asn*	Sar	Ala	D-Lys	mOAsp	D-Asn	3mGlu	Ile	2	16	8
CB-182,575	D-*Asn*	hAsn	Sar	Ala	D-Lys	mOAsp	D-Asn	3mGlu	Ile	2	4	2

Compound										MIC		
CB-183,298	d-Glu	hAsn	Sar	Ala	d-Lys	mOAsp	d-Asn	**Glu**	**Kyn**	1	2	2
CB-182,509	d-Glu	hAsn	Sar	Ala	d-Lys	mOAsp	d-**Ala**	**Glu**	Ile	8	16	2
CB-182,336	d-Glu	hAsn	Sar	Ala	d-Lys	mOAsp	d-**Ser**	**Glu**	Ile	64	128	2
CB-182,350	d-Glu	hAsn	Sar	Ala	d-Lys	**hAsp**	d-Asn	**Glu**	Ile	8	16	2
CB-182,333	d-Glu	hAsn	Sar	Ala	d-Lys	**Asp**	d-Asn	**Glu**	Ile	32	64	2
CB-182,567	d-Glu	hAsn	Sar	Ala	d-**Ala**	mOAsp	d-Asn	**Glu**	Ile	4	8	2
CB-182,532	d-Glu	hAsn	Sar	Ala	d-**Ser**	mOAsp	d-Asn	**Glu**	Ile	8	8	1
CB-182,531	d-Glu	hAsn	Sar	Ala	d-**Asn**	mOAsp	d-Asn	**Glu**	Ile	16	16	1
CB-182,391	d-Glu	hAsn	**Gly**	Ala	d-Lys	mOAsp	d-Asn	**Glu**	Ile	16	32	2
CB-182,325	d-Glu	**Asn**	Sar	Ala	d-Lys	mOAsp	d-Asn	**Glu**	Ile	32	32	1
CB-182,444	d-**Asn**	hAsn	Sar	Ala	d-Lys	mOAsp	d-Asn	**Glu**	Ile	8	8	1
CB-182,597	d-Glu	**Asn**	Sar	Ala	d-Lys	**hAsp**	d-Asn	**3mGlu**	Ile	1	16	16
CB-182,390	d-Glu	**Asn**	Sar	Ala	d-Lys	**Asp**	d-Asn	**3mGlu**	Ile	2	2	1
CB-182,561	d-**Asn**	**Asp**	Sar	Ala	d-Lys	mOAsp	d-Asn	**3mGlu**	Ile	1	2	2
CB-182,349	d-Glu	**Asn**	Sar	Ala	d-Lys	**hAsp**	d-Asn	**Glu**	Ile	32	64	2
CB-182,348	d-Glu	**Asn**	Sar	Ala	d-Lys	**Asp**	d-Asn	**Glu**	Ile	16	32	2
CB-182,560	d-**Asn**	**Asp**	Sar	Ala	d-Lys	mOAsp	d-Asn	**Glu**	Ile	8	16	2

[a] Daptomycin has an *n*-decanoyl side chain. All others compounds have *anteiso*-undecanoyl side chains. MICs against *S. aureus* were determined with and without the addition of 1% bovine sufactant. Data from references 28, 46, 49, 50; Alexander, Rock, He, et al., in preparation; Alexander, Rock, Gu, et al., in preparation; Alexander, Rock, Mascio, et al., in preparation; and Nguyen et al., submitted.

27.5.5. Incorporation of Multiple Fatty Acid Side Chains

All of the manipulations described above are coupled with the natural propensity to initiate lipopeptide assembly with three different preferred fatty acids. During A21978 biosynthesis, the three major fatty acid side chains are *anteiso*-undecanoyl, *iso*-dodecanoyl, and *anteiso*-tridecanoyl, and during A54145 biosynthesis the major fatty acids are *n*-decanoyl, *iso*-decanoyl (8-methylnonanoyl), and *anteiso*-undecanoyl (8-methyldecanoyl) (Fig. 1). The heterologous pathway expression experiments (44, 52; Alexander, Rock, He, et al., in preparation) demonstrated that the fatty acid precursor specificity is inherent in the lipopeptide biosynthetic enzymes and not in the streptomycete host. The daptomycin pathway incorporates preferentially fatty acids that are two carbons longer than those preferred in the A54145 pathway. It is not known whether the specificity resides within the acyl-coenzyme A ligase and acyl carrier proteins (DptE and DptF or LptEF), in the $^{F}C_L$ condensation domain of the Trp$_1$ modules, or in both. The fatty acid preferences can be preempted by feeding high levels of other fatty acids (6, 9, 14, 33), and many more novel compounds can be produced by feeding different fatty acids, thus expanding the numbers achievable by combinatorial biosynthesis.

27.5.6. Properties of Novel Lipopeptides

Table 3 shows the amino acid substitutions and antibacterial properties of a subset of novel lipopeptides produced by genetic engineering and combinatorial biosynthesis in *S. roseosporus* or *S. fradiae*. Antibacterial activities against *S. aureus* were determined with and without 1% bovine surfactant to identify candidates that might be efficacious in *S. pneumoniae* pulmonary infections. For simplicity, only compounds containing *anteiso*-undecanoyl (8-methyldecanoyl) side chains are compared with daptomycin (which has a decanoyl side chain) and CB-181,220 (which has an 8-methyldecanoyl side chain). The in vitro antibacterial activities varied by 256-fold when the core peptides of daptomycin and A54145 were modified, and the magnitude of inhibition by bovine surfactant varied by 128-fold, with daptomycin and CB-181,220 showing the highest level of surfactant inhibition. These data demonstrate that combinatorial biosynthesis can be used to change the fundamental properties of complex cyclic lipopeptide antibiotics. Several observations from Table 3 are worth noting. (i) The presence of Kyn$_{13}$ is important for the antibacterial activity of daptomycin, but it also contributes substantially to surfactant inhibition. (ii) Substitutions for Kyn$_{13}$ reduce surfactant inhibition, but they cause reduced antibacterial activity against *S. aureus*. (iii) Substitution of Glu$_{12}$ for 3mGlu$_{12}$ in the context of daptomycin reduces surfactant inhibition by fourfold, but increases the MIC in the absence of surfactant 16-fold, leading to no net gain. (iv) Substitutions of D-Ala or D-Asn at position 11 in the daptomycin core peptide reduced surfactant inhibition about fourfold, but increased the MIC in the absence of surfactant twofold. (v) A54145 factor E is twofold less active than daptomycin in the absence of surfactant, but twofold more active in the presence of surfactant. (vi) Last, and importantly, A54145 factor D, which has Glu$_{12}$, is fourfold less active than daptomycin without surfactant, but 16-fold more active in the presence of surfactant. Also, the natural A54145 factors containing 3mGlu have higher levels of acute toxicity in mice than those containing Glu (16; Nguyen et al., submitted).

The properties of A54145D suggested that it might be possible to mitigate the surfactant inhibition encountered by daptomycin while maintaining good antibacterial activity and low mammalian toxicity. This prompted the development of molecular engineering tools to carry out combinatorial biosynthesis in *S. fradiae* focused on modifying the A54145 core peptide (section 27.4.2). The following observations were made from these studies. (i) Substitution of D-Asn for D-Glu at position 2 (CB-182,575), or Asp for mOAsp at position 9 (CB-182,443), while maintaining 3mGlu$_{12}$, had the same effect on antibacterial properties as substituting Glu for 3mGlu at position 12. (ii) Disruption of three genes involved in modifying Asn$_3$ and Asp$_9$ generated a very active lipopeptide (CB-182,390) that showed no surfactant inhibition. (iii) Exchanging the D-Asn$_2$ and Asp$_3$ from the daptomycin pathway for D-Glu$_2$ and hAsn$_3$ generated CB-182,561 with MICs of 1 and 2 µg/ml in the absence and presence of surfactant. (iv) Finally, coupling Kyn$_{13}$ with Glu$_{12}$ in the A54145 core peptide generated CB-183,298, with the same improved properties as CB-182,561. The three most active compounds (CB-182,561, CB-182,390, and CB-183,298), with three distinctly different peptide core modifications, had MICs of 2 µg/ml and were 32-fold more active than daptomycin in the presence of 1% bovine surfactant.

27.6. CONCLUSIONS AND PROSPECTS

It is now feasible (i) to clone complex lipopeptide antibiotic biosynthetic pathways in BAC vectors that can be rapidly manipulated in complex ways in *E. coli* by λ Red-mediated recombination (recombineering); (ii) to transfer partial or complete pathways from *E. coli* to homologous or heterologous streptomycete hosts by conjugation; (iii) to insert partial or complete biosynthetic pathways into bacteriophage or insertion element integration/attachment sites in streptomycete chromosomes; and (iv) to express complete pathways from native promoters and partial pathways from native or other promoters in homologous and heterologous streptomycetes. Conjugal transfer from *E. coli* is a key element for successful introduction of very large BAC vectors, and the efficiency is media dependent. AS-1 agar and derivatives work well for conjugal transfer to *S. roseosporus* and *S. ambofaciens*. The concepts summarized above comprise the basic elements for successful genetic engineering and combinatorial biosynthesis in streptomycetes. The specific versatile tools that have facilitated genetic engineering in *S. roseosporus* and *S. fradiae* are (i) pStreptoBAC V, which has *oriT* and φC31 *att/int* functions, Am^r for selection in streptomycetes and *E. coli*, and other features that facilitate the process; (ii) other vectors with *oriT* and *att/int* functions from φC31, φBT1, or IS117 for conjugal transfer and site-specific integration in streptomycetes; (iii) λ Red-mediated recombination in *E. coli*; and (iv) *ermEp** for driving transcription of cloned genes inserted at ectopic locations in streptomycete chromosomes. The methods discussed in this chapter include some that are specific to NRPS engineering, including (i) splicing at intermodule or interpeptide linker sites; (ii) recognizing and exploiting the correct type of C domain when coupling fatty acids to L-amino acids ($^{F}C_L$), D-amino acids to L-amino acids ($^{D}C_L$), or L-amino acids to L-amino acids ($^{L}C_L$); and (iii) maintaining the integrity of T-TE didomains when engineering terminal modules. It is probably important to maintain the integrity of C-A didomains (25, 62, 65), but this was not addressed specifically in the current studies. However, successful examples of transplanting C-A didomains were demonstrated.

This chapter has pointed out that both *S. roseosporus* and *S. fradiae* can be readily manipulated by the genetic engineering methods described above. As such, they may be useful hosts for the expression and engineering of other secondary metabolic pathways, particularly other NRPS pathways. For example, the A54145 gene cluster was expressed successfully in an *S. roseosporus* strain deleted for daptomycin biosynthetic genes. Of particular note was the high-level expression of the complete A54145 biosynthetic pathway in a mutant of *S. ambofaciens* blocked in spiramycin production, yielding nearly 400 mg/liter of A54145 lipopeptides. *S. ambofaciens* BES2074 was specifically selected for stability and high-level spiramycin production (22) followed by the disruption of a key regulatory gene, *srmR* (26, 54), required for spiramycin biosynthesis. This strain is also defective in netropsin biosynthesis, making it a good host for the expression of antibiotic biosynthetic pathways. *S. ambofaciens* is easily manipulated genetically and thus may be a useful strain for the genetic engineering of pathways cloned from genetically recalcitrant strains. It might also be useful for the expression of cryptic pathways identified in genome sequencing projects (7).

27.7. MEDIA

A355 (44)
Per liter:

Glucose	10.0 g
Glycerol	15.0 g
Soya peptone	15.0 g
NaCl	3.0 g
Malt extract	5.0 g
Yeast extract	5.0 g
Tween 80	1.0 g
MOPS* (pH 7.0)	20.0 g

*Morpholinepropanesulfonic acid

A345 (44)
A355 components

MOPS (pH 7.0)	46.0 g

A346 (44)
Per liter:

Glucose	10.0 g
Soluble starch	20.0 g
Yeast extract	5.0 g
N-Z amine	5.0 g
MOPS	23.0 g

Adjust to pH 7.0.

AS-1 agar (40)
Per liter:

Yeast extract	1.0 g
L-Alanine	0.2 g
L-Arginine	0.2 g
L-Asparagine	0.5 g
Soluble starch	5.0 g
NaCl	2.5 g
Na_2SO_4	10.0 g
Agar	20.0 g

Adjust to pH 7.5.

CSM (32)
Per liter:

Trypticase soy broth	30.0 g
Yeast extract	3.0 g
Glucose	5.0 g
Maltose	4.0 g
$MgSO_4$	2.0 g

DSF (14)
Per liter:

Glucose	30.0 g
Soybean flour	25.0 g
Blackstrap molasses	5.0 g
$(NH_4)_2Fe(SO_4)_2 \cdot 6H_2O$	0.6 g
$CaCO_3$	4.0 g

Adjust to pH 7.0.

ISP Medium 4 (Difco)
Per liter:

ISP Medium 4 agar	37.0 g

F10A (44)
Per liter:

Distillers solubles	25.0 g
Soluble starch	25.0 g
Yeast extract	5.0 g
Glucose	2.0 g
Bactopeptone	5.0 g
$CaCO_3$	3.0 g

Medium A9 (44)
Per liter:

Oats	20.0 g
NaCl	5.0 g
Tryptone	7.0 g
Soya peptone	2.0 g
Trace salts	1.0 ml

Medium B (52)
Per liter:

Glycerol	20.0 g
Sucrose	2.5 g
L-Proline	12.0 g
MOPS	15.0 g
K_2HPO_4	0.56 g
NaCl	0.5 g
Tween 80	10.0 g
Trace salts solution	5.0 ml
Vitamin solution	2.0 ml

Trace salts solution
Per liter:

$ZnSO_4 \cdot 7H_2O$	1.722 g
$FeSO_4 \cdot 7H_2O$	1.112 g
$MnSO_4 \cdot 4H_2O$	0.223 g
H_3BO_3	0.062 g
$CuSO_4 \cdot 5H_2O$	0.125 g
$Na_2MoO_4 \cdot 2H_2O$	0.048 g
$CoCl_2 \cdot 6H_2O$	0.048 g
KI	0.083 g
H_2SO_4 (1 M)	1.0 ml

Vitamin solution
Per liter:

Thiamine	0.025 g
Riboflavin	0.025 g
Pantothenate	0.025 g
Nicotinic acid	0.025 g

Pyridoxine .... 0.025 g
Thioctic acid .... 0.025 g
Folic acid .... 0.0025 g
Cyanocobalamin .... 0.0025 g
p-Aminobenzoic acid .... 0.0025 g
Vitamin K_1 .... 0.05 ml
$CaCl_2 \cdot 2H_2O$ (10 mg/liter) 0.02%
$MgSO_4 \cdot 7H_2O$ (10 mg/liter) 0.02%
$FeSO_4 \cdot 7H_2O$ (10 mg/liter) 0.00055%
Tween 80 .... 2.0 ml

Adjust to pH 7.0 and filter sterilize.

R2 modified soft agar (40)
Per liter:

Sucrose .... 103.0 g
$MgCl_2$ .... 10.12 g
$CaCl_2$ .... 2.2 g
TES .... 5.73 g
Agar .... 4.0 g

Adjust to pH 7.2.

mR2YE (34)
Per liter:

K_2SO_4 .... 0.25 g
$MgCl_2 \cdot 6H_2O$ .... 10.12 g
Glucose .... 10.0 g
Difco Casamino Acids .... 0.1 g
Difco yeast extract .... 5.0 g
TES .... 5.73 g
Trace element solution* .... 2.0 ml
Agar .... 22.0 g

*Trace element stock solution
Per liter:

$ZnCl_2$ .... 40 mg
$FeCl_3 \cdot 6H_2O$ .... 200 mg
$CuCl_2 \cdot 2H_2O$ .... 10 mg
$MnCl_2 \cdot 4H_2O$ .... 10 mg
$Na_2B_4O_7 \cdot 10H_2O$ .... 10 mg
$(NH_4)_6Mo_7O_{24} \cdot 4H_2O$ .... 10 mg

At the time of use, add
Per liter:

KH_2PO_4 (0.5%) .... 10 ml
$CaCl_2 \cdot 2H_2O$ (5 M) .... 4 ml
NaOH (1 N) .... 7 ml

NB soft agar (34)
Per liter:

Nutrient broth .... 8.0 g
Agar .... 6.5 g

TSB (BBL)
Per liter:

Trypticase soy broth .... 30.0 g

TY broth (BBL)
Per liter:

Tryptone .... 16 g
Yeast extract B .... 10 g
NaCl .... 5 g

REFERENCES

1. **Alexander, D. C., D. J. Devlin, D. D. Hewitt, A. C. Horan, and T. J. Hosted.** 2003. Development of the *Micromonospora carbonacea* var. *africana* ATCC 39149 bacteriophage pMLP1 integrase for site-specific integration in *Micromonospora* spp. *Microbiology* **149:**2443–2453.
2. **Arbeit, R. D., D. Maki, F. P. Tally, E. Campanaro, and B. I. Eisenstein.** 2004. The safety and efficacy of daptomycin for the treatment of complicated skin and skin-structure infections. *Clin. Infect. Dis.* **38:**1673–1681.
3. **Baltz, R. H.** 1997. Molecular genetic approaches to yield improvement in actinomycetes, p. 49–62. *In* W. R. Strohl (ed.), *Biotechnology of Industrial Antibiotics*, 2nd ed. Marcel Dekker, Inc., New York, NY.
4. **Baltz, R. H.** 2006. Molecular engineering approaches to peptide, polyketide and other antibiotics. *Nat. Biotechnol.* **24:**1533–1540.
5. **Baltz, R. H.** 2006. Combinatorial biosynthesis of novel antibiotics and other secondary metabolites in actinomycetes. *SIM News* **56:**148–160.
6. **Baltz, R. H.** 2008. Biosynthesis and genetic engineering of lipopeptide antibiotics related to daptomycin. *Curr. Top. Med. Chem.* **8:**618–638.
7. **Baltz, R. H.** 2008. Renaissance in antibacterial discovery from actinomycetes. *Curr. Opin. Pharmacol.* **8:**557–563.
8. **Baltz, R. H.** 2009. Daptomycin: mechanisms of action and resistance, and biosynthetic engineering. *Curr. Opin. Chem. Biol.* **13:**1–8.
9. **Baltz, R. H., V. Miao, and S. K. Wrigley.** 2005. Natural products to drugs: daptomycin and related lipopeptide antibiotics. *Nat. Prod. Rep.* **22:**717–741.
10. **Bibb, M. J., G. R. Janssen, and J. M. Ward.** 1985. Cloning and analysis of the promoter region of the erythromycin resistance gene (*ermE*) of *Streptomyces erythraeus*. *Gene* **38:**215–226.
11. **Bibb, M. J., J. White, J. M. Ward, and G. R. Janssen.** 1994. The mRNA for the 23S rRNA methylase encoded by the *ermE* gene of *Saccharopolyspora erythraea* is translated in the absence of a conventional ribosome-binding site. *Mol. Microbiol.* **14:**533–545.
12. **Bierman, M., R. Logan, K. O'Brien, E. T. Seno, R. N. Rao, and B. E. Schoner.** 1992. Plasmid cloning vectors for conjugal transfer from *Escherichia coli* to *Streptomyces* spp. *Gene* **116:**43–49.
13. **Boeck, L. D., H. R. Papiska, R. W. Wetzel, J. S. Mynderse, D. S. Fukuda, F. P. Mertz, and D. M. Berry.** 1990. A54145, a new lipopeptide antibiotic complex: discovery, taxonomy, fermentation and HPLC. *J. Antibiot.* **43:**587–593.
14. **Boeck, L. D., and R. W. Wetzel.** 1990. A54145, a new lipopeptide antibiotic complex: factor control through precursor directed biosynthesis. *J. Antibiot.* **43:**607–615.
15. **Coëffet-Le Gal, M.-F., L. Thurston, P. Rich, V. Miao, and R. H. Baltz.** 2006. Complementation of daptomycin *dptA* and *dptD* deletion mutations *in-trans* and production of hybrid lipopeptide antibiotics. *Microbiology* **152:**2993–3001.
16. **Counter, F. T., N. E. Allen, D. S. Fukuda, J. N. Hobbs, J. Ott, P. W. Ensminger, J. S. Mynderse, D. A. Preston, and C. Y. Wu.** 1990. A54145, a new lipopeptide antibiotic complex: microbiological evaluation. *J. Antibiot.* **43:**616–622.
17. **Cox, K. L., and R. H. Baltz.** 1984. Restriction of bacteriophage plaque formation in *Streptomyces* spp. *J. Bacteriol.* **159:**499–504.
18. **Datsenko, K. A., and B. Wanner.** 2000. One-step inactivation of chromosomal genes in *Escherichia coli* K-12 using PCR products. *Proc. Natl. Acad. Sci. USA* **97:**6640–6645.

19. Debono, M., M Barnhart, C. B. Carrell, J. A. Hoffmann, J. L. Occolowitz, B. J. Abbott, D. S. Fukuda, R. L. Hamill, K. Biemann, and W. C. Herlihy. 1987. A21978C, a complex of new acidic peptide antibiotics: isolation, chemistry, and mass spectral structure elucidation. *J. Antibiot.* **40:**761–777.

20. Doekel, S., M.-F. Coëffet-Le Gal, J.-Q. Gu, M. Chu, R. H. Baltz, and P. Brian. 2008. Non-ribosomal peptide synthetase module fusions to produce derivatives of daptomycin in *Streptomyces roseosporus*. *Microbiology* **154:**2872–2880.

21. Fischbach, M. A., and C. T. Walsh. 2006. Assembly-line enzymology for polyketide and nonribosomal peptide antibiotics: logic, machinery, and mechanisms. *Chem. Rev.* **106:**3468–3496.

22. Ford, L. M., T. E. Eaton, and O. W. Godfrey. 1990. Selection for *Streptomyces ambofaciens* mutants that produce large quantities of spiramycin and determination of optimal conditions for spiramycin production. *Appl. Environ. Microbiol.* **56:**3511–3514.

23. Fowler, V. G., H. W. Boucher, G. R. Corey, E. Abrutyn, A. W. Karchmer, M. E. Rupp, D. P. Levine, H. F. Chambers, F. P. Tally, G. A. Vigliani, C. H. Cabell, A. S. Link, I. DeMeyer, G. S. Filler, M. Zervos, P. Cook, J. Parsonnet, J. M. Bernstein, C. S. Price, G. N. Forrest, G. Fätkenheure, M. Gareca, S. J. Rehm, H. R. Brodt, A. Tice, S. E. Cosgrove, and S. *aureus* Endocarditis and Bacteremia Study Group. 2006. Daptomycin versus standard therapy for bacteremia and endocarditis caused by *Staphylococcus aureus*. *N. Engl. J. Med.* **355:**653–655.

24. Frijters, A. C. J., Z. Zhang, M. van Damme, G.-L. Wang, P. C. Ronald, and R. W. Michelmore. 1997. Construction of a bacterial artificial chromosome library containing large *Eco*RI and *Hin*dIII genome fragments of lettuce. *Theor. Appl. Genet.* **94:**390–399.

25. Frueh, D. P., H. Arthanari, A. Koglin, D. A. Vosburg, A. E. Bennett, C. T. Walsh, and G. Wagner. 2008. Dynamic thiolation-thioesterase structure of a non-ribosomal peptide synthetase. *Nature* **454:**903-906.

26. Geistlich, M., R. Losick, J. R. Turner, and R. N. Rao. 1992. Characterization of a novel regulatory gene governing the expression of a polyketide synthase gene in *Streptomyces ambofaciens*. *Mol. Microbiol.* **6:**2019–2029.

27. Gregory, M. A., R. Till, and M. C. M. Smith. 2003. Integration site for *Streptomyces* phage φBT1 and development of site-specific integrating vectors. *J. Bacteriol.* **185:**5320–5323.

28. Gu, J.-Q., K. T. Nguyen, C. Gandhi, V. Rajgarhia, R. H. Baltz, and M. Chu. 2007. Structural characterization of daptomycin analogues A21978C$_{1-3}$(D-Asn11) produced by a recombinant *Streptomyces roseosporus* strain. *J. Nat. Prod.* **70:**233–240.

29. Gust, B., G. L. Challis, K. Fowler, T. Kieser, and K. F. Chater. 2003. PCR-targeted *Streptomyces* gene replacement identifies a protein domain needed for biosynthesis of the sesquiterpene soil odor geosmin. *Proc. Natl. Acad. Sci. USA* **100:**1541–1546.

30. Hahn, D. R., P. J. Solenberg, and R. H. Baltz. 1991. Tn5099, a *xylE* promoter probe transposon for *Streptomyces* spp. *J. Bacteriol.* **173:**5573–5577.

31. Hojati, Z., C. Milne, B. Harvey, L. Gordon, M. Borg, F. Flett, B. Wilkinson, P. J. Sidebottom, B. A. M. Rudd, M. A. Hayes, C. P. Smith, and J. Michlefield. 2002. Structure, biosynthetic origin, and engineered biosynthesis of calcium-dependent antibiotics from *Streptomyces coelicolor*. *Chem. Biol.* **9:**1175–1187.

32. Hosted, T. J., and R. H. Baltz. 1997. Use of *rpsL* for dominance selection and gene replacement in *Streptomyces roseosporus*. *J. Bacteriol.* **179:**180–186.

33. Huber, F. M., R. L. Pieper, and A. J. Tietz. 1988. The formation of daptomycin by supplying decanoic acid to *Streptomyces roseosporus* cultures producing the antibiotic complex A21978C. *J. Biotechnol.* **7:**283–292.

34. Kieser, T., M. J. Bibb, M. J. Buttner, K. F. Chater, and D. A. Hopwood. 2000. *Practical Streptomyces Genetics.* John Innes Foundation, Norwich, United Kingdom.

35. Kuhstoss, S., and R. N. Rao. 1991. Analysis of the integration function of the streptomycete bacteriophage φC31. *J. Mol. Biol.* **222:**897–908.

36. Landman, D., C. Georgescu, D. A. Martin, and J. Quale. 2008. Polymyxins revisited. *Clin. Microbiol. Rev.* **21:**449–465.

37. Lee, E., D. Yu, J. M. de Velasco, L. Tessarollo, D. A. Swing, D. L. Court, N. A. Jenkins, and N. G. Copeland. 2001. A highly efficient *Escherichia coli*-based chromosome engineering system adapted for recombinogenic targeting and subcloning of BAC DNA. *Genomics* **73:**56–65.

38. Mahlert, C., F. Kopp, J. Thirlway, J. Micklefield, and M. A. Marahiel. 2007. Stereospecific enzymatic transformation of α-ketoglutarate to (2S,3R)-3-methyl glutamate during acidic lipopeptide biosynthesis. *J. Am. Chem. Soc.* **129:**12011–12018.

39. Matsushima, P., and R. H. Baltz. 1985. Efficient plasmid transformation of *Streptomyces ambofaciens* and *Streptomyces fradiae* protoplasts. *J. Bacteriol.* **163:**180–185.

40. Matsushima, P., and R. H. Baltz. 1986. Protoplast fusion, p. 170–183. *In* A. L. Demain and N. A. Solomon (ed.), *Manual of Industrial Microbiology and Biotechnology*. American Society for Microbiology, Washington, DC.

41. McHenney, M. A., and R. H. Baltz. 1988. Transduction of plasmid DNA in *Streptomyces* spp. and related genera by bacteriophage FP43. *J. Bacteriol.* **170:**2276–2282.

42. McHenney, M. A., and R. H. Baltz. 1996. Gene transfer and transposition mutagenesis in *Streptomyces roseosporus*: mapping of insertions that influence daptomycin or pigment production. *Microbiology* **142:**2363–2373.

43. McHenney, M. A., T. J. Hosted, B. S. Dehoff, P. R. Rosteck, Jr., and R. H. Baltz. 1998. Molecular cloning and physical mapping of the daptomycin gene cluster from *Streptomyces roseosporus*. *J. Bacteriol.* **180:**143–151.

44. Miao, V., M.-F. Coëffet-LeGal, P. Brian, R. Brost, J. Penn, A. Whiting, S. Martin, R. Ford, I. Parr, M. Bouchard, C. J. Silva, S. K. Wrigley, and R. H. Baltz. 2005. Daptomycin biosynthesis in *Streptomyces roseosporus*: cloning and analysis of the gene cluster and revision of peptide stereochemistry. *Microbiology* **151:**1507–1523.

45. Miao, V., R. Brost, J. Chapple, M.-F. Coëffet-LeGal, and R. H. Baltz. 2006. The lipopeptide antibiotic A54145 biosynthetic gene cluster from *Streptomyces fradiae*. *J. Ind. Microbiol. Biotechnol.* **33:**129–140.

46. Miao, V., M.-F. Coëffet-LeGal, K. Nguyen, P. Brian, J. Penn, A. Whiting, J. Steele, D. Kau, S. Martin, R. Ford, T. Gibson, M. Bouchard, S. K. Wrigley, and R. H. Baltz. 2006. Genetic engineering in *Streptomyces roseosporus* to produce hybrid lipopeptide antibiotics. *Chem. Biol.* **13:**269–276.

47. Motamedi, H., A. Shafiee, and S. J. Cai. 1995. Integrative vectors for heterologous gene expression in *Streptomyces* spp. *Gene* **160:**25–31.

48. Müller, C., S. Nolden, P. Gebhardt, E. Heinzelmann, C. Lange, O. Puk, K. Wetzel, W. Wohlleben, and D. Schwartz. 2007. Sequencing and analysis of the biosynthetic gene cluster of the lipopeptide antibiotic friulimicin in *Actinoplanes friuliensis*. *Antimicrob. Agents Chemother.* **51:**1028–1037.

49. Nguyen, K. T., D. Kau, J.-Q. Gu, P. Brian, S. W. Wrigley, R. H. Baltz, and V. Miao. 2006. Identification of a

glutamic acid 3-methyltransferase gene by functional analysis of an accessory gene locus important for daptomycin biosynthesis in *Streptomyces roseosporus*. *Mol. Microbiol.* **61:**1294–1307.

50. **Nguyen, K., D. Ritz, J.-Q. Gu, D. Alexander, M. Chu, V. Miao, P. Brian, and R. H. Baltz.** 2006. Combinatorial biosynthesis of lipopeptide antibiotics related to daptomycin. *Proc. Natl. Acad. Sci. USA* **103:**17462–17467.

51. **Ochi, K.** 2006. From microbial differentiation to ribosome engineering. *Biosci. Biotechnol. Biochem.* **71:**1373–1386.

52. **Penn, J., A. Whiting, S. K. Wrigley, M. Latif, T. Gibson, C. J. Silva, X. Li, V. Miao, P. Brian, and R. H. Baltz.** 2006. Heterologous production of daptomycin in *Streptomyces lividans*. *J. Indust. Microbiol. Biotechnol.* **33:**121–128.

53. **Pertel, P. E., P. Bernardo, C. Fogerty, P. Matthews, R. Northland, M. Benvenuto, G. M. Thorne, S. A. Luperchio, R. D. Arbeit, and J. Alder.** 2008. Effects of prior effective therapy on the efficacy of daptomycin and ceftriaxone for the treatment of community-acquired pneumonia. *Clin. Infect. Dis.* **46:**1142–1151.

54. **Richardson, M. A., S. Kuhstoss, M. L. B. Huber, L. Ford, O. Godfrey, J. R. Turner, and R. N. Rao.** 1990. Cloning of spiramycin biosynthetic genes and their use in constructing *Streptomyces ambofaciens* mutants defective in spiramycin biosynthesis. *J. Bacteriol.* **172:**3790–3798.

55. **Sambrook, J., E. F. Fritsch, and T. Maniatis.** 1989. *Molecular Cloning: a Laboratory Manual*, 2nd ed. Cold Spring Harbor Laboratory Press, Cold Spring Harbor, NY.

56. **Sieber, S. A., and M. A. Marahiel.** 2005. Molecular mechanisms underlying nonribosomal peptide synthesis: approaches to new antibiotics. *Chem. Rev.* **105:**715–738.

57. **Silverman, J. A., L. I. Morton, A. D. Vanpraagh, T. Li, and J. Alder.** 2005. Inhibition of daptomycin by pulmonary surfactant: *in vitro* modeling and clinical impact. *J. Infect. Dis.* **191:**2149–2152.

58. **Solenberg, P. J., P. Matsushima, D. R. Stack, S. C. Wilkie, R. C. Thompson, and R. H. Baltz.** 1997. Production of hybrid glycopeptide antibiotics *in vitro* and in *Streptomyces toyocaensis*. *Chem. Biol.* **4:**195–202.

59. **Straus, S. K., and R. E. W. Hancock.** 2006. Mode of action of the new antibiotic for Gram-positive pathogens daptomycin: comparison with cationic antimicrobial peptides and lipopeptides. *Biochim. Biophys. Acta* **1758:**1215–1223.

60. **Streiker, M., F. Kopp, C. Mahlert, L. O. Essen, and M. A. Marahiel.** 2007. Mechanistic and structural basis of stereospecific Cβ-hydroxylation in calcium-dependent antibiotic, a daptomycin-type lipopeptide. *ACS Chem. Biol.* **2:**152–154.

61. **Streiker, M., and M. A. Marahiel.** 2009. The structural diversity of acidic lipopeptide antibiotics. *ChemBiolChem* **10:**607–616.

62. **Tanovic, A., S. S. Samel, L.-O. Essen, and M. A. Marahiel.** 2008. Crystal structure of the termination module of a nonribosomal peptide synthetase. *Science* **321:**659–663.

63. **Wittman, M., U. Linne, V. Pohlmann, and M. A. Marahiel.** 2008. Role of DptE and DptF in the lipidation reaction of daptomycin. *FEBS J.* **275:**5343–5354.

64. **Zaas, A. K.** 2008. Echinocandins: a wealth of choice—how clinically different are they? *Curr. Opin. Infect. Dis.* **21:**426–432.

65. **Zhou, Z., J. R. Lai, and C. T. Walsh.** 2006. Interdomain communication between the thiolation and thioesterase domains of EntF explored by combinatorial mutagenesis and selection. *Chem. Biol.* **13:**869–879.

Genetic Engineering To Regulate Production of Secondary Metabolites in *Streptomyces clavuligerus*

SUSAN E. JENSEN

28

Streptomyces clavuligerus is a filamentous soil bacterium originally isolated from a South American soil sample in a screening program targeting β-lactamase-resistant β-lactam antibiotics (34, 48, 66). The antibiotics for which it was originally isolated, cephamycin C and 7-demethoxycephamycin C (Fig. 1), were soon overshadowed in importance when rescreening of the organism for production of β-lactamase inhibitory compounds led to the discovery of clavulanic acid (12, 76). While cephamycin antibiotics are valuable clinical agents in their own right, clavulanic acid is recognized as the most important product of *S. clavuligerus* today. Clavulanic acid is a powerful irreversible inhibitor of many serine active site β-lactamases. Since these β-lactamases are responsible for much of the clinical β-lactam resistance, clavulanic acid is used in combination with conventional penicillin or cephalosporin-type antibiotics to restore the effectiveness of these agents against otherwise resistant strains.

S. clavuligerus NRRL 3585 (ATCC 27064) is the original, and was for many years the only known, strain of this species. Clavulanic acid for clinical use is still produced only by fermentations using the descendants of this original strain. Very recently, a second strain, *S. clavuligerus* CKD1119, was isolated from a Korean soil sample in a screen for immunosuppresive compounds (51, 52) and found to produce tacrolimus (also known as FK-506). Analyses showed that this strain also produces clavulanic acid, but no comparative information on productivity relative to the original strain is available. No other information is available on this new strain, and so the remainder of this article will refer to studies on the original *S. clavuligerus* NRRL 3585 strain.

In addition to cephamycin C and clavulanic acid, *S. clavuligerus* is known to produce two non-β-lactam metabolites, holomycin and tunicamycins (48), as well as at least four other clavam compounds (13, 75) (Fig. 1). Superficially, these other clavams resemble clavulanic acid in structure, but they differ in the side chains attached at C-2 and in the stereochemistry of the fused ring system. These other clavam metabolites have an S configuration at C-5, and so are referred to as 5S clavams, whereas clavulanic acid has a 5R configuration. β-Lactamase inhibitory activity is associated with the 5R configuration, and so none of the 5S clavams has this property; instead, they display a variety of weak antibacterial and antifungal activities. In the cases of alanylclavam and 2-hydroxymethylclavam, antibacterial activity is manifest only against organisms growing on

defined medium and is due to inhibition of the methionine biosynthetic enzyme homoserine succinyltransferase, resulting in methionine antimetabolite properties (77). In contrast, the antifungal activity of the 5S clavams is not due to this antimetabolite property, but to an effect on RNA synthesis. Clavam-2-carboxylate is toxic, but its mechanism of toxicity has not been described.

As well as small-molecular-weight metabolites, *S. clavuligerus* also produces a β-lactamase inhibitor protein, BLIP, which, unlike clavulanic acid, is a very narrow-range inhibitor affecting only a handful of class A type β-lactamases, but with very high affinity (K_i in the sub-nanomolar range) (22, 87). While there is no obvious connection between the production of clavulanic acid or the 5S clavams and BLIP, mutants of *S. clavuligerus* defective in production of the pathway-specific regulator, CcaR, that regulates both clavulanic acid and cephamycin C production, also produce lower levels of BLIP than does the wild type (93). Furthermore, production of BLIP proteins is quite restricted within *Streptomyces* spp., and so its occurrence in a species that also produces both cephamycin C and clavulanic acid seems unlikely to be just coincidence.

Many early studies on penicillin and cephamycin biosynthesis were carried out in *S. clavuligerus*, but because of the commercial importance of clavulanic acid, most recent studies have focused on that compound. In particular, genetic engineering to generate improved clavulanic acid producer strains has been a main interest. Although the bioactivity of the 5S clavams has no clinical value, the structural variation seen in this family of metabolites offers the possibility of biological routes to structural variants of clavulanic acid once the full biosynthetic details of their production are known. Finally, the cross-regulation seen between the β-lactam metabolites makes understanding the production of this entire family of metabolites of great interest.

28.1. PRACTICAL CONSIDERATIONS FOR GENETIC ENGINEERING OF *S. CLAVULIGERUS*

Genetic engineering of *Streptomyces* spp. in general is made challenging by the slow growth and filamentous nature of the organisms, the limited selection of genetic tools available, and the lack of natural competence in the genus, among other factors. In addition, *S. clavuligerus* has its own particular characteristics which add complexity.

411

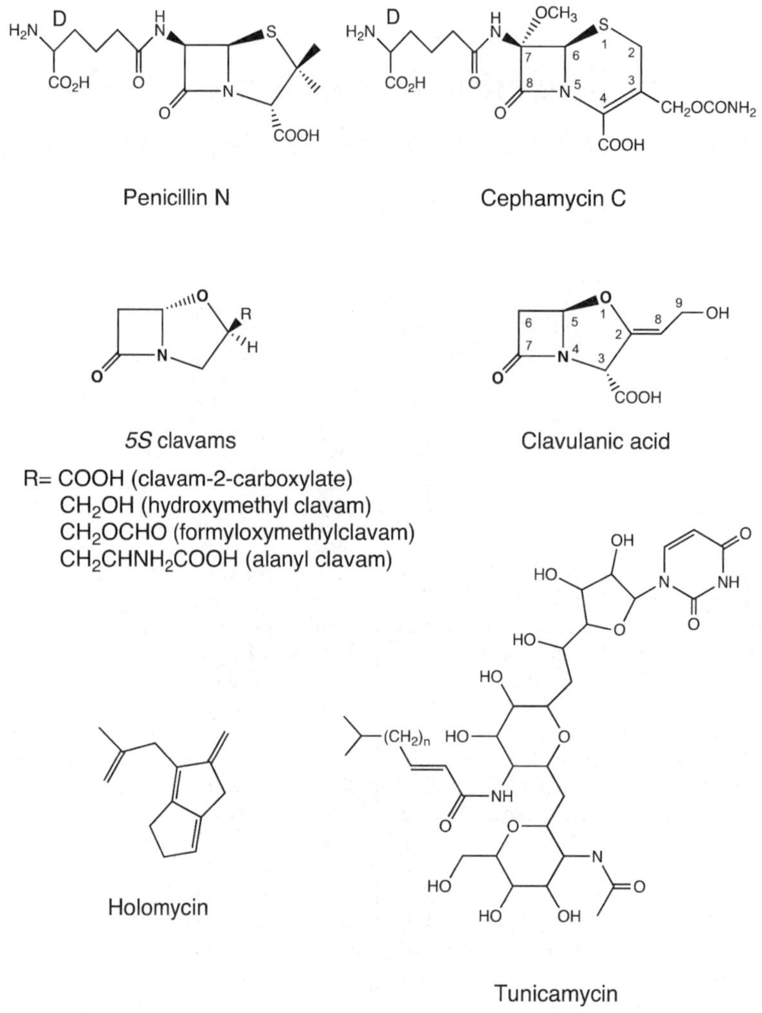

Penicillin N

Cephamycin C

5S clavams

R= COOH (clavam-2-carboxylate)
CH₂OH (hydroxymethyl clavam)
CH₂OCHO (formyloxymethylclavam)
CH₂CHNH₂COOH (alanyl clavam)

Clavulanic acid

Holomycin

Tunicamycin

FIGURE 1 Secondary metabolites produced by *S. clavuligerus.*

28.1.1. Basic Biological Features

S. clavuligerus is unusual among *Streptomyces* spp. in that it cannot use simple sugars such as glucose or sucrose as sole carbon sources (34), but it does grow well on glycerol, starch, maltose, or dextrin. Appropriate substitutions to growth media developed for other *Streptomyces* spp. will typically allow these media, and any genetic protocols with which they are associated, to be applied to *S. clavuligerus*. However, *S. clavuligerus* is fastidious about conditions supporting sporulation and will sporulate poorly if at all on most rich complex media. Good sporulation can usually be obtained on ISP-4 (Difco) or tomato oatmeal agar media. *S. clavuligerus* grows and produces secondary metabolites optimally at 28 to 30°C, but it is unusually sensitive to elevated temperature and will not grow above 31 to 32°C.

Liquid media employed by research groups studying the production of the various types of β-lactam metabolites produced by *S. clavuligerus* vary widely. Both clavulanic acid and cephamycin C can be produced reliably at high levels on a complex soy flour–starch-based medium (Soy medium; 78) as well as on a defined starch–asparagine-based medium (SA medium; 46). Other groups report using Trypticase soy broth for clavulanic acid production (81), but in our hands, production is poor and cultures lyse extensively in stationary

phase, so we use this medium only for staging seed cultures. Similarly, glycerol-sucrose-proline-glutamate (GSPG) medium supports only poor growth in our experience, although others report good results (74). Numerous bioprocess-type studies examining growth medium optimization as a means of improving production of clavulanic acid have appeared in recent years, typically reinforcing the association of soy-based growth medium constituents with high productivity (30, 68, 82, 97).

In contrast to clavulanic acid, reproducible production of 5S clavams is difficult to achieve in any liquid medium for reasons not understood at present. Soy medium is the only liquid medium that we have encountered that supports good 5S clavam production, and even then, variations in culture-to-flask volume ratio and aeration can greatly affect production (unpublished observations). Furthermore, despite careful attention to such details, amounts and relative proportions of the various 5S clavams vary widely from one experiment to the next. Surprisingly, liquid SA medium, which does not support 5S clavam production, will support production when solidified with agar, with metabolites recovered by a "freeze and squeeze" procedure. We have also encountered similar liquid-solid discrepancies with respect to production of 5S clavams on media employing soy protein extracts in place of soy flour (30).

28.1.2. Genetic Features

28.1.2.1. Introduction of Foreign DNA

Despite its lack of natural competence, *S. clavuligerus* is amenable to introduction of foreign DNA by polyethylene glycol-mediated transformation of protoplasts using standard techniques (27, 40, 50), although with some provisos. *S. clavuligerus* protoplasts are unusually sensitive to lysis even in osmotically supportive concentrations of sucrose unless appropriate levels of Ca^{2+} and Mg^{2+} are provided. *S. clavuligerus* protoplasts also regenerate at low rates, and that, combined with their powerful DNA restriction system(s), makes introduction of foreign DNA by polyethylene glycol-mediated transformation inefficient. Transformation with DNA constructs prepared in *Escherichia coli* directly into *S. clavuligerus* is also typically unsuccessful, even if methylation-defective host strains of *E. coli* are employed to prepare the DNA. Passage of *E. coli*-derived plasmid constructs through a nonrestricting *Streptomyces* sp. such as *S. lividans*, and from there into *S. clavuligerus*, can be used to overcome the restriction barrier. However, this precludes the use of integrative vectors that would insert into the intermediate host.

For all of these reasons, intergeneric conjugation has largely replaced transformation in recent years as a means of introducing both integrative and freely replicating plasmid and cosmid vectors into *S. clavuligerus*. Such constructs, so long as they include an origin of transfer such as $oriT_{RK2}$, can be transferred directly from methylation-deficient *E. coli* hosts carrying the necessary transfer functions into *S. clavuligerus*. Standard techniques developed for conjugative transfer from *E. coli* into *S. coelicolor* are generally applicable (50). However, while recovery of *S. clavuligerus* exconjugants using mannitol soy agar plates has been described by others (79), in our experience, conjugation is only successful when exconjugants are recovered on AS-1+Mg agar plates (6).

28.1.2.2. Plasmid Vectors Used in *S. clavuligerus*

Many plasmid vectors have been developed for use in various *Streptomyces* spp., but relatively few have been tested in *S. clavuligerus*, and so the following is not intended to be a complete list of all useful plasmid vectors. Multicopy vectors based on the pIJ101 replicon, such as pIJ486 (41, 64, 98) and pUWL-KS (99), typically work well in *S. clavuligerus* and have been widely used. Unlike the situation in *S. coelicolor*, however, they show segregational instability in *S. clavuligerus* in the absence of antibiotic selection. As a result, sporulation in the absence of antibiotic selection routinely gives near-complete loss of these plasmids, which can be useful for delivery of gene disruption constructs. Resistance to apramycin, encoded by the *apra* cassette and mediated by the *aac(3)IV* gene (50), is particularly useful as a selectable marker because spontaneous resistance is rarely encountered, and resistance can be selected in both *Streptomyces* spp. and *E. coli*. The thiostrepton resistance gene (*thio*; 50) is another widely used marker that gives resistance to thiostrepton with little background of spontaneous resistance in *S. clavuligerus* (although at the unusually low level of 2 to 5 μg/ml), but with the limitation that resistance cannot be selected in *E. coli*. Neomycin resistance due to *neo* (Tn5) is another useful marker (21) that is selectable in both *S. clavuligerus* and *E. coli*, but hampered by a greater tendency to show some spontaneous background resistance. Hygromycin, viomycin, streptomycin, and spectinomycin all have very high natural MIC values in *S. clavuligerus* that limit the usefulness of their resistance genes, although transformants showing hygromycin resistance (65) have been selected using levels of 250 μg/ml.

The segregational instability of pIJ101-based vectors makes them suitable delivery vectors for gene disruption constructs, since they are easily lost from the cell. The partition-defective vector pJV326 (33) also works well in *S. clavuligerus* for delivery of gene disruption constructs. In contrast, the low maximum growth temperature of *S. clavuligerus* precludes the use of temperature-sensitive plasmid vectors for delivery of gene disruption constructs (10).

Providing that antibiotic selection is maintained, plasmid vectors based on the pIJ101 replicon can also be used for complementation or protein expression purposes, but in practice, such high-copy-number vectors can have deleterious effects on cell growth and antibiotic production. Furthermore, in *S. clavuligerus*, attempts to use freely replicating plasmids for complementation or overexpression of native genes can promote integration of the plasmid into the chromosome via homologous recombination. For this reason, integrative plasmids such as pSET152 (8) and pHM8a (65) are preferred for protein overproduction or complementation studies. Furthermore, pSET152 and plasmids derived from it transfer into *S. clavuligerus* with unusually high efficiency by conjugation or transformation, and so they have become widely used vectors for such studies.

E. coli-derived promoters are usually nonfunctional in *Streptomyces* spp., but heterologous promoters from other actinomycete species such as the strong, constitutively expressed *ermE** promoter (83) have proven useful for expression studies in *S. clavuligerus*. Useful regulatable promoters include the thio-inducible *tipA* promoter (19, 50, 81) and the glycerol-inducible $gylP_1P_2$ promoters (35, 41, 94), but the Ni^{2+}-regulatable *sodF1* promoter from *S. coelicolor* proved to be weak and unregulated in *S. clavuligerus* (unpublished observations).

Cosmid libraries based on *E. coli*-derived SuperCos-type vectors (Stratagene) have proven useful, both for general-purpose cloning of *S. clavuligerus* genomic DNA fragments and for gene disruption as part of the REDIRECT© PCR-targeted system (see below). The *E. coli-Streptomyces* shuttle cosmid pOJ436 (8) has also been used to prepare a library of genomic DNA for integration of large fragments of *S. clavuligerus* DNA into the chromosome of other *Streptomyces* spp.

28.1.2.3. Gene Disruption in *S. clavuligerus*

Gene disruption in *S. clavuligerus* is readily achievable, either by preparing gene disruption constructs and delivering them on pIJ101-based plasmid vectors, or by employing PCR-targeting techniques as part of the REDIRECT© system to introduce mutations (31). Of the various cassettes available for use in PCR-targeted mutagenesis, the *apra* cassette carried on pIJ773 is the only one that has been widely used for *S. clavuligerus* because the viomycin and spectinomycin cassettes do not provide useful levels of resistance in this species. Alternative resistance markers are desirable because creation of double mutants typically requires that two different selectable markers be employed. The availability of differently marked single mutants also makes possible their combination into double mutants by the simple expedient of protoplast fusion. Using established techniques for protoplast fusion (39), we were able to fuse BLIP single mutants (*bli::apra*) with BLIP-like protein (BLP) single mutants (*blp::neo*) to generate *bli::apra-blp::neo* double mutants with relative ease (unpublished observations). Alternatively, the REDIRECT© system allows single mutants to be rendered unmarked by using FLP/FLP recombination target (FLP/FRT)-based recombination to remove antibiotic resistance cassettes, making it theoretically possible to create double mutants by sequential use of the *apra* cassette. As originally designed, use of the FLP/FRT system relies on the ability to deliver *E. coli*-derived mutant cosmids by transformation, but the restriction barrier prevents this in *S. clavuligerus*, requiring that alternative techniques be used to remove the resistance cassette. A recently developed *Streptomyces*-based vector carrying a synthetic *flp* gene optimized for expression in *Streptomyces* (25) should streamline this procedure by enabling in vivo excision of the FRT-flanked resistance cassettes in *Streptomyces* hosts, although this has not yet been tested in *S. clavuligerus*. Alternatively, we have just developed a vector which supports expression of the native *flp* gene in *S. clavuligerus* and *S. coelicolor*, allowing direct in vivo excision of the resistance cassette (103).

In vitro transposon mutagenesis of cosmid-based libraries of genomic DNA can be used as an alternative means to introduce mutations into the chromosome of *S. clavuligerus* using the modified Tn5-type transposon Tn5062 (11, 104). However, in our hands, a number of instances (about 20% of genes) were encountered where mutations carried on cosmids could not be successfully transferred into the *S. clavuligerus* chromosome, despite subsequent proof that the genes were not essential (104).

28.2. APPLICATIONS OF GENETIC ENGINEERING TO THE PRODUCTION OF β-LACTAM COMPOUNDS

28.2.1. Genetics of Production of β-Lactam Compounds in *S. clavuligerus*

Like the sequenced *Streptomyces* spp., *S. clavuligerus* is assumed to have a linear chromosome of about 8 to 9 Mb. A complete genome sequence for *S. clavuligerus* is currently not available, but a preliminary draft genome sequence, available through the NCBI site (http://www.ncbi.nlm.nih.gov/sites/entrez?db=genomeprj&cmd=search&term=clavuligerus%20genome), comprising 6.7 Mb of genomic sequence information distributed in 597 contigs and encoding 5,981

proteins, has been obtained by the Broad Institute Genome Sequencing Platform. The Korean Research Institute of Bioscience and Biotechnology continues to work to complete this genome sequence. In addition to its chromosome, *S. clavuligerus* is also known to carry three extrachromosomal linear plasmids, ranging in size from 11.2 kb to greater than 450 kb (67). The plasmids are stably maintained, and no reports of "cured" strains have appeared in the literature.

Genes associated with the production of the β-lactam metabolites are distributed across three separate locations in the genome of *S. clavuligerus*. Studies of the resident linear plasmids have shown them not to carry any genetic information required for the production of the β-lactam metabolites (90). This discontinuous arrangement of the β-lactam-producing genes is unusual and goes against the established paradigm that genes associated with antibiotic production are clustered in the chromosomes of producer bacteria. Clustering would facilitate horizontal gene transfer, which is presumed to be a common route for acquisition of antibiotic production ability, consistent with the sporadic distribution of antibiotic production ability across taxonomically unrelated species.

Clavulanic acid production is limited to a handful of species (43) and is accompanied by the production of cephamycin C, suggesting that the same grouping of clavulanic acid and cephamycin C biosynthetic genes into separate but adjacent gene clusters (the cephamycin C-clavulanic acid supercluster, Fig. 2) may also exist in other clavulanic acid producers. Production of 5S clavam metabolites is a trait shared with a larger number of species (43), but *S. clavuligerus* is unique among 5S clavam producers in its ability to produce both clavulanic acid and 5S clavam metabolites. Because the biosynthetic pathways to clavulanic acid and the 5S clavams overlap, enzymes encoded by genes located in the cephamycin C-clavulanic acid supercluster contribute to the production of the 5S clavams as well as clavulanic acid despite their separate locations. This leads

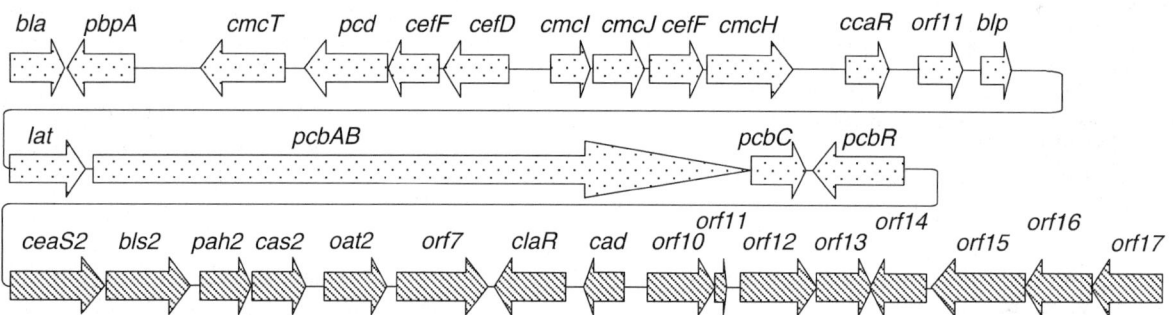

Gene	Proposed function	Gene	Proposed function	Gene	Proposed function
bla	β-lactamase	*orf11*	unknown	**orf7*	peptide binding protein
pbpA	penicillin binding protein	*blp*	BLIP-like protein	**claR*	transcriptional regulator
*cmcT	transmembrane protein	**lat*	lysine aminotransferase	**cad*	clavaldehyde dehydrogenase
*pcd	piperidiene carboxylate dehydrogenase	**pcbAB*	aminoadipyl-cysteinyl-valine synthetase	**orf10*	cytochrome P450
*cefE	deacetoxycephalosporin C synthase	*pcbC*	Isopenicillin N synthase	**orf11*	ferridoxin
*cefD	Isopenicillin N epimerase	*pcbR*	penicillin binding protein -resistance	**orf12*	β-lactamase related protein
*cmcI	C-7 hydroxylase	**ceaS2*	carboxyethylarginine synthase	**orf13*	efflux pump
*cmcJ	C-7 methyltransferase	**bls2*	β-lactam synthetase	**orf14*	acetyltransferase
*cefF	deacetoxycephalosporin C hydroxylase	**pah2*	proclavaminate amidinohydrolase	**orf15*	peptide binding protein
*cmcH	carbamoyltransferase	**cas2*	clavaminate synthase	**orf16*	unknown
*ccaR	transcriptional regulator	*oat2*	ornithine acetyltransferase	**orf17*	glycyltransferase

FIGURE 2 The cephamycin C-clavulanic acid gene supercluster in *S. clavuligerus*. Arrows filled with dots indicate genes for cephamycin C biosynthesis; arrows filled with slanted lines indicate genes for clavulanic acid production. Genes marked with asterisks are essential or important for cephamycin C or clavulanic acid production.

to a complex and overlapping regulatory situation that affects attempts to undertake genetic modifications aimed at improving cephamycin C or clavulanic acid production or modifying the array of metabolites produced.

28.2.2. Cephamycin C-Clavulanic Acid Supercluster

Genes involved in the production of cephalosporins and cephamycins have been studied from several species, and the details of cephamycin C biosynthesis are quite well understood (for reviews see references 20, 44, 55, and 56). In *S. clavuligerus*, biosynthesis begins with the amino acids L-lysine, L-cysteine, and L-valine and proceeds through the synthesis of penicillin and cephalosporin intermediates to give the

final cephamycin product (Fig. 3). In the first step of the pathway, L-lysine is converted to piperideine-6-carboxylate by the enzyme lysine-ε-aminotransferase, and from there to L-α-aminoadipic acid. Once α-aminoadipic acid is formed, it condenses with cysteine and valine through the action of a nonribosomal peptide synthetase, δ-(L-α-aminoadipyl)-L-cysteinyl-D-valine synthetase, and the resulting tripeptide is cyclized by isopenicillin N synthase to give a penicillin intermediate. The thiazolidine ring structure of penicillin N is then expanded by deacetoxycephalosporin C synthase to give the dihydrothiazine ring system characteristic of cephalosporin and cephamycin antibiotics. Other gene products from the cluster modify these intermediates to form the final cephamycin C product. Penicillin, cephalosporin, and cephamycin

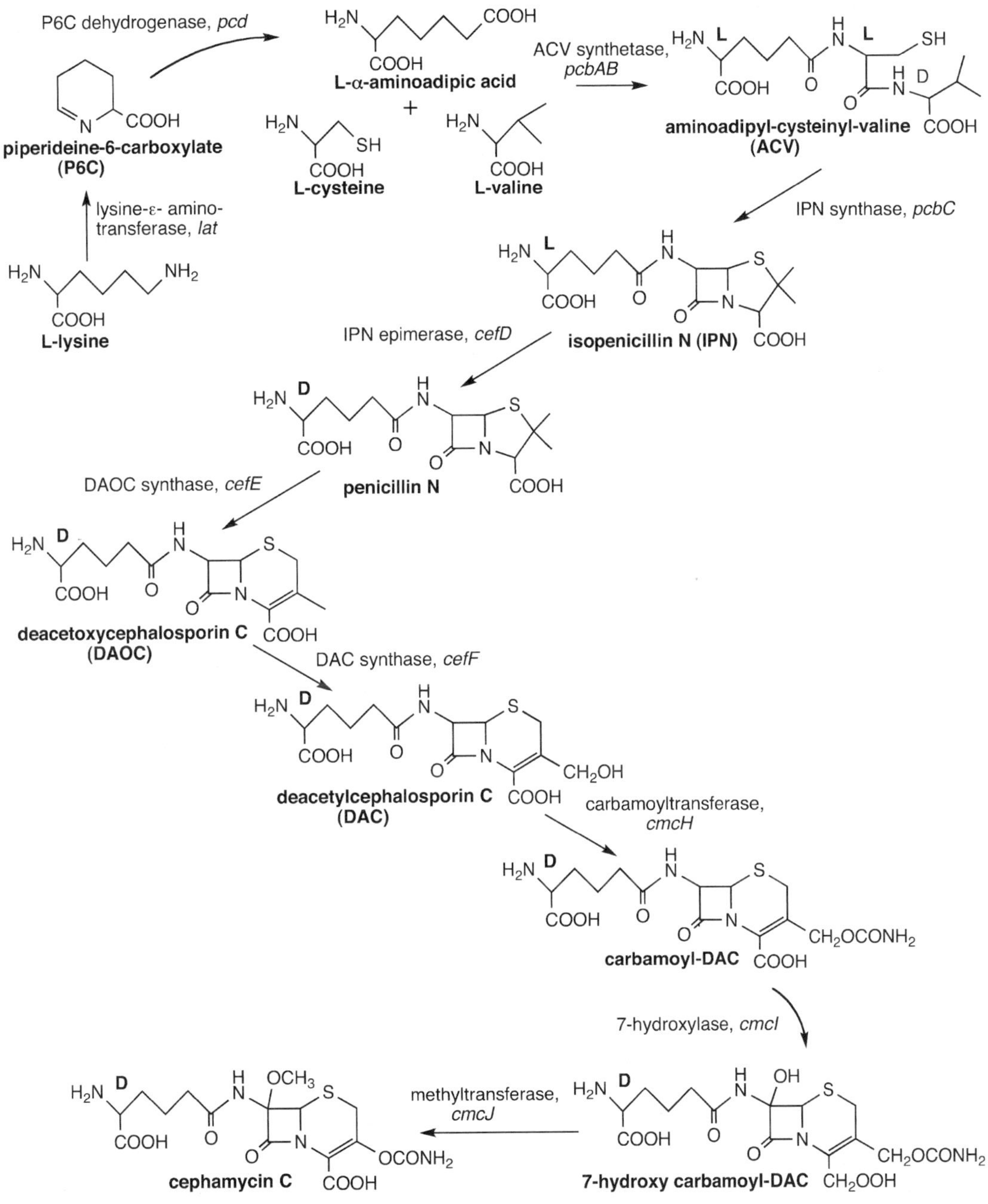

FIGURE 3 The biosynthetic pathway to cephamycin C in *S. clavuligerus*. Enzyme names are followed by the corresponding gene designation shown in italics.

metabolites can therefore all be thought of as cyclized and highly modified nonribosomal peptide antibiotics.

Despite their superficially similar bicyclic β-lactam nuclear structure and adjacently located gene clusters, cephamycin C and clavulanic acid do not share any common biosynthetic elements except for the pathway-specific transcriptional activator, CcaR, encoded by *ccaR*, located within the cephamycin arm of the supercluster (Fig. 2). Clavulanic acid biosynthesis is less well understood than cephamycin C biosynthesis, but the clavulanic acid gene cluster is believed to contain all of the genetic information required for the production of clavulanic acid, with *ceaS2* shown to be the first, and *orf17* the most distant, gene of the cluster essential for clavulanic acid production (2, 36, 41, 45, 54, 59, 61, 62, 70, 71, 73, 74, 84, 98, 102). Mutation of genes beyond *orf17* has only limited or no effect on clavulanic acid production, although some conditional regulators may be encoded there. However, clavulanic acid producing capability has never been transferred to a nonproducer species, and the late steps of clavulanic acid biosynthesis are not understood, making it hard to establish that all of the genes required for these steps have been identified. Therefore, the completeness of the cluster remains an assumption pending further proof.

The early steps of clavulanic acid biosynthesis convert the primary metabolites L-arginine and D-glyceraldehyde-3-PO$_4$ into clavaminic acid, the last known common intermediate for clavulanic acid and 5S clavam biosynthesis (Fig. 4). This conversion is accomplished in six steps by four enzymes, carboxyethylarginine synthase, β-lactam synthetase, proclavaminate amidinohydrolase, and clavaminate synthase (which catalyzes three independent steps). Genes encoding each of these activities make up the first four genes of the clavulanic acid gene cluster. Each of these genes has been expressed in *E. coli*, and the resulting proteins have been characterized and their structures examined by X-ray crystallography (14, 24, 63, 105).

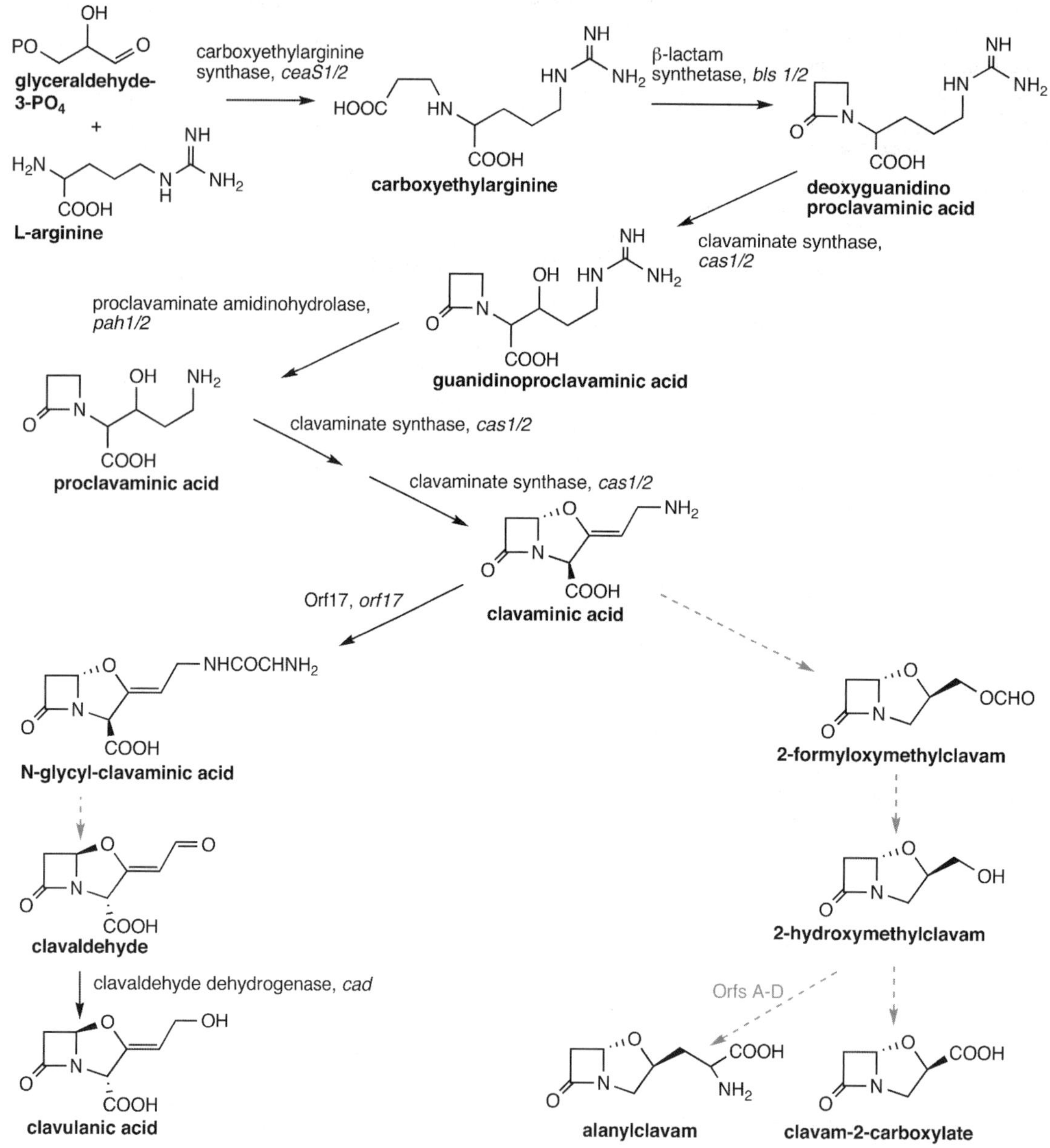

FIGURE 4 The biosynthetic pathway to clavulanic acid and the 5S clavams in *S. clavuligerus*. Enzyme names are followed by the corresponding gene designation shown in italics. Parts of the pathway shown with broken arrows represent pathway steps that are proposed but not yet demonstrated.

While these early steps of clavulanic acid biosynthesis are now well understood, the late steps are not. Twelve more open reading frames (ORFs), *oat2-orf17*, complete the minimal clavulanic acid gene cluster. Gene disruption studies have shown that all of the ORFs in this region, with the exception of *oat2*, are essential or important for clavulanic acid production, although the nature of their involvement is often not clear. *oat2* encodes an ornithine aminotransferase (OAT) shown to catalyze the reversible transfer of acetyl groups from *N*-acetylornithine to glutamate (23, 49, 95). Since this step is part of the arginine biosynthesis pathway, *oat2* may serve to provide additional precursor for clavulanic acid production, a function consistent with its dispensable nature. Alternatively, although OAT acetylates glutamate in vitro, its actual role in clavulanic acid biosynthesis may be to acetylate some other substrate, in much the same way that clavaminate synthase can hydroxylate *N*-acetylarginine in vitro even though that is not its natural substrate in vivo (58). ClaR is a transcriptional activator essential for expression of the genes involved in the late, post-clavaminic acid stages of clavulanic acid biosynthesis, and clavaldehyde dehydrogenase (Cad) catalyzes the final step of the pathway, the reduction of clavaldehyde to clavulanic acid. The only other characterized step in the late pathway is that catalyzed by Orf17, an enzyme that converts clavaminic acid to *N*-glycylclavaminic acid (4). This activity was observed in Orf17 enzyme produced in *E. coli*, and trace amounts of *N*-glycylclavaminic acid were also identified as a metabolite accumulating in *S. clavuligerus* mutants with defects in *orf15* or

orf16 (41), suggesting that production of *N*-glycylclavaminic acid may represent the first step beyond clavaminic acid on the clavulanic acid-specific branch of the pathway. Since this *N*-acylated derivative of clavaminic acid retains the 3S,5S configuration of clavaminic acid, the question of how the double inversion of stereochemistry needed to form clavulanic acid is achieved remains open, as do the roles played by the other unassigned and yet essential ORFs. Beyond *orf17* is a series of ORFs with less clear-cut relationships to clavulanic acid production (41). *orf18* and *orf19* encode apparent penicillin binding proteins, yet their disruption either is impossible (*orf18* is an essential gene) or has no effect on clavulanic acid production or on antibiotic resistance (*orf19*). *orf20* encodes an apparent cytochrome P450, but once again disruption studies suggest no direct involvement in clavulanic acid production. Beyond *orf20* lie three genes encoding apparent regulatory proteins, a sigma factor and the elements of a two-component regulatory system (84). While disruption studies suggest the possibility of conditional effects on clavulanic acid production, these effects are relatively minor compared to disruption of genes central to the cluster.

28.2.3. The Clavam and Paralogue Gene Clusters

From the time of its discovery, the gene encoding clavaminate synthase, *cas2*, from the clavulanic acid gene cluster was known to be only one of a pair of *cas* genes encoding two isozymes of clavaminate synthase. The second of these, *cas1*, is located in a separate region called the clavam gene cluster (Fig. 5) because it is flanked on both sides by genes

Clavam cluster

cvm3 cvm2 cvm1　cas1　cvm4 cvm5　cvm6　　cvm7

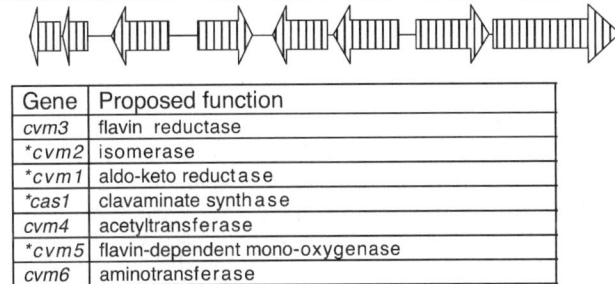

Gene	Proposed function
cvm3	flavin reductase
*cvm2	isomerase
*cvm1	aldo-keto reductase
*cas1	clavaminate synthase
cvm4	acetyltransferase
*cvm5	flavin-dependent mono-oxygenase
cvm6	aminotransferase
cvm7	transcriptional regulator

Paralogue cluster

orfD orfC orfB orfA　ceaS1　bls1　pah1　oat1　c6p　　c7p

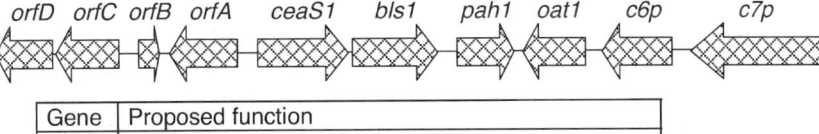

Gene	Proposed function
*orfD	dehydratase
*orfC	aminotransferase
*orfB	regulatory protein
*orfA	hydroxymethyltransferase
*ceaS1	carboxyethylarginine synthase
*bls1	β-lactam synthetase
*pah1	proclavaminate amidinohydrolase
oat1	ornithine acetyltransferase
*c6p	aminotransferase
*c7p	transcriptional regulator

FIGURE 5 The clavam and paralogue gene clusters in *S. clavuligerus*. Genes marked with asterisks are essential or important for 5S clavam production.

involved specifically in the production of 5S clavam metabolites (64, 88). Clavulanic acid and the 5S clavams share a common pathway to the level of clavaminic acid, but the genes of the clavam cluster, except for cas1, are involved in the late, 5S clavam-specific stages of biosynthesis.

In the same way that cas2 has a paralogue, cas1, residing in the clavam cluster, the other early genes of clavulanic acid biosynthesis have paralogues located in yet another region of the chromosome called the paralogue gene cluster (42, 89) (Fig. 5). This cluster includes ceaS1, bls1, pah1, and oat1, paralogues of genes for early shared steps of the pathway, but also includes c6p and c7p, paralogues of the cvm6 and cvm7 genes of the clavam gene cluster. Upstream of ceaS1, the paralogue cluster includes a group of four genes, orfA through orfD, that are specifically involved in the biosynthesis of alanylclavam (104). Mutants with defects in any of these genes fail to produce alanylclavam but continue to produce both clavulanic acid and the other 5S clavams.

28.2.4. Regulatory Systems Influencing Production of β-Lactam Compounds in *S. clavuligerus*

Antibiotic production is not uniform throughout the growth cycle of an organism, and it is usually associated with the late, non-growth stages of a culture. In the model species *S. coelicolor*, a complex network of interacting factors has been shown to control both physiological and morphological differentiation (7). While *S. clavuligerus* has not been investigated to the same extent, elements of a similar regulatory network are apparently involved in this species as well.

The first antibiotic regulatory protein identified in *S. clavuligerus* was CcaR, encoded by the *ccaR* gene in the cephamycin arm of the cephamycin-clavulanic acid gene cluster (3, 9, 53, 72, 81, 95, 96). This protein shows a winged helix structure typical of the *Streptomyces* antibiotic resistance protein (SARP) family of transcriptional regulators that frequently serve as pathway-specific activators of antibiotic production (101). CcaR coregulates the expression of genes for two pathways, cephamycin C and clavulanic acid biosynthesis, and thereby ensures that β-lactamase inhibitor and β-lactam antibiotics are produced simultaneously in this species. Molecular studies to identify direct targets of CcaR have produced contradictory results. SARP-type regulators frequently control expression of most or all of the structural genes of their antibiotic gene clusters, but binding studies have identified few promoter regions that bind CcaR. Confirmed targets include the *ccaR* promoter itself and the bidirectional *cmcI-cefD* promoter controlling expression of genes involved in the middle stages of cephamycin C biosynthesis (81). Data for binding of CcaR to the *lat* promoter, controlling the first step of cephamycin C biosynthesis, are contradictory (53, 56). Although not all promoter regions have been examined, no evidence for binding of CcaR to any promoters of the clavulanic acid gene cluster has yet been obtained. SARP-type transcriptional regulators typically bind characteristic heptameric repeat sequences in the promoter regions of target genes. Such heptameric repeats can be identified in the *cmcI-cefD* intergenic region, but not in the *ccaR* promoter. On the other hand, suggestive, appropriately spaced pentameric repeat sequences are evident in the promoter regions of both the *lat* and *claR* genes despite the lack of clear-cut binding by CcaR.

Whether or not DNA binding has been observed, mutation of *ccaR* does abolish the expression of the tricistronic *lat-pcbAB-pcbC* operon (3) and also prevents expression

from the *ceaS2* promoter in the clavulanic acid gene cluster (89). Furthermore, mutation of *ccaR* blocks production of ClaR, a transcriptional regulator encoded by the *claR* gene contained within the clavulanic acid gene cluster (74). ClaR is a LysR-type transcriptional activator that controls expression of the genes specific for the late (post-clavaminic acid) stages of clavulanic acid biosynthesis (70, 74). Therefore, although an understanding of the molecular mechanism may be lacking, it is clear that CcaR controls expression of genes involved in the biosynthesis of cephamycin C and the early stages of clavulanic acid biosynthesis. It also controls expression of *claR*, which in turn regulates the late stages of clavulanic acid biosynthesis. No other regulatory genes are evident within the cephamycin-clavulanic acid gene cluster.

One possible explanation for the inability to demonstrate direct binding of CcaR to more target genes is that other binding factors may be required to recruit CcaR to the promoters. In this regard, autoregulatory small molecules are commonly produced by *Streptomyces* spp. (92) and *S. clavuligerus* is known to produce IM-2-type γ-butyrolactone compounds (32). Correspondingly, an autoregulator binding protein, ScaR (also called Brp), has been described (38, 80), which has a repressive effect on antibiotic production since disruption of *scaR* increases both cephamycin C and clavulanic acid production. Autoregulatory binding proteins such as ScaR bind to characteristic DNA sequences referred to as autoregulatory elements (ARE), and one such sequence has been identified in the intergenic region upstream of *ccaR* (26, 80). Furthermore, this ARE region upstream of *ccaR* also simultaneously binds a second protein, AreB, a regulatory element related to leucine biosynthesis, and binding of AreB requires the participation of a small molecule, perhaps a γ-butyrolactone (79). This demonstrates how complex the interplay of regulatory factors governing gene expression can get, and explains why failure to observe binding to DNA by a purified regulatory protein in vitro may not rule out the possibility of involvement of the protein in regulation.

Just as CcaR and ClaR are the pathway-specific regulators governing clavulanic acid production, C7p (91) may perform the same function for the 5S clavams. Disruption of C7p (from the paralogue cluster) abolishes production of 5S clavams specifically without affecting clavulanic acid production. No molecular studies have yet been performed on C7p to define its regulons. It is also not clear whether any cross-regulation exists between the regulators of clavulanic acid production and those of the 5S clavams. However, evidence for cross-regulation between antibiotic regulatory pathways has been reported between clavulanic acid and holomycin production (19). While these two classes of compounds are completely unrelated in structure and biosynthesis, mutations in genes for the late steps of the clavulanic acid biosynthetic pathway resulted in strains that overproduced holomycin. Reciprocal experiments have not been done since the elements of the holomycin gene cluster have not yet been identified.

Pathway-specific regulators are themselves generally subject to higher levels of regulation, thus making a cascade or network of regulation. In *S. coelicolor*, an array of higher-level regulators have been identified that control the production of the pathway-specific regulators, including the global "bald" (Bld) regulators that affect both morphological differentiation and antibiotic production. A similar system exists in *S. clavuligerus*, although only a few of the elements have been examined to date. Studies have shown that the *S. clavuligerus* homologue of *bldG* controls both sporulation and antibiotic production in S. *clavuligerus*, just as it does in S. *coelicolor* (9). *bldG* mutants do not sporulate, nor do they express *ccaR* or produce cephamycin C, clavulanic acid, or the 5S clavams. BldG encodes an anti-anti-sigma factor, and while its cognate sigma factor has not been identified, the adjacent anti-sigma factor encoding gene, *orf3*, is presumably part of this partner-switching mechanism, controlling utilization of an alternative sigma factor specific for sporulation and secondary metabolism.

The only other *bld* gene that has been studied in S. *clavuligerus* is *bldA*, encoding the leucyl-tRNA that recognizes the rare UUA anticodon. Genes containing TTA codons are very restricted in high-G+C organisms such as *Streptomyces* spp., and they are distributed asymmetrically. Genes containing TTA codons are overwhelmingly associated with secondary metabolism, sporulation, or utilization of secondary nutrient sources. Genes encoding pathway-specific regulators of antibiotic biosynthesis are commonly found to contain TTA codons, and so when *ccaR* was found to contain a TTA codon, it was fully expected that a *bldA* mutant of S. *clavuligerus* would fail to produce cephamycin C and clavulanic acid. However, antibiotic production was unaffected in the *bldA* mutant and CcaR protein production continued unabated although the mutant was unable to sporulate (94). A proposed explanation for this unexpected result is that the consequences of a *bldA* mutation on expression of TTA-containing genes depend on the context of the TTA codon, with TTA codons followed by pyrimidine nucleotides giving the phenotype of a null mutation, while codons followed by purine nucleotides show little or no effect. Consistent with this theory, in *ccaR* the TTA codon is followed by a purine residue. Interestingly, the *orfA* gene of alanylclavam biosynthesis represents just the second TTA codon-containing gene found in S. *clavuligerus*. In *orfA* the TTA codon is followed by a pyrimidine residue and alanylclavam biosynthesis is abolished in the *bldA* mutant.

Antibiotic production is also regulated at even higher levels in *Streptomyces* spp., and in particular, availability of amino acids, as detected by the stringent response, seems a point of control. At its simplest level, the stringent response is activated in response to an accumulation of uncharged tRNA molecules and is mediated by an accumulation of guanosine polyphosphate [ppGpp(p)] compounds formed by the RelA enzyme. In several *Streptomyces* spp., nutritional shift-down stimulates production of ppGpp and triggers antibiotic production. Recent studies have shown that S. *clavuligerus*, like other *Streptomyces* spp., has two *relA* homologues, *relA* and *rsh*, both encoding enzymes capable of synthesizing ppGpp (47). Mutation of *relA* was found to abolish ppGpp synthesis under conditions of starvation for amino acids, resulting in a bald phenotype and inability to produce either cephamycin C or clavulanic acid, whereas mutation of *rsh* affected only antibiotic production and not sporulation. Thus antibiotic production in S. *clavuligerus* is governed by the stringent response mediated by both RelA and Rsh. How the stringent response interfaces with other elements of the regulatory system, and whether intermediaries such as BldG or the γ-butyrolactone signaling system are involved, remain to be determined.

28.3. GENETIC ENGINEERING TO MODIFY PRODUCTION OF β-LACTAM COMPOUNDS BY *S. CLAVULIGERUS*

As an example of the use of genetic manipulations as described earlier to bring about changes in S. *clavuligerus* metabolite

production, the steps involved in generation of an in-frame *c7p* deletion mutant are described in more detail.

From a library of S. *clavuligerus* genomic DNA fragments carried in the E. *coli* cosmid vector SuperCos, a recombinant cosmid carrying a ~45-kb stretch of genomic DNA encompassing the *c7p* gene was selected (88). This cosmid, 14E10, served as the vehicle for PCR-mediated mutagenesis using the REDIRECT© system (31), which resulted in removal of the *c7p* gene and replacement with an *apra* cassette comprising the *apra* resistance gene and an *oriT*$_{RK2}$ region (supports plasmid transfer by conjugation). The cassette was also flanked on both ends by *frt* sites to allow eventual removal of the resistance marker. As a first step (Fig. 6), two hybrid primers (59 nucleotides [nt] and 58 nt)—one primer consisting of 39 nt from the start codon and upstream of *c7p* fused to 20 nt specific for the 5′ end of the *apra* cassette, and the second primer consisting of 39 nt from the stop codon and downstream of *c7p* fused to 19 nt specific for the 3′ end of the *apra* cassette—were designed and used to amplify the *apra* cassette, with plasmid pIJ773 (31) as template. The resulting PCR product consisted of the *apra* cassette but with 39-nt sequences on each end to target the cassette to the *c7p* gene.

In order to achieve the replacement of the *c7p* gene, first the 14E10 cosmid and then the *apra* cassette were introduced sequentially by electroporation into the recombinogenic E. *coli* host strain BW25113/pIJ790 (31), with selection for kanamycin (*kan*) and ampicillin resistance (from the 14E10 cosmid) and then for apramycin resistance (from the cassette). The *apra* cassette could only persist in E. *coli* to give resistance if it recombined with the 14E10 cosmid, and the *c7p* specific sequences on the ends of the cassette ensured that it recombined to delete the *c7p* gene. The resulting mutant version of the cosmid (14E10Δ*c7p*::*apra*) was transferred into the methylation-deficient E. *coli* host strain ET12567/pUZ8002 (31), which carries the transfer functions necessary to allow this strain to act as donor for interspecies conjugation between E. *coli* and *Streptomyces*. When cells from a mid-log-phase culture of E. *coli* ET12567/pUZ8002/14E10Δ*c7p*::*apra* were combined with heat-shocked spores of wild-type S. *clavuligerus*, and then the conjugation mixture was plated on AS-1+Mg agar plates, transfer of the 14E10Δ*c7p*::*apra* cosmid into S. *clavuligerus* occurred at some frequency. After overnight incubation in the absence of selection, plates were overlaid with nalidixic acid to counterselect the donor E. *coli* strain, and with apramycin to select for the incoming mutant cosmid. Since the SuperCos-based 14E10Δ*c7p*::*apra* cosmid has no origin of replication for *Streptomyces*, it could only persist to give apramycin resistance if it recombined into the S. *clavuligerus* chromosome via the long stretches of S. *clavuligerus* genomic DNA flanking the now-deleted *c7p* gene. A single recombination event would result in insertion of the entire cosmid into the chromosome to give both apramycin and kanamycin (from the cosmid) resistance, but the tandem duplication of the region of the chromosome carried on the cosmid makes the cells prone to a second recombination event. To encourage this second recombination, apramycin-resistant exconjugants were plated onto ISP-4 agar and allowed to sporulate. Resulting spores were then diluted and plated on ISP-4 medium to obtain colonies derived from single spores and then replica-plated to identify the desired cosmid-free apramycin-resistant, kanamycin-susceptible Δ*c7p*::*apra* mutants.

While useful for many purposes, the presence of the *apra* cassette in the Δ*c7p*::*apra* mutants may have polar effects on expression of downstream genes, and it also precludes the use of the *apra* marker for introduction of other complementation

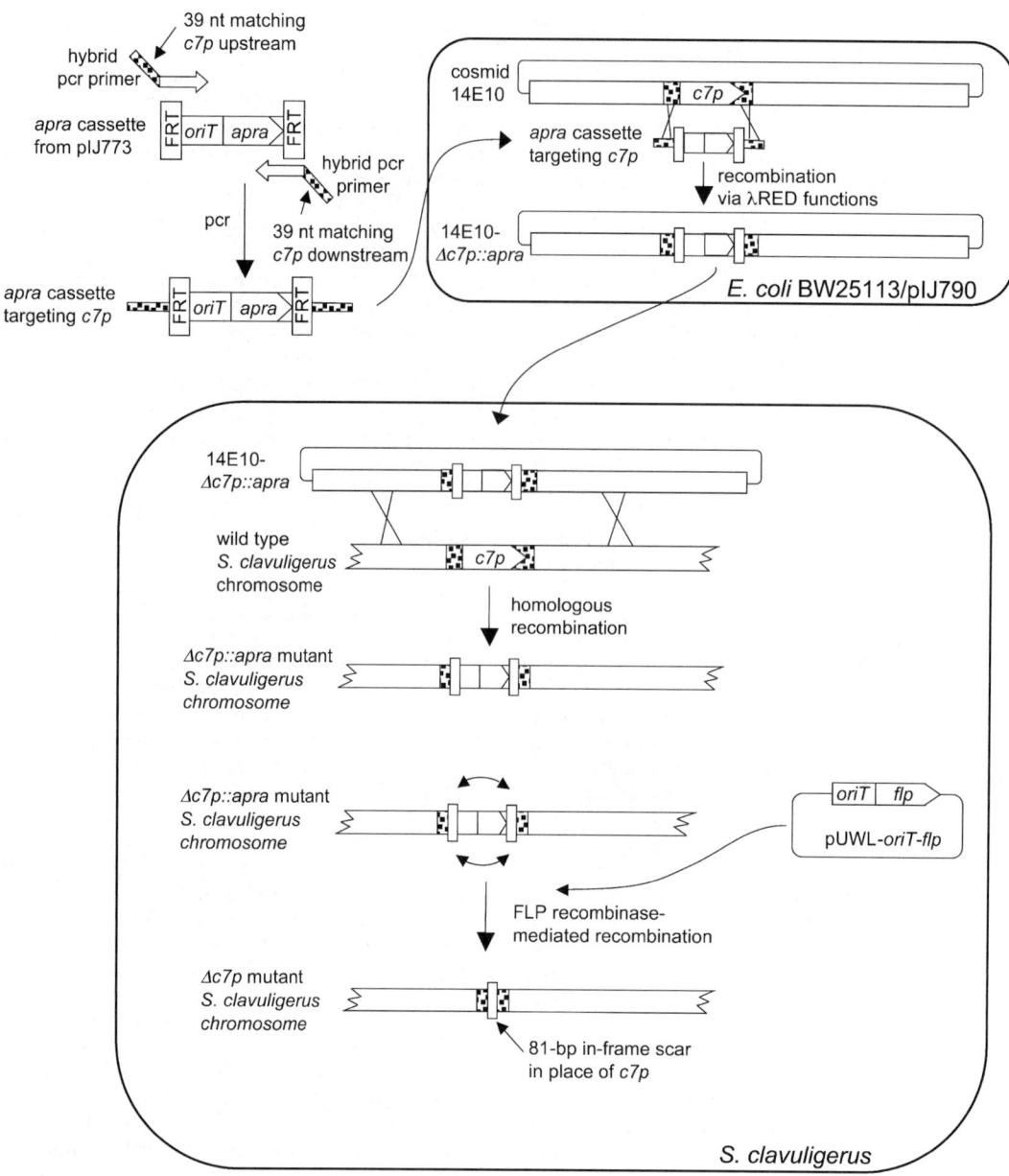

FIGURE 6 Generation of an *S. clavuligerus* Δ*c7p* mutant via PCR-mediated mutagenesis using the REDIRECT© system. An *apra* cassette targeting the *c7p* gene was prepared in vitro via PCR and then electroporated into *E. coli* BW25113/pIJ790 for recombination with cosmid 14E10. The resulting mutant 14E10-Δ*c7p::apra* cosmid was then transferred by conjugation to *S. clavuligerus*, where it exchanged with the resident *c7p* gene to give a Δ*c7p::apra* mutant strain of *S. clavuligerus*. Introduction of plasmid pUWL-*flp-oriT* to the Δ*c7p::apra* mutant resulted in loss of the *apra* cassette, leaving an unmarked Δ*c7p* mutant.

or expression plasmids. To remove the *apra* cassette, the plasmid pUWL-*oriT-flp* (103), carrying the gene encoding FLP recombinase, was conjugated from *E. coli* ET12567/pUZ8002 into the Δ*c7p::apra* mutant with selection for thiostrepton resistance, wherein the FLP protein promoted recombination at the FRT sites to excise the *apra* cassette from the Δ*c7p::apra* mutant. The resulting exconjugants were freed of the pUWL-*oriT-flp* plasmid by sporulation in the absence of antibiotic selection to promote loss of the segregationally unstable pUWL-*oriT-flp* plasmid, followed by replica-plating to identify the desired plasmid-free apramycin-susceptible, thiostrepton-

susceptible Δ*c7p* mutants. In place of the *apra* cassette, the Δ*c7p* mutants retain only an 81-bp in-frame "scar."

Other applications of techniques such as these to modify metabolite production in *S. clavuligerus* are described in the following sections.

28.3.1. Manipulation of Genes Encoding Pathway Enzymes

While *S. clavuligerus* is not used for industrial production of cephamycin C, it does serve as a useful model organism for studies aimed at improving or modifying the production of penicillins and cephalosporins. In particular the demonstrations that isopenicillin N synthase can accept unnatural peptide precursors to form novel penicillins (5, 18), and deacetoxycephalosporin C synthase can ring-expand unnatural penicillins to form high-value cephalosporins (15–17), have attracted interest. While native enzymes carry out these unnatural conversions with low efficiency, site-directed mutagenesis of the genes encoding the enzymes can improve conversion rates. Development of high-throughput assays for these enzymes should facilitate these mutational studies (86). The demonstration that it is possible to convert aromatic penicillins such as penicillin G directly into aromatic cephalosporins via enzymatic processes has been of special interest and has led to intensive mutational studies on *cefE* (1, 57, 100). In addition to studies aimed at altering the native enzyme activity, increasing the dosage of the *lat* gene (encoding the first dedicated enzyme of cephamycin C biosynthesis) was shown to increase cephamycin C production two- to fivefold in *S. clavuligerus* (60).

Although cephamycins can be valuable metabolites, *S. clavuligerus* is primarily of interest as a producer of clavulanic acid, and for this purpose, production of cephamycin C is a liability. Cephamycin C and clavulanic acid production are tightly linked and coregulated in *S. clavuligerus*. Since the two metabolites arise from separate pathways that share no common steps, it was anticipated that clavulanic acid production might be improved by mutating a gene essential for cephamycin C production, thereby diverting metabolic resources into clavulanic acid production. With this in mind, the *lat* gene was disrupted in a wild-type strain of *S. clavuligerus*, resulting in two- to threefold improvement in clavulanic acid production. A similar improvement, although of lesser magnitude, was also noted when the mutation was introduced into a strain of *S. clavuligerus* used for industrial production of clavulanic acid (69). The industrial strain was the product of years of intensive mutation and selection as part of a strain improvement process, demonstrating that targeted mutation through genetic engineering is useful both for initiating a strain improvement process in a wild-type organism and for effecting additional improvements in a strain already exposed to intensive mutation and selection programs.

Just as cephamycin C production is undesirable in a clavulanic acid producer strain, so too production of 5S clavam metabolites represents an unwanted drain of metabolic resources away from clavulanic acid production. Furthermore, the need to ensure complete removal of these structurally similar 5S clavams from the final product can complicate downstream processing. Mutation of the *cvm1* gene (from the clavam cluster) was shown to completely abolish production of all 5S clavams, and when *lat* and *cvm1* mutations were combined in the production strain, a 10% improvement in clavulanic acid production persisted through scale-up to the pilot plant scale with no production of 5S clavams (69).

While gene knockouts can be used to improve antibiotic production by eliminating competing pathways, further improvements can be made by increasing dosage of genes encoding critical enzymes of the biosynthetic pathway. *ceaS2* encodes the first enzyme of the clavulanic acid pathway, and since the early steps of the pathway are shared between clavulanic acid and the 5S clavams, it also contributes to 5S clavam production. *ceaS2* is the lead gene in an operon comprising *ceaS2*, *bls2*, *pah2*, *cas2*, and *oat2*, although internal promoters also drive expression of some genes. When additional copies of *ceaS2* were introduced into *S. clavuligerus* on a multicopy plasmid, earlier formation and higher levels of production of clavulanic acid (73) resulted. Since other genes from the operon or the cluster were not included in

the expression construct, this implies that clavulanic acid production in the wild-type cell must be limited by availability of CeaS2. Consistent with this theory, increasing copy number of the *cad* gene, encoding the enzyme for the last step of clavulanic acid biosynthesis, had no beneficial effect on metabolite production (74). However, other studies have since shown that introduction of an additional copy of the *pah2* gene increased clavulanic acid productivity (85), and similarly, increased copy number of *cas2* was reported to increase clavulanic acid production (37). Since each of these studies increased the dosage of one enzyme only, this raises the question of what the rate-limiting step for this pathway may be. The situation is further confused because of the duplication of these elements of the pathway that already exists naturally. Perhaps increasing copy number of the elements from one set of genes can result in unanticipated effects on expression of the other set of genes, making prediction of effects of increasing gene dosage difficult. However, increased copy number of genes other than *cad*, from the late uncharacterized stages of clavulanic acid biosynthesis (*orf10* through *orf14*), has also been reported to increase clavulanic acid production, and these genes are present at single copy only (62). The different growth conditions employed and different units used to express clavulanic acid productivity in the various studies make it hard to compare clavulanic acid productivity directly from one study to the other, but clearly the matter of the rate-limiting step for clavulanic acid production remains to be determined.

Despite this confusion, it is clear that overproduction of pathway enzymes offers yet another route for increasing productivity of the pathway, particularly once the rate-limiting step can be identified.

28.3.2. Manipulation of Genes Encoding Regulatory Proteins

While improvements in productivity can be achieved by increasing production of pathway enzymes, in the absence of a full understanding of pathway flux this becomes a trial-and-error approach to strain improvement. An alternative approach to achieving the same end is to identify genes encoding critical regulatory proteins and increase their expression levels. Although our understanding of the regulatory networks that affect production of β-lactam compounds by *S. clavuligerus* is still incomplete, manipulation of known regulatory genes can simultaneously affect expression of whole clusters of genes.

Knockout mutation of *ccaR* was found to simultaneously abolish both cephamycin C and clavulanic acid production in *S. clavuligerus* (3, 72), and conversely, introduction of extra copies of *ccaR* improved production of both cephamycin C and clavulanic acid by two- to threefold when they were introduced on a freely replicating multicopy plasmid. In a separate study, the effects of overexpression of *ccaR* alone, or together with the *cas2* gene on an integrative vector with expression driven by the *ermE** promoter, were examined (37). Clavulanic acid production was improved for all of the expression constructs, and in the case of the *ccaR-cas2* expression construct, the resulting strain showed a 24-fold increase in productivity compared to the wild-type background, with clavulanic acid production levels approaching 1 g/liter. No data on effects on cephamycin C production levels were provided.

Increased production of ClaR resulting from overexpression of *claR* on a multicopy plasmid has also been shown to result in a threefold increase in clavulanic acid production, whereas cephamycin C production was decreased (74). Surprisingly, alanylclavam production was also reported to increase in the ClaR-overproducing mutant, a contradictory result since increased production of clavulanic acid would be expected to channel intermediates away from the 5S clavams.

Pathway-specific regulators for the 5S clavam arm of the clavam biosynthetic pathway are just now being identified and so no studies to define their regulons at the molecular level have yet been reported. Gene knockout studies have however established that disruption of *c7p* generates a mutant able to produce only clavulanic acid and cephamycin C, and not any of the 5S clavams (88). *cvm1* and *cvm5* mutants show this same phenotype, but the *c7p* mutation may be a more desirable route to take if abolishing all production of 5S clavams is the goal, rather than targeting only one gene encoding a biosynthetic enzyme. Conversely, if increased production of 5S clavams is desired, a strategy of knocking out *ccaR* and/or *claR* combined with overexpression of *c7p* might be expected to promote increased levels of 5S clavam production. Although impressive improvements in productivity are possible, pathway-specific regulators are typically expressed at relatively low levels in wild-type cells, and so very high expression levels can be potentially deleterious or toxic to the cell, in our experience.

More global approaches to improvement of antibiotic production could also be coupled with strategies aimed at the pathway-specific regulators. Although the γ-butyrolactone-based quorum sensing system of *S. clavuligerus* is not fully characterized, it does affect antibiotic production, and inactivation of the repressor-like product of the *scaR* gene of *S. clavuligerus* resulted in up to threefold increases in production of cephamycin C and clavulanic acid (80). Although no studies have yet been undertaken, improvements in antibiotic production through altered induction of ppGpp synthesis or alteration of the ribosomal proteins with which it interacts (28, 29) should also hold promise.

In contrast, regulation by *bldA* may at first glance suggest only opportunities for reducing rather than increasing antibiotic production. However, the observed nonuniform effect of the presence of TTA codons on gene expression depending on sequence context may offer the means for selectively modifying metabolite productivity. For example, fermentation of a *bldA* mutant specifically favors production of 5S clavams other than alanylclavam, which is specifically eliminated (unpublished observation). In contrast, BldG mutants show loss of productivity of all antibiotics, suggesting that overproduction of BldG may hold promise for improving productivity (9).

28.4. CONCLUSIONS

Genetic engineering of antibiotic pathways to promote production of unnatural products or to increase levels of production of natural end products is being intensively investigated for many classes of antibiotics. These approaches also have demonstrated value in manipulating the production of β-lactam antibiotics, although in the case of clavulanic acid and the 5S clavams, an understanding of the late stages of the biosynthesis of these products is needed to allow a fully rational approach to be taken. Simple overproduction of pathway enzymes has provided improvement in production levels, but without a better understanding of pathway flux to identify rate-limiting steps, this approach can offer only unpredictable results. Effects of greater magnitude are seen when pathway-specific regulators are targeted for overproduction, and combination approaches targeting both global and specific regulators along with individual pathway enzymes for overproduction, together with elimination of competing pathways, will likely yield even greater results. Such approaches offer the promise of yielding vigorous, high-producing strains uncompromised by the effects of repeated rounds of random mutagenesis.

REFERENCES

1. **Adrio, J. L., and A. L. Demain.** 2002. Improvements in the formation of cephalosporins from penicillin G and other penicillins by bioconversion. *Org. Process Res. Dev.* **6:**427–433.
2. **Aidoo, K. A., A. Wong, D. C. Alexander, R. A. Rittammer, and S. E. Jensen.** 1994. Cloning, sequencing and disruption of a gene from *Streptomyces clavuligerus* involved in clavulanic acid biosynthesis. *Gene* **147:**41–46.
3. **Alexander, D. C., and S. E. Jensen.** 1998. Investigation of the *Streptomyces clavuligerus* cephamycin C gene cluster and its regulation by the CcaR protein. *J. Bacteriol.* **180:**4068–4079.
4. **Arulanantham, H., N. J. Kershaw, K. S. Hewitson, C. E. Hughes, J. E. Thirkettle, and C. J. Schofield.** 2006. ORF17 from the clavulanic acid biosynthesis gene cluster catalyzes the ATP-dependent formation of N-glycyl-clavaminic acid. *J. Biol. Chem.* **281:**279–287.
5. **Baldwin, J. E., M. Bradley, S. D. Abbott, and R. M. Adlington.** 1990. Formation of novel unsaturated side chain penicillins with isopenicillin N synthase. *J. Chem. Soc. Chem. Commun.* **1990:**1008–1010.
6. **Baltz, R. H.** 1980. Genetic recombination by protoplast fusion in *Streptomyces. Dev. Indust. Microbiol.* **21:**43–54.
7. **Bibb, M. J.** 2005. Regulation of secondary metabolism in streptomycetes. *Curr. Opin. Microbiol.* **8:**208–215.
8. **Bierman, M., R. Logan, E. T. O'Brien, E. T. Seno, N. Rao, and B. E. Schoner.** 1992. Plasmid cloning vectors for the conjugal transfer of DNA from *Escherichia coli* to *Streptomyces* spp. *Gene* **116:**43–49.
9. **Bignell, D. R. D., K. Tahlan, K. R. Colvin, S. E. Jensen, and B. K. Leskiw.** 2005. Expression of *ccaR*, encoding the positive activator of cephamycin C and clavulanic acid production in *Streptomyces clavuligerus*, is dependent on *bldG. Antimicrob. Agents Chemother.* **49:**1529–1541.
10. **Birch, A. W., and J. Cullum.** 1985. Temperature-sensitive mutants of the *Streptomyces* plasmid pIJ702. *J. Gen. Microbiol.* **131:**1299–1303.
11. **Bishop, A., S. Fielding, P. Dyson, and P. Herron.** 2004. Systematic insertional mutagenesis of a streptomycete genome: a link between osmoadaptation and antibiotic production. *Genome Res.* **14:**893–900.
12. **Brown, A. G., D. Butterworth, M. Cole, G. Hanscombe, J. D. Hood, C. Reading, and G. N. Robinson.** 1976. Naturally occurring beta-lactamase inhibitors with antibacterial activity. *J. Antibiot.* **29:**668–669.
13. **Brown, D., J. R. Evans, and R. A. Fletton.** 1979. Structures of three novel β-lactams isolated from *Streptomyces clavuligerus. J. Chem. Soc. Chem. Commun.* **1979:**282–283.
14. **Caines, M. E. C., J. M. Elkins, K. S. Hewitson, and C. J. Schofield.** 2004. Crystal structure and mechanistic implications of N2-(2-carboxyethyl)arginine synthase, the first enzyme in the clavulanic acid biosynthesis pathway. *J. Biol. Chem.* **279:**5685–5692.
15. **Chin, H. S., K. S. Goo, and T. S. Sim.** 2004. A complete library of amino acid alterations at N304 in *Streptomyces clavuligerus* deacetoxycephalosporin C synthase elucidates the basis for enhanced penicillin analogue conversion. *Appl. Environ. Microbiol.* **70:**607–609.
16. **Chin, H. S., and T. S. Sim.** 2002. C-terminus modification of *Streptomyces clavuligerus* deacetoxycephalosporin C synthase improves catalysis with an expanded substrate specificity. *Biochem. Biophys. Res. Commun.* **295:**55–61.
17. **Chin, H. S., J. Sim, and T. S. Sim.** 2001. Mutation of N304 to leucine in *Streptomyces clavuligerus* deacetoxycephalosporin C synthase creates an enzyme with increased

penicillin analogue conversion. *Biochem. Biophys. Res. Commun.* **287:**507–513.

18. **Cohen, G., D. Shiffman, M. Mevarech, and Y. Aharonowitz.** 1990. Microbial isopenicillin N synthase genes: structure, function, diversity and evolution. *Trends Biotechnol.* **8:**105–111.

19. **de La Fuente, J. L., L. M. Lorenzana, J. F. Martin, and P. Liras.** 2002. Mutants of *Streptomyces clavuligerus* with disruptions in different genes for clavulanic acid biosynthesis produce large amounts of holomycin: possible cross-regulation of two unrelated secondary metabolic pathways. *J. Bacteriol.* **184:**6559–6565.

20. **Demain, A. L.** 2000. Edward P. Abraham, cell-free systems and the fungal biosynthesis of beta-lactams. *J. Antibiot.* **53:**995–1002.

21. **Denis, F., and R. Brzezinski.** 1991. An improved aminoglycoside resistance gene cassette for use in Gram-negative bacteria and *Streptomyces*. *FEMS Microbiol. Lett.* **81:**261–264.

22. **Doran, J. L., B. K. Leskiw, S. Aippersbach, and S. E. Jensen.** 1990. Isolation and characterization of a β-lactamase-inhibitory protein from *Streptomyces clavuligerus* and cloning and analysis of the corresponding gene. *J. Bacteriol.* **172:**4909–4918.

23. **Elkins, J. M., N. J. Kershaw, and C. J. Schofield.** 2005. X-ray crystal structure of ornithine acetyltransferase from the clavulanic acid biosynthesis gene cluster. *Biochem. J.* **385:**565–573.

24. **Elkins, J. M., I. J. Clifton, H. Hernández, L. X. Doan, C. V. Robinson, C. J. Schofield, and K. S. Hewitson.** 2002. Oligomeric structure of proclavaminic acid amidino hydrolase: evolution of a hydrolytic enzyme in clavulanic acid biosynthesis. *Biochem. J.* **366:**423–434.

25. **Fedoryshyn, M., L. Petzke, E. Welle, A. Bechthold, and A. Luzhetskyy.** 2008. Marker removal from actinomycetes genome using Flp recombinase. *Gene* **419:**43–47.

26. **Folcher, M., H. Gaillard, L. T. Nguyen, K. T. Nguyen, P. Lacroix, N. Bamas-Jacques, M. Rinkel, and C. J. Thompson.** 2001. Pleiotropic functions of a *Streptomyces pristinaespiralis* autoregulator receptor in development, antibiotic biosynthesis, and expression of a superoxide dismutase. *J. Biol. Chem.* **276:**44297–44306.

27. **Garcia-Dominguez, M., J. F. Martin, and B. Mahro.** 1987. Efficient plasmid transformation of the β-lactam producer *Streptomyces clavuligerus*. *Appl. Environ. Microbiol.* **53:**1376–1381.

28. **Gomez-Escribano, J. P., J. F. Martín, A. Hesketh, M. J. Bibb, and P. Liras.** 2008. *Streptomyces clavuligerus relA*-null mutants overproduce clavulanic acid and cephamycin C: negative regulation of secondary metabolism by (p)ppGpp. *Microbiology* **154:**744–755.

29. **Gomez-Escribano, J. P., P. Liras, A. Pisabarro, and J. F. Martin.** 2006. An *rplK*^Δ29-PALG-32 mutation leads to reduced expression of the regulatory genes *ccaR* and *claR* and very low transcription of the *ceaS2* gene for clavulanic acid biosynthesis in *Streptomyces clavuligerus*. *Mol. Microbiol.* **61:**758–770.

30. **Gouveia, E. R., A. Baptista-Neto, A. G. Azevedo, A. C. Badino, Jr., and C. O. Hokka.** 1999. Improvement of clavulanic acid production by *Streptomyces clavuligerus* in medium containing soybean derivatives. *World J. Microbiol. Biotechnol.* **15:**623–627.

31. **Gust, B., G. L. Challis, K. Fowler, T. Kieser, and K. F. Chater.** 2003. PCR-targeted *Streptomyces* gene replacement identifies a protein domain needed for biosynthesis of the sesquiterpene soil odor geosmin. *Proc. Natl. Acad. Sci. USA* **100:**1541–1546.

32. **Hashimoto, K., T. Nihira, and Y. Yamada.** 1992. Distribution of virginiae butanolides and IM-2 in the genus *Streptomyces*. *J. Ferment. Bioeng.* **73:**61–65.

33. **He, J., N. Magarvey, M. Piraee, and L. C. Vining.** 2001. The gene cluster for chloramphenicol biosynthesis in *Streptomyces venezuelae* ISP5230 includes novel shikimate pathway homologues and a monomodular non-ribosomal peptide synthetase gene. *Microbiology* **147:**2817–2829.

34. **Higgens, C. E., and R. E. Kastner.** 1971. *Streptomyces clavuligerus* sp. nov., a β-lactam antibiotic producer. *Int. J. Syst. Bacteriol.* **21:**326–331.

35. **Hindle, Z., and C. P. Smith.** 1994. Substrate induction and catabolite repression of the *Streptomyces coelicolor* glycerol operon are mediated through the GylR protein. *Mol. Microbiol.* **12:**737–745.

36. **Hodgson, J. E., A. P. Fosberry, N. S. Rawlinson, H. N. M. Ross, R. J. Neal, J. C. Arnell, A. J. Earl, and E. J. Lawlor.** 1995. Clavulanic acid biosynthesis in *Streptomyces clavuligerus*: gene cloning and characterization. *Gene* **166:**49–55.

37. **Hung, T. V., S. Malla, B. C. Park, K. Liou, H. C. Lee, and J. K. Sohng.** 2007. Enhancement of clavulanic acid by replicative and integrative expression of *ccaR* and *cas2* in *Streptomyces clavuligerus* NRRL3585. *J. Microbiol. Biotechnol.* **17:**1538–1545.

38. **Hyun, S. K., J. L. Yong, K. L. Chang, U. C. Sun, S. H. Yeo, I. H. Yong, S. Y. Tae, H. Kinoshita, and T. Nihira.** 2004. Cloning and characterization of a gene encoding the gamma-butyrolactone autoregulator receptor from *Streptomyces clavuligerus*. *Arch. Microbiol.* **182:**44–50.

39. **Illing, G. T., I. D. Normansell, and J. F. Peberdy.** 1989. Genetic mapping in *Streptomyces clavuligerus* by protoplast fusion. *J. Gen. Microbiol.* **135:**2299–2305.

40. **Illing, G. T., I. D. Normansell, and J. F. Peberdy.** 1989. Protoplast isolation and regeneration in *Streptomyces clavuligerus*. *J. Gen. Microbiol.* **135:**2289–2297.

41. **Jensen, S. E., A. S. Paradkar, R. H. Mosher, C. Anders, P. H. Beatty, M. J. Brumlik, A. Griffin, and B. Barton.** 2004. Five additional genes are involved in clavulanic acid biosynthesis in *Streptomyces clavuligerus*. *Antimicrob. Agents Chemother.* **48:**192–202.

42. **Jensen, S. E., A. Wong, A. Griffin, and B. Barton.** 2004. *Streptomyces clavuligerus* has a second copy of the proclavaminate amidinohydrolase gene. *Antimicrob. Agents Chemother.* **48:**514–520.

43. **Jensen, S. E., and A. S. Paradkar.** 1999. Biosynthesis and molecular genetics of clavulanic acid. *Antonie Leeuwenhoek* **75:**125–133.

44. **Jensen, S. E., and A. L. Demain.** 1995. Beta-lactams, p. 239–268. *In* L. C. Vining and C. Stuttard (ed.), *Genetics and Biochemistry of Antibiotic Production*. Butterworth-Heinemann, Boston, MA.

45. **Jensen, S. E., D. C. Alexander, A. S. Paradkar, and K. A. Aidoo.** 1993. Extending the β-lactam biosynthetic gene cluster in *Streptomyces clavuligerus*, p. 169–176. *In* R. H. Baltz, G. D. Hegeman, and P. L. Skatrud (ed.), *Industrial Microorganisms: Basic and Applied Molecular Genetics*. American Society for Microbiology, Washington, DC.

46. **Jensen, S. E., D. W. Westlake, and S. Wolfe.** 1982. Cyclization of δ-(L-α-aminoadipyl)-L-cysteinyl-D-valine to penicillins by cell-free extracts of *Streptomyces clavuligerus*. *J. Antibiot.* **35:**483–490.

47. **Jin, W., Y. G. Ryu, S. G. Kang, S. K. Kim, N. Saito, K. Ochi, S. H. Lee, and K. J. Lee.** 2004. Two *relA/spoT* homologous genes are involved in the morphological and physiological differentiation of *Streptomyces clavuligerus*. *Microbiology* **150:**1485–1493.

48. **Kenig, M., and C. Reading.** 1979. Holomycin and an antibiotic (MM 19290) related to tunicamycin, metabolites of *Streptomyces clavuligerus*. *J. Antibiot.* **32:**549–554.

49. **Kershaw, N. J., H. J. McNaughton, K. S. Hewitson, H. Hernandez, J. Griffin, C. Hughes, P. Greaves, B. Barton, C. V. Robinson, and C. J. Schofield.** 2002. ORF6 from the clavulanic acid gene cluster of *Streptomyces*

clavuligerus has ornithine acetyltransferase activity. *Eur. J. Biochem.* **269:**2052–2059.

50. **Kieser, T., M. J. Bibb, M. J. Buttner, K. F. Chater, and D. A. Hopwood.** 2000. *Practical* Streptomyces *Genetics.* John Innes Foundation, Norwich, United Kingdom.

51. **Kim, H. S., and Y. I. Park.** 2008. Isolation and identification of a novel microorganism producing the immunosuppressant tacrolimus. *J. Biosci. Bioeng.* **105:**418–421.

52. **Kim, H. S., and Y. I. Park.** 2007. Lipase activity and tacrolimus production in *Streptomyces clavuligerus* CKD 1119 mutant strains. *J. Microbiol. Biotechnol.* **17:**1638–1644.

53. **Kyung, Y. S., W.-S. Hu, and D. S. Sherman.** 2001. Analysis of temporal and spatial expression of the CcaR regulatory element in the cephamycin C biosynthetic pathway using green fluorescent protein. *Mol. Microbiol.* **40:**530–541.

54. **Li, R., N. Khaleeli, and C. A. Townsend.** 2000. Expansion of the clavulanic acid gene cluster: identification and in vivo functional analysis of three new genes required for biosynthesis of clavulanic acid by *Streptomyces clavuligerus.* *J. Bacteriol.* **182:**4087–4095.

55. **Liras, P., J. P. Gomez-Escribano, and I. Santamarta.** 2008. Regulatory mechanisms controlling antibiotic production in *Streptomyces clavuligerus.* *J. Ind. Microbiol. Biotechnol.* **35:**667–676.

56. **Liras, P.** 1999. Biosynthesis and molecular genetics of cephamycins. *Antonie Leeuwenhoek* **75:**109–124.

57. **Lloyd, M. D., S. J. Lipscomb, K. S. Hewitson, C. M. H. Hensgens, J. E. Baldwin, and C. J. Schofield.** 2004. Controlling the substrate selectivity of deacetoxycephalosporin/deacetylcephalosporin C synthase. *J. Biol. Chem.* **279:**15420–15426.

58. **Lloyd, M. D., K. D. Merritt, V. Lee, T. J. Sewell, B. Wha-Son, J. E. Baldwin, C. J. Schofield, S. W. Elson, K. H. Baggaley, and N. H. Nicholson.** 1999. Product-substrate engineering by bacteria: studies on clavaminate synthase, a trifunctional dioxygenase. *Tetrahedron* **55:**10201–10220.

59. **Lorenzana, L. M., R. Pérez-Redondo, I. Santamarta, J. F. Martín, and P. Liras.** 2004. Two oligopeptide-permease-encoding genes in the clavulanic acid cluster of *Streptomyces clavuligerus* are essential for production of the β-lactamase inhibitor. *J. Bacteriol.* **186:**3431–3438.

60. **Malmberg, L. H., W. S. Hu, and D. H. Sherman.** 1993. Precursor flux control through targeted chromosomal insertion of the lysine epsilon-aminotransferase (*lat*) gene in cephamycin C biosynthesis. *J. Bacteriol.* **175:**6916–6924.

61. **Marsh, E. N., M. D. Chang, and C. A. Townsend.** 1992. Two isozymes of clavaminate synthase central to clavulanic acid formation: cloning and sequencing of both genes from *Streptomyces clavuligerus.* *Biochemistry* **31:**12648–12657.

62. **Mellado, E., L. M. Lorenzana, M. Rodriguez-Saiz, B. Diez, P. Liras, and J. L. Barredo.** 2002. The clavulanic acid biosynthetic cluster of *Streptomyces clavuligerus:* genetic organization of the region upstream of the *car* gene. *Microbiology* **149:**1427–1438.

63. **Miller, M. T., B. O. Bachmann, C. A. Townsend, and A. C. Rosenzweig.** 2002. The catalytic cycle of β-lactam synthetase observed by x-ray crystallographic snapshots. *Proc. Natl. Acad. Sci. USA* **99:**14752–14757.

64. **Mosher, R. H., A. S. Paradkar, C. Anders, B. Barton, and S. E. Jensen.** 1999. Genes specific for the biosynthesis of clavam metabolites antipodal to clavulanic acid are clustered with the gene for clavaminate synthase 1 in *Streptomyces clavuligerus.* *Antimicrob. Agents Chemother.* **43:**1215–1224.

65. **Motamedi, H., A. Shafiee, and S. J. Cai.** 1995. Integrative vectors for heterologous gene expression in *Streptomyces* spp. *Gene* **160:**25–31.

66. **Nagarajan, R., L. D. Boeck, M. Gorman, R. I. Hamill, C. E. Higgins, M. M. Hoehn, W. M. Stark, and J. G.** Whitney. 1971. β-Lactam antibiotics from *Streptomyces.* *J. Am. Chem. Soc.* **93:**2308–2310.

67. **Netolitzky, D. J., X. Wu, S. E. Jensen, and K. L. Roy.** 1995. Giant linear plasmids of β-lactam antibiotic producing *Streptomyces.* *FEMS Microbiol. Lett.* **131:**27–34.

68. **Ortiz, S. C. A., C. O. Hokka, and A. C. Badino.** 2007. Utilization of soybean derivatives on clavulanic acid production by *Streptomyces clavuligerus.* *Enzyme Microb. Technol.* **40:**1071–1077.

69. **Paradkar, A. S., R. H. Mosher, C. Anders, A. Griffin, J. Griffin, C. Hughes, P. Greaves, B. Barton, and S. E. Jensen.** 2001. Applications of gene replacement technology to *Streptomyces clavuligerus* strain development for clavulanic acid production. *Appl. Environ. Microbiol.* **67:**2292–2297.

70. **Paradkar, A. S., K. A. Aidoo, and S. E. Jensen.** 1998. A pathway-specific transcriptional activator regulates late steps of clavulanic acid biosynthesis in *Streptomyces clavuligerus.* *Mol. Microbiol.* **27:**831–843.

71. **Paradkar, A. S., and S. E. Jensen.** 1995. Functional analysis of the gene encoding the clavaminate synthase 2 isoenzyme involved in clavulanic acid biosynthesis in *Streptomyces clavuligerus.* *J. Bacteriol.* **177:**1307–1314.

72. **Perez-Llarena, F. J., P. Liras, A. Rodriguez-Garcia, and J. F. Martin.** 1997. A regulatory gene (*ccaR*) required for cephamycin and clavulanic acid production in *Streptomyces clavuligerus:* amplification results in overproduction of both β-lactam compounds. *J. Bacteriol.* **179:**2053–2059.

73. **Perez-Redondo, R., A. Rodriguez-Garcia, J. F. Martin, and P. Liras.** 1999. Deletion of the *pyc* gene blocks clavulanic acid biosynthesis except in glycerol-containing medium: evidence for two different genes in formation of the C3 unit. *J. Bacteriol.* **181:**6922–6928.

74. **Perez-Redondo, R., A. Rodriguez-Garcia, J. F. Martin, and P. Liras.** 1998. The *claR* gene of *Streptomyces clavuligerus,* encoding a LysR-type regulatory protein controlling clavulanic acid biosynthesis, is linked to the clavulanate-9-aldehyde reductase (*car*) gene. *Gene* **211:**311–321.

75. **Pruess, D. L., and M. Kellett.** 1983. Ro-22-5417, a new clavam antibiotic from *Streptomyces clavuligerus.* I. Discovery and biological activity. *J. Antibiot.* **36:**208–212.

76. **Reading, C., and M. Cole.** 1977. Clavulanic acid: a beta-lactamase-inhibiting beta-lactam from *Streptomyces clavuligerus.* *Antimicrob. Agents Chemother.* **11:**852–857.

77. **Rohl, F., J. Rabenhorst, and H. Zahner.** 1987. Biological properties and mode of action of clavams. *Arch. Microbiol.* **147:**315–320.

78. **Salowe, S. P., E. N. Marsh, and C. A. Townsend.** 1990. Purification and characterization of clavaminate synthase from *Streptomyces clavuligerus:* an unusual oxidative enzyme in natural product biosynthesis. *Biochemistry* **29:**6499–6508.

79. **Santamarta, I., M. T. López-García, R. Pérez-Redondo, B. Koekman, J. F. Martín, and P. Liras.** 2007. Connecting primary and secondary metabolism: AreB, an IclR-like protein, binds the ARE*ccaR* sequence of *S. clavuligerus* and modulates leucine biosynthesis and cephamycin C and clavulanic acid production. *Mol. Microbiol.* **66:**511–524.

80. **Santamarta, I., R. Pérez-Redondo, L. M. Lorenzana, J. F. Martín, and P. Liras.** 2005. Different proteins bind to the butyrolactone receptor protein ARE sequence located upstream of the regulatory *ccaR* gene of *Streptomyces clavuligerus.* *Mol. Microbiol.* **56:**824–835.

81. **Santamarta, I., A. Rodriguez-Garcia, R. Perez-Redondo, J. F. Martin, and P. Liras.** 2002. CcaR is an autoregulatory protein that binds to the *ccaR* and *cefD-cmcI* promoters of the cephamycin C-clavulanic acid cluster in *Streptomyces clavuligerus.* *J. Bacteriol.* **184:**3106–3113.

82. **Saudagar, P. S., S. A. Survase, and R. S. Singhal.** 2008. Clavulanic acid: a review. *Biotechnol. Adv.* **26:**335–351.

83. **Schmitt-John, J., and J. W. Engels.** 1992. Promoter constructions for efficient secretion expression in *Streptomyces lividans. Appl. Microbiol. Biotechnol.* **36:**493–498.

84. **Song, J. Y., E. S. Kim, D. W. Kim, S. E. Jensen, and K. J. Lee.** 2009. A gene located downstream of the clavulanic acid gene cluster in *Streptomyces clavuligerus* ATCC 27064 encodes a putative response regulator that affects clavulanic acid production. *J. Ind. Microbiol. Biotechnol.* **36:**301–311.

85. **Song, J. Y., E. S. Kim, D. W. Kim, S. E. Jensen, and K. J. Lee.** 2008. Functional effects of increased copy number of the gene encoding proclavaminate amidino hydrolase on clavulanic acid production in *Streptomyces clavuligerus* ATCC 27064. *J. Microbiol. Biotechnol.* **18:**417–426.

86. **Stok, J. E., and J. E. Baldwin.** 2006. Development of enzyme-linked immunosorbent assays for the detection of deacetoxycephalosporin C and isopenicillin N synthase activity. *Anal. Chim. Acta* **577:**153–162.

87. **Strynadka, N. C., S. E. Jensen, P. M. Alzari, and M. N. James.** 1996. A potent new mode of β-lactamase inhibition revealed by the 1.7 Å X-ray crystallographic structure of the TEM-1-BLIP complex. *Nat. Struct. Biol.* **3:**290–297.

88. **Tahlan, K., C. Anders, A. Wong, R. H. Mosher, P. H. Beatty, M. J. Brumlik, A. Griffin, C. Hughes, J. Griffin, B. Barton, and S. E. Jensen.** 2007. 5S clavam biosynthetic genes are located in both the clavam and paralog gene clusters in *Streptomyces clavuligerus. Chem. Biol.* **14:**131–142.

89. **Tahlan, K., C. Anders, and S. E. Jensen.** 2004. The paralogous pairs of genes involved in clavulanic acid and clavam metabolite biosynthesis are differently regulated in *Streptomyces clavuligerus. J. Bacteriol.* **186:**6286–6297.

90. **Tahlan, K., H. U. Park, and S. E. Jensen.** 2004. Three unlinked gene clusters are involved in clavam metabolite biosynthesis in *Streptomyces clavuligerus. Can. J. Microbiol.* **50:**803–810.

91. **Tahlan, K., H. U. Park, A. Wong, P. H. Beatty, and S. E. Jensen.** 2004. Two sets of paralogous genes encode the enzymes involved in the early stages of clavulanic acid and clavam metabolite biosynthesis in *Streptomyces clavuligerus. Antimicrob. Agents Chemother.* **48:**930–939.

92. **Takano, E.** 2006. γ-Butyrolactones: *Streptomyces* signalling molecules regulating antibiotic production and differentiation. *Curr. Opin. Microbiol.* **9:**287–294.

93. **Thai, W., A. S. Paradkar, and S. E. Jensen.** 2001. Construction and analysis of β-lactamase-inhibitory protein (BLIP) non-producer mutants of *Streptomyces clavuligerus. Microbiology* **147:**325–335.

94. **Trepanier, N. K., S. E. Jensen, D. C. Alexander, and B. K. Leskiw.** 2002. The positive activator of cephamycin C and clavulanic acid production in *Streptomyces clavuligerus* is mistranslated in a *bldA* mutant. *Microbiology* **148:**643–656.

95. **Walters, N. J., B. Barton, and A. J. Earl.** November, 1995. Novel compounds. International patent WO 94/18326-A 1.

96. **Wang, L., K. Tahlan, T. L. Kaziuk, D. C. Alexander, and S. E. Jensen.** 2004. Transcriptional and translational analysis of the *ccaR* gene from *Streptomyces clavuligerus. Microbiology* **150:**4137–4145.

97. **Wang, Y. H., B. Yang, J. Ren, M. L. Dong, D. Liang, and A. L. Xu.** 2005. Optimization of medium composition for the production of clavulanic acid by *Streptomyces clavuligerus. Process Biochem.* **40:**1161–1166.

98. **Ward, J. M., and J. E. Hodgson.** 1993. The biosynthetic genes for clavulanic acid and cephamycin production occur as a "super-cluster" in three *Streptomyces. FEMS Microbiol. Lett.* **110:**239–242.

99. **Wehmeier, U. F.** 1995. New multifunctional *Escherichia coli-Streptomyces* shuttle vectors allowing blue-white screening on XGal plates. *Gene* **165:**149–150.

100. **Wei, C. L., Y. B. Yang, C. H. Deng, W. C. Liu, J. S. Hsu, Y. C. Lin, S. H. Liaw, and Y. C. Tsai.** 2005. Directed evolution of *Streptomyces clavuligerus* deacetoxycephalosporin C synthase for enhancement of penicillin G expansion. *Appl. Environ. Microbiol.* **71:**8873–8880.

101. **Wietzorrek, A., and M. Bibb.** 1997. A novel family of proteins that regulates antibiotic production in streptomycetes appears to contain an OmpR-like DNA-binding fold. *Mol. Microbiol.* **25:**1181–1184.

102. **Wu, T. K., R. W. Busby, T. A. Houston, D. B. McIlwaine, L. A. Egan, and C. A. Townsend.** 1995. Identification, cloning, sequencing, and overexpression of the gene encoding proclavaminate amidino hydrolase and characterization of protein function in clavulanic acid biosynthesis. *J. Bacteriol.* **177:**3714–3720.

103. **Zelyas, N., K. Tahlan, and S. E. Jensen.** 2009. Use of the native *flp* gene to generate in-frame unmarked mutations in *Streptomyces* spp. *Gene* **443:**48–54.

104. **Zelyas, N. J., H. Cai, T. Kwong, and S. E. Jensen.** 2008. Alanylclavam biosynthetic genes are clustered together with one group of clavulanic acid biosynthetic genes in *Streptomyces clavuligerus. J. Bacteriol.* **190:**7957–7965.

105. **Zhang, Z., J. Ren, D. K. Stammers, J. E. Baldwin, K. Harlos, and C. J. Schofield.** 2000. Structural origins of the selectivity of the trifunctional oxygenase clavaminic acid synthase. *Nat. Struct. Biol.* **7:**127–33.

Genetic Engineering of Myxobacterial Natural Product Biosynthetic Genes

BRYAN JULIEN AND EDUARDO RODRIGUEZ

29

29.1. INTRODUCTION

Plants and microbes have been a rich source of bioactive natural products, some of which have been developed for medical or animal health uses (17). The importance of natural products is highlighted by the fact that 34% of all small molecule drugs developed over the last 25 years have been derived from natural products or their semisynthetic derivates (32). Of the microbes, actinomycetes are the largest known producer of natural products. Recently, myxobacteria have been identified as an alternative source of natural products through the work of the group led by Hans Reichenbach at Gesellschaft für Biotechnologische Forschung. From ~7,500 strains of myxobacteria isolated, many novel bioactive compounds have been discovered (35).

Myxobacteria are gram-negative soil bacteria that, under adverse environmental conditions (36), have the ability to form complex fruiting bodies, which can be simple mound structures as seen in *Myxococcus xanthus* or elaborate multibranched, stalked structures as in *Chondromyces crocatus*. These fruiting bodies house dormant spores that are resistant to many environmental stresses.

Natural products produced from myxobacteria have gained wide interest recently. To date, more than 100 new structures and some 500 variants have been identified (3). The majority of myxobacterial natural products have been isolated from strains of *Sorangium cellulosum* (48%), *Myxococcus* (20%), and *Chondromyces* (10%) (12). Most of these natural products are hybrid molecules consisting of polyketides and non-ribosomally made peptides in a linear or cyclic form. Although the addition of modified sugars to natural products from *Streptomyces* is common, they are rarely found on myxobacterial natural products. Less common but noteworthy features of myxobacterial natural products are two compounds with chlorinated amino acids (19, 20), one with a cyclopropane ring (39), and compounds that complex with boron (18). Finally, a few terpenoid compounds have been identified (12).

Of all myxobacterial natural compounds identified, only one compound, the lactam version of epothilone B (Ixempra®) (Fig. 1), is currently marketed as a human drug. The epothilones were originally identified by their antifungal activity against the zygomycete *Mucor hiemalis* (11). They were later rediscovered in a screen to identify compounds that stabilize microtubules, similar to the mechanism of paclitaxel (4). The advantage of epothilones over paclitaxel is their ability to inhibit tumors that exhibit multidrug resistance (28). Ixempra® is currently approved for treatment of metastatic or locally advanced breast cancer.

Because the majority of the natural products from myxobacteria are synthesized by type I polyketide synthases (PKSs) and nonribosomal polypeptide synthetases (NRPSs) (Fig. 1), they are amenable to structure modification through genetic engineering using the information gained over the years on the structure and biochemistry of the polyketide and nonribosomal peptide megasynthases. Genetic engineering of PKSs and NRPSs is aimed at producing novel analogs of natural products for drug development (30, 41, 52).

In order to engineer PKS genes, one must be able to genetically manipulate the producing host. Alternatively, expression and engineering of the genes in a heterologous host is required. However, several challenges with engineering myxobacteria are present. The most prolific producers, *Sorangium* spp., grow relatively slowly, can be difficult to grow as a dispersed culture, and have limited molecular and genetic tools for engineering. Advancements have been made in using heterologous hosts for expression and production of myxobacterial natural products.

In this chapter we summarize progress made with engineering novel analogs of epothilone and ambruticin, an antifungal compound isolated from *S. cellulosum* more than 30 years ago (39). We describe the cloning of the gene clusters from *S. cellulosum*, the development of *M. xanthus* as a heterologous expression host, and the development of tools and techniques to genetically manipulate the natural producer for generating new analogs. The reader is also directed to several reviews of myxobacterial natural products (3, 12, 35, 51, 54).

29.2. HETEROLOGOUS EXPRESSION OF PKS/NRPS GENES IN MYXOBACTERIA

29.2.1. Cloning and Identification of PKS/NRPS Gene Clusters

After identification of a natural product as a lead candidate for drug discovery, limited material supply can be a major obstacle to clinical evaluation. Chemists usually resolve

Epothilone gene cluster

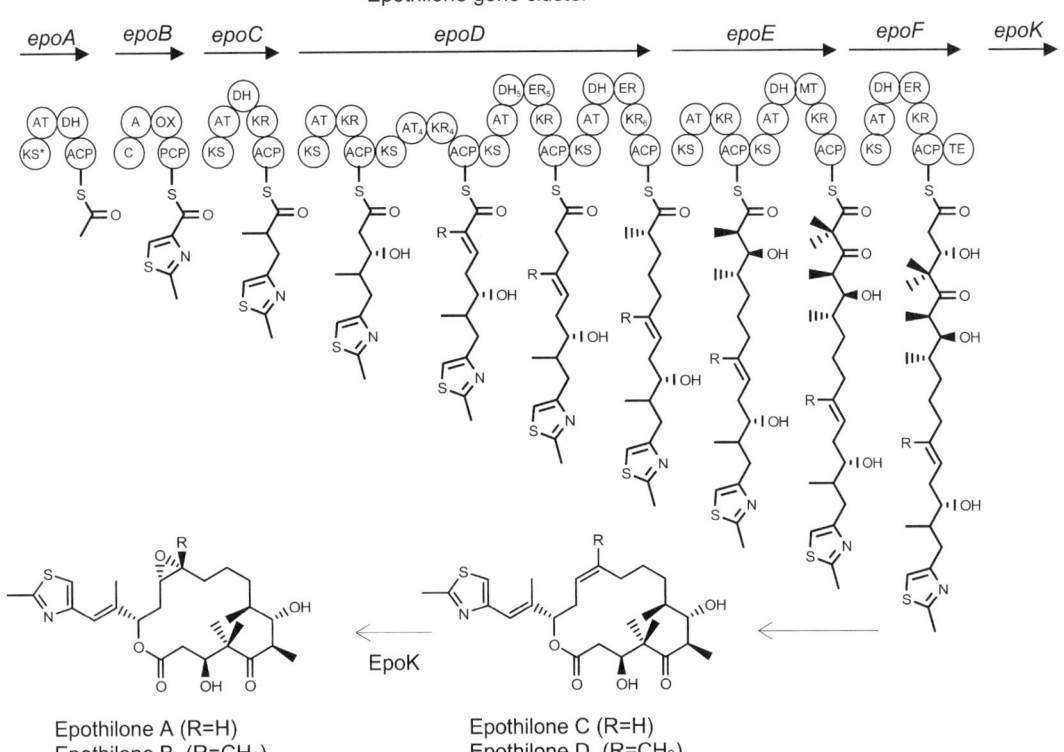

FIGURE 1 The PKS/NRPS assembly line for epothilone biosynthesis. The six polypeptides of the megasynthase are shown at the top (EpoA to EpoF) as arrows (N to C terminus). The domain composition is shown schematically. The numbered domains on the polypeptides were targets for genetic engineering as described in section 29.5.1. The *epoK* gene encodes a cytochrome P450 protein which adds an epoxide to epothilones C and D to produce epothilone A and B, respectively.

this by developing methods for chemical synthesis from commercially available materials. However, for polyketides and peptides, the number of steps required for synthesis excludes this approach as a long-term economical method of production. A good example is epothilone. The Danishefsky and Nicolaou research groups completed the synthesis of epothilone and numbers of analogs (2, 33, 46, 55), but given the complexity of the 20-step synthetic process, chemical synthesis would be more costly than if a final molecule or an intermediate could be made by fermentation and then modified chemically.

Fermentation production of natural products can be costly in the beginning if titers are low. However, a classical strain improvement program involving mutagenesis and screening for higher-producing strains is the proven route to bring fermentation costs down. For example, an *S. cellulosum* strain was improved 500-fold to increase soraphen titers from 3 to 1,500 mg/liter (12), and more recently epothilone B titers were increased 130 times from 0.8 to 104 mg/liter by using a combination of classical mutagenesis and genome shuffling techniques (14).

An alternative method, especially for compounds produced by microorganisms that are difficult to grow at large scale or where techniques are not available for genetic manipulation, is heterologous expression of the natural product biosynthetic genes. The first step in this process is their identification. To identify the epothilone biosynthetic genes, two approaches were taken (22, 43). Both resulted in

generating DNA fragments for screening a library of DNA fragments. Because of the common biosynthetic nature of all type I PKSs, there are conserved protein regions that can be used for making DNA probes. Of these homologous protein regions, the β-ketoacyl acyl carrier protein (ACP) synthase domain (KS) of a PKS, which is found in all extension modules, was used to design degenerate primers which were then used to amplify genomic DNA from the epothilone-producing strain *S. cellulosum* SMP44. The putative KS sequences were cloned, sequenced to verify their identity, and then used to probe a cosmid library of genomic DNA from the producing strain (Fig. 2A). Thirteen cosmids were identified and further analyzed by restriction mapping, hybridization with an enoylreductase probe, and end-sequencing of the DNA cloned into each cosmid. Using the information gained from the analysis of the cosmids, four were identified for sequencing (22). The complete sequence of the cosmids revealed a modular type I PKS, and analysis of the enzymatic functions encoded in each module correlated well with those required for the synthesis of epothilone.

The second approach taken was to generate a more random set of probes for identifying PKS gene clusters. This entailed constructing a library of 1- to 2-kb DNA fragments and sequencing the ends of the cloned fragments. From this library approximately 500 DNA sequences were produced and then analyzed for homology to PKS or NRPS sequences. From these sequences, 16 showed

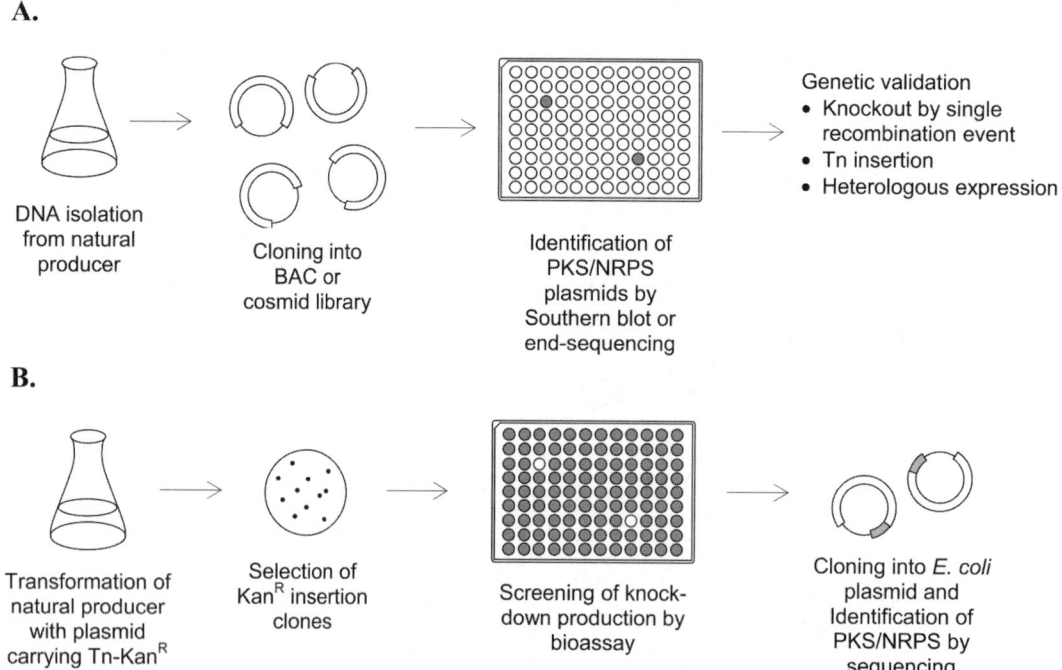

FIGURE 2 (A) Schematic representation of cloning and identification of the PKS/NRPS gene cluster from a *Myxobacterium* natural producer. (B) Schematic steps for using the *mariner* transposon derivative for identification of PKS/NRPS in myxobacteria.

homology to PKS sequences and 4 showed homology to NRPS sequences. Although these PKS/NRPS DNA fragments were never used to probe a library, they did provide the first glimpse at the large number of natural product gene clusters encoded by *S. cellulosum*. The data indicate that approximately 4% of the genome is PKS/NRPS sequence (43); assuming a genome size of approximately 10 Mb (13), then 400 kb of DNA encodes PKS/NRPS sequences. With an average size of approximately 40 to 50 kb per biosynthetic gene cluster, then six to eight different gene clusters are present in SMP44. The large number of natural product gene clusters in myxobacteria have been verified from the genome sequencing of *M. xanthus* DK1622 and *S. cellulosum* So ce56 (13, 44).

Another method of identifying natural product biosynthetic gene clusters in myxobacteria was later developed for isolation of the ambruticin biosynthetic genes. This method uses a transposon to insert and tag a gene cluster of interest. The insertion inactivates the biosynthetic genes, and the loss of production can be screened for by a simple bioassay or by analyzing culture extracts by high-pressure liquid chromatography (HPLC) or liquid chromatography/mass spectroscopy (LC/MS) (Fig. 2B). A detailed protocol for using the transposon is described in section 29.3.1. Using the techniques described for identifying natural product biosynthetic genes, 23 complete gene clusters have been sequenced from myxobacteria (51).

29.2.2. A Bacterial Host for Heterologous Production of Myxobacterial Natural Products

The first myxobacterial natural product to be produced in a heterologous host was epothilone (48). The gene cluster, consisting of nine PKS modules, one NRPS, and a cytochrome P450 spanning 56 kb (Fig. 1), was cloned onto two compat-

ible *Streptomyces* plasmids and transformed into *Streptomyces coelicolor*, a widely used host for expression of polyketide biosynthetic genes from actinomycetes (see chapter 26 by Kealey, this volume). In *S. coelicolor*, epothilones A and B were produced, but at a low level due to an apparent toxicity. This prompted a search for an alternative host.

Such an alternative host would ideally be one similar to *S. cellulosum*. Examination of research on myxobacteria revealed that *M. xanthus* is a good host for several reasons. First, it is well characterized and has the most developed engineering tools among the myxobacteria. It has a shorter doubling time than *S. cellulosum* (5 versus 16 h). Several promoters from other myxobacteria function properly in *M. xanthus*; we have shown the functionality of the myxothiazole promoter from *Stigmatella aurantiaca* and the epothilone and ambruticin promoters from *S. cellulosum* in *M. xanthus* (24; B. Julien, unpublished data). Also, the genome sequence of *M. xanthus* is known. Other requisites, such as precursor supply and posttranslational modifications of PKS/NRPS, have been demonstrated in *M. xanthus*, proving it to be a suitable host for heterologous production of myxobacterial natural products (24).

The epothilone biosynthetic gene cluster was built in the genome of *M. xanthus* in a two-step process due to its large size (24). The first step required introduction of fragments of the epothilone gene cluster for use as recombination sites. Subsequent steps introduced large fragments of the gene cluster from cosmids into the chromosome. Although this work involved time-consuming techniques, it proved the viability of using the myxobacterium *M. xanthus* as a heterologous host for production of myxobacterial natural products, and led the way for combinatorial biosynthesis of novel epothilone analogs (section 29.4). Also, the genetic manipulation of the epothilone PKS cluster allowed an easy method to inactivate the *epoK* gene for production of epothilones C and D in M. *xanthus* (24). The *epoK* gene encodes a cytochrome P450 protein, which adds an epoxide moiety to epothilones C and D to produce epothilones A and B, respectively (Fig. 1) (48).

With the complete epothilone biosynthetic gene introduced into M. *xanthus* host, initial production of epothilone B was 0.3 mg/liter. Unexpectedly, the ratio of epothilone B to epothilone A is significantly higher in M. *xanthus*, 6:1, than in S. *cellulosum*, which has a ratio of 1:2. Through media optimization, this was quickly raised to 25 mg/liter by using a fed-batch process (29). Finally, with a classical strain improvement program, titers over 100 mg/liter were reached, which provided a production host for epothilone D that provided material for phase I and phase II clinical trials. Although *Escherichia coli*, *Streptomyces venezuelae*, and *Pseudomonas putida* have been engineered with the epothilone biosynthetic genes, none has reached the titers obtained from M. *xanthus* (9, 31, 34).

29.3. GENETIC TOOLS FOR MYXOBACTERIA

29.3.1. *mariner*-Based Transposon for *Sorangium* spp.

Because a single strain of myxobacteria can harbor multiple PKS/NRPS gene clusters for natural products, it can be challenging to identify the proper gene cluster for a product of interest. For epothilone, PCR-based amplification of conserved regions in the KS sequence and sequencing of a library of genomic fragments eventually led to the identification of the biosynthetic genes (section 29.2.1). However, this process is time-consuming because it requires the identification of all PKS/NRPS gene clusters, usually by partial sequencing of cosmids, to determine which ones are correct.

To facilitate the identification of the biosynthetic gene cluster in S. *cellulosum*, a transposon based on the *mariner* transposon from eukaryotes was developed. For use in S. *cellulosum*, a conjugative plasmid was constructed that harbored both a *mariner tnp* gene encoding the transposase and the *mariner* inverted repeats flanking the antibiotic resistance gene for bleomycin. Using a conjugation procedure (described in chapter 18, this volume), a transposition frequency of 10^{-4} to 10^{-5} per cell was obtained with S. *cellulosum* So ce90 (25). Additionally, a *tnp* gene harboring two point mutations allowed a fourfold increase in the transposition frequency relative to wild type. Since the transposase only requires the dinucleotide TA to insert into DNA, the modified *mariner* transposon essentially inserts randomly throughout the genome (40).

The utility of such a transposon was demonstrated by isolation of an *epoK* knockout from a library of approximately 12,000 mutants of the S. *cellulosum* strain So ce90 (25). In addition, this transposon was used for mutagenesis of S. *cellulosum* So ce10 and S. *cellulosum* So ce12 for the rapid identification of the ambruticin and disorazole biosynthetic genes (Fig. 2B) (5, 26). For identification of the ambruticin biosynthetic genes, the transposon mutagenesis technique proved superior to the sequencing of random cosmid inserts. Thus, the *mariner* transposon provides a valuable tool for generating random mutations in *Sorangium* spp., and it has been shown to work in other myxobacteria as well (27, 42, 57).

29.3.2. Gene Replacements in *M. xanthus*

The major advantage of using M. *xanthus* over other myxobacteria is the ability to introduce stable genetic modifications into the chromosome by double homologous recombination. This process can be accelerated by the use of a counterselectable marker. For M. *xanthus*, two markers have been used, *sacB* (encoding a levansucrase) and *galK* (encoding a galactokinase) (50). Both genes, when integrated into the M. *xanthus* chromosome, provide selective conditions for their subsequent loss: for *sacB*, growth in the presence of sucrose, and for *galK*, growth in the presence of 2-deoxygalactose. Because 2-deoxygalactose is expensive, it was later found that galactose can be used in the *galK* counterselection system (Julien, unpublished data). This substitution is possible because M. *xanthus* does not metabolize galactose. Expression of *galK* in M. *xanthus* results in the phosphorylation of galactose where there is no pathway to convert or export it. This results in an accumulation of phosphogalactose inside the cell, causing death. Similar results have been seen with E. *coli* mutants that are blocked after the phosphorylation of galactose (56).

29.3.3. Protocol for Performing Gene Replacements in *M. xanthus*

1. In a plasmid carrying the *galK kan*[r] genes, clone both flanking regions of the target gene at each side of the gene. PCR, DNA synthesis, or λRed recombineering technology (54) can be used to construct the DNA. Design of restriction enzyme sites or overlap extension PCR can help to stitch the fragments together.
2. Transform the plasmid into the strain using electroporation or conjugation (see chapter 18, this volume) and select on Casitone-yeast extract (CYE) plates containing 50 μg/ml kanamycin. Verify insertion into target location by PCR or Southern blot.
3. Inoculate the *kan*[r] strain, containing the integrated replacement plasmid, in CYE without kanamycin until cells are in mid-log to early stationary phase.
4. Dilute cells to 10^{-3} and 10^{-4}, add 2.5 ml of CYE top agar, and pour onto CYE plates supplemented with 1% galactose.
5. Incubate plates at 32°C for 7 to 10 days.
6. Restreak colonies and confirm the loss of the *kan*[r] marker by the lack of growth on CYE with 50 μg/ml kanamycin.
7. Confirm presence of the mutation by PCR or Southern blot.

In cases where several replacements need to be tested, as was performed in module 4 replacements of the epothilone gene cluster (section 29.4.2), an additional step in the gene replacement protocol can be included as follows:

1. Replace the target gene with the *galK kan*[r] genes using a plasmid carrying these genes cloned between the two flanking regions of the target gene.
2. Transform the linearized plasmid into the strain using electroporation. Select on CYE plates containing 50 μg/ml kanamycin. Verify the replacement into target location by PCR or Southern blot.
3. Construct a new plasmid containing the bleomycin resistance (*ble*[r]) gene and the same flanking regions of the target gene at both sides of the new version of the gene, and transform into the above *kan*[r] strain, selecting for bleomycin-resistant colonies.
4. Inoculate the *kan*[r] *ble*[r] strain, containing the integrated replacement plasmid, in CYE without kanamycin until cells are in mid-log to early stationary phase and then follow step 4 of the previous protocol.

This additional step generates a nonproducing Kan^r strain which allows the possibility of using a bioassay for screening the positive clones after the replacement.

29.3.3.1. Media

CYE medium: (per liter) 10 g of Casitone, 5 g of yeast extract, 1 g of MgSO$_4$·7H$_2$O. After autoclaving, add 10 ml of 1 M morpholinepropanesulfonic acid (MOPS) adjusted to pH 7.6.
CYE plates: CYE medium containing 15 g of Bacto agar.
CYE top agar: CYE medium containing 7.5 g of Bacto agar.

29.3.4. Integration of Large Fragments of DNA into *M. xanthus* Using Mx9 Site-Specific Integration System

As mentioned above, the introduction of the epothilone biosynthetic genes into *M. xanthus* was time-consuming. However, it was the most efficient method at the time given its size. More efficient methods of introducing large DNAs have been developed. Recently a technique using a transposon to move large pieces of DNA into the chromosome was used for introduction of the epothilone gene cluster into *M. xanthus* and *P. putida* (9, 53). Because of the size of the epothilone gene cluster, the technique required "stitching" large pieces of DNA together on the transposon by Red/ET recombination. Although the authors were successful at building the transposon with the epothilone cluster, the recombination was less efficient, due to the large size of the gene cluster, and there was a high frequency of intermolecular recombination due to the high homology of the DNA sequence encoding each module. Furthermore, the results showed size-dependent transposition frequency: the larger the transposon, the lower the transposition frequency. Thus, very large DNA pieces can be problematic with this technique.

A new method has been developed that enables large fragments of DNA, >100 kb, to be integrated into the chromosome of *M. xanthus*. This technique uses a modified bacterial artificial chromosome (BAC) with large DNA inserts which can be transferred by conjugation and then integrated into the chromosome. This modified BAC contains an *oriT* for conjugation into *M. xanthus* from *E. coli*, and the site-specific recombination system from the myxophage Mx9, to integrate the DNA into the chromosome (Julien, unpublished data). The Mx9 site-specific recombination system was used because integration at the Mx9-*attB* site does not interfere with gene expression, unlike integration at the Mx8 integration sites (23).

This method was validated using a BAC containing a DNA fragment greater than 100 kb which harbored the complete ambruticin biosynthetic gene cluster. Through the use of the *loxP* site for specific recombination carried on the BAC, the *oriT* for conjugal transfer, the bleomycin resistance gene for selection in *M. xanthus*, and the Mx9 *int* gene and Mx9-*attP* site for integration into the *M. xanthus* chromosome were added. The modified BAC was successfully conjugated into six different *M. xanthus* strains, and production of ambruticins varied from 0.1 to 10 mg/liter depending on the *M. xanthus* strain used (Julien, unpublished data).

This technique of integrating large pieces of DNA into *M. xanthus* provides a method to identify new natural product biosynthetic genes from any myxobacteria by first constructing a library of large DNA fragments on this integrating BAC and then conjugating the individual BAC clones into *M. xanthus*. Since the genomes of myxobacteria are in the range of 10 Mb in size, 100 clones from a BAC library with an average insert size of 100 kb should give one genome size worth of coverage.

29.4. ENGINEERING NOVEL ANALOGS OF EPOTHILONE

29.4.1. Inactivation of KR, DH, and ER Domains in the Epothilone Gene Cluster

An important advantage of using *M. xanthus* as a heterologous host is the possibility of applying combinatorial biosynthesis. Thus, the heterologous expression of the epothilone gene cluster in *M. xanthus* provided the opportunity for making novel analogs and understanding novel enzyme mechanisms. This technology makes use of genetic engineering of the assembly line of a PKS through addition, deletion, or mutation of catalytic domains or the reorganization of whole modules aimed at generating novel analogs (Fig. 3).

After the condensation of a carboxylic residue into the carbon skeleton, the β-ketone can be sequentially reduced first by the actions of a ketoreductase (KR), then a dehydratase (DH), and finally an enoylreductase (ER) domain. Thus, one method to alter polyketide structures is through inactivation of one or more of these reductive reactions so that the corresponding step in the synthesis is bypassed, resulting in a novel compound. Our ability to inactivate these domains comes from molecular modeling combined with mutational studies that have led to the identification of key amino acid residues that participate in the catalysis of the KR, DH, and ER domains. In addition, the amino acids required for setting the stereochemistry of the alcohol group by the KR domain were determined (7, 37).

For abolishing the activity of KR domains, it was shown that a conserved active site tyrosine was important for activity, and that changing this amino acid to phenylalanine resulted in a loss of activity (37). In addition, mutation of the conserved glycine in the Rossmann fold important for NADPH binding is predicted to prevent KR activity (7). Using this information, replacement of the corresponding residues in the KR in modules 4 and 6 of the epothilone gene cluster in *M. xanthus* strain was attempted.

For the KR domain in module 4, single amino acid substitutions of the tyrosine or glycine did not inactivate the KR, as evidenced by the production of epothilone D and not the expected 13-oxoepothilone. However, combining both mutations resulted in complete inactivation of the KR domain and production of the expected compound 12,13-dihydro-13-oxoepothilone C, in trace amounts, and the unexpected compound 11,12-dehydro-12,13-dihydro-13-oxoepothilone D as the major product (Fig. 3A) (50). Complete deletion of the KR$_4$ domain and insertion of a linker region resulted in the same products as the double mutants. For KR$_6$ the two point mutations were constructed. This strain produced the expected compound 9-oxoepothilone D and its isomer 8-*epi*-9-oxoepothilone D as the major products (Fig. 3B) (47).

In addition to inactivating KR modules, mutations were made in the DH or ER domains of module 5. To inactivate ER$_5$, the NADPH binding site was targeted, similar to the KR point mutations. Mutation of two adjacent glycine residues to an alanine and serine in the NADPH binding site and introduction into the epothilone D-producing *M. xanthus* strain resulted in the production of the predicted 10,11-didehydroepothilone D (1).

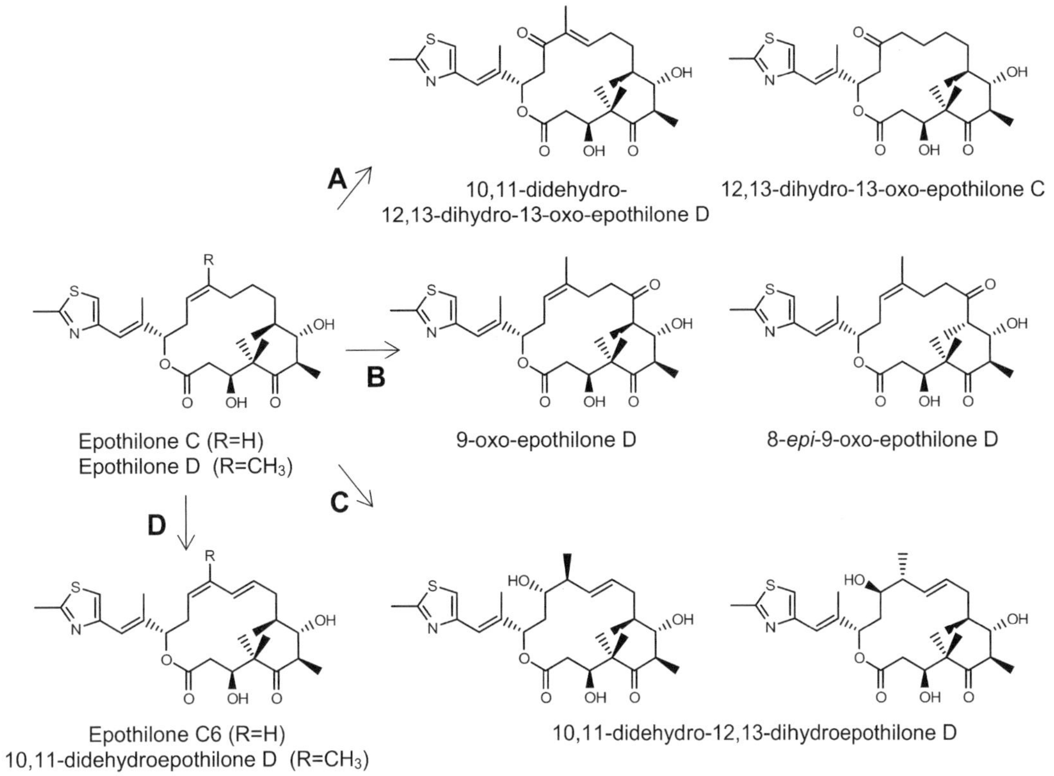

FIGURE 3 Novel epothilone analogs generated by manipulation of the *epo* PKS. Most of the successful changes were produced through inactivation of domains involved in the reductive cycle. (A) KR_4. (B) KR_6. (C) DH_5. (D) ER_5.

In addition to making novel epothilone compounds by engineering the biosynthetic gene cluster, important biochemical information was obtained on the formation of the *cis* double bond between C12 and C13. For this double bond to form, a DH domain is required. However, there is no DH activity encoded in module 4 of the epothilone PKS. Inactivation of DH_5 revealed its role in *cis* double bond formation. Mutations were constructed in DH_5 that replaced eight highly conserved amino acids with three amino acids, or replaced the entire DH/ER/KR segment with a KR domain from another module (49). In both mutants the major product was 10,11-didehydro-12,13-dihydro-13-hydroxyepothilone D (Fig. 3C). The presence of a hydroxyl group at C13 strongly connects the activity of DH_5 with the dehydration and formation of the 12,13 double bond (48). However, the unanticipated 10,11 double bond must have been produced by another DH domain, potentially DH_6. Thus, DH_5 performs an unprecedented series of two reactions, one for *cis* double bond formation and the other for dehydration of the C13 hydroxyl, resulting in the formation of a *trans* double bond (49). Thus, besides being able to make 10,11-didehydroepothilone D in *M. xanthus*, this procedure uncovered the mechanism by which the C12-C13 *cis* double bond is formed (Fig. 3D).

29.4.2. AT Domain Replacement in the Epothilone Gene Cluster

In addition to modification of the reductive domains within the epothilone biosynthetic enzyme, modifications of the specificity of ATs were attempted. The AT of each module recognizes either a malonyl or methylmalonyl extender unit to be added to the growing polyketide chain. Changing the

specificity of an AT will create novel molecules. Modifications of several of the ATs in the epothilone PKS were attempted, but only one yielded a functional replacement (E. Rodriguez and L. Tang, unpublished data). Thus the epothilone PKS is less amenable to changes in the extender unit, likely due to a stricter structure requirement to be recognized by each module as the polyketide chain is lengthened.

The only module that was amenable to changes of specificity of its AT was module 4 (AT_4), which is able to use either malonate or methylmalonate. The two major products made by the epothilone PKS are epothilones A and B. The difference between the two is the presence of a methyl group at C12. Labeling studies have confirmed that this methyl group is derived from propionate (10), the precursor for methylmalonate; thus, to make epothilones A and B, AT_4 must be able to utilize both extender units, malonate or methylmalonate. Because the epothilones with the C12 methyl group are more potent, a strain was engineered to produce only epothilone B or epothilone D in an *epoK* mutant strain.

To engineer this strain, the AT_6 from the soraphen gene cluster from *S. cellulosum* So ce26 and the AT_7 from the epothilone gene cluster were chosen for the replacement, based on their similar domain organization to the module 4 of EpoD protein and their preference for methylmalonyl-coenzyme A (CoA) (E. Rodriguez and B. Julien, unpublished data) (Fig. 4). Three unique junction boundaries between the KS and AT domains in combination with three boundaries between the AT and KR domains were designed based on protein alignments. The gene replacements were performed according to the protocol described in section 29.3.3. The results showed that both the soraphen AT_6 and epothilone AT_7 could be used to increase the ratio of epothilone D to

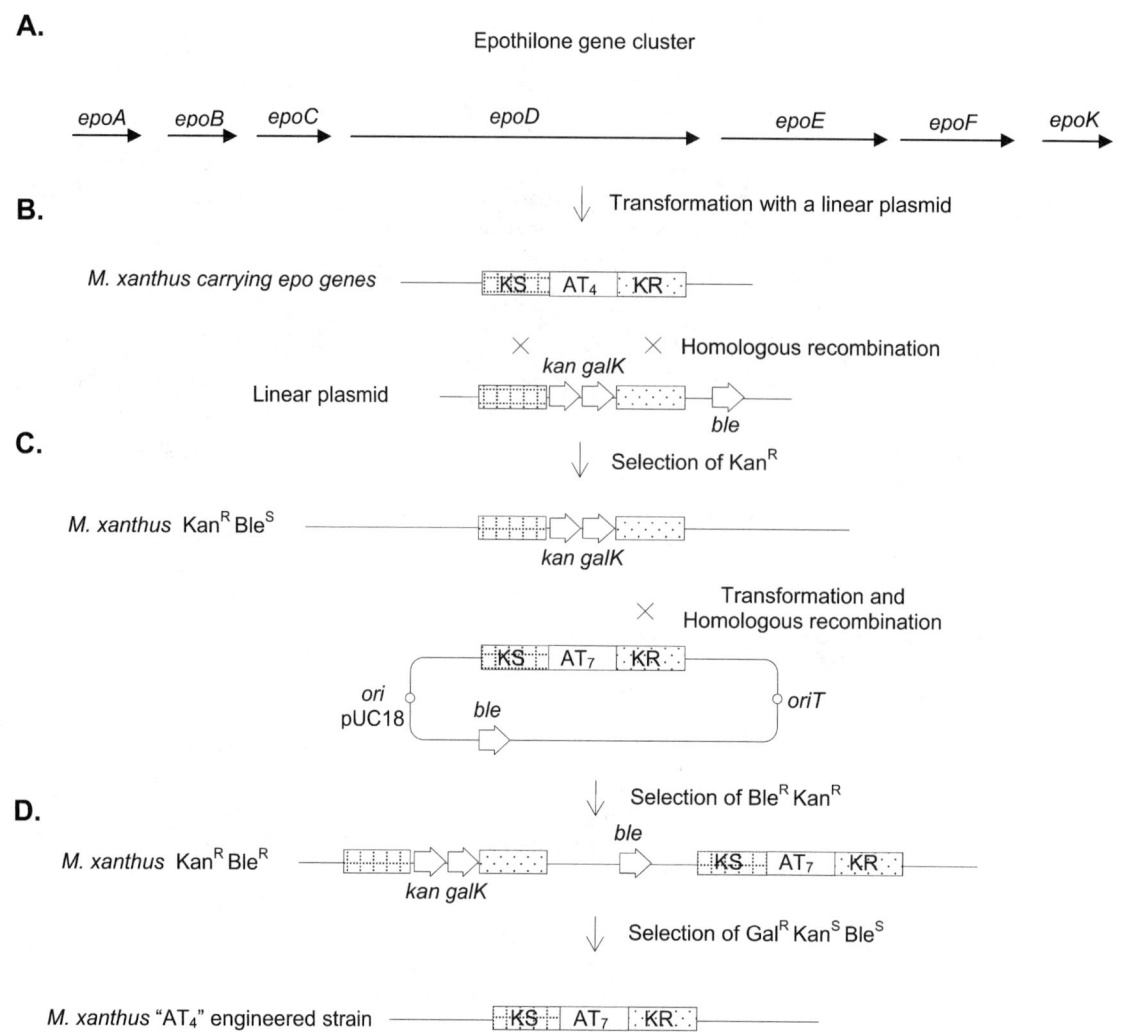

FIGURE 4 Scheme for engineering an AT replacement in the chromosome of M. *xanthus* carrying the epothilone gene cluster. (A) Schematic representation of the epothilone gene cluster. (B) Step 1: replacement of AT$_4$ with the *kan^r galK* genes through double recombination using a linear delivery vector. (C) Step 2: a cassette, consisting of a heterologous AT flanked by upstream and downstream regions, is introduced into the M. *xanthus* strain constructed in step 1. The first crossover can occur with either of the flanking regions. (D) Step 3: the second crossover is selected using plates containing galactose, resulting in the recombination event to remove duplicated regions and segments with the kanamycin resistance and *galK* genes. Fermentation of the Gal^r Kan^s Phle^s strain allows selection of the correct construct.

epothilone C 50-fold in the *epoK* mutant M. *xanthus* strain. These results were obtained using only one of the KS-AT boundaries in combination with any of the three AT-KR boundaries. In addition, a twofold increase in overall titers was obtained by use of the epothilone AT$_7$ domain.

29.5. MODULATION OF EPOTHILONE PRODUCTION IN MYXOBACTERIA

29.5.1. Protocol for Production of Epothilone in M. *xanthus*

M. *xanthus* presents a special challenge for development of a production medium for secondary metabolites. It does not metabolize simple sugars, but relies on amino acids and oils to provide its carbon source. Using amino acids as a carbon source results in the accumulation of ammonium, which

has a negative impact on secondary metabolite production. To circumvent this, the oil methyloleate is used as a carbon source. Because methyloleate disrupts the cell membrane if added at high concentrations, the cells are adapted to the oil before inoculation into production medium. The following protocol has been established for production of secondary metabolites in M. *xanthus* (29).

1. *Inoculum preparation.* Inoculate M. *xanthus* into 3 ml of CYE medium supplemented with 3 μl of methyloleate in a 50-ml glass tube. Incubate at 32°C and 175 rpm until culture is saturated. Slow agitation is critical to prevent too much mixing of the oil and subsequent lysis of the cells. After 3 days, transfer into a 250-ml Erlenmeyer flask containing 50 ml of CYE supplemented with 100 μl of methyloleate, and grow for 2 days at 32°C and 175 rpm.

2. *Production medium.* In a 250-ml Erlenmeyer flask, add 1 g of XAD-16 and enough water to immerse the resin, and autoclave to sterilize. Remove water from sterile flask and add 50 ml of CTS fermentation medium supplemented with 350 μl of methyloleate and 200 μl of trace elements.

3. *Fermentation conditions.* Inoculate medium with 2.5 ml (5%, vol/vol) of the saturated culture. Grow cultures at 32°C. For simple batch fermentations, grow the cultures for 5 days, then remove the XAD-16 resin and extract the compounds. For fed-batch fermentation, which results in increased secondary metabolite production, feed the cultures daily, starting after day 2, with 150 μl of methyloleate and 100 μl of a 1-g/liter solution of Casitone. After 10 days, remove the XAD-16 resin and extract the compound.

4. *Analysis of the products.* To extract the product from the resin, allow the culture to sit for several minutes for the resin to settle to the bottom of the flask, and then pour off the culture broth. Wash the resin twice with several volumes of water, and remove the final few milliliters of water by aspiration. To the resin, add 25 ml of methanol and incubate at room temperature for a couple of hours, preferably with shaking. Analyze the methanol extract by HPLC or LC/MS for desired products.

29.5.1.1. Media

CTS medium: (per liter) 5 g of Casitone, 2 g of $MgSO_4 \cdot 7H_2O$. After autoclaving, add 50 ml of 1 M HEPES (pH 7.6).

Trace element solution: (per liter) 10 ml of concentrated H_2SO_4, 14.6 g of $FeCl_3 \cdot 6H_2O$, 2 g of $ZnCl_3$, 1 g of $MnCl_2 \cdot 4H_2O$, 0.43 g of $CuCl_2 \cdot 2H_2O$, 0.31 g of H_3BO_3, 0.24 g of $CaCl_2 \cdot 6H_2O$, and 0.24 g of $Na_2MO_4 \cdot 2H_2O$.

29.5.2. Altering Epothilone Congener Production through Medium Design

Epothilones A and B are the major fermentation products of *S. cellulosum* So ce90 and of the *M. xanthus* strain engineered to contain and express the epothilone biosynthetic genes. However, the ratios of the compounds produced in the two organisms are dramatically different. For *S. cellulosum*, the ratio of A to B is 2:1, but for *M. xanthus* it is 1:6. The difference in ratio is likely due to the available pools of malonyl- or methylmalonyl-CoA, providing the hypothesis that modulating the levels of these two extender units can result in altering the ratio of the two major products.

To change the ratio of epothilones, the intracellular concentrations of malonyl-CoA and methylmalonyl-CoA were altered by the addition of acetate or propionate to the basal production medium. Supplementation of increasing concentrations of acetate resulted in decreases in the epothilone D:C ratio (8). Conversely, addition of propionate to the basal medium resulted in a higher epothilone D:C ratio. The growth rates and cell densities of these cultures were also markedly decreased as the propionate concentration was increased in the basal medium, indicating a potential toxicity associated with high levels of propionate in the media or high levels of propionate, propionyl-CoA, or methylmalonyl-CoA.

In addition to epothilones A and B, *S. cellulosum* has been shown to make small amounts of 37 additional epothilone variants due to the incorporation of different precursors or alterations in the biosynthetic steps by the epothilone NRPS/PKS enzymes (16). One of the variants contains an oxazole ring in place of the thiazole that results from the utilization of serine rather than cysteine by EpoB, the NRPS. To increase the production of the oxazole-containing compounds by the engineered *M. xanthus* strain, the fermentation medium was supplemented with serine in concentrations ranging from 2.3 mM to 168 mM (8). At a serine concentration of 71 mM, the maximum production of the oxazole-containing compound was achieved, producing roughly equal amounts of the thiazole- and oxazole-containing epothilones. However, the overall epothilone titers were decreased by 75%. Thus, in addition to modification of the PKS enzyme, fermentation conditions can aid in making novel compounds or increasing the production of minor ones.

29.5.3. Metabolic Engineering

In addition to modifications of the PKS enzyme or media optimization to increase production of target molecules, engineering biochemical pathways can prove valuable. One example in *S. cellulosum* is the engineering of the pathway for methylmalonyl-CoA to produce more epothilone B. Because AT_4 is able to utilize either malonyl-CoA or methylmalonyl-CoA, increasing the intracellular pool of methylmalonyl-CoA would favor the production of epothilone B. To achieve this, the propionyl-CoA synthase gene, *prpE*, from *Ralstonia solanacearum* was heterologously expressed in *S. cellulosum* (15). PrpE converts propionate into propionyl-CoA, which then is subsequently converted to methylmalonyl-CoA. This engineered *S. cellulosum* strain exhibited a significant increase in the epothilone B:A ratio of 127:1, which was 100-fold higher than that of the parental strain.

29.6 ENGINEERING NOVEL ANALOGS OF AMBRUTICIN

29.6.1 Inactivation of Post PKS Enzymes

The first myxobacterial natural product was isolated from *S. cellulosum* by researchers at Warner-Lambert in the 1970s (39). The compound, ambruticin S (Fig. 5), exhibited potent activity against many fungal strains, in particular the pathogens *Coccidioides immitis*, *Histoplasma capsulatum*, and *Blastomyces dermatitidis*. It was also shown to be active against *C. immitis* and *H. capsulatum* infections in murine models. However, its development was halted due to its lack of activity against *Candida albicans* (38).

Interest in ambruticins resurfaced as other fungal infections became more problematic, particularly aspergillosis in immunocompromised people. Kosan Biosciences, Inc., began a development program testing new ambruticin analogs that were derived from chemical modification of the basic ambruticin scaffold. Several of these analogs showed efficacy in murine models for coccidioidomycosis and aspergillosis (6, 45).

To make analogs that were inaccessible through facile chemical modification, the biosynthetic genes for ambruticin were identified (26). Analysis of the biosynthesis genes revealed novel enzymatic reactions for PKSs. One is the use of a separate methyltransferase for capturing an enolate to form the first pyran ring. The most intriguing is a Favorskii-like rearrangement that results in the excision of a carbon atom from the growing polyketide chain and the formation of the internal cyclopropane ring. Because of the unique chemistries required for the biosynthesis of

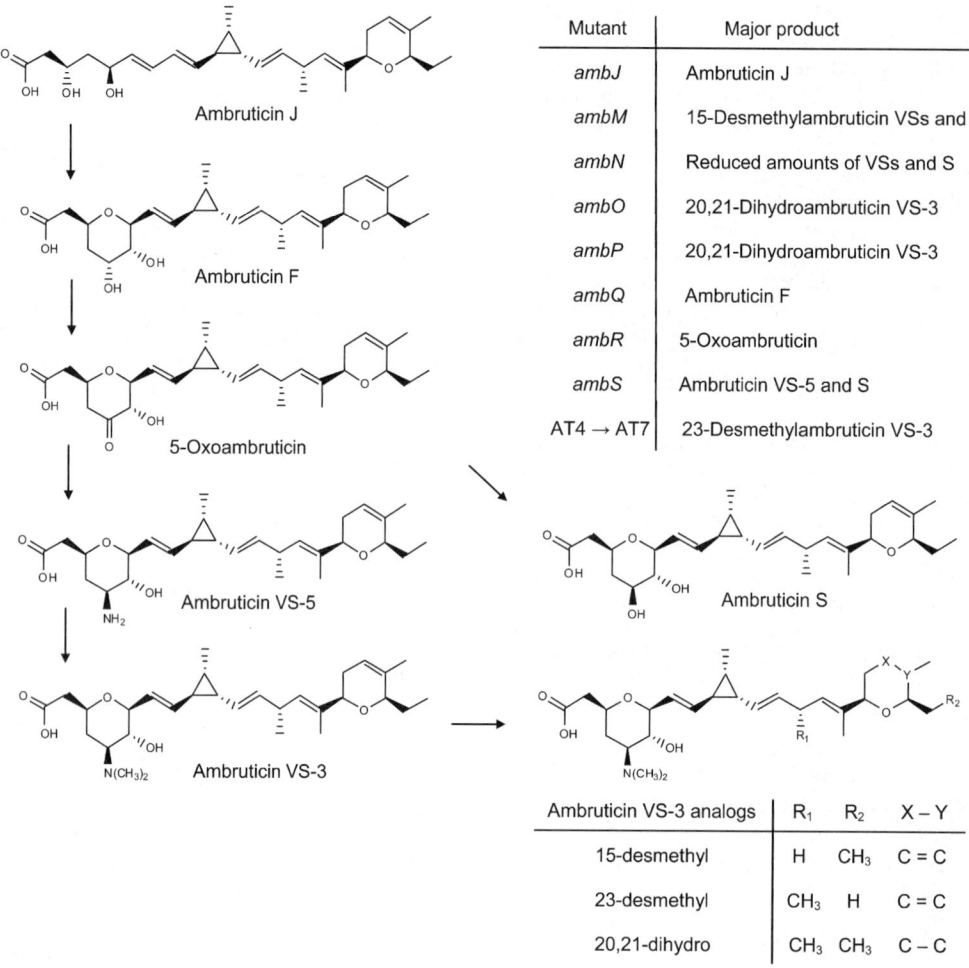

Mutant	Major product
ambJ	Ambruticin J
ambM	15-Desmethylambruticin VSs and S
ambN	Reduced amounts of VSs and S
ambO	20,21-Dihydroambruticin VS-3
ambP	20,21-Dihydroambruticin VS-3
ambQ	Ambruticin F
ambR	5-Oxoambruticin
ambS	Ambruticin VS-5 and S
AT4 → AT7	23-Desmethylambruticin VS-3

Ambruticin VS-3 analogs	R_1	R_2	X – Y
15-desmethyl	H	CH_3	C = C
23-desmethyl	CH_3	H	C = C
20,21-dihydro	CH_3	CH_3	C – C

FIGURE 5 Pathway proposed for the biosynthesis of ambruticin. Inactivation of genes from the ambruticin gene cluster allows the isolation of intermediates in the biosynthesis of ambruticins.

ambruticin, the ability to engineer novel analogs from our current understanding of the PKS is limited. However, eight genes downstream of the last PKS gene, *ambH*, were identified that played a role in the biosynthesis of ambruticin. Inactivation of each of them provided strains that made novel analogs or resulted in strains that made greater quantities of minor products.

A summary of the mutants and the products that are made by each is shown in Fig. 5. Four novel compounds were identified: ambruticin J, 20,21-dihydroambruticin S and VSs, 5-oxoambruticin, and 15-desmethylambruticin S and VSs. Two mutant strains produced previously identified minor compounds as major products; the *ambS* mutant strain made predominantly ambruticin VS5; and the *ambQ* mutant strain made ambruticin F as a major component. Initial testing of these compounds for antifungal activity revealed that all were active (L. Vetcher and B. Julien, unpublished data).

29.6.2. AT Domain Replacement in the Ambruticin Gene Cluster

The classical procedure to make a domain replacement in PKS genes is by homologous recombination, requiring two steps. Because currently it is not possible to perform gene replacement in *Sorangium*, a strategy was devised that allowed for integration of a plasmid by a single homolo-

gous recombination that inactivated the native *ambA* gene of the ambruticin gene cluster and expressed a new, engineered version of *ambA* (Rodriguez and Julien, unpublished data) (Fig. 6).

AmbA initiates the synthesis of ambruticin by decarboxylation of a methylmalonyl unit specified by the loading acyltransferase (AT_0) domain (26). To alter the specificity of AT_0, the AT from module 7 was used because its domain organization is identical to the loading module. The junctions between the KS, AT_0, and ACP were based on the engineering of the AT_4 in the epothilone biosynthetic genes (section 29.5.2) (Rodriguez and Julien, unpublished data).

For this engineering approach, the *ambA* gene was inactivated by truncation of the 5′ end of KS_0. The malonate specific AT_7 domain was cloned along with the truncated KS_0 under the transcriptional control of the *epoA* promoter. Isolation of a recombinant that integrated this engineered loading module within the KS_0 sequence in the chromosome resulted in a recombinant strain that produced 23-desmethylambruticin VS3 (Fig. 6). This result represents the first successful example of polyketide engineering in an *S. cellulosum* strain, showing the potential for engineering this important microorganism. However, this approach can only be employed at the beginning or end regions of the PKS gene cluster.

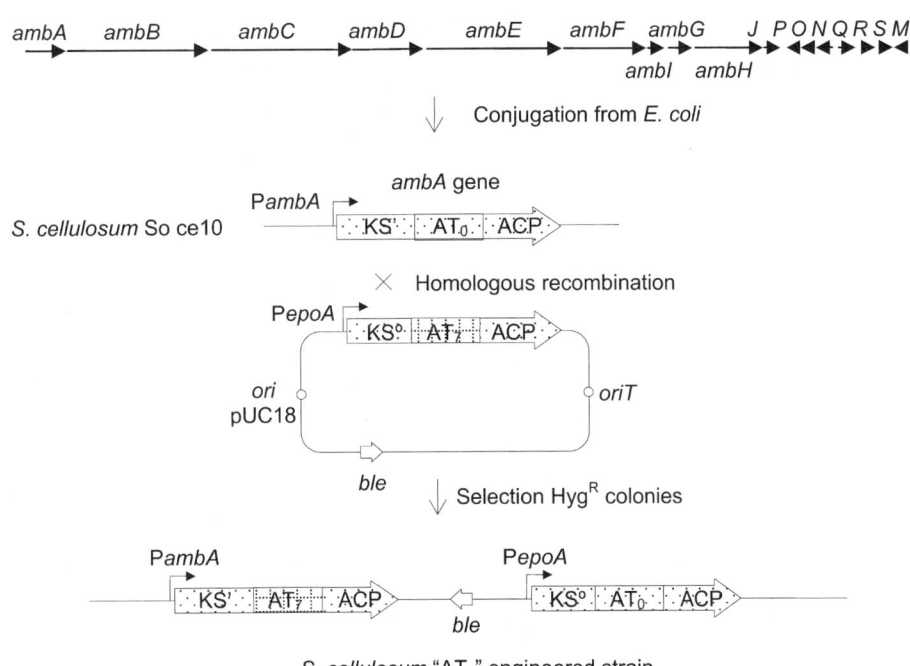

FIGURE 6 PKS engineering in *S. cellulosum* using a single crossover event. The first PKS gene, *ambA*, encodes the loading module consisting of KS′, a methylmalonyl-CoA specific AT (AT$_0$), and an ACP. A plasmid is constructed carrying an in-frame deletion at the 5′ end of the KS (KS^0), an AT domain replaced with one specific for malonyl-CoA (AT$_7$), and an ACP under the expression of a heterologous promoter, P*epoA*. Isolation of a transformant in which crossover has occurred at the KS′ sequence allows the expression of the engineered loading module gene carrying the malonyl-CoA-specific AT$_7$ domain under the P*ambA* promoter, while the expression of remaining downstream PKS genes is under the heterologous promoter *epoA*. The expression of the 5′ truncated loading module gene carrying the AT$_0$ domain is maintained to avoid a polar effect over the downstream PKS genes.

29.6.3. Protocol for Production of Ambruticin from *S. cellulosum* So ce10

Unlike *M. xanthus*, *S. cellulosum* uses a variety of sugars as carbon sources. As its name implies, it is able to metabolize cellulose and starches to their monomeric subunits. The following fermentation protocol has been developed for high-level production in shake flasks and can be scaled into fermentors.

1. Prepare seed culture by inoculating 25 ml of CF9 medium into a 250-ml shake flask.
2. In a 250-ml flask, add 1 g of XAD-1180 (for ambruticin S) or 1 g of XAD-16HP (for VS series of ambruticins) and enough water to immerse the resin, and autoclave to sterilize.
3. Remove water from sterile flask and add 50 ml of Medium F (batch fermentations) or SF1-P medium (fed-batch fermentations).
4. Once the seed culture reaches saturation, use a 10% (vol/vol) inoculum to inoculate the production flask. Incubate the culture at 32°C at 190 rpm. For batch fermentation, grow the cultures for 7 to 8 days. For fed-batch fermentations, initiate feeding on day 3 with the addition of 1 g/liter fructose and continue until harvest on day 7.
5. To extract product from resin, allow the culture to sit for several minutes for the resin to settle to the bottom

of the flask, then pour off the culture broth. Wash the resin twice with several volumes of water, and remove the final few milliliters of water by aspiration.
6. Add 25 ml of methanol to the resin, and incubate at room temperature for about 2 h, preferably with shaking. Analyze the methanol extract by HPLC or LC/MS for desired products.

For large-scale production in fermentors, Medium F is used and the culture is fed with maltodextrin DE18 (1.5 g/liter/day) starting on day 7. The pH is maintained at 7.1 and the dO$_2$ is maintained at 40%. The XAD resin is recovered after 20 days of fermentation.

29.6.3.1. Media

CF9 medium: (per liter) 9 g of Casitone, 6 g of fructose, 0.5 g of CaCl$_2$·2H$_2$O, 0.5 g of MgSO$_4$·7H$_2$O, 0.008 g of ferric citrate, 25 ml of 1.0 M HEPES (pH 7.6). Ferric citrate and HEPES should be filter sterilized and added after autoclaving.

SF1-P medium: (per liter) 3 g of soy peptone, 6 g of fructose, 1 g of CaCl$_2$·2H$_2$O, 1 g of MgSO$_4$·7H$_2$O, 0.008 g of ferric citrate, 50 ml of 1.0 M HEPES (pH 7.6). Ferric citrate and HEPES should be filter sterilized and added after autoclaving.

Medium F: (per liter) 10 g of maltodextrin DE18, 4 g of soy grits, 4 g of skim milk, 5 g of Pharmamedia, 1 g of $CaCl_2 \cdot 2H_2O$, 1 g of $MgSO_4 \cdot 7H_2O$, 4.3 ml of glycerol, 0.008 g of ferric citrate, 50 ml of 1.0 M HEPES (pH 7.6). Ferric citrate and HEPES should be filter sterilized and added after autoclaving.

29.7. SUMMARY

Myxobacteria have been proven to be an alternative source of natural products, with more than 100 novel structures and 500 structural variants. One natural product from *S. cellulosum*, Ixempra®, is currently marketed for treatment of metastatic or locally advanced breast cancer. With 23 complete biosynthetic clusters sequenced, combinatorial biochemistry is possible to generate a variety of new molecules. Using the techniques described in this chapter with the epothilone and ambruticin biosynthetic genes, many new analogs have been made. This technology paves the way for developing more natural products from myxobacteria.

REFERENCES

1. **Arslamian, R. L., L. Tang, S. Blough, W. Ma, R.-G. Qiu, L. Katz, and J. R. Carney.** 2002. A new cytotoxic epothilone from modified polyketide synthases heterologously expressed in *Myxococcus xanthus*. *J. Nat. Prod.* **65**:1061–1064.

2. **Balog, A., D. Meng, T. Kamenecka, P. Bertinato, D-S. Su, E. J. Sorensen, and S. J. Danishefsky.** 1996. Total synthesis of (−)-epothilone A. *Angew. Chem. Int. Ed. Engl.* **35**:2801–2803.

3. **Bode, H. B., and R. J. Müller.** 2006. Analysis of myxobacterial secondary metabolism goes molecular. *J. Ind. Microbiol. Biotechnol.* **33**:577–588.

4. **Bollag, D. M., P. A. McQueney, J. Zhu, O. Hensens, L. Koupal, J. Liesch, M. Goetz, E. Lazarides, and C. M. Woods.** 1995. Epothilones, a new class of microtubule-stabilizing agents with a taxol-like mechanism of action. *Cancer Res.* **55**:2325–2333.

5. **Carvalho, R., R. Reid, N. Viswanathan, H. Gramajo, and B. Julien.** 2005. The biosynthetic genes for disorazoles, potent cytotoxic compounds that disrupt microtubule formation. *Gene* **359**:91–98.

6. **Chian, L. Y., D. E. Ejzykowicz, Z.-Q. Tian, L. Katz, and S. G. Filler.** 2006. Efficacy of ambruticin analogs in a murine model of invasive pulmonary aspergillosis. *Antimicrob. Agents Chemother.* **50**:3464–3466.

7. **Donadio, S., J. B. McAlpine, P. J. Sheldon, M. Jackson, and L. Katz.** 1993. An erythromycin analog produced by reprogramming of polyketide synthesis. *Proc. Natl. Acad. Sci. USA* **90**:7119–7123.

8. **Frykman, S., H. Tsuruta, J. Lau, R. Regentin, S. Ou, C. Reeves, J. Carney, D. Santi, and P. Licari.** 2002. Modulation of epothilone analog production through media design. *J. Ind. Miocrobiol. Biotechnol.* **28**:17–20.

9. **Fu, J., S. C. Wenzel, O. Perlova, J. Wang, F. Gross, Z. Tang, Y. Yin, A. F. Stewart, R. Müller, and Y. Zhang.** 2008. Efficient transfer of two large secondary metabolite pathway gene clusters into heterologous hosts by transposition. *Nucleic Acids Res.* **36**:e113.

10. **Gerth, K., H. Steinmetz, G. Höfle, and H. Reichenbach.** 2001. Studies on the biosynthesis of epothilones: the biosynthetic origin of the carbon skeleton. *J. Antibiot.* **53**:1373–1377.

11. **Gerth, K., N. Bedorf, G. Höfle, H. Irschik, and H. Reichenbach.** 1996. Epothilons A and B: antifungal and cytotoxic compounds from *Sorangium cellulosum* (Myxobacteria). *J. Antibiot.* **49**:560–563.

12. **Gerth, K., S. Pradella, O. Perlova, S. Beyer, and R. Müller.** 2003. Myxobacteria: proficient producers of novel natural products with various biological activities—past and future biotechnological aspects with the focus on the genus *Sorangium*. *J. Biotechnol.* **106**:233–253.

13. **Goldman, B. S., W. C. Nierman, D. Kaiser, S. C. Slater, A. S. Durkin, J. Eisen, C. M. Ronning, W. B. Barbazuk, M. Blanchard, C. Field, C. Halling, G. Hinkle, O. Iartchuk, H. S. Kim, C. Mackenzie, R. Madupu, N. Miller, A. Shvartsbeyn, S. A. Sullivan, M. Vaudin, R. Wiegand, and H. B. Kaplan.** 2006. Evolution of sensory complexity recorded in a myxobacterial genome. *Proc. Natl. Acad. Sci. USA* **103**:15200–15205.

14. **Gong, G.-L., X. Sun, X.-L. Liu, W. Hu, W.-R. Cao, W.-F. Liu, and Y.-Z. Li.** 2007. Mutation and a high-throughput screening method for improving the production of epothilones of *Sorangium*. *J. Ind. Microbiol. Biotechnol.* **34**:615–623.

15. **Han, S. Jong, S. W. Park, B. W. Kim, and S. J. Sim.** 2008. Selective production of epothilone B by heterologous expression of propionyl-CoA synthetase in *Sorangium cellulosum*. *J. Microbiol. Biotechnol.* **18**:135–137.

16. **Hardt, I. H., H. Steinmetz, K. Gerth, F. Sasse, H. Reichenbach, and G. Höfle.** 2001. New natural epothilones from *Sorangium cellulosum*, strains So ce90/B2 and So ce90/D13: isolation, structure elucidation, and SAR studies. *J. Nat. Prod.* **64**:847–856.

17. **Hopwood, D. A., and C. Khosla.** 1992. Genes for polyketide secondary metabolic pathways in microorganisms and plants. *Ciba Found. Symp.* **171**:88–106; discussion 106–112.

18. **Irschik, H., D. Schummer, K. Gerth, G. Höfle, and H. Reichenbach.** 1995. Tartrolons, new boron-containing antibiotics from a myxobacterium, *Sorangium cellulosum*. *J. Antibiot.* **48**:26–30.

19. **Jansen, R. B. K., H. Reichenbach, and G. Höfle.** 1996. Antibiotics from gliding bacteria, LXX chondramides A-D, new cytostatic and antifungal cyclodepsipeptides from *Chondromyces crocatus* (myxobacteria): isolation and structure elucidation. *Liebigs Ann.* **1996:** 285–290.

20. **Jansen, R. B. K., H. Reichenbach, and G. Höfle.** 2003. Chondrochloren A and B, new beta-amino styrenes from *Chondromyces crocatus* (Myxobacteria). *Eur. J. Org. Chem.* **2003:**2684–2689.

21. Reference deleted.

22. **Julien, B., S. Shah, R. Ziermann, R. Goldman, L. Katz, and C. Khosla.** 2000. Isolation and characterization of the epothilone biosynthetic gene cluster from *Sorangium cellulosum*. *Gene* **249**:153–160.

23. **Julien, B.** 2003. Characterization of integrase gene and attachment site for the *Myxococcus xanthus* bacteriophage Mx9. *J. Bacteriol.* **185**:6325–6330.

24. **Julien, B., and S. Shah.** 2002. Heterologous expression of the epothilone biosynthetic genes in *Myxococcus xanthus*. *Antimicrob. Agents Chemother.* **46**:2772–2778.

25. **Julien, B., and R. Fehd.** 2003. Development of a mariner-based transposon for use in *Sorangium cellulosum*. *Appl. Environ. Microbiol.* **69**:6299–6301.

26. **Julien, B., Z.-Q. Tian, R. Reid, and C. D. Reeves.** 2006. Analysis of the ambruticin and jerangolid gene clusters of *Sorangium cellulosum* reveals unusual mechanisms of polyketide biosynthesis. *Chem. Biol.* **13**:1277–1286.

27. **Kopp, M., H. Irschik, F. Gross, O. Perlova, A. Sandmann, K. Gerth, and R. Müller.** 2004. Critical variations of conjugational DNA transfer into secondary metabolite multiproducing *Sorangium cellulosum* strains So ce12 and So ce56: development of a mariner-based transposon mutagenesis system. *J. Biotechnol.* **107**:29–40.

28. **Kowalski, R. J., P. Giannakakou, and E. Hamel.** 1997. Activities of the microtubule-stabilizing agents epothilones A and B with purified tubulin and in cells resistant to paclitaxel (Taxol(R)). *J. Chem. Biol.* **272:**2534–2541.

29. **Lau, J., S. Frykman, R. Regentin, S. Ou, H. Tsuruta, and P. Licari.** 2002. Optimizing the heterologous production of epothilone D in *Myxococcus xanthus. Biotechnol. Bioeng.* **78:**280–288.

30. **Menzella, H. G., and C. D. Reeves.** 2007. Combinatorial biosynthesis for drug development. *Curr. Opin. Microbiol.* **10:**238–245.

31. **Mutka, S. C., J. R. Carney, Y. Liu, and J. Kennedy.** 2006. Heterologous production of epothilone C and D in *Escherichia coli. Biochemistry* **45:**1321–1330.

32. **Newman, D. J., and G. M. Cragg.** 2007. Natural products as sources of new drugs over the last 25 years. *J. Nat. Prod.* **70:**461–477.

33. **Nicolaou, K. C., F. Sarabia, S. Ninkovic, and Z. Yang.** 1997. Total synthesis of epothilone A: the macro-lactonization approach. *Angew. Chem. Int. Edn. Engl.* **36:**525–527.

34. **Park, S. R., J. W. Park, W. S. Jung, A. R. Han, Y.-H. Ban, E. J. Kim, J. K. Shong, S. J. Sim, and Y. J. Yoon.** 2008. Heterologous production of epothilone B and D in *Streptomyces venezuelae. Appl. Microbiol. Biotechnol.* **81:**109–117.

35. **Reichenbach, H.** 2001. Myxobacteria, producers of novel bioactive substances. *J. Ind. Microbiol. Biotechnol.* **27:**149–156.

36. **Reichenbach, H., and M. Dworkin.** 1992. The myxobacteria, p. 3416–3487. *In* A. Balows, H. G. Trüper, M. Dworkin, W. Harder, and K. H. Schleifer (ed.), *The Prokaryotes,* 2nd ed., vol. IV. Springer Verlag, New York, NY.

37. **Reid, R., M. Piagentini, E. Rodriguez, G. Ashley, N. Viswanathan, J. Carney, D. V. Santi, C. R. Hutchinson, and R. McDaniel.** 2003. A model of structure and catalysis for ketoreductase domains in modular polyketide synthases. *Biochemistry* **42:**72–79.

38. **Ringel, S. M.** 1978. In vitro and in vivo stidies of ambruticin (W7783): new class of antifungal antibiotics. *Antimicrob. Agents Chemother.* **13:**762–769.

39. **Ringel, S. M., R. C. Greenough, S. Roemer, D. Connor, A. L. Gutt, B. Blair, G. Kanter, and M. von Strandtmann.** 1977. Ambruticin (W7783), a new antifungal antibiotic. *J. Antibiot.* **30:**371–375.

40. **Robertson, H. M., and D. J. Lampe.** 1995. Recent horizontal transfer of a *mariner* transposable element among and between Diptera and Neuroptera. *Mol. Biol. Evol.* **12:**850–862.

41. **Rodriguez, E., and R. McDaniel.** 2001. Combinatorial biosynthesis of antimicrobials and other natural products. *Curr. Opin. Microbiol.* **4:**526-534.

42. **Sandmann, A., B. Frank, and R. Müller.** 2008. A transposon-based strategy to scale up myxothiazol production in myxobacterial cell factories. *J. Biotechnol.* **135:**255–261.

43. **Santi, D. V., M. A. Siani, B. Julien, D. Kupfer, and B. Roe.** 2000. An approach for obtaining perfect hybridization probes for unknown polyketide synthase genes: a search for the epothilone gene cluster. *Gene* **247:**97–102.

44. **Schneiker, S., O. Perlova, O. Kaiser, K. Gerth, A. Alici, M. O. Altmeyer, D. Bartels, T. Bekel, S. Beyer, E. Bode, H. B. Bode, C. J. Bolten, J. V. Choudhuri, S. Doss, Y. A. Elnakady, B. Frank, L. Gaigalat, A. Goesmann, C. Groeger, F. Gross, L. Jelsbak, L. Jelsbak, J. Kalinowski, C. Kegler, T. Knauber, S. Konietzny, M. Kopp, L. Krause, D. Krug, B. Linke, T. Mahmud, R. Martinez-Arias, A. C. McHardy, M. Merai, F. Meyer, S. Mormann, J. Munoz-Dorado, J. Perez, S. Pradella, S. Rachid, G. Raddatz, F. Rosenau, C. Ruckert, F. Sasse, M. Scharfe, S. C. Schuster, G. Suen, A. Treuner-Lange, G. J. Velicer, F. J. Vorholter, J. K. Weissman, R. D. Welch, S. C. Wenzel, D. E. Whitworth, S. Wilhelm, C. Wittmann, H. Blöcker, A. Pühler, and R. Müller.** 2007. Complete genome sequence of the myxobacterium *Sorangium cellulosum. Nat. Biotechnol.* **25:**1281–1289.

45. **Shubitz, L. F., J. N. Galgiani, Z. Q. Tian, Z. Zhong, P. Timmermans, and L. Katz.** 2006. Efficacy of ambruticin analogs in a murine model of coccidioidomycosis. *Antimicrob. Agents Chemother.* **50:**3467–3469.

46. **Su, D.-S., D. Meng, P. Bertinato, A. Balog, E. J. Sorensen, S. J. Danishefsky, Y.-H. Zheng, T.-C. Chou, L. He, and S. B. Horwitz.** 1997. Total synthesis of (–)-epothilone B: an extension of the Suzuki coupling method and insights into structure-activity relationships of the epothilones. *Angew. Chem. Int. Ed. Engl.* **36:**757–759.

47. **Tang, L., L. Chung, J. R. Carney, C. M. Starks, P. Licari, and L. Katz.** 2005. Generation of new epothilones by genetic engineering of a polyketide synthase in *Myxococcus xanthus. J. Antibiot.* **58:**178–184.

48. **Tang, L., S. Shah, L. Chung, J. Carney, L. Katz, C. Khosla, and B. Julien.** 2000. Cloning and heterologous expression of the epothilone gene cluster. *Science* **287:**640–642.

49. **Tang, L., S. Ward, L. Chung, J. R. Carney, Y. Li, R. Reid, and L. Katz.** 2004. Elucidating the mechanism of *cis* double bond formation in epothilone biosynthesis. *J. Am. Chem. Soc.* **126:**46–47.

50. **Ueki, T., S. Inouye, and M. Inouye.** 1996. Positive-negative KG cassettes for construction of multi-gene deletions using a single drug marker. *Gene* **183:**153–157.

51. **Weissman, K. J., and R. Müller.** 2009. A brief tour of myxobacterial secondary metabolism. *Bioorg. Med. Chem.* **17:**2121–2136.

52. **Weissman, K. J., and P. F. Leadlay.** 2005. Combinatorial biosynthesis of reduced polyketides. *Nat. Rev. Microbiol.* **3:**925–936.

53. **Wenzel, S. C., F. Gross, Y. Zhang, J. Fu, A. F. Stewart, and R. Müller.** 2005. Heterologous expression of a myxobacterial natural products assembly line in pseudomonads via red/ET recombineering. *Chem. Biol.* **12:**349–356.

54. **Wenzel, S., and R. Müller.** 2005. Recent developments towards the heterologous expression of complex bacterial natural product biosynthetic pathways. *Curr. Opin. Biotechnol.* **16:**594–606.

55. **Yang, Z., Y. He, D. Vourloumis, H. Vallberg, and K. C. Nicolaou.** 1997. Total synthesis of epothilone A: the olefin metathesis approach. *Angew. Chem. Int. Edn. Engl.* **36:**166–168.

56. **Yarmolinsky, M. B., H. Wiesmeyer, H. M. Kalckar, and E. Jordan.** 1959. Hereditary defects in galactose metabolism in *Escherichia coli* mutants. II. Galactose-induced sensitivity. *Proc. Natl. Acad. Sci. USA* **45:**1786–1791.

57. **Youderian, P., N. Burke, D. J. White, and P. L. Hartzell.** 2003. Identification of genes required for adventurous gliding motility in *Myxococcus xanthus* with the transposable element mariner. *Mol. Microbiol.* **49:**555–570.

INDUSTRIAL ENZYMES, BIOCATALYSIS, AND ENZYME EVOLUTION

BIOCATALYSIS MAY BE BROADLY DEFINED AS THE USE OF ENZYMES OR WHOLE cells as biocatalysts for industrial synthetic chemistry. Compared to chemical catalysts, biocatalysts generally offer several advantages such as high selectivity, high catalytic efficiency, and mild reaction conditions. Thanks to its association with reduced energy consumption, greenhouse gas emissions, and waste generation, biocatalysis is increasingly used in pharmaceutical, chemical, food, and agricultural industries. For example, glucose isomerase is used to produce high-fructose corn syrup at an industrial scale. Subtilisins and lipases are widely used as additives in laundry detergents. More recently, biocatalysis has become an indispensable and powerful tool for production of biofuels and chemicals from renewable plant materials.

There are a wide variety of biocatalysts capable of catalyzing numerous biological reactions. However, naturally occurring biocatalysts are often not optimal for many specific industrial applications, due to characteristics such as low stability and low activity. Moreover, naturally occurring biocatalysts may not catalyze the reaction with the desired non-natural substrates or produce the desired products. To address these limitations, many molecular techniques have been developed to create improved or novel biocatalysts with altered industrial operating parameters.

In this section, chapter 30 by Yasuhisa Asano describes a number of molecular tools for enzyme discovery. Louis A. Clark emphasizes the synergy between computational protein design tools and directed evolution approaches in chapter 31, whereas Manfred T. Reetz, in chapter 32, provides insightful discussions on various directed evolution methods used in enzyme engineering. In chapter 33, Nikhil U. Nair et al. provide a comprehensive overview of the applications of enzyme catalysts in various industries, including chemical, pharmaceutical, food, textile, and pulp and paper industries, and chapter 34 by Feng Xu provides a comprehensive survey of the enzymes involved in biomass conversion and their applications in biofuels production. Chapter 35, by Elton P. Hudson et al., discusses the use of enzymes for chemical transformations in nonaqueous organic solvents. In chapter 36, Bert Van Loo and Florian Hollfelder discuss enzyme promiscuity and its role in creation of new protein functions, which is not only scientifically interesting but also practically important since novel biocatalysts are highly desirable in biocatalysis. Chapter 37, by Daniel J. Sayut et al., discusses various aspects involved in the high-level production of enzymes in the bacterium *Escherichia coli*, such as gene regulation, vector design, host selection, inducible gene expression, and high-cell-density fermentation. Finally, Lutz Hilterhaus and Andreas Liese deal with various aspects of bioprocess development, such as bioreactor design, process modeling, optimization, scale-up, and product separation and recovery, in chapter 38.

It is clear that the integration of all the above-mentioned tools and approaches will make biocatalysis even more powerful and more broadly applicable.

Tools for Enzyme Discovery

YASUHISA ASANO

30

30.1. INTRODUCTION

Success in microbial transformation has been based on screening for microbial enzymes catalyzing new reactions, or screening known enzymes for an unknown activity with synthetic substrates. Since the advantages of the use of enzymes in organic synthesis have been fully illustrated (2, 23, 44), more and more attention has been paid to the systematic exploitation of microorganisms producing new enzymes, and how to obtain enzymes with desired activities from various enzyme sources and databases (2, 18, 44, 45).

In screening enzymes, it is extremely important to clarify the purpose of the experiment. One would screen microorganisms or databases for an enzyme when (i) nothing is known about the specific reaction, but a homologous enzymatic reaction of the same category is known; (ii) the same enzyme is known only in another biological source; (iii) not enough enzyme is produced by a certain organism for a practical transformation; (iv) improved function, such as better productivity, stability, etc., for practical use is required; and even when (v) almost no information is available for the desired reaction, etc.

One might just like to have an enzyme to see how the reaction proceeds, and further to have a new enzyme. In such a case, one can buy an enzyme from the suppliers or clone the known gene by PCR according to the information given in the database on the Internet (e.g., DBGET Search, http://www.genome.jp/dbget/dbget.html), and express the gene for the enzyme in a versatile heterologous host such as *Escherichia coli*. Knowing the reaction, one might then start further screening. In some cases, a starting material is a new drug intermediate to be converted by known enzymes. In such a case, it would be convenient to have libraries of wide varieties of known enzymes from different sources catalyzing the expected reactions so as to identify the chemical reaction more quickly and selectively, instead of starting random screening from microbial stock cultures (31). If one would like the challenge of establishing an entirely new enzymatic industrial process, no similar case should exist in the world; the enzyme reaction should be unique, without a previous publication or a patent. Interestingly, basic studies such as this result not only in finding new enzymes, but also in discovering unknown biological phenomena and new materials hidden in nature. Thus, a successful screening should focus on what is new in the screening: substrate, product, gene, protein, property, screening source, method, etc.

In this chapter, recent remarkable examples in the screening for new enzymes, biocatalysis, and enzyme evolution are introduced and their merits are discussed. Figure 1 shows a flow chart of the use of tools for discovery and the development of industrial enzymes. The different steps are identified by letters, which are cited below in the text to indicate the various tools being utilized in the process under discussion. Other possibilities of creating proteins by RNA display technique, etc., will be covered in other chapters of this volume.

30.2. SCREENING FOR ENZYME ACTIVITIES

This section will discuss screening for enzymes by enrichment culture technique from stock cultures, followed by enzyme purification and characterization, coupled with gene cloning. Examples will include the discovery of nitrile hydratase and industrial production amides, aldoxime dehydratase and nitrile synthesis, ω-laurolactam hydrolase for enzymatic transcrystallization, and hydroxynitrile lyase from plants. These traditional approaches often yield primary information that can be industrialized successfully.

30.2.1. Nitrile Hydratase in Acrylamide Production (Fig. 1: a, d, i, r, q, s)

Enrichment culture is a technique to isolate microorganisms having special growth characteristics. (Please note: letters in section heads as above correspond to the labels in Fig. 1, indicating the tools used in each section.) Some microorganisms grow faster than others in media with limited nutrients, high temperature, extreme pH, etc. Microorganisms that grow faster than other species become dominant after several transfers of the culture. An acclimation technique is applied when a toxic or unnatural compound is used as a substrate; such a technique is usually run long term to isolate microorganisms that are not easily isolated by enrichment culture. An adaptation to a synthetic medium containing a target compound often results in the isolation of microorganisms having a new enzyme. A typical example of the acclimation procedure is the successful isolation of an acrylonitrile-utilizing bacterium, *Rhodococcus rhodochrous* (formerly *Arthrobacter* sp.) I-9 (K-22, AKU 629) (61). Activated sludge obtained from sewage disposal facilities was suspended in an acclimation

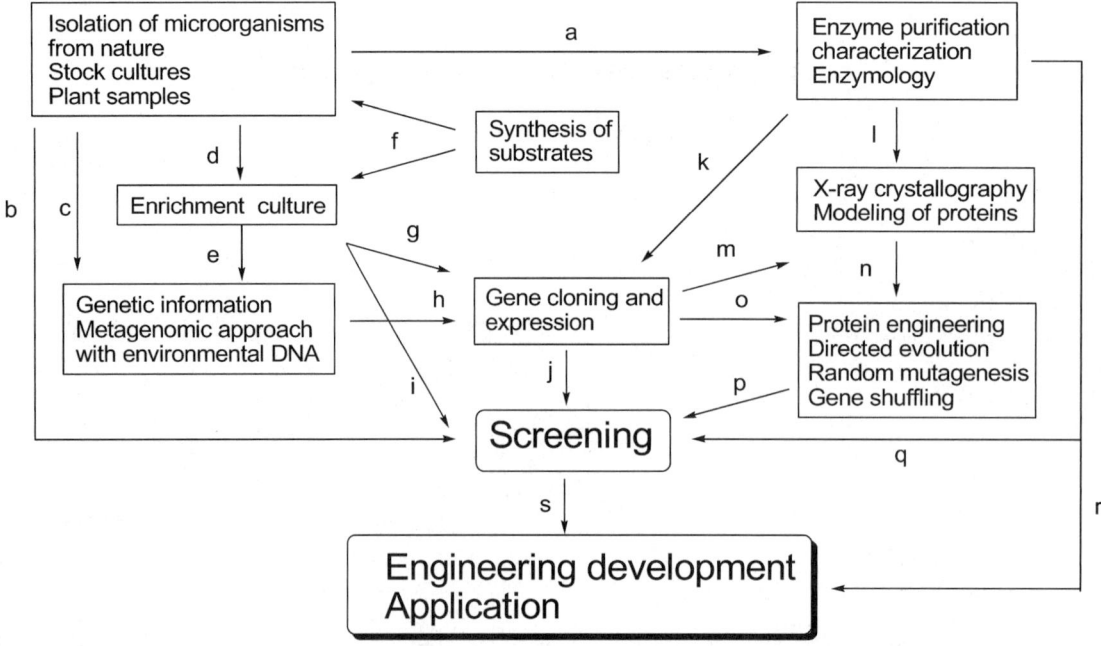

FIGURE 1 Discovery of industrial enzymes, biocatalysis, and enzyme evolution. These various steps in discovery and development are cited by their respective letters throughout this chapter.

medium containing 0.0008% acrylonitrile. After 3 months, *R. rhodochrous* I-9 was isolated for the first time as a degrader of toxic acrylonitrile as carbon and nitrogen source. By the enrichment culture technique, *R. rhodochrous* (formerly *Arthrobacter* sp.) J-1 (12) and *Pseudomonas chlororaphis* B23 (16), etc., were isolated as acetonitrile and isobutyronitrile utilizers, respectively. From *R. rhodochrous* J-1, nitrile hydratase was discovered, purified, named, and characterized (4, 12). Two pathways of nitrile hydrolysis were also discovered to coexist in *R. rhodochrous* J-1; one combines nitrile hydratase and amidase and the other

involves nitrilase (16) (Fig. 2). *P. chlororaphis* B23 was isolated as an industrial acrylamide producer by differential screening from non-acrylonitrile-degrading microorganisms, since acrylonitrile is highly toxic, and the degradation of acrylonitrile by *R. rhodochrous* I-9 was unexpectedly found to be catalyzed by a nitrilase. Nitrile hydratase has become one of the most important industrial biocatalysts, used for the microbial production of acrylamide, nicotinamide, 5-cyanovaleramide, etc., on an industrial scale. Acrylamide synthesis is now the largest enzyme-catalyzed organic synthesis in the world (2, 16, 62).

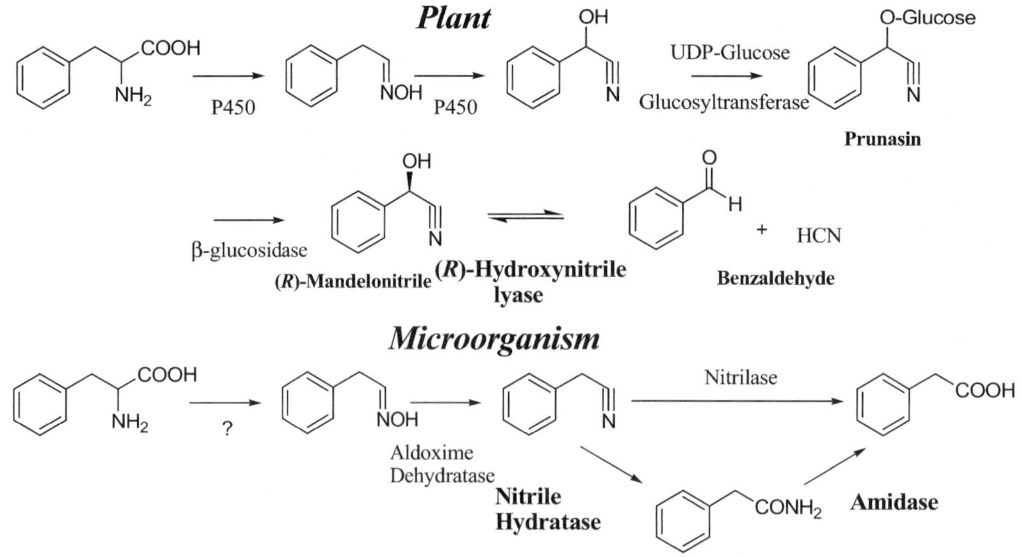

FIGURE 2 Plant and microbial aldoxime-nitrile pathway.

30.2.2. Aldoxime Dehydratase in the First Enzymatic Synthesis of Nitriles (Fig. 1: a, d, g, i, l, s)

There has been no report concerning the microbial biosynthesis of nitriles and the physiological function of such enzymes. It is also not known whether aliphatic and aromatic nitriles are biological compounds or just petrochemicals. Asano and Kato and their colleagues discovered the coexistence of aldoxime dehydratase and nitrile-degrading enzymes and elucidated their role in aldoxime metabolism in microorganisms. Nitrile hydratase has been found to be linked with aldoxime dehydratase both genetically and physiologically.

Asano, Kato, and their coworkers have been successful in the enzymatic synthesis of nitriles from aldoximes by using the new microbial enzyme "aldoxime dehydratase." They acclimated soil samples for about 4 months to isolate *Bacillus* sp. strain OxB-1, a degrader of *Z*-phenylacetaldoxime (6). The strain possessed an inducible enzyme catalyzing the formation of phenylacetonitrile from *Z*-phenylacetaldoxime. Microorganisms converting *E*-pyridine-3-aldoxime to the corresponding nitrile were screened, and *Rhodococcus* sp. strain YH3-3 was also isolated from soil. The cells catalyzed a dehydration reaction of various aryl- and alkyl-aldoximes to form the corresponding nitriles, but did not act on arylalkyl- and substituted alkyl-aldoximes (7). Various nitriles, such as 3-cyanopyridine, 2-cyanopyridine, and cyanopyrazine, were synthesized at preparative scales from 10 to 100 mM aldoximes under optimized reaction conditions (pH 7.0, 30°C). This is the first report on the enzymatic synthesis of nitriles (32, 59).

The *oxd* gene coding for the enzyme was cloned, and an open reading frame coding for 351 amino acid residues was identified (33). A nitrilase, which participates in aldoxime metabolism in the organism, was found to be encoded by the region just upstream from the *oxd* gene. These findings provide genetic evidence for a novel gene cluster that is responsible for aldoxime metabolism in this microorganism. It was found that the occurrence of aldoxime dehydratase is as wide as that for nitrile-degrading enzymes such as nitrile hydratase, amidase, and nitrilase. Most of the nitrile degraders hitherto isolated contain aldoxime dehydratase activities (34). In addition, Asano and Kato and their colleagues proposed that the pathway in which aldoximes are successively degraded via nitrile be termed the "aldoxime-nitrile pathway" (2) (Fig. 2). Thus, such exploratory studies yield richer basic knowledge on enzymology and microbial physiology; they are not just tools for enzyme discovery.

30.2.3. A New Enzymatic Process for the Synthesis of 12-Aminolauric Acid by Enzymatic Transcrystallization of ω-Laurolactam Using ω-Laurolactam Hydrolase (Fig. 1: a, d, f, g, i, k, s)

Asano et al. proposed a new enzymatic method for the production of 12-aminolauric acid by hydrolyzing ω-laurolactam (Fig. 3), a chemical widely used in the synthesis of nylon, adhesives, and hardening agents (5, 25). Although there are some harsh chemical processes for the preparation of 12-aminolauric acid in industry, no report has appeared on mild, e.g., enzymatic production of this chemical. Hydrolysis of the intramolecular amide bond of ω-laurolactam has been possible with 6 N HCl at high temperature.

Recently, Asano et al. (5) isolated from soil samples several ω-laurolactam-degrading microorganisms that were capable of growing in a medium containing ω-laurolactam as a sole source of carbon and nitrogen. Five bacterial strains were selected according to their unique activities: *Cupriavidus* sp. strain T7, *Acidovorax* sp. strain T31, *Cupriavidus* sp. strain U124, *Rhodococcus* sp. strain U224, and *Sphingomonas* sp. strain U238. The primary structure of the enzymes from these five isolates was revealed and shown to be significantly similar to that of 6-aminohexanoate-cyclic-dimer hydrolase from *Arthrobacter* sp. strain KI72 (formerly *Flavobacterium* sp. strain KI72), known as a nylon oligomer-degrading microorganism.

The enzyme from *Acidovorax* sp. strain T31 was most successfully expressed in *E. coli*. Cell-free extract of the recombinant strain was used for the synthesis of 12-aminolauric acid from ω-laurolactam by a process termed "enzymatic transcrystallization," because crystalline ω-laurolactam added into the enzyme solution was converted to crystalline 12-aminolauric acid (>97.3% yield). Under optimum conditions, 208 g/liter of 12-aminolauric acid was produced in 17 h. The synthesized product could be purified by filtration and washed by water; thus, there is no other purification step such as column chromatography (25).

30.2.4. Plant Hydroxynitrile Lyase for Synthesis of Cyanohydrins (Fig. 1: a, b, r, s)

More than 2,000 plants are considered to be cyanogenic, and more than 60 cyanoglycosides made from aliphatic and aromatic cyanohydrins are known. The cyanohydrin aglycones of the cyanoglycosides are considered to be synthesized from hydrophobic amino acids such as L-tyrosine, L-phenylalanine, L-valine, L-isoleucine, etc.

Asano et al. (11) established a simple high-performance liquid chromatography method to determine the activity and stereochemistry of the chiral mandelonitrile synthesized from benzaldehyde and cyanide, and applied it to screen for hydroxynitrile lyase (HNL) activity of plant origin. In a similar manner, in screening for microorganisms with a desirable enzyme activity, they screened plants for new HNLs. A total of 163 species of plants among 74 families were examined for (*R*)- and (*S*)-HNL activities. Asano et al. discovered that the homogenate of leaves of *Baliospermum montanum* shows (*S*)-HNL activity, while leaves and seeds from *Passiflora edulis*, *Eriobotrya japonica*, *Chaenomles sinensis*,

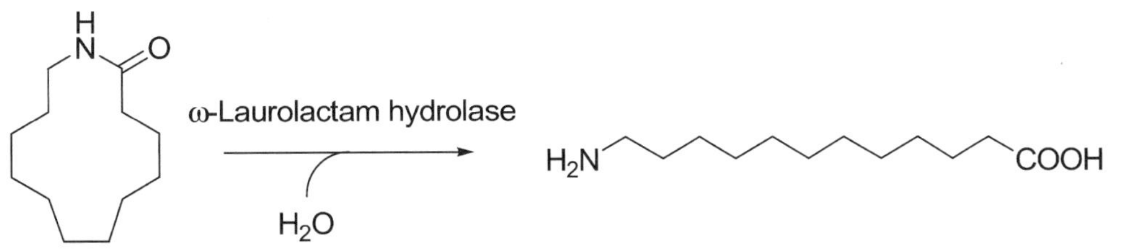

FIGURE 3   Synthesis of 12-aminolauric acid from ω-laurolactam.

Sorbus aucuparia, Prunus mume, and *Prunus persica* show (*R*)-HNL activity. Thus, 30% of HNLs were discovered within a few years in Japan, while others were found over 100 years of history of studies on HNL. Some of the newly found HNLs are used for chiral synthesis of cyanohydrins (43) (Fig. 3).

30.3. APPROACHES WITH MOLECULAR TOOLS

In section 30.2, we explained that the isolation of microorganisms and direct screening for an enzyme in natural organisms are very important steps to initiate the study of the enzymatic transformation of chemicals and to exploit new enzymes, because of the wide biological diversity that exists. In this section we discuss molecular and "omics" approaches, such as databases, whole-genome sequencing, metagenomic approaches, and library construction of genes for industrially important enzymes. There is much more diversity in metagenomes found in the environment, since it is estimated that there are 10 times more species of "unculturable" microorganisms than culturable ones. Since most microorganisms in the environment are difficult to cultivate, scientists have been prompted to clone useful genes directly from environmental metagenomes. An average environmental sample will contain 3,000 to 11,000 genomes per g of soil, with less than 1% of these accessible through cultivation techniques (51). Thus, metagenomes are considered to be a very important

likely source of useful enzymes (19, 21, 51, 53). Chitinases, lipases, esterases, nitrilases, amylases, nucleases, alcohol oxidoreductases, etc., are some examples. Metagenomic approaches include sequence-based screening after shotgun large-scale sequencing of environmental DNA, enzyme activity-based screening of expressed libraries of environmental DNA, and a recent methodology called substrate-induced gene expression screening approaches (SIGEX).

30.3.1. Biodiversity of Nitrilases for Enantioselective Catalysis (Fig. 1: d, e, f, g, h, j, k, s)

Nitrilases are attractive catalysts for forming chiral centers by kinetic resolution reaction. It is reported that about 20 nitrilases have been described. Recently, 137 unique nitrilases, discovered from DNA libraries processed from more than 651 environmental samples collected worldwide from terrestrial and aquatic microenvironments, were characterized (49). The genes were expressed in *E. coli* and tested for the hydrolysis of three nitrile substrates in terms of substrate specificity and stereoselectivity. These enzymes, discovered using a high-throughput, culture-independent method, provide a toolbox for enantioselective hydrolysis of carboxylic acid derivatives from nitriles. The enzymes were tested for their abilities to catalyze stereoselective hydrolysis of three nitrile substrates, 3-hydroxyglutaronitrile, mandelonitrile, and phenylacetaldehyde cyanohydrins (Fig. 4). One of the products of the desymmetrizing hydrolysis on the

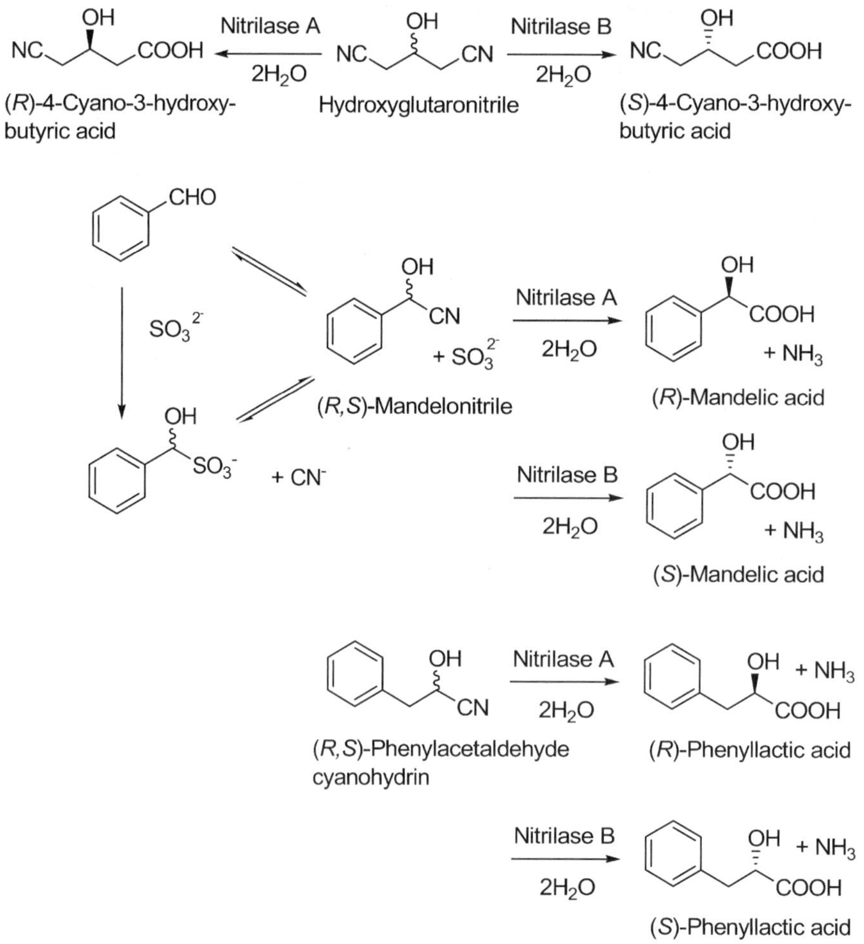

FIGURE 4 Reactions catalyzed by nitrilases.

prochiral substrate 3-hydroxyglutaronitrile, (R)-4-cyano-3-hydroxybutyric acid, is an intermediate in the synthesis of the cholesterol-lowering drug Lipitor. The most selective enzymes generated products with enantiomeric excesses (e.e.s) of 90% for (S)-4-cyano-3-hydroxybutyric acid and 95% for (R)-4-cyano-3-hydroxybutyric acid. On mandelonitrile, the highest e.e. was 99% for (R)-mandelic acid and up to 30% for (S)-mandelic acid. On phenylacetaldehyde cyanohydrins, (S)-specific enzyme (e.e. 77%), and (R)-selective enzyme (e.e. 76%) were found among libraries constructed for genes from all over the world.

It is noteworthy that the desymmetrization of the prochiral substrate 3-hydroxyglutaronitrile was made possible by newly found nitrilases yielding (S)-4-cyano-3-hydroxybutyric acid in 90% e.e., and (R)-4-cyano-3-hydroxybutyric acid in 95% e.e. As for mandelonitrile, it was already known that (S)-specific nitrilase usually acts on mandelonitrile with low e.e. (56). However, the collection of a large and diverse group of sequences and the characterization of their activities provide an opportunity to explore principles of protein structure and evolution and their relationship to catalytic activity. It was surprising that enzymes with opposite stereochemistry have sometimes shared high structural similarities. The desymmetrization of the prochiral substrate 3-hydroxyglutaronitrile has been of interest in enzyme-catalyzed organic synthesis, although there has been limited success in the preparation of (S)-4-cyano-3-hydroxybutyric acid of 22% e.e. by immobilized whole cells of *Rhodococcus* sp. (20). The starting material for Lipitor synthesis, (R)-4-cyano-3-hydroxybutyric acid, has also been found by screening single-site mutagenesis of the known nitrilases, containing 10,528 variants. Ala190 and Phe191 were found to be effective for synthesis of (R)-4-cyano-3-hydroxybutyric acid with increased e.e. (22). It was possible to load up to 3 M of the substrate, yielding the product with 99% e.e. (17).

Thus, this work on high-throughput enzyme discovery from metagenomes has shown very clear benefits both for substantially expanding protein sequence space and for solving problems in industrial catalysis.

30.3.2. Isolation of a Phenol-Induced Enzyme System from Metagenomes by the SIGEX Approach (Fig. 1: c, d, e, g, h, i, s)

Current metagenome approaches use either nucleotide sequence-based or enzyme activity-based screening for isolation of catabolic genes. Nucleotide sequence-based screening by hybridization is highly efficient when PCR is used for direct retrieval of target DNA fragments. However, PCR screening can access only homologs of known genes in the environmental gene pool. Catechol 2,3-dioxygenase genes were isolated from DNA of microbial cultures enriched with phenol and crude oil. In such enzyme activity-based screening, clones expressing a desired enzymatic activity have to be selected from a vast quantity of clones from a metagenome library (48), and it is difficult for some enzymes to be expressed as active forms in cloning hosts such as in *E. coli*.

Considering the limitations associated with current screening methods, Uchiyama and Watanabe introduced a third option for the metagenome approach, the SIGEX method for screening (57). This method takes advantage of the induction of a set of genes by a substrate. In the above-mentioned metagenomic approaches, it has been necessary to establish a method to concentrate useful genes or clones from candidates of genes from a metagenomic library of DNA for the purpose of application, because the conventional microbial enrichment culture technique is very powerful in concentrating desirable microorganisms from a huge number of microorganisms in the environment by their autocatalytic growth ability. The SIGEX method is based on the knowledge that catabolic gene expression is generally induced by substrates and controlled by regulatory elements proximate to catabolic genes. Uchiyama et al. used green fluorescent protein as an indicator of expression detected, and the *E. coli* transformants were screened with a cell sorter. With the SIGEX method, it was possible to clone aromatic hydrocarbon (benzoic acid and naphthalene)-induced genes from a groundwater metagenome library very efficiently. Uchiyama et al. found operons for benzoate and catechol degradation and gene homologs of xenobiotic reductase, tyrosine-phenol lyase, ubiquinone precursor hydroxylase, etc.

30.3.3. Comparison of Conventional and In Silico Screening Approaches: Discoveries of Two New Enzymes for the Production of Functional Dipeptides

30.3.3.1. Conventional Approach (Fig. 1: b, f, s)

Yokozeki and Hara reported efficient enzymatic production of L-Ala-L-Gln—a dipeptide of significant industrial interest by virtue of its widespread use in infusion therapy—from L-Ala methyl ester (acyl donor) and L-Gln (nucleophile) by a new microbial enzyme, such as from *Empedobacter brevis*, screened by a conventional approach from culture collections (65) (Fig. 5a). They eventually succeeded in the industrial production of L-Ala-L-Gln (64). The newly found enzyme purified from *E. brevis* facilitates significantly high production yields of L-Ala-L-Gln in an aqueous solution—more than an 80% yield based on L-Ala methyl ester. The enzyme has wide substrate specificity, both for acyl donors and nucleophiles, and can catalyze peptide-forming reactions, not only to produce various dipeptides, but also to produce various oligopeptides from the corresponding amino acid esters and peptides.

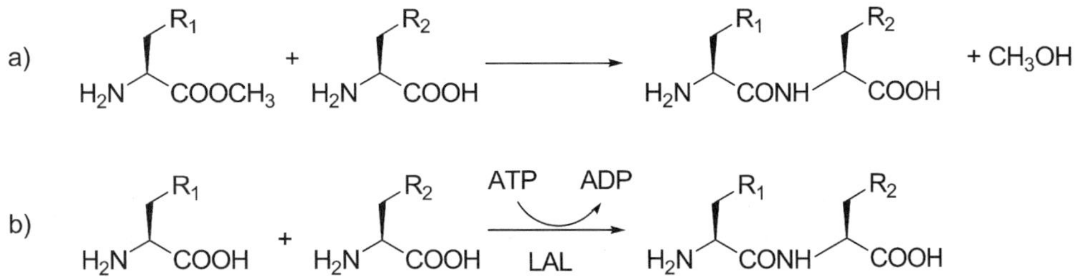

FIGURE 5 Enzyme-catalyzed peptide synthesis. (a) A peptidase from *E. brevis* catalyzing aminolysis reaction. (b) LAL.

30.3.3.2. In Silico Screening Approach (Fig. 1: h, j, s)

On the other hand, dipeptide-forming enzymes in the presence of ATP, as shown in Fig. 5b, may rarely be detected by activity-based screening when activities are weak and products are liable to degradation. One might hesitate to screen for such an enzyme, because no enzyme synthesizing α-dipeptides of L-amino acids by an ATP-dependent manner has been known. Discovery of L-amino acid ligase (LAL) enabled the microbial production of dipeptides such as L-Ala-L-Gln, which is a useful ingredient of infusion.

Tabata and Hashimoto attempted to find such an enzyme by in silico screening based on the consensus sequence of the superfamily, followed by an in vitro assay with purified enzyme to avoid the degradation of the peptide(s) synthesized (55). *ywfE* of *Bacillus subtilis* was found to code for an activity forming L-Ala-L-Gln from L-Ala and L-Gln with hydrolysis of ATP to ADP (Fig. 5). The enzyme accepted a wide variety of L-amino acids. Among 231 combinations of L-amino acids tested, reaction products were obtained for 111 combinations and 44 kinds of α-dipeptides were synthesized; no tripeptide or longer peptide was detected, and the D-amino acids were inert (55, 60).

30.3.4. An Alternative Menaquinone Biosynthetic Pathway Operating in Microorganisms (Fig. 1: h, i, s)

Menaquinone (MK; vitamin K) is an essential vitamin because it is a critical factor of blood coagulation in humans and an obligatory component of the electron transfer pathway in microorganisms, respectively. MK has been shown to be derived from chorismate by seven enzymes, designated MenA, -B, -C, -D, -E, -F, and -G, in *E. coli*. Recently, Dairi et al. ascertained, by bioinformatic analyses

of whole-genome sequences, that some microorganisms, such as *Helicobacter pylori* and *Campylobacter jejuni*, do not have the homologs of most of the *men* genes, although they biosynthesize MK (26). A novel MK biosynthetic pathway was investigated by using *Streptomyces coelicolor* A3(2), which also possesses the novel pathway. The ^{13}C-labeling pattern of MK purified from *S. coelicolor* A3(2) grown on [U-^{13}C]glucose was quite different from that of *E. coli*, suggesting that a novel pathway was operating in the strain. Candidate genes participating in the novel pathway were searched by in silico screening, and the involvement of these genes in the pathway was confirmed by gene-disruption experiments. Mutants that required MK for their growth were also isolated by mutagenesis and used as hosts for shotgun cloning experiments. Metabolites accumulated in culture broth of these mutants were isolated and their structures were determined (52). Taking these results together, the outline of this pathway, which branched at chorismate in a manner similar to that in the known pathway but then followed a completely different pathway, was proposed. Since human cells and some useful intestinal bacteria such as lactobacilli do not have this pathway, it would be an attractive target for the development of novel chemotherapeutics. This biosynthetic pathway is a good source of new enzymes (Fig. 6).

30.4. FINE-TUNING OF ENZYMES BY DIRECTED EVOLUTION

Here we discuss the fine-tuning of enzymes with regard to optimal temperature, substrate specificity, and yield of the reaction. Examples will include the following: mutation of

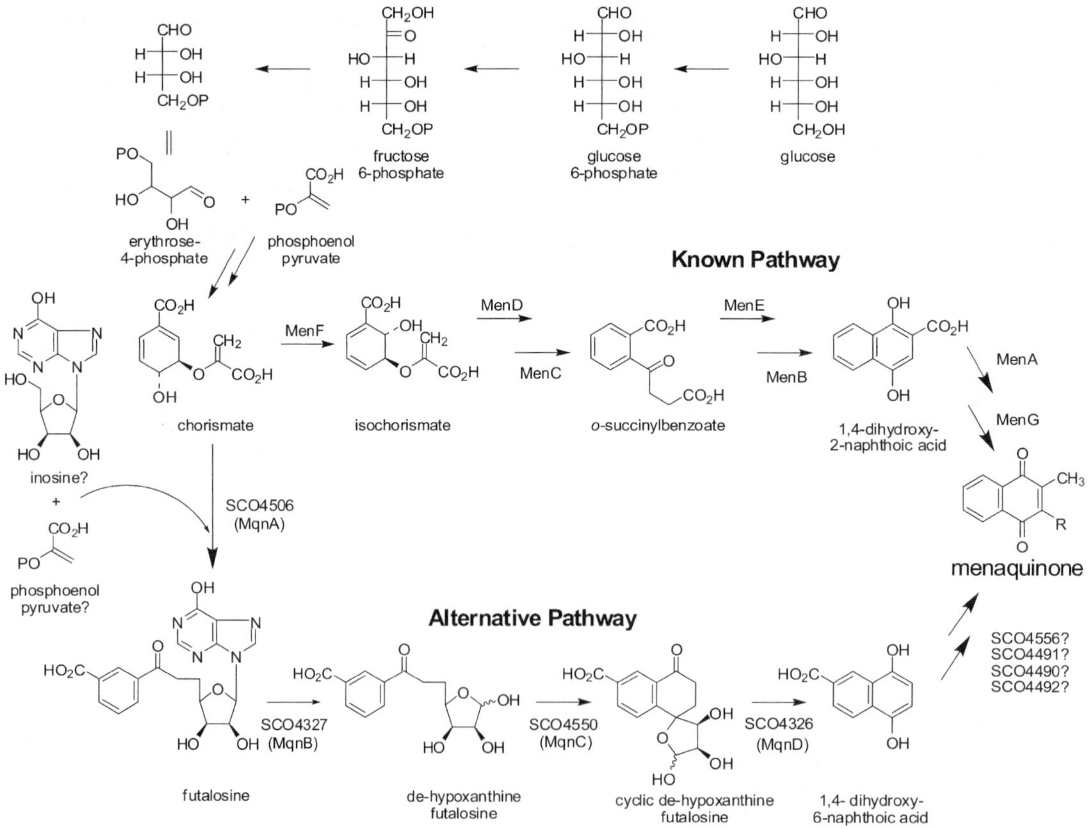

FIGURE 6 Biosynthesis of menaquinone in microorganisms.

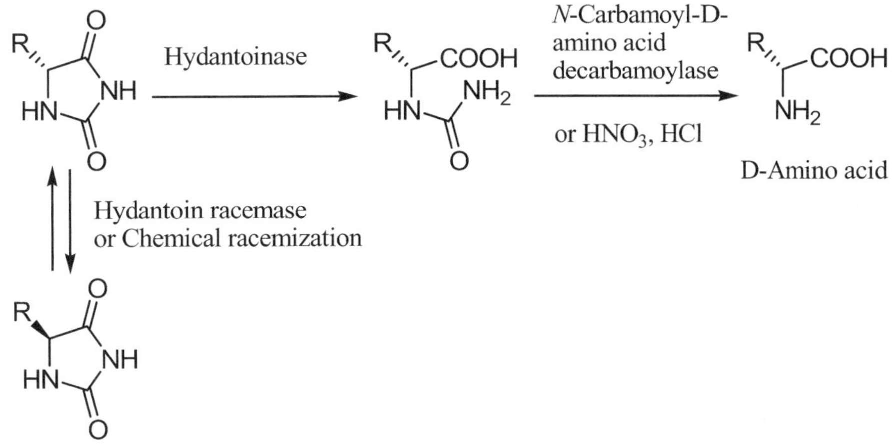

FIGURE 7 Industrial dynamic kinetic resolution of hydantoin by hydantoinase and DCase to produce p-hydroxy-D-phenylglycine.

N-carbamoyl-D-amino acid decarbamoylase (DCase) to a thermostable form for the industrial production of p-hydroxy-D-phenylglycine, and directed evolution of acid phosphatase for the industrial production of 5′-IMP (inosine 5′-monophosphate) and 5′-GMP (guanosine 5′-monophosphate). The use of software for molecular modeling of the enzyme structure, followed by protein engineering, is also described, with creation of D-amino acid dehydrogenase from meso-diaminopimelate dehydrogenase as an example.

30.4.1. DCase in p-Hydroxy-D-Phenylglycine Production (Fig. 1: a, f, k, o, p, s)

D-Amino acids are used for the starting materials for semisynthetic penicillin and bioactive peptides. For example, p-hydroxy-D-phenylglycine is an important D-amino acid produced on a 2,000-tons/year scale by an enzymatic method, as a component of the wide-spectrum β-lactam antibacterial amoxicillin. These studies were begun in 1980 by a collaboration of Kyoto University and Kaneka Corp. (Fig. 7) (44, 45). The reaction involves D-stereospecific hydrolysis of synthetic DL-5-substituted hydantoins to N-carbamoyl-D-amino acid, with spontaneous racemization of hydantoins. Thus, all of the racemic hydantoins will be converted to N-carbamoyl-D-amino acids by the "dynamic kinetic resolution" reaction. N-Carbamoyl-D-amino acid is decarbamoylated to form D-amino acid. The process was improved to a two-step bioreactor reaction by changing the latter chemical decarbamoylation reaction to an enzymatic reaction. By improving DCase to a much more thermostable enzyme, it was possible to change the whole process from a batch to a bioreactor system with immobilized enzymes catalyzing the two-step reactions.

Since the existence of DCase with high stability was not known, soil samples were screened for a wild-type enzyme, and Agrobacterium sp. strain KNK712 was isolated. The enzyme from this organism showed good reactivity toward the substrates, although it was necessary to improve its operational stability. The gene for the enzyme was cloned and subjected to random mutagenesis by chemical treatment with hydroxylamine and $NaNO_2$. The mutants were screened by replication to filter paper, heat treatment, and screening for the remaining activity by a color formation. Of 34,000 colonies, 16 clones were isolated (27). Their gene analyses revealed that three amino acid residues caused the stabilization against heat.

They are His57Tyr, Pro203Leu/Ser, and Val236Ala. Next, saturation mutagenesis was carried out in the three sites, and some other mutants were found to stabilize the enzyme as well. Furthermore, the mutations were combined, with interesting results. A His57Tyr/Pro203Glu mutant caused about 12°C increase in thermostability, His57Tyr/Val236Ala caused about 14°C increase, and Pro203Glu/Val236Ala caused about 17°C increase. His57Tyr/Pro203Glu/Val236Ala showed the highest thermostability at 75°C, approximately 10°C higher than the wild-type enzyme. There was no loss of stereoselectivity, nor a change in the substrate specificity. The K_m increased 50%, and the V_{max} decreased 16%, although these changes were negligible for production (28, 42). The crystal structure of the mutant was analyzed and the cause of the stabilization was rationalized (41).

30.4.2. Creation of D-Amino Acid Dehydrogenase from meso-Diaminopimelate Dehydrogenase (Fig. 1: h, m, n, p, s)

The production of D-amino acids is mainly conducted by enzymatic methods, with the exception of the fermentative production of D-Ala. The enzymatic methods include (i) dynamic kinetic resolution of hydantoin derivatives by hydantoinase (EC 3.5.2.2), followed by DCase reaction; (ii) optical resolution of N-acyl amino acids by N-aminoacylase (EC 3.5.1.14) in the possible presence of N-acyl D-amino acid racemase; (iii) asymmetric synthesis from α-keto acids in the presence of D-amino acid transaminase (EC 2.6.1.21), alanine dehydrogenase, formate dehydrogenase, and alanine racemase; and (iv) dynamic kinetic resolution of amino acid amide by D-aminopeptidase and amino acid amide racemase (3, 14) (Fig. 8).

Novik et al. (58) have created an NAD^+-dependent D-amino acid dehydrogenase with broad substrate specificity by rational and random mutagenesis. By the asymmetric reductive amination reaction from the corresponding α-keto acid and ammonia, as has been well studied in L-amino acid dehydrogenases (13), the enzyme was evolved by three rounds of mutagenesis, and screening was performed on meso-diaminopimelate D-dehydrogenase. The first round targeted the active sites by the previously studied X-ray crystal structure of the wild-type enzyme and produced mutants (Arg196Met, Thr170Ile, and His245Asn) which were active only on a substrate D-lysine, but no longer on meso-diaminopimelate, the original substrate. The screening was

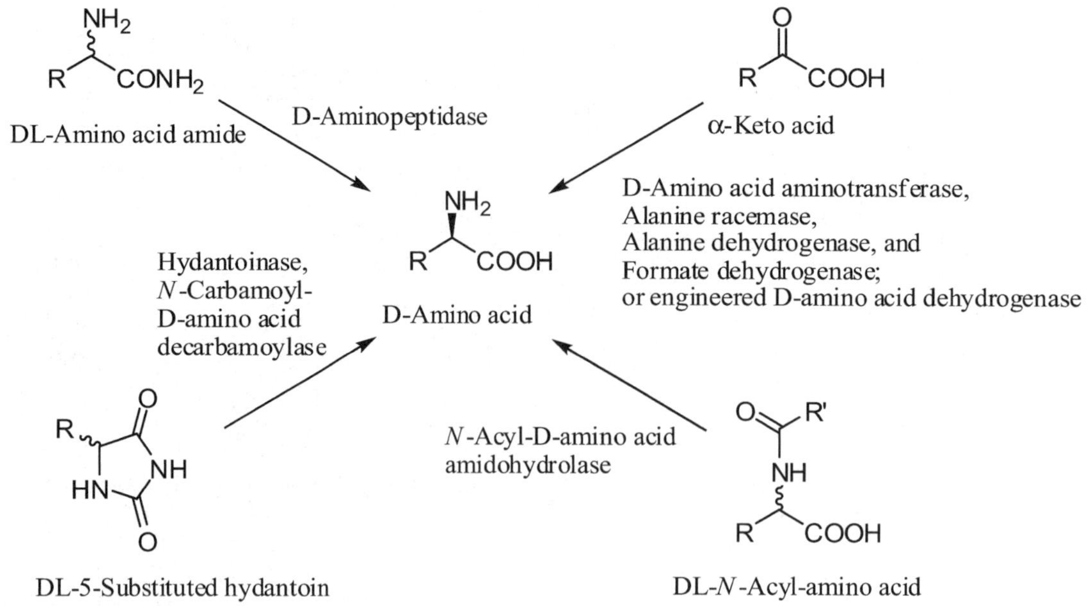

FIGURE 8 Enzymatic synthesis of D-amino acids.

done by monitoring color formation with nitrotetrazolium dye. The second mutation was carried out randomly by error-prone PCR on the entire gene, and a fourth mutation, Gln51Leu, was added. The third round of mutagenesis produced mutants that had an increased substrate range including straight- and branched-aliphatic amino acids and aromatic amino acids, and a fifth mutation, Asp155Gly, was added. Very high selectivity toward the D-enantiomer (95 to >99% e.e.) was shown to be preserved even after the total of five mutations were found in the three rounds of mutagenesis and screening. Reductive amination reaction of α-keto acid to form D-amino acid was made possible with this new enzyme. D-Isoleucine, D-glutamic acid, D-phenylalanine, and D-4-chlorophenylalanine were the substrates on which the wild-type enzyme was completely nonactive.

30.4.3. A New Industrial Enzymatic Method of Selective Phosphorylation of Nucleosides for 5′-IMP and 5′-GMP Production (a, j, k, l, m, n, o, p, s)

5′-IMP and 5′-GMP are important nucleotides because they give "Umami" taste in foods. There is no taste in other isomers such as 2′- and 3′-inosinic acids. Two phosphorylation methods have been reported. One is a chemical phosphorylation process that uses phosphoryl chloride (POCl₃), and the other is an enzymatic process that uses inosine kinase.

A "greener" and newer enzymatic method to produce 5′-IMP was sought. Microorganisms that phosphorylated nucleosides using pyrophosphate (PPᵢ) as the phosphate donor regioselectively at the 5′ position were screened for from 3,000 strains (Fig. 9) (9). Although many of the microorganisms screened were able to phosphorylate inosine, phosphotransferase activity specific to the 5′ position was found to be distributed among the bacteria belonging to the family *Enterobacteriaceae*. *Morganella morganii* NCIMB 10466 was selected as a 5′-IMP producer (10).

A selective nucleoside phosphorylating enzyme was purified to homogeneity from *M. morganii* NCIMB 10466 crude extract. The enzyme appeared to consist of six subunits identical in molecular weight (Mᵣ 25,000). It phosphorylated various nucleosides at the 5′ position to produce nucleoside 5′-monophosphates using PPᵢ as the phosphate source. Energy-rich compounds, such as carbamoyl phosphate and acetyl phosphate, were also very effective phosphate donors. The enzyme also exhibited phosphatase activity and dephosphorylated various phosphate esters, but had a weak effect on nucleoside 3′-monophosphates. The M. *morganii* gene encoding a nucleoside-phosphorylating enzyme was isolated by a shotgun-cloning strategy. It was identical to the M. *morganii* PhoC acid phosphatase gene. With use of the purified enzyme, 32.6 mM 5′-IMP was synthesized from inosine with a 41% molar yield, but the synthesized 5′-IMP was hydrolyzed back to inosine because of its phosphatase activity as the reaction time was extended.

To suppress the dephosphorylation reaction and increase the efficiency of the transphosphorylation reaction, a random mutagenesis approach was used. By error-prone

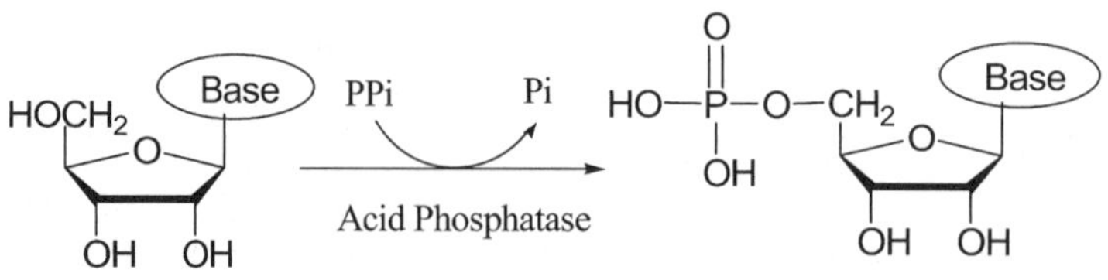

FIGURE 9 Acid phosphatase-catalyzed 5′-monophosphorylation of nucleoside with PPᵢ as a phosphorylating agent.

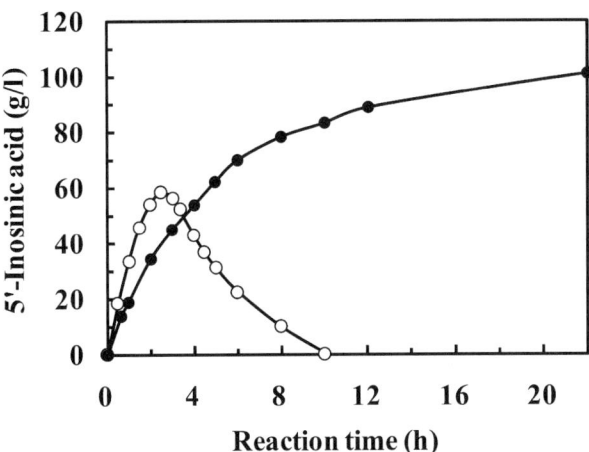

FIGURE 10 Directed evolution of acid phosphatase as an industrial biocatalyst to produce 5′-IMP and 5′-GMP. ○, wild type; •, mutant.

PCR, one mutated acid phosphatase that increased in phosphotransferase reaction yield was obtained (38–40). With the *E. coli* overproducing the mutated acid phosphatase, 101 g/liter (191 mM) of 5′-IMP was synthesized from inosine in an 85% molar yield (Fig. 10). This improvement was achieved with two mutations, Gly to Asp at position 92 and Ile to Thr at position 171. A decreased K_m value for inosine was responsible for the increased productivity.

Ile151Thr caused improvement of the affinity toward inosine, and the mutation at Gly72Asp, together with Ile151Thr, caused not only an increase in affinity toward inosine but also a decrease in the dephosphorylating activity. On the basis of X-ray analysis of the enzyme crystal, two mutation sites are located near the active site, and Thr151 seems to form hydrogen bonds with inosine, enhancing the affinity toward inosine (29, 30, 54). Since 2003, 5′-IMP and 5′-GMP have been produced at multiple thousands of tons per year by Ajinomoto Co. Inc., Japan.

30.5. FINDING NEW FUNCTIONS OF KNOWN ENZYMES BY LABORATORY EXPERIMENTS

An example of obtaining the desired enzyme without screening, or directed evolution, is introduced. Several dynamic kinetic enzymatic resolutions of synthetic substrates have been developed for the industrial production of chiral amino acids. The use of L-α-amino-ε-caprolactam (L-ACL) hydrolase in the presence of α-amino-ε-caprolactam (ACL) racemase is a

typical example of L-lysine production (24). L-Cysteine is produced by hydrolases acting on amino-2-thiazoline-4-carboxylic acid in the presence of a racemase (50). As described in section 30.3, DCase catalyzes the hydrolysis of N-carbamoyl-D-amino acids to form D-amino acids when used in combination with D-specific hydantoinase, either with spontaneous racemization of hydantoin or with hydantoin racemase. N-Acyl-D-amino acid amidohydrolase and N-acyl amino acid racemase are used together to form D-amino acids.

Asano and Yamaguchi (14, 15) discovered amino acid amide racemizing activity in ACL racemase (EC 5.1.1.15) from *Achromobacter obae*. They reported the discoveries, properties, and structures of the new enzymes D-aminopeptidase (EC 3.4.11.19) (8), D-amino acid amidase (EC 3.5.1.–) (35), and alkaline D-peptidase (EC 3.4.11.–) from *Ochrobactrum anthropi* C1-38, *O. anthropi* SV3, and *Bacillus cereus* DF4-B, respectively (3). These D-stereospecific hydrolases can be applied to kinetic resolution of racemic amino acid amides to yield D-amino acids. By this route, D-amino acid may be prepared in three steps from an aldehyde. The amino acids can only be obtained in 50% yield, because the enzyme accepts one of the D-enantiomers as a substrate. If amino acid amide racemase were used together with these D-stereospecific hydrolases, the remaining L-amino acid amide could be racemized and DL-amino acid amide would be completely hydrolyzed to form D-amino acid (36). It would also be possible to synthesize L-amino acids when amino acid amide racemase is used with L-stereospecific hydrolases (37).

Amino acid amide-racemizing activity in ACL racemase (EC 5.1.1.15) from *A. obae* was discovered. Previously, Soda et al. did pioneering work on ACL racemase and reported that the substrates for ACL racemase are only ACL, α-amino-δ-valerolactam, and α-amino-β-thia-ε-caprolactam, but the enzyme does not act on amino acid amide (1). The result of construction of an efficient system to produce D-amino acid from L-amino acid amide by a combination of ACL racemase and D-aminopeptidase is demonstrated in Fig. 11. This is the first report of the enzymatic synthesis of stoichiometric amounts of D-amino acid from L-amino acid amide by a combination of a new amino acid amide-racemizing enzyme and D-stereospecific amino acid amide hydrolase (Fig. 12) (3, 63). The X-ray crystal structures of the enzymes have been solved (46, 47).

30.6. CONCLUSION

In this chapter, I have discussed how new enzymes have been discovered and improved, with the following examples: nitrile hydratase, aldoxime dehydratase, ω-laurolactam hydrolase, L-amino acid ligase, DCase, D-amino acid dehydrogenase, acid phosphatase, amino acid amidases, and amino acid amide racemase. Half of these are industrial

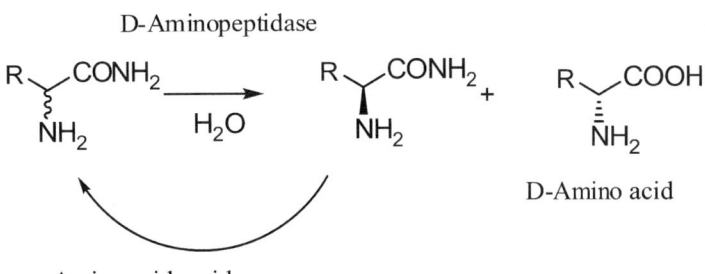

FIGURE 11 Dynamic kinetic resolution of amino acid amide.

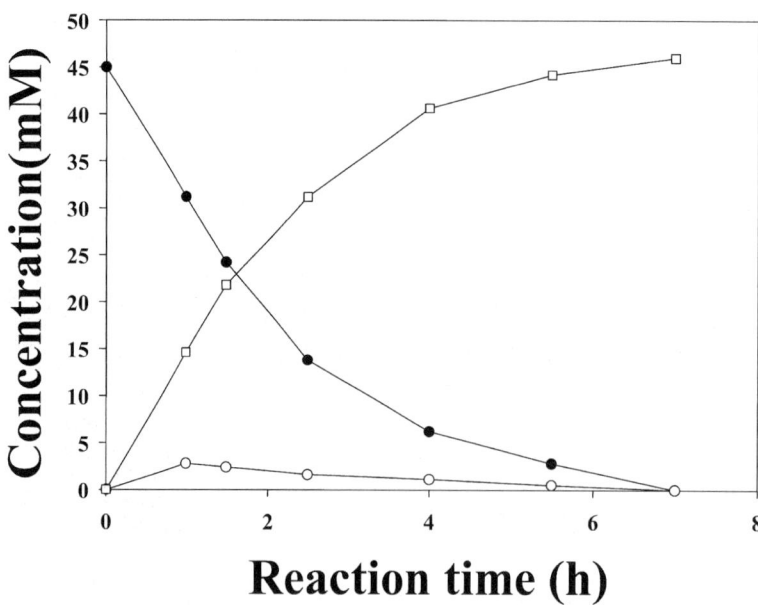

FIGURE 12 Dynamic kinetic resolution of alanine amide with D-aminopeptidase and ACL racemase. ●, L-alanine; ○, D-alanine amide; □, D-alanine.

enzymes. Furthermore, the discovery of a new biosynthetic pathway of MK by in silico screening was introduced.

Traditional approaches, with isolation of microorganisms and characterization of the enzyme, followed by gene cloning, etc., have had a huge impact on current enzyme engineering. Discoveries, however, sometimes needed serendipitous work.

The resource of genome information, including the metagenome, is in our hands and could certainly be utilized more in the future. A new SIGEX metagenome approach was invented to screen inducible enzymes efficiently, since the metagenome approaches are based on higher-fidelity discovering methods such as hybridization and PCR. The discoveries of LAL and the new MK biosynthetic pathway are good examples of in silico screening for enzymes and the biosynthetic pathway. There is no doubt that microbiology will be involved in this search, including metagenomic information, as seen in the successful exploitation of nitrilase genes from environmental DNA.

A tremendous amount of information is available from the databases, and it is now possible to predict the function of enzymes from the similarity of their primary structures. This has helped to accelerate the exploitation of genetic resources, as seen in the discovery of LAL for dipeptide synthesis. The discovery of the new MK biosynthetic pathway also used the database to find missing information. Research on the use of ACL racemase in the dynamic kinetic resolution of amino acid amide was made possible by information from accessible databases, although it has been shown that data accumulation in the "wet" laboratory was the key factor in finding the new function of those enzymes. It should be emphasized that the counterpart enzymes such as D-aminopeptidase and D-amino acid amidases have been discovered by screening from nature by enrichment culture technique (see Fig. 1, steps a, d, g, i, j, k, and s).

The directed evolution of enzymes has proved to be a very powerful technique, as can be seen in the development of the new enzyme D-amino acid dehydrogenase, a thermostable DCase, and industrial production of nucleic acids with umami flavor by the evolved acid phosphatase.

The mutated acid phosphatase is probably one of the first examples of the use of directed evolution in the industrial production of useful materials. It seems that there should be no limit to the evolution of proteins for the benefit of human beings; these tremendous possibilities should be utilized to the fullest.

Looking at the history of the development of new drugs, discoveries have been a major factor. How do the readers of this chapter think about the future of enzyme discovery? Do they judge that the 99% of total microbial DNA from uncul-turable microorganisms is a huge or a modest resource, com-pared with the already known 1%? As we have seen here, many more advances may be made in the development of industrial enzymes, biocatalysis, and enzyme evolution with the aid of rational approaches combining the information from databases with actual experiments in the laboratory.

REFERENCES

1. **Ahmed, S. A., N. Esaki, H. Tanaka, and K. Soda.** 1986. Mechanism of α-amino-ε-caprolactam racemase reaction. *Biochemistry* **25**:385–388.
2. **Asano, Y.** 2002. Overview of screening for new microbial catalysts and their uses in organic synthesis. Selection and optimization of biocatalysts. *J. Biotechnol.* **94**:65–72.
3. **Asano, Y.** 2008. Enzymatic production of D-amino acids, p. 68–77. *In* M. Ueda (ed.), *Microbial Bioconversion and Biopro-duction: Development of White Biotechnology beyond Chemical Synthesis.* CMC Press Inc., Tokyo, Japan. (In Japanese.)
4. **Asano, Y., K. Fujishiro, Y. Tani, and H. Yamada.** 1982. Aliphatic nitrile hydratase from *Arthrobacter* sp. J-1—purification and characterization. *Agric. Biol. Chem.* **46**:1165–1174.
5. **Asano, Y., Y. Fukuta, Y. Yoshida, and H. Komeda.** 2008. A new enzymatic synthesis of 12-aminolauric acid: screening, characterization and the use of ω-laurolactam hydrolase. *Biosci. Biotech. Biochem.* **72**: 2141–2150.
6. **Asano, Y., and Y. Kato.** 1998. Z-Phenylacetaldoxime degradation by a novel aldoxime dehydratase from *Bacillus* sp. strain OxB-1. *FEMS Microbiol. Lett.* **158**:185–190.

7. Asano, Y., and Y. Kato. 1999. A new enzymatic method of nitrile synthesis by *Rhodococcus* sp. strain YH3-3. *J. Mol. Catal. B Enzym.* **6:**249–256.

8. Asano, Y., Y. Kato, A. Yamada, and K. Kondo. 1992. Structural similarity of D-aminopeptidase to carboxypeptidase DD and β-lactamase. *Biochemistry* **31:**2316–2328.

9. Asano, Y., Y. Mihara, and H. Yamada. 1999. A new enzymatic method of selective phosphorylation of nucleosides. *J. Mol. Catal. B Enzym.* **6:**271–277.

10. Asano, Y., Y. Mihara, and H. Yamada. 1999. A novel nucleoside phosphorylating enzyme from *Morganella morganii*. *J. Biosci. Bioeng.* **87:**732–738.

11. Asano, Y., K. Tamura, N. Doi, T. Ueatrongchit, A. H-Kittikun, and T. Ohmiya. 2005. Screening for new hydroxynitrilases from plants. *Biosci. Biotechnol. Biochem.* **69:**2349–2357.

12. Asano, Y., Y. Tani, and H. Yamada. 1980. A new enzyme "nitrile hydratase" which degrades acetonitrile in combination with amidase. *Agric. Biol. Chem.* **44:**2251–2252.

13. Asano, Y., A. Yamada, Y. Kato, K. Yamaguchi, Y. Hibino, K. Hirai, and K. Kondo. 1990. Enantioselective synthesis of (S)-amino acids by phenylalanine dehydrogenase from *Bacillus sphaericus*: use of natural and recombinant enzymes. *J. Org. Chem.* **55:**5567–5571.

14. Asano, Y., and S. Yamaguchi. 2005. Dynamic kinetic resolution of amino acid amide catalyzed by D-aminopeptidase and α-amino-ε-caprolactam racemase. *J. Am. Chem. Soc.* **127:**7696–7697.

15. Asano, Y., and S. Yamaguchi. 2005. Discovery of new substrates amino acid amides for α-amino-ε-caprolactam racemase from *Achromobacter obae*. *J. Mol. Catal. B Enzym.* **36:**22–29.

16. Asano, Y., T. Yasuda, Y. Tani, and H. Yamada. 1982. A new enzymatic method of acrylamide production. *Agric. Biol. Chem.* **46:**1183–1189.

17. Bergeron, S., D. A. Chaplin, J. H. Edwards, B.S.W. Ellis, C. L. Hill, K. Holt-Tiffin, J. R. Knight, T. Mahoney, A. P. Osborne, and G. Ruecroft. 2006. Nitrilase-catalyzed desymmetrization of 3-hydroxyglutaronitrile: preparation of a statin side-chain intermediate. *Org. Process Res. Dev.* **10:**661–665.

18. Bornscheuer, U. T. 2005. Trends and challenges in enzyme technology. *Adv. Biochem. Eng. Biotechnol.* **100:**181–203.

19. Cowan, D., Q. Meyer, W. Stafford, S. Muyanga, R. Cameron, and P. Wittwer. 2005. Metagenomic gene discovery: past, present and future. *Trends Biotechnol.* **23:**321–329.

20. Crosby, J. A., J. S. Parratt, and N. J. Turner. 1992. Enzymic hydrolysis of prochiral dinitriles. *Tetrahedron Asymmetry* **3:**1547–1550.

21. Daniel, R. 2002. The soil metagenome—a rich resource for the discovery of novel natural products. *Curr. Opin. Biotechnol.* **15:**199–204.

22. DeSantis, G., K. Wong, B. Farwell, K. Chatman, Z. Zhu, G. Tomlinson, H. Huang, X. Tan, L. Bibbs, P. Chen, K. Kretz, and M. J. Burk. 2003. Creation of a productive, highly enantioselective nitrilase through gene site saturation mutagenesis (GSSM). *J. Am. Chem. Soc.* **125:**11476–11477.

23. Faber, K. 2004. *Biotransformations in Organic Chemistry*, 5th ed. Springer Verlag, Berlin, Germany.

24. Fukumura, T. 1977. Conversion of D- and DL-α-amino-ε-caprolactam into L-lysine using both yeast cells and bacterial cells. *Agric. Biol. Chem.* **41:**1327–1330.

25. Fukuta, Y., H. Komeda, Y. Yoshida, and Y. Asano. 2009. High yield synthesis of 12-aminolauric acid by "enzymatic transcrystallization" of ω-laurolactam using ω-laurolactam hydrolase from *Acidovorax* sp. T31. *Biosci. Biotech. Biochem.* **73:**980–986.

26. Hiratsuka, T., K. Furihata, J. Ishikawa, H. Yamashita, N. Itoh, H. Seto, and T. Dairi. 2008. An alternative menaquinone biosynthetic pathway operating in microorganisms. *Science* **321:**1670–1673.

27. Ikenaka, Y., H. Nanba, K. Yajima, Y. Yamada, M. Takano, and S. Takahashi. 1998. Increase in thermostability of N-carbamyl-D-amino acid amidohydrolase on amino acid substitutions. *Biosci. Biotechnol. Biochem.* **62:**1668–1671.

28. Ikenaka, Y., H. Nanba, K. Yajima, Y. Yamada, M. Takano, and S. Takahashi. 1998. Relationship between an increase in thermostability and amino acid substitutions in N-carbamyl-D-amino acid amidohydrolase. *Biosci. Biotechnol. Biochem.* **62:**1672–1675.

29. Ishikawa, K., Y. Mihara, K. Gondoh, E. Suzuki, and Y. Asano. 2000. X-ray structures of a novel acid phosphatase from *Escherichia blattae* and its complex with the transition-state analog molybdate. *EMBO J.* **19:** 2412–2423.

30. Ishikawa, K., E. Suzuki, Y. Mihara, T. Utagawa, and Y. Asano. 2002. Enhancement of nucleoside phosphorylation activity in an acid phosphatase. *Protein Eng.* **15:**539–543.

31. Kataoka, M., and S. Shimizu. 2008. Screening of novel microbial enzymes and their application to chiral compound production, p. 355–373. *In* C.-T. Hou and J.-F. Shaw (ed.), *Biocatalysis and Bioenergy*. John Wiley & Sons, Inc., New York, NY.

32. Kato, Y., and Y. Asano. 1999. A new enzymatic method of nitrile synthesis by *Rhodococcus* sp. strain YH3-3. *J. Mol. Catal. B Enzym.* **6:**249–256.

33. Kato, Y., K. Nakamura, H. Sakiyama, S. G. Mayhew, and Y. Asano. 2000. A novel heme-containing lyase, phenylacetaldoxime dehydratase from *Bacillus* sp. strain OxB-1: purification, characterization, and molecular cloning of the gene. *Biochemistry* **39:**800–809.

34. Kato, Y., R. Ooi, and Y. Asano. 2000. Distribution of aldoxime dehydratase in microorganisms. *Appl. Environ. Microbiol.* **66:**2290–2296.

35. Komeda, H., and Y. Asano. 2000. Gene cloning, nucleotide sequencing, and purification and characterization of the D-stereospecific amino acid amidase from *Ochrobactrum anthropi* SV3. *Eur. J. Biochem.* **267:**1–9.

36. Komeda, H., and Y. Asano. 2008. A novel D-stereoselective amino acid amidase from *Brevibacterium iodinum*: gene cloning, expression and characterization. *Enzyme Microb. Technol.* **43:**276–283.

37. Komeda, H., H. Harada, S. Washika, T. Sakamoto, M. Ueda, and Y. Asano. 2004. A novel R-stereoselective amidase from *Pseudomonas* sp. MCI3434 acting on piperazine-2-tert-butylcarboxamide. *Eur. J. Biochem.* **271:** 1580–1590.

38. Mihara, Y., K. Ishikawa, E. Suzuki, and Y. Asano. 2004. Improving the pyrophosphate-inosine phosphotransferase activity of *Escherichia blattae* acid phosphatase by sequential site-directed mutagenesis. *Biosci. Biotech. Biochem.* **68:**1046–1050.

39. Mihara, Y., T. Utagawa, H. Yamada, and Y. Asano. 2000. Phosphorylation of nucleosides by the mutated acid phosphatase from *Morganella morganii*. *Appl. Environ. Microbiol.* **66:**2811–2816.

40. Mihara, Y., T. Utagawa, H. Yamada, and Y. Asano. 2001. Acid phosphatase/phosphotransferases from enteric bacteria. *J. Biosci. Bioeng.* **92:**50–54.

41. Nakai, T., T. Hasegawa, E. Yamashita, M. Yamamoto, T. Kumasaka, T. Ueki, H. Nanba, Y. Ikenaka, S. Takahashi, M. Sato, and T. Tsukihara. 2000. Crystal structure of N-carbamyl-D-amino acid amidohydrolase with a novel catalytic framework common to amidohydrolases. *Structure* **8:**729–737.

42. Nanba, H. 2007. Application of hyperthermophilic enzyme to a bioreactor producing D-amino acid. *Seibutsu Kogaku Kaishi* **85:**403–405. (In Japanese.)

43. **Nanda, S., Y. Kato, and Y. Asano.** 2005. A new (R)-hydroxynitrile lyase from *Prunus mume*: asymmetric synthesis of cyanohydrins. *Tetrahedron* **61:**10908–10916.

44. **Ogawa, J., and S. Shimizu.** 1999. Microbial enzymes: new industrial applications from traditional screening methods. *Trends Biotechnol.* **17:**13–20.

45. **Ogawa, J., and S. Shimizu.** 2002. Industrial microbial enzymes: their discovery by screening and use in large-scale production of useful chemicals in Japan. *Curr. Opin. Biotechnol.* **13:**367–375.

46. **Okazaki, S., A. Suzuki, H. Komeda, Y. Asano, and T. Yamane.** 2007. Crystal structure and function of a D-stereospecific amino acid amidase from *Ochrobactrum anthropi* SV3, a new member of the penicillin-recognizing proteins. *J. Mol. Biol.* **368:**79–91.

47. **Okazaki, S., A. Suzuki, T. Mizushima, T. Kawano, H. Komeda, Y. Asano, and T. Yamane.** 2009. The novel structure of a pyridoxal-5′-phosphate-dependent fold-type I racemase, α-amino-ε-caprolactam racemase from *Achromobacter obae*. *Biochemistry* **48:**941–950.

48. **Okuta, A., K. Ohnishi, and S. Harayama.** 1998. PCR isolation of catechol 2,3-dioxygenase gene fragments from environmental samples and their assembly into functional genes. *Gene* **212:**221–228.

49. **Robertson, D. E., J. A. Chaplin, G. DeSantis, M. Podar, M. Madden, E. Chi, T. Richardson, A. Milan, M. Miller, D. P. Weiner, K. Wong, J. McQuaid, B. Farwell, L. A. Preston, X. Tan, M. A. Snead, M. Keller, E. Mathur, P. L. Kretz, M. J. Burk, and J. M. Short.** 2004. Exploring nitrilase sequence space for enantioselective catalysis. *Appl. Environ. Microbiol.* **70:**2429–2436.

50. **Sano, K., and K. Mitsugi.** 1978. Enzymatic production of L-cysteine from D,L-2-amino-Δ²-thiazoline-4-carboxylic acid by *Pseudomonas thiazolinophilum*: optimal conditions for the enzyme formation and enzymatic reaction. *Agric. Biol. Chem.* **42:**2315–2321.

51. **Schmeisser, C., H. Steele, and W. R. Streit.** 2007. Metagenomics, biotechnology with non-culturable microbes. *Appl. Microbiol. Biotechnol.* **75:**955–962.

52. **Seto, H., Y. Jinnai, T. Hiratsuka, M. Fukawa, K. Furihata, N. Itoh, and T. Dairi.** 2008. An alternative menaquinone biosynthetic pathway operating in microorganisms. *J. Am. Chem. Soc.* **130:**5614–5615.

53. **Steele, H. L., K.-E. Jaeger, R. Daniel, and W. R. Streit.** 2009. Advances in recovery of novel biocatalysts from metagenomes. *J. Mol. Microbiol. Biotechnol.* **16:**25–37.

54. **Suzuki, E., K. Ishikawa, Y. Mihara, N. Shimba, and Y. Asano.** 2007. Structural-based engineering for transferases to improve the industrial production of 5′-nucleotide. *Bull. Chem. Soc. Jpn.* **80:**276–286.

55. **Tabata, K., and S. Hashimoto.** 2007. Fermentative production of L-alanyl-L-glutamine by a metabolically engineered *Escherichia coli* strain expressing L-amino acid α-ligase. *Appl. Environ. Microbiol.* **73:**6378–6385.

56. **Tamura, K., Y. Hirata, Y. Kobayashi, and T. Endo.** 1998. Development of bio-production of optically active α-hydroxy acids, p. 33–44. *In* T. Nakai and T. Ohashi (ed.), *Aspect of Industrial Chiral Technology*. CMC Press Inc., Tokyo, Japan. (In Japanese.)

57. **Uchiyama, T., T. Abe, T. Ikemura, and K. Watanabe.** 2004. Substrate-induced gene-expression screening of environmental metagenome libraries for isolation of catabolic genes. *Nat. Biotechnol.* **23:**88–93.

58. **Vedha-Peters, K., M. Gunawardana, J. D. Rozzell, and S. J. Novick.** 2006. Creation of a broad-range and highly stereoselective D-amino acid dehydrogenase for the one-step synthesis of D-amino acids. *J. Am. Chem. Soc.* **128:**10923–10929.

59. **Xie, S.-X., Y. Kato, and Y. Asano.** 2001. High yield synthesis of nitriles by a new enzyme phenylacetaldoxime dehydratase from *Bacillus* sp. strain OxB-1. *Biosci. Biotech. Biochem.* **65:**2666–2672.

60. **Yagasaki, M., and S. Hashimoto.** 2008. Synthesis and application of dipeptides; current status and perspectives, *Appl. Microbiol. Biotechnol.* **81:**13–22.

61. **Yamada, H., Y. Asano, T. Hino, and Y. Tani.** 1979. Microbial utilization of acrylonitrile. *J. Ferment. Technol.* **57:**8–14.

62. **Yamada, H., and M. Kobayashi.** 1996. Nitrile hydratase and its application to industrial production of acrylamide. *Biosci. Biotechnol. Biochem.* **60:**1391–1400.

63. **Yamaguchi, S., H. Komeda, and Y. Asano.** 2007. New enzymatic method of chiral amino acid synthesis by dynamic kinetic resolution of amino acid amides: use of stereoselective amino acid amidases in the presence of α-amino-ε-caprolactam racemase. *Appl. Environ. Microbiol.* **73:**5370–5373.

64. **Yokozeki, K.** 2007. An enzymatic breakthrough in the industrial production of oligopeptides. *Specialty Chemicals Magazine* **27**(May):44–45.

65. **Yokozeki, K., and S. Hara.** 2005. A novel and efficient enzymatic method for the production of peptides from unprotected starting materials. *J. Biotechnol.* **115:**211–220.

Enzyme Engineering: Combining Computational Approaches with Directed Evolution

LOUIS A. CLARK

31

31.1. INTRODUCTION

It is increasingly being recognized that the most effective way to evolve proteins, and especially enzymes, involves the use of both screening and rational design. This chapter seeks to give a cursory overview of how one might use rational or semirational design techniques to guide an evolution project. The goal is to give scientists who are not experienced in rational protein design, but who may have experience with the necessary experimental techniques, a view of what one can achieve. Emphasis is placed on practical evolution goals and the determination of what is reasonable. Large modifications in substrate range, enantioselectivity, and enzyme stability are routinely, if not easily, achievable. In addition, it now appears possible to switch catalytic mechanism and to design active sites from first principles in some cases. Support for these claims and details of techniques are provided solely by examples from the open literature. There are now many convincing and instructive examples of rational and semirational enzyme design.

There are many aspects of an evolution project that will not be covered here, but which are essential. Perhaps the two most important such aspects are the DNA manipulation techniques and the activity assays. Without effective methods to construct desired variants, rational design cannot be practiced. Likewise, without a fast assay sensitive to any pertinent chiral products, evolution is restricted or impossible (95). It will be argued that the throughput of the assay is the most critical factor in choosing to use rational design and selecting an approach. Beyond making that choice and a few mentions to the literature, strategy aspects will not be discussed, in part because they are best learned through experience.

It is useful to examine some reasonable evolution goals for the industrial usage of enzymes. Substantial progress has been made in incorporating enzymes into pharmaceutical processes where high regioselectivity and stereoselectivity are critical (94). Biocatalysts are typically used as crude lysate and exposed to the substrate under nonphysiological conditions. For the pharmaceutical industry, goals of 100 g/liter substrate, 1 g/liter enzyme, 99% enantiomeric excess, and 98% conversion in less than 24 h of reaction are reasonable and necessary (81). These conditions often require extensive evolution for activity, selectivity, and stability. Emerging uses of exogenous enzymes as part of in vivo systems may be less stringent.

This chapter begins with an overview of the synergy between rational design and screening or selection. The choice of a rational design approach requires consideration of the number of enzyme variants that can be evaluated experimentally. A review of fundamentals of enzyme catalysis from an energy landscape point of view is presented to illustrate why and where rational design could be applied. The bulk of the chapter follows with many examples from the literature of techniques that have worked. Throughout the discussion, mention of basic computational techniques needed to support design is made. It will be seen that it is often quite feasible to adapt an enzyme to many chemical process or organism-engineering needs. From these collective examples, an assessment of what is reasonable to expect from evolution will emerge. Finally, a brief attempt to foresee applications of emerging computational techniques will be made.

31.2. SYNERGY BETWEEN DESIGN AND SCREENING

It is useful to examine the bounds of screening and rational design combinations. If one can make an exhaustive library of all possible sequences and pull the desired phenotype out of it using a selection method, then there is little or no need to bother with design. There is a sole example in the literature of a selection system developing de novo (from the beginning) enzyme activity (105). That being said, such a situation is very rare. It can be very time-consuming to develop an in vivo selection system, and in many cases, it may be impossible. For example, it would be quite difficult to tie production of a nonnatural product with a distinct chirality to organism fitness.

For many cases, a versatile practical approach is to intelligently construct as many enzyme variants as possible and screen them in high throughput. High throughput may mean ten reactions mixed and injected onto a chiral column by hand, or it may be tens of thousands handled only by robotics and assayed spectroscopically. Once an assay is developed, it is foolish not to take advantage of whatever throughput it can offer. The more variants one tests, the more chances of success. It is throughput that determines how many variants one can test and therefore what type of rational design one could use.

Figure 1 attempts to show the range of likely combinations between rational design and screening. For a given

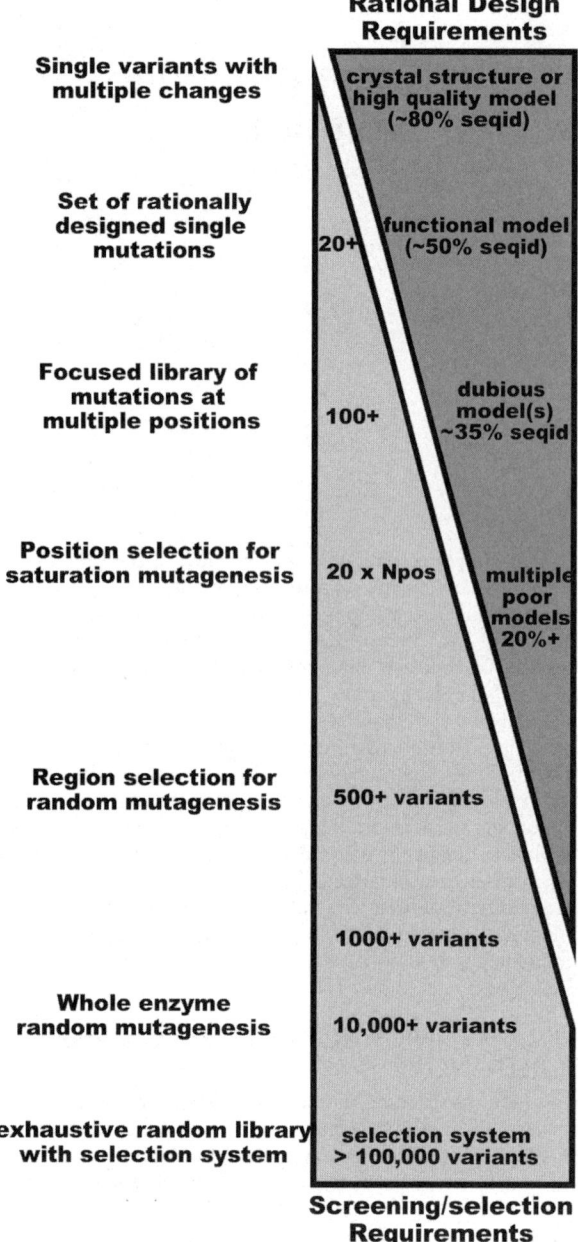

FIGURE 1 Possible choices of rational design methods for a given level of screening capacity. The upper triangle gives approximate requirements for structural information in the form of percentage of sequence identity (seqid) of the homology model template to the enzyme sequence. The lower triangle specifies the corresponding screening requirements for each approach Npos, number of positrons targeted.

screening capacity (lower triangle), a possible design approach is listed in the left column. Corresponding minimum requirements to support that structure-based design approach are given in the upper triangle. Many of these scenarios will be seen in the examples provided in later sections.

There is always a trade-off between incorporating multiple potentially beneficial mutations into a design and risking destabilization of the enzyme. A balance must be made when considering a mix of screening and rational design. Stability can be engineered by selecting mutations that are likely to improve protein-stabilizing interactions or at least be neutral (see later section). The ability to rationally pick such mutations depends in part on the quality of the structural information. If the quality is high, it may be possible to make more than one mutation per tested variant. This is represented in the top level of Fig. 1. Conversely, if structural information is poor or not present, the chance of each mutation decreasing stability is higher and multiple mutations are more likely to waste screening resources.

The quality of structural information is the accuracy at which atoms are present and properly positioned in the computer representation of the structure. High-resolution crystal structures reflect reality best, and models are best if they are based on crystal structures with fewer sequence differences. Models are made by first aligning a crystal structure sequence (the template) with a sequence from the protein to be modeled, then replacing residues that do not match. They are called "homology models" under the assumption that the template is a homologous protein. Various software is available for performing this common procedure. Swiss Model is freely available on the Web (4). MODELLER is one of the most established packages and is free to academic users (102; MODELLER software available at: http://www.salilab.org/modeller/). The most reliable homology models are those made by using a template that has high (ca. 80% or more) sequence identity to the modeled sequence and no gaps in the sequence alignment. Trustworthiness decreases approximately with sequence identity until even the overall fold becomes unreliable (perhaps ca. 20% sequence identity).

31.3. RATIONAL DESIGN IS A COST-EFFECTIVE METHOD

The more difficult it is to develop a screening assay for your enzyme reaction, the more rational design is likely to be useful. If one is limited to testing only a few variants, it makes sense to use any available methods to improve the chance that those variants will be productive. At this extreme, any amount of in silico screening and validation is worthwhile.

Rational design itself costs little to implement. The literature contains much information on enzyme mechanisms, and crystal structures for many common classes of enzymes are available. In most cases, it is possible to make at least a low-quality homology model. With that model and the mechanistic knowledge, it should be possible to identify the catalytic residues and the binding pocket. Even without modeling the substrate in the active site, the scientist can return to the laboratory and focus any mutagenesis experiments at residues near or in the binding pocket. This semirational design scenario is exemplified in the middle of Fig. 1, where many positions are chosen for saturation mutagenesis.

As an example of why directed evolution alone is not ideal, consider random mutagenesis. Even if it were truly random, mutations resulting in amino acids coded by the greatest number of codons would be more favored. Residues coded by a single codon (methionine and tryptophan) would be six times less likely than those coded by six codons (arginine, leucine, and serine). Random mutation introduced by polymerase errors occurs almost exclusively by single base changes. Simple calculations show that it is not possible to access all amino acid types from a given

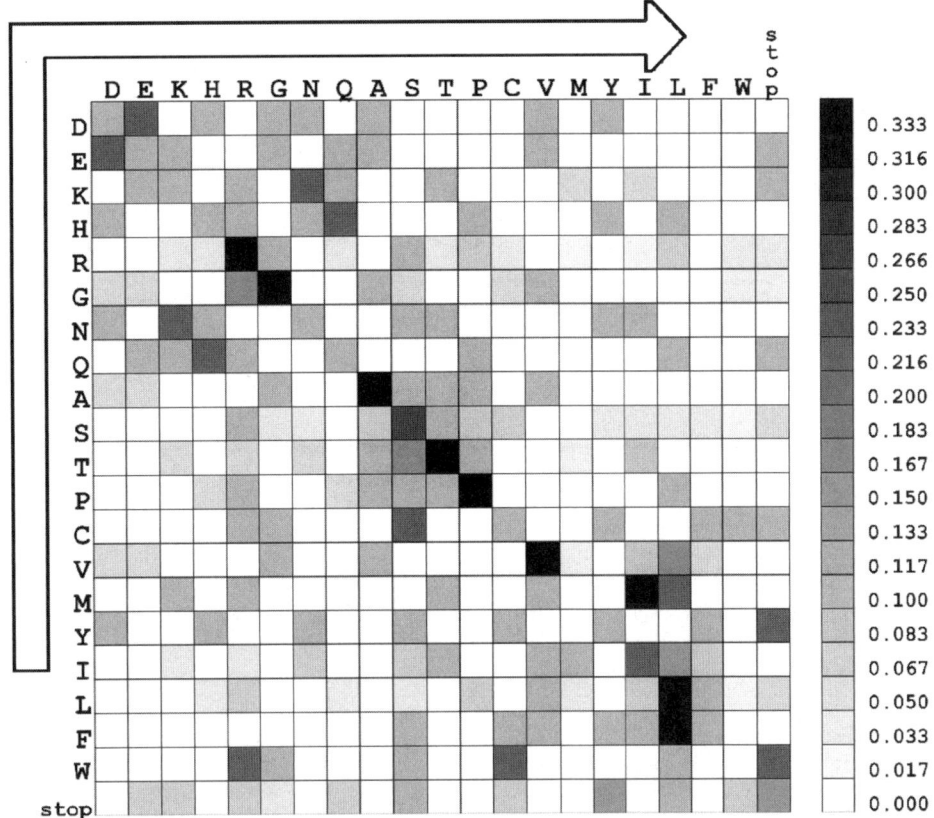

FIGURE 2 Probability of mutating from a given residue type to another during a single-base change. Random mutations are inherently biased on a residue level, and 42% of all residue-to-residue type changes are impossible. Residues are ordered approximately by hydrophobicity. The figure is read from left to right (row index to column index). Example: the probability of mutating from M to I is higher than mutating from I to M.

starting amino acid codon. Figure 2 shows the probability of making a given amino acid type change by random mutagenesis. Forty-two percent of all amino acid type changes are not accessible. In some cases (e.g., phenylalanine), drastic changes in hydrophobicity are impossible or very unlikely. When interfacing rational design with traditional evolution, it may be useful to use this information to suggest mutations that will not occur via random mutagenesis.

31.4. REVIEW OF THE ENERGY LANDSCAPE VIEW OF CATALYSIS

Biocatalysts accelerate and determine which chemical reactions occur. The rate acceleration results in part from reducing the free-energy barrier of the rate-limiting reaction step. From this perspective, enzyme function may appear simple. The relative free energies of the substrates, transition states, intermediates, and products make up the reaction *energy landscape* and determine much of the rate and selectivity of reaction (56). Computational methods can give estimates of these free energies (51, 68, 74). Catalytic effects that do not fall well into a thermodynamic viewpoint, such as vibrations (47) or tunneling (84), are generally smaller, but could also be treated computationally with more development effort. For engineering, it is beneficial to have as much knowledge as possible of the relative energies of these reaction states in the context of

the reaction mechanism. While not all aspects of enzyme catalysis fit well into an energy landscape view, much can be rationalized and it provides a convenient basis for discussion of engineering approaches.

It is useful to partition enzyme catalysis into binding and chemical transformation. Under the Michaelis-Menten description of kinetics, reaction is first order in bound substrate concentration, which is in turn determined by adsorption equilibrium (see reference 50 for greater depth). Stronger binding translates to a lower equilibrium dissociation constant (K_d, which is conceptually related to the Michaelis constant, K_m), a higher concentration of prereactive substrate, and often higher reaction rate. Improving substrate binding is thus a reasonable goal, especially when substrate concentrations in solution are expected to be low. However, substrate binding is just the first step in catalysis.

A simplistic example of an enzymatic reaction energy landscape beginning with binding is given in Fig. 3. In this example, the reaction in the enzyme (black lines) is compared with reaction in solution (gray dotted lines). Catalysis is achieved here by lowering the energy gap between the reactants and the relevant transition state. In solution phase, the reactants (E+S state) must proceed directly to a high-transition state (TS0). This hypothetical enzyme lowers the first energy barrier (TS1 − ES) by providing a more accessible intermediate state. It could be a covalently bound species such as the acyl intermediate formed by the reactive serine in serine proteases (see reference 119 for

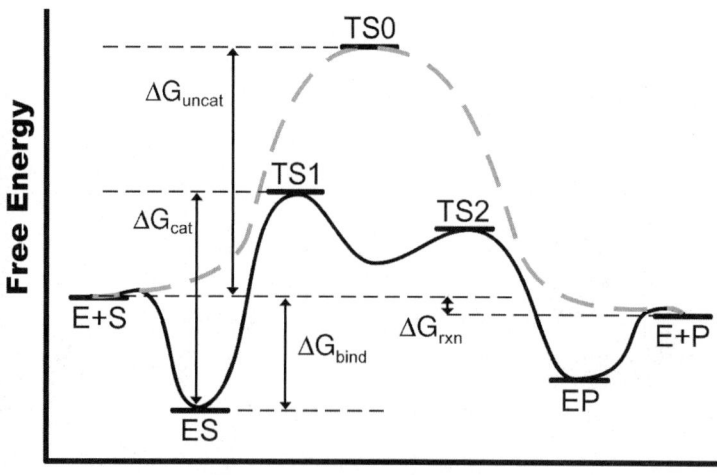

FIGURE 3 A hypothetical energy landscape for an enzyme-catalyzed reaction (black line) compared with the uncatalyzed reaction (gray dotted line). The overall reaction is thermodynamically favorable ($\Delta G_{rxn} < 0$). Notice that, in this case, the catalyzed reaction has only a slightly more favorable activation energy (ΔG_{cat}) than the uncatalyzed reaction (ΔG_{uncat}) due to strong binding to the substrate (ΔG_{bind}). E, ezyme; S, substrate; P, product.

examples of covalent effects). The enzyme may also catalyze by binding the reactants in a configuration that is more likely to lead to reaction. Other theories hypothesize enzyme residue arrangements that provide interactions that preferentially stabilize the transition state, perhaps by electrostatics (10). In any case, the enzyme provides a lower energy barrier path to reaction. If one can improve the energetics of the path or provide an alternative lower energy path by rational mutagenesis, then design is possible.

The reaction path is determined in part by the preference of the enzyme to accept certain substrates (specificity) or produce certain products (selectivity). It results from variation in the relative energy levels for possible species at any of the points at or bordering the rate-determining transition state. In Fig. 3, substrate specificity could manifest as the lack of an ES energy minimum (no binding) for one possible substrate. Alternatively, it could be that one substrate cannot approach the reactive transition state geometry because of steric clashes. In that case, TS1 for that substrate could be very high.

Although it is often assumed that the bond-forming or bond-breaking step in catalysis is rate-limiting, there are plenty of counterexamples. Often a product must be released before the active site is regenerated and able to accept another substrate. If the product is bound too tightly, release could become rate-limiting. Halohydrin dehalogenase HheC from *Agrobacterium radiobacter* provides one example. Tang et al. (110) showed that destabilizing the halide (product)-binding pocket by removing a hydrogen bond could accelerate the reaction rate for some substrates.

31.5. PHYSICOCHEMICAL CONTRIBUTIONS TO FREE ENERGY

A number of basic physicochemical effects provide a convenient framework for thinking about what determines the free energy of a reaction state. Interatomic interactions modify the potential energy of a collection of atoms and are enthalpic in nature. Rather than attempt to review the

computational methods for calculating these various interactions, the reader is directed toward recent reviews (14). In many cases, it is most important to understand what type of interactions could be important. *Electrostatic interactions* are one type. They occur between charged atoms and formally decay only gradually with distance ($1/r^2$). In the absence of dielectric shielding from solvent, and less so from bulk protein, they would be significant well beyond 10 Å. Electrostatic stabilization is often hypothesized as a major contributor to transition state stability and catalysis. *Steric repulsion* results from close approach of two atoms to the point where electron clouds begin to overlap and strongly repel. It is the easiest to calculate and most conceptually powerful interaction. If a substrate cannot be fit in the active site in the necessary reactive configuration without steric clashes, it will not react. If experiment shows that it does react, then one needs to refine the model and rework assumptions.

Hydrogen bonding and *van der Waals forces* are attractive interactions that can be manipulated by protein engineering. As any two atoms approach, there is a relatively minor attraction due to electronic effects that is often called van der Waals forces or London interactions. This electronic effect increases with the number of electrons on the atoms and their polarizability. It is difficult to calculate from first principles and is always treated empirically for systems as complex as proteins. Hydrogen bonding is also a complex electronic effect between a hydrogen atom and an acceptor atom such as oxygen, but it can be reduced to geometric rules for great engineering benefit. At the optimal configuration between donor and acceptor, and in a low dielectric environment, a hydrogen bond can contribute multiple kilocalories per mole of stabilization. Hydrogen bonds are critical for stabilizing proteins and binding reactants.

Solvation effects are the most complex and difficult interactions to compute and are sometimes even difficult to predict qualitatively. The absence of explicit water molecules in many simulations and visualizations of enzyme structure makes solvation effects easy to forget. Despite its comparatively small size, each water molecule can act as donor for two hydrogen bonds and as acceptor for two. Other molecules with less ability to hydrogen bond are, to varying degrees, hydrophobic. Water molecules avoid them to maximize water-water hydrogen bonding and minimize free energy. The hydrophobic effect wherein nonpolar molecules are effectively attracted to each other can be thought of as a consequence of surrounding water molecules squeezing them together. At about room temperature, water molecules can adopt some greater degree of order around a hydrophobe to maintain good hydrogen bonding. Because the partial ordering of the surrounding water decreases the entropy, the hydrophobic effect has a strong entropy component. When two hydrophobes come into contact, the partially ordered water molecules between them are released and the total free energy decreases (30). As the temperature increases, fluctuations increase and the surrounding water is less able to form all the hydrogen bonds present in bulk water. The hydrophobic effect *increases* with temperature. Hydrophobicity is often cited as the main driving force behind protein folding.

31.6. EXAMPLES OF STABILITY ENGINEERING

The resistance of an enzyme to thermal or solvent conditions is an integral part of almost any application. Most natural enzymes do not have ideal stability characteristics. They may not function at pH extremes, form aggregates at higher temperature, or simply lose activity with time under even mild conditions. Stability characteristics are loosely transferable across different challenge conditions. For example, mutations that improve thermostability may also promote solvent stability (111). However, folding stability as measured by the free energy of folding may not translate to solubility (see reference 20 for an example). Protein stability engineering is a broad topic in the literature (46), and only the aspects most relevant for enzyme evolution are mentioned below.

There is growing evidence that higher enzyme stability promotes evolvability. In work directly addressing this question, it was shown that stabilizing mutations in a bacterial P450 enzyme were required before necessary destabilizing active-site mutations could be made (13). Global stability must be maintained to keep the enzyme folded. The active site can be viewed as a necessary pocket of instability in the near-optimal background of bulk protein. In support of this concept, Shoichet, Matthews, and coworkers have shown that enzymes can easily be stabilized at the sacrifice of their enzymatic activity by making mutations in the active site (9, 106). Properties necessary for biocatalysis, such as maintenance of a hydrophobic water-filled cavity or burial of uncompensated charged residues, are detrimental to protein stability.

It is worth asking what the chances are of finding a single mutation in an enzyme that will increase thermostability. Certainly, the answer depends on the origin of the enzyme. Enzymes from mesophilic organisms experience less selective thermostability pressure than those from thermophilic organisms. One can also examine published data giving folding free-energy changes due to mutations. Figure 4 shows the distribution of mutation effect on protein stability for all single mutations with known structure in the Protherm database (8). Although the data set is not a random sample, it is large (ca. 3,000 entries). It can be seen that mutations tend to be destabilizing (1.1 kcal/mol) on average. Mutations on the surface are significantly less (0.4 kcal/mol) destabilizing, and those buried in bulk protein are more destabilizing (1.5 kcal/mol) on average. The more confined and tightly packed the environment is, the

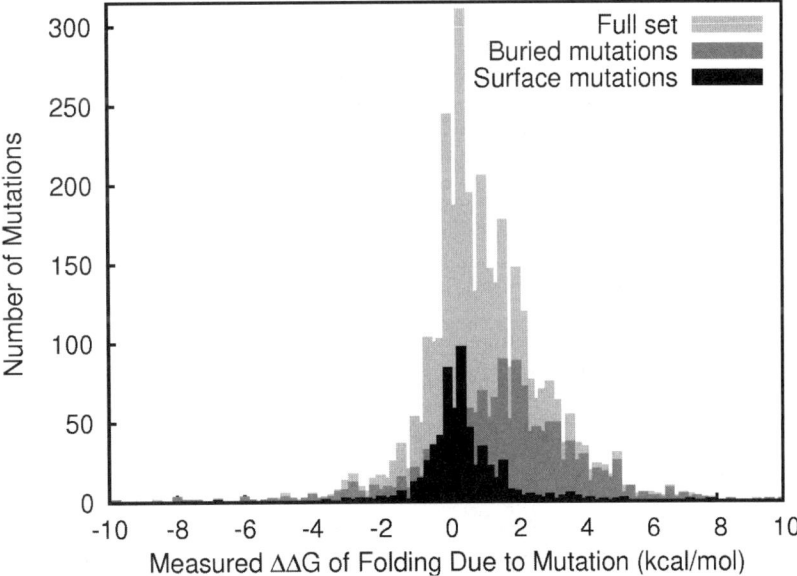

FIGURE 4 Distribution of mutation effect on the ΔG of protein folding. Data are taken from the Protherm4.0 data set and restricted to single mutations where the protein structure is known (8). Buried residues are <20% exposed relative to exposure in an alanine-X-alanine tripeptide. Surface residues are defined as having at least 60% relative exposure.

fewer alternative amino acid types that can be accommodated. This destabilizing effect of pseudorandom mutations and the previously mentioned destabilization of the active site implies that unchecked evolution for only activity may result in destabilization.

While random amino acid substitutions are unlikely on average to lead to stabilization, rational design can improve the chances of making stabilizing changes. A model of the structure and the basic physicochemical principles discussed in the last section can be used for rational selection of stabilizing mutations. Insights into what stabilizes enzymes from hyperthermophiles can also be useful (112). Repacking of side chains in enzyme cores in pursuit of stabilization can be performed using ROSETTA software version 3.0 (http://www.rosettacommons.org/) (36, 76). Other freely available software packages, such as PoPMuSiC (20) and FoldX (58), that use side-chain repacking or less-structure-based approaches have also been used for stabilization.

Mutations that decrease conformational flexibility are often used to stabilize proteins. One may view such constraints as decreasing the entropy of the unfolded state or as hindrances to unfolding. Disulfide bonds are the most effective because they provide covalent constraints on folding and unfolding. Several studies demonstrate that adding disulfide bonds to proteases (73, 109, 110a) and xylanases (49, 66, 114) may increase thermostability, sometimes by more than 10°C (69). Mutations to proline place a kink in the backbone that can also be stabilizing. Matthews et al. (85) made mutations to proline as well as glycine-to-alanine mutations, then used crystal structures to show that backbone conformations were unchanged during stabilization. The procedure of finding proline mutations can be automated by use of side-chain repacking software (29). Readers interested in structural stabilization are directed to a particularly instructive review by DeGrado et al. (37) for further insights on the design of protein secondary structure.

31.7. CONSENSUS APPROACH

The use of similar sequences to suggest mutations during the evolution of an enzyme is one of the more effective and easiest approaches to stabilization of proteins. Although it may also be used for selectivity engineering, its origins are in stability engineering (108), to which it is well suited. The basic idea is that nature has already explored variations on a given enzyme sequence or fold of interest and arrived at a "consensus" sequence that may have good stability properties. Variants similar to the sequence of interest (sometimes called "homologs") are often present in public databases. If the sequences are similar enough (>50% sequence identity), then the fold and many of the local residue environments will be the same. The similar sequences can be aligned to the sequence of interest and, at positions where the sequence of interest differs from the most common residue, change can be made. If the similar sequences are taken from thermophilic organisms, then chances of finding stabilizing mutations are further improved.

There are many applications of the consensus approach in the literature. It has worked well in the hands of many researchers on many different protein scaffolds. The thermostability of commonly used glucose dehydrogenase was increased substantially to allow utilization at 65°C (110c). Consensus-based libraries were used to increase the T_m of β-lactamase by 9.1°C (2). A consensus approach emphasizing local sequence alignment quality increased the T_m of

subtilisin E by 13°C (42). In an extension of the approach, Larson et al. (78) included pair correlation effects and used the method to stabilize an SH3 fold.

31.8. EXAMPLES OF SELECTIVITY ENGINEERING

In contrast to conventional heterogeneous or homogeneous catalysts, high chiral selectivity from enzymes is the rule, not the exception. In fact, it is generally difficult to make an enzyme nonselective. Enantioselectivity is a consequence of the arrangement of the catalytic residues in a geometrically constrained active site. Active sites will often be shaped to accommodate only a certain arrangement of bulky or small substituents at a chiral center, meaning that it is also common for a given enzyme to retain its R or S selectivity, even if the substrate is changed. In some cases, binding site pockets can be altered in size or shape to accommodate different chirality-determining side groups. Substantial changes in selectivity can be made. Directed evolution experiments show that it is possible to invert enantioselectivity (86). The examples here are chosen to demonstrate the range of modifications that can be achieved.

True rational design involves formulation of a hypothesis concerning how the enzyme function could be changed by mutagenesis, followed by construction and testing of the predicted variants. Semirational design uses logic to focus variant construction to fewer, but still many possibilities, which are then screened for activity. This section is divided by the approach taken. The semirational approaches will be covered first, moving roughly from the bottom of Fig. 1 to the top.

31.8.1. Targeted Random Mutagenesis for Selectivity Engineering

Very little or no structural information is required to target random mutagenesis libraries to large sequence regions. For multidomain proteins, the catalytically active domain(s) can often be identified by sequence-based classification. If a homology model can be built, it is likely that the active site can be identified and libraries further focused. There are many examples of selectivity-changing mutations occurring far from the active site, but logic and experimental evidence support targeting mutations for selectivity to the active site (87). Targeted random mutagenesis has been advocated and reviewed by Chica et al. (28). It is the least rational of the approaches in Fig. 1.

Random mutagenesis and rational or semirational mutation strategies can work well together. For example, Cheon et al. (27) followed up on consensus-guided and structure-based mutations at four positions with random mutagenesis of a D-hydantoinase to give a 280-fold shift in selectivity.

When random mutagenesis is focused to a specific residue position, it is called saturation mutagenesis, which refers to the construction and testing of all possible amino acids at a single-residue position. It has long been the favored technique for exhaustive exploration of function arising from small regions of proteins. More recently, effective evolution strategies have been built around repeated application of saturation mutagenesis at the same sites (96) and blanketing the whole enzyme with saturation mutagenesis (41).

There are examples of semirational saturation mutagenesis on several different enzyme classes in the literature. Rui et al. (100) used a library of saturation mutagenesis at four sites to improve both k_{cat} and K_m of an epoxide hydrolase, resulting in a fivefold improvement in k_{cat}/K_m. Park et al. (91) used a homology model of an esterase to pick four active-site positions for individual saturation mutagenesis. They found mutations improving enantioselectivity by as much as fivefold and noted that saturation mutagenesis in the active site produced more selectivity than random mutation in the whole protein.

Although most reports in the literature deal with the construction of only a single set of saturation mutagenesis variants, evolution benefits from multiple rounds. Hill et al. (62) picked two active-site positions for simultaneous saturation mutagenesis by using a crystal structure of a phosphotriesterase, yielding a two-mutation variant with a 100-fold improvement in k_{cat}/K_m. They then followed up with saturation mutagenesis at another active-site position to gain an additional 10-fold.

For some catalytic mechanisms, the accessible transition states can determine the product selectivity of the reaction. Terpene cyclases utilize a high-energy carbocationic intermediate that can easily access many related structures before being quenched. Yoshikuni et al. (117) used product selectivity from saturation mutagenesis at 19 active-site positions in a sesquiterpene cyclase to guide rounds of rational mutations. A homology model was sufficient to pick the active-site residues. The final designed variants had at least 50% selectivity for seven different products.

31.8.2. Rationally Focused Libraries

As the number of library positions to sample increases or the screening resources decrease, it occasionally becomes prudent to explore less than all 20 amino acid types at each position. This type of focused library can explore concerted changes at multiple sites while keeping the number of variants to screen at a minimum. In general, one must suspect that synergistic mutations will be more beneficial than single mutations before this approach becomes effective. Reetz et al. (97) found the technique to be more effective than saturation mutagenesis or random mutagenesis. Schmitzer et al. (103) used rationally chosen degenerate codons at four active-site positions in a library format to achieve a sixfold activity improvement for the native substrate of a dihydrofolate reductase.

A focused library helped Koga et al. (75) invert the enantioselectivity of a lipase. They chose four leucine positions in the substrate-binding pocket by using a homology model and varied their residue identities within a hydrophobic set. The restriction drastically reduced the size of the library (from 20^4 to 7^4) while presumably increasing the fraction of variants with reasonable binding pockets. An interesting cell-free system was used for screening, and the best variants were found to also have beneficial random mutations.

As multiple-site libraries increase in complexity, so do the results from the screening. Truly beneficial mutations may occur only sparsely and in combination with neutral or even detrimental mutations. Random unintentional mutations are a common side effect of many library construction techniques. Statistical methods such as regression analysis can be very helpful to separate the good mutations from the bad (2, 54, 55). Interested readers might also examine the work from Ranganathan and coworkers that seeks to infer mutation energetics from multiple sequence alignments (101).

31.8.3. Single-Variant Design

The design of single variants is the most "rational" of the selectivity change categories. Much of the earliest work on rational enzyme design was done without the assistance of

libraries. Without libraries, other approaches must be taken to increase the chances of maintaining stability and finding improved activity. This is especially true when the variants contain more than one mutation. A number of researchers have used either a consensus approach or structural information to guide mutations.

Proteases were one of the first enzyme classes to receive substantial engineering attention. A minor glycine-to-alanine mutation in trypsin gave a 20-fold change in specificity (34). Carter et al. (21) used a modeled substrate in a subtilisin crystal structure to successfully design replacement of a catalytic histidine with a histidine from the substrate. Ballinger et al. (115) used comparison with eukaryotic homologs and electrostatic arguments to switch the subtilisin protease toward dibasic (6), and tribasic peptides (7). The specificity change was 60,000-fold for a tribasic substrate after three synergistic mutations (7). A single mutation in a hydrophobic pocket of the same enzyme was shown to give a specificity change of 200-fold for a different substrate (98).

Lipases are another class of enzymes that have received much engineering attention. Reetz's group tested several approaches, including random mutagenesis and site saturation mutagenesis on a *Pseudomonas aeruginosa* lipase (80). They later found focused libraries based on previous knowledge to be most effective (97). In a strictly rational approach, Magnusson et al. (83) chose a tryptophan on the base of the binding site of *Candida antarctica* lipase B (CALB) for size reduction. As hypothesized, enlarging the binding pocket allowed the enzyme to accommodate larger alcohols. Docking of the substrates showed steric clashes for the larger alcohols in the wild-type active site and explained specificity changes of more than 5 orders of magnitude for some substrate comparisons. In an opposite approach, Ema et al. (48) increased the size of a hydrophobic residue in the binding pocket of a *Burkholderia cepacia* lipase to constrain reaction of one enantiomer.

If done carefully, it is possible to make more than one rational mutation in a single variant. Multiple mutations will tend to destabilize proteins more than single mutations (see arguments presented with Fig. 4), so it is important to work with good information. The most useful guidance comes from a crystal structure of a related enzyme with the desired function. There are several examples of using a related enzyme structure to transplant function and/or selectivity.

Mutations at four positions designed from direct structure comparison helped convert an isocitrate dehydrogenase to an isopropylmalate dehydrogenase, although it was subsequently shown that random mutagenesis was more effective (43). A homology model was used to pick six active-site residues in an aspartate aminotransferase for substitution from tyrosine aminotransferase. Substitution improved activity on the intended substrate by 1,000-fold, but fell short of donor enzyme activity by 10-fold, providing evidence that effects outside the first sphere of residues are important (89). Structure comparison was used to rationally invert enantioselectivity of a vanillyl-alcohol oxidase with two mutations (110b). With use of a related crystal structure for information, a catalytic residue was moved to the opposite face of the active site. The best resulting enzyme had an enantiomeric excess of 80%, but lost approximately 100-fold in k_{cat}/K_m relative to wild type. Cheon et al. (27) used structure-guided mutations at four positions in a D-hydantoinase followed by random mutagenesis to give a 280-fold shift in specificity.

It is beneficial, but not strictly necessary, to use a related structure for comparison. The same consensus-guided approach used for stability enhancements can be used to sample selectivity variants. However, because it lowers the success rate, this is best done through a library approach. Chen et al. (26) used residue substitutions from related sequences in a consensus-type approach and specific structure-based pocket mutations to change the specificity of strictosidine synthase to accommodate larger substrates.

Successful structure-based enzyme engineering can be done without the aid of a related structure with the desired function. Crystal structure availability for a coenzyme A ligase complexed with a 4-chlorobenzoate substrate enabled rational expansion of substrate specificity (116). An isoleucine residue in the binding pocket was selected for replacement with a smaller alanine residue. After engineering, crystal structure evidence established that the mutation had the desired effect of accommodating another substituent on the substrate ring. A second round of mutations was beneficial; more structure-based mutations were made and a wide variety of substrates were tested, and selectivity changed by as much as 100-fold.

31.8.4. Substrate Docking for Selectivity Engineering

An accurate model of how the substrate fits into the active site and approaches the transition state is critical for rational engineering. In a rare example, Karimäki et al. (72) produced a crystal structure of a nonnative substrate bound to xylose isomerase and used it for redesign. They modeled the L-arabinose into the electron density at the active site and rationally selected two mutations that increased the reaction rate with the smaller C5 sugar by 2.8-fold. One mutation was selected to push the reacting oxygen closer to a catalytic metal. In typical circumstances, one is very unlikely to have crystal structure information for the substrate of interest.

Computational docking of the substrate into a crystal structure or a good homology model is a viable alternative. Gocke et al. (57) used previous modeling results and examination of an R-selective benzoylformate decarboxylase active site to locate and enlarge a small S-substrate pocket. Resulting enzymes were as much as 99% S-selective and a crystal structure of one mutant verified the mutation strategy.

Docking and design around transition state analogs (TSAs) is a logical and successful technique for selectivity design. It is based on the theory that molding the enzyme to best bind the transition state will decrease the activation energy (see Fig. 3). Selection for binding to TSAs has been used experimentally to engineer catalytic activity into antibodies. The technique produces catalytic antibodies, but may be limited because binding affinity rather than activity is directly selected (63). Improving binding to a TSA will decrease the energy level of the transition state, but it may decrease the bound substrate energy level similarly, resulting in no net decrease of the activation energy.

TSA docking has been used to make rational mutations to a carbonic anhydrase that increase the esterase rate for bulky substituents by 3,000-fold (64). It may also be used to rationalize known mutation effects. Bocola et al. (15) used a model TSA taken from more rigorous quantum chemical calculations to carefully explain the function of lipase mutations found by using random mutagenesis and other techniques. They were able to propose a mechanism for the effect of both a mutation to glycine and one to proline. Mutations involving glycine and proline are more likely than others to alter the secondary structure and complicate modeling. Later the same group built on the modeling and used it to guide experiments that further improved the enzyme (96).

Once docking of the substrate or TSA has been performed, it is possible to use computer algorithms to search for side-chain possibilities (rotamers) at selected positions. This repacking of the side chains is typically done on a fixed protein backbone. Side-chain repacking around an alternative DNA substrate was used to redesign the specificity of an endonuclease at a single base (5). In a related example, a DNA-binding antibody was converted to a ribonuclease by strategic introduction of a histidine without repacking, but with the help of a crystal structure (52). Lassila et al. incorporated a rigorous TSA and internal ligand flexibility into the repacking-driven redesign of an *Escherichia coli* chorismate mutase (79). They observed small but measurable improvements in both k_{cat} and K_m on the native substrate.

31.8.5. Grafting from Similar Enzymes

Grafting or change of substantial sequence sections at once produces large jumps in sequence and potentially function space. The downside is that the resulting enzymes tend to be destabilized because of incompatibilities at the interface of new and old sequence (see references 43 and 70 for examples). Destabilization can also result in unexpected effects such as domain swapping (31). However, library-based approaches have shown that swapping of large regions can be successful some fraction of the time (35, 88), especially when breakpoints are chosen to minimize broken or changed contacts (113). Despite the low likelihood of producing a stable functional protein, many interesting examples of loop grafting are available in the literature.

Loop grafting requires high-quality structural information, good modeling, and as much screening throughput as possible. In keeping with the expected power of a selection system, Park et al. (90) showed that it is possible to use extensive consensus-based loop changes to change the mechanism of an enzyme. A glyoxylase was converted to a metallo-β-lactamase using the guidance of a very dissimilar metallo-β-lactamase structure from the same fold family. Library sizes of ca. 10^7 kept the evolution on track while 81 of the 198 amino acids were changed. In contrast to the suggestion in Fig. 1, structural information and rational design was useful in this case even when a selection system was available.

In a successful effort to switch specificity of cofactor binding, Chen et al. (25) grafted a 13-residue stretch from an isocitrate dehydrogenase into an isopropylmalate dehydrogenase. The structures were superimposed, the stretch transplanted into the donor enzyme, and the resulting hybrid enzyme energy minimized to investigate the fit. During this process steric classes were noticed and three mutations were introduced into the final design to eliminate them.

Using a crystal structure of their starting enzyme and sequence from a related enzyme with the same mechanism and different specificity, Ma and Penning (82) found that grafting three long loops produced a stable enzyme with the target specificity. They also noted that making only residue substitutions predicted to contact the substrate was insufficient to change the selectivity. Even long loop grafts can result in stable enzymes if the scaffold is similar enough.

In an another encouraging report, Bocola et al. (16) removed an active-site loop in a phenyl acetone monooxygenase to accommodate a larger substrate. This may be a generally applicable strategy for situations when the desired substrate would make steric clashes with the backbone. Conversely, it may be possible to insert residues into an active site that is too open. Boersma et al. showed that inserting two- or three-residue-long loops from related crystal structures resulted in an inversion of enantioselectivity (17). Further site-directed saturation mutagenesis improved the enantioselectivity to almost 60%.

31.9. EXAMPLES OF COFACTOR ENGINEERING

Enzymatic cofactors enable many important reaction types by providing labile chemical elements such as hydrides and electrons. The type and orientation of an enzymatic cofactor can be critical in determining the activity and selectivity of an enzyme. There have been clear successes in the design of metal binding sites and the alteration of cofactor binding specificity that will be useful to those contemplating rational enzyme design.

31.9.1. Design of Metal Binding Sites

Several examples of spectroscopically verified metal binding sites have been introduced by rational design. An informational wealth of residue type and configurations that support metal binding sites are available in crystal structures and organized in accessible databases (22, 38). The strategy typically involves locating a site in a given protein fold where the backbone positions are compatible with the desired side-chain placement. Copper centers (11, 107) have been introduced by copying known sites. Loop grafting from a homolog also achieved the same goal (59). Iron binding sites have been successfully designed (12), as well as cuboidal iron-sulfur centers, using the DEZYMER program (32).

31.9.2. Interconversion of NADH and NADPH Cofactor Binding

With the emergence of rational enzyme engineering for in vivo applications and the continued importance of cofactor cost in pharmaceutical processes, it can be quite advantageous to control the nicotinamide adenine dinucleotide (NADH) cofactor usage. NADH and NADPH differ by a single phosphate group, which carries a formal -2 charge and provides a handle for recognition. An early example of structure-guided mutations was given by Scrutton et al. (104), who succeeded in switching the cofactor specificity of a glutathione reductase from NADPH to NADH. They used related enzymes to make consensus-style mutations.

Curiously, Flores and Ellington (53) were able to switch the cofactor specificity for a lactate dehydrogenase from NADH to NADPH by using consensus-guided mutations *outside* the phosphate binding pocket. Chen et al. (24) also found two mutations outside the phosphate binding pocket which, in combination with six other mutations, shifted selectivity by $>10^6$-fold from NADPH to NADH. They later used loop grafting to achieve the opposite selectivity change in another enzyme (25).

Kristan et al. (77) used comparative homology modeling to choose mutations in a 17β-hydroxysteroid dehydrogenase. Single mutations gave improvements of twofold to fivefold, and a single Y to D mutation switched cofactor specificity from NADPH to NAD. The most common effective strategy behind all the reports in the literature is the introduction of a negatively charged residue to mimic the phosphate charge or removal of positively charged phosphate binding residues. For the NAD to NADPH change, the opposite is done.

31.9.3. Active-Site de Novo Design

Design of enzyme active sites from first principles is beginning to be possible. Clever researchers have shown that it is sometimes possible to rationally redesign an enzyme to use a related mechanism (see references 19 and 92 for reviews). However, even when they exist, recognizing such engineering opportunities may be difficult. The de novo design work covered here uses generally applicable approaches to introduce catalytic activity. It introduces a number of intriguing possibilities. In its simplest form, it can be used to graft known active sites into a protein scaffold of interest. Still more promising in terms of applications is the possibility of introducing novel or nonnatural catalytic reaction mechanisms that enable useful transformations. It is an area where the benefits of rational design are clear and where almost any amount of effort spent in silico is worthwhile. For now, the primary practical goal of de novo active-site design should be to provide several proteins with measurable starting activity for further evolution.

Active-site design and introduction of de novo activity are difficult. The field has a history of failed validations (1, 33, 44) which seem to stem from difficulties in distinguishing low activity from background activity. Promising structure-based approaches use various search methods to match a desired active-site geometry with possible sites in known protein backbones (60, 118). The approach is flexible because the active-site geometry may include the catalytic residues, cofactors, and even neighboring residues. In this way, the proposed transition state of even nonnatural reactions can be specified.

Some of the earliest de novo design successes involved the introduction of catalytic metal centers. An active-site conformation for an iron superoxide dismutase was derived from known structures and used to redesign a thioredoxin. The single tested design had four mutations, three of them mutations to histidine for metal binding. Superoxide dismutase activity was observed and noted to be approximately 10^4 lower than natural enzymes (93). The same system was later revisited, and five separate superoxide dismutase designs in thioredoxin were characterized more extensively. Apparent correlations between redox potential or surrounding positive charge with activity were observed (12). The researchers also note that the deepest active site had the most shifted pK_a, confirming general expectations for shifts due to hydrophobic burial. These additional characterizations beyond activity are essential to further the rational design field and inform continued design on similar enzyme classes.

Similar successes at introducing known activity into a nonenzyme scaffold have been achieved with dock-and-design methods. Hydrolysis activity was introduced into thioredoxin by docking a TSA near a known histidine and molding the active site around it (18). A number of designs were tested, and two to four mutations were required to accommodate a *para*-nitrophenylacetate substrate. The researchers made use of the solvent accessibility of the TSA to choose designs for construction and testing. Some solvent accessibility is necessary to allow binding, but too much and the active site lacks the order it needs to function. Kaplan and DeGrado (71) took the dock-and-design approach further by applying it to a scaffold that itself had been designed to fold and form a helical bundle. They docked a 4-aminophenol TSA with two catalytic iron atoms and made accommodating mutations in the center of the bundle.

It has recently been demonstrated that active sites catalyzing nonnatural mechanisms can be designed into nonenzyme scaffolds. Researchers from the Baker group have been successful in constructing nonnatural enzymes that catalyze both the Kemp elimination (99) and retro-aldol reactions (67). These designs were made possible by a generalized approach combining hashing methods used in ligand docking with rotamer-based protein design methods (118). Two methods were developed to deal with the problem of side-chain position variability on the fringe of the active-site geometry specification. One matches a large set of variations on the same active site geometry to scaffolds and the other seeks to match independent placements of the catalytic residue rotamers to the transition state geometry.

For each of the two reaction mechanisms, multiple designs were constructed and tested in multiple scaffolds: 20 designs in 4 scaffolds for Kemp elimination and 72 designs in 10 scaffolds for the retro-aldol reaction. This work illustrates the benefit in trying many approaches to the same reaction. Our understanding of what makes a good enzyme is not complete and multiple trials are required to achieve evolution starting points with reasonable starting activity. The researchers also tested variations on the same mechanism involving different catalytic and binding residues: two for the Kemp elimination and four for the retro-aldol reaction. For the retro-aldol reaction, 32 of 72 designed constructs had measurable activity, putting the success rate over 40% for a reaction that requires multiple substrate-enzyme catalytic interactions. Quality assurance was provided by crystal structures of the tested variants and the resulting good superposition with the intended design.

31.10. CONCLUSIONS AND OUTLOOK

It should be clear from the examples provided in this chapter that rational computational-based design can play a strong role in evolving enzyme activity, especially when screening is difficult. As argued in other reviews, there are clear synergies between screening and rational design (3, 23, 28).

From the full set of selectivity engineering examples above, it is possible to draw some conclusions concerning what can be expected from rational and semirational approaches. A wide range of enzymes have been rationally engineered (more than 25 classes in more than 50 examples). Most of the efforts were focused in the active sites and resulted in substantial improvements by as much as several orders of magnitude in specificity or selectivity. Activity improvements were seen with substantial increases over wild type for nonnative substrates and even one example of a sixfold increase for a native substrate. Although little attention has been paid in these primarily academic reports of finishing the evolution for process requirements, it should be clear that these rational design techniques can also support evolution for industrial processes (55).

What is the best way to get started with structure-based design? After reading some of the work cited above, consider downloading a free or low-cost structure viewer such as PyMOL (39) and look at crystal structures related to your enzyme of interest. Pay attention to what the current most accepted mechanism implies for substrate positioning. Learn as much as possible from the literature and model the mutations you think might be beneficial.

31.10.1. Outlook for Future Advances

One area where one can expect improvements are contributions from more sophisticated calculations. The difficulty in achieving activities as high as natural enzymes is an ever-present theme in enzyme engineering. Improvements in potential energy calculations and configuration sampling could address this gap. None of the examples cited in this chapter use mixed quantum mechanical and molecular mechanical methods, molecular dynamics, or high-level ab initio quantum chemistry calculations. The only mention of free-energy perturbation methods was in the postexperimental validation of a design (72). The low usage presumably indicates that these methods are simply not yet easily accessible to the people who do the design, construction, and testing.

The area of de novo design also holds significant promise as both a tool to refine our understanding of enzymes and as a source for starting activity for enzyme evolution. Improvements in conformational sampling that have already been successful in loop design (65) can be expected to make contributions to active-site design. Authors of some recent successes have speculated that improvements in backbone flexibility handling (99) and discrimination during handling of polar networks (67) will be important factors in achieving high reaction rates common to natural enzymes. Jiang et al. (67) noted that their designed enzymes had low binding affinities. More sophisticated free-energy prediction methods, perhaps from the ligand-docking field, may play a useful role. For some reaction mechanisms, it will be necessary to incorporate negative design techniques to avoid side reactions. In any case, it is clear that interested readers can expect significant developments before the next edition of this manual.

REFERENCES

1. **Altamirano, M. M., J. M. Blackburn, C. Aguayo, and A. R. Fersht.** 2002. Retraction. (Altamirano, M. M., J. M. Blackburn, C. Aguayo, and A. R. Fersht. 2000. Directed evolution of new catalytic activity using the alpha/beta-barrel scaffold. *Nature* **403:**617–622). *Nature* **417:**468.

2. **Amin, N., A. D. Liu, S. Ramer, W. Aehle, D. Meijer, M. Metin, S. Wong, P. Gualfetti, and V. Schellenberger.** 2004. Construction of stabilized proteins by combinatorial consensus mutagenesis. *Protein Eng. Des. Sel.* **17:**787–793.

3. **Arnold, F. H.** 2001. Combinatorial and computational challenges for biocatalyst design. *Nature* **409:**253–257.

4. **Arnold, K., L. Bordoli, J. Kopp, and T. Schwede.** 2006. The SWISS-MODEL workspace: a web-based environment for protein structure homology modelling. *Bioinformatics* **22:**195–201.

5. **Ashworth, J., J. J. Havranek, C. M. Duarte, D. Sussman, R. J. J. Monnat, B. L. Stoddard, and D. Baker.** 2006. Computational redesign of endonuclease DNA binding and cleavage specificity. *Nature* **441:**656–659.

6. **Ballinger, M. D., J. Tom, and J. A. Wells.** 1995. Designing subtilisin BPN′ to cleave substrates containing dibasic residues. *Biochemistry* **34:**13312–13319.

7. **Ballinger, M. D., J. Tom, and J. A. Wells.** 1996. Furilisin: a variant of subtilisin BPN′ engineered for cleaving tribasic substrates. *Biochemistry* **35:**13579–13585.

8. **Bava, K. A., M. M. Gromiha, H. Uedaira, K. Kitajima, and A. Sarai.** 2004. ProTherm, version 4.0: thermodynamic database for proteins and mutants. *Nucleic Acids Res.* **32**(Database Issue):D120–D121.

9. **Beadle, B. M., and B. K. Shoichet.** 2002. Structural bases of stability-function tradeoffs in enzymes. *J. Mol. Biol.* **321:**285–296.

10. **Benkovic, S. J., and S. Hammes-Schiffer.** 2003. A perspective on enzyme catalysis. *Science* **301:**1196–1202.

11. **Benson, D. E., A. E. Haddy, and H. W. Hellinga.** 2002. Converting a maltose receptor into a nascent binuclear copper oxygenase by computational design. *Biochemistry* **41:**3262–3269.

12. **Benson, D. E., M. S. Wisz, and H. W. Hellinga.** 2000. Rational design of nascent metalloenzymes. *Proc. Natl. Acad. Sci. USA* **97:**6292–6297.

13. **Bloom, J. D., S. T. Labthavikul, C. R. Otey, and F. H. Arnold.** 2006. Protein stability promotes evolvability. *Proc. Natl. Acad. Sci. USA* **103:**5869–5874.

14. **Boas, F. E., and P. B. Harbury.** 2007. Potential energy functions for protein design. *Curr. Opin. Struct. Biol.* **17:**199–204.

15. **Bocola, M., N. Otte, K. E. Jaeger, M. T. Reetz, and W. Thiel.** 2004. Learning from directed evolution: theoretical investigations into cooperative mutations in lipase enantioselectivity. *Chembiochem* **5:**214–223.

16. **Bocola, M., F. Schulz, F. Leca, A. Vogel, M. W. Fraaije, and M. T. Reetz.** 2005. Converting phenylacetone monooxygenase into phenylcyclohexanone monooxygenase by rational design: towards practical Baeyer-Villiger monooxygenases. *Adv. Synth. Catal.* **347:**979–986.

17. **Boersma, Y. L., T. Pijning, M. S. Bosma, A. M. van der Sloot, L. F. Godinho, M. J. Dröge, R. T. Winter, G. van Pouderoyen, B. W. Dijkstra, and W. J. Quax.** 2008. Loop grafting of *Bacillus subtilis* lipase a: inversion of enantioselectivity. *Chem. Biol.* **15:**782–789.

18. **Bolon, D. N., and S. L. Mayo.** 2001. Enzyme-like proteins by computational design. *Proc. Natl. Acad. Sci. USA* **98:**14274–14279.

19. **Bornscheuer, U. T., and M. Pohl.** 2001. Improved biocatalysts by directed evolution and rational protein design. *Curr. Opin. Chem. Biol.* **5:**137–143.

20. **Cabrita, L. D., D. Gilis, A. L. Robertson, Y. Dehouck, M. Rooman, and S. P. Bottomley.** 2007. Enhancing the stability and solubility of TEV protease using in silico design. *Protein Sci.* **16:**2360–2367.

21. **Carter, P., B. Nilsson, J. P. Burnier, D. Burdick, and J. A. Wells.** 1989. Engineering subtilisin BPN′ for site-specific proteolysis. *Proteins* **6:**240–248.

22. **Castagnetto, J. M., S. W. Hennessy, V. A. Roberts, E. D. Getzoff, J. A. Tainer, and M. E. Pique.** 2002. MDB: the metalloprotein database and browser at the Scripps Research Institute. *Nucleic Acids Res.* **30:**379–382.

23. **Cedrone, F., A. Ménez, and E. Quéméneur.** 2000. Tailoring new enzyme functions by rational redesign. *Curr. Opin. Struct. Biol.* **10:**405–410.

24. **Chen, R., A. Greer, and A. M. Dean.** 1995. A highly active decarboxylating dehydrogenase with rationally inverted coenzyme specificity. *Proc. Natl. Acad. Sci. USA* **92:**11666–11670.

25. **Chen, R., A. Greer, and A. M. Dean.** 1996. Redesigning secondary structure to invert coenzyme specificity in isopropylmalate dehydrogenase. *Proc. Natl. Acad. Sci. USA* **93:**12171–12176.

26. **Chen, S., M. C. Galan, C. Coltharp, and S. E. O'Connor.** 2006. Redesign of a central enzyme in alkaloid biosynthesis. *Chem. Biol.* **13:**1137–1141.

27. **Cheon, Y. H., H. S. Park, J. H. Kim, Y. Kim, and H. S. Kim.** 2004. Manipulation of the active site loops of d-hydantoinase, a (beta/alpha)8-barrel protein, for modulation of the substrate specificity. *Biochemistry* **43:**7413–7420.

28. **Chica, R. A., N. Doucet, and J. N. Pelletier.** 2005. Semi-rational approaches to engineering enzyme activity: combining the benefits of directed evolution and rational design. *Curr. Opin. Biotechnol.* **16:**378–384.

29. **Choi, E. J., and S. L. Mayo.** 2006. Generation and analysis of proline mutants in protein G. *Protein Eng. Des. Sel.* **19:**285–289.

30. **Choudhury, N., and B. M. Pettitt.** 2006. Enthalpy-entropy contributions to the potential of mean force of nanoscopic hydrophobic solutes. *J. Phys. Chem. B* **110:**8459–8463.

31. **Clark, L. A., P. A. Boriack-Sjodin, J. Eldredge, C. Fitch, M. Jarpe, S. Miller, Y. Li, K. Simon, and H. W. van Vlijmen.** 2008. An antibody loop replacement design feasibility study and a loop-swapped dimer structure. *Protein Eng. Des. Sel.* **22:**93–101.

32. **Coldren, C. D., H. W. Hellinga, and J. P. Caradonna.** 1997. The rational design and construction of a cuboidal iron-sulfur protein. *Proc. Natl. Acad. Sci. USA* **94:**6635–6640.

33. **Corey, M. J., and E. Corey.** 1996. On the failure of de novo-designed peptides as biocatalysts. *Proc. Natl. Acad. Sci. USA* **93:**11428–11434.

34. **Craik, C. S., C. Largman, T. Fletcher, S. Roczniak, P. J. Barr, R. Fletterick, and W. J. Rutter.** 1985. Redesigning trypsin: alteration of substrate specificity. *Science* **228:**291–297.

35. **Crameri, A., S. A. Raillard, E. Bermudez, and W. P. Stemmer.** 1998. DNA shuffling of a family of genes from diverse species accelerates directed evolution. *Nature* **391:**288–291.

36. **Dantas, G., B. Kuhlman, D. Callender, M. Wong, and D. Baker.** 2003. A large scale test of computational protein design: folding and stability of nine completely redesigned globular proteins. *J. Mol. Biol.* **332:**449–460.

37. **DeGrado, W. F., C. M. Summa, V. Pavone, F. Nastri, and A. Lombardi.** 1999. De novo design and structural characterization of proteins and metalloproteins. *Annu. Rev. Biochem.* **68:**779–819.

38. **Degtyarenko, K. N., A. C. North, and J. B. Findlay.** 1999. PROMISE: a database of bioinorganic motifs. *Nucleic Acids Res.* **27:**233–236.

39. **DeLano, W.** 2002. The PyMOL molecular graphics system. DeLano Scientific, Palo Alto, CA. http://www.pymol.org.

40. Reference deleted.

41. **Desantis, G., K. Wong, B. Farwell, K. Chatman, Z. Zhu, G. Tomlinson, H. Huang, X. Tan, L. Bibbs, P. Chen, K. Kretz, and M. J. Burk.** 2003. Creation of a productive, highly enantioselective nitrilase through gene site saturation mutagenesis (GSSM). *J. Am. Chem. Soc.* **125:**11476–11477.

42. **Ditursi, M. K., S. J. Kwon, P. J. Reeder, and J. S. Dordick.** 2006. Bioinformatics-driven, rational engineering of protein thermostability. *Protein Eng. Des. Sel.* **19:**517–524.

43. **Doyle, S. A., S. Y. Fung, and D. E. J. Koshland.** 2000. Redesigning the substrate specificity of an enzyme: isocitrate dehydrogenase. *Biochemistry* **39:**14348–14355.

44. **Dwyer, M. A., L. L. Looger, and H. W. Hellinga.** 2008. Retraction (Dwyer, M. A., L. L. Looger, and H. W. Hellinga. 2004. Computational design of a biologically active enzyme. *Science* **304:**1967–1971). *Science* **319:**569.

45. Reference deleted.

46. **Eijsink, V. G., A. Bjork, S. Gaseidnes, R. Sirevag, B. Synstad, B. van den Burg, and G. Vriend.** 2004. Rational engineering of enzyme stability. *J. Biotechnol.* **113:**105–120.

47. **Eisenmesser, E. Z., O. Millet, W. Labeikovsky, D. M. Korzhnev, M. Wolf-Watz, D. A. Bosco, J. J. Skalicky, L. E. Kay, and D. Kern.** 2005. Intrinsic dynamics of an enzyme underlies catalysis. *Nature* **438:**117–121.

48. **Ema, T., T. Fujii, M. Ozaki, T. Korenaga, and T. Sakai.** 2005. Rational control of enantioselectivity of lipase by site-directed mutagenesis based on the mechanism. *Chem. Commun. (Camb.)* **4650:**4650–4651.

49. **Fenel, F., M. Leisola, J. Jänis, and O. Turunen.** 2004. A de novo designed n-terminal disulphide bridge stabilizes the *Trichoderma reesei* endo-1,4-beta-xylanase II. *J. Biotechnol.* **108:**137–143.

50. **Fersht, A.** 1998. *Structure and Mechanism in Protein Science: A Guide to Enzyme Catalysis and Protein Folding.* W. H. Freeman & Co., New York, NY.

51. **Field, M. J.** 2002. Simulating enzyme reactions: challenges and perspectives. *J. Comp. Chem.* **23:**48–58.

52. **Fletcher, M. C., A. Kuderova, M. Cygler, and J. S. Lee.** 1998. Creation of a ribonuclease abzyme through site-directed mutagenesis. *Nat. Biotechnol.* **16:**1065–1067.

53. **Flores, H., and A. D. Ellington.** 2005. A modified consensus approach to mutagenesis inverts the cofactor specificity of *Bacillus stearothermophilus* lactate dehydrogenase. *Protein Eng. Des. Sel.* **18:**369–377.

54. **Fox, R., A. Roy, S. Govindarajan, J. Minshull, C. Gustafsson, J. T. Jones, and R. Emig.** 2003. Optimizing the search algorithm for protein engineering by directed evolution. *Protein Eng.* **16:**589–597.

55. **Fox, R. J., S. C. Davis, E. C. Mundorff, L. M. Newman, V. Gavrilovic, S. K. Ma, L. M. Chung, C. Ching, S. Tam, S. Muley, J. Grate, J. Gruber, J. C. Whitman, R. A. Sheldon, and G. W. Huisman.** 2007. Improving catalytic function by ProSAR-driven enzyme evolution. *Nat. Biotechnol.* **25:**338–344.

56. **Garcia-Viloca, M., J. Gao, M. Karplus, and D. G. Truhlar.** 2004. How enzymes work: analysis by modern rate theory and computer simulations. *Science* **303:**186–195.

57. **Gocke, D., L. Walter, E. Gauchenova, G. Kolter, M. Knoll, C. L. Berthold, G. Schneider, J. Pleiss, M. Müller, and M. Pohl.** 2008. Rational protein design of ThDP-dependent enzymes—engineering stereoselectivity. *Chembiochem* **9:**406–412.

58. **Guerois, R., J. E. Nielsen, and L. Serrano.** 2002. Predicting changes in the stability of proteins and protein complexes: a study of more than 1000 mutations. *J. Mol. Biol.* **320:**369–387.

59. **Hay, M., J. H. Richards, and Y. Lu.** 1996. Construction and characterization of an azurin analog for the purple copper site in cytochrome c oxidase. *Proc. Natl. Acad. Sci. USA* **93:**461–464.

60. **Hellinga, H. W., and F. M. Richards.** 1991. Construction of new ligand binding sites in proteins of known structure. I. Computer-aided modeling of sites with pre-defined geometry. *J. Mol. Biol.* **222:**763–785.

61. Reference deleted.

62. **Hill, C. M., W. S. Li, J. B. Thoden, H. M. Holden, and F. M. Raushel.** 2003. Enhanced degradation of chemical warfare agents through molecular engineering of the phosphotriesterase active site. *J. Am. Chem. Soc.* **125:**8990–8991.

63. **Hilvert, D.** 2000. Critical analysis of antibody catalysis. *Annu. Rev. Biochem.* **69:**751–793.

64. **Höst, G., L. G. Mårtensson, and B. H. Jonsson.** 2006. Redesign of human carbonic anhydrase II for increased esterase activity and specificity towards esters with long acyl chains. *Biochim. Biophys. Acta* **1764:**1601–1606.

65. **Hu, X., H. Wang, H. Ke, and B. Kuhlman.** 2007. High-resolution design of a protein loop. *Proc. Natl. Acad. Sci. USA* **104:**17668–17673.

66. **Jeong, M. Y., S. Kim, C. W. Yun, Y. J. Choi, and S. G. Cho.** 2007. Engineering a de novo internal disulfide bridge to improve the thermal stability of xylanase from *Bacillus stearothermophilus* no. 236. *J. Biotechnol.* **127:**300–309.

67. **Jiang, L., E. A. Althoff, F. R. Clemente, L. Doyle, D. Röthlisberger, A. Zanghellini, J. L. Gallaher, J. L. Betker, F. Tanaka, C. F. Barbas III, D. Hilvert, K. N. Houk, B. L. Stoddard, and D. Baker.** 2008. De novo computational design of retro-aldol enzymes. *Science* **319:**1387–1391.

68. **Jorgensen, W. L.** 1989. Free energy calculations: a breakthrough for modeling organic chemistry in solution. *Acc. Chem. Res.* **22:**184–189.

69. **Kabashima, T., Y. Li, N. Kanada, K. Ito, and T. Yoshimoto.** 2001. Enhancement of the thermal stability of pyroglutamyl peptidase I by introduction of an intersubunit disulfide bond. *Biochim. Biophys. Acta* **1547:**214–220.

70. **Kagami, O., K. Shindo, A. Kyojima, K. Takeda, H. Ikenaga, K. Furukawa, and N. Misawa.** 2008. Protein engineering on biphenyl dioxygenase for conferring activity to convert 7-hydroxyflavone and 5,7-dihydroxyflavone (chrysin). *J. Biosci. Bioeng.* **106:**121–127.

71. **Kaplan, J., and W. F. DeGrado.** 2004. De novo design of catalytic proteins. *Proc. Natl. Acad. Sci. USA* **101:**11566–11570.

72. **Karimäki, J., T. Parkkinen, H. Santa, O. Pastinen, M. Leisola, J. Rouvinen, and O. Turunen.** 2004. Engineering the substrate specificity of xylose isomerase. *Protein Eng. Des. Sel.* **17:**861–869.

73. **Ko, J. H., W. H. Jang, E. K. Kim, H. B. Lee, K. D. Park, J. H. Chung, and O. J. Yoo.** 1996. Enhancement of thermostability and catalytic efficiency of AprP, an alkaline protease from *Pseudomonas* sp., by the introduction of a disulfide bond. *Biochem. Biophys. Res. Commun.* **221:**631–635.

74. **Koch, W., and M. C. Holthausen.** 2000. *A Chemist's Guide to Density Functional Theory.* John Wiley & Sons, Inc., Hoboken, NJ.

75. **Koga, Y., K. Kato, H. Nakano, and T. Yamane.** 2003. Inverting enantioselectivity of *Burkholderia cepacia* KWI-56 lipase by combinatorial mutation and high-throughput screening using single-molecule PCR and in vitro expression. *J. Mol. Biol.* **331:**585–592.

76. **Korkegian, A., M. E. Black, D. Baker, and B. L. Stoddard.** 2005. Computational thermostabilization of an enzyme. *Science* **308:**857–860.

77. **Kristan, K., J. Stojan, J. Adamski, and T. L. Rizner.** 2007. Rational design of novel mutants of fungal 17beta-hydroxysteroid dehydrogenase. *J. Biotechnol.* **129:**123–30.

78. **Larson, S. M., A. A. Di Nardo, and A. R. Davidson.** 2000. Analysis of covariation in an SH3 domain sequence alignment: applications in tertiary contact prediction and the design of compensating hydrophobic core substitutions. *J. Mol. Biol.* **303:**433–446.

79. **Lassila, J. K., J. R. Keeffe, P. Oelschlaeger, and S. L. Mayo.** 2005. Computationally designed variants of *Escherichia coli* chorismate mutase show altered catalytic activity. *Protein Eng. Des. Sel.* **18:**161–163.

80. **Liebeton, K., A. Zonta, K. Schimossek, M. Nardini, D. Lang, B. W. Dijkstra, M. T. Reetz, and K. E. Jaeger.** 2000. Directed evolution of an enantioselective lipase. *Chem. Biol.* **7:**709–718.

81. **Luetz, S., L. Giver, and J. Lalonde.** 2008. Engineered enzymes for chemical production. *Biotechnol. Bioeng.* **101:**647–653.

82. **Ma, H., and T. M. Penning.** 1999. Conversion of mammalian 3alpha-hydroxysteroid dehydrogenase to 20alpha-hydroxysteroid dehydrogenase using loop chimeras: changing specificity from androgens to progestins. *Proc. Natl. Acad. Sci. USA* **96:**11161–11166.

83. **Magnusson, A. O., J. C. Rotticci-Mulder, A. Santagostino, and K. Hult.** 2005. Creating space for large secondary alcohols by rational redesign of *Candida antarctica* lipase B. *Chembiochem* **6:**1051–1056.

84. **Masgrau, L., A. Roujeinikova, L. O. Johannissen, P. Hothi, J. Basran, K. E. Ranaghan, A. J. Mulholland, M. J. Sutcliffe, N. S. Scrutton, and D. Leys.** 2006. Atomic description of an enzyme reaction dominated by proton tunneling. *Science* **312:**237–241.

85. **Matthews, B. W., H. Nicholson, and W. J. Becktel.** 1987. Enhanced protein thermostability from site-directed mutations that decrease the entropy of unfolding. *Proc. Natl. Acad. Sci. USA* **84:**6663–6667.

86. **May, O., P. T. Nguyen, and F. H. Arnold.** 2000. Inverting enantioselectivity by directed evolution of hydantoinase for improved production of L-methionine. *Nat. Biotechnol.* **18:**317–320.

87. **Morley, K. L., and R. J. Kazlauskas.** 2005. Improving enzyme properties: when are closer mutations better? *Trends Biotechnol.* **23:**231–237.

88. Ness, J. E., S. Kim, A. Gottman, R. Pak, A. Krebber, T. V. Borchert, S. Govindarajan, E. C. Mundorff, and J. Minshull. 2002. Synthetic shuffling expands functional protein diversity by allowing amino acids to recombine independently. *Nat. Biotechnol.* 20:1251–1255.

89. Onuffer, J. J., and J. F. Kirsch. 1995. Redesign of the substrate specificity of *Escherichia coli* aspartate aminotransferase to that of *Escherichia coli* tyrosine aminotransferase by homology modeling and site-directed mutagenesis. *Protein Sci.* 4:1750–1757.

90. Park, H. S., S. H. Nam, J. K. Lee, C. N. Yoon, B. Mannervik, S. J. Benkovic, and H. S. Kim. 2006. Design and evolution of new catalytic activity with an existing protein scaffold. *Science* 311:535–538.

91. Park, S., K. L. Morley, G. P. Horsman, M. Holmquist, K. Hult, and R. J. Kazlauskas. 2005. Focusing mutations into the *P. fluorescens* esterase binding site increases enantioselectivity more effectively than distant mutations. *Chem. Biol.* 12:45–54.

92. Penning, T. M., and J. M. Jez. 2001. Enzyme redesign. *Chem. Rev.* 101:3027–3046.

93. Pinto, A. L., H. W. Hellinga, and J. P. Caradonna. 1997. Construction of a catalytically active iron superoxide dismutase by rational protein design. *Proc. Natl. Acad. Sci. USA* 94:5562–5567.

94. Pollard, D. J., and J. M. Woodley. 2007. Biocatalysis for pharmaceutical intermediates: the future is now. *Trends Biotechnol.* 25:66–73.

95. Reetz, M. T. 2004. Changing the enantioselectivity of enzymes by directed evolution. *Methods Enzymol.* 388:238–256.

96. Reetz, M. T., and J. D. Carballeira. 2007. Iterative saturation mutagenesis (ISM) for rapid directed evolution of functional enzymes. *Nat. Protoc.* 2:891–903.

97. Reetz, M. T., S. Wilensek, D. Zha, and K. E. Jaeger. 2001. Directed evolution of an enantioselective enzyme through combinatorial multiple-cassette mutagenesis. *Angew. Chem. Int. Ed. Engl.* 40:3589–3591.

98. Rheinnecker, M., G. Baker, J. Eder, and A. R. Fersht. 1993. Engineering a novel specificity in subtilisin BPN'. *Biochemistry* 32:1199–1203.

99. Röthlisberger, D., O. Khersonsky, A. M. Wollacott, L. Jiang, J. Dechancie, J. Betker, J. L. Gallaher, E. A. Althoff, A. Zanghellini, O. Dym, S. Albeck, K. N. Houk, D. S. Tawfik, and D. Baker. 2008. Kemp elimination catalysts by computational enzyme design. *Nature* 453:190–195.

100. Rui, L., L. Cao, W. Chen, K. F. Reardon, and T. K. Wood. 2004. Active site engineering of the epoxide hydrolase from *Agrobacterium radiobacter* AD1 to enhance aerobic mineralization of cis-1,2-dichloroethylene in cells expressing an evolved toluene ortho-monooxygenase. *J. Biol. Chem.* 279:46810–46817.

101. Russ, W. P., and R. Ranganathan. 2002. Knowledge-based potential functions in protein design. *Curr. Opin. Struct. Biol.* 12:447–452.

102. Sali, A., and T. L. Blundell. 1993. Comparative protein modelling by satisfaction of spatial restraints. *J. Mol. Biol.* 234:779–815.

103. Schmitzer, A. R., F. Lépine, and J. N. Pelletier. 2004. Combinatorial exploration of the catalytic site of a drug-resistant dihydrofolate reductase: creating alternative functional configurations. *Protein Eng. Des. Sel.* 17:809–819.

104. Scrutton, N. S., A. Berry, and R. N. Perham. 1990. Redesign of the coenzyme specificity of a dehydrogenase by protein engineering. *Nature* 343:38–43.

105. Seelig, B., and J. W. Szostak. 2007. Selection and evolution of enzymes from a partially randomized non-catalytic scaffold. *Nature* 448:828–831.

106. Shoichet, B. K., W. A. Baase, R. Kuroki, and B. W. Matthews. 1995. A relationship between protein stability and protein function. *Proc. Natl. Acad. Sci. USA* 92:452–456.

107. Sigman, J. A., B. C. Kwok, and Y. Lu. 2000. From myoglobin to heme-copper oxidase: design and engineering of a CuB center into sperm whale myoglobin. *J. Am. Chem. Soc.* 122:8192–8196.

108. Steipe, B., B. Schiller, A. Plückthun,, and S. Steinbacher. 1994. Sequence statistics reliably predict stabilizing mutations in a protein domain. *J. Mol. Biol.* 240:188–192.

109. Takagi, H., T. Takahashi, H. Momose, M. Inouye, Y. Maeda, H. Matsuzawa, and T. Ohta. 1990. Enhancement of the thermostability of subtilisin e by introduction of a disulfide bond engineered on the basis of structural comparison with a thermophilic serine protease. *J. Biol. Chem.* 265:6874–6878.

110. Tang, L., D. E. T. Pazmino, M. W. Fraaije, R. M. D. Jong, B. W. Dijkstra, and D. B. Janssen. 2005. Improved catalytic properties of halohydrin dehalogenase by modification of the halide-binding site. *Biochemistry* 44:6609–6618.

110a. van den Burg, B., B. W. Dijkstra, B. van der Vinne, B. K. Stulp, V. G. Eijsink, and G. Venema. 1993. Introduction of disulfide bonds into *Bacillus subtilis* neutral protease. *Protein Eng.* 6:521–527.

110b. van den Heuvel, R. H., M. W. Fraaije, M. Ferrer, A. Mattevi, and W. J. van Berkel. 2000. Inversion of stereospecificity of vanillyl-alcohol oxidase. *Proc. Natl. Acad. Sci. USA* 97:9455–9460.

110c. Vázquez-Figueroa, E., J. Chaparro-Riggers, and A. S. Bommarius. 2007. Development of a thermostable glucose dehydrogenase by a structure-guided consensus concept. *Chembiochem* 8:2295–2301.

111. Vazquez-Figueroa, E., V. Yeh, J. M. Broering, J. F. Chaparro-Riggers, and A. S. Bommarius. 2008. Thermostable variants constructed via the structure-guided consensus method also show increased stability in salts solutions and homogeneous aqueous-organic media. *Protein Eng. Des. Sel.* 21:673–680.

112. Vieille, C., and G. J. Zeikus. 2001. Hyperthermophilic enzymes: sources, uses, and molecular mechanisms for thermostability. *Microbiol. Mol. Biol. Rev.* 65:1–43.

113. Voigt, C. A., C. Martinez, Z. G. Wang, S. L. Mayo, and F. H. Arnold. 2002. Protein building blocks preserved by recombination. *Nat. Struct. Biol.* 9:553–558.

114. Wakarchuk, W. W., W. L. Sung, R. L. Campbell, A. Cunningham, D. C. Watson, and M. Yaguchi. 1994. Thermostabilization of the *Bacillus circulans* xylanase by the introduction of disulfide bonds. *Protein Eng.* 7:1379–1386.

115. Wells, J. A., D. B. Powers, R. R. Bott, T. P. Graycar, and D. A. Estell. 1987. Designing substrate specificity by protein engineering of electrostatic interactions. *Proc. Natl. Acad. Sci. USA* 84:1219–1223.

116. Wu, R., A. S. Reger, J. Cao, A. M. Gulick, and D. Dunaway-Mariano. 2007. Rational redesign of the 4-chlorobenzoate binding site of 4-chlorobenzoate: coenzyme A ligase for expanded substrate range. *Biochemistry* 46:14487–14499.

117. Yoshikuni, Y., T. E. Ferrin, and J. D. Keasling. 2006. Designed divergent evolution of enzyme function. *Nature* 440:1078–1082.

118. Zanghellini, A., L. Jiang, A. M. Wollacott, G. Cheng, J. Meiler, E. A. Althoff, D. Rothlisberger, and D. Baker. 2006. New algorithms and an in silico benchmark for computational enzyme design. *Protein Sci.* 15:2785–2794.

119. Zhang, X., and K. N. Houk. 2005. Why enzymes are proficient catalysts: beyond the Pauling paradigm. *Acc. Chem. Res.* 38:379–385.

Enzyme Engineering by Directed Evolution

MANFRED T. REETZ

32

32.1. INTRODUCTION

Enzymes have been used as catalysts in synthetic organic chemistry and in biotechnology for at least a century, but it was not until the past few decades that the chemical community has finally accepted biocatalysis as a truly useful tool in synthetic endeavors. To some researchers it may seem surprising that many enzymes not only accept numerous nonnatural substrates, but also turn them over with high degrees of chemo-, regio-, and/or stereoselectivity (11). Moreover, the discovery that many enzymes are also active in organic solvents (although at much lower rates than in the usual aqueous medium) originally came as a surprise, but is now state of the art. Progress in bioreactor design and process engineering has also been instrumental in making biocatalysis a viable option in industry, including the use of whole cells (26). Nevertheless, until the mid-1990s, many limitations to biocatalysis remained.

1. Numerous (if not the majority of) synthetic substrates of interest to chemists and/or biotechnologists are not accepted by enzymes, or react so slowly that enzyme catalysis is not feasible.
2. Enantioselectivity may well be poor, even if activity is acceptable.
3. Some enzymes, although active and enantioselective in a given transformation, may not be thermally stable enough for practical applications.
4. Many enzymes do not survive well enough in hostile organic solvents to be of practical use.

Traditional approaches to solve these problems include immobilization techniques, posttranslational modification, the use of additives, and/or protein engineering utilizing site-specific mutagenesis based on rational design, but unfortunately these methods are far from general (11, 26). In contrast, directed evolution, meaning Darwinian evolution in the test tube, has emerged during the past 15 years as the method of choice for addressing these problems (1, 21, 28, 40). The roots of directed evolution go back to the contributions of such researchers as Spiegelman and Eigen, but it was not until the mid-1980s that molecular biologists began to develop reliable gene mutagenesis methods required for the overall process.

Today, directed evolution of proteins comprises the following steps. The gene of the wild-type (WT) enzyme is used as a template for producing a library of mutated genes, which is then inserted into an appropriate bacterial host needed for (over)expressing the corresponding enzyme variants. Upon plating out on agar plates, the bacterial colonies containing the transformants are harvested by a colony picker and placed in the wells of microtiter plates containing nutrient broth. Since each colony originates from a single cell, undesired mixtures of mutants never form. Following another step, the enzyme variants are tested as catalysts in a given reaction of interest by using an appropriate high-throughput screen (or selection method). The gene of the best enzyme variant (hit) is then used as a template for another cycle, and the Darwinian process is repeated as often as necessary until the desired degree of catalyst improvement has been achieved (Fig. 1). The most commonly used mutagenesis methods are error-prone polymerase chain reaction (epPCR), saturation mutagenesis, and DNA shuffling (section 32.3).

The problems in putting this form of laboratory evolution into practice revolve around the need to devise optimal gene mutagenesis methods and strategies for probing protein sequence space efficiently (1, 28, 40), and the necessity to develop fast and reliable methods for high-throughput screening or selection (52). Because of the screening problem, laboratory evolution of enantioselective enzymes constitutes a special challenge (40). In any directed evolution study of proteins, it is necessary to ensure the linkage between genotype and phenotype, the former referring to the information stored by the gene in the form of nucleic acids (which can be replicated), and the latter referring to the corresponding protein variants. In the case of nucleic acid libraries of ribozymes, aptamers, and DNAzymes, a well-developed research area (22), genotype and phenotype are the same. However, such biocatalysts are not as versatile or as stable as protein-based enzymes, and nature itself uses the latter in the vast majority of catalytic processes. This chapter treats exclusively directed evolution of proteins, the primary focus being on recent methodology developments.

32.2. SCREENING VERSUS SELECTION

The need to develop high-throughput screening methods (or selection) in directed evolution relates to the vast extent of protein sequence space (39, 52). The number of enzyme variants N at the theoretically maximum degree of diversity is traditionally described by the algorithm $N = 19^M X!/(X - M)!M!$, where M denotes the total number of amino acid substitutions per enzyme molecule

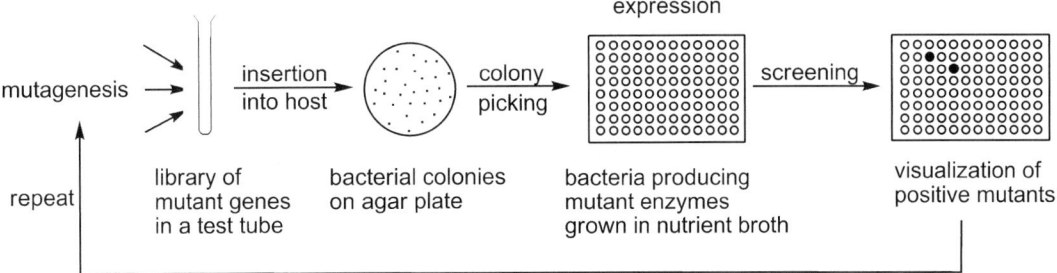

FIGURE 1 Individual steps in directed evolution.

and X is the number of residues in the enzyme molecule. When considering an enzyme composed of 300 residues, for example, 5,700 mutants are possible if one amino acid is substituted randomly, but 16 million in the case of two simultaneous substitutions, and about 30 billion if three amino acids are exchanged simultaneously. In practice, the size of libraries of enzyme variants ranges typically between 10^3 and 10^6 transformants. Screening using some sort of an analytical technique follows the expression of enzyme variants, whereas selection, in the strict sense of the word, relies on the survival of the host organism that harbors an improved enzyme variant, thereby allowing for very large gene libraries. The concepts are sometimes confused in the literature.

In most directed evolution studies some sort of screening is applied, often in combination with a crude pretest that assesses activity and thereby automatically eliminates "dead" variants (52). An example of an efficient pretest is the traditional tributyrin assay for lipase activity in which the agar plates are charged with tributyrin undergoing lipase-catalyzed hydrolysis, active variants being indicated by the appearance of halos on the agar plates. Such tests are important because directed evolution produces enzyme libraries in which the vast majority of variants are inactive.

Many different high-throughput screening systems have been developed (52), including those that assess enantioselectivity (39). Conventional chiral gas chromatography or high-performance liquid chromatography can handle only a few dozen samples per day, but cleverly designed assays based on UV-visible, infrared, or nuclear magnetic resonance spectroscopy, mass spectrometry, fluorescence or circular dichroism have been devised, allowing typically the determination of 800 to 5,000 ee-values per day (ee = enantiomeric excess), depending upon the nature of the substrate. In favorable cases automated gas chromatography or high-performance liquid chromatography can be used to assay up to 400 to 500 samples. Pooling strategies can also be used to alleviate the screening problem (37), which to this day constitutes the bottleneck in directed evolution.

In nature, compartmentalization of genes in cells ensures the genotype-phenotype linkage. Inspired by this phenomenon, droplet-based strategies for in vitro compartmentalization have been developed (16). By use of oil, detergents, and emulsifiers, emulsions with droplets having diameters of approximately 2 μm are easily constructed, "mimicking" natural cells. The members of a gene library are then partitioned into microscopic compartments in such a way that one copy comes to exist in each droplet, and in vitro protein expression is used to synthesize multiple copies of the encoded protein. Since the droplets are so small, large libraries (10^6 to 10^8 members) can be generated and handled. Several variations of this technique have been reported, including systems in which the droplets are the sole connection between genotype and phenotype and DNA display in such droplets using covalent or noncovalent links or even beads (16). It remains to be seen how this technique will develop in the future, as for example in the directed evolution of enantioselective enzymes, and how it compares with other approaches.

Various systems for genotype-phenotype linkages have been reviewed (1, 4, 16, 28), including display techniques such as ribozyme display, phage display, bacterial surface display, and yeast surface display. Sometimes the desired mutants are isolated or enriched directly from a suspension of the corresponding display species, while in other systems analytical instruments such as the fluorescence-activated cell sorting (FACS) need to be invoked. The numbers describing the size of the libraries in all of these systems are high (10^6 to 10^{10}), and, indeed, successful examples of directed evolution of proteins have been reported (1, 4, 16, 28). Nevertheless, the question of generality and ease of performance needs to be addressed. For example, phage display is primarily suited to handle the binding properties of proteins. That is probably why attempts to apply phage display in the directed evolution of enantioselective enzymes (as in the case of a lipase) have not been exceedingly rewarding so far (12). In contrast, a promising approach to yeast surface display for selecting horseradish peroxidase variants as catalysts in the stereoselective dimerization of tyrosinol has been devised, the synthesis of fluorescence-labeled enantiomeric tyrosinol substrates being necessary (27). One was immobilized on the surface of live yeast cells (together with the library of enzyme variants), and the other was supplied in solution and used by the active variants to label those cells that express the active enzyme variants. A library of approximately 2×10^6 horseradish peroxidase variants was generated by saturation mutagenesis at five positions next to the binding pocket. The library was subjected to FACS analysis, once for enantioselectivity favoring D-tyrosinol over the L-enantiomer, and once favoring the reverse. Variants with up to eightfold-altered enantioselectivity toward L/D-tyrosinol were identified, including those with reversed stereoselectivity (27).

A system for single-cell, super-high-throughput screening to identify enantioselective esterase mutants as catalysts in the hydrolytic kinetic resolution of chiral esters also deserves mention (3). The FACS-based concept requires the two enantiomeric esters to be labeled, each with a different fluorescent dye (green/red). Appropriately labeled (R)- and (S)-tyramide esters as substrates were subjected to hydrolytic kinetic resolution, peroxidase-mediated radical formation ensuring the immediate covalent attachment of the reaction products to the surface of an esterase-proficient bacterial cell (*Escherichia coli*) (3). The system allows 10^8 cells, and thus this number of clones is screened within a few hours. The goal was to reverse

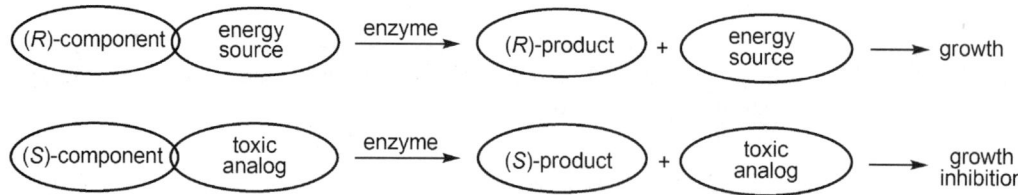

FIGURE 2 Genetic selection system for laboratory evolution of enantioselectivity in a kinetic resolution (43).

the sense of enantioselectivity. In a proof-of-principle study, epPCR was applied at an error rate corresponding to the introduction of two to four amino acid substitutions per enzyme molecule. The best hit was found to display a selectivity factor of $E = 16$ in favor of the (R)-enantiomer [WT: $E = 1.2$ in favor of the (S)-substrate]. More work is required here, especially to see how far enantioselectivity can be boosted with epPCR or with other gene mutagenesis methods.

As already stated, selection systems can handle larger libraries than conventional screening assays, meaning libraries of gene mutants versus libraries of protein variants (21, 52). The question of whether the generation of super-large libraries is really desirable, certainly relative to the use of small but super-high-quality protein libraries, will be addressed in section 32.4. Here, a short discussion regarding selection is in order. Experimental platforms that link microbial survival as detected by growth rate with a catalytic parameter of an enzyme are needed. Growth conditions are required so that, from libraries comprising 10^6 to 10^{10} gene mutants, only those cells that express improved enzyme variants appear on the agar plates. In most successful examples of selection systems, the necessary physiological effects are straightforward and easy to predict (1, 21, 28). In most cases, substrate mimics are used for a defined reaction to be catalyzed in a way that fulfills the metabolic requirement. Further requirements concern the necessity of water solubility of the substrates and the prerequisite that the substrates and reaction products do not interfere with the cellular environment. Many systems regarding the evolution of activity, substrate acceptance, or stability are based on the use of antibiotics. However, this limits the scope of applications.

The question whether selection can be harnessed in the directed evolution of enantioselective enzymes is a particularly difficult one. Why should an organism experience a growth and thus survival advantage just because it harbors an enzyme that is enantioselective for a given substrate? Two recent studies have shown that, in principle, this is possible, although generalization has yet to be achieved. In one approach, a selection system for enantioselective lipase variants of *Bacillus subtilis* was devised using a mutant library in an aspartate auxotroph *E. coli.* (5). This was supplemented with an aspartate ester of a chiral alcohol, namely isopropylidene glycerol in the enantiomeric (S)-form. Such a step alone does not ensure any growth advantage. Therefore, the researchers added an enzyme inhibitor consisting of (R)-isopropylidene glycerol covalently bonded to a phosphonate ester. Since phosphonate esters are known to be lipase inhibitors by undergoing covalent attachment to serine of the catalytic triad Asp/His/Ser, the combined effects were expected to induce selection. Saturation mutagenesis using an appropriate cassette for amino acid positions 132 to 136 was used to construct a library, and after three selection rounds in which the concentration of the inhibitor was successively raised, a lipase variant was indeed identified having improved enantioselectivity (ee = +73.1% at 28.9% conversion versus ee = −29.6% at 23.4% conversion of the WT) (5).

In another study describing a selection system for evolving enantioselectivity, a more general approach was taken that does not require such surrogate diastereomeric substrates or the use of inhibitors (43). The basic concept is to mimic kinetic resolution in such a way that "positive" and "negative" components are used in a single system according to the absolute configuration of the chiral compounds under study. This ensures simultaneous selection for activity and enantioselectivity (Fig. 2). For the evolution of (R)-selectivity, the (R)-substrate needs to contain a positive component that serves as a potential energy source for the host organism, thereby promoting growth following the desired enantioselective cleavage reaction. In contrast, the (S)-substrate is designed so as to contain a negative component, in this case the cleavage reaction generating a toxic component as a poison for the organism. This dual-selection system requires the use of a mixture of isosteric pseudo-enantiomers, the ratio of the two starting substrates serving as a convenient way to optimize selection pressure (43).

The concept was illustrated in the directed evolution of *Candida antarctica* lipase B as a catalyst in the kinetic resolution of the acetate of isopropylidene glycerol (43). The WT is slightly (R)-selective ($E = 1.9$). The (S)-acetate was used as the "positive" component, the cleavage reaction generating acetic acid as an energy source. The enantiomeric (R)-fluoroacetate served as the "negative" component, leading to fluoroacetic acid as the toxic agent, which means that with this choice reversal of enantioselectivity was programmed. After saturation mutagenesis, the use of a 1:100 ratio of this mixture led to the appearance of bacterial colonies that harbor almost exclusively (S)-selective *C. antarctica* lipase B variants [≥90% active variants favoring the (S)-substrate with E-factors up to 8]. In this proof-of-principle study, only a relatively small gene library was generated (43), and it is therefore of interest to see how this and related selection systems for enantioselectivity will perform when studying 10^6- to 10^{10}-sized gene libraries.

32.3. GENE MUTAGENESIS METHODS

During the past 25 years, numerous gene mutagenesis methods have been developed. It is difficult even for experts to make the optimal choice, because comparative studies focusing on relative efficiency are rare. Since the methods have been summarized in previous reviews and monographs (1, 28, 40), only a few basic facets are repeated here. It should be pointed out that although many gene mutagenesis methods were already available in the 1980s, the idea that repetitive cycles of mutagenesis/screening constitute the crucial tool for exerting evolutionary pressure was rarely practiced at this time (25).

32.3.1. epPCR and Related Methods

To this day the mutagenesis method used most often is epPCR (1, 28, 40). By adjusting such parameters as the concentration of $MgCl_2$ (or $MnCl_2$) or using unbalanced concentrations of nucleotides, the error rate can be adjusted empirically so that on average one, two, three, or more amino acid substitutions occur in the protein. epPCR is often designated as random mutagenesis, but this is only a rough first approximation, and, in fact, several sources of serious bias have been reviewed (33, 57, 59). Due to the degeneracy of the genetic code, inter alia, amino acid bias occurs. The full potential diversity is hardly reached, on average, only about one-third (57). It is unlikely that two or even three point mutations will occur at neighboring residues in the protein. Bias can also arise because of the exponential nature of the amplification process (33). A detailed statistical analysis of these limitations has been performed (57).

Some of these problems can be reduced by combining libraries generated by Tag-based PCR and those obtained by application of the Stratagene GeneMorph kit which is characterized by a different bias (33). A very different way to reduce amino acid bias is sequence saturation mutagenesis, comprising four steps (60). It is independent of the mutational bias of DNA polymerases, and also has the advantage that the fragment distribution of a DNA library can be controlled by the use of different concentrations of the individual Sp-dNTPαS or a combination thereof. The mutation rate can be tuned by varying the concentration of NaCl and/or NaOH in the DNA-melting step. It remains to be seen whether the advantages inherent in these alternatives outweigh the increased labor effort relative to the use of standard epPCR.

Despite the limitations of standard epPCR, the introduction of point mutations by this method is easy to perform, and it is particularly appropriate if no structural information is available. In an early publication the stability and activity of aspartase, catalyzing the industrially important addition of ammonia to fumarate with formation of L-aspartic acid, were improved by a notable margin as a result of screening a library of enzyme mutants produced by epPCR (63). Another early study concerns the enhancement of activity of subtilisin E in dimethylformamide as a hostile solvent, which constitutes a landmark contribution, because for the first time sequential rounds of epPCR were shown to be effective in directed evolution, such a strategy providing a means to exert evolutionary pressure (7). Since then, many if not most studies in directed evolution have utilized successive rounds of epPCR.

32.3.2. Saturation Mutagenesis

The generation of focused libraries can be achieved by saturation mutagenesis, which requires some structural knowledge to make an appropriate decision regarding where in the enzyme amino acid randomization should take place (1, 28, 40). In the simplest case a single residue is targeted, meaning the introduction of all of the other 19 amino acids by using the appropriate codon degeneracy. It is also possible to randomize sites composed of two or more amino acid positions, in which case the term "combinatorial cassette mutagenesis" is sometimes used. Since in such a process the individual point mutations may influence each other, the simultaneous introduction of amino acids at more than one position makes synergistic effects possible (not just additivity).

Since the mid-1980s many methods for saturation mutagenesis, usually as variations of the general approach based on the use of appropriate oligodeoxynucleotides, have been reported. Accordingly, primers need to be designed and prepared that carry the genetic information encoding the desired mutational changes. The most widely used procedure is the so-called QuikChange protocol of Stratagene (19), which is based on previous work. It was originally developed for traditional site-specific single-amino-acid introduction, but can also be used for randomization at up to five amino acid positions in the enzyme (Stratagene's QuikChange® Multi Site-Directed Mutagenesis Kit). The general sequence of events required in the QuikChange protocol is summarized in Fig. 3.

Unfortunately, the extended protocol is not general, e.g., problems related to the primer length and design were noted. Therefore, various improvements were proposed, e.g., the use of partially overlapping or even nonoverlapping oligonucleotides, the resulting amplicon being employed as a megaprimer that completes the synthesis of the plasmid in a second PCR (30). Nevertheless, difficulties were still encountered in the case of recalcitrant targets such as large plasmids, as in the case of *P450-bm3* from *Bacillus megaterium* (55). Extending the idea of using nonoverlapping oligonucleotides (23, 66), a highly improved two-stage PCR-based method

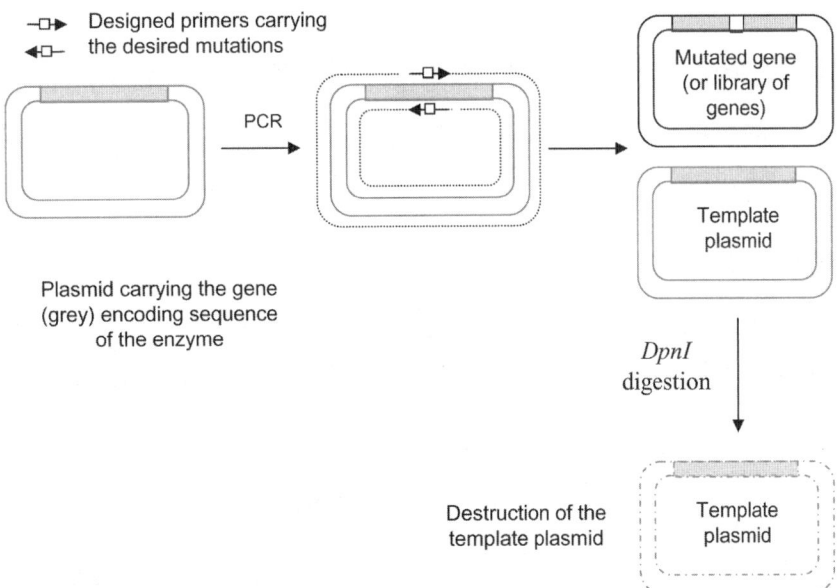

FIGURE 3 Schematic illustration of QuikChange (Stratagene) (19).

for the creation of saturation mutagenesis libraries was recently developed specifically for difficult-to-amplify templates (Fig. 4) (55). In the first stage, both the mutagenic primer and the antiprimer that are not complementary anneal to the template. The amplified sequence is then used in the second stage as a megaprimer. In this straightforward process, sites composed of one or more residues can be randomized in a single PCR reaction, irrespective of their location in the gene sequence. In a carefully performed comparative investigation, the virtues of the new method relative to QuikChange and related protocols were demonstrated by use of four different enzymes (55).

The optimal use of saturation mutagenesis requires the consideration of oversampling (44, 53). Fortunately, algorithms have been developed which are useful in the design of libraries (6, 10, 36), and indeed these have been applied in assessing the degree of oversampling as a function of the nature of the codon degeneracy and of the percentage of coverage of the respective protein sequence space (44). As a result of a systematic study addressing the numbers problem in directed evolution, practical recommendations were made (44). NNK degeneracy is usually applied in directed evolution based on saturation mutagenesis, which means that all

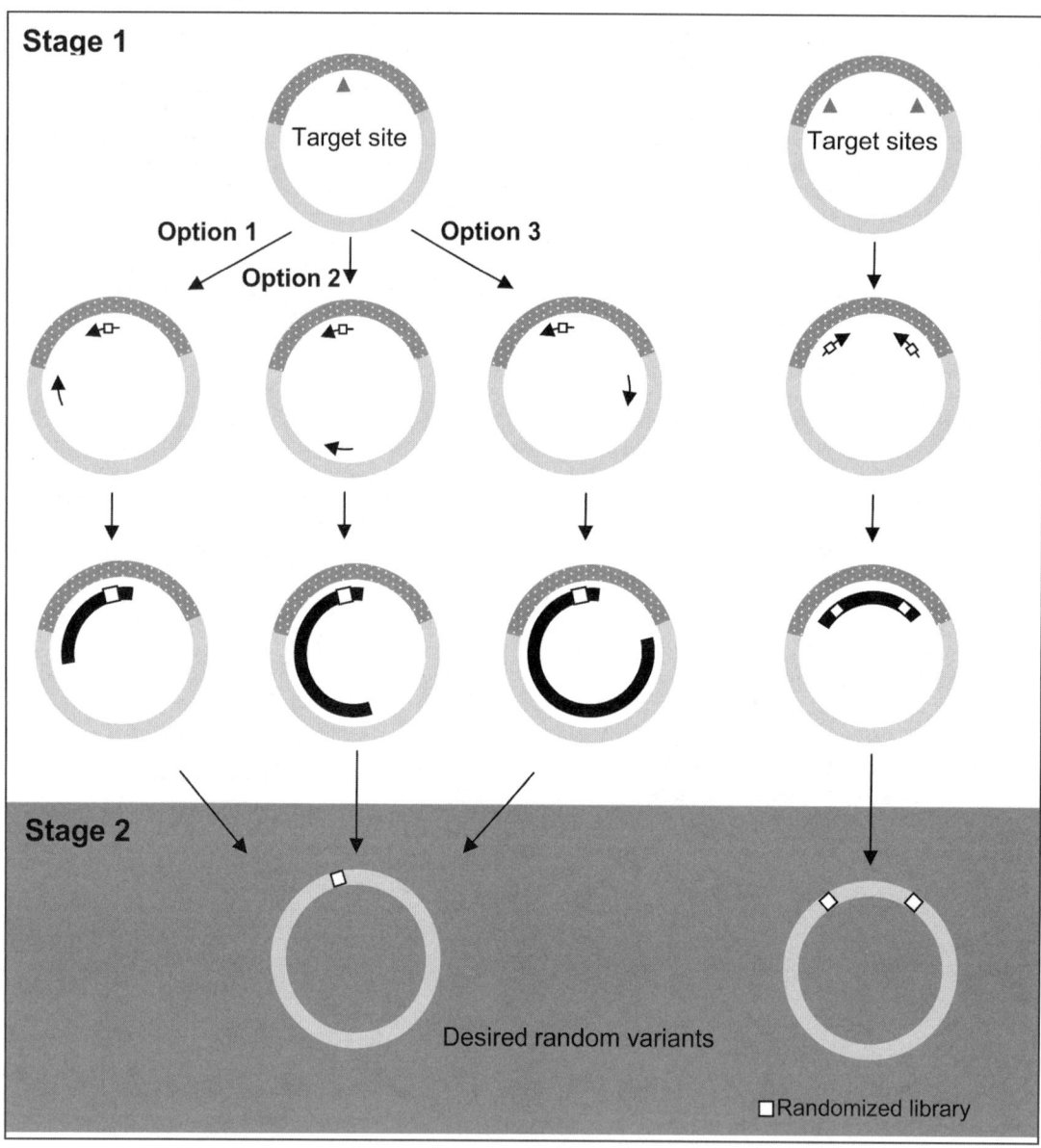

FIGURE 4 Scheme illustrating the improved method for PCR-based saturation mutagenesis useful in the case of difficult-to-amplify templates (55). The gene is represented by the dotted section, the vector backbone is shown in light gray, and the formed megaprimer in black. In the first stage of the PCR both the mutagenic primer (positions randomized represented by a white square) and the antiprimer (or another mutagenic primer, shown to the right) anneal to the template, and the amplified sequence is used as a megaprimer in the second stage. Finally, the template plasmids are digested by use of DpnI and the resulting library is transformed in bacteria. The scheme on the left illustrates the three possible options in the choice of the megaprimer size for a single-site randomization experiment. The scheme to the right represents an experiment with two sites simultaneously randomized.

TABLE 1   Oversampling necessary for 95% coverage as a function of NNK and NDT codon degeneracy assuming the absence of amino acid bias (44)

No.[a]	Codons	NNK transformants needed	Codons	NDT transformants needed
1	32	94	12	34
2	1,028	3,066	144	430
3	32,768	98,163	1,728	5,175
4	1,048,576	3,141,251	20,736	62,118
5	33,554,432	100,520,093	248,832	745,433
6	$>1.0 \times 10^9$	$>3.2 \times 10^9$	$>2.9 \times 10^6$	$>8.9 \times 10^6$
7	$>3.4 \times 10^{10}$	$>1.0 \times 10^{11}$	$>3.5 \times 10^7$	$>1.1 \times 10^8$
8	$>1.0 \times 10^{12}$	$>3.3 \times 10^{12}$	$>4.2 \times 10^8$	$>1.3 \times 10^9$
9	$>3.5 \times 10^{13}$	$>1.0 \times 10^{14}$	$>5.1 \times 10^9$	$>1.5 \times 10^{10}$
10	$>1.1 \times 10^{15}$	$>3.4 \times 10^{15}$	$>6.1 \times 10^{10}$	$>1.9 \times 10^{11}$

[a]Number of amino acid positions at a given site.

20 proteinogenic amino acids are used as building blocks (N, adenine/cytosine/guanine/thymine; K, guanine/thymine). However, reduced amino acid alphabets are also possible by applying the appropriate codon degeneracy, NDT being one of many options (D, adenine/guanine/thymine; T, thymine). In this case only 12 amino acids are encoded (Phe, Leu, Ile, Val, Tyr, His, Asn, Asp, Cys, Arg, Ser, and Gly), reducing structural diversity but having strong ramifications regarding the quality of saturation mutagenesis libraries. In a comparative investigation regarding the relative merits of NNK versus NDT codon degeneracy, the respective oversampling necessary for 95% coverage of the respective libraries was first calculated for randomization sites comprising different numbers of amino acid positions (Table 1) (44). These calculations are only rough approximations since amino acid bias is neglected. Bias can be avoided by the use of appropriate mixtures of oligonucleotides, but such a procedure is more expensive and

time-consuming; however, it is being studied for comparison in the author's laboratory. In the discussion that follows, bias is neglected. For example, in the case of a site composed of three amino acids, NNK requires almost 100,000 clones, whereas NDT needs only approximately 5,000 clones for 95% coverage. This raises the question as to which codon degeneracy is optimal, especially when the screening problem forces the experimenter to restrict the library size, e.g., to 5,000.

The results of the calculations over the whole range of coverage (0 to 95%) for sites composed of one, two, three, four, and five amino acid positions as a function of NNK versus NDT are shown in Fig. 5 and 6 (44). Here again, the pronounced differences in NNK and NDT libraries become apparent. For example, if, in the case of a three-amino-acid site, the experimenter restricts screening to 5,000 transformants for practical and economical reasons, the use of an NDT library correlates with approximately 95% coverage,

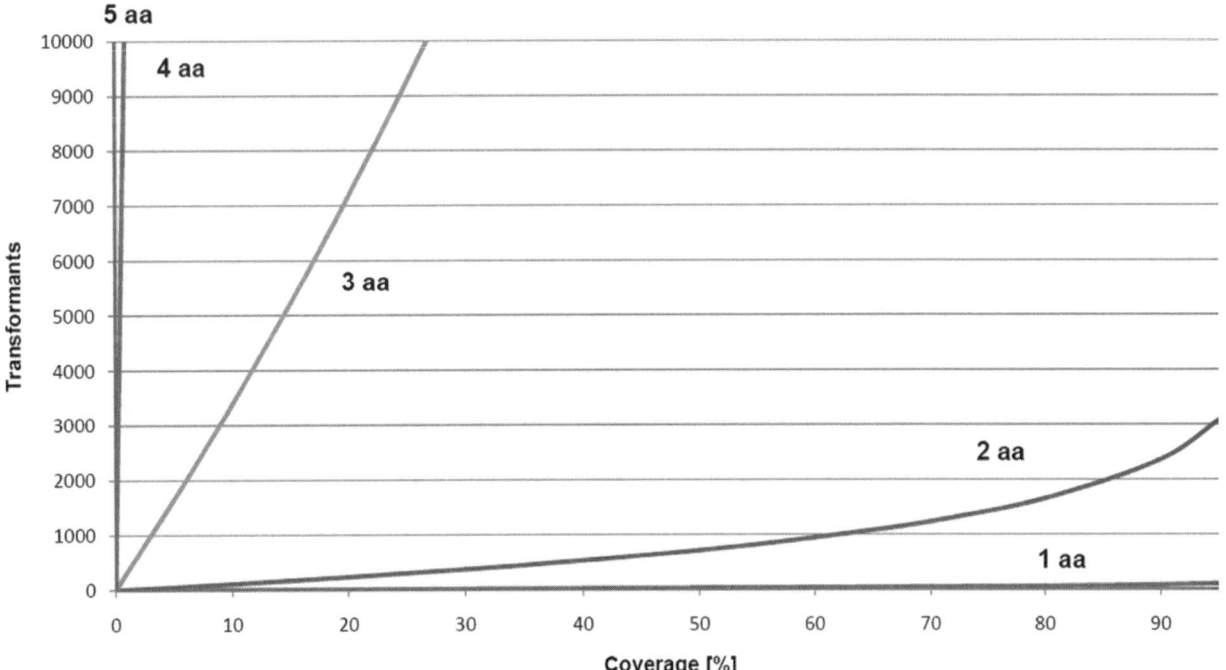

FIGURE 5   Library coverage calculated for NNK codon degeneracy at sites composed of one, two, three, four, and five amino acid positions (aa, amino acids) (44).

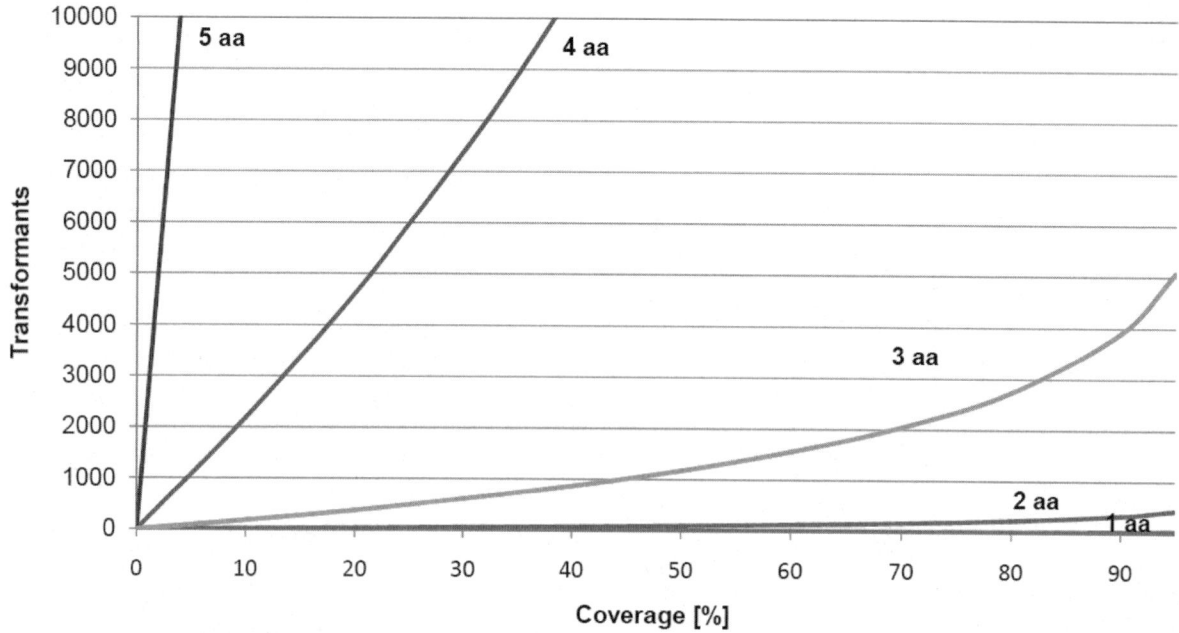

FIGURE 6 Library coverage calculated for NDT degeneracy at sites composed of one, two, three, four, and five amino acid positions (aa, amino acids) (44).

whereas the same number of screened clones in an NNK library allows for only 15% coverage.

Other codon degeneracies, which may be more suitable for a given task, can be analyzed analogously. It was concluded that it is best to choose systems in which the number of codons is equal to the number of amino acids, which helps to reduce the inherent bias and overrepresentation of certain amino acids (44). These statistical analyses were used as the basis for evolving substrate acceptance (rate) and enantioselectivity of an epoxide hydrolase (section 32.4). In particular, it was demonstrated that the quality of a size-limited enzyme library (5,000 transformants) produced by saturation mutagenesis at a site made up of three amino acid positions, using NDT codon degeneracy, is dramatically higher than that of a 5,000-membered library generated on the basis of the conventional NNK codon degeneracy. NDT is not the only alternative to a reduced amino acid alphabet, and indeed the use of more than one codon degeneracy in a given saturation mutagenesis experiment is possible.

The use of focused libraries in directed evolution has been practiced successfully many times (1, 28, 40), as, for example, in the successful quest to enhance the enantioselectivity of a lipase, in which a site composed of four amino acid positions aligning the binding pocket was randomized (49). In a later study it was suggested that whenever substrate acceptance or enantioselectivity need to be improved, sites closer to the active center are liable to be more important than remote positions (20).

It was not until systematization and iterativity were implemented that saturation mutagenesis was really exploited to its full potential, specifically in the form of iterative saturation mutagenesis (ISM) (41, 42, 48). The experimenter systematically identifies all relevant "hot" sites with the help of structural data (X-ray data or homology model), a given site being composed of one or more amino acid positions. After the generation and screening of the respective saturation mutagenesis libraries, the genes of the respective hits are then used as templates for performing further rounds of randomization at the other sites. Consider the case of four sites, A, B, C, and D (Fig. 7) (41, 48). If each site is "visited" only once in a given upward pathway, convergence is reached after preparing

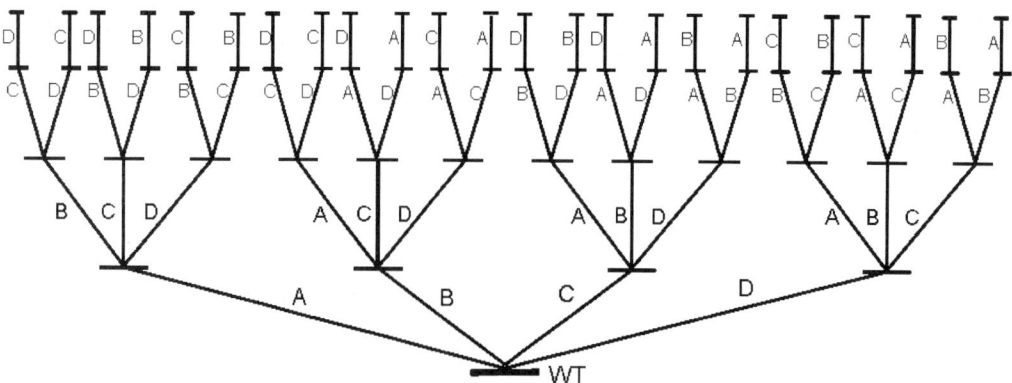

FIGURE 7 ISM using four sites, A, B, C, and D, each site in a given upward pathway in the fitness landscape being visited only once (48).

and screening a total of 64 libraries. As will be seen in section 32.4, complete scanning of such a defined and limited section of protein sequence space is not necessary, i.e., any one of the 24 pathways can be chosen. If a "dead end" as a consequence of a local minimum is encountered, backtracking is possible. In all applications of ISM, it is crucial to develop reliable criteria for choosing the appropriate randomization sites.

Thus far, several enzyme properties important in practical biocatalysis have been addressed when applying ISM, namely, enantioselectivity/substrate acceptance (48) and thermostability (41, 42). In the case of enantioselectivity and/or substrate acceptance (rate), the criterion is defined in the combinatorial active-site saturation test (CAST) (48). All sites next to the binding pocket are systematically identified on the basis of X-ray data or homology model, a site being composed of one or more amino acid residues. Second-sphere residues may also be considered (41). Rather than focusing just on a select site as done in previous saturation mutagenesis experiments (1, 20, 28, 40, 49), CASTing constitutes systematization, setting the stage for iterativity (ISM). It is a relatively new approach, but it has already been applied to several different enzymes in the quest to broaden substrate scope and to enhance enantioselectivity (section 32.4) (2, 24, 48). Such a systematic method is a practical way to reduce the screening effort, which is the bottleneck of directed evolution.

When applying ISM in the enhancement of thermostability of proteins, a different criterion for choosing randomization sites needs to be applied (41, 42). Since it was well known that hyperthermophilic enzymes are more rigid than the mesophilic analogs (13), it seemed reasonable to introduce appropriate mutations at sites displaying high degrees of flexibility. To have a rough guide for identifying such sites, atomic displacement parameters obtained from X-ray data can be used, namely, B factors that reflect smearing of atomic electron densities with respect to equilibrium positions as a result of thermal motion and positional disorder.

On the basis of this physical phenomenon, the B-factor iterative test (B-FIT) was developed, according to which only those sites displaying the highest B factors are considered for saturation mutagenesis, the process being performed iteratively according to Fig. 7 (41, 42). This embodiment of ISM has likewise proved to be successful; unusual degrees of thermostabilization have been achieved (section 32.4). Moreover, B-FIT can also be used to enhance enzyme tolerance toward hostile organic solvents, since a rough correlation between thermostability and stability in organic solvents exists (13).

32.3.4. Recombinant Gene Mutagenesis Methods

Instead of introducing point mutations, as in the methods above, recombination can also be considered, meaning the breaking and rejoining of DNA in new combinations (1, 21, 28, 40). DNA shuffling is the most prominent method (38). In general, one or more genes are first digested with a DNase to yield double-stranded oligonucleotide fragments of 10 to 50 base pairs, which are then amplified in a PCR-like process. Repeating cycles of strand separation and reannealing in the presence of a DNA polymerase followed by a final PCR amplification result in the reassembly of full-length mutant genes. DNA shuffling was first applied to tumor endothelial marker β-lactamase, the goal being to achieve increased antibiotic resistance. DNA shuffling can be performed with one gene, with two or more natural genes, or with mutant genes (Fig. 8). In general, a relatively high degree of homology is necessary (at least 60%). A particularly efficient version is family shuffling (38), in which homologous genes from different species are chosen; such an approach provides high catalyst diversity. In all of these recombinant methods a certain degree of self-hybridization of parental genes occurs, thereby lowering the quality of the mutant libraries. Various improvements have been suggested (1, 21, 28, 40).

An alternative to the above-mentioned in vitro recombination is the so-called staggered extension process (StEP), which is based on cross-hybridization of growing

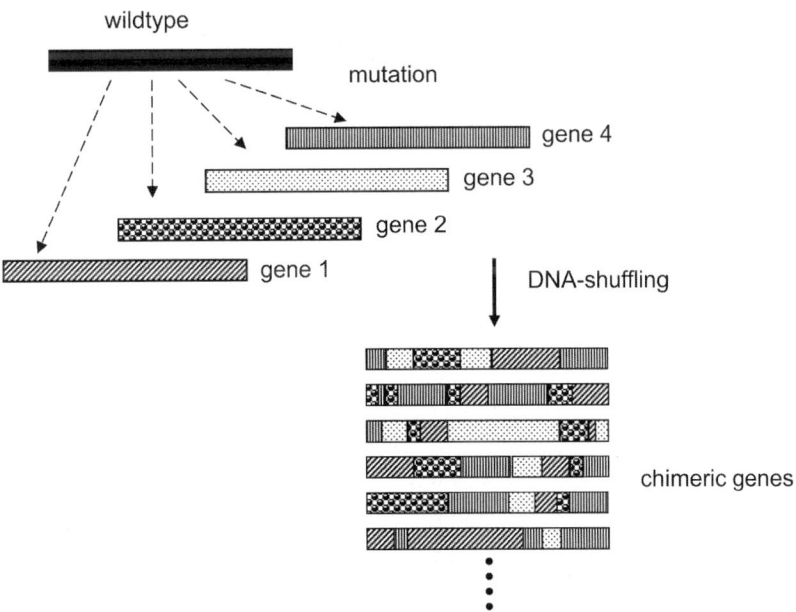

FIGURE 8 Scheme illustrating DNA shuffling for the case in which the parental genes originate from the WT by some sort of mutagenesis.

gene fragments as the DNA polymerase-catalyzed primer extension process (65). Subsequent to denaturation the primers anneal and extend under conditions that limit extension, allowing the primers to reanneal randomly to different parent sequences throughout the multiple cycles. Finally, the recombinant full-length gene products are amplified by PCR.

A number of homology-dependent and independent in vitro recombination methods have been summarized in reviews (1, 21, 28, 40), and only a few recent advancements are highlighted here. A prominent method is random chimeragenesis on transient templates (RACHITT), which constitutes a conceptually distinct alternative to sexual PCR for gene family shuffling (9). Accordingly, thermocycling, strand switching, or staggered extension are not necessary, the method relying on the trimming, gap filling, and ligation of parental gene fragments hybridized on a transient DNA template. This approach was applied successfully to the directed evolution of dibenzothiophene monooxygenase, which catalyzes the first step of the *dszABCD* diesel biodesulfurization pathway. Significantly increased reaction rate and a broadened scope of substrate acceptance were achieved (9).

Another advancement is random strand transfer recombination (RSTR), which is based on the ability of reverse transcriptases to undergo homology-independent template switches during the DNA synthesis (51). The method appears to be fairly general, involving spontaneous base-pairing-dependent recombination at high frequency between genes having low or high sequence homology. A different recombinant method is "biased mutation-assembly," in which a library is generated by overlap extension PCR with DNA fragments from a WT enzyme and phenotypically advantageous mutant genes (17). The number of mutations assembled in the WT gene is controlled stochastically by mixing ratio of the WT fragments to the mutant DNA fragments. In yet another approach, synthetic oligonucleotides are added to a mixture of gene fragments prior to assembly (56), a method that has been optimized and dubbed "incorporating synthetic oligonucleotides via gene reassembly" (ISOR) (18).

Insertion and deletion constitutes a different way to obtain mutant enzymes, but random versions of this approach have not been routinely applied. In addition to the first successful approach called random insertion/deletion (RID), which allows for the deletion of up to 16 consecutive bases at random positions and the insertion of a specific sequence or random sequences of an arbitrary number of bases threat (31), a method based on gene shuffling termed "random insertional-deletional strand exchange" mutagenesis (RAISE) was recently proposed (15). It is composed of three steps, namely random DNA fragmentation by DNase I, attachment of a random short nucleotide sequence to the 3″ terminus of the fragment using TdT, and finally reconstruction of each fragment with a tail of random nucleotides into a full-length sequence by self-priming PCR. The method was first applied to a tumor endothelial marker β-lactamase for improved antibiotic resistance.

In a different approach to recombination, the WT genes normally necessary for digestion are not needed, because the starting points are synthetic DNA fragments that are then simply assembled. Three versions have been reported, namely "degenerate homoduplex recombination" (8), "synthetic shuffling" (32), and "assembly of designed oligonucleotides" (61). These methods tolerate low homology, while undesired self-hybridization of parental genes is minimized. However, they have not been exploited extensively thus far.

32.3.5. Computation-Guided Approaches to Mutagenesis

Some of these mutagenesis methods utilize computer aids, as for example, in the user-friendly design of saturation mutagenesis libraries (10, 41, 42, 44) or in strategies relying on sequence alignments (1, 28). In this section, the focus is on strategies in which computation is the primary tool; two select cases are highlighted. In one approach called SCHEMA, homologous recombination, guided by structure-based computation, is used to create libraries of protein sequences that are not only extensively mutated, but that also have a high likelihood of having the same folding as the parental structures (54). To adapt this to shuffling, the computational algorithm SCHEMA was developed, which allows the prediction of fragments that need to be inherited from the same parent. The algorithm computes all interactions between residues, and then predicts the number of interactions that are disrupted in the process of recombination, i.e., in the formation of new proteins. SCHEMA was originally developed by use of two distantly related β-lactamases, but has since been applied to cytochromes P450 as well (54).

In another in silico approach, recombination-based directed evolution was augmented by incorporating a strategy for the statistical analysis of protein sequence activity relationships (ProSAR) (14), in analogy to the well-known quantitative structure-activity relationships (QSAR) that have been used widely in drug design and peptide optimization. ProSAR is a multivariate optimization process composed of repetitive diversity generation and screening accompanied by a statistical analysis based on linear regression on "training sets" derived from one or more libraries per cycle. After the identification of a hit obtained at the end of a given round, the variant is used as a template for programming diversity in the next round. Statistical learning comes about by applying an algorithm that correlates mutations with protein function. In ProSAR, DNA shuffling and semisynthetic DNA shuffling are used as the mutagenesis methods. The method was applied in the directed evolution of a bacterial halohydrin dehalogenase, the goal being highly improved volumetric productivity (14). In this endeavor, one to four libraries were designed and constructed from a pool of 10 to 50 mutations; approximately 14,000 enzyme variants are typically screened per round. Each round required 3 to 4 weeks. After 18 rounds, a 4,000-fold improvement in activity was indeed achieved in the cyanation process relevant in the synthesis of a chiral intermediate useful in the preparation of the cholesterol-lowering drug Lipitor®. Studies are needed in which the performance of both ProSAR and SCHEMA are compared with other strategies being applied in directed evolution.

32.4. COMPARATIVE ENZYME EVOLUTION STUDIES

In this section a few select studies in directed evolution are reviewed, because they allow conclusions regarding the relative merits of different mutagenesis methods and strategies. However, the results should be viewed as possible trends, because the scarcity of comparative data regarding the vast majority of other studies does not allow for general conclusions.

32.4.1. Transforming a β-Galactosidase into a β-Fucosidase by DNA Shuffling and Saturation Mutagenesis

In a landmark study, single-gene DNA shuffling was applied to turn E. coli β-galactosidase (BGAL) into a β-fucosidase, thereby changing the substrate scope (62). The enzyme hydrolyzes β-galactosyl linkages such as the β(1,4)-linkage in lactose. Following seven cycles of DNA shuffling and screening approximately 10,000 transformants in each round, an enzyme variant was obtained showing a 10-fold improvement in activity as measured by k_{cat}/K_m in the reaction of the model compound p-nitrophenyl-β-D-fucopyranoside (pNP-fuc), while exhibiting a 39-fold decrease in activity with the "native" substrate p-nitrophenyl-β-D-galactopyranoside (pNP-gal). A 2.7-fold preference for pNP-gal over pNP-fuc was maintained.

Later, the same system was used to systematically compare DNA shuffling with saturation mutagenesis (35). Using appropriate primer design, residues Asp201, His540, and Asn604 were randomized by using NNK codon degeneracy. These positions were chosen because the X-ray structure of the E537QBGAL/pNP-gal complex reveals a hydrogen bond between the C6 hydroxyl moiety and residues His540 and Asn604 with a fixed sodium, which is held in place by Asp201. Only approximately 10,000 transformants were screened using a color test that indicated the presence of approximately 250 possibly improved clones, 29 of which showed a particularly prominent color change (blue). These were studied more closely, leading to the discovery of 11 variants that exhibited greater activity and specificity in reactions with pNP-fuc than the WT does with pNP-gal, indicating that substrate specificity had inverted in a single round of saturation mutagenesis/screening. Several highly improved variants were identified, the combination His540Val/Asn604Thr (Asp201 remaining) being particularly efficient. It was shown to have a k_{cat}/K_m value $(s^{-1} M^{-1})$ of 179 for pNP-gal and 2,563 for pNP-fuc, which is an impressive result. The failure to realize that Asp201 is necessary to fix the sodium greatly reduced the efficiency of the combinatorial search, because only approximately 1/32 of the library contained Asp201. This means that a mere 300 transformants had any chance of harboring improved variants. Yet, this strategy outperformed the DNA approach, by far, in terms of the degree of catalyst improvement (kinetic data), labor intensity, and speed (35). The authors of this comparative study offer several explanations for the reason why DNA shuffling does not produce the crucial mutations His540Val and Asn604Thr, but they are careful not to generalize.

32.4.2. Increasing the Thermostability of an Esterase by epPCR

Increasing the thermostability of enzymes is an important endeavor in biotechnology for practical and theoretical reasons. Rational design using site-specific mutagenesis and directed evolution have been used successfully in many studies (13). In early reports, several rounds of epPCR, sometimes alternating with recombinant methods, were shown to be effective in the thermostabilization of such enzymes as esterases and proteases (64). Since then, most studies regarding thermostabilization have relied on epPCR (13), a recent example being highlighted here.

A case study concerns the esterase from Burkholderia gladioli (EstB) which belongs to a family of esterases related to β-lactamases and DD-peptidases (58). It catalyzes the chemoselective hydrolysis of the acetyl group at the

3 position of cephalosporin derivatives, which is of considerable industrial interest. To obtain a more stable variant, epPCR was applied corresponding to approximately five amino acid substitutions per enzyme molecule. By use of a crude but efficient colony filter pretest based on the pH change, approximately 10^6 transformants were screened in a process requiring 2,000 plates. This led to the identification of 11 moderately improved mutants showing ΔT_m values of up to 7.6°C. In the attempt to enforce further improvements, DNA shuffling was applied, which failed. Even after optimizing every stage of the recombinant process, including fragment sizes, elution protocols, DNA concentrations, and the PCR conditions, no product could be obtained after the final reassembly PCR step. The researchers postulated that the high GC content of EstB (74%) imposes a barrier to successful PCR gene assembly (58). Then the mutations of several of the previous variants were combined, followed by another round of epPCR involving 10^5 transformants, which led to the best variant displaying distinctly improved thermostability ($\Delta T_m = 13°C$). High activity toward cephalosporin C was retained, underscoring the success of this study. Nevertheless, it would be interesting to subject this enzyme to directed evolution in a knowledge-guided manner such as the B-FIT method (sections 32.3.2 and 32.4.3).

32.4.3. Increasing the Thermostability of a Lipase by the B-FIT Method

In principle, epPCR and DNA shuffling are independent of structural data, whereas saturation mutagenesis generally needs such knowledge to make a decision regarding the randomization sites. Alternatively, saturation mutagenesis can be applied systematically at every single amino acid position, as was reported in the thermostabilization of a xylanase (34). This process provided nine improved mutants, eight of which showed enzymatic activity similar to the WT, but the thermally most stable variant exhibited only 35% activity. The mutations were subsequently combined, leading to a highly robust variant having $T_m = 95.6°C$. This means an improvement of $\Delta T_m = 35°C$, which constitutes a dramatic increase in thermostability. The best mutant showed similar initial rate/temperature optima at 70°C, but the WT had higher activity at lower temperatures (25 to 50°C). Another recent approach to saturation mutagenesis in the quest to increase thermostability is based on a two-step strategy, as demonstrated in the case of phosphite dehydrogenase (29). First, epPCR was performed with the identification of improved variants, followed by saturation mutagenesis at the observed mutation points. Improved variants displaying ΔT_m values of up to 25°C were obtained, demonstrating that they are potential candidates for practical cofactor regeneration.

In yet another approach, the B-FIT method (section 32.3.2) was applied to the mesophilic lipase from B. subtilis (Lip A) (41, 42). The 10 amino acids with the highest average B factors, as provided by the X-ray data and the B-FIT-TER computer aid, were chosen for saturation mutagenesis: Arg33 (B factor = 50.9), Lys69 (44.1), Gln164 (40.7), Asp34 (39.8), Lys112 (39.6), Lys35 (38.9), Met134 (38.5), Tyr139 (37.9), Ile157 (37.4), and Gly13 (37.0), and these were grouped as eight sites: site A (Gly13), B (Arg33, Asp34, Lys35), C (Lys69), D (Lys112), E (Met134), F (Tyr139), G (Ile157), and H (Gln164).

Thermostability was assessed by measuring the residual activity subsequent to the exposure to high temperature and ascertaining the T_{50} value, which is the temperature required

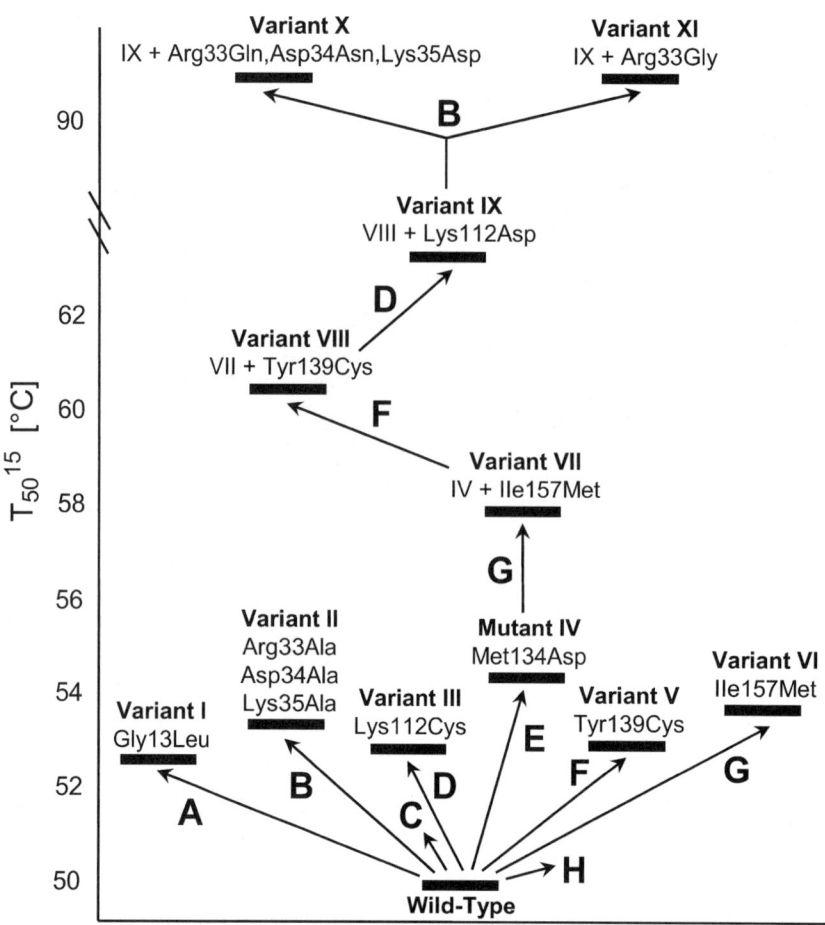

FIGURE 9 Thermostability diagram in the B-FIT-based evolution of Lip A (41, 42).

to reduce the initial enzymatic activity by 50% within a given time (in this case, as measured by the hydrolysis rate of *p*-nitrophenylcaprylate). For rapid screening, a 15-min heat treatment was applied (T_{50}^{15} values), but the final variants were subjected to a heat treatment of 1 h, providing T_{50}^{60} values (which are quite similar to T_{50}^{15}). The WT Lip A has a T_{50}^{60} value of 48°C. Of the many possible pathways, only one was actually tested, namely E→G→F→D→B, which provided the best Lip A hyperthermophilic variants, X and XI, having T_{50}^{60} values of 89 and 93°C, respectively, meaning an increase of 41 and 45°C, respectively, which have no precedence in the literature (Fig. 9). The overall effort required the screening of only 8,000 transformants (41, 42). In further experiments, it was shown that activity and enantioselectivity of variant XI at room temperature are essentially identical to those of the WT. The data speak for the unusual efficacy of the B-FIT method (45), but more case studies using other proteins are needed.

32.4.4. Enhancing the Enantioselectivity of an Epoxide Hydrolase by Iterative CASTing

Subsequent to the first directed evolution study regarding the enantioselectivity of enzymes (50), which involved four cycles of epPCR in the hydrolytic kinetic resolution of a chiral ester catalyzed by a lipase, numerous academic and industrial studies have appeared that contribute to the generalization of this new approach to asymmetric

catalysis (40). In most cases, epPCR, DNA shuffling, and/or saturation mutagenesis at select sites were applied. Efficiency was greatly increased by the development of iterative CASTing (48) (section 32.3.2). In a case study, the epoxide hydrolase from *Aspergillus niger* (ANEH) was used as the catalyst in the hydrolytic kinetic resolution of glycidyl phenyl ether, the WT showing slight (*S*)-selectivity (*E* = 4.6). The CAST analysis based on the X-ray structure of ANEH suggested six sites for saturation mutagenesis, namely A (amino acid positions 193/195/296), B (215/217/219), C (329/330), D (349/350), E (317/318), and F (244/245/249). An arbitrarily chosen pathway, B→C→D→F→E, led to a dramatically improved ANEH variant LW202 exhibiting a selectivity factor of *E* = 115 (48) (Fig. 10). Only 20,000 transformants had to be screened in the overall process, which happens to be the same number required in an earlier study based on epPCR leading to a maximum enantioselectivity of only *E* = 11 (47). This comparison illuminates the difference in efficiency of the two approaches.

Subsequent investigations addressed the reasons for the apparent efficacy of iterative CASTing as a form of ISM (45, 46). A deconvolution strategy was devised in which the five sets of mutations leading from the WT ANEH to the best variant LW202 were used to form all theoretically possible permutational combinations, leading to the construction of a fitness landscape comprising 5! = 120 pathways. Of these 120 trajectories leading from WT ANEH to LW202,

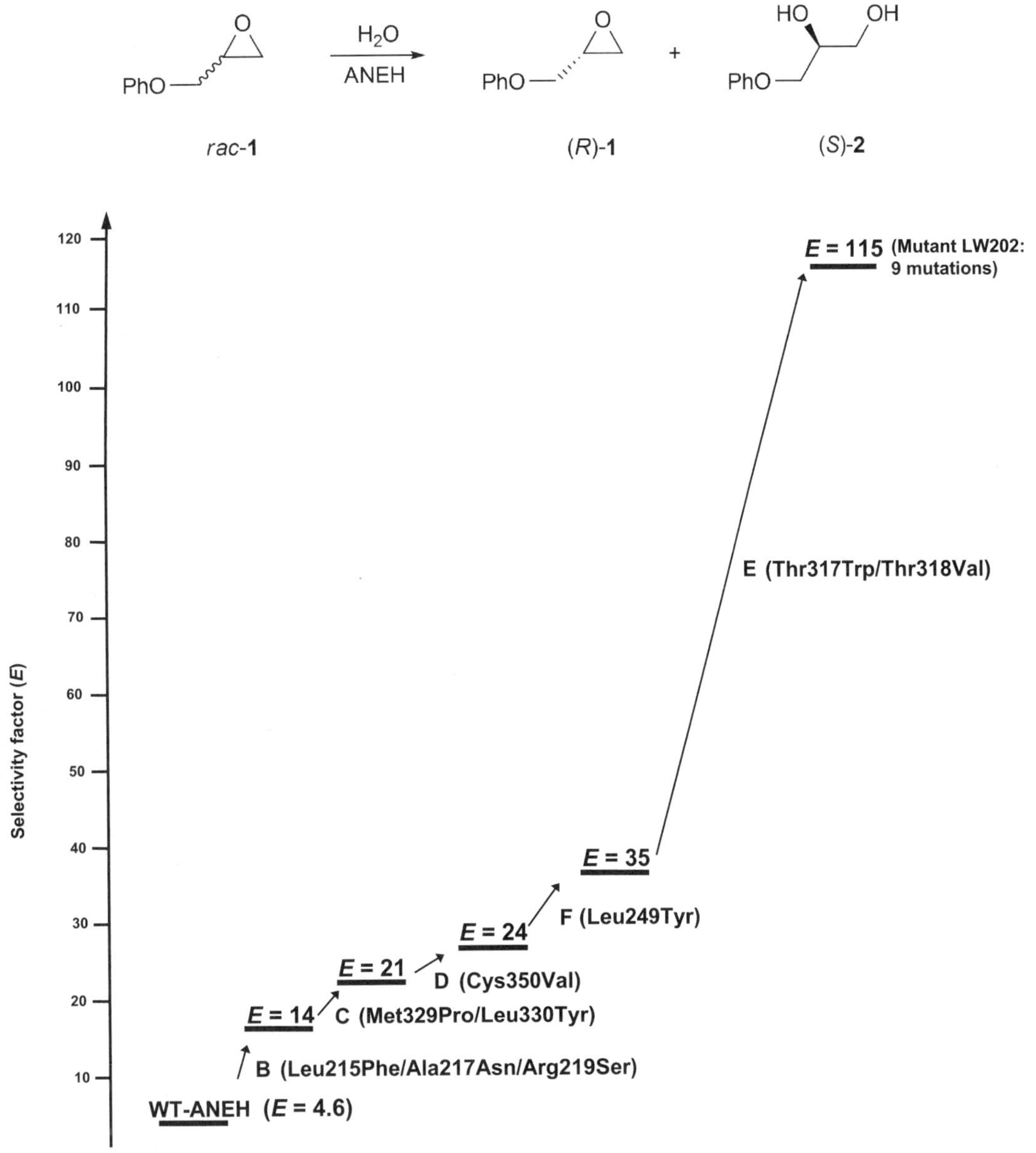

FIGURE 10 Iterative CASTing in the enhancement of enantioselectivity of the hydrolytic kinetic resolution of glycidyl phenyl ether catalyzed by ANEH variants (48).

55 proved to be energetically favored, meaning the absence of any local minima (45, 46). Moreover, strong synergistic effects operating between the sets of mutations were uncovered, likewise shedding light on the nature of the ISM process. This is also an indication that ISM creates an amino acid network in which the residues communicate with one another. Finally, CASTing utilizing a reduced amino acid alphabet for the construction of ANEH libraries constitutes another important tool in directed evolution, because it points the way to handling the numbers problem in directed evolution (44) (section 32.3.2).

32.5. CONCLUSIONS

The past decade has witnessed the generalization of directed evolution of enzymes as a reliable protein-engineering technique for improving such catalytic parameters as thermostability, stability in hostile organic solvents, substrate acceptance, and enantioselectivity. Numerous mutagenesis protocols and strategies for their application were applied successfully in the quest to achieve a practical result and/or to deepen our understanding of enzyme mechanisms. A certain degree of success can always be anticipated, irrespective of the mutagenesis method applied. It is not easy to make the optimal choice regarding the mutagenesis method, although trends are now visible. Fortunately, in recent years one of the important foci of research has centered on the development of user-friendly methods and strategies for probing protein sequence space more efficiently, making fast and reliable directed evolution possible (28, 45). This means that small but high-quality libraries are now accessible, as highlighted in this chapter. Hopefully, in the future more comparative studies focusing on relative efficiency will follow.

REFERENCES

1. **Arndt, K. M., and K. M. Müller.** 2007. *Protein Engineering Protocols (Methods in Molecular Biology).* Humana Press, Totowa, NJ.

2. **Bartsch, S., R. Kourist, and U. T. Bornscheuer.** 2008. Complete inversion of enantioselectivity towards acetylated tertiary alcohols by a double mutant of a *Bacillus subtilis* esterase. *Angew. Chem. Int. Ed.* **47:**1508–1511.

3. **Becker, S., H. Höbenreich, A. Vogel, J. Knorr, S. Wilhelm, F. Rosenau, K.-E. Jaeger, M. T. Reetz, and H. Kolmar.** 2008. Single-cell high-throughput screening to identify enantioselective hydrolytic enzymes. *Angew. Chem. Int. Ed.* **47:** 5085–5088.

4. **Becker, S., H.-U. Schmoldt, T. M. Adams, S. Wilhelm, and H. Kolmar.** 2004. Ultra-high-throughput screening based on cell-surface display and fluorescence-activated cell sorting for the identification of novel biocatalysts. *Curr. Opin. Biotechnol.* **15:**323–329.

5. **Boersma, Y. L., M. J. Dröge, A. M. van der Sloot, T. Pijning, R. H. Cool, B. W. Dijkstra, and W. J. Quax.** 2008. A novel genetic selection system for improved enantioselectivity of *Bacillus subtilis* lipase A. *Chembiochem* **9:**1110–1115.

6. **Bosley, A. D., and M. Ostermeier.** 2005. Mathematical expressions useful in the construction, description and evaluation of protein libraries. *Biomol. Eng.* **22:**57–61.

7. **Chen, K., and F. H. Arnold.** 1993. Tuning the activity of an enzyme for unusual environments: sequential random mutagenesis of subtilisin E for catalysis in dimethylformamide. *Proc. Natl. Acad. Sci. USA* **90:**5618–5622.

8. **Coco, W. M., L. P. Encell, W. E. Levinson, M. J. Crist, A. K. Loomis, L. L. Licato, J. J. Arensdorf, N. Sica, P. T. Pienkos, and D. J. Monticello.** 2002. Growth factor engineering by degenerate homoduplex gene family recombination. *Nat. Biotechnol.* **20:**1246–1250.

9. **Coco, W. M., W. E. Levinson, M. J. Crist, H. J. Hektor, A. Darzins, P. T. Pienkos, C. H. Squires, and D. J. Monticello.** 2001. DNA shuffling method for generating highly recombined genes and evolved enzymes. *Nat. Biotechnol.* **19:**354–359.

10. **Denault, M., and J. N. Pelletier.** 2007. Protein library design and screening: working out the probabilities, p. 127–154. *In* K. M. Arndt and K. M. Müller (ed.), *Protein Engineering Protocols.* Humana Press, Totowa, NJ.

11. **Drauz, K., and H. Waldmann.** 2002. *Enzyme Catalysis in Organic Synthesis: a Comprehensive Handbook.* Wiley-VCH, Weinheim, Germany.

12. **Dröge, M. J., Y. L. Boersma, G. van Pouderoyen, T. E. Vrenken, C. J. Rüggeberg, M. T. Reetz, B. W. Dijkstra, and W. J. Quax.** 2006. Directed evolution of *Bacillus subtilis* lipase a by use of enantiomeric phosphonate inhibitors: crystal structures and phage display selection. *Chembiochem* **7:**149–157.

13. **Eijsink, V. G. H., S. Gåseidnes, T. V. Borchert, and B. van den Burg.** 2005. Directed evolution of enzyme stability. *Biomol. Eng.* **22:**21–30.

14. **Fox, R. J., S. C. Davis, E. C. Mundorff, L. M. Newman, V. Gavrilovic, S. K. Ma, L. M. Chung, C. Ching, S. Tam, S. Muley, J. Grate, J. Gruber, J. C. Whitman, R. A. Sheldon, and G. W. Huisman.** 2007. Improving catalytic function by ProSAR-driven enzyme evolution. *Nat. Biotechnol.* **25:**338–344.

15. **Fujii, R., M. Kitaoka, and K. Hayashi.** 2006. RAISE: a simple and novel method of generating random insertion and deletion mutations. *Nucleic Acids Res.* **34:**e30.

16. **Griffiths, A. D., and D. S. Tawfik.** 2006. Miniaturising the laboratory in emulsion droplets. *Trends Biotechnol.* **24:**395–402.

17. **Hamamatsu, N., T. Aita, Y. Nomiya, H. Uchiyama, M. Nakajima, Y. Husimi, and Y. Shibanaka.** 2005. Biased

mutation-assembling: an efficient method for rapid directed evolution through simultaneous mutation accumulation. *Protein Eng. Des. Sel.* **18:**265–271.

18. **Herman, A., and D. S. Tawfik.** 2007. Incorporating synthetic oligonucleotides via gene reassembly (ISOR): a versatile tool for generating targeted libraries. *Protein Eng. Des. Sel.* **20:**219–226.

19. **Hogrefe, H. H., J. Cline, G. L. Youngblood, and R. M. Allen.** 2002. Creating randomized amino acid libraries with the QuikChange® multi site-directed mutagenesis kit. *BioTechniques* **33:**1158–1165.

20. **Horsman, G. P., A. M. F. Liu, E. Henke, U. T. Bornscheuer, and R. J. Kazlauskas.** 2003. Mutations in distant residues moderately increase the enantioselectivity of *Pseudomonas fluorescens* esterase towards methyl 3-bromo-2-methylpropanoate and ethyl 3-phenylbutyrate. *Chemistry* **9:**1933–1939.

21. **Jäckel, C., P. Kast, and D. Hilvert.** 2008. Protein design by directed evolution. *Annu. Rev. Biophys.* **37:**153–173.

22. **Joyce, G. F.** 2007. Forty years of in vitro evolution. *Angew. Chem. Int. Ed.* **46:**6420–6436.

23. **Kirsch, R. D., and E. Joly.** 1998. An improved PCR-mutagenesis strategy for two-site mutagenesis or sequence swapping between related genes. *Nucleic Acids Res.* **26:**1848–1850.

24. **Liang, L., J. Zhang, and Z. Lin.** 2007. Altering coenzyme specificity of *Pichia stipitis* xylose reductase by the semirational approach CASTing. *Microb. Cell Fact.* **6:**36.

25. **Liao, H., T. McKenzie, and R. Hageman.** 1986. Isolation of a thermostable enzyme variant by cloning and selection in a thermophile. *Proc. Natl. Acad. Sci. USA* **83:**576–580.

26. **Liese, A., K. Seelbach, and C. Wandrey.** 2006. *Industrial Biotransformations,* 2nd ed. Wiley-VCH, Weinheim, Germany.

27. **Lipovšek, D., E. Antipov, K. A. Armstrong, M. J. Olsen, A. M. Klibanov, B. Tidor, and K. D. Wittrup.** 2007. Selection of horseradish peroxidase variants with enhanced enantioselectivity by yeast surface display. *Chem. Biol.* **14:**1176–1185.

28. **Lutz, S., and U. T. Bornscheuer.** 2009. *Protein Engineering Handbook,* vol. 1 and 2. Wiley-VCH, Weinheim, Germany.

29. **McLachlan, M. J., T. W. Johannes, and H. Zhao.** 2008. Further improvement of phosphite dehydrogenase thermostability by saturation mutagenesis. *Biotechnol. Bioeng.* **99:**268–274.

30. **Miyazaki, K., and M. Takenouchi.** 2002. Creating random mutagenesis libraries using megaprimer PCR of whole plasmid. *BioTechniques* **33:**1033–1038.

31. **Murakami, H., T. Hohsaka, and M. Sisido.** 2002. Random insertion and deletion of arbitrary number of bases for codon-based random mutation of DNAs. *Nat. Biotechnol.* **20:**76–81.

32. **Ness, J. E., S. Kim, A. Gottman, R. Pak, A. Krebber, T. V. Borchert, S. Govindarajan, E. C. Mundorff, and J. Minshull.** 2002. Synthetic shuffling expands functional protein diversity by allowing amino acids to recombine independently. *Nat. Biotechnol.* **20:**1251–1255.

33. **Neylon, C.** 2004. Chemical and biochemical strategies for the randomization of protein encoding DNA sequences: library construction methods for directed evolution. *Nucleic Acids Res.* **32:**1448–1459.

34. **Palackal, N., Y. Brennan, W. N. Callen, P. Dupree, G. Frey, F. Goubet, G. P. Hazlewood, S. Healey, Y. E. Kang, K. A. Kretz, E. Lee, X. Tan, G. L. Tomlinson, J. Verruto, V. W. K. Wong, E. J. Mathur, J. M. Short, D. E. Robertson, and B. A. Steer.** 2004. An evolutionary route to xylanase process fitness. *Protein Sci.* **13:**494–503.

35. **Parikh, M. R., and I. Matsumura.** 2005. Site-saturation mutagenesis is more efficient than DNA shuffling for the directed evolution of β-fucosidase from β-galactosidase. *J. Mol. Biol.* **352:**621–628.

36. **Patrick, W. M., and A. E. Firth.** 2005. Strategies and computational tools for improving randomized protein libraries. *Biomol. Eng.* **22:**105–112.

37. **Polizzi, K. M., M. Parikh, C. U. Spencer, I. Matsumura, J. H. Lee, M. J. Realff, and A. S. Bommarius.** 2006. Pooling for improved screening of combinatorial libraries for directed evolution. *Biotechnol. Prog.* **22:**961–967.

38. **Powell, K. A., S. W. Ramer, S. B. del Cardayré, W. P. C. Stemmer, M. B. Tobin, P. F. Longchamp, and G. W. Huisman.** 2001. Directed evolution and biocatalysis. *Angew. Chem. Int. Ed.* **40:**3948–3959.

39. **Reetz, M. T.** 2006. High-throughput screening systems for assaying the enantioselectivity of enzymes, p. 41–76. In J.-L. Reymond (ed.), *Enzyme Assays—High-Throughput Screening, Genetic Selection and Fingerprinting.* Wiley-VCH, Weinheim, Germany.

40. **Reetz, M. T.** 2008. Directed evolution as a means to engineer enantioselective enzymes, p. 21–63. In V. Gotor, I. Alfonso, and E. García-Urdiales (ed.), *Asymmetric Organic Synthesis with Enzymes.* Wiley-VCH, Weinheim, Germany.

41. **Reetz, M. T., and J. D. Carballeira.** 2007. Iterative saturation mutagenesis (ISM) for rapid directed evolution of functional enzymes. *Nat. Protoc.* **2:**891–903.

42. **Reetz, M. T., J. D. Carballeira, and A. Vogel.** 2006. Iterative saturation mutagenesis on the basis of B factors as a strategy for increasing protein thermostability. *Angew. Chem., Int. Ed.* **45:**7745–7751.

43. **Reetz, M. T., H. Höbenreich, P. Soni, and L. Fernández.** 2008. A genetic selection system for evolving enantioselectivity of enzymes. *Chem. Commun.* (Cambridge) **2008:**5502–5504.

44. **Reetz, M. T., D. Kahakeaw, and R. Lohmer.** 2008. Addressing the numbers problem in directed evolution. *Chembiochem* **9:**1797–1804.

45. **Reetz, M. T., D. Kahakeaw, and J. Sanchis.** 2009. Shedding light on the efficacy of laboratory evolution based on iterative saturation mutagenesis. *Mol. Biosyst.* **5:**115–122.

46. **Reetz, M. T., and J. Sanchis.** 2008. Constructing and analyzing the fitness landscape of an experimental evolutionary process. *Chembiochem* **9:**2260–2267.

47. **Reetz, M. T., C. Torre, A. Eipper, R. Lohmer, M. Hermes, B. Brunner, A. Maichele, M. Bocola, M. Arand, A. Cronin, Y. Genzel, A. Archelas, and R. Furstoss.** 2004. Enhancing the enantioselectivity of an epoxide hydrolase by directed evolution. *Org. Lett.* **6:**177–180.

48. **Reetz, M. T., L.-W. Wang, and M. Bocola.** 2006. Directed evolution of enantioselective enzymes: iterative cycles of CASTing for probing protein-sequence space. *Angew. Chem. Int. Ed.* **45:**1236–1241. (Erratum **45:**2494.)

49. **Reetz, M. T., S. Wilensek, D. Zha, and K.-E. Jaeger.** 2001. Directed evolution of an enantioselective enzyme through combinatorial multiple cassette mutagenesis. *Angew. Chem. Int. Ed.* **40:**3589–3591.

50. **Reetz, M. T., A. Zonta, K. Schimossek, K. Liebeton, and K.-E. Jaeger.** 1997. Creation of enantioselective biocatalysts for organic chemistry by in vitro evolution. *Angew. Chem. Int. Ed. Engl.* **36:**2830–2832.

51. **Reiter, B., A. Faschinger, A. Glieder, and H. Schwab.** 2007. Random strand transfer recombination (RSTR) for homology-independent nucleic acid recombination. *J. Biotechnol.* **129:**39–49.

52. **Reymond, J.-L.** 2006. *Enzyme Assays—High-Throughput Screening, Genetic Selection and Fingerprinting.* Wiley-VCH, Weinheim, Germany.

53. **Rui, L., Y. M. Kwon, A. Fishman, K. F. Reardon, and T. K. Wood.** 2004. Saturation mutagenesis of toluene *ortho*-monooxygenase of *Burkholderia cepacia* G4 for enhanced 1-naphthol synthesis and chloroform degradation. *Appl. Environ. Microbiol.* **70:**3246–3252.

54. **Saab-Rincon, G., Y. Li, M. Meyer, M. Carbone, M. Landwehr, and F. H. Arnold.** 2009. Protein engineering by structure-guided SCHEMA recombination, p. 481–492. In S. Lutz and U. T. Bornscheuer (ed.), *Protein Engineering Handbook.* Wiley-VCH, Weinheim, Germany.

55. **Sanchis, J., L. Fernández, J. D. Carballeira, J. Drone, Y. Gumulya, H. Höbenreich, D. Kahakeaw, S. Kille, R. Lohmer, J. J.-P. Peyralans, J. Podtetenieff, S. Prasad, P. Soni, A. Taglieber, S. Wu, F. E. Zilly, and M. T. Reetz.** 2008. Improved PCR method for the creation of saturation mutagenesis libraries in directed evolution: application to difficult-to-amplify templates. *Appl. Microbiol. Biotechnol.* **81:**387–397.

56. **Stemmer, W. P. C.** 1994. DNA shuffling by random fragmentation and reassembly: in vitro recombination for molecular evolution. *Proc. Natl. Acad. Sci. USA* **91:**10747–10751.

57. **Sylvestre, J., H. Chautard, F. Cedrone, and M. Delcourt.** 2006. Directed evolution of biocatalysts. *Org. Process Res. Dev.* **10:**562–571.

58. **Valinger, G., M. Hermann, U. G. Wagner, and H. Schwab.** 2007. Stability and activity improvement of cephalosporin esterase EstB from *Burkholderia gladioli* by directed evolution and structural interpretation of muteins. *J. Biotechnol.* **129:**98–108.

59. **Wong, T. S., D. Roccatano, M. Zacharias, and U. Schwaneberg.** 2006. A statistical analysis of random mutagenesis methods used for directed protein evolution. *J. Mol. Biol.* **355:**858–871.

60. **Wong, T. S., K. L. Tee, B. Hauer, and U. Schwaneberg.** 2004. Sequence saturation mutagenesis (SeSaM): a novel method for directed evolution. *Nucleic Acids Res.* **32:**e26.

61. **Zha, D., A. Eipper, and M. T. Reetz.** 2003. Assembly of designed oligonucleotides as an efficient method for gene recombination: a new tool in directed evolution. *Chembiochem* **4:**34–39.

62. **Zhang, J.-H., G. Dawes, and W. P. C. Stemmer.** 1997. Directed evolution of a fucosidase from a galactosidase by DNA shuffling and screening. *Proc. Natl. Acad. Sci. USA* **94:**4504–4509.

63. **Zhang, H. Y., J. Zhang, L. Lin, W. Y. Du, and J. Lu.** 1993. Enhancement of the stability and activity of aspartase by random and site-directed mutagenesis. *Biochem. Biophys. Res. Commun.* **192:**15–21.

64. **Zhao, H., and F. H. Arnold.** 1999. Directed evolution converts subtilisin E into a functional equivalent of thermitase. *Protein Eng.* **12:**47–53.

65. **Zhao, H., L. Giver, Z. Shao, J. A. Affholter, and F. H. Arnold.** 1998. Molecular evolution by staggered extension process (StEP) in vitro recombination. *Nat. Biotechnol.* **16:**258–261.

66. **Zheng, L., U. Baumann, and J.-L. Reymond.** 2004. An efficient one-step site-directed and site-saturation mutagenesis protocol. *Nucleic Acids Res.* **32:**e115.

Industrial Applications of Enzymes as Catalysts

NIKHIL U. NAIR, WENG LIN TANG, DAWN T. ERIKSEN, AND HUIMIN ZHAO

33

33.1. INTRODUCTION

The industrial use of enzymes exploded in the late 1960s with the development of protease additives to detergents in the detergent industry and the use of a fungal glucoamylase (hydrolase) in the breakdown of starch into glucose in the food industry. Following the trend set by these pioneering efforts, the detergent and food industries have continued to dwarf all other industries in terms of total industrial enzyme usage and sales. Food-manufacturing processes, including starch hydrolysis and bread, cheese, and juice manufacturing, are responsible for an estimated 40 to 45% of all enzyme sales. Detergent additives such as proteases, lipases, and cellulases are responsible for 35 to 40% of total soluble enzyme sales. The growth of the enzyme industry, concomitant with a greater demand for enzymes, has been quite remarkable in the past 40 years, with a 3 orders of magnitude increase in total sales. For example, Novo Industri A/S (now Novozymes), the leading enzyme manufacturer in the 1960s, had enzyme sales of approximately US $1 million in 1965. By 2007 the same metric had reached a value of over US $1 billion.

While the aforementioned two industries may be quite capable of sustaining the burgeoning enzyme industry, applications have extended to other areas, including paper and pulp processing, textile manufacture, pharmaceutical syntheses, and now even the commodity chemicals business. Increased emphasis on sustainability and use of renewable resources is driving this trend further, because biocatalysts align very well with the current trend toward green chemistry (2). The Freedonia Group (Cleveland, OH) estimated that the global demand for enzymes will grow at a rate of 7.6% per year, to reach US $6.1 billion by 2011. However, just because the potential exists does not mean it will be realized. The rate at which biocatalysis or principles of green chemistry will be adopted depends on whether or not it makes economic sense, which in turn greatly depends on the prices of the feedstock—be it oil or biomass.

Irrespective of the starting raw material or the process, biocatalysis offers several significant advantages over chemocatalysts. First, since enzymes function at moderate temperatures and pressures, they require less energy input. Second, enzymes can be highly selective (regio-, diastereo-, enantio-, or chemo-), and thus are able to replace multistep reactions or difficult purification schemes. This can circumvent the need for many blocking and deblocking steps that are required for stereo- or regioselective reactions. Third, enzymes can be extremely fast, increasing reaction rates by 4 to 12 orders of magnitude. In some cases, enzymes can be so fast that their turnover rates are limited only by diffusion of substrate(s) and product(s) to and from the active site. Fourth, since biocatalysts are biodegradable and rarely contain heavy metals, waste streams are much more manageable, requiring minimal treatment. Fifth, since most biocatalysts function in aqueous media, effluent volatile organic compound emissions from production plants can be drastically reduced.

However, the above-mentioned secondary concerns are more relevant to bulk chemical industries where profit margins are low. In pharmaceuticals, or even the closely related fine chemicals, product purity is of greater concern. Selectivities of >95% are difficult to achieve by chemocatalysis and, if essential to synthesis, would require the use of biocatalysis. This can also be relevant when high enantiopurities are required by regulatory authorities. At present, 22 of 38 large-scale asymmetric syntheses have already incorporated biocatalysis (93). In industrial settings, whole cells are usually preferred over purified enzymes, even for single-step transformations. This makes more sense for enzymes such as oxidoreductases and lyases that are used for asymmetric syntheses, because such enzymes require cofactors in stoichiometric quantities. Hydrolases, however, do not require cofactors, yet the industry standard practice is to use whole cells to avoid enzyme purification costs. The following sections describe some of these industrial-scale applications of biocatalysis.

33.2. APPLICATIONS IN THE CHEMICAL INDUSTRY

Biocatalysts have been increasingly used in the chemical industry. However, because of their high overhead of production and purification costs, biocatalysts are usually not competitive for use in synthesis of most bulk-quantity, low-margin chemicals. Therefore, they are employed in a niche set of reactions that are either extremely difficult or uneconomical to perform by chemocatalysis. It is not surprising that manufacturers of fine chemicals that produce compounds with complex stereocenters have embraced biocatalysis. The following examples reveal that optimized biocatalytic processes are quite viable commercially, and provide further precedence for their application outside their forte in fine chemicals and pharmaceuticals.

33.2.1. Regioselective Catalysis

33.2.1.1. Acrylamide

Acrylamide 1 is a commodity chemical used industrially in coagulators, paper treatment, paper sizing, and soil conditioning, as well as for adhesives, polymers, paints, and petroleum recovery agents. The traditional route of producing acrylamide from acrylonitrile uses either an acid or a Raney-copper catalyst for regioselective hydration. While the Raney-copper process is superior to the former, it is plagued with issues such

as catalyst poisoning and leaching of copper ions. This necessitates not only the constant replenishment for fresh catalyst, but also wastewater treatment to remove the toxic heavy metal ions (50, 65). Mitsubishi Rayon Co. (formerly Nitto Chemical Industries, Japan) produces acrylamide by regioselective hydration of acrylonitrile at a scale of approximately 100,000 tons/year by use of biocatalysis (50, 97). The first-generation process commercialized in 1985 used whole cells harboring an iron-containing nitrile hydratase from *Rhodococcus* sp. N-774 (65) (Fig. 1). In 1988, a second-generation process

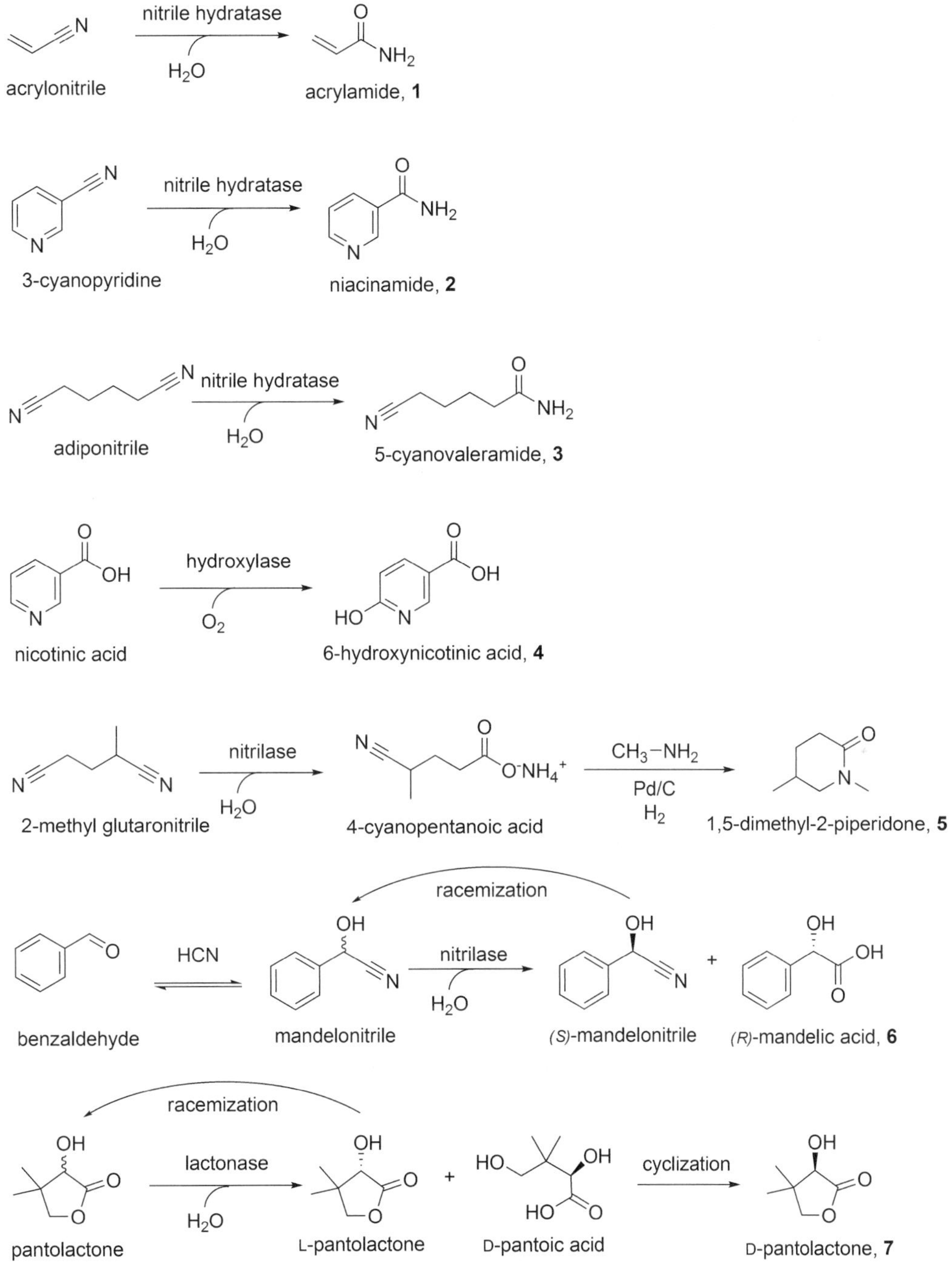

FIGURE 1 Schemes for use of biocatalysis to synthesize industrially important chemicals including acrylamide 1, niacinamide 2, 5-cyanovaleramide 3, 6-hydroxynicotinic acid 4, 1,5-dimethyl-2-piperidone 5, (R)-mandelic acid 6, and D-pantolactone 7.

was developed using another iron-containing nitrile hydratase from *Pseudomonas chlororaphis* B23. The final upgrade in 1991 implemented a copper-containing nitrile hydratase from *Rhodococcus rhodochrous* J1 (64). This third-generation process can produce acrylamide titers of up to 700 g/liter (>50 wt%) with ~99.99% yield in an immobilized-cell bioreactor. In addition to being a highly productive (>7 g of product/g dry cell weight) and environmentally friendly process, the low operating temperature (10°C) ensures lower operating costs than the Raney-copper process and minimizes spontaneous polymerization of acrylamide (1, 38, 94). SNF Floerger (France) and Mitsui Chemicals (Japan) have licensed this process and built several acrylamide plants, each with capacity of 20,000 tons/year. Considering that biocatalysts have been typically viewed as stereoselective catalysts, it is ironic that this nonstereoselective Mitsubishi Rayon process is one of the most well-known industrial syntheses using a biocatalyst.

33.2.1.2. Niacinamide

Another well-known regioselective process was developed by Lonza (Switzerland) for the production of niacinamide **2** (also known as vitamin B3 or nicotinamide), which is used against the skin disease pellagra in humans, and is also an additive to animal feed (6). Commercialized by Lonza Guangzhou Fine Chemicals (China) in 1999, the entire process involves a four-step chemoenzymatic synthesis, with the final step being an enzyme-catalyzed hydration (94) (Fig. 1). The process starts with catalytic conversion of a nylon-6,6 by-product, 3-methyl-1,5-diaminopentane (Dytek®), to 3-picoline by cyclization and dehydrogenation, followed by ammoxidation to 3-cyanopyridine. *R. rhodochrous* J1-immobilized cells expressing a nitrile hydratase are then used for hydration to niacinamide (37). The plant has a capacity of 3,400 tons/year and employs a three-stirred-tank reactor continuously fed with 10 to 20 wt% 3-cyanopyridine. The final amide is produced with >99.3% selectivity, >99.5% purity, and near-complete conversion. Due to the addition of the final enzymatic step, the entire process uses lower energy and has lower emissions than with the caustic hydration step used previously (1, 37, 94). Also, unlike the alkaline hydration, it does not produce any nicotinic acid, an unwanted by-product that causes diarrhea in cattle (6).

33.2.1.3. 5-Cyanovaleramide

Yet another commercialized process using a nitrilase for regioselective hydration is employed by DuPont (United States) for synthesis of 5-cyanovaleramide (5-CVAM) **3**, a precursor to the insecticide Milestone® (azafenidin) (61, 94) (Fig. 1). This process was developed because the previously used manganese dioxide-based catalysis was plagued by several issues. First, nonspecific reactivity of water in the presence of this catalyst resulted in hydration of the product 5-CVAM, yielding unwanted adipamide. Up to 5% of the total reacted adiponitrile was converted to adipamide at a total conversion of only 25%. Second, the separation of the product from the reactant involved a difficult extraction step. Finally, the rapid deactivation of the manganese dioxide catalyst resulted in the production of 1.25 kg of waste/kg 5-CVAM (94). The biocatalytic process used *P. chloraphis* B23, the same organism used in the second-generation Mitsubishi Rayon process, immobilized in calcium alginate beads (6, 32). Using batch reactions with catalyst recycle, 13.6 tons of 5-CVAM were produced at 97% conversion with 93% yield and 96% selectivity (6). The total turnover for the biocatalyst was 3,150 kg of

5-CVAM/kg cell dry weight (6). With the biocatalytic process, the final step was made much more environmentally friendly, produced little waste and by-products, and had minimal purification steps.

33.2.1.4. 6-Hydroxynicotinic Acid

A derivative of niacin, 6-hydroxynicotinic acid **4** is a versatile building block in many modern insecticides such as imidachloprid (40). During chemical synthesis, nonspecific hydroxylations of the pyridine ring lead to side products (76), requiring extensive downstream processing. This renders the entire process prohibitively expensive. The discovery of a regioselective hydroxylase has alleviated many of the issues associated with synthesis. *Achromobacter xylosoxydans* LK1 (DSM 2783) is able to use niacin as its sole nitrogen and carbon source. The first step in niacin metabolism is exactly the regioselective hydroxylation required to produce 6-hydroxynicotinic acid. A unique property of *A. xylosoxydans* LK1 metabolism exploited by Lonza is that, at high concentrations (>1%) of niacin, its assimilation is stalled at 6-hydroxynicotinic acid because of inhibition of the next catabolic enzyme. Therefore, cells are induced with low concentrations of niacin (<1%) to express high levels of the hydroxylase, and then growth is quenched by addition of a high concentration of caustic soda-neutralized niacin. Incubation in well-aerated conditions at 30°C results in yields of >99% (43, 86). Ultrafiltration of the biomass followed by crystallization of product by sulfuric acid addition yields nearly pure precipitated 6-hydroxynicotinic acid crystals. After this downstream processing, the overall yield is >93% and purity is >99%. Lonza has produced more than 10 tons of 6-hydroxynicotinic acid by using this process since its implementation.

33.2.1.5. 1,5-Dimethyl-2-piperidone

Also known as Xolvone®, 1,5-dimethyl-2-piperidone (1,5-DMPD) **5** is a highly water-soluble, nonflammable, biodegradable cleaning solvent with low toxicity. Industrially, it is suitable for application in electronics cleaning, photoresist stripping, degreasing, and metal and resin cleanup (38). It is also used in the formulation of industrial adhesives and inks and as a reaction solvent for production of polymers and chemicals (15, 33). Initial production of 1,5-DMPD was performed by direct hydrogenation of precursor 2-methylglutaronitrile in the presence of methylamine. However, this produced a mixture of 1,5- and 1,3-DMPD. DuPont developed a two-step chemoenzymatic route to overcome the production of unwanted 1,3-DMPD (Fig. 1). Regiospecific hydration of 2-methylglutaronitrile by the nitrilase activity of immobilized *Acidovorax facilis* 72W cells gives 4-cyanopentanoic acid. This is subsequently nonenzymatically converted to the final product by direct hydrogenation in the presence of methylamine (15). The *A. facilis* step is carried out at high loadings (~200 g/liter) and has ~98% selectivity, even after catalyst recycling to give high turnovers (~1,000 kg of product/kg cell dry weight). A further improvement to this process using alginate-immobilized recombinant *Escherichia coli* expressing *A. facilis* 72W nitrilase has been reported (33).

33.2.2. Racemic Resolution

33.2.2.1. (R)-Mandelic Acid

(R)-Mandelic acid **6** [(R)-2-hydroxy-2-phenylacetic acid] has been used in the past as an antibacterial, in particular to treat urinary tract infections (98), but it is now primarily used in cosmetics as an acne or wrinkle treatment (90). Starting with a racemic mixture of mandelonitrile, *E. coli* cells expressing an (R)-specific nitrilase are used to create enantiomerically pure (R)-mandelic acid (38) (Fig. 1). With use of classical chemical hydrocyanation, the unreacted mandelonitrile is racemized via aldehyde in situ. This enzyme-mediated dynamic resolution allows for enantiomerically pure (R)-mandelic acid to be produced on a multiton-per-year scale from a racemic mixture (84).

33.2.2.2. D-Pantolactone

Used in feed for pigs and poultry, D-pantolactone **7** is a chiral-building block for chemical syntheses, particularly for pantothenic acid (vitamin B_5) and coenzyme A (1, 38). A racemic mixture of pantolactones can be achieved by hydrocyanation of the formaldehyde/isobutanal aldol product followed by acidic lactonization. However, separation of the enantiomers is troublesome, and, as a result, a biocatalytic resolution to this problem was developed. Calcium alginate-immobilized *Fusarium oxysporum* AKU302 cells were employed to hydrolyze D-pantolactone to its corresponding acid (Fig. 1). The L-enantiomer is relatively unreactive in the presence of this fungal aldolactonase and is racemized after extraction, whereas the D-pantoic acid is chemically lactonized (81, 87). Since then several other chemical and enzymatic syntheses for both enantiomers have been developed and reviewed in detail for their merits (10).

33.2.2.3. 2,5-Hexanediol

Hexanediols are building block chemicals for synthesis of chiral acetals. Catalytic resolution using Chirazyme L2® (a lipase) in heptane with vinyl propionate esterifies only (R)-hydroxy groups (Fig. 2). This selectivity creates pure (2S,5S)-hexanediol **8**, which is easily extracted into water. The mono- and diesters remaining in the heptane can be rehydrolyzed chemically to (2R,5R)-hexanediol **9** (38). Both diols can be produced at 99% diastereomeric excess (d.e.) and 99% enantiomeric excess (e.e.) and the production has been scaled up to 40-kg batches.

33.2.3. Asymmetric and Enantioselective Synthesis

33.2.3.1. Aspartame

Holland Sweetener Company, a joint venture between Tosoh (Japan) and DSM (Netherlands), produced the low-calorie sweetener aspartame **10** (L-α-aspartyl-L-phenylalanine methyl ester) on a kiloton scale (Fig. 2) until 2006, when they shut down their production facilities. Aspartame is used extensively (12,000 to 15,000 tons per year), however, in soft drinks, salad dressing, and tabletop sweeteners, as well as in pharmaceuticals (31). Starting from fumaric acid and ammonia, aspartase (a lyase) catalyzes the formation of L-aspartic acid. After the addition of a protection group, thermolysin (a metalloprotease from the thermophile *Bacillus thermoproteolyticus*) catalyzes its coupling to L-phenylalanine methyl ester from a D,L-isomeric mixture. The protected dipeptide carboxylate has low solubility and precipitates, driving the equilibrium-disfavored reaction forward. Deprotection of precipitated product is done by traditional chemistry to yield aspartame.

33.2.3.2. Terminal Epoxides

Oxygenation of terminal alkenes to the corresponding epoxides with high enantiomeric purity is difficult by traditional chemistry because of the insufficient activation of the terminal double bond. Such epoxides are particularly useful building blocks for various enantiopure syntheses (22). The alkene monooxygenase from *Nocardia corallina* B276 used by Nippon Mining Co. Ltd. (Japan) is capable of producing predominantly the (R)-enantiomer, starting from a wide range of C_3 to C_{18} alkenes (Fig. 2). Depending on the starting olefin, the bioprocess is carried out in different configurations to cope with the varying levels of toxicity of the resulting epoxide **11**. Short-chain (C_3 to C_5) gaseous epoxides are highly toxic and must be quickly removed by high aeration, followed by an extraction to recover the low amounts of product. Medium-chain epoxides (C_6 to C_{12}) are liquid and less toxic, and are therefore produced in a two-phase bioreactor process. The aqueous phase contains resting cells expressing the monooxygenase, and the organic phase is used to continuously extract the epoxide, driving the reaction forward and also allowing for ease of product recovery. Finally, for long-chain epoxides (C_{13} to C_{18}) that have very low toxicity, growing cells are used in a bioreactor (25–28, 38).

33.2.3.3. (S)-m-Phenoxybenzaldehyde Cyanohydrin

Cyanohydrins are useful building block chemicals, and (S)-m-phenoxybenzaldehyde cyanohydrin in particular is used in synthesis of a highly effective class of insecticides and repellents called pyrethroids. A biphasic system using an oxynitrilase as a catalyst for hydrocyanation has been studied extensively to demonstrate the enzyme's applicability to unnatural systems (19, 30, 44, 85). Of particular interest is the recombinant enzyme from the rubber tree, *Hevea brasiliensis*, which is used to produce (S)-m-phenoxybenzaldehyde cyanohydrin **12** by DSM and Nippon Shokubai (Japan) (Fig. 2).

33.2.3.4. D-Malic Acid

Used as a synthon, resolving agent, or as a ligand in asymmetric synthesis, D-malic acid **13** is a chiral hydroxy dicarboxylic acid produced by DSM. After screening 300 organisms for the ability to produce D-malic acid from maleic acid, permeabilized *Pseudomonas pseudoalcaligenes* NCIMB 9867, expressing the lyase malease, was found to be capable of producing the desired enantiomer with 99.97% purity (96).

33.3. BIOCATALYSIS IN THE PHARMACEUTICAL INDUSTRY

Chirality is a very important factor in the efficacy of many drug products. A survey of major pharmaceutical companies such as GlaxoSmithKline (United Kingdom), AstraZeneca (United Kingdom), and Pfizer (United States) showed that more than half of the drug compounds examined contained one or more chiral centers (11). Furthermore, an enantiomeric purity of at least 99.5% is often necessary to meet regulatory requirements. Recognizing its importance, many fine-chemical and pharmaceutical companies have started to focus on acquiring biocatalysis expertise (Table 1), and those that have already done so are trying to maintain their positions as technological leaders.

33.3.1. Production of Amino Acids

Chiral amino acids are important building blocks for a number of drug compounds. Industrial application of enzymes for production of L-amino acids began almost 40 years

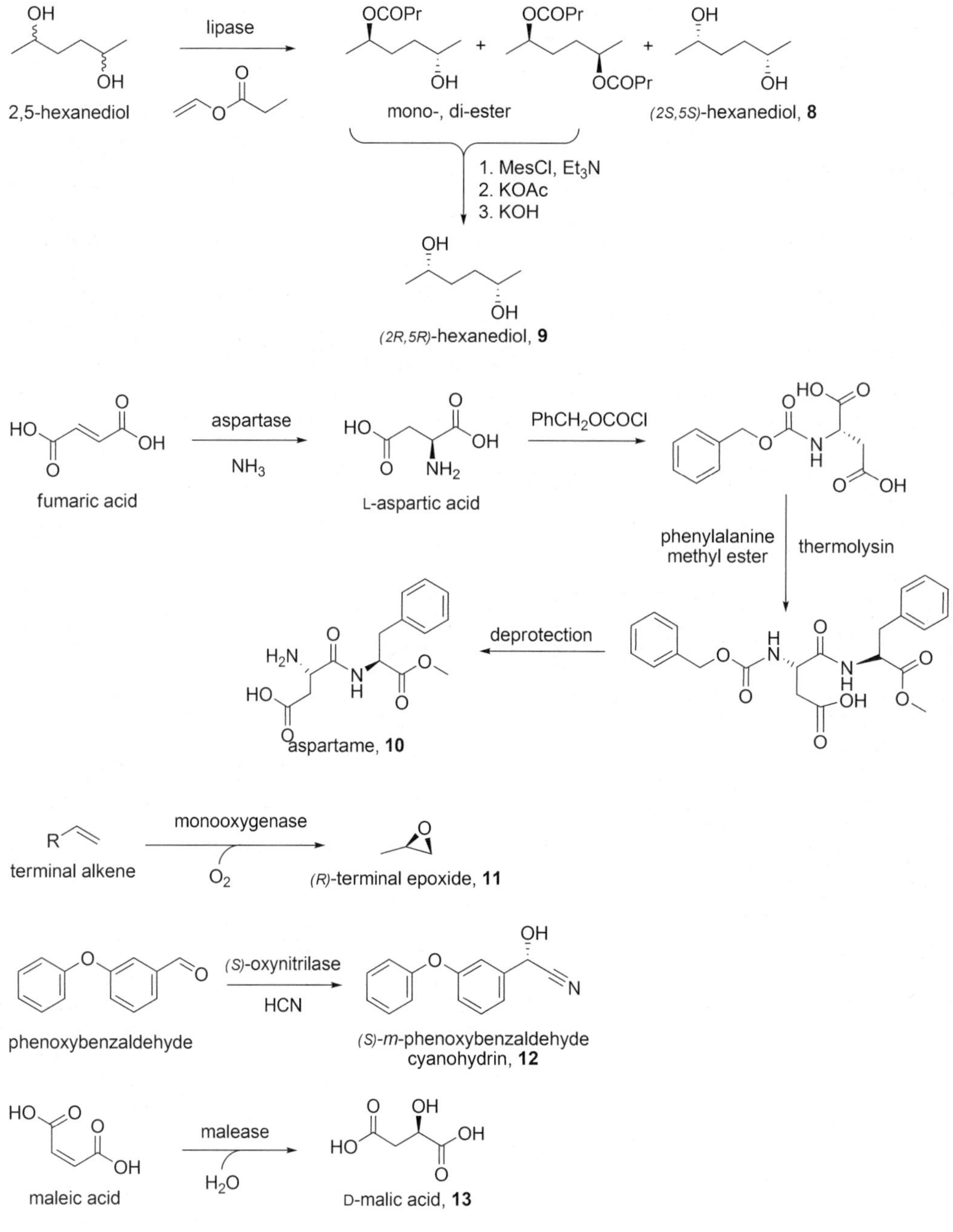

FIGURE 2 Schemes for use of biocatalysis to synthesize industrially important chemicals including (2S,5S)-hexanediol **8**, (2R,5R)-hexanediol **9**, aspartame **10**, (R)-terminal epoxide **11**, (S)-m-phenoxybenzaldehyde cyanohydrin **12**, and D-malic acid **13**.

ago in Japan with the resolution of N-acetyl D,L-amino acids by immobilized acylase (14). Degussa (Germany) has established a large-scale process for the production of D- and L-amino acids via the enantiospecific hydrolysis of N-acyl amino acids by amino acylase I from *Aspergillus oryzae* (101). The broad substrate specificity of A. *oryzae* enables a wide range of proteinogenic and nonproteinogenic N-acetyl and N-chloroacetyl amino acids to be transformed by use of a continuous enzyme membrane reactor, with the enzyme retained by a hollow-fiber ultrafiltration membrane. Established in the early 1980s, this process enabled

the production of hundreds of tons of enantiopure amino acids each year. In more recent years, Degussa established a process which combines the use of the acylase with a novel racemase from *Amycolatopsis orientalis* subsp. *lurida*. Selective in situ racemization of N-acetyl amino acids greatly improved the acylase process by eliminating costly racemization and separation steps (60).

Meanwhile, the DSM process for the production of L-amino acids is based on the resolution of racemic amino acid amides. Although the amidase (aminopeptidase) is L-specific, both the D- and L-amino acids are obtained, and

TABLE 1 Biotransformations developed by the pharmaceutical industry (89)[a]

Target compound	Company	Reaction	Biocatalyst
SCH56592	Schering Plough	Acylation	CALB
β-Lactams	GlaxoSmithKline	Acylation	CALB
Lotrafiban	GlaxoSmithKline	Hydrolysis	CALB
Paclitaxel	Bristol-Myers Squibb	Hydrolysis	*Pseudomonas* lipase AK
			P. cepacia lipase PS-30
HMG-CoA reductase inhibitor	Bristol-Myers Squibb	Acylation	*P. cepacia* lipase PS-30
SCH66336	Schering-Plough	Acylation	*Pseudomonas aeruginosa* lipase
Xemilofiban	Monsanto	Hydrolysis	*E. coli* penicillin acylase
Renin inhibitor	Hoffmann-La Roche	Hydrolysis	*B. licheniformis* subtilisin
Lamivudine	GlaxoSmithKline	Hydrolysis	*E. coli* cytidine deaminase
AG7088	Pfizer	Reduction	*Leuconostoc mesenteroides* D-LDH
			Candida boidinii FDH
ACE inhibitor	Ciba-Geigy	Reduction	*Staphylococcus epidermidis* D-LDH
LY300164	Eli Lilly	Reduction	*Zygosaccharomyces rouxii* dehydrogenase
Omapatrilat	Bristol-Myers Squibb	Reductive amination	*Thermoactinomyces intermedius* PDH
N-Butyl DNJ	Pharmacia	Oxidation	*G. oxydans* SDH
2-Quinoxaline-carboxylic acid	Pfizer	Oxidation	*Absidia repens* MO
HMG-CoA reductase inhibitor	Merck Sharp and Dohme	Oxidation	*Nocardia autotropica* MO
Lobucavir prodrug	Bristol-Myers Squibb	Regioselective acylation	*B. licheniformis* subtilisin

[a]ACE, angiotensin-converting enzyme; CALB, *C. antarctica* lipase B; FDH, formate dehydrogenase; D-LDH, D-lactate dehydrogenase; MO, monooxygenase; PDH, phenylalanine dehydrogenase; SDH, sorbitol dehydrogenase.

the undesired enantiomer is recycled. DSM's biocatalytic amidase toolbox consists of enzymes from *Pseudomonas putida*, *Mycobacterium neoaurum*, and *Ochrobactrum anthropi*, thus enabling a wide range of amino acid amides to be resolved (84).

The hydantoinase/carbamylase system has been used since the 1970s to produce D-phenylglycine and *p*-hydroxy-D-phenylglycine, which are building blocks for the semisynthetic antibiotics ampicillin and amoxicillin. Today, it is commercially applied at a scale of >1,000 tons per year. Using directed evolution, the D-specificity of hydantoinases was switched to L-specificity (59). In addition, Degussa has also established a dynamic kinetic resolution of 5-monosubstituted hydantoins using evolved whole-cell biocatalysts coexpressing an L-carbamoylase, a hydantoin racemase, and a hydantoinase. The hydantoin-converting pathway was optimized by adjusting expression levels of the respective enzymes and by inverting the enantioselectivity of the D-selective hydantoinase by directed evolution. As a result, productivity was improved by 50-fold and biocatalyst cost was significantly reduced (60).

33.3.2. β-Lactam Antibiotics

Based on the condensation of the appropriate D-amino acid derivative with a β-lactam nucleus, processes for large-scale enzymatic production of antibiotics such as ampicillin **14**, amoxicillin **15**, cefaclor, cephalexin **16**, and cefadroxil **17** have been well developed (Fig. 3a). With worldwide capacity exceeding 20,000 tons annually, 6-aminopenicillanic acid (6-APA, **18**), 7-aminocephalosporanic acid (7-ACA), and 7-aminodesacetoxycephalosporanic acid (7-ADCA, **19**) are important precursors for the semisynthetic β-lactam

antibiotics (1). In the 1980s, the chemical preparations of 6-APA and 7-ADCA were replaced with enzymatic routes that were found to be cheaper and environmentally friendly. The enzyme that removes the side chain from penicillin G **20**, penicillin V, and phenacetyl-8-ADCA is penicillin G amidase. A high conversion rate of 98% is reported for the industrial production of 6-APA and 7-ADCA. 7-ACA, on the other hand, is produced in a three-step process using two enzymes. The starting material, cephalosporin C, is subjected to deamination by a D-amino acid oxidase, followed by spontaneous decarboxylation of the α-ketoadipyl-7-ACA into glutaryl-7-ACA. Its side chain is finally hydrolyzed by glutaryl-7-ACA acylase to form 7-ACA.

Penicillin amidases are also used to couple the precursor 7-ADCA with D-phenylglycine methyl ester (D-PGM) or D-phenylglycine amide (D-PGA) in a kinetically controlled enzymatic peptide synthesis to form cephalexin. Similarly, ampicillin is synthesized from 6-APA and excess D-PGA or D-PGM. Amoxicillin, on the other hand, is synthesized using high substrate concentrations by the enzyme-catalyzed reaction of 6-APA and D-4-hydroxyphenylglycine amide (D-HPGA) or D-4-hydroxyphenylglycine methyl ester (D-HPGM) using a semicontinuous reactor system. Despite several drawbacks such as solubility, difficulty in pinpointing an optimal process pH, and chemical degradation issues, the aqueous enzymatic synthesis of β-lactam antibiotics is indeed feasible with yields comparable to or even better than the chemical route. Scientists at DSM research and Gist-Brocades (Netherlands) have developed a biocatalyst that can be recycled many times and also improved the downstream process for better yield (8).

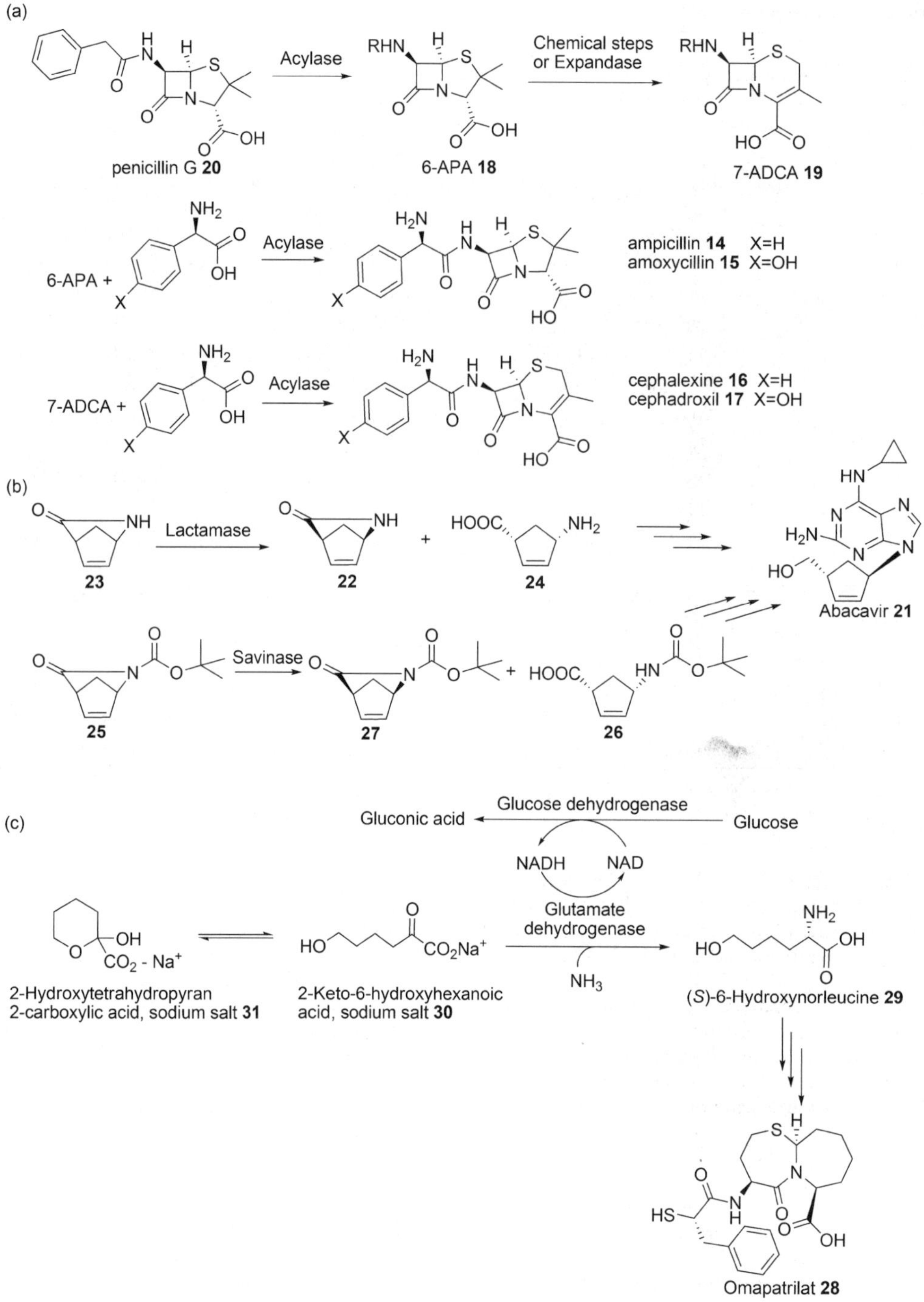

FIGURE 3 Schemes for use of biocatalysis to synthesize chiral intermediates for the production of pharmaceuticals, including β-lactam antibiotics (a), abacavir **21** (b), and omapatrilat **28** (c).

33.3.3. Antiviral Drugs

33.3.3.1. Abacavir

Abacavir **21** (Ziazen™) is a 2-aminopurine nucleoside analog that functions as a selective reverse transcriptase inhibitor for the treatment of human immunodeficiency viruses (HIVs) and hepatitis B viruses. Its intermediate, γ-lactam 2-azabicyclo(2.2.1)hept-5-en-3-one **22**, is produced via the resolution of racemic γ-lactam **23** to

yield the desired product **22** and amino acid **24** using the γ-lactamase-containing organisms *Pseudomonas solanacearum* NCIMB 40249 and *Rhodococcus* NCIMB 40213 (57). An enzymatic process using a commercially available enzyme, Savinase®, to hydrolyze the lactam bond of racemic *tert*-butyl-3-oxo-2-azabicyclo-(2.2.1)hept-5-ene-2-carboxylate **25** has also been developed (56). The hydrolysis of **25** yields the corresponding amino acid **26**, with the desired (1R, 2S)-**27** unreacted (Fig. 3b). The reaction can be carried out at 100 g/liter substrate and gives a reaction yield of 50% and an e.e. of 99%.

33.3.3.2. BMS-186318

Inhibition of HIV protease stops the replication of the virus in vivo. Patel et al. (72) developed an enzymatic process to prepare the chiral intermediate (1S,2R)-[3-chloro-2-hydroxy-1-(phenylmethyl)propyl] carbamic acid, 1,l-dimethylethyl ester for the total synthesis of an HIV protease inhibitor, BMS-186318. The stereoselective reduction of (1S)-[3-chloro-2-oxol-(phenylmethyl)propyl] carbamic acid, 1,1-dimethylethyl ester is carried out using microbial cultures, among which *Streptomyces nodosus* SC 13149 efficiently reduced the ester to its corresponding (1S,2S)-alcohol. A reaction yield of 80% is obtained with product optical purity of 99.8% and diastereomeric purity of 99%.

33.3.3.3. Lobucavir

Lobucavir, a cyclobutyl guanine nucleoside analog, is a potential antiviral agent to treat herpes and hepatitis B. Synthesis of lobucavir L-valine prodrug, BMS 233866, requires regioselective coupling of one of the two hydroxyl groups of lobucavir (BMS 180194) with valine (35). This regioselective aminoacylation is difficult via chemical routes. Hence, an enzymatic route was developed. Selective hydrolysis of di-Cbz-valine ester with lipase M gives *N*-[(phenylmethoxy)carbonyl]-L-valine, [(1R,2R,4S)-2-(2-amino-6-oxo-1*H*-purin-9-yl)-4-(hydroxymethyl)cyclobutyl]methyl ester in 82.5% yield. Meanwhile, L-valine, [(1R,2R,4S)-2-(2-amino-1,6-dihydro-6-oxo-9*H*-purin-9-yl)-4-(hydroxymethyl)cyclobutyl]methyl ester monohydrochloride (87% yield), is obtained by hydrolysis of the divaline ester dihydrochloride with lipase from *Candida cylindracea*. The final intermediate for the lobucavir prodrug, *N*-[(phenylmethyl)carbonyl]-L-valine, [(1S,2R,4R)-3-(2-amino-6-oxo-1*H*-purin-9-yl)-2-(hydroxymethyl)cyclobutyl]methyl ester, is obtained by either transesterification of lobucavir using ChiroCLEC BL™ (crosslinked enzyme crystals of subtilisin Carlsberg) (61% yield), or more selectively by using an immobilized lipase from *Pseudomonas cepacia* (84% yield).

33.3.4. Anticancer Drug: Paclitaxel Semisynthesis

The anticancer drug Taxol® (paclitaxel) is an antimitotic agent that inhibits the depolymerization process of microtubulin during mitosis. It has been used in the treatment of ovarian cancer and metastatic breast cancer with annual sales revenue of US $1 billion (39). Paclitaxel was originally extracted and purified from the bark of the yew *Taxus brevifolia* in very low yield, with approximately 9,000 kg of yew bark (3,000 trees) required to produce 1 kg of purified paclitaxel. Semisynthetic approaches to obtain paclitaxel by coupling baccatin III (paclitaxel without the C-13 side chain) or 10-deacetylbaccatin II ([10-DAB], paclitaxel without the C-13 side chain and the C-10 acetate) to C-13 paclitaxel side chains have also been established. Baccatin III and 10-DAB can be obtained from renewable resources such as needles, shoots, and young *Taxus* cultivars, thus eliminating the cutting down of yew trees.

The C-13 paclitaxel side chain is obtained from the enantioselective hydrolysis of racemic acetate-*cis*-3-(acetoxy)-4-phenyl-2-azetidione to the corresponding (3S)-alcohol and the intact desired (3R)-acetate. The reaction is catalyzed by the lipase PS-30 from *P. cepacia* or Bristol-Myers Squibb (United States) lipase from *Pseudomonas* sp. SC 13856 which was immobilized on Accurel polypropylene

prior to use. Reaction yields of >48% and an e.e. of >99.5% are obtainable for (3R)-acetate. The process has been scaled up to 75 and 150 liters using immobilized enzymes. Finally, the (3R)-acetate is converted into (3R)-alcohol (C-13 paclitaxel side chain) via a chemical route (71).

33.3.5. Anti-inflammatory Drugs

33.3.5.1. Ibuprofen

Ibuprofen, or 2-(4-isobutylphenyl)propionic acid, is a nonsteroidal anti-inflammatory drug that inhibits the binding of arachidonic acid to prostaglandin H2 synthase-l and prevents the synthesis of prostaglandins acting on the inflammatory response. The (S)-ibuprofen molecule is 160 times more potent than the (R)-enantiomer (39). The resolution of racemic ibuprofen is achieved via a lipase-catalyzed esterification reaction with methanol or butanol in organic media. The (S)-ibuprofen is selectively esterified and is easily separated from the (R)-ibuprofen. The (S)-ester is then chemically hydrolyzed to obtain the (S)-ibuprofen. The production of (S)-ibuprofen has also been demonstrated through a lipase-catalyzed hydrolysis of its chemically synthesized ester (46).

33.3.5.2. Naproxen

The racemic resolution of the nonsteroidal anti-inflammatory drug (S)-naproxen or 2-(6-methoxy-2-naphthyl)-propionic acid has been developed by Chirotech (United Kingdom) (77). Widely used as a drug for human connective tissue diseases, the (S)-enantiomer is 30-fold more active than the corresponding (R)-enantiomer (52). Chirotech screened and cloned an efficient *Bacillus* esterase that is specific for the (S)-ester only. The enzyme can accept an initial ester loading of 150 g/liter and the (R)-ester and (S)-acid can be separated easily by centrifugation while the (R)-ester is recycled.

Over the past few years, various enzymatic kinetic resolution reactions have been developed for the preparation of (S)-(+)-naproxen involving *trans*-esterification and hydrolytic methods, using nitrilases (20) as well as lipases and esterases (7, 12, 52, 104). For example, Koul et al. (52) successfully used *Trichosporon* sp. whole cells or its cell-free ester hydrolase to selectively hydrolyze (±)-6-methoxy-α-methyl-2-naphthaleneacetic acid alkyl ester to yield (S)-naproxen (e.e. >99%, E~500). The whole-cell process has been scaled up to a multikilogram level. Further optimization of the process resulted in downstream processing at 80 to 100 g/liter substrate concentration with >90% recovery.

33.3.6. Anticholesterol and Antihypertensive Drugs

33.3.6.1. Lipitor

Atorvastatin is an active ingredient in Pfizer's Lipitor®, a cholesterol-lowering drug with annual sales exceeding US $12 billion per year (91). Atorvastatin belongs to a class of drugs named statins that reduce levels of total cholesterol and low-density lipoprotein. Statins inhibit 3-hydroxy-3-methylglutaryl-coenzyme A (HMG-CoA) reductase, an enzyme that catalyzes the conversion of HMF-CoA to mevalonate. This conversion is one of the initial steps in cholesterol biosynthesis. Therefore, by inhibiting the production of mevalonate, the production of cholesterol is also limited. HN, chemically named ethyl (R)-4-cyano-3-hydroxybutyrate, is the key chiral building block in the synthesis of atorvastatin. Scientists at Codexis (United

States) have developed a green HN process using a three-enzyme system obtained by directed evolution technologies. In the first step, two evolved enzymes, a ketoreductase and a glucose dehydrogenase, are used to catalyze the enantioselective reduction of a prochiral chloroketone (ethyl 4-chloroacetoacetate) by glucose to form an enantiopure chlorohydrin. In the second step, *Agrobacterium radiobacter* halohydrin dehalogenase, expressed in *E. coli*, is used to catalyze the conversion of ethyl (S)-4-chloro-3-hydroxybutyrate to HN at neutral pH, thus avoiding the side reactions associated with chemical cyanation. Using a protein sequence activity relationship-driven strategy, scientists successfully evolved a halohydrin dehalogenase that improves the volumetric productivity of the cyanation process by ~4,000-fold (23). This green HN process is currently being commercialized by Arch Pharmalabs (India), a Codexis production partner, and Lonza.

33.3.6.2. Omapatrilat

Omapatrilat **28** is an antihypertensive drug that inhibits angiotensin-converting enzyme and neutral endopeptidase. It is used not only in the treatment of hypertension but also in the management of congestive heart failure. (S)-6-Hydroxynorleucine **29** is an important intermediate and is synthesized through the complete conversion of 2-keto-6-hydroxyhexanoic acid **30**, in equilibrium with 2-hydroxytetrahydropyran-2-carboxylic acid sodium salt **31** (Fig. 3c), using beef liver glutamate dehydrogenase, in the presence of ammonia and NADH (34). NAD$^+$ produced is recycled to NADH by the oxidation of glucose to gluconic acid by use of glucose dehydrogenase from *Bacillus megaterium*. A reaction yield of 92% and an e.e. of >99% for (S)-6-hydroxynorleucine are obtained from this process.

Alternatively, 2-keto-6-hydroxyhexanoic acid can be prepared by the treatment of racemic 6-hydroxynorleucine with porcine kidney D-amino acid oxidase and beef liver catalase or *Trigonopsis variabilis* whole cells (source of both the oxidase and catalase). Once the e.e. of the remaining (S)-6-hydroxynorleucine has risen to >99%, reductive amination procedure is used to convert the mixture containing 2-keto-6-hydroxyhexanoic acid to (S)-6-hydroxynorleucine with a yield of 97% and e.e. of 98% from racemic 6-hydroxynorleucine (70).

33.3.7. Calcium Channel Blocker: Diltiazem

Diltiazem is a benzothiazepinone calcium channel-blocking agent that inhibits influx of extracellular calcium through L-type voltage-operated calcium channels. Used widely in the treatment of hypertension and angina, diltiazem sales were reported to be $250 million in 2002. Tanabe (Japan) manufactures 50 tons of diltiazem annually. Resolution of racemic epoxyesters is a key step in the production of an important intermediate in the synthesis of diltiazem. Catalyzed by a lipase from *Serratia marcescens*, the enantiospecific hydrolysis yields (2R,3S)-methyl-*p*-methoxyphenylglycidate with an e.e. of >98%. This key diltiazem intermediate has also been successfully produced by Sepracor (United States) in a multiphase membrane bioreactor at a multikilogram scale (39).

In searching for a more potent analog, (*cis*)-3-(acetoxy)-1-[2-(dimethylamino) ethyl]-1,3,4,5-tetrahydro-4-(4-methoxyphenyl)-6-(trifluoromethyl)-2H-1-benzazepin-2-one was identified as a longer lasting and better antihypertensive agent. A key intermediate in the synthesis of this compound is (3R-*cis*)-1,3,4,5-tetrahydro-3-hydroxy-4-(4-methoxyphenyl)-6-(trifluoromethyl)-2H-1-benzazepin-2-one,

which is produced via the enantioselective reduction of 4,5-dihydro-4-(4-methoxyphenyl)-6-(trifluoromethyl)-1H-1-benzazepin-2,3-dione using *Nocardia salmonicolor* SC 6310. A reaction yield of 96% with 99.8% e.e. is obtained for this process (73).

33.3.8. Anti-Alzheimer's Disease Drugs

Several potential anti-Alzheimer's disease drugs that inhibit β-amyloid peptide release or its synthesis have been developed (83). (S)-2-Pentanol, an important intermediate in the synthesis of these drugs, is obtained from the enzymatic resolution of racemic 2-pentanol by lipase B from *Candida antarctica* (69). Various commercially available lipases were screened for stereoselective acylation of racemic alcohols in the presence of vinyl acetate as an acyl donor, and lipase B from *C. antarctica* gave a reaction yield of 43 to 45% with an e.e. for (S)-2-pentanol of >99% at a substrate concentration of 100 g/liter. Racemic 2-pentanol is used both as solvent and substrate. By use of 0.68 mol equivalent of succinic anhydride and 13 g of lipase B per kg of racemic 2-pentanol, an overall yield of 38% with an e.e. of >98% is obtained for (S)-2-pentanol.

In another approach, 2-pentanone is enantioselectively reduced to (S)-2-pentanol by using the 2-ketoreductase from *Gluconobacter oxydans* (66). On a pilot scale, 1.06 kg of (S)-2-pentanol was prepared from 2.3 kg of 2-pentanone by use of Triton X-100-treated *G. oxydans* cells with an e.e. of >99%.

33.3.9. Antituberculosis Drugs

5-Hydroxypyrazine-2-carboxylic acid, a versatile building block for the synthesis of new antituberculous agents, is prepared by a two-step biotransformation of 2-cyanopyrazine using whole cells of *Agrobacterium* sp. DSM 6336 (102). The biotransformation involves the hydrolysis of the nitrile group to pyrazonecarboxylic acid, followed by a regioselective hydroxylation to 5-hydroxypyrazinecarboxylic acid. A product concentration of 286 mM (40 g/liter), corresponding to a total yield of 80%, is achieved. Earlier studies demonstrated that various 5-chloropyrazine-2-carboxylic acid esters, prepared from 5-hydroxypyrazine-2-carboxylic acid, were up to 1,000 times more potent against *Mycobacterium tuberculosis* and other *Mycobacterium* strains than pyrazinamide, a drug commonly used in treating tuberculosis. This result indicated a great potential for 5-hydroxypyrazine-2-carboxylic acid as a building block for new antituberculous agents.

33.3.10. Conversions of Steroids and Oligosaccharides

Biotransformation of steroids for use in contraceptives and other steroid hormone derivatives is a well-established industrial application of biocatalysis. The world demand for these steroid drugs exceeds 1,000 tons per year (84). Basic building blocks for these drugs can be obtained from natural phytosterols from soya (typically a mixture of 40% β-sitosterol, 25% campesterol, and 25% stigmasterol), from conifers (tall-oil: 70% β-sitosterol, 10% campesterol, and 15% β-sitostanol), and from rapeseed (45% β-sitosterol, 35% campesterol, and 12% brassicasterol). Schering (Berlin and Bergkamen, Germany) utilizes these natural resources together with mutants of *Mycobacterium* sp. devoid of steroid-ring degradation activities for the large-scale production of androstenedione and androstadienedione. These steroids are used as basic substrates for the subsequent production of drugs. Among the biocatalytic processes are hydroxylations (e.g., at the 11α or 11β positions with *Curvularia* sp.), dehydrogenations (Δ1-position; hydrocortisone to prednisolone), and reductions (17-keto reduction).

Oligosaccharides play important roles in various types of biochemical recognition processes on the cell surface and are useful pharmaceuticals. Kyowa Hakko Kogyo Co. Ltd. (Japan) has developed a multistep whole-cell biocatalysis for the production of oligosaccharides with high yield, productivity, and product concentration (51). They implemented a large-scale production system of uridine 5′-diphosphogalactose (UDP-Gal) from inexpensive starting materials by using a combination of metabolically engineered recombinant *E. coli* and *Corynebacterium ammoniagenes*. *E. coli* was engineered to express the UDP-Gal biosynthetic genes galactose-1-phosphate uridyltransferase (*galT*), galactokinase (*galK*), glucose-1-phosphate uridyltransferase (*galU*), and pyrophosphatase (*ppa*). *C. ammoniagenes* was engineered to produce uridine 5′-triphosphate (UTP), a substrate for UDP-Gal biosynthesis, from orotic acid. By coupling the UDP-Gal production system with *E. coli* cells that expressed the α1,4-galactosyltransferase gene (*lgtC*) of *Neisseria gonorrhoeae*, 372 mM (188 g/liter) globotriose (Galα1-4Galβ1-4Glc), a trisaccharide portion of the verotoxin receptor, is produced from inexpensive starting materials including orotic acid, galactose, and lactose. No oligosaccharide byproducts are observed in the reaction mixture because of the strict substrate specificity of α1,4-galactosyltransferase. It is interesting that the production of globotriose is much higher than that of UDP-Gal itself (72 mM). This system, combining *C. ammoniagenes* with metabolically engineered recombinant *E. coli* cells that overexpressed sugar nucleotide biosynthetic genes, overcame the drawbacks of enzymatic methods of sugar nucleotide synthesis that require expensive raw materials such as phosphoenolpyruvate and nucleoside 5′-phosphates (42, 103).

33.4. ENZYMES IN FOOD, TEXTILE, AND PULP AND PAPER INDUSTRIES

Enzymes have been used to make cheese, bread, and beer for thousands of years, long before the knowledge of their existence (49, 74). With the recent advancements in recombinant DNA technology and protein engineering, more efficient enzymes are being applied to these processes, as well as others in the food industry. In addition, enzymes are being applied to improve processes in the textile, pulp, and paper industries (Fig. 4). The following sections describe a few representative enzymes that are commonly used in these industries.

33.4.1. Amylases
The amylase family is a starch-degrading enzyme acting on the α-1,4- and α-1,6-O-glycosidic linkages of the

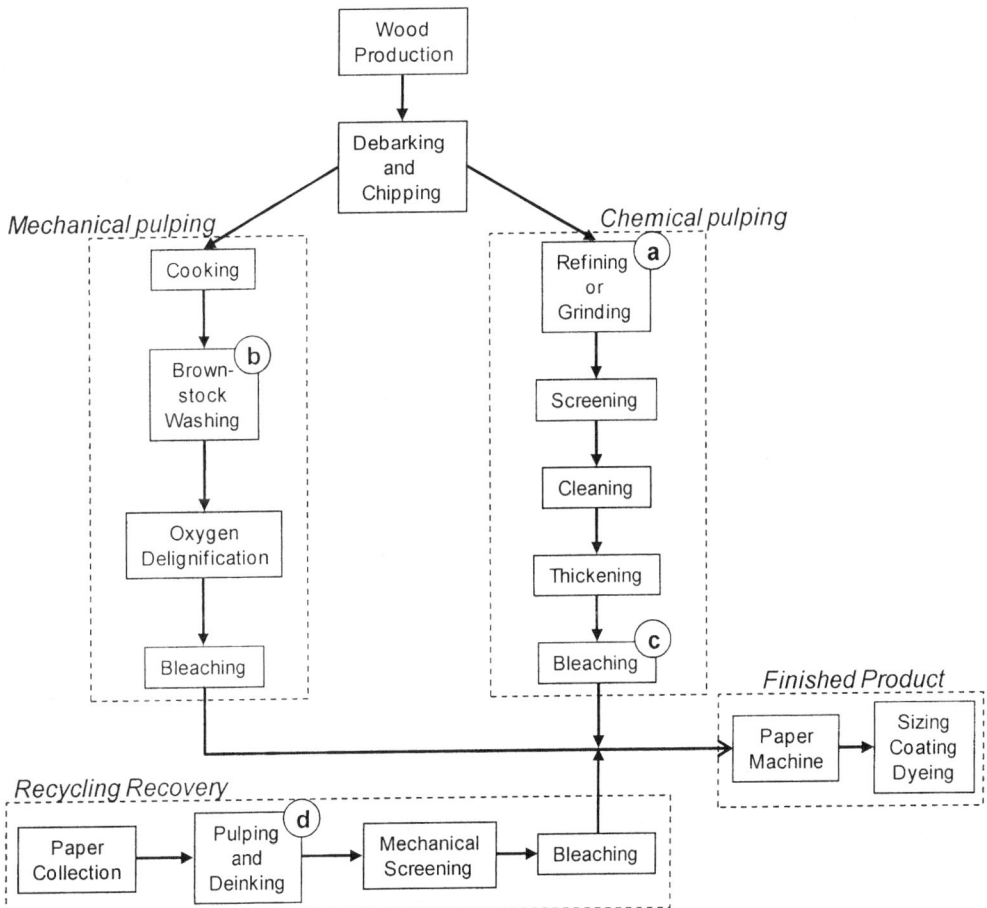

FIGURE 4 Flow diagram of the main processes in the pulp and paper industry, with indication of enzymes applied at an industrial scale. (a) Cellulases, (b) xylanases, (c) lipases, and (d) cellulases.

polysaccharide (68). The largest demand of amylases in the food industry is for the production of high-fructose corn syrups (13). The starch is pressure cooked, and thermostable α-amylases hydrolyze the internal glycosidic linkages of the starch molecule to form maltodextrins. β-Amylases are then applied to produce maltose by hydrolyzing only the nonreducing ends of the molecule, and a final treatment with glucose isomerase produces fructose (80). Protein-engineering techniques were used to produce thermostable enzymes that significantly increased the efficiency of the reaction and thus improved the process (17, 45, 55, 78). The thermostable *Bacillus licheniformis* α-amylase is one of the most widely used amylases in the bioindustry (99). The wild-type enzyme performs optimally at 90°C and a pH of 6, but requires the addition of calcium for high thermostability (100). Machius and coworkers (55) used directed evolution techniques on the protein to introduce five-point mutations (H133V, N190F, A209V, and Q264S), which increased the temperature of the unfolding of the enzyme by 13°C. The single-point mutations had a cumulative effect, resulting in a final mutant that unfolded 32 times slower than the wild type. Richardson et al. (78) used biodiversity studies to identify thermostable amylases from multiple organisms and applied DNA-shuffling techniques to produce protein chimeras. These chimeras not only had a 5°C increase in thermostability and required no additives for stability compared with the wild type from *B. licheniformis*, but were also tolerant to pH as low as 4.5.

Amylases have also been used as an additive to improve the quality of bread (75). During baking, the amylases decrease starch gelatinization, which decreases the dough viscosity (18). This improves the oven spring (ability of bread to rise when heated), thus enlarging the bread volume. Dogan (18) has shown that increased amylase activity improves gas production, which leads to enhanced crumb structure. Amylases are also known to improve the shelf life of baking products through their antistaling properties (29, 58, 62). During staling, a decrease in moisture level causes the amorphous starch structure to retrograde and crystallize. Palacios et al. (67) have shown that addition of thermostable amylases that remain active after baking hydrolyze chain segments in the amorphous regions near crystallites, which reduces crystallization.

Since the 1950s, the textile industry has also applied amylases to improve the process of producing cotton fabrics (3). To protect the product during weaving, the threads are coated with a starch called size that must subsequently be removed before dyeing and finishing. Conventionally, desizing was carried out by treating the fabric with chemicals such as acids, alkali, or oxidizing agents that can have damaging effects on the product. However, amylases have been applied to the process to selectively degrade the starch and not harm the product (41).

33.4.2. Pectinases

A main structural component of plant cell walls called pectins are polysaccharides, consisting of galacturonic acid residues linked by α-(1,4)-glycosidic linkages that are hydrolyzed by pectinases. In the food industry, pectinases from *Aspergillus niger* are used to increase the yield of fruit juice. When fruit tissue is ground, pectin is localized in the liquid phase and increases the viscosity of the juice (47). Pectinases partially hydrolyze the pectins, which causes the particles to aggregate and settle out of solution, increasing the juice yield. To increase efficiency of the process, immobilization of pectinases has been investigated. Specifically,

Sarioglu et al. (82) immobilized a commercially available Pectinex Ultra SP-L, from *Aspergillus aculeatus* produced by Novo Nordisk Ferment in Switzerland, onto an anion-exchange resin particle by electrostatic adsorption. The kinetic constants were found to be V_{max} 0.0091% (wt/vol)/s and K_m 2.172% (wt/vol) compared with the free enzyme V_{max} 0.0046% (wt/vol)/s and K_m 1.137% (wt/vol).

In the production of cotton, one of the most energy- and water-consuming steps is the scouring step: the removal of various remaining cell wall components on the cellulose fibers before dyeing (49). Scouring gives fabric a high and even wetting ability, so that it can be bleached and dyed successfully. The first application of enzymes to this process was by Tzanov et al. (95), who developed a process to treat the fabric with an alkaline bacterial pectinase (Bioprep 3,000L from Novo Nordisk), which reduced the overall cellulose structure with only a 27% decrease in weight compared with chemical scouring. The positive environmental impact of the new process was recognized by the U.S. Presidential Green Chemistry Challenge Award in 2001 (49).

33.4.3. Xylanases

Hemicellulose makes up 20 to 30% of plant matter, and xylans are the major component. This polysaccharide consists of a backbone chain of xylopyranose units that are linked by 1,4-β-D-glycosidic bonds that can be hydrolyzed by xylanases. The ability to degrade the xylans in flour makes the application of xylanases in the food industry desirable to increase the quality of bread and pastry dough (9, 53). Compared with white flour, wheat flour has a higher content of soluble arabinoxylan, which increases viscosity of the dough (16). Degradation of the arabinoxylan by xylanases decreases the dough viscosity, thus increasing volume and also creating finer, more uniform crumbs (9, 79). The most common xylanase applied in the baking industry originated from *A. niger*, but Jiang et al. (43) investigated the potential of the xylanase from the extremophile *Thermomyces lanuginosus*. The enzyme remained active up to 75°C. This increase in activity of the enzyme increased specific volume of the bread by 20% (43).

Xylanases are also used in the pulp and paper industry to degrade and weaken wood fibers before processing. A white and bright product is considered high quality; therefore, the pulp is bleached to produce the desired result. The degradation of the lignin and wood fibers makes the pulp more susceptible to bleaching and is called "bleach boosting" (88). This treatment significantly decreases the amount of bleaching material required and limits the emissions of absorbable organic halides (24). The efficiency of xylanases originating from different organisms has been investigated, and Khandeparkar and Bhosle (48) determined that xylanases isolated from *Arthrobacter* sp. were the most efficient and decreased the chlorine requirement the most compared with other organisms. Previous studies in using the xylanase from *Streptomyces thermoviolaceus* yielded only a 20% reduction in chlorine usage compared with a nonenzymatic treatment, while *Arthrobacter* sp. decreased the chlorine requirement by 28%.

33.4.4. Cellulases

Cellulases are used in both the textile and pulp and paper industries to hydrolyze cellulose, a polysaccharide of 1,4-β-D-linked glucose units. In the textile industry, the quality of cotton can be improved by an enzymatic treatment known as bio-polishing (49). The treatment removes tiny strands of fiber that protrude from the surface of yarn, giving the fabric a smoother, glossier appearance and softer handling. In denim processing, many garments are subjected to a treatment to give them a slightly worn look, called stonewashing. Traditionally, stonewashing involves pumice stones beating the surface of the fabric, which fades the color and loosens the fabric (5). This process actually damages the garment and can break the stonewashing machine. In bio-stonewashing, cellulases loosen the indigo dye on the denim, achieving the same worn look without damaging the garment and giving the option of a variety of finishes. Miettinen-Oinonen et al. (61) compared four cellulase enzymes produced by *Melanocarpus albomyces*, *Myceliophthora thermophila*, *Chaetomium thermophilum*, and *Sporotrichum thermophilum*. They determined that cellulase originating from M. *albomyces* had the highest dye-releasing activity and also had the lowest back-staining (the released dye particles restaining the back side of the fabric or other fabrics).

In the pulp and paper industry, wood chips are physically separated into free fibers. This process can utilize a lot of electricity. Cellulases break down the cellulose in the wood fibers, softening the chips so that the processing time is shortened and electrical usage is diminished (74). Cellulases are also used in the deinking of recycled papers. Conventional deinking requires large amounts of chemicals such as NaOH, NaSO$_3$, and H$_2$O$_2$. The cellulases loosen the fibers of the recycled paper and thus release the ink from the fiber (4). Lee et al. (54) developed a protocol for the enzymatic deinking of laser-printed waste papers with use of cellulases and hemicellulases from A. *niger*. A maximum deinking efficiency of nearly 95% was achieved with a combination of the two enzymes. The bio-deinked recycled papers had quality similar to commercial papers (54).

33.4.5. Transglutaminases

Transglutaminases catalyze an acyl-transfer reaction of γ-carboxamide groups of peptide-bound glutamine residues, which cross-links whey, soya, beef, and poultry meat proteins (106). This can be used to modify the functional properties of the foods, as first shown by Motoki et al. (63). Products such as hamburgers, minced meat, and dumplings have improved elasticity, flavor, and texture as a result of the cross-linking. Using a liver enzyme from guinea pigs, this group developed a novel meat-binding system that was based on the protein cross-linking ability of the transglutaminase to bind fragments of meat such as fish and beef together. For industrial application of the process, microbial transglutaminase was isolated from *Streptomyces mobaraensis* (105). Herrero et al. (36) used Raman spectroscopy to show that treating meats with transglutaminases increased protein β-sheet and -turns structure, including an increase in hardness, springiness, and cohesiveness.

33.5. CONCLUSIONS AND FUTURE PROSPECTS

Biocatalysis has emerged as a powerful tool for the production of many important industrial products including, but not limited to, enantiomerically pure building blocks for pharmaceuticals and agrochemicals, active pharmaceutical ingredients, antibiotics, and food ingredients such as sweeteners and vitamins. With the discovery of powerful enzyme manipulation tools such as directed evolution techniques and the advances in genomics, applications of biocatalysis in industrial synthesis are likely to increase rapidly. The feasibility of applying biocatalysis in the chemical, pharmaceutical, food, textile, and pulp and paper industries depends very much on the economics of each specific process. Biocatalysis provides enormous opportunities for new development, and once fine-chemical and pharmaceutical companies have acquired more experience and the necessary expertise in such processes, these companies will become more confident in developing next-generation bioprocesses.

Biocatalytic processes are heavily focused on the sustainable production of chemicals, materials, and fuels from renewable resources. These processes do not use toxic raw materials, operate under mild reaction conditions, and produce significantly less hazardous waste. Biocatalysis can help shorten complicated syntheses from as many as 12 steps to just three or four, cuts manufacturing costs 40 to 60%, and reduces capital expenditures by more than 25%, while at the same time reducing the environmental footprint (92). All these factors often translate into a reduction of production costs and an increase in competitiveness, especially in highly regulated countries. Development of a new sustainable biocatalysis platform results in job retention and creation, as well as the conservation of fossil resources. In short, social, environmental, and economic benefits of industrial biocatalysis are a self-reinforcing cycle of sustainability; industrial biocatalysis can create new employment and economic values and simultaneously reduce the negative impact on the environment.

The application of biocatalysis in industry has yet to be exploited to its maximum potential. Public databases like BioCatalysis (accelrys.com) and the University of Minnesota Biocatalysis Database (21) list over 35,000 enzyme reactions. This, together with the expanding database of whole genomes and yet unexplored microbial diversity, represents a huge pool of enzymes that are waiting to be exploited as future industrial catalysts. As technologies for developing efficient biocatalysts, such as high-throughput screening, random and recombinatory mutagenesis, metabolic engineering, and enzyme immobilization, continue to advance, many new industrial applications of biocatalysis will soon replace traditional chemical catalysis methods. The challenge is to develop a biocatalyst with high activity, specificity, and stability in the presence of high concentrations of substrate and organic solvents, or at elevated temperatures, to make the industrial process economically viable. Process-engineering solutions such as in situ product removal using a multiphase process and a robust cofactor-recycling/replacement system will help improve productivity and reduce cost. In addition, elucidation of enzyme crystal structures in combination with computer-aided modeling will enable enzymes to be custom-tailored for industrial applications. The enzymes can be evolved to be highly active, to be thermostable, and to have altered substrate specificity. The breadth and depth of knowledge in the biocatalysis area is undoubtedly still limited compared with traditional chemical catalysis. However, in years to come, the expanding biocatalytic and biomolecular toolbox and better understanding of enzymes and cell stability in industrial environments, such as large reactors and organic solvents, will soon make it easier to develop many more efficient industrial bioprocesses.

N.U.N. and W.L.T. contributed equally to this work. We thank the National Institutes of Health (GM077596), the Biotechnology Research and Development Consortium (BRDC) (Project 2-4-121), the British Petroleum Energy Biosciences Institute, and the National Science Foundation for financial support in our biocatalysis and biosynthesis studies. D.T.E. acknowledges support from the National Science Foundation Graduate Research Fellowship Program.

REFERENCES

1. **Aehle, W., H. Waldmann, C. Schultz, H. Gröger, C. Dinkel, and K. Drauz.** 2008. Enzymes. 5. Enzymes in organic synthesis. *Ullmann's Encyclopedia of Industrial Chemistry,* 7th ed. John Wiley & Sons, Inc., Hoboken, NJ.

2. **Anastas, P. T., and J. C. Warner.** 2000. *Green Chemistry: Theory and Practice.* Oxford University Press, New York, NY.

3. **Araujo, R., M. Casal, and A. Cavaco-Paulo.** 2008. Application of enzymes for textile fibres processing. *Biocatal. Biotransform.* **26:**332–349.

4. **Bajpai, P.** 1999. Application of enzymes in the pulp and paper industry. *Biotechnol. Prog.***15:**147–157.

5. **Bhat, M. K.** 2000. Cellulases and related enzymes in biotechnology. *Biotechnol. Adv.* **18:**355–383.

6. **Bommarius, A. S., and B. R. Riebel.** 2004. Application of enzymes as catalysts: basic chemicals, fine chemicals, food, crop protection, bulk pharmaceuticals, p. 159–208. *In* A. S. Bommarius and B. R. Riebel (ed.), *Biocatalysis: Fundamentals and Applications,* vol. 1. Wiley-VCH, Weinheim, Germany.

7. **Brady, D., A. Beeton, J. Zeevaart, C. Kgaje, F. Rantwijk, and R. A. Sheldon.** 2004. Characterisation of nitrilase and nitrile hydratase biocatalytic systems. *Appl. Microbiol. Biotechnol.* **64:**76–85.

8. **Bruggink, A., E. C. Roos, and E. de Vroom.** 1998. Penicillin acylase in the industrial production of β-lactam antibiotics. *Org. Process Res. Dev.* **2:**128–133.

9. **Butt, M. S., M. Tahir-Nadeem, Z. Ahmad, and M. T. Sultan.** 2008. Xylanases and their applications in baking industry. *Food Technol. Biotechnol.* **46:**22–31.

10. **Camps, P., and D. Munoz-Torrero.** 2004. Synthesis and applications of (R)- and (S)-pantolactone as chiral auxiliaries. *Curr. Org. Chem.* **8:**1339–1380.

11. **Carey, J. S., D. Laffan, C. Thomson, and M. T. Williams.** 2006. Analysis of the reactions used for the preparation of drug candidate molecules. *Org. Biomol. Chem.* **4:**2337–2347.

12. **Chang, C. S., and S. W. Tsai.** 1999. Lipase-catalyzed dynamic resolution of naproxen thioester by thiotransesterification in isooctane. *Biochem. Eng. J.* **3:**239–242.

13. **Cherry, J. R., and A. L. Fidantsef.** 2003. Directed evolution of industrial enzymes: an update. *Curr. Opin. Biotechnol.* **14:**438–443.

14. **Chibata, I.** 1978. *Immobilized Enzymes.* John Wiley & Sons Inc., Hoboken, NJ.

15. **Cooling, F. B., S. K. Fager, R. D. Fallon, P. W. Folsom, F. G. Gallagher, J. E. Gavagan, E. C. Hann, F. E. Herkes, R. L. Phillips, A. Sigmund, L. W. Wagner, W. Wu, and R. DiCosimo.** 2001. Chemoenzymatic production of 1,5-dimethyl-2-piperidone. *J. Mol. Catal. B Enzym.* **11:**295–306.

16. **Courtin, C. M., and J. A. Delcour.** 2002. Arabinoxylans and endoxylanases in wheat flour bread-making. *J. Cereal Sci.* **35:**225–243.

17. **Declerck, N., P. Joyet, J. Y. Trosset, J. Garnier, and C. Gaillardin.** 1995. Hyperthermostable mutants of *Bacillus licheniformis* alpha-amylase: multiple amino acid replacements and molecular modelling. *Protein Eng.* **8:**1029–1037.

18. **Dogan, I.** 2002. Dynamic rheological properties of dough as affected by amylases from various sources. *Nahrung/Food* **46:**399–403.

19. **Effenberger, F., S. Forster, and H. Wajant.** 2000. Hydroxynitrile lyases in stereoselective catalysis. *Curr. Opin. Biotechnol.* **11:**532–539.

20. **Effenberger, F., B. W. Graef, and S. Osswald.** 1997. Preparation of (S)-naproxen by enantioselective hydrolysis of racemic naproxen amide with resting cells of *Rhodococcus erythropolis* MP50 in organic solvents. *Tetrahedron Asymmetry* **8:**2749–2755.

21. **Ellis, L. B. M., C. D. Hershberger, E. M. Bryan, and L. P. Wackett.** 2001. The University of Minnesota Biocatalysis/Biodegradation Database: emphasizing enzymes, vol. 29, p. 340–343.

22. **Faber, K., and R. Patel.** 2000. Chemical biotechnology: a happy marriage between chemistry and biotechnology: asymmetric synthesis via green chemistry. *Curr. Opin. Biotechnol.* **11:**517–519.

23. **Fox, R. J., S. C. Davis, E. C. Mundorff, L. M. Newman, V. Gavrilovic, S. K. Ma, L. M. Chung, C. Ching, S. Tam, S. Muley, J. Grate, J. Gruber, J. C. Whitman, R. A. Sheldon, and G. W. Huisman.** 2007. Improving catalytic function by ProSAR-driven enzyme evolution. *Nat. Biotechnol.* **25:**338–344.

24. **Fu, G., A. Chan, and D. Minns.** 2005. Preliminary assessment of the environmental benefits of enzyme bleach boosting for pulp and paper making. *Int. J. Life Cycle Assess.* **10:**136–142.

25. **Furuhashi, K.** 1988. Microbial production of optically-active epoxide. *Nippon Nogeikagaku Kaishi-J. Jpn. Soc. Biosci. Biotechnol. Agrochem.* **62:**772–774.

26. **Furuhashi, K.** 1987. Production of optically-active epoxides by microbial oxidation of olefins. *J. Synthet. Organ. Chem. Jpn.* **45:**162–168.

27. **Furuhashi, K., A. Taoka, S. Uchida, I. Karube, and S. Suzuki.** 1981. Production of 1,2-epoxyalkanes from 1-alkenes by *Nocardia corallina* B-276. *Eur. J. Appl. Microbiol. Biotechnol.* **12:**39–45.

28. **Gallagher, S. C., R. Cammack, and H. Dalton.** 1997. Alkene monooxygenase from *Nocardia corallina* B-276 is a member of the class of dinuclear iron proteins capable of stereospecific epoxygenation reactions. *Eur. J. Biochem.* **247:**635–641.

29. **Gerrard, J. A., D. Every, K. H. Sutton, and M. J. Gilpin.** 1997. The role of maltodextrins in the staling of bread. *J. Cereal Sci.* **26:**201–209.

30. **Griengl, H., H. Schwab, and M. Fechter.** 2000. The synthesis of chiral cyanohydrins by oxynitrilases. *Trends Biotechnol.* **18:**252–256.

31. **Hagen, J.** 2006. Aspartame through enzymatic peptide synthesis, p. 94–95. *In Industrial Catalysis: a Practical Approach.* Wiley-VCH, Weinheim, Germany.

32. **Hann, E. C., A. Eisenberg, S. K. Fager, N. E. Perkins, F. G. Gallagher, S. M. Cooper, J. E. Gavagan, B. Stieglitz, S. M. Hennessey, and R. DiCosimo.** 1999. 5-Cyanovaleramide production using immobilized *Pseudomonas chlororaphis* B23. *Bioorg. Med. Chem.* **7:**2239–2245.

33. **Hann, E. C., A. E. Sigmund, S. M. Hennessey, J. E. Gavagan, D. R. Short, A. Ben-Bassat, S. Chauhan, R. D. Fallon, M. S. Payne, and R. DiCosimo.** 2002. Optimization of an immobilized-cell biocatalyst for production of 4-cyanopentanoic acid. *Org. Process Res. Dev.* **6:**492–496.

34. **Hanson, R. L., M. D. Schwinden, A. Banerjee, D. B. Brzozowski, B. C. Chen, B. P. Patel, C. G. McNamee, G. A. Kodersha, D. R. Kronenthal, R. N. Patel, and L. J. Szarka.** 1999. Enzymatic synthesis of L-6-hydroxynorleucine. *Bioorg. Med. Chem.* **7:**2247–2252.

35. **Hanson, R. L., Z. Shi, D. B. Brzozowski, A. Banerjee, T. P. Kissick, J. Singh, A. J. Pullockaran, J. T. North, J. Fan, J. Howell, S. C. Durand, M. A. Montana, D. R. Kronenthal, R. H. Mueller, and R. N. Patel.** 2000. Regioselective enzymatic aminoacylation of lobucavir to give an intermediate for lobucavir prodrug. *Bioorg. Med. Chem.* **8:**2681–2687.

36. **Herrero, A. M., M. I. Cambero, J. A. Ordonez, L. De la Hoz, and P. Carmona.** 2008. Raman spectroscopy study of the structural effect of microbial transglutaminase on meat systems and its relationship with textural characteristics. *Food Chem.* **109:**25–32.

37. Heveling, J., E. Armbruster, L. Utiger, M. Rohner, H.-R. Dettwiler, and R. J. Chuck. February, 1998. Process for preparing nicotinamide. U.S. patent 5,719,045.

38. Hilterhaus, L., and A. Liese. 2007. Building blocks, p. 133–173. In R. Ulber and D. Sell (ed.), White Biotechnology. Springer, Berlin, Germany.

39. Houde, A., A. Kademi, and D. Leblanc. 2004. Lipases and their industrial applications. Appl. Biochem. Biotechnol. 118:155–170.

40. Hurh, B., M. Ohshima, T. Yamane, and T. Nagasawa. 1994. Microbial production of 6-hydroxynicotinic acid, an important building block for the synthesis of modern insecticides. J. Ferment. Bioeng. 77:382–385.

41. Ibrahim, N. A., M. El-Hossamy, M. S. Morsy, and B. M. Eid. 2004. Optimization and modification of enzymatic desizing of starch-size. Polymer-Plastics Technol. Eng. 43:519–538.

42. Ichikawa, Y., R. Wang, and C. H. Wong. 1994. Regeneration of sugar nucleotide for enzymatic oligosaccharide synthesis. Methods Enzymol. 247:107–127.

43. Jiang, Z. Q., S. Q. Yang, S. S. Tan, L. T. Li, and X. T. Li. 2005. Characterization of a xylanase from the newly isolated thermophilic Thermomyces lanuginosus CAU44 and its application in bread making. Lett. Appl. Microbiol. 41:69–76.

44. Johnson, D. V., A. A. Zabelinskaja-Mackova, and H. Griengl. 2000. Oxynitrilases for asymmetric C-C bond formation. Curr. Opin. Chem. Biol. 4:103–109.

45. Joyet, P., N. Declerck, and C. Gaillardin. 1992. Hyperthermostable variants of a highly thermostable alpha-amylase. Bio-Technology 10:1579–1583.

46. Kademi, A., D. Leblanc, and A. Houde. 2004. Microbial Enzymes: Production and Applications. The Haworth Press, Binghamton, NY.

47. Kashyap, D. R., P. K. Vohra, S. Chopra, and R. Tewari. 2001. Applications of pectinases in the commercial sector: a review. Bioresource Technol. 77:215–227.

48. Khandeparkar, R., and N. B. Bhosle. 2007. Application of thermoalkalophilic xylanase from Arthrobacter sp MTCC 5214 in biobleaching of kraft pulp. Bioresource Technol. 98:897–903.

49. Kirk, O., T. V. Borchert, and C. C. Fuglsang. 2002. Industrial enzyme applications. Curr. Opin. Biotechnol. 13:345–351.

50. Kobayashi, M., T. Nagasawa, and H. Yamada. 1992. Enzymatic synthesis of acrylamide—a success story not yet over. Trends Biotechnol. 10:402–408.

51. Koizumi, S., T. Endo, K. Tabata, and A. Ozaki. 1998. Large-scale production of UDP-galactose and globotriose by coupling metabolically engineered bacteria. Nature Biotechnol. 16:827–850.

52. Koul, S., R. Parshad, S. C. Taneja, and G. N. Qazi. 2003. Enzymatic resolution of naproxen. Tetrahedron Asymmetry 14:2459–2465.

53. Kulkarni, N., A. Shendye, and M. Rao. 1999. Molecular and biotechnological aspects of xylanases. FEMS Microbiol. Rev. 23:411–456.

54. Lee, C. K., I. Darah, and C. O. Ibrahim. 2007. Enzymatic deinking of laser printed office waste papers: some governing parameters on deinking efficiency. Bioresource Technol. 98:1684–1689.

55. Machius, M., N. Declerck, R. Huber, and G. Wiegand. 2003. Kinetic stabilization of Bacillus licheniformis alpha-amylase through introduction of hydrophobic residues at the surface. J. Biol. Chem. 278:11546–11553.

56. Mahmoudian, M. 2000. Biocatalytic production of chiral pharmaceutical intermediates. Biocatal. Biotransform. 18:105–118.

57. Mahmoudian, M., A. Lowdon, M. Jones, M. Dawson, and C. Wallis. 1999. A practical enzymatic procedure for the resolution of N-substituted 2-azabicyclo[2.2.1]hept-5-en-3-one. Tetrahedron Asymmetry 10:1201–1206.

58. Martin, M. L., and R. C. Hoseney. 1991. A mechanism of bread firming. 2. Role of starch hydrolyzing enzymes. Cereal Chem. 68:503–507.

59. May, O., P. T. Nguyen, and F. H. Arnold. 2000. Inverting enantioselectivity by directed evolution of hydantoinase for improved production of l-methionine. Nat. Biotechnol. 18:317–320.

60. May, O., S. Verseck, A. Bommarius, and K. Drauz. 2002. Development of dynamic kinetic resolution processes for biocatalytic production of natural and nonnatural l-amino acids. Org. Process Res. Dev. 6:452–457.

61. Miettinen-Oinonen, A., J. Londesborough, V. Joutsjoki, R. Lantto, J. Vehmaanpera, and Primalco Ltd. Biotec. 2004. Three cellulases from Melanocarpus albomyces for textile treatment at neutral pH. Enzyme Microb. Technol. 34:332–341.

62. Miller, B. S. J., J. A.; Palmer, D. L. 1953. A comparison of cereal, fungal, and bacterial R-amylases as supplements for breadmaking. Food Technol. Biotechnol. 7:38–42.

63. Motoki, M., and N. Nio. 1983. Crosslinking between different food proteins by transglutaminase. J. Food Sci. 48:561–566.

64. Nagasawa, T., and H. Yamada. 1995. Interrelations of chemistry and biotechnology. 6. Microbial-production of commodity chemicals. Pure Appl. Chem. 67:1241–1256.

65. Nagasawa, T., and H. Yamada. 1989. Microbial transformations of nitriles. Trends Biotechnol. 7:153–158.

66. Nanduri, V. B., A. Banerjee, J. M. Howell, D. B. Brzozowski, R. F. Eiring, and R. N. Patel. 2000. Purification of a stereospecific 2-ketoreductase from Gluconobacter oxydans. J. Ind. Microbiol. Biotechnol. 25:171–175.

67. Palacios, H. R., P. B. Schwarz, and B. L. D'Appolonia. 2004. Effect of alpha-amylases from different sources on the retrogradation and recrystallization of concentrated wheat starch gels: relationship to bread staling. J. Agric. Food Chem. 52:5978–5986.

68. Park, K. H., T. J. Kim, T. K. Cheong, J. W. Kim, B. H. Oh, and B. Svensson. 2000. Structure, specificity and function of cyclomaltodextrinase, a multispecific enzyme of the alpha-amylase family. Biochim. Biophys. Acta 1478:165–185.

69. Patel, R., A. Banerjee, V. Nanduri, A. Goswami, and F. Comezoglu. 2000. Enzymatic resolution of racemic secondary alcohols by lipase B from Candida antarctica. J. Am. Oil 'Chem. Soc. 77:1015–1019.

70. Patel, R., R. Hanson, A. Goswami, V. Nanduri, A. Banerjee, M.-J. Donovan, S. Goldberg, R. Johnston, D. Brzozowski, T. Tully, J. Howell, D. Cazzulino, and R. Ko. 2003. Enzymatic synthesis of chiral intermediates for pharmaceuticals. J. Ind. Microbiol. Biotechnol. 30:252.

71. Patel, R. N. 1998. Tour de paclitaxel: biocatalysis for semisynthesis. Annu. Rev. Microbiol. 52:361–395.

72. Patel, R. N., A. Banerjee, C. G. McNamee, D. B. Brzozowski, and L. J. Szarka. 1997. Preparation of chiral synthon for HIV protease inhibitor: stereoselective microbial reduction of N-protected [alpha]-aminochloroketone. Tetrahedron Asymmetry 8:2547–2552.

73. Patel, R. N., R. S. Robison, L. J. Szarka, J. Kloss, J. K. Thottathil, and R. H. Mueller. 1991. Stereospecific microbial reduction of 4,5-dihydro-4-(4-methoxyphenyl)-6-(trifluoromethyl-1H-1)-benzazepin-2-one. Enzyme Microb. Technol. 13:906–912.

74. Pere, J., M. Siika-Aho, and L. Viikari. 2000. Biomechanical pulping with enzymes: response of coarse mechanical pulp to enzymatic modification and secondary refining. Tappi J. 83:1–8.

75. Pritchard, P. E. 1992. Studies on the bread-improving mechanism of fungal alpha-amylase. J. Biol. Educ. 26:12–18.

76. **Quarroz, D.** December, 1985. Process for the production of 2-halopyridine derivatives. E.P. patent 0,084,118 B1.

77. **Rasor, J. P., and E. Voss.** 2001. Enzyme-catalyzed processes in pharmaceutical industry. *Appl. Catal. A Gen.* **221:**145–158.

78. **Richardson, T. H., X. Q. Tan, G. Frey, W. Callen, M. Cabell, D. Lam, J. Macomber, J. M. Short, D. E. Robertson, and C. Miller.** 2002. A novel, high performance enzyme for starch liquefaction—discovery and optimization of a low pH, thermostable alpha-amylase. *J. Biol. Chem.* **277:**26501–26507.

79. **Rouau, X.** 1993. Investigations into the effects of an enzyme preparation for baking on wheat-flour dough pentosans. *J. Cereal Sci.* **18:**145–157.

80. **Rubingh, D. N.** 1997. Protein engineering from a bioindustrial point of view. *Curr. Opin. Biotechnol.* **8:**417–422.

81. **Sakamoto, K., H. Yamada, and S. Shimizu.** January, 1994. Process for the preparation of D-pantolactone. U.S. patent 5,275,949.

82. **Sarioglu, K., N. Demir, J. Acar, and M. Mutlu.** 2001. The use of commercial pectinase in the fruit juice industry. 2: Determination of the kinetic behaviour of immobilized commercial pectinase. *J. Food Eng.* **47:**271–274.

83. **Sauerberg, P. O., and H. Preben.** May, 1995. Heterocyclic compounds and their preparation and use. U.S. patent 5,418,240.

84. **Schmid, A., J. Dordick, B. Hauer, A. Kiener, M. Wubbolts, and B. Witholt.** 2001. Industrial biocatalysis today and tomorrow. *Nature* **409:**258–268.

85. **Schmidt, M., and H. Griengl.** 1999. Oxynitrilases: from cyanogenesis to asymmetric synthesis. *Biocat. Discovery Appl.* **200:**193–226.

86. **Shaw, N. M., K. T. Robins, and A. Kiener.** 2003. Lonza: 20 years of biotransformations. *Adv. Synthesis Catal.* **345:**425–435.

87. **Shimizu, S., and M. Kataoka.** 1996. Optical resolution of pantolactone by a novel fungal enzyme, lactonohydrolase. *Ann. N. Y. Acad. Sci.* **799:**650–658.

88. **Skals, P. B., A. Krabek, P. H. Nielsen, and H. Wenzel.** 2008. Environmental assessment of enzyme assisted processing in pulp and paper industry. *Int. J. Life Cycle Assess.* **13:**124–132.

89. **Straathof, A. J. J., S. Panke, and A. Schmid.** 2002. The production of fine chemicals by biotransformations. *Curr. Opin. Biotechnol.* **13:**548–556.

90. **Taylor, M. B.** 1999. Summary of mandelic acid for the improvement of skin conditions. *Cosmet. Dermatol.* **21:**26–28.

91. **Thayer, A. M.** 2006. Competitors want to get a piece of Lipitor. *Chem. Eng. News* **84:**26–27.

92. **Thayer, A. M.** 2006. Enzymes at work. *Chem. Eng. News* **84:**15–25.

93. **Thoden, J. B., E. A. T. Ringia, J. B. Garrett, J. A. Gerlt, H. M. Holden, and I. Rayment.** 2004. Evolution of enzymatic activity in the enolase superfamily: structural studies of the promiscuous o-succinylbenzoate synthase from Amycolatopsis. *Biochemistry* **43:**5716–5727.

94. **Thomas, S. M., R. DiCosimo, and A. Nagarajan.** 2002. Biocatalysis: applications and potentials for the chemical industry. *Trends Biotechnol.* **20:**238–242.

95. **Tzanov, T., M. Calafell, G. M. Guebitz, and A. Cavaco-Paulo.** 2001. Bio-preparation of cotton fabrics. *Enzyme Microb. Technol.* **29:**357–362.

96. **Vanderwerf, M., W. J. J. Vandentweel, and S. Hartmans.** 1992. Screening for microorganisms producing D-malate from maleate. *Appl. Environ. Microbiol.* **58:**2854–2860.

97. **van Pelt, S., S. Quignard, D. Kubac, Y. S. B. Dimitry, F. van Rantwijk, and R. A. Sheldon.** 2008. Nitrile hydratase CLEAs: the immobilization and stabilization of an industrially important enzyme. *Green Chem.* **10:**395–400.

98. **Vanputten, P. L.** 1979. Mandelic-acid and urinary-tract infections. *Antonie Van Leeuwenhoek J. Microbiol.* **45:**622–623.

99. **Vihinen, M., and P. Mantsala.** 1989. Microbial amylolytic enzymes. *Crit. Rev. Biochem. Mol. Biol.* **24:**329–418.

100. **Violet, M., and J. C. Meunier.** 1989. Kinetic-study of the irreversible thermal-denaturation of *Bacillus licheniformis* alpha-amylase. *Biochem. J.* **263:**665–670.

101. **Wandrey, C., R. Wichmann, W. Leuctenberger, M.-R. Kula, and A. Bueckmann.** December, 1981. A process for the continuous enzymatic change of water soluble α-ketocarboxylic acids into the corresponding amino acids. U.S. patent 4,304,858.

102. **Wieser, M., K. Heinzmann, and A. Kiener.** 1997. Bioconversion of 2-cyanopyrazine to 5-hydroxypyrazine-2-carboxylic acid with *Agrobacterium* sp. DSM 6336. *Appl. Microbiol. Biotechnol.* **48:**174–176.

103. **Wong, C.-H., S. L. Haynie, and G. M. Whitesides.** 1982. Enzyme-catalyzed synthesis of N-acetyllactosamine with in situ regeneration of uridine 5′-diphosphate glucose and uridine 5′-diphosphate galactose. *J. Org. Chem.* **47:**5416–5418.

104. **Xin, J. Y., S. B. Li, X. H. Chen, L. L. Wang, and Y. Xu.** 2000. Improvement of the enantioselectivity of lipase-catalyzed naproxen ester hydrolysis in organic solvent. *Enzyme Microb. Technol.* **26:**137–141.

105. **Yokoyama, K., N. Nio, and Y. Kikuchi.** 2004. Properties and applications of microbial transglutaminase. *Appl. Microbiol. Biotechnol.* **64:**447–454.

106. **Zhu, Y, A. Rinzema, and J. Tramper.** 1995. Microbial transglutaminase—a review of its production and application in food processing. *Appl. Microbiol. Biotechnol.* **44:**277–282.

Biomass-Converting Enzymes and Their Bioenergy Applications

FENG XU

34

34.1. INTRODUCTION

Sustainability of energy sources is of strategic importance for the future advancement of humanity. Among various renewable energy candidates, the ones derived from biological sources, in particular, the so-called bioenergy for transportation fuels, are attracting increasing attention (for recent reviews, see references 48 and 56). Bioethanol production from sugary/starchy biomaterials and biodiesel production from vegetable oils are now viable industries (often referred as "the first generation bioenergy"). However, because of the sheer size of our fuel usage, it is believed that lignocellulosic biomass feedstocks have to be tapped as fermentable carbohydrate sources to meet the projected demand of bioenergy.

Enzymes play key roles in the development of bioenergy. Production of the first-generation bioethanol from starch relies on amylases to transform starch to glucose. To produce the "second generation" of bioethanol from lignocellulosic biomass, the development of efficient cellulases, hemicellulases, and accessory enzymes will be indispensable.

Natural evolution has equipped many organisms with the ability to degrade and feed on biomass substances with an army of enzymes. For instance, cellulolytic aerobic fungi, in general, have extracellular and freely dissociated enzymes, while some cellulolytic anaerobic bacteria have enzymes assembled into cell-bound, supramolecular cellulosomes. Comprehensive studies on biomass-active enzymes have been carried out for many decades by classical microbiological and biochemical means, and recently have been advanced by genetic engineering, genome sequencing, and systems biology (for recent reviews, see references 33, 48, and 96).

To better understand the structure-function relationships of these enzymes, many different yet corroborating and complementing (to a certain extent) classifications have been proposed based on the following.

1. IUBMB Enzyme Nomenclature and Classification (http://www.chem.qmul.ac.uk/iubmb/enzyme/), according to the enzymatic reactions in the EC 3.2.1 glycosidase, EC 4.2.2 lyase, and EC 3.1.1 esterase classes. Because many enzymes can catalyze different types of reactions, the EC-based classification is not always definitive.
2. CAZy database (http://www.cazy.org/), according to the sequence homology and three-dimensional structure in glycoside hydrolase (GH), polysaccharide lyase (PL), and carbohydrate esterase (CE) families (13). Members of the same family have homologous three-dimensional structure and catalytic mechanism. Different families may form clans based on determinant folding features. Enzymes from different families may catalyze the same reaction (or have the same EC designation).
3. Catalytic mechanism, according to the stereochemistry of enzymatic reactions (94). For "retaining" glycoside hydrolases, the anomeric C bearing the target glycosidic bond retains the same configuration after a double-displacement hydrolysis, while for "inverting" hydrolases, the anomeric C inverts its configuration after a single nucleophilic displacement hydrolysis. Many retaining hydrolases also possess significant transglycosylation activity, because of the covalently linked transient intermediate and the anomeric compatibility of transglycosylation products with the hydrolases' active site. Most of these hydrolases rely on two acidic amino acid residues (Glu or Asp), one as proton donor or general acid and another as nucleophile or base as the catalytic nucleophile. Two known exceptions of these canonical mechanisms are with GH4 and 109 glycosidases that use NAD^+ cofactor in a redox mechanism (54, 107).
4. "Supertertiary"/quaternary structure, according to the modular structure or multimolecule assembly of the enzymes. Modular glycoside hydrolases may comprise one or more catalytic modules, carbohydrate-binding modules (CBMs), fibronectin 3-like modules, dockerins, immunoglobulin-like domains, or functionally unknown "X" domains, connected by linkers (48, 74, 96). Many anaerobic bacteria have biomass-active enzymes assembled into cellulosomes, in which different cellulases are anchored onto scaffoldin molecules via cohesin-dockerin interaction (1, 28).

In this chapter, the basic structure, property, and function of biomass-active enzymes are introduced, with emphasis on those most relevant to the development of cellulosic bioenergy, e.g., major cellulases and hemicellulases. Due to limited space, only the most recent reviews and selected reports are cited in this chapter. These and other published articles may be used for more detailed and authoritative referencing. Most of the cited structural and enzymological data come from CAZy (http://www.cazy.org/) and BRENDA (http://www.brenda-enzymes.info/ [15]) databases, where comprehensive information on biomass-active enzymes can be found.

34.2. CELLULASES

In Nature, cellulose [a linear polysaccharide consisting of β(1→4)-linked D-glucopyranosyl (Glc) units] is degraded by a combination of mainly three types of cellulases (Cel): cellobiohydrolase (CBH), endo-1,4-β-D-glucanase (EG), and β-glucosidase (BG), whose catalytic modules belong to ~14 GH families and CBMs belong to ~15 CBM families (15, 35). Because of their different molecular and enzymological properties (in particular, catalytic mechanism and substrate specificity), cellulases can act synergistically on cellulosic biomass substances (for recently studied examples, see references 15, 31, 35, and 74).

34.2.1. CBHs

CBHs (cellulose 1,4-β-cellobiosidase, exo-1,4-β-D-glucanase, or Avicelase; EC 3.2.1.91, EC 3.2.1.−) are able to degrade the crystalline part of cellulose, which is recalcitrant to other types of cellulases. Different CBHs have catalytic modules belonging to the GH5, GH6, GH7, GH9, GH48, and GH74 families. CBHs may possess CBM, which is believed critical for a CBH's binding to and action on crystalline cellulose. Based on either experimental data or structural homology analysis, the stereochemistry of Cel6, Cel9, Cel48, and Cel74 CBHs is of the inverting type, while the stereochemistry of Cel5 and Cel7 CBHs is of the retaining type.

The most significant topological feature of CBH's catalytic module is the tunnel structure, which may cover the entirety (e.g., in the case of Cel7 CBH) or part of the active site (e.g., in the case of Cel48 CBH). Such a tunnel enables the enzyme to hydrolyze cellulose in a unique "processive" manner (94). The movement restriction of CBH brought by the processive dynamics, as well as the insoluble and polymeric nature of the substrate, cause the enzyme's kinetics to deviate significantly from the classical Michaelis-Menten model (44, 65, 96) and become fractal (100).

Hypocrea jecorina Cel7A CBH-I and Cel6A CBH-II are archetypical CBHs. Experimental evidence indicates that Cel7 CBHs (EC 3.2.1.−) are specific to the reducing end of the cellulose chain to initiate their processive hydrolysis. A Cel7 CBH-I may have ten subsites (denoted from −7 to +3 [94]) that involve intricate H-bonding and π-stacking interactions to bind and activate a bound cellulose segment, and a CBM at their C terminus. It is believed that Cel6 CBHs (EC 3.2.1.91) are specific to the nonreducing end of the cellulose chain. In general, a Cel6 CBH-II has five subsites (−2 to +2, +4) and a CBM at its N terminus. A Cel7 CBH and a Cel6 CBH can be synergistic in hydrolyzing cellulose, a phenomenon attributed to their specificities at opposing cellulose chain ends. Cel5, Cel9, Cel48, and Cel74 CBHs may also be differentiated into reducing-end or nonreducing-end specific (5, 105). Many cellulose-degrading microbes secrete more than one CBH, probably to benefit from the synergism between CBHs of different specificity. For instance, *H. jecorina* produces both Cel7A CBH-I and Cel6A CBH-II (with abundance of ~60 and 15% wt/wt of all secreted proteins, respectively) (34, 74).

Compared with enzymes involved in homogenous catalysis, CBHs have quite slow kinetics. With soluble substrates, apparent K_m values on the order of 1 mM and k_{cat} values on the order of 1 turnover s^{-1} are often observed for CBHs, but the kinetic parameters can vary in a wide range (~10^{-2} to 10^2 mM, ~10^{-3} to 10^3 s^{-1}) depending on substrate or CBH. For insoluble substrates, only apparent K_m and k_{cat} in units such as grams of cellulose per liter and grams of cellulose per liter per milligram of enzyme per minute,

respectively, may be obtained after fitting observed kinetic data to the Michaelis-Menten model. Polymeric celluloses (e.g., microcrystalline cellulose Avicel, filter paper, bacterial cellulose, amorphous phosphoric acid-swollen cellulose [PASC]), soluble cellodextrins, and soluble β-glycosides derivatized with chromogenic or fluorogenic functional groups (e.g., p-nitrophenyl cellobioside, 4-methylumberlliferyl lactoside) are commonly used to assay CBH, although the soluble fluoro/chromogenic substrates in general are not active toward Cel6 CBH-II. Many CBHs are active around pH ~5 to 7 and under ~20 to 60°C, and some extremophilic CBHs can extend their activity/stability beyond the pH or temperature range.

In addition to general protein destabilizers and carboxylate-active reagents (able to modify the catalytic Glu or Asp), CBHs may be inactivated by insoluble cellulose via unproductive binding (22), adsorption onto lignin or other noncellulose substances (6, 67, 101), or inhibited by cellobiose product. At high cellobiose concentration, a Cel7 CBH-I may act as transglycosylase, elongating cellodextrin chains instead of degrading them. Such unproductive adsorption, inhibition, or transglycosylation may lead to significant loss of desired CBH activity for bioenergy applications, besides thermal unfolding due to high reaction temperature (36).

34.2.2. EGs

EGs (4-β-D-glucan 4-glucanohydrolase, or CMCase, EC 3.2.1.4) fragment cellulose by hydrolyzing internal β(1→4) glycosidic bonds. Different EGs have catalytic modules belonging to the GH5 to GH9, GH12, GH44, GH45, GH48, GH51, and GH74 families. The stereochemistry of Cel8 and Cel45 EGs is of the inverting type, while that of Cel5, Cel7, Cel12, Cel44, and Cel51 EGs is of the retaining type. The catalytic modules of most EGs have a cleft/groove-shaped active site, which allows the EGs to bind and clip a cellodextrin or a cellulose segment in a somewhat random, on-off manner. However, a few Cel5, Cel9, and Cel48 EGs can act processively (as CBHs do) (16, 61, 68). Because EGs may create new reducing and nonreducing ends in cellulose for CBHs to attack, these two types of cellulases can act synergistically in cellulose degradation.

H. jecorina Cel7B EG-I, Cel5A EG-II, Cel12A EG-III, *Humicola insolens* Cel6B EG-VI, Cel45A EG-V, *Clostridium thermocellum* Cel9R, and *Clostridium cellulolyticum* Cel48F are archetypical EGs. *H. jecorina* Cel7B EG-I has four subsites and a CBM at its C terminus, and prefers internal β(1→4) glycosidic bonds when cellodextrins serve as substrate. *H. jecorina* Cel5A EG-II has seven subsites and a CBM at its N terminus, and prefers glycosidic bonds near the reducing end. *H. jecorina* Cel12A EG-III has six subsites but no CBM. *H. insolens* Cel45A EG-V has six subsites and a C-terminus CBM. *H. insolens* Cel6B EG-VI has six subsites but no CBM, and prefers cellohexaose (over shorter cellodextrins) as well as the glycosidic bonds near the reducing end. Many cellulose-degrading microbes secrete more than one EG. For instance, *H. jecorina* produces at least five EGs: Cel7B EG-I, Cel5A EG-II (together the two EGs can have ~15% wt/wt abundance in secreted *H. jecorina* proteins), Cel12A EG-III, Cel45A EG-V, and CBM-less Cel5B (34). On purified, homogeneous cellulose, no significant synergism among different EGs seems to exist, except between processive and conventional EGs (96). On polymorphous/heterogeneous cellulose and complex biomass substances, however, having different EGs may be beneficial. It is postulated that Cel12 EGs may act first on the amorphous, outer part of cellulose, and CBM-possessing EGs (particularly the processive ones) then act on the more crystalline, inner part of cellulose (74). Cel12 EG is also postulated as having expansin-like activity (108). EGs from different GHs can have different preferences for cellulose regions to bind (due to difference in CBM specificity, multiplicity, or linkage to the catalytic module), length of cellulose segment to activate, or the site of glycosidic bond to cleave (due to difference in the architecture of EG's active site). EGs from different GHs may possess significant "side activity" on noncellulose poly- or oligosaccharides. For instance, H. jecorina Cel7B EG-I, Cel5A EG-II, and Cel12A EG-III can hydrolyze xylan, mannan, and xyloglucan, respectively (50, 70, 83). Thus, different EGs may act in concert, with their different yet complementary or supplementary side activities, toward plant cell wall substrates in which cellulose is entangled with and shielded by hemicellulose components. Brown rots do not have CBH, but their processive EG (in general, the most abundant in their secreted proteins) may act on crystalline cellulose and team up with other EGs and enzymes to constitute a "complete" cellulolytic system (16). In addition to their polysaccharide-degrading activity, some EGs may also function in the generation of inducers for the production of other enzymes by the hosting organisms.

With soluble substrates, K_m values on the order of 1 mM and k_{cat} values on the order of 10 s^{-1} are often observed for EGs, but the kinetic parameters can vary in a range ($\sim 10^{-1}$ to 10^1 mM, $\sim 10^{-2}$ to 10^2 s^{-1}) depending on substrate or EG. Insoluble or polymeric celluloses (e.g., PASC, carboxymethyl cellulose [CMC]), soluble cellodextrins, and soluble derivatized β-glycosides are commonly used to assay EGs, based on either viscosity decrease, reducing-end generation, different cellodextrin distribution, or photometric change. Many EGs are active at pH ~ 4 to 8 and under ~ 20 to 60°C. Some EGs may have significant transglycosylase activity.

"CMCase" (CMC-hydrolyzing) activity is commonly applied to differentiate EG from CBH. Among their side activities, Cel5, Cel7, Cel8, Cel9, Cel12, and Cel74 EGs may prefer mannan (and related hemicelluloses such as galactomannan), xylan (and related hemicelluloses such as arabinoxylan), chitosan, chitin, β-glucan, and xyloglucan, respectively (50, 66, 70, 79, 83, 94a, 98). Such discrimination on noncellulose polysaccharides may be applied to differentiate different EGs.

34.2.3. β-ᴅ-Glucosidases

BGs (β-ᴅ-glucosidase, β-ᴅ-glucoside glucohydrolase, or cellobiase, EC 3.2.1.21) hydrolyze cellobiose or longer, soluble cellodextrins into Glc. Different BGs have catalytic modules belonging to the GH1, GH3, and GH9 families (25). Microbial BGs are, in general, not multimodular; they lack distinct CBM. Based on either experimental data or structural homology analysis, the stereochemistry of Cel1 and Cel3 BGs is of the retaining type, while the stereochemistry of Cel9 BG is of the inverting type. BGs in general have a pocket-shaped active site, which allows them to bind to the nonreducing Glc unit and clip it from cellobiose or cellodextrin. At least for Cel3 BG, only the nonreducing glucosyl is important in BG activity; steric and electronic properties of the "aglycon" part of a glycoside or oligosaccharide (occupying the +1 subsite of BG) only play a minor role (49). Because it degrades cellobiose, a known inhibitor of CBH and EG, BG is indispensable in a complete cellulase system.

H. jecorina Cel1A BG-II and *Aspergillus niger* Cel3A BG-I are archetypical extracellular BGs. Cel1 BGs often have higher tolerance toward Glc (product) inhibition, as well as broader side activity on noncellobiose di- or oligosaccharides, than Cel3 BGs. Such kinetic differences might render it advantageous to have more than one BG in a cellulase system (34).

With cellobiose or other β-glucosides, K_m values on the order of 1 mM and k_{cat} values on the order of 10^3 s^{-1} are often observed for BGs, but the kinetic parameters can vary in a range ($\sim 10^{-2}$ to 10^2 mM, $\sim 10^{-1}$ to 10^4 s^{-1}) depending on substrate or BG. Cellodextrins, β-glucosides, and enzyme-coupled assays (such as those based on glucose oxidase's action on Glc) are commonly used to assay BGs. Many BGs are active at pH ~ 4 to 7 and under ~ 20 to 60°C. BGs may be subjected to substrate inhibition from cellobiose (K_i generally on the order of 1 mM), or product inhibition from Glc (K_i generally on the order of 1 mM, but may vary from 10^{-1} to 10^2 mM, depending on BG). At high Glc concentration and low water activity, BG may act as transglycosylase, condensing Glc and other carbohydrates into dimeric or oligomeric molecules (80).

34.2.4. Cellobiose Phosphorylase

Cellobiose phosphorylase (cellobiose:phosphate β-ᴅ-glucosyltransferase, or cellodextrin phosphorylase, EC 2.4.1.20 or EC 2.4.1.49) cleaves the β(1→4)-bond in cellobiose (or cellulodextrin) by phosphorylating a glucosyl unit while releasing a Glc. Because they are mainly intracellular and occur in anaerobic bacteria, the enzymes belonging to the GH94 family are important for in vivo bacterial cellulose utilization. Potential applications of the enzymes in enzymatic cellulose hydrolysis and H$_2$ production from cellulosic materials have been evaluated (15, 106).

34.2.5. Glucuronan Lyase

A few microbial glucuronan lyases have recently been characterized (19, 45). The enzyme [(1→4)-β-ᴅ-glucuronan lyase, EC 4.2.2.14] cleaves β(1→4)-bonds in polyglucuronan (or cellouranan) by β-elimination, generating a Glc (or reducing end) and a Δ4:5 unsaturated Glc (or corresponding nonreducing end). The lyase is highly specific to polyglucuronan and has no or low activity on polygalacturonan [α(1→4) backbone], hyaluronan [β(1→3, 4) backbone], or alginate [β, α(1→4) backbone]. Based on sequence, glucuronan lyase belongs to the PL20 family. The in vivo function and in vitro application of the enzyme remain to be further investigated.

34.3. HEMICELLULASES

Hemicellulases hydrolyze hemicellulose, a group of complex polysaccharides consisting of different units, linkages, intramolecular architectures, or intermolecular interactions. Common hemicelluloses include β-glucan (not cellulose), xylan, xyloglucan, arabinoxylan, mannan, galactomannan, arabinan, galactan, polygalacturonan, etc., which are degraded in Nature by β-glucanase, xylanase, xyloglucanase, mannanase, arabinase, galactanase, polygalacturonase, glucuronidase, acetyl esterase, and other enzymes (82). Based on the bonds they cleave, hemicellulases may be divided into two groups: glycoside hydrolases/lyases (on glycosidic bonds) and carbohydrate esterases (on ester bonds). Based on their targets, hemicellulases may also be divided into two different groups: one acts on the backbones and the other acts on the side chains of hemicelluloses. Similar to cellulases, many hemicellulases are modular, with catalytic modules belonging to ≥29 GH, 5 PL, and 9 CE families and CBMs belonging to ≥37 CBM families. Many GH families contain numerous cellulases and hemicellulases, indicating likely common evolution ancestry and possible overlapping functions. Most cellulolytic microbes produce an array of hemicellulases along with cellulases. Because of their different molecular and enzymological properties (in particular, catalytic mechanism and substrate specificity), hemicellulases can act synergistically among themselves or with cellulases on complex biomass substances to effectively produce fermentable sugars for bioenergy or other biorefinery applications (30, 48, 73, 74, 81).

34.3.1. Endo-β-Xylanases and β-Xylosidase

Endo-β-xylanases (endo-1,4-β-D-xylan xylanohydrolase, EC 3.2.1.8; endo-1,3-β-D-xylanase, EC 3.2.1.32) cleave backbone glycosidic bonds in xylan, a group of β(1→4)-linked D-xylopyranosyl polysaccharides with different O-substitutions and frequent copresence with cellulose in biomass. Different xylanases (Xyn) have catalytic modules belonging to the GH5, GH8, GH10, GH11, GH26, and GH43 families (17, 69, 72). Most known xylanases are in GH10 and GH11 families: GH11 contains solely xylanases, while GH10 contains a few other glycoside hydrolases. Xyn26 are 1,3-β-xylanases only. β-Xylosidases (EC 3.2.1.37) cleave xylobiose or longer xylooligosaccharides into xylose (Xyl). Different β-xylosidases have catalytic modules belonging to the GH3, GH30, GH39, GH43, GH52, and GH54 families.

The stereochemistry of Xyn8 and Xyn43 xylanases is of the inverting type, while that of Xyn5, Xyn10, Xyn11, and Xyn26 xylanases is of the retaining type. The catalytic modules of most endo-xylanases have a cleft-shaped active site, with four to seven subsites, which allows the enzymes to bind and clip a xylooligosaccharide or a segment of a xylan chain, in a somewhat random, on-off manner. The stereochemistry of most β-xylosidases is of the retaining type, with the exception of Xyn43 members. β-Xylosidases have a pocket-like active site with two subsites, and cleave xylobiose or xylooligosaccharides from the nonreducing end (41).

Cellulomonas fimi Xyn10A Cex, *H. jecorina* Xyn11 xylanase-I, and *H. jecorina* Xyn3 β-xylosidase are archetypical. *C. fimi* Xyn10A has a CBM2a CBM with a high affinity toward crystalline cellulose and chitin (but weak affinity toward amorphous cellulose). *H. jecorina* Xyn11 has no CBM (although many bacterial Xyn11 xylanases do have CBM). Besides molecular shape and size, Xyn10s and Xyn11s differ mainly on their substrate specificity. Compared with Xyn11s, Xyn10s in general degrade xylan into shorter oligosaccharides and have broader side activity on other polysaccharides (69, 92). Xyn10s typically act on the third or second β(1→4)-glycosidic bond at the "nonreducing" side of a substituted Xyl unit or a β(1→3)-bond, respectively (17). Xyn11s typically act only on plain/unsubstituted 1,4-β-xylan or xylooligosaccharides. Xylanases with preference on either the reducing end or nonreducing end of a xylan chain are also known (17). Some Xyn5s and Xyn43s may also act on non-xylan polysaccharides, such as β-glucan and arabinan, respectively.

Xylanases from different GHs may possess not only different specificity on xylan, but also significant side activity on non-xylan substrates. For instance, *H. jecorina* Xyn10 xylanase can hydrolyze cellulose, probably a remnant function from the enzyme's evolution ancestry shared by Cel5 EG belonging to the same GH-A clan. Thus, different

xylanases may act in concert among themselves. It is postulated that Xyn10 may further degrade the large xylooligosaccharides generated by Xyn11, thus achieving greater overall xylan degradation (69). Different xylanases may collaborate with enzymes acting on xylan side chains and cellulases to collectively attack plant cell wall substrates in which cellulose is entangled with and shielded by hemicellulose components. It has been noticed that Xyn11 and Xyn10 collaborate with feruloyl esterase better in releasing feruloyl and 5,5′-diferuloyl, respectively, from ferulated polysaccharides (26). Numerous natural fusion proteins containing xylanase and other biomass-active enzyme catalytic modules are known (17). Given the abundance of xylan in hardwood, xylanases should be an important part of enzyme systems targeting hardwood-derived biomass feedstocks. For other lignocellulosic substrates (even when only residual hemicelluloses exists after pretreatment), the benefit of including xylanase in cellulase mixtures has been observed (7, 16, 27, 30, 73, 81, 89). β-Xylosidase can keep endoxylanase from product inhibition by degrading xylobiose or longer xylooligosaccharides (81). Many β-xylosidases have dual activity on α-L-arabinofuranoside, probably because of the configurational similarity between D-xylopyranose and L-arabinofuranose, making the enzymes useful in removing xylanase-hindering xylan side chains.

For endo-xylanase, K_m values on the order of 1 mM and k_{cat} values on the order of 10^2 s^{-1} are often observed for soluble substrates, but the kinetic parameters can vary in a range ($\sim 10^{-1}$ to 10^2 mM, $\sim 10^{-2}$ to 10^3 s^{-1}) depending on substrate or xylanase. Various xylan and xyloside substrates are used to assay xylanases, although some Xyn5 xylanases may have low reactivity toward short derivatized xylosides. For β-xylosidase, K_m values on the order of 10^{-1} mM and k_{cat} values on the order of 10^{-2} s^{-1} are often observed, with a range ($\sim 10^{-2}$ to 10^1 mM, $\sim 10^{-3}$ to 10^2 s^{-1}) depending on substrate or β-xylosidase.

Many xylanases and β-xylosidases are active at pH ~ 4 to 8 and temperatures from ~ 40 to 70°C. Xylanases from (hyper)thermophiles can extend their activity/stability at much higher temperatures. These xylanases, among the most thermally active glycoside hydrolases known, may hold promise for high-temperature biomass hydrolysis.

Xylanase and β-xylosidase may be inactivated by unproductive adsorption from lignin or inhibition from substrate, product, or analogs (6, 40). They may also be strongly inhibited by plant-derived xylanase-specific proteins (e.g., *Triticum aestivum* xylanase inhibitor I and II) (8). Many retaining xylanases and β-xylosidases have significant transglycosylase activity (74). Such inhibition or transglycosylation may lead to significant xylanase activity loss in bioenergy applications (17, 69).

34.3.2. β-Glucanases

Endo-β-glucanases (non-EG) [1,3-(1,3;1,4)-β-D-glucan 3(4)-glucanohydrolase or laminarinase, EC 3.2.1.6; 1,3-β-D-glucan glucanohydrolase, EC 3.2.1.39; 1,3-1,4-β-D-glucan 4-glucanohydrolase or licheninase, EC 3.2.1.73] cleave backbone linkages in non-cellulose β-glucans [Glc polymers with β(1→3), (1→4), or (1→6) linkages]. Exo-β-glucanases (non-BG) (1,3-β-glucan glucohydrolase, EC 3.2.1.58; 1,6-β-glucan glucohydrolase, EC 3.2.1.75) cleave nonreducing ends of β-glucans to produce mostly Glc and sometimes gentiobiose. Different β-glucanases have catalytic modules belonging to the GH3, GH5, GH12, GH16, GH17, GH55, GH64, and GH81 families. GH16s are (1,3-1,4)-β-D-glucanases only, and GH64's β-glucanases are specific to β(1→3)-D-glucan only. The stereochemistry of GH8, GH64, and GH81 β-glucanases is of the inverting type, while the stereochemistry of GH5, GH12, GH16, and GH17 β-glucanases is of the retaining type. Many microbes produce more than one β-glucanase (mostly β-1,3-glucanases), which may act in concert to degrade complex β-glucans.

For many β-glucanases, K_m values on the order of 10^{-1} mM and k_{cat} values on the order of 10^2 s^{-1} are often observed for soluble substrates, with a range of ~10^{-1} to 10^1 mM and ~10^{-1} to 10^3 s^{-1}, respectively. For non-BG β-glucosidases, K_m values on the order of 1 mM and k_{cat} values on the order of 10 s^{-1} are often observed, with a range of ~10^{-2} to 10^1 mM and ~1 to 10 s^{-1}, respectively. Many β-glucanases are active at temperatures ranging from ~40 to 70°C and show optimal activity between pH 4 and 8.

34.3.3. Endo-β-Mannanase and β-Mannosidase

Mannans consist of β(1→4)-D-mannosyl or manno/glucopyranosyl polymeric backbone with variable β(1→6)-D-galactosyl side chains, as well as O2 and/or O3 acetylations. Endo-mannanases (1,4-β-D-mannan mannanohydrolase, EC 3.2.1.78) cleave backbone β(1→4)-bonds in mannan (21, 62). Different endo-mannanases (Man) have catalytic modules belonging to the GH5, GH26, and GH113 families. β-Mannosidases (β-D-mannoside mannohydrolase, EC 3.2.1.25) cleave mannosyl units from nonreducing ends of mannan or mannooligosaccharides. Different β-mannosidases have catalytic modules belonging to the GH1, GH2, and GH5 families. Man1, Man2, Man5, Man26, and Man113 enzymes all have the retaining stereochemistry.

H. jecorina Man5A, C. fimi Man26A, and Alicyclobacillus acidocaldarius Man113A are archetypical endo-mannanases. Man5 and Man26 have approximately five to six subsites (−4 or −3 to +2), and often need bindings onto at least four subsites to activate a substrate. Because the CBMs in fungal mannanases in general bind to cellulose, but not to mannan, their function is postulated as mainly a locator to bring the enzymes to cellulose-bound mannan. However, many of the CBMs of bacterial mannanases are mannan specific, such as the CBM of C. fimi Man26A (a CBM23 CBM, linked also to an immunoglobulin-like and S-layer module specific to bacterial surface), which might directly assist in mannan binding and activating by the enzymes. Because of their retaining catalytic mechanism, some mannanases have significant transglycosylase activity. Solanum lycopersicum Man1, A. niger Man2A, and Cellvibrio mixtus Man5A are archetypical β-mannosidases, most of them being intracellular or membrane bound.

Many biomass-degrading microbes secrete endo-mannanase and/or β-mannosidase along with various cellulases, xylanases, and other biomass-active enzymes (34). Given the abundance of (galacto)glucomannan in softwood, mannanases could be an important part of enzyme systems targeting softwood-derived biomass feedstocks. For instance, mannanase may act together with xylanase and EG to synergistically degrade sugarcane bagasse (9). BG may further degrade endo-mannanase-produced glucomannooligosaccharide by removing nonreducing end Glc. α-Galactosidase and acetyl mannan esterase may assist endo-mannanase's hydrolysis of substituted (gluco)mannan by removing side chains (62). For enzymatic hydrolysis of complex biomass substrates, however, such side chain removal may have mixed effects, because "polished" homo-mannan may lose water solubility and stick to cellulose, thus impeding the action of cellulases (21).

For endo-mannanase, K_m values on the order of 10^{-2} mM and k_{cat} values on the order of 10^2 s^{-1} are often observed for soluble substrates, but the kinetic parameters can vary in a range (~10^{-3} to 10 mM, ~10^{-1} to 10^3 s^{-1}). For β-mannosidase, K_m values on the order of 1 mM and k_{cat} values on the order of 10^2 s^{-1} are often observed, with a range of ~10^{-2} to 10^1 mM and ~10^{-2} to 10^2 s^{-1}, respectively. Many mannanases and β-mannosidases show optimal activity between pH 3 and 8, and are active at temperatures ranging from ~40 to 80°C.

34.3.4. Xyloglucan Hydrolase and Xyloglucan Endo-Transglycosylase

Xyloglucan is a group of polysaccharides consisting of a 1,4-β-glucan backbone with O6 substitution of α(1→6)-linked Xyl units that themselves are substituted (at O2) by either α(1→2)-L-arabinofuranosyl (Ara) or β(1→2)-D-galactosyl (Gal) units [partially acetylated or substituted (at O2) by α(1→2)-L-fucopyranosyl (Fuc)] (109). Xyloglucan hydrolases [1,4-β-D-glucan glucanohydrolase, EC 3.2.1.151; (1,6-α-D-xylo)-1,4-β-D-glucan exo-glucanohydrolase, EC 3.2.1.155; oligoxyloglucan reducing-end cellobiohydrolase, EC 3.2.1.150] cleave polymeric xyloglucan into oligoxyloglucans. Different xyloglucan hydrolases have catalytic modules belonging to the GH5, GH12, GH16, GH44, and GH74 families. Xyloglucan endo-transglycosylases (xyloglucan:xyloglucan xyloglucanotransferase, EC 2.4.1.207), found mainly in plants, transfer an oligoxyloglucan from a xyloglucan to the O4 of a nonreducing end Glc unit of another (oligo)xyloglucan, and have catalytic modules belonging to the GH12 and GH16 families. Recently, it has been proposed to group xyloglucan hydrolase and xyloglucan endo-transglycosylase into one xyloglucan transferase/hydrolase (XTH) superfamily (77; http://labs.plantbio.cornell.edu/XTH/). XTH may possess CBM, which seems to serve as a locator to bring XTH to specific cell wall regions, rather than an enhancer for the catalytic module's catalysis (38). The stereochemistry of XTH74 xyloglucanases is of the inverting type, while the stereochemistry of XTH5, XTH12, XTH16, and XTH44 is of the retaining type.

Paenibacillus pabuli XTH5, Bacillus licheniformis XTH12, and H. jecorina XTH74A (also noted as EG-VI) are archetypical. Besides molecular shape and size, XTHs may differ on their substrate specificity. For instance, H. jecorina XTH74A can hydrolyze a xyloglucan at substituted Glc units, and near either end of a xyloglucan chain (20). Many XTHs may act on both xyloglucan and unsubstituted β(1→4)-glucan (e.g., cellulose), but their activity on the former is, in general, significantly greater than that on the latter (38), in contrast to Cel7 EG, which also acts on xyloglucan but with an activity significantly less than that on cellulose (94a).

Many biomass-degrading microbes secrete at least one XTH. For instance, both H. jecorina and Phanerochaete chrysosporium produce two XTHs. XTHs from different GH families may possess not only different specificity on xyloglucan, but also side activity on other β-glucans/polysaccharides (38). Arabinofuranosidase and esterase may enhance XTH activity by trimming off Ara-containing or acetylated side chains in xyloglucan, respectively. Given the presence of xyloglucan in plant cell wall, XTH may enhance cellulases' activity on cellulose by degrading away xyloglucan and subsequently exposing more cellulose in a biomass feedstock (3).

Various XTHs are believed to take part in the formation/expansion of plant cell wall (77). These XTHs may be active in loosening rigid lignocellulosic cell wall, by cutting and relinking load-bearing xyloglucan, to allow new growth. Different plant XTHs may have different donor and/or acceptor specificity, in terms of the size (oligo- or polymeric xyloglucan) or fine structure (e.g., Fuc substitution, arrangement of plain and Xyl-substituted backbone Glc units), as well as a different hydrolysis pattern, in terms of being internal or chain end-preferred. The xyloglucan endo-transglycosylase property may be useful in biomass conversion in which disentangling and weakening recalcitrant polysaccharides are vital for effective (hemi)cellulose hydrolysis. Thus, XTH may have a potential of becoming a cellulase enhancer for bioenergy application, although economic production of plant XTH (in fermentative microbes) remains challenging.

For XTH's hydrolase activity, K_m values on the order of 1 mM and k_{cat} values on the order of 10^2 s^{-1} are often observed, with a range of $\sim 10^{-1}$ to 10 mM and 10^{-2} to 10^3 s^{-1}, respectively. For xyloglucan endo-transglycosylase activity, K_m values on the order of 10^{-1} mM and k_{cat} values on the order of 10^{-3} s^{-1} are often observed, with a range of $\sim 10^{-2}$ to 10^1 mM and $\sim 10^{-4}$ to 10^{-3} s^{-1}. Many XTHs are active as hydrolases at pH 5 to 6 and $\sim$40 to 70°C, and as transglycosylases at pH 5 to 6 and $\sim$20 to 40°C.

34.3.5. Polygalacturonan Hydrolases and Lyases

Pectic polysaccharides are complex plant cell wall components consisting of poly-α-galacturonic acids (GalU) or polyrhamnogalacturonic acids with variable methylation/acetylation (on backbone GalU) and carbohydrate side chains (containing mainly Ara and Gal) linked to backbone rhamnosyl units. Based on catalytic mechanism, three groups of pectinolytic enzymes are involved in the degradation of pectic polysaccharides: polygalacturonases or pectinases [cleaving internal backbone $\alpha(1{\rightarrow}4)$-bonds by hydrolysis], pectin/pectate lyases (cleaving internal backbone bonds by β-elimination), and pectin methyl esterase (pectin pectylhydrolase, EC 3.1.1.11; converting pectin to pectate). Polygalacturonases may be further divided into endo-polygalacturonase (EC 3.2.1.15), exo-polygalacturonidase (poly-1,4-α-D-galacturonide galacturonohydrolase, EC 3.2.1.67), and exo-polygalacturonosidase (EC 3.2.1.82). Pectin/pectate lyases cleave the glycosidic bond between glycosidic O and C4, with the help of C6-uronate, and form a Δ4:5 C=C bond at the nonreducing side GalU. The lyases may be further divided into endo-type pectate lyase (1,4-α-D-galacturonan lyase, EC 4.2.2.2), endo-type oligogalacturonide lyase (EC 4.2.2.6), endo-type pectin lyase (1,4-6-O-methyl-α-D-galacturonan lyase, EC 4.2.2.10), and exo-type pectate disaccharide lyase (1,4-α-D-galacturonan reducing-end-disaccharide-lyase, EC 4.2.2.9). Many pectinolytic hydrolases and lyases can act on both pectin (polymethylgalacturonic ester) and pectate (polygalacturonic acid) (39, 90). Based on sequence, polygalacturonases belong to the GH28, pectin methyl esterases belong to the CE8, and pectin/pectate lyases belong to the PL1, PL2, PL3, PL9, and PL10 families. It is known that GH28 polygalacturonases have the inverting stereochemistry.

Exo-polygalacturonases degrade their substrates processively with one dominant single product [GalU for fungal enzymes, but GalU-$\alpha(1{\rightarrow}4)$-GalU for bacterial enzymes], while endo-polygalacturonases cleave internal $\alpha(1{\rightarrow}4)$-bonds randomly, yielding an array of oligogalacturonan products (39, 63). Ca^{2+} is required by pectate lyase, but not pectin lyase, although the latter may be stimulated by the cation. Fungal pectin methyl esterases are believed to act randomly along a pectin chain, but plant pectin methyl esterases may prefer to start at the ends of a pectin chain and continue in a processive manner (39, 71).

Different pectinolytic enzymes may act in concert among themselves. Endo- and exo-polygalacturonases may synergize with each other as EG and CBH do. Polygalacturonases and pectin/pectate lyases with different specificity or tolerance on poly(rhamno)galacturonan or side chains may collectively attack complex pectin or plant cell wall substrates. Pectin methyl esterase may enhance the catalysis of polygalacturonase by removing the methyl groups in pectin. α-L-Arabinofuranosidase, arabinase, galactosidase, and other enzymes may also enhance the catalysis of pectinolytic enzymes by removing the side chains from polyrhamnogalacturonan. Given the abundance of pectic polysaccharides in sugar beet and fruits, pectinolytic enzymes could be an important part of enzyme systems targeting sugar beet pulp, fruit peels, and other related biomass feedstocks.

For polygalacturonase, K_m values on the order of 1 mM and k_{cat} values on the order of 10^2 s^{-1} are often observed for soluble substrates, but the kinetic parameters can vary in a range ($\sim 10^{-2}$ to 10^1 mM, ~ 1 to 10^3 s^{-1}). For pectin/pectate lyases, K_m values on the order of 1 mM and k_{cat} values on the order of 10^2 s^{-1} are often observed, with a range of $\sim 10^{-3}$ to 1 mM and $\sim 10^{-1}$ to 10^2 s^{-1}. For pectin methyl esterases, K_m values on the order of 1 mM and k_{cat} values on the order of 10^2 s^{-1} are often observed, with a range of $\sim 10^{-2}$ to 1 mM and ~ 10 to 10^2 s^{-1}, respectively. Many polygalacturonases, pectin/pectate lyases, and pectin methyl esterases are active at temperatures ranging from $\sim$30 to 60°C, 40 to 70°C, and 30 to 70°C, as well as pH values ranging from 4 to 6, 5 to 10, and 4 to 9, respectively. Polygalacturonases and pectin methyl esterases may be susceptible to plant-derived specific protein inhibitors (71).

34.3.6. α-L-Arabinofuranosidase and Endo-1,5-α-Arabinanase

α-L-Arabinofuranosidase (α-L-arabinofuranoside arabinofuranohydrolase, EC 3.2.1.55) cleaves α-glycosidic bonds linking Ara in many hemicelluloses. Different arabinofuranosidases have catalytic modules belonging to the GH3, GH43, GH51, GH54, and GH62 families. Endo-1,5-α-arabinanase (1,5-α-L-arabinan 1,5-α-L-arabinanohydrolase, EC 3.2.1.99) cleaves internal $\alpha(1{\rightarrow}5)$-glycosidic bonds in arabinan or Ara side chain of some hemicelluloses. The catalytic modules of α-arabinases belong to the GH43 family. The stereochemistry of GH43 arabinofuranosidase and arabinase is of the inverting type, while that of GH3, GH51, and GH54 arabinofuranosidases is of the retaining type.

Arabidopsis thaliana GH3, *Bacillus subtilis* GH43A, *C. thermocellum* GH51A, *Aspergillus kawachii* GH54, and *H. jecorina* GH62 arabinofuranosidases are archetypical. Many arabinofuranosidases contain CBM, whose role may be to anchor the enzymes to a particular region in a biomass substrate rather than to bind the enzymes onto an arabinan (37). Besides molecular shape and size, arabinofuranosidases may differ in their substrate specificity. Some arabinofuranosidases prefer Ara singly linked to either a Xyl's O2, O3, or O5 site in (oligo)xylan or arabinan, while others prefer Ara doubly linked to O2 and O3 sites (55, 64, 82). It is reported that GH51 arabinofuranosidases only act on short Ara-oligosaccharides, while GH54 arabinofuranosidases act on both the oligosaccharides and polymeric arabinoxylan (32).

Many arabinofuranosidases have significant xylosidase, transglycosylase, or other dual or side activities (55). Different arabinofuranosidases may act in concert among themselves when their specificities toward different Ara substitution supplement/synergize each other (85). Arabinofuranosidase and arabinase may also act together, in an exo-endo synergism, to effectively degrade arabinan or Ara side chains of other polysaccharides (103). Arabinofuranosidase's removal of Ara side chains in xylan, pectin, or other polysaccharides may enhance the activity of xylanase, pectinase, or other hemicellulases (85). Given the heavy occurrence of Ara substitution of xylan in softwood, arabinoxylan in cereal and grasses, and pectin in many plant cell walls, arabinofuranosidase could be an important part of enzyme systems targeting feedstocks derived from such biomass materials (75).

For arabinofuranosidases, K_m values on the order of 1 mM and k_{cat} values on the order of 1 s^{-1} are often observed for soluble substrates, but the kinetic parameters can vary in a range ($\sim10^{-2}$ to 10 mM, $\sim10^{-2}$ to 10^2 s^{-1}). For arabinase, K_m values on the order of 1 mM and k_{cat} values on the order of 10 s^{-1} are often observed, with a range of $\sim10^{-1}$ to 10 mM and $\sim10^{-2}$ to 10^2 s^{-1}. Many arabinofuranosidases are active at temperatures ranging from ~30 to 80°C, and some arabinofuranosidases from extremophiles can extend their activity/stability at higher temperatures. Arabinofuranosidases, in general, show optimal activity between pH 3 and 7. Endo-arabinases are active at temperatures ranging from ~50 to 70°C, and pH values ranging from 4 to 6.

34.3.7. α-Galactosidase and Endo-Galactanase

α-Galactosidase (α-D-galactoside galactohydrolase, EC 3.2.1.22) cleaves α-glycosidic bonds linking Gal in many hemicelluloses. Different α-galactosidases have catalytic modules belonging to the GH4, GH27, GH36, GH57, and GH110 families. Endo-galactanase (arabinogalactan 4-β-D-galactanohydrolase, EC 3.2.1.89) cleaves internal β(1→4)-bonds in galactan or Gal side chains of some hemicelluloses. All endo-galactanases belong to the GH53 family. The stereochemistry of GH110 α-galactosidases is of the inverting type, while that of GH4, GH27, GH36, and GH57 α-galactosidases, as well as GH53 endo-galactanases, is of the retaining type. B. subtilis GH4, H. jecorina GH27, Thermotoga maritima GH36, Pyrococcus furiosus GH57, and Bacteroides fragilis GH110A α-galactosidases, as well as H. insolens GH53 endo-galactanase, are archetypical. GH4 α-galactosidases along with other GH4 enzymes have the only mechanism exception in all known glycoside hydrolases: they use an NAD$^+$ cofactor in a redox mechanism to hydrolyze their substrates (107).

Comparative study on microbial α-galactosidase and endo-galactanase remains limited, in contrast to other biomass-active enzymes. Given the heavy occurrence of Gal α(1→6) substitution and/or Gal side chain in galactomannan, pectin, and other hemicelluloses, these enzymes could be a useful part of enzyme systems targeting feedstocks rich in such hemicelluloses.

For α-galactosidases, K_m values on the order of 1 mM and k_{cat} values on the order of 10^2 s^{-1} are often observed for soluble substrates, but the kinetic parameters can vary in a range ($\sim10^{-1}$ to 10 mM, $\sim10^{-2}$ to 10^4 s^{-1}). For endo-galactanase, K_m values on the order of 1 mM and k_{cat} values on the order of 10^2 s^{-1} are often observed, with a range of ~10 to 10^3 s^{-1}. Many α-galactosidases are active at temperatures ranging from ~30 to 70°C, and some from extremophiles can extend their activity/stability at higher temperatures. α-Galactosidases, in general, show optimal

activity between pH 3 and 7. Endo-galactanases are active at temperatures ranging from ~40 to 60°C and pH values ranging from 4 to 8.

34.3.8. α-Glucuronidase

α-Glucuronidases [α-D-glucosiduronate glucuronohydrolase, EC 3.2.1.139; xylan α-D-1,2-(4-O-methyl)glucuronohydrolase, EC 3.2.1.131] cleave α(1→2)-bonds linking (methyl)GlcU units in xylan or other hemicelluloses. Almost all α-glucuronidases belong to the exclusive GH67 family (GH4 contains two T. maritima α-glucuronidases) with an inverting stereochemistry.

Cellvibrio japonicus GH67A and Aspergillus nidulans GH67A are archetypical. Different α-glucuronidases may have different specificity. For instance, some xylan α-glucuronidases prefer glucuroylated xylooligosaccharides, while others prefer polymeric glucuronoxylan (69). α-Glucuronidase may assist xylanase action on complex xylans by removing GlcU side chains.

For α-glucuronidases, K_m values on the order of 1 mM and k_{cat} values on the order of 10^2 s^{-1} are often observed, with a range of $\sim10^{-1}$ to 10 mM and $\sim10^{-2}$ to 10^2 s^{-1}, respectively. Many α-glucuronidases are active at temperatures ranging from ~40 to 80°C and pH values ranging from 3 to 7.

34.3.9. Acetyl Xylan Esterase, Feruloyl Esterase, and Glucuronoyl Esterase

Acetyl xylan esterase (AXE, EC 3.1.1.72) and feruloyl esterase (FAE, 4-hydroxy-3-methoxycinnamoyl-sugar hydrolase or ferulic acid esterase, EC 3.1.1.73) hydrolyze the acetyl and hydroxycinnamoyl ester bonds, respectively, found in xylan or other hemicelluloses. Acetylation occurs in many hemicelluloses' backbone glycosyl units (at the O2 or O3 site), and hydroxycinnamoyl (e.g., feruloyl) esterification occurs in α-L-Ara (O2 or O5 site), β-D-Gal (O6 site), or α-D-Xyl side chains of arabinan/arabinoxylan, rhamnogalacturonan, or xyloglucan. The esterifications are believed to serve the purpose of linking hemicellulose to lignin. Glucuronoyl esterase (GE, EC 3.1.1.-) hydrolyzes the O6 methyl in methyl-esterified GlcU α(1→2)-linked to some backbone Xyl in glucuronoarabinoxylan. Compared with AXE, sequence information on FAE and GE is rather limited (53). The catalytic modules of all FAEs are classified in the CE1 family (although seven subfamilies of FAE may be constructed [4]), all GEs are in the CE15, while AXEs are in the CE1, CE2, CE3, CE4, CE5, CE6, CE7, and CE12 families. CE2, CE3, and CE6 contain only AXE.

Being esterases, many AXEs and FAEs are believed to possess the canonical Ser-His-Asp catalytic triad as well as the "oxy anion hole" for their catalysis, similar to other esterases, lipases, or serine proteases. Some AXEs and FAEs may also use metal ions (e.g., Zn^{2+}) for their activity. Many FAEs are modular and contain CBM. Based on specificity, four types of FAE have been proposed. Type A FAEs (e.g., A. niger FAE-A) prefer to act on esters of methoxylated cinnamoyl; feruloyl α-Ara ester (at O5, but not O2; site in Ara) in xylan; feruloyl β-Gal ester in pectin; and 5,5′-diferuloyl esters. In contrast, type B FAEs (e.g., A. niger FAE-B) prefer to act on esters of hydroxylated cinnamoyl, feruloyl α-Ara ester (at both O2 and O5 sites) in both xylan and pectin, but not diferuloyl esters. Type C (e.g., Sporotrichum thermophile FAE-C) and type D (e.g., Piromyces equi FAE) FAEs can act on esters of both methoxylated and hydroxylated cinnamoyls, but only the latter type FAEs can act on diferuloyl esters (4, 91, 97). Although few PEs have been thoroughly characterized, Schizophyllum commune PE does not seem to be a serine type or metalloesterase (87). Based on cloned sequence, a few GEs have CBM (23).

Different AXEs, FAEs, and GEs may complement/ supplement each other in attacking heterogeneously acetylated, ferulylated, or methyl esterified substrates. For FAEs, FAE-B would suit the need to hydrolyze dicots arabinans in which feruloyls esterify the O2 site of α-L-Ara units, and FAE-A or FAE-D would suit the need to hydrolyze 5,5'-diferuloyl esters. AXE or FAE can assist hemicellulases and/or cellulase, by deacetylating or lignin-delinking hemicelluloses in hydrolyzing complex plant cell wall substrates (30, 89). Other enzymes (including arabinase and α-L-arabinofuranosidase) can also assist AXE or FAE by generating shorter, less hindered oligomeric substrates for AXE or FAE (97). To hydrolyze feruloyl or diferuloyl esters in ferulated biomass feedstocks, FAE prefers collaboration from Xyn11 or Xyn10 xylanase, respectively (26). GE seems to have specificity different from those of AXE or FAE (87), and may assist α-glucuronidase by converting GlcU ester to GlcU. Given the heavy occurrence of acetylation of hardwood xylan and feruloyl esterification of arabinoxylan, AXE and FAE could be an important part, respectively, of enzyme systems targeting hardwood- and grass-derived biomass feedstocks.

For AXE, K_m values on the order of 1 mM and k_{cat} values on the order of 10^2 s^{-1} are often observed for soluble substrates, but the kinetic parameters can vary in a range ($\sim$10^{-1} to 1 mM, $\sim$10 to 10^2 s^{-1}). For FAE, K_m values on the order of 0.1 mM and k_{cat} values on the order of 10^3 s^{-1} are often observed, with a range of $\sim$10^{-2} to 10 mM and $\sim$10^{-1} to 10^4 s^{-1}, respectively. Kinetic data on GE are limited, but *S. commune* PE has a K_m of 0.3 mM and a k_{cat} of 3 s^{-1} on soluble, surrogate substrates (87). Many AXEs and FAEs are active at temperatures ranging from $\sim$40 to 70°C, and pH values from $\sim$4 to 8.

34.4. ACCESSORY ENZYMES, PROTEINS, AND OTHER SUBSTANCES

In Nature, biomass-active organisms often produce other types of enzymes or even noncatalytic proteins, often called "accessory enzymes or proteins," that may assist polysaccharide-degrading enzymes. These enzymes or proteins may exert their effects by degrading lignin, disrupting H-bonds in cellulose, preventing "nonproductive" enzyme adsorption by lignin, interacting with biomass-active enzymes, or functioning in mechanisms yet to be understood. These proteins are of interest, not only for the elucidation of natural biomass conversion processes, but also for the development of enzymes for bioenergy.

34.4.1. Expansin and Swollenin

Believed responsible for pH-dependent in vivo plant wall loosening, expansin is a group of plant proteins able to loosen, but not weaken, plant cell walls (18). Being apparently noncatalytic, expansin may disrupt noncovalent bonds among cell wall polysaccharides, leading to wall relaxation. Expansin has an N-terminal domain with a distant sequence homology to Cel45 EG, and a C-terminal domain homologous to grass pollen allergen. Among the four subfamilies of expansin, only the α-expansin and β-expansin subfamilies are known able to extend cell wall. While the diversity of natural plant expansins continues to be probed, microbial genes homologous to plant expansins have recently been found, and their recombinant protein products have plant cell wall-loosening activity (43). This opens a new venue for finding and characterizing new expansins.

Besides independent expansins, expansin-like modules are found in some microbial proteins, in particular, fungal swollenin. Swollenin is an extracellular protein known to be produced by several cellulolytic fungi, consisting of an expansin-like domain linked to a CBM and/or other modules often found in cellulases. It is reported that swollenin is able to disperse cellulosic cotton fiber, weaken filter paper, or disrupt *Valonia* cell walls, without releasing detectable reducing sugars (degradation products). No significant cellulase activity has been observed for swollenin, indicating a noncatalytic nature of the protein (104). Recently, an expansin-like module has also been found in a putative plant EG, suggesting possible interaction with EG activity (11).

The unique plant cell wall-loosening effect of expansin and swollenin makes the proteins interesting as potential contributors to the enzyme systems for bioenergy application. Indeed, a few studies showed the benefit of adding expansin or swollenin to cellulases for improved lignocellulose hydrolysis, an effect ascribed to the proteins' putative cellulose-disrupting action (18, 43). However, the generality and applicability of such enhancement from expansin or swollenin on biomass-active enzymes remain to be further investigated.

34.4.2. CBM

For modular biomass-active enzymes, CBM is the second most important module after the catalytic module. The amazing specificity, versatility, and efficiency of CBM can be demonstrated by the frequent occurrence in different enzymes of a same or highly homologous CBM linked to different catalytic modules, as well as different CBMs linked to a same or homologous catalytic module. CBM has a large natural diversity, with variable sequence, composition, size, or folding, leading to diverse affinities toward polysaccharides. Based on sequence homology, CBMs are classified into $\sim$53 families (mostly fungal CBMs in CBM1). Based on their interaction with carbohydrate, CBMs can be divided into three general groups. The first group (exemplified by CBM1, CBM2, and CBM3) has planar surfaces with exposed aromatic residues (e.g., Phe, Tyr) suited for the binding of bundled polysaccharide chains (e.g., crystalline cellulose microfibrils), mainly via π-stacking between the aromatic residues and furan/pyranosyl units. The second type (exemplified by CBM4 and CBM6) has clefts with hidden aromatic residues suited for the binding of individual polysaccharide chains or oligosaccharides, similar to many glycoside hydrolases' catalytic modules. The third type (exemplified by CBM42) has pockets suited for the binding of small saccharides, similar to lectins. Many CBMs need Ca^{2+} for folding or carbohydrate binding. Different CBMs have different specificities. For instance, CBM1 and CBM49 prefer crystalline cellulose (but not chitin); CBM2 and CBM37 bind to cellulose, xylan, or chitin; CBM2–CBM5, CBM8–CBM10, CBM16, CBM17, CBM28, CBM30, CBM44, and CBM46 bind to amorphous cellulose; CBM4, CBM6, CBM11, CBM28, CBM39, CBM43, and CBM52 bind to β-glucan; CBM2, CBM6, CBM13, CBM15, CBM22, CBM31, and CBM35–CBM37 bind to xylan; CBM44 binds to xyloglucan; CBM32 binds to pectin; CBM16, CBM23, CBM27, and CBM29 bind to (gluco)mannan; CBM20, CBM21, CBM25, CBM26, CBM34, CBM41, CBM45, and CBM53 bind to starch; CBM41 binds to pullulan; CBM24 binds to α-glucan (mutan); CBM38 binds to inulin (a fructan); CBM2, CBM12, CBM14, CBM18, CBM33, CBM37, and CBM50 bind to chitin; CBM13 binds to Man; CBM33 and CBM51 bind to Gal; CBM42 binds to Ara; and CBM47 binds to Fuc. Most native CBMs are part of various biomass-active enzymes, but some exist as either independent binding proteins or as part of apparently noncatalytic proteins (32, 35, 84).

Two key native roles are postulated for CBM: anchoring enzymes to targeted carbohydrate substrates (and bringing catalytic modules to close proximity of substrates), and/or directing enzymes to specific regions in complex biomass substances (10). In addition, a xylan-specific CBM22 can stabilize a xylanase catalytic module against thermal inactivation (88). A few studies show that free CBM or noncatalytic CBM-containing proteins may even "directly" assist in the catalysis of a glycoside hydrolase (e.g., cellulase), under laboratory conditions, by physically modifying a polysaccharide substrate (e.g., disrupting crystalline cellulose microfibrils) so it becomes more reactive toward the enzyme (74). However, more comparative and systematic studies are needed to further validate and explore this intriguing CBM effect (96).

34.4.3. GH61, Cip, and HbpA

Among *H. jecorina*-secreted proteins, EG-IV is the product of one of three GH61 genes of the organism. Weak EG-like activity from wild-type and recombinant GH61s is reported in some studies, but not others (42, 46). All known GH61s are from cellulolytic fungi with one exception (a single GH61 gene in *Cryptococcus neoformans*), and some of them have CBM. *H. jecorina* GH61B has a fold and metal ion site similar to that of a chitin-binding protein (42). Although GH61 may be noncatalytic on many typical glycoside hydrolase substrates, a recent study showed for the first time that it can enhance the action of cellulases on lignocellulose (12).

H. jecorina secretes a Cip1 protein, consisting of a CBM and a "core" homologous to a putative *Streptomyces coelicolor* hydrolase, which has recently been reported to enhance cellulase activity either alone or together with GH61 and swollenin (B. R. Scott, C. Hill, J. Tomashek, and C. Liu, U.S. patent application 0,061,484). *H. jecorina* also secretes another CBM-bearing protein, Cip2, whose core is identified as a glucuronoyl esterase (53).

C. cellulovorans produces a noncatalytic cellulosomal protein, HbpA, which has a surface layer module and a cohesin. The protein may interact with two dockerin-bearing EGs, bind them to cell surface and to polysaccharides, and stimulate the EGs' hydrolysis of insoluble cellulosic substances (60).

34.4.4. Oxidoreductase

Many cellulolytic microbes (in particular, white and brown rots) secrete oxidoreductases, believed to be important for the organisms' degradation of lignaceous biomass. Lignin is a group of highly heterogeneous, branched, and recalcitrant aromatic polymers that cross-link to, entangle with, and shield hemicellulose or cellulose. Degradation of lignin is crucial for effective enzymatic hydrolysis of many biomass feedstocks, in terms of increased cellulose accessibility as well as decreased lignin inhibition for cellulase/hemicellulase. Natural microbial degradation of lignin involves phenol oxidase (laccase), peroxidase, or H_2O_2-generating carbohydrate/alcohol oxidoreductase, which attack lignin via high-valence oxidative species (52, 59). Laccase (EC 1.10.3.2), lignin peroxidase (EC 1.11.1.14), Mn peroxidase (EC 1.11.1.13), versatile peroxidase (EC 1.11.1.16), cellobiose dehydrogenase (EC 1.1.99.18), and various oxidases (e.g., aryl-alcohol oxidase, EC 1.1.3.7; glyoxal oxidase, EC 1.1.3.-) are of particular

interest as industrial delignifying enzymes and potential contributors to enzymatic systems for bioenergy. Comprehensive structural information on these oxidoreductases can be found in the Fungal Oxidative Lignin Enzymes (FOLy) database (52; http://foly.esil.univ-mrs.fr/).

Almost all of the known delignifying enzymes invoke highly reactive species to oxidatively degrade lignin. However, the oxidants may also oxidatively modify cellulose or hemicelluloses, leading to decreased enzyme digestibility (2; F. Xu, H. Ding, A. Tejirian, and J. Quinlan, presented at the 30th Symposium on Biotechnology for Fuels and Chemicals, 2008). Applying such enzymes may facilitate enzymatic biomass hydrolysis under certain conditions (47) but not under others (89); hence, more research is needed to evaluate both the positive and negative impacts from the oxidoreductases.

34.5. AMYLASES

Amylases are enzymes that hydrolyze starch comprising mainly linear amylose [$\alpha(1{\rightarrow}4)$-linked D-Glc polymer] and branched amylopectin [$\alpha(1{\rightarrow}4, 1{\rightarrow}6)$-linked D-Glc polymer]. In Nature, starch is degraded mainly by a combination of two types of hydrolases, endo-type (e.g., α-amylase) and exo-type (e.g., glucoamylase), with approximately 30 enzymatic specificities (58). In general, amylases are modular enzymes, with catalytic modules and CBMs belonging to eight GH and CBM families, respectively. Starch-binding CBMs are believed to enable amylase activity on raw starch granules, by increasing local amylase concentration on insoluble starch or even by disrupting surface starch structure (76). Because of their different molecular and enzymological properties (in particular, catalytic mechanism and substrate specificity), amylases can act synergistically on starchy substances. Industrial-scale starch hydrolysis (for bioethanol production) generally applies first α-amylase to fragment raw starch into soluble starch or dextrins ("liquefaction"), then glucoamylase and other amylolytic enzymes (e.g., α-amylase, isoamylase) to hydrolyze dextrins into Glc ("saccharification"). Amylolytic enzymes acting on $\alpha(1{\rightarrow}6)$-bonds or branched segments in starch can assist the hydrolysis of the $\alpha(1{\rightarrow}4)$-bond-specific amylases, leading to more effective starch conversion.

34.5.1. α-Amylases

α-Amylase (1,4-α-D-glucan glucanohydrolase, EC 3.2.1.1) is an endo-amylolytic enzyme that cleaves internal $\alpha(1{\rightarrow}4)$-bonds in starch. The catalytic modules of the majority of α-amylases belong to the GH13 family, while those of some prokaryotic α-amylases belong to the GH57 family. The stereochemistry of α-amylases is of the retaining type. A minority (~10%) of α-amylases use one or more starch-binding CBMs (belonging to the CBM20, CBM21, CBM25, CBM26, CBM34, and CBM45 family, commonly linked to the C terminus of catalytic modules), prerequisite for amylase activity on raw starch (57, 58, 76). *Aspergillus oryzae* TAKA α-amylase and barley Amy1 are archetypical Amy13 α-amylases. Many α-amylases are Ca^{2+} dependent (for both activity and stability).

With soluble substrates, K_m values on the order of 1 mM and k_{cat} values on the order of 10^2 s^{-1} are often observed for α-amylases, but the kinetic parameters can vary in a range (~10^{-2} to 10 mM, ~10^{-1} to 10^3 s^{-1}). Insoluble (raw) starch granule, soluble amylose or maltodextrins, and soluble α-glucosides are commonly used to assay α-amylase, based on changes in viscosity, reducing sugar, dextrin distribution, photometry, or other physicochemical properties. Many α-amylases show optimal activity around pH 3 to 10, and are active at temperatures ranging from ~30 to 80°C (hyperthermophilic α-amylases can extend their activity/stability above 100°C). In general, α-amylases may be inhibited by metal chelators such as EDTA, and some α-amylases may be inhibited by proteinaceous inhibitors, such as barley amylase/subtilisin inhibitor. Lacking a CBM may decrease the thermal stability of some α-amylases.

Although producing bioethanol from starch is a viable industry at present, continued optimizations, including that on enzymatic starch hydrolysis, are still of interest. For instance, better liquefying α-amylases are sought in terms of higher thermal activity and stability (which may be achieved with hyperthermophilic amylases), higher raw starch-binding and activity (which may be achieved with single or tandem CBM grafting), or lower dependence on Ca^{2+} concentration.

34.5.2. β-Amylase

β-Amylase (1,4-α-D-glucan maltohydrolase, EC 3.2.1.2) is an exo-amylolytic enzyme that cleaves the second α(1→4)-bond at the nonreducing end of the amylose chain to produce maltose. β-Amylases have catalytic modules belonging to the GH14 family, and act under the inverting stereochemistry. Some bacterial β-amylases have starch-binding CBM (belonging to the CBM20 and CBM25 families). *Bacillus cereus* and soybean β-amylases are archetypical. With soluble substrates, K_m values on the order of 1 mM and k_{cat} values on the order of 10^3 s^{-1} are often observed for β-amylases, but the kinetic parameters can vary (~10^{-2} to 10 mM, ~10^{-1} to 10^4 s^{-1}). Many β-amylases are active at temperatures ranging from ~30 to 80°C and pH values ranging from 4 to 8.

34.5.3. Glucoamylase and α-Glucosidase

Glucoamylase (1,4-α-D-glucan glucohydrolase or amyloglucosidase, EC 3.2.1.3) is an exo-amylolytic enzyme that cleaves Glc from the nonreducing end of an α-glucan chain. The enzyme's catalytic module belongs to the GH15 family, and the enzyme acts under the inverting stereochemistry. Some fungal glucoamylases have starch-binding CBM (belonging to the CBM20 and CBM21 families). *Aspergillus awamori* glucoamylase is archetypical.

α-Glucosidase (α-D-glucoside glucohydrolase or maltase, EC 3.2.1.20) is another exo-amylolytic enzyme that cleaves Glc from the nonreducing end of α-glucooligosaccharides. α-Glucosidases have catalytic modules belonging to the GH4, GH13, GH31, GH63, and GH97 families. GH63 and part of GH97 α-glucosidases have the inverting stereochemistry, while GH4, GH13, GH31, and the other part of GH97 α-glucosidases have the retaining stereochemistry. Unlike other α-glucosidases, GH4 α-glucosidases are expected to use NAD$^+$ cofactor in a redox mechanism (107). Some glucoamylases have starch-binding CBM (belonging to the CBM20 and CBM21 families). *A. awamori* glucoamylase is archetypical.

Although both glucoamylase and α-glucosidase are exo-amylases, there are differences between the two types of enzymes (93). In general, glucoamylases prefer starch (with relatively low k_{cat} and high K_m on maltose, but high k_{cat} and low K_m on maltotetraose, longer maltodextrins, or starch), while α-glucosidases prefer maltodextrins (with approximately the same k_{cat} on maltotetraose and longer maltodextrins, but higher K_m on longer maltodextrins), although a few α-glucosidases may act on raw starch granules.

K_m values on the order of 10 mM and k_{cat} values on the order of 10^2 s^{-1} are common for glucoamylases and α-glucosidases, although the kinetic parameters can vary (~10^{-3} to 10^2 mM, ~10^{-1} to 10^3 s^{-1}). Many glucoamylases and α-glucosidases are active at pH values ranging from 3 to 8 under temperatures ranging from ~20 to 80°C.

34.5.4. Pullulanase and Isoamylase

Pullulanase (pullulan α-1,6-glucanohydrolase, EC 3.2.1.41) is a "debranching" amylolytic enzyme that cleaves α(1→6)-bonds in amylopectin, glycogen, pullulan (α(1→6)-linked maltotriosyl polymer), or branched dextrins. The catalytic modules of pullulanases belong to the GH13, GH49, and GH57 families. The stereochemistry of GH13 and GH57 pullulanases is retaining, while that of GH49 pullulanases is inverting. Some pullulanases have CBM (belonging to the CBM20 and CBM41 families).

Isoamylase (glycogen α-1,6-glucanohydrolase, EC 3.2.1.68) is another debranching amylolytic enzyme that cleaves α(1→6)-bonds in amylopectin or glycogen. The catalytic modules of isoamylases belong to the GH13 family, with a retaining stereochemistry.

Although both pullulanase and isoamylase cleave α(1→6)-bonds, there are differences between the two types of enzymes. Isoamylases act on amylopectin (starch) and glycogen, but not pullulan, while pullulanases act on all three α-glucans. For amylopectin and glycogen, isoamylases can cleave both inner and outer α(1→6)-bonds, but pullulanases prefer those near the ends of α-glucan chains.

K_m values on the order of 1 mM and k_{cat} values on the order of 10^2 s^{-1} are often observed for pullulanases, but the kinetic parameters can vary in a range (~10^{-2} to 1 mM, ~10^2 to 10^3 s^{-1}). Many pullulanases are active at temperatures ranging from ~40 to 90°C (but hyperthermophilic pullulanases can retain activity above 100°C), and pH values ranging from ~5 to 9. K_m values on the order of 10^{-4} mM and k_{cat} values on the order of 10^2 s^{-1} are often observed for isoamylases, with a range of ~10^{-4} to 10^{-2} mM and ~1 to 10^2 s^{-1}, respectively. Many isoamylases are active at temperatures ranging from ~30 to 60°C, and pHs ranging from ~3 to 9.

Pullulanase and isoamylase may, through their specificity on α(1→6)-bonds, act synergistically with α(1→4)-bond-specific amylases in hydrolyzing starch, making them important components of efficient enzymatic starch hydrolysis systems. At high starch consistency, the transglycosylase side activity of pullulanase and isoamylase may become significant enough to negatively impact the starch-to-Glc hydrolysis.

34.5.5. α-Glucan Lyase

1,4-α-Glucan lyase (1,4-α-D-glucan exo-4-lyase [1,5-anhydro-D-fructose-forming], EC 4.2.2.13) cleaves internal α(1→4)-bonds in linear α-glucan by β-elimination. Such lyases are classified to the GH31 family. The glucan lyase is specific to α(1→4)-glycosidic bonds, acts on the nonreducing end of an α-glucan chain, prefers maltodextrins or α-glucan over simple α-glucosides, and has high affinity to starch despite lacking a distinct CBM (51).

For α-glucan lyase, K_m values on the order of 1 mM and k_{cat} values on the order of 10 s^{-1} are often observed for soluble substrates, but the kinetic parameters can vary (~10^{-1} to 10 mM, ~1 to 10^2 s^{-1}). Many α-glucan lyases are active in the temperature range of ~30 to 50°C and pH range of 3 to 7.

Compared with polygalacturonan lyases, α-glucan lyase has a different mechanism and produces a different product. Polygalacturonan lyase cleaves the bond between glycosidic O and C4, and the β-elimination results in a Δ4:5-unsaturated GalU moiety. α-Glucan lyase cleaves the bond between anomeric C1 and glycosidic O, and the β-elimination results in a Δ1:2 unsaturated Glc moiety, which rearranges into a 1,5-anhydro-D-Fru (51).

34.6. ONGOING DEVELOPMENTS AND FUTURE PROSPECTS OF BIOENERGY ENZYMES

Increasingly intensive research is being conducted to understand natural conversion of energy-storing polysaccharides into fermentable carbohydrates, and to discover and improve enzymes for industrial conversion of such polysaccharides from various biomass feedstocks (24, 56, 99). Current strategies include engineering known enzymes (for better specificity, activity, or stability), searching for carbohydrate-active enzymes that can work at elevated temperature (to benefit from potential thermal acceleration on biomass hydrolysis), and exploring synergistic enzyme mixtures (with lowered overall dosage).

34.6.1. Protein Engineering of Biomass-Active Enzymes

Among the bioenergy enzymes actively under study, cellulases have attracted, and will continue to attract, the most attention because of the significance of converting cellulose to Glc in the development of cellulosic biofuels. The increase in solved three-dimensional cellulase structures and decoded cellulase genes has propelled rapid progress in site-directed mutagenesis, directed evolution, module or domain grafting, and other genetic/protein engineerings of cellulases. Many variants of CBH, EG, or BG have been constructed and characterized for improved kinetic or thermal properties. Hemicellulases with improved activities and CBMs with improved or altered specificity have also been studied by various protein engineering methods (29).

Of many improvements on cellulases, enhanced thermal activity and stability are particularly desirable. Two approaches are currently being taken: engineering known but mesophilic cellulases, or finding new thermophilic wild-type cellulases. New cellulases and hemicellulases with enhanced specificity, shifted pH optimum, or increased inhibition resilience are of interest as well.

Constructing designer carbohydrate-active enzymes with selected catalytic modules, CBMs, dockerins, or other modules may result in novel and/or desired functionalities. Hybrid cellulases may have additional copies of a native module or nonnative modules, so as to multiply a native or possess a nonnative but beneficial functionality. Some fusion proteins bring a cellulase at close vicinity with its collaborative partner enzyme (e.g., CBH-EG fusion, cellulase-feruloyl esterase) or accessory protein (e.g., cellulase-expansin, feruloyl esterase-swollenin, cellulase-protease inhibitor fusions), potentially beneficial for not only hydrolysis but also control of enzyme stoichiometry. Some chimeras may have a dockerin linked to a native cellulase or other enzymes to allow their incorporation, along with synergistic partner enzymes, into a tailor-made cellulosome (14). CBM may be engineered for nonnative specificity (29). Optimal combination of the selected modules/enzymes may be achieved by careful selection of linkers (rigidity and length), dockerin-cohesion pairs, or scaffoldins (hosting capacity, cell binding, and substrate binding), although the art is far from perfection. Biomass-active enzymes may even be planted on a surface of selected microbes, via interactions such as that from bacterial S-layer proteins, to facilitate combined reactions or recovery.

34.6.2. Synergistic and Novel Multienzyme Systems

It is believed that a minimal complete cellulase mixture needs four interdependent and/or complementary enzymes: one CBH-I, one CBH-II, one EG, and one BG. In general, cellulolytic organisms have far more than four cellulase genes, and produce more than four cellulases. Multiple enzymes often exist for one basic type of cellulase action (e.g., EG catalysis) in natural cellulase mixtures. Because the cellulase compositions produced by laboratory-grown microbes may well not be optimal with regard to industry-relevant biomass substrates and hydrolysis conditions, they often need to be supplemented for one or more components (99), or complemented with other enzymes that may assist the cellulases and/or convert noncellulose parts of the feedstocks (3, 27, 30, 73, 78, 81, 89). Similar to the case of cellulases, complete hemicellulase mixtures for hydrolyzing hemicelluloses, as well as needs for supplementing/complementing them for improved hydrolysis, also exist (86). Seeking the best stoichiometry and synergism among different cellulases, hemicellulases, other enzymes, or accessory proteins is vital for the development of bioenergy enzymes.

New structural and enzymological studies are expanding or even challenging our current beliefs. For instance, there might be a third type of natural cellulose degradation relying on nonprocessive, noncellulosomal cellulases to fragment cellulose, cellular import of cellodextrins, and intracellular cellodextrin degradation, a process fundamentally different from the two conventional degradations relying on mixtures of dissociated CBH, EG, and BG or on cellulosomes (96). We are in an early stage in the investigation of noncatalytic proteins, coproduced along with biomass-active enzymes by many cellulolytic microbes, about their in vivo function and potential accessory role in enzyme systems for bioenergy (24). The cellulolytic machinery of brown rots is also of interest, because, in addition to hydrolases and hydrolysis, the organisms apparently use oxidoreductases and redox or radical chemistry (59, 95), which might complement bioenergy enzymes produced by cellulolytic organisms.

Robert L. Starnes, Paul Harris, and Claus C. Fuglsang from Novozymes provided critical suggestions.

REFERENCES

1. **Bayer, E. A., J. P. Belaich, Y. Shoham, and R. Lamed.** 2004. The cellulosomes: multienzyme machines for degradation of plant cell wall polysaccharides. *Annu. Rev. Microbiol.* **58:**521–554.
2. **Bendl, R. F., J. M. Kandel, K. D. Amodeo, A. M. Ryder, and E. M. Woolridge.** 2008. Characterization of the oxidative inactivation of xylanase by laccase and a redox mediator. *Enzyme Microb. Technol.* **43:**149–156.
3. **Benko, Z., M. Siika-aho, L. Viikari, and K. Reczey.** 2008. Evaluation of the role of xyloglucanase in the enzymatic hydrolysis of lignocellulosic substrates. *Enzyme Microb. Technol.* **43:**109–114.
4. **Benoit, I., E. G. J. Danchin, R. J. Bleichrodt, and R. P. de Vries.** 2008. Biotechnological applications and potential of fungal feruloyl esterases based on prevalence, classification and biochemical diversity. *Biotechnol. Lett.* **30:**387–396.

5. **Berger, E., D. Zhang, V. V. Zverlov, and W. H. Schwarz.** 2007. Two noncellulosomal cellulases of *Clostridium thermocellum*, Cel9I and Cel48Y, hydrolyse crystalline cellulose synergistically. *FEMS Microbiol. Lett.* **268:**194–201.

6. **Berlin, A., M. Balakshin, N. Gilkes, J. Kadla, V. Maximenko, S. Kubo, and J. Saddler.** 2006. Inhibition of cellulase, xylanase and beta-glucosidase activities by softwood lignin preparations. *J. Biotechnol.* **125:**198–209.

7. **Berlin, A., N. Gilkes, D. Kilburn, R. Bura, A. Markov, A. Skomarovsky, O. Okunev, A. Gusakov, V. Maximenko, D. Gregg, A. Sinitsyn, and J. Saddler.** 2005. Evaluation of novel fungal cellulase preparations for ability to hydrolyze softwood substrates—evidence for the role of accessory enzymes. *Enzyme Microb. Technol.* **37:**175–184.

8. **Berrin, J. G., and N. Juge.** 2008. Factors affecting xylanase functionality in the degradation of arabinoxylans. *Biotechnol. Lett.* **30:**1139–1150.

9. **Beukes, N., H. Chan, R. H. Doi, and B. I. Pletschke.** 2008. Synergistic associations between *Clostridium cellulovorans* enzymes XynA, ManA and EngE against sugarcane bagasse. *Enzyme Microb. Technol.* **42:**492–498.

10. **Blake, A. W., L. McCartney, J. E. Flint, D. N. Bolam, A. B. Boraston, H. J. Gilbert, and J. P. Knox.** 2006. Understanding the biological rationale for the diversity of cellulose-directed carbohydrate-binding modules in prokaryotic enzymes. *J. Biol. Chem.* **281:**29321–29329.

11. **Bouzarelou, D., M. Billini, K. Roumelioti, and V. Sophianopoulou.** 2008. EglD, a putative endoglucanase, with an expansin like domain is localized in the conidial cell wall of *Aspergillus nidulans*. *Fungal Genet. Biol.* **45:**839–850.

12. **Brown, K., P. Harris, E. Zaretsky, E. Re, E. Vlasenko, K. McFarland, and A. Lopez de Leon.** January, 2005. Polypeptide from a cellulolytic fungus having cellulolytic enhancing activity. U.S. patent 7,361,495.

13. **Cantarel, B. L., P. M. Coutinho, C. Rancurel, T. Bernard, V. Lombard, and B. Henrissat.** 2009. The Carbohydrate-Active EnZymes database (CAZy): an expert resource for glycogenomics. *Nucleic Acids Res.* **37:**D233–D238.

14. **Caspi, J., D. Irwin, R. Lamed, Y. C. Li, H. P. Fierobe, D. B. Wilson, and E. A. Bayer.** 2008. Conversion of *Thermobifida fusca* free exoglucanases into cellulosomal components: comparative impact on cellulose-degrading activity. *J. Biotechnol.* **135:**351–357.

15. **Chang, A., M. Scheer, A. Grote, I. Schomburg, and D. Schomburg.** 2009. BRENDA, AMENDA and FRENDA the enzyme information system: new content and tools in 2009. *Nucleic Acids Res.* **37:**D588–D592.

16. **Cohen, R., M. R. Suzuki, and K. E. Hammel.** 2005. Processive endoglucanase active in crystalline cellulose hydrolysis by the brown rot basidiomycete *Gloeophyllum trabeum*. *Appl. Environ. Microbiol.* **71:**2412–2417.

17. **Collins, T., C. Gerday, and G. Feller.** 2005. Xylanases, xylanase families and extremophilic xylanases. *FEMS Microbiol. Rev.* **29:**3–23.

18. **Cosgrove, D.** 2005. Growth of the plant cell wall. *Nat. Rev. Mol. Cell. Biol.* **6:**850–861.

19. **Da Costa, A., P. Michaud, E. Petit, A. Heyraud, P. Colin-Morel, B. Courtois, and J. Courtois.** 2001. Purification and properties of a glucuronan lyase from *Sinorhizobium meliloti* M5N1CS (NCIMB 40472). *Appl. Environ. Microbiol.* **67:**5197–5203.

20. **Desmet, T., T. Cantaert, P. Gualfetti, W. Nerinckx, L. Gross, C. Mitchinson, and K. Piens.** 2007. An investigation of the substrate specificity of the xyloglucanase Cel74A from *Hypocrea jecorina*. *FEBS J.* **274:**356–363.

21. **Dhawan, S., and J. Kaur.** 2007. Microbial mannanases: an overview of production and applications. *Crit. Rev. Biotechnol.* **27:**197–216.

22. **Ding, H., and F. Xu.** 2004. Productive cellulase adsorption on cellulose, p. 154–169. *In* B. C. Saha (ed.), *Lignocellulose Biodegradation (Symposium Series 889).* American Chemical Society, Washington, DC.

23. **Duranova, M., S. Spanikova, H. A. Wosten, P. Biely, and R. P. de Vries.** 2009. Two glucuronoyl esterases of *Phanerochaete chrysosporium.* *Arch. Microbiol.* **191:**133–140.

24. **Eijsink, V. G. H., G. Vaaje-Kolstad, K. M. Varum, and S. J. Horn.** 2008. Towards new enzymes for biofuels: lessons from chitinase research. *Trends Biotechnol.* **26:**228–235.

25. **Eyzaguirre, J., M. Hidalgo, and A. Leschot.** 2005. Beta-glucosidases from filamentous fungi: properties, structure, and applications, p. 645–685. *In* K. J. Yarema (ed.), *Handbook of Carbohydrate Engineering.* CRC Press, Boca Raton, FL.

26. **Faulds, C. B., G. Mandalari, R. B. Lo Curto, G. Bisignano, P. Christakopoulos, and K. W. Waldron.** 2006. Synergy between xylanases from glycoside hydrolase family 10 and family 11 and a feruloyl esterase in the release of phenolic acids from cereal arabinoxylan. *Appl. Microbiol. Biotechnol.* **71:**622–629.

27. **Garcia-Aparicio, M. P., M. Ballesteros, P. Manzanares, I. Ballesteros, A. Gonzalez, and M. J. Negro.** 2007. Xylanase contribution to the efficiency of cellulose enzymatic hydrolysis of barley straw. *Appl. Biochem. Biotechnol.* **137:**353–365.

28. **Gilbert, H. J.** 2007. Cellulosomes: microbial nanomachines that display plasticity in quaternary structure. *Mol. Microbiol.* **63:**1568–1576.

29. **Gunnarsson, L. C., Q. Zhou, C. Montanier, E. N. Karlsson, H. Brumer, and M. Ohlin.** 2006. Engineered xyloglucan specificity in a carbohydrate-binding module. *Glycobiology* **16:**1171–1180.

30. **Gupta, R., T. H. Kim, and Y. Y. Lee.** 2008. Substrate dependency and effect of xylanase supplementation on enzymatic hydrolysis of ammonia-treated biomass. *Appl. Biochem. Biotechnol.* **148:**59–70.

31. **Gusakov, A. V., T. N. Salanovich, A. I. Antonov, B. B. Ustinov, O. N. Okunev, R. Burlingame, M. Emalfarb, M. Baez, and A. P. Sinitsyn.** 2007. Design of highly efficient cellulase mixtures for enzymatic hydrolysis of cellulose. *Biotechnol. Bioeng.* **97:**1028–1038.

32. **Hashimoto, H.** 2006. Recent structural studies of carbohydrate-binding modules. *Cell. Mol. Life Sci.* **63:**2954–2967.

33. **Henrissat, B., G. Sulzenbacher, and Y. Bourne.** 2008. Glycosyltransferases, glycoside hydrolases: surprise, surprise! *Curr. Opin. Struct. Biol.* **18:**527–533.

34. **Herpoel-Gimbert, I., A. Margeot, A. Dolla, G. Jan, D. Molle, S. Lignon, H. Mathis, J.-C. Sigoillot, F. Monot, and M. Asther.** 2008. Comparative secretome analyses of two *Trichoderma reesei* RUT-C30 and CL847 hypersecretory strains. *Biotechnol. Biofuels* **1:**18.

35. **Hilden, L., and G. Johansson.** 2004. Recent developments on cellulases and carbohydrate-binding modules with cellulose affinity. *Biotechnol. Lett.* **26:**1683–1693.

36. **Hodge, D. B., M. N. Karim, D. J. Schell, and J. D. McMillan.** 2008. Soluble and insoluble solids contributions to high-solids enzymatic hydrolysis of lignocellulose. *Bioresour. Technol.* **99:**8940–8948.

37. **Ichinose, H., M. Yoshida, Z. Fujimoto, and S. Kaneko.** 2008. Characterization of a modular enzyme of exo-1,5-alpha-L-arabinofuranosidase and arabinan binding module from *Streptomyces avermitilis* NBRC14893. *Appl. Microbiol. Biotechnol.* **80:**399–408.

38. **Ishida, T., K. Yaoi, A. Hiyoshi, K. Igarashi, and M. Samejima.** 2007. Substrate recognition by glycoside hydrolase family 74 xyloglucanase from the basidiomycete *Phanerochaete chrysosporium.* *FEBS J.* **274:**5727–5736.

39. **Jayani, R. S., S. Saxena, and R. Gupta.** 2005. Microbial pectinolytic enzymes: a review. *Process Biochem.* **40:**2931–2944.

40. **Jordan, D. B., and J. D. Braker.** 2007. Inhibition of the two-subsite beta-D-xylosidase from *Selenomonas ruminantium* by sugars: competitive, noncompetitive,

double binding, and slow binding modes. *Arch. Biochem. Biophys.* **465:**231–246.

41. Jordan, D. B., X. L. Li, C. A. Dunlap, T. R. Whitehead, and M. A. Cotta. 2007. Structure-function relationships of a catalytically efficient beta-D-xylosidase. *Appl. Biochem. Biotechnol.* **141:**51–76.

42. Karkehabadi, S., H. Hansson, S. Kim, K. Piens, C. Mitchinson, and M. Sandgren. 2008. The first structure of a glycoside hydrolase family 61 member, Cel61B from *Hypocrea jecorina*, at 1.6 angstrom resolution. *J. Mol. Biol.* **383:**144–154.

43. Kim, E., H. Lee, W. G. Bang, I. G. Choi, and K. H. Kim. 2009. Functional characterization of a bacterial expansin from *Bacillus subtilis* for enhanced enzymatic hydrolysis of cellulose. *Biotechnol. Bioeng.* **102:**1342–1353.

44. Kipper, K., P. Valjamae, and G. Johansson. 2005. Processive action of cellobiohydrolase Cel7A from *Trichoderma reesei* is revealed as "burst" kinetics on fluorescent polymeric model substrates. *Biochem. J.* **385:**527–535.

45. Konno, N., T. Ishida, K. Igarashi, S. Fushinobu, N. Habu, M. Samejima, and A. Isogai. 2009. Crystal structure of polysaccharide lyase family 20 endo-β-1,4-glucuronan lyase from the filamentous fungus *Trichoderma reesei*. *FEBS Lett.* **583:**1323–1326.

46. Koseki, T., Y. Mese, S. Fushinobu, K. Masaki, T. Fujii, K. Ito, Y. Shiono, T. Murayama, and H. Iefuji. 2008. Biochemical characterization of a glycoside hydrolase family 61 endoglucanase from *Aspergillus kawachii*. *Appl. Microbiol. Biotechnol.* **77:**1279–1285.

47. Krusa, M., H. Lennholm, and G. Henriksson. 2008. Pre-treatment of cellulose by cellobiose dehydrogenase increases the degradation rate by hydrolytic cellulases. *Cell Chem. Technol.* **41:**105–111.

48. Kumar, R., S. Singh, and O. V. Singh. 2008. Bioconversion of lignocellulosic biomass: biochemical and molecular perspectives. *J. Ind. Microbiol. Biotechnol.* **35:**377–391.

49. Langston, J., N. Sheehy, and F. Xu. 2006. Substrate specificity of *Aspergillus oryzae* family 3 β-glucosidase. *Biochim. Biophys. Acta* **1764:**972–978.

50. Lawoko, M., A. Nutt, H. Henriksson, G. Gellerstedt, and G. Henriksson. 2000. Hemicellulase activity of aerobic fungal cellulases. *Holzforschung* **54:**497–500.

51. Lee, S. S., S. Yu, and S. G. Withers. 2005. Mechanism of action of exo-acting α-1,4-glucan lyase: a glycoside hydrolase family 31 enzyme. *Biologia* (Bratislava) **60:**137–148.

52. Levasseur, A., F. Piumi, P. Coutinho, C. Rancurel, M. Asther, M. Delattre, B. Henrissat, P. Pontarotti, M. Asther, and E. Record. 2008. FOLy: an integrated database for the classification and functional annotation of fungal oxidoreductases potentially involved in the degradation of lignin and related aromatic compounds. *Fungal Genet. Biol.* **45:**638–645.

53. Li, X.-L., S. Spanikova, R. P. de Vries, and P. Biely. 2007. Identification of genes encoding microbial glucuronoyl esterases. *FEBS Lett.* **581:**4029–4035.

54. Liu, Q. P., G. Sulzenbacher, H. Yuan, E. P. Bennett, G. Pietz, K. Saunders, J. Spence, E. Nudelman, S. B. Levery, T. White, J. M. Neveu, W. S. Lane, Y. Bourne, M. L. Olsson, B. Henrissat, and H. Clausen. 2007. Bacterial glycosidases for the production of universal red blood cells. *Nat. Biotechnol.* **25:**454–464.

55. Lopez, G., C. Nugier-Chauvin, C. Remond, and M.O'Donohue. 2007. Investigation of the specificity of an alpha-L-arabinofuranosidase using C-2 and C-5 modified alpha-L-arabinofuranosides. *Carbohydr. Res.* **342:**2202–2211.

56. Lynd, L. R., M. S. Laser, D. Bransby, B. E. Dale, B. Davison, R. Hamilton, M. Himmel, M. Keller, J. D. McMillan, J. Sheehan, and C. E. Wyman. 2008. How biotech can transform biofuels. *Nat. Biotechnol.* **26:**169–172.

57. MacGregor, E. A. 2005. An overview of clan GH-H and distantly-related families. *Biologia* (Bratislava) **60:**5–12.

58. Machovic, M., and S. Janecek. 2007. Amylolytic enzymes: types, structures and specificities, p. 3–18. *In* J. Polaina and A. P. MacCabe (ed.), *Industrial Enzymes: Structure, Function and Applications.* Springer, Dordrecht, The Netherlands.

59. Martinez, A. T., M. Speranza, F. J. Ruiz-Dueñas, P. Ferreira, S. Camarero, F. Guillen, M. J. Martinez, A. Gutierrez, and J. C. del Rio. 2005. Biodegradation of lignocellulosics: microbial, chemical, and enzymatic aspects of the fungal attack of lignin. *Int. Microbiol.* **8:**195–204.

60. Matsuoka, S., H. Yukawa, M. Inui, and R. H. Doi. 2007. Synergistic interaction of *Clostridium cellulovorans* cellulosomal cellulases and HbpA. *J. Bacteriol.* **189:**7190–7194.

61. Mejia-Castillo, T., M. E. Hidalgo-Lara, L. G. Brieba, and J. Ortega-Lopez. 2008. Purification, characterization and modular organization of a cellulose-binding protein, CBP105, a processive beta-1,4-endoglucanase from *Cellulomonas flavigena*. *Biotechnol. Lett.* **30:**681–687.

62. Moreira, L., and E. Filho. 2008. An overview of mannan structure and mannan-degrading enzyme systems. *Appl. Microbiol. Biotechnol.* **79:**165–178.

63. Niture, S. K. 2008. Comparative biochemical and structural characterizations of fungal polygalacturonases. *Biologia* (Bratislava) **63:**1–19.

64. Numan, M. T., and N. B. Bhosle. 2006. α-L-Arabinofuranosidases: the potential applications in biotechnology. *J. Ind. Microbiol. Biotechnol.* **33:**247–260.

65. O'Dwyer, J. P., L. Zhu, C. B. Granda, and M. T. Holtzapple. 2007. Enzymatic hydrolysis of lime-pretreated corn stover and investigation of the HCH-1 Model: inhibition pattern, degree of inhibition, validity of simplified HCH-1 Model. *Bioresour. Technol.* **98:**2969–2977.

66. Ogura, J., A. Toyoda, T. Kurosawa, A. L. Chong, S. Chohnan, and T. Masaki. 2006. Purification, characterization, and gene analysis of cellulase (Cel8A) from *Lysobacter* sp IB-9374. *Biosci. Biotechnol. Biochem.* **70:**2420–2428.

67. Pan, X. J. 2008. Role of functional groups in lignin inhibition of enzymatic hydrolysis of cellulose to glucose. *J. Biobased Mater. Bioenergy* **2:**25–32.

68. Parsiegla, G., C. Reverbel, C. Tardif, H. Driguez, and R. Haser. 2008. Structures of mutants of cellulase Ce148F of *Clostridium cellulolyticum* in complex with long hemithiocellooligosaccharides give rise to a new view of the substrate pathway during processive action. *J. Mol. Biol.* **375:**499–510.

69. Pastor, F. I. J., O. Gallardo, J. Sanz-Aparicio, and P. Diaz. 2007. Xylanases: molecular properties and applications, p. 65–82. *In* J. Polaina and A. P. MacCabe (ed.), *Industrial Enzymes*. Springer, Dordrecht, The Netherlands.

70. Pauly, M., L. N. Andersen, S. Kauppinen, L. V. Kofod, W. S. York, P. Albersheim, and A. Darvill. 1999. A xyloglucan-specific endo-β-1,4-glucanase from *Aspergillus aculeatus*: expression cloning in yeast, purification and characterization of the recombinant enzyme. *Glycobiology* **9:**93–100.

71. Payasi, A., R. Sanwal, and G. G. Sanwal. 2009. Microbial pectate lyases: characterization and enzymological properties. *World J. Microbiol. Biotechnol.* **25:**1–4.

72. Polizeli, M. L. T. M., A. C. S. Rizzatti, R. Monti, H. F. Terenzi, J. A. Jorge, and D. S. Amorim. 2005. Xylanases from fungi: properties and industrial applications. *Appl. Microbiol. Biotechnol.* **67:**577–591.

73. Prior, B. A., and D. F. Day. 2008. Hydrolysis of ammonia-pretreated sugar cane bagasse with cellulase, beta-glucosidase, and hemicellulase preparations. *Appl. Biochem. Biotechnol.* **146:**151–164.

74. Rabinovich, M. L., M. S. Melnik, and A. V. Bolobova. 2002. Microbial cellulases (review). *J. Appl. Biochem. Microbiol.* **38:**305–322.

75. Raweesri, P., P. Riangrungrojana, and P. Pinphanichakarn. 2008. alpha-L-Arabinofuranosidase from *Streptomyces* sp PC22: purification, characterization and its synergistic

action with xylanolytic enzymes in the degradation of xylan and agricultural residues. *Bioresour. Technol.* **99:**8981–8986.

76. **Rodriguez-Sanoja, R., N. Oviedo, and S. Sanchez.** 2005. Microbial starch-binding domain. *Curr. Opin. Microbiol.* **8:**260–267.

77. **Rose, J. K. C., J. Braam, S. C. Fry, and K. Nishitani.** 2002. The XTH family of enzymes involved in xyloglucan endotransglucosylation and endohydrolysis: current perspectives and a new unifying nomenclature. *Plant Cell Physiol.* **43:**1421–1435.

78. **Rosgaard, L., S. Pedersen, J. R. Cherry, P. Harris, and A. S. Meyer.** 2006. Efficiency of new fungal cellulase systems in boosting enzymatic degradation of barley straw lignocellulose. *Biotechnol. Prog.* **22:**493–498.

79. **Schagerlöf, U., H. Schagerlöf, D. Momcilovic, G. Brinkmalm, and F. Tjerneld.** 2007. Endoglucanase sensitivity for substituents in methyl cellulose hydrolysis studied using MALDI-TOFMS for oligosaccharide analysis and structural analysis of enzyme active sites. *Biomacromolecules* **8:**2358–2365.

80. **Seidle, H. F., and R. E. Huber.** 2005. Transglucosidic reactions of the *Aspergillus niger* family 3 beta-glucosidase: qualitative and quantitative analyses and evidence that the transglucosidic rate is independent of pH. *Arch. Biochem. Biophys.* **436:**254–264.

81. **Selig, M. J., E. P. Knoshaug, W. S. Adney, M. E. Himmel, and S. R. Decker.** 2008. Synergistic enhancement of cellobiohydrolase performance on pretreated corn stover by addition of xylanase and esterase activities. *Bioresour. Technol.* **99:**4997–5005.

82. **Shallom, D., and Y. Shoham.** 2003. Microbial hemicellulases. *Curr. Opin. Microbiol.* **6:**219–228.

83. **Shimokawa, T., H. Shibuya, M. Nojiri, S. Yoshida, and M. Ishihara.** 2008. Purification, molecular cloning, and enzymatic properties of a family 12 endoglucanase (EG-II) from *Fomitopsis palustris*: role of EG-II in larch holocellulose hydrolysis. *Appl. Environ. Microbiol.* **74:**5857–5861.

84. **Shoseyov, O., Z. Shani, and I. Levy.** 2006. Carbohydrate binding modules: biochemical properties and novel applications. *Microbiol. Mol. Biol. Rev.* **70:**283–295.

85. **Sorensen, H. R., C. T. Jorgensen, C. H. Hansen, C. I. Jorgensen, S. Pedersen, and A. S. Meyer.** 2006. A novel GH43 alpha-L-arabinofuranosidase from *Humicola insolens*: mode of action and synergy with GH51 alpha-L-arabinofuranosidases on wheat arabinoxylan. *Appl. Microbiol. Biotechnol.* **73:**850–861.

86. **Sorensen, H. R., S. Pedersen, C. T. Jorgensen, and A. S. Meyer.** 2007. Enzymatic hydrolysis of wheat arabinoxylan by a recombinant "minimal" enzyme cocktail containing beta-xylosidase and novel endo-1,4-beta-xylanase and alpha-(L)-arabinofuranosidase activities. *Biotechnol. Prog.* **23:**100–107.

87. **Spanikova, S., and P. Biely.** 2006. Glucuronoyl esterase—novel carbohydrate esterase produced by *Schizophyllum commune*. *FEBS Lett.* **580:**4597–4601.

88. **Sunna, A., M. D. Gibbs, and P. L. Bergquist.** 2000. The thermostabilizing domain, XynA, of *Caldibacillus cellulovorans* xylanase is a xylan binding domain. *Biochem. J.* **346:**583–586.

89. **Tabka, M. G., I. Herpoel-Gimbert, F. Monod, M. Asther, and J. C. Sigoillot.** 2006. Enzymatic saccharification of wheat straw for bioethanol production by a combined cellulase xylanase and feruloyl esterase treatment. *Enzyme Microb. Technol.* **39:**897–902.

90. **Tewari, R., R. P. Tewari, and G. S. Hoondal.** 2005. Microbial pectinases, p. 191–208. *In* J. L. Barredo (ed.), *Microbial Enzymes and Biotransformations.* Humana Press, Totowa, NJ.

91. **Topakas, E., C. Vafiadi, and P. Christakopoulos.** 2007. Microbial production, characterization and applications of feruloyl esterases. *Process Biochem.* **42:**497–509.

92. **Ustinov, B. B., A. V. Gusakov, A. I. Antonov, and A. P. Sinitsyn.** 2008. Comparison of properties and mode of

action of six secreted xylanases from *Chrysosporium lucknowense*. *Enzyme Microb. Technol.* **43:**56–65.

93. **Vihinen, M., and P. Mäntsala.** 1989. Microbial amylolytic enzymes. *Crit. Rev. Biochem. Mol. Biol.* **24:**329–418.

94. **Vocadlo, D. J., and G. J. Davies.** 2008. Mechanistic insights into glycosidase chemistry. *Curr. Opin. Chem. Biol.* **12:**539–555.

94a. **Vlasenko, E., M. Schülein, J. Cherry, and F. Xu.** 2010. Substrate specificity of family 5, 6, 7, 9, 12, and 45 endoglucanases. *Bioresour. Technol.* epub before print.

95. **Wang, W., and P. J. Gao.** 2003. Function and mechanism of a low-molecular-weight peptide produced by *Gloeophyllum trabeum* in biodegradation of cellulose. *J. Biotechnol.* **101:**119–130.

96. **Wilson, D. B.** 2008. Three microbial strategies for plant cell wall degradation. *Ann. N. Y. Acad. Sci.* **1125:**289–297.

97. **Wong, D. W. S.** 2006. Feruloyl esterase—a key enzyme in biomass degradation. *Appl. Biochem. Biotechnol.* **133:**87–112.

98. **Xia, W. S., P. Liu, and J. Liu.** 2008. Advance in chitosan hydrolysis by non-specific cellulases. *Bioresour. Technol.* **99:**6751–6762.

99. **Xu, F.** 2004. Enhancing biomass conversion to fermentable sugars: a progress report of a joint government-industrial project, p. 793–804. *In* K. Ohmiya, K. Sakka, S. Karita, T. Kimura, M. Sakka, and Y. Onishi (ed.), *Biotechnology of Lignocellulose Degradation and Biomass Utilization.* Uni Publishers, Tokyo, Japan.

100. **Xu, F., and H. Ding.** 2007. A new kinetic model for heterogeneous (or spatially confined) enzymatic catalysis: contributions from the fractal and jamming (overcrowding) effects. *Appl. Catal. A Gen.* **317:**70–81.

101. **Xu, F., H. Ding, D. Osborn, A. Tejirian, K. Brown, W. Albano, N. Sheehy, and J. Langston.** 2007. Partition of enzymes between the solvent and insoluble substrate during the hydrolysis of lignocellulose by cellulases. *J. Mol. Catal. B. Enzym.* **51:**42–48.

102. **Xu, F., H. Ding, and A. Tejirian.** 2009. Detrimental effect of cellulose oxidation on cellulose hydrolysis by cellulase. *Enzyme Microb. Technol.* **45:**203–209.

103. **Yang, H., H. Ichinose, M. Nakajima, H. Kobayashi, and S. Kaneko.** 2006. Synergy between an alpha-L-arabinofuranosidase from *Aspergillus oryzae* and an endo-arabinanase from *Streptomyces coelicolor* for degradation of arabinan. *Food Sci. Technol. Res.* **12:**43–49.

104. **Yao, Q., T. T. Sun, W. F. Liu, and G. J. Chen.** 2008. Gene cloning and heterologous expression of a novel endoglucanase, swollenin, from *Trichoderma pseudokoningii* S38. *Biosci. Biotechnol. Biochem.* **72:**2799–2805.

105. **Yaoi, K., H. Kondo, A. Hiyoshi, N. Noro, H. Sugimoto, S. Tsuda, Y. Mitsuishi, and K. Miyazaki.** 2007. The structural basis for the exo-mode of action in GH74 oligoxyloglucan reducing end-specific cellobiohydrolase. *J. Mol. Biol.* **370:**53–62.

106. **Ye, X., Y. Wang, R. C. Hopkins, M. W. W. Adams, B. R. Evans, J. R. Mielenz, and Y. H. P. Zhang.** 2009. Spontaneous high-yield production of hydrogen from cellulosic materials and water catalyzed by enzyme cocktails. *ChemSusChem* **2:**149–152.

107. **Yip, V. L. Y., and S. G. Withers.** 2006. Family 4 glycoside hydrolases are special: the first-elimination mechanism amongst glycoside hydrolases. *Biocat. Biotransform.* **24:**167–176.

108. **Yuan, S., Y. Wu, and D. J. Cosgrove.** 2001. A fungal endoglucanase with plant cell wall extension activity. *Plant Physiol.* **127:**324–333.

109. **Zhou, Q., M. J. Baumann, P. S. Piispanen, T. T. Teeri, and H. Brumer.** 2006. Xyloglucan and xyluglucan endo-transglycosylases (XET): tools for *ex vivo* cellulose surface modification. *Biocatal. Biotransform.* **24:**107–120.

The Use of Enzymes for Nonaqueous Organic Transformations

ELTON P. HUDSON, MICHAEL J. LISZKA, AND DOUGLAS S. CLARK

35

The use of enzymes in organic synthesis is becoming an increasingly attractive alternative to traditional chemical routes (39). This chapter focuses on synthetic applications of enzymes in monophasic organic solvents and is intended to illustrate many types of transformations that can be catalyzed by enzymes in organic solvents. Methods to prepare the biocatalyst for optimal activity or stability are also presented, noting that enzymes in organic solvents are the subject of recent monographs and reviews (16, 30). The following section surveys techniques used to prepare and activate enzymes for use in organic solvents.

35.1. PREPARATION OF ENZYMES FOR NONAQUEOUS BIOCATALYSIS

Nonaqueous enzyme systems can be divided into two classes: homogeneous systems in which enzymes are modified to be soluble, and heterogeneous systems in which the catalyst is in an insoluble form. The bulk of the reaction mixture can be a liquid, gas, or supercritical fluid. Heuristics for choosing an appropriate preparation have been developed based on the combination of enzyme and solvent (12, 66). Several types of enzyme preparations are illustrated in Fig. 1.

Although many enzymes remain active in a variety of low-water organic systems, when used as unmodified and insoluble powders (commonly referred to as "as received") they usually exhibit activities that are orders of magnitude lower than in aqueous solutions. The low activity in organic solvents has been attributed to a number of factors, including a decrease in the enzyme's active-site polarity, a loss of conformational mobility, and structural changes during freeze-drying of the biocatalyst. Fortunately, there are well-documented methods to activate enzymes relative to their as-received form and frequently these methodologies are simple to implement. However, in some cases the activation technique can alter an enzyme's reactive properties. A notable example of this comes from Altreuter et al. (5), who found that two preparations of subtilisin, lyophilized with nonbuffer salts and solubilized with a surfactant, exhibited different regioselectivities toward reactive hydroxyl groups of the chemotherapeutic doxorubicin (Fig. 2). This unexpected regioselectivity was recently exploited in generating a library of new doxorubicin analogs (26).

35.1.1. Heterogeneous Preparations

Enzymes are routinely obtained commercially as purified powders or immobilized on an inert support. Solid catalyst preparations have the advantage of easy recoverability for reuse and are often more stable than their soluble counterparts. In the case of enzyme powders, a contributing factor to the low reaction rates observed in organic solvents can be that the powder has been formulated from an aqueous solution at a pH different than the enzyme's optimal pH. Therefore, a typical first step in using enzyme powders is resuspension and freeze-drying (lyophilization) from an aqueous buffer at the optimal pH (144). Further adjustment of the reaction rate can be made through water control, or by doping in small amounts of typically denaturing organic solvents (4). Diffusional limitations may also be an important consideration, although, in at least one case, they were ruled out as a factor affecting overall reaction rates (8). However, the adjustments due to water and pH are often overshadowed by the activation afforded by many of the preparation methods described below.

35.1.1.1. Immobilization

A common form of biocatalyst for use in organic solvents is immobilized on a solid support, either through covalent attachment or adsorption. Most often the enzyme is attached to the support in aqueous solution, and a wealth of literature exists on attachment chemistries and adsorption conditions (for an early but thorough review, see reference 69). Common supports to which enzymes can be covalently attached (usually through amino groups) include activated resins, coated glass, or zeolites. In some cases covalent attachment of enzymes to solid supports can increase both the thermal and solvent stability, and immobilized enzymes often have longer half-lives than their suspended counterparts (17, 19). For example, cutinase immobilized onto an NaY (Faujasite) zeolite showed very little loss in activity after incubation in isooctane for 45 days (42). Among many other examples, subtilisin and thermolysin exhibited increased thermal and solvent stability over aqueous and commercially available immobilized forms when covalently attached to an acrylic support through a polyethylene linker (141).

Popular supports for noncovalent immobilization in organic solvents include porous glass beads and fumed silica, which can be impregnated through saturation with an aqueous enzyme solution and subsequent washings to

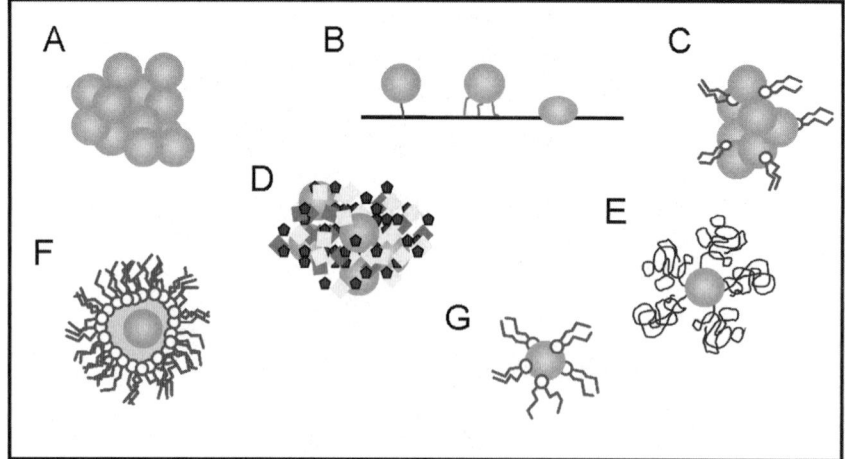

FIGURE 1 Enzyme preparations used in organic solvents. (A) Lyophilized powder. (B) Immobilized enzymes (left to right): single-point covalent attachment, multipoint covalent attachment, and physisorption. (C) Directly solubilized as a cluster via a surfactant. (D) Colyophilized powder containing an excipient. (E) Covalently modified with PEG. (F) Reverse micelle encapsulation with retained water. (G) Surfactant-paired, extracted single-enzyme molecules.

remove nonadsorbed enzyme. Other commercially available resins can serve as ion exchangers for enzymes, in which adsorption is due to strong ionic interactions. Upon enzyme deactivation, these catalysts can be cleansed with solutions of high ionic strength and regenerated. When choosing an immobilization support, it is important to note that the ionization state of both the enzyme *and* the support contribute to the activity of the catalyst (145). In addition, the scale of the material's morphology relative to the enzyme's size may also influence activity. For example, Takahashi et al. (125) investigated the effect of pore size on the activity of horseradish peroxidase adsorbed onto mesoporous silica and found that the highest activity was achieved in materials with a pore size slightly larger than the enzyme diameter. Other nanoscale materials, such as single-walled carbon nanotubes or nanoparticles, have been shown to impart increased thermal and solvent stability (7, 60).

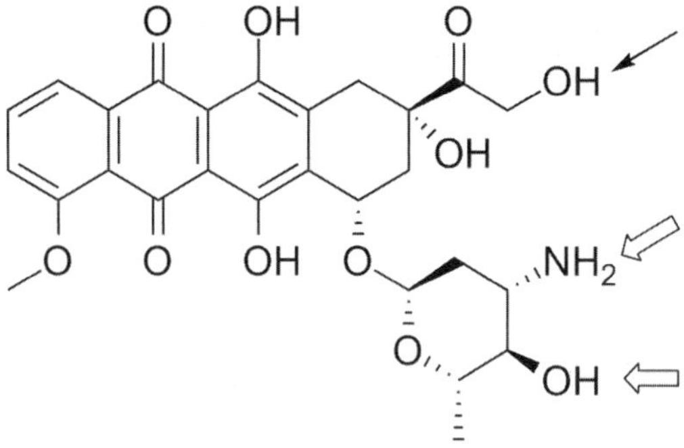

FIGURE 2 Doxorubicin derivatives generated with solubilized and salt-activated lipase and subtilisin preparations. All preparations could acylate the 14-O hydroxyl of doxorubicin (black arrow) in toluene, but only salt-activated subtilisin Carlsberg could modify the amino and hydroxyl groups indicated by the white arrows (5).

35.1.1.2. Lyophilization with Excipients

Colyophilisates for use with enzymes in organic solvents can be categorized according to three primary acting mechanisms: activating salts, molecular imprinting agents/molds, and lyoprotectants. Each of these groups affects the activity and structure of the enzyme in different ways, and the entangled effects can be difficult to separate.

Salt activation has yielded some of the largest improvements in enzyme activity in organic solvents, up to a 35,000-fold increase (78). Serdakowski and Dordick (115) recently reviewed colyophilized enzyme preparations, with a focus on salt activation. The current hypothesis is that chao- and kosmotropic ions order water differently around the enzyme. Both the activity of salt-lyophilized enzymes and the amount of bound water were reported to increase with the kosmotropicity of the salt (110). Deuterium nuclear magnetic resonance studies on the dynamics of water in salt-activated formulations revealed that activity also increases as the mobility of the bound water increases (38).

Recognizing that enzymes are more static in organic solvents, colyophilizing enzymes with ligands or transition state analogs can activate them by "freezing" the enzyme into an active conformation (70). Such imprinted catalysts are created by adding a small amount of ligand to an aqueous enzyme solution, freeze-drying, and washing the resulting powder with an anhydrous solvent to remove the bound ligand (111). Molecular imprinting has more recently found use in nonbiological catalysts, where functional groups are incorporated into a polymer and the system is cured/cross-linked with a ligand present (76).

Other molecules, such as crown ethers (132), sugars, and polymers, are collectively referred to as lyoprotectants (28). Associated with the work by Hofmeister on salt stabilization, these molecules are believed to prevent the enzyme from deactivating during the freeze-drying process by increasing the free energy of transfer of hydrophobic residues.

35.1.1.3. Cross-Linked Enzyme Crystals/Aggregates (CLECs/CLEAs)

Cross-linking of enzymes was originally used for diffraction measurements because cross-linked crystals were found to have similar cell dimensions but improved mechanical stability compared with non-cross-linked protein crystals (104). After an enzyme is crystallized, the crystal is treated with a difunctional cross-linking agent, typically gluteraldehyde. Cross-linked enzyme crystals, or CLECS, have since gained favor as biocatalysts for nonaqueous reactions. Cross-linked subtilisin was shown to be more stable against inactivation in organic solvents; in octane and acetonitrile the half-life of subtilisin was extended 40- and 90-fold, respectively, relative to suspended powders.

Whereas cross-linked crystals exhibit increased stability, their preparation is time-consuming and requires highly pure enzymes. To remedy this, Sheldon et al. (117) developed a procedure where enzymes are instead precipitated first from solution and then cross-linked. Any common precipitant, such as sulfate salts or polymers, may be used providing the enzyme retains an active conformation prior to cross-linking.

35.1.2. Homogeneous Preparations

35.1.2.1. Covalent Modification

Solubilizing enzymes can increase their specific activity over suspended powders (136), but reduces the possibility of recovery and reuse. To this end, a wide array of molecules can be coupled to the surface groups displayed on a protein. Amines, carboxylic acids, and thiols are the most readily available for modification, with amines being the most utilized. Amphiphiles or polyelectrolytes can be attached to provide a variety of functionality, including increased solubility, selective partitioning, or ionic adsorption for immobilization.

Inada et al. (58) introduced a method of creating highly soluble enzymes by polyethylene glycol (PEG) modification, and covalent attachment of PEG has become a common method for solubilizing enzymes in organic solvents. PEG is first functionalized for specific chemistry or multiple attachments, and then coupled to an exposed protein residue. An extensive review of the possible attachment chemistries is provided by Veronese (133). Most reactions can be performed under mild aqueous conditions, and the modified enzyme is purified by repeated dialysis. Purified PEGylated proteins are then lyophilized and dissolved in the solvent of choice. The degree of modification can be adjusted by changing the length of the polymer or the molar ratio of protein to activated PEG. Lipase and chymotrypsin have both shown improvement in activity in a range of solvents because of PEG modification (18, 58). Castillo et al. (20) studied the effect of PEG-modified subtilisin in dioxane, and found that, as the amount of PEG attached to the enzyme increased, both the dynamics (measured by H/D exchange) and the activity increased. When the number of modifiable functional groups is either too high or too low, mutagenesis can be used to remove or add suitable residues.

35.1.2.2. Noncovalent Modification

Surfactants can be used to extract enzymes into organic solvents with highly varying amounts of water and thus highly varying properties. Reverse micelles can segregate a near-native enzyme from the solvent and contain "pools" of water with bulk-like properties. At the other extreme, reverse micelles can be dehydrated to the point of including only structured or enzyme-bound water molecules. At very low surfactant concentrations, the enzyme is directly exposed to the solvent and even less water may be present.

Reverse micelles can form stable microemulsions containing water in millimolar or lower concentrations. A reverse-micelle system is prepared by liquid-liquid extraction of an aqueous enzyme solution with an organic phase containing surfactant. This can typically be accomplished by vigorously mixing the two phases together. In these systems, the amount of water included with the enzyme in the organic phase can be modified by direct addition of water or through techniques discussed in the next section.

The use of surfactants at amounts below the critical micelle concentration solubilizes enzymes without forming reverse micelles. By modulating the surfactant and solvent properties, nearly all of the enzyme in the aqueous phase can be extracted to the organic phase. Paradkar et al. (97) used low concentrations of dioctyl sodium sulfosuccinate (AOT) (2 mM) to extract 93% of aqueous chymotrypsin into isooctane. Hobbs et al. (52) paired a fluorinated surfactant with a fluorinated solvent to extract greater than 99% of aqueous cytochrome c into the organic phase. Lipase PS (Amano Pharm. Ltd.) paired with a nonionic surfactant provided the greatest activation versus either cationic or anionic surfactants (95).

So-called "direct solubilization" of enzyme suspensions was shown to yield highly active, soluble aggregates. In this system, the enzyme powder was added directly to a solvent containing a small amount of surfactant and water. Directly solubilized subtilisin was ca. 70 times more active in tetrahydrofuran (THF) than either the extracted or suspended enzyme system. A similar trend was found for isooctane with a 25-fold increase over the extracted enzyme (3).

35.1.3. Other Considerations

35.1.3.1. Water Control

One of the simplest methods for manipulating reaction rates of enzymes in organic solvents is to control the amount of water present in the system. Water content can be controlled through established techniques of using salt hydrates or saturated salt solutions (72, 142). The water content of a nonpolar organic solvent can be adjusted by mixing the solvent with a salt hydrate. The other components of the system, as well as polar organic solvents, can be equilibrated to a desired water content by exposing them indirectly to a saturated salt solution. In situ control can be achieved by including salt hydrates directly in the reaction mixture to act as a source or sink of water. Care is required when selecting which salt pair to use, because both the cation and anion have been shown to affect the reaction independently (142).

35.1.3.2. pH

Early studies revealed that the pH of the aqueous solution during extraction or lyophilization can have a dramatic effect on the enzyme activity in the organic phase. This "pH memory" usually leads to the same bell-shaped dependence of the activity on transfer pH in the organic phase as in the aqueous phase. An interesting exception to this trend was reported for liver alcohol dehydrogenase, where the pH optimum in heptane was 2.0, far from the aqueous optimum of 7.5 (47). Buffer salts can be used as solid-state pH buffers for control of protonation states of the enzyme in organic solvents, similar to in aqueous solutions. Partridge et al. (98) showed that zwitterionic salts can be used to actively control the protonation state of the protein, and to remove acidic or basic by-products that can inhibit the reaction. Zwitterionic salts were preferred because they have a lower tendency to dissolve (by remaining charged), and they are able to act as both an acid and a base.

35.1.3.3. Solvent Effects

Thermal stability, catalytic activity, and selectivity in many organic-phase enzyme systems are strongly affected by the solvent, and the literature on solvent effects is large and expanding (54, 115). Numerous recent experimental and

molecular simulation studies demonstrate negative solvent effects on enzyme properties thought to be critical to catalysis: loss of protein secondary structure (18, 45, 140), a decrease in active-site polarity (116), and reduced protein dynamics (37, 53, 84). The confluence of these effects can cause enzyme half-life (136), enantioselectivity (40), and catalytic activity (114) to vary by orders of magnitude between solvents.

Many studies have attempted to relate selectivity and catalytic activity to simple solvent properties such as hydrophobicity, polarizability, and size (see references 9

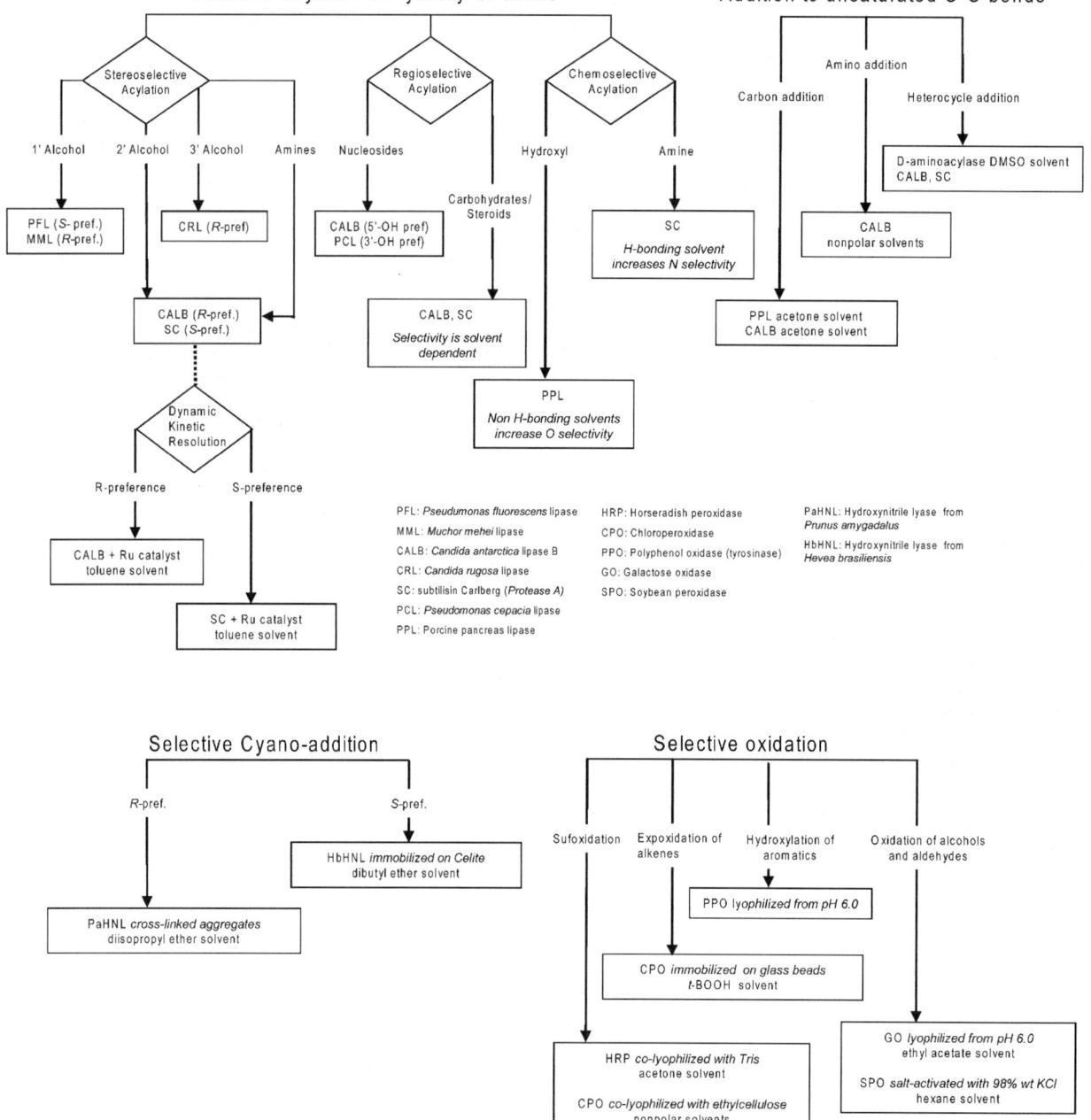

FIGURE 3 Recommended enzymes for catalyzing various reactions in organic solvents. The recommended systems are those shown to give the best performance for a certain reaction (high enzyme stability, reactivity, and selectivity) based on the literature cited in this chapter.

and 14 for recent reviews). Some heuristics are available, which are demonstrated mostly with proteases and lipases. In general, these enzymes are more thermally stable, enantioselective, and catalytically active in nonpolar solvents such as hexane than in polar solvents such as THF or acetone (73). The low dielectric environment provided by nonpolar solvents restricts enzyme conformational dynamics, preventing unfolding (143) and locking the active site into a specific conformation (106). Several recent reports present inverse correlations between protein flexibility and stability (96, 109) as well as flexibility and enantioselectivity (20). In polar solvents, enzyme-bound water molecules are stripped from the protein surface (43), and the resulting solvent-protein interactions alter enzyme structure (53, 136), reducing stability and catalytic activity relative to nonpolar solvents. As mentioned previously, the addition of water can alleviate these effects (2, 114), and can drastically increase enzyme activity, in particular, in polar solvents (101). The previously mentioned activation techniques, such as covalent modification or salt activation, can increase enzyme activity and allow the choice of solvent to be dictated more by substrate solubility than enzyme behavior.

35.2. SYNTHETIC APPLICATIONS OF ENZYMES IN NONAQUEOUS SOLVENTS

This section summarizes many applications of enzymes in organic solvents, with focus placed on three classes of enzymes that have found a wide measure of use in organic solvents: hydrolases (EC 3), lyases (EC 4), and oxidoreductases (EC 1). Seminal examples illustrate the types of transformations that enzymes can catalyze in organic solvents, and when an enzyme may be preferred over a traditional chemical method. Often the particular desired reaction is identified and several possible enzymes are tested for the proper reactivity (see Fig. 3). After finding an enzyme that catalyzes the desired reaction, its reactivity can be increased (or diversified) using the preparation techniques discussed in section 35.1.

35.2.1. Hydrolases

Hydrolase enzymes (EC 3) comprise the predominant biocatalysts for transformations reported in organic solvents, and two groups, lipases and proteases, enjoy widespread use. Hydrolases have traditionally been used to catalyze selective acylations of alcohols and amines using a variety of acyl donors and solvents. Their selectivities have been exploited in kinetic resolution of racemic mixtures (15), combinatorial lead optimization (90), and organic-phase peptide synthesis (80). It has been shown that wild-type proteases and lipases retain their ping-pong mechanisms in organic solvents, and in some cases individual rate constants have been determined (21, 55, 62).

Many hydrolases are available commercially as freeze-dried powders or as specialized preparations, such as immobilized on acrylic resin (Novozyme 435), polypropylene (Accurel), glass beads, or cellulose; or as CLEAs or CLECs. These preparations can be very robust, exhibiting half-lives up to hundreds of hours in a range of organic solvents, even when used as received. The most common lipases used in organic solvents are from *Candida antarctica* (CALB), porcine pancreas (PPL), *Pseudomonas* spp. (PCL and PFL), and *Muchor* spp. (MML and MJL). Commonly used proteases in organic solvents are subtilisins from *Bacillus* spp. (subtilisin Carlsberg and subtilisin BPN′), penicillin

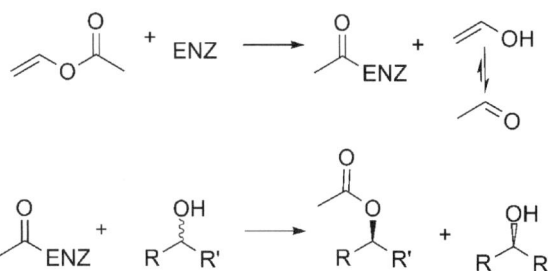

FIGURE 4 A simple scheme for the hydrolase-catalyzed resolution of an alcohol. Vinyl acetate is a common acyl donor in alcohol resolutions. Upon enzyme acylation by vinyl acetate, vinyl alcohol is released, which rapidly tautomerizes to the nonnucleophilic acetaldehyde. The acyl-enzyme intermediate is then attacked preferentially by the (R)- or (S)-alcohol enantiomer.

amidase from *Escherichia coli*, and α-chymotrypsin from bovine pancreas. Appreciable conversions are often achieved for model reactions in a few hours by use of biocatalyst concentrations of 5 to 10 mg/ml, although this heuristic will depend on the specific activity of the enzyme preparation and the solvent used.

35.2.1.1. Enantioselective Resolution of Racemic Alcohols and Amines

In kinetic resolutions the hydrolase enzyme reacts preferentially with one enantiomer, (R)- or (S)-, of a racemic substrate mixture. Figure 4 shows a typical reaction scheme. The enzyme is first acylated by an acyl donor (vinyl acetate is common because of its nonnucleophilic product, acetaldehyde [134]). The acyl-enzyme complex is then attacked preferentially by the (R)- or (S)-alcohol or amine enantiomer. The acylated product can be separated from the solution via chromatography or extraction, enabling resolution of polyfunctionalized alcohols or amines on large scales. For example, Feng and coworkers recently demonstrated the large-scale CALB-catalyzed resolution of several racemic alkyl alcohols with vinyl acetate and succinic anhydride, producing 2.0 g of (S)-2-pentanol at 99% enantiomeric excess (ee %) (135).

An enzyme's enantioselectivity, E, is commonly expressed as the ratio of k_{cat}/K_m for one enantiomer over the other (equation 1). Enantioselectivity is related to the product's enantiomeric excess ee and substrate conversion c (equations 2 and 3), where R and S are the concentrations of (R)- and (S)-forms of the product at a given substrate conversion c. For example, if E = 60, the $ee_{product}$ = 98% at 30% substrate conversion. Enantioselectivities for hydrolases in organic solvent range from E = 1 for a nonselective process to E = 400, but typically fall in the range of 30 to 200. For one-step resolutions, E = 100 is a typical benchmark.

$$E = \frac{\left(\frac{k_{cat}}{K_m}\right)_R}{\left(\frac{k_{cat}}{K_m}\right)_S} \tag{1}$$

$$ee_R\ (\%) = \left(\frac{R-S}{R+S}\right) * 100 \tag{2}$$

$$E = \frac{\ln\left[1 - c(1 + ee_{product})\right]}{\ln\left[1 - c(1 - ee_{product})\right]} \tag{3}$$

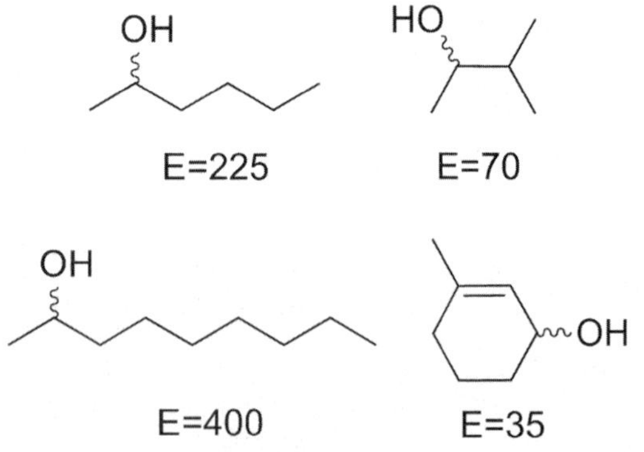

FIGURE 5 These *rac*-alcohols were resolved enantioselectively [(*R*)-preference] by acylation with vinyl acetate in diisopropyl ether by three lipases. *E* values are for CALB. Adapted from reference 135.

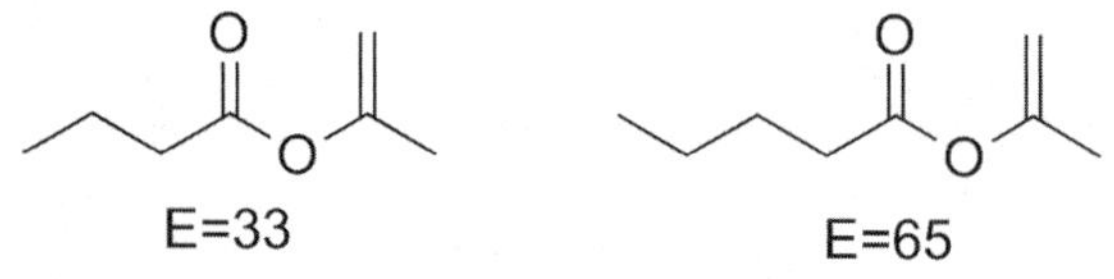

FIGURE 6 Increasing the chain length of acyl donor increased *E* from 33 to 65 in the subtilisin Carlsberg-catalyzed acylation of 1-phenyl ethanol in THF. Adapted from reference 10.

35.2.1.1.1. Alcohols

Hydrolases display the greatest enantioselectivity for secondary alcohols (often $E > 100$). Resolutions of primary alcohols typically exhibit lower *E* values (often $E < 10$). Resolution of tertiary alcohols is rare. Figure 5 presents a sampling of secondary alcohols that can be resolved with high selectivity (135). In general, increasing the chain length of the acyl donor (Fig. 6) (10) and addition of bulky substituents to the acyl acceptor (Fig. 7) (88) can increase lipase enantioselectivities by exacerbating discriminatory steric constraints.

Molecular modeling studies of enzyme-substrate interactions have been used to rationalize observed enantioselectivities for certain racemic substrates (61, 77, 102), as well as regioselectivities for acylations of sugars (105) and nucleosides (74). Molecular modeling can help predict which enzyme will display the highest *E* for a given transformation and can also provide guidance into the rational design of enzymes with new reactivities. In one striking example, CALB was rationally redesigned to exhibit inverted enantioselectivity in a variety of organic solvents (81).

For resolution of secondary alcohols it is known that most lipases react preferentially with the (*R*)-enantiomer [(*R*)-preference] (64), while proteases such as subtilisin BPN' and Carlsberg exhibit (*S*)-preference (10) (Fig. 8). This trend is not applicable to the resolution of primary alcohols (126); for example, the lipase PCL shows opposite preference (*S*-) when acting on primary alcohols (128).

Enantioselectivities of hydrolases toward primary alcohols are typically lower than those observed for secondary alcohols. This is particularly true for simple methyl alkanols, where enantioselectivities were low ($2 < E < 10$) for the lipases PFL and PCL in chloroform (93). PFL- and PCL-catalyzed resolution of primary alcohols with larger substituents such as phenyl rings exhibited enantioselectivities that are an order of magnitude higher ($E > 100$) (94).

There are presently very few examples of hydrolase-catalyzed resolutions of tertiary alcohols. The inability of lipases to accept tertiary alcohols is due to steric exclusion of the substrate from the enzyme's active site. In a recent study, 25 commercial lipases were screened for the ability to *hydrolyze* several acylated *tert*-alcohols. Only *Candida rugosa* lipase (CRL) and *C. antarctica* lipase A (CALA) were reactive toward the substrates (51). A unique flexible loop in the active site of CRL (the crystal structure of CALA has not been solved) allows the enzyme to accommodate bulkier substrates. Recent work has shown that binding-site mutations can enable esterases and proteases to accept tertiary alcohols, but these studies have been limited to aqueous/organic mixtures (50).

It is also noteworthy that hydrolases can resolve racemic substrates with stereocenters far from the reactive OH moiety, although with less fidelity (93). It may therefore be necessary to use multiple enzymes to obtain satisfactory enantiomeric excess. Figure 9 shows a primary alcohol with a stereocenter three carbon atoms removed from the hydroxyl group. This substrate was resolved with two lipases in multiple steps, yielding high enantiomeric excess (ee > 98%) (113). When each lipase was used individually, ee$_{product}$ was less than 60%.

35.2.1.1.2. Amines

Hydrolases have been used to resolve alkyl and aryl amines in a manner similar to that of alcohols, and these reactions have been reviewed recently (131). The preferential enantioselectivity of lipases and proteases follows the same

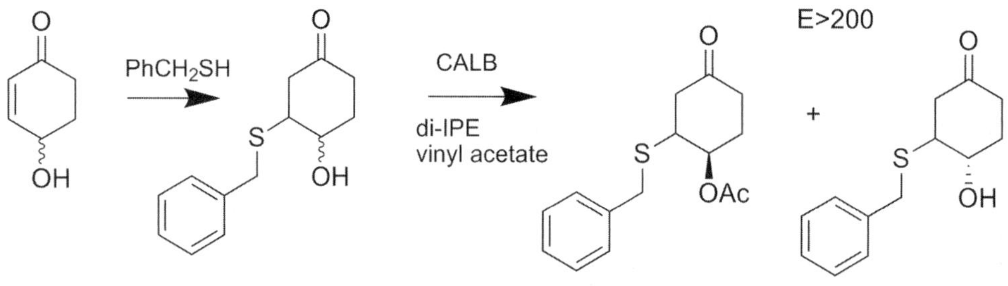

FIGURE 7 To facilitate resolution of *rac*-cyclo-hexenol, the bulky benzylthiol was added. After resolution by CALB in diisopropyl ether (di-IPE), the benzylthiol was removed. Adapted from reference 88.

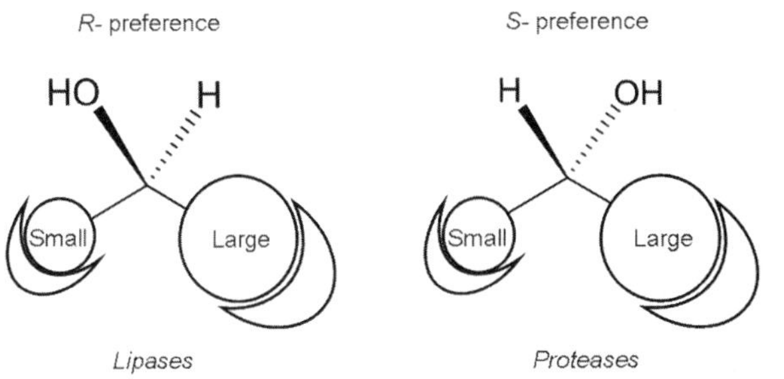

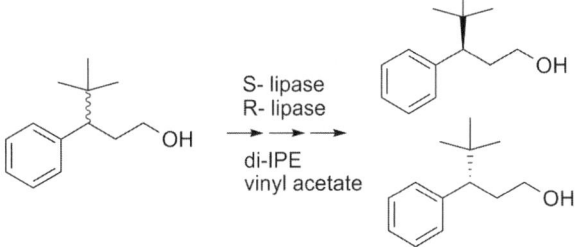

FIGURE 8 Due to different active-site geometries, lipases and proteases show opposite enantiopreference in the resolution of *rac*-secondary alcohols if the substituents differ in size (64).

FIGURE 9 Multistep lipase-catalyzed resolution of a primary alcohol with distant stereocenter. di-IPE, diisopropyl ether. Adapted from reference 113.

trend as with secondary alcohols [most lipases show (R)-preference and most proteases show (S)-preference]. As observed with alcohols, increasing the size of the substituent can substantially increase E (Fig. 10). Amines are more susceptible to noncatalyzed (and nonselective) acylation than alcohols, and this spontaneous reaction occurs faster in nonpolar solvents. Therefore, special care should be taken in choosing the appropriate reaction solvent.

Several commercial hydrolases have been shown to catalyze the formation of peptide bonds in organic solvents as immobilized (86) or soluble (6) biocatalysts. These have been used for small oligomer synthesis and segment condensation. A variety of activating moieties on the acyl donor in the peptide synthesis reaction broaden enzyme specificity and improve coupling yields for amino acids that are typically poor substrates for a given enzyme (85, 87). Enzyme-catalyzed peptide synthesis has been reviewed recently (80).

35.2.1.1.3. Dynamic Kinetic Resolution

In an enantio*specific* resolution of a racemic substrate, only one enantiomer will react, limiting the maximum percentage yield to 50%. Dynamic kinetic resolution can be used to overcome this theoretical limitation. As one substrate enantiomer is selectively modified by the enzyme, it is replenished by the chemically catalyzed (metal catalysts are typically used) racemization of the other substrate enantiomer (Fig. 11). This "one-pot" technique has been optimized through a thorough tuning of conditions, enzymes, and racemization catalysts (33, 82, 100). Toluene affords high racemization rates for ruthenium and palladium catalysts, and high selectivities for

immobilized lipases such as CALB. Dynamic kinetic resolution has been reviewed recently (41).

35.2.1.2. Regio- and Chemoselectivity of Hydrolases

The ability of hydrolases to modify a particular functional group in the presence of more reactive functional groups has been exploited in a variety of transformations on many classes of compounds. This section provides illustrative examples of such selective reactivities for lipases and proteases in neat organic solvents. Similar to the kinetic resolutions described above, most of these reactions involve acyl acceptors that are large and traditionally difficult to modify in a regioselective or chemoselective fashion. By comparison, the acyl donors are usually smaller and are thus able to coexist with the larger acyl acceptor in the enzyme's active site.

The regioselectivity of lipases and proteases in organic solvents is elegantly demonstrated in the acylation patterns of dihydroxy steroids: lipases show preference for A-ring acylations, while proteases such as subtilisin BPN′ show preference for D-ring acylations (27, 108) (Fig. 12A). After acylation of one hydroxyl, the others can be oxidized chemically; subsequent alkaline deacylation thus provides a route to regioselective oxidation of the steroid without an oxidase enzyme. Regioselective acylation and deacylation of steroids have been used recently in the synthesis of several pharmacologically relevant steroid metabolites (119).

Distinct regioselectivity patterns are also observed in hydrolase-catalyzed acylations of a variety of carbohydrates, including polysaccharides (99), carboxymethyl cellulose

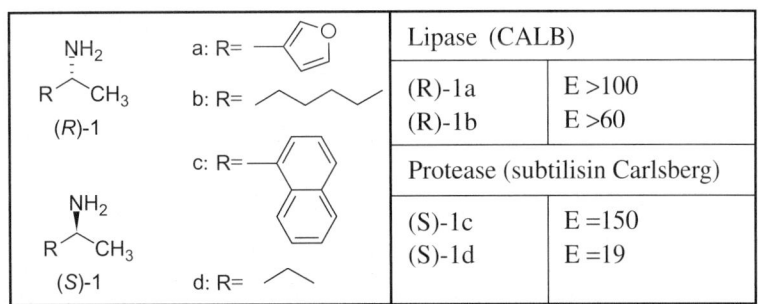

FIGURE 10 Enantioselectivities of suspended CALB and suspended subtilisin Carlsberg against various acyl acceptors in organic solvents. CALB reaction was in neat ethyl acetate, which was also the acyl donor (57). Subtilisin reactions were in 3-methyl-3-pentanol, with trifluoroethyl butyrate as the acyl donor (67).

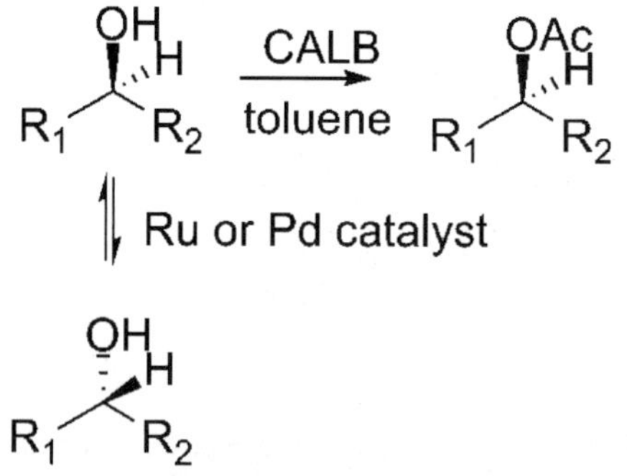

FIGURE 11 A dynamic kinetic resolution scheme. The lipase CALB and ruthenium and palladium catalysts are compatible with resolutions in organic solvents, affording fast racemization and 98% product yield (23).

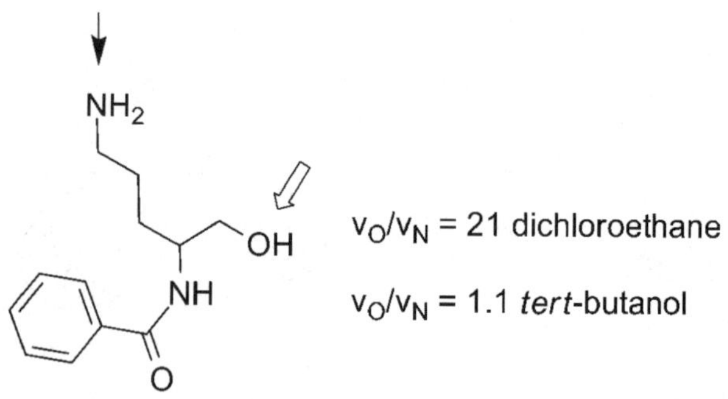

$v_O/v_N = 21$ dichloroethane

$v_O/v_N = 1.1$ *tert*-butanol

FIGURE 13 Loss of chemoselectivity by PCL when the solvent is changed. The enzyme was 20-fold more selective at hydroxyl acylation in dichloroethane than in *t*-butanol (127).

(139), and *N*-acetylhexosamines (120). Hydrolases have been used in the organic synthesis of carbohydrate-peptide conjugates (121) and novel vitamin D and deoxynucleoside derivatives (44).

As with regio- and enantioselectivity, chemoselectivity of hydrolases in organic solvents (e.g., preferred acylation of -OH or -NH$_2$) can be adjusted by solvent selection. In certain cases the solvent can increase the effective nucleophilicity of the OH and NH$_2$ groups by disrupting intramolecular hydrogen bonding (32). In contrast, hydrogen-bonding solvents can preferentially hydrogen bond with a substrate OH group, rendering it less nucleophilic and shifting chemoselectivity toward an NH$_2$ group (Fig. 13).

The ability to use "medium engineering" to bias reactions toward a desired product is a hallmark of nonaqueous biocatalysis. When multiple reaction products are formed,

it is possible to change the product distribution by performing the reaction in a solvent more amenable to the formation of one product over the other. For example, Castillo and coworkers recently demonstrated the ability to adjust the equilibrium distribution of CALB-catalyzed products by changing the polarity of the solvent (103). Nonpolar solvents such as hexane gave low yields of the aminolysis product, due to its limited solubility, and therefore allowed for the accumulation of the more slowly formed Michael addition product (Fig. 14).

These examples of unexpected selectivity raise the possibility of discovering new bioactive compounds by screening various solvents and enzymes against natural products. For example, PPL was used in *tert*-amyl alcohol to selectively diversify hydroxyls on the carboxylic acid and amine components of the four-component Ugi reaction, resulting in a new nine-member

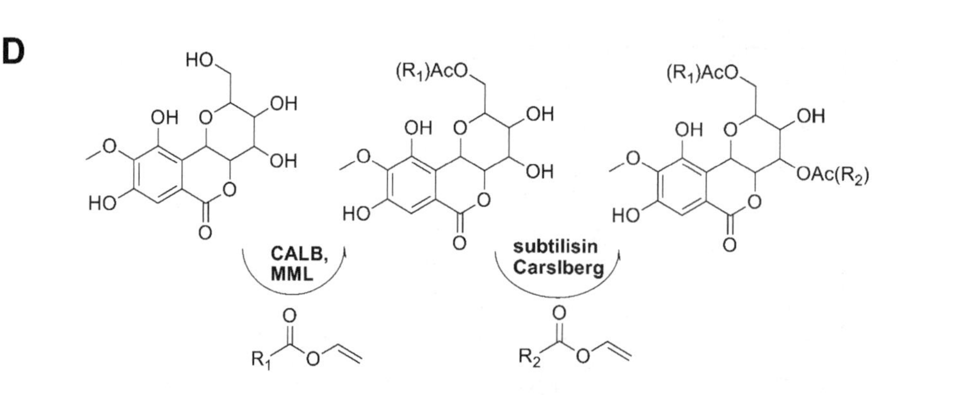

FIGURE 12 Regioselective acylations or alcoholysis of steroids, flavonoids, and nucleosides by hydrolases in organic solvents. (A) A model steroid (108). (B) The flavonoid morin (29). (C) Adenosine (44, 107). (Note: No amino acylations are observed.) (D) The antioxidant bergenin (90).

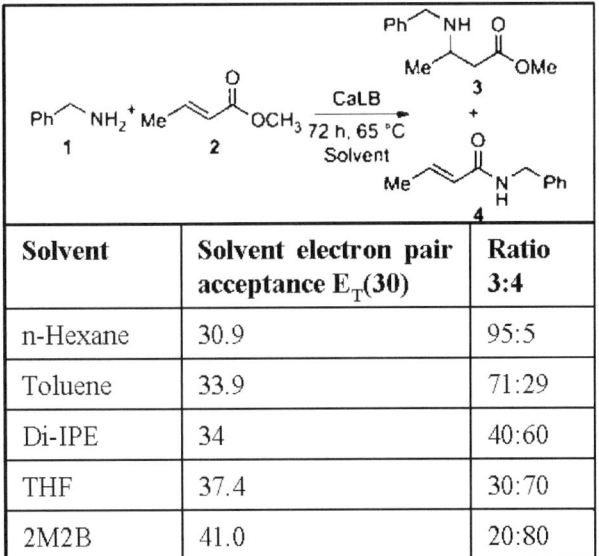

Solvent	Solvent electron pair acceptance $E_T(30)$	Ratio 3:4
n-Hexane	30.9	95:5
Toluene	33.9	71:29
Di-IPE	34	40:60
THF	37.4	30:70
2M2B	41.0	20:80

FIGURE 14 Medium engineering to favor CALB-catalyzed Michael addition (3) over aminolysis (4) in various anhydrous organic solvents. Adapted from reference 103.

Ugi library (79). In addition, new analogs of the flavonoid morin were generated by regioselective modification with several lipases in THF (29, 92) (Fig. 12B), and over 50 commercial hydrolases were screened as biocatalysts en route to generating over 160 analogs of the antioxidant berginin (90) (Fig. 12D). A similar screening strategy was used in preparing taxol derivatives, whereby the zinc protease thermolysin and lipase were key enzymes involved in generating taxol derivatives that were up to 1,000 times more soluble in water (65).

35.2.1.3. Other Hydrolase Reactions

The most common reactions for hydrolases in organic solvents are the alcoholysis and aminolysis reactions described above; however, other reactions have been reported, including Michael additions (PPL and subtilisin Carlsberg [13]), Markovnikov additions (D- and L-amino acylases [137]), and aldol additions (PPL [75]). Two examples are shown in Fig. 15. The very low rates of these promiscuous reactions (it is not uncommon to let the reaction run for several days) have provided impetus for protein engineering of hydrolases

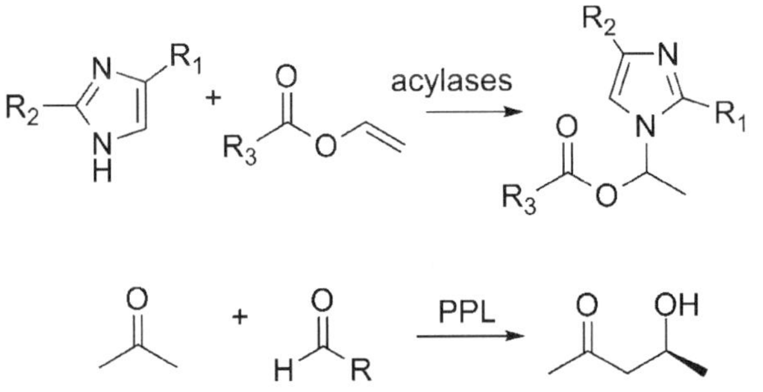

with higher activities for new reactions. Several studies (see reference 56 for a recent review) have used a mutated variant of CALB, in which the catalytic serine is changed to alanine. The mutant's catalytic dyad can activate a number of different nucleophiles, allowing for reactions that cannot be catalyzed with the wild-type enzyme, such as H_2O_2 epoxidations of unsaturated aldehydes (123), carbon-carbon bond formation via Michael addition of unsaturated carbonyl compounds to diketones (124), and aldol addition of aldehydes (11).

35.2.2. Hydroxynitrile Lyases

From the lyase class of enzymes (EC 4), hydroxynitrile lyases (HNLs) have been used for synthetic applications in organic solvents. Under the proper conditions HNLs can be used to asymmetrically add HCN to aldehydes (Fig. 16) and show high selectivities ($E > 100$) for a range of branched and polycyclic substrates (118). Common (R)-selective HNLs are from *Prunus amygdalus* (PaHNL) and *Linum usitatissimum* (LuHNL); common (S)-selective HNLs are from *Hevea brasiliensis* (HbHNL), *Manihot esculenta* (MeHNL), and *Sorghum bicolor* (SbHNL). The (S)-HNLs typically exhibit lower reaction rates (36). Chiral cyanohydrins produced by HNLs in organic solvents have been used as building blocks in the synthesis of various of bioactive compounds, such as epinephrine derivatives (34).

The use of HNLs in organic/aqueous biphasic systems has been reviewed recently (see chapter 9 in reference 16). Monophasic solvent systems are used to improve enzyme stability and to suppress the uncatalyzed, nonselective addition of HCN. The consensus view of the literature is that the best solvents for monophasic reactions are diisopropyl ether and ethyl acetate. These solvents afford very high enantioselectivities ($E = 200$) and afford exceptional stability (half-lives of hundreds of hours have been reported [24]) for crude preparations such as defatted almond meal and HNLs immobilized on solid supports such as Celite and Avicel. However, monophasic solvents can reduce enzyme reaction rates by orders of magnitude relative to aqueous/organic biphasic systems (36).

Several studies aimed at optimizing activity and enantioselectivity of monophasic HNL systems have demonstrated that addition of small amounts of water (0.1 to 1% vol) to an immobilized or crude enzyme system dramatically increases enzyme activity and enantioselectivity (22, 48). For example, Adlercreutz and coworkers (25) measured the effect of adding water on HCN addition to 3-phenylpropionaldehyde catalyzed by HbHNL adsorbed on Celite in dibutylether, and found that 1% vol/vol added water increased the initial rate 30-fold and the product ee_p from 15% to 85%. Higher water contents reduced reaction rates but had no effect on ee_p.

As an alternative to solid-supported enzymes, HNLs can be used as CLEAs (glutaraldehyde and dextran polyaldehyde are common cross-linkers), whereby the enzyme retains enough water to show high activity in monophasic organic

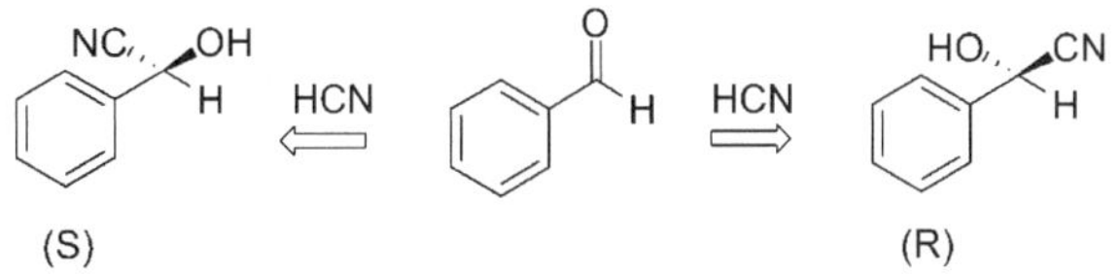

FIGURE 15 The Markovnikov addition of imidazoles to vinyl esters in dimethyl sulfoxide by several acylases (137) (top), and the asymmetric aldol addition of ketones and aldehydes catalyzed by PPL in acetone (75) (bottom). di-IPE, diisopropyl ether.

FIGURE 16 Asymmetric hydrocyanic addition of prochiral benzaldehyde is a popular model reaction for HNLs.

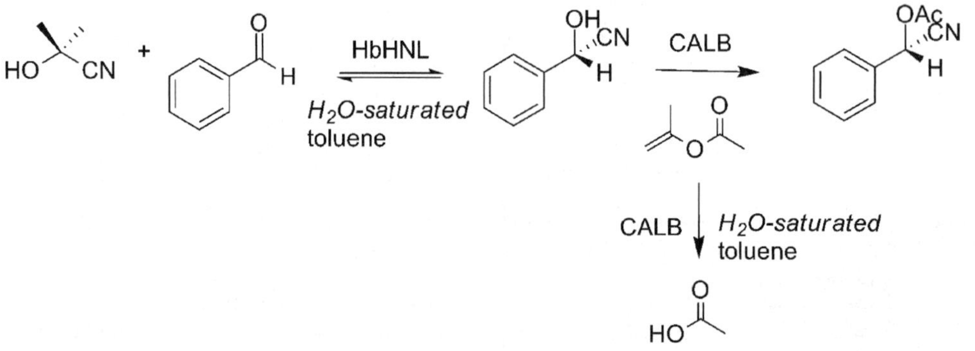

FIGURE 17 A one-pot synthesis involving Celite-HbHNL and CALB in water-saturated toluene. The water present was used by CALB to generate acetic acid, which deactivated the HbHNL. Adapted from reference 49.

solvents. For example, Sheldon and coworkers (130) demonstrated that the CLEA of PaHNL produced high conversions (>95%) and enantiomeric excesses (>95%) with several functionalized aldehydes in diisopropyl ether, and the catalyst could be recycled up to 10 times without loss of activity.

HNLs have also been used in several attractive "one-pot" schemes that combine selective cyanoadditions with further reactions, using either separately suspended enzymes or coenzyme CLEAs (83). While their synthetic utility is apparent, these schemes illustrate the potential incompatibility of HNLs with other enzymes. An attempt to couple HbHNL cyanoaddition with selective acylation by CALB was thwarted by the undesired hydrolysis of vinyl acetate catalyzed by CALB. The released acid deactivated the Celite-immobilized HbHNL (Fig. 17) (49).

Although HNLs provide a direct way for selective cyanoaddition, it is possible to arrive at the same product by use of a different enzyme system. An illustrative example is the production of α-cyano-3-phenoxybenzyl alcohol, a chiral building block for the insecticide pyrethrin (Fig. 18). In one synthesis route, (S)-selective SbHNL was used for direct cyanoaddition at ee_p = 90% and a 90% product yield (35). An alternative route was nonselective, chemically catalyzed cyanoaddition, followed by selective acylation using a lipase, followed by nonselective hydrolysis. In this one-pot scheme, the basic conditions for cyanoaddition also racemize the cyanohydrin in a dynamic kinetic resolution (59). This approach gave ee_p = 90% at 85% product yield.

35.2.3. Oxidoreductases

Examples of enzymes from this class (EC 1) that have been used in monophasic organic solvents include oxidoreductases that use peroxide (horseradish peroxidase [112]; soybean peroxidase [116]; and chloroperoxidase [129]), molecular oxygen (phenol oxidase [63], galactose oxidase [71], and xanthine oxidase [89]), and nicotinamide cofactors (alcohol dehydrogenase, ADH). Dehydrated whole cells of *Rhodococcus ruber* have also been used, catalyzing the oxidation of *sec*-alcohols in anhydrous hexane at 5% the rate in aqueous buffer (122).

In an early, seminal report demonstrating the potential of oxidoreductases in organic solvents, Klibanov and coworkers (63) used polyphenol oxidase immobilized on glass beads to regioselectively oxidize phenols to quinones (Fig. 19). These reactions were difficult in aqueous media because of rapid inactivation of the enzyme and the

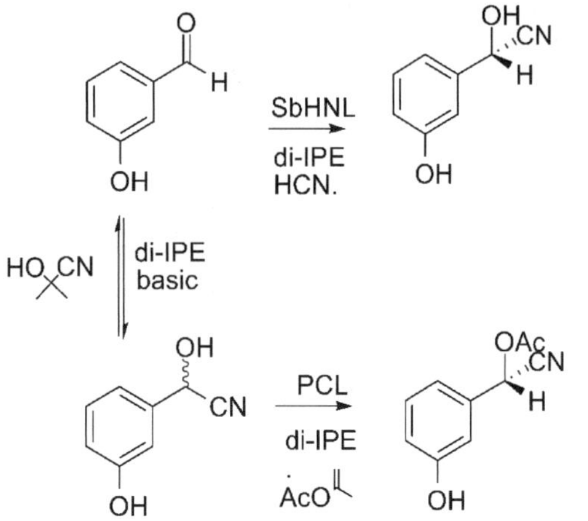

FIGURE 18 Alternative reaction paths to optically active α-cyano-3-phenoxybenzyl alcohol, a building block in pyrethrin insecticides. In the top path, an (S)-selective HNL is used for direct cyanoaddition (35). In the bottom path, nonselective cyanoaddition is coupled with enantioselective lipase acylation in a dynamic kinetic resolution (59). di-IPE, diisopropyl ether.

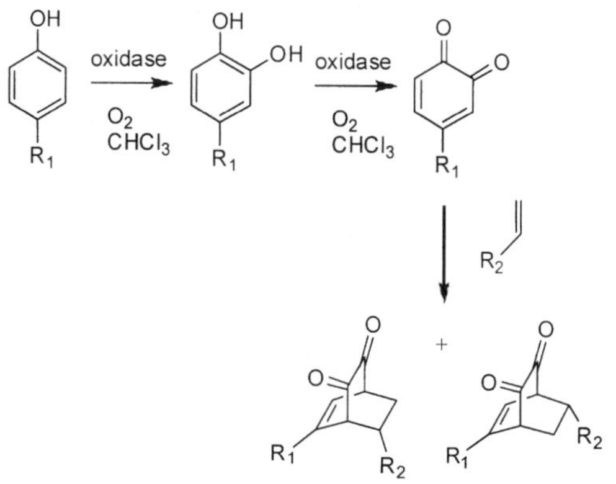

FIGURE 19 Polyphenol oxidase conversion of benzalcohols to diquinones in chloroform. This reaction was difficult in aqueous buffer because both the enzyme and quinone were unstable (63). In a subsequent study, the enzyme-generated quinone was subjected to various olefins to prepare bicyclooctenones (91).

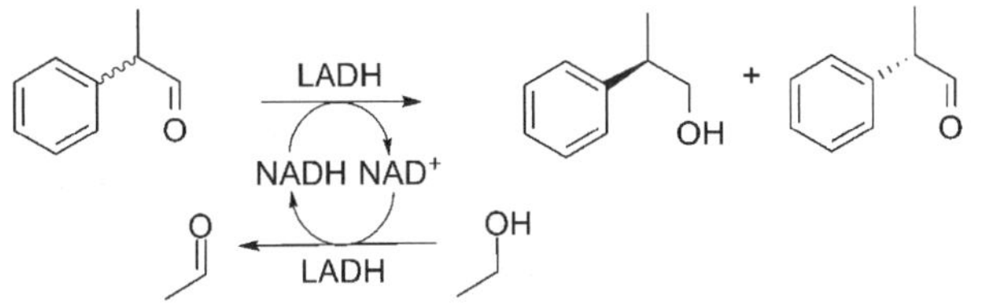

FIGURE 20 Horse liver alcohol dehydrogenase catalyzes the selective reduction of a racemic aldehyde with the consumption of NADH in isopropyl ether. The enzyme and NADH cofactor were coimmobilized on a glass support. The NADH was regenerated by the same enzyme in the oxidation of ethanol to acetaldehyde (46).

tendency of the quinones to polymerize. The reactions also benefitted from the high solubility of O_2 in chloroform. Polyphenol oxidase-generated quinones have been used in one-pot cycloadditions with various olefins to yield useful bicyclooctenones (91).

Horseradish peroxidase and chloroperoxidase have also been used for enantioselective sulfoxidations in organic solvents, where they exhibit increased enantioselectivity relative to the aqueous enzyme preparations (see reference 68 for a recent review). For example, the enantioselectivity of a horseradish peroxidase suspension toward the oxidation of thioanisole with *t*-butyl hydroperoxide was $E = 8.0$ in isopropyl alcohol and 1.8 in aqueous buffer (138).

Cofactor-requiring oxidoreductases such as the $NAD^+/NADH$-requiring ADH must be colyophilized or coprecipitated with the cofactor from an aqueous buffer to ensure enzyme-cofactor contact in organic solvent. Coimmobilization of horse liver ADH with NAD^+ on controlled-pore glass (47), coimmobilization on Sepharose (31), and colyophilization from an aqueous mixture (89) have all been used. Coimmobilized preparations required the addition of small amounts of water (0.1 to 1% vol/vol) for catalytic activity in polar solvents such as acetone and acetonitrile to ensure the proper conformation of the ADH active site (46, 47). However, even with added water, the ADH reactions were very slow, requiring several days for appreciable conversion.

Cofactor regeneration in the ADH $NAD^+/NADH$ system is most easily realized with the "second substrate" approach, in which the same enzyme catalyzes the depletion and regeneration of the cofactor in a cycle with a second substrate (Fig. 20). It has also been demonstrated that coimmobilization of a second enzyme (such as diaphorase) can accept the NADH generated by ADH and convert it to NAD^+ (31). There are many ways to regenerate depleted cofactors in aqueous or organic-aqueous mixtures, typically involving a second enzyme and substrate (see reference 1 for a thorough review).

35.3. CONCLUSION

Enzymes are not only active in organic solvents, they often display high regio-, chemo-, and enantioselectivities, making them particularly suited for the selective modification of complex molecules. Enzymatic reactions are thus often attractive alternatives to complex and inefficient chemical synthesis schemes for the preparation of pharmaceutical intermediates and derivatives. While the hydrolase enzymes are most widely used, enzymes from the lyase and oxidoreductase classes are also prominent, and other enzyme classes will no

doubt receive greater attention in the future. Novel enzyme preparation methods and system conditions, along with multienzyme processes and the ability to engineer enzymes with improved properties for synthesis, will lead to new synthetic routes and ensure an expanding role of organic-phase biocatalysis in the synthetic chemist's repertoire.

This work was supported by the National Science Foundation (BES-0228145) and the National Institutes of Health (GM66712).

REFERENCES

1. **Adlercreutz, P.** 1996. Cofactor regeneration in biocatalysis in organic media. *Biocatal. Biotransform.* **14:**1–30.
2. **Affleck, R., Z. F. Xu, V. Suzawa, K. Focht, D. S. Clark, and J. S. Dordick.** 1992. Enzymatic catalysis and dynamics in low-water environments. *Proc. Natl. Acad. Sci. USA* **89:**1100–1104.
3. **Akbar, U., C. D. Aschenbrenner, M. R. Harper, H. R. Johnson, J. S. Dordick, and D. S. Clark.** 2007. Direct solubilization of enzyme aggregates with enhanced activity in nonaqueous media. *Biotechnol. Bioeng.* **96:**1030–1039.
4. **Almarsson, O., and A. M. Klibanov.** 1996. Remarkable activation of enzymes in nonaqueous media by denaturing organic cosolvents. *Biotechnol. Bioeng.* **49:**87–92.
5. **Altreuter, D. H., J. S. Dordick, and D. S. Clark.** 2002. Nonaqueous biocatalytic synthesis of new cytotoxic doxorubicin derivatives: exploiting unexpected differences in the regioselectivity of salt-activated and solubilized subtilisin. *J. Am. Chem. Soc.* **124:**1871–1876.
6. **Altreuter, D. H., J. S. Dordick, and D. S. Clark.** 2003. Solid-phase peptide synthesis by ion-paired alpha-chymotrypsin in nonaqueous media. *Biotechnol. Bioeng.* **81:**809–817.
7. **Asuri, P., S. S. Karajanagi, A. A. Vertegel, J. S. Dordick, and R. S. Kane.** 2007. Enhanced stability of enzymes adsorbed onto nanoparticles. *J. Nanosci. Nanotechnol.* **7:**1675–1678.
8. **Bedell, B. A., V. V. Mozhaev, D. S. Clark, and J. S. Dordick.** 1998. Testing for diffusion limitations in salt-activated enzyme catalysts operating in organic solvents. *Biotechnol. Bioeng.* **58:**654–657.
9. **Bordusa, F.** 2002. Proteases in organic synthesis. *Chem. Rev.* **102:**4817–4867.
10. **Boren, L., B. Martin-Matute, Y. M. Xu, A. Cordova, and J. E. Backvall.** 2005. (S)-Selective kinetic resolution and chemoenzymatic dynamic kinetic resolution of secondary alcohols. *Chem. Eur. J.* **12:**225–232.
11. **Branneby, C., P. Carlqvist, A. Magnusson, K. Hult, T. Brinck, and P. Berglund.** 2003. Carbon-carbon bonds by hydrolytic enzymes. *J. Am. Chem. Soc.* **125:**874–875.
12. **Butler, L. G.** 1979. Enzymes in non-aqueous solvents. *Enzyme Microb. Technol.* **1:**253–259.
13. **Cai, Y., S. P. Yao, Q. Wu, and X. F. Lin.** 2004. Michael addition of imidazole with acrylates catalyzed by alkaline protease from *Bacillus subtilis* in organic media. *Biotechnol. Lett.* **26:**525–528.
14. **Cainelli, G., P. Galletti, and D. Giacomini.** 2009. Solvent effects on stereoselectivity: more than just an environment. *Chem. Soc. Rev.* **38:**990–1001.
15. **Carrea, G., and S. Riva.** 2000. Properties and synthetic applications of enzymes in organic solvents. *Angew. Chem. Int. Ed.* **39:**2226–2254.
16. **Carrea, G., and S. Riva.** 2008. *Organic Synthesis with Enzymes in Non-Aqueous Media.* Wiley-VCH, Weinheim, Germany.
17. **Castillo, B., V. Bansal, A. Ganesan, P. Halling, F. Secundo, A. Ferrer, K. Griebenow, and G. Barletta.** 2006. On the activity loss of hydrolases in organic solvents. II. A mechanistic study of subtilisin Carlsberg. *BMC Biotechnol.* **6:**13.

18. **Castillo, B., J. Mendez, W. Al-Azzam, G. Barletta, and K. Griebenow.** 2006. On the relationship between the activity and structure of PEG-alpha-chymotrypsin conjugates in organic solvents. *Biotechnol. Bioeng.* **94:**565–574.

19. **Castillo, B., Y. Pacheco, W. Al-Azzam, K. Griebenow, M. Devi, A. Ferrer, and G. Barletta.** 2005. On the activity loss of hydrolases in organic solvents. I. Rapid loss of activity of a variety of enzymes and formulations in a range of organic solvents. *J. Mol. Catal. B Enzym.* **35:**147–153.

20. **Castillo, B., R. J. Sola, A. Ferrer, G. Barletta, and K. Griebenow.** 2008. Effect of PEG modification on subtilisin Carlsberg activity, enantioselectivity, and structural dynamics in 1,4-dioxane. *Biotechnol. Bioeng.* **99:**9–17.

21. **Chatterjee, S., and A. J. Russell.** 1992. Determination of equilibrium and individual rate constants for subtilisin-catalyzed transesterification in anhydrous environments. *Biotechnol. Bioeng.* **40:**1069–1077.

22. **Chmura, A., G. M. van der Kraan, F. Kielar, L. M. van Langen, F. van Rantwijk, and R. A. Sheldon.** 2006. Cross-linked aggregates of the hydroxynitrile lyase from Manihot esculenta: highly active and robust biocatalysts. *Adv. Synth. Catal.* **348:**1655–1661.

23. **Choi, Y. K., M. J. Kim, Y. Ahn, and M. J. Kim.** 2001. Lipase/palladium-catalyzed asymmetric transformations of ketoximes to optically active amines. *Org. Lett.* **3:**4099–4101.

24. **Costes, D., G. Rotcenkovs, E. Wehtje, and P. Adlercreutz.** 2001. Stability and stabilization of hydroxynitrile lyase in organic solvents. *Biocatal. Biotransform.* **19:**119–130.

25. **Costes, D., E. Wehtje, and P. Adlercreutz.** 1999. Hydroxynitrile lyase-catalyzed synthesis of cyanohydrins in organic solvents. Parameters influencing activity and enantiospecificity. *Enzyme Microb. Technol.* **25:**384–391.

26. **Cotterill, I. C., J. O. Rich, M. D. Scholten, L. Mozhaeva, and P. C. Michels.** 2008. Reversible derivatization to enhance enzymatic synthesis: chemoenzymatic synthesis of doxorubicin-14-O-esters. *Biotechnol. Bioeng.* **101:**435–440.

27. **Cruz Silva, M. M., S. Riva, and M. L. Sá e Melo.** 2005. Regioselective enzymatic acylation of vicinal diols of steroids. *Tetrahedron* **61:**3065–3073.

28. **Dabulis, K., and A. M. Klibanov.** 1993. Dramatic enhancement of enzymatic-activity in organic solvents by lyoprotectants. *Biotechnol. Bioeng.* **41:**566–571.

29. **D'Antona, N., D. Lambusta, G. Nicolosi, and P. Bovicelli.** 2008. Preparation of regioprotected morins by lipase-catalysed transesterification. *J. Mol. Catal. B Enzym.* **52-3:**78–81.

30. **Davis, B. G., and V. Borer.** 2001. Biocatalysis and enzymes in organic synthesis. *Nat. Prod. Rep.* **18:**618–640.

31. **Deetz, J. S., and J. D. Rozzell.** 1988. Enzyme-catalyzed reactions in non-aqueous media. *Trends Biotechnol.* **6:**15–19.

32. **Ebert, C., L. Gardossi, P. Linda, R. Vesnaver, and M. Bosco.** 1996. Influence of organic solvents on enzyme chemoselectivity and their role in enzyme-substrate interaction. *Tetrahedron* **52:**4867–4876.

33. **Edin, M., J. Steinreiber, and J. E. Backvall.** 2004. One-pot synthesis of enantiopure syn-1,3-diacetates from racemic syn/anti mixtures of 1,3-diols by dynamic kinetic asymmetric transformation. *Proc. Natl. Acad. Sci USA* **101:**5761–5766.

34. **Effenberger, F., and J. Eichhorn.** 1997. Enzyme catalyzed reactions. 27. Stereoselective synthesis of thienyl and furyl analogues of ephedrine. *Tetrahedron Asymmetry* **8:**469–476.

35. **Effenberger, F., B. Horsch, S. Forster, and T. Ziegler.** 1990. Enzyme-catalyzed reactions. 5. Enzyme-catalyzed synthesis of (S)-cyanohydrins and subsequent hydrolysis to (S)-alpha-hydroxy-carboxylic acids. *Tetrahedron Lett.* **31:**1249–1252.

36. **Effenberger, F., T. Ziegler, and S. Forster.** 1987. Enzyme-catalyzed cyanohydrin synthesis in organic solvents. *Angew. Chem. Int. Ed. Engl.* **26:**458–460.

37. **Eppler, R. K., E. P. Hudson, S. D. Chase, J. S. Dordick, J. A. Reimer, and D. S. Clark.** 2008. Biocatalytic activity in nonaqueous environments correlates with centisecond-range protein motions. *Proc. Natl. Acad. Sci. USA* **105:**15672–15677.

38. **Eppler, R. K., R. S. Komor, J. Huynh, J. S. Dordick, J. A. Reimer, and D. S. Clark.** 2006. Water dynamics and salt-activation of enzymes in organic media: mechanistic implications revealed by NMR spectroscopy. *Proc. Natl. Acad. Sci. USA* **103:**5706–5710.

39. **Faber, K.** 2004. *Biotransformations in Organic Chemistry: a Textbook*, 5th rev. & corr. ed. Springer-Verlag, Berlin, Germany.

40. **Fitzpatrick, P. A., and A. M. Klibanov.** 1991. How can the solvent affect enzyme enantioselectivity. *J. Am. Chem. Soc.* **113:**3166–3171.

41. **Ghanem, A., and H. Y. Aboul-Enein.** 2004. Lipase-mediated chiral resolution of racemates in organic solvents. *Tetrahedron Asymmetry* **15:**3331–3351.

42. **Goncalves, A. P. V., J. M. Lopes, F. Lemos, F. R. Ribeiro, D. M. F. Prazeres, J. M. S. Cabral, and M. R. AiresBarros.** 1997. Effect of the immobilization support on the hydrolytic activity of a cutinase from Fusarium solani pisi. *Enzyme Microb. Technol.* **20:**93–101.

43. **Gorman, L. A. S., and J. S. Dordick.** 1992. Organic solvents strip water off enzymes. *Biotechnol. Bioeng.* **39:**392–397.

44. **Gotor, V.** 2002. Biocatalysis applied to chemoselective transformations on vitamin D and nucleoside derivatives. *J. Mol. Catal. B Enzym.* **19:**21–30.

45. **Griebenow, K., Y. D. Laureano, A. M. Santos, I. M. Clemente, L. Rodriguez, M. W. Vidal, and G. Barletta.** 1999. Improved enzyme activity and enantioselectivity in organic solvents by methyl-beta-cyclodextrin. *J. Am. Chem. Soc.* **121:**8157–8163.

46. **Grunwald, J., B. Wirz, M. P. Scollar, and A. M. Klibanov.** 1986. Asymmetric oxidoreductions catalyzed by alcohol-dehydrogenase in organic solvents. *J. Am. Chem. Soc.* **108:**6732–6734.

47. **Guinn, R. M., P. S. Skerker, P. Kavanaugh, and D. S. Clark.** 1991. Activity and flexibility of alcohol-dehydrogenase in organic solvents. *Biotechnol. Bioeng.* **37:**303–308.

48. **Han, S. Q., G. Q. Lin, and Z. Y. Li.** 1998. Synthesis of (R)-cyanohydrins by crude (R)-oxynitrilase-catalyzed reactions in micro-aqueous medium. *Tetrahedron Asymmetry* **9:**1835–1838.

49. **Hanefeld, U., A. J. J. Straathof, and J. J. Heijnen.** 2001. Enzymatic formation and esterification of (S)-mandelonitrile. *J. Mol. Catal. B Enzym.* **11:**213–218.

50. **Heinze, B., R. Kourist, L. Fransson, K. Hult, and U. T. Bornscheuer.** 2007. Highly enantioselective kinetic resolution of two tertiary alcohols using mutants of an esterase from Bacillus subtilis. *Protein Eng. Des. Sel.* **20:**125–131.

51. **Henke, E., E. Pleiss, and U. T. Bornscheuer.** 2002. Activity of lipases and esterases towards tertiary alcohols: insights into structure-function relationships. *Angew. Chem. Int. Ed.* **41:**3211–3213.

52. **Hobbs, H. R., H. M. Kirke, M. Poliakoff, and N. R. Thomas.** 2007. Homogeneous biocatalysis in both fluorous biphasic and supercritical carbon dioxide systems. *Angew. Chem. Int. Ed.* **46:**7860–7863.

53. **Hudson, E. P., R. K. Eppler, J. M. Beaudoin, J. S. Dordick, J. A. Reimer, and D. S. Clark.** 2009. Active-site motions and polarity enhance catalytic turnover of hydrated subtilisin dissolved in organic solvents. *J. Am. Chem. Soc.* **131:**4294–4300.

54. Hudson, E. P., R. K. Eppler, and D. S. Clark. 2005. Biocatalysis in semi-aqueous and nearly anhydrous conditions. *Curr. Opin. Biotechnol.* **16:**637–643.

55. Hult, K. 1995. Enzyme-kinetics and the design of lipase-catalyzed acyl transfer-reactions. *Abstr. Pap. Am. Chem. Soc.* **209:**155–BIOT.

56. Hult, K., and P. Berglund. 2007. Enzyme promiscuity: mechanism and applications. *Trends Biotechnol.* **25:**231–238.

57. Iglesias, L. E., V. M. Sanchez, F. Rebolledo, and V. Gotor. 1997. Candida antarctica B lipase catalysed resolution of (+/−)-1-(heteroaryl)ethylamines. *Tetrahedron Asymmetry* **8:**2675–2677.

58. Inada, Y., K. Takahashi, T. Yoshimoto, A. Ajima, A. Matsushima, and Y. Saito. 1986. Application of polyethylene glycol-modified enzymes in biotechnological processes—organic solvent-soluble enzymes. *Trends Biotechnol.* **4:**190–194.

59. Inagaki, M., J. Hiratake, T. Nishioka, and J. Oda. 1992. One-pot synthesis of optically-active cyanohydrin acetates from aldehydes via lipase-catalyzed kinetic resolution coupled with insitu formation and racemization of cyanohydrins. *J. Org. Chem.* **57:**5643–5649.

60. Jia, H. F., G. Y. Zhu, and P. Wang. 2003. Catalytic behaviors of enzymes attached to nanoparticles: the effect of particle mobility. *Biotechnol. Bioeng.* **84:**406–414.

61. Kahlow, U. H. M., R. D. Schmid, and J. Pleiss. 2001. A model of the pressure dependence of the enantio selectivity of Candida rugosa lipase towards (+/−)-menthol. *Protein Sci.* **10:**1942–1952.

62. Kanerva, L. T., and A. M. Klibanov. 1989. Hammett analysis of enzyme action in organic solvents. *J. Am. Chem. Soc.* **111:**6864–6865.

63. Kazandjian, R. Z., and A. M. Klibanov. 1985. Regioselective oxidation of phenols catalyzed by polyphenol oxidase in chloroform. *J. Am. Chem. Soc.* **107:**5448–5450.

64. Kazlauskas, R. J., A. N. E. Weissfloch, A. T. Rappaport, and L. A. Cuccia. 1991. A rule to predict which enantiomer of a secondary alcohol reacts faster in reactions catalyzed by cholesterol esterase, lipase from *Pseudomonas cepacia*, and lipase from *Candida rugosa*. *J. Org. Chem.* **56:**2656–2665.

65. Khmelnitsky, Y. L., C. Budde, J. M. Arnold, A. Usyatinsky, D. S. Clark, and J. S. Dordick. 1997. Synthesis of water-soluble paclitaxel derivatives by enzymatic acylation. *J. Am. Chem. Soc.* **119:**11554–11555.

66. Khmelnitsky, Y. L., A. V. Levashov, N. L. Klyachko, and K. Martinek. 1988. Engineering biocatalytic systems in organic media with low water-content. *Enzyme Microb. Technol.* **10:**710–724.

67. Kitaguchi, H., P. A. Fitzpatrick, J. E. Huber, and A. M. Klibanov. 1989. Enzymatic resolution of racemic amines—crucial role of the solvent. *J. Am. Chem. Soc.* **111:**3094–3095.

68. Klibanov, A. M. 2003. Asymmetric enzymatic oxidoreductions in organic solvents. *Curr. Opin. Biotechnol.* **14:**427–431.

69. Klibanov, A. M. 1983. Immobilized enzymes and cells as practical catalysts. *Science* **219:**722–727.

70. Klibanov, A. M. 2001. Improving enzymes by using them in organic solvents. *Nature* **409:**241–246.

71. Klibanov, A. M., B. N. Alberti, and M. A. Marletta. 1982. Stereospecific oxidation of aliphatic-alcohols catalyzed by galactose-oxidase. *Biochem. Biophys. Res. Commun.* **108:**804–808.

72. Kvittingen, L., B. Sjursnes, T. Anthonsen, and P. Halling. 1992. Use of salt hydrates to buffer optimal water level during lipase catalyzed synthesis in organic media. A practical procedure for organic chemists. *Tetrahedron* **48:**2793–2802.

73. Laane, C., S. Boeren, K. Vos, and C. Veeger. 1987. Rules for optimization of biocatalysis in organic solvents. *Biotechnol. Bioeng.* **30:**81–87.

74. Lavandera, W., S. Fernandez, J. Magdalenala, M. Ferrero, H. Grewal, C. K. Savile, R. J. Kazlauskas, and V. Gotor. 2006. Remote interactions explain the unusual regioselectivity of lipase from *Pseudomonas cepacia* toward the secondary hydroxyl of 2′-deoxynucleosides. *Chembiochem* **7:**693–698.

75. Li, C., X. W. Feng, N. Wang, Y. J. Zhou, and X. Q. Yu. 2008. Biocatalytic promiscuity: the first lipase-catalysed asymmetric aldol reaction. *Green Chem.* **10:**616–618.

76. Li, W., and S. J. Li. 2007. Molecular imprinting: a versatile tool for separation, sensors and catalysis. *Adv. Polymer Sci.* **206:**191–210.

77. Li, W. C., B. Yang, Y. H. Wang, D. Q. Wei, C. Whiteley, and X. N. Wang. 2009. Molecular modeling of substrate selectivity of Candida antarctica lipase B and Candida rugosa lipase towards c9, t11-and t10, c12-conjugated linoleic acid. *J. Mol. Catal. B Enzym.* **57:**299–303.

78. Lindsay, J. P., D. S. Clark, and J. S. Dordick. 2004. Combinatorial formulation of biocatalyst preparations for increased activity in organic solvents: salt activation of penicillin amidase. *Biotechnol. Bioeng.* **85:**553–560.

79. Liu, X. C., D. S. Clark, and J. S. Dordick. 2000. Chemoenzymatic construction of a four-component Ugi combinatorial library. *Biotechnol. Bioeng.* **69:**457–460.

80. Lombard, C., J. Saulnier, and J. M. Wallach. 2005. Recent trends in protease-catalyzed peptide synthesis. *Protein Pept. Lett.* **12:**621–629.

81. Magnusson, A. O., M. Takwa, A. Harnberg, and K. Hult. 2005. An S-selective lipase was created by rational redesign and the enantioselectivity increased with temperature. *Angew. Chem. Int. Ed.* **44:**4582–4585.

82. Martin-Matute, B., M. Edin, K. Bogar, and J. E. Backvall. 2004. Highly compatible metal and enzyme catalysts for efficient dynamic kinetic resolution of alcohols at ambient temperature. *Angew. Chem. Int. Ed.* **43:**6535–6539.

83. Mateo, C., A. Chmura, S. Rustler, F. van Rantwijk, A. Stolz, and R. A. Sheldon. 2006. Synthesis of enantiomerically pure (S)-mandelic acid using an oxynitrilase-nitrilase bienzymatic cascade: a nitrilase surprisingly shows nitrile hydratase activity. *Tetrahedron Asymmetry* **17:**320–323.

84. Micaelo, N. M., and C. M. Soares. 2007. Modeling hydration mechanisms of enzymes in nonpolar and polar organic solvents. *FEBS J.* **274:**2424–2436.

85. Miyazawa, T., E. Ensatsu, M. Hiramatsu, R. Yanagihara, and T. Yamada. 2002. alpha-Chymotrypsin-catalysed segment condensations via the kinetically controlled approach using carbamoylmethyl esters as acyl donors in organic media. *J. Chem. Soc. Perkin Trans. 1* **3:**396–401.

86. Miyazawa, T., M. Hiramatsu, T. Murashima, and T. Yamada. 2003. Utilization of proteases from Aspergillus species for peptide synthesis via the kinetically controlled approach. *Biocatal. Biotransform.* **21:**93–100.

87. Miyazawa, T., K. Tanaka, E. Ensatsu, R. Yanagihara, and T. Yamada. 2001. Broadening of the substrate tolerance of alpha-chymotrypsin by using the carbamoylmethyl ester as an acyl donor in kinetically controlled peptide synthesis. *J. Chem. Soc. Perkin Trans. 1* **1:**87–93.

88. Morgan, B. S., D. Hoenner, P. Evans, and S. M. Roberts. 2004. Facile biocatalytic syntheses of optically active 4-hydroxycyclohex-2-enone and 4-benzylthiacyclopent-2-enone. *Tetrahedron Asymmetry* **15:**2807–2809.

89. Morgan, J. A., and D. S. Clark. 2004. Salt-activation of nonhydrolase enzymes for use in organic solvents. *Biotechnol. Bioeng.* **85:**456–459.

90. Mozhaev, V. V., C. L. Budde, J. O. Rich, A. Y. Usyatinsky, P. C. Michels, Y. L. Khmelnitsky, D. S. Clark, and J. S. Dordick. 1998. Regioselective enzymatic acylation as a tool for producing solution-phase combinatorial libraries. *Tetrahedron* **54:**3971–3982.

91. Muller, G. H., A. Lang, D. R. Seithel, and H. Waldmann. 1998. An enzyme-initiated hydroxylation—oxidation carbo Diels-Alder domino reaction. *Chem. Eur. J.* **4:**2513–2522.

92. Natoli, M., G. Nicolosi, and M. Piattelli. 1992. Regioselective alcoholysis of flavonoid acetates with lipase in an organic solvent. *J. Org. Chem.* **57:**5776–5778.

93. Nordin, O., E. Hedenstrom, and H. E. Hogberg. 1994. Enantioselective transesterifications of 2-methyl-1-alcohols catalyzed by lipases from Pseudomonas. *Tetrahedron Asymmetry* **5:**785–788.

94. Nordin, O., B. V. Nguyen, C. Vorde, E. Hedenstrom, and H. E. Hogberg. 2000. Kinetic resolution of primary 2-methyl-substituted alcohols via *Pseudomonas cepacia* lipase-catalysed enantioselective acylation. *J. Chem. Soc. Perkin Trans. 1* **3:**367–376.

95. Okazaki, S. Y., N. Kamiya, K. Abe, M. Goto, and F. Nakashio. 1997. Novel preparation method for surfactant-lipase complexes utilizing water in oil emulsions. *Biotechnol. Bioeng.* **55:**455–460.

96. Pagan, M., R. J. Sola, and K. Griebenow. 2009. On the role of protein structural dynamics in the catalytic activity and thermostability of serine protease subtilisin Carlsberg. *Biotechnol. Bioeng.* **103:**77–84.

97. Paradkar, V. M., and J. S. Dordick. 1994. Aqueous-like activity of alpha-chymotrypsin dissolved in nearly anhydrous organic solvents. *J. Am. Chem. Soc.* **116:**5009–5010.

98. Partridge, J., P. J. Halling, and B. D. Moore. 2000. Solid-state proton/sodium buffers: "chemical pH stats" for biocatalysts in organic solvents. *J. Chem. Soc. Perkin Trans. 2* **3:**465–471.

99. Perez-Victoria, I., and J. C. Morales. 2006. Complementary regioselective esterification of non-reducing oligosaccharides catalyzed by different hydrolases. *Tetrahedron* **62:**878–886.

100. Persson, B. A., A. L. E. Larsson, M. Le Ray, and J. E. Backvall. 1999. Ruthenium- and enzyme-catalyzed dynamic kinetic resolution of secondary alcohols. *J. Am. Chem. Soc.* **121:**1645–1650.

101. Petersson, A. E. V., P. Adlercreutz, and B. Mattiasson. 2007. A water activity control system for enzymatic reactions in organic media. *Biotechnol. Bioengineer.* **97:**235–241.

102. Pleiss, J., H. Scheib, and R. D. Schmid. 2000. The His gap motif in microbial lipases: a determinant of stereoselectivity toward triacylglycerols and analogs. *Biochimie* **82:**1043–1052.

103. Priego, J., C. Ortiz-Nava, M. Carrillo-Morales, A. Lopez-Munguia, J. Escalante, and E. Castillo. 2009. Solvent engineering: an effective tool to direct chemoselectivity in a lipase-catalyzed Michael addition. *Tetrahedron* **65:**536–539.

104. Quiocho, F. A., and F. M. Richards. 1964. Intermolecular cross linking of protein in crystalline state: carboxypeptidase-A. *Proc. Natl. Acad. Sci. USA* **52:**833–839.

105. Rich, J. O., B. A. Bedell, and J. S. Dordick. 1995. Controlling enzyme-catalyzed regioselectivity in sugar ester synthesis. *Biotechnol. Bioeng.* **45:**426–434.

106. Rich, J. O., V. V. Mozhaev, J. S. Dordick, D. S. Clark, and Y. L. Khmelnitsky. 2002. Molecular imprinting of enzymes with water-insoluble ligands for nonaqueous biocatalysis. *J. Am. Chem. Soc.* **124:**5254–5255.

107. Riva, S., J. Chopineau, A. P. G. Kieboom, and A. M. Klibanov. 1988. Protease-catalyzed regioselective esterification of sugars and related compounds in anhydrous dimethylformamide. *J. Am. Chem. Soc.* **110:**584–589.

108. Riva, S., and A. M. Klibanov. 1988. Enzymochemical regioselective oxidation of steroids without oxidoreductases. *J. Am. Chem. Soc.* **110:**3291–3295.

109. Rodriguez-Martinez, J. A., R. J. Sola, B. Castillo, H. R. Cintron-Colon, I. Rivera-Rivera, G. Barletta, and K. Griebenow. 2008. Stabilization of alpha-chymotrypsin upon pegylation correlates with reduced structural dynamics. *Biotechnol. Bioeng.* **101:**1142–1149.

110. Ru, M. T., S. Y. Hirokane, A. S. Lo, J. S. Dordick, J. A. Reimer, and D. S. Clark. 2000. On the salt-induced activation of lyophilized enzymes in organic solvents: effect of salt kosmotropicity on enzyme activity. *J. Am. Chem. Soc.* **122:**1565–1571.

111. Russell, A. J., and A. M. Klibanov. 1988. Inhibitor-induced enzyme activation in organic solvents. *J. Biol. Chem.* **263:**11624–11626.

112. Ryu, K., and J. S. Dordick. 1989. Free-energy relationships of substrate and solvent hydrophobicities with enzymatic catalysis in organic media. *J. Am. Chem. Soc.* **111:**8026–8027.

113. Sabbani, S., E. Hedenstrom, and J. Andersson. 2007. Lipase catalyzed acylation of primary alcohols with remotely located stereogenic centres: the resolution of (+/−)-4,4-dimethyl-3-phenyl-1-pentanol. *Tetrahedron Asymmetry* **18:**1712–1720.

114. Schmitke, J. L., C. R. Wescott, and A. M. Klibanov. 1996. The mechanistic dissection of the plunge in enzymatic activity upon transition from water to anhydrous solvents. *J. Am. Chem. Soc.* **118:**3360–3365.

115. Serdakowski, A. L., and J. S. Dordick. 2008. Enzyme activation for organic solvents made easy. *Trends Biotechnol.* **26:**48–54.

116. Serdakowski, A. L., I. Z. Munir, and J. S. Dordick. 2006. Dramatic solvent and hydration effects on the transition state of soybean peroxidase. *J. Am. Chem. Soc.* **128:**14272–14273.

117. Sheldon, R. A., R. Schoevaart, and L. M. Van Langen. 2005. Cross-linked enzyme aggregates (CLEAs): a novel and versatile method for enzyme immobilization (a review). *Biocatal. Biotransform.* **23:**141–147.

118. Silva, M. M. C., M. Melo, M. Parolin, D. Tessaro, S. Riva, and B. Danieli. 2004. The biocatalyzed stereoselective preparation of polycyclic cyanohydrins. *Tetrahedron Asymmetry* **15:**21–27.

119. Silva, M. M. C., S. Riva, and M. L. S. Melo. 2004. Highly selective lipase-mediated discrimination of diastereomeric 5,6-epoxysteroids. *Tetrahedron Asymmetry* **15:**1173–1179.

120. Simerska, P., A. Pisvejcova, M. Kuzma, P. Sedmera, V. Kren, S. Nicotra, and S. Riva. 2004. Regioselective enzymatic acylation of N-acetylhexosamines. *J. Mol. Cat. B Enz.* **29:**219–225.

121. Somashekar, B. R., and S. Divakar. 2007. Lipase catalyzed synthesis of L-alanyl esters of carbohydrates. *Enzyme Microb. Technol.* **40:**299–309.

122. Stampfer, W., B. Kosjek, W. Kroutil, and K. Faber. 2003. On the organic solvent and thermostability of the biocatalytic redox system of *Rhodococcus ruber* DSM 44541. *Biotechnol. Bioeng.* **81:**865–869.

123. Svedendahl, M., P. Carlqvist, C. Branneby, O. Allner, A. Frise, K. Hult, P. Berglund, and T. Brinck. 2008. Direct epoxidation in *Candida antarctica* lipase B studied by experiment and theory. *Chembiochem* **9:**2443–2451.

124. Svedendahl, M., K. Hult, and P. Berglund. 2005. Fast carbon-carbon bond formation by a promiscuous lipase. *J. Am. Chem. Soc.* **127:**17988–17989.

125. Takahashi, H., B. Li, T. Sasaki, C. Miyazaki, T. Kajino, and S. Inagaki. 2000. Catalytic activity in organic solvents and stability of immobilized enzymes depend on the

pore size and surface characteristics of mesoporous silica. *Chem. Mater.* **12:**3301–3305.

126. **Tanaka, K., H. Osuga, H. Suzuki, Y. Shogase, and Y. Kitahara.** 1998. Synthesis, enzymic resolution and enantiomeric enhancement of bis(hydroxymethyl)[7]thia heterohelicenes. *J. Chem. Soc. Perkin Trans. 1* **5:**935–940.

127. **Tawaki, S., and A. M. Klibanov.** 1992. Inversion of enzyme enantioselectivity mediated by the solvent. *J. Am. Chem. Soc.* **114:**1882–1884.

128. **Tuomi, W. V., and R. J. Kazlauskas.** 1999. Molecular basis for enantioselectivity of lipase from *Pseudomonas cepacia* toward primary alcohols. Modeling, kinetics, and chemical modification of Tyr29 to increase or decrease enantioselectivity. *J. Org. Chem.* **64:**2638–2647.

129. **van de Velde, F., M. Bakker, F. van Rantwijk, and R. A. Sheldon.** 2001. Chloroperoxidase-catalyzed enantioselective oxidations in hydrophobic organic media. *Biotechnol. Bioeng.* **72:**523–529.

130. **van Langen, L. M., R. P. Selassa, F. van Rantwijk, and R. A. Sheldon.** 2005. Cross-linked aggregates of (R)-oxynitrilase: a stable, recyclable biocatalyst for enantioselective hydrocyanation. *Org. Lett.* **7:**327–329.

131. **van Rantwijk, F., and R. A. Sheldon.** 2004. Enantioselective acylation of chiral amines catalysed by serine hydrolases. *Tetrahedron* **60:**501–519.

132. **van Unen, D. J., I. K. Sakodinskaya, J. F. J. Engbersen, and D. N. Reinhoudt.** 1998. Crown ether activation of cross-linked subtilisin Carlsberg crystals in organic solvents. *J. Chem. Soc. Perkin Trans. 1* **2:**3341–3343.

133. **Veronese, F. M.** 2001. Peptide and protein PEGylation: a review of problems and solutions. *Biomaterials* **22:**405–417.

134. **Wang, Y. F., J. J. Lalonde, M. Momongan, D. E. Bergbreiter, and C. H. Wong.** 1988. Lipase-catalyzed irreversible transesterifications using enol esters as acylating reagents: preparative enantioselective and regioselective syntheses of alcohols, glycerol derivatives, sugars, and organometallics. *J. Am. Chem. Soc.* **110:**7200–7205.

135. **Wang, Y. H., R. Wang, Q. S. Li, Z. M. Zhang, and Y. Feng.** 2009. Kinetic resolution of rac-alkyl alcohols via lipase-catalyzed enantioselective acylation using succinic anhydride as acylating agent. *J. Mol. Catal. B Enzym.* **56:**142–145.

136. **Wangikar, P. P., P. C. Michels, D. S. Clark, and J. S. Dordick.** 1997. Structure and function of subtilisin BPN' solubilized in organic solvents. *J. Am. Chem. Soc.* **119:**70–76.

137. **Wu, W. B., J. M. Xu, Q. Wu, D. S. Lv, and X. F. Lin.** 2006. Promiscuous acylases-catalyzed Markovnikov addition of N-heterocycles to vinyl esters in organic media. *Adv. Synth. Catal.* **348:**487–492.

138. **Xie, Y. C., P. K. Das, J. M. M. Caaveiro, and A. M. Klibanov.** 2002. Unexpectedly enhanced stereoselectivity of peroxidase-catalyzed sulfoxidation in branched alcohols. *Biotechnol. Bioeng.* **79:**105–111.

139. **Yang, K., and Y. J. Wang.** 2003. Lipase-catalyzed cellulose acetylation in aqueous and organic media. *Biotechnol. Prog.* **19:**1664–1671.

140. **Yang, L., J. S. Dordick, and S. Garde.** 2004. Hydration of enzyme in nonaqueous media is consistent with solvent dependence of its activity. *Biophys. J.* **87:**812–821.

141. **Yang, Z., A. J. Mesiano, S. Venkatasubramanian, S. H. Gross, J. M. Harris, and A. J. Russell.** 1995. Activity and stability of enzymes incorporated into acrylic polymers. *J. Am. Chem. Soc.* **117:**4843–4850.

142. **Zacharis, E., I. C. Omar, J. Partridge, D. A. Robb, and P. J. Halling.** 1997. Selection of salt hydrate pairs for use in water control in enzyme catalysis in organic solvents. *Biotechnol. Bioeng.* **55:**367–374.

143. **Zaks, A., and A. M. Klibanov.** 1984. Enzymatic catalysis in organic media at 100 degrees C. *Science* **224:**1249–1251.

144. **Zaks, A., and A. M. Klibanov.** 1985. Enzyme-catalyzed processes in organic solvents. *Proc. Natl. Acad. Sci. USA* **82:**3192–3196.

145. **Zhang, X. Z., X. Wang, S. M. Chen, X. Q. Fu, X. X. Wu, and C. H. Li.** 1996. Protease-catalyzed small peptide synthesis in organic media. *Enzyme Microb. Technol.* **19:**538–544.

Enzyme Promiscuity and Evolution of New Protein Functions

BERT VAN LOO AND FLORIAN HOLLFELDER

36

36.1. INTRODUCTION

Promiscuous enzymes can perform additional functions besides their "native" activities, challenging the textbook adage "one enzyme, one activity." Changing the function of enzymes or even creating entirely new catalysts de novo by design has proven difficult (43), despite some exciting recent advances (52, 96). Likewise, directed evolution of an enzyme toward an entirely new function has to overcome the—often large—thermodynamic barriers of the target reaction, until a catalytic effect can be visibly detected. Jensen (51) and later O'Brien and Herschlag (85) recognized that promiscuous activities can provide a solution to this dilemma: catalytic promiscuity equips a protein with a second head-start activity that can be adapted and enhanced when the gene is duplicated without compromising the original activity (2, 48, 61, 85). Typical k_{cat}/K_m values for specialized functions that are considered highly evolved range up to 10^6 to 10^7 M^{-1}s^{-1} (123). However, substantially lower activities could already be sufficient to support a metabolic function. In natural evolution the level of promiscuous activity needed for the new function to provide a selective advantage was estimated to be as low as 0.3 M^{-1} s^{-1} for k_{cat}/K_m, provided the enzyme was expressed at high levels (90). This figure suggests that even rates that one would consider small compared with those of "mature" enzymes can matter in evolution. Likewise, the lower limit for a promiscuous activity assigned in a laboratory experiment will depend on the sensitivity of the assay system, a feature that must be kept in mind for any in vitro evolution studies: if activities fall below the experimental threshold, they cannot be selected for.

In an effort to discover evolutionary starting points, there has been an increase in systematic analysis of enzyme promiscuity in recent years combining not only newly discovered enzymes, but also numerous historic examples from earlier studies in which the phenomenon of promiscuity was not explicitly discussed as such. This chapter introduces the various categories of promiscuity and their possible role in enzyme evolution. We also review some of the well-known examples of enzymes for which the promiscuous function is industrially relevant, leading to conversions of nonnatural compounds. In some cases these promiscuous activities have been improved by site-directed and/or random mutagenesis methods, suggesting that the study of catalytic promiscuity may help identification of new functions by directed evolution. Thus, enzyme promiscuity is becoming an increasingly important phenomenon in the field of biocatalysis, both directly and indirectly: directly, because many desirable conversions are "unnatural," so expansion of the existing functions of protein catalysts is necessary to convert these "promiscuous" substrates; indirectly, through the observation that promiscuous enzymes may be particularly evolvable, opening a wide range of possibilities for engineering these potentially valuable side activities.

36.2. ENZYME PROMISCUITY: DEFINITION AND OVERVIEW

Promiscuity is the ability of an enzyme to catalyze a reaction other than its native one, i.e., active site residues can recognize and turn over alternative substrates. Promiscuity can be divided conceptually into several classes, based on the similarities and/or differences between the two or more substrates that are accepted by one enzyme, as illustrated in Fig. 1. The examples mentioned are discussed in more detail in subsequent paragraphs.

1. *Substrate promiscuity* (Fig. 1A)—sometimes described as "substrate ambiguity" or simply referred to as "substrate range"—is the ability to perform the same chemical reaction on different substrates. For instance, many hydrolases, such as esterases (53) and epoxide hydrolases (3, 101, 116), accept a wide range of substrates with different substituents, but always tackle the same chemical functionality. This type of promiscuity, e.g., in esterases and lipases, is used successfully in fine-chemical synthesis (53). It is often difficult to derive molecular recognition rules from the accidental discovery of this phenomenon because most studies exploring the substrate range of enzymes have been driven by application to industrially relevant compounds. The recent systematic analysis of a comprehensive set of more than 30 promiscuous substrates in a biosynthetic enzyme (TycA, an adenylation domain involved in nonribosomal peptide synthesis) suggests that profiling the active-site properties that determine enzyme-substrate recognition is possible based on an extensive survey of alternative substrates (117).

2. In *product promiscuity*, one substrate is converted into various products (Fig. 1B) by use of similar or identical chemistry. Fascinating examples of this principle are the various sesquiterpene cyclases. This class of enzyme uses the same farnesyl diphosphate substrate that can be

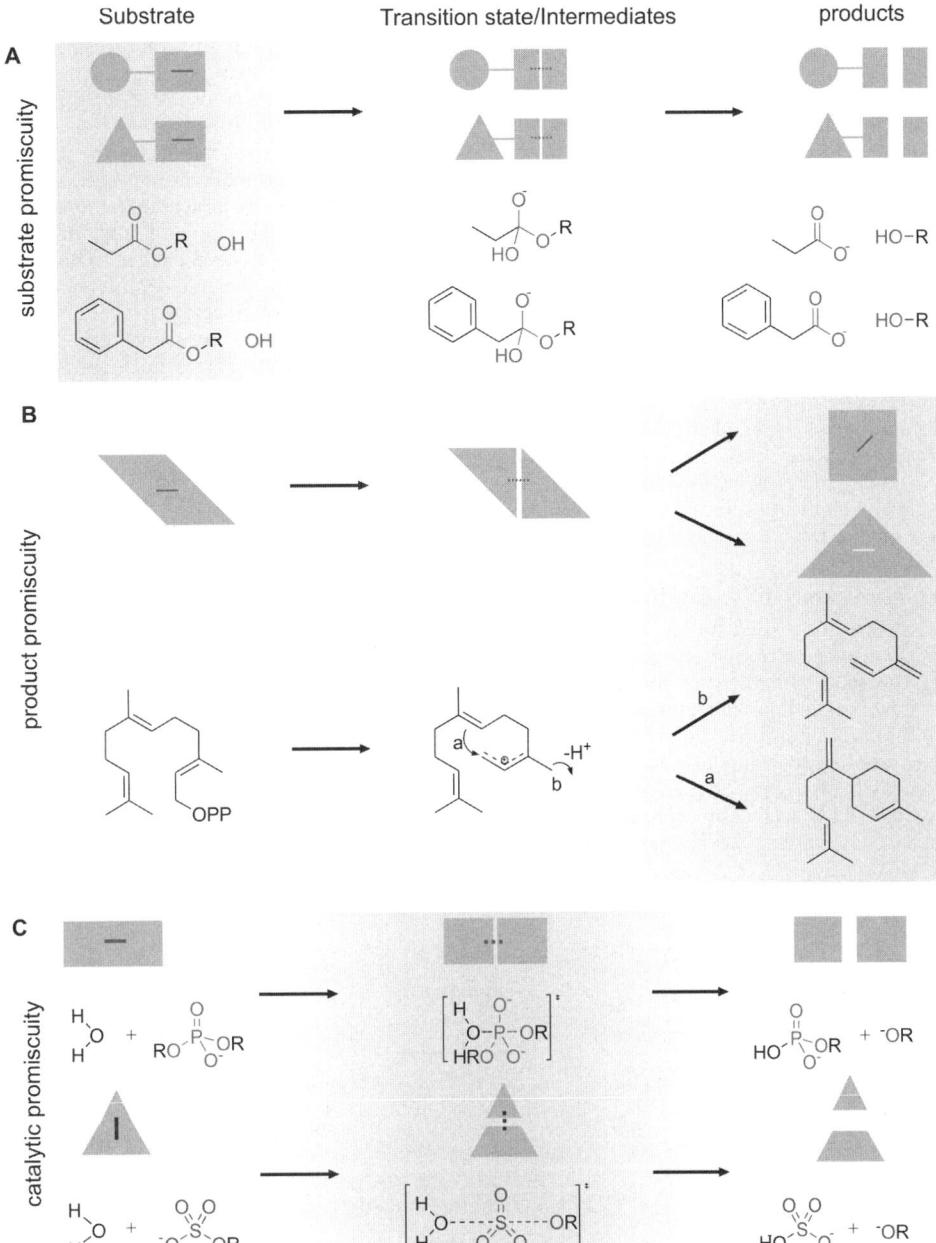

FIGURE 1 Different types of enzyme promiscuity. Promiscuous molecular recognition can take place at different stages of a reaction cycle. (A) In substrate promiscuity, the enzyme is able to recognize substrates with different structural features, but the chemical transformation performed is identical despite different substituents. (B) In product promiscuity, one substrate can be converted to many alternative products. The molecular recognition of the intermediate is conserved, but the folding of the intermediate or rearrangements of the reactive centers on the intermediate, e.g., by steric complementarity of the active-site template, lead to a variety of products onward from this reactive intermediate. (C) To achieve catalytic promiscuity, the tightest interactions—those in the transition state—have to be tolerant enough to recognize different reaction centers and bond-making and -breaking processes around them.

converted into various cyclic products, depending on carbocation rearrangements that take place in the active site (100). Here, the active site provides a template for "folding" a reactive intermediate into a conformation that leads to a transition state resembling the product structure. Depending on the shape of the active site, rerouting of a common reactive intermediate leads to various "promiscuous" alternative products.

3. *Catalytic promiscuity* is the ability of one enzyme to catalyze chemically different conversions (Fig. 1C).

This may be the same class of reaction (e.g., hydrolysis), but the bonds made and broken or the transition states differ between native and promiscuous functions. A prominent example for a structural superfamily of hydrolases catalyzing a range of chemistries is the ability of phosphatases to hydrolyze sulfate ester bonds and vice versa (4, 55, 66, 84, 86, 87, 115a, 124). Alternatively, the promiscuous function facilitates a different class of reaction altogether, using at least part of the same catalytic machinery, as is the case for the carbon-carbon and carbon-oxygen bond-forming reactions that are catalyzed by *Candida antarctica* lipase B (CALB), which normally hydrolyzes carboxylic ester bonds (10, 12, 13, 104, 105).

The following review discusses examples for each of these types of promiscuity with special attention for enzymes of relevance for applied biocatalysis. Several other reviews are available that cover the growing body of evidence in this area and discuss its implication for a variety of research fields (8, 21, 38, 49, 54, 55, 61, 83, 85).

36.2.1. Hydrolase Promiscuity: CALB

CALB is a well-known lipase used for the stereoselective production of a range of enantiopure esters and alcohols (36, 118), using the same chemistry as for its native lipolytic activity. CALB is a typical representative of the α/β-hydrolase fold family of enzymes comprising not only a number of other esterases, but also haloalkane dehalogenases and epoxide hydrolases (46, 88). As for most of its family members the CALB active site contains the Asp-His-Ser catalytic triad characteristic of the serine hydrolases

of this family, and also an oxyanion hole that activates the carbonyl functionality of the substrate. The Asp-His pair activates the serine residue for nucleophilic attack on the carbonyl function of the ester substrate (Fig. 2). The oxyanion hole stabilizes the negative charge that develops on the carbonyl oxygen during the transition state. The acyl-enzyme intermediate that is formed is subsequently hydrolyzed by a water molecule that is activated by the Asp-His pair. During this second step, negative charge again develops on the carbonyl oxygen. The water molecule can be replaced by another nucleophile, resulting in transesterification instead of hydrolysis. The abundance of reactive features in this active site seems to be sufficient to activate both an electrophile and a nucleophile in close proximity to a substrate molecule.

It is therefore perhaps not surprising that both the general base (Asp-His pair) and the oxyanion hole play an important role in a number of synthetic reactions that are catalyzed by CALB, making it also catalytically promiscuous. The oxyanion hole appears to be an intrinsically promiscuous feature that can be used by other reactions that also involve charge accumulation on carbonyl oxygens: CALB can also catalyze aldol condensations (10), Michael additions (13, 105), and epoxidations (104) (Fig. 3B to D). A molecular modeling study suggests that CALB can also catalyze Baeyer-Villiger reactions (12) (Fig. 3E), but no experimental data are available to support this claim. For the Michael additions and the epoxidation reactions the carbonyl group of the electrophilic partner in the bimolecular reaction is recognized and activated by the oxyanion hole. The activation occurs through polarization of the carbonyl group and subsequent stabilization of the developing

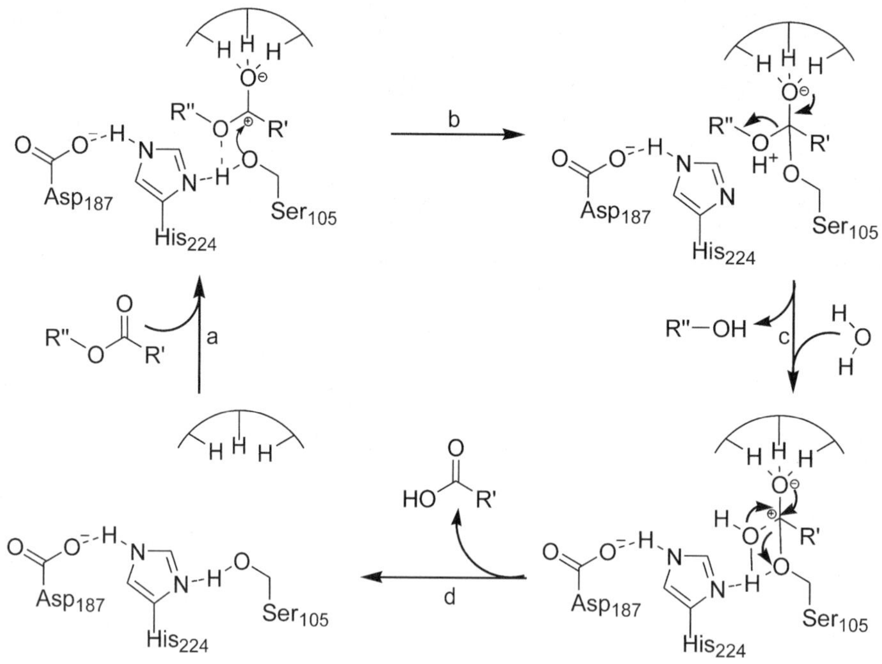

FIGURE 2 Mechanism of the native esterolytic activity of CALB. Upon substrate binding, the carbonyl functionality becomes positioned and polarized by the oxyanion hole (step a). The activated Asp-His pair activates the serine nucleophile, which performs a nucleophilic attack on the carbonyl carbon (b), ultimately resulting in the release of an alcohol and binding of a water molecule (c). The Asp-His pair now activates a water molecule for nucleophilic attack on the carbonyl carbon by general base catalysis, and the oxyanion hole performs the same role as it did for the first step, resulting in hydrolysis of the acyl-enzyme complex and release of the carboxylic acid (d).

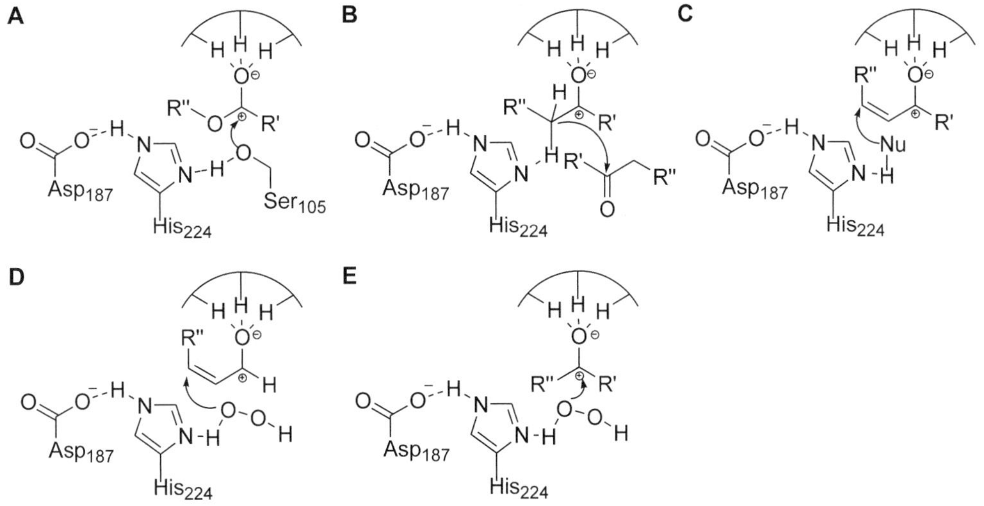

FIGURE 3 Catalytic promiscuity of CALB. The oxyanion hole is a key catalytic feature, polarizing the carbonyl bond of the electrophile in each reaction and offsetting the development of negative charge at oxygen. The reactions catalyzed by CALB are: (A) native carboxylic ester hydrolysis (112); (B) aldol condensation (10); (C) Michael-type addition (13, 105); (D) epoxidation (104); and (E) the predicted Baeyer-Villiger reaction (12). For the promiscuous functions, the catalytic nucleophile does not take part in catalysis. The other parts of the active-site residues play a similar role, and the serine has been replaced by an external nucleophile that is activated by the Asp-His pair, which acts as a general base.

negative charge on this oxygen center, as in its native hydrolytic reaction. The Asp-His pair of the catalytic triad acts as a general base to activate the nucleophile in these promiscuous reactions, again mirroring the mechanism of its native activity. For the aldol condensation, the oxyanion hole and Asp-His pair stabilize an enolate ion. During the formation of the enolate intermediate the histidine picks up a proton. This protonated histidine acts as a general acid in the second step of the reaction, in which the enolate ion reacts with a second carbonyl functionality (10).

The originally catalytic serine nucleophile is not required for all these promiscuous reactions. The mutant CALB Ser105Ala, in which this serine is removed by site-directed mutagenesis, shows improved activity for all promiscuous reactions (10, 13, 104, 105), whereas the native esterolytic activity is severely impaired. This mutant and the wild type seem fairly indiscriminate with respect to the promiscuous substrates: as long as a carbonyl functionality is present in the electrophile, many reactions are possible. Unlike for the native lipase activity, the promiscuous reactions do not involve a covalent intermediate, circumventing the possible formation of a suicide complex between enzyme and substrate.

36.2.2. Promiscuity in Secondary Metabolism: Combinatorial Biosynthesis

Nonribosomal peptide synthases (NRPSs) and polyketide synthases are large multimodular enzymes that are involved in the synthesis of natural products. Many of these natural products have unique bioactivity and are thus of great interest to the pharmaceutical industry. NRPSs and polyketide synthases act as "molecular assembly lines" in which each module of the multienzyme complex incorporates a different building block into the correspondingly modular natural product (11, 62, 122). A temptingly logical approach to produce a variety of altered natural products would be to "mix and match" modules from other enzyme

assembly lines, although transplantation of modules often yields chimeric proteins with impaired activity (82, 97, 98), probably because of the lack of crucial interdomain contacts in the multienzyme complexes (35, 40, 59, 108). Substrate promiscuity of individual modules would greatly facilitate the production of alternative natural products, allowing the incorporation of alternative building blocks while minimizing potentially disruptive module insertion.

One of the most striking examples of the exploitation of substrate promiscuity in such enzymes is the biosynthesis of avermectins, a series of anthelmintic compounds produced by *Streptomyces avermitilis* (Fig. 4). A mutant *S. avermitilis* strain that lacks branched-chain 2-oxo acid dehydrogenase activity is unable to produce avermectin. Production of avermectin can be restored by feeding the appropriate fatty acid to the culture medium (39). Feeding alternative fatty acids to this mutant strain results in avermectin analogs for more than 40 acids that are incorporated (resulting in variation at position R_2 as indicated in Fig. 4), making the loading module for the fatty acid one of the most promiscuous enzymes described. These data also suggest that downstream modules are similarly tolerant: not only the loading module, but also the downstream extension and termination modules must process the alternative intermediates to be able to generate the avermectin framework (28), suggesting that "local" specificity (near the reaction center) is maintained, while "remote" specificity (away from the reaction center) is relaxed. Fusing the avermectin-loading module to the erythromycin elongation and termination modules (comprising some 30 active sites) also resulted in functional antibiotic products, showing that the erythromycin-producing modules are likewise highly tolerant (76). If one carboxylic acid substrate can be fed exclusively—i.e., it is the only available module to be incorporated and thus does not compete against the more efficient incorporation of native substrate—this promiscuity can bring about new natural products and is highly useful. The same pathway-directed evolution

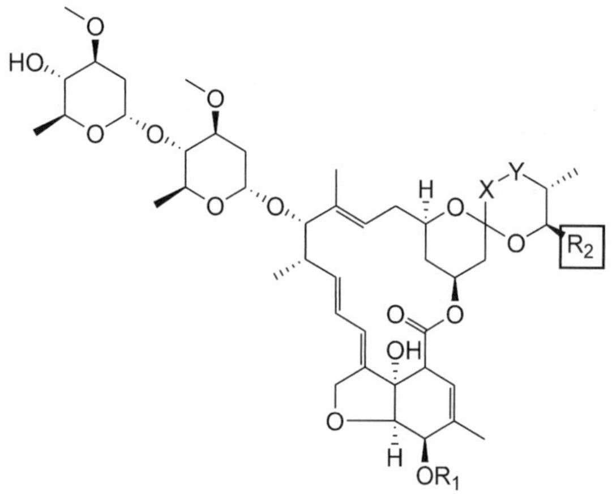

FIGURE 4 Substrate promiscuity allows the combinatorial biosynthesis of avermectin analogs. The R_2 group in the avermectin structure originates from either 2-methylbutyryl-coenzyme A or isobutyryl-coenzyme A, which are generated from L-leucine and L-valine, respectively. In *S. avermitilis* ATCC 53569, a branched-chain α-keto acid dehydrogenase that is essential for the in vivo production of the branched-chain fatty acids is absent (27, 39), rendering this strain unable to produce avermectins. Feeding up 40 different fatty acid compounds (with a variety of substitutions in R_2) to this mutant strain restores the biosynthetic machinery, resulting, in each case, in a different avermectin analog (28). For instance, in doramectin biosynthesis, cyclohexane carboxylic acid is incorporated (102), replacing either methylbutyric acid or isobutyric acid in the wild-type strain (39).

has been used (102, 103) to shift the product ratio in favor of the desired analog doramectin (28, 39), which incorporates a promiscuous substrate, cyclohexane carboxylic acid, to replace the native butyric acid (28, 39). No kinetic data on these transformations are available, so the extent of promiscuity is only qualitatively, not quantitatively characterized.

There are a number of indications that, in principle, NRPS enzymes can also be promiscuous. For example, the composition of peptide antibiotics depends on the amino acid-feeding regime during their production (37, 67, 78). However, in this case, a kinetic analysis is available to inform on the quantitative aspects of the specificity of catalysis, rather than the largely qualitative detection of product. A recent systematic study showed that TycA, the first module

in tyrocidine biosynthesis, can accept 29 different amino acids besides the native phenylalanine substrate (117). This large number of substrates was systematically characterized by their relative catalytic efficiencies (measured by k_{cat}/K_m) highlighting the factors involved in substrate recognition. Although, in principle, many substrates are accepted, the efficiency varies greatly: k_{cat}/K_m values range over 6 orders of magnitude. The hydrophobicity of the substrates (measured by the log P for the substrate side chain) showed a linear correlation, suggesting that matching the hydrophobicity of substrate and active site is a key requirement of achieving promiscuity. In addition, there are other layers of substrate recognition control, namely, size exclusion, that make this enzyme by direct comparison with aminoacyl-tRNA synthetases similarly specific. Aminoacyl-tRNA synthetases catalyze the same chemical reaction, but are involved in safeguarding the fidelity of protein production during translation. The observation of similar discrimination—and, in fact, similar promiscuity—in both enzymes may suggest the generality of these observations. By extension, one may assume that promiscuity is a feature even of enzymes under evolutionary pressure to be specific to some extent and thus a general property of active sites, which could be quantified if substrate collections (and enough time for such painstaking, repetitive experiments) were available. The low activities for the promiscuous reactions in TycA and the observed discrimination suggest that not every module in combinatorial biosynthesis is tolerant. Downstream domains (e.g., those responsible for condensation [5, 29] of amino acid building blocks and processing by cyclization [64, 111]) have also been shown to possess some specificity. Collectively, these results suggest that enzymes involved in secondary metabolism can be nonspecific (as has been suggested [30, 31]), but that the actual promiscuity should be tested on a case-by-case basis.

36.2.3. Tolerance for the Catalytic Nucleophile: Halohydrin Lyases/Haloalcohol Dehalogenases

Halohydrin lyases or haloalcohol dehalogenases are enzymes that catalyze the dehalogenation of vicinal haloalcohols by intramolecular attack of the hydroxyl group that displaces the halide, resulting in an epoxide, a halide ion, and a proton. They are found in various bacteria that are able to use halogenated hydrocarbons as a sole source of carbon and energy (14, 94, 113, 114). In these organisms they are part of the degradation pathways of halo propanols and 1,2-dibromoethane in concert with an epoxide hydrolase (Fig. 5A) and haloalkane dehalogenase (Fig. 5B), respectively (94, 113).

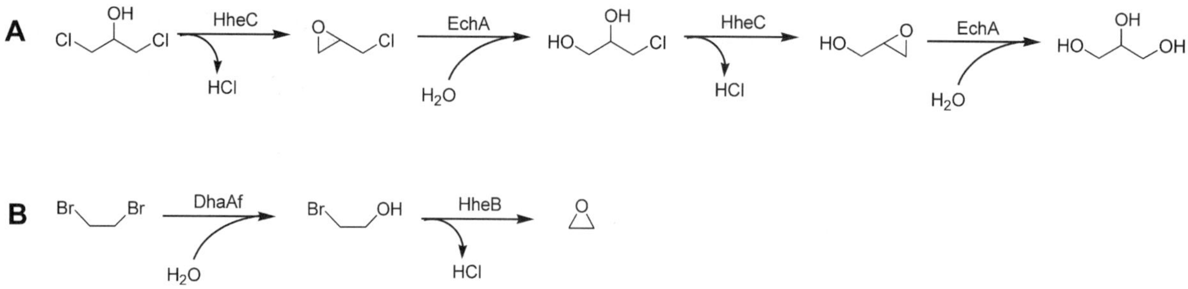

FIGURE 5 Biodegradation routes for halogenated hydrocarbons involving haloalcohol dehalogenases. (A) Degradation of 1,3-dichloro-2-propanol to glycerol in *A. radiobacter* AD1, catalyzed by a haloalcohol dehalogenase (HheC) and an epoxide hydrolase (EchA) (113). (B) Degradation of 1,2-dibromoethane to ethylene oxide in *Mycobacterium* sp. GP1, catalyzed by a haloalkane dehalogenase (DhaAf) and a haloalcohol dehalogenase (HheB) (94).

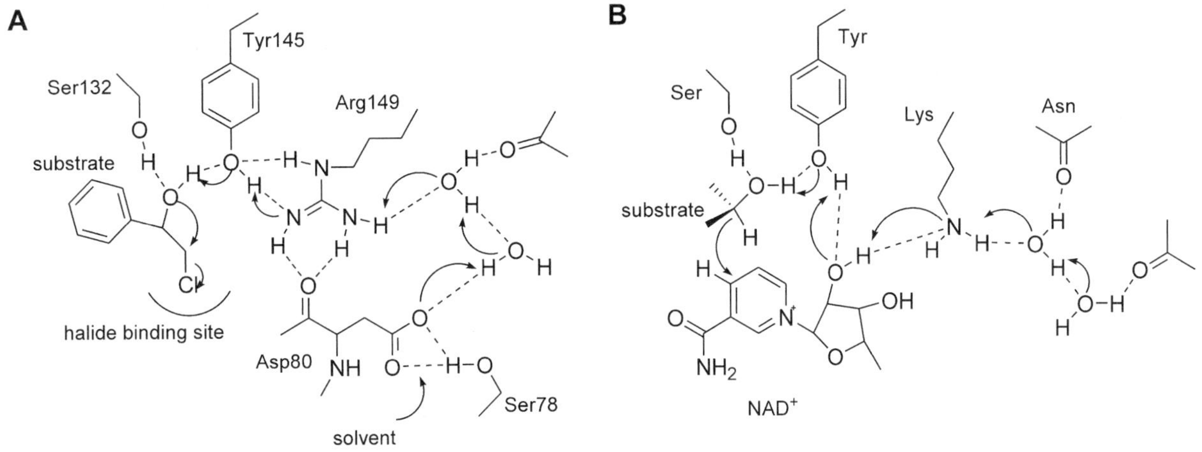

A

B

FIGURE 6 Mechanism of haloalcohol dehalogenases and their likely precursors, the short-chain dehydrogenases/reductases (SDRs). (A) Proposed mechanism of *A. radiobacter* AD1 haloalcohol dehalogenase (HheC)-catalyzed ring closure of a vicinal haloalcohol. The Tyr145/Arg149 pair acts as a general base to abstract a proton from the hydroxyl group, resulting in an intramolecular nucleophilic displacement of the halide, aided by leaving group stabilization by the halide binding site (25). (B) Proposed mechanism of SDR enzymes, in which the proton abstraction from the hydroxyl group is followed by hydrogen abstraction by the NAD$^+$, resulting in the corresponding ketone.

Sequence similarity searches and sequence alignments show that haloalcohol dehalogenases are related to proteins from the short-chain dehydrogenase/reductase (SDR) family of enzymes (115). These NAD(P)H-dependent oxidoreductases catalyze a wide variety of oxidation and reduction reactions. The X-ray structures of two haloalcohol dehalogenases strengthened the conclusion that these dehalogenases are related to SDR proteins (24, 25), and showed that the proposed divergence of the haloalcohol dehalogenases from the SDR precursors occurred twice: one divergence resulted in the A- and C-type haloalcohol dehalogenases, the other in the B type (25). The main difference between SDR proteins and the halohydrin lyases is the absence of a nicotinamide cofactor in the latter. In SDR proteins this cofactor accepts/donates a hydride from/to the alcohol/ketone substrate. In haloalcohol dehalogenases the location at which the sugar phosphate part of the cofactor would bind in SDR proteins is filled up with amino acids, and the space where the nicotinamide ring would bind is now the halide binding site (Fig. 6) (25).

As with the SDR proteins, the reactions that these haloalcohol dehalogenases catalyze are reversible. For vicinal haloalcohols, the equilibrium lies toward the side of the epoxide and the free halide. These enzymes are promiscuous with regard to the nucleophiles they can accept in the ring-opening reaction. This ability was first acknowledged in the early 1990s, when haloalcohol dehalogenases from *Corynebacterium* sp. strain N-1074 were used for the enantioselective ring opening of epoxides with cyanide as the nucleophile (80, 81). More extensive research, especially with the haloalcohol dehalogenase from *Agrobacterium radiobacter* AD1 (HheC), showed that up to nine different nucleophiles are accepted by the enzyme (Fig. 7), depending on which epoxide substrate is used (41, 42, 68, 70–74). For the three different halides (Cl$^-$, Br$^-$, and I$^-$) the equilibrium of the reaction is on the side of the epoxide, as mentioned above, but for all other accepted nucleophiles the equilibrium is shifted toward the vicinal alcohol (42).

The molecular basis for this remarkable promiscuity was suggested to be possible when alternative nucleophiles were known halide mimics (25). A systematic screening of possible nucleophiles for the HheC-catalyzed ring opening of 1,2-epoxybutane showed a preference of the enzyme for relatively small anionic linear nucleophiles (42). Nonanionic nucleophiles, such as primary alcohols and amines, were not accepted and neither were more bulky nonlinear ions such as phosphates, sulfates, and organic carboxylates. To be accepted in the halide binding site a molecule must be small and negatively charged and have "linear" geometry (42), although the binding of several of the accepted nucleophiles may be in a different location from the halide binding site (42, 68). Fluoride ion, the least polarizable halide nucleophile, was not accepted, in line with the extreme recalcitrance of fluorinated hydrocarbons compared with their chlorinated and brominated counterparts. The promiscuous cyanolysis activity appears to be present in most of the haloalcohol dehalogenases that have been characterized so far (71, 80, 81). A systematic understanding of nucleophile acceptance in the different classes of haloalcohol dehalogenases (A, B, and C types) is lacking, and it is not known whether the A and B types can use nucleophiles that are not accepted by HheC.

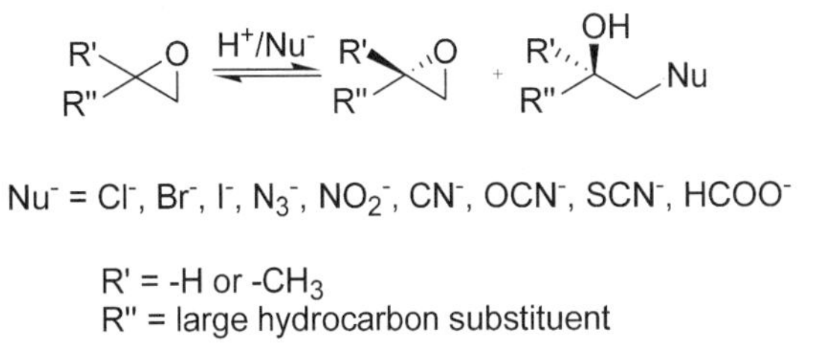

Nu$^-$ = Cl$^-$, Br$^-$, I$^-$, N$_3^-$, NO$_2^-$, CN$^-$, OCN$^-$, SCN$^-$, HCOO$^-$

R' = -H or -CH$_3$
R" = large hydrocarbon substituent

FIGURE 7 Nucleophile promiscuity for the halohydrin dehalogenase-catalyzed epoxide ring opening. The position of the equilibrium is dependent on the nucleophile.

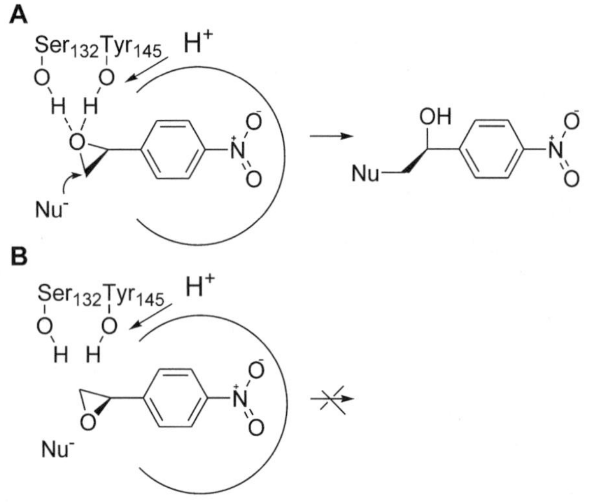

FIGURE 8 Schematic representation of (A) (*R*)- and (B) (*S*)-pNSO in the active site of haloalcohol dehalogenase from *A. radiobacter* AD1 (HheC) (26). The proposed difference in binding modes is based on X-ray structures of HheC in complex with (*R*)- and (*S*)-pNSO, respectively. Both structures were refined with *p*-nitrophenylcyclopropane as the substrate to obtain an unbiased view of the interactions of the Ser132/Tyr145 pair with the epoxide ring. These data show that the refinement data for both structures can be completely superimposed, apart from the epoxide ring, in which the (*R*)-enantiomer comes close enough to the Ser/Tyr pair to form a productive hydrogen bond. The calculated difference in Gibbs binding energy for both structures is also in agreement with the loss of one hydrogen bond.

In contrast to the very promiscuous behavior with respect to the nucleophile in the ring-opening reaction, HheC is often selective with respect to substrate stereochemistry, since the enantioselectivity as represented by the *E* value is often high (*E* > 100) (41, 68–71, 73, 74). The basis for this behavior can be explained based on X-ray structures of HheC in complex with both (*R*)- and (*S*)-*para*-nitrostyrene oxide (pNSO) (Fig. 8), in which the substituent on the epoxide ring guides substrate binding. This means that binding of one of the enantiomers results in an unproductive enzyme-substrate complex since the carbon atom that would normally be susceptible to nucleophilic ring opening is too far away from a water molecule sitting where the nucleophile would normally bind (26). The stereoselectivities toward epoxides with smaller substituents are lower, probably because of higher degrees of freedom with respect to binding

of the side chain, resulting in productive binding modes for both enantiomers. The structure of the A-type haloalcohol dehalogenase from *Arthrobacter* sp. (HheA) (24) revealed a more spacious active site, which was consistent with its lower stereoselectivity compared with HheC (71, 115).

Many of the promiscuous reactions catalyzed by Hhes result in the stereoselective production of a number of interesting compounds. In one particular example, the native and one of the promiscuous activities of HheC W249F, a mutant with improved stereoselectivity (107), were exploited for the one-pot conversion of a racemic haloalcohol to the corresponding (*S*)-cyano alcohol, leaving the (*R*)-haloalcohol largely unreacted (72) (Fig. 9). The first step of the synthesis involved the native activity (haloalcohol ring closure); the second step was one of the promiscuous activities (cyanolysis of an epoxide). Both reactions were enantioselective, and the one-pot approach amplified this phenomenon.

36.2.4. Promiscuity in Cofactor-Dependent Enzymes: Baeyer-Villiger Monooxygenases

The Baeyer-Villiger reaction is the conversion of a ketone into a carboxylic ester using peracid as the oxidative reagent, and it is a reaction of industrial relevance. Baeyer-Villiger monooxygenases (BVMOs) are flavin-dependent monooxygenases that use molecular oxygen as the oxidant. All BVMOs that have been characterized thus far use NADPH to reduce the flavin cofactor during the catalytic cycle (33, 34, 57). The reducing equivalents (H^- and H^+) are ultimately accepted by the remaining oxygen atom to form H_2O (Fig. 10). The first BVMO that was extensively characterized, and for which the gene was cloned, was cyclohexanone monooxygenase from *Acinetobacter* NCIB 9871 (16). Since then various other BVMOs have been cloned and overexpressed (34, 50, 58, 65, 79), and an X-ray structure was solved for phenyl acetone monooxygenase (PAMO), providing insight into the mechanism of BVMOs (75). Based on early work by Walsh and Chen on cyclohexanone monooxygenase (119) and, more recently, by Fraaije and coworkers on structural and kinetic studies with PAMO (75, 109), a mechanism was proposed for the native reaction of BVMOs (Fig. 10). The central feature of these enzymes is the highly reactive peroxyflavin intermediate, which is responsible for catalysis. A strictly conserved arginine was originally thought to be mainly responsible for stabilization of the peroxyflavin intermediate (75), but spectroscopic and kinetic data for an alanine mutant at this conserved position showed that the peroxyflavin intermediate could still be formed and, in fact, had a lower decomposition rate than in the wild type (109).

BVMOs are promiscuous enzymes and have been shown to catalyze epoxidations (17), boron oxidations (9, 22, 23), sulfoxidations (15, 18, 19, 22, 23), seleno oxidations (119),

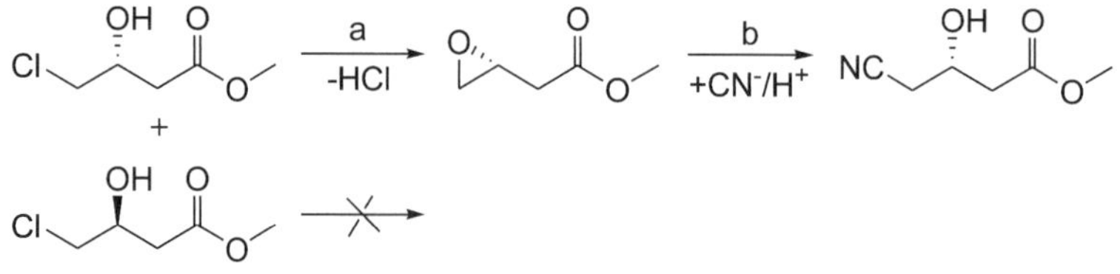

FIGURE 9 One-pot synthesis of (*S*)-4-cyano-3-hydroxybutanoate methyl ester from racemic 4-chloro-3-hydroxybutanoate methylester using a mutant haloalcohol dehalogenase (HheC W249F [107]). This method makes use of both the native activity (step a) and one of the promiscuous activities (step b) of HheC (72).

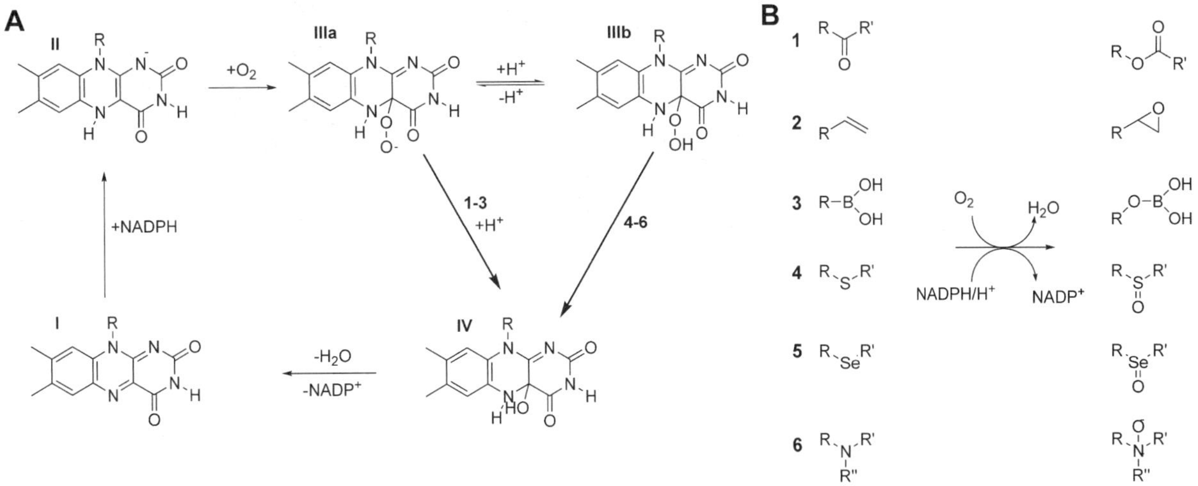

FIGURE 10 Catalytic promiscuity of BVMOs. (A) Proposed reaction cycle for each of the different conversions (57, 109). The reaction cycle starts with reduction of the oxidized flavin (I) by NADPH, resulting in reduced flavin (II) and NADP$^+$, which is released from the enzyme only at the end of the catalytic cycle. The reduced flavin subsequently reacts with molecular oxygen to form the peroxyflavin intermediate (IIIa), which has been suggested to exist also in the protonated state (IIIb). This highly reactive intermediate is stabilized by various interactions with residues in the active site and can react with a variety of different substrates, both electron-rich and electron-deficient substrates (B) (119). The resulting hydroxyflavin (IV) subsequently decomposes into oxidized flavin (I) by eliminating water. Finally, NADP$^+$ is released from the enzyme, allowing the enzyme to start another catalytic cycle.

and amine oxidations (20, 23, 89), besides their native Baeyer-Villiger reaction (Fig. 10). The sulfoxidation activity of a BVMO in *Mycobacterium tuberculosis* (EtaA) (33) is responsible for the activation of thioamide drugs. Mutant strains in which the activity was impaired were therefore resistant to the treatment with these second-line antitubercular drugs. The very promiscuous behavior of these enzymes most likely stems from the intrinsic reactivity of the peroxyflavin intermediate. In solution this intermediate would rapidly decompose, producing hydrogen peroxide. When this is the case for a BVMO it, in principle, should act as an NADPH oxidase.

The enzyme active site provides stabilization of the peroxyflavin and introduces some form of selectivity, since the various BVMOs do differ with respect to the kind of R-groups they can accept (6, 15, 18–20, 22, 23, 34, 57, 77, 119). The detailed mechanisms for the alternative conversions are largely unknown, but some suggestions have been made. The boron oxidations are thought to involve the deprotonated peroxyflavin, as is the case for the native conversion, whereas the heteroatom oxidations (Fig. 10B, reactions 4 to 6) are thought to require the more electrophilic protonated peroxyflavin (57). The pathway for BVMO-catalyzed epoxidation was originally suggested to proceed via the electrophilic peroxyflavin (119), but experimental data obtained by Colonna and coworkers (17) suggest that the nucleophilic peroxyflavin is responsible for this reaction.

The catalytic prowess for each of the different oxidations differs little between the different chemistries. Catalytic efficiency seems at least equally or even more dependent on the carbon skeletons that surround the ketone, sulfide, or amino functionality. An example in support of this idea is the observation that PAMO oxidizes α-methylacetone

(Baeyer-Villiger reaction) and methyl 4-tolylsulfide (sulfoxidation) with equal efficiency ($k_{cat}/K_m = 2.4 \times 10^3$), yet phenylacetone is oxidized with far greater efficiency ($k_{cat}/K_m = 32 \times 10^3$) (34). The reaction pathways for both these reaction types are at least partly similar, since removal of the essential arginine in PAMO abolished both activities (109). In addition, these enzymes are also highly stereoselective for some conversions (22), despite their mechanistic tolerance.

36.3. CROSSWISE CATALYTIC PROMISCUITY IN THE ALKALINE PHOSPHATASE SUPERFAMILY: EFFICIENT PROMISCUOUS HYDROLASES

Perhaps the most systematically studied group of promiscuous enzymes is a group of hydrolytic enzymes related to the familiar textbook enzyme alkaline phosphatase (AP; Fig. 11A). Assignment to the AP superfamily is based on structural homology (despite low sequence homology) and complemented by mechanistic parallels: the active-site residues diverge considerably, but the character of the active-site residues is conserved. Like most metallohydrolases, the proteins in this superfamily benefit from the availability of a nucleophile activated (i.e., deprotonated) by general base catalysis and metal coordination, and from offsetting negative-charge development on the leaving group of phosphoryl/sulfuryl oxygens by Lewis acid catalysis. Figure 11 shows how this general pattern is adapted in individual superfamily members. Remarkably, in the context of this review, the superfamily is also defined by extraordinarily efficient crosswise promiscuity; i.e., the native activity of one family member corresponds to the promiscuous activity of another.

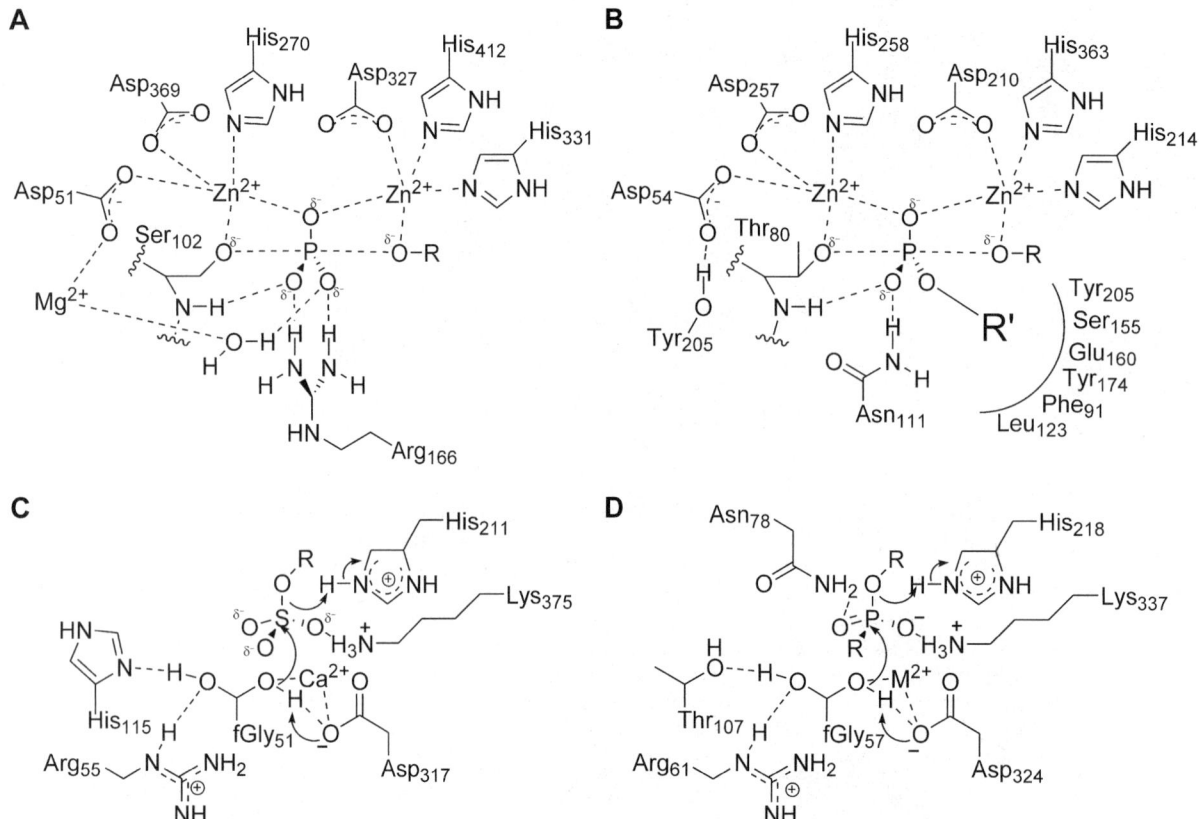

FIGURE 11 Mechanisms of the various hydrolases that belong to the alkaline phosphatase superfamily. (A) *Escherichia coli* AP (99). (B) *Xanthomonas axonopodis* NPP (124). (C) *P. aeruginosa* arylsulfatase (7). (D) Mechanism proposed for the PMH from *Rhizobium leguminosarum* (*Rl*PMH), based on X-ray structural data and the kinetic data of various active-site mutants (56). The active-site residues of PMH from *Burkholderia caryophylli* (*Bc*PMH) superimpose ideally on *Rl*PMH (86% amino acid identity), suggesting that *Bc*PMH uses the same mechanism (115a).

O'Brien and Herschlag pioneered the functional exploration of promiscuity in this superfamily and detected additional promiscuous activities in AP, namely, sulfate monoesterase (84), phosphate diesterase, and phosphonate monoesterase (86) activities in addition to AP's original phosphate monoesterase activity. Promiscuous catalysis occurred with substantial rate accelerations characterized by second-order rate enhancements [$(k_{cat}/K_m)/k_2$] ranging up to 10^{11} (Table 1). A structurally homologous family member, the arylsulfatase from *Pseudomonas aeruginosa* (PAS), exhibits promiscuity mirroring the repertoire of AP. PAS accelerates the hydrolysis of phosphate monoesters by 10^{13} (87) and phosphate diesters by factors between 10^{15} and 10^{18}, depending on the structure of the diester substrate (4).

A further member of the AP superfamily, a phosphonate monoester hydrolase (PMH), shows activity toward the originally assigned cognate substrate phosphonate monoester as well as phosphate mono- and diesters, sulfates, and sulfonates with accelerations ranging from 10^{10} to as high as 10^{18} (Table 1) (115a). PMH thus sums up all activities of other family members, but branches out into new activities, too.

Despite the common net hydrolysis reaction in the AP superfamily, this collection encompasses a range of diverse substrates: their charges range between 0 and -2, the reactions involve transition states of a different nature

(in terms of bond making and breaking) (44, 45, 63), they attack at two different reaction centers (P and S), and they exhibit diverse intrinsic reactivities (with half-lives between 20 and 85,000 years under near-neutral conditions). It is remarkable that one catalytic machine can solve the diverging molecular recognition challenges that are involved in turnover for this set of substrates—with extraordinary rate accelerations.

The data in Table 1 challenge the notion that high rate accelerations can only be observed for specialized enzymes, since PMH is an excellent phosphodiesterase, yet highly promiscuous, and may be adapted further by evolution. One may speculate that the enzymes of this superfamily are derived from a common ancestor that was even less specific, catalyzing all the reactions that are now the extant activities of the enzyme family: this would be the ideal starting point for differentiation of hydrolases by directed evolution.

Promiscuity has been observed in other superfamilies: Glasner et al. (38) used assignment to a superfamily as a clue to identify uncharacterized family members in the enolase family. In a recent review by Poelarends and coworkers (95), the evolutionary relationship between the various members of the tautomerase superfamily, each of which carries an N-terminal proline as part of its active site, was extensively discussed. This proline can act either as a general base or as a general acid. Although there is

TABLE 1 Catalytic promiscuity in the alkaline phosphatase superfamily[a]

Enzyme		$-O{-}\overset{O}{\underset{O^-}{P}}{-}O{-}R$	$R'{-}O{-}\overset{O}{\underset{O^-}{P}}{-}O{-}R$	$R'{-}\overset{O}{\underset{O^-}{P}}{-}O{-}R$	$-O{-}\overset{O\ O}{\underset{}{S}}{-}O{-}R$	$R'{-}O{-}\overset{O}{\underset{O-R''}{P}}{-}O{-}R$	$R'{-}\overset{O\ O}{\underset{}{S}}{-}O{-}R$	Reference(s)
AP	k_{cat}/K_m $(M^{-1}\,s^{-1})$	**$10^{7.5}$**	$10^{-1.3}$	$10^{-1.5}$	$10^{-2.0}$			54, 84, 86
	$(k_{cat}/K_m)/k_2$	**$10^{17.8}$**	$10^{11.58}$	$10^{10.6}$	$10^{9.0}$			
NPP	k_{cat}/K_m $(M^{-1}\,s^{-1})$	$10^{0.04}$	**$10^{3.4}$**		$10^{-4.7}$			66
	$(k_{cat}/K_m)/k_2$	$10^{10.6}$	**$10^{15.9}$**		$10^{6.3}$			
PAS	k_{cat}/K_m $(M^{-1}\,s^{-1})$	$10^{2.9}$	$10^{5.4}$		**$10^{7.7}$**			4, 87
	$(k_{cat}/K_m)/k_2$	$10^{13.3}$	$10^{18.1}$		**$10^{18.7}$**			
BcPMH	k_{cat}/K_m $(M^{-1}\,s^{-1})$	$10^{1.3}$	$10^{4.0}$	**$10^{4.1}$**	$10^{-0.2}$	$10^{-1.7}$	$10^{1.7}$	115a
	$(k_{cat}/K_m)/k_2$	$10^{10.4}$	$10^{18.3}$	**$10^{16.7}$**	$10^{10.5}$	$10^{7.4}$	$10^{11.7}$	

[a] Boldface numbers indicate the primary activity of each family member.

limited evidence of crosswise catalytic promiscuity in this enzyme family with respect to the native substrates (91, 120), a number of family members show hydratase activity (92, 93, 121), unifying these enzymes mechanistically. Furthermore, for this family, the observed promiscuous behavior provided insight into the mechanism of one of the family members (91).

36.4. IMPROVEMENT OF PROMISCUOUS FUNCTIONS BY MUTAGENESIS

Some of the promiscuous functions mentioned above have been improved, either by random methods (32) or by structure-inspired, site-directed mutagenesis (12, 13, 104, 105, 110).

The production of an enantiopure cyano alcohol from the corresponding haloalcohol by HheC, similar to the one depicted in Fig. 9, was improved 4,000-fold (volumetric productivity) by use of protein sequence activity relationships (ProSAR). For this approach diversity was generated by use of various methods such as site-saturation mutagenesis and error-prone PCR, and activity data were obtained for a limited number of variants. This multivariate optimization strategy links the activity of a number of enzyme variants in each round to the mutations they carry, inferring the effect of the contribution of each single mutation. Thereby, detrimental mutations or "genetic hitchhikers" could be sifted out. At the start of each round, new diversity was introduced by mutagenic techniques to maintain diversity. On average, each round resulted in 1.5- to 3-fold improvement of the parent (the best variant from the previous round), which after 18 rounds resulted in 4,000-fold improvement of the activity under production conditions (32). Since the selection includes both the ring closure of a haloalcohol and the ring opening of an epoxide by cyanide, both the native and the promiscuous activity were maintained at a high level in the same catalyst.

The improvement of several of the promiscuous reactions of CALB was obtained by removing the catalytic nucleophile, naturally resulting in decreased performance for the native activity (10, 13, 104, 105). Removal of Ser105

resulted in 2- to 10-fold-increased turnover for all the promiscuous functions compared with wild type. The more spacious active site was apparently more tolerant than wild type for the alternative nucleophiles.

More systematic studies into the role of promiscuous functions for the evolution of new functions were performed with carbonic anhydrase II (which also has weak esterase activity), phosphotriesterase (esterase/lactonase), and PON1 lactonase (phosphotriesterase/esterase) (2). These studies showed that mutations that confer improved promiscuous activities do not necessarily decrease the native activity. This means that a promiscuous function can evolve in an enzyme without compromising its native one, potentially allowing evolution prior to a duplication event. Analysis of the positions of these mutations in the three enzymes under investigation in this study and of a number of literature examples on directed evolution of promiscuous features showed that many of these mutations accumulated in loops and that the active-site residues and residues important for stability were generally unmodified (2).

36.5. CONCLUSIONS

The increasing number of bona fide promiscuous enzymes suggests that their active sites deserve this name in a more general sense than usually applied: apparently the high concentration of functional groups brings about additional activities. New high-throughput screening systems (1, 47, 60, 106) that can characterize multiple activities for entire libraries will no doubt play a role in comprehensively assessing how far this observation can be generalized. The systematic analysis of this phenomenon—either in metagenomic libraries or during cycles of directed evolution—will inform a future strategy for identification and alteration of promiscuous functions.

The question then arising is where promiscuity should be suspected and what the best choice of a starting template for enhancing an existing promiscuous activity by directed evolution is. At the moment, the ultimate answer is not available, but some preliminary rules and requirements can be drawn up to guide the experimentalist to new promiscuous activities, as follows.

1. *The shape of the active site.* At a minimum, alternative substrates have to fit in the active site, to be within reach of the catalytic functionality, so a large, accessible binding pocket is a key requirement for substrate acceptance. For promiscuous hydrolases such as PMH, AP, or PAS, this criterion is sufficient. In addition, conformational flexibility of an active site may promote acceptance of alternative substrates by "induced fit," e.g., based on relatively nonspecific hydrophobic interactions (117).

   For the product-promiscuous enzymes (such as the cationic cyclization reactions carried out by γ-humulene synthase [100]), the rules are somewhat reversed. Here, the shape of the active site is used to favor specificity, and strategically placing them provides access to alternative products.

2. *Reactive active-site groups.* This need for high reactivity is most clearly illustrated by the promiscuous examples that carry reactive cofactors, such as metal ions (in PMH, PAS, and AP [7, 56, 99, 115a]) or flavin (BVMOs [34, 50, 58, 65, 79]), which are not covalently bound to the enzyme. Promiscuous behavior of some of the adjacent amino acid side chains and cofactors can result in promiscuous functions, mainly because of different protonation states, which, depending on the pK_a of any particular group, can exist side by side. For instance, the catalysis carried out by BVMOs is highly dependent on the protonation state of the peroxyflavin intermediate, which makes it react either as a nucleophile (deprotonated) or electrophile (protonated), or, the case of CALB, where His224 can act as a general base and acid in the same reaction (aldol condensation) (10).

3. *Native and promiscuous reactions share some key features.* Stabilization of negative charges that accumulate as the transition state is approached at considerable energetic cost will benefit a range of mechanisms. Groups that could provide such stabilization are, e.g., the oxyanion hole (as in CALB [10, 13, 104, 105]) or metal ions (as in AP, PAS, nucleotide pyrophosphatase/phosphodiesterase [NPP], or PMH [7, 56, 99, 115a, 124]) acting as general acid catalysts and/or electrostatic catalysts. In all reactions catalyzed by these enzymes the buildup of negative charge is part of the mechanism. Similar arguments can be made for the accumulation of positive charges. The positioning of stabilizing groups is important, but—at the cost of efficiency—does not have to match alternative transition states too precisely.

   Fortunately, the observation of degrees of freedom extends to the molecular recognition of transition-state structures for native and promiscuous reactions as defined by the extent of bond making and breaking. For example, both AP and PAS achieve higher rate accelerations for phosphate diester hydrolysis (with an associative transition state) than for their respective promiscuous sulfate or phosphate monoesterase reactions (with a dissociative transition state resembling the transition state of their native reactions) (4, 84, 86, 87) (Table 1). So the nature of the transition state appears to be relatively unimportant, and even for efficient multiple promiscuous catalysis. The same groups must be able to stabilize transition states with different bond-breaking and -making characteristics (115a). The implication is that a reaction mechanism will not easily change its nature and—rather than fitting into the corset of the native reaction—the most favorable interactions with the given active-site functionality will be sought.

4. *Avoiding dead-end complexes.* Covalent intermediates are commonplace in highly evolved enzymes taking advantage of nucleophilic catalysis. This catalytic strategy carries the danger of a dead-end covalent adduct (similar to a suicide substrate), leading to only one turnover and rendering the catalyst a stoichiometric reagent only. Many of the examples of catalytic promiscuity discussed in this review do not involve covalent catalysis, and in the very promiscuous PMH and AS, resolution of the covalent intermediate is unified for all conversions (4, 54, 87, 115a).

For the moment, these emerging rules are consistent with the available body of evidence, but they will have to be revised or refined in response to the discovery of more promiscuous activities. Since biocatalytic application intrinsically involves "nonnatural" conversions, there is a strong commercial driving force behind this effort. Fortunately, even promiscuous enzymes are still chiral reagents, giving access to regio- and stereoselective catalysis (22, 41, 68–71, 73, 74). Thus, enzyme promiscuity remains a useful vehicle for the identification of catalysts for a wider range of green, high-quality, and hopefully efficient biotransformations.

This work was supported by the Medical Research Council. F.H. is an ERC Starting Investigator. We thank Peter Leadlay for helpful suggestions and Stefanie Jonas and Ann Babtie for comments on the manuscript.

REFERENCES

1. **Aharoni, A., A. D. Griffiths, and D. S. Tawfik.** 2005. High-throughput screens and selections of enzyme-encoding genes. *Curr. Opin. Chem. Biol.* **9:**210–216.
2. **Aharoni, A., L. Gaidukov, O. Khersonsky, S. McQ Gould, C. Roodveldt, and D. S. Tawfik.** 2005. The "evolvability" of promiscuous protein functions. *Nat. Genet.* **37:**73–76.
3. **Archelas, A., and R. Furstoss.** 2001. Synthetic applications of epoxide hydrolases. *Curr. Opin. Chem. Biol.* **5:**112–119.
4. **Babtie, A. C., S. Bandyopadhyay, L. F. Olguin, and F. Hollfelder.** 2009. Efficient catalytic promiscuity for chemically distinct reactions. *Angew. Chem. Int. Ed.* **48:**3692–3694.
5. **Belshaw, P. J., C. T. Walsh, and T. Stachelhaus.** 1999. Aminoacyl-CoAs as probes of condensation domain selectivity in nonribosomal peptide synthesis. *Science* **284:**486–489.
6. **Berezina, N., V. Alphand, and R. Furstoss.** 2002. Microbiological transformations. Part 51. The first example of a dynamic kinetic resolution process applied to a microbiological Baeyer-Villiger oxidation. *Tetrahedron Asymmetry* **13:**1953–1955.
7. **Boltes, I., H. Czapinska, A. Kahnert, R. von Bulow, T. Dierks, B. Schmidt, K. von Figura, M. A. Kertesz, and I. Uson.** 2001. 1.3 Å structure of arylsulfatase from *Pseudomonas aeruginosa* establishes the catalytic mechanism of sulfate ester cleavage in the sulfatase family. *Structure* **9:**483–491.
8. **Bornscheuer, U. T., and R. J. Kazlauskas.** 2004. Catalytic promiscuity in biocatalysis: using old enzymes to form new bonds and follow new pathways. *Angew. Chem. Int. Ed.* **43:**6032–6040.
9. **Branchaud, B. P., and C. T. Walsh.** 1985. Functional group diversity in enzymatic oxygenation reactions catalyzed

by bacterial flavin-containing cyclohexanone oxygenase. *J. Am. Chem. Soc.* **107:**2153–2161.

10. **Branneby, C., P. Carlqvist, A. Magnusson, K. Hult, T. Brinck, and P. Berglund.** 2003. Carbon-carbon bonds by hydrolytic enzymes. *J. Am. Chem. Soc.* **125:**874–875.

11. **Cane, D. E., and C. T. Walsh.** 1999. The parallel and convergent universes of polyketide synthases and nonribosomal peptide synthetases. *Chem. Biol.* **6:**R319–R325.

12. **Carlqvist, P., R. Eklund, K. Hult, and T. Brinck.** 2003. Rational design of a lipase to accommodate catalysis of Baeyer-Villiger oxidation with hydrogen peroxide. *J. Mol. Model.* **9:**164–171.

13. **Carlqvist, P., M. Svedendahl, C. Branneby, K. Hult, T. Brinck, and P. Berglund.** 2005. Exploring the active-site of a rationally redesigned lipase for catalysis of Michael-type additions. *Chembiochem* **6:**331–336.

14. **Castro, C. E., and E. W. Bartnicki.** 1968. Biodehalogenation. Epoxidation of halohydrins, epoxide opening and transhalogenation by *Flavobacterium* sp. *Biochemistry* **7:**3213–3218.

15. **Chen, G., M. M. Kayser, M. D. Mihovilovic, M. E. Mrstik, C. A. Martinez, and J. D. Stewart.** 1999. Asymmetric oxidation at sulfur catalyzed by engineered strains that overexpress cyclohexanone monooxygenase. *New J. Chem.* **23:**827–832.

16. **Chen, Y. C., O. P. Peoples, and C. T. Walsh.** 1988. *Acinetobacter* cyclohexanone monooxygenase: gene cloning and sequence determination. *J. Bacteriol.* **170:**781–789.

17. **Colonna, S., N. Gaggero, G. Carrea, G. Ottolina, P. Pasta, and F. Zambianchi.** 2002. First asymmetric epoxidation catalysed by cyclohexanone monooxygenase. *Tetrahedron Lett.* **43:**1797–1799.

18. **Colonna, S., N. Gaggero, G. Carrea, and P. Pasta.** 1998. Oxidation of organic cyclic sulfites to sulfates: a new reaction catalyzed by cyclohexanone monooxygenase. *Chem. Commun.* 415–416.

19. **Colonna, S., N. Gaggero, G. Carrea, P. Pasta, V. Alphand, and R. Furstoss.** 2001. Enantioselective synthesis of *tert*-butyl *tert*-butanethiosulfinate catalyzed by cyclohexanone monooxygenase. *Chirality* **13:**40–42.

20. **Colonna, S., V. Pironti, P. Pasta, and F. Zambianchi.** 2003. Oxidation of amines catalyzed by cyclohexanone monooxygenase. *Tetrahedron Lett.* **44:**869–871.

21. **Copley, S. D.** 2003. Enzymes with extra talents: moonlighting functions and catalytic promiscuity. *Curr. Opin. Chem. Biol.* **7:**265–272.

22. **de Gonzalo, G., D. E. Torres Pazmino, G. Ottolina, M. W. Fraaije, and G. Carrea.** 2006. 4-Hydroxyacetophenone monooxygenase from *Pseudomonas fluorescens* ACB as an oxidative biocatalyst in the synthesis of optically active sulfoxides. *Tetrahedron Asymmetry* **17:**130–135.

23. **de Gonzalo, G., D. E. Torres Pazmino, G. Ottolina, M. W. Fraaije, and G. Carrea.** 2005. Oxidations catalyzed by phenylacetone monooxygenase from *Thermobifida fusca*. *Tetrahedron Asymmetry* **16:**3077–3083.

24. **de Jong, R. M., K. H. Kalk, L. Tang, D. B. Janssen, and B. W. Dijkstra.** 2006. The X-ray structure of the haloalcohol dehalogenase HheA from *Arthrobacter* sp. strain AD2: insight into enantioselectivity and halide binding in the haloalcohol dehalogenase family. *J. Bacteriol.* **188:**4051–4056.

25. **de Jong, R. M., J. J. Tiesinga, H. J. Rozeboom, K. H. Kalk, L. Tang, D. B. Janssen, and B. W. Dijkstra.** 2003. Structure and mechanism of a bacterial haloalcohol dehalogenase: a new variation of the short-chain dehydrogenase/reductase fold without an NAD(P)H binding site. *EMBO J.* **22:**4933–4944.

26. **de Jong, R. M., J. J. Tiesinga, A. Villa, L. Tang, D. B. Janssen, and B. W. Dijkstra.** 2005. Structural basis for the enantioselectivity of an epoxide ring opening reaction catalyzed by halo alcohol dehalogenase HheC. *J. Am. Chem. Soc.* **127:**13338–13343.

27. **Denoya, C. D., R. W. Fedechko, E. W. Hafner, H. A. McArthur, M. R. Morgenstern, D. D. Skinner, K. Stutzman-Engwall, R. G. Wax, and W. C. Wernau.** 1995. A second branched-chain alpha-keto acid dehydrogenase gene cluster (*bkdFGH*) from *Streptomyces avermitilis*: its relationship to avermectin biosynthesis and the construction of a *bkdF* mutant suitable for the production of novel antiparasitic avermectins. *J. Bacteriol.* **177:**3504–3411.

28. **Dutton, C. J., S. P. Gibson, A. C. Goudie, K. S. Holdom, M. S. Pacey, J. C. Ruddock, J. D. Bu'Lock, and M. K. Richards.** 1991. Novel avermectins produced by mutational biosynthesis. *J. Antibiot.* (Tokyo) **44:**357–365.

29. **Ehmann, D. E., J. W. Trauger, T. Stachelhaus, and C. T. Walsh.** 2000. Aminoacyl-SNACs as small-molecule substrates for the condensation domains of nonribosomal peptide synthetases. *Chem. Biol.* **7:**765–772.

30. **Firn, R. D., and C. G. Jones.** 2000. The evolution of secondary metabolism—a unifying model. *Mol. Microbiol.* **37:**989–994.

31. **Fischbach, M. A., and J. Clardy.** 2007. One pathway, many products. *Nat. Chem. Biol.* **3:**353–355.

32. **Fox, R. J., S. C. Davis, E. C. Mundorff, L. M. Newman, V. Gavrilovic, S. K. Ma, L. M. Chung, C. Ching, S. Tam, S. Muley, J. Grate, J. Gruber, J. C. Whitman, R. A. Sheldon, and G. W. Huisman.** 2007. Improving catalytic function by ProSAR-driven enzyme evolution. *Nat. Biotechnol.* **25:**338–344.

33. **Fraaije, M. W., N. M. Kamerbeek, A. J. Heidekamp, R. Fortin, and D. B. Janssen.** 2004. The prodrug activator EtaA from *Mycobacterium tuberculosis* is a Baeyer-Villiger monooxygenase. *J. Biol. Chem.* **279:**3354–3360.

34. **Fraaije, M. W., J. Wu, D. P. Heuts, E. W. van Hellemond, J. H. Spelberg, and D. B. Janssen.** 2005. Discovery of a thermostable Baeyer-Villiger monooxygenase by genome mining. *Appl. Microbiol. Biotechnol.* **66:**393–400.

35. **Frueh, D. P., H. Arthanari, A. Koglin, D. A. Vosburg, A. E. Bennett, C. T. Walsh, and G. Wagner.** 2008. Dynamic thiolation-thioesterase structure of a non-ribosomal peptide synthetase. *Nature* **454:**903–906.

36. **Frykman, H., N. Ohrner, T. Norin, and K. Hult.** 1993. S-Ethyl thiooctanoate as acyl donor in lipase catalysed resolution of secondary alcohols. *Tetrahedron Lett.* **34:**1367–1370.

37. **Galli, G., F. Rodriguez, P. Cosmina, C. Pratesi, R. Nogarotto, F. de Ferra, and G. Grandi.** 1994. Characterization of the surfactin synthetase multi-enzyme complex. *Biochim. Biophys. Acta* **1205:**19–28.

38. **Glasner, M. E., J. A. Gerlt, and P. C. Babbitt.** 2006. Evolution of enzyme superfamilies. *Curr. Opin. Chem. Biol.* **10:**492–497.

39. **Hafner, E. W., B. W. Holley, K. S. Holdom, S. E. Lee, R. G. Wax, D. Beck, H. A. McArthur, and W. C. Wernau.** 1991. Branched-chain fatty acid requirement for avermectin production by a mutant of *Streptomyces avermitilis* lacking branched-chain 2-oxo acid dehydrogenase activity. *J. Antibiot.* (Tokyo) **44:**349–356.

40. **Hans, M., A. Hornung, A. Dziarnowski, D. E. Cane, and C. Khosla.** 2003. Mechanistic analysis of acyl transferase domain exchange in polyketide synthase modules. *J. Am. Chem. Soc.* **125:**5366–5374.

41. **Hasnaoui, G., J. H. Lutje Spelberg, E. J. de Vries, L. Tang, B. Hauer, and D. B. Janssen.** 2005. Nitrite-mediated hydrolysis of epoxides catalyzed by halohydrin dehalogenase from *Agrobacterium radiobacter* AD1: a new tool for the kinetic resolution of epoxides. *Tetrahedron Asymmetry* **16:**1685–1692.

42. **Hasnaoui-Dijoux, G., M. Majeric Elenkov, J. H. Lutje Spelberg, B. Hauer, and D. B. Janssen.** 2008. Catalytic promiscuity of halohydrin dehalogenase and its application

in enantioselective epoxide ring opening. *Chembiochem* **9**:1048–1051.

43. **Hecht, M. H., A. Das, A. Go, L. H. Bradley, and Y. Wei.** 2004. *De novo* proteins from designed combinatorial libraries. *Protein Sci.* **13**:1711–1723.

44. **Hengge, A. C., and I. Onyido.** 2005. Physical organic perspectives on phospho group transfer from phosphates and phosphonates. *Curr. Org. Chem.* **9**:61–74.

45. **Hollfelder, F., and D. Herschlag.** 1995. The nature of the transition state for enzyme-catalyzed phosphoryl transfer. Hydrolysis of O-aryl phosphorothioates by alkaline phosphatase. *Biochemistry* **34**:12255–12264.

46. **Holmquist, M.** 2000. Alpha/beta-hydrolase fold enzymes: structures, functions and mechanisms. *Curr. Protein Pept. Sci.* **1**:209–235.

47. **Huebner, A., S. Sharma, M. Srisa-Art, F. Hollfelder, J. B. Edel, and A. J. Demello.** 2008. Microdroplets: a sea of applications? *Lab. Chip* **8**:1244–1254.

48. **Hughes, A. L.** 1994. The evolution of functionally novel proteins after gene duplication. *Proc. Biol. Sci.* **256**:119–124.

49. **Hult, K., and P. Berglund.** 2007. Enzyme promiscuity: mechanism and applications. *Trends Biotechnol.* **25**:231–238.

50. **Iwaki, H., Y. Hasegawa, S. Wang, M. M. Kayser, and P. C. Lau.** 2002. Cloning and characterization of a gene cluster involved in cyclopentanol metabolism in *Comamonas* sp. strain NCIMB 9872 and biotransformations effected by *Escherichia coli*-expressed cyclopentanone 1,2-monooxygenase. *Appl. Environ. Microbiol.* **68**:5671–5684.

51. **Jensen, R. A.** 1976. Enzyme recruitment in evolution of new function. *Annu. Rev. Microbiol.* **30**:409–425.

52. **Jiang, L., E. A. Althoff, F. R. Clemente, L. Doyle, D. Rothlisberger, A. Zanghellini, J. L. Gallaher, J. L. Betker, F. Tanaka, C. F. Barbas III, D. Hilvert, K. N. Houk, B. L. Stoddard, and D. Baker.** 2008. *De novo* computational design of retro-aldol enzymes. *Science* **319**:1387–1391.

53. **Jing, Q., and R. J. Kazlauskas.** 2008. Determination of absolute configuration of secondary alcohols using lipase-catalyzed kinetic resolutions. *Chirality* **20**:724–735.

54. **Jonas, S., and F. Hollfelder.** 2009. Mechanism and catalytic promiscuity: emerging mechanistic principles for identification and manipulation of catalytically promiscuous enzymes, p. 47–79. *In* S. Lutz and U. T. Bornscheuer (ed.), *Protein Engineering Handbook*, vol. 1. Wiley VCH, Weinheim, Germany.

55. **Jonas, S., and F. Hollfelder.** 2009. Mapping catalytic promiscuity in the alkaline phosphatase superfamily. *Pure Appl. Chem.* **81**:733–744.

56. **Jonas, S., B. van Loo, M. Hyvonen, and F. Hollfelder.** 2008. A new member of the alkaline phosphatase superfamily with a formylglycine nucleophile: structural and kinetic characterisation of a phosphonate monoester hydrolase/ phosphodiesterase from *Rhizobium leguminosarum*. *J. Mol. Biol.* **384**:120–136.

57. **Kamerbeek, N. M., D. B. Janssen, W. J. H. van Berkel, and M. W. Fraaije.** 2003. Baeyer-Villiger monooxygenases, an emerging family of flavin-dependent biocatalysts. *Adv. Synth. Catal.* **345**:667–678.

58. **Kamerbeek, N. M., M. J. Moonen, J. G. Van Der Ven, W. J. Van Berkel, M. W. Fraaije, and D. B. Janssen.** 2001. 4-Hydroxyacetophenone monooxygenase from *Pseudomonas fluorescens* ACB. A novel flavoprotein catalyzing Baeyer-Villiger oxidation of aromatic compounds. *Eur. J. Biochem.* **268**:2547–2557.

59. **Kapur, S., and C. Khosla.** 2008. Biochemistry: fit for an enzyme. *Nature* **454**:832–833.

60. **Kelly, B. T., J. C. Baret, V. Taly, and A. D. Griffiths.** 2007. Miniaturizing chemistry and biology in microdroplets. *Chem. Commun.* 1773–1788.

61. **Khersonsky, O., C. Roodveldt, and D. S. Tawfik.** 2006. Enzyme promiscuity: evolutionary and mechanistic aspects. *Curr. Opin. Chem. Biol.* **10**:498–508.

62. **Khosla, C., S. Kapur, and D. E. Cane.** 2009. Revisiting the modularity of modular polyketide synthases. *Curr. Opin. Chem. Biol.* **13**:135–143.

63. **Kirby, A. J., and W. P. Jencks.** 1965. The reactivity of nucleophilic reagents toward the *p*-nitrophenyl phosphate dianion. *J. Am. Chem. Soc.* **87**:3209–3216.

64. **Kohli, R. M., C. T. Walsh, and M. D. Burkart.** 2002. Biomimetic synthesis and optimization of cyclic peptide antibiotics. *Nature* **418**:658–661.

65. **Kostichka, K., S. M. Thomas, K. J. Gibson, V. Nagarajan, and Q. Cheng.** 2001. Cloning and characterization of a gene cluster for cyclododecanone oxidation in *Rhodococcus ruber* SC1. *J. Bacteriol.* **183**:6478–6486.

66. **Lassila, J. K., and D. Herschlag.** 2008. Promiscuous sulfatase activity and thio-effects in a phosphodiesterase of the alkaline phosphatase superfamily. *Biochemistry* **47**:12853–12859.

67. **Lawen, A., and R. Traber.** 1993. Substrate specificities of cyclosporin synthetase and peptolide SDZ 214-103 synthetase. Comparison of the substrate specificities of the related multifunctional polypeptides. *J. Biol. Chem.* **268**:20452–20465.

68. **Lutje Spelberg, J. H., L. Tang, M. van Gelder, R. M. Kellogg, and D. B. Janssen.** 2002. Exploration of the biocatalytic potential of a halohydrin dehalogenase using chromogenic substrates. *Tetrahedron Asymmetry* **13**:1083–1089.

69. **Lutje Spelberg, J. H., J. E. van Hylckama Vlieg, T. Bosma, R. M. Kellogg, and D. B. Janssen.** 1999. A tandem enzyme reaction to produce optically active halohydrins, epoxides and diols. *Tetrahedron Asymmetry* **10**:2963–2870.

70. **Lutje Spelberg, J. H., J. E. van Hylckama Vlieg, L. Tang, D. B. Janssen, and R. M. Kellogg.** 2001. Highly enantioselective and regioselective biocatalytic azidolysis of aromatic epoxides. *Org. Lett.* **3**:41–43.

71. **Majeric Elenkov, M., B. Hauer, and D. B. Janssen.** 2006. Enantioselective ring opening of epoxides with cyanide catalysed by halohydrin dehalogenase: a new approach to non-racemic β-hydroxy nitriles. *Adv. Synth. Catal.* **348**:579–585.

72. **Majeric Elenkov, M., L. Tang, B. Hauer, and D. B. Janssen.** 2006. Sequential kinetic resolution catalyzed by halohydrin dehalogenase. *Org. Lett.* **8**:4227–4229.

73. **Majeric Elenkov, M., L. Tang, A. Meetsma, B. Hauer, and D. B. Janssen.** 2008. Formation of enantiopure 5-substituted oxazolidinones through enzyme-catalysed kinetic resolution of epoxides. *Org. Lett.* **10**:2417–2420.

74. **Majerić Elenkov, M., H. W. Hoeffken, L. Tang, B. Hauer, and D. B. Janssen.** 2007. Enzyme-catalyzed nucleophilic ring opening of epoxides for the preparation of enantiopure tertiary alcohols. *Adv. Synth. Catal.* **349**:2279–2285.

75. **Malito, E., A. Alfieri, M. W. Fraaije, and A. Mattevi.** 2004. Crystal structure of a Baeyer-Villiger monooxygenase. *Proc. Natl. Acad. Sci. USA* **101**:13157–13162.

76. **Marsden, A. F., B. Wilkinson, J. Cortes, N. J. Dunster, J. Staunton, and P. F. Leadlay.** 1998. Engineering broader specificity into an antibiotic-producing polyketide synthase. *Science* **279**:199–202.

77. **Mihovilovic, M. D., B. Müller, and P. Stanetty.** 2002. Monooxygenase-mediated Baeyer-Villiger oxidations. *Eur. J. Org. Chem.* 3711–3730.

78. **Mootz, H. D., and M. A. Marahiel.** 1997. The tyrocidine biosynthesis operon of *Bacillus brevis*: complete nucleotide sequence and biochemical characterization of functional internal adenylation domains. *J. Bacteriol.* **179**:6843–6850.

79. Morii, S., S. Sawamoto, Y. Yamauchi, M. Miyamoto, M. Iwami, and E. Itagaki. 1999. Steroid monooxygenase of *Rhodococcus rhodochrous*: sequencing of the genomic DNA, and hyperexpression, purification, and characterization of the recombinant enzyme. *J. Biochem.* **126:**624–631.

80. Nakamura, T., T. Nagasawa, F. Yu, I. Watanabe, and H. Yamada. 1991. A new catalytic function of halohydrin hydrogen-halide-lyase, synthesis of β-hydroxynitriles from epoxides and cyanide. *Biochem. Biophys. Res. Commun.* **180:**124–130.

81. Nakamura, T., T. Nagasawa, F. Yu, I. Watanabe, and H. Yamada. 1994. A new enzymatic synthesis of (*R*)-γ-chloro-β-hydroxybutyronitrile. *Tetrahedron* **50:**11821–11826.

82. Nguyen, K. T., D. Ritz, J. Q. Gu, D. Alexander, M. Chu, V. Miao, P. Brian, and R. H. Baltz. 2006. Combinatorial biosynthesis of novel antibiotics related to daptomycin. *Proc. Natl. Acad. Sci. USA* **103:**17462–17467.

83. Nobeli, I., A. D. Favia, and J. M. Thornton. 2009. Protein promiscuity and its implications for biotechnology. *Nat. Biotechnol.* **27:**157–167.

84. O'Brien, P. J., and D. Herschlag. 1998. Sulfatase activity of *E. coli* alkaline phosphatase demonstrates a functional link to arylsulfatases, an evolutionary related enzyme family. *J. Am. Chem. Soc.* **120:**12369–12370.

85. O'Brien, P. J., and D. Herschlag. 1999. Catalytic promiscuity and the evolution of new enzymatic activities. *Chem. Biol.* **6:**R91–R105.

86. O'Brien, P. J., and D. Herschlag. 2001. Functional interrelationships in the alkaline phosphatase superfamily: phosphodiesterase activity of *Escherichia coli* alkaline phosphatase. *Biochemistry* **40:**5691–5699.

87. Olguin, L. F., S. E. Askew, A. C. O'Donoghue, and F. Hollfelder. 2008. Efficient catalytic promiscuity in an enzyme superfamily: an arylsulfatase shows a rate acceleration of 10^{13} for phosphate monoester hydrolysis. *J. Am. Chem. Soc.* **130:**16547–16555.

88. Ollis, D. L., E. Cheah, M. Cygler, B. Dijkstra, F. Frolow, S. M. Franken, M. Harel, S. J. Remington, I. Silman, J. Schrag, J. L. Sussman, K. H. G. Verschueren, and A. Goldman. 1992. The α/β hydrolase fold. *Protein Eng.* **5:**197–211.

89. Ottolina, G., S. Bianchi, B. Belloni, G. Carrea, and B. Danieli. 1999. First asymmetric oxidation of tertiary amines by cyclohexanone monooxygenase. *Tetrahedron Lett.* **40:**8483–8486.

90. Patrick, W. M., and I. Matsumura. 2008. A study in molecular contingency: glutamine phosphoribosylpyrophosphate amidotransferase is a promiscuous and evolvable phosphoribosylanthranilate isomerase. *J. Mol. Biol.* **377:**323–336.

91. Poelarends, G. J., W. H. Johnson, Jr., H. Serrano, and C. P. Whitman. 2007. Phenylpyruvate tautomerase activity of *trans*-3-chloroacrylic acid dehalogenase: evidence for an enol intermediate in the dehalogenase reaction? *Biochemistry* **46:**9596–9604.

92. Poelarends, G. J., H. Serrano, W. H. Johnson, Jr., D. W. Hoffman, and C. P. Whitman. 2004. The hydratase activity of malonate semialdehyde decarboxylase: mechanistic and evolutionary implications. *J. Am. Chem. Soc.* **126:**15658–15659.

93. Poelarends, G. J., H. Serrano, M. D. Person, W. H. Johnson, Jr., A. G. Murzin, and C. P. Whitman. 2004. Cloning, expression, and characterization of a *cis*-3-chloroacrylic acid dehalogenase: insights into the mechanistic, structural, and evolutionary relationship between isomer-specific 3-chloroacrylic acid dehalogenases. *Biochemistry* **43:**759–772.

94. Poelarends, G. J., J. E. van Hylckama Vlieg, J. R. Marchesi, L. M. Freitas Dos Santos, and D. B. Janssen. 1999. Degradation of 1,2-dibromoethane by *Mycobacterium* sp. strain GP1. *J. Bacteriol.* **181:**2050–2058.

95. Poelarends, G. J., V. P. Veetil, and C. P. Whitman. 2008. The chemical versatility of the β-α-β fold: catalytic promiscuity and divergent evolution in the tautomerase superfamily. *Cell. Mol. Life Sci.* **65:**3606–3618.

96. Rothlisberger, D., O. Khersonsky, A. M. Wollacott, L. Jiang, J. DeChancie, J. Betker, J. L. Gallaher, E. A. Althoff, A. Zanghellini, O. Dym, S. Albeck, K. N. Houk, D. S. Tawfik, and D. Baker. 2008. Kemp elimination catalysts by computational enzyme design. *Nature* **453:**190–195.

97. Schneider, A., T. Stachelhaus, and M. A. Marahiel. 1998. Targeted alteration of the substrate specificity of peptide synthetases by rational module swapping. *Mol. Gen. Genet.* **257:**308–318.

98. Stachelhaus, T., A. Schneider, and M. A. Marahiel. 1995. Rational design of peptide antibiotics by targeted replacement of bacterial and fungal domains. *Science* **269:**69–72.

99. Stec, B., K. M. Holtz, and E. R. Kantrowitz. 2000. A revised mechanism for the alkaline phosphatase reaction involving three metal ions. *J. Mol. Biol.* **299:**1303–1311.

100. Steele, C. L., J. Crock, J. Bohlmann, and R. Croteau. 1998. Sesquiterpene synthases from grand fir (*Abies grandis*). Comparison of constitutive and wound-induced activities, and cDNA isolation, characterization, and bacterial expression of delta-selinene synthase and γ-humulene synthase. *J. Biol. Chem.* **273:**2078–2089.

101. Steinreiber, A., and K. Faber. 2001. Microbial epoxide hydrolases for preparative biotransformations. *Curr. Opin. Biotechnol.* **12:**552–558.

102. Stutzman-Engwall, K., S. Conlon, R. Fedechko, F. Kaczmarek, H. McArthur, A. Krebber, Y. Chen, J. Minshull, S. A. Raillard, and C. Gustafsson. 2003. Engineering the *aveC* gene to enhance the ratio of doramectin to its CHC-B2 analogue produced in *Streptomyces avermitilis*. *Biotechnol. Bioeng.* **82:**359–369.

103. Stutzman-Engwall, K., S. Conlon, R. Fedechko, H. McArthur, K. Pekrun, Y. Chen, S. Jenne, C. La, N. Trinh, S. Kim, Y. X. Zhang, R. Fox, C. Gustafsson, and A. Krebber. 2005. Semi-synthetic DNA shuffling of *aveC* leads to improved industrial scale production of doramectin by *Streptomyces avermitilis*. *Metab. Eng.* **7:**27–37.

104. Svedendahl, M., P. Carlqvist, C. Branneby, O. Allner, A. Frise, K. Hult, P. Berglund, and T. Brinck. 2008. Direct epoxidation in *Candida antarctica* lipase B studied by experiment and theory. *Chembiochem* **9:**2443–2451.

105. Svedendahl, M., K. Hult, and P. Berglund. 2005. Fast carbon-carbon bond formation by a promiscuous lipase. *J. Am. Chem. Soc.* **127:**17988–17989.

106. Taly, V., B. T. Kelly, and A. D. Griffiths. 2007. Droplets as microreactors for high-throughput biology. *Chembiochem* **8:**263–272.

107. Tang, L., A. E. van Merode, J. H. Lutje Spelberg, M. W. Fraaije, and D. B. Janssen. 2003. Steady-state kinetics and tryptophan fluorescence properties of halohydrin dehalogenase from *Agrobacterium radiobacter*. Roles of W139 and W249 in the active site and halide-induced conformational change. *Biochemistry* **42:** 14057–14065.

108. Tanovic, A., S. A. Samel, L. O. Essen, and M. A. Marahiel. 2008. Crystal structure of the termination module of a nonribosomal peptide synthetase. *Science* **321:**659–663.

109. Torres Pazmino, D. E., B. J. Baas, D. B. Janssen, and M. W. Fraaije. 2008. Kinetic mechanism of phenylacetone monooxygenase from *Thermobifida fusca*. *Biochemistry* **47:**4082–4093.

110. Torres Pazmino, D. E., R. Snajdrova, D. V. Rial, M. D. Mihovilovic, and M. W. Fraaije. 2007. Altering the substrate specificity and enantioselectivity of phenylacetone

monooxygenase by structure-inspired redesign. *Adv. Synth. Catal.* **349:**1361–1368.

111. **Trauger, J. W., R. M. Kohli, H. D. Mootz, M. A. Marahiel, and C. T. Walsh.** 2000. Peptide cyclization catalysed by the thioesterase domain of tyrocidine synthetase. *Nature* **407:**215–218.

112. **Uppenberg, J., N. Ohrner, M. Norin, K. Hult, G. J. Kleywegt, S. Patkar, V. Waagen, T. Anthonsen, and T. A. Jones.** 1995. Crystallographic and molecular-modeling studies of lipase B from *Candida antarctica* reveal a stereospecificity pocket for secondary alcohols. *Biochemistry* **34:**16838–16851.

113. **van den Wijngaard, A. J., D. B. Janssen, and B. Witholt.** 1989. Degradation of epichlorohydrin and halohydrins by bacterial cultures isolated from freshwater sediment. *J. Gen. Microbiol.* **135:**2199–2208.

114. **van den Wijngaard, A. J., P. T. Reuvekamp, and D. B. Janssen.** 1991. Purification and characterization of halo-alcohol dehalogenase from *Arthrobacter* sp. strain AD2. *J. Bacteriol.* **173:**124–129.

115. **van Hylckama Vlieg, J. E., L. Tang, J. H. Lutje Spelberg, T. Smilda, G. J. Poelarends, T. Bosma, A. E. van Merode, M. W. Fraaije, and D. B. Janssen.** 2001. Halohydrin dehalogenases are structurally and mechanistically related to short-chain dehydrogenases/reductases. *J. Bacteriol.* **183:**5058–5066.

115a.**van Loo, B., S. Jonas, A. C. Babtie, A. Benjdia, O. Berteau, M. Hyvonen, and F. Hollfelder.** 2010. An efficient multiply promiscuous hydrolase in the alkaline phosphatase superfamily. *Proc. Natl. Acad. Sci. USA* doi/10.1073/pnas.0903951107.

116. **van Loo, B., J. Kingma, M. Arand, M. G. Wubbolts, and D. B. Janssen.** 2006. Diversity and biocatalytic potential of epoxide hydrolases identified by genome analysis. *Appl. Environ. Microbiol.* **72:**2905–2917.

117. **Villiers, B., and F. Hollfelder.** 2009. Mapping the limits of substrate specificity of the adenylation domain of TycA. *Chembiochem* **10:**671–682.

118. **Waagen, V., I. Hollingsaeter, V. Partali, O. Thorstad, and T. Anthonsen.** 1993. Enzymatic resolution of butanoic esters of 1-phenyl, 1-phenylmethyl, 1-[2-phenylethyl] and 1-[2-phenoxyethyl] ethers of 3-methoxy-1,2-propanediol. *Tetrahedron Asymmetry* **4:**2265–2274.

119. **Walsh, C. T., and Y. C. J. Chen.** 1988. Enzymic Baeyer-Villiger oxidations by flavin-dependent monooxygenases. *Angew. Chem. Int. Ed.* **27:**333–343.

120. **Wang, S. C., W. H. Johnson, Jr., and C. P. Whitman.** 2003. The 4-oxalocrotonate tautomerase- and YwhB-catalyzed hydration of 3E-haloacrylates: implications for the evolution of new enzymatic activities. *J. Am. Chem. Soc.* **125:**14282–14283.

121. **Wang, S. C., M. D. Person, W. H. Johnson, Jr., and C. P. Whitman.** 2003. Reactions of *trans*-3-chloroacrylic acid dehalogenase with acetylene substrates: consequences of and evidence for a hydration reaction. *Biochemistry* **42:**8762–8773.

122. **Weissman, K. J.** 2009. Introduction to polyketide biosynthesis. *Methods Enzymol.* **459:**3–16.

123. **Wolfenden, R.** 2006. Degrees of difficulty of water-consuming reactions in the absence of enzymes. *Chem. Rev.* **106:**3379–3396.

124. **Zalatan, J. G., T. D. Fenn, A. T. Brunger, and D. Herschlag.** 2006. Structural and functional comparisons of nucleotide pyrophosphatase/phosphodiesterase and alkaline phosphatase: implications for mechanism and evolution. *Biochemistry* **45:**9788–9803.

Enzyme Production in *Escherichia coli*

DANIEL J. SAYUT, PAVAN K. R. KAMBAM, WILLIAM G. HERRICK,
AND LIANHONG SUN

37

37.1. INTRODUCTION

As efficient biocatalysts, enzymes are used widely in the food, pharmaceutical, and chemical industries. In addition to these traditional applications, many enzymes, and in particular those with human origins, are increasingly being used as therapeutic agents to treat a wide range of genetic diseases including lysosomal storage disorders (90). Because of their widespread use, methods for producing enzymes have been developed for a variety of biological systems including bacteria, yeasts, fungi, mammalian cells, insect cells, human or animal tissues, and plants. Despite the availability of these diverse expression systems, industrial enzymes traditionally have been produced commercially by use of overproducing strains of microorganisms from which they are naturally secreted. These overproducing strains are usually generated from the wild-type strains over years of genetic selection/screening. For example, the nitrile-converting strain *Rhodococcus rhodochrous* J1, which is used for production of acrylamide from nitrile at a scale of 4,000 tons/year, is a third-generation biocatalyst developed 6 years after the first large-scale production process using *Rhodococcus* sp. N774 was initialized (60). Depending on the intended application, some industrial enzymes are isolated and purified before being formulated, such as subtilisin, a common protease used in detergents (1). Others are used in the form of whole-cell catalysts, such as nitrile hydratase, the enzyme produced by *R. rhodochrous* for acrylamide production. In contrast, therapeutic proteins are all purified and formulated with procedures approved by the FDA. In addition, once a production method for a therapeutic enzyme has been established, further improvement of the method is generally undesirable because obtaining FDA approval, a prolonged and costly process, is necessary to use the altered method (58).

With the development of recombinant DNA technology, industrial enzymes are increasingly being produced in heterologous hosts, particularly in the bacterium *Escherichia coli* and the yeasts *Saccharomyces cerevisiae* and *Pichia pastoris*. These microbial strains are chosen largely because of their well-understood genetics, established genetic manipulation methods, and the availability of large numbers of expression vectors and strains. Therapeutic enzymes are produced almost exclusively in recombinant forms, resulting in a recombinant therapeutic protein market that has increased steadily during the past few decades (110).

Although mammalian cells are predominately used for the production of recombinant human enzymes, this review will focus on the use of *E. coli* as an alternative host for the production of recombinant enzymes (93).

Developing a cost-effective recombinant protein production method requires the integration of genetic engineering with fermentation process design. At the genetic level, regulatory methods not only have to be selected to control protein expression levels (68), but additional elements that affect protein folding, stability, and transport might also need to be considered to achieve high production levels or to simplify downstream processing (64). The selection of the host strain is critical during this process, because the genetic background and phenotypes of the host can greatly affect gene expression, protein stabilization, and protein secretion. In designing the fermentation process, the preferred method is the use of high-cell-density fermentations in stirred-tank reactors. To this end, novel autoinducing gene regulatory methods that are able to function in fed-batch fermentation processes may represent a strategy superior to conventional inducible gene expression methods. We will discuss these aspects of recombinant protein production in the following sections.

37.2. OVERVIEW OF GENE EXPRESSION REGULATION

Once a protein of interest is identified, the first steps in developing a production method include selection of the host cells to be used for expression, selection of a method for gene expression regulation, and selection of a method for protein targeting (50). Although constitutive gene expression has traditionally been used for the industrial production of recombinant proteins because of its simplicity, inducible gene regulatory systems are desired for high-level recombinant protein production (Table 1). High-level recombinant protein production commonly redirects metabolic flux from cellular growth to the overproduction of the desired proteins, and the resulting increased metabolic burden can hamper host cell growth, resulting in an overall low level of recombinant protein production. In comparison, inducible gene regulation allows for activation of gene expression at a desired cell density or cell growth phase, at which point the large number of cells in the culture can achieve a high level of recombinant protein production despite the presence of

TABLE 1 Some recent examples of producing recombinant proteins in *E. coli*

Product	Host strain	Reactor	Productivity	Reference
Human interferon-α2	DH5α	2 liters, fed-batch	5.2 g/liter	101
Human interferon-γ	BL21(DE3)	2 liters, fed-batch	42.5 g/liter	4
Human tumor necrosis factor-related apoptosis-inducing ligand (TRAIL)	C600	3.7 liters, fed-batch	4.51 g/liter	62
Aminolevulinate synthase	MG1655	2.5 liters, batch	5.2 g/liter	112
Alkaline phosphatase	HB101	6.6 liters, fed-batch	5.2 g/liter	16
Tilapia insulin-like growth factor-2	BL21(DE3)	5 liters, fed-batch	9.69 g/liter	47
Fructose 1,6-bisphosphate aldolase	BL21(DE3)	1.5 liters, batch	1.5 g (70 U/mg)	57
Glycine oxidase	Rosetta(DE3)	5 liters, fed-batch	1,154 U/liter	66

metabolic burden. This delay is particularly important for the large-scale production of toxic recombinant proteins that are able to inhibit cell growth even at low cellular concentrations (88). In addition to the timing of gene expression, inducible gene regulation allows for the fine-tuning of expression levels, which is frequently necessary for efficient protein secretion, correct protein folding, and the formation of disulfide bonds (5, 69). In these cases, optimizing rather than maximizing gene expression levels is desired.

The choice of an expression host strain for recombinant protein production is determined by the expected expression levels, origins of the proteins of interest, presence of post-translational modifications, and methods of gene expression regulation. While most recombinant proteins are expressed in their biologically active forms in industrial production processes, some recombinant proteins, and biopharmaceuticals produced in *E. coli*, in particular, are expressed as inclusion bodies with the biological activities being recovered only after isolation and refolding (70, 93). Isolation and purification of recombinant proteins can be significantly simplified if the proteins can be secreted into culture media or periplasm, because the recovery of cytoplasmic proteins requires disruption of the cell wall to release all cellular components, which complicates downstream purification. For proteins with disulfide bonds, periplasmic expression in *E. coli* is necessary to obtain their active forms. Despite the presence of various protein secretion pathways and the ability to obtain high levels of secretory protein production (18, 52), *E. coli* is not considered the preferred host organism for extracellular protein secretion.

37.3. HOST CELL SELECTION

The gram-negative bacterium *E. coli* is the most widely used prokaryotic organism for recombinant protein production. This is largely due to its long history and ubiquitous use in the laboratory as a model organism for bacteriology and molecular biology research. Extensive characterizations have also generated rich information about its physiology and genetics (72). This information has been used to generate a comprehensive genetic knockout library of *E. coli* mutants that can be readily obtained from culture collections (3), and has led to the creation of efficient genetic manipulation methods, such as DNA transformation and plasmid purification, that are exclusive to *E. coli*. In addition, *E. coli* can be cultivated in a wide variety of defined and rich media with a minimum achievable doubling time of 20 to 30 min, which is substantially faster than other model organisms. During the past several decades, efficient high-cell-density fermentation methods for *E. coli* have been established,

and cell densities as high as 200 g/liter (dry weight) can be achieved (96). Its potential to produce high-level recombinant proteins has been demonstrated by accumulating recombinant proteins in the periplasm at concentrations of approximately 8.5 g/liter, constituting 30% of the total intracellular proteins (52).

The most common *E. coli* strains used for recombinant protein expression are commercially available derivatives of the *E. coli* K-12 and BL strains (Table 2). These strains are engineered to enhance recombinant protein production by eliminating the host proteases encoded by the *lon* and *ompT* genes, which tend to attach to inclusion bodies separated by centrifugation to cause degradation during protein recovery. To minimize the frequency of rearrangements of genes in plasmids with multiple copy numbers, most *E. coli* expression strains are also mutated to be deficient in homologous recombination. In addition to these common mutations, a number of special strains are also available to allow for the enhanced production of proteins that are difficult to produce. These include special strains that are engineered to enable the expression of proteins with rare codons, and others that are engineered to more efficiently express toxic proteins, such as the ABLE C and ABLE K strains, which reduce the copy number of ColE1-derived plasmids to maintain a low basal expression level (12). In addition to these considerations, another important factor in the selection of an *E. coli* strain for recombinant protein production is the type of inducible gene expression system that is to be used. For example, many inducible gene expression systems, including those with the T7, *lacUV5*, and *tac* promoters, require the expression of the LacI repressor in the host strain for the repression of gene expression in the absence of chemical inducers. If the T7 gene expression system is used, expression of the T7 RNA polymerase in the host strain is also necessary to activate gene expression (26). Alternatively, it is not uncommon for repressor proteins to be produced independent of the host strain in *trans*. This is especially true for production from high-copy-number plasmids, because the higher production levels in *trans* ensure that significant amounts of the repressor are produced to completely repress gene expression. This approach is exemplified by the commercially available P_{BAD}-inducible gene expression system (36).

Despite the availability of an array of engineered *E. coli* strains for recombinant protein production, the engineering of an *E. coli* host strain is frequently necessary to achieve high-level recombinant protein production for a specific application. One example is the engineering of strains for the periplasmic production of recombinant

TABLE 2 Genotypes and key features of common *E. coli* strains[a]

Strain	Genotype	Key features
ABLE C	*E. coli* C, *lac*(LacZω⁻), *mcrA*, *mcrCB*, *mcrF*, *mrr*, *hsdR*(r$_{K-}$, m$_{K-}$), F′ [*proAB*, *lacI*^{q}, *lacZ*ΔM15::Tn10] (Kan^r, Tet^r)	Reduced plasmid copy number; tetracycline resistance
BL21	*E. coli* B, F⁻, *ompT*, *hsdS*$_B$(r$_{B-}$, m$_{B-}$), *dcm*, *gal*	Protease deficient; Eco restriction and methylation deficient; galactose metabolism deficient
BL21(DE3)	BL21, λ(DE3)	Lysogene of DE3 containing T7 RNA polymerase regulated by a *lacUV5* promoter
BL21(DE3)pLysS	BL21(DE3), pLysS, Cam^r	Expresses T7 lysozyme for inhibition of T7 RNA polymerase; chloramphenicol resistance
Tuner[b]	BL21, *lacY*	Decreased protein expression heterogeneity caused by removal of lactose permease
Origami B[b]	Tuner, *ahpC*, *gor522*::Tn10, *trxB* (Kan^r, Tet^r)	Facilitates cytoplasmic disulfide bond formation; lactose permease deficient; kanamycin and tetracycline resistance
Origami 2[b]	*E. coli* K-12, Δ(*ara-leu*)7697, Δ*lacX74*, Δ*phoA*, PvuII, *phoR*, *araD139*, *ahpC*, *galE*, *galK*, *rpsL*, F′[*lac*⁺, *lacI*^{q}, *pro*], *gor522*::Tn10 *trxB* (Str^r, Tet^r)	Enhanced cytoplasmic disulfide bond formation; tetracycline and streptomycin resistance
Rosetta[b]	BL21, pRARE (Cam^r)	Enhances expression of genes with rare codons; chloramphenicol resistance
BLR[b]	BL21, Δ(*srl-recA*)306::Tn10 (Tet^r)	Improved plasmid monomer yields; tetracycline resistance
HMS174[b]	*E. coli* K-12, F⁻, *recA1*, *hsdR*(r$_{K12-}$, m$_{K12+}$) (Rif^r)	Improved plasmid monomer yields; rifampin resistance

[a]Data are gathered from literature and suppliers' product catalogs. Strains are available from common commercial resources, such as Invitrogen, Novagene/EMD Biosciences, and Stratagene/Agilent Technologies.
[b]Also available with DE3 or (DE3)pLysS.

proteins, which is considered superior to cytoplasmic production and is necessary for the active expression of target proteins containing disulfide bonds (18, 34). In one such study, the periplasmic production of insulin-like growth factor-I in *E. coli* was doubled by coexpression of DsbA or DsbC, two proteins involved in disulfide bond formation (52). The overexpression of membrane proteins in *E. coli* is also notoriously challenging (25). To incorporate a membrane protein into the membrane to form a functional protein, periplasmic domains of the membrane protein are first translocated across the inner membrane via the Sec secretory pathways and then targeted into the inner membrane primarily via the signal recognition particle (SRP) targeting pathway (22, 40, 61). However, membrane proteins frequently form inclusion bodies during cytoplasmic overexpression, presumably because of fast protein expression and slow protein translocation. This problem has been addressed by Wagner et al. (109), who reported the engineering of *E. coli* strains C41(DE3) and C43(DE3), two derivatives of BL21(DE3), for fine-tuning of the expression levels of membrane proteins. These strains use the expression of T7 lysozyme to inhibit the activity of T7 RNA polymerase, allowing for the expression levels of membrane proteins to be optimized to yield maximum amounts of active protein.

37.4. VECTOR DESIGN
Genes encoding proteins of interest can either be inserted into plasmids or into the *E. coli* chromosomal DNA for gene expression. Integration into the chromosomal DNA has many advantages including enhanced expression stability, easy strain maintenance, and antibiotic-free cultivation. Despite these advantages, the use of chromosomal

integration is limited by the difficulty in achieving a high level of expression because of low gene copy numbers. This problem can be slightly alleviated by introducing multiple copies of the gene into the chromosome (78).

As an alternative to chromosomal insertion, the use of plasmids to harbor genes of interest has proven to be effective for high-level gene expression. Most native plasmids are circular DNA, which can replicate autonomously, and vary from a few to more than 1,000 kilobase pairs in size. Most natural plasmids harbor genes encoding proteins that contribute diverse functions to the host cells, such as antibiotic resistance, conjugation, virulence, and degradation of toxic substances. Plasmids are present in their host cells at a specific number of copies, ranging from one to more than one hundred. During cell division, high-copy-number plasmids are partitioned into daughter cells in a near-random fashion, while low-copy-number plasmids may be distributed via active partition systems to ensure faithful inheritance (9). Wild-type plasmids also take advantage of the bacterial site-specific recombination mechanism to overcome homologous recombination and maintain plasmid monomers (13, 104). Some plasmids contain a toxin-antitoxin system to ensure at least one copy of the plasmid is present in each surviving daughter cell (10, 35). This is best illustrated by the CcdAB toxin-antitoxin system encoded in the *E. coli* F plasmid (19). CcdB is a toxin that specifically targets gyrase to inhibit DNA replication. The CcdA antitoxin binds CcdB to prevent gyrase inhibition, and allows DNA replication to resume. Because CcdA is prone to degradation by the Lon protease, constitutive production of CcdA is necessary to neutralize the CcdB toxicity. If a daughter cell does not acquire an F plasmid during cell division, production of CcdA and CcdB discontinues; however, CcdA inherited from the parental cell is more rapidly degraded than CcdB, resulting in cell death due to blocked DNA replication by CcdB. Based on this principle, a strategy has been devised to engineer *E. coli* for antibiotic-free cell cultivation in a minimal medium by use of plasmids containing a toxin-antitoxin system (21).

Most plasmids used for recombinant protein production in *E. coli* consist of an origin of replication, a promoter, a set of multicloning sites, transcription and translation terminators, and a selection marker (Table 3) (68). The origin of replication determines the copy number of a plasmid, and also plays a critical role in determining the compatibility of the plasmid with other plasmids. Various promoters are available that use a diverse array of inducing agents or methods with the expression levels of recombinant proteins largely determined by promoter strength. Many commercial plasmids also contain N- and C-terminal tags to form fusion proteins that facilitate recombinant protein detection and purification. Common tags include the *myc* tag for protein detection and the *his* tag for protein purification (30, 43, 44). These tags, however, are rarely used in large-scale recombinant protein production. Many different methods have been used to enhance gene expression in a plasmid-based system including using high-copy-number plasmids (38, 99) and strong promoters (51), as well as engineering translation efficiency (11, 80). However, enhanced gene expression inevitably increases the metabolic burden on the host cells, resulting in slower growth rates that ultimately lower production yields (68). As a result, there is no general rule for choosing a "perfect" plasmid to maximize recombinant protein expression, and "trial and error" is still applied to identify a balance between protein production levels and metabolic burden on the host cells.

TABLE 3 Genetic elements that determine gene expression in a typical *E. coli* plasmid

Element	Function
Repressor	Represses or activates gene expression in *trans*; enables inducible gene expression
Promoter	Activates gene expression by recruiting RNA polymerase and transcription factors; fine-tuning of gene expression levels using inducing agents; binary or graded gene expression control
mRNA secondary structure	Fine-tuning of translation efficiency
RBS	Recruits ribosomes; fine-tuning of translation efficiency
Stop codon	Terminates gene translation
Transcription terminator	Terminates transcription; enhances mRNA stability
Antibiotic resistance gene	Confers antibiotic resistance; determines compatibility with other antibiotic-resistance plasmids
Origin of replication	Initiates plasmid DNA replication; determines copy number and incompatibility with other plasmids

37.4.1. Origin of Replication

The copy number of a plasmid is largely determined by the origin of replication. A strong origin of replication is generally preferred for the expression of recombinant proteins so as to achieve high production levels. Commonly used origins of replication include the high-copy-number pUC replicon, which can maintain a couple of hundred copies per cell (113), the ColE1 replicon, which has a medium copy number with 15 to 60 copies (39, 65), and the p15A replicon, with a copy number of 10 to 12 (42). Some origins of replication, such as pSC101, generate an even smaller number of copies per cell, and may be useful for the production of toxic proteins (79).

Origins of replication also determine plasmid incompatibility, which has been used to classify plasmids into incompatibility groups (20, 23, 106). When multiple incompatible plasmids are present in the same host strain simultaneously, not all plasmids will be inherited into the daughter cells because of the absence of selection pressure, a process referred to as plasmid segregation (27). Although many biological processes are involved in determining plasmid incompatibility, stochasticity during cell division is believed to be a significant factor that results in the loss of plasmids with the same origins of replication during the random partitioning process (9, 76). To avoid the plasmid loss, the use of plasmids in different incompatibility groups is preferred when two plasmids are necessary for recombinant protein production, because this eliminates the need to use two antibiotics simultaneously to maintain the plasmids. However, the use of antibiotics to maintain compatible plasmids is necessary, though undesired for increased cost, to prevent the loss of both plasmids during cell division, which generates plasmid-free cells whose faster growth rates may result in the domination of the culture.

37.4.2. Transcriptional Regulation

Promoters and transcription terminators are the primary genetic elements for transcriptional regulation. The essential components of a typical *E. coli* promoter are shown in Table 3. The *E. coli* RNA polymerase specifically recognizes two DNA sequences located 35 bases (-35 site) and 10 bases (-10 site) upstream of the transcription start site; the latter is also known as TATA boxes. These sites have conserved sequences, with the -35 site having a consensus sequence of TTGACA and the -10 site having a consensus sequence of TATAAT (37). Base changes in these regions or length changes to the spacer can significantly reduce gene expression levels. In addition, strong promoters have an additional AT-rich sequence (known as UP element) upstream of the -35 position (86). Finally, many bacterial promoters also contain an operator that interacts with a cognate transcription factor. In the presence of this transcription factor, gene expression from the promoter can be activated or repressed depending on the nature of the transcription factor as well as the presence of an inducing agent. This mode of gene regulation is commonly referred to as inducible gene expression, and is frequently the most important property of a promoter used for recombinant protein production.

Because the number of gene transcripts is the strongest determinant of high-level protein production, transcriptional regulation is the most important consideration when developing protein production methods. In *E. coli*, the rate of transcription from a promoter depends on the strength of the interaction between the promoter and RNA polymerase, with strong promoters having strong interactions. As a result, strong promoters are preferred in recombinant protein production, although using a strong promoter does not always warrant a higher expression level (discussed in the following sections). The basal expression level of a promoter is also an important consideration, in particular, for toxic protein production, because high basal expression levels of toxic proteins can significantly inhibit cell growth, resulting in an overall low production level. Common strategies for reducing the basal expression level of a promoter include the insertion of a DNA sequence into the native promoter that is recognized by a repressor protein, such as the *lac* operator, to which the LacI repressor protein specifically binds (8). This strategy is used in a number of commercial plasmids to ensure low basal expression levels from their regulated promoters, including the strong T7 and P_{BAD} promoters (see the following sections). In addition, this strategy was demonstrated in recent work to reduce the basal expression level of the P_{luxI} promoter (92). After insertion of the *lacI* recognition sequence upstream of the transcriptional start site, the basal expression level of the P_{luxI} promoter was reduced by more than fourfold.

Transcription terminators are also an important determinant of the strength of protein production, because they seem to affect mRNA stability (74, 77). It has been well demonstrated that differential mRNA stability can influence relative gene expression levels (46, 67, 73). As a result, enhancing mRNA stability in a transcriptional regulatory system can be used to further increase gene expression levels. This can be achieved by inserting repeat sequences downstream of the coding sequence to form stable RNA structures or stem loops, which can slow mRNA degradation by preventing the $3' \rightarrow 5'$ exonuclease activity of RNase II and polynucleotide phosphorylase (41). Bacterial transcription terminators are composed of repeat sequences (46). As a result, transcription terminators can be used to prevent continued transcription and to increase mRNA stability and enhance gene expression (111). Because of their importance, most commercial plasmids contain two tandem transcription terminators downstream of the encoding gene, such as the T1 and T2 terminators.

37.4.3. Translational Regulation

Translation is initialized by the binding of a ribosome to a ribosome binding site (RBS) located downstream of the transcription initiation site. An RBS contains a Shine-Dalgarno (SD) sequence that is complementary to nucleotides at the 3′ end of the 16S rRNA, a spacer sequence to the translation start codon, a start codon, and a sequence downstream of the start codon. The actual SD sequence in an RBS varies, but all contain a purine-rich sequence positioned 5 to 12 nucleotides upstream of the start codon (49). It has been demonstrated that the sequence UAAGGAGG is more efficient than AAGGA, and that the length and sequence of the spacer region can also have a significant influence on the translational efficiency (85). Despite extensive exploration of the influence of the SD and the spacer sequences on translational efficiency, the re-engineering of these sequences in a commercial plasmid is rare. Studies have also examined the influence of the start codon and the sequences immediately downstream of the start codon on translation efficiency, and it was found that AUG is the most efficient start codon for translation initiation in *E. coli* (83, 85). It should be noted that the use of the AUG start codon may introduce a nonnative methionine residue as the first amino acid in the protein sequence. If such an addition is undesired, periplasmic production of the protein can be used to completely eliminate this residue during the

proteolytic removal of the leader sequence. Finally, efficient translation also depends on the presence of a stop codon, with the sequence UAA being preferred in *E. coli* and the most efficient termination sequence being UAAU (82).

The secondary structure of the transcribed mRNA is also an important factor in the efficiency of translation, and the engineering of mRNA secondary structures and 5′-untranslated regions (5′-UTR) can be used to alter expression levels. The local secondary structures of mRNA in the RBS are known to have a profound impact on the translation efficiency, and up to a 100-fold increase in expression level has been observed for engineered systems (94). Such a significant change is believed to be caused by competition between the formation of secondary structure and the binding of the ribosome (24). Based on this principle, expression of a porcine P450 enzyme in *E. coli* was enhanced by approximately 100-fold by engineering the secondary structure of the RBS (56). Despite the potential for marked improvements in recombinant protein production, engineering mRNA secondary structures is more commonly used for fine-tuning gene expression levels rather than maximizing protein production. This has been demonstrated nicely by Keasling and coworkers (98, 100), who varied the relative expression levels of two genes expressed in an operon by inserting different DNA sequences that encoded for mRNA secondary structures or RNase sites between the two genes. This strategy was subsequently applied to modulate the expression levels of multiple genes in a synthetic pathway (81). In another study, Isaacs et al. inserted a small DNA sequence between a promoter and an RBS to repress gene expression in *cis* through formation of mRNA secondary structure that blocked ribosome binding. The repression can be completely alleviated by use of a second DNA sequence provided in *trans* which can form a more stable mRNA secondary structure with the introduced DNA fragment to release the RBS and activate gene expression (48), providing a novel method for positive and negative regulation of gene expression.

37.4.4. Protein Secretion

Secretion of recombinant proteins in *E. coli* can refer to delivery of the proteins to the periplasmic space, or the export of proteins across the outer membrane to the culture media. The secretion of proteins across the outer membrane of *E. coli* has been reported occasionally. However, *E. coli* is not the preferred organism for protein export because it has not been evolved to produce large quantities of exported proteins. Protein secretion to the periplasm, in contrast, has been used extensively to produce functional proteins, obtain authentic protein sequences, increase yields, and facilitate protein purification (18, 34). Five types of secretory pathways have been identified in gram-negative bacteria, with the type I and II secretory mechanisms being the most commonly used for recombinant protein secretion (69). Type I secretion transports proteins without formation of a periplasmic intermediate and is capable of exporting target proteins to the culture media (95). In contrast, all type II secretory pathways, including the SecB-dependent secretion pathway (28), the SRP pathway (61), and the twin-arginine translocation (TAT) secretion pathway (6, 7), transport target proteins via a two-step process. The advantages and disadvantages of each secretion pathway are summarized in Table 4. Regardless of the actual secretion mechanisms, an N-terminal peptide sequence, also referred to as a leader sequence, is necessary for targeted proteins to be recognized for secretion (49). It should be noted that the effectiveness

TABLE 4 Secretory systems used in *E. coli* for recombinant protein production

Type	Advantages	Disadvantages
Type I secretion pathway	Single-step secretion; capable of extracellular export	The leader sequence is not removed after secretion; low transport efficiency
SecB-dependent pathway	Used extensively for recombinant protein secretion; rich biochemical and biophysical information; various strategies for improving secretion efficiency	Unable to secrete folded proteins; low transport efficiency for some proteins
PRP	Transport efficiency can be altered by changing the leader sequence	Less commonly used for recombinant protein production
TAT	Capable of secreting folded proteins as a mechanism for protein-folding quality control	Relatively new system; low efficiency; slow transport rates; high energy cost

of a leader sequence for protein secretion is highly protein sequence dependent, and is substantially influenced by the genotype of the host strain (14). Finally, it is known that the fine-tuning of gene expression rates is necessary to obtain increased secretion efficiency, because highly expressed proteins tend to overwhelm the secretion machinery, resulting in the formation of inclusion bodies. Because of this effect, secretion efficiency can be enhanced by coexpression of chaperones to facilitate protein folding and disulfide bond formation in the periplasm (52).

37.5. INDUCIBLE GENE EXPRESSION SYSTEMS

Inducible gene expression systems are commonly used in recombinant protein production. The desired properties for such a system include tight regulation (low basal expression levels), inducibility (activation of gene expression in the presence of inducing agents), low cost of inducer, and large dynamic ranges (fine-tuning of expression levels through a broad range). For therapeutic proteins, any inducing agents also need to be approved for use by the FDA. If the proteins of interest are toxic to *E. coli* hosts, extremely low basal expression levels are critical. Although many inducible gene expression systems have been developed, there are a handful of systems that are predominantly used for recombinant protein production.

The *lac* system is derived from the *E. coli lac* operon and contains a *lac* promoter that is repressed by the binding of the LacI repressor protein to the *lac* operator. LacI can form a complex with lactose or synthetic analogs of lactose that prevents the repressor from interacting with the *lac* operator, resulting in activation of the *lac* promoter. Even though lactose can efficiently activate the *lac* promoter, it is also decomposed by β-galactosidase, an enzyme regulated by the native *lac* operon. As a result, synthetic (gratuitous) inducers, such as isopropyl-β-D-thiogalactoside (IPTG), that are not degraded by β-galactosidase are more commonly used to activate the *lac* promoter. IPTG has the additional advantage in that it can freely diffuse across the cell wall, while LacY (lactose permease) is required to transport lactose into cells (55). Since LacY is regulated by the *lac* promoter of the *E. coli* chromosome, it forms a positive-feedback loop to transport more lactose into the cell upon activation, resulting in heterogeneity of gene expression in a cell population. It should be noted that activation of the *lac* promoter is also subject to catabolite repression, because activation of the promoter requires the efficient formation

of cyclic AMP (cAMP)-cAMP receptor protein complexes. Low concentrations of cAMP are generated in the presence of glucose in the culture medium, and therefore, catabolite repression of the *lac* promoter excludes the use of the *lac* promoter in cells growing in rich media. A variant of the *lac* promoter, *lacUV5*, is a stronger promoter with activation that is less affected by catabolite repression and is more commonly used in production of recombinant proteins. A mutant repressor protein, LacI^q, with improved expression levels is also commonly used as the increased level of LacI reduces basal gene expression.

The T7 gene expression system has become an increasingly popular tool for recombinant protein production (103). The T7 promoter is activated exclusively by the bacteriophage T7 RNA polymerase and cannot be activated by the *E. coli* RNA polymerase because of the lack of −35 and −10 sequences in the promoter. A significant advantage in using the T7 expression system is the highly efficient transcriptional initiation of the T7 RNA polymerase that allows for high-level recombinant protein production. A variety of *E. coli* host strains is commercially available for use with the T7 gene expression system, where the T7 RNA polymerase is encoded on an F plasmid [such as BL21(DE3)] regulated by an engineered *lac* promoter. The basal expression level of the T7 promoter can also be lowered by using a hybrid promoter incorporating the *lac* operator (26), making it possible to express highly toxic proteins in *E. coli* (88). Because of the presence of the *lac* regulatory components, host strains used for T7 expression generally harbor the mutations used by the *lac* gene expression system for increased repression and reduced expression heterogeneity. Further reductions in basal expression levels can be achieved by expressing the inhibitor of the T7 RNA polymerase, T7 lysozyme (strains with pLysS or pLysE plasmid). Additional strains, such as the Tuner strains, can be used for the fine-tuning of gene expression (Table 2).

The pBAD expression system has been used widely for recombinant protein production because it offers tight regulation and high tunability (36, 75). The system is based on the P$_{BAD}$ promoter, which controls the *E. coli* arabinose operon and is regulated by AraC. In contrast to the LacI repressor, which functions as a negative regulator, AraC acts as a both positive and negative regulator. In the absence of arabinose, AraC binds to one set of sites in the P$_{BAD}$ promoter and inhibits gene expression; in the presence of arabinose, AraC binds to a different set of sites in the promoter to activate gene expression. Because AraC is required for both repression and activation of gene expression, it is preferentially provided in *trans* so as to achieve the high levels of cellular AraC that are necessary to regulate the promoter harbored in high-copy-number plasmids. Analogous to the *lac* promoter, heterogeneous expression has also been observed upon induction by arabinose due to the autocatalytic transport of arabinose by the gene products of *araA* and *araFGH*, which are regulated by the *E. coli* P_{BAD} promoter (97). Engineered strains have been developed to allow homogeneous gene expression in the presence of subsaturating concentrations of arabinose, primarily containing mutations of *araA*, *araGFH*, *araBCD* (genes encoding proteins for arabinose degradation), and *lacY* (71). A hybrid P_{BAD} promoter incorporating a *lac* operator has also been engineered, and the resulting hybrid promoter allows tunable gene expression with a dynamic range of up to 1,800-fold by use of a combination of IPTG and arabinose (63).

Inducible gene expression allows production of high levels of recombinant proteins; however, the introduction of inducing agents is undesired in large-scale fermentations, because this requires interruption of the fermentation process and the use of special devices to introduce the chemical inducers. These additional operations increase the cost of manufacturing recombinant proteins, and more importantly, may introduce contaminants that compromise the quality of the whole culture batch. While the use of temperature-sensitive gene expression systems may eliminate the risk of introducing contamination, changing the temperature of a large fermentation vessel is inefficient, and is generally avoided for the production of therapeutic proteins. A more practical alternative is the use of autoinducible gene expression systems. One example of an autoinducible strategy is the use of autoinducing media (102). Such media are composed of a specific amount of glucose in addition to other carbon sources and nutrition. In the presence of glucose, catabolite repression inhibits gene expression from promoters regulated by a *lac* operator. Upon the complete consumption of glucose, the promoters are activated, causing gene expression. By adjusting the starting concentrations of glucose, the induction time can be fine-tuned to optimize recombinant protein production (45). Another strategy to implement autoinduction is based on the bacterial quorum-sensing mechanism (33). Quorum-sensing systems produce an autoinducer that is recognized by its cognate transcription factor to activate gene expression once concentrations of the autoinducer exceed a threshold level. Because autoinducer concentrations largely depend on local population concentrations, quorum sensing enables cell-density-dependent gene expression. In the LuxI-LuxR quorum-sensing system, LuxI catalyzes the synthesis of the autoinducer, *N*-3-oxo-hexanoyl homoserine lactone (OHHL). At concentrations higher than the threshold level, OHHL interacts with the LuxR transcriptional activator, and the LuxR-OHHL complex activates gene expression from the P_{luxI} promoter. Because a quorum-sensing system is self-sufficient, its activation is completely autonomous. An attempt to use the LuxI-LuxR system for recombinant gene production has been reported (107). Because a wild-type quorum-sensing system is activated at a defined cell density, tunable quorum-sensing systems are desired for recombinant protein production so the induction timing can be optimized for each target protein. A method for creating these tunable systems by use of protein engineering has been proposed (89), and mutant quorum-sensing components have been generated by use of directed evolution (53, 54, 91). By use of LuxI and LuxR mutants, a series of engineered LuxI-LuxR systems can be generated for the activation of gene expression at a wide range of unique cell densities. Compared with the use of autoinducing media, quorum sensing can be used in a broad range of culture media, including those designed for high-cell-density fermentations.

37.6. HIGH-CELL-DENSITY FERMENTATION

High-cell-density fermentations of *E. coli* are used to obtain high volumetric productivities of recombinant proteins (17). During the past few decades, various medium compositions and operating conditions have been established to cultivate *E. coli* to densities as high as 200 g/liter (dry cell weight), making it possible to produce recombinant proteins at concentrations above 10 g/liter (59). It should be noted that all nutrients and trace ions become toxic to cells at high concentrations and should be avoided in high-cell-density fermentations. For example, glucose becomes inhibitory if its concentration exceeds 50 g/liter, and the inhibitory concentrations of ammonia and iron are 3 g/liter and 1.15 g/liter, respectively (84). Oxygen levels also commonly exert significant constraints on achievable cell densities, and efficient aeration conditions and the use of pure oxygen are necessary. It has been established that 1.96 g of oxygen is required to grow *E. coli* to 1 g of dry cell weight (84). Cell growth is also significantly inhibited by acetate accumulation at concentrations of 2 g/liter or higher, and control of acetate secretion by *E. coli* is necessary for high-cell-density fermentations (29). Limiting essential nutrient levels, and therefore cell growth rates, is an effective strategy to reduce acetate production. However, because this approach reduces cell growth rates, it is not desired if other options are available.

High-cell-density fermentations are performed in fermentors or stirred-tank reactors, and fed-batch (semibatch) fermentation is more common than batch operation (84). In a fed-batch operation, the culture medium can be simply introduced into a fermentor at a predefined mass flow rate, which can be constant, stepwise (gradual or linear), or exponential. Constant rate feeding is technically simple, but the specific cellular growth rate changes over time. Stepwise feeding enables increased cell growth at the late growth phase and extends the exponential growth phase, which favors recombinant protein production. Exponential rate feeding has a significant advantage in that a constant specific growth rate can be maintained, and methods exist to calculate the exponential feeding rates that are necessary for a specific growth rate (15). By choosing specific growth rates below the threshold level for acetate accumulation (1.5 to 2 g/liter), cell growth inhibition by acetate can be minimized. A specific growth rate can also be maintained if the feeding rate is coupled to feedback-control systems that regulate cell growth conditions (2).

37.7. CONCLUSION

Despite extensive developments in the engineering of *E. coli* strains and the construction of vectors for recombinant protein production, it still remains a challenge to produce any protein of interest profitably at an industrial scale. In particular, the isolation of expressed proteins remains an inefficient process involving multiple steps. On the production side, achieving a balance between protein production levels and the metabolic burden on the cell is a time-consuming process. However, sophisticated biomolecular engineering methods have been developed over the past few decades to address similar issues in metabolic engineering, and they could be used to circumvent some of the challenges present in the high-level production of recombinant proteins in *E. coli*. For example, Farmer and Liao (31) designed a genetic feedback-control circuit to adjust expression levels according to carbon availability. This strategy autonomously responded to the metabolic burden caused by overexpression of recombinant proteins, and resulted in increased production titers. The arresting of *E. coli* cells at the quiescent state, allowing for cells to divert resources typically used for biomass synthesis to recombinant protein production, has also been demonstrated, effectively uncoupling protein overexpression and cellular growth (87). Other approaches have looked at simplifying the purification of produced proteins, typically by engineering of the protein secretory pathways for enhanced secretion efficiency (105). A more novel approach was pursued by Suzuki et al., who developed a method for the degradation of all cellular proteins except the protein of interest by engineering mRNA degradation, a strategy that may significantly facilitate protein purification (105). Finally, a minimal cell harboring exclusively essential genes has been constructed (32, 108). Although such a cell may exhibit undesired vulnerability, it may allow as many resources as possible to be used for recombinant protein production when cultivated in well-controlled environments. While many of these strategies have not been tested extensively enough to determine their actual benefits, they show how our current ability to engineer and program cellular systems has created new methods to enhance the production of recombinant proteins, and show the promise of ever more sophisticated methods.

REFERENCES

1. Aehle, W. 2004. *Enzymes in Industry*, 2nd ed. Wiley-VCH, Weinheim, Germany.
2. Akesson, M., P. Hagander, and J. P. Axelsson. 2001. Probing control of fed-batch cultivations: analysis and tuning. *Control Eng. Practice* 9:709–723.
3. Baba, T., T. Ara, M. Hasegawa, Y. Takai, Y. Okumura, M. Baba, K. A. Datsenko, M. Tomita, B. L. Wanner, and H. Mori. 2006. Construction of *Escherichia coli* K-12 in-frame, single-gene knockout mutants: the Keio collection. *Mol. Syst. Biol.* 2:1–11.
4. Babaeipour, V., S. A. Shojaosadati, S. M. Robatjazi, R. Khalilzadeh, and N. Maghsoudi. 2007. Over-production of human interferon-gamma by HCDC of recombinant *Escherichia coli*. *Process Biochem.* 42:112–117.
5. Baneyx, F., and M. Mujacic. 2004. Recombinant protein folding and misfolding in *Escherichia coli*. *Nat. Biotechnol.* 22:1399–1408.
6. Berks, B. C., T. Palmer, and F. Sargent. 2005. Protein targeting by the bacterial twin-arginine translocation (Tat) pathway. *Curr. Opin. Microbiol.* 8:174–181.
7. Berks, B. C., F. Sargent, and T. Palmer. 2000. The Tat protein export pathway. *Mol. Microbiol.* 35:260–274.
8. Boer, H. A. D., L. J. Comstock, and M. Vasser. 1983. The *tac* promoter: a functional hybrid derived from the *trp* and *lac* promoters. *Proc. Natl. Acad. Sci. USA* 80:21–25.
9. Bouet, J. Y., K. Nordstrom, and D. Lane. 2007. Plasmid partition and incompatibility—the focus shifts. *Mol. Microbiol.* 65:1405–1414.
10. Buts, L., J. Lah, M.-H. Dao-Thi, L. Wyns, and R. Loris. 2005. Toxin-antitoxin modules as bacterial metabolic stress managers. *Trends Biochem. Sci.* 30:672–679.
11. Carrier, T. A., and J. D. Keasling. 1999. Library of synthetic 5′ secondary structures to manipulate mRNA stability in *Escherichia coli*. *Biotechnol. Prog.* 15:58–64.
12. Casali, N. 2003. *E. coli* plasmid vectors, p. 27–48. *In* N. Casali and A. Preston (ed.), *Methods in Molecular Biology*, vol. 235. Humana Press, Totowa, NJ.
13. Chatwin, H. M., and D. K. Summers. 2001. Monomer-dimer control of the ColE1 P-*cer* promoter. *Microbiology* 147:3071–3081.
14. Chen, C., B. Snedecor, J. C. Nishihara, J. C. Joly, N. McFarland, D. C. Andersen, J. E. Battersby, and K. M. Champion. 2004. High-level accumulation of a recombinant antibody fragment in the periplasm of *Escherichia coli* requires a triple-mutant (*degP prc spr*) host strain. *Biotechnol. Bioeng.* 85:463–474.
15. Cheng, L. C., J. Y. Wu, and T. L. Chen. 2002. A pseudo-exponential feeding method for control of specific growth rate in fed-batch cultures. *Biochem. Eng. J.* 10:227–232.
16. Choi, J. H., K. J. Jeong, S. C. Kim, and S. Y. Lee. 2000. Efficient secretory production of alkaline phosphatase by high cell density culture of recombinant *Escherichia coli* using the *Bacillus sp* endoxylanase signal sequence. *Appl. Microbiol. Biotechnol.* 53:640–645.
17. Choi, J. H., K. C. Keum, and S. Y. Lee. 2006. Production of recombinant proteins by high cell density culture of *Escherichia coli*. *Chem. Eng. Sci.* 61:876–885.
18. Choi, J. H., and S. Y. Lee. 2004. Secretory and extracellular production of recombinant proteins using *Escherichia coli*. *Appl. Microbiol. Biotechnol.* 64:625–635.
19. Couturier, M., E. M. Bahassi, and L. Van Melderen. 1998. Bacterial death by DNA gyrase poisoning. *Trends Microbiol.* 6:269–275.
20. Couturier, M., F. Bex, P. L. Bergquist, and W. K. Maas. 1988. Identification and classification of bacterial plasmids. *Microbiol. Rev.* 52:375–395.
21. Cranenburgh, R. M., J. A. Hanak, S. G. Williams, and D. J. Sherratt. 2001. *Escherichia coli* strains that allow antibiotic-free plasmid selection and maintenance by repressor titration. *Nucleic Acids Res.* 29:E26.
22. Dalbey, R. E., and M. Chen. 2004. Sec-translocase mediated membrane protein biogenesis. *Biochim. Biophys. Acta* 1694:37–53.
23. del Solar, G., R. Giraldo, M. J. Ruiz-Echevarria, M. Espinosa, and R. Diaz-Orejas. 1998. Replication and control of circular bacterial plasmids. *Microbiol. Mol. Biol. Rev.* 62:434–464.
24. Desmit, M. H., and J. Vanduin. 1994. Control of translation by messenger RNA secondary structure in *Escherichia coli*—a quantitative analysis of literature data. *J. Mol. Biol.* 244:144–150.
25. Drew, D., L. Froderberg, L. Baars, and J. W. de Gier. 2003. Assembly and overexpression of membrane proteins in *Escherichia coli*. *Biochim. Biophys. Acta* 1610:3–10.
26. Dubendorf, J. W., and F. W. Studier. 1991. Controlling basal expression in an inducible T7 expression system by blocking the target T7 promoter with *lac* repressor. *J. Mol. Biol.* 219:45–59.
27. Ebersbach, G., and K. Gerdes. 2005. Plasmid segregation mechanisms. *Annu. Rev. Genet.* 39:453–479.
28. Economou, A. 1999. Following the leader: bacterial protein export through the Sec pathway. *Trends Microbiol.* 7:315–320.
29. Eiteman, M. A., and E. Altman. 2006. Overcoming acetate in *Escherichia coli* recombinant protein fermentations. *Trends Biotechnol.* 24:530–536.
30. Evan, G. I., G. K. Lewis, G. Ramsay, and J. M. Bishop. 1985. Isolation of monoclonal antibodies specific for human C-Myc Proto-oncogene product. *Mol. Cell. Biol.* 5:3610–3616.
31. Farmer, W. R., and J. C. Liao. 2000. Improving lycopene production in *Escherichia coli* by engineering metabolic control. *Nat. Biotechnol.* 18:533–537.
32. Forster, A. C., and G. M. Church. 2006. Towards synthesis of a minimal cell. *Mol. Syst. Biol.* 2:2–10.
33. Fuqua, C., and E. P. Greenberg. 2002. Listening in on bacteria: acyl-homoserine lactone signaling. *Nat. Rev. Mol. Cell Biol.* 3:685–695.

34. **Georgiou, G., and L. Segatori.** 2005. Preparative expression of secreted proteins in bacteria: status report and future prospects. *Curr. Opin. Biotechnol.* **16:**538–545.

35. **Gerdes, K., F. W. Bech, S. T. Jorgensen, A. Lobnerolesen, P. B. Rasmussen, T. Atlung, L. Boe, O. Karlstrom, S. Molin, and K. Vonmeyenburg.** 1986. Mechanism of postsegregational killing by the *hok* gene-product of the *parB* system of plasmid R1 and its homology with the *relF* gene-product of the *Escherichia coli relB* operon. *EMBO J.* **5:**2023–2029.

36. **Guzman, L. M., D. Belin, M. J. Carson, and J. Beckwith.** 1995. Tight regulation, modulation, and high-level expression by vectors containing the arabinose P-*bad* promoter. *J. Bacteriol.* **177:**4121–4130.

37. **Harley, C. B., and R. P. Reynolds.** 1987. Analysis of E. coli promoter sequences. *Nucleic Acids Res.* **15:**2343–2361.

38. **Herman-Antosiewicz, A., M. Obuchowski, and G. Węgrzyn.** 2001. A plasmid cloning vector with precisely regulatable copy number in *Escherichia coli*. *Mol. Biotechnol.* **17:**193.

39. **Hershfield, V., H. W. Boyer, C. Yanofsky, M. A. Lovett, and D. R. Helinski.** 1974. Plasmid Cole1 as a molecular vehicle for cloning and amplification of DNA. *Proc. Natl. Acad. Sci. USA* **71:**3455–3459.

40. **Herskovits, A. A., E. S. Bochkareva, and E. Bibi.** 2000. New prospects in studying the bacterial signal recognition particle pathway. *Mol. Microbiol.* **38:**927–939.

41. **Higgins, C. F., R. S. McLaren, and S. F. Newbury.** 1988. Repetitive extragenic palindromic sequences, mRNA stability and gene expression: evolution by gene conversion? A review. *Gene* **72:**3–14.

42. **Hiszczynska-Sawicka, E., and J. Kur.** 1997. Effect of *Escherichia coli* IHF mutations on plasmid p15A copy number. *Plasmid* **38:**174–179.

43. **Hochuli, E., W. Bannwarth, H. Dobeli, R. Gentz, and D. Stuber.** 1988. Genetic approach to facilitate purification of recombinant proteins with a novel metal chelate adsorbent. *BioTechniques* **6:**1321–1325.

44. **Hochuli, E., H. Dobeli, and A. Schacher.** 1987. New metal chelate adsorbent selective for proteins and peptides containing neighboring histidine-residues. *J. Chromatogr.* **411:**177–184.

45. **Hoffman, B. J., J. A. Broadwater, P. Johnson, J. Harper, B. G. Fox, and W. R. Kenealy.** 1995. Lactose fed-batch overexpression of recombinant metalloproteins in *Escherichia coli* Bl21(DE3)—process control yielding high-levels of metal-incorporated, soluble protein. *Protein Expr. Purif.* **6:**646–654.

46. **Holmes, W. M., T. Platt, and M. Rosenberg.** 1983. Termination of transcription in *E. coli*. *Cell* **32:**1029–1032.

47. **Hu, S. Y., J. L. Wu, and J. H. Huang.** 2004. Production of tilapia insulin-like growth factor-2 in high cell density cultures of recombinant *Escherichia coli*. *J. Biotechnol.* **107:**161–171.

48. **Isaacs, F. J., D. J. Dwyer, C. M. Ding, D. D. Pervouchine, C. R. Cantor, and J. J. Collins.** 2004. Engineered riboregulators enable post-transcriptional control of gene expression. *Nat. Biotechnol.* **22:**841–847.

49. **Izard, J. W., and D. A. Kendall.** 1994. Signal peptides: exquisitely designed transport promoters. *Mol. Microbiol.* **13:**765–773.

50. **Jana, S., and J. K. Deb.** 2005. Strategies for efficient production of heterologous proteins in *Escherichia coli*. *Appl. Microbiol. Biotechnol.* **67:**289–298.

51. **Jensen, P. R., and K. Hammer.** 1998. The sequence of spacers between the consensus sequences modulates the strength of prokaryotic promoters. *Appl. Environ. Microbiol.* **64:**82–87.

52. **Joly, J. C., W. S. Leung, and J. R. Swartz.** 1998. Overexpression of *Escherichia coli* oxidoreductases increases recombinant insulin-like growth factor-I accumulation. *Proc. Natl. Acad. Sci. USA* **95:**2773–2777.

53. **Kambam, P. K. R., D. T. Eriksen, J. Lajoie, D. J. Sayut, and L. Sun.** 2009. Altering the substrate specificity of RhlI by directed evolution. *Chembiochem* **10:**553–558.

54. **Kambam, P. K. R., D. J. Sayut, Y. Niu, D. T. Eriksen, and L. Sun.** 2008. Directed evolution of LuxI for enhanced OHHL production. *Biotechnol. Bioeng.* **101:**263–272.

55. **Khlebnikov, A., and J. D. Keasling.** 2002. Effect of *lacY* expression on homogeneity of induction from the P-*tac* and P-*trc* promoters by natural and synthetic inducers. *Biotechnol. Progr.* **18:**672–674.

56. **Kimura, S., and T. Iyanagi.** 2003. High-level expression of porcine liver cytochrome P-450 reductase catalytic domain in *Escherichia coli* by modulating the predicted local secondary structure of mRNA. *J. Biochem.* **134:**403–413.

57. **Labbe, G., J. Bezaire, S. de Groot, C. How, T. Rasmusson, J. Yaeck, E. Jervis, G. I. Dmitrienko, and J. G. Guillemette.** 2007. High level production of the *Magnaporthe grisea* fructose 1,6-bisphosphate aldolase enzyme in *Escherichia coli* using a small volume bench-top fermentor. *Protein Expr. Purif.* **51:**110–119.

58. **Leader, B., Q. J. Baca, and D. E. Golan.** 2008. Protein therapeutics: a summary and pharmacological classification. *Nat. Rev. Drug Discov.* **7:**21–39.

59. **Lee, S. Y.** 1996. High cell-density culture of *Escherichia coli*. *Trends Biotechnol.* **14:**98–105.

60. **Liese, A., K. Seelbach, and C. Wandrey.** 2000. *Industrial Biotransformations.* Wiley-VCH, Weinheim, Germany.

61. **Luirink, J., and I. Sinning.** 2004. SRP-mediated protein targeting: structure and function revisited. *Biochim. Biophys. Acta* **1694:**17–35.

62. **Luo, Q. P., Y. L. Shen, D. Z. Wei, and W. Cao.** 2006. Optimization of culture on the overproduction of TRAIL in high-cell-density culture by recombinant *Escherichia coli*. *Appl. Microbiol. Biotechnol.* **71:**184–191.

63. **Lutz, R., and H. Bujard.** 1997. Independent and tight regulation of transcriptional units in *Escherichia coli* via the LacR/O, the TetR/O and AraC/I-1-I-2 regulatory elements. *Nucleic Acids Res.* **25:**1203–1210.

64. **Makrides, S. C.** 1996. Strategies for achieving high-level expression of genes in *Escherichia coli*. *Microbiol. Rev.* **60:**512–538.

65. **Martinez, E., B. Bartolome, and F. Delacruz.** 1988. pACYC184-derived cloning vectors containing the multiple cloning site and *lacZα* reporter gene of pUC8/9 and pUC18/19 plasmids. *Gene* **68:**159–162.

66. **Martinez-Martinez, I., C. Kaiser, A. Rohde, A. Ellert, F. Garcia-Carmona, A. Sanchez-Ferrer, and R. Luttmann.** 2007. High-level production of Bacillus subtilis glycine oxidase by fed-batch cultivation of recombinant *Escherichia coli* rosetta (DE3). *Biotechnol. Prog.* **23:**645–651.

67. **McCarthy, J. E., B. Gerstel, B. Surin, U. Wiedemann, and P. Ziemke.** 1991. Differential gene expression from the *Escherichia coli atp* operon mediated by segmental differences in mRNA stability. *Mol. Microbiol.* **5:**2447–2458.

68. **Mergulhao, F. J. M., G. A. Monteiro, J. M. S. Cabral, and M. A. Taipa.** 2004. Design of bacterial vector systems for the production of recombinant proteins in *Escherichia coli*. *J. Microbiol. Biotechnol.* **14:**1–14.

69. **Mergulhao, F. J. M., D. K. Summers, and G. A. Monteiro.** 2005. Recombinant protein secretion in *Escherichia coli*. *Biotechnol. Adv.* **23:**177–202.

70. **Middelberg, A. P. J.** 1996. Large-scale recovery of recombinant protein inclusion bodies expressed in *Escherichia coli*. *J. Microbiol. Biotechnol.* **6:**225–231.

71. **Morgan-Kiss, R. M., C. Wadler, and J. E. Cronan.** 2002. Long-term and homogeneous regulation of the *Escherichia coli araBAD* promoter by use of a lactose transporter of relaxed specificity. *Proc. Natl. Acad. Sci. USA* **99:**7373–7377.

72. **Mori, H.** 2004. From the sequence to cell modeling: comprehensive functional genomics in *Escherichia coli*. *J. Biochem. Mol. Biol.* **37:**83–92.

73. **Newbury, S. F., N. H. Smith, and C. F. Higgins.** 1987. Differential mRNA stability controls relative gene expression within a polycistronic operon. *Cell* **51:**1131–1143.

74. **Newbury, S. F., N. H. Smith, E. C. Robinson, I. D. Hiles, and C. F. Higgins.** 1987. Stabilization of translationally active mRNA by prokaryotic REP sequences. *Cell* **48:**297–310.

75. **Newman, J. R., and C. Fuqua.** 1999. Broad-host-range expression vectors that carry the L-arabinose-inducible *Escherichia coli araBAD* promoter and the *araC* regulator. *Gene* **227:**197–203.

76. **Novick, R. P.** 1987. Plasmid incompatibility. *Microbiol. Rev.* **51:**381–395.

77. **Nudler, E., and M. E. Gottesman.** 2002. Transcription termination and anti-termination in *E. coli*. *Genes Cells* **7:**755–768.

78. **Peredelchuk, M. Y., and G. N. Bennett.** 1997. A method for construction of *E. coli* strains with multiple DNA insertions in the chromosome. *Gene* **187:**231–238.

79. **Peterson, J., and G. J. Phillips.** 2008. New pSC101-derivative cloning vectors with elevated copy numbers. *Plasmid* **59:**193–201.

80. **Pfleger, B. F., N. J. Fawzi, and J. D. Keasling.** 2005. Optimization of DsRed production in *Escherichia coli*: effect of ribosome binding site sequestration on translation efficiency. *Biotechnol. Bioeng.* **92:**553–558.

81. **Pfleger, B. F., D. J. Pitera, C. D Smolke, and J. D. Keasling.** 2006. Combinatorial engineering of intergenic regions in operons tunes expression of multiple genes. *Nat. Biotechnol.* **24:**1027–1032.

82. **Poole, E. S., C. M. Brown, and W. P. Tate.** 1995. The identity of the base following the stop codon determines the efficiency of *in vivo* translational termination in *Escherichia coli*. *EMBO J.* **14:**151–158.

83. **Qing, G. L., B. Xia, and M. Inouye.** 2003. Enhancement of translation initiation by A/T-rich sequences downstream of the initiation codon in *Escherichia coli*. *J. Mol. Microbiol. Biotechnol.* **6:**133–144.

84. **Riesenberg, D., and R. Guthke.** 1999. High-cell-density cultivation of microorganisms. *Appl. Microbiol. Biotechnol.* **51:**422–430.

85. **Ringquist, S., S. Shinedling, D. Barrick, L. Green, J. Binkley, G. D. Stormo, and L. Gold.** 1992. Translation initiation in *Escherichia coli* sequences within the ribosome binding site. *Mol. Microbiol.* **6:**1219–1229.

86. **Ross, W., K. K. Gosink, J. Salomon, K. Igarashi, C. Zou, A. Ishihama, K. Severinov, and R. L. Gourse.** 1993. A third recognition element in bacterial promoters: DNA binding by the alpha subunit of RNA polymerase. *Science* **262:**1407–1413.

87. **Rowe, D. C. D., and D. K. Summers.** 1999. The quiescent-cell expression system for protein synthesis in *Escherichia coli*. *Appl. Environ. Microbiol.* **65:**2710–2715.

88. **Saida, F., M. Uzan, B. Odaert, and F. Bontems.** 2006. Expression of highly toxic genes in *E. coli*: special strategies and genetic tools. *Curr. Protein Pept. Sci.* **7:**47–56.

89. **Sayut, D. J., P. K. R. Kambam, and L. Sun.** 2007. Engineering and applications of genetic switches. *Mol. Biosyst.* **3:**835–840.

90. **Sayut, D. J., P. K. R. Kambam, and L. Sun.** 2008. Enzyme replacement therapy for lysosomal storage disorders. *Recent Pat. Biomed. Eng.* **1:**141–147.

91. **Sayut, D. J., Y. Niu, and L. Sun.** 2006. Construction and engineering of positive feedback loops. *ACS Chem. Biol.* **1:**692–696.

92. **Sayut, D. J., Y. Niu, and L. Sun.** 2009. Construction and enhancement of a minimal genetic AND logic gate. *Appl. Environ. Microbiol.* **75:**637–642.

93. **Schmidt, F. R.** 2004. Recombinant expression systems in the pharmaceutical industry. *Appl. Environ. Microbiol.* **65:**363–372.

94. **Schulz, V. P., and W. S. Reznikoff.** 1991. Translation initiation of Is50r read-through transcripts. *J. Mol. Biol.* **221:**65–80.

95. **Sheps, J. A., I. Cheung, and V. Ling.** 1995. Hemolysin transport in *Escherichia coli*. *J. Biol. Chem.* **270:**14829–14834.

96. **Shiloach, J., and R. Fass.** 2005. Growing *E.coli* to high cell density—a historical perspective on method development. *Biotechnol. Adv.* **23:**345–357.

97. **Siegele, D. A., and J. C. Hu.** 1997. Gene expression from plasmids containing the *araBAD* promoter at subsaturating inducer concentrations represents mixed populations. *Proc. Natl. Acad. Sci. USA* **94:**8168–8172.

98. **Smolke, C. D., T. A. Carrier, and J. D. Keasling.** 2000. Coordinated, differential expression of two genes through directed mRNA cleavage and stabilization by secondary structures. *Appl. Environ. Microbiol.* **66:**5399–5405.

99. **Smolke, C. D., and J. D. Keasling.** 2002. Effect of copy number and mRNA processing and stabilization on transcript and protein levels from an engineered dual-gene operon. *Biotechnol. Bioeng.* **78:**412–424.

100. **Smolke, C. D., and J. D. Keasling.** 2002. Effect of gene location, mRNA secondary structures, and RNase sites on expression of two genes in an engineered operon. *Biotechnol. Bioeng.* **80:**762–776.

101. **Srivastava, P., P. Bhattacharaya, G. Pandey, and K. J. Mukherjee.** 2005. Overexpression and purification of recombinant human interferon alpha2b in *Escherichia coli*. *Protein Expr. Purif.* **41:**313–322.

102. **Studier, F. W.** 2005. Protein production by autoinduction in high-density shaking cultures. *Protein Expr. Purif.* **41:**207–234.

103. **Studier, F. W., and B. A. Moffatt.** 1986. Use of bacteriophage T7 RNA polymerase to direct selective high-level expression of cloned genes. *J. Mol. Biol.* **189:**113–130.

104. **Summers, D. K., and D. J. Sherratt.** 1984. Multimerization of high copy number plasmids causes instability—Cole1 encodes a determinant essential for plasmid monomerization and stability. *Cell* **36:**1097–1103.

105. **Suzuki, M., J. Zhang, M. Liu, N. A. Woychik, and M. Inouye.** 2005. Single protein production in living cells facilitated by an mRNA interferase. *Mol. Cell* **18:**253–261.

106. **Thomas, C. M., and C. A. Smith.** 1987. Incompatibility group-P plasmids—genetics, evolution, and use in genetic manipulation. *Annu. Rev. Microbiol.* **41:**77–101.

107. **Thomas, M. D., and A. V. Tilburg.** 2000. Overexpression of foreign proteins using the *Vibrio fischeri lux* control system. *Method Enzymol.* **305:**315–329.

108. **Trinh, C. T., P. Unrean, and F. Srienc.** 2008. Minimal *Escherichia coli* cell for the most efficient production of ethanol from hexoses and pentoses. *Appl. Environ. Microbiol.* **74:**3634–3643.

109. **Wagner, S., M. M. Klepsch, S. Schlegel, A. Appel, R. Draheim, M. Tarry, M. Hogbom, K. J. van Wijk, D. J. Slotboom, J. O. Persson, and J. W. de Gier.** 2008. Tuning *Escherichia coli* for membrane protein overexpression. *Proc. Natl. Acad. Sci. USA* **105:**14371–14376.

110. **Walsh, G.** 2006. Biopharmaceutical benchmarks 2006. *Nat. Biotechnol.* **24:**769–776.

111. **Wong, H. C., and S. Chang.** 1986. Identification of a positive retroregulator that stabilizes messenger RNAs in bacteria. *Proc. Natl. Acad. Sci. USA* **83:**3233–3237.

112. **Xie, L., D. Hall, M. A. Eiteman, and E. Altman.** 2003. Optimization of recombinant aminolevulinate synthase production in *Escherichia coli* using factorial design. *Appl. Microbiol. Biotechnol.* **63:**267–273.

113. **Yanisch-Perron, C., J. Vieira, and J. Messing.** 1985. Improved M13 phage cloning vectors and host strains—nucleotide sequences of the M13mp18 and Puc19 vectors. *Gene* **33:**103–119.

Bioprocess Development

LUTZ HILTERHAUS AND ANDREAS LIESE

38

38.1. BACKGROUND

Within bioprocesses, organic compounds are converted by either isolated enzymes or whole-cell biocatalysts. Thus, the enzyme is the elementary operational component in a bioprocess, while the spectrum of bioprocesses ranges from reactions with single purified enzymes to complex cellular systems. Biotransformations can be differentiated into enzymatic and metabolic bioconversions. Enzymatic biotransformations are characterized by a low number of specific reactions, whereas metabolic bioconversions need the metabolic system of living and growing cells. The latter can be bacteria, yeasts, or fungi. In contrast to these, enzymatic biotransformations are carried out by resting cells or isolated proteins in different types of bioreactors (41). The production of fine and bulk chemicals via biocatalyzed processes can be competitive with classical chemical processes when economic advantages are given. Further on, ecological benefits such as a reduced carbon footprint might increase the market potential of biocatalytic products, dependent on the actual pressure exerted by the respective governments and societies (Fig. 1).

In contrast to chemical processes, bioprocesses often apply milder reaction conditions and are more selective. On the one hand, this is advantageous because of fewer purification steps, a lower amount of side products, and lower energy consumption. On the other hand, one has to consider the sensitivity of the biocatalyst in view of substrate and solvent tolerance, temperature, and shear force stability, as well as its storage, retention, and recycling. Advances in biocatalyst improvement by recombinant technologies provide the possibility to alternate the substrate spectra, activity, and stability of enzymes, and the young discipline of systems biology allows the optimization of whole microorganisms with respect to a desired product (42).

In sum, the arguments for the usage of enzymes and whole cells in industrial synthesis of bulk and fine chemicals are crucial when high standards regarding selectivity and purity have to be reached. Therefore, requirements for a successful industrial bioprocess are rational design and development starting from basic reactor types, optimizing the setup with respect to the biocatalyst. Fundamental aspects of bioreaction engineering, which need particular considerations, include enzyme kinetics, immobilization of enzymes, and scale-up. Here, the reader is referred to the respective chapters within this book, whereas this chapter deals with the reactor setup and the bioprocess itself.

38.2. CONCEPT AND GENERAL FEATURES OF BIOPROCESSES

A bioprocess is based on a biological catalyst that is used to conduct a chemical reaction leading to a defined product. The choice of the reaction media, the biocatalyst, or the chemical and physical parameters is the natural scientific starting point to establish and optimize a bioprocess (Fig. 2). However, associated reactors, process control, and modeling are engineering scientific starting points to understand and enhance a bioprocess. Within this section the specific requirements of a biocatalytic reaction from a natural scientist point of view are illustrated and discussed, whereas aspects of process engineering are covered in section 38.3.

38.2.1. Special Challenges of Biocatalysis

In the early times of bioprocess engineering, water was used principally for the reaction medium, because it was assumed that a higher stability and activity can be reached in the natural medium of enzymes. Connected to this is a low solubility of organic substrate and products, entailing low yields and productivities that again necessitate extensive product purification. Next to these parameters, biocatalyzed reactions are limited by other factors which directly influence the applied enzymes and microorganisms. These are temperature, pH value, shear forces, foam formation, organic solvents, and so on. In the view of fermentation processes, these factors are oxygen enrichment, mixing, carbon and nitrogen source, and so forth.

Today, it is a matter of common knowledge that enzymes and whole cells can be applied successfully also in organic solvents (13), ionic liquids (58, 69), supercritical solvents (30), and mixtures thereof, as well as in a solvent-free manner (23, 24). Accordingly, the spectrum of possible applications of biocatalysts was enormously increased (see also section 38.5).

The temperature of bioprocesses is a crucial point when high turnover numbers will be reached. The reaction rate of a biocatalyzed reaction follows the Arrhenius equation to a certain point. Similar to other chemical reactions, the activation energy and frequency coefficient are biocatalyst dependent and lead to a rate constant k at a certain temperature, whereas in biotransformations not k is used but v_{max}.

$$k = b \cdot \exp\left[-\frac{E_A}{RT}\right]$$

This equation can be used only in the case of simple reaction steps and in a certain temperature interval.

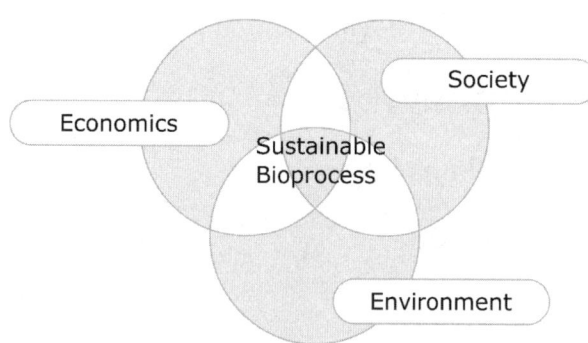

FIGURE 1 Terms of sustainable development.

At an enzyme-dependent temperature, denaturation of the enzyme occurs (Fig. 3) and the reaction rate decreases. Therefore, next to the activity, the enzyme's stability is important, including the half-life time of the biocatalyst.

Regarding the stability of enzymes at increased temperatures, different approaches were carried out to increase the temperature range of bioprocesses. One possibility is using enzymes from extremophilic microorganisms which have been adapted during the natural evolution process to deal with higher temperatures or extreme pH values (47). Another possibility is directed evolution, which can complement the spectrum of natural enzymes by adapting enzymes in the laboratory in view of stability at higher temperature, pH, and solvent tolerance (1). To broaden the applicability at different temperatures, the stability of enzymes can be also improved by immobilization. Next to the effect on the temperature tolerance, immobilization can also help to enable the use of enzymes in different solvents, at extremes of pH and exceptionally high substrate concentrations (18).

However, although the protein literature contains many references to shear denaturation, there are contradictory reports on this phenomenon with respect to solubilized enzymes (28). The negative effect on immobilized enzymes and whole cells is assured. Another force influencing protein denaturation and enzyme inactivation is the surface tension (6). Therefore, during process design, these criteria also have to be considered, especially when foam formation or shear sensitivity are problems known from preliminary experiments.

38.2.2. Advantages of Biocatalysis
Enzymes catalyze several defined reactions, rather than defined classes of reactions; hence, there is a big potential for enzymes in industrial processes. Compared with classical chemical catalysts, enzymatically catalyzed reactions are and have to be performed under mild reaction conditions regarding, e.g., pH value and temperature. Enzymes catalyze exactly one defined reaction; therefore, in most cases no impurities by side reactions occur. The reaction medium consists of a few components, which have mostly a precise composition, so that purification causes relatively low costs. However, one has to watch out not to add too many salts in the form of buffer and metal salts that need to be removed later during downstream processing. By coupling several enzymatic reactions, sequences can be built up to synthesize also more complex molecules, omitting intermediate purification. However, in practice it has to be considered that by increasing the number of reaction steps, the technical implementation gets more complex, demanding integration of process engineering and biocatalysis at an early time. A further advantage compared with classical chemical catalysts, like acids or heavy metal-based ones, is the lower toxicity of biocatalysts. Although it is known that some enzymes cause allergic reactions, in general, enzymes are not acutely toxic.

38.2.3. Systems Biology and Metabolomics
Metabolic bioconversions need the metabolic system of living and growing microorganisms, e.g., bacteria, yeasts, or fungi. The special challenges of these bioprocesses are

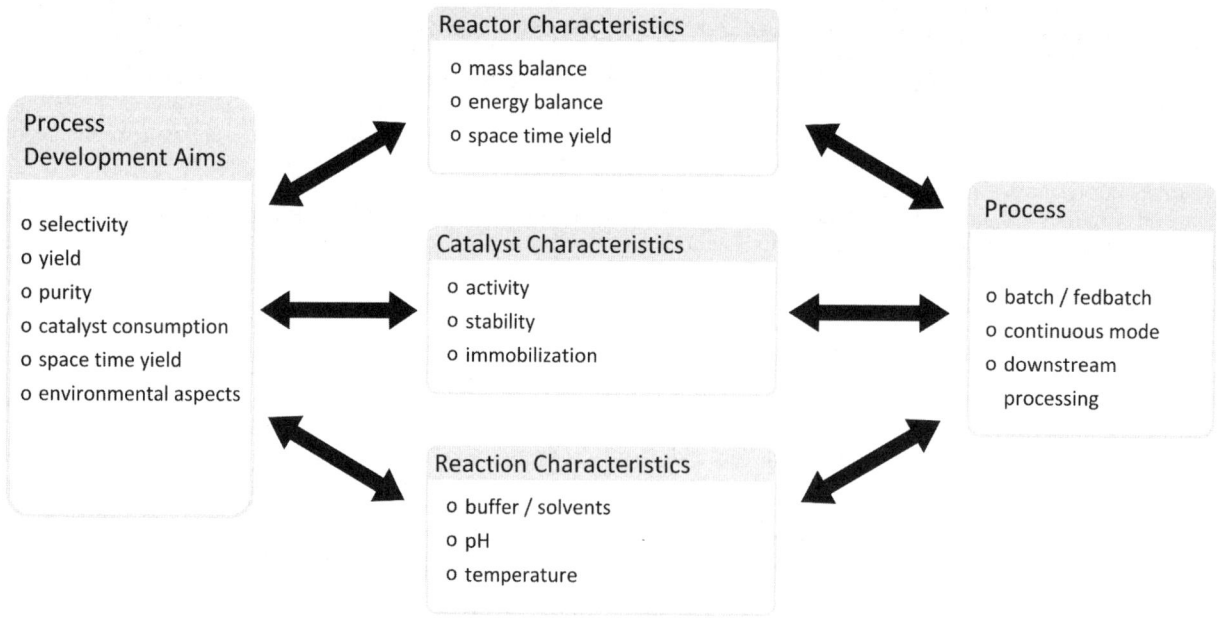

FIGURE 2 Rational process design.

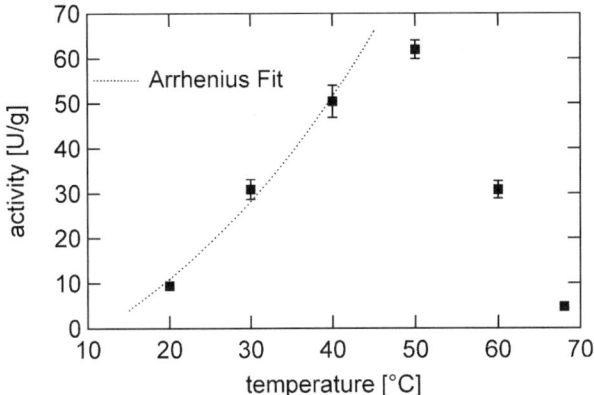

FIGURE 3 Specific activity of immobilized wild-type benzoyl formate decarboxylase at different temperatures (22).

maintaining growing conditions, excluding contaminating organisms, and aerating large volumes of fermentation broth. One disadvantage often found is the extremely low volumetric productivity in combination with the production of undesired side products because of the metabolism of the microorganisms. Therefore, the focus on the overproduction of one specific product by so-called designer bugs is desired.

Traditionally, industrial microorganisms have been developed via multiple rounds of random mutagenesis and selection (2). A goal of systems biology is predictive metabolic engineering, where genes within metabolic pathways are purposefully amplified or deleted based on the consideration of the metabolic network as an entirety (63). However, the application of systems biology to strain development has only just begun to emerge. By combining the results of multiple genome-wide analyses and computational analyses, systems biology can subsequently be used for designing strategies for strain improvement (52). Substantial improvements of analytical hardware allow metabolomics to run routinely now. Data are successfully used to investigate genotype-phenotype relations of strains and mutants. Metabolomics facilitates metabolic engineering to optimize microorganisms for industrial biotechnology and spreads to the investigation of biotransformations (48). Metabolomics allows the construction of designer organisms for process application by use of biotransformation and fermentative approaches making effective use of single enzymes, whole microbial cells, and even eukaryotic cells.

Recently, several amino acid producers have been successfully developed by systems metabolic engineering, where the metabolic engineering procedures were performed within a systems biology framework and entire metabolic networks, including complex regulatory circuits, were engineered in an integrated manner. Park and Lee (51) reviewed the current status of systems metabolic engineering for developing amino acid-producing strains. Here, engineering targets are determined by considering entire metabolic and regulatory networks together with midstream (fermentation) and downstream (recovery and purification) processes. The impact on the entire metabolism of altering these targets is examined to provide feedback, and through the iterative mode of operation of metabolic engineering, the ultimate metabolic phenotype desired can be obtained.

38.3. PROCEDURES OF BIOPROCESS DEVELOPMENT

The industrial application of biotransformations is often limited to the transfer of laboratory bench experiments to the cubic-meter scale. The scale-up implies an increase in the volume handled rather the decrease in the surface-to-volume ratio. The issues involved are, above all, mixing and heat, as well as mass transfer being connected to power consumption and rheology. The basic reactor types and their mode of operation will be discussed with regard to these physical aspects.

38.3.1. Bioprocess Engineering Principles

The increase of mass and heat transfer within the bioreactor and, by this, the increase of the conversion is achieved by a homogenization of the reaction components. Mixing is the uniform distribution of several components in a specific volume leading to minimal differences in chemical composition, physical characteristics (temperature, density, and viscosity), and/or morphology (size and form of bubbles or particles) (8). Both quality of mixing and mixing time are applied to evaluate a mixing process. The specification of the mixing time only has sense when it is combined with the quality of mixing M. This is defined as:

$$M = 1 - \frac{\Delta c}{\Delta c_0}$$

Here, Δc_0 is the difference in concentration at the beginning of the mixing process, whereas Δc is the concentration difference at the time t_c. The reduction of concentration gradient can be described mathematically with adequate accuracy by an exponential function:

$$\frac{\Delta c}{\Delta c_0} = k_0 \cdot \exp\left(-\frac{\Theta}{t_c}\right)$$

It is assumed that differences in density and viscosity can be neglected; hence, mixing time Θ depends on the revolutions per minute n, the stirrer diameter d, and the kinematic viscosity v.

The basics to calculate ideal reactor types (38) and the choice of an appropriate reactor are discussed below. Here, only the change in the amount of substance and not in energy is considered, whereas the equations are valid both for enzymatic and chemical catalysis. Only the three basic types of reactors are presented here. All others are variations or deductions thereof:

- Stirred-tank reactor (STR)
- Continuously operated STR (CSTR)
- Plug flow reactor (PFR)

There are no mass flows in the STR during operation, meaning that a closed system is present. In contrast, CSTR and PFR are continuously operated reactors; hence, in these open systems, ingoing and outgoing mass flows have to be considered (Fig. 4). The key feature of the STR is the time-independent homogenous mixing in the reactor, diminishing all temperature, pressure, or concentration gradients. In contrast to the STR, the PFR is characterized by a plug flow velocity profile in the reactor where all fluid elements show the same velocity. Mixing in the axial direction can therefore be neglected here under ideal conditions.

The conversions attainable in the different reactor types are influenced by enzyme kinetics. However, STR and PFR are described by the same equation:

$$v_{max} \cdot \tau = -K_m \cdot \ln(1 - \chi) + c_{S,in} \cdot \chi$$

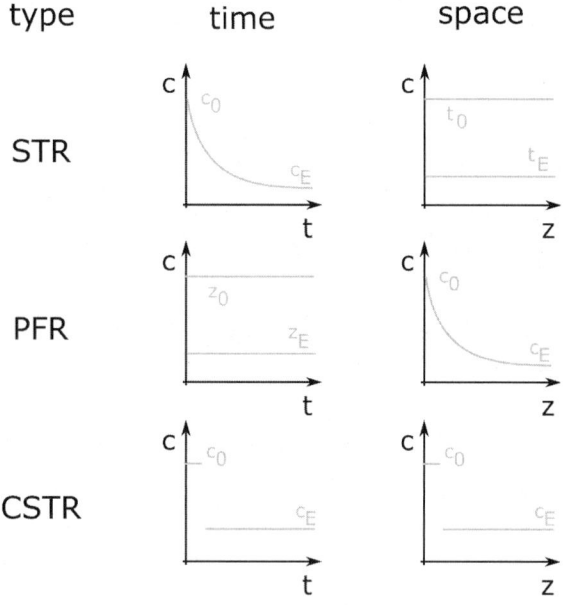

FIGURE 4 Schematic illustration of concentration profiles versus time and space for the different ideal reactor types.

Here, v_{max} is the maximum velocity of the enzyme, τ is the residence time, K_m is the Michaelis constant, χ is the conversion, and $c_{S,in}$ is the substrate concentration entering the reactor. Consequently, both reactors show the same performance considering the enzyme kinetics. Thus, batch reactor performance mainly depends on setup time. Longer setup times in the overall process lead to lower performance of the batch reactor. In the case of neglecting setup times, the performance of the batch reactor and PFR is similar. The same conversion reactions inhibited by excess of substrate, and those inhibited by product in the batch reactor or PFR, can be carried out faster in CSTR. Because efflux conditions are present in the CSTR, low substrate and high product concentrations can be obtained. In contrast, the product concentration in batch reactors increases in the course of the reaction; accordingly, the substrate concentration of the inflow of the PFR is high. On passing through the reactor, the substrate is converted into the product; hence, depending on the axial coordinate, different substrate and product concentrations are found.

Furthermore, all conversions with a reaction order higher than 0 run faster in PFR than in CSTR. At high substrate concentrations, enzymes catalyze a reaction kinetics of zero order and at low concentration of first order. The transition is described by the Michaelis-Menten equation. For this reason, to evaluate a given reactor type, $r_{S_0 K_m}$, the ratio of initial substrate concentration, c_{S_0}, and the K_m value of the enzyme are the parameters of choice:

$$r_{S_0 K_m} = \frac{c_{S_0}}{K_m}$$

For $r_{S_0 K_m} \gg 1$ and $c_{S_0} \gg K_m$ within the whole reaction range, the kinetics are of zero order and identical conversions in CSTR and PFR are obtained. This is the case when low K_m values are present or a high conversion is not sought. In other cases, an identical residence time causes a lower conversion in CSTR compared with the PFR.

The enzyme demand of different reactor types is compared by plotting the ratio e_{CSTR}/e_{PFR} versus the desired conversion. Doubling of enzyme activity v_{max} shows the same effect as doubling residence time τ:

$$\frac{e_{CSTR}}{e_{PFR}} = \frac{\tau_{CSTR}}{\tau_{PFR}}$$

At a given $r_{S_0 K_m}$ value of 100, for example, the reaction has zero-order kinetics over a broad range. Approaching total conversion, the reduced substrate concentration leads to first-order kinetics after passing through a transition phase. Therefore, differences between CSTR and PFR do not occur until very high concentrations. The lower the $r_{S_0 K_m}$ value, the faster the departure from the ratio e_{CSTR}/e_{PFR} of 1 arises at lower conversions. From the very first, the reaction is of first order at a low $r_{S_0 K_m}$ value of 0.1, meaning a higher enzyme demand of the CSTR over the PFR at low conversions to obtain equal conversions.

The CSTR outmatches the PFR up to high conversion if strong inhibition by substrate excess, namely, small K_{iS}/K_m value, is present. Minor inhibition means that at low conversions the CSTR is favored, whereas at high conversions the PFR is favored. In the case of product inhibition, the PFR shows better reaction performance than the CSTR. At high K_{iP}/K_m values, meaning minor inhibition by the product, the kinetic values are described by the Michaelis-Menten equation; rather, the reaction is of first order. At low K_{iP}/K_m values, the PFR is favored at all conversions because of product inhibition.

In general, the reaction can only be carried out efficiently when $K_{iP}/K_m > 1$. Values of $K_{iP}/K_m < 1$ signify a very strong inhibition preventing high conversions (37). Because the kinetic parameters are constants, high conversions can only be obtained when an adequate process design is applied, e.g., in situ product removal (39). This keeps the product concentration low and will be illustrated in sections 38.4.3 and 38.6.2.

38.3.2. Modeling and Assessment

The insight and understanding of a bioprocess can be enhanced by process modeling, leading to potential improvements and identification of possible difficulties. Furthermore, simulation can supplement experiments and broaden the basis for sound decision making (26, 31). While the general approach is similar, typical bioprocesses differ in their kinetics of product formation, process structure, and operating constraints when compared with chemical processes. The general steps in bioprocess modeling (Fig. 5) are process analysis, model creation, and the consideration of uncertainties in the model (19).

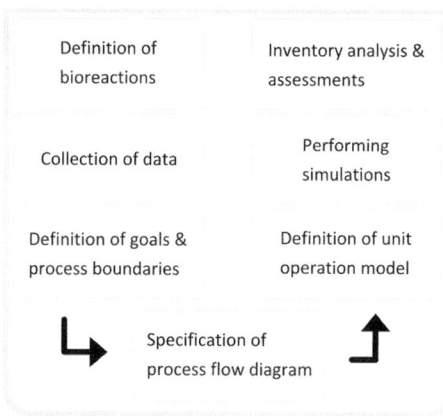

FIGURE 5 The general steps in bioprocess modeling.

For successful modeling, the definition of the modeling goal is crucial; it includes the final product specification and purity, the plant size, the biocatalyst, and the model boundaries. In general, each model has to include all necessary process steps, while keeping complexity at a minimum.

When the goal and the model boundaries are defined, the necessary data have to be collected, which can be derived from one's own experiments or external sources like literature or patents. Parameter values often have to be estimated from different sources or extrapolated from conditions that differ from the expected process, e.g., in scale, process conditions, biocatalyst used, etc. Critical expert assessment of data reliability and applicability is necessary here.

To model a bioreaction, the reaction equations and conditions are derived from the collected data and the general bioprocess knowledge. First, the starting materials needed for the applied biocatalyst are listed. Afterward, parameters like yields, reaction time, final product concentration, by-product formation, etc., are determined. If the reaction data are not known from experiment, a kinetic or a stoichiometric model can be applied to calculate these parameters. Then, the reaction conditions for the process model have to be defined. After this the process flow diagram is assembled and all unit procedures and the process streams of the model have to be specified. Every unit operation has to be described in a model, and the model parameters have to be defined.

The influence of assumptions, estimates, and simplifications on individual steps and the overall performance can be addressed in an uncertainty analysis. Therefore, it is essential to document and explain all assumptions, estimates, and simplifications. This transparent documentation serves as a reference point and enables comprehension and interpretation of the simulation results in the identification of uncertainties. The model is created and finally transferred into suitable software, where simulations are performed. Apart from improving the general understanding of the process, these simulation results are used for further optimization of the bioprocess.

To assess a bioprocess, the understanding of the uncertainties and the identification and quantification of risks and opportunities within a process are crucial. A scenario analysis can compare alternative process flow diagrams, and a sensitivity analysis studies the impact of single input variables, like medium cost or reaction time. In this context, variability is an intrinsic feature of the system and an effect of chance as seen in the actual variation. It cannot be reduced by further studies. Uncertainty is caused by a lack of knowledge about a parameter that does not show variability in reality, but its exact value is not known. Further studies can reduce this type of uncertainty. The variation of model parameters often involves both variability and uncertainty.

Scenario analysis examines variations of the process flow diagram and the process scale. Especially in early process development, there might be a need to compare alternative process flow sheet topologies. For changes in downstream processing, for example, the economic and environmental impact can be derived in a scenario analysis. The base model can be used as a benchmark.

Sensitivity analyses study the impact of a single process parameter on the objective functions of the model. The analysis is usually done within the existing process flow diagram. By comparing the sensitivity of different parameters, the most sensitive can be identified. Special attention must be paid to these parameters in the process development. Sensitivity analysis quantifies the dependence of the objective functions on single parameters; it may not capture nonlinearities between multiple parameters that may vary simultaneously. However, it does not provide any information about the probability of certain values of the examined parameter. For further reading, the book by Heinzle, Biwer, and Cooney is recommended (19).

38.4. TECHNIQUES

38.4.1. Micro- and Minibioreactors

Microbioreactors are characterized by their area of application, by effective heat transfer and ultrafast mixing characteristics, and by their suitability for online monitoring and control (36). Within microprocess engineering, chemical or enzymatic processes are conducted inside small volumes, typically inside channels with diameters of less than 1 mm or other structures with submillimeter dimensions. These processes are usually carried out in continuous-flow mode, allowing a high throughput, and have further benefits like defined residence time setting and highly regular flow profiles.

Microreaction technology is an interdisciplinary field combining natural science and engineering (20, 45). Because there have been demands for innovation in process engineering, particularly for enzymatic reactions, microreaction devices can serve as efficient tools for the development of enzyme processes. The study of individual bioprocess parameters at the microliter scale using either microwell or microfluidic formats can be used to generate quantitative information. Bioprocess design and scale-up based on such broad information can be more rational and effective. Furthermore, automation can enhance experimental throughput, facilitating the parallel evaluation of different process parameters or alternative process flow sheet topologies (section 38.3.2).

Next to the reaction itself, unit operation, e.g., mixing and heat transfer or product purification, can be realized, leading to a complete process on the micrometer level. For further details the reader is referred to Hessel et al. (20).

In contrast, minibioreactors (53) possess volumes below about 100 ml. Different types of these reactors have been constructed:

- Shaken bioreactors (shake flask, microtiter plate)
- Stirred bioreactors (STR, stirred cuvette)
- Other bioreactors (hollow-fiber membrane)

Bioprocess design using micro- or minireactors makes an advanced process control necessary but has a great advantage: the reduction in development time and costs. Minibioreactors can be monitored by special online measurements that allow comprehensive studies with high-priced substrates. Many process parameters can be tested in small-scale devices and later on transferred to large-scale bioreactors. Large numbers of small-scale experiments allow broader screening for appropriate biocatalysts, reaction conditions, and a fast and rational scale-up. Therefore, micro- and minibioreactors offer the potential to speed up the delivery of new products to the market, to reduce development costs, and to increase economic benefit (44).

Besides these advantages in the speed-up of process development, there are some advantages in handling. Microreactors can be applied in those cases where mixing and temperature control are crucial for selectivity and yield. Here, the large surface-to-volume ratio allows an excellent energy and mass transfer. Furthermore, the production of critical components can be advantageous because highly exothermic reactions—leading to high temperature or high pressure—and highly corrosive or toxic reaction components can be processed more safely.

38.4.2. Miniplant

Often, to reduce development costs of bioprocesses, the miniplant technology is suggested (3). This involves establishing a prospective industrial process on the smallest possible scale, including all operation units and the original product. A direct scale-up from laboratory to production facility can thus be made easier in many cases. Accordingly, by mimicking operative unit production, the miniplant technology should help to simplify the scale-up of (bio)processes.

When a new or changed process is to be established, the desired product and reactants, their physical and chemical characteristics, the reaction and downstream processing pathway, and the sequence thereof have to be known. At the beginning, besides mathematical calculations, experiments also have to be done to obtain data not previously available and, hence, to design operable experimental plants. A key factor is the determination of dimensionless key numbers that enable a scale-independent description of the process. To reduce costs, the experimental plant should be as small as possible, leading to a large scale-up factor. These will show the feasibility of the several process steps and deliver data for target-oriented scale-up.

Thus, important characteristics of an optimal miniplant technology are:

- Fast buildup and reconstruction
- High flexibility to react on different experimental conditions
- Long-time operation

The rate at which biotechnology processes are introduced and used in different chemical markets depends on the technical implementation. For economic success,

TABLE 1 Comparison of time requirements (61)

Step	Pilot plant	Miniplant
Planning	1 year	3–4 months
Purchase/setup	1 year	>3 months
Initiation time	3 months	1 month
Test duration	6–9 months	6–9 months
Total time	3 years	1–1.5 years

a fast introduction to the market (short time to market) is necessary. In addition, costs are reduced when shorter development time can be achieved (5). The application of miniplant technology for biocatalytical processes is a helpful approach to attain this goal.

Roughly comparing the time requirement between pilot plant and miniplant, it is obvious that setup and operation predominate (see Table 1), but the development time in sum can be reduced drastically by applying miniplant technology.

Next to time consumption, the costs for process development are crucial. In Fig. 6 the sum of costs for both the pilot plant and the miniplant are illustrated. Here too, the advantage of the miniplant technology can be seen in a faster achievement of the profit zone (11).

38.4.3. Membrane Bioreactors

Membranes have found a broad range of applications, including the production of drinking water, energy generation, tissue repair, pharmaceutical production, food packaging, and the separations needed for the manufacture of chemicals, electronics, and a range of other products (56). In the manufacturing of chemicals by chemical or biotechnological methods, the combination of catalysts and membrane-based reactors is a key to obtaining

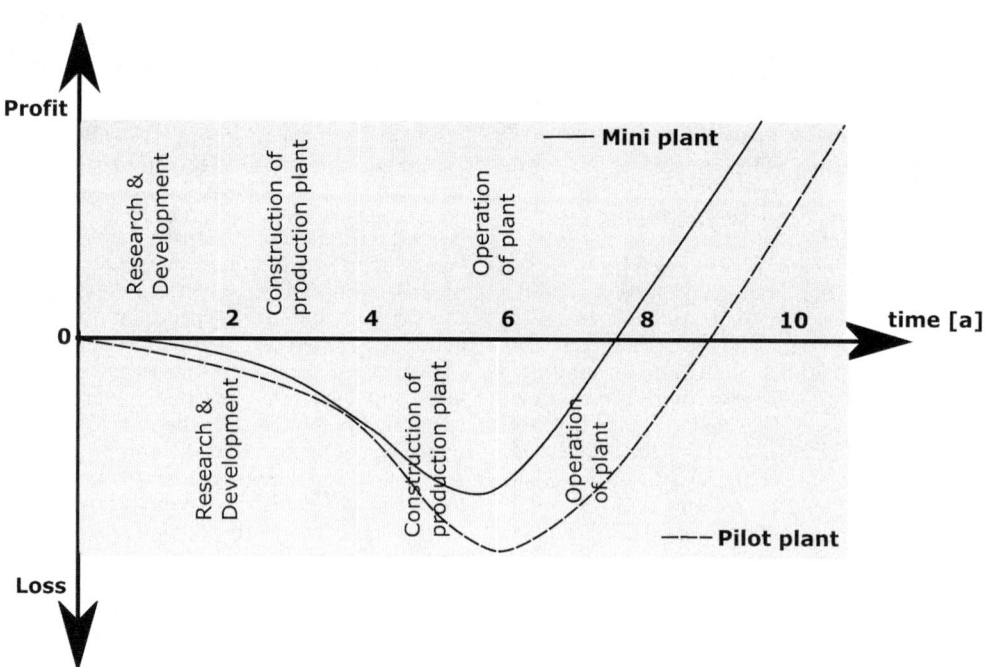

FIGURE 6 Comparison of pilot plant and miniplant: costs for process development versus time (11).

Dead-End-Filtration   **Cross Flow-Filtration**

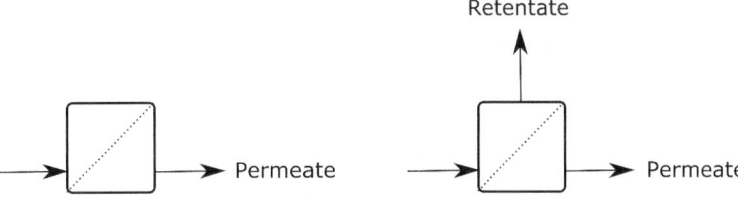

FIGURE 7   Comparison of different membrane process setups.

higher productivities and less waste. Bioprocesses including membrane reactors mimic the biological model of compartmentalization. Here, different reactions and products are separated from each other to minimize interfering reactions and to circumvent by-product formation.

The central point of a membrane bioreactor is the interface, which is formed and stabilized by a thin, partly permeable membrane. This membrane can control the exchange of solvent (conventional or nonconventional), substrates, products, and/or catalysts between two neighboring fluid phases. Here, as in Nature, different forces are found that influence the flow through the membrane. First, there is the viscosity and pressure of the liquid on both sides of the membrane, which influence the flow of the solvent through the membrane. Beyond this, there are forces that limit or allow the passage of solubilized substances. The retention of some of these substances, meaning the selectivity of membranes, is directly influenced by their physical and chemical characteristics. Here, the process engineer has to consider the fluid properties (aqueous phase, organic phase, etc.), the membrane material (hydrophobic, hydrophilic, and selectivity), and the requirement of the process itself.

Different membrane types and driving forces for the filtrations are applied in several processes. The diffusive mass transport via the membrane is mostly very selective but limited by a low flow rate. To increase the flow rate, a higher pressure difference has to be applied. However, porous membranes allow higher permeate flow rates per area because of their membrane structure (46, 60). In view of the conduction of the retentate flow, two operational modes can differentiated: (i) the dead-end filtration, where the retentate flow is directed vertically on the membrane, and (ii) the cross-flow filtration, where the retentate flow is directed tangentially along the membrane surface (Fig. 7).

Permeation of the solvent through the membrane leads to increasing retentate concentration at the inflow side (74). A boundary and a cover layer are formed, whereby a hydraulic resistance is built up (Fig. 8). This resistance incrementally blocks the permeate flow (J). This polarization of concentration can be counteracted by increasing the tangential flow.

For continuous processes with isolated, solubilized enzymes, the enzyme membrane reactor was designed (73). In this system, a membrane retains the enzyme; on industrial scale, a hollow-fiber module is used. Within this process variant, the substrate is fed to the enzyme solution and the product and nonreacted substrates pass through the membrane. Immobilization of the enzyme directly on the membrane surface is also possible, and in the process, the substrate passes the membrane, reaction occurs, and the product can be obtained in the permeate. The big advantage of such reactors is the operation under outflow conditions in the case of CSTR operation mode,

resulting in minimization of specifically substrate surplus inhibition. Thus, enzyme membrane reactors have found application in the synthesis of pharmaceuticals and the building blocks thereof.

The first industrial application of membrane bioreactors was published by the Degussa Company in the 1980s. This process can be used in the production of L-tert-leucine, L-neopentylglycine, or L-5,5-dimethyl-butyl-glycine, which are building blocks for drug synthesis, especially for the synthesis of antitumor or human immunodeficiency virus protease inhibitors, as well as templates in asymmetric synthesis (7). These amino acids are synthesized starting from ammonia and keto acids. High concentrations of the keto acid trimethyl pyruvate lead to inhibition of the enzyme L-leucine-dehydrogenase within the production of L-tert-leucine. Furthermore, the productivity of the process is limited by chemical side reactions. Therefore, a continuously operated process in the mode of a stirred tank with an ultrafiltration step is applied. This enzyme membrane reactor enables low concentrations of the substrates and recycling of the enzyme (73).

A big disadvantage in using pressure-driven membrane processes is membrane fouling, which leads to a loss of performance (4, 46). This effect includes:

- Adhesion of particles on the membrane surface
- Adhesion of particles in the pores of the membrane
- Soaking of the membrane
- Chemical reaction of membrane components with the reaction solution

Special surface modifications can reduce this problem. Changes of chemical surface properties as well as changes of the macroporous structure of the material can improve the performance and stability of the membrane. An overview

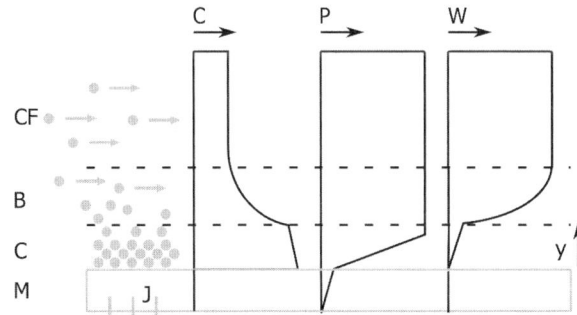

FIGURE 8   Polarization of concentration at a membrane: c, concentration profile; p, pressure profile; w, flow profile; M, membrane; C, cover layer; B, boundary layer; CF, core flow; J, permeate flow; y, membrane distance (57).

about the status quo of membrane bioreactor technology is given in reference 15 and more recently in reference 29. A summary of new applications of enzyme membrane reactor technology is found in references 40, 55, and 75.

38.4.4. Separation Technologies

Many different disciplines are involved in bioprocess engineering, but the separation process is the part that most involves the discipline of chemical engineering. To produce a purified product, an effective separation process has to be carried out. This process can be done after the reaction (classical downstream processing) or during the reaction (hybrid separation process). Within biotechnology, many different separation processes are used; chromatographic methods, membrane technologies, and extraction methods are often applied. A detailed discussion of these technologies will not be done in this subsection; the reader is referred to the literature (25, 62). The focus now should be on the choice of an adequate separation technology in connection with a specific reaction. Because inherent limitations are often exhibited in enzymatic biotransformations, specific separation processes have to be established. Some limitations often found are cofactor regeneration, racemic mixtures, equilibrium conversion, by-product formation, inhibition phenomena by substrate/product, or low solubilities.

For the synthesis of 1-phenyl-2-propanol from 1-phenyl-2-propanone, effective cofactor regeneration necessitates a separation of the cofactors from the products to increase the total turnover number (ttn; mole product/mole cofactor) of the cofactor. Therefore, an STR, an ultrafiltration module, a hollow-fiber membrane module, and distillation process are applied; thus, this enzyme bimembrane reactor consists of three loops (Fig. 9). The CSTR is coupled to two membrane modules (33, 34). The aqueous phase from the reactor is pumped in a loop through the ultrafiltration module. The hydrophilic membrane retains the enzymes, and the filtrate is pumped through the second loop. This aqueous solution, containing products, nonconverted reactants, and cofactors, is passed through the lumen of the extraction module. Here, a microporous, hydrophobic hollow-fiber membrane separates the aqueous and the organic phases and the product and nonconverted reactants are extracted, while the charged cofactors are retained in the aqueous phase. Then, the organic solvent is recycled by continuous distillation and the product remains at the bottom of the distillation column. Besides cofactor regeneration, this process also allows the usage of reactants of low solubility. Here, a ttn of up to 1,350 and a space-time yield (STY) of 64 g liter^{-1} day^{-1} are reached.

The (S)-enantiomer of N-(tert-butoxycarbonyl)-3-hydroxymethylpiperidine is produced by stereospecific esterification of the racemic alcohol with succinic anhydride,

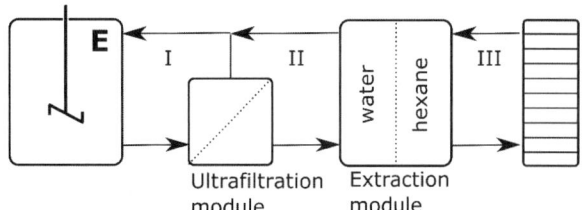

FIGURE 9 The enzyme-bimembrane reactor for the synthesis of 1-phenyl-2-propanol consists of STR, ultrafiltration module, extraction module, and distillation column.

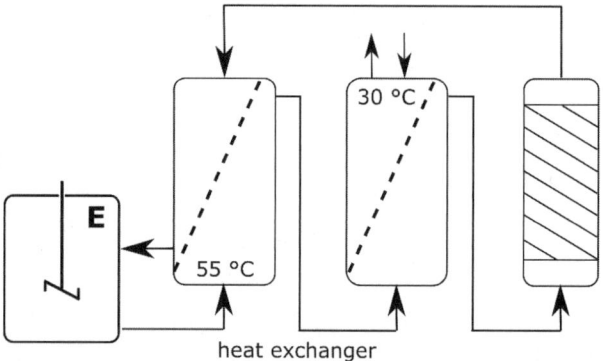

FIGURE 10 Synthesis of α-cyclodextrin by selective adsorption applying a batch process. Here, a sequence of STR, heat exchanger modules, and adsorption step is used.

as carried out by Bristol-Myers-Squibb (16). A sequence of an STR, an extraction module, a second STR, and a second extraction module is applied. The enzymatic resolution is followed by separation of the unreacted (R)-alcohol by extraction with a basic NaHCO$_3$ solution. The subsequent hydrolysis of the (S)-hemisuccinate with NaOH provides the desired (S)-alcohol. To improve the enantiomeric excess and the yield, repeated esterification of the enriched alcohol can be pursued.

Hydrolase-catalyzed reactions lead to limitations regarding the equilibrium in those cases where ester synthesis is carried out, because this condensation reaction produces water. The equilibrium is shifted at high amounts of water toward the reactants; therefore, an efficient separation is necessary to reach high conversions. For this purpose, a process setup with in situ product removal was developed by Unichema Chemie BV (43). The setup is based on azeotropic distillation of the water produced to synthesize isopropyl myristate. This is enabled by a continuous feed of 2-propanol to the reactor, which forms an azeotropic mixture with water. This mixture is distilled, thereby removing the water produced. Afterward, the immobilized enzyme used can be easily separated by filtration. The feasibility of this technique is also illustrated in the synthesis of glucose stearate. A mixture of ethyl methyl ketone and hexane as solvent is used, forming an azeotropic mixture with the water produced (78).

For the case of highly viscous reactants, like polyglycerol and lauric acid, a different process was developed; see section 38.7 (23).

The enzymatic synthesis of cyclodextrins by Mercian Co., Ltd., applying glycosyltransferases, leads to a mixture of α-, β-, and γ-cyclic oligosaccharides and can be carried out in a batch process by use of a sequence of STR, heat exchanger modules, and adsorption step (Fig. 10). To establish economic cyclodextrin production, separation of the cyclodextrins from the reaction medium is essential (67), because the reaction medium contains many by-products and the enzyme is inhibited by a high cyclodextrin concentration. The selective adsorption of α- and β-cyclodextrin on chitosan beads with appropriate ligands enables the separation. To enable effective adsorption, the temperature is lowered to 30°C before the solution is passed through the adsorption column. At this temperature, almost no cyclodextrins are formed; before reentering the reactor, the temperature of the solution is readjusted to 55°C by passing through a heat exchanger. The dissolved enzyme used in this process does not adsorb on the chitosan beads at the given reaction conditions. A further example of product separation can be reached by crystallization and is applied as in situ product removal in several industrial processes (see section 38.6.2).

These examples demonstrate that optimization of a given process can only be reached if an effective product separation is available. The bioprocess engineering must focus on downstream processing in addition to the reaction progress. Starting at the reaction itself, a selective synthesis circumvents complicated downstream processing. The separation and purification steps must be both addressed within classical downstream processing and integrated by reaction engineering. Adapting the reaction parameters with respect to easy downstream processing enables a reduction of production costs. Further examples can be found in reference 21.

38.5. REACTION MEDIA

The effectiveness of a biotransformation in view of reaction rate, yield, and/or selectivity strongly depends on the medium applied. Here, aqueous, organic, ionic, and supercritical solvents have to be differentiated. In addition, one can use solvent-free systems, meaning a mixture of the reactant. Every medium possesses its specific advantages, which are summarized in Table 2 in comparison with the aqueous medium, whose disadvantages were discussed at the beginning of section 38.2.1. A general discussion can be found in reference 35; for more information about the biotransformations in organic solvents see Faber's textbook (13).

To reduce the risks for humans and the environment, "green chemistry" advises minimizing the use of volatile, toxic organic solvents where possible. One approach is to investigate options for replacing such organic solvents with more environmentally compatible solvents. However, it needs to be taken into account that, when discussing nontoxic alternative solvents like supercritical CO_2, an atmosphere of too much CO_2 is lethal (admissible exposition value, 5,000 ppm). The energy necessary to evaporate a given amount of water is far from being environmentally friendly as well. Moreover, solvent-free synthesis can be applied, which is simpler, because there is no separation of the organic solvent necessary. This strategy can save energy, reduces the amount of solvent waste, and circumvents the risks of handling of organic solvents as well as their disposal (65). Organic solvent use accounts for approximately 80% of mass utilization

in typical pharmaceutical and fine-chemical synthetic processes (76). In addition, solvent-free bioprocesses reach higher volumetric productivities compared with syntheses in organic solvents. Solvent-free synthesis also eliminates costs for solvent supply, solvent separation, and waste disposal.

In the area of ester synthesis, a triolein lipase-catalyzed, solvent-free synthesis process was published in 1990 (12). In addition, monoacylglycerols, diacylglycerols, triacylglycerols, and mono- and dithioesters were synthesized by esterification and transesterification of pure substances or glycerolysis (70, 71). However, the absence of a solvent is disadvantageous in these cases, because the viscosity is higher than in solvent-based processes. Here, the solvent secures a fast mass transfer and a reduction of viscosity (14). In this way a faster conversion at a similar energy input is possible. This means that thorough mixing of reactants, a selective reaction by a biocatalyst, and its high stability are the crucial points in solvent-free bioprocesses. When these requirements are satisfied, solvent-free synthesis can outperform classical synthesis thanks to higher STYs and lower energy consumption.

38.6. OPTIMIZATION PROCEDURES

38.6.1. Diffusion Limitations

Heterogeneous biocatalysis on an industrial scale is only lucrative when high STYs are obtained or, alternatively, an easy separation of the biocatalyst is afforded. The application of immobilized enzymes aims to bind a maximal amount of enzyme per carrier, whereas porous materials, because of their large surface, represent a good carrier material (17). Here, the enzymes can also penetrate into the carrier and bind within the pores. The total surface of such porous carriers is much higher compared with solid ones. Hence, a maximal amount of enzyme can be immobilized on a specific volume of carrier material.

However, diffusion limitation is observed in various degrees in such immobilized enzyme systems. This diffusion limitation reduces overall conversion rates, because enzymes immobilized at the inner surface of the carrier are not optimally passed by the reactant flow. This nonuniform distribution of substrate and/or product between the enzyme matrix and the surrounding solution affects the measured (apparent) kinetic constants (Fig. 11). The substrate molecules reach the enzyme within the pores of a heterogeneous carrier material by film and pore diffusion from the solution. This occurs because the substrate must diffuse from the bulk solution up to the immobilized enzyme prior to the reaction. The rate of diffusion relative to the enzyme reaction determines whether there are limitations on the conversion rate. The flow rate at which the substrate passes over the insoluble particle affects the thickness of the diffusion film. The film diffusion and the pore diffusion determine the concentration of the substrate near the enzyme molecule, and thus, the rate of reaction. Diffusion of large molecules will obviously be limited by steric interactions with the matrix. This leads to a decrease of relative activity of bound enzymes toward high-molecular-weight substrates compared with low-molecular-weight substances.

The experimental determination of the diffusion limitation can be easily carried out by operation of two identical PFRs with different bed heights (second reactor has a doubled height, i.e., a doubled amount of enzyme) in parallel. Under non-diffusion-limited conditions the same conversion

TABLE 2 Reaction media for biotransformations

Medium	Advantage	Reference
Organic solvents	Increased enantioselectivity	49
	Increased stability	36
	Reduced risk of microbial contamination	32
Ionic liquids	Very low volatility	54
	Versatile solubilities	68
Supercritical solvents	High diffusion rates	50
	Nontoxic	27

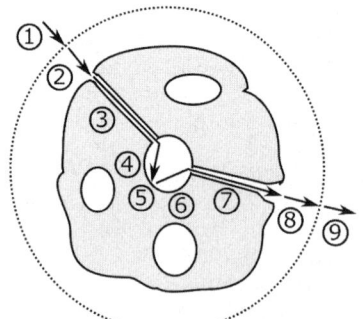

1 mass transport to carrier
2 mass transport through boundary layer
  (film diffusion)
3 pore diffusion of substrates
4 binding of substrate to enzyme
5 enzymatic reaction
6 desorption of products
7 pore diffusion of products
8 mass transport through boundary layer
9 mass transport from carrier

FIGURE 11 Processes within the catalytic reaction at porous carriers.

is reached, if the reactor with the doubled bed volume is operated as well with the doubled flow rate. By this means the same residence time is established in both PFRs.

$$\tau = \frac{V}{F} = \text{const.} \Rightarrow \frac{F_1}{F_2} = \frac{V_1}{V_2}$$

In diffusion limitation, different conversions are reached in both PFRs. The reactor with the higher flow rate will reach a higher conversion at the same residence time, since the flux on the carrier surface is higher, lowering the film diffusion.

38.6.2. In Situ Product Removal

A range of conversion between >80% to total conversion is of high interest for industrial processes (8). Here, great changes in the ratio $e_{\text{CSTR}}/e_{\text{PFR}}$ are found, and thus the choice of an appropriate reactor type strongly influences the effectivity of the process. If the reaction equilibrium is on the side of the substrates or the product has an inhibiting effect on the enzyme activity, a selective removal of product can increase the productivity of the reaction system. This can be realized by application of selective membranes, for example, which retain the product but not the enzyme or substrate. In those cases, the use of continuous processes is favored. The increase of conversion by continuous product removal can be exemplified by a simple enzymatic process carried out in a CSTR. The proposed reaction is an irreversible one-substrate reaction showing competitive product inhibition:

$$v = \frac{v_{\max} \cdot c_S}{K_m \left(1 + \dfrac{c_P}{K_{iP}}\right) + c_S}$$

The selectivities of the product separation step with regard to substrate and product are described by β_S and β_P, respectively. Here c_S and c_R are the substrate and product concentration in the outflow of the reactor:

$$\beta_S = \frac{c_{S,\text{out}}}{c_{S,R}} \quad ; \quad \beta_P = \frac{c_{P,\text{out}}}{c_{P,R}}$$

$\beta_S = \beta_P = 1$ represents a classical STR, where substrates and products are removed in the same degree. In steady state the mass balances are described as follows, where V_R is the reactor volume and $c_{S,\text{in}}$ is the substrate concentration in the inflow:

$$\dot{Q}_{\text{in}} = \dot{Q}_{\text{out}} = \dot{Q}_F$$

$$0 = \dot{Q}_F \cdot c_{S,\text{in}} - \dot{Q}_F \cdot c_{S,\text{out}} + v\big|_{c\,=\,c_{S,R}} \cdot V_R$$

$$0 = -\dot{Q}_F \cdot c_{P,\text{out}} + v\big|_{c\,=\,c_{S,R}} \cdot V_R$$

It follows from the above:

$$c_{P,\text{out}} = c_{S,\text{in}} - c_{S,\text{out}}$$

Transposing leads to reactor concentration $c_{S,R}$ as a function of kinetic constants, selectivities, and residence time:

$$c_{S,R} = \frac{b}{2a} \pm \sqrt{\frac{b^2}{4a^2} - \frac{c}{a}} \quad \text{with } a = \frac{K_m \cdot \beta_S^2}{K_{iP} \cdot \beta_P} - \beta_S$$

$$b = 2\frac{K_m \cdot \beta_S}{K_{iP} \cdot \beta_P} \cdot c_{S,\text{in}} + K_m \cdot \beta_S + v_{\max} \cdot \tau - c_{S,\text{in}}$$

$$c = K_m \cdot c_{S,\text{in}} \cdot \left(1 + \frac{c_{S,\text{in}}}{K_{iP} \cdot \beta_P}\right)$$

Productivity can be increased dramatically by selective product removal, where the increase of conversion depends on the degree of product inhibition.

Besides these product inhibitions in thermodynamically unfavorable reactions, a limit to the achievable product concentration may also be imposed. In situ product removal (ISPR) is a useful strategy to overcome such problems via hybrid separation processes, integrating a one-step reaction with the first step of the downstream processing (77). ISPR may be carried out either internally or externally (Fig. 12). When the reaction media are in contact with a second, product-removing phase that is present in the reactor, the setup used for separation of biocatalyst and by-product is internal. An external setup exists when the reaction media are in contact with a product-removing phase within an external unit. In both cases, the ISPR configuration is intended to reduce the product concentration in the reactor, whereas direct and indirect contact with the product-removing phase can be differentiated in both setups. A process configuration with indirect contact of biocatalyst can be done by ultrafiltration for external setups or by membrane technology for internal setups. In all cases, the biocatalytic reaction is influenced with respect to the chemical equilibrium, enabling higher conversions. The choice of the setup depends on the properties of the by-product and the biocatalyst, the number of phases (liquid/liquid; liquid/gas) concerned, and the type of recovery process in use. The techniques used for ISPR depend on the properties of the target product, the expected benefit that ISPR needs to achieve, and the biocatalyst involved (9).

Depending on the form of the biocatalyst (resting cell, isolated enzyme, or immobilized enzyme), different levels of process integration are possible. A complete integration is reached when the reaction and product recovery steps take place within the reactor. Here, a compromise in the case of the pH value or the temperature must be made. To circumvent such compromise, discrete reaction and product

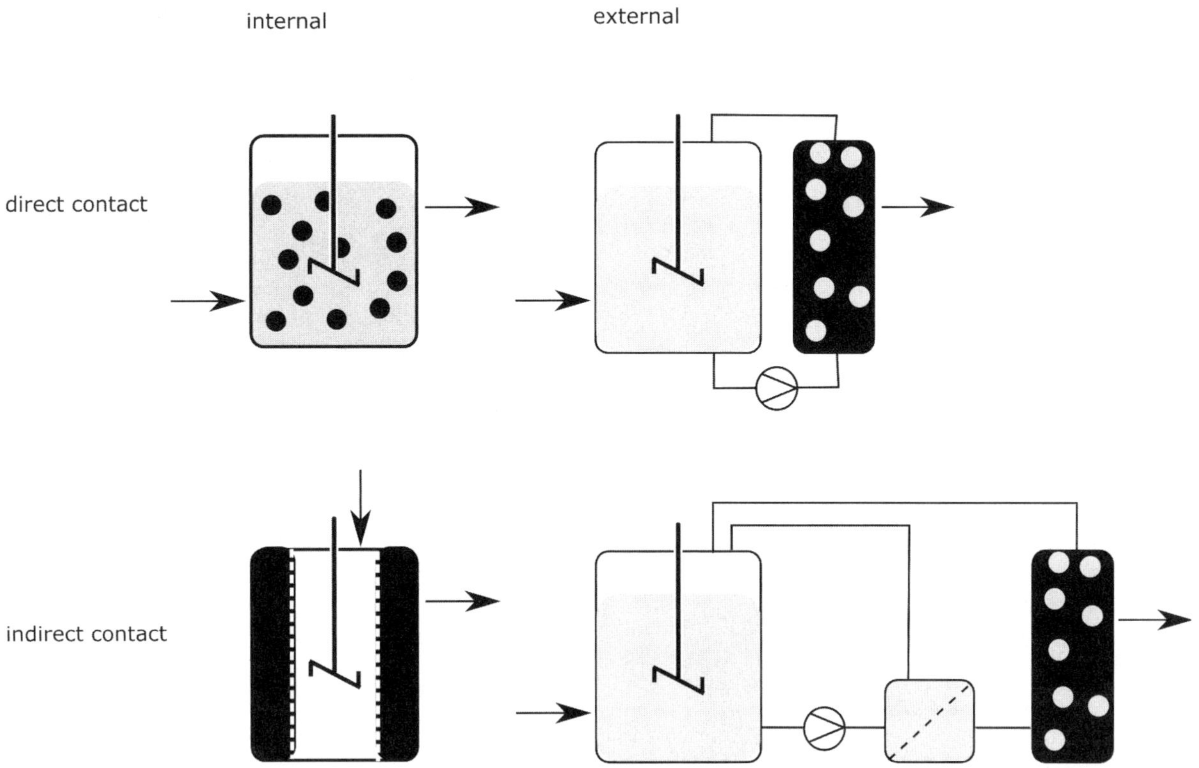

internal   external

direct contact

indirect contact

FIGURE 12 Internal and external modes of ISPR operation with direct and indirect catalyst contact. In the internal setup, the reaction media are in contact with a second, product-removing phase that is present in the reactor, whereas in the external setup, the reaction media are in contact via a loop with a product-removing phase in an external unit. Direct and indirect contact can be differentiated in both cases.

removal steps are linked via a recycle loop, but the biocatalyst must be retained. This may involve a second recycle loop, already illustrated above in the production process of 1-phenyl-2-propanol (section 38.4.4).

The potential of crystallization as an ISPR technique for bioprocesses has found application in fine-chemicals production. In chemical engineering, processes are often referred to as reactive crystallization if an internal configuration (Fig. 11) is used (72). For several enzymatic processes, a crystallization step has been integrated with the reaction step by use of either crude or pure enzyme. The reader will find an interesting example in reference 59. In addition, crystallization might be an attractive option in the production of crystalline compounds using whole cells as biocatalyst (9).

The selection of the appropriate separation method is of key importance. Different separation methods have been illustrated above and applied in ISPR (e.g., chromatographic methods, membrane methods, or crystallization). To choose the optimal process, the important criterion is the product concentration. A low concentration will improve the reaction performance, because enzyme inhibition is minimized and the thermodynamic equilibrium is shifted. In addition, high product concentrations simplify the further downstream processing.

38.7. EXAMPLES OF TYPICAL BIOPROCESSES

The enzymatic synthesis of fatty acid ester-based care specialties is a prime example of a modern, innovative,

and sustainable technology (66). The advantages of this process compared with traditional production methods are lower reaction temperature and higher selectivity leading to fewer by-products, fewer purification steps, less waste, and, therefore, an economically competitive bioprocess. These economic and quality aspects led to the implementation of a fixed-bed reactor system for the synthesis of ester oils at Evonik Industries because the enzymatic esterification could yield specialty esters of high quality. Nevertheless, the field of application is restricted by the availability of enzymes on an industrial scale and their thermal stability. Furthermore, the application of a fixed-bed reactor is restricted to raw materials and products with a melting point below 60°C and low-viscous substances, because otherwise no pumping through the enzymatic fixed bed is possible. Therefore, there was a high interest in establishing an alternative process. To overcome limitations due to the high viscosity in the solvent-free esterification of polyglycerol-3 and related polyols, such as polyethylene glycols (PEGs), an alternative reactor concept was developed (23). The new reactor constitutes a bubble column that prevents mechanical erosion of Novozym 435 (lipase B from *Candida antarctica*) found by mechanical stirring of the mixture. In that way, polyglycerol-3 laurate was synthesized at an STY of 3 kg liter^{-1} day^{-1} and PEG-55-propylene glycol dioleate was synthesized at an STY of 0.7 kg liter^{-1} day^{-1}. To prove the broad application range of this system, low-viscous myristyl myristate was synthesized at an STY of 6.7 kg liter^{-1} day^{-1}, thus outperforming conventional methods such as stirred-tank or fixed-bed.

The newly developed reactor concept is universally applicable to esterification reactions and can be advantageously applied in the synthesis of a broad range of high-quality surfactants.

The product high-fructose corn syrup (HFCS) is used as an alternative sweetener to sucrose or invert sugars in the food and beverage industry (41). HFCS obtained directly by enzymatic isomerization with a content of 42% fructose is used mainly in the baking and dairy industries. This HFCS can be obtained without further downstream processing by a continuously worked fixed-bed reactor setup with a capacity of more than 7,000,000 tons/year. For use in alcoholic beverages, the HFCS is enriched afterward up to 55% fructose by chromatography. The production process of HFCS is based on the isomerization of glucose to fructose by glucose isomerase in the form of immobilized whole cells or enzymes. The biocatalyst is available commercially from different suppliers under various trademarks, e.g., Sweetzyme, Maxazyme, Optisweet, or Sweetase. The reaction enthalpy is slightly endothermic and reversible. The equilibrium conversion is approximately 50% at 55°C; thus, to limit by-product formation, the reaction time has to be kept short. This can be done by using high isomerase concentrations showing half-life times of more than 100 days. Several reactors (typically 20) are operated in parallel containing enzymes of different ages and activities. To maintain productivity the enzyme is replaced when activity is reduced to 12.5% of initial activity. The feed to a single reactor is controlled by the extent of conversion in the reactor. This process setup allows a productivity of 1,000 tons/day.

38.8. DOWNSTREAM PROCESSING

Industrial applications using enzymes as biocatalysts are very diverse. The major market is the use of enzymes within detergents. Other applications are in the food and feed sector, as well as in paper and textile production. Besides this application, there are numerous enzymatic syntheses; e.g., hydrolases are used in the production of bulk chemicals like surfactants and of fine chemicals, whereas lyases like decarboxylases can be used for enantioselective carbon-carbon couplings.

The complexity of the biocatalyst separation is crucial for an effective industrial process. The type of enzyme retention has a strong influence on the costs of the final product, because repetitive use of the enzyme increases the productivity. Within the bioprocess development, there is the question of whether the enzyme will be heterogenized or retained by other approaches (e.g., membranes), whereas the answer is specific for a singular biotransformation. To retain free enzyme, ultrafiltration has to be carried out; in comparison, the filtration of a heterogeneous catalyst is much more simple and allows the use of fixed-bed reactors. The decision about immobilized or solubilized enzymes depends on cost savings resulting from simplified downstream processing by heterogenization. The costs for the production of the enzyme depend on the fermentation scale and yield, as well as the number of purification steps. Nevertheless, these costs have to be considered in both cases; thus, the costs for immobilization depend on the efficiency of the immobilization and the costs for the carrier material. The big advantages of immobilization are the dramatic cost savings because of the repetitive and longer use of the enzyme in synthesis. Here, a long half-life time is afforded to minimize the costs for the final product by increasing the ratio of product mass to mass of catalyst (10, 64).

38.8.1. Integrated Approach and Ideal Reaction Conditions

The classic disciplines of chemistry and biology offer methods for selective catalytic reactions. The task of the process engineer is the transfer of these syntheses from the laboratory scale to the industrial scale, as well as to ensure the feasibility of the bioprocess. Problems often occur that cannot be solved by singular disciplines; therefore, an integration of chemical and biological steps as well as bioprocess engineering is helpful to accomplish the advantages of these disciplines by establishing new reaction sequences. This affords new thinking in process development, especially the realization of hybrid separation methods as illustrated in the section on in situ product removal. The integration of different catalysts necessitates an adequate formulation of the catalyst to establish efficient process management. Furthermore, modeling of these bioprocesses is mandatory to find the bottlenecks and to establish an efficient overall process.

An effective reaction is also required to establish an efficient bioprocess. This is characterized by high end concentrations and high productivity, and at the same time, by low amounts of side products. An early view on the reaction with respect to the product purification can reduce later expenses for product isolation and development and production costs. An additional advantage is the reduced environmental impact because of higher efficiency and therefore less waste.

Effective reactions are characterized by the use of a minimum of reactants but high product amounts. Biocatalytic processes are especially useful to achieve this goal of high atom efficiency. In all bioprocesses, cyclization, retention, and separation of catalyst and product are arbitrative parameters to implement an economically successful bioprocess.

REFERENCES

1. **Arnold, F. H., P. L. Wintrode, K. Miyazaki, and A. Gershenson.** 2001. How enzymes adapt: lessons from directed evolution. *Trends Biochem. Sci.* **26:**100–106.
2. **Bailey, J. E.** 1991. Toward a science of metabolic engineering. *Science* **252:**1668–1675.
3. **Behr, A., V. A. Brehme, C. L. J. Ewers, H. Gron, T. Kimmel, S. Kuppers, and I. Symietz.** 2004. New developments in chemical engineering for the production of drug substances. *Eng. Life Sci.* **4:**15–24.
4. **Belfort, G., R. H. Davis, and A. L. Zydney.** 1994. The behaviour of suspensions and macromolecular solutions in cross-flow microfiltration. *J. Memb. Sci.* **96:**1–58.
5. **Blumenberg, B.** 1994. Development of techniques today. *Nachr. Chem. Tech. Lab.* **5:**480–485.
6. **Bommarius, A. S., and A. Karau.** 2005. Deactivation of formate dehydrogenase (FDH) in solution and at gas-liquid interfaces. *Biotechnol. Prog.* **21:**1663–1672.
7. **Bommarius, A. S., M. Scharm, and K. Drauz.** 1998. Biocatalysis to amino acid-based chiral pharmaceuticals—examples and perspectives. *J. Mol. Catal. B Enzym.* **5:**1–11.
8. **Briechle, S., M. Howaldt, T. Röthig, and A. Liese.** 2006. Enzymatische Prozesse, p. 361–408. *In* H. Chmiel (ed.), *Bioprozesstechnik: Einführung in die Bioverfahrenstechnik.* Elsevier, Spektrum Akad., Munich, Germany.
9. **Buque-Taboada, E. M., A. J. J. Straathof, J. J. Heijnen, and L. A. M. van der Wielen.** 2004. In-situ product removal using crystallization loop in the asymmetric reduction of 4-oxoisophorone by *Saccharomyces cerevisiae*. *Biotechnol. Bioeng.* **86:**795–800.
10. **Cao, L.** 2005. *Carrier-Bound Immobilized Enzymes.* Wiley-VCH, Weinheim, Germany.

11. **Deibele, L., and R. Dohrm (ed.).** 2006. *Miniplant-Technik.* Wiley-VCH, Weinheim, Germany.

12. **Ergan, F., M. Trani, and G. Andre.** 1990. Production of glycerides from glycerol and fatty acid by immobilized lipases in non-aqueous media. *Biotechnol. Bioeng.* **35:**195–200.

13. **Faber, K.** 2004. *Biotransformations in Organic Chemistry: a Textbook,* 4th ed. Springer, Berlin, Germany.

14. **Foresti, M. L., and M. L. Ferreira.** 2005. Solvent-free ethyl oleate synthesis mediated by lipase from *Candida antarctica* B adsorbed on polypropylene powder. *Catal. Today* **107–08:**23–30.

15. **Giorno, L. and E. Drioli.** 2000. Biocatalytic membrane reactors: applications and perspectives. *Trends Biotechnol.* **18:**339–349.

16. **Goswami, A., J. M. Howell, E. Y. Hua, K. D. Mirfakhrae, M. C. Soumeillant, S. Swaminathan, X. H. Qian, F. A. Quiroz, T. C. Vu, X. B. Wang, B. Zheng, D. R. Kronenthal, and R. N. Patel.** 2001. Chemical and enzymatic resolution of (R,S)-N-(tertbutoxycarbonyl)-3-hydroxymethylpiperidine. *Org. Proc. Res. Dev.* **5:**415–420.

17. **Guisan J. M. (ed.).** 2006. *Immobilization of Enzymes and Cells.* Humana Press, New York, NY.

18. **Hanefeld, U., L. Gardossi, and E. Magner.** 2009. Understanding enzyme immobilization. *Chem. Soc. Rev.* **38:**453–468.

19. **Heinzle, E., A. P. Biwer, and C. L. Cooney.** 2006. *Development of Sustainable Bioprocesses: Modeling and Assessment.* Wiley, Chichester, United Kingdom.

20. **Hessel, V., H. Löwe, A. Müller, and G. Kolb.** 2004. *Chemical Micro Process Engineering.* Wiley-VCH, Weinheim, Germany.

21. **Hilterhaus, L., and A. Liese.** 2009. Applications of reaction engineering to industrial biotransformations, p. 65–88. *In* J. Tao, G. Lin, and A. Liese (ed.), *Biocatalysis for the Pharmaceutical Industry.* John Wiley & Sons Asia, Singapore.

22. **Hilterhaus, L., B. Minow, J. Muller, M. Berheide, H. Quitmann, M. Katzer, O. Thum, G. Antranikian, A. P. Zeng, and A. Liese.** 2008. Practical application of different enzymes immobilized on Sepabeads. *Bioproc. Biosys. Eng.* **31:**163–171.

23. **Hilterhaus, L., O. Thum, and A. Liese.** 2008. A reactor concept for lipase-catalyzed solvent-free conversion of highly viscous reactants forming two-phase systems. *Org. Proc. Res. Dev.* **12:**618–625.

24. **Hobbs, H. R., and N. R. Thomas.** 2007. Biocatalysis in supercritical fluids, in fluorous solvents, and under solvent-free conditions. *Chem. Rev.* **107:**2786–2820.

25. **Huang, H.-J., S. Ramaswamy, U. W. Tschirner, and B. V. Ramarao.** 2008. A review of separation technologies in current and future biorefineries. *Sep. Pur. Technol.* **62:**1–21.

26. **Ingham, J., I. J. Dunn, E. Heinzle, J. E. Prenosil, and J. B. Snape.** 2007. *Chemical Engineering Dynamics: an Introduction to Modelling and Computer Simulation,* 3rd ed. Wiley-VCH, Weinheim, Germany.

27. **Jarzebski, A. B., and J. J. Malinoski.** 1995. Potentials and prospects for application of supercritical-fluid technology in bioprocessing. *Proc. Biochem.* **30:**343–352.

28. **Jaspe, J., and S.J. Hagen.** 2006. Do protein molecules unfold in a simple shear flow? *Biophys. J.* **91:**3415–3424.

29. **Judd, S.** 2008. The status of membrane bioreactor technology. *Trends Biotechnol.* **26:**109–116.

30. **Karmee, S. K., L. Casiraghi, and L. Greiner.** 2008. Technical aspects of biocatalysis in non-CO2-based supercritical fluids. *Biotechnol. J.* **3:**104–111.

31. **Keil, F., W. Mackens, H. Voss, and J. Werther.** 1999. *Scientific Computing in Chemical Engineering II.* Springer, Berlin, Germany.

32. **Knezevic, Z., S. Bobic, A. Milutinovic, B. Obradovic, L. Mojovic, and B. Bugarski.** 2002. Alginate-immobilized lipase by electrostatic extrusion for the purpose of palm oil hydrolysis in lecithin/isooctane system. *Proc. Biochem.* **38:**313–318.

33. **Kragl, U., and T. Dwars.** 2001. The development of new methods for the recycling of chiral catalysts. *Trends Biotechnol.* **19:**442–449.

34. **Kragl, U., W. Kruse, W. Hummel, and C. Wandrey.** 1996. Enzyme engineering aspects of biocatalysis: cofactor regeneration as example. *Biotechnol. Bioeng.* **52:**309–319.

35. **Krieger, N., T. Bhatnagar, J. C. Baratti, A. M. Baron, V. M. de Lima, and D. Mitchell.** 2004. Non-aqueous biocatalysis in heterogeneous solvent systems. *Food Technol. Biotechnol.* **42:**279–286.

36. **Kumar, S., C. Wittmann, and E. Heinzle.** 2004. Minibioreactors. *Biotechnol. Lett.* **26:**1–10.

37. **Lee, L. G., and G. M. Whitesides.** 1985. Enzyme-catalyzed organic synthesis: a comparison of strategies for in situ regeneration of NAD from NADH. *J. Am. Chem. Soc.* **107:**6999–7008.

38. **Levenspiel, O.** 1999. *Chemical Reaction Engineering.* Wiley, Hoboken, NJ.

39. **Liese, A., M. Karutz, J. Kamphuis, C. Wandrey, and U. Kragl.** 1996. Enzymatic resolution of 1-phenyl-1,2-ethanediol by enantioselective oxidation: overcoming product inhibition by continuous extraction. *Biotechnol. Bioeng.* **51:**544–550.

40. **Liese, A., U. Kragl, H. Kierkels, and B. Schulze.** 2002. Membrane reactor development for the kinetic resolution of ethyl 2-hydroxy-4-phenylbutyrate. *Enzyme Microb. Technol.* **30:**673–681.

41. **Liese, A., K. Seelbach, and C. Wandrey (ed.).** 2006. *Industrial Biotransformations,* 2nd ed. Wiley-VCH, Weinheim, Germany.

42. **Lutz, S., and U. Bornscheuer (ed.).** 2009. *Protein Engineering Handbook.* Wiley-VCH, Weinheim, Germany.

43. **McCrae, A. R., E.-L. Roehl, and H. M. Brand.** 1990. Bioester. *SOFW J.* **116:**201–205.

44. **Micheletti, M., and G. J. Lye.** 2006. Microscale bioprocess optimisation. *Curr. Opin. Biotechnol.* **17:**611–618.

45. **Miyazaki, M., and H. Maeda.** 2006. Microchannel enzyme reactors and their applications for processing. *Trends Biotechnol.* **24:**463–470.

46. **Mulder, M.** 2000. *Basic Principles of Membrane Technology,* 2nd ed. Kluwer, Dordrecht, The Netherlands.

47. **Niehaus, F., C. Bertoldo, M. Kahler, and G. Antranikian.** 1999. Extremophiles as a source of novel enzymes for industrial application. *Appl. Microbiol. Biotechnol.* **51:**711–729.

48. **Oldiges, M., S. Lutz, S. Pflug, K. Schroer, N. Stein, and C. Wiendahl.** 2007. Metabolomics: current state and evolving methodologies and tools. *Appl. Microbiol. Biotechnol.* **76:**495–511.

49. **Orrenius, C., T. Norin, K. Hult, and G. Carrea.** 1995. The *Candida antarctica* lipase B catalysed kinetic resolution of seudenol in non-aqueous media of controlled water activity. *Tetrahedron Asymmetry* **6:**3023–3030.

50. **Osanai, Y., K. Toshima, and S. Matsumura.** 2006. Enzymatic transformation of aliphatic polyesters into cyclic oligomers using enzyme packed column under continuous flow of supercritical carbon dioxide with toluene. *Sci. Technol. Adv. Mater.* **7:**202–208.

51. **Park, J. H., and S. Y. Lee.** 2008. Towards systems metabolic engineering of microorganisms for amino acid production. *Curr. Opin. Biotechnol.* **19:**454–460.

52. **Park, J. H., S. Y. Lee, T. Y. Kim, and H. U. Kim.** 2008. Application of systems biology for bioprocess development. *Trends Biotechnol.* **26:**404–412.

53. **Puskeiler, R., A. Kusterer, G. T. John, and D. Weuster-Botz.** 2005. Miniature bioreactors for automated high-throughput bioprocess design (HTBD): reproducibility of parallel fed-batch cultivations with *Escherichia coli. Biotechnol. Appl. Biochem.* **42:**227–235.

54. **Reetz, M. T., W. Wiesenhofer, G. Francio, and W. Leitner.** 2003. Continuous flow enzymatic kinetic resolution and enantiomer separation using ionic liquid/supercritical carbon dioxide media. *Adv. Synth. Catal.* **345:**1221–1228.

55. **Rios, G. M., M. P. Belleville, D. Paolucci, and J. Sanchez.** 2004. Progress in enzymatic membrane reactors—a review. *J. Memb. Sci.* **242:**189–196.

56. **Rios, G. M., M.-P. Belleville, and D. Paolucci-Jeanjean.** 2007. Membrane engineering in biotechnology: quo vamus? *Trends Biotechnol.* **25:**242–246.

57. **Rippberger, S.** 1992. *Mikrofiltration mit Membranen: Grundlagen, Verfahren, Anwendungen.* VCH, Weinheim, Germany.

58. **Roosen, C., P. Muller, and L. Greiner.** 2008. Ionic liquids in biotechnology: applications and perspectives for biotransformations. *Appl. Microbiol. Biotechnol.* **81:**607–614.

59. **Schroen, C. G. P. H., V. A. Nierstrasz, R. Bosma, G. J. Kemperman, M. Strubel, L. P. Ooijkaas, H. H. Beeftink, and J. Tramper.** 2002. In situ product removal during enzymatic cephalexin synthesis by complexation. *Enzyme Microb. Technol.* **31:**264–273.

60. **Staude, E.** 1992. *Membranen und Membranprozesse.* Wiley-VCH, Weinheim, Germany.

61. **Steude, H., L. Deibele, and J. Schröter.** 1997. Miniplant engineering—selected examples of equipment design. *Chem. Ing. Tech.* **5:**623–631.

62. **Subramanian, G. (ed.).** 2007. *Bioseparation & Bioprocessing,* vol. 1 and 2. Wiley-VCH, Weinheim, Germany.

63. **Sweetlove, L. J., R. L. Last, and A. R. Fernie.** 2003. Predictive metabolic engineering: a goal for systems biology. *Plant Physiol.* **132:**420–425.

64. **Tanaka, A., T. Tosa, and T. Kobayashi (ed.).** 1993. *Industrial Application of Immobilized Biocatalysts.* Marcel Dekker Inc., New York, NY.

65. **Tanaka, K.** 2003. *Solvent-free Organic Synthesis.* Wiley-VCH, Weinheim, Germany.

66. **Thum, O.** 2004. Enzymatic production of care specialties based on fatty acid esters. *Tens. Surf. Deterg.* **41:**287–290.

67. **Tsuchiyama, Y., K.-I. Yamamoto, T. Asou, M. Okabe, Y. Yagi, and R. Okamoto.** 1991. A novel process of cyclodextrin production by use of specific adsorbents. Part I. Screening of specific adsorbents. *J. Ferment. Bioeng.* **71:**407–412.

68. **van Rantwijk, F., F. Secundo, and R. A. Sheldon.** 2006. Structure and activity of *Candida antarctica* lipase B in ionic liquids. *Green Chem.* **8:**282–286.

69. **Wasserscheid, P., and T. Welton.** 2008. *Ionic Liquids in Synthesis,* 2nd ed. Wiley-VCH, Weinheim, Germany.

70. **Weber, N., E. Klein, K. Vosmann, and K. D. Mukherjee.** 2004. Mono-thioesters and di-thioesters by lipase-catalyzed reactions of alpha,omega-alkanedithiols with palmitic acid or its methyl ester. *Appl. Microbiol. Biotechnol.* **64:**800–805.

71. **Weber, N., and K. D. Mukherjee.** 2004. Solvent-free lipase-catalyzed preparation of diacylglycerols. *J. Agric. Food Chem.* **52:**5347–5353.

72. **Wibowo, C., V. V. Kelkar, K. D. Samant, J. W. Schroer, and K. M. Ng.** 2005. Development of reactive crystallization processes, p. 339–358. *In* K. Sundmacher, A. Kienle, and A. Seidel-Morgenstern (ed.), *Integrated Chemical Processes.* Wiley-VCH, Weinheim, Germany.

73. **Wichmann, R., C. Wandrey, A. F. Bückmann, and M.-R. Kula.** 1981. Continuous enzymatic transformation in an enzyme membrane reactor with simultaneous NAD(H) regeneration. *Biotechnol. Bioeng.* **23:**2789–2796.

74. **Winzeler, H. B.** 1990. Membran-Filtration mit hoher Trennleistung und minimalem Energiebedarf. *Chim. Int. J. Chem.* **44:**288–291.

75. **Wöltinger, J., K. Drauz, and A. S. Bommarius.** 2001. The membrane reactor in the fine chemicals industry. *Appl. Catal. A Gen.* **221:**171–185.

76. **Woodley, J. M.** 2008. New opportunities for biocatalysis: making pharmaceutical processes greener. *Trends Biotechnol.* **26:**321–327.

77. **Woodley, J. M., M. Bisschops, A. J. J. Straathof, and M. Ottens.** 2008. Future directions for in-situ product removal (ISPR). *J. Chem. Technol. Biotechnol.* **83:**121–123.

78. **Yan, Y. C., U. T. Bornscheuer, and R. D. Schmid.** 2002. Efficient water removal in lipase-catalyzed esterifications using a low-boiling-point azeotrope. *Biotechnol. Bioeng.* **78:**31–34.

MICROBIAL FUELS (BIOFUELS) AND FINE CHEMICALS

VI

O NE OF THE MOST SIGNIFICANT RECENT DEVELOPMENTS AFFECTING industrial microbiology has been the marked increase in activity and anticipation related to microbially produced chemicals and, especially, fuels. As detailed elsewhere, a dramatic increase in biofuel and related investment by government, large companies, and venture capitalists occurred during 2005 to 2008 (1–3). At the time of writing, research and commercialization activity still continue at a pace unprecedented prior to 2005.

The chapters in this section address some of the key themes that have emerged since the 2nd edition of the *Manual for Industrial Microbiology and Biotechnology* (MIMB) appeared in 1999. One of these themes is the use of high-throughput methods for accessing biodiversity, which is explored in chapter 39 by Carl Abulencia et al., who draw largely on the experience of Verenium Corp. A second, the subject of chapter 40 by Susan T. L. Harrison et al. of the University of Cape Town, is the use of algae for production of fuels, chemicals, and nutritional products. Metabolic engineering has come of age during this time, with chemicals leading the way to a large extent, as amply illustrated in chapter 41 by Parick Cirino of Penn State University. This period has also seen the development of systems biology tools offering researchers access to unprecedented volumes of information regarding gene sequences, gene expression, and metabolite levels. Jonathan Mielenz and David Hogsett, from Oak Ridge National Laboratory and Mascoma Corporation, respectively, address in chapter 42 the use of systems biology tools to further understand the industrial robustness of microorganisms.

Although not a new theme, the conversion of lignocellulose to fuels and chemicals continues to be a prominent focus of industrial microbiologists and biotechnologists, appears close to commercial fruition after many decades of effort, and involves many important microbiological topics that still have much to contribute through yet-to-be-realized advances. Charles Abbas et al. from ADM address some important techniques and practical considerations in chapter 43. In chapter 44, Alexandru Dumitrache and Gideon Wolfaardt of Ryerson University, with Lee Lynd from Dartmouth, consider the surface microbiology of cellulolytic bacteria. Also from Dartmouth, Nicolai Panikov and Lee Lynd address physiological and methodological aspects of cellulolytic microbial cultures in chapter 45.

Further progress and change are sure to characterize industrial microbiology and biotechnology when the 4th edition of MIMB is prepared in the future. Until then, and it can be imagined thereafter, it is hoped that practitioners in the field will find the chapters herein to be informative and useful.

REFERENCES

1. **Herrera, S.** 2006. Bonkers about biofuels. *Nature Biotech.* **24:**755–760.
2. **Lynd, L. R., M. S. Laser, D. Bransby, B. E. Dale, B. Davison, R. Hamilton, M. Himmel, M. Keller, J.** McMillan, J. Sheehan, and C. E. Wyman. 2008. How biotech can transform biofuels. *Nat. Biotechnol.* **26:** 169–172.
3. **Service, R. F.** 2007. Biofuel researchers prepare to reap a new harvest. *Science* **315:**1488–1491.

Accessing Microbial Communities Relevant to Biofuels Production

CARL B. ABULENCIA, STEVEN M. WELLS, KEVIN A. GRAY, MARTIN KELLER, AND JOEL A. KREPS

39

39.1. INTRODUCTION

Microbiology, a field pioneered by Ferdinand Cohn, Louis Pasteur, and Robert Koch in the late 1800s, has seen a renaissance through novel cultivation methods and insight into the tremendous diversity of microorganisms. Ferdinand Cohn discovered heat-resistant endospores of bacilli; Louis Pasteur disproved the theory of spontaneous generation, leading to effective sterilization procedures; and Robert Koch introduced media solidified with agar and the concept of isolating pure cultures of microorganisms. Koch's discovery was a precursor to the development of bacterial taxonomy, genetics, and general isolation of pure cultures (36). The use of solidified media is still the most common technique to isolate pure cultures, as well as to isolate strains from the environment. The development of molecular techniques, focusing on the isolation, cloning, and sequencing of 16S rRNA, a commonly used phylogenetic marker, demonstrated that prokaryotes represent a large portion of the earth's biota. When Carl Woese and colleagues described the main bacterial phyla almost 20 years ago, 11 bacterial phyla were described (36). Two decades of environmental 16S rRNA gene sequencing from many habitats has shown that most of the 54 described bacterial phyla have only a few representatives, and 13 phyla have no cultured representatives (36). Some groups, such as *Acidobacteria* and *Verrucomicrobia*, represent significant proportions of examined ecosystems, with only a handful of isolates obtained in pure culture. The metabolic capabilities of these uncultured phyla are unknown. However, based on the abundance of cellulose within many environments, it could be speculated that genes for cellulose degradation are present in many of these unknown phyla.

39.2. CELLULOSE DEGRADATION IN NATURE

Due to the insolubility of lignocellulosic material, microbes developed various methods of accessing cellulosic plant matter. In general, cellulose degradation and utilization are carried out not by a pure microbial culture but by microbial consortia representing both cellulolytic and noncellulolytic organisms, with CO_2 in aerobic environments, and methane and CO_2 in anaerobic environments, as final end products (46). In soil the major source of cellulose is decaying plant material, rich in lignin and generally low in moisture, often favoring fungi as dominant cellulolytic biota (45). In aquatic environments, plant material is usually low in lignin and degraded within the anoxic zone by a consortium of anaerobic bacteria (46).

In ruminant animals, the retention time of plant material is long enough to allow fiber utilization through fibrolytic microbial populations and degradation of cellulose. In addition, the microbial protein produced during growth of these microorganisms, followed by lysis within the ruminant's abomasum, contributes significantly to the protein nutrition of the ruminant (30, 53).

Insects are the largest group of organisms on earth that host diverse microbial populations (13). Many insects, such as termites or Coleoptera larvae, specialize on a diet rich in lignocellulosic biomass. Until recently, it was unclear whether the enzymes required for cellulose degradation were produced by the termite or the microorganisms present within its hindgut. Tokuda and Watanabe showed a direct role of the symbiotic bacteria in the gut of the termite in cellulose and xylan hydrolysis (71). An extensive metagenomic analysis of the P3 segment of the hindgut of the "higher" termite species, *Nasutitermes*, demonstrated the presence of a large, diverse set of bacterial genes for cellulose and xylan degradation (78). Cook et al. studied the larvae of *Tipula abdominalis*, an aquatic crane fly found in aquatic riparian environments (13). While the larvae feed on coarse particulate organic matter, primarily leaf litter, the larvae themselves do not synthesize cellulolytic enzymes (13). It was proposed that the larvae may benefit nutritionally from fermentation products produced by microorganisms during digestion of plant material. Larval hindgut bacterial isolates demonstrated the ability to degrade model plant polymers such as carboxymethylcellulose, xylan, and others (13). However, all isolations were performed aerobically on model substrates. It would be of great interest to include isolation strategies including anaerobes enriched on the original larva diet, such as leaf litter obtained directly from the aquatic riparian environment.

Lynd et al. described a general trend of increasing growth rate on crystalline cellulose as a function of temperature with an upper limit of about 75°C (46). In general, thermophiles such as *Anaerocellum thermophilum*, isolated from the plant residues of a hot spring in Russia, degrade crystalline cellulose through free primary cellulases, which are distinct from other multienzyme complexes associated with the cell wall and found in organisms such as *Clostridium thermocellum* (5). Some hyperthermophiles, defined as organisms with a growth temperature optimum at 80°C or above, are able to degrade α- and β-linked glucans and hemicelluloses such as xylan and mannan. None of these organisms, however, effectively degrades crystalline cellulose (5). It is of great scientific interest to discover hyperthermophiles that have the ability to degrade crystalline cellulose.

39.3. MAN-MADE CELLULOSE-DEGRADING ENVIRONMENTS

Man-made environments with a high load of crystalline cellulose are found in compost piles of animal waste, backyard clippings, and wastewater treatment plants. *C. thermocellum* was originally isolated from a cotton bale (18). Many microbial ecology studies have been performed on these types of environments to explore microorganisms performing various steps during lignin and cellulose digestion (44). Shiratori et al. reported the isolation of a novel *Clostridium* sp. that performs effective cellulosic waste digestion from a thermophilic methanogenic bioreactor fed with shredded office paper for methane production (64). In this environment, the conversion of cellulose to methane is mediated by four microbial populations: cellulolytic and noncellulolytic saccharolytic microorganisms, synthrophic hydrogen-generating bacteria, and methane-producing *Archaea*, with the hydrolysis of cellulose being, in general, the rate-limiting step (64).

39.4. PHYLOGENY OF CELLULOSE-DEGRADING MICROORGANISMS IN RELATION TO CELLULOSE DEGRADATION

Using 16S rRNA and 18S rRNA as the most commonly used molecules to classify *Bacteria* and *Eucarya*, the ability to digest cellulose is widely distributed. So far, however, no members of the *Archaea* have been identified with the ability to digest crystalline cellulose (5).

Within the *Bacteria*, there is a clear concentration of cellulolytic organisms within the phyla *Actinobacteria* and *Firmicutes*. Most of the representatives within the *Actinobacteria* are aerobic, while the organisms within the *Firmicutes* are anaerobes. Within the *Eucarya*, cellulolytic organisms are distributed across the fungal kingdom, spanning from Ascomycota to Basidiomycota.

Among the bacteria, there is a distinct difference in cellulolytic strategies between the aerobic and anaerobic groups of microorganisms (46). With very few exceptions, anaerobes degrade cellulose primarily via complexed cellulose systems, while aerobic cellulose degraders utilize cellulose through the production of substantial amounts of extracellular cellulases secreted into the culture supernatant, leading to two primary strategies for crystalline cellulose utilization. Aerobic bacteria and fungi do not adhere to cellulose, produce noncomplexed cellulases, and oxidize hydrolytic products to CO_2 and water (46). Anaerobic bacteria and fungi attach to cellulose, produce primarily complexed cellulases, and secrete a variety of fermentation end products. This fundamental difference between anaerobic and aerobic cellulose-degrading microorganisms has significant impact on isolation and cultivation strategies for various microorganisms. The anaerobic degradation of crystalline cellulose to various fermentation end products, such as small organic acids or alcohols, can lead to product inhibition and changes in medium conditions, inhibiting growth. In addition, the adhesion of many of these anaerobic bacteria to insoluble substrates such as crystalline cellulose complicates the traditional cultivation of microorganisms and led to the use of soluble model substrates.

In nature, the degradation of lignocellulosic materials is not performed by a single microbial species. In contrast, many cellulolytic species coexist with noncellulolytic microorganisms which grow on the side products of metabolic intermediates, leading to a synergy between cellulolytic and noncellulolytic microorganisms. This synergy has been described in many publications (46).

39.5. QUANTIFICATION OF MICROBIAL CELLULOSE UTILIZATION

To study microbial physiology it is essential to obtain physiology data such as rate of cell growth, substrate utilization and product formation, and cell yields. For microbial utilization of soluble substrates such as glucose or cellobiose these data are available, and the exploration of microbial physiology has shifted to more molecular-level studies (46). However, these types of data are, in general, not available for microbial utilization of cellulose. Growth on insoluble substrates coupled with the growth characteristic of cellulose-degrading microorganisms makes the collection of these types of data challenging. For soluble substrates, cells can easily be quantified through various methods such as direct cell count or optical measurements like light scattering. To determine the cell count of cells grown and adhered to insoluble substrate requires the enumeration of cells on three-dimensional particles. In general, accurate cell counts are normally not possible for cellulosic substrates by conventional methods. Alternative and more indirect methods such as measurement of total protein or DNA are required and complicate the determination of cell growth. Other methods were explored as described by Lynd et al. (46).

The accurate determination of cellulase concentration and the measurement of specific activity are also challenging. Cellulases can be bound to the insoluble substrate, and it is unclear how much of the total cellulase activity is bound to the substrate, bound to the cell, or free within the supernatant. In addition, cellulase systems are composed of multiple proteins that interact by performing complementary functions and in some cases forming multienzyme complexes. These multiprotein complexes are often tethered to the bacterial cell wall and also bind to the insoluble substrate, making quantification extremely challenging. Determination of exact cell counts in conjunction with measuring specific activity of cellulase, cellulose-cellulase, and cellulose-cellulase-cell complexes will be critical to compare novel strains isolated from the environment (46).

39.6. NEW CULTIVATION METHODS

To further explore the diversity of novel cellulose-degrading microorganisms, novel cultivation methods need to be explored. Most of the current strains capable of degrading crystalline cellulose were isolated using traditional techniques. However, it could be assumed that these traditional methods are not suited to isolate strains specialized in degrading crystalline cellulose present in lignocellulosic plant material. New cultivation methods need to be applied without the use of the soluble model substrates used in traditional cultivation methods. Ideally, all steps of isolation, starting with the primary enrichment and finishing with the purification of the strain, should be performed on relevant insoluble substrates. Relevant substrates include those that mimic natural environments or are used in biorefineries.

The development of novel cultivation methods targeting previously uncultured microbes gained importance with the goal of obtaining information from these uncultured phyla. In general, many of these methods mimic conditions found in nature. So far, only a few of these novel techniques have been applied to isolate lignocellulosic-degrading microorganisms. These techniques could be adapted to lignocellulosic degraders and might serve as platforms for further developments.

The general method to ensure cultivation of a pure strain is to start cultivation from a single cell, resulting in an actively growing colony created by clonal growth. This method dates back to Robert Koch and the plating of microbial cultures on solid surfaces such as agar (37). Even now, in the era of molecular tools and methods such as RNA profiling, metabolomics, and proteomics, the establishment of a pure culture is obligatory for detailed physiological characterization, is important for patent issues, and is critical in the search for multienzyme protein complexes, which are often not easily expressed in recombinant hosts. These goals have been, and are still, the driving force behind the development of novel techniques for the isolation of a pure culture derived from a single bacterial or eukaryotic cell. For cultures grown on soluble substrates the easiest approach, besides plating on agar plates, is serial dilution to extinction. Normally, these dilution series are repeated multiple times. Overall, this method cannot easily be used for the separation and purification of cells grown on insoluble substrates, since cells must first be sheared off the substrate and then separated from the remaining substrates before the dilution series can be generated.

Many of the traditional cultivation methods are not directly applicable and need to be optimized to be applied to insoluble substrates. Soluble model substrates could be used; however, they present their own issues, as discussed above. Alternatively, microbial cell separation methods have been developed and could help to overcome many of these cultivation challenges.

39.7. BACTERIAL CELL SEPARATION AND MANIPULATION

39.7.1. Cell Separation by Cell Manipulator Technologies

Cell separation technologies use optical instrumentation to visualize the specific microorganisms in conjunction with cell manipulation methods. Cell manipulation is dominated by two principal methods: laser manipulation and mechanical manipulation. Ashkin et al. described the use of infrared laser beams for trapping and manipulating biological specimens, such as single cells of *Escherichia coli* (3). The laser manipulation method was further improved by Huber et al. (27). They used a focused laser beam to capture and move archaeal cells within a capillary filled with fresh medium, from one side of the capillary to the other. The neodymium laser is focused by a microscope objective with a computer-controlled microscope stage. A rectangular glass capillary with a predetermined breaking point is used as the separation chamber, which is filled with fresh medium on one end and the mixed microbial population on the other. Cells are visualized through an inverted microscope, enabling the selection of specific cells based on morphology and also

phylogenetic criteria if combined with fluorescent in situ hybridization (FISH). A single selected cell is fixed with the laser beam and separated from the mixed culture by moving the microscope stage. The capillary is cut to remove the front end filled with the mixed culture. The end with the individual cell can be used for cultivation or, theoretically, for DNA isolation (27). Schutze et al. described a laser pressure catapulting method, which uses a laser for the microdissection and transfer of single cells (62). This method has been successfully applied for the isolation of single cells from human tissue and the dissection of various cell types. The specimens are spread on a sheet of thin polyethylene membrane, and the single cell of choice is circumscribed with the high photonic energy of a focused nitrogen laser. The selected cell, together with a small surrounding strip of the polyethylene membrane, is cut out; the selected cell adheres to the polyethylene membrane. The laser is then focused below the microdissected target cell, and the microdissected sample is catapulted into a common microcentrifuge tube positioned above the sample with a laser shot of increased energy (62). This method can be applied to any cell size. The laser can separate parts of cells in addition to having the capability to kill unwanted cells. However, this method is limited by the porosity of the polyethylene membrane. At high magnification, necessary to visualize individual microorganisms, the membrane shows a granular structure which lowers the optical resolution to a point that certain bacteria are not visible. Further research and the use of alternative membranes would make this a very powerful cell separation technology. Modern commercial microdissection microscopes are computer controlled and allow for automated cutting of selected particles. Xie et al. reported a reagentless method for identification and discrimination of single bacterial cells in aqueous solutions using a combination of laser tweezers and confocal Raman spectroscopy (79). The optical trapping enables the capture of individual bacteria in aqueous solution in the focus of the laser beam, levitating the captured cell well off the cover plate to minimize unwanted background from the cover plate and environment. Raman spectral patterns excited by a near-infrared laser beam provide intrinsic molecular information. Bacterial cells cultured at stationary phase could be discriminated by principal component analyses as well as hierarchical cluster analysis of their Raman spectra (79). This method, applied to biofilms formed on insoluble substrates, should be able to determine if the biofilm is formed from different species. Combination with an optical tweezer might allow separation of different species for subsequent isolation.

Another cell separation approach is based on microinjectors in combination with the precision of a servo-powered micromanipulator, enabling the easy handling of a single microbial cell (20). Fröhlich and König applied such a system using prefabricated Bacto-tips to isolate and cultivate single cells from the natural environment (19). For the isolation of microbial cells, a commercial micromanipulator equipped with a pressure device and mounted onto an inverse phase-contrast microscope can be used. For the isolation of bacteria, a sterile capillary tube (Bacto-tip) was used, sealed at the posterior end with a droplet of sterile oil. Luttermann et al. (cited in reference 20) described a micromanipulation method for transferring bacteria from agar plates using microcapillary tubes (31). An angled capillary tube was positioned between the condenser and the objective. The agar plate with the selected bacteria was moved below the opening of the capillary tube using the microscope stage. The aspirated bacterium was placed on the surface of a solid medium or in liquid medium in a microtiter plate. Further experiments would be required to evaluate if bacteria could be removed from solid substrates and placed on fresh lignocellulosic material. Ishoy et al. further improved this technology and developed a system which is based on an inverted microscope, a microinjector, and a micromanipulator (31). The isolated cell was captured in a microcapillary pipette and transferred into an appropriate growth medium. They demonstrated that the micromanipulator system was able to isolate a single cell from a mixed culture without contamination by any of the nontarget cells from the surrounding medium. However, micromanipulation of individual microorganisms can be challenging. Thomsen et al. used the combination of micromanipulation of FISH-labeled filaments or microcolonies in activated sludge samples with reverse transcriptase PCR for further identification (70). This method was more successful on filamentous bacteria, as they were easier to separate from the floc material. However, it was not possible to isolate microcolonies consisting of only a single bacterial species due to the presence of several different species within a microcolony or the contamination of the selected microcolony by surrounding free bacteria (70). One would expect that similar issues would be encountered during separation of microorganisms from the surface of insoluble substrates. A biofilm formed by various microorganisms might make the selection and separation of an individual microorganism challenging. Luo et al. further combined a microfluidic device with a microwell array and an optical tweezer (44). This method was used to manipulate yeast cells, wherein cells were captured by the optical tweezer and separated into microwells for further cultivation. This combined technology might allow the parallel cultivation of many separated cells within the microfluidic microwells. Further research will be required to adapt this technology to the cultivation of cellulose-degrading microorganisms.

39.7.2. Separation of Cells by Flow Cytometry

Another technology widely used for cell separation and cell analysis is flow cytometry. This technology is commonly used in many medical applications to analyze and separate eukaryotic cells, but it is gaining more and more popularity for bacterial applications. Flow cytometry coupled with cell sorting is a powerful fluorescence-based cell diagnostic and separation tool that enables the rapid analysis of entire cell populations on the basis of single cell characteristics. Characteristics can be analyzed simultaneously, including cell count, cell size or content, and responses to fluorescent probes as diagnostics of cell function. Cells in a liquid sample are passed individually in front of an intense light source (laser or laser diode), and data on light scattering and fluorescence are collected (14, 77). Bacteria can be separated and sorted based on the forward- and side-scatter effects alone, even without the use of any fluorescent dyes. We have successfully demonstrated that this technology can be applied to the isolation of single bacteria grown on insoluble substrates (unpublished data). Cells attached to insoluble substrates can be sheared off by vigorous shaking and separated from the insoluble substrates with the remainder of attached cells, by slow centrifugation (unpublished data). The supernatant is applied to cell sorting, and single individual cells are sorted into fresh medium containing insoluble substrates.

Flow cytometry, used in conjunction with FISH, combines identification of specific cells in a mixed culture with cell separation (35, 63). Raghunathan et al. used flow cytometry to separate single, unstained *E. coli* cells and subsequently amplified the isolated DNA 5×10^9-fold (54). Podar et al. combined flow cytometry with FISH and multiple displacement amplification to amplify DNA from five sorted bacterial cells belonging to the candidate division TM7, an uncultured division of microorganisms commonly found in different soils (52).

Recently, nanotechnology has made significant progress and miniaturized many standard laboratory procedures. This technology, in combination with rapid development in laser technology, led to the miniaturization of flow cytometry and the development of an integrated microfabricated cell sorter using multilayer soft lithography (21, 22). This integrated cell sorter can be incorporated with various microfluidic functionalities, including peristaltic pumps, dampers, switch valves, and input and output wells, to perform cell sorting in a coordinated and automated fashion with extremely low fluidic volumes. Inexpensive materials and a simple fabrication process allow the production of these devices as disposables, eliminating the danger of cross-contamination from previous runs (22). In addition, the cell sorting device can be directly combined with chemical or enzymatic reactions, such as cell lysis and DNA amplification methods. This approach has recently been used to characterize a different ecotype of TM7 phylum bacteria from the human subgingival microbiota (47). Large, rod-shaped cells, resembling oral TM7 bacteria previously characterized by FISH (29), were separated and lysed in a microchamber, followed by DNA amplification in nanoliter-scale volume to confirm identity and provide access to the genome of the isolated cell. However, the sorting speed of these devices is much lower than that of traditional flow cytometers. At this stage it has not been demonstrated whether these devices can be used for the cultivation of microorganisms in the presence of insoluble substrates. Also, insoluble substrate particles have various sizes which might obstruct the channels of the microfluidic devices.

39.7.3. Novel Microbial Cultivation Methods

The traditional cultivation methods, based on plating of mixed populations of microorganisms on solidified medium such as agar, are selective and biased toward growth of specific microorganisms (15, 29). Traditional media strongly select for microbes that are fast growing, grow to high density, are resistant to high concentrations of nutrients, and are able to grow in isolation. Most microbial cells studied directly in the environment are viable but do not form visible colonies on solid agar plates when cultured. Therefore, it could be argued that these traditional culturing strategies use conditions that are completely different from the normal growth habitat of many microbes and are a major contributing factor to the failure to cultivate most microorganisms in pure culture (15, 72, 73). Janssen and coworkers addressed the issue of inhibition by high concentrations of nutrients in cultivation media and investigated the cultivability of soil microorganisms using diluted nutrient broth (33). Extended incubation times, of at least 10 weeks, were required to allow maximum colony development leading to the isolation of previously uncultured microorganisms (11, 61). Other studies investigated the effect of signal compounds such as cyclic AMP on the cultivation efficiency of aquatic bacteria, showing, in general, an increase in culturability (8, 9, 25). Button et al. developed a dilution cultivation method for growing bacteria from oligotrophic aquatic environments (10). These findings suggested that most marine bacteria (60%) are viable but reach stationary phase before attaining visible turbidity and are therefore not detectable with traditional techniques. Kaeberlein et al. used natural chemical components to enrich microorganisms directly from the environment (34). Marine organisms were placed in a chamber, separated through a membrane, and deposited into a natural environment. They could demonstrate growth of novel, previously uncultured microorganisms (34). Gavrish et al. further expanded this technology to cultivate filamentous actinobacteria from soil samples based on the ability of these bacteria to form hyphae and penetrate solid environments (23). A similar idea was applied to the cultivation of previously uncultured soil bacteria using a soil substrate membrane system. The system used a polycarbonate membrane as a solid support for growth and soil extract as the substrate. Microcolonies were visualized using total bacterial staining combined with FISH (16).

Connon and Giovannoni described the development of a high-throughput cultivation method enabling the identification of a large number of extinction cultures (12). They combined the concept of extinction culturing in natural medium with the use of microtiter dishes in combination with an automated cell array and imaging process. This combination enabled them to increase the detection sensitivity of cells, leading to a significant shortening in incubation time of cells with low growth rates (10, 12). Zengler et al. developed a high-throughput cultivation approach based on the encapsulation of single microorganisms in microcapsules followed by flow cytometry for the detection of cell growth and cell separation (80). Members of the microbial community, represented in single encapsulated cells, are grown in columns under very-low-nutrient flux conditions using mineral medium supplemented with low concentrations of nutrients extracted from the sampling site. Over time, each cell capable of growth under the conditions in the column forms a microcolony within its enclosure. Using this technology, it was demonstrated that novel, previously uncultured organisms can form microcolonies within the microcapsules

and can generate a pure culture if separated and sorted into microtiter plates (80).

39.8. CULTURE-INDEPENDENT ENZYME DISCOVERY

Culture-independent methods enable the discovery of genes and proteins from microbial species that are uncultivable. Microbes are capable of growth in a variety of environments and conditions, and the stable and active enzymes of microbes growing at the extremes of pH and temperature could have utility in industrial applications. These microbes, however, have proven to be difficult to cultivate (2). Indeed, the overwhelming majority of all bacteria remain uncultivated (56), and the potential of tapping into this majority for industrial enzymes is vast. Explorations of the metagenome—the collective genomes within a microbial community (26) obtained directly from the environment without cultivation—have increased in the last decade (28). Novel enzymes have been discovered (68), and one metagenomic sequencing effort nearly doubled the number of identified proteins (60).

Screening of the metagenome for novel genes and proteins can be accomplished by first cloning the DNA into a library, and also by interrogating the DNA directly. Methods include PCR screening, DNA hybridization, metagenome sequencing and data mining, and genomic library expression screening (Table 1).

39.8.1. DNA Extraction

Culture-independent screening for genes or enzymes is based on the extraction and analysis of a sample community metagenome. The ideal method is to extract the metagenome in its entirety to retain unbiased representation of all microbes present. Methods to extract DNA from soils and sediments are generally classified as direct or indirect extractions (55). The methods differ in the amount of sample required, the size of DNA extracted, and the separation of cells from their surrounding environmental matrix, which affects species representation. Given these differences, the downstream

TABLE 1 Metagenome screening

Method	Benefits	Limitations
Sequencing	High throughput	Hits may be incomplete ORFs
	Library construction not required	Can find only known domains
		High tech
PCR	Works with small amounts of DNA	Can find only known domains
	Can find similar sequences	Low throughput
	Low tech	
	Library construction not required	
DNA hybridization	Can find similar sequences	Can find only known domains
	Low tech	Hits may be incomplete ORFs
		Medium throughput
		Library construction required
Activity	High throughput	Lower success rate
	Hits are active clones	Host must be able to produce active enzyme
	Can find enzymes with novel domains	Library construction required
		High tech

analytical application must be considered when choosing an extraction method.

39.8.1.1. Direct Extraction

In the direct DNA extraction process the cells are lysed directly within the sample matrix. The DNA is then isolated from the matrix and cell debris (49, 74). The methods for cell lysis include thermal shock, bead mill homogenization, bead beating, microwave heating, and ultrasonication (55). These are preferred to chemical or enzymatic lysis methods because they allow for disruption of the surrounding matrix, allowing for more complete, unbiased lysis of all cells. The liberated DNA is then purified using methods such as silica-based binding, agarose gel electrophoresis, precipitation, and gel filtration. A disadvantage of the direct extraction method is that since the sample is extracted as a whole, any inhibitors present in the matrix, especially humic acids commonly found in soils, accompany the DNA. These substances can inhibit downstream molecular processes such as PCR. Also, DNA from all cells within the sample, including eukaryotic cells, is extracted. Due to the generally larger genomes of eukaryotes and the presence of noncoding sequences, eukaryotic DNA contamination may be undesired. The main consideration of direct extraction is the amount of DNA shearing that takes place with the mechanical disruption of cells and the surrounding matrix, resulting in smaller fragments of genomic DNA. This limits the insert size of genomic clone libraries and precludes the use of fosmid or bacterial artificial chromosome (BAC) libraries.

39.8.1.2. Indirect Extraction

The indirect DNA extraction process separates cells from the sample matrix before lysis (4, 81). Samples are typically blended in a disruption buffer to separate the cells from the matrix while keeping them intact. The cells are then separated from the sample debris by differential centrifugation. The cells are further purified by density gradient centrifugation (4) or the use of a cation-exchange resin (32). The combination of the centrifugations also helps to eliminate the eukaryotic cells from the sample, resulting in a pure prokaryotic metagenome. The purified cells are encased in low-melting-point agarose and then lysed by enzymatic or chemical methods (69). During the lysis and washing steps the immobilization of the DNA in agarose eliminates DNA shearing and preserves high-molecular-weight DNA. The DNA is then released by melting and digesting the agarose with an agarase.

As with direct extraction, there are advantages and disadvantages to indirect extraction. The extraction of high-molecular-weight DNA allows for the construction of larger insert libraries (fosmids and BACs). These libraries enable the screening of longer stretches of DNA, such as those encoding metabolic pathways or multienzyme protein complexes. They also aid in the recovery of full-length open reading frames (ORFs) by increasing the chances of finding a full-length ORF within a clone insert. Since the harsh mechanical methods of direct extraction are not employed, a portion of the cells may not be released from the sample matrix. This leads to overall lower coverage of the community, with rare species possibly unrepresented. The DNA yields from indirect extraction are typically lower than direct extraction. Thus, the sample amounts needed to obtain enough DNA for library construction need to be as much as 10 times higher, if not more, than the amount needed for direct extraction.

39.8.2. Environmental Genomic DNA Screening Methods

39.8.2.1. Sequence-Based Discovery Using Genomic DNA

39.8.2.1.1. PCR Screen

PCR screening is based on amplifying target gene fragments from metagenome DNA templates using degenerate primers that anneal to conserved regions. Gene sequence databases are quite large, and many protein functional domains have well-defined primary amino acid sequences. Thus, a highly conserved target sequence can be defined for many enzyme activities of interest. Once a target is chosen, e.g., a cellulase, the known, sequenced cellulases are aligned to determine any conserved regions. In designing degenerate primers, the primers must be specific enough to amplify only cellulases but general enough to amplify the variety of cellulases that may be present in any sample. Annealing temperatures are typically low, to decrease the stringency of the PCR and to aid in capturing novel genes. The PCR product is a collection of the different cellulases that were amplified in the sample. The product is cloned, and clone inserts are sequenced—with sequence depth determined by the complexity of the metagenome and the number of unique cellulases amplified by PCR.

Since most conserved regions are not found at the start and stop regions of a gene, the amplified products are only gene fragments. These PCR tags can then be used as starting points for complete cloning of the full ORF using thermal asymmetric interlaced PCR or other PCR recovery techniques. Activity of the protein encoded by the ORF must then be confirmed by activity assays. Another limitation to the PCR screening approach is that the only genes that can be amplified are those encoding already known domains with sufficient conservation to allow for degenerate primer annealing.

The main benefit of PCR screening is sample coverage. As long as the sample metagenome DNA is free of PCR inhibitors, 1 ng of template DNA in a reaction is greater than 1,000 times the coverage of a sample containing 1,000 genomes. Relative abundance plays a large role. To discover genes from rare species, and to limit the dominance of genes from abundant species, the lowest number of PCR cycles should be employed and the greater number of clones should be sequenced.

39.8.2.1.2. Hybridization Screen

Another traditional approach for isolating novel coding sequences is to screen a DNA library at low stringency with a DNA probe. While this approach can work with environmental DNA samples, it does have limitations. Recent analysis of a soil sample led to an estimate of 3,555 bacterial operational taxonomic units in that sample (48). If one assumes a typical prokaryotic genome size of 4 Mb, then there would be 1.4×10^{10} bp of sequence information in that sample. A typical small insert library with an average insert size of 5 kb would require screening 1.4×10^{7} clones. If one can screen 5,000 colonies per plate, it would require 2,800 plates to screen all the information contained in the plasmid library at 5× coverage. Thus, any standard DNA hybridization probe approach will cover only a very small percentage of the potential within a typical metagenomic library. Such a screen is also limited to finding only sequences related to previously identified sequences and will miss truly novel sequences.

39.8.2.2. Metagenomic Sequencing and Data Mining

Since the first large-scale sequencing of metagenomes (75, 76), a method of data mining metagenome sequences to discover novel proteins has emerged. Numerous studies offering comparisons of metagenomes and functional insights into microbial communities have been published (28, 40, 59) and will continue to be invaluable. In collaboration with the Joint Genome Institute and the California Institute of Technology, Verenium Corp. discovered novel genes encoding lignocellulosic-degrading enzymes from the metagenome of hindgut microbiota of wood-eating higher termites (71). These enzymes have potential applications in the lignocellulosic ethanol industry. With ever-increasing sequencing capacity and technology, data mining for discovery and application of novel proteins for the biofuel, pharmaceutical, chemical, and industrial fields continues to gain ground.

39.8.2.2.1. Library End Sequencing

Traditional dideoxynucleotide chain termination sequencing, or Sanger sequencing, has been the approach to obtain metagenome sequences for the past three decades. This method has also been used for gene pathway and whole-genome sequencing. The metagenomic DNA of a community is randomly sheared or digested and then cloned into a library, with clone insert sizes in the range of 1 to 10 kb for a plasmid library, 32 to 47 kb for a fosmid or cosmid library, and 150 to 350 kb for a BAC library. The clones are plated, and individual colonies are picked and grown. The cloned DNA is extracted from each clone separately, and the end of each clone insert is sequenced. Two sequencing reactions are run for each clone using a forward and a reverse primer, producing a forward and reverse "end read." The priming sites are situated on the vector sequence nearest to the insert, to minimize known vector sequence in the read. Typical end reads are 500 to 1,000 bp long. Software is used to assemble any overlapping reads to form contigs. Depending on the diversity and abundance of species within a community, contigs may or may not form. Unlike a single genome sequencing effort, the goal of metagenome sequencing is not to "close" the metagenome, or sequence it in its entirety, but rather to gather a sampling of all the genomes present. Data mining of metagenome sequences has been used to discover novel glycosyl hydrolases (7, 78).

The advantage of metagenome data mining is that the genes discovered are not encumbered by the necessity of hybridization to conserved regions, as is the case for PCR and hybridization screens. More genes may be discovered without these constraints. The disadvantages are limited coverage and the relative abundance of species. The amount of metagenome sequenced in the termite gut study and the Sargasso Sea was only a fraction of the total metagenomes present (76, 78). However, from these sequencing efforts numerous novel genes were discovered. In any given community the distribution of species is not equal; there are over- and underrepresented species present. Metagenome sequencing will reveal a greater amount of data from the abundant species, with very little from the rare species. This bias can be overcome by sequencing to a greater depth, if cost is not a factor.

The main disadvantage of metagenome sequence data mining for novel genes encoding active proteins is the paucity of full-length genes discovered. The vast amount of data generated by metagenome sequencing is subject to gene prediction and annotation using software programs that employ different gene prediction methods (40). In most cases the genes discovered will be gene fragments or genes containing partial ORFs. These partial ORFs do not contain—on one read or contig—the start and stop codons of the gene. The utility of these partial sequences is invaluable for comparative analyses, where genes can be predicted from partial sequence, but limited for obtaining novel proteins from full-length ORFs. When the full-length ORF does exist within the sequence, primers can be designed to the start and stop of the gene and PCR amplified from the genomic DNA or library. The gene may also be created synthetically.

39.8.2.2.2. Full-Length Gene Recovery

The candidate genes discovered by data mining are all hypothetical, subject to vetting in subsequent functional characterization experiments. Many unknowns exist as to the activity of the proteins encoded by sequence data-mined genes. To be active, a gene must contain no debilitating mutations or deletions, contain functional start and stop sites, be capable of expression and proper folding in the expression host cell, and have activity on the chosen substrate. Inactive proteins and their genes may possibly be fine-tuned to produce activity. Indeed, the same issues also apply to genes and proteins in the function-driven approach. However, only those proteins that are active will be discovered through the screen; the number of inactive clones does not affect the workload. In the case of sequence data mining, any gene of interest must be subcloned and tested for expression. Some will be inactive, for any number of the reasons listed above (78). It should be assumed, when embarking on a sequence mining discovery effort, that some portion of the genes discovered will be inactive. When only a partial ORF is discovered, there are a number of methods to obtain the missing sequences—the start or stop codons, or both. The power of library end sequencing is that there is a potentially inexhaustible supply of clone DNA to go back to and obtain the portion that neighbors the end sequence. The clone insert from which a partial end sequence was discovered may contain the entire ORF. A new primer is designed toward the end of the known sequence. It is used to "primer-walk," or continue sequencing from the known sequence. When the full-length ORF is contained within a clone insert, this is the simplest and least labor-intensive method.

If a partial gene is truncated at the end of a clone insert, other more laborious methods are needed to obtain its full ORF. These methods depend on the use of metagenomic DNA, which may be scarce. The DNA must also be clean and free of PCR inhibitors to work effectively. Environmental samples, especially soils, often contain inhibitors that make these methods of full-length recovery inefficient.

39.8.2.2.3. Next-Generation Sequencing

Next-generation sequencing methods such as pyrosequencing, reverse termination, and amplification by ligation (51) can yield much greater amounts of sequence data than Sanger sequencing. The new technologies have provided substantial increases in predicted gene sequences and the resulting predicted enzyme sequences (7). However, given the relative shortness in the sequencing reads (<400 bp) from all the next-generation DNA sequencing technologies, the sequence data tend to cover only portions of a complete ORF. These short reads limit investigators to discussing sequences in terms of environmental gene tags (7, 39). The methods do enable sequencing directly from extracted metagenomic DNA, bypassing the need to create genomic clone libraries. However, for the purpose of functional gene and protein discovery, the absence of actual library clones underlying the DNA sequence precludes the relatively easy subsequent clone sequencing efforts to complete partial sequences.

The obvious advantage to highly processive new sequencing technologies is greater coverage of the metagenome, which would yield greater numbers of, albeit partial, genes discovered. Depending on the complexity and relative abundance of a sample, the increased coverage may yield overlapping sequences and contigs, which may include full-length genes. Without contigs, however, the laborious PCR methods mentioned above may be the only way to recover ORFs, using the metagenomic DNA as a template. The large differences in per-base sequencing costs dictate that the new methods will be used, and improvements in average read lengths will lead to larger contigs and better ORF calls (reviewed in reference 40).

In a previous effort that used Sanger-based sequencing methods to identify new cellulases (78), scientists at Verenium Corp. were able to follow up on partial genes and sequence actual clones, express them, and confirm the enzyme activity of 35 of the 44 genes tested. It follows that to most effectively discover novel functional proteins by data mining metagenome sequences, the combination of new and old technologies is needed. This can be done by building up large data sets with next-generation sequencing and then associating these sequences and contigs with actual library clone ends generated by Sanger sequencing.

39.8.2.3. Genomic Clone Libraries and Expression Screening

Direct sequencing of environmental DNA can lead to the identification of novel enzymes but has limitations, as mentioned above. Also, sequence-based discovery still requires subsequent testing of the cloned or synthesized DNA to confirm that the enzyme predicted to be encoded by the gene in question does have the activity of interest. Thus, it is important to have expression systems for testing putative ORFs as well as using the environmental DNA as source material for building libraries in expression vectors suitable for high-throughput activity (24).

39.8.2.3.1. Library Construction

39.8.2.3.1.1. Characteristics of different bacterial cloning. Researchers may now choose from a large variety of bacterial cloning systems available for use in an assortment of *E. coli* strains. These cloning systems cover a full range of potential insert sizes from less than 1 kb to greater than 100 kb.

Each of the available cloning strategies has advantages and disadvantages.

Lambda cloning vectors provide significant advantages over plasmid vectors, such as lack of size bias, easier amplification and screening, and more complete representation because lytic systems have a higher tolerance for toxic proteins. Early versions of lambda vectors were difficult to manipulate in downstream applications such as subcloning and sequencing due to their unwieldy size, nearly 50 kb. However, this issue has been addressed by newer generations of lambda vectors such as Stratagene's Lambda Zap.

Lambda insertion vectors typically accept inserts of 0 to 12 kb, depending on the vector. Stratagene's Lambda Zap vectors combine the high efficiency of a lambda vector system with the versatility of a plasmid system by allowing in vivo excision of cloned fragments into a phagemid

vector. The plasmid form of the clone is recovered in the pBluescript plasmid through a simple in vivo process of coinfecting the appropriate *E. coli* strain with the phagemid and a helper phage.

Cosmid vectors have larger cloning capacities than lambda insertion vectors, up to 45 kb. This has the advantage of allowing large genes or operons to be isolated within a single clone. However, nearly all large-fragment cloning systems have the potential of generating cloning artifacts such as chimeras and can also give rise to a percentage of clones that are unstable and rearrange or delete inserts during cell replication.

39.8.2.3.1.2. Average insert size and representation of the genome. For environmental genomic cloning strategies, microbial sources and preparations can sometimes be directed to enrich for targets of interest. When constructing environmental libraries intended for multigenome-wide studies, the usefulness of the library depends on maximizing the proportion of entire genomes present in the library. The two defining characteristics of library quality are clone insert size and total number of clones containing inserts.

Insert size relates directly to the overall genome coverage of an environmental library, which may include several to hundreds of genomes. Inserts must also be of sufficient size to be useful for downstream analysis such as sequencing. As described above, even larger inserts may be required to identify complex pathways; however, their large size also presents challenges to procedures such as sequencing.

The number of clones or library size also relates directly to the probability of finding clones of interest. As library size increases, the probability of finding the desired segment of any genome within the library increases. This also applies to the probability of finding overlapping clones which make up pathways. One must keep in mind that while larger insert libraries have inherent advantages with genome analysis, they also have disadvantages in downstream manipulation. Consequently, how large does a library need to be to theoretically guarantee that a target will be present? "Genome equivalent" is a term used to describe the depth of genome coverage. Equivalents are calculated as the ratio of the amount of genomic DNA in the library to the amount of DNA in the genome, or the genome redundancy. The calculation of an environmental library's genome coverage involves five factors: (i) genome size of the organisms, (ii) total number of clones in the library, (iii) number of background clones, (iv) average insert size of contributing clones, and (v) with environmental libraries, the number of different genomes present. For example, a soil sample may contain hundreds to thousands of different genomes. A library with 5× genome coverage is generally adequate for finding one or more clones of interest, since more than 99% of the genome should be present.

39.8.2.3.1.3. Amount and quality of DNA required. The quantity of DNA needed to construct a library varies depending on numerous factors, including the cloning system chosen, average size of the genomic DNA prepared, and genome coverage needed. Depending on the experience of the user, multiple rounds of optimization may be needed to achieve an acceptable initial DNA preparation. The ideal preparation of environmentally sourced DNA consists of both sufficient quantity and quality of size-selected DNA for library construction. Regardless of the vector type chosen for library construction, it is helpful to have available environmental source DNA which is at least five times the size of the desired insert size. DNA of five times the size will allow for random shearing or partial enzyme digestion of the DNA with a high probability of gene representation in the library. For example, lambda insertion libraries which accept inserts up to 12 kb would require 60 kb of source DNA and cosmid libraries with 40-kb inserts would require 200 kb of source DNA.

DNA prepared in liquid format is more susceptible to shearing forces and breakage, although it can be of sufficient length to be suitable for cloning into lambda, cosmid, and fosmid vectors.

39.8.2.3.1.4. The bacterial host strain.

There are a variety of challenges to maintaining foreign genomic DNA in a bacterial host. Environmental sources will likely contain both prokaryotic and eukaryotic genomes, adding to the overall complexity of maintaining library representation. Wild-type *E. coli* has evolved numerous restriction, modification, and recombination systems to combat introduction of foreign DNA. Fortunately, a variety of host strains with mutant genotypes is now available for library construction, and these strains minimize such effects.

For instance, disabling *recA* and other related genes in the host strain eliminates most recombination related to repetitive sequences in the genomes of higher organisms. Another common host strain modification includes mutations to restriction systems that degrade foreign DNA. Selection of a suitable host strain is simplified by the general availability of library construction kits which include host strains (Table 2).

39.8.2.3.2. Activity Screening Methods

Expression libraries made from environmental DNA can be screened in several different ways, each differing in the format for keeping individual clones separate and exposing each clone to the substrate. The basic distinctions are between solid agar plate formats and liquid-based microtiter plate formats. Each has its pros and cons. Solid agar plate formats are simple to run and have reasonable throughput but are limited to assays that can be run at lower temperatures and are more amenable to colorimetric substrates that allow direct visual readout of the activity. Liquid-based screens, using microtiter plates, can result in higher throughput because of the use of photonic readouts (fluorescence and luminescence) and robotic systems (24, 65). Liquid-based screens also allow for assay conditions that may not be amenable to host cell growth or agar stability, such as higher temperatures or organic solvent-requiring substrates.

In all the cases, primary screens are conducted with highly complex libraries to identify a clone expressing the enzyme activity of interest, or hit (66, 67). These primary hits are then subsequently rescreened at much lower complexity to confirm the presence of a clone with the desired activity and to isolate individual clones. In some cases there may be a third or even fourth round of screening at low density to ensure that single clones with the desired activity have been isolated. The final clone is then ready for DNA sequencing. The resulting DNA sequence is scanned for potential coding sequences. The translations of the coding sequences are then searched for similarity to previously identified enzymes or domains using, for example, BLAST (1) and Pfam (17).

39.8.2.3.2.1. Solid-medium screening.

Activity-based screening of environmental DNA libraries on solid media such as agar plates is best suited to assays where a colorimetric substrate is available (67). An azo dye-linked xylan was used at Verenium (6) to screen libraries from insect gut contents to discover novel xylanases. Palackal et al. (50), also at Verenium, used the same substrate to isolate a novel cellulase from bovine ruminant gut bacteria. In both cases, bacteriophage lambda plaques were screened by plating the phage in top agar containing the dye-linked substrate. A thermostable α-amylase that is used in the grain ethanol industry was discovered at Verenium using a similar method with dye-linked starch (57). In this case, a nitrocellulose lift was done to replicate the plaques onto a solid support and allow for screening at high temperature (~75°C).

39.8.2.3.2.2. Liquid-based screening.

Another method for screening by activity is to use a liquid medium. A library is propagated in plasmid form, and bacterial clones are grown and assayed for the novel activity (65). Typically the library is plated out and colonies are picked into microtiter plates for culturing. The substrate is subsequently added, and the microtiter plates are tested for generation of the expected product. Colorimetric substrates are useful because the products can be detected with a basic spectrometer; however, they can have problems with background noise, low sensitivity, and linearity. Fluorometric assays are also used and have the benefits of higher sensitivity, better linearity, and lower noise. Kim et al. (38) used a fluorogenic compound to screen environmental DNA from a soil sample to find a β-glucosidase. They used 4-methylumbelliferyl β-D-cellobioside as a substrate to screen a metagenomic expression library arrayed into 96-well microtiter plates, visualizing the product using UV light. Robertson et al. (58), from Verenium, used a variant of this method to select for the clone of interest. They developed a strategy to find nitrilases by growing the cells expressing the environmental

TABLE 2 Bacterial cloning systems

Vector	Max capacity (kb)	Avg insert size (kb)	Growth format on agar plates	No. of clones required for 10× representation[a]	Typical no. of clones in a library constructed from 1 µg of size-selected DNA (under good conditions)	Source
ZAP Express	12	5	Plaques or colonies if excised	400,000	2,000,000	Stratagene
Uni-ZAP XR	10	5	Plaques or colonies if excised	400,000	2,000,000	Stratagene
pWEB	45	40	Colonies	50,000	20,000	Epicentre
EpiFos	45	40	Colonies	50,000	20,000	Epicentre

[a]Environmental source containing 100 genomes = 200,000 kb (assuming 2-Mb genome).

DNA library in a medium where the only nitrogen source was a nitrile. After up to 5 weeks of culturing, several flasks were identified that contained actively growing cultures. They were subsequently streaked for single-colony isolation and confirmation of the activity, yielding 137 novel nitrilases. The efficiency of microtiter plates for screening has increased from the original standard of 96 wells up to 1,536 wells per plate in the same area as occupied by the 96-well version. The GigaMatrix® technology (42, 43), invented at Verenium (41), dramatically increased that efficiency by using a microtiter plate that contains up to 1,000,000 wells in the same footprint. GigaMatrix® has been used to isolate proteases, cellulases, β-xylosidases, and α-L-arabinofuranosidase (42).

REFERENCES

1. **Altschul, S. F., W. Gish, W. Miller, E. W. Myers, and D. J. Lipman.** 1990. Basic local alignment search tool. *J. Mol. Biol.* **215:**403–410.

2. **Amann, R. I., W. Ludwig, and K. H. Schleifer.** 1995. Phylogenetic identification and in situ detection of individual microbial cells without cultivation. *Microbiol. Rev.* **59:**143–169.

3. **Ashkin, A., J. M. Dziedzic, and T. Yamane.** 1987. Optical trapping and manipulation of single cells using infrared laser beams. *Nature* **330:**769–771.

4. **Berry, A. E., C. Chiocchini, T. Selby, M. Sosio, and E. M. Wellington.** 2003. Isolation of high molecular weight DNA from soil for cloning into BAC vectors. *FEMS Microbiol. Lett.* **223:**15–20.

5. **Blumer-Schuette, S. E., I. Kataeva, J. Westpheling, M. W. Adams, and R. M. Kelly.** 2008. Extremely thermophilic microorganisms for biomass conversion: status and prospects. *Curr. Opin. Biotechnol.* **19:**210–217.

6. **Brennan, Y., W. N. Callen, L. Christoffersen, P. Dupree, F. Goubet, S. Healey, M. Hernandez, M. Keller, K. Li, N. Palackal, A. Sittenfeld, G. Tamayo, S. Wells, G. P. Hazlewood, E. J. Mathur, J. M. Short, D. E. Robertson, and B. A. Steer.** 2004. Unusual microbial xylanases from insect guts. *Appl. Environ. Microbiol.* **70:**3609–3617.

7. **Brulc, J. M., D. A. Antonopoulos, M. E. Miller, M. K. Wilson, A. C. Yannarell, E. A. Dinsdale, R. E. Edwards, E. D. Frank, J. B. Emerson, P. Wacklin, P. M. Coutinho, B. Henrissat, K. E. Nelson, and B. A. White.** 2009. Gene-centric metagenomics of the fiber-adherent bovine rumen microbiome reveals forage specific glycoside hydrolases. *Proc. Natl. Acad. Sci. USA* **106:**1948–1953.

8. **Bruns, A., H. Hoffelner, and J. Overmann.** 2003. A novel approach for high throughput cultivation assays and the isolation of planktonic bacteria. *FEMS Microbiol. Ecol.* **45:**161–171.

9. **Bruns, A., U. Nubel, H. Cypionka, and J. Overmann.** 2003. Effect of signal compounds and incubation conditions on the culturability of freshwater bacterioplankton. *Appl. Environ. Microbiol.* **69:**1980–1989.

10. **Button, D. K., F. Schut, P. Quang, R. Martin, and B. R. Robertson.** 1993. Viability and isolation of marine bacteria by dilution culture: theory, procedures, and initial results. *Appl. Environ. Microbiol.* **59:**881–891.

11. **Chin, K. J., D. Hahn, U. Hengstmann, W. Liesack, and P. H. Janssen.** 1999. Characterization and identification of numerically abundant culturable bacteria from the anoxic bulk soil of rice paddy microcosms. *Appl. Environ. Microbiol.* **65:**5042–5049.

12. **Connon, S. A., and S. J. Giovannoni.** 2002. High-throughput methods for culturing microorganisms in very-low-nutrient media yield diverse new marine isolates. *Appl. Environ. Microbiol.* **68:**3878–3885.

13. **Cook, D. M., E. DeCrescenzo Henriksen, R. Upchurch, and J. B. Doran Peterson.** 2007. Isolation of polymer-degrading bacteria and characterization of the hindgut bacterial community from the detritus-feeding larvae of *Tipula abdominalis* (Diptera: Tipulidae). *Appl. Environ. Microbiol.* **73:**5683–5686.

14. **Davey, H. M., and D. B. Kell.** 1996. Flow cytometry and cell sorting of heterogeneous microbial populations: the importance of single-cell analyses. *Microbiol. Rev.* **60:**641–696.

15. **Eilers, H., J. Pernthaler, F. O. Glockner, and R. Amann.** 2000. Culturability and in situ abundance of pelagic bacteria from the North Sea. *Appl. Environ. Microbiol.* **66:**3044–3051.

16. **Ferrari, B. C., T. Winsley, M. Gillings, and S. Binnerup.** 2008. Cultivating previously uncultured soil bacteria using a soil substrate membrane system. *Nat. Protoc.* **3:**1261–1269.

17. **Finn, R. D., J. Tate, J. Mistry, P. C. Coggill, S. J. Sammut, H. R. Hotz, G. Ceric, K. Forslund, S. R. Eddy, E. L. Sonnhammer, and A. Bateman.** 2008. The Pfam protein families database. *Nucleic Acids Res.* **36:**D281–D288.

18. **Freier, D., C. P. Mothershed, and J. Wiegel.** 1988. Characterization of *Clostridium thermocellum* JW20. *Appl. Environ. Microbiol.* **54:**204–211.

19. **Fröhlich, J., and H. König.** 1999. Rapid isolation of single microbial cells from mixed natural and laboratory populations with the aid of a micromanipulator. *Syst. Appl. Microbiol.* **22:**249–257.

20. **Fröhlich, J., and H. König.** 2000. New techniques for isolation of single prokaryotic cells. *FEMS Microbiol. Rev.* **24:**567–572.

21. **Fu, A. Y., H. P. Chou, C. Spence, F. H. Arnold, and S. R. Quake.** 2002. An integrated microfabricated cell sorter. *Anal. Chem.* **74:**2451–2457.

22. **Fu, A. Y., C. Spence, A. Scherer, F. H. Arnold, and S. R. Quake.** 1999. A microfabricated fluorescence-activated cell sorter. *Nat. Biotechnol.* **17:**1109–1111.

23. **Gavrish, E., A. Bollmann, S. Epstein, and K. Lewis.** 2008. A trap for in situ cultivation of filamentous actinobacteria. *J. Microbiol. Methods* **72:**257–262.

24. **Gray, K. A., T. H. Richardson, D. E. Robertson, P. E. Swanson, and M. V. Subramanian.** 2003. Soil-based gene discovery: a new technology to accelerate and broaden biocatalytic applications. *Adv. Appl. Microbiol.* **52:**1–27.

25. **Guan, L. L., H. Onuki, and K. Kamino.** 2000. Bacterial growth stimulation with exogenous siderophore and synthetic N-acyl homoserine lactone autoinducers under iron-limited and low-nutrient conditions. *Appl. Environ. Microbiol.* **66:**2797–2803.

26. **Handelsman, J., M. R. Rondon, S. F. Brady, J. Clardy, and R. M. Goodman.** 1998. Molecular biological access to the chemistry of unknown soil microbes: a new frontier for natural products. *Chem. Biol.* **5:**R245–R249.

27. **Huber, R., S. Burggraf, T. Mayer, S. M. Barns, P. Rossnagel, and K. O. Stetter.** 1995. Isolation of a hyperthermophilic archaeum predicted by in situ RNA analysis. *Nature* **376:**57–58.

28. **Hugenholtz, P., and G. W. Tyson.** 2008. Microbiology: metagenomics. *Nature* **455:**481–483.

29. **Hugenholtz, P., G. W. Tyson, R. I. Webb, A. M. Wagner, and L. L. Blackall.** 2001. Investigation of candidate division TM7, a recently recognized major lineage of the domain *Bacteria* with no known pure-culture representatives. *Appl. Environ. Microbiol.* **67:**411–419.

30. **Irwin, D. M., and A. C. Wilson.** 1990. Concerted evolution of ruminant stomach lysozymes. Characterization of lysozyme cDNA clones from sheep and deer. *J. Biol. Chem.* **265:**4944–4952.

31. Ishoy, T., T. Kvist, P. Westermann, and B. K. Ahring. 2006. An improved method for single cell isolation of prokaryotes from meso-, thermo- and hyperthermophilic environments using micromanipulation. *Appl. Microbiol. Biotechnol.* **69:**510–514.

32. Jacobsen, C. S., and O. F. Rasmussen. 1992. Development and application of a new method to extract bacterial DNA from soil based on separation of bacteria from soil with cation-exchange resin. *Appl. Environ. Microbiol.* **58:**2458–2462.

33. Janssen, P. H., P. S. Yates, B. E. Grinton, P. M. Taylor, and M. Sait. 2002. Improved culturability of soil bacteria and isolation in pure culture of novel members of the divisions *Acidobacteria, Actinobacteria, Proteobacteria,* and *Verrucomicrobia. Appl. Environ. Microbiol.* **68:**2391–2396.

34. Kaeberlein, T., K. Lewis, and S. S. Epstein. 2002. Isolating "uncultivable" microorganisms in pure culture in a simulated natural environment. *Science* **296:**1127–1129.

35. Kalyuzhnaya, M. G., R. Zabinsky, S. Bowerman, D. R. Baker, M. E. Lidstrom, and L. Chistoserdova. 2006. Fluorescence in situ hybridization-flow cytometry-cell sorting-based method for separation and enrichment of type I and type II methanotroph populations. *Appl. Environ. Microbiol.* **72:**4293–4301.

36. Keller, M., and K. Zengler. 2004. Tapping into microbial diversity. *Nat. Rev. Microbiol.* **2:**141–150.

37. Keller, M., K. Zengler, G. Toledo, J. Elkins, E. Mathur, and J. Short. 2002. Microbiology: what is next? *Geochim. Cosmochim. Acta* **66:**A390.

38. Kim, S. J., C. M. Lee, M. Y. Kim, Y. S. Yeo, S. H. Yoon, H. C. Kang, and B. S. Koo. 2007. Screening and characterization of an enzyme with beta-glucosidase activity from environmental DNA. *J. Microbiol. Biotechnol.* **17:**905–912.

39. Krause, L., N. N. Diaz, A. Goesmann, S. Kelley, T. W. Nattkemper, F. Rohwer, R. A. Edwards, and J. Stoye. 2008. Phylogenetic classification of short environmental DNA fragments. *Nucleic Acids Res.* **36:**2230–2239.

40. Kunin, V., A. Copeland, A. Lapidus, K. Mavromatis, and P. Hugenholtz. 2008. A bioinformatician's guide to metagenomics. *Microbiol. Mol. Biol. Rev.* **72:**557–578.

41. Lafferty, M. March 2006. GigaMatrix holding tray having through-hole wells. U.S. patent 7,019,827.

42. Lafferty, M., and M. J. Dycaico. 2004. GigaMatrix™: an ultra high-throughput tool for accessing biodiversity. *J. Assoc. Lab. Automation* **9:**200–208.

43. Lafferty, M., and M. J. Dycaico. 2004. GigaMatrix: a novel ultrahigh throughput protein optimization and discovery platform. *Methods Enzymol.* **388:**119–134.

44. Luo, C., H. Li, C. Xiong, X. Peng, Q. Kou, Y. Chen, H. Ji, and Q. Ouyang. 2007. The combination of optical tweezers and microwell array for cells physical manipulation and localization in microfluidic device. *Biomed. Microdevices* **9:**573–578.

45. Lynch, J. M. 1988. The terrestrial environment, p. 103–131. *In* J. M. Lynch and J. E. Hobbie (ed.), *Microorganisms in Action: Concepts and Applications in Microbial Ecology.* Blackwell Scientific Publishers, Oxford, United Kingdom.

46. Lynd, L. R., P. J. Weimer, W. H. van Zyl, and I. S. Pretorius. 2002. Microbial cellulose utilization: fundamentals and biotechnology. *Microbiol. Mol. Biol. Rev.* **66:**506–577.

47. Marcy, Y., C. Ouverney, E. M. Bik, T. Losekann, N. Ivanova, H. G. Martin, E. Szeto, D. Platt, P. Hugenholtz, D. A. Relman, and S. R. Quake. 2007. Dissecting biological "dark matter" with single-cell genetic analysis of rare and uncultivated TM7 microbes from the human mouth. *Proc. Natl. Acad. Sci. USA* **104:**11889–11894.

48. Morales, S. E., T. F. Cosart, J. V. Johnson, and W. E. Holben. 2009. Extensive phylogenetic analysis of a soil bacterial community illustrates extreme taxon evenness and the effects of amplicon length, degree of coverage,

and DNA fractionation on classification and ecological parameters. *Appl. Environ. Microbiol.* **75:**668–675.

49. Ogram, A., G. Sayler, and T. Barkay. 1987. The extraction and purification of microbial DNA from sediments. *J. Microbiol. Methods* **7:**57–66.

50. Palackal, N., C. S. Lyon, S. Zaidi, P. Luginbuhl, P. Dupree, F. Goubet, J. L. Macomber, J. M. Short, G. P. Hazlewood, D. E. Robertson, and B. A. Steer. 2007. A multifunctional hybrid glycosyl hydrolase discovered in an uncultured microbial consortium from ruminant gut. *Appl. Microbiol. Biotechnol.* **74:**113–124.

51. Pettersson, E., J. Lundeberg, and A. Ahmadian. 2009. Generations of sequencing technologies. *Genomics* **93:**105–111.

52. Podar, M., C. B. Abulencia, M. Walcher, D. Hutchison, K. Zengler, J. A. Garcia, T. Holland, D. Cotton, L. Hauser, and M. Keller. 2007. Targeted access to the genomes of low-abundance organisms in complex microbial communities. *Appl. Environ. Microbiol.* **73:**3205–3214.

53. Prieur, D. J. 1986. Tissue specific deficiency of lysozyme in ruminants. *Comp. Biochem. Physiol. B* **85:**349–353.

54. Raghunathan, A., H. R. Ferguson, Jr., C. J. Bornarth, W. Song, M. Driscoll, and R. S. Lasken. 2005. Genomic DNA amplification from a single bacterium. *Appl. Environ. Microbiol.* **71:**3342–3347.

55. Rajendhran, J., and P. Gunasekaran. 2008. Strategies for accessing soil metagenome for desired applications. *Biotechnol. Adv.* **26:**576–590.

56. Rappe, M. S., and S. J. Giovannoni. 2003. The uncultured microbial majority. *Annu. Rev. Microbiol.* **57:**369–394.

57. Richardson, T. H., X. Tan, G. Frey, W. Callen, M. Cabell, D. Lam, J. Macomber, J. M. Short, D. E. Robertson, and C. Miller. 2002. A novel, high performance enzyme for starch liquefaction. Discovery and optimization of a low pH, thermostable alpha-amylase. *J. Biol. Chem.* **277:**26501–26507.

58. Robertson, D. E., J. A. Chaplin, G. DeSantis, M. Podar, M. Madden, E. Chi, T. Richardson, A. Milan, M. Miller, D. P. Weiner, K. Wong, J. McQuaid, B. Farwell, L. A. Preston, X. Tan, M. A. Snead, M. Keller, E. Mathur, P. L. Kretz, M. J. Burk, and J. M. Short. 2004. Exploring nitrilase sequence space for enantioselective catalysis. *Appl. Environ. Microbiol.* **70:**2429–2436.

59. Rodriguez-Brito, B., F. Rohwer, and R. A. Edwards. 2006. An application of statistics to comparative metagenomics. *BMC Bioinformatics* **7:**162.

60. Rusch, D. B., A. L. Halpern, G. Sutton, K. B. Heidelberg, S. Williamson, S. Yooseph, D. Wu, J. A. Eisen, J. M. Hoffman, K. Remington, K. Beeson, B. Tran, H. Smith, H. Baden-Tillson, C. Stewart, J. Thorpe, J. Freeman, C. Andrews-Pfannkoch, J. E. Venter, K. Li, S. Kravitz, J. F. Heidelberg, T. Utterback, Y. H. Rogers, L. I. Falcon, V. Souza, G. Bonilla-Rosso, L. E. Eguiarte, D. M. Karl, S. Sathyendranath, T. Platt, E. Bermingham, V. Gallardo, G. Tamayo-Castillo, M. R. Ferrari, R. L. Strausberg, K. Nealson, R. Friedman, M. Frazier, and J. C. Venter. 2007. The Sorcerer II Global Ocean Sampling expedition: northwest Atlantic through eastern tropical Pacific. *PLoS Biol.* **5:**e77.

61. Sait, M., P. Hugenholtz, and P. H. Janssen. 2002. Cultivation of globally distributed soil bacteria from phylogenetic lineages previously only detected in cultivation-independent surveys. *Environ. Microbiol.* **4:**654–666.

62. Schutze, K., H. Posl, and G. Lahr. 1998. Laser micromanipulation systems as universal tools in cellular and molecular biology and in medicine. *Cell. Mol. Biol.* (Noisy-le-grand) **44:**735–746.

63. Sekar, R., B. M. Fuchs, R. Amann, and J. Pernthaler. 2004. Flow sorting of marine bacterioplankton after fluorescence in situ hybridization. *Appl. Environ. Microbiol.* **70:**6210–6219.

64. **Shiratori, H., H. Ikeno, S. Ayame, N. Kataoka, A. Miya, K. Hosono, T. Beppu, and K. Ueda.** 2006. Isolation and characterization of a new *Clostridium* sp. that performs effective cellulosic waste digestion in a thermophilic methanogenic bioreactor. *Appl. Environ. Microbiol.* **72:**3702–3709.

65. **Short, J. M.** January 2001. Screening methods for enzymes and enzyme kits. U.S. patent 6,168,919.

66. **Short, J. M.** January 2004. Protein activity screening of clones having DNA from uncultivated microorganisms. U.S. patent 6,677,115.

67. **Short, J. M.** September 1999. Protein activity screening of clones having DNA from uncultivated microorganisms. U.S. patent 5,958,672.

68. **Steele, H. L., K. E. Jaeger, R. Daniel, and W. R. Streit.** 2009. Advances in recovery of novel biocatalysts from metagenomes. *J. Mol. Microbiol. Biotechnol.* **16:**25–37.

69. **Stein, J. L., T. L. Marsh, K. Y. Wu, H. Shizuya, and E. F. DeLong.** 1996. Characterization of uncultivated prokaryotes: isolation and analysis of a 40-kilobase-pair genome fragment from a planktonic marine archaeon. *J. Bacteriol.* **178:**591–599.

70. **Thomsen, T. R., J. L. Nielsen, N. B. Ramsing, and P. H. Nielsen.** 2004. Micromanipulation and further identification of FISH-labelled microcolonies of a dominant denitrifying bacterium in activated sludge. *Environ. Microbiol.* **6:**470–479.

71. **Tokuda, G., and H. Watanabe.** 2007. Hidden cellulases in termites: revision of an old hypothesis. *Biol. Lett.* **3:**336–339.

72. **Torsvik, V., and L. Ovreas.** 2002. Microbial diversity and function in soil: from genes to ecosystems. *Curr. Opin. Microbiol.* **5:**240–245.

73. **Torsvik, V., L. Ovreas, and T. F. Thingstad.** 2002. Prokaryotic diversity—magnitude, dynamics, and controlling factors. *Science* **296:**1064–1066.

74. **Tsai, Y. L., and B. H. Olson.** 1991. Rapid method for direct extraction of DNA from soil and sediments. *Appl. Environ. Microbiol.* **57:**1070–1074.

75. **Tyson, G. W., J. Chapman, P. Hugenholtz, E. E. Allen, R. J. Ram, P. M. Richardson, V. V. Solovyev, E. M. Rubin, D. S. Rokhsar, and J. F. Banfield.** 2004. Community structure and metabolism through reconstruction of microbial genomes from the environment. *Nature* **428:**37–43.

76. **Venter, J. C., K. Remington, J. F. Heidelberg, A. L. Halpern, D. Rusch, J. A. Eisen, D. Wu, I. Paulsen, K. E. Nelson, W. Nelson, D. E. Fouts, S. Levy, A. H. Knap, M. W. Lomas, K. Nealson, O. White, J. Peterson, J. Hoffman, R. Parsons, H. Baden-Tillson, C. Pfannkoch, Y. H. Rogers, and H. O. Smith.** 2004. Environmental genome shotgun sequencing of the Sargasso Sea. *Science* **304:**66–74.

77. **Vives-Rego, J., P. Lebaron, and G. Nebe-von Caron.** 2000. Current and future applications of flow cytometry in aquatic microbiology. *FEMS Microbiol. Rev.* **24:**429–448.

78. **Warnecke, F., P. Luginbuhl, N. Ivanova, M. Ghassemian, T. H. Richardson, J. T. Stege, M. Cayouette, A. C. McHardy, G. Djordjevic, N. Aboushadi, R. Sorek, S. G. Tringe, M. Podar, H. G. Martin, V. Kunin, D. Dalevi, J. Madejska, E. Kirton, D. Platt, E. Szeto, A. Salamov, K. Barry, N. Mikhailova, N. C. Kyrpides, E. G. Matson, E. A. Ottesen, X. Zhang, M. Hernandez, C. Murillo, L. G. Acosta, I. Rigoutsos, G. Tamayo, B. D. Green, C. Chang, E. M. Rubin, E. J. Mathur, D. E. Robertson, P. Hugenholtz, and J. R. Leadbetter.** 2007. Metagenomic and functional analysis of hindgut microbiota of a wood-feeding higher termite. *Nature* **450:**560–565.

79. **Xie, C., J. Mace, M. A. Dinno, Y. Q. Li, W. Tang, R. J. Newton, and P. J. Gemperline.** 2005. Identification of single bacterial cells in aqueous solution using confocal laser tweezers Raman spectroscopy. *Anal. Chem.* **77:**4390–4397.

80. **Zengler, K., G. Toledo, M. Rappe, J. Elkins, E. J. Mathur, J. M. Short, and M. Keller.** 2002. Cultivating the uncultured. *Proc. Natl. Acad. Sci. USA* **99:**15681–15686.

81. **Zhou, J., M. A. Bruns, and J. M. Tiedje.** 1996. DNA recovery from soils of diverse composition. *Appl. Environ. Microbiol.* **62:**316–322.

Microalgal Culture as a Feedstock for Bioenergy, Chemicals, and Nutrition

SUSAN T. L. HARRISON, MELINDA J. GRIFFITHS, NICHOLAS LANGLEY,
CARYN VENGADAJELLUM, AND ROBERT P. VAN HILLE

40

40.1. INTRODUCTION

40.1.1. Benefit of Algal Culture for Renewable Products

Interest in algal culture stems back to the 1940s (15, 48) and formed a key component in the early developments of the bioprocess engineering discipline. Recently, renewed interest in algal research has centered on its potential role as a renewable feedstock for energy and commodity products while utilizing solar energy to fix CO_2. Algae also have a role in the production of niche fine chemicals and nutraceuticals.

Algal culture can be used to extend resources beyond those conventionally exploited by agricultural systems owing to algal culture using marginal land areas and water bodies. This prevents direct competition with food crops. While the theoretical efficiency of photosynthesis has been estimated at about 10% (19), plant crops typically yield an efficiency of 1 to 3%, while algal systems can approach the 10% level under low light conditions ($<1,000$ µmol photons m^{-2} s^{-1}) (9). This underlies the improved productivities of microalgal culture. In the biotechnology context, this can be demonstrated by comparing algal oil productivities with those of seed crops. Typical values for seed crops range from 0.6 liter m^{-2} year^{-1} (palm oil) through 0.1 to 0.2 liter m^{-2} year^{-1} (rapeseed, sunflower, and jatropha) to 0.04 liter m^{-2} year^{-1} (soya) (44). By contrast, algal oil productivities of 4.7 to 14 liters m^{-2} year^{-1} have been reported (45). Furthermore, ready access of the unicellular systems to water and nutrients as well as continuous harvesting of algae contributes to these superior productivities.

40.1.2. Potential Algal Product Range

Commercial interest in algal culture originated with low-value products, focusing on biomass as a source of protein or a source of energy. The large-scale application has been restricted by the cost of production of algal biomass, which ranges from US$15 to $20 kg^{-1} where open culture is feasible to US$50 to $600 kg^{-1} for selective cultivation of fastidious algae in closed systems (12). The current product range has been expanded to include fine chemicals and secondary metabolites as products of higher value as well as bioenergy products, the latter typically coupled to wastewater treatment or using carbon credits to offset costs.

The range of high-value algal products includes health foods, food additives, aquaculture supplements, and growth regulators. Examples are given in Table 1. In these cases, product value affords opportunity for controlled growth in sophisticated bioreactor systems.

While commodity products such as protein, algal oil, and bioenergy products remain of great interest, these require improvements in productivities to be economically feasible as independent products. Currently they are produced as byproducts or coproducts where economically feasible. Here the potential value of algal biomass as an energy source should be noted, as it has a lower heating value of 21 MJ kg^{-1} (Phyllis database for biomass and waste [http://www.ecn.nl/Phyllis]), comparing favorably with conventional biomass resources.

40.2. ALGAL RESOURCES

40.2.1. Overview of Algal Types With Respect to Product Range and Requirements

The term "algae" is loosely used to describe a huge variety of prokaryotic (strictly termed cyanobacteria) and eukaryotic organisms with wide-ranging morphologies and phylogenies. Microalgae are generally small, photosynthetic, heterotrophic or, occasionally, phagotrophic, and unicellular or colonial aquatic plants. The term is typically extended colloquially to include the cyanobacteria. Macroalgae are larger (can be seen without the aid of a microscope), multicellular, and eukaryotic and often show some form of cellular specialization. Some of the more well-known microalgal types are described in Table 2.

40.2.2. A Decision Framework for Algal Species Selection for Desired Product

The choice of algal species is a primary critical decision in any algal enterprise. Success requires selection of a species with both the appropriate productivity and the right characteristics for specific culture conditions. To assist the decision-making process, a decision-making framework is useful. Data on candidate algal species are collated for comparison and assessment. Product value is key in determining the relative weighting of traits of interest in algal culture and product recovery (e.g., capital cost tolerated, need for sterile processing, harvesting, and product liberation) and their influence on process economics (for a good review on this, see reference 11).

For large-scale outdoor pond culture typical for low-value products, one or more of the following traits are desirable in

TABLE 1 Examples of currently commercially available algal products[a]

Product metabolite	Commercial application(s)	Microorganism	Production system(s)	Geographical location(s)
β-Carotene	Pigment, vitamin A precursor	*Dunaliella salina*	Open ponds and raceways	Australia, China, India, Chile, United States, Israel
Astaxanthin	Pigment	*Chlorella*, *Haematococcus pluvialis*	Circular ponds	Taiwan, Japan, United States, Thailand
Phycocyanin	Blue colorant	*Spirulina*	Open ponds, natural lakes	
Eicosapentaenoic acid	Medical use	*Phaeodactylum tricornutum*		
Docosahexaenoic acid	Functional food	*Isochrysis galbana*, *Crypthecodinium*	Heterotrophic fermentation	United States
Fatty acids	Animal nutrition	*Isochrysis galbana*	Open ponds, PBR[b]	
Biomass	Health food	*Spirulina*	Open ponds	

[a]See reference 15.
[b]PBR, photobioreactor.

the algal species: fast growing, robust, optimized for the local culture conditions, able to thrive under conditions providing a niche habitat (thereby minimizing contamination), tolerant of a range of temperatures, and tolerant of a range of light intensities. Growth in an extreme environment (e.g., high temperature, pH, or salt) reduces contamination and predation, as a restricted spectrum of organisms grow under such conditions. This has largely contributed to the successful culture of *Dunaliella salina* (high salt tolerance) and *Spirulina* (high pH and salt). In smaller-scale, closed photobioreactors, the environment can be better controlled to minimize contamination.

Productivity, the product of growth rate and product content or concentration, has been demonstrated as an

TABLE 2 Characteristics of some algal groups relevant to microalgal biotechnology

Group (common name)	Description	Morphology	Dominant storage product	Well-known organisms	Applications[a]
Chlorophyta (green algae)	Diverse group. Origin of higher plants. Usually photoautotrophic but can be heterotrophic.	Large range of cellular structures—unicellular flagellates to complex multicellular arrangements. Cell walls contain cellulose.	Starch (α-1,4-glucan)	*Chlamydomonas*, *Chlorella*, *Scenedesmus*, *Spirogyra*, *Volvox*	*Chlorella* is commercially grown as nutraceutical and aquaculture feed. *Haematococcus pluvialis* is used to produce astaxanthin. *Dunaliella salina* is grown for β-carotene.
Cyanobacteria (blue-green algae)	Nonmotile, gram-negative, prokaryotic eubacteria. Most widely distributed algal group. Dominate particularly in oceans. Important component of picoplankton.	Can be unicellular, filamentous, or colonial.	Starch and cyanophycin (arginine and asparagine polymer)	*Spirulina* (*Arthrospira*), *Anabaena*, *Oscillatoria*	*Spirulina* biomass sold commercially to health food and nutraceutical market.
Dinophyta (dinoflagellates)	Important components of microplankton. Nutritionally diverse. Half the known species are obligate heterotrophs.	Typically unicellular flagellates. Have armor-like cell covering beneath cell membrane.	Starch (α-1,4-glucan)	*Gymnodinium*, *Crypthecodinium cohnii*	Known for blooms and toxin production; many exhibit bioluminescence. Responsible for red tides. Some species produce DHA.

(*Continued on next page*)

TABLE 2 (*Continued*)

Euglenophyta (euglenas)	Occur in freshwater, brackish water, and marine environments, mostly in soils and mud, especially in highly heterotrophic environments. Obligate mixotrophic, as require B vitamins. Colorless. Species are phagotrophic. More closely related to trypanosomes than any other algal group.	Unicellular or colonial flagellates.	Paramylon (β-1,3-glucan)	*Euglena gracilis*	Unique cellular and biochemical features; may have pharmaceutical applications.
Haptophyta (also known as Prymnesiophyta)	Generally marine. Mostly photosynthetic but can be heterotrophic or phagotrophic.	Largely motile unicells. The best-known haptophytes are cocco-lithophores, which have an exoskeleton of calcareous plates called coccoliths.	Chrysolaminaran (β-1,3-glucan)	*Pavlova lutheri, Isochrysis galbana, Prymnesium parvum*	Several species grown as food for fish and other aquaculture organisms such as bivalves and abalone.
Heterokontophyta (includes brown algae, golden algae, and diatoms)	Large group containing the *Chrysophyceae, Xanthophyceae, Eustigmatophyceae,* and *Bacillariophyceae* (diatoms). Largely marine, but some freshwater varieties.	Cells with two different flagella (as opposed to iso-kont—two the same). *Bacillariophyceae* are unicellular, brown cells with a silica cell wall (unique type of casing made of two frustules that fit together like a lid on a box).	Chrysolaminaran (β-1,3-glucan)	*Amphora, Nitzschia, Thalassiosira pseudonana, Phaeodactylum tricornutum, Nannochloropsis*	Several species grown as aquaculture feed; others known for EPA production.
Rhodophyta (red algae)	One of oldest and largest groups of algae. The accessory pigments phycobiliproteins give them their red color.	Free-living, unicellular.	Starch (α-1,4-glucan)	*Porphyridium purpureum*	Some species secrete calcium carbonate—could be used in CO_2 sequestration. Most economically important macroalgae are from this family, e.g., dulse (*Palmaria palmata*), nori (*Porphyra*), and species used to make agar, carrageenans, and other food additives.

[a]DHA, docosahexaenoic acid; EPA, eicosapentaenoic acid.

important characteristic which should be considered in preference to growth rate, biomass concentration, and product content individually (23). Biomass harvesting represents a significant capital and operating cost in the algal process owing to the typically low biomass concentration and associated large culture volumes. The alga may be selected for properties to simplify harvesting, e.g., filamentous morphology, large cell size, high specific gravity, and cell surface properties that promote flotation or autoflocculation. The liberation of intracellular products is facilitated by a species that can be easily disrupted. This is influenced by cell wall type and size. Typically, algal species are more difficult to disrupt than bacteria or yeasts (5), a notable exception being *Dunaliella*. Other desirable characteristics include tolerance of shear force, the ability to outcompete contaminants, the production of autoinhibitors, and a high rate of CO_2 uptake (particularly for CO_2 sequestration applications).

The desirable algal characteristics should be ranked relative to other species, ideally by numerical quantification. These features are weighted depending on their importance and relevance to issues such as cost and impact on the large-scale feasibility of establishing a process for the product under consideration. The composite weighted score for each species allows direct comparison. The use of published information to compare species can present challenges. For example, data are often presented using different units of measurement, without sufficient information for their interconversion unless broad assumptions are made (e.g., conversion between areal and volumetric productivities cannot be made without information on culture depth). Studies may have been conducted to meet distinct objectives, using different strains (not always identified) and degrees of optimization of the culture conditions. Furthermore, species may have different inherent optimum nutrient levels and pH, temperature, and light requirements, and they may not reach their maximum potential under standard conditions.

40.2.3. Acquiring Algae from Culture Collections and Natural Habitats

Microalgae grow in almost every habitat in the world. They can be isolated from rivers, lagoons, dams, pools, salt lakes, oceans, springs, soils, rocks, animals, plants, and even snow and ice. There are two main types: planktonic (floating) and benthic (attached to a solid substrate). Samples should be labeled with locality, date of collection, and as much information on the environment as possible, e.g., whether the water is salt, brackish, or fresh; temperature; and whether the algae are free-floating or attached.

Algal samples can be stored in glass or plastic bottles, vials, or bags. Wide shallow containers are better than tall narrow ones, as they facilitate gas exchange. Samples should be examined under the microscope for initial identification and choice of isolation method. The isolation techniques described below can be used to eliminate predators and obtain monocultures of algae or, preferably, axenic cultures, i.e., single algal species also free of bacteria and other microorganisms.

Single colonies can be isolated by streaking samples onto agar plates, incubated in the light. These isolated colonies may then be used to inoculate liquid media. However, not all algal species grow on agar media. Further, solid media may also facilitate the growth of bacteria and fungi, so colonies should be transferred with care to fresh plates as soon as they appear (days to weeks). An alternative method to obtain pure culture, not utilizing solid media, is serial dilution. Cell concentration is determined by direct counting

and appropriately diluted such that a single algal cell can be transferred into a test tube, flask, or multiwell plate. Several tubes or flasks should be seeded, as some may contain no cells, while others will contain more than one. On the basis of probability, some should contain a single cell and hence yield an axenic culture containing a single algal species. Nutrient enrichment of the culture medium can be used to create an environment where certain species grow in preference to others. These may include the addition of ammonia, selenium, soil-water extract, macronutrients, organic substrates, or trace metals. Isolation by micropipette can be achieved using a glass pipette or fine capillary tube. This fine tip is used to isolate single cells from a drop of liquid to a multiwell plate, agar, or microscope slide. Transfer through a series of sterile droplets is used to dilute the sample until a single cell can be selected. Alternatively, unwanted cells can be removed from the droplet, leaving the target cell free of contaminants. The diameter of the pipette opening should be at least twice that of the cell to prevent shear damage. To remove adhering material from filamentous algae, a hook can be prepared on the end of a Pasteur pipette and used to drag the filament into and through an agar plate. Flow cytometry with cell sorting is recognized as a high-throughput automated technique for isolation of single cells. Density centrifugation can be used to separate the species of interest from contaminants of different size or density. Certain flagellates are known to swim toward or away from light. This can be used to isolate them by application of a focused light source to one end of a channeled culture, through phototaxis. Chemical agents such as germanium dioxide can be used to kill diatoms contaminating cultures of other algal species. Similarly, antibiotics are often effective at eliminating cyanobacteria. A combination of antibiotics or other antimicrobial agents can be used to reduce the number of bacteria in an algal culture without killing all the algae. Exposing the culture to UV light can also be used to eliminate bacteria, based on their different response to UV light from algae. The lower end of the UV-B spectrum and UV-C light are most effective, and exposure times vary from 1 to 30 min.

Typically, a selection protocol composed of the following steps, in the order given, is used for isolation and purification of microalgae to achieve a virus-free, axenic culture (adapted from reference 27).

1. Examination for predators and debris
2. Size/density separation
3. General enrichment (most algae) or selective enrichment (certain species)
4. Isolation by agar streaks, micropipetting, dilution, flow cytometry, centrifugation, etc., to achieve an algal monoculture
5. Antibiotic or UV treatment to achieve axenic culture
6. Further treatment to remove viruses

Pure algal monocultures can be ordered from culture collections around the world. The site http://en.wikipedia.org/wiki/List_of_algal_culture_collections provides a list of appropriate culture collections. Popular ones include the CSIRO culture collection in Australia (http://www.marine.csiro.au) and the collections of the University of Göttingen (http://www.epsag.uni-goettingen.de) and University of Cologne (http://www.ccac.uni-koeln.de) in Germany, the University of Toronto in Canada (http://www.botany.utoronto.ca/utcc), Algobank in France (http://www.unicaen.fr/ufr/ibfa/algobank), the National Institute for Environmental Studies in Japan (http://mcc.nies.go.jp), the Culture Collection of Algae and Protozoa in the United Kingdom (http://www.ccap.ac.uk), and UTEX in the United States (http://web.biosci.utexas.edu/utex/).

40.2.4. Culture Maintenance

The most common maintenance method for microalgal cultures is routine serial subculturing, also known as routine serial transfer (39). A culture in late log or stationary phase is subcultured into fresh growth medium (35). The newly transferred cultures are usually placed on lighted shelves until a desired colony size or culture density is achieved, and then they are stored away from light until needed. Indirect natural daylight or solely artificial light can be used for maintaining cultures. In some instances, a day/night cycle of 12 h each is used (1). The transfer time varies with species from 2 weeks to 6 months. Serial transfer is used to maintain healthy physiologically, morphologically, and genetically representative populations. Therefore, a balance is required between sustaining active stock cultures and extending the period between transfers. The serial transfer method is labor-intensive and expensive for maintaining a large number of strains. Other limitations include possible loss of important morphological features or physical traits and the possibilities of contamination and mutation (35, 39). Despite these limitations, a number of the large international culture collections such as the Culture Collection of Algae and Protozoa in the United Kingdom, Canadian Phycology Culture Centre in Canada, and UTEX in the United States still employ this method. Microalgal strains which grow readily on agar are maintained on agar slants. Long-term culture storage methods such as encapsulation and cryopreservation have been investigated in order to prevent genetic drift (18). The encapsulation method entraps microalgal cells in a calcium alginate gel to extend transfer time to 12 to 36 months; however, it is very labor-intensive and not suitable for large numbers of strains (18, 39). Under cryopreservation, algae are held at an ultra-low temperature while remaining capable of survival upon thawing. Most cryopreservation techniques include two steps: slow cooling at $1°C\ min^{-1}$ in a $-86°C$ freezer, followed by submersion in liquid nitrogen. This has been used successfully for a range of cyanobacteria and unicellular microalgae (1, 18). However, many large and complex algae cannot be cryopreserved, as the mechanism of cell injury during freeze-thaw cycles is not fully understood (18). Microalgal strains which grow readily on agar can be cryopreserved directly on agar. The most commonly used cryoprotectant for microalgae is methanol, not glycerol, which is used conventionally for bacteria. The viability of strains after cryopreservation varies, and it is recognized that the technique of cryopreservation will benefit from further development.

Recently, a new method of maintaining microalgal cultures was proposed, employing dual-layer 96-well microtiter plates (39). The cultures were successfully maintained by immobilization onto membrane filters fitted to individual wells (source layer), allowing passage of nutrients from the underlying growth medium (substrate layer) housed in the bottom of the chamber. Numerous axenic cultures were maintained in this manner.

40.3. BASIC REQUIREMENTS OF ALGAL CULTURE

40.3.1. Identifying Key Culture Variables

While optimal microalgal culturing systems vary greatly with application, the following characteristics should be

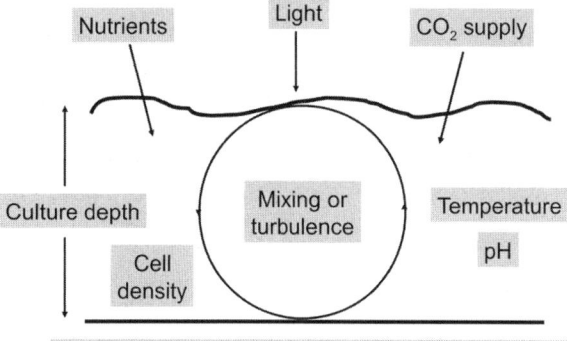

FIGURE 1 Parameters that can be manipulated in algal culture. (Adapted from reference 25.)

considered for culture system design and operation: provision of CO_2 and removal of O_2, provision of nutrients, temperature, pH control, salinity, light provision, mixing, mass and heat transfer, and hydrodynamics. The interaction of these variables is illustrated in Fig. 1. Their optimal and tolerated ranges tend to be species specific.

40.3.2. Light—Provision and Its Intensity

Sufficient light is vital for photosynthesis, the sole source of energy and carbon for autotrophic algae. As the depth in the cell culture increases, the penetration of light through the system decreases. This creates two effective lighting regions in a culture. In the photic volume, light is sufficient for photosynthesis, while in the dark region light intensity is below the threshold value for photosynthesis and photolimitation occurs. The cycling of cells between these volumes can increase algal productivity (42). By increasing the ratios of surface area to volume in the culture system, the exposure of the culture to light is increased. Furthermore, penetration of light through the culture decreases exponentially with increasing cell concentration (42) and is a decreasing function of liquid depth and absorbance capacity of the suspending medium and an increasing function of irradiance intensity. The combination of these parameters alters the ratio of photic and dark regions and is recognized as important for achieving high algal growth rates (15, 42).

Increasing the incident irradiance on the system to overcome photolimitation in dense algal cultures improves light penetration through the photic volume, increasing the algal growth rate (22) until the threshold value is surpassed beyond which photoinhibition occurs (9, 24). This threshold value for photoinhibition is highly dependent on algal species, previous culturing conditions, and the hydrodynamic regime within the system, the last affecting light cycling. An example of this is illustrated in Fig. 2. Many algal species cannot tolerate direct sunlight. Light intensities generally employed in the laboratory are 100 to 200 $\mu mol\ m^{-2}\ s^{-1}$ (about 5 to 10% full sunlight, 2,000 $\mu mol\ m^{-2}\ s^{-1}$). Many algae have been reported to grow better under a light:dark cycle (between 16:8 h and 12:12 h) than continuous light; however, contradictory reports are also found.

Most green algae rely primarily on the pigments chlorophyll a and chlorophyll b to harvest light for photosynthesis. The absorption spectra of chlorophylls a and b show strong absorbance in the blue (410 to 500 nm) and red

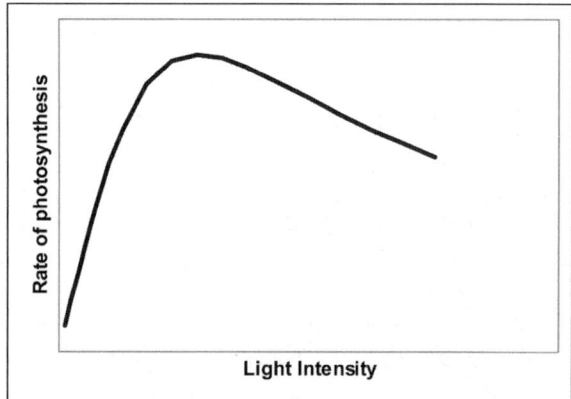

FIGURE 2 Dependence of photosynthetic rate on light intensity, illustrating limitation and inhibition.

(620 to 700 nm) regions and very low absorbance in the green region (500 to 600 nm) (31). Another well-studied pigment, β-carotene, exists in nearly all algal strains and absorbs light between 400 and 500 nm (31). The efficiency at which light is harvested by a certain pigment varies with the wavelength of light supplied (31). Although studies undertaken focus on the quantity of light supplied to cultures, the quality of this light should also be of concern.

Measurements of solar irradiance show a relatively even distribution of radiation across the visible spectrum (31). Most types of lightbulbs, on the other hand, emit light in a more confined region of the visible spectrum (22). Some of the most commonly suggested bulbs for algal cultivation on a laboratory scale are fluorescent tubes, metal halide bulbs, standard incandescent bulbs, and halogen lamps (10, 14, 22). Both standard incandescent bulbs and halogen lamps emit small amounts of blue light, while metal halide bulbs emit small amounts of red light. Fluorescent tubes offer the most practical lighting option, as they put out small amounts of heat. In addition, fluorescent tubes have become available with modified phosphor coating emitting a more evenly distributed spectrum similar to natural daylight; however, they have a relatively low light output (22). More work is needed to establish to what extent the quality of light affects algal growth.

40.3.3. Hydrodynamics—Mixing and Mass Transfer

When a microalgal culturing system is designed, it is important that the hydrodynamic regimes in the system be considered, primarily to prevent settling, expose algae to light, and enhance mass and heat transfer, especially gas-liquid mass transfer. Dead zones in the photobioreactor can result in undesirable settling of the cells (15). Appropriate mixing and circulation times are required, as is optimum cycling between light and dark zones. Care must be taken not to subject shear-sensitive cells to excessive shear stress, as this can cause decreased growth rate, metabolic activity, and even cell disruption (53).

40.3.4. CO₂ Provision and Gas-Liquid Mass Transfer

The rate of transfer of CO_2, the primary carbon source in gas-sparged autotrophic systems, from the gas into the liquid frequently limits the rate of growth of algal biomass (48). New culturing systems designed to facilitate CO_2 mass transfer at a sufficient rate to satisfy the cell growth rate are required. The rate of CO_2 mass transfer (CTR), given

in equation 1, from gas to liquid is dependent on the concentration driving force between the saturation (C_s) and instantaneous (C_L) CO_2 concentrations in the liquid and the volumetric mass transfer coefficient (k_La) (31).

$$\text{CTR} = k_La.(C_s - C_L) \tag{1}$$

The volumetric mass transfer coefficient describes the CO_2 mass transfer capability of a particular system and is a strong function of bubble size, agitation rate, temperature, and superficial gas velocity through the system (31). It can be estimated from a number of empirical correlations (52). Typical values reported for algal photobioreactors range from 0.002 to 0.020 s^{-1} (48). The saturation concentration is directly proportional to the concentration of CO_2 in the gas stream and can be calculated using Henry's law (31). This indicates that mass transfer limitations can be partially alleviated by using CO_2-enriched gas. It must be noted that the recovery of CO_2 from the gas stream is incomplete owing to the need for a driving force for mass transfer. Further, where CO_2-enriched gas is used, the fractional recovery of CO_2 from the gas stream may be expected to decrease despite improved gas transfer rates. The extent of the recovery is a function of the residence time of the gas stream in the reactor.

40.3.5. Nutrient Provision—Medium Requirements

Microalgal medium is composed of macronutrients, microelements, and vitamins. Macronutrients are required in relatively large amounts and include nitrogen, phosphate, and silicon, the last required only for diatoms, silicoflagellates, and some chrysophytes. Nitrogen is typically provided as nitrates, ammonia, or urea, while phosphorus is provided as potassium phosphate. Silicon, added as $Na_2SiO_3 \cdot 9H_2O$, enhances precipitation and hence should be minimized. Macronutrients are generally required in a ratio of 16N:16Si:1P for diatoms, although typically a single macronutrient is set as limiting (3).

Microelements are required in trace amounts. They are usually prepared as a stock solution of chloride or sulfate salts of zinc, cobalt, manganese, selenium, and nickel and kept in solution by a chelator such as EDTA. Iron is also added to marine culture media, usually as a separate solution.

Three vitamins are generally used for algal culture: B_1 (thiamine), B_{12} (cyanocobalamin), and H (biotin) (51). Many algae require only one or two of these, although all three are typically supplied at the laboratory scale. Stock solutions of vitamins should be filter sterilized and added to media after autoclaving, as they are destroyed by heating.

Comprehensive listings of media, trace elements, and growth factors are given by Andersen (3). Frequently used laboratory media for freshwater and marine algae are listed in Table 3.

Freshwater media, tailored to specific algal needs, typically include macronutrients, a micronutrient mix, and distilled water, with a buffer if necessary. The pH of the

TABLE 3 Frequently used general-purpose media for microalgae[a]

Freshwater	Marine
Bristol medium	F/2 medium
Bold's basal medium	Enriched seawater medium
Proteose medium	Erdschreiber's medium
Soil extract medium	Artificial seawater medium

[a]Recipes available on the UTEX website: http://web.biosci.utexas.edu/utex/.

media is adjusted to specification prior to sterilization by heat or filtration. Several media include a soil extract, made by heating, boiling, or autoclaving soil in water and then filtering. This contains a complex mixture of nutrients and is useful when precise knowledge of medium composition is not necessary.

Two main types of marine culture media can be identified, natural and artificial. Natural media use seawater, while artificial seawater recipes attempt to mimic the composition of natural seawater using laboratory chemicals. Natural or simulated seawater forms the basis of the media, to which nutrients, microelements, and vitamins are added.

Natural seawater is collected offshore to avoid pollution and variation in composition. The salinity and nutritional value of collected seawater may vary depending on season and collection point and should be tested. Seawater should be filtered as soon as possible (using 0.45- or 0.22-μm-pore-size filters) and stored at 4°C in the dark. If the exact composition of the medium is important, artificial seawater should be used.

Seawater is often heat sterilized at 121°C. Precipitation of salts may occur during this process, rendering key nutrients unavailable to the algae. The removal of CO_2 during heating with a concomitant pH increase to above pH 10 causes precipitation. Adding acid or a buffer or saturating the medium with CO_2 before autoclaving can reduce precipitation. Rapid cooling of medium also reduces precipitation. Alternatively, other sterilization techniques such as filtration, pasteurization, or microwaving can be used.

Several algal species can be grown heterotrophically, obtaining carbon and energy from an organic carbon source instead of, or in addition to, photosynthesis. This is advantageous, as it reduces and in some cases eliminates the requirement for light. A list of commercially important algal species that can be grown on organic carbon sources such as glucose, acetate, ethanol, glycerol, yeast extract, and various amino acids is given by Lee (33).

40.3.6. Physiological Culture Conditions: Temperature, pH, and Salinity

The optimal temperature for algal species is often related to the environment in which they naturally occur. Temperatures between 16 and 35°C are most common, although polar algae may grow at less than 10°C and growth of thermophilic algae is reported to 62°C, depending on the species (14, 29). Deep-ocean species (e.g., picoplankton) generally prefer lower temperatures than freshwater pond species (e.g., *Chlorella*).

The pH range of most algae lies between pH 7 and 9. CO_2 supply and consumption largely determines the pH of a culture owing to its speciation on dissolution.

$$CO_2 + H_2O \leftrightarrow H_2CO_3$$
$$H_2CO_3 \leftrightarrow HCO_3^- + H^+ \quad (2)$$
$$HCO_3^- \leftrightarrow CO_3^{2-} + H^+$$

While salinity may be used as a selection pressure for halophilic algae of interest, it may also inhibit the growth of nonhalophilic, freshwater algal species. Control of salinity is important in systems where evaporative losses are significant and the makeup water contains dissolved solutes (seawater, brackish water, wastewater, and brines). Marine algae are reasonably tolerant to changes in salinity. Optimal levels are generally slightly below that of seawater, 20 to 24 g NaCl liter^{-1}. Halophilic species such as *Dunaliella* undergo lysis in the absence of salt.

40.4. ALGAL CULTURE SYSTEMS

40.4.1. Laboratory Culture Systems

Many types of photobioreactors are suitable for laboratory culturing of algae. Their suitability depends on application. Shake flasks and sparged bottles provide very simple systems for algal culturing but generally result in low rates of mass transfer. Algae can be grown in bubble columns; however, external or internal loop airlift reactors are more effective as photobioreactors for algal growth owing to the development of defined flow patterns with a concomitant increase in mass and heat transfer (30). Flat plate reactors increase the ratios of surface area to volume and result in high volumetric algal productivities (15). They remain impractical in small spaces. If larger-scale laboratory cultivation is required, open raceway ponds can be used where sterilization is not required.

40.4.2. Large-Scale Culture Systems

Large-scale cultivation systems fall into two main categories: outdoor ponding systems and closed photobioreactors. Commercial-scale operations for low-value products typically require the ability to produce tons of algal biomass. Based on productivity values reported for most species, this requires culture volumes ranging from 10,000 to in excess of a million liters (13). For this reason, almost all commercial microalgal enterprises employ relatively low-cost open ponding systems. While naturally occurring shallow ponds may be used, in most cases ponds are constructed. Constructed ponds vary in complexity from large unlined, unmixed ponds, frequently used for *Dunaliella* cultivation, to mechanically agitated, circular or raceway ponds. Rectangular ponds, up to 1 ha in size, divided by a central baffle and agitated with a single paddle wheel represent the most efficient and widely utilized design. These have been applied at a scale of 5,000 tons of algal biomass per year. Photobioreactors are typically closed, or mostly closed, systems which facilitate a far greater degree of control, resulting in significantly enhanced productivities. Increased capital and operating costs, influenced by the construction materials and energy requirements, currently limit their operation to high-added-value compounds. In Table 4, the benefits and disadvantages of open ponding systems and closed reactor systems for algal growth are compared.

The range of reactor configurations includes tubular, flat plate, and cylindrical systems. Example productivities reported in these systems are given as a function of size and light path across the systems in Table 5.

A number of design considerations are relevant to both types of large-scale culture systems. These include light provision, algal circulation, regulation of pH and temperature, and minimization of contamination. The efficient provision of CO_2 and removal of O_2 are more critical for photobioreactor systems, given the enhanced productivities (8).

Chlorophyll is characterized by a high extinction coefficient, meaning that it is very efficient at absorbing light (31). As a result, light penetration is limited at high algal concentrations. Consequently, the effective depth of open ponds is limited to a maximum of 20 to 30 cm for the majority of commercially important species, with this depth depending on appropriate mixing. Mixing is critical in open ponds to prevent biomass settling, temperature and oxygen stratification, and cyclic exposure to light. Biomass productivities up to 10-fold higher have been recorded in well-mixed systems relative to similar unmixed ones. The most

TABLE 4 Comparison of open and closed reactor systems for algal growth[a]

Reactor type	Advantages	Disadvantages
Open ponds	Lower cost to establish Low operating cost Suitable for mass cultivation	Low biomass concentrations (0.1–0.5 g liter^{-1}) Poor productivity Limited control of culture conditions Weather dependent Easily contaminated Occupy large land areas High evaporative losses
Closed photobioreactors	Improved control of culture conditions Reduced contamination Increased biomass concentration (2–8 g liter^{-1}) Increased productivity Reduced land area required Reduced evaporative losses	Increased cost of manufacture Buildup of dissolved oxygen which may cause inhibition Fouling of walls by biomass limits light penetration, hence requiring cleaning Increased shear stress

[a]Compiled from references 41 and 48.

TABLE 5 Variation in biomass productivity with photobioreactor systems reported for algal culture[a]

Photobioreactor type	Alga used	Reactor vol (liters)	Length of light path (cm)	Max volumetric biomass productivity reported (g liter^{-1} day^{-1})
Open pond	Spirulina platensis		13–15	0.18
	Chlorella sp.		1	2.50
Tubular photobioreactor, horizontal	Spirulina maxima		12.3	0.25
	Spirulina platensis		2.5	1.60
	Isochrysis galbana		2.6	0.32
	Phaeodactylum		6.0	2.02
	Phaeodactylum		3.0	2.76
Tubular photobioreactor, inclined	Chlorella pyrenoidosa		2.5	2.90
	Chlorella pyrenoidosa		1.2	3.64
	Chlorella sorokiniana	6		1.47
Tubular photobioreactor, airlift	Porphyridium cruentum	200		1.50
	Phaeodactylum tricornutum	200		1.20
	Phaeodactylum tricornutum	200		1.90
Helical coil photobioreactor	Phaeodactylum tricornutum	75		1.40
	Tetraselmis chuii		24	1.20
Column photobioreactor, vertical	Haematococcus pluvialis	55		0.06
	Phaeodactylum		20	0.69
	Isochrysis galbana		2.6	1.60
Flat plate bioreactor	Spirulina platensis		10.4	0.30
	Spirulina platensis		1.3	4.30
	Spirulina platensis		3.2	0.80
	Nannochloropsis sp.	440		0.27

[a]Adapted from references 33 and 48.

important engineering considerations related to mixing in open ponding systems are reviewed by Borowitzka (13). Photobioreactors are often illuminated using artificial light sources, and most designs are configured to reduce light path length to a minimum. Mixing is most often achieved using the airlift principle. Control of pH and temperature is possible in photobioreactors but is limited in open systems, where evaporation and subsequent changes in salinity are more important.

Contamination and predation are important considerations for open systems, where maintenance of axenic cultures is impossible. Most of the species currently cultivated at a commercial scale are grown under highly selective conditions of high pH (*Spirulina*), salinity (*Dunaliella*), or nutrient loading (*Chlorella* and *Nannochloropsis*). Cultivation of genera such as *Haematococcus*, or lipid-producing species, under more benign conditions has been prone to contamination in pilot-scale operations. Predation by shrimp, rotifers, and amoebae can significantly reduce algal productivity, but in most cases these can be controlled by short-term adjustment of pH or elevation of salinity levels (7). Closed photobioreactors are less prone to gross contamination, although care must be taken to avoid contamination with competing algal species.

Prevailing climatic conditions are an important consideration for large-scale open ponding systems, particularly where drying of the biomass is required. Excessive humidity prevents efficient solar drying, while areas with high rainfall are unsuitable due to dilution, particularly in the case of halophilic species. Arid regions may appear attractive due to increased sunlight and a longer growing season; however, evaporative water loss from shallow ponds may make these systems unsustainable unless cultivated on a readily available, nonpotable water source. Further, where saline water is used, evaporation may cause concentration of salts.

40.5. MEASURING BIOMASS AND PRODUCT FORMATION

40.5.1. Measuring Biomass Concentration

Microalgal biomass concentration is typically presented as either cell concentration (cells per milliliter) or biomass concentration, normally in terms of cell dry weight per volume (kilograms per cubic meter). The most direct method of quantification remains direct cell counting, using specialized cell counting chambers and an optical microscope. The counting chamber consists of a microscope slide into which a grid representing a specific volume has been etched. Cells are counted within a predetermined number of fields on the grid, and the concentration is calculated using the specific volume of the fields. A stain, such as Lugol's solution, may be added to assist visualization and reduce motility. Direct counting is time-consuming, can be subjective (particularly when concentration or dilution of the sample is required), and may not be appropriate for colonial or filamentous algae. Some of these disadvantages can be overcome using automated particle counters, typically Coulter counters, or flow cytometry. Both of these approaches are increasingly used.

Gravimetric methods of biomass quantification are relatively simple but can be time-consuming. Cells are separated from the liquid medium (typically by centrifugation or filtration), washed to remove dissolved solids, dried, and weighed. Biomass is calculated by subtracting the mass of the preweighed tube or filter. This method is most suitable for axenic cultures, and replicate samples should be taken to account for inherent variability. Its accuracy is limited where dilute algal cultures are used.

In order to interconvert algal cell concentrations on a number and mass basis, it is necessary to determine a mass per cell for the particular species and culture conditions of interest through the determination of the cell concentration by both approaches. The typical absence of such data in the literature restricts overall data comparison.

As both direct counting and gravimetric methods are time-consuming, it is often desirable to use a proxy method, quantitatively correlated by means of a standard curve. Popular proxy methods include measuring optical density or extracted pigment concentration. Optical density is measured at 580 nm to minimize the impact of pigment content of the cells. Chlorophyll pigments in suspensions are typically quantified at 450 nm rather than 680 nm, avoiding impact of turbidity at 660 nm, while carotenoids are measured between 480 and 520 nm, as shown in Fig. 3. The correlation of these methods must be evaluated on a species-by-species basis. Bacterial contamination, precipitation, and medium turbidity interfere with absorbance, while physiological status and life cycle can affect pigment content.

Recent developments in the field of molecular biology have allowed quantification of microbial species using quantitative PCR (28). The process requires complete DNA extraction from all component species and relies on specific amplification of a portion of a selected gene, typically the 16S or 18S rRNA gene or intergenic spacer region. A fluorescent DNA binding dye or reporter probe is added to the PCR mixture, and the increase in signal intensity with successive cycles can be used to quantify the template concentration. While the technique is powerful, allowing simultaneous quantification of several species, and is not affected by bacterial contamination of the culture, it is expensive. Further, it is effective only if the gene sequence of the target organisms is known and specific primers or probes have been designed (34).

40.5.2. Algal Kinetics and Productivity

Where asexual cell division leads to exponential growth, the algal growth may be described by the Malthus equation (equation 3). The rate of growth is then reported as doubling time, T_d (in hours or days), or specific growth rate, μ (per day). These are related through equation 4.

$$dN/dt = \mu N \qquad (3a)$$

$$N_t = N_0 \, e^{\mu t} \qquad (3b)$$

$$\mu = \ln(2)/T_d \qquad (4)$$

N_0 is the initial cell concentration (as cell number; similarly, we can write the equation in terms of X, the cell concentration on a mass basis), and N_t is the cell concentration at time t. Where the culture is exposed to continual lighting, these equations may be used in the same manner as heterotrophic bacterial growth. However, where a diurnal pattern is used, data collection for determination of growth kinetic constants must take this pattern into account, sampling at the same point in each photoperiod.

Volumetric biomass productivity, Q_x (in grams per liter per day), provides a measure of space-time effects and is the

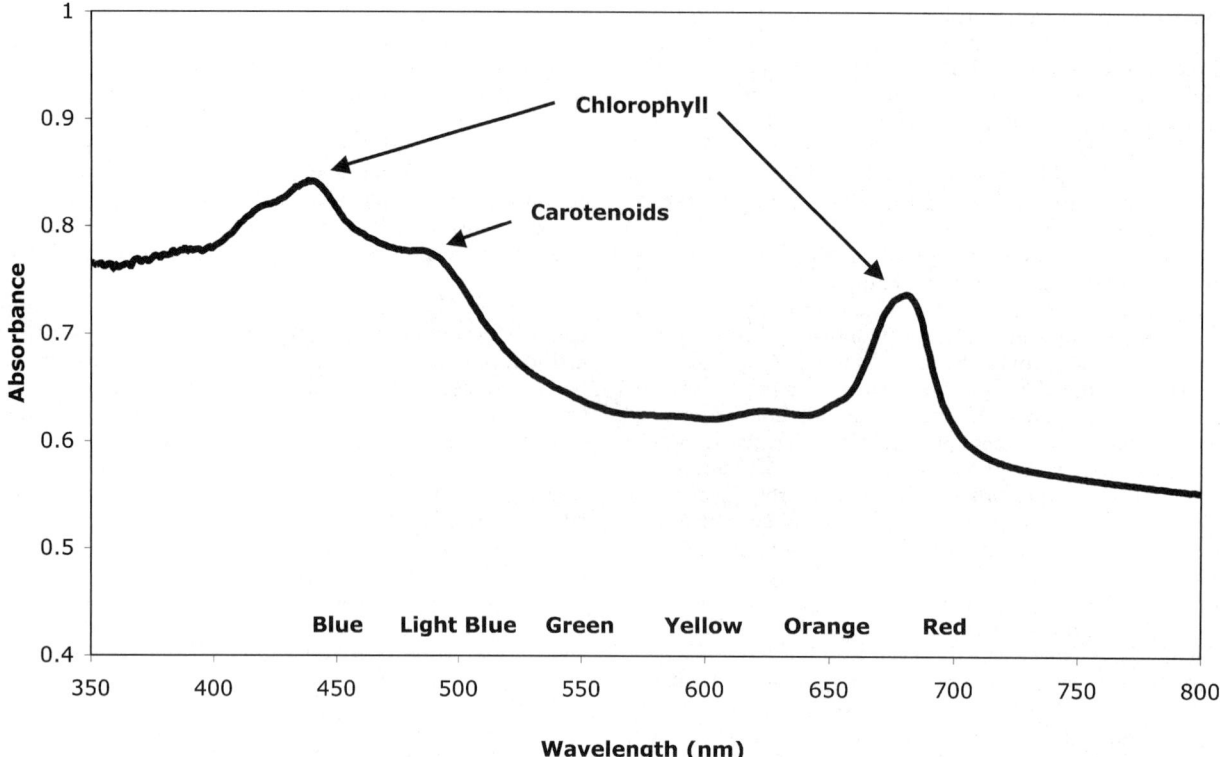

FIGURE 3 Average absorbance across the range of visible light wavelengths for five microalgal species (*Chlorella*, *Scenedesmus*, *Nannochloropsis*, *Isochrysis*, and *Phaeodactylum*).

product of the specific growth rate, μ (per day), and biomass concentration, X (in grams per liter):

$$Q_x = \mu X \qquad (5)$$

Biomass productivity may be reported on an areal (grams per square meter per day) or volumetric (grams per liter per day) basis. The choice of unit depends on the purpose of the study (e.g., optimizing areal footprint versus water requirements), but for purposes of comparison across reactor types, volumetric productivity seems the most appropriate. In order to convert areal productivity to volumetric productivity, culture depth is required; however, this is frequently not reported.

Where a product other than the algal biomass is of interest, product productivity is an important characteristic. This is the product of biomass productivity and product content. Use of specific growth rate, biomass concentration, and product content individually are of limited value

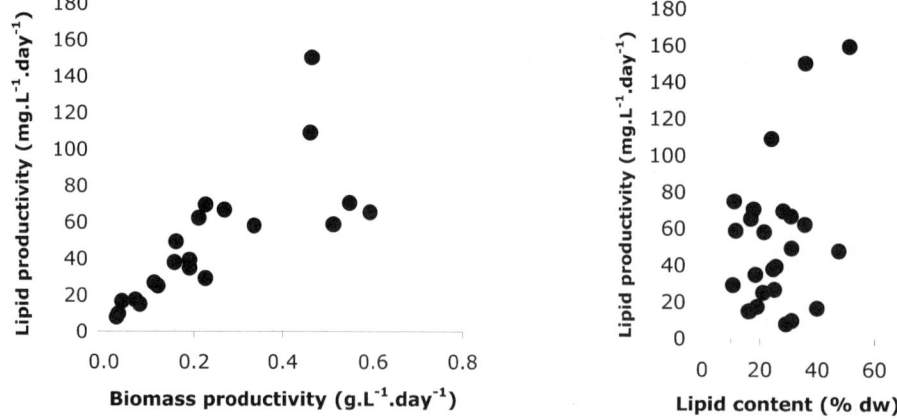

FIGURE 4 Correlation between algal oil productivity, biomass productivity, and oil content under nutrient-replete conditions (23).

TABLE 6 Typical algal growth rates and example product contents and productivities[a]

Algal class (no. of species considered)	Doubling time (h)	Specific growth rate (day^{-1})	Biomass productivity (g liter^{-1} day^{-1})	Lipid content (%, dry wt)	Lipid productivity (mg liter^{-1} day^{-1})
Green algae (17 species)	24 (6–72)	0.69	0.24	23 (13–31)	66
Cyanobacteria (5 species)	17 (7–32)	0.96	No data	8 (5–13)	No data
Other taxa, including Dinophyta, Heterokontophyta, Haptophyta, Rhodophyta, and Euglenophyta (33 species)	18 (5–44)	0.92	0.22	25 (11–51)	44
Freshwater spp. (22 species)	20 (5–72)	0.83	0.26	22 (5–36)	54
Marine spp. (33 species)	19 (7–44)	0.88	0.21	24 (7–51)	41

[a]Results are given as average of species number indicated and as range, the latter in parentheses. Adapted from reference 23.

as indicators of algal productivity. For example, a culture with a high product content but low growth rate could yield less product per unit time than a faster-growing species with a lower product content. These relationships have been explored for algal oil production as a feedstock for biodiesel production through literature review by Griffiths and Harrison (23) and are summarized in Fig. 4.

A selection of typical growth rates, lipid contents, and productivities are given in Table 6 for illustration. A review of these on a species basis is presented by Griffiths and Harrison (23).

40.6. MAXIMIZING CO$_2$ FIXATION

The equilibrium concentration of CO_2 in water at 25°C is approximately 10 μM. This is significantly lower than the K_m (CO_2) of Rubisco, the primary CO_2-fixing enzyme of most microalgae. Many cyanobacteria have K_m values in the region of 200 μM. Furthermore, oxygen is a competitive inhibitor of the carboxylase activity of the enzyme. As a result, most microalgae and cyanobacteria grow under CO_2-limiting conditions. Algae contain a carbon concentrating mechanism, built around the enzyme carbonic anhydrase (CA), which catalyzes the reversible decomposition of bicarbonate (38). External CA elevates the CO_2 concentration at the cell surface, facilitating diffusion into the cell, where the intracellular pH promotes bicarbonate formation. Internal CA catalyzes the same reaction and elevates the CO_2 concentration around the Rubisco, enhancing CO_2 fixation. The carbon concentrating mechanism is particularly well developed in organisms adapted to grow under alkaline conditions where the inorganic carbon speciation strongly favors bicarbonate.

In closed photobioreactor systems, the equilibrium concentration can be increased by using CO_2-enriched air (see item 40.3.4). Increasing the dissolved-CO_2 concentration leads to acidification of the medium. This can be compensated for by the alkalinity generated from nitrate accumulation (8) or controlled by coupling CO_2 addition to an automated feedback system. Achieving this in open ponding systems is more difficult due to reduced gas-liquid contact time. The most commonly employed strategies are improving passive mass transfer by increasing the contact area between the atmosphere and pond surface and active injection. The use of diffusers to reduce bubble size, spoilers to increase turbulence around the injection point, and airlifts to induce defined circulation patterns may further improve mass transfer. The decoupling of CO_2 dissolution

from the algal reactor through the use of scrubbers is of increasing interest for algal applications motivated by CO_2 sequestration or cycling.

40.7. MAXIMIZING PRODUCT FORMATION—ALGAL LIPID AS AN EXAMPLE

The aim of many microalgal biotechnology applications is to maximize productivity, which is influenced by the biology of the organism as well as the culture conditions and operating mode. The goal of optimizing culture conditions is to reach the maximum production biologically possible by the species of interest while considering ease of product recovery. Further improvement requires genetic modification.

Successful optimization of product formation requires an understanding of the flow of material from carbon and energy sources to biomass and product formation at a biochemical level. Metabolic flux analysis can be used to model the movement of materials through metabolic pathways. Metabolic pathways are complex interacting networks; however, they can be simplified by identifying the key enzymes, branch points, and bottlenecks through assessing the time scale of the reactions and identifying nodes at which biological control is exerted most strongly. The simplified pathway, typically representing about 30 reactions, is subjected to integrated kinetic modeling. Metabolic modeling can help to identify the important controlling factors in a pathway of interest and hence to design interventions that influence the flux toward desired products (for more information, see reference 47).

As an example, the optimization of lipid productivity in algae is considered. Here the goal is to maximize the flow of photosynthetic products to lipids rather than carbohydrates, pigments, proteins, and other products. This comes at the expense of cell growth and maintenance. By understanding the flux through these pathways and the partitioning of precursor metabolites such as pyruvate and acetyl coenzyme A (acetyl-CoA) at branch points, the desired pathway can be balanced to provide sufficient growth while optimizing lipid production.

Lipid synthesis is an energy-intensive process and relies on carbon compounds generated from CO_2 by photosynthesis, as well as energy and reducing power [in the forms of ATP and NAD(P)H, respectively]. There are several branch points at which metabolic intermediates are partitioned between synthesis of lipids and other products, such as carbohydrates and proteins, as illustrated in Fig. 5.

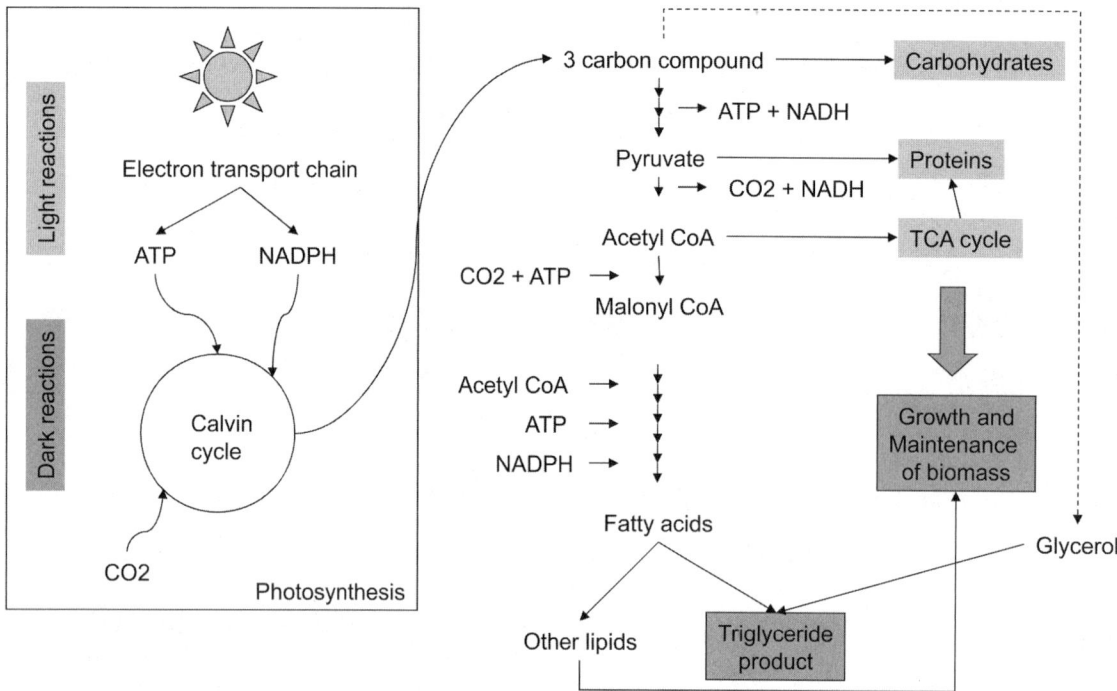

FIGURE 5 Simplified metabolic pathway illustrating CO_2 fixation and metabolism to biomass and triacylglycerides. TCA, tricarboxylic acid.

Acetyl-CoA is a central metabolite for lipid synthesis as well as the entry point into the tricarboxylic acid cycle for generation of energy and biosynthetic precursors for proteins and nucleic acids. Light determines the amount of CO_2 taken up and hence the total carbon available for anabolism. Introduction of an organic carbon source into the medium can enhance lipid formation and reduce dependence on light in heterotrophic species. At the other end of metabolism, rates can be accelerated by removing the product, e.g., designing algal species that secrete lipid.

Nitrogen deprivation is well known to enhance lipid formation in certain algae by limiting protein formation, thereby reducing growth. By reducing the carbon and energy flux into proteins, these resources are channeled into lipid and carbohydrate production. The extent and timing of nitrogen removal must be carefully balanced to maximize the number of cells as well as their lipid content. This illustrates the value of a comprehensive understanding of metabolic pathways and their control for application of the rational design approach to product optimization through both the physiological and molecular routes.

40.8. CHARACTERISTICS OF ALGAL SUSPENSIONS AND THEIR INFLUENCE ON HARVESTING BIOMASS FOR PRODUCT RECOVERY

Algal cultures are typically restricted to a low biomass density to facilitate light penetration. Typical volumetric biomass concentrations reported in large-scale processes range from 0.3 kg m^{-3} to some 5 kg m^{-3}. This algal biomass is harvested and concentrated by solid-liquid separation prior to use of the biomass as product or extraction of the product. Owing to the dilute nature of the biomass, the limited driving force for separation, and the large volumes

to be processed, this separation can contribute as much as 20 to 30% of the total cost of producing the biomass (26). The value of the end product therefore influences the choice of harvesting method. Furthermore, harvesting is critical to the overall dewatering step. Short-sighted savings in harvesting could result in higher drying and end-product extraction costs later in the process train.

Physical properties of the algal cell that can be exploited for solid-liquid separation include cell density, size, charge, and surface properties. Algal cells range in size from 3 to 30 μm. Their size is influenced by growth conditions (2, 50). Example cell sizes are given in Table 7. Filamentous algae such as *Spirulina* present a larger effective size, facilitating collection by filtering. As typically found for microbial biomass, limited density difference exists with respect to water, with typical densities of 1,030 to 1,050 kg m^{-3}. Cell surface properties, such as zeta potential and hydrophobicity, influence flocculation and flotation and may be a function of growth conditions.

Size-based separation by dead-end filtration is infrequently used for algal separation (37). Where filtration is used, cross-flow filtration shows the greatest potential, as it prevents fouling of the filter with cells and is optimized as a function of pressure. In the harvesting of filamentous algae, such as *Spirulina*, screen filtration is used with larger pore diameters than the cellulosic or polycarbonate membranes used for bacterial harvesting (46). Therefore, the effectiveness of filtration is strongly influenced by the morphology of the species being cultivated.

Cell density coupled with cell size determines the settling velocity of the cell, which can be used in either gravity or centrifugal sedimentation. Gravity sedimentation in settling tanks is suitable for processing large volumes of fluid, and the units are relatively inexpensive to construct and operate. The resulting concentrate retains a relatively high

TABLE 7 Typical algal cell size[a]

Algal species	Cell diam (μm)	Cell vol (μm³)
Chlamydomonas		1,080–6,200
Chlorella		34
Dunaliella tertiolecta	9–10	490
Dunaliella sp.	3–5	
Isochrysis galbana	3–5	
Prorocentrum minimum	13.5	1,400
Rhodomonas salina	10–15	
Scenedesmus		48
Synechococcus	1–2	1.3
Tetraselmis	12–14	1,275

[a]Data from references 2, 3, and 50.

fluid content, and the tank itself takes up a large area. Centrifuges enable rapid separation and produce a concentrate of lower moisture content but are energy intensive. Guidelines for recovery of microalgal biomass via centrifugation are provided in reference 17.

For both filtration and settling, recovery can be enhanced by flocculation. Algae carry a negative surface charge, which keeps them apart (37). A multivalent cation or bioflocculant (40) can be used to neutralize this charge, allowing algal flocculation and increasing their effective size. The effect of flocculant addition on further downstream processing and toxicity should be considered (40). Flocculation can also be induced by altering the pH to the isoelectric point of the cell surface. Care must be taken to avoid cell lysis and the release of intracellular products during cell harvesting (37, 40).

Flotation provides an alternative harvesting method. This uses the cell surface properties in terms of both hydrophobic properties and charge of the algae cell wall to attach selectively to rising air bubbles. These surface properties may be altered through manipulating growth conditions or the addition of collectors. Flotation has been investigated as a method for removal of algae in water treatment. The efficiency of separation is dependent on bubble size and contact time. Flotation of algae has been demonstrated through autoflotation (using oxygen evolved on algal culture), dispersed air flotation, and dissolved air flotation (16, 43, 49, 54).

40.8. FUTURE CHALLENGES

Algal biotechnology presents a valuable range of products, ranging from energy products through commodity products to high-value products for medical use. Currently the exploitation of algal biotechnology is in its infancy, with a great range of biodiversity remaining to be explored. While the production of high-value products using photobioreactor systems is economically feasible, exploitation of algal systems for commodity products is limited to extreme environments providing a competitive advantage. Using algae to produce commodity products requires improve-

ment of algal productivity through overcoming mass transfer and light limitations. Further to this, the significant downstream processing costs incurred require emphasis on both improving product concentration and identifying low-energy unit operations suited to the efficient processing of large volumes of algal suspensions.

REFERENCES

1. **Acreman, J. C.** 2004. The University of Toronto Culture Collection of Algae and Cyanobacteria (UTCC): a Canadian phycological research centre. *Nova Hedwigia* **79:**135–144.
2. **Agusti, S., and J. Kalff.** 1989. The influence of growth conditions on size dependence of maximal algal density and biomass. *Limnol. Oceanogr.* **34:**1104–1108.
3. **Andersen, R. A. (ed.).** 2005. *Algal Culturing Techniques.* Elsevier Academic Press, San Diego, CA.
4. **Apt, K. E., and P. W. Behrens.** 1999. Commercial developments in microalgal biotechnology. *J. Phycol.* **35:**215–226.
5. **Barnes, M., A. D. Ansell, and R. N. Gibson.** 1988. *Oceanography and Marine Biology, an Annual Review,* vol. 26. Aberdeen University Press, Aberdeen, United Kingdom.
6. **Barsanti, L., and P. Gualtieri.** 2006. *Algae: Aanatomy, Biochemistry and Biotechnology.* CRC Press, Taylor and Francis Group, Boca Raton, FL.
7. **Becker, E. W.** 1994. *Microalgae: Biotechnology and Microbiology.* Cambridge University Press, Cambridge, United Kingdom.
8. **Behrens, P. W.** 2005. Photobioreactors and fermentors: the light and dark side of growing algae, p. 189–203. *In* R. A. Andersen (ed.), *Algal Culturing Techniques.* Elsevier Academic Press, San Diego, CA.
9. **Benneman, J. R., and W. J. Oswald.** 1996. Systems and economic analysis of microalgae ponds for conversion of CO₂ to biomass. Report DOE/PC/93204-T5. Department of Energy, Washington, DC.
10. **Benson, B. C., and K. A. Rusch.** 2006. Investigation of the light dynamics and their impact on algal growth in a hydraulically integrated serial turbidostat algal reactor (HISTAR). *Aquacultural Eng.* **35:**122–134.
11. **Borowitzka, M. A.** 1992. Algal biotechnology products and processes—matching science and economics. *J. Appl. Phycol.* **4:**267–279.
12. **Borowitzka, M. A.** 1997. Microalgae for aquaculture: opportunities and constraints. *J. Appl. Phycol.* **9:**393–401.
13. **Borowitzka, M. A.** 2005. Culturing microalgae in outdoor ponds, p. 205–218. *In* R. A. Andersen (ed.), *Algal Culturing Techniques.* Elsevier Academic Press, San Diego, CA.
14. **Bouterfas, R., M. Belkoura, and A. Dauta.** 2002. Light and temperature effects on the growth rate of three freshwater algae from a eutrophic lake. *Hydrobiologia* **489:**207–217.
15. **Carvalho, A. P., L. A. Meireles, and F. X. Malcata.** 2006. Microalgal reactors: a review of enclosed system designs and performances. *Biotechnol. Progr.* **22:**1490–1506.
16. **Chen, Y., J. Liu, and Y. Ju.** 1998. Flotation removal of algae from water. *Colloids Surfaces B* **12:**49–55.
17. **Chisti, Y., and M. Moo-Young.** 1991. Fermentation technology, bioprocessing, scale-up and manufacture, p. 167–209. *In* M. Moses and R. E. Cape (ed.), *Biotechnology: the Science and the Business.* Harwood Academic Publishers, New York, NY.
18. **Day, J. G., and J. J. Brand.** 2005. Cryopreservation methods for maintaining microalgal cultures, p. 165–188. *In* R. A. Andersen (ed.), *Algal Culturing Techniques.* Elsevier Academic Press, San Diego, CA.
19. **Dimitrov, K.** March 2007. *GreenFuel Technologies: a Case Study for Industrial Photosynthetic Energy Capture.* http://www.nanostring.net/Algae/CaseStudy.pdf.

20. **El-Mansi, E. M. T., C. F. A. Bryce, A. L. Demain, and A. R. Allman.** 2007. *Fermentation Microbiology and Biotechnology*, 2nd ed. CRC Press, Taylor and Francis Group, Boca Raton, FL.

21. **Fell, D.** 1997. *Understanding the Control of Metabolism.* Portland Press, London, United Kingdom.

22. **Geider, R. J., and B. A. Osbourne.** 1992. *Algal Photosynthesis: the Measurement of Algal Gas Exchange.* Chapman and Hall, New York, NY.

23. **Griffiths, M., and S. T. L. Harrison.** 2009. Lipid productivity as a key characteristic for choosing algal species for biodiesel production. *J. Appl. Phycol.* **21:**493–507.

24. **Grima, E. M., J. M. F. Sevilla, J. A. S. Perez, and F. G. Camacho.** 1996. A study on the simultaneous photolimitation and photoinhibition in dense microalgal cultures taking into account incident and averaged irradiances. *J. Biotechnol.* **45:**59–69.

25. **Grobbelaar, J. U.** 2000. Physiological and technological considerations for optimising mass algal cultures. *J. Appl. Phycol.* **12:**201–206.

26. **Gudin, C., and C. Therpenier.** 1986. Bioconversion of solar energy into organic chemicals by microalgae. *Adv. Biotechnol. Processes* **6:**73–110.

27. **Guillard, R. R., and S. L. Morton.** 2003. Culture methods, p. 77–97. *In* G. M. Hallegraeff, D. M. Anderson, and A. D. Cembella (ed.), *Manual on Harmful Marine Microalgae.* UNESCO, Paris, France.

28. **Handy, S. M., E. Demir, D. A. Hutchins, K. J. Portune, E. B. Whereat, C. E. Hare, J. M. Rose, M. Warner, M. Farestad, S. C. Cary, and K. J. Coyne.** 2008. Using quantitative real-time PCR to study competition and community dynamics among Delaware Inland Bays harmful algae in field and laboratory studies. *Harmful Algae* **7:**599–613.

29. **Hsueh, H. T., H. Chu, and S. T. Yu.** 2007. A batch study on the bio-fixation of carbon dioxide in the absorbed solution from a chemical wet scrubber by hot spring and marine algae. *Chemosphere* **66:**878–886.

30. **Kaewpintong, K., A. Shotipruk, S. Powtongsook, and P. Pavasant.** 2007. Photoautotrophic high-density cultivation of vegetative cells of *Haematococcus pluvialis* in airlift bioreactor. *Bioresource Technol.* **98:**288–295.

31. **Kirk, J. T. O.** 1994. *Light and Photosynthesis in Aquatic Ecosystems*, 2nd ed. Cambridge University Press, Cambridge, United Kingdom.

32. **Kirk, R. E., D. F. Othmer, M. Grayson, D. Eckroth, et al.** 1978–1984. *Kirk-Othmer Encyclopedia of Chemical Technology*, 3rd ed., vol. 26, p. 153–163. Wiley, New York, NY.

33. **Lee, Y. K.** 2001. Microalgal mass culture systems and methods: their limitation and potential. *J. Appl. Phycol.* **13:**307–315.

34. **Logan, J., K. Edwards, and N. Saunders (ed.).** 2009. *Real-Time PCR: Current Technology and Applications.* Caister Academic Press, Norwich, United Kingdom.

35. **Lorenz, M., T. Friedel, and J. G. Day.** 2005. Perpetual maintenance of actively metabolizing microalgal cultures, p. 145–156. *In* R. A. Andersen (ed.), *Algal Culturing Techniques.* Elsevier Academic Press, San Diego, CA.

36. **Millamena, O. M., E. J. Aajero, and I. G. Borlongan.** 1990. Techniques on algae harvesting and preservation for use in culture as larval food. *Aquaculture Eng.* **9:**295–304.

37. **Molina Grima, E., E. H. Belarbi, F. G. Acién Fernández, A. Robles Medina, and Y. Chisti.** 2003. Recovery of microalgal biomass and metabolites, process options and economics. *Biotechnol. Adv.* **20:**491–515.

38. **Moroney, J. V., and A. Somanchi.** 1999. How do algae concentrate CO_2 to increase the efficiency of photosynthetic carbon fixation? *Plant Physiol.* **119:**9–16.

39. **Nowack, E. C. M., B. Podola, and M. Melkonian.** 2005. A novel approach in the cultivation of microalgae—96 well twin layer system. *Protist* **156:**239–251.

40. **Oh, H. M., S. J. Lee, M.-H. Park, H.-S. Kim, H.-C. Kim, J.-H. Yoon, G.-S. Kwon, and B.-D. Yoon.** 2001. Harvesting of *Chlorella vulgaris* using a bioflocculant from *Paenibacillus sp.* AM49. *Biotechnol. Lett.* **23:**1229–1234.

41. **Pulz, O.** 2004. Valuable products from biotechnology of microalgae. *Appl. Microbiol. Biotechnol.* **65:**635–648.

42. **Richmond, A.** 2004. Principles for attaining maximal microalgal productivity in photobioreactors: an overview. *Hydrobiologia* **512:**33–37.

43. **Roh, S., D. Kwak, H. Jung, K. Hwang, I. Baek, Y. Chun, S. Kim, and J. Lee.** 2008. Simultaneous removal of algae and their secondary algal metabolites from water by hybrid system of DAF and PAC adsorption. *Separation Sci. Technol.* **43:**113–131.

44. **Sazdanoff, N.** 2006. Honors thesis. Ohio State University, Columbus.

45. **Sheehan, J., T. Dunahay, J. Benemann, and P. Roessler.** 1998. *A Look Back at the U.S. Department of Energy's Aquatic Species Program: Biodiesel from Algae.* Close-out report. Report no. NREL/TP-580-24190. National Renewable Energy Lab, Department of Energy, Golden, CO.

46. **Spolaore, P., C. Joannis-Cassan, E. Duran, and A. Isambert.** 2006. Commercial applications of microalgae: review. *J. Biosci. Bioeng.* **101:**87–96.

47. **Stephanopoulos, G. N., A. A. Aristidou, and J. N. Nielson.** 1998. *Metabolic Engineering, Principles and Methodologies.* Academic Press, San Diego, CA.

48. **Ugwu, C. U., H. Aoyagi, and H. Uchiyama.** 2008. Photobioreactors for mass cultivation of algae. *Bioresource Technol.* **99:**4021–4028.

49. **Van Puffelen, J., P. Buijs, P. Nuhn, and W. Hijnen.** 1995. Dissolved air flotation in potable water treatment: the Dutch experience. *Water Sci. Technol.* **31:**149–157.

50. **Verity, P. G., C. Y. Robertson, C. R. Tronzo, M. G. Andrews, J. R. Nelson, and M. E. Sieracki.** 1992. Relationships between cell volume and the carbon and nitrogen content of marine photosynthetic nanoplankton. *Limnol. Oceanogr.* **37:**1434–1446.

51. **Watanabe, M. M.** 2005. Freshwater culture media, p. 13–21. *In* R. A. Andersen (ed.), *Algal Culturing Techniques.* Elsevier Academic Press, San Diego, CA.

52. **Welty, J. R., C. E. Wicks, R. E. Wilson, and G. Rorrer.** 2001. *Fundamentals of Momentum, Heat and Mass Transfer*, 4th ed. Wiley, New York, NY.

53. **Wu, X., and J. C. Merchuk.** 2004. Simulation of algae growth in a bench scale internal loop airlift reactor. *Chem. Eng. Sci.* **59:**2899–2912.

54. **Yan, Y., and G. Jameson.** 2004. Application of the Jameson cell technology for algae and phosphorous removal from maturation ponds. *Int. J. Mineral Processing* **73:**23–28.

Metabolic Engineering Strategies for Production of Commodity and Fine Chemicals: *Escherichia coli* as a Platform Organism

PATRICK C. CIRINO

41.1. INTRODUCTION

Metabolic engineering encompasses a dynamic field of research that increasingly influences the development of microbial strains for industrial applications. Metabolic engineering efforts pertaining to industrial microorganisms are typically intended to improve (i) productivity and/or the yield and titer of a desired product, (ii) ability to utilize inexpensive raw materials, and (iii) robustness under production conditions. Achieving these goals is facilitated by increasing understanding of cellular metabolism, gene regulation, enzyme kinetics, and allosteric regulation. Advances in molecular biology methods and functional genomics tools promote more efficient manipulation of these biological processes.

This chapter presents a summary of examples in which various metabolic engineering strategies were employed to improve microbial production of small molecules having a range of commercial value (commodity and fine chemicals). Specifically, notable achievements in engineering the biotechnological workhorse organism *Escherichia coli* for chemical production are described. While metabolic engineering approaches have been applied to a great variety of microorganisms (e.g., *Corynebacterium*, lactic acid bacteria [LAB], *Zymomonas mobilis*, *Saccharomyces cerevisiae*, *Pichia pastoris*, etc.) (12, 13, 17, 31, 33, 34, 61, 62, 70, 78), *E. coli*'s metabolic versatility, genetic amenability, fast growth, and track record of engineering success (49, 66) have made this organism the most popular host for both proof-of-concept demonstrations and successful applications of metabolic engineering, often resulting in dramatic improvements in strain performance.

Described below are examples in which *E. coli* metabolism was engineered for overproduction of acetate, pyruvate, D- and L-lactate, succinate, L-alanine, L-valine, L-threonine, shikimic acid, *cis,cis*-muconic acid (*cc*MA), carotenoids, mevalonate, and amorphadiene. The genetic techniques employed can be summarized as chromosomal gene deletions, plasmid-based or chromosomal insertion-based heterologous gene expression, adaptive strain evolution of growth-coupled product formation, targeted enzyme evolution for improved flux through a nonnative pathway, optimization of pathway expression in response to the flux capacity of the cell, and fine-tuning the relative expression levels of genes in a synthetic operon. While not comprehensive, the selected examples of metabolic engineering illustrate the diversity of tools available, the variety of accessible transformations and compounds, and the versatility of *E. coli* as a platform organism and as a system for developing metabolic engineering strategies that pave the way for future developments. Many reviews describe additional metabolic engineering efforts toward the microbially based industrial production of commodity and fine chemicals such as indigo, 2-keto-L-gulonic acid (precursor to ascorbic acid), and the exemplary case of 1,3-propanediol production from glucose (13, 49, 86).

41.2. ORGANIC ACIDS

In terms of commodity chemicals, metabolic engineering has found the most application and success toward the production of small organic acids and amino acids (perhaps with the exception of engineering *E. coli* to produce 1,3-propanediol from glucose), owing largely to the natural coupling between growth and production of these compounds through fermentation. Recent progress in microbial production of acetate, pyruvate, lactate, and succinate is first described here, followed by examples of increased amino acid production. These examples represent a category of metabolic engineering applications in which the host organism is made to overproduce a native metabolite (rather than inserting heterologous pathways for foreign compounds). These may appear to be relatively simple demonstrations of targeted strain improvement, since the host organisms naturally produce the compound of interest. Genetic modifications primarily involve redistribution of carbon flux through increased expression of enzymes which may be bottlenecks in product formation, as well as deletion of pathways that compete for carbon and result in by-products. However, these examples also address the often challenging tasks of overcoming native mechanisms of feedback inhibition and regulation in response to metabolite accumulation and of functionally expressing foreign genes.

Figure 1 depicts the major pathways in *E. coli* central carbon metabolism involved in production of the metabolites described. Many of the corresponding genes shown in Fig. 1 are targets for deletion or amplification in the examples described.

41.2.1. Acetic Acid

Acetic acid (CH_3COOH) is produced commercially primarily for use in polymer production and as a solvent.

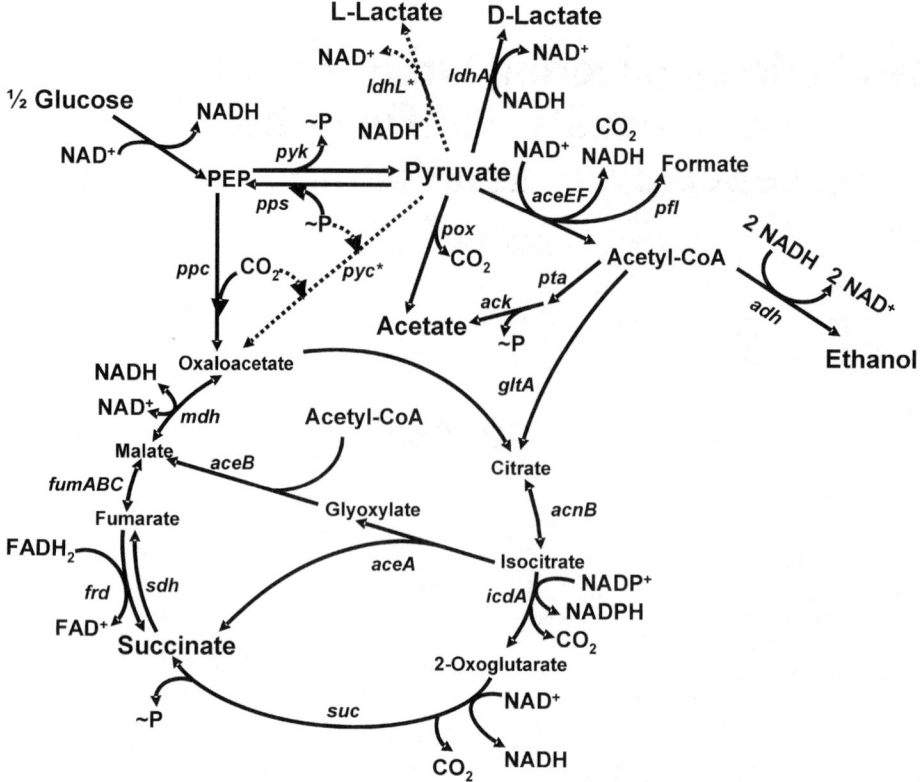

FIGURE 1 *E. coli* primary metabolic pathways involved in production of acetate, pyruvate, lactate, and succinate. Relevant genes are listed with the corresponding transformation catalyzed. ~P represents utilization or production of ATP. Genes encoding pyruvate carboxylase (*pyc*) and L-specific lactate dehydrogenase (*ldhL*) are labeled with an asterisk to indicate that they are not native to *E. coli*.

Commercial biological production of acetate is currently limited to food use (e.g., as vinegar or as a food acidity regulator), although microbial production has the potential to displace petrochemical routes as these processes are made more efficient and petroleum supplies diminish. Acetate production is commonly an undesired by-product resulting from excess glycolytic flux during microbial culturing for production of proteins or higher-value metabolites (80, 82). Conversion of glucose to two acetic acids eliminates the availability of pyruvate and acetyl coenzyme A (acetyl-CoA) as electron acceptors, as they are in anaerobic fermentation pathways. Homoacetate production therefore requires another route for NADH oxidation. Natural clostridial homoacetate producers are capable of reducing formate or CO_2, although these strains suffer from limitations that include low volumetric productivity or inability to use glucose directly (29, 71). Current microbial production of acetate uses yeast for fermentation of sugars to ethanol, followed by ethanol oxidation to acetate by *Acetobacter*.

Ingram and coworkers sought to engineer *E. coli* for homoacetate production from glucose, with oxygen serving as the required electron acceptor for NAD^+ regeneration (9). Their approach combines attributes of fermentative and oxidative metabolism and provides new opportunities for high-level

microbial production of commodity chemicals with higher oxidation states than fermentation products. To prevent loss of carbon to fermentation by-products, the authors eliminated native fermentation pathways by deleting the genes encoding lactate dehydrogenase (*ldhA*), pyruvate formate lyase (*pfl*), fumarate reductase (*frd*), and alcohol/aldehyde dehydrogenase (*adh*). The presence of oxygen enables respiration and would promote further oxidation of acetyl-CoA through the tricarboxylic acid (TCA) cycle, rather than conversion to acetate. To prevent this potentially dramatic loss of carbon as CO_2, the TCA cycle was interrupted by deleting part of the 2-oxoglutarate dehydrogenase complex (*suc*). The deletion resulted in an auxotrophic requirement for succinate (Succ⁻), which was subsequently alleviated through serial transfer of the auxotrophic strain in glucose minimal medium containing decreasing amounts of succinate. While the genetic modification leading to the Succ⁺ phenotype was not characterized, the authors speculated that upregulation of the glyoxylate shunt or citrate lyase may have been responsible.

The availability of respiratory metabolism for NADH oxidation resulted in increased ATP yields that caused high levels of growth and loss of carbon to cell mass. In order to reduce the ATP/ADP ratio and limit cell mass production, the authors introduced a mutation in the $(F_1F_0)H^+$-ATP synthase complex (*atpFH*) that released the (still functional) ATP synthase from the cell membrane, not only uncoupling proton translocation from ATP production but also promoting cytoplasmic ATP hydrolysis. This provided the added benefit of increased glycolytic flux and acetate production rates as a result of the low ATP yields. Strain TC36 resulting from these efforts is devoid of plasmids, foreign genes, and antibiotic resistance markers and grows well in mineral salts medium without complex nutrients (9). Acetate was produced as the sole product, with up to 86% of the theoretical maximum from glucose (0.67 g acetate/g glucose). In a fed-batch fermentation supplied with 5 % dissolved O_2, TC36 produced 878 mM acetate (~53 g/liter), with a volumetric productivity of ~1.4 g/liter · h (notably much higher than values reported for native acetogens) (9, 29).

41.2.2. Pyruvic Acid

Pyruvic acid ($CH_3COCOOH$) serves as the biological precursor to many bio-based commodity chemicals (e.g., ethanol, lactate, 1,3-propanediol). In addition to its commercial use as a synthetic precursor to several amino acids, including L-3,4-dihydroxyphenylalanine (40), pyruvate offers potential as a platform chemical for production of a variety of renewable fuels and chemicals. Pyruvate is also marketed as a weight control supplement. Traditionally, pyruvate has been synthesized chemically from tartaric acid (85). While microbially based processes increasingly gain a competitive edge, the general requirement for complex and expensive culture ingredients has hampered their commercial success. Recent metabolic engineering efforts similar to those applied for acetate production have proven useful for overproduction of pyruvate from glucose by *E. coli* grown aerobically in inexpensive, mineral salts medium.

Ingram and coworkers constructed a strain similar to TC36 (see above) but with additional deletions in the genes encoding pyruvate:quinine oxidoreductase (*pox*) and acetate kinase (*ack*), the two primary sources of pyruvate conversion to acetate (8). The resulting strain, TC44, produced up to 749 mM pyruvate (~66 g/liter) in fed-batch culture and achieved a maximum volumetric productivity of 1.21 g/liter · h and a yield of 0.76 g/g glucose. Up to 62 mM of contaminating acetate also accumulated in the broth.

A similar approach to pyruvate overproduction was reported by Takors and coworkers, who constructed a strain containing gene deletions that disabled conversion of pyruvate to lactate (*ldhA*), acetyl-CoA (*aceEF* and *pfl*), phosphoenolpyruvate (PEP) (*pps*), and acetate (*pox*) (84). The strain grows in minimal glucose medium supplemented with acetate (it is acetate auxotrophic). A maximum molar yield of 1.78 mol pyruvate/mol glucose (0.87 g/g) was reported (85), along with a maximum pyruvate titer of 62 g/liter and a volumetric productivity of 1.75 g/liter · h (84).

In a final and again similar strategy, Altman and coworkers combined the same mutations as for the acetate-auxotrophic strain (*ldhA*, *pfl*, *pox*, *pps*, and *aceEF*), with deletions in the (F_1F_0)H^+-ATP synthase complex (*atpFH*) and in the redox-sensitive global regulatory protein ArcA, which is known to control gene expression of respiratory pathways (91). The resulting strain, ALS1059, achieved 90 g/liter pyruvate, with an overall productivity of 2.1 g/liter · h and a yield of 0.68 g/g glucose in the presence of 5 mM betaine as an osmoprotectant and under acetate-limited growth (91).

41.2.3. Lactic Acid

Lactic acid [$C_3H_6O_3$; $CH_3CH(OH)COOH$] finds a wide range of applications in the food and chemical industries (67). Polylactic acid is a biodegradable polymer finding increasing applications and replacing petrochemical products, resulting in growing demand for lactic acid production. Physical properties of polylactic acid are dependent on the relative amounts of D-lactic acid and L-lactic acid in the polymer, so it is important to provide these isomers in the desired ratio. The environmentally friendly solvent ethyl lactate is also gaining popularity. LAB are Generally Regarded As Safe (GRAS) and are the primary source of lactic acid from sugar fermentation (17, 66, 67). Disadvantages to the use of LAB for commodity chemical production include their requirement for expensive complex nutrients and their inability or poor ability to ferment pentose sugars. In addition, these organisms typically produce racemic mixtures of lactic acid and suffer from significant inhibition as the product accumulates in the fermentation broth. Microbially based commercial production of lactic acid would ideally involve the efficient conversion of a variety of sugars derived from hemicellulose to a single isomer with high volumetric productivity and yield. While many improved mutant strains of LAB have been developed over the years and directed metabolic engineering efforts have been increasingly successful toward LAB as new tools and knowledge are furnished (17), the unprecedented wealth of metabolic knowledge and genetic engineering techniques available for *E. coli*, along with its native ability to metabolize both pentose and hexose sugars, has made this organism a widely studied alternative host for homolactate production, as discussed below (note that engineered yeasts have also been considered for lactate production [13]).

Dien et al. constructed an *E. coli* strain carrying deletions in genes encoding pyruvate formate lyase (*pfl*) and the native D-lactate-producing lactate dehydrogenase (*ldhA*), resulting in strain NZN111, which is unable to grow fermentatively (19). They then used plasmid-based heterologous gene expression to introduce an L-specific lactate dehydrogenase (*ldhL*) gene from *Streptococcus bovis* (81) into this strain, resulting in the ability to grow anaerobically and ferment either glucose or xylose to L-lactic acid. *E. coli* preferentially metabolizes glucose over other sugars through a complex mechanism known as catabolite repression, which is at least partly controlled by the PEP-dependent glucose phosphotransferase system (PTS) (16). A more desirable property for industrial fermentation of biomass-derived sugars is the ability to simultaneously utilize pentose and hexose sugars. The authors constructed a mutant of NZN111 in which the gene (*ptsG*) encoding IIBC^{Glc} of the PTS is disrupted, in an effort to partially alleviate glucose-dependent catabolite repression (18). The resulting strain (ND19) had a yield of 0.77 g L-lactic acid/g sugar when fermenting an equal mixture of 10% glucose and xylose in complex medium (containing tryptone and yeast extract).

Ingram and coworkers constructed a series of *E. coli* strains that efficiently ferment glucose and sucrose into optically pure D- or L-lactic acid (27, 90). Their efforts included directed gene deletions to eliminate loss of carbon to by-products (*pfl*, *frd*, *adh*, and *ack*), heterologous expression of an L-specific lactate dehydrogenase from *Pediococcus acidilactici* (*ldhL*) that was chromosomally inserted in place of the native *ldhA* gene (for the case of L-lactate production)

(89), and many rounds of growth-based metabolic evolution in which serial subculturing was used to enrich for spontaneous mutations that both increased growth and improved volumetric productivity and lactate production. Addition of 1 mM betaine as a protective osmolyte significantly improved lactate production and tolerance to high concentrations of lactate and sugars in mineral salts medium (88). Flux through the methylglyoxal bypass was identified as the source of low levels of chirally impure lactic acid (both isomers can be produced via methylglyoxal), and deletion of *mgsA* (encoding methylglyoxal synthase) ensured chiral purity of both D-lactate (using the native *ldhA*) and L-lactate (with *ldhL* in place of *ldhA*) (27).

The culmination of the Ingram group's lactate strain development, which includes the accumulation of as-yet-uncharacterized random mutations resulting from adaptive growth evolution at multiple stages of strain design, is strain TG108, which produced >1.2 M L-lactic acid with a maximum volumetric productivity of ~2.7 g/liter · h, and strain TG114, capable of producing >1.2 M D-lactic acid with a maximum volumetric productivity of ~3.2 g/liter · h (27). Cultures of these strains completely utilized 10 to 12% (wt/vol) glucose in 48 h using mineral salts medium containing glucose and 1 mM betaine, with yields of up to 98% and chiral purities of >99%. An important distinction between the adaptive evolution strategy used in the Ingram studies and the more traditional microbial evolution experiments applied for industrial strain optimization is that in the former case, detailed understanding of the central metabolic pathways leading to straightforward strain engineering steps was necessary to constrain the metabolic network and ensure coupling between cell growth and lactic acid production. The increasing availability of affordable genome sequencing in the postgenomic era provides tremendous opportunities for understanding the mechanisms of strain adaptation and further improving subsequent strains through more rational approaches.

In a final example, Zhu et al. developed a strain having deletions in all major genes responsible for pyruvate metabolism, other than *ldhA* (i.e., *aceEF*, *pfl*, *pox*, and *pps*) (92). This strain requires acetate for aerobic growth due to inactivation of both pyruvate dehydrogenase and pyruvate formate lyase and is unable to grow under anaerobic conditions. The authors used an aerobic growth stage to generate the biocatalyst, followed by a nongrowing anaerobic stage for conversion of glucose to D-lactate. Residual acetate following the growth stage was determined to be the primary source of contaminating succinate in the fermentation broth, and inactivation of fumarate reductase by deleting *frd* genes helped to reduce succinate formation. By minimizing residual acetate at the end of the growth stage, their final strain produced 138 g/liter lactate representing 97% of all carbon products, with a yield of 0.99 g/g glucose and a productivity of 6.3 g/liter · h during the anaerobic stage (92).

41.2.4. Succinic Acid

Succinic acid ($C_4H_6O_4$) serves as a precursor to a variety of other commodity chemicals such as butanediol, tetrahydrofuran, and γ-butyrolactone; can be used as a monomer toward biodegradable polymer synthesis; and finds applications in the food, pharmaceutical, and cosmetics industries (47, 78, 79). Commercial demand for succinic acid is far from mature and is expected to increase dramatically as petrochemical prices increase and as more efficient routes to renewable succinate production are developed. A variety of naturally occurring organisms produce succinate as a major

but not solo fermentation product, and some strain development efforts have focused on improving yield and purity from these organisms (47, 69). Noteworthy are the efforts of Lee and coworkers, who isolated the high-succinate-producing bacterium "*Mannheimia succiniciproducens*" and subsequently performed genomic, transcriptomic, and proteomic analyses and developed genetic tools enabling metabolic engineering of this organism (37). However, as is the case with so many target products, *E. coli* has served as the primary host for metabolic engineering efforts toward efficient conversion of sugars to succinate. Again, the ability to perform this conversion using inexpensive, minimal medium ingredients is essential.

The theoretical yield of succinate from glucose depends on whether additional CO_2 or external reductants are also supplied. Regardless of the pathways used, CO_2 fixation is required to achieve theoretical maximum yields, since all glucose is first converted to the three-carbon metabolite PEP (Fig. 1). Glucose-derived carbon can yield 1.5 succinate per glucose, while CO_2 supplementation increases the theoretical yield to 1.7 mol/mol. A yield of 2 succinate per glucose is theoretically possible, provided the organism is appropriately engineered to balance CO_2 fixation and trafficking of reducing equivalents from an external source such as H_2 toward succinate production, resulting in the redox-balanced reaction

$$C_6H_{12}O_6 + 2\ CO_2 + 2\ H_2 \rightarrow 2\ C_4H_6O_4 + 2\ H_2O$$

E. coli has been engineered to produce succinate under both aerobic and anaerobic conditions. A mutant of strain NZN111 (*pfl ldhA* double deletion strain described above for L-lactate production) showing dramatically enhanced anaerobic growth on glucose and producing high levels of succinate (0.92 mol/mol glucose in rich medium) (20) was found to carry a mutation in the PTS system gene *ptsG* (11). While the mechanisms leading to fast anaerobic growth and succinate production are not well understood, altered PEP levels, the consequential regulatory effects, and flux through PEP carboxylase clearly play important roles in this mutant strain (named AFP111). That initial discovery prompted a series of subsequent studies with *ptsG* mutant strains. To increase flux of three-carbon intermediates to succinate via the reductive branch of the TCA cycle (i.e., oxaloacetate), Altman and coworkers expressed *pyc*, encoding the pyruvate carboxylase enzyme from *Rhizobium etli*, in AFP111 (Fig. 1) (74, 75). Succinate production was compared under exclusively anaerobic growth and dual-phase conditions (aerobic growth followed by anaerobic production). Using dual-phase, fed-batch fermentation with complex medium and supplementation with $MgCO_3$ (source of CO_2), succinate was produced to a titer of 841 mM (~98 g/liter), with an average volumetric productivity of 1.31 g/liter · h and a yield of 1.68 mol/mol glucose (75).

San and coworkers constructed a strain for aerobic succinate production by preventing oxidation of succinate to fumarate (*sdh* genes were deleted) and preventing acetate secretion (*ack*, *pta*, and *pox* were deleted) (41–43). To prevent loss of carbon through the oxidative branch of the TCA cycle, aerobic flux to succinate should be routed through the glyoxylate bypass. The gene *iclR*, encoding the repressor of the glyoxylate bypass operon, was therefore deleted, promoting expression of the genes encoding isocitrate lyase (*aceA*) and malate synthase (*aceB*). Finally, a PEP carboxylase from *Sorghum vulgare* (*ppc*) (catalyzing conversion of PEP to oxaloacetate) was overexpressed, and PEP consumption during glucose transport through the PTS was prevented by deleting *ptsG* (43). The resulting strain produced ~500 mM succinate (58 g/liter) during aerobic growth in complex medium, with an average volumetric productivity of ~1.1 g/liter · h and a yield of ~0.9 mol/mol glucose (41). The same laboratory also developed similar strains for anaerobic succinate production (63). Strain SBS550MG contains deletions in *adh*, *ldhA*, *ack*, and *pta* (to prevent fermentation to ethanol, lactate, and acetate) and contains the *iclR* deletion for activating the glyoxylate bypass. Plasmid-based expression of pyruvate carboxylase (*pyc*, this time from *Lactococcus lactis*) resulted in the anaerobic production of 340 mM succinate (~40 g/liter) from glucose with CO_2 supplementation using a rich growth medium, with a final yield of 1.6 mol/mol glucose, near the theoretical maximum of 1.7 mol/mol (63). This strain apparently balances fluxes through the reductive branch of the TCA cycle (where oxaloacetate is replenished via carboxylation of PEP and pyruvate) and the glyoxalate bypass, such that redox balance is achieved through succinate production, with net consumption of CO_2.

To avoid scale-up complications related to poor growth in the absence of expensive growth nutrients, Jantama et al. focused on the use of a mineral salts-based defined medium for conversion of glucose to succinate (30). They combined direct genetic modifications (similar to those used for the previously described succinate-producing strains, but without addition of plasmids or foreign genes) with a metabolic evolution strategy similar to that described above for lactic acid-producing strains. The experiments started from strain KJ012, having deletions in *ldhA*, *adh*, and *ack*. While a moderate succinate producer when growing in rich medium, this strain grew much slower and produced much less succinate than the wild-type strain (*E. coli* C in this case) when using minimal medium (6 mM succinate for KJ012, compared to 49 mM for the wild type). Metabolic evolution via serial dilution was then employed, with culture conditions and transfer times appropriately changed throughout the process to accommodate the adapting strain phenotype. It is important to note that as was the case for growth-based evolution of lactate producers, rational metabolic engineering was first required to constrain the metabolic network such that growth is coupled to succinate production. The result of this first set of transfers, strain KJ017, showed much higher growth and succinate production in the minimal medium but also produced formate. Formate production was next eliminated by deleting *pfl*, resulting in acetate-auxotrophic strain KJ032. Another round of serial dilutions with gradual decreases in acetate used to supplement the culture medium eventually yielded strain KJ060, having no acetate requirements. Subsequent modifications included deletion of *mgsA* (encoding methylglyoxal synthase) to eliminate residual lactate production and *pox*, followed by further rounds of metabolic evolution to improve growth.

Throughout this work, several unexpected metabolic changes were noted, such as increased malate production due to the *mgsA* deletion and reduced malate secretion with the *pox* deletion. Most noteworthy from these studies is the drastic difference in strain performance in minimal versus complex media, and the power of adaptive strain evolution to overcome these metabolically complex (and not understood) limitations. In the end, the best succinate-producing biocatalyst produced 733 mM succinate (0.90 g/liter · h, 1.4 mol/mol glucose), with 39 mM malate and 250 mM acetate produced as by-products, while a high-malate-producing strain produced 516 mM malate, with significant succinate, pyruvate, and acetate as by-products (30).

Elimination of coproducts for higher-purity succinate or malate production from minimal medium cultures presents opportunities for further strain optimization.

41.3. AMINO ACIDS

Microorganisms have been used for the production of amino acids for many decades. A variety of amino acids find applications in industries including food, animal feed, pharmaceuticals, and cosmetics (38). While *Corynebacterium glutamicum* strains gained early popularity for their ability to produce large quantities of glutamate and lysine, *E. coli* strains have also proven effective hosts for overproduction of many amino acids (38, 78). Whereas microbes were traditionally improved for amino acid production using strain adaptation techniques (random mutation and auxotrophic selection procedures) (52), recently, metabolic engineering and systems biology approaches have been effectively applied to *C. glutamicum* (for lysine, serine, tryptophan, isoleucine, and threonine), LAB (for alanine), and *E. coli* (for threonine, phenylalanine, alanine, and valine) (14, 17, 38, 52, 53, 62, 70, 78). Considering the low titers of amino acids obtained from wild-type *E. coli* cultures, improvements of this host through metabolic engineering have been impressive, as described below.

41.3.1. Alanine

Just as the research teams of Altman and coworkers and Ingram and coworkers reported similar approaches to engineer *E. coli* for pyruvate and lactate overproduction (as described above), these two groups also took similar approaches for overproduction of L-alanine from pyruvate, by the action of a nonnative, heterologously expressed NADH-linked alanine dehydrogenase (*alaD*) (Fig. 2).

The Altman approach was to eliminate all major metabolic routes of aerobic and anaerobic pyruvate assimilation. Pyruvate-overproducing strain ALS929 carries deletions in *pfl*, *pps*, *pox*, *ldhA*, and *aceEF* and has an acetate-auxotrophic requirement for aerobic growth (68). In a two-phase production strategy similar to that described above by the same laboratory, strain ALS929 expressing plasmid-borne *alaD* from *Bacillus sphaericus* was first grown in complex medium under aerobic conditions, with glucose plus acetate feeding to control the growth rate. Following this biocatalyst production stage (~25 h), the culture was switched to anaerobic conditions and additional glucose was fed to the system for conversion to alanine. A 23-h production phase was sufficient to deplete all glucose and produce 88 g/liter D,L-alanine (not optically pure), with a yield of ~1 g/g glucose (68).

Starting from lactate-overproducing strain SZ194 (*pfl*, *frd*, *adh*, and *ack* deletions) (89), Zhang et al. replaced the native *E. coli ldhA* gene (encoding lactate dehydrogenase) with the *alaD* gene from *Geobacillus stearothermophilus*, such that *alaD* was regulated by the native *ldhA* promoter (87). Deletion of the methylglyoxal synthase gene (*mgsA*) eliminated low levels of lactate production (refer to above description for construction of lactate strain TG108), and the catabolic alanine racemase gene (*dadX*) was deleted to prevent conversion of L-alanine to D-alanine (87). Since alanine production replaces the role of lactate dehydrogenase in NADH oxidation, the supply of ATP for anaerobic cell growth from glycolysis is obligately coupled to alanine production. Therefore, as in the previous studies from this group described above, strains that were rationally engineered to ensure coupling between growth and product formation were subsequently subjected to metabolic evolution by selecting for faster-growing mutants on minimal glucose

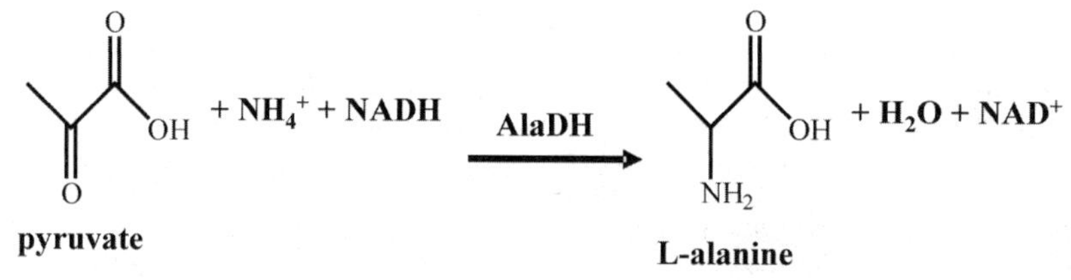

FIGURE 2 L-Alanine production from pyruvate by alanine dehydrogenase (*alaD*).

medium through serial transfers. Resultant strain XZ132 produced <1.3 M L-alanine from 120 g/liter glucose in batch fermentation in mineral salts medium, with an average volumetric productivity of <2.4 g/liter · h (87). The yield was 0.95 g L-alanine/g glucose, with >99.5% chiral purity. Results from this work appear superior to other attempts at microbial production of L-alanine. Chromosomal expression of the foreign *alaD* gene was sufficient for supporting the high levels of glycolytic flux, so this strain did not require plasmid maintenance or antibiotics. The authors also point out that L-alanine production by XZ132 exceeded the performance of their previously reported pyruvate-producing strain (TC44, described above), suggesting an alternate route to pyruvate production via enzymatic hydrolysis of L-alanine.

41.3.2. Valine

Amino acid metabolism is known to be tightly regulated in *E. coli*, such that supply and demand are tightly coupled. This regulation was avoided in the alanine studies described above because a heterologous and unregulated pathway was used for the nonnative conversion of pyruvate to alanine. In contrast to that approach, Lee and coworkers sought to engineer *E. coli* for L-valine and L-threonine production via their native pathways (36, 51). In addition to the common strategy of deleting competing pathways, their efforts required alleviation of natural regulatory mechanisms designed to prevent loss of carbon and energy through amino acid overproduction. To identify the correct genetic targets, a variety of systems biology and metabolic modeling tools were combined with in-depth knowledge of amino acid metabolism to methodically construct L-valine- and L-threonine-overproducing *E. coli* strains.

In their first example, known mechanisms of regulation of L-valine production, such as feedback inhibition and gene repression, were eliminated by genetic modification (51). Acetohydroxy acid synthase (AHAS) catalyzes the conversion of pyruvate to 2-acetolactate in the L-valine biosynthetic pathway. Previous research identified point mutations in AHAS that relieve L-valine-mediated feedback inhibition, and these mutations were added to the *E. coli* chromosome. The expression of AHAS isoenzymes was enhanced by replacing native promoters known to be subject to transcriptional attenuation with a strong constitutive promoter (*tac*), as well as by amplifying AHAS activity through plasmid-based expression. Pathways known to be competing with L-valine biosynthesis were next blocked by gene deletion (*ilvA*, *leuA*, and *panB*). The transcriptomes of the wild-type and intermediate valine-producing strains were compared to identify additional genes whose upregulation may further boost valine production. The genes selected for amplification encoded a leucine-responsive global regulator (Lrp) and an L-valine exporter (YgaZH). Finally,

in silico gene knockout simulations using a stoichiometric model of *E. coli* metabolism (adapted from reference 60) were used to study the effects of various genetic modifications on valine production. This model-driven approach identified three additional gene deletions (*aceF*, *mdh*, and *pfkA*) to further enhance valine production. The final strain resulting from these systematic efforts produced 7.55 g/liter L-valine, with a yield of 0.378 g/g glucose in batch fermentation (51).

41.3.3. Threonine

A similar systems-based metabolic engineering approach was taken for L-threonine production (36). Collectively, relief of built-in negative regulation by point mutation and promoter replacement, deletion of genes involved in threonine degradation and biosynthetic precursor competition, transcriptome profiling of wild-type and intermediate threonine-producing strains to identify gene amplification targets, and in silico flux response analysis to help optimize amplified gene expression levels resulted in a genetically defined *E. coli* mutant able to produce 82.4 g/liter threonine in fed-batch culture, with the high yield of 0.39 g/g glucose (36).

41.3.4. Unnatural Amino Acids

Unnatural amino acids are attractive synthetic building blocks for pharmaceuticals and for incorporation into proteins for structure and function studies and for added stability or to introduce unique functionality (77). Whereas unnatural amino acids are typically produced by chemical synthesis, Maier developed an *E. coli* fermentation-based process for production of unnatural L-α-amino acids by metabolic engineering of the cysteine biosynthetic pathway (44). As depicted in Fig. 3, O-acetylserine sulfhydrylases encoded by *cysK* and *cysM* catalyze the final step of cysteine biosynthesis using sulfide as the nucleophile and ultimate source of cysteine sulfur. The known relaxed specificity of these enzymes for the nucleophilic substrate enables the use of externally supplied substrates (thiols, selenols, and N-heterocycles), leading to the synthesis of unnatural amino acids. O acetylation of serine by the serine acetyltransferase (encoded by *cysE*) is the committed step in L-cysteine synthesis and is subject to L-cysteine (product) inhibition. A feedback inhibition-resistant mutant of *cysE* was therefore isolated (*cysE*^{fbr}) and overexpressed during nucleophile feeding. Using this strategy, along with optimized biotransformation procedures that included a two-stage process to avoid feeding of toxic nucleophiles during growth, the author was able to produce a wide variety of unnatural amino acids (up to 13 g/liter), including the valuable S-phenyl-L-cysteine which serves as a building block for a commercially important human immunodeficiency virus protease inhibitor (32, 44).

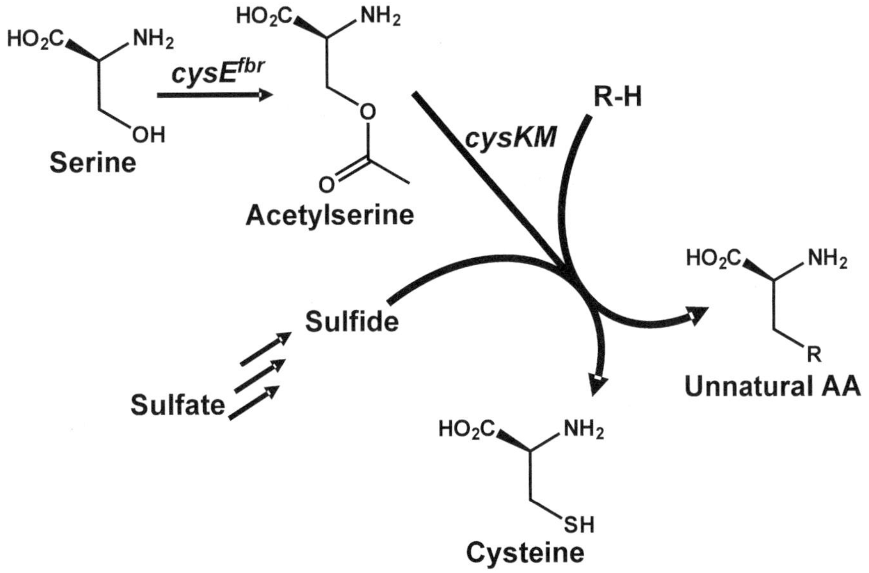

FIGURE 3 Engineering the cysteine biosynthetic pathway (starting from serine) for production of unnatural amino acids (AA). The O-acetylserine sulfhydrylases encoded by *cysK* and *cysM* utilize sulfide as a cosubstrate to produce cysteine, but the enzymes have relaxed specificity such that addition of select, nonnative nucleophiles (R-H) results in production of unnatural amino acids. *cysE^{fbr}* encodes a feedback-resistant variant of serine acetyltransferase.

41.4. PRODUCTS OF THE AROMATIC AMINO ACID PATHWAY

Frost and coworkers have devoted two decades of research toward engineering the aromatic amino acid biosynthetic pathways of *E. coli*. Their efforts have resulted in the biological production of a variety of aromatic compounds, products that otherwise require petroleum-derived aromatics such as benzene for chemical synthesis, and other valuable metabolites that are difficult to isolate from natural resources (1, 6, 21, 26, 28, 39, 50, 58). Of primary importance has been shikimic acid, which is a key starting material for the synthesis of the neuraminidase inhibitor Tamiflu®, an orally effective anti-influenza agent (35). Isolation of shikimic acid from plants has been considered commercially cost-prohibitive, while high-level production from bacterial culture is cost-effective.

41.4.1. Shikimic Acid and Quinic Acid

As shown in Fig. 4, shikimic acid is produced by *E. coli* as an intermediate toward the production of aromatic amino acids and aromatic vitamins. The genes *aroK* and *aroL*, encoding enzymes downstream of shikimic acid, were deleted (21). The reaction catalyzed by 3-dehydroquinate (DHQ) synthase (encoded by *aroB*) was identified as a rate-limiting step in shikimic acid production (15), so an extra copy of *aroB* was integrated into the *E. coli* chromosome. The aromatic amino acid pathway starts with the condensation of PEP and erythrose-4-phosphate (E4P) to 3-deoxy-heptulosonate-7-phosphate (DAHP), catalyzed by DAHP synthase (encoded by *aroF*), and this reaction is tightly regulated (allosterically) through feedback inhibition by aromatic amino acids. Using a strategy similar to that described above for amino acid overproduction, a feedback-resistant mutant of *aroF* was overexpressed in the

strain, along with an additional gene encoding shikimate dehydrogenase (*aroE*). As a result of these modifications, the *E. coli* strain converted glucose to 27 g/liter shikimic acid in fed-batch fermentation with defined medium (21). Quinic acid was produced as a significant by-product (12.6 g/liter), as a result of a side reaction between shikimate dehydrogenase and DHQ. By changing the glucose feeding strategy during fermentation, the authors were able to shift the shikimate/quinate molar ratio to 11.8:1 (although with a net lower yield of shikimate), thus allowing for easier recovery of shikimic acid by crystallization.

Quinic acid (shown in Fig. 4) is also a high-value pharmaceutical precursor, and high levels of this product without shikimic acid by-product were obtained by deleting the gene *aroD* in the strain described above, resulting in production of 60 g/liter quinic acid plus 2.6 g/liter DHQ in 60 h from a fed-batch fermentation of glucose (21).

Further improvements in flux through the engineered aromatic amino acid synthesis pathway focused on increased availability of upstream intermediates. Overexpression of the pentose phosphate pathway enzyme transketolase (*tktA*) increases E4P availability (22), causing metabolic competition for PEP to become a limiting factor in DAHP synthesis. A major source of PEP competition is PEP-dependent glucose uptake and phosphorylation via the PTS. To improve PEP availability, PTS uptake of glucose was eliminated by *pts* gene deletions, and PEP-independent glucose uptake and phosphorylation were enhanced by heterologous expression of the *Z. mobilis glf*-encoded glucose facilitator protein (for which glucose diffusion is the mechanism of transport) and *glk*-encoded glucokinase (for ATP-dependent glucose phosphorylation) (10). These efforts resulted in production of 87 g/liter shikimic acid from glucose in fed-batch culture supplemented with yeast extract.

As an alternate approach to relieving the demand for PEP, a novel enzymatic step toward DAHP was introduced into the

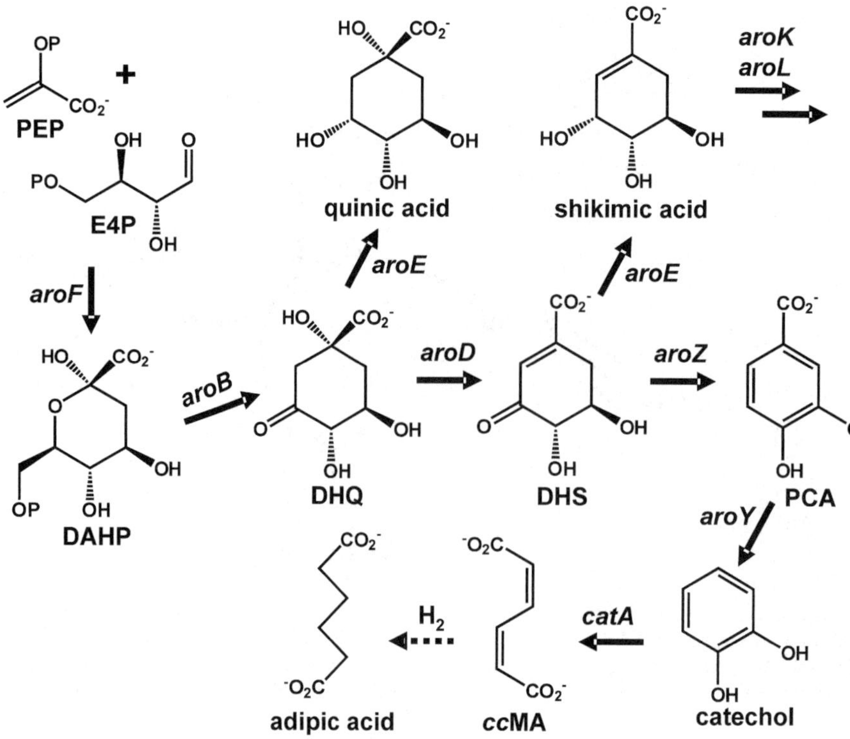

FIGURE 4 Engineered biosynthesis of shikimic acid, quinic acid, and ccMA from the *E. coli* aromatic amino acid synthesis pathway. PCA, protocatechuic acid.

shikimate pathway (57, 59). Ran et al. sought to replace PEP with pyruvate in an enzyme-catalyzed condensation with E4P to synthesize DAHP. 2-Keto-3-deoxy-phosphogalactonate (KDPGal) aldolase (DgoA) naturally catalyzes the reversible cleavage of KDPGal to pyruvate and glyceraldehyde-3-phosphate (G3P). Directed evolution of DgoA activity toward E4P was accomplished by screening mutants based on their ability to improve growth of DAHP synthase-deficient *E. coli*. Successful recovery of novel DgoA variants resulted in an effective bypass route to DAHP, which could prove useful for improved production of shikimic acid as well as a variety of other compounds derived from the pathway, including adipic acid, catechol, vanillin, and pyrogallol (59).

41.4.2. Adipic Acid

Adipic acid (shown in Fig. 4) is used in the manufacture of nylon and is traditionally synthesized from benzene, with release of harmful N_2O as a by-product. Adipic acid can also be easily synthesized chemically from ccMA (shown in Fig. 4). Niu et al. modified the *E. coli* aromatic amino acid biosynthetic pathway for the conversion of the native intermediate 3-dehydroshikimic acid (DHS) to ccMA (50). This was accomplished by expressing the genes *aroZ* from *Klebsiella pneumoniae* (encoding DHS dehydratase), *aroY* from *K. pneumoniae* (encoding protocatechuic acid decarboxylase), and *catA* from *Acinetobacter calcoaceticus* (encoding catechol 1,2-dioxygenase) in a strain lacking *aroE* and carrying extra copies of *aroB* and *tktA*. The reactions catalyzed by these enzymes are depicted in Fig. 4. These strain modifications resulted in production of 37 g/liter ccMA from glucose in 48 h (50). Additional efforts from the Frost laboratory include *E. coli*-based conversion of glucose to *p*-hydroxybenzoic acid (12 g/liter) (6) via the aromatic amino acid pathway.

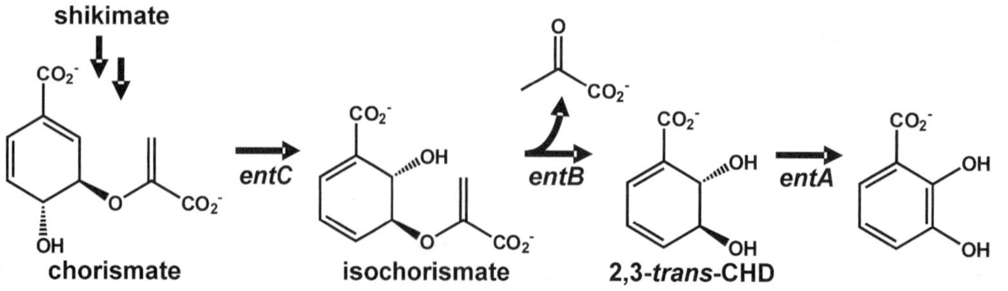

FIGURE 5 Biosynthesis of 2,3-*trans*-CHD starting from chorismate. The genes *entC*, *entB*, and *entA* encode isochorismate synthase, isochorismatase, and 2,3-dihydroxybenzoate synthase, respectively, which catalyze the corresponding reactions depicted.

41.4.3. 2,3-*trans*-CHD

In a final example of engineering the *E. coli* aromatic amino acid pathway, Franke et al. sought to overproduce (*S,S*)-2,3-dihydroxy-2,3-dihydrobenzoic acid (2,3-*trans*-CHD), which serves as a valuable chiral intermediate toward synthesis of a variety of biologically active natural products (25). As depicted in Fig. 5, 2,3-*trans*-CHD was produced by deleting *entA*, encoding 2,3-dihydroxybenzoate synthase, while overexpressing the genes *entB* and *entC*, encoding isochorismatase and isochorismate synthase. The resulting strain produced 2,3-*trans*-CHD in concentrations up to 4.6 g/liter (25).

41.5. ISOPRENOIDS

Isoprenoids (or terpenoids) represent a wide variety of specialty and pharmaceutical compounds synthesized biologically by the assembly of two or more branched, five-carbon isopentenyl pyrophosphate (IPP) subunits (7). As depicted in Fig. 6a, IPP is synthesized through two distinct routes involving either the synthesis of D-1-deoxy-xylulose-5-P (DXP) from pyruvate and G3P or the synthesis of mevalonic acid from acetyl-CoA. IPP is then isomerized to dimethylallyl pyrophosphate (DMAPP) by the enzyme IPP isomerase (*idi*), initiating chain elongation via condensation reactions

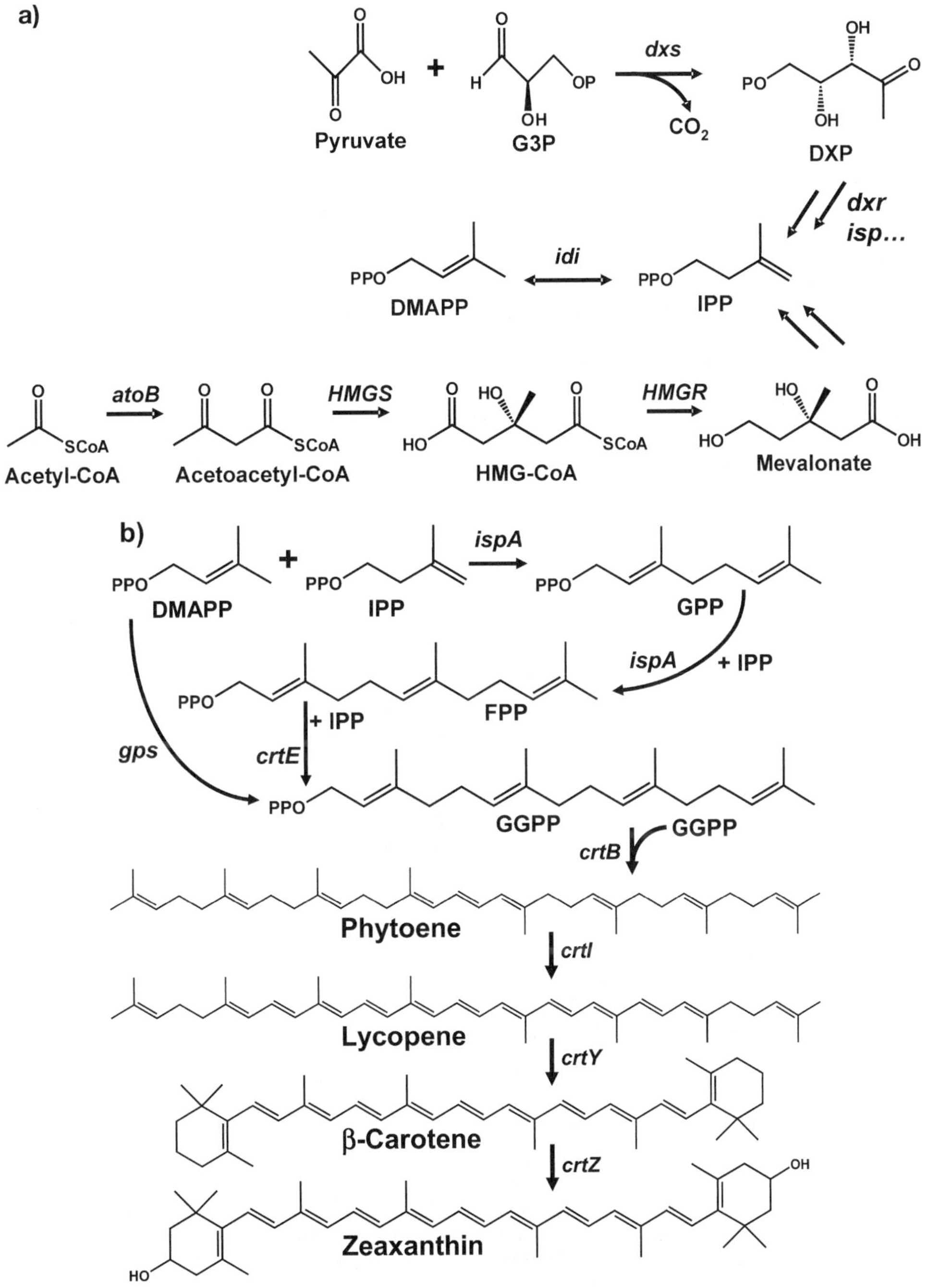

FIGURE 6 (a) Two different pathways to IPP; (b) pathways to carotenoids.

between DMAPP and IPP. As depicted in Fig. 6b, a single condensation yields the 10-carbon intermediate geranyl pyrophosphate (GPP), which can be transformed into a variety of 10-carbon natural products known as monoterpenes. Addition of another IPP to GPP yields the 15-carbon farnesyl pyrophosphate (FPP), which can be transformed into products such as quinines, sterols, and sesquiterpenes. Another IPP addition to FPP yields the 20-carbon intermediate geranylgeranyl pyrophosphate (GGPP), which can be converted to diterpenes and carotenoids.

41.5.1. Carotenoids

Carotenoids are a class of typically tetraterpenoid (40-carbon) pigments that act as light-harvesting and light-protecting agents, antioxidants, and anticancer compounds and serve as valuable synthetic precursors (48). Early metabolic engineering efforts to develop microbial strains for isoprenoid production focused on production of carotenoids, as reviewed by Barkovich and Liao (7) and Misawa and Shimada (48). In general, these involved heterologous expression of bacterial *crt* genes from the naturally carotenogenic *Erwinia* species for production of lycopene, β-carotene, and zeaxanthin in *E. coli* and other bacteria (Fig. 6b). *E. coli* has limited supplies of IPP, DMAPP, GPP, FPP, and GGPP, and strategies to improve carotenoid production using the native DXP pathway in this organism have focused on increasing the supply of IPP and DMAPP building blocks by increasing expression of native genes (*idi*, *dxs*, or *dxr*), as well as expressing the multifunctional GGPP synthase (*gps*) from the archaebacterium *Archaeoglobus fulgidus* and optimizing heterologous expression of the various downstream *crt* genes (7, 46, 48, 64, 76). The optical properties of carotenoids make them easy targets for high-throughput screening, and as a result, various enzymes involved in carotenoid synthesis have been engineered using combinatorial directed evolution techniques, leading to improved production as well as synthesis of novel natural compounds in *E. coli* (65, 73, 76).

Synthesis of IPP in *E. coli* requires G3P and pyruvate (Fig. 6a). Farmer and Liao showed that lycopene production in *E. coli* overexpressing *dxs*, *idi*, *gps*, and *crtBI* could be improved by altering the flux distribution between pyruvate and G3P (24). Increasing the availability of PEP by overexpressing PEP synthase (*pps*) or by deleting pyruvate kinase (*pyk*) led to improved lycopene production by directing excess pyruvate to G3P and balancing the availabilities of the two precursors for IPP synthesis. One drawback to the continuous expression of heterologous pathways is the metabolic burden imposed on the cells. For example, continuously overexpressing the genes involved in lycopene synthesis is likely to diminish the availability of important precursor metabolites that serve as substrates to the same pathway (e.g., pyruvate and G3P). The expression levels of pathway enzymes may at times be far in excess of the pathway's metabolic flux capacity. To improve the efficiency of lycopene production in *E. coli*, Farmer and Liao designed a dynamic genetic control system, such that the expression of rate-controlling enzymes involved in isoprenoid synthesis (*idi* and *pps*) is regulated in response to the availability of precursors required for synthesis of IPP and, ultimately, lycopene (23). Acetyl phosphate (AcP) concentrations provide a measure of excess glucose flux (overflow metabolism), and AcP was chosen as the molecule to signal availability of energetic precursors derived from glucose metabolism. The Ntr regulon of *E. coli*, which normally plays a role in adapting to nitrogen depletion, was modified to control

gene expression in response to AcP. In the native system, the nitrogen sensor protein NRII controls phosphorylation of the regulatory protein NRI, whose phosphorylated form (NRI-P) activates expression from promoter *glnAp2*. However, in the absence of NRII, the phosphorylation state of NRI depends on AcP levels. Therefore, by placing expression of *idi* and *pps* under the control of *glnAp2* in an NRII-deficient strain, expression of these genes is activated in response to precursor availability. Using this approach, the authors were able to drastically improve final lycopene titers from batch cultures compared to control strains in which *idi* and *pps* were continuously expressed from a *tac* promoter. Whereas cells without the dynamic controller secreted large amounts of acetate, and constitutive expression of *pps* caused a growth deficit, AcP-dependent expression of *idi* and *pps* alleviated metabolic imbalance and growth retardation. The sum of their efforts was a >18-fold increase in the lycopene carbon yield on glucose and an ~3-fold decrease in acetate secretion in batch cultures containing 1.5% (wt/vol) glucose. The productivity was 0.16 g/liter · h, with a final lycopene titer of ~150 mg/liter.

Stephanopoulos and coworkers used a global stoichiometric network model of *E. coli* metabolism, modified to include the lycopene pathway (*crtEBI* operon), to identify single and multiple gene deletion targets that would improve lycopene production (2). Using this model-driven, optimization-based design approach, a total of seven single or multiple gene deletions were predicted and experimentally validated to increase lycopene production by increasing the supply of precursors and cofactors important for lycopene synthesis. An alternate approach used to identify gene knockout targets involved visual screening for increased lycopene production by cells carrying transposons randomly inserted into their chromosomes (thereby randomly disrupting genes in the genome). This strategy yielded three additional gene knockout targets (4). They next screened for further improvements in lycopene production from strains carrying all combinations of the two sets of knockout targets identified from these two different strategies (64 unique knockout combinations), resulting in new gene deletion combinations representing "global maximum" strains with twofold-improved lycopene production over the parental strain (4). High-density fermentations led to production of ~220 mg/liter lycopene from ~27 g (dry cell weight)/liter (3).

41.5.2. MVA Pathway Engineering for Artemisinin Production

To achieve an adequate supply of IPP to a heterologous isoprenoid pathway in *E. coli*, as an alternative to relying on the native DXP-dependent pathway, which likely suffers from limitations owing to native feedback control mechanisms, Keasling and coworkers expressed the foreign mevalonate-dependent (MVA) pathway in this organism (45). An eight-gene biosynthetic pathway combining genes from *E. coli* and *S. cerevisiae* was divided into a "top" operon that transforms acetyl-CoA to mevalonate in three enzymatic steps and a "bottom" operon that converts mevalonate to IPP, DMAPP, and/or FPP (depending on the construct). Coexpressing these operons in a mutant *E. coli* strain carrying a lethal deletion in an essential gene in the endogenous isoprenoid biosynthetic pathway (*ispC*) complemented the deletion, demonstrating a functional MVA pathway. However, the resulting excessive accumulation of intermediates produced through the heterologous biosynthetic pathway was toxic. The top and bottom operons were next coupled to expression of a codon-optimized variant of amorphadiene synthase, which catalyzes conversion of FPP to amorphadiene, the sesquiterpene olefin precursor to the antimalarial drug artemisinin. This alleviated the growth inhibition and resulted in high-level production of amorphadiene from rich medium supplemented with glycerol. Peak amorphadiene production from the complete MVA pathway was 36- and 10-fold improved over peak production by similar strains using the native DXP pathway and an engineered DXP pathway, respectively (45).

Following this initial study, Keasling and coworkers reported a variety of additional efforts to further enhance flux through the MVA pathway for production of amorphadiene. While supply of mevalonate was identified as a limiting step in amorphadiene production, simply increasing expression of the top operon (to increase mevalonate production) led to reduced growth and productivity (56). Accumulation of the mevalonate precursor 3-hydroxy-3-methylglutaryl-CoA (HMG-CoA; synthesized through the top operon by HMG-CoA synthase) was identified as a source of inhibition. Mevalonate production was improved threefold by amplifying the expression of HMG-CoA reductase (which converts HMG-CoA to mevalonate) relative to the expression levels of the upstream enzymes that produce the intermediate from acetyl-CoA (56).

In an alternate approach to optimizing the expression levels of individual enzymes in the MVA pathway, Pfleger et al. inserted between the top operon genes combinatorial libraries of tunable intergenic regions containing genetic control elements that include mRNA secondary structures, RNase cleavage sites, and ribosome binding site sequestering sequences (55). Intergenic region libraries were first screened for their ability to satisfy a mevalonate-auxotrophic requirement in engineered *E. coli* cells (54). Combinations of intergenic regions conferring enhanced mevalonate production were then identified using a fluorescence screen that involved constitutive green fluorescent protein expression from a mevalonate auxotroph. This novel strategy of optimizing expression of multiple genes in a biosynthesis pathway resulted in approximately 10-fold-improved mevalonate production (55). Further improvements have included optimizing promoter strengths and plasmid copy numbers of genes in the bottom pathway (5), replacing the top operon genes encoding HMG-CoA synthase and HMG-CoA reductase from *S. cerevisiae* with the genes from *Staphylococcus aureus* (72), and optimizing fermentation conditions, resulting in production of >25 g/liter amorphadiene by fermentation in defined glucose medium (72).

Prior to these efforts, artemisinin production relied on extraction from plants, with low yields. The ability to synthesize large quantities of the artemisinin precursor amorphadiene from inexpensive carbon sources using microbial culture presents revolutionary opportunities to treat malaria in the developing world at drastically reduced costs. Moreover, similar efforts following these examples should result in microbial production of other terpene-based natural products having important biological activity and potential to treat disease but traditionally obtained in low yields from plants. As a final example, the heterologous MVA pathway in *E. coli* has been applied to the production of β-carotene, resulting in 465 mg/liter β-carotene from batch culture supplemented with glycerol (83).

41.6. CONCLUSIONS

Since the successful engineering of *E. coli* metabolism for conversion of glucose to the nonnative product 1,3-propanediol (13, 49), a commodity chemical now used in the industrial manufacture of renewable polymers, the field has seen tremendous advances in genomics, systems biology tools and information, and synthetic biology tools that facilitate genetic engineering. As a result, microbial metabolic engineering continues to gain momentum, with increasing frequency of success. This chapter has highlighted successful applications of metabolic engineering toward the goal of microbially based industrial chemical production and has demonstrated the diversity of strain modification strategies that can be applied to achieve overproduction of native and nonnative metabolites by *E. coli*. From these examples, it is clear that the likelihood of success increases with awareness of natural diversity (e.g., accessible enzyme functions), rate-limiting steps and pathway bottlenecks, and mechanisms of inhibition. Noteworthy are the power of combining rational and evolutionary approaches and the value of molecular tools that enable fine-tuning of gene expression.

REFERENCES

1. **Achkar, J., M. Xian, H. Zhao, and J. W. Frost.** 2005. Biosynthesis of phloroglucinol. *J. Am. Chem. Soc.* **127:**5332–5333.
2. **Alper, H., Y. S. Jin, J. F. Moxley, and G. Stephanopoulos.** 2005. Identifying gene targets for the metabolic engineering of lycopene biosynthesis in *Escherichia coli*. *Metab. Eng.* **7:**155–164.
3. **Alper, H., K. Miyaoku, and G. Stephanopoulos.** 2006. Characterization of lycopene-overproducing *E. coli* strains in high cell density fermentations. *Appl. Microbiol. Biotechnol.* **72:**968–974.
4. **Alper, H., K. Miyaoku, and G. Stephanopoulos.** 2005. Construction of lycopene-overproducing *E. coli* strains by combining systematic and combinatorial gene knockout targets. *Nat. Biotechnol.* **23:**612–616.
5. **Anthony, J. R., L. C. Anthony, F. Nowroozi, G. Kwon, J. D. Newman, and J. D. Keasling.** 2009. Optimization of the mevalonate-based isoprenoid biosynthetic pathway in *Escherichia coli* for production of the anti-malarial drug precursor amorpha-4,11-diene. *Metab. Eng.* **11:**13–19.
6. **Barker, J. L., and J. W. Frost.** 2001. Microbial synthesis of p-hydroxybenzoic acid from glucose. *Biotechnol. Bioeng.* **76:**376–390.
7. **Barkovich, R., and J. C. Liao.** 2001. Metabolic engineering of isoprenoids. *Metab. Eng.* **3:**27–39.
8. **Causey, T. B., K. T. Shanmugam, L. P. Yomano, and L. O. Ingram.** 2004. Engineering *Escherichia coli* for efficient conversion of glucose to pyruvate. *Proc. Natl. Acad. Sci. USA* **101:**2235–2240.
9. **Causey, T. B., S. Zhou, K. T. Shanmugam, and L. O. Ingram.** 2003. Engineering the metabolism of *Escherichia coli* W3110 for the conversion of sugar to redox-neutral and oxidized products: homoacetate production. *Proc. Natl. Acad. Sci. USA* **100:**825–832.
10. **Chandran, S. S., J. Yi, K. M. Draths, R. von Daeniken, W. Weber, and J. W. Frost.** 2003. Phosphoenolpyruvate availability and the biosynthesis of shikimic acid. *Biotechnol. Progr.* **19:**808–814.
11. **Chatterjee, R., C. S. Millard, K. Champion, D. P. Clark, and M. I. Donnelly.** 2001. Mutation of the *ptsG* gene results in increased production of succinate in fermentation of glucose by *Escherichia coli*. *Appl. Environ. Microbiol.* **67:**148–154.
12. **Chatterjee, R., and L. Yuan.** 2006. Directed evolution of metabolic pathways. *Trends Biotechnol.* **24:**28–38.
13. **Chotani, G., T. Dodge, A. Hsu, M. Kumar, R. LaDuca, D. Trimbur, W. Weyler, and K. Sanford.** 2000. The commercial production of chemicals using pathway engineering. *Biochim. Biophys. Acta* **1543:**434–455.

14. **de Graaf, A. A., L. Eggeling, and H. Sahm.** 2001. Metabolic engineering for L-lysine production by *Corynebacterium glutamicum*. *Adv. Biochem. Eng. Biotechnol.* **73**:9–29.

15. **Dell, K. A., and J. W. Frost.** 1993. Identification and removal of impediments to biocatalytic synthesis of aromatics from D-glucose: rate-limiting enzymes in the common pathway of aromatic amino-acid biosynthesis. *J. Am. Chem. Soc.* **115**:11581–11589.

16. **Deutscher, J., C. Francke, and P. W. Postma.** 2006. How phosphotransferase system-related protein phosphorylation regulates carbohydrate metabolism in bacteria. *Microbiol. Mol. Biol. Rev.* **70**:939–1031.

17. **de Vos, W. M., and J. Hugenholtz.** 2004. Engineering metabolic highways in lactococci and other lactic acid bacteria. *Trends Biotechnol.* **22**:72–79.

18. **Dien, B. S., N. N. Nichols, and R. J. Bothast.** 2002. Fermentation of sugar mixtures using *Escherichia coli* catabolite repression mutants engineered for production of L-lactic acid. *J. Ind. Microbiol. Biotechnol.* **29**:221–227.

19. **Dien, B. S., N. N. Nichols, and R. J. Bothast.** 2001. Recombinant *Escherichia coli* engineered for production of L-lactic acid from hexose and pentose sugars. *J. Ind. Microbiol. Biotechnol.* **27**:259–264.

20. **Donnelly, M. I., C. S. Millard, D. P. Clark, M. J. Chen, and J. W. Rathke.** 1998. A novel fermentation pathway in an *Escherichia coli* mutant producing succinic acid, acetic acid, and ethanol. *Appl. Biochem. Biotechnol.* **70–72**:187–198.

21. **Draths, K. M., D. R. Knop, and J. W. Frost.** 1999. Shikimic acid and quinic acid: replacing isolation from plant sources with recombinant microbial biocatalysis. *J. Am. Chem. Soc.* **121**:1603–1604.

22. **Draths, K. M., D. L. Pompliano, D. L. Conley, J. W. Frost, A. Berry, G. L. Disbrow, R. J. Staversky, and J. C. Lievense.** 1992. Biocatalytic synthesis of aromatics from D-glucose—the role of transketolase. *J. Am. Chem. Soc.* **114**:3956–3962.

23. **Farmer, W. R., and J. C. Liao.** 2000. Improving lycopene production in *Escherichia coli* by engineering metabolic control. *Nat. Biotechnol.* **18**:533–537.

24. **Farmer, W. R., and J. C. Liao.** 2001. Precursor balancing for metabolic engineering of lycopene production in *Escherichia coli*. *Biotechnol. Prog.* **17**:57–61.

25. **Franke, D., V. Lorbach, S. Esser, C. Dose, G. A. Sprenger, M. Halfar, J. Thommes, R. Muller, R. Takors, and M. Muller.** 2003. (S,S)-2,3-Dihydroxy-2,3-dihydrobenzoic acid: microbial access with engineered cells of *Escherichia coli* and application as starting material in natural-product synthesis. *Chem. Eur. J.* **9**:4188–4196.

26. **Gibson, J. M., P. S. Thomas, J. D. Thomas, J. L. Barker, S. S. Chandran, M. K. Harrup, K. M. Draths, and J. W. Frost.** 2001. Benzene-free synthesis of phenol. *Angew. Chem. Int. Ed. Engl.* **40**:1945–1948.

27. **Grabar, T. B., S. Zhou, K. T. Shanmugam, L. P. Yomano, and L. O. Ingram.** 2006. Methylglyoxal bypass identified as source of chiral contamination in L(+) and D(−)-lactate fermentations by recombinant *Escherichia coli*. *Biotechnol. Lett.* **28**:1527–1535.

28. **Guo, J., and J. W. Frost.** 2004. Synthesis of aminoshikimic acid. *Org. Lett.* **6**:1585–1588.

29. **Huang, Y. L., K. Mann, J. M. Novak, and S. T. Yang.** 1998. Acetic acid production from fructose by *Clostridium formicoaceticum* immobilized in a fibrous-bed bioreactor. *Biotechnol. Prog.* **14**:800–806.

30. **Jantama, K., M. J. Haupt, S. A. Svoronos, X. Zhang, J. C. Moore, K. T. Shanmugam, and L. O. Ingram.** 2008. Combining metabolic engineering and metabolic evolution to develop nonrecombinant strains of *Escherichia coli* C that produce succinate and malate. *Biotechnol. Bioeng.* **99**:1140–1153.

31. **Jeffries, T. W.** 2006. Engineering yeasts for xylose metabolism. *Curr. Opin. Biotechnol.* **17**:320–326.

32. **Kaldor, S. W., V. J. Kalish, J. F. Davies II, B. V. Shetty, J. E. Fritz, K. Appelt, J. A. Burgess, K. M. Campanale, N. Y. Chirgadze, D. K. Clawson, B. A. Dressman, S. D. Hatch, D. A. Khalil, M. B. Kosa, P. P. Lubbehusen, M. A. Muesing, A. K. Patick, S. H. Reich, K. S. Su, and J. H. Tatlock.** 1997. Viracept (nelfinavir mesylate, AG1343): a potent, orally bioavailable inhibitor of HIV-1 protease. *J. Med. Chem.* **40**:3979–3985.

33. **Kern, A., E. Tilley, I. S. Hunter, M. Legisa, and A. Glieder.** 2007. Engineering primary metabolic pathways of industrial micro-organisms. *J. Biotechnol.* **129**:6–29.

34. **Koffas, M., C. Roberge, K. Lee, and G. Stephanopoulos.** 1999. Metabolic engineering. *Annu. Rev. Biomed. Eng.* **1**:535–557.

35. **Kramer, M., J. Bongaerts, R. Bovenberg, S. Kremer, U. Muller, S. Orf, M. Wubbolts, and L. Raeven.** 2003. Metabolic engineering for microbial production of shikimic acid. *Metab. Eng.* **5**:277–283.

36. **Lee, K. H., J. H. Park, T. Y. Kim, H. U. Kim, and S. Y. Lee.** 2007. Systems metabolic engineering of *Escherichia coli* for L-threonine production. *Mol. Syst. Biol.* **3**:149.

37. **Lee, S. Y., J. M. Kim, H. Song, J. W. Lee, T. Y. Kim, and Y. S. Jang.** 2008. From genome sequence to integrated bioprocess for succinic acid production by *Mannheimia succiniciproducens*. *Appl. Microbiol. Biotechnol.* **79**:11–22.

38. **Leuchtenberger, W., K. Huthmacher, and K. Drauz.** 2005. Biotechnological production of amino acids and derivatives: current status and prospects. *Appl. Microbiol. Biotechnol.* **69**:1–8.

39. **Li, W., D. Xie, and J. W. Frost.** 2005. Benzene-free synthesis of catechol: interfacing microbial and chemical catalysis. *J. Am. Chem. Soc.* **127**:2874–2882.

40. **Li, Y., J. Chen, and S. Y. Lun.** 2001. Biotechnological production of pyruvic acid. *Appl. Microbiol. Biotechnol.* **57**:451–459.

41. **Lin, H., G. N. Bennett, and K. Y. San.** 2005. Fed-batch culture of a metabolically engineered *Escherichia coli* strain designed for high-level succinate production and yield under aerobic conditions. *Biotechnol. Bioeng.* **90**:775–779.

42. **Lin, H., G. N. Bennett, and K. Y. San.** 2005. Genetic reconstruction of the aerobic central metabolism in *Escherichia coli* for the absolute aerobic production of succinate. *Biotechnol. Bioeng.* **89**:148–156.

43. **Lin, H., G. N. Bennett, and K. Y. San.** 2005. Metabolic engineering of aerobic succinate production systems in *Escherichia coli* to improve process productivity and achieve the maximum theoretical succinate yield. *Metab. Eng.* **7**:116–127.

44. **Maier, T. H.** 2003. Semisynthetic production of unnatural L-alpha-amino acids by metabolic engineering of the cysteine-biosynthetic pathway. *Nat. Biotechnol.* **21**:422–427.

45. **Martin, V. J., D. J. Pitera, S. T. Withers, J. D. Newman, and J. D. Keasling.** 2003. Engineering a mevalonate pathway in *Escherichia coli* for production of terpenoids. *Nat. Biotechnol.* **21**:796–802.

46. **Matthews, P. D., and E. T. Wurtzel.** 2000. Metabolic engineering of carotenoid accumulation in *Escherichia coli* by modulation of the isoprenoid precursor pool with expression of deoxyxylulose phosphate synthase. *Appl. Microbiol. Biotechnol.* **53**:396–400.

47. **McKinlay, J. B., C. Vieille, and J. G. Zeikus.** 2007. Prospects for a bio-based succinate industry. *Appl. Microbiol. Biotechnol.* **76**:727–740.

48. **Misawa, N., and H. Shimada.** 1998. Metabolic engineering for the production of carotenoids in non-carotenogenic bacteria and yeasts. *J. Biotechnol.* **59**:169–181.

49. Nakamura, C. E., and G. M. Whited. 2003. Metabolic engineering for the microbial production of 1,3-propanediol. *Curr. Opin. Biotechnol.* **14:**454–459.

50. Niu, W., K. M. Draths, and J. W. Frost. 2002. Benzene-free synthesis of adipic acid. *Biotechnol. Prog.* **18:**201–211.

51. Park, J. H., K. H. Lee, T. Y. Kim, and S. Y. Lee. 2007. Metabolic engineering of *Escherichia coli* for the production of L-valine based on transcriptome analysis and *in silico* gene knockout simulation. *Proc. Natl. Acad. Sci. USA* **104:**7797–7802.

52. Park, J. H., and S. Y. Lee. 2008. Towards systems metabolic engineering of microorganisms for amino acid production. *Curr. Opin. Biotechnol.* **19:**454–460.

53. Peters-Wendisch, P., M. Stolz, H. Etterich, N. Kennerknecht, H. Sahm, and L. Eggeling. 2005. Metabolic engineering of *Corynebacterium glutamicum* for L-serine production. *Appl. Environ. Microbiol.* **71:**7139–7144.

54. Pfleger, B. F., D. J. Pitera, J. D. Newman, V. J. Martin, and J. D. Keasling. 2007. Microbial sensors for small molecules: development of a mevalonate biosensor. *Metab. Eng.* **9:**30–38.

55. Pfleger, B. F., D. J. Pitera, C. D. Smolke, and J. D. Keasling. 2006. Combinatorial engineering of intergenic regions in operons tunes expression of multiple genes. *Nat. Biotechnol.* **24:**1027–1032.

56. Pitera, D. J., C. J. Paddon, J. D. Newman, and J. D. Keasling. 2007. Balancing a heterologous mevalonate pathway for improved isoprenoid production in *Escherichia coli*. *Metab. Eng.* **9:**193–207.

57. Ran, N., K. M. Draths, and J. W. Frost. 2004. Creation of a shikimate pathway variant. *J. Am. Chem. Soc.* **126:**6856–6857.

58. Ran, N., D. R. Knop, K. M. Draths, and J. W. Frost. 2001. Benzene-free synthesis of hydroquinone. *J. Am. Chem. Soc.* **123:**10927–10934.

59. Ran, N. Q., and J. W. Frost. 2007. Directed evolution of 2-keto-3-deoxy-6-phosphogalactonate aldolase to replace 3-deoxy-D-arabino-heptulosonic acid 7-phosphate synthase. *J. Am. Chem. Soc.* **129:**6130–6139.

60. Reed, J. L., T. D. Vo, C. H. Schilling, and B. O. Palsson. 2003. An expanded genome-scale model of *Escherichia coli* K-12 (iJR904 GSM/GPR). *Genome Biol.* **4:**R54.

61. Saha, B. C. 2003. Commodity chemicals production by fermentation: an overview. *Am. Chem. Soc. Symp. Ser.* **862:**3–17.

62. Sahm, H., L. Eggeling, and A. A. de Graaf. 2000. Pathway analysis and metabolic engineering in *Corynebacterium glutamicum*. *Biol. Chem.* **381:**899–910.

63. Sanchez, A. M., G. N. Bennett, and K. Y. San. 2005. Novel pathway engineering design of the anaerobic central metabolic pathway in *Escherichia coli* to increase succinate yield and productivity. *Metab. Eng.* **7:**229–239.

64. Sandmann, G., M. Albrecht, G. Schnurr, O. Knorzer, and P. Boger. 1999. The biotechnological potential and design of novel carotenoids by gene combination in *Escherichia coli*. *Trends Biotechnol.* **17:**233–237.

65. Schmidt-Dannert, C., D. Umeno, and F. H. Arnold. 2000. Molecular breeding of carotenoid biosynthetic pathways. *Nat. Biotechnol.* **18:**750–753.

66. Shanmugam, K. T., and L. O. Ingram. 2008. Engineering biocatalysts for production of commodity chemicals. *J. Mol. Microbiol. Biotechnol.* **15:**8–15.

67. Singh, S. K., S. U. Ahmed, and A. Pandey. 2006. Metabolic engineering approaches for lactic acid production. *Process Biochem.* **41:**991–1000.

68. Smith, G. M., S. A. Lee, K. C. Reilly, M. A. Eiteman, and E. Altman. 2006. Fed-batch two-phase production of alanine by a metabolically engineered *Escherichia coli*. *Biotechnol. Lett.* **28:**1695–1700.

69. Song, H., and S. Y. Lee. 2006. Production of succinic acid by bacterial fermentation. *Enzyme Microb. Technol.* **39:**352–361.

70. Stephanopoulos, G., A. A. Aristidou, and J. Nielsen. 1998. *Metabolic Engineering: Principles and Methodologies*. Academic Press, San Diego, CA.

71. Tammali, R., G. Seenayya, and G. Reddy. 2003. Fermentation of cellulose to acetic acid by *Clostridium lentocellum* SG6: induction of sporulation and effect of buffering agent on acetic acid production. *Lett. Appl. Microbiol.* **37:**304–308.

72. Tsuruta, H., C. J. Paddon, D. Eng, J. R. Lenihan, T. Horning, L. C. Anthony, R. Regentin, J. D. Keasling, N. S. Renninger, and J. D. Newman. 2009. High-level production of amorpha-4,11-diene, a precursor of the antimalarial agent artemisinin, in *Escherichia coli*. *PLoS ONE* **4:**e4489.

73. Umeno, D., A. V. Tobias, and F. H. Arnold. 2005. Diversifying carotenoid biosynthetic pathways by directed evolution. *Microbiol. Mol. Biol. Rev.* **69:**51–78.

74. Vemuri, G. N., M. A. Eiteman, and E. Altman. 2002. Effects of growth mode and pyruvate carboxylase on succinic acid production by metabolically engineered strains of *Escherichia coli*. *Appl. Environ. Microbiol.* **68:**1715–1727.

75. Vemuri, G. N., M. A. Eiteman, and E. Altman. 2002. Succinate production in dual-phase *Escherichia coli* fermentations depends on the time of transition from aerobic to anaerobic conditions. *J. Ind. Microbiol. Biotechnol.* **28:**325–332.

76. Wang, C., M. K. Oh, and J. C. Liao. 2000. Directed evolution of metabolically engineered *Escherichia coli* for carotenoid production. *Biotechnol. Prog.* **16:**922–926.

77. Wang, L., J. Xie, and P. G. Schultz. 2006. Expanding the genetic code. *Annu. Rev. Biophys. Biomol. Struct.* **35:**225–249.

78. Wendisch, V. F., M. Bott, and B. J. Eikmanns. 2006. Metabolic engineering of *Escherichia coli* and *Corynebacterium glutamicum* for biotechnological production of organic acids and amino acids. *Curr. Opin. Microbiol.* **9:**268–274.

79. Werpy, T., and G. Petersen (ed.). 2004. *Value Added Chemicals from Biomass*, vol. I—*Results of Screening for Potential Candidates from Sugars and Synthesis Gas*. National Renewable Energy Laboratory, Washington, DC.

80. Wolfe, A. J. 2005. The acetate switch. *Microbiol. Mol. Biol. Rev.* **69:**12–50.

81. Wyckoff, H. A., J. Chow, T. R. Whitehead, and M. A. Cotta. 1997. Cloning, sequence, and expression of the L-(+) lactate dehydrogenase of *Streptococcus bovis*. *Curr. Microbiol.* **34:**367–373.

82. Xu, B., M. Jahic, G. Blomsten, and S. O. Enfors. 1999. Glucose overflow metabolism and mixed-acid fermentation in aerobic large-scale fed-batch processes with *Escherichia coli*. *Appl. Microbiol. Biotechnol.* **51:**564–571.

83. Yoon, S. H., S. H. Lee, A. Das, H. K. Ryu, H. J. Jang, J. Y. Kim, D. K. Oh, J. D. Keasling, and S. W. Kim. 2009. Combinatorial expression of bacterial whole mevalonate pathway for the production of beta-carotene in *E. coli*. *J. Biotechnol.* **140:**218–226.

84. Zelic, B., T. Gerharz, M. Bott, D. Vasic-Racki, C. Wandrey, and R. Takors. 2003. Fed-batch process for pyruvate production by recombinant *Escherichia coli* YYC202 strain. *Eng. Life Sci.* **3:**299–305.

85. Zelic, B., S. Gostovic, K. Vuorilehto, D. Vasic-Racki, and R. Takors. 2004. Process strategies to enhance pyruvate production with recombinant *Escherichia coli*: from repetitive fed-batch to in situ product recovery with fully integrated electrodialysis. *Biotechnol. Bioeng.* **85:**638–646.

86. Zeng, A. P., and H. Biebl. 2002. Bulk chemicals from biotechnology: the case of 1,3-propanediol production and the new trends. *Adv. Biochem. Eng. Biotechnol.* **74:**239–259.

87. **Zhang, X., K. Jantama, J. C. Moore, K. T. Shanmugam, and L. O. Ingram.** 2007. Production of L-alanine by metabolically engineered *Escherichia coli. Appl. Microbiol. Biotechnol.* **77:**355–366.

88. **Zhou, S., T. B. Grabar, K. T. Shanmugam, and L. O. Ingram.** 2006. Betaine tripled the volumetric productivity of D(−)-lactate by *Escherichia coli* strain SZ132 in mineral salts medium. *Biotechnol. Lett.* **28:**671–676.

89. **Zhou, S., K. T. Shanmugam, and L. O. Ingram.** 2003. Functional replacement of the *Escherichia coli* D-(−)-lactate dehydrogenase gene (*ldhA*) with the L-(+)-lactate dehydrogenase gene (*ldhL*) from *Pediococcus acidilactici. Appl. Environ. Microbiol.* **69:**2237–2244.

90. **Zhou, S., K. T. Shanmugam, L. P. Yomano, T. B. Grabar, and L. O. Ingram.** 2006. Fermentation of 12% (w/v) glucose to 1.2 M lactate by *Escherichia coli* strain SZ194 using mineral salts medium. *Biotechnol. Lett.* **28:**663–670.

91. **Zhu, Y., M. A. Eiteman, R. Altman, and E. Altman.** 2008. High glycolytic flux improves pyruvate production by a metabolically engineered *Escherichia coli* strain. *Appl. Environ. Microbiol.* **74:**6649–6655.

92. **Zhu, Y., M. A. Eiteman, K. DeWitt, and E. Altman.** 2007. Homolactate fermentation by metabolically engineered *Escherichia coli* strains. *Appl. Environ. Microbiol.* **73:**456–464.

Improving Microbial Robustness Using Systems Biology

JONATHAN R. MIELENZ AND DAVID A. HOGSETT

42

42.1. MICROBIAL ROBUSTNESS IN INDUSTRIAL PROCESSES

What is robustness regarding microorganisms and microbial processes? We define robustness to be a collection of properties a microorganism has that allow it to grow, conduct biochemical processes, and survive in a changing environment better than less robust microorganisms. Robustness requires the microbe to be able to rapidly adjust to different stresses derived from temperature and pH changes, nutritional limitations, presence of chemical toxins, and pathogen attacks such as by viruses and phages, while both surviving and continuing to function (Table 1). Indeed, these adaptations theoretically could result in a microbial strain that is less competitive under nonstressed conditions. Still, these characteristics are sought for industrial and environmental processes such as large-scale fermentations or bioremediation applications. Robustness is not necessarily a requirement for highly productive microorganisms that are developed for well-controlled environments, although robustness is nevertheless quite beneficial there too. Therefore, robust microbial processes are sought for all microbial processes used in the medical, food, feed nutrition, cleaning, chemical, bioenergy, and environmental industries. In this chapter, we aim to outline aspects of microbial robustness desirable for specific industries and processes and examine the current status and potential benefits of application of systems biology analysis tools to further improve these processes.

42.1.1. Biological Robustness

While robustness is a key recognized property of biological systems, the term "robustness" is frequently applied across a broad range of biological attributes, ranging in scale from genes to biological populations. For example, an enzyme can be described as robust if it performs its function across a wide range of conditions, such as pH and temperature, although optimal (peak) activity is defined by a narrow range of pH and temperature, and stability is temperature dependent (Mosier and Ladish, 2009). Likewise, the genetic makeup of an organism can be regarded as robust if it contains redundant genes and pathways for key metabolic activities or incorporates sophisticated metabolic control schemes. Additionally, "robustness" is used to describe both the ability of biological systems to respond rapidly to perturbations and their ability to adapt to evolutionary pressures

on a much longer timescale. Such broad use of the term underscores the significance of understanding biological robustness and has motivated several attempts to more precisely define the term (117).

In the context of industrial microbiology, the value of robustness can be seen across processes, organisms, and biological scales. It is a topic of both fundamental and applied value, the centrality of which has been essentially rediscovered with the development of systems biology tools. For example, robustness of even very simple biochemical networks has been shown to be important to bacterial chemotaxis, circadian rhythms, and the yeast cell cycle (10, 13, 116). Such analyses draw heavily from the network theories and computational tools applied previously to physical networks and incorporate consideration of dynamic stability, control theory, and biochemical "noise" (66, 69, 80).

This application of systems biology and network theory to biological robustness has generated valuable insights that may potentially be useful in identifying or developing industrial organisms with improved robustness. At the genetic level, latent metabolic pathways have been shown to be as important as excess metabolic capacity when organisms adapt to the loss of key metabolic enzymes (31). Perhaps not surprisingly, such metabolic flexibility often exists around core metabolic pathways that synthesize or utilize metabolites essential for biomass formation under all environmental conditions (5, 63). For example, high-throughput, integrated "omics" tools were used to evaluate *Escherichia coli* strains with disrupted glycolytic and pentose phosphate pathways. Despite the alterations to these central metabolic networks, the organism was able to reroute fluxes and exhibited only minor changes in gene transcription and protein expression (57).

In many organisms, genotypic variation far exceeds phenotypic variation, and a great many genes and gene variants can be considered inconsequential in terms of normal function. However, such variants may impart phenotypic or evolutionary capacitance, which enables the organism to behave in a normal phenotypic manner even when environmental conditions or mutations would otherwise make the system less robust (11, 84). While such phenotypic capacitance is not yet well understood, it may play a significant role in determining the robustness of industrial organisms designed around so-called "minimal" genomes and metabolic functionality. In a practical example of

TABLE 1 Preferred characteristics of a robust microorganism

Able to use inexpensive organic and inorganic sources of nitrogen and other nutrients

Able to utilize multiple complex carbohydrates as substrates for carbon and energy

Able to withstand a temperature or pH shock with minimal loss in viability

Multiple restriction modification systems to minimize virus and other external genetic attacks

Presence of metabolic and genetic redundancy for core biological processes

Able to enter dormant state to survive inhospitable environments or nutrient limitations

Ability to withstand brief oxygen deprivation or exposure for aerobes or anaerobes, respectively

Able to multiply rapidly and respond appropriately to changing environments

Reaches and maintains very high viable cell density

Able to withstand high levels of metabolic products and input inhibitors

robustness of enzymes, substitution of several amino acid residues was sufficient to enhance the stability of an alkaline protease 10-fold (43). Hence, robustness comes from a range of potential biochemical, metabolic, and genetic characteristics.

Robustness gained through genetic or metabolic redundancy is not without trade-offs, as such redundancy requires that an organism have additional genetic complexity, that it utilize more sophisticated regulatory networks, and that it carry a greater metabolic burden. These extra burdens exert pressure on biological systems to find an efficient and effective balance between redundancy, adaptability, and robustness. One possible means for accomplishing this balance is reliance upon a few simple yet complementary strategies for responding to predictable environmental stresses, such as nutrient limitation. Elementary node analysis of *E. coli* central metabolism, when applied in such a case, indicates that while only a very small fraction of permissible metabolic pathways are ecologically competitive, they represent enough diversity to account for the observed range of metabolic responses (16). In this way, a robust response results from the combined application of only a few metabolic strategies.

42.1.2. Genetic Adaptability

A key aspect of attaining robustness is a population's ability to adapt genetically to perturbations to the environment to not only survive but also possibly thrive to improved fitness. Evolutionary geneticists distinguish adaptation as different from natural selection primarily in that adaptation includes mutations that in and of themselves do not result in large genetic changes but individually move an organism to a more fit optimum (101). Such a process may have been involved in evolution of the very successful variety of beverage yeast, which is discussed below, that developed over time through genetic paths that will remain unknown. An ongoing difficulty is that beneficial, adaptive mutations are difficult to study due to low frequencies (on the order of 4×10^{-9} per cell and generation), and they are difficult to identify compared to deleterious mutations (auxotrophy and loss of other critical functions) (52). Using phage evolution in the laboratory, Wichman et al. determined that fitness can be the result of multiple mutations, many defined as adaptive yet resulting in a similar improved organism or, in this case, a phage genome (130). In their study, they examined the adaptation of φX174 to high temperature and an altered host. They found that two different lines had 15 mutations and documented the specific amino acid changed within five proteins. Most interestingly, they determined

that seven of these mutations were precisely identical as to amino acid affected and the nature of substitution, but critically, the order of their appearance was very different, suggesting that no single change or order of adaptation was decisive regarding ability to survive in a stressful environment. In the case of bacterial evolution, a laboratory 3-liter fermentor of *E. coli* approaching stationary phase ($\sim 2 \times 10^9$ cells/ml) should contain at least one mutation in every one of the $\sim$4,600 genes in the genome. Using this microorganism, two different studies analyzed adaptation of *E. coli* across hundreds of generations under specific adaptive pressures. The authors concluded that many of the genomic changes are highly reproducible although not identical as with the aforementioned study, improving *E. coli* with many mutations that had small impacts and few "silver bullets" resulting in sudden fitness (52, 141). So such a population will likely contain many silent or nearly silent mutations from an evolutionary standpoint, many of which have the potential to improve the fitness when responding to metabolic challenges or perturbations.

42.2. SYSTEMS BIOLOGY TOOLS

Systems biology tools include measurement of the level of transcripts, proteins, and metabolites in a biological system at any moment in time. With rapid sampling, these tools can be used to paint a picture of the basal operation of cellular functions in a temporal context, including gene and protein expression plus carbon and metabolic flux. Similarly, the cellular response to an external change such as environmental alterations, metabolic limitations, or exterior assaults by pathogens or infective agents can be measured and compared with respect to the basal metabolism to detect genetic and metabolic responses to the new situation. Consequently, these tools are ideal to begin to further define the nature of "robustness" in microorganisms, and cellular and genetic characteristics that comprise this beneficial survivability. Certainly, a metabolic model of such robust microorganisms should have components not found in many "lesser" microorganisms. Systems biology tools have been the subject of a number of excellent reviews (1, 61, 64, 66) and so are reviewed here only briefly.

42.2.1. Genomic Analysis and Resequencing

The genetic sequence of the organism under analysis is a fundamental requirement for systems biology tools. Fortunately, the technology for sequencing is continuing to evolve rapidly from the "ancient" gel-based sequencing to fluorescent-tag-based sequencing to the evolving Roche

454 Genome Sequencer FLX™ and FLX Titanium™ and the Applied Biosystems SOLiD™ instruments. Interestingly, large numbers of repeated sequences can compromise assembly of contiguous sequences (contigs), especially for newer short-contig sequencing technologies. Even with the acquisition of genomic data, fully assembled, the function of many genes will remain unknown or hypothetical within all genomes, including that of very-well-studied organisms such as *E. coli*, due to the lack of homologous genes found in other organisms. Therefore, gene annotation remains one of the most limiting factors for genome analysis. At this writing, there are approximately 761 bacterial genome sequences, 55 archaeal microorganism genomes complete, and 100 animal and plant genomes complete. Table 2 highlights a sampling of the numerous microorganisms with fully sequenced and annotated genomes, along with areas of research that motivated their genomic analysis. Each of these microorganism's genomic data fulfill a fundamental requirement for application of systems biology tools to investigate these many important research topics.

Genome resequencing refers to determination of changes to a previously known genome sequence after a designed modification and/or evolution of the genome. An obvious example is a mutagenesis and selection procedure, but also interesting is detection of changes after multiple generations under mild selection pressure, including continuous culture. With proper genome sequence knowledge completed at the outset of a designed selection, as opposed to a genome sequence available from public data banks, there is a good likelihood a relatively limited number of genetic changes, whether they are single nucleotide polymorphisms or deletions or genome rearrangements, will be identified, providing valuable information regarding genetic changes relative to phenotypic changes. For example, the power of genome resequencing is described in the amino acid production section. In short, the genome sequence permitted specific rebuilding of a production strain, avoiding existing unintended potentially deleterious genetic changes found in the conventionally developed amino acid producer.

42.2.2. Transcriptomics for Gene Expression

Given availability of a genomic sequence, microarray gene expression analysis has made it possible to determine differences in gene expression under two or more conditions. Briefly, a microarray contains presynthesized oligonucleotide segments of selected regions of the genome, usually all the open reading frames/genes in the genome. The mRNA is isolated from an organism and converted into labeled DNA segments by enzymatic synthesis, producing genetic probes that represent the genes active under the selected condition. These labeled DNA segments derived from the mRNA are used to challenge the microarray by hybridization. If a gene is active, the mRNA is present as its labeled DNA equivalent, which will hybridize to the microarray oligonucleotide segment representing that gene. Since the locations of all selected genes are known on the microarray, hybridization and therefore expression of each of the genes can be determined. As an example, using a bacterial culture, genes expressed by a wild-type strain can be compared to a mutant to detect those genes not expressed or those whose expression has changed due to the mutation. Additionally, microarrays are very valuable to determine temporal expression of genes, for example, during the course of fermentation. Since thousands of genes are active during normal metabolism of an organism, a very large number of signals will be detected by the microarray.

The best microarray data come from differential expression of clearly definable differences, as with the example mentioned above, and completed with multiple biological and technical replicates to provide statistical validation of the results. As the technology matures, the number of arrays and replicates of arrays that can be placed on one analysis slide has grown considerably, providing tens of thousands of gene spots per microarray, yielding multiple gene coverage for simpler genomes with one microarray.

An additional tool, serial analysis of gene expression (SAGE), is used to evaluate gene expression patterns in a comprehensive manner (125). The technology relies upon the uniqueness of an ~14-bp segment (called a tag) from a gene obtained from expressed mRNA. The multiple tags are amplified and linked together for sequencing. The key advantage is that genes can be identified by these small sequences that can be linked together, allowing a sequencing run to identify dozens of genes expressed at a particular time. Using homology searching, the cells that are being examined do not require a genomic sequence because sequences of related (micro)organisms or homologous genes from anywhere could be used to identify the tagged gene. This technology has been useful for analysis of *Saccharomyces cerevisiae* and other yeasts (125, 126), as well as other cellular systems, including cancerous cells. Like all new techniques, improvements have been made with longer tags (25 to 27 bp) called super-SAGE, allowing better identification of gene location (90), and other variants aimed at microRNA expression quantitation, especially for cancer research (140).

42.2.3. Proteomics for Protein Synthesis

Proteomic analysis involves the examination of the proteins present in an organism. While early work was based on two-dimensional gels, technological improvements in genomics and mass spectrometry (MS) have permitted identification of the majority of proteins present in simple microorganisms, far more than before. For example, gel-based techniques required sufficient protein to form a detectable spot and required its structural characteristics to permit it to be within the pH and molecular weight range separable in the gel. While shotgun proteomics requires the partial purification of proteins, the technology permits many more proteins to be identified than by gel-based methods. The key to the technique is the ability to cleave the proteins specifically by site-specific proteases that generate peptides of predictable molecular weight. The peptides are separated by high-performance liquid chromatography (HPLC) and detected by either matrix-assisted laser desorption/ionization or electrospray ionization possibly coupled with time-of-flight technology (2). Computer analysis is required to connect the hundreds of peptides of specific molecular weight to the protein encoded by the genome of the targeted organism. MS protein identification can be challenging when there is extensive posttranslational modification of proteins by phosphorylation, glycosylation, transferases, or other modifications that increase the peptide molecular weight, making the computational identification of those peptides impossible without additional treatment of the protein sample. Interestingly, coupled with genomic data, experience has shown that proteins expressed from microorganisms resident in a complex matrix such as soil or biomass can still be identified in the presence of these environmental "contaminating" proteins, making proteomic technology very powerful for dissection of complex biological materials (108, 132).

TABLE 2 Examples of industrial organisms with completed genome sequence and potential areas of applied use[a]

Organism	Use(s)
Acidithiobacillus ferrooxidans	Bioleaching, bioremediation
Acidothermus cellulolyticus	Bioremediation, environmental
Actinobacillus succinogenes	Succinic acid production
Anaerocellum thermophilum	Energy production
Anaeromyxobacter sp.	Bioremediation, environmental
Arabidopsis thaliana[b]	Model plant species
Arthrobacter chlorophenolicus	Bioremediation, environmental
Aspergillus niger[b]	Citric acid production, enzyme production
Aspergillus oryzae[b]	Beverage alcohol, food industry
Bacillus amyloliquefaciens	Antibiotic production, suppresses plant pathogens
Bacillus licheniformis	Biotechnological, enzyme production
Bacillus subtilis	Biotechnological, enzyme production
Bacillus thuringiensis	Insect pathogen
Bradyrhizobium japonicum	Soybean nitrogen-fixing symbiote
Caldicellulosiruptor saccharolyticus	Cellulases, biofuels
Cellvibrio japonicus	Cellulases, biofuels
Chlamydomonas reinhardtii	Model plant species
Clostridium acetobutylicum	Acetone, butanol, ethanol production
Clostridium beijerinckii	Acetone, butanol, ethanol production
Clostridium cellulolyticum	Cellulases, biofuels
Clostridium phytofermentans	Cellulases, biofuels
Clostridium thermocellum	Cellulases, biofuels
Corynebacterium efficiens	Amino acid production
Corynebacterium glutamicum	Amino acid production, food industry
Desulfovibrio vulgaris	Bioremediation, environmental
Escherichia coli	Biomaterials, biofuels, biochemicals
Geobacter sp.	Bioremediation, microbial fuel cells
Gluconobacter oxydans	Vitamin C production
Hydrogenobaculum sp.	Hydrogen production
Klebsiella oxytoca	Biofuels
Lactobacillus casei	Lactic acid production
Lactobacillus delbrueckii subsp. *bulgaricus*	Lactic acid production
Lactobacillus plantarum	Lactic acid production, food industry
Lactococcus lactis	Cheese production
Mannheimia succiniciproducens	Succinic acid production
Methanocaldococcus jannaschii	Methane production
Methanosarcina barkeri	Methane production
Methanospirillum hungatei	Methane production
Nostoc (Anabaena)[b]	Alga model organism
Pichia stipitis[b]	Energy production, biofuels
Populus balsamifera subsp. *trichocarpa*[b]	Biomass source
Ralstonia eutropha	Bioplastic producer
Rhizobium leguminosarum	Bean nitrogen-fixing symbiote
Rhodobacter sphaeroides	Bioenergy, biohydrogen
Rhodopseudomonas palustris	Hydrogen production
Saccharomyces cerevisiae[b]	Biofuel production, food industry
Sinorhizobium meliloti	Alfalfa nitrogen-fixing symbiote
Streptococcus thermophilus	Yogurt production
Sphingomonas (Pseudomonas) elodea	Biogum production

continued on next page

TABLE 2 *(continued)*

Organism	Use(s)
Streptomyces avermitilis	Antibiotic production
Streptomyces coelicolor	Antibiotic production
Thermoanaerobacter ethanolicus	Biofuels
Thermoanaerobacter pseudethanolicus	Biofuels
Thermobifida fusca	Cellulases, biofuels
Trichoderma reesei[b]	Enzyme production
Vitis vinifera L.[b]	Grapes and wine
Yarrowia lipolytica[b]	Citric acid production
Xanthomonas campestris	Biogum production
Zymomonas mobilis	Biofuels

[a]Data gathered from www.genomesonline.org; www.genolevures.org.
[b]Eukaryote.

42.2.4. Metabolomics

Biological systems are constructed of chemical molecules either obtained from the environment (i.e., food) or synthesized from portions of these input materials by cellular metabolism. While much can be learned from examination of gene expression and the resulting protein expression, identification of the chemical molecules that comprise internal cellular metabolism can be critical to understanding regulation of cellular gene and protein/enzyme expression. Metabolomics aims to accomplish this by use of advanced separation and detection tools to identify small metabolic molecules. Typically, a sample is taken and quickly extracted to remove the small-molecule metabolites by use of very cold solvents, such as methanol, to halt metabolic activity as quickly as possible. Special approaches are needed for charged molecules or those not readily solubilized by cold solvents. As metabolism is dynamic, temporal sampling of a biological system in rapid succession can detect changes in metabolic processes which can confer robustness after the organism is challenged by inhibitors or other stressful or selective conditions. These small molecules are typically separated by chromatography (gas chromatography, HPLC, or capillary) and detected by either nuclear magnetic resonance spectroscopy or MS, which are particularly suited to small molecules, with modifications to separate charged molecules such as phosphorylated molecules common in cellular processes. Unlike with the previous "omics," the genome sequence is not required to delve deeply into the metabolism of biological systems using metabolomics (89).

42.2.5. Fluxomics for Metabolite Flow Analysis

Analysis of the flow of metabolites within an organism is termed fluxomics and is a natural extension of metabolomic analysis. Indeed, temporal analysis of the metabolome comprises the fluxome, with flow of cellular small molecules delineated in a linked metabolic pathway, permitting analysis of the flow of materials, as shown in *Saccharomyces cerevisiae* (14). Pathway engineering in combination with fluxomic analysis provides a powerful approach by which to understand the changes engineered into a metabolic chain of reactions. Such understanding can be critical to understanding an organism's ability to survive stresses and is therefore closely aligned to understanding differences between robust (micro)organisms and more sensitive relatives. Fluxomics has been helpful for production of small molecules such as amino acids and is being used to analyze production of ethanol production from both corn starch and cellulosic feedstocks.

42.2.6. Metabolic Model Development

Metabolic models are valuable tools to begin to structure and test understanding of cellular processes. Models can have different degrees of complexity and can attempt to model the whole cell or, more often, specific aspects of cellular processes such as carbon flow, DNA replication, or bacterial sporulation. As computational power has grown, the possibility to model hundreds of genes and enzymes and link them to the fluxome has become realistic. Models provide platforms to test interactions of different cellular processes using multiple approaches including all the "omics" tools described briefly above. Metabolic modeling is particularly useful for fermentation chemicals such as organic acids, alcohols, and amino acids, as outlined below.

42.3. INDUSTRIAL ENZYME PROCESSING AIDS AND PRODUCTION STRAINS

Industrial enzyme production is a mature industry that contributes to numerous products and processes, including food production, improved cleaning products, pulp and paper processes, textiles, animal feed, and niche markets such as chemical detectors, research enzymes, and chiral chemical processes. Due to their complex structure, enzymes are synthesized exclusively by biological processes. Microorganisms are most commonly used for enzyme production, although there are examples of enzymes isolated from plant and animal sources. Since enzyme production is largely a microbial process, highly productive fermentation processes have been developed yielding enzymes in excess of 50 g of protein per liter, which is now required in many instances to be commercially competitive.

Bringing an enzyme to the commercial market requires a number of steps. Initially an enzyme must have catalytic and stability characteristics that meet the requirements of the application for which they are destined. Such functional criteria are typically very stringent, especially now with the many enzymes that have been developed over the last three decades (106). For example, early commercial enzyme producers quickly realized that process knowledge was essential, consistent with the formation of the strategic partnership between Procter & Gamble and Genencor International developed in the 1980s for development of detergent enzymes. This partnership permitted Genencor to understand the very stringent requirements needed for a successful and superior detergent and processing enzymes needed in Procter & Gamble product lines. As is discussed in some detail below, Henkel KGaA (Düsseldorf, Germany), Staley/Tate & Lyle, and CPC

International are three of a small number of companies that developed in-house capability for enzymes used in detergent and sweetener applications. Proprietary details for "doing the laundry" or production of high-fructose corn syrup were readily available to in-house microbiologists/geneticists seeking improved enzymes for their processes. This proprietary process knowledge has been one of the significant hurdles facing nascent enzyme companies, such as Diversa/Verenium, Dyadic, Direvo, and Maxygen, whose products must compete with and hopefully displace carefully crafted existing enzyme products that are commercially successful using existing processes with attendant depreciated capital.

Selection of an enzyme that provides the best match to existing or anticipated process needs is critical, as it provides the springboard for both enzyme improvement and high-level production. Enzyme production has been accomplished through the use of convenient, reasonably productive original host microorganisms such as *Bacillus* or *Streptomyces*, as well as filamentous fungi such as *Aspergillus niger*. However, it was understood early on that improved production can be accomplished better by transferring the enzyme structural gene to a new, more suitable host, such as genetically modifiable fungi or bacteria, initially with associated production improvement by simple gene dosage. Such gene cloning opened the doors to genetic improvement by numerous approaches, including random gene mutagenesis, site-directed modification, gene and nucleotide shuffling, and gene recruiting. Productivity improvements were gained by modification or recruitment of regulatory sequences, most prominently a paired promoter/Shine-Dalgarno sequence with a strong terminator, but more sophisticated improvements continue to be developed (113).

Selection of the best microbial production host organism is extremely critical, since follow-on genetic and fermentation process development will be built around this organism. Approaches have taken and continue to take two general forms. The first is the development of a process around a microorganism that can easily be genetically modified, with the expectation that deficient characteristics can be genetically improved. The alternate approach is the selection of a robust host microorganism that has a known track record of functioning in an industrial process, and struggling with the usually difficult task of genetically modifying a microorganism for which the requisite tools are not developed. Both approaches have merit, and their selection often hinges upon the value of the product as well as the complexity of the protein regarding required posttranslational modification. For example, *E. coli* K-12 was chosen by Eli Lilly as the host for production of human insulin even with its lack of protein transport and potential for endotoxins because the high value of the product supports more expensive, highly regimented regulation-driven reproducible production and downstream processing. By contrast, industrial carbohydrases, such as glucoamylases, and cellulases continue to be produced by filamentous fungi, even with the associated complexity of production due to required extensive posttranslational modification and since their applications do not require highly purified enzyme products.

42.3.1. Detergent Enzymes and Production Strains

As mentioned previously, robust production strains are highly beneficial economically because their performance in industrial fermentors, 50 m^3 and larger, is predictable and dependable. Henkel KGaA, as the largest producer of laundry detergent in Europe, was faced with the need to produce an improved detergent protease to replace their in-house-produced *Bacillus licheniformis* alkaline protease

P300 or to purchase a protease from an enzyme manufacturer such as Novo Nordisk at a higher price. The current production host had been subjected to multiple rounds of random chemical mutagenesis, which yielded a strain able to produce active protease at economically competitive levels. In-house research identified a more effective protease from *Bacillus lentus* which passed required stability tests in company applications laboratories, which included the required perborate bleach and temperature stability needed for European laundry practices. The structural gene for the protease was cloned, and production in *Bacillus subtilis* rose modestly but did not match the productivity of the older enzyme, P300 (133), being produced on a large scale from *B. licheniformis*, a well-characterized high-yielding enzyme producer. To take advantage of this production strain, the *B. lentus* protease gene was cloned behind the P300 protease promoter and Shine-Dalgarno ribosome binding sequence (133). However, it was not known whether the signal sequence for the *B. lentus* enzyme would function as efficiently in the new industrial host, so the new protease gene was either left linked to the resident homologous secretion sequence or attached to the *B. licheniformis* P300 secretion sequence prior to cloning into the robust, dependable production host, *B. licheniformis*. Results showed that the secretion sequence homologous to the host and the new structural gene was significantly more productive, permitting the well-characterized production process to now produce a superior alkaline protease at protein concentrations obtained previously with the inferior enzyme. Subsequently, the genetic construction was inserted in the chromosome after multiple rounds of site-directed mutagenesis, yielding one of the best alkaline proteases for detergent applications, all being produced by the original superior production host strain (43, 44).

While production of alkaline proteases, such as described above for Henkel KGaA, involves quite mature technology where possibly only small incremental improvements remain, much can be learned from analysis of the production process with common heterologous host-enzyme combinations using systems biology tools. For example, sequencing of the genome of the production strain will yield valuable hints as to why this host organism is so effective in protein production under industrial conditions. Furthermore, with the genome sequence available, it could be determined if the protein production proceeds consistently throughout the fermentation run using both transcriptomic and proteomic approaches, although the latter will be a challenge in the presence of very high levels of protease. Certainly protein synthesis has a high energy demand, and potential limitations in synthetic capability may become evident with gene expression analysis. For example, a detergent protease expression was modeled for mass flux looking at enzyme synthesis (15). Furthermore, Henkel, as an oleochemical company, could benefit from this information were they to want to produce lipases or other enzymes for detergent, chemical, or personal-care businesses with a well-understood host organism.

42.4. FERMENTED BEVERAGE INDUSTRY

One of the oldest uses of microorganisms is production of fermented beverages by yeasts and other microorganisms, which has evolved from a spontaneous process to the current well-developed art and science of production of wine, beer, and processed fermentation beverages such as various liquors. Indeed, alcoholic fermentation was used by early societies to preserve beverages from fresh fruits and other sources of fermentable sugars. For example, *Zymomonas* has long been used for production of pulque from fermented agave sap and has appeared as a troublesome contaminant in wine making. *S. cerevisiae* and other yeasts have been used for wine and beer production since antiquity (9, 92). Currently the technology is very advanced, with dramatic improvements in quality and quantity due to improved processing equipment and selection and development of improved fermentation yeasts. Indeed, studies using systems biology tools are beginning to delineate differences between laboratory yeast strains and those developed through the ages as robust fermentative strains that permit the production of high-quality alcoholic beverages.

Laboratory strains of *S. cerevisiae* have been the subject of intense analysis since the genome sequence was determined in 1996 (45, 134); such analysis includes generation of a metabolic network model for a portion of the intracellular metabolism, including over 700 genes/open reading frames, or about 16% of the genome (32). For example, the linkage of transcriptomic analysis with the metabolome during a sudden increase in carbon source (glucose) in *S. cerevisiae* showed that within 5 min, the organism can respond to addition of glucose both with increased intracellular metabolites and with activation of genes associated with purine synthesis due to the sudden demand for ATP (71). Interestingly, the demand for the energy carrier's metabolites was severe enough that the mRNA degradation rate increased to provide purine building blocks, giving a clue with respect to the extent of cellular metabolism reprogramming expected due to a sudden change in the external environment. Finally, a comprehensive multiomic study correlating the gene expression (transcriptome), protein expression (proteome) and the external and internal metabolome with a chemostat-grown *S. cerevisiae* strain permitted the delineation of some of the details of growth control of this laboratory yeast (19). Genes associated with growth under nutrient-limited and -sufficient conditions were identified, as well as stress- and stimulus-related genes, and provided a guide for integrated systems biology analysis of a complex biological system and possible approaches to understand how robustness can be detected and quantified in a cellular system.

Regarding robustness, laboratory yeasts cannot successfully compete with commercial wine and fermentation yeasts, which have been selected to be very hardy and survive a very stressful environment while rapidly fermenting metabolizable sugars to ethanol. When a yeast is pitched into a wine must, it is suddenly exposed to very high sugar levels (i.e., >22% sugar by weight) with associated high osmotic pressure, plus low pH (pH 3 to 4) due to high organic acid levels, especially for white wines. Relatively low nitrogen levels and high levels of sulfur dioxide used to kill indigenous yeasts and bacteria present with incoming grapes plus the onset of anaerobiosis (105) provide other stresses. Beer production exposes the beer yeast to a similar environment, with the exception of minimal sulfur dioxide levels, but can include limitations in expression of genes associated with membrane synthesis that require small levels of oxygen (106). As a result of this stressful and inhospitable environment, yeasts have presumably evolved that are quite different from laboratory *S. cerevisiae*.

Transcriptome and proteome tools require a genome sequence for gene identification. Wine and beer yeasts, including *Saccharomyces bayanus*, *Saccharomyces kudriavzevii*, *Saccharomyces mikatae*, *Saccharomyces paradoxus*, and *Saccharomyces pastorianus* (*Saccharomyces carlsbergensis*), are among the approximately 117 *Saccharomyces* genomes being sequenced which will greatly facilitate systems biological analysis (www.genomesonline.org; www.genolevures.org).

Wine production often uses *S. cerevisiae* or *S. bayanus*, while beer yeasts comprise a wider group of *Saccharomyces* species. Initial analysis has shown considerable genome and chromosome rearrangement, and there is considerable divergence between beer yeasts for top-fermenting ales and bottom-fermenting lagers (62, 70, 115), providing for different characteristics among fermentation strains likely to affect the resulting product. The availability of a large and growing number of genome sequences, with multiple strains being sequenced from the aforementioned species, will greatly broaden the potential for understanding the value of the genetic differences found in industrial yeasts.

During wine production there is no need for starter cultures, as dry wine yeasts are typically pitched into the must, where they have to quickly adapt to very different environments upon rehydration. Having been produced in a rich and rapidly fermenting aerobic fermentor, harvested, and spray dried, the starter yeasts find themselves now in a high-sugar, acidic, anaerobic, rather toxic milieu comprising the wine must. Surprisingly, a few studies have failed to detect activation of classical stress response genes, which might be expected under these conditions. The rehydrated yeasts undertake activation of anaerobiosis genes, glycolytic genes, and catabolism repression due to the high sugar levels (98, 111), suggesting that these yeasts have been selected to avoid an energy-consuming stress response in favor of quickly adapting to the environment of an industrial wine yeast, i.e., are able to undertake fermentation in the environment and not become the classical "stuck" fermentation common to laboratory yeasts when facing high sugar levels. As the fermentation proceeds, an additional stress is self-introduced as the ethanol levels rise due to fermentation. Indeed, levels above 25% (by weight) sugar are common with wine production yielding over 12% ethanol, a level that is highly toxic to laboratory yeasts and nearly all bacteria. One common response is accumulation of the sugar trehalose, as well as expression of genes related to membrane fitness and electron transport and, finally, generalized stress response genes (110, 124). Cellular metabolism slows down in late stationary phase due to minimal sugar and nitrogen and high levels of ethanol in readiness for harvesting of the product by the enologist.

By comparison, a true wild-type yeast isolated from a spontaneous grape fermentation caused by a yeast resident on the grape exterior was analyzed for protein changes during fermentation (121). During the fermentation, when the glucose level dropped significantly, there was a dramatic decrease in protein synthesis for about 50 proteins and evidence of proteolytic activity. Also noted was upregulation of classical stress response genes, a response not seen in selected wine yeast strains, indicating that the natural yeast indigenous to vineyards was not as robust or resilient as selected commercial wine yeasts.

Beer yeasts are more diverse and, as mentioned above, have multiple fermentation modes divided at least between top- and bottom-fermenting yeasts plus other proprietary fermentation approaches. Analysis of brewing yeasts has shown chromosomal diversity regarding structure and number, as first noted by Casey (17). Using comparative competitive genomic hybridization techniques, various laboratories have found that wine and lager yeast chromosomal structures have distinct site-specific changes in a number of genes at or near the mating type (MAT) loci (12, 53, 70). These hybridization studies have identified regions of *S. cerevisiae*-like and non-*S. cerevisiae*-like genes, suggesting interspecies mating and hybridization, which lead generally to two situations. First, an addition of large regions of chromosomes after mating clearly accelerates evolution of traits that can be advantageous if selection pressures are imposed. Additionally, analysis of industrial yeast strains has shown them to have poor mating and sporulation ability, providing genetic isolation eventually. This provides for stability and maintenance of traits such as robustness in the face of the challenges of alcoholic beverage manufacture, for example, but makes conventional strain improvement via natural gene transfer difficult. The use of genetic engineering to further improve yeast strains for beverage fermentation can, therefore, be limited to modern genetic engineering approaches, raising concerns regarding genetically modified organism (GMO) application for products destined for human consumption. The application of systems biology tools such as those underway with laboratory strains (19, 32) can be applied to wine and especially lager beer production to build an integrated metabolic model of gene response and metabolite flows as the fermentation process proceeds. With such a model, non-GMO approaches can be used after identification of genes that should be either removed (knockout) or overexpressed to improve further the robust industrial nature that the brewmaster has selected through time.

42.5. AMINO ACID PRODUCTION

Amino acid production is a mature industry, with very large volumes of selected amino acids being produced primarily for food and feed applications. For example, lysine is deficient in corn-based rations for animals, so a number of companies such as Ajinomoto, ADM, Cargill, Kyowa Hakko Kogyo, and DSM produce lysine as well as glutamic acid by microbial fermentation with additional amino acids, including threonine and phenylalanine. *Corynebacterium glutamicum* has been developed for very-large-scale fermentations (>50 m^3) using low-cost substrates with current production of glutamic acid reaching 1.5 tonnes annually, primarily for food use, and lysine production of more than 600,000 tonnes as a feed additive.

Historically *C. glutamicum* was identified as a glutamic acid producer, and this discovery started the continued development of this microorganism for industrial processes, with many years of conventional mutagenesis and selection. Early work included the use of amino acid analogues to select for overproduction (115) and improved fermentation yields were improved with time. Fermentation conditions were developed that took advantage of the flexible nature of *Corynebacterium*. For example, many corynebacteria do not demonstrate diauxic lag for use of multiple carbohydrates beside glucose and as gram-positive microorganisms are associated with a very broad and diverse substrate utilization potential. Due to these advantages, such microorganisms, which include *Brevibacterium*, have been developed as producers of multiple amino acids, with the advantage of employing similar fermentation regimens for different amino acid products.

C. glutamicum was first used for glutamic acid production in the late 1950s, and work to increase yields was begun using fermentation process development and strain improvement, the latter with conventional mutagenesis and screening with amino acid analogues. Strain improvement continued as gene cloning and expression contributed to productivity, which included early flux analysis and substrate pathway modification for higher yields (51, 65). While the productivity of these conventional and genetically improved *C. glutamicum* and related *Brevibacterium* organisms provided sufficient economic productivity, the advent of genomic sequencing opened additional investigative avenues of strain improvement using

systems biology tools (60). For example, initial comparative analysis of related genomes revealed that the ability of *Corynebacterium efficiens* to produce amino acids at higher temperatures was due to both higher G+C content and altered codon usage, providing a beneficial change to production at higher temperatures, and likely improved stability to temperature spikes, an important aspect of robustness.

Industry and academia have used microarray technology that followed the emergence of genome sequences. Microarrays generate hundreds to thousands of signals, often leading to data overload and potentially reducing the perceived value of this tool. However, with well-designed experiments, a specific metabolic quandary was solved by microarray analysis involving valine production. The sequenced *C. glutamicum* strain ATCC 13032 was genetically modified to overproduce valine, and unexpectedly it was discovered to be inhibited by exogenously added valine (as was a related lysine producer). Microarray analysis comparison of the overproducing valine strain with the wild type detected an isoleucine limitation overcome by adding isoleucine, thus identifying an actionable outcome for further genetic engineering (74). As another example, microbial genome sequencing permitted an additional unique approach to strain building. It is well known that conventional mutagenesis followed by screening permitted significant progress in developing industrial fermentation strains, but it often resulted in unrelated deleterious mutations causing poor growth and stability. Whole comparative genomic analysis of a *C. glutamicum* production strain was used to examine the lysine pathway, along with transport and supportive genes. Initial work was limited to specific genes, but further research used the full-genome analysis, which detected many mutations in the conventionally derived strain, along with the identification of key genetic changes that caused overproduction of amino acids. These specific mutations predicted to be critical to amino acid production were recruited for "genome-based strain reconstruction" in which the critical mutations were added to an unmutated *C. glutamicum* strain, which resulted in a much more robust rapidly fermenting production strain with no known detrimental mutations (88, 99).

In addition to genetic analysis, analysis of the intermediate metabolites is progressing with advanced analytical tools with increasing sensitivity. For example, an automated method for gas chromatography-MS analysis of *C. glutamicum* metabolites able to detect over 1,000 compounds with conclusive identification of a selected 164 of these has led to improved understanding of internal metabolism (118).

These examples highlight the growing potential of the "omics" toolbox that can be beneficially applied to industrial microbes that were developed before the advent of genetic engineering. Competing with these efficient industrial coryneform bacteria for amino acid production is *E. coli*, which was investigated as a preferred host for phenylalanine production by GD Searle in the 1980s for NutraSweet production. Apparently it was decided that *E. coli* was advantageous due to its ease of genetic modification and that strain development research could overcome economic advantages of the competitor's *Corynebacterium* and *Brevibacterium* strains. This approach discounted any apparent benefits from robustness of the gram-positive microorganism, likely due to patent concern, with the conclusion that *E. coli* could be developed for high levels of phenylalanine production quickly. Interestingly, GD Searle successfully developed *E. coli* as a phenylalanine producer from two different directions. The first included conventional mutagenesis and selection with amino acid analogues at their Harbor Beach, MI, facility. Subsequently the unpublished cloning of the entire phenylalanine pathway was completed by a research team led by S. Primrose in the Searle High Wycombe, England, research facility (33), leading to another *E. coli* production strain for phenylalanine. The initial nonrecombinant *E. coli* strain went into production, providing important knowledge regarding large-scale *E. coli* phenylalanine production. Among the knowledge gained from this work is the lack of robustness of *E. coli* regarding phage infections, as early fermentations were plagued by viral attacks due to a nearby waste treatment facility, until additional engineering modifications were added to ensure sterility of incoming air. Indeed, this is a real concern for any bacterial fermentation process that is potentially worse for *E. coli* hosts due to ubiquitous coliform-containing waste management facilities. This new information was undoubtedly included in facility designs as the fully recombinant strain went into production by Searle in their Georgia facility. This example demonstrates that nonindustrial microorganisms can be harnessed successfully and compete with robust counterparts if proper facility and process practices are included.

42.6. XANTHAN PRODUCTION

The origins of the industrial production of bacterially derived biogums, such as xanthan, can be traced back to work done in the 1950s at the U.S. Department of Agriculture Laboratory in Peoria, IL. Current production of xanthan gum for use in foods, pharmaceuticals, cosmetics, oil drilling, and industrial processing amounts to more than 50,000 tonnes per annum globally. Xanthan production leaders include companies such as CP Kelco, ADM, Cargill, Danisco, Jungbunzlauer, and Deosen, the majority of whom have been shifting production capacity to China to remain competitive in this growing but cost-sensitive market (25).

The production of xanthan gum by *Xanthomonas campestris* has been extensively studied and is the subject of several good reviews (30, 41). Produced aerobically via submerged fermentation, xanthan is a large, anionic, water-soluble heteropolysaccharide with a primary structure containing repeated pentasaccharide units of glucose, mannose, and glucuronic acid. Even early mucoid isolates of *X. campestris* were capable of producing >50 g/liter xanthan, so much of the development of the xanthan process has focused on industrial scale-up and application development, especially as a food ingredient. The development of techniques for large-scale xanthan production have focused on overcoming the challenges inherent in the highly viscous fermentation, including controlling dissolved oxygen, temperature, and pH as the acidic polysaccharide is produced (87). Xanthan gum fermentations exhibit a strong dependence on oxygen transfer rate, as it affects cell growth and xanthan production rates, as well as the mean molar mass of the xanthan produced. As the fermentations are viscous, typically >100 cP, various carbon and nitrogen feeding strategies are used to maintain productivity during the course of fermentation (30, 41). As the commercially relevant properties of the xanthan gum differ among strains, cultivation techniques, and purification schemes, these have also been investigated in detail (30, 37, 112).

The genomes of three strains of *Xanthomonas campestris* pv. *campestris* have been published since 2002 (24, 107, 128). While the genomic information has been primarily interrogated to better understand the pathogenicity of the microbe, which is the causal agent for black rot disease in crucifers, the genome sequences have also been utilized to reconstruct the metabolic pathways for synthesis of xanthan and its precursors, as well as to perform comparative genomic analyses (128). The reconstruction of the xanthan pathway was nontrivial, for while the genes responsible for xanthan gum production were first reported in the 1980s, the number of cyclic metabolic pathways involved in xanthan synthesis complicates stoichiometry. In addition to providing insight into xanthan synthesis, the genome provides evidence for a large number of carbohydrate import systems. Following publication of the genome, the links between the import systems and specific carbohydrates have been further elucidated using transcriptomic data (114).

While structured kinetic models of *X. campestris* growth and xanthan production have been developed and shown to be useful in predicting bioprocess performance, the development of a genome-scale metabolic model has not yet been reported (39, 40, 83). Metabolic flux analyses have been developed and utilized to evaluate an advanced reactor configuration, to evaluate the energetics of xanthan synthesis, and to increase understanding of the relationship between nutrient addition and product attributes (30, 50, 81). Proteomic analysis has also been utilized with *X. campestris*, although the focus has been largely restricted to its role as a phytopathogen (20, 129). The presence of copious amounts of exopolysaccharide creates challenges in the "omic" investigation of *X. campestris*, as it can interfere with standard techniques for proteomic, transcriptomic, and metabolomic sampling and analysis. To overcome this issue, researchers have utilized modified methods or nonmucoid strains (20, 82). To make real progress with "omics" tools, method development aimed at extraction of genetic transcripts and microbial proteins for transcriptomic and proteomic analysis, respectively, must be accomplished. Once these methods are available, a full "omic" analysis of gum synthesis will be possible and potentially yield a metabolic model with hopefully actionable outcomes to further improve a very unique microbial fermentation process.

42.7. CELLULOSIC ETHANOL PRODUCTION

The conversion of cheap, abundant cellulosic biomass into alcohol fuels has been pursued for more than a century. Like ethanol derived from cane and corn, ethanol derived from cellulose will require low-cost operations possible only in very large, simply designed fermentation equipment. However, while relatively inexpensive, cellulosic feedstocks are much more difficult and costly to process than feedstocks providing starch or sucrose. Taken as a whole, the robustness challenges of cellulosic ethanol production are numerous (Table 3).

Cellulosic biomass is composed mainly of cellulose, hemicellulose, and lignin. While precise compositions vary significantly across types of cellulosic biomass, such as grasses and trees, typical dry biomass is >50% carbohydrate by weight, with lignin comprising between 20 and 35%. While the major portion of biomass is typically cellulose, composed of $\beta(1{\rightarrow}4)$-linked D-glucose units, a significant portion is composed of pentose-rich polysaccharides such as xylan and arabinan, or hexose-rich polysaccharides such as galactan and mannan. In part due to the presence of lignin, which sheathes cellulose fibrils and is cross-linked within the hemicellulose, the biological breakdown of lignocellulosic biomass is challenging. To attain the conversion rates and yields required of industrial fermentation processes, this recalcitrance to biological decomposition must be overcome.

TABLE 3 Robustness challenges in biomass ethanol production

Type of challenge	Description
Substrate	
Concentration	Processes employing prehydrolysis require rapid growth at high initial sugar concentrations.
	Processes featuring simultaneous saccharification release sugars slowly, typically over several days.
	Process economics favor very high solid concentrations, often >20% (by wt), resulting in limited free water, difficult mixing, and potential for local accumulation of sugars and/or ethanol.
Heterogeneity	Biomass feedstocks exhibit considerable compositional variability.
	Processes require the use of both pentose and hexose sugars.
Environment	
pH, temp, nutrients	Low-cost, large-scale fermentation systems have minimal mechanical mixing, resulting in potentially large spatial variations in pH, temperature, and nutrient availability.
Product accumulation	Process economics favor ethanol titers in excess of 6% (vol/vol).
	High concentrations of dissolved carbon dioxide can be anticipated due to equipment scale and mixing.
Infection	Scale, feedstock, and equipment limitations increase the likelihood of bacterial and phage infections.
Biomass-derived inhibitors	
Cellulose	Many pretreatments convert a portion of the hexose sugars into HMF, levulinic acid, and formic acid.
Hemicellulose	Many pretreatments convert a portion of the pentose sugars into furfural; acetyl groups are converted to acetic acid.
Lignin	Many pretreatments convert a portion of the lignin polyphenolics into vanillin, 4-hydroxybenzoic acid, 4-hydroxybenzaldehyde, and syringaldehyde.

As early as 1900, concentrated acid was employed to hydrolyze cellulose into glucose. However, low yields and high cost led researchers to look for other means of liberating sugars from cellulosic biomass. Most current research is focused on processes that employ pretreatments which disrupt the lignocellulosic structure prior to biological conversion to ethanol. A wide variety of pretreatment approaches have been proposed, including processes that operate at low pH (dilute acid and sulfur dioxide), high pH (ammonia and lime), and near neutral pH (liquid hot water and steam explosion). In other processes, such as wet oxidation, pH is a matter of choice (103). In addition to disrupting the cellulosic structure, the low-pH pretreatments hydrolyze the hemicellulose fraction and yield monomer sugars such as xylose. Pretreatments at high pH tend to solubilize lignin, without hydrolyzing the hemicellulose to a significant degree. The various approaches to pretreatment and their performance trade-offs are the subject of several good reviews (36, 49, 95). Pretreatments vary in terms of their ability to enhance enzymatic digestibility of different cellulosic feedstocks, but for most cellulosic feedstocks of interest, one or more pretreatments have been shown to be effective at rendering the cellulosic biomass digestible (135, 136).

While the various pretreatments improve biomass digestibility, many of them also make the biomass hydrolysates inhibitory to the growth of the fermentative microorganisms. Growth inhibition by biomass-derived inhibitors has been a recognized challenge since the earliest days of concentrated acid hydrolysis (79). Various means of detoxification have been tried, including overliming and activated carbon treatment of liquid hydrolysates and neutralization and washing of the insoluble materials. While several of the methods have been shown to be effective at reducing inhibition, their cost has motivated efforts to both minimize the production of inhibitors during pretreatment and identify or develop robust organisms capable of tolerating biomass-derived inhibitors (76, 94).

The exact identity and mechanism of the biomass-derived inhibitors are not well understood. Degradation products of glucose and xylose, such as 5-hydroxymethylfurfural (HMF) and furfural, have been a focus of attention for many years. The same is true of lignin degradation products such as vanillin and syringaldehyde. However, recently developed analytical methods capable of quantifying a wide range of potentially inhibitory water-soluble compounds have made it clear that dilute-acid-pretreated biomass hydrolysates have many more potential inhibitors (21, 22). The inhibitor profile following alkaline pretreatments differs considerably from that following pretreatments at lower pH and can result in pretreated biomass with low toxicity. For example, both wheat straw pretreated by alkaline wet oxidation and corn stover pretreated using ammonia fiber expansion have been shown to exert little or no inhibition on *S. cerevisiae* (67, 77). For ammonia fiber expansion and other mild types of pretreatments, such as liquid hot water (93), carboxylic acids such as acetic and formic acids are the primary biomass-derived inhibitors.

Among the potential inhibitors, furfural and HMF have received the most attention. Examination of furfural and HMF toxicity in yeasts led to the recognition that both compounds are converted to their corresponding alcohols during the course of the fermentation (86, 97, 127). While the sensitivities and responses to furfural and HMF vary widely among yeast strains, under anaerobic conditions, cell growth is typically inhibited until the conversions to the less toxic alcohol derivatives are complete (38, 93, 102, 119). In *S. cerevisiae* the reduction of the aldehyde group is catalyzed by one or more alcohol dehydrogenases, using both NADH and NADPH as cofactors (85). Researchers at Lund University utilized mRNA expression profiles to identify *adh6* as the gene responsible for conveying NADPH-dependent HMF conversion. Later, the same group identified a second, NADH-dependent alcohol dehydrogenase using electrospray ionization-MS and demonstrated that overexpression of the two alcohol dehydrogenases in a laboratory strain of

S. cerevisiae markedly improved anaerobic fermentation in the presence of HMF and furfural (6, 73, 104). The bacterial ethanologens *Zymomonas mobilis*, *E. coli* KO11, and *Klebsiella oxytoca* P2 have also been evaluated with respect to furfural and HMF toxicity and conversion. As with the yeast, the bacteria appear to detoxify the aldehydes by converting them to their respective alcohols (47, 109, 139).

While the appearance of HMF and furfural in pretreated biomass is associated with the degradation of otherwise fermentable carbohydrates, the presence of acetic acid is an unavoidable outcome of the depolymerization and hydrolysis of hemicellulose. Microbial sensitivity to acetic acid is strongly pH dependent, as toxicity is well correlated to the protonated form of acetic acid. Acetic acid toxicity has been shown to be significant for *E. coli* and *Z. mobilis*, as well as *S. cerevisiae* (38, 78, 102, 109). Interestingly, a gene that provides sodium acetate resistance in *Z. mobilis* has been identified by resequencing the genome of a mutant strain, in conjunction with transcriptomic profiling and genetic studies (S. Brown, Oak Ridge National Laboratory, personal communication). Microarray analysis of *S. cerevisiae* grown in glucose-limited chemostats showed hundreds of up- and downregulated transcripts when exposed to acetic ethanol at sublethal levels (1). Analysis of *S. cerevisiae* strains adapted to acetic acid, utilizing either whole-genome microarrays or lipidomics, highlights the adaptive importance of membrane composition and ion pumps (42, 137).

Microbial inhibition due to lignin-derived inhibitors such as vanillin, 4-hydroxybenzoic acid, 4-hydroxybenzaldehyde, and syringaldehyde has also been shown to be significant (26, 68, 75, 139). While genetic elements responsible for tolerance to these compounds have not been reported to date, this can be expected to change. For example, recent analysis of an *S. cerevisiae* deletion library identified 76 mutants with increased sensitivity to vanillin. The analysis indicated a key role for ergosterol biosynthesis, as well as for genes in the functional categories for chromatin remodeling and vesicle transport (29).

In addition to the robustness challenges posed by inhibitors deriving from pretreated biomass, microorganisms to be used in cellulosic ethanol production also need to be able to withstand high levels of ethanol and dissolved carbon dioxide. In cellulosic ethanol production, like starch- and sugar-based processes, attaining high ethanol concentrations has a significant impact on production costs. Existing commercial ethanol processes often achieve ethanol yields in excess of 10% (vol/vol), with much of the ethanol production occurring after cell growth ceases. High partial pressures of CO_2 accompany high ethanol titers, especially in large-scale ethanol production processes where economic considerations include very large (>500,000-gal) fermentation vessels with very limited agitation.

The effects of dissolved carbon dioxide on microorganisms have been studied at length, including its impact on aerobic and anaerobic yeast cultivations (28, 58). High levels of dissolved CO_2 have been shown to influence cell growth and protein production in aerobic systems as well as cell growth and product formation in anaerobic ethanol production processes (72, 96, 120). Analysis of these systems using microarrays has led to new insight into the underlying genetic mechanisms of CO_2 inhibition at the regulatory and transcriptional levels (3, 96). The physiological effect of dissolved carbon dioxide has also been examined in continuous cultures of the bacterial ethanol producers, including *Z. mobilis* and *Clostridium thermocellum*, although without the benefit of systems biology tools.

Ethanol tolerance has been of applied and fundamental interest since before the modern era of biotechnology, especially in wine making. Several good reviews have been written on ethanol tolerance in yeasts and other industrial organisms, which highlight the impact of ethanol on cell membranes, glycolysis, end production distribution, and cell growth (18, 23, 54, 56). Despite years of study, the biological mechanisms involved in ethanol tolerance, and how they are affected by temperature, medium composition, and genetics, remain the subject of some debate.

Genetic-level understanding of alcohol tolerance in yeasts has been significantly advanced through the use of systems biology. In 2006, separate groups published studies in which *S. cerevisiae* deletion libraries were screened for ethanol tolerance. While dozens of genes were correlated with ethanol sensitivity, genes encoding proteins involved in vacuolar function were identified by both groups as being critical in ethanol tolerance (35, 123). Since the initial reports, screening of more comprehensive libraries has expanded the list of genes which potentially influence ethanol tolerance (138).

Microarrays have also been employed in examining the response of yeast to ethanol. Alexandre et al. showed that more than 3% of the genes in the yeast genome are upregulated within 30 min of an ethanol challenge (4). Similar challenge experiments with a variety of alcohols showed significant upregulation in genes associated with stress response, glycolysis, metabolism of energy reserves, and synthesis of amino acids (34). Transcriptional profiling has also been applied to evaluate the difference between ethanol-adapted and parental yeast strains during growth in the presence of elevated ethanol. Genes related to ribosomal proteins were highly upregulated in the ethanol-tolerant strain. In addition, when cultivated under ethanol stress, both parental and adapted strains showed upregulation of genes associated with mitochondrial ATP generation and oxidative stress response (27). The number of transcriptional differences seen in response to short- and long-term ethanol exposure underscores the complexity inherent in yeast fermentations. Further evidence of this can be seen in research showing that modifications in global transcriptional machinery can impart significant improvements in overall ethanol and stress resistance (7, 8).

Systems biology tools have also been applied to evaluation of ethanol-adapted and parental of bacteria, such as *C. thermocellum* and *E. coli*, engineered for ethanol production. When performing transcriptional profiling to compare ethanol-adapted and parental strains of *E. coli*, researchers found that gene expression levels for more than 5% of the genome differed significantly. Among the affected genes were those involved in aromatic amino acid transport and synthesis, cell structure, energy metabolism, and stress response (46). In proteomic analysis of a *C. thermocellum* strain adapted to grow in 5% ethanol, the majority of membrane proteins were shown to be differentially expressed. Most of the expression difference was the result of downregulation of membrane proteins, especially those related to carbohydrate transport and metabolism (131). Subsequent resequencing, using 454 pyrosequencing and microarray-based comparative genome sequencing, revealed hundreds of mutations between the ethanol-adapted strain and the ATCC type strain from which it was derived (Brown, personal communication). Undoubtedly, additional systems biology investigations will reveal additional clues regarding the response of ethanologens to this important industrial chemical.

The development of processes for biomass ethanol production will require strains that tolerate the combined effect of ethanol, acetic acid, and degradation products arising from biomass pretreatment. While the impact of individual inhibitors on growth and product formation may not severely impair performance, the effect of the combined stresses may be more than additive (59, 100, 102). So the quintessential multiple-inhibitor-resistant super-robust ethanologen has yet to be obtained.

In biomass ethanol production, economic considerations necessitate the use of organisms capable of fermenting both cellulose- and hemicellulose-derived carbohydrates, including cellobiose, glucose, xylose, arabinose, galactose, and mannose. In practice, this requires broadening the range of sugars fermented by proven ethanol-producing organisms, enhancing ethanol production in organisms natively capable of using numerous biomass-derived sugars, or both. While metabolic engineering to obtain these improvements falls beyond the scope of this chapter, it is the subject of several good reviews and an area of active, ongoing research that includes multiple examples of successful application of systems biology tools (48, 55, 91, 122).

42.8. SUMMARY

Development of a vigorous, rugged microorganism by adaptation or as the result of directed selection is of selective value to microbes and economic value to industrial microbiologists. This robustness will have many characteristics derived from a combination of selection/adaptation and genetic changes, resulting in better survivability under stressful conditions. One of the best examples of robustness is yeast strains derived for beer and wine manufacture through the centuries which resulted in many phenotypic and genetic changes in the yeasts, leading to very beneficial survival responses during alcoholic fermentation. Similarly, engineered microorganisms for enzyme or other products are being developed to be very rugged and potent, which is part of the definition of robustness, but this microbial soundness requires a protected, engineered production environment to maintain and demonstrate their hardiness. Still, in either situation, the microorganism's response to the environment, however controlled, can be dissected and defined in ever-increasing detail by employing systems biology tools. This process starts with identification of gene and genome changes, and resultant data are used for analysis of both expression of genes and attendant proteins and their impact on the flux of metabolite in the cell and environment. As these differences become known, new or expanded modification strategies can be planned to further improve the microbe's robustness as needed for each industry.

REFERENCES

1. **Abbott, D. A., T. A. Knijnenburg, L. M. I. de Poorter, M. J. T. Reinders, J. T. Pronk, and A. J. A. van Maris.** 2007. Generic and specific transcriptional responses to different weak organic acids in anaerobic chemostat cultures of *Saccharomyces cerevisiae*. *FEMS Yeast Res.* **7:**819–833.

2. **Aebersold, R., and M. Mann.** 2003. Mass spectrometry-based proteomics. *Nature* **422:**198–207.

3. **Aguilera, J., T. Petit, J. H. Winde, and J. T. Pronk.** 2005. Physiological and genome-wide transcriptional responses of *Saccharomyces cerevisiae* to high carbon dioxide concentrations. *FEMS Yeast Res.* **5:**579–593.

4. **Alexandre, H., V. Ansanay-Galeote, S. Dequin, and B. Blondin.** 2001. Global gene expression during short-term ethanol stress in *Saccharomyces cerevisiae*. *FEBS Lett.* **498:**98–103.

5. **Almaas, E.** 2007. Biological impacts and context of network theory. *J. Exp. Biol.* **210:**1548–1558.

6. **Almeida, J. R. M., A. Röder, T. Modig, B. Laadan, G. Lidén, and M.-F. Gorwa-Grauslund.** 2008. NADH- vs NADPH-coupled reduction of 5-hydroxymethyl furfural (HMF) and its implications on product distribution in *Saccharomyces cerevisiae*. *Appl. Microbiol. Biotechnol.* **78:**939–945.

7. **Alper, H., J. Moxley, E. Nevoigt, G. R. Fink, and G. Stephanopoulos.** 2006. Engineering yeast transcription machinery for improved ethanol tolerance and production. *Science* **314:**1565–1568.

8. **Alper, H., and G. Stephanopoulos.** 2007. Global transcription machinery engineering: a new approach for improving cellular phenotype. *Metab. Eng.* **9:**258–267.

9. **Arnold, J. P.** 1911. *Origin and History of Beer and Brewing from Prehistoric Times to the Beginning of Brewing Science and Technology.* Alumni Association of the Wahl-Henius Institute of Fermentology, Chicago, IL.

10. **Barkai, N., and S. Leibler.** 1997. Robustness in simple biochemical networks. *Nature* **387:**913–917.

11. **Bergman, A., and M. L. Siegal.** 2003. Evolutionary capacitance as a general feature of complex gene networks. *Nature* **424:**549–552.

12. **Bond, U., C. Neal, D. Donnelly, and T. C. James.** 2004. Aneuploidy and copy number breakpoints in the genome of lager yeasts mapped by microarray hybridisation. *Curr. Genet.* **45:**360–370.

13. **Braunewell, S., and S. Bornholdt.** 2007. Superstability of the yeast cell-cycle dynamics: ensuring causality in the presence of biochemical stochasticity. *J. Theoret. Biol.* **245:**638–643.

14. **Çakir, T., B. Kirdar, and K. Ö. Ülgen.** 2004. Metabolic pathway analysis of yeast strengthens the bridge between transcriptomics and metabolic networks. *Biotechnol. Bioeng.* **86:**251–260.

15. **Calik, P., and T. H. Ozdamar.** 1999. Mass flux balance-based model and metabolic pathway engineering analysis for serine alkaline protease synthesis by *Bacillus licheniformis*. *Enzyme Microb. Technol.* **24:**621–635.

16. **Carlson, R. P.** 2009. Decomposition of complex microbial behaviors into resource-based stress responses. *Bioinformatics* **25:**90–97.

17. **Casey, G. P.** 1986. Molecular and genetic analysis of chromosomes X in *Saccharomyces carlsbergensis*. *Carlsberg Res. Commun.* **51:**343–362.

18. **Casey, G. P., and W. M. Ingledew.** 1986. Ethanol tolerance in yeasts. *Crit. Rev. Microbiol.* **13:**219–280.

19. **Castrillo, J. I., L. A. Zeef, D. C. Hoyle, N. Zhang, A. Hayes, D. C. Gardner, M. J. Cornell, J. Petty, L. Hakes, L. Wardleworth, M. B. Bharat Rash, W. B. Dunn, D. Broadhurst, K. O'Donoghue, S. S. Hester, T. P. Dunkley, S. R. Hart, N. Swainston, P. Li, S. J. Gaskell, N. W. Paton, K. S. Lilley, D. B. Kell, and S. G. Oliver.** 2007. Growth control of the eukaryote cell: a systems biology study in yeast. *J. Biol.* **6(2):**4.

20. **Chang, W.-H., M.-C. Lee, M.-T. Yang, and Y.-H. Tseng.** 2005. Expression of heat-shock genes groESL in *Xanthomonas campestris* is upregulated by CLP in an indirect manner. *FEMS Microbiol. Lett.* **243:**365–372.

21. **Chen, S.-F., R. A. Mowery, V. A. Castleberry, G. P. van Walsum, and C. K. Chambliss.** 2006. High-performance liquid chromatography method for simultaneous determination of aliphatic acid, aromatic acid and neutral degradation products in biomass pretreatment hydrolysates. *J. Chromatogr. A* **1104:**54–61.

22. **Chen, S.-F., R. A. Mowery, C. J. Scarlata, and C. K. Chambliss.** 2007. Compositional analysis of water-

soluble materials in corn stover. *J. Agric. Food Chem.* **55:**5912–5918.

23. **D'Amore, T., C. J. Panchal, I. Russell, and G. G. Stewart.** 1990. A study of ethanol tolerance in yeast. *Crit. Rev. Biotechnol.* **9:**287–304.

24. **da Silva, A. C. R., J. A. Ferro, F. C. Reinach, C. S. Farah, L. R. Furlan, R. B. Quaggio, C. B. Monteiro-Vitorello, M. A. Van Sluys, N. F. Almeida, L. M. C. Alves, A. M. do Amaral, M. C. Bertolini, L. E. A. Camargo, G. Camarotte, F. Cannavan, J. Cardozo, F. Chambergo, L. P. Ciapina, R. M. B. Cicarelli, L. L. Coutinho, J. R. Cursino-Santos, H. El-Dorry, J. B. Faria, A. J. S. Ferreira, R. C. C. Ferreira, M. I. T. Ferro, E. F. Formighieri, M. C. Franco, C. C. Greggio, A. Gruber, A. M. Katsuyama, L. T. Kishi, R. P. Leite, E. G. M. Lemos, M. V. F. Lemos, E. C. Locali, M. A. Machado, A. M. B. N. Madeira, N. M. Martinez-Rossi, E. C. Martins, J. Meidanis, C. F. M. Menck, C. Y. Miyaki, D. H. Moon, L. M. Moreira, M. T. M. Novo, V. K. Okura, M. C. Oliveira, V. R. Oliveira, H. A. Pereira, A. Rossi, J. A. D. Sena, C. Silva, R. F. de Souza, L. A. F. Spinola, M. A. Takita, R. E. Tamura, E. C. Teixeira, R. I. D. Tezza, M. Trindade dos Santos, D. Truffi, S. M. Tsai, F. F. White, J. C. Setubal, and J. P. Kitajima.** 2002. Comparison of the genomes of two *Xanthomonas* pathogens with differing host specificities. *Nature* **417:**459–463.

25. **de Guzman, D.** 2006. Xanthan gum supply thickens. *Chem.Market Rep.* **270:**26.

26. **Delgenes, J. P., R. Moletta, and J. M. Navarro.** 1996. Effects of lignocellulose degradation products on ethanol fermentations of glucose and xylose by *Saccharomyces cerevisiae, Zymomonas mobilis, Pichia stipitis,* and *Candida shehatae. Enzyme Microb. Technol.* **19:**220–225.

27. **Dinh, T., K. Nagahisa, K. Yoshikawa, T. Hirasawa, C. Furusawa, and H. Shimizu.** 2009. Analysis of adaptation to high ethanol concentration in *Saccharomyces cerevisiae* using DNA microarray. *Bioprocess Biosyst. Eng.* **32:**681–688.

28. **Dixon, N. M., and D. B. Kell.** 1989. The inhibition by CO_2 of the growth and metabolism of microorganisms. *J. Appl. Microbiol.* **67:**109–136.

29. **Endo, A., T. Nakamura, A. Ando, K. Tokuyasu, and J. Shima.** 2008. Genome-wide screening of the genes required for tolerance to vanillin, which is a potential inhibitor of bioethanol fermentation, in *Saccharomyces cerevisiae. Biotechnol. Biofuels* **1:**3.

30. **Flores Candia, J.-L., and W.-D. Deckwer.** 1999. Effect of the nitrogen source on pyruvate content and rheological properties of xanthan. *Biotechnol. Prog.* **15:**446–452.

31. **Fong, S. S., A. Nanchen, B. O. Palsson, and U. Sauer.** 2006. Latent pathway activation and increased pathway capacity enable *Escherichia coli* adaptation to loss of key metabolic enzymes. *J. Biol. Chem.* **281:**8024–8033.

32. **Forster, J., I. Famili, P. Fu, B. O. Palsson, and J. Nielsen.** 2003. Genome-scale reconstruction of the *Saccharomyces cerevisiae* metabolic network. *Genome Res.* **13:**244–253.

33. **Fotheringham, I. G., S. A. Dacey, P. P. Taylor, T. J. Smith, M. G. Hunter, M. E. Finlay, S. B. Primrose, D. M. Parker, and R. M. Edwards.** 1986. The cloning and sequence analysis of the *aspC* and *tyrB* genes from *Escherichia coli* K12—comparison of the primary structures of the aspartate aminotransferase and aromatic aminotransferase of *E. coli* with those of the pig aspartate aminotransferase isoenzymes. *Biochem. J.* **234:**593–604.

34. **Fujita, K., A. Matsuyama, Y. Kobayashi, and H. Iwahashi.** 2004. Comprehensive gene expression analysis of the response to straight-chain alcohols in *Saccharomyces cerevisiae* using cDNA microarray. *J. Appl. Microbiol.* **97:**57–67.

35. **Fujita, K., A. Matsuyama, Y. Kobayashi, and H. Iwahashi.** 2006. The genome-wide screening of yeast deletion mutants to identify the genes required for tolerance to ethanol and other alcohols. *FEMS Yeast Res.* **6:**744–750.

36. **Galbe, M., and G. Zacchi.** 2007. Pretreatment of lignocellulosic materials for efficient bioethanol production. *Adv. Biochem. Eng. Biotechnol.* **108:**41–65.

37. **Galindo, E., and V. N. Albiter.** 1996. High-yield recovery of xanthan by precipitation with isopropyl alcohol in a stirred tank. *Biotechnol. Prog.* **12:**540–547.

38. **Garay-Arroyo, A., A. A. Covarrubias, I. Clark, I. Niño, G. Gosset, and A. Martinez.** 2004. Response to different environmental stress conditions of industrial and laboratory *Saccharomyces cerevisiae* strains. *Appl. Microbiol. Biotechnol.* **63:**734–741.

39. **Garcia-Ochoa, F., V. E. Santos, and A. Alcon.** 2004. Structured kinetic model for *Xanthomonas campestris* growth. *Enzyme Microb. Technol.* **34:**583–594.

40. **García-Ochoa, F., V. E. Santos, and A. Alcón.** 1998. Metabolic structured kinetic model for xanthan production. *Enzyme Microb. Technol.* **23:**75–82.

41. **García-Ochoa, F., V. E. Santos, J. A. Casas, and E. Gómez.** 2000. Xanthan gum: production, recovery, and properties. *Biotechnol. Adv.* **18:**549–579.

42. **Gilbert, A., D. P. Sangurdekar, and F. Srienc.** 2009. Rapid strain improvement through optimized evolution in the cytostat. *Biotechnol. Bioeng.* **103:**500–512.

43. **Goddette, D., T. Christianson, B. Ladin, M. Lau, J. Mielenz, C. Paech, R. Reynolds, S. Yang, and C. Wilson.** 1993. Strategy and implementation of a system for protein engineering. *J. Biotechnol.* **28:**41–54.

44. **Goddette, D. W., C. Paech, S. S. Yang, J. R. Mielenz, C. Bystroff, M. E. Wilke, and R. J. Fletterick.** 1992. The crystal structure of the *Bacillus lentus* alkaline protease, subtilisin BL, at 1.4 A resolution. *J. Mol. Biol.* **228:**580–595.

45. **Goffeau, A., B. G. Barrell, H. Bussey, R. W. Davis, B. Dujon, H. Feldmann, F. Galibert, J. D. Hoheisel, C. Jacq, M. Johnston, E. J. Louis, H. W. Mewes, Y. Murakami, P. Philippsen, H. Tettelin, and S. G. Oliver.** 1996. Life with 6000 genes. *Science* **274:**546, 563–567.

46. **Gonzalez, R., H. Tao, J. E. Purvis, S. W. York, K. T. Shanmugam, and L. O. Ingram.** 2003. Gene array-based identification of changes that contribute to ethanol tolerance in ethanologenic *Escherichia coli*: comparison of KO11 (parent) to LY01 (resistant mutant). *Biotechnol. Prog.* **19:**612–623.

47. **Gutiérrez, T., M. L. Buszko, L. O. Ingram, and J. F. Preston.** 2002. Reduction of furfural to furfuryl alcohol by ethanologenic strains of bacteria and its effect on ethanol production from xylose. *Appl. Biochem. Biotechnol.* **98–100:**327–340.

48. **Hahn-Hägerdal, B., K. Karhumaa, C. Fonseca, I. Spencer-Martins, and M. Gorwa-Grauslund.** 2007. Towards industrial pentose-fermenting yeast strains. *Appl. Microbiol. Biotechnol.* **74:**937–953.

49. **Hendriks, A. T. W. M., and G. Zeeman.** 2009. Pretreatments to enhance the digestibility of lignocellulosic biomass. *Bioresour. Technol.* **100:**10–18.

50. **Hsu, C.-H., and Y. M. Lo.** 2003. Characterization of xanthan gum biosynthesis in a centrifugal, packed-bed reactor using metabolic flux analysis. *Process Biochem.* **38:**1617–1625.

51. **Ikeda, M., and R. Katsumata.** 1999. Hyperproduction of tryptophan by *Corynebacterium glutamicum* with the modified pentose phosphate pathway. *Appl. Environ. Microbiol.* **65:**2497–2502.

52. **Imhof, M., and C. Schlotterer.** 2001. Fitness effects of advantageous mutations in evolving *Escherichia coli* populations. *Proc. Natl. Acad. Sci. USA* **98:**1113–1117.

53. **Infante, J. J., K. M. Dombek, L. Rebordinos, J. M. Cantoral, and E. T. Young.** 2003. Genome-wide amplifications caused by chromosomal rearrangements play a major role in the adaptive evolution of natural yeast. *Genetics* **165:**1745–1759.

54. **Ingram, L. O.** 1990. Ethanol tolerance in bacteria. *Crit. Rev. Biotechnol.* **9:**305–319.

55. **Ingram, L. O., P. F. Gomez, X. Lai, M. Moniruzzaman, B. E. Wood, L. P. Yomano, and S. W. York.** 1998. Metabolic engineering of bacteria for ethanol production. *Biotechnol. Bioeng.* **58:**204–214.

56. **Ingram, L. O. N., T. M. Buttke, A. H. Rose, and D. W. Tempest.** 1984. Effects of alcohols on micro-organisms. *Adv. Microb. Physiol.* **25:**253–300.

57. **Ishii, N., K. Nakahigashi, T. Baba, M. Robert, T. Soga, A. Kanai, T. Hirasawa, M. Naba, K. Hirai, A. Hoque, P. Y. Ho, Y. Kakazu, K. Sugawara, S. Igarashi, S. Harada, T. Masuda, N. Sugiyama, T. Togashi, M. Hasegawa, Y. Takai, K. Yugi, K. Arakawa, N. Iwata, Y. Toya, Y. Nakayama, T. Nishioka, K. Shimizu, H. Mori, and M. Tomita.** 2007. Multiple high-throughput analyses monitor the response of *E. coli* to perturbations. *Science* **316:**593–597.

58. **Jones, R. P., and P. F. Greenfield.** 1982. Effect of carbon dioxide on yeast growth and fermentation. *Enzyme Microb. Technol.* **4:**210–223.

59. **Kádár, Z., S. F. Maltha, Z. Szengyel, K. Réczey, and W. de Laat.** 2007. Ethanol fermentation of various pretreated and hydrolyzed substrates at low initial pH. *Appl. Biochem. Biotechnol.* **137–140:**847–858.

60. **Kalinowski, J., B. Bathe, D. Bartels, N. Bischoff, M. Bott, A. Burkovski, N. Dusch, L. Eggeling, B. J. Eikmanns, L. Gaigalat, A. Goesmann, M. Hartmann, K. Huthmacher, R. Kramer, B. Linke, A. C. McHardy, F. Meyer, B. Mockel, W. Pfefferle, A. Puhler, D. A. Rey, C. Ruckert, O. Rupp, H. Sahm, V. F. Wendisch, I. Wiegrabe, and A. Tauch.** 2003. The complete *Corynebacterium glutamicum* ATCC 13032 genome sequence and its impact on the production of L-aspartate-derived amino acids and vitamins. *J. Biotechnol.* **104:**5–25.

61. **Kell, D. B.** 2004. Metabolomics and systems biology: making sense of the soup. *Curr. Opin. Microbiol.* **7:**296–307.

62. **Kielland-Brandt, M. C., T. Nilsson-Tillgren, C. Gjermansen, S. Holmberg, and M. B. Pedersen.** 1995. Genetics of brewing yeast, p. 233–253. *In* A. E. Wheals, A. H. Rose, and J. S. Harrison (ed.), *The Yeasts*. Academic Press, London, United Kingdom.

63. **Kim, P.-J., D.-Y. Lee, T. Y. Kim, K. H. Lee, H. Jeong, S. Y. Lee, and S. Park.** 2007. Metabolite essentiality elucidates robustness of *Escherichia coli* metabolism. *Proc. Natl. Acad. Sci. USA* **104:**13638–13642.

64. **Kim, T., S. Sohn, H. Kim, and S. Lee.** 2008. Strategies for systems-level metabolic engineering. *Biotechnol. J.* **3:**612–623.

65. **Kirchner, O., and A. Tauch.** 2003. Tools for genetic engineering in the amino acid-producing bacterium *Corynebacterium glutamicum*. *J. Biotechnol.* **104:**287–299.

66. **Kitano, H.** 2002. Systems biology: a brief overview. *Science* **295:**1662–1664.

67. **Klinke, H. B., L. Olsson, A. B. Thomsen, and B. K. Ahring.** 2003. Potential inhibitors from wet oxidation of wheat straw and their effect on ethanol production of *Saccharomyces cerevisiae*: wet oxidation and fermentation by yeast. *Biotechnol. Bioeng.* **81:**738–747.

68. **Klinke, H. B., A. B. Thomsen, and B. K. Ahring.** 2004. Inhibition of ethanol-producing yeast and bacteria by degradation products produced during pre-treatment of biomass. *Appl. Microbiol. Biotechnol.* **66:**10–26.

69. **Klipp, E., R. Herwig, A. Kowald, C. Wierling, and H. Lehrach.** 2005. *Systems Biology in Practice: Concepts, Implementation and Application*. Wiley-VCH Verlag GmbH & Co. KGaA, Weinheim, Germany.

70. **Kodama, Y., M. Kielland-Brandt, and J. Hansen.** 2006. Lager brewing yeast, p. 145–164. *In* P. Sunnerhagen and J. Piskur (ed.), *Topics in Current Genetics*, vol. 15. *Comparative Genomics*. Springer, Berlin, Germany.

71. **Kresnowati, M., W. A. van Winden, M. J. H. Almering, A. ten Pierick, C. Ras, T. A. Knijnenburg, P. Daran-Lapujade, J. T. Pronk, J. J. Heijnen, and J. M. Daran.** 2006. When transcriptome meets metabolome: fast cellular responses of yeast to sudden relief of glucose limitation. *Mol. Syst. Biol.* **2:**49.

72. **Kuriyama, H., W. Mahakarnchanakul, S. Matsui, and H. Kobayashi.** 1993. The effects of pCO_2 on yeast growth and metabolism under continuous fermentation. *Biotechnol. Lett.* **15:**189–194.

73. **Laadan, B., J. R. M. Almeida, P. Rådström, B. Hahn-Hägerdal, and M. Gorwa-Grauslund.** 2008. Identification of an NADH-dependent 5-hydroxymethylfurfural-reducing alcohol dehydrogenase in *Saccharomyces cerevisiae*. *Yeast* **25:**191–198.

74. **Lange, C., D. Rittmann, V. F. Wendisch, M. Bott, and H. Sahm.** 2003. Global expression, profiling, and physiological characterization of *Corynebacterium glutamicum* grown in the presence of L-valine. *Appl. Environ. Microbiol.* **69:**2521–2532.

75. **Larsson, S., A. Quintana-Sáinz, A. Reimann, N.-O. Nilvebrant, and L. Jönsson.** 2000. Influence of lignocellulose-derived aromatic compounds on oxygen-limited growth and ethanolic fermentation by *Saccharomyces cerevisiae*. *Appl. Biochem. Biotechnol.* **84–86:**617–632.

76. **Larsson, S., A. Reimann, N.-O. Nilvebrant, and L. Jönsson.** 1999. Comparison of different methods for the detoxification of lignocellulose hydrolysates of spruce. *Appl. Biochem. Biotechnol.* **77:**91–103.

77. **Lau, M. W., and B. E. Dale.** 2009. Cellulosic ethanol production from AFEX-treated corn stover using *Saccharomyces cerevisiae* 424A(LNH-ST). *Proc. Natl. Acad. Sci. USA* **106:**1368–1373.

78. **Lawford, H. G., and J. D. Rousseau.** 1993. Effects of pH and acetic acid on glucose and xylose metabolism by a genetically engineered ethanologenic *Escherichia coli*. *Appl. Biochem. Biotechnol.* **39–40:**301–322.

79. **Leonard, R. H., and G. J. Hajny.** 1945. Fermentation of wood sugars to ethyl alcohol. *Ind. Eng. Chem.* **37:**390–395.

80. **Lesne, A.** 2008. Robustness: confronting lessons from physics and biology. *Biol. Rev. Camb. Philos. Soc.* **83:**509–532.

81. **Letisse, F., P. Chevallereau, J.-L. Simon, and N. Lindley.** 2002. The influence of metabolic network structures and energy requirements on xanthan gum yields. *J. Biotechnol.* **99:**307–317.

82. **Letisse, F., and N. D. Lindley.** 2000. An intracellular metabolite quantification technique applicable to polysaccharide-producing bacteria. *Biotechnol. Lett.* **22:**1673–1677.

83. **Letisse, F., N. D. Lindley, and G. Roux.** 2002. Development of a phenomenological modeling approach for prediction of growth and xanthan gum production using *Xanthomonas campestris*. *Biotechnol. Prog.* **19:**822–827.

84. **Levy, S. F., and M. L. Siegal.** 2008. Network hubs buffer environmental variation in *Saccharomyces cerevisiae*. *PLoS Biol.* **6:**e264.

85. **Liu, Z. L., J. Moon, B. J. Andersh, P. J. Slininger, and S. Weber.** 2008. Multiple gene-mediated NAD(P)H-dependent aldehyde reduction is a mechanism of in situ detoxification of furfural and 5-hydroxymethylfurfural by *Saccharomyces cerevisiae*. *Appl. Microbiol. Biotechnol.* **81:**743–753.

86. **Liu, Z. L., P. J. Slininger, B. S. Dien, M. A. Berhow, C. P. Kurtzman, and S. W. Gorsich.** 2004. Adaptive response of yeasts to furfural and 5-hydroxymethylfurfural and new chemical evidence for HMF conversion to 2,5-bis-hydroxymethylfuran. *J. Ind. Microbiol. Biotechnol.* **31:**345–352.

87. **Margaritis, A., and J. E. Zajic.** 1978. Mixing, mass transfer, and scale-up of polysaccharide fermentations. *Biotechnol. Bioeng.* **20:**939–1001.

88. **Masato, I., O. Junko, H. Mikiro, and M. Satoshi.** 2006. A genome-based approach to create a minimally mutated

Corynebacterium glutamicum strain for efficient L-lysine production. *J. Ind. Microbiol. Biotechnol.* **33**:610–615.

89. **Mashego, M. R., K. Rumbold, M. De Mey, E. Vandamme, W. Soetaert, and J. J. Heijnen.** 2007. Microbial metabolomics: past, present and future methodologies. *Biotechnol. Lett.* **29**:1–16.

90. **Matsumura, H., A. Ito, H. Saitoh, P. Winter, G. Kahl, M. Reuter, D. H. Kruger, and R. Terauchi.** 2005. SuperSAGE. *Cell. Microbiol.* **7**:11–18.

91. **Matsushika, A., H. Inoue, T. Kodaki, and S. Sawayama.** 2009. Ethanol production from xylose in engineered *Saccharomyces cerevisiae* strains: current state and perspectives. *Appl. Microbiol. Biotechnol.* **84**:37–53.

92. **McGovern, P., S. Fleming, and S. Katz.** 1997. *The Origins and Ancient History of Wine.* Routledge, Abingdon, United Kingdom.

93. **Modig, T., J. R. M. Almeida, M. F. Gorwa-Grauslund, and G. Lidén.** 2008. Variability of the response of *Saccharomyces cerevisiae* strains to lignocellulose hydrolysate. *Biotechnol. Bioeng.* **100**:423–429.

94. **Mohagheghi, A., M. Ruth, and D. J. Schell.** 2006. Conditioning hemicellulose hydrolysates for fermentation: effects of overliming pH on sugar and ethanol yields. *Process Biochem.* **41**:1806–1811.

94a. **Mosier, N., and M. Ladisch.** 2009. *Modern Biotechnology: Connecting Innovations in Microbiology and Biochemistry to Engineering Fundamentals.* John Wiley & Sons, Inc., Hoboken, NJ.

95. **Mosier, N., C. Wyman, B. Dale, R. Elander, Y. Y. Lee, M. Holtzapple, and M. Ladisch.** 2005. Features of promising technologies for pretreatment of lignocellulosic biomass. *Bioresour. Technol.* **96**:673–686.

96. **Nagahisa, K., T. Nakajima, K. Yoshikawa, T. Hirasawa, Y. Katakura, C. Furusawa, S. Shioya, and H. Shimizu.** 2005. DNA microarray analysis on *Saccharomyces cerevisiae* under high carbon dioxide concentration in fermentation process. *Biotechnol. Bioprocess Eng.* **10**:451–461.

97. **Nilsson, A., M. F. Gorwa-Grauslund, B. Hahn-Hägerdal, and G. Lidén.** 2005. Cofactor dependence in furan reduction by *Saccharomyces cerevisiae* in fermentation of acid-hydrolyzed lignocellulose. *Appl. Environ. Microbiol.* **71**:7866–7871.

98. **Novo, M., G. Beltran, N. Rozes, J. M. Guillamon, S. Sokol, V. Leberre, J. Francois, and A. Mas.** 2007. Early transcriptional response of wine yeast after rehydration: osmotic shock and metabolic activation. *FEMS Yeast Res.* **7**:304–316.

99. **Ohnishi, J., S. Mitsuhashi, M. Hayashi, S. Ando, H. Yokoi, K. Ochiai, and M. Ikeda.** 2002. A novel methodology employing *Corynebacterium glutamicum* genome information to generate a new L-lysine-producing mutant. *Appl. Microbiol. Biotechnol.* **58**:217–223.

100. **Oliva, J. M., I. Ballesteros, M. J. Negro, P. Manzanares, A. Cabañas, and M. Ballesteros.** 2004. Effect of binary combinations of selected toxic compounds on growth and fermentation of *Kluyveromyces marxianus*. *Biotechnol. Prog.* **20**:715–720.

101. **Orr, H. A.** 2005. The genetic theory of adaptation: a brief history. *Nat. Rev. Genet.* **6**:119–127.

102. **Palmqvist, E., H. Grage, N. Q. Meinander, and B. Hahn-Hägerdal.** 1999. Main and interaction effects of acetic acid, furfural, and p-hydroxybenzoic acid on growth and ethanol productivity of yeasts. *Biotechnol. Bioeng.* **63**:46–55.

103. **Palonen, H., A. Thomsen, M. Tenkanen, A. Schmidt, and L. Viikari.** 2004. Evaluation of wet oxidation pretreatment for enzymatic hydrolysis of softwood. *Appl. Biochem. Biotechnol.* **117**:1–17.

104. **Petersson, A., J. R. M. Almeida, T. Modig, K. Karhumaa, B. Hahn-Hägerdal, M. F. Gorwa-Grauslund, and G. Lidén.** 2006. A 5-hydroxymethyl furfural reducing

105. **Pizarro, F., F. A. Vargas, and E. Agosin.** 2007. A systems biology perspective of wine fermentations. *Yeast* **24**:977–991.

106. **Polaina, J., and A. P. E. MacCabe.** 2007. *Industrial Enzymes: Structure, Function and Applications.* Springer, Dordrecht, The Netherlands.

107. **Qian, W., Y. Jia, S.-X. Ren, Y.-Q. He, J.-X. Feng, L.-F. Lu, Q. Sun, G. Ying, D.-J. Tang, H. Tang, W. Wu, P. Hao, L. Wang, B.-L. Jiang, S. Zeng, W.-Y. Gu, G. Lu, L. Rong, Y. Tian, Z. Yao, G. Fu, B. Chen, R. Fang, B. Qiang, Z. Chen, G.-P. Zhao, J.-L. Tang, and C. He.** 2005. Comparative and functional genomic analyses of the pathogenicity of phytopathogen *Xanthomonas campestris* pv. campestris. *Genome Res.* **15**:757–767.

108. **Raman, B., C. Pan, G. B. Hurst, M. Rodriguez, Jr., C. K. McKeown, P. K. Lankford, N. F. Samatova, and J. R. Mielenz.** 2009. Impact of pretreated Switchgrass and biomass carbohydrates on *Clostridium thermocellum* ATCC 27405 cellulosome composition: a quantitative proteomic analysis. *PLoS ONE* **4**(4):e5271.

109. **Ranatunga, T., J. Jervis, R. Helm, J. McMillan, and C. Hatzis.** 1997. Identification of inhibitory components toxic toward *Zymomonas mobilis* CP4(pZB5) xylose fermentation. *Appl. Biochem. Biotechnol.* **67**:185–198.

110. **Rossignol, T., L. Dulau, A. Julien, and B. Blondin.** 2003. Genome-wide monitoring of wine yeast gene expression during alcoholic fermentation. *Yeast* **20**:1369–1385.

111. **Rossignol, T., O. Postaire, J. Storai, and B. Blondin.** 2006. Analysis of the genomic response of a wine yeast to rehydration and inoculation. *Appl. Microbiol. Biotechnol.* **71**:699–712.

112. **Rottava, I., G. Batesini, M. Fernandes Silva, L. Lerin, D. de Oliveira, F. F. Padilha, G. Toniazzo, A. Mossi, R. L. Cansian, M. Di Luccio, and H. Treichel.** 2008. Xanthan gum production and rheological behavior using different strains of *Xanthomonas* sp. *Carbohydrate Polymers* **77**:65–71.

113. **Schallmey, M., A. Singh, and O. P. Ward.** 2004. Developments in the use of *Bacillus* species for industrial production. *Can. J. Microbiol.* **50**:1–17.

114. **Serrania, J., F.-J. Vorhölter, K. Niehaus, A. Pühler, and A. Becker.** 2008. Identification of *Xanthomonas campestris* pv. campestris galactose utilization genes from transcriptome data. *J. Biotechnol.* **135**:309–317.

115. **Shio, I., H. Sato, and M. Nakagawa.** 1972. L-Tryptophan production by 5-methyltryptophan-resistant mutants of glutamate-producing bacteria. *Agric. Biol. Chem.* **36**:2315–2322.

116. **Stelling, J., E. D. Gilles, and F. J. Doyle.** 2004. Robustness properties of circadian clock architectures. *Proc. Natl. Acad. Sci. USA* **101**:13210–13215.

117. **Stelling, J., U. Sauer, Z. Szallasi, F. J. Doyle, and J. Doyle.** 2004. Robustness of cellular functions. *Cell* **118**:675–685.

118. **Strelkov, S., M. von Elstermann, and D. Schomburg.** 2004. Comprehensive analysis of metabolites in *Corynebacterium glutamicum* by gas chromatography/mass spectrometry. *Biol. Chem.* **385**:853–861.

119. **Taherzadeh, M. J., L. Gustafsson, C. Niklasson, and G. Lidén.** 2000. Physiological effects of 5-hydroxymethylfurfural on *Saccharomyces cerevisiae*. *Appl. Microbiol. Biotechnol.* **53**:701–708.

120. **Thibault, J., A. Leduy, and F. Côté.** 1987. Production of ethanol by *Saccharomyces cerevisiae* under high-pressure conditions. *Biotechnol. Bioeng.* **30**:74–80.

121. **Trabalzini, L., A. Paffetti, A. Scaloni, F. Talamo, E. Ferro, G. Coratza, L. Bovalini, P. Lusini, P. Martelli, and A. Santucci.** 2003. Proteomic response to physiological fermentation stresses in a wild-type wine strain of *Saccharomyces cerevisiae*. *Biochem. J.* **370**(Pt.1):35–46.

122. van Maris, A., D. Abbott, E. Bellissimi, J. van den Brink, M. Kuyper, M. Luttik, H. Wisselink, W. Scheffers, J. van Dijken, and J. Pronk. 2006. Alcoholic fermentation of carbon sources in biomass hydrolysates by *Saccharomyces cerevisiae*: current status. *Antonie van Leeuwenhoek* **90:**391–418.

123. van Voorst, F., J. Houghton-Larsen, L. Jønson, M. C. Kielland-Brandt, and A. Brandt. 2006. Genome-wide identification of genes required for growth of *Saccharomyces cerevisiae* under ethanol stress. *Yeast* **23:**351–359.

124. Varela, C., J. Cardenas, F. Melo, and E. Agosin. 2005. Quantitative analysis of wine yeast gene expression profiles under winemaking conditions. *Yeast* **22:**369–383.

125. Velculescu, V. E., L. Zhang, B. Vogelstein, and K. W. Kinzler. 1995. Serial analysis of gene expression. *Science* **270:**484–487.

126. Velculescu, V. E., L. Zhang, W. Zhou, J. Vogelstein, M. A. Basrai, D. E. Bassett, P. Hieter, B. Vogelstein, and K. W. Kinzler. 1997. Characterization of the yeast transcriptome. *Cell* **88:**243–251.

127. Villa, G. P., R. Bartroli, R. Lopez, M. Guerra, M. Enrique, M. Penas, E. Rodriquez, D. Redondo, I. Jglesias, and M. Diaz. 1992. Microbial transformation of furfural to furfuryl alcohol by *Saccharomyces cerevisiae*. *Acta Biotechnol.* **12:**509–512.

128. Vorhölter, F.-J., S. Schneiker, A. Goesmann, L. Krause, T. Bekel, O. Kaiser, B. Linke, T. Patschkowski, C. Rückert, J. Schmid, V. K. Sidhu, V. Sieber, A. Tauch, S. A. Watt, B. Weisshaar, A. Becker, K. Niehaus, and A. Pühler. 2008. The genome of *Xanthomonas campestris* pv. campestris B100 and its use for the reconstruction of metabolic pathways involved in xanthan biosynthesis. *J. Biotechnol.* **134:**33–45.

129. Watt, S. A., A. Wilke, T. Patschkowski, and K. Niehaus. 2005. Comprehensive analysis of the extracellular proteins from *Xanthomonas campestris* pv. campestris B100. *Proteomics* **5:**153–167.

130. Wichman, H. A., M. R. Badgett, L. A. Scott, C. M. Boulianne, and J. J. Bull. 1999. Different trajectories of parallel evolution during viral adaptation. *Science* **285:**422–424.

131. Williams, T. I., J. C. Combs, B. C. Lynn, and H. J. Strobel. 2007. Proteomic profile changes in membranes of ethanol-tolerant *Clostridium thermocellum*. *Appl. Microbiol. Biotechnol.* **74:**422–432.

132. Wilmes, P., and P. L. Bond. 2006. Metaproteomics: studying functional gene expression in microbial ecosystems. *Trends Microbiol.* **14:**92–97.

133. Wilson, C. R., B. F. Ladin, J. R. Mielenz, S. S. M. Hom, D. Hansen, R. B. Reynolds, N. C. T. Kennedy, J. Schindler, M. Bahn, R. Schmid, M. Markgraf, C. Paech, and K. Maurer. 1994. Alkaline proteolytic enzyme and method of production. U.S. patent 5,352,604.

134. Winzeler, E. A., and R. W. Davis. 1997. Functional analysis of the yeast genome. *Curr. Opin. Genet. Dev.* **7:**771–776.

135. Wyman, C. E., B. E. Dale, R. T. Elander, M. Holtzapple, M. R. Ladisch, and Y. Y. Lee. 2005. Coordinated development of leading biomass pretreatment technologies. *Bioresour. Technol.* **96:**1959–1966.

136. Wyman, C. E., B. E. Dale, R. T. Elander, M. Holtzapple, M. R. Ladisch, Y. Y. Lee, C. Mitchinson, and J. N. Saddler. 2009. Comparative sugar recovery and fermentation data following pretreatment of poplar wood by leading technologies. *Biotechnol. Prog.* **25:**333–339.

137. Xia, J.-M., and Y.-J. Yuan. 2009. Comparative lipidomics of four strains of *Saccharomyces cerevisiae* reveals different responses to furfural, phenol, and acetic acid. *J. Agric. Food Chem.* **57:**99–108.

138. Yoshikawa, K., T. Tanaka, C. Furusawa, K. Nagahisa, T. Hirasawa, and H. Shimizu. 2009. Comprehensive phenotypic analysis for identification of genes affecting growth under ethanol stress in *Saccharomyces cerevisiae*. *FEMS Yeast Res.* **9:**32–44.

139. Zaldivar, J., A. Martinez, and L. O. Ingram. 1999. Effect of selected aldehydes on the growth and fermentation of ethanologenic *Escherichia coli*. *Biotechnol. Bioeng.* **65:**24–33.

140. Zhang, L., W. Zhou, V. E. Velculescu, S. E. Kern, R. H. Hruban, S. R. Hamilton, B. Vogelstein, and K. W. Kinzler. 1997. Gene expression profiles in normal and cancer cells. *Science* **276:**1268–1272.

141. Zhong, S. B., A. Khodursky, D. E. Dykhuizen, and A. M. Dean. 2004. Evolutionary genomics of ecological specialization. *Proc. Natl. Acad. Sci. USA* **101:**11719–11724.

Bioethanol Production from Lignocellulosics: Some Process Considerations and Procedures

CHARLES A. ABBAS, WU LI BAO, KYLE E. BEERY, PAMELA CORRINGTON, CONSUELO CRUZ, LUCAS LOVELESS, MARTIN SPARKS, AND KELLI TREI

43

43.1. INTRODUCTION

The current emphasis on environmentally friendly liquid transportation fuels has caused a renewed interest in bioethanol and, in particular, lignocellulosic feedstocks for the production of bioethanol. As a result, research into the development of an economic process for the bioconversion of lignocellulosics has accelerated (21). The primary areas of research crucial to improved process economics consist of feedstock selection, pretreatment, hydrolysis of the pretreated feedstocks, and developing an optimized fermentation process that uses newly engineered ethanologens capable of utilizing various biomass-derived sugars (1). Faced with the above challenges, biotechnologists and engineers are addressing these issues in an integrated fashion by combining many of the above steps, thereby reducing the need for costly processing scenarios (52). For those new to this area of research, this chapter addresses in a stepwise and practical manner some of the general considerations, procedures, and approaches that can be used in process optimization.

43.2. FEEDSTOCK SELECTION

Choosing a lignocellulosic biomass feedstock is an important step in the biochemical conversion of biomass to ethanol. Feedstock composition dictates the choice of pretreatments. The hemicellulose, cellulose, lignin, silica/ash, starch, and moisture content all need to be taken into consideration.

A feedstock high in lignin, such as wood or corn stover, requires a more severe pretreatment and also requires a higher enzyme usage, as lignin inhibits cellulase activity by adsorbing the enzymes (49, 55, 56). A feedstock high in silica, such as rice straw or rice hulls, needs to be discounted, as the silica will be inert, at best, throughout the pretreatment. A feedstock high in starch, such as corn fiber hulls, will require a less severe pretreatment to avoid excessive degradation of the glucose from the easily hydrolyzable starch. Another factor is moisture, and agricultural processing by-products such as corn fiber hulls and sugarcane bagasse will already contain a high content of water. The initial moisture content and chip size have both been shown to have an impact on the efficiency of the bioconversion of softwood lignocellulosics (17). This needs to be factored into the pretreatment step, and it can aid specific types of pretreatment processes, such as dilute acid, ammonia recycle percolation, and steam explosion, among others. Total carbohydrate content is important, as that is the only portion of the feedstock that can be converted to ethanol biochemically. However, the lignin in the biomass surrounds the carbohydrate to protect the carbohydrate from degradation and to offer structural rigidity to the plant; therefore, lignin composition is also important. More important than all of the compositional considerations is economics. Feedstocks such as wheat straw have use as animal bedding, while other higher-protein, ruminant-digestible feedstocks, such as corn fiber hulls, soybean hulls, or wheat middlings, are used as animal feed. These types of feedstocks will have a much higher value than wood chips, corn stover, or switchgrass; however, the total carbohydrate content is also higher than that of wood or corn stover, so more ethanol production is possible per ton of biomass feedstock. Table 1 shows the approximate compositions of various biomass feedstocks, including grains, oilseeds, agricultural processing by-products, agricultural residues, energy crops, and wood.

43.3. PRETREATMENT SELECTION

After the selection of feedstock, the selection of the pretreatment method is the next difficult choice. There are many types of pretreatment, some severe and some mild, ranging from hot-water treatment to thermochemical, organosolv, or ionic liquid treatments. The pretreatment may be expected to release fermentable sugars by the chemical treatment alone or in combination with an enzyme hydrolysis step. Generally, the treatments are aimed at either disrupting or solubilizing the hemicellulose, cellulose, lignin, or a combination of the three (Table 2). The following sections detail these approaches.

The selection of the pretreatment step also depends upon the composition of the feedstock in question. Various levels of cellulose, hemicellulose, lignin, ash, silica, nitrogen, and moisture will require various pretreatment conditions and chemicals to ensure the most optimum pretreatment for that feedstock.

Typically, the first step in feedstock pretreatment aims to increase surface area. The larger the surface area of the feedstock, the more area which is available for chemical reactions. Particle size is one determinant of surface area, but the situation is complicated because internal surface area is much larger than external surface area for many pretreated feedstocks, and accessible surface area is affected by the size

TABLE 1 Compositions of various biomass feedstocks

Feedstock	% of indicated ingredient						
	Cellulose	Hemicellulose	Starch	Fat	Ash	Protein	Lignin
Corn	2.0	7.6	72.0	4.4	1.6	9.1	1.0
Corn fiber hulls	16.0	40.0	18.0	3.0	3.0	11.0	4.0
Corn stover	38.0	25.0	NA[a]	0.8	6.1	4.0	17.5
Corn cobs	32.0	35.0	NA	0.4	1.8	2.5	20.0
DDG/DDGS[b]	22–26	24–28	NA	8–12	2.5	26–29	4.0
Corn gluten feed	13.0	25.1	23.0	3.3	8.2	23.9	NA
Soybean	2.0	5.0	NA	18.8	5.5	42.8	NA
Soybean hulls	46.0	18.0	NA	2.5	5.0	12.0	2.0
Wheat	8.0	4.0	70.0	2.2	1.6	12.2	2.0
Wheat straw	35.0	24.0	NA	NA	6.0	4.0	25.0
Wheat hulls	10.5	29.5	34.0	NA	NA	13.5	5.0
Brown rice	1.0	2.0	74.4	2.6	1.6	8.5	NA
Rice hulls	30.0	20.0	NA	0.8	16.3	3.2	21.4
Sorghum grain	5.5	7.0	55.5	6.4	2.6	11.0	1.0
Sweet sorghum, whole stalk	32.9	21.4	2.0	0.9	5.6	5.0	6.4
Oat	11.0	15.0	44.7	5.4	3.4	13.3	2.7
Oat hulls	30.0	34.0	NA	1.6	6.1	3.6	13.2
Barley	5.0	13.0	62.8	2.2	2.2	12.8	2.0
Barley hulls	33.6	37.2	NA	NA	3.6	NA	19.3
Cocoa shells	13.7	7.1	NA	8.3	15.3	NA	3.4
Cottonseed hulls	59.0	NA	NA	1.7	2.8	NA	24.0
Miscanthus	48.7	26.4	NA	0.9	5.2	2.8	11.7
Pine	50.0	15–25	NA	NA	NA	NA	15–30
Sugarcane bagasse	41.70	28.60	NA	4.40	1.8	NA	23.1

[a]NA, not applicable.
[b]DDG/DDGS, distillers dried grains/distillers dried grains with solubles.

of the adsorbing species. Grinding to reduce particle size is very costly from an energy standpoint (7, 27).

Another important factor to consider is the length of time for the pretreatment. In general, times that are too short result in solids that are not sufficiently reactive, while times that are too long result in unwanted reactions such as degradation of sugars and formation of inhibitors. The various pretreatments range from seconds for dilute acid, ammonia fiber explosion (AFEX), and steam explosion to days and weeks for lime pretreatment (28, 34, 36, 63).

43.3.1. Pretreatments with Acids

Acid pretreatments include both processes featuring the addition of acids and processes without added acids in which the pH is lowered due to release of acetic acid in the course of pretreatment. This family of pretreatments brings about decrystallization and swelling of the cellulose structure as well as hydrolysis of hemicellulose polysaccharides (and, to a much lesser extent, cellulose). Although little or no lignin is solubilized by acid pretreatment, the state of lignin is significantly altered in the course of melting and recondensation. Hydrolysis of hemicellulose leads to soluble oligosaccharides and free sugars being generated (2), while the hydrolysis of the cellulose leads to swelling of the cellulose matrix, but very few soluble maltooligosaccharides, due to the highly crystalline nature of cellulose (23). The acid pretreatments also disrupt the lignin, which allows for enzymes to effectively hydrolyze the cellulose (70).

These pretreatments can be characterized and compared by calculating the severity factor (R_0) for the pretreatment (40). The formula for severity factor is

$$R_0 = t \times \exp[\frac{(T_H - T_R)}{14.75}]$$

where t is reaction time in minutes, T_H is hydrolysis temperature in degrees Celsius, and T_R is a reference temperature, most often 100°C.

If acid is used in the pretreatment, a modification of the severity factor is used: the combined severity factor $CS = \log_{10}(R_0) - pH$.

Generally the range for pretreatment by dilute acid, hot water, or steam explosion is 150 to 200°C, with 0.15 to 3% acid for 1 to 30 min. Other studies have examined pretreating biomass feedstocks under autoclave conditions (24, 53). Figure 1 shows the relationship between severity factor and amount of solids solubilized and monomeric arabinose yield for corn fiber. It can be seen that as the severity factor is increased, the monomeric arabinose yield (the hemicellulose side chain sugar) and the total solids solubilized both increase linearly over the range of severity factors shown.

43.3.2. Pretreatments with Bases

Pretreatments using bases, such as AFEX (61), lime pretreatment (34), and ammonia recycle percolation alone or followed by a successive treatment with hydrogen peroxide (32, 35, 41),

TABLE 2 Pretreatment selection and impact of pretreatment on lignocellulosics

Process	Solids loading (%)	Cellulose	Hemicellulose	Lignin
Dilute acid	10–40	Some depolymerization	80–100% solubilization to monomers	Little or no solubilization but extensive redistribution
Steam explosion at high solids	20–40	Some depolymerization	80–100% solubilization to monomers or oligomers	Little or no solubilization but extensive redistribution
Hydrothermal	10–30	Some depolymerization	80–100% solubilization to >50% oligomers	Partial solubilization (20–50%)
Organic solvents with water (organosolv)	10–30	Some depolymerization	Substantial solubilization to near completion	Substantial solubilization to near completion
Wet oxidation	6–15	Some decrystallization	30–90% solubilization depending on severity	20–50% solubilization depending on severity
AFEX	60–90	Substantial swelling	Solubilization from 0–60% depending on moisture with >90% oligomers	Some solubilization (10–20%)
Sodium hydroxide or ammonia	10–30	Substantial swelling	Substantial solubilization, often to >50%	Substantial solubilization, often >50%
Lime pretreatment	10–30	Substantial swelling	Substantial solubilization (>30%) under some conditions	Partial solubilization (40%)
Concentrated acid	10–30	Substantial solubilization	80–100% solubilization	Extensive redistribution

work mainly by swelling the biomass and modifying or solubilizing the hemicellulose and lignin (70). By removing the bonds between the lignin, hemicellulose, and cellulose, the cellulase enzymes can effectively hydrolyze the cellulose (34).

43.3.3. Other Pretreatments
Wet oxidation, organosolv, and ionic-liquid-based pretreatments modify the biomass in different ways than the acid

and base pretreatments. Wet oxidation works by solubilizing the hemicellulose and lignin and cleaving the bonds between the three fiber components. This works best when wet oxidation takes place under alkaline conditions (44).

Organosolv removes lignin nearly completely, by first hydrolyzing the alkyl aryl ether bonds in the lignin and then solubilizing the remaining lignin fragments (16). The lignin can be separated from the residual solids and

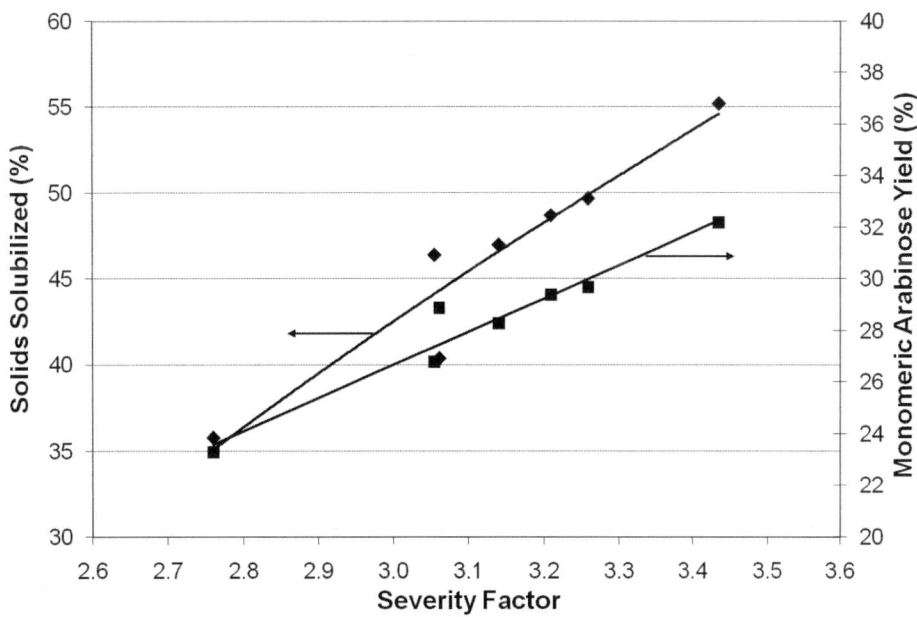

FIGURE 1 Severity factor impact on solubilization and arabinose release from the corn fiber hull hemicellulose arabinoxylan.

precipitated from the solvent/water system. This allows the enzymes to attack the remaining carbohydrates.

Ionic liquids are a recent development in the lignocellulosic biomass conversion arena. This technology works by completely dissolving the carbohydrate portion of the biomass and leaving the lignin as a solid. The soluble carbohydrates can be chemically hydrolyzed in the ionic liquid to create water-soluble oligosaccharides for use as a chemical or fermentation feedstock (57). The difficulty of this type of pretreatment is recovery of the expensive, and often environmentally hazardous, ionic liquid.

43.4. ENZYME SELECTION AND HYDROLYSIS

Following pretreatment of a lignocellulosic feedstock, an enzymatic step can be employed to further hydrolyze the solubilized material and residue into simple sugars. These sugars are then, in turn, used by the microbes in fermentation to produce a desired biobased product such as ethanol. The enzymatic hydrolysis step is one of the major obstacles to the industrial production of bioethanol from a process efficiency and economic standpoint (1). Therefore, optimizing the enzyme hydrolysis blend for a given feedstock is critical to an industrial process. Choosing and optimizing the right blend of enzymes for an industrial process depends on many factors, such as feedstock composition, pretreatment method used, overall process design, enzyme reaction efficiency, and cost of enzymes (1, 72).

43.4.1. Substrate Preparation

The substrates of the enzymatic reactions (the pretreated feedstocks) are derived from diverse sources of materials which undergo different methods of pretreatments as described in section 43.3. The diversity in feedstock and pretreatment methods generate substrates for enzymatic hydrolysis with different physical and chemical properties (1). Properties such as particle size, solid and liquid and/or moisture content, degree of crystallinity, chemical composition, polymer size or length of fiber filaments and degree of substitution, and the specific type of chemical bonds all dictate the kind of enzymes and dose of each enzyme required to hydrolyze the substrates (17). The slurry generated from pretreatment usually includes a variety of carbohydrates, such as cellulose fibers, soluble oligosaccharides, and some simple free sugars, as well as other compounds that are derived from lignin (1). Hydrolysis of the cellulose and oligosaccharides with different compositions and chemical bonds requires specific cellulosic enzyme cocktails, as these enzymes act in a synergistic fashion. Optimizing the cocktail or components of enzymes for a given substrate is necessary to quantitatively depolymerize the carbohydrates at an acceptable rate (72). When lignin is present in high concentrations, as in the case of wood-derived feedstocks or agricultural residues (see Table 1), inhibition of cellulases and hemicellulases is a common occurrence (10). Some differences in lignin inhibition of enzyme activity following feedstock pretreatment have been shown (10). The addition of nonionic surfactants has been shown to have a positive impact on enzymatic conversion of softwood lignocellulose by reducing the amount of enzyme necessary and by enhancing the rate of conversion (13, 37). Screening for weak lignin-binding cellulases is one approach that has been proposed to deal with the interference of this polymer so as to improve the activity of cellulases on lignocellulosics (12).

The whole slurry after pretreatment can be fractionated into a cellulose stream and a hemicellulose stream prior to selective enzyme treatment (hemicellulases versus cellulases) to improve the effectiveness of the hydrolysis step. This requires the use of different enzyme blends for the different fractions. Fractionation can consist of a liquid-solid separation step such as filtration or centrifugation. If filtration is employed, the cake or solid retentate contains mostly fractured and cellulosic fibers. An enzyme cocktail consisting mostly of cellulases should be considered for the hydrolysis of this residual solid material. The filtrate or permeate will consist mostly of soluble oligosaccharides of hemicellulose. A blend with mostly hemicellulases should be selected for this fraction. In some cases where the hydrolysate is derived from agricultural residues or processing by-products that contain starch such as corn fiber and wheat middlings, a glucanase-type enzyme such as glucoamylase is also essential to hydrolyze the residual starch present in the soluble fraction.

43.4.2. Enzyme Selection Criteria

Enzyme samples to be tested can be obtained from enzyme suppliers such as Novozymes, Genencor (Danisco), and other smaller companies. These companies have developed several commercial cellulase blends (Table 3) that can be evaluated for a given pretreated feedstock. In general these blends contain high cellulase activity and are produced by fermentation using the fungus *Trichoderma reesei* (46, 59). By comparison, hemicellulases are derived from different fungi and bacteria (8, 38, 47, 68). In addition to cellulase and xylanase activities, several secondary activities may also be produced during fermentation, which may be necessary for optimum cellulase hydrolysis of a pretreated lignocellulosic feedstock (11, 47). Variability in the enzyme blends is due to the choice of microorganisms and differences in fermentation parameters. Therefore, dissimilar protein compositions are observed among different commercial xylanase blends (Fig. 2). For this reason, it is crucial to ensure that all needed activities are present in the chosen blend.

A cellulase blend usually contains endo-β-glucanases and exo-β-glucanases (47, 72). The combined activities can hydrolyze the solid cellulose fibers into glucose in a stepwise fashion. The endoglucanase reacts first with the cellulose chain to cleave an internal β-glucosidic bond, thus creating two new chain ends, one reducing and one nonreducing. The exoglucanases then cut the polymer from the reducing end or nonreducing end to make cellobiose, which is eventually hydrolyzed by the β-glucosidase into glucose (6, 43, 48). The inhibition of cellulases and β-glucosidases by increased glucose content during enzymatic treatment has been well described (71). The cellulase complex from *T. reesei* has been well characterized by several research institutes and is produced by several manufacturers, listed in Table 3. A variety of cellulase enzymes from other sources, especially those from thermostable microbes, have also been evaluated for the ability to hydrolyze cellulosic fibers (67). While the list of commercially available blends is limited, there are many more cellulases and hemicellulases in development that can be considered for evaluation on these substrates.

Commercial hemicellulase mixtures can have multiple activities and are not as well defined as the cellulase mixtures. Most pretreatment methods employed to specifically release hemicellulose create soluble oligosaccharides and do not release all the hemicellulose. This hemicellulose is still attached to the lignin component of the cellulose

TABLE 3 Some commercial enzyme products available for testing

Enzyme name	Function	Source	Temp (°C)	Best pH	Supplier
Cellic Ctec	Cellulase	*T. reesei*	50	5	Novozymes
Celluclast 1.5L	Cellulase	*T. reesei*	Up to 70	3.5–5.5	Novozymes
GC220 (C2)	Cellulase	*T. reesei*	Up to 70	3.5–5.5	Genencor
Multifect GC	Cellulase	*T. reesei*	Up to 70	4	Genencor
Spezyme CP	Cellulase	*T. reesei*	50	4	Genencor
Accellerase 1000	Cellulase	*T. reesei*	50	5	Genencor
ENM series	Cellulase	*Aspergillus*	30–55	3–6	Amano
UltraFlo L	Hemicellulase	*Humicola*	Up to 70	5	Novozymes
Viscoenzyme L	Hemicellulase	*Aspergillus*	25–50	3.3–5.5	Novozymes
Cellic Htec	Xylanase	?	35–55	4.5–6.0	Novozymes
Shearzyme Plus	Xylanase	*Aspergillus*	30–50	4–5.5	Novozymes
Viscostar L	Xylanase	*Chrysosporium*	50	5	Dyadic
Multifect xylanase	Xylanase	*T. reesei*	50–55	5	Genencor

fiber matrix. The hemicellulase blend is then necessary to disintegrate those soluble and insoluble polymers into usable carbohydrates. The hemicellulose is composed of many different monomers, such as xylose, arabinose, mannose, galactose, and other C_5 or C_6 chemicals (67). These sugars are linked together by a variety of chemical bonds, and these bonds may require application of different enzymes. A typical hemicellulase blend contains exo- and endoxylanases and arabinofuranosidase. Because of the highly substituted and branched nature of hemicellulose, other enzymes, such as mannanase, polygalactouronase, rhamnosidase, and ferulic esterase or other esterases, could sometimes be required for different substrates.

In addition to the carbohydrate hydrolytic enzymes, there are several other types of enzymes that could play a role in the process and have a synergistic effect. Examples of this are the wood rot fungus-derived oxidases that have been proposed to have a function in the degradation of lignin compounds. The activities of these oxidases by in vitro assays on the intact lignocellulose material are not easily detectable (25). Evaluating the function of these enzymes on the pretreated substrates can be useful in assessing the impact of pretreatment on a lignocellulosic feedstock (1). Also, an acid protease has been shown to have some synergistic effect on the hydrolysis of some selected polysaccharides from a variety of feedstocks (W. L. Bao, unpublished data).

43.4.3. Enzyme Mixture Optimization

Due to the recalcitrant nature of lignocellulosic feedstocks, the compositional variation within the feedstock, and differences in enzyme sources, enzyme mixture optimization is necessary to the processing of each pretreated substrate. For

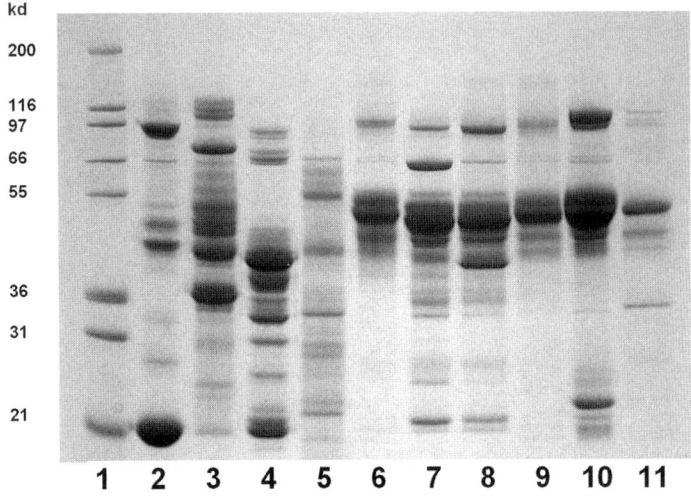

FIGURE 2 Sodium dodecyl sulfate-polyacrylamide gel electrophoresis of commercially available cellulosic enzymes. Lane 1, molecular mass marker; lanes 2 to 6, commercial xylanases; lanes 7 to 11, commercial cellulases.

the downstream processes, separate hydrolysis and fermentation and simultaneous saccharification and fermentation, the enzyme blend should have high catalytic efficiency. A long reaction time contributes to additional cost of the process and increases the chance of contamination. Therefore, the ideal enzyme blend should hydrolyze most of the substrates with minimum enzyme concentration at a high rate. According to published reports and due in part to the U.S. Department of Energy's funding of cellulase development work, the cost of production of commercial cellulosic enzymes has dropped significantly in recent years (72). In spite of this, the cost of these enzymes is still a prohibitive factor in the process. To achieve additional cost reduction, further research and development work and enzyme blend optimization are needed. Blend optimization is one of the ways that end users can minimize the amount of enzyme in processing of lignocellulosics and thereby decrease cost.

Blend optimization includes finding the right enzymes for the substrate and fine-tuning the amount of each enzyme in the blend. This can be achieved by utilizing tools that are available for experimental designs, statistical analysis, and computer modeling (15, 66). Approaches include testing different enzyme sources in order to find the right mix for the substrate. Once the right mix is determined, different compositions of the blend should be tested for optimization. The hydrolysis data can be plotted on a graph to establish the dosage effect of each enzyme. By using this method, the optimum blend can be determined (Fig. 3).

43.4.4. Detection of Enzyme Products

Several analytical methods can be used to test the catalytic efficiency of enzymatic hydrolysis of cellulosic substrates, some of which are described below. The enzyme products can be analyzed on silica thin-layer chromatography plates, where sugars are visualized by sulfuric acid charring at high temperature. The glucose can also be determined by the Yellow Springs Instrument Company's (Yellow Springs, OH) biochemical analyzer, which is an instrument that utilizes glucose oxidase to measure the amount of glucose in a solution. The phenol-sulfuric method can be used for the total soluble carbohydrate analysis (20), as can enzymatic assays using hexokinase and glucose 6-phosphate to measure glucose concentration. Finally, high-performance liquid chromatography (HPLC) or gas chromatographic (GC) methods can be used for quantitative carbohydrate determination (3).

43.4.5. Other Considerations

The cost of enzymes can be reduced by modification of the enzymes or fermentation organisms, such as enzyme display on the membrane surface of yeasts and transformation of expressed enzymes into yeasts (19, 29, 39, 64, 65). Current research includes the use of membrane separation to recycle the biomass-degrading enzymes and on enzyme immobilization to extend the lifetime of enzymes for the hydrolysis of cellulosic material (9, 26, 58, 62). Enhancement of enzymatic hydrolysis has also been shown using simultaneous ball milling in combination with added cellulases (42). Other developments in protein engineering include the development of artificial enzymes or high-throughput screening platforms. These methods hold some promise for improving the catalytic activity of enzymes or the rapid discovery of better enzymes (22, 51). Whether these new approaches can be applied successfully remains to be demonstrated.

43.5. TESTING PROCEDURES FOR ETHANOLOGENS ON HYDROLYSATES

Improved ethanologens can be created by classical mutagenesis and/or genetic engineering approaches. Ethanologens can also be obtained from numerous sources, which include proprietary and public culture collections. A partial list of ethanologens of interest is provided in Table 4. These ethanologens can be tested for performance on feedstocks prepared as described in sections 43.3 and 43.4. Laboratory testing of ethanologens on hydrolysates proceeds by experiments in

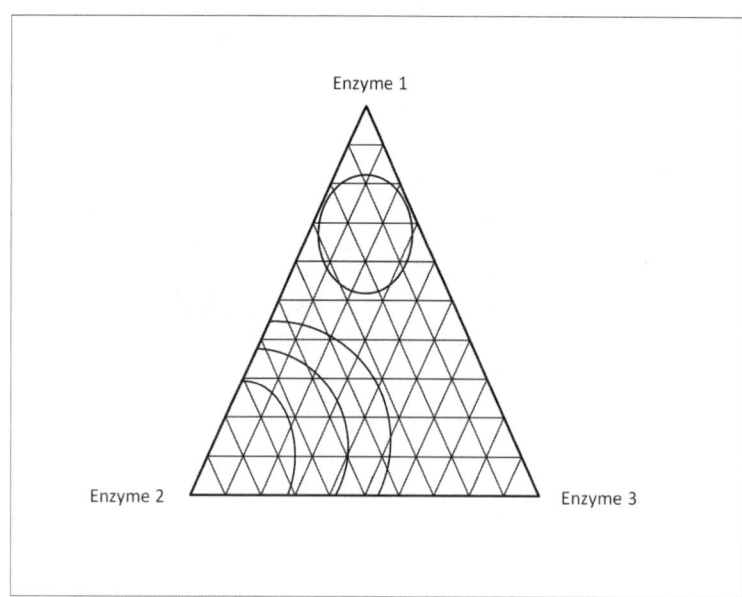

FIGURE 3 Ternary plot for enzyme blend optimization. Three enzymes, each with 10 concentrations, can be tested and plotted in the same experiment. Product (glucose or other sugars) concentration is plotted on the graph, and each line represents a concentration range.

TABLE 4 Properties of leading candidate microorganisms for industrial production of ethanol from xylose

Organism (wild type)	Growth on xylose	Anaerobic growth on glucose	Homo-ethanolic fermentation	Growth at pH 4.0	Ethanol tolerance (% [vol/vol])
Escherichia coli	+	+	−	−	<5
Clostridium spp.	+	+	−	−	<5
Zymomonas spp.	−	+	+	−	>10
Pichia stipitis	+	−	+	(+)	<5
Saccharomyces cerevisiae	−	+	+	+	>15

shake flasks and fermentors. Shake flasks are often used as a prescreening method before moving organisms to bench-scale fermentors. Successful ethanologen strains will proceed to the commercial-scale plant. Some benefits of using shake flask experiments to screen ethanologens include the following.

- Simplicity of planning
- Fast results
- Minimal cost due to the small scale and the simple equipment required
- Sufficient flexibility to run in batch format or as cascades which mimic continuous processes

Some of the issues to consider in choosing an ethanologen include the ability to ferment completely feedstock components, resistance to inhibitors, and the ability to thrive at temperatures and pH levels compatible with enzyme activity (4, 5, 14, 19, 30). Table 5 lists some of these considerations.

43.5.1. Shake Flask Protocol

Successful fermentations depend on the inoculation of a sufficient number of healthy cells into a fermentation broth. This is achieved by growing the inoculum in an enriched seed medium before starting the fermentation. The fermentation itself can use either an enriched medium with the addition of cellulosic material (33, 50) or a minimal medium with the addition of particular carbohydrates. Antimicrobials and antifoaming agents may be added as deemed necessary. The preparation of a seed medium for shake flask evaluation proceeds as follows.

1. Ethanologens should be grown in an enriched medium such as YPD broth for yeast (usually 1% yeast extract, 2% peptone, and 2% dextrose) (18) and tryptic soy broth or RM broth (yeast extract, KH_2PO_4, glucose, and other sugars) for bacteria. Cultures are incubated at an optimal temperature and agitation for the specific organisms.
2. Determine broth cell counts using a hemacytometer or automated cell counter, and determine cell mass by spectrophotometer. Adjust inocula so that all flasks have comparable cell counts or masses.

After the seed stage, fermentation tests proceed as outlined below.

1. Add the desired amount of cellulosic hydrolysate to YP broth (1% [wt/vol] yeast extract and 2% [wt/vol] peptone) or other medium with or without dextrose supplementation, depending upon the content of the cellulosic hydrolysate, in order to achieve a targeted broth concentration of dextrose.

2. Adjust the pH of the fermentation medium using ammonium hydroxide, potassium hydroxide, and calcium hydroxide or sodium hydroxide for an acid hydrolysate or sulfuric acid for an alkaline hydrolysate to a level which is acceptable to the chosen ethanologen and the required enzymes.
3. Add enzymes to hydrolyze any complex carbohydrates that have not been hydrolyzed by pretreatment protocols.
4. Inoculate enriched fermentation medium with the seed medium using comparable cell concentrations in each flask. Wash cells with sterile water several times before inoculation if the fermentation broth is a minimal medium.
5. Incubate flasks at the appropriate temperature and agitation. This agitation will be typically slower than that used when culturing the seeds because the intent now is to produce ethanol rather than cell mass.

During the fermentation, samples are taken daily and analyzed by HPLC or GC. One HPLC method employed by Agblevor et al. (3) analyzes sugars using an acetonitrile-water mixture as mobile phase, a column packed with a proprietary bonding material, and both a low-temperature evaporative light-scattering detector (ELSD-LT) and a refractive index detector. Figures 4 and 5 show chromatograms of standard sugar mixtures and a pretreated corn stover liquid fraction, respectively, analyzed by this method. Ethanol can be analyzed by centrifuging and filtering samples before injecting them into a GC system (50). The measurement of dry solids in the fermentation medium can be accomplished rapidly using a moisture balance set at 110°C. For these measurements, 1- to 2-g samples are

TABLE 5 Evaluation scheme for improved ethanologens for scale-up testing

Organism of choice is obtained by genetic engineering, mutagenesis, and selection or a combination of two using high-throughput screening approaches.

Single isolates or clones that meet a set of criteria that range from robustness, tolerance to hydrolysate, and fermentation of sugars of interest are further screened in shake flasks.

The most promising isolates or clones are evaluated further in bench-scale fermentors to better define process parameters for scale-up purposes.

The pool of single isolates or clones is then limited further to a small pool that can be tested in pilot plant fermentors.

A single isolate and/or clone that meets overall desirable process criteria is moved on to a production facility.

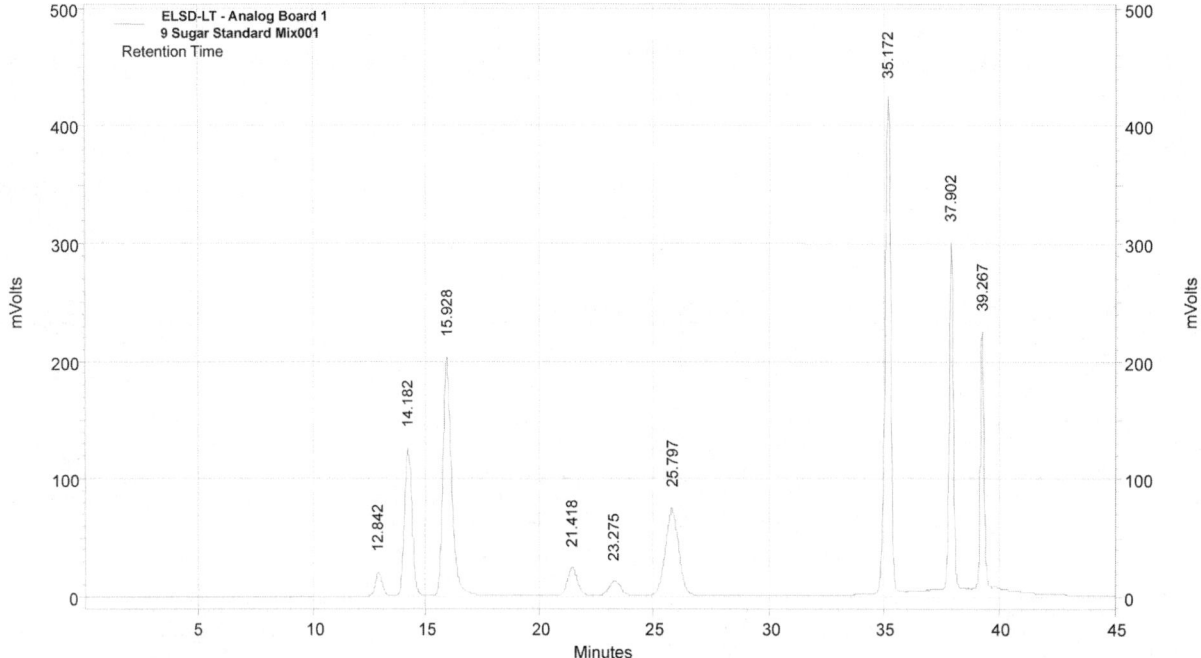

FIGURE 4 Chromatogram of seven standard sugar mixtures and internal standard showing their retention times (RT) in minutes. RT 12.84 = arabinose; RT 14.18 = xylose; RT 15.92 = fructose; RT 21.41 = mannose; RT 23.27 = galactose; RT 25.79 = glucose; RT 35.17 = inositol (internal standard); RT 37.90 = sucrose; and RT 39.27 = cellobiose. (Courtesy of F. Agblevor from reference 3.)

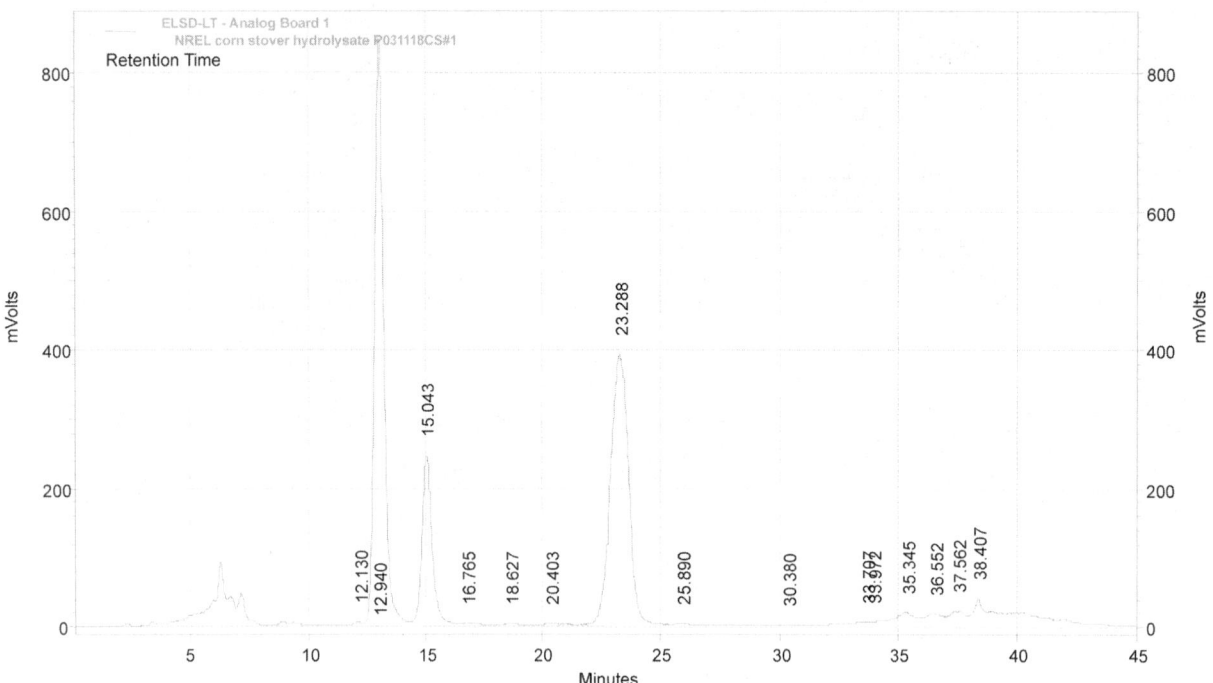

FIGURE 5 Chromatogram of pretreated corn stover liquid fraction showing three regions of unknown compounds (0 to 10 min), monomeric sugars (10 to 30 min), and oligomeric sugars (30 to 45 min). (Courtesy of F. Agblevor from reference 3.)

tared on the balance pan, the lid is closed, and measurement is made by pushing a start button and waiting several minutes until the reading stabilizes. An alternate method that provides greater accuracy uses an oven. In this method, preweighed and oven-dried disposable sample pans are used and the oven temperature is set in the range of 75 to 85°C. After determining the weight of the disposable sample pan containing a glass fiber pad, a sample is added and the weight is determined. The pan is placed in an oven until the weight of the dried sample remains unchanged. The pan is then removed from the oven and cooled in the desiccator. The weight of the pan and dried material is measured and dry solids are calculated as follows:

$$\{1 - [\frac{(\text{wet sample mass} - \text{dry sample mass})}{(\text{wet sample mass})}]\} \times 100$$

Dry solids are typically determined for fermentation medium at the outset and at the end of the fermentation. When using a pretreated feedstock that is high in suspended solids, total dry and soluble solids can be determined before and after enzyme hydrolysis. This is often done for material balance determinations.

43.5.2. Small-Scale Fermentor Protocols

Small-scale fermentation is carried out in fermentors varying in size from a few hundred milliliters to 6 liters. Small-scale fermentation is generally the next step in testing a strain or process conditions before testing at a pilot plant scale. Due to variances in agitation and aeration, difficulty in pH control in shake flasks, osmotic pressure, and nutrient distribution, in many cases strains perform differently in fermentation vessels than in shake flasks.

The goal in fermentation experiments in bench-scale vessels is similar to that in shake flasks. Fermentors are used to establish performance of selected strains on hydrolysate media without the limitations encountered in shake flasks. To accomplish this purpose, we test various media with different levels of hydrolysate to determine the performance of an ethanologen under difficult growth conditions in the presence of increasing levels of inhibitors. The following steps are usually followed.

1. Grow the organism in seed medium from a freezer vial lot representing plant media with a small percentage of hydrolysate for 18 to 21 h.
2. Measure seeds for optical density and viable cell count in order to inoculate the main fermentors to about 10^7 cells.
3. Prepare main media and add α-amylase if starch is present prior to heat sterilization in an autoclave.
4. Prior to inoculation, do the following.
   a. Set target temperature for experiment in fermentors.
   b. Set target pH (using the above-mentioned acids, preferably those used in the plant) in fermentors. Measure external pH to calibrate the pH probes in fermentors.
   c. Set target agitation in fermentors.
5. Take initial sample before inoculation of fermentation media and add glucoamylase, cellulases, hemicellulases, and other enzymes; incubate overnight at the same fermentation temperature.
6. Inoculate the fermentors.
   a. Start pH control to maintain set pH.
   b. Start temperature control to maintain set temperature.
7. Add enzyme blend to fermentors.
8. Take a postinoculation sample to measure initial ethanol that is carried over from the seeds as well as to monitor sugar consumption.

9. Sample the fermentors at least every 12 h, if not more frequently.
   a. Read external pH.
   b. Read biochemical analyzer for dextrose.
   c. Read Brix measurement. (Degrees Brix is a measurement of the dissolved sugar-to-water mass ratio of a liquid. It is measured with a refractometer.)
   d. Spin down samples and freeze supernatant for analytical submission.
10. At 48 h, measure the final volume of the fermentation.
11. Submit all samples, including the seed samples and the initial uninoculated control sample, for the following analysis: ethanol, glycerol, acetic acid, lactic acid, glucose, xylose, arabinose, mannose, fructose, galactose, and any other sugars or inhibitors that are likely to be present in the media or fermented by the strain being tested.

43.6. FERMENTATION METHODS

Once the appropriate organism and feedstock are selected, the method of fermentation must be chosen. Lignocellulosic hydrolysate fermentation or bioconversion may be achieved by either solid-state or submerged-liquid fermentation (4, 19, 31).

Solid-state fermentation consists of the direct inoculation of dry hydrolysate cake or partially digested precipitate. This method holds tremendous potential for the production of enzymes and various other fermented products (60). Tray, packed-bed, fluidized-bed, and rotary-drum fermentors are common vehicles of solid-state fermentation.

Although there are economic advantages to the use of solid-state fermentation, e.g., the cost of water and extraction are minimized, process limitations in the current methods of solid-state fermentation may favor submerged-liquid-broth fermentation for hydrolysates. Due to the high solids of the cake and reduced surface area created by the tray and packed-bed methods of solid-state fermentation, microbial growth and bioconversion occur mainly on the substrate's surface. Aeration and homogenous distribution of pH, nutrients, moisture, and temperature may pose an obstacle to adequate fermentation (54).

Carbohydrates in hydrolysates may be metabolized using the four main forms of liquid fermentation: batch, fed-batch, semicontinuous, and continuous. Continuous fermentation can be carried out in a single reactor, also referred to as a chemostat reactor, as described by Brandberg et al. (14), or in a cascade mode as detailed below. Tolerance of ethanologens to inhibitory compounds in lignocellulosics can be improved by cell recycling (30, 45).

43.6.1. Batch Method

During batch fermentation, pretreated substrate/slurry and supplemental nutrients are combined in the initial media. Possible growth inhibitors formed during the pretreatment (acids, hydroxymethyl furfural, and furfural) may also be present at levels higher than during continuous fermentation. It may be necessary to increase cell inoculation dosages and adjust the pH when fermenting pretreated biomass using the batch method. The main advantage of the batch process is that cleaning of the equipment can be done after each fermentation; therefore, process vulnerability to yeast and enzyme contaminants is reduced. The main disadvantage is the increased downtime necessary to clean and sterilize fermentors.

43.6.2. Fed-Batch Method

The fed-batch method is similar initially to the batch method; however, pretreated solids/slurries and other nutrients are

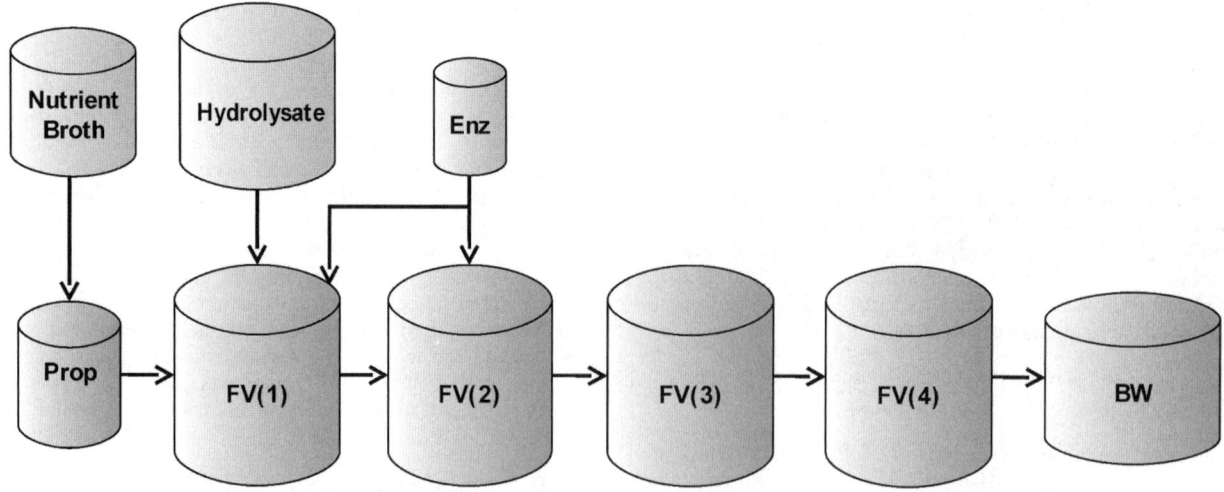

FIGURE 6 Cascade testing: illustration of a continuous setup for testing of ethanologens on a hydrolysate stream derived from a lignocellulosic. Prop, propagation; Enz, enzyme; BW, beer well; FV, fermentation vessel.

fed continually to prevent a plateau in the growth curve. The feed rate greatly influences cell viability. The pH must be adjusted based on the inclusion rate of the hydrolysate. An alkaline solution may be used to adjust the pH for acid hydrolysates or an acid solution for alkaline hydrolysates to a range of 3.5 to 7, depending on the fermentation organism.

43.6.3. Continuous-Cascade Method

Fermentation in the continuous-cascade configuration consists of a series of tanks (or flasks, depending on experimental scale) into which are pumped pretreated lignocellulose and nutrients at crucial stages in the fermentation process. An example of cascade design follows. The first tank typically serves as a propagator and is inoculated with the ethanologen culture being tested. Overflow from this tank is pumped into a second tank as illustrated in Fig. 6. The second tank is also fed hydrolysate and a blend of enzymes to aid in the gradual release of available sugars. The third tank receives overflow from the second tank as well as additional enzymes. The fourth tank holds overflow from the third tank. Surplus broth from the fourth is pumped to the fifth. Overflow from the fifth tank is pumped into the beer well and is recycled back to the propagator. If necessary, and in order to maintain a high cell viability count in the cascade, an overflow loop from the final collection tank (beer well) with supplemental nutrient solution can be fed to the propagator.

In the bench-scale experimental setup illustrated in Fig. 5, the temperature is controlled using a circulating water bath and the flow rate through the cascade is adjusted using peristaltic pumps calibrated for the residence time sought. Stir plates maintain medium to low agitation within each flask during bench-scale experiments to circulate the yeast and hydrolysate media. The main advantages to the cascade process are as follows.

- Less downtime for maximum use of equipment. Maximum heat load can be targeted to specific fermentors for cost savings.

- Easily scaled down to a bench-scale experiment in which variables such as flow rates, feeds, etc., are evaluated at a relatively low cost (relative to production scale).

The primary disadvantages are as follows.

- Cleaning the equipment may be problematic.
- The system is vulnerable to shifts in external temperature.
- More vigilance is required by operators to monitor potential problems in the system.
- The system is highly vulnerable to yeast and enzyme contaminants.

43.6.4. Semicontinuous-Cascade Method

Semicontinuous fermentation is similar to continuous-cascade fermentation; however, a fraction of the culture is replaced with fresh medium at regular intervals (69). The greatest disadvantage is the removal of cells when culture is replaced with medium.

43.6.5. Other Specific Considerations

- Stover hydrolysate exhibits properties of a non-Newtonian fluid; this should be considered when selecting tubing and pump methods.
- Fermentation vessels (flasks, tubing, etc.) should possess resistance to chemical erosion due to content of the hydrolysate.
- The optimal hydrolysate concentration is dependent on the organism's tolerance of inhibitors.

43.7. CONCLUDING REMARKS

In the past decade, significant advances have been made in the pretreatment and hydrolysis of a wide range of lignocellulosic feedstocks. Development of improved biocatalysts (enzymes and ethanologens) and the pace of discovery of new ones have also increased. When combined with improvements in design and engineering of processing facilities, these advances have set the stage for the construction of pilot- and demonstration-scale plants to validate this new technology. Beyond these developments, significant reduction in processing costs must be made to ensure that the cost of producing cellulosic ethanol is competitive with that of bioethanol produced from conventional feedstocks such as corn, sweet sugar beets, and sugarcane. Improving the logistics and sourcing of lignocellulosic supplies is still necessary for future large-scale production and expansion of this emerging industry.

REFERENCES

1. **Abbas, C. A.** 2003. Lignocellulosics to ethanol: meeting ethanol demand in the future, p. 41–57. In K. A. Jacques, T. P. Lyons, and D. R. Kelsall (ed.), The Alcohol Texbook, 4th ed. Nottingham University Press, Nottingham, United Kingdom.

2. **Abbas, C. A., K. E. Beery, E. K. Dennison, and P. M. Corrington.** 2004. Thermochemical treatment, separation and conversion of corn fiber to ethanol. ACS Symp. Ser. **889:**84–97.

3. **Agblevor, F. A., B. R. Hames, D. Schell, and H. L. Chum.** 2007. Analysis of biomass sugars using a novel HPLC method. Appl. Biochem. Biotechnol. **136:**309–326.

4. **Alkasrawi, M., A. Rudolf, G. Liden, and G. Zacchi.** 2006. Influence of strain and cultivation procedure on the performance of simultaneous saccharification and fermentation of steam pretreated spruce. Enzyme Microb. Technol. **38:**279–286.

5. **Almeida, J. R. M., M. Bertilsson, M. F. Gorwa-Grauslund, S. Gorsich, and G. Lidén.** 2009. Metabolic effects of furaldehydes and impacts on biotechnological processes. Appl. Microbiol. Biol. **82:**625–638.

6. **Baker, J. O., C. L. Ehrman, W. S. Adney, S. R. Thomas, and M. E. Himmel.** 1998. Hydrolysis of cellulose using ternary mixtures of purified cellulases. Appl. Biochem. Biotechnol. **70–72:**395–403.

7. **Ballesteros, I., J. M. Olivia, A. A. Navarro, A. Gonzalez, J. Carrasco, and M. Ballesteros.** 2000. Effect of chip size on steam explosion pretreatment of softwood. Appl. Biochem. Biotechnol. **84–86:**97–110.

8. **Battan, B., J. Sharma, and R. C. Kuhad.** 2006. High-level xylanase production by alkaliphilic Bacillus pumilus ASH under solid-state fermentation. World J. Microbiol. Biotechnol. **22:**1281–1287.

9. **Bélafi-Bakó, K., A. Koutinas, N. Nemestóthy, L. Gubicza, and C. Webb.** 2006. Continuous enzymatic cellulose hydrolysis in a tubular membrane bioreactor. Enzyme Microb. Technol. **38:**155–161.

10. **Berlin, A., M. Balaskshin, N. Gilkes, J. Kadla, V. Maximenko, S. Kubo, and J. Saddler.** 2006. Inhibition of cellulase, xylanase, and β-glucosidase activities by softwood lignin preparations. J. Biotechnol. **125:**198–209.

11. **Berlin, A., N. Gilkes, D. Kilburn, R. Bura, A. Markov, A. Skomarosky, O. Okunev, A. Gusakov, V. Maximenko, D. Gregg, A. Sinitsyn, and J. Saddler.** 2005. Evaluation of novel fungal preparations for ability to hydrolyze softwood substrates—evidence for the role of accessory enzymes. Enzyme Microb. Technol. **37:**175–184.

12. **Berlin, A., N. Gilkes, A. Kurabi, R. Bura, M. Tu, D. Kilburn, and J. Saddler.** 2005. Weak lignin-binding enzymes: a novel approach to improve activity of cellulases for hydrolysis of lignocellulosics. Appl. Biochem. Biotechnol. **121–124:**163–170.

13. **Borjesson, J., R. Peterson, and F. Tjerneld.** 2007. Enhanced enzymatic conversion of softwood lignocellulose by poly(ethylene glycol) addition. Enzyme Microb. Technol. **40:**754–762.

14. **Brandberg, T., N. Sanandaji, L. Gustafsson, and C. J. Franzen.** 2005. Continuous fermentation of undetoxified dilute acid lignocellulose hydrolysate by Saccharomyces cerevisiae ATCC 96581 using cell recirculation. Biotechnol. Prog. **21:**1093–1101.

15. **Caminal, G., J. López-Santín, and C. Sola.** 1985. Kinetic modeling of the enzymatic hydrolysis of pretreated cellulose. Biotechnol. Bioeng. **27:**1282–1290.

16. **Chum, H. L., S. K. Black, D. K. Johnson, K. V. Sarkanen, and D. Robert.** 1999. Organosolv pretreatment for enzymatic hydrolysis of poplars: isolation and quantitative structural studies of lignins. Clean Products Processes **1:**187–198.

17. **Cullis, I. F., J. N. Saddler, and S. D. Mansfield.** 2004. Effect of initial moisture content and chip size on the bioconversion efficiency of softwood lignocellulosics. Biotechnol. Bioeng. **85:**413–421.

18. **Degelmann, A.** 2002. Methods, p. 286–288. In G. Gellissen (ed.), Hansenula polymorpha: Biology and Applications. Wiley-VCH Verlag GmbH, Weinheim, Germany.

19. **Den Haan, R., S. H. Rose, L. R. Lynd, and W. H. van Zyl.** 2007. Hydrolysis and fermentation of amorphous cellulose by recombinant Saccharomyces cerevisiae. Metabolic Eng. **9:**87–94.

20. **Dubois, M., K. A. Gilles, J. K. Hamilton, P. A. Rebers, and F. Smith.** 1956. Colorimetric method for determination of sugars and related substances. Anal. Chem. **28:**350–356.

21. **Galbe, M., and G. Zacchi.** 2002. A review of the production of ethanol from softwood. Appl. Microbiol. Biotechnol. **59:**618–628.

22. **Ghirlanda, G.** 2008. Old enzymes, new tricks. Nature **453:**164–166.

23. **Gray, K. A., L. Zhao, and M. Emptage.** 2006. Bioethanol. Curr. Opin. Chem. Biol. **10:**141–146.

24. **Grohmann, K., and R. J. Bothast.** 1997. Saccharification of corn fibre by combined treatment with dilute sulphuric acid and enzymes. Process Biochem. **32:**405–415.

25. **Hammel, K. E., and D. Cullen.** 2008. Role of fungal peroxidases in biological ligninolysis. Curr. Opin. Plant Biol. **11:**349–355.

26. **Ho, K. M., X. Mao, L. Gu, and P. Li.** 2008. Facile route to enzyme immobilization: core-shell nanoenzyme particles consisting of well-defined poly(methyl methacrylate) cores and cellulase shells. Langmuir **24:**11036–11042.

27. **Jin, S., and H. Chen.** 2006. Superfine grinding of steam-exploded rice straw and its enzymatic hydrolysis. Biochem. Eng. J. **30:**225–230.

28. **Kabel, M. A., G. Bos, J. Zeevalking, A. G. J. Voragen, and H. A. Schols.** 2007. Effect of pretreatment severity on xylan solubility and enzymatic breakdown of the remaining cellulose from wheat straw. Bioresource Technol. **98:**2034–2042.

29. **Katahira, S., Y. Fujita, A. Mizuike, H. Fukuda, and A. Kondo.** 2004. Construction of a xylan-fermenting yeast strain through codisplay of xylanolytic enzymes on the surface of xylose-utilizing Saccharomyces cerevisiae cells. Appl. Environ. Microbiol. **70:**5407–5414.

30. **Keating, J. D., C. Panganiban, and S. D. Mansfield.** 2006. Tolerance and adaptation of ethanologenic yeasts to lignocellulosic inhibitory compounds. Biotechnol. Bioeng. **93:**1196–1206.

31. **Keating, J. D., J. Robinson, M. A. Cotta, J. N. Saddler, and S. D. Mansfield.** 2004. An ethanologenic yeast exhibiting unusual metabolism in the fermentation of lignocellulosic hexose sugars. J. Ind. Microbiol. Biotechnol. **31:**235–244.

32. **Kim, J. K., Y. Y. Lee, and S. C. Park.** 2000. Pretreatment of wastepaper and pulp mill sludge by aqueous ammonia and hydrogen peroxide. Appl. Biochem. Biotechnol. **84–86:**129–139.

33. **Kim, J. S., K. K. Oh, S. W. Kim, Y. S. Jeong, and S. I. Hong.** 1999. Ethanol production from lignocellulosic biomass by simultaneous saccharification and fermentation employing the reuse of yeast and enzyme. J. Microbiol. Biotechnol. **9:**297–302.

34. **Kim, S., and M. T. Holtzapple.** 2005. Lime pretreatment and enzymatic hydrolysis of corn stover. Bioresource Technol. **96:**1994–2006.

35. Kim, T. H., and Y. Y. Lee. 2005. Pretreatment and fractionation of corn stover by ammonia recycle percolation process. *Bioresource Technol.* **96:**2007–2013.

36. Kim, Y., R. Hendrickson, N. S. Mosier, M. R. Ladisch, B. Bals, V. Balan, and B. E. Dale. 2008. Enzyme hydrolysis and ethanol fermentation of liquid hot water and AFEX pretreated distillers' grains at high-solids loadings. *Bioresource Technol.* **99:**5206–5215.

37. Kristensen, J. B., J. Borjesson, M. H. Bruun, F. Tjerneld, and H. Jorgensen. 2007. Use of surface active additives in enzymatic hydrolysis of wheat straw lignocellulose. *Enzyme Microb. Technol.* **40:**888–895.

38. Leskinen, S., A. Mäntylä, R. Fagerström, J. Vehmaanperä, R. Lantto, M. Paloheimo, and P. Suominen. 2005. Thermostable xylanases, Xyn10A and Xyn11A, from the actinomycete *Nonomuraea flexuosa*: isolation of the genes and characterization of recombinant Xyn11A polypeptides produced in *Trichoderma reesei*. *Appl. Microbiol. Biotechnol.* **67:**495–505.

39. Liu, G., X. Tang, S-L. Tian, X. Deng, and M. Xing. 2006. Improvement of the cellulolytic activity of *Trichoderma reesei* endoglucanase IV with an additional catalytic domain. *World J. Microb. Biotechnol.* **22:**1301–1305.

40. Lloyd, T. A., and C. E. Wyman. 2005. Combined sugar yields for dilute sulfuric acid pretreatment of corn stover followed by enzymatic hydrolysis of the remaining solids. *Bioresource Technol.* **96:**1967–1977.

41. Maekawa, E. 1996. On an available pretreatment for the enzymatic saccharification of lignocellulosic materials. *Wood Sci. Technol.* **30:**133–139.

42. Mais, U., A. R. Esteghlalian, J. N. Saddler, and S. D. Mansfield. 2002. Enhancing the enzymatic hydrolysis of cellulosic materials using simultaneous ball milling. *Appl. Biochem. Biotechnol.* **98–100:**815–832.

43. Markov, A. V., A. V. Gusakov, E. G. Kondratyeva, O. N. Okunev, A. O. Bekkarevich, and A. P. Sinitsyn. 2005. New effective method for analysis of the component composition of enzyme complexes from *Trichoderma reesei*. *Biochemistry* (Moscow) **70:**657–663.

44. Martín, C., H. B. Klinke, and A. B. Thomsen. 2007. Wet oxidation as a pretreatment method for enhancing the enzymatic convertibility of sugarcane bagasse. *Enzyme Microb. Technol.* **40:**426–432.

45. Martin, C., M. Marcet, O. Almazan, and L. J. Jonsson. 2007. Adaptation of a recombinant xylose-utilizing *Saccharomyces cerevisiae* strain to a sugarcane bagasse hydrolysate with high content of fermentation inhibitors. *Bioresource Technol.* **98:**1767–1773.

46. Muthuvelayudham, R., and T. Viruthagiri. 2006. Fermentative production and kinetics of cellulase protein production on *Trichoderma reesei* using sugarcane bagasse and rice straw. *African J. Biotechnol.* **5:**1873–1881.

47. Nieves, R. A., C. I. Ehrman, W. S. Adney, R. T. Elander, and M. E. Himmel. 1998. Technical communication: survey and analysis of commercial cellulase preparations suitable for biomass conversion to ethanol. *World J. Microbiol. Biotechnol.* **14:**301–304.

48. Nutt, A., V. Sild, G. Pettersson, and G. Johansson. 1998. Progress curves—a mean for functional classification of cellulases. *Eur. J. Biochem.* **258:**200–206.

49. O'Dwyer, J. 2005. Ph.D. thesis. Texas A & M, College Station.

50. Patle, S., and B. Lal. 2008. Investigation of the potential of agro-industrial material as low cost substrate for ethanol production by using *Candida tropicalis* and *Zymomonas mobilis*. *Biomass Bioenergy* **32:**596–602.

51. Pohn, B., J. Gerlach, M. Scheider, H. Katz, M. Uray, H. Bischof, I. Klimant, and H. Schwab. 2007. Micro-colony array based high throughput platform for enzyme library screening. *J. Biotechnol.* **129:**162–170.

52. Ragauskas, A. J., C. K. Williams, B. H. Davison, G. Britovsek, J. Carney, C. E. Eckert, W. J. Fredrick, J. P. Hallett, D. J. Leak, C. L. Liotta, J. R. Mielenz, R. Murphy, R. Templer, and T. Tschaplinski. 2006. The path forward for biofuels and biomaterials. *Science* **311:**484–489.

53. Saha, B. C., and R. J. Bothast. 1999. Pretreatment and enzymatic saccharification of corn fiber. *Appl. Biochem. Biotechnol.* **76:**65–77.

54. Sato, K. 1999. Small-scale solid-state fermentations, p. 61–73. *In* A. Demain and J. Davies (ed.), *Manual of Industrial Microbiology and Biotechnology*, 2nd ed. ASM Press, Washington, DC.

55. Sewalt, V. J. H., K. A. Beauchemin, L. M. Rode, S. Acharya, and V. S. Baron. 1997. Lignin impact on fiber degradation. IV. Enzymatic saccharification and in vitro digestibility of alfalfa and grasses following selective solvent delignification. *Bioresource Technol.* **61:**199–206.

56. Sewalt, V. J. H., W. G. Glasser, and K. A. Beauchemin. 1997. Lignin impact on fiber degradation. 3. Reversal of inhibition of enzymatic hydrolysis by chemical modification of lignin and by additives. *J. Agric. Food Chem.* **45:**1823–1828.

57. Sievers, C., M. B.Valenzuela-Olarte, T. Marzialetti, I. Musin, P. K. Agrawal, and C. W. Jones. 2009. Ionic-liquid-phase hydrolysis of pine wood. *Ind. Eng. Chem. Res.* **48:**1277–1286.

58. Steele, B., S. Raj, J. Nghiem, and M. Stowers. 2005. Enzyme recovery and recycling following hydrolysis of ammonia fiber explosion-treated corn stover. *Appl. Biochem. Biotechnol.* **121–124:**901–910.

59. Szengyel, Z., G. Zacchi, A. Varga, and K. Reczey. 2000. Cellulase production of *Trichoderma reesei* Rut C 30 using steam-pretreated spruce. *Appl. Biochem. Biotechnol.* **84–86:**679–691.

60. Tengerdy, R. P. 1998. Solid state fermentation for enzyme production, p. 13–16. *In* A. Pandey (ed.), *Advances in Biotechnology*. Educational Publishers and Distributors, New Delhi, India.

61. Teymouri, F., L. Laureano-Perez, H. Alizadeh, and B. E. Dale. 2005. Optimization of the ammonia fiber explosion (AFEX) treatment parameters for enzymatic hydrolysis of corn stover. *Bioresource Technol.* **96:**2014–2018.

62. Tu, M., X. Zhang, A. Kurabi, N. Gilkes, W. Mabee, and J. Saddler. 2006. Immobilization of beta-glucosidase on Eupergit C for lignocellulose hydrolysis. *Biotechnol. Lett.* **28:**151–156.

63. Tucker, M. P., K. H. Kim, M. M. Newman, and Q. A. Nguyen. 2003. Effects of temperature and moisture on dilute-acid steam explosion pretreatment of corn stover and cellulase enzyme digestibility. *Appl. Biochem. Biotechnol.* **105:**1–3.

64. Ueda, M., and A. Tanaka. 2000. Cell surface engineering of yeast: construction of arming yeast with biocatalyst. *J. Biosci. Bioeng.* **90:**125–136.

65. Uryu, T., M. Sugie, S. Ishida, S. Konoma, H. Kato, K. Katsuraya, K. Okuyama, G. Borjihan, K. Iwashita, and H. Iefuji. 2006. Chemo-enzymatic production of fuel ethanol from cellulosic materials utilizing yeast expressing β-glucosidases. *Appl. Biochem. Biotechnol.* **135:**15–31.

66. Vital-Lopez, F. G., A. Armaou, E. V. Nikolaev, and C. D. Maranas. 2006. A computational procedure for optimal engineering interventions using kinetic models of metabolism. *Biotechnol. Prog.* **22:**1507–1517.

67. Vries, R. P., and J. Visser. 2001. *Aspergillus* enzymes involved in degradation of plant cell wall polysaccharides. *Microbiol. Mol. Biol. Rev.* **65:**497–522.

68. Wagschal, K., D. Franqui-Espiet, C. C. Lee, R. E. Kibblewhite-Accinelli, G. H. Robertson, and D. W. S. Wong. 2007. Genetic and biochemical characterization of an α-L-arabinofuranosidase isolated from a compost starter mixture. *Enzyme Microb. Technol.* **40:**747–753.

69. **Westgate, P., and A. Emery.** 1990. Approximation of continuous fermentation by semicontinuous cultures. *Biotechnol. Bioeng.* **35:**437–453.

70. **Wyman, C. E., B. E. Dale, R. T. Elander, M. Holtzapple, M. R. Ladisch, and Y. Y. Lee.** 2005. Coordinated development of leading biomass pretreatment technologies. *Bioresource Technol.* **96:**1959–1966.

71. **Xiao, Z., X. Zhang, D. J. Gregg, and J. N. Saddler.** 2004. Effects of sugar inhibition on cellulases and β-glucosidase during enzymatic hydrolysis of softwood substrates. *Appl. Biochem. Biotechnol.* **113–116:**1115–1126.

72. **Zhang, Y. H., M. E. Himmel, and J. R. Mielenz.** 2006. Outlook for cellulase improvement: screening and selection strategies. *Biotechnol. Adv.* **24:**452–481.

Surface Microbiology of Cellulolytic Bacteria

ALEXANDRU DUMITRACHE, GIDEON M. WOLFAARDT, AND LEE R. LYND

44

44.1. INTRODUCTION

Cell-enzyme-substrate interactions may be major determinants for the efficiency of cellulose hydrolysis during biomass conversion. These associations, where the substratum serves as the attachment support and carbon source for the cells as well as the substrate for saccharolytic enzymes, differs substantially from those in biofilms typically studied on inert interfaces. While the microbiology associated with bioprocess technology traditionally focused on suspended populations, there is evidence that biofilms may be central elements of cellulolytic microorganisms' life cycles. Delineating the underlying fundamentals of such biofilms should advance bioconversion technology and specifically contribute to realizing the potential of cellulosic biomass as a feedstock for renewable fuel production. In this chapter, biofilm structure and cell behavior in fully hydrated, undisturbed biofilms of *Clostridium thermocellum* are considered as a means to study biofilm formation by these anaerobic thermophiles.

44.2. CELLULOSE CONVERSION FROM A BIOFILM PERSPECTIVE: A NEW FRONTIER

44.2.1. The Biofilm Mode of Microbial Existence

Bacteria are often studied in a planktonic mode as suspended cells growing in a liquid culture; however, in most natural environments, up to 95% of bacteria are present in surface-associated biofilms (38). Sessile bacteria thrive in multicellular communities associated with phase interfaces, which convey advantages over the planktonic life form, most notably protection against environmental factors such as extreme conditions, host immune response, predation, and antimicrobial stress. The classical biofilm is a microbial assemblage of adherent cells, situated (most commonly) at a liquid-solid interface and encased within a matrix of extracellular polymeric material. It may be noted that this definition is applicable to microbial aggregates and floccules (7). Biofilm translocation to new niches is generally understood to occur when cells become planktonic and then recolonize new interfaces or, in the case of some flagellated bacteria, by a massive coordinated group behavior called swarming motility (54). Other surface-associated methods of translocation have been described (16). Taken to the extreme, the planktonic mode of bacterial life can therefore be viewed as

primary mechanisms of cell translocation from one surface to another (55); however, from a holistic perspective, both biofilm and planktonic phenotypes are seen as integrated components of the prokaryotic lifestyle (51). Many studies suggest that the profiles of gene transcription are significantly different for adhered and nonadhered cells (45) and that cell-to-cell signaling is a required condition for biofilm formation (9, 46).

44.2.2. Cell-Cellulose Interactions in Nature

Close associations between cellulolytic organisms and their substrates have been reported in the literature. For example, a survey of 13 cellulolytic and 10 noncellulolytic strains of *Ruminococcus albus* showed that most cellulolytic strains bind to the substrate, whereas most noncellulolytic strains do not; however, the extent of bacterial adhesion was not well correlated with culture cellulolytic activity (36). Recent evidence demonstrates the strong correlation between first-order hydrolysis rates of cellulose with the concentration of sessile bacteria rather than with the concentration of total or planktonic biomass after inoculation with enriched leachate and rumen fluid (20).

In the case of *C. thermocellum*, cellulolytic enzyme complexes, defined as cellulosomes, were found to form large clusters embedded in the cell surface layer, appearing as protuberances in electron micrographs. Mutants isolated based on their inability to bind cellulose were found to lack such protuberances of cell-bound cellulosomes and to display reduced cellulose hydrolysis rates (1–3). Further study led to a comprehensive description of the macromolecular cellulosome architecture. A key feature of this architecture is a noncellulolytic scaffoldin protein including a cellulose-binding domain, dockerin domains to which catalytic components bind, and a type II cohesin region that mediates binding to the cell surface (4). Other investigations using electron microscopy revealed that such cell surface "protuberances" are present on the surfaces of a variety of cellulolytic bacteria (gram positive or negative, thermophilic or mesophilic, and aerobic or anaerobic) but are not found in noncellulolytic bacteria or on cellulolytic bacteria grown under conditions that are known to repress cellulase synthesis (17, 25).

Rumen bacteria are known to strongly attach to grass particles and model cellulosic substrates. Electron micrographs of the cellulolytic rumen bacterium *Fibrobacter succinogenes* show tight adherence to the cellulose fibers, with erosion of the substrate visible beneath the cells (12). Glycosylated residues of cellulose-binding proteins, some of which have endoglucanase activity, isolated from the cell envelope of *Fibrobacter intestinalis* DR7 were suggested to have an important role in the adhesion of cells to cellulose (35), and electron micrographs of adherent *Ruminococcus flavefaciens* cells have shown the presence of a discernible exopolysaccharide (EPS) matrix, presumably responsible for nonspecific adhesion (24). A glycocalyx, thought to be responsible for substrate adherence through its constitutive polymeric elements, has also been identified in the anaerobic cellulolytic *R. albus* 7. Variable-pressure scanning electron microscopy revealed the formation of thin cellular extensions that mediate substrate attachment, followed by the development of a ramifying network that connects individual cells to one another (56). Additionally, in the gram-positive strain *R. albus* 8, a novel type of cellulose-binding protein (CbcC) belonging to the Pil protein family has been isolated with similarity to the type 4 fimbrial proteins of gram-negative pathogens. *R. albus* 8 was found to employ a fimbria-mediated attachment to cellulose (37).

Adhesion to the substrate and the formation of a cellulose-enzyme-microbe ternary complex has been suggested to provide a series of advantages to the cellulolytic organism, which include enzyme concentration at the cellulose surface, fast sequestration of oligomeric hydrolysis products, protection from predation (grazing protozoa) or bacteriophage attack, and, in the case of ruminal bacteria, protection from rumen proteases (30). The benefits seem to outweigh the extra energy demand for the biosynthesis of glycocalyx precursors and of cellulose binding factors, which, for example, in the case of the nonprotein fraction of EPS, has been shown to represent ≤4% of the total anabolic ATP expenditure of *C. thermocellum* (56).

Overall, substrate adherence by anaerobic cellulolytic species is primarily attributed to (i) the cellulase complex (cellulosome) through the carbohydrate binding module, (ii) the glycosylated moieties of catalytic or noncatalytic cellulose-binding proteins in the cell envelope, (iii) the carbohydrate epitopes of the bacterial glycocalyx, and (iv) fimbriae or other proteinaceous appendages. The actual strategies employed by individual species may be a combination of the above mechanisms, as suggested for adherent rumen cellulolytic bacteria (34); however, further research is needed to extend this knowledge to other naturally occurring anaerobic bacteria.

44.3. CONCEPTUAL AND POTENTIAL METHODOLOGICAL OVERLAP BETWEEN BIOFILM RESEARCH AND BIOMASS PROCESSING

44.3.1. Tools for Discovery and Current State of Biofilm Research

Modern biofilm research has relied primarily on a few major approaches: genetic screening of mutants (22, 23), the application of confocal laser scanning microscopy in conjunction with fluorescent molecular probes and digital image analysis (26, 33), and, to a lesser extent, biophysical studies that measure adhesion kinetics and cohesive strength (e.g., the quartz crystal microbalance technique, atomic force microscopy, or optical tweezers) (42). Other approaches with demonstrated or suggested utility for biofilm research include measurement of biofilm respiratory activity (50); quantification of cell mass and numbers

(19); and biochemical, spectroscopic, and chromatographic analyses of cells and EPS composition and structure (47).

Biofilm formation has been proposed as a model system for the study of microbial development. Reviewing a large number of studies from the preceding 10 years, O'Toole et al. (41) concluded that biofilms are a stable point in a biological cycle that includes initiation, maturation, maintenance, and dissolution and that most microorganisms appear to be able to make the transition to a surface-associated life, regardless of their physiological capabilities. The number of biofilm-related studies has indeed grown at an impressive rate during the last two decades in diverse fields such as medical, natural, and engineered systems. A recent keyword search revealed that the number of articles published in the scientific literature containing the term "biofilm" in the abstract steadily increased from less than 200 per year in 1993 to more than 2,000 per year in 2008. Research that started as primarily microscopic observations of attached cells now involves a range of tools and techniques to analyze spatial relationships, gene expression, and metabolic activity, and as a result, the structure-function relationships of biofilms are increasingly better understood.

The same is not true for cellulolytic biofilm research, for which there are only a few reports that describe the role of biofilms in cellulose hydrolysis, and among those, the focus is primarily on natural processes and degradation of cellulose-based waste. Adhesion by *C. cellulolyticum* to cellulose in order to overcome carbon starvation was addressed by Gehin et al. (13), who demonstrated higher levels of adhesion by cells in the exponential growth phase than those in the stationary phase. They further suggested that spores may play a role in increased adhesion towards the late stationary phase. The positive correlation between cell activity and adhesion to cellulose in the exponential growth phase suggests that biofilm formation facilitates cellulose degradation, which is in line with the phenomenon of enzyme-microbe synergy described by Lu et al. (28). In the latter, the effectiveness of a cellulase system is significantly enhanced when it functions as part of a cellulose-enzyme-microbe complex. Rotter et al. (48) developed a model to evaluate the importance of parameters and processes in the modeling of cellulose degradation. Although this model focused on landfill and anaerobic digesters, their assumptions and conclusions may be relevant to cellulolytic processes in general. The authors' major concern with the most commonly used approaches for modeling cellulose-utilizing bacteria is that hydrolysis is typically described as a zero- or first-order reaction that is independent of the planktonic biomass, implying that the impact of bacterial growth is negligible and that enzyme-producing cells are in excess. It is argued that these assumptions represent an oversimplification, as there is no concrete evidence that the cellulolytic cells indeed have direct access to cellulose particles. Furthermore, referencing Schwarz (49) and Lynd et al. (30), the authors indicated that cell-enzyme-cellulose binding may indeed be essential for the microbially mediated cellulose hydrolysis process to proceed. Although subsequent models incorporated cell-cellulose binding, the authors indicated that these models do not take spatial separation between bacterial cells and cellulose into account, nor do they consider interaction between the attached and suspended populations. In response to these potential shortfalls, the model developed by Rotter et al. (48) incorporates the potential for colonization of the cellulose substrate by suspended bacteria, which then become the primary agent of hydrolysis, including controlling the rate of hydrolysis. At this point it should be emphasized that this model focuses on landfill and anaerobic digesters, thus with an ultimate goal of complete hydrolysis of biomass to simple products such as CO_2 and CH_4, mediated by complex microbial communities with a high degree of cooperation, cometabolism, and often competition, which is in contrast to bioprocesses where biomass conversion is perceived through the activity of single species, native or engineered, or potentially through synergy of well-defined microbial consortia, in some cases supplemented with selected hydrolytic enzymes. It therefore remains to be determined to what degree models that are based on biofilm characteristics generally attributed to degradative microbial communities can account for or predict cellulose degradation by *C. thermocellum* and other cellulolytic strains with industrial potential—the need for cooperation and sequential degradation that is typically described for the biofilm mode of existence may not be relevant for cellulolytic organisms where it appears that specialized species have evolved independently to overcome the challenge of growth on a single refractory polymer.

McAllister et al. (31) summarized findings by various groups that showed distinct populations in the rumen ecosystem: the nonattached organisms in the ruminal fluid and two subpopulations attached loosely (can be removed by gentle washing) and firmly to feed particles. While these attached subpopulations accounted for 70 to 80% of the microbial matter in the rumen, and were responsible for up to 80% of endoglucanase, 70% of amylase, and 75% of the protease activity in the rumen, the subpopulation associated with the ruminal fluid was linked with the important role of colonizing newly ingested feed particles. The demonstration that hemicellulase and cellulase activities were notably higher in the particulate fraction of ruminal contents than in the fluid suggested that biofilm populations are responsible for the majority of feed digestion in the rumen.

A study of the colonization potential of naturally occurring, mixed species in the anaerobic layers of a landfill site made use of an adapted aerobic flow cell to promote strict anaerobic conditions (21). The culture chamber, designated as an anaerobic continuous-culture microscopy unit, was a miniature, self-sealing, oxygen-impermeable, transparent reactor that could be fit on a microscope stage. The investigators used inocula of enriched collections from the anaerobic layers of a landfill site, which were injected at the outflow port of the flow chamber to avoid passive transport of organisms by the media flow and to ensure that any observed attachment upstream was due to some active motility or transposition event. The colonization process on biodegradable and nonbiodegradable materials was monitored over time by capturing phase-contrast micrographs and calculating the rate at which the biofilm front advanced. Fluorescence microscopy and the natural coenzyme F_{420} autofluorescence were used to track and quantify the localization of methanogenic bacteria. It was found that wood, cotton, and polyester fabric remained poorly colonized or not colonized after weeks of incubations, whereas the nonbiodegradable glass, plastic, and polyethane were readily and heavily colonized. However, the study acknowledged that enrichment for butyrate- and hexanoate-degrading methanogens could have excluded cellulolytic organisms from the flow cell inocula, which could explain the lack of cellulose colonization.

The use of flow cells to grow pure cultures of the oral bacterium *Porphyromonas gingivalis* under anaerobic and microaerophilic conditions was described by Hansen et al. (15). Their findings show poor biofilm formation after 18 h

of growth under low oxygen concentrations (1 to 2 ppm) and improved growth in anaerobic culturing. To ensure anaerobic culture conditions, an inflatable glove chamber was used to isolate the medium flask (kept under continuous $N:CO_2$ sparging) and the flow cell from the normal atmosphere environment. Biofilm formation was visualized with fluorescence tagging and confocal laser microscopy. Their extensive literature review revealed that relatively few attempts have been made to grow anaerobic biofilms, with little evidence that the most commonly used anaerobic chemostat technology had been extended to obtain in situ observations.

A recent study in which microbial attachment and cellulose degradation in leachate and rumen systems were evaluated in situ and in real time acknowledges the potential and utility of using flow cell concepts in the study of anaerobic biofilms (39). In their system, a flow cell, which contained immobilized cellulose fibers on the underside of the glass coverslips, was connected to an anaerobic medium flask (containing crystalline cellulose) in a closed loop so that its content was continuously recirculated through the flow chamber. The medium flask was inoculated with rumen fluid or anaerobic leachate from a municipal solid-waste digester and during the incubation the extent of cellulose solubilization was estimated by nitrogen/chemical oxygen demand balance. Concurrently, phase-contrast microscopy was used to estimate the extent of biofilm formation at the cellulose interface inside the flow cell. The study aimed to correlate the imaging of cellulolytic biofilms with process performance data in real time, and the results suggested that the rumen culture formed thicker and more stable biofilms than the inoculum from digester leachate, which were consistent with the higher rates of cellulose solubilization in the rumen reactors. It was also acknowledged that the presence of multilayer biofilms could be partly due to the interactions between the species within the microbial consortia and that close interactions between bacteria and cellulose particles are required for efficient cellulose degradation.

44.3.2. Biofilm Equilibrium: Reactive Versus Inert Attachment Substrates

Studies on biofilms have overwhelmingly focused on the association of cells with inert surfaces, where the nutrients are obtained from the bulk liquid phase (thus supply to the basal biofilm layers, i.e., at the biofilm attachment interface, is diffusion dependent). This model is representative for engineered systems such as industrial bioreactors for the treatment of waste effluents, natural systems such as biofilm deposits on a riverbed, or biofilms with relevance to medical applications such as those developing on implants and prosthetics. Biofilms on inert surfaces typically have multiple layers of cells, and their form and function are dependent on the large variability in concentration and bioavailability of nutrients (29). It is generally expected that the nature of the attachment substrate, whether it is biotic or abiotic—reactive or inert—has an impact on biofilm form and function, and therefore on the overall metabolic activity of the associated microbial communities. However, there is a paucity of comparative analyses between bacterial colonization of inert and reactive surfaces, with a few studies suggesting an overlap between the mechanisms contributing to biofilm formation at these interfaces. For instance, studies of the mutualistic interaction between plant roots and *Pseudomonas putida* showed the presence of genetic determinants that similarly influence adhesion on plant surfaces and inert abiotic material (58). Insertional mutations affecting flagellar genes or the large adhesin LapA decreased adhesion to corn seeds as well as abiotic surfaces. In contrast, a mutant of the cytochrome oxidase assembly has deficient bacterium-plant interactions only, while a lipopolysaccharide biosynthesis-impaired phenotype presents reduced adhesion in the rhizosphere but initiates attachment to plastic faster than the wild type. These examples suggest that bacterial adaptation to the sessile state can be dependent on the physicochemical properties of the attachment interface and be governed by surface-specific or global molecular mechanisms. Microbially influenced corrosion represents a good example of biofilm formation at a reactive interface where sulfate-reducing bacteria have been shown to play a primary role in anaerobic environments by metabolically producing sulfide ions to form sulfide iron deposits and by secreting hydrogenase enzymes that adsorb on steel surfaces. These elements are responsible for the catalysis of cathodic proton reduction, which is generally understood as microbial corrosion (6, 8). Recently, corrosion induced by direct microbe-cathode interactions has been linked to *Geobacter sulfurreducens*, which, when grown in a medium with low electron donor concentrations, was found to interact directly with the attachment surface by extracting electrons required for its metabolism from the cathode through membrane-bound redox compounds (32). Other organisms, termed electricigens, are well known for their potential to oxidize organic matter and to transfer the electrons produced directly to solid electrodes (27).

Cellulolytic organisms occupy a special niche where (the primarily) insoluble cellulose serves both as the carbon source and as biofilm solid support, theoretically providing optimal conditions for biomass accumulation with preferential access to the primary substrate. It can be expected that continuous-flow cultures of planktonic cellulolytic cells would have reduced hydrolysis with lower retention rates which are inversely proportional to cell and substrate dilution. Bacterial populations attached to inert surfaces that hydrolyze the cellulosic substrate at distance would suffer from hydrolytic end product loss by dilution. Attachment to and proliferation at the cellulosic source are thought to be beneficial due to reduced microbial biomass loss and the potential capture of the enzymatic hydrolysis products. In *C. thermocellum*, polycellulosomal structures have been demonstrated to exist at the cell surface, forming multiple protuberances (2). Electron micrographs of antibody-labeled cells adherent to cellulose indicated that the protuberances undergo structural conformations to form fibrous contact corridors connecting the cell and the substrate at distances of more than 500 nm (this being, to our knowledge, the only reference of measured cell-substrate gap in the literature). These protracted protuberances were proposed to assist in channeling the products of cellulose hydrolysis to the cell to be taken up via the appropriate transport mechanisms (4), potentially providing a competitive advantage for the adherent bacterial cells.

44.3.3. Microbial Dynamics: Cooperation Versus Competition

In evolutionary ecology, Darwin's concept of selection is applied to individual species. However, within a community, the biological complexity at the different levels of organization pose notable challenges to selection theory. Caldwell et al. (5) proposed that communities are considered as units of proliferation (hence units of evolution) to offer alternative or additional community-level testable hypotheses. They introduced a proliferation hypothesis as an evolutionary

model, which recognized the possibility of simultaneous propagation and reproductive success at different levels of organization (e.g., genes, cells, communities, and ecosystems) rather than at the level of individual organisms. In line with classical evolution, this theory involves the assumption that organisms are more successful by adapting through genetic mutation and recombination. However, it also assumes that adaptation is enhanced by association and copropagation; i.e., self-replicating molecules associated as macromolecules, macromolecules as prokaryotic cells, eukaryotic and prokaryotic cells as communities through behavioral changes, or communities propagated more efficiently as a result of association to form ecosystems. The symbiotic relations in the microbial consortium of the rumen ecosystems are exemplary for the evolutionary success with adaptation at multiple levels of organization. The rate of fiber digestion in the rumen content, rich in organic matter at pH 5.5 to 6.9 with low redox potential, is affected by many factors which include the nature of the species of anaerobic and facultative anaerobic organisms and their degree of adaptation to the diet.

In line with the proliferation theory described above, species which carry traits unnecessary for their survival in isolation, but beneficial for a consortium, are considered an evolutionary success for reproduction. Here the focus is not solely at the organismal level, as implied by the selection theory, but rather at the community level. For example, in the case of *C. thermocellum*, the enzymatic subunits of the cellulosome are complex proteins containing one or more catalytic domains in the form of glycosyl hydrolases, which include cellulases, hemicellulases (xylanase, mannose, chitinase, and lichenase), and carbohydrate esterases (4). The hydrolytic products of hemicellulolytic activity are not utilized by *C. thermocellum* and thus suggest that these enzymes' function may be to facilitate a better exposure of cellulose in the complex cell wall, or that these cellulosomal elements are the coevolutionary result of the symbiotic interaction in mixed cellulolytic communities. These examples demonstrate the complex nature of microbial interactions. This complexity poses numerous challenges to bioprocessing aimed at cellulose conversion but at the same time offers a broad range of possibilities. Realization of this potential will depend on the development of strategies that will make it possible to manipulate microbial cell-enzyme-substrate interactions; studying these processes from a biofilm perspective may be an important first step towards this goal.

44.4. BIOFILM FORMATION BY *C. THERMOCELLUM*

44.4.1. Experimental System

While studies on microbial cellulose degradation are numerous, the principal approach to understanding organism behavior and cellulose colonization has been by direct visualization, in which slurry samples are collected from fermentation reactors and subjected to light and fluorescence microscopy, fluorescence in situ hybridization, and electron microscopy (4, 40, 52), or indirectly by molecular and biochemical probing of batch cultures (14). Arguably, these methods have a destructive nature associated with sample preparation and can skew the associated observations on the complexity of adherent structures. A major limitation remains the lack of detail about colonization patterns or spatial relationships. The use of flow cells involves in situ observation of biofilms that can be made in real time under flow conditions. The flow chamber allows the growth of microbial biofilms on solid surfaces under controlled conditions and their microscopic visualization through a transparent window under fully hydrated conditions and in a nondestructive manner, with the possibility of repeated observation over long cultivation periods and the extrapolation of behavioral changes (44).

We have successfully used flow cells (57) with modified chamber design (Fig. 1) to study associative behavior during microbial cellulolose utilization by C. thermocellum. For this purpose, solid cellulose (cotton fibers) was immobilized inside flow chambers and served as the single, insoluble source of cellulosic carbon for the bacteria. Continuous-flow experiments were carried out under an N₂ atmosphere and under a normal oxygen atmosphere. Growth in flow cells kept under a nitrogen atmosphere was possible with the use of an anaerobic cabinet equipped with an incubator and was suited for experiments where occasional observation and evaluation were desired or where sample collection and flow cell analysis over time were not limited by the confined chamber space. The continuous-flow system was composed of a vessel containing RM medium (43, 53), silicone tubing, a flow cell, a peristaltic pump, and an effluent waste container. The complex synthetic RM medium, lacking Avicel, was prepared by a modified Hungate technique (18) and stored for up to 48 h to allow the photocatalytic reduction of oxygen by ʟ-cysteine HCl. Exposure of media to a higher light intensity (11) hastened the reduction time. The tubing system and the flow chamber were sterilized by pumping a 10% sodium hypochlorite solution at approximately 40 ml/h for 30 to 60 min and was then washed out with sterile distilled water overnight at 10 ml/h. Sterile medium was pumped in at 40 ml/h for approximately 2 h with the flow chamber kept in the incubator at 60°C. Cultures of C. thermocellum were transferred (10 to 30% [vol/vol] inocula) into the flow chamber via sterile syringes and needles, and the medium was kept stationary for 60 min to allow potential cell-surface interactions to occur. Thereafter, flow was initiated and maintained at constant rates for the duration of the experiments at 10 ml/h. The cultures were examined at various stages of growth ranging from 24 to 120 h for microscopic investigations. Sample preparation and staining procedures are detailed below.

Growth of C. thermocellum in flow cells maintained outside of an anaerobic chamber was possible with the use of a modified continuous-flow system design (Fig. 2) and was suited for experiments where sample collection and repeated flow cell measurements required the use of devices not fitted for an anaerobic chamber (e.g., real-time microscopy, CO₂ measurement, and flow cytometry). Under ambient atmosphere, the system described above was modified to accommodate an oxygen-free medium source in a unidirectional flow system and a set of sample collection devices on the effluent side. External ports were protected with 0.2-μm-pore-size filters to ensure sterility (not shown in Fig. 2), and the silicone tubing was replaced with Norprene because of its much lower gas permeability.

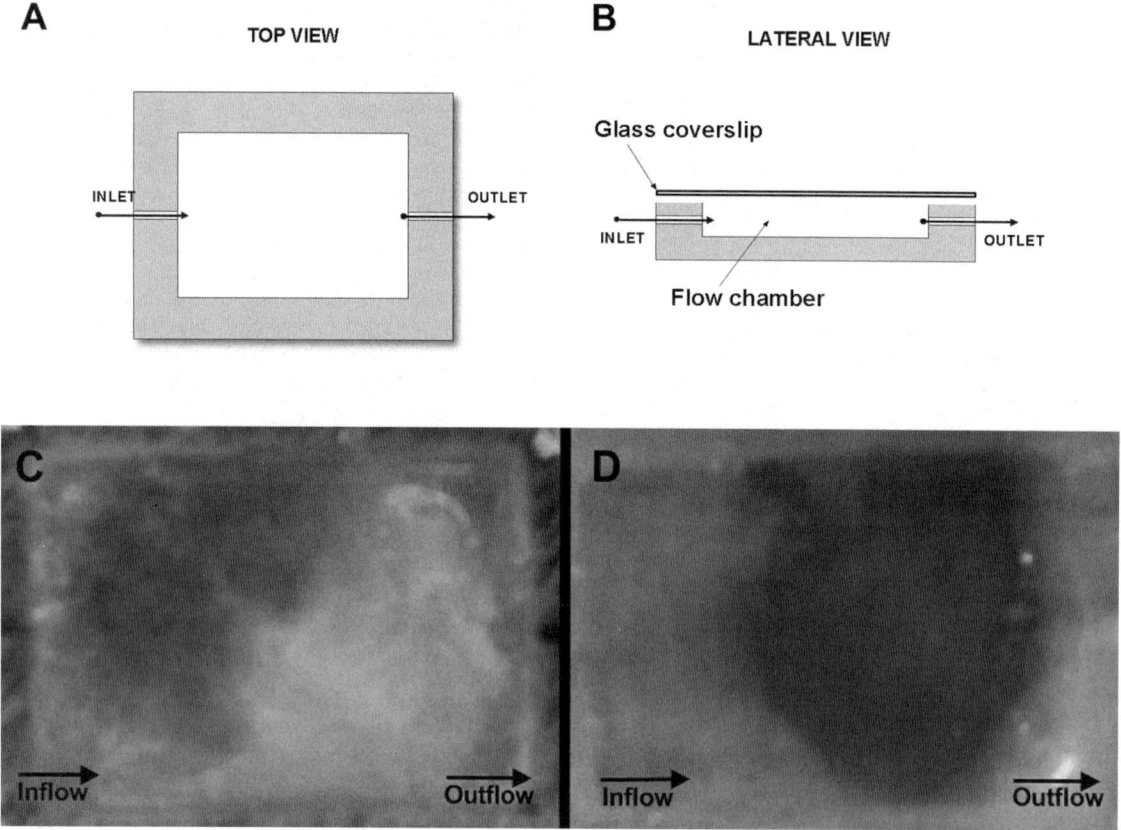

FIGURE 1 Schematic flow cell design showing top view (A) and lateral view (B) and typical C. thermocellum continuous-flow culture growth within flow cells showing the initial growth (dark areas) pattern around the inoculation site (C) with subsequent spreading to uncolonized regions and the depletion of its only cellulosic carbon source, which serves as the attachment substratum (D).

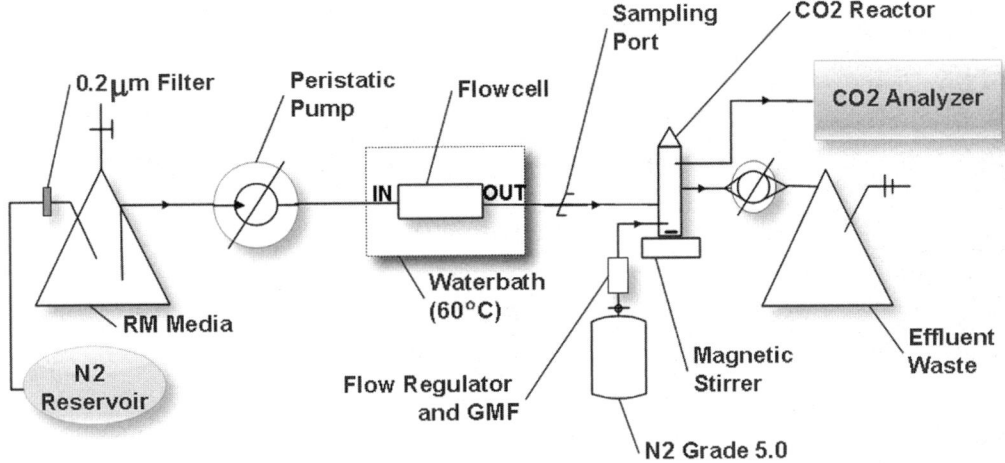

FIGURE 2 Schematic diagram of a continuous-flow system used under normal atmosphere for growth of anaerobic, thermophilic C. *thermocellum* cultures. A sampling port allows the aseptic collection of liquid effluent samples, and a "CO_2 exchange reactor" purged with N_2 gas and coupled to an infrared analyzer allows real-time measurements of dissolved CO_2 in the effluent. GFM, gas mass flowmeter.

Continuous-flow cultures of C. *thermocellum* were analyzed in situ with the confocal laser scanning microscope by fitting the flow cell directly onto the microscope stage. Stain (fluorescent probe) stocks and the working solutions derived from them were prepared with ultrapure sterile deionized water as detailed in Table 1.

For the preparation of culture samples, volumes of stain working solutions were transferred very slowly via needles directly into the flow cell while the medium flow was halted.

The amount of stain injected was dependent on the flow chamber size. The working solutions were tested across the range of concentrations provided in Table 1, but typically a middle-range value was used. All the stains were tested separately and in combinations of two or more. A typical combination of two consisted of a nucleic acid stain supplemented with a secondary stain targeted against carbohydrates or lipids. Noninvasive imaging of the undisturbed anaerobic cultures was performed through the thin glass coverslip of the

TABLE 1 List of stain stocks and working volumes used on continuous-flow cultures of C. *thermocellum* in confocal laser scanning microscopy[a]

Stain[b]	Stock concn	Working solution (μl/1 ml H_2O)	Target(s)
Syto9	3.34 mM	2–5	Nucleic acids
Propidium iodide	20 mM	2	Nucleic acids
FM 1-43	1 mg/ml	4–20	Cell membrane (outer leaflet)
FM 4-64	2 mg/ml	4–20	Cell membrane (outer leaflet)
Conjugated lectins	1 mg/ml	4–20	
ConA-FITC + TRITC			α-Glucose and α-mannose
PSA-FITC			α-Mannose
DBA-FITC			N-Acetylgalactosamine
PNA-FITC			Galactose
UEA-FITC			Fucose
LEA-FITC			N-Acetylglucosamine
WGA-TRITC	5 mg/ml	1–4	N-Acetylglucosamine
Acridine orange	0.25 mg/ml	10	Nucleic acids
Ethidium bromide	10 mg/ml	2–10	Nucleic acids
Nile red	1 mg/ml	100	Intracellular lipids

[a]Final volumes of the working solutions are represented as the solvent volume.
[b]Abbreviations: ConA, concanavalin A; FITC, fluorescein isothiocyanate; PSA, *Pisum sativum* agglutinin; DBA, *Dolichos biflorus* agglutinin; PNA, peanut agglutinin; UEA, *Ulex europaeus* agglutinin; LEA, *Lycopersicon esculentum* agglutinin; WGA, wheat germ agglutinin.

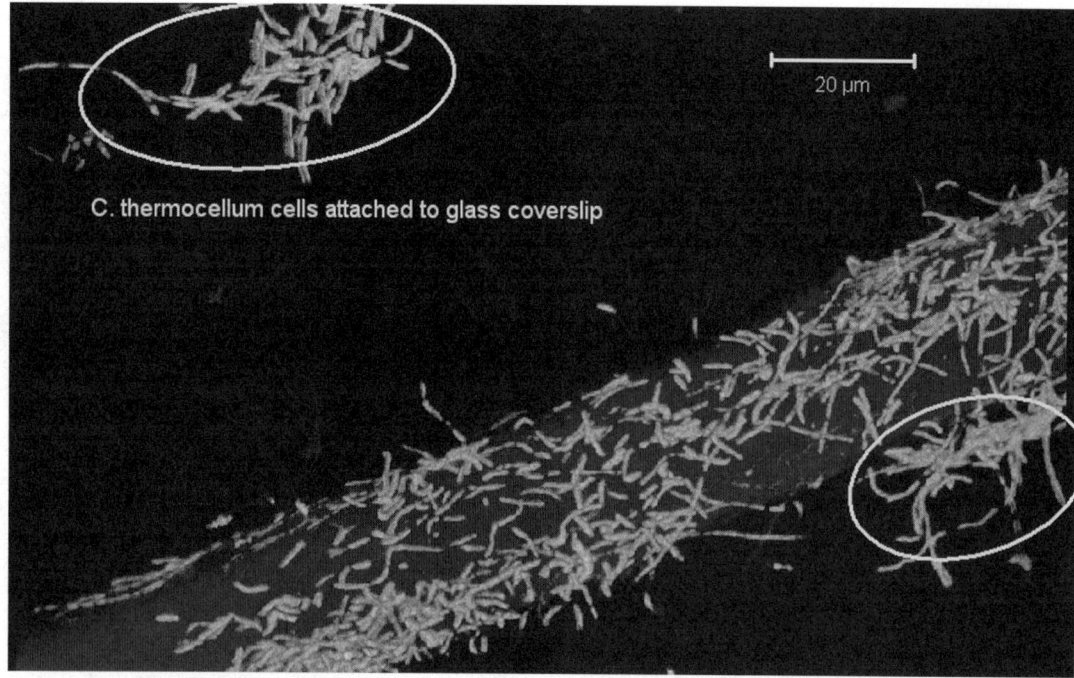

FIGURE 3 Confocal laser scanning micrographs of C. *thermocellum* biofilms growing on solid cellulosic substratum. Attachment to abiotic surfaces was recorded when the solid cellulosic support had been considerably depleted. Cells were stained with Syto9, and cellulose was stained with wheat germ agglutinin-TRITC.

flow cell. The optical properties of the objectives and the glass cover thickness determined the vertical working distance within the chamber. The confocal microscopy optical slicing feature along the z axis allowed the postacquisition reconstruction of orthographic projections of the scanned fields and high-resolution three-dimensional volume rendering.

44.4.2. Cell Organization

Growth in flow chambers was found to begin at the site of inoculation and then progressively spread and colonize the available adjacent cellulosic surface areas, with concomitant depletion of the solid substratum. Biofilms were observed at various degrees of cellulose colonization, from scattered cell groups to regions with very high cell densities, covering the entire surface area of the fibers. Occasionally, attachment to inert glass surfaces was recorded in late growth stages, primarily around substrate depletion zones (Fig. 3).

Typical C. *thermocellum* biofilm morphology was characterized by accumulation of terminal endospore and rod-shaped vegetative cells at the cellulose surface forming a single layer of cells with various degrees of density and with multiple cell orientations. Cells with terminal endospores invariably exhibited an end-on attachment with spores away from the cellulose surface (Fig. 4), and vegetative cells during division appeared to have a parallel orientation to the substrate. Cells divided along their longitudinal axis while attached, thus forming the typical long chains of cells described in previous work (10), which were also encountered in the flow cell effluent. Each individual cell appeared to be in close proximity to the cellulosic interface, as shown in the orthographic projections along the xz and yz planes. A typical biofilm matrix (consisting of multiple cell layers and an extensive EPS matrix) was absent. Notably, cell yield to the suspended phase (typically $\geq 10^5$ cells/ml after 24 h) occurred soon after initial attachment and persisted through all stages of biofilm development.

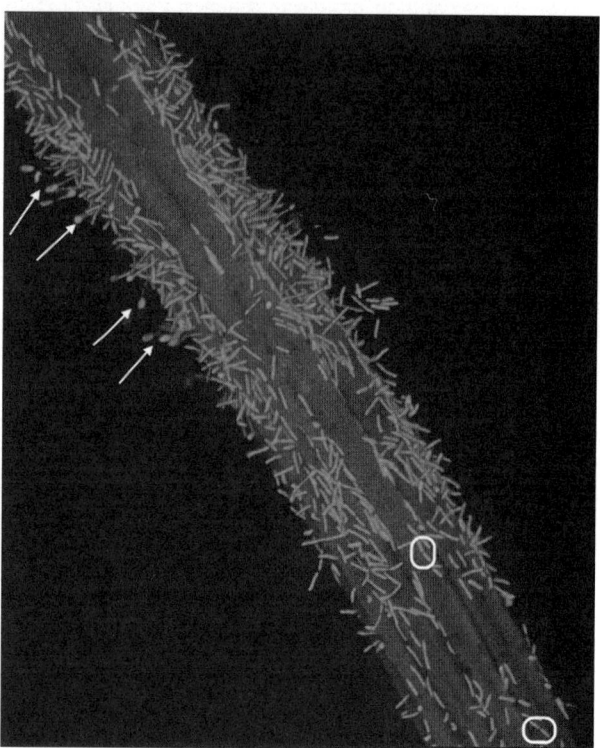

FIGURE 4 Confocal laser scanning micrograph of C. *thermocellum* cells colonizing a cellulose substrate (fiber) showing spores (arrows) with distinct end-on attachment on the nonsporulating side. Cells were stained with Syto9, and cellulose was stained with wheat germ agglutinin-TRITC. The circles indicate dividing cells with parallel orientation to the attachment interface. Objective, 63 × 1.2 water immersion.

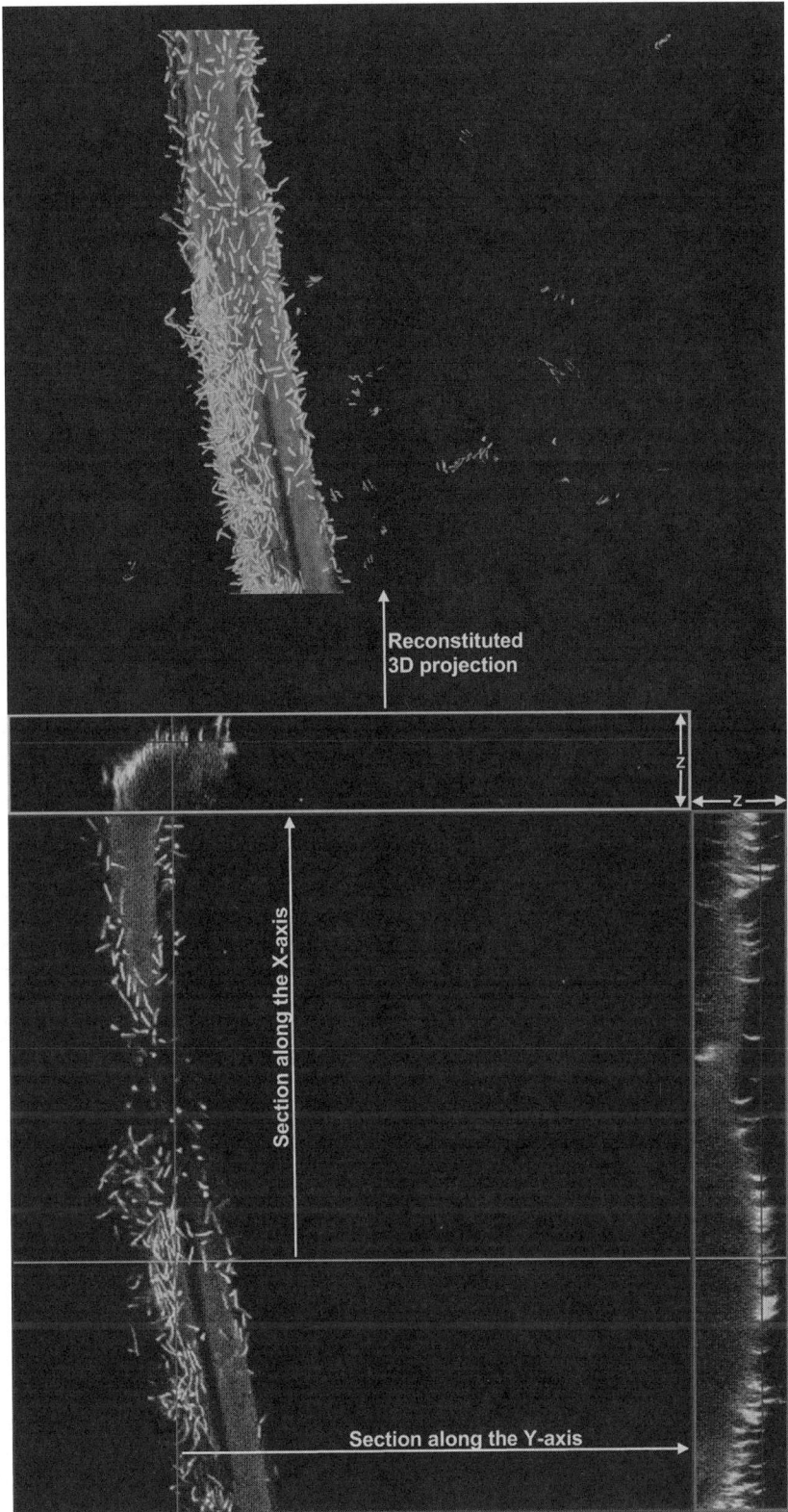

FIGURE 5 Confocal laser scanning micrograph of C. *thermocellum* biofilm developing on a cellulose fiber, seen from the top-down view of a three-dimensional projection (top); sectioning through the projection along the x and y axes in the z plane (bottom) revealed the distance between cells and substrate, which was recorded to be lower than the 0.44-μm z-scaling limit of the scanning microscope. Cells were stained with Syto9, and cellulose was stained with wheat germ agglutinin-TRITC.

44.4.3. Form-Function Relationship

In situ microscopic imaging of fully hydrated cultures of *C. thermocellum* growing on cellulose revealed biofilm clusters where each cellulose fiber was colonized with various degrees of cell density and distribution. Each flow cell thus hosts independent biofilms in different stages of development. Confocal laser scanning and postacquisition image analysis allowed us to observe development over time to obtain a better understanding of biofilm formation by *C. thermocellum*: following initial attachment, cells divide along the longitudinal axis, thereby increasing the biofilm cell density laterally but not orthogonally relative to the surface. It further appeared that biofilm development was coupled with the continuous recruitment of cells from the planktonic mode. The fact that adherent cells were always visualized in monolayers, with various orientations to the attachment substratum, supports the idea that the cellulolytic organisms benefit from a close proximity to the cellulosic source for efficient substrate utilization, as opposed to the acquisition of diffuse hydrolysis products from the bulk liquid phase (or the leaching of hydrolytic products from enzymatic cellulolysis). Invariably, cellulose colonization progresses from localized patches of adherent cells to the complete coverage of the entire surface area. Many of the cells during the initial stages of colonization have spores, suggesting that these cells might be involved in biofilm translocation to uncolonized regions and that sporulation may be regulated by surface conditions or by cell-to-cell interactions in the biofilm layer. Carbohydrate- and protein-targeted staining in continuous-flow biofilms failed to reveal a typical EPS matrix interconnecting the adherent cells. However, tetramethyl rhodamine isocyanate (TRITC)-conjugated concanavalin A, with primary sugar specificity for α-glucose and α-mannose, did bind to the outer regions of biofilm cells, which suggests the presence of a glycocalyx. Previous electron microscopy findings suggest the existence of protracted polycellulosome protuberances at the surface of the adherent cells, mediating the contact between cellulose and the microbe (2). We propose that a glycocalyx layer is also involved in the cell-substrate interaction, as evidenced by observations of microbe adherence to inert surfaces in regions where cellulose was believed to have been in contact with the inert interface. Attachment to inert material suggests the existence of secondary, nonspecific binding mechanisms mediated, for example, by the constituents of a glycocalyx. Orthographic projections (Fig. 5) along the xz and yz planes of colonized cellulose fibers scanned along the vertical axis with a scaling of 0.44 μm between sequential images in the xy plane revealed no visible separation between the microbes and the interface, confirming a cell-substrate gap no larger than 0.44 μm.

The role of spores in translocation, cellular division, and biofilm lateral sprawling along the liquid-cellulose interface with subsequent detachment events and substrate depletion can be further understood by the use of time-lapse microscopy, a technique not previously applied in the study of thermophilic anaerobic organisms. The formation of monolayer biofilms of strongly adherent cellulolytic cells, emphasizing the intimate association with the carbon substrate, can be further investigated in biofilms formed by microbial consortia, where targeted staining techniques have to be developed to enable the in situ localization of cellulolytic microbes in the consortial structures, which have previously been reported to form multilayer biofilms (39).

C. thermocellum is one of the potential candidates for strain development through recombinant and classical strategies for application in consolidated bioprocessing aimed at large-scale conversion of renewable biomass into biofuels and other products. The numerous examples of biofilm formation in the literature, and the observed association between *C. thermocellum* cells and substrate, make this organism a suitable model to delineate cell-cellulosic substratum associations and to optimize the tools to study this phenomenon in other cellulolytic cultures and consortia with potential for application in biomass conversion.

REFERENCES

1. **Bayer, E. A., R. Kenig, and R. Lamed.** 1983. Adherence of *Clostridium thermocellum* to cellulose. *J. Bacteriol.* **156:**818–827.
2. **Bayer, E. A., and R. Lamed.** 1986. Ultrastructure of the cell surface cellulosome of *Clostridium thermocellum* and its interaction with cellulose. *J. Bacteriol.* **167:**828–836.
3. **Bayer, E. A., E. Setter, and R. Lamed.** 1985. Organization and distribution of the cellulosome in *Clostridium thermocellum*. *J. Bacteriol.* **163:**552–559.
4. **Bayer, E. A., L. J. W. Shimon, Y. Shoham, and R. Lamed.** 1998. Cellulosomes—structure and ultrastructure. *J. Struct. Biol.* **124:**221–234.
5. **Caldwell, D. E., G. M. Wolfaardt, D. R. Korber, and J. R. Lawrence.** 1997. Do bacterial communities transcend Darwinism? *Adv. Microb. Ecol.* **15:**105–191.
6. **Cheng, S., J. Tian, S. Chen, Y. Lei, X. Chang, T. Liu, and Y. Yin.** 2009. Microbially influenced corrosion of stainless steel by marine bacterium *Vibrio natriegens*: (I) corrosion behavior. *Materials Sci. Eng. C* **29:**751–755.
7. **Costerton, J. W.** 1995. Overview of microbial biofilms. *J. Ind. Microbiol.* **15:**137–140.
8. **Da Silva, S., R. Basséguy, and A. Bergel.** 2002. The role of hydrogenases in the anaerobic microbiologically influenced corrosion of steels. *Bioelectrochemistry* **56:**77–79.
9. **Davies, D. G., M. R. Parsek, J. P. Pearson, B. H. Iglewski, J. W. Costerton, and E. P. Greenberg.** 1998. The involvement of cell-to-cell signals in the development of a bacterial biofilm. *Science* **280:**295–298.
10. **Freier, D., C. P. Mothershed, and J. Wiegel.** 1988. Characterization of *Clostridium thermocellum* JW20. *Appl. Environ. Microbiol.* **54:**204–211.
11. **Fukushima, R. S., P. J. Weimer, and D. A. Kunz.** 2003. Use of photocatalytic reduction to hasten preparation of culture media for saccharolytic *Clostridium* species. *Braz. J. Microbiol.* **34:**22–26.
12. **Gaudet, G., and B. Gaillard.** 1987. Vesicle formation and cellulose degradation in *Bacteroides succinogenes* cultures: ultrastructural aspects. *Arch. Microbiol.* **148:**150–154.
13. **Gehin, A., E. Gelhaye, G. Raval, and H. Petitdemange.** 1995. *Clostridium cellulolyticum* viability and sporulation under cellobiose starvation conditions. *Appl. Environ. Microbiol.* **61:**868–871.
14. **Gelhaye, E., H. Petitdemange, and R. Gay.** 1992. Characteristics of cellulose colonization by a mesophilic, cellulolytic *Clostridium* (strain C401). *Res. Microbiol.* **143:**891–895.
15. **Hansen, M. C., R. J. Palmer Jr., and D. C. White.** 2000. Flowcell culture of *Porphyromonas gingivalis* biofilms under anaerobic conditions. *J. Microbiol. Methods* **40:**233–239.
16. **Henrichsen, J.** 1972. Bacterial surface translocation: a survey and a classification. *Bacteriol. Rev.* **36:**478–503.
17. **Hostalka, F., A. Moultrie, and F. Stutzenberger.** 1992. Influence of carbon source on cell surface topology of *Thermomonospora curvata*. *J. Bacteriol.* **174:**7048–7052.
18. **Hungate, R. E.** 1969. A roll tube method for cultivation of strict anaerobes. *Methods Microbiol.* **3B:**117–132.

19. Jensen, P. D., M. T. Hardin, and W. P. Clarke. 2008. Measurement and quantification of sessile and planktonic microbial populations during the anaerobic digestion of cellulose. *Water Sci. Technol.* **57:**465–469.

20. Jensen, P. D., M. T. Hardin, and W. P. Clarke. 2009. Effect of biomass concentration and inoculum source on the rate of anaerobic cellulose solubilization. *Bioresour. Technol.* **100:**5219–5225.

21. Jones, L. R., I. A. Watson-Craik, and E. Senior. 1997. Image analysis for the study of the development of anaerobic biofilms on materials characteristic of landfilled refuse. *Water Sci. Technol.* **36:**485–492.

22. Kim, T.-J., B. M. Young, and G. M. Young. 2008. Effect of flagellar mutations on *Yersinia enterocolitica* biofilm formation. *Appl. Environ. Microbiol.* **74:**5466–5474.

23. Kobayashi, K. 2008. SlrR/SlrA controls the initiation of biofilm formation in *Bacillus subtilis. Mol. Microbiol.* **69:**1399–1410.

24. Kudo, H., K.-J. Cheng, and J. W. Costerton. 1987. Electron microscopic study of the methylcellulose-mediated detachment of cellulolytic rumen bacteria from cellulose fibers. *Can. J. Microbiol.* **33:**267–272.

25. Lamed, R., J. Naimark, E. Morgenstern, and E. A. Bayer. 1987. Specialized cell surface structures in cellulolytic bacteria. *J. Bacteriol.* **169:**3792–3800.

26. Larsen, P., B. H. Olesen, P. H. Nielsen, and J. L. Nielsen. 2008. Quantification of lipids and protein in thin biofilms by fluorescence staining. *Biofouling* **24:**241–250.

27. Lovley, D. R. 2008. The microbe electric: conversion of organic matter to electricity. *Curr. Opin. Biotechnol.* **19:**564–571.

28. Lu, Y., Y. P. Zhang, and L. R. Lynd. 2006. Enzyme-microbe synergy during cellulose hydrolysis by *Clostridium thermocellum. Proc. Natl. Acad. Sci. USA* **103:**16165–16169.

29. Lynd, L. R., P. J. Weimer, G. Wolfaardt, and Y.-P. Zhang. 2006. Cellulose hydrolysis by *Clostridium thermocellum:* a microbial perspective, p. 95–117. *In* I. A. Kataeva (ed.), *Cellulosome.* Nova Science Publishers, Inc., Hauppauge, NY.

30. Lynd, L. R., P. J. Weimer, W. H. Van Zyl, and I. S. Pretorius. 2002. Microbial cellulose utilization: fundamentals and biotechnology. *Microbiol. Mol. Biol. Rev.* **66:**506–577.

31. McAllister, T. A., H. D. Bae, G. A. Jones, and K. J. Cheng. 1994. Microbial attachment and feed digestion in the rumen. *J. Anim. Sci.* **72:**3004–3018.

32. Mehanna, M., R. Basseguy, M.-L. Delia, and A. Bergel. 2008. Role of direct microbial electron transfer in corrosion of steels. *Electrochem. Commun.* **11:**568–571.

33. Merod, R. T., J. E. Warren, H. McCaslin, and S. Wuertz. 2007. Toward automated analysis of biofilm architecture: bias caused by extraneous confocal laser scanning microscopy images. *Appl. Environ. Microbiol.* **73:**4922–4930.

34. Miron, J., D. Ben-Ghedalia, and M. Morrison. 2001. Invited review: adhesion mechanisms of rumen cellulolytic bacteria. *J. Dairy Sci.* **84:**1294–1309.

35. Miron, J., and C. W. Forsberg. 1999. Characterisation of cellulose-binding proteins that are involved in the adhesion mechanism of *Fibrobacter intestinalis* DR7. *Appl. Microbiol. Biotechnol.* **51:**491–497.

36. Morris, E. J., and O. J. Cole. 1987. Relationship between cellulolytic activity and adhesion to cellulose in *Ruminococcus albus. J. Gen. Microbiol.* **133:**1023–1032.

37. Morrison, M., and J. Miron. 2000. Adhesion to cellulose by *Ruminococcus albus:* a combination of cellulosomes and Pil-proteins? *FEMS Microbiol. Lett.* **185:**109–115.

38. Nikolaev, Y. A., and V. K. Plakunov. 2007. Biofilm—"city of microbes" or an analogue of multicellular organisms? *Microbiology* **76:**125–138.

39. O'Sullivan, C., P. C. Burrell, M. Pasmore, W. P. Clarke, and L. L. Blackall. 2008. Application of flowcell technology for monitoring biofilm development and cellulose degradation in leachate and rumen systems. *Bioresour. Technol.* **100:**492–496.

40. O'Sullivan, C. A., P. C. Burrell, W. P. Clarke, and L. L. Blackall. 2005. Structure of a cellulose degrading bacterial community during anaerobic digestion. *Biotechnol. Bioeng.* **92:**871–878.

41. O'Toole, G., H. B. Kaplan, and R. Kolter. 2000. Biofilm formation as microbial development. *Annu. Rev. Microbiol.* **54:**49–79.

42. Otto, K. 2008. Biophysical approaches to study the dynamic process of bacterial adhesion. *Res. Microbiol.* **159:**415–422.

43. Ozkan, M., S. G. Desai, Y. Zhang, D. M. Stevenson, J. Beane, E. A. White, M. L. Guerinot, and L. R. Lynd. 2001. Characterization of 13 newly isolated strains of anaerobic, cellulolytic, thermophilic bacteria. *J. Ind. Microbiol. Biotechnol.* **27:**275–280.

44. Palmer, R. J., Jr. 1999. Microscopy flowcells: perfusion chambers for real-time study of biofilms. *Methods Enzymol.* **310:**160–166.

45. Prigent-Combaret, C., O. Vidal, C. Dorel, and P. Lejeune. 1999. Abiotic surface sensing and biofilm-dependent regulation of gene expression in *Escherichia coli. J. Bacteriol.* **181:**5993–6002.

46. Purevdorj, B., J. W. Costerton, and P. Stoodley. 2002. Influence of hydrodynamics and cell signaling on the structure and behavior of *Pseudomonas aeruginosa* biofilms. *Appl. Environ. Microbiol.* **68:**4457–4464.

47. Rau, U., A. Kuenz, V. Wray, M. Nimtz, J. Wrenger, and H. Cicek. 2008. Production and structural analysis of the polysaccharide secreted by *Trametes (Coriolus) versicolor* ATCC 200801. *Appl. Microbiol. Biotechnol.* **81:**827–837.

48. Rotter, B. E., D. A. Barry, J. I. Gerhard, and J. S. Small. 2008. Parameter and process significance in mechanistic modeling of cellulose hydrolysis. *Bioresour. Technol.* **99:**5738–5748.

49. Schwarz, W. H. 2001. The cellulosome and cellulose degradation by anaerobic bacteria. *Appl. Microbiol. Biotechnol.* **56:**634–649.

50. Simões, M., S. Cleto, M. O. Pereira, and M. J. Vieira. 2007. Influence of biofilm composition on the resistance to detachment. *Water Sci. Technol.* **55:**473–480.

51. Stoodley, P., K. Sauer, D. G. Davies, and J. W. Costerton. 2002. Biofilms as complex differentiated communities. *Annu. Rev. Microbiol.* **56:**187–209.

52. Syutsubo, K., Y. Nagaya, S. Sakai, and A. Miya. 2005. Behavior of cellulose-degrading bacteria in thermophilic anaerobic digestion process. *Water Sci. Technol.* **52:**79–84.

53. Tsoi, T. V., N. A. Chuvil'skaia, I. I. Atakishieva, T. T. Dzhavakhishvili, and V. K. Akimenko. 1987. *Clostridium thermocellum*—a new object of genetic studies. *Mol. Genet. Mikrobiol. Virusol.* **1987**(11):18–23.

54. Verstraeten, N., K. Braeken, B. Debkumari, M. Fauvart, J. Fransaer, J. Vermant, and J. Michiels. 2008. Living on a surface: swarming and biofilm formation. *Trends Microbiol.* **16:**496–506.

55. Watnick, P., and R. Kolter. 2000. Biofilm, city of microbes. *J. Bacteriol.* **182:**2675–2679.

56. Weimer, P. J., N. P. J. Price, O. Kroukamp, L.-M. Joubert, G. M. Wolfaardt, and W. H. Van Zyl. 2006. Studies of the extracellular glycocalyx of the anaerobic cellulolytic bacterium *Ruminococcus albus* 7. *Appl. Environ. Microbiol.* **72:**7559–7566.

57. Wolfaardt, G. M., J. R. Lawrence, R. D. Robarts, S. J. Caldwell, and D. E. Caldwell. 1994. Multicellular organization in a degradative biofilm community. *Appl. Environ. Microbiol.* **60:**434–446.

58. Yousef-Coronado, F., M. L. Travieso, and M. Espinosa-Urgel. 2008. Different, overlapping mechanisms for colonization of abiotic and plant surfaces by *Pseudomonas putida. FEMS Microbiol. Lett.* **288:**118–124.

Physiological and Methodological Aspects of Cellulolytic Microbial Cultures

NICOLAI S. PANIKOV AND LEE R. LYND

45

45.1. INTRODUCTION

Microbial cellulose utilization is a key part of the global cycle, important for the production of ruminants and ensiling, and promising for new industrial processes for production of fuels and chemicals from solar energy captured in chemical form by photosynthesis. In all of these contexts, isolation, cultivation, and characterization of cellulolytic microbes is of interest. Yet physiological understanding of microbial cellulose utilization substantially lags behind that of utilization of soluble substrates. This is due both to the complexity of cellulose utilization and to the methodological challenges associated with its study (15). To illustrate these points, we note the following.

- Cell concentration and growth rate measurement are routine for microbial cultures grown on soluble substrates, but this is not the case for cultures grown on cellulose.
- We know of no controlled comparisons of the rate of cellulose utilization by cellulolytic microorganisms. Whereas highly standardized methods such as the filter paper assay are developed for measuring the specific activity of cellulose-solubilizing enzymes (9), there are no such standardized methods for cellulose-solubilizing microbes.
- The widely used Monod model for kinetics of microbial utilization of soluble substrates was proposed in 1942 (18). By contrast, kinetic models for enzymatic hydrolysis of cellulose are in a nascent stage of development (31), and incorporation of enzyme-level models into kinetic models for microbial cellulose utilization has yet to be attempted. Moreover, foundational concepts associated with soluble substrate utilization such as exponential growth and a substrate concentration-independent affinity constant have seldom been demonstrated experimentally for microbial cellulose utilization and are questionable on mechanistic grounds (15).

Fermentation technique specifically developed for cellulose fermentation has been covered in several extensive reviews (6, 15, 34) and numerous experimental studies (1–4, 7, 11, 13, 20, 24, 25, 28, 29, 33). Drawing from both this prior work and our own experience, we address in this chapter three aspects of cultivating cellulose-utilizing microbes: (i) growth media, (ii) fermentation equipment for small-scale laboratory cultivation, and (iii) methods to monitor cellulose consumption and fermentation dynamics with online and offline instrumental techniques. Data relevant to item iii are presented for *Clostridium thermocellum*, an anaerobic, thermophilic bacterium with among the highest rates of cellulose utilization reported (15).

45.2. GROWTH MEDIA

45.2.1. General Principles

As discussed elsewhere (22), growth media used to culture heterotrophic microorganisms for fundamentally oriented studies should contain a carbon and energy source, sources of all inorganic elements present in cells (macro- and microelements), and essential organic growth factors (vitamins and amino acids) which the organism to be cultivated is unable to synthesize from simple precursors. The temperature, pH, and redox potential must also be maintained in a range compatible with the organism of interest.

The concentration of the ith nutrients in growth media (s^0_i) can be found from a mass balance equation if the desired cell density (x^*) of cultured organisms is specified:

$$s^0_i = s_i + (x^* - x_0)/Y_i$$

where x_0 is the initial cell density (inoculum concentration), s_i is residual concentration of the ith nutrient when growth stops, and Y_i is respective cell yield (grams of cells/grams of nutrient). Although it is usually desired that s_i be close to zero for economic reasons, the value of s_i should be chosen such as to prevent growth rate limitation. Typical cell yield values for energy sources are 0.1 g of cellular biomass synthesized/g of cellulose consumed, although higher apparent yield values can be observed under some conditions for cellulolytic anaerobes (15, 26, 32). For nutrients which are not energy sources, such as sources of N, P, K, S, Mg, Ca, Fe, and trace elements, the consumed elements are incorporated into cell mass; therefore, the Y_i value can be calculated from published data on elemental composition of the organism of interest or a similar organism:

$$Y_i = 1/\sigma_i$$

where σ_i is intracellular content of the respective element per unit of cell dry weight. For example, the N content (σ_N) in *C. thermocellum* is about 0.1 g of N/g of cell dry weight; therefore, $Y_N = 1/0.10 = 10.0$ g of cell dry weight/g of N. To produce 2 g/liter of cell mass, one has to use medium containing at least $2/10.0 = 2*0.10 = 0.2$ g of N/liter, which is equivalent to 0.94 g/liter of $(NH_4)_2SO_4$. In practice, this minimal amount should be at least doubled to ensure lack of growth rate limitation.

Common undesirable secondary reactions accompanying preparation of chemically defined media are precipitation of insoluble salts formed after mixing of soluble ingredients, e.g., formation of NH_4MgPO_4, $Ca_3(PO_4)_2$, $Fe_3(PO_4)_2$, etc.; caramelization (Shiff base formation between sugars and NH_2 group); and spontaneous oxidation of reduced medium components, e.g., $Fe^{2+} \rightarrow Fe^{3+}$. One strategy to avoid these reactions is to separately sterilize reactive medium components. Precipitation of di- and trivalent metals by PO_4^{3-} and SO_4^{2-} can be prevented by addition of chelating agents, mainly citric or nitroloacetic acid. EDTA is less appropriate because of its inhibitory effect on cellulases (5). Autoxidation is eliminated by purging of media with O_2-free gases (CO_2 and N_2) in the presence of reducing agents.

The choice of cellulose-containing substrate is a key aspect of medium formulation for cellulolytic microorganisms. Model substrates containing essentially pure cellulose, e.g., Avicel (FMC Corp.) and Sigmacel (Sigma Chemical), have significant merits for fundamental studies in that they have reproducible properties and are commonly accessible. Solka Floc (International Fiber Corp.) is one of a few commercially available model lignocellulosic substrates. Forage samples are readily obtained for studies related to animal feeding for which a real-world substrate is desired, although reproducibility between samples is likely to be an issue. For studies aimed at conversion of pretreated lignocellulosic materials, both access and reproducibility are significant issues, as such materials are not commercially available to our knowledge.

45.2.2. Published Growth Medium Recipes and Protocols for *C. thermocellum*

Table 1 lists six nutrient media for *C. thermocellum* including brief protocols. It may be seen that medium composition varies widely dependent upon the specific purposes of cultivation experiments. In addition to their relevance for *C. thermocellum* and the particular media listed, aspects of the following discussion may be of more general interest for other organisms and media.

MTC and MJM were designed to support high cell densities accompanying growth at high cellulose concentrations (30 to 100 g/liter). These media (especially MTC) are the most concentrated of those considered, and they include up to 3 g/liter of citric acid to prevent precipitation of Mg, Ca, Fe, and trace elements. MJM is simpler and does not contain yeast extract (YE), $NaHCO_3$, or Na_2SO_4; as a result, its use is recommended for "clean" experiments in instrumentally controlled bioreactors when uncertain effects of undefined growth supplements or high ionic strength are not desirable (experiments aimed at quantitative analysis, isolation and purification of extracellular metabolites, uptake kinetics, off-gas analysis, spectrum of C and N sources, etc.). On the other hand, MTC is more appropriate for experiments aimed at obtaining high cell densities either in pH-controlled fermentors or in serum vials without pH control due to the strong buffering effect of bicarbonates. Very dilute MM and MDM media do not contain chelating agents; therefore, they are simpler to prepare, less expensive, and good for strains sensitive to salts. On the other hand, these media can be expected to fail to support high growth rates in continuous culture or complete utilization of more than 3 g/liter of cellulose.

Trace elements can sometimes be omitted without any noticeable effect on cell growth, and such effects are often only observed when ultrapure chemicals, water, and gases are employed. Many investigators choose to use a standard mixture of trace elements to prevent unintentional growth limitation.

Recipes for trace element stock solutions (Table 1) are based on either chlorides or sulfates of trace metals. For pure cultures that do not reduce sulfate, there is likely to be little difference between chlorides and sulfates. For experiments involving enrichment of mixed microbial consortia, chloride ions are expected to be relatively inert, whereas sulfate provides an electron acceptor for sulfate-reducing bacteria. Stock solutions of trace elements (usually concentrated 1,000-fold) are typically stabilized by acidification to pH 1.5 by HCl or by addition of nitriloacetic acid, which prevents precipitation of hydroxides.

Defined mixtures of vitamins (e.g., as developed for *C. thermocellum* by Johnson et al. [10]) are often preferred relative to undefined growth factors such as YE for several reasons: (i) elimination of variation from one batch of YE to another (we have observed, for example, that some batches of YE contain glucose, whereas others do not); (ii) a much smaller background of organic constituents, which is good for experiments involving quantification and characterization of cells, enzymes, and metabolites; and (iii) avoidance of complications due to metabolism of organic growth factors. The concentration of vitamins varies in different medium recipes and is 10 to 100 times higher in MTC than in the other media listed. Even at such high concentrations, added vitamins do not significantly affect the mass balance of microbial growth.

To maintain anoxic conditions, both cultivation vessels and growth medium reservoirs (for continuous culture) should be purged with O_2-free gases: He, Ar, N_2, or CO_2. Use of He is justified when gas phase is continuously monitored by mass spectrometry in a SIM mode (single ion monitor); otherwise, N_2 and CO_2 are preferred, as they are cheaper and more nearly replicate the gas phase composition that would be observed under most conditions of interest in nature or industrial processes. Within our experience, use of a CO_2 gas phase frequently encourages growth and shortens the lag phase, although continuous use of CO_2 flushing makes it impossible to monitor CO_2, which may be desired. A good compromise can be to use 100% CO_2 at startup and then switch to an N_2 purge stream. We have found that if continuous CO_2 monitoring is not desired, a purge gas is not required for *C. thermocellum* and many other known cellulolytic microbes. It should be noted that the H_2 partial pressure can influence the ratio of fermentation products produced (1, 13).

The order of mixing of individual ingredients is an important consideration when preparing media. For instance, addition of $FeSO_4$ to even slightly buffered neutral solution before citric acid can cause precipitation, but if citric acid is added first, then precipitation does not occur. Filtration can be a good substitute for autoclaving because it prevents numerous side reactions. Millipore disposable filters provide a high rate of filter sterilization, e.g., less than 10 min for a 20-liter carboy in our experience.

45.3. CONTROLLED CULTIVATION OF BACTERIA ON MEDIA CONTAINING SUSPENDED CELLULOSE

45.3.1. General Problems of Using Suspended Media

Methodological challenges associated with microbial cultivation on suspended substrates (cellulose and pretreated wood) have been reviewed (15) and investigated experimentally (14, 26). Key technical obstacles associated with microbial cultivation on insoluble cellulose particles include the following.

- Solid particles make difficult or even impossible the direct assessment of microbial growth by all of the best-developed

TABLE 1 Summary of nutrient media used to culture *Clostridium thermocellum* and related organisms

Function (unit of concn in medium)	Ingredient	Concn in medium (reference)					
		MTC[a] (30)	MJ[b] (10)	MM[c] (8)	MM[d] (28)	MDM[e] (25)	MJM[f] (this publication)
C source (g/liter)	Cellobiose or cellulose	30–100		3–20	4	4	3–20
Inorganic macroconstituents (g/liter)	KH_2PO_4	1	1.5	1.5	1.5	0.9	0
	K_2HPO_4	0	2.9	0	0	0	4.4
	$Na_2HPO_4 \cdot 12H_2O$	0	0	4.2	1.5	0	0
	NH_4Cl	5	0	0.5	0.5	0.73	0
	$(NH_4)_2SO_4$	0	0	0.5	0.5	0	2.1
	Urea	0	2.1	0	0	0	0
	$MgCl_2 \cdot 6H_2O$	1	1	0.09	0.09	0.085	1
	$MgSO_4 \cdot 7H_2O$	0	0	0.015	0	0	0
	$CaCl_2 \cdot 2H_2O$	0.2	0.15	0.03	0.03	0.066	0.15
	$FeCl_2 \cdot 4H_2O$	0.1	0	0	0	0	0
	$FeSO_4 \cdot 6H_2O$		0.00125	0.0005	0	0.02	0.005
	Na_2SO_4	1	0	0	0	0	0
	$NaHCO_3$	2.5	0	0.5	4	0	0
	Na_2CO_3	0	0	0	0	3.2	0
	NaCl	0	0	0.005	0	0.9	0
Chelating agent (g/liter)	Potassium citrate	2	0	0	0	0	0
	Sodium citrate $\cdot 2H_2O$	0	3	0	0	0	3
	Citric acid $C_6H_8O_7 \cdot H_2O$	1.25	0	0	0	0	0
	Nitriloacetic acid	0	0	0.0075	0	0	0
Buffer (g/liter)	MOPS	0	10	0	0	0	0
Vitamins and growth factors (mg/liter)	YE	10,000	0	2,000	2,000	500	0
	Pyridoxamine dihydrochloride	20	2	0	20	0	2
	Pyridoxine hydrochloride	0	0	0.1	0	0	0
	p-Aminobenzoic acid	4	0.4	0.05	1	0	0.4
	D-Biotin	2	0.2	0.02	0.5	0	0.2
	Vitamin B_{12}	2	0.2	0.001	0.5	0	0.2
	Thiamine	4	0	0.05	20	0	0.4
	Folic acid	0	0	0.02	0.5	0.125	0
	Nicotinic acid	0	0	0.05	0	0	0
	Nicotinamide	0	0	0	20	0	0
	Calcium pantothenate	0	0	0.05	20	0	0
	Riboflavin	0	0	0.005	20	0	0
	Thioctic acid	0	0	0.05	0	0	0
	Lipoic acid	0	0	0	10	0	0
	$MnCl_2$	1.25	0	0	0	0.028	1.25
	$MnSO_4 \cdot H_2O$	0	0	2.5	2.5	0	0
	$ZnCl_2$	0.5	0	0	0	0.01	0.5

(Continued on next page)

TABLE 1 (*Continued*)

Function (unit of concn in medium)	Ingredient	Concn in medium (reference)					
		MTC[a] (30)	MJ[b] (10)	MM[c] (8)	MM[d] (28)	MDM[e] (25)	MJM[f] (this publication)
	$CoCl_2$	0.125	0	0	0	0.002	0.125
	$Co(NO_3)_2 \cdot 6H_2O$	0	0	0.5	0.5	0	0
	$NiCl_2$	0.125	0	0.25	0.25	0	0.125
	$CuSO_4$	0.125	0	0.05	0.05	0	0.125
	H_3BO_3	0.125	0	0.05	0.05	0	0.125
	Na_2MoO_4	0.125	0	0.05	0.05	0	0.125
	$AlK_2(SO_4)_3$	0	0	0.05	0.05	0	0
	$Na_2WO_4 \cdot 2H_2O$	0	0	0.05	0.05	0	0
	Na_2SeO_3	0	0	0.005	0.005	0	0
Reducing agent (mg/liter)	L-Cysteine HCl	1,000	1,000	1,000	500	1,000	1,000
Redox indicator (mg/liter)	Resazurin	2	2	2	1	0. 4	2

[a]MTC. Autoclave separately sources of C; N; Mg, Ca, Fe, and L-Cys; and the rest of ingredients. The vitamins and trace elements are filter sterilized and combined with the sources of C; N; Mg, Ca, Fe, and L-Cys; and the rest of the ingredients.

[b]MJ-medium. Cellobiose and Mg-Ca-Fe salts mixture are autoclaved separately as 10-fold-concentrated solutions and then mixed with filter-sterilized vitamins.

[c]MM, minimal medium. CO_2, 100% in gas phase.

[d]MM, minimal medium. The basal medium, adjusted to pH 6.7, is autoclaved for 10 min to remove dissolved gases and purged with CO_2. The cooled solution is mixed anaerobically with Na_2CO_3, dispensed into bottles, and autoclaved for another 20 min. A total of 4 g/liter of cellobiose is added after autoclaving.

[e]The modified Dehority medium. Autoclave for 15 min separately YE, L-Cys, and the rest of ingredients, and then add YE and L-Cys into the rest of the ingredients aseptically and expose to bright light to accelerate the action of reducing agent.

[f]MJM, the modified JM medium developed at Dartmouth College. A carboy with 5 to 20 liters of deionized water, resazurin, and cellulose powder is autoclaved for 60 min. Solutions containing (i) Mg, Ca, Fe, and citrate and (ii) the rest of ingredients are adjusted to pH 6.8 and pumped to the carboy through the filter.

and most widely used techniques for soluble substrates: microscopic counts, a Coulter Counter, turbidity, and dry-weight measurement.

- The presence of solids gives rise to significant culture heterogeneity even in the presence of intensive agitation. In contrast to soluble substrates, microbial utilization of cellulose by *C. thermocellum* and other cellulolytic anaerobes is accompanied by formation of biofilms (see chapter 44, this volume) and aggregates containing cellulose particles and cells in various ratios and arrangements. Growth conditions are different for planktonic cells and cellulose-adhered cells, and potentially on the inside and outside of substrate-cell aggregates. Thus, the aggregated behavior of a cellulose-utilizing microbial culture is the weighted sum of the behavior of different interconvertible subpopulations which experience different environments and are in different physiological states.

- Sedimentation of solids makes it hard to maintain consistent delivery of substrate and removal of substrate and cells and can cause operational difficulties such as clogging of lines. Nonrepresentative solids removal, either in sampling or in effluent removal for continuous culture, can easily occur and greatly complicates and confounds interpretation of data.

Avoiding sedimentation and related difficulties is addressed in detail below.

45.3.2. Basic Fermentation System for Batch and Continuous Cultivation of Cellulolytic Bacteria

Figure 1 gives an overview of a fermentation system based on the BIOSTAT® Aplus bioreactor (Sartorius Stedim,

http://www.sartorius-stedim.com/index.php?id=9196). The system described could also be based on other small-scale (0.5- to 5-liter) bioreactors, e.g., from Applikon, New Brunswick Scientific, Dasgip, and other manufacturers. Bioreactors from all these manufacturers have built-in computer control of pH, agitation speed, temperature, volume of cultural liquid (foam/level control), and pO_2. For anaerobic microorganisms, measurement of O_2 is not needed, and the vacant port could be used to accommodate another probe, for example, the optical near-infrared (NIR) sensor DO4 (Dasgip, Jülich, Germany). In addition, systems from the manufacturers listed all contain an acid pump to neutralize excessive alkalinity. In our experience, cellulose fermentation almost never causes alkalization, and two active pumps (for base and acid) can create undesired situations such as activation of the acid pump to compensate base overshoot. We have thus found it preferable to disable acid addition functionality and utilize the respective built-in pump for other purposes such as automatic sampling of the culture for offline analyses (see below).

Four environmental parameters (temperature, pH, mixing rate, and level of liquid) are controlled as described by the manufacturer (e.g., http://www.sartorius-stedim.com/biostat-aplus/documents.html). Brief additional comments are given below.

- We have found temperature sensing using a regular thermistor and warming by a resistive "heating blanket" to be quite satisfactory. A water-jacketed heat exchanger (available in BIOSTAT® Qplus reactor) is more expensive but is better suited for quick visual inspection of culture (color, aggregation, and possible film growth). Neither a heating blanket nor water jacket is fast enough to prevent cooling during repetitive batch culture (see

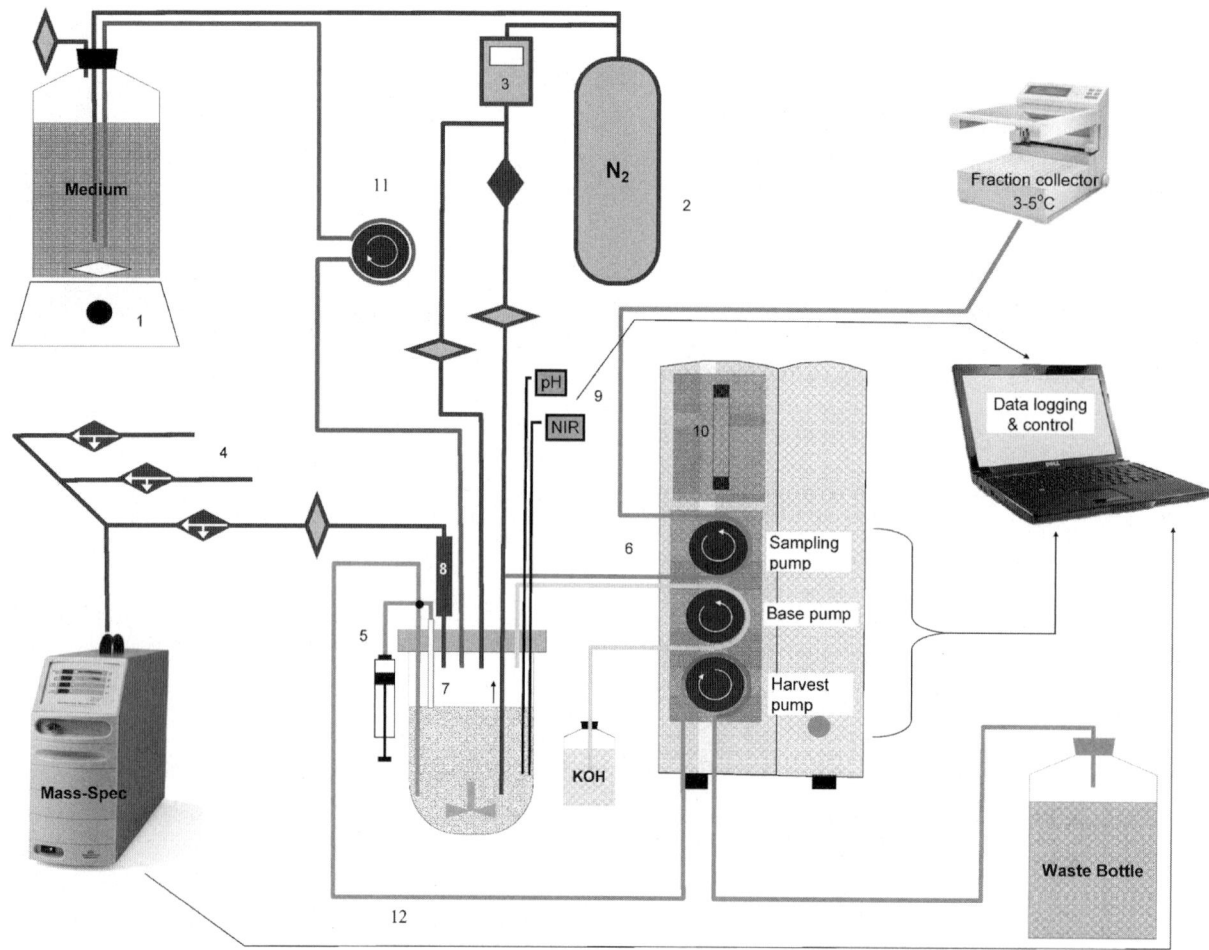

FIGURE 1 Scheme of bioreactor designed for quantitative physiological studies of cellulose fermentation. 1, Medium carboy continuously stirred with magnetic stirrer to maintain microcrystalline cellulose in suspended state; 2, N_2 tank equipped with pressure regulator; 3, mass flow controller maintaining constant gas flow (40 to 200 ml/min) through reactor headspace (FMA5512; Omega); 4, manifold of three-way solenoid valves directing gas flow from several (6 to 12; 1 is shown) individual bioreactors to a QMS100 mass spectrometer (high-pressure gas analyzer; Stanford Research Systems); 5, manual sampling port, standard 60-cc syringe connected to neoprene tubing via female Luer connector; 6, automated sampling device consisting of solenoid valve and pump delivering 2- to 15-ml samples at regular intervals to the set of test tubes placed in refrigerated fraction collector (Bio-Rad); 7, liquid level sensor, based on conductivity principle, is used to maintain constant volume of the bioreactor during continuous cultivation; when the volume reaches the threshold level and the sensor's tip touches liquid, the conductivity in the circuit jumps and activates the harvesting pump, pushing excessive cultural liquid to the waste bottle; 8, the condenser functions as a barrier for intensive medium evaporation; 9, pH probe and online optical NIR sensor (OD4, Dasgip); 10, Sartorius-Stedim BIOSTAT APlus controlling unit; 11, optional peristaltic pump; 12, tubing line used to discard cells which can be replaced by tangential filtration unit.

below), for which an additional heat exchanger is required if near-constant temperature is to be maintained.

- During autoclaving, the tip of the pH probe should stay submerged in liquid. The user can change the quality of pH control by changing the time delay between signal and base pump response as well as the tolerance limit (allowable mismatch between actual and preset pH values).
- Mixing control is advisable to regulate dependent upon growth stage of the culture. Immediately after inoculation, mixing should be low or eliminated to encourage colonization of cellulose particles by bacteria. Growing

fermenting culture should be agitated with a speed of 150 to 300 rpm. More rigorous mixing (up to 500 rpm or higher) appears to be tolerated and supports high mass transfer rates, which are needed in kinetic studies when metabolic activity is measured based on gas production. Another justification for a high mixing rate can be removal of H_2, which is a strong regulator of fermentation stoichiometry (see above). Except for these situations, intensive mixing is not in general necessary for anaerobic cellulose-utilizing microbes, in contrast to the situation with aerobic microorganisms, which are easily limited by insufficient oxygen mass transfer.

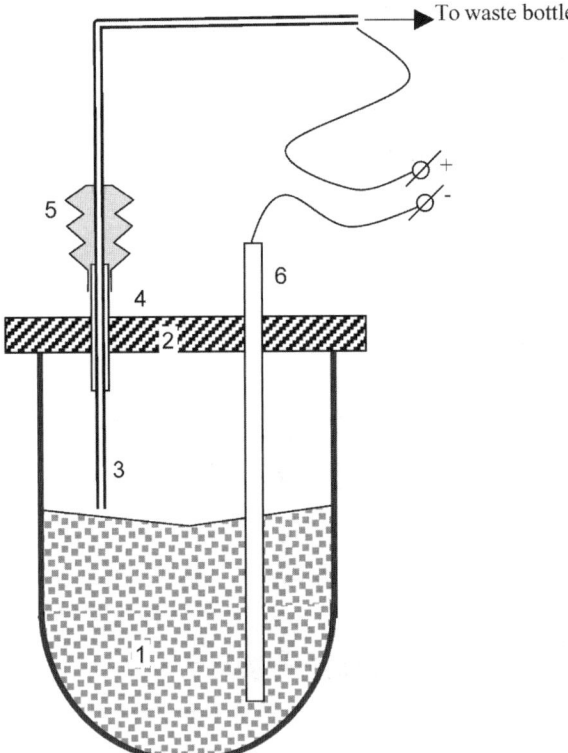

FIGURE 2 Modification of the level control of a Sartorius-Stedim bioreactor (Dartmouth design). 1, Cultural liquid (electroconductive); 2, metal headplate; 3, stainless steel pipe used as level sensor and harvest tubing; 4, Teflon pipe (insulator); 5, flexible rubber tubing which allows changing the position of the pipe by moving it up and down without compromising the sterility of the fermentor; 6, temperature probe (electroconductive). When the liquid level is below the sensor, the electric circuit is open, the current is zero, and the pump is idle. At the instant of contact, the circuit is closed and the harvesting pump is activated.

• Level control is used to keep a constant volume in continuous culture (chemostat or auxostat mode of cultivation; see below). Sartorius provides a short (~3-in. length) conductivity sensor which activates the harvest pump instantly when the sensor tip touches culture liquid. We constructed a level sensor of varying length (Fig. 2); it can be pushed down into the vessel to decrease the volume (taking care to cover newly exposed sensor surfaces with a flexible rubber sleeve) or withdrawn to increase the culture volume.

45.3.3. Modification of the Basic Fermentation System To Meet Requirements of Suspended Solid (Cellulose and Pretreated Wood) Fermentation

It is desired that solids be fully suspended in the bioreactor, in the medium reservoir (for continuous culture), and during medium delivery and withdrawal). However, significant deviation from uniformity in spatial distribution of particles can occur in narrow tubing carrying slurry from the medium reservoir to the bioreactor and in vertically positioned pipes inside both vessels (Fig. 3). In these unmixed parts (stagnation zones), we always observe some sedimentation. While the use of large-diameter tubing decreases

the risk of clogging, it also tends to exacerbate nonuniform distribution of particles because fluid velocities are lower.

To prevent particle sedimentation in practice, Weimer et al. (26) successfully applied a "segmented gas/liquid delivery system" entailing dividing feed slurries into small (~20-µl) discrete liquid segments separated by intervening gas bubbles. We have successfully employed a simpler approach consisting of a pump activated for 7 s (flow rate, ~10 ml/min) and then turned off for 40 to 100 s depending on the residence time desired. The on/off cycling is repeated ad infinitum using cheap and robust electronic time relays (e.g., NIST traceable lab controller; Cole-Parmer). We have found that this pulsed method completely prevents sedimentation even in tubing with multiple bends. Although technically intermittent, it can be considered practically continuous substrate delivery because the cycle period (~1 min) is much shorter than bacterial generation time, varying in the chemostat from 2 to 30 h.

There are several other general rules which help to achieve uniform distribution of suspended cellulose during continuous cultivation. The tubing connecting the reservoir with the fermentor should be as short and straight as possible, with predominantly vertical or downward-sloped orientation and avoiding folds and loops. It is strongly recommended to avoid diameter changes, as these tend to cause particle accumulation and tubing blockage (Fig. 3B). To stop solids sedimentation within vertical pipes inside the nutrient reservoir (Fig. 3C), we have found it effective to curve upward the end of the pipe by 5 to 10 mm to prevent exiting of particles back to the carboy. To prevent unintended return of cells and residual cellulose to the fermentation vessel (Fig. 3D), the open end of the vertical harvesting pipe should be positioned at the gas-liquid interface rather than below the liquid surface; this condition is satisfied when the harvesting pipe is used to control the level in the fermentor (Fig. 3). Uniform substrate dispersion can be verified by running tests with continuous culture for 1 to 2 days before inoculation by cellulolytic microorganisms. The concentration of solids is determined by dry-weight or total organic carbon (TOC) analysis (see below) in periodically withdrawn subsamples; the uniformity is confirmed if the solid concentration remains the same in the reservoir, culture vessel, and product bottle. Another test is to establish whether the solid concentration is uniform with respect to height in the carboy or fermentor (16).

45.3.4. Diversity of Cultivation Techniques as Applied to Cellulose Fermentation

The fermentation system shown in Fig. 1 can be used for diverse cultivation regimens: batch, repetitive batch, simple chemostat and chemostat with cell recycle, pH auxostat, CO_2 auxostat, D-stat, and others. Definitions and purposes of various cultivation methods together with respective mass balance equations have been reviewed elsewhere (20, 21). Table 2 lists six cultivation techniques (batch and repetitive batch, chemostat, cell recycle, pH auxostat, CO_2 auxostat, and D-stat) in terms of substrate and cell fluxes (Fig. 4). Long-term continuous cultures are accompanied by autoselection, in which spontaneous mutants with changed properties replace the original inoculum strain. The outcome of autoselection depends on the type of continuous cultivation. The "Goals" column of Table 2 lists anticipated results of autoselection and summarizes its potential biotechnological benefits.

45.3.4.1. Batch and Repetitive Batch Culture

When bioreactors are run in batch mode, it is desirable to prepare the nutrient reservoir with medium sufficient for several cycles. The first cycle is usually a preliminary

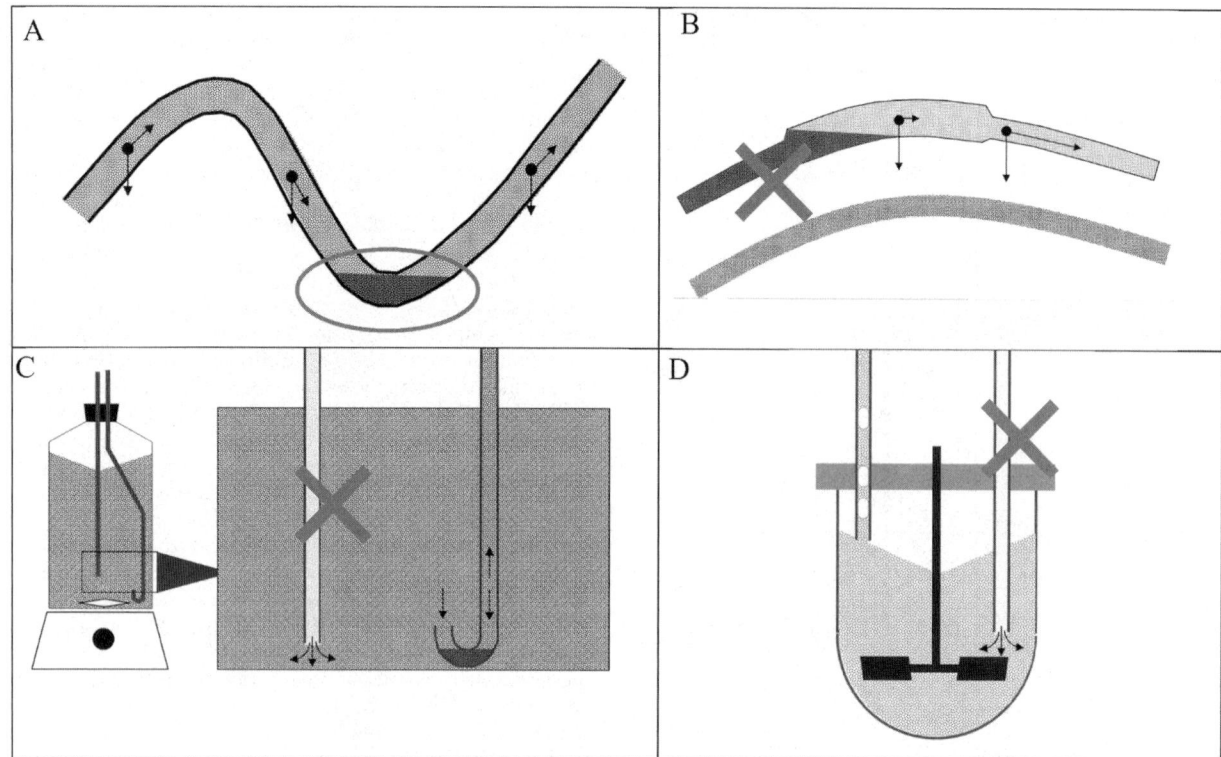

FIGURE 3 Managing suspended solids in continuous culture. (A) Schematic illustration of forces affecting solid particles moving through narrow tubing: advective forces (vectors parallel to flow direction) are interacting with gravitational forces (vertical vectors), and as a result, solids sediment at the bottom of tube folding (enclosed area). (B) Sedimentation is maximal at junction point of wide and narrow tubing. (C) Wrong (crossed) and correct shape of the vertical pipe used to pump out substrate slurry. With a straight pipe inside a reservoir, the stirring of the carboy does not prevent sedimentation of particulate material back to the reservoir (arrows). The circular end of pipe completely stops sedimentation. (D) Wrong (crossed) and correct way of solids recovery (cells plus unused cellulose) from the fermentation vessel. Unintentional recycling of solids occurs because of their sedimentation in a vertical harvesting pipe (the tip of this pipe is positioned below the gas-liquid interface). Recycling is completely prevented if the harvesting pipe is positioned at the gas-liquid interface: sedimentation is blocked by gas bubbles, and the harvesting pipe serves as a level sensor (see also Fig. 2).

cultivation to "wake up" inoculum material stored frozen under glycerol. The duration of the lag phase is not fully predictable and can last up to a week in our experience. Once the culture is growing rapidly, it can be diluted by fresh medium to initiate the second cycle of batch fermentation. Multiply repetitive batch cultures can be used to monitor autoselection, evolution of microbial communities, etc. Fresh medium is normally kept at room temperature. For thermophilic species like C. thermocellum, it should be warmed up to 55 to 60°C immediately before addition to the fermentor. This can be accomplished with a heat exchanger operating along the tubing line from the reservoir to the culture vessel. A standard chemical condenser of appropriate internal diameter can be inserted in the tubing line with hot water (65 to 75°C) circulated through the condenser's water jacket during medium delivery.

All other elements of the cultivation system shown in Fig. 1 can be used for repetitive batch culture without major modification. The peristaltic pump (Fig. 1, item 11) is turned on only at the end of the growth cycle to replace spent medium. The range of online and offline parameters measured during cultivation is optional.

Protocol for Startup of Batch Culture

1. Assemble, clean, and sterilize fermentation system shown in Fig. 1.
2. Prepare sterile medium according to one of the protocols outlined in Table 1.
3. Connect the sterilized medium reservoir to the fermentor via a Luer connector and start purging the fermentor and medium reservoir with N_2 at a rate of 100 ml/min or higher. Fill the vessel with the required amount of medium (typically 1 liter). Turn on the control panel of the BIOSTAT Aplus; start agitation (200 rpm), pH, and temperature control.
4. When the medium in the fermentor is completely reduced (resazurin turns colorless), the temperature is 60°C, and the pH is around 6.8, inoculate the fermentor with exponentially grown cells. Use a sterile disposable syringe and a Luer port on the fermentor.
5. Reduce the purging rate to ~20 ml of N_2/min in the medium reservoir and fermentor. Stop agitation and leave the culture overnight. It is recommended to start off-gas recording immediately for early detection of metabolic activity.

TABLE 2   Various cultivation techniques and their possible goals

Cultivation method	Operation conditions	Goals
Chemostat	F = constant, $x = x_{out}$, $p = p_{out}$	Continuous conversion of cellulose to fermentation products, minimizing residual concentration of substrate
Batch and repetitive batch	$F = \begin{cases} 0 & if & 0 \leq t \leq T_1 \\ A > 0 & if & T_1 < t \leq T_2 \end{cases}$ where A is liquid replacement rate (100–200 ml/min) at the end of growth cycle, T_1 (20–50 h): removal of 90–95% of spent liquid with following addition of fresh medium to 100%, the replacement time $(T_1 - T_2)$ is ~0.5 h.	Discontinuous or pulsed fermentation (in the case of repetitive batch); selection pressure minimizes lag phase and maximizes growth and colonization of cellulose particles
Cell recycle	F = constant, $x \gg x_{out}$, $s \geq s_{out}$, $p = p_{out}$; cells and residual cellulose are returned to the fermentor by tangential filtration, while fermentation products are continuously washed out.	Continuous fermentation with deep substrate conversion in the absence of severe product inhibition
pH auxostat	$x = x_{out}$, $p = p_{out}$; F is generally not constant but rapidly merges to steady state under absence of perturbations. Fresh medium is used to titrate metabolic acidity to keep pH constant.	Continuous fermentation at intermediate degree of substrate conversion. Increase of buffer strength results in higher degree of cellulose conversion. Selection of fast growers.
CO_2 auxostat	$x = x_{out}$, $p = p_{out}$; F is generally not constant but tends to steady state, similar to pH auxostat. The medium flow is instrumentally stopped if the CO_2 level is below preset value. CO_2 formation rate is constant.	Continuous fermentation favoring fast-growing cells. Degree of substrate conversion is proportional to selected CO_2 level. The multistability is possible with two or more steady states at the same CO_2 level and different x_{out} and s_{out}.
D-stat	F = constant, $x = x_{out}$, $p = p_{out}$; added medium contains inhibitor (in the main reservoir or in the second small carboy). Its concentration varies stepwise or smoothly as time gradient.	Continuous fermentation aimed at selection of product-resistant strains or strains tolerant to exogenous toxins (e.g., wood pretreatment by-products)

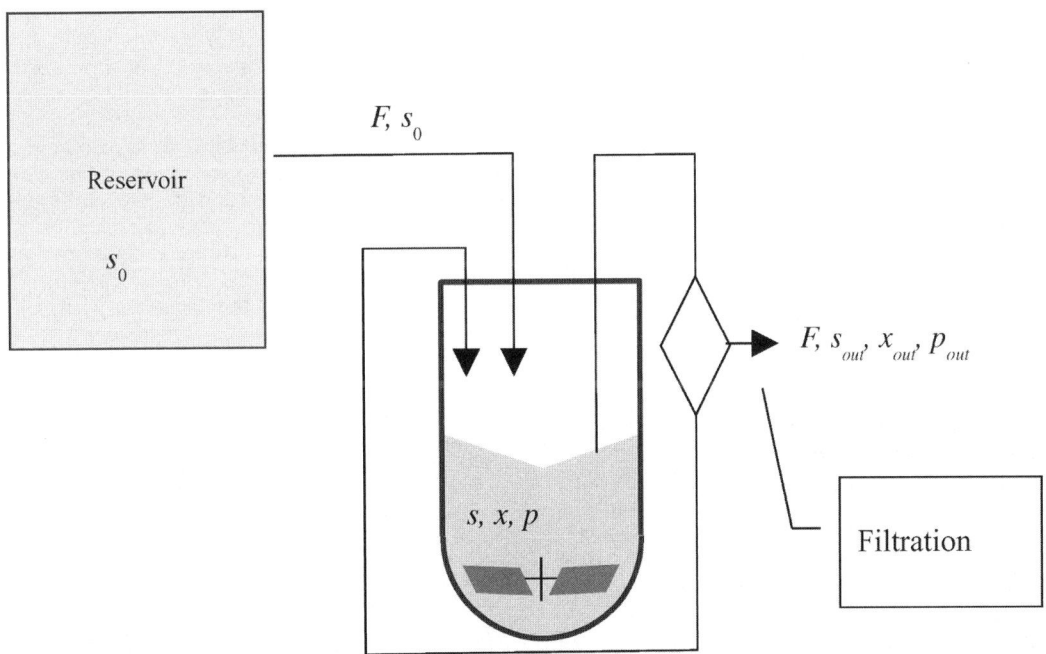

FIGURE 4   Schematic representation of cultivation system listed in Table 2. F, medium flow rate; s_0, feed substrate concentration; s, residual substrate concentration; x, cell mass concentration in fermentor; x_{out}, cell concentration in outflow liquid; p, metabolic products in fermentor; p_{out}, products in outflow liquid.

6. Check growth the next day: CO_2 and H_2 level above background level, acidification (pH drop down and/or amount of 1 N KOH automatically added), changes in color, and formation of slimy aggregates. Start agitation and increase the purging rate to 50 to 100 ml/min if growth has started. Start measuring the optical density with an online probe.

7. When the culture is growing vigorously (close to maximal rate of gas production and KOH titration rate), dilute it with fresh medium to initiate the second cycle. Increase the purging rate to 200 ml of N_2/min, and turn on the harvest pump (50 to 100 ml/min), remove 90% of cultural liquid, and add fresh medium with passage through a heat exchanger. Make sure that the temperature for the culture does not drop below 55°C. Agitation, temperature, and pH control is continued during feeding, ensuring that the tip of the pH probe remains submerged.

8. Monitor growth as required, including offline analysis and sampling with the fraction collector (see below). Start the next cycle before or after substrate depletion as indicated by constancy of optical density and slight increase of pH usually observed upon substrate exhaustion.

45.3.4.2. Chemostat Culture

Chemostat culture is carried out similarly to batch culture, except that the pump delivers suspended medium continuously or in short pulses as described above. To determine medium flow rate, one can use a 60-cc syringe permanently connected via a T-piece in the tubing leading from the feed reservoir to the pump. Growth medium may be drawn into the syringe, and the time required for the feed pump to empty the syringe may be determined under conditions where no flow is coming from the medium reservoir. More advanced automated determination of flow rate may be carried out by recording the volume of culture effluent, a built-in function of the BIOSTAT Aplus. The logged time series is numerically differentiated (e.g., with "SLOPE" or "LINEST" function in Microsoft Excel) to obtain the flow rate. Correction must be taken for any manually withdrawn samples or evaporation (usually a small fraction if the condenser runs at 5 to 10°C). A third method is to weigh periodically the carboy with medium or take a reading of the residual volume. All three methods should give reasonable compatibility.

45.3.4.3. pH Auxostat

The main operational difference between a pH auxostat and a chemostat is that fresh medium is delivered by the base pump rather than the peristaltic pump (Fig. 1, item 11). The bottle with 1 N KOH is disconnected or might be used only at startup. The pH auxostat keeps the pH at a constant level preset by the experimenter. However, the biomass level and degree of cellulose conversion are not simple functions of the pH. In practice, one can start pH auxostat cultivation by preliminary batch experiment without autotitration but with instrumental recording of the pH and biomass concentration (x). The growth rate progressively declines because of medium acidification. On the plot "x vs pH," select the required x level and then use the respective pH as the preset value. If a higher biomass is desirable, then increase the buffer strength of the medium by increasing the K_2HPO_4 concentration or adding nonassimilable buffer (morpholinepropanesulfonic acid [MOPS]) (Table 1).

45.3.4.4. CO_2 Auxostat

A robust and relatively inexpensive CO_2 analyzer (e.g., LI-800; LI-Cor, Lincoln, NE) can be attached to the fermentor exhaust gas line (Fig. 1) instead of or in addition to a mass spectrometer. This CO_2 sensor contains built-in "low-alarm" and "high-alarm" outlets generating an electrical signal at 5 V when the current CO_2 levels are, respectively, below and above the preset value. The high alarm (excessive CO_2 production) turns on a relay which activates the pump delivering medium from the reservoir. Dilution of the culture decreases the CO_2 level until the preset value is restored. An appropriate National Instrument data acquisition system can be used to log the amount of pumped liquid to calculate the medium flow rate (F_s) and then dilution rate ($D = F_s/V$, where V is culture volume). N_2 purging of the fermentor's headspace must be perfectly constant (we use the FMA5512 Omega mass flow controller) with a residence time of about 1 min (e.g., a purging rate of 200 ml/min in a fermentor with a headspace of 200 cc). The CO_2 auxostat is more easily adjusted to the desired level of steady-state cell mass and cellulose conversion than is the pH auxostat because CO_2 is stoichiometrically related to cell yield and substrate consumption.

45.3.4.5. Biomass Recycling

To recycle microbial biomass, the fermentation system (Fig. 1) may include a tangential filter (Sigma) as shown schematically in the flow diagram in Table 2. Initially the culture is run in a chemostat mode, the peristaltic pump (Fig. 1, item 11) continuously adds medium, and the harvest pump removes excessive liquid as regulated by the level sensor. To continue the process with cell recycle, the tubing line (Fig. 1, item 12) is closed, and instead the cultural liquid is directed to the loop containing the tangential filter. Solids, including cells and unutilized cellulose, are returned to the fermentor, while clear filtrate flows to the effluent collection vessel. With low substrate flow rates (~50 ml/h), a standard harvest pump (e.g., from a BIOSTAT Aplus) can be used for filtration activated by the level sensor in the same way as it is done in simple chemostat mode.

45.3.4.6. D-Stat

A D-stat maintains a constant dilution rate in the presence of changes (either stepwise or continuous) in any environmental factor (pH or temperature) or inhibitor concentration. This technique has been described in the literature (12, 21). Although we have not tested this technique in our laboratory at Dartmouth College, we can propose its straightforward realization as follows. The fermentation system (Fig. 1) is appended with a second medium reservoir containing the same medium plus inhibitor (ethanol, acetate, lignin derivatives from wood pretreatment, etc.) equipped with an additional pump. Two pumps work concurrently to keep a total flow constant while their ratio is varied; it would be reasonable to increase the inhibitor concentration gradually to avoid undesirable washout of cells. Another option is to activate the second pump in response to a CO_2 or pH sensor only when the culture retains an adequate level of metabolic activity. In this case, we would have a hybrid of a chemostat, auxostat, and D-stat.

45.4. MONITORING OF GROWTH

45.4.1. Online Methods

45.4.1.1. Gases and Volatile Compounds

Analytical chemistry of gaseous and volatile products is well developed, with several gas analyzers on the market. We have positive experience in using the portable quadrupole mass spectrometer QMS100 (high-pressure gas analyzer; Stanford Research Systems, Sunnyvale, CA) and infrared analyzers LI-800 and LI-840 (LI-Cor). Gas exhaust from several fermentors may be fed through a solenoid multiplexer into one master gas analyzer, and data can be logged to a computer using either custom or standard software. A one-out-of-eight gas line serves as background (N_2-flushing line). The difference in gas exhaust versus background multiplied by N_2 flow rate gives the formation rate of the respective gaseous product. Although fermentation generates a wide spectrum of volatiles (ethanol and other alcohols, aldehydes, and volatile fatty acids), their direct detection with mass spectrometry is not a trivial task. The main obstacle is low vapor pressure and binding of these compounds to the walls of capillary tubes. However, detection of true gases—H_2, CO_2, CH_4, and CO—is achieved satisfactorily even without gas chromatographic separation based on unique m/z values or their unique combinations. Example data are shown in Fig. 5 for a *C. thermocellum* batch culture producing CO_2 and H_2 from cellulose.

45.4.1.2. Titration Rate

The cumulative amount of alkali used to neutralize metabolic acidity may be recorded by standard MFCS software. Differentiation of the logged data gives rates of base addition (milliliters of KOH per hour) which provide a measure of the instantaneous metabolic activity of growing microorganisms. It should be realized that identifying particular biochemical processes responsible for acidification can be nontrivial. The common assumption that metabolic acidity is proportional to formation of acidic fermentation products is often wrong, as indicated by the observation of similar pH drops for cultures producing neutral products (ethanol, propanol, and glycerol) and organic acids (lactate, acetate, and propionate). The global ionic balance of aerobic growth indicates that NH_4^+ uptake is the main reaction causing acidification (16, 17), primarily because of high requirements in N of all bacteria and the high cell yield of aerobic microorganisms. In practice, titration rates often correlate closely with growth and fermentation rates (Fig. 5; compare titration rate with rates of gas production), particularly when the pH is controlled. Titration rate is probably the most inexpensive and easiest way of online monitoring of the fermentation dynamics.

45.4.1.3. Optical Probes

NIR optical probes (absorbance at 800 to 1,000 nm) are commonly used in fermentation research to record turbidity in solid-free media and are used to infer cell density. In our laboratory, we have used Optek and Dasgip NIR probes for recording turbidity dynamics in cellulose-containing media. During batch fermentation, the decline in turbidity caused by solubilization and uptake of cellulose is four to five times stronger than the increase in turbidity by growing cells. Therefore, NIR probes originally designed for cell mass recording can be applied for measurement of residual cellulose (and probably pretreated wood) after calibration.

45.4.2. Offline Methods

45.4.2.1. Autosampling

Since batch fermentation of cellulose commonly requires 20 to 50 h, manual sampling is very tedious. To address this, we have adapted a system based on a fraction collector (e.g., BioFrac fraction collector; Bio-Rad, Hercules, CA) which preserves sterility and does not require prior flushing of tubes to remove material from previous sampling. The method is illustrated by Fig. 6. The fermentation vessel is continuously purged by N_2, which enters into the culture

through two lines, one of which is periodically closed by a solenoid valve for 20 to 40 s (NIST electronic time relay); during this period an auxiliary pump is activated to deliver 4 to 10 ml of sample into a sample tube in a refrigerated collector. The valve is open but the pump keeps going for several seconds to force out the residual liquid and clean the tubing for the next sample. Samples can be collected every 1 to 2 h and are kept at 3 to 5°C for 1 to 2 days in tightly capped plastic tubes before analysis.

45.4.2.2. Cell Mass and Residual Cellulose C

The mass concentration of microbial biomass in cellulose-containing slurries can be estimated by analysis of some specific cellular constituent: DNA, membrane phospholipids, proteins, particulate N, muramic acid, etc. The conversion factor from these biochemical proxies to cell mass is generally not constant, the highest variability being reported for phospholipids and cell wall constituents, while DNA, cell proteins, and cell N display the lowest degree of variation. Below we outline protocols for cell N measurement designed to be amenable to a moderately high throughput (a few minutes per sample).

Since the mass fraction of nitrogen in cells (e.g., 10%) is much higher than in most cell-free cellulosic substrates, nitrogen can be used as a proxy of microbial biomass in studies of cellulose-utilizing microorganisms (14). Soluble N-containing compounds (NH_4^+, extracellular amino acids and proteins, and products of cell lysis) can be removed by washing pellets. If it is desired to differentiate microbial biomass from cellulase, a direct measurement of cellulase concentration (e.g., enzyme-linked immunosorbent assay and quantitative proteomic analysis) is required as developed by Lynd and Zhang (15a) and Zhang and Lynd (30).

Protocol for Cell N and Residual Cellulose C Analysis

1. One milliliter of culture is centrifuged and the pellet is washed twice with deionized water. We find that vacuum suction via a stainless steel needle removes the liquid phase as fully as possible, leaving an intact pellet.
2. The pellet is quantitatively transferred to 20 ml of water and placed in the autosampler of a TOC/N analyzer (e.g., Shimadzu OCT-2) with suspension maintained via a magnetic stirrer. The Shimadzu TOC/N analyzer (model TOC-V combustion analyzer with TNM-1 total nitrogen option) performs full catalytic combustion of organic matter to CO_2 and N_2O, which are then quantified, respectively, by NDIR and chemoluminescence detectors.
3. Calculation of the total pellet C (TPC) and total pellet N (TPN) in suspension is done from peak area versus standard solution of glycine (1,000.0 mg/liter, acidified to pH 1.5 by H_3PO_4). For a cellulosic substrate containing no nitrogen, and cells with mass fractions of 0.46 and 0.12 for carbon and nitrogen, respectively, calculations are as follows: cell mass concentration = TPN/0.12. Residual cellulose C = TPC − TPN · 0.46/0.12. An example of fermentation dynamics inferred based on TPC and TPN is shown in Fig. 5.

45.4.2.3. Other Offline Analytical Methods

We omit here explicit description of several other standard procedures which should be highly recommended as additional valuable characteristics of bacterial growth on cellulose:

- High-performance liquid chromatographic analysis of cellulose degradation products (cellobiose, glucose, and cellodextrin[s]) and fermentation products—ethanol, lactate, glycerol, acetate, formate, and other fatty acids (see example in Fig. 5).

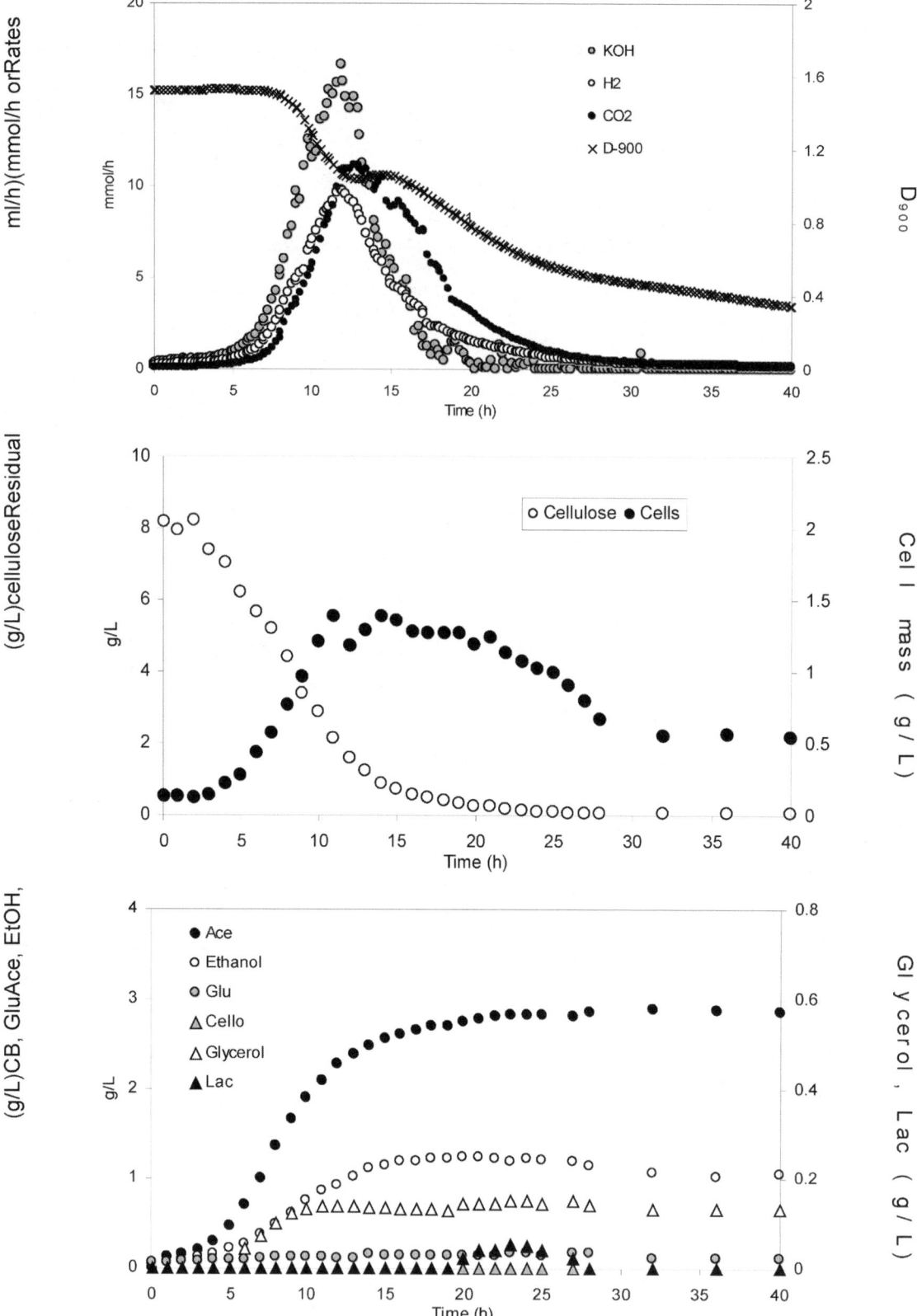

FIGURE 5 *C. thermocellum* batch growth on Avicel, 10 g/liter. (Top) Example of online monitoring of gas production, titration rate, and turbidity; (middle) residual cellulose and cell biomass based on analyses of pellet C and N (example of offline growth monitoring in samples taken with fraction collector); (bottom) high-performance liquid chromatography data for the same experiment.

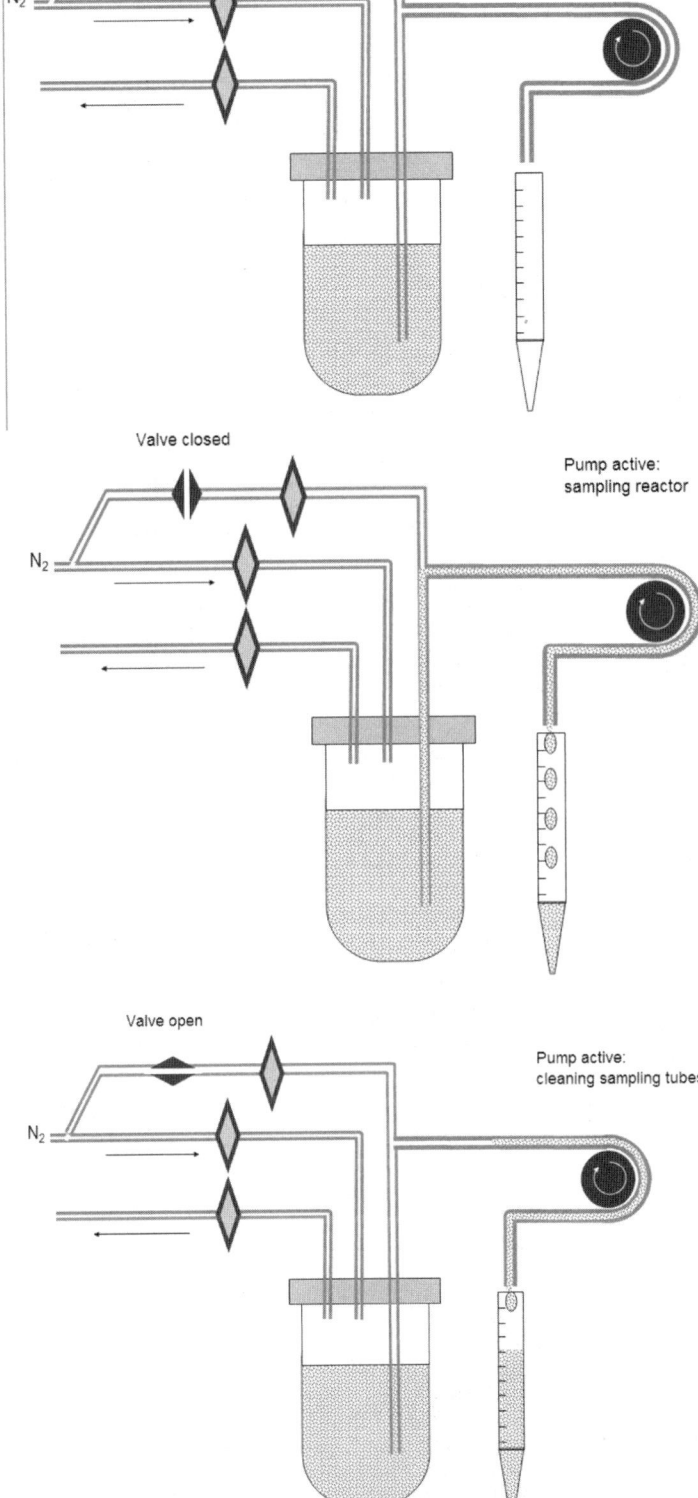

Valve open

Pump idle

N_2

Valve closed

Pump active:
sampling reactor

N_2

Valve open

Pump active:
cleaning sampling tubes

N_2

FIGURE 6 Automated sampling which requires no preliminary tubing flushing. (Top) Configuration of the system between samplings (sampling frequency is typically every 1 to 2 h); the regulating solenoid valve is open and pump is idle. (Middle) The first sampling phase: the valve is closed and the pump is active. Cultural liquid is delivered to a tube kept in a refrigerated fraction collector. (Bottom) The second sampling phase: the valve is open and N_2 gets in and interrupts the flow of cultural liquid; however, the pump keeps going for 5 to 10 s and cleans up the peripheral part of the sampling loop. Unidirectional flow of liquid and positive pressure gradient guarantee the sterility of the entire sampling line. This device can be used in a manual mode as an alternative to traditional sampling with disposable syringes or vials. In this case, the pump is activated manually and N_2 flow is stopped with a clamp instead of a solenoid.

- Pellet proteins with bovine serum albumin as standard by the Lowry and BCA methods (BCA Protein Assay Reagent [bicinchoninic acid], Thermo Scientific)
- Total carbohydrate content by quantitative saccharification as glucose released by differential acidic hydrolysis

Part of the experimental studies described in this chapter were supported by DOE (BESC), NSF (award 0732966, support of development of gas analysis) and Mascoma.

REFERENCES

1. **Bothun, G. D., B. L. Knutson, J. A. Berberich, H. J. Strobel, and S. E. Nokes.** 2004. Metabolic selectivity and growth of *Clostridium thermocellum* in continuous culture under elevated hydrostatic pressure. *Appl. Microbiol. Biotechnol.* **65:**149–157.
2. **Bothun, G. D., B. L. Knutson, H. J. Strobel, and S. E. Nokes.** 2005. Molecular and phase toxicity of compressed and supercritical fluids in biphasic continuous cultures of *Clostridium thermocellum*. *Biotechnol. Bioeng.* **89:**32–41.
3. **Brener, D., and B. F. Johnson.** 1984. Relationship between substrate concentration and fermentation product ratios in *Clostridium thermocellum* cultures. *Appl. Environ. Microbiol.* **47:**1126–1129.
4. **Chinn, M. S., S. E. Nokeset, and H. J. Strobel.** 2007. Influence of process conditions on end product formation from *Clostridium thermocellum* 27405 in solid substrate cultivation on paper pulp sludge. *Bioresour. Technol.* **98:**2184–2193.
5. **Choi, S. K., and L. G. Ljungdahl.** 1996. Dissociation of the cellulosome of *Clostridium thermocellum* in the presence of ethylenediaminetetraacetic acid occurs with the formation of truncated polypeptides. *Biochemistry* **35:**4897–4905.
6. **Demain, A. L., M. Newcombet, and J. H. D. Wu.** 2005. Cellulase, clostridia, and ethanol. *Microbiol. Mol. Biol. Rev.* **69:**124–154.
7. **Dror, T. W., E. Morag, A. Rolider, E. A. Bayer, R. Lamedet, and Y. Shoham.** 2003. Regulation of the cellulosomal CelS (cel48A) gene of *Clostridium thermocellum* is growth rate dependent. *J. Bacteriol.* **185:**3042–3048.
8. **Freier, D., C. P. Mothershedet, and J. Wiegel.** 1988. Characterization of *Clostridium thermocellum* JW20. *Appl. Environ. Microbiol.* **54:**204–211.
9. **Ghose, T. K.** 1987. Measurement of cellulase activities. *Pure Appl. Chem.* **59:**257–268.
10. **Johnson, E. A., A. Madiaet, and A.L. Demain.** 1981. Chemically defined minimal medium for growth of the anaerobic cellulolytic thermophile *Clostridium thermocellum*. *Appl. Environ. Microbiol.* **41:**1060–1062.
11. **Johnson, E. A., M. Sakajoh, G. Halliwell, A. Madiaet, and A. L. Demain.** 1982. Saccharification of complex cellulosic substrates by the cellulase system from *Clostridium thermocellum*. *Appl. Environ. Microbiol.* **43:**1125–1132.
12. **Kasemets, K., M. Drews, I. Nisamedtinov, K. Adamberget, and T. Paalme.** 2003. Modification of A-stat for the characterization of microorganisms. *J. Microbiol. Methods* **55:**187–200.
13. **Lamed, R. J., J. H. Loboset, and T. M. Su.** 1988. Effects of stirring and hydrogen on fermentation products of *Clostridium thermocellum*. *Appl. Environ. Microbiol.* **54:**1216–1221.
14. **Lynd, L. R., H. E. Grethleinet, and R. H. Wolkin.** 1989. Fermentation of cellulosic substrates in batch and continuous culture by *Clostridium thermocellum*. *Appl. Environ. Microbiol.* **55:**3131–3139.
15. **Lynd, L. R., P. J. Weimer, W. H. van Zylet, and I. S. Pretorius.** 2002. Microbial cellulose utilization: fundamentals and biotechnology. *Microbiol. Mol. Biol. Rev.* **66:**506–577.
15a. **Lynd, L. R., and Y. Zhang.** 2002. Quantitative determination of cellulase concentration as distinct from cell concentration in studies of microbial cellulose utilization: analytical framework and methodological approach. *Biotechnol. Bioeng.* **77:**467–475.
16. **Minkevich, I. G.** 1979. Ionic balance during cultivation of cell populations in spatial homogenous medium. *Biophysics* (Moscow) **24:**712–716.
17. **Minkevich, I. G., A. Y. Krynitskayaet, and V. K. Eroshin.** 1989. Bistat—a novel method of continuous cultivation. *Biotechnol. Bioeng.* **33:**1157–1161.
18. **Monod, J.** 1942. *Recherches sur la croissance des cultures bacteriennes.* Hermann, Paris, France.
19. **Nolte, A., and A. Holzenburg.** 1990. Studies on the anaerobic degradation of crystalline cellulose by *Clostridium thermocellum* using a new assay. *FEMS Microbiol. Lett.* **72:**201–207.
20. **Panikov, N. S.** 1995. *Microbial Growth Kinetics.* Chapman & Hall, London, England.
21. **Panikov, N. S.** 2008. Kinetics, microbial growth, p. 1513–1543. *In* M. C. Flickinger and S. W. Drew (ed.), *Encyclopedia of Bioprocess Technology: Fermentation, Biocatalysts and Bioseparation.* John Wiley & Sons, Inc., New York, NY.
22. **Pirt, S. J.** 1975. *Principles of Microbe and Cell Cultivation.* Blackwell Scientific, Oxford, United Kingdom.
23. **Stevenson, D. M., and P. J. Weimer.** 2005. Expression of 17 genes in *Clostridium thermocellum* ATCC 27405 during fermentation of cellulose or cellobiose in continuous culture. *Appl. Environ. Microbiol.* **71:**4672–4678.
24. **Tailliez, P., H. Girard, R. Longin, P. Beguinet, and J. Millet.** 1989. Cellulose fermentation by an asporogenous mutant and an ethanol-tolerant mutant of *Clostridium thermocellum*. *Appl. Environ. Microbiol.* **55:**203–206.
25. **Weimer, P. J., A. H. Conneret, and L. F. Lorenz.** 2003. Solid residues from *Ruminococcus* cellulose fermentations as components of wood adhesive formulations. *Appl. Microbiol. Biotechnol.* **63:**29–34.
26. **Weimer, P. J., Y. Shiet, and C. L. Odt.** 1991. A segmented gas/liquid delivery system for continuous culture of microorganisms on insoluble substrates and its use for growth of *Ruminococcus flavefaciens* on cellulose. *Appl. Microbiol. Biotechnol.* **36:**178–183.
27. **Wiegel, J., and M. Dykstra.** 1984. *Clostridium thermocellum*: adhesion and sporulation while adhered to cellulose and hemicellulose. *Appl. Microbiol. Biotechnol.* **20:**59–65.
28. **Williams, T. I., J. C. Combs, B. C. Lynnet, and H. J. Strobel.** 2007. Proteomic profile changes in membranes of ethanol-tolerant *Clostridium thermocellum*. *Appl. Microbiol. Biotechnol.* **74:**422–432.
29. **Zertuche, L., and R. R. Zall.** 1982. A study of producing ethanol from cellulose using *Clostridium thermocellum*. *Biotechnol. Bioeng.* **24:**57–68.
30. **Zhang, Y., and L. R. Lynd.** 2003. Quantification of cell and cellulase mass concentrations during anaerobic cellulose fermentation: development of an enzyme-linked immunosorbent assay-based method with application to *Clostridium thermocellum* batch cultures. *Anal. Chem.* **75:**219–227.
31. **Zhang, Y. H., and L. R. Lynd.** 2004. Toward an aggregated understanding of enzymatic hydrolysis of cellulose: noncomplexed cellulase systems. *Biotechnol. Bioeng.* **88:**797–824.
32. **Zhang, Y. H., and L. R. Lynd.** 2005. Cellulose utilization by *Clostridium thermocellum*: bioenergetics and hydrolysis product assimilation. *Proc. Natl. Acad. Sci. USA* **102:**7321–7325.
33. **Zhang, Y. H., and L. R. Lynd.** 2006. Biosynthesis of radiolabeled cellodextrins by the *Clostridium thermocellum* cellobiose and cellodextrin phosphorylases for measurement of intracellular sugars. *Appl. Microbiol. Biotechnol.* **70:**123–129.
34. **Zverlov, V. V., and W. H. Schwarz.** 2008. Bacterial cellulose hydrolysis in anaerobic environmental subsystems—*Clostridium thermocellum* and *Clostridium stercorarium*, thermophilic plant-fiber degraders. *Ann. N. Y. Acad. Sci.* **1125:**298–307.

BIOLOGICAL ENGINEERING AND SCALE-UP OF INDUSTRIAL PROCESSES

T HIS SECTION CONTAINS CHAPTERS THAT FOCUS ON THE MAJOR ELEMENTS of biological engineering and scale-up that are considered key for industrial process success. The chapters are authored by a mix of academic and industrial practitioners, noted for their extensive experience and valuable insights as well as valuable direct "hands-on" perspectives.

The section begins where most bioprocess development efforts start. The selection of raw materials and subsequent medium development for industrial fermentation processes has been, and continues to be, the foundation of sound bioprocess performance. Samun K. Dahod, Randolph Greasham, and Max Kennedy include recent best practices relating to raw-material quality and performance in their chapter 46. As of this writing, there are some significant examples of raw-material issues that have disrupted biopharmaceutical processing at more than one company, indicating the importance of achieving robustness as well as productivity from raw material selection.

Bioreactors are the main focus of the next two chapters. Microbial fermentation scale-up is described by Xaoming Yang in chapter 47. Wan-Seop Kim describes cell culture bioreactors in chapter 48. Both of these industrial authors include material that focuses on the importance of scaling down as well as scaling up bioprocesses with fidelity of process performance. As Quality by Design efforts continue to emphasize acquiring process understanding at multiple scales, bioprocess practitioners need to appropriately identify scale-dependent and scale-independent process parameters for further study.

Continuous-culture principles are the subject of chapter 49, written by An-Ping Zeng and Jibin Sun. This chapter remains particularly timeless since the pharmaceutical industry in general, and the biopharmaceutical industry in particular, continue to wrestle with the question of whether batch or continuous processing is more effective. Continuous culture also has remained one key method to efficiently study and optimize process parameter set points or feed compositions.

Precise measurement and consistent control of the cellular environment have been the goal of bioprocess engineers for decades. In chapter 50, Adeyma Arroyo of Genentech describes advances in sensor and sampling technologies, and David Hopkins, Melissa St. Amand, and Jack Prior describe bioreactor automation in chapter 51. These industrial authors describe the wide variety of sensor and automation philosophies currently available for implementation to match bioprocessing monitoring demands. The biopharmaceutical industry and academics working in this field consistently have been effective partners with equipment and instrument manufacturers. The end result has been a continued, decades-long, extensive selection of reliable sensor offerings.

This section ends with chapter 52, on purification and characterization of proteins, authored by Ulrich Strych and Richard C. Willson. The quality of the product produced is perhaps the most important aspect of a bioprocess and the key to evaluating bioprocess success. Strych and Willson describe the available process and analytical methods appropriate for use as the protein becomes progressively more purified.

In summary, the field of bioprocessing is continuing to evolve, but at vastly different speeds depending on the specific area being examined. Although the fundamentals have remained relatively constant, their wide application to different process elements remains astonishing. The goal of the authors contributing to this updated section was to assist readers in making these applications successfully. In turn, it is hoped that the readers of this third edition of the *Manual of Industrial Microbiology and Biotechnology* can pave the way for continued enhancement of these applications and progression of these fundamentals for use by future readers.

Raw Materials Selection and Medium Development for Industrial Fermentation Processes

SAMUN K. DAHOD, RANDOLPH GREASHAM, AND MAX KENNEDY

46

46.1. INTRODUCTION

The fermentation medium forms the environment in which the fermentation microorganisms live, reproduce, and carry out their specific metabolic reactions to produce useful products. The importance of this environment cannot be overemphasized when it comes to the development of a productive fermentation process. Over the years, substantial progress has been made in developing fermentation medium design as a systematic science. However, experienced industrial microbiologists and biochemical engineers will be the first to point out that this field is as much an art as it is a science. In most industrial fermentations, where the product is something other than the cell mass itself, there are two distinct biological requirements for medium design. First, nutrients have to be supplied to establish the growth of the organism. Second, after growth is established, proper nutritional conditions have to be provided to maximize product formation. Besides these obvious biological requirements, one needs to worry about selection of nutrient components that are cost-effective, readily available, and consistent from lot to lot. In recent years, as integrated approaches to fermentation and downstream processing have been developed, it has also been recognized that the fermentation medium should not unduly hinder the downstream processing and, if possible, should even facilitate downstream processing. For new fermentation processes brought up from microbiology laboratories, considerable flexibility and latitude in medium design are possible. The process is not locked into a fixed set of raw materials (for example, due to a Food and Drug Administration [FDA] filing), and the medium components can be freely selected for the sole purpose of maximizing the product yield and minimizing the cost. For an established fermentation process, the choice of medium components may be limited by such factors as FDA filing, the cost structure for the product, and the requirements of downstream processing. In spite of these limitations, continued medium development remains a necessity so that an established product retains its competitive edge in the marketplace.

While literature reports for medium development in specific fermentation processes are plentiful, a general treatment of broad principles involved in fermentation medium development is comparatively harder to find. Readers may find the reviews by Kennedy and Krouse (9) and Corbett (4) informative.

This chapter focuses primarily on raw materials and medium development for microbial fermentation processes. Although general principles also apply to it, mammalian cell culture will not be emphasized. This chapter is not intended to provide a literature search or a review of specific medium types used in specific fermentation processes. Rather, it is designed to provide practicing microbiologists and biochemical engineers with a rational basis for medium development and improvement. At the start of the chapter, chemically defined fermentation media are considered. Then, various commercially available ingredients for key nutrient components of traditional complex fermentation media are described in generic terms. This discussion is followed by a discussion of general considerations and a set of guidelines for medium development and improvement. The information provided is derived from experiences in the fermentation industry, and little effort has been made to cite references for specific examples and dicta mentioned in the chapter, even though similar information may also be presented in the literature.

46.2. CHEMICALLY DEFINED FERMENTATION MEDIA

Chemically defined media have been used routinely in the laboratory to study the microbial biosynthesis of primary and secondary metabolites. A key characteristic that has made these media desirable is consistency of performance. Although chemically defined media may be used to study the biosynthesis of the metabolites, complex media are usually used to produce them. Today, most fermentation processes employ complex media. However, chemically defined media are becoming popular where process consistency is very important, such as the production of biologics. Frequently, the process for making the biologic is considered part of the product definition. In addition to enhancing process consistency, chemically defined media have several other favorable characteristics, including better control and monitoring of the fermentation process, rapid scale-up, reduced sensitivity to large-scale sterilization conditions, and simplification of downstream processing. Concerns associated with these media include medium cost and development time as well as cell growth and production levels. Some of these concerns are being favorably addressed by the availability of more-sophisticated analytical tools to quickly analyze initial complex medium components, the use of statistical design of experiments to rapidly develop and optimize media, and the use of the well-established technique of nutrient feeding to circumvent nutrient toxicity at high concentrations and to enhance productivity.

There are several approaches one may employ to develop chemically defined media. One is to perform a thorough search of the literature for chemically defined media that support growth of the microorganism of choice or a closely related one. Once identified, it may be optimized using statistical design of experiments. Another approach is to simulate complex medium ingredients using modern analytical tools. For example, in the medium supporting biotin production, histidine and methionine were found to replace the complex component casein hydrolysate. In general, monosodium glutamate and ammonium sulfate have proven to be good replacements for complex nitrogen medium ingredients. A third approach is to develop the initial medium composition based on the typical elemental formula of the cell being $C_6H_{11}NO_3$ with a molecular weight of 161 including ash (the cell dry matter is 90% organic and 10% ash) [18] and the elemental composition of the cell as presented in Table 1 [20].

For example, to estimate the amount of glucose (carbon source) and ammonium sulfate (nitrogen source) required to support a yeast dry cell weight (DCW) of 100 g/liter, the following calculations may be used. For the amount of glucose (for cell growth): (100 g of DCW/liter)(0.48 g of C/g of DCW)[(180 g of glucose/mol of glucose)/(72 g of C/mol of glucose)] = 120 g of glucose/liter. However, since glucose is also an energy source for cell maintenance, the total amount of glucose required for both growth and maintenance is estimated as follows: cell yield = grams of DCW/gram of glucose consumed = 0.51 [17]. Thus, for 100 g of DCW/liter, 100/0.51 = 196 g of glucose per liter. For the amount of nitrogen: (100 g of DCW/liter)(0.075 g of N/g of DCW){[132 g of $(NH_4)_2SO_4$/mol of $(NH_4)_2SO_4$]/[28 g of N/mol of $(NH_4)_2SO_4$]} = 35.4 g of $(NH_4)_2SO_4$ per liter.

Similar calculations are performed for the remaining elements of the microbe. If a growth factor required by the microbe is known (e.g., yeasts require biotin), it should be added. If the requirement for growth factors is unknown, a small amount of yeast extract (0.001 to 0.05%) may be added initially.

Care must be taken when using this approach to calculate the initial concentration of medium ingredients, since they may be growth inhibitory. For example, glucose at a concentration of 50 g/liter inhibits the growth of *Escherichia coli* and can be detrimental even at much lower concentrations. To circumvent this inhibition and achieve high cell densities, appropriate glucose feeding is usually employed. For example, a chemically defined medium used to produce recombinant human interferon-α_1 by *E. coli* is presented in Table 2 [19]. As shown, glucose was present initially at a noninhibitory concentration of 30 g/liter. Subsequent glucose feeding was computer controlled, maintaining glucose below its inhibitory concentration throughout most (14 days) of the 16-day fermentation cycle.

By calculating the cell mass supported by each of the elements listed in the medium in Table 2 (such as carbon, nitrogen, phosphorus, sulfur, etc.), ammonium sulfate was identified as the growth-limiting nutrient. Based on the nitrogen level, 34 g of ammonium sulfate was calculated to support a maximum cell mass of 58 g of DCW per liter—the cell density achieved with the actual fermentation process.

In addition to the production of biologics, chemically defined media are proving to be profitable for producing secondary metabolites at manufacturing scale; a good example is the production of penicillin. Examples of chemically defined media that have proven to be economically successful for secondary metabolites are presented in Table 3 [8, 21].

When the chemically defined medium was compared with the initial, optimized complex medium for producing the secondary metabolite by *Streptomyces*, the chemically defined medium reduced the medium cost by 4.5-fold and increased the titer by 80% at the 800-liter scale.

TABLE 1 Typical elemental composition of microbes[a]

Element	Composition (% of DCW) in:		
	Bacteria	Yeast	Fungi
Carbon	48	48	48
Nitrogen	12.5	7.5	6
Phosphorus	2.5	1.7	2.5
Sulfur	0.6	0.13	0.3
Potassium	2.8	2.5	1.4
Magnesium	0.3	0.3	0.2
Sodium	0.8	0.06	0.26
Calcium	0.56	0.2	0.75
Iron	0.11	0.26	0.15
Copper	0.02	0.006	
Manganese	0.006	0.004	
Molybdenum		0.0002	

[a]Adapted from reference 20.

TABLE 2 Production of recombinant human interferon-α_1 by *E. coli*[a]

Component	Initial medium	Feeding solution
KH_2PO_4	3.0 g/liter	
K_2HPO_4	5.0 g/liter	
$(NH_4)_2SO_4$	4.0 g/liter	30.0 g/liter
$MgSO_4 \cdot 7H_2O$	2.0 g/liter	5.0 g/liter
Vitamin B_1	0.1 g/liter	2.0 g/liter
Trace metal	3.5 ml	
Glucose	30.0 g/liter	500 g/liter
Antifoam	0.5 g/liter	
pH	7.0	

[a]Adapted from reference 19.

46.3. COMPONENTS OF INDUSTRIAL FERMENTATION MEDIA

As noted above, most industrial fermentation media are complex formulations containing poorly defined ingredients. Often these ingredients contain multiple nutrients for the growth of fermentation microorganisms. However, for the purposes of medium development, a given ingredient is thought to provide primarily a single nutrient. For example, soy flour is used primarily to supply complex nitrogen or protein for the growth of a microorganism. However, soy flour also contains substantial amounts of metabolizable carbohydrate and minerals. In the discussion below, the medium ingredients are classified according to their primary role in the fermentation process. On this basis, we can classify the fermentation raw materials in four broad nutrient categories: materials used primarily as sources of carbon, nitrogen, or minerals, and materials used for special purposes.

TABLE 3 Chemically defined media for representatives of actinomycetes and filamentous fungi

Medium component	Composition for production in:	
	Streptomyces	*Gliocladium*
(A) Medium		
Glucose	112.5 g/liter	
Sucrose		170.0 g/liter
$(NH_4)_2SO_4$	16.5 g/liter	8.9 g/liter
K_2HPO_4	1.5 g/liter	1.73 g/liter
Monosodium glutamate	7.5 g/liter	66.4 g/liter
$CaCl_2$	1.0 g/liter	
$CaCO_3$		1.73 g/liter
$MgSO_4 \cdot 7H_2O$		1.12 g/liter
Biotin		0.17 g/liter
Salt solution (see part B)	20 ml/liter	2.6 ml/liter
P-2000	2 ml/liter	1.7 ml/liter
(B) Salt solution		
$MgSO_4 \cdot 7H_2O$	28.9 g/liter	
$ZnSO_4 \cdot 7H_2O$	0.5 g/liter	5.0 g/liter
$CuSO_4 \cdot 5H_2O$	0.05 g/liter	0.5 g/liter
$FeSO_4 \cdot 7H_2O$	0.5 g/liter	5.0 g/liter
$MnSO_4 \cdot H_2O$	0.1 g/liter	1.0 g/liter
$CoCl_2 \cdot 6H_2O$	0.04 g/liter	0.4 g/liter

46.3.1. Carbon Sources

46.3.1.1. Carbohydrates

Glucose is the most frequently used carbohydrate in the fermentation industry. In the United States, it is derived from the corn-processing industry. Two types of products are in use, dextrose monohydrate and hydrolyzed corn syrups containing glucose at a level greater than 95% (called DE95 or dextrose equivalent of 95%). While dextrose monohydrate comes in the form of easy-to-handle crystalline material, it is more expensive. This material is used primarily in small-scale applications as in seed fermentors and when consistency is of the utmost importance. For the bulk of the glucose needs, such as for large-scale fermentations and for in-process feeding, the hydrolysate is the more economical material. If the fermentation microorganism is able to hydrolyze low-molecular-weight saccharides, less expensive corn syrups of various lower degrees of hydrolysis can be used. Industrial fermentation processes such as those for the production of penicillin can readily utilize hydrolysates with a dextrose equivalent as low as 20 (DE20). In fact, some processes give higher yields with these higher-molecular-weight saccharides than they do with pure glucose. The next level of complexity in these glucose-based carbohydrates comes in the form of various dextrins. These are primarily cornstarch products with just enough hydrolysis carried out to make them soluble in the fermentation medium. The dextrins, cornstarch, other starches (such as potato starch), and solid substrates in general are rarely used for in-process feeding. They are generally used as batched-in carbon sources for initial growth of the organism or as carbon sources that are gradually assimilated by the microorganism during the product synthesis phase. In the United States, the crudest and the cheapest source of complex carbohydrate is corn flour. This product is primarily starch but also contains about 5% protein. An important cost reduction strategy used by many fermentation companies is to use crude starch or corn flour in the batch along with the commercially available enzyme amylase. The amylase breaks down starch molecules to generate more readily utilizable carbohydrates. In many fermentation processes, the primary carbon source (the most readily utilizable, such as glucose) is metabolized rapidly during the growth phase and a secondary carbon source (such as oils) is utilized during production.

Sucrose is often used in fermentation processes. In its crystalline form, sucrose is available as table sugar of various degrees of refinement. The white crystalline sucrose is generally used in small-scale applications and in seed fermentors. However, it can also be used as a gradually utilized carbon source in some fermentations in which the organism has a limited ability for metabolizing sucrose. The use of disaccharides is often explored as alternate carbon sources for monosaccharides (such as glucose) when catabolite repression is encountered. The crudest form of sucrose comes as molasses, which contains anywhere from 3 to 10% protein. In some fermentations (for example, glutamic acid fermentation), this product gives excellent results as a combined carbon-nitrogen feed.

In the early days of penicillin fermentations, the carbon source of choice was lactose. This sugar is gradually metabolized by the penicillin-producing organism and hence can be batched into the medium from the beginning of the process. However, since the advent of controlled feeding of glucose, the importance of lactose in the fermentation industry has decreased. Lactose is available in granular form for small-scale applications, and it is still used in some fermentations, especially in Europe, where it is more readily available than dextrose and corn syrups. The most economical source of lactose is derived from the cheese industry by-product cheese whey. This product is available in a spray-dried form and is an excellent source of protein and minerals besides being a source of lactose.

Other sugars that are used less frequently in the fermentation industry include maltose, mannitol, sorbitol, and xylose. All of these are generally used in their purified forms. A related carbon source for the fermentation industry is glycerol. It is useful in many processes as a gradually metabolized carbon source. Additionally, organic acids, such as acetic acid, may be used on rare occasions as combination pH control agents and carbon nutrients. Minoda (13) has reported on the potential uses of other unusual carbon sources for amino acid fermentations.

46.3.1.2. Oils

Various oils are widely used as carbon sources in the fermentation industry, especially in antibiotic fermentations. Oils can supply both the energy and the growth carbon needs of the organism. In many antibiotic fermentations, where the antibiotic backbone is synthesized from low-molecular-weight fatty acids, the oils make ideal carbon sources since they gradually supply these fatty acids during the fermentation process. The oils are used both as batched-in ingredients and as continuous feeds. In some fermentations, oils play an important auxiliary role even when they are not actively metabolized by the fermentation microorganism. The yield-enhancing effect of the oil when it is not metabolized is not well understood. However, oxygen diffusivity in oils is higher than in water, which may have a beneficial effect on oxygen transfer from bubbles. It is possible that the oil provides protection to cells from excessive shear forces or that it makes a key micronutrient from the complex medium more available to the microorganism in the form of micelles. In fermentations in which oils can be utilized as carbon feeds, they offer important benefits. First, the caloric content and the corresponding energy availability per unit volume of feed are appreciably higher for oils than for carbohydrates. One liter of vegetable oil has more than twice the utilizable energy as 1 liter of a 55% solution of glucose. This high energy density allows for lower feed rates and smaller feed vessels. Consequently, the fermentor volume management for long-cycle fermentations is easier with oil-fed fermentations than with sugar-fed fermentations. This enhanced management is true not only because less feed is introduced into the fermentor but also because the metabolism of oil does not produce as much water as the metabolism of sugars. The antifoaming property of the oils is also beneficial for most fermentation processes. Before the advent of synthetic defoamers, oils were used for foam control in many fermentation processes even when the carbon source of choice was a sugar. However, the oil added for foam control is metabolized by the organism, and continuous addition is required to control foam. The synthetic defoamers are more effective because they are not readily degraded by the fermenting microorganism and they are cost-effective. In special cases, when the presence of synthetic defoamer interferes with the downstream processing, oils are still used as defoamers. The antifoaming properties of several natural oils are reviewed by Vardar-Sukan (16).

The most important oil in the U.S. fermentation industry is soybean oil. It is abundant and relatively inexpensive. Other oils that are often used are lard oil, fish oil, and oils of other plants such as corn, cottonseed, peanut, sunflower, and safflower. One specialty oil product that is synthetically made and has found application in the fermentation industry is methyl oleate. Methyl oleate is often used as a supplemental feed in conjunction with another feed such as soybean oil. The fatty acid contents of various oils vary according to their source, and there may be a theoretical basis for one type of oil to perform better than another type. However, the choice of oil in a given fermentation is generally determined empirically. The oil that is used in the shake flask fermentations during screening of the producing strains very often also gives better results in large-scale fermentations.

46.3.2. Sources of Organic Nitrogen or Protein

There are principally three classes of raw materials available to supply the organic nitrogen or protein requirement of a fermentation process: (i) those derived from agricultural products, (ii) those derived from brewery industry by-products, and (iii) those derived from meat and fish by-products. All of these products supply other important fermentation nutrients in addition to organic nitrogen.

46.3.2.1. Nitrogen Sources Derived from Agricultural Products

The sources derived from agricultural products are the workhorse ingredients of the fermentation industry. They include the products of commodities such as various grains and soybean. The soybean flours, meals, and grits head the list of applications in antibiotic fermentations. The popularity of the soy products is based on the fact that after the soy oil is extracted from the soybeans, the residue is about 50% protein, which is readily available for cell growth. In addition, soy flour, meals, and grits contain up to 30% utilizable carbohydrates. Most minerals required for microbial growth are also present in soy-based products. In many seed medium applications, where growth is the primary consideration, all that is required in the medium is soy flour along with salts such as magnesium sulfate and potassium phosphate. A product that is processed very similarly to soy flour is cottonseed flour. The protein in the cottonseed flour is less readily available and thus makes a good slow-releasing nitrogen source. Corn gluten meal is another readily available product that is suitable as a slow-releasing nitrogen source. Corn steep liquor, a by-product of the corn milling industry, was very extensively used in the early years of the antibiotic fermentation industry. In recent years, though, due to the variability in the product quality, the liquid form of corn steep liquor has fallen out of favor. Spray-dried corn steep liquor is now available and is used in many antibiotic fermentations because it is less variable. Other agricultural commodities used as nitrogen sources in the fermentation industry include peanut meal, linseed meal, wheat flour, barley meal, and rice meal. Plant or animal hydrolysates produced by the degrading action of enzymes, usually proteases, are widely used. Should a plant source be required, soybean hydrolysates can be used, or if an animal source is acceptable, then casein or whey protein hydrolysates may be used.

46.3.2.2. Nitrogen Sources Derived from Brewery Industry By-Products

The brewing industry is an important source of fermentation raw materials. The principal product is the yeast left over after beer fermentation. The suitability of the yeast by-product for a given fermentation depends upon the method of drying. The yeast may be drum dried or spray dried. It is also sold as a paste produced by water evaporation in an industrial evaporator. All of these products have found applications in the fermentation industry as sources of nitrogen. However, the yeast is never used as the primary source of nitrogen. Instead, it is thought of as a nitrogen supplement with additional beneficial nutrients that are not available from grain-based nitrogen sources. Generally, these additional nutrients are organic phosphorus and unknown micronutrients. Brewery yeast is also refined into yeast extracts of different water solubilities, which are more expensive and used in smaller quantities. Yeast extract is often the single undefined component used in so-called semidefined fermentation media to provide micronutrients. The brewing and distilling industries supply two other by-products that are sometimes used in the fermentation industry: distillers' solubles, in the form of a concentrate or spray-dried powder, and leftover grains from the brewing process.

46.3.2.3. Nitrogen Sources Derived from Meat and Fish By-Products

Meat and fish products are very rich in protein. So are the by-products of these industries. The primary meat-based product is generically known as spray-dried lard water. This is a by-product of lard processing. The animal bones and tissues are boiled in water, sometimes in the presence of proteases, to free the fat. The resulting liquor is separated into fat and water layers. The water part is rich in proteins and peptides. This water, when spray dried, gives a product with a protein content of 80% or greater. The lard water can be obtained with different degrees of chemical or enzymatic hydrolysis. Hydrolyzed lard water products are sold as meat peptones under various brand names. A parallel line of products labeled fish meals and fish hydrolysates is derived from heat and enzymatic treatment of fish wastes. These products are generally about 70% protein.

46.3.3. Minerals

Minerals are used in fermentation media to serve many purposes, e.g., as major nutrients, as trace metal suppliers, as ionic strength-balancing agents, as precursors for secondary-metabolite synthesis, as buffering agents, as pH control agents, and as reactants to remove specific inhibitory nutrients from the medium. The nitrogen-containing salts (e.g., ammonium sulfate, ammonium nitrate, sodium nitrate, and potassium nitrate) can provide a substantial portion of the nitrogen requirement for cell growth when combined with organic nitrogen. When salts are used as nitrogen nutrients, their metabolism invariably results in pH changes in the medium. For example, when ammonium sulfate is utilized by the organism, the pH tends to fall, and when sodium nitrate is utilized, the pH tends to rise. Therefore, it is very important that adequate buffering or pH control be provided to counterbalance these pH effects. Ammonia used for pH control has the advantage of regulating pH while replenishing ammonium nitrogen used up from ammonium sulfate in the medium.

Another major nutrient supplied as inorganic salt is phosphorus in the form of phosphate salts. Phosphorus from soluble phosphate salts is more readily available to the organism than the phosphorus derived from organic nutrients such as yeast. As a result, it is possible to control the rate of growth by balancing organic phosphorus against inorganic phosphorus salts.

Although most organic nitrogen sources such as grain meals and yeast extracts contain many of the minerals required for growth, the fermentation medium is often supplemented with salts that provide elements that are required in greater than trace quantities. For example, magnesium and potassium salts and the salts containing sulfate are generally included in the medium if they have not already been included for other purposes. Trace elements such as iron, zinc, manganese, copper, cobalt, and molybdenum are generally not included in fermentation media containing high concentrations of complex ingredients unless they serve specific purposes in metabolism. For example, if product synthesis is known to be carried out by an enzyme complex containing cobalt, this element will be included in the medium at a concentration of a few parts per million to ensure that it is not scarce. When a medium contains low concentrations of complex ingredients, it is important to include a trace element mixture in the fermentation medium.

In fermentations where the ionic strength has to be relatively high, sodium chloride or sodium sulfate is included in the medium. The insoluble salt calcium carbonate is added to prevent the fermentation pH from falling below 6.0. As the pH drops below 6, calcium carbonate dissolves in the medium, raising its pH. Phosphate salts are rarely used for buffering in fermentation media because the phosphorus balance has to be based on the metabolism rather than on the buffering needs. The soluble calcium salts such as calcium chloride and calcium acetate are often used to precipitate out soluble phosphate (in the form of calcium phosphate) from the media of fermentations in which product synthesis is strongly inhibited by phosphate. Minerals also serve as precursors in antibiotic fermentations. In penicillin and cephalosporin fermentations, sufficient sulfate salts have to be included in the medium to supply the sulfur required for the syntheses of these sulfur-containing antibiotics. Similarly, chloride salts must be included in the medium for vancomycin fermentation since the vancomycin molecule contains several chlorine atoms.

46.3.4. Specialty Chemicals

Several types of specialty chemicals are added to large-scale fermentation media. The most important of these chemicals are the defoamers. The defoamers reduce the interfacial surface tension between air and water to facilitate bubble coalescence. In the fermentation industry, silicone and polyol-based defoamers have largely replaced vegetable oils as defoamers. The advantages of the synthetic defoamers are that they are cost-effective and very slowly metabolized and do not have appreciable metabolic side effects. The two most popular defoamers in use in the fermentation industry are polypropylene glycol and silicone emulsion. The defoamers are generally batched with the starting medium. In many fermentations, however, it is necessary to supply defoamer throughout the fermentation cycle to control foam and to control air holdup. Emulsifiers used in fermentations (such as Tween and Span) play a role opposite to that of defoamers. They are added to stabilize small droplets of oily nutrients by increasing the surface tension between oil and water. The small droplets have a dramatically increased surface area and thus allow oily substrates to be more readily utilized by the fermentation organism. Metal-chelating agents such as EDTA are often included in fermentation media. The chelating agents have two diametrically opposed effects. On the one hand, they can tie up metal ions that are toxic to the microorganism. On the other hand, they can prevent the precipitation of a required trace metal by

forming a soluble complex. The availability of the metal to the fermenting microorganism depends upon whether the microorganism can effectively compete with the complexing agent for the required metal.

An important class of specialty products used in the fermentation industry is made up of various enzyme preparations. Crude preparations of enzymes such as amylase, protease, and cellulase are used to precondition the medium. Invariably, these enzymes are used at the mixing stage before medium sterilization. A partial breakdown of the starch of medium components such as corn flour can be achieved by the addition of amylase. The cellulase complex can be used to reduce the viscosity of a medium containing a high concentration of ingredients such as soy or cottonseed flour. Proteases can predigest the medium proteins before sterilization. Enzymatic pretreatment of a fermentation medium thus allows a crude and cheaper raw material to be substituted for a more refined and expensive raw material. Significant efforts are currently under way to investigate the pretreatment of cheaper raw materials to enable their economic fermentation to ethanol or other biofuels.

46.3.5. Sources of Information on Fermentation Raw Materials and Microbial Composition

The best source of information on a given class of fermentation raw material is the industry in which it is generated. Information about such things as the protein, fat, carbohydrate, and mineral contents of various raw materials is readily available from the supplier of the raw materials. However, this information is not necessarily generated for the use of the fermentation industry. It is generated for the benefit of the primary users, which in most cases are the animal feed and food industries. As a result, interpretation of the information for fermentation use is up to the fermentation scientist. For example, while the total nitrogen value of a grain-based product may be meaningful from the point of view of a weight gain calculation when the product is fed to a farm animal, it may not necessarily have the same meaning as the nitrogen available for the fermentation microorganism to grow on. For the same reason, the carbohydrate value provided by the manufacturer of one product may be higher than the value provided for a second product, and yet the second product could have more available carbon for a particular fermentation microorganism. The information provided by the manufacturer is a good approximation for the initial evaluation and for preliminary cost calculations. Actual fermentation experiments are necessary in all cases to justify a change of raw material. In recent years, some of the raw materials suppliers have taken it upon themselves to evaluate their products for various fermentation processes and publish the results in their own manuals or in scientific journals.

A list or database of fermentation raw materials is essential for the fermentation medium designer. Such a database should include medium composition (both molecular and elemental), but could also include such things as price, supplier, and availability. Useful lists of such data include the following.

- Miller and Churchill (12) lists many fermentation raw materials by their trade names along with their applications in various types of fermentation processes.
- Atkinson and Mavituna (1) is a useful source of data for a range of fermentation situations and has a list of compositions of fermentation medium components.

- Solomons (15), an older but classic practical fermentation guide, has a chapter on constituents of fermentation culture media.
- Kennedy and Reader (10) may have general applicability, although this paper describes only raw materials available in New Zealand.

Many such lists are prepared by the suppliers of fermentation media, such as the *Traders' Guide to Fermentation Media Formulation* (20). A similar information booklet relating to soy products, titled *Soy Protein Products in Fermentation*, is supplied by Cargill, Inc. (3). Knowledge of suppliers reveals trends in composition and availability and is particularly useful when it comes to using agricultural by-products for fermentation media.

In the case of some commodity products where there is intense price pressure on raw-material costs, or when the goal of the fermentation is to add value to an otherwise low-value by-product stream, the list of possible medium components is quite constricted. Usually this means incorporating agricultural products in the fermentation medium in a very crude state. Data on agricultural products are best found from the relevant trade organization or from food nutrition databases. Many countries maintain databases of food composition, including the United States, Denmark, Australia, and New Zealand, which profile the main agricultural products of the country concerned. These are usually conveniently Web searchable; see, for example, those provided by the U.S. Department of Agriculture (www.ars.usda.gov/Aboutus/docs.htm?docid=6300) and Plant and Food Research (www.crop.cri.nz/home/products-services/nutrition/foodcompdata/fcd-products/fcd-food-comp-tables.php). Another good source of data on agricultural by-products is the *Feed Industry Red Book* by Goihl and McEllhiney (6).

There are also compilations of media that can be useful for selecting a medium specific for a certain situation. Most of these compilations are laboratory media, rather than industrial fermentation media, and thus limited in use for scale-up. They do, however, provide a useful starting point. Examples of already specified media sources include *Handbook of Culture Media for Food Microbiology* (5), *Difco & BBL Manual: Manual of Microbiological Culture Media* (2), and *The Oxoid Manual of Culture Media, Ingredients and Other Laboratory Services* (14).

Most practitioners of fermentation medium design have favorite compositions for specific microorganisms, so soliciting colleagues for a suggested medium for a particular microorganism is often enlightening.

As discussed earlier in this chapter, the composition of the microorganism (molecular and elemental) is also important in medium design for the purpose of calculating potential biomass and product yield, via mass balance. For accurate data, it is best to measure the composition for the microorganism in question.

46.4. GENERAL CONSIDERATIONS FOR INDUSTRIAL (COMPLEX) MEDIUM DEVELOPMENT OR IMPROVEMENT

46.4.1. Rationale for Improving a Fermentation Medium

Designing an improved fermentation medium can be laborious, expensive, open-ended, and time-consuming, involving a large number of experiments. Consequently, it is important that the effort is justified and, most importantly,

that the target of the optimization is identified. The main reasons the fermentation industry designs improved fermentation media are to (i) improve product yield, (ii) decrease the cost of the medium, (iii) ease product separation, (iv) avoid the microbe making undesired by-products, (v) reduce waste treatment costs, and (vi) improve robustness (the ability of a medium to repeatedly perform well at large scale). It is very important at the start of a medium design campaign to clearly identify which of these reasons is the primary target of the effort.

In a typical industrial antibiotic fermentation, improving product yield always has a larger impact on the overall process cost than does simple medium cost reduction. Very often, yield improvement not only improves the economy of the fermentation process itself but also has beneficial effects on downstream processing. The product-to-impurity ratio increases as the fermentation yield increases, making the recovery process more efficient. In some mature fermentations, productivity improvement beyond a certain level is difficult to attain due to genetic limitations or the inability of the microorganism to tolerate increasingly higher concentration of the product. In such cases, fermentation raw-material cost reduction alone can be the major component of the overall cost reduction efforts. The value of the final product and the volume of the product produced are other important considerations. First, consider the final product value relative to the cost of the raw materials used. In the fermentation industry, the contribution of fermentation raw materials to the overall production cost may vary from as little as 5% (for example, the production of high-value biological agents such as interferon or the production of steroids) to as much as 50% (for example, the production of commodities such as ethanol). The scientist working on the former type of product has much greater flexibility in selecting raw materials, since the overall production cost is not appreciably increased by introduction of a relatively costly raw material. The goal here is to reduce the overall cost by increasing the fermentation yield. In the latter case, however, the incremental cost increase due to the introduction of a new raw material has to be more than compensated by the increase in yield and product quality. The agricultural commodity products and by-products from the brewery and corn wet-milling industries are the typical raw materials used in fermentation processes for low- and medium-value products such as organic acids and well-established antibiotics. On the other hand, exotic raw materials such as refined yeast extracts and exotic growth factors can be cost-effective in fermentation processes of high-value products such as biological peptides. The usage rate of a given raw material and the overall volume of the fermentation broth processed also have to be taken into account for medium development decisions. If an ingredient is used at a few parts per million, its unit cost does not significantly affect the overall process cost. If the volume of the fermentation broth is very large, however, the overall cost may still be significantly affected.

The availability of a given raw material in a given geographic location is another consideration. Should a specific material be shipped long distance, or should the medium formulation be changed so that a readily available material can be used in its place? This depends largely upon how sensitive the fermentation yield is to the type of material used. While a readily available raw material may give a somewhat reduced yield, in the long run it may be more cost-effective to standardize the medium with that material than to depend upon a material that gives higher yield but may be subject to supply disruption. On the large scale, supply disruption is a crucial issue. Access to at least two suppliers of any particular component in the medium is recommended. This recommendation prevents a large disruption should one supplier suddenly no longer supply a given component or go out of business. In the case of some agricultural products, adverse weather can disturb supply. Agricultural products are also susceptible to price fluctuations. A rapid rise in cost of one component may mean its removal from the fermentation and substitution with another. One recent example of such price fluctuations was caused by the rapid rise in the use of biofuels, disturbing supply and prices of agricultural commodities. It is a useful exercise to rank each substrate on a price/kilogram of carbon and price/kilogram of nitrogen basis for comparison. In this way critical prices, where an alternative substrate becomes more cost-effective, can be identified and market prices tracked. A new component validation procedure ensures that new components can quickly be given an end-use test to confirm their performance and robustness.

Other factors to consider are whether the quality of the material will be adversely affected during long-distance shipping and/or prolonged storage. Raw materials such as yeast paste and corn steep liquor are not stable enough for prolonged storage. On the other hand, raw materials with low moisture content such as cottonseed meal, soy flour, and spray-dried yeast are reasonably stable over long periods of storage.

46.4.2. Nature of Fermentation Raw Materials

Most raw materials used in the fermentation industries are not designed for that use. They are generally designed to supply commodities for the food and feed industry. Thus, soy meals, cottonseed meals, and corn gluten meals are designed primarily as animal feed protein sources. Various yeast products are designed for both human food and animal feed applications. Corn syrups of different levels of hydrolysis are made for application in the food-processing industry. Since the fermentation industry is not the primary user of these raw materials, the industry does not have much control over their processing and the resulting quality from the point of view of their use in fermentation processes. Also, agricultural products are subject to variation due to growing seasons, soil conditions, and storage conditions. In short, raw-material variability is the rule rather than the exception. In medium design, then, it is necessary to use multiple sources of the same class of nutrient to reduce process variability. Thus, including two complex nitrogen sources in the medium formulation is more desirable than depending upon a single ingredient. It is also recommended that several lots of the same raw material be tested before settling on a given medium formulation. If the product yield varies excessively due to lot-to-lot variability, it is better to avoid that raw material in the medium formulation altogether. Crude complex raw materials are more likely to exhibit greater variability in composition compared to synthetic medium components.

At this point, it should be noted that water used to prepare fermentation medium is the major component of the medium. In large fermentation plants, this water is usually not distilled or deionized water, as may be the case in the laboratory. As a result, certain metal ions and organic components that come dissolved in the water as impurities become part of the fermentation medium. These impurities and their concentrations may vary on a seasonal basis. In addition, the profile of inorganic and organic components that come with the water may vary when the municipal water treatment plant experiences upsets in its operations. Many fermentation plants use readily available water from adjacent water sources such as lakes, rivers, or deep wells with minimal pretreatment. These water sources are also subject to seasonal variability. Water quality is an important variable when fermentation processes are scaled up from the laboratory, where deionized or distilled water may be used. The water quality is also an important consideration when fermentations involving identical raw materials perform differently at differing physical plant locations. Most fermentation plants monitor the water quality only superficially, and it is seldom known which water quality parameters are important for a given fermentation process.

46.5. GENERAL GUIDELINES FOR FERMENTATION MEDIUM DEVELOPMENT

46.5.1. Seed Medium and Product Synthesis Medium

Generally, the purpose of the seed culture is to grow cells as fast as possible on the basis of predefined criteria such as dissolved-oxygen level, oxygen uptake rate, or centrifuged cell volume. This can be readily achieved by supplying the required nutrient for growth without regard to the product formation needs. When developing a medium for the production stage of a process, the selection of medium components and optimization of their concentrations in the medium are more involved. The objective is not only to develop cell mass but also to synthesize the product at the highest rate possible. The cell density attained, the growth rate during the cell growth, the fermentation time, and the subsequent maintenance metabolism are all important factors in maximizing product formation. The rate of cell growth often can be controlled by controlling the level of readily available nutrients such as glucose, amino acids, and soluble phosphate and by controlling the growth temperature. The slow-growth and maintenance metabolism during the product synthesis phase of the fermentation process is generally controlled by supplying additional nutrients slowly. This controlled nutrient feed is usually composed of glucose or vegetable oil. In some cases, ammonia or complex nitrogen sources are also supplied during this phase. Another way of controlling the slow-growth and maintenance phase of the fermentation process is to include in the medium a carbon (or nitrogen) source that is only gradually utilized by the organism. Often, carbohydrates such as lactose or starch are used for this purpose. The organism being cultivated must produce specific enzymes such as β-galactosidase or amylase to be able to utilize these carbon sources. Various oils are frequently added as the source of carbon that is gradually consumed by microorganisms exhibiting lipase activity. Coarse raw materials such as soybean grits and corn gluten meal are used to supply slow-releasing nitrogen. Enzymes such as cellulase and protease must be induced for the organism to utilize these coarse nitrogen sources. In addition to maintenance nutrients, some secondary-metabolite fermentations require the addition of precursor compounds. For example, the precursors phenylacetic acid and uracil are added to fermentations of the antibiotics penicillin and nikkomycin, respectively.

46.5.2. Using Laboratory Fermentation Medium as the Starting Point

Industrial fermentation organisms are generally highly mutated organisms that are developed in strain development laboratories over many years. The fermentation conditions under which these organisms have been selected must be taken into consideration during medium development work. If a strain has been selected with a laboratory fermentation medium that is based on cottonseed meal as the primary nitrogen source, it may not perform well in a medium based on corn steep liquor as the primary nitrogen source. This is not to say that more closely related medium ingredients such as soy flour and peanut meal may not give yield improvements. The relationship between the laboratory carbon source and the carbon source used in the large-scale fermentation is often not straightforward. Because in shake flask fermentations external pH control is not possible, a readily utilized carbon source such as glucose is very seldom used unless the medium is heavily buffered. Typically, a carbon source such as sucrose, lactose, dextrin, or starch is used to maintain the pH in a reasonable range. When these processes are scaled up, similar complex carbohydrates are initially used in a batch mode. As the process is developed further, however, they are often replaced with an external feed of a readily utilizable carbon source such as glucose, accompanied by pH control. A similar situation can also arise with regard to simple nitrogen sources. Although nitrates or amino acids have to be used in a laboratory fermentation medium for the purpose of pH balancing and slow nitrogen release, they can be replaced in large-scale fermentations with more readily available and cheaper materials such as ammonium sulfate or ammonia, with appropriate control mechanisms. The overall efficiency of nutrient utilization may also change when the fermentation process is scaled up to large fermentors, in which the agitation and aeration conditions are more intense than those in shake flasks. More often than not, the nutrient requirement increases when going from shake flasks to large fermentors.

46.5.3. Considerations of the Fermentation Medium as a Whole

A fermentation medium is typically prepared by dissolving or suspending various raw materials in water. Before the medium is inoculated with the desired microorganism, it is heat sterilized. The batch sterilization involves heating the medium to over 121°C for a period ranging from 30 to 60 min. Continuous sterilization is carried out by rapidly bringing up the temperature to 145 to 155°C and holding it at that temperature for 5 to 10 min. This heat sterilization of a mixture of ingredients in water has a profound effect on the resulting fermentation medium. A number of chemical and physical changes occur during sterilization. Insoluble ingredients such as grain flours and meals are partially solubilized. Macromolecules such as proteins and starch are partly degraded to more soluble and readily metabolizable lower-molecular-weight peptides and oligosaccharides. The inorganic components of the medium react among themselves and with organic components to give new compounds. For example, various metal ions complex with protein molecules to alter protein solubility, organic phosphorus compounds release phosphate into the medium, dissolved phosphorus is precipitated as insoluble metal phosphates, etc. In some cases, the heat sterilization generates toxic chemicals from relatively benign medium ingredients. A well-known example of this toxicity is the Maillard reaction between reducing sugars and amino compounds to give growth-inhibiting amino sugars. To prevent this reaction, reducing sugars such as glucose are sterilized separately from the medium containing amino acids and ammonia. The two components of the medium are mixed after they are cooled to about 40°C. Since various medium components interact during sterilization, it is important to examine the effect of an ingredient being added or removed on the overall chemistry of the medium. The organism may not require calcium salt for growth or for product formation. However, calcium may play a critical role by precipitating out excess phosphate from the medium in the form of insoluble calcium phosphate and allowing a phosphate-regulated product to be synthesized. The elimination of soluble phosphate will also change the medium's buffering capacity. The pH of the medium during sterilization is important because the chemical reactions occurring in an aqueous medium are affected by pH. The pH can have an effect on both the rates of reactions and the equilibrium composition. For this reason, it is generally necessary to experiment with sterilization pH to optimize the performance of the medium under development. It is well known that by manipulating sterilization pH, one can increase or decrease protein solubilization from a medium containing insoluble protein sources such as grain flours and meals.

Sterilization heat damage to a medium can have a significant impact on fermentation medium performance. Unfortunately, heat input to a fermentation medium during sterilization can vary considerably on scale-up, depending on fermentor geometry. One way to quantitatively track the heat input as scale is varied is the use of the del factor (or Ro) calculation (18). Del factor can be used to track variations in heat input between sterilization batches, should a medium be unavoidably sensitive to such damage.

Some components of the medium may have an indirect effect even in the absence of heat sterilization. For example, seemingly inert oils and defoamers may create micelles in the broth that solubilize proteinaceous components and fats that may otherwise be unavailable for metabolism. Some surface-active agents have no metabolic effect but may have substantial effects on the oxygen transfer characteristics of the fermentation broth by changing the surface tension at the air-liquid interface. Many fermentation media containing complex proteins tend to foam heavily during sterilization, and addition of defoamer may be necessary even though the fermentation process itself does not require foam control chemicals. On rare occasions, the order of addition of various ingredients when the medium is prepared, the temperature at which the presterilized medium is prepared, and the length of time the medium is held before sterilization will affect the performance of the fermentation process.

Particle size can have a significant effect on fermentation medium performance. Large particles have a smaller surface area per unit volume than small particles, and this means that they may dissolve or get degraded at a slower rate during the fermentation, affecting fermentation kinetics. Calcium carbonate is particularly susceptible to this effect. Large particles also require a longer time to sterilize, meaning more heat damage to other components in the medium. Solid substrates also interfere with the common DCW cell mass assay, and one of the advantages of a synthetic medium is the ability to conduct meaningful DCW cell mass assays.

Regulatory preferences also play a part in fermentation medium design. Some companies may wish to avoid using all animal-derived products due to the potential contamination by self-replicating proteins referred to as prions. Strong evidence supports prions' being responsible for causing bovine spongiform encephalopathy, a transmissible spongiform encephalopathy, in cattle. Other companies may accept the use of milk-based products, e.g., casein hydrolysate, or fish-derived products, e.g., fish meal.

When designing a fermentation medium, it is worthwhile doing an elemental mass balance, not only to see that sufficient components are present to achieve a desired yield but also to inspect the medium for overdosing. Too high a concentration of some minerals may be toxic. Copper is an example of a mineral of which it is easy to add too much. The exact amount that a microbe will tolerate can be strain specific. Corrosion of equipment may inadvertently add minerals to toxic levels. For this reason, copper pipes contacting the medium are to be avoided. Corrosion may also present an issue in the special case of growing marine microbes. Some marine microbes need high levels of sodium chloride in the fermentation medium, which can provide a significant corrosion challenge for stainless steel equipment.

Often forgotten in fermentation medium design is the fact that oxygen is a metabolic requirement and that carbon dioxide can dissolve in the medium and become inhibitory. While developing a medium at laboratory level, poor results may be due to poor oxygen transfer rather than the components in the medium. Another trap when developing a medium at small scale is the appearance of wall growth. This will invalidate results, as product formation (or lack thereof) in this wall growth is not representative of what will happen in a large-scale fermentor (where wall growth, if present at all, will be only a small fraction of the total fermentation volume). Wall growth on the small scale can be greatly reduced by adding a small amount of a growth dispersion agent for filamentous microorganisms, such as Junlon polyacrylic acid (7).

Because the fermentation medium after sterilization (and hence after the chemical and physical changes have taken place) is the real medium in which the organism of interest is to be grown, it is important to characterize the sterilized medium. Certain overall indices such as soluble nitrogen, reducing-sugar equivalent, and soluble phosphate are often used to characterize the sterilized fermentation medium. However, these indices give only a gross measure of the properties of the medium. Only by understanding the chemical and physical phenomena taking place in the medium during batching, sterilization, and the fermentation process itself can one truly master the art of fermentation medium development.

Last, it is important to ask, when are medium design efforts sufficient? Usually some form of medium design continues during the life of the product due to the introduction of new strains from a strain improvement program, the necessity of further yield improvements, or the replacement of components for some reason. The time to stop is when resources run out or when no changes seem to improve the best medium. It is important not to overinvest in medium design, and one tool useful in assessing this is to plot the number of media tested versus the maximum performance to date. Typically such a curve follows asymptotic behavior quite quickly (especially if statistical experimental design is used), which indicates visually when the point of diminishing returns is reached (11).

Fermentation medium design is both an art form and a logical science, and this is what makes it a challenge.

REFERENCES

1. **Atkinson, K., and F. Mavituna.** 1991. *Biochemical Engineering and Biotechnology Handbook,* 2nd ed. Stockton Press, New York, NY.
2. **Becton, Dickinson and Company.** *Difco & BBL Manual: Manual of Microbiological Culture Media.* Becton, Dickinson and Company, Franklin Lakes, NJ.
3. **Cargill, Inc.** *Soy Protein Products in Fermentation.* Cargill, Inc., Cedar Rapids, IA.
4. **Corbett, K.** 1985. Design, preparation and sterilization of fermentation media, p. 127–139. *In* A. T. Bull and H. Dalton (ed.), *Comprehensive Biotechnology,* vol. 1. *The Principles of Biotechnology: Scientific Fundamentals.* Pergamon Press, Inc., New York, NY.
5. **Corry, J. E. L., G. D. W. Curtis, and R. M. Baird.** 2003. *Handbook of Culture Media for Food Microbiology,* 2nd ed. (Progress in Industrial Microbiology Series). Elsevier, Amsterdam, The Netherlands.
6. **Goihl, J. H., and R. R. McEllhiney.** 1994. *Feed Industry Red Book.* Comm. Marketing, Inc., Eden Prairie, MN.
7. **Hobbs, G., C. M. Frazer, D. C. J. Gardner, J. A. Cullum, and S. G. Oliver.** 1998. Dispersed growth of *Streptomyces* in liquid culture. *Appl. Microbiol. Biotechnol.* **31:**272–277.
8. **Junker, B., J. Zhang, Z. Mann, J. Reddy, and R. Greasham.** 2001. Scale-up studies on a defined medium process for pilot plant production of illicicolin by *Gliocladium roseum. Biotechnol. Prog.* **17:**278–286.
9. **Kennedy, M. J., and D. Krouse.** 1999. Strategies for improving fermentation medium performance: a review. *J. Ind. Microbiol. Biotechnol.* **23:**456–475.
10. **Kennedy, M. J., and S. L. Reader.** 1991. Industrial fermentation substrates available in New Zealand and a strategy for industrial fermentation medium formulation. *Aust. Biotechnol.* **1:**116–120.
11. **Kennedy, M. J., S. L. Reader, and R. J. Davies.** 1994. The kinetics of developing fermentation media. *Proc. Biochem.* **29:**529–534.
12. **Miller, T. L., and B. W. Churchill.** 1986. Substrates for large scale fermentations, p. 122–136. *In* A. L. Demain and N. A. Solomon (ed.), *Manual of Industrial Microbiology and Biotechnology.* American Society for Microbiology, Washington, DC.
13. **Minoda, Y.** 1986. Raw materials for amino acid fermentation—culture medium C—source development. *Prog. Ind. Microbiol.* **24:**51–66.
14. **Oxoid Ltd.** 1982. *The Oxoid Manual of Culture Media, Ingredients and Other Laboratory Services,* 5th ed. Oxoid Ltd., London, United Kingdom.
15. **Solomons, G.** 1969. *Materials and Methods in Fermentation.* Academic Press, New York, NY.
16. **Vardar-Sukan, F.** 1988. Efficiency of natural oils as antifoaming agents in bioprocesses. *J. Chem. Technol. Biotechnol.* **43:**39–47.
17. **Verduyn, C.** 1991. Physiology of yeast in relation to biomass yields. *Antonie Van Leeuwenhoek* **60:**325–353.
18. **Wang, D. I. C., C. L. Cooney, A. L. Demain, P. Dunnill, A. E. Humphrey, and M. D. Lilly.** 1979. *Fermentation and Enzyme Technology.* John Wiley & Sons, New York, NY.
19. **Yang, X. M., L. Xu, and L. Eppstein.** 1992. Production of recombinant human interferon-alpha 1 by *Escherichia coli* using a computer-controlled cultivation process. *J. Biotechnol.* **23:**291–301.
20. **Zabriskie, D. W., W. B. Armiger, D. H. Phillips, and P. A. Albano.** 1999. *Traders' Guide to Fermentation Media Formulation.* Traders' Protein, Memphis, TN.
21. **Zhang, J., and R. Greasham.** 1999. Chemically defined media for commercial fermentations. *Appl. Microbiol. Biotechnol.* **51:**407–421.

Scale-Up of Microbial Fermentation Process

XIAOMING YANG

47.1. INTRODUCTION

47.1.1. Scope

Microbial fermentation processes play a critical role in many important industrial applications. In various applications, fermentation processes are utilized to produce a final product, convert substrates, catalyze reactions, remediate environmentally toxic materials, or simply make mass biological materials. Process scale-up, in a broad sense, is a critical activity that enables a fermentation process achieved in research and development to operate at a commercially viable scale for manufacturing. A successful scale-up involves many aspects of successful preparation and planning beyond pure process scale-up technology. These include setting up clear goals and expectations, timelines and milestones, resources and organization, facility fit considerations, and quality and specifications. This chapter covers a broad scope of issues related to microbial fermentation process scale-up and describes specific steps to plan and execute a scale-up project.

47.1.2. Plan

The first step is to define the project goals, targets, facility, and timeline for completion. Common project goals and targets include the following parameters.

- Annual output of products (in grams or kilograms), based on an analysis of market demand and manufacturing capacity
- Productivity targets (such as grams/day, grams/liter/day, or lot/number of days, etc.), based on process capability and facility capability
- Yield per lot (grams or kilograms/lot), based on process capability, facility scale, and operational capability
- Process cycle time (days/lot), based on throughput, process, and operational constraints
- Cost targets (cost/gram, cost/lot, etc.), based on process capability, product price, market demand, and facility operating cost.

There may be other special targets as well. Ideally, one may have a strong desire to optimize all the targets for maximum benefit of a process. In reality, facility utilization and availability sometimes place a critical constraint on process scale-up planning. A firm may have to use an existing facility that confines how a new process may be scaled up to fit into the facility. In this case, it may not be possible to optimize all of the above targets, and they have to be balanced and prioritized. Setup of the right objectives and targets is a cross-functional activity with business input and technology input.

With clear targets, the next step is to organize a team with all the necessary expertise. A detailed analysis of technical risks and challenges may lead to the appropriate level of resources and a realistic timeline. A project timeline with verifiable milestones is essential at the beginning of a project.

47.2. FERMENTATION PROCESS SCALE-UP

47.2.1. Cell Banking Strategy for Large-Scale Manufacturing

A commercial fermentation process requires a stable cell bank that ensures consistent and predictable production in the long term. A common cell bank strategy for large-scale production is to maintain a master cell bank (MCB), from which a working cell bank (WCB) is generated for routine production. This strategy allows the original source of genetically stable MCB to be preserved for an extensive period of time. Cell banks are usually stored in −70°C liquid nitrogen or a −80°C freezer.

The number of vials in the MCB and WCB depends on the frequency of production and the stability profile of the cell bank. It is common to have hundreds of vials for both cell banks. Many cell banks should be able to last for many years. The volume of each cell bank vial depends on the production inoculation needs. It is common to have 1 ml to tens of milliliters of cells in microbial cell banks.

The cell bank quality and stability need to be adequately tested and monitored. General tests of a cell bank include:

1. Cell recovery upon freeze and thaw: growth test
2. Contamination or purity tests: plating for colony morphology, genetic discrimination tests, phenotype tests, etc.
3. Real-time stability tests: time points for growth, genetic stability, and product expression stability
4. Production qualification: use of a small-scale production model process to check the cell bank's suitability for production
5. A comparability study of the MCB and WCB, recommended to ensure the WCB's performance

47.2.2. Seed Train Scale-Up Strategy

The purpose of a seed train is to propagate cells to a desired mass for inoculation into the production bioreactor. The traditional seed train includes thawing a vial and inoculating into shaker flasks for a certain number of stages with increasing flask sizes, and may include stainless steel reactors. The guiding principle for a seed train scale-up is to minimize the number of stages for the entire seed train, maintain robust growth, and finally prevent any negative impact of extended generations to productivity and product quality.

Typical studies of seed train scale-up may include:

1. *Vial thaw conditions.* This may include temperature and time between thaw and inoculation.
2. *Inoculation ratio.* It is always desirable to have fewer steps for the seed train for simplicity of operations. Therefore, a minimum inoculation ratio (vol%) needs to be defined at every step. For microbes, 0.1 to 5% is commonly used. However, overall process robustness needs to be demonstrated with the inoculation ratios.
3. *Seed train media.* Usually, seed train media are optimized to support growth, rather than product expression, in a batch mode. For operational simplicity, it is recommended to use a simple recipe for the seed train media.
4. *Genetic stability and production stability during the seed train process.* This requires studies with an extended duration, beyond that estimated to be required for generation at the production scale.

In practical operations, a manufacturing plant may carry a parallel seed train as a backup in case the main seed train fails, usually due to contamination. A typical seed train for fermentation is shown in Fig. 1.

47.2.3. Production Fermentation Scale-Up

The essence of scale-up of a fermentation process is to demonstrate fermentation production at large scale resulting in the same productivity and quality as that developed at small scale. One of the outcomes of a process scale-up is to finalize a detailed large-scale process description with settings of all operational parameters and their ranges at scale. The scaled-up fermentation has to be demonstrated with a certain number of runs for consistency and statistical significance. To achieve these goals, one may take the following steps.

1. Identify linear scale-up parameters.
   a. *Temperature.* Temperature should be maintained the same at all scales. Temperatures of microbial processes range widely, mostly from 20 to 45°C. For large scale, consideration should be given to the mixing time, heat transfer rates, and control precision. Challenges

to meet temperature control requirements may be due to limited cooling capacity of a large bioreactor. To maintain temperature, a large-scale bioreactor needs to provide the same heat transfer rate (heat/volume/time) as the small-scale bioreactor does throughout the entire course of a process. Many high-cell-density cultures require extremely high heat removal rates. Good examples for various *Escherichia coli* and yeast fermentations can be found in the literature (1, 5).

   b. *pH.* pH provides an important cell physiological condition. It is to be maintained the same at large scale. Automated pH control needs to be tuned to provide an adequate control regimen for pH. This may be influenced by acid/base concentration and delivery system (pump, pressure, valves, etc.).

   c. *Pressure.* The pressure of a fermentation may remain the same across scales. It may be necessary to adjust the pressure set point from small scale to large scale by taking into account the higher hydrostatic pressure at large scale. Another factor for a potential adjustment arises when a small-scale process is developed at a location with a different elevation, and thus the atmospheric pressure is significantly different from that in the large-scale region. Pressure may also be adjusted as a strategy or tool to help improve gas transfer at larger scales, where a high agitation rate is difficult to reach.

   d. *Dissolved-oxygen (DO) level.* To keep the culture in the same respiration condition, the DO set point is kept consistent across scales. An automated DO control mechanism for a large-scale bioreactor needs to be defined. Commonly, DO is controlled in a cascade mode, in which agitation is increased based on culture oxygen demand and then airflow rate is increased or pure oxygen is introduced after the agitation reaches its maximum. In some cases, pressure may be increased to control DO.

   e. *Airflow rate.* Usually, the airflow rate is scaled up by a constant volumetric flow rate per fermentation broth volume. For a 1-liter fermentation with an airflow rate of 0.2 liter/min, this is 0.2 vvm. At the 100-liter scale, the flow rate would be 20 liter/min to maintain 0.2 vvm. To obtain an accurate airflow rate measurement at any scale, a mass flow device is needed. Due to pressure differences at various scales, the volumetric flow rate readings need to be calibrated via the mass flow measurement. In some cases, successful scale-up is conducted with superficial gas velocity (gas flow per unit cross-sectional area of reactor) kept constant. This approach addresses concerns of gas holdup due to the difference in geometry of fermentors at different scales.

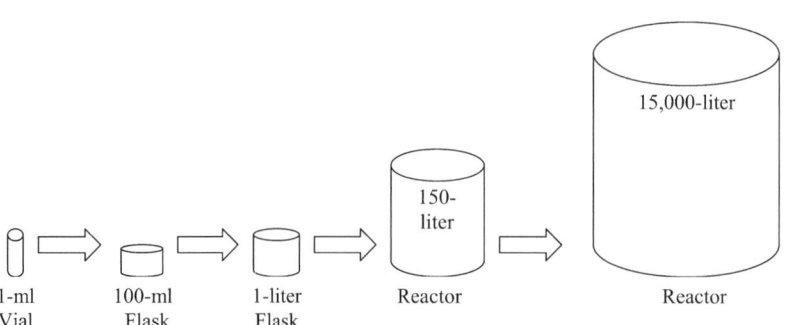

FIGURE 1 A typical microbial seed train.

f. *Final fermentation volume and initial (starting) batch volume.* Usually, the ratio of the final fermentation volume versus the initial batch volume is kept constant in scale-up. The final fermentation volume is determined by the maximum working volume, usually about 70% of the total vessel volume, and with consideration of headspace needed to contain potential foaming and keeping the volume in a well-mixed region.

g. *Nutrient concentrations.* Nutrient concentrations are maintained at large scale. The batch medium volume is scaled up proportionally. For fed-batch processes, nutrient feeding rates can be calculated based on the initial batch medium volume. For example, a feeding rate starts at 5 g/min at 2.5-liter batch volume in a small scale (5-liter) process. When scaling up to 10,000-liter fermentation, the feeding rate should start at 10,000 g/min at 5,000-liter batch volume. Evaporation may occur at any given scale. It needs to be accounted for in the design of nutrient feed rate and concentrations.

2. Identify nonlinear scale-up parameters.
   a. *Agitation.* This parameter is related to mixing time and oxygen transfer. It is described in detail in section 47.2.4.
   b. *Nonlinear feeding rates.* Often glucose and some nutrients are fed exponentially to match the cell growth profile. It is assumed that the exponential feeding of nutrient feed allows the nutrient concentration or uptake rate for each cell to be constant.

3. Perform fermentation preparation scale-up.
   a. *Media preparation.* At large scale, media preparation procedure, vessels, and conditions must be defined. Large-scale media preparation may be a challenge because of the delivery of large amounts of materials, the slow dissolution of ingredients, and the relatively longer hold time prior to sterilization and inoculation. Media solubility and stability should be studied prior to implementation. Sometimes media need to be filtered prior to use.

   Media sterilization methods include heat sterilization and filter sterilization. Generally, components that are sensitive to heat degradation are filter sterilized and components that are not sensitive to heat degradation are heat sterilized for large-scale fermentation. For filter sterilization it is important to consider absorption of critical trace nutrients of fermentation. A small-scale filtration model study is important to determine the impact.

   b. *Media hold.* Certain media will be made ahead of time and be fed during the fermentation process. One or more medium hold vessels and their conditions, such as temperature and mixing, need to be studied and defined.

   c. *Acid and base for pH control.* It is very important to plan ahead to identify the types of acid and base that are suitable for large-scale fermentation. Large-scale acid/base vessels are not always suitable for various types of acid and base used in small-scale studies since acid/base storage vessels have to meet corrosion requirements.

   d. *Medium and acid/base delivery systems.* Usually, automated valves and pressure or pump systems are used for the delivery of nutrient feeds, acid, and base. The control accuracy of the delivery system should match in proportion to the small-scale system to ensure process performance across scale. Sometimes the apparatus needs to be changed or retuned to fit a new process.

47.2.4. Bioreactor/Fermentor Scale-Up

In this section, we discuss specifically the scale-up of the bioreactor operational parameters such as agitation, airflow rate, and cooling capability based on small-scale data. These parameters determine a bioreactor's ability to provide a well-mixed nutrient environment, maintain DO for the organism, and keep a desired temperature for the culture.

To scale up a bioreactor, several criteria are used (2–4):

1. Constant agitation power input (power/volume)
2. Constant oxygen transfer coefficient ($k_L a$) (h^{-1})
3. Constant mixing time (s)
4. Constant agitation impeller tip speed (m/s)
5. Constant heat transfer rate (heat/volume)
6. Constant gas volumetric flow rate (vvm)
7. Constant gas superficial velocity (m/s)

Criteria 1, 2, 5, and/or 6 are most common and successfully applied for large-scale microbial fermentation. Criteria 3, 4, and 7 are useful for special applications that are sensitive to mixing time, shear, and gas flooding, respectively. However, scale-up cannot be done while keeping all of these parameters constant at the same time due to the bioreactor's physical limitation. It is suggested to study and select appropriate scale-up parameters with a series of experiments from small scale to pilot scale.

A fermentor's agitation power input value (power consumption/volume) drives oxygen transfer and mixing time, so it is a simple and practical way to evaluate scale-up (3, 4).

$$P = N_{p0}\rho n^3 d_l^5 \qquad (1)$$

or

$$N_{p0} = P/\rho n^3 d_l^5 \qquad (2)$$

where P is power input (kg m^2 s^{-2} = W), N_{p0} is a dimensionless power number, n is agitation speed, ρ is density of the medium, and d_l is impeller diameter. The dimensionless power number can usually be obtained from standard impeller references and impeller vendors.

Scale-up of microbial fermentation is usually conducted by first assessing if adequate oxygen transfer rate (OTR) can be provided at a large scale. OTR is limited by maximum agitation speed, airflow rate, pressure, and pure oxygen supply.

$$\text{OTR} = k_L a(C_0 - C) \qquad (3)$$

where C_0 (mM) is the saturation concentration of oxygen in fermentation broth, influenced by the oxygen content in the inlet airflow and vessel pressure; C (mM) is the DO concentration of a culture at a given time; and $k_L a$ (h^{-1}) is the oxygen transfer coefficient, which is directly related to the fermentor's configuration, agitation speed, impeller design, and airflow rate. Therefore, $k_L a$ is a good indicator of a fermentor's oxygen transfer capability.

In general, OTRs decrease as fermentor scales increase. The maximum OTRs (in mM O_2/liter/h) are 300 for a bench fermentor, 100 to 150 for a pilot fermentor (>500-liter volume), and 50 to 100 for a large-scale fermentor (>10,000-liter volume) (6). The comparison of these OTR values can be used for scale-up considerations.

When using oxygen gas supplementation, C_0 is enhanced. Therefore, OTR can be proportionally increased according to equation 3. In industry, pure oxygen supplementation is a common approach to increase oxygen transfer capability at large scale.

High-density microbial processes usually generate a significant amount of heat. Cooling capacity is important at large-scale production since the ratio of jacket surface area to volume decreases as processes scale up. Heat transfer rate (HTR) is a measure of the cooling capability of a fermentor.

$$HTR = hA(T - T_j) \qquad (4)$$

where h is the heat transfer coefficient, related to the specific heat transfer efficiency of the jacket material and configuration; A is the vessel jacket area; T is the culture temperature; and T_j is the temperature of coolant inside the jacket.

As indicated by equation 4, low coolant temperature and large jacket area will improve heat transfer capability. To ensure a successful large-scale manufacturing process, HTR data must be collected during process optimization. The maximum heat transfer capacity of a large fermentor system must be evaluated along with the coolant system that supplies the jacket water.

47.2.5. Evaluation of Scale-Up Success

For a fermentation scale-up, the following parameters can be used to evaluate if the scale-up is a success.

1. *Product formation rate.* First of all, the final product titer or yield at large scale should equal that of small-scale studies. In addition, product expression profiles during the course of the fermentation at different scales should trend similarly.
2. *Cell growth profile.* A successful scale-up results in a comparable growth profile between the large-scale fermentation and the small-scale fermentation. In addition, nutrient consumption rates, acid and base addition rates, and important metabolic profiles throughout the fermentation should be comparable.
3. *Operational aspects.* The success of a scale-up is also judged by how efficiently and effectively a new operation team at the receiving facility can execute the scaled-up process, including media preparation, media transfer, and process documents.
4. *Product quality.* Product quality should be examined to confirm a successful scale-up. This may require further purification of the product in order to assess its quality.

Table 1 provides an example of the parameters considered for the scale-up of a 15-liter fermentation to a 15,000-liter fermentation.

TABLE 1 An example of the parameters considered in fermentation scale-up

Parameters	Scale-up ratio	Initial fermentation weight (kg)	
		Small scale (15-liter)	Large scale (15,000-liter)
Start volume		6.8	5,000
Seed flasks			
No. of flasks		1	1
Flask volume (ml)		2,000	2,000
Per-flask medium volume (ml)		500	500
Per-flask inoculum volume (µl)		150	150
Inoculum %		0.03	0.03
Shaker speed (rpm)		275	275
pH		7	7
Temp (°C)		30	30
Time (h)		14	14
Optical density (OD) specification		0.7–1.2	0.7–1.2
Seed fermentation			
Start volume (liters)		—[a]	60
Inoculum volume (ml)		—	500
Inoculum %		—	1
Aeration rate (standard liters/min)		—	30
Pressure (psig)		—	5
DO set point (%)		—	>30
pH		—	7
Temp (°C)		—	30
Time (h)		—	6
OD specification		—	2.0–7.0
Additions			
For pH control		—	

(Continued on next page)

TABLE 1 *(Continued)*

Parameters	Scale-up ratio	Initial fermentation weight (kg)	
		Small scale (15-liter)	Large scale (15,000-liter)
Acid addition (expected) (g)		—	162
Base addition (expected) (g)		—	54
Production fermentation[b]			
Start volume (liters) (after inoculation)		6.8	5,000
Condensate pickup during sterilization (kg)		0.5	0[c]
Inoculum volume (liters)	1.0%	0.07	50.0
Agitation speed (rpm)	Constant k_La	350	125
Aeration rate (standard liters/min)	1 vvm	6.8	5,000
Pressure (psig)		12	10
DO set point (%)	Constant	>40	>40
pH	Constant	7.0	7.0
Temp (°C)	Constant	30.0	30.0
Additions	Density (g/ml)		
Glucose (kg)	1.26	0.09	178.13
Salt1 (kg)	1.10	0.20	382.98
Trace metals solution (kg)	1.00	0.02	29.69
Acid (expected) (kg)	1.00	0.24	457.20
Acid feed (prepared) (kg)	1.00	0.31	593.76
Base (expected) (kg)	1.00	0.13	255.32
Base (prepared) (kg)		0.22	415.63
Antifoam (expected) (kg)	1.00	0.01	24.34
Antifoam (prepared) (kg)		0.14	267.19
Feed 1 (expected) (kg)	1.26	1.46	2,766.93
Feed 1 (prepared) (kg)		1.77	3,354.75
Feed 2 (expected) (kg)	1.14	2.48	4,696.65
Feed 2 (prepared) (kg)		2.77	5,248.85
Inducer (ml) (0.5 mg/ml)		12.5	23,750.4
Volume of components added		4.00	7,595.67
Final volume (expected) (liters)		**10.80**	**12,595.67**
Fermentation feed parameters			
Feed 1 exponential feeding parameters[d]			
Feed 1 density (kg/liter)		1.26	1.26
Start feed 1 (OD)	Constant	6 ± 1	6 ± 1
Initial feed (K1, g/liter/h)	Constant	10	10
Growth rate (K2, h^{-1})	Constant	0.4	0.4
Feed 2 feeding parameters			
Feed 2 density (kg/liter)	1.14		
Start feed 2 (OD)		29 ± 2.0	29 ± 2.0
Feed 2 rate	23.08 g/h/initial wt	156.7 g/h	115.4 kg/h

[a]—, not applicable.
[b]All parameters are scaled from initial fermentation weight.
[c]The volume of "Condensate pickup during sterilization" can vary significantly in large-scale fermentors based on how the sterilization is carried out and the fermentor design. In some instances, this can be a negative number due to evaporation during sterilization.
[d]Feed 1 rate = K1 × EXP(K2 × time).

47.2.6. Product Intermediate Stability and Storage

Product generated during fermentation can be unstable. Studies are needed to define how long and under what conditions the fermentation product intermediate can be held in the broth before it is recovered or purified from the fermentation broth. This also depends largely on the form in which the product is produced, such as secreted, cytoplasmic, soluble, or insoluble. More vulnerable products are susceptible to degradation; therefore, less hold time is possible. It is very common that cooling down of a fermentor broth at the end of fermentation is required to slow down or stop product degradation. For large scale, it is also common to have a harvest hold tank available to allow efficient turnaround of the production fermentor.

47.2.7. Product Quality Considerations

It is not a surprise that scaling up a process may result in product quality differences. Large-scale fermentation may provide a different environment for microbial cells compared to small-scale fermentation. The differences are usually related to nonhomogeneous environments at large scale. At large scale, mixing time, gas distribution, nutrient concentration, and temperature distribution may not precisely match those at small scale. These conditions may affect product formation or transport in microbial cells. Therefore, it is prudent to adequately analyze product quality during process scale-up.

47.2.8. Waste Streams and Treatment

For a given process scale, process waste outputs and treatment requirements must be carefully calculated and considered as a part of a project plan and facility design. This is a requirement by environmental agencies. The waste output can be calculated based on material throughput at each step of the process as detailed in the process description. The required treatment of each waste stream should be evaluated according to regulations. In many cases, waste biological materials or organisms may present potential hazards to the environment; a waste collection tank and a treatment system including chemical or heat treatment often are required for large facilities.

47.3. RAW-MATERIAL SOURCING, HANDLING, AND CONTROL AT LARGE SCALE

For large-scale commercial production, one needs to consider and ensure that all the raw materials can be sourced reliably and with consistent quality. An important study for scale-up is to assess raw-material lot-to-lot variability. Fermentation processes often use raw materials that are less defined and tend to vary in nutrient content from lot to lot and from time to time. A quality or performance test plan for raw materials used in fermentation is highly recommended.

At large scale, large amounts of bulk raw materials are expected. Their transportation, delivery, handling, and storage are to be considered and planned prior to scale-up.

47.4. EQUIPMENT QUALIFICATION AND MODIFICATIONS

Scale-up projects often require implementation of new equipment or modification of existing equipment. This can be time-consuming and costly; however, it is essential. This activity may follow these steps to ensure a satisfactory outcome.

1. Define user requirements early.
2. Draft engineering drawings for functional review.
3. Monitor fabrication timeline.
4. Conduct factory acceptance testing before shipping to installation site.
5. Implement installation plan and wet testing.
6. Review documentation for completeness.

Equipment preparation is critical to ensure the overall success of scale-up.

47.5. SPECIAL CONSIDERATIONS FOR BIOPHARMACEUTICAL PROCESS SCALE-UP

In the biopharmaceutical industry, regulations are much more detailed and strictly enforced to ensure public safety and product efficacy. Therefore, special considerations are given during scale-up of a biological product fermentation. The following sections briefly describe a few major regulatory requirements applied to biological process scale-up.

47.5.1. Equipment Validation

Equipment validation is to document that equipment meets all predefined functionality criteria per process requirements with a high assurance. It is conducted in three stages: installation qualification, operational qualification, and performance qualification. Installation qualification documents how the equipment is fabricated and installed and whether the equipment meets the design specifications. Operational qualification documents satisfactory operational parameters per specification. It usually uses water instead of actual product for the testing. Performance qualification is an exercise that uses actual product to demonstrate performance as required. All the requirements need to be predefined and documented before any execution of operation and testing. In addition, a meaningful number of repeats have to be performed to validate the equipment.

47.5.2. Documentation

Unlike many other industries, the biopharmaceutical industry emphasizes documentation of detailed information about process, materials, analytical procedures, equipment, and staff training. For fermentation process scale-up, documentation may include process descriptions at small scale and large scale, material qualification and testing, analytical methods qualification, process intermediate stability, process hold steps, and scale-up production campaign summaries.

47.5.3. Process Validation

A unique requirement for commercial biopharmaceutical process scale-up is process validation. Process validation usually specifies a set of predefined criteria for each process step based on the purpose of the step, as well as for the final product quality. This is documented in a process validation plan or protocol. Three to five successful runs of the process need to be performed, and the results must meet all the predefined criteria. All important process parameters and product quality attributes of the validation runs will be discussed, analyzed, and then documented in a validation campaign report as a proof of process validation (see the guidelines of the International Conference on Harmonisation of Technical Requirements for Registration of Pharmaceuticals for Human Use at www.ich.org).

47.5.4. Product Comparability

For pharmaceutical applications, product quality attributes must be comparable at all scales from development to commercial. Biological product quality attributes include product purity, impurities such as aggregates, clipped product and oxidized product, process-related impurities such as process chemicals, and contaminants such as other microbes. For fermentation scale-up, samples are purified for product quality analysis. The product quality attributes must be analyzed and compared to those of the small-scale process in order to determine if product is comparable.

47.6. CONCLUSION

With the tremendous diversity of organisms and products in industrial fermentation, a precise and one-size-fits-all strategy for scale-up is not practical. However, the general principles of scale-up of microbial fermentation as described in this chapter are well established and widely practiced in industry with reasonable success. For each individual product and organism, there will always be a specific model weighted by critical parameters based on experimental knowledge.

Scale-up activity is complex, including cell biological factors, nutrient and metabolic factors, and processing-related factors. In addition, its success is highly dependent on product quality requirements and project emphasis. It is true that not all parameters and goals of a fermentation process are equal. Therefore, approaches and models in scale-up may vary due to project emphasis. The keys for scale-up are to achieve project goals, produce desired product with

acceptable quality and yield, and fully understand and control operational risks at large-scale manufacturing.

I thank Swapnil Bhargava, Mark Berge, and Henry Lin, who provided initial review and suggestions. Thomas Seewoester and Sam Guhan are appreciated for their time in critical review and valuable comments. Beth Junker gave me many good suggestions from the beginning of this chapter planning.

REFERENCES

1. **Berge, M., S. Bhargava, R. Georgescu, and X. Yang.** 2008. Leveraging fermentation heat transfer data to better understand metabolic activity. *BioPharm Int.* **21:**2–7.
2. **Garcia-Ochoa, F., and E. Gomez.** 2009. Bioreactor scale-up and oxygen transfer rate in microbial processes: an overview. *Biotechnol. Adv.* **27:**153–176.
3. **Junker, B. H.** 2004. Scale-up methodologies for *Escherichia coli* and yeast fermentation process. *J. Biosci. Bioeng.* **97:**347–364.
4. **Schmidt, F. R.** 2005. Optimization and scale up of industrial fermentation processes. *Appl. Microbiol. Biotechnol.* **68:**425–435.
5. **Yang, X.** 2008. Large-scale microbial production technology for human therapeutic products, p. 329–362. *In* E. S. Langer (ed.), *Advances in Large-Scale Biopharmaceutical Manufacturing and Scale-Up Production,* 2nd ed. ASM Press and BioPlan Associates, Washington, DC.
6. **Yang, X., and L. Cui.** 2004. Microbial process development for biopharmaceuticals. *BioProcess Int.* **2:**32–39.

Cell Culture Bioreactors: Controls, Measurements, and Scale-Down Model

WAN-SEOP KIM

48

48.1. INTRODUCTION

Cell culture is an animal cell cultivation process to produce a protein of interest from genetically engineered cells in various culture vessels. Cells are cultivated under the optimized conditions for process parameters such as temperature, pH, and dissolved oxygen (DO). The protein of interest is produced into the extracellular culture media for the subsequent purification process step. A culture vessel can be a small (<100-ml) culture flask up to a large-scale (25,000-liter) vessel known as a bioreactor. In this chapter, types of bioreactors, bioreactor controls for process parameters, measurement of cell growth and metabolites, and a scale-down model using small- and pilot-scale culture vessels used in the biopharmaceutical industry are described.

48.2. TYPES OF BIOREACTORS

The term "bioreactor" is used for the cell culture vessel which cultivates animal cells to produce biological products, compared to a chemical reactor, which controls chemical reactions. The term "fermentor" is also often used, but it is more appropriate to describe the microbial (i.e., bacteria or yeast) fermentation vessel. The major types of bioreactors used for cell culture processes in the biopharmaceutical industry today are stirred and airlift bioreactors. Recently, disposable bioreactors have been utilized for their convenience. Each type of bioreactor can be used for batch, fed-batch, and continuous cultures. Table 1 provides an overview of each of the bioreactor types used for the cell culture process.

48.2.1. Stirred Bioreactor

The stirred bioreactor is the most widely used at the small (1- to 20-liter), pilot (50- to 1,000-liter), and large manufacturing (up to 25,000-liter) scales. There are no universal definitions of sizes for small, pilot, and large manufacturing scales. The stated scale ranges are representative scales that are generally used in the industry. The operation of the stirred bioreactor is relatively simple; cells and nutrient components are mixed by agitation and oxygen is supplied for aeration.

Most cell culture processes in industry today are developed with small- and pilot-scale stirred bioreactors. Because of the popularity of stirred bioreactors in process development and in-depth experience in the operation of large-scale stirred fermentors, most large manufacturing facilities for cell culture processes use large-scale stirred bioreactors for commercial production runs.

The advantage of using the same type of bioreactor for small-scale through large-scale runs is that new cell culture processes can be more easily scaled up for commercial manufacturing. Also, large-scale manufacturing processes can be scaled down for investigation of cell growth, productivity, and quality-of-product issues in small-scale model systems.

A major concern with stirred bioreactors is the mechanical-shear effect on fragile cells caused by a high agitation rate. Especially in large-scale bioreactor runs, a proper agitation rate should be used to mix cells, nutrient components, and oxygen adequately without locally insufficient mixing. Use of low-shear impellers can reduce the shear damage of cells at a high agitation rate. In addition, recent developments in robust cell lines and new culture media containing a shear-protective agent will allow a higher agitation rate to be used to improve mixing of cells, nutrient components, and DO in the large-scale runs without damaging cells.

The second concern is microbial contamination by leaking at the mechanical agitator area. Since the agitator is a moving part that is operated with a motor external to the bioreactor, microbial contamination by leaking should be well prevented. The leaking can be avoided by using a double mechanical seal with a sterile barrier fluid such as clean steam condensate.

48.2.2. Airlift Bioreactor

The airlift bioreactor is the second choice of bioreactor for cell culture processes. The airlift system uses aeration from the bottom to mix cells and nutrient components instead of using an agitator as in the stirred bioreactor.

The advantages of the airlift bioreactor are that it does not produce mechanical-shear effects by the agitator and is a much more closed system, which results in a lower chance of contamination due to the lack of agitator moving parts. However, the airlift system needs a much higher aeration rate and pressure to avoid locally insufficient mixing within the whole bioreactor system.

Process development areas and pilot plants are typically equipped with small-scale stirred bioreactors. If large-scale airlift bioreactors are planned to be installed, small- and pilot-scale airlift bioreactors would be additionally required for new process development and scale-down support of large-scale manufacturing runs.

TABLE 1 Types of bioreactors

Type of bioreactor	Type of mixing	Use of bioreactor	Advantages and disadvantages
Stirred bioreactor	Agitation by impeller	Widely used in small-scale through large-scale (~25,000-liter) manufacturing	Well-known mixing technology Widely used at different scales High shear damage to cells—can be reduced by a low-shear impeller, robust cell line, or use of a shear-protective agent in the culture media Contamination due to agitator seal failure
Airlift bioreactor	Bottom airflow without agitation by impeller	Used much less than stirred bioreactors at manufacturing scale	No shear damage by agitation Closed system with no moving parts High aeration rate and pressure required Insufficient mixing may occur locally
Disposable bioreactor	Mostly rocking or shaking; magnetic agitation	Used for seed culture and small- and pilot-scale production runs	Ease of use and installation—no cleaning and sterilization required Closed system Currently available at up to 2,000-liter volume Low installation cost, but high operating cost

48.2.3. Disposable Bioreactor

A disposable bioreactor is another type of bioreactor that uses a sterilized disposable bag. The stirred bioreactor system using the disposable bag has recently become available in up to 2,000-liter volumes from Xcellerex (Marlborough, MA). The disposable bag is placed inside the vessel and the internal mixing of the bag is performed by a magnetic drive agitation. The pH, DO, and temperature controls of disposable bioreactors are similar to those of conventional stirred bioreactors.

Another disposable bioreactor called the Wave bioreactor is available from GE Healthcare (Piscataway, NJ). The mixing in the Wave bioreactor is typically performed by rocking or shaking the disposable bag on a stainless steel platform. The platform is used to hold and shake the bag and also to control temperature. Air and carbon dioxide gases are continuously supplied to the bag headspace to control the DO concentration and pH of the culture, respectively. The disposable bag is maintained as an inflated bag to create a large air-liquid interface for gas transfer.

The use of a disposable bag eliminates the requirement for bioreactor cleaning and sterilization procedures before and after runs, which creates much simpler product changeover procedures.

Since disposable bioreactors are available at scales of up to only about 2,000 liters, the disposable bioreactor is being used mostly as a seed bioreactor in large-scale manufacturing instead of the production bioreactor step. However, considering the advantages of disposable bags for convenience of use and the trend of higher productivity in recent cell culture process development, the use of the disposable bioreactor in pilot-scale production runs will significantly increase in the next 10 years.

48.3. BIOREACTOR CONTROL

Bioreactor setup and control of process parameters are the major factors in cell culture processes that consistently control cell growth and produce products according to their predefined quality attributes.

The control of process parameters and the associated concerns described below are based on the use of the most widely used, stirred bioreactors. Table 2 provides an overview of the controls of process parameters and concerns. While the control of process parameters is described in detail, bioreactor setup (i.e., cleaning-in-place [CIP], leak test, and sterilization-in-place [SIP]), culture medium addition, and inoculation are only briefly described here because they are well-known, common processes.

48.3.1. Setup and Operation of Bioreactors

Bioreactor operation involves multiple steps of operations as follows.

1. Bioreactor setup
   a. Preuse CIP, leak test, and preuse SIP
   b. Standardization of pH and DO probes
2. Addition of culture medium and preparation of feed medium
3. Inoculation of seed cells and maintenance of control parameters: pH, DO, temperature, etc.
4. Sampling and culture maintenance with feeding and controlling probes
5. End of the run and postuse CIP

48.3.1.1. Setup, Media Addition, and Inoculation

CIP is a cleaning method for the internal area of bioreactors and pipes without disassembly to remove cell debris, culture media, remains of product, and impurities and residues of cleaning agents. Generally, it is automatically performed for the cleaning of pilot- and large-scale bioreactors before and after the runs.

A leak test is performed to ensure that the bioreactors maintain sterilized condition during the cultivation. Leakage is evaluated by measuring the pressure drop during a certain time interval after pressurization or by detecting helium gas leak after helium gas is supplied to the bioreactor.

SIP is a process to sterilize bioreactors by using steam prior to the addition of culture medium to cultivate cells of interest only. After the SIP is completed, culture medium is added into the bioreactor vessel by 0.2- and 0.1-μm-pore-size filtration. Prior to the filtration, the culture medium can be treated with a high-temperature, short-time sterilizer for viral inactivation. The temperature used is typically about 100°C for about 10 s. Temperature, agitation, aeration, and pH

TABLE 2 Controls of process parameters

Control parameter	Range of control	Method of control	Comments
Temperature	32–37°C in fixed or shift modes	Steam and cool system	Failure of gaskets and control valves if preventive maintenance is not adequate
pH	pH 6.8–7.6 in fixed or shift modes; pH 6.9–7.2 commonly used	CO_2 gas and caustic chemical	The performance of pH probes should be verified with offline pH measurements
			A dead band can be considered if better pH control is needed, but it alters the control pH of culture if its range is wide
DO and aeration	DO 10–50% saturation; DO 25–40% commonly used	Overlaying air, sparging air, and pure oxygen gas	A proper strategy for the use of air-sparging rate and pure oxygen needs to be defined in high-oxygen-demanding cultures
			Internal bioreactor pressure and hydrostatic pressure should be considered for the standardization of the DO probes in pilot- and large-scale runs
			A foaming issue with a high sparging aeration rate
Agitation	Variable per scale	Agitation with Rushton, pitch-, or marine-type blades	Shear damage of cells by a high agitation rate
			Low agitation can reduce bulk mixing and affect pH, temperature, and dissolved gas control
Dissolved carbon dioxide	Less than 200 mm Hg; maximum of 150 mm Hg commonly used	Aeration and agitation	High dissolved carbon dioxide observed in large-scale runs; proper reduction of carbon dioxide is necessary

controls start after culture medium is added, and then cells from a seed culture are transferred to inoculate the culture medium for seed culture expansion or production runs.

48.3.2. Control of Process Parameters

48.3.2.1. pH
The control of pH is one of the most critical process controls during cell culture. It significantly affects cell growth, productivity, and quality of product, e.g., glycosylation (1, 3). The most common pH range in cell culture is 6.8 to 7.6, but typically the production process requires a narrower pH control range (6.9 to 7.2) for consistent quality and productivity. The control of pH is usually achieved by regulating CO_2 gas and base addition through a proportional-integral-derivative (PID) controller with feedback from the pH probe. Addition of CO_2 gas to the culture decreases the pH level of culture media, and base addition (e.g., sodium hydroxide) raises the pH.

While a small-scale bioreactor usually uses a single pH probe to maintain a target pH level for the entire culture duration, pilot- and large-scale manufacturing bioreactors typically utilize two pH probes to properly control the culture's pH level. One of the two pH probes is assigned as a control probe for the active control of culture pH, and the other probe serves as a backup probe. If the control pH probe shows continuous pH drifts or continuously significant differences compared to offline pH measurements, the control probe is replaced with the backup probe to ensure acceptable pH control.

During the cultivation of cells, the readings of two probes are monitored and compared with offline pH measurements made using a benchtop pH meter or a biochemical analyzer. If the pH difference between pH probe readings and offline pH measurements is out of the predetermined acceptable range, a 1-point pH correction is performed. A maximum pH difference of 0.1 is generally used to determine if the 1-point correction needs to be carried out.

The set point of pH can be controlled with or without a dead band. Due to the significant impact of pH on productivity and quality of product, the pH in a cell culture process should be tightly controlled. A dead band is not required or can be applied in a narrow range to control the pH at the target value. If pH perturbations are observed by adding excess CO_2 gas or base, the pH controller is not properly tuned or the agitation rate is not sufficient to mix the culture well. The pH control can be improved by widening the dead band or raising the agitation rate. However, widening of the dead band changes the actual control point of pH and may adversely affect productivity and quality of product. When a dead band for pH is applied to control pH at the beginning of culture, the pH is maintained at the upper limit of the pH set range with the dead band because aeration to a bioreactor increases the pH of culture by reducing dissolved carbon dioxide in the culture. So, CO_2 gas is mostly used to reduce and control the culture's pH. After cells start growing and produce lactate and carbon dioxide, the pH level drops and is maintained at the bottom limit of the pH set range with the dead band, primarily by use of base.

Different pH set points or pH shifts are often used during the cultivation of cells. The pH set point can be manually or automatically changed at a certain elapsed culture time. Alternatively, a dead band can be used instead of changing the pH set point if the set point needs to be shifted from high to low pH values.

48.3.2.2. DO and Aeration
DO concentration in the culture also significantly affects cell growth, productivity, and quality of product (8, 13, 20). A DO set point can be considered in the range of 20 to 50% air saturation. The most common DO set point is within 25 to 40% air saturation. Oxygen is supplied primarily by overlaying and sparging clean air from the top and bottom of the bioreactor. Additional pure oxygen or oxygen-enriched air is required to provide sufficient oxygen to the cells in high-density cell cultures.

Pilot- and large-scale manufacturing bioreactors typically use two DO probes to control the DO level at the set point. As described above for the use of two pH probes, one of the two DO probes is used to control the DO level in the culture and the other is to back up the control probe.

Standardization of the DO probes is usually performed by supplying nitrogen gas and clean air to the bioreactor. A DO probe in a small-scale bioreactor is usually standardized with clean air for the 100% level, without considering bioreactor pressure and hydrostatic pressure as is done for pilot- and large-scale bioreactor runs. The consideration of bioreactor and hydrostatic pressures during DO probe calibration allows for the maintenance of the DO concentration at the same level as small-scale bioreactors can maintain.

Air overlaying at the top of the culture medium replaces dissolved carbon dioxide in the culture with oxygen and also maintains the internal positive pressure of the bioreactor after completion of SIP. Air sparging from the bottom of the bioreactor also replaces dissolved carbon dioxide in the culture with oxygen, and is provided to the culture gradually or at a fixed rate until the end of the culture. The maximum aeration rate during the scale-up of culture is generally determined to maintain a constant volume of gas per volume of liquid per minute (vvm).

A gradual increase of the air-sparging rate, as the number of viable cells increases, efficiently provides oxygen to maintain the DO level at the set point. However, the gradual increase of the air-sparging rate itself at the peak time of the oxygen requirements may not provide sufficient oxygen to high-density cell cultures. When very high air-sparging rates are required, excess foaming could be observed. These issues can be addressed by sparging pure oxygen gas. A major disadvantage of a high fixed rate of air sparging at the early stage of the culture is that pH is increased by lowering dissolved carbon dioxide and then CO_2 gas is supplied, which results in pH fluctuations by both excess CO_2 gas and excess base. Foaming is also observed due to both excessively high aeration rate and CO_2 gas supply.

Therefore, the initial sparging aeration rate should be low at the beginning of culture to balance dissolved carbon dioxide concentration in the culture medium, and then the sparging aeration rate can be increased gradually or in stepwise fashion as required by cells to maintain a DO level at the set point. To avoid excessively high aeration rates, the maximum airflow rate can be set, and pure oxygen can be additionally supplemented if the aeration rate reaches a maximum set point and the DO set point cannot be achieved.

The use of antifoam negatively affects the oxygen transfer rate of the culture medium (12). Thus, in cases where antifoam is used, a higher oxygen concentration in air/oxygen gas and a lower DO set point (higher driving force) are required to transfer oxygen sufficiently to the cells. A high agitation rate improves the oxygen transfer rate and also minimizes the occurrence of local anaerobic conditions in the large-scale bioreactor. Due to shear damage of cells, a high agitation rate is not recommended, but recent trends including robust cell line development and use of shear-protective agents (15, 16) such as Pluronic F-68 allow the use of higher agitation rates for better mixing of oxygen and nutrient components.

48.3.2.3. Agitation

Agitation rate and impeller type should be well studied with small-scale and pilot-scale bioreactors at the process development stage so as to successfully scale up the cell culture process to the large-scale manufacturing runs (18).

Rushton impellers are often considered for microbial fermentation and not for cell culture processes due to their high shear impact on cells. Pitch-blade and marine-blade impellers are used for shear-sensitive cell culture to mix cells and nutrient components in axial and radial directions with lower shear rates. Recent robust cell line development allows for the use of any Rushton, pitch-, or marine-blade impellers, which are already installed in the bioreactor, to provide sufficient mixing of cells and nutrient components.

A low agitation rate may cause insufficient mixing of culture at the pilot- and large-scale manufacturing bioreactor runs, which results in locally uncontrolled pH, temperature, DO, dissolved carbon dioxide, and nutrient levels. A higher agitation rate improves mixing but generates a high shear rate that may damage cells.

After an agitation rate in small-scale process development is defined, scale-up of the agitation rate for pilot- and large-scale manufacturing bioreactors is determined by scale-up parameters: power per volume, volumetric oxygen transfer coefficient, tip speed, mixing time, etc. (18). Regardless of what scale-up parameter is used, the most important aspect of scale-up of the agitation rate is to achieve a similar metabolic condition of cells in culture at small scale through large-scale manufacturing runs. Measurement of metabolites during the culture is one of the most common ways to understand the metabolic condition of cells.

48.3.2.4. Dissolved Carbon Dioxide

Accumulation of high levels of dissolved carbon dioxide (>200 mm Hg) is one of the major concerns in large-scale cell culture because it is known to negatively affect cell growth, productivity, and quality of product (6, 17). In pilot- and large-scale runs, dissolved carbon dioxide levels of less than 150 mm Hg are generally maintained.

In culture medium containing bicarbonate, high concentrations of dissolved carbon dioxide can be reduced to the predetermined level (usually about 100 mm Hg) by gassing air from the top of the bioreactor. As the number of cells increases, production of carbon dioxide by cells increases and the dissolved carbon dioxide concentration can be elevated due to insufficient gas exchange in large-scale bioreactor runs. Increase of the agitation rate or air-sparging rate can reduce the level of dissolved carbon dioxide in the culture. It is not recommended to use very high air-sparging levels, which generate excess foaming and pH fluctuations by excess removal of dissolved carbon dioxide. During scale-up/scale-down of the cell culture process, the level of carbon dioxide is one of the indicators for proper agitation and aeration.

48.3.2.5. Temperature

It is also known that temperature significantly affects cell growth and product quality (11, 22). The temperature range of 36 to 37°C is used for the optimum growth of Chinese hamster ovary (CHO) cells and NS0 cells. Lower temperatures down to 32°C are often applied during the late stage of culture. A low temperature extends the duration of culture by maintaining higher viability of cells until the end of culture and also reduces enzymatic activities that negatively affect the quality of glycoproteins. The impact of temperature shift is dependent on the processes, cell lines, and products.

Temperature of culture is maintained by steam and cooling systems in pilot- and manufacturing-scale bioreactors. Small-scale bioreactors do not use steam, and the temperature is maintained with an electric heater or hot

TABLE 3   Cell growth and metabolite measurement

Parameter	Method of measurement	Comments
Cell growth: VCD and viability	Manual cell counting	Counting variation due to cell clumping—good mixing is needed
	Automatic cell counting	Selection of viable cells in the image with an automatic cell counter should be optimized
	Offline automatic cell counter	
	In situ optical light absorption probe	The optical light probe measures all cell mass including cell debris
	In situ RFI probe	The RFI probe measures mass of viable cells only
		In situ probes for cell growth measurement cannot measure viability
Metabolites	Offline biochemical analyzer	A biochemical analyzer is relatively reliable and widely used due to its convenience
	In situ probes	In situ probes are being tested widely for future use in cell culture processes and for PAT application
	In situ probe for carbon dioxide measurement	
	In situ near-infrared region probe	

jacket. It is necessary to equip an efficient cooling system to maintain low temperature for large-scale bioreactors and harvest tanks if the quality of product is affected during harvest duration. Steam gaskets should be well maintained to avoid temperature control excursions by their malfunction. SIP procedures and the use of steam for temperature control can damage the gaskets or generate residues that cause the gaskets to fail. The interval for preventive maintenance programs should be carefully determined to prevent failure of the gaskets.

48.4.   MEASUREMENT OF CELL GROWTH AND METABOLITES

An overview of cell growth and metabolite measurements is presented in Table 3. Since changes in cell growth may alter productivity and quality of product, cell growth in cell culture processes is monitored daily or at predetermined intervals during the entire culture duration. Sampling at regular intervals is performed to measure viable cells and viability as well as metabolites such as glucose, lactate,

ammonia, and glutamine. In fed-batch cultures, viable cell density (VCD) is often used to determine feeding amount and other process parameter changes.

VCD is generally expressed as viable cell numbers per unit volume. A peak VCD is typically observed at the range of 5×10^6 to 12×10^6 cells/ml in commercial manufacturing runs. Viability is determined by dividing viable cells by total cells and is often used to determine harvest time since it indicates the metabolic status of cells and degree of debris that may affect the harvest step. Integral viable cell density (IVCD) and growth rate are also used to monitor cell growth. IVCD is the integral of VCD over time during the entire culture duration. The increase of IVCD may indicate increased product titer. Specific productivity is calculated by dividing product titer by the IVCD, which represents the product amount produced per cell. Growth rate of cells is the slope of cell density over time. Changes in the control of process parameters such as pH and temperature affect the cell growth rate, which can be observed by cell growth measurement. Representative cell growth (VCD and IVCD) and titer curves are shown in Fig. 1.

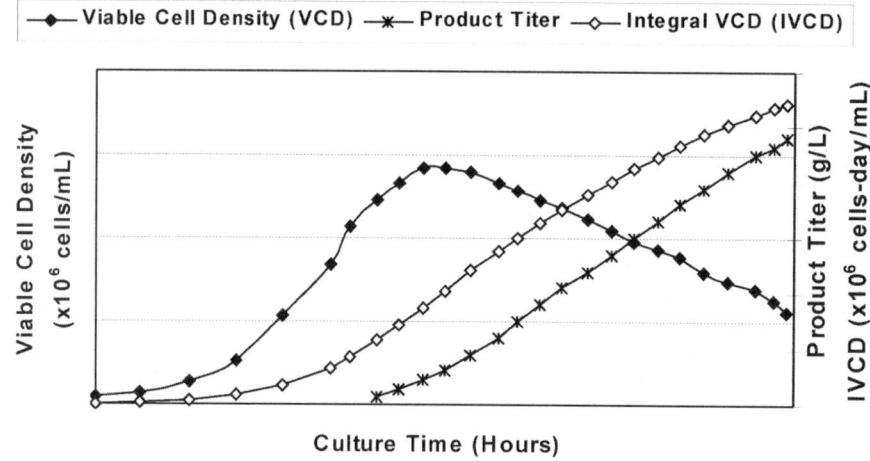

FIGURE 1   Representative cell growth (VCD and IVCD) and product titer profiles.

Metabolite levels are often used to calculate the amount of feeding in fed-batch processes and also to determine the shift time of process parameters such as pH and temperature. While glucose and glutamine are used as energy resources to produce cell mass, lactate and ammonia are produced as by-products from glucose and other nutrients and negatively affect cell growth and productivity (14). Generally, the by-products are more severely accumulated in fed-batch cultures with extended culture duration in large-scale bioreactors. When mixing is optimized, lactate level is typically increased as cells grow exponentially and then drops or remains constant during the stationary and death phases of culture. A continuous increase of lactate level is often observed until the end of culture when the process is improperly scaled up to the large-scale bioreactor runs. Considering that lactate production occurs owing to local anaerobic conditions, increasing agitation rate and/or air sparging helps to reduce the accumulation of lactate. Also, feeding medium composition and feeding strategy of the fed-batch culture should be optimized to reduce the lactate level. Maintaining glucose at a lower level will also help to reduce lactate accumulation. Representative glucose and lactate curves are shown in Fig. 2. Similar levels of by-products should be observed at small scale through large-scale manufacturing runs if the process is properly scaled up or down. Slightly higher lactate levels can be observed as the culture scale increases.

Since amino acid concentrations in culture media also affect cell growth, productivity, and quality of product, it may be necessary to measure amino acid levels in culture media to maintain cell growth and productivity consistently during the culture (5). Proper concentrations of amino acids can be maintained by feeding or further development of basal and feeding medium compositions.

48.4.1. VCD and Viability Measurement

The number of viable cells during the cultivation is measured by manual cell counting or automatic cell counters. The Vi-CELL™ cell viability analyzer from Beckman Coulter (Fullerton, CA) and the Cedex counter from Innovatis (Malvern, PA) are widely used for automatic cell counting.

A trypan blue exclusion method is generally used for both manual and automatic cell counting. The trypan blue exclusion method is used to determine the numbers of viable and dead cells by counting white (viable) and blue (dead) cells. The trypan blue dye does not penetrate intact cell membranes of viable cells, which will remain white under a microscope. However, the trypan blue dye passes the damaged cell membranes of dead cells and stains the cytoplasm of dead cells in blue. Classical manual cell counting using a hemacytometer is still widely used in manufacturing runs, but recently mechanical, automatic cell counters have been more often applied to newly developed processes and current manufacturing processes. Due to varying operator techniques and cell clumping issues, count variations are often observed in manual cell counting. It is important to mix culture samples with a trypan blue dye sufficiently prior to cell counting in order to separate cells from the cell clumps and stain the individual cells with the dye. The automatic cell counter mechanically mixes cells with the trypan blue dye very well to separate and stain cells in the clumps. Cell numbers and viability in the automatic cell counter are affected by incorrectly counting debris and particles that meet predetermined photometric parameter ranges such as size, brightness, sharpness, and circularity of particle. The optimum range of each parameter in the automatic cell counter should be well determined to distinguish viable cells from dead cells and to avoid the counting of debris, which is mostly observed in low-viability cultures. Thus, the optimization of each parameter for each cell culture process and cell line should be performed prior to use in commercial runs. The default settings provided from the manufacturer can be the starting point for optimization. Considering convenience of use and reliability after the parameter settings are optimized, the automatic cell counter has recently become the first choice for cell counting in cell culture.

There are also in situ probes available for cell mass measurement. Total cell mass can be monitored by classical optical measurement using light absorption. It is widely used for microbial fermentation but not appropriate for cell culture because the probe does not distinguish between the viable cells and dead cells. Also, there is inaccuracy of measurement because the probe recognizes cell debris as cell mass, which is mostly observed in low-viability cell cultures. Another in situ probe, using a radio frequency

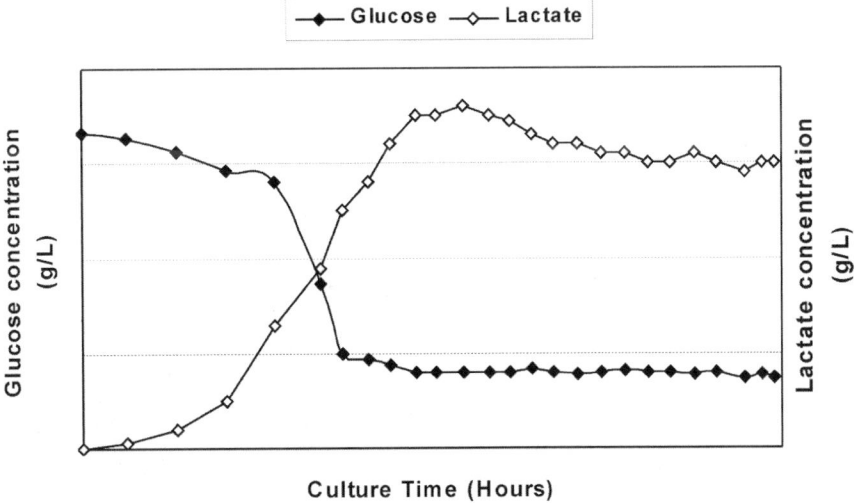

FIGURE 2 Representative glucose and lactate profiles.

impedance (RFI) method, is available for microbial fermentation and cell culture processes (4). However, this in situ probe can measure viable cell mass only. As with the classical optical measurement probe, the use of the RFI in situ probe should be limited due to lack of viability measurement.

48.4.2. Metabolite Measurement

Metabolites (glucose, glutamine, lactate, ammonia, etc.) are usually measured by sampling and analysis with offline biochemical analyzers during cell cultivation.

Biochemical analyzers are the first choice for the measurement of metabolites because of ease of use and reliability. The BioProfile® analyzer from Nova Biomedical (Waltham, MA) and YSI biochemistry analyzer from YSI Life Sciences (Yellow Springs, OH) are widely used for metabolite measurement. Most metabolites are simultaneously measured by analyzing one sample only. The combination of automatic sampling, measurement, and data collection provided by the biochemical analyzers supports the application of process analytical technology (PAT) to evaluate the impact of process changes (9, 10).

An in situ probe for carbon dioxide measurement by sensing a fluorescence signal is available for cell culture processes (19). An in situ probe, using absorbance of light in the near-infrared region (700 to 2,500 nm) of the electromagnetic spectrum, is also available to monitor metabolic conditions during cell culture in real time (2) and is widely applied to PAT projects by monitoring metabolites to control process parameters in a timely manner (21). Metabolites such as glucose, glutamine, lactate, ammonia, and amino acids can be monitored with an advantage of real-time measurements without sampling for in situ analysis of culture metabolites in order to maintain predetermined ranges of each metabolite in the fed-batch culture. The in situ probes will be widely used in the future due to the ease of real-time measurements and lack of sampling.

48.5. SCALE-DOWN MODEL

During the development of a new process, small- and pilot-scale runs generate data to develop future process control strategies in large-scale manufacturing runs by determining in-process control ranges and obtaining product quality information. The new cell culture process, developed in small-scale runs, is usually scaled up to the pilot plant scale prior to transferring the process to the large-scale manufacturing plant. The cost of a large-scale run is very high, so it is not feasible to test multiple process conditions in the large-scale bioreactors. Small- and pilot-scale bioreactors are used for new process development and troubleshooting for the large-scale runs. The processes should be scalable to make successful large-scale manufacturing runs. After the cell culture process is commercialized, it is often observed that the growth, productivity, and quality of product in the large-scale runs are inconsistent with those in small- and pilot-scale runs due to improper scale-up or process control issues in large-scale bioreactor runs. In order to solve unexpected issues in large-scale runs, a scale-down model to reproduce performance of the large-scale runs is required. After a scale-down model at small and pilot scales is developed and qualified by reproducing the performance of the large-tank runs in growth, productivity, and quality of product, the performance issues in

large-scale runs can be investigated by changing process parameters in the scale-down model. Also, these models are used to develop a design space for process parameters in a quality-by-design application to ensure predefined product quality (10).

48.5.1. Points of Consideration

There is no universal method to scale down the cell culture process. The type and scale of bioreactor and controls of the process parameters should be considered in designing the scale-down model. The typical parameters considered for scale-down models are as follows.

- Type of bioreactor and scale difference
- Process parameters: mixing by agitation, dissolved oxygen and carbon dioxide, aeration, and pH
- Metabolites: glucose, lactate, ammonia, and amino acids

The most reliable scale-down model for large-scale runs could be a pilot-scale model with the same type of bioreactors, because the difference between scales is minimized and because the design of the bioreactors is similar to produce consistent hydrodynamic mixing. In the process development area, culture scales of 1 to 15 liters are typically used for scale-down models to investigate issues in large-scale runs. The scale of the small-scale model can be shake flask or miniature-scale bioreactor scales of less than 10-ml culture volume as long as the small-scale runs reproduce the performance of large-scale runs (7).

Due to large differences in culture volume between scales, the mixing effect in the large-scale runs should be a primary consideration. Thus, the determination of an agitation rate for the small-scale model should be made primarily based on equal power per volume, volumetric oxygen transfer coefficient, shear rate, or mixing time (18). The details of the scale-up/scale-down parameters are not described here, but are covered in chapter 47 and in a reference indicated above (18). By testing different agitation rates, small- or pilot-scale runs may reproduce the large-scale runs by showing comparable growth, productivity, and quality of product. Once the agitation rate for the small-scale model is defined, the root cause of altered productivity or quality of product can be determined by changing agitation rates or by reproducing process control levels with the qualified scale-down model.

The control of process parameters in the large-scale bioreactor runs also needs to be reviewed and applied to the scale-down model to evaluate the impact of the different controls between different scales.

The level of dissolved carbon dioxide can be too high at the beginning of culture or build up during the cultivation of cells when gas exchange is insufficient in the large-scale runs. The high dissolved carbon dioxide level in the large-scale runs can be reproduced by reducing the sparging aeration rate, adding bicarbonate to the medium, or sparging CO_2 gas with the scale-down models.

Differences in the DO concentration between different scales can be observed, due to improper DO probe standardization by failure to consider bioreactor pressure or hydrostatic pressure in the large-scale bioreactors. The impact of the difference in oxygen concentration between different scales can be evaluated by changing a DO set point in the scale-down model. Excess DO fluctuation in large-scale runs with well-turned gas controllers would occur due to insufficient mixing of the culture. A small-scale run may reproduce DO fluctuations in the large-scale runs by changing the agitation rate.

The pH profile in large-scale runs may be different compared to small- and pilot plant-scale runs. Improperly tuned pH controllers may generate pH perturbation by supplying excess or insufficient base or CO_2 gas into the culture. A low agitation rate at the large scale also causes local pH perturbations. With the scale-down model in small and pilot scales, this pH perturbation may be reproduced by lowering the agitation rate to evaluate the impact of the local pH perturbation on the performance of the cell culture process. In the small-scale runs, offline pH measurements are generally consistent with online readings due to proper mixing and well-turned pH controllers. However, offline pH measurements often show significantly higher or lower pH values than online pH readings in the large-scale runs due to local pH perturbation by improper mixing. The impact of the pH difference could be evaluated using the scale-down model by widening a dead band or the pH 1-point correction range, or by setting different pH set points which are observed in the large-scale bioreactor runs.

The changes in productivity and quality of product between different scales are often indicated by metabolite profile changes. Major changes in the metabolite levels are observed with lactate and ammonia, which are considered as by-products of culture, in addition to the changes in glucose, glutamine, and amino acid concentrations. If metabolite profiles in the scale-down model are similar to the profiles observed in the large-scale runs, the scale-down model may represent the large-scale runs in order to investigate the root cause of productivity and quality issues in large-scale manufacturing runs by changing control parameters.

48.6. CONCLUSION

This chapter describes the types of bioreactors, bioreactor control for process parameters, measurements of cell growth and metabolites, and a scale-down model at the small scale through large-scale cell culture. The contents of this chapter could be used by cell culture scientists or engineers who work in process development or at pilot- or large-scale manufacturing areas for the installation and operation of cell culture bioreactors.

The types of bioreactors for cell culture processes are introduced to provide the advantages and disadvantages of each type. The proper control of process parameters such as pH, DO, agitation, and aeration is the most important consideration in the operation of bioreactors to consistently produce good cell growth, productivity, and quality of product. The points of consideration for the control of each parameter are described to explain how to maintain predefined set points for each parameter consistently from the small scale through large-scale cell culture runs. Also, offline and in situ probe measurement of process parameters, cell growth, and metabolites during the cultivation allows for these parameters to be monitored and maintained in predefined ranges. Process control with in situ probe measurement is widely applied to the PAT projects that were initiated by the FDA.

Upon scale-up to the large scale, processes often show "out-of-trend" results for productivity or quality of product. Small- or pilot-scale bioreactors are mostly used to investigate out-of-trend results. The pilot-scale runs are the most reliable for reproducing the performance of large-scale runs because of relatively comparable hydrodynamic mixing between scales. The scale-up parameters, such as equal power per volume, volumetric oxygen transfer coefficient, shear rate, or mixing times, are also applied to determine the agitation rate for the scale-down model. The scale-down model should show comparable metabolite levels, cell growth, productivity, and quality of product to those observed at the large scale. The differences in the control of process parameters and metabolite levels between scales should be reviewed and applied to the scale-down model to investigate the impact on the large-scale runs. The scale-down model is also applied to the quality-by-design applications to develop a design space for process parameters to ensure predefined product quality. The PAT and quality-by-design guidance and application are further described in chapter 50 in this volume.

REFERENCES

1. **Arathoon, W. R., and J. R. Birch.** 1986. Large-scale cell culture in biotechnology. *Science* **232:**1390–1395.
2. **Arnold, S. A., J. Crowley, N. Woods, L. M. Harvey, and B. McNeil.** 2003. In-situ near infrared spectroscopy to monitor key analytes in mammalian cell cultivation. *Biotechnol. Boeng.* **84:**13–19.
3. **Borys, M. C., D. I. Linzer, and E. T. Papoutsakis.** 1993. Culture pH affects expression rates and glycosylation of recombinant mouse placental lactogen proteins by Chinese hamster ovary (CHO) cells. *Bio/Technology* **11:**720–724.
4. **Carvell, J. P., and J. E. Dowd.** 2006. On-line measurements and control of viable cell density in cell culture manufacturing processes using radio-frequency impedance. *Cytotechnology* **50:**35–48.
5. **Chen, P., and S. Harcum.** 2005. Effects of amino acid additions on ammonium stressed CHO cells. *J. Biotechnol.* **117:** 277–286.
6. **deZengotita V. M., R. Kimura, and W. M. Miller.** 1998. Effects of CO_2 and osmolality on hybridoma cells: growth, metabolism, and monoclonal antibody production. *Cytotechnology* **28:**213–227.
7. **Diao, J., L. Young, P. Zhou, and M. L. Schuler.** 2007. An actively mixed mini-bioreactor for protein production from suspended animal cells. *Biotechnol. Bioeng.* **100:**72–81.
8. **Emery, A. N., D. C. Jan, and M. al-Rubeai.** 1995. Oxygenation of intensive cell-culture system. *Appl. Microbiol. Biotechnol.* **43:**1028–1033.
9. **Food and Drug Administration.** 2004. Guidance for industry: PAT—a framework for innovative pharmaceutical development, manufacturing, and quality assurance. Food and Drug Administration, Rockville, MD.
10. **Food and Drug Administration.** 2004. Innovation and continuous improvement in pharmaceutical manufacturing: pharmaceutical CGMPs for the 21st century. Food and Drug Administration, Rockville, MD.
11. **Furukawa, K., and K. Ohsuye.** 1998. Effect of culture temperature on a recombinant CHO cell producing a C-terminal α-amidating enzyme. *Cytotechnology* **26:**153–164.
12. **Kawase, Y., and M. Moo-Young.** 1990. The effect of antifoam agents on mass transfer in bioreactors. *Bioprocess Eng.* **5:**169–173.
13. **Konz, J. O., J. King, and C. L. Cooney.** 1998. Effects of oxygen on recombinant protein expression. *Biotechnol. Prog.* **14:**393–409.
14. **Kurano, N., C. Leist, F. Messi, S. Kurano, and A. Fiechter.** 1990. Growth behavior of Chinese hamster ovary cells in a compact loop bioreactor. 2. Effects of medium components and waste products. *J. Biotechnol.* **15:**113–128.
15. **Ma, N., J. J. Chalmers, J. G. Aunins, W. Zhou, and L. Xie.** 2004. Quantitative studies of cell-bubble interactions and cell damage at different Pluronic F-68 and cell concentrations. *Biotechnol. Prog.* **20:**1183–1191.

16. **Marquis, C. P., K. S. Low, J. P. Barford, and C. Harbour.** 1989. Agitation and aeration effects in suspension mammalian cell cultures. *Cytotechnology* **2:**163–170.

17. **Mostafa, S. S., and X. Gu.** 2003. Strategies for improved dCO_2 removal in large-scale fed-batch cultures. *Biotechnol. Prog.* **19:**45–51.

18. **Oldshue, J. Y.** 1966. Fermentation mixing scale-up techniques. *Biotechnol. Bioeng.* **8:**3–24.

19. **Pattison, R. N., J. Swamy, B. Mendenhall, C. Hwang, and B. T. Frohlich.** 2000. Measurement and control of dissolved carbon dioxide in mammalian cell culture processes using an in situ fiber optic chemical sensor. *Biotechnol. Prog.* **16:**769–774.

20. **Restelli, V., M. D. Wang, N. Huzel, M. Ethier, H. Perreault, and M. Butler.** 2006. The effect of dissolved oxygen on the production and the glycosylation profile of recombinant human erythropoietin produced from CHO cells. *Biotechnol. Bioeng.* **94:**481–494.

21. **Roggo, Y., P. Chalus, L. Maurer, C. Lema-Martinez, A. Edmond, and N. Jent.** 2007. A review of near infrared spectroscopy and chemometrics in pharmaceutical technologies. *J. Pharm. Biomed. Anal.* **44:**683–700.

22. **Yoon, S. K., J. Y. Song, and G. M. Lee.** 2003. Effect of low culture temperature on specific productivity, transcription level, and heterogeneity of erythropoietin in Chinese hamster ovary cells. *Biotechnol. Bioeng.* **82:**289–298.

Continuous Culture

AN-PING ZENG AND JIBIN SUN

49

49.1. INTRODUCTION

Since the fundamental work of Monod (50), Novick and Szilard (53), and Herbert et al. (30) on the theory of continuous culture a half-century ago, the advantages and potential of this type of cultivation have been widely recognized. From an application point of view, these include high volumetric productivity, savings in labor and energy costs (e.g., inoculum preparation, reactor cleaning, and sterilization), uniform product quality, better automation and process control, and the use of more efficient and economical methods of medium preparation and downstream processing.

As a research tool, continuous culture provides well-defined cultivation conditions for genetic, biochemical, and physiological characterizations of cells (13, 26, 76, 81). It allows independent variation of growth parameters, enabling reliable kinetic studies of cell growth and metabolism for process optimization (27, 72). Competition and/or interaction of different species can be studied more easily in continuous mixed cultures than in batch cultures (26). Some phenomena, such as synchronized growth and oscillation, which are hardly observed in batch cultures, can be "trapped" and reproduced in continuous culture (7, 28, 45). The transition behavior of a continuous culture upon shift-up or shift-down of a variable is also a powerful tool to study the regulation of growth and metabolism (4, 6, 15, 32, 44). This kind of transient behavior can also be used for an improved medium design and optimization (22). In addition, continuous culture is also a useful means for selecting strains and/or subclones with improved growth characteristics and productivity (19, 46). On the other hand, mutation and instability of some producing strains due to the selection pressure in continuous culture is a major obstacle for the industrial application of this cultivation technique, particularly in the case of genetically modified cells (8, 38, 65, 75). Another obstacle is the difficulty in maintaining monoseptic conditions over a long period of operation at a large scale. Cell adhesion on the reactor wall and sensors is sometimes a problem, preventing long-term operation of a culture.

With the rapid development in ultrafast genome sequencing, functional genomics, and other "-omics" technologies, which can quantify cellular components at the system and molecular levels, continuous culture has received renewed interest in recent years. Unlike batch or fed-batch culture—wherein environmental factors and cellular regulation and components submit to continuous change, and therefore the cellular state is an integrative outcome of all these cross-talking factors—continuous culture offers controllable and reproducible steady states wherein all cellular components remain unchanged for a long time. Combined with well-designed experiments, this provides a unique opportunity to elucidate the regulatory mechanism of cellular response to a single environmental or genetic perturbation. Microarray is among the most established tools to quantify the mRNA expression profile for this purpose. The reader is referred to the excellent review articles published recently (16, 21, 33). By applying the next-generation genome sequencing technology, one can identify the industrially or academically interesting mutations corresponding to adaptation or selection of the organisms or communities to the given environment. This knowledge can be used to engineer the organisms for further improvement of cellular properties—a strategy called "reverse metabolic engineering" or "evolutionary engineering" (3, 10, 16, 62).

The use of continuous culture received great interest in the 1960s for basic research in biology and cell physiology and led to a renaissance in the 1970s and early 1980s, mainly because of its potential industrial application for the production of single-cell protein, alcohols, solvents, and food-related products and for wastewater treatment. In today's biotechnological applications, continuous operation is fully established in wastewater treatment, production of some primary metabolites (e.g., ethanol and organic acids), and fermented foods and for some reactions catalyzed by enzymes. It is also used for the production of monoclonal antibodies and recombinant proteins by animal cell cultures. The major applications of continuous culture are, however, still found in fundamental studies and process optimization at laboratory scale. A number of new biological processes, such as animal cell cultures and microbial cultures with genetically modified cells, have been quantitatively studied with the help of continuous culture (5, 8, 23, 24, 29, 30, 43, 49, 52, 70, 72). Continuous culture has also been used for bioreactor characterization, control, and scale-up (12, 27, 72). Variations of the continuous culture (e.g., the auxostat) have been introduced and exploited (25, 48, 60, 61). These variations overcome some of the disadvantages associated with chemostat culture and open up new ways of continuous cultivation.

This chapter gives a brief introduction to the general concept and theory of different types of continuous culture. The design and operation of equipment and experiments are discussed from an application point of view. For more detailed treatment and special aspects of continuous culture, the reader is referred to several excellent textbooks and review articles (2, 13, 16, 21, 22, 26, 28, 39, 51, 56, 69, 75, 76).

49.2. GENERAL CONCEPT AND THEORY OF CONTINUOUS CULTURE

Depending on the control parameter and the operation mode, continuous culture can be classified into four general types (Fig. 1). A common feature of these cultures is that they consist of one or more culture vessels into which fresh medium or culture from a preceding vessel is continuously introduced at a rate, F, expressed in liters per hour, and that the culture volume, V, expressed in liters, is kept constant by continuous removal of the culture. The general concept and theory of these four types of continuous culture are described in more detail as follows.

49.2.1. Chemostat

The chemostat (Fig. 1A) is defined as a continuous culture system in which the feed rate is set externally and cell growth is limited by a selected nutrient. The second condition means that the specific growth rate, μ (hour^{-1}), of the organism is a function of a single growth-limiting nutrient. However, this definition may be relaxed to include continuous culture limited simultaneously by more than one nutrient component (26). Continuous cultures that are fed with an inhibitory nutrient or that form toxic products can be limited by growth inhibition under the condition of excess of all nutrients. Strictly speaking, they cannot be referred to as chemostat cultures.

A chemostat is usually started as a batch culture. Before a nutrient becomes limiting, the nutrient feed is started. Cells grow until the chosen nutrient becomes limiting. After this, cell growth is limited by the rate of addition of medium. The specific growth rate of a chemostat culture can be determined from a material balance for biomass: net increase in biomass = biomass in incoming medium + growth − output − death. In mathematical form this is

$$\frac{dX}{dt} = \frac{F}{V}X_0 + \mu X - \frac{F}{V}X - \alpha X \quad (1)$$

where dX/dt is the rate of accumulation of biomass per unit of time and per volume (g of dry cell weight [DCW] per h per liter); X_0 is the biomass concentration (g of DCW per liter) in the incoming medium; X is the biomass concentration (g of DCW per liter) in the bioreactor; and α is specific death rate of cells (h^{-1}). This expression can be simplified if one assumes that the chemostat is at a steady state, that there are no cells in the incoming medium, and that cell death is negligible. The assumption of steady state implies that there is no net accumulation of cells in the bioreactor, so the left-hand side of the equation is equal to zero. Furthermore, if the dilution rate, D, is introduced for F/V, equation 1 becomes

$$\mu X = DX \quad (2)$$

Hence

$$\mu = D \quad (3)$$

Thus, by maintaining a constant volume and changing the nutrient feed rate, one can precisely control the specific growth rate of a culture over a range up to the maximum specific growth rate (μ_{max}) and let the system come to a steady state. This is one of the most important properties

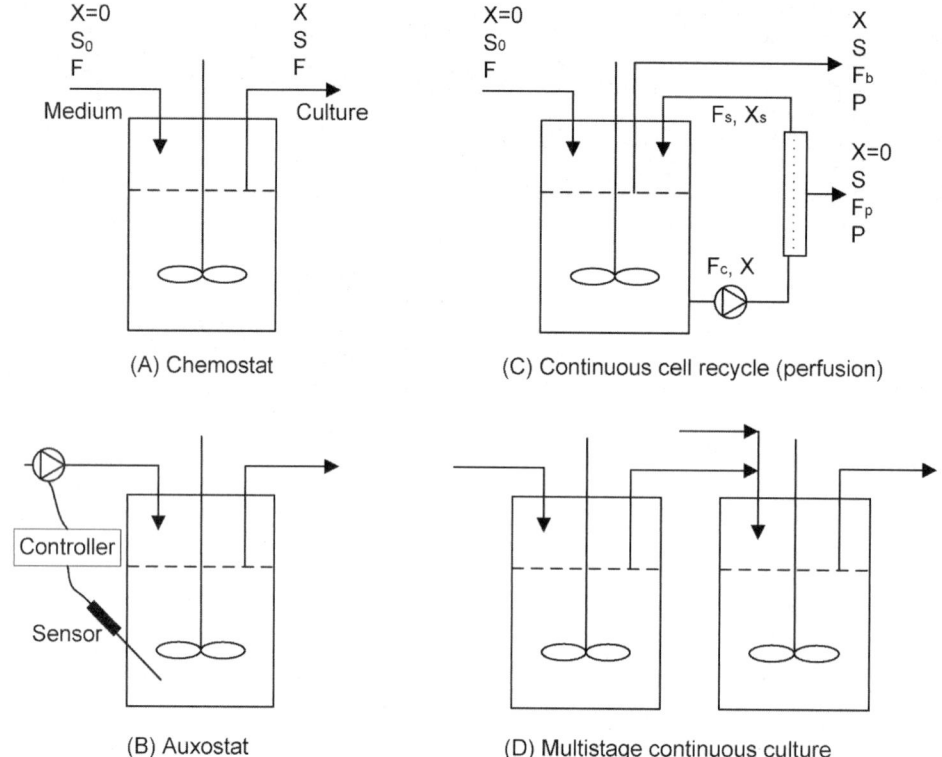

FIGURE 1 Schematic diagram of four types of continuous culture.

of a chemostat. Note that the following additional assumptions are made in the derivation of equations 1 to 3.

1. Cells are distributed randomly in the bioreactor; i.e., the cells do not adhere to each other or to the walls of the reactor, and the suspension is well mixed.
2. The population is nonsegregated and consists of physiologically identical cells.
3. The population density, X, is a continuous variable; i.e., the number of cells is sufficiently high and the size of each cell is sufficiently small for the discreteness of biomass to be ignored.
4. The volume occupied by the cells is negligible compared to the total volume of the culture.

Furthermore, assuming that as soon as medium enters the bioreactor it is instantly distributed uniformly throughout the culture, a material balance on the growth-limiting substrate can be formulated as net increase = input − output − substrate consumed by cells. For an infinitely small time interval, dt, this balance for the whole culture is

$$\frac{dS}{dt} = DS_0 - DS - q_s X \qquad (4)$$

where S_0 is the concentration (g per liter) of substrate in feed, S is the concentration (g per liter) of substrate leaving the reactor, and q_s is the specific consumption rate of substrate (g per g of DCW per h). Equation 4 can also be written in terms of growth yield:

$$\frac{dS}{dt} = DS_0 - DS - \frac{\mu X}{Y_{x/s}^{app}} \qquad (5)$$

where $Y_{x/s}^{app}$ is the apparent yield of biomass on substrate (g of DCW per g of substrate), which is defined as

$$Y_{x/s}^{app} = \frac{\mu}{q_s} \qquad (6)$$

In general, both q_s and $Y_{x/s}^{app}$ are functions of μ and can also be affected by a number of other growth parameters, such as the nature of the limitation and the level of substrate concentration (79, 80). To illustrate the basic characteristics of a chemostat, $Y_{x/s}^{app}$ is assumed here to be constant. Thus, under steady-state conditions, equation 5 reduces to

$$DS_0 - D\overline{S} - \frac{\mu \overline{X}}{Y_{x/s}^{app}} = 0 \qquad (7)$$

where $\overline{S}$ and $\overline{X}$ denote steady-state values. For a chemostat in which growth inhibition by product formation and excess of other nutrients does not occur, the well-known Monod kinetic (50) may be used to describe the relationship between specific cell growth rate, μ, and the concentration of a limiting nutrient in culture:

$$\mu = \mu_{max} \frac{S}{S + K_s} \qquad (8)$$

where μ_{max} is the maximal specific growth rate (h^{-1}) and K_s is a saturation constant corresponding to the substrate concentration at which the maximum specific growth rate is reduced by one-half. Substituting μ for D in equation 8, we obtain for a steady state

$$\overline{S} = K_s \frac{D}{\mu_{max} - D} \qquad (9)$$

and from equations 3 and 7, the following expression for the steady-state biomass concentration is obtained:

$$\overline{X} = Y_{x/s}^{app} (S_0 - \overline{S}) = Y_{x/s}^{app} \left(S_0 - K_s \frac{D}{\mu_{max} - D} \right) \qquad (10)$$

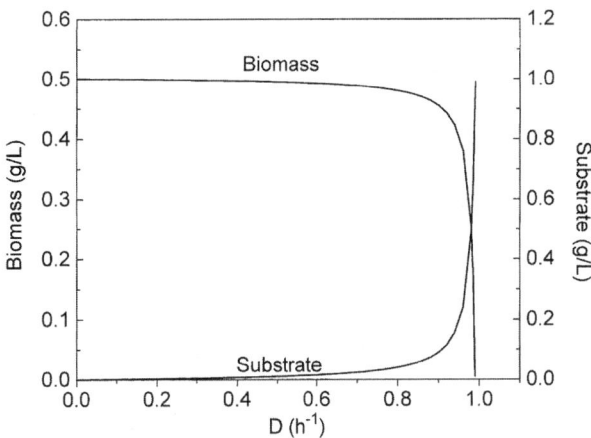

FIGURE 2 Steady-state values of biomass and growth-limiting substrate concentration in a chemostat according to equations 9 and 10. Parameters: $\mu_{max} = 1.0$ h^{-1}; $K_s = 0.01$ g/liter; $Y_{x/s}^{app} = 0.5$ g/g.

The characteristics of a chemostat can be best viewed by examining the behavior of the governing equations derived above. Figure 2 shows plots of $\overline{S}$ and $\overline{X}$ against dilution rate, D, with typical parameter values. The residual substrate concentration is only a function of the dilution rate. It increases slowly with D at low values but very rapidly as D approaches μ_{max}. The biomass concentration depends on the feed substrate concentration. At a given S_0, biomass is almost constant at low values of D but rapidly declines at dilution rates close to μ_{max}.

49.2.2. Deviations from the Ideal Chemostat

The suitability of the simple chemostat theory has been demonstrated for a number of organisms, including bacteria, yeasts, and plant and animal cell cultures. In most cases, these cultures were carried out at very low feed concentrations of nutrient, in small-scale bioreactors, and/or in a relatively narrow range of dilution rate. Under these conditions, the assumptions made in the derivation of the above equations may be approximately fulfilled. Nevertheless, deviations from chemostat theory have been frequently observed. Figure 3 shows typical biomass profiles as a function of dilution rate in continuous culture that depart from chemostat behavior. The deviations may originate from interactions of cells with the equipment and reactor hydrodynamics, such as wall growth and imperfect mixing. Wall growth is caused by the adhesion of cells to glass and metal surfaces. As shown in Fig. 3A, wall growth can increase the steady-state biomass concentration and enable an operation to be carried out at dilution rates higher than the maximum specific growth rate (71). Brown et al. (11) studied the effect of imperfect mixing (stagnant zone) on the performance of a continuous culture (Fig. 3B). Depending on the relative size of the stagnant zone and the velocity of liquid exchange between the well-mixed zone(s) and the stagnant zone, the biomass concentration can be markedly reduced at a certain range of the dilution rate. Similarly, as in the case of wall growth, the operational dilution rate can be higher than μ_{max}. Simple chemostat theory assumes perfect mixing of medium; i.e., a drop of medium entering the vessel should instantly be distributed uniformly throughout the culture. Quantitatively, this means that the time constant required for mixing should be smaller

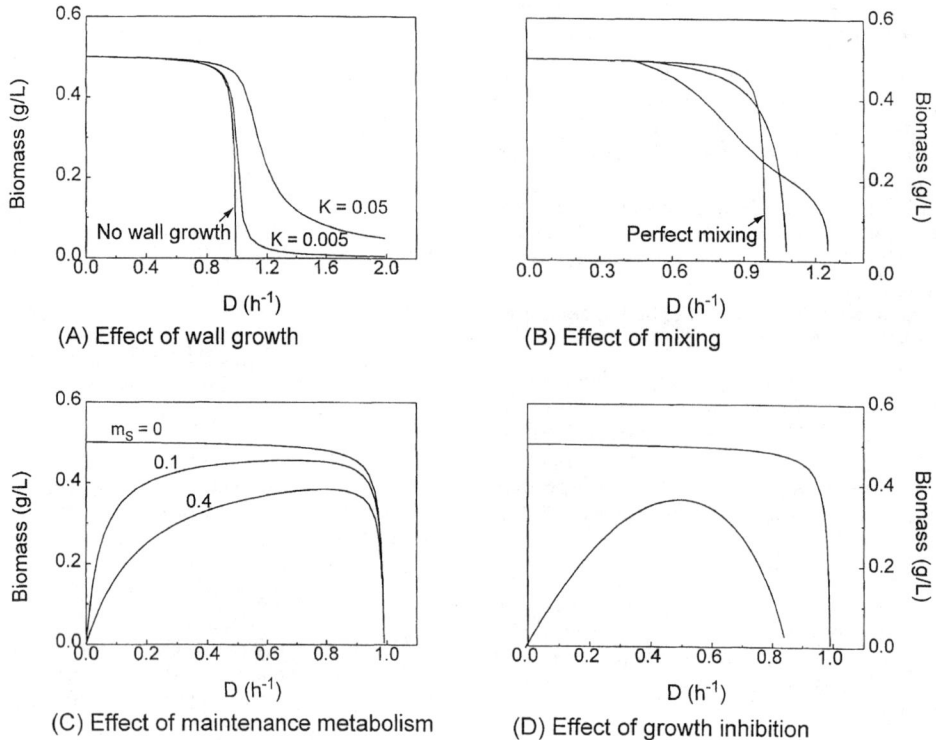

FIGURE 3 Typical profiles and reasons for continuous cultures deviating from chemostat behavior. The biomass curve of a corresponding chemostat culture (see Fig. 2) is also shown for comparison. (A) Effect of wall growth. K is the amount of growing biomass attached to the reactor surface per unit volume of culture. (B) Effect of imperfect mixing. (C) Effect of maintenance metabolism. m_s is the maintenance requirement for substrate (g/g · h). (D) Effect of growth inhibition by product(s) and/or substrate.

than the time constants for mass transfer and bioreactions (83). In laboratory-scale reactors with a culture volume of a few liters, nearly perfect mixing can be obtained at proper agitation rates unless the reactor contains zones that screen the culture from the agitation or the culture becomes very viscous. In large-scale reactors, however, imperfect mixing is inevitably encountered.

The deviations observed in laboratory-scale reactors are quite often caused by biological effects. Among others, these include maintenance requirement (57), variations of growth yield due to metabolic overflow (67, 79, 80), growth inhibition by products and/or substrate (27, 78), and regulatory and segregating effects at low and high growth rates (4, 44). As shown in Fig. 3C and 3D, both the effects of maintenance metabolism and growth inhibition tend to decrease the steady-state biomass concentration, particularly at a low dilution rate. Under conditions of strong growth inhibition, the operation of a continuous culture may be only possible at dilution rates significantly lower than μ_{max}. The factors mentioned above may be simultaneously involved in a culture system. In such a case, the relationship between the biomass concentration and the dilution rate will be more complicated.

49.2.3. Auxostat

An auxostat is a continuous culture in which a growth-dependent parameter is kept constant by adjusting the feeding rate of medium (Fig. 1B). The dilution rate and hence the specific growth rate of the culture adjust accordingly. The choice of the feedback parameter for an auxostat is quite broad. It includes cell density (turbidity), pH, dissolved oxygen concentration, CO_2 in effluent gas, and concentrations of nutrients and products (25). Sometimes the term "nutristat" is also used to refer to auxostats using a nutrient concentration as the feedback growth parameter. Of the different kinds of auxostats, the turbidostat and the pH-auxostat have so far found the most applications (9, 25, 48, 60).

In a turbidostat, the biomass (cell density) is used as a control parameter. A sensor detecting the biomass density gives a signal to a pump to add more medium when the biomass density rises above a chosen level. By means of turbidostat control, therefore, the biomass density is set and the dilution rate adjusts itself to the steady-state value, in contrast to the chemostat, in which the dilution rate is fixed and the biomass concentration adjusts itself to the steady-state level. The governing equations of a turbidostat can be derived from the general equations 1, 5, and 8. Under steady-state conditions, we have

$$\overline{S} = S_0 - \frac{\overline{X}}{Y^{app}_{x/s}} \tag{11}$$

$$D = \mu_{max} \frac{\overline{S}}{\overline{S} + K_s} \tag{12}$$

Here $\overline{S}$ is linearly dependent on both $\overline{X}$ and $\overline{S}_0$, while D is a hyperbolic function of these two parameters. These relationships are illustrated in Fig. 4.

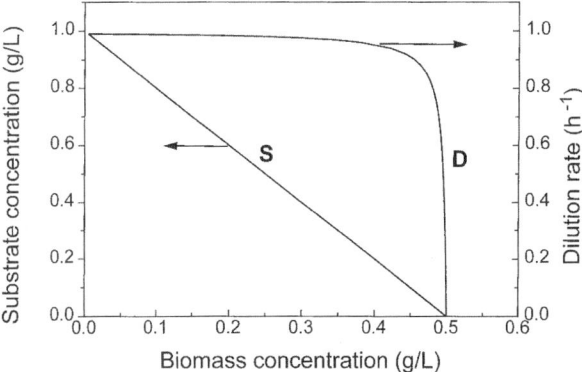

FIGURE 4 Steady-state limiting-nutrient concentration, $\bar{S}$, and dilution rate, D, as functions of biomass concentration in a turbidostat culture. Growth parameters are as given in the legend to Fig. 2.

In principle, turbidostat culture is suitable for the cultivation range in which the biomass concentration varies significantly with change in dilution rate, such as near the critical dilution rate (Fig. 4). This is of particular interest because operation near the maximum growth rate can be very unstable in a chemostat. The turbidostat also provides a means of maintaining cultures in a constant environment with excess of substrate. Since the growth rate is not fixed, the system will select for fast-growing organisms. Increase in the maximum growth rate will result from both selection of genetically different organisms (or mutants) and an adaptation of cellular metabolism. In the past, a photoelectric cell was the main sensor used to monitor cell density. This technique is limited to unicellular organisms and is difficult for long-term cultures, largely because of adhesion of organisms to the surface of the sensor. With the development of optical sensors based on laser or infrared light transmission/scattering, problems with long-term monitoring of biomass may be partly overcome.

In a pH-auxostat, the addition of fresh medium is coupled to pH control. As the pH drifts from a given set point, fresh medium is added to bring the pH back to the set point. The rate of addition of medium is determined by the buffering capacity and the concentration of nutrients in the medium. Buffering capacity is defined as the equivalents of titrant required to change the medium pH to the culture pH. The governing equation is the mass balance on the H$^+$ ion concentration, expressed as milliequivalents (meq) per liter, in the bioreactor (25):

$$\frac{dH^+}{dt} = D(H_F^+ - H^+) + \mu X h - D(BC) \quad (13)$$

where H_F^+ is the H$^+$ ion concentration in feed (meq per liter), h is the acid production yield of cells (meq per g of biomass), and BC is the buffering capacity of medium (meq per liter).

By using fresh medium for pH control, the limiting nutrient level in the reactor can be manipulated by adjusting the buffering capacity and/or limiting nutrient concentration in the feed. For cultures producing high amounts of acid or alkali, the buffering capacity of the medium may not be sufficient. In this case, separate feeding lines for titrant and medium can be used (9). The concentration of nutrient determines the amount of ionic species produced and ions released during nutrient uptake. Assuming that the difference between feed and bioreactor H$^+$ concentrations is very small, the following relationships can be obtained from equation 13 and substrate balance under steady state:

$$\bar{X} = \frac{BC}{h} \quad (14)$$

$$\bar{S} = S_0 - \frac{BC}{hY_{x/s}^{app}} \quad (15)$$

The specific growth rate achieved in a pH-auxostat is positively correlated with the concentration of the nutrient, S_0, but inversely correlated with the buffering capacity, BC, in the feed medium. Roughly speaking, the specific growth rate achieved in a system with low buffering capacity can be about 10 times greater than with a well-buffered system (60). The biomass concentration is mainly determined by the buffering capacity of the medium.

The pH change of a culture is a good indication of cell growth and metabolic activity. However, the exact cause of pH change varies among organisms. It represents the summation of the production of different ionic species and ion release during nutrient uptake. The pH of a culture is normally reduced by the production of organic acids and by the uptake of ammonium. However, for microorganisms growing on protein- or amino acid-rich media, the pH will rise with growth because of the release of excess ammonia.

Generally speaking, auxostats have the following advantages over conventional chemostats. First, auxostats permit stable operation in the "high-gain" areas near the maximum growth rate. Second, they reach steady state more rapidly at high dilution rates than the open-loop chemostat. Third, population selection pressures in an auxostat lead to cultures that grow rapidly. Finally, it is possible to design a dual set point auxostat that controls two growth parameters (e.g., concentrations of two nutrients) simultaneously (25). This kind of dual set point auxostat permits control of nutrient concentration ratios with ease, which tends to be very difficult with a chemostat. On the other hand, the experimental setup and operation of an auxostat are somewhat complicated.

49.2.4. Continuous Culture with Cell Recycle (Perfusion)
Cell recycle is a useful means for increasing the concentrations of biomass and product in a continuous culture (5, 31, 41, 82). This system can be operated at a dilution rate higher than the maximum specific growth rate, leading to a much higher output of the reactor. Another property is that the dilution rate is almost independent of the growth rate. Recycle of biomass can also protect against shock loading with an inhibitory substrate, because the critical dilution rate is raised. It is particularly advantageous in the following cases: (i) the growth-limiting substrate is unavoidably dilute, for example, in the treatment of effluent; (ii) the substrate has a low solubility, such as when a gaseous substrate is used; (iii) the concentration of growth-limiting substrate has to be limited because of the formation of inhibitory product(s); (iv) product formation is not associated with growth; and (v) the product is unstable or may be degraded, e.g., by proteases in the culture with a long residence time.

Several methods can be used for the retention of biomass, such as filtration, sedimentation, centrifugation, and immobilization. Depending on the position of the separation device inside or outside of the reactor, the methods can be further divided into internal and external systems (56). The concentration of culture effluent outside of the reactor by means of membrane filtration has so far found the most frequent application, except for the biological treatment of wastewater, in which the recycling of sludge after sedimentation has been used for a long time.

Figure 1C schematically shows a continuous culture with membrane cell recycle. The culture is circulated at an outflow rate F_c (liters per h) by a pump over a microfiltration membrane from which a cell-free filtrate is withdrawn at a rate of F_p (liters per h). In addition, a portion of the liquid culture containing biomass is withdrawn from the reactor at a bleed rate of F_b (liters per h).

Note that in a membrane bioreactor, the circulation rate of liquid culture, F_c, is normally much higher than the filtration rate, F_p, so that the concentration factor, $\alpha = X_s/X$, is very close to 1:

$$\alpha = \frac{X_s}{X} = \frac{F_c}{F_s} = \frac{F_c}{F_c - F_p} \approx 1 \qquad (16)$$

Where X_s and F_s are the biomass concentration and flow rate at the outlet of the filtration module, respectively. Thus, the position of biomass withdrawal (bleed) does not affect the governing equations for the membrane bioreactor, which can be written as follows:

$$\frac{dX}{dt} = (\mu - B)X \qquad (17)$$

$$\frac{dS}{dt} = D(S_0 - S) - q_sX = (B + f)(S_0 - S) - q_sX \qquad (18)$$

where B is the bleed rate (h^{-1}), defined as $B = F_b/V = 1/\tau$; τ is the mean cell residence time (h); and f is the filtration rate, defined as $f = F_p/V$ (h^{-1}). The dilution rate, D, is the sum of rates of bleed and filtration. Assuming a constant apparent biomass yield (equation 6) and Monod kinetics (equation 8), the following steady-state solution of equations 17 and 18 is obtained:

$$\mu = B \qquad (19)$$

$$\overline{S} = K_s \frac{B}{\mu_{max} - B} \qquad (20)$$

$$\overline{X} = \left(1 + \frac{f}{B}\right)Y_{x/s}^{app}\left(S_0 - K_s \frac{B}{\mu_{max} - B}\right) \qquad (21)$$

Thus, cell growth rate in a chemostat with cell recycle is determined by the bleed rate, B. From equation 17, it is obvious that in a system of total cell recycle ($B = 0$), the biomass would build up to reach a point where the cultivation becomes inoperable owing to high viscosity and difficulties in mixing and nutrient supply. Thus, a controlled bleed of biomass is necessary to maintain an active biomass. In some cases, cell death and lysis at high biomass concentration may be considerable so that total cell retention is possible. A comparison of equation 21 with equation 10 reveals that at a same growth rate, the chemostat culture with cell recycle can increase the biomass concentration of a simple chemostat by a factor of f/B. In practice, a value of f/B in the range 10 to 100 is feasible. The increase of final product concentration is normally much lower because of the reduction of productivity of cells at high concentrations.

49.2.5. Multistage Continuous Culture

The two-stage continuous culture system shown in Fig. 1D can extend the range of application of continuous culture. For example, the second stage may be used to extend the growth rate downward to zero, and the first stage may be used to achieve stable conditions with maximum growth rate, both of which conditions may be desired in certain cases but are impossible in the simple chemostat (56). The latter property is particularly useful when the substrate is

also a growth inhibitor. In the production of secondary metabolites and enzymes by continuous culture, the second stage may be used to provide a nongrowing situation in which product formation occurs. For products of foreign gene expression, the second stage can be used for induction of expression (70). Another useful application of the two-stage continuous culture is for reactor scale-down studies (12). The two-stage culture system can be extended to include more stages and more feeding streams with or without biomass recycle. For more information on these systems, the reader is referred to Pirt (56) and Moser (51).

49.3. EQUIPMENT SETUP AND DESIGN

49.3.1. General Setup of a Continuous Culture

Figure 5 shows a typical setup of a bench-scale reactor system for continuous cultivation. A continuous reactor system consists of at least three parts:

1. Reaction vessel, i.e., a bioreactor (at center) equipped with elements for aeration, mixing of liquid (medium) and solid (cells or insoluble substrate) phases, sampling, and control of temperature, pH, dissolved oxygen, foaming, and reaction volume. Equipment for online effluent gas measurement is also depicted in Fig. 5. Some of these control units may be omitted, depending on the type of fermentation.

2. Storage vessels for medium and solutions for control of pH and foam. They should be suitable for sterilization. Normally, the entire homogeneous, mixed medium is stored therein, containing the substrate as carbon and energy source (e.g., glucose) and other nutrients at given concentrations. More vessels are necessary if heat-sensitive components are used, because they need separate sterilization. Components may be mixed together before being fed into the reactor by a metering device (e.g., a pump) or may have separate connecting lines to the reactor.

3. Broth withdrawal and collection vessel (at right). The collection vessel is not mandatory but is useful for measuring the effluent rate of the culture and avoiding the release of organisms into the environment.

For some applications, such as selection of strains (or mutants), much simpler configurations of continuous culture can be used (39).

49.3.2. Equipment Selection and Design

Stafford (66) provides an excellent overview of potential problems and solutions concerning selection, design, and operation of individual equipment used in continuous culture. The following discussion is mainly based on Stafford and includes some updated information and the authors' experience with continuous culture. In addition, necessary equipment and its operation for continuous culture with cell recycle are addressed. As pointed out by Stafford (66), the type of fermentation may dictate the choice of equipment. Medium composition (chemically defined or complex) and organism morphology (filamentous or unicellular) set restrictions on the type of equipment that can be used. Therefore, the suggestions below serve only as a guideline and may need to be modified or even replaced according to the actual requirements.

49.3.2.1. Reactor

The scale of reactor is an important factor in selection of equipment and setup of a continuous culture. The con-

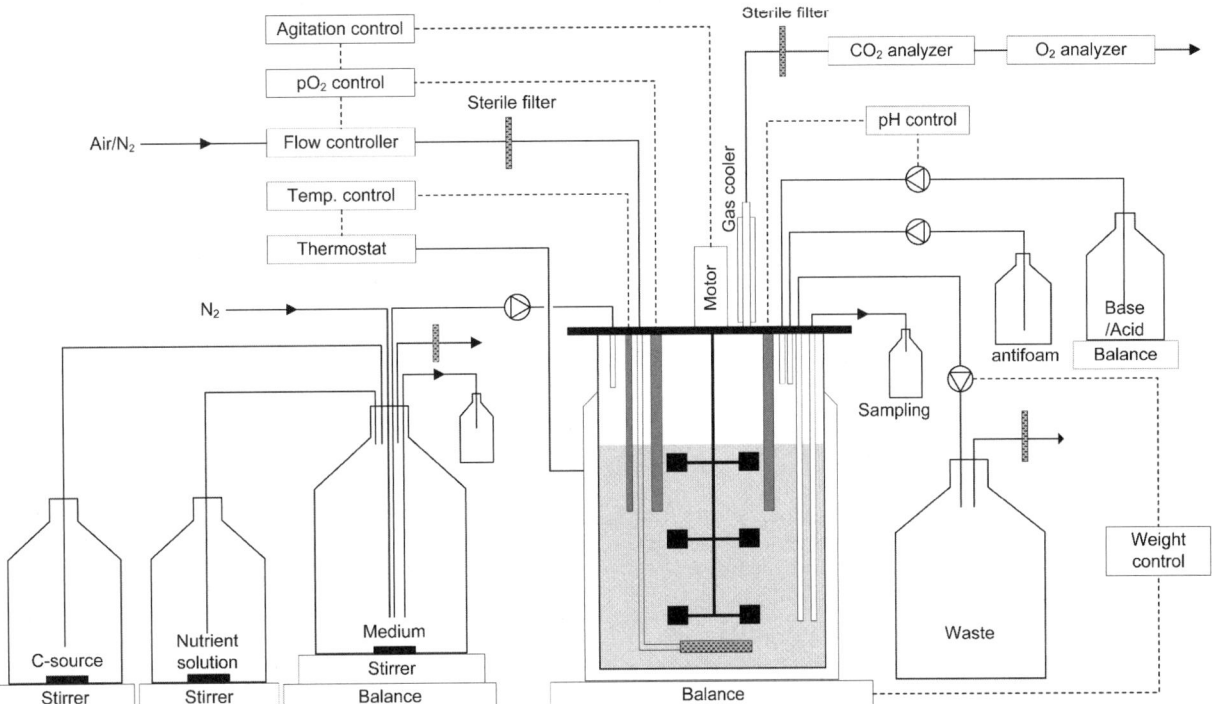

FIGURE 5 A typical setup of a bench-scale continuous-culture system. The feeding rate and the culture volume are controlled by a weight control unit.

sumption of medium and the need for other peripheral equipment are largely determined by the working volume of the reactor. For research purposes, a benchtop-scale reactor is convenient to operate. In addition, investment and space requirements are modest, and peristaltic pumps of adequate reliability are available. A benchtop reactor is easily dismantled for cleaning.

Small-scale culture vessels have some unique problems, however. When operated at volumes of less than 500 ml, sampling can significantly reduce the volume, affecting the dilution rate and leading to fluctuation of the system. Flow rates for obtaining low dilution rates can be very difficult to control, and feeding a complex medium at low flow rates can result in settling out of insoluble medium components in feed lines, causing plugging. It is often difficult to implement an online gas analysis owing to low back-pressure and low aeration rate. In addition, wall growth may represent a considerable fraction of the total population in a small vessel. For microbial cultures, a working volume of 1 to 5 liters (total volume of 2 to 7 liters) is in most cases appropriate. A lower working volume (e.g., 0.5 to 2 liters) may be desired for animal cell culture owing to the relatively high cost of medium. The alternative use of a reactor for batch or fed-batch cultures should also be considered.

Stirred-tank reactors are usually used in benchtop scale, although other reactor types, such as airlift and fluidized-bed reactors, are also available in small scale. The choice of a reactor type depends very much on the goal of the research. For small-scale stirred-tank reactors, there are also several variations in design, including flat-bottom and round-bottom vessels. The flat bottom eases handling on the bench and in the autoclave. On the other hand, round- or dished-bottom vessels ensure better mixing and are mostly used for culturing shear-sensitive cells. Jacketed reactors are also

available. The glass jacket gives better temperature control than the use of a heat exchanger inside the reactor. The jacketed reactor is more suitable for heat-sensitive cultures.

49.3.2.2. Nutrient Feed Reservoir and Auxiliaries

Nutrient reservoirs for continuous culture should have ports for feeding, addition and/or mixing of heat-labile nutrients and substrate, venting, and sparging of the medium (Fig. 5). The feed and sparging lines should be stainless steel tubes going to the bottom of the vessel. The venting line should be attached with a sterile filter. A magnetic stirring bar must be included to suspend any particulates and mix nutrients that are added after sterilization of the bulk medium. A separate sampling line is also useful for taking samples to determine the exact substrate concentration after refilling. A carboy or glass flask with a medium volume large enough for at least one steady state may be used as an intermediate reservoir. This can be calibrated by volume or placed on a weighing device so that the nutrient feed rate can be checked over several hours or days. It is also flexible enough for adjusting the nutrient concentration if different growth limitations are to be used.

Silicone rubber tubing or special gas-impermeable tubing should be used when tubing is required. It can be sterilized repeatedly, retains its resilience, and has excellent chemical resistance. The method of connecting tubing should be aseptic, fast, and reliable. One should be aware that the flow rate through a peristaltic pump is dependent on the resilience of the tubing. Because silicone rubber tubing loses its resilience with use, a long piece of tubing should be used so that the pump can be moved every other day to a new section. For extended operation, replacement of the tubing may be necessary.

The medium inlet line may be connected with a drip tube very close to the reactor. A drip tube provides a barrier

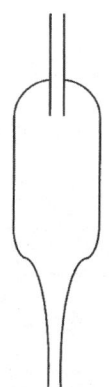

FIGURE 6   Diagram of a drip tube.

in the medium flow path and prevents microorganisms from growing into the feed line. It can be purchased from fermentor manufacturers or constructed as shown in Fig. 6. The drip tube may fail if pressure builds up in the reactor.

49.3.2.3.   Broth Removal and Level Control

To achieve an accurate and constant control of the dilution rate, the working volume of the culture must be determined and controlled. The working volume of a culture is normally defined as the ungassed liquid volume. Two parameters can be used to control the ungassed liquid volume: gassed liquid level (assuming constant gas holdup) and weight of the reactor content. Because gassed volume is easier to control, it is widely used. However, constant gas holdup is difficult to achieve. Alternatively, weight control can be more accurate if properly installed, but it is much more expensive to implement on the bench scale of operation. The working volume

of the reactor system shown in Fig. 5 is kept constant by a weight control unit. The signal from the balance is used to control the pump for culture removal.

There are basically two ways of controlling gassed liquid level. The simplest method is to make use of the overpressure in the reactor from aeration. The effluent gas outlet can be placed at the broth surface. The effluent gas is directed into a waste reservoir that is vented aseptically to the atmosphere. When the culture volume increases above the effluent gas outlet, broth is forced out into the reservoir with the gas. Sometimes a pump is used to withdraw broth instead of using effluent gas. Both of these systems suffer from the problem that the cell concentration at the air/liquid interface can be quite different from that in the bulk liquid, especially if foam is formed. This can cause a nonrepresentative withdrawal of culture from the reactor.

Another method is to use a liquid level controller. Although this technique requires more equipment, it allows withdrawal of broth from below the liquid surface. A number of liquid level controllers are available. Most of them are based on a sensor within the vessel that makes physical contact with the culture fluid. Because the sensor is located within the reactor, failure of the probe will end the process. Depending on the type of sensor used, inaccuracies in volume can be caused by variations in agitation, aeration, foam generation, and wall growth. Ultrasonic level controllers are available that can be attached to the outside of a glass or steel vessel. They can be attached after autoclaving, and if a sensor malfunctions, it can be replaced without violating sterility.

Stafford (66) proposed the use of a redundant control scheme as shown in Fig. 7. In this system, a liquid level controller is backed up by the effluent gas system. The culture exit tube is located just below the liquid level, while the effluent gas exit is located just above the liquid

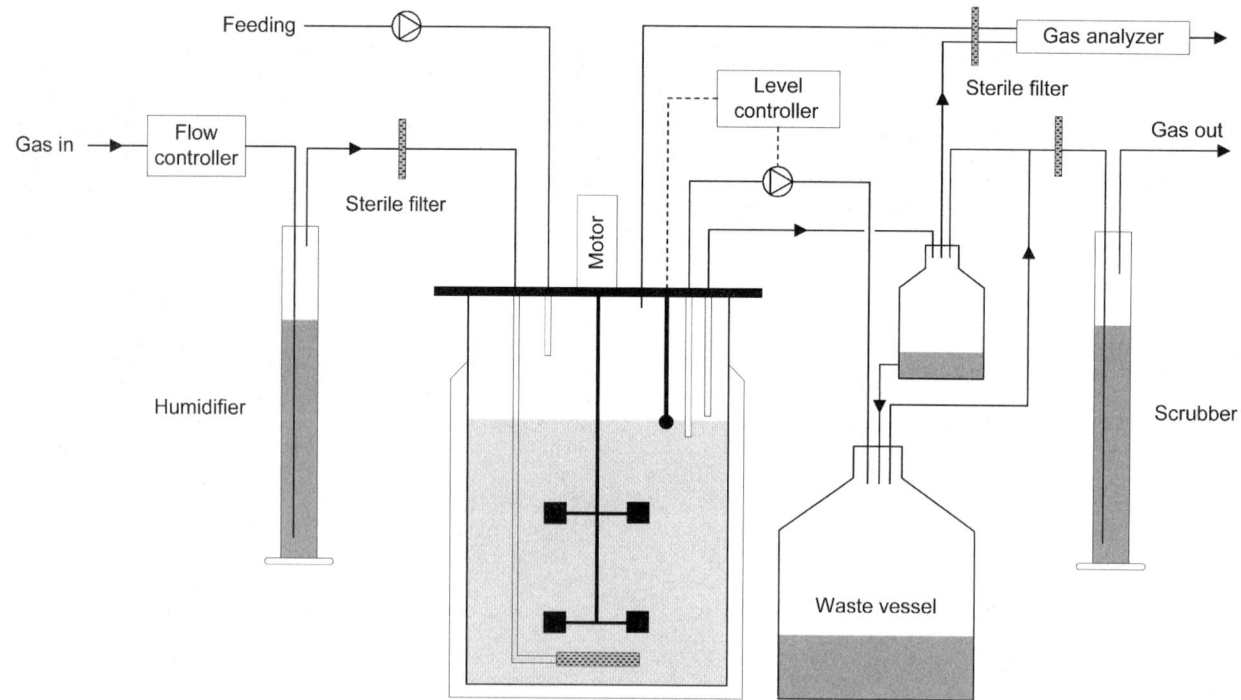

FIGURE 7   Control scheme for the reactor volume of a continuous culture that uses a liquid level controller backed up by an effluent gas system. (Modified from reference 66.)

level. If the level controller fails and remains in the "on" position, the waste broth pump cannot reduce the liquid level below the culture take-off line. If the level controller fails and remains in the "off" position, or the pump stops or the tubing becomes clogged, then the liquid level cannot rise above the effluent gas exit tubing. However, problems caused by the change of gas holdup and foam formation still cannot be fully eliminated. If an antifoam agent is added during operation, fluctuation in the liquid level may result. To avoid error in determining the dilution rate, the real ungassed liquid volume may need to be checked from time to time by a brief switch-off of aeration and agitation (in case of strong foaming), for example, after a steady state has been achieved. Nevertheless, this measure may lead to other negative influences on cellular metabolism owing to the shock of oxygen limitation.

A stable continuous culture requires accurate and reliable pumps. The pump heads and the pumping lines should be sterilizable. Most important, it should be capable of long-term aseptic operation. On the benchtop scale, peristaltic pumps meet these criteria. As fermentation scale increases, they become impractical. Silicone tubing should be replaced by piping. Generally, in situ steam sterilization is required for reactors greater than the 10-liter scale. Diaphragm pumps can be used that provide reliable, adjustable flow rates. These pumps have several advantages over centrifugal and gear pumps. First, they lack a mechanical seal. Over time, mechanical seals wear out and can be a source of contamination. Second, the pump head and the lines leading to and from the head can be sterilized by steam. Finally, the flow rate can be maintained over a wide range of pressure. This means that changes in the back-pressure of the vessel can be accomplished without substantially changing the pump flow rate.

49.3.2.4. Gas Inlet and Exit

An example of a gas supply and effluent system is shown in Fig. 7. Humidification of the gas prevents evaporation of medium from the vessel and improves the estimation of dilution rate if the culture effluent rate is measured in the outlet line. This may be significant when operating at a relatively high temperature (e.g., above 35°C), when using compressed air that is very dry, or when operating at dilution rates of less than $0.1\ h^{-1}$. Sterilization of the inlet gas can be done by a membrane filter with proper pore size or a packed glass-wool filter. Two filters connected by a tee are recommended as a precaution, especially when using humidifiers. In the event that one filter becomes plugged, the other filter can be unclamped and used immediately. A regulated pressure of gas supply is important for safety. In case the gas exit becomes plugged, the gas pressure should be sufficiently low to prevent the vessel from exploding.

The effluent gas should be directed to an overflow reservoir (safety vessel). In this way, foam-out can be contained. The overflow reservoir is then connected to a graduated cylinder filled with 0.5% bleach. If an online gas measurement is desired, a bypass gas line from the reservoir can be used. The rest of the effluent gas is sparged through the liquid in the cylinder and vented through a sterilizing filter. The cylinder serves several functions. First, it gives a certain degree of back-pressure to the vessel, which is needed for effluent gas analyzers. It also facilitates sampling. Second, it provides a means to visually check whether the gas flow rate is high enough for the gas analyzer and if the reactor system is well sealed. Absence of bubbling may indicate loss of containment, which is of particular concern if pathogenic or recombinant organisms are being cultured. Finally, it acts as a scrubber to reduce unpleasant odors in the laboratory.

49.3.2.5. Cell Recycle

At laboratory scale, cell recycle is easily achieved by using a steam-sterilizable membrane microfiltration module. Figure 8 shows a simplified structure of such a module. It is composed of microporous membrane capillaries. The membrane is usually made of polypropylene or similar materials and is originally hydrophobic. After wetting with ethanol, it becomes hydrophilic. The capillaries normally have a diameter of about 8 mm and a pore diameter of about 0.2 μm. Membrane modules with different diameters, pore sizes, and filtration areas are available.

An often observed problem of microfiltration is the decrease of filtration rate during operation. This is mainly caused by the formation of biofilms and adsorption of proteinaceous materials on the membrane surface (fouling) and/or loss of hydrophilicity. Factors affecting membrane fouling include concentrations and properties of cells, circulation rate, and properties of the broth and filtration rate. The use of antifoam may also significantly affect

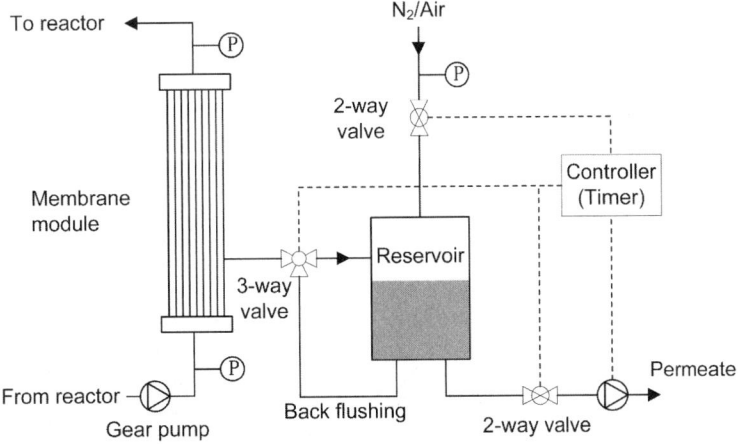

FIGURE 8 Schematic diagram of a back-flushing configuration for a cell recycle reactor with membrane microfiltration. The permeate in the reservoir is back-flushed through the membrane module by using pressurized nitrogen gas or air and controlled by a timer.

membrane performance. To reduce membrane fouling, a minimal broth circulation velocity of about 0.5 to 1 m/s should be maintained, normally by a gear pump. This may impose problems for shear-sensitive cells. The filtration module should also be periodically back-flushed with permeate. Such a back-flushing configuration is shown in Fig. 8 (82). A small permeate reservoir is used for the back-flushing. It is controlled by a computer system or by a timer. When back-flushed, the gas (N_2) line is opened and the permeate line and pump are switched off. The permeate is forced back into the membrane module through a back-flushing line. The back-flushing pressure (e.g., 10^5 Pa) should not exceed the maximum operation pressure of the module given by the manufacturer (about 2×10^5 to 3×10^5 Pa for most modules). For long-term operation, the module may be back-washed daily with sterilized distilled water. A previous cleaning with a NaOH solution or wetting with ethanol may substantially recover the filtration performance of the module. Care should be taken to prevent NaOH and/or ethanol from running into the reactor. It is advisable to use two parallel filtration modules; while one module is being cleaned, the other remains in operation.

In recent years, ultrafiltration has been increasingly used for cell retention. An ultrafiltration module is made of a membrane with a pore size in the range of 0.1 to 0.01 μm. In addition to cells, macromolecules such as proteins and fats can also be retained. This can simplify the downstream processing in some cases. Present industrial applications of ultrafiltration include food and beverage processing; pharmaceutical production such as purification of monoclonal antibodies; and fractionation of plasma, whey, and egg proteins. In these applications, viruses and pathogens can be also retained and removed.

In mammalian cell cultures used for production of pharmaceuticals (e.g., factor XIII), an inclined sedimentation device with controlled vibration has been used in industrial scale. Hydroclone has also been examined for selective retention of mammalian cells.

49.4. EXPERIMENTAL DESIGN AND OPERATION

49.4.1. Experimental Design

In designing a continuous culture for long-term operation, one must be aware of the possibility of spontaneous mutations because of selective pressure. The rate of spontaneous mutation is estimated at about 1 in 10^6 cell divisions (66). Sooner or later, cells with a competitive advantage will take over the original strain in the reactor. Thus, if one is interested in studying the physiology of a homogeneous cell population, the operation of a culture should not be too long in order to distinguish between metabolic responses of cells to changes in environmental conditions and genetic changes due to mutation. Some cultures form aggregates or exhibit strong wall growth after prolonged cultivation (e.g., after about 2 to 3 weeks for some *Enterobacter* strains) under metabolic stress (e.g., substrate excess, limitation by growth factors, product inhibition). For these kinds of culture, length of operation is also limited. Care should be taken when interpreting data obtained under conditions of aggregate formation and/or wall growth. If a long-term operation is planned, one should consider the necessity of replacing tubes (damage due to pumping and actions of acid and alkaline solutions) and filters for the incoming and effluent gases used during the process.

49.4.2. Medium Formulation

The basis for medium design is the nutritional requirement for cell growth and product synthesis. In addition to the identification of all components that are necessary for balanced growth or that may interfere with the kinetics of growth and product formation, one has also to establish the appropriate concentrations of the components in a continuous culture, especially if a certain kind of growth limitation is desired. As more and more genetically modified organisms are used for the production of high-value products, chemically defined media and tailored feeding are finding more applications. Considerations of stoichiometry and regulation of metabolism are important for these purposes (43).

The stability of medium formulation and interactions of nutrient components should be considered to avoid precipitation. Precipitation of medium components may be caused by high concentrations of metal ions (e.g., Fe, Ca, and Mg), high temperature during autoclaving, and high pH. This may result in the loss of metal ions and some other trace elements, impairing cell growth and product formation. To achieve a high cell density, the concentrations of metal ions and trace elements must be carefully controlled (or reduced). The use of chelating agents such as EDTA and citrate, particularly in combination, can effectively reduce precipitation. The proper selection of metal ions is also important, because the formation of insoluble salts, such as those associated with phosphate and magnesium, is often a problem.

49.4.3. Medium and Equipment Sterilization

The sterilization of a continuous-culture system is somewhat more complicated than that for a batch-culture system because of the necessity of having more and larger reservoirs for feed medium and acid and alkali solutions and the need to refill or replace them during the process (Fig. 5). Some components of the medium or the acid and alkali solutions may need to be sterilized separately. The lines connecting the various reservoirs and the waste broth vessel to the reactor can become sources of contamination and therefore should be sterilized and connected carefully. The connection should be done under aseptic conditions, e.g., on a clean bench if the reactor system is relatively small. For reactors larger than 5 liters, in situ sterilization may be necessary. This can be done with steam at 121°C, either indirectly through the double wall or by a heat exchanger inside the reactor, or directly by means of steam injection into the reactor. Steam injection is faster and easier but will cause an increase of medium volume by 10 to 15% owing to condensation. The piping lines for large reactors are normally sterilized in situ by using steam. Piping lines should be as short as possible, straight, and laid with a downward slope to ensure the outflow by gravity. T connections should be avoided if possible. To keep the danger of contamination to a minimum, all piping in the sterile stage of a reactor may need to be welded, with no detachable connections. This is very cost-intensive and is therefore only used for special processes (e.g., cell culture processes). In the practical operation of bioreactors, detachable connections are desired to achieve a high degree of flexibility. In this case, only flanges designed in accordance with sterilization requirements should be employed for all connections. The cross sections of piping should be neither reduced nor increased at the flange connection points to avoid accumulation of dirt. The two most popular flange systems are the so-called quick-connection clamping ring system and the screw-on system (17). A common feature of these flanges is that a sterile connection is achieved by "O" rings. O rings are automatic sealing elements, the total sealing pressure of which increases with increasing pressure in the system.

49.4.4. Environmental Control

Temperature, pH, and dissolved oxygen (pO_2) measurement and control are similar in batch and continuous cultures. However, pH and pO_2 probes may present problems that are not usually encountered in short fermentations. Drifts in the pH and pO_2 probes can occur because of biofilm formation or adsorption of proteins and other materials on the tips of the probes. Whereas the pH probe can be roughly recalibrated by briefly changing the culture pH values and comparing with the offline measurements, this cannot be done with the oxygen probe. Whenever possible, polarographic oxygen probes should be used that have much better stability than galvanic probes after autoclaving, and the electronic zero feature can be used to zero them. For more detailed information on the environmental control of bioreactors, the reader is referred to Schügerl (63).

49.4.5. Start-Up

Continuous cultivation of microorganisms is started as a batch culture. It is normally switched to continuous culture during the exponential growth phase before substrate is completely consumed. The transition of a culture after the switch-over can display very different dynamic behaviors, depending on the nature of the culture and the reactor conditions at the time of switch-over (7, 18). Oscillatory transitions may occur as a result of sudden changes of environmental conditions, particularly during transition from substrate limitation to substrate excess (7, 18, 28, 45). If the culture is suddenly exposed to a very high concentration of toxic substrate, the oscillation may become so severe that cells are washed out. If a system can display multiplicity, the timing and operation mode of start-up of a continuous culture may also lead to different transition behaviors and different steady states (36, 77).

In most cases, the use of a half-strength medium for the initial batch growth can give a smooth transition from batch to continuous cultivation. Nutrient feed is started when the cell concentration is about half of that expected for steady state. In this way, the culture initially approaches a nutrient-limited fed batch after the switch-over. The specific growth rate declines as the cell concentration increases until the steady-state concentration is achieved. At this time, the level controller is turned on and the continuous fermentation begins.

49.4.6. Dynamic Behavior and Determination of Steady State

Steady state is defined as the condition of a continuous culture in which changes in the process parameters (e.g., concentrations of substrate, biomass, and product) and the physiological state of cells are no longer detectable. Normally, only the process parameters that can be measured either online or offline are used to judge if a culture has reached a steady state. In this case, as many process parameters as possible should be included. If product(s) is formed, particular attention should be given to the concentration of the product(s). The product usually reaches constant values later than the concentrations of biomass and substrate. The physiological state of cells reaches a steady state even more slowly, especially if long-term adaptations take place (55, 58).

In an ideal chemostat culture governed by Monod kinetics, the time required to achieve a steady state is about three to four times of volume replacement after a condition changes (56). The volume replacement time (residence time) is given by $1/D$. In reality, a much longer time is often needed to achieve steady state, depending on the dynamic behavior of the culture and the magnitude of changes (30, 45, 58). Figure 9 gives some examples of transient behavior of a continuous culture after different step changes of substrate concentration in the feed medium at a constant dilution rate (77). In an extreme case, where the new feed concentration (y_0) increases subtly (from 0.334 to 0.335), the new steady state is reached after almost 40 residence times. The parameter change is visible only after about 24 residence times. On the other hand, if y_0 is increased by a large step, the new steady state is quickly reached after about nine residence times. Overshoot and underwing of

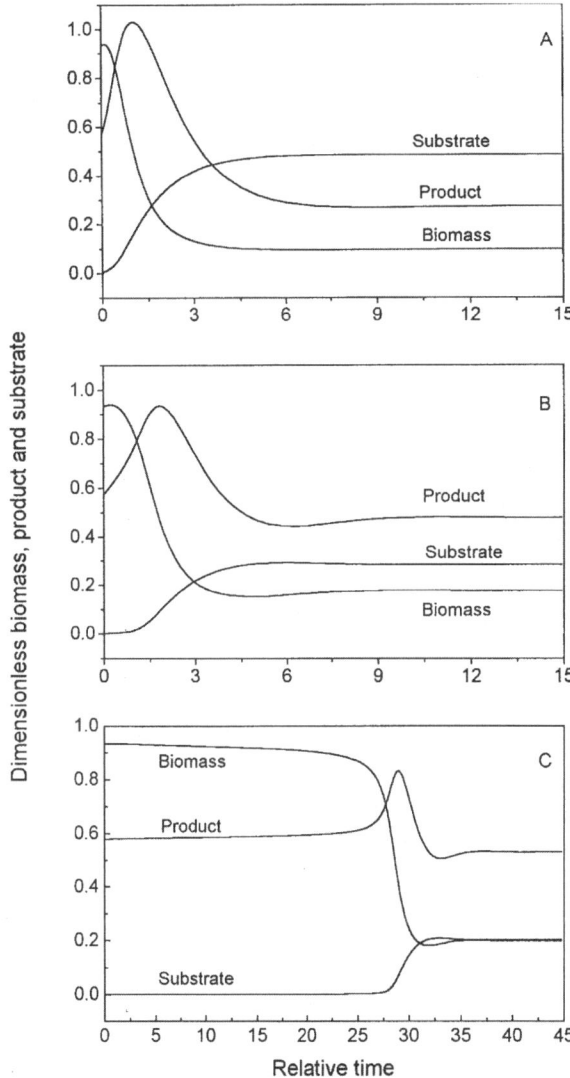

FIGURE 9 Transient behavior of a steady state after step changes in a continuous culture operated at a constant dilution rate. The culture is subjected to metabolic overflow and growth inhibition by substrate and products. The substrate concentration in the feed (y_0) is increased from (A) 0.334 to 0.550 (in dimensionless units), (B) 0.334 to 0.400, and (C) 0.334 to 0.335. The relative time is numerically equal to the number of reactor volume exchanges (residence times) (64).

biomass and product are observed in this culture system. More profound oscillatory transient behaviors are observed under certain conditions (6, 7, 22, 28, 45, 58).

In some culture systems, multiplicity of steady states can occur (1, 36, 45). Depending on the operation mode, two markedly different stable steady states can be obtained under the same cultivation conditions. Xiu et al. (77) presented a theoretical and mathematical analysis of multiplicity and instability of microorganisms in continuous culture and compared them with recent experimental results. Figure 10

depicts results for a culture operated at a constant feed substrate concentration. For cells grown on one substrate without limitation, the combined effects of product inhibition and enhanced formation rate of product under substrate excess appear to be the main reasons for the multiplicity and hysteresis phenomena so far experimentally observed. These combined effects can also lead to unusual dynamic behaviors, such as a prolonged time to reach a steady state, oscillatory transition from one steady state to another, and sustained oscillations (Fig. 9).

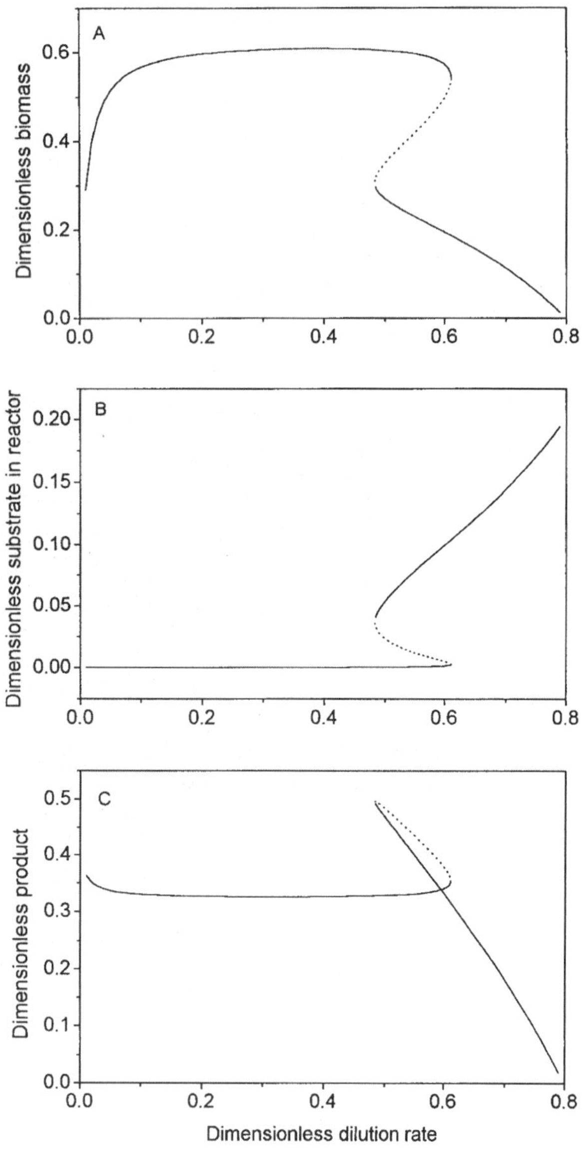

49.5. RECENT DEVELOPMENTS AND FUTURE TRENDS

Whereas the theory of continuous culture is well established, there have been several new developments in the use of this technology. In particular, recent developments in implementing continuous culture in microfluidics or microdevices are worth mentioning and briefly summarized below. With the maturation of microfabrication techniques, continuous culture in microfluidics is drawing increasing interest (35, 37, 40, 59, 64, 84, 85). Microfluidics deals with the behavior, precise control, and manipulation of fluids that are geometrically constrained to a small, typically submillimeter, scale. Microfluidic devices offer many unique properties, such as high surface-to-volume ratio, mass, and heat transfer rates, that are not accessible in conventional bioreactors. Combined with continuous culture, they can be used for a wide range of new applications. Their potential applications include (i) high-throughput screening of medium and parameter optimization; (ii) high-throughput screening of cell lines, drugs, and toxicity testing; (iii) investigation of cellular physiology, in particular at the single-cell level; (iv) controlled study of cellular differentiation and interactions; and (v) cell-based micro biosensors.

For example, fetal human hepatocytes and human hepatocarcinoma cells (Hep-G2) growing in a microchamber showed significantly higher production of albumin compared to standard cultivations in dishes. The higher production of albumin characterizes the higher activity of the cells and therefore the enhanced differentiation, which argues for close in vivo conditions (40). Reduction of the medium volume on top of the cells was considered one of the important reasons (73, 74). By running in perfusion mode, depletion of nutrients and oxygen and accumulation of waste products can be successfully prevented (40, 59). Moreover, continuous culture can be implemented in fully automated biosensor systems by continuous removal and analysis of perfusate. This type of biosensor is a cell-based assay chip wherein the cells act as a transducer that can be used, for example, for drug screenings. Shackman et al. (64) created a biosensor to study the response of Langerhans islet cells to step change of glucose. Various sensors for different cells have been built so far (37, 68). The extension of the throughput of biosensors can be realized by parallelization as arrays of different dimensions, e.g., 8 by 8, 10 by 10, or higher (34, 35, 42, 85). The patterning technologies for fabrication and the potential applications as biosensors are described in detail by Park and Shuler (54).

Microchambers can be integrated with other submicrometer elements such as fluid-handling devices and sensors, becoming sophisticated devices that integrate one or several laboratory functions on a single chip, often called "lab on a chip," or known as micro total analysis systems (μTAS) when used for analysis purposes. A completed system would include cultivation, treatment, selection, lysis,

FIGURE 10 Steady-state concentrations of biomass (A), substrate in reactor (B), and product (C) in a continuous culture at a constant feed substrate concentration but varied dilution rates (dimensionless feed substrate concentration, $y_0 = 0.2$). This culture is subject to metabolic overflow and growth inhibition by product. Multiple steady states are found in a given range of dilution rates. Depending on the operation mode, different steady states can be obtained under identical operation conditions (64).

separation, analysis, etc. (20). Zhang et al. (84) presented a chip on which a chemostat culture was established for microbial cells with the capability of monitoring dissolved oxygen, pH, and optical density. Polydimethylsiloxane, used as a material for the fabrication of the chips, can offer the advantages of direct microscopic visibility of the culture and also biocompatibility. Polydimethylsiloxane is gas permeable and therefore very suitable for aerobic cultivation (14). The next challenge in the technology of on-chip cultivation systems is to adopt standardized interfaces. Not yet available is compatibility with labor equipment that would result in easy-to-use microdevices even for researchers who are not familiar with microfluidic technologies. Attempts are being made to overcome this issue by, for example, employing devices compatible with standard pipettes (47, 85).

As exemplified above by the combined use of continuous culture with microfluids, continuous culture may experience a renaissance again when combined with other enabling technologies or new concepts. With the development of more stable recombinant strains and cell lines and more reliable and accurate on-line measurement and control techniques, continuous culture will receive greater acceptance in the future, as a tool both for basic research and for industrial applications. In particular, its increased use in industrial biotechnology for bioconversion of biomass is expected. Bioconversion processes typically aim at large-scale production of chemicals and fuels from renewable biomass. A continuous operation of such processes is crucial to reduce the production cost. To realize the promise of the new concept, biorefinery continuous cascade or multistep processes are desired. Principles of operation and equipment setup and design for this type of continuous culture deserve further study.

We thank Benedikt Schöpke for providing information on continuous culture in microfluidics.

REFERENCES

1. **Axelsson, J. P., T. Münch, and B. Sonnleitner.** 1992. Multiple steady states in continuous cultivation of yeast, p. 383–386. *In* M. N. Karim and G. Stephanopoulos (ed.), *Proceedings of ICCAFT5/IFAC Modeling and Control of Biotechnical Processes*, Keystone, CO. Pergamon Press, Oxford, UK.
2. **Bailey, J. E., and D. F. Ollis.** 1986. *Biochemical Engineering Fundamentals.* McGraw-Hill, New York, NY.
3. **Bailey, J. E., A. Sburlati, V. Hatzimanikatis, K. Lee, W. A. Renner, and P. S. Tsai.** 2002. Inverse metabolic engineering: a strategy for directed genetic engineering of useful phenotypes. *Biotechnol. Bioeng.* **79:**568–579.
4. **Baloo, S., and D. Ramkrishna.** 1991. Metabolic regulation in bacterial continuous cultures. I. *Biotechnol. Bioeng.* **38:**1337–1352.
5. **Banik, G. G., and C. A. Heath.** 1995. Hybridoma growth and antibody production as a function of cell density and specific growth rate in perfusion culture. *Biotechnol. Bioeng.* **48:**289–300.
6. **Barford, J. P., J. H. Johnston, and P. K. Mwesigye.** 1995. Continuous culture study of transient behaviour of *Saccharomyces cerevisiae* growing on sucrose and fructose. *J. Ferment. Bioeng.* **79:**158–162.
7. **Benthin, S., J. Nielsen, and J. Villadsen.** 1993. Two uptake systems for fructose in *Lactococcus lactis* subsp. *cremoris* FD1 produce glycolytic and gluconeogenic fructose phosphates and induce oscillations in growth and lactic acid formation. *Appl. Environ. Microbiol.* **59:**3206–3211.
8. **Bentley, W., and D. S. Kompala.** 1990. Stability in continuous cultures of recombinant bacteria: a metabolic approach. *Biotechnol. Lett.* **12:**329–334.
9. **Biebl, H.** 1991. Glycerol fermentation to 1,3-propanediol by *Clostridium butyricum*. Measurement of product inhibition by use of a pH-auxostat. *Appl. Microbiol. Biotechnol.* **35:**701–705.
10. **Bro, C., and J. Nielsen.** 2004. Impact of 'ome' analyses on inverse metabolic engineering. *Metab. Eng.* **6:**204–211.
11. **Brown, D. E., D. J. Halsted, and C. G. Sinclair.** 1979. Application of an aerated mixing model to a continuous culture system. *Biotechnol. Lett.* **1:**159–164.
12. **Byun, T. G., A.-P. Zeng, and W.-D. Deckwer.** 1994. Reactor comparison and scale-up for the microaerobic production of 2,3-butanediol by *Enterobacter aerogenes* at constant oxygen transfer rate. *Bioprocess Eng.* **11:**167–175.
13. **Calcott, P. H.** 1981. *Continuous Culture of Cells*, vol. 1 and vol. 2. CRC Press, Inc., Boca Raton, FL.
14. **Charati, S., and S. Stern.** 1998. Diffusion of gases in silicone polymers: molecular dynamics simulation. *Macromolecules* **31:**5529–5535.
15. **Cooney, C. L., H. M. Koplove, and M. Häggstrom.** 1981. Transient phenomena in continuous culture, p. 143–168. *In* P. H. Calcott (ed.), *Continuous Culture of Cells*, vol. 2. CRC Press, Inc., Boca Raton, FL.
16. **Daran-Lapujade, P., J. M. Daran, A. J. van Maris, J. H. de Winde, and J. T. Pronk.** 2009. Chemostat-based micro-array analysis in baker's yeast. *Adv. Microb. Physiol.* **54:**257–311.
17. **Deckwer, W.-D., R. Luttmann, H. G. Reng, and S. Yonsel.** 1987. *Bioreaktoren: Ein Leitfaden für Anwender.* GBF Texte 4, Braunschweig-Stöckheim, Germany.
18. **Dunn, I. J., and J. R. Mor.** 1975. Variable volume continuous cultivation. *Biotechnol. Bioeng.* **17:**1805–1822.
19. **Dykhuizen, D. E., and D. L. Hartl.** 1983. Selection in chemostats. *Microbiol. Rev.* **47:**150–168.
20. **El-Ali, J., P. K. Sorger, and K. F. Jensen.** 2006. Cells on chips. *Nature* **442:**403–411.
21. **Ferenci, T.** 2008. Bacterial physiology, regulation and mutational adaptation in a chemostat environment. *Adv. Microb. Physiol.* **53:**169–229.
22. **Fiechter, A.** 1981. Batch and continuous culture of microbial, plant and animal cells, p. 454–505. *In* H. J. Rehm and G. Reed (ed.), *Biotechnology: a Comprehensive Treatise*, vol. 1. Verlag Chemie, Weinheim, Germany.
23. **Frame, K. K., and W.-S. Hu.** 1991. Kinetic study of hybridoma cell growth in continuous culture. II. Behavior of producers and comparison to non-producers. *Biotechnol. Bioeng.* **38:**1020–1028.
24. **Frame, K. K., and W.-S. Hu.** 1991. Kinetic study of hybridoma cell growth in continuous culture. I. A model for non-producing cells. *Biotechnol. Bioeng.* **37:**55–64.
25. **Gostomski, P., M. Mühlemann, Y.-H. Lin, R. Mormino, and H. Bungay.** 1994. Auxostats for continuous culture research. *J. Biotechnol.* **37:**167–177.
26. **Gottschal, J. C.** 1992. Continuous culture, p. 559–572. *In* J. Lederberg (ed.), *Encyclopedia of Microbiology*, vol. 1. Academic Press, Inc., New York, NY.
27. **Gregory, M. E., M. Bulmer, I. D. L. Bogle, and N. Titcherner-Hooker.** 1996. Optimising enzyme production by bakers yeast in continuous culture: physiological knowledge useful for process design and control. *Bioprocess Eng.* **15:**239–245.
28. **Harrison, D. E. F., and H. H. Topiwala.** 1974. Transient and oscillatory states of continuous culture, p. 168–219. *In* T. H. Ghose and A. Fiechter (ed.), *Advanced Biochemical Engineering*, vol. 3. Springer-Verlag, Berlin, Germany.
29. **Hellmuth, K., D. J. Korz, E. A. Sanders, and W.-D. Deckwer.** 1994. Effect of growth rate on stability and gene expression of recombinant plasmids during continuous

and high cell density cultivation of *Escherichia coli* TG1. *J. Biotechnol.* **32:**289–298.

30. **Herbert, D., R. Elsworth, and R. C. Telling.** 1956. The continuous culture of bacteria; a theoretical and experimental study. *J. Gen. Microbiol.* **14:**601–622.

31. **Hiller, G. W., D. S. Clark, and H. W. Blanch.** 1993. Cell retention-chemostat studies of hybridoma cells: analysis of hybridoma growth and metabolism in continuous suspension culture in serum-free medium. *Biotechnol. Bioeng.* **42:**185–195.

32. **Hiller, G. W., D. S. Clark, and H. W. Blanch.** 1994. Transient responses of hybridoma cells in continuous culture to step changes in amino acid and vitamin concentrations. *Biotechnol. Bioeng.* **44:**303–321.

33. **Hoskisson, P. A., and G. Hobbs.** 2005. Continuous culture—making a comeback? *Microbiology* **151**(Pt. 10):3153–3159.

34. **Hung, P. J., P. J. Lee, P. Sabounchi, N. Aghdam, R. Lin, and L. P. Lee.** 2005. A novel high aspect ratio microfluidic design to provide a stable and uniform microenvironment for cell growth in a high throughput mammalian cell culture array. *Lab Chip* **5:**44–48.

35. **Hung, P. J., P. J. Lee, P. Sabounchi, R. Lin, and L. P. Lee.** 2005. Continuous perfusion microfluidic cell culture array for high-throughput cell-based assays. *Biotechnol. Bioeng.* **89:**1–8.

36. **Imanaka, T., T. Kaieda, K. Sato, and H. Taguchi.** 1972. Optimization of α-galactosidase production by mold. I. α-Galactosidase production in batch and continuous culture and a kinetic model for enzyme production. *J. Ferment. Technol.* **50:**633–646.

37. **Jang, K., K. Sato, K. Igawa, U. I. Chung, and T. Kitamori.** 2008. Development of an osteoblast-based 3D continuous-perfusion microfluidic system for drug screening. *Anal. Bioanal. Chem.* **390:**825–832.

38. **Kiss, R. D., and G. Stephanopoulos.** 1992. Culture instability of auxotrophic amino acid producers. *Biotechnol. Bioeng.* **40:**75–85.

39. **Kubitschek, H. E.** 1970. *Introduction to Research with Continuous Culture.* Prentice Hall, Englewood Cliffs, NJ.

40. **Leclerc, E., Y. Sakai, and T. Fujii.** 2004. Perfusion culture of fetal human hepatocytes in microfluidic environments. *Biochem. Eng. J.* **20:**143–148.

41. **Lee, C. W., and H. N. Chang.** 1987. Kinetics of ethanol fermentations in membrane cell recycle fermentors. *Biotechnol. Bioeng.* **29:**1105–1112.

42. **Lee, P. J., P. J. Hung, V. M. Rao, and L. P. Lee.** 2006. Nanoliter scale microbioreactor array for quantitative cell biology. *Biotechnol. Bioeng.* **94:**5–14.

43. **Linz, M., A.-P. Zeng, R. Wagner, and W.-D. Deckwer.** 1997. Stoichiometry, kinetics, and regulation of glucose and amino acid metabolism of a recombinant BHK cell line in batch and continuous cultures. *Biotechnol. Prog.* **13:**453–463.

44. **Marr, A. G.** 1991. Growth rate of *Escherichia coli. Microbiol. Rev.* **55:**316–333.

45. **Menzel, K., A.-P. Zeng, H. Biebl, and W.-D. Deckwer.** 1996. Kinetic, dynamic, and pathway studies of glycerol metabolism by *Klebsiella pneumoniae* in anaerobic continuous culture. Part I. The phenomena and characterization of oscillation and hysteresis. *Biotechnol. Bioeng.* **52:**549–560.

46. **Merten, O. W., D. Moeurs, H. Keller, M. Leno, G. E. Palfi, L. Cabanie, and E. Couve.** 1994. Modified monoclonal antibody production kinetics, kappa/gamma mRNA levels, and metabolic activities in a murine hybridoma selected by continuous culture. *Biotechnol. Bioeng.* **44:**753–764.

47. **Meyvantsson, I., J. W. Warrick, S. Hayes, A. Skoien, and D. J. Beebe.** 2008. Automated cell culture in high density tubeless microfluidic device arrays. *Lab Chip* **8:**717–724.

48. **Minkevich, I. G., A. Yu Krynitskaya, and V. K. Eroshin.** 1989. Bistat: a novel method of continuous cultivation. *Biotechnol. Bioeng.* **33:**1157–1161.

49. **Mohan, S. B., S. R. Chohan, J. Eade, and A. Lyddiatt.** 1993. Molecular integrity of monoclonal antibodies produced by hybridoma cells in batch culture and in continuous-flow culture with integrated product recovery. *Biotechnol. Bioeng.* **42:**974–986.

50. **Monod, J.** 1950. La technique de culture continue: theorie et application. *Ann. Inst. Pasteur (Paris)* **79:**390.

51. **Moser, A.** 1985. Continuous cultivation, p. 285–309. *In* H. J. Rehm and G. Reed (ed.), *Biotechnology: a Comprehensive Treatise,* vol. 2. Verlag Chemie, Weinheim, Germany.

52. **Nancib, N., and J. Boudrant.** 1992. Effect of growth rate on stability and gene expression of a recombinant plasmid during continuous culture of *Escherichia coli* in a nonselective medium. *Biotechnol. Lett.* **14:**643–648.

53. **Novick, A., and L. Szilard.** 1950. Description of the chemostat. *Science* **112:**715–716.

54. **Park, T. H., and M. L. Shuler.** 2003. Integration of cell culture and microfabrication technology. *Biotechnol. Prog.* **19:**243–253.

55. **Petrik, M., O. Käppeli, and A. Fiechter.** 1983. An expanded concept for the glucose effect in the yeast *Saccharomyces uvarum:* involvement of short- and long-term regulation. *J. Gen. Microbiol.* **129:**43–49.

56. **Pirt, S. J.** 1975. *Principles of Microbe and Cell Cultivation.* Blackwell Scientific Publications, Oxford, UK.

57. **Pirt, S. J.** 1965. The maintenance energy of bacteria in growing cultures. *Proc. R. Soc. Lond. Ser. B* **163:**305–314.

58. **Postma, E., C. Verduyn, W. A. Scheffers, and J. P. van Dijken.** 1989. Enzymic analysis of the Crabtree effect in glucose-limited chemostat cultures of *Saccharomyces cerevisiae. Appl. Environ. Microbiol.* **55:**468–477.

59. **Powers, M. J., K. Domansky, M. R. Kaazempur-Mofrad, A. Kalezi, A. Capitano, A. Upadhyaya, P. Kurzawski, K. E. Wack, D. B. Stolz, R. Kamm, and L. G. Griffith.** 2002. A microfabricated array bioreactor for perfused 3D liver culture. *Biotechnol. Bioeng.* **78:**257–269.

60. **Rice, C. W., and W. P. Hempfling.** 1985. Nutrient-limited continuous culture in the pH-auxostat. *Biotechnol. Bioeng.* **27:**187–191.

61. **Sanchez, O., H. van Gemerden, and J. Mas.** 1996. Description of a redox-controlled sulfidostat for the growth of sulfide-oxidizing phototrophs. *Appl. Environ. Microbiol.* **62:**3640–3645.

62. **Sauer, U.** 2001. Evolutionary engineering of industrially important microbial phenotypes. *Adv. Biochem. Eng. Biotechnol.* **73:**129–169.

63. **Schügerl, K.** 1991. Common instruments for process analysis and control, p. 5–25. *In* H. J. Rehm and G. Reed (ed.), *Biotechnology: a Comprehensive Treatise,* vol. 4. Verlag Chemie, Weinheim, Germany.

64. **Shackman, J. G., G. M. Dahlgren, J. L. Peters, and R. T. Kennedy.** 2005. Perfusion and chemical monitoring of living cells on a microfluidic chip. *Lab Chip* **5:**56–63.

65. **Shoham, Y., and A. L. Demain.** 1991. Kinetics of loss of a recombinant plasmid in *Bacillus subtilis. Biotechnol. Bioeng.* **37:**927–935.

66. **Stafford, K.** 1986. Continuous fermentation, p. 137–151. *In* A. L. Demain and N. A. Solomon (ed.), *Manual of Industrial Microbiology and Biotechnology.* American Society for Microbiology, Washington, DC.

67. **Tempest, D. W., and O. M. Neijssel.** 1992. Physiological and energetic aspects of bacterial metabolite overproduction. *FEMS Microbiol. Lett.* **79:**169–176.

68. **Thompson, D. M., K. R. King, K. J. Wieder, M. Toner, M. L. Yarmush, and A. Jayaraman.** 2004. Dynamic gene expression profiling using a microfabricated living cell array. *Anal. Chem.* **76:**4098–4103.

69. Toda, K. 2003. Theoretical and methodological studies of continuous microbial bioreactors. *J. Gen. Appl. Microbiol.* **49:**219–233.

70. Togna, A. P., J. Fu, and M. L. Shuler. 1993. Use of a simple mathematical model to predict the behavior of *Escherichia coli* overproducing β-lactamase within continuous single- and two-stage reactor systems. *Biotechnol. Bioeng.* **42:**557–570.

71. Topiwala, H. H., and G. Hamer. 1971. Effect of wall growth in steady-state continuous cultures. *Biotechnol. Bioeng.* **13:**919–922.

72. Vierheller, C., A. Goel, M. Peterson, M. M. Domach, and M. M. Ataai. 1995. Sustained and constitutive high levels of protein production in continuous cultures of *Bacillussubtilis. Biotechnol. Bioeng.* **47:**520–524.

73. Walker, G. M., M. S. Ozers, and D. J. Beebe. 2002. Insect cell culture in microfluidic channels. *Biomed. Microdevices* **4:**161–166.

74. Walker, G. M., H. C. Zeringue, and D. J. Beebe. 2004. Microenvironment design considerations for cellular scale studies. *Lab Chip* **4:**91–97.

75. Werner, R. G., F. Walz, W. Noe, and A. Konrad. 1992. Safety and economic aspects of continuous mammalian cell culture. *J. Biotechnol.* **22:**51–68.

76. Weusthuis, R. A., J. T. Pronk, P. J. van den Broek, and J. P. van Dijken. 1994. Chemostat cultivation as a tool for studies on sugar transport in yeasts. *Microbiol. Rev.* **58:**616–630.

77. Xiu, Z.-L., A.-P. Zeng, and W.-D. Deckwer. 1998. Multiplicity and stability analysis of microorganisms in continuous culture: effects of metabolic overflow and growth inhibition. *Biotechnol. Bioeng.* **57:**251–261.

78. Yang, R. D., and A. E. Humphrey. 1975. Dynamic and steady state studies of phenol biodegradation in pure and mixed cultures. *Biotechnol. Bioeng.* **17:**1211–1235.

79. Zeng, A.-P., and W.-D. Deckwer. 1995. A kinetic model for substrate and energy consumption of microbial growth under substrate-sufficient conditions. *Biotechnol. Prog.* **11:**71–79.

80. Zeng, A.-P., and W.-D. Deckwer. 1995. Mathematical modeling and analysis of glucose and glutamine utilization and regulation in cultures of continuous mammalian cells. *Biotechnol. Bioeng.* **47:**334–346.

81. Zeng, A.-P., H. Biebl, and W.-D. Deckwer. 1990. 2,3-Butanediol production by *Enterobacter aerogenes* in continuous culture: role of oxygen supply. *Appl. Microbiol. Biotechnol.* **33:**264–268.

82. Zeng, A.-P., H. Biebl, and W.-D. Deckwer. 1991. Production of 2,3-butanediol in a membrane bioreactor. *Appl. Microbiol. Biotechnol.* **34:**463–468.

83. Zeng, A.-P., and W.-D. Deckwer. 1996. Bioreaction techniques under low oxygen tension and oxygen limitation: from molecular level to pilot plant reactor. *Chem. Eng. Sci.* **51:**2305–2314.

84. Zhang, Z., P. Boccazzi, H. G. Choi, G. Perozziello, A. J. Sinskey, and K. F. Jensen. 2006. Microchemostat-microbial continuous culture in a polymer-based, instrumented microbioreactor. *Lab Chip* **6:**906–913.

85. Zhu, X., C. L. Yi, B. H. Chueh, M. Shen, B. Hazarika, N. Phadke, and S. Takayama. 2004. Arrays of horizontally-oriented mini-reservoirs generate steady microfluidic flows for continuous perfusion cell culture and gradient generation. *Analyst* **129:**1026–1031.

Advances in Sensor and Sampling Technologies in Fermentation and Mammalian Cell Culture

ADEYMA Y. ARROYO

50

50.1. INTRODUCTION: PROCESS ANALYTICAL TECHNOLOGY AND NEW SENSOR TECHNOLOGY

Significant progress has been made in the past 2 decades in understanding the impact of the cellular environment on culture physiology and its subsequent impact on productivity, product quality, and downstream processes. In September 2004, the Food and Drug Administration (FDA) published the *Guidance for Industry PAT—a Framework for Innovative Pharmaceutical Development, Manufacturing, and Quality Assurance* (20). In it, the FDA describes process analytical technology (PAT) as a mechanism to design, analyze, and control pharmaceutical manufacturing processes through the measurement of critical process parameters which affect critical quality attributes. The document notes that the term PAT includes "chemical, physical, microbiological, mathematical, and risk analysis conducted in an integrated manner" as a series of tools designed to increase process understanding while meeting regulatory requirements. Therefore, the PAT framework includes at-line and on-line sensing or analytical technologies that along with design of experiments (DOE), chemometrics, information technology, and multivariate analysis allow scientists and researchers to assess the quality of the product. Through the PAT initiative, the Office of Pharmaceutical Sciences at the FDA encourages a better understanding of the manufacturing process, identifying relationships between process parameters and reproducibility, culture performance, and product quality attributes. Although originally conceptualized for small-molecule manufacturing processes, it is increasingly being extended to biological processes, which are inherently more complex and generally less understood processes.

Since the publication of this guidance, and leading up to the time of this writing, research in this area has significantly increased. Industry and regulators have come together in conferences and meetings to invest in PAT, generating improvements in existing technologies and resulting in the evaluation and implementation of new tools for greater process understanding. Application of PAT tools across the industry also facilitates quality-by-design (QBD) efforts. QBD is a systematic approach to determining the critical quality attributes of a product. A QBD approach utilizes DOE to establish a design space or a process variability range within which the product attributes can vary without affecting product quality. The ability to measure process data in real time using PAT tools, i.e., sensor technologies, data analysis, and process understanding, would enable manufacturing flexibility, efficiency, and, potentially, real-time release. This chapter reviews research on existing and new sampling and sensor technologies for monitoring fermentations and mammalian cell culture in the last decade. New technologies are discussed, as well as advances in existing technologies. Each one of these technologies contributes to enabling PAT and increased process understanding.

50.2. IMPLEMENTING A NEW SENSOR TECHNOLOGY

In order to understand the acceptability of any new measurement technology, it is important to consider both the accuracy and precision of a measurement. Measurement accuracy reflects how close a measurement is to the "true" value. For example, in the case of autosampling technologies, a researcher is typically introducing a new automated technology to replace an existing manual method. In this case, an accurate measurement is defined as one that gives values comparable to those obtained with the established measurement technology. Based on this, when an automated cell counting method is introduced to replace a manual method, it is important to understand how the cell density and viability measurements compare between the two methods. Understanding the comparability is important to allow useful comparison between old and new data. Without this understanding, a researcher can lose the value of many years of process development and/or manufacturing data. Further, the comparability data are a regulatory requirement if a new sample analysis technology is introduced into a commercial manufacturing process. Demonstrating comparability can be problematic, particularly if the newer technology is actually better at delivering measurements closer to the true value of the variable. In cases where there is an offset between the established and new measurements, it is important to characterize this offset and appropriate correlations so that data from the old and new measurements can be suitably compared.

Precision measures the ability of an analysis method to be reproducible. Understanding precision is part of evaluating any new sensor technology. Sensor reproducibility may need to be measured over a wide range of conditions to understand the performance. Typical conditions studied are precision over time, with different sensors (including different vendor lot numbers), with different cell lines, at different cell densities and viabilities, using different media, and under various conditions of temperature, pH, and osmolality.

Satisfying the accuracy and precision requirements should be strongly considered when evaluating any new technology. Manufacturing and development considerations must be carefully evaluated and established early in the process. For manufacturing implementation, a regulatory strategy may be required near the beginning stages of the evaluation to ensure that the appropriate comparability data are gathered in a timely manner.

50.3. AUTOSAMPLING TECHNOLOGIES

Cell culture and bacterial processes continue to heavily rely on manual sample and off-line analyses for process monitoring. Manual sampling and analysis have several disadvantages, including risk of contaminating cultures, high labor costs, and lack of reproducibility. Given these issues, there has been increased interest in automated sample handling and analysis technologies in recent years. As mentioned above, the introduction of automated sample handling and analysis technologies requires demonstration that the automated technology provides results that are acceptable for the intended use.

50.4. CELL CULTURE PROCESS AUTOSAMPLING TECHNOLOGIES

In cell culture processes, a few technologies have been developed to automate sampling and analysis from bioreactors. The most explored technologies include BioProfile FLEX (Nova Biomedical, Waltham, MA) and the Automated Reactor Sampling (ARS) 800 (Groton Biosystems, Boxborough, MA) systems. Table 1 lists the process variables measured by the current systems.

50.4.1. BioProfile FLEX Autosampler

The BioProfile FLEX autosampler is an off-line cell culture monitoring technology that combines three off-line analyzers, namely, a nutrient metabolite analyzer, a cell counter, and an osmometer (14). The system has a pumping module and 10 reactor valve modules (RVM). The pumping module drives all fluid movement within the autosampler system via a syringe pump. Each RVM is designed to open only when the pressure gradient, determined by a built-in pressure sensor, is such that flow will be out of the bioreactor, thus serving to protect the sterility of the culture. The nutrient metabolite module analyzer provides gas, electrolyte, nutrient, and metabolite data. The cell counter measures viable and total cell counts, percent viability, and live cell diameter, while the osmometer measures osmolality. The autosampler is a modular system, with each module capable of working individually or in combination with others. If all modules are selected, a sample volume of about 1 ml is required with an analysis time of 7 to 7.5 min. This autosampler can be connected up to 10 bioreactors at the same time, with the flexibility of analyzing all or just selected bioreactors. The modular design and the single automation system allow for all process data to be collected under one system. Figure 1 shows the analyzer, module layout and general setup.

TABLE 1 Variables measured using autosampling technologies[a]

Process variables	Analyte	Standard off-line analyzer
Gas, electrolytes[b]	pH	Nova BioProfile 400
	pO_2	
	pCO_2	
	Ammonium ion	
	Potassium ion	
	Sodium ion	
Nutrient, metabolite	Glucose	Nova BioProfile 400
	Lactate	
	Glutamine	
	Glutamate	
Cell density	Viable cell density	Beckman Coulter Vi-Cell AS
	Viability	
	Live cell diameter	
Osmometer	Osmolality	Advanced Instruments 3600

[a]Adapted from reference 14.
[b]Nova FLEX can also measure calcium ion.

Comparability between the BioProfile FLEX and traditional off-line analyzers has been demonstrated (14). Results of the automated versus manual sampling method showed acceptable accuracy: 97 to 100% of the samples within 25% of each other and 81 to 100% of all samples within 10% of each other, with one exception, dissolved oxygen (DO) concentration. Oxygen levels measured as partial pressure (pO_2) can be highly variable, probably due to off-gassing and other phenomena during sampling handling. The data are based on runs with 13 cell lines and over 20 operators. The inter-instrument precision was also found to be acceptable, with most parameters showing minimum variability. However, there were some exceptions in the nutrient metabolite module when the concentrations tested were in the low end of the analytical range.

50.4.2. Groton Biosystems' ARS Systems

The ARS system from Groton Biosystems can sample from up to eight bioreactors and deliver the sample to up to four different analyzers. Remote valve interfaces (RVI) are used to sample from the bioreactors with no backflow into the reactor. The sample is pulled through the RVI into the sampling valve and then out to an analytical instrument. A cleaning procedure cleans all parts of the unit up to each RVI, including the sampling line to the analytical instrument. Unlike the modular design of the BioProfile FLEX, the Groton Biosystems ARS is an independent unit that requires software development to allow for communications between the sampling unit and each one of the analytical instruments, generally via object linking and embedding for process control protocols. The challenge of implementation relies heavily in the automation resources required for communications to the different analytical instruments and the automation support from each one of the vendors. Figure 2 shows the basic layout

A

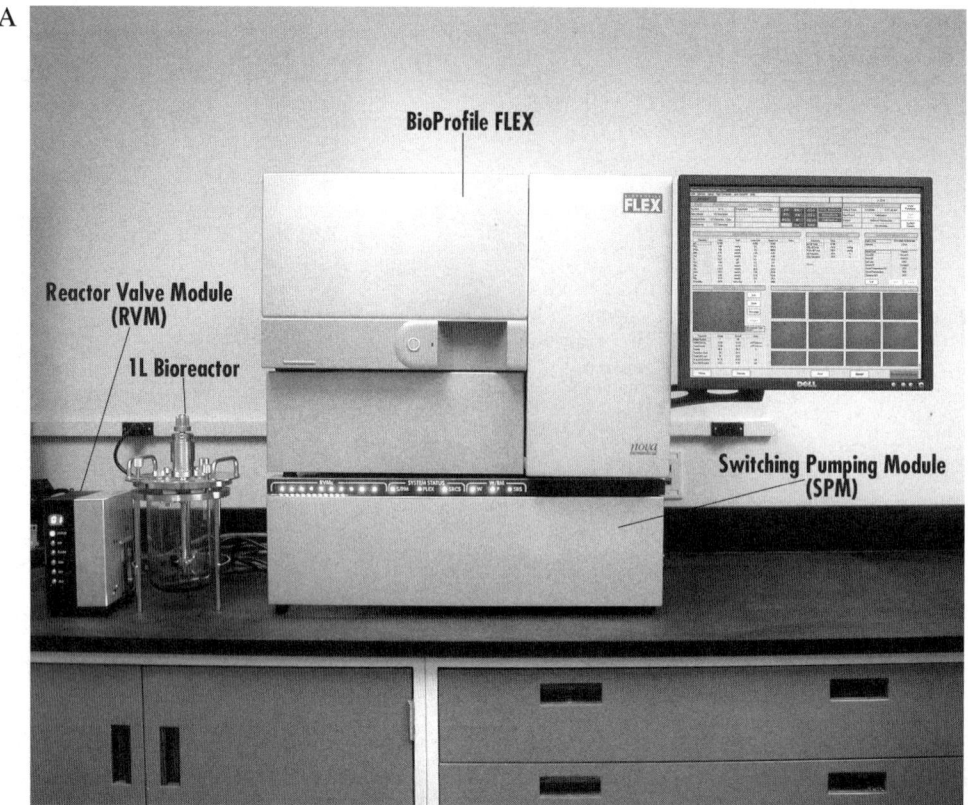

B

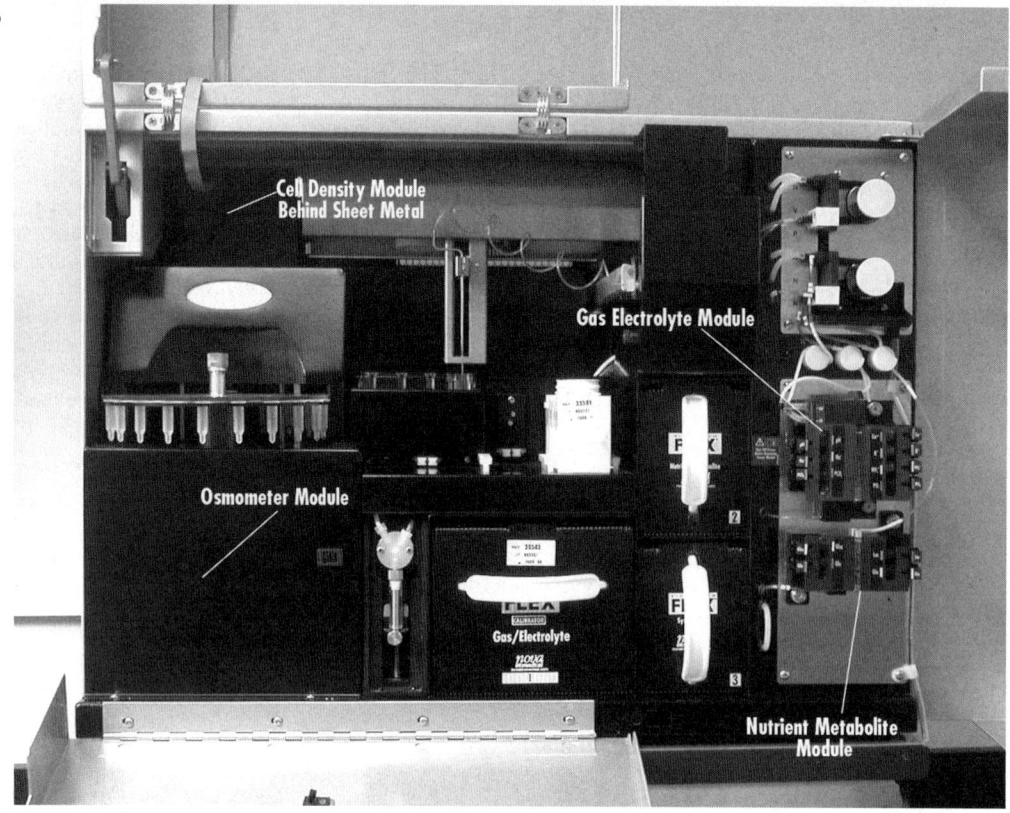

FIGURE 1 Nova BioProfile FLEX system. (A) BioProfile FLEX unit (courtesy of John Balsavich, Nova Biomedical). (B) Internal modules (courtesy of John Balsavich, Nova Biomedical). (*continued*)

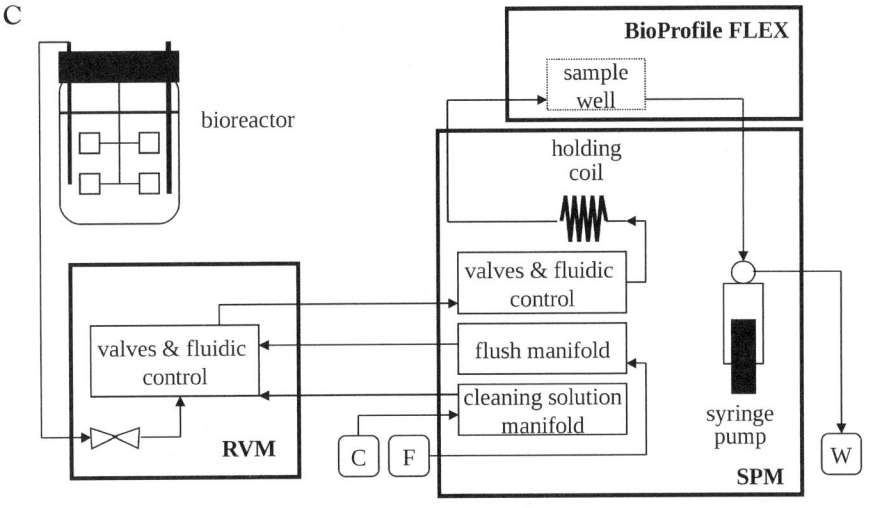

FIGURE 1 (*continued*) (C) Schematic of fluid flow paths through system. The sample flows from the bioreactor to the RVM to the switching pumping module (SPM) before reaching the BioProfile FLEX. Waste flows from the BioProfile FLEX to the SPM to waste container (W). Cleaning solution (C) and saline flush (F) flow from their respective containers to SPM manifolds to RVM fluidics, following the sample path through the system to the waste container. (Adapted from reference 14.).

of the system and a picture of the system connected to four bioreactors.

Several studies showed comparability between the manual and automated sampling methods (70, 83). Automated sampling showed pH results comparable with those of manual sampling, with 80% of the samples showing values within 0.05 pH unit and 100% within 0.1 pH unit. In the case of glucose, automated sampling values were within 25% of manual sampling values. All other metabolites showed results within 10% of those obtained with the manual sampling method. pCO_2 and pO_2 were variable, as in the BioProfile FLEX, probably due to similar off-gassing and sampling issues. Finally, automation and sampling and flow rate optimization were required for cell density and viability measurements (70). A different study showed successful interface with Nova BioProfile, Cedex, a fraction collector, and a high-performance liquid chromatography (HPLC) system for amino acid analysis (83).

50.5. BACTERIAL FERMENTATION AUTOSAMPLING TECHNOLOGIES

In bacterial fermentations, implementation of autosampling systems is challenging. High sample viscosity and cleaning requirements have to be overcome to ensure sample flow and sterility. Despite the challenges of automated sampling, there have been recent advances in automated sample analysis using robotics and a liquid handling system for sample management and analysis (V. Saucedo, unpublished data). This solution is generally customized, and it involves a partnership between users, researchers, and automation engineers to ensure that the design and implementation results in accurate measurements of the process variables. Several companies with expertise in custom liquid handling systems, automation, and robotics can be explored, including Tecan Group Ltd. (Männedorf, Switzerland), Beckman Coulter, Inc. (Fullerton, CA), and Hamilton Company (Reno, NV).

50.6. FLOW INJECTION ANALYSIS

In addition to autosampling systems, another technology that can be used to deliver samples from bioreactors for analysis is flow injection analysis (FIA). The sample is taken using in situ filtration ceramic probes (Flownamics, Madison, WI) to a detection system based on amperometric, potentiometric, fluorescence, chemiluminescence, turbidity, or absorption measurements. The application of FIA to monitor cell culture processes has been described in the literature. Ammonia and glutamine were simultaneously monitored on-line in a hollow-fiber reactor containing hybridoma cells (71). Glucose and glutamine were monitored in serum-free hybridoma media using a split-stream FIA (52). One advantage of FIA systems is that the sensors employed do not have to be operated in a sterile manner and are therefore not subjected to the same stringent requirements as for in situ sensors. Although FIA techniques can be employed to measure a large number of cell culture components of interest, they have not been widely adopted due to concerns of introducing additional contamination risks, multiple detection systems, and calibration requirements. A review of this and spectroscopic methods (discussed next) can be found in the 2001 article by Schügerl (65).

50.7. IN SITU PROCESS MONITORING USING SPECTROSCOPY

Autosampling technologies can minimize the disadvantages associated with manual sampling. However, automated

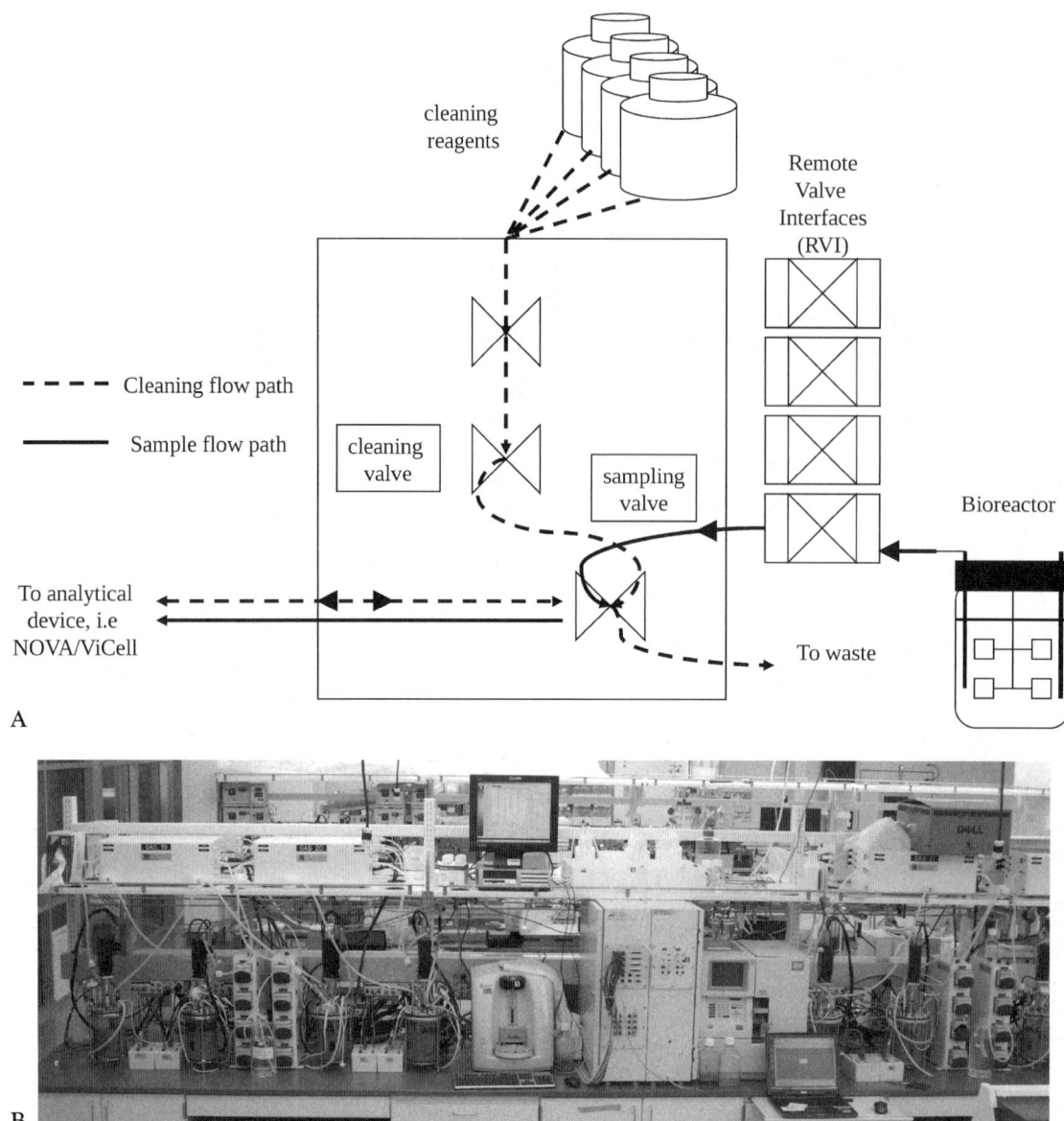

FIGURE 2 Groton ARS System. (A) Schematic of the Groton ARS. Cleaning flow cleans all parts of the unit up to each RVI, including the sampling line to the analytical device. The sample is pulled through the RVI into the sampling valve and then out to an analytical instrument. Adapted from reference 70. (B) Picture of the ARS system connected to six bioreactors. (Courtesy of Daniel Abramzon, Genentech, Inc.)

methods continue to pose a sterility risk throughout multiple cycles of sampling and cleaning procedures. They also add complexity associated with the sampling software and hardware. In addition, the frequency of an at-line measurement (automated or not) can be low due to available volume and/ or long analysis times. In situ sensors offer the advantages of maintaining the sterile envelope during culture, relatively simple hardware, measuring well-mixed samples, and high measurement frequency. In situ bioreactor sensor requirements include (i) ability to be sterilized, (ii) reliability and robustness, (iii) high specificity to the measured parameter, (iv) fast response time, and (v) minimal drift. In addition, the sensor must require minimal maintenance and must not increase the risk of contamination. Spectroscopic probes

are an established in situ technology in other industries, but they have not traditionally been used in bioreactor applications. One major challenge with these sensors is the complexity of the spectrum gathered in a bioreactor. Interpreting the spectra in a way that provides accurate, precise measurement of one or more variables is a complex challenge requiring chemometric tools to correlate spectra to a specific variable.

Spectroscopic methods have improved significantly in the last decade. Several companies have explored the use of near-infrared spectroscopy (NIRS), Raman spectroscopy, and mid-infrared spectroscopy (MIRS) for on-line bioreactor monitoring in bacterial fermentations and mammalian cell culture. Spectroscopy, particularly using

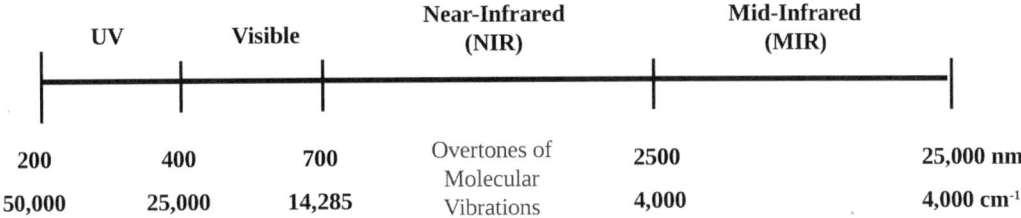

FIGURE 3 Regions of the spectra and their wavelengths.

NIR or MIR, is an attractive alternative for the measurement of cell culture components (including substrates, waste products, and essential and nonessential amino acids), cell concentration, and viability. Sample preparation is not required, and this technique can be used to measure a large number of parameters simultaneously.

50.7.1. NIRS

The IR region of the electromagnetic spectrum is composed of several sections: the MIR, the NIR, and the far IR, based on their proximity to the visible light (Fig. 3). NIRS is an optical method of gathering measurements between the visible and MIR regions of the electromagnetic spectrum (800 to 2,500 nm). The measurements are based on overtones of molecular vibrations. NIRS captures the contributions of all sample constituents, resulting in a complex spectrum with absorbance at multiple wavelengths. Data interpretation is thus dependent on multivariate data analysis (MVDA) and strong calibration methods (74).

In applications to cell culture processes or bacterial fermentations, a light source is introduced to the sample via a steam-sterilizable probe mounted using standard bioreactor ports. The resulting transmittance measurement can be correlated to individual analytes at various concentrations in a sample matrix. Chemometric methods are employed for data deconvolution and analysis. Initially the NIRS scans are trained using off-line measurements of cell counts, osmolality, metabolites, and amino acids or other cell culture or bacterial fermentation medium components. This training results in a calibration model that is then used to predict unknown spectral scans in independent bioreactor cultures.

A number of studies have been reported in the last decade using NIRS for bioprocess monitoring in cell cultures and bacterial fermentations. Some studies have been focused on investigating the tools required for interpreting the spectral information (74). Others have focused on real-time monitoring of specific analytes in animal cell culture or bacterial fermentations using in situ probes (3, 53, 61) or at-line analysis (49). In spite of this, NIRS is rarely applied in cell culture processes (even in a laboratory setting). This may be due to the fact that the spectroscopic measurements (position and intensity of spectrum) are sensitive to changes in process parameters such as pH, ionic strength, temperature, and the analytical ranges used when developing the models. Furthermore, depending on the application, multiple mathematical treatments may be required for different analytes in the same process. For example, the second derivative can be used to deconvolute overlapping peaks and resolve baseline drift, whereas the standard normal variant can be used to clean up noise or interference (3). In general, a robust application would require a model based on a comprehensive design of experiments that incorporates the analyte ranges, process changes or excursions, and the appropriate mathematical treatments for each analyte.

Despite the challenges, the use of in situ sterilizable probes continues to be evaluated, as vendors have improved their design to withstand bioreactor operations. Several vendors are developing NIRS solutions, including Foss NIR Systems (Laurel, MD), which has a commercially available unit capable of connecting probes in up to nine bioreactors (nine channels). Successful implementation of these probes requires extensive testing, including probe design and integrity across multiple autoclaving or steam-in-place cycles, signal optimization (path length evaluation), bioreactor condition evaluations (aeration levels, fouling, temperature shifts, etc.), and the effect of sensor position in bioreactors of different scales. In addition, application of this promising technology for cell culture monitoring will require the establishment of robust calibration models that can be implemented routinely by nonexpert operators. Several studies using in situ probes are summarized in Table 2.

50.7.2. Raman Spectroscopy

Raman spectroscopy is a technique based on inelastic scattering. When a sample is illuminated by light, inelastic scattering is observed and a shift in energy (wavelength) can be detected. This energy shift is specific to a chemical bond, thus providing information about the sample chemistry or composition.

In the last several years, Raman spectroscopy has seen tremendous advances due to instrumentation and data analysis tool improvements. Applications of this technique to bioprocess monitoring have been limited. At-line or off-line monitoring using Raman spectroscopy has been explored in bacterial fermentations for the analysis of glucose, lactic acid, and optical density (69), and in fungal fermentations for the analysis of gibberellins, important agricultural products (51). Real-time or in situ monitoring has also been evaluated for carotenoid production in yeast cultures (6) and for glucose, acetate, formate, lactate, and phenylalanine in bacterial fermentations (43). As in the case of NIRS, robust mathematical algorithms and identification of the best "Raman shift" for a particular analyte are required.

The results of these studies suggest that Raman spectroscopy, though promising, is still evolving for bioprocess monitoring. Several aspects need to be evaluated or considered during the evaluation. These include attenuation of light scatter due to bubbles and high biomass, calibration models, probe window design, analyte concentration and variability (complex mixtures), and exposure time. Studies using in situ probes are summarized in Table 3.

TABLE 2 Summary of NIRS in situ probe monitoring applications

Type of culture	Analyte(s) monitored	Probes used	Model(s) used	Major findings and challenges	Reference
Mammalian cell culture					
CHO-K1	Glucose, lactate, glutamine, ammonia	In situ Foss NIR Systems probe Path length, 1.2 mm NIR range, 400–2,500 nm	2nd derivative, standard normal variant and combination of both	Large data sets were used to develop models for glucose, lactate, glutamine, and ammonia. The models were developed in a fed-batch bioprocess involving a temperature change.	3
CHO	Glucose, lactate	Seven in situ Bruker Matrix F FT-NIR system	PLS Multiplicative scatter correction	Multiplexing 5 to 6 probes in multiple bioreactors Sources of variance for multiple probes include spectrophotometer channels, probe construction variability, and mirror optical properties.	61
Bacterial fermentation					
Vibrio cholerae	Biomass, glucose, acetate	In situ Foss NIR Systems probe Path length, 0.5 mm NIR range, 400–2,500 nm	PLS	Control strategies improved using NIR data and electronic noses or chemical gas sensors Challenges include aeration disturbances, gas phase effects, probe fouling, and position of the probe within the reactor in relation to agitator and sparger.	53
E. coli	Biomass	In situ Foss NIR Systems probe Path length, 0.5 mm NIR range, 400–2,500 nm	PLS, 2nd derivative	At-line data were compared to on-line or in situ data. Temperature, agitation, aeration, and feeding affect the in situ results. Path length optimization was performed.	4

50.7.3. MIRS

The MIR range is the region past the NIR spectrum, between 2,500 and 25,000 nm (Fig. 3). There have been a number of studies exploring MIRS for fermentation monitoring. As in the case of NIRS, the spectra can provide specific information for different molecules and thus provide the possibility of real-time monitoring. One advantage of MIRS can be a higher selectivity and signal-to-noise ratio than with NIRS for some analytes. Comparisons between Fourier transform MIRS (FT-MIRS), NIRS, and FT-Raman spectroscopy in lactic fermentations showed FT-MIRS to be superior to the other spectroscopic methods in bacterial fermentations for at-line applications (69).

Multiple studies have been performed using MIRS for in situ monitoring and real-time analysis (15, 40, 41, 60). In all the studies, the most common probe used was the ReactIR 1000 (ASI Applied Systems, Millersville, MD). As with all other in situ applications, in situ monitoring results

are highly dependent on the design, performance, and reliability of the probe as well as the calibration models used for spectrum analysis. In situ MIRS studies are summarized in Table 4.

50.8. ADVANCES IN EXISTING SENSOR TECHNOLOGIES

Advances in traditional bioreactor sensor technologies are important. Some of these include glucose, biomass, and pH sensors.

50.8.1. Glucose

Glucose is an essential carbon and energy source in cell culture and bacterial fermentations. A real-time in situ glucose sensor would allow further exploration of feed control strategies coupled with optimization of growth and protein production. The implementation of an in situ glucose sensor

TABLE 3 Summary of Raman spectroscopy in situ probe monitoring applications[a]

Type of culture	Analyte(s) monitored	Probe(s) used	Models constructed	Major findings and challenges	Reference
Bacterial fermentation, *E. coli*	Glucose, phenylalanine, acetate, formate, lactate	InPhotonics Raman probe	PLS, PCA, artificial neural networks	Light scattering from air bubbles and biomass needed correction. Errors in the physical model, instead of noise, limited the accuracy of the reading. Improved physical modeling and window materials with stable Raman spectrum can improve detection limits and sensitivity.	43
Yeast fermentation, *Phaffia rhodozyma*	Carotenoid	Process analyzer System 100 (Renishaw, Gloucestershire, United Kingdom) coupled to a 12.5-mm probe (EIC Laboratories, Inc., Norwood, MA)	PLS, PCR	Complicated multivariate techniques are not necessary when the target component has a strong signal, as in the case of carotenoids in this process.	6

[a]Publications of in-situ applications are limited.

TABLE 4 Summary of MIRS in situ probe monitoring applications

Type of culture	Analytes monitored	Probes used	Models constructed	Major findings and challenges	Reference
Bacterial fermentation					
Gluconacetobacter xylinus	Glucose, acetate, ethanol, glucoacetan, fructose	ReactIR 1000 (ASI Applied Systems) 4,000–400 cm^{-1} (spectrum analysis between 1,500 and 950 cm^{-1})	PLS; specifics of model development not discussed.	General use of these techniques is challenging and highly dependent on the calibration models. Model is strengthened by the addition of synthetic standards of the target analytes.	40
G. xylinus	Acetate, glucoacetan, fructose	ReactIR 1000 (ASI Applied Systems) 4,000–400 cm^{-1} (spectrum analysis between 1,500 and 950 cm^{-1})	PLS regression	Calibration models are improved throughout the fermentation by the addition of new standards collected *in process*.	41
E. coli	Glucose, acetic acid	ReactIR 1000 (ASI Applied Systems) 4,000–500 cm^{-1}	PLS	Selecting the appropriate calibration data to generate the best model can be challenging. Sterilization can have an effect in medium components and affect spectra. The extent of the effect is different per sterilization cycle.	15
Cell culture processes					
CHO/SF3 cells	Glucose, lactate	ReactIR 1000 (ASI Applied Systems) 4,000–500 cm^{-1}	PLS regression	Sensitivity and selectivity of the sensor are important tests in establishing the sensor capabilities.	60

has been challenging. Current measurements are based on off-line sample analysis using a metabolite analyzer or an HPLC. An in situ sensor would need to withstand sterilization and chemical cleaning (as needed, since sensors may be removed and cleaned if unable to withstand chemicals) and measure glucose concentrations in the presence of a variety of analytes over potentially long durations (up to 2 to 3 weeks for mammalian cell culture).

Current glucose sensing technologies in bioprocessing utilize the reaction of glucose oxidase (GOx) and oxygen. The reaction converts glucose to gluconic acid and hydrogen peroxide as shown in equation 1:

$$\text{Glucose} + O_2 \quad \rightarrow \quad \text{gluconate}^- + H^+ + H_2O_2 \qquad \textbf{(1)}$$

The search for an in situ sensor has been explored using this reaction. The approach has been to immobilize the enzyme in a matrix, membrane, or another surface where it is allowed to react with the process fluid. Glucose concentration can be correlated to oxygen consumption or hydrogen peroxide generation. In one study, a multilayered GOx-containing membrane was exposed to oxygen via on-line sampling (82). The generated hydrogen peroxide was measured amperometrically on a platinum working electrode as shown in equations 2 and 3, and the output of the sensor was converted to glucose concentration. In another study, an optical fiber biosensor was developed using a matrix for oxygen-sensitive complexes and GOx reaction (66). Oxygen consumption was determined by measuring the fluorescence lifetime of the complexes which are quenched by the oxygen.

$$\text{Platinum anode} \quad H_2O_2 \quad \rightarrow \quad 2H^+ + 2e^- + O_2 \qquad \textbf{(2)}$$

$$\text{Platinum cathode} \quad 4H^+ + 4e^- + O_2 \quad \rightarrow \quad H_2O_2 \qquad \textbf{(3)}$$

Biosensors have also been used in several studies. Glucose binding protein labeled with a fluorescent probe can be used as a sensor by detecting the conformational change as a result of the glucose binding (23). Commercially available biosensors such as SIRE® (sensors based on injection of the recognition element) in conjuction with GOx have also been investigated (45). In this study, the glucose diffuses through a membrane and reacts with the GOx enzyme inside the SIRE® amperometric biosensor chamber.

In general, glucose measurements in cell culture processes and bacterial fermentations rely on off-line measurements. Commercial availability of an in situ probe will depend on sterilization capabilities, stability of the immobilized enzyme, activity without degradation, and other factors such as temperature and interference by other analytes over the course of the culture or fermentation. Spectrocospic methods discussed earlier may provide a solution before any of these approaches materialize into a reliable commercial sensor in a production environment.

50.8.2. Biomass

Biomass is traditionally measured by off-line methods. Systems like the Beckman Coulter Vi-Cell AS can be used to measure viable and total cell counts, percent viability, and cell size distribution in mammalian cell cultures. In bacterial fermentations, dry cell weight measurements are often used. All of these technologies require sampling and manual manipulations, which tend to be labor-intensive. In situ sensors for biomass measurement have been explored for a number of years. Some of the issues encountered include calibration, maintenance, or inability to measure very high optical densities, as in some bacterial fermentation cases (dynamic range limitation). However, on-line sensor

technology for biomass measurements has advanced in the last decade. Several sensing principles have been evaluated and further developed, including dielectric spectroscopy, optical density, IRS, and fluorescence.

50.8.2.1 Dielectric Spectroscopy

A review of the last 10 years of development shows that the main technologies used today are dielectric spectroscopy and optical methods (37). Dielectric spectroscopy measures the dielectric properties of a medium as a function of frequency. The use of dielectric spectroscopy as a tool to measure biomass has been clearly described (33). The biomass sensor produces an electrical field in the culture medium, resulting in charge separation or polarization. This polarization is the result of ion movement in the ionic culture fluid, with restriction across the intact cellular membrane. Each cell acts as a capacitor, and the charge separation is measured as capacitance. The capacitance of the cell suspension gives a measure of the extent of the field-induced polarization. A scanning frequency from 0.01 to 10 MHz is applied, and from the resulting β dispersion, it is possible to determine the permittivity due to media alone and media containing cells. The difference between these readings is related to the viable cell volume or concentration, provided the cell diameter does not change appreciably. The measurement is dependent on the integrity of the cell membrane. Thus, only viable cells are detected, and the capacitance signal can be correlated with viable cell count. Lysed cells or solid particles do not significantly affect the variation in capacitance.

There are several in situ dielectric spectroscopes available today. Aber Instruments (Aberystwyth, United Kingdom) biomass sensors have been documented in academic research (7, 16, 18, 38). A Fogale (Fogale Nanotech, Nimes, France) biomass system has also been evaluated in bacterial and yeast fermentations as well as mammalian cell cultures. Bacterial fermentation data showed good reproducibility in cultures with high conductivity and low cell concentration (5). In bacterial applications with sporulation, some challenges were observed, as cell structure changes during spore liberation and germination (63). In mammalian cell culture, monitoring and control (57) and viable cell number measurements (58) have been evaluated.

Implementation of dielectric spectroscopy sensors in a particular application will require testing, as their performance may be affected by the culture conditions (conductivity of the culture, stirring, and aeration conditions), probe geometry, or probe-to-probe variability, depending on the vendor and specific model (38). In any application, successful measurements can be used to calculate mean cell size and size distribution using off-line calibration methods (2), enable feedback control of feeds, automate batch refeed cultures commonly used for cell maintenance, and allow culture transfers based on on-line cell density targets.

50.8.2.2. Optical Sensors

Optical sensor systems are widely used in bioprocess monitoring (50). Optical density is the most common method for on-line biomass monitoring. The probes measure turbidity by either transmission or backscatter. Specific selection is dependent on the turbidity of the sample solution. There are multiple sensors on the market for mammalian cell culture and bacterial and yeast fermentations, and several of them have been evaluated in the literature (32, 79). These sensors are simple and do not require extensive calibration or mathematical models, although calibrations against an off-line analyzer may be required. However,

TABLE 5 In situ optical density probes in bacterial fermentations and mammalian cell culture applications

Sensor	Vendor	Mechanism	Application	Challenge(s)	Reference(s)
Aquasant AF 44 SIR	Aquasant AG, Wettingen, Switzerland, www.aquasant.com	Backscattering	Mammalian cell culture	Contamination and viability decrease (as a result of process changes) affected predicted cell densities.	56, 79
BE2100	BugLab, LLC., Danville, CA, http://www.buglab.com	Backscattering	Bacterial fermentation	Calibration required for each particular molecule to be monitored. Effective at high optical density.	76
InPro 8100	Mettler-Toledo Ingold, Bedford, MA, http://us.mt.com	Backscattering	Bacterial fermentation	Effective at high cell density. Calibration procedures may need optimization.	J. Wakeem, unpublished data
Optek (specific model omitted)	Optek-Danulat, Inc., Germantown, WI, http://www.optek.com	Forward scattering	Not specified	Contains hybrid LED light source emitting in NIR zone that proved to be more accurate than other wavelength regions.	75
BT65 cell growth sensor	Endress+Hauser Conducta Inc. (formerly Wedgewood Analytical, Inc.), Anaheim, CA, http://www.wedgewood-analytical.com	Transmission	Mammalian cell culture	Not as sensitive as the Aquasant to high cell density	56, 79
MAX™ cell mass sensor	Cerex, Inc., Wellesley, MA, www.cerexinc.com	90° scattering and absorbance	Bacterial fermentations	Effective calibration required to minimize sensing errors. Probe design modifications are required to address mechanical difficulties.	32, 79

optical density measurements are affected by cell concentrations, cell morphology (dynamic range nonlinearity), and bioreactor conditions such as agitation rates and bubble sizes. A summary of the different sensors and applications is provided in Table 5.

50.8.2.3. In Situ Microscope

In the last decade, several studies have shown that cell number can also be measured using microscopy. An in situ microscope acquires images in a given sample volume during the fermentation process and performs cell counts. The system is essentially an in situ hemacytometer mounted on the bioreactor (Fig. 4). Comparisons between the hemacytometer hand counts, Cedex (biomass analyzer similar to the Vi-Cell AS), and an in situ microscope in mammalian cell cultures showed very good correlations (30). The system was also successfully tested in yeast fermentations, in which good correlations of cell density and viability were demonstrated (77). Additionally, it has been successful in measuring the level of colonization of fibroblasts on microcarrier surfaces during cultivation (62). The advantages of this technology include the ability to monitor cell morphology, cell density, and viability in real time without the need for reagents. Challenges for implementation include the verification and/or development of image processing software for the specific applications (cell culture, yeast fermentation, etc.), cleaning and maintenance of the sampling window, sample zone optimization for a range of cell densities, and sampling zone consistency (reproducibility and accuracy testing).

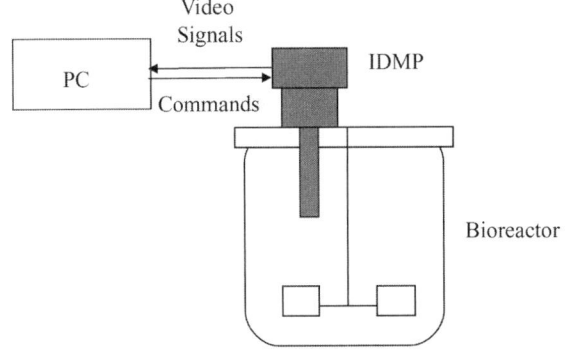

FIGURE 4 Bioreactor with in situ dark-field microscopy probe (IDMP). Adapted from reference 77.

50.8.4. pH Sensors in Mammalian Cell Cultures

pH is a measure of hydrogen concentration in a solution. The pH value is calculated as minus the decimal logarithm of the hydrogen ion activity in an aqueous solution. Because of its algorithmic nature, a small change in pH units implies a large change in the concentration of hydrogen ions in the process medium. Mammalian cell cultures are significantly affected by changes in the pH of the culture. Very small variations, as small as 0.1 or 0.2 pH unit, can significantly affect culture growth and metabolism, in particular glucose consumption and lactate production (80). It is important to note that the pH set point and control strategy (e.g., dead band) are intimately linked to levels of dissolved carbon dioxide (dCO_2), base consumption for pH control, and, therefore, osmolality. These considerations make the accurate measurement of pH even more imperative.

Measuring and controlling pH are regularly performed during cell culture operations. Glass sensors have traditionally been used across the biotechnology industry for this measurement. This technology has been established for years and is generally reliable. A typical pH glass sensor design consists of a tube within a tube. The bottom of the inner tube has a bulb-shaped glass membrane of a specific pH. The inner glass tube body is filled with a buffer solution called inner reference electrolyte, containing Cl^- ions usually with a pH value of 7. The outer side of the glass tube body is filled with a KCl solution at a high concentration. A reference electrode is immersed in the inner reference electrolyte and equilibrates with the Cl^- ions in it. A reference electrode is also immersed in the outer reference electrolyte. The outer reference electrolyte is electrically in contact with the solution of interest via a reference junction. In essence, the pH probe is an electrochemical cell between the inner reference or measuring electrode and the outer reference electrode.

Some issues related to the use of pH probes for extended culture duration include drift and reduced sensitivity. To protect against on-line sensor drift, on-line pH measurements are usually compared to off-line analyzers or bench-scale meters to ensure that the correct value is maintained in the culture. Literature on these differences is limited, but disparity between on-line and off-line pH measurements has been reported (17). Comparisons between the standard glass sensor technology and an off-line analyzer and bench meter confirmed a downward drift in the on-line measurement. In this study, the phenomenon was clearly detected but a root cause was not identified. In searching for this root cause, an evaluation of the pH performance of different technologies in cell culture was performed (64). The results showed that glass technologies are the most accurate, but variability does exist across different vendors. This evaluation investigated the effect of process conditions (agitation and gassing) on pH measurements and the performance of different pH technologies in cell culture processes through several sterilization cycles. Technologies included glass, solid state, and optical. Further evaluation of the glass sensor technology showed that the main cause for drift in cell culture processes is the effect of steam sterilization on the glass membranes of the pH sensors (Saucedo, unpublished data).

50.8.5. dCO_2

dCO_2 concentrations are important to mammalian cell growth. High dCO_2 levels can inhibit cell growth, affect protein production, and result in product quality issues. dCO_2 is also an important process variable which can accumulate to inhibitory levels at values of >120 to 150 mm Hg (21, 25, 34) and/or influence product quality at high cell density cultivations or scale-up due to inadequate CO_2 stripping. Generally, cell culture processes monitor dCO_2 using off-line analyzers such as the NovaFLEX and BioProfile, both from Nova Biomedical. This technology requires sampling and off-line analysis. On-line dCO_2 measurement would enable the implementation of control strategies for processes that require accurate measurements and close monitoring.

In situ probes have been developed since 1958 (67), and a number of evaluations have followed. Two of them are worth noting: the YSI 8500 (YSI Inc., Yellow Springs, OH) and the InPro 5000 (Mettler Toledo, Bedford, MA). The YSI 8500 is a fiber-optic dCO_2 sensor that has shown acceptable maintenance, response time, and sensitivity in cell culture processes (59). The InPro 5000 is a membrane-based probe for fermentation and cell culture applications. The CO_2 diffuses through the membrane, where a change in pH can be monitored by an electrode. Extensive evaluation in large-scale manufacturing of either probe was not available. Maintenance, probe-to-probe variability, and long-term performance may have to be evaluated for complete implementation of this measurement as well as control strategies. Until an in situ probe can be implemented, the use of the NovaFLEX autosampler system, discussed earlier, would allow for on-line data and enable control strategies to be implemented.

50.9. DISPOSABLE SENSORS

The need for disposable sensors has significantly increased in the last decade. A variety of companies have developed disposable bioreactors for bacterial fermentations and mammalian cell cultures, including GE Healthcare, Xcellerex, Thermo Scientific HyClone, and Applikon Biotechnology. In the same way, there has been a significant increase in the development of miniaturization and high-throughput technology for bioreactors. A number of companies have developed disposable-sensor technologies to provide sensing technology capabilities to these systems, including PreSens Precision Sensing GmbH (Regensburg, Germany), Polestar Technologies (Needham Heights, MA), Finesse Solutions (Santa Clara, CA), and Fluorometrix Corporation (Baltimore, MD). Disposable sensors have been developed and are currently available to measure pH, DO, dCO_2, and, in some cases, dual variables, e.g., pH/DO. In these sensors, a fluorescent dye sensitive to the analyte of interest, e.g., pH, O_2, or CO_2, is immobilized in a matrix or patch. The patch can be mounted inside a microplate well or a miniature bioreactor (MBR), the bottom of a shake flask, or the end of fiber-optic cable, depending on the application (Fig. 5). A light-emitting diode (LED) source and a photodetector for light detection are required. The measurement is based on either fluorescence intensity or fluorescence decay.

Disposable sensors have been used in a number of bioreactor applications, including shake flasks, microplates, MBRs, and single-use bioreactors. Shake flasks are one of the most widely used tools for process development. However, sensor data have been challenging to obtain due to the large size of existing probes relative to the flask sizes. Outfitting shake flasks with patches has enabled on-line measurement and improved monitoring in these culture devices in a number of publications (27, 78). Quantification of DO in flasks was performed by immobilizing a thin-layer sensor, or spot with a 10-mm diameter and a thickness of

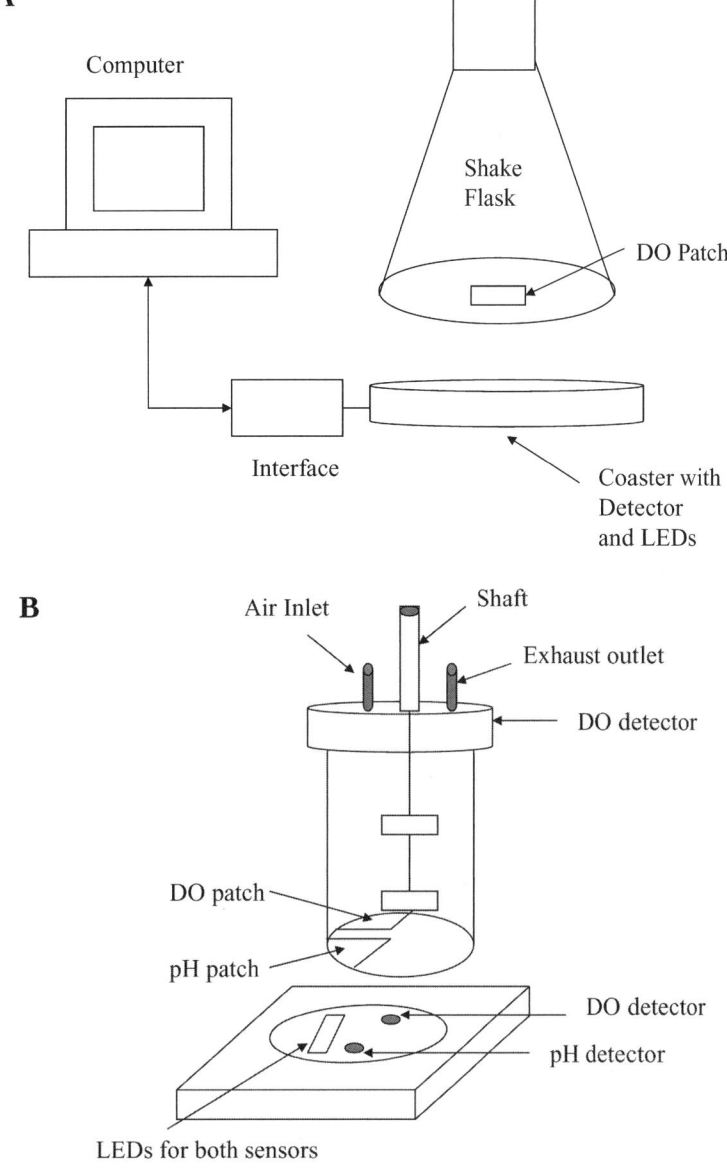

FIGURE 5 Disposable-sensor applications. (A) Shake flasks (adapted from reference 78); (B) microplates (adapted from reference 24).

less than 100 μm, at the bottom of the shake flasks (78). Oxygen mass transfer coefficients were compared in shake flask and stirred tank reactor *Escherichia coli* fermentations when using noninvasive sensors (27).

Sensor patches in microplate systems have also been evaluated using products from different manufacturers. Optical pH and DO were evaluated in microplates and mammalian cell culture using a high-throughput bioreactor system (24). Consistency and calibration across the different plates showed acceptable results. Dual-sensor capabilities have also been explored in microplates. This is of particular importance in microplate applications due to volume and surface area limitations. A dual-sensor application for pH and DO was evaluated in bacterial fermentations using a 24-well format, showing good reproducibility with measurements every 30 min (39). Dual sensors have also been used in MBRs. In one study, a 24-well plate MBR

system was assessed for *Saccharomyces cerevisiae*, *E. coli*, and *Pichia pastoris* cultivations (29). Each 10-ml MBR was equipped with optically isolated pH and DO fluorescence-emitting patches and sensors (PreSens-Precision Sensing GmbH). The results showed high interwell reproducibility for on-line temperature, pH, and DO. In a second study, the same system was assessed for cell culture applications using a Chinese hamster ovary (CHO) cell line from Genentech, Inc., also showing good results (8).

A few studies have also been published showing comparisons between optical disposable sensors and traditional electrochemical sensors. The results suggest that photobleaching control or optical sampling rate (how often the patch is excited) is important in establishing calibration and operation limits for a particular sensor (28, 46, 47). Additionally, depending on the vendor, software development may be required to allow for in-process calibrations and

TABLE 6 Disposable-sensor evaluations

Parameter	Company	Application	Reference(s)
pH	PreSens	Single-use bioreactors	46
	Fluorometrix	Microplates	24
DO	Fluorometrix	Shake flasks	27, 73
	PreSens	Shake flasks	78
		Microplates	24, 39
		Single-use bioreactors	46, 47
	Custom made for research	Perfusion bioreactors	22
Dual sensor applications: pH and O_2	PreSens	Microplates	39
		MBRs	8, 29

integration to control platforms. In these studies, the data showed good agreement between electrochemical probes and optical sensors used for pH and DO measurements in 5-liter bioreactors and 250-liter single-use systems.

Disposable sensors provide a number of advantages, including high sensitivity (specific chemistry identified for a given application), ease of miniaturization, and, in some cases, noninvasiveness. However, several factors need to be considered for implementation, including signal variations as a result of photobleaching, light intensity variability, sensor positioning and orientation, potential irregularities in the patches, detector sensitivity, fluorescence of the sample, and software requirements. Various studies are summarized in Table 6.

50.10. PEPTIDE AND PROTEIN ANALYSIS MONITORING

In the last few years, there have been advances in other sensing technologies, in addition to sensors for the process variables discussed above, i.e., biomass, pH, DO, and dCO_2. At-line or on-line protein quantity and quality monitoring can provide significant advantages to process understanding. The current process usually requires sampling and analysis by the quality control department or protein analytical laboratories, depending on whether the sample is from manufacturing or development. Sample gathering, delivery, and assay completion can take hours to days. The implementation of rapid and accurate on-line or at-line analytical technologies for monitoring product quantity and quality during processing would improve large-scale manufacturing and process development. Having near-real-time data in large-scale manufacturing could enable control loops to optimize productivity and maintain the process within the design space or allowable process variability range. Implementation of high-throughput testing during process development would increase the ability to monitor a large number of conditions and establish the design space.

Several chromatographic methods have been used to monitor product concentration and product quality, including HPLC and mass spectrometry (31, 48, 65), among others. These technologies are generally well established and can be relatively easy to operate, given appropriate training. However, on-line implementation can be complex. Samples must be transferred from the reactor to the system regularly, and a cell removal device is required. Additionally, a strong program to support the maintenance requirements of these systems is very important to ensure column life and the

accuracy of the measurements. Most recently, the lab-on-a-chip (LOC) technology has become more attractive due to its high-throughput capabilities.

50.10.1. LOC

LOC generally refers to a device that allows the performance of one or multiple analyses on the microscale level. LOC sensors take advantage of microfluidic technology. Microfluidics refers to the control and manipulation of fluids in submilliliter scales where very small volumes, as low as picoliters, are required. The low volumes and small size of the device and system allow for samples to be analyzed faster and potentially at the point of sampling rather than in special laboratories. This can be of significant advantage for process monitoring in development laboratories, pilot plants, or manufacturing facilities. High-throughput applications of protein quality analyses in these environments can provide very valuable information about the process in almost real time.

The application of LOC technology to process monitoring, specifically protein quality, has become more popular in the last few years. The literature is limited, but a few studies have been presented or published on the analysis of protein quality in mammalian cell culture and fermentation products. The most common systems in the literature include the LabChip® 90 (Caliper Life Sciences, Hopkinton, MA) and the Agilent 2100 Bioanalyzer (Agilent Technologies, Palo Alto, CA).

The LabChip® 90 was used to perform a capillary electrophoresis-sodium dodecyl sulfate (CE-SDS) assay in high-throughput mode to monitor monoclonal antibody protein quality (11). Several assays were developed and optimized, including an assay to resolve nonglycosylated and glycosylated heavy chains and a glycan typing assay. The developed assays were much faster than the standard CE-SDS, with one of the assays taking 41 s for the microchip, versus 50 min for the standard assay. The LabChip® 90 has also been used in high-throughput mode to support *E. coli* process development (72). In this study, samples from 2-ml 96-well plates were purified using automated purification on a Tecan Genesis robot (Tecan, Research Triangle Park, NC). Purified samples were prepared, also by the Tecan Genesis robot, and processed by the LabChip® 90 system for protein quantification and identification. The fast sample handling and analysis enabled process optimization via a DOE approach. In the case of the Agilent 2100 Bioanalyzer, this technology has been evaluated to determine protein size (molecular weight distribution) and concentration in cell

TABLE 7 Multivariable analysis review

Method(s) used	Culture type	Application	Reference
PCA	Fed-batch cell culture	Fault detection and diagnosis of abnormal process conditions in pilot plant 2-liter bioreactors	26
PCA and AANN	Yeast fermentation	Fault detection by two different multivariate analysis methods: linear (PCA) and nonlinear (AANN)	68
PFA	Cell culture	Identification of threshold values of repressing metabolites (lactate, ammonia, osmolality, and carbon dioxide) in 5,000-liter bioreactors	81
PCA and PLS	Yeast fermentation	Use of seed quality information from manufacturing data to estimate final productivity	13
MVDA	Cell culture	Investigation of MVDA as a tool for assessing scale-up and comparability of the cell culture process	35
MVDA	Cell culture	Using MVDA as a root cause analysis tool for identifying scale-up differences and parameter interactions affecting process performance	36

culture samples (55). In this study, the Protein 200 LabChip assay kit using the Agilent 2100 Bioanalyzer was compared to the standard SDS-polyacrylamide gel electrophoresis tests. The results showed no statistical differences and a high level of reproducibility. Both technologies have the capability of providing protein quality information during production and look very promising as at-line protein quality monitoring technologies.

50.11. DATA TECHNOLOGY

Process understanding via data analysis has also become a powerful tool for bioprocess monitoring in the last decade. Companies are generally investing more in their data infrastructure and management to enhance their ability to capture the appropriate data and perform appropriate analysis. Several approaches have been evaluated in the literature, including hybrid modeling, multivariate modeling, and soft sensors. Table 7 summarizes a few studies since 1998 using different tools to evaluate data from cell culture and bacterial and yeast fermentation processes.

50.11.1. Hybrid Modeling

Hybrid modeling utilizes artificial intelligence, previous process knowledge, and reaction kinetics to predict fermentation process performance. Mass balance equations, process description, and available measurements are combined with neural networks to predict specific parameters. Implementation of these models can be challenging, as it requires appropriate data infrastructure and automation systems support, in addition to process experts. A few publications on fermentation applications are available. Some studies focus on comparisons between conventional models using mass balances and hybrid models utilizing artificial neural networks and process knowledge (19). Others discuss a process for the identification of hybrid models for bioprocesses (9). These studies illustrate significant challenges and testing required for implementation of this technology.

50.11.2. Multivariate Modeling

In an effort to increase process understanding, the biopharmaceutical industry is increasingly investing in multivariate analysis models that allow process optimization and contin-

uous process improvement. Applications in cell culture and bacterial fermentations are discussed in a number of articles reporting the use of different techniques to analyze multiple sets of measurements or variables (Table 7). Some of these techniques include principal-component analysis (PCA), auto-associated neural networks (AANN), partial least squares (PLS), principal-factor analysis (PFA), and MVDA. PCA has been used to detect abnormal process conditions in 2-liter cell culture bioreactors using both on-line and off-line data (26). In another evaluation, PCA was used to assess seed quality using data gathered from the manufacturing plant (13). These data were then used to evaluate the final productivity estimation of an antibiotic fermentation process. PCA and AANN techniques were used in yeast fermentations for the detection of faulty temperature sensors and plasmid instability (68). PFA multivariate analysis was used to identify inhibitory threshold values of specific metabolites to cell culture and protein quality in 5,000-liter bioreactors (81). The feasibility of using MVDA was evaluated for assessing scale-up performance and comparability of the cell culture manufacturing process (35). MVDA was also evaluated as a tool for root cause analysis in the identification of scale-up differences and parameter interactions affecting a cell culture process (36).

50.11.3. Soft Sensors

A soft sensor or software sensor can be described as the combination of hardware and software to derive variables that cannot be directly measured. The hardware is a sensor or a combination of sensors providing output about the process conditions in the bioreactor. The software is a series of algorithms or mathematical models correlating the hardware output to the unmeasurable variable of interest. Thus, the success of this sensor is highly dependent on the reliability of the on-line process variable measurements (hardware), the mathematical models (software) used, and, most importantly, the knowledge of the measurement and the process (12). An important factor in soft sensors' success is the use of artificial neural networks and recurrent neural networks, which combine information and recognize patterns to model and control dynamic systems (10).

Multiple software sensors have been explored in bioprocessing in the last few years, including biomass, acetate formation, and specific growth rate. Recurrent neural networks were used in combination with feed rate, liquid volume, and DO to predict biomass concentration in yeast fermentations within 11% of the off-line value (10). Acetate formation from *E. coli* cultures was predicted using DO responses to feed transients (1), and glucose feed rate to enable sustained high specific growth rates was optimized in response to inferred glucose demand rate based on similar glucose/DO transients (84). Specific growth rate has also been predicted in a yeast fermentation using on-line DO concentration and mass balances (54). Growth rate was used in a control strategy to maximize protein production.

The use of soft sensors in practice has seen limited success. The complex data manipulation required makes it difficult to obtain a satisfactory measurement over a range of process conditions. In addition, noise and variability in traditional sensor measurements add to the complexity of designing a robust software algorithm. These challenges are important, but the reward is also potentially quite significant. The ability to obtain additional process information without implementing new sensor hardware continues to make soft sensors an attractive area for technology development. Rapid advances in computing technology can be expected to improve mathematical modeling capabilities and thus the robustness of soft sensors.

50.12. SENSORS IN DOWNSTREAM APPLICATIONS

The sensors discussed in this chapter allow for increased process understanding in upstream applications, i.e., cell culture and fermentation processes. New sensor technologies in downstream processing have also been explored in the last decade. As in the case of upstream sensors, downstream applications focus on data throughput and in situ sensors used in purification development and manufacturing operations.

There are several new technologies that have been explored and developed in the last decade. These technologies allow for high-throughput or on-the-floor testing of process variables such as endotoxin levels, total organic carbon concentrations, and at-line protein concentration measurements without dilutions. Protein quality using LOC technology, mentioned above, is also used to monitor product quality in downstream applications.

New developments in monitoring host cell protein clearance (HCP) in downstream processing are also important. In general, removal of CHO cell-derived proteins from the biopharmaceutical product is monitored using multiproduct immunoassays (42). However, these assays require intensive sample preparation and the use of robotics to facilitate high throughput. An example of a new technology involves the miniaturization of an affinity capture column to generate a sandwich-based immunoassay integrated into a compact disk microlaboratory for high-throughput HCP measurements (44). However, most of the newer technologies are for at-line measurements. On-line or in situ probes for downstream sensors have the same requirements as mentioned above, i.e., (i) ability to be sterilized (chemical sanitization only may apply to downstream sensors), (ii) reliability and robustness, (iii) high specificity to the measured parameter, (iv) fast response time, and (v) minimal drift. Extensive development and testing are required to achieve these requirements. Examples of downstream technologies are listed in Table 8.

50.13. IMPLEMENTATION OF A NEW PAT

Implementing PAT in development or manufacturing will require extensive testing to ensure acceptable performance and required comparability. Equally important to implementation is the infrastructure support required to manage the technology life cycle. Intensive vendor, instrument services, and automation support may be required to ensure reliability and robustness during testing and after imple-

TABLE 8 Examples of new technologies in downstream applications

Commercial name	Application[a]	Vendor information
Endosafe®-PTS™	Rapid, real-time endotoxin testing	Charles River Laboratories International, Inc., Wilmington, MA, http://www.criver.com
Sievers® 900 Portable TOC Analyzer	Portable TOC analyzer for at-line grab samples, or connected to an autosampler for high-volume applications	GE Analytical Instruments, Inc., Boulder, CO, http://www.geinstruments.com
NanoDrop™ (models 2000 and 2000c)	UV/Vis spectrophotometer for rapid quantification of protein samples without the need for dilution	NanoDrop, Wilmington, DE, http://www.nanodrop.com
NanoPhotometer™	UV/Vis spectrophotometer for rapid quantification of protein samples without the need for dilution	Implen, Inc., Calabasas, CA, http://www.implen.de
NanoVue™	UV/Vis spectrophotometer for rapid quantification of protein samples without the need for dilution	GE Healthcare Bio-Sciences AB, Uppsala, Sweden, http://www.gelifesciences.com
Gyrolab Bioaffy 200	Quantification of monoclonal IgG and HCP for the entire purification process from cell supernatant to purified product	Gyros AB, Uppsala, Sweden, http://www.gyros.com

[a]Abbreviations: TOC, total organic carbon; Vis, visual; IgG, immunoglobulin G.

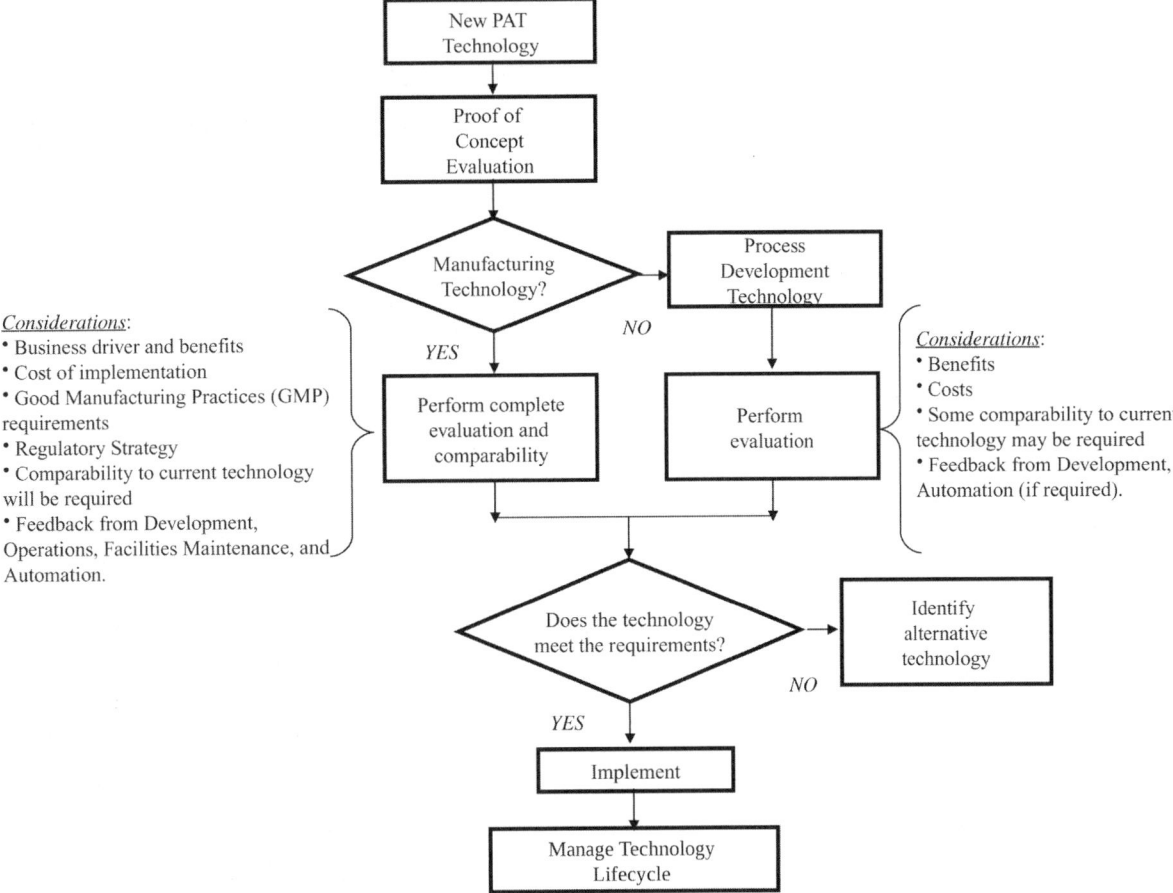

FIGURE 6 Potential steps required for the evaluation and implementation of a new PAT.

mentation. In addition, manufacturing implementation requires support from the regulatory organization within the company. Regulatory representatives can establish the strategy of implementation required for a particular technology early during development. Discussions with the FDA may be recommended in the technology development phase. This becomes more important if the implementation is complex or across multiple products, multiple sites, and multiple countries.

Depending on the evaluation challenges and the robustness of the technology, implementation of a new PAT may require several years. Ideally, technologies should be tested in development laboratories or pilot plants where data and knowledge can be gathered before implementation in a manufacturing environment. Figure 6 shows a potential path for the evaluation and implementation of a new PAT.

50.14. SUMMARY

It is clear from the literature that there have been tremendous efforts in the development and evaluation of new sensors and tools to enable PAT implementation in cell culture and fermentation processes. Universities, vendors, and companies are working together to advance sensor and sampling technology research and development and ensure process understanding and product quality. These advances have also been encouraged by the multiple QBD sessions

and workshops, increasingly common in conferences with pharmaceutical and biotechnology topics. This chapter covers some of the numerous technologies that have been developed or improved in the last decade.

Autosampling technologies that enable real-time data and potentially feedback control will become more common in processing facilities. Advances in in situ spectroscopic sensors could facilitate real-time monitoring of multiple analytes and process parameters. New developments in biomass and glucose sensing technologies could allow for feed optimization and maximum product yield. High-throughput product quality technologies such as LOC could allow real-time product quality monitoring during production. These technologies along with strong data analysis tools will ensure strong process understanding and allow for real-time release in the future.

I acknowledge Tina M. Larson, Robert Kiss, and Ashraf Amanullah, Genentech, Inc., South San Francisco, CA, for their review and helpful comments.

REFERENCES

1. **Åkesson, M., E. N. Karlsson, P. Hagander, J. P. Axelsson, and A. Tocaj.** 1999. On-line detection of acetate formation in *Escherichia coli* cultures using dissolved oxygen responses to feed transients. *Biotechnol. Bioeng.* **64:**590–598.
2. **Ansorge, S., G. Esteban, and G. Schmid.** 2007. On-line monitoring of infected Sf-9 insect cell cultures by scanning

permittivity measurements and comparison with off-line biovolume measurements. *Cytotechnology* **55**:115–124.

3. **Arnold, S. A., J. Crowley, N. Woods, L. M. Harvey, and B. McNeil.** 2003. *In situ* near infrared spectroscopy to monitor key analytes in mammalian cell cultivation. *Biotechnol. Bieng.* **84**:13–19.

4. **Arnold, S. A., R. Gaensakoo, L. M. Harvey, and B. McNeil.** 2002. Use of at-line and *in situ* near-infrared spectroscopy to monitor biomass in an industrial fed-batch *Escherichia coli* process. *Biotechnol. Bioeng.* **80**:405–413.

5. **Arnoux, A. S., L. Preziosi-Belloy, G. Esteban, P. Teissier, and C. Ghommidh.** 2005. Lactic acid bacteria biomass monitoring in highly conductive media by permittivity measurements. *Biotechnol. Lett.* **27**:1551–1557.

6. **Cannizzaro, C., M. Rhiel, I. Marison, and U. von Stockar.** 2003. On-line monitoring of *Phaffia rhodozyma* fed-batch process with *in situ* dispersive Raman spectroscopy. *Biotechnol. Bioeng.* **83**:668–680.

7. **Cannizzaro, C., R. Gugerli, I. Marison, and U. von Stockar.** 2003. On-line biomass monitoring of CHO perfusion culture with scanning dielectric spectroscopy. *Biotechnol. Bioeng.* **84**:597–610.

8. **Chen, A., R. Chitta, D. Chang, and A. Amanullah.** 2009. Twenty-four well plate miniature bioreactor system as a scale-down model for cell culture process development. *Biotechnol. Bioeng.* **102**:148–160.

9. **Chen, L., O. Bernard, G. Bastin, and P. Angelov.** 2000. Hybrid modeling of biotechnological processes using neural networks. *Contr. Eng. Pract.* **8**:821–827.

10. **Chen, L. Z., S. K. Nguang, X. M. Li, and X. D. Chen.** 2004. Soft sensors for on-line biomass measurements. *Bioprocess Biosyst. Eng.* **26**:191–195.

11. **Chen, X., K. Tang, M. Lee, and G. C. Flynn.** 2008. Microchip assays for screening monoclonal antibody product quality. *Electrophoresis* **29**:4993–5002.

12. **Chéruy, A.** 1997. Software sensors in bioprocess engineering. *J. Biotechnol.* **52**:193–199.

13. **Cunha, C. C. F., J. Glassey, G. A. Montague, S. Albert, and P. Mohan.** 2002. An assessment of seed quality and its influence on productivity estimation in an industrial antibiotic fermentation. *Biotechnol. Bioeng.* **78**:658–669.

14. **Derfus, G. E., D. Abramzon, M. Tung, D. Chang, R. Kiss, and A. Amanullah.** 2009. Cell culture monitoring via an auto-sampler and an integrated multi-functional off-line analyzer. *Biotechnol. Prog.* 2009 Nov 13 (Epub ahead of print).

15. **Doak, D. L., and J. Phillips.** 1999. In-situ monitoring of an *Escherichia coli* fermentation using a diamond composition ATR probe and mid-infrared spectroscopy. *Biotechnol. Prog.* **15**:529–539.

16. **Ducommun, P., A. Kadouri, U. Von Stockar, and I. W. Marison.** 2002. On-line determination of animal cell concentration in two industrial high-density culture processes by dielectric spectroscopy. *Biotechnol. Bioeng.* **77**:316–323.

17. **Evans, H., and T. Larson.** 2006. Dealing with disparity in on-line and off-line pH measurements. IFPAC Meet., 21 to 23 February 2006, Arlington, VA.

18. **Ferreira, A. P., L. M. Vieira, J. P. Cardoso, and J. C. Menezes.** 2005. Evaluation of a new annular capacitance probe for biomass monitoring in industrial pilot-scale fermentations. *J. Biotechnol.* **116**:403–409.

19. **Feyo de Azevedo, S., B. Dahm, and F. R. Oliveira.** 1997. Hybrid modeling of biochemical processes: a comparison with the conventional approach. *Comput. Chem. Eng.* **21**:S751–S756.

20. **Food and Drug Administration.** 2004. *Guidance for Industry PAT—a Framework for Innovative Pharmaceutical Development, Manufacturing, and Quality Assurance.* Food and Drug Administration, Silver Spring, MD. (Online.) http://www.fda.gov/cder/guidance/6419fnl.pdf.

21. **Gamier, A., R. Voyer, R. Tom, S. Perret, B. Jardin, and A. Kamen.** 1996. Dissolved carbon dioxide accumulation in a large scale and high-density production of TGF receptor with baculovirus infected Sf-9 cells. *Cytotechnology* **22**:53–63.

22. **Gao, F., A. S. Jeevarajan, and M. M. Anderson.** 2004. Long-term continuous monitoring of dissolved oxygen in cell culture medium for perfused bioreactors using optical oxygen sensors. *Biotechnol. Bioeng.* **86**:425–433.

23. **Ge, X., G. Rao, and L. Tolosa.** 2008. On the possibility of real-time monitoring of glucose in cell culture by microdialysis using a fluorescent glucose binding protein sensor. *Biotechnol. Prog.* **24**:691–697.

24. **Ge, X., M. Hanson, H. Shen, Y. Kostov, K. A. Brorson, D. D. Frey, A. R. Moreira, and G. Rao.** 2006. Validation of an optical sensor-based high-throughput bioreactor system for mammalian cell culture. *J. Biotechnol.* **122**:293–306.

25. **Gray, D. R., S. Chen, W. Howarth, D. Inlow, and B. L. Maiorella.** 1996. CO_2 in large-scale and high-density CHO cell perfusion culture. *Cytotechnology* **22**:65–78.

26. **Gunther, J. C., J. S. Conner, and D. E. Seborg.** 2007. Fault detection and diagnosis in an industrial fed-batch cell culture process. *Biotechnol. Prog.* **23**:851–857.

27. **Gupta, A. and G. Rao.** 2003. A study of oxygen transfer in shake flasks using a non-invasive oxygen sensor. *Biotechnol. Bioeng.* **84**:351–358.

28. **Hanson, M. A., X. Ge, Y. Kostov, K. A. Brorson, A. R. Moreira, and G. Rao.** 2007. Comparisons of optical pH and dissolved oxygen sensors with traditional electrochemical probes during mammalian cell culture. *Biotechnol. Bioeng.* **97**:833–841.

29. **Isett, K., H. George, W. Herber, and A. Amanullah.** 2007. Twenty-four-well plate miniature bioreactor high-throughput system: assessment for microbial cultivations. *Biotechnol. Bioeng.* **98**:1017–1028.

30. **Joeris, K., J. G. Frerichs, K. Konstantinov, and T. Sheper.** 2002. *In situ* microscopy: on-line process monitoring of mammalian cell cultures. *Cytotechnology* **38**:129–134.

31. **Jones, J. J., C. L. Wilkins, Y. Cai, R. R. Beitle, R. Liyanage, and J. O. Lay, Jr.** 2005. Real-time monitoring of recombinant bacterial proteins by mass spectrometry. *Biotechnol. Prog.* **21**:1754–1758.

32. **Junker, B. H., J. Reddy, K. Gbewonyo, and R. Greasham.** 1994. On-line and *in situ* monitoring technology for cell density measurement in microbial and animal cell cultures. *Bioprocess Eng.* **10**:195–207.

33. **Kell, D. B., G. H. Markx, C. L. Davey, and R. W. Todd.** 1990. Real-time monitoring of cellular biomass: methods and applications. *Trends Anal. Chem.* **9**:190–194.

34. **Kimura, R., and W. M. Miller.** 1996. Effects of elevated pCO_2 and/or osmolality on the growth and recombinant tPA production of CHO cells. *Biotechnol. Bioeng.* **52**:152–160.

35. **Kirdar, A. O., K. D. Green, and A. S. Rathore.** 2007. Application of multivariate analysis toward biotech processes: case study of a cell-culture unit operation. *Biotechnol. Prog.* **23**:61–67.

36. **Kirdar, A. O., K. D. Green, and A. S. Rathore.** 2008. Application of multivariate data analysis for identification and successful resolution of a root cause for a bioprocessing application. *Biotechnol. Prog.* **24**:720–726.

37. **Kiviharju, K., K. Salonen, and U. Moilanen.** 2008. Biomass measurement on-line: the performance of in situ measurements and software sensors. *J. Ind. Microbiol. Biotechnol.* **35**:657–665.

38. **Kiviharju, K., K. Salonen, U. Moilanen, and E. Meskanen.** 2007. On-line biomass measurements in bioreactor cultivations: comparison study of two on-line probes. *J. Ind. Microbiol. Biotechnol.* **34**:561–566.

39. **Kocincová, A. S., S. Nagl, S. Arain, C. Krause, S. M. Borisov, M. Arnold, and O. S. Wolfbeis.** 2008. Multi-

plex bacterial growth monitoring in 24-well microplates using a dual optical sensor for dissolved oxygen and pH. *Biotechnol. Bioeng.* **100:**430–438.

40. **Kornmann, H., M. Rhiel, C. Cannizzaro, I. Marison, and U. von Stockar.** 2003. Methodology for real-time, multianalyte monitoring of fermentations using an *in-situ* mid-infrared sensor. *Biotechnol. Bioeng.* **82:**702–709.

41. **Kornmann, H., S. Valentinotti, I. Marison, and U. Von Stockar.** 2004. Real-time update of calibration model for better monitoring of batch processes using spectroscopy. *Biotechnol. Bioeng.* **87:**593–601.

42. **Krawitz, D. C., W. Forrest, G. T. Moreno, J. Kittleson, and K. M. Champion.** 2006. Proteomic studies support the use of multi-product immunoassays to monitor host cell protein impurities. *Proteomics* **6:**94–110.

43. **Lee, H. L. T., P. Boccazzi, N. Gorret, R. J. Ram, and A. J. Sinskey.** 2004. *In situ* bioprocess monitoring of *Escherichia coli* bioreactions using Raman spectroscopy. *Vibrat. Spectrosc.* **35:**131–137.

44. **Lehtonen, P., F. Sandegren, and M. Inganäs.** 2007. Improved analytical tools for quantification of monoclonal IgG and host cell proteins during biopharmaceutical process development. Well Characterized Biotechnol. Pharm. (WCBP), 29 to 31 January 2007, Washington, DC.

45. **Lidgren, L., O. Lilja, M. Krook, and D. Kriz.** 2006. Automatic fermentation control based on real-time in situ SIRE® biosensor regulated glucose feed. *Biosensors Bioelectronics* **21:**2010–2013.

46. **Lorenz, C., and J. Long.** 2008. Optical, digital, and disposables: measurements without the mess. WilBio Single-Use BioProcessing Components Syst., 28 to 30 July 2008, San Francisco, CA.

47. **Lorenz, C., T. Matthews, J. Long, and B. Wolk.** 2007. Moving toward a completely disposable solution: evaluating single-use pH and dissolved oxygen sensors in disposable bioreactors. Single-Use Bioprocess Syst. Appl. IBC, 4 to 6 June 2007, La Jolla, CA.

48. **Lundstrom, H., M. Brobjer, B. Osterlof, and T. Moks.** 1990. A completely automated system for on-line monitoring of the production of a growth factor secreted during fermentation of *Escherichia coli*. *Biotechnol. Bioeng.* **36:**1056–1062.

49. **Macaloney, G., J. W. Hall, M. J. Rollins, I. Draper, K. B. Anderson, J. Preston, B. G. Thompson, and B. McNeil.** 1997. The utility and performance of near-infrared spectroscopy in simultaneous monitoring of multiple components in a high cell density recombinant *Escherichia coli* production process. *Bioprocess Eng.* **17:**157–167.

50. **Marose, S., C. Lindemann, R. Ulber, and T. Sheper.** 1999. Optical sensor systems for bioprocessing monitoring. *Trends Biotechnol.* **17:**30–34.

51. **McGovern, A. C., C. D. Broadhurst, J. Taylor, N. Kaderbhai, M. K. Winson, D. A. Small, J. J. Rowland, D. B. Kell, and R. Goodacre.** 2002. Monitoring of complex industrial bioprocesses for metabolite concentrations using modern spectroscopies and machine learning: application to gibberellic acid production. *Biotechnol. Bioeng.* **78:**527–538.

52. **Meyerhoff, M. E., M. Trojanowicz, and B. O. Palsson.** 1993. Simultaneous enzymatic/electrochemical determination of glucose and L-glutamine in hybridoma media by flow-injection analysis. *Biotechnol. Bioeng.* **41:**964–969.

53. **Návratil, M., A. Norberg, L. Lembrén, and C. F. Mandenius.** 2004. On-line multi-analyzer monitoring biomass, glucose and acetate for growth rate control of a *Vibrio cholerae* fed-batch cultivation. *J. Biotechnol.* **115:**67–79.

54. **Nor, Z. M., M. I. Tamer, J. M. Scharer, M. Moo-Young, and E. J. Jervis.** 2001. Automated fed-batch culture of *Kluyveromyces fragilis* based on a novel method for on-line estimation of cell specific growth rate. *Biochem. Eng. J.* **9:**221–231.

55. **Ohashi, R., J. M. Otero, A. Chwistek, and J.-F. P. Hamel.** 2002. Determination of monoclonal antibody production in cell culture using novel microfluidic and traditional assays. *Electrophoresis* **23:**3623–3629.

56. **Olsson, L., and J. Nielsen.** 1997. On-line and in situ monitoring of biomass in submerged cultivations. *Trends Biotechnol.* **15:**517–522.

57. **Opel, C., and A. Amanullah.** 2008. Monitoring and control of cell culture bioreactors using on-line scanning dielectric spectroscopy. AIChE Annu. Meet., 16 to 21 November 2008, Philadelphia, PA.

58. **Opel, C., F. Li, and A. Amanullah.** 2008. On-line viable cell density measurements using scanning dielectric spectroscopy. Cell Culture Eng. XI, 13 to 18 April 2008, Coolum, Australia.

59. **Pattison, R. N., J. Swamy, B. Mendenhall, C. Hwang, and B. T. Frohlich.** 2000. Measurement and control of dissolved carbon dioxide in mammalian cell culture processes using an in situ fiber optic chemical sensor. *Biotechnol. Prog.* **16:**769–774.

60. **Rhiel, M., P. Ducommun, I. Bolzonella, I. Marison, and U. Von Stockar.** 2002. Real-time *in situ* monitoring of freely suspended and immobilized cell cultures based on mid-infrared spectroscopic measurements. *Biotechnol. Bioeng.* **77:**174–185.

61. **Roychoudhury, P., R. O'Kennedy, B. McNeil, and L. M. Harvey.** 2007. Multiplexing fibre optic near infrared (NIR) spectroscopy as an emerging technology to monitor industrial bioprocesses. *Anal. Chim. Acta* **590:**110–117.

62. **Rudolph, G., P. Lindner, A. Gierse, A. Bluma, G. Martinez, B. Hitzmann, and T. Scheper.** 2008. On-line monitoring of microcarrier based fibroblast cultivations with *in situ* microscopy. *Biotechnol. Bioeng.* **99:**136–145.

63. **Sarrafzadeh, M. H., L. Belloy, G. Esteban, J. M. Navarro, and C. Ghommidh.** 2005. Dielectric monitoring of growth and sporulation of *Bacillus thuringiensis*. *Biotechnol. Lett.* **27:**511–517.

64. **Saucedo, V., B. Wolk, and A. Arroyo.** 2008. Evaluating the pH performance of different technologies in cell culture. IFPAC Annu. Meet., 27 to 30 January 2008, Baltimore, MD.

65. **Schügerl, K.** 2001. Progress in monitoring, modeling and control of bioprocesses during the last 20 years. *J. Biotechnol.* **85:**149–173.

66. **Scully, P. J., L. Betancor, J. Bolyo, S. Dzyadevych, J. M. Guisan, R. Fernandez-Lafuente, N. Jaffrezic-Renault, G. Kuncová, V. Matejec, B. O'Kennedy, O. Podrazky, K. Rose, L. Sasek, and J. S. Young.** 2007. Optical fiber biosensors using enzymatic transducers to monitor glucose. *Meas. Sci. Technol.* **18:**3177–3186.

67. **Severinghaus, J. W., and A. F. Bradley.** 1958. Electrodes for blood pO_2, and pCO_2, determination. *J. Appl. Physiol.* **13:**515–520.

68. **Shimizu, H., K. Yasuoka, K. Uchiyama, and S. Shioya.** 1998. Bioprocess fault detection by nonlinear multivariate analysis: application of an artificial autoassociative neural network and wavelet filter bank. *Biotechnol. Prog.* **14:**79–87.

69. **Sivakesava, S., J. Irudayaraj, and D. Ali.** 2001. Simultaneous determination of multiple components in lactic acid fermentation using FT-MIR, NIR, and FT-Raman spectroscopic techniques. *Process Biochem.* **37:**371–378.

70. **Speciner, L., J. Perez, C. Grimaldi, R. Reineke, and A. Arroyo.** 2007. Implementing an automated reactor sampling system for monitoring cell culture bioreactors. AIChE Annu. Meet., 4 to 9 November 2007, Salt Lake City, UT.

71. **Stoll, T. S., P. A. Ruffieux, M. Schneider, U. von Stockar, and I. W. Marison.** 1993. On-line simultaneous monitoring of ammonia and glutamine in a hollow-

fiber reactor using flow injection analysis. *J. Biotechnol.* **51**:27–35.

72. **Swalley, S. E., J. R. Fulghum, and S. P. Chambers.** 2006. Screening factors effecting a response in soluble protein expression: formalized approach using design of experiments. *Anal. Biochem.* **351**:122–127.

73. **Tolosa, L., K. Yordan, P. Harms, and G. Rao.** 2002. Noninvasive measurement of dissolved oxygen in shake flasks. *Biotechnol. Bioeng.* **80**:594–597.

74. **Vaidyayanathan, S., S. A. Arnold, L. Matheson, P. Mohan, B. McNeil, and L. M. Harvey.** 2001. Assessment of near-infrared spectral information for rapid monitoring of bioprocess quality. *Biotechnol. Bioeng.* **74**:376–388.

75. **Vojinović, V., J. M. S. Cabral, and L. P. Fonseca.** 2006. Real-time bioprocess monitoring. Part I. In situ sensors. *Sensors Actuators B* **114**:1083–1091.

76. **Wakeem, J., S. Tolani, and A. Arroyo.** 2009. Evaluation of bug eye on-line sensor for measuring optical density in *E. coli* processes. IFPAC Annu. Meet., 25 to 28 January 2009, Baltimore, MD.

77. **Wei, N., J. You, K. Friehs, E. Flashel, and T. W. Nattkemper.** 2007. An *in situ* probe for on-line monitoring of cell density and viability on the basis of dark field microscopy in conjunction with image processing and supervised machine learning. *Biotechnol. Bioeng.* **97**:1489–1500.

78. **Wittman, C., H. M. Kim, G. John, and E. Heinzle.** 2003. Characterization and application of an optical sensor for quantification of dissolved O_2 in shake-flasks. *Biotechnol. Lett.* **25**:377–380.

79. **Wu, P., S. Ozturk, J. D. Blackie, J. C. Thrift, C. Figueroa, and D. Naveh.** 1995. Evaluation and applications of optical cell density probes in mammalian cell bioreactors. *Biotechnol. Bioeng.* **45**:495–502.

80. **Xie, L., W. Pilbrough, C. Metallo, T. Zhong, L. Pikus, J. Leung, J. G. Auniņš, and W. Zhou.** 2002. Serum-free suspension cultivation of PER.C6 cells and recombinant adenovirus production under different pH conditions. *Biotechnol. Bioeng.* **80**:569–579.

81. **Xing, Z., Z. Li, V. Chow, and S. S. Lee.** 2008. Identifying inhibitory threshold values of repressing metabolites in CHO cell culture using multivariate analysis methods. *Biotechnol. Prog.* **24**:675–683.

82. **Xu, Y., A. S. Jeevarajan, J. M. Fay, T. D. Taylor, and M. Andersen.** 2002. On-line measurement of glucose in a rotating wall perfused vessel bioreactor using an amperometric glucose sensor. *J. Electrochem. Soc.* **149**: H103–H106.

83. **Yu, E. W., K. Davis, M. Meacham, G. Barringer, and S. Casnocha.** 2008. Performance evaluation of an automated bioreactor sampling system for mammalian cell cultivation. ACS Natl. Meet., 17 to 21 August 2008, Philadelphia, PA.

84. **Zawada, J., and J. Swartz.** 2005. Maintaining rapid growth in moderate density *Escherichia coli* fermentations. *Biotechnol. Bioeng.* **89**:407–415.

Bioreactor Automation

DAVID HOPKINS, MELISSA ST. AMAND,
AND JACK PRIOR

51

51.1. INTRODUCTION

Although successful growth of prokaryotic or eukaryotic cells in bioreactors can be achieved with little or no automation, increased monitoring and control of bioreactors can often improve process reliability, consistency, and understanding. The goal of bioreactor control is to manage the cells' environment and metabolism to maximize productivity and product quality. The complexity of an automation system will vary depending on a number of factors including process requirements, production costs, product value, quality requirements, regulatory constraints, and data acquisition requirements. This chapter introduces basic automation terminology and concepts as well as the rationale for implementing various levels of automation.

51.1.1. Brief History of Bioreactor Automation

Modern control theory was first applied to biotechnology in the 1970s (24) as developments in bioreactor sensor technology and increased availability of computer hardware in the years leading up to this era set the stage for more advanced bioreactor control implementations. Particularly useful were the commercial availability of the modern pH probe in 1945 (19, 23) and the membrane-covered dissolved oxygen (DO) probe in 1956 (7). Later, in 1964, Johnson et al. (13) introduced steam-sterilizable DO probes, which enabled practical use of the probes in bioreactors.

In the early 1970s, Humphrey and Nyiri demonstrated computer control of DO by varying agitation and airflow rates (20, 21). Other pioneers of bioreactor monitoring and computer control include Lim, Weigand, Zabriskie, N. S. Wang, Stephanopoulos, Cooney, H. Y. Wang, and D. I. C. Wang (8, 9, 18, 24, 29). An automated in-line sampling system which maintained sterility was introduced by Leisola and Kauppinen in 1978 (16). Later in the mid-1980s, new types of biosensors became commercially available for measuring components for glucose, amino acids, antibiotics, and macromolecules (14). Since the mid-1980s, the increased availability of low-cost computer and data storage hardware has significantly increased the amount of data collected for bioreactor processes, with data historians frequently being employed to facilitate archiving and retrieving data for analysis and troubleshooting.

51.1.2. Process Requirements

Before designing or purchasing an automation system, the full range of process operations and requirements must be understood. Major operations associated with a given batch typically consist of equipment preparation and assembly, sterilization, inoculation, cell growth and production, and cleaning.

While these operations can often be performed manually for laboratory and smaller pilot-scale bioreactors (~500 liters or smaller), large-scale reactors require automation for most operations. For example, small-scale reactors can often be moved to cleaning/steaming areas or be sterilized with autoclaves, whereas larger bioreactors must be dealt with by cleaning in place (CIP) and sterilization in place (SIP). Larger bioreactors also pose a financial risk if a mistake or unusual event occurs that affects product quality or productivity. This often justifies additional automation to facilitate consistent operation. Some form of automation is typically used at all scales during inoculation, cell growth, and production in order to maintain proper culture conditions.

51.1.2.1. Automated Reactor Operations

Sterilization is required after reactor assembly and prior to inoculation. The sterilization process begins by heating the gas and vent filters with clean steam, followed by a vessel heat-up via a heated jacket and/or with steam injected directly into the reactor. Air in the bioreactor is eventually replaced with clean steam as the air exits through the vent and steam traps. The vessel is then pressurized to allow the temperature to rise to a minimum of 121°C, where it is held until a temperature- and time-based criterion is met. Steam is then shut off. While the steam condenses, a vacuum is prevented by the introduction of sterile air. The pressure is then slowly reduced from the sterilization pressure to the operating pressure. The bioreactor is also cooled by introducing water into the jacket, if present.

After sterilization and prior to inoculation, unstable or reactive media components may be added aseptically. In-situ sensors to detect variables such as cell density, DO, and pH may require standardization or adjustment prior to inoculation with the seed culture. During cell growth and production, additional transfers of nutrients, acid or base, and antifoam as well as periodic sampling may be necessary. For continuous or semibatch operations, harvest transfers for further downstream processing occur during the production phase. Care must be taken to ensure that the bioreactor remains sterile while adjoining lines are cleaned and sterilized.

Once production is complete, larger bioreactors are typically connected to a pump or a CIP skid that circulates cleaning solutions through spray-balls to clean the interior as well as each port. Valves must be manipulated as the vessel and each port are cleaned and rinsed. Temperature control may be active to maintain the desired cleaning solution temperature. Automation can be used to coordinate valve cycling on the bioreactor in conjunction with the cleaning equipment.

Sensors, control strategies, and automation systems capable of meeting these process requirements will be discussed in the following sections.

51.2. INSTRUMENTATION REQUIREMENTS

Bioreactor measurements can be categorized by the physical location of the sensor relative to the process they are measuring. "In-situ" or "in-line" sensors are exposed directly to the process fluid. "At-line" measurements are conducted on samples locally after they are aseptically withdrawn from the bioreactor in order to maximize accuracy or their timely utilization. Living cells in the noncontrolled sample environment may influence the sample itself. Dissolved gases can quickly change in concentration and, in the case of dissolved CO_2, can affect the pH of the sample. "Off-line" measurements are less time sensitive and are usually performed on aseptically stored supernatant within hours or days of sampling. Supernatant samples may be frozen or refrigerated to minimize continued biological or chemical reactions.

In a manufacturing setting, data from these three types of measurements—in-line, at-line, and off-line—may be obtained electronically or gathered manually by manufacturing or quality control personnel. Each type records data in separate systems, and provisions must be made for recombining these data electronically for review and analysis.

Desirable sensor traits include high precision and accuracy, signal linearity across the full operational range, minimal drift over time, insensitivity to other process conditions such as temperature, and robustness against failure for all process conditions. Sensors susceptible to drift must be verified by periodic at-line analysis of samples. The transmitter or automation system can then be adjusted as required. Measurements which are a function of temperature, such as conductivity and pH, require compensation if the batch temperature varies over time. This compensation may be incorporated into the sensor apparatus itself or into the automation system.

The materials of construction for the instrumentation must withstand the pressure and temperature requirements of sterilization and not leach material into the process. Typical materials acceptable for product contact include 316L stainless steel electropolished to an average roughness (R_a) of 10 to 20, ethylene propylene diene M-class rubber, Viton, silicone, Teflon, and glass.

Identifying advanced instruments that achieve these goals while monitoring cell density or metabolism can be more challenging than for sensors which directly measure physical and chemical properties. Bioreactors also have a limited number of ports for sensors. All sensors involve cost and risks that must be balanced with the operating benefit they provide.

51.2.1. Culture Conditions

Typical in-line measurements of reactor conditions include pH, DO, dissolved carbon dioxide, pressure, temperature, weight/level, foam level, and agitation speed.

51.2.1.1. pH

pH was one of the first in-line measurements of cell culture state and is a critical process control variable because cell growth, enzyme activity, and other biological processes are affected by pH (5, 14). Choosing a pH set point (SP) or trajectory for a particular process often involves striking a balance between the optimal pH for growth and that for protein production (26).

pH electrodes measure free H^+ ions in the surrounding medium. In bioreactors, pH probes are typically inserted via a sterile connection port in the sidewall to allow direct contact with the culture medium. The pH probe should be placed in a well-mixed location and must be designed to withstand frequent sterilization and prolonged exposure to cell culture fluid (5, 26).

51.2.1.2. DO

Dissolved oxygen (DO) is a critical substrate during aerobic cell cultivation since cellular respiration requires oxygen. Gaseous O_2 is typically supplied to the culture medium by sparging air or oxygen into the reactor vessel. DO probes measure oxygen partial pressure (pO_2) in the culture medium rather than direct oxygen concentration. The probe should be located in a well-mixed region away from spargers or solids that may interfere with the measurement (26).

DO probes are typically galvanic or polarographic. Both types use an electrode system in which DO reacts with the cathode to produce an electrical current. Galvanic probes are generally preferred because they require less maintenance.

51.2.1.3. Dissolved Carbon Dioxide

Carbon dioxide partial pressure (pCO_2) is a measure of the dissolved CO_2 content of the culture medium. Carbon dioxide is often added to the headspace above the culture at small scale to maintain minimum pCO_2 levels. At large scale, reduced mass transfer rates can result in accumulation of pCO_2 in the medium as a result of cellular respiration. If not monitored and controlled, it can adversely affect cellular metabolism by lowering the cells' internal pH. Cell growth, product formation, and even glycosylation patterns have been shown to be negatively affected by high pCO_2 concentrations in culture medium (22).

Off-line blood-gas analyzers can be used to measure pCO_2. However, careful sample handling is necessary to ensure accurate readings. If the culture produces little acid, pCO_2 may be indirectly estimated based on pH (27). Recently, steam-sterilizable pCO_2 probes have been developed which operate as a modified pH electrode (22).

51.2.1.4. Pressure

Bioreactors are maintained at positive pressure to ensure aseptic operation. Bioreactor pressure also influences the partial pressure of dissolved gases. Higher vessel pressure increases the driving force for O_2 transfer and increases the capacity of the reactor to support higher cell densities. Conversely, higher pCO_2 pressures can have adverse effects on the culture. Diaphragm sensors are typically used to measure pressure (26).

51.2.1.5. Temperature

Living cells can be highly sensitive to temperature. As with pH, the optimal temperature for growth may differ from that for protein production. At temperatures exceeding the optimal range, cells will generally undergo thermal death. Insertable thermocouples, resistance thermometers, or thermistors are used to measure bioreactor temperature (26).

51.2.1.6. Volume

Level sensors based on conductivity, capacitance, or differential pressure measurements may be used to determine the bioreactor volume. Mass sensors such as load cells or scales may also be used, but require flexible connections between the bioreactor and the adjoining platform to perform correctly.

51.2.1.7. Foam

Foam formation can adversely affect a fermentation process by causing cell death and protein denaturation, and can also interfere with level measurements and exhaust filter operation. Various features of a bioreactor such as impeller blades and the position of additive inlet ports are designed to minimize foaming. Conductance and capacitance probes are typically used to monitor foam level.

51.2.1.8. Agitator Speed

Agitator speed influences several phenomena inside the reactor vessel and is thus important to monitor and control. Minimal agitation levels are required to ensure off-bottom mixing and sufficient mass transfer rate of gases to the culture. Conversely, high agitation rates can lead to shear damage to the cells and can induce foam formation. Agitation rate is typically measured by a frequency counter signal converted to rotations per minute and controlled via a variable frequency drive (VFD).

51.2.1.9. Nutrient Feed and Gas Flow

Gas flow is typically measured and controlled by mass flow controllers or volumetric rotameters. Liquid flow rate of feed or titrant solutions may be measured by source tank level, magnetic flow meters, or mass flow meters.

5.1.2.2. Cell Growth and Metabolism

Historically, various measurements at-line or off-line are used to monitor cell growth and metabolism of the cell culture. However, technology innovations are increasingly enabling the use of in-line measurements for bioreactor control.

51.2.2.1. OD (Turbidity)

Optical density (OD), or turbidity, measurements can be correlated with biomass concentrations. These measurements are typically performed off-line, but in-line OD sensors are also available. OD sensors rely on the transmittance, absorbance, and/or scatter of light. The wavelength of incident light, the angle of detection, interference from sample bubbling, and extraneous particulate matter are all important considerations for OD measurements (15, 17, 27). Cell viability and aggregation must be considered when translating OD measurements to cell concentrations.

51.2.2.2. Capacitance

Capacitance probes such as the ABER Biomass Monitor (ABER Instruments, Aberystwyth, United Kingdom) can determine viable cell density in-line. Capacitance measurements are similar in principle to trypan blue staining—in which only viable cells remain unstained due to intact membranes—with the plasma membrane of the cell acting as an insulator and preventing low-frequency electrical currents from charging the intracellular fluid. The resulting dielectric dispersion is proportional to cell radius and volume. Since only living cells have intact plasma membranes, many of the typical interferences such as lysed cells and media components are minimized with capacitance measurements (15).

51.2.2.3. NIR

Near-infrared spectroscopy (NIR) is another form of in-line biomass measurement. NIR is based on electromagnetic absorption between 700 and 2,500 nm, which has overtones and combination bands of the fundamental bending and stretching vibrations (11, 28). Weaker than mid-infrared, NIR enables direct sample analysis. NIR measurements are also possible at longer path lengths than traditional spectroscopic techniques, making it a suitable real-time method for biomass measurement (1, 28).

51.2.2.4. Exit Gas Composition

Measurements of exit gas composition and gas flow rates can be used to monitor the rate of cellular metabolism by enabling the calculation of oxygen uptake and carbon dioxide evolution rates. CO_2 in the gas phase is typically measured by infrared absorbance at <15 μm or by acoustic means. O_2 analysis is based on paramagnetic properties. Photoacoustic spectroscopy and magnetoacoustic technology are also used for CO_2 and O_2 analysis. Time delay and sample pretreatment issues must be addressed for these measurements (27). A centralized mass spectrometer can be used to monitor gas streams multiplexed from several reactors and can provide measurements of O_2, CO_2, and N_2 (6).

51.2.2.5. Redox Potential

Redox potential refers to the overall availability of electrons in the culture broth that result from reduced and oxidized forms of various media and culture components (2). Redox potential measurements are influenced by multiple factors, making interpretation of results difficult. However, redox measurements are useful when pO_2 sensor measurements become inaccurate (i.e., <10 mmol O_2 /liter) (27). Redox potential is generally measured potentiometrically with electrodes that are similar in construction to pH electrodes (24).

51.3. CONTROL METHODS

This section describes various control objectives and methodologies for bioreactor operation.

51.3.1. Bioreactor Automation Design Goals

A bioreactor system has several levels of control, distinguished by their primary objectives as well as the frequency of execution. The system design should address each of the layers of control that are described in the following sections. Only safety, process control, and process optimization are necessary for most laboratory-scale bioreactor systems. Large-scale production facilities can benefit from the addition of procedural and plant floor control applications.

51.3.1.1. Personnel, Equipment, and Process Safety

Protection of personnel, equipment, and process should be considered during all phases of bioreactor design and operation. At the hardware specification phase, for example, the bioreactor pressure rating should be set above the associated utility supply pressures to prevent catastrophic bioreactor failure. A second layer of physical control is added by placing a rupture disk on the inner vessel and a pressure relief valve on the jacket of the vessel. This layer protects the bioreactor regardless of external component failures (e.g., pressure regulator).

Rupture disk sensors are monitored by the automation system in order to bring the bioreactor to a safe, nonenergized state in the event of a failure. Typical failure-state adjustments include shutting valves to isolate the bioreactor from entering gases and fluids, and halting associated agitators and pumps. Operators must also be able to completely shut down a bioreactor with a main power switch (typical for small bioreactors) or emergency shutdown ("E-Stop") switch (typical for large bioreactors).

Equipment can also be protected by direct feedback from a sensor. For example, a pressure switch placed directly after a positive displacement pump and wired to the pump motor controller can directly shut down the pump if the pressure exceeds the switch's limit, thus protecting the pump and downstream equipment from damage. Note that these examples of physical and hard-wired protection are always active. They have a response time on the order of milliseconds and collectively provide the last line of defense against failures not handled by the automation system.

When protection is not required under all conditions, software automation is used to enable interlocks between sensors and energized equipment, such as motors and valves, at the appropriate time. An example of a bioreactor process interlock is the automatic closure of an addition valve when the titrant tank is empty, to avoid gas from the titrant tank flowing into the bioreactor. The interlock overrides the pH controller and prevents this condition from occurring, while activating an alarm to notify the operator of the abnormal condition. This interlock is enabled only when pH control is enabled and is not otherwise active. Software interlocks typically react on the order of tenths of seconds to seconds.

51.3.1.2. Basic Process Control

The purpose of a standard process controller is to maintain a process variable within a specified range, or as close to an SP as possible, by adjusting one or more process control devices. The frequency of adjustment is typically on the order of seconds to minutes. Process control algorithms are discussed in more detail in section 51.3.2.

51.3.1.3. Process Optimization

Control strategies can be implemented to optimize cell metabolism for cell growth, production, and/or product quality by periodically adjusting SPs of standard process controllers. The frequency of adjustment is typically on the order of seconds to days. The wide time scale is due to the fact that data used to compute the controller's output may require evaluating several days of in-line and off-line data. This subject is discussed in more detail in section 51.3.3.

51.3.1.4. Phase and Recipe Automation

Bioreactor systems require procedural control of the sequences of process steps necessary to complete an operation. Each step requires some criteria to be met before the next can begin. Examples include completing a hold timer, reaching a target process value, receiving feedback that equipment is in the correct position, confirmation of successful communications between automation subsystems, or combinations thereof. Conditions are typically monitored on a once-per-second basis, with changes made to process controllers ranging from seconds to days. Procedural control includes phase, recipe, or unit batch control.

51.3.1.5. Plant Floor Management

Plant floor control involves the coordination and scheduling of several production systems within a manufacturing area. It has little to do with direct process control, but often manages communication between higher-level corporate systems and process control systems. A typical scenario is for a production order to be issued by a corporate Enterprise Resource Planning system instructing the plant floor system to select the appropriate area recipe. Recipe parameters are selected and the area recipe is then started. As the area recipe is executed, equipment is selected and parameters are passed to each unit procedure, which is executed in a sequential, coordinated fashion. Key process data are collected from the selected equipment and related manufacturing computer systems as the area recipe is executed. Recipe steps can take anywhere from minutes to days to complete. These systems are typically known as manufacturing execution systems (MESs) or area batch control systems and are discussed further in section 51.4.

51.3.2. Process Control Algorithms

During process control, a process variable, termed the controlled variable (CV), is maintained at a desired SP. Most bioreactor automation uses a single-input, single-output controller, which consists of a sensor input, a controller, and a control device/actuator.

The controller calculates an output to send to the control device(s) based on the current and historical values obtained from the CV relative to the desired SP. The output of the controller is sent to one or more manipulated variables (MVs), or to another controller, which will cause physical changes in the process. These manipulations should cause the CV to approach the SP. During operation, disturbance variables are variables which affect the CV but cannot be controlled. They represent the natural fluctuation of the process. Figure 1 displays some of the more common attributes of an industrial-sized bioreactor. The CV, MV, and controller types corresponding to this figure are listed in Table 1.

Two key criteria for choosing the type of controller are the process sensitivity to fluctuations around the SP and whether the MV adjustments are to be performed periodically or continuously.

If the process is insensitive to some variance, and only periodic adjustments in the MV are desired, then basic on/off control is the controller of choice. If the CV must be precisely controlled or continuous adjustment to the MV is desired, then proportional-integral-derivative (PID) control is the controller of choice.

51.3.2.1. On/Off Control

On/off controllers are simple to understand and adjust. A control device, such as a pump, solenoid, or valve, is energized based on an on/off signal from a sensor or automation system. A digital sensor, such as a pressure switch, can directly drive equipment on/off. An analog sensor can be compared to a threshold value by an automation system to drive equipment on/off (e.g., "if level < 100 then open feed valve").

The triggering events can be set to work in only one direction or both directions. For example, if the goal of an interlock is to place the equipment in a safe, nonenergized state when an unsafe condition is detected, it may not be desired to restart until an operator confirms that the unsafe condition has been addressed. A level controller, on the other hand, may open a feed valve when the level falls below a target value, and then close when the level reaches the target. Control can continue as the level rises and falls, as the actions to both open and close the valve are based on logically opposite conditions.

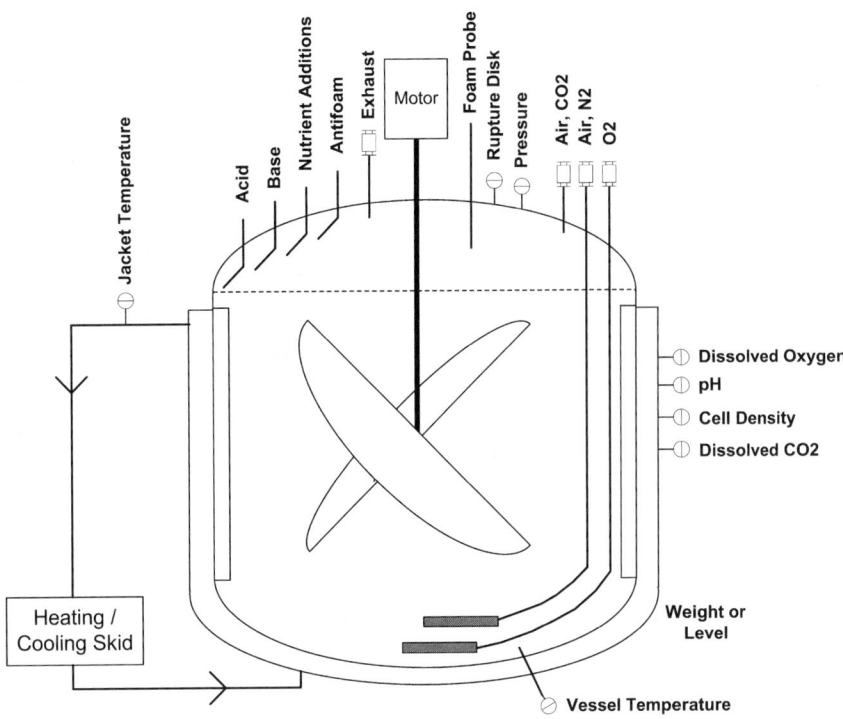

FIGURE 1   Typical bioreactor components.

A region of inactivity (deadband) can be introduced to avoid frequent switches between the on and off position and thereby reduce equipment wear. This region should be large enough to distinguish between a real process change and sensor noise. In the example shown in Fig. 2, the control equipment is energized when the CV falls below the lower action limit and remains on until the SP is reached. Once the SP is reached, the control equipment is de-energized and remains off until the process causes the CV to fall below the action limit again.

A common use of this type of controller is to maintain pH. Catabolism of carbohydrates produces acids that must be neutralized to maintain pH for optimal cell growth and production. The goal is to make these adjustments infrequently and minimize the reactor volume increase over the course of a batch. This can be facilitated by using concentrated solutions, but care must be taken to ensure that mixing is sufficient to avoid local regions of extreme pH detrimental to culture health. Figure 3 shows a hypothetical example of a pH profile using this type of control. As the cell concentration increases exponentially over time, the rate at which the pH falls, and therefore the frequency of base additions, also increases.

Some cultures may require pH adjustment in both directions over the course of the culture due to changes in metabolism, dissolved CO_2 concentration, and cell density. Figure 4 shows how this controller can be modified for acid and base addition. As shown, the deadband region between the lower and upper action limits prevents overshoot from an acid or base addition from activating the other. The risk of overshoot is typically highest at the start of a run when the cell density is lowest.

For cultures where the cell density increases by several orders of magnitude, the frequency of addition may reach a point where the titrant is added constantly and pH cannot be maintained. This is known as controller saturation and can be corrected by adding adaptive features. In addition to

TABLE 1   Typical bioreactor control loops

CV	MV	Loop type
Vessel temperature	Heating/cooling	
Fluid flow	Cascaded PID	
Pressure	Exhaust flow	PID
pH	Acid/base pump	PID, on/off
pH	Headspace air/CO_2 flow	PID
DO	Sparge air or O_2 flow	PID, on/off
Dissolved CO_2	Sparge N_2 flow	PID, on/off
Foam	Antifoam pump	On/off
Gas inlet/outlet composition	Nutrient addition rate	PID
Agitator rpm	Agitator motor speed	PID
Level/weight	Nutrient/harvest	
Pump rates	PID	

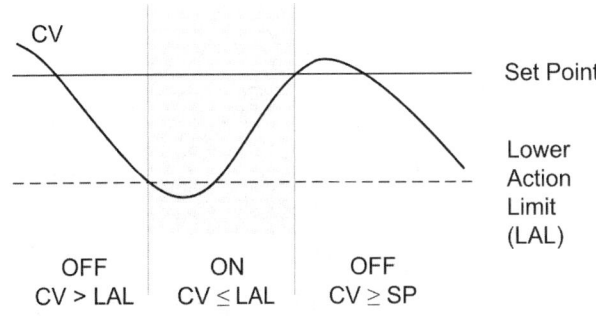

FIGURE 2   On/off range control.

FIGURE 3 On/off pH range control.

turning a base addition pump on and off, the speed of the pump can be stepped up or down to extend the sensitivity and range of the controller. If the pH drops well below the lower action limit (see Fig. 5), a response limit can be set to double the pump speed, allowing the controller to return the pH to SP. If the reverse occurs and the demand for base drops so low that the smallest addition causes the pH to overshoot the SP, an overshoot limit can be set to reduce the pump speed by half. A delay is required to ensure that the speed is not adjusted repeatedly, and to allow both the controller and the process time to adjust to the new pump setting.

51.3.2.2. PID Controller

If the control equipment is continuously active, then PID controllers are typically used. These controllers respond to three components of the error measurement over time (see Fig. 6).

Although manufacturers of automation equipment vary the implementation of PID control, a typical form of the PID equation is:

Output =

$$\text{Output}_{SS} + K\left[\text{Error} + \left(\frac{1}{\tau_I}\right)\int \text{Error}\,dt + \tau_D\left(\frac{d\,\text{Error}}{dt}\right)\right] \quad (1)$$

where Output_{SS} is the steady-state output required to maintain SP; Error is SP − CV; K is the proportional control constant; τ_I is the integral control constant and τ_D is the derivative control constant; $\int \text{Error}\,dt$ is the integral of error with respect to time; and $d\,\text{Error}/dt$ is the derivative of Error with respect to time (i.e., $\Delta\text{Error}/\Delta\text{time}$).

If the CV is maintained at the SP, all three error terms equal zero and the output equals the steady-state value, Output_{SS}. As process demands change, or as sporadic disturbances occur, the error becomes nonzero and the various terms come into play. The proportional (P) term, $K * \text{Error}$, adjusts for temporary disturbances but cannot completely return the process to SP if a step change in the process occurs. When the process or SP changes, Output = $\text{Output}_{SS} + K * \text{Error}$. If CV = SP, the error is zero, but Output_{SS} no longer represents the output required to maintain SP. Therefore, as shown in Fig. 7, the process will settle into a new steady state with a constant offset from SP if only proportional control is used.

The integral (I) term, $K * (1/\tau_I)\int \text{Error}\,dt$, accumulates error over time and will continue to increase in magnitude while error persists. The integral term will therefore

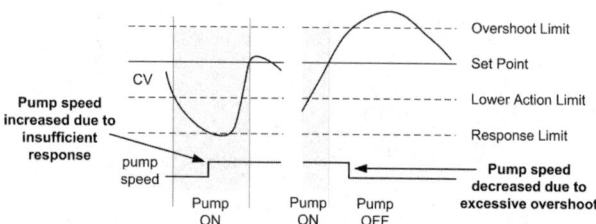

FIGURE 5 Adaptive on/off control.

compensate for the offset problem seen with proportional-only control. If the integral constant τ_I is too large, excessive oscillation around the SP can occur, as shown in Fig. 7.

The derivative (D) term, $K * \tau_D (d\,\text{Error}/dt)$, is seldom necessary, but can be useful for countering rapid disturbance changes more quickly than with proportional and integral control alone. This reduces SP overshoot and allows the controller to reach SP more quickly, as shown in Fig. 7. The disadvantage of derivative control is that noise in the CV is magnified and can translate into erratic control. To avoid this issue, one can simply use PI control, or smooth the derivative error data using various methods before calculating the derivative term.

Most modern automation systems have a built-in auto-tune feature to determine appropriate K, τ_I, and τ_D parameters. If this feature is not available, Seborg et al. (25) reviews several methods for PID parameter estimation and provides a good introduction to control theory and advanced PID controller issues.

It is particularly important to test bioreactor PID controller performance throughout the full process range to ensure that the controller can maintain SP without excessive oscillations and does not become unstable at any point in the process. The output required to maintain steady state can change exponentially over the course of a batch, and some cases may require more than one set of PID parameters to correctly control during periods of low as well as high demand. Figure 8 shows that PID parameter set 1 is used initially and switched to set 2 when the high output limit is reached. If the demand slows and the output drops, set 1 is reinstated when the low output limit is reached. The range between low and high output limits acts as a buffer zone to ensure that the PID controller is not constantly disrupted by changes between PID parameter sets 1 and 2.

Another common issue is that a control device with the capability to deliver the output required to maintain control over the full process range may not have the precise response needed to control smoothly during periods of low demand. This can result in near on/off control of the MV, resulting in large oscillations of the CV. In this case, the output range may be split between small-scale and large-scale control equipment, each with its own PID controller, to provide precise control throughout the full range of the process. Figure 9 illustrates how control is passed between the two separate PID controllers in split range PID control, using the output of the currently active controller to determine when control should be switched to the other controller. The buffer range between the low and high output limits prevents frequent

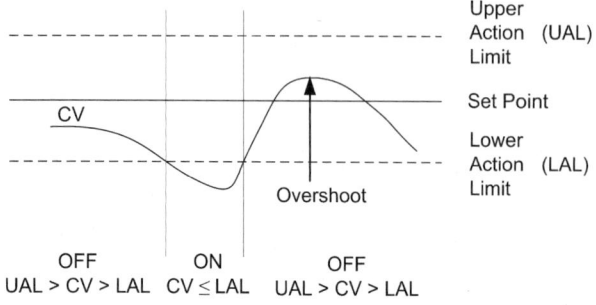

FIGURE 4 On/off control with deadband.

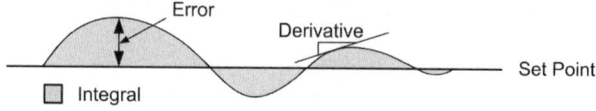

FIGURE 6 PID measurements.

Proportional Response

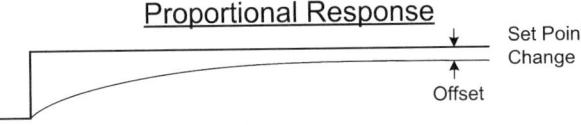

Proportional + Integral (PI) Response

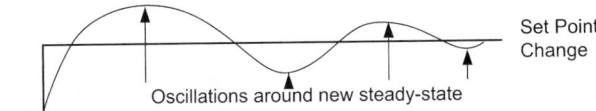

PID Response

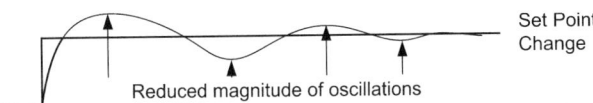

FIGURE 7　PID term contributions.

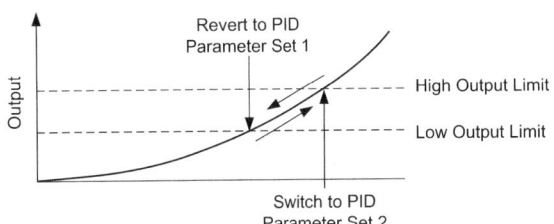

FIGURE 8　Adaptive PID control.

switching between the high- and low-range controllers. The integral term of the inactive PID controller must also be reset before control is passed to it to ensure a smooth transfer.

51.3.2.3.　Cascade Control

Cascade control connects an outer (master) controller to an inner (slave) controller by passing the output from the master to the SP of the slave. This is useful when the physical control of one subsystem has a direct impact on another. Bioreactor temperature control is typically performed by circulating a fluid through a jacket that surrounds the side and bottom of the vessel (see Fig. 1). The jacket fluid is heated or cooled as necessary, which in turn transfers heat to or from the vessel. The jacket controller is set to cascade mode to allow its SP to be fed remotely by the vessel temperature controller's output. The jacket temperature controller is said to be a slave to the master vessel temperature controller. It is not important for a slave controller to actually achieve its SP, but simply to respond in the correct direction to changes in the SP. For this reason, slave PID controllers can operate successfully using proportional-only control.

51.3.2.4.　Split Range Output

A split range output takes a single analog controller value and separates it into two separate values that can be used to drive two individual control devices. To continue with the temperature control example, the jacket

controller output requires a single output value to drive both heating and cooling actions. Typical controllers range from 0 to 100%, with the control equipment at rest at 0%. In contrast, a jacket temperature controller is at rest at 50%, which allows 50 to 100% output to be translated to 0–100% heating and 50 to 0% output to be translated to 0–100% cooling (see Fig. 10). Note that the controller responds equally in magnitude to positive and negative error. Therefore, the MVs should respond similarly for the controller to have a balanced effect on the process.

In our example, 0% PID output corresponds to full cooling and should transfer an equal but opposite amount as that delivered by a PID output of 100%, which corresponds to full heating. A deadband can be implemented to provide a buffer zone where no output is active, in order to reduce oscillations and equipment wear by preventing small variations near the SP from causing a counter-response as the temperature crosses the SP.

51.3.2.5.　TPO

When a controller with an analog output is used, such as a PID controller, and the MV is digital, such as an open/close valve or an on/off pump, then a time proportional output (TPO) must be used to convert the PID percent output to a digital signal that is active for a proportional fraction of the specified cycle time.

For example, if the cycle time is 10 s and the controller output is 10%, then the valve is opened for 1 s and closed for 9 s. The calculation is repeated at the end of each cycle. The TPO scan interval is the time period at which the valve position is determined. Therefore, the total time a valve can be open is a multiple of the scan interval. The ratio of the scan interval over the cycle time determines the resolution of the TPO. If the cycle time is 10 s and the scan interval is 1 s, then only controller output changes in multiples of 10% will cause an actual change to the time a valve is open. The scan interval should be set short enough to prevent controller overshoot during the most responsive portion of the process, but not shorter than the time it takes for the valve to physically open.

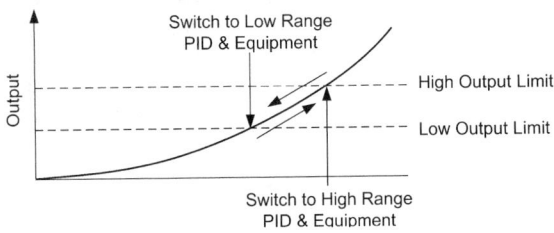

FIGURE 9　Split range PID control.

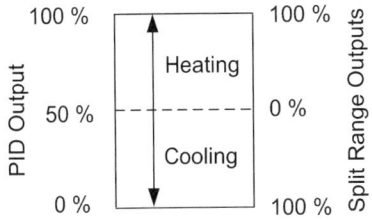

FIGURE 10　Split range output.

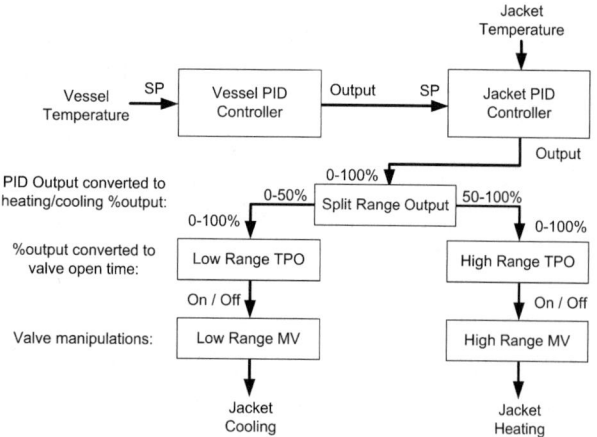

FIGURE 11 Temperature control example.

Figure 11 summarizes how these control elements are integrated in our temperature control example. The vessel PID controller output is transferred to the jacket PID SP. The jacket PID output is split between heating and cooling values, which are in turn converted to valve open/close signals with the TPO modules.

51.3.3. Process Optimization

Process optimization involves monitoring cell growth and metabolic state and making appropriate changes in the SPs of the process control loops discussed in the previous section to maximize production. Often off-line, at-line, and in-line data must be combined and evaluated, with adjustments performed manually. In some cases, the process changes too quickly for this to be a viable option, and inferences from the available data are made to make adjustments automatically (3, 4, 10, 18).

51.3.3.1. Process Monitoring

Section 51.2 lists several sensors that are available for bioreactors. Widely used measurements include pH, DO, temperature, pressure, agitation rate, overlay and sparge gas flows, titrant addition volume, and substrate addition volume. Good-quality sensors which measure inlet and outlet gas composition, cell density, foam detection, redox potential, and pCO_2 are also available, but are not as prevalent and not necessary to properly control cell growth and production in all bioreactor processes.

Several methods can be used to assess data quality from these in-line sensors before process conditions are calculated and used for feedback control. Dual sensors for pH, DO, and temperature are often installed to provide redundancy. In this case, the average value is used for control and the difference between the two values is monitored. If the difference becomes large, an alarm is issued so a sample can be taken to determine which sensor is correct. The correct sensor is then selected, rather than the average value, for subsequent control.

Some smart sensors run their own diagnostics and can alert the automation system of failure. Noise from fluctuations in the bioreactor, sensor signal, transmitter circuitry, and analog-to-digital conversion are typically dampened with simple averaging or filter functions within the transmitter or the control system. Excessive filtering should be avoided as it introduces delays between the observed value and the actual event when real disturbances occur.

At-line and off-line data are available infrequently, which can slow the process considerably. Biochemical assays of product, media components, cell quantities, and various metabolite concentrations are often somewhat inaccurate. It is important to standardize these instruments often and to perform repeated, randomized sample analysis so that an unbiased average value can be determined. Also, comparison of direct measurements to values estimated from other measurements can detect outliers, which can then be rejected.

It can be difficult to completely capture these data reconciliation steps within an automation system's programming, and thus these steps are typically performed by experienced manufacturing and process support staff. If at-line and off-line data are used for automatic process optimization, alarm limits should be put in place to alert the staff of unusual values or unusual new process control SPs. The alarm should also prevent the process optimization system from making process changes until the problem is corrected.

Key parameters for process optimization often include cell density, growth rate, oxygen uptake rate, carbon dioxide evolution rate, base or nutrient demand, and by-product formation.

51.3.3.2. Batch Optimization

Simple batch processes basically allow a freshly inoculated culture to grow while maintaining culture conditions. Optimization of the culture is therefore limited to choosing the time to terminate the batch and to making adjustments to SPs based on culture age, cell density, or changes in metabolism. Although infrequent additions of titrants and limiting nutrients may occur, batch processes tend to optimize production by making controller adjustments as expected milestones are reached, rather than implementing regulatory controls. A simple example is a temperature SP change to induce production of a protein by a recombinant organism once a target cell mass is achieved. This shifts the bulk of the cell's machinery from cell growth to protein production. Other common adjustments required during batch processes are related to equipment limitations and are discussed in section 51.3.3.5, as they are common to all modes of operation.

51.3.3.3. Fed-Batch Optimization

The goal of fed-batch and semicontinuous feeding strategies is to extend cell growth and production by controlling the state of cell metabolism. An example of a fed-batch process includes using a defined medium to stochiometrically match the elemental content of the cell and ensure that inhibitory concentrations of feed components are not reached during late stages of the culture. The initial medium contains all nutrients required for the entire batch with the exception of the key substrate. The concentrated substrate solution is fed to maintain the proper concentration in the bioreactor. Maintaining the concentration also ensures that the cell metabolic state remains stable. A desired metabolic state has a corresponding yield of cell mass on substrate, for which the appropriate feed rate can be calculated using equation 2, based on current estimates or measurements of cell density, growth rate, and bioreactor volume.

$$\text{Feed rate} = \frac{\mu XV}{(YS_0)})$$ (2)

where feed rate is the substrate flow rate (liters/hour), μ is the current growth rate (1/hour), X is the current cell density (grams/liter), V is the current bioreactor volume (liters), Y is the yield of cells on substrate (grams of cells/gram of substrate), and S_0 is the feed solution substrate concentration (grams/liter).

In some processes, ammonia can be used both for pH control and as a nitrogen source. To maintain a stochiometric ratio of carbon to nitrogen (C/N), the total batch substrate volume required to maintain the desired molar C/N ratio is calculated throughout the batch by multiplying the total volume of ammonia fed as a result of pH control by a conversion constant after each addition and directing the substrate addition controller to add substrate until the new target total batch substrate volume is reached. An advantage of this approach is that total batch volumes are used to calculate additions, rather than multiple instantaneous calculations, which accumulate error over time.

51.3.3.4. Semicontinuous Optimization

Semicontinuous processes begin with a batch growth stage, followed by a continuous feed and harvest stage. This strategy can provide nearly continuous, high-product-concentration material and therefore is the most efficient use of a bioreactor. The choice between fed-batch and semicontinuous feeding strategies is often dependent on productivity, product stability, and the selected downstream process.

Chemostats feed media with all components in excess except for a single limiting substrate while withdrawing culture broth to maintain a constant bioreactor volume. The feed rate is chosen to optimize product concentration and quality, cell density, growth rate, and by-product concentrations, which all become constant as the process approaches steady state.

Perfusion processes feed media at a fixed rate, with all components in excess of that required for the cell mass in the bioreactor. Cells either are returned to the bioreactor after separation from the harvest stream, or are retained within the bioreactor. Changes in cell density can be compensated for by modifying the feed flow rate to ensure that sufficient nutrient concentrations are maintained. Process control adjustments of semicontinuous processes are minimal, with the majority of changes occurring in the transition between growth and continuous harvest modes.

51.3.3.5. Equipment Constraint Management

Bioreactor heat and mass transfer limitations often constrain the optimization of high-cell-density processes. The oxygen mass transfer requirements and metabolic heat load are proportional to the product of growth rate and cell density (μX). Viscosity, which increases with cell density, also has an adverse effect on mass and heat transfer.

To maintain temperature, the bioreactor design must provide sufficient jacket surface area and cooling systems capable of dropping the jacket temperature as low as needed. If the physical design of the bioreactor cannot control temperature, then the substrate flow rate can be reduced to maintain temperature. This will cause the growth rate to be reduced, thus preventing the temperature from rising to a point where the cell death rate increases rapidly. Cultures are typically terminated in a controlled fashion prior to reaching this physical limitation.

To maintain DO throughout the full range of demand, mass transfer from sparged gas can be managed by automatically switching the control strategy between agitation, sparge gas flow, sparge gas O_2 composition, and pressure as the maximum desired output of each controller is successively reached.

Agitation reduces bubble size and thus increases surface area, while also improving mixing within the bioreactor. Agitation is often the first controller to be used for DO control. Agitation also increases shear and is therefore limited to low rotations per minute (rpm) values for shear-sensitive insect and mammalian cells.

Sparge flow rate can be increased to provide a higher liquid-gas interface area to allow more oxygen to be transferred. Increasing O_2 composition of the sparged gas increases the driving force from the gas to the liquid. For bacterial fermentations, the process typically involves increasing airflow first and then increasing the O_2 sparge gas concentration by supplementing the air with pure oxygen, preferably using separate sparger lines (Fig. 1). Increasing sparge flow rates eventually becomes ineffective as the agitator is "flooded" with gas and cannot mix the liquid and gas as efficiently. Agitator power consumption will decrease as this occurs. Cells sensitive to gas-liquid interfaces are sparged with pure O_2 to minimize the flow rate of sparged gas.

Cultures with very high O_2 demand may also require the pressure of the bioreactor to be increased, which causes the concentration of all dissolved gases to increase. This step is not performed for cells sensitive to high pCO_2.

Finally, if the DO controller cannot maintain SP, either the substrate feed can be slowed or the temperature can be reduced to decrease growth rate.

51.4. AUTOMATION SYSTEMS

A variety of commercial automation packages are available to provide the various levels of control discussed in section 51.3. Some common examples of automation hardware and software are discussed in this section.

51.4.1. Sensor Signal Processing

In-line sensors typically send millivolt (mV), milliamp (mA), or optical signals to an associated transmitter which converts the signal to a current, a voltage, or a numeric value, which is then communicated to the automation system. The raw signal is amplified and sometimes filtered to stabilize the output. Common transmitter output voltage ranges are 0 to 10 V direct current (VDC), 1 to 5 VDC, and 0 to 1 VDC. Common transmitter output currents include 0 to 20 mA and 4 to 20 mA. Digital transmission of numeric values by "smart devices" via ethernet or other protocols is increasingly replacing analog transmission.

The transmitter often has a display and the ability to calibrate the signal so that the displayed engineering units and output signal range match actual process conditions. The transmitter may also provide other signal conditioning functions, such as compensation for temperature effects or signal linearization. The transmitter may exist in a stand-alone device, as a card that slides into a personal computer or rack system, or as a module that can be mounted in a panel (see Fig. 12).

51.4.2. Analog and Digital Signal Conversion

Sensor signals and transmitter outputs can also be connected to microprocessor-based control systems such as programmable logic controllers (PLCs), supervisory control and data acquisition (SCADA) systems, and distributed control systems (DCSs).

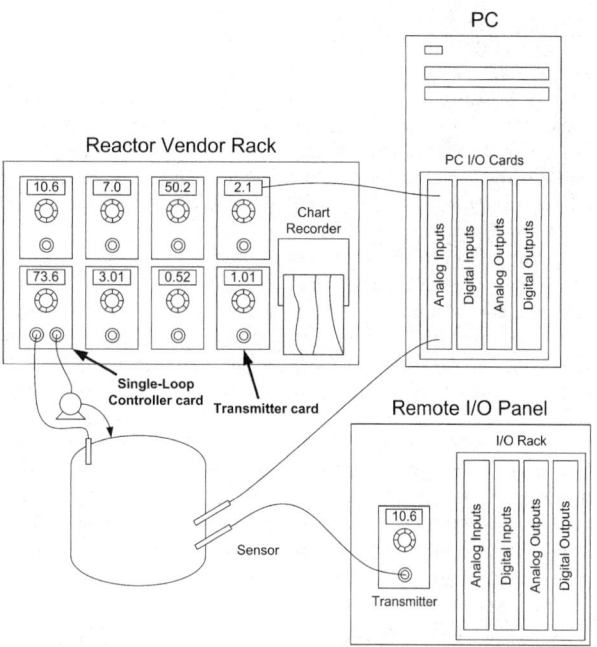

FIGURE 12 Automation I/O hardware examples.

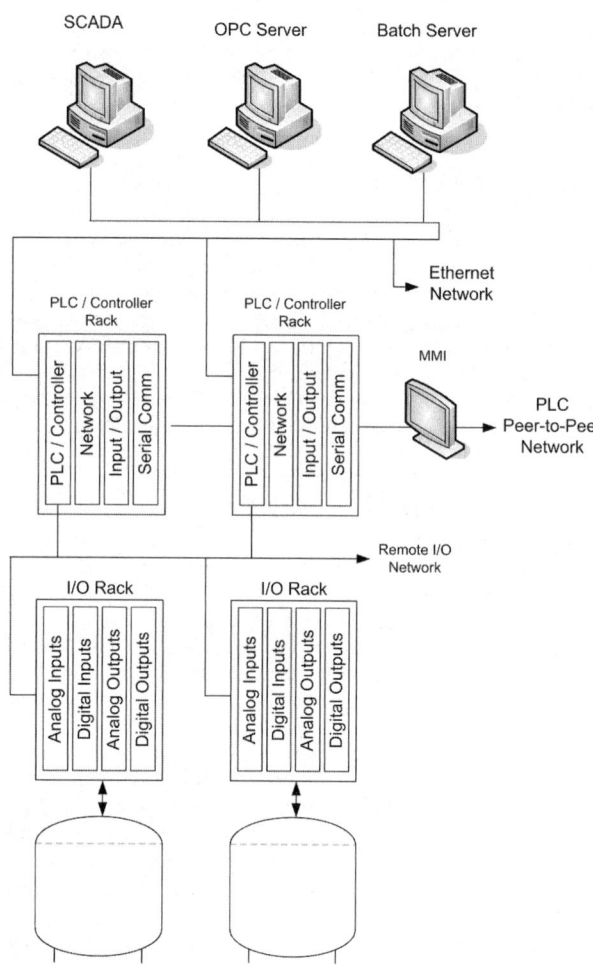

FIGURE 13 PLC/SCADA network examples.

SCADA systems are software applications installed on standard computers or servers that allow data to be transferred between various input and output devices as well as other computer-based process data systems. One form of interfacing is to use input/output (I/O) cards inserted directly into the SCADA computer. These cards convert analog and digital sensor input signals to numeric values stored in the SCADA database. Similarly, numeric values within the SCADA database can be converted to analog and digital output signals which cause a response in the control equipment. SCADA vendors supply software drivers specific to the I/O cards used, to allow the SCADA system to pass data between the SCADA database and the I/O cards.

PLC and DCS systems have their own proprietary remote I/O racks where I/O cards are installed and connected to sensors and control equipment. These systems also provide I/O cards for common equipment. For example, resistance temperature detector cards take thermocouple inputs directly and convert them to temperature without the need for intermediate transmitters. 24-VDC output cards are used to drive solenoids directly. The remote I/O racks are connected to PLCs, or a DCS equivalent, via proprietary digital communication networks (Fig. 13).

Smart devices are microprocessor controlled and provide their own signal conversion. They may contain other functions such as self-calibration, sensor fault diagnosis, and digital communication with control systems via standard protocols such as Foundation Fieldbus, Profibus, HART, RS-232, and RS-485.

The analog and digital signals are eventually converted to numeric values which are stored in the control system's memory. These numeric values are no longer subject to signal interference and can be precisely used for further calculations or communicated to other microprocessor-based systems. The precision of the conversion is determined by the number of bits used by the converter, which must be adequate for the application.

51.4.3. Actuation

Bioreactor control primarily involves the manipulation of gas and liquid flows. On/off flow control is achieved with a variety of valves. Solenoids are driven by 24-VDC or 120-VAC signals and are used for non-process contact applications. Example uses include process gases and jacket heating/cooling temperature control. Solenoids are also used to allow air pressure actuation (typically at 90 to 110 psig) of diaphragm valves, which are designed for aseptic process-contact applications such as culture, titrant, and media transfers. Diaphragm valves are spring-loaded to position the valve in the fail-safe condition. When actuated, the valve moves to the opposite position. For the most part, the fail-safe condition is the closed position, but bioreactors will often have a gas supply valve, such as sparge air, and a vent valve which fail to the open position so the culture has a chance of survival during a short power outage.

Flow rate control is performed with a pump or a flow control valve. Peristaltic pumps are low-flow-range positive-displacement pumps that can be operated aseptically. They respond linearly to a control system's output signal and are relatively inexpensive. Examples of positive-displacement pumps capable of larger flow rates include stainless steel rotary lobe and centrifugal pumps. Rotary lobe pumps cause less cell shear and are therefore useful for cell culture transfers. Centrifugal pumps are less expensive and can be used for non–cell contact applications such as bioreactor cleaning procedures. Metering (or dosing) pumps and valves are also available which very precisely control rotations per minute and position, respectively, by digitally positioning toothed gears. Metering pumps are used for short, precise volume additions.

Flow control valves require the source vessel to be pressurized at a roughly constant pressure above the destination vessel. The control system's output signal drives a current-to-pressure (I/P) device which converts 4–20 mA to 3–15 psig air pressure. The increasing air pressure positions the diaphragm from the spring-closed, fail-safe position to the fully actuated, open position. Flow control valves are susceptible to some hysteresis as the mechanism wears and loosens.

Agitation rate is controlled by a VFD. The control system output is converted to increasing frequencies within coils in three-phase motors which rotate the agitator. VFDs are often microprocessor-driven devices which may store fault codes and display power usage. VFDs often fault when asked to rotate the motor at less than 5 to 10% of full speed, as power requirements increase quickly in this range. VFDs are also used to drive pumps requiring variable flow rates.

51.4.4. PLC/SCADA Systems

As discussed, SCADA systems use I/O software drivers to communicate to internal I/O cards or external equipment such as PLCs or smart devices. Current I/O values are stored in the SCADA database. Various function blocks are available to manipulate the data, provide control logic, and send output values to control equipment. There is often an option to write custom code in a proprietary macro language, or in common programming languages such as Visual Basic or C. These custom code options provide additional flexibility that can be useful for process optimization control.

Operator interface screens can be quickly developed using templates which connect to SCADA database values. Templates can include simple value display, value entry, shapes that can change size according to process or calculated values, buttons that run code or select options, and alarm displays. Function block templates may also be available which display all block parameters and settings for a particular instance of a function block. These screens allow the user to visualize and control the process. In addition to operator-initiated actions, the SCADA system is capable of making requests of the operator, such as alarm or prompt acknowledgments which may be required before the automation system takes further action.

SCADA systems also provide a means for storing the data of interest and communicating with other computer-based data systems such as data historians, relational databases, batch control systems, and MES systems. For larger distributed systems, a main SCADA server will communicate with various I/O sources, execute function block logic, and update the SCADA database, while remote SCADA workstations will simply provide operator interface functionality. This allows the SCADA server to focus on communication and control and not be disrupted by operator actions.

PLCs are robust microprocessor-based systems designed specifically for process control. PLCs have no user interface and communicate with I/O equipment on isolated networks via vendor-specific protocols. They are programmed in various low-level languages, which allow them to provide control responses on the order of milliseconds. PLCs can communicate with each other and with man-machine interfaces (MMIs) over a peer-to-peer network (see Fig. 13). MMIs contain basic screen templates, similar to those described earlier for SCADA systems, which are connected directly to PLC memory locations. Keypad or touch-screen selections on the MMI change PLC memory values, which the PLC then acts on. MMIs have no control capabilities, nor do they store or process data in any way. They are simply PLC interfaces.

SCADA servers are not as robust or as efficient as PLCs. Combining the SCADA with PLCs is a more expensive option than using SCADA alone, but provides a solid automation platform which is preferred for large-scale industrial bioreactors where automation failures can have a large financial impact. PLC design should include all information necessary for direct process control so if the SCADA system should become unavailable, control can continue without interruption. SCADA design should focus on providing a view of the process to the operator and handling all operator and external data system requests. In this way, a SCADA system can provide a single interface point for a large production area.

PLCs are connected to the SCADA server via a separate network, typically an ethernet connection (see Fig. 13). PLCs therefore improve data integrity and communication speed by segregating I/O, PLC/MMI, and SCADA traffic among three different networks. SCADA servers may also contain two network cards so that the PLC-SCADA network can be separated from SCADA communication with other computers. Separating communications in this way and splitting functionality between MMIs, PLCs, and SCADA results in an efficient, robust automation system.

51.4.5. Batch Control Systems

Batch control systems provide recipe-based sequential process control and are specifically based on the ISA S88 standard (12). These systems are useful for large production facilities where the equipment is frequently reassigned to produce various products. They are not as useful for pilot plants where the process is still under development and frequent changes are made.

51.4.6. DCS

DCSs (distributed control systems) essentially combine the functionality of PLC, SCADA, and batch systems under one umbrella. DCS systems are designed for larger production facilities and are not often used for benchtop or pilot-scale bioreactor systems.

51.4.7. MES

MESs (manufacturing execution systems) integrate all plant floor activities and provide an electronic record for each batch of material produced. In practice, the scope of MESs can vary widely. At one extreme, the MES simply records electronically what was once written manually on a paper batch record. These are known as paper-on-glass systems. At the other extreme, the MES fully integrates the batch record with the automation system and all the business systems that drive or support plant floor activities. Such business systems can include those which schedule operations, purchase and track raw materials, verify personnel training documentation, track equipment maintenance, schedule instrument calibration, provide electronic copies of Standard Operating Procedures (SOPs), and provide bioreactor sample assay results. The MES also tracks the status of all plant floor equipment and in-process materials. With all of this information, quality checks can be performed before the process is allowed to proceed, thereby reducing errors and waste. Building these quality checks into a validated system also minimizes the amount of review required. Only exceptions, such as observations recorded by the operator and significant process alarms and events recorded by the MES and automation systems, need to be reviewed. All data are available electronically, minimizing batch record review time and reducing batch release time.

The key to choosing the proper balance of MES complexity versus risk reduction is to perform a risk evaluation. Reviewing quality records that resulted in lost bioreactor batches and conducting a formal risk analysis with engineering, manufacturing, and quality departments will help identify the highest-priority problem areas. The MES design can then focus on eliminating the major issues. Organizations must also balance the long-term benefits associated with a given design with the investment and potential time delays associated with starting up and validating a complex system.

REFERENCES

1. **Arnold, A., R. Gaensakoo, L. M. Harvey, and B. McNeil.** 2002. Use of at-line and in-situ near infrared spectroscopy to monitor biomass in an industrial fed-batch *Escherichia coli* process. *Biotechnol. Bioeng.* **80:**405–413.
2. **Azevedo, S. F., R. Oliviera, and B. Sonnleitner.** 2001. New methodologies for multiphase bioreactors 3: data acquisition, modelling and control, p. 57–84. *In* M. Joaquim, S. Cabral, M. Mota, and J. Tramper (ed.), *Multiphase Bioreactor Design.* Taylor & Francis Inc., New York, NY.
3. **Bailey, J. E., and D. F. Ollis.** 1986. *Biochemical Engineering Fundamentals*, 2nd ed. McGraw-Hill, Inc., New York, NY.
4. **Bastin, G., and D. Dochain.** 1990. *In-Line Estimation and Adaptive Control of Bioreactors.* Elsevier Science Publishers, Amsterdam, The Netherlands.
5. **Buckbee, G., and J. Alford.** 2008. *Automation Applications in Bio-Pharmaceuticals*, p. 28–34. ISA, Research Triangle Park, NC.
6. **Buckland, B., T. Brix, H. Fastert, K. Gbewonyo, G. Hunt, and D. Jain.** 1985. Fermentation exhaust gas analysis using mass spectrometry. *Bio/Technology* **3:**982–988.
7. **Clark, L. C., Jr.** 1956. Monitor and control of blood and tissue oxygenation. *Trans. Am. Soc. Artif. Intern. Org.* **2:**41–45.
8. **Cooney, C. L., D. I. C. Wang, and R. I. Mateles.** 1968. Measurement of heat evolution and correlation with oxygen consumption during microbial growth. *Biotechnol. Bioeng.* **11:**269–281.
9. **Cooney, C. L., H. Y. Wang, and D. I. C. Wang.** 1977. Computer-aided material balancing for prediction of fermentation parameters. *Biotechnol. Bioeng.* **19:**55–67.
10. **El-Mansi, E. M. T., C. F. A. Bryce, A. L. Demain, and A. R. Allman.** 2007. *Fermentation Microbiology and Biotechnology.* CRC Press, Taylor & Francis Group, Boca Raton, FL.
11. **Hall, J. W., and A. Pollard.** 1992. Near-infrared spectrophotometry: a new dimension in clinical chemistry. *Clin. Chem.* **38:**1623–1631.
12. **International Society of Automation.** 2006. *Batch Control Part 1: Models and Terminology.* ISA-88.01-1995 (R2006). http://www.isa.org/Content/Microsites165/SP18,_Instrument_Signals_and_Alarms/Home163/ISA_Standards_for_Committee_Use/S_8801.PDF. (Online.)
13. **Johnson, M. J., J. Borowski, and C. Engblom.** 1964. Steam sterilizable probes for dissolved oxygen measurement. *Biotechnol. Bioeng.* **6:**457–468.
14. **Junker, B. H., and H. Y. Wang.** 2006. Bioprocess monitoring and computer control: key roots of the current PAT initiative. *Biotechnol. Bioeng.* **95:**226–261.
15. **Junker, B. H., J. Reddy, K. Gbewonyo, and R. Greasham.** 1994. In-line and in-situ monitoring technology for cell density measurement in microbial and animal cell cultures. *Bioproc. Eng.* **10:**195–207.
16. **Leisola, M., and V. Kauppinen.** 1978. Automatic assay of cellulase activity during fermentation. *Biotechnol. Bioeng.* **20:**837–846.
17. **Li, J., and A. E. Humphrey.** 1990. Use of fluorometry for monitoring and control of a bioreactor. *Biotechnol. Bioeng.* **37:**1043–1049.
18. **Lim, H., and K.-S. Lee.** 1991. Control of bioreactor systems, p. 509–560. *In* K. Schugerl (ed.), *Measuring, Modelling and Control.* VCH, Weinheim, Germany.
19. **MacInnes, D. A., and M. Dole.** 1930. The behavior of glass electrodes of different compositions. *J. Am. Chem. Soc.* **52:**29–36.
20. **Nyiri, L. K., G. M. Toth, and M. Charles.** 1975. Application of computers to the analysis and control of microbiological processes. *Biotechnol. Bioeng. Symp.* **4:**613–628.
21. **Nyiri, L. K., J. D. Wilson, A. E. Humphrey, and C. S. Harmes.** December, 1975. Method and apparatus for control of biochemical processes. U.S. patent no. 3,926,738.
22. **Pattison, R. N., J. Swamy, B. Mendenhall, C. Hwang, and B. T. Frohlich.** 2000. Measurement and control of dissolved carbon dioxide in mammalian cell culture processes using an in situ fiber optic chemical sensor. *Biotechnol. Prog.* **16:**769–774.
23. **Perley, G. A.** December, 1945. Improvements in the manufacture of glass electrodes. British patent GB574029.
24. **Schügerl, K.** 2000. Development of bioreactor engineering, p. 49–51. *In* T. Scheper, A. Fiechter, W. Babel, H. W. Blanch, C. L. Cooney, I. Endo, S.-O. Enfors, K.-E. L. Eriksson, M. Hoare, B. Mattiasson, H. J. Rehm, P. L. Rogers, H. Sahm, K. Schügerl, G. Stephanopoulos, G. T. Tsao, K. Venkat, J. Villadsen, U. von Stockar, and C. Wandrey (ed.), *Advances in Biochemical Engineering Biotechnology: History of Modern Biotechnology II.* Springer, Berlin, Germany.
25. **Seborg, D. E., T. F. Edgar, and D. A. Mellichamp.** 2004. *Process Dynamics and Control*, 2nd ed. John Wiley & Sons, Inc., New York, NY.
26. **Shuler, M. L., and F. Kargi.** 2002. *Bioprocess Engineering Basic Concepts*, p. 307–311. Prentice-Hall, New Delhi, India.
27. **Sonnleitner, B.** 2000. Instrumentation of biotechnological processes, p. 3–21. *In* T. Scheper, B. Sonnleitner, W. Babel, H. W. Blanch, C. L. Cooney, S.-O. Enfors, K.-E. L. Eriksson, A. Fiechter, A. M. Klibanov, B. Mattiasson, S. B. Primrose, H. J. Rehm, P. L. Rogers, H. Sahm, K. Schügerl, G. T. Tsao, K. Venkat, J. Villadsen, U. von Stockar, and C. Wandrey (ed.), *Advances in Biochemical Engineering Biotechnology: Bioanalysis and Biosensors for Bioprocess Monitoring.* Springer, Berlin, Germany.
28. **Tosi, S., R. Maddalena, E. Tamburini, G. Vaccari, A. Amaretti, and D. Matteuzzi.** 2003. Assessment of in-line near infrared spectroscopy for continuous monitoring of fermentation processes. *Biotechnol. Prog.* **19:**1816–1821.
29. **Wang, N. S., and G. N. Stephanopoulos.** 1984. Computer application to fermentation processes. *CRC Crit. Rev. Biotechnol.* **2:**1–90.

Purification and Characterization of Proteins

ULRICH STRYCH AND RICHARD C. WILLSON

52

52.1. BACKGROUND

This chapter provides an introduction to the fascinating, useful, and sometimes dreaded world of protein purification and characterization. It is intended to provide an introduction to the most commonly used laboratory-scale methods, along with pointers to detailed references for further reading.

The goals of purification vary with the intended use of the purified protein. The amount of protein needed can range from micrograms for microsequencing or the raising of antiserum, through milligrams for activity characterization or structural studies, to kilograms or even tons for pharmaceutical products. Purity is determined by the general level of other protein contaminants, but also by the absence of contaminants of special interest, which can include endotoxins, highly antigenic proteins, viruses, high levels of salts, and competing enzymes or biological activities. With glycosylated proteins, in particular, the presence of various glycoforms of the protein might also negatively affect certain applications.

Protein purification can generally be divided into five broad stages, which do not all need to occur sequentially: (i) preparation of the source, (ii) gathering of all available information about the protein's properties, (iii) development of an assay, (iv) primary isolation, and (v) final purification (Fig. 1). Each of these steps can often be made easier, or even avoided, by effectively searching the literature. Only rarely is the protein of interest unrelated to any that have previously been purified. Hints and tricks from previous work can greatly simplify development of a novel purification strategy; it sometimes may even be possible to "purify by phone" if a previously isolated protein will serve the intended purpose and can be obtained in sufficient quantities.

52.2. PREPARATION OF THE SOURCE

Preparation begins with selection of the raw material (or the system) from which the protein will be isolated. The source is here assumed to be microbial or a cultured metazoan cell line, although transgenic animals and plants may well be of interest for future applications.

It is important to start with ample quantities of target protein in the raw material. Overall purification yields (defined as the ratio of the amount of target protein finally purified to the amount of the target protein in the raw material) for optimized pharmaceutical protein processes often exceed 80%, but the first few exploratory purifications of a poorly characterized target expressed at low levels may achieve levels closer to 1%, or even less. It is thus prudent to start with an abundant amount (at least 10 to 20 times as much as needed) of protein in the raw material; a common mistake is to start with raw material that already contains less target protein than eventually desired either because purification yields were unexpectedly low or the measurement of the original concentration in the presence of a large excess of host cell proteins was inaccurate.

Abundant supplies are also desirable to permit greater evaluation of alternative purification strategies and will allow each step to be optimized for a greater degree of purification, at the expense of yield.

Protein supplies can be increased by increasing the cultivation volume, by increasing the cell density (e.g., by computer-controlled fermentation with pH control, enhanced oxygen transfer, and regulated feeding of substrate), or by producing more of the desired protein per cell (8). Expression levels can be increased by screening for high-producing cell lines; by cloning and overexpression; by use of nonmetabolizable antibiotics to better maintain selection pressures (e.g., switching away from ampicillin, which is degraded by resistant cells); by use of tightly regulated, strong promoters such as T7 (*Escherichia coli*, pET-System [91]) or NICE (*Lactococcus*, nisin-controlled expression [60]); and often by expression at lower temperatures, as low as 10°C for *E. coli* (74).

In addition to its quantity, the *quality* of the raw material is also subject to optimization. Improvements can include elimination of key contaminants (e.g., proteases, competing enzymatic activities, and serum components), optimization of posttranslational processes, maximization of extracellular secretion or correct folding, and the genetic addition of purification tags such as polyhistidines or glutathione-S-transferase, as discussed below.

In some cases it will be impossible to express and purify a particular protein in a living system, because the protein either has a toxic effect on the cells or it is too rapidly degraded by enzymes of the host cell. Here cell-free translation systems can alleviate the problem by providing all the necessary components for the transcription and translation of a given gene in a premixed cocktail of nucleotides, amino

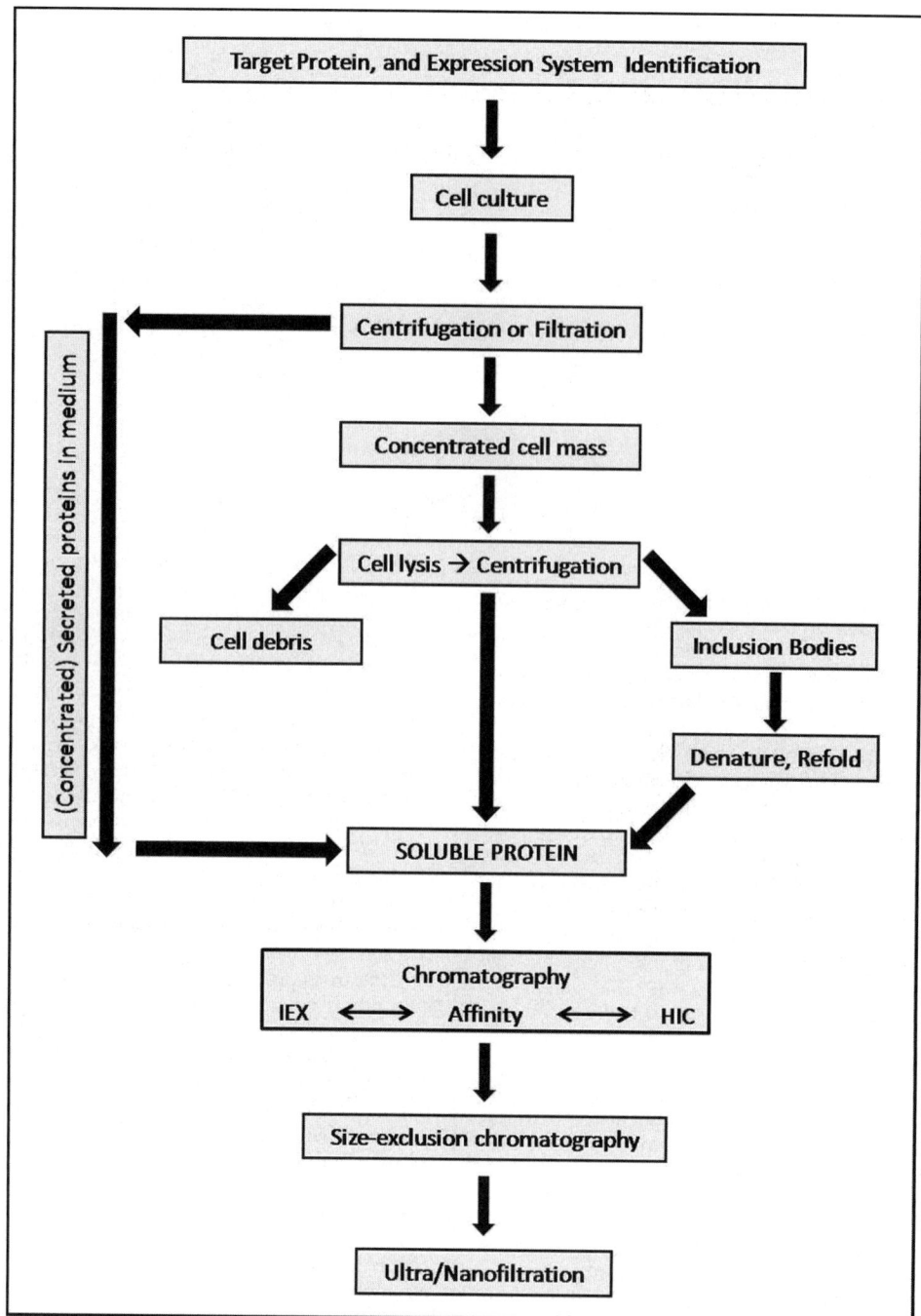

FIGURE 1 Overview of a typical protein purification protocol. After being expressed in the host, the protein of interest will be found in the supernatant or will remain inside the cells, either soluble or as inclusion bodies. HIC, hydrophobic interaction chromatography; IEX, ion-exchange chromatography.

acids (natural and/or unnatural), tRNAs, enzymes, and ribosomes (83). Three in vitro translation systems are most commonly used: *E. coli* lysates (53), wheat germ extracts (57), and rabbit reticulocytes (18). Expressing proteins in cell-free systems does come at high expense and traditionally was a matter of last resort if in vivo systems had failed, but this approach is becoming increasingly popular. Novel cell-free expression systems allow the rapid screening of large numbers of different proteins (100), can incorporate

nonnatural amino acids (38) and introduce disulfide bonds (37), and, overall, have reached a stage where their industrial application appears to be imminent (93).

Recombinant proteins are still most commonly expressed in the enterobacterium *E. coli*. While the bias towards this system might may be partly historical (it was one of the first bacteria described in the late 19th century), its ability to grow rapidly on inexpensive nutrients, both aerobically and anaerobically, is an advantage over other bacterial strains in the laboratory environment. It is the best understood microbial system, with the widest array of genomic and proteomic tools available for its manipulation. *E. coli* strains have been genomically optimized to encode fewer nucleases and proteases, and/or to possess additional genes for the expression of, for instance, highly processive bacteriophage polymerases (91) or rare tRNAs (20). Last but not least, pathogenicity factors present in some wild-type *E. coli* isolates have been removed from laboratory strains.

In recent years an autoinduction medium originally developed by Studier (90) and now distributed by EMD Chemicals has gained an increased following in the protein expression community. The system employs the common *lacUV* promoter-controlled T7 polymerase-dependent bacterial strains and plasmids, but after growing uninduced to a high cell density and metabolizing all carbon sources but lactose, the cells then spontaneously *autoinduce* high-level protein expression, without the need to add a costly synthetic inducer such as IPTG (isopropyl-β-D-thiogalactopyranoside). Not only does this constitute an economical advantage, but the well-defined and optimized medium also increases cell density and target protein yield.

As noted above, some genes are not expressed well in *E. coli*; in particular, eukaryotic genes whose products require posttranslational modifications are better expressed in alternative systems. These include insect cells that are infected with genetically engineered baculovirus particles, yeasts such as *Saccharomyces* sp. and *Pichia pastoris* (13, 70), and other fungi (62) that have found application as hosts for the industrial-scale production of recombinant proteins. Because of their fastidious nature, mammalian cell lines were not historically the first choice for protein production, but as mammalian expression systems and cell culture technology have improved, they have become more and more competitive. There also is increasing scientific and medical interest in complex and/or glycosylated proteins (notably antibodies) best produced in mammalian cells (48). Nevertheless, establishing and maintaining a stable mammalian cell line expressing recombinant proteins remains a technically challenging, lengthy, and quite costly endeavor. In one remarkable improvement of this process, transient transfection or transient gene expression (3), large amounts of relatively inexpensive plasmid DNA are transfected into eukaryotic cells, rapidly but temporarily turning transfected cells into protein production engines. Although the recombinant genetic material is lost from the culture after mitosis, by using optimized vectors and improved transfection protocols and by coexpressing helper proteins, cell densities obtained through transient transfection/transient gene expression can be equivalent to those obtained with stable cell lines (2). So, just as with the above-mentioned *E. coli* systems, these protein production hosts are constantly being optimized in order to yield larger quantities of the target protein at lower cost (21, 98).

52.3. KNOWLEDGE OF THE PROTEIN'S PROPERTIES

It is essential to collect as much information on the protein as possible. Knowledge of almost any characteristic of the protein may find application in the development of a purification process. Much of this information can be obtained using publicly available Internet resources such as Entrez (http://www.ncbi.nlm.nih.gov/sites/gquery) or ExPASy (http://www.expasy.ch/). The properties of interest are summarized as follows.

- Source: cell type, expression level, intra/extracellular location, folding state, presence of proteases/glycosidases
- Tolerance of: endogenous proteases, temperature range, pH range, ionic strength, hydrophobic surfaces, cofactor or metal ion loss; freeze-thaw stability (for storage, during and after purification)
- Aggregation tendencies
- Size: molecular weight, amino acid sequence, multimeric state (monomer or homo/heteromultimer), hydrodynamic radius
- Charge: isoelectric point, titration curve, electrophoretic mobility
- Binding partners: substrates and cofactors (and analogs), screening-derived binding agents (antibodies, peptides, aptamers), metal affinity; binding to any of these in immobilized forms for affinity chromatography; stabilizing effects of binding partners

52.4. ASSAY

While a method to monitor the presence of the target protein during the purification protocol is essential, an assay for the desired activity of the protein is also highly desirable. General assays for use in purification must be convenient and rapid; precision is less important. Typical methods include colorimetric assays (Bradford Assay [10]), UV spectrophotometry, visible absorbance of proteins with some cofactors, sodium dodecyl sulfate-polyacrylamide gel electrophoresis (SDS-PAGE), immunological assays (enzyme-linked immunosorbent assay, dot-blot), gel-shift assays, and enzyme activity assays, as well as biosensor testing of fractions (SPR, Biacore, etc.; see below). While in principle these assays have not changed much over the last few decades, improved instrumentation and increased availability of prepackaged chemical detection systems (kits) have made these assays faster and more reproducible and have significantly reduced the amount of sample that has to be sacrificed.

52.5. RECOVERY AND INITIAL ISOLATION

Recovery and initial isolation steps separate the product from the majority of the water in the cultivation medium and from the majority of the host cell components. Recovery of both extracellular and intracellular proteins thus usually begins with the separation of the soluble culture medium from the insoluble cell pellet by centrifugation or filtration.

52.5.1. Concentration

The secretion of the target protein into the extracellular medium is generally desirable, but is rare with *E. coli*. It may require genetic signals to be placed on the target protein, or manipulations of the complex cell secretion machinery. Secreted proteins are naturally separated from most contaminating proteins (i.e., the cell), making purification easier, but requiring concentration of the target protein from a large volume of cell-free culture medium. This concentration step can be achieved by ultrafiltration or adsorption on chromatographic media. Ultrafiltration can be carried out at small scale in centrifugally driven cells, or at larger scale in stirred cells or tangential-flow units (59, 97). Adsorption can employ any of the chemistries noted below, although expensive and/or delicate adsorbents (e.g., immobilized antibodies) are normally only used for initial capture on the small scale, or for highly unstable proteins. An increasingly popular approach at larger scale is direct protein capture in fluidized beds of dense adsorbent particles from culture media and cell lysates, thus reducing the number of early processing steps.

52.5.2. Cell Lysis

Total lysis of cells is required for the liberation of cytoplasmic proteins. Microbial lysis often employs chemical agents such as chloroform, toluene, SDS, or EDTA. In addition, freeze-thaw cycling and mechanical stress will also aid in this crucial first step of the purification. Enzymes such as lysozyme or nuclease (e.g., Benzonase™) are used to degrade the cell wall (especially of gram-negative organisms) and reduce the viscosity caused by nucleic acids.

Chemical, enzymatic, and freeze-thaw methods are often less than completely effective and may best be used on smaller scales. Mechanical methods include sonication (for small scale), grinding, glass bead mills, the French press, and the Manton-Gaulin homogenizer. Each of these methods affects proteins as well as cells and requires optimization for the optimum recovery of the target protein, but each can give reliable, satisfactory results. The French press in particular is a standard tool for lysis of some tens of grams of wet cell paste by repeatedly forcing cells through a small orifice at high pressure (10,000 to 40,000 lb/in²). Primary isolation yields may benefit from cooling, or from the addition of protease or glycosidase inhibitors. These measures can be particularly valuable with cell lysis using mechanical methods, which tend to heat samples.

Periplasmic proteins can be selectively liberated by osmotic shock, which ruptures only the outer membrane by sudden transfer from hyper- to hypotonic medium. Successful performance and scale-up of this operation can be tricky and may benefit from measurement of release of enzyme activities known to be cytoplasmic (e.g., β-galactosidase) or periplasmic (e.g., alkaline phosphatase).

Cell lysates are unattractive substances, rich in degradative activities, sticky lipids, and viscous nucleic acids along with thousands of protein contaminants. Soluble proteins are often recovered from cell lysates by using precipitation or (less commonly) liquid partitioning, as these methods are tolerant of viscosity, particulates, and fouling. Ribosomes and nucleic acids can first be precipitated with streptomycin, protamine, spermine, or polyethyleneimine (22). The desired protein is often concentrated and partially purified by precipitation with ammonium sulfate or polyethylene glycol (80, 92). The stepwise addition of optimized levels of precipitant can be used for the controlled precipitation of first some contaminants and then, using a higher concentration, the desired protein. Additional contaminants remain in the final supernatant. Exact results depend on the rate of precipitant addition, temperature, mixing, and aging for precipitate formation, but these are generally robust laboratory-scale methods from which partially purified proteins can be recovered at high yields.

52.5.3. Refolding Proteins from Inclusion Bodies

Recombinant proteins expressed in *E. coli* often misfold to form dense, insoluble aggregates of inactive protein (19). Depending on the overall protein concentration, these aggregations might be immediately obvious in the form of visible protein inclusion bodies in the cell, or they can be detected by SDS-PAGE of the soluble and insoluble fractions of a cell lysate. Generally, protein in these inclusion bodies is inactive and not useful for any kind of enzymatic, mechanistic, or structural investigation. Thus researchers aspire to minimize the formation of inclusion bodies through, for instance, systematic site-directed mutagenesis of the protein sequence, the coexpression of chaperones (folding helper proteins), or the reduction of the temperature at the time of expression, or they undertake to refold the protein to its native conformation (96). Although misfolded, inactive protein in the form of inclusion bodies usually is not directly useful, proteins in inclusion bodies can be harvested rather easily by centrifugation at purities exceeding 70%, and can often be renatured to their native tertiary structure. While there are alternative methods under development (16), the first step in this process is commonly the solubilization of the inclusion bodies in strong chaotropic solutions such as 6 M urea or guanidine hydrochloride (65). After this rapid and reliable first step, the dissolved, denatured protein is then allowed to slowly renature to its native conformation by successively removing the denaturant through dialysis, dilution, or chromatographic separation. In the renaturation step, unimolecular refolding processes must compete with multimolecular reaggregation reactions, so immobilized or highly dilute protein molecules must be used. The development of a process that quantitatively recovers the target protein in its native form can be challenging and is always unpredictable. In many cases, though, a low refolding yield is acceptable because the protein is abundantly available from inclusion bodies in the first place. Thus, large-scale purification protocols for inclusion bodies of commercially attractive proteins, such as human epidermal growth factor (82) or thrombolytic agents (28), are still employed.

Several rules of thumb for refolding have emerged (27, 66):

1. Small, disulfide-free, single-peptide-chain proteins are easiest to refold, but proteins of all types have been refolded.
2. Refold at low protein concentrations.
3. If possible, immobilize the refolding protein by adsorption to prevent multimolecular reactions which promote reaggregation.
4. Control redox potential (through dithiothreitol and β-mercaptoethanol addition) and regulate oxygenation.
5. Add additives such as arginine, polyethylene glycol, or cyclodextrins to suppress aggregation during protein refolding (56, 94).
6. Reduce denaturant concentration in multiple small steps as opposed to one single drastic reduction step.
7. Allow up to 2 to 10 days for refolding.
8. Recycle any reaggregated protein back to the refolding process.

52.6. PROTEIN PURIFICATION

Chromatography is the most common method of preparing highly purified, active proteins (54). Most forms of chromatography involve selective adsorption of proteins on the surface of porous particles, through interactions that can primarily be classified as ion-exchange, hydrophobic, and affinity. Chromatographic operations are also classified as low-pressure, medium-pressure (including GE Healthcare's popular ÄKTA), and high-pressure (high-pressure liquid chromatography [HPLC]), depending on the pressure used to force liquid through the packed bed of adsorbent particles (Table 1). In general, high pressures imply higher costs but also finer adsorbent particles and better resolving power.

TABLE 1 Types of liquid chromatography

Feature	Low-pressure	Medium-pressure	High-pressure (HPLC)
Particle size	40–150 µm	10–75 µm	2–15 µm
Flow driver	Gravity, peristaltic	Piston or syringe	Positive displacement
Run time	40–1,000 min	15–60 min	0.5–30 min
Apparatus cost	Low	Medium–high	High
Resolving power	Lowest	Intermediate	Highest
Particulate tolerance	Low	Very low	Lowest

Samples loaded onto high- and medium-pressure columns must be rigorously free of particulates.

In addition to the adsorbent matrix, an apparatus for chromatography includes a column into which the particles are packed (glass, plastic, or steel, depending on operating pressure), a pump or height differential to drive liquid flow, some method for introducing the sample into the flow before the column (a switching valve, or manual pipetting onto the top of the packed bed), and a fraction collector that deposits the emerging, fractionated proteins into different vessels. Possible enhancements include UV and visual absorbance detectors for monitoring the emergence of the various proteins from the column, a conductivity monitor for tracking changes in salt concentration used to elute proteins from the column (especially in ion-exchange chromatography), and a computer for control and evaluation of the results. Large-scale operations frequently use step changes in elution conditions, and valves to direct individual fractions to separate vessels.

Not all adsorptive separations involve particles in a column (71). Particles can be used for batch adsorption by gently mixing them into a solution containing proteins to be captured or by suspending dense particles in upflowing liquid in a fluidized bed. The resolution obtained in this manner is low, but the method is tolerant of fouling and fine particulates and can easily be applied to large volumes of solution. Nonparticulate supports, notably membranes, are available in many of the surface chemistries discussed below. These yield rapid adsorption and desorption kinetics, albeit at reduced capacities which may be balanced by rapid cycling (9, 77).

There are a variety of excellent references for the methodological details of chromatographic separation of proteins (see below); the present discussion will address general characteristics, along with some guidelines for method development.

52.6.1. Ion-Exchange Chromatography

Ion-exchange chromatography (1) is the most common high-resolution method for preparative separation of proteins and is used in most protocols (Table 2). It will be discussed in particular detail because of its importance, and also to illustrate the general issues involved in the development of chromatographic separations. Adsorbents with fixed positive charges (immobilized amines, e.g., DEAE, quaternary methyl-ammonium) are called anion exchangers and are employed at pH values above the isoelectric point of the protein to be adsorbed so that the protein's net charge will be of opposite sign to that of the adsorbent. Matrices bearing negative charges (carboxylates and sulfonates, e.g., carboxymethyl and sulfo-propyl) are called cation exchangers and are used at lower pH values. In practice, the protein of interest is usually adsorbed rather than passed through the column, as higher resolution and some degree of concentration can be achieved. Ion-exchange adsorption depends strongly on the ionic strength (I), equal to half the sum of the concentrations of all ions present, each multiplied by the square of the ion's charge. For 1 M NaCl, I = 1 M; for 1 M $(NH_4)_2SO_4$, I = 3 M. Initial trials might employ ionic strengths of 10 to 100 mM for loading and a gradient increasing to 1 M for elution.

The combinations of pH and matrix type (anion or cation exchange) at which the protein will be adsorbed can be predicted with some confidence from its isoelectric point (pI), although the detailed distribution of charges on the protein's surface plays a role (72), as do nonspecific interactions with the adsorbent backbone. If the pI is not known, it can sometimes be measured through isoelectric focusing (IEF) electrophoresis, or estimated from the sequence using services widely available online, for instance at the ExPASy Proteomics Server (http://www.expasy.ch/). Otherwise, a sample containing the protein is loaded (at low ionic strength) on an anion-exchange column at the lowest pH at which it is known to be stable, or on a cation-exchange

TABLE 2 Modes of adsorptive chromatography

Feature	Ion-exchange	Hydrophobic	Metal chelate	Biospecific
Adsorbent	Carboxyl, amine, sulfonate	Propyl, butyl, phenyl	Chelator, loaded with Cu^{2+}, Ni^{2+}, Zn^{2+}	Antibody, cofactor, receptor
Selectivity	Moderate–high	Moderate–high	Moderate–high	High–very high
Capacity	High	High	Moderate–high	Moderate–high
Matrix cost	Low	Low	Moderate	High
Elution	High salt, pH	Low salt	pH, imidazole	pH, chaotrope
Initial salt	Low	High	Indifferent	Often indifferent

column at its highest stable pH. The column is then washed with 2 to 3 volumes of the starting buffer, or until the initial peak of nonbinding protein has fallen to baseline. Most proteins will be retained on the column and can be eluted with a gradient of increasing salt concentration up to 1 to 2 M applied over a volume 5 to 10 times that of the column packing volume. Failure to quantitatively bind all protein can be due to overloading of the column (5 to 10 mg of total protein per ml of column packing is a conservative starting point); high loading ionic strength; failure to regenerate the column sufficiently at low ionic strength after the previous elution (this can require more than 5 column volumes of low-salt buffer and is ideally verified using a conductivity monitor); instability of the protein while adsorbed (this can improve at lower temperatures, or with slightly increased loading ionic strengths); or instability at too extreme a loading pH.

Optimization of the method involves increasing the fraction of the protein recovered in active form, the purity of the recovered material, and the rate at which purified protein is isolated (through loading of larger samples or shorter separation times). Steps to optimize process development often include loading under less permissive conditions (higher salt, pH closer to neutrality) to cause more contaminants to wash through unadsorbed; higher operating temperatures; faster liquid flow rates (to the limits specified by the adsorbent manufacturer); and testing of shorter gradients or gradients tailored to be shallow only near the ionic strength at which the protein of interest is eluted.

52.6.2. Hydrophobic Interaction Chromatography

Hydrophobic interaction chromatography (HIC) (35) exploits the presence of exposed hydrophobic groups on the surfaces of proteins, which can interact with immobilized nonpolar moieties such as short alkyl chains and phenyl rings (i.e., immobilized propane and benzene). Adsorption is promoted by high ionic strength, and elution often employs a salt gradient of decreasing concentration. As burial of nonpolar groups is a major contributor to protein stability, there is reason for concern that exposure to hydrophobic surfaces may be denaturing. Many very-hydrophobic polymer surfaces are in fact rapidly coated with denatured protein, making this a major constraint on the selection of support materials for chromatographic matrices. Isolated hydrophobic groups sparsely immobilized on polar supports such as polysaccharides and acrylates, however, are not nearly as denaturing as bulk materials, and recoveries of many proteins from optimized hydrophobic separations can approach 100%. The cost and capacity of HIC matrices are attractive, as is the method's tolerance of high salt concentrations, which can allow immediate processing of proteins eluted in high salt from a previous ion-exchange

chromatography or ammonium sulfate precipitation step. HIC is widely used in industrial processes and is probably underutilized in academic laboratories.

52.6.3. Affinity Chromatography

The remaining adsorptive techniques are affinity methods, which exploit selective interactions characteristic of biological systems. The normal biological functions of many proteins depend on their ability to associate with substrates, cofactors, etc., and proteins often will bind selectively to immobilized forms of these partners. Alternatively, an affinity agent such as an antibody can be used to selectively capture the protein of interest. Finally, it has become increasingly popular to append DNA encoding a "purification handle" protein with affinity for an available ligand to the gene encoding a recombinant protein of interest. The resulting fusion protein can then be purified by its affinity for this ligand. The purification tag sequence can often be removed posttranslationally by proteolysis using site-specific proteases, or left in place for many applications. In one particular system, intein and a chitin binding tag are fused to the protein of interest. The recombinant protein is then adsorbed to a chitin column, and intein-mediated self-cleavage of the tag occurs upon addition of thiols (e.g., dithiothreitol), releasing the tag-free protein into the eluate (14). Typical purification handles are listed in Table 3.

The most commonly used type of affinity chromatography is probably immobilized-metal affinity chromatography, or IMAC. It exploits the affinity of certain chelated metal ions (e.g., Ni^{2+} and Zn^{2+}) for the amino acid side chains of cysteine and especially histidine. Elution is achieved by pH gradient or the addition of competitive metal-chelating compounds such as imidazole. IMAC adsorbent matrices are relatively inexpensive and robust, and selectivity can be high. The method is also tolerant of variations in ionic strength.

Biospecific affinity chromatography describes separations based on molecular recognition interactions normally found in nature, such as antibody/antigen, enzyme/cofactor, and hormone/receptor affinity (89). Many widely applicable premade affinity matrices (at least 50 types) are commercially available. Among the most popular affinity methods are Cibacron Blue dye affinity for purification of enzymes binding ATP or NAD(P)(H), to which this textile dye bears a coincidental structural resemblance, and chromatography on protein A of *Staphylococcus aureus* for purification of antibodies (32, 43, 50). There are also many "activated" matrices on the market that have been chemically derivatized for facile coupling of affinity ligands through surface amines and carbohydrate groups.

The selectivity of affinity separations can be enormously high. This advantage is somewhat balanced in certain cases

TABLE 3 Affinity chromatography using protein tags

Protein tag	Affinity medium	Eluent	Reference(s)
Polyhistidine	(Cu^{2+}, Ni^{2+}, Zn^{2+})-charged agarose	Imidazole	4, 45
Glutathione-S-transferase	Glutathione agarose	Glutathione (reduced)	86
Maltose binding protein	Dextrin agarose	Maltose	75
Strep(II)-peptide	StrepTactin agarose	Desthiobiotin	79
F_c (antibody constant region)	Protein A/G	Citric acid, glycine (pH ≤ 3)	5, 26
Intein-chitin binding domain	Chitin	Thiols	14

by the cost and delay of obtaining and coupling suitable ligands. Also, the harsh conditions that may be required for elution of the bound protein can damage both target protein and affinity column. An active area of development in purification technology is the screening of large numbers of potential ligands, including phage-displayed peptides, aptamers, and small organic molecules, for use as improved affinity separation ligands (76).

From a commercial aspect, it appears that the purification of antibodies is attracting particularly strong interest. Human immunoglobulin (polyclonal immunoglobulin G [IgG]) for the treatment of immune deficiencies is prepared from human plasma through cold ethanol precipitation and purified from contaminating molecules such as IgM by anion-exchange chromatography (44). In addition to IgG, there are currently 23 other antibody-based drugs on the market (49), and an expansion of this arsenal is likely. These therapeutic monoclonal antibodies are produced recombinantly in large (>10,000-liter) Chinese hamster or mouse cell cultures. After the monoclonal antibodies are separated from the host cells by centrifugation and filtration, most remaining contaminants (>98%) are removed by protein A affinity chromatography (103). Commonly, one or more additional chromatography steps, such as ion-exchange chromatography, are then employed to generate the final product. A crucial element in all pharmaceutical purification protocols is the removal of viral particles. Virus inactivation is achieved by heat, nanofiltration (64), or UV light treatment (78). Historically, lipid-enveloped viruses (e.g., human immunodeficiency virus and hepatitis type C virus) have also been removed from therapeutic protein preparations by solvent/detergent treatment (31).

52.6.4. Size-Exclusion Chromatography

Size-exclusion chromatography (SEC; also known as gel filtration or molecular sieve chromatography) differs from the absorptive methods in that protein interactions with the chromatographic matrix are not exploited, but minimized (34). SEC media display pores of varying sizes, and differential access to these pores is the basis of separation. Large molecules run faster, as they enter fewer pores and flow through an effectively smaller volume. All molecules above a certain size run together, "excluded," as they enter very few pores. Small molecules below the "included" size limit run together, exploring nearly all pores. Useful resolving power is confined to a range of hydrodynamic radii (not molecular weights) between the effective sizes of the largest and smallest pores; matrices of differing average pore size are used to purify proteins of different sizes. The inclusion and exclusion limits of a given matrix are given by the manufacturer. It is important to remember that larger molecules run faster (the opposite of a sieve or filter) and that separation is on the basis of molecular radius, not molecular weight. This implies that an elongated protein runs more rapidly than a compact globular protein of the same molecular weight, and both run much more slowly than a high-aspect-ratio nucleic acid or random coil carbohydrate of the same molecular weight as the proteins.

SEC has relatively low sample capacity, and the packing and sample loading of long, narrow SEC columns is more demanding than preparation of relatively squat columns of adsorptive media. It is well worth testing the quality of a newly packed SEC column using a tracer such as acetone or blue dextran. SEC is a gentle, nondenaturing method, and in contrast to most methods is capable of removing aggregated forms from nearly pure protein samples. An SEC column can be "calibrated" by running a set of proteins differing in molecular weight, but all of similar shape (in usual practice, all globular). The retention times will ideally give a linear dependence on the logarithm of the molecular weight of the protein (deviations often indicate interactions with the matrix, which can be suppressed by adding salts or traces of organic solvents to the running buffer). The retention time of the unknown protein indicates its largest dimension. This information is especially valuable if it can be compared with the molecular weight obtained through (denaturing) SDS-PAGE, as large differences in the results of the two methods can often be the first sign of aggregation or multimerization.

52.6.5. Field-Flow Fractionation

Field-flow fractionation is a means to separate complex mixtures of proteins (and particles) based upon their difference in the diffusion coefficient, which itself is determined by their hydrodynamic size (36). In this method, proteins are separated in a field-flow fractionation channel in lieu of a chromatographic column, so no specific adsorbents such as gel matrixes are required. In addition, proteins can be analyzed under physiological conditions. Today, field-flow fractionation is often coupled with further, more sensitive analytical methods such as mass spectrometry (MS) or capillary isoelectric focusing (73).

52.6.6. High-Throughput Protein Purification

The recent improvements in DNA sequencing have given rise to an exponential increase in the amount of information on genomes and transcriptomes from all kingdoms. So-called "next-generation sequencing" platforms generate gigabytes of nucleic acid information per day, and the technology is advancing rapidly. Many of these nucleic acid sequences may in silico appear to encode promising new protein products, but unless they are actually translated and the protein is purified and characterized, no final decision on the real value of any individual protein can be made. Traditional protein purification methods often constitute the bottleneck in this process, delaying the discovery of potential new protein products and the advancement of biological understanding.

In an effort to address this problem, in recent years more and more automated protein purification platforms have entered the marketplace. The inherent diversity of protein molecules makes the parallel processing of a mixed batch very challenging, if not impossible. This problem is circumvented by the (genetic) addition of tags such as polyhistidine or protein A to all proteins of interest. The resulting pool of proteins can then be purified by a single type of affinity chromatography (e.g., IMAC or IgG affinity) using, for instance, magnetic particles. Using a microtiter plate format, for example, 96 individual recombinant *E. coli* colonies each expressing a different open reading frame can be grown in parallel and chemically lysed. The crude extract is purified in a mobile solid-phase approach using paramagnetic affinity particles. During the purification, the bead-bound protein of interest can easily be separated from the nonmagnetic proteins of the host cell by applying a magnetic field. The field can be removed for washes and elutions. Eventually the protein of interest can be eluted from the beads. Robotic pipetting systems such as Qiagen's BioSprint or Beckman Coulter's Biomek® make this parallel and exploratory approach even more reliable and rapidly yield purified proteins.

Other instruments such as BioRad's Profinia™ protein purification system employ the same principal approach of purifying tagged proteins by affinity chromatography on a somewhat larger scale. The purification process is automated and monitored by UV absorption, and the protein of interest is automatically separated from the background. While these systems do not allow the parallel processing of more than a few proteins at a time, they yield much larger quantities of the target protein than the microtiter plate purification described above.

Probably the most commonly found instruments in the protein purification laboratory and increasingly also in large-scale protein purification facilities are GE Healthcare's ÄKTA™ automated liquid chromatography systems. These devices use the same chromatographic media as in standard chromatography, but can conduct automatic screening for optimized purification conditions, provide better buffer control, and allow the combination of multiple purification steps into one automated purification protocol. While this family of instruments comprises platforms for purification from bench to industrial scale, the relatively high initial investment may make them cost-prohibitive for smaller laboratories.

52.6.7. Industrial-Scale Protein Purification

While the above-mentioned methods are all generally applicable to large, industrial-scale purification processes, the focus in these protocols is slightly shifted. For instance, scaling up the initial steps of a protein isolation protocol from an Erlenmeyer flask to a fermentor requires a much stronger emphasis on highly efficient protein concentration methods that can process volumes of 10,000 to 300,000 liters. Thus, these large volumes are almost always concentrated by various cell-centrifugation and filtration processes before any kind of chromatographic purification is undertaken (for some examples of industrial-scale protein purifications see references 45, 50, 54, and 85).

Given the inherently diverse nature of proteins—every protein purification protocol is probably to some extent unique, and often requires the specific adaptation of a particular set of methods—a few general rules for protein purification can be summarized as follows.

1. Tailor the process to the intended purpose. Do not over-purify, and focus on the contaminants that matter in the planned application.
2. Establish a convenient assay.
3. Select the right source of raw material, and use sufficient amounts.
4. Work quickly, or in the cold, or both.
5. Eliminate proteases early through host selection, separation, or inhibition.
6. Identify stable intermediate storage forms.
7. Avoid freeze-thaw cycles.
8. Use columns of appropriate capacity.
9. Build on previous work.

52.7. PRODUCT CHARACTERIZATION

Complete characterization of the purified product involves study of the idiosyncratic activities of each protein, which are beyond the scope of this chapter. There are also, however, general structural properties that are subject to determination by generic methods (13, 16, 26).

52.7.1. Electrophoresis

Probably the most common method of protein characterization is SDS-PAGE, which gives the molecular weight(s) of the peptide chain(s) of the protein as well as a good indication of the purity of the sample (33). SDS converts nearly all protein into regular rodlike forms of constant charge per unit mass, so that electrophoretic mobility is independent of amino acid composition. The polyacrylamide gel serves to suppress mixing induced by temperature gradients arising from the heating associated with current flow, which would reduce resolution. SDS-PAGE can use commercial precast gels, or gels can be cast at the time of use. Related electrophoretic methods include IEF (39, 40), for measuring isoelectric point; native gel electrophoresis, in which proteins are not denatured with SDS; and two-dimensional gel electrophoresis (12, 47), in which IEF separation by charge is combined with orthogonal separation by mass using SDS-PAGE. Native electrophoresis and IEF also show promise as preparative methods (30, 81).

Capillary electrophoresis is conducted in tubing (or chip microchannels, e.g., the popular Agilent Bioanalyzer) of very small diameter (<100 mm) using high voltages (e.g., 30 kV). Capillary electrophoresis can achieve very high resolution in analytical separations, requiring only a few minutes (33).

52.7.2. Peptide Sequencing

The classic Edman degradation (25), albeit in an automated way, is still being used to identify the N-terminal amino acid residues of peptides (87). Most workers will choose to have this procedure performed by a specialist core facility. This sequence information can be used to confirm the identity of the protein; to establish the degree of removal of N-terminal methionines, purification handles, or secretion leader peptides; and to detect exoproteolytic degradation. Protein sequencing depends on sequential, stepwise removal of N-terminal amino acids in the form of labeled derivatives that are separated by HPLC and identified by their characteristic retention times. Blocked N termini are not uncommon and require removal of the blockage, isolation and sequencing of internal peptides, or use of MS sequencing as discussed below. Many applications of protein sequencing are now being replaced by MS (88).

52.7.3. Tryptic Mapping

Tryptic mapping, in which small peptides derived from a protein by endoprotease (e.g., trypsin) action are separated by high-resolution reverse-phase HPLC or MS, allows sensitive detection of variations in protein (especially pharmaceutical) preparations (51, 84). Individual peptides can be subjected to sequencing as described above (e.g., avoiding problems with blocked N termini) or, more commonly nowadays, MS to confirm the identity of a given peptide, to develop information for DNA probe development, or even to establish the entire sequence of the protein, often with the aid of peptide fragment databases, which are increasingly derived from genome sequences.

52.7.4. Analytical Ultracentrifugation

In protein purification, analytical ultracentrifugation is a powerful tool for understanding protein association in solution under physiological conditions (6). This technique, which requires considerable investment in equipment and training, allows measurement of a variety of properties of a protein sample, including solution molecular weight, self-association, interaction with other molecules, and sample homogeneity. In principle, samples are analyzed either by sedimentation velocity, a hydrodynamic method, or sedimentation equilibrium analysis, a thermodynamic method (46). Sedimentation velocity, a high-speed centrifugation method, is used to determine the mass of a sample, detect aggregates, verify the homogeneity of a sample, and allow some estimation of the shape of protein molecules. Sedimentation equilibrium, a low-speed centrifugation method, is also used to measure protein molecular masses in solution, but is most suitable to determine the stoichiometry between different proteins, for instance, between a receptor and its ligand. With modern instruments, samples can now be analyzed in real time using either absorbance, interference, or fluorescence (17).

52.7.5. Spectroscopy

Circular dichroism gives an indication of the fraction of the polypeptide that is composed of specific secondary structural features, such as α-helix and β-sheet. Raman spectroscopy has been used (less widely) for similar purposes and especially for characterizing metal-containing protein cofactors (41, 58).

52.7.6. Dynamic Light Scattering

Small particles such as globular proteins move in solution in a random pattern described as Brownian motion. In dynamic light scattering (DLS), a laser light of a known wavelength is directed at a protein solution and the shift in light frequency caused by the particles in the solution is measured by a fast photon counter (7). The shift in frequency is inversely proportional to the hydrodynamic or Stokes radius of the protein. Since DLS only provides an averaged value for the whole solution, the sample under investigation has to be pure and cannot contain heterogeneous mixtures of different proteins. The advantages of DLS are that it is nondestructive, and it can be performed in solution, in real time. This provides for immediate results (as opposed to, for instance, lengthy analytical centrifugations) and allows the direct monitoring of protein binding events when two defined samples are mixed in the DLS reaction cuvette.

52.7.7. Biosensors

Biosensor devices such as GE Healthcare's BIAcore, Neo-Sensors' IASys instrument, and Sensata's Spreeta sensor, are now widely used for detecting and characterizing intermolecular interactions. These devices allow the label-free, continuous detection of proteins, cells, or nucleic acids present in a narrow region adjacent to a gold-coated detector surface that is permanently targeted by a laser light. After the molecules of interest (the *ligands*) have been chemically immobilized on the sensor, their putative binding partners (the *analytes*) can be passed over the sensor. The chemistries involved are derived from those of affinity chromatography and are now generally well characterized by the instrument manufacturers. For a signal to be produced, a meaningful amount of mass must be captured by the interaction.

For example, in studying the interaction of an antibody with a small hapten or antigen, it would be better to immobilize the hapten, i.e. make it the ligand, and capture the much larger antibody, as capturing the hapten might not produce a useful signal. The binding of the analyte to the ligand changes the refractive index of the surface, proportional to the molecular weight of the analyte, thus allowing the determination of the concentration of the analyte. Because of their sensitivity, low material consumption, and convenience, these devices are very useful for tracking the desired protein in the course of purification.

They have an equally important role in characterizing the purified protein, helping to identify specific binding pairs such as antibodies and antigens or receptors and their agonists or antagonists (for an excellent, topical review on current applications, see reference 23). Since all binding events are detected in real time, complex kinetic studies are often conducted using surface plasmon resonance instruments. It is necessary, however, to guard against the often severe effects of mass transfer limitations (15). This becomes an issue when the transfer of the analyte from the bulk solution to the sensor surface is slower than its actual binding to the ligand, resulting in a shortage of analyte at the sensor surface. These and other errors (rebinding, avidity effects of multivalency, e.g., with antibodies) have been reduced as users have become more familiar with these devices.

52.7.8. Analysis of Glycosylation

Analysis of glycosylation is an actively developing field, in particular because the comprehensive study of all glycoproteins in an organism (*glycomics*) has recently attracted heightened interest. The characterization of glycoproteins, however, remains quite challenging. Many eukaryotic (and a few prokaryotic) proteins contain sugars, which are sometimes essential to the protein's function. For instance, we now know that glycosylated proteins are involved in a wide variety of crucial molecular recognition events. Available methods include stains and blots for detection of glycosylation, deglycosylating enzymes, sugar-specific lectins for blotting and separations (42), and nuclear magnetic resonance and MS methods of characterizing individual side chains (61, 102).

52.7.9. Mass Spectrometry

With the tremendous growth in the field of proteomics, MS has blossomed in recent years into an indispensable tool for the identification of complex protein samples, with market growth driving continuing advances in instrumentation and methods of sample introduction (24, 52). While a family of related MS techniques exists, the present discussion will introduce the uses of MS as a class of methods, all of which are driven by the ability to determine the mass of proteins and peptides with remarkable precision, of the order of a few hydrogen atoms in a typical protein. The technique, therefore, can detect and characterize almost any important modification or variation in a protein's covalent structure. These include posttranslational modifications, such as glycosylation, N-terminal methionine addition or removal, methionine oxidation, proteolytic trimming, and many sequence mutations and deamidation events. Because these modifications are detected without being specifically looked for, MS is an excellent way to characterize any newly purified protein.

As mentioned above, in the era of proteomics (and postproteomics) MS has advanced into new applications. These include protein sequencing, on-line coupling to HPLC and capillary electrophoresis, and identification of "hits" from combinatorial libraries. A new application driven by the availability of genome sequences and cDNA libraries is the identification of unknown proteins separated by electrophoresis or HPLC from a diseased tissue, or a cell line exposed to an agent thought to alter expression of one or more genes. MS determination of the molecular weight of a protein showing interesting behavior (or, if digestion can be done, of one or more peptides derived from the protein) gives mass values that can then be searched against the (large) database of all proteins/peptides encoded by that genome or cDNA bank. This type of search can often be successful and in many cases will avoid the need to purify the protein in question at all.

52.7.10. Calorimetry

Calorimetry is a label-free method that is unique in that it allows the direct measurements of thermodynamic binding constants, reaction stoichiometry, enthalpy (ΔH), and entropy (ΔS) during a protein-protein interaction. Isothermal titration calorimetry (ITC) (69) is the most quantitative means available for directly measuring the thermodynamic properties of a protein-protein interaction. An ITC instrument measures the heat evolved on association of a ligand with its binding partner. A typical ITC instrument thus contains two reaction chambers, a sample and a reference cell, whose temperatures are monitored continuously. Any tiny change in temperature between sample and reference cell is compensated by a heating device in the ITC instrument. The power necessary to perform this compensation provides a measurable value that can be used to infer the reaction kinetics.

In differential scanning calorimetry (11), a protein is heated at a constant rate. When the protein becomes thermally denatured there is a measurable heat change. The instrument thus reports the enthalpy (ΔH) of unfolding due to heat denaturation. A characteristic reporting value from a differential scanning calorimetry experiment is the *thermal transition midpoint* (*Tm*), where 50% of the biomolecules are unfolded. The more stabilizing the intra- and intermolecular forces within a given protein molecule, the higher the *Tm*.

52.7.11. Protein Structure Determination

The ultimate way to characterize and understand a protein and its interactions with its substrate, cofactors, or other proteins is through nuclear magnetic resonance (NMR) spectroscopy and X-ray crystallography. NMR (63, 68, 99) is a particularly powerful tool in understanding the dynamic interaction of proteins with ligands in solution (101). This nondestructive method was originally limited to small molecules but, in certain cases, has been adapted to analyze molecules as large as 900 kDa (29, 95). While NMR can be used at many stages of the protein characterization process, it has a found a distinctly strong role in screening potential protein ligands during the drug discovery process (67).

X-ray crystallography has long had the reputation of being an inherently slow process that is more of an art than a scientific method. However, in recent years remarkable advances in this technology have been achieved. This is probably most evident by the increase of protein structures deposited in the Protein Database (55). Nearly every step of X-ray crystallography can now be automated. Robots perform the parallel expression and purification of proteins. Each protein is then screened for crystal formation in a sparse matrix approach using approximately 100 conditions of various buffers, precipitants, and additives. Once again, robots perform the pipetting steps, and cameras screen the reaction compartments for crystallization products. However, even with these recent advances, most X-ray structures still require time-consuming hands-on optimization of the crystallization conditions and interpretation of the diffraction patterns. Realistically, as shown by a 2007 survey of structural genomics centers worldwide (55), still only a small fraction (13%) of all soluble proteins can be crystallized right away.

REFERENCES

1. **Amersham-Biosciences.** 2004. *Ion Exchange Chromatography: Principles and Methods.* Amersham-Biosciences, Uppsala, Sweden.
2. **Backliwal, G., M. Hildinger, S. Chenuet, S. Wulhfard, M. De Jesus, and F. M. Wurm.** 2008. Rational vector design and multi-pathway modulation of HEK 293E cells yield recombinant antibody titers exceeding 1 g/l by transient transfection under serum-free conditions. *Nucleic Acids Res.* **36:**e96.
3. **Baldi, L., D. L. Hacker, M. Adam, and F. M. Wurm.** 2007. Recombinant protein production by large-scale transient gene expression in mammalian cells: state of the art and future perspectives. *Biotechnol. Lett.* **29:**677–684.
4. **Belew, M., T. T. Yip, L. Andersson, and R. Ehrnstrom.** 1987. High-performance analytical applications of immobilized metal ion affinity chromatography. *Anal. Biochem.* **164:**457–465.
5. **Bergmann-Leitner, E. S., R. M. Mease, E. H. Duncan, F. Khan, J. Waitumbi, and E. Angov.** 2008. Evaluation of immunoglobulin purification methods and their impact on quality and yield of antigen-specific antibodies. *Malaria J.* **7:**129.
6. **Berkowitz, S. A.** 2006. Role of analytical ultracentrifugation in assessing the aggregation of protein biopharmaceuticals. *AAPS J.* **8:**E590–605.
7. **Berne, B. J., and R. Pecora.** 2000. *Dynamic Light Scattering: with Applications to Chemistry, Biology, and Physics.* Dover Publications, Mineola, NY.
8. **Blanch, H. W., and D. Clark.** 1997. *Biochemical Engineering.* Marcel Dekker, New York, NY.
9. **Boi, C.** 2007. Membrane adsorbers as purification tools for monoclonal antibody purification. *J. Chromatogr. B* **848:**19–27.
10. **Bradford, M. M.** 1976. A rapid and sensitive method for the quantitation of microgram quantities of protein utilizing the principle of protein-dye binding. *Anal. Biochem.* **72:**248–254.
11. **Bruylants, G., J. Wouters, and C. Michaux.** 2005. Differential scanning calorimetry in life science: thermodynamics, stability, molecular recognition and application in drug design. *Curr. Med. Chem.* **12:**2011–2020.
12. **Carrette, O., P. R. Burkhard, J. C. Sanchez, and D. F. Hochstrasser.** 2006. State-of-the-art two-dimensional gel electrophoresis: a key tool of proteomics research. *Nat. Protoc.* **1:**812–823.
13. **Cereghino, G. P., J. L. Cereghino, C. Ilgen, and J. M. Cregg.** 2002. Production of recombinant proteins in fermenter cultures of the yeast *Pichia pastoris. Curr. Opin. Biotechnol.* **13:**329–332.
14. **Chong, S., F. B. Mersha, D. G. Comb, M. E. Scott, D. Landry, L. M. Vence, F. B. Perler, J. Benner, R. B. Kucera, C. A. Hirvonen, J. J. Pelletier, H. Paulus, and M. Q. Xu.** 1997. Single-column purification of free recombinant proteins using a self-cleavable affinity tag derived from a protein splicing element. *Gene* **192:**271–281.
15. **Christensen, L. L.** 1997. Theoretical analysis of protein concentration determination using biosensor technology under conditions of partial mass transport limitation. *Anal. Biochem.* **249:**153–164.
16. **Clark, E. D.** 2001. Protein refolding for industrial processes. *Curr. Opin. Biotechnol.* **12:**202–207.
17. **Cole, J. L., J. W. Lary, T. Moody, and T. M. Laue.** 2008. Analytical ultracentrifugation: sedimentation velocity and sedimentation equilibrium. *Methods Cell Biol.* **84:**143–179.

18. Craig, D., M. T. Howell, C. L. Gibbs, T. Hunt, and R. J. Jackson. 1992. Plasmid cDNA-directed protein synthesis in a coupled eukaryotic in vitro transcription-translation system. *Nucleic Acids Res.* **20**:4987–4995.

19. de Groot, N. S., A. Espargaro, M. Morell, and S. Ventura. 2008. Studies on bacterial inclusion bodies. *Future Microbiol.* **3**:423–435.

20. Del Tito, B. J., Jr., J. M. Ward, J. Hodgson, C. J. Gershater, H. Edwards, L. A. Wysocki, F. A. Watson, G. Sathe, and J. F. Kane. 1995. Effects of a minor isoleucyl tRNA on heterologous protein translation in *Escherichia coli*. *J. Bacteriol.* **177**:7086–7091.

21. Demain, A. L., and J. L. Adrio. 2008. Strain improvement for production of pharmaceuticals and other microbial metabolites by fermentation. *Prog. Drug Res.* **65**:251, 253–289.

22. DeWalt, B. W., J. C. Murphy, G. E. Fox, and R. C. Willson. 2003. Compaction agent clarification of microbial lysates. *Protein Expr. Purif.* **28**:220–223.

23. Doern, G. V., S. S. Richter, A. Miller, N. Miller, C. Rice, K. Heilmann, and S. Beekmann. 2005. Antimicrobial resistance among *Streptococcus pneumoniae* in the United States: have we begun to turn the corner on resistance to certain antimicrobial classes? *Clin. Infect. Dis.* **41**:139–148.

24. Domon, B., and R. Aebersold. 2006. Mass spectrometry and protein analysis. *Science* **312**:212–217.

25. Edman, K. A. 1950. The action of ouabain on heart actomyosin. *Acta Physiol. Scand.* **21**:230–237.

26. Eliasson, M., A. Olsson, E. Palmcrantz, K. Wiberg, M. Inganas, B. Guss, M. Lindberg, and M. Uhlen. 1988. Chimeric IgG-binding receptors engineered from staphylococcal protein A and streptococcal protein G. *J. Biol. Chem.* **263**:4323–4327.

27. Fahnert, B., H. Lilie, and P. Neubauer. 2004. Inclusion bodies: formation and utilisation. *Adv. Biochem. Eng. Biotechnol.* **89**:93–142.

28. Fan, X., D. Xu, B. Lu, J. Xia, and D. Wei. 2009. Refolding and purification of rhNTA protein by chromatography. *Biomed. Chromatogr.* **23**:257–266.

29. Fiaux, J., E. B. Bertelsen, A. L. Horwich, and K. Wuthrich. 2002. NMR analysis of a 900K GroEL GroES complex. *Nature* **418**:207–211.

30. Fountoulakis, M., and P. Dimitraki. 2008. Protein fractionation by preparative electrophoresis. *Methods Mol. Biol.* **424**:301–313.

31. Fricke, W. A., and M. A. Lamb. 1993. Viral safety of clotting factor concentrates. *Semin. Thromb. Hemost.* **19**:54–61.

32. Gagnon, P. 1996. *Purification Tools for Monoclonal Antibodies*. Validated Biosystems, Tucson, AZ.

33. Gallagher, S. R. 2007. One-dimensional SDS gel electrophoresis of proteins. *Curr. Protoc. Cell Biol.* **Chapter 6**:Unit 6.1.

34. GE-Healthcare-Bio-Sciences-AB. 2006. *Gel Filtration: Principles and Methods*. GE-Healthcare-Bio-Sciences-AB, Uppsala, Sweden.

35. GE-Healthcare-Bio-Sciences-AB. 2006. *Hydrophobic Interaction and Reversed Phase Chromatography*. GE-Healthcare-Bio-Sciences-AB, Uppsala, Sweden.

36. Giddings, J. C. 1993. Field-flow fractionation: analysis of macromolecular, colloidal, and particulate materials. *Science* **260**:1456–1465.

37. Goerke, A. R., and J. R. Swartz. 2008. Development of cell-free protein synthesis platforms for disulfide bonded proteins. *Biotechnol. Bioeng.* **99**:351–367.

38. Goerke, A. R., and J. R. Swartz. 2009. High-level cell-free synthesis yields of proteins containing site-specific non-natural amino acids. *Biotechnol. Bioeng.* **102**:400–416.

39. Gorg, A., C. Obermaier, G. Boguth, A. Harder, B. Scheibe, R. Wildgruber, and W. Weiss. 2000. The current state of two-dimensional electrophoresis with immobilized pH gradients. *Electrophoresis* **21**:1037–1053.

40. Gorg, A., W. Weiss, and M. J. Dunn. 2004. Current two-dimensional electrophoresis technology for proteomics. *Proteomics* **4**:3665–3685.

41. Havel, H. A. 1996. *Spectroscopic Methods for Determining Protein Structure in Solution*. VCH Publishers, New York, NY.

42. Hirabayashi, J., and K. Kasai. 2002. Separation technologies for glycomics. *J. Chromatogr. B* **771**:67–87.

43. Hober, S., K. Nord, and M. Linhult. 2007. Protein A chromatography for antibody purification. *J. Chromatogr. B* **848**:40–47.

44. Hooper, J. A. 2008. Intravenous immunoglobulins: evolution of commercial IVIG preparations. *Immunol. Allergy Clin. North Am.* **28**:765–778, viii.

45. Hoskins, J., W. E. Alborn, Jr., J. Arnold, L. C. Blaszczak, S. Burgett, B. S. DeHoff, S. T. Estrem, L. Fritz, D. J. Fu, W. Fuller, C. Geringer, R. Gilmour, J. S. Glass, H. Khoja, A. R. Kraft, R. E. Lagace, D. J. LeBlanc, L. N. Lee, E. J. Lefkowitz, J. Lu, P. Matsushima, S. M. McAhren, M. McHenney, K. McLeaster, C. W. Mundy, T. I. Nicas, F. H. Norris, M. O'Gara, R. B. Peery, G. T. Robertson, P. Rockey, P. M. Sun, M. E. Winkler, Y. Yang, M. Young-Bellido, G. Zhao, C. A. Zook, R. H. Baltz, S. R. Jaskunas, P. R. Rosteck, Jr., P. L. Skatrud, and J. I. Glass. 2001. Genome of the bacterium *Streptococcus pneumoniae* strain R6. *J. Bacteriol.* **183**:5709–5717.

46. Howlett, G. J., A. P. Minton, and G. Rivas. 2006. Analytical ultracentrifugation for the study of protein association and assembly. *Curr. Opin. Chem. Biol.* **10**:430–436.

47. Issaq, H., and T. Veenstra. 2008. Two-dimensional polyacrylamide gel electrophoresis (2D-PAGE): advances and perspectives. *Biotechniques* **44**:697–698, 700.

48. Jelkmann, W. 2007. Recombinant EPO production—points the nephrologist should know. *Nephrol. Dial. Transplant.* **22**:2749–2753.

49. Kamoda, S., and K. Kakehi. 2008. Evaluation of glycosylation for quality assurance of antibody pharmaceuticals by capillary electrophoresis. *Electrophoresis* **29**:3595–3604.

50. Kelley, B. 2007. Very large scale monoclonal antibody purification: the case for conventional unit operations. *Biotechnol. Prog.* **23**:995–1008.

51. Kellner, R., F. Lottspeich, and H. E. Meyer. 1994. *Microcharacterization of Proteins*. VCH, New York, NY.

52. Khalsa-Moyers, G., and W. H. McDonald. 2006. Developments in mass spectrometry for the analysis of complex protein mixtures. *Brief Funct. Genomic Proteomic* **5**:98–111.

53. Kim, D. M., T. Kigawa, C. Y. Choi, and S. Yokoyama. 1996. A highly efficient cell-free protein synthesis system from *Escherichia coli*. *Eur. J. Biochem.* **239**:881–886.

54. Ladisch, M. R. 2001. *Bioseparations Engineering: Principles, Practice, and Economics*. Wiley, New York, NY.

55. Levitt, M. 2007. Growth of novel protein structural data. *Proc. Natl. Acad. Sci. USA* **104**:3183–3188.

56. Liu, Y. D., J. J. Li, F. W. Wang, J. Chen, P. Li, and Z. G. Su. 2007. A newly proposed mechanism for arginine-assisted protein refolding—not inhibiting soluble oligomers although promoting a correct structure. *Protein Expr. Purif.* **51**:235–242.

57. Madin, K., T. Sawasaki, T. Ogasawara, and Y. Endo. 2000. A highly efficient and robust cell-free protein synthesis system prepared from wheat embryos: plants apparently contain a suicide system directed at ribosomes. *Proc. Natl. Acad. Sci. USA* **97**:559–564.

58. Mantsch, H. H., and D. Chapman. 1996. *Infrared Spectroscopy of Biomolecules*. Wiley-Liss, New York, NY.

59. McGregor, W. C. 1986. *Membrane Separations in Biotechnology*. Dekker, New York, NY.

60. Mierau, I., and M. Kleerebezem. 2005. 10 years of the nisin-controlled gene expression system (NICE) in *Lactococcus lactis*. *Appl. Microbiol. Biotechnol.* **68**:705–717.

61. **Morelle, W., and J. C. Michalski.** 2005. Glycomics and mass spectrometry. *Curr. Pharm. Des.* **11:**2615–2645.

62. **Nevalainen, K. M., V. S. Te'o, and P. L. Bergquist.** 2005. Heterologous protein expression in filamentous fungi. *Trends Biotechnol.* **23:**468–474.

63. **Nietlispach, D., H. R. Mott, K. M. Stott, P. R. Nielsen, A. Thiru, and E. D. Laue.** 2004. Structure determination of protein complexes by NMR. *Methods Mol. Biol.* **278:**255–288.

64. **Omar, A., and C. Kempf.** 2002. Removal of neutralized model parvoviruses and enteroviruses in human IgG solutions by nanofiltration. *Transfusion* **42:**1005–1010.

65. **Panda, A. K.** 2003. Bioprocessing of therapeutic proteins from the inclusion bodies of *Escherichia coli. Adv. Biochem. Eng. Biotechnol.* **85:**43–93.

66. **Panda, A. K.** 2005. High-throughput recovery of therapeutic proteins from the inclusion bodies of *Escherichia coli,* p. 155–161. *In* C. M. Smales and D. C. James (ed.), *Therapeutic Proteins: Methods and Protocols.* Humana Press, Totowa, NJ.

67. **Pellecchia, M., I. Bertini, D. Cowburn, C. Dalvit, E. Giralt, W. Jahnke, T. L. James, S. W. Homans, H. Kessler, C. Luchinat, B. Meyer, H. Oschkinat, J. Peng, H. Schwalbe, and G. Siegal.** 2008. Perspectives on NMR in drug discovery: a technique comes of age. *Nat. Rev. Drug Discov.* **7:**738–745.

68. **Pellecchia, M., D. S. Sem, and K. Wuthrich.** 2002. NMR in drug discovery. *Nat. Rev. Drug Discov.* **1:**211–219.

69. **Pierce, M. M., C. S. Raman, and B. T. Nall.** 1999. Isothermal titration calorimetry of protein-protein interactions. *Methods* **19:**213–221.

70. **Porro, D., M. Sauer, P. Branduardi, and D. Mattanovich.** 2005. Recombinant protein production in yeasts. *Mol. Biotechnol.* **31:**245–259.

71. **Przybycien, T. M., N. S. Pujar, and L. M. Steele.** 2004. Alternative bioseparation operations: life beyond packed-bed chromatography. *Curr. Opin. Biotechnol.* **15:**469–478.

72. **Ramsden, J. J., D. J. Roush, D. S. Gill, R. Kurrat, and R. C. Willson.** 1995. Protein adsorption kinetics drastically altered by repositioning a single charge. *J. Am. Chem. Soc.* **119:**8511–8516.

73. **Reschiglian, P., and M. H. Moon.** 2008. Flow field-flow fractionation: a pre-analytical method for proteomics. *J. Proteomics* **71:**265–276.

74. **Reznikoff, W. S., and L. Gold.** 1986. *Maximizing Gene Expression.* Butterworths, Boston, MA.

75. **Riggs, P.** 2000. Expression and purification of recombinant proteins by fusion to maltose-binding protein. *Mol. Biotechnol.* **15:**51–63.

76. **Roque, A. C., C. S. Silva, and M. A. Taipa.** 2007. Affinity-based methodologies and ligands for antibody purification: advances and perspectives. *J. Chromatogr. A* **1160:**44–55.

77. **Saxena, A., B. P. Tripathi, M. Kumar, and V. K. Shahi.** 2009. Membrane-based techniques for the separation and purification of proteins: an overview. *Adv. Colloid Interface Sci.* **145:**1–22.

78. **Schmidt, S.** 2006. Powerful ability of UVC irradiation to inactivate viruses, especially the most critical small non-enveloped viruses (e.g. parvovirus). *Biologicals* **34:**237–238.

79. **Schmidt, T. G., and A. Skerra.** 2007. The Strep-tag system for one-step purification and high-affinity detection or capturing of proteins. *Nat. Protoc.* **2:**1528–1535.

80. **Scopes, R. K.** 1994. *Protein Purification: Principles and Practice,* 3rd ed. Springer Verlag, New York, NY.

81. **Seelert, H., and F. Krause.** 2008. Preparative isolation of protein complexes and other bioparticles by elution from polyacrylamide gels. *Electrophoresis* **29:**2617–2636.

82. **Sharma, K., P. V. Babu, P. Sasidhar, V. K. Srinivas, V. K. Mohan, and E. Krishna.** 2008. Recombinant human epidermal growth factor inclusion body solubilization and refolding at large scale using expanded-bed adsorption chromatography from *Escherichia coli. Protein Expr. Purif.* **60:**7–14.

83. **Shimizu, Y., Y. Kuruma, B. W. Ying, S. Umekage, and T. Ueda.** 2006. Cell-free translation systems for protein engineering. *FEBS J.* **273:**4133–4140.

84. **Shively, J. E.** 1986. *Methods of Protein Microcharacterization: a Practical Handbook.* Humana Press, Clifton, NJ.

85. **Smales, C. M., and D. C. James.** 2005. *Therapeutic Proteins: Methods and Protocols.* Humana Press, Totowa, NJ.

86. **Smith, D. B., and K. S. Johnson.** 1988. Single-step purification of polypeptides expressed in *Escherichia coli* as fusions with glutathione S-transferase. *Gene* **67:**31–40.

87. **Smith, J. B.** 2001. Peptide sequencing by Edman degradation. *In Encyclopedia of Life Sciences.* John Wiley & Sons, Ltd., Chichester, United Kingdom. (Online.) DOI: 10.1038/npg.els.0002688.

88. **Steen, H., and M. Mann.** 2004. The ABC's (and XYZ's) of peptide sequencing. *Nat. Rev. Mol. Cell Biol.* **5:**699–711.

89. **Street, G.** 1994. *Highly Selective Separations in Biotechnology.* Blackie Academic & Professional, London, United Kingdom.

90. **Studier, F. W.** 2005. Protein production by auto-induction in high density shaking cultures. *Protein Expr. Purif.* **41:**207–234.

91. **Studier, F. W., and B. A. Moffatt.** 1986. Use of bacteriophage T7 RNA polymerase to direct selective high-level expression of cloned genes. *J. Mol. Biol.* **189:**113–130.

92. **Sumner, J. B.** 1926. The isolation and crystallization of the enzyme urease. *J. Biol. Chem.* **69:**435–441.

93. **Swartz, J.** 2006. Developing cell-free biology for industrial applications. *J. Ind. Microbiol. Biotechnol.* **33:**476–485.

94. **Tsumoto, K., M. Umetsu, I. Kumagai, D. Ejima, J. S. Philo, and T. Arakawa.** 2004. Role of arginine in protein refolding, solubilization, and purification. *Biotechnol. Prog.* **20:**1301–1308.

95. **Tzakos, A. G., C. R. Grace, P. J. Lukavsky, and R. Riek.** 2006. NMR techniques for very large proteins and RNAs in solution. *Annu. Rev. Biophys. Biomol. Struct.* **35:**319–342.

96. **Ventura, S., and A. Villaverde.** 2006. Protein quality in bacterial inclusion bodies. *Trends Biotechnol.* **24:**179–185.

97. **Vieth, W. R.** 1988. *Membrane Systems: Analysis and Design. Applications in Biotechnology, Biomedicine and Polymer Science.* Hanser Publishers, New York, NY.

98. **Villaverde, A., and D. Mattanovich.** 2007. Recombinant protein production in the new Millennium. *Microb. Cell Fact.* **6:**33.

99. **Widmer, H., and W. Jahnke.** 2004. Protein NMR in biomedical research. *Cell. Mol. Life Sci.* **61:**580–599.

100. **Woodrow, K. A., and J. R. Swartz.** 2007. A sequential expression system for high-throughput functional genomic analysis. *Proteomics* **7:**3870–3879.

101. **Wuthrich, K.** 1990. Protein structure determination in solution by NMR spectroscopy. *J. Biol. Chem.* **265:**22059–22062.

102. **Zaia, J.** 2008. Mass spectrometry and the emerging field of glycomics. *Chem. Biol.* **15:**881–892.

103. **Zhou, J. X., T. Tressel, X. Yang, and T. Seewoester.** 2008. Implementation of advanced technologies in commercial monoclonal antibody production. *Biotechnol. J.* **3:**1185–1200.

Author Index

Subject Index